COMPREHENSIVE
MEDICINAL CHEMISTRY

IN 6 VOLUMES

COMPREHENSIVE MEDICINAL CHEMISTRY

The Rational Design, Mechanistic Study & Therapeutic Application of Chemical Compounds

Chairman of the Editorial Board
CORWIN HANSCH
Pomona College, Claremont, CA, USA

Joint Executive Editors
PETER G. SAMMES
Brunel University of West London, Uxbridge, UK

JOHN B. TAYLOR
Rhône-Poulenc Ltd, Dagenham, UK

Volume 6
CUMULATIVE SUBJECT INDEX & DRUG COMPENDIUM

Volume Editor
COLIN J. DRAYTON
Pergamon Press, Oxford, UK

PERGAMON PRESS

Member of Maxwell Macmillan Pergamon Publishing Corporation
OXFORD • NEW YORK • BEIJING • FRANKFURT
SÃO PAULO • SYDNEY • TOKYO • TORONTO

U.K.	Pergamon Press plc, Headington Hill Hall, Oxford OX3 0BW, England
U.S.A.	Pergamon Press, Inc., Maxwell House, Fairview Park, Elmsford, New York 10523, U.S.A.
PEOPLE'S REPUBLIC OF CHINA	Pergamon Press, Room 4037, Qianmen Hotel, Beijing, People's Republic of China
FEDERAL REPUBLIC OF GERMANY	Pergamon Press GmbH, Hammerweg 6, D-6242 Kronberg, Federal Republic of Germany
BRAZIL	Pergamon Editora Ltda, Rua Eça de Queiros, 346, CEP 04011, Paraiso, São Paulo, Brazil
AUSTRALIA	Pergamon Press Australia Pty Ltd., P.O. Box 544, Potts Point, N.S.W. 2011, Australia
JAPAN	Pergamon Press, 5th Floor, Matsuoka Central Building, 1-7-1 Nishishinjuku, Shinjuku-ku, Tokyo 160, Japan
CANADA	Pergamon Press Canada Ltd., Suite No. 241, 253 College Street, Toronto, Ontario, Canada M5T 1R5

First edition 1990

Library of Congress Cataloging in Publication Data

Comprehensive medicinal chemistry: the rational design, mechanistic study & therapeutic application of chemical compounds/ chairman of the editorial board, Corwin Hansch; joint executive editors, Peter G. Sammes, John B. Taylor. — 1st ed.
p. cm.
Includes index.
1. Pharmaceutical chemistry. I. Hansch, Corwin. II. Sammes, P. G. (Peter George) III. Taylor, J. B. (John Bodenham), 1939–
[DNLM: 1. Chemistry, Pharmaceutical. QV 744 C737]
RS402.C65
615'.19—dc20
DNLM/DLC 89–16329

British Library Cataloguing in Publication Data

Hansch, Corwin
Comprehensive medicinal chemistry
1. Pharmaceutics
I. Title
615'.19

ISBN 0–08–037062–4 (Vol. 6)
ISBN 0–08–032530–0 (set)

Printed in Great Britain by
BPCC Hazell Books Ltd, Aylesbury, Bucks, England

Contents

Preface

Medicinal chemistry is a subject which has seen enormous growth in the past decade. Traditionally accepted as a branch of organic chemistry, and the near exclusive province of the organic chemist, the subject has reached an enormous level of complexity today. The science now employs the most sophisticated developments in technology and instrumentation, including powerful molecular graphics systems with 'drug design' software, all aspects of high resolution spectroscopy, and the use of robots. Moreover, the medicinal chemist (very much a new breed of organic chemist) works in very close collaboration and mutual understanding with a number of other specialists, notably the molecular biologist, the genetic engineer, and the biopharmacist, as well as traditional partners in biology.

Current books on medicinal chemistry inevitably reflect traditional attitudes and approaches to the field and cover unevenly, if at all, much of modern thinking in the field. In addition, such works are largely based on a classical organic structure and therapeutic grouping of biologically active molecules. The aim of *Comprehensive Medicinal Chemistry* is to present the subject, the modern role of which is the understanding of structure–activity relationships and drug design from the mechanistic viewpoint, as a field in its own right, integrating with its central chemistry all the necessary ancillary disciplines.

To ensure that a broad coverage is obtained at an authoritative level, more than 250 authors and editors from 15 countries have been enlisted. The contributions have been organized into five major themes. Thus Volume 1 covers general principles, Volume 2 deals with enzymes and other molecular targets, Volume 3 describes membranes and receptors, Volume 4 covers quantitative drug design, and Volume 5 discusses biopharmaceutics. As well as a cumulative subject index, Volume 6 contains a unique drug compendium containing information on over 5500 compounds currently on the market. All six volumes are being published simultaneously, to provide a work that covers all major topics of interest.

Because of the mechanistic approach adopted, Volumes 1–5 do not discuss those drugs whose modes of action are unknown, although they will be included in the compendium in Volume 6. The mechanisms of action of such agents remain a future challenge for the medicinal chemist.

We should like to acknowledge the way in which the staff at the publisher, particularly Dr Colin Drayton (who initially proposed the project), Dr Helen McPherson and their editorial team, have supported the editors and authors in their endeavour to produce a work of reference that is both complete and up-to-date.

Comprehensive Medicinal Chemistry is a milestone in the literature of the subject in terms of coverage, clarity and a sustained high level of presentation. We are confident it will appeal to academic and industrial researchers in chemistry, biology, medicine and pharmacy, as well as teachers of the subject at all levels.

CORWIN HANSCH
Claremont, USA

PETER G. SAMMES
Uxbridge, UK

JOHN B. TAYLOR
Dagenham, UK

Contributors to Volume 6

Dr P. N. Craig
4931 Mariners Drive, Shady Side, MD 20764, USA

Dr J. Newton
David John Services Ltd, 221 Wentworth Avenue, Farnham Royal, Slough, Berks SL2 2AP, UK

Contents of All Volumes

Volume 2 Enzymes and Other Molecular Targets

Cumulative Subject Index

JOHN NEWTON

David John Services Ltd, Maidenhead, Berks, UK

B

C

ceftazidime synthesis from, **2**, 643
enzyme inhibition, **2**, 104
hydrophobicity
    crypticity, **2**, 98
introduction, **1**, 43
nephrotoxicity, **2**, 618
synthesis, **2**, 624
Cephalosporanate, 7-amino-3-pyridiniummethyl-
    synthesis, **2**, 643
Cephalosporanic acid, 7-amino-
    cephem ring system, **2**, 609
    synthesis, **2**, 623
Cephalosporin C, **1**, 43
    action
        cell walls, **1**, 171
    activity, **2**, 623
    biosynthesis, **2**, 610
    synthesis, **2**, 612, 623
Cephalosporin, amino-
    oral absorption, **5**, 15
    synthesis, **2**, 638
Cephalosporin, 7α-chlorosulfoxide
    inhibition
        porcine pancreatic elastase, **2**, 81
Cephalosporin C, deacetoxy-
    biosynthesis, **2**, 610
Cephalosporin, 3-exomethylene-
    ozonolysis, **2**, 645
Cephalosporin, 7α-formamido-
    activity, **2**, 642
Cephalosporin, 7α-methoxy-
    stability, **2**, 629
    synthesis, **2**, 631
Cephalosporin, 7α-methoxy-7β-amino-
    synthesis, **2**, 631
Cephalosporin, methoxyiminoaminothiazolyl-
    preparation, **2**, 636
Cephalosporin, oximino-
    properties, **2**, 631
Cephalosporin, ureido-
    activity, **2**, 642
Cephalosporinases
    classification, **2**, 104
Cephalosporin ester
    elastase inhibitor, **2**, 670
Cephalosporins
    antibacterial activity, **2**, 625, 630
    antipseudomonal, **2**, 642
        antibacterial activity, **2**, 643
    binding proteins
        Escherichia coli, **2**, 615
    biosynthesis, **2**, 610
    cell wall agents, **2**, 609–650
    classification, **1**, 255
    delivery systems
        oral, **5**, 617
    drug–drug interactions
        tubular active transport, **5**, 180
    elastase inhibitors, **2**, 426
    enzyme inhibition, **2**, 104
    gastrointestinal tract
        absorption, **5**, 618
    history, **2**, 623
    inhibitors
        transpeptidases, **2**, 81
    isolation, **1**, 43
    β-lactamase stable, **2**, 629
    manufacture, **1**, 335
    marketing
        USA, **1**, 631
    microbiological assay, **5**, 418
    nuclear analogs, **2**, 677
    oral
        antibacterial activity, **2**, 645

partitioning, **4**, 263
structure–activity relationships, **2**, 627
susceptibility, **2**, 92
therapeutic index, **5**, 611
third generation, **2**, 635
    antibacterial activity, **2**, 636
toxicity testing, **1**, 585
Cephalosporium acremonium
cephalins, **1**, 43
cephalosporin biosynthesis, **2**, 610
Cephalothin
activity, **2**, 641
hydrophobicity
    crypticity, **2**, 97
introduction, **1**, 43
metabolism
    kidneys, **5**, 120
nephrotoxicity, **5**, 542
synthesis, **2**, 623
total synthesis, **2**, 612
Cephalothin, deacetyl-
synthesis, **2**, 623
Cephams, 3-hetero-
properties, **2**, 691
Cephamycin C
structure, **2**, 629
Cephamycins
biosynthesis, **2**, 610
detection, **2**, 662
discovery, **2**, 655
Cephapirin
metabolism
    kidneys, **5**, 120
Cephems
4,7-bicyclic system
    activity, **2**, 692
conformation, **2**, 613
nuclear modification, **2**, 677
ring system
    β-lactam antibiotics, **2**, 609
Cephems, 3-hetero-
properties, **2**, 691
Cephradine
activity, **2**, 626, 644
introduction, **1**, 45
oral absorption, **2**, 627
Ceramides
structure, **3**, 9
Cerebellum
brain, **1**, 155, 156
Cerebral aqueduct
nervous system, **1**, 158
Cerebral cortex
brain, **1**, 155
Cerebral vasodilating activity
benzyldiphenylmethylpiperazines
    QSAR, **4**, 546
Cerebrosides
structure, **3**, 9
Cerebroside sulfate
opioid receptor
    model, **3**, 836
Cerebrospinal fluid
nervous system, **1**, 158
Cerebrum
brain, **1**, 155
Certificate of need
health care
    USA, **1**, 539
Certification marks
trade marks, **1**, 700
Ceruloplasmin
copper poisoning, **2**, 184
Cetamolol

Coca leaf
  alkaloids, **1**, 17
Cocitation clustering
  biological information retrieval, **1**, 731
Codamine
  partial synthesis, **1**, 1
Codecarboxylase — *see* Coenzyme B₆
Codeine, **1**, 11, 12
  analgesic properties, **1**, 2
  cartels, **1**, 92
  coated dosage forms
    ion-exchange resins, **5**, 639
  crystal forms
    bioavailability, **5**, 555
  isolation, **1**, 11
  opioid activity, **3**, 823
  partial structures
    amino alcohol derivatives, **1**, 1
Codes of conduct
  pharmaceutical industry
    promotion, **1**, 489
Cod liver oil
  emulsions, **5**, 572
Codons
  genes, **1**, 413
    synthesis, **1**, 413
  RNA, **1**, 365
  translation efficiency, **1**, 413
Coenzyme A, acetyl-
  Krebs cycle, **1**, 185
Coenzyme B₆ (codecarboxylase; pyridoxal 5′- phos-
      phate)
  history, **2**, 214
  mechanism, **2**, 214
  stereochemistry, **2**, 214
  structure, **2**, 214
Coenzyme NAD
  Krebs cycle, **1**, 185
Coenzymes
  specificity, **2**, 83
Coformycin
  adenosine deaminase inhibitor, **2**, 449
  antiviral agent, **2**, 320
Coformycin, 2′-chloro-
  adenosine deaminase inhibition, **2**, 73
Coformycin, 2′-chloro-2′-deoxy-
  adenosine deaminase inhibitor, **2**, 450
Coformycin, 2′-deoxy- (pentostatin)
  adenosine deaminase inhibitor, **2**, 73, 449; **3**, 603
    animals, **2**, 452
  antiparasitic agent, **2**, 302
  antiviral agent, **2**, 320
  clinical studies, **2**, 452
  immunosuppressive agent
    rats, **2**, 452
  malaria, **2**, 475
  organ transplantation, **2**, 474
  purine antagonist, **2**, 309
  T-lymphoblastic leukemia
    remission, **2**, 453
  toxicities, **2**, 453
Cohesive energy density
  definition, **4**, 247
Cointegrate
  Tn3 transposition, **1**, 386
Colaspase
  introduction, **1**, 59
Colchicine
  anticancer therapy, **1**, 188
  classification, **1**, 254
  gout, **2**, 469
  isolation, **1**, 9, 14
  malabsorption syndrome, **5**, 168
  mitotic inhibitor, **2**, 784

Colchicum
  alkaloids, **1**, 14
Colchicum autumnale
  alkaloids, **1**, 9
Colcynth
  Egyptian medicine, **1**, 9
Cold and heat syndromes
  traditional Chinese medicine, **1**, 103
Colestipol
  drug interactions
    hydrochlorothiazide, **5**, 173
  HMGCoA reductase
    inhibition, **2**, 338
Coleus forskolii
  Ayurvedic medicine, **1**, 128
Colicin E3
  ribosomes
    inactivation, **2**, 817
Colicins
  potassium channels
    membranes, **3**, 1048
Colistins
  cell membrane disorganization, **2**, 590
Collagenase
  concentration
    hepatocytes, **5**, 456
  fetal liver cells
    isolation, **5**, 460
  hepatocytes, **5**, 455
    isolation, **5**, 456
  inhibitors, **2**, 394, 412
  isolated cells
    drug metabolism, **5**, 454
  isolated renal fragments, **5**, 458
  metalloproteases, **2**, 71
  parathyroid hormone
    receptors, **3**, 1034
  substrates
    molecular modeling, **2**, 412
Collander equation
  drug transport, **4**, 396
  partition coefficients, **4**, 296
    estimation, **4**, 276
Collecting tubule
  kidney, **1**, 149
Colley's anemia
  iron chelator, **2**, 190
Colliculi
  brain, **1**, 157
Collinearity
  QSAR, **4**, 655
Colloidal dispersions
  definition, **5**, 557
Colon
  average transit time
    drug absorption, **5**, 7
  digestion, **1**, 149
  drug delivery, **5**, 639
  perfusion, **5**, 452
  physiology, **5**, 5
Colony hybridization
  cloned genes, **1**, 406
Colony stimulating factors
  assay, **3**, 1113
  biological effects, **3**, 1111
  biosynthesis, **3**, 1105
  cellular immunity, **1**, 231
  cellular sources, **3**, 1105
  clinical trials, **3**, 1113
  distribution, **3**, 1105
  genetic engineering, **1**, 467
  localization, **3**, 1105
  mechanism of action, **3**, 1111
  sites of action, **3**, 1105

D

D600
  calcium channels, **3**, 1078
DA 4643
  clinical tests, **3**, 412
Dactinomycin — *see* Actinomycin D
DADLE
  opioid
    δ-selective agonist, **3**, 813
    structure–activity relationships, **3**, 818
DAGO
  opioid receptors
    binding studies, **3**, 811
  μ-opioid receptor selectivity, **3**, 828
Daidzein
  traditional Chinese medicine, **1**, 107
Daidzin
  traditional Chinese medicine, **1**, 107
Dale's Law
  receptor theory, **3**, 75
Danazol
  coagulation factors, **2**, 491
Dantrolene
  calcium channels, **3**, 1085
  muscular rigidity, **5**, 265
DAPA
  thrombin inhibitor, **2**, 490
Dapsone, **4**, 6
  acetylation
    metabolic pathways, **5**, 156
  analogs
    folate synthesis inhibition, **4**, 525, 527
  antibacterial action, **2**, 260
  introduction, **1**, 74
  leukotriene B4 receptors
    antagonist, **3**, 791
  resistance, **2**, 265
  structure, **2**, 258
DARC, **4**, 46, 49, 61
  connection tables
    chemical structure, **1**, 721
  information storage and retrieval, **4**, 36
DARC/PELCO approach
  antibacterial activity
    alcohols, **4**, 636
  Free–Wilson analysis, **4**, 636
DARC-RMS
  chemical reaction database, **1**, 735, 737
DARC-SMS
  ChemQuest database, **1**, 743
Darvas
  graphical method, **4**, 19
Data banks
  biological activity data, **4**, 93
  structures, **4**, 154
Databases
  companies, **4**, 40
  computers, **4**, 36, 47
  connection tables, **4**, 38
  data input, **4**, 51
  data security, **4**, 52
  definition, **4**, 51
  development tools, **4**, 38
  hardware, **4**, 52
  inhouse, **4**, 49
  interfaces
    computers, **4**, 38
  internal
    development, **4**, 49

  maintenance, **4**, 51
  management systems, **4**, 47
  online, **1**, 713; **4**, 47
  reactions, **4**, 51
  training, **4**, 51
  update, **4**, 51
Data confidentiality
  computers, **4**, 38
Data distribution
  normal
    characteristics, **4**, 353
Data processing
  chromatography
    drugs, **5**, 408
Data security
  computers, **4**, 38
Data space
  pattern recognition, **4**, 694
Data structure
  pattern recognition, **4**, 700
Data-Star
  database vendor, **1**, 713
Datura metel
  drugs
    Ayurvedic medicine, **1**, 122
Daunorubicin (daunomycin)
  antitumor drug, **2**, 187, 727
  conjugation
    monoclonal antibodies, **5**, 673
  DNA complex
    structure, **4**, 468
  DNA intercalator, **1**, 207
  DNA polymerases
    inhibition, **2**, 780
  DNA strand-breakage, **2**, 736
  DNA synthesis
    inhibition, **2**, 714
  DNA topoisomerase, **2**, 779
    inhibition, **2**, 779
  drug targeting, **5**, 675
  erythrocytes
    drug targeting, **5**, 694
  intercalating agent
    DNA, **2**, 704
  isolation, **1**, 48
  toxicity, **2**, 722
Dazmegrel
  microalbuminuria, **2**, 163
Dazoxiben
  thromboxane synthase
    inhibition, **2**, 163
DBH
  catecholamines
    release, **3**, 136
  traditional Chinese medicine, **1**, 107
DDT
  breast milk
    toxicity, **5**, 310
  cytochromes *P*-450 induction
    toxicity, **5**, 329
  metabolism
    resistance, **5**, 313
  selectivity, **1**, 240
  sodium channels
    structure–activity relationships, **3**, 1062
  solubilization, **4**, 256
  toxicity

E

Ebastine
  non-sedative
    antihistamine, **3**, 376
Ebers papyrus
  Egyptian medicine, **1**, 9
Ebselen
  glucuronide conjugation
    metabolic pathways, **5**, 153
Eccrine sweat glands
  human body, **1**, 136
ECETOC
  QSAR, **4**, 663
Echinocandin B
  sterols, **2**, 361
Echinocandins
  antifungal activity, **2**, 588
  glucan synthase inhibitors, **2**, 587
Echinomycin
  intercalating agent
    DNA, **2**, 704, 718
Ecogenetics
  drug response, **5**, 254
Econazole
  fungal resistance, **2**, 114
Economics
  pharmaceutical industry, **1**, 485
ECO-R1
  DNA restriction enzyme complex
    structure, **4**, 468
Ecotoxicology
  computers, **4**, 35
  selectivity, **1**, 240
Ecto-5′-nucleotidase
  biochemistry and immunology, **2**, 456
  inhibitors, **2**, 457
Edema
  prostaglandins, **2**, 151
  prostanoids, **3**, 695
EDLFIA — *see* Dissociation-enhanced lanthanide
Edman degradation
  proteins, **2**, 17
EDTA
  Gram-negative bacteria, **2**, 561
  lead incorporation, **2**, 185
  microsomal fractions
    isolation, **5**, 467
  postmitochondrial supernatant
    isolation, **5**, 463
  rectal drug absorption, **5**, 21
    enhancer, **5**, 599
  subcellular fractions, **5**, 463
Education system
  pharmaceutical industry, **1**, 485
Effect compartment
  pharmacodynamics, **5**, 348
Effective filtration pressure
  kidneys, **1**, 151
Efferent arteriole
  kidney, **1**, 149
Efficacy
  ligands
    receptors, **3**, 48
  pharmacodynamics
    definition, **5**, 336
  receptor–ligand interactions, **3**, 51
Effluents
  biological treatment
    pharmaceutical manufacture, **1**, 351

disposal
  pharmaceutical manufacture, **1**, 351
pharmaceuticals
  manufacture, **1**, 327, 333, 336
Egestion
  digestive system, **1**, 145
Eglin
  inhibitor
    emphysema, **2**, 71
EGTA
  hepatocytes
    calcium removal, **5**, 455
    isolation, **5**, 456
  isolated cells
    drug metabolism, **5**, 454
Egyptian medicine
  isolated cells, **1**, 9
EHNA — *see* Adenine, *erythro*-9-(2-hydroxy-3-
    nonyl)-
Ehrlich, Paul
  chemotherapy, **1**, 86
  drug regulations
    history, **1**, 87
EHT — *see* Extended Hückel theory
Eicosanoids
  from arachidonic acid, **2**, 149
Eicosapentaenoic acid
  arachidonic acid
    metabolism, **2**, 168
  hyperlipidemia, **2**, 169
  inflammatory disease, **2**, 169
Eicosatetraenoic acid, 6,8-*trans*-10,14-*cis*-
  leukotriene B$_4$ receptors, **3**, 783
Eicosatetraenoic acid, 6,10-*trans*-8,14-*cis*-
  leukotriene B$_4$ receptors
    antagonist, **3**, 783
Eicosatetraenoic acid, 5-hydroxy- (5-HETE)
  platelet activating factor
    neutrophils, **3**, 723
Eicosatrienoic acid, epoxy-
  arachidonic acid
    metabolism, **2**, 160
Elases
  inhibition
    boranes, **2**, 71
Elastase
  binding, **2**, 70
  complexes
    structure, **4**, 465
  immune system, **1**, 227
  inhibition, **2**, 71, 79, 422
    benzoxazinones, **2**, 79
  peptidyl inhibitors, **2**, 422
  release
    platelet activating factor, **3**, 739
  X-ray structure, **4**, 460
Elastatinal
  binding, **2**, 70
Elastic arteries
  human body, **1**, 141
Elastin
  elastase substrate
    structure, **2**, 422
Elastinal
  elastase inhibitor, **2**, 423
Elcatonin
  calcitonin
    therapeutic applications, **3**, 1030

Eterylate
  radiolabeled
    absorption, **5**, 372
Ethacridine
  introduction, **1**, 72
Ethacrynic acid
  introduction, **1**, 60
  metabolism
    gastrointestinal tract, **5**, 117
  ototoxicity, **5**, 542
Ethambutol
  introduction, **1**, 75
Ethane
  partition coefficient
    calculation, **4**, 305
Ethane, 1,2-dibromo-
  glutathione conjugation
    metabolic pathways, **5**, 159
Ethane, 1,2-diethoxy-
  partition coefficient
    calculation, **4**, 309
Ethane, 1,2-diphenyl-
  partition coefficient
    calculation, **4**, 305
Ethane, diphenoxy-
  partition coefficient, **4**, 309
Ethane, tetrachloro-
  toxicity
    structure, **5**, 319
Ethanethiol
  penem metabolite, **1**, 272
  prodrug
    odor, **5**, 127
Ethanol
  drug dependence, **2**, 115
  drug disposition
    twins, **5**, 268
  drug interactions
    propantheline, **5**, 168
  elimination
    age, **5**, 271
  metoclopramide interaction, **5**, 166
  pharmacogenetics, **5**, 265
  transdermal drug delivery
    penetration enhancement, **5**, 655
Ethanol, 1-aryl-2-(alkylamino)-
  antimalarials, **4**, 227
Ethanol, dimethylamino-
  methylation
    metabolic pathways, **5**, 156
Ethanol, 4-hydroxy-3-methoxyphenyl-
  sulfation
    brain, **5**, 122
Ethanol, phenanthryl-
  antimalarials, **4**, 227
Ethanol, trichloro-
  conjugation
    glucuronides, **5**, 270
  introduction, **1**, 53
Ethanolamines, aryl-
  α_1-adrenergic receptor antagonists
    structure–activity relationships, **3**, 166
Ethaverine
  introduction, **1**, 13
Ethchlorvynol
  introduction, **1**, 65
Ether
  introduction, **1**, 53, 55
  isolated perfused liver, **5**, 448
  malignant hyperthermia, **5**, 265
  selectivity, **1**, 241
Ethereal glucuronides
  hydrolysis, **5**, 482
Ethers

blood–brain barrier, **4**, 402
  halogenated ethyl methyl
    toxicity and QSAR, **4**, 578
Ethics Committee
  UK
    clinical trials, **1**, 597
Ethidium bromide
  DNA binding agent, **2**, 767
  DNA dyes, **1**, 407
  intercalating agent
    DNA, **2**, 709, 712
Ethinamate
  introduction, **1**, 65
Ethionamide
  reduction
    metabolic pathways, **5**, 150
Ethisterone, **1**, 30, 65
  introduction, **1**, 29
Ethoglucid
  introduction, **1**, 69
Ethosuximide
  centrifugal analyzers, **5**, 437
  introduction, **1**, 55
Ethoxyquine
  enzyme induction, **5**, 330
Ethylamine, 3,4-dimethoxyphenyl-
  acetylation
    brain, **5**, 122
Ethylamine, haloaryl-
  adrenergic blocking activity, **4**, 514
Ethylamine, N^α-methyl-2-pyridyl-
  histamine
    agonist, **3**, 349
Ethylamine, phenyl-
  acetylation
    brain, **5**, 114
  dopamine
    agonist, **3**, 244
Ethylamine, 3-pyrazolyl-
  histamine
    agonist, **3**, 347
Ethylamine, 2-pyridyl-
  active tautomers
    histamine receptors, **3**, 352
  histamine
    agonist, **3**, 347
  histamine receptors
    chiral agonist, **3**, 367
    pharmacological characterization, **3**, 364
Ethylamine, 2-thiazolyl-
  active tautomers
    histamine receptors, **3**, 352, 364
  dication
    histamine receptors, **3**, 351
  histamine
    agonist, **3**, 347
Ethylamine, 3-(1,2,4-triazolyl)-
  dication
    histamine receptors, **3**, 351
  histamine agonist, **3**, 350
Ethylammonium nitrate
  hydrogen bonds, **4**, 253
Ethyl biscoumacetate
  introduction, **1**, 21
Ethylene, **1**, 67
  anesthetic, **1**, 67
  inactivation
    cytochrome P-450, **4**, 121
Ethylene, 1,2-dichloro-
  partition coefficients
    calculation, **4**, 317
Ethylene, tetrachloro-
  toxicity
    glutathione conjugation, **5**, 312

F

Facilitated diffusion
  drug absorption
    gastrointestinal tract, **5**, 165
Facilitated transport
  cell membranes, **3**, 29
Factor V
  plasminogen activators, **1**, 464
Factor VIII
  genetic engineering, **1**, 468
  plasminogen activators, **1**, 464
Factor IX
  genetic engineering, **1**, 468
Factor IXa
  molecular modeling
    computer graphics, **4**, 425
Factor Xa
  molecular modeling
    computer graphics, **4**, 425
Factor analysis, **4**, 15
  QSAR, **4**, 580
  test series
    Topliss manual method, **4**, 567
Factorial design
  process optimization
    pharmaceutical manufacture, **1**, 338
  test series
    compound selection, **4**, 569
Faith healing
  history, **1**, 626
Falintolol
  activity, **3**, 205
  glaucoma, **3**, 217
Fallopian tubes
  smooth muscle
    prostanoid receptors, **3**., 686
Familial hyperbilirubinemia
  glucuronide conjugation, **5**, 270
Familial hypercholesterolemia
  LDL receptors, **2**, 334
Family Practitioner Committees
  UK health care, **1**, 532
Family Practitioner Services
  UK National Health Service, **1**, 524
Famotidine
  clinical tests, **3**, 411
  histamine receptors, **1**, 275
    antagonist, **3**, 384
  lead structure
    development, **1**, 275
Farben, I.G.
  history, **1**, 93
Fast excitatory postsynaptic potential, **3**, 426
Fasting
  microsomal composition, **5**, 464
Fast protein liquid chromatography
  enzyme purification
    drug metabolism, **5**, 472
Fats
  small intestine, **1**, 148
Fatty acid desaturase
  cell membranes, **3**, 22
Fatty acid esters
  bioerodible
    drug absorption, **5**, 623
Fatty acids
  biosynthesis, **3**, 24
  drug absorption, **5**, 18
  polyunsaturated

drug toxicity, **5**, 326
  unsaturated
    cell membranes, **3**, 11
    phospholipase A$_2$, inhibition, **2**, 526
Fatty acid synthase
  fatty acids
    biosynthesis, **3**, 24
Fatty acyl chain distribution
  cell membranes, **3**, 11
Favism
  glucose-6-phosphate dehydrogenase deficiency, **5**, 262
Fazadinium
  neuromuscular-blocking agent, **3**, 477, 478
FCD
  chemical information database, **4**, 50
FCE 22101
  synthesis, **2**, 676
FCE 22509
  IP receptors
    agonist, **3**, 668
Febrifugine
  introduction, **1**, 64
Federal Food and Drug Act (1906)
  United States of America
    history, **1**, 89
Feeding
  brain, **1**, 156
Fenbufen
  adverse drug reactions, **1**, 642–647
  prescription–event monitoring, **1**, 642
Fendiline
  calcium channels, **3**, 1072
Fenethazine
  antihistamine, **3**, 281
Fenoctimine
  proton pump blockers, **2**, 205
Fenofibrate
  radiolabeled
    absorption, **5**, 372
Fenofibric acid
  solvent extraction, **5**, 483
Fenoldopam
  dopamine receptor agonists, **3**, 254, 308
    properties, **3**, 287
  dopamine receptor antagonists, **3**, 313
  food–drug interactions, **5**, 174
Fenoldopam mesylate
  properties, **3**, 316
Fenoprofen
  introduction, **1**, 22
  stereoselectivity
    inversion, **5**, 196
  triglycerides
    toxicity, **5**, 313
Fenoterol
  activity, **3**, 196, 211
  bronchodilator, **3**, 214
  clinical studies, **3**, 214
  introduction, **1**, 25
  β-receptor assay, **3**, 193
Fenoximone
  enzyme inhibition, **1**, 205
Fenpiprane
  introduction, **1**, 17
Fenprostalene
  FP receptors
    agonist, **3**, 665

G

G proteins — *see* Protein G
GABA — *see* Butanoic acid, 4-amino-
GABA aminotransferase
  glial uptake, **3**, 499
  neuronal uptake, **3**, 499
  receptors
    binding, **3**, 498
Gabaculine
  amino acid transaminase
    inhibitors, **2**, 244
  GABA-T
    inhibitors, **2**, 82, 242; **3**, 497
  inhibitor
    bacterial cell wall biosynthesis, **2**, 564
D-Galactal
  galactosidase inhibition, **2**, 74
D-Galactitol, 1,5-dideoxy-1,5-imino-
  glycosidases
    inhibition, **2**, 371
Galactocerebroside
  cell membranes, **1**, 175, 177
Galactosamine
  cofactor depletion, **5**, 133
β-Galactosidase
  encoding
    lacZ gene, **1**, 392
  enzyme immunoassay, **5**, 426
  fusion
    somatostatin gene, **1**, 425
  inhibition
    alkylation, **2**, 78
    cyclohexene polyol epoxides, **2**, 76
    D-galactal, **2**, 74
  insulin genes, **1**, 425
β-Galactoside permease
  encoding
    lacA gene, **1**, 392
β-Galactoside transacetylase
  encoding
    lacY gene, **1**, 392
Galegine
  introduction, **1**, 58
Gallamine
  introduction, **1**, 18
  M₂ receptors
    blocking agent, **3**, 440
  neuromuscular-blocking agent, **3**, 441, 477, 478
Gallic acid
  isolation, **1**, 11
Gametes
  drug metabolism, **5**, 239
Ganciclovir
  antiviral drug, **1**, 672
  DNA polymerases
    inhibition, **2**, 762
Ganglion-blocking agents
  neuromuscular-blocking agent, **3**, 452, 476
Gangliosides
  cell membranes, **1**, 177
  protein kinase C
    antagonist, **2**, 545
  structure, **3**, 9
Ganoderma lucidum
  traditional Chinese medicinal material, **1**, 101
Gap junctions
  epithelial cells, **3**, 4
Gardnerella vaginalis
  metronidazole, **2**, 728

Garlic
  platelet aggregation inhibition
    phospholipase A₂, **2**, 525
Gas chromatography
  automation, **5**, 435
  columns
    drug metabolites, **5**, 488
  derivatization
    drug metabolites, **5**, 490
  detectors
    drug metabolites, **5**, 489
  drug metabolism, **5**, 488, 491
  electron capture detectors
    drug metabolites, **5**, 489
  flame ionization detectors
    drug metabolites, **5**, 489
  nitrogen–phosphorus detectors
    drug metabolites, **5**, 490
  solid support materials
    drug metabolites, **5**, 488
  stationary phases
    drug metabolites, **5**, 488
  thermal conductivity detectors
    drug metabolites, **5**, 489
Gas exchange
  respiratory system, **1**, 142, 143
Gas–liquid chromatography
  drugs, **4**, 259; **5**, 395
  infrared spectroscopy
    drugs, **5**, 407
    mass spectrometry, **5**, 407
  mass spectrometry
    drugs, **5**, 407
Gas–liquid partitioning
  standard states, **4**, 246
Gassing mixture
  isolated perfused liver, **5**, 447
Gas transport
  respiratory system, **1**, 144
Gastric acid
  H⁺,K⁺-ATPase
    role in secretion, **2**, 194
  secretion
    inhibitors, **2**, 195
    prostaglandins, **2**, 151
Gastric balloons
  bioavailability, **5**, 628
Gastric emptying
  bioavailability, **5**, 165
  circadian variation, **5**, 286
  drug absorption, **5**, 11
  drug interactions, **5**, 166
  drug toxicity, **5**, 327
  food
    drug interactions, **5**, 174
  gastric retention devices, **5**, 628
  matrix dosage forms, **5**, 625
  pH, **5**, 169
  posture
    drug absorption, **5**, 12
  rate, **5**, 165
Gastric juice
  digestion, **1**, 147
Gastric lipase
  stomach, **1**, 147
Gastric retention devices
  bioavailability, **5**, 627
Gastric secretion

Glycine
  agonists, **3**, 519
  antagonists, **3**, 519
  biosynthesis, **3**, 518
  conjugation
    metabolic pathways, **5**, 157
    species differences, **5**, 233
  GABA interactions, **3**, 520
  GABA neurotransmitter, **3**, 518
  high affinity uptake, **3**, 519
  nervous system, **1**, 154
  pathophysiological role, **3**, 518
  physiological role, **3**, 518
  receptors, **3**, 519
  spinal cord, **3**, 519
Glycine, aminoethoxyvinyl-
  1-amino-1-carboxycyclopropane synthetase
    inhibition, **2**, 248
Glycine, *N*-chloroacetyl-*N*-phenyl-
  ethyl esters
    herbicidal activity, **4**, 547
Glycine, ethynyl-
  alanine racemase
    inhibitor, **2**, 246
Glycine, glutamyl-
  EAA receptors
    inhibitor, **3**, 525
Glycine, 3-methoxyvinyl-
  alanine transaminase
    inhibitor, **2**, 241
  glutamate decarboxylase
    inhibitor, **2**, 240
Glycine, propargyl-
  alanine transaminase
    inhibitor, **2**, 241
  amino acid transaminase
    inhibitor, **2**, 244
  cystathione γ-synthetase
    inhibitor, **2**, 248
  pyridoxal phosphate enzymes
    inhibitor, **2**, 82
Glycine, vinyl-
  amino acid transaminase
    inhibitors, **2**, 244
  reactions at C-4, **2**, 238
Glyciphosphoramide
  traditional Chinese medicine, **1**, 104
Glycocalyx
  cell membranes, **1**, 178
  cell walls, **2**, 556
Glycogen
  metabolism
    castanospermine, **2**, 378
Glycogenolysis
  histamine, **3**, 333
Glycogen phosphorylase
  glucose complex
    structure, **4**, 464
  pyridoxal-dependent systems, **2**, 239
Glycogen storage disease
  hyperuricemia, **2**, 468
Glycolic acid, phenylcyclohexyl-
  esters
    SAR at muscarinic receptors, **3**, 468
Glycolipids
  cell membranes, **1**, 175, 177; **3**, 5
Glycollate oxidase
  inhibition
    2-hydrobut-3-ynoate, **2**, 83
Glycols
  drug absorption, **5**, 27
Glycopeptides
  radiolabeled
    sulfur-35, **5**, 363

Glycophorin
  cell membranes, **3**, 22
Glycoproteins
  castanospermine, **2**, 376
  cell membranes, **3**, 5
  deoxynojirimycin, **2**, 376
  drug binding
    drug distribution, **5**, 80
  gene expression, **1**, 433
  hybrids
    swainsonine, **2**, 373
  processing
    neutral glycosidases, **2**, 366
  processing inhibition
    swainsonine, **2**, 372
  variant surface
    gene switching, **1**, 401
  viral
    swainsonine, **2**, 373
Glycopyrrolate
  antimuscarinic agent, **3**, 482
  peptic ulcer treatment, **3**, 481
  physiological effects, **3**, 481
Glycosaminoglycan
  sulfation
    Golgi apparatus, **1**, 184
Glycosidases
  hydrolysis, **2**, 366
  inhibitors, **2**, 68, 365
    biochemistry, **2**, 371
    occurrence, **2**, 369
  in metabolism, **2**, 366
  neutral
    glycoprotein processing, **2**, 366
    occurrence, **2**, 366
Glycosides
  drugs, **1**, 21
Glycosomes
  cell microbodies, **1**, 187
Glycosylation
  endoplasmic reticulum, **3**, 26
  gene expression, **1**, 433
  Golgi, **1**, 421
  proteins
    eukaryotes, **1**, 421
  species differences, **5**, 230
Glycosyl-phosphatidylinositol anchors
  structure and function, **3**, 125
Gnidimacrin
  protein kinase C, **2**, 544
Gnotobiotic animals
  animal experimentation, **1**, 661
  drug metabolism, **5**, 117
Gold compounds
  introduction, **1**, 51
  rheumatoid arthritis, **2**, 181, 468
Goldman–Hodgkin–Katz constant-field theory
  ion selectivity, **3**, 1050
Gold sodium thiosulfate
  antirheumatic agent, **1**, 51
Golgi apparatus
  cell structure, **1**, 183
  coated vesicles, **3**, 28
  structure, **3**, 3
Golgi I neurons
  nervous system, **1**, 153
Golgi II neurons
  nervous system, **1**, 153
Gonadocrinins
  placenta, **3**, 851
Gonadotrophin-releasing hormone, **1**, 287, 288,
    304–307, 311; **4**, 133
  associated peptide
    action, **1**, 287, 288, 304, 307

H

I

Ibopamine
  activity, **3**, 194
  congestive heart failure, **3**, 216
  dopamine receptor agonists
    prodrugs, **3**, 304
  properties, **3**, 317
Ibotenic acid
  NMDA agonist, **3**, 525, 527
Ibuprofen
  arachidonic acid
    metabolism, **2**, 160
  classification, **1**, 257
  clinical aspects
    stereoselectivity, **5**, 197
  drug absorption
    gastrointestinal tract, **3**, 632
  introduction, **1**, 22
  stereoselectivity
    pharmacokinetics, **5**, 196
  triglycerides
    toxicity, **3**, 313
Ice
  melting, **4**, 252
  structure, **4**, 252
Iceberg theory
  evidence against, **4**, 268
  water structure, **4**, 252
ICI 15995
  TP receptors
    antagonist, **3**, 678
ICI 80205
  EP receptors
    agonist, **3**, 662
  PGE_2 analog, **3**, 661
ICI 89406
  activity, **3**, 200, 209
ICI 118551
  activity, **3**, 205, 211
  tremor, **3**, 217
ICI 118587
  activity, **3**, 210
ICI 154129
  opioid antagonism, **3**, 835
ICI 162846
  clinical tests, **3**, 411
ICI 174864
  opioid
    δ-selective antagonist, **3**, 811, 835
ICI 176334
  androgen
    receptor, **3**, 1210
  antiandrogen, **3**, 1213
ICI 180080
  TP receptors
    antagonist, **3**, 678
ICI 185282
  TP receptors
    antagonist, **3**, 678
ICI 198615
  leukotriene receptors
    agonist, **3**, 766
    antagonist, **3**, 778
ICI 204219
  leukotriene receptors
    antagonist, **3**, 779
Icotidine
  activity, **3**, 397
  clinical tests, **3**, 406

structure–activity studies, **3**, 377
ICS 205-930
  5-HT receptors
    antagonist, **3**, 596
ICS 205-930, *N*-methyl-
  5-HT receptors, **3**, 594
Idazoxan
  α-adrenergic receptor agonist, **3**, 144
  α_2-adrenergic receptor antagonists, **3**, 139, 169
  antidepressant drug, **3**, 176
Ideal liquids
  solubility, **4**, 244
Ideal solubility
  definition, **4**, 246
Ideal solutions
  behavior, **4**, 243
  definition, **4**, 244
Idiotypes
  immune system, **1**, 227
Ido-swainsonine
  synthesis, **2**, 375
Idoxuridine
  antiviral agent, **2**, 315
  introduction, **1**, 75
IGF-I — *see* Somatomedin C
Ileostomy
  drug absorption, **5**, 14
Ilkovic equation
  polarography, **5**, 390
Iloprost
  EP receptors
    agonist, **3**, 664
  IP receptors
    agonist, **3**, 666
Image analysis
  autoradiography, **5**, 363
  whole-body autoradiography, **5**, 367
Imaging techniques
  drug distribution, **5**, 97
  monoclonal antibodies
    radionuclide conjugates, **5**, 673
Imazodan
  inhibitor
    phosphodiesterases, **2**, 507, 510, 511
Imidazole
  classification, **1**, 255
  2,3-disubstituted
    cimetidine synthesis, **1**, 323
  drug oxidation
    inhibitors, **5**, 215
  fungal resistance, **2**, 114
  radicals
    DNA strand-breakage, **2**, 741
  susceptibility, **2**, 92
Imidazole, nitro-
  classification, **1**, 256
  DNA damage, **2**, 726
  DNA strand-breakage, **2**, 739
  reduction
    hydrogenosomes, **1**, 187
Imidazole, 2-nitro-
  antitumor drugs, **2**, 727
  coulometry
    DNA damage, **2**, 740
  nitro radical anion, **2**, 741
  radiosensitizing drugs, **2**, 740
  ruthenium complexes
    radiosensitization, **2**, 743

J, K

Jaundice
  oral contraceptives, **5**, 267
Jet lag, **1**, 316
Joint Commission on Prescription Drug Use
  USA
    post-marketing surveillance, **1**, 637
Journal of Synthetic Methods
  chemical information database, **4**, 50

K-252a
  inhibitor
    protein kinases, **2**, 539
K-252b
  inhibitor
    protein kinase C, **2**, 545
K-351
  activity, **3**, 207
Kabikinase
  genetic engineering, **1**, 465
Kadsurenone
  platelet activating factor
    antagonist, **3**, 743, 747, 751
Kadsurin A
  platelet activating factor
    antagonist, **3**, 744
Kadsurin B
  platelet activating factor
    antagonist, **3**, 744
Kainic acid
  KAIN receptors, **3**, 524
  receptor agonist, **3**, 529
Kairin
  introduction, **1**, 14
Kairoline A
  introduction, **1**, 14
Kaiser–Permanente Group
  USA
    drug monitoring, **1**, 635
Kallidin
  biosynthesis, **3**, 976
Kanamycin
  minimum inhibitory concentrations, **5**, 415
  resistance, **2**, 92, 107
Kaolin
  chloroquine
    absorption, **5**, 172
  drug absorption, **5**, 16, 173
  drug interactions
    quinidine, **5**, 173
  gastrointestinal tract
    oral drug administration, **5**, 595
Kassinin
  activity
    tissue preparations, **3**, 1005
  isolation, **3**, 1001
Kasugamycin
  drug dependence, **2**, 116
  mode of action, **2**, 822
  resistance, **2**, 100, 101, 107, 822
  ribosomal RNA
    methylated residues, **2**, 820
Kelatorphan
  aminopeptidases
    inhibition, **3**, 808
  enkephalinase inhibitor, **2**, 411
Kernicterus
  glucuronides
    conjugation, **5**, 271

chemical reaction information, **1**, 734
current awareness, **1**, 750
online database, **1**, 742
REACCS database, **1**, 742
Juvenile diabetes
  autoimmune disease, **2**, 468
  cyclosporin, **2**, 466

Ketamine
  general anesthetic, **3**, 1092
  monitored release, **1**, 637
Ketanserin
  α_1-adrenergic receptor antagonists, **3**, 167
  5-HT receptors, **3**, 588
    agonist, **3**, 588
    antagonist, **3**, 592, 593
Ketanserin, 6-azido-
  histamine receptors
    photoaffinity studies, **3**, 339
Ketanserin, 7-azido-
  histamine receptors
    photoaffinity studies, **3**, 339
Ketanserin, 7-azido-8-iodo-
  histamine receptors
    photoaffinity studies, **3**, 339
Ketazocine
  opiate agonist, **3**, 826
  properties, **3**, 809
Ketazocine, ethyl-
  opioid receptor
    model, **3**, 836
  κ-selective opioid ligands, **3**, 830
Kethoxal
  chemical footprinting
    ribosomal binding sites, **2**, 830
Ketoconazole
  antifungal agent, **2**, 140, 354, 356–358
  cholesterol
    inhibition, **2**, 340
  drug interactions
    toxicity, **5**, 328
  enzyme inhibition, **5**, 328
  fungal resistance, **2**, 114
  gastrointestinal tract
    dissolution, **5**, 618
  metabolism, **5**, 315
  steroid hormones
    inhibition, **2**, 348
Ketocyclazocine, ethyl-
  opioid
    κ-agonist, **3**, 813
  opioid receptors
    binding studies, **3**, 810
Ketones, β-amino-
  dopamine
    agonists, **3**, 267
Ketones, chloromethyl
  binding, **2**, 70
Ketones, α-diazomethyl
  pyroglutamyl peptide hydrolase
    inhibition, **2**, 76
Ketones, α-halomethyl
  enzyme inhibition
    alkylation, **2**, 78
Ketones, methyl ethyl
  drug absorption, **5**, 27
Ketones, peptidyl halomethyl

L

M

M₁ receptors
M_1 receptors
  antagonists, **3**, 473
  CNS, **3**, 438
  coupling to second messengers, **3**, 439
  electrophysiology, **3**, 439
  heterogeneity, **3**, 439, 445
  location, **3**, 438
  peripheral ganglia, **3**, 438
  presynaptic location, **3**, 438
  SAR for agonists, **3**, 465
M_2 receptors
  CNS, **3**, 440
  coupling to second messengers, **3**, 440
  electrophysiology, **3**, 440
  heterogeneity, **3**, 440
  location, **3**, 440
  peripheral effector organs, **3**, 440
  presynaptic location, **3**, 440
  subtypes, **3**, 442
  sympathetic ganglia, **3**, 440
M-7
  α_2-adrenergic receptor agonists, **3**, 163
M13 bacteriophage
  DNA chain, **1**, 192
Mabuterol
  activity, **3**, 197
MACCS
  chemical information system, **4**, 46
  ChemQuest database, **1**, 743
  substructure searching, **4**, 651
Macrolide antibiotics
  cytochrome P-450
    inhibition, **2**, 135
  susceptibility, **2**, 92
MACROMODEL
  molecular modeling
    software, **4**, 42
Macromolecules
  structure
    molecular graphics, **4**, 459–494
  targets
    drug action, **1**, 195–208
Macrophage colony stimulating factor
  genetic engineering, **1**, 467
Macrophages
  drug targeting, **5**, 665, 681
  immune system, **1**, 224, 227
  liver
    drug targeting, **5**, 674
  oxidative burst
    platelet activating factor, **3**, 724
  platelet activating factor, **3**, 723
  structure, **3**, 3
  superoxide generation
    platelet activating factor, **3**, 724
  surface antigens, **1**, 224
Macrotetralides
  cell membrane
    permeability, **2**, 598
Madrid Agreement (1891)
  trade marks, **1**, 709
Maduramicin
  coccidostatic activity, **2**, 602
Magnesium
  biochemical competition, **2**, 178
  tetracyline absorption, **5**, 172
Magnesium carbonate
  drug interactions

quinidine, **5**, 173
Magnesium hydroxide
  drug absorption, **5**, 173
  drug interactions
    nitrofurantoin, **5**, 173
    penicillamine, **5**, 174
    phenytoin, **5**, 173
    quinidine, **5**, 173
Magnesium oxide
  drug interactions
    phenytoin, **5**, 173
Magnesium perhydrol
  digoxin bioavailability, **5**, 172
Magnesium stearate
  capsules, **5**, 578
  lubricant
    drug absorption, **5**, 17
Magnesium trisilicate
  chloroquine
    absorption, **5**, 172
  drug absorption, **5**, 173
  drug interactions
    nitrofurantoin, **5**, 173
    phenytoin, **5**, 173
    quinidine, **5**, 173
Magnetism
  microspheres
    drug targeting, **5**, 691
Magnetite
  microspheres
    drug targeting, **5**, 691
Mahalanobis metrics
  pattern recognition
    colinear data, **4**, 713
Mail order pharmacy
  USA, **1**, 540
Malabsorption syndromes
  drug interactions, **5**, 168
Malaria
  purine metabolism, **2**, 474
  vaccines
    genetic engineering, **1**, 471
Malate dehydrogenase
  cytoplasmic
    inhibitors, **4**, 573
L-Malate hydrolase
  nomenclature, **2**, 36
Malathion
  metabolism
    blood, **5**, 122
  neurotoxicity, **5**, 306
  toxicity
    species differences, **5**, 322
  toxicokinetics, **5**, 306
Maleates
  partition coefficients
    calculation, **4**, 317
Maleic acid
  amino acid transaminase
    inhibitor, **2**, 240
  partition coefficients
    calculation, **4**, 317
Maleic anhydride/divinyl ether copolymer
  immunomodulatory drug, **1**, 237
Maleimide, N-ethyl-
  adenosine receptors
    assay, **3**, 612
Malic acid

CSA and drug design, **4**, 570
depression, **5**, 266
electronic effects, **4**, 225
introduction, **1**, 63
properties, **1**, 313
role of cyclopropane ring in, **1**, 2, 207
type A, **2**, 125
5-HT metabolism, **3**, 569
type B, **2**, 125
Monobactams, **2**, 656
detection, **2**, 658
discovery, **2**, 656
hetero-substituted, **2**, 660
resistance, **2**, 114
structure–activity relationships, **2**, 659
Monobactams, 4-alkyl-
synthesis, **2**, 659
Monocarbams
activity, **2**, 661
Monoclonal antibodies
chemotherapy, **2**, 326
conjugation, **5**, 673
drug targeting, **5**, 668
enzyme purification, **5**, 472
hormones, **1**, 289
immunoaffinity chromatography
membrane receptors, **3**, 86
immunoassay, **5**, 428
immunoprecipitation, **3**, 86
immunotherapy, **1**, 236
membrane receptors, **3**, 93
radionuclide conjugates, **5**, 673
receptors, **3**, 76
Monocrotaline
hepatotoxicity, **5**, 330
pulmonary toxicity, **5**, 137
Monocytes
immune system, **1**, 224
Monokines
genetic engineering, **1**, 466
Monooxygenases
neonatal, **5**, 245
Monophosphams
activity, **2**, 661
Monopole bond polarizability
drug–receptor interactions, **4**, 329
Monosaccharides
small intestine, **1**, 148
Monosulfactams
synthesis, **2**, 661
Monte Carlo methods
conformation, **4**, 132
molecular structures, **4**, 179
Monthly Index of Medical Specialties
medical research
UK, **1**, 495
Moprolol
stereoselectivity
pharmokinetics, **5**, 190
MORGAN algorithm
information storage and retrieval, **4**, 36
Morin rearrangement
penicillins, **2**, 626
Moroxella spp.
P-450 isoenzymes, **5**, 223
Morphiceptin
conformation
receptor-bound, **4**, 451
opioid activity, **3**, 822
Morphinan
analgesic activity
predictions, **4**, 616
derivatives
structure–activity relationships, **3**, 826

Morphinan, *N*-methyl-
introduction, **1**, 12
Morphine, **1**, 12, 16, 287; **4**, 67, 69, 90
addiction
role in development of medicinal chemistry, **1**, 1
analgesic effect
Hill equation, **5**, 341
link model, **5**, 348
analogs
analgesic activity, **4**, 500
binding, **4**, 94
buccal absorption, **5**, 32
chemical modification, **1**, 1
clinical properties, **3**, 838
congeners
structure–activity relationships, **3**, 823
conjugation
metabolic pathways, **5**, 154
derivatives
opioid activity, **3**, 823
dimers
opioid activity, **3**, 823
drug–receptor dissociation time, **5**, 351, 353
electrostatic potentials, **4**, 117
epidural administration, **5**, 597
ethyl ether
introduction, **1**, 11
extraction, **5**, 484
fluorescence immunoassay, **5**, 425
histamine releaser, **3**, 329
ion pair extraction, **5**, 485
isolation, **1**, 9, 11
metabolism
blood, **5**, 122
kidneys, **5**, 120
nervous system, **1**, 159
μ-opioid receptor selectivity, **3**, 828
pharmacological response
two-compartment models, **5**, 515
properties, **3**, 806, 809
radioimmunoassay kits, **5**, 373
steric models, **1**, 2
toxicity, **4**, 3
Morphine, diacetyl-
introduction, **1**, 11
Morphine, ethyl-
metabolism
gastrointestinal tract, **5**, 114
kidneys, **5**, 120
Morphinone, methyldihydro-
introduction, **1**, 11
Morson, Thomas
history, **1**, 83
Mortality
toxicity testing, **1**, 574
Motilin, **1**, 314, 315
secretion, **1**, 306
Motivation
personnel
good manufacturing practice, **1**, 556
Motor impulses
nervous system, **1**, 153
Mouth
digestion and absorption, **1**, 145
Moxalactam
activity, **2**, 641
resistance, **2**, 102, 114
side effects, **2**, 642, 685
MR 2033
opiate antagonist, **3**, 839
MR 2266
κ-selective opioid ligands, **3**, 830
MSA — *see* Molecular shape analysis
MSD — *see* Minimal steric difference

N

N-0057
 DP receptors
  antagonist, **3**, 671
N-0161
 DP receptors
  antagonist, **3**, 671
N-0164
 DP receptors
  antagonist, **3**, 671
 EP receptors
  antagonist, **3**, 673
 PGD$_2$ antagonist, **3**, 660
 TP receptors
  antagonist, **3**, 675
N-0434
 dopamine
  agonist, **3**, 247
N-0437
 dopamine agonist, **3**, 247
  properties, **3**, 285
 radiolabeled
  dopamine assay, **3**, 238
N-0500
 dopamine agonist, **3**, 249
  properties, **3**, 285
  stereochemistry, **3**, 272
Nabilone
 introduction, **1**, 23
 polymorphic forms
  drug dosage, **5**, 555
Nachschlag phenomenon, **3**, 427
NAD
 molecular modeling, **4**, 438
NADH
 Krebs cycle, **1**, 185
NADH-cytochrome b_5 reductase
 cytochrome P-450
  reconstitution, **5**, 472
Nadolol
 clinical studies, **3**, 213
NADP
 drug toxicity, **5**, 326
NADPH
 binding
  dihydrofolate reductase, **2**, 284
 microsomal fraction, **5**, 469
 S9, **5**, 464
NADPH cytochrome P-450 reductase
 cytochrome P-450
  reconstitution, **5**, 472
 drug reduction
  metabolic pathways, **5**, 150
 free radicals
  generation, **5**, 134
 gut wall activity, **5**, 14
Nafcillin
 resistance, **2**, 103
Nafidimide
 intercalating agent
  DNA, **2**, 704
Naftifine
 squalene oxidase
  inhibition, **2**, 359
Nagana red
 introduction, **1**, 70
Nalbuphine
 buccal absorption, **5**, 32
 introduction, **1**, 12

Nalidixic acid
 adverse drug reactions, **1**, 630
 antibacterial activity, **2**, 773
  QSAR, **4**, 549
 DNA topoisomerases
  inhibition, **1**, 374
 microbiological assay, **5**, 418
 resistance, **2**, 101
Nalorphine
 introduction, **1**, 12
 opioid antagonist, **3**, 831
Naloxone
 binding affinity
  opioid receptors, **3**, 837
 opioid antagonist, **3**, 58, 832
 μ-opioid receptor selectivity, **3**, 828
Naltrexamine
 opioid antagonist activity, **3**, 823
Naltrexone
 opioid antagonist, **3**, 832
Nantradol, deacetyl-
 DP receptors
  antagonist, **3**, 671
 EP receptors
  antagonist, **3**, 674
Naphazoline
 α-adrenergic receptor agonists
  structure–activity relationships, **3**, 151
 introduction, **1**, 26
Naphthalene, **4**, 26
 introduction, **1**, 59
 partition coefficients
  calculation, **4**, 311
Naphthalene, 2-amino-5,6-dihydroxy-1,2,3,4-tetra-
   hydro-
 dopamine receptor agonists
  structure–activity relationships, **3**, 302
Naphthalene, 2-amino-6,7-dihydroxy-1,2,3,4-tetra-
   hydro-
 dopamine receptor agonists, **3**, 246
  structure–activity relationships, **3**, 302
Naphthalene-1,2-diol, 6-amino-1,2,3,4-tetra-
   hydro-
 dopamine receptor agonists
  structure–activity relationships, **3**, 302
2-Naphthaleneethanamine, N,N-dimethyl-4-(2-
   fluorophenyl)-
 extraction, **5**, 487
Naphthalenesulfonamides
 inhibitors
  protein kinases, **2**, 533, 538
2-Naphthoflavone
 drug metabolism
  blastocysts, **5**, 240
 enzyme induction, **5**, 212
1-Naphthoic acid
 acetate
  metabolism, blood, **5**, 122
1-Naphthol
 metabolism
  gastrointestinal tract, **5**, 117
Naphthoquinones
 antitumor drugs, **2**, 728
Naphthoxazine
 stereochemistry, **3**, 272
Naphthoxazines, hexahydro-
 dopamine agonists
  properties, **3**, 285

O

Obesity
  β-agonists, **3**, 210
Obofluorin
  activity, **2**, 694
Occipital lobe
  brain, **1**, 155
Occlusive vascular disease
  prostanoids, **3**, 698
Occupational Safety and Health Administration
  funding of medical research
    USA, **1**, 511
Ocimum sanctum
  Ayurvedic medicine, **1**, 126
OCSS
  chemical reaction library, **1**, 743
Octanoates
  blood–brain barrier, **4**, 402
Octanoic acid
  metabolism
    kidneys, **5**, 120
Octanol
  membranes
    model, **4**, 287
  partitioning, **4**, 284
  solvent
    partitioning, **4**, 284, 287
Octapeptins
  cell membrane disorganization, **2**, 590
  resistance, **2**, 110
  structure, **2**, 590
Octoclothepin
  dopamine
    agonist, **3**, 262
  stereochemistry, **2**, 273
Octyl glucoside
  membrane receptors
    solubilization, **2**, 82
  protein solubilization, **2**, 20
Oculomucocutaneous syndrome
  practolol, **1**, 570, 571
Odor
  prodrugs, **5**, 126
Office of Health Economics
  UK health care, **1**, 530
Office of Population Censuses and Surveys
  UK
    drug monitoring scheme, **1**, 639
Ofloxacin
  DNA gyrase
    inhibitor, **2**, 774
  resistance, **2**, 101
Ointments
  formulation, **5**, 574
  manufacture
    validation, **1**, 566
OK-432
  clinical trials, **3**, 1120
Okazaki fragments
  DNA polymerases, **2**, 757
  synthesis
    DNA primase, **2**, 757
OKY-1581
  myocardial infarction, **2**, 163
Old yellow enzyme
  nomenclature, **2**, 34
Oleandomycin
  resistance, **2**, 109
Oleandomycin, triacetyl-

isozyme *P*-450 III induction, **5**, 329
Oleandrin
  gastrointestinal irritability
    prodrugs, **5**, 127
Oleic acid
  protein kinase C, **2**, 544
  transdermal drug delivery
    penetration enhancement, **5**, 656
Olfactory compounds
  QSAR, **4**, 539
Oligodendrocytes
  nervous system, **1**, 154
Oligonucleoside methylphosphonates
  antiviral agents, **2**, 327
Oligonucleotides
  site-specific mutagenesis
    DNA, **1**, 441
  synthesis
    chemotherapy, **2**, 326
Oligosaccharides
  anticoagulant, **2**, 490
Olivanic acids
  activity, **2**, 671
  biosynthesis, **2**, 671
  detection, **2**, 662, 671
  history, **2**, 656
Olive oil
  cell membrane
    partitioning, **4**, 284
  lipid bilayer
    partitioning, **4**, 284
  partitioning, **4**, 284
Olivomycin
  nonintercalating agent
    DNA, **2**, 709
Omeprazole
  H$^+$,K$^+$-ATPase
    inhibition, **2**, 198
  lead structure
    development, **1**, 272
  mode of action, **2**, 198
  structure–activity relationships, **2**, 202
  therapeutic activity, **2**, 204
Oncogenes
  gene expression, **1**, 439
  growth factors
    receptors, **1**, 439
Online searching
  medicinal chemistry information, **1**, 752
ONO 3708
  thromboxane A$_2$
    binding sites, **3**, 657
  TP receptors
    antagonist, **3**, 678
ONO 11120
  TP receptors
    antagonist, **3**, 678
ONO-RS-347
  leukotriene receptors
    antagonist, **3**, 778
ONO-RS-411
  leukotriene receptors
    antagonist, **3**, 778
Oocyte maturation inhibitor, **1**, 311
Operators
  gene expression
    eukaryotes, **1**, 422
    prokaryotes, **1**, 415

P

P388 mouse leukemia, **1**, 280–283
  anticancer activity
    screening, **1**, 279
  Venn diagrams
    drug development, **4**, 569
P4S
  GABA aminotransferase
    agonists, **3**, 504, 509
  GABA-A receptors
    benzodiazepine interactions, **3**, 514
  pharmacokinetics, **3**, 515
PABA — *see* Benzoic acid, *p*-amino-
Pachystermines
  discovery, **2**, 655
Packed columns
  gas–liquid chromatography
    drugs, **5**, 395
PADAC
  chromogenic cephalosporin, **2**, 626
Paget's disease
  bone
    calcitonin, **3**, 1030
Pain
  gastric emptying
    drug absorption, **5**, 12
  injection site
    prodrugs, **5**, 127
  prostanoids, **3**, 696
Palladium on charcoal
  pharmaceutical manufacture
    safety, **1**, 348
Palmitic acid
  metabolism
    kidneys, **5**, 120
PAM —*see* Pyridinium-2-carbaldoxime chloride, *N*-methyl-
Pamaquin
  introduction, **1**, 71
Pamphlets
  drug marketing, **1**, 91
Panax ginseng
  traditional Chinese medicinal material, **1**, 100, 101, 109
Pancreas
  extracts, **1**, 26
  insulin, **1**, 293
  physiology, **1**, 162
Pancreastatin, **1**, 287, 314
  properties, **1**, 293
Pancreatic juice
  control, **1**, 148
  digestion, **1**, 148
Pancreatic polypeptide, **1**, 314
Pancreozymin
  discovery, **3**, 929
Pancuronium
  introduction, **1**, 19
  muscle twitch
    link model, **5**, 348
  neuromuscular-blocking agent, **3**, 441, 477, 478
  pharmacodynamics
    Hill equation, **5**, 340
Pantothenate
  absorption, **5**, 372
Papain
  amide binding
    Free–Wilson analysis, **4**, 628
  binding, **2**, 70

computer graphics, **4**, 25
  inhibition, **2**, 431
    QSAR, **4**, 494
  ligand interactions
    Free–Wilson analysis, **4**, 628
  molecular modeling
    computer graphics, **4**, 423
  nomenclature, **2**, 34
  QSAR, **4**, 14
    computer graphics, **4**, 482
Papaverine, **1**, 12, 17
  absorption
    perfused intestine, **5**, 452
  adenosine
    inhibitor, **3**, 603
  inhibitor
    phosphodiesterases, **2**, 510
  isolation, **1**, 12
  partial synthesis, **1**, 1
Papaver somniferum
  alkaloids, **1**, 9
Papillary duct
  kidney, **1**, 149
PAPP
  5-HT receptors, **3**, 571
Papulacandins
  activity
    Candida albicans, **2**, 589
  antifungal activity, **2**, 588
  glucan synthase inhibitors, **2**, 587
Paracetamol (acetaminophen)
  absorption, **5**, 8
    circadian rhythm, **5**, 288
    gastric emptying, **5**, 327
  classification, **1**, 252
  cyclooxygenase
    inhibition, **2**, 161
  covalent binding, **5**, 133
  detoxication, **5**, 314
  dissolution, **4**, 376
    absorption, **5**, 564
  drug interactions, **5**, 168
    propantheline, **5**, 168
  gastric emptying
    drug absorption, **5**, 11
  glucuronidation, **5**, 208
  hepatotoxicity
    species differences, **5**, 322
  high performance liquid chromatography, **5**, 493
  introduction, **1**, 59
  metabolic activation, **5**, 329
  metabolism, **5**, 132
    circadian rhythm, **5**, 290
    cigarette smoking, **5**, 213
    isolated perfused kidney, **5**, 451
    kidneys, **5**, 120
  metoclopramide interaction, **5**, 166
  partition coefficient
    drug transport, **4**, 387
    routes of administration, **4**, 392
  QSAR
    double peaked, **4**, 394
  radiolabeled
    glutathione hepatic depletion, **5**, 371
  toxicity, **5**, 214
    alcohol induced, **5**, 213
    prodrug design, **5**, 132
    species differences, **5**, 322

154

Piperidine-4-sulfonic acid
  GABA aminotransferase
    receptor binding, **3**, 499
Piperidolate
  introduction, **1**, 16
Piperonyl butoxide
  delayed neurotoxicity
    species differences, **5**, 317
Piperoxan
  α-adrenergic receptor antagonists
    structure–activity relationships, **3**, 164
  histamine receptors
    antagonist, **3**, 369
  introduction, **1**, 26, 35
Pirbuterol
  activity, **3**, 197
  β-adrenoceptor partial agonist, **1**, 202
  congestive heart failure, **3**, 216
  introduction, **1**, 25
Pirdonium bromide
  histamine receptors
    antagonist, **3**, 370
Pirenperone
  5-HT receptors, **3**, 588
    antagonist, **3**, 592
Pirenzepine
  antimuscarinic activity, **3**, 437, 472
  M_1 blockers
    gastric acid secretion, **2**, 196
  M_1 receptor antagonist, **3**, 481
  muscarinic antagonist
    bioselectivity, **1**, 216
Piretanide
  introduction, **1**, 60
Piribedil
  dopamine
    agonist, **3**, 254
Piritramide
  activity, **1**, 17
Piritrexem
  dihydrofolate reductase
    inhibitor, **2**, 283, 286, 288
Piroxicam, **1**, 268
  adverse drug reactions, **2**, 642–648
  lead structure
    discovery, **2**, 268
  prescription–event monitoring, **2**, 642
Pitrazepin
  GABA aminotransferase
    antagonist, **3**, 509
Pituitary
  brain, **1**, 156
  extracts, **1**, 27
  function
    diagnosis, **1**, 305
  hormones, **1**, 285, 288, 307
    release, **1**, 303
  physiology, **1**, 160
  portal system
    physiology, **1**, 160
Pituitary gland
  dopamine
    receptors, **3**, 237
  gonadotropin releasing hormone
    receptors, **3**, 850
Pivampicillin
  introduction, **1**, 42
  metabolism
    gastrointestinal tract, **5**, 114
  prodrug, **4**, 406
  synthesis, **2**, 622
Pivmecillinam
  oral absorption, **2**, 629
Pizotifen

5-HT receptors
  antagonist, **3**, 591
PK
  chemical information database, **4**, 50
pK_a
  calculation
    computers, **4**, 44
  skin
    drug absorption, **5**, 27
Placenta
  drug metabolism, **5**, 112
  drug transfer, **5**, 310
    whole-body autoradiography, **5**, 367
  glucuronidation, **5**, 244
  hormones, **1**, 288, 292, 311
  protein 12, **1**, 316
  protein 14, **1**, 316
Planning
  health care
    USA, **1**, 539
Plant acids
  antiseptics and anticoagulants, **1**, 20
Plant collection
  Ayurvedic medicine
    drugs, **1**, 121
Plant growth regulators
  aryloxypropionic acids, **4**, 191
Plants
  active principles, **1**, 9, 11
Plasma
  biological assay, **5**, 418
  drug oxidation, **5**, 460
  human body, **1**, 163, 164
  unbound drug concentration, **5**, 353
Plasma membrane
  drug metabolism, **5**, 471
  fractions
    preparation, **3**, 5
  function, **3**, 81
Plasma proteins
  drug binding
    circadian rhythm, **5**, 288
    food–drug interactions, **5**, 174
Plasma thromboplastin
  genetic engineering, **1**, 468
Plasmalogens
  structure, **3**, 9
Plasmids
  cloning, **1**, 405
  DNA, **1**, 404
    isolation, **1**, 407
    supercoiling, **1**, 374
  drug resistance, **2**, 116
  β-lactamases
    coding, **2**, 614
  penicillin resistance, **2**, 629
  protein expression
    bacteria, **1**, 410
Plasmin
  derivation from plasminogen, **2**, 492
  fibrinolysis, **2**, 492
  inhibitors, **2**, 494
    benzamidines, **4**, 617
    drugs, **2**, 496
    Free–Wilson analysis, **4**, 616
Plasminogen activators, **2**, 492
  action on plasminogen, **2**, 492
  biosynthesis, **2**, 496
  drugs, **2**, 496
  genetic engineering, **1**, 464
  immunochemical classification, **2**, 493
  inhibitors, **2**, 494
  modified activators, **2**, 496
  prourokinase

PR-879-3177A
 clinical trials, **3**, 1118
Practolol
 activity, **3**, 209, 211
 acute myocardial infarction, **3**, 216
 adverse drug reactions, **1**, 628, 636
  characteristic symptoms, **1**, 629
  underreporting, **1**, 629
 cardioselective β-antagonist, **3**, 201
 introduction, **1**, 25
 marketing
  USA, **1**, 631
 oculomucocutaneous syndrome, **1**, 571
 post-marketing surveillance
  effectiveness, **1**, 636
 side effects, **3**, 218
 toxicology, **1**, 570
Prasinomycins
 phosphoglycolipid antibiotic, **2**, 578
Prazosin
 α₁-adrenergic receptor antagonists
  structure–activity relationships, **3**, 165
 α-adrenergic receptors
  agonist, **3**, 144
  inhibition, **3**, 146
 antihypertensive agent, **4**, 551
 binding, **4**, 366
Prealbumin
 binding, **4**, 136
 molecular modeling
  computer graphics, **4**, 425, 426
  solvation energy, **4**, 21
 thyroid hormone interactions
  computer graphics, **4**, 474
 transport, **3**, 1149
Precipitation
 membrane receptors, **3**, 87
Prednisolone
 drug distribution
  single-compartment model, **5**, 504
 first-pass metabolism
  prodrugs, **5**, 129
 introduction, **1**, 33
 metabolism
  diurnal variation, **5**, 313
  gut microflora, **5**, 118
 pharmacodynamics
  tissue compartments, **5**, 348
 solubility
  prodrugs, **5**, 128
Prednisolone, methyl-
 introduction, **1**, 33
 manufacture, **1**, 357
Prednisone
 drug disposition
  liver disease, **5**, 274
 introduction, **1**, 32
Predoctoral training
 funding
  USA, **1**, 512
Preferred Provider Organization
 role in health care
  USA, **1**, 538
Prefertilization
 drug metabolism, **5**, 239
Pregnancy
 drug metabolism, **5**, 239, 247
 jaundice, **5**, 267
α₁-Pregnancy-associated endometrial protein, **1**, 316
Pregnanedione
 anesthetic, **1**, 68
Pregn-4-ene-3,20-diones
 glucocorticoid activity
  Free–Wilson analysis, **4**, 612

 Fujita–Ban analysis, **4**, 614
Pregnenolone-16α-carbonitriles
 drug metabolism
  sex differences, **5**, 246
 isozyme *P*450 III induction, **5**, 329
Preimplantation embryos
 drug metabolism, **5**, 240
Premenstrual tension
 endogenous opioid production, **1**, 310
Premises
 siting
  good manufacturing practice, **1**, 559
 structure
  good manufacturing practice, **1**, 559
Premutagens
 drug metabolism, **5**, 239
Prenalterol
 activity, **3**, 209, 211
 β-adrenoceptor activation, **1**, 213
 agonist/antagonist, **1**, 196
 congestive heart failure, **3**, 216
 intrinsic efficacy, **1**, 209
 β-receptors
  agonism, **3**, 192
Prenylamine
 calcium channels, **3**, 1072
 introduction, **1**, 17
PREP automated sample processor
 drug analysis
  biological fluids, **5**, 435
Preprocholecystokinin
 biosynthesis, **3**, 929
Preproglucagon
 biosynthesis, **3**, 909
 human gene
  sequence, **3**, 909
Preproinsulin
 insulin production, **1**, 457
 structure, **3**, 902
Preproparathyroid hormone
 biosynthesis, **3**, 1031
 structure, **3**, 1035
Preprosecretin
 structure, **3**, 947
Prescription–event monitoring
 drugs, **1**, 617
 questionnaires
  response, **1**, 642
 UK
  methodology, **1**, 640
  post-marketing surveillance, **1**, 640
Prescription Pricing Authority
 UK
  adverse drug reactions, **1**, 635
  drug monitoring scheme, **1**, 638
Prescriptions
 UK National Health Service, **1**, 495
 USA, **1**, 546
Preservatives
 solution formulations, **5**, 569
 suspensions, **5**, 571
Pressure
 control
  pilot plant, **1**, 347
 pharmaceuticals
  manufacture, **1**, 337
Preswell dilution technique
 erythrocytes
  drug targeting, **5**, 693
Presynaptic receptors
 axon terminals, **3**, 137
Prices
 pharmaceutical industry, **1**, 487, 490
Pridinol

Q

R

R 2061
  progestin
    receptor, **3**, 1201
R 5020
  progestin, **3**, 1211
  receptor
    molecular mechanisms, **3**, 1206
R 5135
  GABA aminotransferase
    antagonist, **3**, 509
R 28935
  α_1-adrenergic receptor antagonists, **3**, 168
R 56413
  5-HT receptors
    antagonist, **3**, 592
Rabbits
  drug absorption
    determination, **5**, 35
  isolated rectococcygens muscle
    dopamine receptors, **3**, 242
  splenic artery
    dopamine receptors, **3**, 242
RAC 109-I
  sodium channels
    structure–activity relationships, **3**, 1066
RAC 109-II
  sodium channels
    structure–activity relationships, **3**, 1066
Racemases
  classification, **2**, 36
  inhibitors, **2**, 246
Racemates
  pharmacokinetics, **5**, 188
Raclopride
  dopamine
    agonist, **3**, 270
β-Radiation
  analysis, **5**, 360
γ-Radiation
  analysis, **5**, 360
Radiation exposure
  analysis, **5**, 361
Radical reactions
  drugs, **4**, 214
Radioactive derivatizing agents
  use, **5**, 372
Radioactive probes
  gene location, **1**, 406
Radiochemical purity
  analysis, **5**, 361, 363
Radioimmunoassay, **5**, 372
  accuracy, **5**, 361
  centrifugal analyzers, **5**, 437
  drugs, **5**, 422
  glibenlamide, **5**, 427
  hormones, **1**, 289, 295
  human chorionic gonadotrophin, **1**, 297
  kits, **5**, 373
  optimization, **5**, 424
  second antibody method, **5**, 425
  steroids, **1**, 296
  tritium, **5**, 425
Radioisotopes
  analysis, **5**, 360
    advantages, **5**, 361
    disadvantages, **5**, 361
  labeled compounds, **5**, 360
    synthesis, **5**, 361

  measurement, **5**, 363
Radiolabeled drugs, **5**, 423
  distribution, **5**, 97
  drug metabolism
    isolation, **5**, 481
  physiology, **5**, 372
Radioligands
  binding studies, **4**, 362
    membrane receptors, **3**, 82
    receptors, **3**, 58
  histamine receptors, **3**, 336
Radioreceptor assays, **5**, 373, 420
  advantages and disadvantages, **5**, 421
  benzodiazepines, **5**, 421
Radiotherapy
  monoclonal antibodies
    radionuclide conjugates, **5**, 674
Radix *Aconiti carmichaeli*
  traditional Chinese medicinal material, **1**, 100, 101
Radix *Aconiti kusnezoffii*
  traditional Chinese medicinal material, **1**, 102
Radix *Angelicae sinensis*
  traditional Chinese medicinal material, **1**, 109
Radix *Astragali membranacei*
  traditional Chinese medicinal material, **1**, 109
Radix *Codonopsis pilosulae*
  traditional Chinese medicinal material, **1**, 109
Radix *Puerariae lobatae*
  traditional Chinese medicinal material, **1**, 107
Radix *Rhei palmati*
  traditional Chinese medicinal material, **1**, 103
Radix *Salviae miltiorrhizae*
  traditional Chinese medicinal material, **1**, 108
Raman spectroscopy
  proteins
    secondary structure, **2**, 22
Ramixotidine
  clinical tests, **3**, 407
Randic index
  biological phenomena, **4**, 22
  Kier and Hall problems, **4**, 23
  QSAR, **4**, 22
Random selection
  screening
    anticancer activity, **1**, 280
Random walk
  concept, **4**, 9
Raney nickel
  pharmaceutical manufacture
    safety, **1**, 348
RANGEX
  biosteric substituent selection, **4**, 570
Ranitidine, **1**, 273, 275
  blood–brain barrier, **3**, 404
  clinical tests, **3**, 397, 406
  drug interactions
    antipeptic agent, **5**, 328
  histamine H_2 receptor antagonist, **1**, 198; **3**, 384
  indexing
    databases, **1**, 725
  introduction, **1**, 35
  lead structure
    development, **1**, 275
  manufacture
    intermediates, **1**, 328
    thermal hazards, **1**, 348
  peptic ulcers, **2**, 204
  post-marketing surveillance, **1**, 649

S

S9
  drug metabolism, **5**, 463
    kinetics, **5**, 464
  isolation, **5**, 463
S-1623
  fibrinolysis, **2**, 497
Saccharin, *N*-acyl-
  enzyme acylation, **2**, 79
Saccharomyces
  genetic engineering, **1**, 457
Saccharomyces cerevisiae
  cell envelope
    drug susceptibility, **2**, 97
  gene expression
    protein production, **1**, 418
  P-450 isoenzymes, **5**, 223
  protein production
    gene expression, **1**, 418
Safety
  animal experimentation, **1**, 665
  data sheets
    pharmaceuticals, **1**, 351
  medical research, **1**, 502
  pharmaceuticals
    manufacture, **1**, 325, 341
  QSAR, **4**, 13
  testing
    pilot plant, **1**, 349
Safrole
  sulfation
    metabolic pathways, **5**, 155
Safrole, dihydro-
  topological descriptors, **4**, 181
Salbutamol
  activity, **3**, 197, 211
  asthma, **5**, 594
  bronchodilator, **3**, 214
  clinical studies, **3**, 214
  congestive heart failure, **3**, 216
  inhibitor
    platelet activating factor, **3**, 731
  introduction, **1**, 24
  profile, **3**, 219
  pulmonary administration, **5**, 33, 604
  β-receptors
    selectivity, **3**, 192
Salicin, **1**, 59
  cartels, **1**, 92
  isolation, **1**, 22
Salicylamide
  drug absorption
    rectum, **5**, 20
  intestinal enzymes
    inhibition, **5**, 14
  metabolism
    gastrointestinal tract, **5**, 114
    toxicity, **5**, 314
  sulfoconjugation
    dogs, **5**, 231
Salicylates
  absorption
    charcoal, **5**, 173
  classification, **1**, 252
  conjugation
    glucuronides, **5**, 270
  drug disposition
    twins, **5**, 268
  inhibitor

dehydrogenase NADH, **2**, 63
  metabolism
    kidneys, **5**, 120
  therapeutic index, **5**, 610
Salicylic acid
  absorption
    antacids, **5**, 173
    skin, **5**, 28
  drug interactions, **5**, 328
  high performance liquid chromagraphy, **5**, 493
  introduction, **1**, 20, 22, 59
  molecular orbital indices, **4**, 118
  permeability
    skin, **5**, 27
Salicylic acid, *p*-amino-
  acetylation, **5**, 256
  phenytoin intoxication, **5**, 259
Saligenin
  activity, **3**, 197
Salinity
  selectivity, **1**, 242
Saliva
  digestion, **1**, 146
  drug transfer, **5**, 96
  secretion
    control, **1**, 146
Salmo gairdneri
  P-450 isoenzymes
    drug metabolism, **5**, 223
Salmonella typhi
  drug resistance, **2**, 93
Salmonella typhimurium
  antibiotic resistance, **2**, 92
  cell envelope
    drug susceptibility, **2**, 96
Salol
  introduction, **1**, 22
Salts
  small intestine, **1**, 148
Salvage pathways
  ribonucleotides, **2**, 301
Salvarsan
  history, **1**, 87, 90, 94; **2**, 90
Samples
  end-product
    testing, **1**, 552
  testing
    good manufacturing practice, **1**, 564
Sampling
  medicines, **1**, 553
  pharmaceuticals
    manufacture, **1**, 353
  tables
    quality control, **1**, 551
Sandoz
  history, **1**, 93
SANDRA
  Beilstein
    structure searching, **1**, 725
Sanger method
  DNA
    sequence analysis, **2**, 19
Sangivamycin
  purine antagonist, **2**, 310
Sanitation
  history, **1**, 626
SANSS
  activity information database

T

U

V

W

X

Y

Y-590
  inhibitor
    phosphodiesterase, **2**, 507
Yang
  modulators
    traditional Chinese medicine, **1**, 109
  tonifying drug
    effects, **1**, 103
  traditional Chinese medicine, **1**, 100
Yeasts
  cell walls, **1**, 174
  DNA topoisomerase
    function, **2**, 772
  genetic engineering, **1**, 457
  mating type
    gene control, **1**, 399
  secretion, **1**, 421
Yellow card system
  drug monitoring, **1**, 628
    value, **1**, 630
Yin
  modulators

traditional Chinese medicine, **1**, 109
nourishing agent
  effects, **1**, 103
traditional Chinese medicine, **1**, 100
Yingzhaosu A
  traditional Chinese medicine, **1**, 105
YM 09151-2
  dopamine agonist, **3**, 269
  dopamine antagonist
    properties, **3**, 286
Yohimbine
  α_1-adrenergic receptor antagonists
    structure–activity relationships, **3**, 165
  α_2-adrenergic receptors
    affinity, **3**, 139
    antagonists, **3**, 168
  calcium channels, **3**, 1073
  composite dose–response curves, **5**, 344
  multiple actions, **1**, 209
Yukawa–Tsuno equations
  substituent effects
    drugs, **4**, 210

Z

Drug Compendium

PAUL N. CRAIG

National Library of Medicine, Washington, DC, USA

---

1 INTRODUCTION

This Compendium of Drugs is intended to serve several roles: first, as a comprehensive source of drug data, an intensive effort has been made to include as many as possible of the compounds which have been used or studied as medicinal agents in man (over 5500 are included); these have been augmented by more than 200 chemicals used as pharmacological agents; second, as an aid in research, this compilation was prepared by use of the ChemBase program,[1] which not only provides good quality structures but also permits sub-structure and text searching of the database which was generated in the preparation of this compendium; third, as a unique source of unpublished data, more than 4500 measured or estimated log P values ($\log_{10}$ of the *n*-octanol/water partition coefficient) and 516 pK_a values from the Pomona College Medicinal Chemistry Project databases were added.[2] In many cases both estimated and measured log P values are reported for the same compound.

The complete Compendium database will be available from Molecular Design Ltd. as an electronic three-dimensional (3D) database.[3] Called CMC-3D, the new database will be searchable based on 3D properties using MACCS-3D, the module to MDL's graphics-based MACCS-II system that provides a means to store, search and retrieve 3D models and model-related data. It is expected that CMC-3D will enable pharmaceutical researchers to examine correlations between biological activity, geometric properties, log P values, *etc.* as an important aid in rational drug design.

2 SOURCES

The major source for compounds in this compendium was the inclusion of most of the drugs identified by either USAN[4] or INN[4] generic names, since these include practically all medicinal agents or compounds intended for clinical study in the major countries of the world. All radioactive diagnostic drugs were excluded, as were drugs of unknown structure. Some 155 anticancer agents not in USAN, for which the National Cancer Institute (USA) has prepared New Drug Application submissions,[5] were included; 64 chemicals were identified for inclusion from the 'Merck Index, 10th edition',[6] and other compounds not elsewhere identified were obtained from the 'The National Formulary',[7] 'Annual Reports in Medicinal Chemistry'[8] for 1988 and 1989, 'The American Drug Index'[9] and 'Martindale: The Extra Pharmacopoeia'.[10]

The compendium includes selected references. Where readily available, a reference to a patent is given, with the priority given to a US or British patent when patents from several countries are available. One or two references to the preparation and to pharmacological or clinical findings are included for about 80% of the entries. These references were obtained from the following sources: 'The Merck Index', 10th edition;[6] 'Martindale: The Extra Pharmacopoeia', 28th edition;[10] 'Dictionary of Antibiotics and Related Substances';[11] 'Clarke's Isolation and Identification of Drugs';[12] 'Registry of Toxic Effects of Chemical Substances';[13] and 'Unlisted Drugs'.[14]

3 PRESENTATION OF DATA

Use of the ChemBase program to generate the structures and to create the record format for final printing required a balance to be struck between the space available for the structure and the accompanying data elements. The form finally arrived at is illustrated in Figure 1 and contains the following data elements.

3.1 Chemical Structure [(1) in Figure 1]

The structures displayed were generated by the ChemBase program.[1] These are two-dimensional, with stereobond indicators as used in ChemBase version 1.2. With few exceptions the parent structure

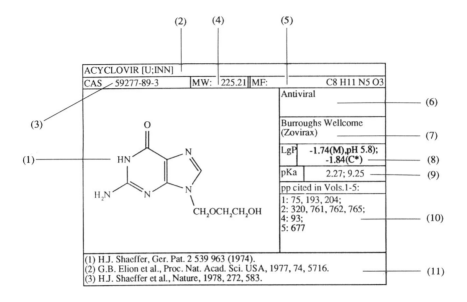

Figure 1

is given, rather than a salt or hydrate which may be used as the actual drug (see discussion under Section 3.3 below).

3.2 Generic Name [(2) in Figure 1]

The presence of the drug in USAN and/or INN compilations is shown by 'U' and/or 'INN' within brackets following the generic name. If the drug is not listed in either USAN or INN, then the National Cancer Institute (NCI), Merck Index (MI), National Formulary (NF), Annual Reports in Medicinal Chemistry (ARMC), American Drug Index (ADI), or Martindale is given.

The presence of a drug in USAN and INN is always indicated. If a drug is not listed in USAN as an INN or USAN entry, only the first-encountered source is given. If the drug is not listed in USAN or INN, the name given is not an 'approved' generic name, and in a few instances a name was coined to fit the available space (this is often true for the NCI compounds). Space limitations prevented the inclusion of a systematic chemical name, but because of the complexity of many drugs their long systematic names are of little use for identification in a printed index. If one is not certain that a drug name is a USAN or INN generic name, use of the CAS RN Index is recommended for locating structures, rather than the Name Index.

3.3 Chemical Abstracts Registry Number (RN) [(3) in Figure 1]

This is the unique identification number assigned by Chemical Abstracts Service[15] to each compound. Because different numbers are assigned to each salt or hydrate of a parent compound, the number and structure for the parent compound have been used, rather than those for the form used commercially in marketed drugs. Thus there often is not a direct correspondence between the CAS RN used in this Compendium and the RN of the marketed form of a drug. This approach simplifies the identification (or location) of a structure in this Compendium, by use of the CAS RN Index or the Molecular Formula Index.

3.4   Molecular Weight (MW) [(4) in Figure 1]

This is the molecular weight for the structure entered into ChemBase, and is algorithmically calculated from the structure displayed in (1). For certain complex structures and for all polymeric compounds the correspondence is not correct, and varies depending on the structural approach used to show the complex nature of the compound. These molecular weights have been removed from the printed records to avoid confusion.

3.5   Molecular Formula (MF) [(5) in Figure 1]

The molecular formula corresponding to the structure is displayed here. However, polymeric compounds (for which a repeating unit is shown) do not have the correct MF values. The MF is the total sum of all atoms displayed in the structure field (1), which must be considered for those drugs which are combinations of two or more moieties (*e.g.* Dimenhydrinate). Compounds which form hydrates are included only as the anhydrous form in this compendium.

3.6   Classification [(6) in Figure 1; index on p. 966]

The drug activity classifications given in USAN have usually been used, with some necessary abbreviations due to space limitations. For compounds not in USAN, an attempt has been made to conform with the USAN classifications, but it has proven difficult to distinguish between, for example, antiasthmatic and bronchodilator, antispasmodic and anticholinergic, *etc.*, without performing an in-depth study. Classifications have not been assigned to such compounds even if they appear to be typical antihistamines or antispasmodics, and class data were only used if found in the listed sources. Because the INN listings in the USAN publication do not contain classification data, there are gaps in this data field, even after all the listed sources were examined.

3.7   Company [(7) in Figure 1]

This field identifies the first organization which prepared and/or studied the drug. In most cases this was a simple decision, but for some drugs several organizations have reported the same agent at almost the same time. It was decided to report here the organization which obtained the earliest US patent covering a given drug, but the reader is strongly cautioned not to consider this field as a definitive assignment of priority.

The situation is further complicated by numerous licensing and cross-licensing agreements, in which area the drug industry has been extremely active, on an international level, for more than 40 years. In many cases where it was possible (resorting to those sources mentioned in Section 2 above) to assign the first preparation or study of a drug to an organization which later licensed it to others, both organizations have been mentioned in this field. An attempt has been made to list a company from the USA, the EEC or Japan wherever a marketed product has been identified. In many cases the trade name of the drug is included in parentheses immediately following the name of the company. The reader is cautioned that this information is not to be considered as complete; these data are intended only to provide additional information which may assist the reader in beginning an in-depth study of the literature.

The company is assumed to be located in the USA unless another country is listed immediately following the company name. Several companies have research laboratories located in two or more countries. In some cases the wrong country may be listed, because the sources used did not permit an accurate determination of the originating location to be made.

3.8   Log *P* Data [(8) in Figure 1]

This field presents *n*-octanol/water partition coefficient data in log form. These data are from the Medicinal Chemistry Project at Pomona College and give both measured and calculated values which were estimated using CLOGP versions 3.54 and 3.55.[2] The following parenthetical notes further qualify the log *P* data:

(M) = An experimentally determined value.
(C) = Calculated using CLOGP version 3.54.
(C*) = Calculated using CLOGP version 3.55.
(1) = Low confidence; measured value probably would be higher.

(2) = Low confidence; measured value probably would be lower.
(3) = Low confidence; may be either high or low.
(4) = Cannot calculate; either a missing fragment value or an uncommon structural feature.
(5) = Values of log P > 7 should be used with caution, if at all, in QSAR work.
(6) = Manually corrected by A. J. Leo.
(7) = Values of log P < –3.5 should be used with caution, if at all, in QSAR studies.

3.9 Ionization Constant pK_a [(9) in Figure 1]

The bulk (516) of the pK_a values are from the database of the Medicinal Chemistry Project[2] at Pomona College. These were augmented by several values obtained from 'Clarke's Isolation and Identification of Drugs' and 'The Merck Index', 10th edition.

3.10 Citations to Volumes 1–5 [(10) in Figure 1]

Citations to important discussions or data about a drug which are contained in Volumes 1–5 of this series are given in this field. These citations take the form 'volume number: page number', *e.g.* '4: 879'.

3.11 Key References [(11) in Figure 1]

The sources listed in Section 2 were used to locate the earliest references to the preparation and/or report of pharmacological or clinical data for the drug. For older drugs, such as acetylsalicylic acid or phenobarbital, a reference to a modern preparation, a review or a book may be given instead.

The earliest US patent listed in the sources used was also given; if no US patent was located, a British, French, West German or other EEC country or Japanese patent was reported, if located.

A reference to a review article or a book is often given for those drugs which have been well developed. Some 80% of the compounds have at least one literature reference. These references may lack the authors' names depending on how the references were listed in the sources used. They have also been shortened to fit the limited space available; thus many do not conform to the use of recommended journal abbreviations. The journal citations take the form 'year, volume number, page number', *e.g.* 1983, 55, 395. Only the first-named author is listed, with rare exceptions. No effort was made to verify the references located in the sources used; hence incorrect citations which might have been included in those sources may have been repeated here. The CAS registry numbers should be used to initiate more exhaustive literature sources for compounds of interest.

4 CHEMICAL STRUCTURES

The structures shown were drawn using the ChemBase software (version 1.2) of Molecular Design Ltd.[1] ChemBase allows considerable variation in the presentation of the structures, including the ability to use Me, Et, Bu, *etc.* in place of the skeletal structures, if desired. It was decided to use only a few of the possible abbreviations, such as Me, Et, n- and i-Pr, n-, i-, s- and t-Bu, COOH, COOEt, $CONH_2$, and to print the structure of the phenyl ring, rather than use Ph, wherever space constraints permitted. The resulting structures give a better sense of the relative size and polarity for each portion of the molecule than is conveyed by the use of most abbreviations. This gives one an intuitive feeling of how the log P values are derived by the CLOGP programs, as well as more closely presenting the relative sizes of portions of the molecules (within the limitations of two-dimensional structures). One exception to this general rule is the need to abbreviate long alkyl chains, such as octyl and higher, for reasons of space.

5 ACKNOWLEDGEMENTS

The cooperation and support of Drs. Corwin Hansch, Albert Leo and Colin Drayton in planning with me the scope and presentation of the data is gratefully acknowledged. The staff of Molecular Design Ltd. were most helpful, especially the telephone Software Support team.

The author wishes to express his particular appreciation to Dr. Albert J. Leo, who not only furnished partition and pK_a data but provided valuable proof-reading assistance for the compendium. In addition, my grandson David Craig and Michael Medlin (Al Leo's grandson) input much of the numerical data with exceptional accuracy, and Michael also assisted in the proof reading. The author also wishes to thank Drs.

Glenn Ullyot and Carl Kaiser for proof reading portions of the Compendium. The Research Library staff at Smith Kline Beecham and the Specialized Information Services Division at the National Library of Medicine kindly permitted me access to many of the sources of information mentioned above.

Last, but not least, I wish to thank my wife for graciously accepting but little help from her husband for the more than 18 months of full-time effort which was required for this task.

6 REFERENCES

1. ChemBase is the name given to computer software programs developed by Molecular Design Ltd., San Leandro, California. ChemBase provides the capability of drawing two-dimensional structures, generating the molecular formulas and molecular weights which correspond to the structure, and permits the recording of various (searchable) data elements which are linked to the structure. All of these capabilities can be performed on a microcomputer.
2. Dr. Albert J. Leo, Director, Medicinal Chemistry Project, Chemistry Department, Pomona College, Claremont, CA 91711, USA. The Project maintains computer records of partition coefficient data and pK_a values, among other activities.
3. Contact Dr. Philip J. McHale, Molecular Design Ltd., 2132 Farallon Drive, San Leandro, CA 94577, USA, for details concerning the availability of this database and of the ChemBase version.
4. USAN is used for 'USAN and the USP Dictionary of Drug Names', published by the United States Pharmacopeial Convention, Inc., 12601 Twinbrook Parkway, Rockville, MD 20852, USA. It is a compilation of United States Adopted Names (USAN) and also contains international nonproprietary names (INN) which are published by the World Health Organization.
5. These compounds were taken from a compilation of 391 Investigational New Drug Applications (INDAs) that have been filed with the Food and Drug Administration for anticancer drugs, from 1945–1988. The compilation is entitled 'Chemicals for Which Investigational New Drug Applications have been Filed by NCI', 4 January 1988, and was obtained from Dr. Frank Quinn, Information Technology Branch, Division of Cancer Treatment, National Cancer Institute, Bethesda, MD 20892, USA.
6. 'The Merck Index', 10th edition, published by Merck & Co., Inc., Rahway, NJ 07065, USA, 1983 (the 11th edition appeared late in 1989).
7. 'The National Formulary', American Pharmaceutical Association, Washington, DC, 1974.
8. Published annually by the Division of Medicinal Chemistry, American Chemical Society, 1155 N.W. Sixteenth St., Washington, DC 20036, USA.
9. 'The American Drug Index', ed. N. F. Billups, Lippincott, Philadelphia, published annually.
10. 'Martindale: The Extra Pharmacopoeia', 28th edition, ed. J. E. F. Reynolds, The Pharmaceutical Press, London, 1982 (the 29th edition appeared in 1989).
11. 'Dictionary of Antibiotics and Related Substances', ed. B. W. Bycroft, Chapman & Hall, London, 1988.
12. 'Clarke's Isolation and Identification of Drugs', ed. A. C. Moffat, The Pharmaceutical Press, London, 1986.
13. 'Registry of Toxic Effects of Chemical Substances', 1985–86 edition, ed. D. V. Sweet, US Department of Health and Human Services, Public Health Service, Centers for Disease Control, National Institute for Occupational Safety and Health, DHHS (NIOSH) Publ. No. 87-114, US Government Printing Office, Washington, DC 20402, 1988.
14. 'Unlisted Drugs', Pharmaceutical Division, Special Libraries Association, PO Box 429, Chatham, NJ 07928, USA.
15. Chemical Abstracts Service (CAS), 2540 Olentangy River Rd., Columbus, OH 43210, USA. CAS assigns a registry number (RN) to every newly published chemical substance.

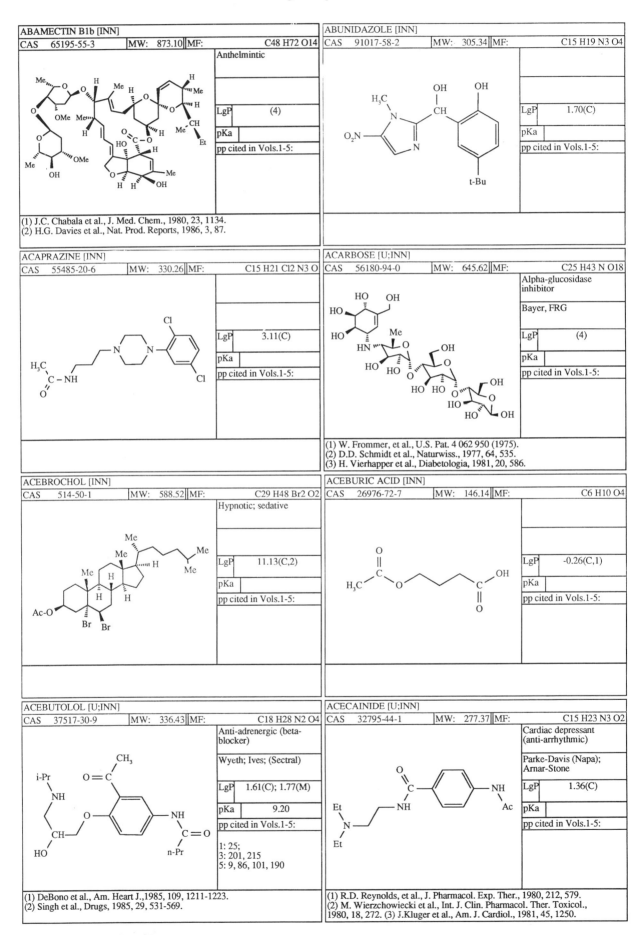

ABAMECTIN B1b [INN]

CAS	65195-55-3	MW: 873.10	MF:	C48 H72 O14

Anthelmintic

LgP	(4)
pKa	
pp cited in Vols.1-5:	

(1) J.C. Chabala et al., J. Med. Chem., 1980, 23, 1134.
(2) H.G. Davies et al., Nat. Prod. Reports, 1986, 3, 87.

ABUNIDAZOLE [INN]

CAS	91017-58-2	MW: 305.34	MF:	C15 H19 N3 O4

LgP	1.70(C)
pKa	
pp cited in Vols.1-5:	

ACAPRAZINE [INN]

CAS	55485-20-6	MW: 330.26	MF:	C15 H21 Cl2 N3 O

LgP	3.11(C)
pKa	
pp cited in Vols.1-5:	

ACARBOSE [U;INN]

CAS	56180-94-0	MW: 645.62	MF:	C25 H43 N O18

Alpha-glucosidase inhibitor

Bayer, FRG

LgP	(4)
pKa	
pp cited in Vols.1-5:	

(1) W. Frommer, et al., U.S. Pat. 4 062 950 (1975).
(2) D.D. Schmidt et al., Naturwiss., 1977, 64, 535.
(3) H. Vierhapper et al., Diabetologia, 1981, 20, 586.

ACEBROCHOL [INN]

CAS	514-50-1	MW: 588.52	MF:	C29 H48 Br2 O2

Hypnotic; sedative

LgP	11.13(C,2)
pKa	
pp cited in Vols.1-5:	

ACEBURIC ACID [INN]

CAS	26976-72-7	MW: 146.14	MF:	C6 H10 O4

LgP	-0.26(C,1)
pKa	
pp cited in Vols.1-5:	

ACEBUTOLOL [U;INN]

CAS	37517-30-9	MW: 336.43	MF:	C18 H28 N2 O4

Anti-adrenergic (beta-blocker)

Wyeth; Ives; (Sectral)

LgP	1.61(C); 1.77(M)
pKa	9.20
pp cited in Vols.1-5:	

1: 25;
3: 201, 215
5: 9, 86, 101, 190

(1) DeBono et al., Am. Heart J.,1985, 109, 1211-1223.
(2) Singh et al., Drugs, 1985, 29, 531-569.

ACECAINIDE [U;INN]

CAS	32795-44-1	MW: 277.37	MF:	C15 H23 N3 O2

Cardiac depressant (anti-arrhythmic)

Parke-Davis (Napa); Arnar-Stone

LgP	1.36(C)
pKa	
pp cited in Vols.1-5:	

(1) R.D. Reynolds, et al., J. Pharmacol. Exp. Ther., 1980, 212, 579.
(2) M. Wierzchowiecki et al., Int. J. Clin. Pharmacol. Ther. Toxicol., 1980, 18, 272. (3) J.Kluger et al., Am. J. Cardiol., 1981, 45, 1250.

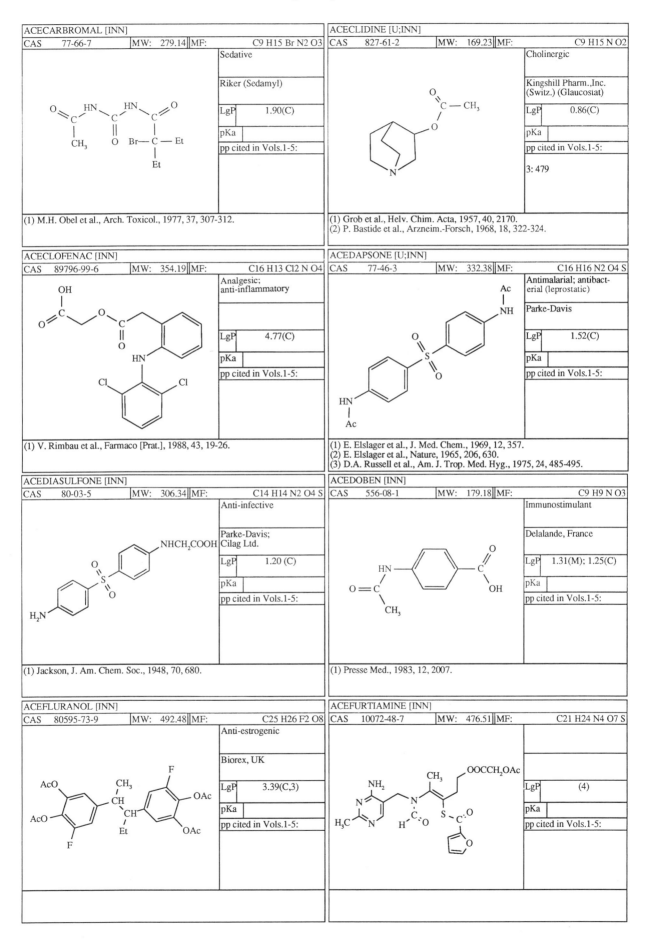

ACECARBROMAL [INN]

CAS	77-66-7	MW:	279.14	MF:	C9 H15 Br N2 O3

Sedative

Riker (Sedamyl)

LgP	1.90(C)
pKa	
pp cited in Vols.1-5:	

(1) M.H. Obel et al., Arch. Toxicol., 1977, 37, 307-312.

ACECLIDINE [U;INN]

CAS	827-61-2	MW:	169.23	MF:	C9 H15 N O2

Cholinergic

Kingshill Pharm.,Inc. (Switz.) (Glaucostat)

LgP	0.86(C)
pKa	
pp cited in Vols.1-5:	
3: 479	

(1) Grob et al., Helv. Chim. Acta, 1957, 40, 2170.
(2) P. Bastide et al., Arzneim.-Forsch, 1968, 18, 322-324.

ACECLOFENAC [INN]

CAS	89796-99-6	MW:	354.19	MF:	C16 H13 Cl2 N O4

Analgesic; anti-inflammatory

LgP	4.77(C)
pKa	
pp cited in Vols.1-5:	

(1) V. Rimbau et al., Farmaco [Prat.], 1988, 43, 19-26.

ACEDAPSONE [U;INN]

CAS	77-46-3	MW:	332.38	MF:	C16 H16 N2 O4 S

Antimalarial; antibacterial (leprostatic)

Parke-Davis

LgP	1.52(C)
pKa	
pp cited in Vols.1-5:	

(1) E. Elslager et al., J. Med. Chem., 1969, 12, 357.
(2) E. Elslager et al., Nature, 1965, 206, 630.
(3) D.A. Russell et al., Am. J. Trop. Med. Hyg., 1975, 24, 485-495.

ACEDIASULFONE [INN]

CAS	80-03-5	MW:	306.34	MF:	C14 H14 N2 O4 S

Anti-infective

Parke-Davis; Cilag Ltd.

LgP	1.20 (C)
pKa	
pp cited in Vols.1-5:	

(1) Jackson, J. Am. Chem. Soc., 1948, 70, 680.

ACEDOBEN [INN]

CAS	556-08-1	MW:	179.18	MF:	C9 H9 N O3

Immunostimulant

Delalande, France

LgP	1.31(M); 1.25(C)
pKa	
pp cited in Vols.1-5:	

(1) Presse Med., 1983, 12, 2007.

ACEFLURANOL [INN]

CAS	80595-73-9	MW:	492.48	MF:	C25 H26 F2 O8

Anti-estrogenic

Biorex, UK

LgP	3.39(C,3)
pKa	
pp cited in Vols.1-5:	

ACEFURTIAMINE [INN]

CAS	10072-48-7	MW:	476.51	MF:	C21 H24 N4 O7 S

LgP	(4)
pKa	
pp cited in Vols.1-5:	

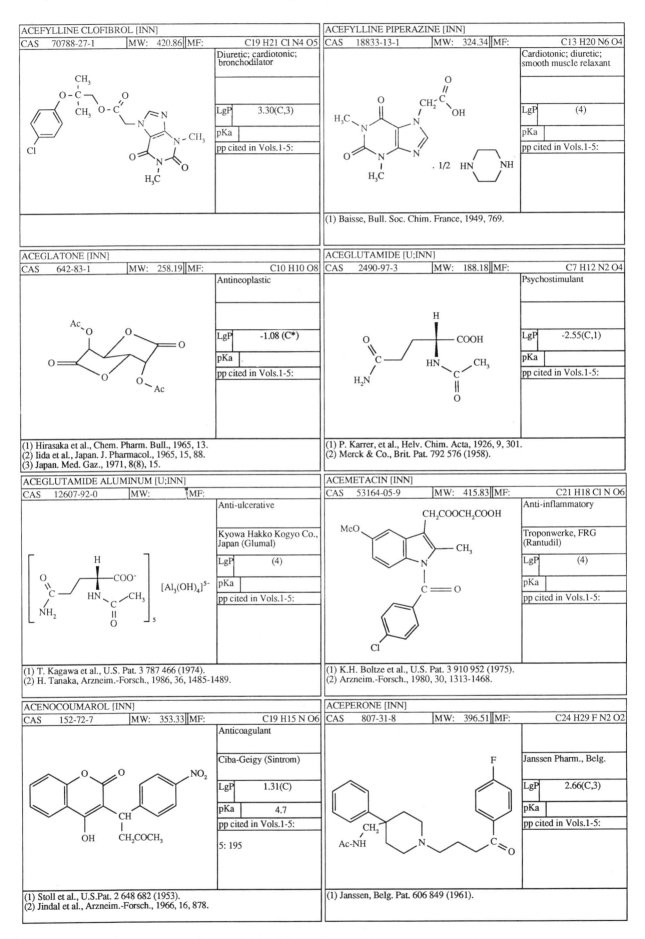

ACEFYLLINE CLOFIBROL [INN]

CAS	70788-27-1	MW: 420.86	MF:	C19 H21 Cl N4 O5

Diuretic; cardiotonic; bronchodilator

LgP	3.30(C,3)
pKa	
pp cited in Vols.1-5:	

ACEFYLLINE PIPERAZINE [INN]

CAS	18833-13-1	MW: 324.34	MF:	C13 H20 N6 O4

Cardiotonic; diuretic; smooth muscle relaxant

LgP	(4)
pKa	
pp cited in Vols.1-5:	

. 1/2 HN NH

(1) Baisse, Bull. Soc. Chim. France, 1949, 769.

ACEGLATONE [INN]

CAS	642-83-1	MW: 258.19	MF:	C10 H10 O8

Antineoplastic

LgP	-1.08 (C*)
pKa	
pp cited in Vols.1-5:	

(1) Hirasaka et al., Chem. Pharm. Bull., 1965, 13.
(2) Iida et al., Japan. J. Pharmacol., 1965, 15, 88.
(3) Japan. Med. Gaz., 1971, 8(8), 15.

ACEGLUTAMIDE [U;INN]

CAS	2490-97-3	MW: 188.18	MF:	C7 H12 N2 O4

Psychostimulant

LgP	-2.55(C,1)
pKa	
pp cited in Vols.1-5:	

(1) P. Karrer, et al., Helv. Chim. Acta, 1926, 9, 301.
(2) Merck & Co., Brit. Pat. 792 576 (1958).

ACEGLUTAMIDE ALUMINUM [U;INN]

CAS	12607-92-0	MW:	MF:	

Anti-ulcerative

Kyowa Hakko Kogyo Co., Japan (Glumal)

LgP	(4)
pKa	
pp cited in Vols.1-5:	

$[Al_3(OH)_4]^{5-}$

(1) T. Kagawa et al., U.S. Pat. 3 787 466 (1974).
(2) H. Tanaka, Arzneim.-Forsch., 1986, 36, 1485-1489.

ACEMETACIN [INN]

CAS	53164-05-9	MW: 415.83	MF:	C21 H18 Cl N O6

CH_2COOCH_2COOH

Anti-inflammatory

Troponwerke, FRG (Rantudil)

LgP	(4)
pKa	
pp cited in Vols.1-5:	

(1) K.H. Boltze et al., U.S. Pat. 3 910 952 (1975).
(2) Arzneim.-Forsch., 1980, 30, 1313-1468.

ACENOCOUMAROL [INN]

CAS	152-72-7	MW: 353.33	MF:	C19 H15 N O6

Anticoagulant

Ciba-Geigy (Sintrom)

LgP	1.31(C)
pKa	4.7
pp cited in Vols.1-5:	
	5: 195

(1) Stoll et al., U.S.Pat. 2 648 682 (1953).
(2) Jindal et al., Arzneim.-Forsch., 1966, 16, 878.

ACEPERONE [INN]

CAS	807-31-8	MW: 396.51	MF:	C24 H29 F N2 O2

Janssen Pharm., Belg.

LgP	2.66(C,3)
pKa	
pp cited in Vols.1-5:	

(1) Janssen, Belg. Pat. 606 849 (1961).

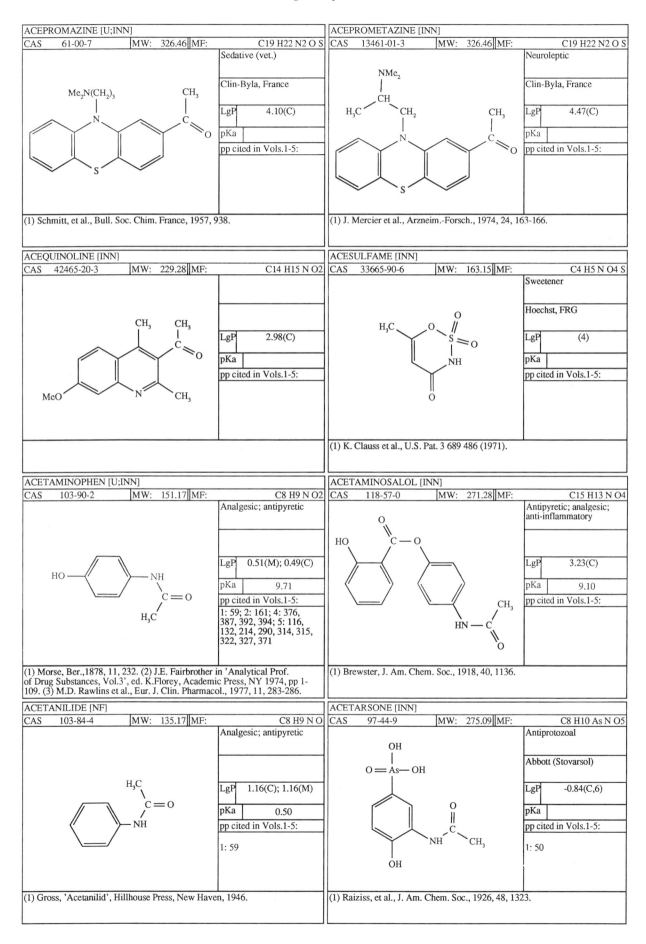

ACEPROMAZINE [U;INN]			
CAS 61-00-7	MW: 326.46	MF:	C19 H22 N2 O S

Sedative (vet.)

Clin-Byla, France

LgP	4.10(C)
pKa	
pp cited in Vols.1-5:	

(1) Schmitt, et al., Bull. Soc. Chim. France, 1957, 938.

ACEPROMETAZINE [INN]			
CAS 13461-01-3	MW: 326.46	MF:	C19 H22 N2 O S

Neuroleptic

Clin-Byla, France

LgP	4.47(C)
pKa	
pp cited in Vols.1-5:	

(1) J. Mercier et al., Arzneim.-Forsch., 1974, 24, 163-166.

ACEQUINOLINE [INN]			
CAS 42465-20-3	MW: 229.28	MF:	C14 H15 N O2

LgP	2.98(C)
pKa	
pp cited in Vols.1-5:	

ACESULFAME [INN]			
CAS 33665-90-6	MW: 163.15	MF:	C4 H5 N O4 S

Sweetener

Hoechst, FRG

LgP	(4)
pKa	
pp cited in Vols.1-5:	

(1) K. Clauss et al., U.S. Pat. 3 689 486 (1971).

ACETAMINOPHEN [U;INN]			
CAS 103-90-2	MW: 151.17	MF:	C8 H9 N O2

Analgesic; antipyretic

LgP	0.51(M); 0.49(C)
pKa	9.71
pp cited in Vols.1-5:	

1: 59; 2: 161; 4: 376, 387, 392, 394; 5: 116, 132, 214, 290, 314, 315, 322, 327, 371

(1) Morse, Ber.,1878, 11, 232. (2) J.E. Fairbrother in 'Analytical Prof. of Drug Substances, Vol.3', ed. K.Florey, Academic Press, NY 1974, pp 1-109. (3) M.D. Rawlins et al., Eur. J. Clin. Pharmacol., 1977, 11, 283-286.

ACETAMINOSALOL [INN]			
CAS 118-57-0	MW: 271.28	MF:	C15 H13 N O4

Antipyretic; analgesic; anti-inflammatory

LgP	3.23(C)
pKa	9.10
pp cited in Vols.1-5:	

(1) Brewster, J. Am. Chem. Soc., 1918, 40, 1136.

ACETANILIDE [NF]			
CAS 103-84-4	MW: 135.17	MF:	C8 H9 N O

Analgesic; antipyretic

LgP	1.16(C); 1.16(M)
pKa	0.50
pp cited in Vols.1-5:	

1: 59

(1) Gross, 'Acetanilid', Hillhouse Press, New Haven, 1946.

ACETARSONE [INN]			
CAS 97-44-9	MW: 275.09	MF:	C8 H10 As N O5

Antiprotozoal

Abbott (Stovarsol)

LgP	-0.84(C,6)
pKa	
pp cited in Vols.1-5:	

1: 50

(1) Raiziss, et al., J. Am. Chem. Soc., 1926, 48, 1323.

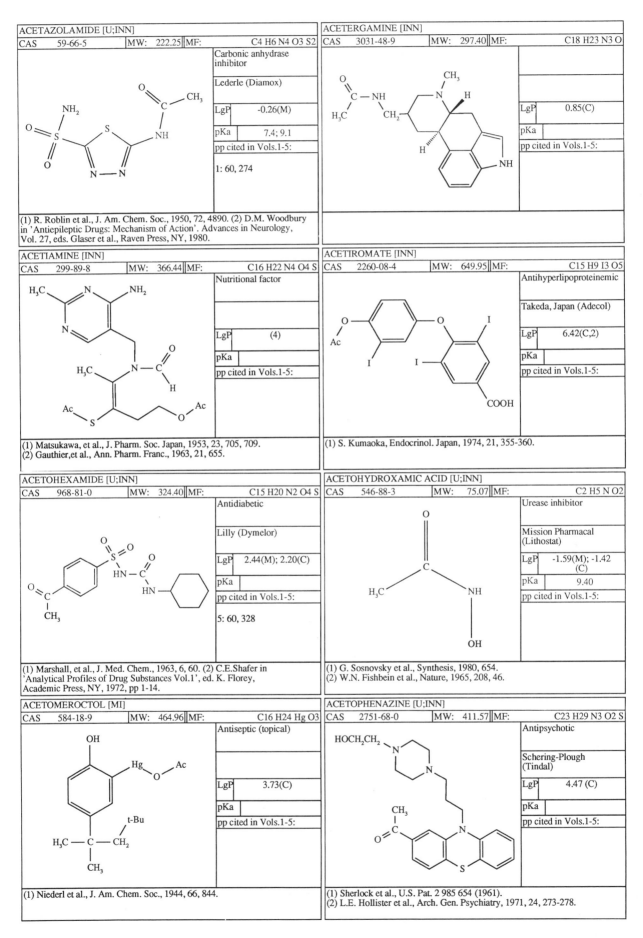

ACETAZOLAMIDE [U;INN]			
CAS 59-66-5	MW: 222.25	MF:	C4 H6 N4 O3 S2

Carbonic anhydrase inhibitor

Lederle (Diamox)

LgP	-0.26(M)
pKa	7.4; 9.1
pp cited in Vols.1-5:	

1: 60, 274

(1) R. Roblin et al., J. Am. Chem. Soc., 1950, 72, 4890. (2) D.M. Woodbury in 'Antiepileptic Drugs: Mechanism of Action'. Advances in Neurology, Vol. 27, eds. Glaser et al., Raven Press, NY, 1980.

ACETERGAMINE [INN]			
CAS 3031-48-9	MW: 297.40	MF:	C18 H23 N3 O

LgP	0.85(C)
pKa	
pp cited in Vols.1-5:	

ACETIAMINE [INN]			
CAS 299-89-8	MW: 366.44	MF:	C16 H22 N4 O4 S

Nutritional factor

LgP	(4)
pKa	
pp cited in Vols.1-5:	

(1) Matsukawa, et al., J. Pharm. Soc. Japan, 1953, 23, 705, 709.
(2) Gauthier, et al., Ann. Pharm. Franc., 1963, 21, 655.

ACETIROMATE [INN]			
CAS 2260-08-4	MW: 649.95	MF:	C15 H9 I3 O5

Antihyperlipoproteinemic

Takeda, Japan (Adecol)

LgP	6.42(C,2)
pKa	
pp cited in Vols.1-5:	

(1) S. Kumaoka, Endocrinol. Japan, 1974, 21, 355-360.

ACETOHEXAMIDE [U;INN]			
CAS 968-81-0	MW: 324.40	MF:	C15 H20 N2 O4 S

Antidiabetic

Lilly (Dymelor)

LgP	2.44(M); 2.20(C)
pKa	
pp cited in Vols.1-5:	

5: 60, 328

(1) Marshall, et al., J. Med. Chem., 1963, 6, 60. (2) C.E.Shafer in 'Analytical Profiles of Drug Substances Vol.1', ed. K. Florey, Academic Press, NY, 1972, pp 1-14.

ACETOHYDROXAMIC ACID [U;INN]			
CAS 546-88-3	MW: 75.07	MF:	C2 H5 N O2

Urease inhibitor

Mission Pharmacal (Lithostat)

LgP	-1.59(M); -1.42 (C)
pKa	9.40
pp cited in Vols.1-5:	

(1) G. Sosnovsky et al., Synthesis, 1980, 654.
(2) W.N. Fishbein et al., Nature, 1965, 208, 46.

ACETOMEROCTOL [MI]			
CAS 584-18-9	MW: 464.96	MF:	C16 H24 Hg O3

Antiseptic (topical)

LgP	3.73(C)
pKa	
pp cited in Vols.1-5:	

(1) Niederl et al., J. Am. Chem. Soc., 1944, 66, 844.

ACETOPHENAZINE [U;INN]			
CAS 2751-68-0	MW: 411.57	MF:	C23 H29 N3 O2 S

Antipsychotic

Schering-Plough (Tindal)

LgP	4.47 (C)
pKa	
pp cited in Vols.1-5:	

(1) Sherlock et al., U.S. Pat. 2 985 654 (1961).
(2) L.E. Hollister et al., Arch. Gen. Psychiatry, 1971, 24, 273-278.

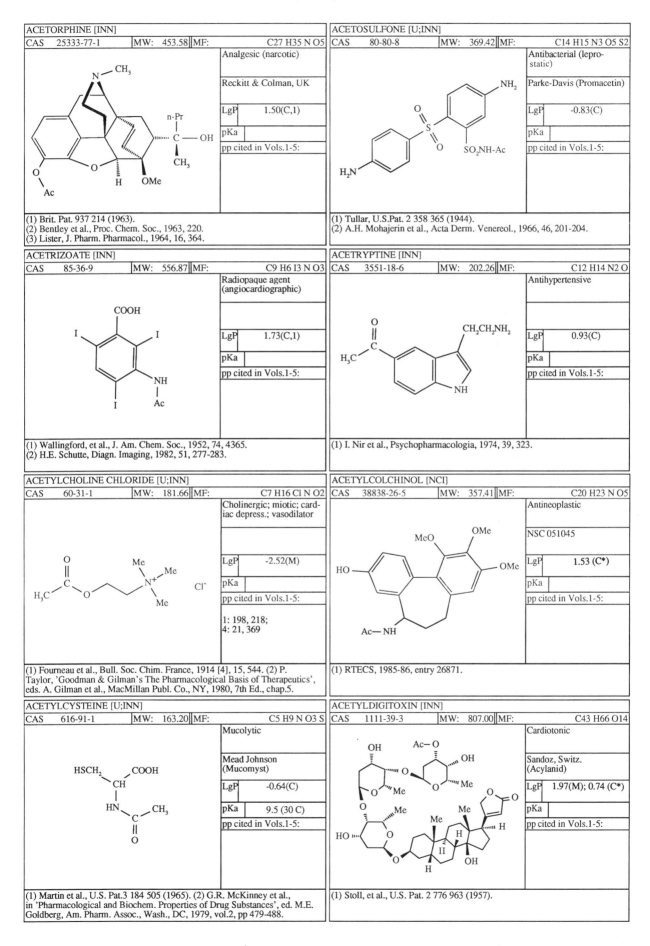

ACETORPHINE [INN]

CAS	25333-77-1	MW: 453.58	MF:	C27 H35 N O5

Analgesic (narcotic)

Reckitt & Colman, UK

LgP	1.50(C,1)
pKa	
pp cited in Vols.1-5:	

(1) Brit. Pat. 937 214 (1963).
(2) Bentley et al., Proc. Chem. Soc., 1963, 220.
(3) Lister, J. Pharm. Pharmacol., 1964, 16, 364.

ACETOSULFONE [U;INN]

CAS	80-80-8	MW: 369.42	MF:	C14 H15 N3 O5 S2

Antibacterial (leprostatic)

Parke-Davis (Promacetin)

LgP	-0.83(C)
pKa	
pp cited in Vols.1-5:	

(1) Tullar, U.S.Pat. 2 358 365 (1944).
(2) A.H. Mohajerin et al., Acta Derm. Venereol., 1966, 46, 201-204.

ACETRIZOATE [INN]

CAS	85-36-9	MW: 556.87	MF:	C9 H6 I3 N O3

Radiopaque agent (angiocardiographic)

LgP	1.73(C,1)
pKa	
pp cited in Vols.1-5:	

(1) Wallingford, et al., J. Am. Chem. Soc., 1952, 74, 4365.
(2) H.E. Schutte, Diagn. Imaging, 1982, 51, 277-283.

ACETRYPTINE [INN]

CAS	3551-18-6	MW: 202.26	MF:	C12 H14 N2 O

Antihypertensive

LgP	0.93(C)
pKa	
pp cited in Vols.1-5:	

(1) I. Nir et al., Psychopharmacologia, 1974, 39, 323.

ACETYLCHOLINE CHLORIDE [U;INN]

CAS	60-31-1	MW: 181.66	MF:	C7 H16 Cl N O2

Cholinergic; miotic; cardiac depress.; vasodilator

LgP	-2.52(M)
pKa	
pp cited in Vols.1-5:	

1: 198, 218;
4: 21, 369

(1) Fourneau et al., Bull. Soc. Chim. France, 1914 [4], 15, 544. (2) P. Taylor, 'Goodman & Gilman's The Pharmacological Basis of Therapeutics', eds. A. Gilman et al., MacMillan Publ. Co., NY, 1980, 7th Ed., chap.5.

ACETYLCOLCHINOL [NCI]

CAS	38838-26-5	MW: 357.41	MF:	C20 H23 N O5

Antineoplastic

NSC 051045

LgP	1.53 (C*)
pKa	
pp cited in Vols.1-5:	

(1) RTECS, 1985-86, entry 26871.

ACETYLCYSTEINE [U;INN]

CAS	616-91-1	MW: 163.20	MF:	C5 H9 N O3 S

Mucolytic

Mead Johnson (Mucomyst)

LgP	-0.64(C)
pKa	9.5 (30 C)
pp cited in Vols.1-5:	

(1) Martin et al., U.S. Pat.3 184 505 (1965). (2) G.R. McKinney et al., in 'Pharmacological and Biochem. Properties of Drug Substances', ed. M.E. Goldberg, Am. Pharm. Assoc., Wash., DC, 1979, vol.2, pp 479-488.

ACETYLDIGITOXIN [INN]

CAS	1111-39-3	MW: 807.00	MF:	C43 H66 O14

Cardiotonic

Sandoz, Switz. (Acylanid)

LgP	1.97(M); 0.74 (C*)
pKa	
pp cited in Vols.1-5:	

(1) Stoll, et al., U.S. Pat. 2 776 963 (1957).

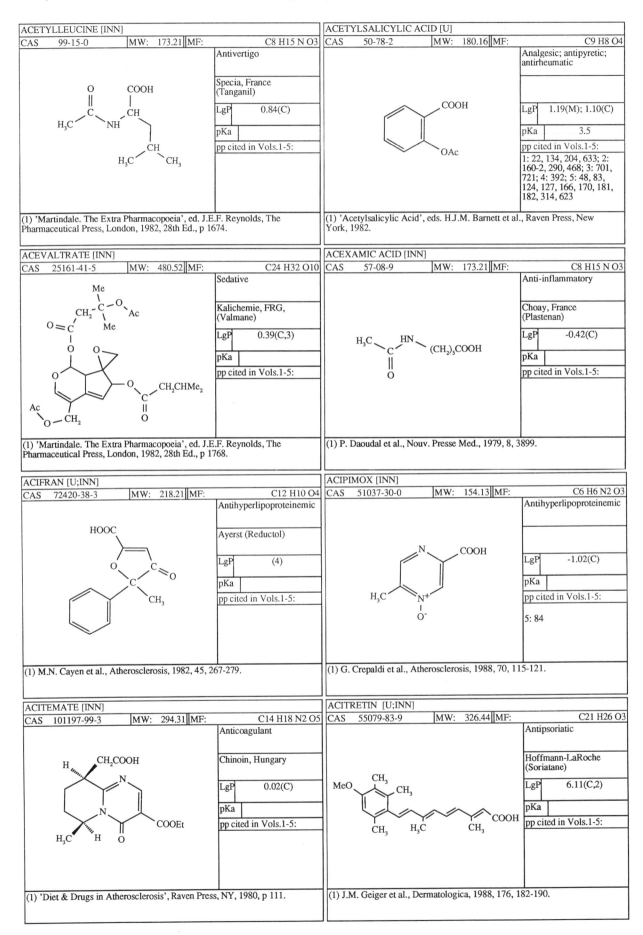

ACETYLLEUCINE [INN]			
CAS 99-15-0	MW: 173.21	MF:	C8 H15 N O3

Antivertigo

Specia, France (Tanganil)

LgP	0.84(C)

| pKa | |

pp cited in Vols.1-5:

(1) 'Martindale. The Extra Pharmacopoeia', ed. J.E.F. Reynolds, The Pharmaceutical Press, London, 1982, 28th Ed., p 1674.

ACETYLSALICYLIC ACID [U]			
CAS 50-78-2	MW: 180.16	MF:	C9 H8 O4

Analgesic; antipyretic; antirheumatic

LgP	1.19(M); 1.10(C)

| pKa | 3.5 |

pp cited in Vols.1-5:
1: 22, 134, 204, 633; 2: 160-2, 290, 468; 3: 701, 721; 4: 392; 5: 48, 83, 124, 127, 166, 170, 181, 182, 314, 623

(1) 'Acetylsalicylic Acid', eds. H.J.M. Barnett et al., Raven Press, New York, 1982.

ACEVALTRATE [INN]			
CAS 25161-41-5	MW: 480.52	MF:	C24 H32 O10

Sedative

Kalichemie, FRG, (Valmane)

LgP	0.39(C,3)

| pKa | |

pp cited in Vols.1-5:

(1) 'Martindale. The Extra Pharmacopoeia', ed. J.E.F. Reynolds, The Pharmaceutical Press, London, 1982, 28th Ed., p 1768.

ACEXAMIC ACID [INN]			
CAS 57-08-9	MW: 173.21	MF:	C8 H15 N O3

Anti-inflammatory

Choay, France (Plastenan)

LgP	-0.42(C)

| pKa | |

pp cited in Vols.1-5:

(1) P. Daoudal et al., Nouv. Presse Med., 1979, 8, 3899.

ACIFRAN [U;INN]			
CAS 72420-38-3	MW: 218.21	MF:	C12 H10 O4

Antihyperlipoproteinemic

Ayerst (Reductol)

LgP	(4)

| pKa | |

pp cited in Vols.1-5:

(1) M.N. Cayen et al., Atherosclerosis, 1982, 45, 267-279.

ACIPIMOX [INN]			
CAS 51037-30-0	MW: 154.13	MF:	C6 H6 N2 O3

Antihyperlipoproteinemic

LgP	-1.02(C)

| pKa | |

pp cited in Vols.1-5:

5: 84

(1) G. Crepaldi et al., Atherosclerosis, 1988, 70, 115-121.

ACITEMATE [INN]			
CAS 101197-99-3	MW: 294.31	MF:	C14 H18 N2 O5

Anticoagulant

Chinoin, Hungary

LgP	0.02(C)

| pKa | |

pp cited in Vols.1-5:

(1) 'Diet & Drugs in Atherosclerosis', Raven Press, NY, 1980, p 111.

ACITRETIN [U;INN]			
CAS 55079-83-9	MW: 326.44	MF:	C21 H26 O3

Antipsoriatic

Hoffmann-LaRoche (Soriatane)

LgP	6.11(C,2)

| pKa | |

pp cited in Vols.1-5:

(1) J.M. Geiger et al., Dermatologica, 1988, 176, 182-190.

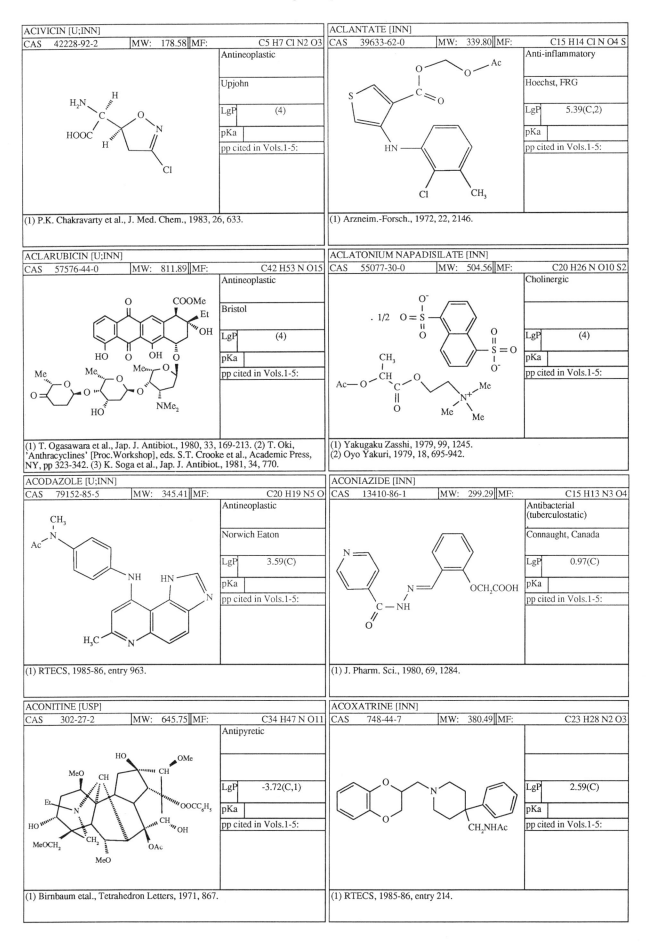

ACIVICIN [U;INN]

CAS	42228-92-2	MW:	178.58	MF:	C5 H7 Cl N2 O3

Antineoplastic

Upjohn

LgP (4)

pKa

pp cited in Vols.1-5:

(1) P.K. Chakravarty et al., J. Med. Chem., 1983, 26, 633.

ACLANTATE [INN]

CAS	39633-62-0	MW:	339.80	MF:	C15 H14 Cl N O4 S

Anti-inflammatory

Hoechst, FRG

LgP 5.39(C,2)

pKa

pp cited in Vols.1-5:

(1) Arzneim.-Forsch., 1972, 22, 2146.

ACLARUBICIN [U;INN]

CAS	57576-44-0	MW:	811.89	MF:	C42 H53 N O15

Antineoplastic

Bristol

LgP (4)

pKa

pp cited in Vols.1-5:

(1) T. Ogasawara et al., Jap. J. Antibiot., 1980, 33, 169-213. (2) T. Oki, 'Anthracyclines' [Proc.Workshop], eds. S.T. Crooke et al., Academic Press, NY, pp 323-342. (3) K. Soga et al., Jap. J. Antibiot., 1981, 34, 770.

ACLATONIUM NAPADISILATE [INN]

CAS	55077-30-0	MW:	504.56	MF:	C20 H26 N O10 S2

Cholinergic

LgP (4)

pKa

pp cited in Vols.1-5:

(1) Yakugaku Zasshi, 1979, 99, 1245.
(2) Oyo Yakuri, 1979, 18, 695-942.

ACODAZOLE [U;INN]

CAS	79152-85-5	MW:	345.41	MF:	C20 H19 N5 O

Antineoplastic

Norwich Eaton

LgP 3.59(C)

pKa

pp cited in Vols.1-5:

(1) RTECS, 1985-86, entry 963.

ACONIAZIDE [INN]

CAS	13410-86-1	MW:	299.29	MF:	C15 H13 N3 O4

Antibacterial (tuberculostatic)

Connaught, Canada

LgP 0.97(C)

pKa

pp cited in Vols.1-5:

(1) J. Pharm. Sci., 1980, 69, 1284.

ACONITINE [USP]

CAS	302-27-2	MW:	645.75	MF:	C34 H47 N O11

Antipyretic

LgP -3.72(C,1)

pKa

pp cited in Vols.1-5:

(1) Birnbaum etal., Tetrahedron Letters, 1971, 867.

ACOXATRINE [INN]

CAS	748-44-7	MW:	380.49	MF:	C23 H28 N2 O3

LgP 2.59(C)

pKa

pp cited in Vols.1-5:

(1) RTECS, 1985-86, entry 214.

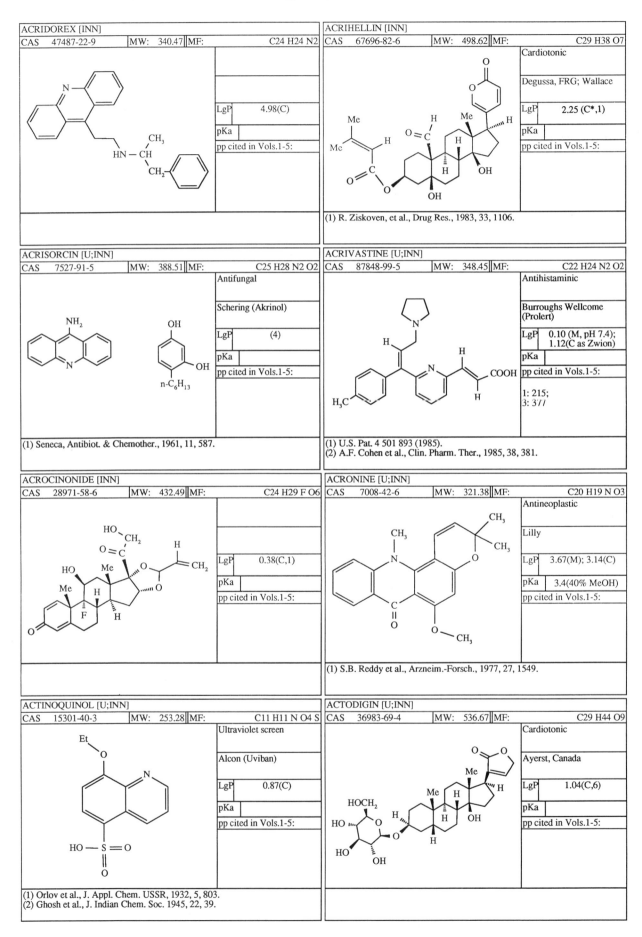

ACRIDOREX [INN]

CAS 47487-22-9	MW: 340.47	MF: C24 H24 N2

LgP	4.98(C)
pKa	
pp cited in Vols.1-5:	

ACRIHELLIN [INN]

CAS 67696-82-6	MW: 498.62	MF: C29 H38 O7

Cardiotonic

Degussa, FRG; Wallace

LgP	2.25 (C*,1)
pKa	
pp cited in Vols.1-5:	

(1) R. Ziskoven, et al., Drug Res., 1983, 33, 1106.

ACRISORCIN [U;INN]

CAS 7527-91-5	MW: 388.51	MF: C25 H28 N2 O2

Antifungal

Schering (Akrinol)

LgP	(4)
pKa	
pp cited in Vols.1-5:	

(1) Seneca, Antibiot. & Chemother., 1961, 11, 587.

ACRIVASTINE [U;INN]

CAS 87848-99-5	MW: 348.45	MF: C22 H24 N2 O2

Antihistaminic

Burroughs Wellcome (Prolert)

LgP	0.10 (M, pH 7.4); 1.12(C as Zwion)
pKa	
pp cited in Vols.1-5:	
	1: 215; 3: 377

(1) U.S. Pat. 4 501 893 (1985).
(2) A.F. Cohen et al., Clin. Pharm. Ther., 1985, 38, 381.

ACROCINONIDE [INN]

CAS 28971-58-6	MW: 432.49	MF: C24 H29 F O6

LgP	0.38(C,1)
pKa	
pp cited in Vols.1-5:	

ACRONINE [U;INN]

CAS 7008-42-6	MW: 321.38	MF: C20 H19 N O3

Antineoplastic

Lilly

LgP	3.67(M); 3.14(C)
pKa	3.4(40% MeOH)
pp cited in Vols.1-5:	

(1) S.B. Reddy et al., Arzneim.-Forsch., 1977, 27, 1549.

ACTINOQUINOL [U;INN]

CAS 15301-40-3	MW: 253.28	MF: C11 H11 N O4 S

Ultraviolet screen

Alcon (Uviban)

LgP	0.87(C)
pKa	
pp cited in Vols.1-5:	

(1) Orlov et al., J. Appl. Chem. USSR, 1932, 5, 803.
(2) Ghosh et al., J. Indian Chem. Soc. 1945, 22, 39.

ACTODIGIN [U;INN]

CAS 36983-69-4	MW: 536.67	MF: C29 H44 O9

Cardiotonic

Ayerst, Canada

LgP	1.04(C,6)
pKa	
pp cited in Vols.1-5:	

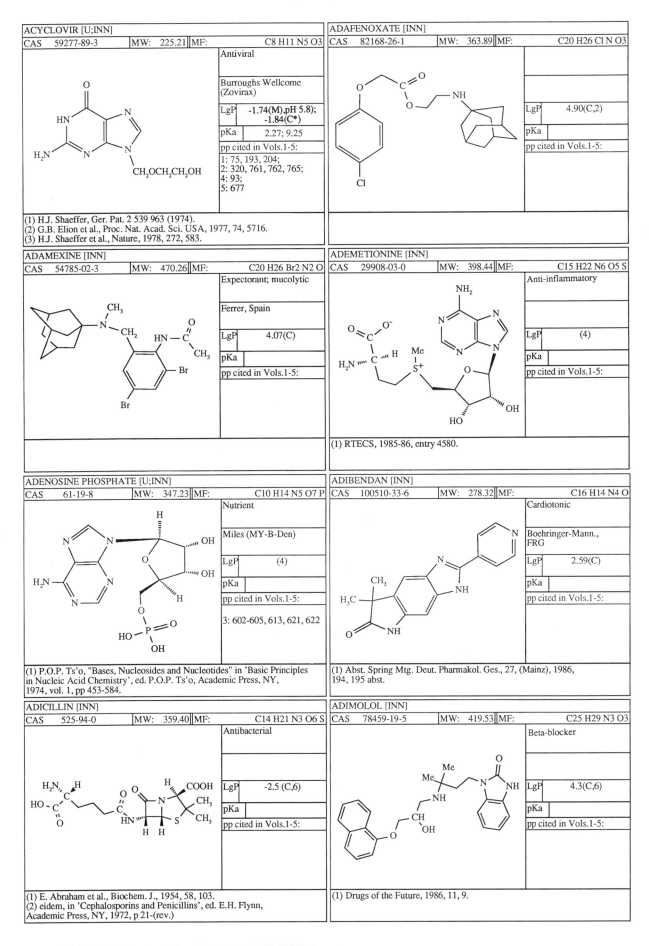

ACYCLOVIR [U;INN]
CAS 59277-89-3 MW: 225.21 MF: C8 H11 N5 O3
Antiviral
Burroughs Wellcome (Zovirax)
LgP -1.74(M),pH 5.8); -1.84(C*)
pKa 2.27; 9.25
pp cited in Vols.1-5:
1: 75, 193, 204;
2: 320, 761, 762, 765;
4: 93;
5: 677
(1) H.J. Shaeffer, Ger. Pat. 2 539 963 (1974).
(2) G.B. Elion et al., Proc. Nat. Acad. Sci. USA, 1977, 74, 5716.
(3) H.J. Shaeffer et al., Nature, 1978, 272, 583.

ADAFENOXATE [INN]
CAS 82168-26-1 MW: 363.89 MF: C20 H26 Cl N O3
LgP 4.90(C,2)
pKa
pp cited in Vols.1-5:

ADAMEXINE [INN]
CAS 54785-02-3 MW: 470.26 MF: C20 H26 Br2 N2 O
Expectorant; mucolytic
Ferrer, Spain
LgP 4.07(C)
pKa
pp cited in Vols.1-5:

ADEMETIONINE [INN]
CAS 29908-03-0 MW: 398.44 MF: C15 H22 N6 O5 S
Anti-inflammatory
LgP (4)
pKa
pp cited in Vols.1-5:
(1) RTECS, 1985-86, entry 4580.

ADENOSINE PHOSPHATE [U;INN]
CAS 61-19-8 MW: 347.23 MF: C10 H14 N5 O7 P
Nutrient
Miles (MY-B-Den)
LgP (4)
pKa
pp cited in Vols.1-5:
3: 602-605, 613, 621, 622
(1) P.O.P. Ts'o, "Bases, Nucleosides and Nucleotides" in 'Basic Principles in Nucleic Acid Chemistry', ed. P.O.P. Ts'o, Academic Press, NY, 1974, vol. 1, pp 453-584.

ADIBENDAN [INN]
CAS 100510-33-6 MW: 278.32 MF: C16 H14 N4 O
Cardiotonic
Boehringer-Mann., FRG
LgP 2.59(C)
pKa
pp cited in Vols.1-5:
(1) Abst. Spring Mtg. Deut. Pharmakol. Ges., 27, (Mainz), 1986, 194, 195 abst.

ADICILLIN [INN]
CAS 525-94-0 MW: 359.40 MF: C14 H21 N3 O6 S
Antibacterial
LgP -2.5 (C,6)
pKa
pp cited in Vols.1-5:
(1) E. Abraham et al., Biochem. J., 1954, 58, 103.
(2) eidem, in 'Cephalosporins and Penicillins', ed. E.H. Flynn, Academic Press, NY, 1972, p 21-(rev.)

ADIMOLOL [INN]
CAS 78459-19-5 MW: 419.53 MF: C25 H29 N3 O3
Beta-blocker
LgP 4.3(C,6)
pKa
pp cited in Vols.1-5:
(1) Drugs of the Future, 1986, 11, 9.

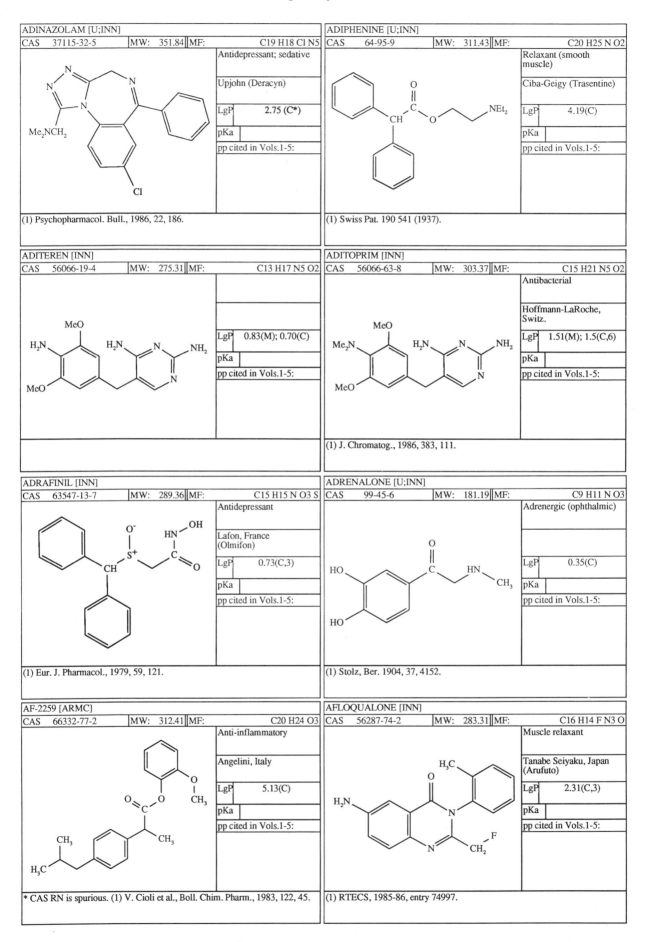

ADINAZOLAM [U;INN]

CAS	37115-32-5	MW:	351.84	MF:	C19 H18 Cl N5

Antidepressant; sedative

Upjohn (Deracyn)

LgP	2.75 (C*)
pKa	
pp cited in Vols.1-5:	

Me$_2$NCH$_2$

Cl

(1) Psychopharmacol. Bull., 1986, 22, 186.

ADIPHENINE [U;INN]

CAS	64-95-9	MW:	311.43	MF:	C20 H25 N O2

Relaxant (smooth muscle)

Ciba-Geigy (Trasentine)

LgP	4.19(C)
pKa	
pp cited in Vols.1-5:	

NEt$_2$

(1) Swiss Pat. 190 541 (1937).

ADITEREN [INN]

CAS	56066-19-4	MW:	275.31	MF:	C13 H17 N5 O2

LgP	0.83(M); 0.70(C)
pKa	
pp cited in Vols.1-5:	

MeO

H$_2$N

H$_2$N

NH$_2$

MeO

ADITOPRIM [INN]

CAS	56066-63-8	MW:	303.37	MF:	C15 H21 N5 O2

Antibacterial

Hoffmann-LaRoche, Switz.

LgP	1.51(M); 1.5(C,6)
pKa	
pp cited in Vols.1-5:	

MeO

Me$_2$N

H$_2$N

NH$_2$

MeO

(1) J. Chromatog., 1986, 383, 111.

ADRAFINIL [INN]

CAS	63547-13-7	MW:	289.36	MF:	C15 H15 N O3 S

Antidepressant

Lafon, France (Olmifon)

LgP	0.73(C,3)
pKa	
pp cited in Vols.1-5:	

O$^-$ HN OH

S$^+$

CH

(1) Eur. J. Pharmacol., 1979, 59, 121.

ADRENALONE [U;INN]

CAS	99-45-6	MW:	181.19	MF:	C9 H11 N O3

Adrenergic (ophthalmic)

LgP	0.35(C)
pKa	
pp cited in Vols.1-5:	

HO

HN CH$_3$

HO

(1) Stolz, Ber. 1904, 37, 4152.

AF-2259 [ARMC]

CAS	66332-77-2	MW:	312.41	MF:	C20 H24 O3

Anti-inflammatory

Angelini, Italy

LgP	5.13(C)
pKa	
pp cited in Vols.1-5:	

O CH$_3$

CH$_3$

CH$_3$

H$_3$C

* CAS RN is spurious. (1) V. Cioli et al., Boll. Chim. Pharm., 1983, 122, 45.

AFLOQUALONE [INN]

CAS	56287-74-2	MW:	283.31	MF:	C16 H14 F N3 O

Muscle relaxant

Tanabe Seiyaku, Japan (Arufuto)

LgP	2.31(C,3)
pKa	
pp cited in Vols.1-5:	

H$_3$C

O

H$_2$N

N

N CH$_2$ F

(1) RTECS, 1985-86, entry 74997.

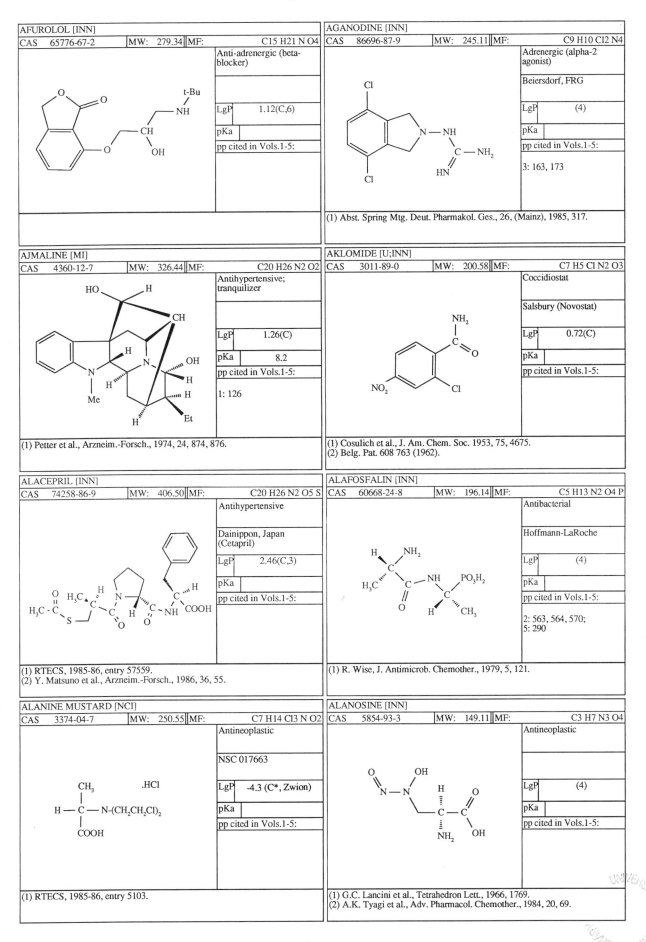

AFUROLOL [INN]

CAS	65776-67-2	MW:	279.34	MF:	C15 H21 N O4

Anti-adrenergic (beta-blocker)

LgP	1.12(C,6)
pKa	
pp cited in Vols.1-5:	

AGANODINE [INN]

CAS	86696-87-9	MW:	245.11	MF:	C9 H10 Cl2 N4

Adrenergic (alpha-2 agonist)

Beiersdorf, FRG

LgP	(4)
pKa	
pp cited in Vols.1-5:	

3: 163, 173

(1) Abst. Spring Mtg. Deut. Pharmakol. Ges., 26, (Mainz), 1985, 317.

AJMALINE [MI]

CAS	4360-12-7	MW:	326.44	MF:	C20 H26 N2 O2

Antihypertensive; tranquilizer

LgP	1.26(C)
pKa	8.2
pp cited in Vols.1-5:	

1: 126

(1) Petter et al., Arzneim.-Forsch., 1974, 24, 874, 876.

AKLOMIDE [U;INN]

CAS	3011-89-0	MW:	200.58	MF:	C7 H5 Cl N2 O3

Coccidiostat

Salsbury (Novostat)

LgP	0.72(C)
pKa	
pp cited in Vols.1-5:	

(1) Cosulich et al., J. Am. Chem. Soc. 1953, 75, 4675.
(2) Belg. Pat. 608 763 (1962).

ALACEPRIL [INN]

CAS	74258-86-9	MW:	406.50	MF:	C20 H26 N2 O5 S

Antihypertensive

Dainippon, Japan (Cetapril)

LgP	2.46(C,3)
pKa	
pp cited in Vols.1-5:	

(1) RTECS, 1985-86, entry 57559.
(2) Y. Matsuno et al., Arzneim.-Forsch., 1986, 36, 55.

ALAFOSFALIN [INN]

CAS	60668-24-8	MW:	196.14	MF:	C5 H13 N2 O4 P

Antibacterial

Hoffmann-LaRoche

LgP	(4)
pKa	
pp cited in Vols.1-5:	

2: 563, 564, 570;
5: 290

(1) R. Wise, J. Antimicrob. Chemother., 1979, 5, 121.

ALANINE MUSTARD [NCI]

CAS	3374-04-7	MW:	250.55	MF:	C7 H14 Cl3 N O2

Antineoplastic

NSC 017663

LgP	-4.3 (C*, Zwion)
pKa	
pp cited in Vols.1-5:	

(1) RTECS, 1985-86, entry 5103.

ALANOSINE [INN]

CAS	5854-93-3	MW:	149.11	MF:	C3 H7 N3 O4

Antineoplastic

LgP	(4)
pKa	
pp cited in Vols.1-5:	

(1) G.C. Lancini et al., Tetrahedron Lett., 1966, 1769.
(2) A.K. Tyagi et al., Adv. Pharmacol. Chemother., 1984, 20, 69.

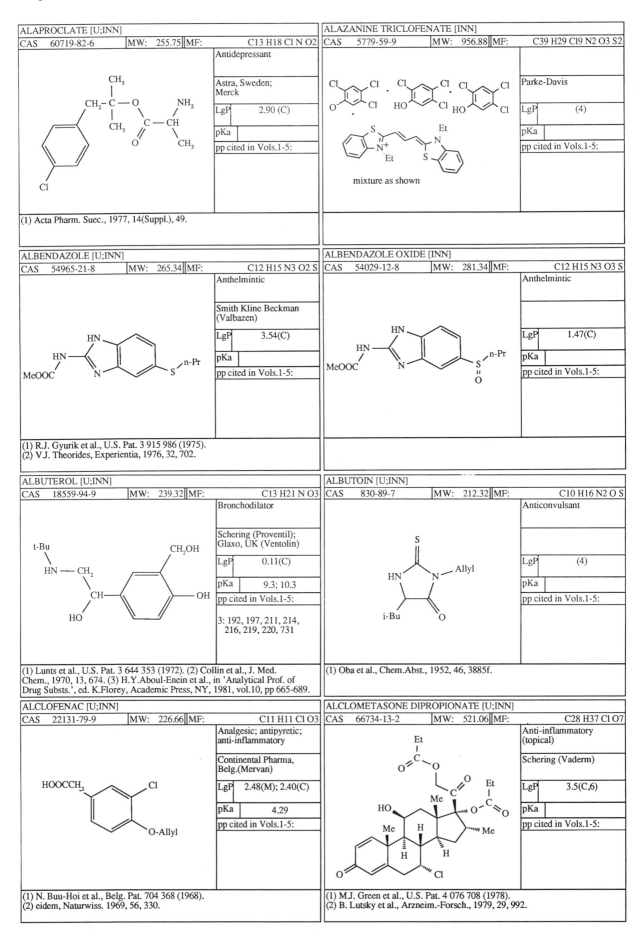

ALAPROCLATE [U;INN]

CAS	60719-82-6	MW:	255.75	MF:	C13 H18 Cl N O2

Antidepressant

Astra, Sweden; Merck

LgP	2.90 (C)
pKa	
pp cited in Vols.1-5:	

(1) Acta Pharm. Suec., 1977, 14(Suppl.), 49.

ALAZANINE TRICLOFENATE [INN]

CAS	5779-59-9	MW:	956.88	MF:	C39 H29 Cl9 N2 O3 S2

Parke-Davis

LgP	(4)
pKa	
pp cited in Vols.1-5:	

mixture as shown

ALBENDAZOLE [U;INN]

CAS	54965-21-8	MW:	265.34	MF:	C12 H15 N3 O2 S

Anthelmintic

Smith Kline Beckman (Valbazen)

LgP	3.54(C)
pKa	
pp cited in Vols.1-5:	

(1) R.J. Gyurik et al., U.S. Pat. 3 915 986 (1975).
(2) V.J. Theorides, Experientia, 1976, 32, 702.

ALBENDAZOLE OXIDE [INN]

CAS	54029-12-8	MW:	281.34	MF:	C12 H15 N3 O3 S

Anthelmintic

LgP	1.47(C)
pKa	
pp cited in Vols.1-5:	

ALBUTEROL [U;INN]

CAS	18559-94-9	MW:	239.32	MF:	C13 H21 N O3

Bronchodilator

Schering (Proventil); Glaxo, UK (Ventolin)

LgP	0.11(C)
pKa	9.3; 10.3
pp cited in Vols.1-5:	

3: 192, 197, 211, 214, 216, 219, 220, 731

(1) Lunts et al., U.S. Pat. 3 644 353 (1972). (2) Collin et al., J. Med. Chem., 1970, 13, 674. (3) H.Y.Aboul-Enein et al., in 'Analytical Prof. of Drug Substs.', ed. K.Florey, Academic Press, NY, 1981, vol.10, pp 665-689.

ALBUTOIN [U;INN]

CAS	830-89-7	MW:	212.32	MF:	C10 H16 N2 O S

Anticonvulsant

LgP	(4)
pKa	
pp cited in Vols.1-5:	

(1) Oba et al., Chem.Abst., 1952, 46, 3885f.

ALCLOFENAC [U;INN]

CAS	22131-79-9	MW:	226.66	MF:	C11 H11 Cl O3

Analgesic; antipyretic; anti-inflammatory

Continental Pharma, Belg.(Mervan)

LgP	2.48(M); 2.40(C)
pKa	4.29
pp cited in Vols.1-5:	

(1) N. Buu-Hoi et al., Belg. Pat. 704 368 (1968).
(2) eidem, Naturwiss. 1969, 56, 330.

ALCLOMETASONE DIPROPIONATE [U;INN]

CAS	66734-13-2	MW:	521.06	MF:	C28 H37 Cl O7

Anti-inflammatory (topical)

Schering (Vaderm)

LgP	3.5(C,6)
pKa	
pp cited in Vols.1-5:	

(1) M.J. Green et al., U.S. Pat. 4 076 708 (1978).
(2) B. Lutsky et al., Arzneim.-Forsch., 1979, 29, 992.

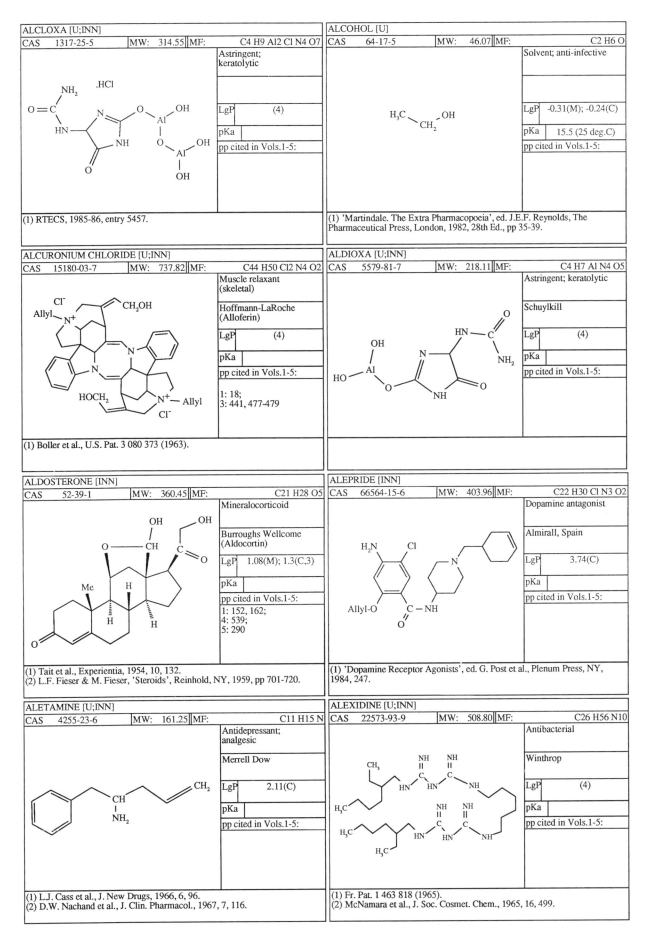

ALCLOXA [U;INN]

CAS	1317-25-5	MW:	314.55	MF:	C4 H9 Al2 Cl N4 O7

Astringent; keratolytic

LgP   (4)

pKa

pp cited in Vols.1-5:

(1) RTECS, 1985-86, entry 5457.

ALCOHOL [U]

CAS	64-17-5	MW:	46.07	MF:	C2 H6 O

Solvent; anti-infective

LgP   -0.31(M); -0.24(C)

pKa   15.5 (25 deg.C)

pp cited in Vols.1-5:

(1) 'Martindale. The Extra Pharmacopoeia', ed. J.E.F. Reynolds, The Pharmaceutical Press, London, 1982, 28th Ed., pp 35-39.

ALCURONIUM CHLORIDE [U;INN]

CAS	15180-03-7	MW:	737.82	MF:	C44 H50 Cl2 N4 O2

Muscle relaxant (skeletal)

Hoffmann-LaRoche (Alloferin)

LgP   (4)

pKa

pp cited in Vols.1-5:

1: 18;
3: 441, 477-479

(1) Boller et al., U.S. Pat. 3 080 373 (1963).

ALDIOXA [U;INN]

CAS	5579-81-7	MW:	218.11	MF:	C4 H7 Al N4 O5

Astringent; keratolytic

Schuylkill

LgP   (4)

pKa

pp cited in Vols.1-5:

ALDOSTERONE [INN]

CAS	52-39-1	MW:	360.45	MF:	C21 H28 O5

Mineralocorticoid

Burroughs Wellcome (Aldocortin)

LgP   1.08(M); 1.3(C,3)

pKa

pp cited in Vols.1-5:

1: 152, 162;
4: 539;
5: 290

(1) Tait et al., Experientia, 1954, 10, 132.
(2) L.F. Fieser & M. Fieser, 'Steroids', Reinhold, NY, 1959, pp 701-720.

ALEPRIDE [INN]

CAS	66564-15-6	MW:	403.96	MF:	C22 H30 Cl N3 O2

Dopamine antagonist

Almirall, Spain

LgP   3.74(C)

pKa

pp cited in Vols.1-5:

(1) 'Dopamine Receptor Agonists', ed. G. Post et al., Plenum Press, NY, 1984, 247.

ALETAMINE [U;INN]

CAS	4255-23-6	MW:	161.25	MF:	C11 H15 N

Antidepressant; analgesic

Merrell Dow

LgP   2.11(C)

pKa

pp cited in Vols.1-5:

(1) L.J. Cass et al., J. New Drugs, 1966, 6, 96.
(2) D.W. Nachand et al., J. Clin. Pharmacol., 1967, 7, 116.

ALEXIDINE [U;INN]

CAS	22573-93-9	MW:	508.80	MF:	C26 H56 N10

Antibacterial

Winthrop

LgP   (4)

pKa

pp cited in Vols.1-5:

(1) Fr. Pat. 1 463 818 (1965).
(2) McNamara et al., J. Soc. Cosmet. Chem., 1965, 16, 499.

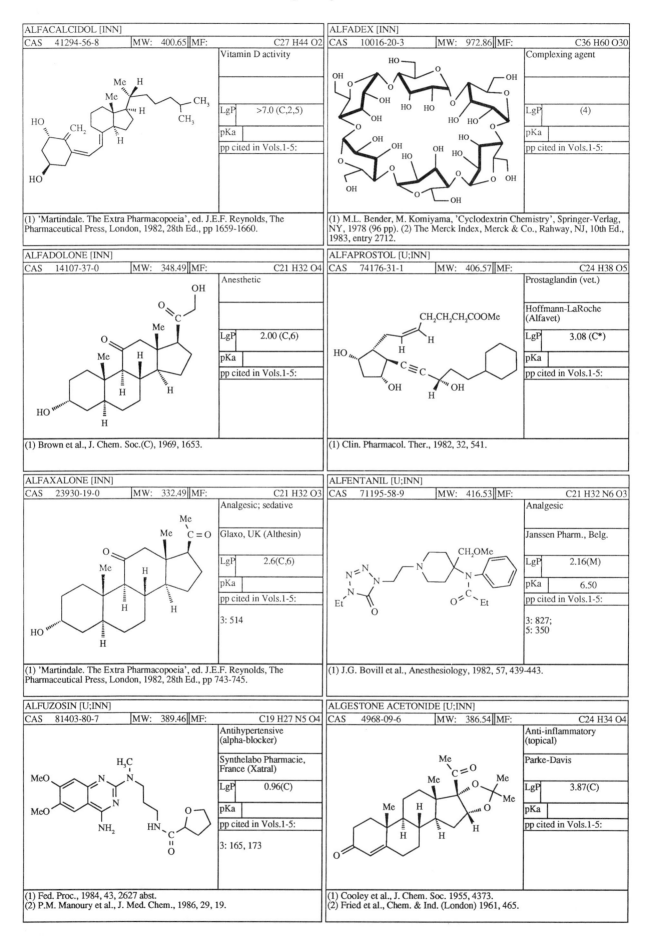

ALFACALCIDOL [INN]

CAS 41294-56-8	MW: 400.65	MF: C27 H44 O2

Vitamin D activity

LgP	>7.0 (C,2,5)
pKa	
pp cited in Vols.1-5:	

(1) 'Martindale. The Extra Pharmacopoeia', ed. J.E.F. Reynolds, The Pharmaceutical Press, London, 1982, 28th Ed., pp 1659-1660.

ALFADEX [INN]

CAS 10016-20-3	MW: 972.86	MF: C36 H60 O30

Complexing agent

LgP	(4)
pKa	
pp cited in Vols.1-5:	

(1) M.L. Bender, M. Komiyama, 'Cyclodextrin Chemistry', Springer-Verlag, NY, 1978 (96 pp). (2) The Merck Index, Merck & Co., Rahway, NJ, 10th Ed., 1983, entry 2712.

ALFADOLONE [INN]

CAS 14107-37-0	MW: 348.49	MF: C21 H32 O4

Anesthetic

LgP	2.00 (C,6)
pKa	
pp cited in Vols.1-5:	

(1) Brown et al., J. Chem. Soc.(C), 1969, 1653.

ALFAPROSTOL [U;INN]

CAS 74176-31-1	MW: 406.57	MF: C24 H38 O5

Prostaglandin (vet.)

Hoffmann-LaRoche (Alfavet)

LgP	3.08 (C*)
pKa	
pp cited in Vols.1-5:	

(1) Clin. Pharmacol. Ther., 1982, 32, 541.

ALFAXALONE [INN]

CAS 23930-19-0	MW: 332.49	MF: C21 H32 O3

Analgesic; sedative

Glaxo, UK (Althesin)

LgP	2.6(C,6)
pKa	
pp cited in Vols.1-5:	
3: 514	

(1) 'Martindale. The Extra Pharmacopoeia', ed. J.E.F. Reynolds, The Pharmaceutical Press, London, 1982, 28th Ed., pp 743-745.

ALFENTANIL [U;INN]

CAS 71195-58-9	MW: 416.53	MF: C21 H32 N6 O3

Analgesic

Janssen Pharm., Belg.

LgP	2.16(M)
pKa	6.50
pp cited in Vols.1-5:	
3: 827;	
5: 350	

(1) J.G. Bovill et al., Anesthesiology, 1982, 57, 439-443.

ALFUZOSIN [U;INN]

CAS 81403-80-7	MW: 389.46	MF: C19 H27 N5 O4

Antihypertensive (alpha-blocker)

Synthelabo Pharmacie, France (Xatral)

LgP	0.96(C)
pKa	
pp cited in Vols.1-5:	
3: 165, 173	

(1) Fed. Proc., 1984, 43, 2627 abst.
(2) P.M. Manoury et al., J. Med. Chem., 1986, 29, 19.

ALGESTONE ACETONIDE [U;INN]

CAS 4968-09-6	MW: 386.54	MF: C24 H34 O4

Anti-inflammatory (topical)

Parke-Davis

LgP	3.87(C)
pKa	
pp cited in Vols.1-5:	

(1) Cooley et al., J. Chem. Soc. 1955, 4373.
(2) Fried et al., Chem. & Ind. (London) 1961, 465.

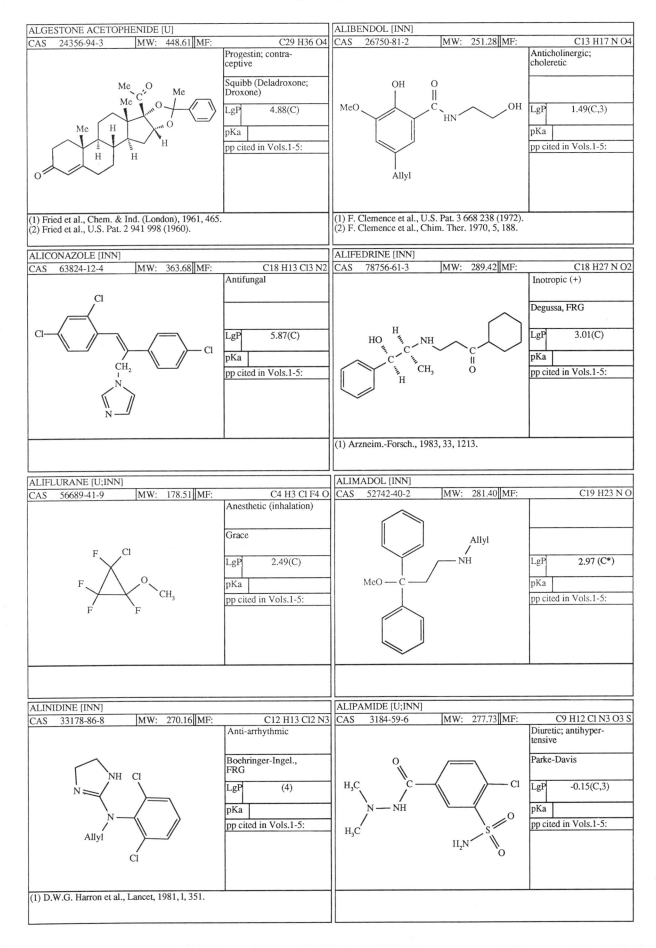

ALGESTONE ACETOPHENIDE [U]			
CAS 24356-94-3	MW: 448.61	MF:	C29 H36 O4

Progestin; contraceptive

Squibb (Deladroxone; Droxone)

LgP	4.88(C)
pKa	
pp cited in Vols.1-5:	

(1) Fried et al., Chem. & Ind. (London), 1961, 465.
(2) Fried et al., U.S. Pat. 2 941 998 (1960).

ALIBENDOL [INN]			
CAS 26750-81-2	MW: 251.28	MF:	C13 H17 N O4

Anticholinergic; choleretic

LgP	1.49(C,3)
pKa	
pp cited in Vols.1-5:	

(1) F. Clemence et al., U.S. Pat. 3 668 238 (1972).
(2) F. Clemence et al., Chim. Ther. 1970, 5, 188.

ALICONAZOLE [INN]			
CAS 63824-12-4	MW: 363.68	MF:	C18 H13 Cl3 N2

Antifungal

LgP	5.87(C)
pKa	
pp cited in Vols.1-5:	

ALIFEDRINE [INN]			
CAS 78756-61-3	MW: 289.42	MF:	C18 H27 N O2

Inotropic (+)

Degussa, FRG

LgP	3.01(C)
pKa	
pp cited in Vols.1-5:	

(1) Arzneim.-Forsch., 1983, 33, 1213.

ALIFLURANE [U;INN]			
CAS 56689-41-9	MW: 178.51	MF:	C4 H3 Cl F4 O

Anesthetic (inhalation)

Grace

LgP	2.49(C)
pKa	
pp cited in Vols.1-5:	

ALIMADOL [INN]			
CAS 52742-40-2	MW: 281.40	MF:	C19 H23 N O

LgP	2.97 (C*)
pKa	
pp cited in Vols.1-5:	

ALINIDINE [INN]			
CAS 33178-86-8	MW: 270.16	MF:	C12 H13 Cl2 N3

Anti-arrhythmic

Boehringer-Ingel., FRG

LgP	(4)
pKa	
pp cited in Vols.1-5:	

(1) D.W.G. Harron et al., Lancet, 1981, 1, 351.

ALIPAMIDE [U;INN]			
CAS 3184-59-6	MW: 277.73	MF:	C9 H12 Cl N3 O3 S

Diuretic; antihypertensive

Parke-Davis

LgP	-0.15(C,3)
pKa	
pp cited in Vols.1-5:	

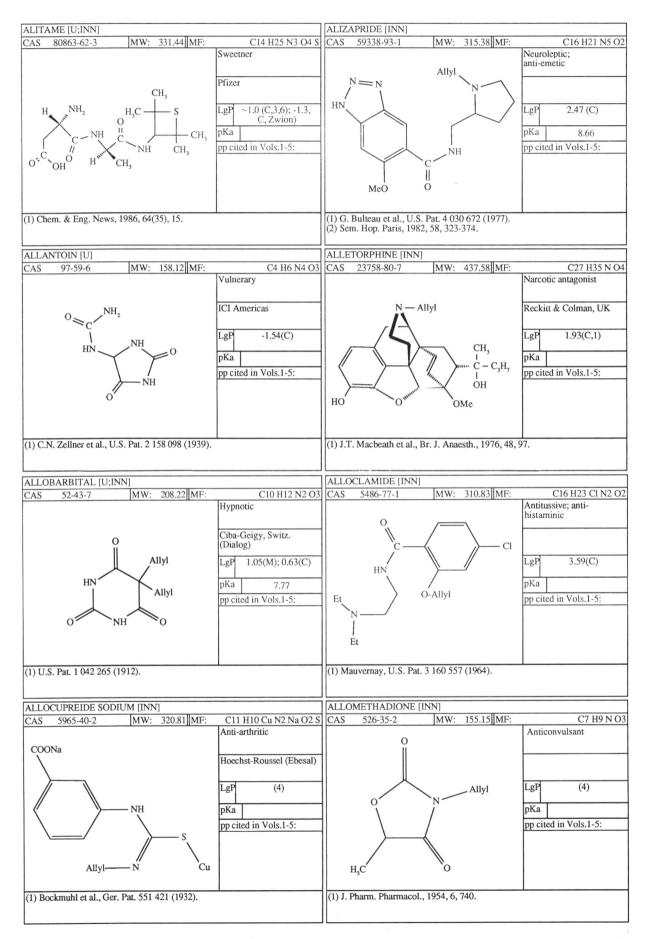

ALITAME [U;INN]

CAS 80863-62-3   MW: 331.44   MF: C14 H25 N3 O4 S

Sweetner

Pfizer

LgP ~1.0 (C,3,6); -1.3, C, Zwion)

pKa

pp cited in Vols.1-5:

(1) Chem. & Eng. News, 1986, 64(35), 15.

ALIZAPRIDE [INN]

CAS 59338-93-1   MW: 315.38   MF: C16 H21 N5 O2

Neuroleptic; anti-emetic

LgP 2.47 (C)

pKa 8.66

pp cited in Vols.1-5:

(1) G. Bulteau et al., U.S. Pat. 4 030 672 (1977).
(2) Sem. Hop. Paris, 1982, 58, 323-374.

ALLANTOIN [U]

CAS 97-59-6   MW: 158.12   MF: C4 H6 N4 O3

Vulnerary

ICI Americas

LgP -1.54(C)

pKa

pp cited in Vols.1-5:

(1) C.N. Zellner et al., U.S. Pat. 2 158 098 (1939).

ALLETORPHINE [INN]

CAS 23758-80-7   MW: 437.58   MF: C27 H35 N O4

Narcotic antagonist

Reckitt & Colman, UK

LgP 1.93(C,1)

pKa

pp cited in Vols.1-5:

(1) J.T. Macbeath et al., Br. J. Anaesth., 1976, 48, 97.

ALLOBARBITAL [U;INN]

CAS 52-43-7   MW: 208.22   MF: C10 H12 N2 O3

Hypnotic

Ciba-Geigy, Switz. (Dialog)

LgP 1.05(M); 0.63(C)

pKa 7.77

pp cited in Vols.1-5:

(1) U.S. Pat. 1 042 265 (1912).

ALLOCLAMIDE [INN]

CAS 5486-77-1   MW: 310.83   MF: C16 H23 Cl N2 O2

Antitussive; anti-histaminic

LgP 3.59(C)

pKa

pp cited in Vols.1-5:

(1) Mauvernay, U.S. Pat. 3 160 557 (1964).

ALLOCUPREIDE SODIUM [INN]

CAS 5965-40-2   MW: 320.81   MF: C11 H10 Cu N2 Na O2 S

Anti-arthritic

Hoechst-Roussel (Ebesal)

LgP (4)

pKa

pp cited in Vols.1-5:

(1) Bockmuhl et al., Ger. Pat. 551 421 (1932).

ALLOMETHADIONE [INN]

CAS 526-35-2   MW: 155.15   MF: C7 H9 N O3

Anticonvulsant

LgP (4)

pKa

pp cited in Vols.1-5:

(1) J. Pharm. Pharmacol., 1954, 6, 740.

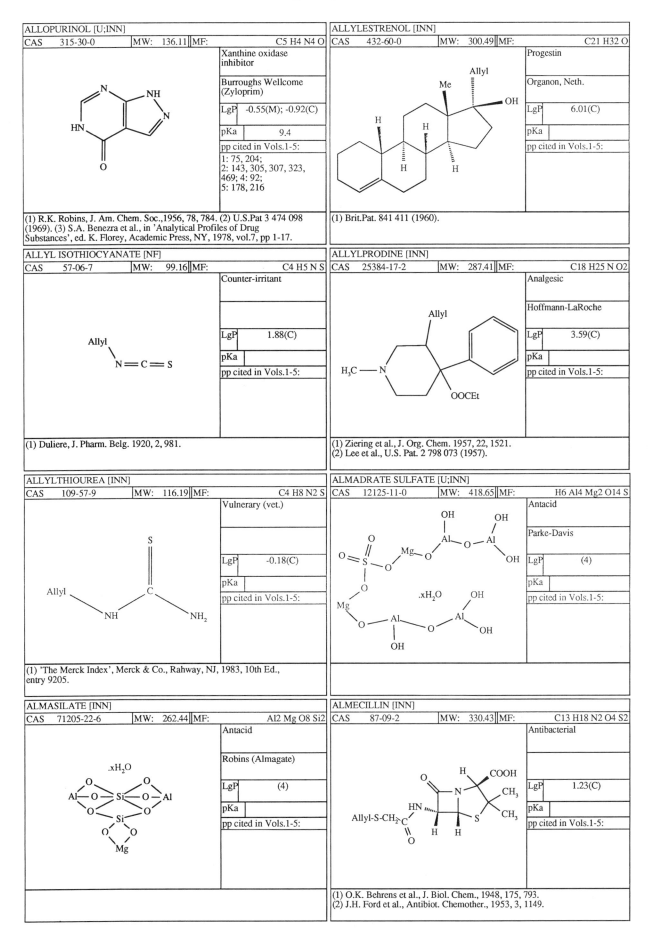

ALLOPURINOL [U;INN]

CAS	315-30-0	MW:	136.11	MF:	C5 H4 N4 O

Xanthine oxidase inhibitor

Burroughs Wellcome (Zyloprim)

LgP	-0.55(M); -0.92(C)

pKa	9.4

pp cited in Vols.1-5:

1: 75, 204;
2: 143, 305, 307, 323, 469; 4: 92;
5: 178, 216

(1) R.K. Robins, J. Am. Chem. Soc.,1956, 78, 784. (2) U.S.Pat 3 474 098 (1969). (3) S.A. Benezra et al., in 'Analytical Profiles of Drug Substances', ed. K. Florey, Academic Press, NY, 1978, vol.7, pp 1-17.

ALLYLESTRENOL [INN]

CAS	432-60-0	MW:	300.49	MF:	C21 H32 O

Progestin

Organon, Neth.

LgP	6.01(C)

pKa	

pp cited in Vols.1-5:

(1) Brit.Pat. 841 411 (1960).

ALLYL ISOTHIOCYANATE [NF]

CAS	57-06-7	MW:	99.16	MF:	C4 H5 N S

Counter-irritant

LgP	1.88(C)

pKa	

pp cited in Vols.1-5:

(1) Duliere, J. Pharm. Belg. 1920, 2, 981.

ALLYLPRODINE [INN]

CAS	25384-17-2	MW:	287.41	MF:	C18 H25 N O2

Analgesic

Hoffmann-LaRoche

LgP	3.59(C)

pKa	

pp cited in Vols.1-5:

(1) Ziering et al., J. Org. Chem. 1957, 22, 1521.
(2) Lee et al., U.S. Pat. 2 798 073 (1957).

ALLYLTHIOUREA [INN]

CAS	109-57-9	MW:	116.19	MF:	C4 H8 N2 S

Vulnerary (vet.)

LgP	-0.18(C)

pKa	

pp cited in Vols.1-5:

(1) 'The Merck Index', Merck & Co., Rahway, NJ, 1983, 10th Ed., entry 9205.

ALMADRATE SULFATE [U;INN]

CAS	12125-11-0	MW:	418.65	MF:	H6 Al4 Mg2 O14 S

Antacid

Parke-Davis

LgP	(4)

pKa	

pp cited in Vols.1-5:

.xH2O

ALMASILATE [INN]

CAS	71205-22-6	MW:	262.44	MF:	Al2 Mg O8 Si2

Antacid

Robins (Almagate)

LgP	(4)

pKa	

pp cited in Vols.1-5:

.xH2O

ALMECILLIN [INN]

CAS	87-09-2	MW:	330.43	MF:	C13 H18 N2 O4 S2

Antibacterial

LgP	1.23(C)

pKa	

pp cited in Vols.1-5:

(1) O.K. Behrens et al., J. Biol. Chem., 1948, 175, 793.
(2) J.H. Ford et al., Antibiot. Chemother., 1953, 3, 1149.

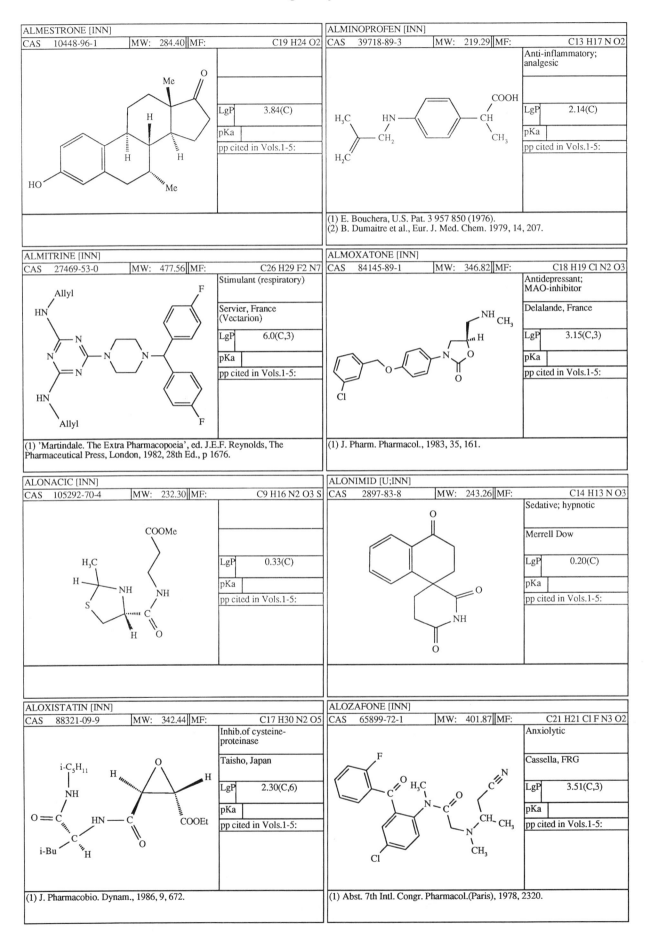

ALMESTRONE [INN]			
CAS 10448-96-1	MW: 284.40	MF:	C19 H24 O2
		LgP	3.84(C)
		pKa	
		pp cited in Vols.1-5:	

ALMINOPROFEN [INN]			
CAS 39718-89-3	MW: 219.29	MF:	C13 H17 N O2
		Anti-inflammatory; analgesic	
		LgP	2.14(C)
		pKa	
		pp cited in Vols.1-5:	

(1) E. Bouchera, U.S. Pat. 3 957 850 (1976).
(2) B. Dumaitre et al., Eur. J. Med. Chem. 1979, 14, 207.

ALMITRINE [INN]			
CAS 27469-53-0	MW: 477.56	MF:	C26 H29 F2 N7
		Stimulant (respiratory)	
		Servier, France (Vectarion)	
		LgP	6.0(C,3)
		pKa	
		pp cited in Vols.1-5:	

(1) 'Martindale. The Extra Pharmacopoeia', ed. J.E.F. Reynolds, The
Pharmaceutical Press, London, 1982, 28th Ed., p 1676.

ALMOXATONE [INN]			
CAS 84145-89-1	MW: 346.82	MF:	C18 H19 Cl N2 O3
		Antidepressant; MAO-inhibitor	
		Delalande, France	
		LgP	3.15(C,3)
		pKa	
		pp cited in Vols.1-5:	

(1) J. Pharm. Pharmacol., 1983, 35, 161.

ALONACIC [INN]			
CAS 105292-70-4	MW: 232.30	MF:	C9 H16 N2 O3 S
		LgP	0.33(C)
		pKa	
		pp cited in Vols.1-5:	

ALONIMID [U;INN]			
CAS 2897-83-8	MW: 243.26	MF:	C14 H13 N O3
		Sedative; hypnotic	
		Merrell Dow	
		LgP	0.20(C)
		pKa	
		pp cited in Vols.1-5:	

ALOXISTATIN [INN]			
CAS 88321-09-9	MW: 342.44	MF:	C17 H30 N2 O5
		Inhib.of cysteine-proteinase	
		Taisho, Japan	
		LgP	2.30(C,6)
		pKa	
		pp cited in Vols.1-5:	

(1) J. Pharmacobio. Dynam., 1986, 9, 672.

ALOZAFONE [INN]			
CAS 65899-72-1	MW: 401.87	MF:	C21 H21 Cl F N3 O2
		Anxiolytic	
		Cassella, FRG	
		LgP	3.51(C,3)
		pKa	
		pp cited in Vols.1-5:	

(1) Abst. 7th Intl. Congr. Pharmacol.(Paris), 1978, 2320.

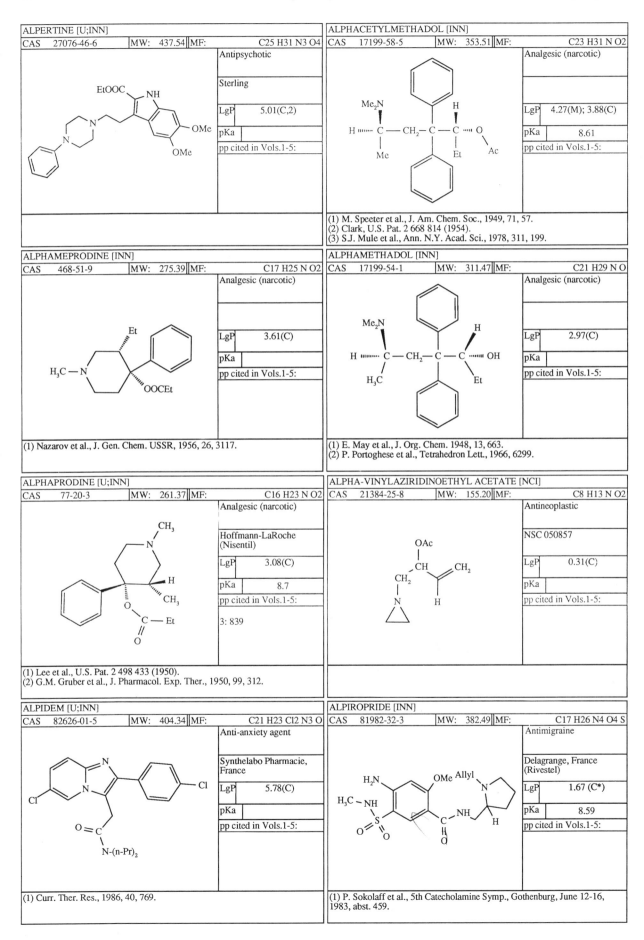

ALPERTINE [U;INN]

CAS 27076-46-6	MW: 437.54	MF:	C25 H31 N3 O4

Antipsychotic

Sterling

LgP	5.01(C,2)
pKa	
pp cited in Vols.1-5:	

EtOOC

OMe
OMe

ALPHACETYLMETHADOL [INN]

CAS 17199-58-5	MW: 353.51	MF:	C23 H31 N O2

Analgesic (narcotic)

LgP	4.27(M); 3.88(C)
pKa	8.61
pp cited in Vols.1-5:	

(1) M. Speeter et al., J. Am. Chem. Soc., 1949, 71, 57.
(2) Clark, U.S. Pat. 2 668 814 (1954).
(3) S.J. Mule et al., Ann. N.Y. Acad. Sci., 1978, 311, 199.

ALPHAMEPRODINE [INN]

CAS 468-51-9	MW: 275.39	MF:	C17 H25 N O2

Analgesic (narcotic)

LgP	3.61(C)
pKa	
pp cited in Vols.1-5:	

(1) Nazarov et al., J. Gen. Chem. USSR, 1956, 26, 3117.

ALPHAMETHADOL [INN]

CAS 17199-54-1	MW: 311.47	MF:	C21 H29 N O

Analgesic (narcotic)

LgP	2.97(C)
pKa	
pp cited in Vols.1-5:	

(1) E. May et al., J. Org. Chem. 1948, 13, 663.
(2) P. Portoghese et al., Tetrahedron Lett., 1966, 6299.

ALPHAPRODINE [U;INN]

CAS 77-20-3	MW: 261.37	MF:	C16 H23 N O2

Analgesic (narcotic)

Hoffmann-LaRoche (Nisentil)

LgP	3.08(C)
pKa	8.7
pp cited in Vols.1-5:	

3: 839

(1) Lee et al., U.S. Pat. 2 498 433 (1950).
(2) G.M. Gruber et al., J. Pharmacol. Exp. Ther., 1950, 99, 312.

ALPHA-VINYLAZIRIDINOETHYL ACETATE [NCI]

CAS 21384-25-8	MW: 155.20	MF:	C8 H13 N O2

Antineoplastic

NSC 050857

LgP	0.31(C)
pKa	
pp cited in Vols.1-5:	

ALPIDEM [U;INN]

CAS 82626-01-5	MW: 404.34	MF:	C21 H23 Cl2 N3 O

Anti-anxiety agent

Synthelabo Pharmacie, France

LgP	5.78(C)
pKa	
pp cited in Vols.1-5:	

(1) Curr. Ther. Res., 1986, 40, 769.

ALPIROPRIDE [INN]

CAS 81982-32-3	MW: 382.49	MF:	C17 H26 N4 O4 S

Antimigraine

Delagrange, France (Rivestel)

LgP	1.67 (C*)
pKa	8.59
pp cited in Vols.1-5:	

(1) P. Sokolaff et al., 5th Catecholamine Symp., Gothenburg, June 12-16, 1983, abst. 459.

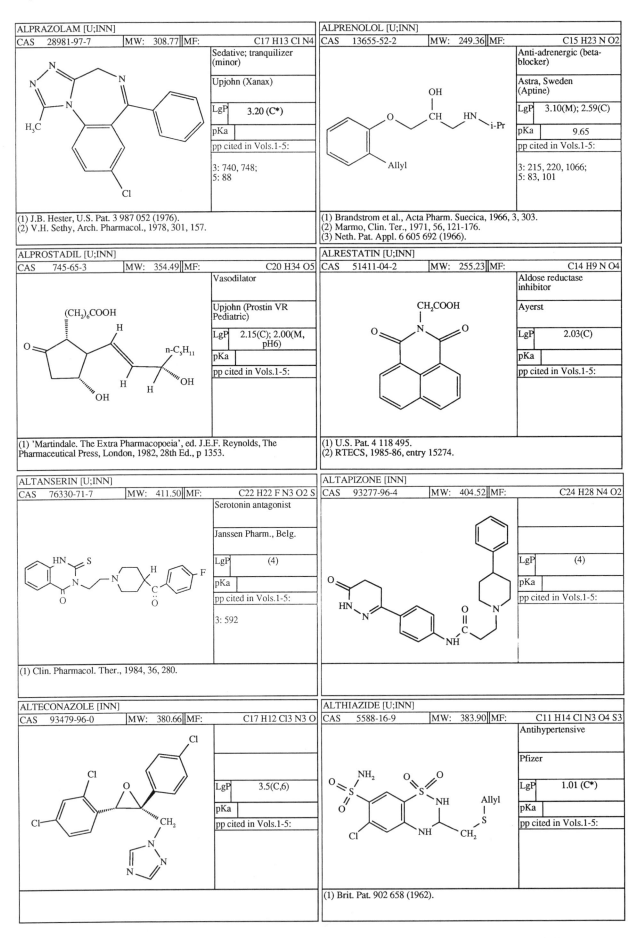

ALPRAZOLAM [U;INN]

CAS	28981-97-7	MW:	308.77	MF:	C17 H13 Cl N4

Sedative; tranquilizer (minor)

Upjohn (Xanax)

LgP	3.20 (C*)
pKa	
pp cited in Vols.1-5:	

3: 740, 748;
5: 88

(1) J.B. Hester, U.S. Pat. 3 987 052 (1976).
(2) V.H. Sethy, Arch. Pharmacol., 1978, 301, 157.

ALPRENOLOL [U;INN]

CAS	13655-52-2	MW:	249.36	MF:	C15 H23 N O2

Anti-adrenergic (beta-blocker)

Astra, Sweden (Aptine)

LgP	3.10(M); 2.59(C)
pKa	9.65
pp cited in Vols.1-5:	

3: 215, 220, 1066;
5: 83, 101

(1) Brandstrom et al., Acta Pharm. Suecica, 1966, 3, 303.
(2) Marmo, Clin. Ter., 1971, 56, 121-176.
(3) Neth. Pat. Appl. 6 605 692 (1966).

ALPROSTADIL [U;INN]

CAS	745-65-3	MW:	354.49	MF:	C20 H34 O5

Vasodilator

Upjohn (Prostin VR Pediatric)

LgP	2.15(C); 2.00(M, pH6)
pKa	
pp cited in Vols.1-5:	

(1) 'Martindale. The Extra Pharmacopoeia', ed. J.E.F. Reynolds, The Pharmaceutical Press, London, 1982, 28th Ed., p 1353.

ALRESTATIN [U;INN]

CAS	51411-04-2	MW:	255.23	MF:	C14 H9 N O4

Aldose reductase inhibitor

Ayerst

LgP	2.03(C)
pKa	
pp cited in Vols.1-5:	

(1) U.S. Pat. 4 118 495.
(2) RTECS, 1985-86, entry 15274.

ALTANSERIN [U;INN]

CAS	76330-71-7	MW:	411.50	MF:	C22 H22 F N3 O2 S

Serotonin antagonist

Janssen Pharm., Belg.

LgP	(4)
pKa	
pp cited in Vols.1-5:	

3: 592

(1) Clin. Pharmacol. Ther., 1984, 36, 280.

ALTAPIZONE [INN]

CAS	93277-96-4	MW:	404.52	MF:	C24 H28 N4 O2

LgP	(4)
pKa	
pp cited in Vols.1-5:	

ALTECONAZOLE [INN]

CAS	93479-96-0	MW:	380.66	MF:	C17 H12 Cl3 N3 O

LgP	3.5(C,6)
pKa	
pp cited in Vols.1-5:	

ALTHIAZIDE [U;INN]

CAS	5588-16-9	MW:	383.90	MF:	C11 H14 Cl N3 O4 S3

Antihypertensive

Pfizer

LgP	1.01 (C*)
pKa	
pp cited in Vols.1-5:	

(1) Brit. Pat. 902 658 (1962).

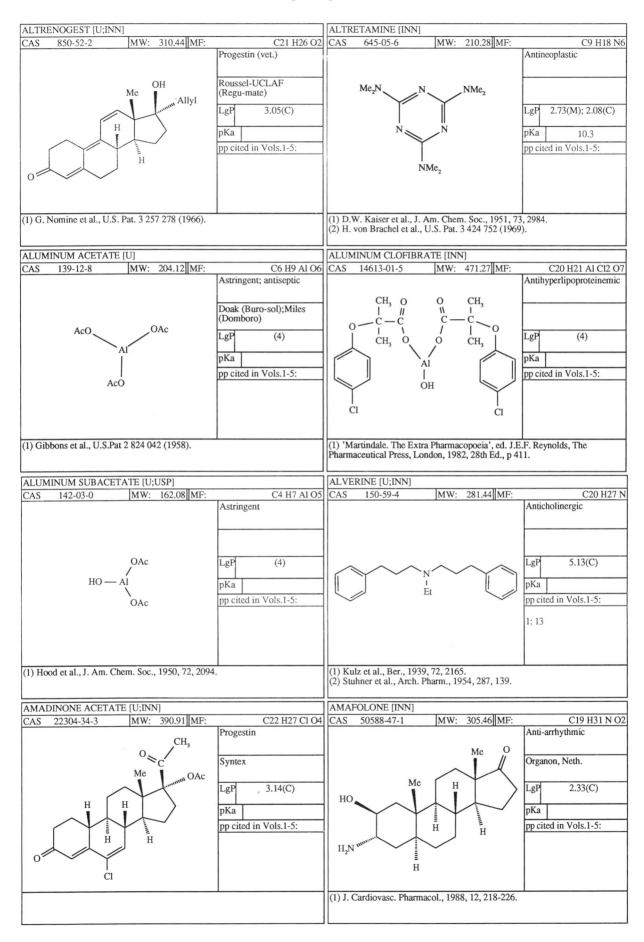

ALTRENOGEST [U;INN]			
CAS 850-52-2	MW: 310.44	MF:	C21 H26 O2

Progestin (vet.)

Roussel-UCLAF (Regu-mate)

LgP	3.05(C)
pKa	
pp cited in Vols.1-5:	

(1) G. Nomine et al., U.S. Pat. 3 257 278 (1966).

ALTRETAMINE [INN]			
CAS 645-05-6	MW: 210.28	MF:	C9 H18 N6

Antineoplastic

LgP	2.73(M); 2.08(C)
pKa	10.3
pp cited in Vols.1-5:	

(1) D.W. Kaiser et al., J. Am. Chem. Soc., 1951, 73, 2984.
(2) H. von Brachel et al., U.S. Pat. 3 424 752 (1969).

ALUMINUM ACETATE [U]			
CAS 139-12-8	MW: 204.12	MF:	C6 H9 Al O6

Astringent; antiseptic

Doak (Buro-sol);Miles (Domboro)

LgP	(4)
pKa	
pp cited in Vols.1-5:	

(1) Gibbons et al., U.S.Pat 2 824 042 (1958).

ALUMINUM CLOFIBRATE [INN]			
CAS 14613-01-5	MW: 471.27	MF:	C20 H21 Al Cl2 O7

Antihyperlipoproteinemic

LgP	(4)
pKa	
pp cited in Vols.1-5:	

(1) 'Martindale. The Extra Pharmacopoeia', ed. J.E.F. Reynolds, The Pharmaceutical Press, London, 1982, 28th Ed., p 411.

ALUMINUM SUBACETATE [U;USP]			
CAS 142-03-0	MW: 162.08	MF:	C4 H7 Al O5

Astringent

LgP	(4)
pKa	
pp cited in Vols.1-5:	

(1) Hood et al., J. Am. Chem. Soc., 1950, 72, 2094.

ALVERINE [U;INN]			
CAS 150-59-4	MW: 281.44	MF:	C20 H27 N

Anticholinergic

LgP	5.13(C)
pKa	
pp cited in Vols.1-5:	
1: 13	

(1) Kulz et al., Ber., 1939, 72, 2165.
(2) Stuhner et al., Arch. Pharm., 1954, 287, 139.

AMADINONE ACETATE [U;INN]			
CAS 22304-34-3	MW: 390.91	MF:	C22 H27 Cl O4

Progestin

Syntex

LgP	3.14(C)
pKa	
pp cited in Vols.1-5:	

AMAFOLONE [INN]			
CAS 50588-47-1	MW: 305.46	MF:	C19 H31 N O2

Anti-arrhythmic

Organon, Neth.

LgP	2.33(C)
pKa	
pp cited in Vols.1-5:	

(1) J. Cardiovasc. Pharmacol., 1988, 12, 218-226.

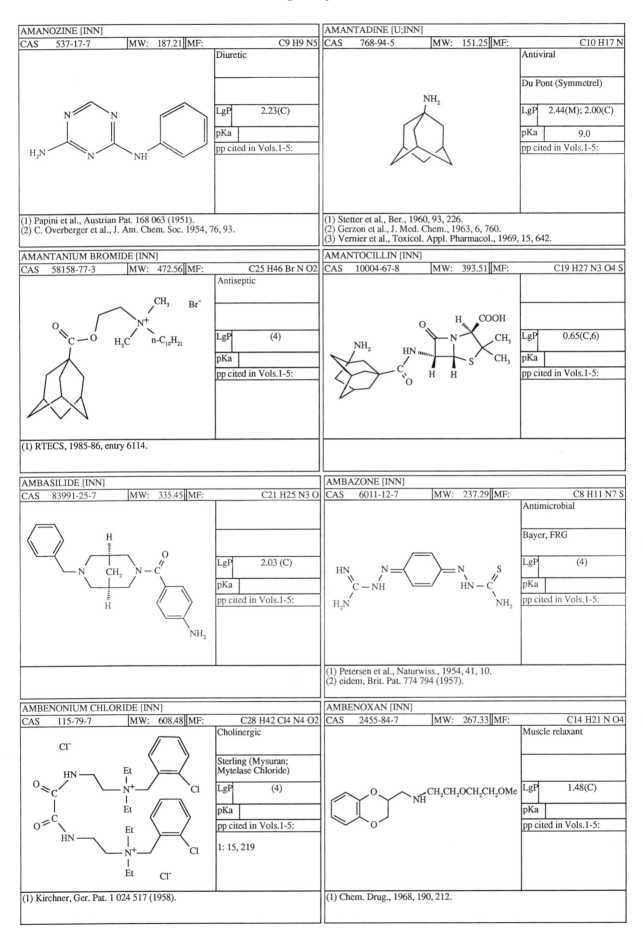

AMANOZINE [INN]			
CAS 537-17-7	MW: 187.21	MF:	C9 H9 N5

Diuretic

LgP	2.23(C)
pKa	
pp cited in Vols.1-5:	

(1) Papini et al., Austrian Pat. 168 063 (1951).
(2) C. Overberger et al., J. Am. Chem. Soc. 1954, 76, 93.

AMANTADINE [U;INN]			
CAS 768-94-5	MW: 151.25	MF:	C10 H17 N

Antiviral

Du Pont (Symmetrel)

LgP	2.44(M); 2.00(C)
pKa	9.0
pp cited in Vols.1-5:	

(1) Stetter et al., Ber., 1960, 93, 226.
(2) Gerzon et al., J. Med. Chem., 1963, 6, 760.
(3) Vernier et al., Toxicol. Appl. Pharmacol., 1969, 15, 642.

AMANTANIUM BROMIDE [INN]			
CAS 58158-77-3	MW: 472.56	MF:	C25 H46 Br N O2

Antiseptic

LgP	(4)
pKa	
pp cited in Vols.1-5:	

(1) RTECS, 1985-86, entry 6114.

AMANTOCILLIN [INN]			
CAS 10004-67-8	MW: 393.51	MF:	C19 H27 N3 O4 S

LgP	0.65(C,6)
pKa	
pp cited in Vols.1-5:	

AMBASILIDE [INN]			
CAS 83991-25-7	MW: 335.45	MF:	C21 H25 N3 O

LgP	2.03 (C)
pKa	
pp cited in Vols.1-5:	

AMBAZONE [INN]			
CAS 6011-12-7	MW: 237.29	MF:	C8 H11 N7 S

Antimicrobial

Bayer, FRG

LgP	(4)
pKa	
pp cited in Vols.1-5:	

(1) Petersen et al., Naturwiss., 1954, 41, 10.
(2) eidem, Brit. Pat. 774 794 (1957).

AMBENONIUM CHLORIDE [INN]			
CAS 115-79-7	MW: 608.48	MF:	C28 H42 Cl4 N4 O2

Cholinergic

Sterling (Mysuran; Mytelase Chloride)

LgP	(4)
pKa	
pp cited in Vols.1-5:	

1: 15, 219

(1) Kirchner, Ger. Pat. 1 024 517 (1958).

AMBENOXAN [INN]			
CAS 2455-84-7	MW: 267.33	MF:	C14 H21 N O4

Muscle relaxant

LgP	1.48(C)
pKa	
pp cited in Vols.1-5:	

(1) Chem. Drug., 1968, 190, 212.

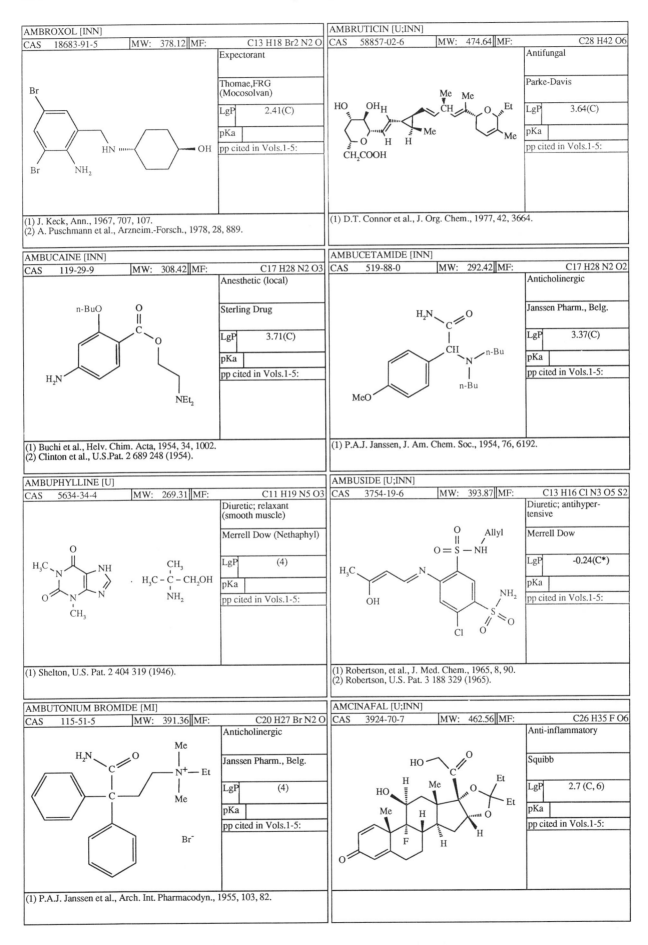

AMBROXOL [INN]

CAS	18683-91-5	MW:	378.12	MF:	C13 H18 Br2 N2 O

Expectorant

Thomae,FRG
(Mocosolvan)

LgP	2.41(C)
pKa	
pp cited in Vols.1-5:	

(1) J. Keck, Ann., 1967, 707, 107.
(2) A. Puschmann et al., Arzneim.-Forsch., 1978, 28, 889.

AMBRUTICIN [U;INN]

CAS	58857-02-6	MW:	474.64	MF:	C28 H42 O6

Antifungal

Parke-Davis

LgP	3.64(C)
pKa	
pp cited in Vols.1-5:	

(1) D.T. Connor et al., J. Org. Chem., 1977, 42, 3664.

AMBUCAINE [INN]

CAS	119-29-9	MW:	308.42	MF:	C17 H28 N2 O3

Anesthetic (local)

Sterling Drug

LgP	3.71(C)
pKa	
pp cited in Vols.1-5:	

(1) Buchi et al., Helv. Chim. Acta, 1954, 34, 1002.
(2) Clinton et al., U.S.Pat. 2 689 248 (1954).

AMBUCETAMIDE [INN]

CAS	519-88-0	MW:	292.42	MF:	C17 H28 N2 O2

Anticholinergic

Janssen Pharm., Belg.

LgP	3.37(C)
pKa	
pp cited in Vols.1-5:	

(1) P.A.J. Janssen, J. Am. Chem. Soc., 1954, 76, 6192.

AMBUPHYLLINE [U]

CAS	5634-34-4	MW:	269.31	MF:	C11 H19 N5 O3

Diuretic; relaxant
(smooth muscle)

Merrell Dow (Nethaphyl)

LgP	(4)
pKa	
pp cited in Vols.1-5:	

(1) Shelton, U.S. Pat. 2 404 319 (1946).

AMBUSIDE [U;INN]

CAS	3754-19-6	MW:	393.87	MF:	C13 H16 Cl N3 O5 S2

Diuretic; antihyper-
tensive

Merrell Dow

LgP	-0.24(C*)
pKa	
pp cited in Vols.1-5:	

(1) Robertson, et al., J. Med. Chem., 1965, 8, 90.
(2) Robertson, U.S. Pat. 3 188 329 (1965).

AMBUTONIUM BROMIDE [MI]

CAS	115-51-5	MW:	391.36	MF:	C20 H27 Br N2 O

Anticholinergic

Janssen Pharm., Belg.

LgP	(4)
pKa	
pp cited in Vols.1-5:	

(1) P.A.J. Janssen et al., Arch. Int. Pharmacodyn., 1955, 103, 82.

AMCINAFAL [U;INN]

CAS	3924-70-7	MW:	462.56	MF:	C26 H35 F O6

Anti-inflammatory

Squibb

LgP	2.7 (C, 6)
pKa	
pp cited in Vols.1-5:	

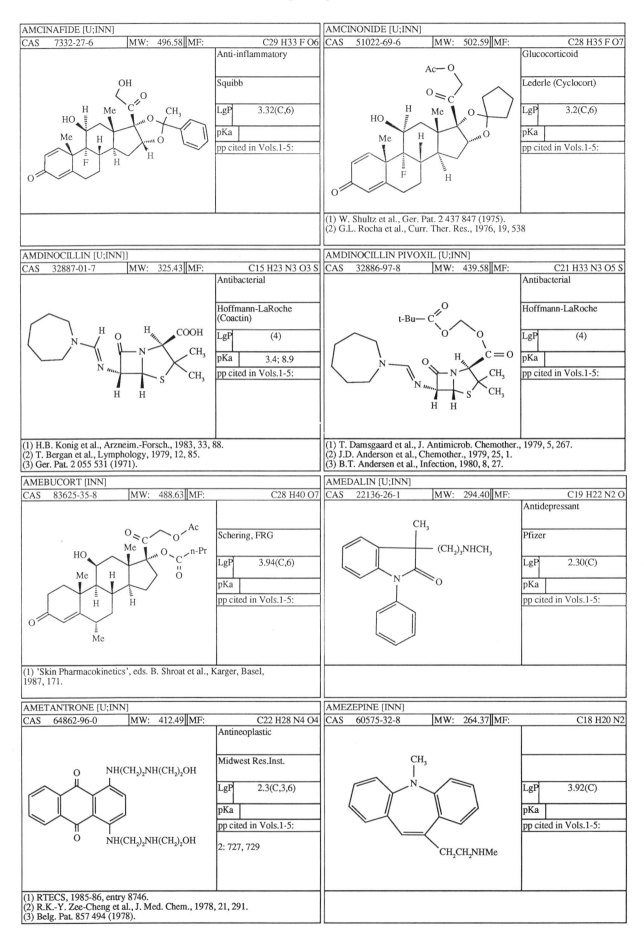

AMCINAFIDE [U;INN]				
CAS 7332-27-6	MW: 496.58	MF:		C29 H33 F O6

Anti-inflammatory

Squibb

LgP	3.32(C,6)
pKa	
pp cited in Vols.1-5:	

AMCINONIDE [U;INN]				
CAS 51022-69-6	MW: 502.59	MF:		C28 H35 F O7

Glucocorticoid

Lederle (Cyclocort)

LgP	3.2(C,6)
pKa	
pp cited in Vols.1-5:	

(1) W. Shultz et al., Ger. Pat. 2 437 847 (1975).
(2) G.L. Rocha et al., Curr. Ther. Res., 1976, 19, 538

AMDINOCILLIN [U;INN]]				
CAS 32887-01-7	MW: 325.43	MF:		C15 H23 N3 O3 S

Antibacterial

Hoffmann-LaRoche (Coactin)

LgP	(4)
pKa	3.4; 8.9
pp cited in Vols.1-5:	

(1) H.B. Konig et al., Arzneim.-Forsch., 1983, 33, 88.
(2) T. Bergan et al., Lymphology, 1979, 12, 85.
(3) Ger. Pat. 2 055 531 (1971).

AMDINOCILLIN PIVOXIL [U;INN]				
CAS 32886-97-8	MW: 439.58	MF:		C21 H33 N3 O5 S

Antibacterial

Hoffmann-LaRoche

LgP	(4)
pKa	
pp cited in Vols.1-5:	

(1) T. Damsgaard et al., J. Antimicrob. Chemother., 1979, 5, 267.
(2) J.D. Anderson et al., Chemother., 1979, 25, 1.
(3) B.T. Andersen et al., Infection, 1980, 8, 27.

AMEBUCORT [INN]				
CAS 83625-35-8	MW: 488.63	MF:		C28 H40 O7

Schering, FRG

LgP	3.94(C,6)
pKa	
pp cited in Vols.1-5:	

(1) 'Skin Pharmacokinetics', eds. B. Shroat et al., Karger, Basel, 1987, 171.

AMEDALIN [U;INN]				
CAS 22136-26-1	MW: 294.40	MF:		C19 H22 N2 O

Antidepressant

Pfizer

LgP	2.30(C)
pKa	
pp cited in Vols.1-5:	

AMETANTRONE [U;INN]				
CAS 64862-96-0	MW: 412.49	MF:		C22 H28 N4 O4

Antineoplastic

Midwest Res.Inst.

LgP	2.3(C,3,6)
pKa	
pp cited in Vols.1-5:	
2: 727, 729	

(1) RTECS, 1985-86, entry 8746.
(2) R.K.-Y. Zee-Cheng et al., J. Med. Chem., 1978, 21, 291.
(3) Belg. Pat. 857 494 (1978).

AMEZEPINE [INN]				
CAS 60575-32-8	MW: 264.37	MF:		C18 H20 N2

LgP	3.92(C)
pKa	
pp cited in Vols.1-5:	

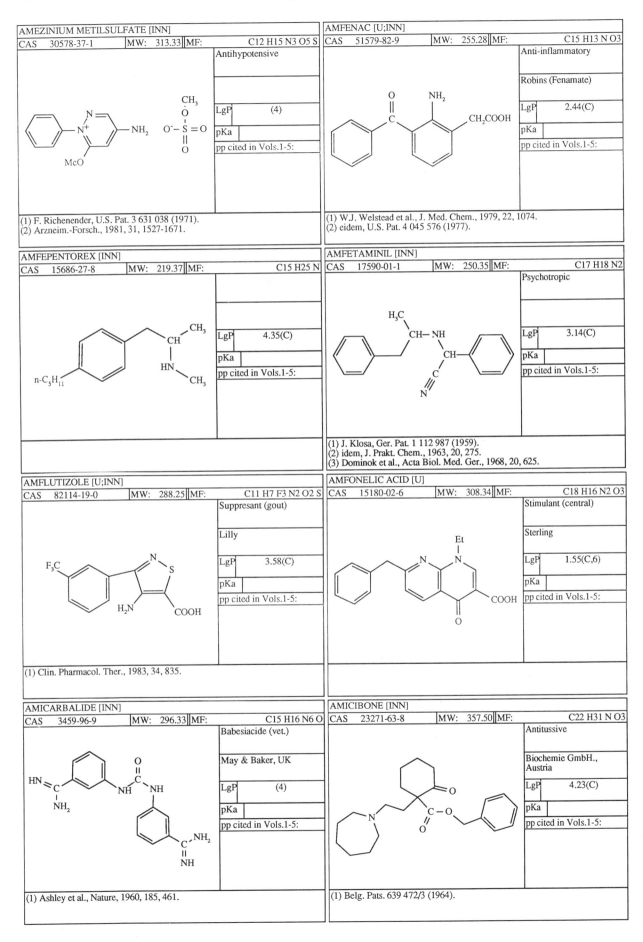

AMEZINIUM METILSULFATE [INN]

CAS	30578-37-1	MW:	313.33	MF:	C12 H15 N3 O5 S

Antihypotensive

LgP	(4)
pKa	
pp cited in Vols.1-5:	

(1) F. Richenender, U.S. Pat. 3 631 038 (1971).
(2) Arzneim.-Forsch., 1981, 31, 1527-1671.

AMFENAC [U;INN]

CAS	51579-82-9	MW:	255.28	MF:	C15 H13 N O3

Anti-inflammatory

Robins (Fenamate)

LgP	2.44(C)
pKa	
pp cited in Vols.1-5:	

(1) W.J. Welstead et al., J. Med. Chem., 1979, 22, 1074.
(2) eidem, U.S. Pat. 4 045 576 (1977).

AMFEPENTOREX [INN]

CAS	15686-27-8	MW:	219.37	MF:	C15 H25 N

LgP	4.35(C)
pKa	
pp cited in Vols.1-5:	

AMFETAMINIL [INN]

CAS	17590-01-1	MW:	250.35	MF:	C17 H18 N2

Psychotropic

LgP	3.14(C)
pKa	
pp cited in Vols.1-5:	

(1) J. Klosa, Ger. Pat. 1 112 987 (1959).
(2) idem, J. Prakt. Chem., 1963, 20, 275.
(3) Dominok et al., Acta Biol. Med. Ger., 1968, 20, 625.

AMFLUTIZOLE [U;INN]

CAS	82114-19-0	MW:	288.25	MF:	C11 H7 F3 N2 O2 S

Suppresant (gout)

Lilly

LgP	3.58(C)
pKa	
pp cited in Vols.1-5:	

(1) Clin. Pharmacol. Ther., 1983, 34, 835.

AMFONELIC ACID [U]

CAS	15180-02-6	MW:	308.34	MF:	C18 H16 N2 O3

Stimulant (central)

Sterling

LgP	1.55(C,6)
pKa	
pp cited in Vols.1-5:	

AMICARBALIDE [INN]

CAS	3459-96-9	MW:	296.33	MF:	C15 H16 N6 O

Babesiacide (vet.)

May & Baker, UK

LgP	(4)
pKa	
pp cited in Vols.1-5:	

(1) Ashley et al., Nature, 1960, 185, 461.

AMICIBONE [INN]

CAS	23271-63-8	MW:	357.50	MF:	C22 H31 N O3

Antitussive

Biochemie GmbH., Austria

LgP	4.23(C)
pKa	
pp cited in Vols.1-5:	

(1) Belg. Pats. 639 472/3 (1964).

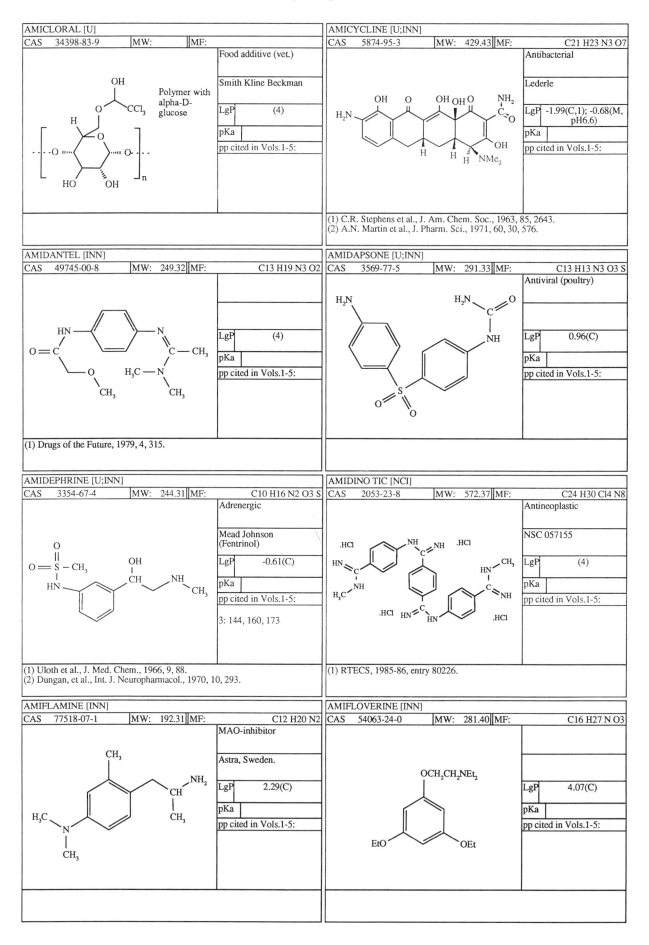

AMICLORAL [U]		
CAS 34398-83-9	MW:	MF:

Polymer with alpha-D-glucose

Food additive (vet.)

Smith Kline Beckman

LgP	(4)
pKa	
pp cited in Vols.1-5:	

AMICYCLINE [U;INN]		
CAS 5874-95-3	MW: 429.43	MF: C21 H23 N3 O7

Antibacterial

Lederle

LgP	-1.99(C,1); -0.68(M, pH6.6)
pKa	
pp cited in Vols.1-5:	

(1) C.R. Stephens et al., J. Am. Chem. Soc., 1963, 85, 2643.
(2) A.N. Martin et al., J. Pharm. Sci., 1971, 60, 30, 576.

AMIDANTEL [INN]		
CAS 49745-00-8	MW: 249.32	MF: C13 H19 N3 O2

LgP	(4)
pKa	
pp cited in Vols.1-5:	

(1) Drugs of the Future, 1979, 4, 315.

AMIDAPSONE [U;INN]		
CAS 3569-77-5	MW: 291.33	MF: C13 H13 N3 O3 S

Antiviral (poultry)

LgP	0.96(C)
pKa	
pp cited in Vols.1-5:	

AMIDEPHRINE [U;INN]		
CAS 3354-67-4	MW: 244.31	MF: C10 H16 N2 O3 S

Adrenergic

Mead Johnson (Fentrinol)

LgP	-0.61(C)
pKa	
pp cited in Vols.1-5:	

3: 144, 160, 173

(1) Uloth et al., J. Med. Chem., 1966, 9, 88.
(2) Dungan, et al., Int. J. Neuropharmacol., 1970, 10, 293.

AMIDINO TIC [NCI]		
CAS 2053-23-8	MW: 572.37	MF: C24 H30 Cl4 N8

Antineoplastic

NSC 057155

LgP	(4)
pKa	
pp cited in Vols.1-5:	

(1) RTECS, 1985-86, entry 80226.

AMIFLAMINE [INN]		
CAS 77518-07-1	MW: 192.31	MF: C12 H20 N2

MAO-inhibitor

Astra, Sweden.

LgP	2.29(C)
pKa	
pp cited in Vols.1-5:	

AMIFLOVERINE [INN]		
CAS 54063-24-0	MW: 281.40	MF: C16 H27 N O3

LgP	4.07(C)
pKa	
pp cited in Vols.1-5:	

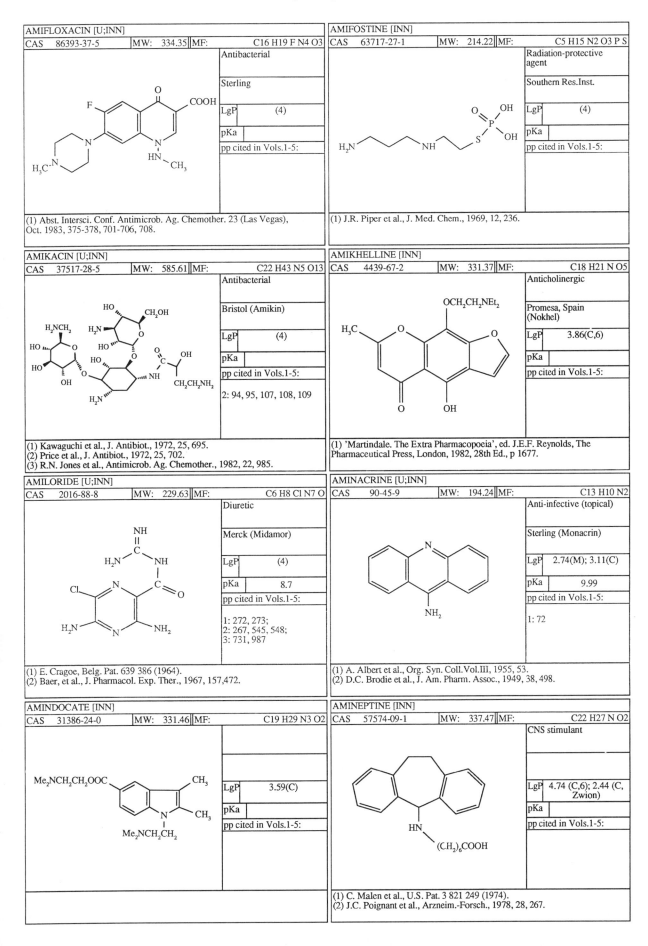

AMIFLOXACIN [U;INN]

CAS	86393-37-5	MW:	334.35	MF:	C16 H19 F N4 O3

Antibacterial

Sterling

LgP	(4)
pKa	
pp cited in Vols.1-5:	

(1) Abst. Intersci. Conf. Antimicrob. Ag. Chemother. 23 (Las Vegas), Oct. 1983, 375-378, 701-706, 708.

AMIFOSTINE [INN]

CAS	63717-27-1	MW:	214.22	MF:	C5 H15 N2 O3 P S

Radiation-protective agent

Southern Res.Inst.

LgP	(4)
pKa	
pp cited in Vols.1-5:	

(1) J.R. Piper et al., J. Med. Chem., 1969, 12, 236.

AMIKACIN [U;INN]

CAS	37517-28-5	MW:	585.61	MF:	C22 H43 N5 O13

Antibacterial

Bristol (Amikin)

LgP	(4)
pKa	
pp cited in Vols.1-5:	

2: 94, 95, 107, 108, 109

(1) Kawaguchi et al., J. Antibiot., 1972, 25, 695.
(2) Price et al., J. Antibiot., 1972, 25, 702.
(3) R.N. Jones et al., Antimicrob. Ag. Chemother., 1982, 22, 985.

AMIKHELLINE [INN]

CAS	4439-67-2	MW:	331.37	MF:	C18 H21 N O5

Anticholinergic

Promesa, Spain (Nokhel)

LgP	3.86(C,6)
pKa	
pp cited in Vols.1-5:	

(1) 'Martindale. The Extra Pharmacopoeia', ed. J.E.F. Reynolds, The Pharmaceutical Press, London, 1982, 28th Ed., p 1677.

AMILORIDE [U;INN]

CAS	2016-88-8	MW:	229.63	MF:	C6 H8 Cl N7 O

Diuretic

Merck (Midamor)

LgP	(4)
pKa	8.7
pp cited in Vols.1-5:	

1: 272, 273;
2: 267, 545, 548;
3: 731, 987

(1) E. Cragoe, Belg. Pat. 639 386 (1964).
(2) Baer, et al., J. Pharmacol. Exp. Ther., 1967, 157,472.

AMINACRINE [U;INN]

CAS	90-45-9	MW:	194.24	MF:	C13 H10 N2

Anti-infective (topical)

Sterling (Monacrin)

LgP	2.74(M); 3.11(C)
pKa	9.99
pp cited in Vols.1-5:	

1: 72

(1) A. Albert et al., Org. Syn. Coll.Vol.III, 1955, 53.
(2) D.C. Brodie et al., J. Am. Pharm. Assoc., 1949, 38, 498.

AMINDOCATE [INN]

CAS	31386-24-0	MW:	331.46	MF:	C19 H29 N3 O2

LgP	3.59(C)
pKa	
pp cited in Vols.1-5:	

AMINEPTINE [INN]

CAS	57574-09-1	MW:	337.47	MF:	C22 H27 N O2

CNS stimulant

LgP	4.74 (C,6); 2.44 (C, Zwion)
pKa	
pp cited in Vols.1-5:	

(1) C. Malen et al., U.S. Pat. 3 821 249 (1974).
(2) J.C. Poignant et al., Arzneim.-Forsch., 1978, 28, 267.

AMINOBENZOIC ACID [U]		
CAS 150-13-0	MW: 137.14	MF: C7 H7 N O2

Ultraviolet screen; anti-rickettsial; analgesic

LgP	0.83(M); 1.00(C)
pKa	2.50; 4.87
pp cited in Vols.1-5:	

(1) Nielson et al., J. Chem. Soc., 1962, 371.
(2) Cronheim, Fed. Proc., 1952, 10, 289.

AMINOCAPROIC ACID [U;INN]		
CAS 60-32-2	MW: 131.18	MF: C6 H13 N O2

Hemostatic

Lederle (Amicar)

$H_2N-(CH_2)_5 - COOH$

LgP	-2.95(M); -2.30(C, zwion.)
pKa	4.43; 10.75
pp cited in Vols.1-5:	

(1) Galat et al., J. Am. Chem. Soc., 1946, 68, 2729.
(2) McNichol et al., J. Lab. Clin. Med., 1962, 59, 15.

AMINOETHYL NITRATE [INN]		
CAS 646-02-6	MW: 106.08	MF: C2 H6 N2 O3

Vasodilator

Pharmacia,Swed. (Itramin)

$H_2NCH_2CH_2\text{-}O\text{-}NO_2$

LgP	-0.10(C)
pKa	
pp cited in Vols.1-5:	

(1) Swedish Pat. 168 308 (1959).

AMINOGLUTETHIMIDE [U;INN]		
CAS 125-84-8	MW: 232.28	MF: C13 H16 N2 O2

Adrenocortical suppressant; antineoplastic

Ciba-Geigy (Cytadren)

LgP	0.67(C)
pKa	
pp cited in Vols.1-5:	
1: 64; 2: 138, 347	

(1) Hoffmann et al., U.S.Pat 2 848 455 (1958).
(2) ' Pharmanual (Basel)' eds. R.J. Sante, I.C. Henderson, Karger, Basel, 1981, vol.2, 160 pp.

AMINOHIPPURIC ACID [U]		
CAS 61-78-9	MW: 194.19	MF: C9 H10 N2 O3

Renal function diagnosis

Merck

LgP	-0.25(C)
pKa	3.83
pp cited in Vols.1-5:	

(1) Shimizu et al., J. Pharm. Soc. Japan, 1953, 73, 523.

AMINOMETRADINE [INN]		
CAS 642-44-4	MW: 195.22	MF: C9 H13 N3 O2

Diuretic

Searle (Mictine)

LgP	0.52(C,3)
pKa	
pp cited in Vols.1-5:	

(1) Papesch et al., U.S. Pat. 2 650 922 (1953).

AMINOPENTAMIDE [INN]		
CAS 60-46-8	MW: 296.42	MF: C19 H24 N2 O

Anticholinergic

Bristol-Myers (Centrine)

LgP	2.02(C,1)
pKa	
pp cited in Vols.1-5:	

(1) Cheyney et al., J. Org. Chem., 1952, 17, 770.
(2) W. Wheatley et al., J. Org. Chem., 1954, 19, 794.

AMINOPHYLLINE [U;INN]		
CAS 317-34-0	MW: 420.43	MF: C16 H24 N10 O4

Relaxant (smooth muscle)

LgP	(4)
pKa	
pp cited in Vols.1-5:	
3: 606, 731; 5: 579	

(1) Gruter, U.S.Pat. 919 161 (1909). (2) K.D.Thacker et al., in 'Analytical Profiles of Drug Substances', ed. K. Florey, Academic Press, 1982, vol.11, pp 1-44.

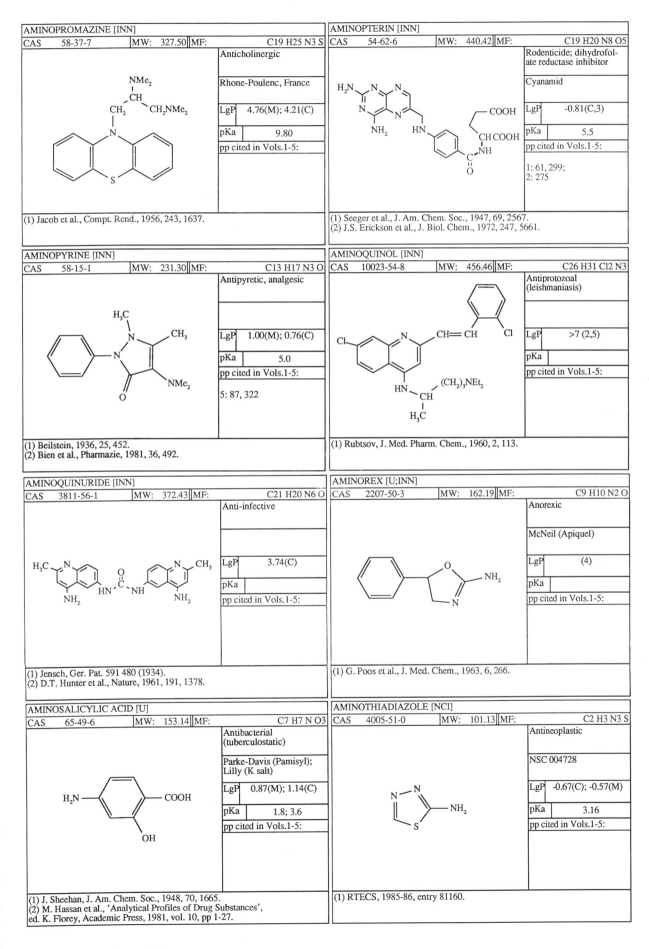

AMINOPROMAZINE [INN]

CAS 58-37-7	MW: 327.50	MF:	C19 H25 N3 S

Anticholinergic

Rhone-Poulenc, France

LgP	4.76(M); 4.21(C)
pKa	9.80
pp cited in Vols.1-5:	

(1) Jacob et al., Compt. Rend., 1956, 243, 1637.

AMINOPYRINE [INN]

CAS 58-15-1	MW: 231.30	MF:	C13 H17 N3 O

Antipyretic, analgesic

LgP	1.00(M); 0.76(C)
pKa	5.0
pp cited in Vols.1-5:	

5: 87, 322

(1) Beilstein, 1936, 25, 452.
(2) Bien et al., Pharmazie, 1981, 36, 492.

AMINOQUINURIDE [INN]

CAS 3811-56-1	MW: 372.43	MF:	C21 H20 N6 O

Anti-infective

LgP	3.74(C)
pKa	
pp cited in Vols.1-5:	

(1) Jensch, Ger. Pat. 591 480 (1934).
(2) D.T. Hunter et al., Nature, 1961, 191, 1378.

AMINOSALICYLIC ACID [U]

CAS 65-49-6	MW: 153.14	MF:	C7 H7 N O3

Antibacterial (tuberculostatic)

Parke-Davis (Pamisyl); Lilly (K salt)

LgP	0.87(M); 1.14(C)
pKa	1.8; 3.6
pp cited in Vols.1-5:	

(1) J. Sheehan, J. Am. Chem. Soc., 1948, 70, 1665.
(2) M. Hassan et al., 'Analytical Profiles of Drug Substances',
ed. K. Florey, Academic Press, 1981, vol. 10, pp 1-27.

AMINOPTERIN [INN]

CAS 54-62-6	MW: 440.42	MF:	C19 H20 N8 O5

Rodenticide; dihydrofolate reductase inhibitor

Cyanamid

LgP	-0.81(C,3)
pKa	5.5
pp cited in Vols.1-5:	

1: 61, 299;
2: 275

(1) Seeger et al., J. Am. Chem. Soc., 1947, 69, 2567.
(2) J.S. Erickson et al., J. Biol. Chem., 1972, 247, 5661.

AMINOQUINOL [INN]

CAS 10023-54-8	MW: 456.46	MF:	C26 H31 Cl2 N3

Antiprotozoal (leishmaniasis)

LgP	>7 (2,5)
pKa	
pp cited in Vols.1-5:	

(1) Rubtsov, J. Med. Pharm. Chem., 1960, 2, 113.

AMINOREX [U;INN]

CAS 2207-50-3	MW: 162.19	MF:	C9 H10 N2 O

Anorexic

McNeil (Apiquel)

LgP	(4)
pKa	
pp cited in Vols.1-5:	

(1) G. Poos et al., J. Med. Chem., 1963, 6, 266.

AMINOTHIADIAZOLE [NCI]

CAS 4005-51-0	MW: 101.13	MF:	C2 H3 N3 S

Antineoplastic

NSC 004728

LgP	-0.67(C); -0.57(M)
pKa	3.16
pp cited in Vols.1-5:	

(1) RTECS, 1985-86, entry 81160.

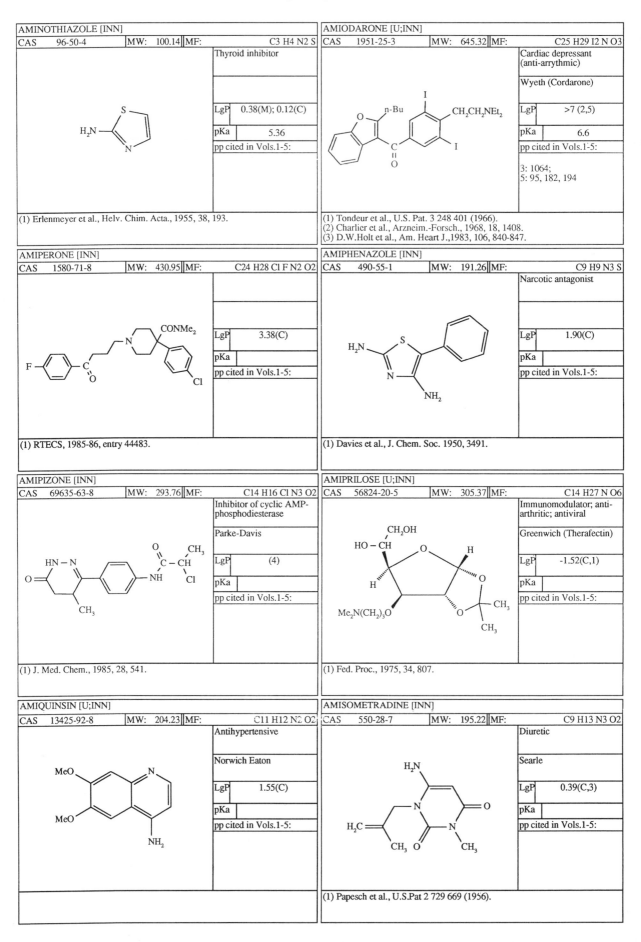

AMINOTHIAZOLE [INN]		
CAS 96-50-4	MW: 100.14 MF:	C3 H4 N2 S

Thyroid inhibitor

LgP	0.38(M); 0.12(C)
pKa	5.36
pp cited in Vols.1-5:	

(1) Erlenmeyer et al., Helv. Chim. Acta, 1955, 38, 193.

AMIODARONE [U;INN]		
CAS 1951-25-3	MW: 645.32 MF:	C25 H29 I2 N O3

Cardiac depressant
(anti-arrythmic)

Wyeth (Cordarone)

LgP	>7 (2,5)
pKa	6.6
pp cited in Vols.1-5:	

3: 1064;
5: 95, 182, 194

(1) Tondeur et al., U.S. Pat. 3 248 401 (1966).
(2) Charlier et al., Arzneim.-Forsch., 1968, 18, 1408.
(3) D.W.Holt et al., Am. Heart J.,1983, 106, 840-847.

AMIPERONE [INN]		
CAS 1580-71-8	MW: 430.95 MF:	C24 H28 Cl F N2 O2

LgP	3.38(C)
pKa	
pp cited in Vols.1-5:	

(1) RTECS, 1985-86, entry 44483.

AMIPHENAZOLE [INN]		
CAS 490-55-1	MW: 191.26 MF:	C9 H9 N3 S

Narcotic antagonist

LgP	1.90(C)
pKa	
pp cited in Vols.1-5:	

(1) Davies et al., J. Chem. Soc. 1950, 3491.

AMIPIZONE [INN]		
CAS 69635-63-8	MW: 293.76 MF:	C14 H16 Cl N3 O2

Inhibitor of cyclic AMP-
phosphodiesterase

Parke-Davis

LgP	(4)
pKa	
pp cited in Vols.1-5:	

(1) J. Med. Chem., 1985, 28, 541.

AMIPRILOSE [U;INN]		
CAS 56824-20-5	MW: 305.37 MF:	C14 H27 N O6

Immunomodulator; anti-
arthritic; antiviral

Greenwich (Therafectin)

LgP	-1.52(C,1)
pKa	
pp cited in Vols.1-5:	

(1) Fed. Proc., 1975, 34, 807.

AMIQUINSIN [U;INN]		
CAS 13425-92-8	MW: 204.23 MF:	C11 H12 N2 O2

Antihypertensive

Norwich Eaton

LgP	1.55(C)
pKa	
pp cited in Vols.1-5:	

AMISOMETRADINE [INN]		
CAS 550-28-7	MW: 195.22 MF:	C9 H13 N3 O2

Diuretic

Searle

LgP	0.39(C,3)
pKa	
pp cited in Vols.1-5:	

(1) Papesch et al., U.S.Pat 2 729 669 (1956).

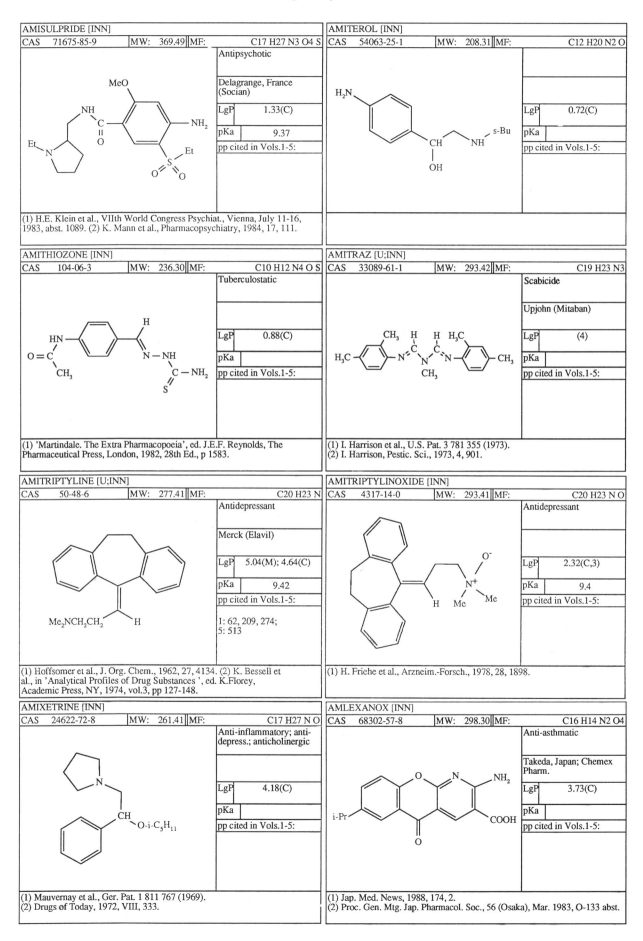

AMISULPRIDE [INN]
CAS 71675-85-9 | MW: 369.49 | MF: C17 H27 N3 O4 S
Antipsychotic
Delagrange, France (Socian)
LgP 1.33(C)
pKa 9.37
pp cited in Vols.1-5:
(1) H.E. Klein et al., VIIth World Congress Psychiat., Vienna, July 11-16, 1983, abst. 1089. (2) K. Mann et al., Pharmacopsychiatry, 1984, 17, 111.

AMITEROL [INN]
CAS 54063-25-1 | MW: 208.31 | MF: C12 H20 N2 O
LgP 0.72(C)
pKa
pp cited in Vols.1-5:

AMITHIOZONE [INN]
CAS 104-06-3 | MW: 236.30 | MF: C10 H12 N4 O S
Tuberculostatic
LgP 0.88(C)
pKa
pp cited in Vols.1-5:
(1) 'Martindale. The Extra Pharmacopoeia', ed. J.E.F. Reynolds, The Pharmaceutical Press, London, 1982, 28th Ed., p 1583.

AMITRAZ [U;INN]
CAS 33089-61-1 | MW: 293.42 | MF: C19 H23 N3
Scabicide
Upjohn (Mitaban)
LgP (4)
pKa
pp cited in Vols.1-5:
(1) I. Harrison et al., U.S. Pat. 3 781 355 (1973). (2) I. Harrison, Pestic. Sci., 1973, 4, 901.

AMITRIPTYLINE [U;INN]
CAS 50-48-6 | MW: 277.41 | MF: C20 H23 N
Antidepressant
Merck (Elavil)
LgP 5.04(M); 4.64(C)
pKa 9.42
pp cited in Vols.1-5: 1: 62, 209, 274; 5: 513
(1) Hoffsomer et al., J. Org. Chem., 1962, 27, 4134. (2) K. Bessell et al., in 'Analytical Profiles of Drug Substances ', ed. K.Florey, Academic Press, NY, 1974, vol.3, pp 127-148.

AMITRIPTYLINOXIDE [INN]
CAS 4317-14-0 | MW: 293.41 | MF: C20 H23 N O
Antidepressant
LgP 2.32(C,3)
pKa 9.4
pp cited in Vols.1-5:
(1) H. Friehe et al., Arzneim.-Forsch., 1978, 28, 1898.

AMIXETRINE [INN]
CAS 24622-72-8 | MW: 261.41 | MF: C17 H27 N O
Anti-inflammatory; anti-depress.; anticholinergic
LgP 4.18(C)
pKa
pp cited in Vols.1-5:
(1) Mauvernay et al., Ger. Pat. 1 811 767 (1969). (2) Drugs of Today, 1972, VIII, 333.

AMLEXANOX [INN]
CAS 68302-57-8 | MW: 298.30 | MF: C16 H14 N2 O4
Anti-asthmatic
Takeda, Japan; Chemex Pharm.
LgP 3.73(C)
pKa
pp cited in Vols.1-5:
(1) Jap. Med. News, 1988, 174, 2. (2) Proc. Gen. Mtg. Jap. Pharmacol. Soc., 56 (Osaka), Mar. 1983, O-133 abst.

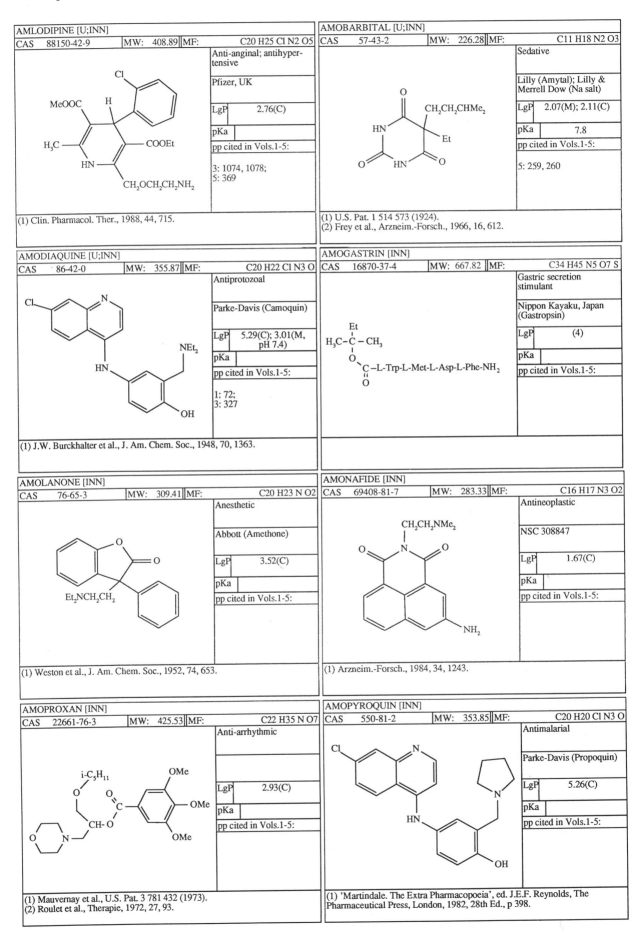

AMLODIPINE [U;INN]

CAS 88150-42-9 | MW: 408.89 | MF: C20 H25 Cl N2 O5

Anti-anginal; antihypertensive

Pfizer, UK

LgP 2.76(C)

pKa

pp cited in Vols.1-5:

3: 1074, 1078; 5: 369

(1) Clin. Pharmacol. Ther., 1988, 44, 715.

AMOBARBITAL [U;INN]

CAS 57-43-2 | MW: 226.28 | MF: C11 H18 N2 O3

Sedative

Lilly (Amytal); Lilly & Merrell Dow (Na salt)

LgP 2.07(M); 2.11(C)

pKa 7.8

pp cited in Vols.1-5:

5: 259, 260

(1) U.S. Pat. 1 514 573 (1924).
(2) Frey et al., Arzneim.-Forsch., 1966, 16, 612.

AMODIAQUINE [U;INN]

CAS 86-42-0 | MW: 355.87 | MF: C20 H22 Cl N3 O

Antiprotozoal

Parke-Davis (Camoquin)

LgP 5.29(C); 3.01(M, pH 7.4)

pKa

pp cited in Vols.1-5:

1: 72; 3: 327

(1) J.W. Burckhalter et al., J. Am. Chem. Soc., 1948, 70, 1363.

AMOGASTRIN [INN]

CAS 16870-37-4 | MW: 667.82 | MF: C34 H45 N5 O7 S

Gastric secretion stimulant

Nippon Kayaku, Japan (Gastropsin)

LgP (4)

pKa

pp cited in Vols.1-5:

AMOLANONE [INN]

CAS 76-65-3 | MW: 309.41 | MF: C20 H23 N O2

Anesthetic

Abbott (Amethone)

LgP 3.52(C)

pKa

pp cited in Vols.1-5:

(1) Weston et al., J. Am. Chem. Soc., 1952, 74, 653.

AMONAFIDE [INN]

CAS 69408-81-7 | MW: 283.33 | MF: C16 H17 N3 O2

Antineoplastic

NSC 308847

LgP 1.67(C)

pKa

pp cited in Vols.1-5:

(1) Arzneim.-Forsch., 1984, 34, 1243.

AMOPROXAN [INN]

CAS 22661-76-3 | MW: 425.53 | MF: C22 H35 N O7

Anti-arrhythmic

LgP 2.93(C)

pKa

pp cited in Vols.1-5:

(1) Mauvernay et al., U.S. Pat. 3 781 432 (1973).
(2) Roulet et al., Therapie, 1972, 27, 93.

AMOPYROQUIN [INN]

CAS 550-81-2 | MW: 353.85 | MF: C20 H20 Cl N3 O

Antimalarial

Parke-Davis (Propoquin)

LgP 5.26(C)

pKa

pp cited in Vols.1-5:

(1) 'Martindale. The Extra Pharmacopoeia', ed. J.E.F. Reynolds, The Pharmaceutical Press, London, 1982, 28th Ed., p 398.

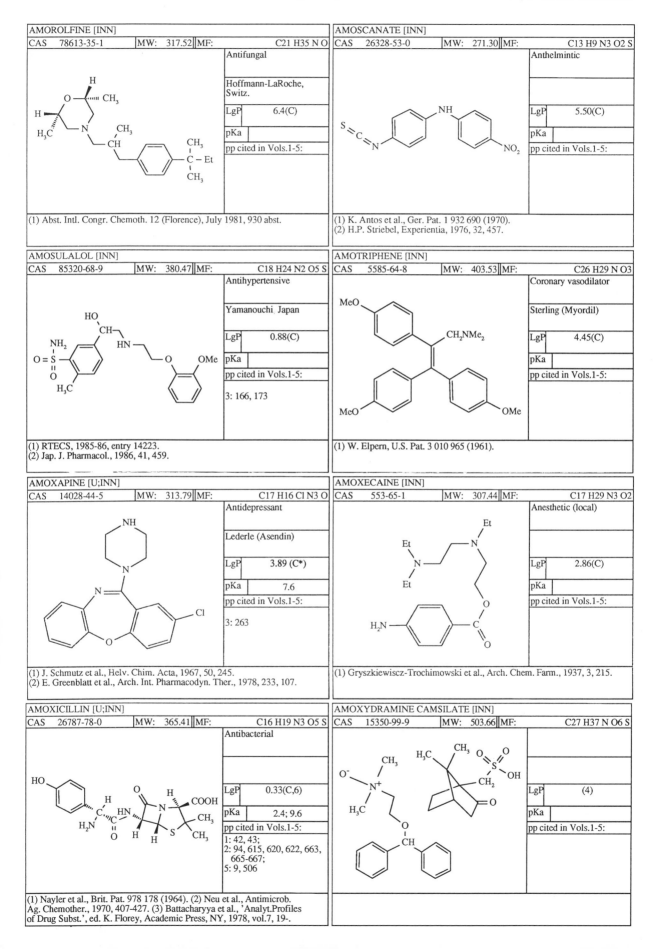

AMOROLFINE [INN]

CAS 78613-35-1	MW: 317.52	MF:	C21 H35 N O

Antifungal

Hoffmann-LaRoche, Switz.

LgP	6.4(C)
pKa	
pp cited in Vols.1-5:	

(1) Abst. Intl. Congr. Chemoth. 12 (Florence), July 1981, 930 abst.

AMOSCANATE [INN]

CAS 26328-53-0	MW: 271.30	MF:	C13 H9 N3 O2 S

Anthelmintic

LgP	5.50(C)
pKa	
pp cited in Vols.1-5:	

(1) K. Antos et al., Ger. Pat. 1 932 690 (1970).
(2) H.P. Striebel, Experientia, 1976, 32, 457.

AMOSULALOL [INN]

CAS 85320-68-9	MW: 380.47	MF:	C18 H24 N2 O5 S

Antihypertensive

Yamanouchi, Japan

LgP	0.88(C)
pKa	
pp cited in Vols.1-5:	
3: 166, 173	

(1) RTECS, 1985-86, entry 14223.
(2) Jap. J. Pharmacol., 1986, 41, 459.

AMOTRIPHENE [INN]

CAS 5585-64-8	MW: 403.53	MF:	C26 H29 N O3

Coronary vasodilator

Sterling (Myordil)

LgP	4.45(C)
pKa	
pp cited in Vols.1-5:	

(1) W. Elpern, U.S. Pat. 3 010 965 (1961).

AMOXAPINE [U;INN]

CAS 14028-44-5	MW: 313.79	MF:	C17 H16 Cl N3 O

Antidepressant

Lederle (Asendin)

LgP	3.89 (C*)
pKa	7.6
pp cited in Vols.1-5:	
3: 263	

(1) J. Schmutz et al., Helv. Chim. Acta, 1967, 50, 245.
(2) E. Greenblatt et al., Arch. Int. Pharmacodyn. Ther., 1978, 233, 107.

AMOXECAINE [INN]

CAS 553-65-1	MW: 307.44	MF:	C17 H29 N3 O2

Anesthetic (local)

LgP	2.86(C)
pKa	
pp cited in Vols.1-5:	

(1) Gryszkiewiscz-Trochimowski et al., Arch. Chem. Farm., 1937, 3, 215.

AMOXICILLIN [U;INN]

CAS 26787-78-0	MW: 365.41	MF:	C16 H19 N3 O5 S

Antibacterial

LgP	0.33(C,6)
pKa	2.4; 9.6
pp cited in Vols.1-5:	
1: 42, 43; 2: 94, 615, 620, 622, 663, 665-667; 5: 9, 506	

(1) Nayler et al., Brit. Pat. 978 178 (1964). (2) Neu et al., Antimicrob. Ag. Chemother., 1970, 407-427. (3) Battacharyya et al., 'Analyt.Profiles of Drug Subst.', ed. K. Florey, Academic Press, NY, 1978, vol.7, 19-.

AMOXYDRAMINE CAMSILATE [INN]

CAS 15350-99-9	MW: 503.66	MF:	C27 H37 N O6 S

LgP	(4)
pKa	
pp cited in Vols.1-5:	

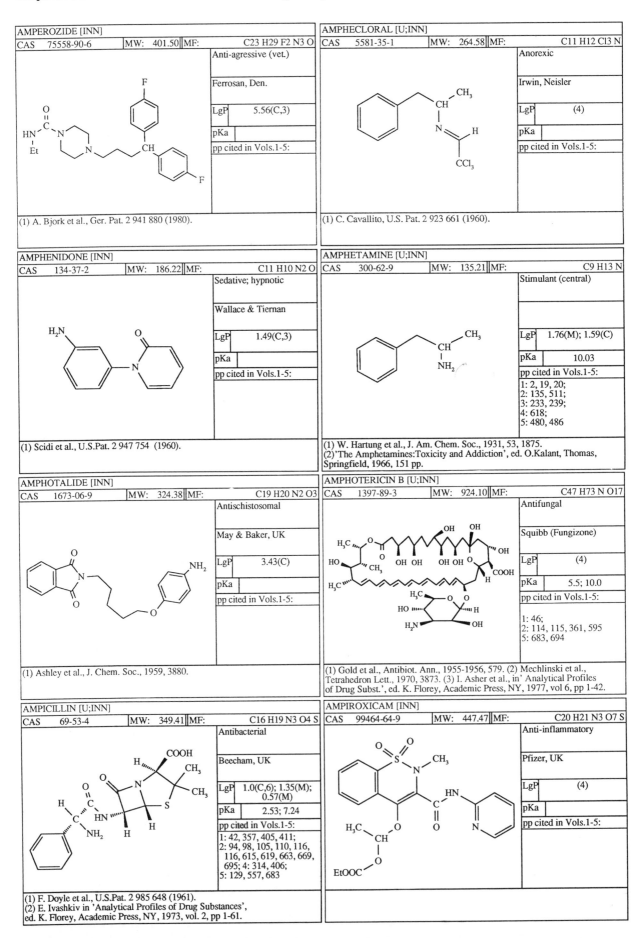

AMPEROZIDE [INN]

CAS	75558-90-6	MW: 401.50	MF:	C23 H29 F2 N3 O

Anti-agressive (vet.)

Ferrosan, Den.

LgP	5.56(C,3)
pKa	
pp cited in Vols.1-5:	

(1) A. Bjork et al., Ger. Pat. 2 941 880 (1980).

AMPHECLORAL [U;INN]

CAS	5581-35-1	MW: 264.58	MF:	C11 H12 Cl3 N

Anorexic

Irwin, Neisler

LgP	(4)
pKa	
pp cited in Vols.1-5:	

(1) C. Cavallito, U.S. Pat. 2 923 661 (1960).

AMPHENIDONE [INN]

CAS	134-37-2	MW: 186.22	MF:	C11 H10 N2 O

Sedative; hypnotic

Wallace & Tiernan

LgP	1.49(C,3)
pKa	
pp cited in Vols.1-5:	

(1) Scidi et al., U.S.Pat. 2 947 754 (1960).

AMPHETAMINE [U;INN]

CAS	300-62-9	MW: 135.21	MF:	C9 H13 N

Stimulant (central)

LgP	1.76(M); 1.59(C)
pKa	10.03
pp cited in Vols.1-5:	

1: 2, 19, 20;
2: 135, 511;
3: 233, 239;
4: 618;
5: 480, 486

(1) W. Hartung et al., J. Am. Chem. Soc., 1931, 53, 1875.
(2)'The Amphetamines:Toxicity and Addiction', ed. O.Kalant, Thomas, Springfield, 1966, 151 pp.

AMPHOTALIDE [INN]

CAS	1673-06-9	MW: 324.38	MF:	C19 H20 N2 O3

Antischistosomal

May & Baker, UK

LgP	3.43(C)
pKa	
pp cited in Vols.1-5:	

(1) Ashley et al., J. Chem. Soc., 1959, 3880.

AMPHOTERICIN B [U;INN]

CAS	1397-89-3	MW: 924.10	MF:	C47 H73 N O17

Antifungal

Squibb (Fungizone)

LgP	(4)
pKa	5.5; 10.0
pp cited in Vols.1-5:	

1: 46;
2: 114, 115, 361, 595
5: 683, 694

(1) Gold et al., Antibiot. Ann., 1955-1956, 579. (2) Mechlinski et al., Tetrahedron Lett., 1970, 3873. (3) I. Asher et al., in' Analytical Profiles of Drug Subst.', ed. K. Florey, Academic Press, NY, 1977, vol 6, pp 1-42.

AMPICILLIN [U;INN]

CAS	69-53-4	MW: 349.41	MF:	C16 H19 N3 O4 S

Antibacterial

Beecham, UK

LgP	1.0(C,6); 1.35(M); 0.57(M)
pKa	2.53; 7.24
pp cited in Vols.1-5:	

1: 42, 357, 405, 411;
2: 94, 98, 105, 110, 116, 116, 615, 619, 663, 669, 695; 4: 314, 406;
5: 129, 557, 683

(1) F. Doyle et al., U.S.Pat. 2 985 648 (1961).
(2) E. Ivashkiv in 'Analytical Profiles of Drug Substances', ed. K. Florey, Academic Press, NY, 1973, vol. 2, pp 1-61.

AMPIROXICAM [INN]

CAS	99464-64-9	MW: 447.47	MF:	C20 H21 N3 O7 S

Anti-inflammatory

Pfizer, UK

LgP	(4)
pKa	
pp cited in Vols.1-5:	

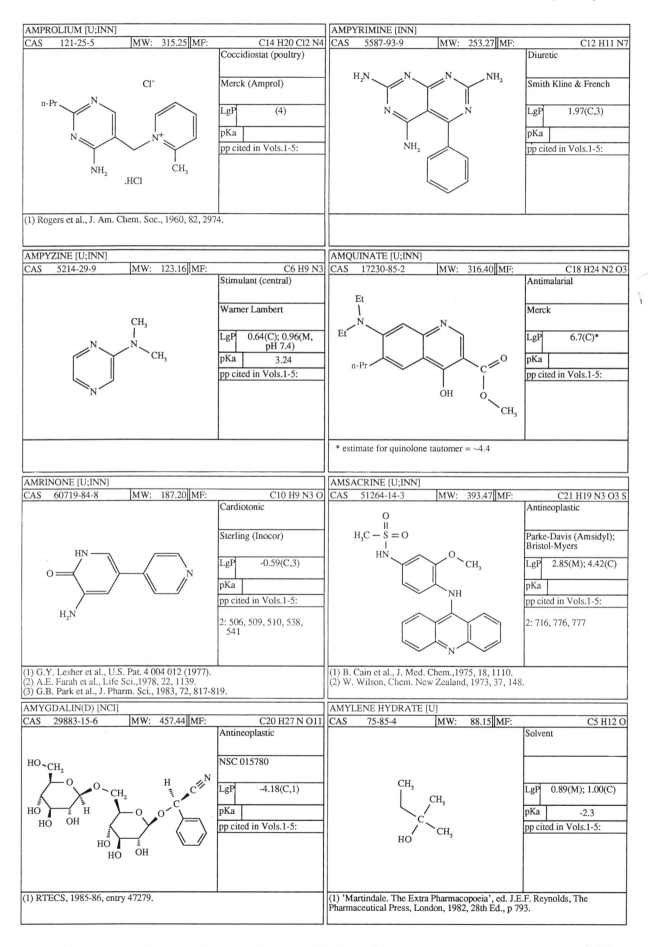

AMPROLIUM [U;INN]

CAS	121-25-5	MW:	315.25	MF:	C14 H20 Cl2 N4

Coccidiostat (poultry)

Merck (Amprol)

LgP	(4)
pKa	
pp cited in Vols.1-5:	

(1) Rogers et al., J. Am. Chem. Soc., 1960, 82, 2974.

AMPYRIMINE [INN]

CAS	5587-93-9	MW:	253.27	MF:	C12 H11 N7

Diuretic

Smith Kline & French

LgP	1.97(C,3)
pKa	
pp cited in Vols.1-5:	

AMPYZINE [U;INN]

CAS	5214-29-9	MW:	123.16	MF:	C6 H9 N3

Stimulant (central)

Warner Lambert

LgP	0.64(C); 0.96(M, pH 7.4)
pKa	3.24
pp cited in Vols.1-5:	

AMQUINATE [U;INN]

CAS	17230-85-2	MW:	316.40	MF:	C18 H24 N2 O3

Antimalarial

Merck

LgP	6.7(C)*
pKa	
pp cited in Vols.1-5:	

* estimate for quinolone tautomer = ~4.4

AMRINONE [U;INN]

CAS	60719-84-8	MW:	187.20	MF:	C10 H9 N3 O

Cardiotonic

Sterling (Inocor)

LgP	-0.59(C,3)
pKa	
pp cited in Vols.1-5:	
	2: 506, 509, 510, 538, 541

(1) G.Y. Lesher et al., U.S. Pat. 4 004 012 (1977).
(2) A.E. Farah et al., Life Sci.,1978, 22, 1139.
(3) G.B. Park et al., J. Pharm. Sci., 1983, 72, 817-819.

AMSACRINE [U;INN]

CAS	51264-14-3	MW:	393.47	MF:	C21 H19 N3 O3 S

Antineoplastic

Parke-Davis (Amsidyl); Bristol-Myers

LgP	2.85(M); 4.42(C)
pKa	
pp cited in Vols.1-5:	
	2: 716, 776, 777

(1) B. Cain et al., J. Med. Chem.,1975, 18, 1110.
(2) W. Wilson, Chem. New Zealand, 1973, 37, 148.

AMYGDALIN(D) [NCI]

CAS	29883-15-6	MW:	457.44	MF:	C20 H27 N O11

Antineoplastic

NSC 015780

LgP	-4.18(C,1)
pKa	
pp cited in Vols.1-5:	

(1) RTECS, 1985-86, entry 47279.

AMYLENE HYDRATE [U]

CAS	75-85-4	MW:	88.15	MF:	C5 H12 O

Solvent

LgP	0.89(M); 1.00(C)
pKa	-2.3
pp cited in Vols.1-5:	

(1) 'Martindale. The Extra Pharmacopoeia', ed. J.E.F. Reynolds, The Pharmaceutical Press, London, 1982, 28th Ed., p 793.

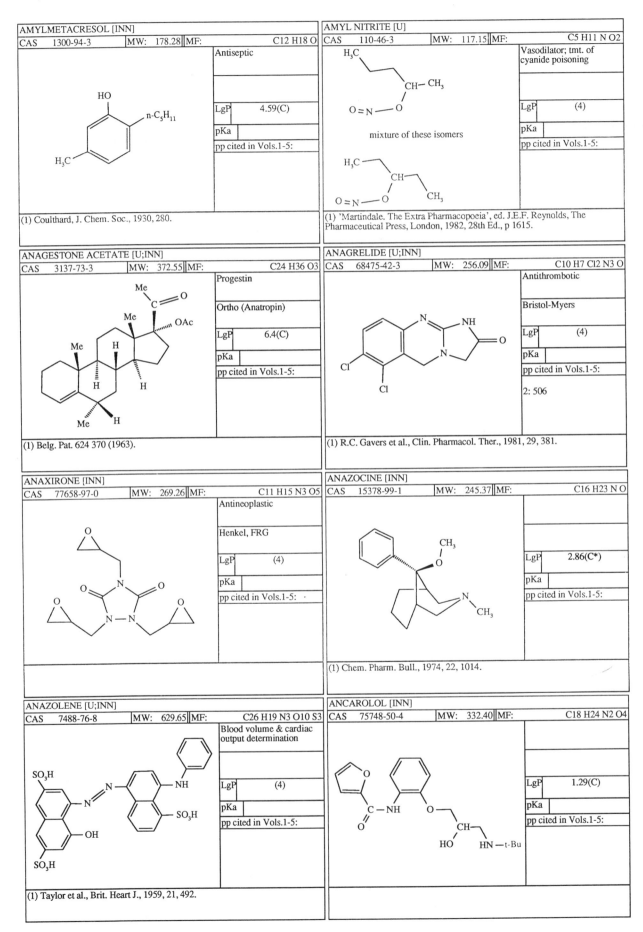

AMYLMETACRESOL [INN]

CAS	1300-94-3	MW:	178.28	MF:	C12 H18 O

Antiseptic

LgP 4.59(C)

pKa

pp cited in Vols.1-5:

(1) Coulthard, J. Chem. Soc., 1930, 280.

AMYL NITRITE [U]

CAS	110-46-3	MW:	117.15	MF:	C5 H11 N O2

Vasodilator; tmt. of cyanide poisoning

LgP (4)

pKa

pp cited in Vols.1-5:

mixture of these isomers

(1) 'Martindale. The Extra Pharmacopoeia', ed. J.E.F. Reynolds, The Pharmaceutical Press, London, 1982, 28th Ed., p 1615.

ANAGESTONE ACETATE [U;INN]

CAS	3137-73-3	MW:	372.55	MF:	C24 H36 O3

Progestin

Ortho (Anatropin)

LgP 6.4(C)

pKa

pp cited in Vols.1-5:

(1) Belg. Pat. 624 370 (1963).

ANAGRELIDE [U;INN]

CAS	68475-42-3	MW:	256.09	MF:	C10 H7 Cl2 N3 O

Antithrombotic

Bristol-Myers

LgP (4)

pKa

pp cited in Vols.1-5:

2: 506

(1) R.C. Gavers et al., Clin. Pharmacol. Ther., 1981, 29, 381.

ANAXIRONE [INN]

CAS	77658-97-0	MW:	269.26	MF:	C11 H15 N3 O5

Antineoplastic

Henkel, FRG

LgP (4)

pKa

pp cited in Vols.1-5: ·

ANAZOCINE [INN]

CAS	15378-99-1	MW:	245.37	MF:	C16 H23 N O

LgP 2.86(C*)

pKa

pp cited in Vols.1-5:

(1) Chem. Pharm. Bull., 1974, 22, 1014.

ANAZOLENE [U;INN]

CAS	7488-76-8	MW:	629.65	MF:	C26 H19 N3 O10 S3

Blood volume & cardiac output determination

LgP (4)

pKa

pp cited in Vols.1-5:

(1) Taylor et al., Brit. Heart J., 1959, 21, 492.

ANCAROLOL [INN]

CAS	75748-50-4	MW:	332.40	MF:	C18 H24 N2 O4

LgP 1.29(C)

pKa

pp cited in Vols.1-5:

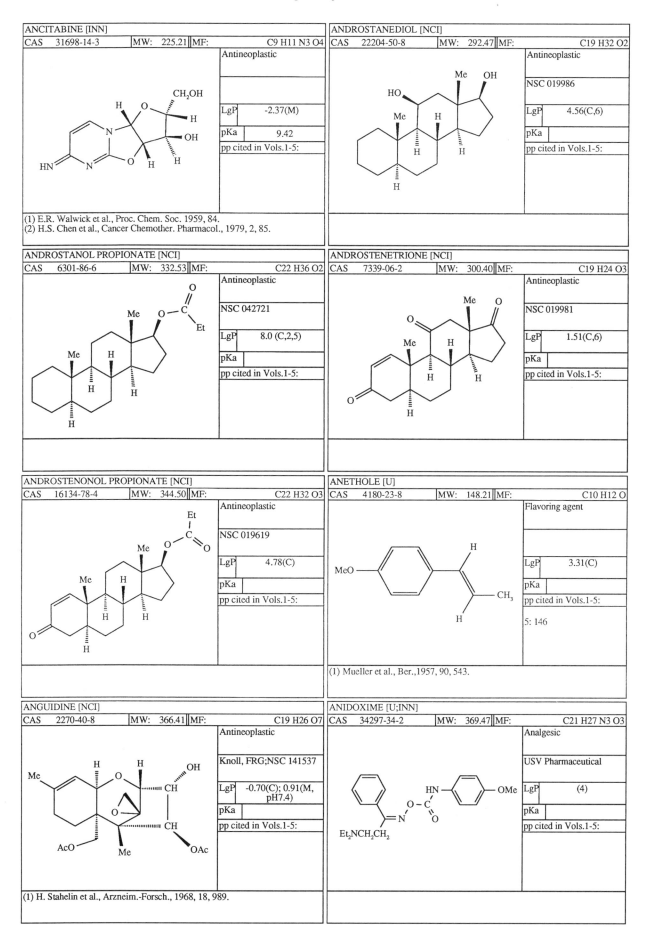

ANCITABINE [INN]

CAS	31698-14-3	MW:	225.21	MF:	C9 H11 N3 O4

Antineoplastic

LgP	-2.37(M)
pKa	9.42

pp cited in Vols.1-5:

(1) E.R. Walwick et al., Proc. Chem. Soc. 1959, 84.
(2) H.S. Chen et al., Cancer Chemother. Pharmacol., 1979, 2, 85.

ANDROSTANEDIOL [NCI]

CAS	22204-50-8	MW:	292.47	MF:	C19 H32 O2

Antineoplastic

NSC 019986

LgP	4.56(C,6)
pKa	

pp cited in Vols.1-5:

ANDROSTANOL PROPIONATE [NCI]

CAS	6301-86-6	MW:	332.53	MF:	C22 H36 O2

Antineoplastic

NSC 042721

LgP	8.0 (C,2,5)
pKa	

pp cited in Vols.1-5:

ANDROSTENETRIONE [NCI]

CAS	7339-06-2	MW:	300.40	MF:	C19 H24 O3

Antineoplastic

NSC 019981

LgP	1.51(C,6)
pKa	

pp cited in Vols.1-5:

ANDROSTENONOL PROPIONATE [NCI]

CAS	16134-78-4	MW:	344.50	MF:	C22 H32 O3

Antineoplastic

NSC 019619

LgP	4.78(C)
pKa	

pp cited in Vols.1-5:

ANETHOLE [U]

CAS	4180-23-8	MW:	148.21	MF:	C10 H12 O

Flavoring agent

LgP	3.31(C)
pKa	

pp cited in Vols.1-5:

5: 146

(1) Mueller et al., Ber.,1957, 90, 543.

ANGUIDINE [NCI]

CAS	2270-40-8	MW:	366.41	MF:	C19 H26 O7

Antineoplastic

Knoll, FRG;NSC 141537

LgP	-0.70(C); 0.91(M, pH7.4)
pKa	

pp cited in Vols.1-5:

(1) H. Stahelin et al., Arzneim.-Forsch., 1968, 18, 989.

ANIDOXIME [U;INN]

CAS	34297-34-2	MW:	369.47	MF:	C21 H27 N3 O3

Analgesic

USV Pharmaceutical

LgP	(4)
pKa	

pp cited in Vols.1-5:

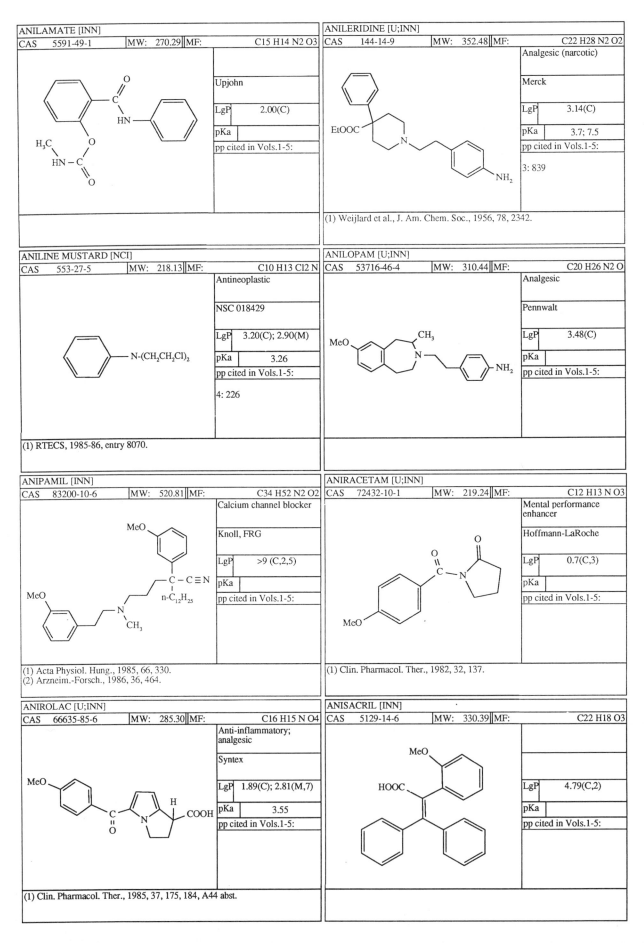

ANILAMATE [INN]

CAS	5591-49-1	MW: 270.29	MF:	C15 H14 N2 O3

Upjohn

LgP	2.00(C)
pKa	

pp cited in Vols.1-5:

ANILERIDINE [U;INN]

CAS	144-14-9	MW: 352.48	MF:	C22 H28 N2 O2

Analgesic (narcotic)

Merck

LgP	3.14(C)
pKa	3.7; 7.5

pp cited in Vols.1-5:

3: 839

(1) Weijlard et al., J. Am. Chem. Soc., 1956, 78, 2342.

ANILINE MUSTARD [NCI]

CAS	553-27-5	MW: 218.13	MF:	C10 H13 Cl2 N

Antineoplastic

NSC 018429

LgP	3.20(C); 2.90(M)
pKa	3.26

pp cited in Vols.1-5:

4: 226

(1) RTECS, 1985-86, entry 8070.

ANILOPAM [U;INN]

CAS	53716-46-4	MW: 310.44	MF:	C20 H26 N2 O

Analgesic

Pennwalt

LgP	3.48(C)
pKa	

pp cited in Vols.1-5:

ANIPAMIL [INN]

CAS	83200-10-6	MW: 520.81	MF:	C34 H52 N2 O2

Calcium channel blocker

Knoll, FRG

LgP	>9 (C,2,5)
pKa	

pp cited in Vols.1-5:

(1) Acta Physiol. Hung., 1985, 66, 330.
(2) Arzneim.-Forsch., 1986, 36, 464.

ANIRACETAM [U;INN]

CAS	72432-10-1	MW: 219.24	MF:	C12 H13 N O3

Mental performance enhancer

Hoffmann-LaRoche

LgP	0.7(C,3)
pKa	

pp cited in Vols.1-5:

(1) Clin. Pharmacol. Ther., 1982, 32, 137.

ANIROLAC [U;INN]

CAS	66635-85-6	MW: 285.30	MF:	C16 H15 N O4

Anti-inflammatory; analgesic

Syntex

LgP	1.89(C); 2.81(M,7)
pKa	3.55

pp cited in Vols.1-5:

(1) Clin. Pharmacol. Ther., 1985, 37, 175, 184, A44 abst.

ANISACRIL [INN]

CAS	5129-14-6	MW: 330.39	MF:	C22 H18 O3

LgP	4.79(C,2)
pKa	

pp cited in Vols.1-5:

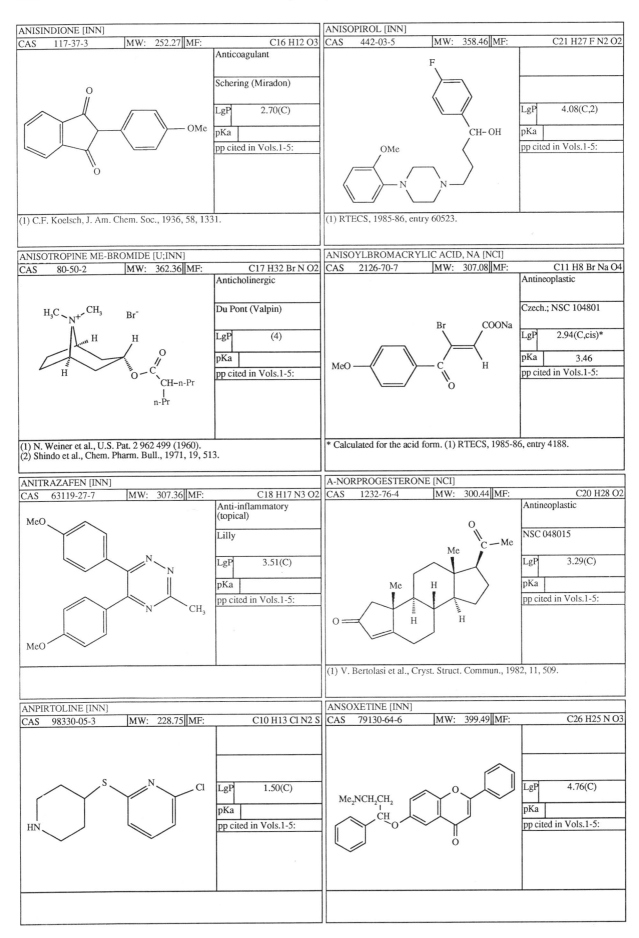

ANISINDIONE [INN]			
CAS 117-37-3	MW: 252.27	MF:	C16 H12 O3

Anticoagulant

Schering (Miradon)

LgP	2.70(C)
pKa	
pp cited in Vols.1-5:	

(1) C.F. Koelsch, J. Am. Chem. Soc., 1936, 58, 1331.

ANISOPIROL [INN]			
CAS 442-03-5	MW: 358.46	MF:	C21 H27 F N2 O2

LgP	4.08(C,2)
pKa	
pp cited in Vols.1-5:	

(1) RTECS, 1985-86, entry 60523.

ANISOTROPINE ME-BROMIDE [U;INN]			
CAS 80-50-2	MW: 362.36	MF:	C17 H32 Br N O2

Anticholinergic

Du Pont (Valpin)

LgP	(4)
pKa	
pp cited in Vols.1-5:	

(1) N. Weiner et al., U.S. Pat. 2 962 499 (1960).
(2) Shindo et al., Chem. Pharm. Bull., 1971, 19, 513.

ANISOYLBROMACRYLIC ACID, NA [NCI]			
CAS 2126-70-7	MW: 307.08	MF:	C11 H8 Br Na O4

Antineoplastic

Czech.; NSC 104801

LgP	2.94(C,cis)*
pKa	3.46
pp cited in Vols.1-5:	

* Calculated for the acid form. (1) RTECS, 1985-86, entry 4188.

ANITRAZAFEN [INN]			
CAS 63119-27-7	MW: 307.36	MF:	C18 H17 N3 O2

Anti-inflammatory (topical)

Lilly

LgP	3.51(C)
pKa	
pp cited in Vols.1-5:	

A-NORPROGESTERONE [NCI]			
CAS 1232-76-4	MW: 300.44	MF:	C20 H28 O2

Antineoplastic

NSC 048015

LgP	3.29(C)
pKa	
pp cited in Vols.1-5:	

(1) V. Bertolasi et al., Cryst. Struct. Commun., 1982, 11, 509.

ANPIRTOLINE [INN]			
CAS 98330-05-3	MW: 228.75	MF:	C10 H13 Cl N2 S

LgP	1.50(C)
pKa	
pp cited in Vols.1-5:	

ANSOXETINE [INN]			
CAS 79130-64-6	MW: 399.49	MF:	C26 H25 N O3

LgP	4.76(C)
pKa	
pp cited in Vols.1-5:	

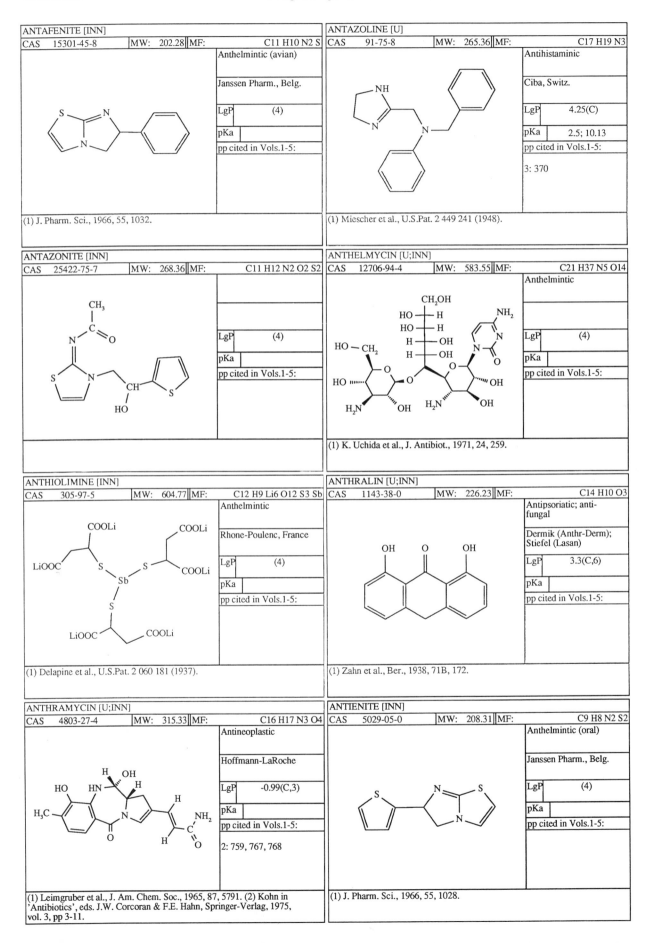

ANTAFENITE [INN]		
CAS 15301-45-8	MW: 202.28	MF: C11 H10 N2 S

Anthelmintic (avian)

Janssen Pharm., Belg.

LgP	(4)
pKa	
pp cited in Vols.1-5:	

(1) J. Pharm. Sci., 1966, 55, 1032.

ANTAZOLINE [U]		
CAS 91-75-8	MW: 265.36	MF: C17 H19 N3

Antihistaminic

Ciba, Switz.

LgP	4.25(C)
pKa	2.5; 10.13
pp cited in Vols.1-5:	

3: 370

(1) Miescher et al., U.S.Pat. 2 449 241 (1948).

ANTAZONITE [INN]		
CAS 25422-75-7	MW: 268.36	MF: C11 H12 N2 O2 S2

LgP	(4)
pKa	
pp cited in Vols.1-5:	

ANTHELMYCIN [U;INN]		
CAS 12706-94-4	MW: 583.55	MF: C21 H37 N5 O14

Anthelmintic

LgP	(4)
pKa	
pp cited in Vols.1-5:	

(1) K. Uchida et al., J. Antibiot., 1971, 24, 259.

ANTHIOLIMINE [INN]		
CAS 305-97-5	MW: 604.77	MF: C12 H9 Li6 O12 S3 Sb

Anthelmintic

Rhone-Poulenc, France

LgP	(4)
pKa	
pp cited in Vols.1-5:	

(1) Delapine et al., U.S.Pat. 2 060 181 (1937).

ANTHRALIN [U;INN]		
CAS 1143-38-0	MW: 226.23	MF: C14 H10 O3

Antipsoriatic; anti-fungal

Dermik (Anthr-Derm); Stiefel (Lasan)

LgP	3.3(C,6)
pKa	
pp cited in Vols.1-5:	

(1) Zahn et al., Ber., 1938, 71B, 172.

ANTHRAMYCIN [U;INN]		
CAS 4803-27-4	MW: 315.33	MF: C16 H17 N3 O4

Antineoplastic

Hoffmann-LaRoche

LgP	-0.99(C,3)
pKa	
pp cited in Vols.1-5:	

2: 759, 767, 768

(1) Leimgruber et al., J. Am. Chem. Soc., 1965, 87, 5791. (2) Kohn in 'Antibiotics', eds. J.W. Corcoran & F.E. Hahn, Springer-Verlag, 1975, vol. 3, pp 3-11.

ANTIENITE [INN]		
CAS 5029-05-0	MW: 208.31	MF: C9 H8 N2 S2

Anthelmintic (oral)

Janssen Pharm., Belg.

LgP	(4)
pKa	
pp cited in Vols.1-5:	

(1) J. Pharm. Sci., 1966, 55, 1028.

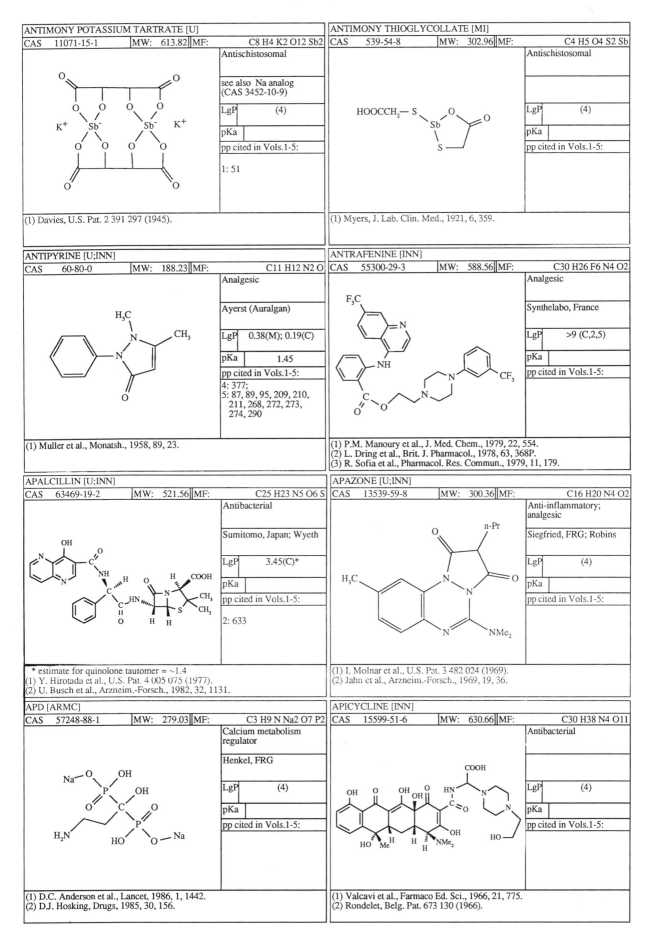

ANTIMONY POTASSIUM TARTRATE [U]

CAS	11071-15-1	MW:	613.82	MF:	C8 H4 K2 O12 Sb2

Antischistosomal

see also Na analog
(CAS 3452-10-9)

LgP	(4)
pKa	
pp cited in Vols.1-5:	

1: 51

(1) Davies, U.S. Pat. 2 391 297 (1945).

ANTIMONY THIOGLYCOLLATE [MI]

CAS	539-54-8	MW:	302.96	MF:	C4 H5 O4 S2 Sb

Antischistosomal

LgP	(4)
pKa	
pp cited in Vols.1-5:	

(1) Myers, J. Lab. Clin. Med., 1921, 6, 359.

ANTIPYRINE [U;INN]

CAS	60-80-0	MW:	188.23	MF:	C11 H12 N2 O

Analgesic

Ayerst (Auralgan)

LgP	0.38(M); 0.19(C)
pKa	1.45
pp cited in Vols.1-5:	

4: 377;
5: 87, 89, 95, 209, 210,
211, 268, 272, 273,
274, 290

(1) Muller et al., Monatsh., 1958, 89, 23.

ANTRAFENINE [INN]

CAS	55300-29-3	MW:	588.56	MF:	C30 H26 F6 N4 O2

Analgesic

Synthelabo, France

LgP	>9 (C,2,5)
pKa	
pp cited in Vols.1-5:	

(1) P.M. Manoury et al., J. Med. Chem., 1979, 22, 554.
(2) L. Dring et al., Brit. J. Pharmacol., 1978, 63, 368P.
(3) R. Sofia et al., Pharmacol. Res. Commun., 1979, 11, 179.

APALCILLIN [U;INN]

CAS	63469-19-2	MW:	521.56	MF:	C25 H23 N5 O6 S

Antibacterial

Sumitomo, Japan; Wyeth

LgP	3.45(C)*
pKa	
pp cited in Vols.1-5:	

2: 633

* estimate for quinolone tautomer = ~1.4
(1) Y. Hirotada et al., U.S. Pat. 4 005 075 (1977).
(2) U. Busch et al., Arzneim.-Forsch., 1982, 32, 1131.

APAZONE [U;INN]

CAS	13539-59-8	MW:	300.36	MF:	C16 H20 N4 O2

Anti-inflammatory;
analgesic

Siegfried, FRG; Robins

LgP	(4)
pKa	
pp cited in Vols.1-5:	

(1) I. Molnar et al., U.S. Pat. 3 482 024 (1969).
(2) Jahn et al., Arzneim.-Forsch., 1969, 19, 36.

APD [ARMC]

CAS	57248-88-1	MW:	279.03	MF:	C3 H9 N Na2 O7 P2

Calcium metabolism
regulator

Henkel, FRG

LgP	(4)
pKa	
pp cited in Vols.1-5:	

(1) D.C. Anderson et al., Lancet, 1986, 1, 1442.
(2) D.J. Hosking, Drugs, 1985, 30, 156.

APICYCLINE [INN]

CAS	15599-51-6	MW:	630.66	MF:	C30 H38 N4 O11

Antibacterial

LgP	(4)
pKa	
pp cited in Vols.1-5:	

(1) Valcavi et al., Farmaco Ed. Sci., 1966, 21, 775.
(2) Rondelet, Belg. Pat. 673 130 (1966).

APOMORPHINE [U]

CAS	58-00-4	MW:	267.33	MF:	C17 H17 N O2

Emetic

LgP	2.71(C); 1.76(M, pH7.4)
pKa	7.20; 8.92
pp cited in Vols.1-5:	

3: 245, 271, 284, 285, 303, 304, 308-311, 318;
4: 134;
5: 344, 351, 352, 353, 355

(1) L. Small et al., J. Org. Chem., 1940, 5, 344.
(2) G. Dichiara et al., Adv. Pharmacol. Chemother., 1978, 15, 87-160.

APOVINCAMINE [INN]

CAS	4880-92-6	MW:	336.44	MF:	C21 H24 N2 O2

Hypotensive

LgP	4.48(C,3)
pKa	
pp cited in Vols.1-5:	

(1) Acta Physiol. Acad. Sci. Hung., 1969, 36, 311.

APRACLONIDINE [ARMC]

CAS	66711-21-5	MW:	245.11	MF:	C9 H10 Cl2 N4

Antiglaucoma; adrenergic, (a-2 agonist)

Boehringer-Ingel., FRG (Lopidine)

LgP	0.85(C,2)
pKa	
pp cited in Vols.1-5:	

(1) C.P. Robinson, Drugs Today, 1988, 24, 557.

APRAMYCIN [U;INN]

CAS	37321-09-8	MW:	539.59	MF:	C21 H41 N5 O11

Antibacterial

Lilly

LgP	(7)
pKa	
pp cited in Vols.1-5:	

2: 107, 108

(1) W.M. Stark, U.S. Pat. 3 691 279 (1972).
(2) R. Ryden et al., J. Antimicrob. Chemother., 1977, 3, 609.

APRINDINE [U;INN]

CAS	37640-71-4	MW:	322.50	MF:	C22 H30 N2

Cardiac depressant (anti-arrhythmic)

Lilly (Fibocil)

LgP	4.9 (C)
pKa	
pp cited in Vols.1-5:	

(1) Vanhoff et al., Ger. Pat. 2 060 721 (1971).
(2) Georges et al., Arzneim.-Forsch., 1973, 23, 519.

APROBARBITAL [INN]

CAS	77-02-1	MW:	210.23	MF:	C10 H14 N2 O3

Sedative; hypnotic

Hoffmann-LaRoche, Switz. (Alurate)

LgP	1.15(M); 1.04(C)
pKa	7.99
pp cited in Vols.1-5:	

(1) U.S. Pat. 1 444 802 (1923).

APROFENE [INN]

CAS	3563-01-7	MW:	325.45	MF:	C21 H27 N O2

Analgesic; anticholinergic

LgP	4.58(C)
pKa	
pp cited in Vols.1-5:	

(1) H. Zaugg et al., J. Am. Chem. Soc., 1950, 72, 3004.
(2) Brit. Pat. 641 573 (1950).

APTAZAPINE [U;INN]

CAS	71576-40-4	MW:	253.35	MF:	C16 H19 N3

Antidepressant

Ciba-Geigy

LgP	3.06(C)
pKa	
pp cited in Vols.1-5:	

(1) Fed. Proc., 1982, 41, 4656 (Abst).

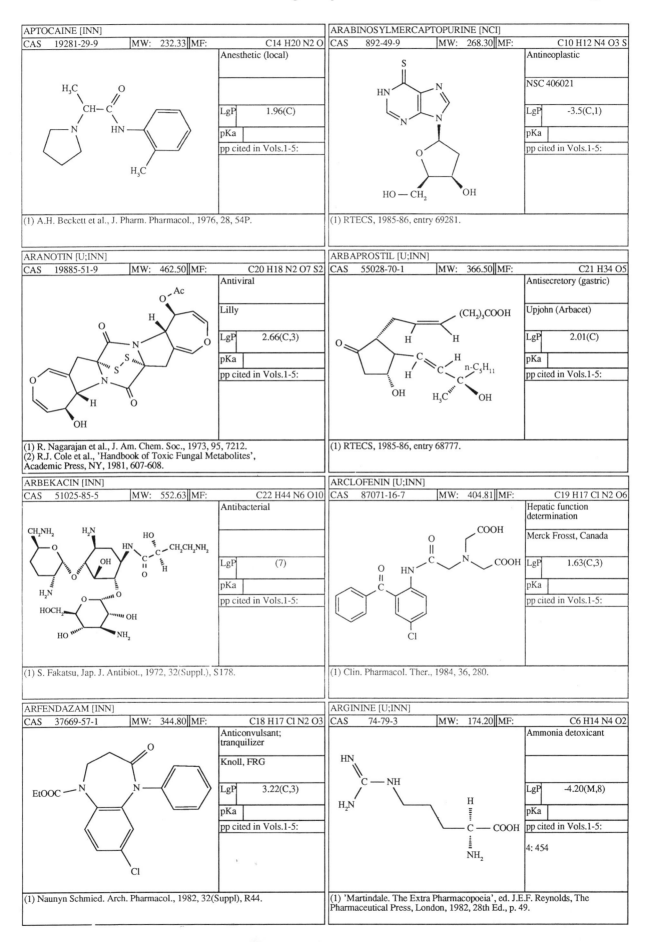

APTOCAINE [INN]

| CAS | 19281-29-9 | MW: | 232.33 | MF: | C14 H20 N2 O |

Anesthetic (local)

LgP	1.96(C)
pKa	
pp cited in Vols.1-5:	

(1) A.H. Beckett et al., J. Pharm. Pharmacol., 1976, 28, 54P.

ARABINOSYLMERCAPTOPURINE [NCI]

| CAS | 892-49-9 | MW: | 268.30 | MF: | C10 H12 N4 O3 S |

Antineoplastic

NSC 406021

LgP	-3.5(C,1)
pKa	
pp cited in Vols.1-5:	

(1) RTECS, 1985-86, entry 69281.

ARANOTIN [U;INN]

| CAS | 19885-51-9 | MW: | 462.50 | MF: | C20 H18 N2 O7 S2 |

Antiviral

Lilly

LgP	2.66(C,3)
pKa	
pp cited in Vols.1-5:	

(1) R. Nagarajan et al., J. Am. Chem. Soc., 1973, 95, 7212.
(2) R.J. Cole et al., 'Handbook of Toxic Fungal Metabolites',
Academic Press, NY, 1981, 607-608.

ARBAPROSTIL [U;INN]

| CAS | 55028-70-1 | MW: | 366.50 | MF: | C21 H34 O5 |

Antisecretory (gastric)

Upjohn (Arbacet)

LgP	2.01(C)
pKa	
pp cited in Vols.1-5:	

(1) RTECS, 1985-86, entry 68777.

ARBEKACIN [INN]

| CAS | 51025-85-5 | MW: | 552.63 | MF: | C22 H44 N6 O10 |

Antibacterial

LgP	(7)
pKa	
pp cited in Vols.1-5:	

(1) S. Fakatsu, Jap. J. Antibiot., 1972, 32(Suppl.), S178.

ARCLOFENIN [U;INN]

| CAS | 87071-16-7 | MW: | 404.81 | MF: | C19 H17 Cl N2 O6 |

Hepatic function determination

Merck Frosst, Canada

LgP	1.63(C,3)
pKa	
pp cited in Vols.1-5:	

(1) Clin. Pharmacol. Ther., 1984, 36, 280.

ARFENDAZAM [INN]

| CAS | 37669-57-1 | MW: | 344.80 | MF: | C18 H17 Cl N2 O3 |

Anticonvulsant; tranquilizer

Knoll, FRG

LgP	3.22(C,3)
pKa	
pp cited in Vols.1-5:	

(1) Naunyn Schmied. Arch. Pharmacol., 1982, 32(Suppl), R44.

ARGININE [U;INN]

| CAS | 74-79-3 | MW: | 174.20 | MF: | C6 H14 N4 O2 |

Ammonia detoxicant

LgP	-4.20(M,8)
pKa	
pp cited in Vols.1-5:	
4: 454	

(1) 'Martindale. The Extra Pharmacopoeia', ed. J.E.F. Reynolds, The
Pharmaceutical Press, London, 1982, 28th Ed., p. 49.

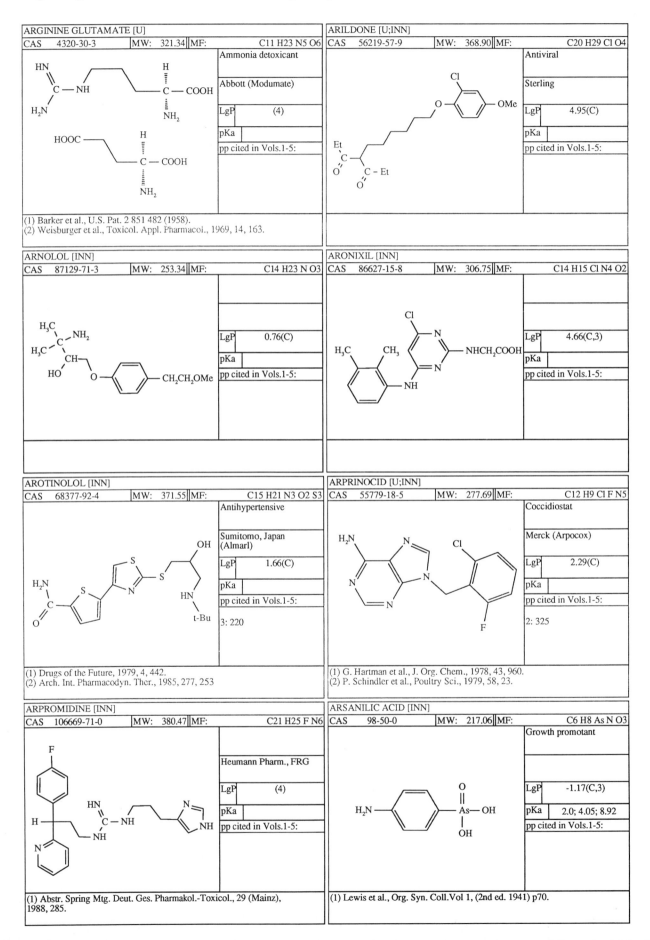

ARGININE GLUTAMATE [U]

| CAS | 4320-30-3 | MW: | 321.34 | MF: | C11 H23 N5 O6 |

Ammonia detoxicant

Abbott (Modumate)

LgP	(4)
pKa	
pp cited in Vols.1-5:	

(1) Barker et al., U.S. Pat. 2 851 482 (1958).
(2) Weisburger et al., Toxicol. Appl. Pharmacol., 1969, 14, 163.

ARILDONE [U;INN]

| CAS | 56219-57-9 | MW: | 368.90 | MF: | C20 H29 Cl O4 |

Antiviral

Sterling

LgP	4.95(C)
pKa	
pp cited in Vols.1-5:	

ARNOLOL [INN]

| CAS | 87129-71-3 | MW: | 253.34 | MF: | C14 H23 N O3 |

LgP	0.76(C)
pKa	
pp cited in Vols.1-5:	

ARONIXIL [INN]

| CAS | 86627-15-8 | MW: | 306.75 | MF: | C14 H15 Cl N4 O2 |

LgP	4.66(C,3)
pKa	
pp cited in Vols.1-5:	

AROTINOLOL [INN]

| CAS | 68377-92-4 | MW: | 371.55 | MF: | C15 H21 N3 O2 S3 |

Antihypertensive

Sumitomo, Japan (Almarl)

LgP	1.66(C)
pKa	
pp cited in Vols.1-5:	
3: 220	

(1) Drugs of the Future, 1979, 4, 442.
(2) Arch. Int. Pharmacodyn. Ther., 1985, 277, 253

ARPRINOCID [U;INN]

| CAS | 55779-18-5 | MW: | 277.69 | MF: | C12 H9 Cl F N5 |

Coccidiostat

Merck (Arpocox)

LgP	2.29(C)
pKa	
pp cited in Vols.1-5:	
2: 325	

(1) G. Hartman et al., J. Org. Chem., 1978, 43, 960.
(2) P. Schindler et al., Poultry Sci., 1979, 58, 23.

ARPROMIDINE [INN]

| CAS | 106669-71-0 | MW: | 380.47 | MF: | C21 H25 F N6 |

Heumann Pharm., FRG

LgP	(4)
pKa	
pp cited in Vols.1-5:	

(1) Abstr. Spring Mtg. Deut. Ges. Pharmakol.-Toxicol., 29 (Mainz), 1988, 285.

ARSANILIC ACID [INN]

| CAS | 98-50-0 | MW: | 217.06 | MF: | C6 H8 As N O3 |

Growth promotant

LgP	-1.17(C,3)
pKa	2.0; 4.05; 8.92
pp cited in Vols.1-5:	

(1) Lewis et al., Org. Syn. Coll.Vol 1, (2nd ed. 1941) p70.

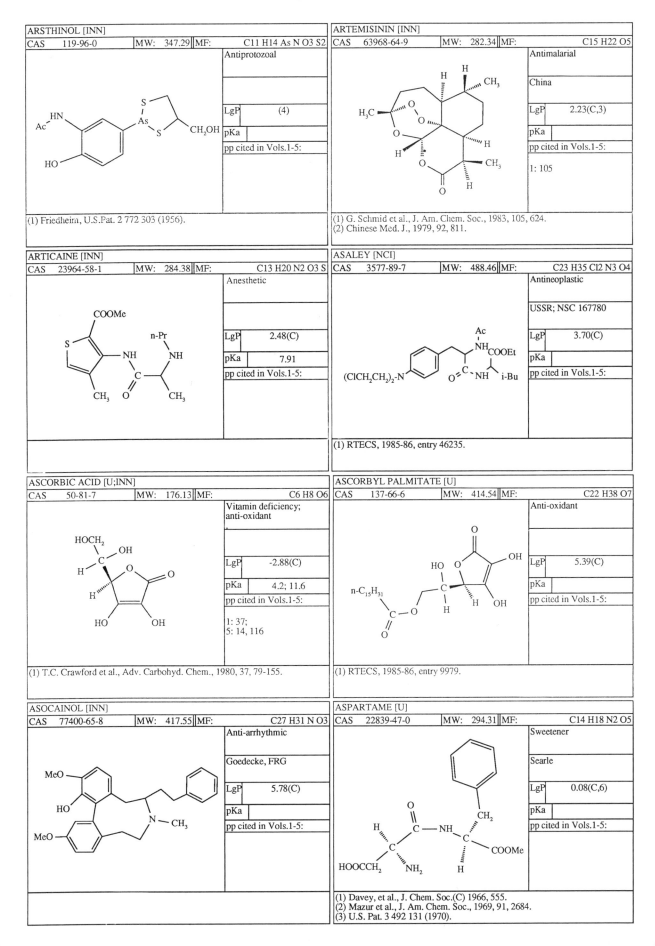

ARSTHINOL [INN]			
CAS 119-96-0	MW: 347.29	MF:	C11 H14 As N O3 S2

Antiprotozoal

LgP	(4)
pKa	
pp cited in Vols.1-5:	

(1) Friedheim, U.S.Pat. 2 772 303 (1956).

ARTEMISININ [INN]			
CAS 63968-64-9	MW: 282.34	MF:	C15 H22 O5

Antimalarial

China

LgP	2.23(C,3)
pKa	
pp cited in Vols.1-5:	

1: 105

(1) G. Schmid et al., J. Am. Chem. Soc., 1983, 105, 624.
(2) Chinese Med. J., 1979, 92, 811.

ARTICAINE [INN]			
CAS 23964-58-1	MW: 284.38	MF:	C13 H20 N2 O3 S

Anesthetic

LgP	2.48(C)
pKa	7.91
pp cited in Vols.1-5:	

ASALEY [NCI]			
CAS 3577-89-7	MW: 488.46	MF:	C23 H35 Cl2 N3 O4

Antineoplastic

USSR; NSC 167780

LgP	3.70(C)
pKa	
pp cited in Vols.1-5:	

(1) RTECS, 1985-86, entry 46235.

ASCORBIC ACID [U;INN]			
CAS 50-81-7	MW: 176.13	MF:	C6 H8 O6

Vitamin deficiency; anti-oxidant

LgP	-2.88(C)
pKa	4.2; 11.6
pp cited in Vols.1-5:	

1: 37;
5: 14, 116

(1) T.C. Crawford et al., Adv. Carbohyd. Chem., 1980, 37, 79-155.

ASCORBYL PALMITATE [U]			
CAS 137-66-6	MW: 414.54	MF:	C22 H38 O7

Anti-oxidant

LgP	5.39(C)
pKa	
pp cited in Vols.1-5:	

(1) RTECS, 1985-86, entry 9979.

ASOCAINOL [INN]			
CAS 77400-65-8	MW: 417.55	MF:	C27 H31 N O3

Anti-arrhythmic

Goedecke, FRG

LgP	5.78(C)
pKa	
pp cited in Vols.1-5:	

ASPARTAME [U]			
CAS 22839-47-0	MW: 294.31	MF:	C14 H18 N2 O5

Sweetener

Searle

LgP	0.08(C,6)
pKa	
pp cited in Vols.1-5:	

(1) Davey, et al., J. Chem. Soc.(C) 1966, 555.
(2) Mazur et al., J. Am. Chem. Soc., 1969, 91, 2684.
(3) U.S. Pat. 3 492 131 (1970).

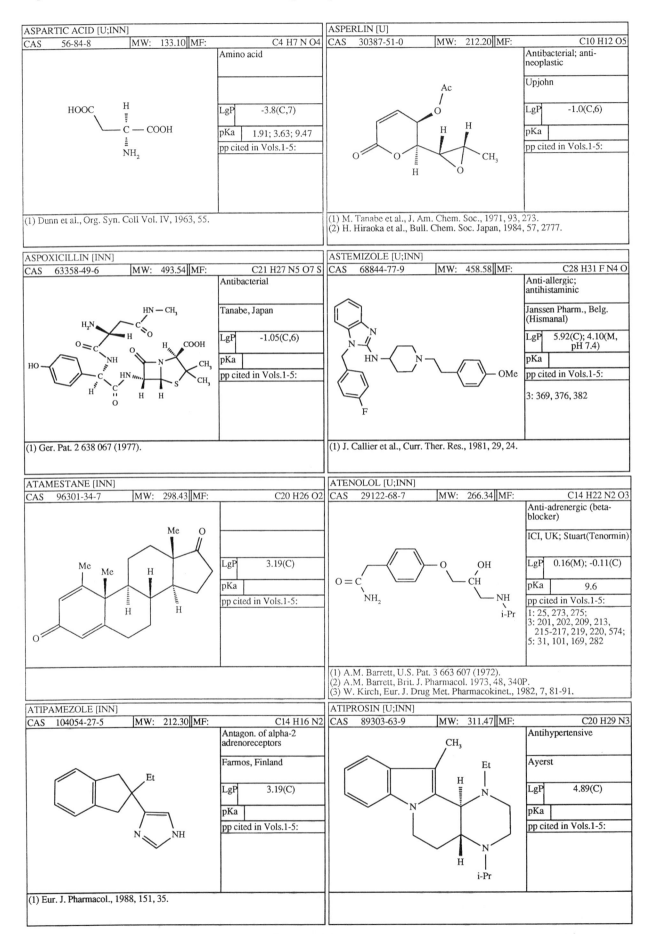

ASPARTIC ACID [U;INN]

CAS	56-84-8	MW: 133.10	MF:	C4 H7 N O4

Amino acid

LgP	-3.8(C,7)
pKa	1.91; 3.63; 9.47
pp cited in Vols.1-5:	

(1) Dunn et al., Org. Syn. Coll Vol. IV, 1963, 55.

ASPERLIN [U]

CAS	30387-51-0	MW: 212.20	MF:	C10 H12 O5

Antibacterial; anti-neoplastic

Upjohn

LgP	-1.0(C,6)
pKa	
pp cited in Vols.1-5:	

(1) M. Tanabe et al., J. Am. Chem. Soc., 1971, 93, 273.
(2) H. Hiraoka et al., Bull. Chem. Soc. Japan, 1984, 57, 2777.

ASPOXICILLIN [INN]

CAS	63358-49-6	MW: 493.54	MF:	C21 H27 N5 O7 S

Antibacterial

Tanabe, Japan

LgP	-1.05(C,6)
pKa	
pp cited in Vols.1-5:	

(1) Ger. Pat. 2 638 067 (1977).

ASTEMIZOLE [U;INN]

CAS	68844-77-9	MW: 458.58	MF:	C28 H31 F N4 O

Anti-allergic; antihistaminic

Janssen Pharm., Belg. (Hismanal)

LgP	5.92(C); 4.10(M, pH 7.4)
pKa	
pp cited in Vols.1-5:	

3: 369, 376, 382

(1) J. Callier et al., Curr. Ther. Res., 1981, 29, 24.

ATAMESTANE [INN]

CAS	96301-34-7	MW: 298.43	MF:	C20 H26 O2

LgP	3.19(C)
pKa	
pp cited in Vols.1-5:	

ATENOLOL [U;INN]

CAS	29122-68-7	MW: 266.34	MF:	C14 H22 N2 O3

Anti-adrenergic (beta-blocker)

ICI, UK; Stuart(Tenormin)

LgP	0.16(M); -0.11(C)
pKa	9.6
pp cited in Vols.1-5:	

1: 25, 273, 275;
3: 201, 202, 209, 213, 215-217, 219, 220, 574;
5: 31, 101, 169, 282

(1) A.M. Barrett, U.S. Pat. 3 663 607 (1972).
(2) A.M. Barrett, Brit. J. Pharmacol. 1973, 48, 340P.
(3) W. Kirch, Eur. J. Drug Met. Pharmacokinet., 1982, 7, 81-91.

ATIPAMEZOLE [INN]

CAS	104054-27-5	MW: 212.30	MF:	C14 H16 N2

Antagon. of alpha-2 adrenoreceptors

Farmos, Finland

LgP	3.19(C)
pKa	
pp cited in Vols.1-5:	

(1) Eur. J. Pharmacol., 1988, 151, 35.

ATIPROSIN [U;INN]

CAS	89303-63-9	MW: 311.47	MF:	C20 H29 N3

Antihypertensive

Ayerst

LgP	4.89(C)
pKa	
pp cited in Vols.1-5:	

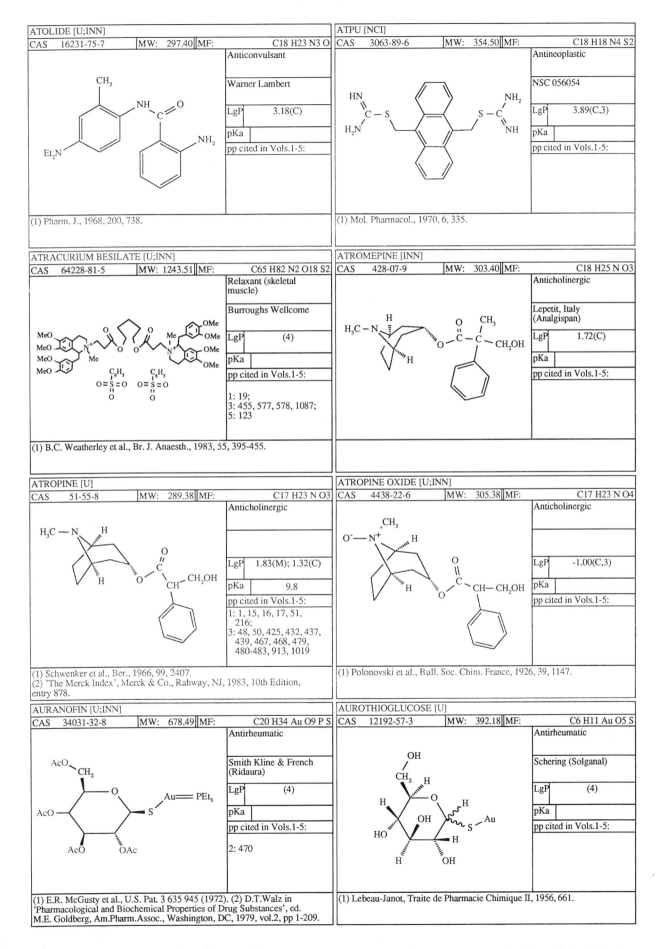

ATOLIDE [U;INN]

CAS 16231-75-7	MW: 297.40	MF:	C18 H23 N3 O

Anticonvulsant

Warner Lambert

LgP	3.18(C)
pKa	
pp cited in Vols.1-5:	

(1) Pharm. J., 1968, 200, 738.

ATPU [NCI]

CAS 3063-89-6	MW: 354.50	MF:	C18 H18 N4 S2

Antineoplastic

NSC 056054

LgP	3.89(C,3)
pKa	
pp cited in Vols.1-5:	

(1) Mol. Pharmacol., 1970, 6, 335.

ATRACURIUM BESILATE [U;INN]

CAS 64228-81-5	MW: 1243.51	MF:	C65 H82 N2 O18 S2

Relaxant (skeletal muscle)

Burroughs Wellcome

LgP	(4)
pKa	
pp cited in Vols.1-5:	

1: 19;
3: 455, 577, 578, 1087;
5: 123

(1) B.C. Weatherley et al., Br. J. Anaesth., 1983, 55, 395-455.

ATROMEPINE [INN]

CAS 428-07-9	MW: 303.40	MF:	C18 H25 N O3

Anticholinergic

Lepetit, Italy
(Analgispan)

LgP	1.72(C)
pKa	
pp cited in Vols.1-5:	

ATROPINE [U]

CAS 51-55-8	MW: 289.38	MF:	C17 H23 N O3

Anticholinergic

LgP	1.83(M); 1.32(C)
pKa	9.8
pp cited in Vols.1-5:	

1: 1, 15, 16, 17, 51,
216;
3: 48, 50, 425, 432, 437,
439, 467, 468, 479,
480-483, 913, 1019

(1) Schwenker et al., Ber., 1966, 99, 2407.
(2) 'The Merck Index', Merck & Co., Rahway, NJ, 1983, 10th Edition, entry 878.

ATROPINE OXIDE [U;INN]

CAS 4438-22-6	MW: 305.38	MF:	C17 H23 N O4

Anticholinergic

LgP	-1.00(C,3)
pKa	
pp cited in Vols.1-5:	

(1) Polonovski et al., Bull. Soc. Chim. France, 1926, 39, 1147.

AURANOFIN [U;INN]

CAS 34031-32-8	MW: 678.49	MF:	C20 H34 Au O9 P S

Antirheumatic

Smith Kline & French
(Ridaura)

LgP	(4)
pKa	
pp cited in Vols.1-5:	

2: 470

(1) E.R. McGusty et al., U.S. Pat. 3 635 945 (1972). (2) D.T.Walz in
'Pharmacological and Biochemical Properties of Drug Substances', ed.
M.E. Goldberg, Am.Pharm.Assoc., Washington, DC, 1979, vol.2, pp 1-209.

AUROTHIOGLUCOSE [U]

CAS 12192-57-3	MW: 392.18	MF:	C6 H11 Au O5 S

Antirheumatic

Schering (Solganal)

LgP	(4)
pKa	
pp cited in Vols.1-5:	

(1) Lebeau-Janot, Traite de Pharmacie Chimique II, 1956, 661.

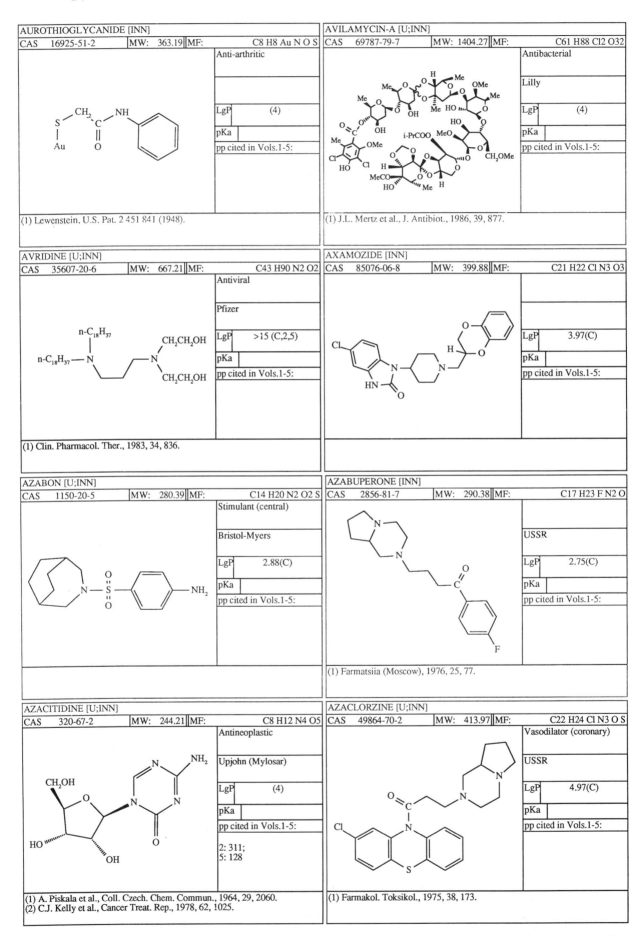

AUROTHIOGLYCANIDE [INN]			
CAS 16925-51-2	MW: 363.19	MF:	C8 H8 Au N O S

Anti-arthritic

LgP	(4)
pKa	
pp cited in Vols.1-5:	

(1) Lewenstein, U.S. Pat. 2 451 841 (1948).

AVILAMYCIN-A [U;INN]			
CAS 69787-79-7	MW: 1404.27	MF:	C61 H88 Cl2 O32

Antibacterial

Lilly

LgP	(4)
pKa	
pp cited in Vols.1-5:	

(1) J.L. Mertz et al., J. Antibiot., 1986, 39, 877.

AVRIDINE [U;INN]			
CAS 35607-20-6	MW: 667.21	MF:	C43 H90 N2 O2

Antiviral

Pfizer

LgP	>15 (C,2,5)
pKa	
pp cited in Vols.1-5:	

(1) Clin. Pharmacol. Ther., 1983, 34, 836.

AXAMOZIDE [INN]			
CAS 85076-06-8	MW: 399.88	MF:	C21 H22 Cl N3 O3

LgP	3.97(C)
pKa	
pp cited in Vols.1-5:	

AZABON [U;INN]			
CAS 1150-20-5	MW: 280.39	MF:	C14 H20 N2 O2 S

Stimulant (central)

Bristol-Myers

LgP	2.88(C)
pKa	
pp cited in Vols.1-5:	

AZABUPERONE [INN]			
CAS 2856-81-7	MW: 290.38	MF:	C17 H23 F N2 O

USSR

LgP	2.75(C)
pKa	
pp cited in Vols.1-5:	

(1) Farmatsiia (Moscow), 1976, 25, 77.

AZACITIDINE [U;INN]			
CAS 320-67-2	MW: 244.21	MF:	C8 H12 N4 O5

Antineoplastic

Upjohn (Mylosar)

LgP	(4)
pKa	
pp cited in Vols.1-5:	
	2: 311; 5: 128

(1) A. Piskala et al., Coll. Czech. Chem. Commun., 1964, 29, 2060.
(2) C.J. Kelly et al., Cancer Treat. Rep., 1978, 62, 1025.

AZACLORZINE [U;INN]			
CAS 49864-70-2	MW: 413.97	MF:	C22 H24 Cl N3 O S

Vasodilator (coronary)

USSR

LgP	4.97(C)
pKa	
pp cited in Vols.1-5:	

(1) Farmakol. Toksikol., 1975, 38, 173.

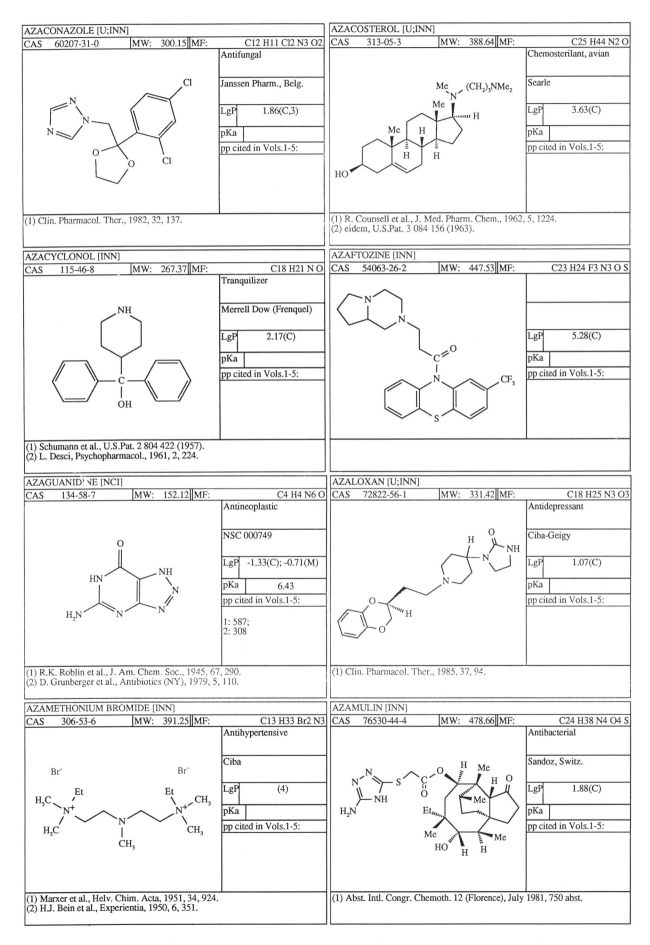

AZACONAZOLE [U;INN]

CAS	60207-31-0	MW:	300.15	MF:	C12 H11 Cl2 N3 O2

Antifungal

Janssen Pharm., Belg.

LgP	1.86(C,3)
pKa	
pp cited in Vols.1-5:	

(1) Clin. Pharmacol. Ther., 1982, 32, 137.

AZACOSTEROL [U;INN]

CAS	313-05-3	MW:	388.64	MF:	C25 H44 N2 O

Chemosterilant, avian

Searle

LgP	3.63(C)
pKa	
pp cited in Vols.1-5:	

(1) R. Counsell et al., J. Med. Pharm. Chem., 1962, 5, 1224.
(2) eidem, U.S.Pat. 3 084 156 (1963).

AZACYCLONOL [INN]

CAS	115-46-8	MW:	267.37	MF:	C18 H21 N O

Tranquilizer

Merrell Dow (Frenquel)

LgP	2.17(C)
pKa	
pp cited in Vols.1-5:	

(1) Schumann et al., U.S.Pat. 2 804 422 (1957).
(2) L. Desci, Psychopharmacol., 1961, 2, 224.

AZAFTOZINE [INN]

CAS	54063-26-2	MW:	447.53	MF:	C23 H24 F3 N3 O S

LgP	5.28(C)
pKa	
pp cited in Vols.1-5:	

AZAGUANIDINE [NCI]

CAS	134-58-7	MW:	152.12	MF:	C4 H4 N6 O

Antineoplastic

NSC 000749

LgP	-1.33(C); -0.71(M)
pKa	6.43
pp cited in Vols.1-5:	

1: 587;
2: 308

(1) R.K. Roblin et al., J. Am. Chem. Soc., 1945, 67, 290.
(2) D. Grunberger et al., Antibiotics (NY), 1979, 5, 110.

AZALOXAN [U;INN]

CAS	72822-56-1	MW:	331.42	MF:	C18 H25 N3 O3

Antidepressant

Ciba-Geigy

LgP	1.07(C)
pKa	
pp cited in Vols.1-5:	

(1) Clin. Pharmacol. Ther., 1985, 37, 94.

AZAMETHONIUM BROMIDE [INN]

CAS	306-53-6	MW:	391.25	MF:	C13 H33 Br2 N3

Antihypertensive

Ciba

LgP	(4)
pKa	
pp cited in Vols.1-5:	

(1) Marxer et al., Helv. Chim. Acta, 1951, 34, 924.
(2) H.J. Bein et al., Experientia, 1950, 6, 351.

AZAMULIN [INN]

CAS	76530-44-4	MW:	478.66	MF:	C24 H38 N4 O4 S

Antibacterial

Sandoz, Switz.

LgP	1.88(C)
pKa	
pp cited in Vols.1-5:	

(1) Abst. Intl. Congr. Chemoth. 12 (Florence), July 1981, 750 abst.

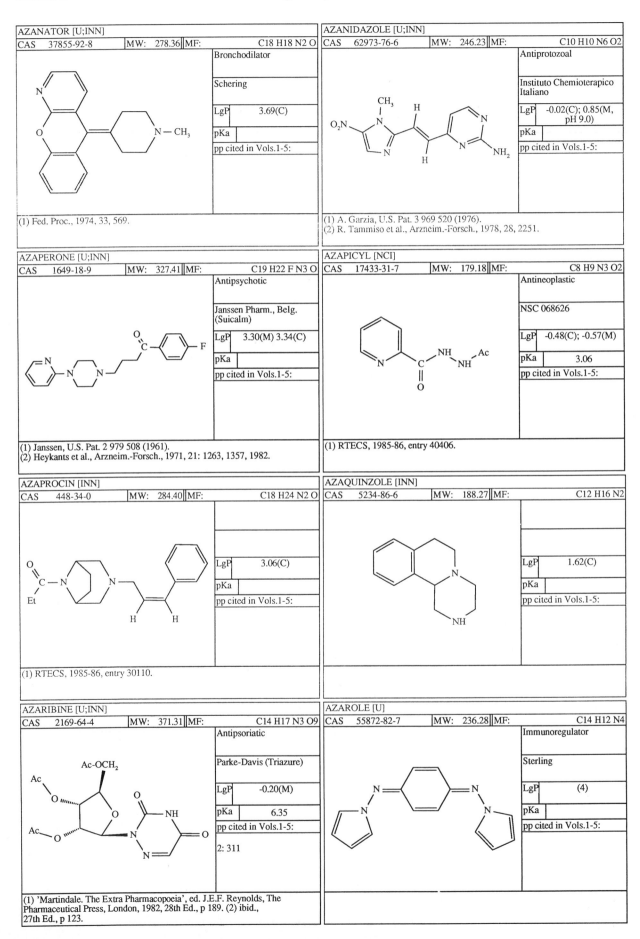

AZANATOR [U;INN]				
CAS	37855-92-8	MW: 278.36	MF:	C18 H18 N2 O

Bronchodilator

Schering

LgP	3.69(C)
pKa	
pp cited in Vols.1-5:	

(1) Fed. Proc., 1974, 33, 569.

AZANIDAZOLE [U;INN]				
CAS	62973-76-6	MW: 246.23	MF:	C10 H10 N6 O2

Antiprotozoal

Instituto Chemioterapico Italiano

LgP	-0.02(C); 0.85(M, pH 9.0)
pKa	
pp cited in Vols.1-5:	

(1) A. Garzia, U.S. Pat. 3 969 520 (1976).
(2) R. Tammiso et al., Arzneim.-Forsch., 1978, 28, 2251.

AZAPERONE [U;INN]				
CAS	1649-18-9	MW: 327.41	MF:	C19 H22 F N3 O

Antipsychotic

Janssen Pharm., Belg. (Suicalm)

LgP	3.30(M) 3.34(C)
pKa	
pp cited in Vols.1-5:	

(1) Janssen, U.S. Pat. 2 979 508 (1961).
(2) Heykants et al., Arzneim.-Forsch., 1971, 21: 1263, 1357, 1982.

AZAPICYL [NCI]				
CAS	17433-31-7	MW: 179.18	MF:	C8 H9 N3 O2

Antineoplastic

NSC 068626

LgP	-0.48(C); -0.57(M)
pKa	3.06
pp cited in Vols.1-5:	

(1) RTECS, 1985-86, entry 40406.

AZAPROCIN [INN]				
CAS	448-34-0	MW: 284.40	MF:	C18 H24 N2 O

LgP	3.06(C)
pKa	
pp cited in Vols.1-5:	

(1) RTECS, 1985-86, entry 30110.

AZAQUINZOLE [INN]				
CAS	5234-86-6	MW: 188.27	MF:	C12 H16 N2

LgP	1.62(C)
pKa	
pp cited in Vols.1-5:	

AZARIBINE [U;INN]				
CAS	2169-64-4	MW: 371.31	MF:	C14 H17 N3 O9

Antipsoriatic

Parke-Davis (Triazure)

LgP	-0.20(M)
pKa	6.35
pp cited in Vols.1-5:	
2: 311	

(1) 'Martindale. The Extra Pharmacopoeia', ed. J.E.F. Reynolds, The Pharmaceutical Press, London, 1982, 28th Ed., p 189. (2) ibid., 27th Ed., p 123.

AZAROLE [U]				
CAS	55872-82-7	MW: 236.28	MF:	C14 H12 N4

Immunoregulator

Sterling

LgP	(4)
pKa	
pp cited in Vols.1-5:	

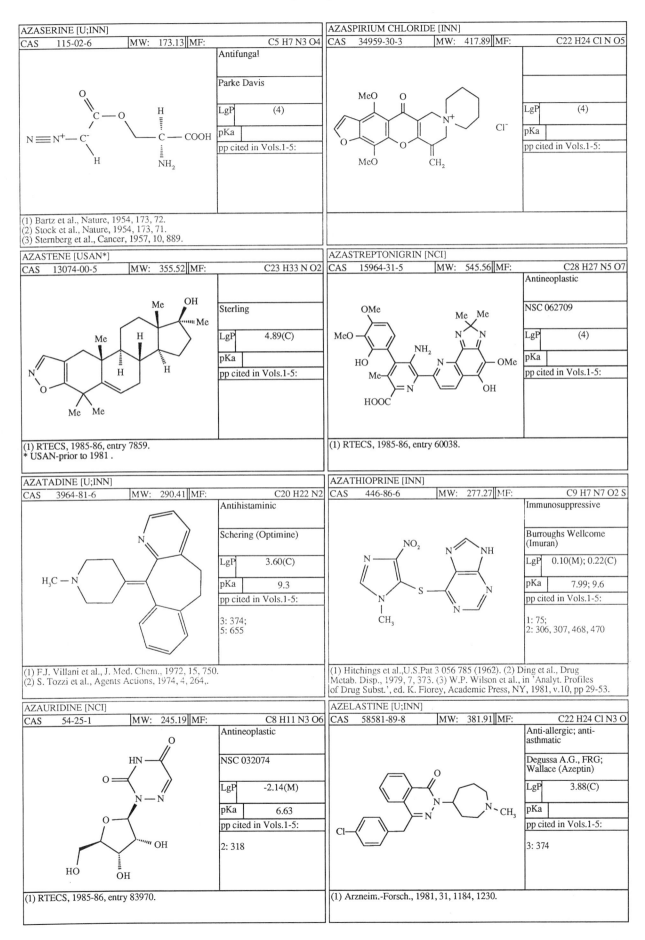

AZASERINE [U;INN]

| CAS | 115-02-6 | MW: | 173.13 | MF: | C5 H7 N3 O4 |

Antifungal

Parke Davis

LgP	(4)
pKa	
pp cited in Vols.1-5:	

(1) Bartz et al., Nature, 1954, 173, 72.
(2) Stock et al., Nature, 1954, 173, 71.
(3) Sternberg et al., Cancer, 1957, 10, 889.

AZASPIRIUM CHLORIDE [INN]

| CAS | 34959-30-3 | MW: | 417.89 | MF: | C22 H24 Cl N O5 |

LgP	(4)
pKa	
pp cited in Vols.1-5:	

AZASTENE [USAN*]

| CAS | 13074-00-5 | MW: | 355.52 | MF: | C23 H33 N O2 |

Sterling

LgP	4.89(C)
pKa	
pp cited in Vols.1-5:	

(1) RTECS, 1985-86, entry 7859.
* USAN-prior to 1981 .

AZASTREPTONIGRIN [NCI]

| CAS | 15964-31-5 | MW: | 545.56 | MF: | C28 H27 N5 O7 |

Antineoplastic

NSC 062709

LgP	(4)
pKa	
pp cited in Vols.1-5:	

(1) RTECS, 1985-86, entry 60038.

AZATADINE [U;INN]

| CAS | 3964-81-6 | MW: | 290.41 | MF: | C20 H22 N2 |

Antihistaminic

Schering (Optimine)

LgP	3.60(C)
pKa	9.3
pp cited in Vols.1-5:	

3: 374;
5: 655

(1) F.J. Villani et al., J. Med. Chem., 1972, 15, 750.
(2) S. Tozzi et al., Agents Actions, 1974, 4, 264,.

AZATHIOPRINE [INN]

| CAS | 446-86-6 | MW: | 277.27 | MF: | C9 H7 N7 O2 S |

Immunosuppressive

Burroughs Wellcome (Imuran)

LgP	0.10(M); 0.22(C)
pKa	7.99; 9.6
pp cited in Vols.1-5:	

1: 75;
2: 306, 307, 468, 470

(1) Hitchings et al.,U.S.Pat 3 056 785 (1962). (2) Ding et al., Drug
Metab. Disp., 1979, 7, 373. (3) W.P. Wilson et al., in 'Analyt. Profiles
of Drug Subst.', ed. K. Florey, Academic Press, NY, 1981, v.10, pp 29-53.

AZAURIDINE [NCI]

| CAS | 54-25-1 | MW: | 245.19 | MF: | C8 H11 N3 O6 |

Antineoplastic

NSC 032074

LgP	-2.14(M)
pKa	6.63
pp cited in Vols.1-5:	

2: 318

(1) RTECS, 1985-86, entry 83970.

AZELASTINE [U;INN]

| CAS | 58581-89-8 | MW: | 381.91 | MF: | C22 H24 Cl N3 O |

Anti-allergic; anti-asthmatic

Degussa A.G., FRG;
Wallace (Azeptin)

LgP	3.88(C)
pKa	
pp cited in Vols.1-5:	

3: 374

(1) Arzneim.-Forsch., 1981, 31, 1184, 1230.

AZEPEXOLE [INN]			
CAS 36067-73-9	MW: 181.24	MF:	C9 H15 N3 O

Antihypertensive

Boehringer-Ingel., FRG

LgP	-0.23(C); 0.05(M, pH7.4;37 deg C)
pKa	
pp cited in Vols.1-5:	

3: 140, 162, 173

(1) J. Cardiov. Pharmacol., 1981, 3, 269.

AZEPINDOLE [U;INN]			
CAS 26304-61-0	MW: 186.26	MF:	C12 H14 N2

Antidepressant

McNeil

LgP	2.11(C)
pKa	
pp cited in Vols.1-5:	

AZETEPA [U]			
CAS 125-45-1	MW: 259.27	MF:	C8 H14 N5 O P S

Antineoplastic

Lederle

LgP	(4)
pKa	
pp cited in Vols.1-5:	

AZIDAMFENICOL [INN]			
CAS 13838-08-9	MW: 295.26	MF:	C11 H13 N5 O5

Antimicrobial

LgP	-0.49(C,3)
pKa	
pp cited in Vols.1-5:	

(1) Meiser et al., U.S. Pat. 2 882 275 (1959).

AZIDOCILLIN [INN]			
CAS 17243-38-8	MW: 375.41	MF:	C16 H17 N5 O4 S

Antibacterial

Beecham, UK

LgP	1.69(C,3)
pKa	
pp cited in Vols.1-5:	

(1) Ekstrom et al., Acta Chem. Scand., 1965, 19, 281.
(2) Hanssen et al., Antimicrob. Ag. Chemother., 1967, 560, 568, 573.

AZIMEXON [INN]			
CAS 64118-86-1	MW: 194.24	MF:	C9 H14 N4 O

Antineoplastic

LgP	0.22(C,1)
pKa	
pp cited in Vols.1-5:	

3: 1115

(1) Drugs of the Future, 1980, 5, 174.

AZINTAMIDE [INN]			
CAS 1830-32-6	MW: 259.76	MF:	C10 H14 Cl N3 O S

Choleretic

LgP	1.26(C)
pKa	
pp cited in Vols.1-5:	

(1) Kloimstein et al., Arzneim.-Forsch.,1964, 14, 261.

AZIPRAMINE [U;INN]			
CAS 58503-82-5	MW: 366.51	MF:	C26 H26 N2

Antidepressant

Pierrel S.p.A., Italy

LgP	6.5(C,2)
pKa	
pp cited in Vols.1-5:	

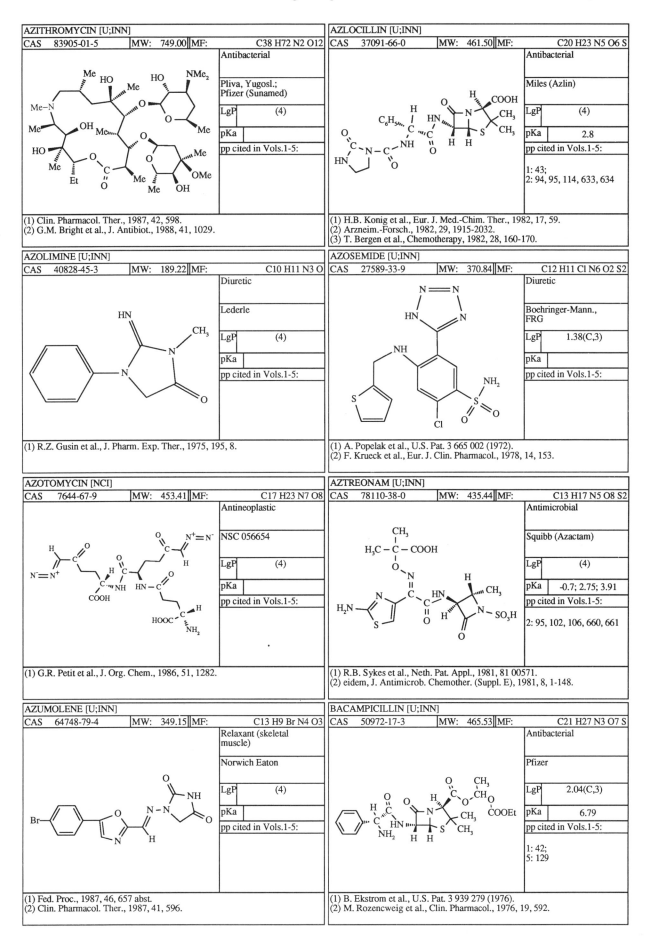

AZITHROMYCIN [U;INN]

CAS	83905-01-5	MW:	749.00	MF:	C38 H72 N2 O12

Antibacterial

Pliva, Yugosl.; Pfizer (Sunamed)

LgP	(4)
pKa	
pp cited in Vols.1-5:	

(1) Clin. Pharmacol. Ther., 1987, 42, 598.
(2) G.M. Bright et al., J. Antibiot., 1988, 41, 1029.

AZLOCILLIN [U;INN]

CAS	37091-66-0	MW:	461.50	MF:	C20 H23 N5 O6 S

Antibacterial

Miles (Azlin)

LgP	(4)
pKa	2.8
pp cited in Vols.1-5:	

1: 43;
2: 94, 95, 114, 633, 634

(1) H.B. Konig et al., Eur. J. Med.-Chim. Ther., 1982, 17, 59.
(2) Arzneim.-Forsch., 1982, 29, 1915-2032.
(3) T. Bergen et al., Chemotherapy, 1982, 28, 160-170.

AZOLIMINE [U;INN]

CAS	40828-45-3	MW:	189.22	MF:	C10 H11 N3 O

Diuretic

Lederle

LgP	(4)
pKa	
pp cited in Vols.1-5:	

(1) R.Z. Gusin et al., J. Pharm. Exp. Ther., 1975, 195, 8.

AZOSEMIDE [U;INN]

CAS	27589-33-9	MW:	370.84	MF:	C12 H11 Cl N6 O2 S2

Diuretic

Boehringer-Mann., FRG

LgP	1.38(C,3)
pKa	
pp cited in Vols.1-5:	

(1) A. Popelak et al., U.S. Pat. 3 665 002 (1972).
(2) F. Krueck et al., Eur. J. Clin. Pharmacol., 1978, 14, 153.

AZOTOMYCIN [NCI]

CAS	7644-67-9	MW:	453.41	MF:	C17 H23 N7 O8

Antineoplastic

NSC 056654

LgP	(4)
pKa	
pp cited in Vols.1-5:	

(1) G.R. Petit et al., J. Org. Chem., 1986, 51, 1282.

AZTREONAM [U;INN]

CAS	78110-38-0	MW:	435.44	MF:	C13 H17 N5 O8 S2

Antimicrobial

Squibb (Azactam)

LgP	(4)
pKa	-0.7; 2.75; 3.91
pp cited in Vols.1-5:	

2: 95, 102, 106, 660, 661

(1) R.B. Sykes et al., Neth. Pat. Appl., 1981, 81 00571.
(2) eidem, J. Antimicrob. Chemother. (Suppl. E), 1981, 8, 1-148.

AZUMOLENE [U;INN]

CAS	64748-79-4	MW:	349.15	MF:	C13 H9 Br N4 O3

Relaxant (skeletal muscle)

Norwich Eaton

LgP	(4)
pKa	
pp cited in Vols.1-5:	

(1) Fed. Proc., 1987, 46, 657 abst.
(2) Clin. Pharmacol. Ther., 1987, 41, 596.

BACAMPICILLIN [U;INN]

CAS	50972-17-3	MW:	465.53	MF:	C21 H27 N3 O7 S

Antibacterial

Pfizer

LgP	2.04(C,3)
pKa	6.79
pp cited in Vols.1-5:	

1: 42;
5: 129

(1) B. Ekstrom et al., U.S. Pat. 3 939 279 (1976).
(2) M. Rozencweig et al., Clin. Pharmacol., 1976, 19, 592.

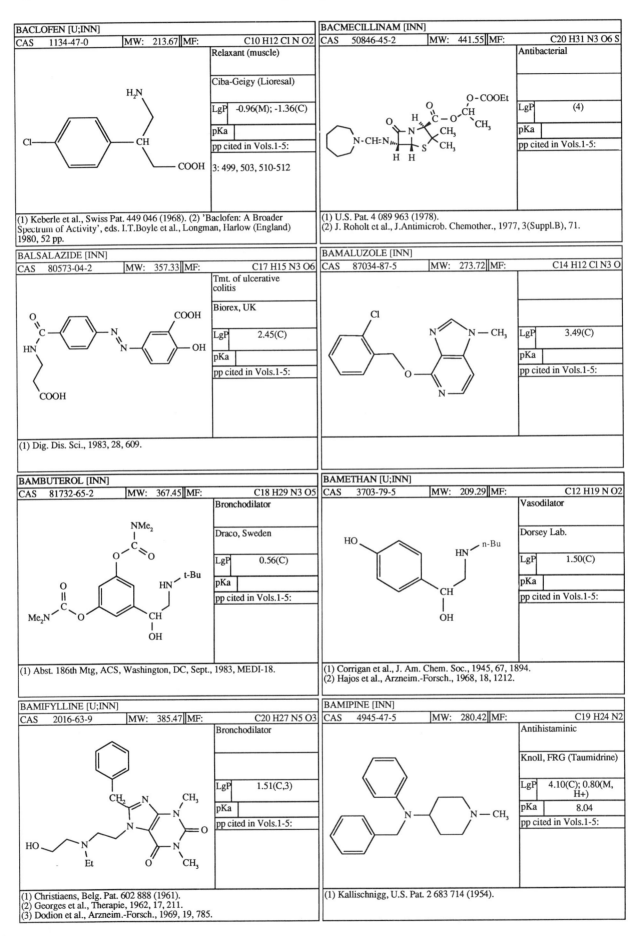

BACLOFEN [U;INN]

CAS	1134-47-0	MW:	213.67	MF:	C10 H12 Cl N O2

Relaxant (muscle)

Ciba-Geigy (Lioresal)

LgP	-0.96(M); -1.36(C)
pKa	
pp cited in Vols.1-5:	3: 499, 503, 510-512

(1) Keberle et al., Swiss Pat. 449 046 (1968). (2) 'Baclofen: A Broader Spectrum of Activity', eds. I.T.Boyle et al., Longman, Harlow (England) 1980, 52 pp.

BALSALAZIDE [INN]

CAS	80573-04-2	MW:	357.33	MF:	C17 H15 N3 O6

Tmt. of ulcerative colitis

Biorex, UK

LgP	2.45(C)
pKa	
pp cited in Vols.1-5:	

(1) Dig. Dis. Sci., 1983, 28, 609.

BAMBUTEROL [INN]

CAS	81732-65-2	MW:	367.45	MF:	C18 H29 N3 O5

Bronchodilator

Draco, Sweden

LgP	0.56(C)
pKa	
pp cited in Vols.1-5:	

(1) Abst. 186th Mtg, ACS, Washington, DC, Sept., 1983, MEDI-18.

BAMIFYLLINE [U;INN]

CAS	2016-63-9	MW:	385.47	MF:	C20 H27 N5 O3

Bronchodilator

LgP	1.51(C,3)
pKa	
pp cited in Vols.1-5:	

(1) Christiaens, Belg. Pat. 602 888 (1961).
(2) Georges et al., Therapie, 1962, 17, 211.
(3) Dodion et al., Arzneim.-Forsch., 1969, 19, 785.

BACMECILLINAM [INN]

CAS	50846-45-2	MW:	441.55	MF:	C20 H31 N3 O6 S

Antibacterial

LgP	(4)
pKa	
pp cited in Vols.1-5:	

(1) U.S. Pat. 4 089 963 (1978).
(2) J. Roholt et al., J.Antimicrob. Chemother., 1977, 3(Suppl.B), 71.

BAMALUZOLE [INN]

CAS	87034-87-5	MW:	273.72	MF:	C14 H12 Cl N3 O

LgP	3.49(C)
pKa	
pp cited in Vols.1-5:	

BAMETHAN [U;INN]

CAS	3703-79-5	MW:	209.29	MF:	C12 H19 N O2

Vasodilator

Dorsey Lab.

LgP	1.50(C)
pKa	
pp cited in Vols.1-5:	

(1) Corrigan et al., J. Am. Chem. Soc., 1945, 67, 1894.
(2) Hajos et al., Arzneim.-Forsch., 1968, 18, 1212.

BAMIPINE [INN]

CAS	4945-47-5	MW:	280.42	MF:	C19 H24 N2

Antihistaminic

Knoll, FRG (Taumidrine)

LgP	4.10(C); 0.80(M, H+)
pKa	8.04
pp cited in Vols.1-5:	

(1) Kallischnigg, U.S. Pat. 2 683 714 (1954).

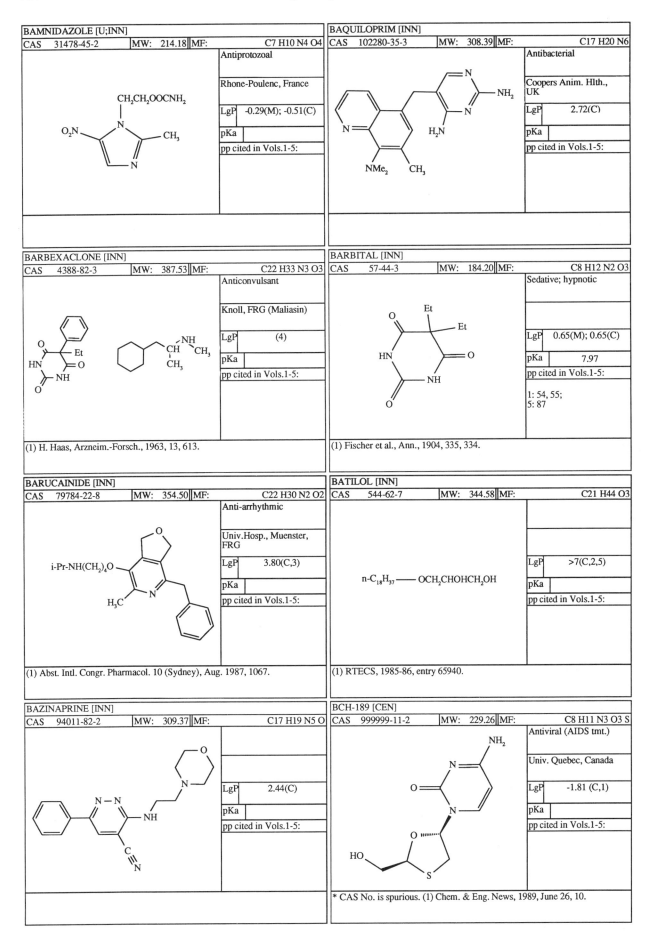

BAMNIDAZOLE [U;INN]

CAS	31478-45-2	MW:	214.18	MF:	C7 H10 N4 O4

Antiprotozoal

Rhone-Poulenc, France

LgP	-0.29(M); -0.51(C)
pKa	
pp cited in Vols.1-5:	

BAQUILOPRIM [INN]

CAS	102280-35-3	MW:	308.39	MF:	C17 H20 N6

Antibacterial

Coopers Anim. Hlth., UK

LgP	2.72(C)
pKa	
pp cited in Vols.1-5:	

BARBEXACLONE [INN]

CAS	4388-82-3	MW:	387.53	MF:	C22 H33 N3 O3

Anticonvulsant

Knoll, FRG (Maliasin)

LgP	(4)
pKa	
pp cited in Vols.1-5:	

(1) H. Haas, Arzneim.-Forsch., 1963, 13, 613.

BARBITAL [INN]

CAS	57-44-3	MW:	184.20	MF:	C8 H12 N2 O3

Sedative; hypnotic

LgP	0.65(M); 0.65(C)
pKa	7.97
pp cited in Vols.1-5:	

1: 54, 55;
5: 87

(1) Fischer et al., Ann., 1904, 335, 334.

BARUCAINIDE [INN]

CAS	79784-22-8	MW:	354.50	MF:	C22 H30 N2 O2

Anti-arrhythmic

Univ.Hosp., Muenster, FRG

LgP	3.80(C,3)
pKa	
pp cited in Vols.1-5:	

(1) Abst. Intl. Congr. Pharmacol. 10 (Sydney), Aug. 1987, 1067.

BATILOL [INN]

CAS	544-62-7	MW:	344.58	MF:	C21 H44 O3

LgP	>7(C,2,5)
pKa	
pp cited in Vols.1-5:	

(1) RTECS, 1985-86, entry 65940.

BAZINAPRINE [INN]

CAS	94011-82-2	MW:	309.37	MF:	C17 H19 N5 O

LgP	2.44(C)
pKa	
pp cited in Vols.1-5:	

BCH-189 [CEN]

CAS	999999-11-2	MW:	229.26	MF:	C8 H11 N3 O3 S

Antiviral (AIDS tmt.)

Univ. Quebec, Canada

LgP	-1.81 (C,1)
pKa	
pp cited in Vols.1-5:	

* CAS No. is spurious. (1) Chem. & Eng. News, 1989, June 26, 10.

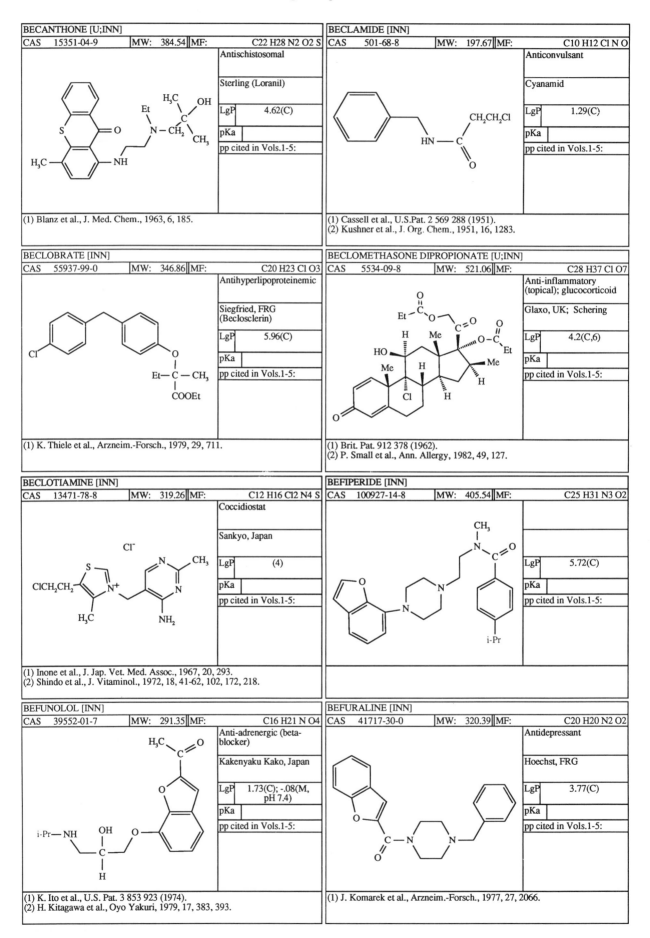

BECANTHONE [U;INN]

CAS	15351-04-9	MW:	384.54	MF:	C22 H28 N2 O2 S

Antischistosomal

Sterling (Loranil)

LgP	4.62(C)
pKa	
pp cited in Vols.1-5:	

(1) Blanz et al., J. Med. Chem., 1963, 6, 185.

BECLAMIDE [INN]

CAS	501-68-8	MW:	197.67	MF:	C10 H12 Cl N O

Anticonvulsant

Cyanamid

LgP	1.29(C)
pKa	
pp cited in Vols.1-5:	

(1) Cassell et al., U.S.Pat. 2 569 288 (1951).
(2) Kushner et al., J. Org. Chem., 1951, 16, 1283.

BECLOBRATE [INN]

CAS	55937-99-0	MW:	346.86	MF:	C20 H23 Cl O3

Antihyperlipoproteinemic

Siegfried, FRG
(Beclosclerin)

LgP	5.96(C)
pKa	
pp cited in Vols.1-5:	

(1) K. Thiele et al., Arzneim.-Forsch., 1979, 29, 711.

BECLOMETHASONE DIPROPIONATE [U;INN]

CAS	5534-09-8	MW:	521.06	MF:	C28 H37 Cl O7

Anti-inflammatory
(topical); glucocorticoid

Glaxo, UK; Schering

LgP	4.2(C,6)
pKa	
pp cited in Vols.1-5:	

(1) Brit. Pat. 912 378 (1962).
(2) P. Small et al., Ann. Allergy, 1982, 49, 127.

BECLOTIAMINE [INN]

CAS	13471-78-8	MW:	319.26	MF:	C12 H16 Cl2 N4 S

Coccidiostat

Sankyo, Japan

LgP	(4)
pKa	
pp cited in Vols.1-5:	

(1) Inone et al., J. Jap. Vet. Med. Assoc., 1967, 20, 293.
(2) Shindo et al., J. Vitaminol., 1972, 18, 41-62, 102, 172, 218.

BEFIPERIDE [INN]

CAS	100927-14-8	MW:	405.54	MF:	C25 H31 N3 O2

LgP	5.72(C)
pKa	
pp cited in Vols.1-5:	

BEFUNOLOL [INN]

CAS	39552-01-7	MW:	291.35	MF:	C16 H21 N O4

Anti-adrenergic (beta-blocker)

Kakenyaku Kako, Japan

LgP	1.73(C); -.08(M, pH 7.4)
pKa	
pp cited in Vols.1-5:	

(1) K. Ito et al., U.S. Pat. 3 853 923 (1974).
(2) H. Kitagawa et al., Oyo Yakuri, 1979, 17, 383, 393.

BEFURALINE [INN]

CAS	41717-30-0	MW:	320.39	MF:	C20 H20 N2 O2

Antidepressant

Hoechst, FRG

LgP	3.77(C)
pKa	
pp cited in Vols.1-5:	

(1) J. Komarek et al., Arzneim.-Forsch., 1977, 27, 2066.

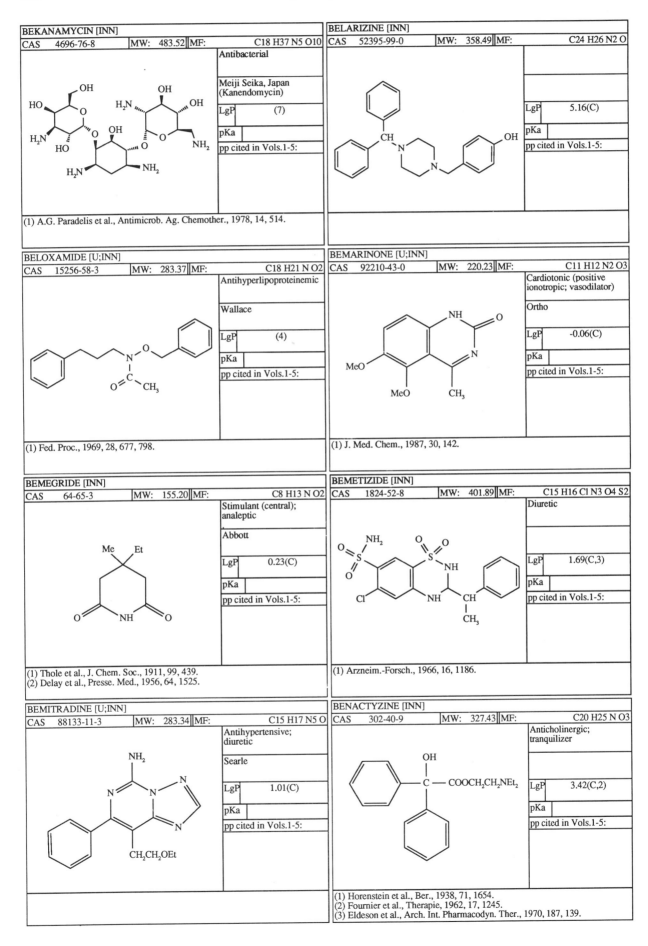

BEKANAMYCIN [INN]

CAS	4696-76-8	MW: 483.52	MF:	C18 H37 N5 O10

Antibacterial

Meiji Seika, Japan (Kanendomycin)

LgP	(7)

pKa	

pp cited in Vols.1-5:

(1) A.G. Paradelis et al., Antimicrob. Ag. Chemother., 1978, 14, 514.

BELARIZINE [INN]

CAS	52395-99-0	MW: 358.49	MF:	C24 H26 N2 O

LgP	5.16(C)

pKa	

pp cited in Vols.1-5:

BELOXAMIDE [U;INN]

CAS	15256-58-3	MW: 283.37	MF:	C18 H21 N O2

Antihyperlipoproteinemic

Wallace

LgP	(4)

pKa	

pp cited in Vols.1-5:

(1) Fed. Proc., 1969, 28, 677, 798.

BEMARINONE [U;INN]

CAS	92210-43-0	MW: 220.23	MF:	C11 H12 N2 O3

Cardiotonic (positive ionotropic; vasodilator)

Ortho

LgP	-0.06(C)

pKa	

pp cited in Vols.1-5:

(1) J. Med. Chem., 1987, 30, 142.

BEMEGRIDE [INN]

CAS	64-65-3	MW: 155.20	MF:	C8 H13 N O2

Stimulant (central); analeptic

Abbott

LgP	0.23(C)

pKa	

pp cited in Vols.1-5:

(1) Thole et al., J. Chem. Soc., 1911, 99, 439.
(2) Delay et al., Presse. Med., 1956, 64, 1525.

BEMETIZIDE [INN]

CAS	1824-52-8	MW: 401.89	MF:	C15 H16 Cl N3 O4 S2

Diuretic

LgP	1.69(C,3)

pKa	

pp cited in Vols.1-5:

(1) Arzneim.-Forsch., 1966, 16, 1186.

BEMITRADINE [U;INN]

CAS	88133-11-3	MW: 283.34	MF:	C15 H17 N5 O

Antihypertensive; diuretic

Searle

LgP	1.01(C)

pKa	

pp cited in Vols.1-5:

BENACTYZINE [INN]

CAS	302-40-9	MW: 327.43	MF:	C20 H25 N O3

Anticholinergic; tranquilizer

LgP	3.42(C,2)

pKa	

pp cited in Vols.1-5:

(1) Horenstein et al., Ber., 1938, 71, 1654.
(2) Fournier et al., Therapie, 1962, 17, 1245.
(3) Eldeson et al., Arch. Int. Pharmacodyn. Ther., 1970, 187, 139.

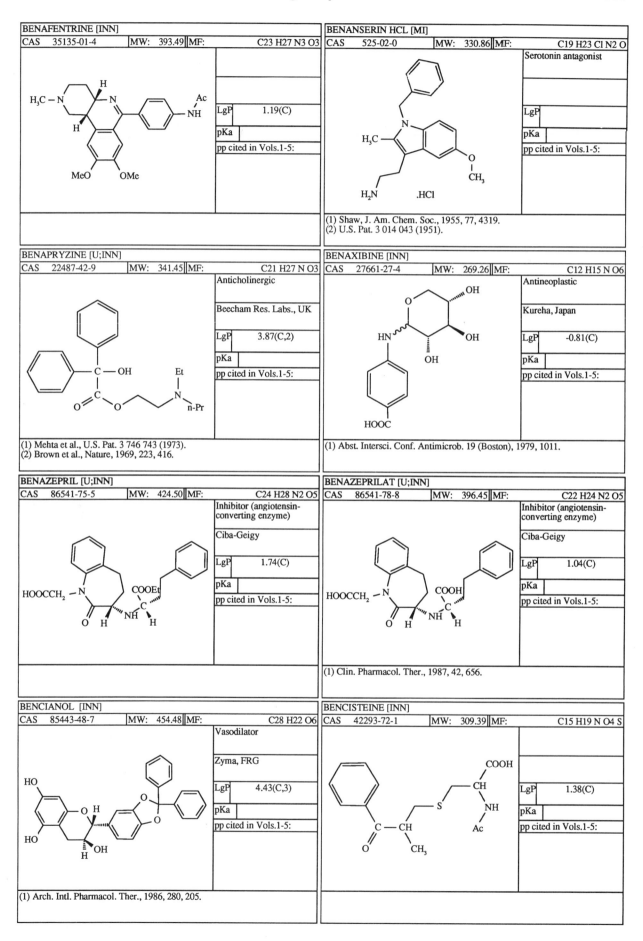

BENAFENTRINE [INN]

CAS	35135-01-4	MW:	393.49	MF:	C23 H27 N3 O3

LgP	1.19(C)
pKa	
pp cited in Vols.1-5:	

BENAPRYZINE [U;INN]

CAS	22487-42-9	MW:	341.45	MF:	C21 H27 N O3

Anticholinergic

Beecham Res. Labs., UK

LgP	3.87(C,2)
pKa	
pp cited in Vols.1-5:	

(1) Mehta et al., U.S. Pat. 3 746 743 (1973).
(2) Brown et al., Nature, 1969, 223, 416.

BENAZEPRIL [U;INN]

CAS	86541-75-5	MW:	424.50	MF:	C24 H28 N2 O5

Inhibitor (angiotensin-converting enzyme)

Ciba-Geigy

LgP	1.74(C)
pKa	
pp cited in Vols.1-5:	

BENCIANOL [INN]

CAS	85443-48-7	MW:	454.48	MF:	C28 H22 O6

Vasodilator

Zyma, FRG

LgP	4.43(C,3)
pKa	
pp cited in Vols.1-5:	

(1) Arch. Intl. Pharmacol. Ther., 1986, 280, 205.

BENANSERIN HCL [MI]

CAS	525-02-0	MW:	330.86	MF:	C19 H23 Cl N2 O

Serotonin antagonist

LgP	
pKa	
pp cited in Vols.1-5:	

(1) Shaw, J. Am. Chem. Soc., 1955, 77, 4319.
(2) U.S. Pat. 3 014 043 (1951).

BENAXIBINE [INN]

CAS	27661-27-4	MW:	269.26	MF:	C12 H15 N O6

Antineoplastic

Kureha, Japan

LgP	-0.81(C)
pKa	
pp cited in Vols.1-5:	

(1) Abst. Intersci. Conf. Antimicrob. 19 (Boston), 1979, 1011.

BENAZEPRILAT [U;INN]

CAS	86541-78-8	MW:	396.45	MF:	C22 H24 N2 O5

Inhibitor (angiotensin-converting enzyme)

Ciba-Geigy

LgP	1.04(C)
pKa	
pp cited in Vols.1-5:	

(1) Clin. Pharmacol. Ther., 1987, 42, 656.

BENCISTEINE [INN]

CAS	42293-72-1	MW:	309.39	MF:	C15 H19 N O4 S

LgP	1.38(C)
pKa	
pp cited in Vols.1-5:	

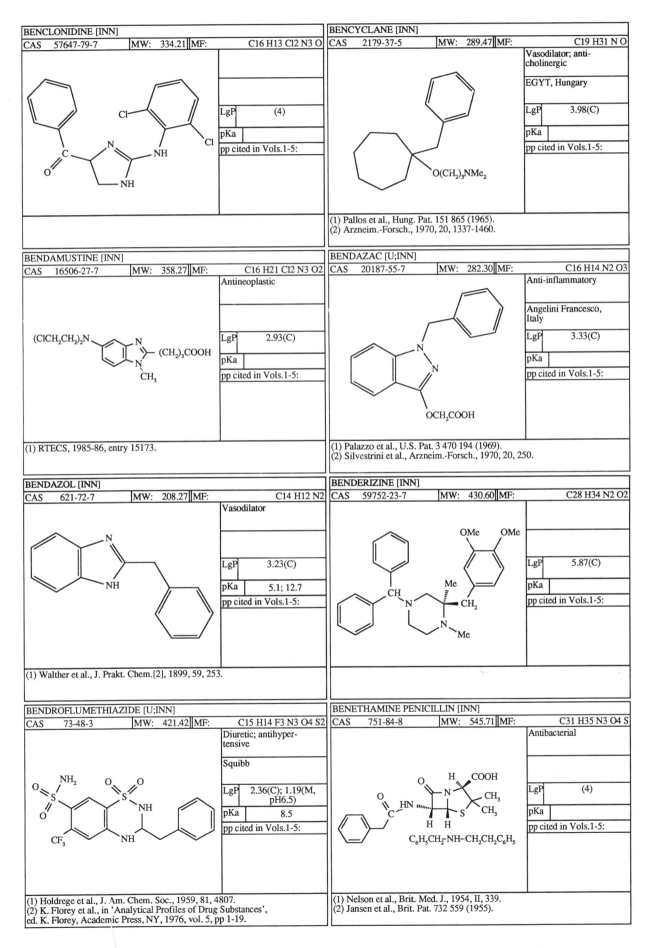

BENCLONIDINE [INN]

CAS 57647-79-7	MW: 334.21	MF:	C16 H13 Cl2 N3 O

	LgP	(4)
	pKa	
	pp cited in Vols.1-5:	

BENCYCLANE [INN]

CAS 2179-37-5	MW: 289.47	MF:	C19 H31 N O

Vasodilator; anti-cholinergic
EGYT, Hungary

LgP	3.98(C)
pKa	
pp cited in Vols.1-5:	

(1) Pallos et al., Hung. Pat. 151 865 (1965).
(2) Arzneim.-Forsch., 1970, 20, 1337-1460.

BENDAMUSTINE [INN]

CAS 16506-27-7	MW: 358.27	MF:	C16 H21 Cl2 N3 O2

Antineoplastic

LgP	2.93(C)
pKa	
pp cited in Vols.1-5:	

(1) RTECS, 1985-86, entry 15173.

BENDAZAC [U;INN]

CAS 20187-55-7	MW: 282.30	MF:	C16 H14 N2 O3

Anti-inflammatory
Angelini Francesco, Italy

LgP	3.33(C)
pKa	
pp cited in Vols.1-5:	

(1) Palazzo et al., U.S. Pat. 3 470 194 (1969).
(2) Silvestrini et al., Arzneim.-Forsch., 1970, 20, 250.

BENDAZOL [INN]

CAS 621-72-7	MW: 208.27	MF:	C14 H12 N2

Vasodilator

LgP	3.23(C)
pKa	5.1; 12.7
pp cited in Vols.1-5:	

(1) Walther et al., J. Prakt. Chem.[2], 1899, 59, 253.

BENDERIZINE [INN]

CAS 59752-23-7	MW: 430.60	MF:	C28 H34 N2 O2

LgP	5.87(C)
pKa	
pp cited in Vols.1-5:	

BENDROFLUMETHIAZIDE [U;INN]

CAS 73-48-3	MW: 421.42	MF:	C15 H14 F3 N3 O4 S2

Diuretic; antihypertensive
Squibb

LgP	2.36(C); 1.19(M, pH6.5)
pKa	8.5
pp cited in Vols.1-5:	

(1) Holdrege et al., J. Am. Chem. Soc., 1959, 81, 4807.
(2) K. Florey et al., in 'Analytical Profiles of Drug Substances', ed. K. Florey, Academic Press, NY, 1976, vol. 5, pp 1-19.

BENETHAMINE PENICILLIN [INN]

CAS 751-84-8	MW: 545.71	MF:	C31 H35 N3 O4 S

Antibacterial

LgP	(4)
pKa	
pp cited in Vols.1-5:	

$C_6H_5CH_2-NH-CH_2CH_2C_6H_5$

(1) Nelson et al., Brit. Med. J., 1954, II, 339.
(2) Jansen et al., Brit. Pat. 732 559 (1955).

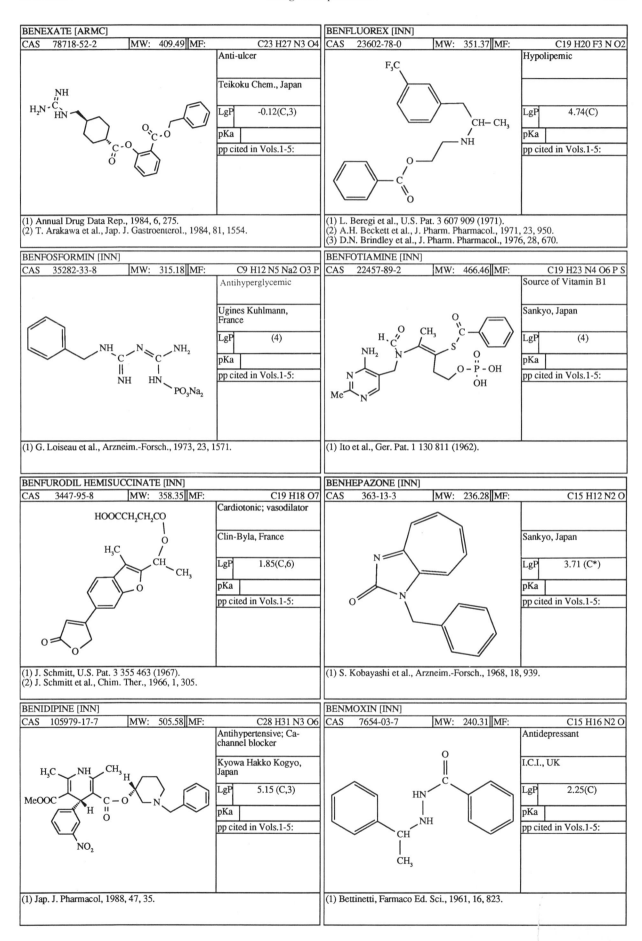

BENEXATE [ARMC]

CAS	78718-52-2	MW: 409.49	MF:	C23 H27 N3 O4

Anti-ulcer

Teikoku Chem., Japan

LgP	-0.12(C,3)
pKa	
pp cited in Vols.1-5:	

(1) Annual Drug Data Rep., 1984, 6, 275.
(2) T. Arakawa et al., Jap. J. Gastroenterol., 1984, 81, 1554.

BENFLUOREX [INN]

CAS	23602-78-0	MW: 351.37	MF:	C19 H20 F3 N O2

Hypolipemic

LgP	4.74(C)
pKa	
pp cited in Vols.1-5:	

(1) L. Beregi et al., U.S. Pat. 3 607 909 (1971).
(2) A.H. Beckett et al., J. Pharm. Pharmacol., 1971, 23, 950.
(3) D.N. Brindley et al., J. Pharm. Pharmacol., 1976, 28, 670.

BENFOSFORMIN [INN]

CAS	35282-33-8	MW: 315.18	MF:	C9 H12 N5 Na2 O3 P

Antihyperglycemic

Ugines Kuhlmann, France

LgP	(4)
pKa	
pp cited in Vols.1-5:	

(1) G. Loiseau et al., Arzneim.-Forsch., 1973, 23, 1571.

BENFOTIAMINE [INN]

CAS	22457-89-2	MW: 466.46	MF:	C19 H23 N4 O6 P S

Source of Vitamin B1

Sankyo, Japan

LgP	(4)
pKa	
pp cited in Vols.1-5:	

(1) Ito et al., Ger. Pat. 1 130 811 (1962).

BENFURODIL HEMISUCCINATE [INN]

CAS	3447-95-8	MW: 358.35	MF:	C19 H18 O7

Cardiotonic; vasodilator

Clin-Byla, France

LgP	1.85(C,6)
pKa	
pp cited in Vols.1-5:	

(1) J. Schmitt, U.S. Pat. 3 355 463 (1967).
(2) J. Schmitt et al., Chim. Ther., 1966, 1, 305.

BENHEPAZONE [INN]

CAS	363-13-3	MW: 236.28	MF:	C15 H12 N2 O

Sankyo, Japan

LgP	3.71 (C*)
pKa	
pp cited in Vols.1-5:	

(1) S. Kobayashi et al., Arzneim.-Forsch., 1968, 18, 939.

BENIDIPINE [INN]

CAS	105979-17-7	MW: 505.58	MF:	C28 H31 N3 O6

Antihypertensive; Ca-channel blocker

Kyowa Hakko Kogyo, Japan

LgP	5.15 (C,3)
pKa	
pp cited in Vols.1-5:	

(1) Jap. J. Pharmacol, 1988, 47, 35.

BENMOXIN [INN]

CAS	7654-03-7	MW: 240.31	MF:	C15 H16 N2 O

Antidepressant

I.C.I., UK

LgP	2.25(C)
pKa	
pp cited in Vols.1-5:	

(1) Bettinetti, Farmaco Ed. Sci., 1961, 16, 823.

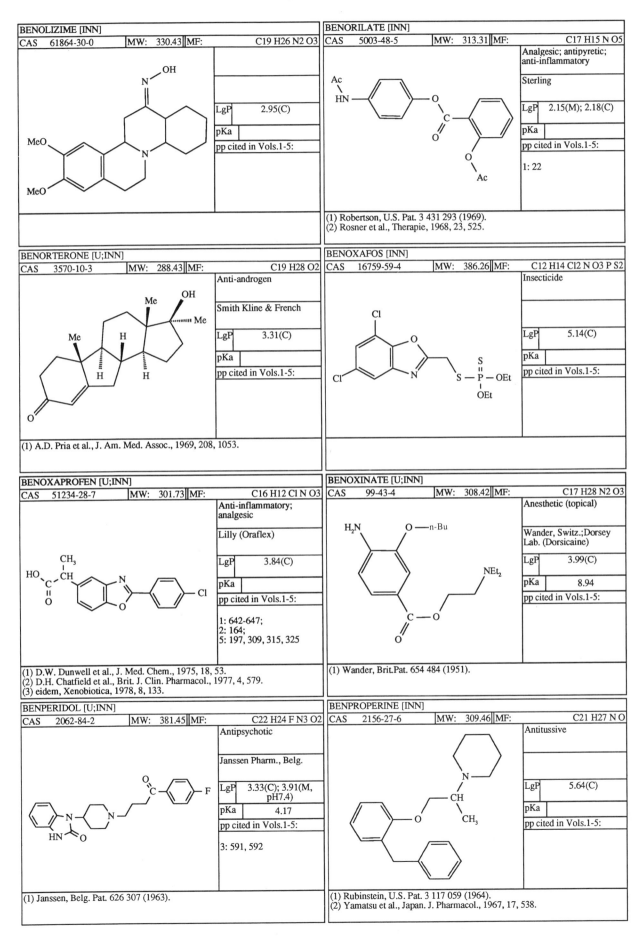

BENOLIZIME [INN]				
CAS 61864-30-0	MW: 330.43	MF:		C19 H26 N2 O3
		LgP	2.95(C)	
		pKa		
		pp cited in Vols.1-5:		

BENORILATE [INN]				
CAS 5003-48-5	MW: 313.31	MF:		C17 H15 N O5
		Analgesic; antipyretic; anti-inflammatory		
		Sterling		
		LgP	2.15(M); 2.18(C)	
		pKa		
		pp cited in Vols.1-5:		
		1: 22		

(1) Robertson, U.S. Pat. 3 431 293 (1969).
(2) Rosner et al., Therapie, 1968, 23, 525.

BENORTERONE [U;INN]				
CAS 3570-10-3	MW: 288.43	MF:		C19 H28 O2
		Anti-androgen		
		Smith Kline & French		
		LgP	3.31(C)	
		pKa		
		pp cited in Vols.1-5:		

(1) A.D. Pria et al., J. Am. Med. Assoc., 1969, 208, 1053.

BENOXAFOS [INN]				
CAS 16759-59-4	MW: 386.26	MF:		C12 H14 Cl2 N O3 P S2
		Insecticide		
		LgP	5.14(C)	
		pKa		
		pp cited in Vols.1-5:		

BENOXAPROFEN [U;INN]				
CAS 51234-28-7	MW: 301.73	MF:		C16 H12 Cl N O3
		Anti-inflammatory; analgesic		
		Lilly (Oraflex)		
		LgP	3.84(C)	
		pKa		
		pp cited in Vols.1-5:		
		1: 642-647; 2: 164; 5: 197, 309, 315, 325		

(1) D.W. Dunwell et al., J. Med. Chem., 1975, 18, 53.
(2) D.H. Chatfield et al., Brit. J. Clin. Pharmacol., 1977, 4, 579.
(3) eidem, Xenobiotica, 1978, 8, 133.

BENOXINATE [U;INN]				
CAS 99-43-4	MW: 308.42	MF:		C17 H28 N2 O3
		Anesthetic (topical)		
		Wander, Switz.;Dorsey Lab. (Dorsicaine)		
		LgP	3.99(C)	
		pKa	8.94	
		pp cited in Vols.1-5:		

(1) Wander, Brit.Pat. 654 484 (1951).

BENPERIDOL [U;INN]				
CAS 2062-84-2	MW: 381.45	MF:		C22 H24 F N3 O2
		Antipsychotic		
		Janssen Pharm., Belg.		
		LgP	3.33(C); 3.91(M, pH7.4)	
		pKa	4.17	
		pp cited in Vols.1-5:		
		3: 591, 592		

(1) Janssen, Belg. Pat. 626 307 (1963).

BENPROPERINE [INN]				
CAS 2156-27-6	MW: 309.46	MF:		C21 H27 N O
		Antitussive		
		LgP	5.64(C)	
		pKa		
		pp cited in Vols.1-5:		

(1) Rubinstein, U.S. Pat. 3 117 059 (1964).
(2) Yamatsu et al., Japan. J. Pharmacol., 1967, 17, 538.

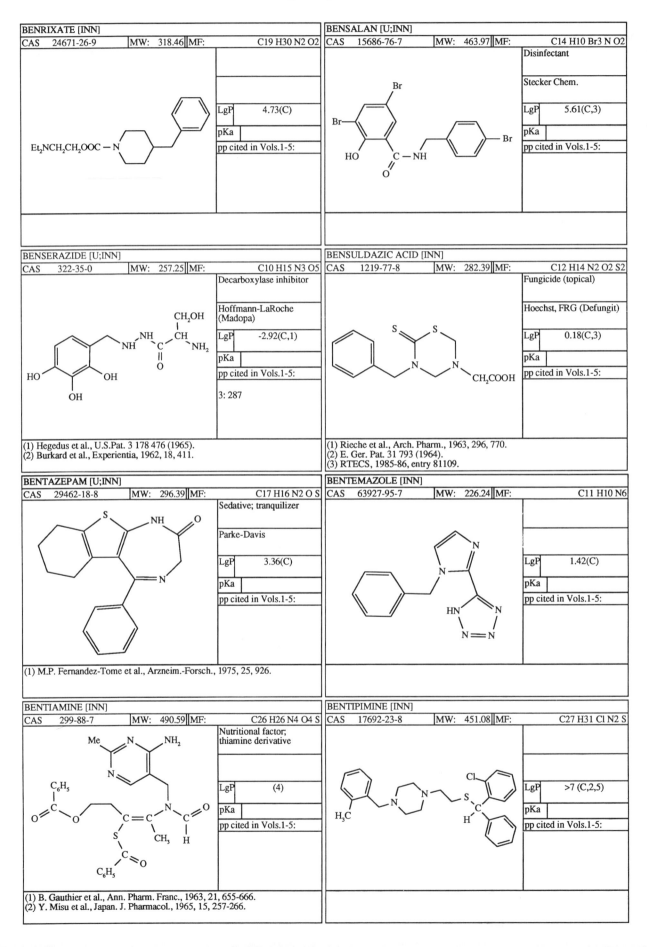

BENRIXATE [INN]

CAS 24671-26-9	MW: 318.46	MF: C19 H30 N2 O2

LgP	4.73(C)
pKa	
pp cited in Vols.1-5:	

Et$_2$NCH$_2$CH$_2$OOC — N (piperidine, 4-benzyl)

BENSALAN [U;INN]

CAS 15686-76-7	MW: 463.97	MF: C14 H10 Br3 N O2

Disinfectant

Stecker Chem.

LgP	5.61(C,3)
pKa	
pp cited in Vols.1-5:	

BENSERAZIDE [U;INN]

CAS 322-35-0	MW: 257.25	MF: C10 H15 N3 O5

Decarboxylase inhibitor

Hoffmann-LaRoche (Madopa)

LgP	-2.92(C,1)
pKa	
pp cited in Vols.1-5:	
3: 287	

(1) Hegedus et al., U.S.Pat. 3 178 476 (1965).
(2) Burkard et al., Experientia, 1962, 18, 411.

BENSULDAZIC ACID [INN]

CAS 1219-77-8	MW: 282.39	MF: C12 H14 N2 O2 S2

Fungicide (topical)

Hoechst, FRG (Defungit)

LgP	0.18(C,3)
pKa	
pp cited in Vols.1-5:	

(1) Rieche et al., Arch. Pharm., 1963, 296, 770.
(2) E. Ger. Pat. 31 793 (1964).
(3) RTECS, 1985-86, entry 81109.

BENTAZEPAM [U;INN]

CAS 29462-18-8	MW: 296.39	MF: C17 H16 N2 O S

Sedative; tranquilizer

Parke-Davis

LgP	3.36(C)
pKa	
pp cited in Vols.1-5:	

(1) M.P. Fernandez-Tome et al., Arzneim.-Forsch., 1975, 25, 926.

BENTEMAZOLE [INN]

CAS 63927-95-7	MW: 226.24	MF: C11 H10 N6

LgP	1.42(C)
pKa	
pp cited in Vols.1-5:	

BENTIAMINE [INN]

CAS 299-88-7	MW: 490.59	MF: C26 H26 N4 O4 S

Nutritional factor; thiamine derivative

LgP	(4)
pKa	
pp cited in Vols.1-5:	

(1) B. Gauthier et al., Ann. Pharm. Franc., 1963, 21, 655-666.
(2) Y. Misu et al., Japan. J. Pharmacol., 1965, 15, 257-266.

BENTIPIMINE [INN]

CAS 17692-23-8	MW: 451.08	MF: C27 H31 Cl N2 S

LgP	>7 (C,2,5)
pKa	
pp cited in Vols.1-5:	

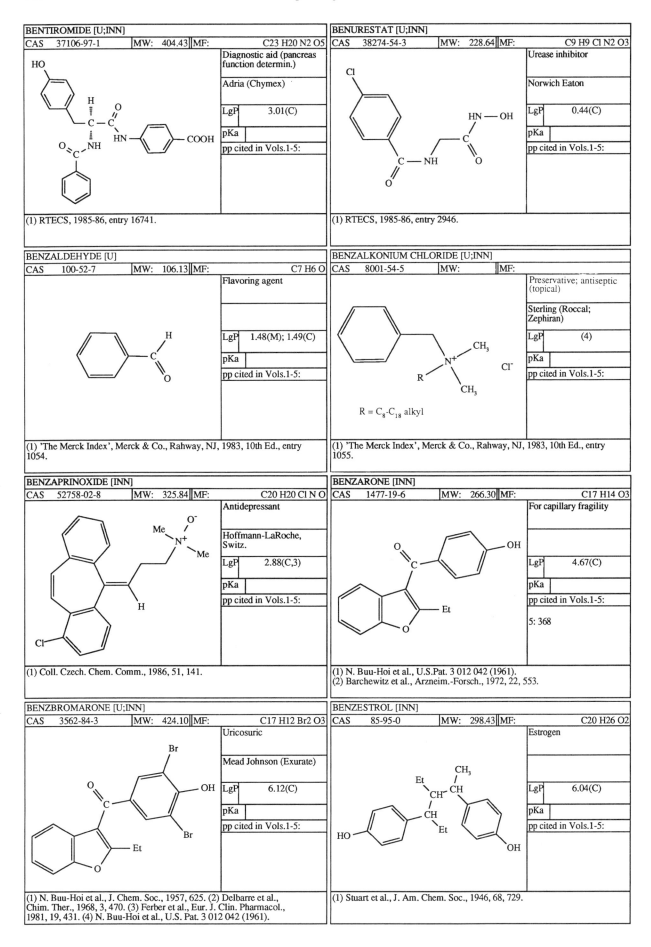

BENTIROMIDE [U;INN]

CAS	37106-97-1	MW:	404.43	MF:	C23 H20 N2 O5

Diagnostic aid (pancreas function determin.)

Adria (Chymex)

LgP	3.01(C)

pKa	

pp cited in Vols.1-5:

(1) RTECS, 1985-86, entry 16741.

BENURESTAT [U;INN]

CAS	38274-54-3	MW:	228.64	MF:	C9 H9 Cl N2 O3

Urease inhibitor

Norwich Eaton

LgP	0.44(C)

pKa	

pp cited in Vols.1-5:

(1) RTECS, 1985-86, entry 2946.

BENZALDEHYDE [U]

CAS	100-52-7	MW:	106.13	MF:	C7 H6 O

Flavoring agent

LgP	1.48(M); 1.49(C)

pKa	

pp cited in Vols.1-5:

(1) 'The Merck Index', Merck & Co., Rahway, NJ, 1983, 10th Ed., entry 1054.

BENZALKONIUM CHLORIDE [U;INN]

CAS	8001-54-5	MW:		MF:	

Preservative; antiseptic (topical)

Sterling (Roccal; Zephiran)

LgP	(4)

pKa	

pp cited in Vols.1-5:

R = C$_8$-C$_{18}$ alkyl

(1) 'The Merck Index', Merck & Co., Rahway, NJ, 1983, 10th Ed., entry 1055.

BENZAPRINOXIDE [INN]

CAS	52758-02-8	MW:	325.84	MF:	C20 H20 Cl N O

Antidepressant

Hoffmann-LaRoche, Switz.

LgP	2.88(C,3)

pKa	

pp cited in Vols.1-5:

(1) Coll. Czech. Chem. Comm., 1986, 51, 141.

BENZARONE [INN]

CAS	1477-19-6	MW:	266.30	MF:	C17 H14 O3

For capillary fragility

LgP	4.67(C)

pKa	

pp cited in Vols.1-5:

5: 368

(1) N. Buu-Hoi et al., U.S.Pat. 3 012 042 (1961).
(2) Barchewitz et al., Arzneim.-Forsch., 1972, 22, 553.

BENZBROMARONE [U;INN]

CAS	3562-84-3	MW:	424.10	MF:	C17 H12 Br2 O3

Uricosuric

Mead Johnson (Exurate)

LgP	6.12(C)

pKa	

pp cited in Vols.1-5:

(1) N. Buu-Hoi et al., J. Chem. Soc., 1957, 625. (2) Delbarre et al., Chim. Ther., 1968, 3, 470. (3) Ferber et al., Eur. J. Clin. Pharmacol., 1981, 19, 431. (4) N. Buu-Hoi et al., U.S. Pat. 3 012 042 (1961).

BENZESTROL [INN]

CAS	85-95-0	MW:	298.43	MF:	C20 H26 O2

Estrogen

LgP	6.04(C)

pKa	

pp cited in Vols.1-5:

(1) Stuart et al., J. Am. Chem. Soc., 1946, 68, 729.

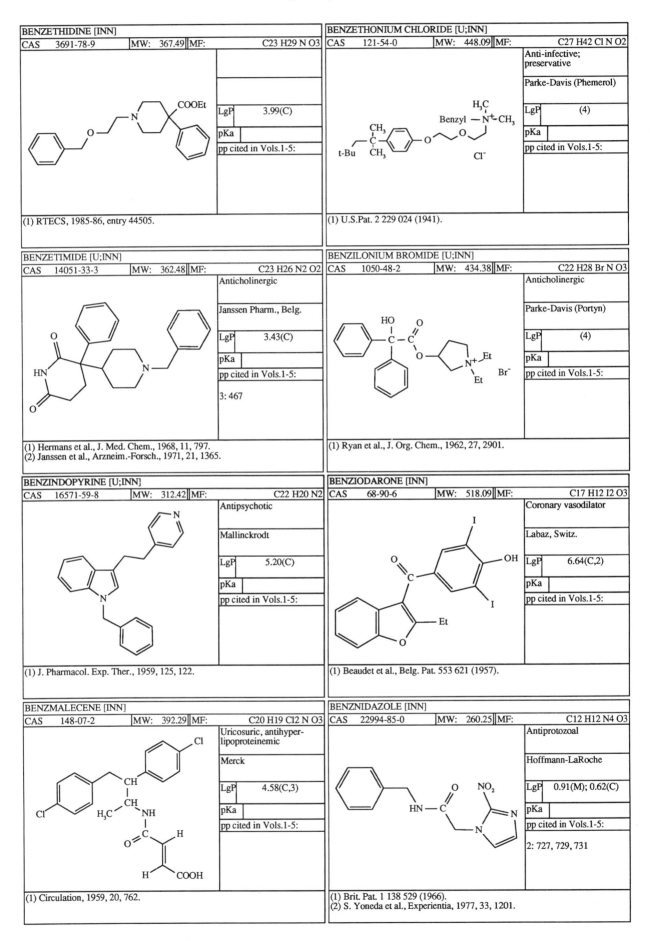

BENZETHIDINE [INN]

CAS	3691-78-9	MW: 367.49	MF:	C23 H29 N O3

LgP	3.99(C)	
pKa		
pp cited in Vols.1-5:		

(1) RTECS, 1985-86, entry 44505.

BENZETHONIUM CHLORIDE [U;INN]

CAS	121-54-0	MW: 448.09	MF:	C27 H42 Cl N O2

Anti-infective; preservative

Parke-Davis (Phemerol)

LgP	(4)
pKa	
pp cited in Vols.1-5:	

(1) U.S.Pat. 2 229 024 (1941).

BENZETIMIDE [U;INN]

CAS	14051-33-3	MW: 362.48	MF:	C23 H26 N2 O2

Anticholinergic

Janssen Pharm., Belg.

LgP	3.43(C)
pKa	
pp cited in Vols.1-5:	

3: 467

(1) Hermans et al., J. Med. Chem., 1968, 11, 797.
(2) Janssen et al., Arzneim.-Forsch., 1971, 21, 1365.

BENZILONIUM BROMIDE [U;INN]

CAS	1050-48-2	MW: 434.38	MF:	C22 H28 Br N O3

Anticholinergic

Parke-Davis (Portyn)

LgP	(4)
pKa	
pp cited in Vols.1-5:	

(1) Ryan et al., J. Org. Chem., 1962, 27, 2901.

BENZINDOPYRINE [U;INN]

CAS	16571-59-8	MW: 312.42	MF:	C22 H20 N2

Antipsychotic

Mallinckrodt

LgP	5.20(C)
pKa	
pp cited in Vols.1-5:	

(1) J. Pharmacol. Exp. Ther., 1959, 125, 122.

BENZIODARONE [INN]

CAS	68-90-6	MW: 518.09	MF:	C17 H12 I2 O3

Coronary vasodilator

Labaz, Switz.

LgP	6.64(C,2)
pKa	
pp cited in Vols.1-5:	

(1) Beaudet et al., Belg. Pat. 553 621 (1957).

BENZMALECENE [INN]

CAS	148-07-2	MW: 392.29	MF:	C20 H19 Cl2 N O3

Uricosuric, antihyper-lipoproteinemic

Merck

LgP	4.58(C,3)
pKa	
pp cited in Vols.1-5:	

(1) Circulation, 1959, 20, 762.

BENZNIDAZOLE [INN]

CAS	22994-85-0	MW: 260.25	MF:	C12 H12 N4 O3

Antiprotozoal

Hoffmann-LaRoche

LgP	0.91(M); 0.62(C)
pKa	
pp cited in Vols.1-5:	

2: 727, 729, 731

(1) Brit. Pat. 1 138 529 (1966).
(2) S. Yoneda et al., Experientia, 1977, 33, 1201.

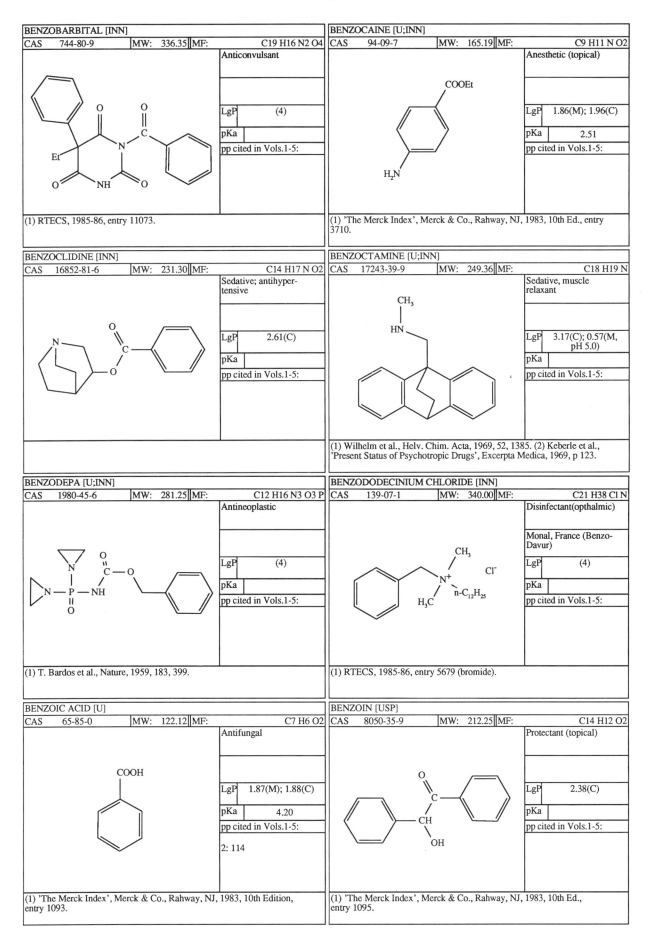

BENZOBARBITAL [INN]

CAS	744-80-9	MW:	336.35	MF:	C19 H16 N2 O4

Anticonvulsant

LgP	(4)
pKa	
pp cited in Vols.1-5:	

(1) RTECS, 1985-86, entry 11073.

BENZOCAINE [U;INN]

CAS	94-09-7	MW:	165.19	MF:	C9 H11 N O2

Anesthetic (topical)

LgP	1.86(M); 1.96(C)
pKa	2.51
pp cited in Vols.1-5:	

(1) 'The Merck Index', Merck & Co., Rahway, NJ, 1983, 10th Ed., entry 3710.

BENZOCLIDINE [INN]

CAS	16852-81-6	MW:	231.30	MF:	C14 H17 N O2

Sedative; antihypertensive

LgP	2.61(C)
pKa	
pp cited in Vols.1-5:	

BENZOCTAMINE [U;INN]

CAS	17243-39-9	MW:	249.36	MF:	C18 H19 N

Sedative, muscle relaxant

LgP	3.17(C); 0.57(M, pH 5.0)
pKa	
pp cited in Vols.1-5:	

(1) Wilhelm et al., Helv. Chim. Acta, 1969, 52, 1385. (2) Keberle et al., 'Present Status of Psychotropic Drugs', Excerpta Medica, 1969, p 123.

BENZODEPA [U;INN]

CAS	1980-45-6	MW:	281.25	MF:	C12 H16 N3 O3 P

Antineoplastic

LgP	(4)
pKa	
pp cited in Vols.1-5:	

(1) T. Bardos et al., Nature, 1959, 183, 399.

BENZODODECINIUM CHLORIDE [INN]

CAS	139-07-1	MW:	340.00	MF:	C21 H38 Cl N

Disinfectant(opthalmic)

Monal, France (Benzo-Davur)

LgP	(4)
pKa	
pp cited in Vols.1-5:	

(1) RTECS, 1985-86, entry 5679 (bromide).

BENZOIC ACID [U]

CAS	65-85-0	MW:	122.12	MF:	C7 H6 O2

Antifungal

LgP	1.87(M); 1.88(C)
pKa	4.20
pp cited in Vols.1-5:	
2: 114	

(1) 'The Merck Index', Merck & Co., Rahway, NJ, 1983, 10th Edition, entry 1093.

BENZOIN [USP]

CAS	8050-35-9	MW:	212.25	MF:	C14 H12 O2

Protectant (topical)

LgP	2.38(C)
pKa	
pp cited in Vols.1-5:	

(1) 'The Merck Index', Merck & Co., Rahway, NJ, 1983, 10th Ed., entry 1095.

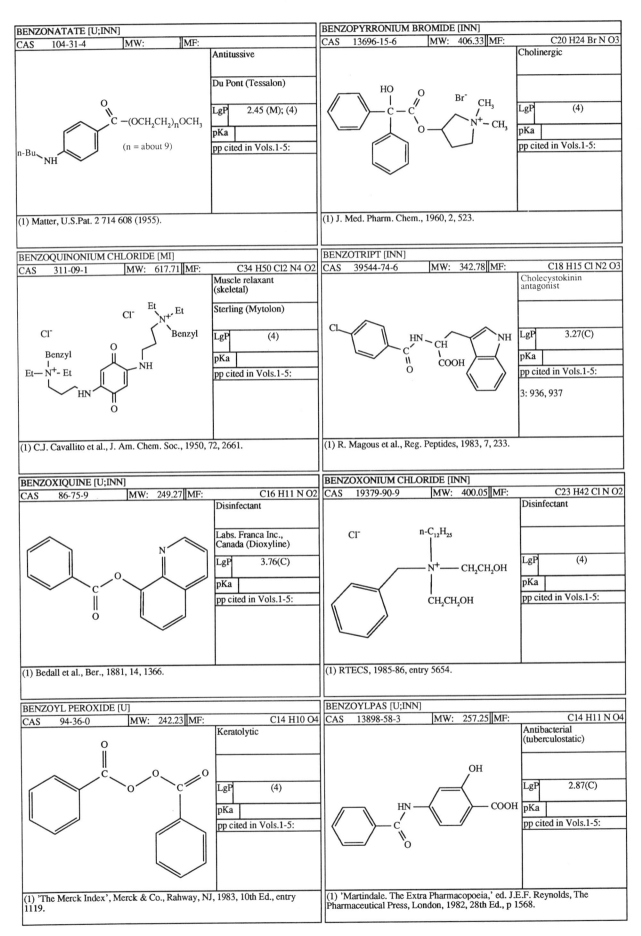

BENZONATATE [U;INN]

CAS	104-31-4	MW:	MF:	

Antitussive

Du Pont (Tessalon)

LgP	2.45 (M); (4)
pKa	
pp cited in Vols.1-5:	

(n = about 9)

(1) Matter, U.S.Pat. 2 714 608 (1955).

BENZOPYRRONIUM BROMIDE [INN]

CAS	13696-15-6	MW:	406.33	MF:	C20 H24 Br N O3

Cholinergic

LgP	(4)
pKa	
pp cited in Vols.1-5:	

(1) J. Med. Pharm. Chem., 1960, 2, 523.

BENZOQUINONIUM CHLORIDE [MI]

CAS	311-09-1	MW:	617.71	MF:	C34 H50 Cl2 N4 O2

Muscle relaxant (skeletal)

Sterling (Mytolon)

LgP	(4)
pKa	
pp cited in Vols.1-5:	

(1) C.J. Cavallito et al., J. Am. Chem. Soc., 1950, 72, 2661.

BENZOTRIPT [INN]

CAS	39544-74-6	MW:	342.78	MF:	C18 H15 Cl N2 O3

Cholecystokinin antagonist

LgP	3.27(C)
pKa	
pp cited in Vols.1-5:	

3: 936, 937

(1) R. Magous et al., Reg. Peptides, 1983, 7, 233.

BENZOXIQUINE [U;INN]

CAS	86-75-9	MW:	249.27	MF:	C16 H11 N O2

Disinfectant

Labs. Franca Inc., Canada (Dioxyline)

LgP	3.76(C)
pKa	
pp cited in Vols.1-5:	

(1) Bedall et al., Ber., 1881, 14, 1366.

BENZOXONIUM CHLORIDE [INN]

CAS	19379-90-9	MW:	400.05	MF:	C23 H42 Cl N O2

Disinfectant

LgP	(4)
pKa	
pp cited in Vols.1-5:	

(1) RTECS, 1985-86, entry 5654.

BENZOYL PEROXIDE [U]

CAS	94-36-0	MW:	242.23	MF:	C14 H10 O4

Keratolytic

LgP	(4)
pKa	
pp cited in Vols.1-5:	

(1) 'The Merck Index', Merck & Co., Rahway, NJ, 1983, 10th Ed., entry 1119.

BENZOYLPAS [U;INN]

CAS	13898-58-3	MW:	257.25	MF:	C14 H11 N O4

Antibacterial (tuberculostatic)

LgP	2.87(C)
pKa	
pp cited in Vols.1-5:	

(1) 'Martindale. The Extra Pharmacopoeia,' ed. J.E.F. Reynolds, The Pharmaceutical Press, London, 1982, 28th Ed., p 1568.

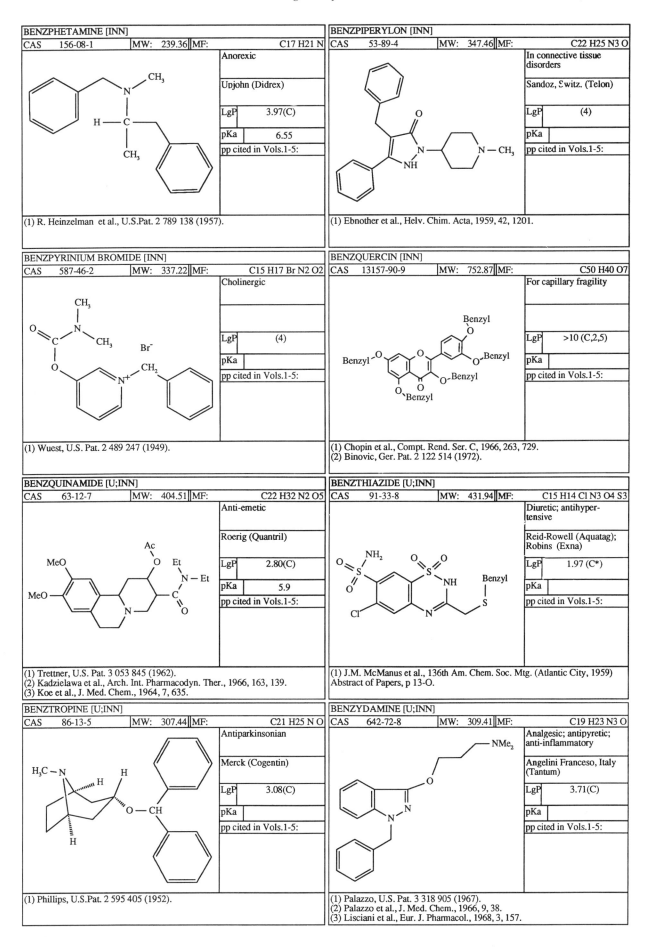

BENZPHETAMINE [INN]

CAS	156-08-1	MW: 239.36	MF:	C17 H21 N

Anorexic

Upjohn (Didrex)

LgP	3.97(C)
pKa	6.55
pp cited in Vols.1-5:	

(1) R. Heinzelman et al., U.S.Pat. 2 789 138 (1957).

BENZPIPERYLON [INN]

CAS	53-89-4	MW: 347.46	MF:	C22 H25 N3 O

In connective tissue disorders

Sandoz, Switz. (Telon)

LgP	(4)
pKa	
pp cited in Vols.1-5:	

(1) Ebnother et al., Helv. Chim. Acta, 1959, 42, 1201.

BENZPYRINIUM BROMIDE [INN]

CAS	587-46-2	MW: 337.22	MF:	C15 H17 Br N2 O2

Cholinergic

LgP	(4)
pKa	
pp cited in Vols.1-5:	

(1) Wuest, U.S. Pat. 2 489 247 (1949).

BENZQUERCIN [INN]

CAS	13157-90-9	MW: 752.87	MF:	C50 H40 O7

For capillary fragility

LgP	>10 (C,2,5)
pKa	
pp cited in Vols.1-5:	

(1) Chopin et al., Compt. Rend. Ser. C, 1966, 263, 729.
(2) Binovic, Ger. Pat. 2 122 514 (1972).

BENZQUINAMIDE [U;INN]

CAS	63-12-7	MW: 404.51	MF:	C22 H32 N2 O5

Anti-emetic

Roerig (Quantril)

LgP	2.80(C)
pKa	5.9
pp cited in Vols.1-5:	

(1) Trettner, U.S. Pat. 3 053 845 (1962).
(2) Kadzielawa et al., Arch. Int. Pharmacodyn. Ther., 1966, 163, 139.
(3) Koe et al., J. Med. Chem., 1964, 7, 635.

BENZTHIAZIDE [U;INN]

CAS	91-33-8	MW: 431.94	MF:	C15 H14 Cl N3 O4 S3

Diuretic; antihypertensive

Reid-Rowell (Aquatag); Robins (Exna)

LgP	1.97 (C*)
pKa	
pp cited in Vols.1-5:	

(1) J.M. McManus et al., 136th Am. Chem. Soc. Mtg. (Atlantic City, 1959) Abstract of Papers, p 13-O.

BENZTROPINE [U;INN]

CAS	86-13-5	MW: 307.44	MF:	C21 H25 N O

Antiparkinsonian

Merck (Cogentin)

LgP	3.08(C)
pKa	
pp cited in Vols.1-5:	

(1) Phillips, U.S.Pat. 2 595 405 (1952).

BENZYDAMINE [U;INN]

CAS	642-72-8	MW: 309.41	MF:	C19 H23 N3 O

Analgesic; antipyretic; anti-inflammatory

Angelini Franceso, Italy (Tantum)

LgP	3.71(C)
pKa	
pp cited in Vols.1-5:	

(1) Palazzo, U.S. Pat. 3 318 905 (1967).
(2) Palazzo et al., J. Med. Chem., 1966, 9, 38.
(3) Lisciani et al., Eur. J. Pharmacol., 1968, 3, 157.

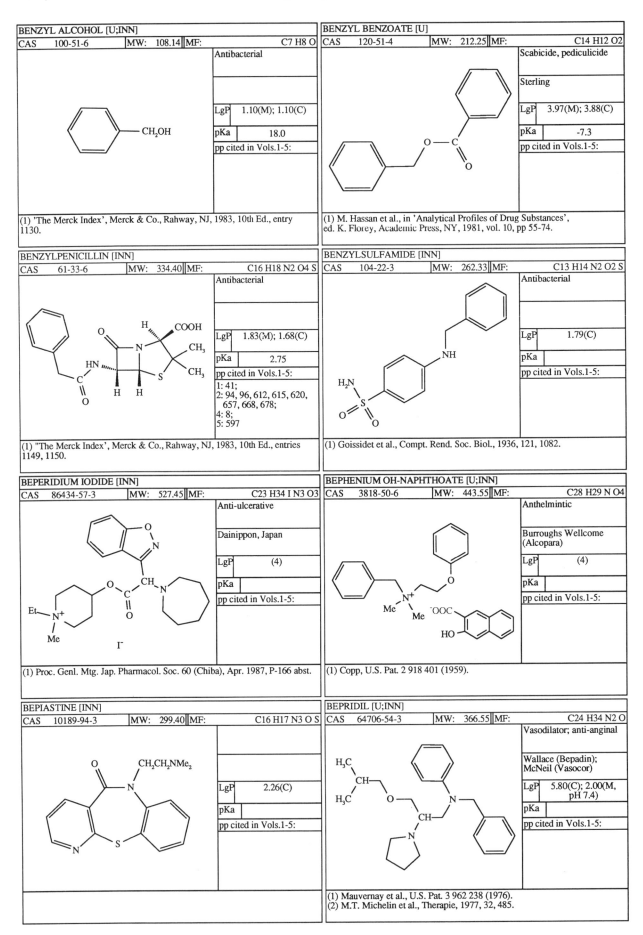

BENZYL ALCOHOL [U;INN]

CAS	100-51-6	MW:	108.14	MF:	C7 H8 O

Antibacterial

LgP	1.10(M); 1.10(C)
pKa	18.0
pp cited in Vols.1-5:	

(1) 'The Merck Index', Merck & Co., Rahway, NJ, 1983, 10th Ed., entry 1130.

BENZYL BENZOATE [U]

CAS	120-51-4	MW:	212.25	MF:	C14 H12 O2

Scabicide, pediculicide

Sterling

LgP	3.97(M); 3.88(C)
pKa	-7.3
pp cited in Vols.1-5:	

(1) M. Hassan et al., in 'Analytical Profiles of Drug Substances', ed. K. Florey, Academic Press, NY, 1981, vol. 10, pp 55-74.

BENZYLPENICILLIN [INN]

CAS	61-33-6	MW:	334.40	MF:	C16 H18 N2 O4 S

Antibacterial

LgP	1.83(M); 1.68(C)
pKa	2.75
pp cited in Vols.1-5:	

1: 41;
2: 94, 96, 612, 615, 620, 657, 668, 678;
4: 8;
5: 597

(1) "The Merck Index', Merck & Co., Rahway, NJ, 1983, 10th Ed., entries 1149, 1150.

BENZYLSULFAMIDE [INN]

CAS	104-22-3	MW:	262.33	MF:	C13 H14 N2 O2 S

Antibacterial

LgP	1.79(C)
pKa	
pp cited in Vols.1-5:	

(1) Goissidet et al., Compt. Rend. Soc. Biol., 1936, 121, 1082.

BEPERIDIUM IODIDE [INN]

CAS	86434-57-3	MW:	527.45	MF:	C23 H34 I N3 O3

Anti-ulcerative

Dainippon, Japan

LgP	(4)
pKa	
pp cited in Vols.1-5:	

(1) Proc. Genl. Mtg. Jap. Pharmacol. Soc. 60 (Chiba), Apr. 1987, P-166 abst.

BEPHENIUM OH-NAPHTHOATE [U;INN]

CAS	3818-50-6	MW:	443.55	MF:	C28 H29 N O4

Anthelmintic

Burroughs Wellcome (Alcopara)

LgP	(4)
pKa	
pp cited in Vols.1-5:	

(1) Copp, U.S. Pat. 2 918 401 (1959).

BEPIASTINE [INN]

CAS	10189-94-3	MW:	299.40	MF:	C16 H17 N3 O S

LgP	2.26(C)
pKa	
pp cited in Vols.1-5:	

BEPRIDIL [U;INN]

CAS	64706-54-3	MW:	366.55	MF:	C24 H34 N2 O

Vasodilator; anti-anginal

Wallace (Bepadin); McNeil (Vasocor)

LgP	5.80(C); 2.00(M, pH 7.4)
pKa	
pp cited in Vols.1-5:	

(1) Mauvernay et al., U.S. Pat. 3 962 238 (1976).
(2) M.T. Michelin et al., Therapie, 1977, 32, 485.

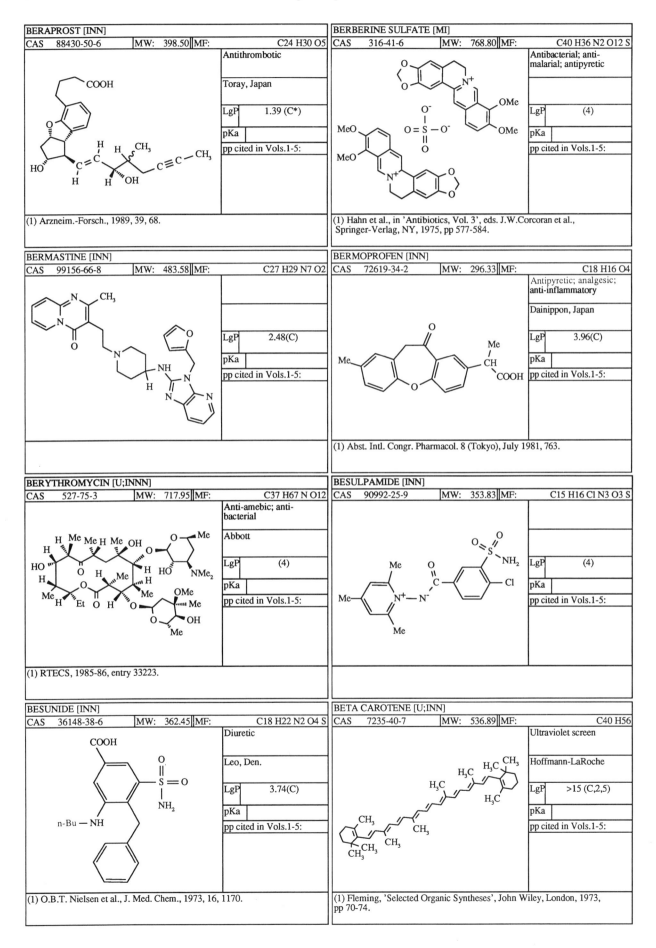

BERAPROST [INN]

CAS	88430-50-6	MW:	398.50	MF:	C24 H30 O5

Antithrombotic

Toray, Japan

LgP	1.39 (C*)
pKa	
pp cited in Vols.1-5:	

(1) Arzneim.-Forsch., 1989, 39, 68.

BERMASTINE [INN]

CAS	99156-66-8	MW:	483.58	MF:	C27 H29 N7 O2

LgP	2.48(C)
pKa	
pp cited in Vols.1-5:	

BERYTHROMYCIN [U;INN]

CAS	527-75-3	MW:	717.95	MF:	C37 H67 N O12

Anti-amebic; anti-bacterial

Abbott

LgP	(4)
pKa	
pp cited in Vols.1-5:	

(1) RTECS, 1985-86, entry 33223.

BESUNIDE [INN]

CAS	36148-38-6	MW:	362.45	MF:	C18 H22 N2 O4 S

Diuretic

Leo, Den.

LgP	3.74(C)
pKa	
pp cited in Vols.1-5:	

(1) O.B.T. Nielsen et al., J. Med. Chem., 1973, 16, 1170.

BERBERINE SULFATE [MI]

CAS	316-41-6	MW:	768.80	MF:	C40 H36 N2 O12 S

Antibacterial; anti-malarial; antipyretic

LgP	(4)
pKa	
pp cited in Vols.1-5:	

(1) Hahn et al., in 'Antibiotics, Vol. 3', eds. J.W.Corcoran et al., Springer-Verlag, NY, 1975, pp 577-584.

BERMOPROFEN [INN]

CAS	72619-34-2	MW:	296.33	MF:	C18 H16 O4

Antipyretic; analgesic; anti-inflammatory

Dainippon, Japan

LgP	3.96(C)
pKa	
pp cited in Vols.1-5:	

(1) Abst. Intl. Congr. Pharmacol. 8 (Tokyo), July 1981, 763.

BESULPAMIDE [INN]

CAS	90992-25-9	MW:	353.83	MF:	C15 H16 Cl N3 O3 S

LgP	(4)
pKa	
pp cited in Vols.1-5:	

BETA CAROTENE [U;INN]

CAS	7235-40-7	MW:	536.89	MF:	C40 H56

Ultraviolet screen

Hoffmann-LaRoche

LgP	>15 (C,2,5)
pKa	
pp cited in Vols.1-5:	

(1) Fleming, 'Selected Organic Syntheses', John Wiley, London, 1973, pp 70-74.

BETACETYLMETHADOL [INN]

CAS	17199-59-6	MW:	353.51	MF:	C23 H31 N O2

Analgesic (narcotic)

LgP	4.27(M); 3.88(C)
pKa	8.61
pp cited in Vols.1-5:	

(1) 'The Merck Index', Merck & Co., Rahway, NJ, 1983, 10th Ed., entry 5800.

BETAHISTINE [U;INN]

CAS	5638-76-6	MW:	136.20	MF:	C8 H12 N2

Vasodilator

Unimed (Serc)

LgP	-0.05(C)
pKa	3.46; 9.78
pp cited in Vols.1-5:	

3: 349, 364, 412

(1) Walter et al., J. Am. Chem. Soc., 1941, 63, 2771.

BETAINE HCl [U]

CAS	590-46-5	MW:	153.61	MF:	C5 H12 Cl N O2

Replenisher adjunct (electrolyte)

Sterling (Acidol-pepsin)

LgP	(4)
pKa	
pp cited in Vols.1-5:	

2: 666

(1) Kuhn et al., Ber.,1950, 83, 420.

BETAMEPRODINE [INN]

CAS	468-50-8	MW:	275.39	MF:	C17 H25 N O2

Analgesic (narcotic)

LgP	3.61(M)
pKa	
pp cited in Vols.1-5:	

BETAMETHADOL [INN]

CAS	17199-55-2	MW:	311.47	MF:	C21 H29 N O

Analgesic (narcotic)

LgP	4.27(M); 2.97(C)
pKa	
pp cited in Vols.1-5:	

(1) 'The Merck Index', Merck & Co., Rahway, NJ, 1983, 10th Ed., entry 3197.

BETAMETHASONE [U;INN]

CAS	378-44-9	MW:	392.47	MF:	C22 H29 F O5

Glucocorticoid

Schering (Celestone)

LgP	1.83(M); 2.21(C,6)
pKa	
pp cited in Vols.1-5:	

1: 33

(1) Taub, et al., J. Am. Chem. Soc., 1958, 80, 4435.

BETAMETHASONE ACETATE [U]

CAS	987-24-6	MW:	434.51	MF:	C24 H31 F O6

Glucocorticoid

Schering

LgP	2.91(M); 2.7(C,6)
pKa	
pp cited in Vols.1-5:	

(1) Oliveto et al., J. Am. Chem. Soc., 1960, 80, 6688.

BETAMETHASONE ACIBUTATE [INN]

CAS	5534-05-4	MW:	504.60	MF:	C28 H37 F O7

Glucocorticoid

LgP	4.05(C,6)
pKa	
pp cited in Vols.1-5:	

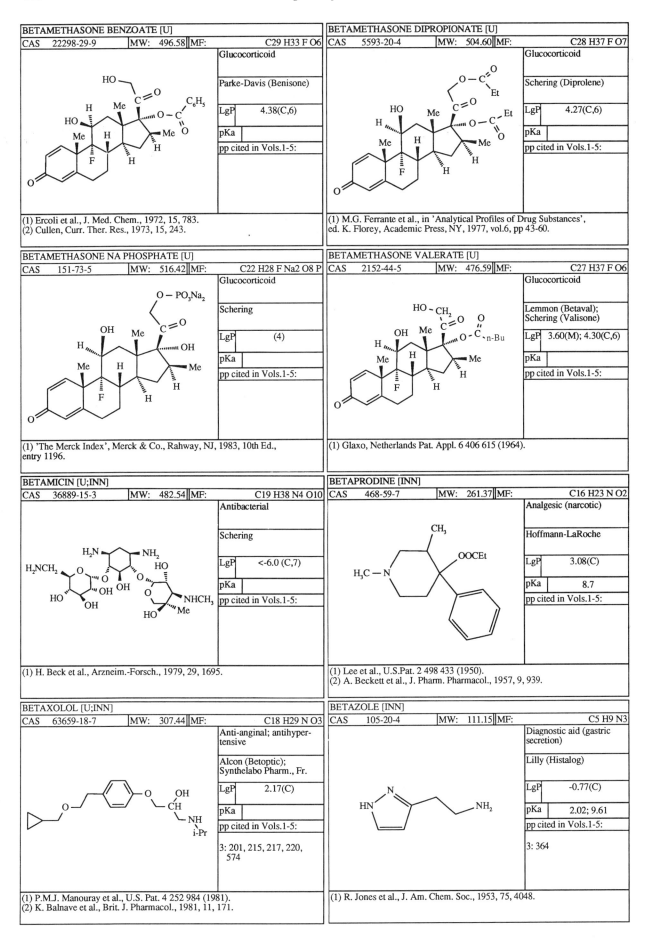

BETAMETHASONE BENZOATE [U]

CAS	22298-29-9	MW:	496.58	MF:	C29 H33 F O6

Glucocorticoid

Parke-Davis (Benisone)

LgP	4.38(C,6)
pKa	
pp cited in Vols.1-5:	

(1) Ercoli et al., J. Med. Chem., 1972, 15, 783.
(2) Cullen, Curr. Ther. Res., 1973, 15, 243.

BETAMETHASONE DIPROPIONATE [U]

CAS	5593-20-4	MW:	504.60	MF:	C28 H37 F O7

Glucocorticoid

Schering (Diprolene)

LgP	4.27(C,6)
pKa	
pp cited in Vols.1-5:	

(1) M.G. Ferrante et al., in 'Analytical Profiles of Drug Substances',
ed. K. Florey, Academic Press, NY, 1977, vol.6, pp 43-60.

BETAMETHASONE NA PHOSPHATE [U]

CAS	151-73-5	MW:	516.42	MF:	C22 H28 F Na2 O8 P

Glucocorticoid

Schering

LgP	(4)
pKa	
pp cited in Vols.1-5:	

(1) 'The Merck Index', Merck & Co., Rahway, NJ, 1983, 10th Ed.,
entry 1196.

BETAMETHASONE VALERATE [U]

CAS	2152-44-5	MW:	476.59	MF:	C27 H37 F O6

Glucocorticoid

Lemmon (Betaval);
Schering (Valisone)

LgP	3.60(M); 4.30(C,6)
pKa	
pp cited in Vols.1-5:	

(1) Glaxo, Netherlands Pat. Appl. 6 406 615 (1964).

BETAMICIN [U;INN]

CAS	36889-15-3	MW:	482.54	MF:	C19 H38 N4 O10

Antibacterial

Schering

LgP	<-6.0 (C,7)
pKa	
pp cited in Vols.1-5:	

(1) H. Beck et al., Arzneim.-Forsch., 1979, 29, 1695.

BETAPRODINE [INN]

CAS	468-59-7	MW:	261.37	MF:	C16 H23 N O2

Analgesic (narcotic)

Hoffmann-LaRoche

LgP	3.08(C)
pKa	8.7
pp cited in Vols.1-5:	

(1) Lee et al., U.S.Pat. 2 498 433 (1950).
(2) A. Beckett et al., J. Pharm. Pharmacol., 1957, 9, 939.

BETAXOLOL [U;INN]

CAS	63659-18-7	MW:	307.44	MF:	C18 H29 N O3

Anti-anginal; antihyper-
tensive

Alcon (Betoptic);
Synthelabo Pharm., Fr.

LgP	2.17(C)
pKa	
pp cited in Vols.1-5:	

3: 201, 215, 217, 220,
574

(1) P.M.J. Manouray et al., U.S. Pat. 4 252 984 (1981).
(2) K. Balnave et al., Brit. J. Pharmacol., 1981, 11, 171.

BETAZOLE [INN]

CAS	105-20-4	MW:	111.15	MF:	C5 H9 N3

Diagnostic aid (gastric
secretion)

Lilly (Histalog)

LgP	-0.77(C)
pKa	2.02; 9.61
pp cited in Vols.1-5:	

3: 364

(1) R. Jones et al., J. Am. Chem. Soc., 1953, 75, 4048.

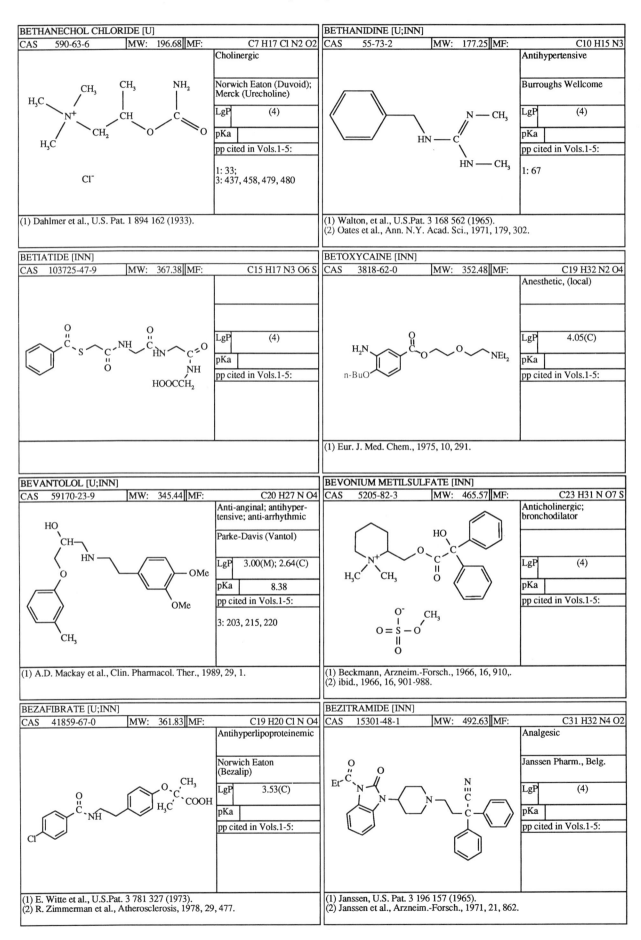

BETHANECHOL CHLORIDE [U]			
CAS 590-63-6	MW: 196.68	MF:	C7 H17 Cl N2 O2

Cholinergic

Norwich Eaton (Duvoid); Merck (Urecholine)

LgP (4)

pKa

pp cited in Vols.1-5:

1: 33;
3: 437, 458, 479, 480

(1) Dahlmer et al., U.S. Pat. 1 894 162 (1933).

BETHANIDINE [U;INN]			
CAS 55-73-2	MW: 177.25	MF:	C10 H15 N3

Antihypertensive

Burroughs Wellcome

LgP (4)

pKa

pp cited in Vols.1-5:

1: 67

(1) Walton, et al., U.S.Pat. 3 168 562 (1965).
(2) Oates et al., Ann. N.Y. Acad. Sci., 1971, 179, 302.

BETIATIDE [INN]			
CAS 103725-47-9	MW: 367.38	MF:	C15 H17 N3 O6 S

LgP (4)

pKa

pp cited in Vols.1-5:

BETOXYCAINE [INN]			
CAS 3818-62-0	MW: 352.48	MF:	C19 H32 N2 O4

Anesthetic, (local)

LgP 4.05(C)

pKa

pp cited in Vols.1-5:

(1) Eur. J. Med. Chem., 1975, 10, 291.

BEVANTOLOL [U;INN]			
CAS 59170-23-9	MW: 345.44	MF:	C20 H27 N O4

Anti-anginal; antihypertensive; anti-arrhythmic

Parke-Davis (Vantol)

LgP 3.00(M); 2.64(C)

pKa 8.38

pp cited in Vols.1-5:

3: 203, 215, 220

(1) A.D. Mackay et al., Clin. Pharmacol. Ther., 1989, 29, 1.

BEVONIUM METILSULFATE [INN]			
CAS 5205-82-3	MW: 465.57	MF:	C23 H31 N O7 S

Anticholinergic; bronchodilator

LgP (4)

pKa

pp cited in Vols.1-5:

(1) Beckmann, Arzneim.-Forsch., 1966, 16, 910,.
(2) ibid., 1966, 16, 901-988.

BEZAFIBRATE [U;INN]			
CAS 41859-67-0	MW: 361.83	MF:	C19 H20 Cl N O4

Antihyperlipoproteinemic

Norwich Eaton (Bezalip)

LgP 3.53(C)

pKa

pp cited in Vols.1-5:

(1) E. Witte et al., U.S.Pat. 3 781 327 (1973).
(2) R. Zimmerman et al., Atherosclerosis, 1978, 29, 477.

BEZITRAMIDE [INN]			
CAS 15301-48-1	MW: 492.63	MF:	C31 H32 N4 O2

Analgesic

Janssen Pharm., Belg.

LgP (4)

pKa

pp cited in Vols.1-5:

(1) Janssen, U.S. Pat. 3 196 157 (1965).
(2) Janssen et al., Arzneim.-Forsch., 1971, 21, 862.

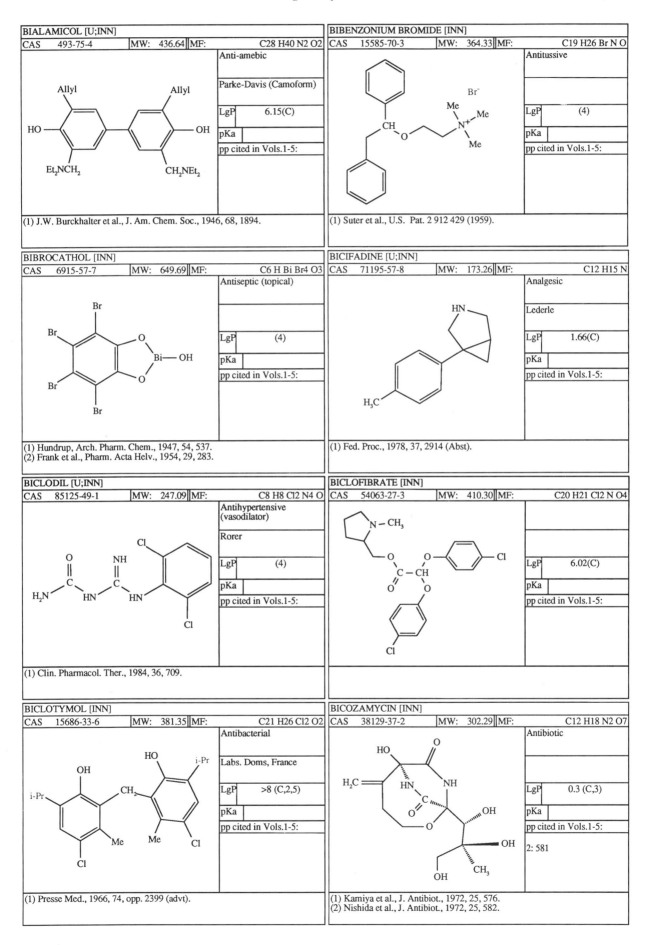

BIALAMICOL [U;INN]

CAS	493-75-4	MW:	436.64	MF:	C28 H40 N2 O2

Anti-amebic

Parke-Davis (Camoform)

LgP	6.15(C)
pKa	
pp cited in Vols.1-5:	

(1) J.W. Burckhalter et al., J. Am. Chem. Soc., 1946, 68, 1894.

BIBENZONIUM BROMIDE [INN]

CAS	15585-70-3	MW:	364.33	MF:	C19 H26 Br N O

Antitussive

LgP	(4)
pKa	
pp cited in Vols.1-5:	

(1) Suter et al., U.S. Pat. 2 912 429 (1959).

BIBROCATHOL [INN]

CAS	6915-57-7	MW:	649.69	MF:	C6 H Bi Br4 O3

Antiseptic (topical)

LgP	(4)
pKa	
pp cited in Vols.1-5:	

(1) Hundrup, Arch. Pharm. Chem., 1947, 54, 537.
(2) Frank et al., Pharm. Acta Helv., 1954, 29, 283.

BICIFADINE [U;INN]

CAS	71195-57-8	MW:	173.26	MF:	C12 H15 N

Analgesic

Lederle

LgP	1.66(C)
pKa	
pp cited in Vols.1-5:	

(1) Fed. Proc., 1978, 37, 2914 (Abst).

BICLODIL [U;INN]

CAS	85125-49-1	MW:	247.09	MF:	C8 H8 Cl2 N4 O

Antihypertensive (vasodilator)

Rorer

LgP	(4)
pKa	
pp cited in Vols.1-5:	

(1) Clin. Pharmacol. Ther., 1984, 36, 709.

BICLOFIBRATE [INN]

CAS	54063-27-3	MW:	410.30	MF:	C20 H21 Cl2 N O4

LgP	6.02(C)
pKa	
pp cited in Vols.1-5:	

BICLOTYMOL [INN]

CAS	15686-33-6	MW:	381.35	MF:	C21 H26 Cl2 O2

Antibacterial

Labs. Doms, France

LgP	>8 (C,2,5)
pKa	
pp cited in Vols.1-5:	

(1) Presse Med., 1966, 74, opp. 2399 (advt).

BICOZAMYCIN [INN]

CAS	38129-37-2	MW:	302.29	MF:	C12 H18 N2 O7

Antibiotic

LgP	0.3 (C,3)
pKa	
pp cited in Vols.1-5:	
2: 581	

(1) Kamiya et al., J. Antibiot., 1972, 25, 576.
(2) Nishida et al., J. Antibiot., 1972, 25, 582.

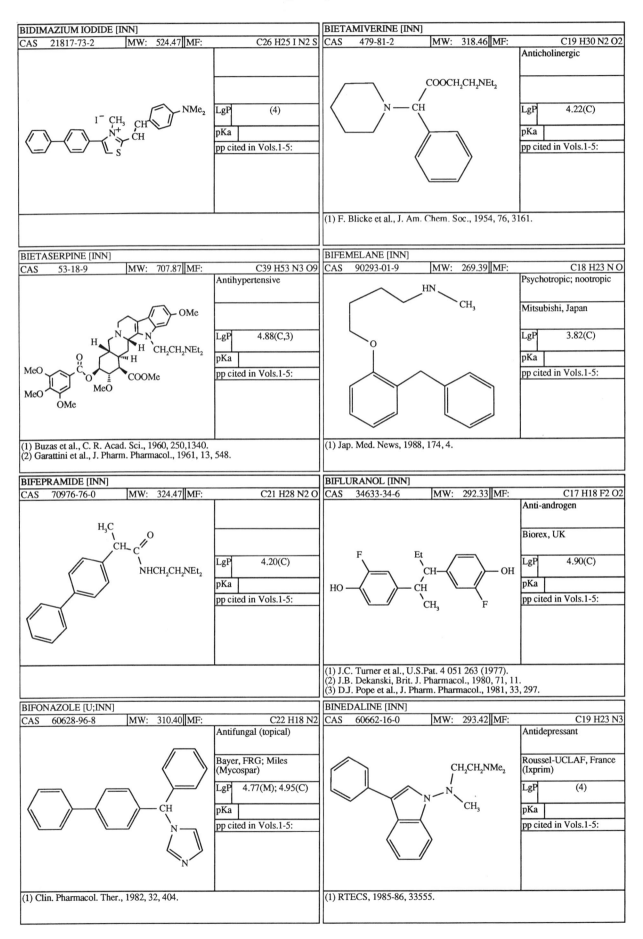

BIDIMAZIUM IODIDE [INN]				
CAS	21817-73-2	MW: 524.47	MF:	C26 H25 I N2 S
			LgP	(4)
			pKa	
			pp cited in Vols.1-5:	

BIETAMIVERINE [INN]				
CAS	479-81-2	MW: 318.46	MF:	C19 H30 N2 O2
			Anticholinergic	
			LgP	4.22(C)
			pKa	
			pp cited in Vols.1-5:	

(1) F. Blicke et al., J. Am. Chem. Soc., 1954, 76, 3161.

BIETASERPINE [INN]				
CAS	53-18-9	MW: 707.87	MF:	C39 H53 N3 O9
			Antihypertensive	
			LgP	4.88(C,3)
			pKa	
			pp cited in Vols.1-5:	

(1) Buzas et al., C. R. Acad. Sci., 1960, 250,1340.
(2) Garattini et al., J. Pharm. Pharmacol., 1961, 13, 548.

BIFEMELANE [INN]				
CAS	90293-01-9	MW: 269.39	MF:	C18 H23 N O
			Psychotropic; nootropic	
			Mitsubishi, Japan	
			LgP	3.82(C)
			pKa	
			pp cited in Vols.1-5:	

(1) Jap. Med. News, 1988, 174, 4.

BIFEPRAMIDE [INN]				
CAS	70976-76-0	MW: 324.47	MF:	C21 H28 N2 O
			LgP	4.20(C)
			pKa	
			pp cited in Vols.1-5:	

BIFLURANOL [INN]				
CAS	34633-34-6	MW: 292.33	MF:	C17 H18 F2 O2
			Anti-androgen	
			Biorex, UK	
			LgP	4.90(C)
			pKa	
			pp cited in Vols.1-5:	

(1) J.C. Turner et al., U.S.Pat. 4 051 263 (1977).
(2) J.B. Dekanski, Brit. J. Pharmacol., 1980, 71, 11.
(3) D.J. Pope et al., J. Pharm. Pharmacol., 1981, 33, 297.

BIFONAZOLE [U;INN]				
CAS	60628-96-8	MW: 310.40	MF:	C22 H18 N2
			Antifungal (topical)	
			Bayer, FRG; Miles (Mycospar)	
			LgP	4.77(M); 4.95(C)
			pKa	
			pp cited in Vols.1-5:	

(1) Clin. Pharmacol. Ther., 1982, 32, 404.

BINEDALINE [INN]				
CAS	60662-16-0	MW: 293.42	MF:	C19 H23 N3
			Antidepressant	
			Roussel-UCLAF, France (Ixprim)	
			LgP	(4)
			pKa	
			pp cited in Vols.1-5:	

(1) RTECS, 1985-86, 33555.

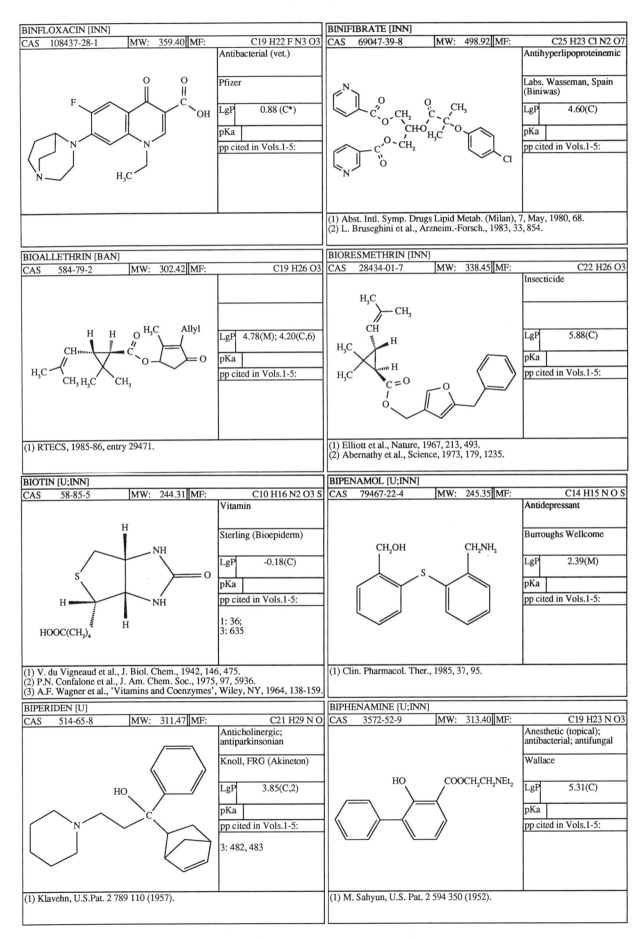

BINFLOXACIN [INN]

CAS	108437-28-1	MW:	359.40	MF:	C19 H22 F N3 O3

Antibacterial (vet.)

Pfizer

LgP	0.88 (C*)
pKa	
pp cited in Vols.1-5:	

BIOALLETHRIN [BAN]

CAS	584-79-2	MW:	302.42	MF:	C19 H26 O3

LgP	4.78(M); 4.20(C,6)
pKa	
pp cited in Vols.1-5:	

(1) RTECS, 1985-86, entry 29471.

BIOTIN [U;INN]

CAS	58-85-5	MW:	244.31	MF:	C10 H16 N2 O3 S

Vitamin

Sterling (Bioepiderm)

LgP	-0.18(C)
pKa	
pp cited in Vols.1-5:	

1: 36;
3: 635

(1) V. du Vigneaud et al., J. Biol. Chem., 1942, 146, 475.
(2) P.N. Confalone et al., J. Am. Chem. Soc., 1975, 97, 5936.
(3) A.F. Wagner et al., 'Vitamins and Coenzymes', Wiley, NY, 1964, 138-159.

BIPERIDEN [U]

CAS	514-65-8	MW:	311.47	MF:	C21 H29 N O

Anticholinergic;
antiparkinsonian

Knoll, FRG (Akineton)

LgP	3.85(C,2)
pKa	
pp cited in Vols.1-5:	

3: 482, 483

(1) Klavehn, U.S.Pat. 2 789 110 (1957).

BINIFIBRATE [INN]

CAS	69047-39-8	MW:	498.92	MF:	C25 H23 Cl N2 O7

Antihyperlipoproteinemic

Labs. Wasseman, Spain
(Biniwas)

LgP	4.60(C)
pKa	
pp cited in Vols.1-5:	

(1) Abst. Intl. Symp. Drugs Lipid Metab. (Milan), 7, May, 1980, 68.
(2) L. Bruseghini et al., Arzneim.-Forsch., 1983, 33, 854.

BIORESMETHRIN [INN]

CAS	28434-01-7	MW:	338.45	MF:	C22 H26 O3

Insecticide

LgP	5.88(C)
pKa	
pp cited in Vols.1-5:	

(1) Elliott et al., Nature, 1967, 213, 493.
(2) Abernathy et al., Science, 1973, 179, 1235.

BIPENAMOL [U;INN]

CAS	79467-22-4	MW:	245.35	MF:	C14 H15 N O S

Antidepressant

Burroughs Wellcome

LgP	2.39(M)
pKa	
pp cited in Vols.1-5:	

(1) Clin. Pharmacol. Ther., 1985, 37, 95.

BIPHENAMINE [U;INN]

CAS	3572-52-9	MW:	313.40	MF:	C19 H23 N O3

Anesthetic (topical);
antibacterial; antifungal

Wallace

LgP	5.31(C)
pKa	
pp cited in Vols.1-5:	

(1) M. Sahyun, U.S. Pat. 2 594 350 (1952).

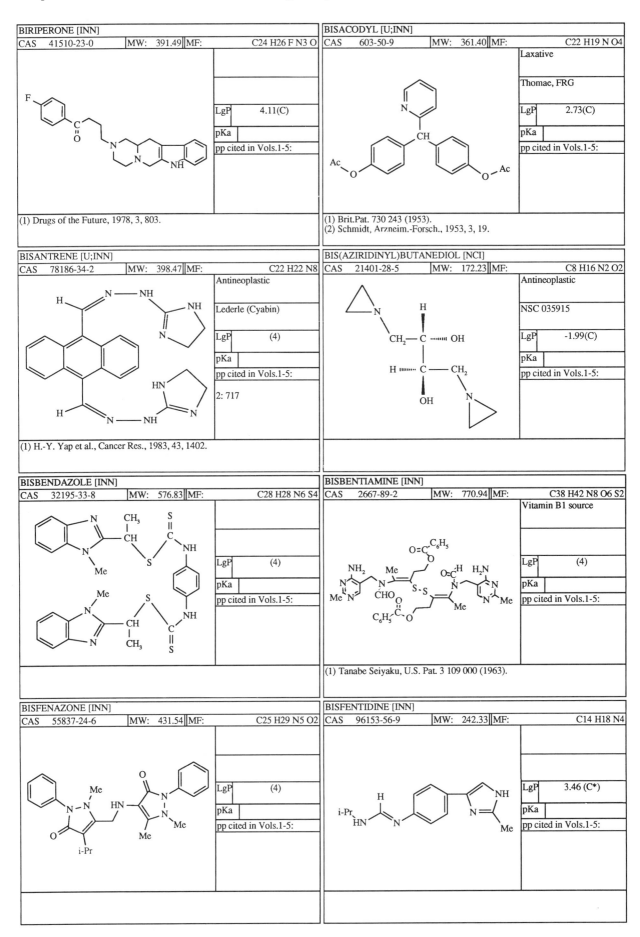

BIRIPERONE [INN]			
CAS 41510-23-0	MW: 391.49	MF:	C24 H26 F N3 O
		LgP	4.11(C)
		pKa	
		pp cited in Vols.1-5:	

(1) Drugs of the Future, 1978, 3, 803.

BISACODYL [U;INN]			
CAS 603-50-9	MW: 361.40	MF:	C22 H19 N O4
		Laxative	
		Thomae, FRG	
		LgP	2.73(C)
		pKa	
		pp cited in Vols.1-5:	

(1) Brit.Pat. 730 243 (1953).
(2) Schmidt, Arzneim.-Forsch., 1953, 3, 19.

BISANTRENE [U;INN]			
CAS 78186-34-2	MW: 398.47	MF:	C22 H22 N8
		Antineoplastic	
		Lederle (Cyabin)	
		LgP	(4)
		pKa	
		pp cited in Vols.1-5:	
		2: 717	

(1) H.-Y. Yap et al., Cancer Res., 1983, 43, 1402.

BIS(AZIRIDINYL)BUTANEDIOL [NCI]			
CAS 21401-28-5	MW: 172.23	MF:	C8 H16 N2 O2
		Antineoplastic	
		NSC 035915	
		LgP	-1.99(C)
		pKa	
		pp cited in Vols.1-5:	

BISBENDAZOLE [INN]			
CAS 32195-33-8	MW: 576.83	MF:	C28 H28 N6 S4
		LgP	(4)
		pKa	
		pp cited in Vols.1-5:	

BISBENTIAMINE [INN]			
CAS 2667-89-2	MW: 770.94	MF:	C38 H42 N8 O6 S2
		Vitamin B1 source	
		LgP	(4)
		pKa	
		pp cited in Vols.1-5:	

(1) Tanabe Seiyaku, U.S. Pat. 3 109 000 (1963).

BISFENAZONE [INN]			
CAS 55837-24-6	MW: 431.54	MF:	C25 H29 N5 O2
		LgP	(4)
		pKa	
		pp cited in Vols.1-5:	

BISFENTIDINE [INN]			
CAS 96153-56-9	MW: 242.33	MF:	C14 H18 N4
		LgP	3.46 (C*)
		pKa	
		pp cited in Vols.1-5:	

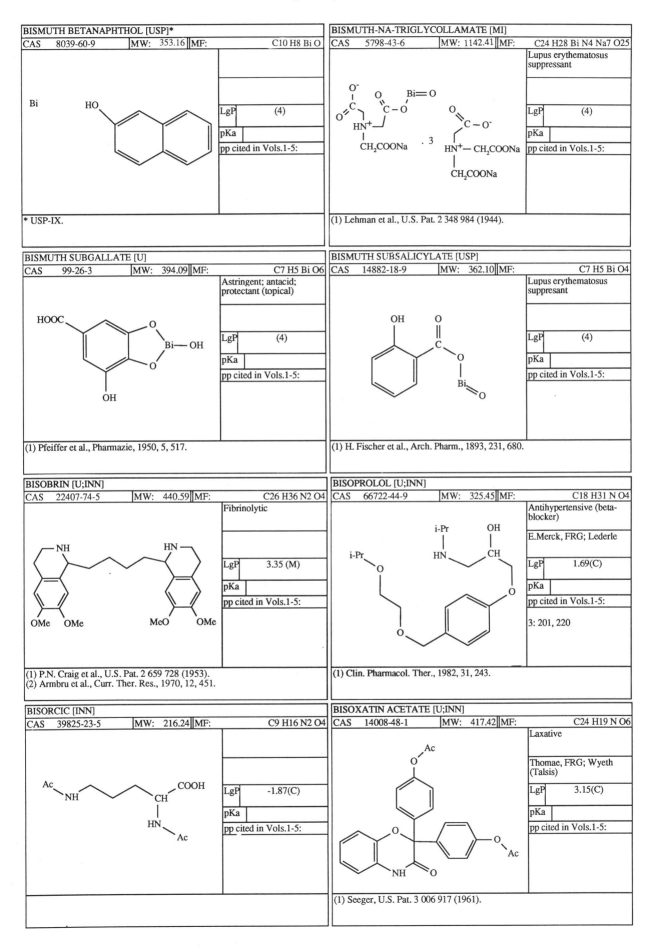

BISMUTH BETANAPHTHOL [USP]*

CAS	8039-60-9	MW:	353.16	MF:	C10 H8 Bi O

LgP	(4)
pKa	
pp cited in Vols.1-5:	

* USP-IX.

BISMUTH SUBGALLATE [U]

CAS	99-26-3	MW:	394.09	MF:	C7 H5 Bi O6

Astringent; antacid; protectant (topical)

LgP	(4)
pKa	
pp cited in Vols.1-5:	

(1) Pfeiffer et al., Pharmazie, 1950, 5, 517.

BISOBRIN [U;INN]

CAS	22407-74-5	MW:	440.59	MF:	C26 H36 N2 O4

Fibrinolytic

LgP	3.35 (M)
pKa	
pp cited in Vols.1-5:	

(1) P.N. Craig et al., U.S. Pat. 2 659 728 (1953).
(2) Armbru et al., Curr. Ther. Res., 1970, 12, 451.

BISORCIC [INN]

CAS	39825-23-5	MW:	216.24	MF:	C9 H16 N2 O4

LgP	-1.87(C)
pKa	
pp cited in Vols.1-5:	

BISMUTH-NA-TRIGLYCOLLAMATE [MI]

CAS	5798-43-6	MW:	1142.41	MF:	C24 H28 Bi N4 Na7 O25

Lupus erythematosus suppressant

LgP	(4)
pKa	
pp cited in Vols.1-5:	

(1) Lehman et al., U.S. Pat. 2 348 984 (1944).

BISMUTH SUBSALICYLATE [USP]

CAS	14882-18-9	MW:	362.10	MF:	C7 H5 Bi O4

Lupus erythematosus suppressant

LgP	(4)
pKa	
pp cited in Vols.1-5:	

(1) H. Fischer et al., Arch. Pharm., 1893, 231, 680.

BISOPROLOL [U;INN]

CAS	66722-44-9	MW:	325.45	MF:	C18 H31 N O4

Antihypertensive (beta-blocker)

E.Merck, FRG; Lederle

LgP	1.69(C)
pKa	
pp cited in Vols.1-5:	

3: 201, 220

(1) Clin. Pharmacol. Ther., 1982, 31, 243.

BISOXATIN ACETATE [U;INN]

CAS	14008-48-1	MW:	417.42	MF:	C24 H19 N O6

Laxative

Thomae, FRG; Wyeth (Talsis)

LgP	3.15(C)
pKa	
pp cited in Vols.1-5:	

(1) Seeger, U.S. Pat. 3 006 917 (1961).

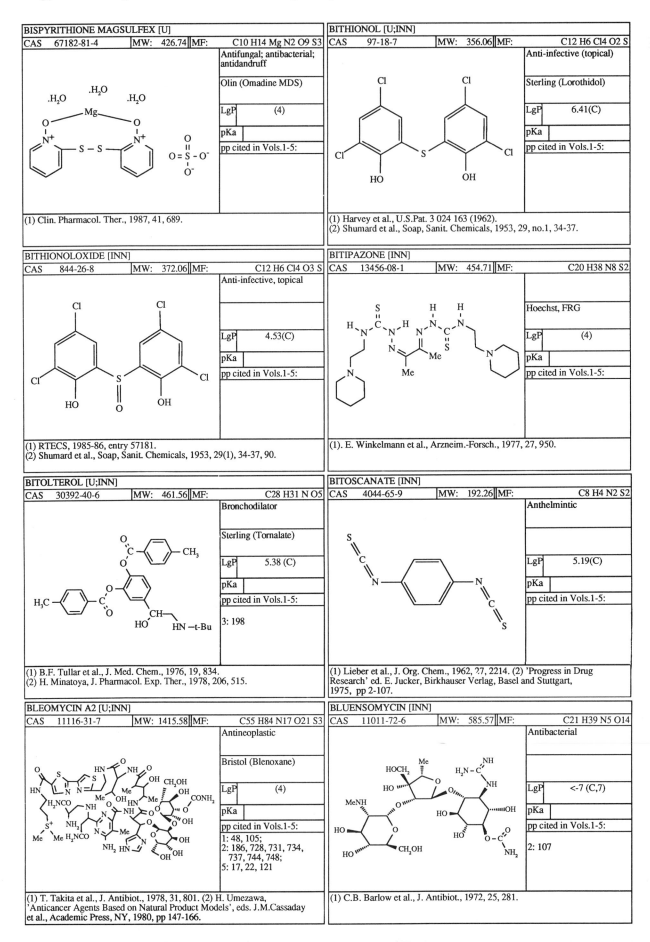

BISPYRITHIONE MAGSULFEX [U]

CAS	67182-81-4	MW:	426.74	MF:	C10 H14 Mg N2 O9 S3

Antifungal; antibacterial; antidandruff

Olin (Omadine MDS)

LgP	(4)
pKa	
pp cited in Vols.1-5:	

.H$_2$O .H$_2$O .H$_2$O

(1) Clin. Pharmacol. Ther., 1987, 41, 689.

BITHIONOL [U;INN]

CAS	97-18-7	MW:	356.06	MF:	C12 H6 Cl4 O2 S

Anti-infective (topical)

Sterling (Lorothidol)

LgP	6.41(C)
pKa	
pp cited in Vols.1-5:	

(1) Harvey et al., U.S.Pat. 3 024 163 (1962).
(2) Shumard et al., Soap, Sanit. Chemicals, 1953, 29, no.1, 34-37.

BITHIONOLOXIDE [INN]

CAS	844-26-8	MW:	372.06	MF:	C12 H6 Cl4 O3 S

Anti-infective, topical

LgP	4.53(C)
pKa	
pp cited in Vols.1-5:	

(1) RTECS, 1985-86, entry 57181.
(2) Shumard et al., Soap, Sanit. Chemicals, 1953, 29(1), 34-37, 90.

BITIPAZONE [INN]

CAS	13456-08-1	MW:	454.71	MF:	C20 H38 N8 S2

Hoechst, FRG

LgP	(4)
pKa	
pp cited in Vols.1-5:	

(1). E. Winkelmann et al., Arzneim.-Forsch., 1977, 27, 950.

BITOLTEROL [U;INN]

CAS	30392-40-6	MW:	461.56	MF:	C28 H31 N O5

Bronchodilator

Sterling (Tornalate)

LgP	5.38 (C)
pKa	
pp cited in Vols.1-5:	
	3: 198

(1) B.F. Tullar et al., J. Med. Chem., 1976, 19, 834.
(2) H. Minatoya, J. Pharmacol. Exp. Ther., 1978, 206, 515.

BITOSCANATE [INN]

CAS	4044-65-9	MW:	192.26	MF:	C8 H4 N2 S2

Anthelmintic

LgP	5.19(C)
pKa	
pp cited in Vols.1-5:	

(1) Lieber et al., J. Org. Chem., 1962, 27, 2214. (2) 'Progress in Drug Research' ed. E. Jucker, Birkhauser Verlag, Basel and Stuttgart, 1975, pp 2-107.

BLEOMYCIN A2 [U;INN]

CAS	11116-31-7	MW:	1415.58	MF:	C55 H84 N17 O21 S3

Antineoplastic

Bristol (Blenoxane)

LgP	(4)
pKa	
pp cited in Vols.1-5:	
	1: 48, 105;
	2: 186, 728, 731, 734, 737, 744, 748;
	5: 17, 22, 121

(1) T. Takita et al., J. Antibiot., 1978, 31, 801. (2) H. Umezawa, 'Anticancer Agents Based on Natural Product Models', eds. J.M.Cassaday et al., Academic Press, NY, 1980, pp 147-166.

BLUENSOMYCIN [INN]

CAS	11011-72-6	MW:	585.57	MF:	C21 H39 N5 O14

Antibacterial

LgP	<-7 (C,7)
pKa	
pp cited in Vols.1-5:	
	2: 107

(1) C.B. Barlow et al., J. Antibiot., 1972, 25, 281.

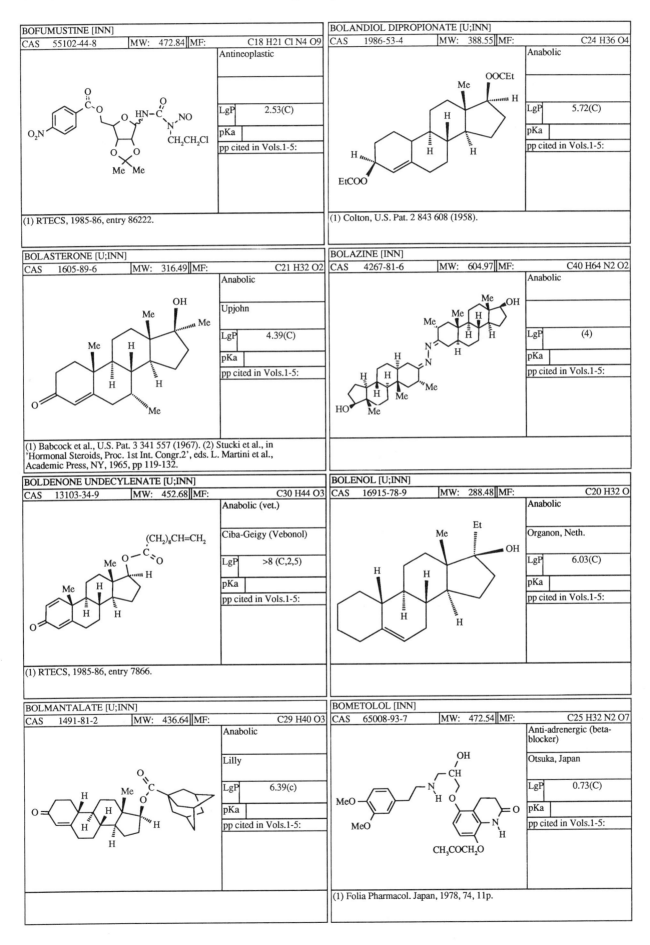

BOFUMUSTINE [INN]

| CAS | 55102-44-8 | MW: | 472.84 | MF: | C18 H21 Cl N4 O9 |

Antineoplastic

LgP	2.53(C)
pKa	
pp cited in Vols.1-5:	

(1) RTECS, 1985-86, entry 86222.

BOLANDIOL DIPROPIONATE [U;INN]

| CAS | 1986-53-4 | MW: | 388.55 | MF: | C24 H36 O4 |

Anabolic

LgP	5.72(C)
pKa	
pp cited in Vols.1-5:	

(1) Colton, U.S. Pat. 2 843 608 (1958).

BOLASTERONE [U;INN]

| CAS | 1605-89-6 | MW: | 316.49 | MF: | C21 H32 O2 |

Anabolic

Upjohn

LgP	4.39(C)
pKa	
pp cited in Vols.1-5:	

(1) Babcock et al., U.S. Pat. 3 341 557 (1967). (2) Stucki et al., in 'Hormonal Steroids, Proc. 1st Int. Congr.2', eds. L. Martini et al., Academic Press, NY, 1965, pp 119-132.

BOLAZINE [INN]

| CAS | 4267-81-6 | MW: | 604.97 | MF: | C40 H64 N2 O2 |

Anabolic

LgP	(4)
pKa	
pp cited in Vols.1-5:	

BOLDENONE UNDECYLENATE [U;INN]

| CAS | 13103-34-9 | MW: | 452.68 | MF: | C30 H44 O3 |

Anabolic (vet.)

Ciba-Geigy (Vebonol)

LgP	>8 (C,2,5)
pKa	
pp cited in Vols.1-5:	

(1) RTECS, 1985-86, entry 7866.

BOLENOL [U;INN]

| CAS | 16915-78-9 | MW: | 288.48 | MF: | C20 H32 O |

Anabolic

Organon, Neth.

LgP	6.03(C)
pKa	
pp cited in Vols.1-5:	

BOLMANTALATE [U;INN]

| CAS | 1491-81-2 | MW: | 436.64 | MF: | C29 H40 O3 |

Anabolic

Lilly

LgP	6.39(c)
pKa	
pp cited in Vols.1-5:	

BOMETOLOL [INN]

| CAS | 65008-93-7 | MW: | 472.54 | MF: | C25 H32 N2 O7 |

Anti-adrenergic (beta-blocker)

Otsuka, Japan

LgP	0.73(C)
pKa	
pp cited in Vols.1-5:	

(1) Folia Pharmacol. Japan, 1978, 74, 11p.

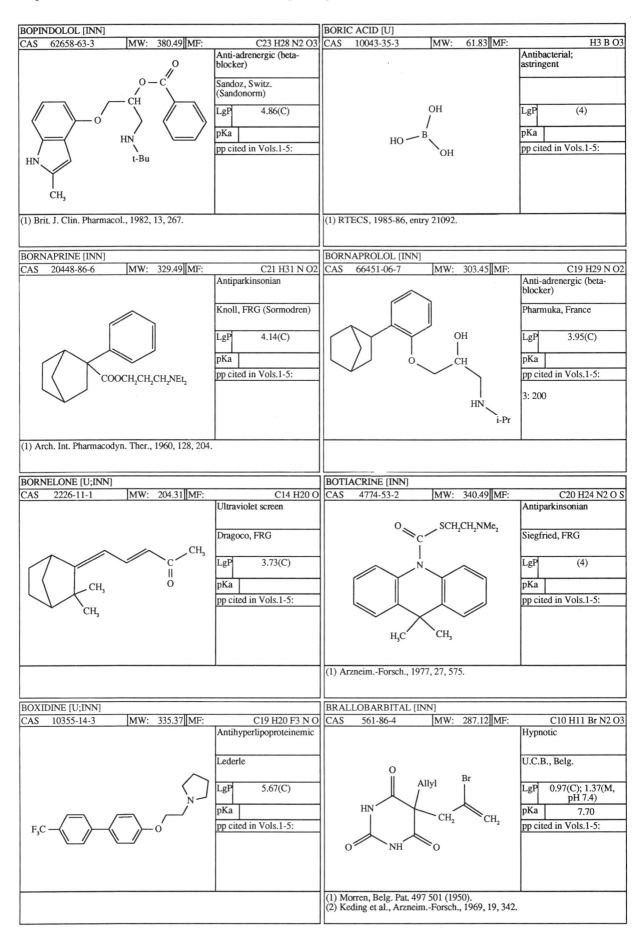

BOPINDOLOL [INN]			
CAS 62658-63-3	MW: 380.49	MF:	C23 H28 N2 O3

Anti-adrenergic (beta-blocker)

Sandoz, Switz. (Sandonorm)

LgP	4.86(C)
pKa	
pp cited in Vols.1-5:	

(1) Brit. J. Clin. Pharmacol., 1982, 13, 267.

BORIC ACID [U]			
CAS 10043-35-3	MW: 61.83	MF:	H3 B O3

Antibacterial; astringent

LgP	(4)
pKa	
pp cited in Vols.1-5:	

(1) RTECS, 1985-86, entry 21092.

BORNAPRINE [INN]			
CAS 20448-86-6	MW: 329.49	MF:	C21 H31 N O2

Antiparkinsonian

Knoll, FRG (Sormodren)

LgP	4.14(C)
pKa	
pp cited in Vols.1-5:	

(1) Arch. Int. Pharmacodyn. Ther., 1960, 128, 204.

BORNAPROLOL [INN]			
CAS 66451-06-7	MW: 303.45	MF:	C19 H29 N O2

Anti-adrenergic (beta-blocker)

Pharmuka, France

LgP	3.95(C)
pKa	
pp cited in Vols.1-5:	
3: 200	

BORNELONE [U;INN]			
CAS 2226-11-1	MW: 204.31	MF:	C14 H20 O

Ultraviolet screen

Dragoco, FRG

LgP	3.73(C)
pKa	
pp cited in Vols.1-5:	

BOTIACRINE [INN]			
CAS 4774-53-2	MW: 340.49	MF:	C20 H24 N2 O S

Antiparkinsonian

Siegfried, FRG

LgP	(4)
pKa	
pp cited in Vols.1-5:	

(1) Arzneim.-Forsch., 1977, 27, 575.

BOXIDINE [U;INN]			
CAS 10355-14-3	MW: 335.37	MF:	C19 H20 F3 N O

Antihyperlipoproteinemic

Lederle

LgP	5.67(C)
pKa	
pp cited in Vols.1-5:	

BRALLOBARBITAL [INN]			
CAS 561-86-4	MW: 287.12	MF:	C10 H11 Br N2 O3

Hypnotic

U.C.B., Belg.

LgP	0.97(C); 1.37(M, pH 7.4)
pKa	7.70
pp cited in Vols.1-5:	

(1) Morren, Belg. Pat. 497 501 (1950).
(2) Keding et al., Arzneim.-Forsch., 1969, 19, 342.

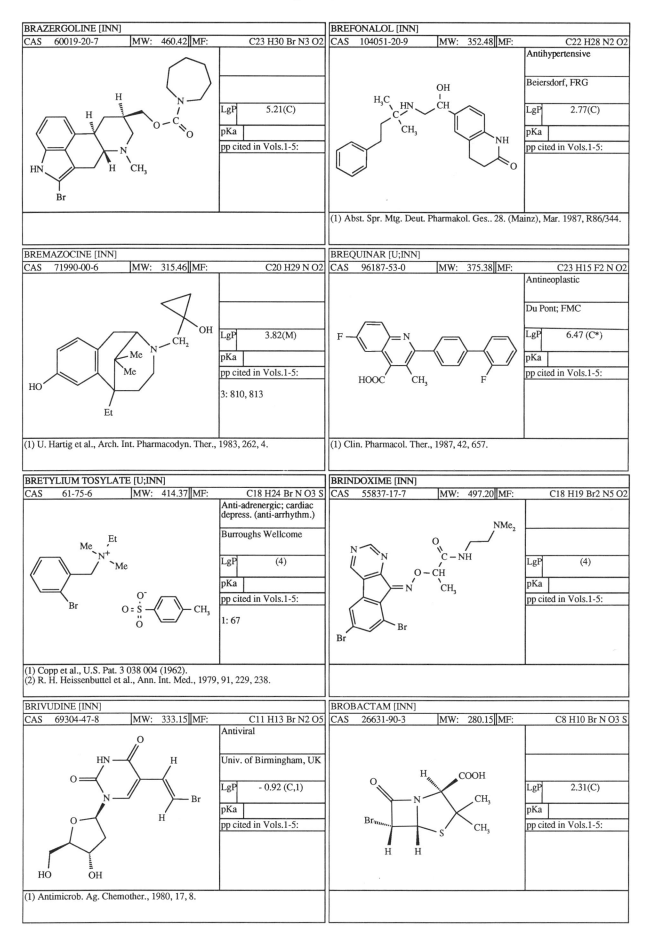

BRAZERGOLINE [INN]

CAS	60019-20-7	MW:	460.42	MF:	C23 H30 Br N3 O2

LgP	5.21(C)
pKa	
pp cited in Vols.1-5:	

BREFONALOL [INN]

CAS	104051-20-9	MW:	352.48	MF:	C22 H28 N2 O2

Antihypertensive

Beiersdorf, FRG

LgP	2.77(C)
pKa	
pp cited in Vols.1-5:	

(1) Abst. Spr. Mtg. Deut. Pharmakol. Ges.. 28. (Mainz), Mar. 1987, R86/344.

BREMAZOCINE [INN]

CAS	71990-00-6	MW:	315.46	MF:	C20 H29 N O2

LgP	3.82(M)
pKa	
pp cited in Vols.1-5:	
3: 810, 813	

(1) U. Hartig et al., Arch. Int. Pharmacodyn. Ther., 1983, 262, 4.

BREQUINAR [U;INN]

CAS	96187-53-0	MW:	375.38	MF:	C23 H15 F2 N O2

Antineoplastic

Du Pont; FMC

LgP	6.47 (C*)
pKa	
pp cited in Vols.1-5:	

(1) Clin. Pharmacol. Ther., 1987, 42, 657.

BRETYLIUM TOSYLATE [U;INN]

CAS	61-75-6	MW:	414.37	MF:	C18 H24 Br N O3 S

Anti-adrenergic; cardiac depress. (anti-arrhythm.)

Burroughs Wellcome

LgP	(4)
pKa	
pp cited in Vols.1-5:	
1: 67	

(1) Copp et al., U.S. Pat. 3 038 004 (1962).
(2) R. H. Heissenbuttel et al., Ann. Int. Med., 1979, 91, 229, 238.

BRINDOXIME [INN]

CAS	55837-17-7	MW:	497.20	MF:	C18 H19 Br2 N5 O2

LgP	(4)
pKa	
pp cited in Vols.1-5:	

BRIVUDINE [INN]

CAS	69304-47-8	MW:	333.15	MF:	C11 H13 Br N2 O5

Antiviral

Univ. of Birmingham, UK

LgP	- 0.92 (C,1)
pKa	
pp cited in Vols.1-5:	

(1) Antimicrob. Ag. Chemother., 1980, 17, 8.

BROBACTAM [INN]

CAS	26631-90-3	MW:	280.15	MF:	C8 H10 Br N O3 S

LgP	2.31(C)
pKa	
pp cited in Vols.1-5:	

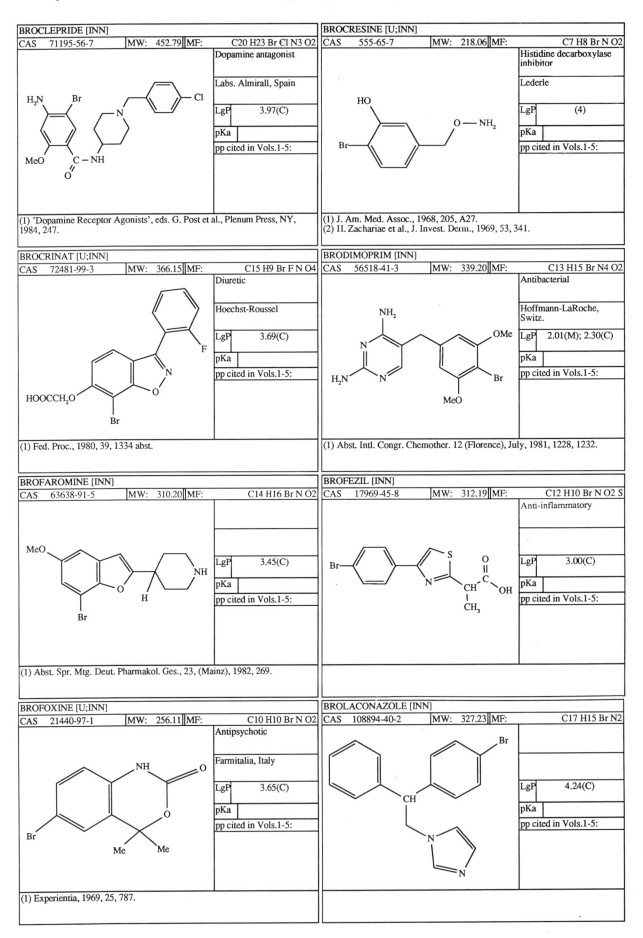

BROCLEPRIDE [INN]				
CAS	71195-56-7	MW: 452.79	MF:	C20 H23 Br Cl N3 O2

Dopamine antagonist

Labs. Almirall, Spain

LgP	3.97(C)
pKa	
pp cited in Vols.1-5:	

(1) 'Dopamine Receptor Agonists', eds. G. Post et al., Plenum Press, NY, 1984, 247.

BROCRESINE [U;INN]				
CAS	555-65-7	MW: 218.06	MF:	C7 H8 Br N O2

Histidine decarboxylase inhibitor

Lederle

LgP	(4)
pKa	
pp cited in Vols.1-5:	

(1) J. Am. Med. Assoc., 1968, 205, A27.
(2) II. Zachariae et al., J. Invest. Derm., 1969, 53, 341.

BROCRINAT [U;INN]				
CAS	72481-99-3	MW: 366.15	MF:	C15 H9 Br F N O4

Diuretic

Hoechst-Roussel

LgP	3.69(C)
pKa	
pp cited in Vols.1-5:	

(1) Fed. Proc., 1980, 39, 1334 abst.

BRODIMOPRIM [INN]				
CAS	56518-41-3	MW: 339.20	MF:	C13 H15 Br N4 O2

Antibacterial

Hoffmann-LaRoche, Switz.

LgP	2.01(M); 2.30(C)
pKa	
pp cited in Vols.1-5:	

(1) Abst. Intl. Congr. Chemother. 12 (Florence), July, 1981, 1228, 1232.

BROFAROMINE [INN]				
CAS	63638-91-5	MW: 310.20	MF:	C14 H16 Br N O2

LgP	3.45(C)
pKa	
pp cited in Vols.1-5:	

(1) Abst. Spr. Mtg. Deut. Pharmakol. Ges., 23, (Mainz), 1982, 269.

BROFEZIL [INN]				
CAS	17969-45-8	MW: 312.19	MF:	C12 H10 Br N O2 S

Anti-inflammatory

LgP	3.00(C)
pKa	
pp cited in Vols.1-5:	

BROFOXINE [U;INN]				
CAS	21440-97-1	MW: 256.11	MF:	C10 H10 Br N O2

Antipsychotic

Farmitalia, Italy

LgP	3.65(C)
pKa	
pp cited in Vols.1-5:	

(1) Experientia, 1969, 25, 787.

BROLACONAZOLE [INN]				
CAS	108894-40-2	MW: 327.23	MF:	C17 H15 Br N2

LgP	4.24(C)
pKa	
pp cited in Vols.1-5:	

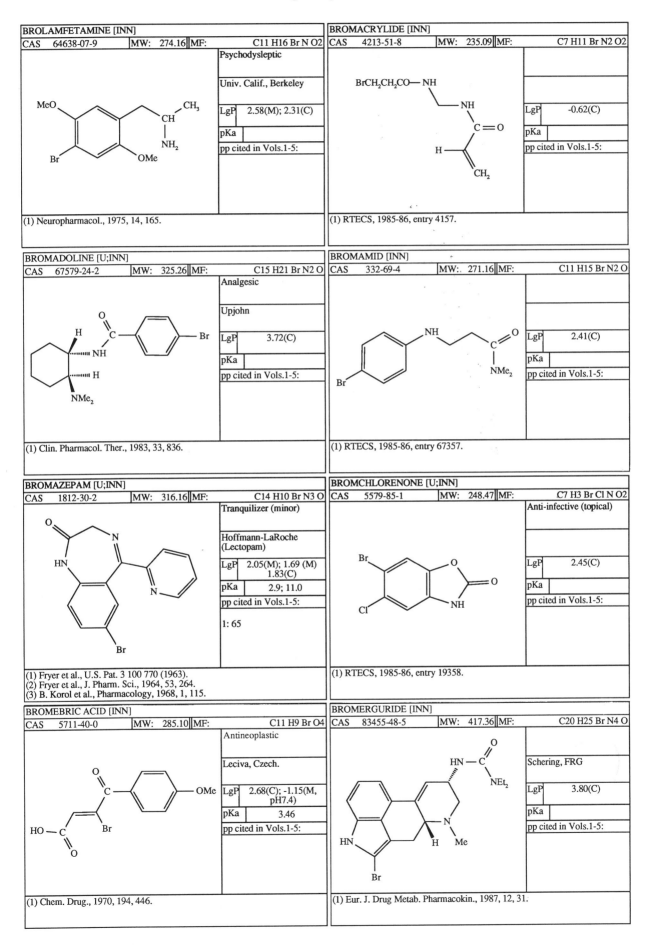

BROLAMFETAMINE [INN]

| CAS | 64638-07-9 | MW: | 274.16 | MF: | C11 H16 Br N O2 |

Psychodysleptic

Univ. Calif., Berkeley

LgP	2.58(M); 2.31(C)
pKa	
pp cited in Vols.1-5:	

(1) Neuropharmacol., 1975, 14, 165.

BROMACRYLIDE [INN]

| CAS | 4213-51-8 | MW: | 235.09 | MF: | C7 H11 Br N2 O2 |

LgP	-0.62(C)
pKa	
pp cited in Vols.1-5:	

(1) RTECS, 1985-86, entry 4157.

BROMADOLINE [U;INN]

| CAS | 67579-24-2 | MW: | 325.26 | MF: | C15 H21 Br N2 O |

Analgesic

Upjohn

LgP	3.72(C)
pKa	
pp cited in Vols.1-5:	

(1) Clin. Pharmacol. Ther., 1983, 33, 836.

BROMAMID [INN]

| CAS | 332-69-4 | MW: | 271.16 | MF: | C11 H15 Br N2 O |

LgP	2.41(C)
pKa	
pp cited in Vols.1-5:	

(1) RTECS, 1985-86, entry 67357.

BROMAZEPAM [U;INN]

| CAS | 1812-30-2 | MW: | 316.16 | MF: | C14 H10 Br N3 O |

Tranquilizer (minor)

Hoffmann-LaRoche (Lectopam)

LgP	2.05(M); 1.69 (M) 1.83(C)
pKa	2.9; 11.0
pp cited in Vols.1-5:	
	1: 65

(1) Fryer et al., U.S. Pat. 3 100 770 (1963).
(2) Fryer et al., J. Pharm. Sci., 1964, 53, 264.
(3) B. Korol et al., Pharmacology, 1968, 1, 115.

BROMCHLORENONE [U;INN]

| CAS | 5579-85-1 | MW: | 248.47 | MF: | C7 H3 Br Cl N O2 |

Anti-infective (topical)

LgP	2.45(C)
pKa	
pp cited in Vols.1-5:	

(1) RTECS, 1985-86, entry 19358.

BROMEBRIC ACID [INN]

| CAS | 5711-40-0 | MW: | 285.10 | MF: | C11 H9 Br O4 |

Antineoplastic

Leciva, Czech.

LgP	2.68(C); -1.15(M, pH7.4)
pKa	3.46
pp cited in Vols.1-5:	

(1) Chem. Drug., 1970, 194, 446.

BROMERGURIDE [INN]

| CAS | 83455-48-5 | MW: | 417.36 | MF: | C20 H25 Br N4 O |

Schering, FRG

LgP	3.80(C)
pKa	
pp cited in Vols.1-5:	

(1) Eur. J. Drug Metab. Pharmacokin., 1987, 12, 31.

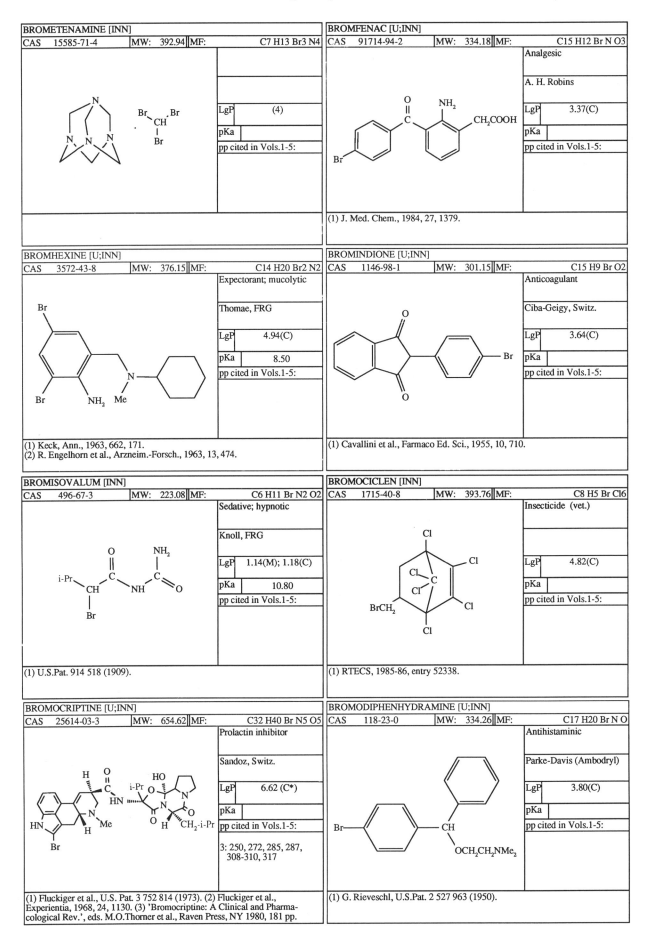

BROMETENAMINE [INN]

CAS	15585-71-4	MW:	392.94	MF:	C7 H13 Br3 N4

LgP	(4)
pKa	
pp cited in Vols.1-5:	

BROMFENAC [U;INN]

CAS	91714-94-2	MW:	334.18	MF:	C15 H12 Br N O3

Analgesic

A. H. Robins

LgP	3.37(C)
pKa	
pp cited in Vols.1-5:	

(1) J. Med. Chem., 1984, 27, 1379.

BROMHEXINE [U;INN]

CAS	3572-43-8	MW:	376.15	MF:	C14 H20 Br2 N2

Expectorant; mucolytic

Thomae, FRG

LgP	4.94(C)
pKa	8.50
pp cited in Vols.1-5:	

(1) Keck, Ann., 1963, 662, 171.
(2) R. Engelhorn et al., Arzneim.-Forsch., 1963, 13, 474.

BROMINDIONE [U;INN]

CAS	1146-98-1	MW:	301.15	MF:	C15 H9 Br O2

Anticoagulant

Ciba-Geigy, Switz.

LgP	3.64(C)
pKa	
pp cited in Vols.1-5:	

(1) Cavallini et al., Farmaco Ed. Sci., 1955, 10, 710.

BROMISOVALUM [INN]

CAS	496-67-3	MW:	223.08	MF:	C6 H11 Br N2 O2

Sedative; hypnotic

Knoll, FRG

LgP	1.14(M); 1.18(C)
pKa	10.80
pp cited in Vols.1-5:	

(1) U.S.Pat. 914 518 (1909).

BROMOCICLEN [INN]

CAS	1715-40-8	MW:	393.76	MF:	C8 H5 Br Cl6

Insecticide (vet.)

LgP	4.82(C)
pKa	
pp cited in Vols.1-5:	

(1) RTECS, 1985-86, entry 52338.

BROMOCRIPTINE [U;INN]

CAS	25614-03-3	MW:	654.62	MF:	C32 H40 Br N5 O5

Prolactin inhibitor

Sandoz, Switz.

LgP	6.62 (C*)
pKa	
pp cited in Vols.1-5:	
	3: 250, 272, 285, 287, 308-310, 317

(1) Fluckiger et al., U.S. Pat. 3 752 814 (1973). (2) Fluckiger et al.,
Experientia, 1968, 24, 1130. (3) 'Bromocriptine: A Clinical and Pharmacological Rev.', eds. M.O.Thorner et al., Raven Press, NY 1980, 181 pp.

BROMODIPHENHYDRAMINE [U;INN]

CAS	118-23-0	MW:	334.26	MF:	C17 H20 Br N O

Antihistaminic

Parke-Davis (Ambodryl)

LgP	3.80(C)
pKa	
pp cited in Vols.1-5:	

(1) G. Rieveschl, U.S.Pat. 2 527 963 (1950).

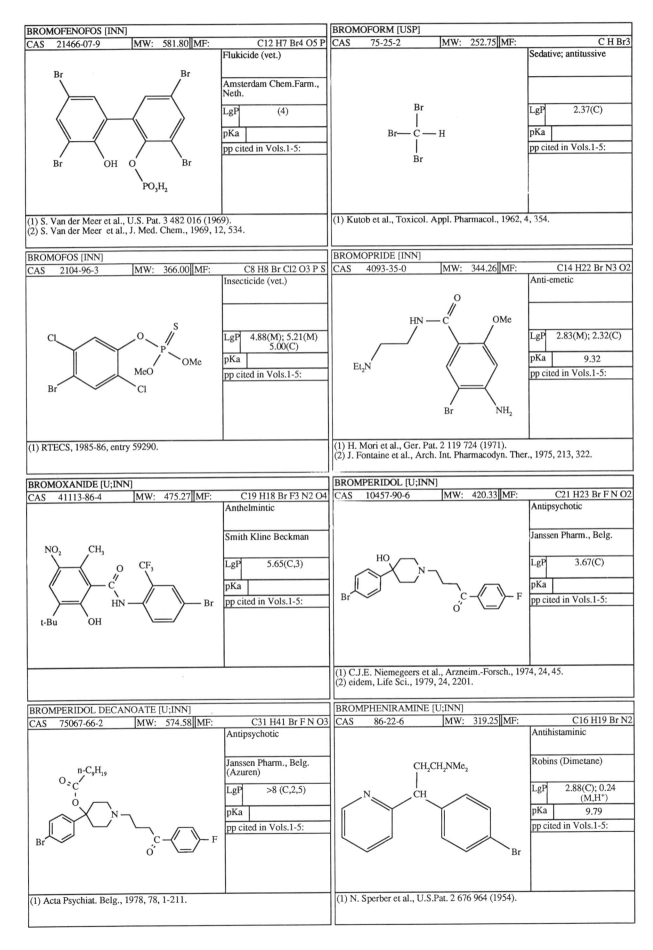

BROMOFENOFOS [INN]

CAS	21466-07-9	MW:	581.80	MF:	C12 H7 Br4 O5 P

Flukicide (vet.)

Amsterdam Chem.Farm., Neth.

LgP	(4)
pKa	
pp cited in Vols.1-5:	

(1) S. Van der Meer et al., U.S. Pat. 3 482 016 (1969).
(2) S. Van der Meer et al., J. Med. Chem., 1969, 12, 534.

BROMOFORM [USP]

CAS	75-25-2	MW:	252.75	MF:	C H Br3

Sedative; antitussive

LgP	2.37(C)
pKa	
pp cited in Vols.1-5:	

(1) Kutob et al., Toxicol. Appl. Pharmacol., 1962, 4, 354.

BROMOFOS [INN]

CAS	2104-96-3	MW:	366.00	MF:	C8 H8 Br Cl2 O3 P S

Insecticide (vet.)

LgP	4.88(M); 5.21(M) 5.00(C)
pKa	
pp cited in Vols.1-5:	

(1) RTECS, 1985-86, entry 59290.

BROMOPRIDE [INN]

CAS	4093-35-0	MW:	344.26	MF:	C14 H22 Br N3 O2

Anti-emetic

LgP	2.83(M); 2.32(C)
pKa	9.32
pp cited in Vols.1-5:	

(1) H. Mori et al., Ger. Pat. 2 119 724 (1971).
(2) J. Fontaine et al., Arch. Int. Pharmacodyn. Ther., 1975, 213, 322.

BROMOXANIDE [U;INN]

CAS	41113-86-4	MW:	475.27	MF:	C19 H18 Br F3 N2 O4

Anthelmintic

Smith Kline Beckman

LgP	5.65(C,3)
pKa	
pp cited in Vols.1-5:	

BROMPERIDOL [U;INN]

CAS	10457-90-6	MW:	420.33	MF:	C21 H23 Br F N O2

Antipsychotic

Janssen Pharm., Belg.

LgP	3.67(C)
pKa	
pp cited in Vols.1-5:	

(1) C.J.E. Niemegeers et al., Arzneim.-Forsch., 1974, 24, 45.
(2) eidem, Life Sci., 1979, 24, 2201.

BROMPERIDOL DECANOATE [U;INN]

CAS	75067-66-2	MW:	574.58	MF:	C31 H41 Br F N O3

Antipsychotic

Janssen Pharm., Belg. (Azuren)

LgP	>8 (C,2,5)
pKa	
pp cited in Vols.1-5:	

(1) Acta Psychiat. Belg., 1978, 78, 1-211.

BROMPHENIRAMINE [U;INN]

CAS	86-22-6	MW:	319.25	MF:	C16 H19 Br N2

Antihistaminic

Robins (Dimetane)

LgP	2.88(C); 0.24 (M,H+)
pKa	9.79
pp cited in Vols.1-5:	

(1) N. Sperber et al., U.S.Pat. 2 676 964 (1954).

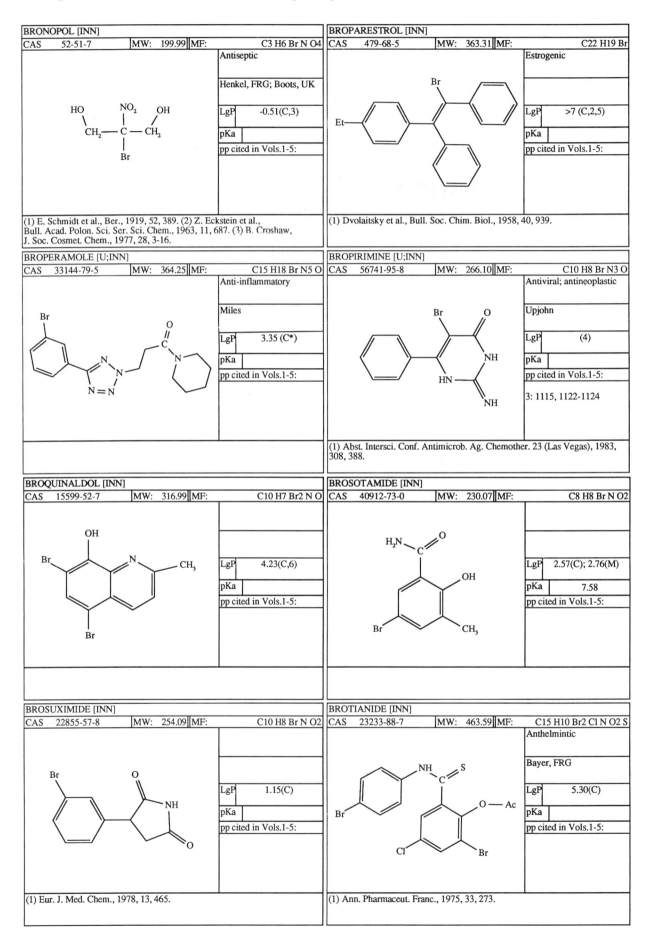

BRONOPOL [INN]

CAS	52-51-7	MW:	199.99	MF:	C3 H6 Br N O4

Antiseptic

Henkel, FRG; Boots, UK

LgP	-0.51(C,3)
pKa	
pp cited in Vols.1-5:	

(1) E. Schmidt et al., Ber., 1919, 52, 389. (2) Z. Eckstein et al., Bull. Acad. Polon. Sci. Ser. Sci. Chem., 1963, 11, 687. (3) B. Croshaw, J. Soc. Cosmet. Chem., 1977, 28, 3-16.

BROPARESTROL [INN]

CAS	479-68-5	MW:	363.31	MF:	C22 H19 Br

Estrogenic

LgP	>7 (C,2,5)
pKa	
pp cited in Vols.1-5:	

(1) Dvolaitsky et al., Bull. Soc. Chim. Biol., 1958, 40, 939.

BROPERAMOLE [U;INN]

CAS	33144-79-5	MW:	364.25	MF:	C15 H18 Br N5 O

Anti-inflammatory

Miles

LgP	3.35 (C*)
pKa	
pp cited in Vols.1-5:	

BROPIRIMINE [U;INN]

CAS	56741-95-8	MW:	266.10	MF:	C10 H8 Br N3 O

Antiviral; antineoplastic

Upjohn

LgP	(4)
pKa	
pp cited in Vols.1-5:	
3: 1115, 1122-1124	

(1) Abst. Intersci. Conf. Antimicrob. Ag. Chemother. 23 (Las Vegas), 1983, 308, 388.

BROQUINALDOL [INN]

CAS	15599-52-7	MW:	316.99	MF:	C10 H7 Br2 N O

LgP	4.23(C,6)
pKa	
pp cited in Vols.1-5:	

BROSOTAMIDE [INN]

CAS	40912-73-0	MW:	230.07	MF:	C8 H8 Br N O2

LgP	2.57(C); 2.76(M)
pKa	7.58
pp cited in Vols.1-5:	

BROSUXIMIDE [INN]

CAS	22855-57-8	MW:	254.09	MF:	C10 H8 Br N O2

LgP	1.15(C)
pKa	
pp cited in Vols.1-5:	

(1) Eur. J. Med. Chem., 1978, 13, 465.

BROTIANIDE [INN]

CAS	23233-88-7	MW:	463.59	MF:	C15 H10 Br2 Cl N O2 S

Anthelmintic

Bayer, FRG

LgP	5.30(C)
pKa	
pp cited in Vols.1-5:	

(1) Ann. Pharmaceut. Franc., 1975, 33, 273.

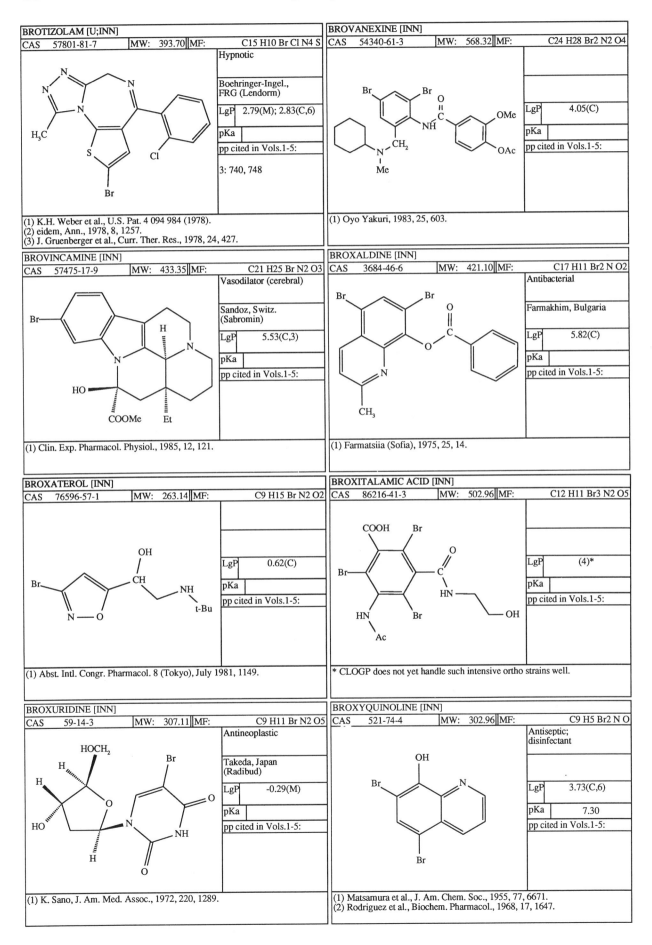

BROTIZOLAM [U;INN]

CAS 57801-81-7	MW: 393.70	MF:	C15 H10 Br Cl N4 S

Hypnotic

Boehringer-Ingel., FRG (Lendorm)

LgP	2.79(M); 2.83(C,6)
pKa	
pp cited in Vols.1-5:	

3: 740, 748

(1) K.H. Weber et al., U.S. Pat. 4 094 984 (1978).
(2) eidem, Ann., 1978, 8, 1257.
(3) J. Gruenberger et al., Curr. Ther. Res., 1978, 24, 427.

BROVANEXINE [INN]

CAS 54340-61-3	MW: 568.32	MF:	C24 H28 Br2 N2 O4

LgP	4.05(C)
pKa	
pp cited in Vols.1-5:	

(1) Oyo Yakuri, 1983, 25, 603.

BROVINCAMINE [INN]

CAS 57475-17-9	MW: 433.35	MF:	C21 H25 Br N2 O3

Vasodilator (cerebral)

Sandoz, Switz. (Sabromin)

LgP	5.53(C,3)
pKa	
pp cited in Vols.1-5:	

(1) Clin. Exp. Pharmacol. Physiol., 1985, 12, 121.

BROXALDINE [INN]

CAS 3684-46-6	MW: 421.10	MF:	C17 H11 Br2 N O2

Antibacterial

Farmakhim, Bulgaria

LgP	5.82(C)
pKa	
pp cited in Vols.1-5:	

(1) Farmatsiia (Sofia), 1975, 25, 14.

BROXATEROL [INN]

CAS 76596-57-1	MW: 263.14	MF:	C9 H15 Br N2 O2

LgP	0.62(C)
pKa	
pp cited in Vols.1-5:	

(1) Abst. Intl. Congr. Pharmacol. 8 (Tokyo), July 1981, 1149.

BROXITALAMIC ACID [INN]

CAS 86216-41-3	MW: 502.96	MF:	C12 H11 Br3 N2 O5

LgP	(4)*
pKa	
pp cited in Vols.1-5:	

* CLOGP does not yet handle such intensive ortho strains well.

BROXURIDINE [INN]

CAS 59-14-3	MW: 307.11	MF:	C9 H11 Br N2 O5

Antineoplastic

Takeda, Japan (Radibud)

LgP	-0.29(M)
pKa	
pp cited in Vols.1-5:	

(1) K. Sano, J. Am. Med. Assoc., 1972, 220, 1289.

BROXYQUINOLINE [INN]

CAS 521-74-4	MW: 302.96	MF:	C9 H5 Br2 N O

Antiseptic; disinfectant

LgP	3.73(C,6)
pKa	7.30
pp cited in Vols.1-5:	

(1) Matsamura et al., J. Am. Chem. Soc., 1955, 77, 6671.
(2) Rodriguez et al., Biochem. Pharmacol., 1968, 17, 1647.

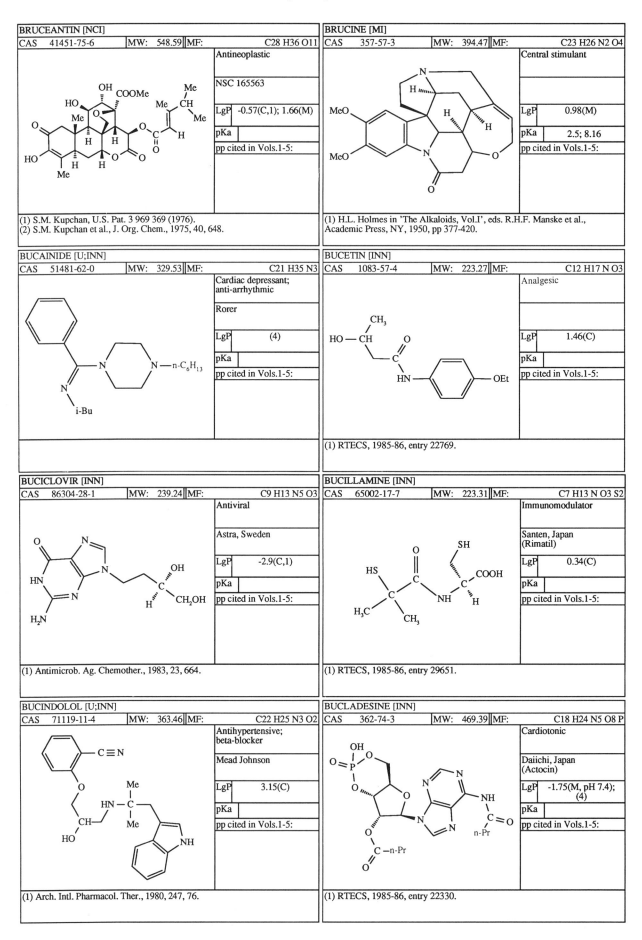

BRUCEANTIN [NCI]

CAS	41451-75-6	MW:	548.59	MF:	C28 H36 O11

Antineoplastic

NSC 165563

LgP	-0.57(C,1); 1.66(M)

pKa	

pp cited in Vols.1-5:

(1) S.M. Kupchan, U.S. Pat. 3 969 369 (1976).
(2) S.M. Kupchan et al., J. Org. Chem., 1975, 40, 648.

BRUCINE [MI]

CAS	357-57-3	MW:	394.47	MF:	C23 H26 N2 O4

Central stimulant

LgP	0.98(M)

pKa	2.5; 8.16

pp cited in Vols.1-5:

(1) H.L. Holmes in 'The Alkaloids, Vol.I', eds. R.H.F. Manske et al., Academic Press, NY, 1950, pp 377-420.

BUCAINIDE [U;INN]

CAS	51481-62-0	MW:	329.53	MF:	C21 H35 N3

Cardiac depressant; anti-arrhythmic

Rorer

LgP	(4)

pKa	

pp cited in Vols.1-5:

BUCETIN [INN]

CAS	1083-57-4	MW:	223.27	MF:	C12 H17 N O3

Analgesic

LgP	1.46(C)

pKa	

pp cited in Vols.1-5:

(1) RTECS, 1985-86, entry 22769.

BUCICLOVIR [INN]

CAS	86304-28-1	MW:	239.24	MF:	C9 H13 N5 O3

Antiviral

Astra, Sweden

LgP	-2.9(C,1)

pKa	

pp cited in Vols.1-5:

(1) Antimicrob. Ag. Chemother., 1983, 23, 664.

BUCILLAMINE [INN]

CAS	65002-17-7	MW:	223.31	MF:	C7 H13 N O3 S2

Immunomodulator

Santen, Japan (Rimatil)

LgP	0.34(C)

pKa	

pp cited in Vols.1-5:

(1) RTECS, 1985-86, entry 29651.

BUCINDOLOL [U;INN]

CAS	71119-11-4	MW:	363.46	MF:	C22 H25 N3 O2

Antihypertensive; beta-blocker

Mead Johnson

LgP	3.15(C)

pKa	

pp cited in Vols.1-5:

(1) Arch. Intl. Pharmacol. Ther., 1980, 247, 76.

BUCLADESINE [INN]

CAS	362-74-3	MW:	469.39	MF:	C18 H24 N5 O8 P

Cardiotonic

Daiichi, Japan (Actocin)

LgP	-1.75(M, pH 7.4); (4)

pKa	

pp cited in Vols.1-5:

(1) RTECS, 1985-86, entry 22330.

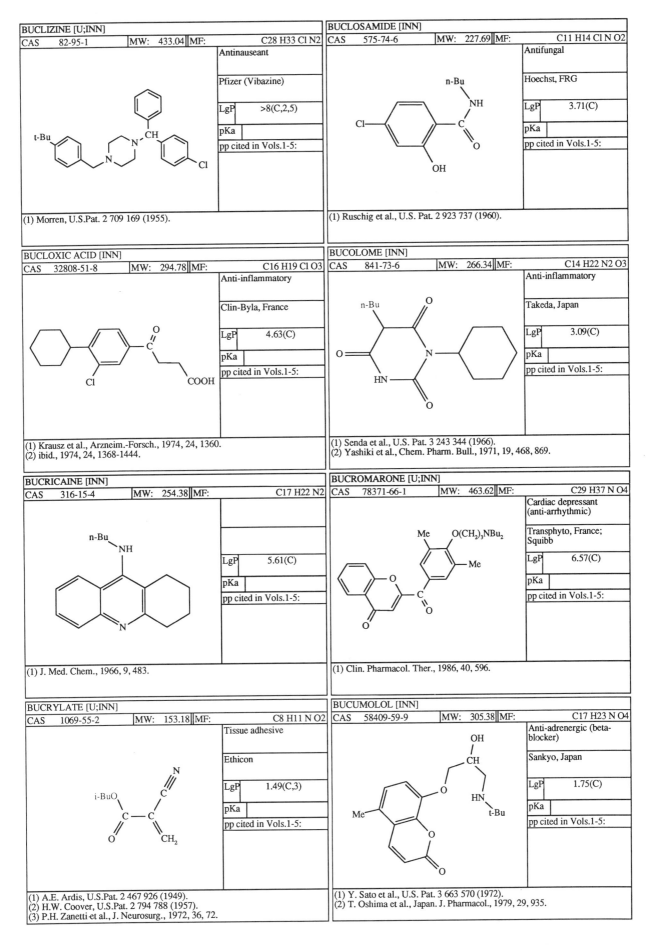

BUCLIZINE [U;INN]

CAS	82-95-1	MW:	433.04	MF:	C28 H33 Cl N2

Antinauseant

Pfizer (Vibazine)

LgP	>8(C,2,5)
pKa	
pp cited in Vols.1-5:	

(1) Morren, U.S.Pat. 2 709 169 (1955).

BUCLOSAMIDE [INN]

CAS	575-74-6	MW:	227.69	MF:	C11 H14 Cl N O2

Antifungal

Hoechst, FRG

LgP	3.71(C)
pKa	
pp cited in Vols.1-5:	

(1) Ruschig et al., U.S. Pat. 2 923 737 (1960).

BUCLOXIC ACID [INN]

CAS	32808-51-8	MW:	294.78	MF:	C16 H19 Cl O3

Anti-inflammatory

Clin-Byla, France

LgP	4.63(C)
pKa	
pp cited in Vols.1-5:	

(1) Krausz et al., Arzneim.-Forsch., 1974, 24, 1360.
(2) ibid., 1974, 24, 1368-1444.

BUCOLOME [INN]

CAS	841-73-6	MW:	266.34	MF:	C14 H22 N2 O3

Anti-inflammatory

Takeda, Japan

LgP	3.09(C)
pKa	
pp cited in Vols.1-5:	

(1) Senda et al., U.S. Pat. 3 243 344 (1966).
(2) Yashiki et al., Chem. Pharm. Bull., 1971, 19, 468, 869.

BUCRICAINE [INN]

CAS	316-15-4	MW:	254.38	MF:	C17 H22 N2

LgP	5.61(C)
pKa	
pp cited in Vols.1-5:	

(1) J. Med. Chem., 1966, 9, 483.

BUCROMARONE [U;INN]

CAS	78371-66-1	MW:	463.62	MF:	C29 H37 N O4

Cardiac depressant (anti-arrhythmic)

Transphyto, France; Squibb

LgP	6.57(C)
pKa	
pp cited in Vols.1-5:	

(1) Clin. Pharmacol. Ther., 1986, 40, 596.

BUCRYLATE [U;INN]

CAS	1069-55-2	MW:	153.18	MF:	C8 H11 N O2

Tissue adhesive

Ethicon

LgP	1.49(C,3)
pKa	
pp cited in Vols.1-5:	

(1) A.E. Ardis, U.S.Pat. 2 467 926 (1949).
(2) H.W. Coover, U.S.Pat. 2 794 788 (1957).
(3) P.H. Zanetti et al., J. Neurosurg., 1972, 36, 72.

BUCUMOLOL [INN]

CAS	58409-59-9	MW:	305.38	MF:	C17 H23 N O4

Anti-adrenergic (beta-blocker)

Sankyo, Japan

LgP	1.75(C)
pKa	
pp cited in Vols.1-5:	

(1) Y. Sato et al., U.S. Pat. 3 663 570 (1972).
(2) T. Oshima et al., Japan. J. Pharmacol., 1979, 29, 935.

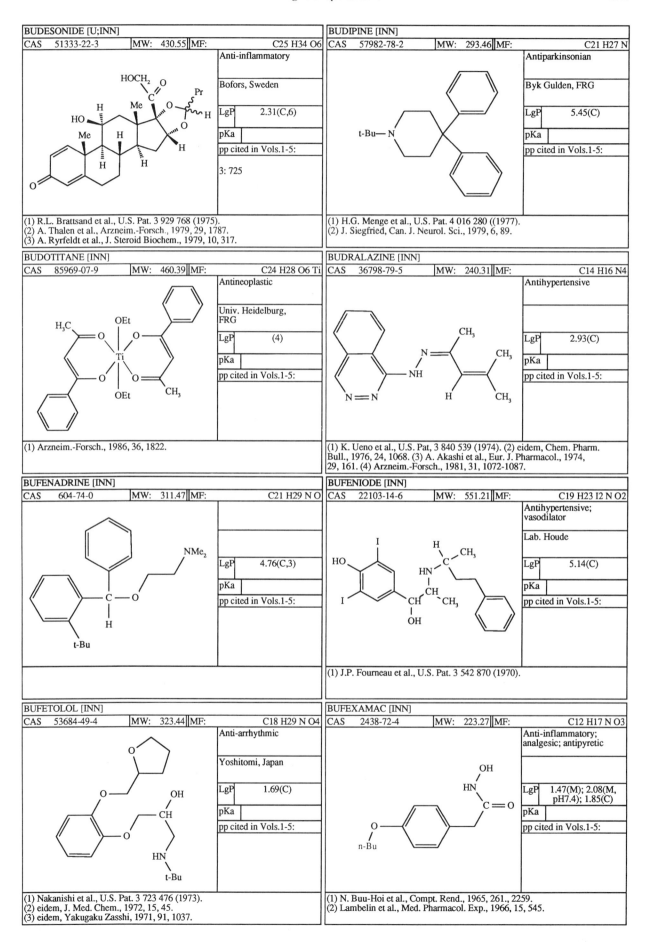

BUDESONIDE [U;INN]

CAS	51333-22-3	MW:	430.55	MF:	C25 H34 O6

Anti-inflammatory

Bofors, Sweden

LgP	2.31(C,6)
pKa	
pp cited in Vols.1-5:	

3: 725

(1) R.L. Brattsand et al., U.S. Pat. 3 929 768 (1975).
(2) A. Thalen et al., Arzneim.-Forsch., 1979, 29, 1787.
(3) A. Ryrfeldt et al., J. Steroid Biochem., 1979, 10, 317.

BUDIPINE [INN]

CAS	57982-78-2	MW:	293.46	MF:	C21 H27 N

Antiparkinsonian

Byk Gulden, FRG

LgP	5.45(C)
pKa	
pp cited in Vols.1-5:	

(1) H.G. Menge et al., U.S. Pat. 4 016 280 ((1977).
(2) J. Siegfried, Can. J. Neurol. Sci., 1979, 6, 89.

BUDOTITANE [INN]

CAS	85969-07-9	MW:	460.39	MF:	C24 H28 O6 Ti

Antineoplastic

Univ. Heidelburg, FRG

LgP	(4)
pKa	
pp cited in Vols.1-5:	

(1) Arzneim.-Forsch., 1986, 36, 1822.

BUDRALAZINE [INN]

CAS	36798-79-5	MW:	240.31	MF:	C14 H16 N4

Antihypertensive

LgP	2.93(C)
pKa	
pp cited in Vols.1-5:	

(1) K. Ueno et al., U.S. Pat, 3 840 539 (1974). (2) eidem, Chem. Pharm. Bull., 1976, 24, 1068. (3) A. Akashi et al., Eur. J. Pharmacol., 1974, 29, 161. (4) Arzneim.-Forsch., 1981, 31, 1072-1087.

BUFENADRINE [INN]

CAS	604-74-0	MW:	311.47	MF:	C21 H29 N O

LgP	4.76(C,3)
pKa	
pp cited in Vols.1-5:	

BUFENIODE [INN]

CAS	22103-14-6	MW:	551.21	MF:	C19 H23 I2 N O2

Antihypertensive; vasodilator

Lab. Houde

LgP	5.14(C)
pKa	
pp cited in Vols.1-5:	

(1) J.P. Fourneau et al., U.S. Pat. 3 542 870 (1970).

BUFETOLOL [INN]

CAS	53684-49-4	MW:	323.44	MF:	C18 H29 N O4

Anti-arrhythmic

Yoshitomi, Japan

LgP	1.69(C)
pKa	
pp cited in Vols.1-5:	

(1) Nakanishi et al., U.S. Pat. 3 723 476 (1973).
(2) eidem, J. Med. Chem., 1972, 15, 45.
(3) eidem, Yakugaku Zasshi, 1971, 91, 1037.

BUFEXAMAC [INN]

CAS	2438-72-4	MW:	223.27	MF:	C12 H17 N O3

Anti-inflammatory; analgesic; antipyretic

LgP	1.47(M); 2.08(M, pH7.4); 1.85(C)
pKa	
pp cited in Vols.1-5:	

(1) N. Buu-Hoi et al., Compt. Rend., 1965, 261, 2259.
(2) Lambelin et al., Med. Pharmacol. Exp., 1966, 15, 545.

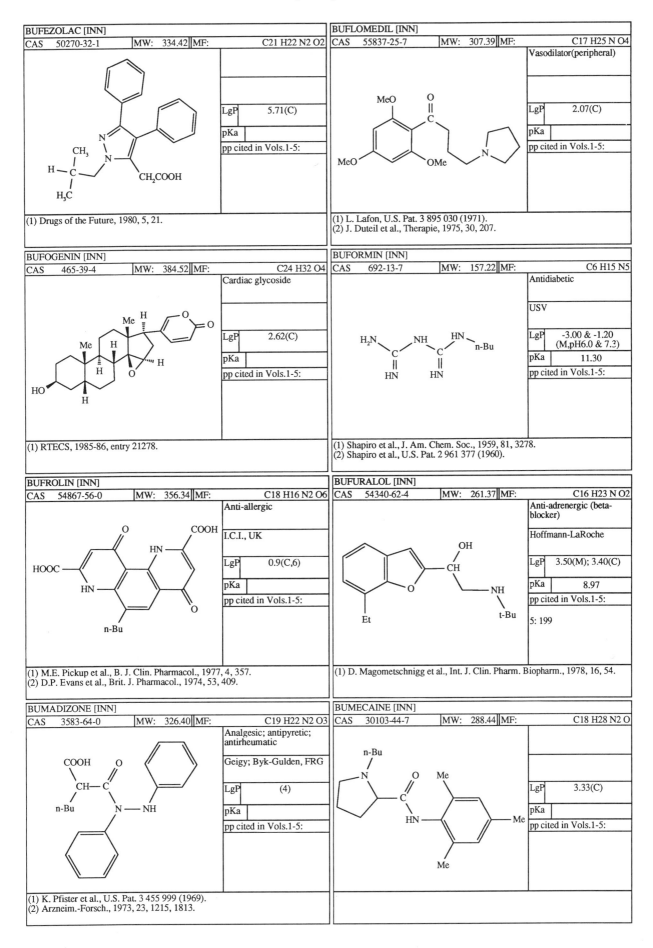

BUFEZOLAC [INN]			
CAS 50270-32-1	MW: 334.42	MF:	C21 H22 N2 O2

LgP	5.71(C)
pKa	
pp cited in Vols.1-5:	

(1) Drugs of the Future, 1980, 5, 21.

BUFLOMEDIL [INN]			
CAS 55837-25-7	MW: 307.39	MF:	C17 H25 N O4

Vasodilator(peripheral)

LgP	2.07(C)
pKa	
pp cited in Vols.1-5:	

(1) L. Lafon, U.S. Pat. 3 895 030 (1971).
(2) J. Duteil et al., Therapie, 1975, 30, 207.

BUFOGENIN [INN]			
CAS 465-39-4	MW: 384.52	MF:	C24 H32 O4

Cardiac glycoside

LgP	2.62(C)
pKa	
pp cited in Vols.1-5:	

(1) RTECS, 1985-86, entry 21278.

BUFORMIN [INN]			
CAS 692-13-7	MW: 157.22	MF:	C6 H15 N5

Antidiabetic

USV

LgP	-3.00 & -1.20 (M,pH6.0 & 7.3)
pKa	11.30
pp cited in Vols.1-5:	

(1) Shapiro et al., J. Am. Chem. Soc., 1959, 81, 3278.
(2) Shapiro et al., U.S. Pat. 2 961 377 (1960).

BUFROLIN [INN]			
CAS 54867-56-0	MW: 356.34	MF:	C18 H16 N2 O6

Anti-allergic

I.C.I., UK

LgP	0.9(C,6)
pKa	
pp cited in Vols.1-5:	

(1) M.E. Pickup et al., B. J. Clin. Pharmacol., 1977, 4, 357.
(2) D.P. Evans et al., Brit. J. Pharmacol., 1974, 53, 409.

BUFURALOL [INN]			
CAS 54340-62-4	MW: 261.37	MF:	C16 H23 N O2

Anti-adrenergic (beta-blocker)

Hoffmann-LaRoche

LgP	3.50(M); 3.40(C)
pKa	8.97
pp cited in Vols.1-5:	
5: 199	

(1) D. Magometschnigg et al., Int. J. Clin. Pharm. Biopharm., 1978, 16, 54.

BUMADIZONE [INN]			
CAS 3583-64-0	MW: 326.40	MF:	C19 H22 N2 O3

Analgesic; antipyretic; antirheumatic

Geigy; Byk-Gulden, FRG

LgP	(4)
pKa	
pp cited in Vols.1-5:	

(1) K. Pfister et al., U.S. Pat. 3 455 999 (1969).
(2) Arzneim.-Forsch., 1973, 23, 1215, 1813.

BUMECAINE [INN]			
CAS 30103-44-7	MW: 288.44	MF:	C18 H28 N2 O

LgP	3.33(C)
pKa	
pp cited in Vols.1-5:	

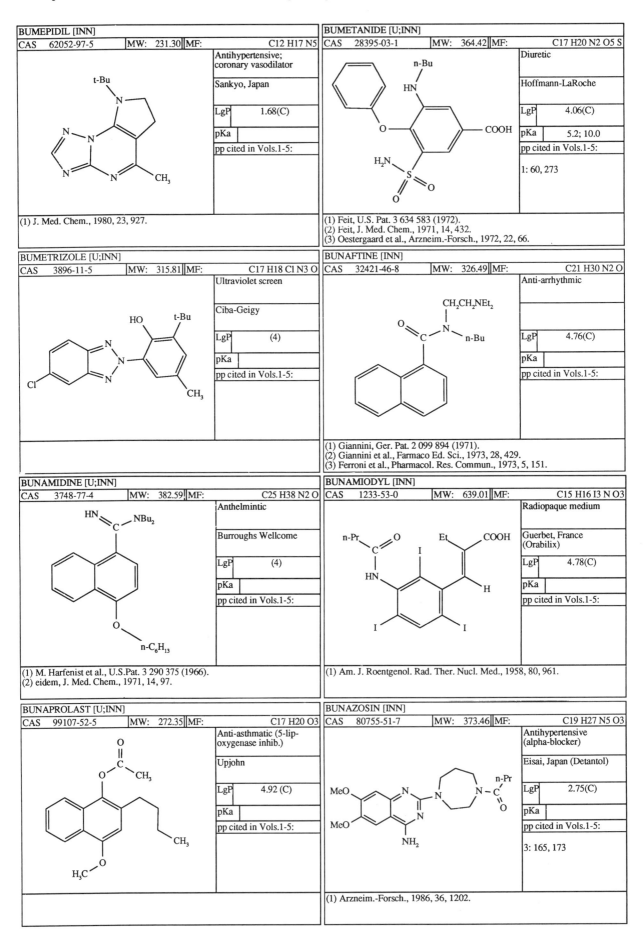

BUMEPIDIL [INN]

CAS 62052-97-5	MW: 231.30	MF:	C12 H17 N5

Antihypertensive; coronary vasodilator

Sankyo, Japan

LgP	1.68(C)

pKa	

pp cited in Vols.1-5:

(1) J. Med. Chem., 1980, 23, 927.

BUMETANIDE [U;INN]

CAS 28395-03-1	MW: 364.42	MF:	C17 H20 N2 O5 S

Diuretic

Hoffmann-LaRoche

LgP	4.06(C)

pKa	5.2; 10.0

pp cited in Vols.1-5:

1: 60, 273

(1) Feit, U.S. Pat. 3 634 583 (1972).
(2) Feit, J. Med. Chem., 1971, 14, 432.
(3) Oestergaard et al., Arzneim.-Forsch., 1972, 22, 66.

BUMETRIZOLE [U;INN]

CAS 3896-11-5	MW: 315.81	MF:	C17 H18 Cl N3 O

Ultraviolet screen

Ciba-Geigy

LgP	(4)

pKa	

pp cited in Vols.1-5:

BUNAFTINE [INN]

CAS 32421-46-8	MW: 326.49	MF:	C21 H30 N2 O

Anti-arrhythmic

LgP	4.76(C)

pKa	

pp cited in Vols.1-5:

(1) Giannini, Ger. Pat. 2 099 894 (1971).
(2) Giannini et al., Farmaco Ed. Sci., 1973, 28, 429.
(3) Ferroni et al., Pharmacol. Res. Commun., 1973, 5, 151.

BUNAMIDINE [U;INN]

CAS 3748-77-4	MW: 382.59	MF:	C25 H38 N2 O

Anthelmintic

Burroughs Wellcome

LgP	(4)

pKa	

pp cited in Vols.1-5:

(1) M. Harfenist et al., U.S.Pat. 3 290 375 (1966).
(2) eidem, J. Med. Chem., 1971, 14, 97.

BUNAMIODYL [INN]

CAS 1233-53-0	MW: 639.01	MF:	C15 H16 I3 N O3

Radiopaque medium

Guerbet, France (Orabilix)

LgP	4.78(C)

pKa	

pp cited in Vols.1-5:

(1) Am. J. Roentgenol. Rad. Ther. Nucl. Med., 1958, 80, 961.

BUNAPROLAST [U;INN]

CAS 99107-52-5	MW: 272.35	MF:	C17 H20 O3

Anti-asthmatic (5-lip-oxygenase inhib.)

Upjohn

LgP	4.92 (C)

pKa	

pp cited in Vols.1-5:

BUNAZOSIN [INN]

CAS 80755-51-7	MW: 373.46	MF:	C19 H27 N5 O3

Antihypertensive (alpha-blocker)

Eisai, Japan (Detantol)

LgP	2.75(C)

pKa	

pp cited in Vols.1-5:

3: 165, 173

(1) Arzneim.-Forsch., 1986, 36, 1202.

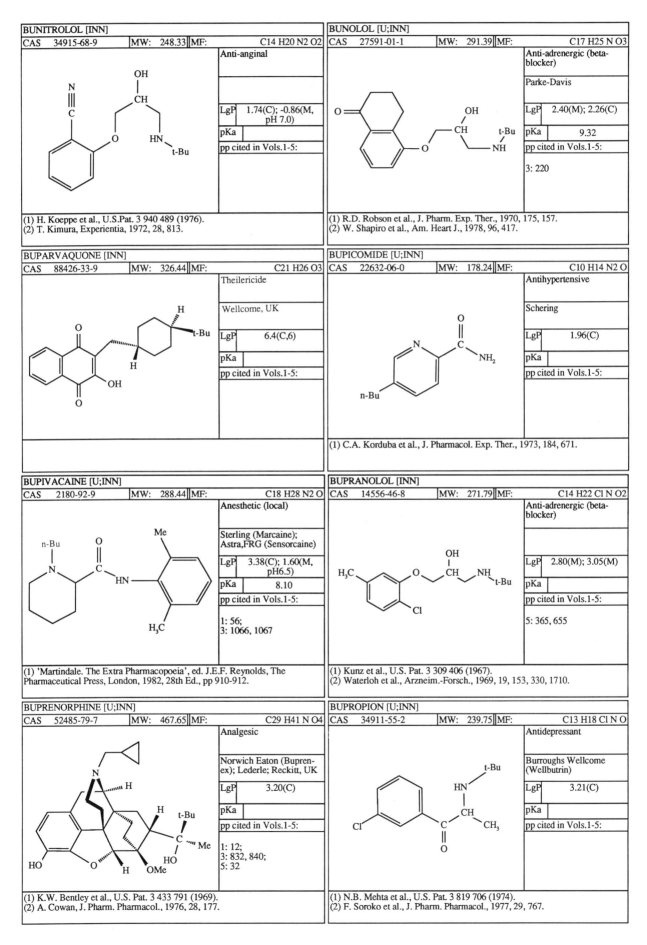

BUNITROLOL [INN]

CAS	34915-68-9	MW:	248.33	MF:	C14 H20 N2 O2

Anti-anginal

LgP	1.74(C); -0.86(M, pH 7.0)
pKa	
pp cited in Vols.1-5:	

(1) H. Koeppe et al., U.S.Pat. 3 940 489 (1976).
(2) T. Kimura, Experientia, 1972, 28, 813.

BUNOLOL [U;INN]

CAS	27591-01-1	MW:	291.39	MF:	C17 H25 N O3

Anti-adrenergic (beta-blocker)

Parke-Davis

LgP	2.40(M); 2.26(C)
pKa	9.32
pp cited in Vols.1-5:	

3: 220

(1) R.D. Robson et al., J. Pharm. Exp. Ther., 1970, 175, 157.
(2) W. Shapiro et al., Am. Heart J., 1978, 96, 417.

BUPARVAQUONE [INN]

CAS	88426-33-9	MW:	326.44	MF:	C21 H26 O3

Theilericide

Wellcome, UK

LgP	6.4(C,6)
pKa	
pp cited in Vols.1-5:	

BUPICOMIDE [U;INN]

CAS	22632-06-0	MW:	178.24	MF:	C10 H14 N2 O

Antihypertensive

Schering

LgP	1.96(C)
pKa	
pp cited in Vols.1-5:	

(1) C.A. Korduba et al., J. Pharmacol. Exp. Ther., 1973, 184, 671.

BUPIVACAINE [U;INN]

CAS	2180-92-9	MW:	288.44	MF:	C18 H28 N2 O

Anesthetic (local)

Sterling (Marcaine); Astra,FRG (Sensorcaine)

LgP	3.38(C); 1.60(M, pH6.5)
pKa	8.10
pp cited in Vols.1-5:	

1: 56;
3: 1066, 1067

(1) 'Martindale. The Extra Pharmacopoeia', ed. J.E.F. Reynolds, The Pharmaceutical Press, London, 1982, 28th Ed., pp 910-912.

BUPRANOLOL [INN]

CAS	14556-46-8	MW:	271.79	MF:	C14 H22 Cl N O2

Anti-adrenergic (beta-blocker)

LgP	2.80(M); 3.05(M)
pKa	
pp cited in Vols.1-5:	

5: 365, 655

(1) Kunz et al., U.S. Pat. 3 309 406 (1967).
(2) Waterloh et al., Arzneim.-Forsch., 1969, 19, 153, 330, 1710.

BUPRENORPHINE [U;INN]

CAS	52485-79-7	MW:	467.65	MF:	C29 H41 N O4

Analgesic

Norwich Eaton (Buprenex); Lederle; Reckitt, UK

LgP	3.20(C)
pKa	
pp cited in Vols.1-5:	

1: 12;
3: 832, 840;
5: 32

(1) K.W. Bentley et al., U.S. Pat. 3 433 791 (1969).
(2) A. Cowan, J. Pharm. Pharmacol., 1976, 28, 177.

BUPROPION [U;INN]

CAS	34911-55-2	MW:	239.75	MF:	C13 H18 Cl N O

Antidepressant

Burroughs Wellcome (Wellbutrin)

LgP	3.21(C)
pKa	
pp cited in Vols.1-5:	

(1) N.B. Mehta et al., U.S. Pat. 3 819 706 (1974).
(2) F. Soroko et al., J. Pharm. Pharmacol., 1977, 29, 767.

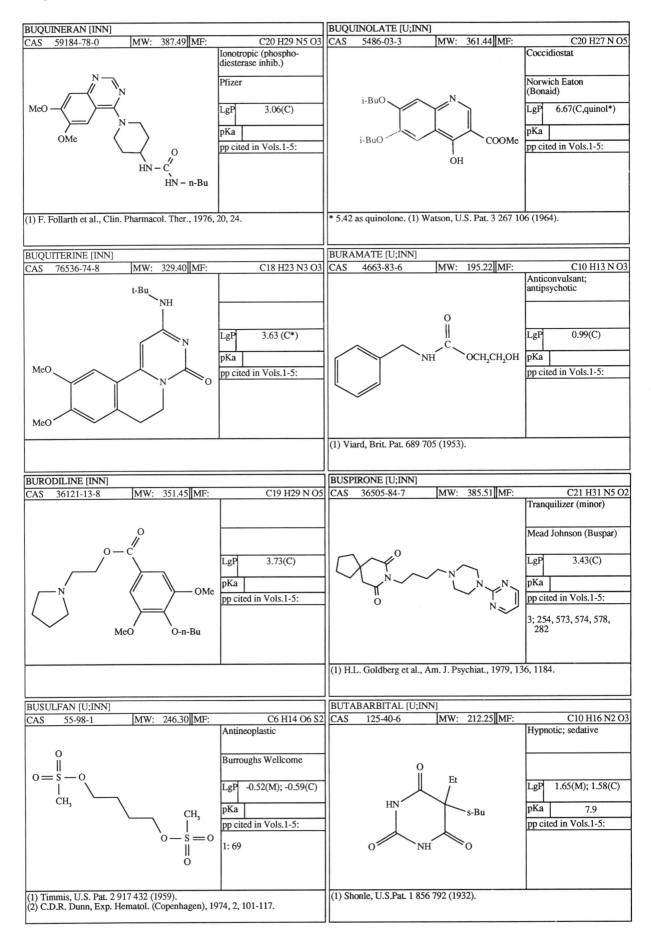

BUQUINERAN [INN]

CAS 59184-78-0	MW: 387.49	MF:	C20 H29 N5 O3

Ionotropic (phospho-diesterase inhib.)

Pfizer

LgP	3.06(C)
pKa	
pp cited in Vols.1-5:	

MeO, OMe on quinazoline; piperidine N; HN–C(=O)–HN–n-Bu

(1) F. Follarth et al., Clin. Pharmacol. Ther., 1976, 20, 24.

BUQUINOLATE [U;INN]

CAS 5486-03-3	MW: 361.44	MF:	C20 H27 N O5

Coccidiostat

Norwich Eaton (Bonaid)

LgP	6.67(C,quinol*)
pKa	
pp cited in Vols.1-5:	

i-BuO, i-BuO; COOMe; OH

* 5.42 as quinolone. (1) Watson, U.S. Pat. 3 267 106 (1964).

BUQUITERINE [INN]

CAS 76536-74-8	MW: 329.40	MF:	C18 H23 N3 O3

LgP	3.63 (C*)
pKa	
pp cited in Vols.1-5:	

t-Bu–NH; MeO, MeO

BURAMATE [U;INN]

CAS 4663-83-6	MW: 195.22	MF:	C10 H13 N O3

Anticonvulsant; antipsychotic

LgP	0.99(C)
pKa	
pp cited in Vols.1-5:	

NH–C(=O)–OCH₂CH₂OH

(1) Viard, Brit. Pat. 689 705 (1953).

BURODILINE [INN]

CAS 36121-13-8	MW: 351.45	MF:	C19 H29 N O5

LgP	3.73(C)
pKa	
pp cited in Vols.1-5:	

OMe, MeO, O-n-Bu; pyrrolidine

BUSPIRONE [U;INN]

CAS 36505-84-7	MW: 385.51	MF:	C21 H31 N5 O2

Tranquilizer (minor)

Mead Johnson (Buspar)

LgP	3.43(C)
pKa	
pp cited in Vols.1-5:	

3; 254, 573, 574, 578, 282

(1) H.L. Goldberg et al., Am. J. Psychiat., 1979, 136, 1184.

BUSULFAN [U;INN]

CAS 55-98-1	MW: 246.30	MF:	C6 H14 O6 S2

Antineoplastic

Burroughs Wellcome

LgP	-0.52(M); -0.59(C)
pKa	
pp cited in Vols.1-5:	

1: 69

(1) Timmis, U.S. Pat. 2 917 432 (1959).
(2) C.D.R. Dunn, Exp. Hematol. (Copenhagen), 1974, 2, 101-117.

BUTABARBITAL [U;INN]

CAS 125-40-6	MW: 212.25	MF:	C10 H16 N2 O3

Hypnotic; sedative

LgP	1.65(M); 1.58(C)
pKa	7.9
pp cited in Vols.1-5:	

Et, s-Bu

(1) Shonle, U.S.Pat. 1 856 792 (1932).

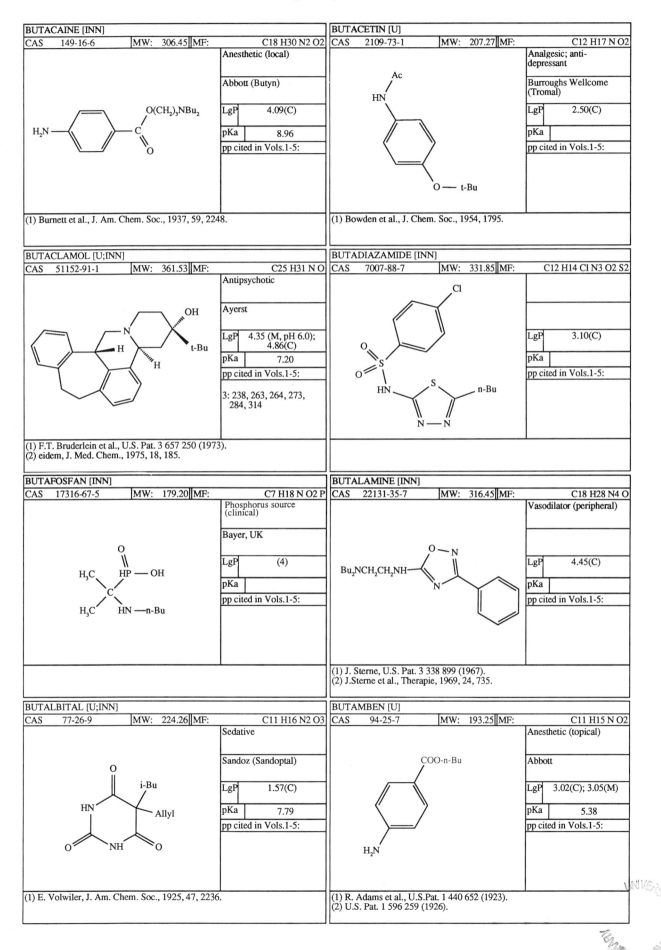

BUTACAINE [INN]

CAS	149-16-6	MW: 306.45	MF:	C18 H30 N2 O2

Anesthetic (local)

Abbott (Butyn)

LgP	4.09(C)
pKa	8.96

pp cited in Vols.1-5:

(1) Burnett et al., J. Am. Chem. Soc., 1937, 59, 2248.

BUTACETIN [U]

CAS	2109-73-1	MW: 207.27	MF:	C12 H17 N O2

Analgesic; anti-depressant

Burroughs Wellcome (Tromal)

LgP	2.50(C)
pKa	

pp cited in Vols.1-5:

(1) Bowden et al., J. Chem. Soc., 1954, 1795.

BUTACLAMOL [U;INN]

CAS	51152-91-1	MW: 361.53	MF:	C25 H31 N O

Antipsychotic

Ayerst

LgP	4.35 (M, pH 6.0); 4.86(C)
pKa	7.20

pp cited in Vols.1-5:

3: 238, 263, 264, 273, 284, 314

(1) F.T. Bruderlein et al., U.S. Pat. 3 657 250 (1973).
(2) eidem, J. Med. Chem., 1975, 18, 185.

BUTADIAZAMIDE [INN]

CAS	7007-88-7	MW: 331.85	MF:	C12 H14 Cl N3 O2 S2

LgP	3.10(C)
pKa	

pp cited in Vols.1-5:

BUTAFOSFAN [INN]

CAS	17316-67-5	MW: 179.20	MF:	C7 H18 N O2 P

Phosphorus source (clinical)

Bayer, UK

LgP	(4)
pKa	

pp cited in Vols.1-5:

BUTALAMINE [INN]

CAS	22131-35-7	MW: 316.45	MF:	C18 H28 N4 O

Vasodilator (peripheral)

$Bu_2NCH_2CH_2NH$

LgP	4.45(C)
pKa	

pp cited in Vols.1-5:

(1) J. Sterne, U.S. Pat. 3 338 899 (1967).
(2) J.Sterne et al., Therapie, 1969, 24, 735.

BUTALBITAL [U;INN]

CAS	77-26-9	MW: 224.26	MF:	C11 H16 N2 O3

Sedative

Sandoz (Sandoptal)

LgP	1.57(C)
pKa	7.79

pp cited in Vols.1-5:

(1) E. Volwiler, J. Am. Chem. Soc., 1925, 47, 2236.

BUTAMBEN [U]

CAS	94-25-7	MW: 193.25	MF:	C11 H15 N O2

Anesthetic (topical)

COO-n-Bu

Abbott

LgP	3.02(C); 3.05(M)
pKa	5.38

pp cited in Vols.1-5:

(1) R. Adams et al., U.S.Pat. 1 440 652 (1923).
(2) U.S. Pat. 1 596 259 (1926).

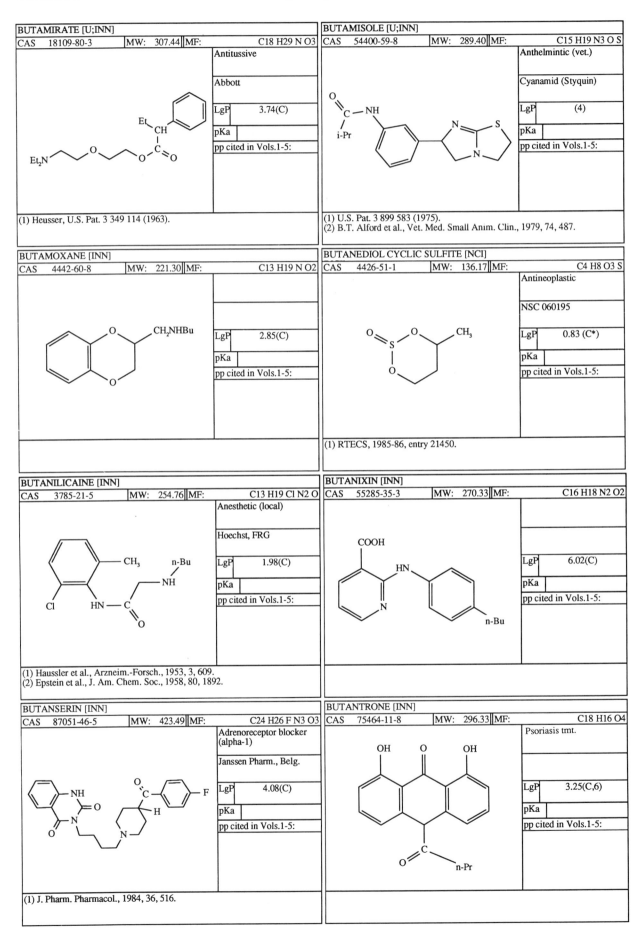

BUTAMIRATE [U;INN]

CAS	18109-80-3	MW:	307.44	MF:	C18 H29 N O3

Antitussive

Abbott

LgP	3.74(C)
pKa	
pp cited in Vols.1-5:	

(1) Heusser, U.S. Pat. 3 349 114 (1963).

BUTAMISOLE [U;INN]

CAS	54400-59-8	MW:	289.40	MF:	C15 H19 N3 O S

Anthelmintic (vet.)

Cyanamid (Styquin)

LgP	(4)
pKa	
pp cited in Vols.1-5:	

(1) U.S. Pat. 3 899 583 (1975).
(2) B.T. Alford et al., Vet. Med. Small Anim. Clin., 1979, 74, 487.

BUTAMOXANE [INN]

CAS	4442-60-8	MW:	221.30	MF:	C13 H19 N O2

LgP	2.85(C)
pKa	
pp cited in Vols.1-5:	

BUTANEDIOL CYCLIC SULFITE [NCI]

CAS	4426-51-1	MW:	136.17	MF:	C4 H8 O3 S

Antineoplastic

NSC 060195

LgP	0.83 (C*)
pKa	
pp cited in Vols.1-5:	

(1) RTECS, 1985-86, entry 21450.

BUTANILICAINE [INN]

CAS	3785-21-5	MW:	254.76	MF:	C13 H19 Cl N2 O

Anesthetic (local)

Hoechst, FRG

LgP	1.98(C)
pKa	
pp cited in Vols.1-5:	

(1) Haussler et al., Arzneim.-Forsch., 1953, 3, 609.
(2) Epstein et al., J. Am. Chem. Soc., 1958, 80, 1892.

BUTANIXIN [INN]

CAS	55285-35-3	MW:	270.33	MF:	C16 H18 N2 O2

LgP	6.02(C)
pKa	
pp cited in Vols.1-5:	

BUTANSERIN [INN]

CAS	87051-46-5	MW:	423.49	MF:	C24 H26 F N3 O3

Adrenoreceptor blocker (alpha-1)

Janssen Pharm., Belg.

LgP	4.08(C)
pKa	
pp cited in Vols.1-5:	

(1) J. Pharm. Pharmacol., 1984, 36, 516.

BUTANTRONE [INN]

CAS	75464-11-8	MW:	296.33	MF:	C18 H16 O4

Psoriasis tmt.

LgP	3.25(C,6)
pKa	
pp cited in Vols.1-5:	

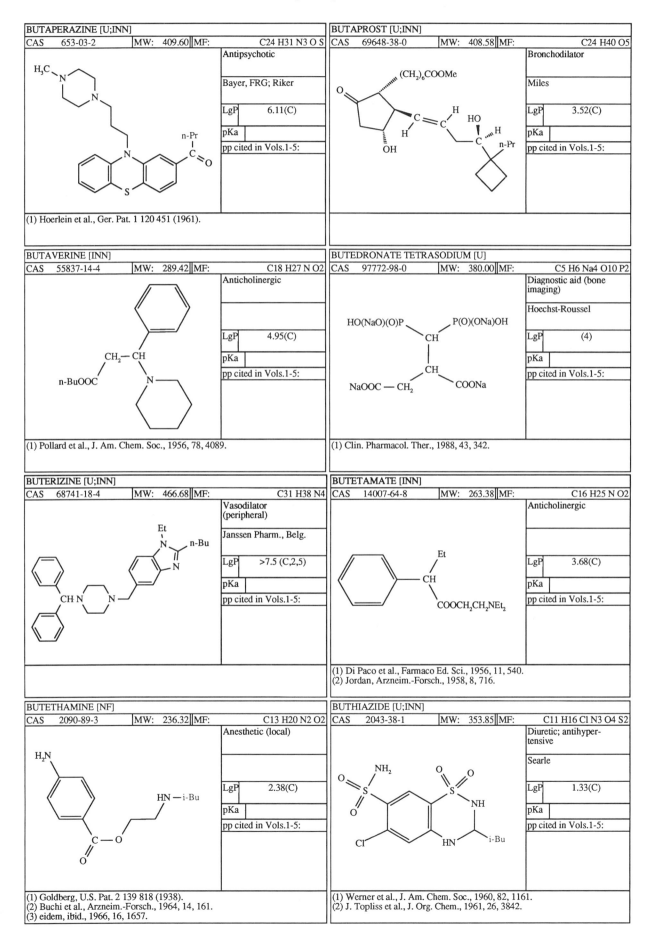

BUTAPERAZINE [U;INN]

CAS	653-03-2	MW:	409.60	MF:	C24 H31 N3 O S

Antipsychotic

Bayer, FRG; Riker

LgP	6.11(C)
pKa	
pp cited in Vols.1-5:	

(1) Hoerlein et al., Ger. Pat. 1 120 451 (1961).

BUTAPROST [U;INN]

CAS	69648-38-0	MW:	408.58	MF:	C24 H40 O5

Bronchodilator

Miles

LgP	3.52(C)
pKa	
pp cited in Vols.1-5:	

BUTAVERINE [INN]

CAS	55837-14-4	MW:	289.42	MF:	C18 H27 N O2

Anticholinergic

LgP	4.95(C)
pKa	
pp cited in Vols.1-5:	

(1) Pollard et al., J. Am. Chem. Soc., 1956, 78, 4089.

BUTEDRONATE TETRASODIUM [U]

CAS	97772-98-0	MW:	380.00	MF:	C5 H6 Na4 O10 P2

Diagnostic aid (bone imaging)

Hoechst-Roussel

LgP	(4)
pKa	
pp cited in Vols.1-5:	

(1) Clin. Pharmacol. Ther., 1988, 43, 342.

BUTERIZINE [U;INN]

CAS	68741-18-4	MW:	466.68	MF:	C31 H38 N4

Vasodilator (peripheral)

Janssen Pharm., Belg.

LgP	>7.5 (C,2,5)
pKa	
pp cited in Vols.1-5:	

BUTETAMATE [INN]

CAS	14007-64-8	MW:	263.38	MF:	C16 H25 N O2

Anticholinergic

LgP	3.68(C)
pKa	
pp cited in Vols.1-5:	

(1) Di Paco et al., Farmaco Ed. Sci., 1956, 11, 540.
(2) Jordan, Arzneim.-Forsch., 1958, 8, 716.

BUTETHAMINE [NF]

CAS	2090-89-3	MW:	236.32	MF:	C13 H20 N2 O2

Anesthetic (local)

LgP	2.38(C)
pKa	
pp cited in Vols.1-5:	

(1) Goldberg, U.S. Pat. 2 139 818 (1938).
(2) Buchi et al., Arzneim.-Forsch., 1964, 14, 161.
(3) eidem, ibid., 1966, 16, 1657.

BUTHIAZIDE [U;INN]

CAS	2043-38-1	MW:	353.85	MF:	C11 H16 Cl N3 O4 S2

Diuretic; antihypertensive

Searle

LgP	1.33(C)
pKa	
pp cited in Vols.1-5:	

(1) Werner et al., J. Am. Chem. Soc., 1960, 82, 1161.
(2) J. Topliss et al., J. Org. Chem., 1961, 26, 3842.

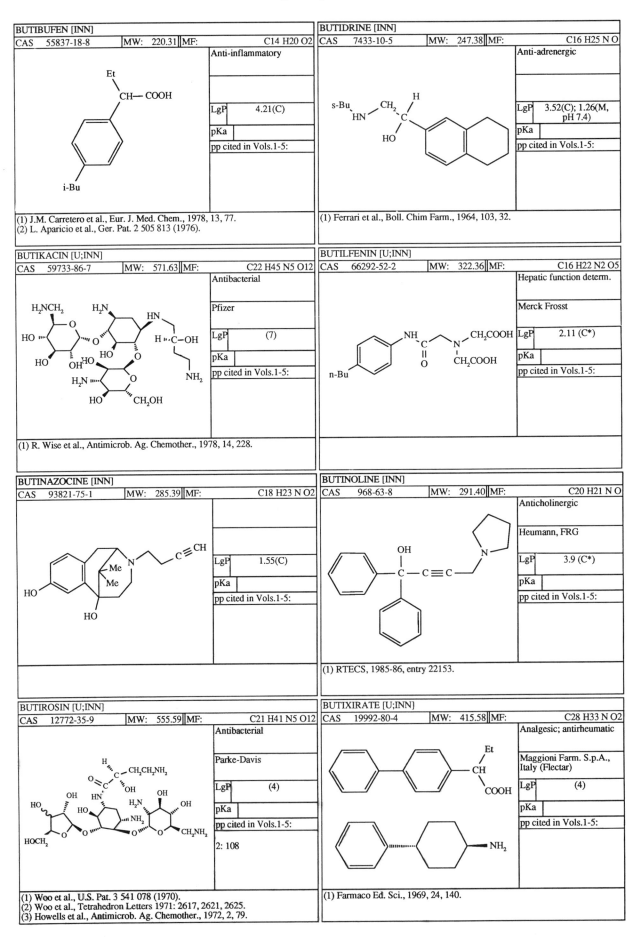

BUTIBUFEN [INN]

CAS	55837-18-8	MW:	220.31	MF:	C14 H20 O2

Anti-inflammatory

LgP	4.21(C)
pKa	
pp cited in Vols.1-5:	

(1) J.M. Carretero et al., Eur. J. Med. Chem., 1978, 13, 77.
(2) L. Aparicio et al., Ger. Pat. 2 505 813 (1976).

BUTIDRINE [INN]

CAS	7433-10-5	MW:	247.38	MF:	C16 H25 N O

Anti-adrenergic

LgP	3.52(C); 1.26(M, pH 7.4)
pKa	
pp cited in Vols.1-5:	

(1) Ferrari et al., Boll. Chim Farm., 1964, 103, 32.

BUTIKACIN [U;INN]

CAS	59733-86-7	MW:	571.63	MF:	C22 H45 N5 O12

Antibacterial

Pfizer

LgP	(7)
pKa	
pp cited in Vols.1-5:	

(1) R. Wise et al., Antimicrob. Ag. Chemother., 1978, 14, 228.

BUTILFENIN [U;INN]

CAS	66292-52-2	MW:	322.36	MF:	C16 H22 N2 O5

Hepatic function determ.

Merck Frosst

LgP	2.11 (C*)
pKa	
pp cited in Vols.1-5:	

BUTINAZOCINE [INN]

CAS	93821-75-1	MW:	285.39	MF:	C18 H23 N O2

LgP	1.55(C)
pKa	
pp cited in Vols.1-5:	

BUTINOLINE [INN]

CAS	968-63-8	MW:	291.40	MF:	C20 H21 N O

Anticholinergic

Heumann, FRG

LgP	3.9 (C*)
pKa	
pp cited in Vols.1-5:	

(1) RTECS, 1985-86, entry 22153.

BUTIROSIN [U;INN]

CAS	12772-35-9	MW:	555.59	MF:	C21 H41 N5 O12

Antibacterial

Parke-Davis

LgP	(4)
pKa	
pp cited in Vols.1-5:	
2: 108	

(1) Woo et al., U.S. Pat. 3 541 078 (1970).
(2) Woo et al., Tetrahedron Letters 1971: 2617, 2621, 2625.
(3) Howells et al., Antimicrob. Ag. Chemother., 1972, 2, 79.

BUTIXIRATE [U;INN]

CAS	19992-80-4	MW:	415.58	MF:	C28 H33 N O2

Analgesic; antirheumatic

Maggioni Farm. S.p.A., Italy (Flectar)

LgP	(4)
pKa	
pp cited in Vols.1-5:	

(1) Farmaco Ed. Sci., 1969, 24, 140.

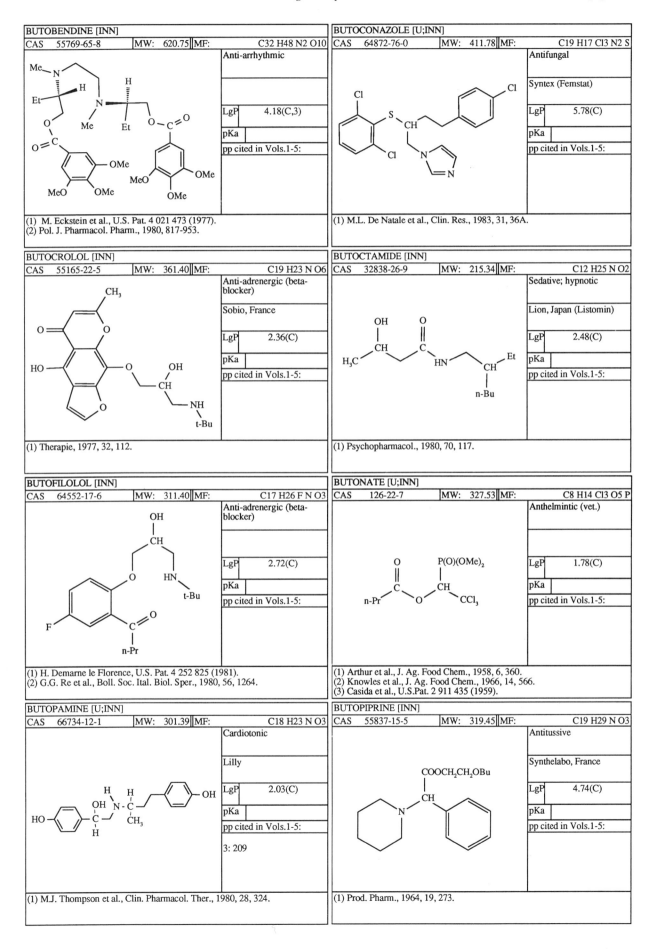

BUTOBENDINE [INN]

CAS	55769-65-8	MW: 620.75	MF:	C32 H48 N2 O10

Anti-arrhythmic

LgP 4.18(C,3)

pKa

pp cited in Vols.1-5:

(1) M. Eckstein et al., U.S. Pat. 4 021 473 (1977).
(2) Pol. J. Pharmacol. Pharm., 1980, 817-953.

BUTOCROLOL [INN]

CAS	55165-22-5	MW: 361.40	MF:	C19 H23 N O6

Anti-adrenergic (beta-blocker)

Sobio, France

LgP 2.36(C)

pKa

pp cited in Vols.1-5:

(1) Therapie, 1977, 32, 112.

BUTOFILOLOL [INN]

CAS	64552-17-6	MW: 311.40	MF:	C17 H26 F N O3

Anti-adrenergic (beta-blocker)

LgP 2.72(C)

pKa

pp cited in Vols.1-5:

(1) H. Demarne le Florence, U.S. Pat. 4 252 825 (1981).
(2) G.G. Re et al., Boll. Soc. Ital. Biol. Sper., 1980, 56, 1264.

BUTOPAMINE [U;INN]

CAS	66734-12-1	MW: 301.39	MF:	C18 H23 N O3

Cardiotonic

Lilly

LgP 2.03(C)

pKa

pp cited in Vols.1-5:

3: 209

(1) M.J. Thompson et al., Clin. Pharmacol. Ther., 1980, 28, 324.

BUTOCONAZOLE [U;INN]

CAS	64872-76-0	MW: 411.78	MF:	C19 H17 Cl3 N2 S

Antifungal

Syntex (Femstat)

LgP 5.78(C)

pKa

pp cited in Vols.1-5:

(1) M.L. De Natale et al., Clin. Res., 1983, 31, 36A.

BUTOCTAMIDE [INN]

CAS	32838-26-9	MW: 215.34	MF:	C12 H25 N O2

Sedative; hypnotic

Lion, Japan (Listomin)

LgP 2.48(C)

pKa

pp cited in Vols.1-5:

(1) Psychopharmacol., 1980, 70, 117.

BUTONATE [U;INN]

CAS	126-22-7	MW: 327.53	MF:	C8 H14 Cl3 O5 P

Anthelmintic (vet.)

LgP 1.78(C)

pKa

pp cited in Vols.1-5:

(1) Arthur et al., J. Ag. Food Chem., 1958, 6, 360.
(2) Knowles et al., J. Ag. Food Chem., 1966, 14, 566.
(3) Casida et al., U.S.Pat. 2 911 435 (1959).

BUTOPIPRINE [INN]

CAS	55837-15-5	MW: 319.45	MF:	C19 H29 N O3

Antitussive

Synthelabo, France

LgP 4.74(C)

pKa

pp cited in Vols.1-5:

(1) Prod. Pharm., 1964, 19, 273.

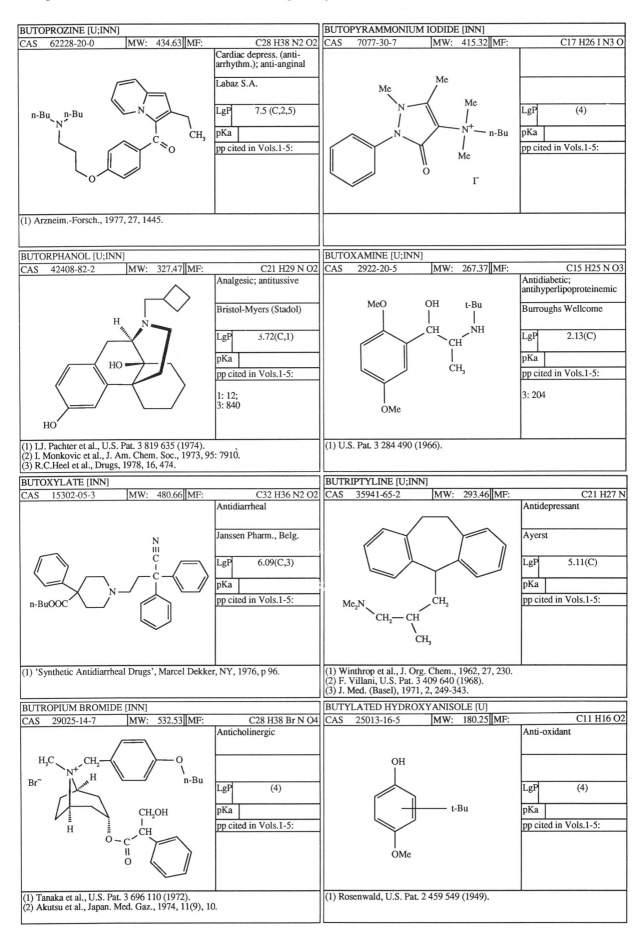

BUTOPROZINE [U;INN]		
CAS 62228-20-0	MW: 434.63	MF: C28 H38 N2 O2
	Cardiac depress. (anti-arrhythm.); anti-anginal	
	Labaz S.A.	
	LgP 7.5 (C,2,5)	
	pKa	
	pp cited in Vols.1-5:	

(1) Arzneim.-Forsch., 1977, 27, 1445.

BUTOPYRAMMONIUM IODIDE [INN]		
CAS 7077-30-7	MW: 415.32	MF: C17 H26 I N3 O
	LgP (4)	
	pKa	
	pp cited in Vols.1-5:	

BUTORPHANOL [U;INN]		
CAS 42408-82-2	MW: 327.47	MF: C21 H29 N O2
	Analgesic; antitussive	
	Bristol-Myers (Stadol)	
	LgP 3.72(C,1)	
	pKa	
	pp cited in Vols.1-5:	
	1: 12; 3: 840	

(1) I.J. Pachter et al., U.S. Pat. 3 819 635 (1974).
(2) I. Monkovic et al., J. Am. Chem. Soc., 1973, 95: 7910.
(3) R.C.Heel et al., Drugs, 1978, 16, 474.

BUTOXAMINE [U;INN]		
CAS 2922-20-5	MW: 267.37	MF: C15 H25 N O3
	Antidiabetic; antihyperlipoproteinemic	
	Burroughs Wellcome	
	LgP 2.13(C)	
	pKa	
	pp cited in Vols.1-5:	
	3: 204	

(1) U.S. Pat. 3 284 490 (1966).

BUTOXYLATE [INN]		
CAS 15302-05-3	MW: 480.66	MF: C32 H36 N2 O2
	Antidiarrheal	
	Janssen Pharm., Belg.	
	LgP 6.09(C,3)	
	pKa	
	pp cited in Vols.1-5:	

(1) 'Synthetic Antidiarrheal Drugs', Marcel Dekker, NY, 1976, p 96.

BUTRIPTYLINE [U;INN]		
CAS 35941-65-2	MW: 293.46	MF: C21 H27 N
	Antidepressant	
	Ayerst	
	LgP 5.11(C)	
	pKa	
	pp cited in Vols.1-5:	

(1) Winthrop et al., J. Org. Chem., 1962, 27, 230.
(2) F. Villani, U.S. Pat. 3 409 640 (1968).
(3) J. Med. (Basel), 1971, 2, 249-343.

BUTROPIUM BROMIDE [INN]		
CAS 29025-14-7	MW: 532.53	MF: C28 H38 Br N O4
	Anticholinergic	
	LgP (4)	
	pKa	
	pp cited in Vols.1-5:	

(1) Tanaka et al., U.S. Pat. 3 696 110 (1972).
(2) Akutsu et al., Japan. Med. Gaz., 1974, 11(9), 10.

BUTYLATED HYDROXYANISOLE [U]		
CAS 25013-16-5	MW: 180.25	MF: C11 H16 O2
	Anti-oxidant	
	LgP (4)	
	pKa	
	pp cited in Vols.1-5:	

(1) Rosenwald, U.S. Pat. 2 459 549 (1949).

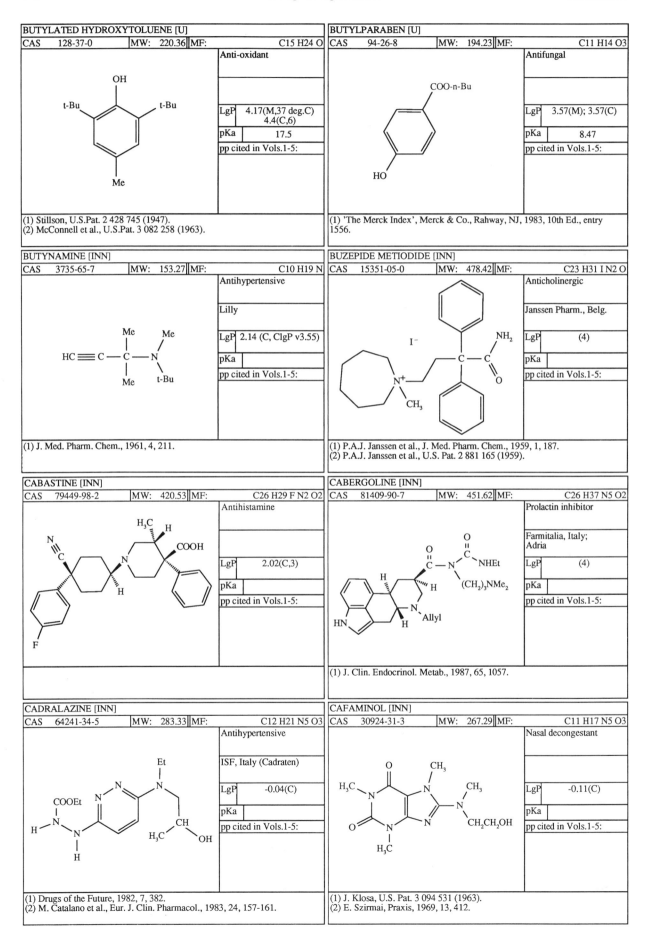

BUTYLATED HYDROXYTOLUENE [U]

CAS	128-37-0	MW:	220.36	MF:	C15 H24 O

Anti-oxidant

LgP	4.17(M,37 deg.C) 4.4(C,6)
pKa	17.5
pp cited in Vols.1-5:	

(1) Stillson, U.S.Pat. 2 428 745 (1947).
(2) McConnell et al., U.S.Pat. 3 082 258 (1963).

BUTYLPARABEN [U]

CAS	94-26-8	MW:	194.23	MF:	C11 H14 O3

Antifungal

LgP	3.57(M); 3.57(C)
pKa	8.47
pp cited in Vols.1-5:	

(1) 'The Merck Index', Merck & Co., Rahway, NJ, 1983, 10th Ed., entry 1556.

BUTYNAMINE [INN]

CAS	3735-65-7	MW:	153.27	MF:	C10 H19 N

Antihypertensive

Lilly

LgP	2.14 (C, ClgP v3.55)
pKa	
pp cited in Vols.1-5:	

(1) J. Med. Pharm. Chem., 1961, 4, 211.

BUZEPIDE METIODIDE [INN]

CAS	15351-05-0	MW:	478.42	MF:	C23 H31 I N2 O

Anticholinergic

Janssen Pharm., Belg.

LgP	(4)
pKa	
pp cited in Vols.1-5:	

(1) P.A.J. Janssen et al., J. Med. Pharm. Chem., 1959, 1, 187.
(2) P.A.J. Janssen et al., U.S. Pat. 2 881 165 (1959).

CABASTINE [INN]

CAS	79449-98-2	MW:	420.53	MF:	C26 H29 F N2 O2

Antihistamine

LgP	2.02(C,3)
pKa	
pp cited in Vols.1-5:	

CABERGOLINE [INN]

CAS	81409-90-7	MW:	451.62	MF:	C26 H37 N5 O2

Prolactin inhibitor

Farmitalia, Italy; Adria

LgP	(4)
pKa	
pp cited in Vols.1-5:	

(1) J. Clin. Endocrinol. Metab., 1987, 65, 1057.

CADRALAZINE [INN]

CAS	64241-34-5	MW:	283.33	MF:	C12 H21 N5 O3

Antihypertensive

ISF, Italy (Cadraten)

LgP	-0.04(C)
pKa	
pp cited in Vols.1-5:	

(1) Drugs of the Future, 1982, 7, 382.
(2) M. Catalano et al., Eur. J. Clin. Pharmacol., 1983, 24, 157-161.

CAFAMINOL [INN]

CAS	30924-31-3	MW:	267.29	MF:	C11 H17 N5 O3

Nasal decongestant

LgP	-0.11(C)
pKa	
pp cited in Vols.1-5:	

(1) J. Klosa, U.S. Pat. 3 094 531 (1963).
(2) E. Szirmai, Praxis, 1969, 13, 412.

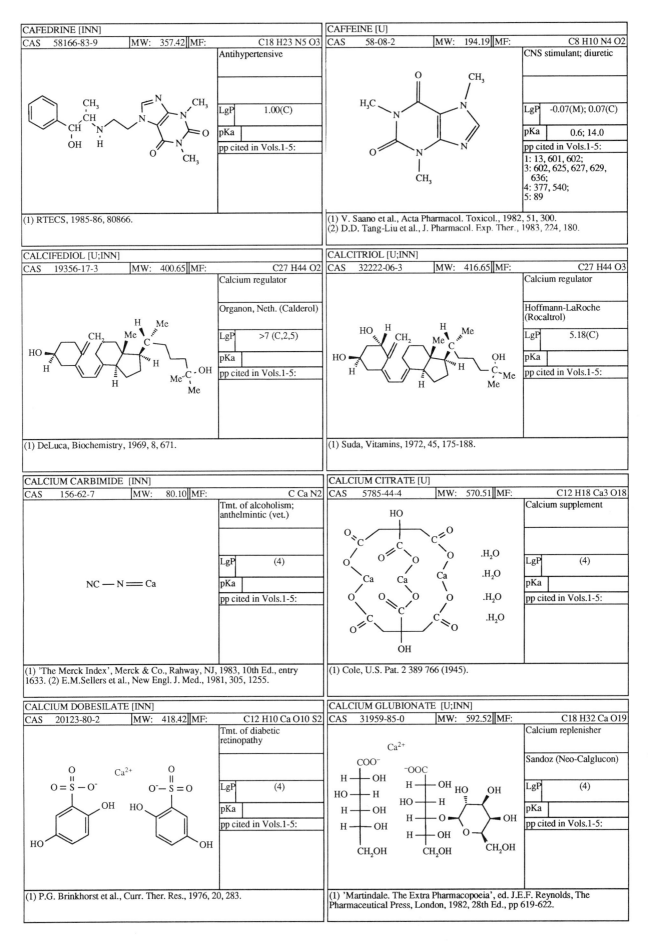

CAFEDRINE [INN]		
CAS 58166-83-9	MW: 357.42	MF: C18 H23 N5 O3
	Antihypertensive	
	LgP	1.00(C)
	pKa	
	pp cited in Vols.1-5:	

(1) RTECS, 1985-86, 80866.

CAFFEINE [U]		
CAS 58-08-2	MW: 194.19	MF: C8 H10 N4 O2
	CNS stimulant; diuretic	
	LgP	-0.07(M); 0.07(C)
	pKa	0.6; 14.0
	pp cited in Vols.1-5:	
	1: 13, 601, 602; 3: 602, 625, 627, 629, 636; 4: 377, 540; 5: 89	

(1) V. Saano et al., Acta Pharmacol. Toxicol., 1982, 51, 300.
(2) D.D. Tang-Liu et al., J. Pharmacol. Exp. Ther., 1983, 224, 180.

CALCIFEDIOL [U;INN]		
CAS 19356-17-3	MW: 400.65	MF: C27 H44 O2
	Calcium regulator	
	Organon, Neth. (Calderol)	
	LgP	>7 (C,2,5)
	pKa	
	pp cited in Vols.1-5:	

(1) DeLuca, Biochemistry, 1969, 8, 671.

CALCITRIOL [U;INN]		
CAS 32222-06-3	MW: 416.65	MF: C27 H44 O3
	Calcium regulator	
	Hoffmann-LaRoche (Rocaltrol)	
	LgP	5.18(C)
	pKa	
	pp cited in Vols.1-5:	

(1) Suda, Vitamins, 1972, 45, 175-188.

CALCIUM CARBIMIDE [INN]		
CAS 156-62-7	MW: 80.10	MF: C Ca N2
	Tmt. of alcoholism; anthelmintic (vet.)	
	LgP	(4)
	pKa	
	pp cited in Vols.1-5:	

(1) 'The Merck Index', Merck & Co., Rahway, NJ, 1983, 10th Ed., entry 1633. (2) E.M.Sellers et al., New Engl. J. Med., 1981, 305, 1255.

CALCIUM CITRATE [U]		
CAS 5785-44-4	MW: 570.51	MF: C12 H18 Ca3 O18
	Calcium supplement	
	LgP	(4)
	pKa	
	pp cited in Vols.1-5:	

(1) Cole, U.S. Pat. 2 389 766 (1945).

CALCIUM DOBESILATE [INN]		
CAS 20123-80-2	MW: 418.42	MF: C12 H10 Ca O10 S2
	Tmt. of diabetic retinopathy	
	LgP	(4)
	pKa	
	pp cited in Vols.1-5:	

(1) P.G. Brinkhorst et al., Curr. Ther. Res., 1976, 20, 283.

CALCIUM GLUBIONATE [U;INN]		
CAS 31959-85-0	MW: 592.52	MF: C18 H32 Ca O19
	Calcium replenisher	
	Sandoz (Neo-Calglucon)	
	LgP	(4)
	pKa	
	pp cited in Vols.1-5:	

(1) 'Martindale. The Extra Pharmacopoeia', ed. J.E.F. Reynolds, The Pharmaceutical Press, London, 1982, 28th Ed., pp 619-622.

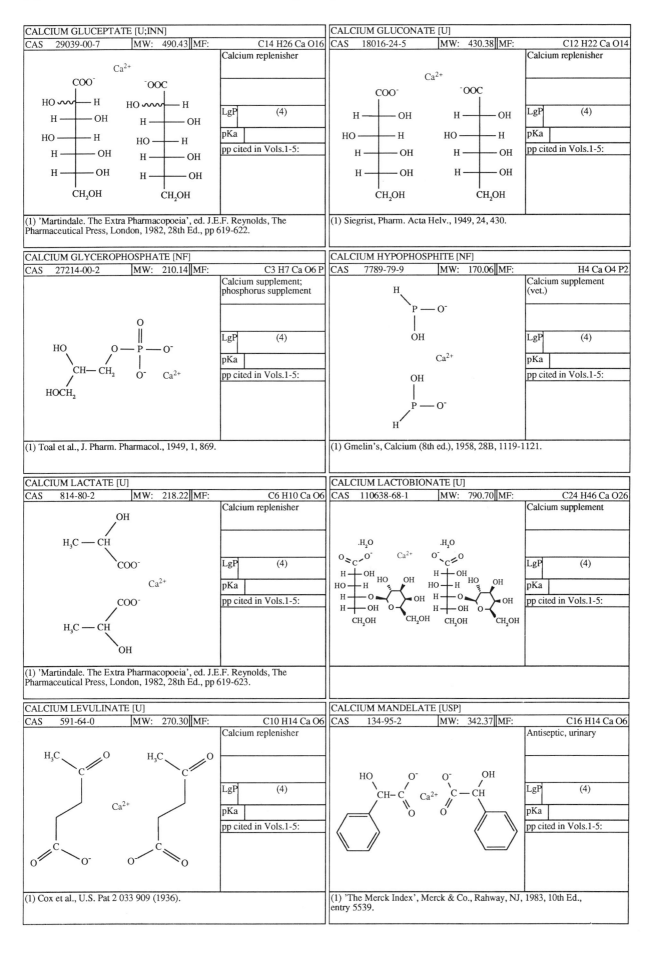

CALCIUM GLUCEPTATE [U;INN]

CAS	29039-00-7	MW:	490.43	MF:	C14 H26 Ca O16

Calcium replenisher

LgP (4)

pKa

pp cited in Vols.1-5:

Ca^{2+}

(1) 'Martindale. The Extra Pharmacopoeia', ed. J.E.F. Reynolds, The Pharmaceutical Press, London, 1982, 28th Ed., pp 619-622.

CALCIUM GLUCONATE [U]

CAS	18016-24-5	MW:	430.38	MF:	C12 H22 Ca O14

Calcium replenisher

LgP (4)

pKa

pp cited in Vols.1-5:

Ca^{2+}

(1) Siegrist, Pharm. Acta Helv., 1949, 24, 430.

CALCIUM GLYCEROPHOSPHATE [NF]

CAS	27214-00-2	MW:	210.14	MF:	C3 H7 Ca O6 P

Calcium supplement; phosphorus supplement

LgP (4)

pKa

pp cited in Vols.1-5:

Ca^{2+}

(1) Toal et al., J. Pharm. Pharmacol., 1949, 1, 869.

CALCIUM HYPOPHOSPHITE [NF]

CAS	7789-79-9	MW:	170.06	MF:	H4 Ca O4 P2

Calcium supplement (vet.)

LgP (4)

pKa

pp cited in Vols.1-5:

Ca^{2+}

(1) Gmelin's, Calcium (8th ed.), 1958, 28B, 1119-1121.

CALCIUM LACTATE [U]

CAS	814-80-2	MW:	218.22	MF:	C6 H10 Ca O6

Calcium replenisher

LgP (4)

pKa

pp cited in Vols.1-5:

Ca^{2+}

(1) 'Martindale. The Extra Pharmacopoeia', ed. J.E.F. Reynolds, The Pharmaceutical Press, London, 1982, 28th Ed., pp 619-623.

CALCIUM LACTOBIONATE [U]

CAS	110638-68-1	MW:	790.70	MF:	C24 H46 Ca O26

Calcium supplement

LgP (4)

pKa

pp cited in Vols.1-5:

Ca^{2+}

CALCIUM LEVULINATE [U]

CAS	591-64-0	MW:	270.30	MF:	C10 H14 Ca O6

Calcium replenisher

LgP (4)

pKa

pp cited in Vols.1-5:

Ca^{2+}

(1) Cox et al., U.S. Pat 2 033 909 (1936).

CALCIUM MANDELATE [USP]

CAS	134-95-2	MW:	342.37	MF:	C16 H14 Ca O6

Antiseptic, urinary

LgP (4)

pKa

pp cited in Vols.1-5:

Ca^{2+}

(1) 'The Merck Index', Merck & Co., Rahway, NJ, 1983, 10th Ed., entry 5539.

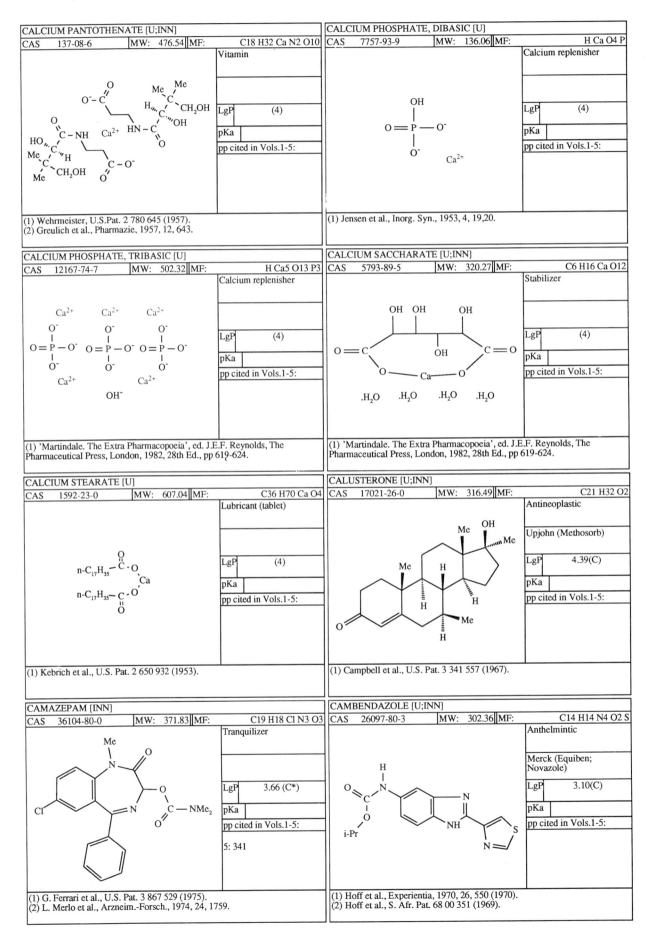

CALCIUM PANTOTHENATE [U;INN]

CAS 137-08-6 | MW: 476.54 | MF: C18 H32 Ca N2 O10

Vitamin

LgP (4)

pKa

pp cited in Vols.1-5:

(1) Wehrmeister, U.S.Pat. 2 780 645 (1957).
(2) Greulich et al., Pharmazie, 1957, 12, 643.

CALCIUM PHOSPHATE, DIBASIC [U]

CAS 7757-93-9 | MW: 136.06 | MF: H Ca O4 P

Calcium replenisher

LgP (4)

pKa

pp cited in Vols.1-5:

(1) Jensen et al., Inorg. Syn., 1953, 4, 19,20.

CALCIUM PHOSPHATE, TRIBASIC [U]

CAS 12167-74-7 | MW: 502.32 | MF: H Ca5 O13 P3

Calcium replenisher

LgP (4)

pKa

pp cited in Vols.1-5:

(1) 'Martindale. The Extra Pharmacopoeia', ed. J.E.F. Reynolds, The
Pharmaceutical Press, London, 1982, 28th Ed., pp 619-624.

CALCIUM SACCHARATE [U;INN]

CAS 5793-89-5 | MW: 320.27 | MF: C6 H16 Ca O12

Stabilizer

LgP (4)

pKa

pp cited in Vols.1-5:

(1) 'Martindale. The Extra Pharmacopoeia', ed. J.E.F. Reynolds, The
Pharmaceutical Press, London, 1982, 28th Ed., pp 619-624.

CALCIUM STEARATE [U]

CAS 1592-23-0 | MW: 607.04 | MF: C36 H70 Ca O4

Lubricant (tablet)

LgP (4)

pKa

pp cited in Vols.1-5:

(1) Kebrich et al., U.S. Pat. 2 650 932 (1953).

CALUSTERONE [U;INN]

CAS 17021-26-0 | MW: 316.49 | MF: C21 H32 O2

Antineoplastic

Upjohn (Methosorb)

LgP 4.39(C)

pKa

pp cited in Vols.1-5:

(1) Campbell et al., U.S. Pat. 3 341 557 (1967).

CAMAZEPAM [INN]

CAS 36104-80-0 | MW: 371.83 | MF: C19 H18 Cl N3 O3

Tranquilizer

LgP 3.66 (C*)

pKa

pp cited in Vols.1-5:

5: 341

(1) G. Ferrari et al., U.S.Pat. 3 867 529 (1975).
(2) L. Merlo et al., Arzneim.-Forsch., 1974, 24, 1759.

CAMBENDAZOLE [U;INN]

CAS 26097-80-3 | MW: 302.36 | MF: C14 H14 N4 O2 S

Anthelmintic

Merck (Equiben; Novazole)

LgP 3.10(C)

pKa

pp cited in Vols.1-5:

(1) Hoff et al., Experientia, 1970, 26, 550 (1970).
(2) Hoff et al., S. Afr. Pat. 68 00 351 (1969).

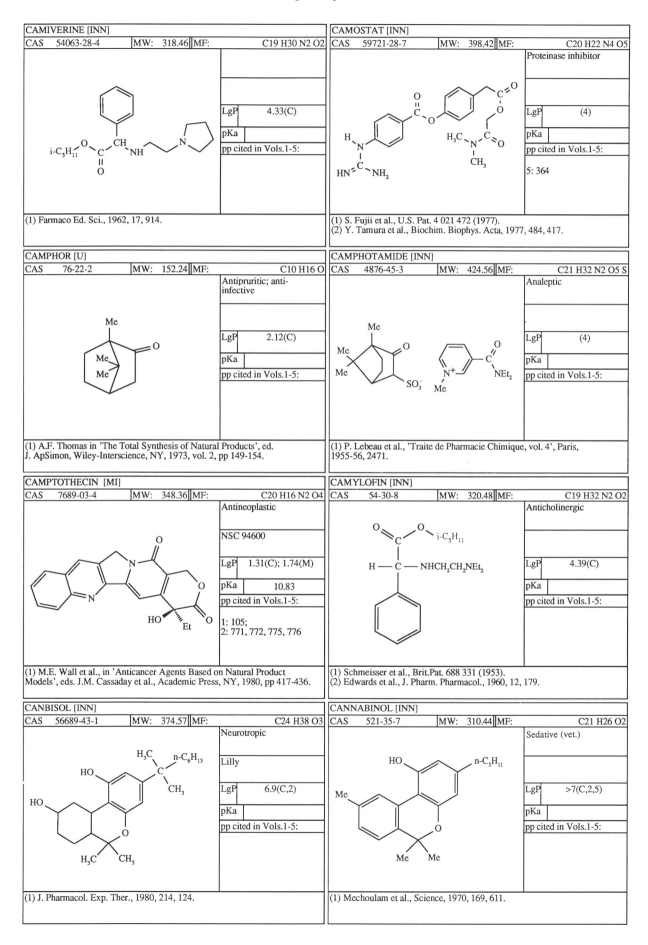

CAMIVERINE [INN]

CAS	54063-28-4	MW:	318.46	MF:	C19 H30 N2 O2

LgP	4.33(C)
pKa	
pp cited in Vols.1-5:	

(1) Farmaco Ed. Sci., 1962, 17, 914.

CAMOSTAT [INN]

CAS	59721-28-7	MW:	398.42	MF:	C20 H22 N4 O5

Proteinase inhibitor

LgP	(4)
pKa	
pp cited in Vols.1-5:	

5: 364

(1) S. Fujii et al., U.S. Pat. 4 021 472 (1977).
(2) Y. Tamura et al., Biochim. Biophys. Acta, 1977, 484, 417.

CAMPHOR [U]

CAS	76-22-2	MW:	152.24	MF:	C10 H16 O

Antipruritic; anti-infective

LgP	2.12(C)
pKa	
pp cited in Vols.1-5:	

(1) A.F. Thomas in 'The Total Synthesis of Natural Products', ed.
J. ApSimon, Wiley-Interscience, NY, 1973, vol. 2, pp 149-154.

CAMPHOTAMIDE [INN]

CAS	4876-45-3	MW:	424.56	MF:	C21 H32 N2 O5 S

Analeptic

LgP	(4)
pKa	
pp cited in Vols.1-5:	

(1) P. Lebeau et al., 'Traite de Pharmacie Chimique, vol. 4', Paris,
1955-56, 2471.

CAMPTOTHECIN [MI]

CAS	7689-03-4	MW:	348.36	MF:	C20 H16 N2 O4

Antineoplastic

NSC 94600

LgP	1.31(C); 1.74(M)
pKa	10.83
pp cited in Vols.1-5:	

1: 105;
2: 771, 772, 775, 776

(1) M.E. Wall et al., in 'Anticancer Agents Based on Natural Product
Models', eds. J.M. Cassaday et al., Academic Press, NY, 1980, pp 417-436.

CAMYLOFIN [INN]

CAS	54-30-8	MW:	320.48	MF:	C19 H32 N2 O2

Anticholinergic

LgP	4.39(C)
pKa	
pp cited in Vols.1-5:	

(1) Schmeisser et al., Brit.Pat. 688 331 (1953).
(2) Edwards et al., J. Pharm. Pharmacol., 1960, 12, 179.

CANBISOL [INN]

CAS	56689-43-1	MW:	374.57	MF:	C24 H38 O3

Neurotropic

Lilly

LgP	6.9(C,2)
pKa	
pp cited in Vols.1-5:	

(1) J. Pharmacol. Exp. Ther., 1980, 214, 124.

CANNABINOL [INN]

CAS	521-35-7	MW:	310.44	MF:	C21 H26 O2

Sedative (vet.)

LgP	>7(C,2,5)
pKa	
pp cited in Vols.1-5:	

(1) Mechoulam et al., Science, 1970, 169, 611.

CANRENOIC ACID [U]	
CAS 4138-96-9	MW: 358.48	MF: C22 H30 O4

Diuretic (aldosterone antagonist)

LgP	3.3(C,6)
pKa	
pp cited in Vols.1-5:	

(1) 'Martindale. The Extra Pharmacopoeia', ed. J.E.F. Reynolds, The Pharmaceutical Press, London, 1982, 28th Ed., p 587.

CANRENONE [U;INN]	
CAS 976-71-6	MW: 340.47	MF: C22 H28 O3

Aldosterone antagonist

Searle

LgP	2.68(M); 2.9(C,6)
pKa	
pp cited in Vols.1-5:	

5: 177

(1) Cella et al., J. Org. Chem., 1959, 24, 1109.
(2) Cella et al., U.S. Pat. 2 900 383 (1959).

CANTHARIDINE [MI]	
CAS 56-25-7	MW: 196.20	MF: C10 H12 O4

Vesicant

LgP	0.0 (C,3)
pKa	
pp cited in Vols.1-5:	

(1) R.B. Woodward et al., J. Am. Chem. Soc., 1941, 63, 3167.
(2) W.G. Dauben et al., ibid., 1980, 102, 6893.

CAPOBENIC ACID [U;INN]	
CAS 21434-91-3	MW: 325.36	MF: C16 H23 N O6

Cardiac depressant (anti-arrhythmic)

Instituto Chemioter. Italiano

LgP	0.73(C)
pKa	
pp cited in Vols.1-5:	

(1) A. Garzia, U.S. Pat. 3 697 563 (1972). (2) Bottazzi et al., Riv. Farmacol. Ter., 1971, 11, 215. (3) Razzaboni et al., Boll. Soc. Ital. Biol. Sper., 1968, 44, 1783 (1968).

CAPREOMYCIN 1B [U;INN]	
CAS 33490-33-4	MW: 652.72	MF: C25 H44 N14 O7

Antibacterial (tuberculostatic)

Lilly (Caprocin)

LgP	(4)
pKa	
pp cited in Vols.1-5:	

(1) Herr et al., U.S. Pat. 3 143 468 (1964).
(2) Sutton et al., Ann. N.Y. Acad.Sci., 1966, 135, 947.
(3) T.Shiba et al., Tetrahedron Lett., 1976, 3907.

CAPROXAMINE [INN]	
CAS 24047-16-3	MW: 263.39	MF: C15 H25 N3 O

Antidepressant

Phillips Duphar, Neth.

LgP	3.82(C,3)
pKa	
pp cited in Vols.1-5:	

CAPSAICINE [MI]	
CAS 404-86-4	MW: 305.42	MF: C18 H27 N O3

Post-hepatic neuralgia; biological research tool

Genderm Labs.

LgP	3.41 (C)
pKa	
pp cited in Vols.1-5:	

3: 1018

(1) D.J. Bennet et al., J. Chem. Soc. C, 1968, 442.
(2) Molnar et al., Acta Physiol., 1969, 35, 369.

CAPTAMINE [U;INN]	
CAS 108-02-1	MW: 105.20	MF: C4 H11 N S

Depigmentor

$HSCH_2CH_2-NMe_2$

LgP	0.47(C)
pKa	
pp cited in Vols.1-5:	

(1) H.M. Swartz et al., in 'Radiation Protection Sensitization', Proc. Int.Sympos., 2nd., 1969, ed. H.Moroson, Taylor and Francis, Ltd., London, 1970.

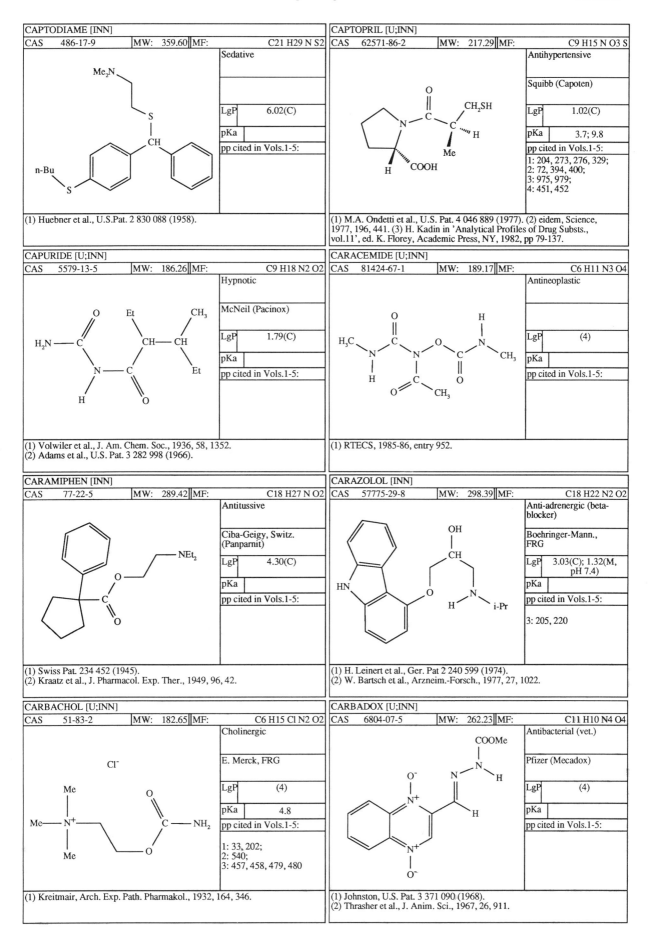

CAPTODIAME [INN]			
CAS 486-17-9	MW: 359.60	MF:	C21 H29 N S2

Sedative

LgP	6.02(C)
pKa	

pp cited in Vols.1-5:

(1) Huebner et al., U.S.Pat. 2 830 088 (1958).

CAPTOPRIL [U;INN]			
CAS 62571-86-2	MW: 217.29	MF:	C9 H15 N O3 S

Antihypertensive

Squibb (Capoten)

LgP	1.02(C)
pKa	3.7; 9.8

pp cited in Vols.1-5:
1: 204, 273, 276, 329;
2: 72, 394, 400;
3: 975, 979;
4: 451, 452

(1) M.A. Ondetti et al., U.S. Pat. 4 046 889 (1977). (2) eidem, Science, 1977, 196, 441. (3) H. Kadin in 'Analytical Profiles of Drug Substs., vol.11', ed. K. Florey, Academic Press, NY, 1982, pp 79-137.

CAPURIDE [U;INN]			
CAS 5579-13-5	MW: 186.26	MF:	C9 H18 N2 O2

Hypnotic

McNeil (Pacinox)

LgP	1.79(C)
pKa	

pp cited in Vols.1-5:

(1) Volwiler et al., J. Am. Chem. Soc., 1936, 58, 1352.
(2) Adams et al., U.S. Pat. 3 282 998 (1966).

CARACEMIDE [U;INN]			
CAS 81424-67-1	MW: 189.17	MF:	C6 H11 N3 O4

Antineoplastic

LgP	(4)
pKa	

pp cited in Vols.1-5:

(1) RTECS, 1985-86, entry 952.

CARAMIPHEN [INN]			
CAS 77-22-5	MW: 289.42	MF:	C18 H27 N O2

Antitussive

Ciba-Geigy, Switz. (Panparnit)

LgP	4.30(C)
pKa	

pp cited in Vols.1-5:

(1) Swiss Pat. 234 452 (1945).
(2) Kraatz et al., J. Pharmacol. Exp. Ther., 1949, 96, 42.

CARAZOLOL [INN]			
CAS 57775-29-8	MW: 298.39	MF:	C18 H22 N2 O2

Anti-adrenergic (beta-blocker)

Boehringer-Mann., FRG

LgP	3.03(C); 1.32(M, pH 7.4)
pKa	

pp cited in Vols.1-5:

3: 205, 220

(1) H. Leinert et al., Ger. Pat 2 240 599 (1974).
(2) W. Bartsch et al., Arzneim.-Forsch., 1977, 27, 1022.

CARBACHOL [U;INN]			
CAS 51-83-2	MW: 182.65	MF:	C6 H15 Cl N2 O2

Cholinergic

E. Merck, FRG

LgP	(4)
pKa	4.8

pp cited in Vols.1-5:

1: 33, 202;
2: 540;
3: 457, 458, 479, 480

(1) Kreitmair, Arch. Exp. Path. Pharmakol., 1932, 164, 346.

CARBADOX [U;INN]			
CAS 6804-07-5	MW: 262.23	MF:	C11 H10 N4 O4

Antibacterial (vet.)

Pfizer (Mecadox)

LgP	(4)
pKa	

pp cited in Vols.1-5:

(1) Johnston, U.S. Pat. 3 371 090 (1968).
(2) Thrasher et al., J. Anim. Sci., 1967, 26, 911.

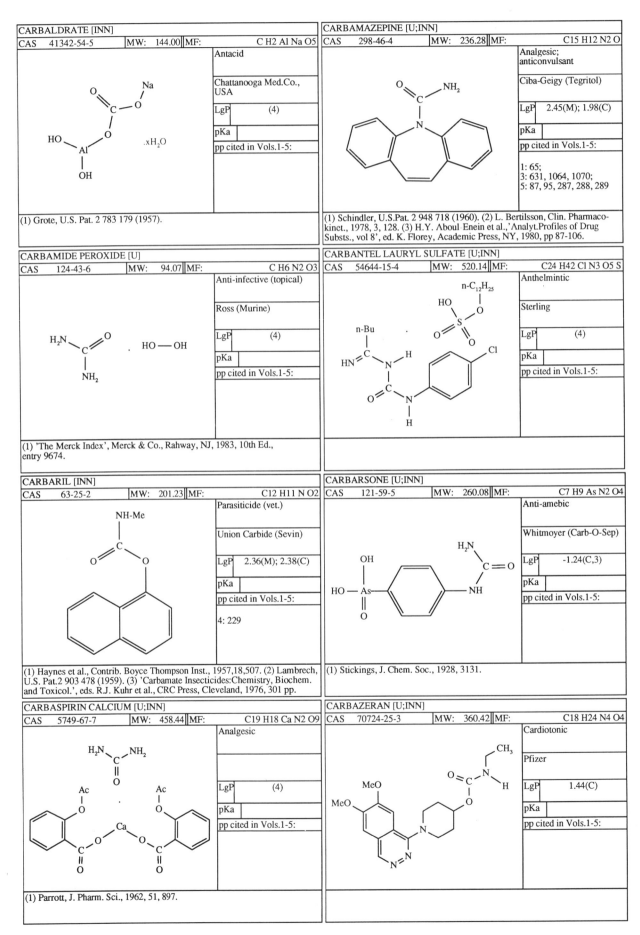

CARBALDRATE [INN]

CAS	41342-54-5	MW:	144.00	MF:	C H2 Al Na O5

Antacid

Chattanooga Med.Co., USA

LgP	(4)

pKa	

pp cited in Vols.1-5:

(1) Grote, U.S. Pat. 2 783 179 (1957).

CARBAMIDE PEROXIDE [U]

CAS	124-43-6	MW:	94.07	MF:	C H6 N2 O3

Anti-infective (topical)

Ross (Murine)

LgP	(4)

pKa	

pp cited in Vols.1-5:

(1) 'The Merck Index', Merck & Co., Rahway, NJ, 1983, 10th Ed., entry 9674.

CARBARIL [INN]

CAS	63-25-2	MW:	201.23	MF:	C12 H11 N O2

Parasiticide (vet.)

Union Carbide (Sevin)

LgP	2.36(M); 2.38(C)

pKa	

pp cited in Vols.1-5:

4: 229

(1) Haynes et al., Contrib. Boyce Thompson Inst., 1957,18,507. (2) Lambrech, U.S. Pat.2 903 478 (1959). (3) 'Carbamate Insecticides:Chemistry, Biochem. and Toxicol.', eds. R.J. Kuhr et al., CRC Press, Cleveland, 1976, 301 pp.

CARBASPIRIN CALCIUM [U;INN]

CAS	5749-67-7	MW:	458.44	MF:	C19 H18 Ca N2 O9

Analgesic

LgP	(4)

pKa	

pp cited in Vols.1-5:

(1) Parrott, J. Pharm. Sci., 1962, 51, 897.

CARBAMAZEPINE [U;INN]

CAS	298-46-4	MW:	236.28	MF:	C15 H12 N2 O

Analgesic; anticonvulsant

Ciba-Geigy (Tegritol)

LgP	2.45(M); 1.98(C)

pKa	

pp cited in Vols.1-5:

1: 65;
3: 631, 1064, 1070;
5: 87, 95, 287, 288, 289

(1) Schindler, U.S.Pat. 2 948 718 (1960). (2) L. Bertilsson, Clin. Pharmaco-kinet., 1978, 3, 128. (3) H.Y. Aboul-Enein et al.,'Analyt.Profiles of Drug Substs., vol 8', ed. K. Florey, Academic Press, NY, 1980, pp 87-106.

CARBANTEL LAURYL SULFATE [U;INN]

CAS	54644-15-4	MW:	520.14	MF:	C24 H42 Cl N3 O5 S

Anthelmintic

Sterling

LgP	(4)

pKa	

pp cited in Vols.1-5:

CARBARSONE [U;INN]

CAS	121-59-5	MW:	260.08	MF:	C7 H9 As N2 O4

Anti-amebic

Whitmoyer (Carb-O-Sep)

LgP	-1.24(C,3)

pKa	

pp cited in Vols.1-5:

(1) Stickings, J. Chem. Soc., 1928, 3131.

CARBAZERAN [U;INN]

CAS	70724-25-3	MW:	360.42	MF:	C18 H24 N4 O4

Cardiotonic

Pfizer

LgP	1.44(C)

pKa	

pp cited in Vols.1-5:

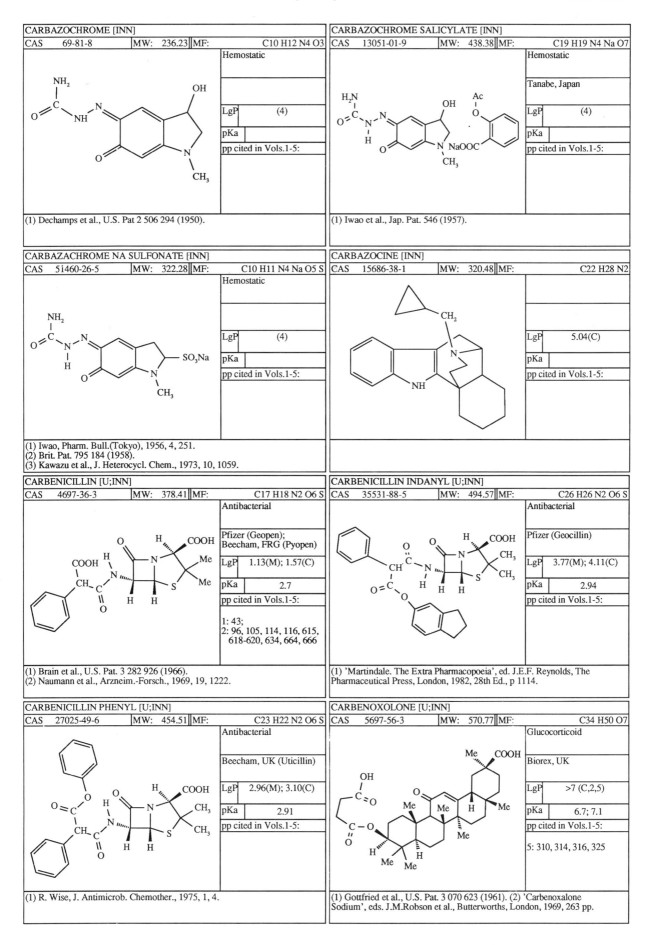

CARBAZOCHROME [INN]			
CAS 69-81-8	MW: 236.23	MF:	C10 H12 N4 O3

Hemostatic

LgP	(4)
pKa	
pp cited in Vols.1-5:	

(1) Dechamps et al., U.S. Pat 2 506 294 (1950).

CARBAZOCHROME SALICYLATE [INN]			
CAS 13051-01-9	MW: 438.38	MF:	C19 H19 N4 Na O7

Hemostatic

Tanabe, Japan

LgP	(4)
pKa	
pp cited in Vols.1-5:	

(1) Iwao et al., Jap. Pat. 546 (1957).

CARBAZACHROME NA SULFONATE [INN]			
CAS 51460-26-5	MW: 322.28	MF:	C10 H11 N4 Na O5 S

Hemostatic

LgP	(4)
pKa	
pp cited in Vols.1-5:	

(1) Iwao, Pharm. Bull.(Tokyo), 1956, 4, 251.
(2) Brit. Pat. 795 184 (1958).
(3) Kawazu et al., J. Heterocycl. Chem., 1973, 10, 1059.

CARBAZOCINE [INN]			
CAS 15686-38-1	MW: 320.48	MF:	C22 H28 N2

LgP	5.04(C)
pKa	
pp cited in Vols.1-5:	

CARBENICILLIN [U;INN]			
CAS 4697-36-3	MW: 378.41	MF:	C17 H18 N2 O6 S

Antibacterial

Pfizer (Geopen);
Beecham, FRG (Pyopen)

LgP	1.13(M); 1.57(C)
pKa	2.7
pp cited in Vols.1-5:	

1: 43;
2: 96, 105, 114, 116, 615,
618-620, 634, 664, 666

(1) Brain et al., U.S. Pat. 3 282 926 (1966).
(2) Naumann et al., Arzneim.-Forsch., 1969, 19, 1222.

CARBENICILLIN INDANYL [U;INN]			
CAS 35531-88-5	MW: 494.57	MF:	C26 H26 N2 O6 S

Antibacterial

Pfizer (Geocillin)

LgP	3.77(M); 4.11(C)
pKa	2.94
pp cited in Vols.1-5:	

(1) 'Martindale. The Extra Pharmacopoeia', ed. J.E.F. Reynolds, The Pharmaceutical Press, London, 1982, 28th Ed., p 1114.

CARBENICILLIN PHENYL [U;INN]			
CAS 27025-49-6	MW: 454.51	MF:	C23 H22 N2 O6 S

Antibacterial

Beecham, UK (Uticillin)

LgP	2.96(M); 3.10(C)
pKa	2.91
pp cited in Vols.1-5:	

(1) R. Wise, J. Antimicrob. Chemother., 1975, 1, 4.

CARBENOXOLONE [U;INN]			
CAS 5697-56-3	MW: 570.77	MF:	C34 H50 O7

Glucocorticoid

Biorex, UK

LgP	>7 (C,2,5)
pKa	6.7; 7.1
pp cited in Vols.1-5:	

5: 310, 314, 316, 325

(1) Gottfried et al., U.S. Pat. 3 070 623 (1961). (2) 'Carbenoxalone Sodium', eds. J.M.Robson et al., Butterworths, London, 1969, 263 pp.

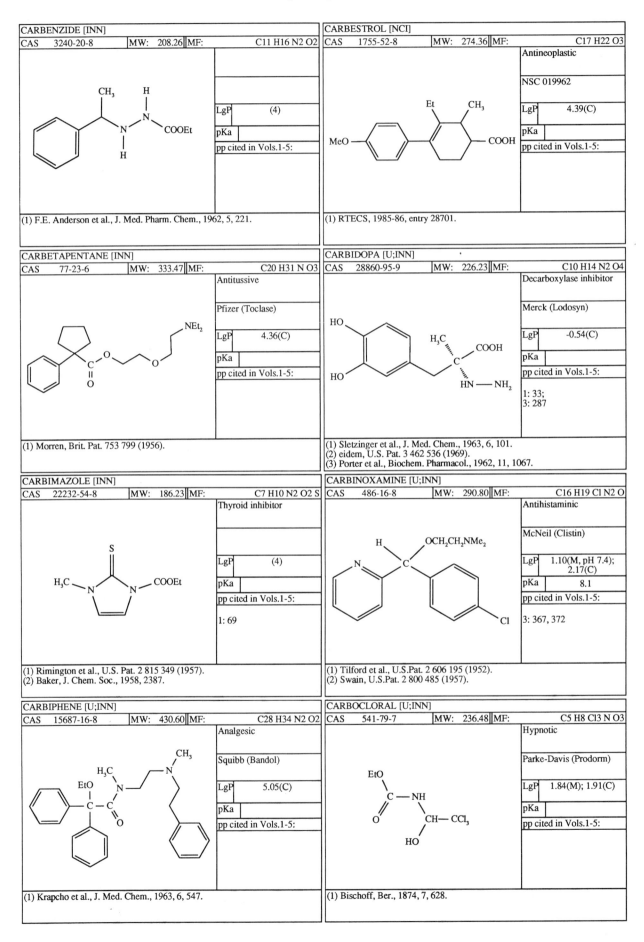

CARBENZIDE [INN]

CAS	3240-20-8	MW: 208.26	MF:	C11 H16 N2 O2

LgP	(4)
pKa	
pp cited in Vols.1-5:	

(1) F.E. Anderson et al., J. Med. Pharm. Chem., 1962, 5, 221.

CARBESTROL [NCI]

CAS	1755-52-8	MW: 274.36	MF:	C17 H22 O3

Antineoplastic

NSC 019962

LgP	4.39(C)
pKa	
pp cited in Vols.1-5:	

(1) RTECS, 1985-86, entry 28701.

CARBETAPENTANE [INN]

CAS	77-23-6	MW: 333.47	MF:	C20 H31 N O3

Antitussive

Pfizer (Toclase)

LgP	4.36(C)
pKa	
pp cited in Vols.1-5:	

(1) Morren, Brit. Pat. 753 799 (1956).

CARBIDOPA [U;INN]

CAS	28860-95-9	MW: 226.23	MF:	C10 H14 N2 O4

Decarboxylase inhibitor

Merck (Lodosyn)

LgP	-0.54(C)
pKa	
pp cited in Vols.1-5:	
1: 33;	
3: 287	

(1) Sletzinger et al., J. Med. Chem., 1963, 6, 101.
(2) eidem, U.S. Pat. 3 462 536 (1969).
(3) Porter et al., Biochem. Pharmacol., 1962, 11, 1067.

CARBIMAZOLE [INN]

CAS	22232-54-8	MW: 186.23	MF:	C7 H10 N2 O2 S

Thyroid inhibitor

LgP	(4)
pKa	
pp cited in Vols.1-5:	
1: 69	

(1) Rimington et al., U.S. Pat. 2 815 349 (1957).
(2) Baker, J. Chem. Soc., 1958, 2387.

CARBINOXAMINE [U;INN]

CAS	486-16-8	MW: 290.80	MF:	C16 H19 Cl N2 O

Antihistaminic

McNeil (Clistin)

LgP	1.10(M, pH 7.4); 2.17(C)
pKa	8.1
pp cited in Vols.1-5:	
3: 367, 372	

(1) Tilford et al., U.S.Pat. 2 606 195 (1952).
(2) Swain, U.S.Pat. 2 800 485 (1957).

CARBIPHENE [U;INN]

CAS	15687-16-8	MW: 430.60	MF:	C28 H34 N2 O2

Analgesic

Squibb (Bandol)

LgP	5.05(C)
pKa	
pp cited in Vols.1-5:	

(1) Krapcho et al., J. Med. Chem., 1963, 6, 547.

CARBOCLORAL [U;INN]

CAS	541-79-7	MW: 236.48	MF:	C5 H8 Cl3 N O3

Hypnotic

Parke-Davis (Prodorm)

LgP	1.84(M); 1.91(C)
pKa	
pp cited in Vols.1-5:	

(1) Bischoff, Ber., 1874, 7, 628.

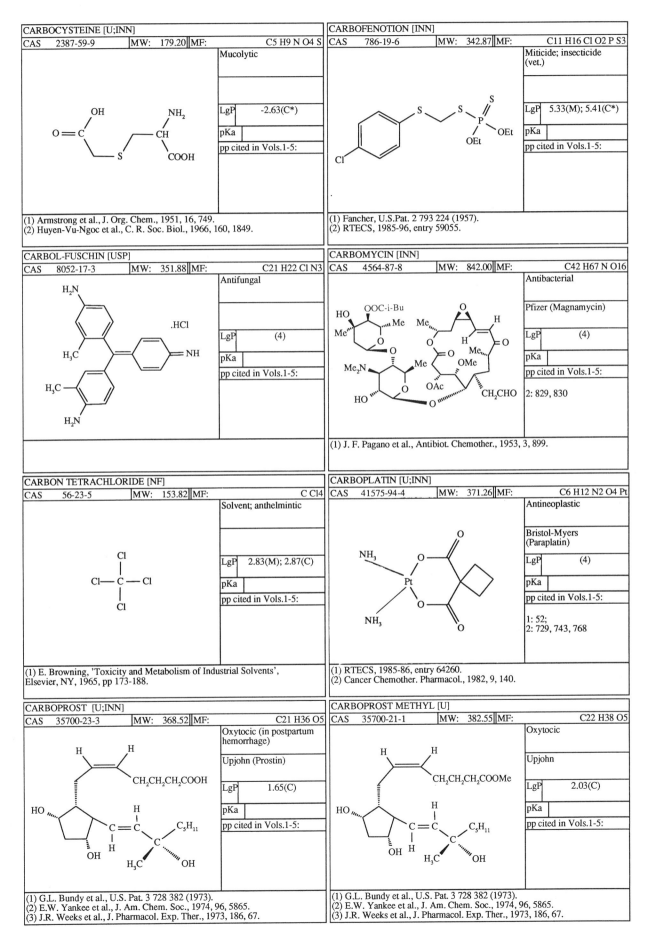

CARBOCYSTEINE [U;INN]

CAS	2387-59-9	MW:	179.20	MF:	C5 H9 N O4 S

Mucolytic

LgP	-2.63(C*)
pKa	
pp cited in Vols.1-5:	

(1) Armstrong et al., J. Org. Chem., 1951, 16, 749.
(2) Huyen-Vu-Ngoc et al., C. R. Soc. Biol., 1966, 160, 1849.

CARBOFENOTION [INN]

CAS	786-19-6	MW:	342.87	MF:	C11 H16 Cl O2 P S3

Miticide; insecticide (vet.)

LgP	5.33(M); 5.41(C*)
pKa	
pp cited in Vols.1-5:	

(1) Fancher, U.S.Pat. 2 793 224 (1957).
(2) RTECS, 1985-96, entry 59055.

CARBOL-FUSCHIN [USP]

CAS	8052-17-3	MW:	351.88	MF:	C21 H22 Cl N3

Antifungal

.HCl

LgP	(4)
pKa	
pp cited in Vols.1-5:	

CARBOMYCIN [INN]

CAS	4564-87-8	MW:	842.00	MF:	C42 H67 N O16

Antibacterial

Pfizer (Magnamycin)

LgP	(4)
pKa	
pp cited in Vols.1-5:	

2: 829, 830

(1) J. F. Pagano et al., Antibiot. Chemother., 1953, 3, 899.

CARBON TETRACHLORIDE [NF]

CAS	56-23-5	MW:	153.82	MF:	C Cl4

Solvent; anthelmintic

LgP	2.83(M); 2.87(C)
pKa	
pp cited in Vols.1-5:	

(1) E. Browning, 'Toxicity and Metabolism of Industrial Solvents', Elsevier, NY, 1965, pp 173-188.

CARBOPLATIN [U;INN]

CAS	41575-94-4	MW:	371.26	MF:	C6 H12 N2 O4 Pt

Antineoplastic

Bristol-Myers (Paraplatin)

LgP	(4)
pKa	
pp cited in Vols.1-5:	

1: 52;
2: 729, 743, 768

(1) RTECS, 1985-86, entry 64260.
(2) Cancer Chemother. Pharmacol., 1982, 9, 140.

CARBOPROST [U;INN]

CAS	35700-23-3	MW:	368.52	MF:	C21 H36 O5

Oxytocic (in postpartum hemorrhage)

Upjohn (Prostin)

LgP	1.65(C)
pKa	
pp cited in Vols.1-5:	

(1) G.L. Bundy et al., U.S. Pat. 3 728 382 (1973).
(2) E.W. Yankee et al., J. Am. Chem. Soc., 1974, 96, 5865.
(3) J.R. Weeks et al., J. Pharmacol. Exp. Ther., 1973, 186, 67.

CARBOPROST METHYL [U]

CAS	35700-21-1	MW:	382.55	MF:	C22 H38 O5

Oxytocic

Upjohn

LgP	2.03(C)
pKa	
pp cited in Vols.1-5:	

(1) G.L. Bundy et al., U.S. Pat. 3 728 382 (1973).
(2) E.W. Yankee et al., J. Am. Chem. Soc., 1974, 96, 5865.
(3) J.R. Weeks et al., J. Pharmacol. Exp. Ther., 1973, 186, 67.

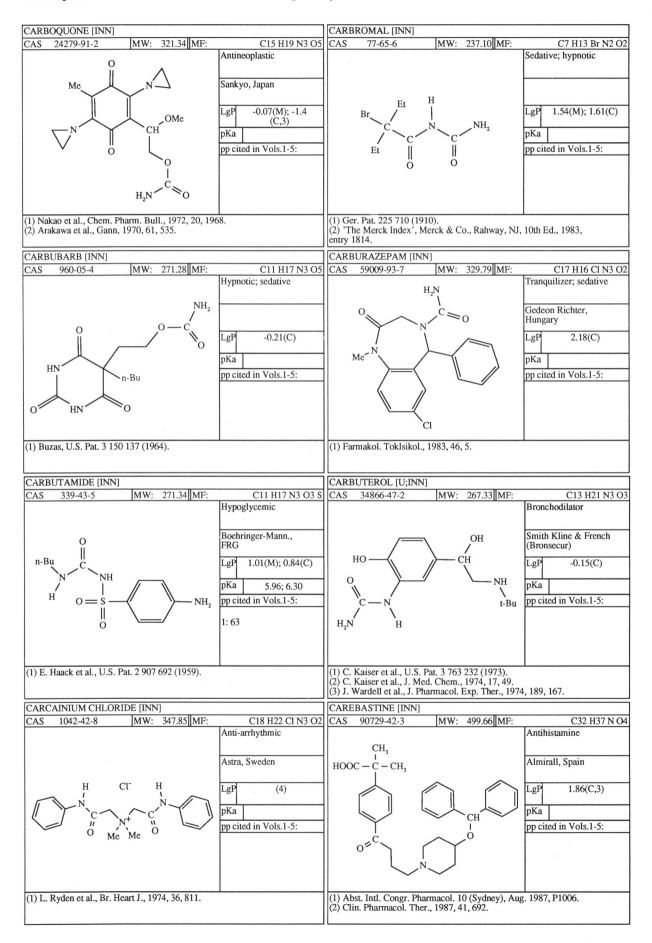

CARBOQUONE [INN]		
CAS 24279-91-2	MW: 321.34	MF: C15 H19 N3 O5

Antineoplastic

Sankyo, Japan

LgP	-0.07(M); -1.4 (C,3)
pKa	
pp cited in Vols.1-5:	

(1) Nakao et al., Chem. Pharm. Bull., 1972, 20, 1968.
(2) Arakawa et al., Gann, 1970, 61, 535.

CARBROMAL [INN]		
CAS 77-65-6	MW: 237.10	MF: C7 H13 Br N2 O2

Sedative; hypnotic

LgP	1.54(M); 1.61(C)
pKa	
pp cited in Vols.1-5:	

(1) Ger. Pat. 225 710 (1910).
(2) 'The Merck Index', Merck & Co., Rahway, NJ, 10th Ed., 1983, entry 1814.

CARBUBARB [INN]		
CAS 960-05-4	MW: 271.28	MF: C11 H17 N3 O5

Hypnotic; sedative

LgP	-0.21(C)
pKa	
pp cited in Vols.1-5:	

(1) Buzas, U.S. Pat. 3 150 137 (1964).

CARBURAZEPAM [INN]		
CAS 59009-93-7	MW: 329.79	MF: C17 H16 Cl N3 O2

Tranquilizer; sedative

Gedeon Richter, Hungary

LgP	2.18(C)
pKa	
pp cited in Vols.1-5:	

(1) Farmakol. Toklsikol., 1983, 46, 5.

CARBUTAMIDE [INN]		
CAS 339-43-5	MW: 271.34	MF: C11 H17 N3 O3 S

Hypoglycemic

Boehringer-Mann., FRG

LgP	1.01(M); 0.84(C)
pKa	5.96; 6.30
pp cited in Vols.1-5:	
1: 63	

(1) E. Haack et al., U.S. Pat. 2 907 692 (1959).

CARBUTEROL [U;INN]		
CAS 34866-47-2	MW: 267.33	MF: C13 H21 N3 O3

Bronchodilator

Smith Kline & French (Bronsecur)

LgP	-0.15(C)
pKa	
pp cited in Vols.1-5:	

(1) C. Kaiser et al., U.S. Pat. 3 763 232 (1973).
(2) C. Kaiser et al., J. Med. Chem., 1974, 17, 49.
(3) J. Wardell et al., J. Pharmacol. Exp. Ther., 1974, 189, 167.

CARCAINIUM CHLORIDE [INN]		
CAS 1042-42-8	MW: 347.85	MF: C18 H22 Cl N3 O2

Anti-arrhythmic

Astra, Sweden

LgP	(4)
pKa	
pp cited in Vols.1-5:	

(1) L. Ryden et al., Br. Heart J., 1974, 36, 811.

CAREBASTINE [INN]		
CAS 90729-42-3	MW: 499.66	MF: C32 H37 N O4

Antihistamine

Almirall, Spain

LgP	1.86(C,3)
pKa	
pp cited in Vols.1-5:	

(1) Abst. Intl. Congr. Pharmacol. 10 (Sydney), Aug. 1987, P1006.
(2) Clin. Pharmacol. Ther., 1987, 41, 692.

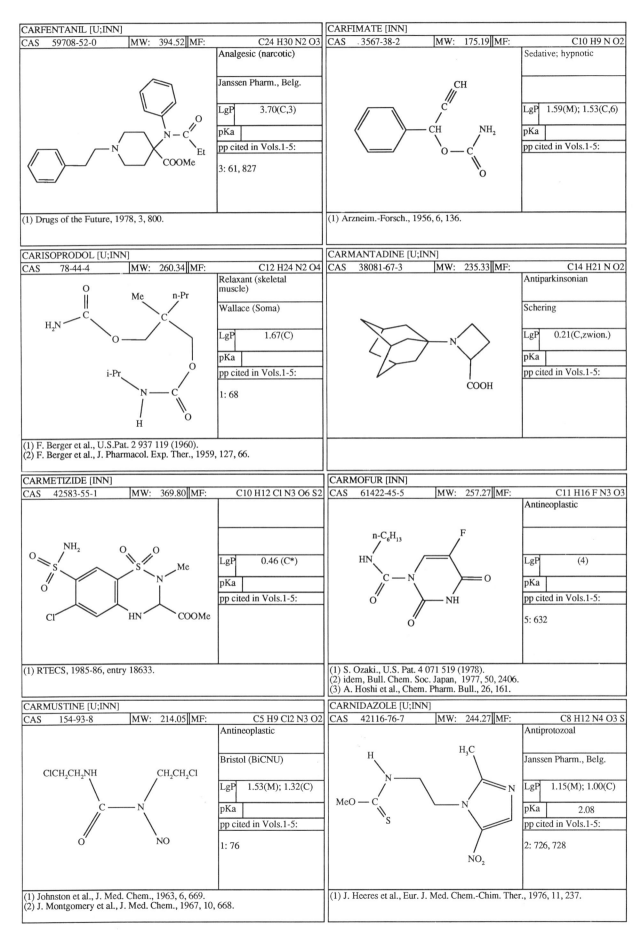

CARFENTANIL [U;INN]			
CAS 59708-52-0	MW: 394.52	MF:	C24 H30 N2 O3

Analgesic (narcotic)

Janssen Pharm., Belg.

LgP	3.70(C,3)
pKa	
pp cited in Vols.1-5:	

3: 61, 827

(1) Drugs of the Future, 1978, 3, 800.

CARFIMATE [INN]			
CAS . 3567-38-2	MW: 175.19	MF:	C10 H9 N O2

Sedative; hypnotic

LgP	1.59(M); 1.53(C,6)
pKa	
pp cited in Vols.1-5:	

(1) Arzneim.-Forsch., 1956, 6, 136.

CARISOPRODOL [U;INN]			
CAS 78-44-4	MW: 260.34	MF:	C12 H24 N2 O4

Relaxant (skeletal muscle)

Wallace (Soma)

LgP	1.67(C)
pKa	
pp cited in Vols.1-5:	

1: 68

(1) F. Berger et al., U.S.Pat. 2 937 119 (1960).
(2) F. Berger et al., J. Pharmacol. Exp. Ther., 1959, 127, 66.

CARMANTADINE [U;INN]			
CAS 38081-67-3	MW: 235.33	MF:	C14 H21 N O2

Antiparkinsonian

Schering

LgP	0.21(C,zwion.)
pKa	
pp cited in Vols.1-5:	

CARMETIZIDE [INN]			
CAS 42583-55-1	MW: 369.80	MF:	C10 H12 Cl N3 O6 S2

LgP	0.46 (C*)
pKa	
pp cited in Vols.1-5:	

(1) RTECS, 1985-86, entry 18633.

CARMOFUR [INN]			
CAS 61422-45-5	MW: 257.27	MF:	C11 H16 F N3 O3

Antineoplastic

LgP	(4)
pKa	
pp cited in Vols.1-5:	

5: 632

(1) S. Ozaki., U.S. Pat. 4 071 519 (1978).
(2) idem, Bull. Chem. Soc. Japan, 1977, 50, 2406.
(3) A. Hoshi et al., Chem. Pharm. Bull., 26, 161.

CARMUSTINE [U;INN]			
CAS 154-93-8	MW: 214.05	MF:	C5 H9 Cl2 N3 O2

Antineoplastic

Bristol (BiCNU)

LgP	1.53(M); 1.32(C)
pKa	
pp cited in Vols.1-5:	

1: 76

(1) Johnston et al., J. Med. Chem., 1963, 6, 669.
(2) J. Montgomery et al., J. Med. Chem., 1967, 10, 668.

CARNIDAZOLE [U;INN]			
CAS 42116-76-7	MW: 244.27	MF:	C8 H12 N4 O3 S

Antiprotozoal

Janssen Pharm., Belg.

LgP	1.15(M); 1.00(C)
pKa	2.08
pp cited in Vols.1-5:	

2: 726, 728

(1) J. Heeres et al., Eur. J. Med. Chem.-Chim. Ther., 1976, 11, 237.

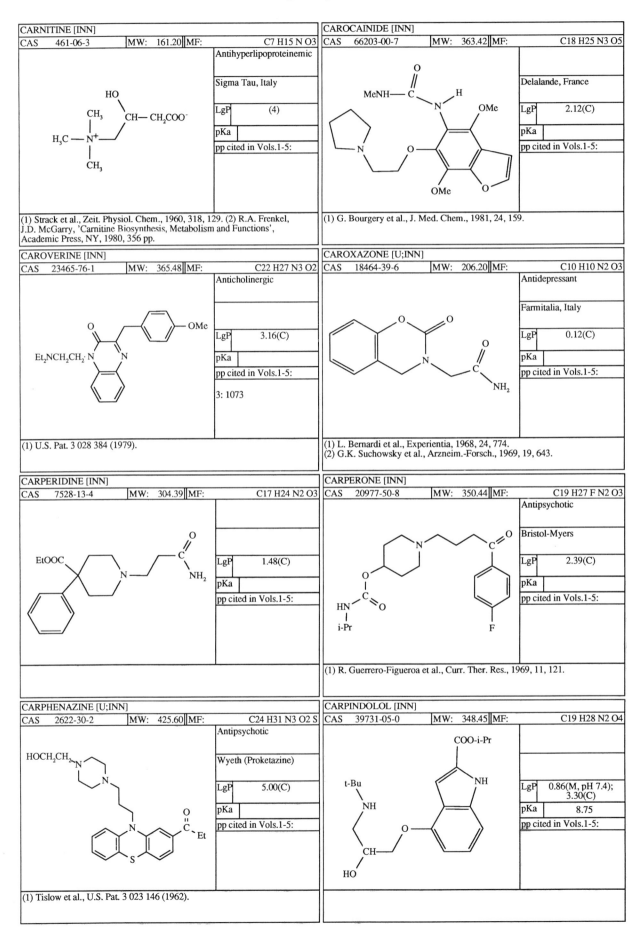

CARNITINE [INN]

CAS 461-06-3	MW: 161.20	MF: C7 H15 N O3

Antihyperlipoproteinemic

Sigma Tau, Italy

LgP	(4)
pKa	
pp cited in Vols.1-5:	

(1) Strack et al., Zeit. Physiol. Chem., 1960, 318, 129. (2) R.A. Frenkel, J.D. McGarry, 'Carnitine Biosynthesis, Metabolism and Functions', Academic Press, NY, 1980, 356 pp.

CAROCAINIDE [INN]

CAS 66203-00-7	MW: 363.42	MF: C18 H25 N3 O5

Delalande, France

LgP	2.12(C)
pKa	
pp cited in Vols.1-5:	

(1) G. Bourgery et al., J. Med. Chem., 1981, 24, 159.

CAROVERINE [INN]

CAS 23465-76-1	MW: 365.48	MF: C22 H27 N3 O2

Anticholinergic

LgP	3.16(C)
pKa	
pp cited in Vols.1-5:	
	3: 1073

(1) U.S. Pat. 3 028 384 (1979).

CAROXAZONE [U;INN]

CAS 18464-39-6	MW: 206.20	MF: C10 H10 N2 O3

Antidepressant

Farmitalia, Italy

LgP	0.12(C)
pKa	
pp cited in Vols.1-5:	

(1) L. Bernardi et al., Experientia, 1968, 24, 774.
(2) G.K. Suchowsky et al., Arzneim.-Forsch., 1969, 19, 643.

CARPERIDINE [INN]

CAS 7528-13-4	MW: 304.39	MF: C17 H24 N2 O3

LgP	1.48(C)
pKa	
pp cited in Vols.1-5:	

CARPERONE [INN]

CAS 20977-50-8	MW: 350.44	MF: C19 H27 F N2 O3

Antipsychotic

Bristol-Myers

LgP	2.39(C)
pKa	
pp cited in Vols.1-5:	

(1) R. Guerrero-Figueroa et al., Curr. Ther. Res., 1969, 11, 121.

CARPHENAZINE [U;INN]

CAS 2622-30-2	MW: 425.60	MF: C24 H31 N3 O2 S

Antipsychotic

Wyeth (Proketazine)

LgP	5.00(C)
pKa	
pp cited in Vols.1-5:	

(1) Tislow et al., U.S. Pat. 3 023 146 (1962).

CARPINDOLOL [INN]

CAS 39731-05-0	MW: 348.45	MF: C19 H28 N2 O4

LgP	0.86(M, pH 7.4); 3.30(C)
pKa	8.75
pp cited in Vols.1-5:	

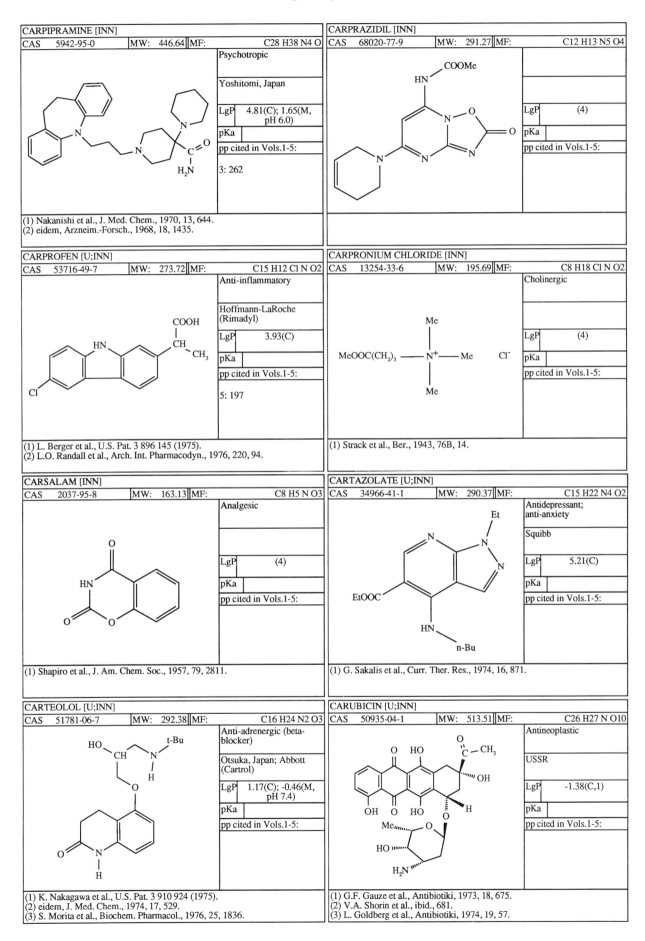

CARPIPRAMINE [INN]

CAS	5942-95-0	MW:	446.64	MF:	C28 H38 N4 O

Psychotropic

Yoshitomi, Japan

LgP	4.81(C); 1.65(M, pH 6.0)
pKa	
pp cited in Vols.1-5:	

3: 262

(1) Nakanishi et al., J. Med. Chem., 1970, 13, 644.
(2) eidem, Arzneim.-Forsch., 1968, 18, 1435.

CARPRAZIDIL [INN]

CAS	68020-77-9	MW:	291.27	MF:	C12 H13 N5 O4

LgP	(4)
pKa	
pp cited in Vols.1-5:	

CARPROFEN [U;INN]

CAS	53716-49-7	MW:	273.72	MF:	C15 H12 Cl N O2

Anti-inflammatory

Hoffmann-LaRoche (Rimadyl)

LgP	3.93(C)
pKa	
pp cited in Vols.1-5:	

5: 197

(1) L. Berger et al., U.S. Pat. 3 896 145 (1975).
(2) L.O. Randall et al., Arch. Int. Pharmacodyn., 1976, 220, 94.

CARPRONIUM CHLORIDE [INN]

CAS	13254-33-6	MW:	195.69	MF:	C8 H18 Cl N O2

Cholinergic

LgP	(4)
pKa	
pp cited in Vols.1-5:	

$MeOOC(CH_2)_3 \longrightarrow N^+ \longrightarrow Me \quad Cl^-$

(1) Strack et al., Ber., 1943, 76B, 14.

CARSALAM [INN]

CAS	2037-95-8	MW:	163.13	MF:	C8 H5 N O3

Analgesic

LgP	(4)
pKa	
pp cited in Vols.1-5:	

(1) Shapiro et al., J. Am. Chem. Soc., 1957, 79, 2811.

CARTAZOLATE [U;INN]

CAS	34966-41-1	MW:	290.37	MF:	C15 H22 N4 O2

Antidepressant; anti-anxiety

Squibb

LgP	5.21(C)
pKa	
pp cited in Vols.1-5:	

(1) G. Sakalis et al., Curr. Ther. Res., 1974, 16, 871.

CARTEOLOL [U;INN]

CAS	51781-06-7	MW:	292.38	MF:	C16 H24 N2 O3

Anti-adrenergic (beta-blocker)

Otsuka, Japan; Abbott (Cartrol)

LgP	1.17(C); -0.46(M, pH 7.4)
pKa	
pp cited in Vols.1-5:	

(1) K. Nakagawa et al., U.S. Pat. 3 910 924 (1975).
(2) eidem, J. Med. Chem., 1974, 17, 529.
(3) S. Morita et al., Biochem. Pharmacol., 1976, 25, 1836.

CARUBICIN [U;INN]

CAS	50935-04-1	MW:	513.51	MF:	C26 H27 N O10

Antineoplastic

USSR

LgP	-1.38(C,1)
pKa	
pp cited in Vols.1-5:	

(1) G.F. Gauze et al., Antibiotiki, 1973, 18, 675.
(2) V.A. Shorin et al., ibid., 681.
(3) L. Goldberg et al., Antibiotiki, 1974, 19, 57.

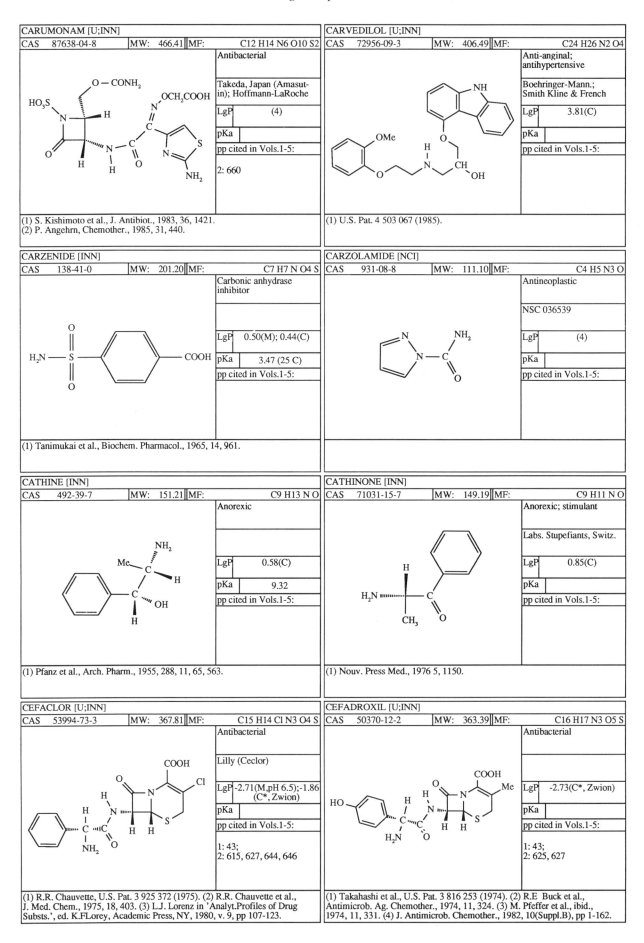

CARUMONAM [U;INN]

CAS	87638-04-8	MW:	466.41	MF:	C12 H14 N6 O10 S2

Antibacterial

Takeda, Japan (Amasutin); Hoffmann-LaRoche

LgP	(4)
pKa	
pp cited in Vols.1-5:	
2: 660	

(1) S. Kishimoto et al., J. Antibiot., 1983, 36, 1421.
(2) P. Angehrn, Chemother., 1985, 31, 440.

CARVEDILOL [U;INN]

CAS	72956-09-3	MW:	406.49	MF:	C24 H26 N2 O4

Anti-anginal; antihypertensive

Boehringer-Mann.; Smith Kline & French

LgP	3.81(C)
pKa	
pp cited in Vols.1-5:	

(1) U.S. Pat. 4 503 067 (1985).

CARZENIDE [INN]

CAS	138-41-0	MW:	201.20	MF:	C7 H7 N O4 S

Carbonic anhydrase inhibitor

LgP	0.50(M); 0.44(C)
pKa	3.47 (25 C)
pp cited in Vols.1-5:	

(1) Tanimukai et al., Biochem. Pharmacol., 1965, 14, 961.

CARZOLAMIDE [NCI]

CAS	931-08-8	MW:	111.10	MF:	C4 H5 N3 O

Antineoplastic

NSC 036539

LgP	(4)
pKa	
pp cited in Vols.1-5:	

CATHINE [INN]

CAS	492-39-7	MW:	151.21	MF:	C9 H13 N O

Anorexic

LgP	0.58(C)
pKa	9.32
pp cited in Vols.1-5:	

(1) Pfanz et al., Arch. Pharm., 1955, 288, 11, 65, 563.

CATHINONE [INN]

CAS	71031-15-7	MW:	149.19	MF:	C9 H11 N O

Anorexic; stimulant

Labs. Stupefiants, Switz.

LgP	0.85(C)
pKa	
pp cited in Vols.1-5:	

(1) Nouv. Press Med., 1976 5, 1150.

CEFACLOR [U;INN]

CAS	53994-73-3	MW:	367.81	MF:	C15 H14 Cl N3 O4 S

Antibacterial

Lilly (Ceclor)

LgP	-2.71(M,pH 6.5);-1.86 (C*, Zwion)
pKa	
pp cited in Vols.1-5:	
1: 43; 2: 615, 627, 644, 646	

(1) R.R. Chauvette, U.S. Pat. 3 925 372 (1975). (2) R.R. Chauvette et al., J. Med. Chem., 1975, 18, 403. (3) L.J. Lorenz in 'Analyt.Profiles of Drug Substs.', ed. K.F.Lorey, Academic Press, NY, 1980, v. 9, pp 107-123.

CEFADROXIL [U;INN]

CAS	50370-12-2	MW:	363.39	MF:	C16 H17 N3 O5 S

Antibacterial

LgP	-2.73(C*, Zwion)
pKa	
pp cited in Vols.1-5:	
1: 43; 2: 625, 627	

(1) Takahashi et al., U.S. Pat. 3 816 253 (1974). (2) R.E Buck et al., Antimicrob. Ag. Chemother., 1974, 11, 324. (3) M. Pfeffer et al., ibid., 1974, 11, 331. (4) J. Antimicrob. Chemother., 1982, 10(Suppl.B), pp 1-162.

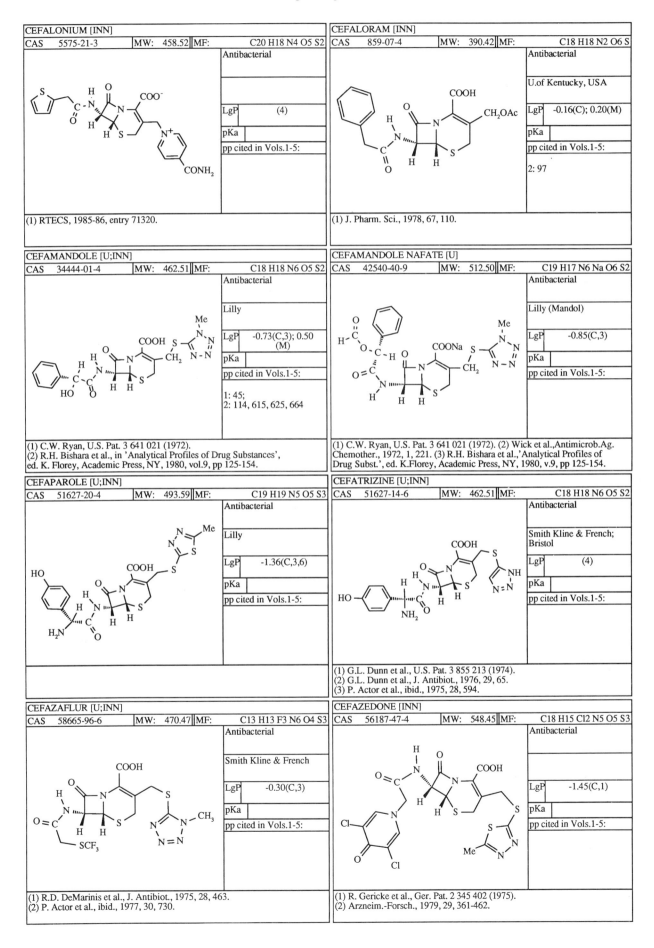

CEFALONIUM [INN]

CAS	5575-21-3	MW:	458.52	MF:	C20 H18 N4 O5 S2

Antibacterial

LgP	(4)
pKa	
pp cited in Vols.1-5:	

(1) RTECS, 1985-86, entry 71320.

CEFALORAM [INN]

CAS	859-07-4	MW:	390.42	MF:	C18 H18 N2 O6 S

Antibacterial

U.of Kentucky, USA

LgP	-0.16(C); 0.20(M)
pKa	
pp cited in Vols.1-5:	

2: 97

(1) J. Pharm. Sci., 1978, 67, 110.

CEFAMANDOLE [U;INN]

CAS	34444-01-4	MW:	462.51	MF:	C18 H18 N6 O5 S2

Antibacterial

Lilly

LgP	-0.73(C,3); 0.50 (M)
pKa	
pp cited in Vols.1-5:	

1: 45;
2: 114, 615, 625, 664

(1) C.W. Ryan, U.S. Pat. 3 641 021 (1972).
(2) R.H. Bishara et al., in 'Analytical Profiles of Drug Substances',
ed. K. Florey, Academic Press, NY, 1980, vol.9, pp 125-154.

CEFAMANDOLE NAFATE [U]

CAS	42540-40-9	MW:	512.50	MF:	C19 H17 N6 Na O6 S2

Antibacterial

Lilly (Mandol)

LgP	-0.85(C,3)
pKa	
pp cited in Vols.1-5:	

(1) C.W. Ryan, U.S. Pat. 3 641 021 (1972). (2) Wick et al.,Antimicrob.Ag.
Chemother., 1972, 1, 221. (3) R.H. Bishara et al.,'Analytical Profiles of
Drug Subst.', ed. K.Florey, Academic Press, NY, 1980, v.9, pp 125-154.

CEFAPAROLE [U;INN]

CAS	51627-20-4	MW:	493.59	MF:	C19 H19 N5 O5 S3

Antibacterial

Lilly

LgP	-1.36(C,3,6)
pKa	
pp cited in Vols.1-5:	

CEFATRIZINE [U;INN]

CAS	51627-14-6	MW:	462.51	MF:	C18 H18 N6 O5 S2

Antibacterial

Smith Kline & French;
Bristol

LgP	(4)
pKa	
pp cited in Vols.1-5:	

(1) G.L. Dunn et al., U.S. Pat. 3 855 213 (1974).
(2) G.L. Dunn et al., J. Antibiot., 1976, 29, 65.
(3) P. Actor et al., ibid., 1975, 28, 594.

CEFAZAFLUR [U;INN]

CAS	58665-96-6	MW:	470.47	MF:	C13 H13 F3 N6 O4 S3

Antibacterial

Smith Kline & French

LgP	-0.30(C,3)
pKa	
pp cited in Vols.1-5:	

(1) R.D. DeMarinis et al., J. Antibiot., 1975, 28, 463.
(2) P. Actor et al., ibid., 1977, 30, 730.

CEFAZEDONE [INN]

CAS	56187-47-4	MW:	548.45	MF:	C18 H15 Cl2 N5 O5 S3

Antibacterial

LgP	-1.45(C,1)
pKa	
pp cited in Vols.1-5:	

(1) R. Gericke et al., Ger. Pat. 2 345 402 (1975).
(2) Arzneim.-Forsch., 1979, 29, 361-462.

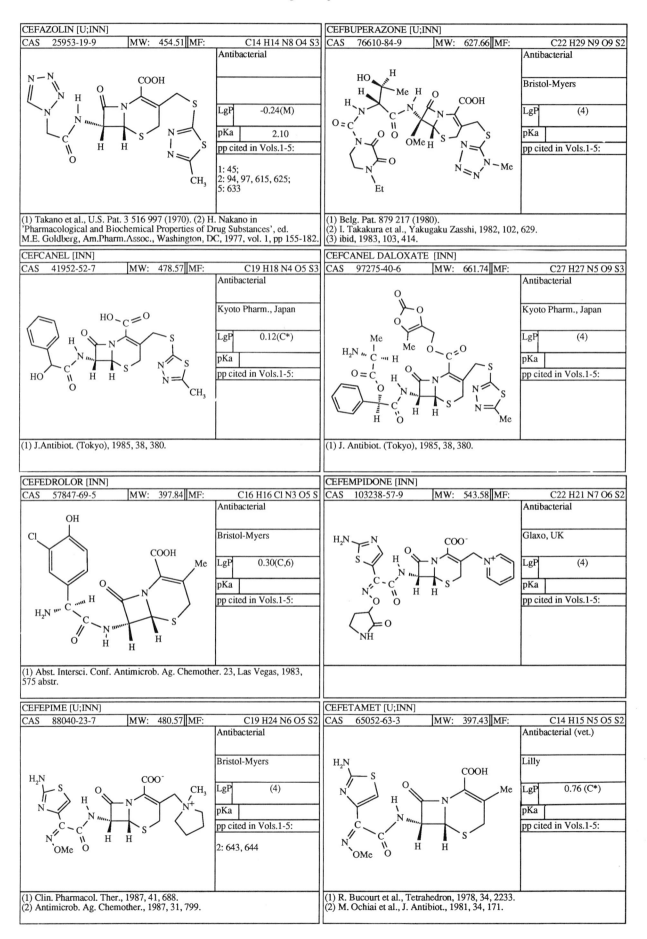

CEFAZOLIN [U;INN]			
CAS 25953-19-9	MW: 454.51	MF:	C14 H14 N8 O4 S3

Antibacterial

LgP -0.24(M)

pKa 2.10

pp cited in Vols.1-5:

1: 45;
2: 94, 97, 615, 625;
5: 633

(1) Takano et al., U.S. Pat. 3 516 997 (1970). (2) H. Nakano in
'Pharmacological and Biochemical Properties of Drug Substances', ed.
M.E. Goldberg, Am.Pharm.Assoc., Washington, DC, 1977, vol. 1, pp 155-182.

CEFBUPERAZONE [U;INN]			
CAS 76610-84-9	MW: 627.66	MF:	C22 H29 N9 O9 S2

Antibacterial

Bristol-Myers

LgP (4)

pKa

pp cited in Vols.1-5:

(1) Belg. Pat. 879 217 (1980).
(2) I. Takakura et al., Yakugaku Zasshi, 1982, 102, 629.
(3) ibid, 1983, 103, 414.

CEFCANEL [INN]			
CAS 41952-52-7	MW: 478.57	MF:	C19 H18 N4 O5 S3

Antibacterial

Kyoto Pharm., Japan

LgP 0.12(C*)

pKa

pp cited in Vols.1-5:

(1) J.Antibiot. (Tokyo), 1985, 38, 380.

CEFCANEL DALOXATE [INN]			
CAS 97275-40-6	MW: 661.74	MF:	C27 H27 N5 O9 S3

Antibacterial

Kyoto Pharm., Japan

LgP (4)

pKa

pp cited in Vols.1-5:

(1) J. Antibiot. (Tokyo), 1985, 38, 380.

CEFEDROLOR [INN]			
CAS 57847-69-5	MW: 397.84	MF:	C16 H16 Cl N3 O5 S

Antibacterial

Bristol-Myers

LgP 0.30(C,6)

pKa

pp cited in Vols.1-5:

(1) Abst. Intersci. Conf. Antimicrob. Ag. Chemother. 23, Las Vegas, 1983,
575 abstr.

CEFEMPIDONE [INN]			
CAS 103238-57-9	MW: 543.58	MF:	C22 H21 N7 O6 S2

Antibacterial

Glaxo, UK

LgP (4)

pKa

pp cited in Vols.1-5:

CEFEPIME [U;INN]			
CAS 88040-23-7	MW: 480.57	MF:	C19 H24 N6 O5 S2

Antibacterial

Bristol-Myers

LgP (4)

pKa

pp cited in Vols.1-5:

2: 643, 644

(1) Clin. Pharmacol. Ther., 1987, 41, 688.
(2) Antimicrob. Ag. Chemother., 1987, 31, 799.

CEFETAMET [U;INN]			
CAS 65052-63-3	MW: 397.43	MF:	C14 H15 N5 O5 S2

Antibacterial (vet.)

Lilly

LgP 0.76 (C*)

pKa

pp cited in Vols.1-5:

(1) R. Bucourt et al., Tetrahedron, 1978, 34, 2233.
(2) M. Ochiai et al., J. Antibiot., 1981, 34, 171.

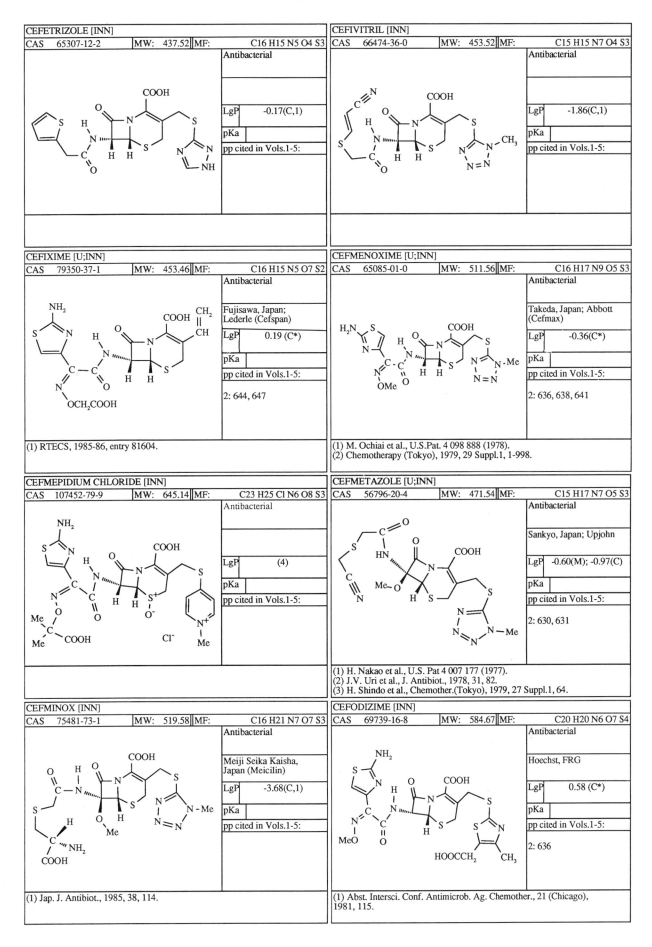

CEFETRIZOLE [INN]

CAS	65307-12-2	MW:	437.52	MF:	C16 H15 N5 O4 S3

Antibacterial

LgP   -0.17(C,1)

pKa

pp cited in Vols.1-5:

CEFIVITRIL [INN]

CAS	66474-36-0	MW:	453.52	MF:	C15 H15 N7 O4 S3

Antibacterial

LgP   -1.86(C,1)

pKa

pp cited in Vols.1-5:

CEFIXIME [U;INN]

CAS	79350-37-1	MW:	453.46	MF:	C16 H15 N5 O7 S2

Antibacterial

Fujisawa, Japan; Lederle (Cefspan)

LgP   0.19 (C*)

pKa

pp cited in Vols.1-5:

2: 644, 647

(1) RTECS, 1985-86, entry 81604.

CEFMENOXIME [U;INN]

CAS	65085-01-0	MW:	511.56	MF:	C16 H17 N9 O5 S3

Antibacterial

Takeda, Japan; Abbott (Cefmax)

LgP   -0.36(C*)

pKa

pp cited in Vols.1-5:

2: 636, 638, 641

(1) M. Ochiai et al., U.S.Pat. 4 098 888 (1978).
(2) Chemotherapy (Tokyo), 1979, 29 Suppl.1, 1-998.

CEFMEPIDIUM CHLORIDE [INN]

CAS	107452-79-9	MW:	645.14	MF:	C23 H25 Cl N6 O8 S3

Antibacterial

LgP   (4)

pKa

pp cited in Vols.1-5:

CEFMETAZOLE [U;INN]

CAS	56796-20-4	MW:	471.54	MF:	C15 H17 N7 O5 S3

Antibacterial

Sankyo, Japan; Upjohn

LgP   -0.60(M); -0.97(C)

pKa

pp cited in Vols.1-5:

2: 630, 631

(1) H. Nakao et al., U.S. Pat 4 007 177 (1977).
(2) J.V. Uri et al., J. Antibiot., 1978, 31, 82.
(3) H. Shindo et al., Chemother.(Tokyo), 1979, 27 Suppl.1, 64.

CEFMINOX [INN]

CAS	75481-73-1	MW:	519.58	MF:	C16 H21 N7 O7 S3

Antibacterial

Meiji Seika Kaisha, Japan (Meicilin)

LgP   -3.68(C,1)

pKa

pp cited in Vols.1-5:

(1) Jap. J. Antibiot., 1985, 38, 114.

CEFODIZIME [INN]

CAS	69739-16-8	MW:	584.67	MF:	C20 H20 N6 O7 S4

Antibacterial

Hoechst, FRG

LgP   0.58 (C*)

pKa

pp cited in Vols.1-5:

2: 636

(1) Abst. Intersci. Conf. Antimicrob. Ag. Chemother., 21 (Chicago), 1981, 115.

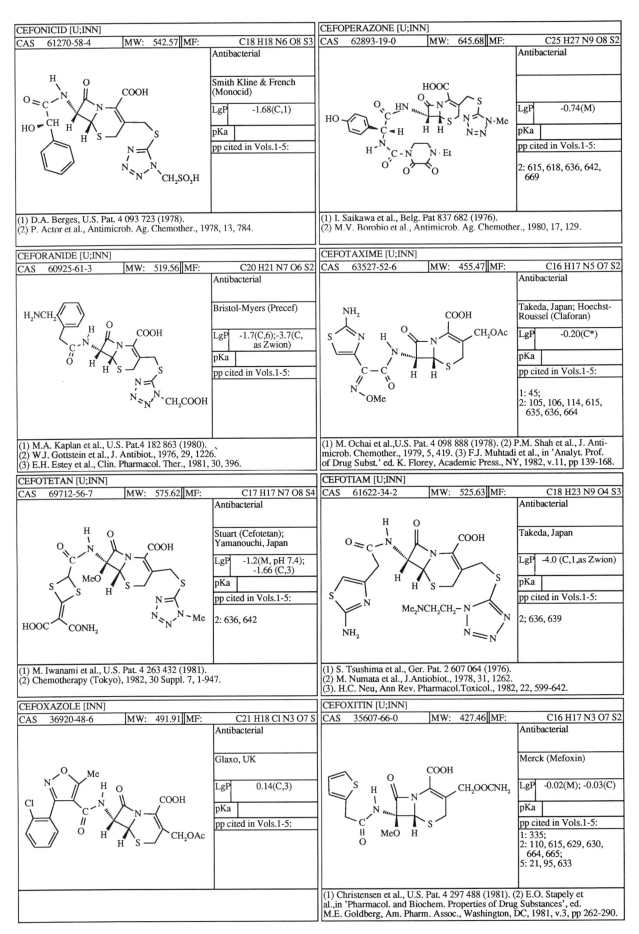

CEFONICID [U;INN]		
CAS 61270-58-4	MW: 542.57	MF: C18 H18 N6 O8 S3

Antibacterial

Smith Kline & French (Monocid)

LgP	-1.68(C,1)
pKa	
pp cited in Vols.1-5:	

(1) D.A. Berges, U.S. Pat. 4 093 723 (1978).
(2) P. Actor et al., Antimicrob. Ag. Chemother., 1978, 13, 784.

CEFOPERAZONE [U;INN]		
CAS 62893-19-0	MW: 645.68	MF: C25 H27 N9 O8 S2

Antibacterial

LgP	-0.74(M)
pKa	
pp cited in Vols.1-5:	
	2: 615, 618, 636, 642, 669

(1) I. Saikawa et al., Belg. Pat 837 682 (1976).
(2) M.V. Borobio et al., Antimicrob. Ag. Chemother., 1980, 17, 129.

CEFORANIDE [U;INN]		
CAS 60925-61-3	MW: 519.56	MF: C20 H21 N7 O6 S2

Antibacterial

Bristol-Myers (Precef)

LgP	-1.7(C,6);-3.7(C, as Zwion)
pKa	
pp cited in Vols.1-5:	

(1) M.A. Kaplan et al., U.S. Pat.4 182 863 (1980).
(2) W.J. Gottstein et al., J. Antibiot., 1976, 29, 1226.
(3) E.H. Estey et al., Clin. Pharmacol. Ther., 1981, 30, 396.

CEFOTAXIME [U;INN]		
CAS 63527-52-6	MW: 455.47	MF: C16 H17 N5 O7 S2

Antibacterial

Takeda, Japan; Hoechst-Roussel (Claforan)

LgP	-0.20(C*)
pKa	
pp cited in Vols.1-5:	
	1: 45; 2: 105, 106, 114, 615, 635, 636, 664

(1) M. Ochai et al.,U.S. Pat. 4 098 888 (1978). (2) P.M. Shah et al., J. Antimicrob. Chemother., 1979, 5, 419. (3) F.J. Muhtadi et al., in 'Analyt. Prof. of Drug Subst.' ed. K. Florey, Academic Press., NY, 1982, v.11, pp 139-168.

CEFOTETAN [U;INN]		
CAS 69712-56-7	MW: 575.62	MF: C17 H17 N7 O8 S4

Antibacterial

Stuart (Cefotetan); Yamanouchi, Japan

LgP	-1.2(M, pH 7.4); -1.66 (C,3)
pKa	
pp cited in Vols.1-5:	
	2: 636, 642

(1) M. Iwanami et al., U.S. Pat. 4 263 432 (1981).
(2) Chemotherapy (Tokyo), 1982, 30 Suppl. 7, 1-947.

CEFOTIAM [U;INN]		
CAS 61622-34-2	MW: 525.63	MF: C18 H23 N9 O4 S3

Antibacterial

Takeda, Japan

LgP	-4.0 (C,1,as Zwion)
pKa	
pp cited in Vols.1-5:	
	2; 636, 639

(1) S. Tsushima et al., Ger. Pat. 2 607 064 (1976).
(2) M. Numata et al., J.Antiobiot., 1978, 31, 1262.
(3). H.C. Neu, Ann Rev. Pharmacol.Toxicol., 1982, 22, 599-642.

CEFOXAZOLE [INN]		
CAS 36920-48-6	MW: 491.91	MF: C21 H18 Cl N3 O7 S

Antibacterial

Glaxo, UK

LgP	0.14(C,3)
pKa	
pp cited in Vols.1-5:	

CEFOXITIN [U;INN]		
CAS 35607-66-0	MW: 427.46	MF: C16 H17 N3 O7 S2

Antibacterial

Merck (Mefoxin)

LgP	-0.02(M); -0.03(C)
pKa	
pp cited in Vols.1-5:	
	1: 335; 2: 110, 615, 629, 630, 664, 665; 5: 21, 95, 633

(1) Christensen et al., U.S. Pat. 4 297 488 (1981). (2) E.O. Stapely et al.,in 'Pharmacol. and Biochem. Properties of Drug Substances', ed. M.E. Goldberg, Am. Pharm. Assoc., Washington, DC, 1981, v.3, pp 262-290.

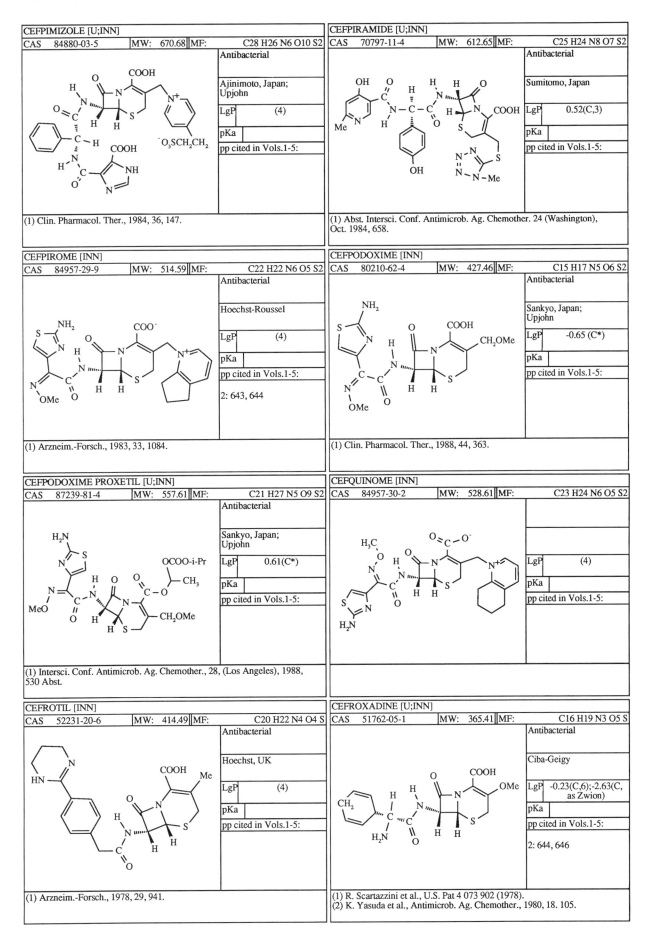

CEFPIMIZOLE [U;INN]			
CAS 84880-03-5	MW: 670.68	MF:	C28 H26 N6 O10 S2

Antibacterial

Ajinimoto, Japan; Upjohn

LgP	(4)
pKa	
pp cited in Vols.1-5:	

(1) Clin. Pharmacol. Ther., 1984, 36, 147.

CEFPIRAMIDE [U;INN]			
CAS 70797-11-4	MW: 612.65	MF:	C25 H24 N8 O7 S2

Antibacterial

Sumitomo, Japan

LgP	0.52(C,3)
pKa	
pp cited in Vols.1-5:	

(1) Abst. Intersci. Conf. Antimicrob. Ag. Chemother. 24 (Washington), Oct. 1984, 658.

CEFPIROME [INN]			
CAS 84957-29-9	MW: 514.59	MF:	C22 H22 N6 O5 S2

Antibacterial

Hoechst-Roussel

LgP	(4)
pKa	
pp cited in Vols.1-5:	
2: 643, 644	

(1) Arzneim.-Forsch., 1983, 33, 1084.

CEFPODOXIME [INN]			
CAS 80210-62-4	MW: 427.46	MF:	C15 H17 N5 O6 S2

Antibacterial

Sankyo, Japan; Upjohn

LgP	-0.65 (C*)
pKa	
pp cited in Vols.1-5:	

(1) Clin. Pharmacol. Ther., 1988, 44, 363.

CEFPODOXIME PROXETIL [U;INN]			
CAS 87239-81-4	MW: 557.61	MF:	C21 H27 N5 O9 S2

Antibacterial

Sankyo, Japan; Upjohn

LgP	0.61(C*)
pKa	
pp cited in Vols.1-5:	

(1) Intersci. Conf. Antimicrob. Ag. Chemother., 28, (Los Angeles), 1988, 530 Abst.

CEFQUINOME [INN]			
CAS 84957-30-2	MW: 528.61	MF:	C23 H24 N6 O5 S2

LgP	(4)
pKa	
pp cited in Vols.1-5:	

CEFROTIL [INN]			
CAS 52231-20-6	MW: 414.49	MF:	C20 H22 N4 O4 S

Antibacterial

Hoechst, UK

LgP	(4)
pKa	
pp cited in Vols.1-5:	

(1) Arzneim.-Forsch., 1978, 29, 941.

CEFROXADINE [U;INN]			
CAS 51762-05-1	MW: 365.41	MF:	C16 H19 N3 O5 S

Antibacterial

Ciba-Geigy

LgP	-0.23(C,6);-2.63(C, as Zwion)
pKa	
pp cited in Vols.1-5:	
2: 644, 646	

(1) R. Scartazzini et al., U.S. Pat 4 073 902 (1978).
(2) K. Yasuda et al., Antimicrob. Ag. Chemother., 1980, 18. 105.

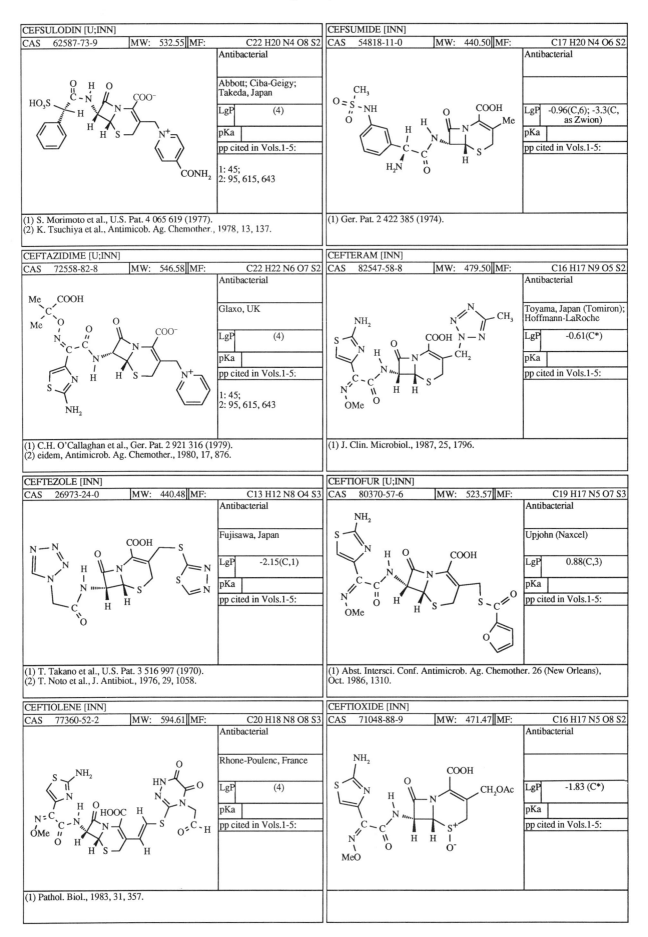

CEFSULODIN [U;INN]			
CAS 62587-73-9	MW: 532.55	MF:	C22 H20 N4 O8 S2

Antibacterial

Abbott; Ciba-Geigy; Takeda, Japan

LgP	(4)
pKa	
pp cited in Vols.1-5:	

1: 45;
2: 95, 615, 643

(1) S. Morimoto et al., U.S. Pat. 4 065 619 (1977).
(2) K. Tsuchiya et al., Antimicob. Ag. Chemother., 1978, 13, 137.

CEFSUMIDE [INN]			
CAS 54818-11-0	MW: 440.50	MF:	C17 H20 N4 O6 S2

Antibacterial

LgP	-0.96(C,6); -3.3(C, as Zwion)
pKa	
pp cited in Vols.1-5:	

(1) Ger. Pat. 2 422 385 (1974).

CEFTAZIDIME [U;INN]			
CAS 72558-82-8	MW: 546.58	MF:	C22 H22 N6 O7 S2

Antibacterial

Glaxo, UK

LgP	(4)
pKa	
pp cited in Vols.1-5:	

1: 45;
2: 95, 615, 643

(1) C.H. O'Callaghan et al., Ger. Pat. 2 921 316 (1979).
(2) eidem, Antimicrob. Ag. Chemother., 1980, 17, 876.

CEFTERAM [INN]			
CAS 82547-58-8	MW: 479.50	MF:	C16 H17 N9 O5 S2

Antibacterial

Toyama, Japan (Tomiron); Hoffmann-LaRoche

LgP	-0.61(C*)
pKa	
pp cited in Vols.1-5:	

(1) J. Clin. Microbiol., 1987, 25, 1796.

CEFTEZOLE [INN]			
CAS 26973-24-0	MW: 440.48	MF:	C13 H12 N8 O4 S3

Antibacterial

Fujisawa, Japan

LgP	-2.15(C,1)
pKa	
pp cited in Vols.1-5:	

(1) T. Takano et al., U.S. Pat. 3 516 997 (1970).
(2) T. Noto et al., J. Antibiot., 1976, 29, 1058.

CEFTIOFUR [U;INN]			
CAS 80370-57-6	MW: 523.57	MF:	C19 H17 N5 O7 S3

Antibacterial

Upjohn (Naxcel)

LgP	0.88(C,3)
pKa	
pp cited in Vols.1-5:	

(1) Abst. Intersci. Conf. Antimicrob. Ag. Chemother. 26 (New Orleans), Oct. 1986, 1310.

CEFTIOLENE [INN]			
CAS 77360-52-2	MW: 594.61	MF:	C20 H18 N8 O8 S3

Antibacterial

Rhone-Poulenc, France

LgP	(4)
pKa	
pp cited in Vols.1-5:	

(1) Pathol. Biol., 1983, 31, 357.

CEFTIOXIDE [INN]			
CAS 71048-88-9	MW: 471.47	MF:	C16 H17 N5 O8 S2

Antibacterial

LgP	-1.83 (C*)
pKa	
pp cited in Vols.1-5:	

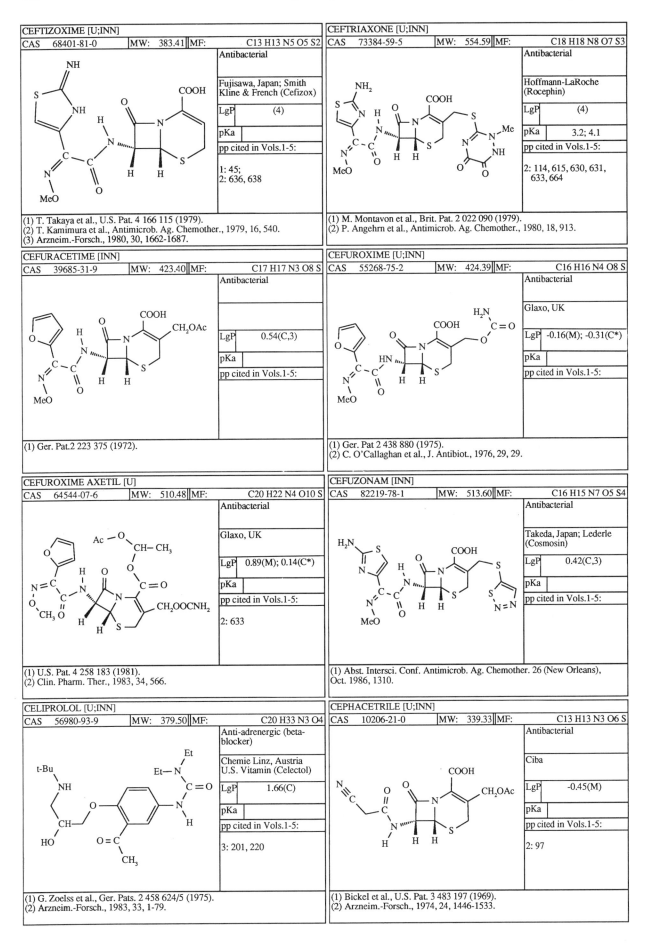

CEFTIZOXIME [U;INN]

CAS 68401-81-0 | MW: 383.41 | MF: | C13 H13 N5 O5 S2

Antibacterial

Fujisawa, Japan; Smith Kline & French (Cefizox)

LgP	(4)
pKa	
pp cited in Vols.1-5:	
1: 45;
2: 636, 638	

(1) T. Takaya et al., U.S. Pat. 4 166 115 (1979).
(2) T. Kamimura et al., Antimicrob. Ag. Chemother., 1979, 16, 540.
(3) Arzneim.-Forsch., 1980, 30, 1662-1687.

CEFTRIAXONE [U;INN]

CAS 73384-59-5 | MW: 554.59 | MF: | C18 H18 N8 O7 S3

Antibacterial

Hoffmann-LaRoche (Rocephin)

LgP	(4)
pKa	3.2; 4.1
pp cited in Vols.1-5:	
2: 114, 615, 630, 631,
633, 664	

(1) M. Montavon et al., Brit. Pat. 2 022 090 (1979).
(2) P. Angehrn et al., Antimicrob. Ag. Chemother., 1980, 18, 913.

CEFURACETIME [INN]

CAS 39685-31-9 | MW: 423.40 | MF: | C17 H17 N3 O8 S

Antibacterial

LgP	0.54(C,3)
pKa	
pp cited in Vols.1-5:	

(1) Ger. Pat.2 223 375 (1972).

CEFUROXIME [U;INN]

CAS 55268-75-2 | MW: 424.39 | MF: | C16 H16 N4 O8 S

Antibacterial

Glaxo, UK

LgP	-0.16(M); -0.31(C*)
pKa	
pp cited in Vols.1-5:	

(1) Ger. Pat 2 438 880 (1975).
(2) C. O'Callaghan et al., J. Antibiot., 1976, 29, 29.

CEFUROXIME AXETIL [U]

CAS 64544-07-6 | MW: 510.48 | MF: | C20 H22 N4 O10 S

Antibacterial

Glaxo, UK

LgP	0.89(M); 0.14(C*)
pKa	
pp cited in Vols.1-5:	
2: 633	

(1) U.S. Pat. 4 258 183 (1981).
(2) Clin. Pharm. Ther., 1983, 34, 566.

CEFUZONAM [INN]

CAS 82219-78-1 | MW: 513.60 | MF: | C16 H15 N7 O5 S4

Antibacterial

Takeda, Japan; Lederle (Cosmosin)

LgP	0.42(C,3)
pKa	
pp cited in Vols.1-5:	

(1) Abst. Intersci. Conf. Antimicrob. Ag. Chemother. 26 (New Orleans),
Oct. 1986, 1310.

CELIPROLOL [U;INN]

CAS 56980-93-9 | MW: 379.50 | MF: | C20 H33 N3 O4

Anti-adrenergic (beta-blocker)

Chemie Linz, Austria
U.S. Vitamin (Celectol)

LgP	1.66(C)
pKa	
pp cited in Vols.1-5:	
3: 201, 220	

(1) G. Zoelss et al., Ger. Pats. 2 458 624/5 (1975).
(2) Arzneim.-Forsch., 1983, 33, 1-79.

CEPHACETRILE [U;INN]

CAS 10206-21-0 | MW: 339.33 | MF: | C13 H13 N3 O6 S

Antibacterial

Ciba

LgP	-0.45(M)
pKa	
pp cited in Vols.1-5:	
2: 97	

(1) Bickel et al., U.S. Pat. 3 483 197 (1969).
(2) Arzneim.-Forsch., 1974, 24, 1446-1533.

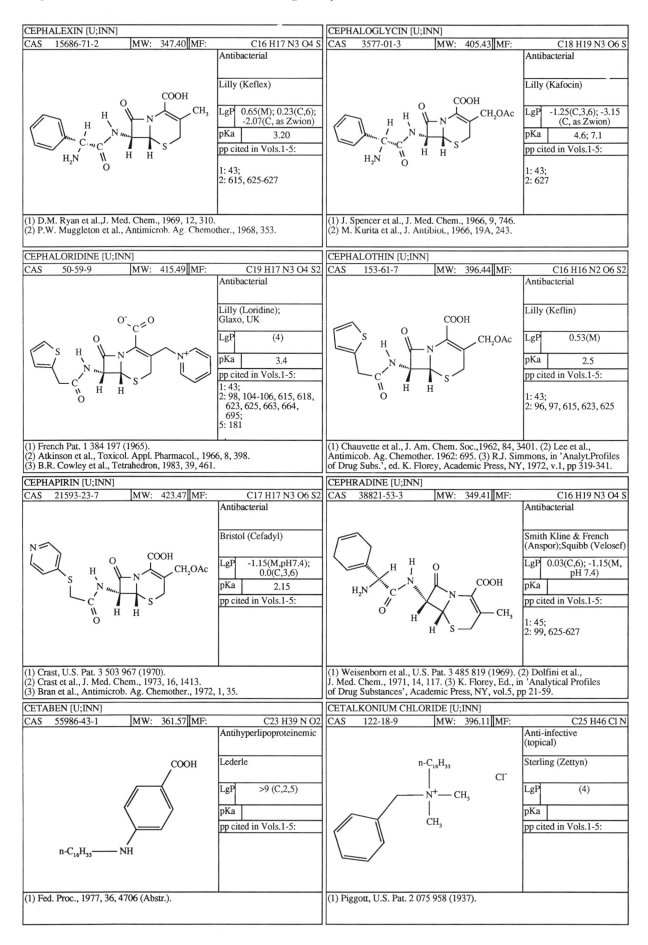

CEPHALEXIN [U;INN]			
CAS 15686-71-2	MW: 347.40	MF:	C16 H17 N3 O4 S

Antibacterial

Lilly (Keflex)

LgP	0.65(M); 0.23(C,6); -2.07(C, as Zwion)
pKa	3.20
pp cited in Vols.1-5:	1: 43; 2: 615, 625-627

(1) D.M. Ryan et al.,J. Med. Chem., 1969, 12, 310.
(2) P.W. Muggleton et al., Antimicrob. Ag. Chemother., 1968, 353.

CEPHALOGLYCIN [U;INN]			
CAS 3577-01-3	MW: 405.43	MF:	C18 H19 N3 O6 S

Antibacterial

Lilly (Kafocin)

LgP	-1.25(C,3,6); -3.15 (C, as Zwion)
pKa	4.6; 7.1
pp cited in Vols.1-5:	1: 43; 2: 627

(1) J. Spencer et al., J. Med. Chem., 1966, 9, 746.
(2) M. Kurita et al., J. Antibiot., 1966, 19A, 243.

CEPHALORIDINE [U;INN]			
CAS 50-59-9	MW: 415.49	MF:	C19 H17 N3 O4 S2

Antibacterial

Lilly (Loridine); Glaxo, UK

LgP	(4)
pKa	3.4
pp cited in Vols.1-5:	1: 43; 2: 98, 104-106, 615, 618, 623, 625, 663, 664, 695; 5: 181

(1) French Pat. 1 384 197 (1965).
(2) Atkinson et al., Toxicol. Appl. Pharmacol., 1966, 8, 398.
(3) B.R. Cowley et al., Tetrahedron, 1983, 39, 461.

CEPHALOTHIN [U;INN]			
CAS 153-61-7	MW: 396.44	MF:	C16 H16 N2 O6 S2

Antibacterial

Lilly (Keflin)

LgP	0.53(M)
pKa	2.5
pp cited in Vols.1-5:	1: 43; 2: 96, 97, 615, 623, 625

(1) Chauvette et al., J. Am. Chem. Soc.,1962, 84, 3401. (2) Lee et al., Antimicob. Ag. Chemother. 1962: 695. (3) R.J. Simmons, in 'Analyt.Profiles of Drug Subs.', ed. K. Florey, Academic Press, NY, 1972, v.1, pp 319-341.

CEPHAPIRIN [U;INN]			
CAS 21593-23-7	MW: 423.47	MF:	C17 H17 N3 O6 S2

Antibacterial

Bristol (Cefadyl)

LgP	-1.15(M,pH7.4); 0.0(C,3,6)
pKa	2.15
pp cited in Vols.1-5:	

(1) Crast, U.S. Pat. 3 503 967 (1970).
(2) Crast et al., J. Med. Chem., 1973, 16, 1413.
(3) Bran et al., Antimicrob. Ag. Chemother., 1972, 1, 35.

CEPHRADINE [U;INN]			
CAS 38821-53-3	MW: 349.41	MF:	C16 H19 N3 O4 S

Antibacterial

Smith Kline & French (Anspor);Squibb (Velosef)

LgP	0.03(C,6); -1.15(M, pH 7.4)
pKa	
pp cited in Vols.1-5:	1: 45; 2: 99, 625-627

(1) Weisenborn et al., U.S. Pat. 3 485 819 (1969). (2) Dolfini et al., J. Med. Chem., 1971, 14, 117. (3) K. Florey, Ed., in 'Analytical Profiles of Drug Substances', Academic Press, NY, vol.5, pp 21-59.

CETABEN [U;INN]			
CAS 55986-43-1	MW: 361.57	MF:	C23 H39 N O2

Antihyperlipoproteinemic

Lederle

LgP	>9 (C,2,5)
pKa	
pp cited in Vols.1-5:	

(1) Fed. Proc., 1977, 36, 4706 (Abstr.).

CETALKONIUM CHLORIDE [U;INN]			
CAS 122-18-9	MW: 396.11	MF:	C25 H46 Cl N

Anti-infective (topical)

Sterling (Zettyn)

LgP	(4)
pKa	
pp cited in Vols.1-5:	

(1) Piggott, U.S. Pat. 2 075 958 (1937).

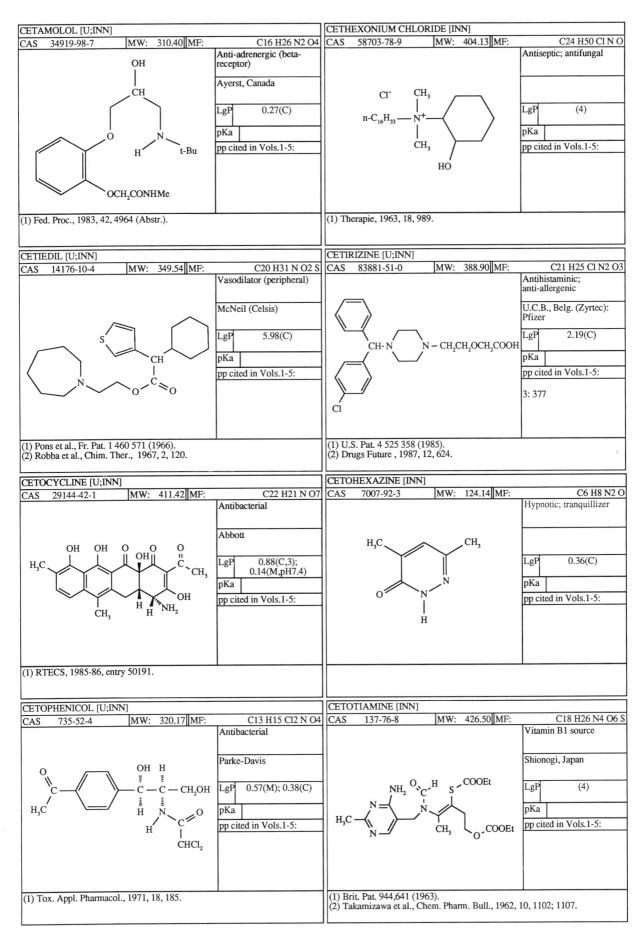

CETAMOLOL [U;INN]

| CAS | 34919-98-7 | MW: | 310.40 | MF: | C16 H26 N2 O4 |

Anti-adrenergic (beta-receptor)

Ayerst, Canada

LgP	0.27(C)
pKa	
pp cited in Vols.1-5:	

(1) Fed. Proc., 1983, 42, 4964 (Abstr.).

CETHEXONIUM CHLORIDE [INN]

| CAS | 58703-78-9 | MW: | 404.13 | MF: | C24 H50 Cl N O |

Antiseptic; antifungal

LgP	(4)
pKa	
pp cited in Vols.1-5:	

(1) Therapie, 1963, 18, 989.

CETIEDIL [U;INN]

| CAS | 14176-10-4 | MW: | 349.54 | MF: | C20 H31 N O2 S |

Vasodilator (peripheral)

McNeil (Celsis)

LgP	5.98(C)
pKa	
pp cited in Vols.1-5:	

(1) Pons et al., Fr. Pat. 1 460 571 (1966).
(2) Robba et al., Chim. Ther., 1967, 2, 120.

CETIRIZINE [U;INN]

| CAS | 83881-51-0 | MW: | 388.90 | MF: | C21 H25 Cl N2 O3 |

Antihistaminic; anti-allergenic

U.C.B., Belg. (Zyrtec): Pfizer

LgP	2.19(C)
pKa	
pp cited in Vols.1-5:	
	3: 377

(1) U.S. Pat. 4 525 358 (1985).
(2) Drugs Future , 1987, 12, 624.

CETOCYCLINE [U;INN]

| CAS | 29144-42-1 | MW: | 411.42 | MF: | C22 H21 N O7 |

Antibacterial

Abbott

LgP	0.88(C,3); 0.14(M,pH7.4)
pKa	
pp cited in Vols.1-5:	

(1) RTECS, 1985-86, entry 50191.

CETOHEXAZINE [INN]

| CAS | 7007-92-3 | MW: | 124.14 | MF: | C6 H8 N2 O |

Hypnotic; tranquillizer

LgP	0.36(C)
pKa	
pp cited in Vols.1-5:	

CETOPHENICOL [U;INN]

| CAS | 735-52-4 | MW: | 320.17 | MF: | C13 H15 Cl2 N O4 |

Antibacterial

Parke-Davis

LgP	0.57(M); 0.38(C)
pKa	
pp cited in Vols.1-5:	

(1) Tox. Appl. Pharmacol., 1971, 18, 185.

CETOTIAMINE [INN]

| CAS | 137-76-8 | MW: | 426.50 | MF: | C18 H26 N4 O6 S |

Vitamin B1 source

Shionogi, Japan

LgP	(4)
pKa	
pp cited in Vols.1-5:	

(1) Brit. Pat. 944,641 (1963).
(2) Takamizawa et al., Chem. Pharm. Bull., 1962, 10, 1102; 1107.

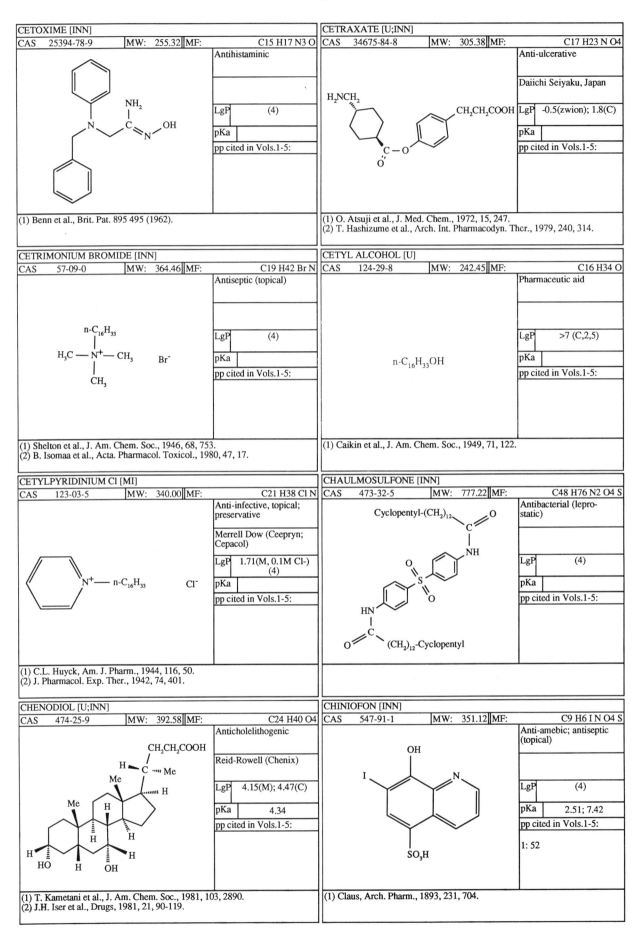

CETOXIME [INN]

CAS	25394-78-9	MW: 255.32	MF:	C15 H17 N3 O

Antihistaminic

LgP	(4)
pKa	
pp cited in Vols.1-5:	

(1) Benn et al., Brit. Pat. 895 495 (1962).

CETRAXATE [U;INN]

CAS	34675-84-8	MW: 305.38	MF:	C17 H23 N O4

Anti-ulcerative

Daiichi Seiyaku, Japan

LgP	-0.5(zwion); 1.8(C)
pKa	
pp cited in Vols.1-5:	

(1) O. Atsuji et al., J. Med. Chem., 1972, 15, 247.
(2) T. Hashizume et al., Arch. Int. Pharmacodyn. Ther., 1979, 240, 314.

CETRIMONIUM BROMIDE [INN]

CAS	57-09-0	MW: 364.46	MF:	C19 H42 Br N

Antiseptic (topical)

LgP	(4)
pKa	
pp cited in Vols.1-5:	

(1) Shelton et al., J. Am. Chem. Soc., 1946, 68, 753.
(2) B. Isomaa et al., Acta. Pharmacol. Toxicol., 1980, 47, 17.

CETYL ALCOHOL [U]

CAS	124-29-8	MW: 242.45	MF:	C16 H34 O

Pharmaceutic aid

$n\text{-}C_{16}H_{33}OH$

LgP	>7 (C,2,5)
pKa	
pp cited in Vols.1-5:	

(1) Caikin et al., J. Am. Chem. Soc., 1949, 71, 122.

CETYLPYRIDINIUM Cl [MI]

CAS	123-03-5	MW: 340.00	MF:	C21 H38 Cl N

Anti-infective, topical; preservative

Merrell Dow (Ceepryn; Cepacol)

LgP	1.71(M, 0.1M Cl-) (4)
pKa	
pp cited in Vols.1-5:	

(1) C.L. Huyck, Am. J. Pharm., 1944, 116, 50.
(2) J. Pharmacol. Exp. Ther., 1942, 74, 401.

CHAULMOSULFONE [INN]

CAS	473-32-5	MW: 777.22	MF:	C48 H76 N2 O4 S

Antibacterial (leprostatic)

LgP	(4)
pKa	
pp cited in Vols.1-5:	

CHENODIOL [U;INN]

CAS	474-25-9	MW: 392.58	MF:	C24 H40 O4

Anticholelithogenic

Reid-Rowell (Chenix)

LgP	4.15(M); 4.47(C)
pKa	4.34
pp cited in Vols.1-5:	

(1) T. Kametani et al., J. Am. Chem. Soc., 1981, 103, 2890.
(2) J.H. Iser et al., Drugs, 1981, 21, 90-119.

CHINIOFON [INN]

CAS	547-91-1	MW: 351.12	MF:	C9 H6 I N O4 S

Anti-amebic; antiseptic (topical)

LgP	(4)
pKa	2.51; 7.42
pp cited in Vols.1-5:	
	1: 52

(1) Claus, Arch. Pharm., 1893, 231, 704.

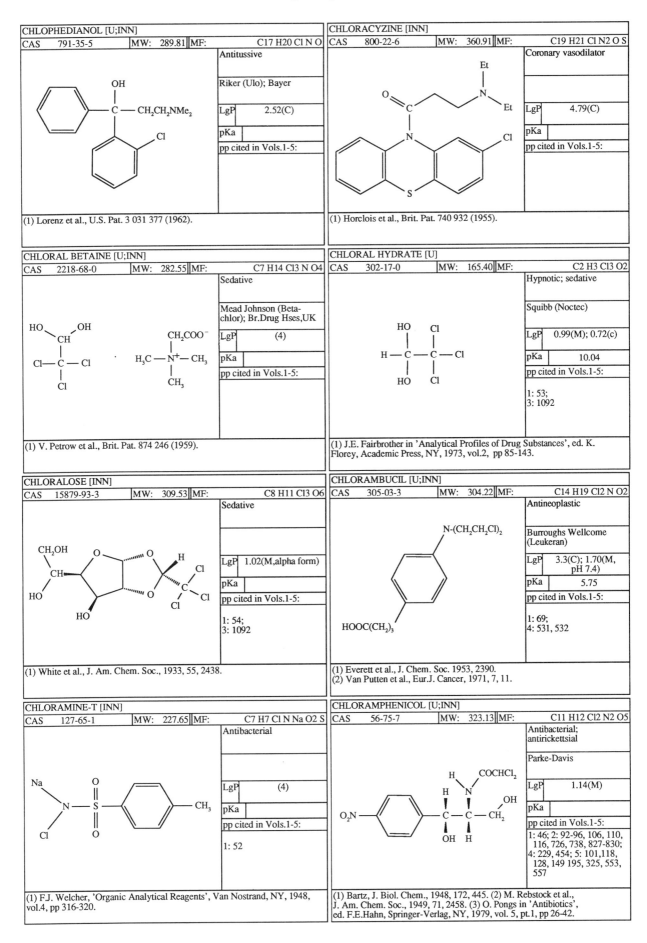

CHLOPHEDIANOL [U;INN]

CAS	791-35-5	MW: 289.81	MF:	C17 H20 Cl N O

Antitussive

Riker (Ulo); Bayer

LgP	2.52(C)
pKa	
pp cited in Vols.1-5:	

(1) Lorenz et al., U.S. Pat. 3 031 377 (1962).

CHLORACYZINE [INN]

CAS	800-22-6	MW: 360.91	MF:	C19 H21 Cl N2 O S

Coronary vasodilator

LgP	4.79(C)
pKa	
pp cited in Vols.1-5:	

(1) Horclois et al., Brit. Pat. 740 932 (1955).

CHLORAL BETAINE [U;INN]

CAS	2218-68-0	MW: 282.55	MF:	C7 H14 Cl3 N O4

Sedative

Mead Johnson (Beta-chlor); Br.Drug Hses,UK

LgP	(4)
pKa	
pp cited in Vols.1-5:	

(1) V. Petrow et al., Brit. Pat. 874 246 (1959).

CHLORAL HYDRATE [U]

CAS	302-17-0	MW: 165.40	MF:	C2 H3 Cl3 O2

Hypnotic; sedative

Squibb (Noctec)

LgP	0.99(M); 0.72(c)
pKa	10.04
pp cited in Vols.1-5:	
	1: 53; 3: 1092

(1) J.E. Fairbrother in 'Analytical Profiles of Drug Substances', ed. K. Florey, Academic Press, NY, 1973, vol.2, pp 85-143.

CHLORALOSE [INN]

CAS	15879-93-3	MW: 309.53	MF:	C8 H11 Cl3 O6

Sedative

LgP	1.02(M,alpha form)
pKa	
pp cited in Vols.1-5:	
	1: 54; 3: 1092

(1) White et al., J. Am. Chem. Soc., 1933, 55, 2438.

CHLORAMBUCIL [U;INN]

CAS	305-03-3	MW: 304.22	MF:	C14 H19 Cl2 N O2

Antineoplastic

Burroughs Wellcome (Leukeran)

LgP	3.3(C); 1.70(M, pH 7.4)
pKa	5.75
pp cited in Vols.1-5:	
	1: 69; 4: 531, 532

(1) Everett et al., J. Chem. Soc. 1953, 2390.
(2) Van Putten et al., Eur.J. Cancer, 1971, 7, 11.

CHLORAMINE-T [INN]

CAS	127-65-1	MW: 227.65	MF:	C7 H7 Cl N Na O2 S

Antibacterial

LgP	(4)
pKa	
pp cited in Vols.1-5:	
	1: 52

(1) F.J. Welcher, 'Organic Analytical Reagents', Van Nostrand, NY, 1948, vol.4, pp 316-320.

CHLORAMPHENICOL [U;INN]

CAS	56-75-7	MW: 323.13	MF:	C11 H12 Cl2 N2 O5

Antibacterial; antirickettsial

Parke-Davis

LgP	1.14(M)
pKa	
pp cited in Vols.1-5:	
	1: 46; 2: 92-96, 106, 110, 116, 726, 738, 827-830; 4: 229, 454; 5: 101,118, 128, 149 195, 325, 553, 557

(1) Bartz, J. Biol. Chem., 1948, 172, 445. (2) M. Rebstock et al., J. Am. Chem. Soc., 1949, 71, 2458. (3) O. Pongs in 'Antibiotics', ed. F.E.Hahn, Springer-Verlag, NY, 1979, vol. 5, pt.1, pp 26-42.

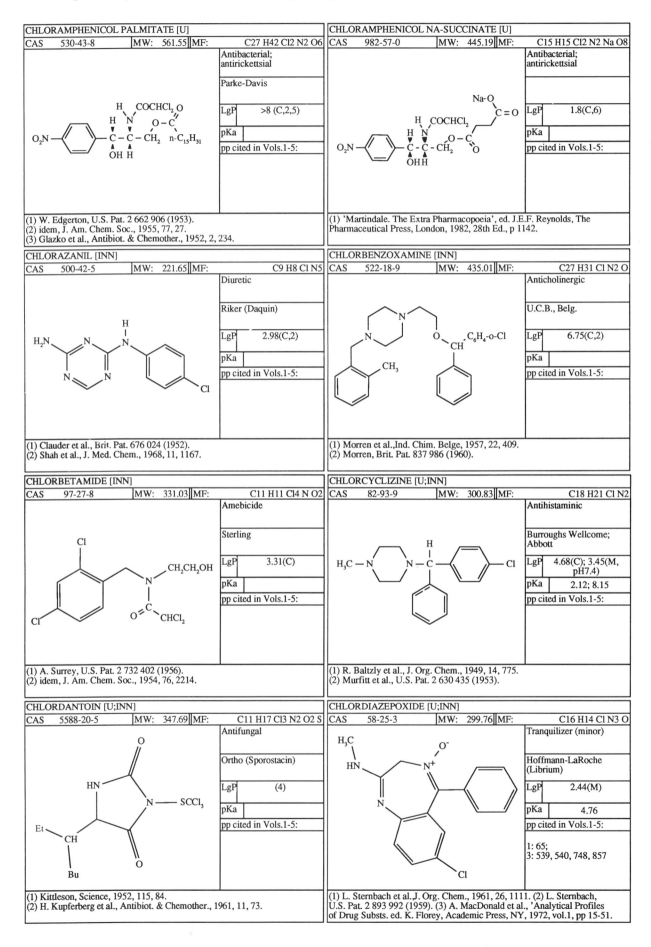

CHLORAMPHENICOL PALMITATE [U]			
CAS 530-43-8	MW: 561.55	MF:	C27 H42 Cl2 N2 O6

Antibacterial; antirickettsial	
Parke-Davis	
LgP	>8 (C,2,5)
pKa	
pp cited in Vols.1-5:	

(1) W. Edgerton, U.S. Pat. 2 662 906 (1953).
(2) idem, J. Am. Chem. Soc., 1955, 77, 27.
(3) Glazko et al., Antibiot. & Chemother., 1952, 2, 234.

CHLORAMPHENICOL NA-SUCCINATE [U]			
CAS 982-57-0	MW: 445.19	MF:	C15 H15 Cl2 N2 Na O8

Antibacterial; antirickettsial	
LgP	1.8(C,6)
pKa	
pp cited in Vols.1-5:	

(1) 'Martindale. The Extra Pharmacopoeia', ed. J.E.F. Reynolds, The Pharmaceutical Press, London, 1982, 28th Ed., p 1142.

CHLORAZANIL [INN]			
CAS 500-42-5	MW: 221.65	MF:	C9 H8 Cl N5

Diuretic	
Riker (Daquin)	
LgP	2.98(C,2)
pKa	
pp cited in Vols.1-5:	

(1) Clauder et al., Brit. Pat. 676 024 (1952).
(2) Shah et al., J. Med. Chem., 1968, 11, 1167.

CHLORBENZOXAMINE [INN]			
CAS 522-18-9	MW: 435.01	MF:	C27 H31 Cl N2 O

Anticholinergic	
U.C.B., Belg.	
LgP	6.75(C,2)
pKa	
pp cited in Vols.1-5:	

(1) Morren et al.,Ind. Chim. Belge, 1957, 22, 409.
(2) Morren, Brit. Pat. 837 986 (1960).

CHLORBETAMIDE [INN]			
CAS 97-27-8	MW: 331.03	MF:	C11 H11 Cl4 N O2

Amebicide	
Sterling	
LgP	3.31(C)
pKa	
pp cited in Vols.1-5:	

(1) A. Surrey, U.S. Pat. 2 732 402 (1956).
(2) idem, J. Am. Chem. Soc., 1954, 76, 2214.

CHLORCYCLIZINE [U;INN]			
CAS 82-93-9	MW: 300.83	MF:	C18 H21 Cl N2

Antihistaminic	
Burroughs Wellcome; Abbott	
LgP	4.68(C); 3.45(M, pH7.4)
pKa	2.12; 8.15
pp cited in Vols.1-5:	

(1) R. Baltzly et al., J. Org. Chem., 1949, 14, 775.
(2) Murfitt et al., U.S. Pat. 2 630 435 (1953).

CHLORDANTOIN [U;INN]			
CAS 5588-20-5	MW: 347.69	MF:	C11 H17 Cl3 N2 O2 S

Antifungal	
Ortho (Sporostacin)	
LgP	(4)
pKa	
pp cited in Vols.1-5:	

(1) Kittleson, Science, 1952, 115, 84.
(2) H. Kupferberg et al., Antibiot. & Chemother., 1961, 11, 73.

CHLORDIAZEPOXIDE [U;INN]			
CAS 58-25-3	MW: 299.76	MF:	C16 H14 Cl N3 O

Tranquilizer (minor)	
Hoffmann-LaRoche (Librium)	
LgP	2.44(M)
pKa	4.76
pp cited in Vols.1-5:	
1: 65; 3: 539, 540, 748, 857	

(1) L. Sternbach et al.J. Org. Chem., 1961, 26, 1111. (2) L. Sternbach, U.S. Pat. 2 893 992 (1959). (3) A. MacDonald et al., 'Analytical Profiles of Drug Substs. ed. K. Florey, Academic Press, NY, 1972, vol.1, pp 15-51.

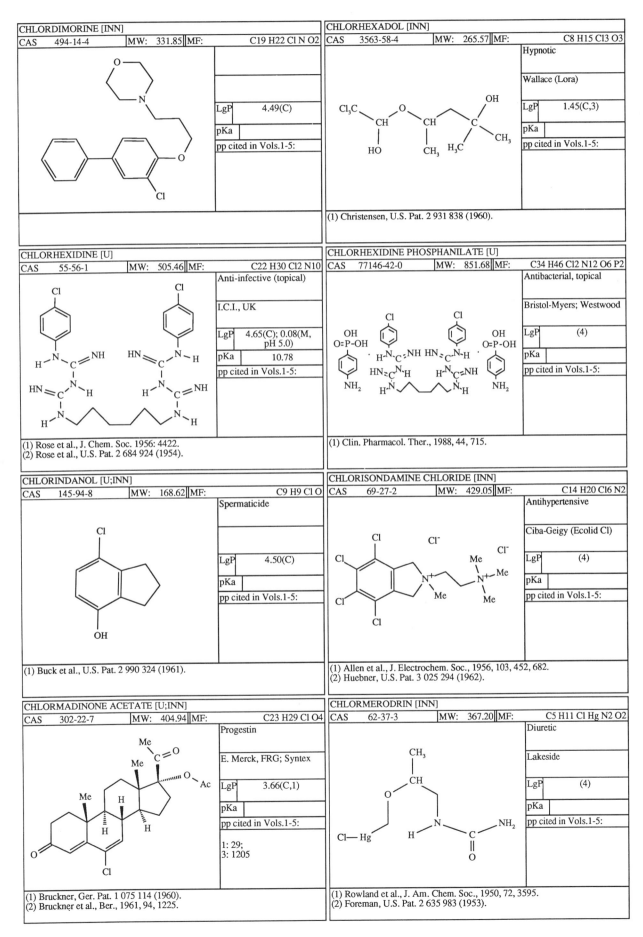

CHLORDIMORINE [INN]

CAS	494-14-4	MW:	331.85	MF:	C19 H22 Cl N O2

LgP	4.49(C)
pKa	
pp cited in Vols.1-5:	

CHLORHEXADOL [INN]

CAS	3563-58-4	MW:	265.57	MF:	C8 H15 Cl3 O3

Hypnotic

Wallace (Lora)

LgP	1.45(C,3)
pKa	
pp cited in Vols.1-5:	

(1) Christensen, U.S. Pat. 2 931 838 (1960).

CHLORHEXIDINE [U]

CAS	55-56-1	MW:	505.46	MF:	C22 H30 Cl2 N10

Anti-infective (topical)

I.C.I., UK

LgP	4.65(C); 0.08(M, pH 5.0)
pKa	10.78
pp cited in Vols.1-5:	

(1) Rose et al., J. Chem. Soc. 1956: 4422.
(2) Rose et al., U.S. Pat. 2 684 924 (1954).

CHLORHEXIDINE PHOSPHANILATE [U]

CAS	77146-42-0	MW:	851.68	MF:	C34 H46 Cl2 N12 O6 P2

Antibacterial, topical

Bristol-Myers; Westwood

LgP	(4)
pKa	
pp cited in Vols.1-5:	

(1) Clin. Pharmacol. Ther., 1988, 44, 715.

CHLORINDANOL [U;INN]

CAS	145-94-8	MW:	168.62	MF:	C9 H9 Cl O

Spermaticide

LgP	4.50(C)
pKa	
pp cited in Vols.1-5:	

(1) Buck et al., U.S. Pat. 2 990 324 (1961).

CHLORISONDAMINE CHLORIDE [INN]

CAS	69-27-2	MW:	429.05	MF:	C14 H20 Cl6 N2

Antihypertensive

Ciba-Geigy (Ecolid Cl)

LgP	(4)
pKa	
pp cited in Vols.1-5:	

(1) Allen et al., J. Electrochem. Soc., 1956, 103, 452, 682.
(2) Huebner, U.S. Pat. 3 025 294 (1962).

CHLORMADINONE ACETATE [U;INN]

CAS	302-22-7	MW:	404.94	MF:	C23 H29 Cl O4

Progestin

E. Merck, FRG; Syntex

LgP	3.66(C,1)
pKa	
pp cited in Vols.1-5:	

1: 29;
3: 1205

(1) Bruckner, Ger. Pat. 1 075 114 (1960).
(2) Bruckner et al., Ber., 1961, 94, 1225.

CHLORMERODRIN [INN]

CAS	62-37-3	MW:	367.20	MF:	C5 H11 Cl Hg N2 O2

Diuretic

Lakeside

LgP	(4)
pKa	
pp cited in Vols.1-5:	

(1) Rowland et al., J. Am. Chem. Soc., 1950, 72, 3595.
(2) Foreman, U.S. Pat. 2 635 983 (1953).

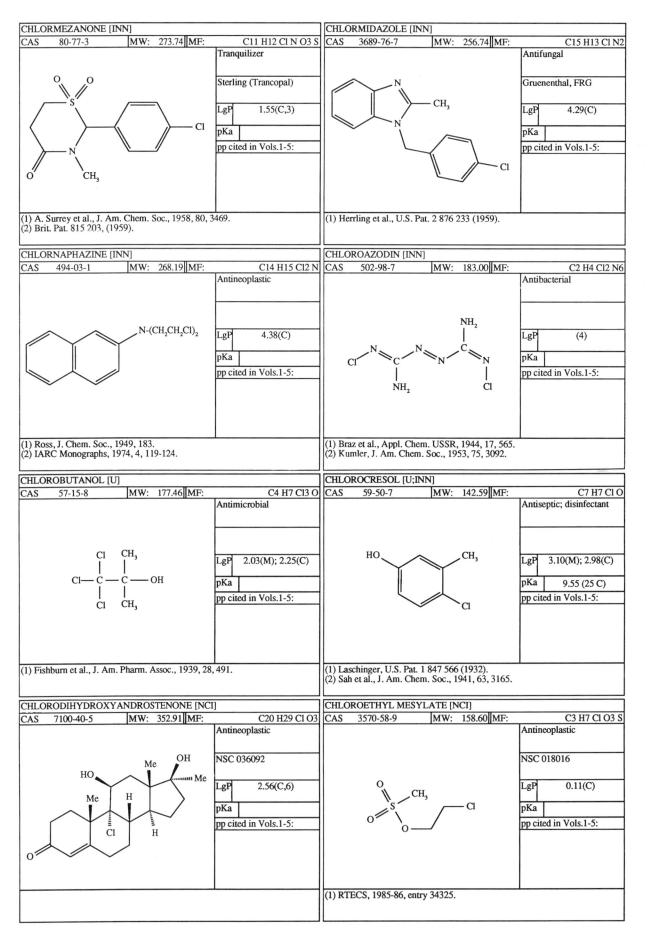

CHLORMEZANONE [INN]

CAS	80-77-3	MW:	273.74	MF:	C11 H12 Cl N O3 S

Tranquilizer

Sterling (Trancopal)

LgP	1.55(C,3)
pKa	
pp cited in Vols.1-5:	

(1) A. Surrey et al., J. Am. Chem. Soc., 1958, 80, 3469.
(2) Brit. Pat. 815 203, (1959).

CHLORMIDAZOLE [INN]

CAS	3689-76-7	MW:	256.74	MF:	C15 H13 Cl N2

Antifungal

Gruenenthal, FRG

LgP	4.29(C)
pKa	
pp cited in Vols.1-5:	

(1) Herrling et al., U.S. Pat. 2 876 233 (1959).

CHLORNAPHAZINE [INN]

CAS	494-03-1	MW:	268.19	MF:	C14 H15 Cl2 N

Antineoplastic

LgP	4.38(C)
pKa	
pp cited in Vols.1-5:	

(1) Ross, J. Chem. Soc., 1949, 183.
(2) IARC Monographs, 1974, 4, 119-124.

CHLOROAZODIN [INN]

CAS	502-98-7	MW:	183.00	MF:	C2 H4 Cl2 N6

Antibacterial

LgP	(4)
pKa	
pp cited in Vols.1-5:	

(1) Braz et al., Appl. Chem. USSR, 1944, 17, 565.
(2) Kumler, J. Am. Chem. Soc., 1953, 75, 3092.

CHLOROBUTANOL [U]

CAS	57-15-8	MW:	177.46	MF:	C4 H7 Cl3 O

Antimicrobial

LgP	2.03(M); 2.25(C)
pKa	
pp cited in Vols.1-5:	

(1) Fishburn et al., J. Am. Pharm. Assoc., 1939, 28, 491.

CHLOROCRESOL [U;INN]

CAS	59-50-7	MW:	142.59	MF:	C7 H7 Cl O

Antiseptic; disinfectant

LgP	3.10(M); 2.98(C)
pKa	9.55 (25 C)
pp cited in Vols.1-5:	

(1) Laschinger, U.S. Pat. 1 847 566 (1932).
(2) Sah et al., J. Am. Chem. Soc., 1941, 63, 3165.

CHLORODIHYDROXYANDROSTENONE [NCI]

CAS	7100-40-5	MW:	352.91	MF:	C20 H29 Cl O3

Antineoplastic

NSC 036092

LgP	2.56(C,6)
pKa	
pp cited in Vols.1-5:	

CHLOROETHYL MESYLATE [NCI]

CAS	3570-58-9	MW:	158.60	MF:	C3 H7 Cl O3 S

Antineoplastic

NSC 018016

LgP	0.11(C)
pKa	
pp cited in Vols.1-5:	

(1) RTECS, 1985-86, entry 34325.

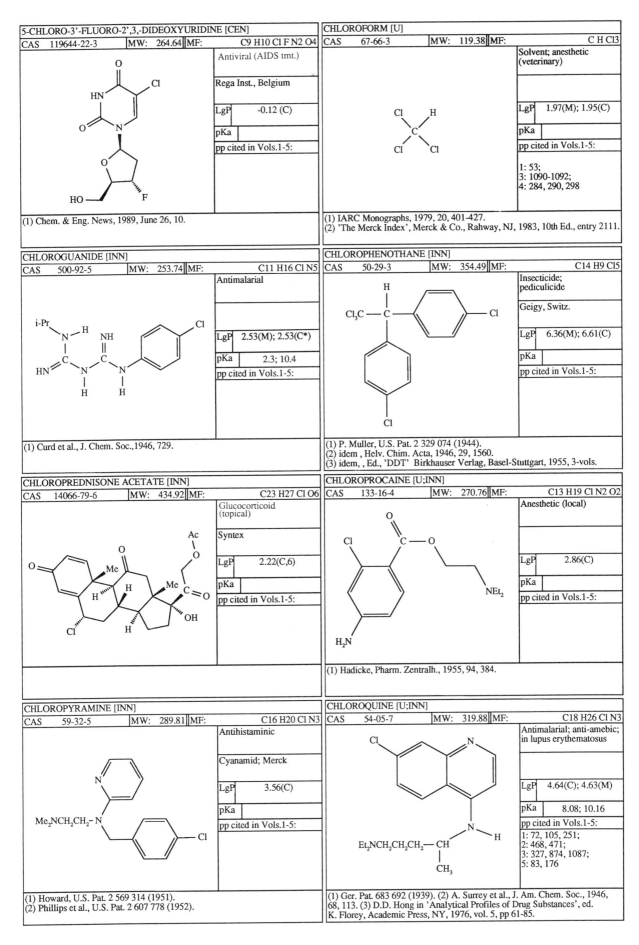

5-CHLORO-3'-FLUORO-2',3,-DIDEOXYURIDINE [CEN]

| CAS | 119644-22-3 | MW: | 264.64 | MF: | C9 H10 Cl F N2 O4 |

Antiviral (AIDS tmt.)

Rega Inst., Belgium

LgP	-0.12 (C)
pKa	
pp cited in Vols.1-5:	

(1) Chem. & Eng. News, 1989, June 26, 10.

CHLOROFORM [U]

| CAS | 67-66-3 | MW: | 119.38 | MF: | C H Cl3 |

Solvent; anesthetic (veterinary)

LgP	1.97(M); 1.95(C)
pKa	
pp cited in Vols.1-5:	

1: 53;
3: 1090-1092;
4: 284, 290, 298

(1) IARC Monographs, 1979, 20, 401-427.
(2) 'The Merck Index', Merck & Co., Rahway, NJ, 1983, 10th Ed., entry 2111.

CHLOROGUANIDE [INN]

| CAS | 500-92-5 | MW: | 253.74 | MF: | C11 H16 Cl N5 |

Antimalarial

LgP	2.53(M); 2.53(C*)
pKa	2.3; 10.4
pp cited in Vols.1-5:	

(1) Curd et al., J. Chem. Soc.,1946, 729.

CHLOROPHENOTHANE [INN]

| CAS | 50-29-3 | MW: | 354.49 | MF: | C14 H9 Cl5 |

Insecticide; pediculicide

Geigy, Switz.

LgP	6.36(M); 6.61(C)
pKa	
pp cited in Vols.1-5:	

(1) P. Muller, U.S. Pat. 2 329 074 (1944).
(2) idem , Helv. Chim. Acta, 1946, 29, 1560.
(3) idem, , Ed., 'DDT' Birkhauser Verlag, Basel-Stuttgart, 1955, 3-vols.

CHLOROPREDNISONE ACETATE [INN]

| CAS | 14066-79-6 | MW: | 434.92 | MF: | C23 H27 Cl O6 |

Glucocorticoid (topical)

Syntex

LgP	2.22(C,6)
pKa	
pp cited in Vols.1-5:	

CHLOROPROCAINE [U;INN]

| CAS | 133-16-4 | MW: | 270.76 | MF: | C13 H19 Cl N2 O2 |

Anesthetic (local)

LgP	2.86(C)
pKa	
pp cited in Vols.1-5:	

(1) Hadicke, Pharm. Zentralh., 1955, 94, 384.

CHLOROPYRAMINE [INN]

| CAS | 59-32-5 | MW: | 289.81 | MF: | C16 H20 Cl N3 |

Antihistaminic

Cyanamid; Merck

LgP	3.56(C)
pKa	
pp cited in Vols.1-5:	

(1) Howard, U.S. Pat. 2 569 314 (1951).
(2) Phillips et al., U.S. Pat. 2 607 778 (1952).

CHLOROQUINE [U;INN]

| CAS | 54-05-7 | MW: | 319.88 | MF: | C18 H26 Cl N3 |

Antimalarial; anti-amebic; in lupus erythematosus

LgP	4.64(C); 4.63(M)
pKa	8.08; 10.16
pp cited in Vols.1-5:	

1: 72, 105, 251;
2: 468, 471;
3: 327, 874, 1087;
5: 83, 176

(1) Ger. Pat. 683 692 (1939). (2) A. Surrey et al., J. Am. Chem. Soc., 1946, 68, 113. (3) D.D. Hong in 'Analytical Profiles of Drug Substances', ed. K. Florey, Academic Press, NY, 1976, vol. 5, pp 61-85.

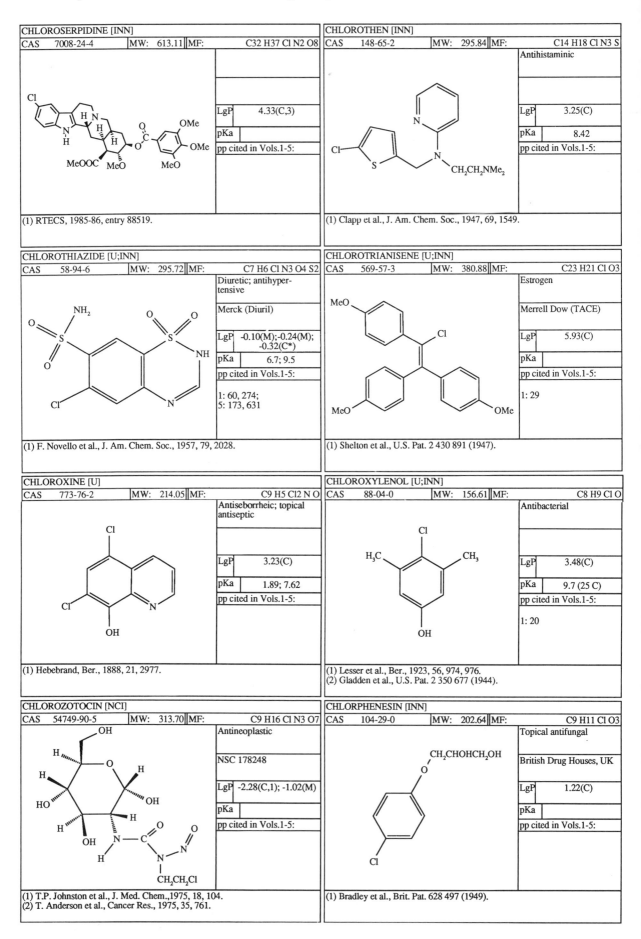

CHLOROSERPIDINE [INN]

CAS	7008-24-4	MW: 613.11	MF:	C32 H37 Cl N2 O8

LgP	4.33(C,3)
pKa	
pp cited in Vols.1-5:	

(1) RTECS, 1985-86, entry 88519.

CHLOROTHEN [INN]

CAS	148-65-2	MW: 295.84	MF:	C14 H18 Cl N3 S

Antihistaminic

LgP	3.25(C)
pKa	8.42
pp cited in Vols.1-5:	

(1) Clapp et al., J. Am. Chem. Soc., 1947, 69, 1549.

CHLOROTHIAZIDE [U;INN]

CAS	58-94-6	MW: 295.72	MF:	C7 H6 Cl N3 O4 S2

Diuretic; antihypertensive

Merck (Diuril)

LgP	-0.10(M);-0.24(M); -0.32(C*)
pKa	6.7; 9.5
pp cited in Vols.1-5:	1: 60, 274; 5: 173, 631

(1) F. Novello et al., J. Am. Chem. Soc., 1957, 79, 2028.

CHLOROTRIANISENE [U;INN]

CAS	569-57-3	MW: 380.88	MF:	C23 H21 Cl O3

Estrogen

Merrell Dow (TACE)

LgP	5.93(C)
pKa	
pp cited in Vols.1-5:	1: 29

(1) Shelton et al., U.S. Pat. 2 430 891 (1947).

CHLOROXINE [U]

CAS	773-76-2	MW: 214.05	MF:	C9 H5 Cl2 N O

Antiseborrheic; topical antiseptic

LgP	3.23(C)
pKa	1.89; 7.62
pp cited in Vols.1-5:	

(1) Hebebrand, Ber., 1888, 21, 2977.

CHLOROXYLENOL [U;INN]

CAS	88-04-0	MW: 156.61	MF:	C8 H9 Cl O

Antibacterial

LgP	3.48(C)
pKa	9.7 (25 C)
pp cited in Vols.1-5:	1: 20

(1) Lesser et al., Ber., 1923, 56, 974, 976.
(2) Gladden et al., U.S. Pat. 2 350 677 (1944).

CHLOROZOTOCIN [NCI]

CAS	54749-90-5	MW: 313.70	MF:	C9 H16 Cl N3 O7

Antineoplastic

NSC 178248

LgP	-2.28(C,1); -1.02(M)
pKa	
pp cited in Vols.1-5:	

(1) T.P. Johnston et al., J. Med. Chem.,1975, 18, 104.
(2) T. Anderson et al., Cancer Res., 1975, 35, 761.

CHLORPHENESIN [INN]

CAS	104-29-0	MW: 202.64	MF:	C9 H11 Cl O3

Topical antifungal

British Drug Houses, UK

LgP	1.22(C)
pKa	
pp cited in Vols.1-5:	

(1) Bradley et al., Brit. Pat. 628 497 (1949).

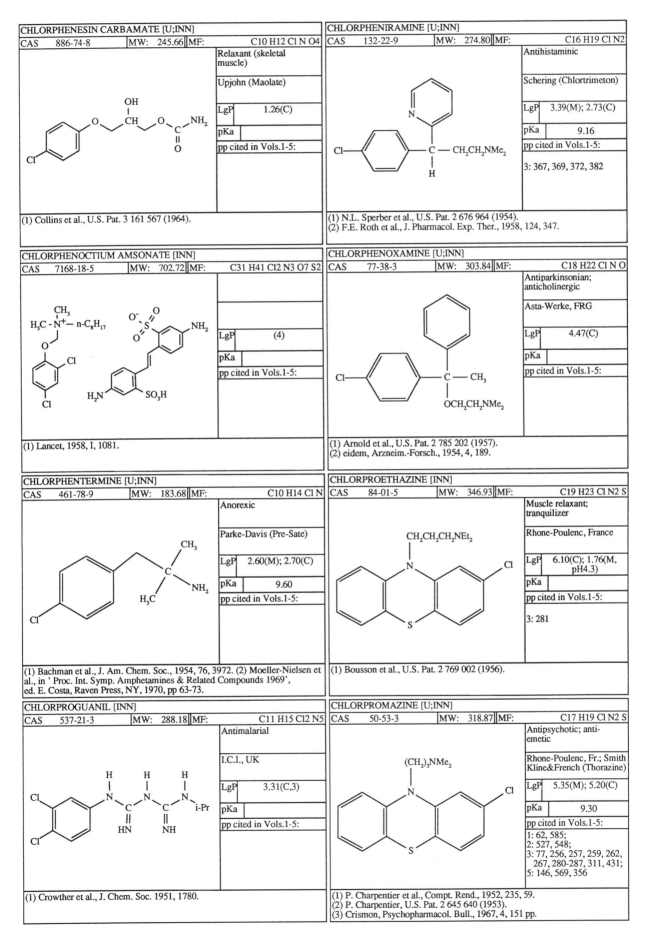

CHLORPHENESIN CARBAMATE [U;INN]

CAS	886-74-8	MW:	245.66	MF:	C10 H12 Cl N O4

	Relaxant (skeletal muscle)
	Upjohn (Maolate)
LgP	1.26(C)
pKa	
pp cited in Vols.1-5:	

(1) Collins et al., U.S. Pat. 3 161 567 (1964).

CHLORPHENOCTIUM AMSONATE [INN]

CAS	7168-18-5	MW:	702.72	MF:	C31 H41 Cl2 N3 O7 S2

LgP	(4)
pKa	
pp cited in Vols.1-5:	

(1) Lancet, 1958, I, 1081.

CHLORPHENTERMINE [U;INN]

CAS	461-78-9	MW:	183.68	MF:	C10 H14 Cl N

	Anorexic
	Parke-Davis (Pre-Sate)
LgP	2.60(M); 2.70(C)
pKa	9.60
pp cited in Vols.1-5:	

(1) Bachman et al., J. Am. Chem. Soc., 1954, 76, 3972. (2) Moeller-Nielsen et al., in ' Proc. Int. Symp. Amphetamines & Related Compounds 1969', ed. E. Costa, Raven Press, NY, 1970, pp 63-73.

CHLORPROGUANIL [INN]

CAS	537-21-3	MW:	288.18	MF:	C11 H15 Cl2 N5

	Antimalarial
	I.C.I., UK
LgP	3.31(C,3)
pKa	
pp cited in Vols.1-5:	

(1) Crowther et al., J. Chem. Soc. 1951, 1780.

CHLORPHENIRAMINE [U;INN]

CAS	132-22-9	MW:	274.80	MF:	C16 H19 Cl N2

	Antihistaminic
	Schering (Chlortrimeton)
LgP	3.39(M); 2.73(C)
pKa	9.16
pp cited in Vols.1-5:	
	3: 367, 369, 372, 382

(1) N.L. Sperber et al., U.S. Pat. 2 676 964 (1954).
(2) F.E. Roth et al., J. Pharmacol. Exp. Ther., 1958, 124, 347.

CHLORPHENOXAMINE [U;INN]

CAS	77-38-3	MW:	303.84	MF:	C18 H22 Cl N O

	Antiparkinsonian; anticholinergic
	Asta-Werke, FRG
LgP	4.47(C)
pKa	
pp cited in Vols.1-5:	

(1) Arnold et al., U.S. Pat. 2 785 202 (1957).
(2) eidem, Arzneim.-Forsch., 1954, 4, 189.

CHLORPROETHAZINE [INN]

CAS	84-01-5	MW:	346.93	MF:	C19 H23 Cl N2 S

	Muscle relaxant; tranquilizer
	Rhone-Poulenc, France
LgP	6.10(C); 1.76(M, pH4.3)
pKa	
pp cited in Vols.1-5:	
	3: 281

(1) Bousson et al., U.S. Pat. 2 769 002 (1956).

CHLORPROMAZINE [U;INN]

CAS	50-53-3	MW:	318.87	MF:	C17 H19 Cl N2 S

	Antipsychotic; anti-emetic
	Rhone-Poulenc, Fr.; Smith Kline&French (Thorazine)
LgP	5.35(M); 5.20(C)
pKa	9.30
pp cited in Vols.1-5:	
	1: 62, 585; 2: 527, 548; 3: 77, 256, 257, 259, 262, 267, 280-287, 311, 431; 5: 146, 569, 356

(1) P. Charpentier et al., Compt. Rend., 1952, 235, 59.
(2) P. Charpentier, U.S. Pat. 2 645 640 (1953).
(3) Crismon, Psychopharmacol. Bull., 1967, 4, 151 pp.

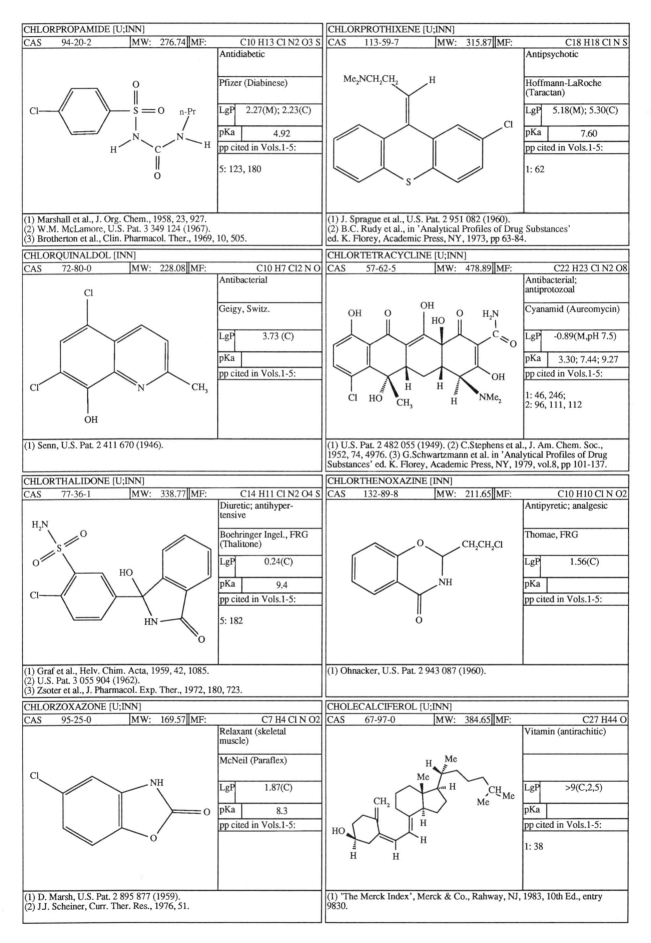

CHLORPROPAMIDE [U;INN]		
CAS 94-20-2	MW: 276.74 MF:	C10 H13 Cl N2 O3 S

Antidiabetic

Pfizer (Diabinese)

LgP	2.27(M); 2.23(C)
pKa	4.92
pp cited in Vols.1-5:	

5: 123, 180

(1) Marshall et al., J. Org. Chem., 1958, 23, 927.
(2) W.M. McLamore, U.S. Pat. 3 349 124 (1967).
(3) Brotherton et al., Clin. Pharmacol. Ther., 1969, 10, 505.

CHLORPROTHIXENE [U;INN]		
CAS 113-59-7	MW: 315.87 MF:	C18 H18 Cl N S

Antipsychotic

Hoffmann-LaRoche (Taractan)

LgP	5.18(M); 5.30(C)
pKa	7.60
pp cited in Vols.1-5:	

1: 62

(1) J. Sprague et al., U.S. Pat. 2 951 082 (1960).
(2) B.C. Rudy et al., in 'Analytical Profiles of Drug Substances'
ed. K. Florey, Academic Press, NY, 1973, pp 63-84.

CHLORQUINALDOL [INN]		
CAS 72-80-0	MW: 228.08 MF:	C10 H7 Cl2 N O

Antibacterial

Geigy, Switz.

LgP	3.73 (C)
pKa	
pp cited in Vols.1-5:	

(1) Senn, U.S. Pat. 2 411 670 (1946).

CHLORTETRACYCLINE [U;INN]		
CAS 57-62-5	MW: 478.89 MF:	C22 H23 Cl N2 O8

Antibacterial; antiprotozoal

Cyanamid (Aureomycin)

LgP	-0.89(M,pH 7.5)
pKa	3.30; 7.44; 9.27
pp cited in Vols.1-5:	

1: 46, 246;
2: 96, 111, 112

(1) U.S. Pat. 2 482 055 (1949). (2) C.Stephens et al., J. Am. Chem. Soc.,
1952, 74, 4976. (3) G.Schwartzmann et al. in 'Analytical Profiles of Drug
Substances' ed. K. Florey, Academic Press, NY, 1979, vol.8, pp 101-137.

CHLORTHALIDONE [U;INN]		
CAS 77-36-1	MW: 338.77 MF:	C14 H11 Cl N2 O4 S

Diuretic; antihypertensive

Boehringer Ingel., FRG (Thalitone)

LgP	0.24(C)
pKa	9.4
pp cited in Vols.1-5:	

5: 182

(1) Graf et al., Helv. Chim. Acta, 1959, 42, 1085.
(2) U.S. Pat. 3 055 904 (1962).
(3) Zsoter et al., J. Pharmacol. Exp. Ther., 1972, 180, 723.

CHLORTHENOXAZINE [INN]		
CAS 132-89-8	MW: 211.65 MF:	C10 H10 Cl N O2

Antipyretic; analgesic

Thomae, FRG

LgP	1.56(C)
pKa	
pp cited in Vols.1-5:	

(1) Ohnacker, U.S. Pat. 2 943 087 (1960).

CHLORZOXAZONE [U;INN]		
CAS 95-25-0	MW: 169.57 MF:	C7 H4 Cl N O2

Relaxant (skeletal muscle)

McNeil (Paraflex)

LgP	1.87(C)
pKa	8.3
pp cited in Vols.1-5:	

(1) D. Marsh, U.S. Pat. 2 895 877 (1959).
(2) J.J. Scheiner, Curr. Ther. Res., 1976, 51.

CHOLECALCIFEROL [U;INN]		
CAS 67-97-0	MW: 384.65 MF:	C27 H44 O

Vitamin (antirachitic)

LgP	>9(C,2,5)
pKa	
pp cited in Vols.1-5:	

1: 38

(1) 'The Merck Index', Merck & Co., Rahway, NJ, 1983, 10th Ed., entry
9830.

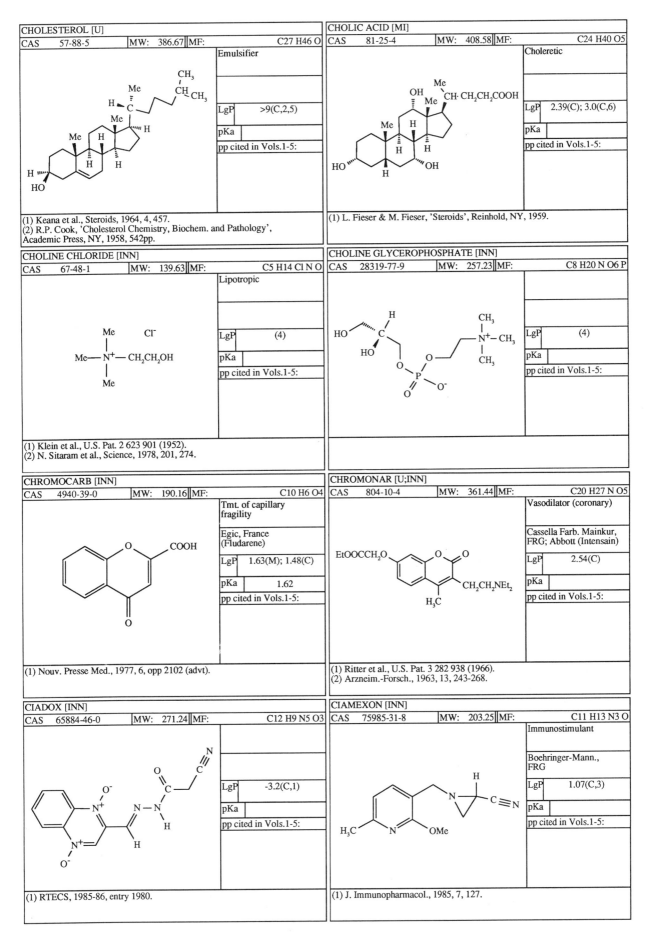

CHOLESTEROL [U]

CAS	57-88-5	MW: 386.67	MF:	C27 H46 O

Emulsifier

LgP	>9(C,2,5)
pKa	
pp cited in Vols.1-5:	

(1) Keana et al., Steroids, 1964, 4, 457.
(2) R.P. Cook, 'Cholesterol Chemistry, Biochem. and Pathology',
Academic Press, NY, 1958, 542pp.

CHOLINE CHLORIDE [INN]

CAS	67-48-1	MW: 139.63	MF:	C5 H14 Cl N O

Lipotropic

LgP	(4)
pKa	
pp cited in Vols.1-5:	

(1) Klein et al., U.S. Pat. 2 623 901 (1952).
(2) N. Sitaram et al., Science, 1978, 201, 274.

CHROMOCARB [INN]

CAS	4940-39-0	MW: 190.16	MF:	C10 H6 O4

Tmt. of capillary fragility

Egic, France
(Fludarene)

LgP	1.63(M); 1.48(C)
pKa	1.62
pp cited in Vols.1-5:	

(1) Nouv. Presse Med., 1977, 6, opp 2102 (advt).

CIADOX [INN]

CAS	65884-46-0	MW: 271.24	MF:	C12 H9 N5 O3

LgP	-3.2(C,1)
pKa	
pp cited in Vols.1-5:	

(1) RTECS, 1985-86, entry 1980.

CHOLIC ACID [MI]

CAS	81-25-4	MW: 408.58	MF:	C24 H40 O5

Choleretic

LgP	2.39(C); 3.0(C,6)
pKa	
pp cited in Vols.1-5:	

(1) L. Fieser & M. Fieser, 'Steroids', Reinhold, NY, 1959.

CHOLINE GLYCEROPHOSPHATE [INN]

CAS	28319-77-9	MW: 257.23	MF:	C8 H20 N O6 P

LgP	(4)
pKa	
pp cited in Vols.1-5:	

CHROMONAR [U;INN]

CAS	804-10-4	MW: 361.44	MF:	C20 H27 N O5

Vasodilator (coronary)

Cassella Farb. Mainkur,
FRG; Abbott (Intensain)

LgP	2.54(C)
pKa	
pp cited in Vols.1-5:	

(1) Ritter et al., U.S. Pat. 3 282 938 (1966).
(2) Arzneim.-Forsch., 1963, 13, 243-268.

CIAMEXON [INN]

CAS	75985-31-8	MW: 203.25	MF:	C11 H13 N3 O

Immunostimulant

Boehringer-Mann.,
FRG

LgP	1.07(C,3)
pKa	
pp cited in Vols.1-5:	

(1) J. Immunopharmacol., 1985, 7, 127.

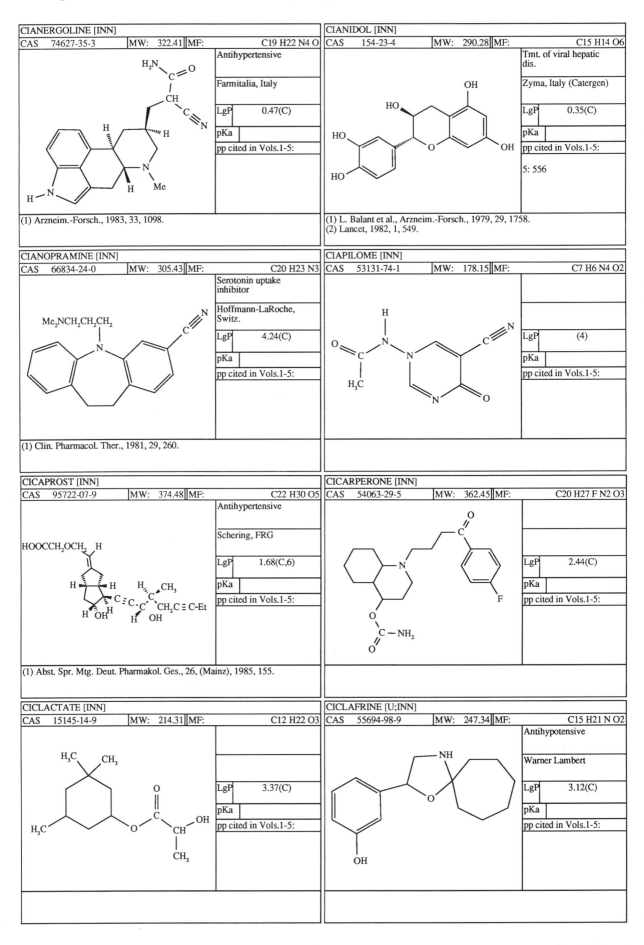

CIANERGOLINE [INN]

| CAS | 74627-35-3 | MW: | 322.41 | MF: | C19 H22 N4 O |

Antihypertensive

Farmitalia, Italy

LgP	0.47(C)
pKa	
pp cited in Vols.1-5:	

(1) Arzneim.-Forsch., 1983, 33, 1098.

CIANIDOL [INN]

| CAS | 154-23-4 | MW: | 290.28 | MF: | C15 H14 O6 |

Tmt. of viral hepatic dis.

Zyma, Italy (Catergen)

LgP	0.35(C)
pKa	
pp cited in Vols.1-5:	

5: 556

(1) L. Balant et al., Arzneim.-Forsch., 1979, 29, 1758.
(2) Lancet, 1982, 1, 549.

CIANOPRAMINE [INN]

| CAS | 66834-24-0 | MW: | 305.43 | MF: | C20 H23 N3 |

Serotonin uptake inhibitor

Hoffmann-LaRoche, Switz.

LgP	4.24(C)
pKa	
pp cited in Vols.1-5:	

(1) Clin. Pharmacol. Ther., 1981, 29, 260.

CIAPILOME [INN]

| CAS | 53131-74-1 | MW: | 178.15 | MF: | C7 H6 N4 O2 |

LgP	(4)
pKa	
pp cited in Vols.1-5:	

CICAPROST [INN]

| CAS | 95722-07-9 | MW: | 374.48 | MF: | C22 H30 O5 |

Antihypertensive

Schering, FRG

LgP	1.68(C,6)
pKa	
pp cited in Vols.1-5:	

(1) Abst. Spr. Mtg. Deut. Pharmakol. Ges., 26, (Mainz), 1985, 155.

CICARPERONE [INN]

| CAS | 54063-29-5 | MW: | 362.45 | MF: | C20 H27 F N2 O3 |

LgP	2.44(C)
pKa	
pp cited in Vols.1-5:	

CICLACTATE [INN]

| CAS | 15145-14-9 | MW: | 214.31 | MF: | C12 H22 O3 |

LgP	3.37(C)
pKa	
pp cited in Vols.1-5:	

CICLAFRINE [U;INN]

| CAS | 55694-98-9 | MW: | 247.34 | MF: | C15 H21 N O2 |

Antihypotensive

Warner Lambert

LgP	3.12(C)
pKa	
pp cited in Vols.1-5:	

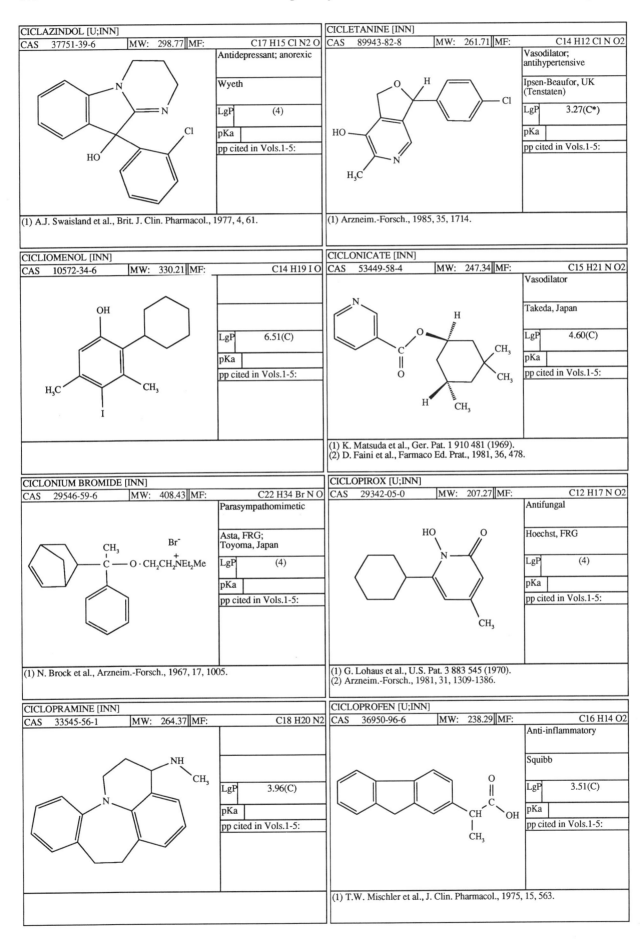

CICLAZINDOL [U;INN]

CAS	37751-39-6	MW:	298.77	MF:	C17 H15 Cl N2 O

Antidepressant; anorexic

Wyeth

LgP	(4)
pKa	
pp cited in Vols.1-5:	

(1) A.J. Swaisland et al., Brit. J. Clin. Pharmacol., 1977, 4, 61.

CICLETANINE [INN]

CAS	89943-82-8	MW:	261.71	MF:	C14 H12 Cl N O2

Vasodilator; antihypertensive

Ipsen-Beaufor, UK (Tenstaten)

LgP	3.27(C*)
pKa	
pp cited in Vols.1-5:	

(1) Arzneim.-Forsch., 1985, 35, 1714.

CICLIOMENOL [INN]

CAS	10572-34-6	MW:	330.21	MF:	C14 H19 I O

LgP	6.51(C)
pKa	
pp cited in Vols.1-5:	

CICLONICATE [INN]

CAS	53449-58-4	MW:	247.34	MF:	C15 H21 N O2

Vasodilator

Takeda, Japan

LgP	4.60(C)
pKa	
pp cited in Vols.1-5:	

(1) K. Matsuda et al., Ger. Pat. 1 910 481 (1969).
(2) D. Faini et al., Farmaco Ed. Prat., 1981, 36, 478.

CICLONIUM BROMIDE [INN]

CAS	29546-59-6	MW:	408.43	MF:	C22 H34 Br N O

Parasympathomimetic

Asta, FRG; Toyoma, Japan

LgP	(4)
pKa	
pp cited in Vols.1-5:	

(1) N. Brock et al., Arzneim.-Forsch., 1967, 17, 1005.

CICLOPIROX [U;INN]

CAS	29342-05-0	MW:	207.27	MF:	C12 H17 N O2

Antifungal

Hoechst, FRG

LgP	(4)
pKa	
pp cited in Vols.1-5:	

(1) G. Lohaus et al., U.S. Pat. 3 883 545 (1970).
(2) Arzneim.-Forsch., 1981, 31, 1309-1386.

CICLOPRAMINE [INN]

CAS	33545-56-1	MW:	264.37	MF:	C18 H20 N2

LgP	3.96(C)
pKa	
pp cited in Vols.1-5:	

CICLOPROFEN [U;INN]

CAS	36950-96-6	MW:	238.29	MF:	C16 H14 O2

Anti-inflammatory

Squibb

LgP	3.51(C)
pKa	
pp cited in Vols.1-5:	

(1) T.W. Mischler et al., J. Clin. Pharmacol., 1975, 15, 563.

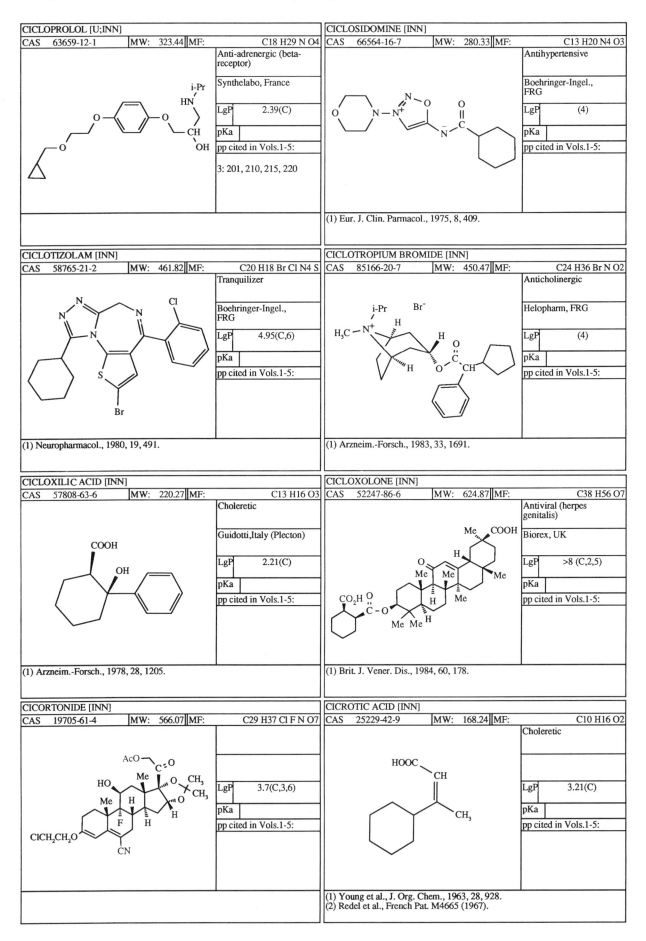

CICLOPROLOL [U;INN]

| CAS | 63659-12-1 | MW: | 323.44 | MF: | C18 H29 N O4 |

Anti-adrenergic (beta-receptor)

Synthelabo, France

LgP 2.39(C)

pKa

pp cited in Vols.1-5:

3: 201, 210, 215, 220

CICLOSIDOMINE [INN]

| CAS | 66564-16-7 | MW: | 280.33 | MF: | C13 H20 N4 O3 |

Antihypertensive

Boehringer-Ingel., FRG

LgP (4)

pKa

pp cited in Vols.1-5:

(1) Eur. J. Clin. Parmacol., 1975, 8, 409.

CICLOTIZOLAM [INN]

| CAS | 58765-21-2 | MW: | 461.82 | MF: | C20 H18 Br Cl N4 S |

Tranquilizer

Boehringer-Ingel., FRG

LgP 4.95(C,6)

pKa

pp cited in Vols.1-5:

(1) Neuropharmacol., 1980, 19, 491.

CICLOTROPIUM BROMIDE [INN]

| CAS | 85166-20-7 | MW: | 450.47 | MF: | C24 H36 Br N O2 |

Anticholinergic

Helopharm, FRG

LgP (4)

pKa

pp cited in Vols.1-5:

(1) Arzneim.-Forsch., 1983, 33, 1691.

CICLOXILIC ACID [INN]

| CAS | 57808-63-6 | MW: | 220.27 | MF: | C13 H16 O3 |

Choleretic

Guidotti,Italy (Plecton)

LgP 2.21(C)

pKa

pp cited in Vols.1-5:

(1) Arzneim.-Forsch., 1978, 28, 1205.

CICLOXOLONE [INN]

| CAS | 52247-86-6 | MW: | 624.87 | MF: | C38 H56 O7 |

Antiviral (herpes genitalis)

Biorex, UK

LgP >8 (C,2,5)

pKa

pp cited in Vols.1-5:

(1) Brit. J. Vener. Dis., 1984, 60, 178.

CICORTONIDE [INN]

| CAS | 19705-61-4 | MW: | 566.07 | MF: | C29 H37 Cl F N O7 |

LgP 3.7(C,3,6)

pKa

pp cited in Vols.1-5:

CICROTIC ACID [INN]

| CAS | 25229-42-9 | MW: | 168.24 | MF: | C10 H16 O2 |

Choleretic

LgP 3.21(C)

pKa

pp cited in Vols.1-5:

(1) Young et al., J. Org. Chem., 1963, 28, 928.
(2) Redel et al., French Pat. M4665 (1967).

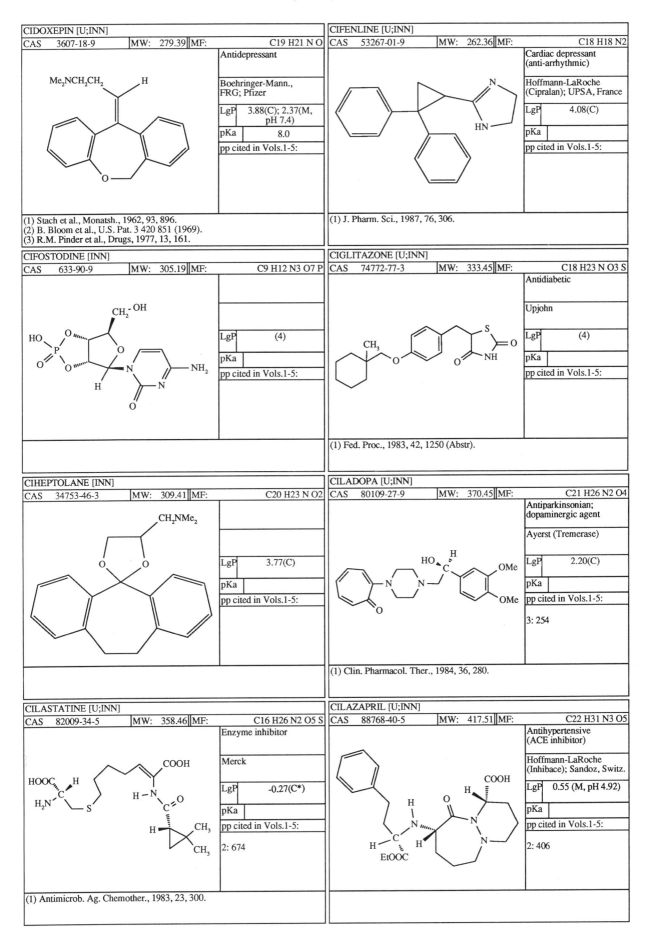

CIDOXEPIN [U;INN]				
CAS 3607-18-9		MW: 279.39	MF:	C19 H21 N O

Antidepressant

Boehringer-Mann., FRG; Pfizer

LgP	3.88(C); 2.37(M, pH 7.4)
pKa	8.0
pp cited in Vols.1-5:	

(1) Stach et al., Monatsh., 1962, 93, 896.
(2) B. Bloom et al., U.S. Pat. 3 420 851 (1969).
(3) R.M. Pinder et al., Drugs, 1977, 13, 161.

CIFENLINE [U;INN]				
CAS 53267-01-9		MW: 262.36	MF:	C18 H18 N2

Cardiac depressant (anti-arrhythmic)

Hoffmann-LaRoche (Cipralan); UPSA, France

LgP	4.08(C)
pKa	
pp cited in Vols.1-5:	

(1) J. Pharm. Sci., 1987, 76, 306.

CIFOSTODINE [INN]				
CAS 633-90-9		MW: 305.19	MF:	C9 H12 N3 O7 P

LgP	(4)
pKa	
pp cited in Vols.1-5:	

CIGLITAZONE [U;INN]				
CAS 74772-77-3		MW: 333.45	MF:	C18 H23 N O3 S

Antidiabetic

Upjohn

LgP	(4)
pKa	
pp cited in Vols.1-5:	

(1) Fed. Proc., 1983, 42, 1250 (Abstr).

CIHEPTOLANE [INN]				
CAS 34753-46-3		MW: 309.41	MF:	C20 H23 N O2

LgP	3.77(C)
pKa	
pp cited in Vols.1-5:	

CILADOPA [U;INN]				
CAS 80109-27-9		MW: 370.45	MF:	C21 H26 N2 O4

Antiparkinsonian; dopaminergic agent

Ayerst (Tremerase)

LgP	2.20(C)
pKa	
pp cited in Vols.1-5:	
3: 254	

(1) Clin. Pharmacol. Ther., 1984, 36, 280.

CILASTATINE [U;INN]				
CAS 82009-34-5		MW: 358.46	MF:	C16 H26 N2 O5 S

Enzyme inhibitor

Merck

LgP	-0.27(C*)
pKa	
pp cited in Vols.1-5:	
2: 674	

(1) Antimicrob. Ag. Chemother., 1983, 23, 300.

CILAZAPRIL [U;INN]				
CAS 88768-40-5		MW: 417.51	MF:	C22 H31 N3 O5

Antihypertensive (ACE inhibitor)

Hoffmann-LaRoche (Inhibace); Sandoz, Switz.

LgP	0.55 (M, pH 4.92)
pKa	
pp cited in Vols.1-5:	
2: 406	

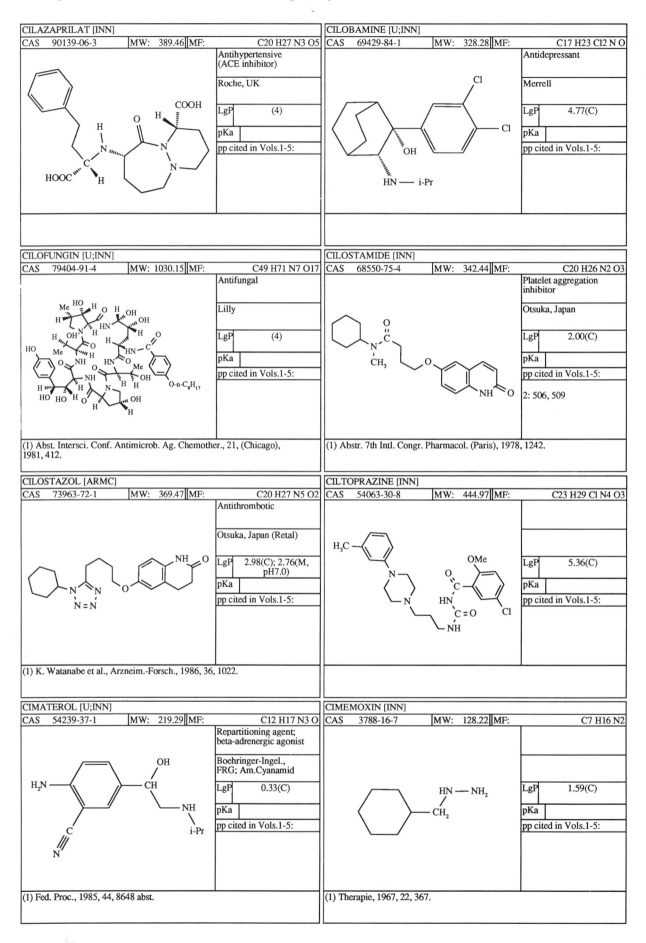

CILAZAPRILAT [INN]		
CAS 90139-06-3	MW: 389.46	MF: C20 H27 N3 O5

Antihypertensive (ACE inhibitor)

Roche, UK

LgP	(4)
pKa	
pp cited in Vols.1-5:	

CILOBAMINE [U;INN]		
CAS 69429-84-1	MW: 328.28	MF: C17 H23 Cl2 N O

Antidepressant

Merrell

LgP	4.77(C)
pKa	
pp cited in Vols.1-5:	

CILOFUNGIN [U;INN]		
CAS 79404-91-4	MW: 1030.15	MF: C49 H71 N7 O17

Antifungal

Lilly

LgP	(4)
pKa	
pp cited in Vols.1-5:	

(1) Abst. Intersci. Conf. Antimicrob. Ag. Chemother., 21, (Chicago), 1981, 412.

CILOSTAMIDE [INN]		
CAS 68550-75-4	MW: 342.44	MF: C20 H26 N2 O3

Platelet aggregation inhibitor

Otsuka, Japan

LgP	2.00(C)
pKa	
pp cited in Vols.1-5:	

2: 506, 509

(1) Abstr. 7th Intl. Congr. Pharmacol. (Paris), 1978, 1242.

CILOSTAZOL [ARMC]		
CAS 73963-72-1	MW: 369.47	MF: C20 H27 N5 O2

Antithrombotic

Otsuka, Japan (Retal)

LgP	2.98(C); 2.76(M, pH7.0)
pKa	
pp cited in Vols.1-5:	

(1) K. Watanabe et al., Arzneim.-Forsch., 1986, 36, 1022.

CILTOPRAZINE [INN]		
CAS 54063-30-8	MW: 444.97	MF: C23 H29 Cl N4 O3

LgP	5.36(C)
pKa	
pp cited in Vols.1-5:	

CIMATEROL [U;INN]		
CAS 54239-37-1	MW: 219.29	MF: C12 H17 N3 O

Repartitioning agent; beta-adrenergic agonist

Boehringer-Ingel., FRG; Am.Cyanamid

LgP	0.33(C)
pKa	
pp cited in Vols.1-5:	

(1) Fed. Proc., 1985, 44, 8648 abst.

CIMEMOXIN [INN]		
CAS 3788-16-7	MW: 128.22	MF: C7 H16 N2

LgP	1.59(C)
pKa	
pp cited in Vols.1-5:	

(1) Therapie, 1967, 22, 367.

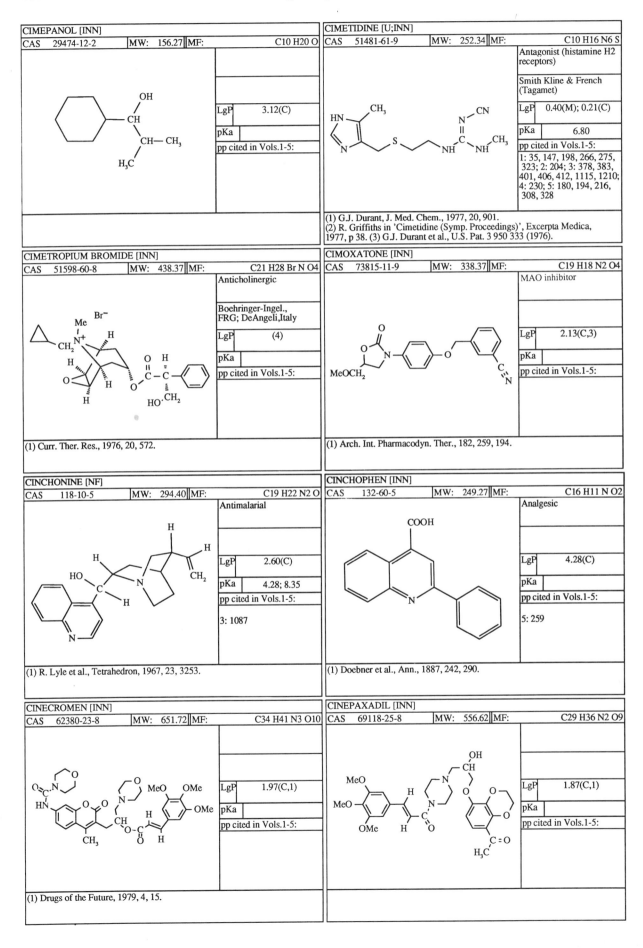

CIMEPANOL [INN]

CAS	29474-12-2	MW:	156.27	MF:	C10 H20 O

LgP	3.12(C)
pKa	
pp cited in Vols.1-5:	

CIMETIDINE [U;INN]

CAS	51481-61-9	MW:	252.34	MF:	C10 H16 N6 S

Antagonist (histamine H2 receptors)

Smith Kline & French (Tagamet)

LgP	0.40(M); 0.21(C)
pKa	6.80
pp cited in Vols.1-5:	

1: 35, 147, 198, 266, 275, 323; 2: 204; 3: 378, 383, 401, 406, 412, 1115, 1210; 4: 230; 5: 180, 194, 216, 308, 328

(1) G.J. Durant, J. Med. Chem., 1977, 20, 901.
(2) R. Griffiths in 'Cimetidine (Symp. Proceedings)', Excerpta Medica, 1977, p 38. (3) G.J. Durant et al., U.S. Pat. 3 950 333 (1976).

CIMETROPIUM BROMIDE [INN]

CAS	51598-60-8	MW:	438.37	MF:	C21 H28 Br N O4

Anticholinergic

Boehringer-Ingel., FRG; DeAngeli,Italy

LgP	(4)
pKa	
pp cited in Vols.1-5:	

(1) Curr. Ther. Res., 1976, 20, 572.

CIMOXATONE [INN]

CAS	73815-11-9	MW:	338.37	MF:	C19 H18 N2 O4

MAO inhibitor

LgP	2.13(C,3)
pKa	
pp cited in Vols.1-5:	

(1) Arch. Int. Pharmacodyn. Ther., 182, 259, 194.

CINCHONINE [NF]

CAS	118-10-5	MW:	294.40	MF:	C19 H22 N2 O

Antimalarial

LgP	2.60(C)
pKa	4.28; 8.35
pp cited in Vols.1-5:	

3: 1087

(1) R. Lyle et al., Tetrahedron, 1967, 23, 3253.

CINCHOPHEN [INN]

CAS	132-60-5	MW:	249.27	MF:	C16 H11 N O2

Analgesic

LgP	4.28(C)
pKa	
pp cited in Vols.1-5:	

5: 259

(1) Doebner et al., Ann., 1887, 242, 290.

CINECROMEN [INN]

CAS	62380-23-8	MW:	651.72	MF:	C34 H41 N3 O10

LgP	1.97(C,1)
pKa	
pp cited in Vols.1-5:	

(1) Drugs of the Future, 1979, 4, 15.

CINEPAXADIL [INN]

CAS	69118-25-8	MW:	556.62	MF:	C29 H36 N2 O9

LgP	1.87(C,1)
pKa	
pp cited in Vols.1-5:	

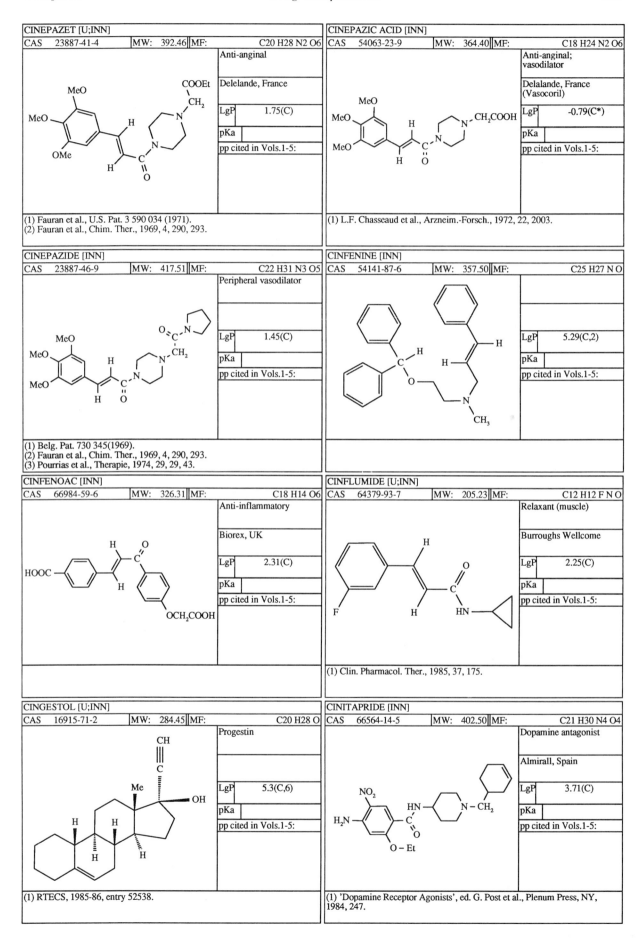

CINEPAZET [U;INN]		
CAS 23887-41-4	MW: 392.46	MF: C20 H28 N2 O6

Anti-anginal

Delelande, France

LgP	1.75(C)
pKa	
pp cited in Vols.1-5:	

(1) Fauran et al., U.S. Pat. 3 590 034 (1971).
(2) Fauran et al., Chim. Ther., 1969, 4, 290, 293.

CINEPAZIC ACID [INN]		
CAS 54063-23-9	MW: 364.40	MF: C18 H24 N2 O6

Anti-anginal;
vasodilator

Delalande, France
(Vasocoril)

LgP	-0.79(C*)
pKa	
pp cited in Vols.1-5:	

(1) L.F. Chasseaud et al., Arzneim.-Forsch., 1972, 22, 2003.

CINEPAZIDE [INN]		
CAS 23887-46-9	MW: 417.51	MF: C22 H31 N3 O5

Peripheral vasodilator

LgP	1.45(C)
pKa	
pp cited in Vols.1-5:	

(1) Belg. Pat. 730 345(1969).
(2) Fauran et al., Chim. Ther., 1969, 4, 290, 293.
(3) Pourrias et al., Therapie, 1974, 29, 29, 43.

CINFENINE [INN]		
CAS 54141-87-6	MW: 357.50	MF: C25 H27 N O

LgP	5.29(C,2)
pKa	
pp cited in Vols.1-5:	

CINFENOAC [INN]		
CAS 66984-59-6	MW: 326.31	MF: C18 H14 O6

Anti-inflammatory

Biorex, UK

LgP	2.31(C)
pKa	
pp cited in Vols.1-5:	

CINFLUMIDE [U;INN]		
CAS 64379-93-7	MW: 205.23	MF: C12 H12 F N O

Relaxant (muscle)

Burroughs Wellcome

LgP	2.25(C)
pKa	
pp cited in Vols.1-5:	

(1) Clin. Pharmacol. Ther., 1985, 37, 175.

CINGESTOL [U;INN]		
CAS 16915-71-2	MW: 284.45	MF: C20 H28 O

Progestin

LgP	5.3(C,6)
pKa	
pp cited in Vols.1-5:	

(1) RTECS, 1985-86, entry 52538.

CINITAPRIDE [INN]		
CAS 66564-14-5	MW: 402.50	MF: C21 H30 N4 O4

Dopamine antagonist

Almirall, Spain

LgP	3.71(C)
pKa	
pp cited in Vols.1-5:	

(1) 'Dopamine Receptor Agonists', ed. G. Post et al., Plenum Press, NY, 1984, 247.

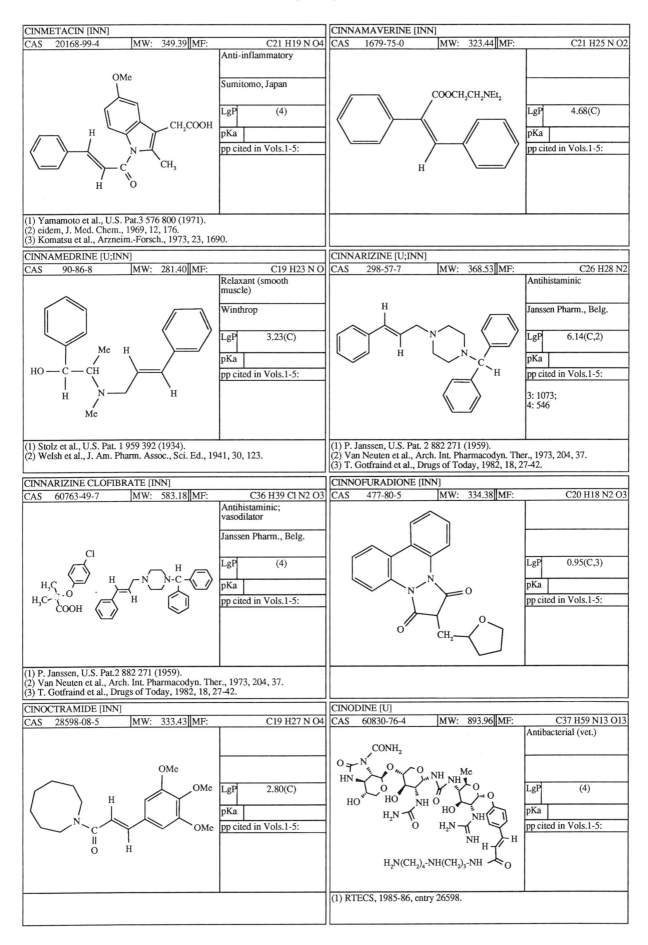

CINMETACIN [INN]

| CAS | 20168-99-4 | MW: | 349.39 | MF: | C21 H19 N O4 |

Anti-inflammatory

Sumitomo, Japan

LgP (4)

pKa

pp cited in Vols.1-5:

(1) Yamamoto et al., U.S. Pat.3 576 800 (1971).
(2) eidem, J. Med. Chem., 1969, 12, 176.
(3) Komatsu et al., Arzneim.-Forsch., 1973, 23, 1690.

CINNAMAVERINE [INN]

| CAS | 1679-75-0 | MW: | 323.44 | MF: | C21 H25 N O2 |

LgP 4.68(C)

pKa

pp cited in Vols.1-5:

CINNAMEDRINE [U;INN]

| CAS | 90-86-8 | MW: | 281.40 | MF: | C19 H23 N O |

Relaxant (smooth muscle)

Winthrop

LgP 3.23(C)

pKa

pp cited in Vols.1-5:

(1) Stolz et al., U.S. Pat. 1 959 392 (1934).
(2) Welsh et al., J. Am. Pharm. Assoc., Sci. Ed., 1941, 30, 123.

CINNARIZINE [U;INN]

| CAS | 298-57-7 | MW: | 368.53 | MF: | C26 H28 N2 |

Antihistaminic

Janssen Pharm., Belg.

LgP 6.14(C,2)

pKa

pp cited in Vols.1-5:

3: 1073;
4: 546

(1) P. Janssen, U.S. Pat. 2 882 271 (1959).
(2) Van Neuten et al., Arch. Int. Pharmacodyn. Ther., 1973, 204, 37.
(3) T. Gotfraind et al., Drugs of Today, 1982, 18, 27-42.

CINNARIZINE CLOFIBRATE [INN]

| CAS | 60763-49-7 | MW: | 583.18 | MF: | C36 H39 Cl N2 O3 |

Antihistaminic; vasodilator

Janssen Pharm., Belg.

LgP (4)

pKa

pp cited in Vols.1-5:

(1) P. Janssen, U.S. Pat.2 882 271 (1959).
(2) Van Neuten et al., Arch. Int. Pharmacodyn. Ther., 1973, 204, 37.
(3) T. Gotfraind et al., Drugs of Today, 1982, 18, 27-42.

CINNOFURADIONE [INN]

| CAS | 477-80-5 | MW: | 334.38 | MF: | C20 H18 N2 O3 |

LgP 0.95(C,3)

pKa

pp cited in Vols.1-5:

CINOCTRAMIDE [INN]

| CAS | 28598-08-5 | MW: | 333.43 | MF: | C19 H27 N O4 |

LgP 2.80(C)

pKa

pp cited in Vols.1-5:

CINODINE [U]

| CAS | 60830-76-4 | MW: | 893.96 | MF: | C37 H59 N13 O13 |

Antibacterial (vet.)

LgP (4)

pKa

pp cited in Vols.1-5:

(1) RTECS, 1985-86, entry 26598.

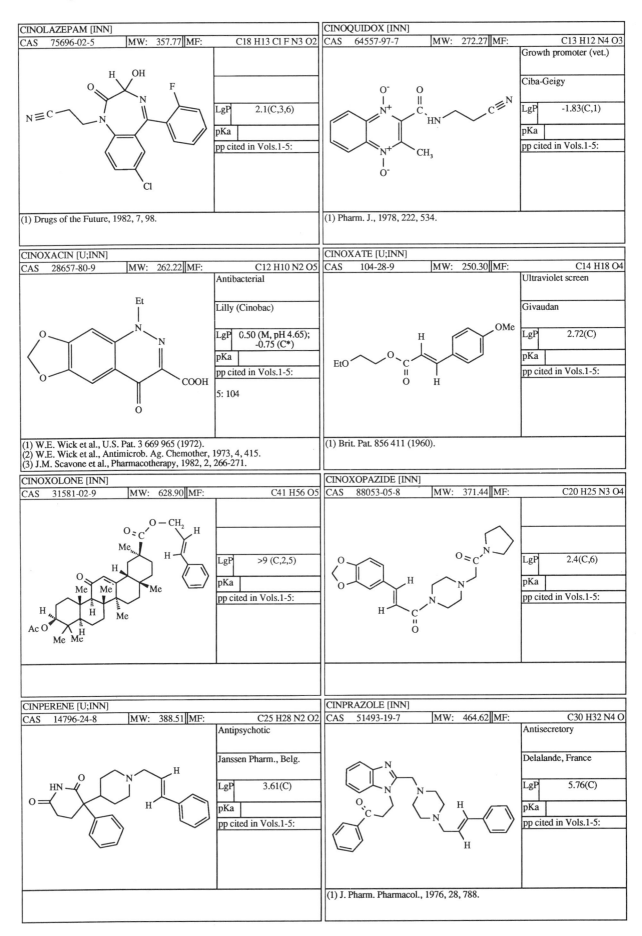

CINOLAZEPAM [INN]

| CAS | 75696-02-5 | MW: | 357.77 | MF: | C18 H13 Cl F N3 O2 |

LgP	2.1(C,3,6)
pKa	
pp cited in Vols.1-5:	

(1) Drugs of the Future, 1982, 7, 98.

CINOQUIDOX [INN]

| CAS | 64557-97-7 | MW: | 272.27 | MF: | C13 H12 N4 O3 |

Growth promoter (vet.)

Ciba-Geigy

LgP	-1.83(C,1)
pKa	
pp cited in Vols.1-5:	

(1) Pharm. J., 1978, 222, 534.

CINOXACIN [U;INN]

| CAS | 28657-80-9 | MW: | 262.22 | MF: | C12 H10 N2 O5 |

Antibacterial

Lilly (Cinobac)

LgP	0.50 (M, pH 4.65); -0.75 (C*)
pKa	
pp cited in Vols.1-5:	5: 104

(1) W.E. Wick et al., U.S. Pat. 3 669 965 (1972).
(2) W.E. Wick et al., Antimicrob. Ag. Chemother, 1973, 4, 415.
(3) J.M. Scavone et al., Pharmacotherapy, 1982, 2, 266-271.

CINOXATE [U;INN]

| CAS | 104-28-9 | MW: | 250.30 | MF: | C14 H18 O4 |

Ultraviolet screen

Givaudan

LgP	2.72(C)
pKa	
pp cited in Vols.1-5:	

(1) Brit. Pat. 856 411 (1960).

CINOXOLONE [INN]

| CAS | 31581-02-9 | MW: | 628.90 | MF: | C41 H56 O5 |

LgP	>9 (C,2,5)
pKa	
pp cited in Vols.1-5:	

CINOXOPAZIDE [INN]

| CAS | 88053-05-8 | MW: | 371.44 | MF: | C20 H25 N3 O4 |

LgP	2.4(C,6)
pKa	
pp cited in Vols.1-5:	

CINPERENE [U;INN]

| CAS | 14796-24-8 | MW: | 388.51 | MF: | C25 H28 N2 O2 |

Antipsychotic

Janssen Pharm., Belg.

LgP	3.61(C)
pKa	
pp cited in Vols.1-5:	

CINPRAZOLE [INN]

| CAS | 51493-19-7 | MW: | 464.62 | MF: | C30 H32 N4 O |

Antisecretory

Delalande, France

LgP	5.76(C)
pKa	
pp cited in Vols.1-5:	

(1) J. Pharm. Pharmacol., 1976, 28, 788.

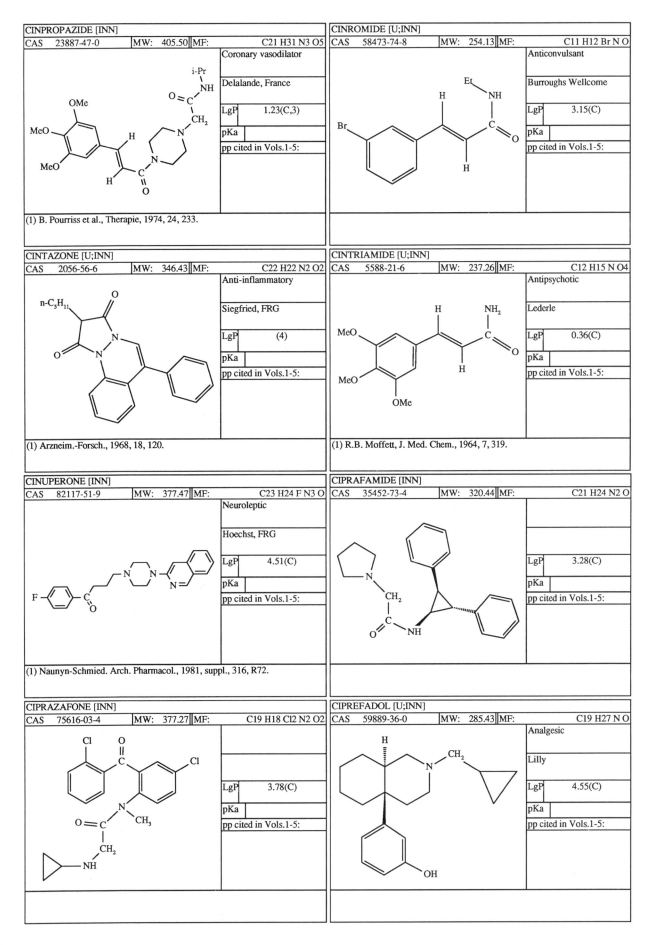

CINPROPAZIDE [INN]

CAS 23887-47-0 | MW: 405.50 | MF: C21 H31 N3 O5

Coronary vasodilator

Delalande, France

LgP 1.23(C,3)

pKa

pp cited in Vols.1-5:

(1) B. Pourriss et al., Therapie, 1974, 24, 233.

CINROMIDE [U;INN]

CAS 58473-74-8 | MW: 254.13 | MF: C11 H12 Br N O

Anticonvulsant

Burroughs Wellcome

LgP 3.15(C)

pKa

pp cited in Vols.1-5:

CINTAZONE [U;INN]

CAS 2056-56-6 | MW: 346.43 | MF: C22 H22 N2 O2

Anti-inflammatory

Siegfried, FRG

LgP (4)

pKa

pp cited in Vols.1-5:

(1) Arzneim.-Forsch., 1968, 18, 120.

CINTRIAMIDE [U;INN]

CAS 5588-21-6 | MW: 237.26 | MF: C12 H15 N O4

Antipsychotic

Lederle

LgP 0.36(C)

pKa

pp cited in Vols.1-5:

(1) R.B. Moffett, J. Med. Chem., 1964, 7, 319.

CINUPERONE [INN]

CAS 82117-51-9 | MW: 377.47 | MF: C23 H24 F N3 O

Neuroleptic

Hoechst, FRG

LgP 4.51(C)

pKa

pp cited in Vols.1-5:

(1) Naunyn-Schmied. Arch. Pharmacol., 1981, suppl., 316, R72.

CIPRAFAMIDE [INN]

CAS 35452-73-4 | MW: 320.44 | MF: C21 H24 N2 O

LgP 3.28(C)

pKa

pp cited in Vols.1-5:

CIPRAZAFONE [INN]

CAS 75616-03-4 | MW: 377.27 | MF: C19 H18 Cl2 N2 O2

LgP 3.78(C)

pKa

pp cited in Vols.1-5:

CIPREFADOL [U;INN]

CAS 59889-36-0 | MW: 285.43 | MF: C19 H27 N O

Analgesic

Lilly

LgP 4.55(C)

pKa

pp cited in Vols.1-5:

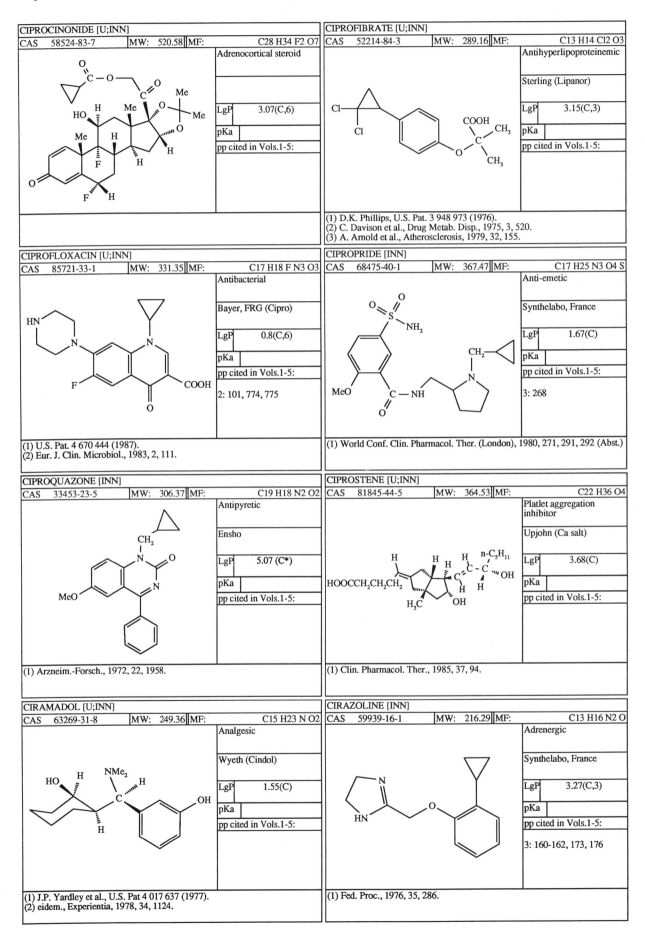

CIPROCINONIDE [U;INN]

CAS	58524-83-7	MW:	520.58	MF:	C28 H34 F2 O7

Adrenocortical steroid

LgP	3.07(C,6)
pKa	
pp cited in Vols.1-5:	

CIPROFIBRATE [U;INN]

CAS	52214-84-3	MW:	289.16	MF:	C13 H14 Cl2 O3

Antihyperlipoproteinemic

Sterling (Lipanor)

LgP	3.15(C,3)
pKa	
pp cited in Vols.1-5:	

(1) D.K. Phillips, U.S. Pat. 3 948 973 (1976).
(2) C. Davison et al., Drug Metab. Disp., 1975, 3, 520.
(3) A. Arnold et al., Atherosclerosis, 1979, 32, 155.

CIPROFLOXACIN [U;INN]

CAS	85721-33-1	MW:	331.35	MF:	C17 H18 F N3 O3

Antibacterial

Bayer, FRG (Cipro)

LgP	0.8(C,6)
pKa	
pp cited in Vols.1-5:	2: 101, 774, 775

(1) U.S. Pat. 4 670 444 (1987).
(2) Eur. J. Clin. Microbiol., 1983, 2, 111.

CIPROPRIDE [INN]

CAS	68475-40-1	MW:	367.47	MF:	C17 H25 N3 O4 S

Anti-emetic

Synthelabo, France

LgP	1.67(C)
pKa	
pp cited in Vols.1-5:	3: 268

(1) World Conf. Clin. Pharmacol. Ther. (London), 1980, 271, 291, 292 (Abst.)

CIPROQUAZONE [INN]

CAS	33453-23-5	MW:	306.37	MF:	C19 H18 N2 O2

Antipyretic

Ensho

LgP	5.07 (C*)
pKa	
pp cited in Vols.1-5:	

(1) Arzneim.-Forsch., 1972, 22, 1958.

CIPROSTENE [U;INN]

CAS	81845-44-5	MW:	364.53	MF:	C22 H36 O4

Platelet aggregation inhibitor

Upjohn (Ca salt)

LgP	3.68(C)
pKa	
pp cited in Vols.1-5:	

(1) Clin. Pharmacol. Ther., 1985, 37, 94.

CIRAMADOL [U;INN]

CAS	63269-31-8	MW:	249.36	MF:	C15 H23 N O2

Analgesic

Wyeth (Cindol)

LgP	1.55(C)
pKa	
pp cited in Vols.1-5:	

(1) J.P. Yardley et al., U.S. Pat 4 017 637 (1977).
(2) eidem., Experientia, 1978, 34, 1124.

CIRAZOLINE [INN]

CAS	59939-16-1	MW:	216.29	MF:	C13 H16 N2 O

Adrenergic

Synthelabo, France

LgP	3.27(C,3)
pKa	
pp cited in Vols.1-5:	3: 160-162, 173, 176

(1) Fed. Proc., 1976, 35, 286.

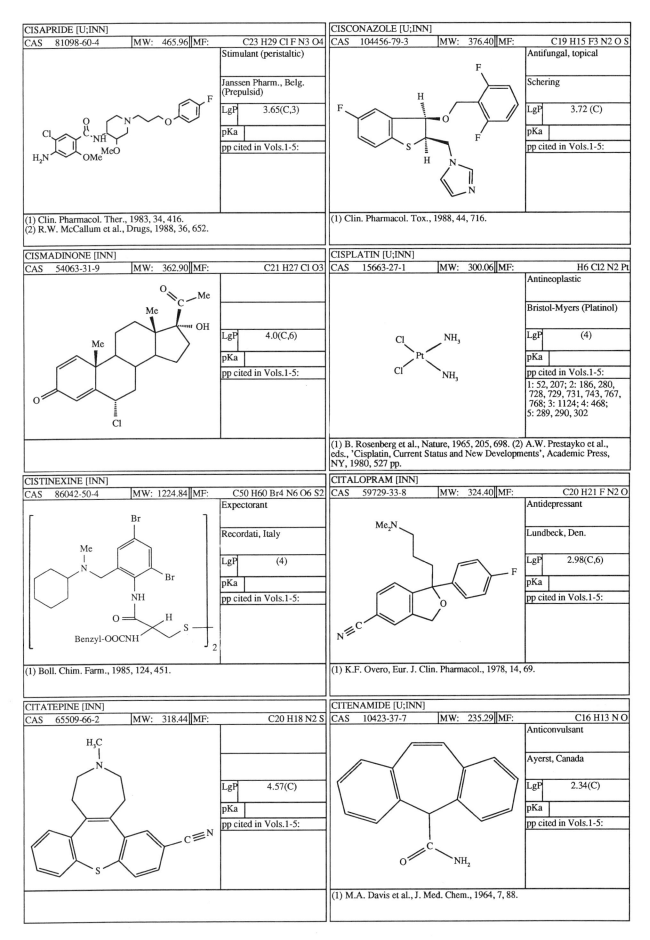

CISAPRIDE [U;INN]

CAS	81098-60-4	MW:	465.96	MF:	C23 H29 Cl F N3 O4

Stimulant (peristaltic)

Janssen Pharm., Belg. (Prepulsid)

LgP	3.65(C,3)
pKa	
pp cited in Vols.1-5:	

(1) Clin. Pharmacol. Ther., 1983, 34, 416.
(2) R.W. McCallum et al., Drugs, 1988, 36, 652.

CISCONAZOLE [U;INN]

CAS	104456-79-3	MW:	376.40	MF:	C19 H15 F3 N2 O S

Antifungal, topical

Schering

LgP	3.72 (C)
pKa	
pp cited in Vols.1-5:	

(1) Clin. Pharmacol. Tox., 1988, 44, 716.

CISMADINONE [INN]

CAS	54063-31-9	MW:	362.90	MF:	C21 H27 Cl O3

LgP	4.0(C,6)
pKa	
pp cited in Vols.1-5:	

CISPLATIN [U;INN]

CAS	15663-27-1	MW:	300.06	MF:	H6 Cl2 N2 Pt

Antineoplastic

Bristol-Myers (Platinol)

LgP	(4)
pKa	
pp cited in Vols.1-5:	
1: 52, 207; 2: 186, 280, 728, 729, 731, 743, 767, 768; 3: 1124; 4: 468; 5: 289, 290, 302	

(1) B. Rosenberg et al., Nature, 1965, 205, 698. (2) A.W. Prestayko et al., eds., 'Cisplatin, Current Status and New Developments', Academic Press, NY, 1980, 527 pp.

CISTINEXINE [INN]

CAS	86042-50-4	MW:	1224.84	MF:	C50 H60 Br4 N6 O6 S2

Expectorant

Recordati, Italy

LgP	(4)
pKa	
pp cited in Vols.1-5:	

(1) Boll. Chim. Farm., 1985, 124, 451.

CITALOPRAM [INN]

CAS	59729-33-8	MW:	324.40	MF:	C20 H21 F N2 O

Antidepressant

Lundbeck, Den.

LgP	2.98(C,6)
pKa	
pp cited in Vols.1-5:	

(1) K.F. Overo, Eur. J. Clin. Pharmacol., 1978, 14, 69.

CITATEPINE [INN]

CAS	65509-66-2	MW:	318.44	MF:	C20 H18 N2 S

LgP	4.57(C)
pKa	
pp cited in Vols.1-5:	

CITENAMIDE [U;INN]

CAS	10423-37-7	MW:	235.29	MF:	C16 H13 N O

Anticonvulsant

Ayerst, Canada

LgP	2.34(C)
pKa	
pp cited in Vols.1-5:	

(1) M.A. Davis et al., J. Med. Chem., 1964, 7, 88.

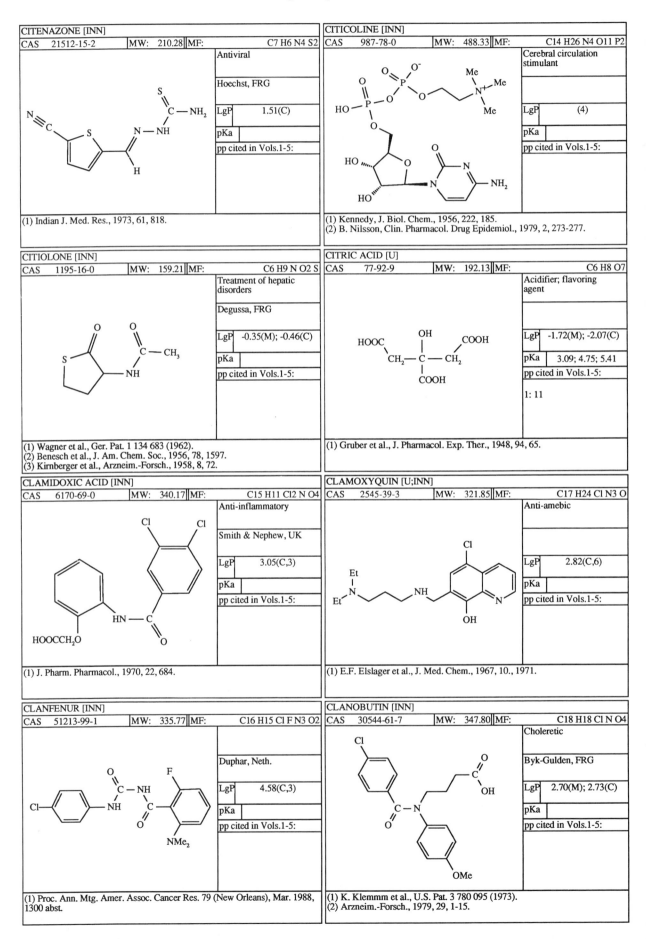

CITENAZONE [INN]		
CAS 21512-15-2	MW: 210.28	MF: C7 H6 N4 S2

Antiviral

Hoechst, FRG

LgP	1.51(C)
pKa	
pp cited in Vols.1-5:	

(1) Indian J. Med. Res., 1973, 61, 818.

CITICOLINE [INN]		
CAS 987-78-0	MW: 488.33	MF: C14 H26 N4 O11 P2

Cerebral circulation stimulant

LgP	(4)
pKa	
pp cited in Vols.1-5:	

(1) Kennedy, J. Biol. Chem., 1956, 222, 185.
(2) B. Nilsson, Clin. Pharmacol. Drug Epidemiol., 1979, 2, 273-277.

CITIOLONE [INN]		
CAS 1195-16-0	MW: 159.21	MF: C6 H9 N O2 S

Treatment of hepatic disorders

Degussa, FRG

LgP	-0.35(M); -0.46(C)
pKa	
pp cited in Vols.1-5:	

(1) Wagner et al., Ger. Pat. 1 134 683 (1962).
(2) Benesch et al., J. Am. Chem. Soc., 1956, 78, 1597.
(3) Kirnberger et al., Arzneim.-Forsch., 1958, 8, 72.

CITRIC ACID [U]		
CAS 77-92-9	MW: 192.13	MF: C6 H8 O7

Acidifier; flavoring agent

LgP	-1.72(M); -2.07(C)
pKa	3.09; 4.75; 5.41
pp cited in Vols.1-5:	
1: 11	

(1) Gruber et al., J. Pharmacol. Exp. Ther., 1948, 94, 65.

CLAMIDOXIC ACID [INN]		
CAS 6170-69-0	MW: 340.17	MF: C15 H11 Cl2 N O4

Anti-inflammatory

Smith & Nephew, UK

LgP	3.05(C,3)
pKa	
pp cited in Vols.1-5:	

(1) J. Pharm. Pharmacol., 1970, 22, 684.

CLAMOXYQUIN [U;INN]		
CAS 2545-39-3	MW: 321.85	MF: C17 H24 Cl N3 O

Anti-amebic

LgP	2.82(C,6)
pKa	
pp cited in Vols.1-5:	

(1) E.F. Elslager et al., J. Med. Chem., 1967, 10., 1971.

CLANFENUR [INN]		
CAS 51213-99-1	MW: 335.77	MF: C16 H15 Cl F N3 O2

Duphar, Neth.

LgP	4.58(C,3)
pKa	
pp cited in Vols.1-5:	

(1) Proc. Ann. Mtg. Amer. Assoc. Cancer Res. 79 (New Orleans), Mar. 1988, 1300 abst.

CLANOBUTIN [INN]		
CAS 30544-61-7	MW: 347.80	MF: C18 H18 Cl N O4

Choleretic

Byk-Gulden, FRG

LgP	2.70(M); 2.73(C)
pKa	
pp cited in Vols.1-5:	

(1) K. Klemmm et al., U.S. Pat. 3 780 095 (1973).
(2) Arzneim.-Forsch., 1979, 29, 1-15.

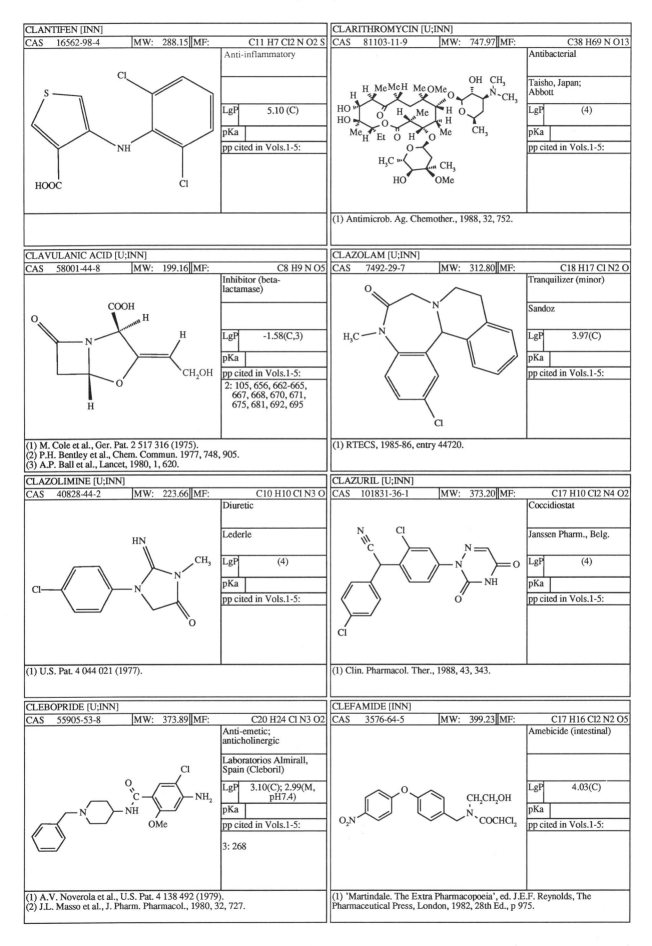

CLANTIFEN [INN]

CAS	16562-98-4	MW:	288.15	MF:	C11 H7 Cl2 N O2 S

Anti-inflammatory

LgP	5.10 (C)
pKa	
pp cited in Vols.1-5:	

CLAVULANIC ACID [U;INN]

CAS	58001-44-8	MW:	199.16	MF:	C8 H9 N O5

Inhibitor (beta-lactamase)

LgP	-1.58(C,3)
pKa	
pp cited in Vols.1-5:	
2: 105, 656, 662-665, 667, 668, 670, 671, 675, 681, 692, 695	

(1) M. Cole et al., Ger. Pat. 2 517 316 (1975).
(2) P.H. Bentley et al., Chem. Commun. 1977, 748, 905.
(3) A.P. Ball et al., Lancet, 1980, 1, 620.

CLAZOLIMINE [U;INN]

CAS	40828-44-2	MW:	223.66	MF:	C10 H10 Cl N3 O

Diuretic

Lederle

LgP	(4)
pKa	
pp cited in Vols.1-5:	

(1) U.S. Pat. 4 044 021 (1977).

CLEBOPRIDE [U;INN]

CAS	55905-53-8	MW:	373.89	MF:	C20 H24 Cl N3 O2

Anti-emetic; anticholinergic

Laboratorios Almirall, Spain (Cleboril)

LgP	3.10(C); 2.99(M, pH7.4)
pKa	
pp cited in Vols.1-5:	
3: 268	

(1) A.V. Noverola et al., U.S. Pat. 4 138 492 (1979).
(2) J.L. Masso et al., J. Pharm. Pharmacol., 1980, 32, 727.

CLARITHROMYCIN [U;INN]

CAS	81103-11-9	MW:	747.97	MF:	C38 H69 N O13

Antibacterial

Taisho, Japan; Abbott

LgP	(4)
pKa	
pp cited in Vols.1-5:	

(1) Antimicrob. Ag. Chemother., 1988, 32, 752.

CLAZOLAM [U;INN]

CAS	7492-29-7	MW:	312.80	MF:	C18 H17 Cl N2 O

Tranquilizer (minor)

Sandoz

LgP	3.97(C)
pKa	
pp cited in Vols.1-5:	

(1) RTECS, 1985-86, entry 44720.

CLAZURIL [U;INN]

CAS	101831-36-1	MW:	373.20	MF:	C17 H10 Cl2 N4 O2

Coccidiostat

Janssen Pharm., Belg.

LgP	(4)
pKa	
pp cited in Vols.1-5:	

(1) Clin. Pharmacol. Ther., 1988, 43, 343.

CLEFAMIDE [INN]

CAS	3576-64-5	MW:	399.23	MF:	C17 H16 Cl2 N2 O5

Amebicide (intestinal)

LgP	4.03(C)
pKa	
pp cited in Vols.1-5:	

(1) 'Martindale. The Extra Pharmacopoeia', ed. J.E.F. Reynolds, The Pharmaceutical Press, London, 1982, 28th Ed., p 975.

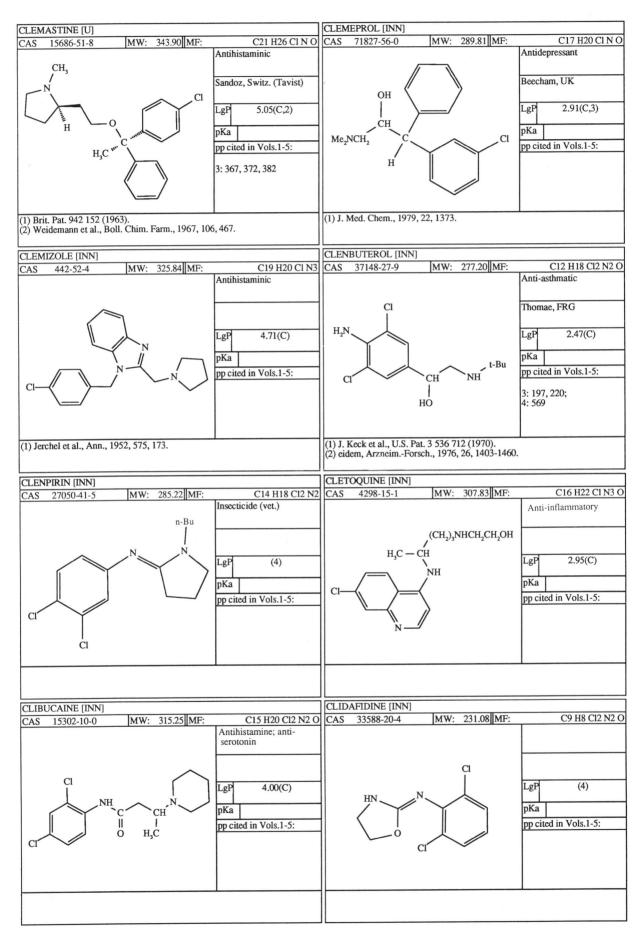

CLEMASTINE [U]			
CAS 15686-51-8	MW: 343.90	MF:	C21 H26 Cl N O

Antihistaminic

Sandoz, Switz. (Tavist)

LgP	5.05(C,2)
pKa	
pp cited in Vols.1-5:	

3: 367, 372, 382

(1) Brit. Pat. 942 152 (1963).
(2) Weidemann et al., Boll. Chim. Farm., 1967, 106, 467.

CLEMEPROL [INN]			
CAS 71827-56-0	MW: 289.81	MF:	C17 H20 Cl N O

Antidepressant

Beecham, UK

LgP	2.91(C,3)
pKa	
pp cited in Vols.1-5:	

(1) J. Med. Chem., 1979, 22, 1373.

CLEMIZOLE [INN]			
CAS 442-52-4	MW: 325.84	MF:	C19 H20 Cl N3

Antihistaminic

LgP	4.71(C)
pKa	
pp cited in Vols.1-5:	

(1) Jerchel et al., Ann., 1952, 575, 173.

CLENBUTEROL [INN]			
CAS 37148-27-9	MW: 277.20	MF:	C12 H18 Cl2 N2 O

Anti-asthmatic

Thomae, FRG

LgP	2.47(C)
pKa	
pp cited in Vols.1-5:	

3: 197, 220;
4: 569

(1) J. Keck et al., U.S. Pat. 3 536 712 (1970).
(2) eidem, Arzneim.-Forsch., 1976, 26, 1403-1460.

CLENPIRIN [INN]			
CAS 27050-41-5	MW: 285.22	MF:	C14 H18 Cl2 N2

Insecticide (vet.)

LgP	(4)
pKa	
pp cited in Vols.1-5:	

CLETOQUINE [INN]			
CAS 4298-15-1	MW: 307.83	MF:	C16 H22 Cl N3 O

Anti-inflammatory

LgP	2.95(C)
pKa	
pp cited in Vols.1-5:	

CLIBUCAINE [INN]			
CAS 15302-10-0	MW: 315.25	MF:	C15 H20 Cl2 N2 O

Antihistamine; anti-
serotonin

LgP	4.00(C)
pKa	
pp cited in Vols.1-5:	

CLIDAFIDINE [INN]			
CAS 33588-20-4	MW: 231.08	MF:	C9 H8 Cl2 N2 O

LgP	(4)
pKa	
pp cited in Vols.1-5:	

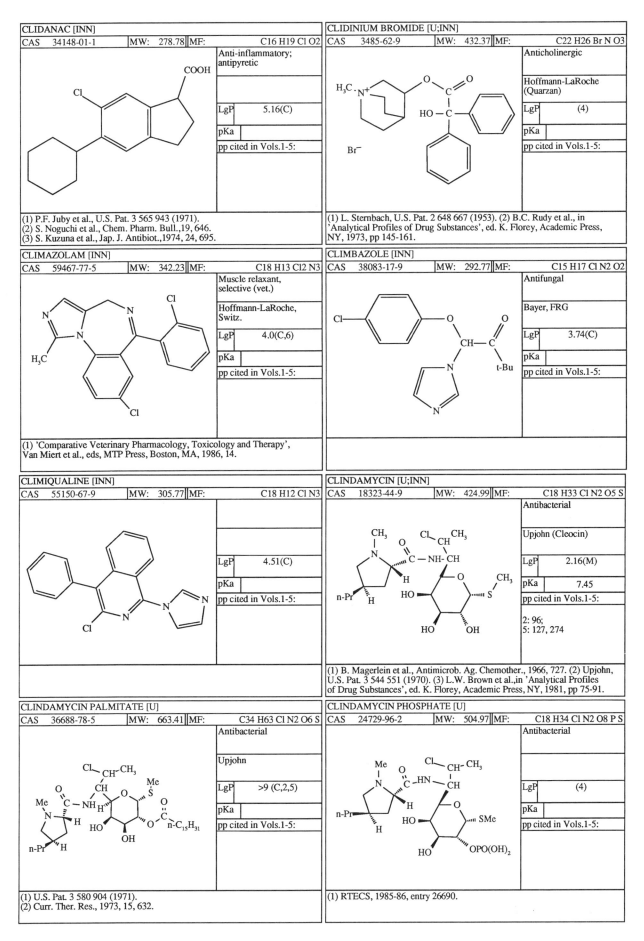

CLIDANAC [INN]

CAS	34148-01-1	MW:	278.78	MF:	C16 H19 Cl O2

Anti-inflammatory; antipyretic

LgP	5.16(C)
pKa	
pp cited in Vols.1-5:	

(1) P.F. Juby et al., U.S. Pat. 3 565 943 (1971).
(2) S. Noguchi et al., Chem. Pharm. Bull.,19, 646.
(3) S. Kuzuna et al., Jap. J. Antibiot.,1974, 24, 695.

CLIDINIUM BROMIDE [U;INN]

CAS	3485-62-9	MW:	432.37	MF:	C22 H26 Br N O3

Anticholinergic

Hoffmann-LaRoche (Quarzan)

LgP	(4)
pKa	
pp cited in Vols.1-5:	

(1) L. Sternbach, U.S. Pat. 2 648 667 (1953). (2) B.C. Rudy et al., in 'Analytical Profiles of Drug Substances', ed. K. Florey, Academic Press, NY, 1973, pp 145-161.

CLIMAZOLAM [INN]

CAS	59467-77-5	MW:	342.23	MF:	C18 H13 Cl2 N3

Muscle relaxant, selective (vet.)

Hoffmann-LaRoche, Switz.

LgP	4.0(C,6)
pKa	
pp cited in Vols.1-5:	

(1) 'Comparative Veterinary Pharmacology, Toxicology and Therapy', Van Miert et al., eds, MTP Press, Boston, MA, 1986, 14.

CLIMBAZOLE [INN]

CAS	38083-17-9	MW:	292.77	MF:	C15 H17 Cl N2 O2

Antifungal

Bayer, FRG

LgP	3.74(C)
pKa	
pp cited in Vols.1-5:	

CLIMIQUALINE [INN]

CAS	55150-67-9	MW:	305.77	MF:	C18 H12 Cl N3

LgP	4.51(C)
pKa	
pp cited in Vols.1-5:	

CLINDAMYCIN [U;INN]

CAS	18323-44-9	MW:	424.99	MF:	C18 H33 Cl N2 O5 S

Antibacterial

Upjohn (Cleocin)

LgP	2.16(M)
pKa	7.45
pp cited in Vols.1-5:	

2: 96;
5: 127, 274

(1) B. Magerlein et al., Antimicrob. Ag. Chemother., 1966, 727. (2) Upjohn, U.S. Pat. 3 544 551 (1970). (3) L.W. Brown et al.,in 'Analytical Profiles of Drug Substances', ed. K. Florey, Academic Press, NY, 1981, pp 75-91.

CLINDAMYCIN PALMITATE [U]

CAS	36688-78-5	MW:	663.41	MF:	C34 H63 Cl N2 O6 S

Antibacterial

Upjohn

LgP	>9 (C,2,5)
pKa	
pp cited in Vols.1-5:	

(1) U.S. Pat. 3 580 904 (1971).
(2) Curr. Ther. Res., 1973, 15, 632.

CLINDAMYCIN PHOSPHATE [U]

CAS	24729-96-2	MW:	504.97	MF:	C18 H34 Cl N2 O8 P S

Antibacterial

LgP	(4)
pKa	
pp cited in Vols.1-5:	

(1) RTECS, 1985-86, entry 26690.

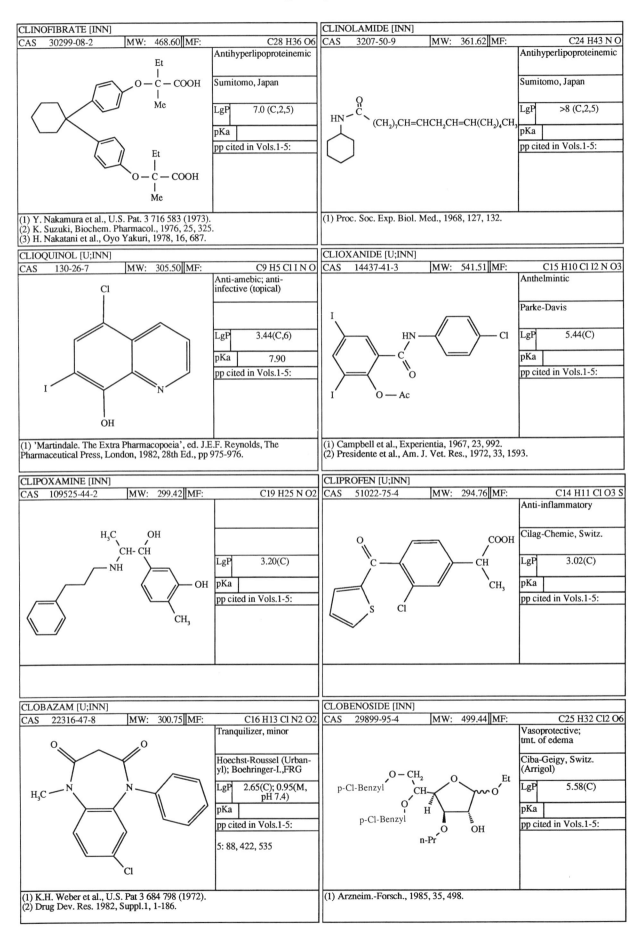

CLINOFIBRATE [INN]

CAS 30299-08-2 | MW: 468.60 | MF: C28 H36 O6

Antihyperlipoproteinemic

Sumitomo, Japan

LgP 7.0 (C,2,5)

pKa

pp cited in Vols.1-5:

(1) Y. Nakamura et al., U.S. Pat. 3 716 583 (1973).
(2) K. Suzuki, Biochem. Pharmacol., 1976, 25, 325.
(3) H. Nakatani et al., Oyo Yakuri, 1978, 16, 687.

CLINOLAMIDE [INN]

CAS 3207-50-9 | MW: 361.62 | MF: C24 H43 N O

Antihyperlipoproteinemic

Sumitomo, Japan

LgP >8 (C,2,5)

pKa

pp cited in Vols.1-5:

(1) Proc. Soc. Exp. Biol. Med., 1968, 127, 132.

CLIOQUINOL [U;INN]

CAS 130-26-7 | MW: 305.50 | MF: C9 H5 Cl I N O

Anti-amebic; anti-infective (topical)

LgP 3.44(C,6)

pKa 7.90

pp cited in Vols.1-5:

(1) 'Martindale. The Extra Pharmacopoeia', ed. J.E.F. Reynolds, The Pharmaceutical Press, London, 1982, 28th Ed., pp 975-976.

CLIOXANIDE [U;INN]

CAS 14437-41-3 | MW: 541.51 | MF: C15 H10 Cl I2 N O3

Anthelmintic

Parke-Davis

LgP 5.44(C)

pKa

pp cited in Vols.1-5:

(1) Campbell et al., Experientia, 1967, 23, 992.
(2) Presidente et al., Am. J. Vet. Res., 1972, 33, 1593.

CLIPOXAMINE [INN]

CAS 109525-44-2 | MW: 299.42 | MF: C19 H25 N O2

LgP 3.20(C)

pKa

pp cited in Vols.1-5:

CLIPROFEN [U;INN]

CAS 51022-75-4 | MW: 294.76 | MF: C14 H11 Cl O3 S

Anti-inflammatory

Cilag-Chemie, Switz.

LgP 3.02(C)

pKa

pp cited in Vols.1-5:

CLOBAZAM [U;INN]

CAS 22316-47-8 | MW: 300.75 | MF: C16 H13 Cl N2 O2

Tranquilizer, minor

Hoechst-Roussel (Urbanyl); Boehringer-I.,FRG

LgP 2.65(C); 0.95(M, pH 7.4)

pKa

pp cited in Vols.1-5:

5: 88, 422, 535

(1) K.H. Weber et al., U.S. Pat 3 684 798 (1972).
(2) Drug Dev. Res. 1982, Suppl.1, 1-186.

CLOBENOSIDE [INN]

CAS 29899-95-4 | MW: 499.44 | MF: C25 H32 Cl2 O6

Vasoprotective; tmt. of edema

Ciba-Geigy, Switz. (Arrigol)

LgP 5.58(C)

pKa

pp cited in Vols.1-5:

(1) Arzneim.-Forsch., 1985, 35, 498.

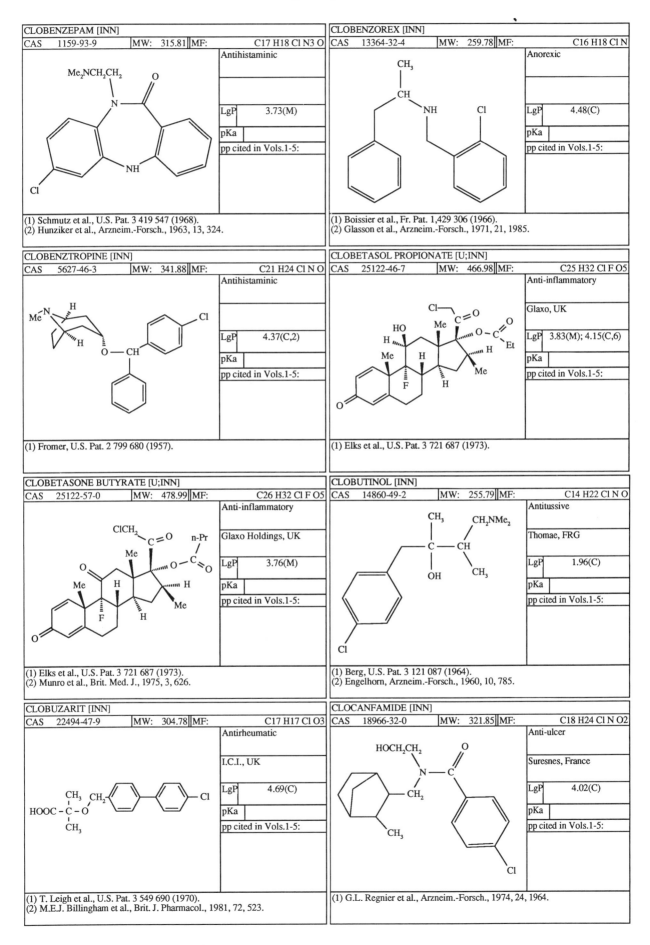

CLOBENZEPAM [INN]

CAS	1159-93-9	MW:	315.81	MF:	C17 H18 Cl N3 O

Antihistaminic

LgP | 3.73(M)
pKa |
pp cited in Vols.1-5:

(1) Schmutz et al., U.S. Pat. 3 419 547 (1968).
(2) Hunziker et al., Arzneim.-Forsch., 1963, 13, 324.

CLOBENZOREX [INN]

CAS	13364-32-4	MW:	259.78	MF:	C16 H18 Cl N

Anorexic

LgP | 4.48(C)
pKa |
pp cited in Vols.1-5:

(1) Boissier et al., Fr. Pat. 1,429 306 (1966).
(2) Glasson et al., Arzneim.-Forsch., 1971, 21, 1985.

CLOBENZTROPINE [INN]

CAS	5627-46-3	MW:	341.88	MF:	C21 H24 Cl N O

Antihistaminic

LgP | 4.37(C,2)
pKa |
pp cited in Vols.1-5:

(1) Fromer, U.S. Pat. 2 799 680 (1957).

CLOBETASOL PROPIONATE [U;INN]

CAS	25122-46-7	MW:	466.98	MF:	C25 H32 Cl F O5

Anti-inflammatory

Glaxo, UK

LgP | 3.83(M); 4.15(C,6)
pKa |
pp cited in Vols.1-5:

(1) Elks et al., U.S. Pat. 3 721 687 (1973).

CLOBETASONE BUTYRATE [U;INN]

CAS	25122-57-0	MW:	478.99	MF:	C26 H32 Cl F O5

Anti-inflammatory

Glaxo Holdings, UK

LgP | 3.76(M)
pKa |
pp cited in Vols.1-5:

(1) Elks et al., U.S. Pat. 3 721 687 (1973).
(2) Munro et al., Brit. Med. J., 1975, 3, 626.

CLOBUTINOL [INN]

CAS	14860-49-2	MW:	255.79	MF:	C14 H22 Cl N O

Antitussive

Thomae, FRG

LgP | 1.96(C)
pKa |
pp cited in Vols.1-5:

(1) Berg, U.S. Pat. 3 121 087 (1964).
(2) Engelhorn, Arzneim.-Forsch., 1960, 10, 785.

CLOBUZARIT [INN]

CAS	22494-47-9	MW:	304.78	MF:	C17 H17 Cl O3

Antirheumatic

I.C.I., UK

LgP | 4.69(C)
pKa |
pp cited in Vols.1-5:

(1) T. Leigh et al., U.S. Pat. 3 549 690 (1970).
(2) M.E.J. Billingham et al., Brit. J. Pharmacol., 1981, 72, 523.

CLOCANFAMIDE [INN]

CAS	18966-32-0	MW:	321.85	MF:	C18 H24 Cl N O2

Anti-ulcer

Suresnes, France

LgP | 4.02(C)
pKa |
pp cited in Vols.1-5:

(1) G.L. Regnier et al., Arzneim.-Forsch., 1974, 24, 1964.

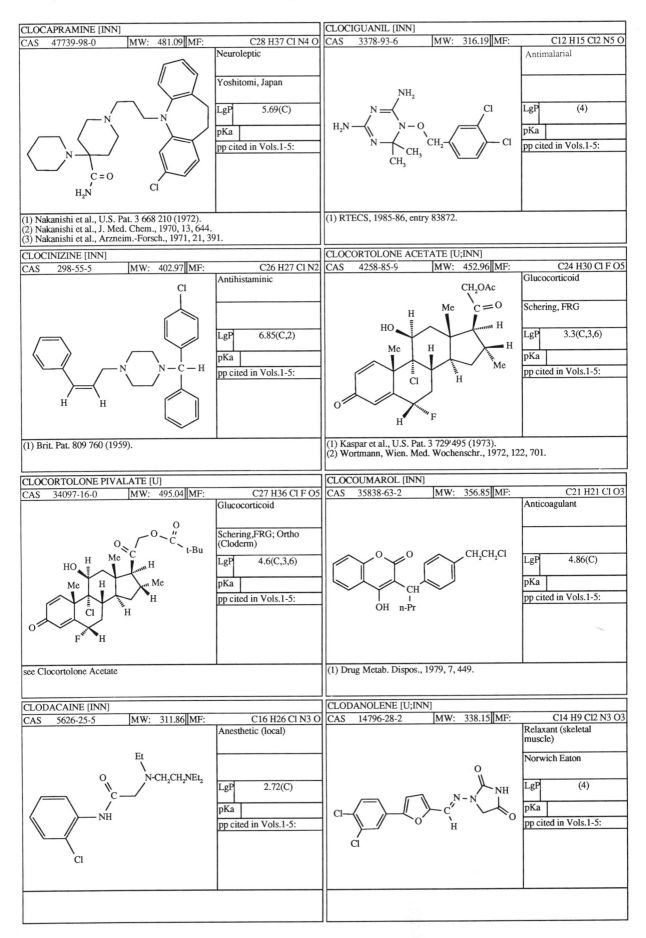

CLOCAPRAMINE [INN]

CAS 47739-98-0	MW: 481.09	MF: C28 H37 Cl N4 O

Neuroleptic

Yoshitomi, Japan

LgP	5.69(C)
pKa	
pp cited in Vols.1-5:	

(1) Nakanishi et al., U.S. Pat. 3 668 210 (1972).
(2) Nakanishi et al., J. Med. Chem., 1970, 13, 644.
(3) Nakanishi et al., Arzneim.-Forsch., 1971, 21, 391.

CLOCIGUANIL [INN]

CAS 3378-93-6	MW: 316.19	MF: C12 H15 Cl2 N5 O

Antimalarial

LgP	(4)
pKa	
pp cited in Vols.1-5:	

(1) RTECS, 1985-86, entry 83872.

CLOCINIZINE [INN]

CAS 298-55-5	MW: 402.97	MF: C26 H27 Cl N2

Antihistaminic

LgP	6.85(C,2)
pKa	
pp cited in Vols.1-5:	

(1) Brit. Pat. 809 760 (1959).

CLOCORTOLONE ACETATE [U;INN]

CAS 4258-85-9	MW: 452.96	MF: C24 H30 Cl F O5

Glucocorticoid

Schering, FRG

LgP	3.3(C,3,6)
pKa	
pp cited in Vols.1-5:	

(1) Kaspar et al., U.S. Pat. 3 729 495 (1973).
(2) Wortmann, Wien. Med. Wochenschr., 1972, 122, 701.

CLOCORTOLONE PIVALATE [U]

CAS 34097-16-0	MW: 495.04	MF: C27 H36 Cl F O5

Glucocorticoid

Schering,FRG; Ortho (Cloderm)

LgP	4.6(C,3,6)
pKa	
pp cited in Vols.1-5:	

see Clocortolone Acetate

CLOCOUMAROL [INN]

CAS 35838-63-2	MW: 356.85	MF: C21 H21 Cl O3

Anticoagulant

LgP	4.86(C)
pKa	
pp cited in Vols.1-5:	

(1) Drug Metab. Dispos., 1979, 7, 449.

CLODACAINE [INN]

CAS 5626-25-5	MW: 311.86	MF: C16 H26 Cl N3 O

Anesthetic (local)

LgP	2.72(C)
pKa	
pp cited in Vols.1-5:	

CLODANOLENE [U;INN]

CAS 14796-28-2	MW: 338.15	MF: C14 H9 Cl2 N3 O3

Relaxant (skeletal muscle)

Norwich Eaton

LgP	(4)
pKa	
pp cited in Vols.1-5:	

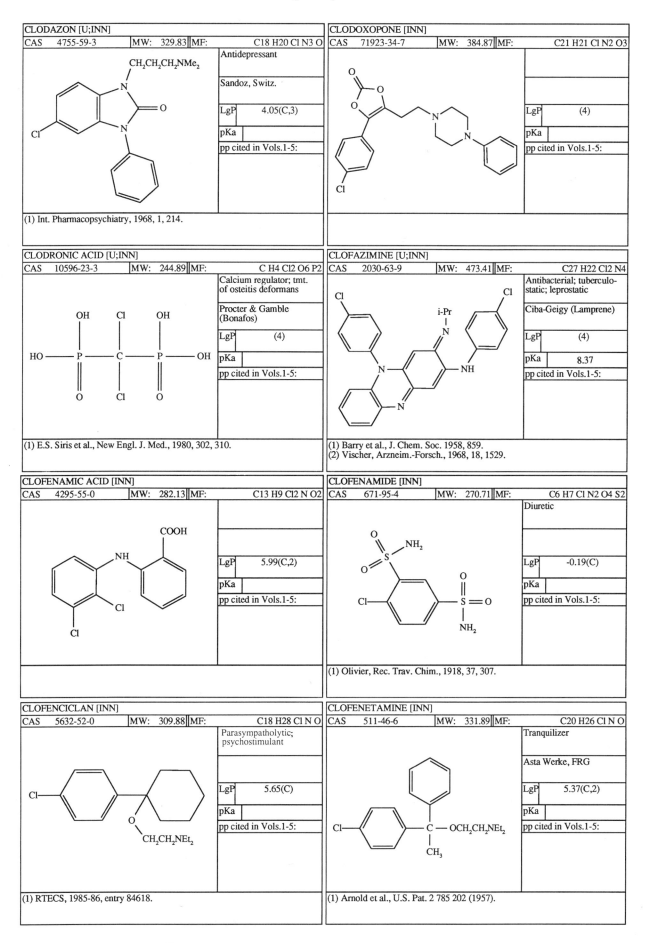

CLODAZON [U;INN]

CAS	4755-59-3	MW:	329.83	MF:	C18 H20 Cl N3 O

Antidepressant

Sandoz, Switz.

LgP	4.05(C,3)
pKa	
pp cited in Vols.1-5:	

CH₂CH₂CH₂NMe₂

(1) Int. Pharmacopsychiatry, 1968, 1, 214.

CLODOXOPONE [INN]

CAS	71923-34-7	MW:	384.87	MF:	C21 H21 Cl N2 O3

LgP	(4)
pKa	
pp cited in Vols.1-5:	

CLODRONIC ACID [U;INN]

CAS	10596-23-3	MW:	244.89	MF:	C H4 Cl2 O6 P2

Calcium regulator; tmt. of osteitis deformans

Procter & Gamble (Bonafos)

LgP	(4)
pKa	
pp cited in Vols.1-5:	

(1) E.S. Siris et al., New Engl. J. Med., 1980, 302, 310.

CLOFAZIMINE [U;INN]

CAS	2030-63-9	MW:	473.41	MF:	C27 H22 Cl2 N4

Antibacterial; tuberculostatic; leprostatic

Ciba-Geigy (Lamprene)

LgP	(4)
pKa	8.37
pp cited in Vols.1-5:	

(1) Barry et al., J. Chem. Soc. 1958, 859.
(2) Vischer, Arzneim.-Forsch., 1968, 18, 1529.

CLOFENAMIC ACID [INN]

CAS	4295-55-0	MW:	282.13	MF:	C13 H9 Cl2 N O2

LgP	5.99(C,2)
pKa	
pp cited in Vols.1-5:	

CLOFENAMIDE [INN]

CAS	671-95-4	MW:	270.71	MF:	C6 H7 Cl N2 O4 S2

Diuretic

LgP	-0.19(C)
pKa	
pp cited in Vols.1-5:	

(1) Olivier, Rec. Trav. Chim., 1918, 37, 307.

CLOFENCICLAN [INN]

CAS	5632-52-0	MW:	309.88	MF:	C18 H28 Cl N O

Parasympatholytic; psychostimulant

LgP	5.65(C)
pKa	
pp cited in Vols.1-5:	

CH₂CH₂NEt₂

(1) RTECS, 1985-86, entry 84618.

CLOFENETAMINE [INN]

CAS	511-46-6	MW:	331.89	MF:	C20 H26 Cl N O

Tranquilizer

Asta Werke, FRG

LgP	5.37(C,2)
pKa	
pp cited in Vols.1-5:	

(1) Arnold et al., U.S. Pat. 2 785 202 (1957).

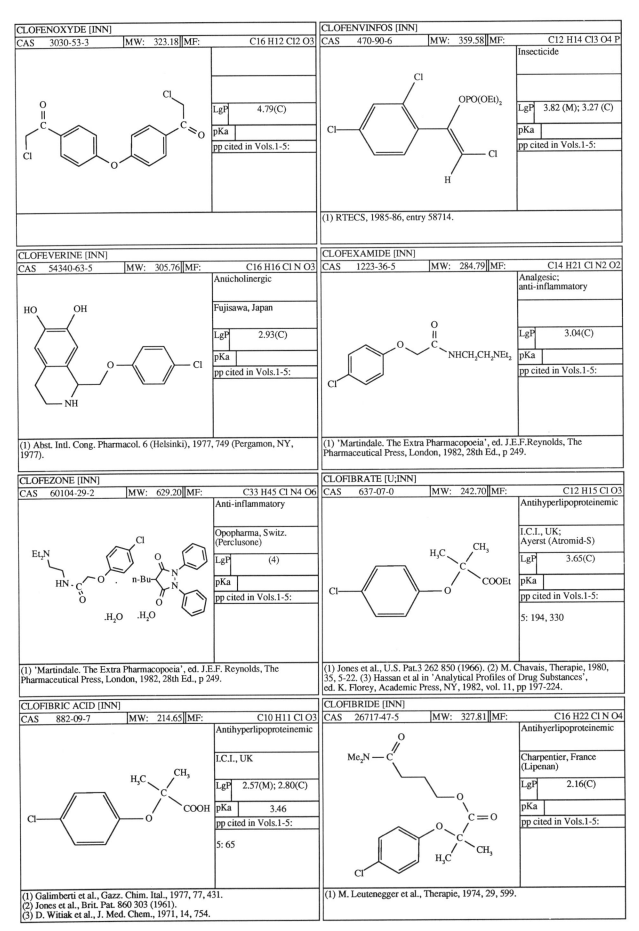

CLOFENOXYDE [INN]

CAS	3030-53-3	MW:	323.18	MF:	C16 H12 Cl2 O3

LgP	4.79(C)
pKa	
pp cited in Vols.1-5:	

CLOFEVERINE [INN]

CAS	54340-63-5	MW:	305.76	MF:	C16 H16 Cl N O3

Anticholinergic

Fujisawa, Japan

LgP	2.93(C)
pKa	
pp cited in Vols.1-5:	

(1) Abst. Intl. Cong. Pharmacol. 6 (Helsinki), 1977, 749 (Pergamon, NY, 1977).

CLOFEZONE [INN]

CAS	60104-29-2	MW:	629.20	MF:	C33 H45 Cl N4 O6

Anti-inflammatory

Opopharma, Switz.
(Perclusone)

LgP	(4)
pKa	
pp cited in Vols.1-5:	

.H₂O .H₂O

(1) 'Martindale. The Extra Pharmacopoeia', ed. J.E.F. Reynolds, The Pharmaceutical Press, London, 1982, 28th Ed., p 249.

CLOFIBRIC ACID [INN]

CAS	882-09-7	MW:	214.65	MF:	C10 H11 Cl O3

Antihyperlipoproteinemic

I.C.I., UK

LgP	2.57(M); 2.80(C)
pKa	3.46
pp cited in Vols.1-5:	

5: 65

(1) Galimberti et al., Gazz. Chim. Ital., 1977, 77, 431.
(2) Jones et al., Brit. Pat. 860 303 (1961).
(3) D. Witiak et al., J. Med. Chem., 1971, 14, 754.

CLOFENVINFOS [INN]

CAS	470-90-6	MW:	359.58	MF:	C12 H14 Cl3 O4 P

Insecticide

LgP	3.82 (M); 3.27 (C)
pKa	
pp cited in Vols.1-5:	

(1) RTECS, 1985-86, entry 58714.

CLOFEXAMIDE [INN]

CAS	1223-36-5	MW:	284.79	MF:	C14 H21 Cl N2 O2

Analgesic;
anti-inflammatory

LgP	3.04(C)
pKa	
pp cited in Vols.1-5:	

(1) 'Martindale. The Extra Pharmacopoeia', ed. J.E.F.Reynolds, The Pharmaceutical Press, London, 1982, 28th Ed., p 249.

CLOFIBRATE [U;INN]

CAS	637-07-0	MW:	242.70	MF:	C12 H15 Cl O3

Antihyperlipoproteinemic

I.C.I., UK;
Ayerst (Atromid-S)

LgP	3.65(C)
pKa	
pp cited in Vols.1-5:	

5: 194, 330

(1) Jones et al., U.S. Pat.3 262 850 (1966). (2) M. Chavais, Therapie, 1980, 35, 5-22. (3) Hassan et al in 'Analytical Profiles of Drug Substances', ed. K. Florey, Academic Press, NY, 1982, vol. 11, pp 197-224.

CLOFIBRIDE [INN]

CAS	26717-47-5	MW:	327.81	MF:	C16 H22 Cl N O4

Antihyerlipoproteinemic

Charpentier, France
(Lipenan)

LgP	2.16(C)
pKa	
pp cited in Vols.1-5:	

(1) M. Leutenegger et al., Therapie, 1974, 29, 599.

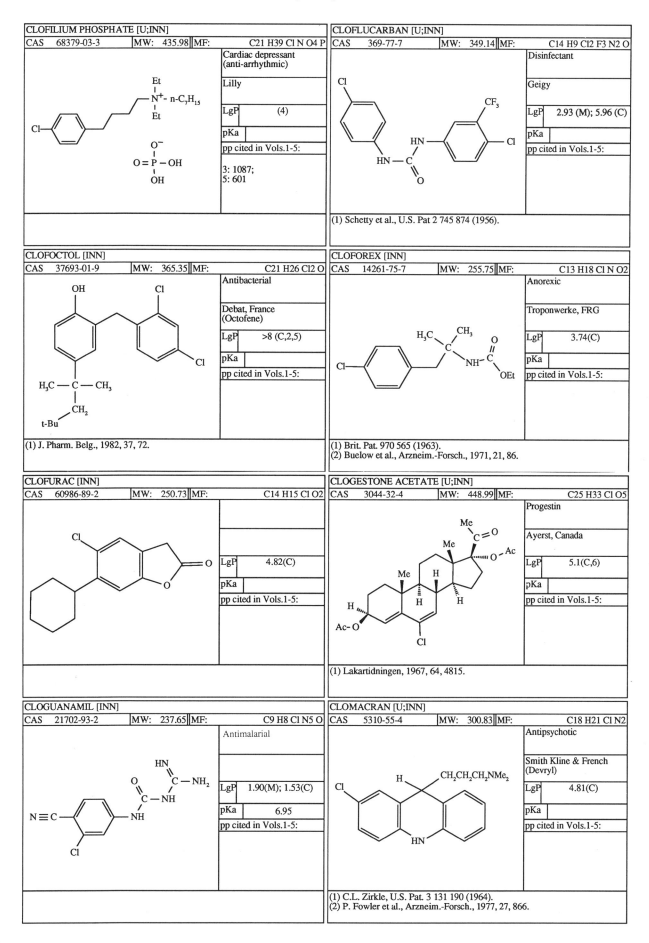

CLOFILIUM PHOSPHATE [U;INN]

CAS 68379-03-3	MW: 435.98	MF:	C21 H39 Cl N O4 P

Cardiac depressant (anti-arrhythmic)

Lilly

LgP	(4)
pKa	
pp cited in Vols.1-5:	

3: 1087;
5: 601

CLOFLUCARBAN [U;INN]

CAS 369-77-7	MW: 349.14	MF:	C14 H9 Cl2 F3 N2 O

Disinfectant

Geigy

LgP	2.93 (M); 5.96 (C)
pKa	
pp cited in Vols.1-5:	

(1) Schetty et al., U.S. Pat 2 745 874 (1956).

CLOFOCTOL [INN]

CAS 37693-01-9	MW: 365.35	MF:	C21 H26 Cl2 O

Antibacterial

Debat, France (Octofene)

LgP	>8 (C,2,5)
pKa	
pp cited in Vols.1-5:	

(1) J. Pharm. Belg., 1982, 37, 72.

CLOFOREX [INN]

CAS 14261-75-7	MW: 255.75	MF:	C13 H18 Cl N O2

Anorexic

Troponwerke, FRG

LgP	3.74(C)
pKa	
pp cited in Vols.1-5:	

(1) Brit. Pat. 970 565 (1963).
(2) Buelow et al., Arzneim.-Forsch., 1971, 21, 86.

CLOFURAC [INN]

CAS 60986-89-2	MW: 250.73	MF:	C14 H15 Cl O2

LgP	4.82(C)
pKa	
pp cited in Vols.1-5:	

CLOGESTONE ACETATE [U;INN]

CAS 3044-32-4	MW: 448.99	MF:	C25 H33 Cl O5

Progestin

Ayerst, Canada

LgP	5.1(C,6)
pKa	
pp cited in Vols.1-5:	

(1) Lakartidningen, 1967, 64, 4815.

CLOGUANAMIL [INN]

CAS 21702-93-2	MW: 237.65	MF:	C9 H8 Cl N5 O

Antimalarial

LgP	1.90(M); 1.53(C)
pKa	6.95
pp cited in Vols.1-5:	

CLOMACRAN [U;INN]

CAS 5310-55-4	MW: 300.83	MF:	C18 H21 Cl N2

Antipsychotic

Smith Kline & French (Devryl)

LgP	4.81(C)
pKa	
pp cited in Vols.1-5:	

(1) C.L. Zirkle, U.S. Pat. 3 131 190 (1964).
(2) P. Fowler et al., Arzneim.-Forsch., 1977, 27, 866.

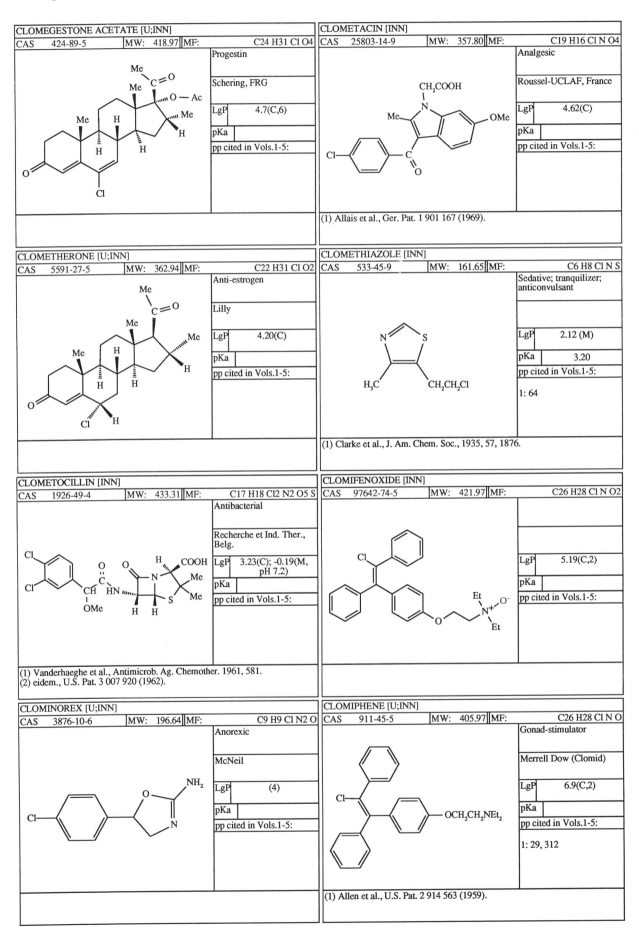

CLOMEGESTONE ACETATE [U;INN]

CAS	424-89-5	MW:	418.97	MF:	C24 H31 Cl O4

Progestin

Schering, FRG

LgP	4.7(C,6)
pKa	
pp cited in Vols.1-5:	

CLOMETACIN [INN]

CAS	25803-14-9	MW:	357.80	MF:	C19 H16 Cl N O4

Analgesic

Roussel-UCLAF, France

LgP	4.62(C)
pKa	
pp cited in Vols.1-5:	

(1) Allais et al., Ger. Pat. 1 901 167 (1969).

CLOMETHERONE [U;INN]

CAS	5591-27-5	MW:	362.94	MF:	C22 H31 Cl O2

Anti-estrogen

Lilly

LgP	4.20(C)
pKa	
pp cited in Vols.1-5:	

CLOMETHIAZOLE [INN]

CAS	533-45-9	MW:	161.65	MF:	C6 H8 Cl N S

Sedative; tranquilizer; anticonvulsant

LgP	2.12 (M)
pKa	3.20
pp cited in Vols.1-5:	
1: 64	

(1) Clarke et al., J. Am. Chem. Soc., 1935, 57, 1876.

CLOMETOCILLIN [INN]

CAS	1926-49-4	MW:	433.31	MF:	C17 H18 Cl2 N2 O5 S

Antibacterial

Recherche et Ind. Ther., Belg.

LgP	3.23(C); -0.19(M, pH 7.2)
pKa	
pp cited in Vols.1-5:	

(1) Vanderhaeghe et al., Antimicrob. Ag. Chemother. 1961, 581.
(2) eidem., U.S. Pat. 3 007 920 (1962).

CLOMIFENOXIDE [INN]

CAS	97642-74-5	MW:	421.97	MF:	C26 H28 Cl N O2

LgP	5.19(C,2)
pKa	
pp cited in Vols.1-5:	

CLOMINOREX [U;INN]

CAS	3876-10-6	MW:	196.64	MF:	C9 H9 Cl N2 O

Anorexic

McNeil

LgP	(4)
pKa	
pp cited in Vols.1-5:	

CLOMIPHENE [U;INN]

CAS	911-45-5	MW:	405.97	MF:	C26 H28 Cl N O

Gonad-stimulator

Merrell Dow (Clomid)

LgP	6.9(C,2)
pKa	
pp cited in Vols.1-5:	
1: 29, 312	

(1) Allen et al., U.S. Pat. 2 914 563 (1959).

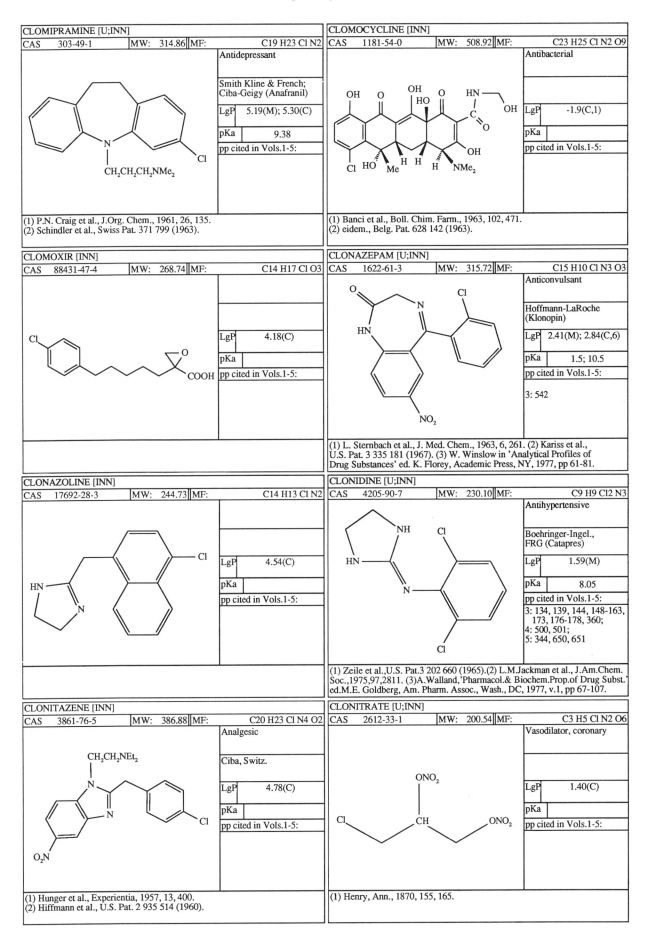

CLOMIPRAMINE [U;INN]

CAS 303-49-1 | MW: 314.86 | MF: C19 H23 Cl N2

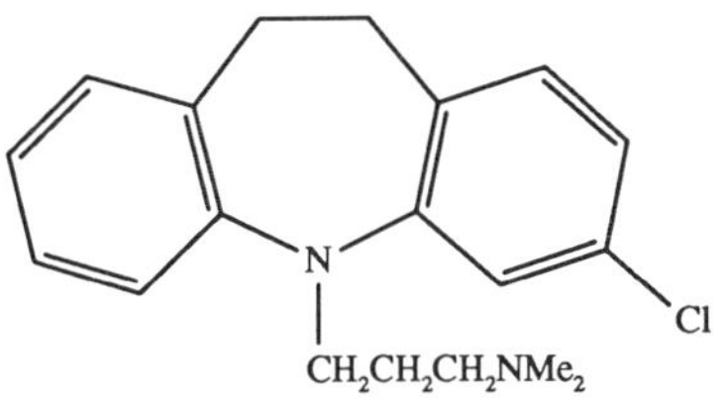

N
Cl
CH2CH2CH2NMe2

Antidepressant	
Smith Kline & French; Ciba-Geigy (Anafranil)	
LgP	5.19(M); 5.30(C)
pKa	9.38
pp cited in Vols.1-5:	

(1) P.N. Craig et al., J.Org. Chem., 1961, 26, 135.
(2) Schindler et al., Swiss Pat. 371 799 (1963).

CLOMOCYCLINE [INN]

CAS 1181-54-0 | MW: 508.92 | MF: C23 H25 Cl N2 O9

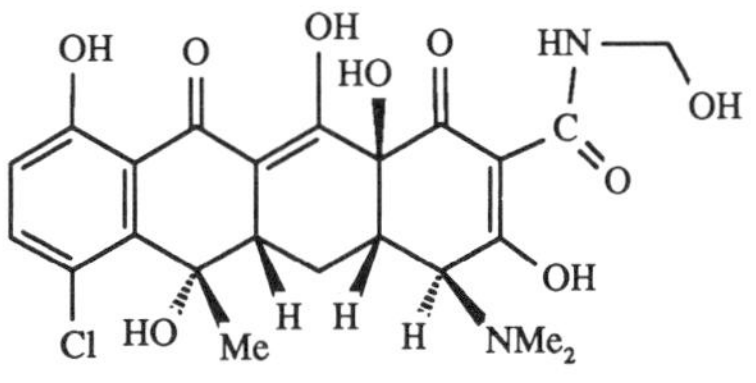

OH O OH O HN
HO
OH
C
O
OH
Cl HO Me H H H NMe2

Antibacterial	
LgP	-1.9(C,1)
pKa	
pp cited in Vols.1-5:	

(1) Banci et al., Boll. Chim. Farm., 1963, 102, 471.
(2) eidem., Belg. Pat. 628 142 (1963).

CLOMOXIR [INN]

CAS 88431-47-4 | MW: 268.74 | MF: C14 H17 Cl O3

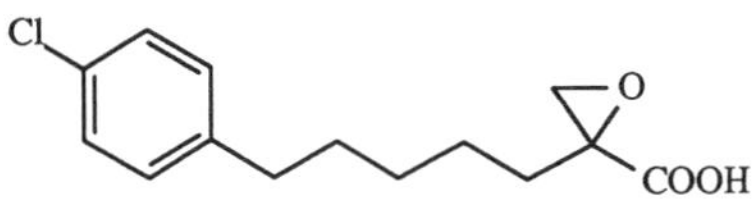

Cl
O
COOH

LgP	4.18(C)
pKa	
pp cited in Vols.1-5:	

CLONAZEPAM [U;INN]

CAS 1622-61-3 | MW: 315.72 | MF: C15 H10 Cl N3 O3

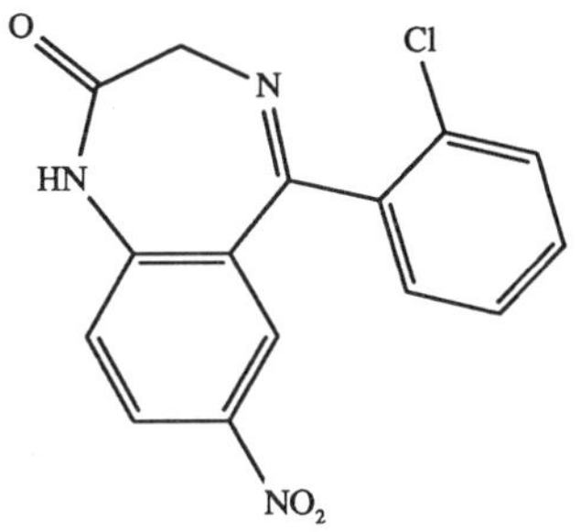

O
Cl
N
HN
NO2

Anticonvulsant	
Hoffmann-LaRoche (Klonopin)	
LgP	2.41(M); 2.84(C,6)
pKa	1.5; 10.5
pp cited in Vols.1-5:	
3: 542	

(1) L. Sternbach et al., J. Med. Chem., 1963, 6, 261. (2) Kariss et al., U.S. Pat. 3 335 181 (1967). (3) W. Winslow in 'Analytical Profiles of Drug Substances' ed. K. Florey, Academic Press, NY, 1977, pp 61-81.

CLONAZOLINE [INN]

CAS 17692-28-3 | MW: 244.73 | MF: C14 H13 Cl N2

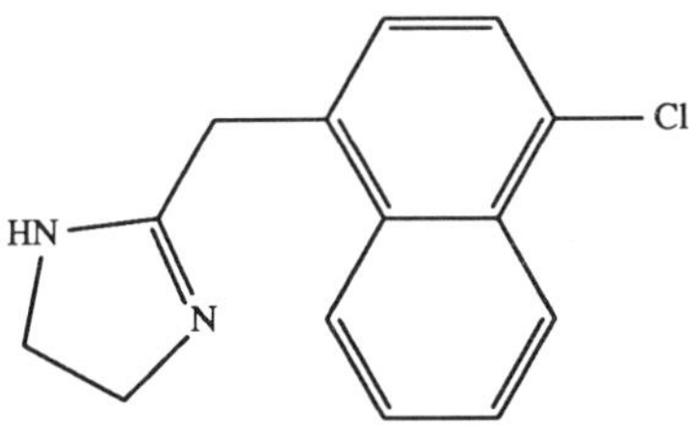

Cl
HN
N

LgP	4.54(C)
pKa	
pp cited in Vols.1-5:	

CLONIDINE [U;INN]

CAS 4205-90-7 | MW: 230.10 | MF: C9 H9 Cl2 N3

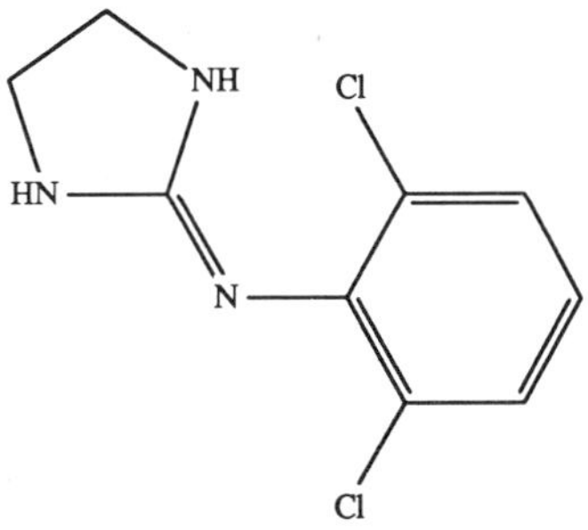

NH
Cl
HN
N
Cl

Antihypertensive	
Boehringer-Ingel., FRG (Catapres)	
LgP	1.59(M)
pKa	8.05
pp cited in Vols.1-5:	
3: 134, 139, 144, 148-163, 173, 176-178, 360; 4: 500, 501; 5: 344, 650, 651	

(1) Zeile et al.,U.S. Pat.3 202 660 (1965).(2) L.M.Jackman et al., J.Am.Chem. Soc.,1975,97,2811. (3)A.Walland,'Pharmacol.& Biochem.Prop.of Drug Subst.' ed.M.E. Goldberg, Am. Pharm. Assoc., Wash., DC, 1977, v.1, pp 67-107.

CLONITAZENE [INN]

CAS 3861-76-5 | MW: 386.88 | MF: C20 H23 Cl N4 O2

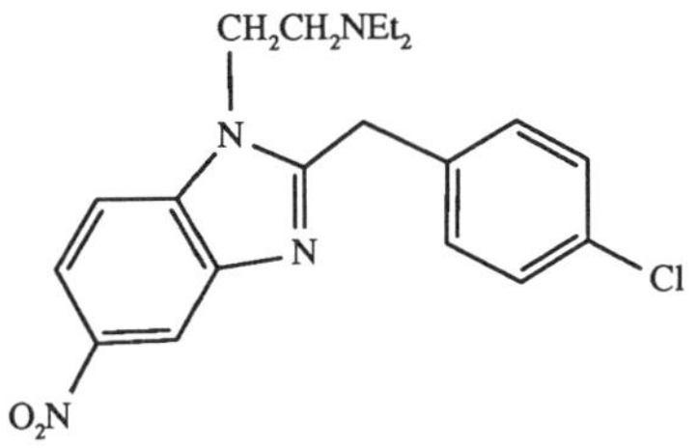

CH2CH2NEt2
N
N
Cl
O2N

Analgesic	
Ciba, Switz.	
LgP	4.78(C)
pKa	
pp cited in Vols.1-5:	

(1) Hunger et al., Experientia, 1957, 13, 400.
(2) Hiffmann et al., U.S. Pat. 2 935 514 (1960).

CLONITRATE [U;INN]

CAS 2612-33-1 | MW: 200.54 | MF: C3 H5 Cl N2 O6

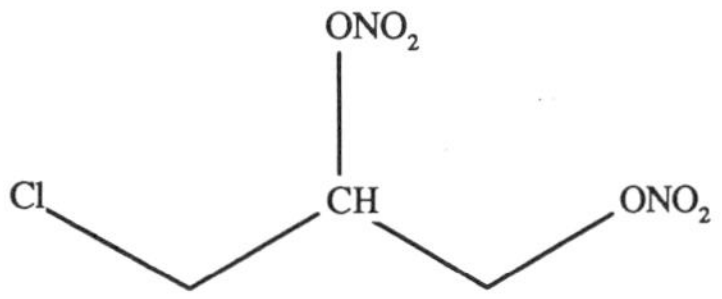

ONO2
Cl
CH
ONO2

Vasodilator, coronary	
LgP	1.40(C)
pKa	
pp cited in Vols.1-5:	

(1) Henry, Ann., 1870, 155, 165.

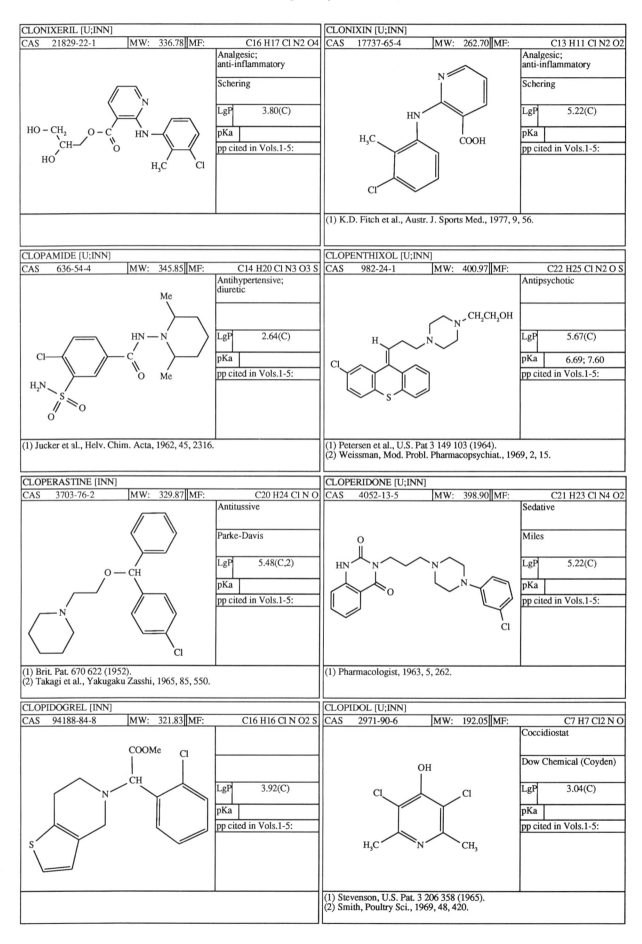

CLONIXERIL [U;INN]			
CAS 21829-22-1	MW: 336.78	MF:	C16 H17 Cl N2 O4

Analgesic; anti-inflammatory

Schering

LgP	3.80(C)
pKa	
pp cited in Vols.1-5:	

CLONIXIN [U;INN]			
CAS 17737-65-4	MW: 262.70	MF:	C13 H11 Cl N2 O2

Analgesic; anti-inflammatory

Schering

LgP	5.22(C)
pKa	
pp cited in Vols.1-5:	

(1) K.D. Fitch et al., Austr. J. Sports Med., 1977, 9, 56.

CLOPAMIDE [U;INN]			
CAS 636-54-4	MW: 345.85	MF:	C14 H20 Cl N3 O3 S

Antihypertensive; diuretic

LgP	2.64(C)
pKa	
pp cited in Vols.1-5:	

(1) Jucker et al., Helv. Chim. Acta, 1962, 45, 2316.

CLOPENTHIXOL [U;INN]			
CAS 982-24-1	MW: 400.97	MF:	C22 H25 Cl N2 O S

Antipsychotic

LgP	5.67(C)
pKa	6.69; 7.60
pp cited in Vols.1-5:	

(1) Petersen et al., U.S. Pat 3 149 103 (1964).
(2) Weissman, Mod. Probl. Pharmacopsychiat., 1969, 2, 15.

CLOPERASTINE [INN]			
CAS 3703-76-2	MW: 329.87	MF:	C20 H24 Cl N O

Antitussive

Parke-Davis

LgP	5.48(C,2)
pKa	
pp cited in Vols.1-5:	

(1) Brit. Pat. 670 622 (1952).
(2) Takagi et al., Yakugaku Zasshi, 1965, 85, 550.

CLOPERIDONE [U;INN]			
CAS 4052-13-5	MW: 398.90	MF:	C21 H23 Cl N4 O2

Sedative

Miles

LgP	5.22(C)
pKa	
pp cited in Vols.1-5:	

(1) Pharmacologist, 1963, 5, 262.

CLOPIDOGREL [INN]			
CAS 94188-84-8	MW: 321.83	MF:	C16 H16 Cl N O2 S

LgP	3.92(C)
pKa	
pp cited in Vols.1-5:	

CLOPIDOL [U;INN]			
CAS 2971-90-6	MW: 192.05	MF:	C7 H7 Cl2 N O

Coccidiostat

Dow Chemical (Coyden)

LgP	3.04(C)
pKa	
pp cited in Vols.1-5:	

(1) Stevenson, U.S. Pat. 3 206 358 (1965).
(2) Smith, Poultry Sci., 1969, 48, 420.

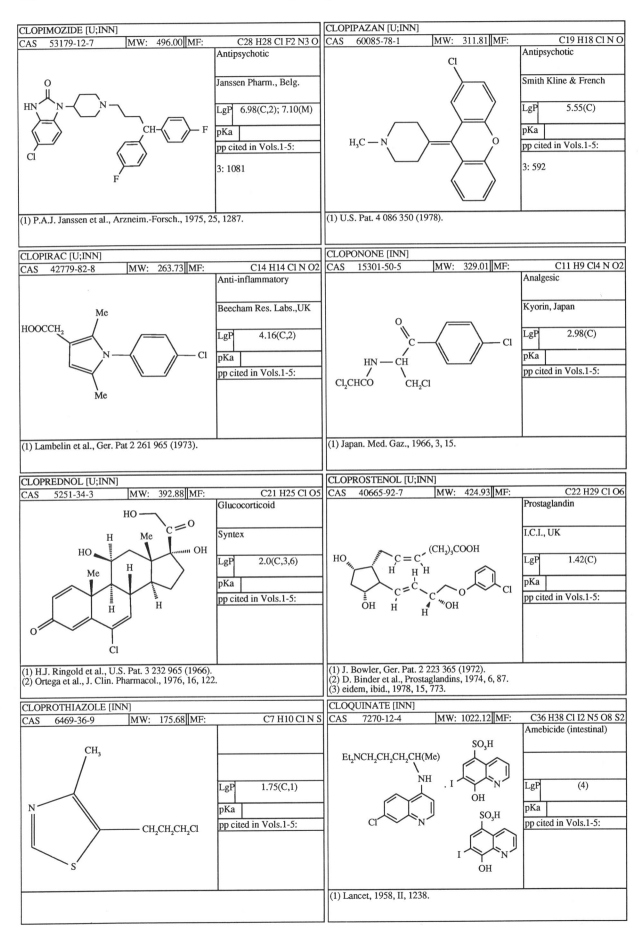

CLOPIMOZIDE [U;INN]

CAS 53179-12-7 | MW: 496.00 | MF: C28 H28 Cl F2 N3 O

Antipsychotic

Janssen Pharm., Belg.

LgP 6.98(C,2); 7.10(M)

pKa

pp cited in Vols.1-5:

3: 1081

(1) P.A.J. Janssen et al., Arzneim.-Forsch., 1975, 25, 1287.

CLOPIPAZAN [U;INN]

CAS 60085-78-1 | MW: 311.81 | MF: C19 H18 Cl N O

Antipsychotic

Smith Kline & French

LgP 5.55(C)

pKa

pp cited in Vols.1-5:

3: 592

(1) U.S. Pat. 4 086 350 (1978).

CLOPIRAC [U;INN]

CAS 42779-82-8 | MW: 263.73 | MF: C14 H14 Cl N O2

Anti-inflammatory

Beecham Res. Labs.,UK

LgP 4.16(C,2)

pKa

pp cited in Vols.1-5:

(1) Lambelin et al., Ger. Pat 2 261 965 (1973).

CLOPONONE [INN]

CAS 15301-50-5 | MW: 329.01 | MF: C11 H9 Cl4 N O2

Analgesic

Kyorin, Japan

LgP 2.98(C)

pKa

pp cited in Vols.1-5:

(1) Japan. Med. Gaz., 1966, 3, 15.

CLOPREDNOL [U;INN]

CAS 5251-34-3 | MW: 392.88 | MF: C21 H25 Cl O5

Glucocorticoid

Syntex

LgP 2.0(C,3,6)

pKa

pp cited in Vols.1-5:

(1) H.J. Ringold et al., U.S. Pat. 3 232 965 (1966).
(2) Ortega et al., J. Clin. Pharmacol., 1976, 16, 122.

CLOPROSTENOL [U;INN]

CAS 40665-92-7 | MW: 424.93 | MF: C22 H29 Cl O6

Prostaglandin

I.C.I., UK

LgP 1.42(C)

pKa

pp cited in Vols.1-5:

(1) J. Bowler, Ger. Pat. 2 223 365 (1972).
(2) D. Binder et al., Prostaglandins, 1974, 6, 87.
(3) eidem, ibid., 1978, 15, 773.

CLOPROTHIAZOLE [INN]

CAS 6469-36-9 | MW: 175.68 | MF: C7 H10 Cl N S

LgP 1.75(C,1)

pKa

pp cited in Vols.1-5:

CLOQUINATE [INN]

CAS 7270-12-4 | MW: 1022.12 | MF: C36 H38 Cl I2 N5 O8 S2

Amebicide (intestinal)

LgP (4)

pKa

pp cited in Vols.1-5:

(1) Lancet, 1958, II, 1238.

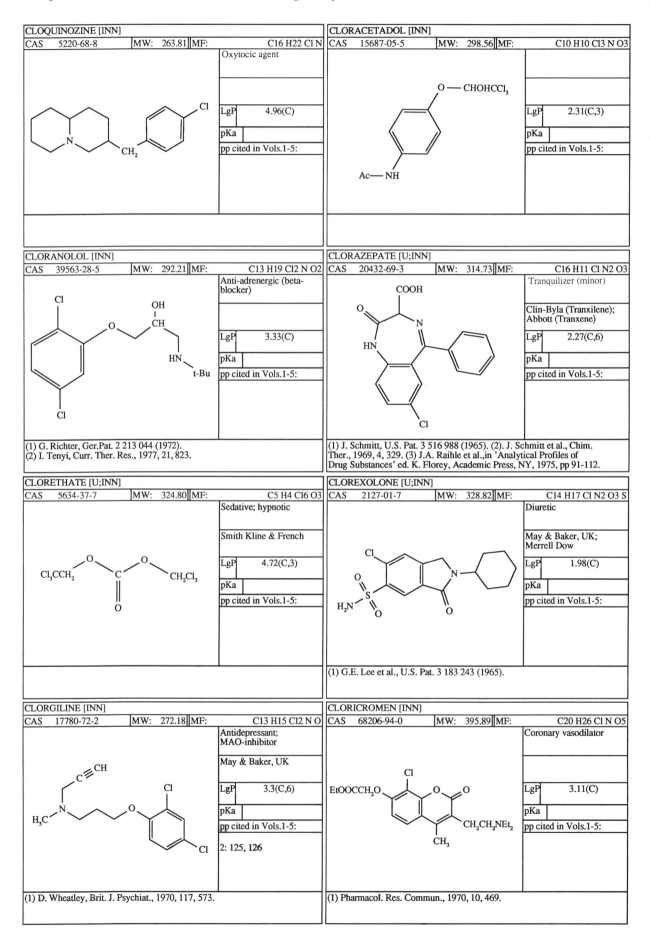

CLOQUINOZINE [INN]

CAS	5220-68-8	MW:	263.81	MF:	C16 H22 Cl N

Oxytocic agent

LgP	4.96(C)
pKa	
pp cited in Vols.1-5:	

CLORACETADOL [INN]

CAS	15687-05-5	MW:	298.56	MF:	C10 H10 Cl3 N O3

LgP	2.31(C,3)
pKa	
pp cited in Vols.1-5:	

CLORANOLOL [INN]

CAS	39563-28-5	MW:	292.21	MF:	C13 H19 Cl2 N O2

Anti-adrenergic (beta-blocker)

LgP	3.33(C)
pKa	
pp cited in Vols.1-5:	

(1) G. Richter, Ger.Pat. 2 213 044 (1972).
(2) I. Tenyi, Curr. Ther. Res., 1977, 21, 823.

CLORAZEPATE [U;INN]

CAS	20432-69-3	MW:	314.73	MF:	C16 H11 Cl N2 O3

Tranquilizer (minor)

Clin-Byla (Tranxilene); Abbott (Tranxene)

LgP	2.27(C,6)
pKa	
pp cited in Vols.1-5:	

(1) J. Schmitt, U.S. Pat. 3 516 988 (1965). (2). J. Schmitt et al., Chim. Ther., 1969, 4, 329. (3) J.A. Raihle et al.,in 'Analytical Profiles of Drug Substances' ed. K. Florey, Academic Press, NY, 1975, pp 91-112.

CLORETHATE [U;INN]

CAS	5634-37-7	MW:	324.80	MF:	C5 H4 Cl6 O3

Sedative; hypnotic

Smith Kline & French

LgP	4.72(C,3)
pKa	
pp cited in Vols.1-5:	

CLOREXOLONE [U;INN]

CAS	2127-01-7	MW:	328.82	MF:	C14 H17 Cl N2 O3 S

Diuretic

May & Baker, UK; Merrell Dow

LgP	1.98(C)
pKa	
pp cited in Vols.1-5:	

(1) G.E. Lee et al., U.S. Pat. 3 183 243 (1965).

CLORGILINE [INN]

CAS	17780-72-2	MW:	272.18	MF:	C13 H15 Cl2 N O

Antidepressant; MAO-inhibitor

May & Baker, UK

LgP	3.3(C,6)
pKa	
pp cited in Vols.1-5:	
2: 125, **126**	

(1) D. Wheatley, Brit. J. Psychiat., 1970, 117, 573.

CLORICROMEN [INN]

CAS	68206-94-0	MW:	395.89	MF:	C20 H26 Cl N O5

Coronary vasodilator

LgP	3.11(C)
pKa	
pp cited in Vols.1-5:	

(1) Pharmacol. Res. Commun., 1970, 10, 469.

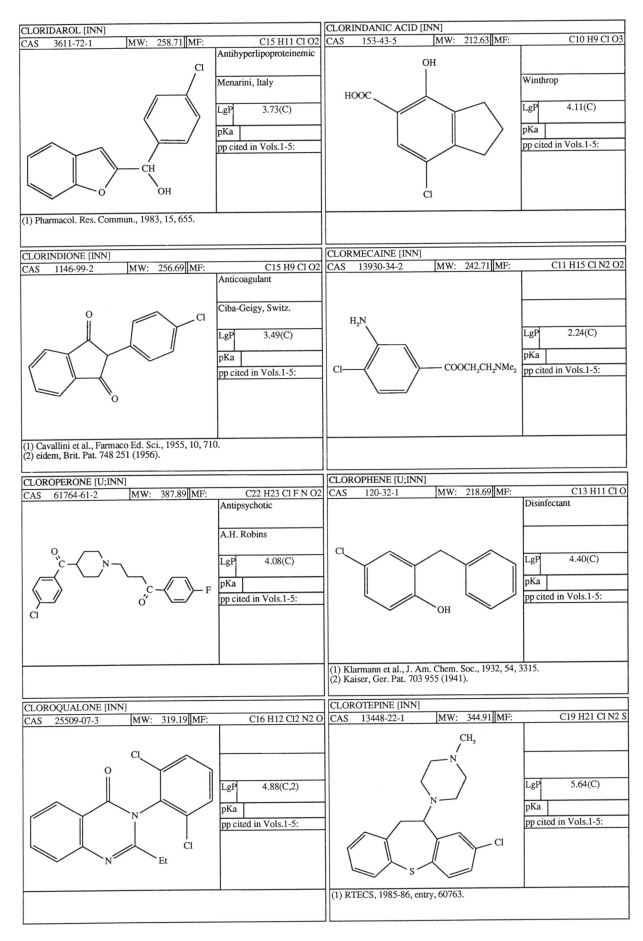

CLORIDAROL [INN]

CAS	3611-72-1	MW: 258.71	MF:	C15 H11 Cl O2

Antihyperlipoproteinemic

Menarini, Italy

LgP	3.73(C)
pKa	
pp cited in Vols.1-5:	

(1) Pharmacol. Res. Commun., 1983, 15, 655.

CLORINDANIC ACID [INN]

CAS	153-43-5	MW: 212.63	MF:	C10 H9 Cl O3

Winthrop

LgP	4.11(C)
pKa	
pp cited in Vols.1-5:	

CLORINDIONE [INN]

CAS	1146-99-2	MW: 256.69	MF:	C15 H9 Cl O2

Anticoagulant

Ciba-Geigy, Switz.

LgP	3.49(C)
pKa	
pp cited in Vols.1-5:	

(1) Cavallini et al., Farmaco Ed. Sci., 1955, 10, 710.
(2) eidem, Brit. Pat. 748 251 (1956).

CLORMECAINE [INN]

CAS	13930-34-2	MW: 242.71	MF:	C11 H15 Cl N2 O2

LgP	2.24(C)
pKa	
pp cited in Vols.1-5:	

CLOROPERONE [U;INN]

CAS	61764-61-2	MW: 387.89	MF:	C22 H23 Cl F N O2

Antipsychotic

A.H. Robins

LgP	4.08(C)
pKa	
pp cited in Vols.1-5:	

CLOROPHENE [U;INN]

CAS	120-32-1	MW: 218.69	MF:	C13 H11 Cl O

Disinfectant

LgP	4.40(C)
pKa	
pp cited in Vols.1-5:	

(1) Klarmann et al., J. Am. Chem. Soc., 1932, 54, 3315.
(2) Kaiser, Ger. Pat. 703 955 (1941).

CLOROQUALONE [INN]

CAS	25509-07-3	MW: 319.19	MF:	C16 H12 Cl2 N2 O

LgP	4.88(C,2)
pKa	
pp cited in Vols.1-5:	

CLOROTEPINE [INN]

CAS	13448-22-1	MW: 344.91	MF:	C19 H21 Cl N2 S

LgP	5.64(C)
pKa	
pp cited in Vols.1-5:	

(1) RTECS, 1985-86, entry, 60763.

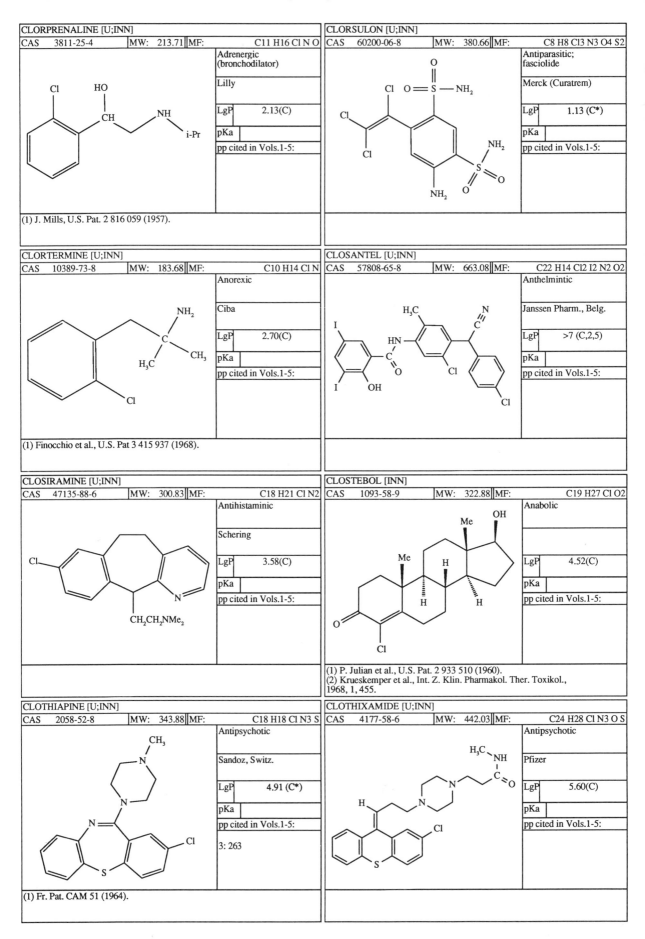

CLORPRENALINE [U;INN]

CAS	3811-25-4	MW: 213.71	MF:	C11 H16 Cl N O

Adrenergic (bronchodilator)

Lilly

LgP	2.13(C)

pKa

pp cited in Vols.1-5:

(1) J. Mills, U.S. Pat. 2 816 059 (1957).

CLORSULON [U;INN]

CAS	60200-06-8	MW: 380.66	MF:	C8 H8 Cl3 N3 O4 S2

Antiparasitic; fasciolide

Merck (Curatrem)

LgP	1.13 (C*)

pKa

pp cited in Vols.1-5:

CLORTERMINE [U;INN]

CAS	10389-73-8	MW: 183.68	MF:	C10 H14 Cl N

Anorexic

Ciba

LgP	2.70(C)

pKa

pp cited in Vols.1-5:

(1) Finocchio et al., U.S. Pat 3 415 937 (1968).

CLOSANTEL [U;INN]

CAS	57808-65-8	MW: 663.08	MF:	C22 H14 Cl2 I2 N2 O2

Anthelmintic

Janssen Pharm., Belg.

LgP	>7 (C,2,5)

pKa

pp cited in Vols.1-5:

CLOSIRAMINE [U;INN]

CAS	47135-88-6	MW: 300.83	MF:	C18 H21 Cl N2

Antihistaminic

Schering

LgP	3.58(C)

pKa

pp cited in Vols.1-5:

CLOSTEBOL [INN]

CAS	1093-58-9	MW: 322.88	MF:	C19 H27 Cl O2

Anabolic

LgP	4.52(C)

pKa

pp cited in Vols.1-5:

(1) P. Julian et al., U.S. Pat. 2 933 510 (1960).
(2) Krueskemper et al., Int. Z. Klin. Pharmakol. Ther. Toxikol., 1968, 1, 455.

CLOTHIAPINE [U;INN]

CAS	2058-52-8	MW: 343.88	MF:	C18 H18 Cl N3 S

Antipsychotic

Sandoz, Switz.

LgP	4.91 (C*)

pKa

pp cited in Vols.1-5:

3: 263

(1) Fr. Pat. CAM 51 (1964).

CLOTHIXAMIDE [U;INN]

CAS	4177-58-6	MW: 442.03	MF:	C24 H28 Cl N3 O S

Antipsychotic

Pfizer

LgP	5.60(C)

pKa

pp cited in Vols.1-5:

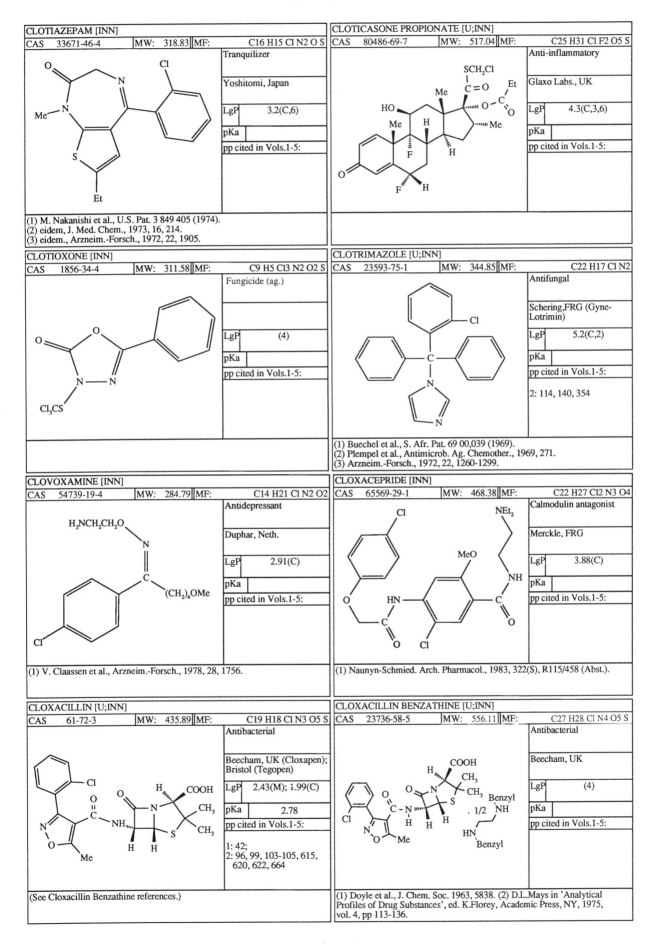

CLOTIAZEPAM [INN]

| CAS | 33671-46-4 | MW: | 318.83 | MF: | C16 H15 Cl N2 O S |

Tranquilizer

Yoshitomi, Japan

LgP	3.2(C,6)
pKa	
pp cited in Vols.1-5:	

(1) M. Nakanishi et al., U.S. Pat. 3 849 405 (1974).
(2) eidem, J. Med. Chem., 1973, 16, 214.
(3) eidem., Arzneim.-Forsch., 1972, 22, 1905.

CLOTICASONE PROPIONATE [U;INN]

| CAS | 80486-69-7 | MW: | 517.04 | MF: | C25 H31 Cl F2 O5 S |

Anti-inflammatory

Glaxo Labs., UK

LgP	4.3(C,3,6)
pKa	
pp cited in Vols.1-5:	

CLOTIOXONE [INN]

| CAS | 1856-34-4 | MW: | 311.58 | MF: | C9 H5 Cl3 N2 O2 S |

Fungicide (ag.)

LgP	(4)
pKa	
pp cited in Vols.1-5:	

CLOTRIMAZOLE [U;INN]

| CAS | 23593-75-1 | MW: | 344.85 | MF: | C22 H17 Cl N2 |

Antifungal

Schering,FRG (Gyne-Lotrimin)

LgP	5.2(C,2)
pKa	
pp cited in Vols.1-5:	

2: 114, 140, 354

(1) Buechel et al., S. Afr. Pat. 69 00,039 (1969).
(2) Plempel et al., Antimicrob. Ag. Chemother., 1969, 271.
(3) Arzneim.-Forsch., 1972, 22, 1260-1299.

CLOVOXAMINE [INN]

| CAS | 54739-19-4 | MW: | 284.79 | MF: | C14 H21 Cl N2 O2 |

Antidepressant

Duphar, Neth.

LgP	2.91(C)
pKa	
pp cited in Vols.1-5:	

(1) V. Claassen et al., Arzneim.-Forsch., 1978, 28, 1756.

CLOXACEPRIDE [INN]

| CAS | 65569-29-1 | MW: | 468.38 | MF: | C22 H27 Cl2 N3 O4 |

Calmodulin antagonist

Merckle, FRG

LgP	3.88(C)
pKa	
pp cited in Vols.1-5:	

(1) Naunyn-Schmied. Arch. Pharmacol., 1983, 322(S), R115/458 (Abst.).

CLOXACILLIN [U;INN]

| CAS | 61-72-3 | MW: | 435.89 | MF: | C19 H18 Cl N3 O5 S |

Antibacterial

Beecham, UK (Cloxapen); Bristol (Tegopen)

LgP	2.43(M); 1.99(C)
pKa	2.78
pp cited in Vols.1-5:	

1: 42;
2: 96, 99, 103-105, 615, 620, 622, 664

(See Cloxacillin Benzathine references.)

CLOXACILLIN BENZATHINE [U;INN]

| CAS | 23736-58-5 | MW: | 556.11 | MF: | C27 H28 Cl N4 O5 S |

Antibacterial

Beecham, UK

LgP	(4)
pKa	
pp cited in Vols.1-5:	

(1) Doyle et al., J. Chem. Soc. 1963, 5838. (2) D.L.Mays in 'Analytical Profiles of Drug Substances', ed. K.Florey, Academic Press, NY, 1975, vol. 4, pp 113-136.

CLOXAZOLAM [INN]

CAS	24166-13-0	MW:	349.22	MF:	C17 H14 Cl2 N2 O2

Tranquilizer (minor)

Sankyo, Japan

LgP	3.9(C,6)
pKa	
pp cited in Vols.1-5:	

(1) Tachikawa et al., U.S. Pat. 3 772 371 (1973).
(2) Kamioka et al., Arzneim.-Forsch., 1972, 22, 884.
(3) Kamioka et al., J. Med. Chem., 1971, 14, 520.

CLOXESTRADIOL [INN]

CAS	54063-33-1	MW:	419.78	MF:	C20 H25 Cl3 O3

LgP	5.9(C)
pKa	
pp cited in Vols.1-5:	

CLOXIMATE [INN]

CAS	58832-68-1	MW:	298.77	MF:	C14 H19 Cl N2 O3

Anti-inflammatory

Phillips-Duphar, Neth.

LgP	3.28(C)
pKa	
pp cited in Vols.1-5:	

(1) J. van Dijk et al., J. Med. Chem., 1977, 20, 1199.
(2) U.S. Pat. 3 853 955 (1974).

CLOXOTESTOSTERONE [INN]

CAS	53608-96-1	MW:	435.82	MF:	C21 H29 Cl3 O3

LgP	5.48(C)
pKa	
pp cited in Vols.1-5:	

CLOXYPENDYL [INN]

CAS	15311-77-0	MW:	404.97	MF:	C20 H25 Cl N4 O S

Chemie-werke Homburg, FRG

LgP	4.77(C)
pKa	
pp cited in Vols.1-5:	

(1) A. Grosz et al., Arzneim.-Forsch., 1968, 18, 435.

CLOXYQUIN [U;INN]

CAS	130-16-5	MW:	179.61	MF:	C9 H6 Cl N O

Antibacterial

LgP	2.88(M); 2.71(C)
pKa	3.8; 8.7
pp cited in Vols.1-5:	

(1) 'Martindale. The Extra Pharmacopoeia', ed. J.E.F. Reynolds, The Pharmaceutical Press, London, 1982, 28th Ed., pp 980 & 1696.

CLOZAPINE [U;INN]

CAS	5786-21-0	MW:	326.83	MF:	C18 H19 Cl N4

Sedative; antipsychotic

Sandoz, Switz. (Clozaril)

LgP	3.62(M); 2.99(M, pH7.4); 4.27(C)
pKa	8.0
pp cited in Vols.1-5:	

1: 271;
3: 139, 140, 263;
4: 117

(1) Schmutz et al., U.S. Pat. 3 539 573 (1970). (2) eidem, Chim. Ther., 1967, 2, 427. (3) A.C. Sayers et al., in 'Pharmacol.& Biochem.Props.Drug Subst.'ed. M.E. Goldberg, Am. Pharm. Assoc., Wash.,DC, 1977, v.1,pp 1-31.

COBAMIDE [INN]

CAS	13870-90-1	MW:	1579.62	MF:	C72 H100 Co N18 O17 P

Hematopoietic vitamin

LgP	(4)
pKa	
pp cited in Vols.1-5:	

(1) Smith, 'Vitamin B12', Methuen & Co., London, 3rd ed., 1965.

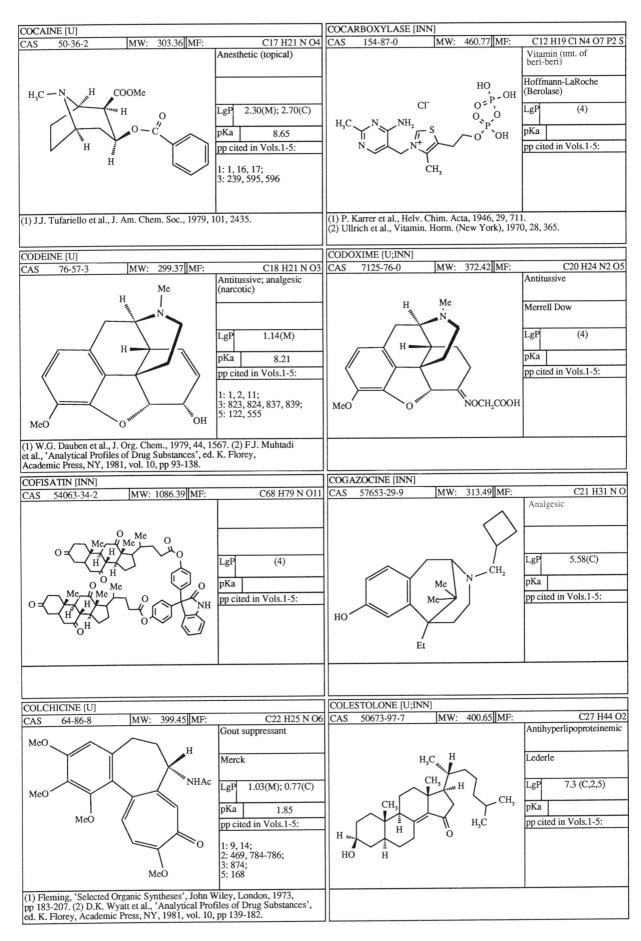

COCAINE [U]

CAS	50-36-2	MW:	303.36	MF:	C17 H21 N O4

Anesthetic (topical)

LgP	2.30(M); 2.70(C)
pKa	8.65

pp cited in Vols.1-5:

1: 1, 16, 17;
3: 239, 595, 596

(1) J.J. Tufariello et al., J. Am. Chem. Soc., 1979, 101, 2435.

COCARBOXYLASE [INN]

CAS	154-87-0	MW:	460.77	MF:	C12 H19 Cl N4 O7 P2 S

Vitamin (tmt. of beri-beri)

Hoffmann-LaRoche (Berolase)

LgP	(4)
pKa	

pp cited in Vols.1-5:

(1) P. Karrer et al., Helv. Chim. Acta, 1946, 29, 711.
(2) Ullrich et al., Vitamin. Horm. (New York), 1970, 28, 365.

CODEINE [U]

CAS	76-57-3	MW:	299.37	MF:	C18 H21 N O3

Antitussive; analgesic (narcotic)

LgP	1.14(M)
pKa	8.21

pp cited in Vols.1-5:

1: 1, 2, 11;
3: 823, 824, 837, 839;
5: 122, 555

(1) W.G. Dauben et al., J. Org. Chem., 1979, 44, 1567. (2) F.J. Muhtadi et al., 'Analytical Profiles of Drug Substances', ed. K. Florey, Academic Press, NY, 1981, vol. 10, pp 93-138.

CODOXIME [U;INN]

CAS	7125-76-0	MW:	372.42	MF:	C20 H24 N2 O5

Antitussive

Merrell Dow

LgP	(4)
pKa	

pp cited in Vols.1-5:

COFISATIN [INN]

CAS	54063-34-2	MW:	1086.39	MF:	C68 H79 N O11

LgP	(4)
pKa	

pp cited in Vols.1-5:

COGAZOCINE [INN]

CAS	57653-29-9	MW:	313.49	MF:	C21 H31 N O

Analgesic

LgP	5.58(C)
pKa	

pp cited in Vols.1-5:

COLCHICINE [U]

CAS	64-86-8	MW:	399.45	MF:	C22 H25 N O6

Gout suppressant

Merck

LgP	1.03(M); 0.77(C)
pKa	1.85

pp cited in Vols.1-5:

1: 9, 14;
2: 469, 784-786;
3: 874;
5: 168

(1) Fleming, 'Selected Organic Syntheses', John Wiley, London, 1973, pp 183-207. (2) D.K. Wyatt et al., 'Analytical Profiles of Drug Substances', ed. K. Florey, Academic Press, NY, 1981, vol. 10, pp 139-182.

COLESTOLONE [U;INN]

CAS	50673-97-7	MW:	400.65	MF:	C27 H44 O2

Antihyperlipoproteinemic

Lederle

LgP	7.3 (C,2,5)
pKa	

pp cited in Vols.1-5:

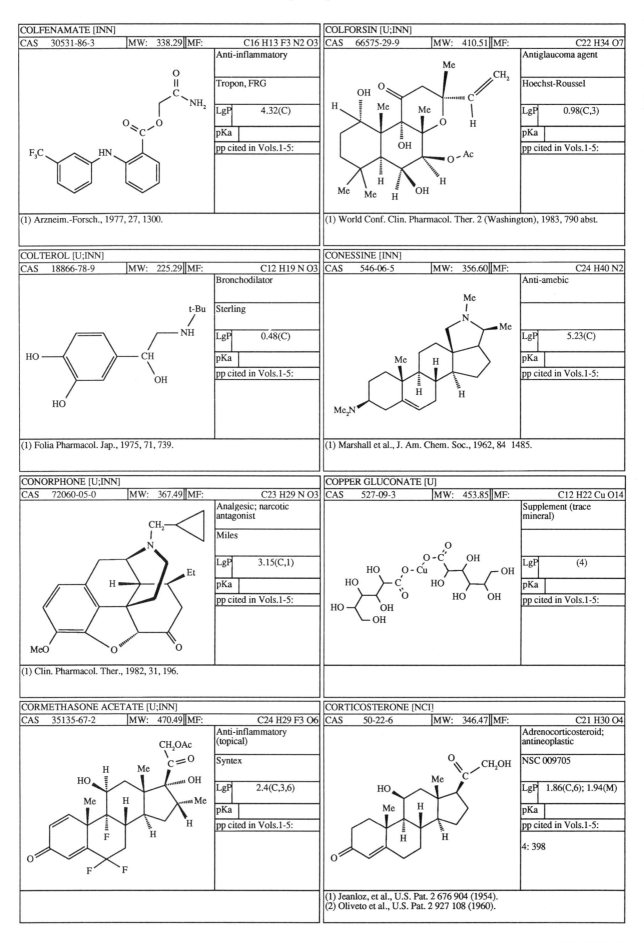

COLFENAMATE [INN]		
CAS 30531-86-3	MW: 338.29	MF: C16 H13 F3 N2 O3

	Anti-inflammatory
	Tropon, FRG
	LgP 4.32(C)
	pKa
	pp cited in Vols.1-5:

(1) Arzneim.-Forsch., 1977, 27, 1300.

COLFORSIN [U;INN]		
CAS 66575-29-9	MW: 410.51	MF: C22 H34 O7

	Antiglaucoma agent
	Hoechst-Roussel
	LgP 0.98(C,3)
	pKa
	pp cited in Vols.1-5:

(1) World Conf. Clin. Pharmacol. Ther. 2 (Washington), 1983, 790 abst.

COLTEROL [U;INN]		
CAS 18866-78-9	MW: 225.29	MF: C12 H19 N O3

	Bronchodilator
	Sterling
	LgP 0.48(C)
	pKa
	pp cited in Vols.1-5:

(1) Folia Pharmacol. Jap., 1975, 71, 739.

CONESSINE [INN]		
CAS 546-06-5	MW: 356.60	MF: C24 H40 N2

	Anti-amebic
	LgP 5.23(C)
	pKa
	pp cited in Vols.1-5:

(1) Marshall et al., J. Am. Chem. Soc., 1962, 84 1485.

CONORPHONE [U;INN]		
CAS 72060-05-0	MW: 367.49	MF: C23 H29 N O3

	Analgesic; narcotic antagonist
	Miles
	LgP 3.15(C,1)
	pKa
	pp cited in Vols.1-5:

(1) Clin. Pharmacol. Ther., 1982, 31, 196.

COPPER GLUCONATE [U]		
CAS 527-09-3	MW: 453.85	MF: C12 H22 Cu O14

	Supplement (trace mineral)
	LgP (4)
	pKa
	pp cited in Vols.1-5:

CORMETHASONE ACETATE [U;INN]		
CAS 35135-67-2	MW: 470.49	MF: C24 H29 F3 O6

	Anti-inflammatory (topical)
	Syntex
	LgP 2.4(C,3,6)
	pKa
	pp cited in Vols.1-5:

CORTICOSTERONE [NCI]		
CAS 50-22-6	MW: 346.47	MF: C21 H30 O4

	Adrenocorticosteroid; antineoplastic
	NSC 009705
	LgP 1.86(C,6); 1.94(M)
	pKa
	pp cited in Vols.1-5:
	4: 398

(1) Jeanloz, et al., U.S. Pat. 2 676 904 (1954).
(2) Oliveto et al., U.S. Pat. 2 927 108 (1960).

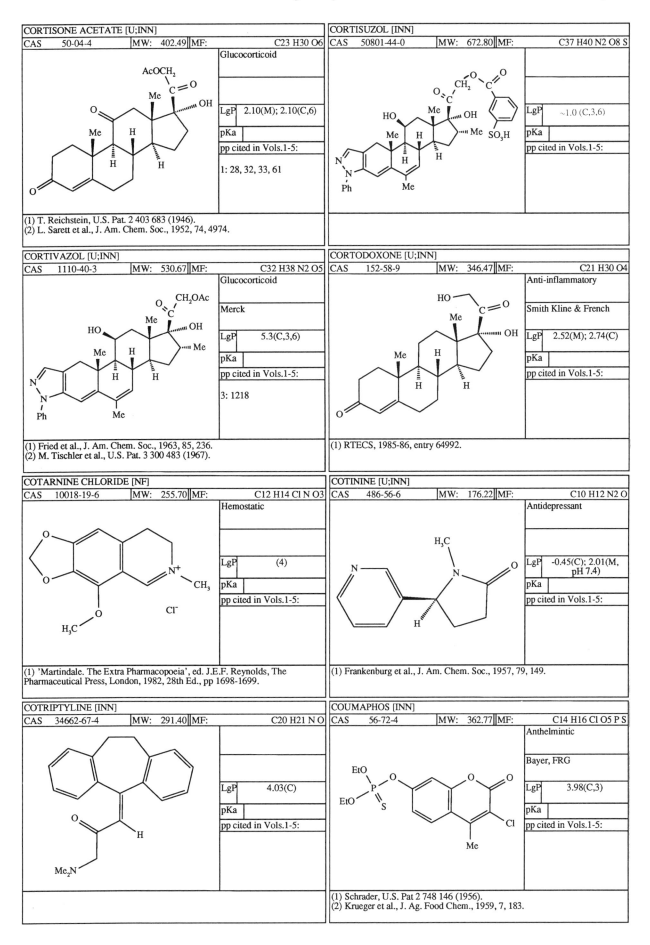

CORTISONE ACETATE [U;INN]

CAS	50-04-4	MW: 402.49	MF:	C23 H30 O6

Glucocorticoid

LgP	2.10(M); 2.10(C,6)
pKa	
pp cited in Vols.1-5:	

1: 28, 32, 33, 61

(1) T. Reichstein, U.S. Pat. 2 403 683 (1946).
(2) L. Sarett et al., J. Am. Chem. Soc., 1952, 74, 4974.

CORTISUZOL [INN]

CAS	50801-44-0	MW: 672.80	MF:	C37 H40 N2 O8 S

LgP	~1.0 (C,3,6)
pKa	
pp cited in Vols.1-5:	

CORTIVAZOL [U;INN]

CAS	1110-40-3	MW: 530.67	MF:	C32 H38 N2 O5

Glucocorticoid

Merck

LgP	5.3(C,3,6)
pKa	
pp cited in Vols.1-5:	

3: 1218

(1) Fried et al., J. Am. Chem. Soc., 1963, 85, 236.
(2) M. Tischler et al., U.S. Pat. 3 300 483 (1967).

CORTODOXONE [U;INN]

CAS	152-58-9	MW: 346.47	MF:	C21 H30 O4

Anti-inflammatory

Smith Kline & French

LgP	2.52(M); 2.74(C)
pKa	
pp cited in Vols.1-5:	

(1) RTECS, 1985-86, entry 64992.

COTARNINE CHLORIDE [NF]

CAS	10018-19-6	MW: 255.70	MF:	C12 H14 Cl N O3

Hemostatic

LgP	(4)
pKa	
pp cited in Vols.1-5:	

(1) 'Martindale. The Extra Pharmacopoeia', ed. J.E.F. Reynolds, The Pharmaceutical Press, London, 1982, 28th Ed., pp 1698-1699.

COTININE [U;INN]

CAS	486-56-6	MW: 176.22	MF:	C10 H12 N2 O

Antidepressant

LgP	-0.45(C); 2.01(M, pH 7.4)
pKa	
pp cited in Vols.1-5:	

(1) Frankenburg et al., J. Am. Chem. Soc., 1957, 79, 149.

COTRIPTYLINE [INN]

CAS	34662-67-4	MW: 291.40	MF:	C20 H21 N O

LgP	4.03(C)
pKa	
pp cited in Vols.1-5:	

COUMAPHOS [INN]

CAS	56-72-4	MW: 362.77	MF:	C14 H16 Cl O5 P S

Anthelmintic

Bayer, FRG

LgP	3.98(C,3)
pKa	
pp cited in Vols.1-5:	

(1) Schrader, U.S. Pat 2 748 146 (1956).
(2) Krueger et al., J. Ag. Food Chem., 1959, 7, 183.

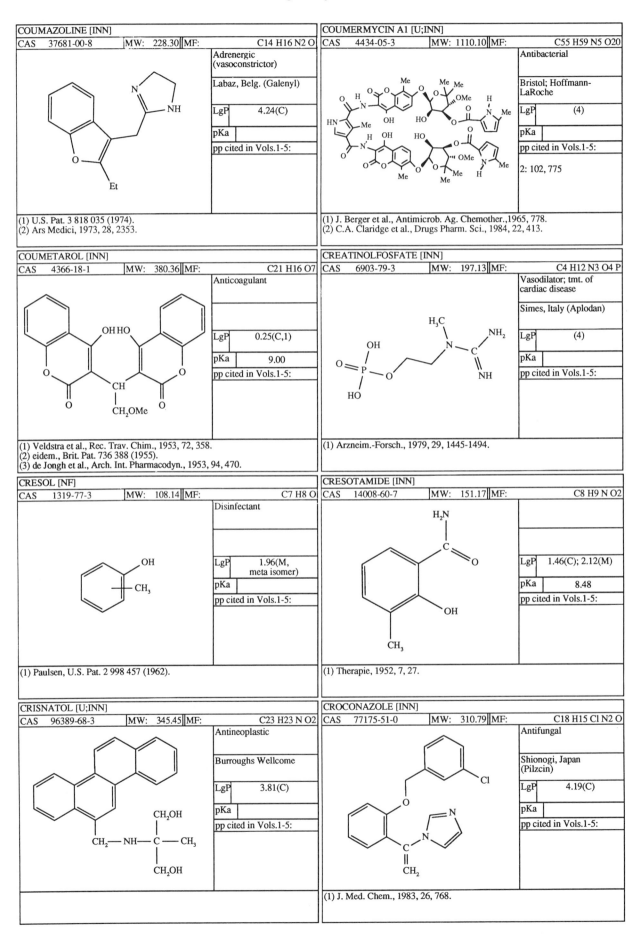

COUMAZOLINE [INN]

CAS	37681-00-8	MW:	228.30	MF:	C14 H16 N2 O

Adrenergic (vasoconstrictor)

Labaz, Belg. (Galenyl)

LgP	4.24(C)
pKa	

pp cited in Vols.1-5:

(1) U.S. Pat. 3 818 035 (1974).
(2) Ars Medici, 1973, 28, 2353.

COUMERMYCIN A1 [U;INN]

CAS	4434-05-3	MW:	1110.10	MF:	C55 H59 N5 O20

Antibacterial

Bristol; Hoffmann-LaRoche

LgP	(4)
pKa	

pp cited in Vols.1-5:

2: 102, 775

(1) J. Berger et al., Antimicrob. Ag. Chemother.,1965, 778.
(2) C.A. Claridge et al., Drugs Pharm. Sci., 1984, 22, 413.

COUMETAROL [INN]

CAS	4366-18-1	MW:	380.36	MF:	C21 H16 O7

Anticoagulant

LgP	0.25(C,1)
pKa	9.00

pp cited in Vols.1-5:

(1) Veldstra et al., Rec. Trav. Chim., 1953, 72, 358.
(2) eidem., Brit. Pat. 736 388 (1955).
(3) de Jongh et al., Arch. Int. Pharmacodyn., 1953, 94, 470.

CREATINOLFOSFATE [INN]

CAS	6903-79-3	MW:	197.13	MF:	C4 H12 N3 O4 P

Vasodilator; tmt. of cardiac disease

Simes, Italy (Aplodan)

LgP	(4)
pKa	

pp cited in Vols.1-5:

(1) Arzneim.-Forsch., 1979, 29, 1445-1494.

CRESOL [NF]

CAS	1319-77-3	MW:	108.14	MF:	C7 H8 O

Disinfectant

LgP	1.96(M, meta isomer)
pKa	

pp cited in Vols.1-5:

(1) Paulsen, U.S. Pat. 2 998 457 (1962).

CRESOTAMIDE [INN]

CAS	14008-60-7	MW:	151.17	MF:	C8 H9 N O2

LgP	1.46(C); 2.12(M)
pKa	8.48

pp cited in Vols.1-5:

(1) Therapie, 1952, 7, 27.

CRISNATOL [U;INN]

CAS	96389-68-3	MW:	345.45	MF:	C23 H23 N O2

Antineoplastic

Burroughs Wellcome

LgP	3.81(C)
pKa	

pp cited in Vols.1-5:

CROCONAZOLE [INN]

CAS	77175-51-0	MW:	310.79	MF:	C18 H15 Cl N2 O

Antifungal

Shionogi, Japan (Pilzcin)

LgP	4.19(C)
pKa	

pp cited in Vols.1-5:

(1) J. Med. Chem., 1983, 26, 768.

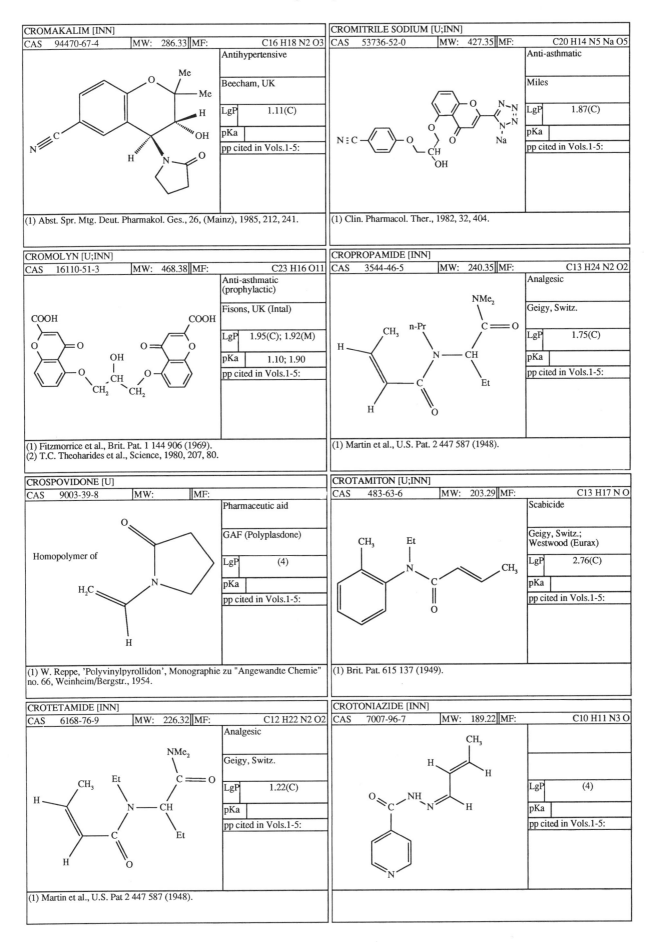

CROMAKALIM [INN]

| CAS | 94470-67-4 | MW: 286.33 | MF: | C16 H18 N2 O3 |

Antihypertensive

Beecham, UK

LgP	1.11(C)
pKa	
pp cited in Vols.1-5:	

(1) Abst. Spr. Mtg. Deut. Pharmakol. Ges., 26, (Mainz), 1985, 212, 241.

CROMITRILE SODIUM [U;INN]

| CAS | 53736-52-0 | MW: 427.35 | MF: | C20 H14 N5 Na O5 |

Anti-asthmatic

Miles

LgP	1.87(C)
pKa	
pp cited in Vols.1-5:	

(1) Clin. Pharmacol. Ther., 1982, 32, 404.

CROMOLYN [U;INN]

| CAS | 16110-51-3 | MW: 468.38 | MF: | C23 H16 O11 |

Anti-asthmatic (prophylactic)

Fisons, UK (Intal)

LgP	1.95(C); 1.92(M)
pKa	1.10; 1.90
pp cited in Vols.1-5:	

(1) Fitzmorrice et al., Brit. Pat. 1 144 906 (1969).
(2) T.C. Theoharides et al., Science, 1980, 207, 80.

CROPROPAMIDE [INN]

| CAS | 3544-46-5 | MW: 240.35 | MF: | C13 H24 N2 O2 |

Analgesic

Geigy, Switz.

LgP	1.75(C)
pKa	
pp cited in Vols.1-5:	

(1) Martin et al., U.S. Pat. 2 447 587 (1948).

CROSPOVIDONE [U]

| CAS | 9003-39-8 | MW: | MF: | |

Pharmaceutic aid

GAF (Polyplasdone)

LgP	(4)
pKa	
pp cited in Vols.1-5:	

Homopolymer of

(1) W. Reppe, 'Polyvinylpyrollidon', Monographie zu "Angewandte Chemie" no. 66, Weinheim/Bergstr., 1954.

CROTAMITON [U;INN]

| CAS | 483-63-6 | MW: 203.29 | MF: | C13 H17 N O |

Scabicide

Geigy, Switz.; Westwood (Eurax)

LgP	2.76(C)
pKa	
pp cited in Vols.1-5:	

(1) Brit. Pat. 615 137 (1949).

CROTETAMIDE [INN]

| CAS | 6168-76-9 | MW: 226.32 | MF: | C12 H22 N2 O2 |

Analgesic

Geigy, Switz.

LgP	1.22(C)
pKa	
pp cited in Vols.1-5:	

(1) Martin et al., U.S. Pat 2 447 587 (1948).

CROTONIAZIDE [INN]

| CAS | 7007-96-7 | MW: 189.22 | MF: | C10 H11 N3 O |

LgP	(4)
pKa	
pp cited in Vols.1-5:	

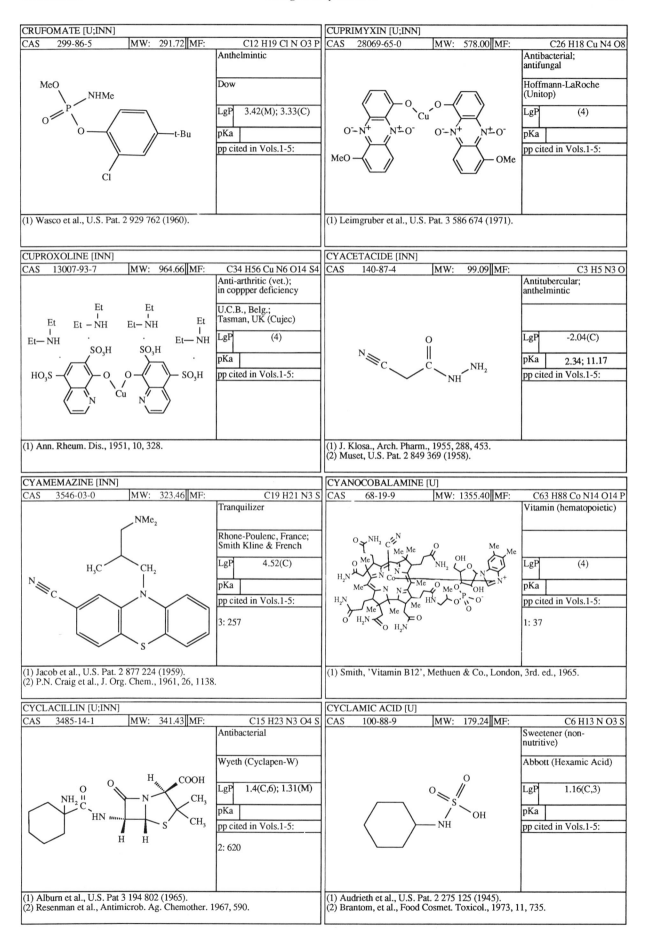

CRUFOMATE [U;INN]		
CAS 299-86-5	MW: 291.72	MF: C12 H19 Cl N O3 P

Anthelmintic

Dow

LgP 3.42(M); 3.33(C)

pKa

pp cited in Vols.1-5:

(1) Wasco et al., U.S. Pat. 2 929 762 (1960).

CUPRIMYXIN [U;INN]		
CAS 28069-65-0	MW: 578.00	MF: C26 H18 Cu N4 O8

Antibacterial; antifungal

Hoffmann-LaRoche (Unitop)

LgP (4)

pKa

pp cited in Vols.1-5:

(1) Leimgruber et al., U.S. Pat. 3 586 674 (1971).

CUPROXOLINE [INN]		
CAS 13007-93-7	MW: 964.66	MF: C34 H56 Cu N6 O14 S4

Anti-arthritic (vet.); in coppper deficiency

U.C.B., Belg.; Tasman, UK (Cujec)

LgP (4)

pKa

pp cited in Vols.1-5:

(1) Ann. Rheum. Dis., 1951, 10, 328.

CYACETACIDE [INN]		
CAS 140-87-4	MW: 99.09	MF: C3 H5 N3 O

Antitubercular; anthelmintic

LgP -2.04(C)

pKa 2.34; 11.17

pp cited in Vols.1-5:

(1) J. Klosa., Arch. Pharm., 1955, 288, 453.
(2) Muset, U.S. Pat. 2 849 369 (1958).

CYAMEMAZINE [INN]		
CAS 3546-03-0	MW: 323.46	MF: C19 H21 N3 S

Tranquilizer

Rhone-Poulenc, France; Smith Kline & French

LgP 4.52(C)

pKa

pp cited in Vols.1-5:

3: 257

(1) Jacob et al., U.S. Pat. 2 877 224 (1959).
(2) P.N. Craig et al., J. Org. Chem., 1961, 26, 1138.

CYANOCOBALAMINE [U]		
CAS 68-19-9	MW: 1355.40	MF: C63 H88 Co N14 O14 P

Vitamin (hematopoietic)

LgP (4)

pKa

pp cited in Vols.1-5:

1: 37

(1) Smith, 'Vitamin B12', Methuen & Co., London, 3rd. ed., 1965.

CYCLACILLIN [U;INN]		
CAS 3485-14-1	MW: 341.43	MF: C15 H23 N3 O4 S

Antibacterial

Wyeth (Cyclapen-W)

LgP 1.4(C,6); 1.31(M)

pKa

pp cited in Vols.1-5:

2: 620

(1) Alburn et al., U.S. Pat 3 194 802 (1965).
(2) Resenman et al., Antimicrob. Ag. Chemother. 1967, 590.

CYCLAMIC ACID [U]		
CAS 100-88-9	MW: 179.24	MF: C6 H13 N O3 S

Sweetener (non-nutritive)

Abbott (Hexamic Acid)

LgP 1.16(C,3)

pKa

pp cited in Vols.1-5:

(1) Audrieth et al., U.S. Pat. 2 275 125 (1945).
(2) Brantom, et al., Food Cosmet. Toxicol., 1973, 11, 735.

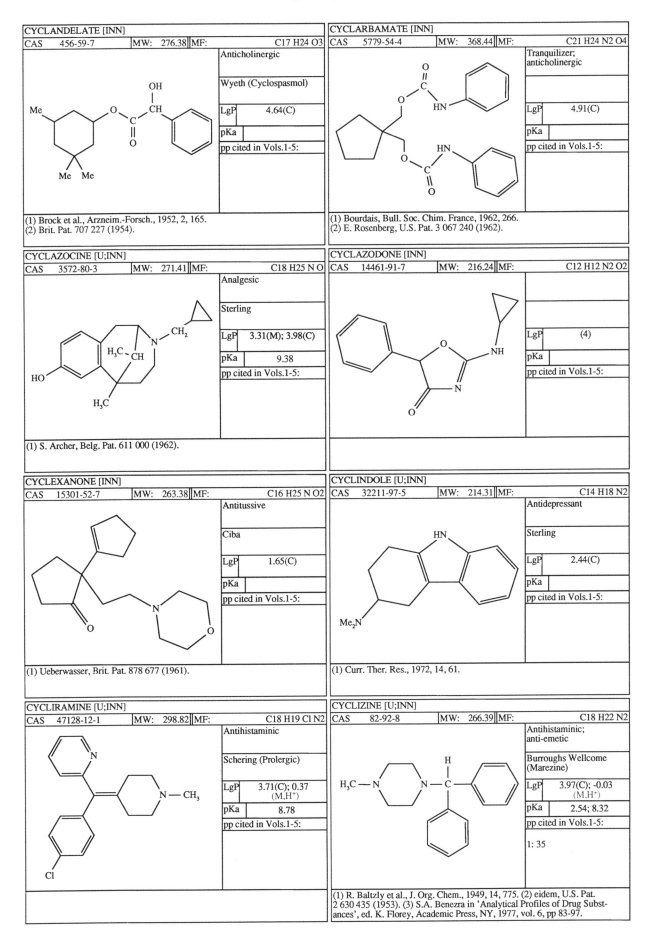

CYCLANDELATE [INN]

CAS	456-59-7	MW:	276.38	MF:	C17 H24 O3

Anticholinergic

Wyeth (Cyclospasmol)

LgP	4.64(C)
pKa	
pp cited in Vols.1-5:	

(1) Brock et al., Arzneim.-Forsch., 1952, 2, 165.
(2) Brit. Pat. 707 227 (1954).

CYCLARBAMATE [INN]

CAS	5779-54-4	MW:	368.44	MF:	C21 H24 N2 O4

Tranquilizer; anticholinergic

LgP	4.91(C)
pKa	
pp cited in Vols.1-5:	

(1) Bourdais, Bull. Soc. Chim. France, 1962, 266.
(2) E. Rosenberg, U.S. Pat. 3 067 240 (1962).

CYCLAZOCINE [U;INN]

CAS	3572-80-3	MW:	271.41	MF:	C18 H25 N O

Analgesic

Sterling

LgP	3.31(M); 3.98(C)
pKa	9.38
pp cited in Vols.1-5:	

(1) S. Archer, Belg. Pat. 611 000 (1962).

CYCLAZODONE [INN]

CAS	14461-91-7	MW:	216.24	MF:	C12 H12 N2 O2

LgP	(4)
pKa	
pp cited in Vols.1-5:	

CYCLEXANONE [INN]

CAS	15301-52-7	MW:	263.38	MF:	C16 H25 N O2

Antitussive

Ciba

LgP	1.65(C)
pKa	
pp cited in Vols.1-5:	

(1) Ueberwasser, Brit. Pat. 878 677 (1961).

CYCLINDOLE [U;INN]

CAS	32211-97-5	MW:	214.31	MF:	C14 H18 N2

Antidepressant

Sterling

LgP	2.44(C)
pKa	
pp cited in Vols.1-5:	

(1) Curr. Ther. Res., 1972, 14, 61.

CYCLIRAMINE [U;INN]

CAS	47128-12-1	MW:	298.82	MF:	C18 H19 Cl N2

Antihistaminic

Schering (Prolergic)

LgP	3.71(C); 0.37 (M,H$^+$)
pKa	8.78
pp cited in Vols.1-5:	

CYCLIZINE [U;INN]

CAS	82-92-8	MW:	266.39	MF:	C18 H22 N2

Antihistaminic; anti-emetic

Burroughs Wellcome (Marezine)

LgP	3.97(C); -0.03 (M,H$^+$)
pKa	2.54; 8.32
pp cited in Vols.1-5:	

1: 35

(1) R. Baltzly et al., J. Org. Chem., 1949, 14, 775. (2) eidem, U.S. Pat. 2 630 435 (1953). (3) S.A. Benezra in 'Analytical Profiles of Drug Substances', ed. K. Florey, Academic Press, NY, 1977, vol. 6, pp 83-97.

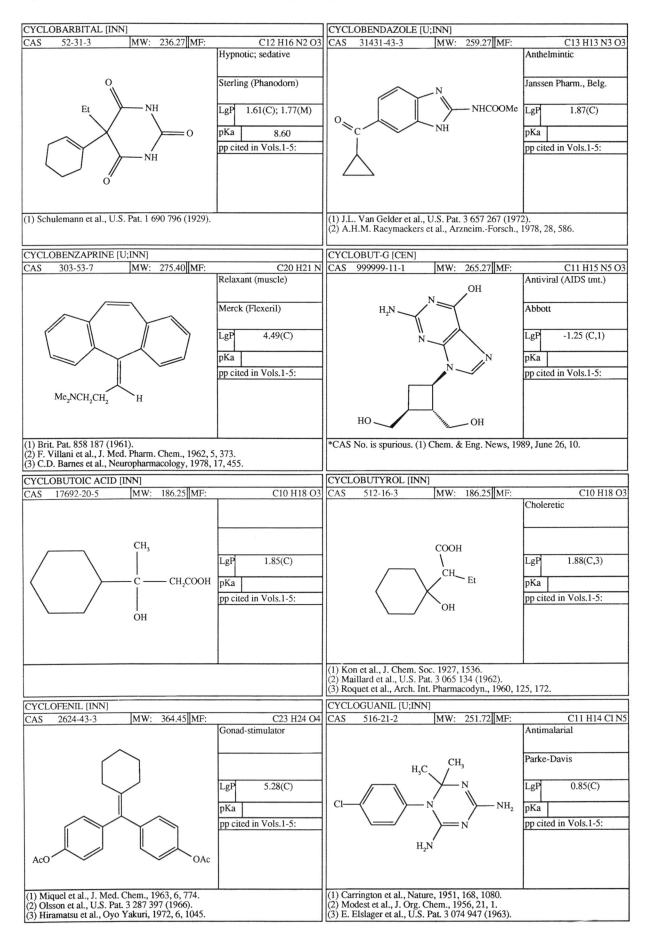

CYCLOBARBITAL [INN]

| CAS | 52-31-3 | MW: 236.27 | MF: | C12 H16 N2 O3 |

Hypnotic; sedative

Sterling (Phanodorn)

LgP	1.61(C); 1.77(M)
pKa	8.60
pp cited in Vols.1-5:	

(1) Schulemann et al., U.S. Pat. 1 690 796 (1929).

CYCLOBENDAZOLE [U;INN]

| CAS | 31431-43-3 | MW: 259.27 | MF: | C13 H13 N3 O3 |

Anthelmintic

Janssen Pharm., Belg.

LgP	1.87(C)
pKa	
pp cited in Vols.1-5:	

(1) J.L. Van Gelder et al., U.S. Pat. 3 657 267 (1972).
(2) A.H.M. Raeymaekers et al., Arzneim.-Forsch., 1978, 28, 586.

CYCLOBENZAPRINE [U;INN]

| CAS | 303-53-7 | MW: 275.40 | MF: | C20 H21 N |

Relaxant (muscle)

Merck (Flexeril)

LgP	4.49(C)
pKa	
pp cited in Vols.1-5:	

(1) Brit. Pat. 858 187 (1961).
(2) F. Villani et al., J. Med. Pharm. Chem., 1962, 5, 373.
(3) C.D. Barnes et al., Neuropharmacology, 1978, 17, 455.

CYCLOBUT-G [CEN]

| CAS | 999999-11-1 | MW: 265.27 | MF: | C11 H15 N5 O3 |

Antiviral (AIDS tmt.)

Abbott

LgP	-1.25 (C,1)
pKa	
pp cited in Vols.1-5:	

*CAS No. is spurious. (1) Chem. & Eng. News, 1989, June 26, 10.

CYCLOBUTOIC ACID [INN]

| CAS | 17692-20-5 | MW: 186.25 | MF: | C10 H18 O3 |

LgP	1.85(C)
pKa	
pp cited in Vols.1-5:	

CYCLOBUTYROL [INN]

| CAS | 512-16-3 | MW: 186.25 | MF: | C10 H18 O3 |

Choleretic

LgP	1.88(C,3)
pKa	
pp cited in Vols.1-5:	

(1) Kon et al., J. Chem. Soc. 1927, 1536.
(2) Maillard et al., U.S. Pat. 3 065 134 (1962).
(3) Roquet et al., Arch. Int. Pharmacodyn., 1960, 125, 172.

CYCLOFENIL [INN]

| CAS | 2624-43-3 | MW: 364.45 | MF: | C23 H24 O4 |

Gonad-stimulator

LgP	5.28(C)
pKa	
pp cited in Vols.1-5:	

(1) Miquel et al., J. Med. Chem., 1963, 6, 774.
(2) Olsson et al., U.S. Pat. 3 287 397 (1966).
(3) Hiramatsu et al., Oyo Yakuri, 1972, 6, 1045.

CYCLOGUANIL [U;INN]

| CAS | 516-21-2 | MW: 251.72 | MF: | C11 H14 Cl N5 |

Antimalarial

Parke-Davis

LgP	0.85(C)
pKa	
pp cited in Vols.1-5:	

(1) Carrington et al., Nature, 1951, 168, 1080.
(2) Modest et al., J. Org. Chem., 1956, 21, 1.
(3) E. Elslager et al., U.S. Pat. 3 074 947 (1963).

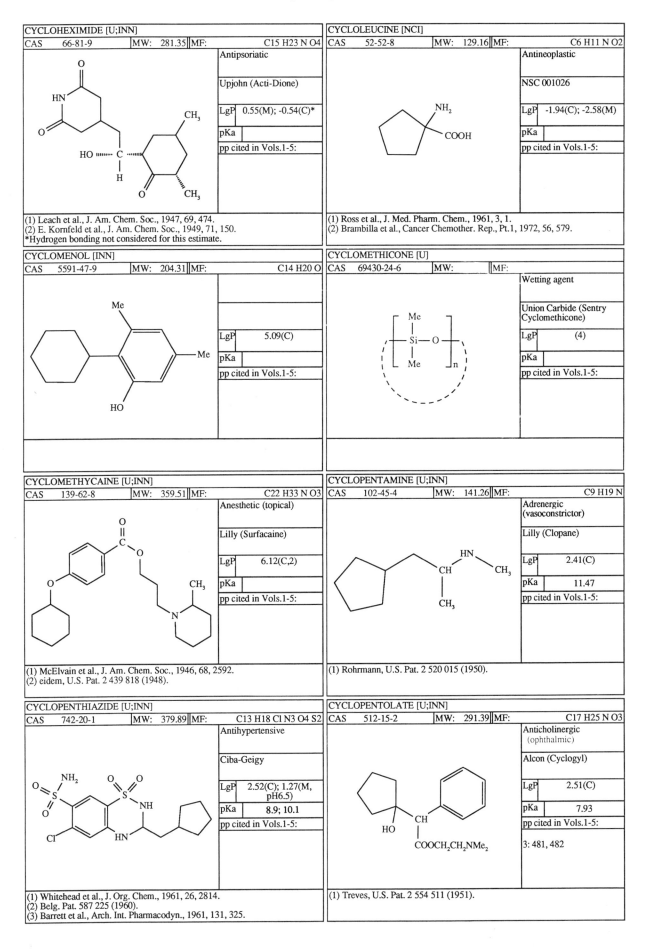

CYCLOHEXIMIDE [U;INN]

CAS	66-81-9	MW:	281.35	MF:	C15 H23 N O4

Antipsoriatic

Upjohn (Acti-Dione)

LgP	0.55(M); -0.54(C)*
pKa	
pp cited in Vols.1-5:	

(1) Leach et al., J. Am. Chem. Soc., 1947, 69, 474.
(2) E. Kornfeld et al., J. Am. Chem. Soc., 1949, 71, 150.
*Hydrogen bonding not considered for this estimate.

CYCLOLEUCINE [NCI]

CAS	52-52-8	MW:	129.16	MF:	C6 H11 N O2

Antineoplastic

NSC 001026

LgP	-1.94(C); -2.58(M)
pKa	
pp cited in Vols.1-5:	

(1) Ross et al., J. Med. Pharm. Chem., 1961, 3, 1.
(2) Brambilla et al., Cancer Chemother. Rep., Pt.1, 1972, 56, 579.

CYCLOMENOL [INN]

CAS	5591-47-9	MW:	204.31	MF:	C14 H20 O

LgP	5.09(C)
pKa	
pp cited in Vols.1-5:	

CYCLOMETHICONE [U]

CAS	69430-24-6	MW:		MF:	

Wetting agent

Union Carbide (Sentry Cyclomethicone)

LgP	(4)
pKa	
pp cited in Vols.1-5:	

CYCLOMETHYCAINE [U;INN]

CAS	139-62-8	MW:	359.51	MF:	C22 H33 N O3

Anesthetic (topical)

Lilly (Surfacaine)

LgP	6.12(C,2)
pKa	
pp cited in Vols.1-5:	

(1) McElvain et al., J. Am. Chem. Soc., 1946, 68, 2592.
(2) eidem, U.S. Pat. 2 439 818 (1948).

CYCLOPENTAMINE [U;INN]

CAS	102-45-4	MW:	141.26	MF:	C9 H19 N

Adrenergic (vasoconstrictor)

Lilly (Clopane)

LgP	2.41(C)
pKa	11.47
pp cited in Vols.1-5:	

(1) Rohrmann, U.S. Pat. 2 520 015 (1950).

CYCLOPENTHIAZIDE [U;INN]

CAS	742-20-1	MW:	379.89	MF:	C13 H18 Cl N3 O4 S2

Antihypertensive

Ciba-Geigy

LgP	2.52(C); 1.27(M, pH6.5)
pKa	8.9; 10.1
pp cited in Vols.1-5:	

(1) Whitehead et al., J. Org. Chem., 1961, 26, 2814.
(2) Belg. Pat. 587 225 (1960).
(3) Barrett et al., Arch. Int. Pharmacodyn., 1961, 131, 325.

CYCLOPENTOLATE [U;INN]

CAS	512-15-2	MW:	291.39	MF:	C17 H25 N O3

Anticholinergic (ophthalmic)

Alcon (Cyclogyl)

LgP	2.51(C)
pKa	7.93
pp cited in Vols.1-5:	
	3: 481, 482

(1) Treves, U.S. Pat. 2 554 511 (1951).

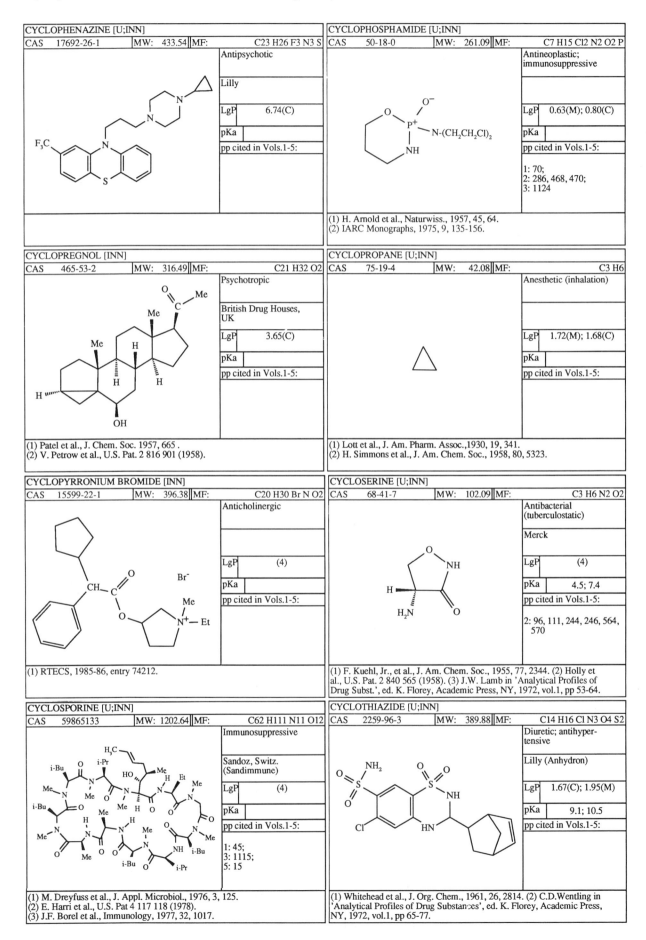

CYCLOPHENAZINE [U;INN]

CAS	17692-26-1	MW: 433.54	MF:	C23 H26 F3 N3 S

Antipsychotic

Lilly

LgP	6.74(C)
pKa	

pp cited in Vols.1-5:

CYCLOPHOSPHAMIDE [U;INN]

CAS	50-18-0	MW: 261.09	MF:	C7 H15 Cl2 N2 O2 P

Antineoplastic; immunosuppressive

LgP	0.63(M); 0.80(C)
pKa	

pp cited in Vols.1-5:

1: 70;
2: 286, 468, 470;
3: 1124

(1) H. Arnold et al., Naturwiss., 1957, 45, 64.
(2) IARC Monographs, 1975, 9, 135-156.

CYCLOPREGNOL [INN]

CAS	465-53-2	MW: 316.49	MF:	C21 H32 O2

Psychotropic

British Drug Houses, UK

LgP	3.65(C)
pKa	

pp cited in Vols.1-5:

(1) Patel et al., J. Chem. Soc. 1957, 665 .
(2) V. Petrow et al., U.S. Pat. 2 816 901 (1958).

CYCLOPROPANE [U;INN]

CAS	75-19-4	MW: 42.08	MF:	C3 H6

Anesthetic (inhalation)

LgP	1.72(M); 1.68(C)
pKa	

pp cited in Vols.1-5:

(1) Lott et al., J. Am. Pharm. Assoc.,1930, 19, 341.
(2) H. Simmons et al., J. Am. Chem. Soc., 1958, 80, 5323.

CYCLOPYRRONIUM BROMIDE [INN]

CAS	15599-22-1	MW: 396.38	MF:	C20 H30 Br N O2

Anticholinergic

LgP	(4)
pKa	

pp cited in Vols.1-5:

(1) RTECS, 1985-86, entry 74212.

CYCLOSERINE [U;INN]

CAS	68-41-7	MW: 102.09	MF:	C3 H6 N2 O2

Antibacterial (tuberculostatic)

Merck

LgP	(4)
pKa	4.5; 7.4

pp cited in Vols.1-5:

2: 96, 111, 244, 246, 564, 570

(1) F. Kuehl, Jr., et al., J. Am. Chem. Soc., 1955, 77, 2344. (2) Holly et al., U.S. Pat. 2 840 565 (1958). (3) J.W. Lamb in 'Analytical Profiles of Drug Subst.', ed. K. Florey, Academic Press, NY, 1972, vol.1, pp 53-64.

CYCLOSPORINE [U;INN]

CAS	59865133	MW: 1202.64	MF:	C62 H111 N11 O12

Immunosuppressive

Sandoz, Switz. (Sandimmune)

LgP	(4)
pKa	

pp cited in Vols.1-5:

1: 45;
3: 1115;
5: 15

(1) M. Dreyfuss et al., J. Appl. Microbiol., 1976, 3, 125.
(2) E. Harri et al., U.S. Pat 4 117 118 (1978).
(3) J.F. Borel et al., Immunology, 1977, 32, 1017.

CYCLOTHIAZIDE [U;INN]

CAS	2259-96-3	MW: 389.88	MF:	C14 H16 Cl N3 O4 S2

Diuretic; antihypertensive

Lilly (Anhydron)

LgP	1.67(C); 1.95(M)
pKa	9.1; 10.5

pp cited in Vols.1-5:

(1) Whitehead et al., J. Org. Chem., 1961, 26, 2814. (2) C.D.Wentling in 'Analytical Profiles of Drug Substances', ed. K. Florey, Academic Press, NY, 1972, vol.1, pp 65-77.

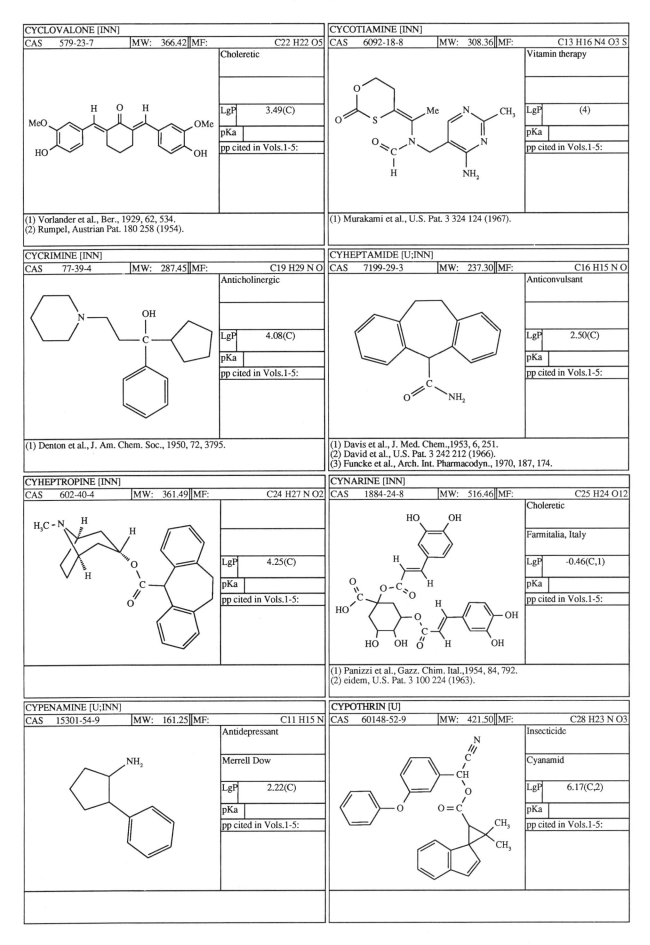

CYCLOVALONE [INN]

CAS	579-23-7	MW:	366.42	MF:	C22 H22 O5

Choleretic

LgP	3.49(C)
pKa	
pp cited in Vols.1-5:	

MeO, H, O, H, OMe, HO, OH

(1) Vorlander et al., Ber., 1929, 62, 534.
(2) Rumpel, Austrian Pat. 180 258 (1954).

CYCOTIAMINE [INN]

CAS	6092-18-8	MW:	308.36	MF:	C13 H16 N4 O3 S

Vitamin therapy

LgP	(4)
pKa	
pp cited in Vols.1-5:	

(1) Murakami et al., U.S. Pat. 3 324 124 (1967).

CYCRIMINE [INN]

CAS	77-39-4	MW:	287.45	MF:	C19 H29 N O

Anticholinergic

LgP	4.08(C)
pKa	
pp cited in Vols.1-5:	

(1) Denton et al., J. Am. Chem. Soc., 1950, 72, 3795.

CYHEPTAMIDE [U;INN]

CAS	7199-29-3	MW:	237.30	MF:	C16 H15 N O

Anticonvulsant

LgP	2.50(C)
pKa	
pp cited in Vols.1-5:	

(1) Davis et al., J. Med. Chem.,1953, 6, 251.
(2) David et al., U.S. Pat. 3 242 212 (1966).
(3) Funcke et al., Arch. Int. Pharmacodyn., 1970, 187, 174.

CYHEPTROPINE [INN]

CAS	602-40-4	MW:	361.49	MF:	C24 H27 N O2

LgP	4.25(C)
pKa	
pp cited in Vols.1-5:	

CYNARINE [INN]

CAS	1884-24-8	MW:	516.46	MF:	C25 H24 O12

Choleretic

Farmitalia, Italy

LgP	-0.46(C,1)
pKa	
pp cited in Vols.1-5:	

(1) Panizzi et al., Gazz. Chim. Ital.,1954, 84, 792.
(2) eidem, U.S. Pat. 3 100 224 (1963).

CYPENAMINE [U;INN]

CAS	15301-54-9	MW:	161.25	MF:	C11 H15 N

Antidepressant

Merrell Dow

LgP	2.22(C)
pKa	
pp cited in Vols.1-5:	

CYPOTHRIN [U]

CAS	60148-52-9	MW:	421.50	MF:	C28 H23 N O3

Insecticide

Cyanamid

LgP	6.17(C,2)
pKa	
pp cited in Vols.1-5:	

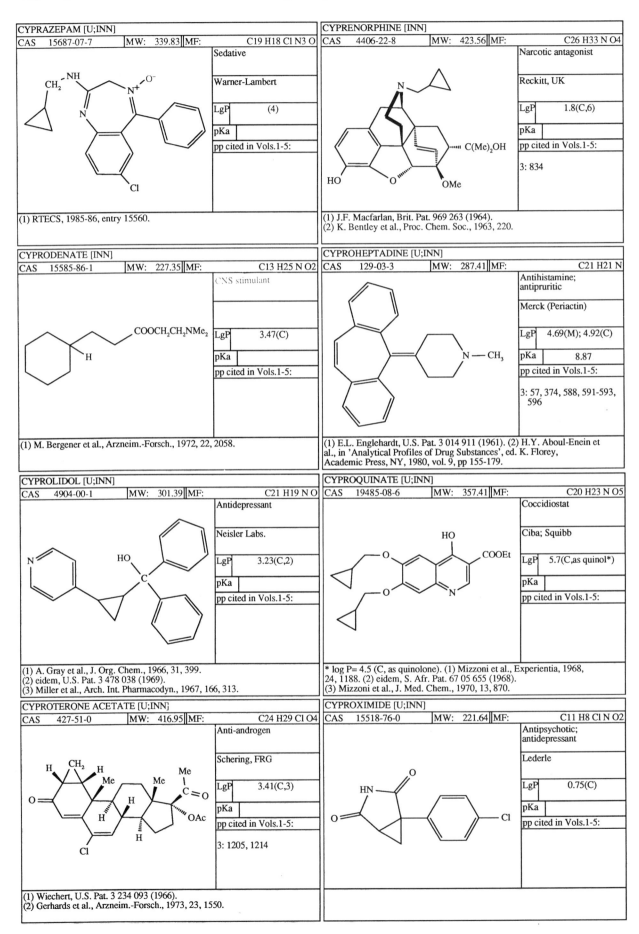

CYPRAZEPAM [U;INN]			
CAS 15687-07-7	MW: 339.83	MF:	C19 H18 Cl N3 O

Sedative

Warner-Lambert

LgP	(4)
pKa	
pp cited in Vols.1-5:	

(1) RTECS, 1985-86, entry 15560.

CYPRENORPHINE [INN]			
CAS 4406-22-8	MW: 423.56	MF:	C26 H33 N O4

Narcotic antagonist

Reckitt, UK

LgP	1.8(C,6)
pKa	
pp cited in Vols.1-5:	

3: 834

(1) J.F. Macfarlan, Brit. Pat. 969 263 (1964).
(2) K. Bentley et al., Proc. Chem. Soc., 1963, 220.

CYPRODENATE [INN]			
CAS 15585-86-1	MW: 227.35	MF:	C13 H25 N O2

CNS stimulant

LgP	3.47(C)
pKa	
pp cited in Vols.1-5:	

(1) M. Bergener et al., Arzneim.-Forsch., 1972, 22, 2058.

CYPROHEPTADINE [U;INN]			
CAS 129-03-3	MW: 287.41	MF:	C21 H21 N

Antihistamine;
antipruritic

Merck (Periactin)

LgP	4.69(M); 4.92(C)
pKa	8.87
pp cited in Vols.1-5:	

3: 57, 374, 588, 591-593, 596

(1) E.L. Englehardt, U.S. Pat. 3 014 911 (1961). (2) H.Y. Aboul-Enein et al., in 'Analytical Profiles of Drug Substances', ed. K. Florey, Academic Press, NY, 1980, vol. 9, pp 155-179.

CYPROLIDOL [U;INN]			
CAS 4904-00-1	MW: 301.39	MF:	C21 H19 N O

Antidepressant

Neisler Labs.

LgP	3.23(C,2)
pKa	
pp cited in Vols.1-5:	

(1) A. Gray et al., J. Org. Chem., 1966, 31, 399.
(2) eidem, U.S. Pat. 3 478 038 (1969).
(3) Miller et al., Arch. Int. Pharmacodyn., 1967, 166, 313.

CYPROQUINATE [U;INN]			
CAS 19485-08-6	MW: 357.41	MF:	C20 H23 N O5

Coccidiostat

Ciba; Squibb

LgP	5.7(C,as quinol*)
pKa	
pp cited in Vols.1-5:	

* log P= 4.5 (C, as quinolone). (1) Mizzoni et al., Experientia, 1968, 24, 1188. (2) eidem, S. Afr. Pat. 67 05 655 (1968).
(3) Mizzoni et al., J. Med. Chem., 1970, 13, 870.

CYPROTERONE ACETATE [U;INN]			
CAS 427-51-0	MW: 416.95	MF:	C24 H29 Cl O4

Anti-androgen

Schering, FRG

LgP	3.41(C,3)
pKa	
pp cited in Vols.1-5:	

3: 1205, 1214

(1) Wiechert, U.S. Pat. 3 234 093 (1966).
(2) Gerhards et al., Arzneim.-Forsch., 1973, 23, 1550.

CYPROXIMIDE [U;INN]			
CAS 15518-76-0	MW: 221.64	MF:	C11 H8 Cl N O2

Antipsychotic;
antidepressant

Lederle

LgP	0.75(C)
pKa	
pp cited in Vols.1-5:	

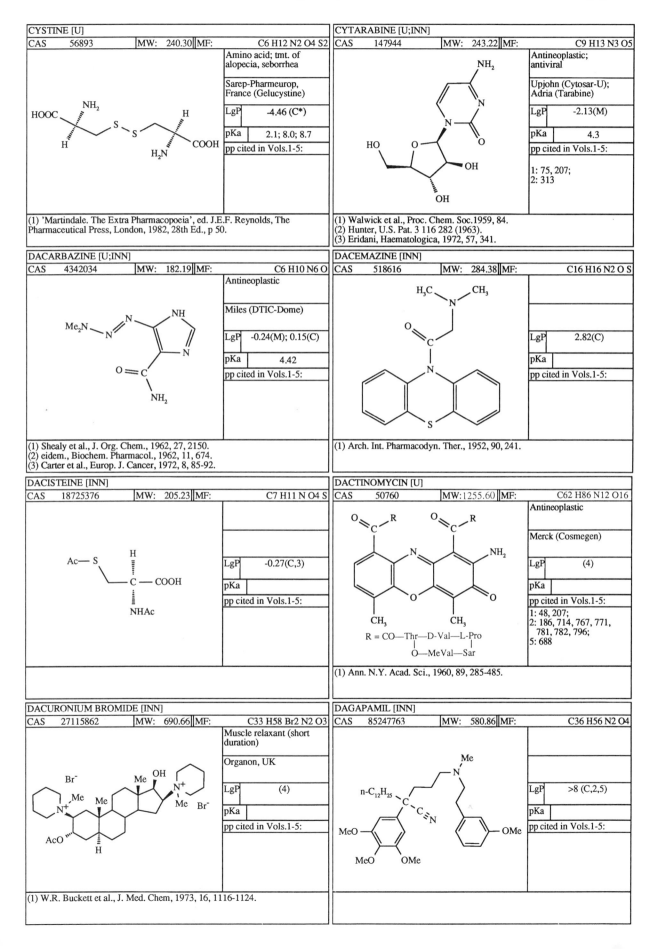

CYSTINE [U]			
CAS 56893	MW: 240.30	MF:	C6 H12 N2 O4 S2

Amino acid; tmt. of alopecia, seborrhea

Sarep-Pharmeurop, France (Gelucystine)

LgP	-4.46 (C*)
pKa	2.1; 8.0; 8.7
pp cited in Vols.1-5:	

(1) 'Martindale. The Extra Pharmacopoeia', ed. J.E.F. Reynolds, The Pharmaceutical Press, London, 1982, 28th Ed., p 50.

CYTARABINE [U;INN]			
CAS 147944	MW: 243.22	MF:	C9 H13 N3 O5

Antineoplastic; antiviral

Upjohn (Cytosar-U); Adria (Tarabine)

LgP	-2.13(M)
pKa	4.3
pp cited in Vols.1-5:	

1: 75, 207;
2: 313

(1) Walwick et al., Proc. Chem. Soc.1959, 84.
(2) Hunter, U.S. Pat. 3 116 282 (1963).
(3) Eridani, Haematologica, 1972, 57, 341.

DACARBAZINE [U;INN]			
CAS 4342034	MW: 182.19	MF:	C6 H10 N6 O

Antineoplastic

Miles (DTIC-Dome)

LgP	-0.24(M); 0.15(C)
pKa	4.42
pp cited in Vols.1-5:	

(1) Shealy et al., J. Org. Chem., 1962, 27, 2150.
(2) eidem., Biochem. Pharmacol., 1962, 11, 674.
(3) Carter et al., Europ. J. Cancer, 1972, 8, 85-92.

DACEMAZINE [INN]			
CAS 518616	MW: 284.38	MF:	C16 H16 N2 O S

LgP	2.82(C)
pKa	
pp cited in Vols.1-5:	

(1) Arch. Int. Pharmacodyn. Ther., 1952, 90, 241.

DACISTEINE [INN]			
CAS 18725376	MW: 205.23	MF:	C7 H11 N O4 S

LgP	-0.27(C,3)
pKa	
pp cited in Vols.1-5:	

DACTINOMYCIN [U]			
CAS 50760	MW:1255.60	MF:	C62 H86 N12 O16

Antineoplastic

Merck (Cosmegen)

LgP	(4)
pKa	
pp cited in Vols.1-5:	

1: 48, 207;
2: 186, 714, 767, 771, 781, 782, 796;
5: 688

R = CO—Thr—D-Val—L-Pro
          | |
      O—MeVal—Sar

(1) Ann. N.Y. Acad. Sci., 1960, 89, 285-485.

DACURONIUM BROMIDE [INN]			
CAS 27115862	MW: 690.66	MF:	C33 H58 Br2 N2 O3

Muscle relaxant (short duration)

Organon, UK

LgP	(4)
pKa	
pp cited in Vols.1-5:	

(1) W.R. Buckett et al., J. Med. Chem, 1973, 16, 1116-1124.

DAGAPAMIL [INN]			
CAS 85247763	MW: 580.86	MF:	C36 H56 N2 O4

LgP	>8 (C,2,5)
pKa	
pp cited in Vols.1-5:	

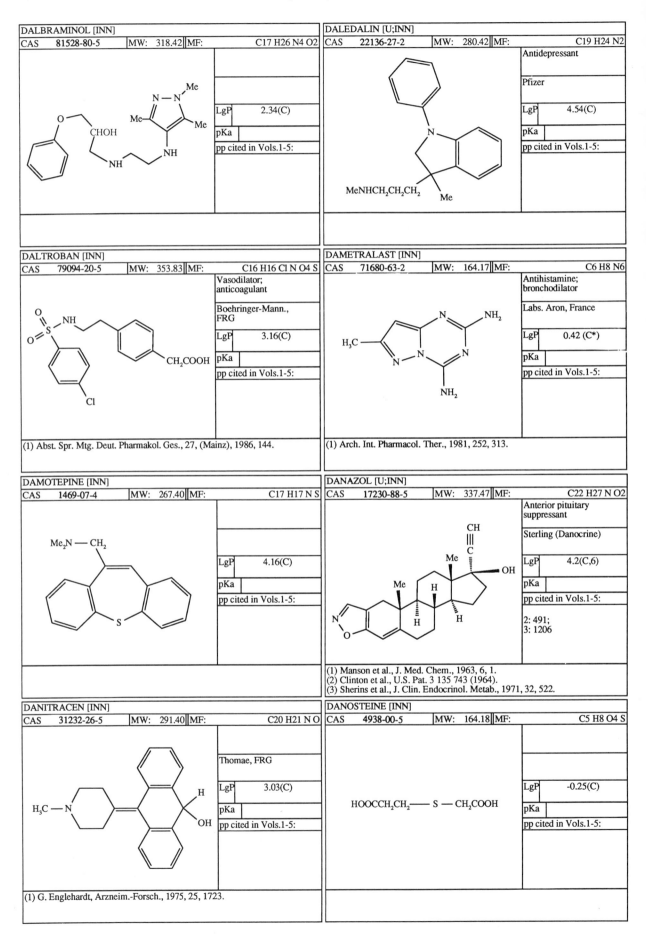

DALBRAMINOL [INN]		
CAS 81528-80-5	MW: 318.42 MF:	C17 H26 N4 O2
	LgP	2.34(C)
	pKa	
	pp cited in Vols.1-5:	

DALEDALIN [U;INN]		
CAS 22136-27-2	MW: 280.42 MF:	C19 H24 N2
	Antidepressant	
	Pfizer	
	LgP	4.54(C)
	pKa	
	pp cited in Vols.1-5:	

DALTROBAN [INN]		
CAS 79094-20-5	MW: 353.83 MF:	C16 H16 Cl N O4 S
	Vasodilator; anticoagulant	
	Boehringer-Mann., FRG	
	LgP	3.16(C)
	pKa	
	pp cited in Vols.1-5:	

(1) Abst. Spr. Mtg. Deut. Pharmakol. Ges., 27, (Mainz), 1986, 144.

DAMETRALAST [INN]		
CAS 71680-63-2	MW: 164.17 MF:	C6 H8 N6
	Antihistamine; bronchodilator	
	Labs. Aron, France	
	LgP	0.42 (C*)
	pKa	
	pp cited in Vols.1-5:	

(1) Arch. Int. Pharmacol. Ther., 1981, 252, 313.

DAMOTEPINE [INN]		
CAS 1469-07-4	MW: 267.40 MF:	C17 H17 N S
	LgP	4.16(C)
	pKa	
	pp cited in Vols.1-5:	

DANAZOL [U;INN]		
CAS 17230-88-5	MW: 337.47 MF:	C22 H27 N O2
	Anterior pituitary suppressant	
	Sterling (Danocrine)	
	LgP	4.2(C,6)
	pKa	
	pp cited in Vols.1-5:	
		2: 491; 3: 1206

(1) Manson et al., J. Med. Chem., 1963, 6, 1.
(2) Clinton et al., U.S. Pat. 3 135 743 (1964).
(3) Sherins et al., J. Clin. Endocrinol. Metab., 1971, 32, 522.

DANITRACEN [INN]		
CAS 31232-26-5	MW: 291.40 MF:	C20 H21 N O
	Thomae, FRG	
	LgP	3.03(C)
	pKa	
	pp cited in Vols.1-5:	

(1) G. Englehardt, Arzneim.-Forsch., 1975, 25, 1723.

DANOSTEINE [INN]		
CAS 4938-00-5	MW: 164.18 MF:	C5 H8 O4 S
	LgP	-0.25(C)
	pKa	
	pp cited in Vols.1-5:	

HOOCCH$_2$CH$_2$ — S — CH$_2$COOH

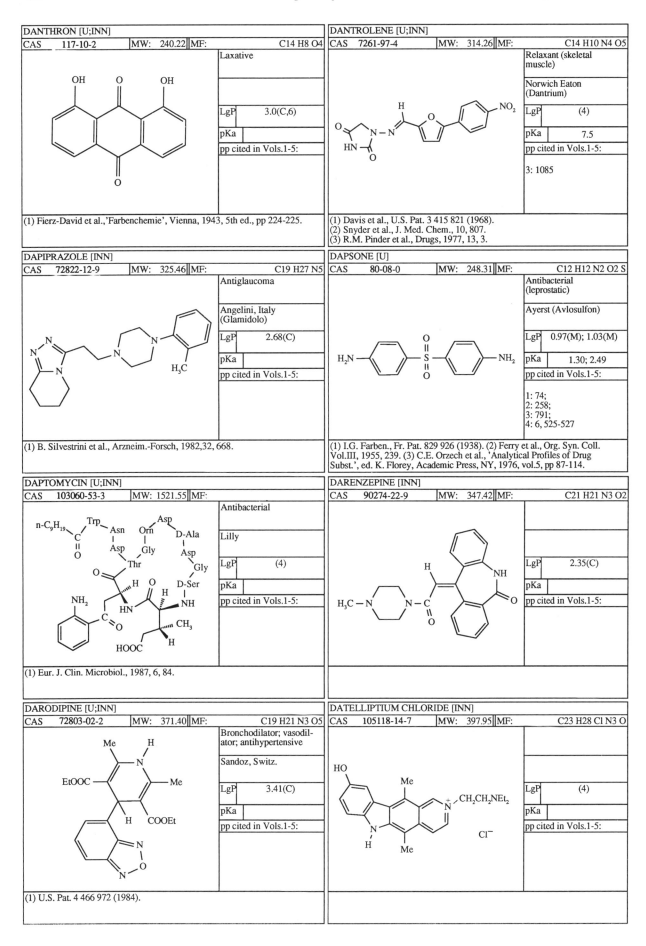

DANTHRON [U;INN]

CAS	117-10-2	MW:	240.22	MF:	C14 H8 O4

Laxative

LgP	3.0(C,6)
pKa	
pp cited in Vols.1-5:	

(1) Fierz-David et al.,'Farbenchemie', Vienna, 1943, 5th ed., pp 224-225.

DANTROLENE [U;INN]

CAS	7261-97-4	MW:	314.26	MF:	C14 H10 N4 O5

Relaxant (skeletal muscle)

Norwich Eaton (Dantrium)

LgP	(4)
pKa	7.5
pp cited in Vols.1-5:	

3: 1085

(1) Davis et al., U.S. Pat. 3 415 821 (1968).
(2) Snyder et al., J. Med. Chem., 10, 807.
(3) R.M. Pinder et al., Drugs, 1977, 13, 3.

DAPIPRAZOLE [INN]

CAS	72822-12-9	MW:	325.46	MF:	C19 H27 N5

Antiglaucoma

Angelini, Italy (Glamidolo)

LgP	2.68(C)
pKa	
pp cited in Vols.1-5:	

(1) B. Silvestrini et al., Arzneim.-Forsch, 1982,32, 668.

DAPSONE [U]

CAS	80-08-0	MW:	248.31	MF:	C12 H12 N2 O2 S

Antibacterial (leprostatic)

Ayerst (Avlosulfon)

LgP	0.97(M); 1.03(M)
pKa	1.30; 2.49
pp cited in Vols.1-5:	

1: 74;
2: 258;
3: 791;
4: 6, 525-527

(1) I.G. Farben., Fr. Pat. 829 926 (1938). (2) Ferry et al., Org. Syn. Coll. Vol.III, 1955, 239. (3) C.E. Orzech et al., 'Analytical Profiles of Drug Subst.', ed. K. Florey, Academic Press, NY, 1976, vol.5, pp 87-114.

DAPTOMYCIN [U;INN]

CAS	103060-53-3	MW:	1521.55	MF:	

Antibacterial

Lilly

LgP	(4)
pKa	
pp cited in Vols.1-5:	

(1) Eur. J. Clin. Microbiol., 1987, 6, 84.

DARENZEPINE [INN]

CAS	90274-22-9	MW:	347.42	MF:	C21 H21 N3 O2

LgP	2.35(C)
pKa	
pp cited in Vols.1-5:	

DARODIPINE [U;INN]

CAS	72803-02-2	MW:	371.40	MF:	C19 H21 N3 O5

Bronchodilator; vasodilator; antihypertensive

Sandoz, Switz.

LgP	3.41(C)
pKa	
pp cited in Vols.1-5:	

(1) U.S. Pat. 4 466 972 (1984).

DATELLIPTIUM CHLORIDE [INN]

CAS	105118-14-7	MW:	397.95	MF:	C23 H28 Cl N3 O

LgP	(4)
pKa	
pp cited in Vols.1-5:	

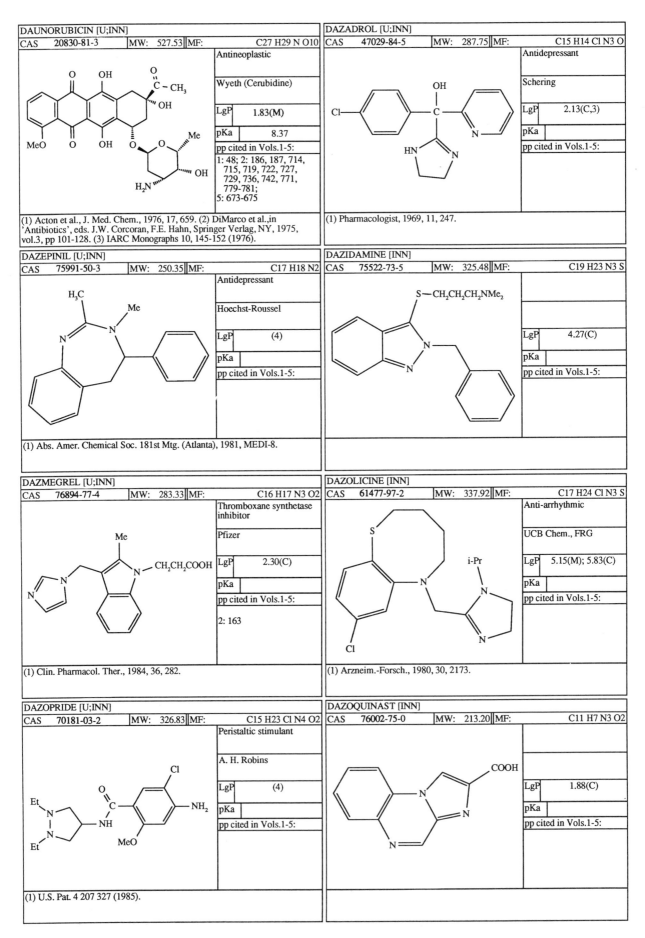

DAUNORUBICIN [U;INN]

CAS	20830-81-3	MW:	527.53	MF:	C27 H29 N O10

Antineoplastic

Wyeth (Cerubidine)

LgP	1.83(M)
pKa	8.37

pp cited in Vols.1-5:
1: 48; 2: 186, 187, 714, 715, 719, 722, 727, 729, 736, 742, 771, 779-781; 5: 673-675

(1) Acton et al., J. Med. Chem., 1976, 17, 659. (2) DiMarco et al.,in 'Antibiotics', eds. J.W. Corcoran, F.E. Hahn, Springer Verlag, NY, 1975, vol.3, pp 101-128. (3) IARC Monographs 10, 145-152 (1976).

DAZADROL [U;INN]

CAS	47029-84-5	MW:	287.75	MF:	C15 H14 Cl N3 O

Antidepressant

Schering

LgP	2.13(C,3)
pKa	

pp cited in Vols.1-5:

(1) Pharmacologist, 1969, 11, 247.

DAZEPINIL [U;INN]

CAS	75991-50-3	MW:	250.35	MF:	C17 H18 N2

Antidepressant

Hoechst-Roussel

LgP	(4)
pKa	

pp cited in Vols.1-5:

(1) Abs. Amer. Chemical Soc. 181st Mtg. (Atlanta), 1981, MEDI-8.

DAZIDAMINE [INN]

CAS	75522-73-5	MW:	325.48	MF:	C19 H23 N3 S

LgP	4.27(C)
pKa	

pp cited in Vols.1-5:

DAZMEGREL [U;INN]

CAS	76894-77-4	MW:	283.33	MF:	C16 H17 N3 O2

Thromboxane synthetase inhibitor

Pfizer

LgP	2.30(C)
pKa	

pp cited in Vols.1-5:
2: 163

(1) Clin. Pharmacol. Ther., 1984, 36, 282.

DAZOLICINE [INN]

CAS	61477-97-2	MW:	337.92	MF:	C17 H24 Cl N3 S

Anti-arrhythmic

UCB Chem., FRG

LgP	5.15(M); 5.83(C)
pKa	

pp cited in Vols.1-5:

(1) Arzneim.-Forsch., 1980, 30, 2173.

DAZOPRIDE [U;INN]

CAS	70181-03-2	MW:	326.83	MF:	C15 H23 Cl N4 O2

Peristaltic stimulant

A. H. Robins

LgP	(4)
pKa	

pp cited in Vols.1-5:

(1) U.S. Pat. 4 207 327 (1985).

DAZOQUINAST [INN]

CAS	76002-75-0	MW:	213.20	MF:	C11 H7 N3 O2

LgP	1.88(C)
pKa	

pp cited in Vols.1-5:

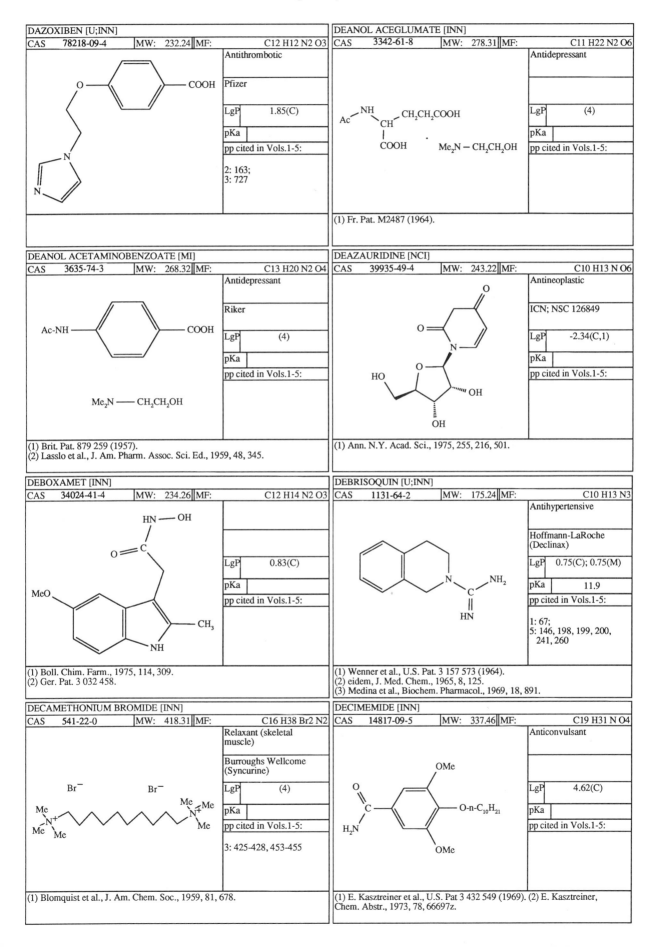

DAZOXIBEN [U;INN]

CAS	78218-09-4	MW: 232.24	MF:	C12 H12 N2 O3

Antithrombotic

Pfizer

LgP	1.85(C)
pKa	
pp cited in Vols.1-5:	

2: 163;
3: 727

DEANOL ACEGLUMATE [INN]

CAS	3342-61-8	MW: 278.31	MF:	C11 H22 N2 O6

Antidepressant

LgP	(4)
pKa	
pp cited in Vols.1-5:	

(1) Fr. Pat. M2487 (1964).

DEANOL ACETAMINOBENZOATE [MI]

CAS	3635-74-3	MW: 268.32	MF:	C13 H20 N2 O4

Antidepressant

Riker

LgP	(4)
pKa	
pp cited in Vols.1-5:	

$Me_2N — CH_2CH_2OH$

(1) Brit. Pat. 879 259 (1957).
(2) Lasslo et al., J. Am. Pharm. Assoc. Sci. Ed., 1959, 48, 345.

DEAZAURIDINE [NCI]

CAS	39935-49-4	MW: 243.22	MF:	C10 H13 N O6

Antineoplastic

ICN; NSC 126849

LgP	-2.34(C,1)
pKa	
pp cited in Vols.1-5:	

(1) Ann. N.Y. Acad. Sci., 1975, 255, 216, 501.

DEBOXAMET [INN]

CAS	34024-41-4	MW: 234.26	MF:	C12 H14 N2 O3

LgP	0.83(C)
pKa	
pp cited in Vols.1-5:	

(1) Boll. Chim. Farm., 1975, 114, 309.
(2) Ger. Pat. 3 032 458.

DEBRISOQUIN [U;INN]

CAS	1131-64-2	MW: 175.24	MF:	C10 H13 N3

Antihypertensive

Hoffmann-LaRoche
(Declinax)

LgP	0.75(C); 0.75(M)
pKa	11.9
pp cited in Vols.1-5:	

1: 67;
5: 146, 198, 199, 200,
241, 260

(1) Wenner et al., U.S. Pat. 3 157 573 (1964).
(2) eidem, J. Med. Chem., 1965, 8, 125.
(3) Medina et al., Biochem. Pharmacol., 1969, 18, 891.

DECAMETHONIUM BROMIDE [INN]

CAS	541-22-0	MW: 418.31	MF:	C16 H38 Br2 N2

Relaxant (skeletal
muscle)

Burroughs Wellcome
(Syncurine)

LgP	(4)
pKa	
pp cited in Vols.1-5:	

3: 425-428, 453-455

(1) Blomquist et al., J. Am. Chem. Soc., 1959, 81, 678.

DECIMEMIDE [INN]

CAS	14817-09-5	MW: 337.46	MF:	C19 H31 N O4

Anticonvulsant

LgP	4.62(C)
pKa	
pp cited in Vols.1-5:	

(1) E. Kasztreiner et al., U.S. Pat 3 432 549 (1969). (2) E. Kasztreiner,
Chem. Abstr., 1973, 78, 66697z.

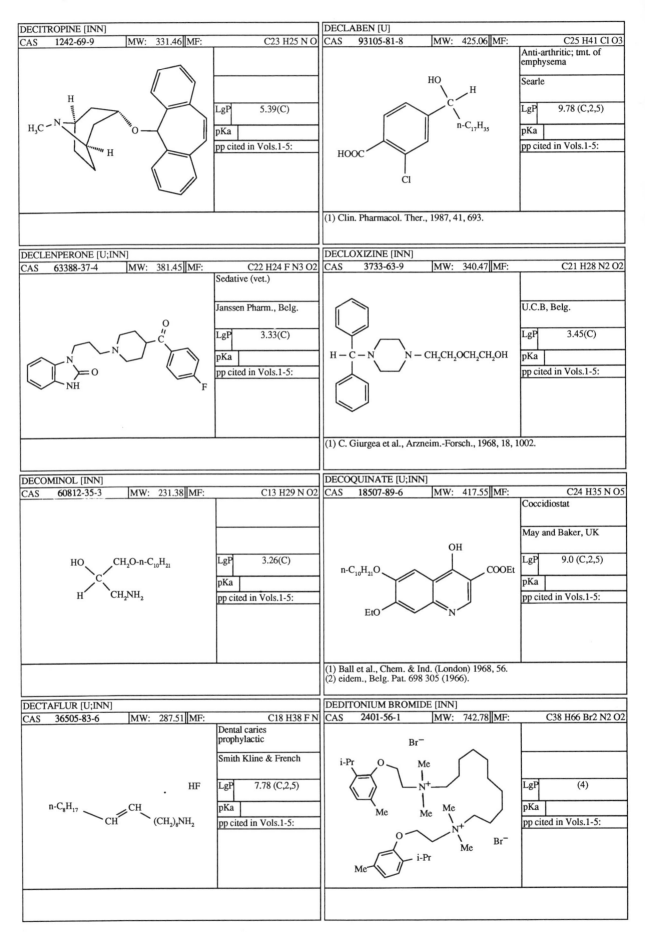

DECITROPINE [INN]

CAS	1242-69-9	MW:	331.46	MF:	C23 H25 N O

LgP	5.39(C)
pKa	
pp cited in Vols.1-5:	

DECLABEN [U]

CAS	93105-81-8	MW:	425.06	MF:	C25 H41 Cl O3

Anti-arthritic; tmt. of emphysema

Searle

LgP	9.78 (C,2,5)
pKa	
pp cited in Vols.1-5:	

(1) Clin. Pharmacol. Ther., 1987, 41, 693.

DECLENPERONE [U;INN]

CAS	63388-37-4	MW:	381.45	MF:	C22 H24 F N3 O2

Sedative (vet.)

Janssen Pharm., Belg.

LgP	3.33(C)
pKa	
pp cited in Vols.1-5:	

DECLOXIZINE [INN]

CAS	3733-63-9	MW:	340.47	MF:	C21 H28 N2 O2

U.C.B, Belg.

LgP	3.45(C)
pKa	
pp cited in Vols.1-5:	

(1) C. Giurgea et al., Arzneim.-Forsch., 1968, 18, 1002.

DECOMINOL [INN]

CAS	60812-35-3	MW:	231.38	MF:	C13 H29 N O2

LgP	3.26(C)
pKa	
pp cited in Vols.1-5:	

DECOQUINATE [U;INN]

CAS	18507-89-6	MW:	417.55	MF:	C24 H35 N O5

Coccidiostat

May and Baker, UK

LgP	9.0 (C,2,5)
pKa	
pp cited in Vols.1-5:	

(1) Ball et al., Chem. & Ind. (London) 1968, 56.
(2) eidem., Belg. Pat. 698 305 (1966).

DECTAFLUR [U;INN]

CAS	36505-83-6	MW:	287.51	MF:	C18 H38 F N

Dental caries prophylactic

Smith Kline & French

LgP	7.78 (C,2,5)
pKa	
pp cited in Vols.1-5:	

DEDITONIUM BROMIDE [INN]

CAS	2401-56-1	MW:	742.78	MF:	C38 H66 Br2 N2 O2

LgP	(4)
pKa	
pp cited in Vols.1-5:	

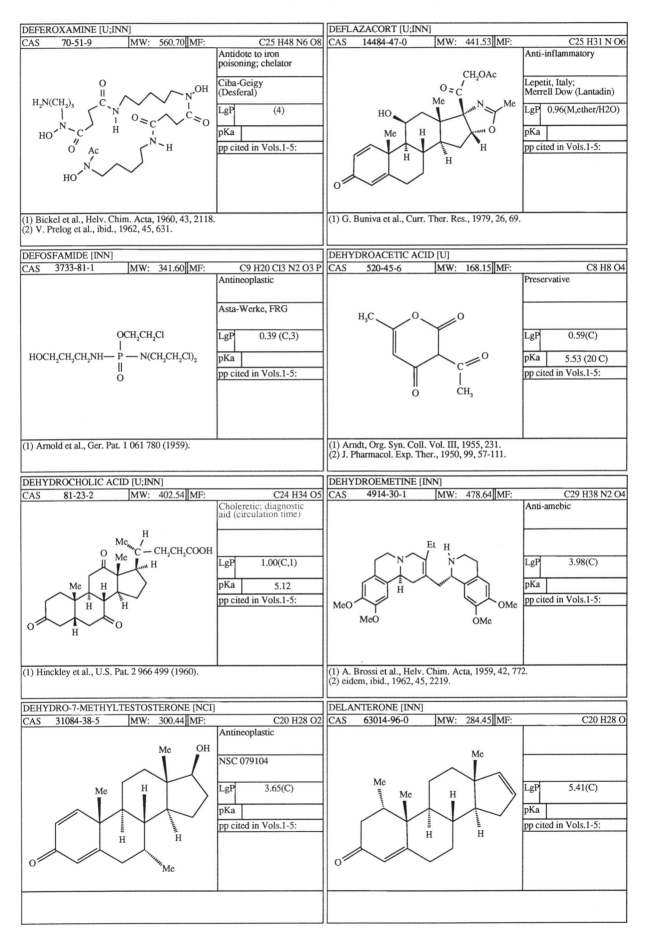

DEFEROXAMINE [U;INN]

CAS	70-51-9	MW:	560.70	MF:	C25 H48 N6 O8

Antidote to iron poisoning; chelator

Ciba-Geigy (Desferal)

LgP	(4)
pKa	
pp cited in Vols.1-5:	

(1) Bickel et al., Helv. Chim. Acta, 1960, 43, 2118.
(2) V. Prelog et al., ibid., 1962, 45, 631.

DEFOSFAMIDE [INN]

CAS	3733-81-1	MW:	341.60	MF:	C9 H20 Cl3 N2 O3 P

Antineoplastic

Asta-Werke, FRG

LgP	0.39 (C,3)
pKa	
pp cited in Vols.1-5:	

(1) Arnold et al., Ger. Pat. 1 061 780 (1959).

DEHYDROCHOLIC ACID [U;INN]

CAS	81-23-2	MW:	402.54	MF:	C24 H34 O5

Choleretic; diagnostic aid (circulation time)

LgP	1.00(C,1)
pKa	5.12
pp cited in Vols.1-5:	

(1) Hinckley et al., U.S. Pat. 2 966 499 (1960).

DEHYDRO-7-METHYLTESTOSTERONE [NCI]

CAS	31084-38-5	MW:	300.44	MF:	C20 H28 O2

Antineoplastic

NSC 079104

LgP	3.65(C)
pKa	
pp cited in Vols.1-5:	

DEFLAZACORT [U;INN]

CAS	14484-47-0	MW:	441.53	MF:	C25 H31 N O6

Anti-inflammatory

Lepetit, Italy; Merrell Dow (Lantadin)

LgP	0.96(M,ether/H2O)
pKa	
pp cited in Vols.1-5:	

(1) G. Buniva et al., Curr. Ther. Res., 1979, 26, 69.

DEHYDROACETIC ACID [U]

CAS	520-45-6	MW:	168.15	MF:	C8 H8 O4

Preservative

LgP	0.59(C)
pKa	5.53 (20 C)
pp cited in Vols.1-5:	

(1) Arndt, Org. Syn. Coll. Vol. III, 1955, 231.
(2) J. Pharmacol. Exp. Ther., 1950, 99, 57-111.

DEHYDROEMETINE [INN]

CAS	4914-30-1	MW:	478.64	MF:	C29 H38 N2 O4

Anti-amebic

LgP	3.98(C)
pKa	
pp cited in Vols.1-5:	

(1) A. Brossi et al., Helv. Chim. Acta, 1959, 42, 772.
(2) eidem, ibid., 1962, 45, 2219.

DELANTERONE [INN]

CAS	63014-96-0	MW:	284.45	MF:	C20 H28 O

LgP	5.41(C)
pKa	
pp cited in Vols.1-5:	

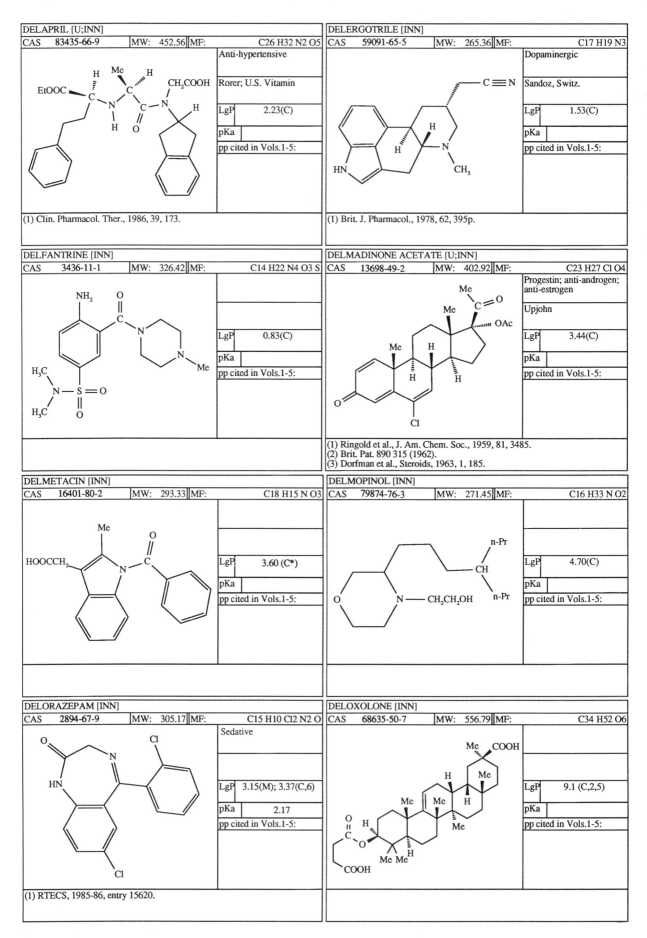

DELAPRIL [U;INN]

CAS	83435-66-9	MW:	452.56	MF:	C26 H32 N2 O5

Anti-hypertensive

Rorer; U.S. Vitamin

LgP	2.23(C)
pKa	
pp cited in Vols.1-5:	

(1) Clin. Pharmacol. Ther., 1986, 39, 173.

DELERGOTRILE [INN]

CAS	59091-65-5	MW:	265.36	MF:	C17 H19 N3

Dopaminergic

Sandoz, Switz.

LgP	1.53(C)
pKa	
pp cited in Vols.1-5:	

(1) Brit. J. Pharmacol., 1978, 62, 395p.

DELFANTRINE [INN]

CAS	3436-11-1	MW:	326.42	MF:	C14 H22 N4 O3 S

LgP	0.83(C)
pKa	
pp cited in Vols.1-5:	

DELMADINONE ACETATE [U;INN]

CAS	13698-49-2	MW:	402.92	MF:	C23 H27 Cl O4

Progestin; anti-androgen; anti-estrogen

Upjohn

LgP	3.44(C)
pKa	
pp cited in Vols.1-5:	

(1) Ringold et al., J. Am. Chem. Soc., 1959, 81, 3485.
(2) Brit. Pat. 890 315 (1962).
(3) Dorfman et al., Steroids, 1963, 1, 185.

DELMETACIN [INN]

CAS	16401-80-2	MW:	293.33	MF:	C18 H15 N O3

LgP	3.60 (C*)
pKa	
pp cited in Vols.1-5:	

DELMOPINOL [INN]

CAS	79874-76-3	MW:	271.45	MF:	C16 H33 N O2

LgP	4.70(C)
pKa	
pp cited in Vols.1-5:	

DELORAZEPAM [INN]

CAS	2894-67-9	MW:	305.17	MF:	C15 H10 Cl2 N2 O

Sedative

LgP	3.15(M); 3.37(C,6)
pKa	2.17
pp cited in Vols.1-5:	

(1) RTECS, 1985-86, entry 15620.

DELOXOLONE [INN]

CAS	68635-50-7	MW:	556.79	MF:	C34 H52 O6

LgP	9.1 (C,2,5)
pKa	
pp cited in Vols.1-5:	

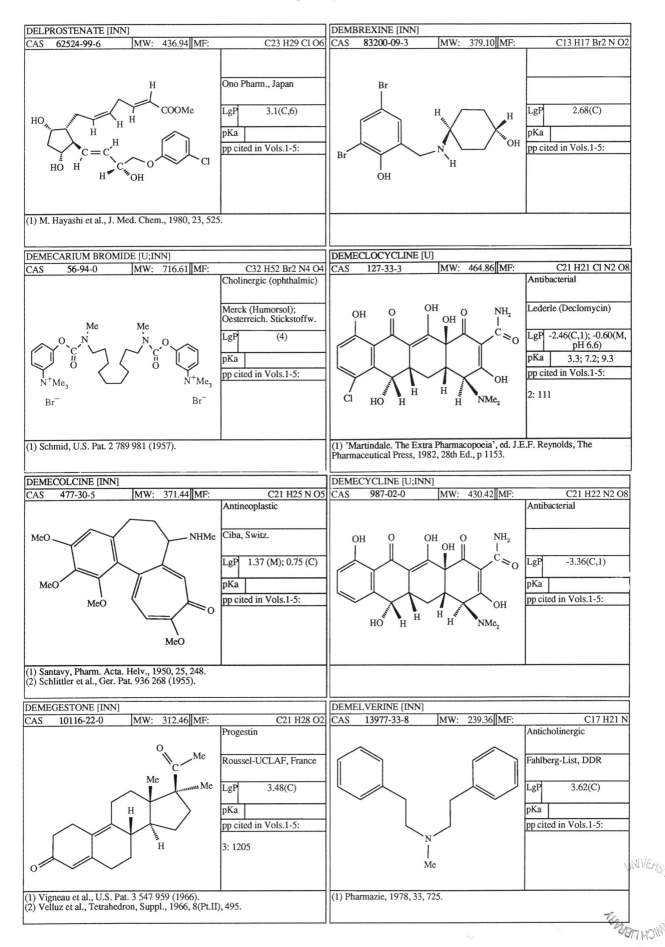

DELPROSTENATE [INN]

CAS	62524-99-6	MW:	436.94	MF:	C23 H29 Cl O6

	Ono Pharm., Japan
	LgP: 3.1(C,6)
	pKa:
	pp cited in Vols.1-5:

(1) M. Hayashi et al., J. Med. Chem., 1980, 23, 525.

DEMBREXINE [INN]

CAS	83200-09-3	MW:	379.10	MF:	C13 H17 Br2 N O2

	LgP: 2.68(C)
	pKa:
	pp cited in Vols.1-5:

DEMECARIUM BROMIDE [U;INN]

CAS	56-94-0	MW:	716.61	MF:	C32 H52 Br2 N4 O4

	Cholinergic (ophthalmic)
	Merck (Humorsol); Oesterreich. Stickstoffw.
	LgP: (4)
	pKa:
	pp cited in Vols.1-5:

(1) Schmid, U.S. Pat. 2 789 981 (1957).

DEMECLOCYCLINE [U]

CAS	127-33-3	MW:	464.86	MF:	C21 H21 Cl N2 O8

	Antibacterial
	Lederle (Declomycin)
	LgP: -2.46(C,1); -0.60(M, pH 6.6)
	pKa: 3.3; 7.2; 9.3
	pp cited in Vols.1-5:
	2: 111

(1) 'Martindale. The Extra Pharmacopoeia', ed. J.E.F. Reynolds, The Pharmaceutical Press, 1982, 28th Ed., p 1153.

DEMECOLCINE [INN]

CAS	477-30-5	MW:	371.44	MF:	C21 H25 N O5

	Antineoplastic
	Ciba, Switz.
	LgP: 1.37 (M); 0.75 (C)
	pKa:
	pp cited in Vols.1-5:

(1) Santavy, Pharm. Acta. Helv., 1950, 25, 248.
(2) Schlittler et al., Ger. Pat. 936 268 (1955).

DEMECYCLINE [U;INN]

CAS	987-02-0	MW:	430.42	MF:	C21 H22 N2 O8

	Antibacterial
	LgP: -3.36(C,1)
	pKa:
	pp cited in Vols.1-5:

DEMEGESTONE [INN]

CAS	10116-22-0	MW:	312.46	MF:	C21 H28 O2

	Progestin
	Roussel-UCLAF, France
	LgP: 3.48(C)
	pKa:
	pp cited in Vols.1-5:
	3: 1205

(1) Vigneau et al., U.S. Pat. 3 547 959 (1966).
(2) Velluz et al., Tetrahedron, Suppl., 1966, 8(Pt.II), 495.

DEMELVERINE [INN]

CAS	13977-33-8	MW:	239.36	MF:	C17 H21 N

	Anticholinergic
	Fahlberg-List, DDR
	LgP: 3.62(C)
	pKa:
	pp cited in Vols.1-5:

(1) Pharmazie, 1978, 33, 725.

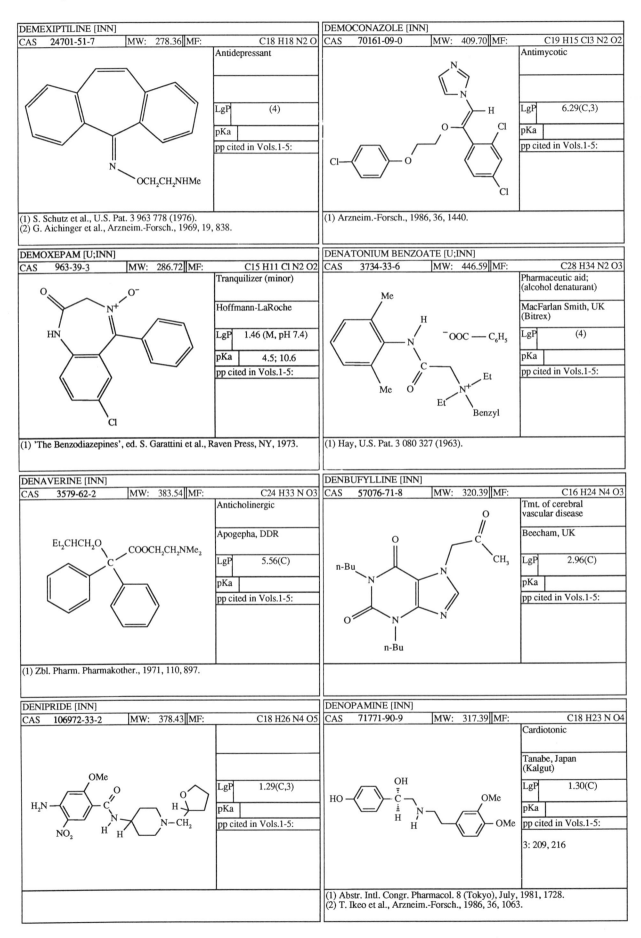

DEMEXIPTILINE [INN]

CAS	24701-51-7	MW:	278.36	MF:	C18 H18 N2 O

Antidepressant

LgP	(4)
pKa	
pp cited in Vols.1-5:	

(1) S. Schutz et al., U.S. Pat. 3 963 778 (1976).
(2) G. Aichinger et al., Arzneim.-Forsch., 1969, 19, 838.

DEMOCONAZOLE [INN]

CAS	70161-09-0	MW:	409.70	MF:	C19 H15 Cl3 N2 O2

Antimycotic

LgP	6.29(C,3)
pKa	
pp cited in Vols.1-5:	

(1) Arzneim.-Forsch., 1986, 36, 1440.

DEMOXEPAM [U;INN]

CAS	963-39-3	MW:	286.72	MF:	C15 H11 Cl N2 O2

Tranquilizer (minor)

Hoffmann-LaRoche

LgP	1.46 (M, pH 7.4)
pKa	4.5; 10.6
pp cited in Vols.1-5:	

(1) 'The Benzodiazepines', ed. S. Garattini et al., Raven Press, NY, 1973.

DENATONIUM BENZOATE [U;INN]

CAS	3734-33-6	MW:	446.59	MF:	C28 H34 N2 O3

Pharmaceutic aid;
(alcohol denaturant)

MacFarlan Smith, UK
(Bitrex)

LgP	(4)
pKa	
pp cited in Vols.1-5:	

(1) Hay, U.S. Pat. 3 080 327 (1963).

DENAVERINE [INN]

CAS	3579-62-2	MW:	383.54	MF:	C24 H33 N O3

Anticholinergic

Apogepha, DDR

LgP	5.56(C)
pKa	
pp cited in Vols.1-5:	

(1) Zbl. Pharm. Pharmakother., 1971, 110, 897.

DENBUFYLLINE [INN]

CAS	57076-71-8	MW:	320.39	MF:	C16 H24 N4 O3

Tmt. of cerebral
vascular disease

Beecham, UK

LgP	2.96(C)
pKa	
pp cited in Vols.1-5:	

DENIPRIDE [INN]

CAS	106972-33-2	MW:	378.43	MF:	C18 H26 N4 O5

LgP	1.29(C,3)
pKa	
pp cited in Vols.1-5:	

DENOPAMINE [INN]

CAS	71771-90-9	MW:	317.39	MF:	C18 H23 N O4

Cardiotonic

Tanabe, Japan
(Kalgut)

LgP	1.30(C)
pKa	
pp cited in Vols.1-5:	
	3: 209, 216

(1) Abstr. Intl. Congr. Pharmacol. 8 (Tokyo), July, 1981, 1728.
(2) T. Ikeo et al., Arzneim.-Forsch., 1986, 36, 1063.

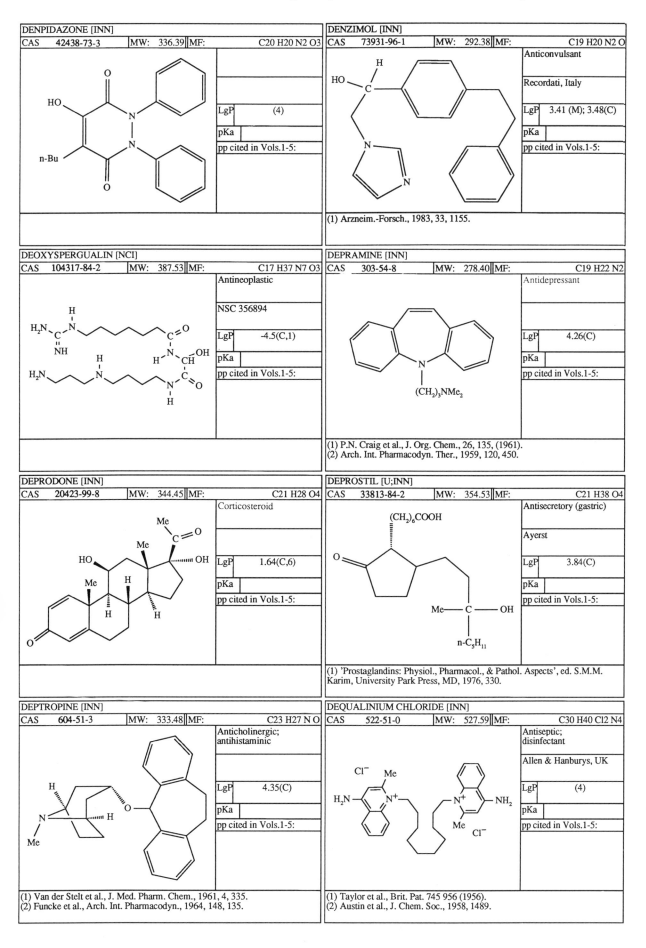

DENPIDAZONE [INN]

| CAS | 42438-73-3 | MW: | 336.39 | MF: | C20 H20 N2 O3 |

LgP	(4)
pKa	
pp cited in Vols.1-5:	

DENZIMOL [INN]

| CAS | 73931-96-1 | MW: | 292.38 | MF: | C19 H20 N2 O |

Anticonvulsant

Recordati, Italy

LgP	3.41 (M); 3.48(C)
pKa	
pp cited in Vols.1-5:	

(1) Arzneim.-Forsch., 1983, 33, 1155.

DEOXYSPERGUALIN [NCI]

| CAS | 104317-84-2 | MW: | 387.53 | MF: | C17 H37 N7 O3 |

Antineoplastic

NSC 356894

LgP	-4.5(C,1)
pKa	
pp cited in Vols.1-5:	

DEPRAMINE [INN]

| CAS | 303-54-8 | MW: | 278.40 | MF: | C19 H22 N2 |

Antidepressant

LgP	4.26(C)
pKa	
pp cited in Vols.1-5:	

(1) P.N. Craig et al., J. Org. Chem., 26, 135, (1961).
(2) Arch. Int. Pharmacodyn. Ther., 1959, 120, 450.

DEPRODONE [INN]

| CAS | 20423-99-8 | MW: | 344.45 | MF: | C21 H28 O4 |

Corticosteroid

LgP	1.64(C,6)
pKa	
pp cited in Vols.1-5:	

DEPROSTIL [U;INN]

| CAS | 33813-84-2 | MW: | 354.53 | MF: | C21 H38 O4 |

Antisecretory (gastric)

Ayerst

LgP	3.84(C)
pKa	
pp cited in Vols.1-5:	

(1) 'Prostaglandins: Physiol., Pharmacol., & Pathol. Aspects', ed. S.M.M. Karim, University Park Press, MD, 1976, 330.

DEPTROPINE [INN]

| CAS | 604-51-3 | MW: | 333.48 | MF: | C23 H27 N O |

Anticholinergic; antihistaminic

LgP	4.35(C)
pKa	
pp cited in Vols.1-5:	

(1) Van der Stelt et al., J. Med. Pharm. Chem., 1961, 4, 335.
(2) Funcke et al., Arch. Int. Pharmacodyn., 1964, 148, 135.

DEQUALINIUM CHLORIDE [INN]

| CAS | 522-51-0 | MW: | 527.59 | MF: | C30 H40 Cl2 N4 |

Antiseptic; disinfectant

Allen & Hanburys, UK

LgP	(4)
pKa	
pp cited in Vols.1-5:	

(1) Taylor et al., Brit. Pat. 745 956 (1956).
(2) Austin et al., J. Chem. Soc., 1958, 1489.

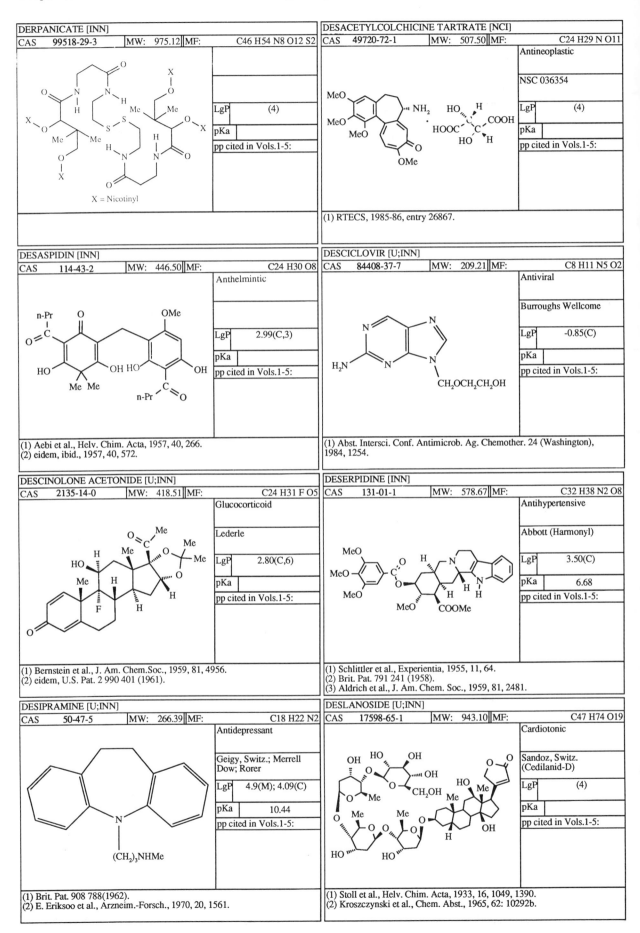

DERPANICATE [INN]

CAS	99518-29-3	MW:	975.12	MF:	C46 H54 N8 O12 S2

LgP	(4)
pKa	
pp cited in Vols.1-5:	

X = Nicotinyl

DESACETYLCOLCHICINE TARTRATE [NCI]

CAS	49720-72-1	MW:	507.50	MF:	C24 H29 N O11

Antineoplastic

NSC 036354

LgP	(4)
pKa	
pp cited in Vols.1-5:	

(1) RTECS, 1985-86, entry 26867.

DESASPIDIN [INN]

CAS	114-43-2	MW:	446.50	MF:	C24 H30 O8

Anthelmintic

LgP	2.99(C,3)
pKa	
pp cited in Vols.1-5:	

(1) Aebi et al., Helv. Chim. Acta, 1957, 40, 266.
(2) eidem, ibid., 1957, 40, 572.

DESCICLOVIR [U;INN]

CAS	84408-37-7	MW:	209.21	MF:	C8 H11 N5 O2

Antiviral

Burroughs Wellcome

LgP	-0.85(C)
pKa	
pp cited in Vols.1-5:	

(1) Abst. Intersci. Conf. Antimicrob. Ag. Chemother. 24 (Washington), 1984, 1254.

DESCINOLONE ACETONIDE [U;INN]

CAS	2135-14-0	MW:	418.51	MF:	C24 H31 F O5

Glucocorticoid

Lederle

LgP	2.80(C,6)
pKa	
pp cited in Vols.1-5:	

(1) Bernstein et al., J. Am. Chem.Soc., 1959, 81, 4956.
(2) eidem, U.S. Pat. 2 990 401 (1961).

DESERPIDINE [INN]

CAS	131-01-1	MW:	578.67	MF:	C32 H38 N2 O8

Antihypertensive

Abbott (Harmonyl)

LgP	3.50(C)
pKa	6.68
pp cited in Vols.1-5:	

(1) Schlittler et al., Experientia, 1955, 11, 64.
(2) Brit. Pat. 791 241 (1958).
(3) Aldrich et al., J. Am. Chem. Soc., 1959, 81, 2481.

DESIPRAMINE [U;INN]

CAS	50-47-5	MW:	266.39	MF:	C18 H22 N2

Antidepressant

Geigy, Switz.; Merrell Dow; Rorer

LgP	4.9(M); 4.09(C)
pKa	10.44
pp cited in Vols.1-5:	

(CH2)3NHMe

(1) Brit. Pat. 908 788(1962).
(2) E. Eriksoo et al., Arzneim.-Forsch., 1970, 20, 1561.

DESLANOSIDE [U;INN]

CAS	17598-65-1	MW:	943.10	MF:	C47 H74 O19

Cardiotonic

Sandoz, Switz. (Cedilanid-D)

LgP	(4)
pKa	
pp cited in Vols.1-5:	

(1) Stoll et al., Helv. Chim. Acta, 1933, 16, 1049, 1390.
(2) Kroszczynski et al., Chem. Abst., 1965, 62: 10292b.

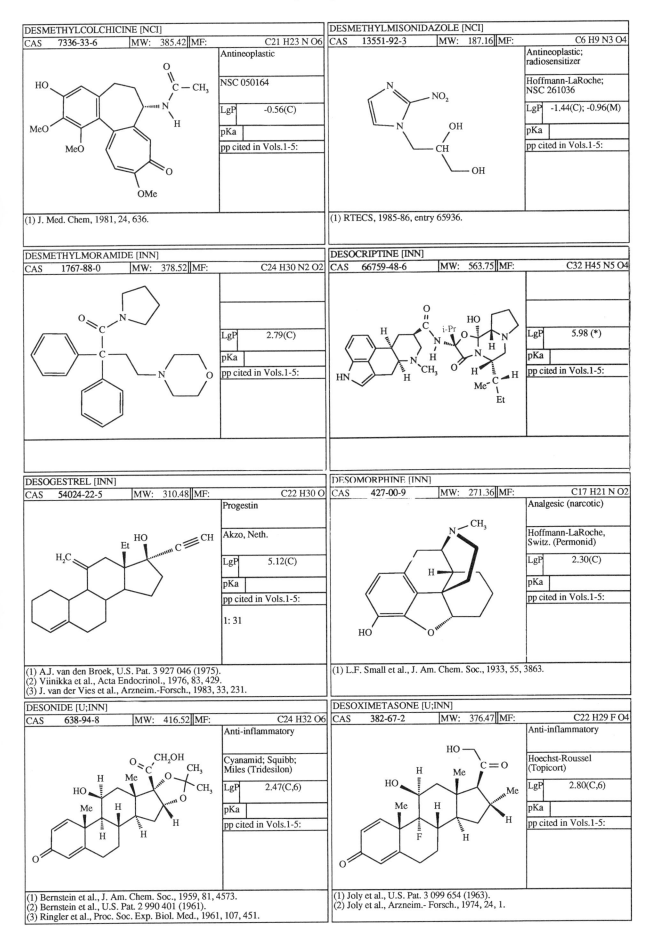

DESMETHYLCOLCHICINE [NCI]

CAS	7336-33-6	MW:	385.42	MF:	C21 H23 N O6

Antineoplastic

NSC 050164

LgP	-0.56(C)
pKa	
pp cited in Vols.1-5:	

(1) J. Med. Chem, 1981, 24, 636.

DESMETHYLMISONIDAZOLE [NCI]

CAS	13551-92-3	MW:	187.16	MF:	C6 H9 N3 O4

Antineoplastic; radiosensitizer

Hoffmann-LaRoche; NSC 261036

LgP	-1.44(C); -0.96(M)
pKa	
pp cited in Vols.1-5:	

(1) RTECS, 1985-86, entry 65936.

DESMETHYLMORAMIDE [INN]

CAS	1767-88-0	MW:	378.52	MF:	C24 H30 N2 O2

LgP	2.79(C)
pKa	
pp cited in Vols.1-5:	

DESOCRIPTINE [INN]

CAS	66759-48-6	MW:	563.75	MF:	C32 H45 N5 O4

LgP	5.98 (*)
pKa	
pp cited in Vols.1-5:	

DESOGESTREL [INN]

CAS	54024-22-5	MW:	310.48	MF:	C22 H30 O

Progestin

Akzo, Neth.

LgP	5.12(C)
pKa	
pp cited in Vols.1-5:	

1: 31

(1) A.J. van den Broek, U.S. Pat. 3 927 046 (1975).
(2) Viinikka et al., Acta Endocrinol., 1976, 83, 429.
(3) J. van der Vies et al., Arzneim.-Forsch., 1983, 33, 231.

DESOMORPHINE [INN]

CAS	427-00-9	MW:	271.36	MF:	C17 H21 N O2

Analgesic (narcotic)

Hoffmann-LaRoche, Switz. (Permonid)

LgP	2.30(C)
pKa	
pp cited in Vols.1-5:	

(1) L.F. Small et al., J. Am. Chem. Soc., 1933, 55, 3863.

DESONIDE [U;INN]

CAS	638-94-8	MW:	416.52	MF:	C24 H32 O6

Anti-inflammatory

Cyanamid; Squibb; Miles (Tridesilon)

LgP	2.47(C,6)
pKa	
pp cited in Vols.1-5:	

(1) Bernstein et al., J. Am. Chem. Soc., 1959, 81, 4573.
(2) Bernstein et al., U.S. Pat. 2 990 401 (1961).
(3) Ringler et al., Proc. Soc. Exp. Biol. Med., 1961, 107, 451.

DESOXIMETASONE [U;INN]

CAS	382-67-2	MW:	376.47	MF:	C22 H29 F O4

Anti-inflammatory

Hoechst-Roussel (Topicort)

LgP	2.80(C,6)
pKa	
pp cited in Vols.1-5:	

(1) Joly et al., U.S. Pat. 3 099 654 (1963).
(2) Joly et al., Arzneim.- Forsch., 1974, 24, 1.

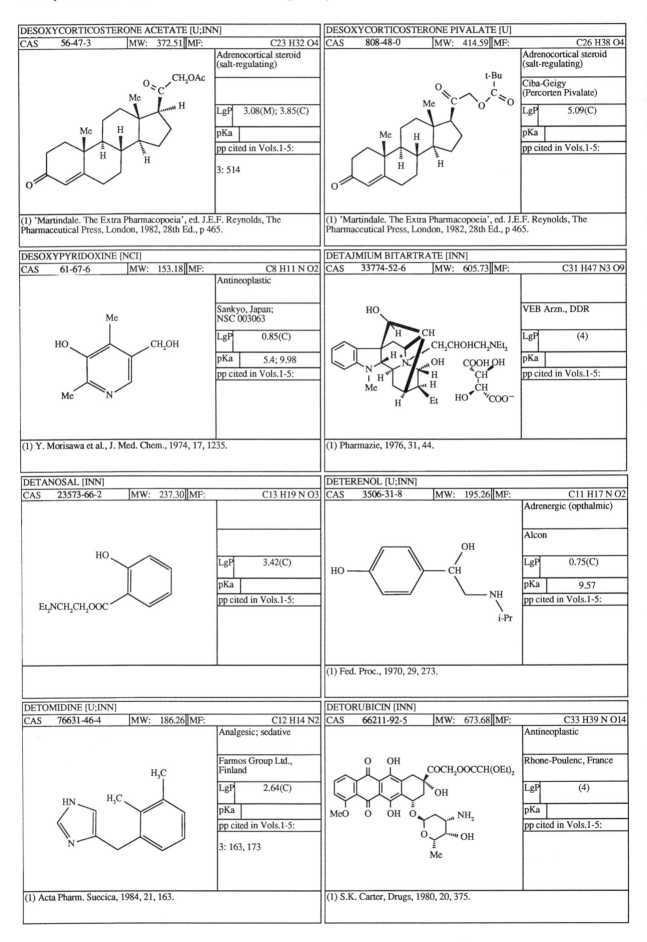

DESOXYCORTICOSTERONE ACETATE [U;INN]		
CAS 56-47-3	MW: 372.51	MF: C23 H32 O4

Adrenocortical steroid (salt-regulating)

LgP	3.08(M); 3.85(C)
pKa	
pp cited in Vols.1-5:	

3: 514

(1) 'Martindale. The Extra Pharmacopoeia', ed. J.E.F. Reynolds, The Pharmaceutical Press, London, 1982, 28th Ed., p 465.

DESOXYCORTICOSTERONE PIVALATE [U]		
CAS 808-48-0	MW: 414.59	MF: C26 H38 O4

Adrenocortical steroid (salt-regulating)

Ciba-Geigy (Percorten Pivalate)

LgP	5.09(C)
pKa	
pp cited in Vols.1-5:	

(1) 'Martindale. The Extra Pharmacopoeia', ed. J.E.F. Reynolds, The Pharmaceutical Press, London, 1982, 28th Ed., p 465.

DESOXYPYRIDOXINE [NCI]		
CAS 61-67-6	MW: 153.18	MF: C8 H11 N O2

Antineoplastic

Sankyo, Japan; NSC 003063

LgP	0.85(C)
pKa	5.4; 9.98
pp cited in Vols.1-5:	

(1) Y. Morisawa et al., J. Med. Chem., 1974, 17, 1235.

DETAJMIUM BITARTRATE [INN]		
CAS 33774-52-6	MW: 605.73	MF: C31 H47 N3 O9

VEB Arzn., DDR

LgP	(4)
pKa	
pp cited in Vols.1-5:	

(1) Pharmazie, 1976, 31, 44.

DETANOSAL [INN]		
CAS 23573-66-2	MW: 237.30	MF: C13 H19 N O3

LgP	3.42(C)
pKa	
pp cited in Vols.1-5:	

DETERENOL [U;INN]		
CAS 3506-31-8	MW: 195.26	MF: C11 H17 N O2

Adrenergic (opthalmic)

Alcon

LgP	0.75(C)
pKa	9.57
pp cited in Vols.1-5:	

(1) Fed. Proc., 1970, 29, 273.

DETOMIDINE [U;INN]		
CAS 76631-46-4	MW: 186.26	MF: C12 H14 N2

Analgesic; sedative

Farmos Group Ltd., Finland

LgP	2.64(C)
pKa	
pp cited in Vols.1-5:	

3: 163, 173

(1) Acta Pharm. Suecica, 1984, 21, 163.

DETORUBICIN [INN]		
CAS 66211-92-5	MW: 673.68	MF: C33 H39 N O14

Antineoplastic

Rhone-Poulenc, France

LgP	(4)
pKa	
pp cited in Vols.1-5:	

(1) S.K. Carter, Drugs, 1980, 20, 375.

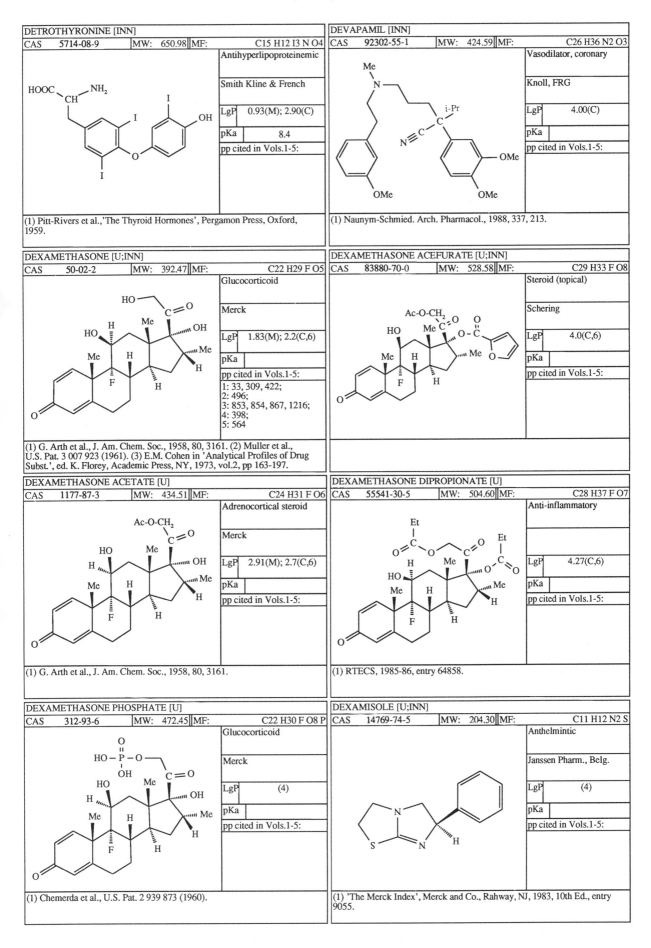

DETROTHYRONINE [INN]

CAS	5714-08-9	MW:	650.98	MF:	C15 H12 I3 N O4

Antihyperlipoproteinemic

Smith Kline & French

LgP	0.93(M); 2.90(C)
pKa	8.4
pp cited in Vols.1-5:	

(1) Pitt-Rivers et al.,'The Thyroid Hormones', Pergamon Press, Oxford, 1959.

DEVAPAMIL [INN]

CAS	92302-55-1	MW:	424.59	MF:	C26 H36 N2 O3

Vasodilator, coronary

Knoll, FRG

LgP	4.00(C)
pKa	
pp cited in Vols.1-5:	

(1) Naunym-Schmied. Arch. Pharmacol., 1988, 337, 213.

DEXAMETHASONE [U;INN]

CAS	50-02-2	MW:	392.47	MF:	C22 H29 F O5

Glucocorticoid

Merck

LgP	1.83(M); 2.2(C,6)
pKa	
pp cited in Vols.1-5:	
	1: 33, 309, 422; 2: 496; 3: 853, 854, 867, 1216; 4: 398; 5: 564

(1) G. Arth et al., J. Am. Chem. Soc., 1958, 80, 3161. (2) Muller et al., U.S. Pat. 3 007 923 (1961). (3) E.M. Cohen in 'Analytical Profiles of Drug Subst.', ed. K. Florey, Academic Press, NY, 1973, vol.2, pp 163-197.

DEXAMETHASONE ACEFURATE [U;INN]

CAS	83880-70-0	MW:	528.58	MF:	C29 H33 F O8

Steroid (topical)

Schering

LgP	4.0(C,6)
pKa	
pp cited in Vols.1-5:	

DEXAMETHASONE ACETATE [U]

CAS	1177-87-3	MW:	434.51	MF:	C24 H31 F O6

Adrenocortical steroid

Merck

LgP	2.91(M); 2.7(C,6)
pKa	
pp cited in Vols.1-5:	

(1) G. Arth et al., J. Am. Chem. Soc., 1958, 80, 3161.

DEXAMETHASONE DIPROPIONATE [U]

CAS	55541-30-5	MW:	504.60	MF:	C28 H37 F O7

Anti-inflammatory

LgP	4.27(C,6)
pKa	
pp cited in Vols.1-5:	

(1) RTECS, 1985-86, entry 64858.

DEXAMETHASONE PHOSPHATE [U]

CAS	312-93-6	MW:	472.45	MF:	C22 H30 F O8 P

Glucocorticoid

Merck

LgP	(4)
pKa	
pp cited in Vols.1-5:	

(1) Chemerda et al., U.S. Pat. 2 939 873 (1960).

DEXAMISOLE [U;INN]

CAS	14769-74-5	MW:	204.30	MF:	C11 H12 N2 S

Anthelmintic

Janssen Pharm., Belg.

LgP	(4)
pKa	
pp cited in Vols.1-5:	

(1) 'The Merck Index', Merck and Co., Rahway, NJ, 1983, 10th Ed., entry 9055.

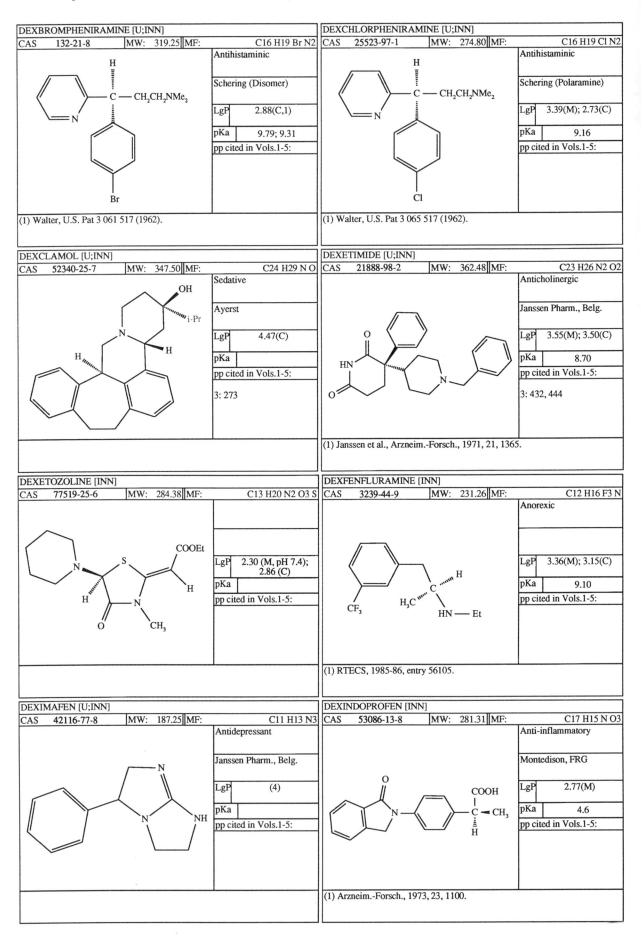

DEXBROMPHENIRAMINE [U;INN]

CAS	132-21-8	MW:	319.25	MF:	C16 H19 Br N2

Antihistaminic

Schering (Disomer)

LgP	2.88(C,1)
pKa	9.79; 9.31
pp cited in Vols.1-5:	

(1) Walter, U.S. Pat 3 061 517 (1962).

DEXCHLORPHENIRAMINE [U;INN]

CAS	25523-97-1	MW:	274.80	MF:	C16 H19 Cl N2

Antihistaminic

Schering (Polaramine)

LgP	3.39(M); 2.73(C)
pKa	9.16
pp cited in Vols.1-5:	

(1) Walter, U.S. Pat 3 065 517 (1962).

DEXCLAMOL [U;INN]

CAS	52340-25-7	MW:	347.50	MF:	C24 H29 N O

Sedative

Ayerst

LgP	4.47(C)
pKa	
pp cited in Vols.1-5:	
3: 273	

DEXETIMIDE [U;INN]

CAS	21888-98-2	MW:	362.48	MF:	C23 H26 N2 O2

Anticholinergic

Janssen Pharm., Belg.

LgP	3.55(M); 3.50(C)
pKa	8.70
pp cited in Vols.1-5:	
3: 432, 444	

(1) Janssen et al., Arzneim.-Forsch., 1971, 21, 1365.

DEXETOZOLINE [INN]

CAS	77519-25-6	MW:	284.38	MF:	C13 H20 N2 O3 S

LgP	2.30 (M, pH 7.4); 2.86 (C)
pKa	
pp cited in Vols.1-5:	

DEXFENFLURAMINE [INN]

CAS	3239-44-9	MW:	231.26	MF:	C12 H16 F3 N

Anorexic

LgP	3.36(M); 3.15(C)
pKa	9.10
pp cited in Vols.1-5:	

(1) RTECS, 1985-86, entry 56105.

DEXIMAFEN [U;INN]

CAS	42116-77-8	MW:	187.25	MF:	C11 H13 N3

Antidepressant

Janssen Pharm., Belg.

LgP	(4)
pKa	
pp cited in Vols.1-5:	

DEXINDOPROFEN [INN]

CAS	53086-13-8	MW:	281.31	MF:	C17 H15 N O3

Anti-inflammatory

Montedison, FRG

LgP	2.77(M)
pKa	4.6
pp cited in Vols.1-5:	

(1) Arzneim.-Forsch., 1973, 23, 1100.

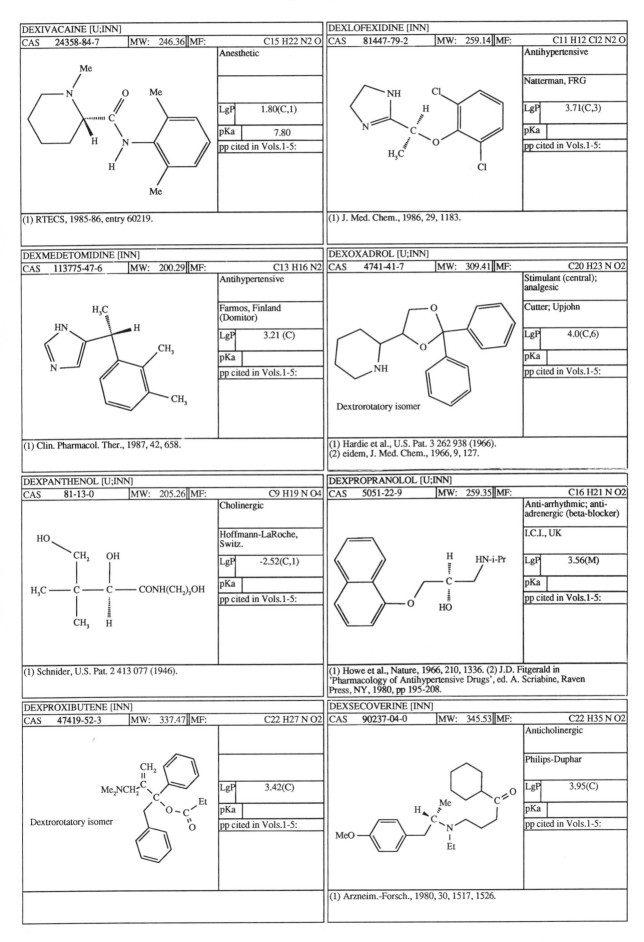

DEXIVACAINE [U;INN]

CAS	24358-84-7	MW:	246.36	MF:	C15 H22 N2 O

Anesthetic

LgP	1.80(C,1)
pKa	7.80
pp cited in Vols.1-5:	

(1) RTECS, 1985-86, entry 60219.

DEXLOFEXIDINE [INN]

CAS	81447-79-2	MW:	259.14	MF:	C11 H12 Cl2 N2 O

Antihypertensive

Natterman, FRG

LgP	3.71(C,3)
pKa	
pp cited in Vols.1-5:	

(1) J. Med. Chem., 1986, 29, 1183.

DEXMEDETOMIDINE [INN]

CAS	113775-47-6	MW:	200.29	MF:	C13 H16 N2

Antihypertensive

Farmos, Finland (Domitor)

LgP	3.21 (C)
pKa	
pp cited in Vols.1-5:	

(1) Clin. Pharmacol. Ther., 1987, 42, 658.

DEXOXADROL [U;INN]

CAS	4741-41-7	MW:	309.41	MF:	C20 H23 N O2

Stimulant (central); analgesic

Cutter; Upjohn

LgP	4.0(C,6)
pKa	
pp cited in Vols.1-5:	

Dextrorotatory isomer

(1) Hardie et al., U.S. Pat. 3 262 938 (1966).
(2) eidem, J. Med. Chem., 1966, 9, 127.

DEXPANTHENOL [U;INN]

CAS	81-13-0	MW:	205.26	MF:	C9 H19 N O4

Cholinergic

Hoffmann-LaRoche, Switz.

LgP	-2.52(C,1)
pKa	
pp cited in Vols.1-5:	

(1) Schnider, U.S. Pat. 2 413 077 (1946).

DEXPROPRANOLOL [U;INN]

CAS	5051-22-9	MW:	259.35	MF:	C16 H21 N O2

Anti-arrhythmic; anti-adrenergic (beta-blocker)

I.C.I., UK

LgP	3.56(M)
pKa	
pp cited in Vols.1-5:	

(1) Howe et al., Nature, 1966, 210, 1336. (2) J.D. Fitgerald in 'Pharmacology of Antihypertensive Drugs', ed. A. Scriabine, Raven Press, NY, 1980, pp 195-208.

DEXPROXIBUTENE [INN]

CAS	47419-52-3	MW:	337.47	MF:	C22 H27 N O2

LgP	3.42(C)
pKa	
pp cited in Vols.1-5:	

Dextrorotatory isomer

DEXSECOVERINE [INN]

CAS	90237-04-0	MW:	345.53	MF:	C22 H35 N O2

Anticholinergic

Philips-Duphar

LgP	3.95(C)
pKa	
pp cited in Vols.1-5:	

(1) Arzneim.-Forsch., 1980, 30, 1517, 1526.

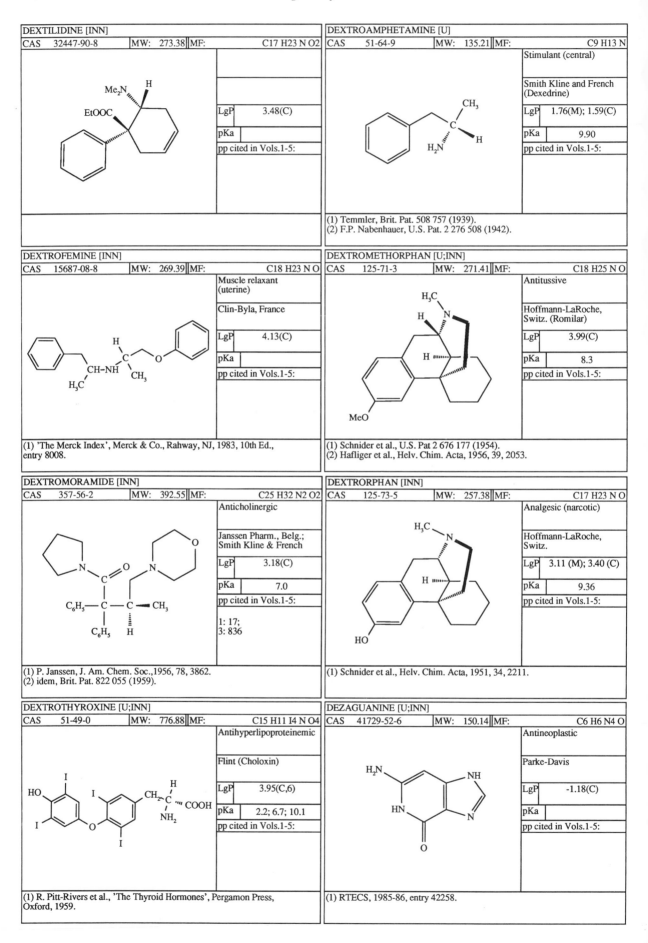

DEXTILIDINE [INN]

CAS	32447-90-8	MW:	273.38	MF:	C17 H23 N O2

LgP	3.48(C)
pKa	
pp cited in Vols.1-5:	

DEXTROAMPHETAMINE [U]

CAS	51-64-9	MW:	135.21	MF:	C9 H13 N

Stimulant (central)

Smith Kline and French (Dexedrine)

LgP	1.76(M); 1.59(C)
pKa	9.90
pp cited in Vols.1-5:	

(1) Temmler, Brit. Pat. 508 757 (1939).
(2) F.P. Nabenhauer, U.S. Pat. 2 276 508 (1942).

DEXTROFEMINE [INN]

CAS	15687-08-8	MW:	269.39	MF:	C18 H23 N O

Muscle relaxant (uterine)

Clin-Byla, France

LgP	4.13(C)
pKa	
pp cited in Vols.1-5:	

(1) 'The Merck Index', Merck & Co., Rahway, NJ, 1983, 10th Ed., entry 8008.

DEXTROMETHORPHAN [U;INN]

CAS	125-71-3	MW:	271.41	MF:	C18 H25 N O

Antitussive

Hoffmann-LaRoche, Switz. (Romilar)

LgP	3.99(C)
pKa	8.3
pp cited in Vols.1-5:	

(1) Schnider et al., U.S. Pat 2 676 177 (1954).
(2) Hafliger et al., Helv. Chim. Acta, 1956, 39, 2053.

DEXTROMORAMIDE [INN]

CAS	357-56-2	MW:	392.55	MF:	C25 H32 N2 O2

Anticholinergic

Janssen Pharm., Belg.; Smith Kline & French

LgP	3.18(C)
pKa	7.0
pp cited in Vols.1-5:	1: 17; 3: 836

(1) P. Janssen, J. Am. Chem. Soc.,1956, 78, 3862.
(2) idem, Brit. Pat. 822 055 (1959).

DEXTRORPHAN [INN]

CAS	125-73-5	MW:	257.38	MF:	C17 H23 N O

Analgesic (narcotic)

Hoffmann-LaRoche, Switz.

LgP	3.11 (M); 3.40 (C)
pKa	9.36
pp cited in Vols.1-5:	

(1) Schnider et al., Helv. Chim. Acta, 1951, 34, 2211.

DEXTROTHYROXINE [U;INN]

CAS	51-49-0	MW:	776.88	MF:	C15 H11 I4 N O4

Antihyperlipoproteinemic

Flint (Choloxin)

LgP	3.95(C,6)
pKa	2.2; 6.7; 10.1
pp cited in Vols.1-5:	

(1) R. Pitt-Rivers et al., 'The Thyroid Hormones', Pergamon Press, Oxford, 1959.

DEZAGUANINE [U;INN]

CAS	41729-52-6	MW:	150.14	MF:	C6 H6 N4 O

Antineoplastic

Parke-Davis

LgP	-1.18(C)
pKa	
pp cited in Vols.1-5:	

(1) RTECS, 1985-86, entry 42258.

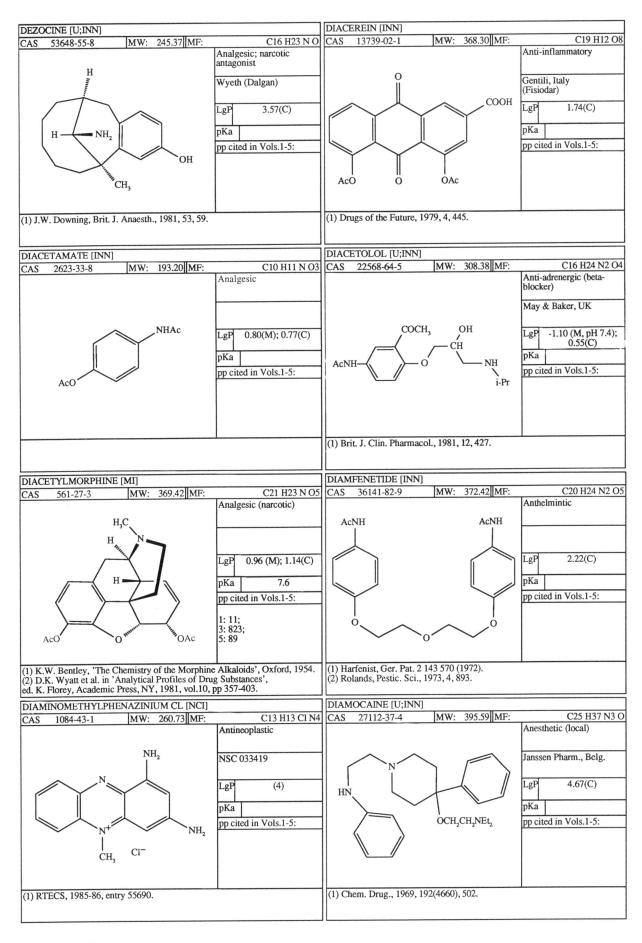

DEZOCINE [U;INN]

CAS	53648-55-8	MW:	245.37	MF:	C16 H23 N O

Analgesic; narcotic antagonist

Wyeth (Dalgan)

LgP	3.57(C)
pKa	
pp cited in Vols.1-5:	

(1) J.W. Downing, Brit. J. Anaesth., 1981, 53, 59.

DIACEREIN [INN]

CAS	13739-02-1	MW:	368.30	MF:	C19 H12 O8

Anti-inflammatory

Gentili, Italy (Fisiodar)

LgP	1.74(C)
pKa	
pp cited in Vols.1-5:	

(1) Drugs of the Future, 1979, 4, 445.

DIACETAMATE [INN]

CAS	2623-33-8	MW:	193.20	MF:	C10 H11 N O3

Analgesic

LgP	0.80(M); 0.77(C)
pKa	
pp cited in Vols.1-5:	

DIACETOLOL [U;INN]

CAS	22568-64-5	MW:	308.38	MF:	C16 H24 N2 O4

Anti-adrenergic (beta-blocker)

May & Baker, UK

LgP	-1.10 (M, pH 7.4); 0.55(C)
pKa	
pp cited in Vols.1-5:	

(1) Brit. J. Clin. Pharmacol., 1981, 12, 427.

DIACETYLMORPHINE [MI]

CAS	561-27-3	MW:	369.42	MF:	C21 H23 N O5

Analgesic (narcotic)

LgP	0.96 (M); 1.14(C)
pKa	7.6
pp cited in Vols.1-5:	

1: 11;
3: 823;
5: 89

(1) K.W. Bentley, 'The Chemistry of the Morphine Alkaloids', Oxford, 1954.
(2) D.K. Wyatt et al. in 'Analytical Profiles of Drug Substances', ed. K. Florey, Academic Press, NY, 1981, vol.10, pp 357-403.

DIAMFENETIDE [INN]

CAS	36141-82-9	MW:	372.42	MF:	C20 H24 N2 O5

Anthelmintic

LgP	2.22(C)
pKa	
pp cited in Vols.1-5:	

(1) Harfenist, Ger. Pat. 2 143 570 (1972).
(2) Rolands, Pestic. Sci., 1973, 4, 893.

DIAMINOMETHYLPHENAZINIUM CL [NCI]

CAS	1084-43-1	MW:	260.73	MF:	C13 H13 Cl N4

Antineoplastic

NSC 033419

LgP	(4)
pKa	
pp cited in Vols.1-5:	

(1) RTECS, 1985-86, entry 55690.

DIAMOCAINE [U;INN]

CAS	27112-37-4	MW:	395.59	MF:	C25 H37 N3 O

Anesthetic (local)

Janssen Pharm., Belg.

LgP	4.67(C)
pKa	
pp cited in Vols.1-5:	

(1) Chem. Drug., 1969, 192(4660), 502.

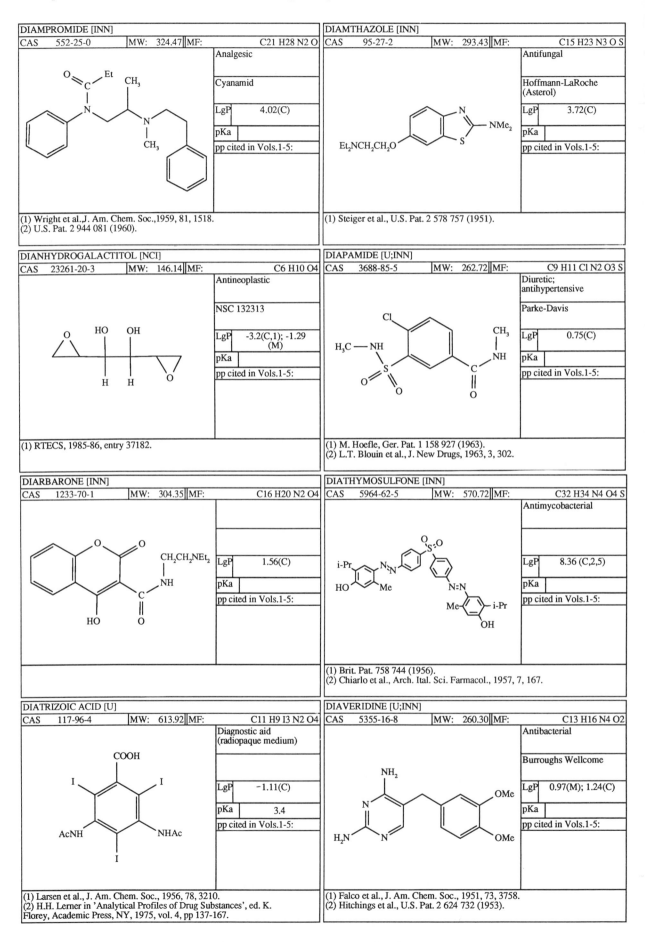

DIAMPROMIDE [INN]

CAS	552-25-0	MW:	324.47	MF:	C21 H28 N2 O

Analgesic

Cyanamid

LgP	4.02(C)
pKa	
pp cited in Vols.1-5:	

(1) Wright et al., J. Am. Chem. Soc.,1959, 81, 1518.
(2) U.S. Pat. 2 944 081 (1960).

DIAMTHAZOLE [INN]

CAS	95-27-2	MW:	293.43	MF:	C15 H23 N3 O S

Antifungal

Hoffmann-LaRoche
(Asterol)

LgP	3.72(C)
pKa	
pp cited in Vols.1-5:	

(1) Steiger et al., U.S. Pat. 2 578 757 (1951).

DIANHYDROGALACTITOL [NCI]

CAS	23261-20-3	MW:	146.14	MF:	C6 H10 O4

Antineoplastic

NSC 132313

LgP	-3.2(C,1); -1.29 (M)
pKa	
pp cited in Vols.1-5:	

(1) RTECS, 1985-86, entry 37182.

DIAPAMIDE [U;INN]

CAS	3688-85-5	MW:	262.72	MF:	C9 H11 Cl N2 O3 S

Diuretic;
antihypertensive

Parke-Davis

LgP	0.75(C)
pKa	
pp cited in Vols.1-5:	

(1) M. Hoefle, Ger. Pat. 1 158 927 (1963).
(2) L.T. Blouin et al., J. New Drugs, 1963, 3, 302.

DIARBARONE [INN]

CAS	1233-70-1	MW:	304.35	MF:	C16 H20 N2 O4

LgP	1.56(C)
pKa	
pp cited in Vols.1-5:	

DIATHYMOSULFONE [INN]

CAS	5964-62-5	MW:	570.72	MF:	C32 H34 N4 O4 S

Antimycobacterial

LgP	8.36 (C,2,5)
pKa	
pp cited in Vols.1-5:	

(1) Brit. Pat. 758 744 (1956).
(2) Chiarlo et al., Arch. Ital. Sci. Farmacol., 1957, 7, 167.

DIATRIZOIC ACID [U]

CAS	117-96-4	MW:	613.92	MF:	C11 H9 I3 N2 O4

Diagnostic aid
(radiopaque medium)

LgP	-1.11(C)
pKa	3.4
pp cited in Vols.1-5:	

(1) Larsen et al., J. Am. Chem. Soc., 1956, 78, 3210.
(2) H.H. Lerner in 'Analytical Profiles of Drug Substances', ed. K. Florey, Academic Press, NY, 1975, vol. 4, pp 137-167.

DIAVERIDINE [U;INN]

CAS	5355-16-8	MW:	260.30	MF:	C13 H16 N4 O2

Antibacterial

Burroughs Wellcome

LgP	0.97(M); 1.24(C)
pKa	
pp cited in Vols.1-5:	

(1) Falco et al., J. Am. Chem. Soc., 1951, 73, 3758.
(2) Hitchings et al., U.S. Pat. 2 624 732 (1953).

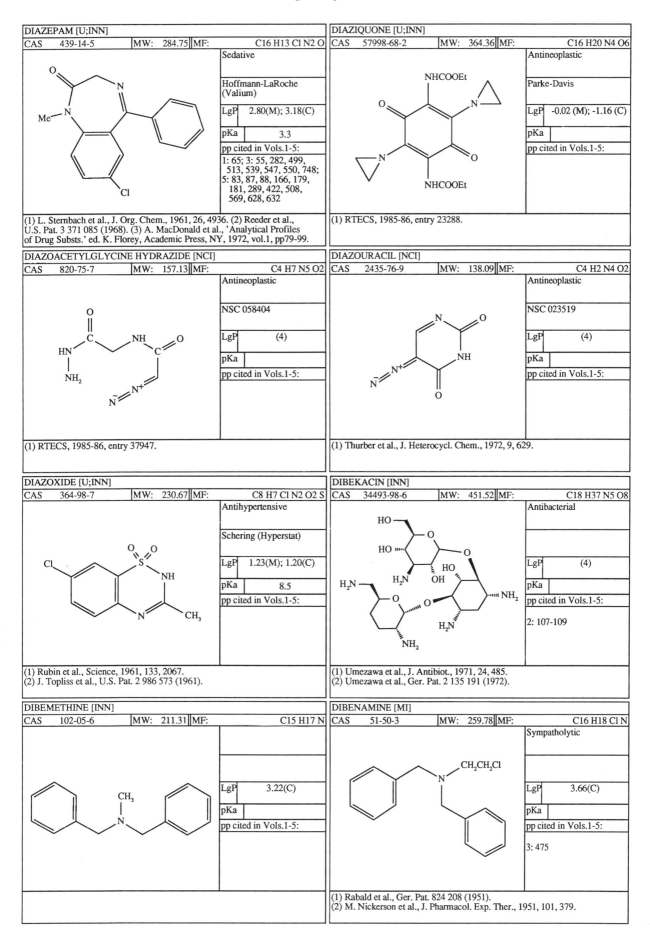

DIAZEPAM [U;INN]

CAS	439-14-5	MW:	284.75	MF:	C16 H13 Cl N2 O

Sedative

Hoffmann-LaRoche (Valium)

LgP	2.80(M); 3.18(C)
pKa	3.3

pp cited in Vols.1-5:
1: 65; 3: 55, 282, 499, 513, 539, 547, 550, 748; 5: 83, 87, 88, 166, 179, 181, 289, 422, 508, 569, 628, 632

(1) L. Sternbach et al., J. Org. Chem., 1961, 26, 4936. (2) Reeder et al., U.S. Pat. 3 371 085 (1968). (3) A. MacDonald et al., 'Analytical Profiles of Drug Substs.' ed. K. Florey, Academic Press, NY, 1972, vol.1, pp79-99.

DIAZIQUONE [U;INN]

CAS	57998-68-2	MW:	364.36	MF:	C16 H20 N4 O6

Antineoplastic

Parke-Davis

LgP	-0.02 (M); -1.16 (C)
pKa	

pp cited in Vols.1-5:

(1) RTECS, 1985-86, entry 23288.

DIAZOACETYLGLYCINE HYDRAZIDE [NCI]

CAS	820-75-7	MW:	157.13	MF:	C4 H7 N5 O2

Antineoplastic

NSC 058404

LgP	(4)
pKa	

pp cited in Vols.1-5:

(1) RTECS, 1985-86, entry 37947.

DIAZOURACIL [NCI]

CAS	2435-76-9	MW:	138.09	MF:	C4 H2 N4 O2

Antineoplastic

NSC 023519

LgP	(4)
pKa	

pp cited in Vols.1-5:

(1) Thurber et al., J. Heterocycl. Chem., 1972, 9, 629.

DIAZOXIDE [U;INN]

CAS	364-98-7	MW:	230.67	MF:	C8 H7 Cl N2 O2 S

Antihypertensive

Schering (Hyperstat)

LgP	1.23(M); 1.20(C)
pKa	8.5

pp cited in Vols.1-5:

(1) Rubin et al., Science, 1961, 133, 2067.
(2) J. Topliss et al., U.S. Pat. 2 986 573 (1961).

DIBEKACIN [INN]

CAS	34493-98-6	MW:	451.52	MF:	C18 H37 N5 O8

Antibacterial

LgP	(4)
pKa	

pp cited in Vols.1-5:
2: 107-109

(1) Umezawa et al., J. Antibiot., 1971, 24, 485.
(2) Umezawa et al., Ger. Pat. 2 135 191 (1972).

DIBEMETHINE [INN]

CAS	102-05-6	MW:	211.31	MF:	C15 H17 N

LgP	3.22(C)
pKa	

pp cited in Vols.1-5:

DIBENAMINE [MI]

CAS	51-50-3	MW:	259.78	MF:	C16 H18 Cl N

Sympatholytic

LgP	3.66(C)
pKa	

pp cited in Vols.1-5:
3: 475

(1) Rabald et al., Ger. Pat. 824 208 (1951).
(2) M. Nickerson et al., J. Pharmacol. Exp. Ther., 1951, 101, 379.

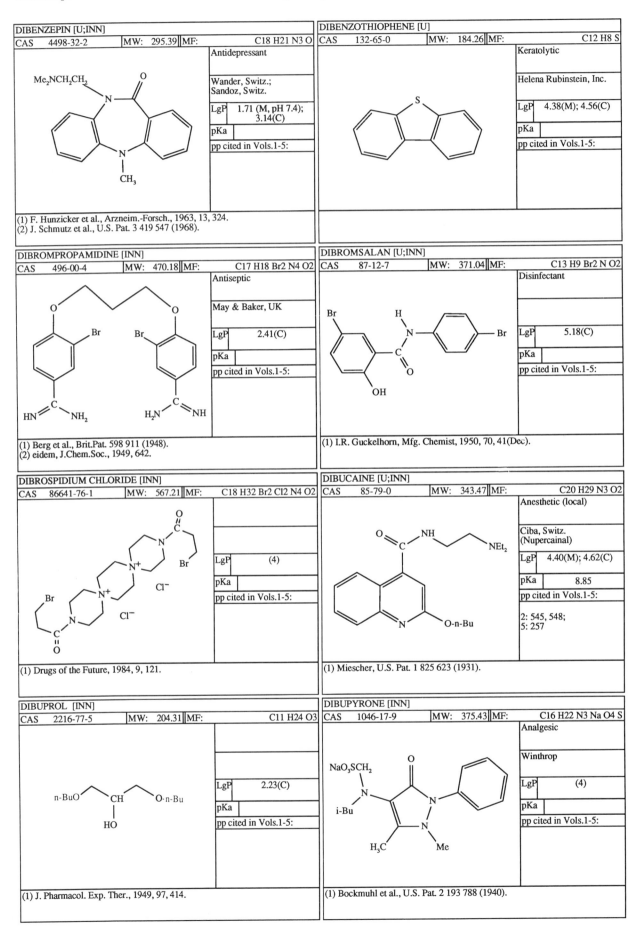

DIBENZEPIN [U;INN]

CAS	4498-32-2	MW:	295.39	MF:	C18 H21 N3 O

Antidepressant

Wander, Switz.; Sandoz, Switz.

LgP	1.71 (M, pH 7.4); 3.14(C)
pKa	
pp cited in Vols.1-5:	

Me$_2$NCH$_2$CH$_2$ — CH$_3$

(1) F. Hunzicker et al., Arzneim.-Forsch., 1963, 13, 324.
(2) J. Schmutz et al., U.S. Pat. 3 419 547 (1968).

DIBENZOTHIOPHENE [U]

CAS	132-65-0	MW:	184.26	MF:	C12 H8 S

Keratolytic

Helena Rubinstein, Inc.

LgP	4.38(M); 4.56(C)
pKa	
pp cited in Vols.1-5:	

DIBROMPROPAMIDINE [INN]

CAS	496-00-4	MW:	470.18	MF:	C17 H18 Br2 N4 O2

Antiseptic

May & Baker, UK

LgP	2.41(C)
pKa	
pp cited in Vols.1-5:	

(1) Berg et al., Brit.Pat. 598 911 (1948).
(2) eidem, J.Chem.Soc., 1949, 642.

DIBROMSALAN [U;INN]

CAS	87-12-7	MW:	371.04	MF:	C13 H9 Br2 N O2

Disinfectant

LgP	5.18(C)
pKa	
pp cited in Vols.1-5:	

(1) I.R. Guckelhorn, Mfg. Chemist, 1950, 70, 41(Dec).

DIBROSPIDIUM CHLORIDE [INN]

CAS	86641-76-1	MW:	567.21	MF:	C18 H32 Br2 Cl2 N4 O2

LgP	(4)
pKa	
pp cited in Vols.1-5:	

(1) Drugs of the Future, 1984, 9, 121.

DIBUCAINE [U;INN]

CAS	85-79-0	MW:	343.47	MF:	C20 H29 N3 O2

Anesthetic (local)

Ciba, Switz. (Nupercainal)

LgP	4.40(M); 4.62(C)
pKa	8.85
pp cited in Vols.1-5:	

2: 545, 548;
5: 257

(1) Miescher, U.S. Pat. 1 825 623 (1931).

DIBUPROL [INN]

CAS	2216-77-5	MW:	204.31	MF:	C11 H24 O3

LgP	2.23(C)
pKa	
pp cited in Vols.1-5:	

n-BuO — CH — O-n-Bu
            HO

(1) J. Pharmacol. Exp. Ther., 1949, 97, 414.

DIBUPYRONE [INN]

CAS	1046-17-9	MW:	375.43	MF:	C16 H22 N3 Na O4 S

Analgesic

Winthrop

LgP	(4)
pKa	
pp cited in Vols.1-5:	

(1) Bockmuhl et al., U.S. Pat. 2 193 788 (1940).

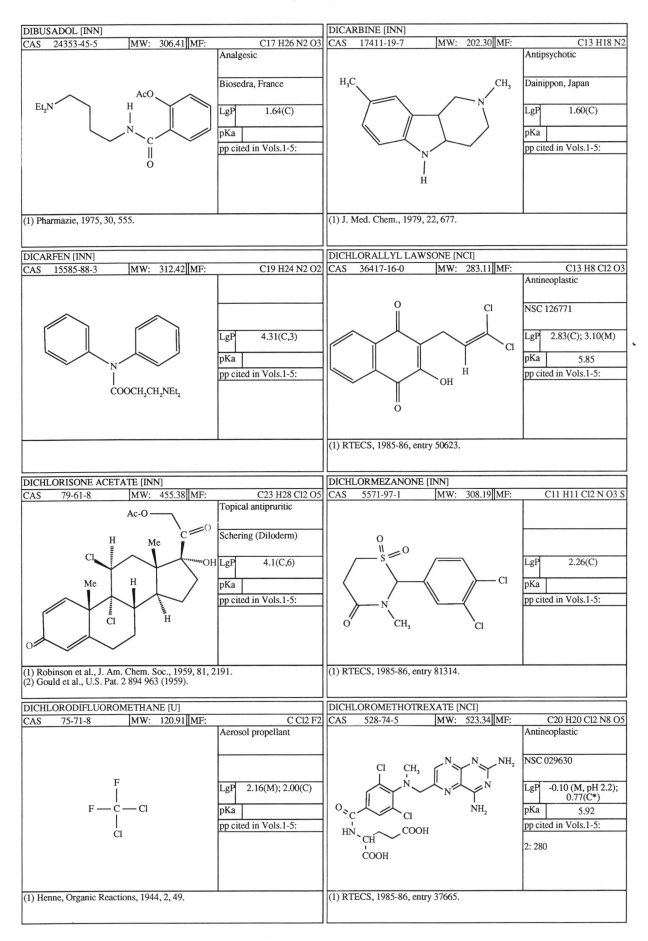

DIBUSADOL [INN]

CAS	24353-45-5	MW:	306.41	MF:	C17 H26 N2 O3

Analgesic

Biosedra, France

LgP	1.64(C)
pKa	
pp cited in Vols.1-5:	

(1) Pharmazie, 1975, 30, 555.

DICARBINE [INN]

CAS	17411-19-7	MW:	202.30	MF:	C13 H18 N2

Antipsychotic

Dainippon, Japan

LgP	1.60(C)
pKa	
pp cited in Vols.1-5:	

(1) J. Med. Chem., 1979, 22, 677.

DICARFEN [INN]

CAS	15585-88-3	MW:	312.42	MF:	C19 H24 N2 O2

LgP	4.31(C,3)
pKa	
pp cited in Vols.1-5:	

DICHLORALLYL LAWSONE [NCI]

CAS	36417-16-0	MW:	283.11	MF:	C13 H8 Cl2 O3

Antineoplastic

NSC 126771

LgP	2.83(C); 3.10(M)
pKa	5.85
pp cited in Vols.1-5:	

(1) RTECS, 1985-86, entry 50623.

DICHLORISONE ACETATE [INN]

CAS	79-61-8	MW:	455.38	MF:	C23 H28 Cl2 O5

Topical antipruritic

Schering (Diloderm)

LgP	4.1(C,6)
pKa	
pp cited in Vols.1-5:	

(1) Robinson et al., J. Am. Chem. Soc., 1959, 81, 2191.
(2) Gould et al., U.S. Pat. 2 894 963 (1959).

DICHLORMEZANONE [INN]

CAS	5571-97-1	MW:	308.19	MF:	C11 H11 Cl2 N O3 S

LgP	2.26(C)
pKa	
pp cited in Vols.1-5:	

(1) RTECS, 1985-86, entry 81314.

DICHLORODIFLUOROMETHANE [U]

CAS	75-71-8	MW:	120.91	MF:	C Cl2 F2

Aerosol propellant

LgP	2.16(M); 2.00(C)
pKa	
pp cited in Vols.1-5:	

(1) Henne, Organic Reactions, 1944, 2, 49.

DICHLOROMETHOTREXATE [NCI]

CAS	528-74-5	MW:	523.34	MF:	C20 H20 Cl2 N8 O5

Antineoplastic

NSC 029630

LgP	-0.10 (M, pH 2.2); 0.77(C*)
pKa	5.92
pp cited in Vols.1-5:	
2: 280	

(1) RTECS, 1985-86, entry 37665.

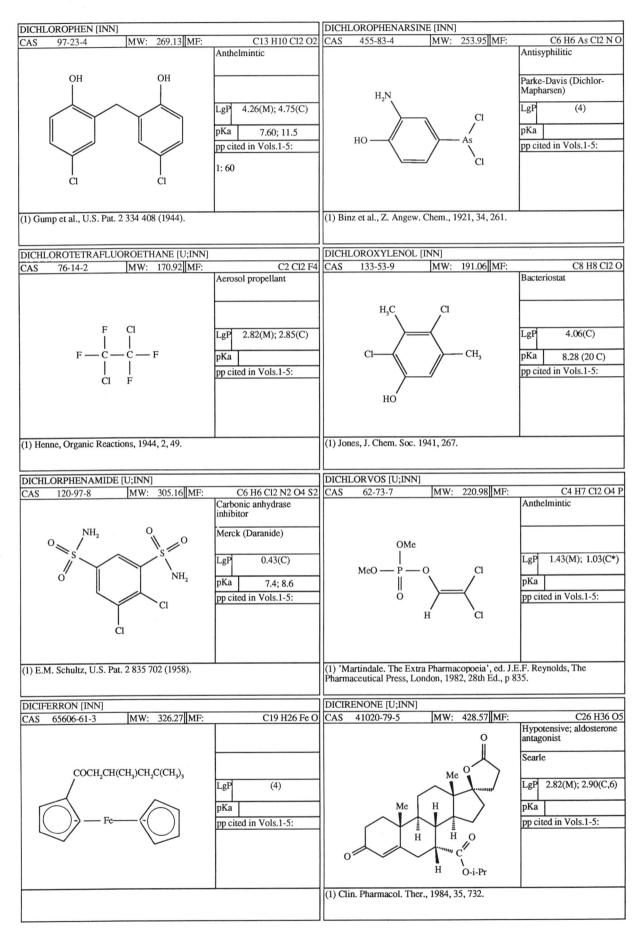

DICHLOROPHEN [INN]

CAS 97-23-4 | MW: 269.13 | MF: C13 H10 Cl2 O2

Anthelmintic

LgP 4.26(M); 4.75(C)

pKa 7.60; 11.5

pp cited in Vols.1-5:

1: 60

(1) Gump et al., U.S. Pat. 2 334 408 (1944).

DICHLOROPHENARSINE [INN]

CAS 455-83-4 | MW: 253.95 | MF: C6 H6 As Cl2 N O

Antisyphilitic

Parke-Davis (Dichlor-Mapharsen)

LgP (4)

pKa

pp cited in Vols.1-5:

(1) Binz et al., Z. Angew. Chem., 1921, 34, 261.

DICHLOROTETRAFLUOROETHANE [U;INN]

CAS 76-14-2 | MW: 170.92 | MF: C2 Cl2 F4

Aerosol propellant

LgP 2.82(M); 2.85(C)

pKa

pp cited in Vols.1-5:

(1) Henne, Organic Reactions, 1944, 2, 49.

DICHLOROXYLENOL [INN]

CAS 133-53-9 | MW: 191.06 | MF: C8 H8 Cl2 O

Bacteriostat

LgP 4.06(C)

pKa 8.28 (20 C)

pp cited in Vols.1-5:

(1) Jones, J. Chem. Soc. 1941, 267.

DICHLORPHENAMIDE [U;INN]

CAS 120-97-8 | MW: 305.16 | MF: C6 H6 Cl2 N2 O4 S2

Carbonic anhydrase inhibitor

Merck (Daranide)

LgP 0.43(C)

pKa 7.4; 8.6

pp cited in Vols.1-5:

(1) E.M. Schultz, U.S. Pat. 2 835 702 (1958).

DICHLORVOS [U;INN]

CAS 62-73-7 | MW: 220.98 | MF: C4 H7 Cl2 O4 P

Anthelmintic

LgP 1.43(M); 1.03(C*)

pKa

pp cited in Vols.1-5:

(1) 'Martindale. The Extra Pharmacopoeia', ed. J.E.F. Reynolds, The Pharmaceutical Press, London, 1982, 28th Ed., p 835.

DICIFERRON [INN]

CAS 65606-61-3 | MW: 326.27 | MF: C19 H26 Fe O

COCH2CH(CH3)CH2C(CH3)3

LgP (4)

pKa

pp cited in Vols.1-5:

DICIRENONE [U;INN]

CAS 41020-79-5 | MW: 428.57 | MF: C26 H36 O5

Hypotensive; aldosterone antagonist

Searle

LgP 2.82(M); 2.90(C,6)

pKa

pp cited in Vols.1-5:

(1) Clin. Pharmacol. Ther., 1984, 35, 732.

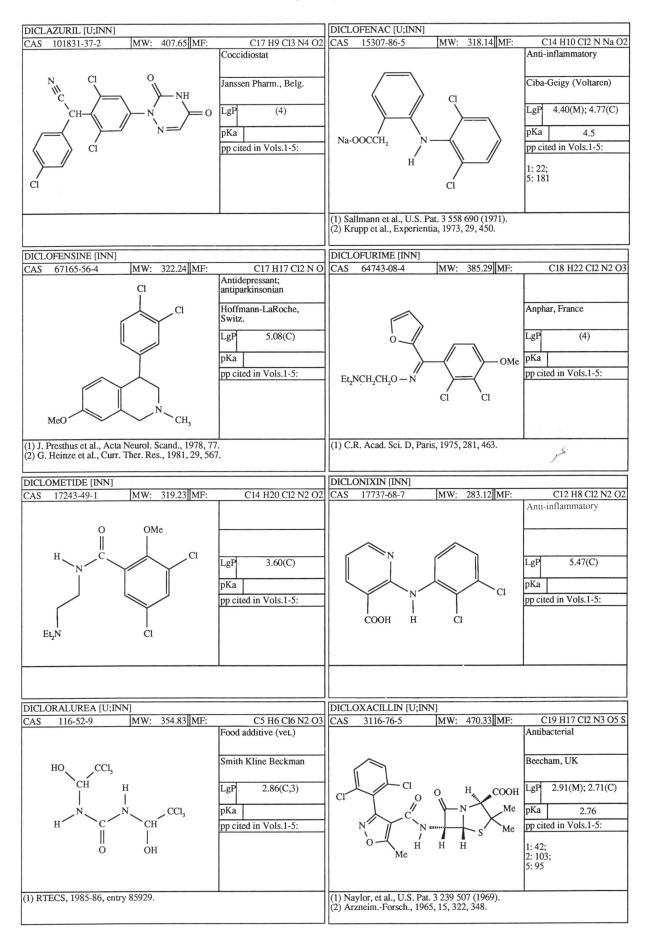

DICLAZURIL [U;INN]

CAS	101831-37-2	MW:	407.65	MF:	C17 H9 Cl3 N4 O2

Coccidiostat

Janssen Pharm., Belg.

LgP	(4)
pKa	
pp cited in Vols.1-5:	

DICLOFENAC [U;INN]

CAS	15307-86-5	MW:	318.14	MF:	C14 H10 Cl2 N Na O2

Anti-inflammatory

Ciba-Geigy (Voltaren)

LgP	4.40(M); 4.77(C)
pKa	4.5
pp cited in Vols.1-5:	

1: 22;
5: 181

(1) Sallmann et al., U.S. Pat. 3 558 690 (1971).
(2) Krupp et al., Experientia, 1973, 29, 450.

DICLOFENSINE [INN]

CAS	67165-56-4	MW:	322.24	MF:	C17 H17 Cl2 N O

Antidepressant; antiparkinsonian

Hoffmann-LaRoche, Switz.

LgP	5.08(C)
pKa	
pp cited in Vols.1-5:	

(1) J. Presthus et al., Acta Neurol. Scand., 1978, 77.
(2) G. Heinze et al., Curr. Ther. Res., 1981, 29, 567.

DICLOFURIME [INN]

CAS	64743-08-4	MW:	385.29	MF:	C18 H22 Cl2 N2 O3

Anphar, France

LgP	(4)
pKa	
pp cited in Vols.1-5:	

(1) C.R. Acad. Sci. D, Paris, 1975, 281, 463.

DICLOMETIDE [INN]

CAS	17243-49-1	MW:	319.23	MF:	C14 H20 Cl2 N2 O2

LgP	3.60(C)
pKa	
pp cited in Vols.1-5:	

DICLONIXIN [INN]

CAS	17737-68-7	MW:	283.12	MF:	C12 H8 Cl2 N2 O2

Anti-inflammatory

LgP	5.47(C)
pKa	
pp cited in Vols.1-5:	

DICLORALUREA [U;INN]

CAS	116-52-9	MW:	354.83	MF:	C5 H6 Cl6 N2 O3

Food additive (vet.)

Smith Kline Beckman

LgP	2.86(C,3)
pKa	
pp cited in Vols.1-5:	

(1) RTECS, 1985-86, entry 85929.

DICLOXACILLIN [U;INN]

CAS	3116-76-5	MW:	470.33	MF:	C19 H17 Cl2 N3 O5 S

Antibacterial

Beecham, UK

LgP	2.91(M); 2.71(C)
pKa	2.76
pp cited in Vols.1-5:	

1: 42;
2: 103;
5: 95

(1) Naylor, et al., U.S. Pat. 3 239 507 (1969).
(2) Arzneim.-Forsch., 1965, 15, 322, 348.

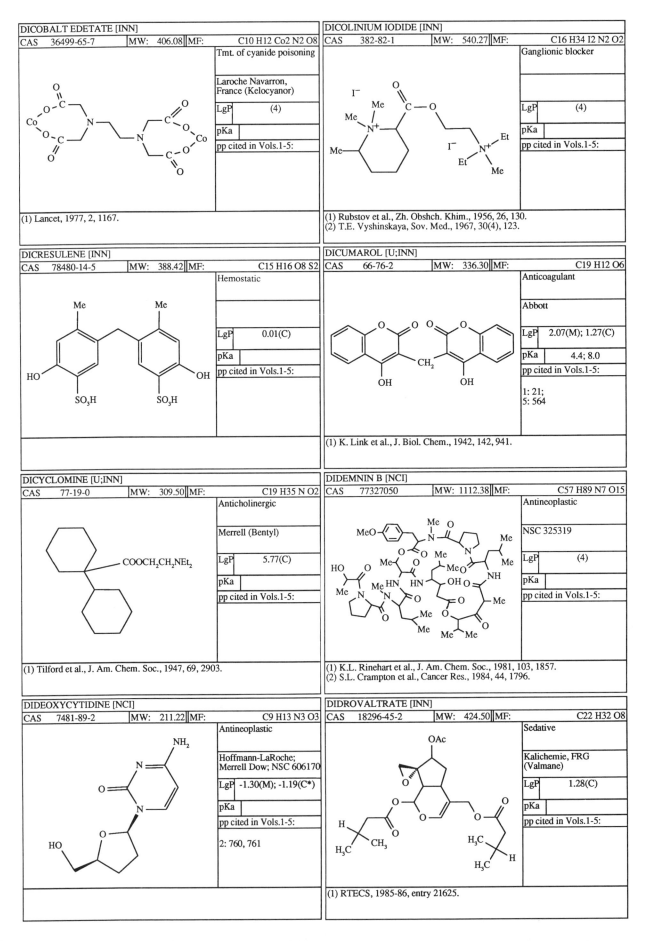

DICOBALT EDETATE [INN]

| CAS | 36499-65-7 | MW: | 406.08 | MF: | C10 H12 Co2 N2 O8 |

Tmt. of cyanide poisoning

Laroche Navarron, France (Kelocyanor)

LgP	(4)
pKa	
pp cited in Vols.1-5:	

(1) Lancet, 1977, 2, 1167.

DICOLINIUM IODIDE [INN]

| CAS | 382-82-1 | MW: | 540.27 | MF: | C16 H34 I2 N2 O2 |

Ganglionic blocker

LgP	(4)
pKa	
pp cited in Vols.1-5:	

(1) Rubstov et al., Zh. Obshch. Khim., 1956, 26, 130.
(2) T.E. Vyshinskaya, Sov. Med., 1967, 30(4), 123.

DICRESULENE [INN]

| CAS | 78480-14-5 | MW: | 388.42 | MF: | C15 H16 O8 S2 |

Hemostatic

LgP	0.01(C)
pKa	
pp cited in Vols.1-5:	

DICUMAROL [U;INN]

| CAS | 66-76-2 | MW: | 336.30 | MF: | C19 H12 O6 |

Anticoagulant

Abbott

LgP	2.07(M); 1.27(C)
pKa	4.4; 8.0
pp cited in Vols.1-5:	

1: 21;
5: 564

(1) K. Link et al., J. Biol. Chem., 1942, 142, 941.

DICYCLOMINE [U;INN]

| CAS | 77-19-0 | MW: | 309.50 | MF: | C19 H35 N O2 |

Anticholinergic

Merrell (Bentyl)

LgP	5.77(C)
pKa	
pp cited in Vols.1-5:	

(1) Tilford et al., J. Am. Chem. Soc., 1947, 69, 2903.

DIDEMNIN B [NCI]

| CAS | 77327050 | MW: | 1112.38 | MF: | C57 H89 N7 O15 |

Antineoplastic

NSC 325319

LgP	(4)
pKa	
pp cited in Vols.1-5:	

(1) K.L. Rinehart et al., J. Am. Chem. Soc., 1981, 103, 1857.
(2) S.L. Crampton et al., Cancer Res., 1984, 44, 1796.

DIDEOXYCYTIDINE [NCI]

| CAS | 7481-89-2 | MW: | 211.22 | MF: | C9 H13 N3 O3 |

Antineoplastic

Hoffmann-LaRoche; Merrell Dow; NSC 606170

LgP	-1.30(M); -1.19(C*)
pKa	
pp cited in Vols.1-5:	

2: 760, 761

DIDROVALTRATE [INN]

| CAS | 18296-45-2 | MW: | 424.50 | MF: | C22 H32 O8 |

Sedative

Kalichemie, FRG (Valmane)

LgP	1.28(C)
pKa	
pp cited in Vols.1-5:	

(1) RTECS, 1985-86, entry 21625.

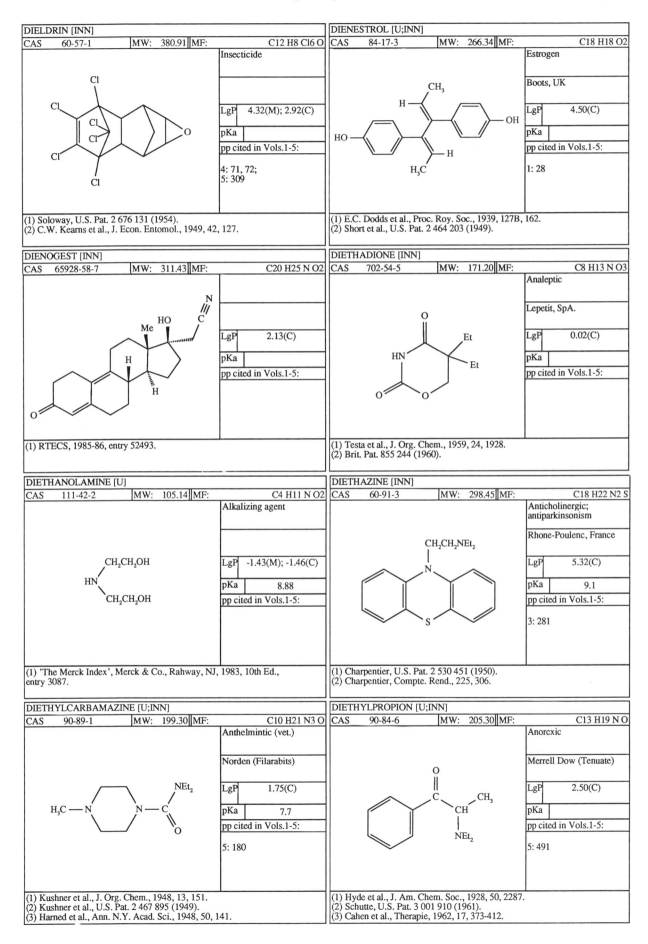

DIELDRIN [INN]		
CAS 60-57-1	MW: 380.91	MF: C12 H8 Cl6 O

Insecticide

LgP 4.32(M); 2.92(C)

pKa

pp cited in Vols.1-5:

4: 71, 72;
5: 309

(1) Soloway, U.S. Pat. 2 676 131 (1954).
(2) C.W. Kearns et al., J. Econ. Entomol., 1949, 42, 127.

DIENESTROL [U;INN]		
CAS 84-17-3	MW: 266.34	MF: C18 H18 O2

Estrogen

Boots, UK

LgP 4.50(C)

pKa

pp cited in Vols.1-5:

1: 28

(1) E.C. Dodds et al., Proc. Roy. Soc., 1939, 127B, 162.
(2) Short et al., U.S. Pat. 2 464 203 (1949).

DIENOGEST [INN]		
CAS 65928-58-7	MW: 311.43	MF: C20 H25 N O2

LgP 2.13(C)

pKa

pp cited in Vols.1-5:

(1) RTECS, 1985-86, entry 52493.

DIETHADIONE [INN]		
CAS 702-54-5	MW: 171.20	MF: C8 H13 N O3

Analeptic

Lepetit, SpA.

LgP 0.02(C)

pKa

pp cited in Vols.1-5:

(1) Testa et al., J. Org. Chem., 1959, 24, 1928.
(2) Brit. Pat. 855 244 (1960).

DIETHANOLAMINE [U]		
CAS 111-42-2	MW: 105.14	MF: C4 H11 N O2

Alkalizing agent

LgP -1.43(M); -1.46(C)

pKa 8.88

pp cited in Vols.1-5:

(1) 'The Merck Index', Merck & Co., Rahway, NJ, 1983, 10th Ed.,
entry 3087.

DIETHAZINE [INN]		
CAS 60-91-3	MW: 298.45	MF: C18 H22 N2 S

Anticholinergic;
antiparkinsonism

Rhone-Poulenc, France

LgP 5.32(C)

pKa 9.1

pp cited in Vols.1-5:

3: 281

(1) Charpentier, U.S. Pat. 2 530 451 (1950).
(2) Charpentier, Compte. Rend., 225, 306.

DIETHYLCARBAMAZINE [U;INN]		
CAS 90-89-1	MW: 199.30	MF: C10 H21 N3 O

Anthelmintic (vet.)

Norden (Filarabits)

LgP 1.75(C)

pKa 7.7

pp cited in Vols.1-5:

5: 180

(1) Kushner et al., J. Org. Chem., 1948, 13, 151.
(2) Kushner et al., U.S. Pat. 2 467 895 (1949).
(3) Harned et al., Ann. N.Y. Acad. Sci., 1948, 50, 141.

DIETHYLPROPION [U;INN]		
CAS 90-84-6	MW: 205.30	MF: C13 H19 N O

Anorexic

Merrell Dow (Tenuate)

LgP 2.50(C)

pKa

pp cited in Vols.1-5:

5: 491

(1) Hyde et al., J. Am. Chem. Soc., 1928, 50, 2287.
(2) Schutte, U.S. Pat. 3 001 910 (1961).
(3) Cahen et al., Therapie, 1962, 17, 373-412.

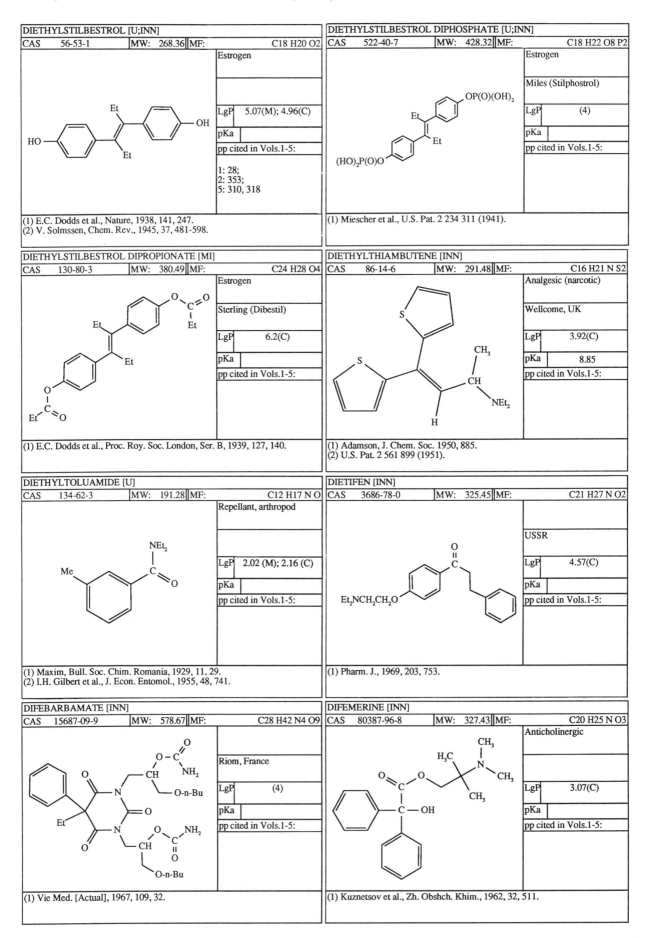

DIETHYLSTILBESTROL [U;INN]

CAS	56-53-1	MW: 268.36	MF:	C18 H20 O2

Estrogen

LgP 5.07(M); 4.96(C)

pKa

pp cited in Vols.1-5:
1: 28;
2: 353;
5: 310, 318

(1) E.C. Dodds et al., Nature, 1938, 141, 247.
(2) V. Solmssen, Chem. Rev., 1945, 37, 481-598.

DIETHYLSTILBESTROL DIPHOSPHATE [U;INN]

CAS	522-40-7	MW: 428.32	MF:	C18 H22 O8 P2

Estrogen

Miles (Stilphostrol)

LgP (4)

pKa

pp cited in Vols.1-5:

(1) Miescher et al., U.S. Pat. 2 234 311 (1941).

DIETHYLSTILBESTROL DIPROPIONATE [MI]

CAS	130-80-3	MW: 380.49	MF:	C24 H28 O4

Estrogen

Sterling (Dibestil)

LgP 6.2(C)

pKa

pp cited in Vols.1-5:

(1) E.C. Dodds et al., Proc. Roy. Soc. London, Ser. B, 1939, 127, 140.

DIETHYLTHIAMBUTENE [INN]

CAS	86-14-6	MW: 291.48	MF:	C16 H21 N S2

Analgesic (narcotic)

Wellcome, UK

LgP 3.92(C)

pKa 8.85

pp cited in Vols.1-5:

(1) Adamson, J. Chem. Soc. 1950, 885.
(2) U.S. Pat. 2 561 899 (1951).

DIETHYLTOLUAMIDE [U]

CAS	134-62-3	MW: 191.28	MF:	C12 H17 N O

Repellant, arthropod

LgP 2.02 (M); 2.16 (C)

pKa

pp cited in Vols.1-5:

(1) Maxim, Bull. Soc. Chim. Romania, 1929, 11, 29.
(2) I.H. Gilbert et al., J. Econ. Entomol., 1955, 48, 741.

DIETIFEN [INN]

CAS	3686-78-0	MW: 325.45	MF:	C21 H27 N O2

USSR

LgP 4.57(C)

pKa

pp cited in Vols.1-5:

(1) Pharm. J., 1969, 203, 753.

DIFEBARBAMATE [INN]

CAS	15687-09-9	MW: 578.67	MF:	C28 H42 N4 O9

Riom, France

LgP (4)

pKa

pp cited in Vols.1-5:

(1) Vie Med. [Actual], 1967, 109, 32.

DIFEMERINE [INN]

CAS	80387-96-8	MW: 327.43	MF:	C20 H25 N O3

Anticholinergic

LgP 3.07(C)

pKa

pp cited in Vols.1-5:

(1) Kuznetsov et al., Zh. Obshch. Khim., 1962, 32, 511.

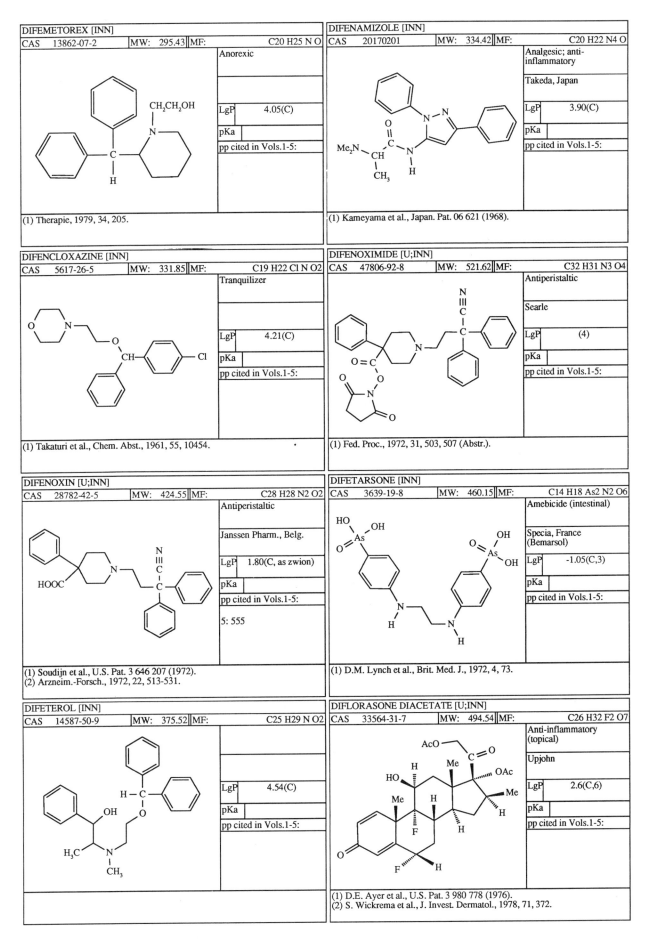

DIFEMETOREX [INN]

CAS 13862-07-2	MW: 295.43	MF:	C20 H25 N O

Anorexic

LgP	4.05(C)
pKa	
pp cited in Vols.1-5:	

(1) Therapie, 1979, 34, 205.

DIFENCLOXAZINE [INN]

CAS 5617-26-5	MW: 331.85	MF:	C19 H22 Cl N O2

Tranquilizer

LgP	4.21(C)
pKa	
pp cited in Vols.1-5:	

(1) Takaturi et al., Chem. Abst., 1961, 55, 10454.

DIFENOXIN [U;INN]

CAS 28782-42-5	MW: 424.55	MF:	C28 H28 N2 O2

Antiperistaltic

Janssen Pharm., Belg.

LgP	1.80(C, as zwion)
pKa	
pp cited in Vols.1-5:	

5: 555

(1) Soudijn et al., U.S. Pat. 3 646 207 (1972).
(2) Arzneim.-Forsch., 1972, 22, 513-531.

DIFETEROL [INN]

CAS 14587-50-9	MW: 375.52	MF:	C25 H29 N O2

LgP	4.54(C)
pKa	
pp cited in Vols.1-5:	

DIFENAMIZOLE [INN]

CAS 20170201	MW: 334.42	MF:	C20 H22 N4 O

Analgesic; anti-inflammatory

Takeda, Japan

LgP	3.90(C)
pKa	
pp cited in Vols.1-5:	

(1) Kameyama et al., Japan. Pat. 06 621 (1968).

DIFENOXIMIDE [U;INN]

CAS 47806-92-8	MW: 521.62	MF:	C32 H31 N3 O4

Antiperistaltic

Searle

LgP	(4)
pKa	
pp cited in Vols.1-5:	

(1) Fed. Proc., 1972, 31, 503, 507 (Abstr.).

DIFETARSONE [INN]

CAS 3639-19-8	MW: 460.15	MF:	C14 H18 As2 N2 O6

Amebicide (intestinal)

Specia, France (Bemarsol)

LgP	-1.05(C,3)
pKa	
pp cited in Vols.1-5:	

(1) D.M. Lynch et al., Brit. Med. J., 1972, 4, 73.

DIFLORASONE DIACETATE [U;INN]

CAS 33564-31-7	MW: 494.54	MF:	C26 H32 F2 O7

Anti-inflammatory (topical)

Upjohn

LgP	2.6(C,6)
pKa	
pp cited in Vols.1-5:	

(1) D.E. Ayer et al., U.S. Pat. 3 980 778 (1976).
(2) S. Wickrema et al., J. Invest. Dermatol., 1978, 71, 372.

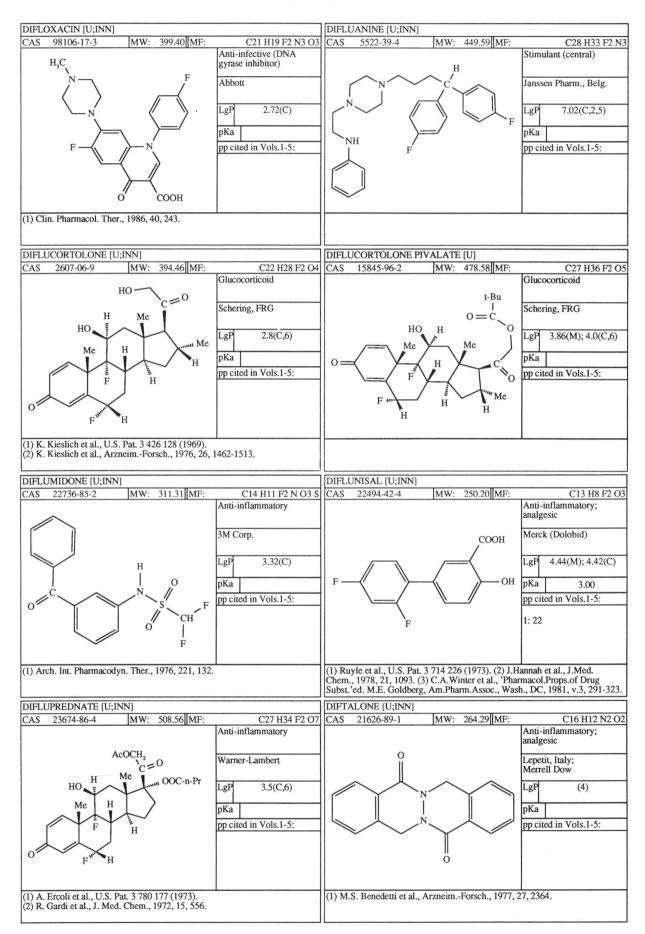

DIFLOXACIN [U;INN]		
CAS 98106-17-3	MW: 399.40	MF: C21 H19 F2 N3 O3

Anti-infective (DNA gyrase inhibitor)

Abbott

LgP	2.72(C)
pKa	
pp cited in Vols.1-5:	

(1) Clin. Pharmacol. Ther., 1986, 40, 243.

DIFLUANINE [U;INN]		
CAS 5522-39-4	MW: 449.59	MF: C28 H33 F2 N3

Stimulant (central)

Janssen Pharm., Belg.

LgP	7.02(C,2,5)
pKa	
pp cited in Vols.1-5:	

DIFLUCORTOLONE [U;INN]		
CAS 2607-06-9	MW: 394.46	MF: C22 H28 F2 O4

Glucocorticoid

Schering, FRG

LgP	2.8(C,6)
pKa	
pp cited in Vols.1-5:	

(1) K. Kieslich et al., U.S. Pat. 3 426 128 (1969).
(2) K. Kieslich et al., Arzneim.-Forsch., 1976, 26, 1462-1513.

DIFLUCORTOLONE PIVALATE [U]		
CAS 15845-96-2	MW: 478.58	MF: C27 H36 F2 O5

Glucocorticoid

Schering, FRG

LgP	3.86(M); 4.0(C,6)
pKa	
pp cited in Vols.1-5:	

DIFLUMIDONE [U;INN]		
CAS 22736-85-2	MW: 311.31	MF: C14 H11 F2 N O3 S

Anti-inflammatory

3M Corp.

LgP	3.32(C)
pKa	
pp cited in Vols.1-5:	

(1) Arch. Int. Pharmacodyn. Ther., 1976, 221, 132.

DIFLUNISAL [U;INN]		
CAS 22494-42-4	MW: 250.20	MF: C13 H8 F2 O3

Anti-inflammatory; analgesic

Merck (Dolobid)

LgP	4.44(M); 4.42(C)
pKa	3.00
pp cited in Vols.1-5:	
1: 22	

(1) Ruyle et al., U.S. Pat. 3 714 226 (1973). (2) J.Hannah et al., J.Med. Chem., 1978, 21, 1093. (3) C.A.Winter et al., 'Pharmacol.Props.of Drug Subst.'ed. M.E. Goldberg, Am.Pharm.Assoc., Wash., DC, 1981, v.3, 291-323.

DIFLUPREDNATE [U;INN]		
CAS 23674-86-4	MW: 508.56	MF: C27 H34 F2 O7

Anti-inflammatory

Warner-Lambert

LgP	3.5(C,6)
pKa	
pp cited in Vols.1-5:	

(1) A. Ercoli et al., U.S. Pat. 3 780 177 (1973).
(2) R. Gardi et al., J. Med. Chem., 1972, 15, 556.

DIFTALONE [U;INN]		
CAS 21626-89-1	MW: 264.29	MF: C16 H12 N2 O2

Anti-inflammatory; analgesic

Lepetit, Italy; Merrell Dow

LgP	(4)
pKa	
pp cited in Vols.1-5:	

(1) M.S. Benedetti et al., Arzneim.-Forsch., 1977, 27, 2364.

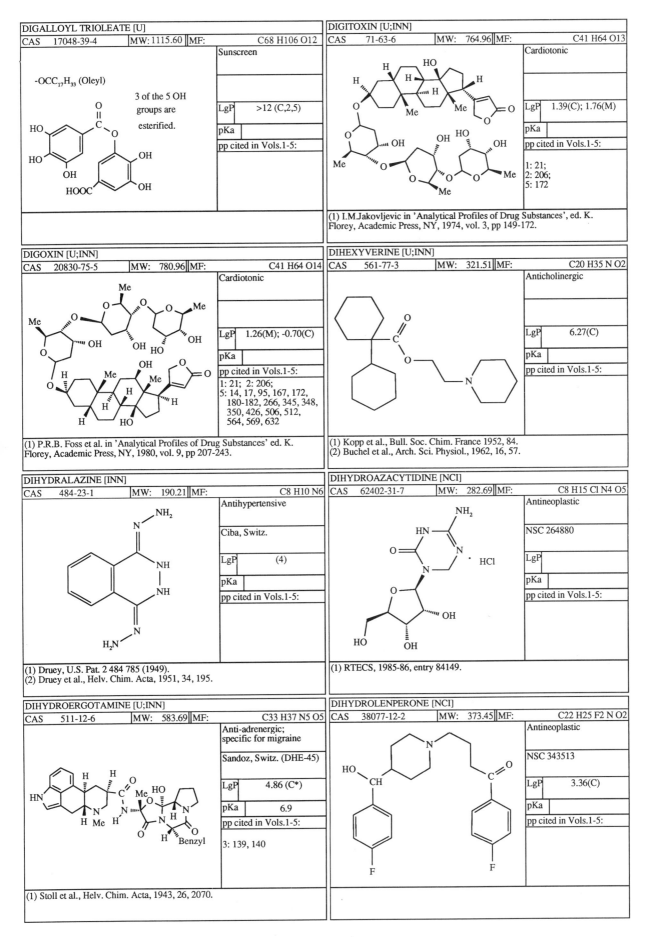

DIGALLOYL TRIOLEATE [U]

CAS	17048-39-4	MW: 1115.60	MF:	C68 H106 O12

Sunscreen

-OCC₁₇H₃₃ (Oleyl)

3 of the 5 OH groups are esterified.

LgP	>12 (C,2,5)
pKa	

pp cited in Vols.1-5:

DIGITOXIN [U;INN]

CAS	71-63-6	MW: 764.96	MF:	C41 H64 O13

Cardiotonic

LgP	1.39(C); 1.76(M)
pKa	

pp cited in Vols.1-5:

1: 21;
2: 206;
5: 172

(1) I.M.Jakovljevic in 'Analytical Profiles of Drug Substances', ed. K. Florey, Academic Press, NY, 1974, vol. 3, pp 149-172.

DIGOXIN [U;INN]

CAS	20830-75-5	MW: 780.96	MF:	C41 H64 O14

Cardiotonic

LgP	1.26(M); -0.70(C)
pKa	

pp cited in Vols.1-5:

1: 21; 2: 206;
5: 14, 17, 95, 167, 172,
180-182, 266, 345, 348,
350, 426, 506, 512,
564, 569, 632

(1) P.R.B. Foss et al. in 'Analytical Profiles of Drug Substances' ed. K. Florey, Academic Press, NY, 1980, vol. 9, pp 207-243.

DIHEXYVERINE [U;INN]

CAS	561-77-3	MW: 321.51	MF:	C20 H35 N O2

Anticholinergic

LgP	6.27(C)
pKa	

pp cited in Vols.1-5:

(1) Kopp et al., Bull. Soc. Chim. France 1952, 84.
(2) Buchel et al., Arch. Sci. Physiol., 1962, 16, 57.

DIHYDRALAZINE [INN]

CAS	484-23-1	MW: 190.21	MF:	C8 H10 N6

Antihypertensive

Ciba, Switz.

LgP	(4)
pKa	

pp cited in Vols.1-5:

(1) Druey, U.S. Pat. 2 484 785 (1949).
(2) Druey et al., Helv. Chim. Acta, 1951, 34, 195.

DIHYDROAZACYTIDINE [NCI]

CAS	62402-31-7	MW: 282.69	MF:	C8 H15 Cl N4 O5

Antineoplastic

NSC 264880

LgP	
pKa	

pp cited in Vols.1-5:

(1) RTECS, 1985-86, entry 84149.

DIHYDROERGOTAMINE [U;INN]

CAS	511-12-6	MW: 583.69	MF:	C33 H37 N5 O5

Anti-adrenergic; specific for migraine

Sandoz, Switz. (DHE-45)

LgP	4.86 (C*)
pKa	6.9

pp cited in Vols.1-5:

3: 139, 140

(1) Stoll et al., Helv. Chim. Acta, 1943, 26, 2070.

DIHYDROLENPERONE [NCI]

CAS	38077-12-2	MW: 373.45	MF:	C22 H25 F2 N O2

Antineoplastic

NSC 343513

LgP	3.36(C)
pKa	

pp cited in Vols.1-5:

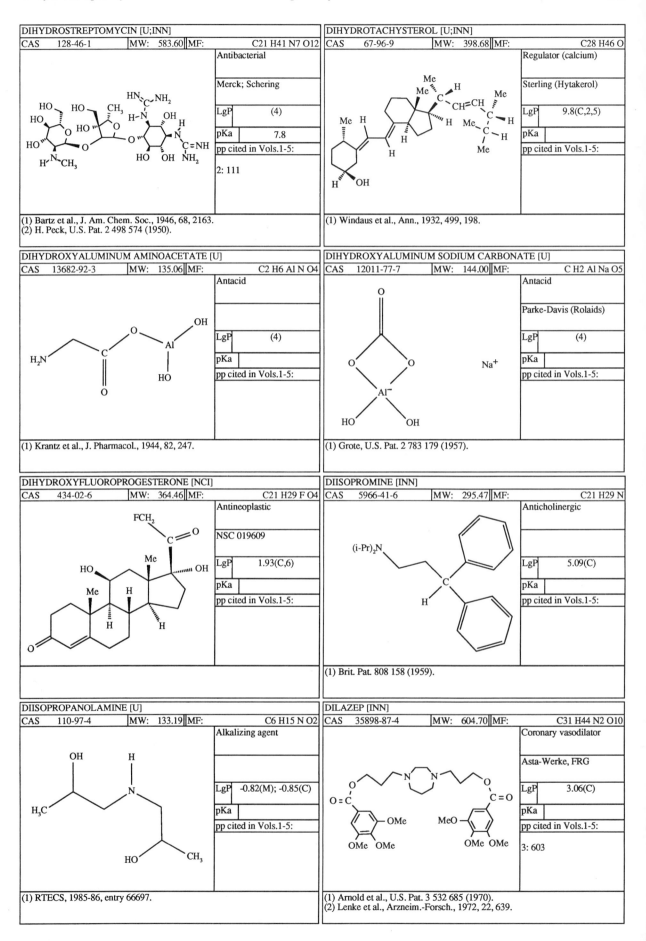

DIHYDROSTREPTOMYCIN [U;INN]

CAS	128-46-1	MW:	583.60	MF:	C21 H41 N7 O12

Antibacterial

Merck; Schering

LgP	(4)
pKa	7.8
pp cited in Vols.1-5:	

2: 111

(1) Bartz et al., J. Am. Chem. Soc., 1946, 68, 2163.
(2) H. Peck, U.S. Pat. 2 498 574 (1950).

DIHYDROTACHYSTEROL [U;INN]

CAS	67-96-9	MW:	398.68	MF:	C28 H46 O

Regulator (calcium)

Sterling (Hytakerol)

LgP	9.8(C,2,5)
pKa	
pp cited in Vols.1-5:	

(1) Windaus et al., Ann., 1932, 499, 198.

DIHYDROXYALUMINUM AMINOACETATE [U]

CAS	13682-92-3	MW:	135.06	MF:	C2 H6 Al N O4

Antacid

LgP	(4)
pKa	
pp cited in Vols.1-5:	

(1) Krantz et al., J. Pharmacol., 1944, 82, 247.

DIHYDROXYALUMINUM SODIUM CARBONATE [U]

CAS	12011-77-7	MW:	144.00	MF:	C H2 Al Na O5

Antacid

Parke-Davis (Rolaids)

LgP	(4)
pKa	
pp cited in Vols.1-5:	

(1) Grote, U.S. Pat. 2 783 179 (1957).

DIHYDROXYFLUOROPROGESTERONE [NCI]

CAS	434-02-6	MW:	364.46	MF:	C21 H29 F O4

Antineoplastic

NSC 019609

LgP	1.93(C,6)
pKa	
pp cited in Vols.1-5:	

DIISOPROMINE [INN]

CAS	5966-41-6	MW:	295.47	MF:	C21 H29 N

Anticholinergic

LgP	5.09(C)
pKa	
pp cited in Vols.1-5:	

(1) Brit. Pat. 808 158 (1959).

DIISOPROPANOLAMINE [U]

CAS	110-97-4	MW:	133.19	MF:	C6 H15 N O2

Alkalizing agent

LgP	-0.82(M); -0.85(C)
pKa	
pp cited in Vols.1-5:	

(1) RTECS, 1985-86, entry 66697.

DILAZEP [INN]

CAS	35898-87-4	MW:	604.70	MF:	C31 H44 N2 O10

Coronary vasodilator

Asta-Werke, FRG

LgP	3.06(C)
pKa	
pp cited in Vols.1-5:	

3: 603

(1) Arnold et al., U.S. Pat. 3 532 685 (1970).
(2) Lenke et al., Arzneim.-Forsch., 1972, 22, 639.

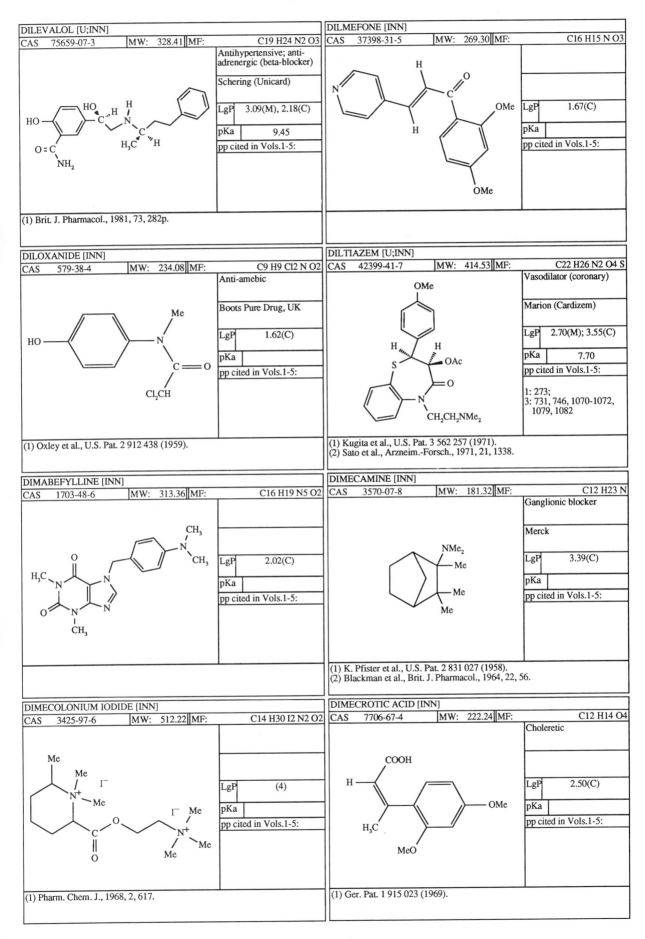

DILEVALOL [U;INN]

CAS	75659-07-3	MW: 328.41	MF:	C19 H24 N2 O3

Antihypertensive; anti-adrenergic (beta-blocker)

Schering (Unicard)

LgP	3.09(M), 2.18(C)
pKa	9.45
pp cited in Vols.1-5:	

(1) Brit. J. Pharmacol., 1981, 73, 282p.

DILMEFONE [INN]

CAS	37398-31-5	MW: 269.30	MF:	C16 H15 N O3

LgP	1.67(C)
pKa	
pp cited in Vols.1-5:	

DILOXANIDE [INN]

CAS	579-38-4	MW: 234.08	MF:	C9 H9 Cl2 N O2

Anti-amebic

Boots Pure Drug, UK

LgP	1.62(C)
pKa	
pp cited in Vols.1-5:	

(1) Oxley et al., U.S. Pat. 2 912 438 (1959).

DILTIAZEM [U;INN]

CAS	42399-41-7	MW: 414.53	MF:	C22 H26 N2 O4 S

Vasodilator (coronary)

Marion (Cardizem)

LgP	2.70(M); 3.55(C)
pKa	7.70
pp cited in Vols.1-5:	

1: 273;
3: 731, 746, 1070-1072, 1079, 1082

(1) Kugita et al., U.S. Pat. 3 562 257 (1971).
(2) Sato et al., Arzneim.-Forsch., 1971, 21, 1338.

DIMABEFYLLINE [INN]

CAS	1703-48-6	MW: 313.36	MF:	C16 H19 N5 O2

LgP	2.02(C)
pKa	
pp cited in Vols.1-5:	

DIMECAMINE [INN]

CAS	3570-07-8	MW: 181.32	MF:	C12 H23 N

Ganglionic blocker

Merck

LgP	3.39(C)
pKa	
pp cited in Vols.1-5:	

(1) K. Pfister et al., U.S. Pat. 2 831 027 (1958).
(2) Blackman et al., Brit. J. Pharmacol., 1964, 22, 56.

DIMECOLONIUM IODIDE [INN]

CAS	3425-97-6	MW: 512.22	MF:	C14 H30 I2 N2 O2

LgP	(4)
pKa	
pp cited in Vols.1-5:	

(1) Pharm. Chem. J., 1968, 2, 617.

DIMECROTIC ACID [INN]

CAS	7706-67-4	MW: 222.24	MF:	C12 H14 O4

Choleretic

LgP	2.50(C)
pKa	
pp cited in Vols.1-5:	

(1) Ger. Pat. 1 915 023 (1969).

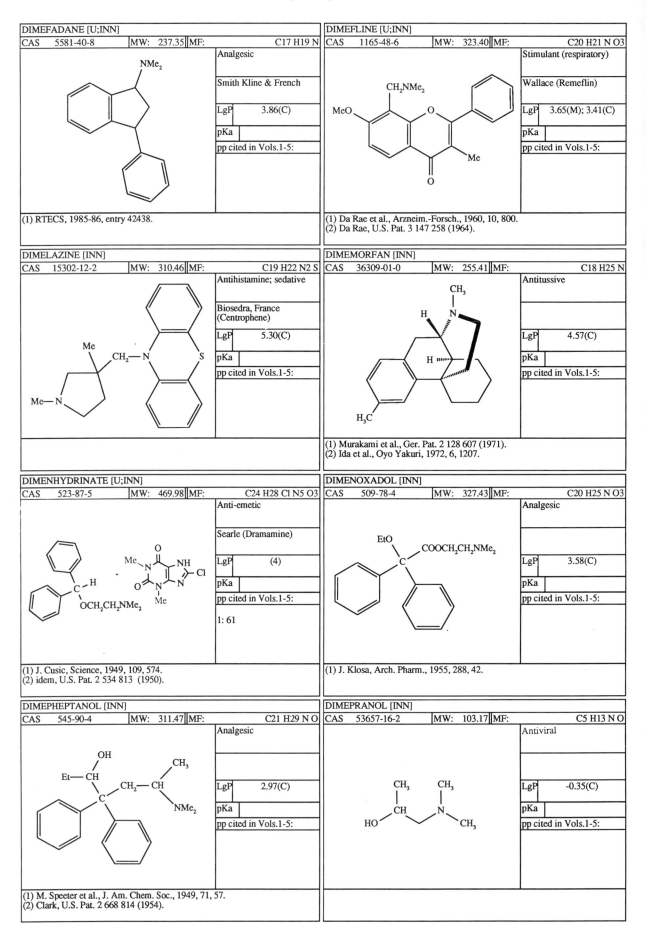

DIMEFADANE [U;INN]	
CAS 5581-40-8　MW: 237.35　MF: C17 H19 N	Analgesic
	Smith Kline & French
	LgP 3.86(C)
	pKa
	pp cited in Vols.1-5:

(1) RTECS, 1985-86, entry 42438.

DIMEFLINE [U;INN]	
CAS 1165-48-6　MW: 323.40　MF: C20 H21 N O3	Stimulant (respiratory)
	Wallace (Remeflin)
	LgP 3.65(M); 3.41(C)
	pKa
	pp cited in Vols.1-5:

(1) Da Rae et al., Arzneim.-Forsch., 1960, 10, 800.
(2) Da Rae, U.S. Pat. 3 147 258 (1964).

DIMELAZINE [INN]	
CAS 15302-12-2　MW: 310.46　MF: C19 H22 N2 S	Antihistamine; sedative
	Biosedra, France (Centrophene)
	LgP 5.30(C)
	pKa
	pp cited in Vols.1-5:

DIMEMORFAN [INN]	
CAS 36309-01-0　MW: 255.41　MF: C18 H25 N	Antitussive
	LgP 4.57(C)
	pKa
	pp cited in Vols.1-5:

(1) Murakami et al., Ger. Pat. 2 128 607 (1971).
(2) Ida et al., Oyo Yakuri, 1972, 6, 1207.

DIMENHYDRINATE [U;INN]	
CAS 523-87-5　MW: 469.98　MF: C24 H28 Cl N5 O3	Anti-emetic
	Searle (Dramamine)
	LgP (4)
	pKa
	pp cited in Vols.1-5:
	1: 61

(1) J. Cusic, Science, 1949, 109, 574.
(2) idem, U.S. Pat. 2 534 813 (1950).

DIMENOXADOL [INN]	
CAS 509-78-4　MW: 327.43　MF: C20 H25 N O3	Analgesic
	LgP 3.58(C)
	pKa
	pp cited in Vols.1-5:

(1) J. Klosa, Arch. Pharm., 1955, 288, 42.

DIMEPHEPTANOL [INN]	
CAS 545-90-4　MW: 311.47　MF: C21 H29 N O	Analgesic
	LgP 2.97(C)
	pKa
	pp cited in Vols.1-5:

(1) M. Speeter et al., J. Am. Chem. Soc., 1949, 71, 57.
(2) Clark, U.S. Pat. 2 668 814 (1954).

DIMEPRANOL [INN]	
CAS 53657-16-2　MW: 103.17　MF: C5 H13 N O	Antiviral
	LgP -0.35(C)
	pKa
	pp cited in Vols.1-5:

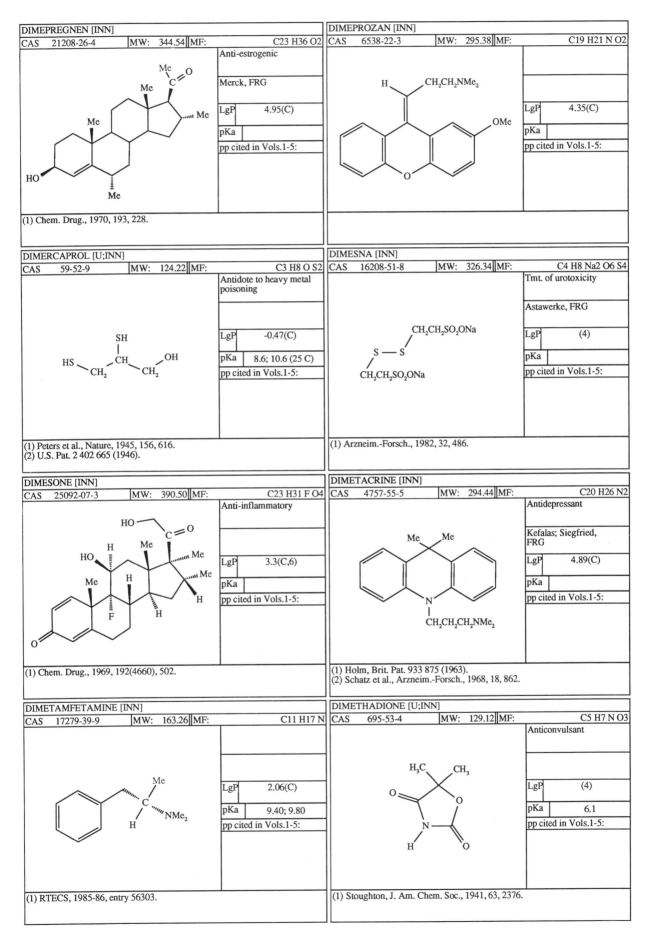

DIMEPREGNEN [INN]

| CAS | 21208-26-4 | MW: | 344.54 | MF: | C23 H36 O2 |

Anti-estrogenic

Merck, FRG

LgP	4.95(C)
pKa	
pp cited in Vols.1-5:	

(1) Chem. Drug., 1970, 193, 228.

DIMEPROZAN [INN]

| CAS | 6538-22-3 | MW: | 295.38 | MF: | C19 H21 N O2 |

LgP	4.35(C)
pKa	
pp cited in Vols.1-5:	

DIMERCAPROL [U;INN]

| CAS | 59-52-9 | MW: | 124.22 | MF: | C3 H8 O S2 |

Antidote to heavy metal poisoning

LgP	-0.47(C)
pKa	8.6; 10.6 (25 C)
pp cited in Vols.1-5:	

(1) Peters et al., Nature, 1945, 156, 616.
(2) U.S. Pat. 2 402 665 (1946).

DIMESNA [INN]

| CAS | 16208-51-8 | MW: | 326.34 | MF: | C4 H8 Na2 O6 S4 |

Tmt. of urotoxicity

Astawerke, FRG

LgP	(4)
pKa	
pp cited in Vols.1-5:	

(1) Arzneim.-Forsch., 1982, 32, 486.

DIMESONE [INN]

| CAS | 25092-07-3 | MW: | 390.50 | MF: | C23 H31 F O4 |

Anti-inflammatory

LgP	3.3(C,6)
pKa	
pp cited in Vols.1-5:	

(1) Chem. Drug., 1969, 192(4660), 502.

DIMETACRINE [INN]

| CAS | 4757-55-5 | MW: | 294.44 | MF: | C20 H26 N2 |

Antidepressant

Kefalas; Siegfried, FRG

LgP	4.89(C)
pKa	
pp cited in Vols.1-5:	

(1) Holm, Brit. Pat. 933 875 (1963).
(2) Schatz et al., Arzneim.-Forsch., 1968, 18, 862.

DIMETAMFETAMINE [INN]

| CAS | 17279-39-9 | MW: | 163.26 | MF: | C11 H17 N |

LgP	2.06(C)
pKa	9.40; 9.80
pp cited in Vols.1-5:	

(1) RTECS, 1985-86, entry 56303.

DIMETHADIONE [U;INN]

| CAS | 695-53-4 | MW: | 129.12 | MF: | C5 H7 N O3 |

Anticonvulsant

LgP	(4)
pKa	6.1
pp cited in Vols.1-5:	

(1) Stoughton, J. Am. Chem. Soc., 1941, 63, 2376.

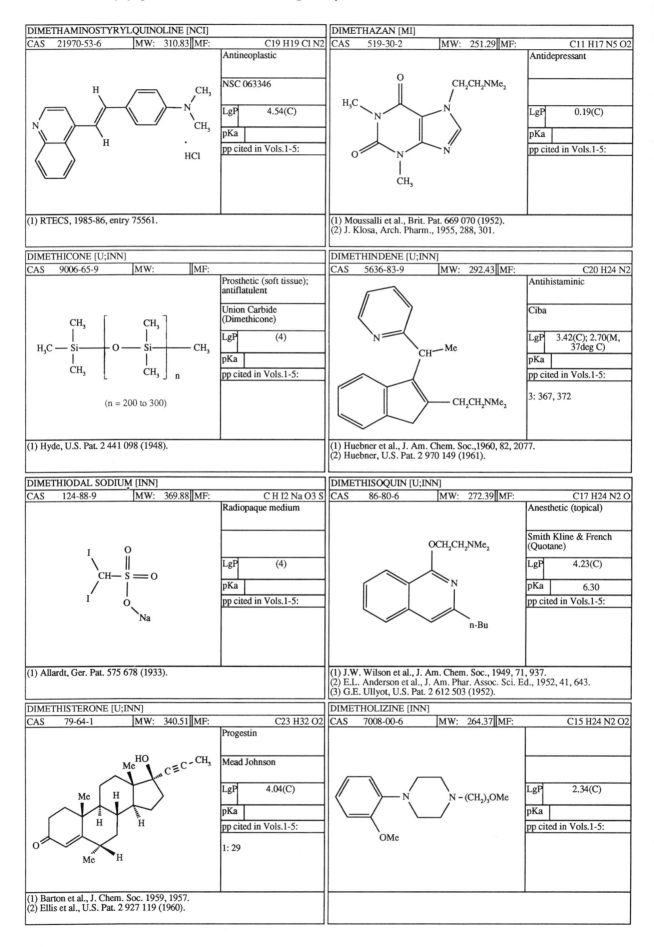

DIMETHAMINOSTYRYLQUINOLINE [NCI]			
CAS 21970-53-6	MW: 310.83	MF:	C19 H19 Cl N2

Antineoplastic

NSC 063346

LgP	4.54(C)

pKa	

pp cited in Vols.1-5:

(1) RTECS, 1985-86, entry 75561.

DIMETHAZAN [MI]			
CAS 519-30-2	MW: 251.29	MF:	C11 H17 N5 O2

Antidepressant

LgP	0.19(C)

pKa	

pp cited in Vols.1-5:

(1) Moussalli et al., Brit. Pat. 669 070 (1952).
(2) J. Klosa, Arch. Pharm., 1955, 288, 301.

DIMETHICONE [U;INN]			
CAS 9006-65-9	MW:	MF:	

Prosthetic (soft tissue);
antiflatulent

Union Carbide
(Dimethicone)

LgP	(4)

pKa	

pp cited in Vols.1-5:

(n = 200 to 300)

(1) Hyde, U.S. Pat. 2 441 098 (1948).

DIMETHINDENE [U;INN]			
CAS 5636-83-9	MW: 292.43	MF:	C20 H24 N2

Antihistaminic

Ciba

LgP	3.42(C); 2.70(M, 37deg C)

pKa	

pp cited in Vols.1-5:

3: 367, 372

(1) Huebner et al., J. Am. Chem. Soc.,1960, 82, 2077.
(2) Huebner, U.S. Pat. 2 970 149 (1961).

DIMETHIODAL SODIUM [INN]			
CAS 124-88-9	MW: 369.88	MF:	C H I2 Na O3 S

Radiopaque medium

LgP	(4)

pKa	

pp cited in Vols.1-5:

(1) Allardt, Ger. Pat. 575 678 (1933).

DIMETHISOQUIN [U;INN]			
CAS 86-80-6	MW: 272.39	MF:	C17 H24 N2 O

Anesthetic (topical)

Smith Kline & French
(Quotane)

LgP	4.23(C)

pKa	6.30

pp cited in Vols.1-5:

(1) J.W. Wilson et al., J. Am. Chem. Soc., 1949, 71, 937.
(2) E.L. Anderson et al., J. Am. Phar. Assoc. Sci. Ed., 1952, 41, 643.
(3) G.E. Ullyot, U.S. Pat. 2 612 503 (1952).

DIMETHISTERONE [U;INN]			
CAS 79-64-1	MW: 340.51	MF:	C23 H32 O2

Progestin

Mead Johnson

LgP	4.04(C)

pKa	

pp cited in Vols.1-5:

1: 29

(1) Barton et al., J. Chem. Soc. 1959, 1957.
(2) Ellis et al., U.S. Pat. 2 927 119 (1960).

DIMETHOLIZINE [INN]			
CAS 7008-00-6	MW: 264.37	MF:	C15 H24 N2 O2

LgP	2.34(C)

pKa	

pp cited in Vols.1-5:

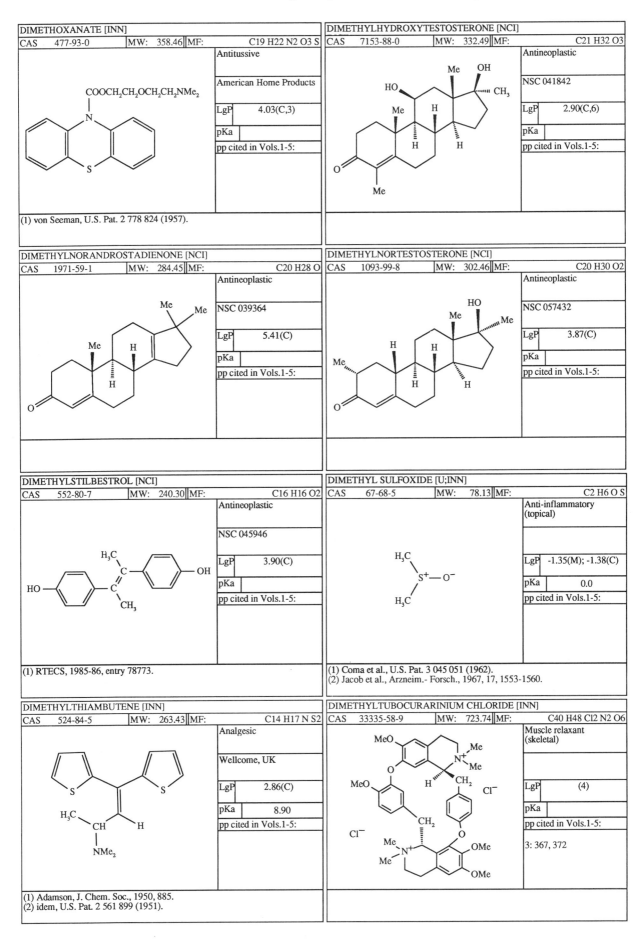

DIMETHOXANATE [INN]

CAS	477-93-0	MW:	358.46	MF:	C19 H22 N2 O3 S

Antitussive

American Home Products

LgP	4.03(C,3)
pKa	

pp cited in Vols.1-5:

(1) von Seeman, U.S. Pat. 2 778 824 (1957).

DIMETHYLHYDROXYTESTOSTERONE [NCI]

CAS	7153-88-0	MW:	332.49	MF:	C21 H32 O3

Antineoplastic

NSC 041842

LgP	2.90(C,6)
pKa	

pp cited in Vols.1-5:

DIMETHYLNORANDROSTADIENONE [NCI]

CAS	1971-59-1	MW:	284.45	MF:	C20 H28 O

Antineoplastic

NSC 039364

LgP	5.41(C)
pKa	

pp cited in Vols.1-5:

DIMETHYLNORTESTOSTERONE [NCI]

CAS	1093-99-8	MW:	302.46	MF:	C20 H30 O2

Antineoplastic

NSC 057432

LgP	3.87(C)
pKa	

pp cited in Vols.1-5:

DIMETHYLSTILBESTROL [NCI]

CAS	552-80-7	MW:	240.30	MF:	C16 H16 O2

Antineoplastic

NSC 045946

LgP	3.90(C)
pKa	

pp cited in Vols.1-5:

(1) RTECS, 1985-86, entry 78773.

DIMETHYL SULFOXIDE [U;INN]

CAS	67-68-5	MW:	78.13	MF:	C2 H6 O S

Anti-inflammatory (topical)

LgP	-1.35(M); -1.38(C)
pKa	0.0

pp cited in Vols.1-5:

(1) Coma et al., U.S. Pat. 3 045 051 (1962).
(2) Jacob et al., Arzneim.- Forsch., 1967, 17, 1553-1560.

DIMETHYLTHIAMBUTENE [INN]

CAS	524-84-5	MW:	263.43	MF:	C14 H17 N S2

Analgesic

Wellcome, UK

LgP	2.86(C)
pKa	8.90

pp cited in Vols.1-5:

(1) Adamson, J. Chem. Soc., 1950, 885.
(2) idem, U.S. Pat. 2 561 899 (1951).

DIMETHYLTUBOCURARINIUM CHLORIDE [INN]

CAS	33335-58-9	MW:	723.74	MF:	C40 H48 Cl2 N2 O6

Muscle relaxant (skeletal)

LgP	(4)
pKa	

pp cited in Vols.1-5:

3: 367, 372

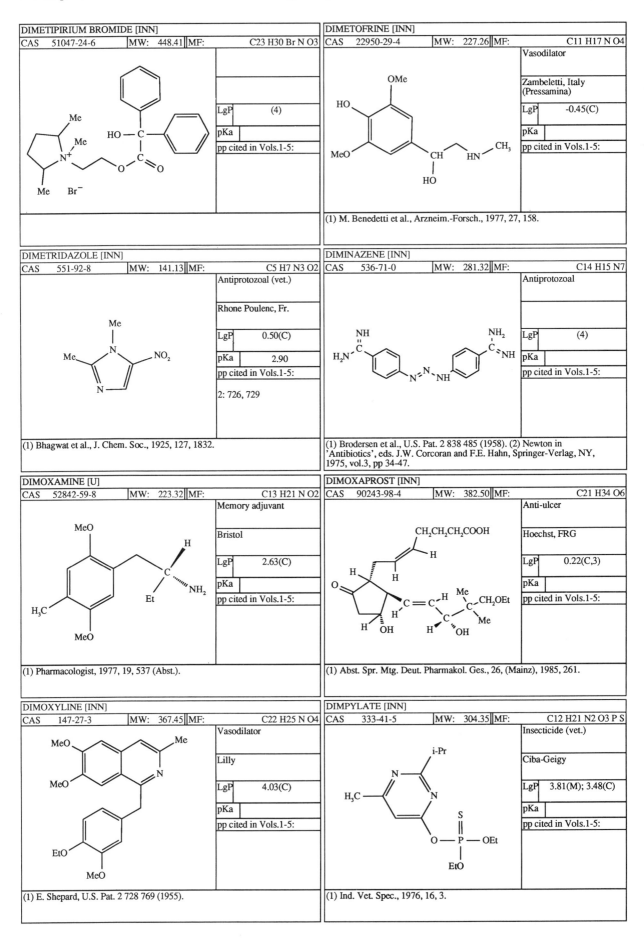

DIMETIPIRIUM BROMIDE [INN]

CAS	51047-24-6	MW: 448.41	MF:	C23 H30 Br N O3

LgP	(4)
pKa	
pp cited in Vols.1-5:	

DIMETOFRINE [INN]

CAS	22950-29-4	MW: 227.26	MF:	C11 H17 N O4

Vasodilator

Zambeletti, Italy
(Pressamina)

LgP	-0.45(C)
pKa	
pp cited in Vols.1-5:	

(1) M. Benedetti et al., Arzneim.-Forsch., 1977, 27, 158.

DIMETRIDAZOLE [INN]

CAS	551-92-8	MW: 141.13	MF:	C5 H7 N3 O2

Antiprotozoal (vet.)

Rhone Poulenc, Fr.

LgP	0.50(C)
pKa	2.90
pp cited in Vols.1-5:	

2: 726, 729

(1) Bhagwat et al., J. Chem. Soc., 1925, 127, 1832.

DIMINAZENE [INN]

CAS	536-71-0	MW: 281.32	MF:	C14 H15 N7

Antiprotozoal

LgP	(4)
pKa	
pp cited in Vols.1-5:	

(1) Brodersen et al., U.S. Pat. 2 838 485 (1958). (2) Newton in
'Antibiotics', eds. J.W. Corcoran and F.E. Hahn, Springer-Verlag, NY,
1975, vol.3, pp 34-47.

DIMOXAMINE [U]

CAS	52842-59-8	MW: 223.32	MF:	C13 H21 N O2

Memory adjuvant

Bristol

LgP	2.63(C)
pKa	
pp cited in Vols.1-5:	

(1) Pharmacologist, 1977, 19, 537 (Abst.).

DIMOXAPROST [INN]

CAS	90243-98-4	MW: 382.50	MF:	C21 H34 O6

Anti-ulcer

Hoechst, FRG

LgP	0.22(C,3)
pKa	
pp cited in Vols.1-5:	

(1) Abst. Spr. Mtg. Deut. Pharmakol. Ges., 26, (Mainz), 1985, 261.

DIMOXYLINE [INN]

CAS	147-27-3	MW: 367.45	MF:	C22 H25 N O4

Vasodilator

Lilly

LgP	4.03(C)
pKa	
pp cited in Vols.1-5:	

(1) E. Shepard, U.S. Pat. 2 728 769 (1955).

DIMPYLATE [INN]

CAS	333-41-5	MW: 304.35	MF:	C12 H21 N2 O3 P S

Insecticide (vet.)

Ciba-Geigy

LgP	3.81(M); 3.48(C)
pKa	
pp cited in Vols.1-5:	

(1) Ind. Vet. Spec., 1976, 16, 3.

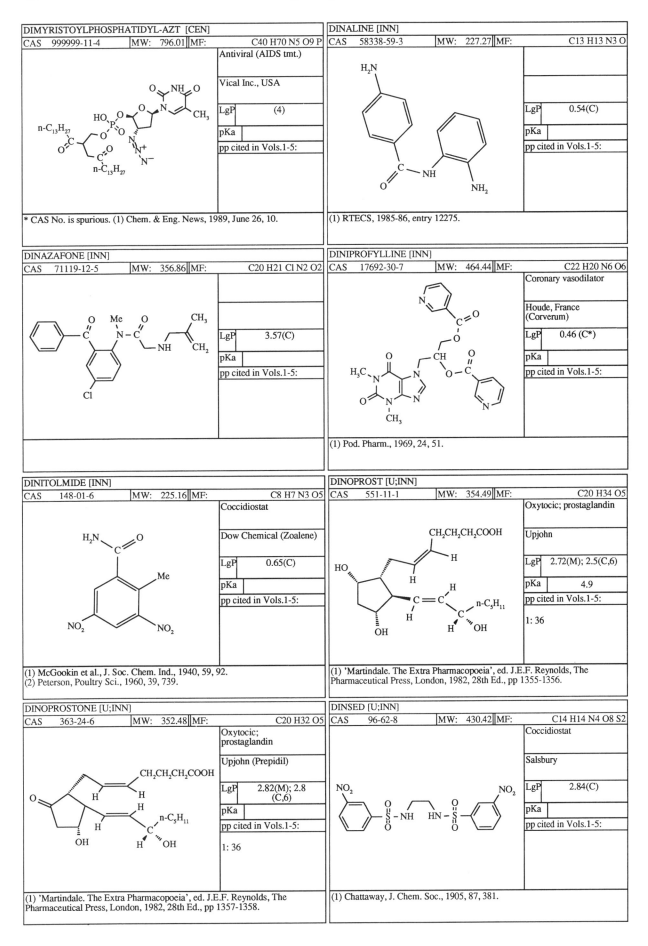

DIMYRISTOYLPHOSPHATIDYL-AZT [CEN]

CAS 999999-11-4	MW: 796.01	MF:	C40 H70 N5 O9 P

Antiviral (AIDS tmt.)

Vical Inc., USA

LgP	(4)
pKa	
pp cited in Vols.1-5:	

* CAS No. is spurious. (1) Chem. & Eng. News, 1989, June 26, 10.

DINALINE [INN]

CAS 58338-59-3	MW: 227.27	MF:	C13 H13 N3 O

LgP	0.54(C)
pKa	
pp cited in Vols.1-5:	

(1) RTECS, 1985-86, entry 12275.

DINAZAFONE [INN]

CAS 71119-12-5	MW: 356.86	MF:	C20 H21 Cl N2 O2

LgP	3.57(C)
pKa	
pp cited in Vols.1-5:	

DINIPROFYLLINE [INN]

CAS 17692-30-7	MW: 464.44	MF:	C22 H20 N6 O6

Coronary vasodilator

Houde, France
(Corverum)

LgP	0.46 (C*)
pKa	
pp cited in Vols.1-5:	

(1) Pod. Pharm., 1969, 24, 51.

DINITOLMIDE [INN]

CAS 148-01-6	MW: 225.16	MF:	C8 H7 N3 O5

Coccidiostat

Dow Chemical (Zoalene)

LgP	0.65(C)
pKa	
pp cited in Vols.1-5:	

(1) McGookin et al., J. Soc. Chem. Ind., 1940, 59, 92.
(2) Peterson, Poultry Sci., 1960, 39, 739.

DINOPROST [U;INN]

CAS 551-11-1	MW: 354.49	MF:	C20 H34 O5

Oxytocic; prostaglandin

Upjohn

LgP	2.72(M); 2.5(C,6)
pKa	4.9
pp cited in Vols.1-5:	

1: 36

(1) 'Martindale. The Extra Pharmacopoeia', ed. J.E.F. Reynolds, The Pharmaceutical Press, London, 1982, 28th Ed., pp 1355-1356.

DINOPROSTONE [U;INN]

CAS 363-24-6	MW: 352.48	MF:	C20 H32 O5

Oxytocic; prostaglandin

Upjohn (Prepidil)

LgP	2.82(M); 2.8 (C,6)
pKa	
pp cited in Vols.1-5:	

1: 36

(1) 'Martindale. The Extra Pharmacopoeia', ed. J.E.F. Reynolds, The Pharmaceutical Press, London, 1982, 28th Ed., pp 1357-1358.

DINSED [U;INN]

CAS 96-62-8	MW: 430.42	MF:	C14 H14 N4 O8 S2

Coccidiostat

Salsbury

LgP	2.84(C)
pKa	
pp cited in Vols.1-5:	

(1) Chattaway, J. Chem. Soc., 1905, 87, 381.

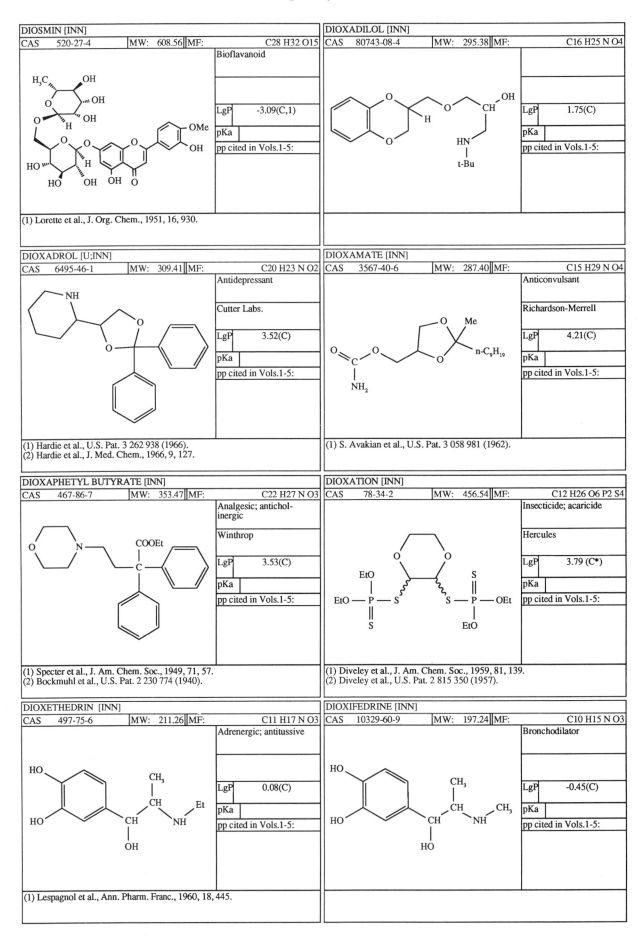

DIOSMIN [INN]

CAS	520-27-4	MW:	608.56	MF:	C28 H32 O15

Bioflavanoid

LgP	-3.09(C,1)
pKa	
pp cited in Vols.1-5:	

(1) Lorette et al., J. Org. Chem., 1951, 16, 930.

DIOXADILOL [INN]

CAS	80743-08-4	MW:	295.38	MF:	C16 H25 N O4

LgP	1.75(C)
pKa	
pp cited in Vols.1-5:	

DIOXADROL [U;INN]

CAS	6495-46-1	MW:	309.41	MF:	C20 H23 N O2

Antidepressant

Cutter Labs.

LgP	3.52(C)
pKa	
pp cited in Vols.1-5:	

(1) Hardie et al., U.S. Pat. 3 262 938 (1966).
(2) Hardie et al., J. Med. Chem., 1966, 9, 127.

DIOXAMATE [INN]

CAS	3567-40-6	MW:	287.40	MF:	C15 H29 N O4

Anticonvulsant

Richardson-Merrell

LgP	4.21(C)
pKa	
pp cited in Vols.1-5:	

(1) S. Avakian et al., U.S. Pat. 3 058 981 (1962).

DIOXAPHETYL BUTYRATE [INN]

CAS	467-86-7	MW:	353.47	MF:	C22 H27 N O3

Analgesic; antichol-inergic

Winthrop

LgP	3.53(C)
pKa	
pp cited in Vols.1-5:	

(1) Specter et al., J. Am. Chem. Soc., 1949, 71, 57.
(2) Bockmuhl et al., U.S. Pat. 2 230 774 (1940).

DIOXATION [INN]

CAS	78-34-2	MW:	456.54	MF:	C12 H26 O6 P2 S4

Insecticide; acaricide

Hercules

LgP	3.79 (C*)
pKa	
pp cited in Vols.1-5:	

(1) Diveley et al., J. Am. Chem. Soc., 1959, 81, 139.
(2) Diveley et al., U.S. Pat. 2 815 350 (1957).

DIOXETHEDRIN [INN]

CAS	497-75-6	MW:	211.26	MF:	C11 H17 N O3

Adrenergic; antitussive

LgP	0.08(C)
pKa	
pp cited in Vols.1-5:	

(1) Lespagnol et al., Ann. Pharm. Franc., 1960, 18, 445.

DIOXIFEDRINE [INN]

CAS	10329-60-9	MW:	197.24	MF:	C10 H15 N O3

Bronchodilator

LgP	-0.45(C)
pKa	
pp cited in Vols.1-5:	

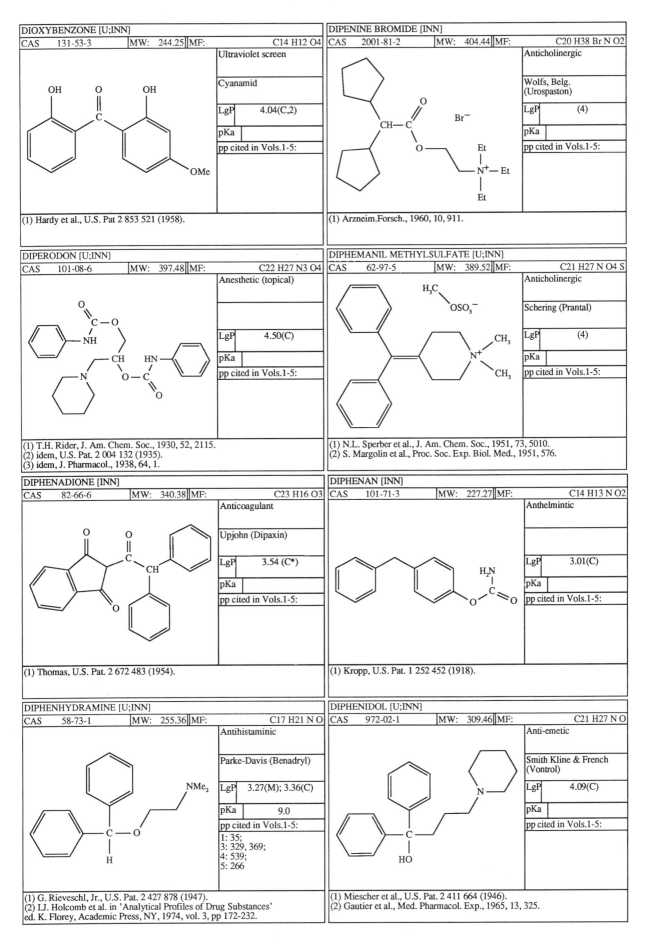

DIOXYBENZONE [U;INN]

CAS	131-53-3	MW:	244.25	MF:	C14 H12 O4

Ultraviolet screen

Cyanamid

LgP	4.04(C,2)
pKa	
pp cited in Vols.1-5:	

(1) Hardy et al., U.S. Pat 2 853 521 (1958).

DIPENINE BROMIDE [INN]

CAS	2001-81-2	MW:	404.44	MF:	C20 H38 Br N O2

Anticholinergic

Wolfs, Belg.
(Urospaston)

LgP	(4)
pKa	
pp cited in Vols.1-5:	

(1) Arzneim.Forsch., 1960, 10, 911.

DIPERODON [U;INN]

CAS	101-08-6	MW:	397.48	MF:	C22 H27 N3 O4

Anesthetic (topical)

LgP	4.50(C)
pKa	
pp cited in Vols.1-5:	

(1) T.H. Rider, J. Am. Chem. Soc., 1930, 52, 2115.
(2) idem, U.S. Pat. 2 004 132 (1935).
(3) idem, J. Pharmacol., 1938, 64, 1.

DIPHEMANIL METHYLSULFATE [U;INN]

CAS	62-97-5	MW:	389.52	MF:	C21 H27 N O4 S

Anticholinergic

Schering (Prantal)

LgP	(4)
pKa	
pp cited in Vols.1-5:	

(1) N.L. Sperber et al., J. Am. Chem. Soc., 1951, 73, 5010.
(2) S. Margolin et al., Proc. Soc. Exp. Biol. Med., 1951, 576.

DIPHENADIONE [INN]

CAS	82-66-6	MW:	340.38	MF:	C23 H16 O3

Anticoagulant

Upjohn (Dipaxin)

LgP	3.54 (C*)
pKa	
pp cited in Vols.1-5:	

(1) Thomas, U.S. Pat. 2 672 483 (1954).

DIPHENAN [INN]

CAS	101-71-3	MW:	227.27	MF:	C14 H13 N O2

Anthelmintic

LgP	3.01(C)
pKa	
pp cited in Vols.1-5:	

(1) Kropp, U.S. Pat. 1 252 452 (1918).

DIPHENHYDRAMINE [U;INN]

CAS	58-73-1	MW:	255.36	MF:	C17 H21 N O

Antihistaminic

Parke-Davis (Benadryl)

LgP	3.27(M); 3.36(C)
pKa	9.0
pp cited in Vols.1-5:	

1: 35;
3: 329, 369;
4: 539;
5: 266

(1) G. Rieveschl, Jr., U.S. Pat. 2 427 878 (1947).
(2) I.J. Holcomb et al. in 'Analytical Profiles of Drug Substances'
ed. K. Florey, Academic Press, NY, 1974, vol. 3, pp 172-232.

DIPHENIDOL [U;INN]

CAS	972-02-1	MW:	309.46	MF:	C21 H27 N O

Anti-emetic

Smith Kline & French
(Vontrol)

LgP	4.09(C)
pKa	
pp cited in Vols.1-5:	

(1) Miescher et al., U.S. Pat. 2 411 664 (1946).
(2) Gautier et al., Med. Pharmacol. Exp., 1965, 13, 325.

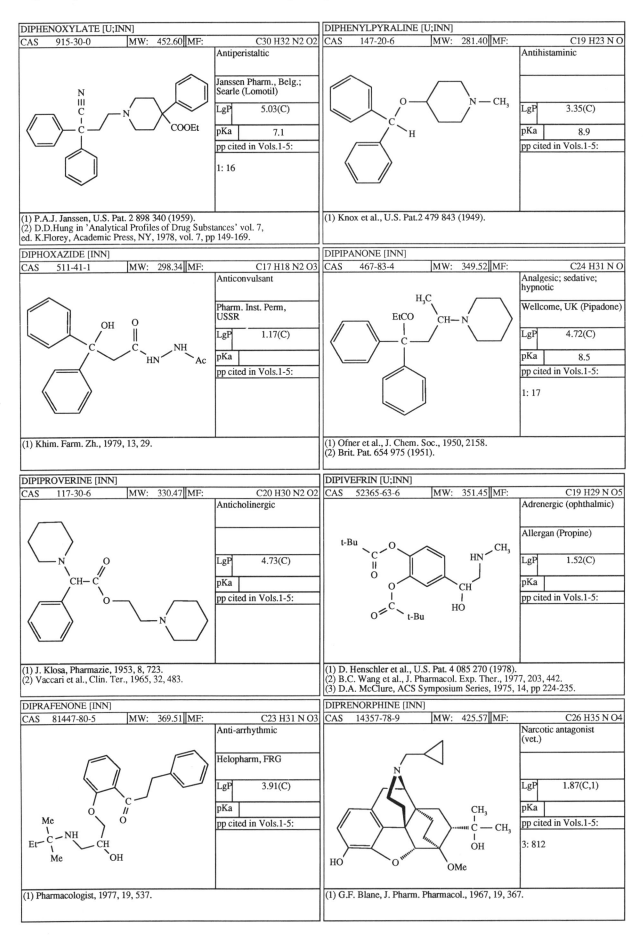

DIPHENOXYLATE [U;INN]

CAS	915-30-0	MW:	452.60	MF:	C30 H32 N2 O2

Antiperistaltic

Janssen Pharm., Belg.; Searle (Lomotil)

LgP	5.03(C)
pKa	7.1

pp cited in Vols.1-5:

1: 16

(1) P.A.J. Janssen, U.S. Pat. 2 898 340 (1959).
(2) D.D.Hung in 'Analytical Profiles of Drug Substances' vol. 7, ed. K.Florey, Academic Press, NY, 1978, vol. 7, pp 149-169.

DIPHENYLPYRALINE [U;INN]

CAS	147-20-6	MW:	281.40	MF:	C19 H23 N O

Antihistaminic

LgP	3.35(C)
pKa	8.9

pp cited in Vols.1-5:

(1) Knox et al., U.S. Pat.2 479 843 (1949).

DIPHOXAZIDE [INN]

CAS	511-41-1	MW:	298.34	MF:	C17 H18 N2 O3

Anticonvulsant

Pharm. Inst. Perm, USSR

LgP	1.17(C)
pKa	

pp cited in Vols.1-5:

(1) Khim. Farm. Zh., 1979, 13, 29.

DIPIPANONE [INN]

CAS	467-83-4	MW:	349.52	MF:	C24 H31 N O

Analgesic; sedative; hypnotic

Wellcome, UK (Pipadone)

LgP	4.72(C)
pKa	8.5

pp cited in Vols.1-5:

1: 17

(1) Ofner et al., J. Chem. Soc., 1950, 2158.
(2) Brit. Pat. 654 975 (1951).

DIPIPROVERINE [INN]

CAS	117-30-6	MW:	330.47	MF:	C20 H30 N2 O2

Anticholinergic

LgP	4.73(C)
pKa	

pp cited in Vols.1-5:

(1) J. Klosa, Pharmazie, 1953, 8, 723.
(2) Vaccari et al., Clin. Ter., 1965, 32, 483.

DIPIVEFRIN [U;INN]

CAS	52365-63-6	MW:	351.45	MF:	C19 H29 N O5

Adrenergic (ophthalmic)

Allergan (Propine)

LgP	1.52(C)
pKa	

pp cited in Vols.1-5:

(1) D. Henschler et al., U.S. Pat. 4 085 270 (1978).
(2) B.C. Wang et al., J. Pharmacol. Exp. Ther., 1977, 203, 442.
(3) D.A. McClure, ACS Symposium Series, 1975, 14, pp 224-235.

DIPRAFENONE [INN]

CAS	81447-80-5	MW:	369.51	MF:	C23 H31 N O3

Anti-arrhythmic

Helopharm, FRG

LgP	3.91(C)
pKa	

pp cited in Vols.1-5:

(1) Pharmacologist, 1977, 19, 537.

DIPRENORPHINE [INN]

CAS	14357-78-9	MW:	425.57	MF:	C26 H35 N O4

Narcotic antagonist (vet.)

LgP	1.87(C,1)
pKa	

pp cited in Vols.1-5:

3: 812

(1) G.F. Blane, J. Pharm. Pharmacol., 1967, 19, 367.

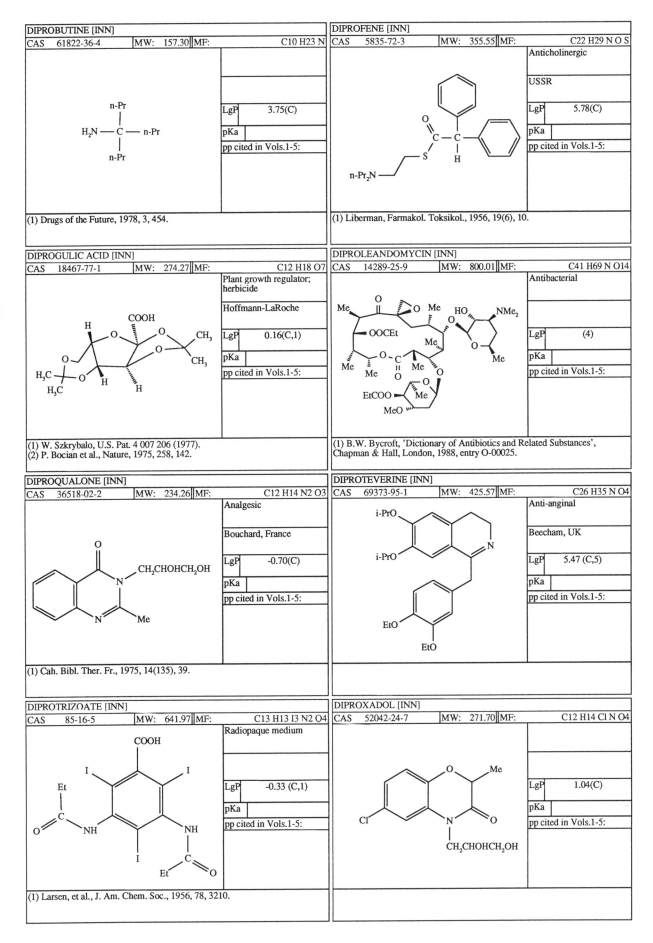

DIPROBUTINE [INN]

CAS 61822-36-4	MW: 157.30	MF:	C10 H23 N
		LgP	3.75(C)
		pKa	
		pp cited in Vols.1-5:	

(1) Drugs of the Future, 1978, 3, 454.

DIPROGULIC ACID [INN]

CAS 18467-77-1	MW: 274.27	MF:	C12 H18 O7
	Plant growth regulator; herbicide		
	Hoffmann-LaRoche		
	LgP	0.16(C,1)	
	pKa		
	pp cited in Vols.1-5:		

(1) W. Szkrybalo, U.S. Pat. 4 007 206 (1977).
(2) P. Bocian et al., Nature, 1975, 258, 142.

DIPROQUALONE [INN]

CAS 36518-02-2	MW: 234.26	MF:	C12 H14 N2 O3
	Analgesic		
	Bouchard, France		
	LgP	-0.70(C)	
	pKa		
	pp cited in Vols.1-5:		

(1) Cah. Bibl. Ther. Fr., 1975, 14(135), 39.

DIPROTRIZOATE [INN]

CAS 85-16-5	MW: 641.97	MF:	C13 H13 I3 N2 O4
	Radiopaque medium		
	LgP	-0.33 (C,1)	
	pKa		
	pp cited in Vols.1-5:		

(1) Larsen, et al., J. Am. Chem. Soc., 1956, 78, 3210.

DIPROFENE [INN]

CAS 5835-72-3	MW: 355.55	MF:	C22 H29 N O S
	Anticholinergic		
	USSR		
	LgP	5.78(C)	
	pKa		
	pp cited in Vols.1-5:		

(1) Liberman, Farmakol. Toksikol., 1956, 19(6), 10.

DIPROLEANDOMYCIN [INN]

CAS 14289-25-9	MW: 800.01	MF:	C41 H69 N O14
	Antibacterial		
	LgP	(4)	
	pKa		
	pp cited in Vols.1-5:		

(1) B.W. Bycroft, 'Dictionary of Antibiotics and Related Substances',
Chapman & Hall, London, 1988, entry O-00025.

DIPROTEVERINE [INN]

CAS 69373-95-1	MW: 425.57	MF:	C26 H35 N O4
	Anti-anginal		
	Beecham, UK		
	LgP	5.47 (C,5)	
	pKa		
	pp cited in Vols.1-5:		

DIPROXADOL [INN]

CAS 52042-24-7	MW: 271.70	MF:	C12 H14 Cl N O4
		LgP	1.04(C)
		pKa	
		pp cited in Vols.1-5:	

DIPYRIDAMOLE [U;INN]

CAS	58-32-2	MW: 504.64	MF:	C24 H40 N8 O4

Vasodilator (coronary)

Thomae, FRG; Boehringer-Ingel., FRG (Persantin)

LgP	2.13(C,3)
pKa	6.4
pp cited in Vols.1-5:	

3: 603

(1) Brit. Pat. 807 826 (1959).
(2) Saraf et al., Indian J. Physiol. Pharmacol., 1971, 15, 135.

DIPYRITHIONE [U;INN]

CAS	3696-28-4	MW: 252.32	MF:	C10 H8 N2 O2 S2

Antibacterial; antifungal

LgP	-3.19(C,1)
pKa	
pp cited in Vols.1-5:	

(1) Shaw et al., J. Am. Chem. Soc., 1950, 72, 4362.

DIPYROCETYL [INN]

CAS	486-79-3	MW: 238.20	MF:	C11 H10 O6

Platelet aggregation inhibitor

U.C.B., Belg. (Morivine)

LgP	0.12(C)
pKa	
pp cited in Vols.1-5:	

(1) 'Agents & Actions, Suppl.1 (Aspirin & Uses), Birkhauser, Basel, 1977, 54.

DIPYRONE [U;INN]

CAS	68-89-3	MW: 333.34	MF:	C13 H16 N3 Na O4 S

Analgesic; antipyretic

Hoechst,FRG; Farmitalia, Italy (Diprofarn)

LgP	(4)
pKa	
pp cited in Vols.1-5:	

(1) Ger. Pat. 259 577 (1911).

DIRITHROMYCIN [INN]

CAS	62013-04-1	MW: 835.09	MF:	C42 H78 N2 O14

Antibacterial

Thomae, FRG

LgP	(4)
pKa	
pp cited in Vols.1-5:	

(1) Therapie, 1987, 42, 263.

DISIQUONIUM CHLORIDE [U;INN]

CAS	68959-20-6	MW: 510.32	MF:	C27 H60 Cl N O3 Si

Antiseptic

Sanitized

LgP	(4)
pKa	
pp cited in Vols.1-5:	

(1) Clin. Pharmacol. Ther., 1987, 41, 690.

DISOBUTAMIDE [U;INN]

CAS	68284-69-5	MW: 408.03	MF:	C23 H38 Cl N3 O

Cardiac depressant (anti-arrhythmic)

Searle

LgP	4.78 (M); 3.21 (C)
pKa	8.2; 9.7
pp cited in Vols.1-5:	

(1) Fed. Proc., 1980, 39, 1431 (Abst.).

DISOFENIN [U;INN]

CAS	65717-97-7	MW: 350.42	MF:	C18 H26 N2 O5

Carrier agent

N. E. Nuclear

LgP	1.88 (C*, zwion)
pKa	
pp cited in Vols.1-5:	

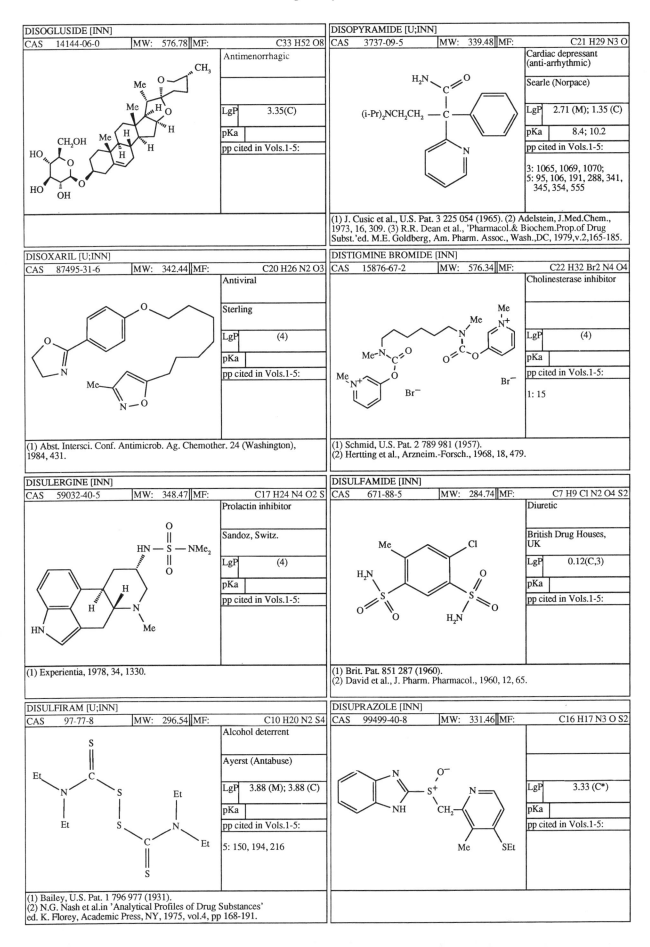

DISOGLUSIDE [INN]

CAS	14144-06-0	MW:	576.78	MF:	C33 H52 O8

Antimenorrhagic

LgP	3.35(C)
pKa	
pp cited in Vols.1-5:	

DISOPYRAMIDE [U;INN]

CAS	3737-09-5	MW:	339.48	MF:	C21 H29 N3 O

Cardiac depressant (anti-arrhythmic)

Searle (Norpace)

LgP	2.71 (M); 1.35 (C)
pKa	8.4; 10.2
pp cited in Vols.1-5:	

3: 1065, 1069, 1070;
5: 95, 106, 191, 288, 341, 345, 354, 555

(1) J. Cusic et al., U.S. Pat. 3 225 054 (1965). (2) Adelstein, J.Med.Chem., 1973, 16, 309. (3) R.R. Dean et al., 'Pharmacol.& Biochem.Prop.of Drug Subst.'ed. M.E. Goldberg, Am. Pharm. Assoc., Wash.,DC, 1979,v.2,165-185.

DISOXARIL [U;INN]

CAS	87495-31-6	MW:	342.44	MF:	C20 H26 N2 O3

Antiviral

Sterling

LgP	(4)
pKa	
pp cited in Vols.1-5:	

(1) Abst. Intersci. Conf. Antimicrob. Ag. Chemother. 24 (Washington), 1984, 431.

DISTIGMINE BROMIDE [INN]

CAS	15876-67-2	MW:	576.34	MF:	C22 H32 Br2 N4 O4

Cholinesterase inhibitor

LgP	(4)
pKa	
pp cited in Vols.1-5:	

1: 15

(1) Schmid, U.S. Pat. 2 789 981 (1957).
(2) Hertting et al., Arzneim.-Forsch., 1968, 18, 479.

DISULERGINE [INN]

CAS	59032-40-5	MW:	348.47	MF:	C17 H24 N4 O2 S

Prolactin inhibitor

Sandoz, Switz.

LgP	(4)
pKa	
pp cited in Vols.1-5:	

(1) Experientia, 1978, 34, 1330.

DISULFAMIDE [INN]

CAS	671-88-5	MW:	284.74	MF:	C7 H9 Cl N2 O4 S2

Diuretic

British Drug Houses, UK

LgP	0.12(C,3)
pKa	
pp cited in Vols.1-5:	

(1) Brit. Pat. 851 287 (1960).
(2) David et al., J. Pharm. Pharmacol., 1960, 12, 65.

DISULFIRAM [U;INN]

CAS	97-77-8	MW:	296.54	MF:	C10 H20 N2 S4

Alcohol deterrent

Ayerst (Antabuse)

LgP	3.88 (M); 3.88 (C)
pKa	
pp cited in Vols.1-5:	

5: 150, 194, 216

(1) Bailey, U.S. Pat. 1 796 977 (1931).
(2) N.G. Nash et al.in 'Analytical Profiles of Drug Substances' ed. K. Florey, Academic Press, NY, 1975, vol.4, pp 168-191.

DISUPRAZOLE [INN]

CAS	99499-40-8	MW:	331.46	MF:	C16 H17 N3 O S2

LgP	3.33 (C*)
pKa	
pp cited in Vols.1-5:	

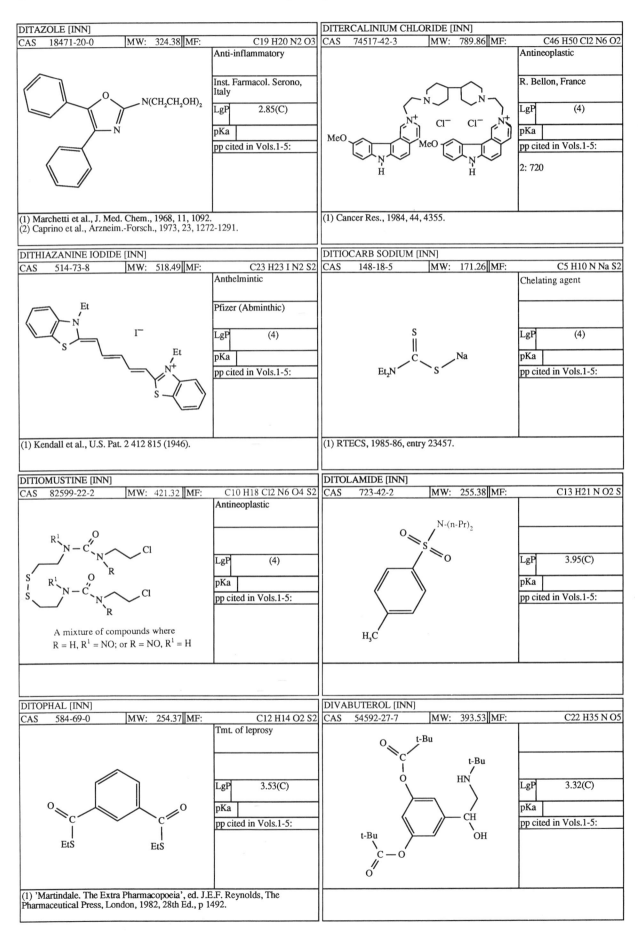

DITAZOLE [INN]

CAS	18471-20-0	MW:	324.38	MF:	C19 H20 N2 O3

Anti-inflammatory

Inst. Farmacol. Serono, Italy

LgP	2.85(C)
pKa	
pp cited in Vols.1-5:	

(1) Marchetti et al., J. Med. Chem., 1968, 11, 1092.
(2) Caprino et al., Arzneim.-Forsch., 1973, 23, 1272-1291.

DITERCALINIUM CHLORIDE [INN]

CAS	74517-42-3	MW:	789.86	MF:	C46 H50 Cl2 N6 O2

Antineoplastic

R. Bellon, France

LgP	(4)
pKa	
pp cited in Vols.1-5:	

2: 720

(1) Cancer Res., 1984, 44, 4355.

DITHIAZANINE IODIDE [INN]

CAS	514-73-8	MW:	518.49	MF:	C23 H23 I N2 S2

Anthelmintic

Pfizer (Abminthic)

LgP	(4)
pKa	
pp cited in Vols.1-5:	

(1) Kendall et al., U.S. Pat. 2 412 815 (1946).

DITIOCARB SODIUM [INN]

CAS	148-18-5	MW:	171.26	MF:	C5 H10 N Na S2

Chelating agent

LgP	(4)
pKa	
pp cited in Vols.1-5:	

(1) RTECS, 1985-86, entry 23457.

DITIOMUSTINE [INN]

CAS	82599-22-2	MW:	421.32	MF:	C10 H18 Cl2 N6 O4 S2

Antineoplastic

LgP	(4)
pKa	
pp cited in Vols.1-5:	

A mixture of compounds where
R = H, R¹ = NO; or R = NO, R¹ = H

DITOLAMIDE [INN]

CAS	723-42-2	MW:	255.38	MF:	C13 H21 N O2 S

LgP	3.95(C)
pKa	
pp cited in Vols.1-5:	

DITOPHAL [INN]

CAS	584-69-0	MW:	254.37	MF:	C12 H14 O2 S2

Tmt. of leprosy

LgP	3.53(C)
pKa	
pp cited in Vols.1-5:	

(1) 'Martindale. The Extra Pharmacopoeia', ed. J.E.F. Reynolds, The Pharmaceutical Press, London, 1982, 28th Ed., p 1492.

DIVABUTEROL [INN]

CAS	54592-27-7	MW:	393.53	MF:	C22 H35 N O5

LgP	3.32(C)
pKa	
pp cited in Vols.1-5:	

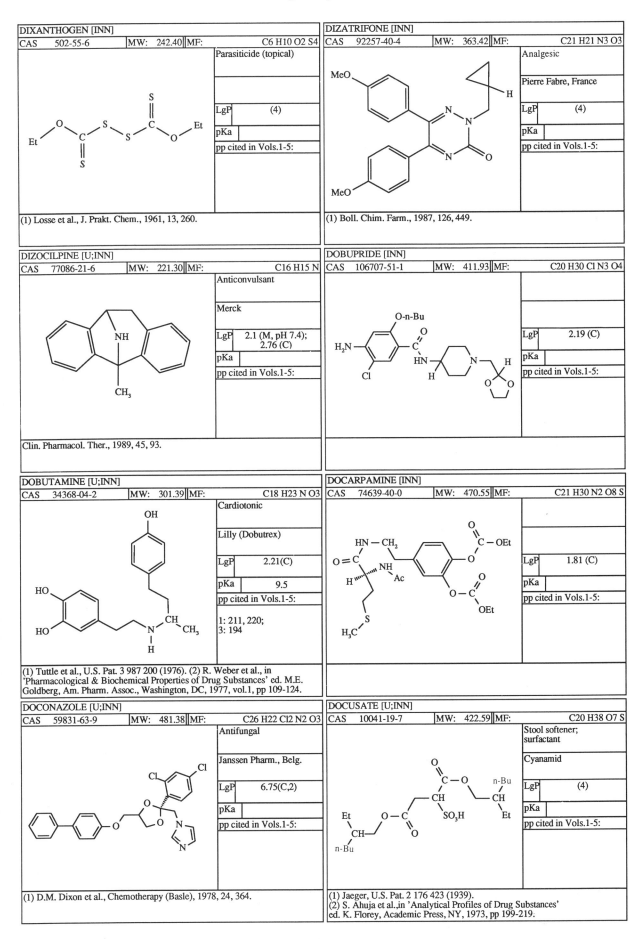

DIXANTHOGEN [INN]

CAS 502-55-6	MW: 242.40	MF: C6 H10 O2 S4

Parasiticide (topical)

LgP (4)

pKa

pp cited in Vols.1-5:

(1) Losse et al., J. Prakt. Chem., 1961, 13, 260.

DIZOCILPINE [U;INN]

CAS 77086-21-6	MW: 221.30	MF: C16 H15 N

Anticonvulsant

Merck

LgP 2.1 (M, pH 7.4); 2.76 (C)

pKa

pp cited in Vols.1-5:

Clin. Pharmacol. Ther., 1989, 45, 93.

DOBUTAMINE [U;INN]

CAS 34368-04-2	MW: 301.39	MF: C18 H23 N O3

Cardiotonic

Lilly (Dobutrex)

LgP 2.21(C)

pKa 9.5

pp cited in Vols.1-5:

1: 211, 220;
3: 194

(1) Tuttle et al., U.S. Pat. 3 987 200 (1976). (2) R. Weber et al., in 'Pharmacological & Biochemical Properties of Drug Substances' ed. M.E. Goldberg, Am. Pharm. Assoc., Washington, DC, 1977, vol.1, pp 109-124.

DOCONAZOLE [U;INN]

CAS 59831-63-9	MW: 481.38	MF: C26 H22 Cl2 N2 O3

Antifungal

Janssen Pharm., Belg.

LgP 6.75(C,2)

pKa

pp cited in Vols.1-5:

(1) D.M. Dixon et al., Chemotherapy (Basle), 1978, 24, 364.

DIZATRIFONE [INN]

CAS 92257-40-4	MW: 363.42	MF: C21 H21 N3 O3

Analgesic

Pierre Fabre, France

LgP (4)

pKa

pp cited in Vols.1-5:

(1) Boll. Chim. Farm., 1987, 126, 449.

DOBUPRIDE [INN]

CAS 106707-51-1	MW: 411.93	MF: C20 H30 Cl N3 O4

LgP 2.19 (C)

pKa

pp cited in Vols.1-5:

DOCARPAMINE [INN]

CAS 74639-40-0	MW: 470.55	MF: C21 H30 N2 O8 S

LgP 1.81 (C)

pKa

pp cited in Vols.1-5:

DOCUSATE [U;INN]

CAS 10041-19-7	MW: 422.59	MF: C20 H38 O7 S

Stool softener; surfactant

Cyanamid

LgP (4)

pKa

pp cited in Vols.1-5:

(1) Jaeger, U.S. Pat. 2 176 423 (1939).
(2) S. Ahuja et al.,in 'Analytical Profiles of Drug Substances' ed. K. Florey, Academic Press, NY, 1973, pp 199-219.

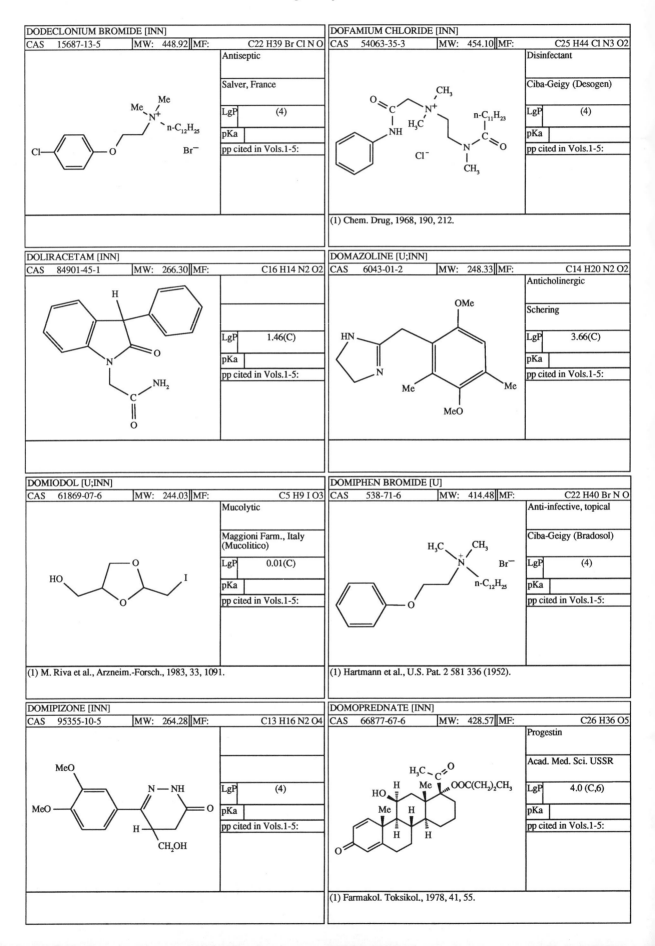

DODECLONIUM BROMIDE [INN]

| CAS | 15687-13-5 | MW: | 448.92 | MF: | C22 H39 Br Cl N O |

Antiseptic

Salver, France

LgP	(4)
pKa	
pp cited in Vols.1-5:	

DOFAMIUM CHLORIDE [INN]

| CAS | 54063-35-3 | MW: | 454.10 | MF: | C25 H44 Cl N3 O2 |

Disinfectant

Ciba-Geigy (Desogen)

LgP	(4)
pKa	
pp cited in Vols.1-5:	

(1) Chem. Drug, 1968, 190, 212.

DOLIRACETAM [INN]

| CAS | 84901-45-1 | MW: | 266.30 | MF: | C16 H14 N2 O2 |

LgP	1.46(C)
pKa	
pp cited in Vols.1-5:	

DOMAZOLINE [U;INN]

| CAS | 6043-01-2 | MW: | 248.33 | MF: | C14 H20 N2 O2 |

Anticholinergic

Schering

LgP	3.66(C)
pKa	
pp cited in Vols.1-5:	

DOMIODOL [U;INN]

| CAS | 61869-07-6 | MW: | 244.03 | MF: | C5 H9 I O3 |

Mucolytic

Maggioni Farm., Italy (Mucolitico)

LgP	0.01(C)
pKa	
pp cited in Vols.1-5:	

(1) M. Riva et al., Arzneim.-Forsch., 1983, 33, 1091.

DOMIPHEN BROMIDE [U]

| CAS | 538-71-6 | MW: | 414.48 | MF: | C22 H40 Br N O |

Anti-infective, topical

Ciba-Geigy (Bradosol)

LgP	(4)
pKa	
pp cited in Vols.1-5:	

(1) Hartmann et al., U.S. Pat. 2 581 336 (1952).

DOMIPIZONE [INN]

| CAS | 95355-10-5 | MW: | 264.28 | MF: | C13 H16 N2 O4 |

LgP	(4)
pKa	
pp cited in Vols.1-5:	

DOMOPREDNATE [INN]

| CAS | 66877-67-6 | MW: | 428.57 | MF: | C26 H36 O5 |

Progestin

Acad. Med. Sci. USSR

LgP	4.0 (C,6)
pKa	
pp cited in Vols.1-5:	

(1) Farmakol. Toksikol., 1978, 41, 55.

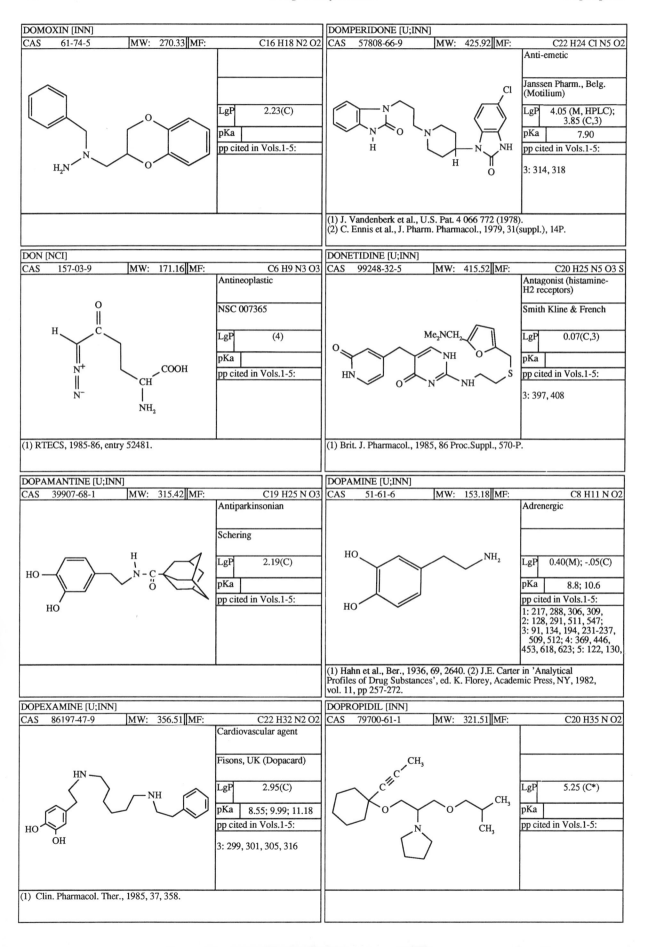

DOMOXIN [INN]

CAS	61-74-5	MW: 270.33	MF:	C16 H18 N2 O2

LgP	2.23(C)
pKa	
pp cited in Vols.1-5:	

DOMPERIDONE [U;INN]

CAS	57808-66-9	MW: 425.92	MF:	C22 H24 Cl N5 O2

Anti-emetic

Janssen Pharm., Belg. (Motilium)

LgP	4.05 (M, HPLC); 3.85 (C,3)
pKa	7.90
pp cited in Vols.1-5:	

3: 314, 318

(1) J. Vandenberk et al., U.S. Pat. 4 066 772 (1978).
(2) C. Ennis et al., J. Pharm. Pharmacol., 1979, 31(suppl.), 14P.

DON [NCI]

CAS	157-03-9	MW: 171.16	MF:	C6 H9 N3 O3

Antineoplastic

NSC 007365

LgP	(4)
pKa	
pp cited in Vols.1-5:	

(1) RTECS, 1985-86, entry 52481.

DONETIDINE [U;INN]

CAS	99248-32-5	MW: 415.52	MF:	C20 H25 N5 O3 S

Antagonist (histamine-H2 receptors)

Smith Kline & French

LgP	0.07(C,3)
pKa	
pp cited in Vols.1-5:	

3: 397, 408

(1) Brit. J. Pharmacol., 1985, 86 Proc.Suppl., 570-P.

DOPAMANTINE [U;INN]

CAS	39907-68-1	MW: 315.42	MF:	C19 H25 N O3

Antiparkinsonian

Schering

LgP	2.19(C)
pKa	
pp cited in Vols.1-5:	

DOPAMINE [U;INN]

CAS	51-61-6	MW: 153.18	MF:	C8 H11 N O2

Adrenergic

LgP	0.40(M); -.05(C)
pKa	8.8; 10.6
pp cited in Vols.1-5:	

1: 217, 288, 306, 309,
2: 128, 291, 511, 547;
3: 91, 134, 194, 231-237,
   509, 512; 4: 369, 446,
453, 618, 623; 5: 122, 130,

(1) Hahn et al., Ber., 1936, 69, 2640. (2) J.E. Carter in 'Analytical Profiles of Drug Substances', ed. K. Florey, Academic Press, NY, 1982, vol. 11, pp 257-272.

DOPEXAMINE [U;INN]

CAS	86197-47-9	MW: 356.51	MF:	C22 H32 N2 O2

Cardiovascular agent

Fisons, UK (Dopacard)

LgP	2.95(C)
pKa	8.55; 9.99; 11.18
pp cited in Vols.1-5:	

3: 299, 301, 305, 316

(1) Clin. Pharmacol. Ther., 1985, 37, 358.

DOPROPIDIL [INN]

CAS	79700-61-1	MW: 321.51	MF:	C20 H35 N O2

LgP	5.25 (C*)
pKa	
pp cited in Vols.1-5:	

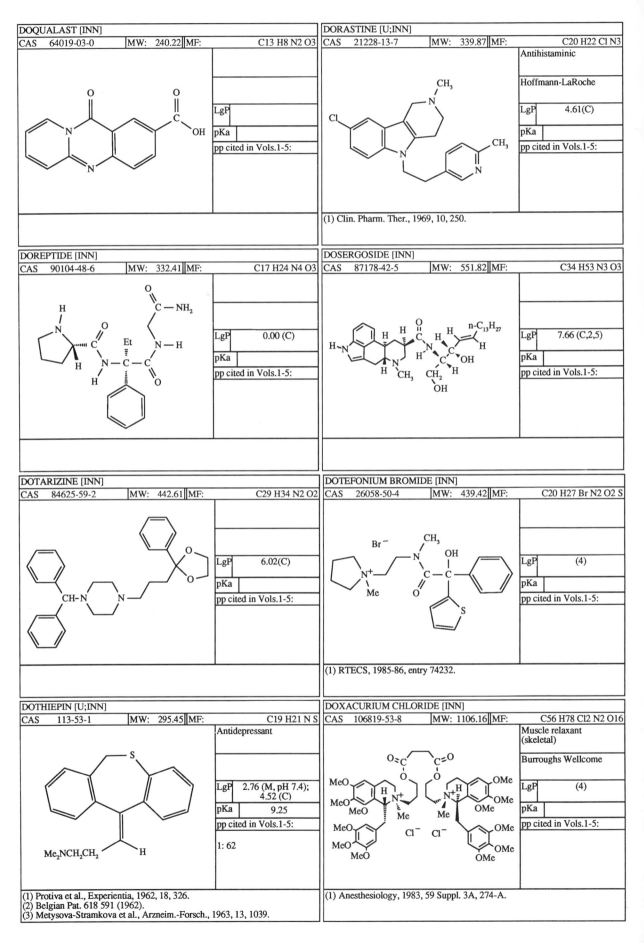

DOQUALAST [INN]

CAS	64019-03-0	MW:	240.22	MF:	C13 H8 N2 O3

LgP	
pKa	
pp cited in Vols.1-5:	

DORASTINE [U;INN]

CAS	21228-13-7	MW:	339.87	MF:	C20 H22 Cl N3

Antihistaminic

Hoffmann-LaRoche

LgP	4.61(C)
pKa	
pp cited in Vols.1-5:	

(1) Clin. Pharm. Ther., 1969, 10, 250.

DOREPTIDE [INN]

CAS	90104-48-6	MW:	332.41	MF:	C17 H24 N4 O3

LgP	0.00 (C)
pKa	
pp cited in Vols.1-5:	

DOSERGOSIDE [INN]

CAS	87178-42-5	MW:	551.82	MF:	C34 H53 N3 O3

LgP	7.66 (C,2,5)
pKa	
pp cited in Vols.1-5:	

DOTARIZINE [INN]

CAS	84625-59-2	MW:	442.61	MF:	C29 H34 N2 O2

LgP	6.02(C)
pKa	
pp cited in Vols.1-5:	

DOTEFONIUM BROMIDE [INN]

CAS	26058-50-4	MW:	439.42	MF:	C20 H27 Br N2 O2 S

LgP	(4)
pKa	
pp cited in Vols.1-5:	

(1) RTECS, 1985-86, entry 74232.

DOTHIEPIN [U;INN]

CAS	113-53-1	MW:	295.45	MF:	C19 H21 N S

Antidepressant

LgP	2.76 (M, pH 7.4); 4.52 (C)
pKa	9.25
pp cited in Vols.1-5:	
1: 62	

(1) Protiva et al., Experientia, 1962, 18, 326.
(2) Belgian Pat. 618 591 (1962).
(3) Metysova-Stramkova et al., Arzneim.-Forsch., 1963, 13, 1039.

DOXACURIUM CHLORIDE [INN]

CAS	106819-53-8	MW:	1106.16	MF:	C56 H78 Cl2 N2 O16

Muscle relaxant (skeletal)

Burroughs Wellcome

LgP	(4)
pKa	
pp cited in Vols.1-5:	

(1) Anesthesiology, 1983, 59 Suppl. 3A, 274-A.

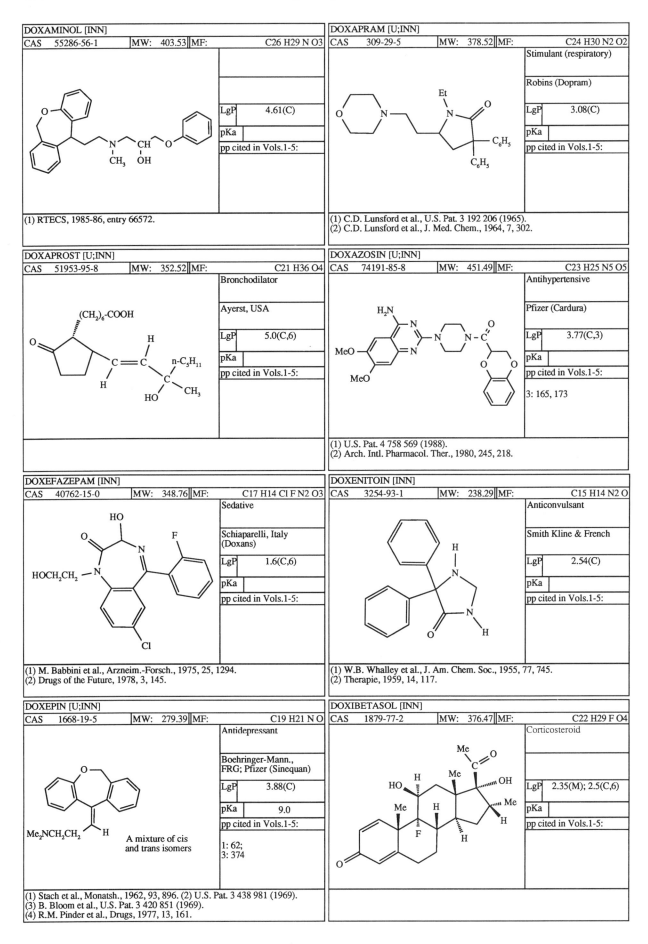

DOXAMINOL [INN]

CAS	55286-56-1	MW:	403.53	MF:	C26 H29 N O3

LgP	4.61(C)
pKa	
pp cited in Vols.1-5:	

(1) RTECS, 1985-86, entry 66572.

DOXAPRAM [U;INN]

CAS	309-29-5	MW:	378.52	MF:	C24 H30 N2 O2

Stimulant (respiratory)

Robins (Dopram)

LgP	3.08(C)
pKa	
pp cited in Vols.1-5:	

(1) C.D. Lunsford et al., U.S. Pat. 3 192 206 (1965).
(2) C.D. Lunsford et al., J. Med. Chem., 1964, 7, 302.

DOXAPROST [U;INN]

CAS	51953-95-8	MW:	352.52	MF:	C21 H36 O4

Bronchodilator

Ayerst, USA

LgP	5.0(C,6)
pKa	
pp cited in Vols.1-5:	

DOXAZOSIN [U;INN]

CAS	74191-85-8	MW:	451.49	MF:	C23 H25 N5 O5

Antihypertensive

Pfizer (Cardura)

LgP	3.77(C,3)
pKa	
pp cited in Vols.1-5:	
3: 165, 173	

(1) U.S. Pat. 4 758 569 (1988).
(2) Arch. Intl. Pharmacol. Ther., 1980, 245, 218.

DOXEFAZEPAM [INN]

CAS	40762-15-0	MW:	348.76	MF:	C17 H14 Cl F N2 O3

Sedative

Schiaparelli, Italy (Doxans)

LgP	1.6(C,6)
pKa	
pp cited in Vols.1-5:	

(1) M. Babbini et al., Arzneim.-Forsch., 1975, 25, 1294.
(2) Drugs of the Future, 1978, 3, 145.

DOXENITOIN [INN]

CAS	3254-93-1	MW:	238.29	MF:	C15 H14 N2 O

Anticonvulsant

Smith Kline & French

LgP	2.54(C)
pKa	
pp cited in Vols.1-5:	

(1) W.B. Whalley et al., J. Am. Chem. Soc., 1955, 77, 745.
(2) Therapie, 1959, 14, 117.

DOXEPIN [U;INN]

CAS	1668-19-5	MW:	279.39	MF:	C19 H21 N O

Antidepressant

Boehringer-Mann., FRG; Pfizer (Sinequan)

LgP	3.88(C)
pKa	9.0
pp cited in Vols.1-5:	
1: 62; 3: 374	

A mixture of cis and trans isomers

(1) Stach et al., Monatsh., 1962, 93, 896. (2) U.S. Pat. 3 438 981 (1969).
(3) B. Bloom et al., U.S. Pat. 3 420 851 (1969).
(4) R.M. Pinder et al., Drugs, 1977, 13, 161.

DOXIBETASOL [INN]

CAS	1879-77-2	MW:	376.47	MF:	C22 H29 F O4

Corticosteroid

LgP	2.35(M); 2.5(C,6)
pKa	
pp cited in Vols.1-5:	

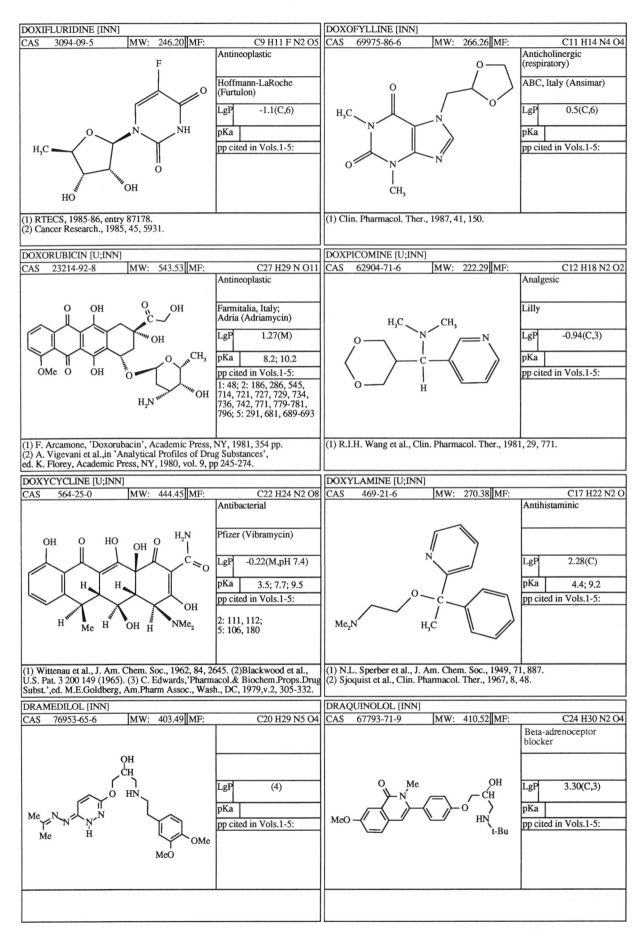

DOXIFLURIDINE [INN]			
CAS 3094-09-5	MW: 246.20	MF:	C9 H11 F N2 O5

Antineoplastic

Hoffmann-LaRoche (Furtulon)

LgP	-1.1(C,6)
pKa	
pp cited in Vols.1-5:	

(1) RTECS, 1985-86, entry 87178.
(2) Cancer Research., 1985, 45, 5931.

DOXOFYLLINE [INN]			
CAS 69975-86-6	MW: 266.26	MF:	C11 H14 N4 O4

Anticholinergic (respiratory)

ABC, Italy (Ansimar)

LgP	0.5(C,6)
pKa	
pp cited in Vols.1-5:	

(1) Clin. Pharmacol. Ther., 1987, 41, 150.

DOXORUBICIN [U;INN]			
CAS 23214-92-8	MW: 543.53	MF:	C27 H29 N O11

Antineoplastic

Farmitalia, Italy; Adria (Adriamycin)

LgP	1.27(M)
pKa	8.2; 10.2
pp cited in Vols.1-5:	1: 48; 2: 186, 286, 545, 714, 721, 727, 729, 734, 736, 742, 771, 779-781, 796; 5: 291, 681, 689-693

(1) F. Arcamone, 'Doxorubacin', Academic Press, NY, 1981, 354 pp.
(2) A. Vigevani et al.,in 'Analytical Profiles of Drug Substances',
ed. K. Florey, Academic Press, NY, 1980, vol. 9, pp 245-274.

DOXPICOMINE [U;INN]			
CAS 62904-71-6	MW: 222.29	MF:	C12 H18 N2 O2

Analgesic

Lilly

LgP	-0.94(C,3)
pKa	
pp cited in Vols.1-5:	

(1) R.I.H. Wang et al., Clin. Pharmacol. Ther., 1981, 29, 771.

DOXYCYCLINE [U;INN]			
CAS 564-25-0	MW: 444.45	MF:	C22 H24 N2 O8

Antibacterial

Pfizer (Vibramycin)

LgP	-0.22(M,pH 7.4)
pKa	3.5; 7.7; 9.5
pp cited in Vols.1-5:	2: 111, 112; 5: 106, 180

(1) Wittenau et al., J. Am. Chem. Soc., 1962, 84, 2645. (2)Blackwood et al.,
U.S. Pat. 3 200 149 (1965). (3) C. Edwards,'Pharmacol.& Biochem.Props.Drug
Subst.',ed. M.E.Goldberg, Am.Pharm Assoc., Wash., DC, 1979,v.2, 305-332.

DOXYLAMINE [U;INN]			
CAS 469-21-6	MW: 270.38	MF:	C17 H22 N2 O

Antihistaminic

LgP	2.28(C)
pKa	4.4; 9.2
pp cited in Vols.1-5:	

(1) N.L. Sperber et al., J. Am. Chem. Soc., 1949, 71, 887.
(2) Sjoquist et al., Clin. Pharmacol. Ther., 1967, 8, 48.

DRAMEDILOL [INN]			
CAS 76953-65-6	MW: 403.49	MF:	C20 H29 N5 O4

LgP	(4)
pKa	
pp cited in Vols.1-5:	

DRAQUINOLOL [INN]			
CAS 67793-71-9	MW: 410.52	MF:	C24 H30 N2 O4

Beta-adrenoceptor blocker

LgP	3.30(C,3)
pKa	
pp cited in Vols.1-5:	

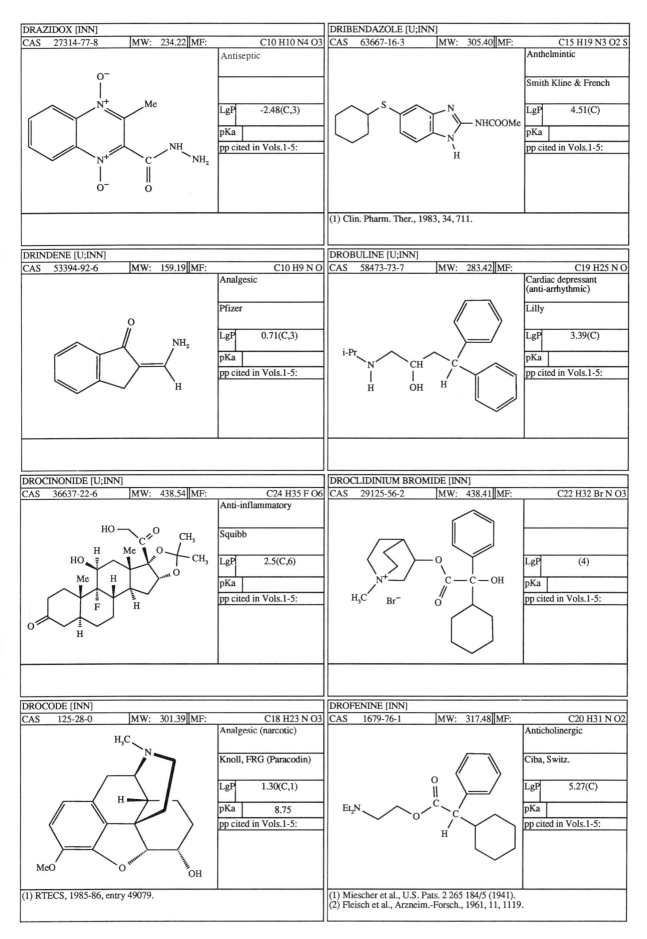

DRAZIDOX [INN]

CAS	27314-77-8	MW:	234.22	MF:	C10 H10 N4 O3

Antiseptic

LgP	-2.48(C,3)
pKa	
pp cited in Vols.1-5:	

DRIBENDAZOLE [U;INN]

CAS	63667-16-3	MW:	305.40	MF:	C15 H19 N3 O2 S

Anthelmintic

Smith Kline & French

LgP	4.51(C)
pKa	
pp cited in Vols.1-5:	

(1) Clin. Pharm. Ther., 1983, 34, 711.

DRINDENE [U;INN]

CAS	53394-92-6	MW:	159.19	MF:	C10 H9 N O

Analgesic

Pfizer

LgP	0.71(C,3)
pKa	
pp cited in Vols.1-5:	

DROBULINE [U;INN]

CAS	58473-73-7	MW:	283.42	MF:	C19 H25 N O

Cardiac depressant (anti-arrhythmic)

Lilly

LgP	3.39(C)
pKa	
pp cited in Vols.1-5:	

DROCINONIDE [U;INN]

CAS	36637-22-6	MW:	438.54	MF:	C24 H35 F O6

Anti-inflammatory

Squibb

LgP	2.5(C,6)
pKa	
pp cited in Vols.1-5:	

DROCLIDINIUM BROMIDE [INN]

CAS	29125-56-2	MW:	438.41	MF:	C22 H32 Br N O3

LgP	(4)
pKa	
pp cited in Vols.1-5:	

DROCODE [INN]

CAS	125-28-0	MW:	301.39	MF:	C18 H23 N O3

Analgesic (narcotic)

Knoll, FRG (Paracodin)

LgP	1.30(C,1)
pKa	8.75
pp cited in Vols.1-5:	

(1) RTECS, 1985-86, entry 49079.

DROFENINE [INN]

CAS	1679-76-1	MW:	317.48	MF:	C20 H31 N O2

Anticholinergic

Ciba, Switz.

LgP	5.27(C)
pKa	
pp cited in Vols.1-5:	

(1) Miescher et al., U.S. Pats. 2 265 184/5 (1941).
(2) Fleisch et al., Arzneim.-Forsch., 1961, 11, 1119.

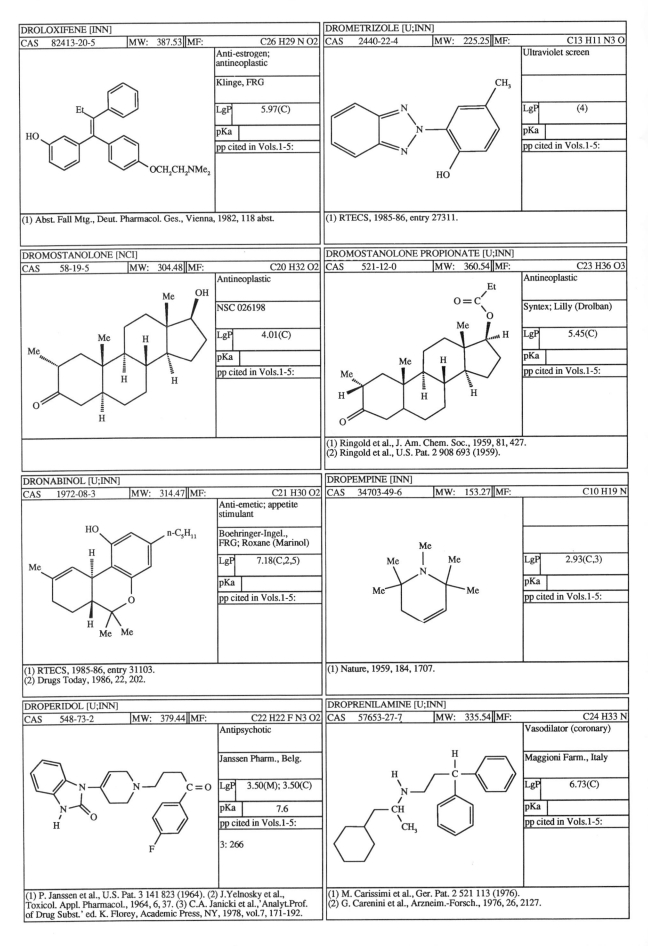

DROLOXIFENE [INN]

| CAS | 82413-20-5 | MW: | 387.53 | MF: | C26 H29 N O2 |

Anti-estrogen; antineoplastic

Klinge, FRG

LgP	5.97(C)
pKa	
pp cited in Vols.1-5:	

OCH₂CH₂NMe₂ → OCH$_2$CH$_2$NMe$_2$

(1) Abst. Fall Mtg., Deut. Pharmacol. Ges., Vienna, 1982, 118 abst.

DROMETRIZOLE [U;INN]

| CAS | 2440-22-4 | MW: | 225.25 | MF: | C13 H11 N3 O |

Ultraviolet screen

LgP	(4)
pKa	
pp cited in Vols.1-5:	

(1) RTECS, 1985-86, entry 27311.

DROMOSTANOLONE [NCI]

| CAS | 58-19-5 | MW: | 304.48 | MF: | C20 H32 O2 |

Antineoplastic

NSC 026198

LgP	4.01(C)
pKa	
pp cited in Vols.1-5:	

DROMOSTANOLONE PROPIONATE [U;INN]

| CAS | 521-12-0 | MW: | 360.54 | MF: | C23 H36 O3 |

Antineoplastic

Syntex; Lilly (Drolban)

LgP	5.45(C)
pKa	
pp cited in Vols.1-5:	

(1) Ringold et al., J. Am. Chem. Soc., 1959, 81, 427.
(2) Ringold et al., U.S. Pat. 2 908 693 (1959).

DRONABINOL [U;INN]

| CAS | 1972-08-3 | MW: | 314.47 | MF: | C21 H30 O2 |

Anti-emetic; appetite stimulant

Boehringer-Ingel., FRG; Roxane (Marinol)

LgP	7.18(C,2,5)
pKa	
pp cited in Vols.1-5:	

n-C$_5$H$_{11}$

(1) RTECS, 1985-86, entry 31103.
(2) Drugs Today, 1986, 22, 202.

DROPEMPINE [INN]

| CAS | 34703-49-6 | MW: | 153.27 | MF: | C10 H19 N |

LgP	2.93(C,3)
pKa	
pp cited in Vols.1-5:	

(1) Nature, 1959, 184, 1707.

DROPERIDOL [U;INN]

| CAS | 548-73-2 | MW: | 379.44 | MF: | C22 H22 F N3 O2 |

Antipsychotic

Janssen Pharm., Belg.

LgP	3.50(M); 3.50(C)
pKa	7.6
pp cited in Vols.1-5:	

3: 266

(1) P. Janssen et al., U.S. Pat. 3 141 823 (1964). (2) J.Yelnosky et al., Toxicol. Appl. Pharmacol., 1964, 6, 37. (3) C.A. Janicki et al.,'Analyt.Prof. of Drug Subst.' ed. K. Florey, Academic Press, NY, 1978, vol.7, 171-192.

DROPRENILAMINE [U;INN]

| CAS | 57653-27-7 | MW: | 335.54 | MF: | C24 H33 N |

Vasodilator (coronary)

Maggioni Farm., Italy

LgP	6.73(C)
pKa	
pp cited in Vols.1-5:	

(1) M. Carissimi et al., Ger. Pat. 2 521 113 (1976).
(2) G. Carenini et al., Arzneim.-Forsch., 1976, 26, 2127.

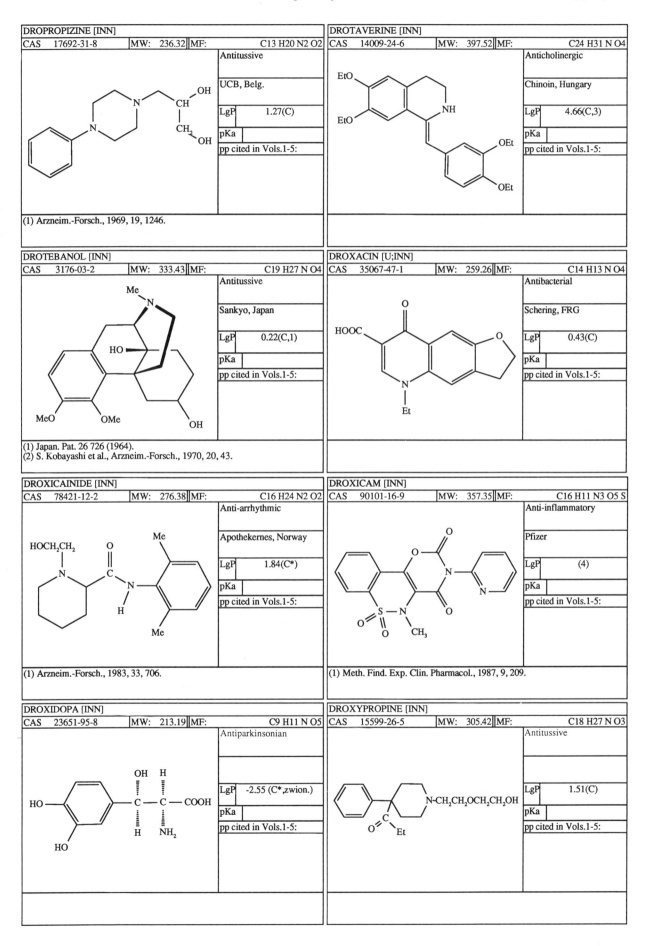

DROPROPIZINE [INN]

CAS	17692-31-8	MW: 236.32	MF:	C13 H20 N2 O2

Antitussive

UCB, Belg.

LgP	1.27(C)

pKa	

pp cited in Vols.1-5:

(1) Arzneim.-Forsch., 1969, 19, 1246.

DROTAVERINE [INN]

CAS	14009-24-6	MW: 397.52	MF:	C24 H31 N O4

Anticholinergic

Chinoin, Hungary

LgP	4.66(C,3)

pKa	

pp cited in Vols.1-5:

DROTEBANOL [INN]

CAS	3176-03-2	MW: 333.43	MF:	C19 H27 N O4

Antitussive

Sankyo, Japan

LgP	0.22(C,1)

pKa	

pp cited in Vols.1-5:

(1) Japan. Pat. 26 726 (1964).
(2) S. Kobayashi et al., Arzneim.-Forsch., 1970, 20, 43.

DROXACIN [U;INN]

CAS	35067-47-1	MW: 259.26	MF:	C14 H13 N O4

Antibacterial

Schering, FRG

LgP	0.43(C)

pKa	

pp cited in Vols.1-5:

DROXICAINIDE [INN]

CAS	78421-12-2	MW: 276.38	MF:	C16 H24 N2 O2

Anti-arrhythmic

Apothekernes, Norway

LgP	1.84(C*)

pKa	

pp cited in Vols.1-5:

(1) Arzneim.-Forsch., 1983, 33, 706.

DROXICAM [INN]

CAS	90101-16-9	MW: 357.35	MF:	C16 H11 N3 O5 S

Anti-inflammatory

Pfizer

LgP	(4)

pKa	

pp cited in Vols.1-5:

(1) Meth. Find. Exp. Clin. Pharmacol., 1987, 9, 209.

DROXIDOPA [INN]

CAS	23651-95-8	MW: 213.19	MF:	C9 H11 N O5

Antiparkinsonian

LgP	-2.55 (C*,zwion.)

pKa	

pp cited in Vols.1-5:

DROXYPROPINE [INN]

CAS	15599-26-5	MW: 305.42	MF:	C18 H27 N O3

Antitussive

LgP	1.51(C)

pKa	

pp cited in Vols.1-5:

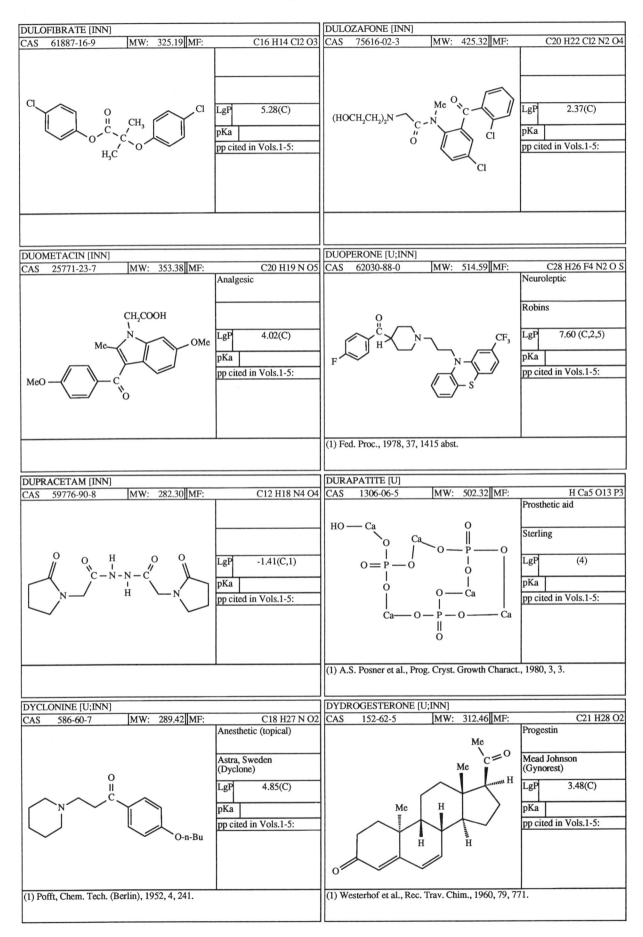

DULOFIBRATE [INN]

| CAS | 61887-16-9 | MW: | 325.19 | MF: | C16 H14 Cl2 O3 |

LgP	5.28(C)
pKa	
pp cited in Vols.1-5:	

DULOZAFONE [INN]

| CAS | 75616-02-3 | MW: | 425.32 | MF: | C20 H22 Cl2 N2 O4 |

LgP	2.37(C)
pKa	
pp cited in Vols.1-5:	

DUOMETACIN [INN]

| CAS | 25771-23-7 | MW: | 353.38 | MF: | C20 H19 N O5 |

Analgesic

LgP	4.02(C)
pKa	
pp cited in Vols.1-5:	

DUOPERONE [U;INN]

| CAS | 62030-88-0 | MW: | 514.59 | MF: | C28 H26 F4 N2 O S |

Neuroleptic

Robins

LgP	7.60 (C,2,5)
pKa	
pp cited in Vols.1-5:	

(1) Fed. Proc., 1978, 37, 1415 abst.

DUPRACETAM [INN]

| CAS | 59776-90-8 | MW: | 282.30 | MF: | C12 H18 N4 O4 |

LgP	-1.41(C,1)
pKa	
pp cited in Vols.1-5:	

DURAPATITE [U]

| CAS | 1306-06-5 | MW: | 502.32 | MF: | H Ca5 O13 P3 |

Prosthetic aid

Sterling

LgP	(4)
pKa	
pp cited in Vols.1-5:	

(1) A.S. Posner et al., Prog. Cryst. Growth Charact., 1980, 3, 3.

DYCLONINE [U;INN]

| CAS | 586-60-7 | MW: | 289.42 | MF: | C18 H27 N O2 |

Anesthetic (topical)

Astra, Sweden (Dyclone)

LgP	4.85(C)
pKa	
pp cited in Vols.1-5:	

(1) Pofft, Chem. Tech. (Berlin), 1952, 4, 241.

DYDROGESTERONE [U;INN]

| CAS | 152-62-5 | MW: | 312.46 | MF: | C21 H28 O2 |

Progestin

Mead Johnson (Gynorest)

LgP	3.48(C)
pKa	
pp cited in Vols.1-5:	

(1) Westerhof et al., Rec. Trav. Chim., 1960, 79, 771.

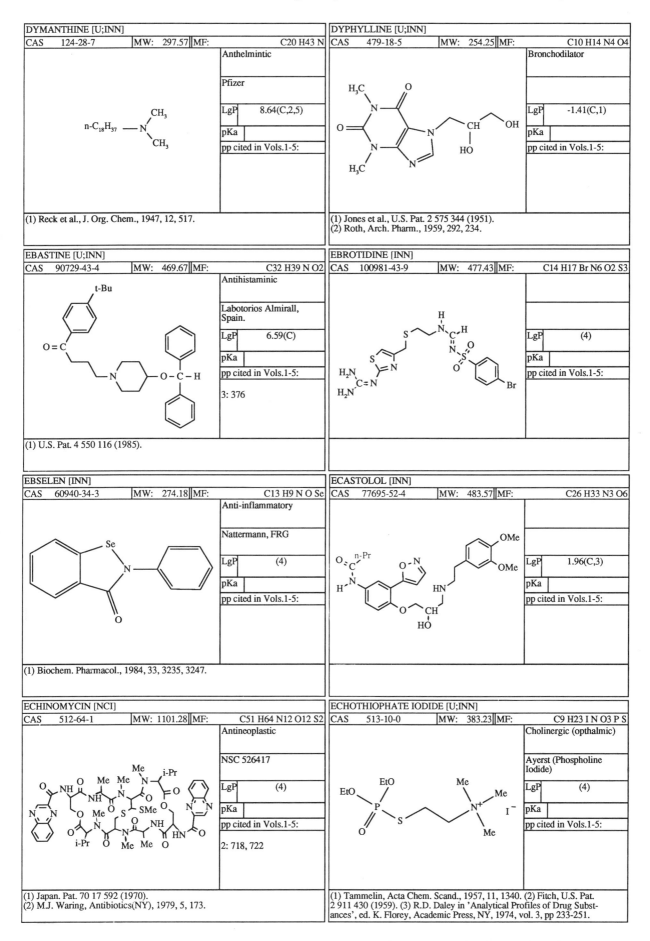

DYMANTHINE [U;INN]

CAS	124-28-7	MW:	297.57	MF:	C20 H43 N

Anthelmintic

Pfizer

LgP	8.64(C,2,5)
pKa	
pp cited in Vols.1-5:	

(1) Reck et al., J. Org. Chem., 1947, 12, 517.

DYPHYLLINE [U;INN]

CAS	479-18-5	MW:	254.25	MF:	C10 H14 N4 O4

Bronchodilator

LgP	-1.41(C,1)
pKa	
pp cited in Vols.1-5:	

(1) Jones et al., U.S. Pat. 2 575 344 (1951).
(2) Roth, Arch. Pharm., 1959, 292, 234.

EBASTINE [U;INN]

CAS	90729-43-4	MW:	469.67	MF:	C32 H39 N O2

Antihistaminic

Labotorios Almirall, Spain.

LgP	6.59(C)
pKa	
pp cited in Vols.1-5:	
3: 376	

(1) U.S. Pat. 4 550 116 (1985).

EBROTIDINE [INN]

CAS	100981-43-9	MW:	477.43	MF:	C14 H17 Br N6 O2 S3

LgP	(4)
pKa	
pp cited in Vols.1-5:	

EBSELEN [INN]

CAS	60940-34-3	MW:	274.18	MF:	C13 H9 N O Se

Anti-inflammatory

Nattermann, FRG

LgP	(4)
pKa	
pp cited in Vols.1-5:	

(1) Biochem. Pharmacol., 1984, 33, 3235, 3247.

ECASTOLOL [INN]

CAS	77695-52-4	MW:	483.57	MF:	C26 H33 N3 O6

LgP	1.96(C,3)
pKa	
pp cited in Vols.1-5:	

ECHINOMYCIN [NCI]

CAS	512-64-1	MW:	1101.28	MF:	C51 H64 N12 O12 S2

Antineoplastic

NSC 526417

LgP	(4)
pKa	
pp cited in Vols.1-5:	
2: 718, 722	

(1) Japan. Pat. 70 17 592 (1970).
(2) M.J. Waring, Antibiotics(NY), 1979, 5, 173.

ECHOTHIOPHATE IODIDE [U;INN]

CAS	513-10-0	MW:	383.23	MF:	C9 H23 I N O3 P S

Cholinergic (opthalmic)

Ayerst (Phospholine Iodide)

LgP	(4)
pKa	
pp cited in Vols.1-5:	

(1) Tammelin, Acta Chem. Scand., 1957, 11, 1340. (2) Fitch, U.S. Pat.
2 911 430 (1959). (3) R.D. Daley in 'Analytical Profiles of Drug Substances', ed. K. Florey, Academic Press, NY, 1974, vol. 3, pp 233-251.

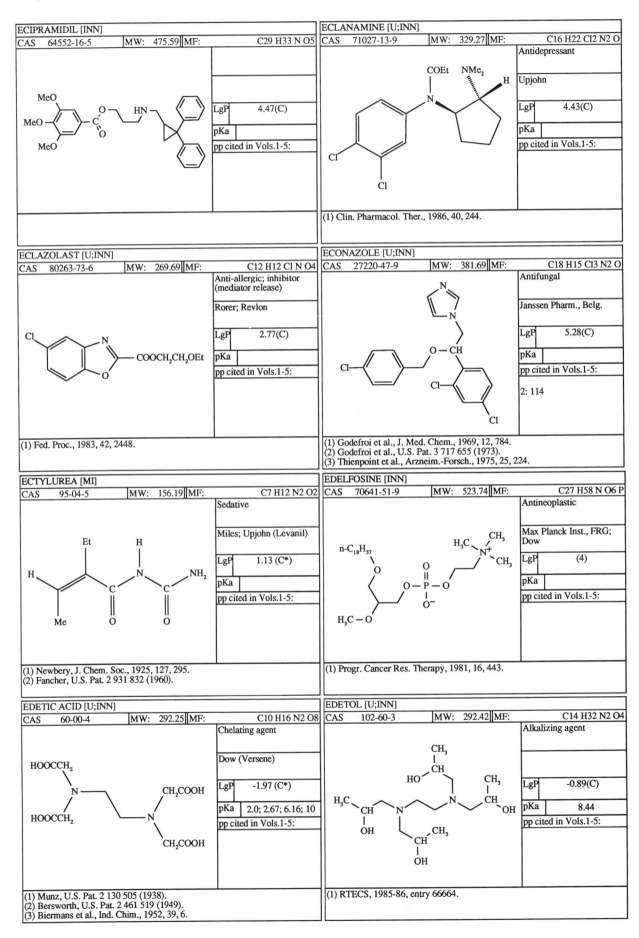

ECIPRAMIDIL [INN]

| CAS | 64552-16-5 | MW: | 475.59 | MF: | C29 H33 N O5 |

LgP	4.47(C)
pKa	
pp cited in Vols.1-5:	

ECLANAMINE [U;INN]

| CAS | 71027-13-9 | MW: | 329.27 | MF: | C16 H22 Cl2 N2 O |

Antidepressant

Upjohn

LgP	4.43(C)
pKa	
pp cited in Vols.1-5:	

(1) Clin. Pharmacol. Ther., 1986, 40, 244.

ECLAZOLAST [U;INN]

| CAS | 80263-73-6 | MW: | 269.69 | MF: | C12 H12 Cl N O4 |

Anti-allergic; inhibitor (mediator release)

Rorer; Revlon

LgP	2.77(C)
pKa	
pp cited in Vols.1-5:	

COOCH₂CH₂OEt

(1) Fed. Proc., 1983, 42, 2448.

ECONAZOLE [U;INN]

| CAS | 27220-47-9 | MW: | 381.69 | MF: | C18 H15 Cl3 N2 O |

Antifungal

Janssen Pharm., Belg.

LgP	5.28(C)
pKa	
pp cited in Vols.1-5:	
2: 114	

(1) Godefroi et al., J. Med. Chem., 1969, 12, 784.
(2) Godefroi et al., U.S. Pat. 3 717 655 (1973).
(3) Thienpoint et al., Arzneim.-Forsch., 1975, 25, 224.

ECTYLUREA [MI]

| CAS | 95-04-5 | MW: | 156.19 | MF: | C7 H12 N2 O2 |

Sedative

Miles; Upjohn (Levanil)

LgP	1.13 (C*)
pKa	
pp cited in Vols.1-5:	

(1) Newbery, J. Chem. Soc., 1925, 127, 295.
(2) Fancher, U.S. Pat. 2 931 832 (1960).

EDELFOSINE [INN]

| CAS | 70641-51-9 | MW: | 523.74 | MF: | C27 H58 N O6 P |

Antineoplastic

Max Planck Inst., FRG; Dow

LgP	(4)
pKa	
pp cited in Vols.1-5:	

(1) Progr. Cancer Res. Therapy, 1981, 16, 443.

EDETIC ACID [U;INN]

| CAS | 60-00-4 | MW: | 292.25 | MF: | C10 H16 N2 O8 |

Chelating agent

Dow (Versene)

LgP	-1.97 (C*)
pKa	2.0; 2.67; 6.16; 10
pp cited in Vols.1-5:	

(1) Munz, U.S. Pat. 2 130 505 (1938).
(2) Bersworth, U.S. Pat. 2 461 519 (1949).
(3) Biermans et al., Ind. Chim., 1952, 39, 6.

EDETOL [U;INN]

| CAS | 102-60-3 | MW: | 292.42 | MF: | C14 H32 N2 O4 |

Alkalizing agent

LgP	-0.89(C)
pKa	8.44
pp cited in Vols.1-5:	

(1) RTECS, 1985-86, entry 66664.

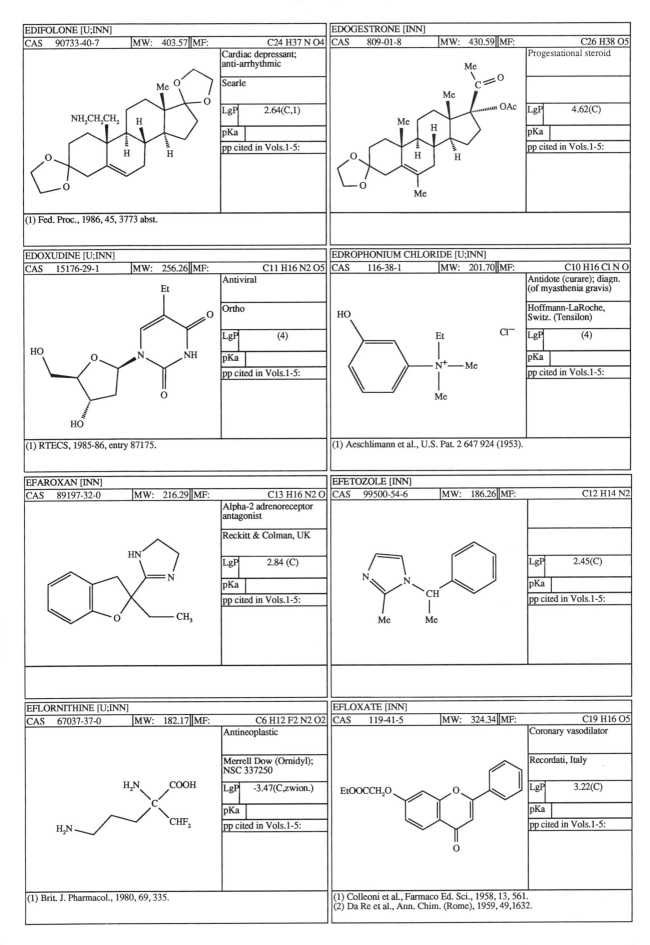

EDIFOLONE [U;INN]

CAS 90733-40-7	MW: 403.57	MF: C24 H37 N O4

Cardiac depressant; anti-arrhythmic

Searle

LgP	2.64(C,1)
pKa	
pp cited in Vols.1-5:	

(1) Fed. Proc., 1986, 45, 3773 abst.

EDOGESTRONE [INN]

CAS 809-01-8	MW: 430.59	MF: C26 H38 O5

Progestational steroid

LgP	4.62(C)
pKa	
pp cited in Vols.1-5:	

EDOXUDINE [U;INN]

CAS 15176-29-1	MW: 256.26	MF: C11 H16 N2 O5

Antiviral

Ortho

LgP	(4)
pKa	
pp cited in Vols.1-5:	

(1) RTECS, 1985-86, entry 87175.

EDROPHONIUM CHLORIDE [U;INN]

CAS 116-38-1	MW: 201.70	MF: C10 H16 Cl N O

Antidote (curare); diagn. (of myasthenia gravis)

Hoffmann-LaRoche, Switz. (Tensilon)

LgP	(4)
pKa	
pp cited in Vols.1-5:	

(1) Aeschlimann et al., U.S. Pat. 2 647 924 (1953).

EFAROXAN [INN]

CAS 89197-32-0	MW: 216.29	MF: C13 H16 N2 O

Alpha-2 adrenoreceptor antagonist

Reckitt & Colman, UK

LgP	2.84 (C)
pKa	
pp cited in Vols.1-5:	

EFETOZOLE [INN]

CAS 99500-54-6	MW: 186.26	MF: C12 H14 N2

LgP	2.45(C)
pKa	
pp cited in Vols.1-5:	

EFLORNITHINE [U;INN]

CAS 67037-37-0	MW: 182.17	MF: C6 H12 F2 N2 O2

Antineoplastic

Merrell Dow (Ornidyl); NSC 337250

LgP	-3.47(C,zwion.)
pKa	
pp cited in Vols.1-5:	

(1) Brit. J. Pharmacol., 1980, 69, 335.

EFLOXATE [INN]

CAS 119-41-5	MW: 324.34	MF: C19 H16 O5

Coronary vasodilator

Recordati, Italy

LgP	3.22(C)
pKa	
pp cited in Vols.1-5:	

(1) Colleoni et al., Farmaco Ed. Sci., 1958, 13, 561.
(2) Da Re et al., Ann. Chim. (Rome), 1959, 49,1632.

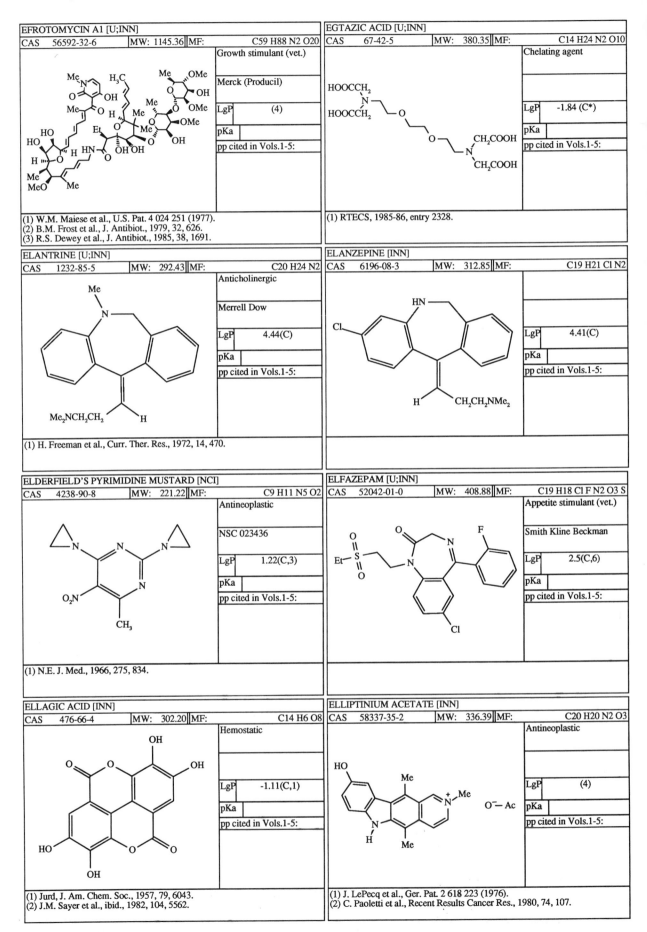

EFROTOMYCIN A1 [U;INN]

CAS	56592-32-6	MW:	1145.36	MF:	C59 H88 N2 O20

Growth stimulant (vet.)

Merck (Producil)

LgP	(4)
pKa	
pp cited in Vols.1-5:	

(1) W.M. Maiese et al., U.S. Pat. 4 024 251 (1977).
(2) B.M. Frost et al., J. Antibiot., 1979, 32, 626.
(3) R.S. Dewey et al., J. Antibiot., 1985, 38, 1691.

EGTAZIC ACID [U;INN]

CAS	67-42-5	MW:	380.35	MF:	C14 H24 N2 O10

Chelating agent

LgP	-1.84 (C*)
pKa	
pp cited in Vols.1-5:	

(1) RTECS, 1985-86, entry 2328.

ELANTRINE [U;INN]

CAS	1232-85-5	MW:	292.43	MF:	C20 H24 N2

Anticholinergic

Merrell Dow

LgP	4.44(C)
pKa	
pp cited in Vols.1-5:	

(1) H. Freeman et al., Curr. Ther. Res., 1972, 14, 470.

ELANZEPINE [INN]

CAS	6196-08-3	MW:	312.85	MF:	C19 H21 Cl N2

LgP	4.41(C)
pKa	
pp cited in Vols.1-5:	

ELDERFIELD'S PYRIMIDINE MUSTARD [NCI]

CAS	4238-90-8	MW:	221.22	MF:	C9 H11 N5 O2

Antineoplastic

NSC 023436

LgP	1.22(C,3)
pKa	
pp cited in Vols.1-5:	

(1) N.E. J. Med., 1966, 275, 834.

ELFAZEPAM [U;INN]

CAS	52042-01-0	MW:	408.88	MF:	C19 H18 Cl F N2 O3 S

Appetite stimulant (vet.)

Smith Kline Beckman

LgP	2.5(C,6)
pKa	
pp cited in Vols.1-5:	

ELLAGIC ACID [INN]

CAS	476-66-4	MW:	302.20	MF:	C14 H6 O8

Hemostatic

LgP	-1.11(C,1)
pKa	
pp cited in Vols.1-5:	

(1) Jurd, J. Am. Chem. Soc., 1957, 79, 6043.
(2) J.M. Sayer et al., ibid., 1982, 104, 5562.

ELLIPTINIUM ACETATE [INN]

CAS	58337-35-2	MW:	336.39	MF:	C20 H20 N2 O3

Antineoplastic

LgP	(4)
pKa	
pp cited in Vols.1-5:	

(1) J. LePecq et al., Ger. Pat. 2 618 223 (1976).
(2) C. Paoletti et al., Recent Results Cancer Res., 1980, 74, 107.

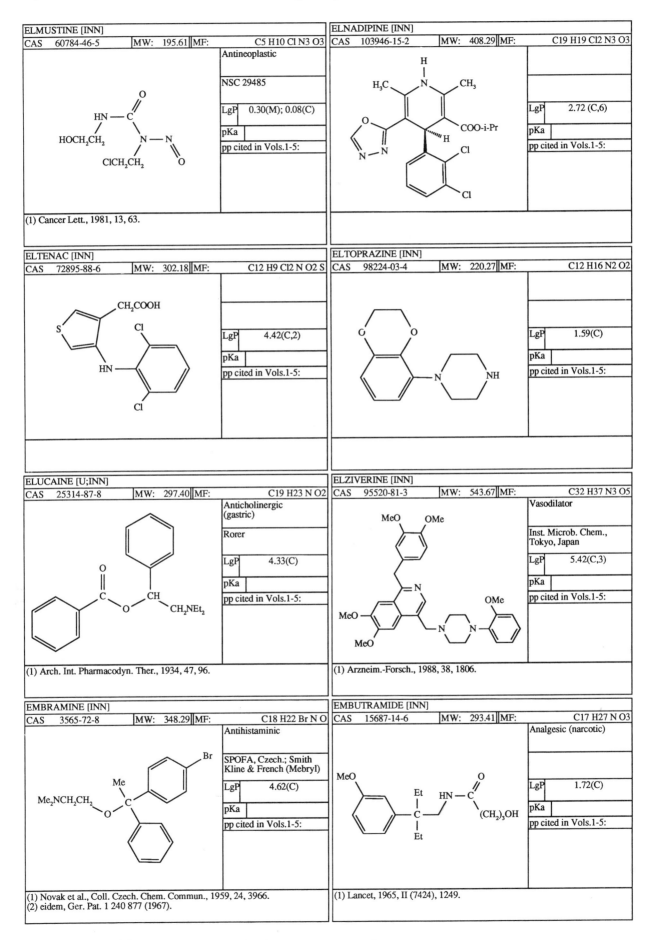

ELMUSTINE [INN]

CAS	60784-46-5	MW:	195.61	MF:	C5 H10 Cl N3 O3

Antineoplastic

NSC 29485

LgP	0.30(M); 0.08(C)
pKa	
pp cited in Vols.1-5:	

(1) Cancer Lett., 1981, 13, 63.

ELNADIPINE [INN]

CAS	103946-15-2	MW:	408.29	MF:	C19 H19 Cl2 N3 O3

LgP	2.72 (C,6)
pKa	
pp cited in Vols.1-5:	

ELTENAC [INN]

CAS	72895-88-6	MW:	302.18	MF:	C12 H9 Cl2 N O2 S

LgP	4.42(C,2)
pKa	
pp cited in Vols.1-5:	

ELTOPRAZINE [INN]

CAS	98224-03-4	MW:	220.27	MF:	C12 H16 N2 O2

LgP	1.59(C)
pKa	
pp cited in Vols.1-5:	

ELUCAINE [U;INN]

CAS	25314-87-8	MW:	297.40	MF:	C19 H23 N O2

Anticholinergic (gastric)

Rorer

LgP	4.33(C)
pKa	
pp cited in Vols.1-5:	

(1) Arch. Int. Pharmacodyn. Ther., 1934, 47, 96.

ELZIVERINE [INN]

CAS	95520-81-3	MW:	543.67	MF:	C32 H37 N3 O5

Vasodilator

Inst. Microb. Chem., Tokyo, Japan

LgP	5.42(C,3)
pKa	
pp cited in Vols.1-5:	

(1) Arzneim.-Forsch., 1988, 38, 1806.

EMBRAMINE [INN]

CAS	3565-72-8	MW:	348.29	MF:	C18 H22 Br N O

Antihistaminic

SPOFA, Czech.; Smith Kline & French (Mebryl)

LgP	4.62(C)
pKa	
pp cited in Vols.1-5:	

(1) Novak et al., Coll. Czech. Chem. Commun., 1959, 24, 3966.
(2) eidem, Ger. Pat. 1 240 877 (1967).

EMBUTRAMIDE [INN]

CAS	15687-14-6	MW:	293.41	MF:	C17 H27 N O3

Analgesic (narcotic)

LgP	1.72(C)
pKa	
pp cited in Vols.1-5:	

(1) Lancet, 1965, II (7424), 1249.

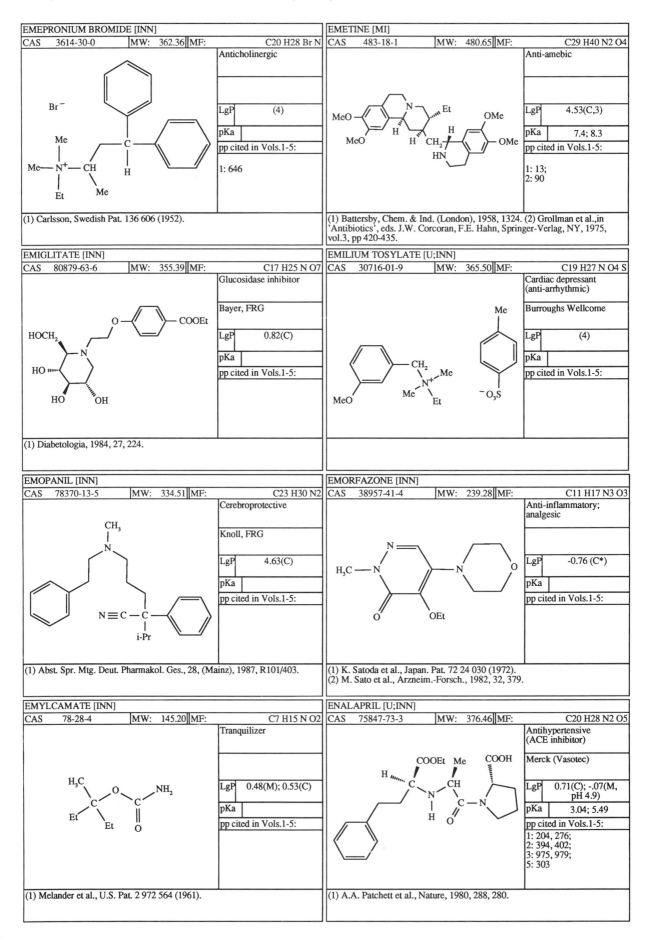

EMEPRONIUM BROMIDE [INN]

CAS 3614-30-0	MW: 362.36	MF:	C20 H28 Br N

Anticholinergic

LgP	(4)
pKa	
pp cited in Vols.1-5:	

1: 646

(1) Carlsson, Swedish Pat. 136 606 (1952).

EMETINE [MI]

CAS 483-18-1	MW: 480.65	MF:	C29 H40 N2 O4

Anti-amebic

LgP	4.53(C,3)
pKa	7.4; 8.3
pp cited in Vols.1-5:	

1: 13;
2: 90

(1) Battersby, Chem. & Ind. (London), 1958, 1324. (2) Grollman et al.,in 'Antibiotics', eds. J.W. Corcoran, F.E. Hahn, Springer-Verlag, NY, 1975, vol.3, pp 420-435.

EMIGLITATE [INN]

CAS 80879-63-6	MW: 355.39	MF:	C17 H25 N O7

Glucosidase inhibitor

Bayer, FRG

LgP	0.82(C)
pKa	
pp cited in Vols.1-5:	

(1) Diabetologia, 1984, 27, 224.

EMILIUM TOSYLATE [U;INN]

CAS 30716-01-9	MW: 365.50	MF:	C19 H27 N O4 S

Cardiac depressant (anti-arrhythmic)

Burroughs Wellcome

LgP	(4)
pKa	
pp cited in Vols.1-5:	

EMOPANIL [INN]

CAS 78370-13-5	MW: 334.51	MF:	C23 H30 N2

Cerebroprotective

Knoll, FRG

LgP	4.63(C)
pKa	
pp cited in Vols.1-5:	

(1) Abst. Spr. Mtg. Deut. Pharmakol. Ges., 28, (Mainz), 1987, R101/403.

EMORFAZONE [INN]

CAS 38957-41-4	MW: 239.28	MF:	C11 H17 N3 O3

Anti-inflammatory; analgesic

LgP	-0.76 (C*)
pKa	
pp cited in Vols.1-5:	

(1) K. Satoda et al., Japan. Pat. 72 24 030 (1972).
(2) M. Sato et al., Arzneim.-Forsch., 1982, 32, 379.

EMYLCAMATE [INN]

CAS 78-28-4	MW: 145.20	MF:	C7 H15 N O2

Tranquilizer

LgP	0.48(M); 0.53(C)
pKa	
pp cited in Vols.1-5:	

(1) Melander et al., U.S. Pat. 2 972 564 (1961).

ENALAPRIL [U;INN]

CAS 75847-73-3	MW: 376.46	MF:	C20 H28 N2 O5

Antihypertensive (ACE inhibitor)

Merck (Vasotec)

LgP	0.71(C); -.07(M, pH 4.9)
pKa	3.04; 5.49
pp cited in Vols.1-5:	

1: 204, 276;
2: 394, 402;
3: 975, 979;
5: 303

(1) A.A. Patchett et al., Nature, 1980, 288, 280.

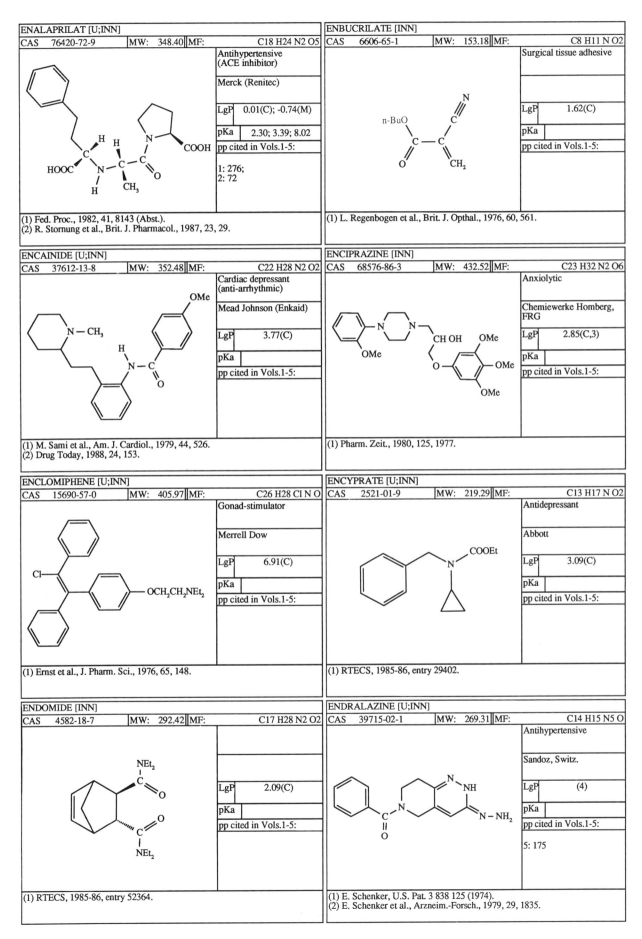

ENALAPRILAT [U;INN]

CAS	76420-72-9	MW:	348.40	MF:	C18 H24 N2 O5

Antihypertensive (ACE inhibitor)	
Merck (Renitec)	
LgP	0.01(C); -0.74(M)
pKa	2.30; 3.39; 8.02
pp cited in Vols.1-5:	
1: 276; 2: 72	

(1) Fed. Proc., 1982, 41, 8143 (Abst.).
(2) R. Stornung et al., Brit. J. Pharmacol., 1987, 23, 29.

ENBUCRILATE [INN]

CAS	6606-65-1	MW:	153.18	MF:	C8 H11 N O2

Surgical tissue adhesive	
LgP	1.62(C)
pKa	
pp cited in Vols.1-5:	

(1) L. Regenbogen et al., Brit. J. Opthal., 1976, 60, 561.

ENCAINIDE [U;INN]

CAS	37612-13-8	MW:	352.48	MF:	C22 H28 N2 O2

Cardiac depressant (anti-arrhythmic)	
Mead Johnson (Enkaid)	
LgP	3.77(C)
pKa	
pp cited in Vols.1-5:	

(1) M. Sami et al., Am. J. Cardiol., 1979, 44, 526.
(2) Drug Today, 1988, 24, 153.

ENCIPRAZINE [INN]

CAS	68576-86-3	MW:	432.52	MF:	C23 H32 N2 O6

Anxiolytic	
Chemiewerke Homberg, FRG	
LgP	2.85(C,3)
pKa	
pp cited in Vols.1-5:	

(1) Pharm. Zeit., 1980, 125, 1977.

ENCLOMIPHENE [U;INN]

CAS	15690-57-0	MW:	405.97	MF:	C26 H28 Cl N O

Gonad-stimulator	
Merrell Dow	
LgP	6.91(C)
pKa	
pp cited in Vols.1-5:	

(1) Ernst et al., J. Pharm. Sci., 1976, 65, 148.

ENCYPRATE [U;INN]

CAS	2521-01-9	MW:	219.29	MF:	C13 H17 N O2

Antidepressant	
Abbott	
LgP	3.09(C)
pKa	
pp cited in Vols.1-5:	

(1) RTECS, 1985-86, entry 29402.

ENDOMIDE [INN]

CAS	4582-18-7	MW:	292.42	MF:	C17 H28 N2 O2

LgP	2.09(C)
pKa	
pp cited in Vols.1-5:	

(1) RTECS, 1985-86, entry 52364.

ENDRALAZINE [U;INN]

CAS	39715-02-1	MW:	269.31	MF:	C14 H15 N5 O

Antihypertensive	
Sandoz, Switz.	
LgP	(4)
pKa	
pp cited in Vols.1-5:	
5: 175	

(1) E. Schenker, U.S. Pat. 3 838 125 (1974).
(2) E. Schenker et al., Arzneim.-Forsch., 1979, 29, 1835.

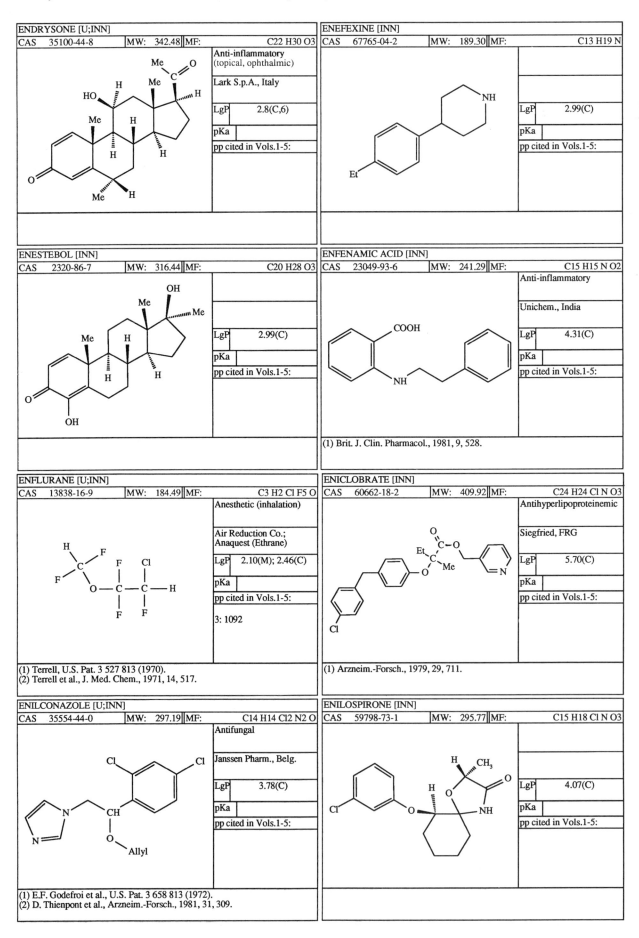

ENDRYSONE [U;INN]

CAS	35100-44-8	MW:	342.48	MF:	C22 H30 O3

Anti-inflammatory
(topical, ophthalmic)

Lark S.p.A., Italy

LgP	2.8(C,6)
pKa	
pp cited in Vols.1-5:	

ENEFEXINE [INN]

CAS	67765-04-2	MW:	189.30	MF:	C13 H19 N

LgP	2.99(C)
pKa	
pp cited in Vols.1-5:	

ENESTEBOL [INN]

CAS	2320-86-7	MW:	316.44	MF:	C20 H28 O3

LgP	2.99(C)
pKa	
pp cited in Vols.1-5:	

ENFENAMIC ACID [INN]

CAS	23049-93-6	MW:	241.29	MF:	C15 H15 N O2

Anti-inflammatory

Unichem., India

LgP	4.31(C)
pKa	
pp cited in Vols.1-5:	

(1) Brit. J. Clin. Pharmacol., 1981, 9, 528.

ENFLURANE [U;INN]

CAS	13838-16-9	MW:	184.49	MF:	C3 H2 Cl F5 O

Anesthetic (inhalation)

Air Reduction Co.;
Anaquest (Ethrane)

LgP	2.10(M); 2.46(C)
pKa	
pp cited in Vols.1-5:	
3: 1092	

(1) Terrell, U.S. Pat. 3 527 813 (1970).
(2) Terrell et al., J. Med. Chem., 1971, 14, 517.

ENICLOBRATE [INN]

CAS	60662-18-2	MW:	409.92	MF:	C24 H24 Cl N O3

Antihyperlipoproteinemic

Siegfried, FRG

LgP	5.70(C)
pKa	
pp cited in Vols.1-5:	

(1) Arzneim.-Forsch., 1979, 29, 711.

ENILCONAZOLE [U;INN]

CAS	35554-44-0	MW:	297.19	MF:	C14 H14 Cl2 N2 O

Antifungal

Janssen Pharm., Belg.

LgP	3.78(C)
pKa	
pp cited in Vols.1-5:	

(1) E.F. Godefroi et al., U.S. Pat. 3 658 813 (1972).
(2) D. Thienpont et al., Arzneim.-Forsch., 1981, 31, 309.

ENILOSPIRONE [INN]

CAS	59798-73-1	MW:	295.77	MF:	C15 H18 Cl N O3

LgP	4.07(C)
pKa	
pp cited in Vols.1-5:	

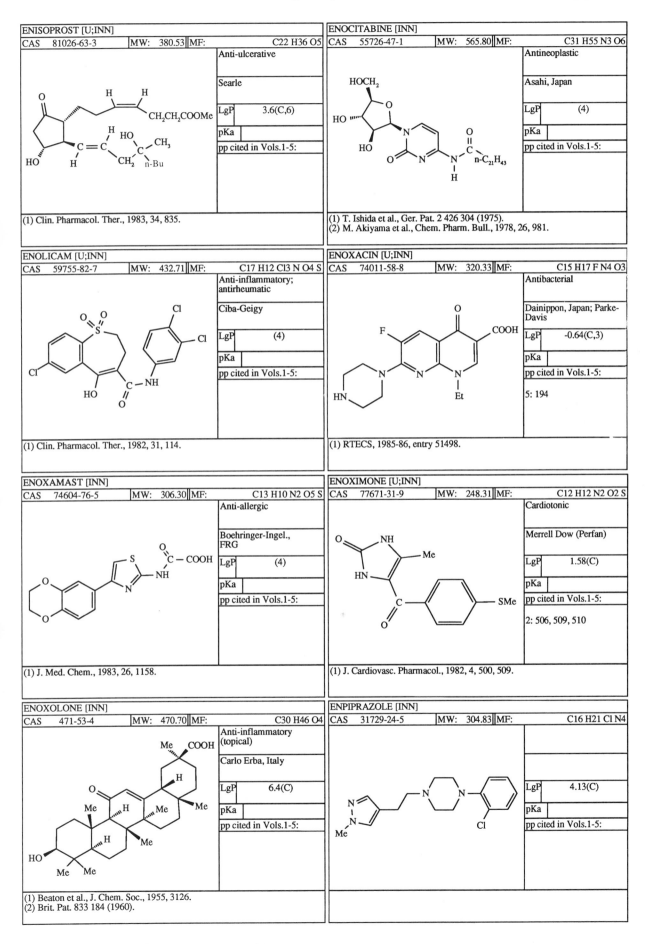

ENISOPROST [U;INN]

CAS	81026-63-3	MW: 380.53	MF:	C22 H36 O5

Anti-ulcerative

Searle

LgP	3.6(C,6)
pKa	
pp cited in Vols.1-5:	

(1) Clin. Pharmacol. Ther., 1983, 34, 835.

ENOCITABINE [INN]

CAS	55726-47-1	MW: 565.80	MF:	C31 H55 N3 O6

Antineoplastic

Asahi, Japan

LgP	(4)
pKa	
pp cited in Vols.1-5:	

(1) T. Ishida et al., Ger. Pat. 2 426 304 (1975).
(2) M. Akiyama et al., Chem. Pharm. Bull., 1978, 26, 981.

ENOLICAM [U;INN]

CAS	59755-82-7	MW: 432.71	MF:	C17 H12 Cl3 N O4 S

Anti-inflammatory; antirheumatic

Ciba-Geigy

LgP	(4)
pKa	
pp cited in Vols.1-5:	

(1) Clin. Pharmacol. Ther., 1982, 31, 114.

ENOXACIN [U;INN]

CAS	74011-58-8	MW: 320.33	MF:	C15 H17 F N4 O3

Antibacterial

Dainippon, Japan; Parke-Davis

LgP	-0.64(C,3)
pKa	
pp cited in Vols.1-5:	
5: 194	

(1) RTECS, 1985-86, entry 51498.

ENOXAMAST [INN]

CAS	74604-76-5	MW: 306.30	MF:	C13 H10 N2 O5 S

Anti-allergic

Boehringer-Ingel., FRG

LgP	(4)
pKa	
pp cited in Vols.1-5:	

(1) J. Med. Chem., 1983, 26, 1158.

ENOXIMONE [U;INN]

CAS	77671-31-9	MW: 248.31	MF:	C12 H12 N2 O2 S

Cardiotonic

Merrell Dow (Perfan)

LgP	1.58(C)
pKa	
pp cited in Vols.1-5:	
2: 506, 509, 510	

(1) J. Cardiovasc. Pharmacol., 1982, 4, 500, 509.

ENOXOLONE [INN]

CAS	471-53-4	MW: 470.70	MF:	C30 H46 O4

Anti-inflammatory (topical)

Carlo Erba, Italy

LgP	6.4(C)
pKa	
pp cited in Vols.1-5:	

(1) Beaton et al., J. Chem. Soc., 1955, 3126.
(2) Brit. Pat. 833 184 (1960).

ENPIPRAZOLE [INN]

CAS	31729-24-5	MW: 304.83	MF:	C16 H21 Cl N4

LgP	4.13(C)
pKa	
pp cited in Vols.1-5:	

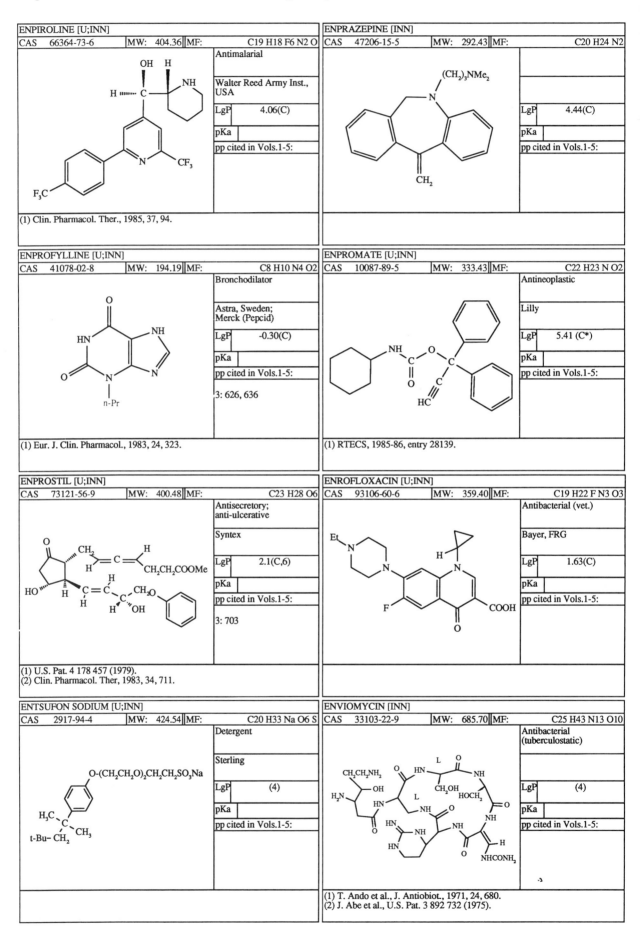

ENPIROLINE [U;INN]

CAS 66364-73-6	MW: 404.36	MF:	C19 H18 F6 N2 O

Antimalarial

Walter Reed Army Inst., USA

LgP	4.06(C)
pKa	
pp cited in Vols.1-5:	

(1) Clin. Pharmacol. Ther., 1985, 37, 94.

ENPRAZEPINE [INN]

CAS 47206-15-5	MW: 292.43	MF:	C20 H24 N2

LgP	4.44(C)
pKa	
pp cited in Vols.1-5:	

ENPROFYLLINE [U;INN]

CAS 41078-02-8	MW: 194.19	MF:	C8 H10 N4 O2

Bronchodilator

Astra, Sweden; Merck (Pepcid)

LgP	-0.30(C)
pKa	
pp cited in Vols.1-5:	

3: 626, 636

(1) Eur. J. Clin. Pharmacol., 1983, 24, 323.

ENPROMATE [U;INN]

CAS 10087-89-5	MW: 333.43	MF:	C22 H23 N O2

Antineoplastic

Lilly

LgP	5.41 (C*)
pKa	
pp cited in Vols.1-5:	

(1) RTECS, 1985-86, entry 28139.

ENPROSTIL [U;INN]

CAS 73121-56-9	MW: 400.48	MF:	C23 H28 O6

Antisecretory; anti-ulcerative

Syntex

LgP	2.1(C,6)
pKa	
pp cited in Vols.1-5:	

3: 703

(1) U.S. Pat. 4 178 457 (1979).
(2) Clin. Pharmacol. Ther, 1983, 34, 711.

ENROFLOXACIN [U;INN]

CAS 93106-60-6	MW: 359.40	MF:	C19 H22 F N3 O3

Antibacterial (vet.)

Bayer, FRG

LgP	1.63(C)
pKa	
pp cited in Vols.1-5:	

ENTSUFON SODIUM [U;INN]

CAS 2917-94-4	MW: 424.54	MF:	C20 H33 Na O6 S

Detergent

Sterling

LgP	(4)
pKa	
pp cited in Vols.1-5:	

ENVIOMYCIN [INN]

CAS 33103-22-9	MW: 685.70	MF:	C25 H43 N13 O10

Antibacterial (tuberculostatic)

LgP	(4)
pKa	
pp cited in Vols.1-5:	

(1) T. Ando et al., J. Antiobiot., 1971, 24, 680.
(2) J. Abe et al., U.S. Pat. 3 892 732 (1975).

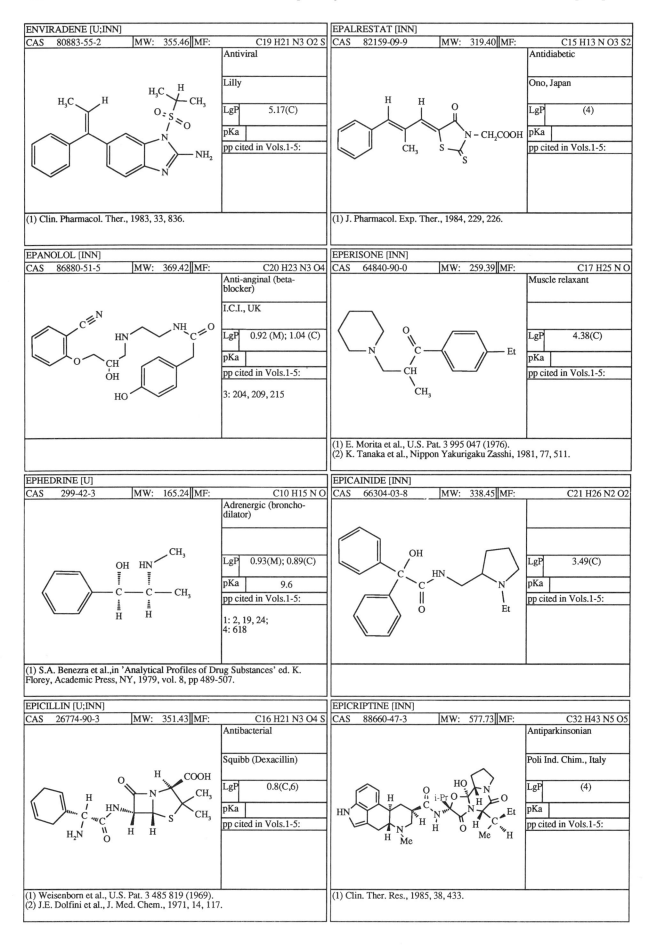

ENVIRADENE [U;INN]			
CAS 80883-55-2	MW: 355.46	MF:	C19 H21 N3 O2 S

Antiviral

Lilly

LgP   5.17(C)

pKa

pp cited in Vols.1-5:

(1) Clin. Pharmacol. Ther., 1983, 33, 836.

EPALRESTAT [INN]			
CAS 82159-09-9	MW: 319.40	MF:	C15 H13 N O3 S2

Antidiabetic

Ono, Japan

LgP   (4)

pKa

pp cited in Vols.1-5:

(1) J. Pharmacol. Exp. Ther., 1984, 229, 226.

EPANOLOL [INN]			
CAS 86880-51-5	MW: 369.42	MF:	C20 H23 N3 O4

Anti-anginal (beta-blocker)

I.C.I., UK

LgP   0.92 (M); 1.04 (C)

pKa

pp cited in Vols.1-5:

3: 204, 209, 215

EPERISONE [INN]			
CAS 64840-90-0	MW: 259.39	MF:	C17 H25 N O

Muscle relaxant

LgP   4.38(C)

pKa

pp cited in Vols.1-5:

(1) E. Morita et al., U.S. Pat. 3 995 047 (1976).
(2) K. Tanaka et al., Nippon Yakurigaku Zasshi, 1981, 77, 511.

EPHEDRINE [U]			
CAS 299-42-3	MW: 165.24	MF:	C10 H15 N O

Adrenergic (broncho-dilator)

LgP   0.93(M); 0.89(C)

pKa   9.6

pp cited in Vols.1-5:

1: 2, 19, 24;
4: 618

(1) S.A. Benezra et al.,in 'Analytical Profiles of Drug Substances' ed. K. Florey, Academic Press, NY, 1979, vol. 8, pp 489-507.

EPICAINIDE [INN]			
CAS 66304-03-8	MW: 338.45	MF:	C21 H26 N2 O2

LgP   3.49(C)

pKa

pp cited in Vols.1-5:

EPICILLIN [U;INN]			
CAS 26774-90-3	MW: 351.43	MF:	C16 H21 N3 O4 S

Antibacterial

Squibb (Dexacillin)

LgP   0.8(C,6)

pKa

pp cited in Vols.1-5:

(1) Weisenborn et al., U.S. Pat. 3 485 819 (1969).
(2) J.E. Dolfini et al., J. Med. Chem., 1971, 14, 117.

EPICRIPTINE [INN]			
CAS 88660-47-3	MW: 577.73	MF:	C32 H43 N5 O5

Antiparkinsonian

Poli Ind. Chim., Italy

LgP   (4)

pKa

pp cited in Vols.1-5:

(1) Clin. Ther. Res., 1985, 38, 433.

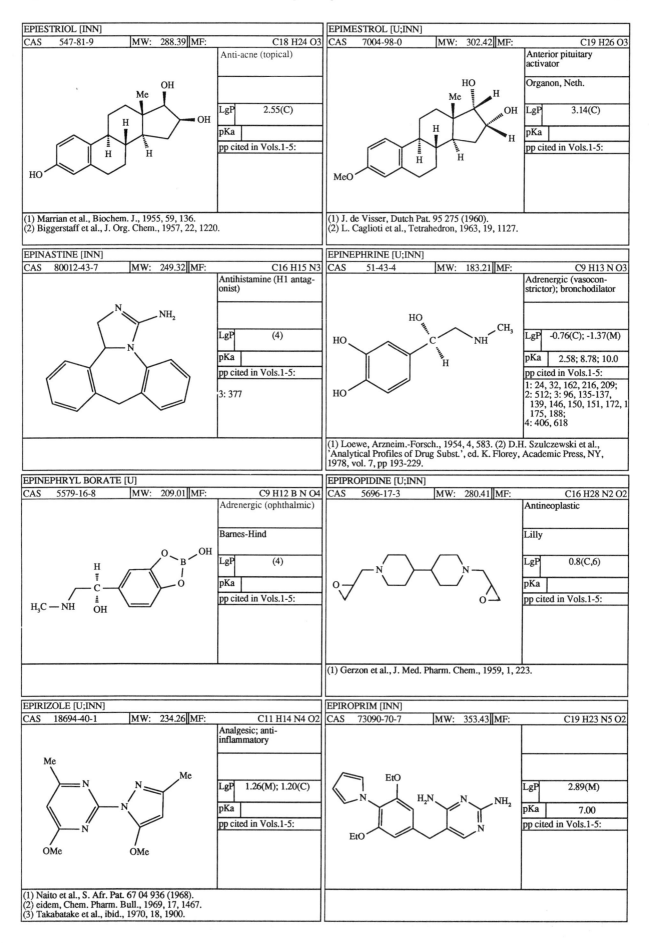

EPIESTRIOL [INN]

CAS	547-81-9	MW:	288.39	MF:	C18 H24 O3

Anti-acne (topical)

LgP	2.55(C)
pKa	
pp cited in Vols.1-5:	

(1) Marrian et al., Biochem. J., 1955, 59, 136.
(2) Biggerstaff et al., J. Org. Chem., 1957, 22, 1220.

EPIMESTROL [U;INN]

CAS	7004-98-0	MW:	302.42	MF:	C19 H26 O3

Anterior pituitary activator

Organon, Neth.

LgP	3.14(C)
pKa	
pp cited in Vols.1-5:	

(1) J. de Visser, Dutch Pat. 95 275 (1960).
(2) L. Caglioti et al., Tetrahedron, 1963, 19, 1127.

EPINASTINE [INN]

CAS	80012-43-7	MW:	249.32	MF:	C16 H15 N3

Antihistamine (H1 antagonist)

LgP	(4)
pKa	
pp cited in Vols.1-5:	
3: 377	

EPINEPHRINE [U;INN]

CAS	51-43-4	MW:	183.21	MF:	C9 H13 N O3

Adrenergic (vasoconstrictor); bronchodilator

LgP	-0.76(C); -1.37(M)
pKa	2.58; 8.78; 10.0
pp cited in Vols.1-5:	

1: 24, 32, 162, 216, 209;
2: 512; 3: 96, 135-137, 139, 146, 150, 151, 172, 1 175, 188;
4: 406, 618

(1) Loewe, Arzneim.-Forsch., 1954, 4, 583. (2) D.H. Szulczewski et al., 'Analytical Profiles of Drug Subst.', ed. K. Florey, Academic Press, NY, 1978, vol. 7, pp 193-229.

EPINEPHRYL BORATE [U]

CAS	5579-16-8	MW:	209.01	MF:	C9 H12 B N O4

Adrenergic (ophthalmic)

Barnes-Hind

LgP	(4)
pKa	
pp cited in Vols.1-5:	

EPIPROPIDINE [U;INN]

CAS	5696-17-3	MW:	280.41	MF:	C16 H28 N2 O2

Antineoplastic

Lilly

LgP	0.8(C,6)
pKa	
pp cited in Vols.1-5:	

(1) Gerzon et al., J. Med. Pharm. Chem., 1959, 1, 223.

EPIRIZOLE [U;INN]

CAS	18694-40-1	MW:	234.26	MF:	C11 H14 N4 O2

Analgesic; anti-inflammatory

LgP	1.26(M); 1.20(C)
pKa	
pp cited in Vols.1-5:	

(1) Naito et al., S. Afr. Pat. 67 04 936 (1968).
(2) eidem, Chem. Pharm. Bull., 1969, 17, 1467.
(3) Takabatake et al., ibid., 1970, 18, 1900.

EPIROPRIM [INN]

CAS	73090-70-7	MW:	353.43	MF:	C19 H23 N5 O2

LgP	2.89(M)
pKa	7.00
pp cited in Vols.1-5:	

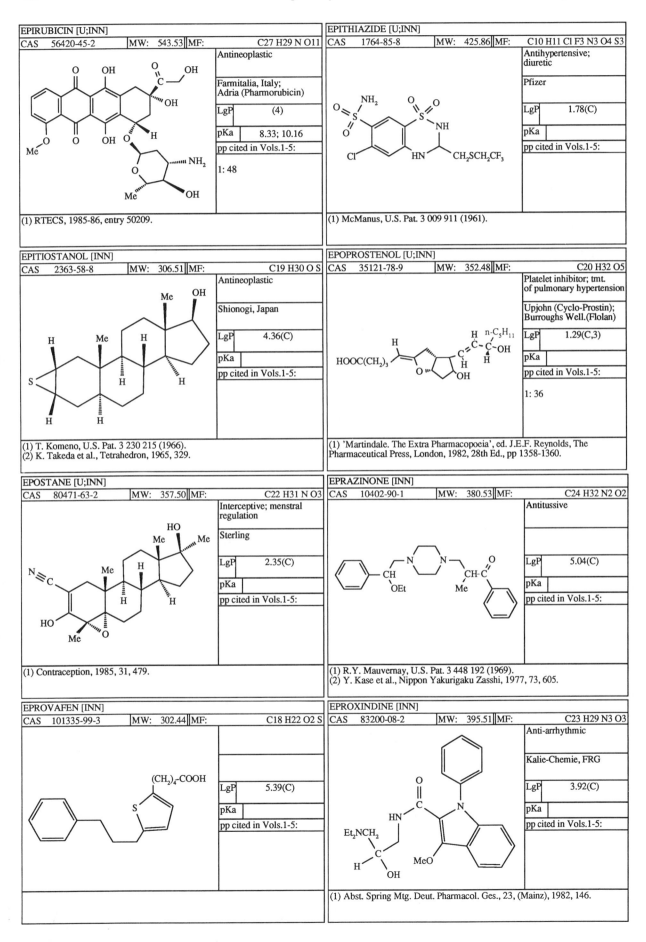

EPIRUBICIN [U;INN]

| CAS | 56420-45-2 | MW: | 543.53 | MF: | C27 H29 N O11 |

Antineoplastic

Farmitalia, Italy; Adria (Pharmorubicin)

LgP	(4)
pKa	8.33; 10.16
pp cited in Vols.1-5:	1: 48

(1) RTECS, 1985-86, entry 50209.

EPITHIAZIDE [U;INN]

| CAS | 1764-85-8 | MW: | 425.86 | MF: | C10 H11 Cl F3 N3 O4 S3 |

Antihypertensive; diuretic

Pfizer

LgP	1.78(C)
pKa	
pp cited in Vols.1-5:	

(1) McManus, U.S. Pat. 3 009 911 (1961).

EPITIOSTANOL [INN]

| CAS | 2363-58-8 | MW: | 306.51 | MF: | C19 H30 O S |

Antineoplastic

Shionogi, Japan

LgP	4.36(C)
pKa	
pp cited in Vols.1-5:	

(1) T. Komeno, U.S. Pat. 3 230 215 (1966).
(2) K. Takeda et al., Tetrahedron, 1965, 329.

EPOPROSTENOL [U;INN]

| CAS | 35121-78-9 | MW: | 352.48 | MF: | C20 H32 O5 |

Platelet inhibitor; tmt. of pulmonary hypertension

Upjohn (Cyclo-Prostin); Burroughs Well.(Flolan)

LgP	1.29(C,3)
pKa	
pp cited in Vols.1-5:	1: 36

(1) 'Martindale. The Extra Pharmacopoeia', ed. J.E.F. Reynolds, The Pharmaceutical Press, London, 1982, 28th Ed., pp 1358-1360.

EPOSTANE [U;INN]

| CAS | 80471-63-2 | MW: | 357.50 | MF: | C22 H31 N O3 |

Interceptive; menstral regulation

Sterling

LgP	2.35(C)
pKa	
pp cited in Vols.1-5:	

(1) Contraception, 1985, 31, 479.

EPRAZINONE [INN]

| CAS | 10402-90-1 | MW: | 380.53 | MF: | C24 H32 N2 O2 |

Antitussive

LgP	5.04(C)
pKa	
pp cited in Vols.1-5:	

(1) R.Y. Mauvernay, U.S. Pat. 3 448 192 (1969).
(2) Y. Kase et al., Nippon Yakurigaku Zasshi, 1977, 73, 605.

EPROVAFEN [INN]

| CAS | 101335-99-3 | MW: | 302.44 | MF: | C18 H22 O2 S |

LgP	5.39(C)
pKa	
pp cited in Vols.1-5:	

EPROXINDINE [INN]

| CAS | 83200-08-2 | MW: | 395.51 | MF: | C23 H29 N3 O3 |

Anti-arrhythmic

Kalie-Chemie, FRG

LgP	3.92(C)
pKa	
pp cited in Vols.1-5:	

(1) Abst. Spring Mtg. Deut. Pharmacol. Ges., 23, (Mainz), 1982, 146.

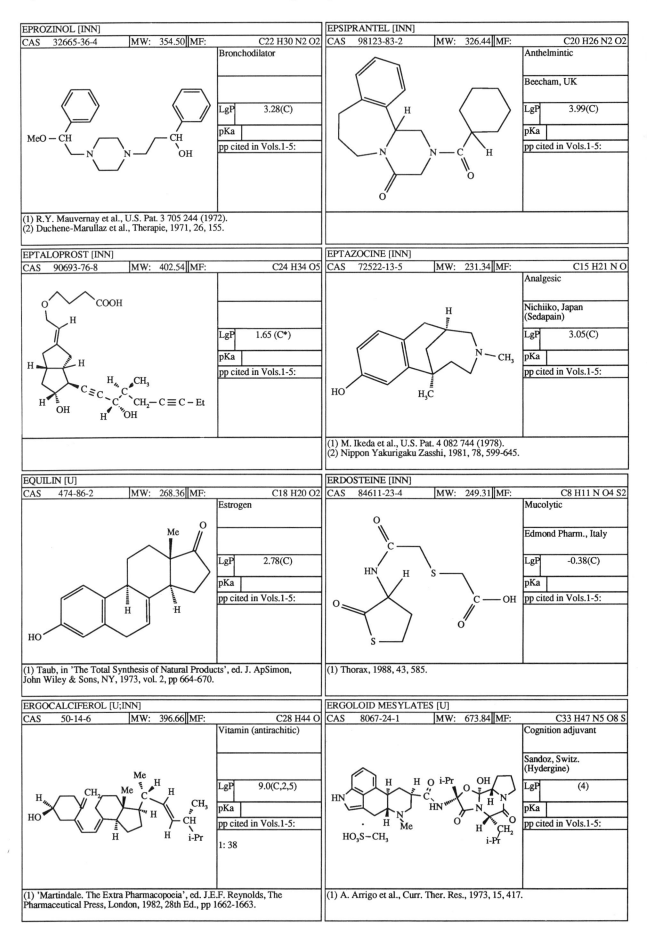

EPROZINOL [INN]

CAS	32665-36-4	MW:	354.50	MF:	C22 H30 N2 O2

Bronchodilator

LgP	3.28(C)
pKa	
pp cited in Vols.1-5:	

MeO — CH

OH

(1) R.Y. Mauvernay et al., U.S. Pat. 3 705 244 (1972).
(2) Duchene-Marullaz et al., Therapie, 1971, 26, 155.

EPSIPRANTEL [INN]

CAS	98123-83-2	MW:	326.44	MF:	C20 H26 N2 O2

Anthelmintic

Beecham, UK

LgP	3.99(C)
pKa	
pp cited in Vols.1-5:	

EPTALOPROST [INN]

CAS	90693-76-8	MW:	402.54	MF:	C24 H34 O5

LgP	1.65 (C*)
pKa	
pp cited in Vols.1-5:	

COOH

EPTAZOCINE [INN]

CAS	72522-13-5	MW:	231.34	MF:	C15 H21 N O

Analgesic

Nichiiko, Japan
(Sedapain)

LgP	3.05(C)
pKa	
pp cited in Vols.1-5:	

N — CH₃

(1) M. Ikeda et al., U.S. Pat. 4 082 744 (1978).
(2) Nippon Yakurigaku Zasshi, 1981, 78, 599-645.

EQUILIN [U]

CAS	474-86-2	MW:	268.36	MF:	C18 H20 O2

Estrogen

LgP	2.78(C)
pKa	
pp cited in Vols.1-5:	

(1) Taub, in 'The Total Synthesis of Natural Products', ed. J. ApSimon, John Wiley & Sons, NY, 1973, vol. 2, pp 664-670.

ERDOSTEINE [INN]

CAS	84611-23-4	MW:	249.31	MF:	C8 H11 N O4 S2

Mucolytic

Edmond Pharm., Italy

LgP	-0.38(C)
pKa	
pp cited in Vols.1-5:	

(1) Thorax, 1988, 43, 585.

ERGOCALCIFEROL [U;INN]

CAS	50-14-6	MW:	396.66	MF:	C28 H44 O

Vitamin (antirachitic)

LgP	9.0(C,2,5)
pKa	
pp cited in Vols.1-5:	
1: 38	

(1) 'Martindale. The Extra Pharmacopoeia', ed. J.E.F. Reynolds, The Pharmaceutical Press, London, 1982, 28th Ed., pp 1662-1663.

ERGOLOID MESYLATES [U]

CAS	8067-24-1	MW:	673.84	MF:	C33 H47 N5 O8 S

Cognition adjuvant

Sandoz, Switz.
(Hydergine)

LgP	(4)
pKa	
pp cited in Vols.1-5:	

HO₃S — CH₃

(1) A. Arrigo et al., Curr. Ther. Res., 1973, 15, 417.

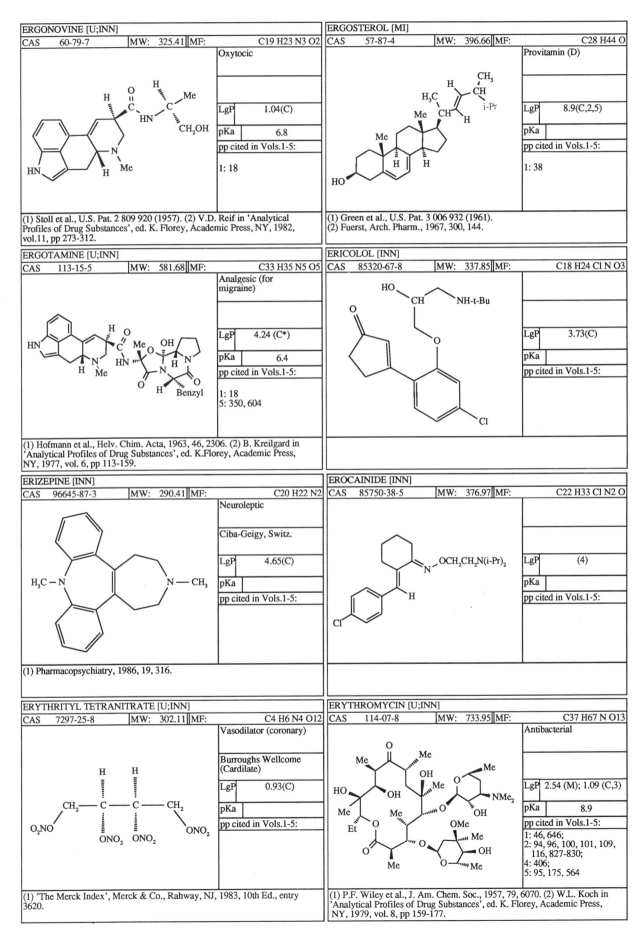

ERGONOVINE [U;INN]

CAS	60-79-7	MW: 325.41	MF:	C19 H23 N3 O2

Oxytocic

LgP	1.04(C)
pKa	6.8

pp cited in Vols.1-5:

1: 18

(1) Stoll et al., U.S. Pat. 2 809 920 (1957). (2) V.D. Reif in 'Analytical Profiles of Drug Substances', ed. K. Florey, Academic Press, NY, 1982, vol.11, pp 273-312.

ERGOSTEROL [MI]

CAS	57-87-4	MW: 396.66	MF:	C28 H44 O

Provitamin (D)

LgP	8.9(C,2,5)
pKa	

pp cited in Vols.1-5:

1: 38

(1) Green et al., U.S. Pat. 3 006 932 (1961).
(2) Fuerst, Arch. Pharm., 1967, 300, 144.

ERGOTAMINE [U;INN]

CAS	113-15-5	MW: 581.68	MF:	C33 H35 N5 O5

Analgesic (for migraine)

LgP	4.24 (C*)
pKa	6.4

pp cited in Vols.1-5:

1: 18
5: 350, 604

(1) Hofmann et al., Helv. Chim. Acta, 1963, 46, 2306. (2) B. Kreilgard in 'Analytical Profiles of Drug Substances', ed. K.Florey, Academic Press, NY, 1977, vol. 6, pp 113-159.

ERICOLOL [INN]

CAS	85320-67-8	MW: 337.85	MF:	C18 H24 Cl N O3

LgP	3.73(C)
pKa	

pp cited in Vols.1-5:

ERIZEPINE [INN]

CAS	96645-87-3	MW: 290.41	MF:	C20 H22 N2

Neuroleptic

Ciba-Geigy, Switz.

LgP	4.65(C)
pKa	

pp cited in Vols.1-5:

(1) Pharmacopsychiatry, 1986, 19, 316.

EROCAINIDE [INN]

CAS	85750-38-5	MW: 376.97	MF:	C22 H33 Cl N2 O

LgP	(4)
pKa	

pp cited in Vols.1-5:

ERYTHRITYL TETRANITRATE [U;INN]

CAS	7297-25-8	MW: 302.11	MF:	C4 H6 N4 O12

Vasodilator (coronary)

Burroughs Wellcome (Cardilate)

LgP	0.93(C)
pKa	

pp cited in Vols.1-5:

(1) 'The Merck Index', Merck & Co., Rahway, NJ, 1983, 10th Ed., entry 3620.

ERYTHROMYCIN [U;INN]

CAS	114-07-8	MW: 733.95	MF:	C37 H67 N O13

Antibacterial

LgP	2.54 (M); 1.09 (C,3)
pKa	8.9

pp cited in Vols.1-5:

1: 46, 646;
2: 94, 96, 100, 101, 109, 116, 827-830;
4: 406;
5: 95, 175, 564

(1) P.F. Wiley et al., J. Am. Chem. Soc., 1957, 79, 6070. (2) W.L. Koch in 'Analytical Profiles of Drug Substances', ed. K. Florey, Academic Press, NY, 1979, vol. 8, pp 159-177.

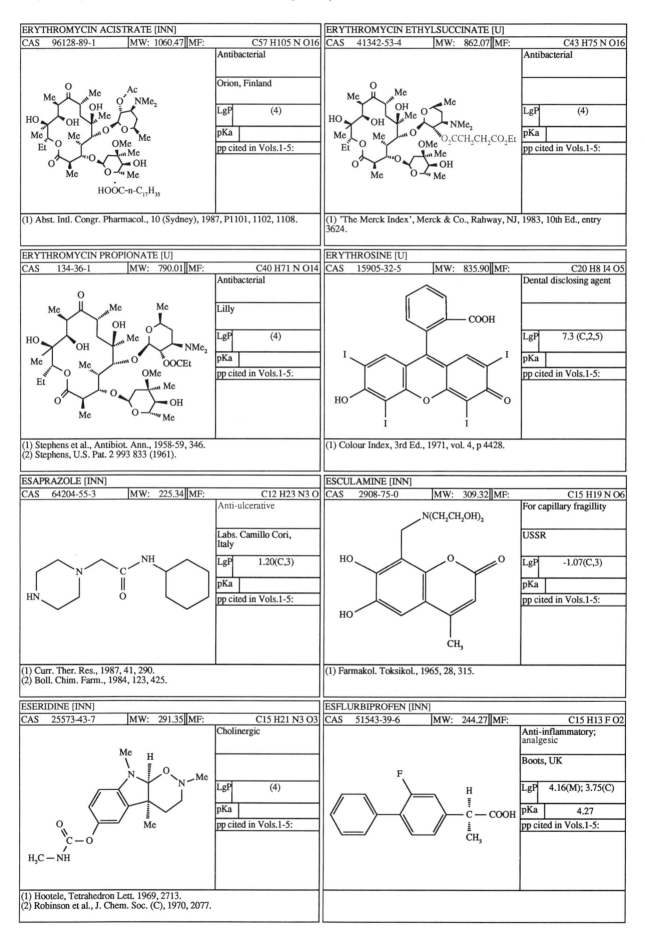

ERYTHROMYCIN ACISTRATE [INN]

CAS 96128-89-1	MW: 1060.47	MF: C57 H105 N O16

Antibacterial

Orion, Finland

LgP	(4)
pKa	
pp cited in Vols.1-5:	

(1) Abst. Intl. Congr. Pharmacol., 10 (Sydney), 1987, P1101, 1102, 1108.

ERYTHROMYCIN ETHYLSUCCINATE [U]

CAS 41342-53-4	MW: 862.07	MF: C43 H75 N O16

Antibacterial

LgP	(4)
pKa	
pp cited in Vols.1-5:	

(1) 'The Merck Index', Merck & Co., Rahway, NJ, 1983, 10th Ed., entry 3624.

ERYTHROMYCIN PROPIONATE [U]

CAS 134-36-1	MW: 790.01	MF: C40 H71 N O14

Antibacterial

Lilly

LgP	(4)
pKa	
pp cited in Vols.1-5:	

(1) Stephens et al., Antibiot. Ann., 1958-59, 346.
(2) Stephens, U.S. Pat. 2 993 833 (1961).

ERYTHROSINE [U]

CAS 15905-32-5	MW: 835.90	MF: C20 H8 I4 O5

Dental disclosing agent

LgP	7.3 (C,2,5)
pKa	
pp cited in Vols.1-5:	

(1) Colour Index, 3rd Ed., 1971, vol. 4, p 4428.

ESAPRAZOLE [INN]

CAS 64204-55-3	MW: 225.34	MF: C12 H23 N3 O

Anti-ulcerative

Labs. Camillo Cori, Italy

LgP	1.20(C,3)
pKa	
pp cited in Vols.1-5:	

(1) Curr. Ther. Res., 1987, 41, 290.
(2) Boll. Chim. Farm., 1984, 123, 425.

ESCULAMINE [INN]

CAS 2908-75-0	MW: 309.32	MF: C15 H19 N O6

For capillary fragillity

USSR

LgP	-1.07(C,3)
pKa	
pp cited in Vols.1-5:	

(1) Farmakol. Toksikol., 1965, 28, 315.

ESERIDINE [INN]

CAS 25573-43-7	MW: 291.35	MF: C15 H21 N3 O3

Cholinergic

LgP	(4)
pKa	
pp cited in Vols.1-5:	

(1) Hootele, Tetrahedron Lett. 1969, 2713.
(2) Robinson et al., J. Chem. Soc. (C), 1970, 2077.

ESFLURBIPROFEN [INN]

CAS 51543-39-6	MW: 244.27	MF: C15 H13 F O2

Anti-inflammatory; analgesic

Boots, UK

LgP	4.16(M); 3.75(C)
pKa	4.27
pp cited in Vols.1-5:	

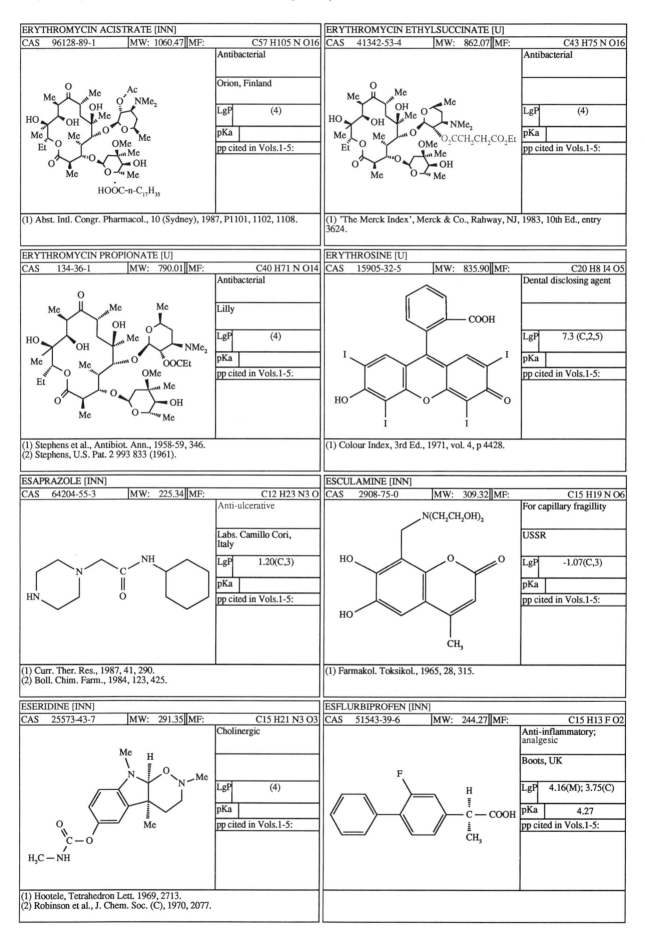

ERYTHROMYCIN ACISTRATE [INN]
CAS 96128-89-1 — MW: 1060.47 — MF: C57 H105 N O16
Antibacterial — Orion, Finland — LgP (4) — pKa — pp cited in Vols.1-5:
(1) Abst. Intl. Congr. Pharmacol., 10 (Sydney), 1987, P1101, 1102, 1108.

ERYTHROMYCIN ETHYLSUCCINATE [U]
CAS 41342-53-4 — MW: 862.07 — MF: C43 H75 N O16
Antibacterial — LgP (4) — pKa — pp cited in Vols.1-5:
(1) 'The Merck Index', Merck & Co., Rahway, NJ, 1983, 10th Ed., entry 3624.

ERYTHROMYCIN PROPIONATE [U]
CAS 134-36-1 — MW: 790.01 — MF: C40 H71 N O14
Antibacterial — Lilly — LgP (4) — pKa — pp cited in Vols.1-5:
(1) Stephens et al., Antibiot. Ann., 1958-59, 346.
(2) Stephens, U.S. Pat. 2 993 833 (1961).

ERYTHROSINE [U]
CAS 15905-32-5 — MW: 835.90 — MF: C20 H8 I4 O5
Dental disclosing agent — LgP 7.3 (C,2,5) — pKa — pp cited in Vols.1-5:
(1) Colour Index, 3rd Ed., 1971, vol. 4, p 4428.

ESAPRAZOLE [INN]
CAS 64204-55-3 — MW: 225.34 — MF: C12 H23 N3 O
Anti-ulcerative — Labs. Camillo Cori, Italy — LgP 1.20(C,3) — pKa — pp cited in Vols.1-5:
(1) Curr. Ther. Res., 1987, 41, 290.
(2) Boll. Chim. Farm., 1984, 123, 425.

ESCULAMINE [INN]
CAS 2908-75-0 — MW: 309.32 — MF: C15 H19 N O6
For capillary fragillity — USSR — LgP -1.07(C,3) — pKa — pp cited in Vols.1-5:
(1) Farmakol. Toksikol., 1965, 28, 315.

ESERIDINE [INN]
CAS 25573-43-7 — MW: 291.35 — MF: C15 H21 N3 O3
Cholinergic — LgP (4) — pKa — pp cited in Vols.1-5:
(1) Hootele, Tetrahedron Lett. 1969, 2713.
(2) Robinson et al., J. Chem. Soc. (C), 1970, 2077.

ESFLURBIPROFEN [INN]
CAS 51543-39-6 — MW: 244.27 — MF: C15 H13 F O2
Anti-inflammatory; analgesic — Boots, UK — LgP 4.16(M); 3.75(C) — pKa 4.27 — pp cited in Vols.1-5:

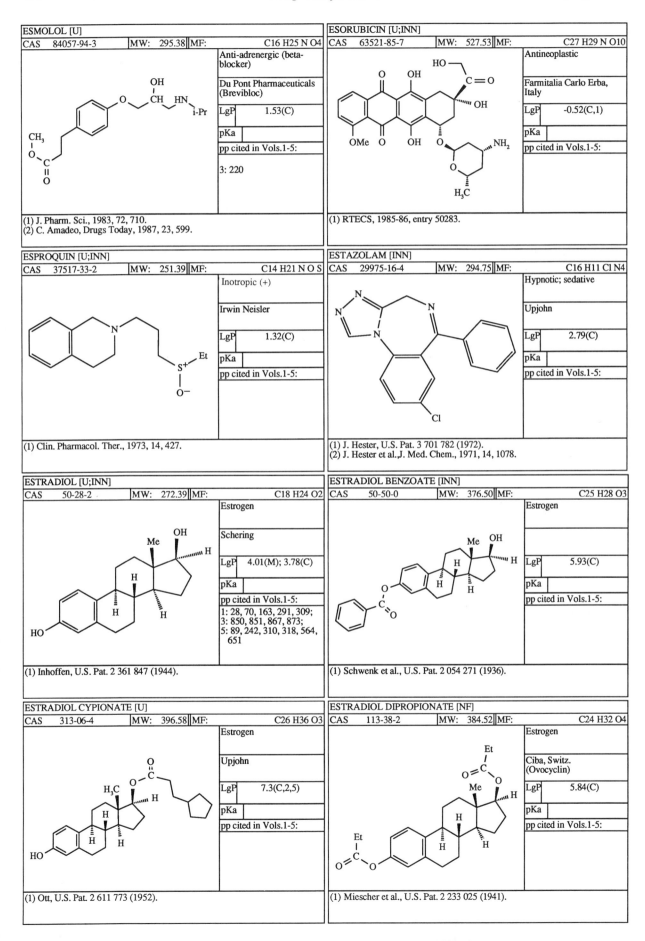

ESMOLOL [U]

CAS	84057-94-3	MW: 295.38	MF:	C16 H25 N O4

Anti-adrenergic (beta-blocker)

Du Pont Pharmaceuticals (Brevibloc)

LgP 1.53(C)

pKa

pp cited in Vols.1-5:
3: 220

(1) J. Pharm. Sci., 1983, 72, 710.
(2) C. Amadeo, Drugs Today, 1987, 23, 599.

ESORUBICIN [U;INN]

CAS	63521-85-7	MW: 527.53	MF:	C27 H29 N O10

Antineoplastic

Farmitalia Carlo Erba, Italy

LgP -0.52(C,1)

pKa

pp cited in Vols.1-5:

(1) RTECS, 1985-86, entry 50283.

ESPROQUIN [U;INN]

CAS	37517-33-2	MW: 251.39	MF:	C14 H21 N O S

Inotropic (+)

Irwin Neisler

LgP 1.32(C)

pKa

pp cited in Vols.1-5:

(1) Clin. Pharmacol. Ther., 1973, 14, 427.

ESTAZOLAM [INN]

CAS	29975-16-4	MW: 294.75	MF:	C16 H11 Cl N4

Hypnotic; sedative

Upjohn

LgP 2.79(C)

pKa

pp cited in Vols.1-5:

(1) J. Hester, U.S. Pat. 3 701 782 (1972).
(2) J. Hester et al., J. Med. Chem., 1971, 14, 1078.

ESTRADIOL [U;INN]

CAS	50-28-2	MW: 272.39	MF:	C18 H24 O2

Estrogen

Schering

LgP 4.01(M); 3.78(C)

pKa

pp cited in Vols.1-5:
1: 28, 70, 163, 291, 309;
3: 850, 851, 867, 873;
5: 89, 242, 310, 318, 564, 651

(1) Inhoffen, U.S. Pat. 2 361 847 (1944).

ESTRADIOL BENZOATE [INN]

CAS	50-50-0	MW: 376.50	MF:	C25 H28 O3

Estrogen

LgP 5.93(C)

pKa

pp cited in Vols.1-5:

(1) Schwenk et al., U.S. Pat. 2 054 271 (1936).

ESTRADIOL CYPIONATE [U]

CAS	313-06-4	MW: 396.58	MF:	C26 H36 O3

Estrogen

Upjohn

LgP 7.3(C,2,5)

pKa

pp cited in Vols.1-5:

(1) Ott, U.S. Pat. 2 611 773 (1952).

ESTRADIOL DIPROPIONATE [NF]

CAS	113-38-2	MW: 384.52	MF:	C24 H32 O4

Estrogen

Ciba, Switz. (Ovocyclin)

LgP 5.84(C)

pKa

pp cited in Vols.1-5:

(1) Miescher et al., U.S. Pat. 2 233 025 (1941).

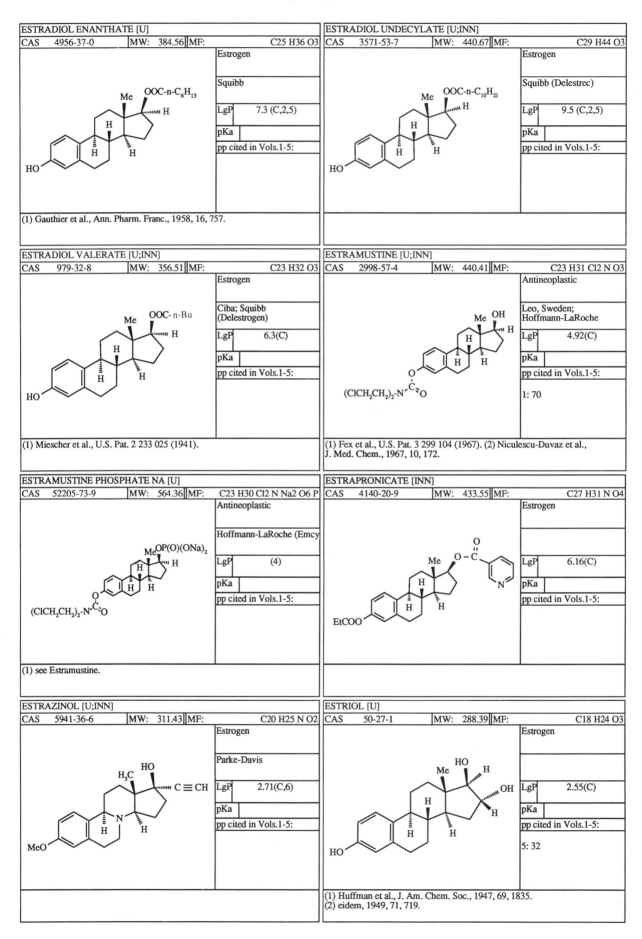

ESTRADIOL ENANTHATE [U]

| CAS | 4956-37-0 | MW: | 384.56 | MF: | C25 H36 O3 |

Estrogen

Squibb

LgP	7.3 (C,2,5)
pKa	
pp cited in Vols.1-5:	

(1) Gauthier et al., Ann. Pharm. Franc., 1958, 16, 757.

ESTRADIOL UNDECYLATE [U;INN]

| CAS | 3571-53-7 | MW: | 440.67 | MF: | C29 H44 O3 |

Estrogen

Squibb (Delestrec)

LgP	9.5 (C,2,5)
pKa	
pp cited in Vols.1-5:	

ESTRADIOL VALERATE [U;INN]

| CAS | 979-32-8 | MW: | 356.51 | MF: | C23 H32 O3 |

Estrogen

Ciba; Squibb (Delestrogen)

LgP	6.3(C)
pKa	
pp cited in Vols.1-5:	

(1) Miescher et al., U.S. Pat. 2 233 025 (1941).

ESTRAMUSTINE [U;INN]

| CAS | 2998-57-4 | MW: | 440.41 | MF: | C23 H31 Cl2 N O3 |

Antineoplastic

Leo, Sweden; Hoffmann-LaRoche

LgP	4.92(C)
pKa	
pp cited in Vols.1-5:	

1: 70

(1) Fex et al., U.S. Pat. 3 299 104 (1967). (2) Niculescu-Duvaz et al., J. Med. Chem., 1967, 10, 172.

ESTRAMUSTINE PHOSPHATE NA [U]

| CAS | 52205-73-9 | MW: | 564.36 | MF: | C23 H30 Cl2 N Na2 O6 P |

Antineoplastic

Hoffmann-LaRoche (Emcy

LgP	(4)
pKa	
pp cited in Vols.1-5:	

(1) see Estramustine.

ESTRAPRONICATE [INN]

| CAS | 4140-20-9 | MW: | 433.55 | MF: | C27 H31 N O4 |

Estrogen

LgP	6.16(C)
pKa	
pp cited in Vols.1-5:	

ESTRAZINOL [U;INN]

| CAS | 5941-36-6 | MW: | 311.43 | MF: | C20 H25 N O2 |

Estrogen

Parke-Davis

LgP	2.71(C,6)
pKa	
pp cited in Vols.1-5:	

ESTRIOL [U]

| CAS | 50-27-1 | MW: | 288.39 | MF: | C18 H24 O3 |

Estrogen

LgP	2.55(C)
pKa	
pp cited in Vols.1-5:	

5: 32

(1) Huffman et al., J. Am. Chem. Soc., 1947, 69, 1835.
(2) eidem, 1949, 71, 719.

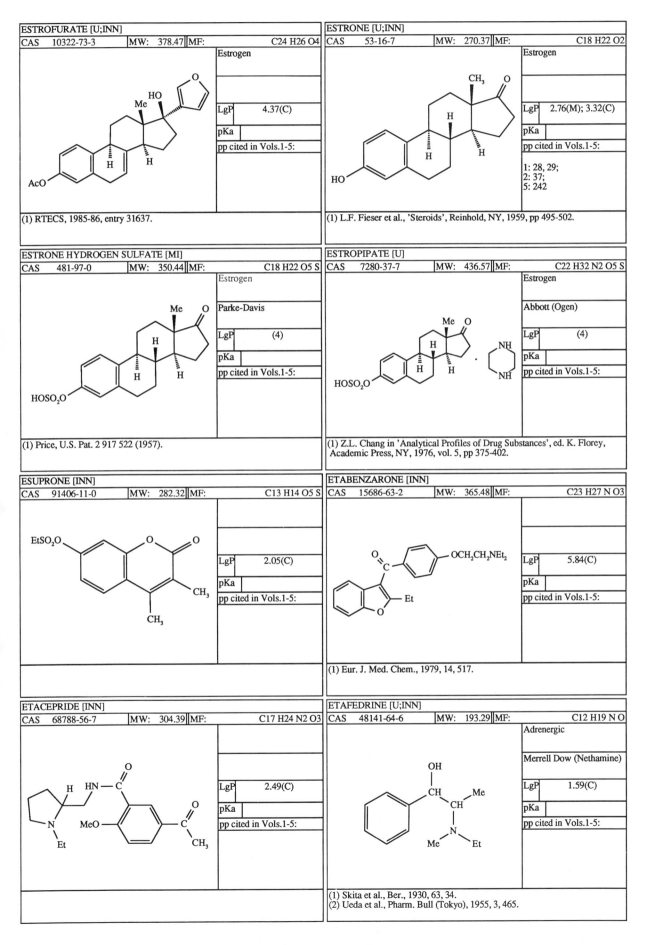

ESTROFURATE [U;INN]

CAS	10322-73-3	MW:	378.47	MF:	C24 H26 O4

Estrogen

LgP	4.37(C)
pKa	
pp cited in Vols.1-5:	

(1) RTECS, 1985-86, entry 31637.

ESTRONE [U;INN]

CAS	53-16-7	MW:	270.37	MF:	C18 H22 O2

Estrogen

LgP	2.76(M); 3.32(C)
pKa	
pp cited in Vols.1-5:	

1: 28, 29;
2: 37;
5: 242

(1) L.F. Fieser et al., 'Steroids', Reinhold, NY, 1959, pp 495-502.

ESTRONE HYDROGEN SULFATE [MI]

CAS	481-97-0	MW:	350.44	MF:	C18 H22 O5 S

Estrogen

Parke-Davis

LgP	(4)
pKa	
pp cited in Vols.1-5:	

(1) Price, U.S. Pat. 2 917 522 (1957).

ESTROPIPATE [U]

CAS	7280-37-7	MW:	436.57	MF:	C22 H32 N2 O5 S

Estrogen

Abbott (Ogen)

LgP	(4)
pKa	
pp cited in Vols.1-5:	

(1) Z.L. Chang in 'Analytical Profiles of Drug Substances', ed. K. Florey, Academic Press, NY, 1976, vol. 5, pp 375-402.

ESUPRONE [INN]

CAS	91406-11-0	MW:	282.32	MF:	C13 H14 O5 S

LgP	2.05(C)
pKa	
pp cited in Vols.1-5:	

ETABENZARONE [INN]

CAS	15686-63-2	MW:	365.48	MF:	C23 H27 N O3

LgP	5.84(C)
pKa	
pp cited in Vols.1-5:	

(1) Eur. J. Med. Chem., 1979, 14, 517.

ETACEPRIDE [INN]

CAS	68788-56-7	MW:	304.39	MF:	C17 H24 N2 O3

LgP	2.49(C)
pKa	
pp cited in Vols.1-5:	

ETAFEDRINE [U;INN]

CAS	48141-64-6	MW:	193.29	MF:	C12 H19 N O

Adrenergic

Merrell Dow (Nethamine)

LgP	1.59(C)
pKa	
pp cited in Vols.1-5:	

(1) Skita et al., Ber., 1930, 63, 34.
(2) Ueda et al., Pharm. Bull (Tokyo), 1955, 3, 465.

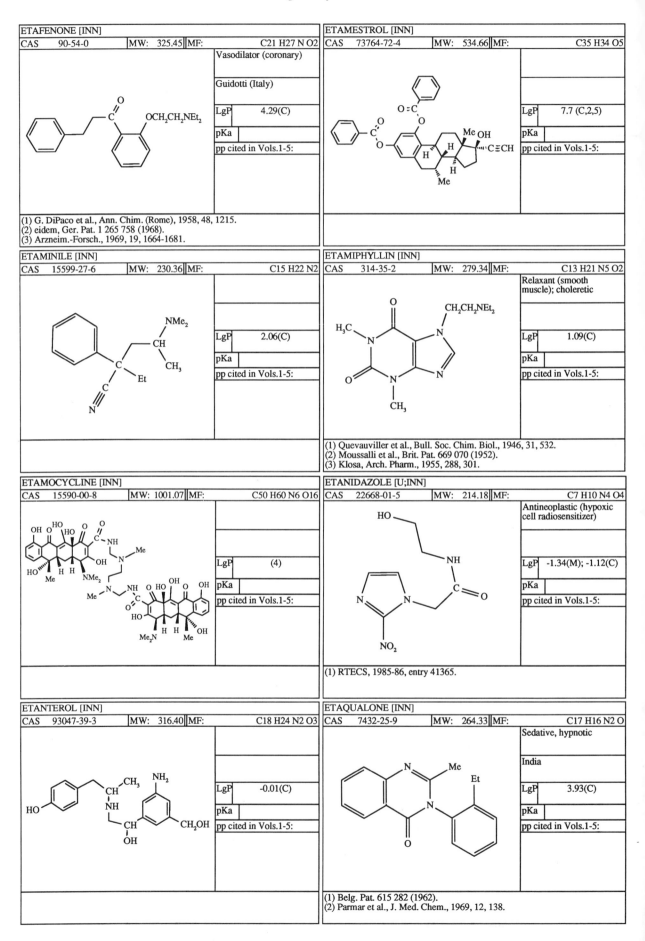

ETAFENONE [INN]

CAS	90-54-0	MW: 325.45	MF:	C21 H27 N O2

Vasodilator (coronary)

Guidotti (Italy)

LgP	4.29(C)
pKa	
pp cited in Vols.1-5:	

(1) G. DiPaco et al., Ann. Chim. (Rome), 1958, 48, 1215.
(2) eidem, Ger. Pat. 1 265 758 (1968).
(3) Arzneim.-Forsch., 1969, 19, 1664-1681.

ETAMESTROL [INN]

CAS	73764-72-4	MW: 534.66	MF:	C35 H34 O5

LgP	7.7 (C,2,5)
pKa	
pp cited in Vols.1-5:	

ETAMINILE [INN]

CAS	15599-27-6	MW: 230.36	MF:	C15 H22 N2

LgP	2.06(C)
pKa	
pp cited in Vols.1-5:	

ETAMIPHYLLIN [INN]

CAS	314-35-2	MW: 279.34	MF:	C13 H21 N5 O2

Relaxant (smooth muscle); choleretic

LgP	1.09(C)
pKa	
pp cited in Vols.1-5:	

(1) Quevauviller et al., Bull. Soc. Chim. Biol., 1946, 31, 532.
(2) Moussalli et al., Brit. Pat. 669 070 (1952).
(3) Klosa, Arch. Pharm., 1955, 288, 301.

ETAMOCYCLINE [INN]

CAS	15590-00-8	MW: 1001.07	MF:	C50 H60 N6 O16

LgP	(4)
pKa	
pp cited in Vols.1-5:	

ETANIDAZOLE [U;INN]

CAS	22668-01-5	MW: 214.18	MF:	C7 H10 N4 O4

Antineoplastic (hypoxic cell radiosensitizer)

LgP	-1.34(M); -1.12(C)
pKa	
pp cited in Vols.1-5:	

(1) RTECS, 1985-86, entry 41365.

ETANTEROL [INN]

CAS	93047-39-3	MW: 316.40	MF:	C18 H24 N2 O3

LgP	-0.01(C)
pKa	
pp cited in Vols.1-5:	

ETAQUALONE [INN]

CAS	7432-25-9	MW: 264.33	MF:	C17 H16 N2 O

Sedative, hypnotic

India

LgP	3.93(C)
pKa	
pp cited in Vols.1-5:	

(1) Belg. Pat. 615 282 (1962).
(2) Parmar et al., J. Med. Chem., 1969, 12, 138.

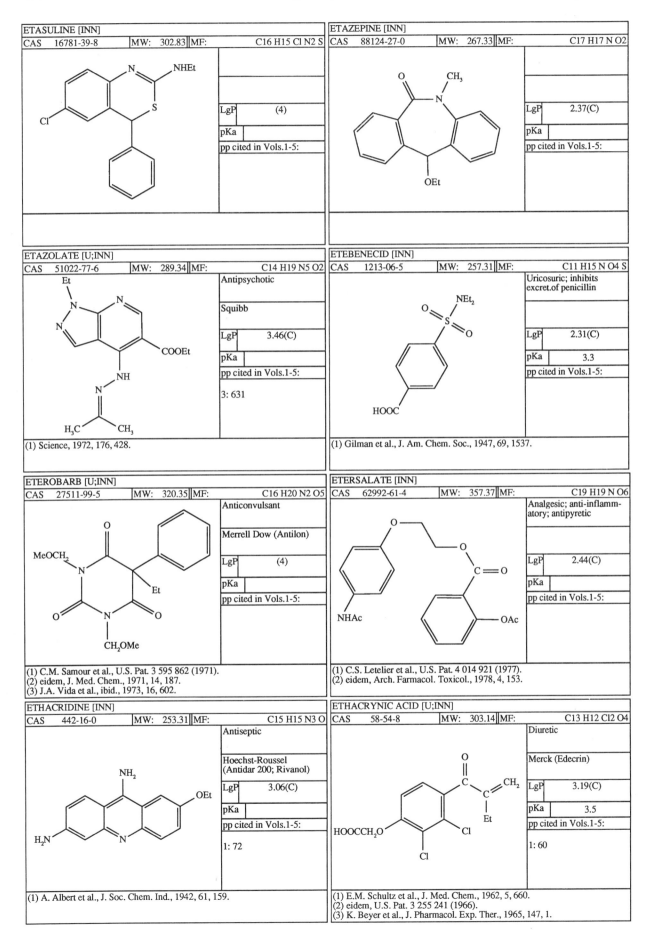

ETASULINE [INN]

CAS	16781-39-8	MW:	302.83	MF:	C16 H15 Cl N2 S

LgP	(4)
pKa	
pp cited in Vols.1-5:	

ETAZEPINE [INN]

CAS	88124-27-0	MW:	267.33	MF:	C17 H17 N O2

LgP	2.37(C)
pKa	
pp cited in Vols.1-5:	

ETAZOLATE [U;INN]

CAS	51022-77-6	MW:	289.34	MF:	C14 H19 N5 O2

Antipsychotic

Squibb

LgP	3.46(C)
pKa	
pp cited in Vols.1-5:	
3: 631	

(1) Science, 1972, 176, 428.

ETEBENECID [INN]

CAS	1213-06-5	MW:	257.31	MF:	C11 H15 N O4 S

Uricosuric; inhibits excret.of penicillin

LgP	2.31(C)
pKa	3.3
pp cited in Vols.1-5:	

(1) Gilman et al., J. Am. Chem. Soc., 1947, 69, 1537.

ETEROBARB [U;INN]

CAS	27511-99-5	MW:	320.35	MF:	C16 H20 N2 O5

Anticonvulsant

Merrell Dow (Antilon)

LgP	(4)
pKa	
pp cited in Vols.1-5:	

(1) C.M. Samour et al., U.S. Pat. 3 595 862 (1971).
(2) eidem, J. Med. Chem., 1971, 14, 187.
(3) J.A. Vida et al., ibid., 1973, 16, 602.

ETERSALATE [INN]

CAS	62992-61-4	MW:	357.37	MF:	C19 H19 N O6

Analgesic; anti-inflammatory; antipyretic

LgP	2.44(C)
pKa	
pp cited in Vols.1-5:	

(1) C.S. Letelier et al., U.S. Pat. 4 014 921 (1977).
(2) eidem, Arch. Farmacol. Toxicol., 1978, 4, 153.

ETHACRIDINE [INN]

CAS	442-16-0	MW:	253.31	MF:	C15 H15 N3 O

Antiseptic

Hoechst-Roussel (Antidar 200; Rivanol)

LgP	3.06(C)
pKa	
pp cited in Vols.1-5:	
1: 72	

(1) A. Albert et al., J. Soc. Chem. Ind., 1942, 61, 159.

ETHACRYNIC ACID [U;INN]

CAS	58-54-8	MW:	303.14	MF:	C13 H12 Cl2 O4

Diuretic

Merck (Edecrin)

LgP	3.19(C)
pKa	3.5
pp cited in Vols.1-5:	
1: 60	

(1) E.M. Schultz et al., J. Med. Chem., 1962, 5, 660.
(2) eidem, U.S. Pat. 3 255 241 (1966).
(3) K. Beyer et al., J. Pharmacol. Exp. Ther., 1965, 147, 1.

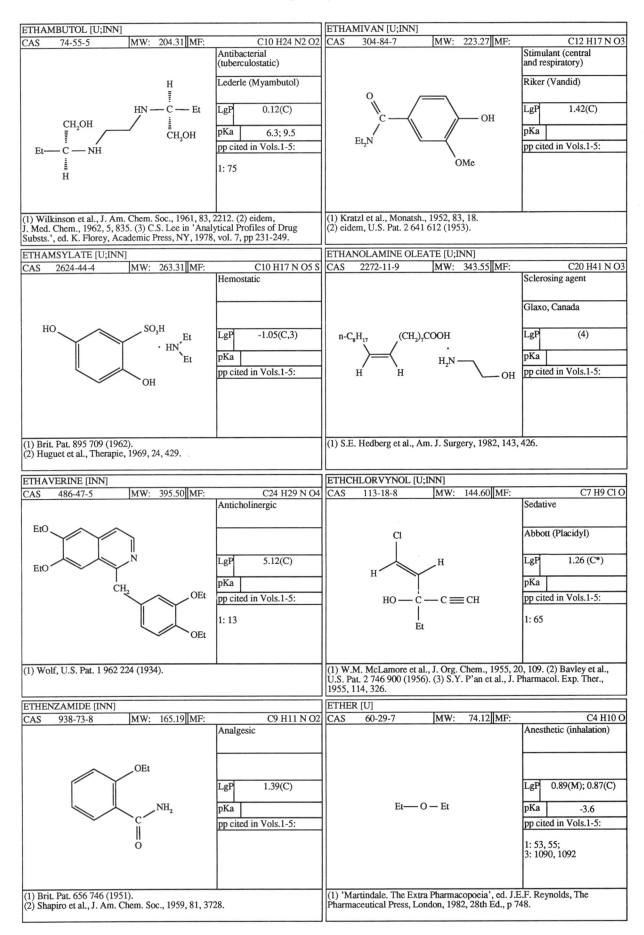

ETHAMBUTOL [U;INN]

CAS	74-55-5	MW: 204.31	MF:	C10 H24 N2 O2

Antibacterial (tuberculostatic)

Lederle (Myambutol)

LgP	0.12(C)
pKa	6.3; 9.5
pp cited in Vols.1-5:	

1: 75

(1) Wilkinson et al., J. Am. Chem. Soc., 1961, 83, 2212. (2) eidem, J. Med. Chem., 1962, 5, 835. (3) C.S. Lee in 'Analytical Profiles of Drug Substs.', ed. K. Florey, Academic Press, NY, 1978, vol. 7, pp 231-249.

ETHAMIVAN [U;INN]

CAS	304-84-7	MW: 223.27	MF:	C12 H17 N O3

Stimulant (central and respiratory)

Riker (Vandid)

LgP	1.42(C)
pKa	
pp cited in Vols.1-5:	

(1) Kratzl et al., Monatsh., 1952, 83, 18.
(2) eidem, U.S. Pat. 2 641 612 (1953).

ETHAMSYLATE [U;INN]

CAS	2624-44-4	MW: 263.31	MF:	C10 H17 N O5 S

Hemostatic

LgP	-1.05(C,3)
pKa	
pp cited in Vols.1-5:	

(1) Brit. Pat. 895 709 (1962).
(2) Huguet et al., Therapie, 1969, 24, 429.

ETHANOLAMINE OLEATE [U;INN]

CAS	2272-11-9	MW: 343.55	MF:	C20 H41 N O3

Sclerosing agent

Glaxo, Canada

LgP	(4)
pKa	
pp cited in Vols.1-5:	

(1) S.E. Hedberg et al., Am. J. Surgery, 1982, 143, 426.

ETHAVERINE [INN]

CAS	486-47-5	MW: 395.50	MF:	C24 H29 N O4

Anticholinergic

LgP	5.12(C)
pKa	
pp cited in Vols.1-5:	

1: 13

(1) Wolf, U.S. Pat. 1 962 224 (1934).

ETHCHLORVYNOL [U;INN]

CAS	113-18-8	MW: 144.60	MF:	C7 H9 Cl O

Sedative

Abbott (Placidyl)

LgP	1.26 (C*)
pKa	
pp cited in Vols.1-5:	

1: 65

(1) W.M. McLamore et al., J. Org. Chem., 1955, 20, 109. (2) Bavley et al., U.S. Pat. 2 746 900 (1956). (3) S.Y. P'an et al., J. Pharmacol. Exp. Ther., 1955, 114, 326.

ETHENZAMIDE [INN]

CAS	938-73-8	MW: 165.19	MF:	C9 H11 N O2

Analgesic

LgP	1.39(C)
pKa	
pp cited in Vols.1-5:	

(1) Brit. Pat. 656 746 (1951).
(2) Shapiro et al., J. Am. Chem. Soc., 1959, 81, 3728.

ETHER [U]

CAS	60-29-7	MW: 74.12	MF:	C4 H10 O

Anesthetic (inhalation)

LgP	0.89(M); 0.87(C)
pKa	-3.6
pp cited in Vols.1-5:	

1: 53, 55;
3: 1090, 1092

(1) 'Martindale. The Extra Pharmacopoeia', ed. J.E.F. Reynolds, The Pharmaceutical Press, London, 1982, 28th Ed., p 748.

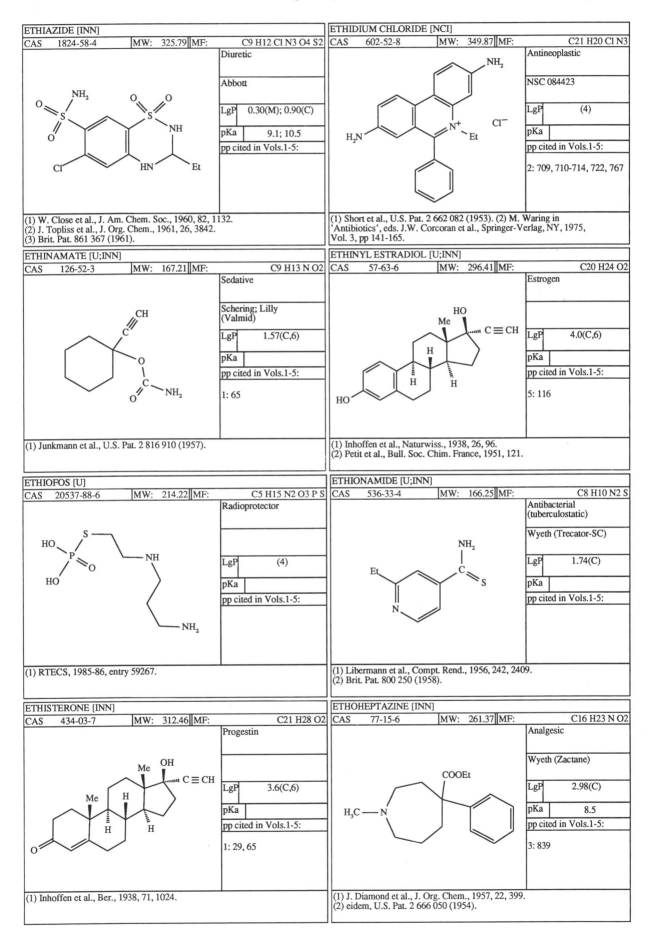

ETHIAZIDE [INN]

CAS	1824-58-4	MW: 325.79	MF:	C9 H12 Cl N3 O4 S2

Diuretic

Abbott

LgP	0.30(M); 0.90(C)
pKa	9.1; 10.5
pp cited in Vols.1-5:	

(1) W. Close et al., J. Am. Chem. Soc., 1960, 82, 1132.
(2) J. Topliss et al., J. Org. Chem., 1961, 26, 3842.
(3) Brit. Pat. 861 367 (1961).

ETHIDIUM CHLORIDE [NCI]

CAS	602-52-8	MW: 349.87	MF:	C21 H20 Cl N3

Antineoplastic

NSC 084423

LgP	(4)
pKa	
pp cited in Vols.1-5:	

2: 709, 710-714, 722, 767

(1) Short et al., U.S. Pat. 2 662 082 (1953). (2) M. Waring in
'Antibiotics', eds. J.W. Corcoran et al., Springer-Verlag, NY, 1975,
Vol. 3, pp 141-165.

ETHINAMATE [U;INN]

CAS	126-52-3	MW: 167.21	MF:	C9 H13 N O2

Sedative

Schering; Lilly
(Valmid)

LgP	1.57(C,6)
pKa	
pp cited in Vols.1-5:	

1: 65

(1) Junkmann et al., U.S. Pat. 2 816 910 (1957).

ETHINYL ESTRADIOL [U;INN]

CAS	57-63-6	MW: 296.41	MF:	C20 H24 O2

Estrogen

LgP	4.0(C,6)
pKa	
pp cited in Vols.1-5:	

5: 116

(1) Inhoffen et al., Naturwiss., 1938, 26, 96.
(2) Petit et al., Bull. Soc. Chim. France, 1951, 121.

ETHIOFOS [U]

CAS	20537-88-6	MW: 214.22	MF:	C5 H15 N2 O3 P S

Radioprotector

LgP	(4)
pKa	
pp cited in Vols.1-5:	

(1) RTECS, 1985-86, entry 59267.

ETHIONAMIDE [U;INN]

CAS	536-33-4	MW: 166.25	MF:	C8 H10 N2 S

Antibacterial
(tuberculostatic)

Wyeth (Trecator-SC)

LgP	1.74(C)
pKa	
pp cited in Vols.1-5:	

(1) Libermann et al., Compt. Rend., 1956, 242, 2409.
(2) Brit. Pat. 800 250 (1958).

ETHISTERONE [INN]

CAS	434-03-7	MW: 312.46	MF:	C21 H28 O2

Progestin

LgP	3.6(C,6)
pKa	
pp cited in Vols.1-5:	

1: 29, 65

(1) Inhoffen et al., Ber., 1938, 71, 1024.

ETHOHEPTAZINE [INN]

CAS	77-15-6	MW: 261.37	MF:	C16 H23 N O2

Analgesic

Wyeth (Zactane)

LgP	2.98(C)
pKa	8.5
pp cited in Vols.1-5:	

3: 839

(1) J. Diamond et al., J. Org. Chem., 1957, 22, 399.
(2) eidem, U.S. Pat. 2 666 050 (1954).

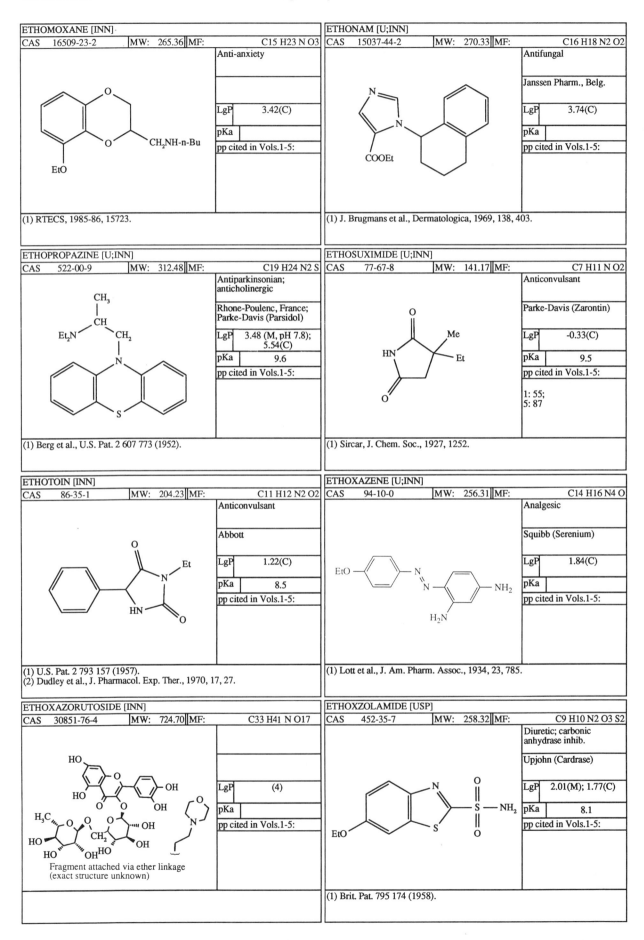

ETHOMOXANE [INN]

| CAS | 16509-23-2 | MW: | 265.36 | MF: | C15 H23 N O3 |

Anti-anxiety

LgP	3.42(C)
pKa	
pp cited in Vols.1-5:	

(1) RTECS, 1985-86, 15723.

ETHONAM [U;INN]

| CAS | 15037-44-2 | MW: | 270.33 | MF: | C16 H18 N2 O2 |

Antifungal

Janssen Pharm., Belg.

LgP	3.74(C)
pKa	
pp cited in Vols.1-5:	

(1) J. Brugmans et al., Dermatologica, 1969, 138, 403.

ETHOPROPAZINE [U;INN]

| CAS | 522-00-9 | MW: | 312.48 | MF: | C19 H24 N2 S |

Antiparkinsonian; anticholinergic

Rhone-Poulenc, France; Parke-Davis (Parsidol)

LgP	3.48 (M, pH 7.8); 5.54(C)
pKa	9.6
pp cited in Vols.1-5:	

(1) Berg et al., U.S. Pat. 2 607 773 (1952).

ETHOSUXIMIDE [U;INN]

| CAS | 77-67-8 | MW: | 141.17 | MF: | C7 H11 N O2 |

Anticonvulsant

Parke-Davis (Zarontin)

LgP	-0.33(C)
pKa	9.5
pp cited in Vols.1-5:	

1: 55;
5: 87

(1) Sircar, J. Chem. Soc., 1927, 1252.

ETHOTOIN [INN]

| CAS | 86-35-1 | MW: | 204.23 | MF: | C11 H12 N2 O2 |

Anticonvulsant

Abbott

LgP	1.22(C)
pKa	8.5
pp cited in Vols.1-5:	

(1) U.S. Pat. 2 793 157 (1957).
(2) Dudley et al., J. Pharmacol. Exp. Ther., 1970, 17, 27.

ETHOXAZENE [U;INN]

| CAS | 94-10-0 | MW: | 256.31 | MF: | C14 H16 N4 O |

Analgesic

Squibb (Serenium)

LgP	1.84(C)
pKa	
pp cited in Vols.1-5:	

(1) Lott et al., J. Am. Pharm. Assoc., 1934, 23, 785.

ETHOXAZORUTOSIDE [INN]

| CAS | 30851-76-4 | MW: | 724.70 | MF: | C33 H41 N O17 |

LgP	(4)
pKa	
pp cited in Vols.1-5:	

Fragment attached via ether linkage
(exact structure unknown)

ETHOXZOLAMIDE [USP]

| CAS | 452-35-7 | MW: | 258.32 | MF: | C9 H10 N2 O3 S2 |

Diuretic; carbonic anhydrase inhib.

Upjohn (Cardrase)

LgP	2.01(M); 1.77(C)
pKa	8.1
pp cited in Vols.1-5:	

(1) Brit. Pat. 795 174 (1958).

ETHYBENZTROPINE [U;INN]

CAS	524-83-4	MW:	321.47	MF:	C22 H27 N O

Anticholinergic; antiparkinsonian

Sandoz, Switz.

LgP	4.19(C)
pKa	
pp cited in Vols.1-5:	

(1) RTECS, 1985-86, 52613.

ETHYL BISCOUMACETATE [INN]

CAS	548-00-5	MW:	408.37	MF:	C22 H16 O8

Anticoagulant

Ciba-Geigy (Tromexan)

LgP	0.90(C)
pKa	7.5
pp cited in Vols.1-5:	

1: 21

(1) Stahmann et al., J. Am. Chem. Soc., 1943, 65, 2285.
(2) U.S. Pat. 2 482 512 (1949).

ETHYL CARFLUZEPATE [INN]

CAS	65400-85-3	MW:	417.83	MF:	C20 H17 Cl F N3 O4

LgP	(4)
pKa	
pp cited in Vols.1-5:	

ETHYL CARTRIZOATE [INN]

CAS	5714-09-0	MW:	700.01	MF:	C15 H15 I3 N2 O6

Radiopaque medium

LgP	0.08(C,3)
pKa	
pp cited in Vols.1-5:	

ETHYL CHLORIDE [U]

CAS	75-00-3	MW:	64.52	MF:	C2 H5 Cl

Anesthetic (topical)

Et— Cl

LgP	1.43(M); 1.47(C)
pKa	
pp cited in Vols.1-5:	

(1) 'The Merck Index', Merck & Co., Rahway, NJ, 1983, 10th Ed., entry 3729. (2) 'Martindale. The Extra Pharmacopoeia', ed. J.E.F. Reynolds, The Pharmaceutical Press, London, 1982, 28th Ed., pp 748-749.

ETHYL DIBUNATE [U;INN]

CAS	5560-69-0	MW:	348.51	MF:	C20 H28 O3 S

Antitussive

Merrell Dow (Neodyne)

LgP	6.17(C)
pKa	
pp cited in Vols.1-5:	

(1) Menard et al., Can. J. Chem., 1961, 39, 729.
(2) Shemano, Arch. Int. Pharmacodyn. Ther., 1967, 165, 410.

ETHYL DIRAZEPATE [INN]

CAS	23980-14-5	MW:	377.23	MF:	C18 H14 Cl2 N2 O3

LgP	3.5(C,6)
pKa	
pp cited in Vols.1-5:	

ETHYLENEDIAMINE [U]

CAS	107-15-3	MW:	60.10	MF:	C2 H8 N2

Urinary acidifier (vet.)

H₂N-CH₂CH₂NH₂

LgP	-2.04(M); -1.22(C)
pKa	7.2; 10.0
pp cited in Vols.1-5:	

(1) 'The Merck Index', Merck & Co., Rahway, NJ, 1983, 10th Ed., entry 3741.

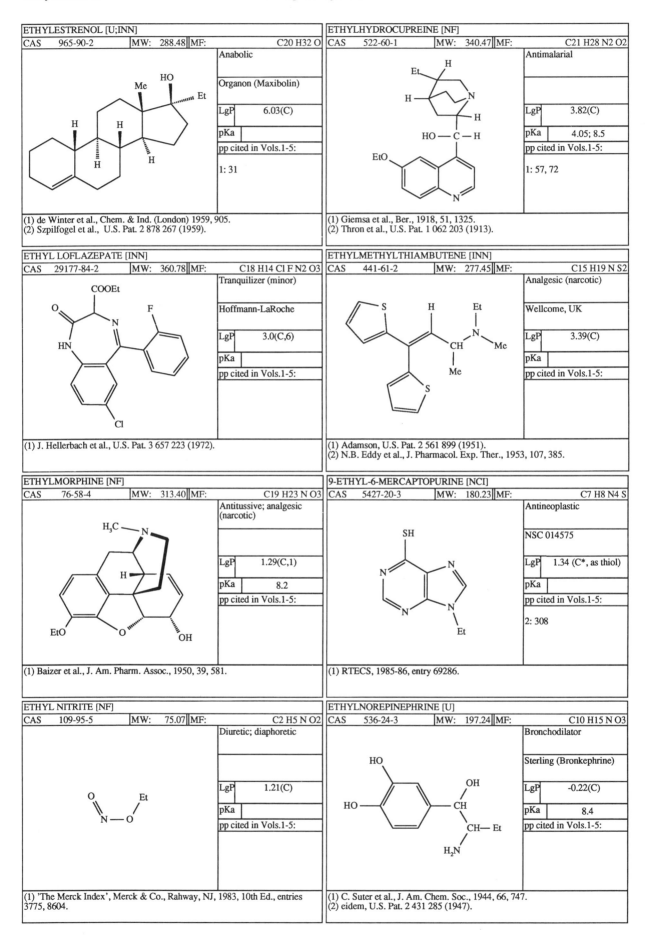

ETHYLESTRENOL [U;INN]			
CAS	965-90-2	MW: 288.48	MF: C20 H32 O

Anabolic

Organon (Maxibolin)

LgP 6.03(C)

pKa

pp cited in Vols.1-5:

1: 31

(1) de Winter et al., Chem. & Ind. (London) 1959, 905.
(2) Szpilfogel et al., U.S. Pat. 2 878 267 (1959).

ETHYLHYDROCUPREINE [NF]			
CAS	522-60-1	MW: 340.47	MF: C21 H28 N2 O2

Antimalarial

LgP 3.82(C)

pKa 4.05; 8.5

pp cited in Vols.1-5:

1: 57, 72

(1) Giemsa et al., Ber., 1918, 51, 1325.
(2) Thron et al., U.S. Pat. 1 062 203 (1913).

ETHYL LOFLAZEPATE [INN]			
CAS	29177-84-2	MW: 360.78	MF: C18 H14 Cl F N2 O3

Tranquilizer (minor)

Hoffmann-LaRoche

LgP 3.0(C,6)

pKa

pp cited in Vols.1-5:

(1) J. Hellerbach et al., U.S. Pat. 3 657 223 (1972).

ETHYLMETHYLTHIAMBUTENE [INN]			
CAS	441-61-2	MW: 277.45	MF: C15 H19 N S2

Analgesic (narcotic)

Wellcome, UK

LgP 3.39(C)

pKa

pp cited in Vols.1-5:

(1) Adamson, U.S. Pat. 2 561 899 (1951).
(2) N.B. Eddy et al., J. Pharmacol. Exp. Ther., 1953, 107, 385.

ETHYLMORPHINE [NF]			
CAS	76-58-4	MW: 313.40	MF: C19 H23 N O3

Antitussive; analgesic (narcotic)

LgP 1.29(C,1)

pKa 8.2

pp cited in Vols.1-5:

(1) Baizer et al., J. Am. Pharm. Assoc., 1950, 39, 581.

9-ETHYL-6-MERCAPTOPURINE [NCI]			
CAS	5427-20-3	MW: 180.23	MF: C7 H8 N4 S

Antineoplastic

NSC 014575

LgP 1.34 (C*, as thiol)

pKa

pp cited in Vols.1-5:

2: 308

(1) RTECS, 1985-86, entry 69286.

ETHYL NITRITE [NF]			
CAS	109-95-5	MW: 75.07	MF: C2 H5 N O2

Diuretic; diaphoretic

LgP 1.21(C)

pKa

pp cited in Vols.1-5:

(1) 'The Merck Index', Merck & Co., Rahway, NJ, 1983, 10th Ed., entries 3775, 8604.

ETHYLNOREPINEPHRINE [U]			
CAS	536-24-3	MW: 197.24	MF: C10 H15 N O3

Bronchodilator

Sterling (Bronkephrine)

LgP -0.22(C)

pKa 8.4

pp cited in Vols.1-5:

(1) C. Suter et al., J. Am. Chem. Soc., 1944, 66, 747.
(2) eidem, U.S. Pat. 2 431 285 (1947).

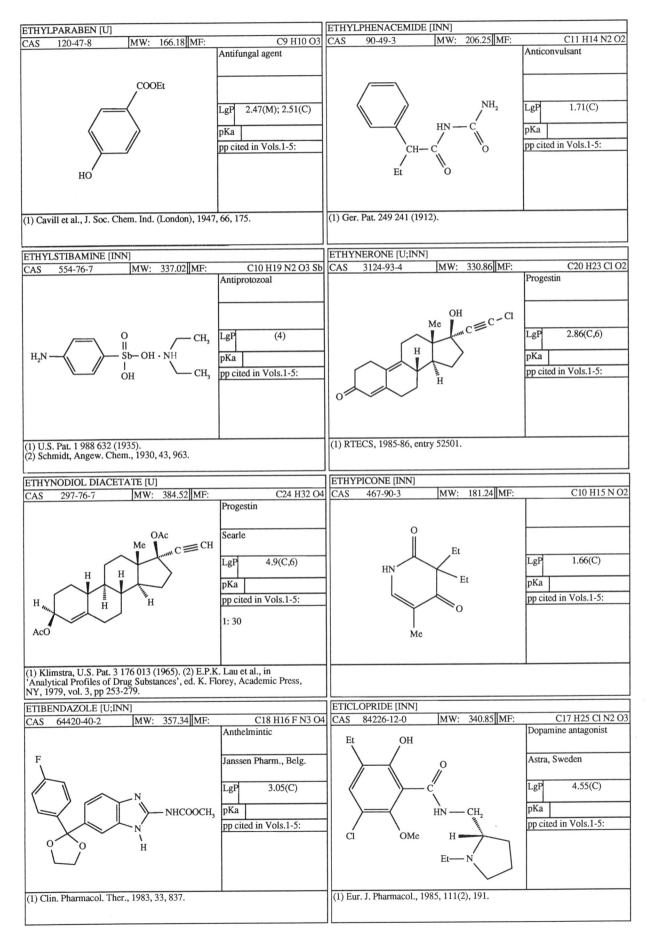

ETHYLPARABEN [U]

CAS	120-47-8	MW:	166.18	MF:	C9 H10 O3

Antifungal agent

LgP	2.47(M); 2.51(C)
pKa	
pp cited in Vols.1-5:	

(1) Cavill et al., J. Soc. Chem. Ind. (London), 1947, 66, 175.

ETHYLSTIBAMINE [INN]

CAS	554-76-7	MW:	337.02	MF:	C10 H19 N2 O3 Sb

Antiprotozoal

LgP	(4)
pKa	
pp cited in Vols.1-5:	

(1) U.S. Pat. 1 988 632 (1935).
(2) Schmidt, Angew. Chem., 1930, 43, 963.

ETHYNODIOL DIACETATE [U]

CAS	297-76-7	MW:	384.52	MF:	C24 H32 O4

Progestin

Searle

LgP	4.9(C,6)
pKa	
pp cited in Vols.1-5:	
1: 30	

(1) Klimstra, U.S. Pat. 3 176 013 (1965). (2) E.P.K. Lau et al., in 'Analytical Profiles of Drug Substances', ed. K. Florey, Academic Press, NY, 1979, vol. 3, pp 253-279.

ETIBENDAZOLE [U;INN]

CAS	64420-40-2	MW:	357.34	MF:	C18 H16 F N3 O4

Anthelmintic

Janssen Pharm., Belg.

LgP	3.05(C)
pKa	
pp cited in Vols.1-5:	

(1) Clin. Pharmacol. Ther., 1983, 33, 837.

ETHYLPHENACEMIDE [INN]

CAS	90-49-3	MW:	206.25	MF:	C11 H14 N2 O2

Anticonvulsant

LgP	1.71(C)
pKa	
pp cited in Vols.1-5:	

(1) Ger. Pat. 249 241 (1912).

ETHYNERONE [U;INN]

CAS	3124-93-4	MW:	330.86	MF:	C20 H23 Cl O2

Progestin

LgP	2.86(C,6)
pKa	
pp cited in Vols.1-5:	

(1) RTECS, 1985-86, entry 52501.

ETHYPICONE [INN]

CAS	467-90-3	MW:	181.24	MF:	C10 H15 N O2

LgP	1.66(C)
pKa	
pp cited in Vols.1-5:	

ETICLOPRIDE [INN]

CAS	84226-12-0	MW:	340.85	MF:	C17 H25 Cl N2 O3

Dopamine antagonist

Astra, Sweden

LgP	4.55(C)
pKa	
pp cited in Vols.1-5:	

(1) Eur. J. Pharmacol., 1985, 111(2), 191.

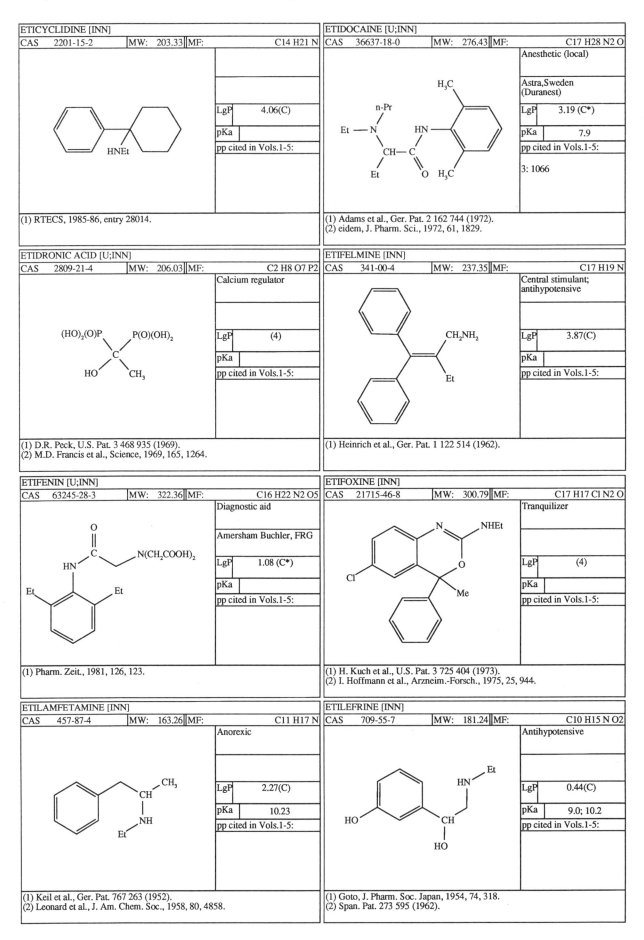

ETICYCLIDINE [INN]

CAS	2201-15-2	MW: 203.33	MF:	C14 H21 N

LgP	4.06(C)
pKa	
pp cited in Vols.1-5:	

(1) RTECS, 1985-86, entry 28014.

ETIDRONIC ACID [U;INN]

CAS	2809-21-4	MW: 206.03	MF:	C2 H8 O7 P2

Calcium regulator

LgP	(4)
pKa	
pp cited in Vols.1-5:	

(1) D.R. Peck, U.S. Pat. 3 468 935 (1969).
(2) M.D. Francis et al., Science, 1969, 165, 1264.

ETIFENIN [U;INN]

CAS	63245-28-3	MW: 322.36	MF:	C16 H22 N2 O5

Diagnostic aid

Amersham Buchler, FRG

LgP	1.08 (C*)
pKa	
pp cited in Vols.1-5:	

(1) Pharm. Zeit., 1981, 126, 123.

ETILAMFETAMINE [INN]

CAS	457-87-4	MW: 163.26	MF:	C11 H17 N

Anorexic

LgP	2.27(C)
pKa	10.23
pp cited in Vols.1-5:	

(1) Keil et al., Ger. Pat. 767 263 (1952).
(2) Leonard et al., J. Am. Chem. Soc., 1958, 80, 4858.

ETIDOCAINE [U;INN]

CAS	36637-18-0	MW: 276.43	MF:	C17 H28 N2 O

Anesthetic (local)

Astra,Sweden
(Duranest)

LgP	3.19 (C*)
pKa	7.9
pp cited in Vols.1-5:	

3: 1066

(1) Adams et al., Ger. Pat. 2 162 744 (1972).
(2) eidem, J. Pharm. Sci., 1972, 61, 1829.

ETIFELMINE [INN]

CAS	341-00-4	MW: 237.35	MF:	C17 H19 N

Central stimulant;
antihypotensive

LgP	3.87(C)
pKa	
pp cited in Vols.1-5:	

(1) Heinrich et al., Ger. Pat. 1 122 514 (1962).

ETIFOXINE [INN]

CAS	21715-46-8	MW: 300.79	MF:	C17 H17 Cl N2 O

Tranquilizer

LgP	(4)
pKa	
pp cited in Vols.1-5:	

(1) H. Kuch et al., U.S. Pat. 3 725 404 (1973).
(2) I. Hoffmann et al., Arzneim.-Forsch., 1975, 25, 944.

ETILEFRINE [INN]

CAS	709-55-7	MW: 181.24	MF:	C10 H15 N O2

Antihypotensive

LgP	0.44(C)
pKa	9.0; 10.2
pp cited in Vols.1-5:	

(1) Goto, J. Pharm. Soc. Japan, 1954, 74, 318.
(2) Span. Pat. 273 595 (1962).

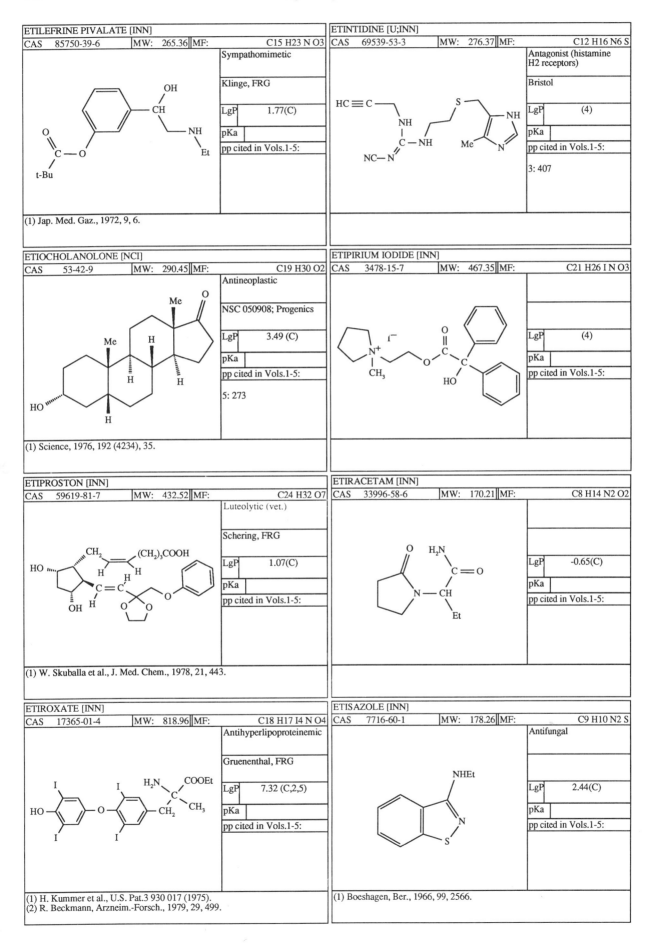

ETILEFRINE PIVALATE [INN]

CAS	85750-39-6	MW:	265.36	MF:	C15 H23 N O3

Sympathomimetic

Klinge, FRG

LgP	1.77(C)
pKa	
pp cited in Vols.1-5:	

(1) Jap. Med. Gaz., 1972, 9, 6.

ETINTIDINE [U;INN]

CAS	69539-53-3	MW:	276.37	MF:	C12 H16 N6 S

Antagonist (histamine H2 receptors)

Bristol

LgP	(4)
pKa	
pp cited in Vols.1-5:	

3: 407

ETIOCHOLANOLONE [NCI]

CAS	53-42-9	MW:	290.45	MF:	C19 H30 O2

Antineoplastic

NSC 050908; Progenics

LgP	3.49 (C)
pKa	
pp cited in Vols.1-5:	

5: 273

(1) Science, 1976, 192 (4234), 35.

ETIPIRIUM IODIDE [INN]

CAS	3478-15-7	MW:	467.35	MF:	C21 H26 I N O3

LgP	(4)
pKa	
pp cited in Vols.1-5:	

ETIPROSTON [INN]

CAS	59619-81-7	MW:	432.52	MF:	C24 H32 O7

Luteolytic (vet.)

Schering, FRG

LgP	1.07(C)
pKa	
pp cited in Vols.1-5:	

(1) W. Skuballa et al., J. Med. Chem., 1978, 21, 443.

ETIRACETAM [INN]

CAS	33996-58-6	MW:	170.21	MF:	C8 H14 N2 O2

LgP	-0.65(C)
pKa	
pp cited in Vols.1-5:	

ETIROXATE [INN]

CAS	17365-01-4	MW:	818.96	MF:	C18 H17 I4 N O4

Antihyperlipoproteinemic

Gruenenthal, FRG

LgP	7.32 (C,2,5)
pKa	
pp cited in Vols.1-5:	

(1) H. Kummer et al., U.S. Pat.3 930 017 (1975).
(2) R. Beckmann, Arzneim.-Forsch., 1979, 29, 499.

ETISAZOLE [INN]

CAS	7716-60-1	MW:	178.26	MF:	C9 H10 N2 S

Antifungal

LgP	2.44(C)
pKa	
pp cited in Vols.1-5:	

(1) Boeshagen, Ber., 1966, 99, 2566.

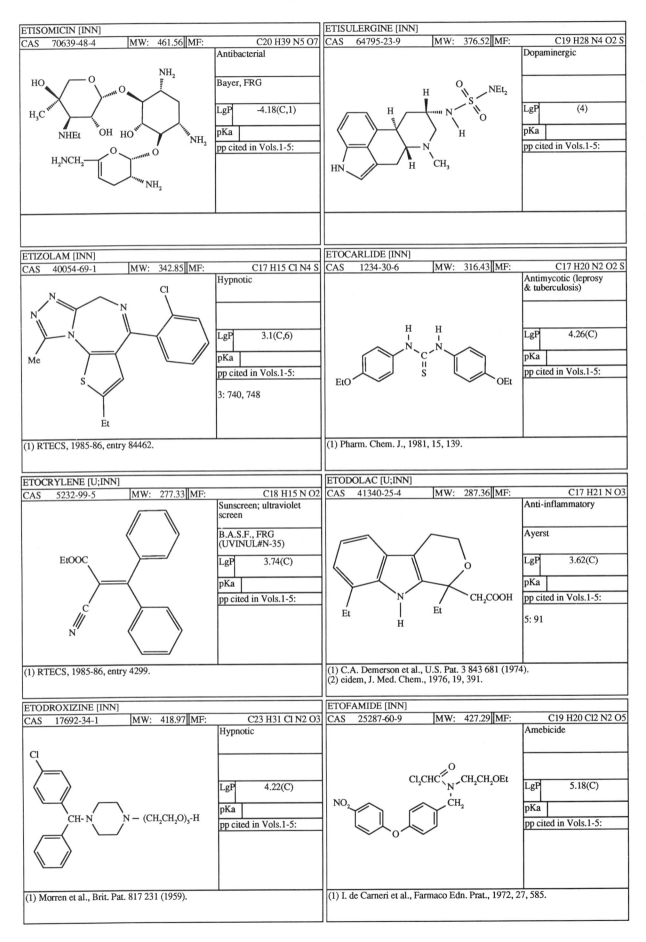

ETISOMICIN [INN]

CAS 70639-48-4 | MW: 461.56 | MF: C20 H39 N5 O7

Antibacterial

Bayer, FRG

LgP -4.18(C,1)

pKa

pp cited in Vols.1-5:

ETISULERGINE [INN]

CAS 64795-23-9 | MW: 376.52 | MF: C19 H28 N4 O2 S

Dopaminergic

LgP (4)

pKa

pp cited in Vols.1-5:

ETIZOLAM [INN]

CAS 40054-69-1 | MW: 342.85 | MF: C17 H15 Cl N4 S

Hypnotic

LgP 3.1(C,6)

pKa

pp cited in Vols.1-5:

3: 740, 748

(1) RTECS, 1985-86, entry 84462.

ETOCARLIDE [INN]

CAS 1234-30-6 | MW: 316.43 | MF: C17 H20 N2 O2 S

Antimycotic (leprosy & tuberculosis)

LgP 4.26(C)

pKa

pp cited in Vols.1-5:

(1) Pharm. Chem. J., 1981, 15, 139.

ETOCRYLENE [U;INN]

CAS 5232-99-5 | MW: 277.33 | MF: C18 H15 N O2

Sunscreen; ultraviolet screen

B.A.S.F., FRG (UVINUL#N-35)

LgP 3.74(C)

pKa

pp cited in Vols.1-5:

(1) RTECS, 1985-86, entry 4299.

ETODOLAC [U;INN]

CAS 41340-25-4 | MW: 287.36 | MF: C17 H21 N O3

Anti-inflammatory

Ayerst

LgP 3.62(C)

pKa

pp cited in Vols.1-5:

5: 91

(1) C.A. Demerson et al., U.S. Pat. 3 843 681 (1974).
(2) eidem, J. Med. Chem., 1976, 19, 391.

ETODROXIZINE [INN]

CAS 17692-34-1 | MW: 418.97 | MF: C23 H31 Cl N2 O3

Hypnotic

LgP 4.22(C)

pKa

pp cited in Vols.1-5:

(1) Morren et al., Brit. Pat. 817 231 (1959).

ETOFAMIDE [INN]

CAS 25287-60-9 | MW: 427.29 | MF: C19 H20 Cl2 N2 O5

Amebicide

LgP 5.18(C)

pKa

pp cited in Vols.1-5:

(1) I. de Carneri et al., Farmaco Edn. Prat., 1972, 27, 585.

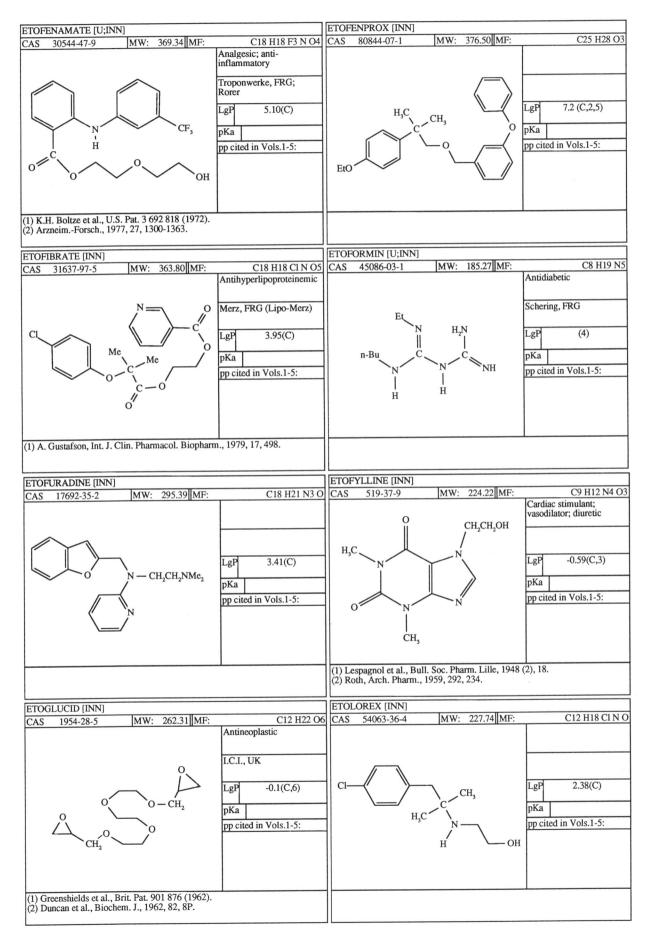

ETOFENAMATE [U;INN]

| CAS | 30544-47-9 | MW: | 369.34 | MF: | C18 H18 F3 N O4 |

Analgesic; anti-inflammatory

Troponwerke, FRG; Rorer

LgP	5.10(C)
pKa	
pp cited in Vols.1-5:	

(1) K.H. Boltze et al., U.S. Pat. 3 692 818 (1972).
(2) Arzneim.-Forsch., 1977, 27, 1300-1363.

ETOFENPROX [INN]

| CAS | 80844-07-1 | MW: | 376.50 | MF: | C25 H28 O3 |

LgP	7.2 (C,2,5)
pKa	
pp cited in Vols.1-5:	

ETOFIBRATE [INN]

| CAS | 31637-97-5 | MW: | 363.80 | MF: | C18 H18 Cl N O5 |

Antihyperlipoproteinemic

Merz, FRG (Lipo-Merz)

LgP	3.95(C)
pKa	
pp cited in Vols.1-5:	

(1) A. Gustafson, Int. J. Clin. Pharmacol. Biopharm., 1979, 17, 498.

ETOFORMIN [U;INN]

| CAS | 45086-03-1 | MW: | 185.27 | MF: | C8 H19 N5 |

Antidiabetic

Schering, FRG

LgP	(4)
pKa	
pp cited in Vols.1-5:	

ETOFURADINE [INN]

| CAS | 17692-35-2 | MW: | 295.39 | MF: | C18 H21 N3 O |

LgP	3.41(C)
pKa	
pp cited in Vols.1-5:	

ETOFYLLINE [INN]

| CAS | 519-37-9 | MW: | 224.22 | MF: | C9 H12 N4 O3 |

Cardiac stimulant; vasodilator; diuretic

LgP	-0.59(C,3)
pKa	
pp cited in Vols.1-5:	

(1) Lespagnol et al., Bull. Soc. Pharm. Lille, 1948 (2), 18.
(2) Roth, Arch. Pharm., 1959, 292, 234.

ETOGLUCID [INN]

| CAS | 1954-28-5 | MW: | 262.31 | MF: | C12 H22 O6 |

Antineoplastic

I.C.I., UK

LgP	-0.1(C,6)
pKa	
pp cited in Vols.1-5:	

(1) Greenshields et al., Brit. Pat. 901 876 (1962).
(2) Duncan et al., Biochem. J., 1962, 82, 8P.

ETOLOREX [INN]

| CAS | 54063-36-4 | MW: | 227.74 | MF: | C12 H18 Cl N O |

LgP	2.38(C)
pKa	
pp cited in Vols.1-5:	

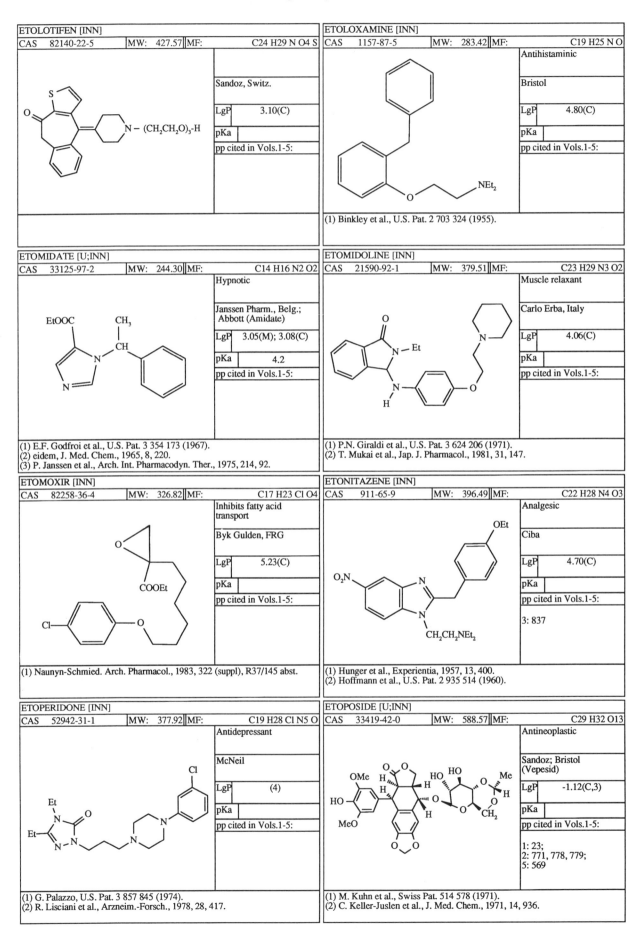

ETOLOTIFEN [INN]

CAS 82140-22-5 | MW: 427.57 | MF: C24 H29 N O4 S

Sandoz, Switz.

LgP 3.10(C)

pKa

pp cited in Vols.1-5:

O=... thieno ...—N—(CH₂CH₂O)₃-H

ETOLOXAMINE [INN]

CAS 1157-87-5 | MW: 283.42 | MF: C19 H25 N O

Antihistaminic

Bristol

LgP 4.80(C)

pKa

pp cited in Vols.1-5:

—O—CH₂CH₂—NEt₂

(1) Binkley et al., U.S. Pat. 2 703 324 (1955).

ETOMIDATE [U;INN]

CAS 33125-97-2 | MW: 244.30 | MF: C14 H16 N2 O2

Hypnotic

Janssen Pharm., Belg.; Abbott (Amidate)

LgP 3.05(M); 3.08(C)

pKa 4.2

pp cited in Vols.1-5:

EtOOC— ... CH₃ ... CH ...

(1) E.F. Godfroi et al., U.S. Pat. 3 354 173 (1967).
(2) eidem, J. Med. Chem., 1965, 8, 220.
(3) P. Janssen et al., Arch. Int. Pharmacodyn. Ther., 1975, 214, 92.

ETOMIDOLINE [INN]

CAS 21590-92-1 | MW: 379.51 | MF: C23 H29 N3 O2

Muscle relaxant

Carlo Erba, Italy

LgP 4.06(C)

pKa

pp cited in Vols.1-5:

N—Et ... NH ... O—

(1) P.N. Giraldi et al., U.S. Pat. 3 624 206 (1971).
(2) T. Mukai et al., Jap. J. Pharmacol., 1981, 31, 147.

ETOMOXIR [INN]

CAS 82258-36-4 | MW: 326.82 | MF: C17 H23 Cl O4

Inhibits fatty acid transport

Byk Gulden, FRG

LgP 5.23(C)

pKa

pp cited in Vols.1-5:

Cl— ... —O— ... COOEt

(1) Naunyn-Schmied. Arch. Pharmacol., 1983, 322 (suppl), R37/145 abst.

ETONITAZENE [INN]

CAS 911-65-9 | MW: 396.49 | MF: C22 H28 N4 O3

Analgesic

Ciba

LgP 4.70(C)

pKa

pp cited in Vols.1-5:

3: 837

O₂N— ... OEt ... CH₂CH₂NEt₂

(1) Hunger et al., Experientia, 1957, 13, 400.
(2) Hoffmann et al., U.S. Pat. 2 935 514 (1960).

ETOPERIDONE [INN]

CAS 52942-31-1 | MW: 377.92 | MF: C19 H28 Cl N5 O

Antidepressant

McNeil

LgP (4)

pKa

pp cited in Vols.1-5:

Et ... Et ... O ... N—N ... Cl

(1) G. Palazzo, U.S. Pat. 3 857 845 (1974).
(2) R. Lisciani et al., Arzneim.-Forsch., 1978, 28, 417.

ETOPOSIDE [U;INN]

CAS 33419-42-0 | MW: 588.57 | MF: C29 H32 O13

Antineoplastic

Sandoz; Bristol (Vepesid)

LgP -1.12(C,3)

pKa

pp cited in Vols.1-5:

1: 23;
2: 771, 778, 779;
5: 569

OMe ... HO ... HO ... Me ... MeO ... O—O

(1) M. Kuhn et al., Swiss Pat. 514 578 (1971).
(2) C. Keller-Juslen et al., J. Med. Chem., 1971, 14, 936.

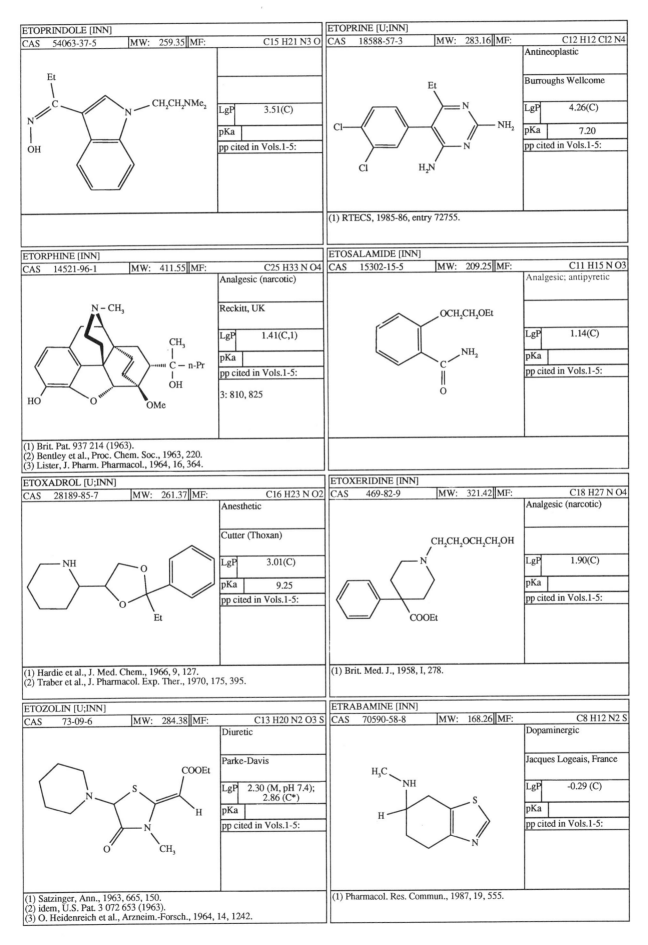

ETOPRINDOLE [INN]

CAS 54063-37-5 | MW: 259.35 | MF: C15 H21 N3 O

LgP 3.51(C)

pKa

pp cited in Vols.1-5:

ETOPRINE [U;INN]

CAS 18588-57-3 | MW: 283.16 | MF: C12 H12 Cl2 N4

Antineoplastic

Burroughs Wellcome

LgP 4.26(C)

pKa 7.20

pp cited in Vols.1-5:

(1) RTECS, 1985-86, entry 72755.

ETORPHINE [INN]

CAS 14521-96-1 | MW: 411.55 | MF: C25 H33 N O4

Analgesic (narcotic)

Reckitt, UK

LgP 1.41(C,1)

pKa

pp cited in Vols.1-5:

3: 810, 825

(1) Brit. Pat. 937 214 (1963).
(2) Bentley et al., Proc. Chem. Soc., 1963, 220.
(3) Lister, J. Pharm. Pharmacol., 1964, 16, 364.

ETOSALAMIDE [INN]

CAS 15302-15-5 | MW: 209.25 | MF: C11 H15 N O3

Analgesic; antipyretic

LgP 1.14(C)

pKa

pp cited in Vols.1-5:

ETOXADROL [U;INN]

CAS 28189-85-7 | MW: 261.37 | MF: C16 H23 N O2

Anesthetic

Cutter (Thoxan)

LgP 3.01(C)

pKa 9.25

pp cited in Vols.1-5:

(1) Hardie et al., J. Med. Chem., 1966, 9, 127.
(2) Traber et al., J. Pharmacol. Exp. Ther., 1970, 175, 395.

ETOXERIDINE [INN]

CAS 469-82-9 | MW: 321.42 | MF: C18 H27 N O4

Analgesic (narcotic)

LgP 1.90(C)

pKa

pp cited in Vols.1-5:

(1) Brit. Med. J., 1958, I, 278.

ETOZOLIN [U;INN]

CAS 73-09-6 | MW: 284.38 | MF: C13 H20 N2 O3 S

Diuretic

Parke-Davis

LgP 2.30 (M, pH 7.4); 2.86 (C*)

pKa

pp cited in Vols.1-5:

(1) Satzinger, Ann., 1963, 665, 150.
(2) idem, U.S. Pat. 3 072 653 (1963).
(3) O. Heidenreich et al., Arzneim.-Forsch., 1964, 14, 1242.

ETRABAMINE [INN]

CAS 70590-58-8 | MW: 168.26 | MF: C8 H12 N2 S

Dopaminergic

Jacques Logeais, France

LgP -0.29 (C)

pKa

pp cited in Vols.1-5:

(1) Pharmacol. Res. Commun., 1987, 19, 555.

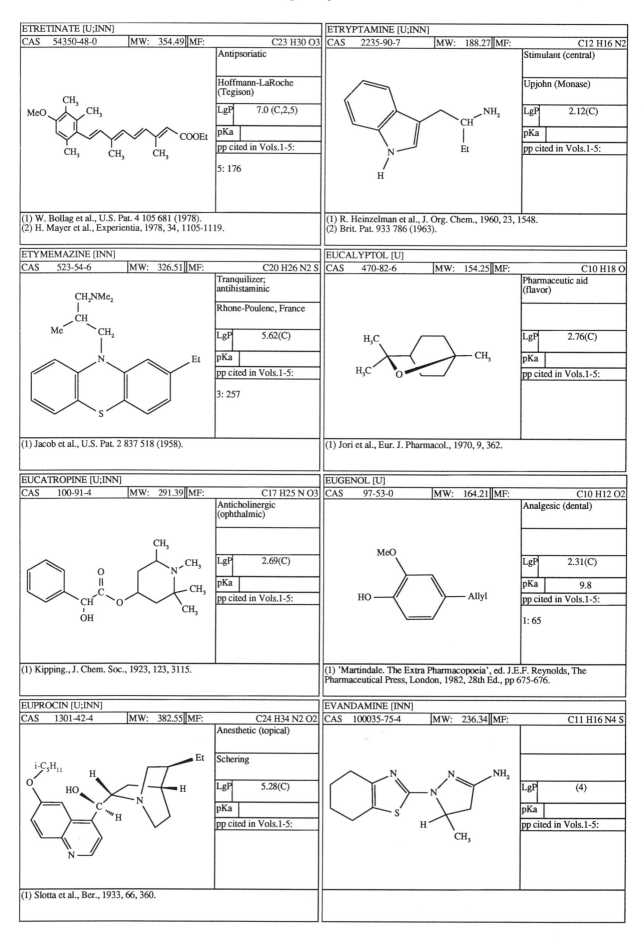

ETRETINATE [U;INN]

CAS	54350-48-0	MW: 354.49	MF:	C23 H30 O3

Antipsoriatic

Hoffmann-LaRoche (Tegison)

LgP	7.0 (C,2,5)
pKa	
pp cited in Vols.1-5:	

5: 176

(1) W. Bollag et al., U.S. Pat. 4 105 681 (1978).
(2) H. Mayer et al., Experientia, 1978, 34, 1105-1119.

ETRYPTAMINE [U;INN]

CAS	2235-90-7	MW: 188.27	MF:	C12 H16 N2

Stimulant (central)

Upjohn (Monase)

LgP	2.12(C)
pKa	
pp cited in Vols.1-5:	

(1) R. Heinzelman et al., J. Org. Chem., 1960, 23, 1548.
(2) Brit. Pat. 933 786 (1963).

ETYMEMAZINE [INN]

CAS	523-54-6	MW: 326.51	MF:	C20 H26 N2 S

Tranquilizer; antihistaminic

Rhone-Poulenc, France

LgP	5.62(C)
pKa	
pp cited in Vols.1-5:	

3: 257

(1) Jacob et al., U.S. Pat. 2 837 518 (1958).

EUCALYPTOL [U]

CAS	470-82-6	MW: 154.25	MF:	C10 H18 O

Pharmaceutic aid (flavor)

LgP	2.76(C)
pKa	
pp cited in Vols.1-5:	

(1) Jori et al., Eur. J. Pharmacol., 1970, 9, 362.

EUCATROPINE [U;INN]

CAS	100-91-4	MW: 291.39	MF:	C17 H25 N O3

Anticholinergic (ophthalmic)

LgP	2.69(C)
pKa	
pp cited in Vols.1-5:	

(1) Kipping., J. Chem. Soc., 1923, 123, 3115.

EUGENOL [U]

CAS	97-53-0	MW: 164.21	MF:	C10 H12 O2

Analgesic (dental)

LgP	2.31(C)
pKa	9.8
pp cited in Vols.1-5:	

1: 65

(1) 'Martindale. The Extra Pharmacopoeia', ed. J.E.F. Reynolds, The Pharmaceutical Press, London, 1982, 28th Ed., pp 675-676.

EUPROCIN [U;INN]

CAS	1301-42-4	MW: 382.55	MF:	C24 H34 N2 O2

Anesthetic (topical)

Schering

LgP	5.28(C)
pKa	
pp cited in Vols.1-5:	

(1) Slotta et al., Ber., 1933, 66, 360.

EVANDAMINE [INN]

CAS	100035-75-4	MW: 236.34	MF:	C11 H16 N4 S

LgP	(4)
pKa	
pp cited in Vols.1-5:	

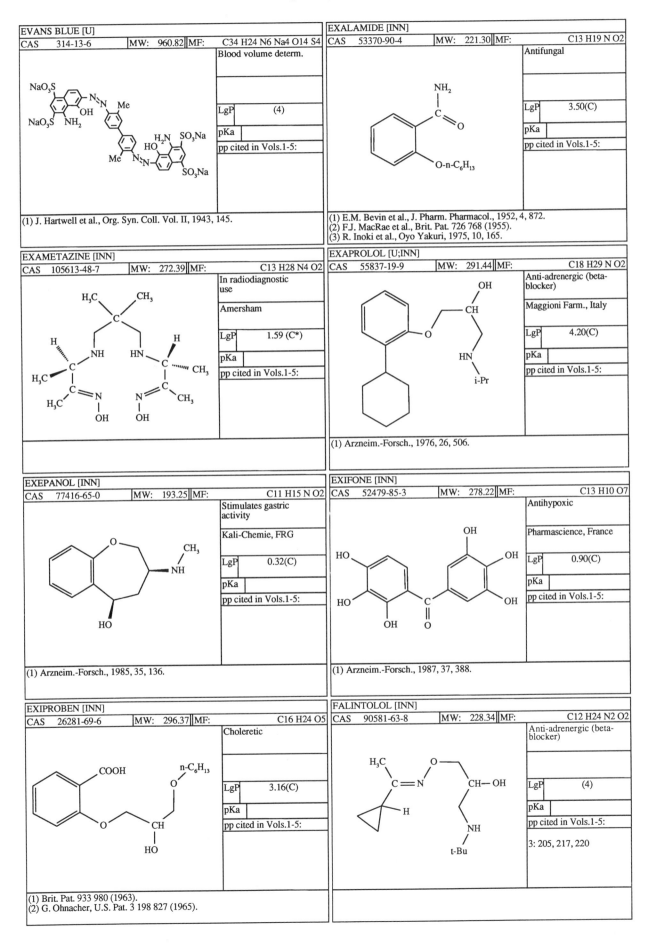

EVANS BLUE [U]

CAS	314-13-6	MW:	960.82	MF:	C34 H24 N6 Na4 O14 S4

Blood volume determ.

LgP	(4)
pKa	
pp cited in Vols.1-5:	

(1) J. Hartwell et al., Org. Syn. Coll. Vol. II, 1943, 145.

EXALAMIDE [INN]

CAS	53370-90-4	MW:	221.30	MF:	C13 H19 N O2

Antifungal

LgP	3.50(C)
pKa	
pp cited in Vols.1-5:	

(1) E.M. Bevin et al., J. Pharm. Pharmacol., 1952, 4, 872.
(2) F.J. MacRae et al., Brit. Pat. 726 768 (1955).
(3) R. Inoki et al., Oyo Yakuri, 1975, 10, 165.

EXAMETAZINE [INN]

CAS	105613-48-7	MW:	272.39	MF:	C13 H28 N4 O2

In radiodiagnostic use

Amersham

LgP	1.59 (C*)
pKa	
pp cited in Vols.1-5:	

EXAPROLOL [U;INN]

CAS	55837-19-9	MW:	291.44	MF:	C18 H29 N O2

Anti-adrenergic (beta-blocker)

Maggioni Farm., Italy

LgP	4.20(C)
pKa	
pp cited in Vols.1-5:	

(1) Arzneim.-Forsch., 1976, 26, 506.

EXEPANOL [INN]

CAS	77416-65-0	MW:	193.25	MF:	C11 H15 N O2

Stimulates gastric activity

Kali-Chemie, FRG

LgP	0.32(C)
pKa	
pp cited in Vols.1-5:	

(1) Arzneim.-Forsch., 1985, 35, 136.

EXIFONE [INN]

CAS	52479-85-3	MW:	278.22	MF:	C13 H10 O7

Antihypoxic

Pharmascience, France

LgP	0.90(C)
pKa	
pp cited in Vols.1-5:	

(1) Arzneim.-Forsch., 1987, 37, 388.

EXIPROBEN [INN]

CAS	26281-69-6	MW:	296.37	MF:	C16 H24 O5

Choleretic

LgP	3.16(C)
pKa	
pp cited in Vols.1-5:	

(1) Brit. Pat. 933 980 (1963).
(2) G. Ohnacher, U.S. Pat. 3 198 827 (1965).

FALINTOLOL [INN]

CAS	90581-63-8	MW:	228.34	MF:	C12 H24 N2 O2

Anti-adrenergic (beta-blocker)

LgP	(4)
pKa	
pp cited in Vols.1-5:	

3: 205, 217, 220

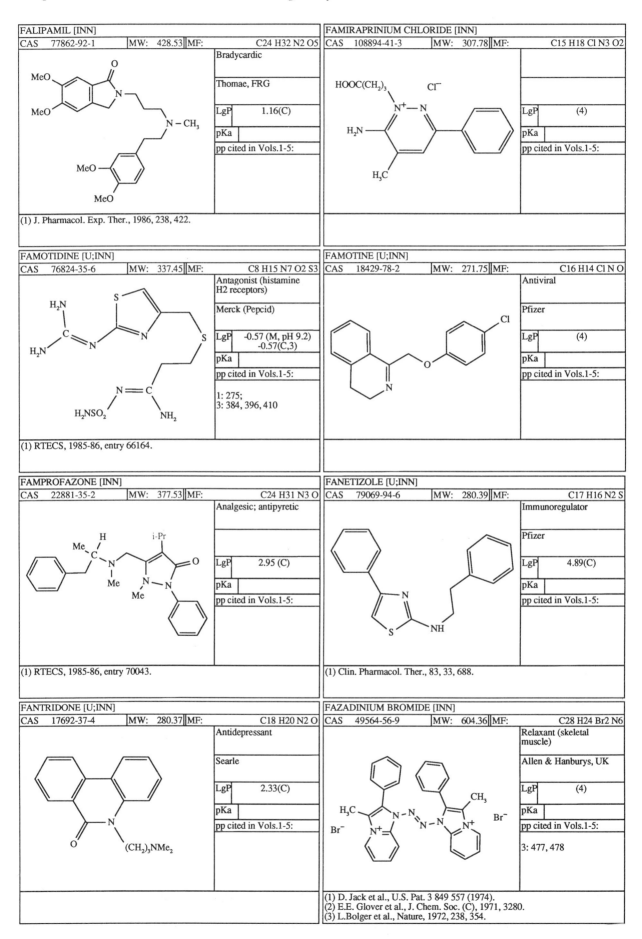

FALIPAMIL [INN]

CAS	77862-92-1	MW:	428.53	MF:	C24 H32 N2 O5

Bradycardic

Thomae, FRG

LgP	1.16(C)
pKa	
pp cited in Vols.1-5:	

(1) J. Pharmacol. Exp. Ther., 1986, 238, 422.

FAMIRAPRINIUM CHLORIDE [INN]

CAS	108894-41-3	MW:	307.78	MF:	C15 H18 Cl N3 O2

LgP	(4)
pKa	
pp cited in Vols.1-5:	

FAMOTIDINE [U;INN]

CAS	76824-35-6	MW:	337.45	MF:	C8 H15 N7 O2 S3

Antagonist (histamine H2 receptors)

Merck (Pepcid)

LgP	-0.57 (M, pH 9.2) -0.57(C,3)
pKa	
pp cited in Vols.1-5:	

1: 275;
3: 384, 396, 410

(1) RTECS, 1985-86, entry 66164.

FAMOTINE [U;INN]

CAS	18429-78-2	MW:	271.75	MF:	C16 H14 Cl N O

Antiviral

Pfizer

LgP	(4)
pKa	
pp cited in Vols.1-5:	

FAMPROFAZONE [INN]

CAS	22881-35-2	MW:	377.53	MF:	C24 H31 N3 O

Analgesic; antipyretic

LgP	2.95 (C)
pKa	
pp cited in Vols.1-5:	

(1) RTECS, 1985-86, entry 70043.

FANETIZOLE [U;INN]

CAS	79069-94-6	MW:	280.39	MF:	C17 H16 N2 S

Immunoregulator

Pfizer

LgP	4.89(C)
pKa	
pp cited in Vols.1-5:	

(1) Clin. Pharmacol. Ther., 83, 33, 688.

FANTRIDONE [U;INN]

CAS	17692-37-4	MW:	280.37	MF:	C18 H20 N2 O

Antidepressant

Searle

LgP	2.33(C)
pKa	
pp cited in Vols.1-5:	

FAZADINIUM BROMIDE [INN]

CAS	49564-56-9	MW:	604.36	MF:	C28 H24 Br2 N6

Relaxant (skeletal muscle)

Allen & Hanburys, UK

LgP	(4)
pKa	
pp cited in Vols.1-5:	

3: 477, 478

(1) D. Jack et al., U.S. Pat. 3 849 557 (1974).
(2) E.E. Glover et al., J. Chem. Soc. (C), 1971, 3280.
(3) L.Bolger et al., Nature, 1972, 238, 354.

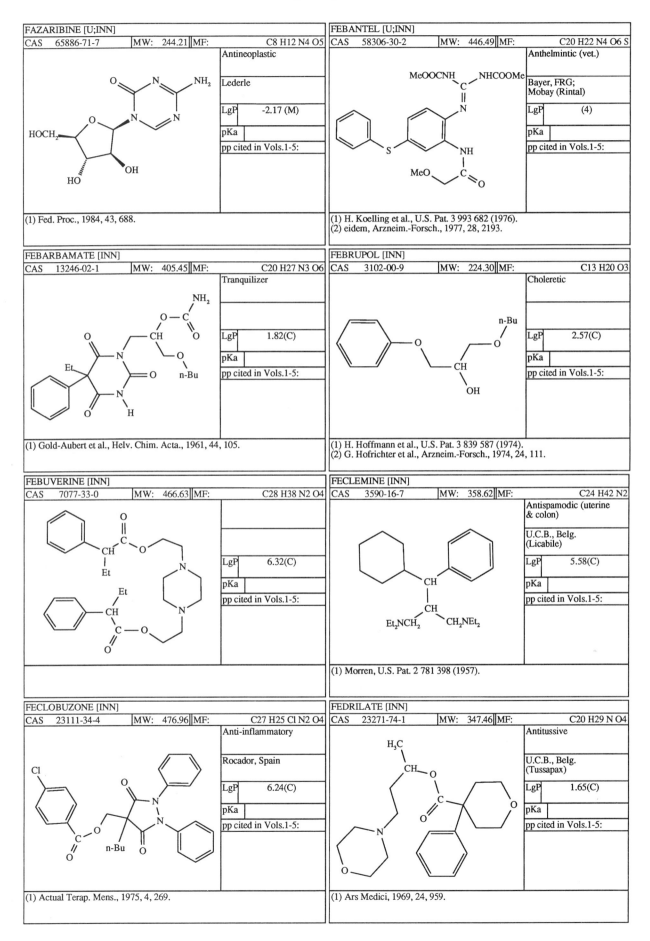

FAZARIBINE [U;INN]

CAS	65886-71-7	MW: 244.21	MF:	C8 H12 N4 O5

Antineoplastic

Lederle

LgP	-2.17 (M)

pKa	

pp cited in Vols.1-5:

(1) Fed. Proc., 1984, 43, 688.

FEBANTEL [U;INN]

CAS	58306-30-2	MW: 446.49	MF:	C20 H22 N4 O6 S

Anthelmintic (vet.)

Bayer, FRG;
Mobay (Rintal)

LgP	(4)

pKa	

pp cited in Vols.1-5:

(1) H. Koelling et al., U.S. Pat. 3 993 682 (1976).
(2) eidem, Arzneim.-Forsch., 1977, 28, 2193.

FEBARBAMATE [INN]

CAS	13246-02-1	MW: 405.45	MF:	C20 H27 N3 O6

Tranquilizer

LgP	1.82(C)

pKa	

pp cited in Vols.1-5:

(1) Gold-Aubert et al., Helv. Chim. Acta., 1961, 44, 105.

FEBRUPOL [INN]

CAS	3102-00-9	MW: 224.30	MF:	C13 H20 O3

Choleretic

LgP	2.57(C)

pKa	

pp cited in Vols.1-5:

(1) H. Hoffmann et al., U.S. Pat. 3 839 587 (1974).
(2) G. Hofrichter et al., Arzneim.-Forsch., 1974, 24, 111.

FEBUVERINE [INN]

CAS	7077-33-0	MW: 466.63	MF:	C28 H38 N2 O4

LgP	6.32(C)

pKa	

pp cited in Vols.1-5:

FECLEMINE [INN]

CAS	3590-16-7	MW: 358.62	MF:	C24 H42 N2

Antispamodic (uterine
& colon)

U.C.B., Belg.
(Licabile)

LgP	5.58(C)

pKa	

pp cited in Vols.1-5:

(1) Morren, U.S. Pat. 2 781 398 (1957).

FECLOBUZONE [INN]

CAS	23111-34-4	MW: 476.96	MF:	C27 H25 Cl N2 O4

Anti-inflammatory

Rocador, Spain

LgP	6.24(C)

pKa	

pp cited in Vols.1-5:

(1) Actual Terap. Mens., 1975, 4, 269.

FEDRILATE [INN]

CAS	23271-74-1	MW: 347.46	MF:	C20 H29 N O4

Antitussive

U.C.B., Belg.
(Tussapax)

LgP	1.65(C)

pKa	

pp cited in Vols.1-5:

(1) Ars Medici, 1969, 24, 959.

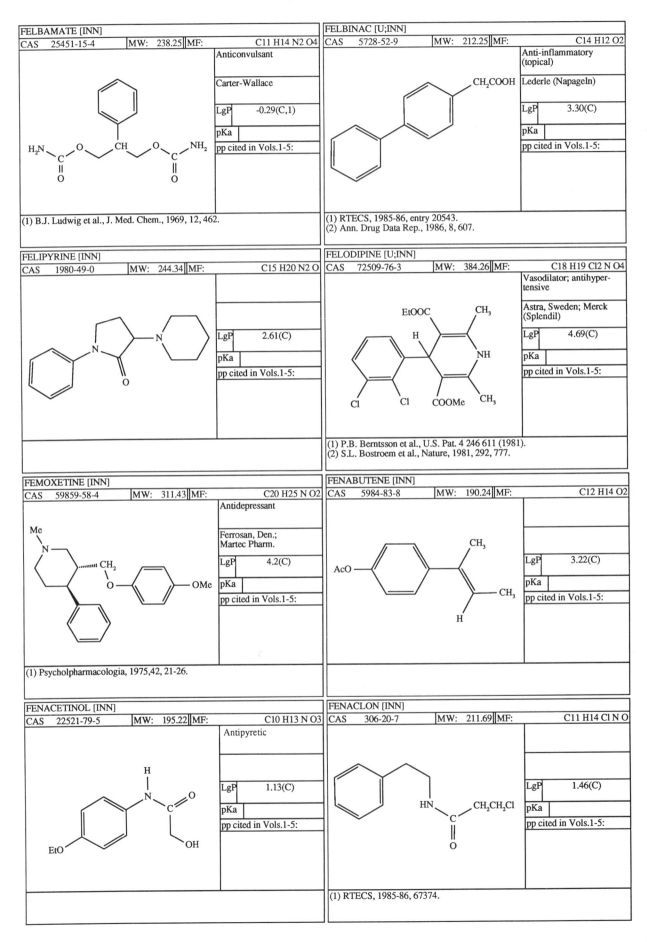

FELBAMATE [INN]

CAS 25451-15-4   MW: 238.25   MF: C11 H14 N2 O4

Anticonvulsant

Carter-Wallace

LgP	-0.29(C,1)
pKa	
pp cited in Vols.1-5:	

H_2N ... O ... CH ... O ... NH_2

(1) B.J. Ludwig et al., J. Med. Chem., 1969, 12, 462.

FELBINAC [U;INN]

CAS 5728-52-9   MW: 212.25   MF: C14 H12 O2

Anti-inflammatory (topical)

Lederle (Napageln)

CH_2COOH

LgP	3.30(C)
pKa	
pp cited in Vols.1-5:	

(1) RTECS, 1985-86, entry 20543.
(2) Ann. Drug Data Rep., 1986, 8, 607.

FELIPYRINE [INN]

CAS 1980-49-0   MW: 244.34   MF: C15 H20 N2 O

LgP	2.61(C)
pKa	
pp cited in Vols.1-5:	

FELODIPINE [U;INN]

CAS 72509-76-3   MW: 384.26   MF: C18 H19 Cl2 N O4

Vasodilator; antihypertensive

Astra, Sweden; Merck (Splendil)

EtOOC CH_3
H
NH
Cl   Cl   COOMe CH_3

LgP	4.69(C)
pKa	
pp cited in Vols.1-5:	

(1) P.B. Berntsson et al., U.S. Pat. 4 246 611 (1981).
(2) S.L. Bostroem et al., Nature, 1981, 292, 777.

FEMOXETINE [INN]

CAS 59859-58-4   MW: 311.43   MF: C20 H25 N O2

Antidepressant

Ferrosan, Den.; Martec Pharm.

LgP	4.2(C)
pKa	
pp cited in Vols.1-5:	

Me
N
CH_2
O ... OMe

(1) Psycholpharmacologia, 1975,42, 21-26.

FENABUTENE [INN]

CAS 5984-83-8   MW: 190.24   MF: C12 H14 O2

CH_3
AcO
CH_3
H

LgP	3.22(C)
pKa	
pp cited in Vols.1-5:	

FENACETINOL [INN]

CAS 22521-79-5   MW: 195.22   MF: C10 H13 N O3

Antipyretic

H
N   O
EtO   OH

LgP	1.13(C)
pKa	
pp cited in Vols.1-5:	

FENACLON [INN]

CAS 306-20-7   MW: 211.69   MF: C11 H14 Cl N O

HN CH_2CH_2Cl
C
O

LgP	1.46(C)
pKa	
pp cited in Vols.1-5:	

(1) RTECS, 1985-86, 67374.

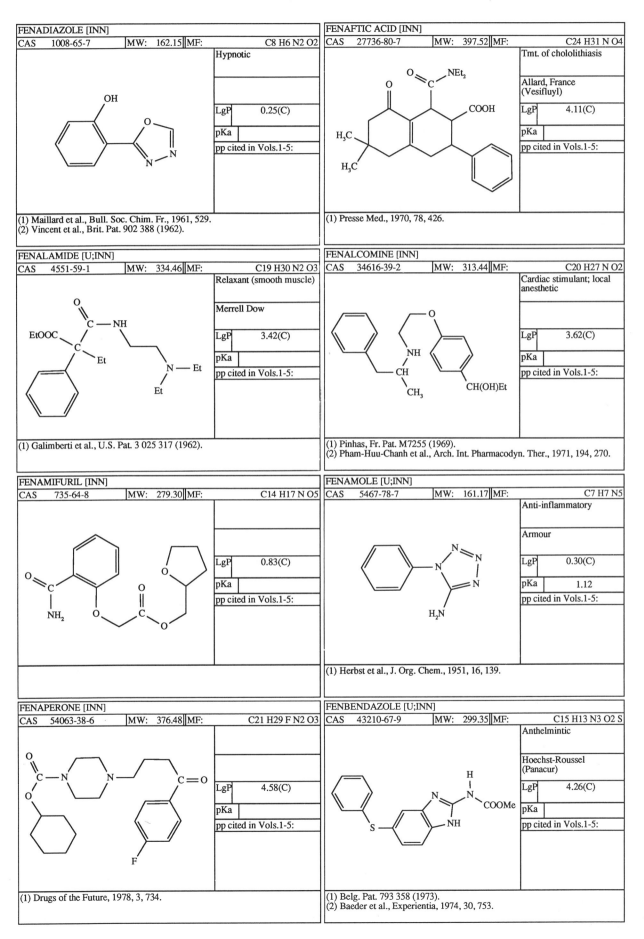

FENADIAZOLE [INN]

CAS	1008-65-7	MW:	162.15	MF:	C8 H6 N2 O2

Hypnotic

LgP	0.25(C)
pKa	
pp cited in Vols.1-5:	

(1) Maillard et al., Bull. Soc. Chim. Fr., 1961, 529.
(2) Vincent et al., Brit. Pat. 902 388 (1962).

FENAFTIC ACID [INN]

CAS	27736-80-7	MW:	397.52	MF:	C24 H31 N O4

Tmt. of chololithiasis

Allard, France
(Vesifluyl)

LgP	4.11(C)
pKa	
pp cited in Vols.1-5:	

(1) Presse Med., 1970, 78, 426.

FENALAMIDE [U;INN]

CAS	4551-59-1	MW:	334.46	MF:	C19 H30 N2 O3

Relaxant (smooth muscle)

Merrell Dow

LgP	3.42(C)
pKa	
pp cited in Vols.1-5:	

(1) Galimberti et al., U.S. Pat. 3 025 317 (1962).

FENALCOMINE [INN]

CAS	34616-39-2	MW:	313.44	MF:	C20 H27 N O2

Cardiac stimulant; local anesthetic

LgP	3.62(C)
pKa	
pp cited in Vols.1-5:	

(1) Pinhas, Fr. Pat. M7255 (1969).
(2) Pham-Huu-Chanh et al., Arch. Int. Pharmacodyn. Ther., 1971, 194, 270.

FENAMIFURIL [INN]

CAS	735-64-8	MW:	279.30	MF:	C14 H17 N O5

LgP	0.83(C)
pKa	
pp cited in Vols.1-5:	

FENAMOLE [U;INN]

CAS	5467-78-7	MW:	161.17	MF:	C7 H7 N5

Anti-inflammatory

Armour

LgP	0.30(C)
pKa	1.12
pp cited in Vols.1-5:	

(1) Herbst et al., J. Org. Chem., 1951, 16, 139.

FENAPERONE [INN]

CAS	54063-38-6	MW:	376.48	MF:	C21 H29 F N2 O3

LgP	4.58(C)
pKa	
pp cited in Vols.1-5:	

(1) Drugs of the Future, 1978, 3, 734.

FENBENDAZOLE [U;INN]

CAS	43210-67-9	MW:	299.35	MF:	C15 H13 N3 O2 S

Anthelmintic

Hoechst-Roussel
(Panacur)

LgP	4.26(C)
pKa	
pp cited in Vols.1-5:	

(1) Belg. Pat. 793 358 (1973).
(2) Baeder et al., Experientia, 1974, 30, 753.

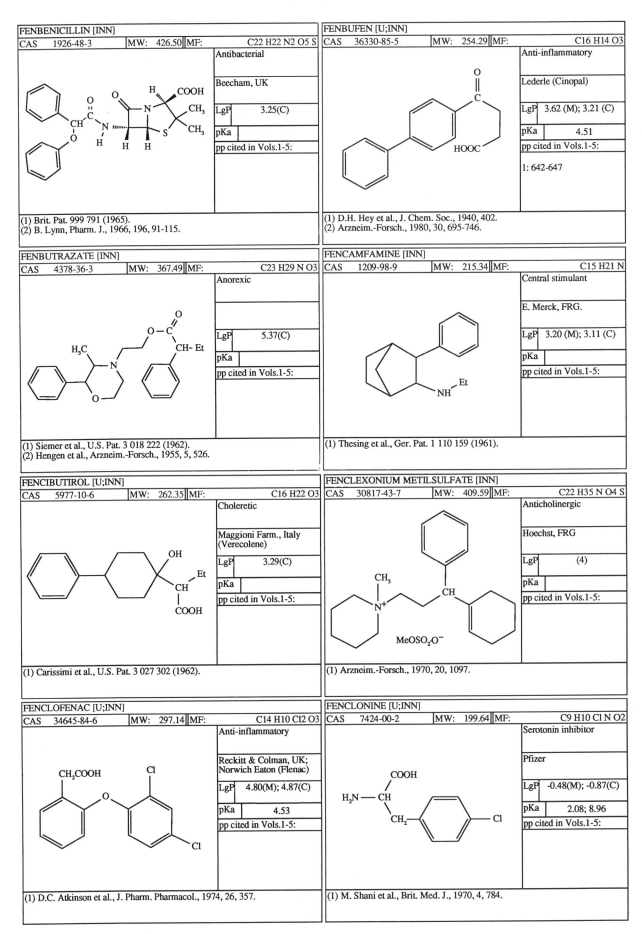

FENBENICILLIN [INN]

CAS	1926-48-3	MW:	426.50	MF:	C22 H22 N2 O5 S

Antibacterial

Beecham, UK

LgP	3.25(C)
pKa	
pp cited in Vols.1-5:	

(1) Brit. Pat. 999 791 (1965).
(2) B. Lynn, Pharm. J., 1966, 196, 91-115.

FENBUFEN [U;INN]

CAS	36330-85-5	MW:	254.29	MF:	C16 H14 O3

Anti-inflammatory

Lederle (Cinopal)

LgP	3.62 (M); 3.21 (C)
pKa	4.51
pp cited in Vols.1-5:	

1: 642-647

(1) D.H. Hey et al., J. Chem. Soc., 1940, 402.
(2) Arzneim.-Forsch., 1980, 30, 695-746.

FENBUTRAZATE [INN]

CAS	4378-36-3	MW:	367.49	MF:	C23 H29 N O3

Anorexic

LgP	5.37(C)
pKa	
pp cited in Vols.1-5:	

(1) Siemer et al., U.S. Pat. 3 018 222 (1962).
(2) Hengen et al., Arzneim.-Forsch., 1955, 5, 526.

FENCAMFAMINE [INN]

CAS	1209-98-9	MW:	215.34	MF:	C15 H21 N

Central stimulant

E. Merck, FRG.

LgP	3.20 (M); 3.11 (C)
pKa	
pp cited in Vols.1-5:	

(1) Thesing et al., Ger. Pat. 1 110 159 (1961).

FENCIBUTIROL [U;INN]

CAS	5977-10-6	MW:	262.35	MF:	C16 H22 O3

Choleretic

Maggioni Farm., Italy (Verecolene)

LgP	3.29(C)
pKa	
pp cited in Vols.1-5:	

(1) Carissimi et al., U.S. Pat. 3 027 302 (1962).

FENCLEXONIUM METILSULFATE [INN]

CAS	30817-43-7	MW:	409.59	MF:	C22 H35 N O4 S

Anticholinergic

Hoechst, FRG

LgP	(4)
pKa	
pp cited in Vols.1-5:	

$MeOSO_2O^-$

(1) Arzneim.-Forsch., 1970, 20, 1097.

FENCLOFENAC [U;INN]

CAS	34645-84-6	MW:	297.14	MF:	C14 H10 Cl2 O3

Anti-inflammatory

Reckitt & Colman, UK; Norwich Eaton (Flenac)

LgP	4.80(M); 4.87(C)
pKa	4.53
pp cited in Vols.1-5:	

(1) D.C. Atkinson et al., J. Pharm. Pharmacol., 1974, 26, 357.

FENCLONINE [U;INN]

CAS	7424-00-2	MW:	199.64	MF:	C9 H10 Cl N O2

Serotonin inhibitor

Pfizer

LgP	-0.48(M); -0.87(C)
pKa	2.08; 8.96
pp cited in Vols.1-5:	

(1) M. Shani et al., Brit. Med. J., 1970, 4, 784.

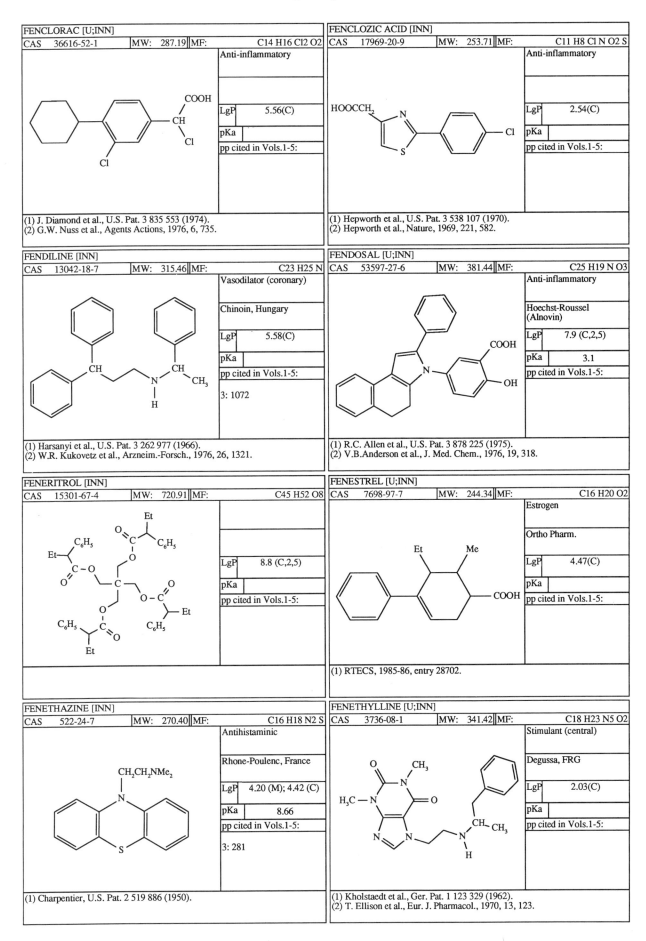

FENCLORAC [U;INN]

CAS 36616-52-1 | MW: 287.19 | MF: C14 H16 Cl2 O2

Anti-inflammatory

LgP 5.56(C)

pKa

pp cited in Vols.1-5:

(1) J. Diamond et al., U.S. Pat. 3 835 553 (1974).
(2) G.W. Nuss et al., Agents Actions, 1976, 6, 735.

FENCLOZIC ACID [INN]

CAS 17969-20-9 | MW: 253.71 | MF: C11 H8 Cl N O2 S

Anti-inflammatory

LgP 2.54(C)

pKa

pp cited in Vols.1-5:

(1) Hepworth et al., U.S. Pat. 3 538 107 (1970).
(2) Hepworth et al., Nature, 1969, 221, 582.

FENDILINE [INN]

CAS 13042-18-7 | MW: 315.46 | MF: C23 H25 N

Vasodilator (coronary)

Chinoin, Hungary

LgP 5.58(C)

pKa

pp cited in Vols.1-5:

3: 1072

(1) Harsanyi et al., U.S. Pat. 3 262 977 (1966).
(2) W.R. Kukovetz et al., Arzneim.-Forsch., 1976, 26, 1321.

FENDOSAL [U;INN]

CAS 53597-27-6 | MW: 381.44 | MF: C25 H19 N O3

Anti-inflammatory

Hoechst-Roussel (Alnovin)

LgP 7.9 (C,2,5)

pKa 3.1

pp cited in Vols.1-5:

(1) R.C. Allen et al., U.S. Pat. 3 878 225 (1975).
(2) V.B.Anderson et al., J. Med. Chem., 1976, 19, 318.

FENERITROL [INN]

CAS 15301-67-4 | MW: 720.91 | MF: C45 H52 O8

LgP 8.8 (C,2,5)

pKa

pp cited in Vols.1-5:

FENESTREL [U;INN]

CAS 7698-97-7 | MW: 244.34 | MF: C16 H20 O2

Estrogen

Ortho Pharm.

LgP 4.47(C)

pKa

pp cited in Vols.1-5:

(1) RTECS, 1985-86, entry 28702.

FENETHAZINE [INN]

CAS 522-24-7 | MW: 270.40 | MF: C16 H18 N2 S

Antihistaminic

Rhone-Poulenc, France

LgP 4.20 (M); 4.42 (C)

pKa 8.66

pp cited in Vols.1-5:

3: 281

(1) Charpentier, U.S. Pat. 2 519 886 (1950).

FENETHYLLINE [U;INN]

CAS 3736-08-1 | MW: 341.42 | MF: C18 H23 N5 O2

Stimulant (central)

Degussa, FRG

LgP 2.03(C)

pKa

pp cited in Vols.1-5:

(1) Kholstaedt et al., Ger. Pat. 1 123 329 (1962).
(2) T. Ellison et al., Eur. J. Pharmacol., 1970, 13, 123.

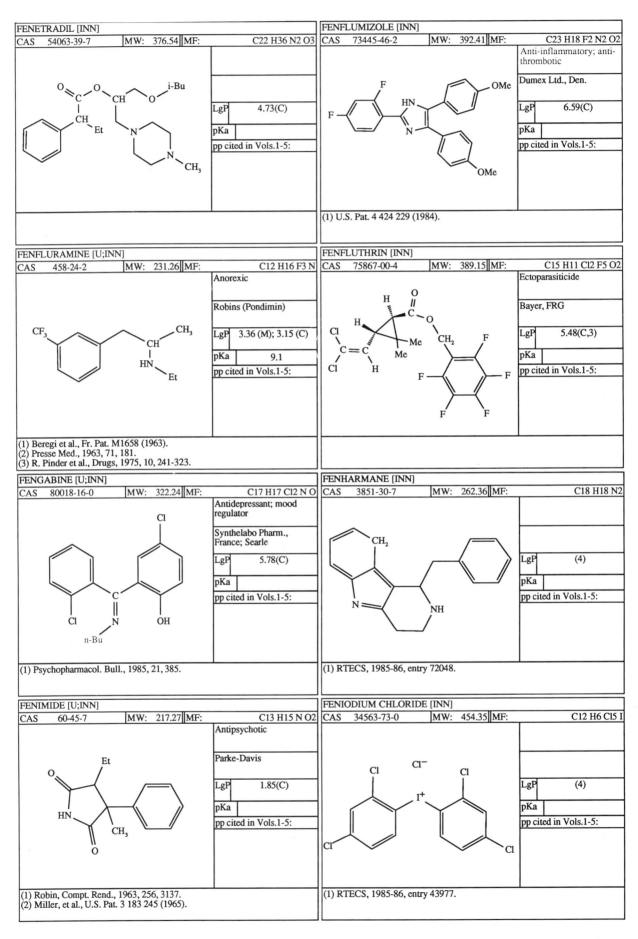

FENETRADIL [INN]

CAS	54063-39-7	MW:	376.54	MF:	C22 H36 N2 O3

LgP	4.73(C)
pKa	
pp cited in Vols.1-5:	

FENFLUMIZOLE [INN]

CAS	73445-46-2	MW:	392.41	MF:	C23 H18 F2 N2 O2

Anti-inflammatory; anti-thrombotic

Dumex Ltd., Den.

LgP	6.59(C)
pKa	
pp cited in Vols.1-5:	

(1) U.S. Pat. 4 424 229 (1984).

FENFLURAMINE [U;INN]

CAS	458-24-2	MW:	231.26	MF:	C12 H16 F3 N

Anorexic

Robins (Pondimin)

LgP	3.36 (M); 3.15 (C)
pKa	9.1
pp cited in Vols.1-5:	

(1) Beregi et al., Fr. Pat. M1658 (1963).
(2) Presse Med., 1963, 71, 181.
(3) R. Pinder et al., Drugs, 1975, 10, 241-323.

FENFLUTHRIN [INN]

CAS	75867-00-4	MW:	389.15	MF:	C15 H11 Cl2 F5 O2

Ectoparasiticide

Bayer, FRG

LgP	5.48(C,3)
pKa	
pp cited in Vols.1-5:	

FENGABINE [U;INN]

CAS	80018-16-0	MW:	322.24	MF:	C17 H17 Cl2 N O

Antidepressant; mood regulator

Synthelabo Pharm., France; Searle

LgP	5.78(C)
pKa	
pp cited in Vols.1-5:	

(1) Psychopharmacol. Bull., 1985, 21, 385.

FENHARMANE [INN]

CAS	3851-30-7	MW:	262.36	MF:	C18 H18 N2

LgP	(4)
pKa	
pp cited in Vols.1-5:	

(1) RTECS, 1985-86, entry 72048.

FENIMIDE [U;INN]

CAS	60-45-7	MW:	217.27	MF:	C13 H15 N O2

Antipsychotic

Parke-Davis

LgP	1.85(C)
pKa	
pp cited in Vols.1-5:	

(1) Robin, Compt. Rend., 1963, 256, 3137.
(2) Miller, et al., U.S. Pat. 3 183 245 (1965).

FENIODIUM CHLORIDE [INN]

CAS	34563-73-0	MW:	454.35	MF:	C12 H6 Cl5 I

LgP	(4)
pKa	
pp cited in Vols.1-5:	

(1) RTECS, 1985-86, entry 43977.

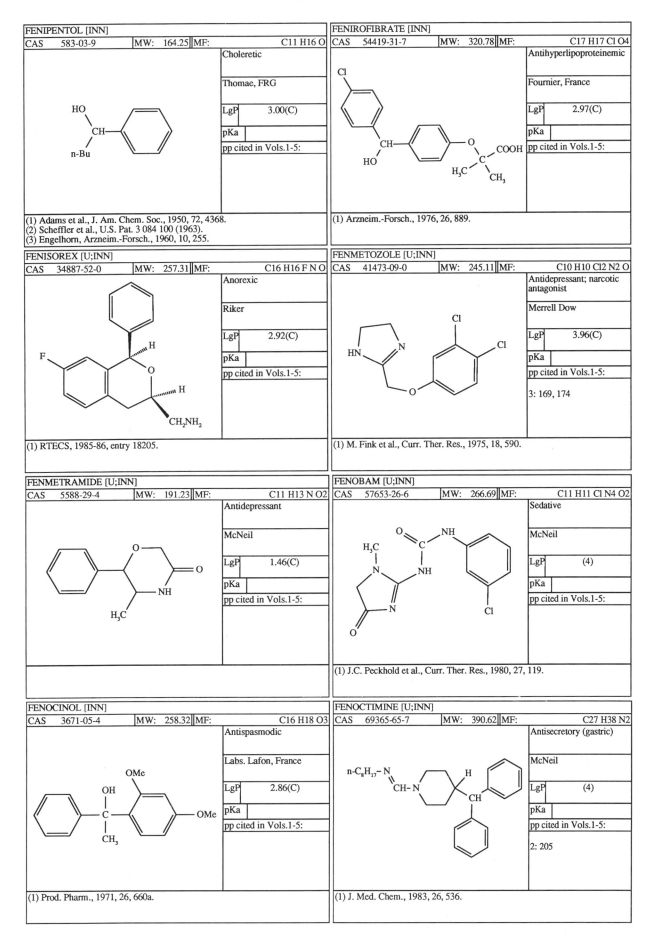

FENIPENTOL [INN]

CAS	583-03-9	MW:	164.25	MF:	C11 H16 O

Choleretic

Thomae, FRG

LgP	3.00(C)

pKa	

pp cited in Vols.1-5:

(1) Adams et al., J. Am. Chem. Soc., 1950, 72, 4368.
(2) Scheffler et al., U.S. Pat. 3 084 100 (1963).
(3) Engelhorn, Arzneim.-Forsch., 1960, 10, 255.

FENIROFIBRATE [INN]

CAS	54419-31-7	MW:	320.78	MF:	C17 H17 Cl O4

Antihyperlipoproteinemic

Fournier, France

LgP	2.97(C)

pKa	

pp cited in Vols.1-5:

(1) Arzneim.-Forsch., 1976, 26, 889.

FENISOREX [U;INN]

CAS	34887-52-0	MW:	257.31	MF:	C16 H16 F N O

Anorexic

Riker

LgP	2.92(C)

pKa	

pp cited in Vols.1-5:

(1) RTECS, 1985-86, entry 18205.

FENMETOZOLE [U;INN]

CAS	41473-09-0	MW:	245.11	MF:	C10 H10 Cl2 N2 O

Antidepressant; narcotic antagonist

Merrell Dow

LgP	3.96(C)

pKa	

pp cited in Vols.1-5:

3: 169, 174

(1) M. Fink et al., Curr. Ther. Res., 1975, 18, 590.

FENMETRAMIDE [U;INN]

CAS	5588-29-4	MW:	191.23	MF:	C11 H13 N O2

Antidepressant

McNeil

LgP	1.46(C)

pKa	

pp cited in Vols.1-5:

FENOBAM [U;INN]

CAS	57653-26-6	MW:	266.69	MF:	C11 H11 Cl N4 O2

Sedative

McNeil

LgP	(4)

pKa	

pp cited in Vols.1-5:

(1) J.C. Peckhold et al., Curr. Ther. Res., 1980, 27, 119.

FENOCINOL [INN]

CAS	3671-05-4	MW:	258.32	MF:	C16 H18 O3

Antispasmodic

Labs. Lafon, France

LgP	2.86(C)

pKa	

pp cited in Vols.1-5:

(1) Prod. Pharm., 1971, 26, 660a.

FENOCTIMINE [U;INN]

CAS	69365-65-7	MW:	390.62	MF:	C27 H38 N2

Antisecretory (gastric)

McNeil

LgP	(4)

pKa	

pp cited in Vols.1-5:

2: 205

(1) J. Med. Chem., 1983, 26, 536.

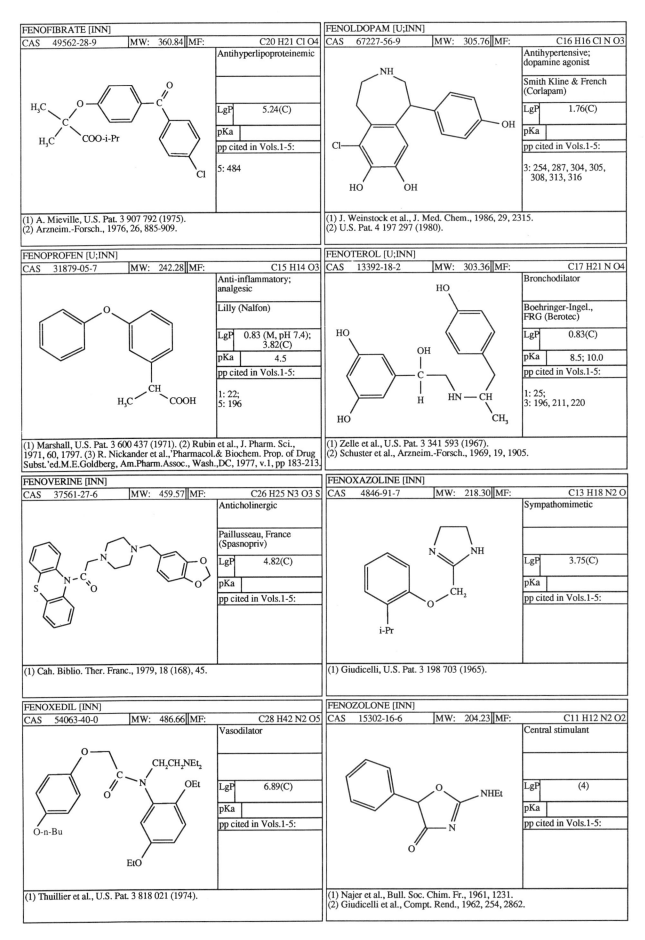

FENOFIBRATE [INN]

| CAS | 49562-28-9 | MW: | 360.84 | MF: | C20 H21 Cl O4 |

Antihyperlipoproteinemic

| LgP | 5.24(C) |
| pKa | |
| pp cited in Vols.1-5: |
| 5: 484 |

(1) A. Mieville, U.S. Pat. 3 907 792 (1975).
(2) Arzneim.-Forsch., 1976, 26, 885-909.

FENOLDOPAM [U;INN]

| CAS | 67227-56-9 | MW: | 305.76 | MF: | C16 H16 Cl N O3 |

Antihypertensive;
dopamine agonist

Smith Kline & French
(Corlapam)

| LgP | 1.76(C) |
| pKa | |
| pp cited in Vols.1-5: |
| 3: 254, 287, 304, 305, 308, 313, 316 |

(1) J. Weinstock et al., J. Med. Chem., 1986, 29, 2315.
(2) U.S. Pat. 4 197 297 (1980).

FENOPROFEN [U;INN]

| CAS | 31879-05-7 | MW: | 242.28 | MF: | C15 H14 O3 |

Anti-inflammatory;
analgesic

Lilly (Nalfon)

| LgP | 0.83 (M, pH 7.4); 3.82(C) |
| pKa | 4.5 |
| pp cited in Vols.1-5: |
| 1: 22; 5: 196 |

(1) Marshall, U.S. Pat. 3 600 437 (1971). (2) Rubin et al., J. Pharm. Sci., 1971, 60, 1797. (3) R. Nickander et al.,'Pharmacol.& Biochem. Prop. of Drug Subst.'ed.M.E.Goldberg, Am.Pharm.Assoc., Wash.,DC, 1977, v.1, pp 183-213,

FENOTEROL [U;INN]

| CAS | 13392-18-2 | MW: | 303.36 | MF: | C17 H21 N O4 |

Bronchodilator

Boehringer-Ingel.,
FRG (Berotec)

| LgP | 0.83(C) |
| pKa | 8.5; 10.0 |
| pp cited in Vols.1-5: |
| 1: 25; 3: 196, 211, 220 |

(1) Zelle et al., U.S. Pat. 3 341 593 (1967).
(2) Schuster et al., Arzneim.-Forsch., 1969, 19, 1905.

FENOVERINE [INN]

| CAS | 37561-27-6 | MW: | 459.57 | MF: | C26 H25 N3 O3 S |

Anticholinergic

Paillusseau, France
(Spasnopriv)

| LgP | 4.82(C) |
| pKa | |
| pp cited in Vols.1-5: |

(1) Cah. Biblio. Ther. Franc., 1979, 18 (168), 45.

FENOXAZOLINE [INN]

| CAS | 4846-91-7 | MW: | 218.30 | MF: | C13 H18 N2 O |

Sympathomimetic

| LgP | 3.75(C) |
| pKa | |
| pp cited in Vols.1-5: |

(1) Giudicelli, U.S. Pat. 3 198 703 (1965).

FENOXEDIL [INN]

| CAS | 54063-40-0 | MW: | 486.66 | MF: | C28 H42 N2 O5 |

Vasodilator

| LgP | 6.89(C) |
| pKa | |
| pp cited in Vols.1-5: |

(1) Thuillier et al., U.S. Pat. 3 818 021 (1974).

FENOZOLONE [INN]

| CAS | 15302-16-6 | MW: | 204.23 | MF: | C11 H12 N2 O2 |

Central stimulant

| LgP | (4) |
| pKa | |
| pp cited in Vols.1-5: |

(1) Najer et al., Bull. Soc. Chim. Fr., 1961, 1231.
(2) Giudicelli et al., Compt. Rend., 1962, 254, 2862.

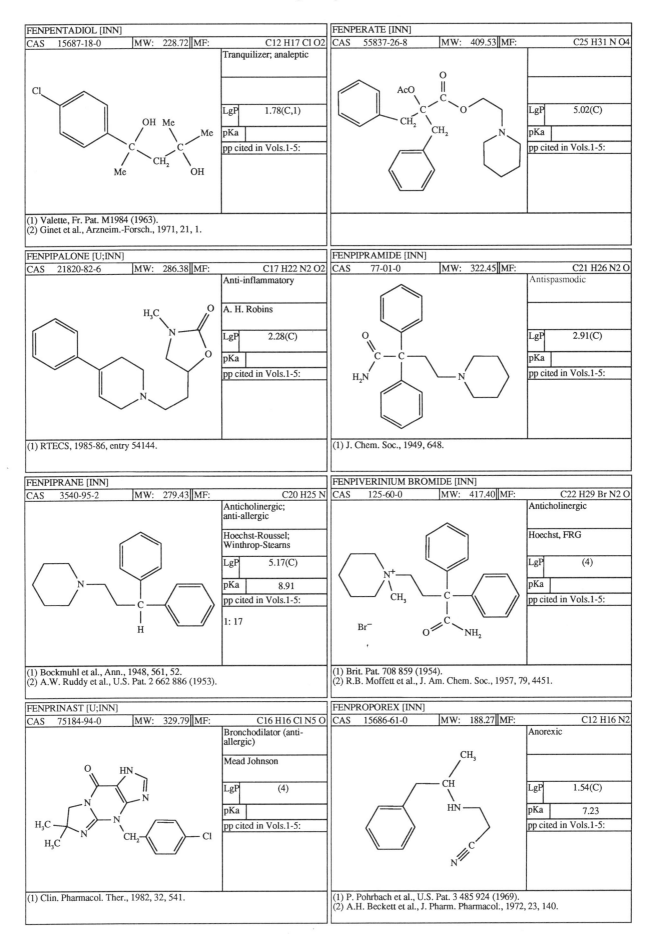

FENPENTADIOL [INN]

CAS 15687-18-0	MW: 228.72	MF:	C12 H17 Cl O2

Tranquilizer; analeptic

LgP	1.78(C,1)
pKa	
pp cited in Vols.1-5:	

(1) Valette, Fr. Pat. M1984 (1963).
(2) Ginet et al., Arzneim.-Forsch., 1971, 21, 1.

FENPERATE [INN]

CAS 55837-26-8	MW: 409.53	MF:	C25 H31 N O4

LgP	5.02(C)
pKa	
pp cited in Vols.1-5:	

FENPIPALONE [U;INN]

CAS 21820-82-6	MW: 286.38	MF:	C17 H22 N2 O2

Anti-inflammatory

A. H. Robins

LgP	2.28(C)
pKa	
pp cited in Vols.1-5:	

(1) RTECS, 1985-86, entry 54144.

FENPIPRAMIDE [INN]

CAS 77-01-0	MW: 322.45	MF:	C21 H26 N2 O

Antispasmodic

LgP	2.91(C)
pKa	
pp cited in Vols.1-5:	

(1) J. Chem. Soc., 1949, 648.

FENPIPRANE [INN]

CAS 3540-95-2	MW: 279.43	MF:	C20 H25 N

Anticholinergic; anti-allergic

Hoechst-Roussel; Winthrop-Stearns

LgP	5.17(C)
pKa	8.91
pp cited in Vols.1-5:	
	1: 17

(1) Bockmuhl et al., Ann., 1948, 561, 52.
(2) A.W. Ruddy et al., U.S. Pat. 2 662 886 (1953).

FENPIVERINIUM BROMIDE [INN]

CAS 125-60-0	MW: 417.40	MF:	C22 H29 Br N2 O

Anticholinergic

Hoechst, FRG

LgP	(4)
pKa	
pp cited in Vols.1-5:	

(1) Brit. Pat. 708 859 (1954).
(2) R.B. Moffett et al., J. Am. Chem. Soc., 1957, 79, 4451.

FENPRINAST [U;INN]

CAS 75184-94-0	MW: 329.79	MF:	C16 H16 Cl N5 O

Bronchodilator (anti-allergic)

Mead Johnson

LgP	(4)
pKa	
pp cited in Vols.1-5:	

(1) Clin. Pharmacol. Ther., 1982, 32, 541.

FENPROPOREX [INN]

CAS 15686-61-0	MW: 188.27	MF:	C12 H16 N2

Anorexic

LgP	1.54(C)
pKa	7.23
pp cited in Vols.1-5:	

(1) P. Pohrbach et al., U.S. Pat. 3 485 924 (1969).
(2) A.H. Beckett et al., J. Pharm. Pharmacol., 1972, 23, 140.

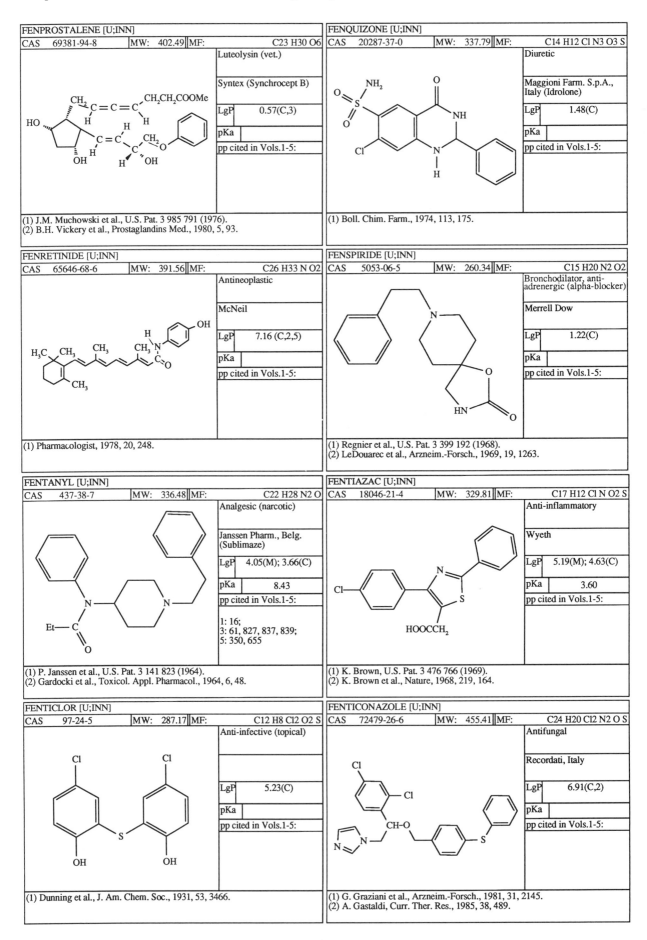

FENPROSTALENE [U;INN]

CAS	69381-94-8	MW:	402.49	MF:	C23 H30 O6

Luteolysin (vet.)

Syntex (Synchrocept B)

LgP	0.57(C,3)
pKa	
pp cited in Vols.1-5:	

(1) J.M. Muchowski et al., U.S. Pat. 3 985 791 (1976).
(2) B.H. Vickery et al., Prostaglandins Med., 1980, 5, 93.

FENQUIZONE [U;INN]

CAS	20287-37-0	MW:	337.79	MF:	C14 H12 Cl N3 O3 S

Diuretic

Maggioni Farm. S.p.A., Italy (Idrolone)

LgP	1.48(C)
pKa	
pp cited in Vols.1-5:	

(1) Boll. Chim. Farm., 1974, 113, 175.

FENRETINIDE [U;INN]

CAS	65646-68-6	MW:	391.56	MF:	C26 H33 N O2

Antineoplastic

McNeil

LgP	7.16 (C,2,5)
pKa	
pp cited in Vols.1-5:	

(1) Pharmacologist, 1978, 20, 248.

FENSPIRIDE [U;INN]

CAS	5053-06-5	MW:	260.34	MF:	C15 H20 N2 O2

Bronchodilator, anti-adrenergic (alpha-blocker)

Merrell Dow

LgP	1.22(C)
pKa	
pp cited in Vols.1-5:	

(1) Regnier et al., U.S. Pat. 3 399 192 (1968).
(2) LeDouarec et al., Arzneim.-Forsch., 1969, 19, 1263.

FENTANYL [U;INN]

CAS	437-38-7	MW:	336.48	MF:	C22 H28 N2 O

Analgesic (narcotic)

Janssen Pharm., Belg. (Sublimaze)

LgP	4.05(M); 3.66(C)
pKa	8.43
pp cited in Vols.1-5:	

1: 16;
3: 61, 827, 837, 839;
5: 350, 655

(1) P. Janssen et al., U.S. Pat. 3 141 823 (1964).
(2) Gardocki et al., Toxicol. Appl. Pharmacol., 1964, 6, 48.

FENTIAZAC [U;INN]

CAS	18046-21-4	MW:	329.81	MF:	C17 H12 Cl N O2 S

Anti-inflammatory

Wyeth

LgP	5.19(M); 4.63(C)
pKa	3.60
pp cited in Vols.1-5:	

(1) K. Brown, U.S. Pat. 3 476 766 (1969).
(2) K. Brown et al., Nature, 1968, 219, 164.

FENTICLOR [U;INN]

CAS	97-24-5	MW:	287.17	MF:	C12 H8 Cl2 O2 S

Anti-infective (topical)

LgP	5.23(C)
pKa	
pp cited in Vols.1-5:	

(1) Dunning et al., J. Am. Chem. Soc., 1931, 53, 3466.

FENTICONAZOLE [U;INN]

CAS	72479-26-6	MW:	455.41	MF:	C24 H20 Cl2 N2 O S

Antifungal

Recordati, Italy

LgP	6.91(C,2)
pKa	
pp cited in Vols.1-5:	

(1) G. Graziani et al., Arzneim.-Forsch., 1981, 31, 2145.
(2) A. Gastaldi, Curr. Ther. Res., 1985, 38, 489.

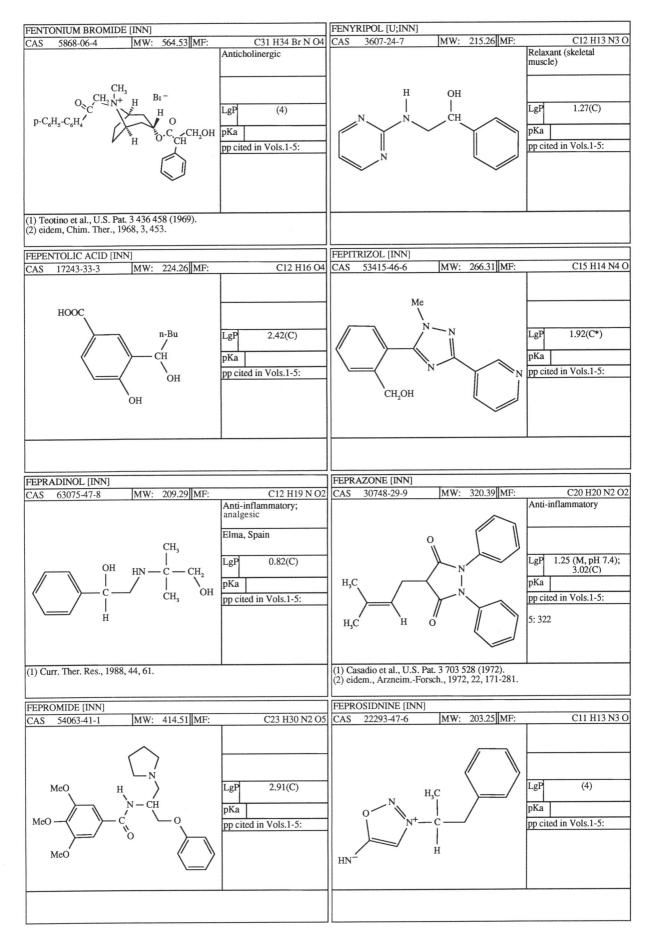

FENTONIUM BROMIDE [INN]

| CAS | 5868-06-4 | MW: | 564.53 | MF: | C31 H34 Br N O4 |

Anticholinergic

LgP	(4)
pKa	
pp cited in Vols.1-5:	

(1) Teotino et al., U.S. Pat. 3 436 458 (1969).
(2) eidem, Chim. Ther., 1968, 3, 453.

FENYRIPOL [U;INN]

| CAS | 3607-24-7 | MW: | 215.26 | MF: | C12 H13 N3 O |

Relaxant (skeletal muscle)

LgP	1.27(C)
pKa	
pp cited in Vols.1-5:	

FEPENTOLIC ACID [INN]

| CAS | 17243-33-3 | MW: | 224.26 | MF: | C12 H16 O4 |

LgP	2.42(C)
pKa	
pp cited in Vols.1-5:	

FEPITRIZOL [INN]

| CAS | 53415-46-6 | MW: | 266.31 | MF: | C15 H14 N4 O |

LgP	1.92(C*)
pKa	
pp cited in Vols.1-5:	

FEPRADINOL [INN]

| CAS | 63075-47-8 | MW: | 209.29 | MF: | C12 H19 N O2 |

Anti-inflammatory; analgesic

Elma, Spain

LgP	0.82(C)
pKa	
pp cited in Vols.1-5:	

(1) Curr. Ther. Res., 1988, 44, 61.

FEPRAZONE [INN]

| CAS | 30748-29-9 | MW: | 320.39 | MF: | C20 H20 N2 O2 |

Anti-inflammatory

LgP	1.25 (M, pH 7.4); 3.02(C)
pKa	
pp cited in Vols.1-5:	
5: 322	

(1) Casadio et al., U.S. Pat. 3 703 528 (1972).
(2) eidem., Arzneim.-Forsch., 1972, 22, 171-281.

FEPROMIDE [INN]

| CAS | 54063-41-1 | MW: | 414.51 | MF: | C23 H30 N2 O5 |

LgP	2.91(C)
pKa	
pp cited in Vols.1-5:	

FEPROSIDNINE [INN]

| CAS | 22293-47-6 | MW: | 203.25 | MF: | C11 H13 N3 O |

LgP	(4)
pKa	
pp cited in Vols.1-5:	

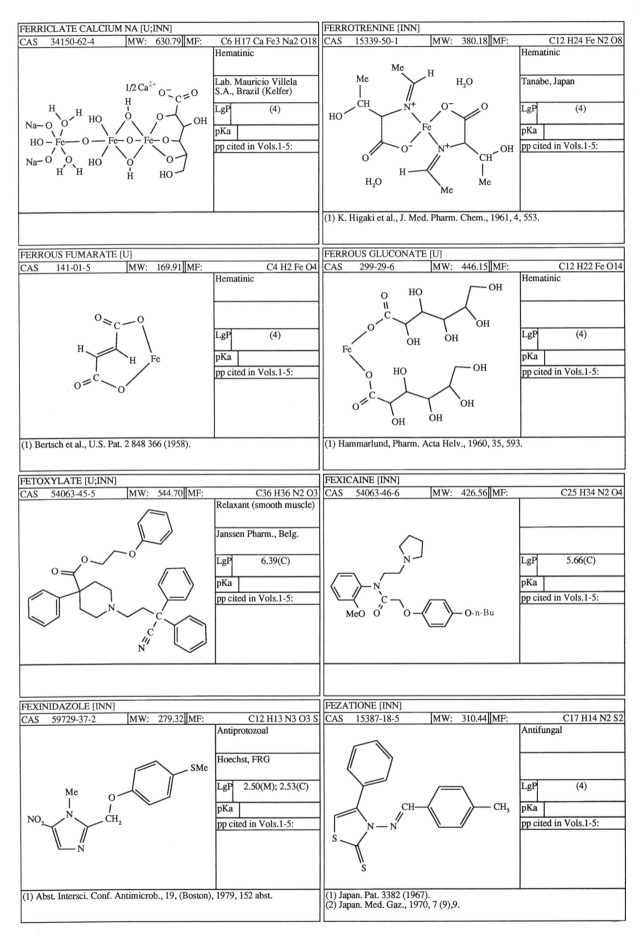

FERRICLATE CALCIUM NA [U;INN]

| CAS | 34150-62-4 | MW: | 630.79 | MF: | C6 H17 Ca Fe3 Na2 O18 |

Hematinic

Lab. Mauricio Villela S.A., Brazil (Kelfer)

LgP	(4)
pKa	
pp cited in Vols.1-5:	

FERROTRENINE [INN]

| CAS | 15339-50-1 | MW: | 380.18 | MF: | C12 H24 Fe N2 O8 |

Hematinic

Tanabe, Japan

LgP	(4)
pKa	
pp cited in Vols.1-5:	

(1) K. Higaki et al., J. Med. Pharm. Chem., 1961, 4, 553.

FERROUS FUMARATE [U]

| CAS | 141-01-5 | MW: | 169.91 | MF: | C4 H2 Fe O4 |

Hematinic

LgP	(4)
pKa	
pp cited in Vols.1-5:	

(1) Bertsch et al., U.S. Pat. 2 848 366 (1958).

FERROUS GLUCONATE [U]

| CAS | 299-29-6 | MW: | 446.15 | MF: | C12 H22 Fe O14 |

Hematinic

LgP	(4)
pKa	
pp cited in Vols.1-5:	

(1) Hammarlund, Pharm. Acta Helv., 1960, 35, 593.

FETOXYLATE [U;INN]

| CAS | 54063-45-5 | MW: | 544.70 | MF: | C36 H36 N2 O3 |

Relaxant (smooth muscle)

Janssen Pharm., Belg.

LgP	6.39(C)
pKa	
pp cited in Vols.1-5:	

FEXICAINE [INN]

| CAS | 54063-46-6 | MW: | 426.56 | MF: | C25 H34 N2 O4 |

LgP	5.66(C)
pKa	
pp cited in Vols.1-5:	

FEXINIDAZOLE [INN]

| CAS | 59729-37-2 | MW: | 279.32 | MF: | C12 H13 N3 O3 S |

Antiprotozoal

Hoechst, FRG

LgP	2.50(M); 2.53(C)
pKa	
pp cited in Vols.1-5:	

(1) Abst. Intersci. Conf. Antimicrob., 19, (Boston), 1979, 152 abst.

FEZATIONE [INN]

| CAS | 15387-18-5 | MW: | 310.44 | MF: | C17 H14 N2 S2 |

Antifungal

LgP	(4)
pKa	
pp cited in Vols.1-5:	

(1) Japan. Pat. 3382 (1967).
(2) Japan. Med. Gaz., 1970, 7 (9),9.

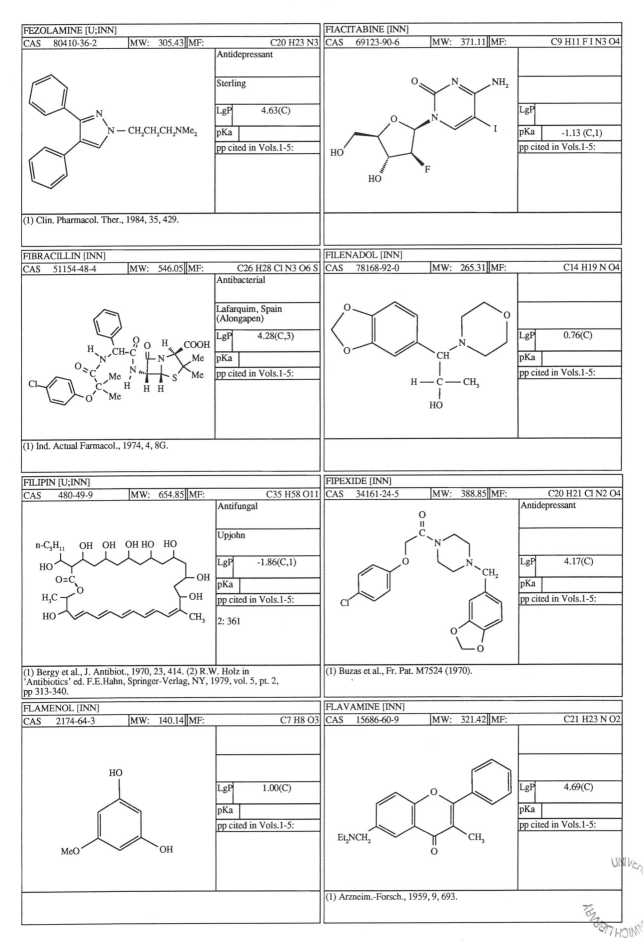

FEZOLAMINE [U;INN]

CAS	80410-36-2	MW:	305.43	MF:	C20 H23 N3

Antidepressant

Sterling

LgP	4.63(C)
pKa	
pp cited in Vols.1-5:	

(1) Clin. Pharmacol. Ther., 1984, 35, 429.

FIACITABINE [INN]

CAS	69123-90-6	MW:	371.11	MF:	C9 H11 F I N3 O4

LgP	
pKa	-1.13 (C,1)
pp cited in Vols.1-5:	

FIBRACILLIN [INN]

CAS	51154-48-4	MW:	546.05	MF:	C26 H28 Cl N3 O6 S

Antibacterial

Lafarquim, Spain
(Alongapen)

LgP	4.28(C,3)
pKa	
pp cited in Vols.1-5:	

(1) Ind. Actual Farmacol., 1974, 4, 8G.

FILENADOL [INN]

CAS	78168-92-0	MW:	265.31	MF:	C14 H19 N O4

LgP	0.76(C)
pKa	
pp cited in Vols.1-5:	

FILIPIN [U;INN]

CAS	480-49-9	MW:	654.85	MF:	C35 H58 O11

Antifungal

Upjohn

LgP	-1.86(C,1)
pKa	
pp cited in Vols.1-5:	
2: 361	

(1) Bergy et al., J. Antibiot., 1970, 23, 414. (2) R.W. Holz in
'Antibiotics' ed. F.E.Hahn, Springer-Verlag, NY, 1979, vol. 5, pt. 2,
pp 313-340.

FIPEXIDE [INN]

CAS	34161-24-5	MW:	388.85	MF:	C20 H21 Cl N2 O4

Antidepressant

LgP	4.17(C)
pKa	
pp cited in Vols.1-5:	

(1) Buzas et al., Fr. Pat. M7524 (1970).

FLAMENOL [INN]

CAS	2174-64-3	MW:	140.14	MF:	C7 H8 O3

LgP	1.00(C)
pKa	
pp cited in Vols.1-5:	

FLAVAMINE [INN]

CAS	15686-60-9	MW:	321.42	MF:	C21 H23 N O2

LgP	4.69(C)
pKa	
pp cited in Vols.1-5:	

(1) Arzneim.-Forsch., 1959, 9, 693.

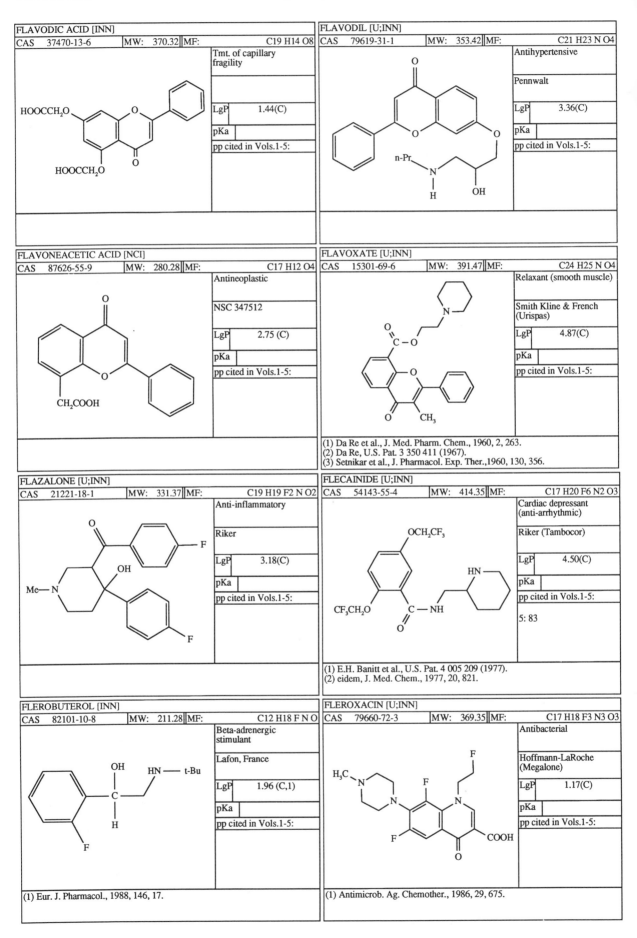

FLAVODIC ACID [INN]

CAS 37470-13-6 | MW: 370.32 | MF: | C19 H14 O8

Tmt. of capillary fragility

LgP 1.44(C)

pKa

pp cited in Vols.1-5:

FLAVODIL [U;INN]

CAS 79619-31-1 | MW: 353.42 | MF: | C21 H23 N O4

Antihypertensive

Pennwalt

LgP 3.36(C)

pKa

pp cited in Vols.1-5:

FLAVONEACETIC ACID [NCI]

CAS 87626-55-9 | MW: 280.28 | MF: | C17 H12 O4

Antineoplastic

NSC 347512

LgP 2.75 (C)

pKa

pp cited in Vols.1-5:

FLAVOXATE [U;INN]

CAS 15301-69-6 | MW: 391.47 | MF: | C24 H25 N O4

Relaxant (smooth muscle)

Smith Kline & French (Urispas)

LgP 4.87(C)

pKa

pp cited in Vols.1-5:

(1) Da Re et al., J. Med. Pharm. Chem., 1960, 2, 263.
(2) Da Re, U.S. Pat. 3 350 411 (1967).
(3) Setnikar et al., J. Pharmacol. Exp. Ther.,1960, 130, 356.

FLAZALONE [U;INN]

CAS 21221-18-1 | MW: 331.37 | MF: | C19 H19 F2 N O2

Anti-inflammatory

Riker

LgP 3.18(C)

pKa

pp cited in Vols.1-5:

FLECAINIDE [U;INN]

CAS 54143-55-4 | MW: 414.35 | MF: | C17 H20 F6 N2 O3

Cardiac depressant (anti-arrhythmic)

Riker (Tambocor)

LgP 4.50(C)

pKa

pp cited in Vols.1-5:

5: 83

(1) E.H. Banitt et al., U.S. Pat. 4 005 209 (1977).
(2) eidem, J. Med. Chem., 1977, 20, 821.

FLEROBUTEROL [INN]

CAS 82101-10-8 | MW: 211.28 | MF: | C12 H18 F N O

Beta-adrenergic stimulant

Lafon, France

LgP 1.96 (C,1)

pKa

pp cited in Vols.1-5:

(1) Eur. J. Pharmacol., 1988, 146, 17.

FLEROXACIN [U;INN]

CAS 79660-72-3 | MW: 369.35 | MF: | C17 H18 F3 N3 O3

Antibacterial

Hoffmann-LaRoche (Megalone)

LgP 1.17(C)

pKa

pp cited in Vols.1-5:

(1) Antimicrob. Ag. Chemother., 1986, 29, 675.

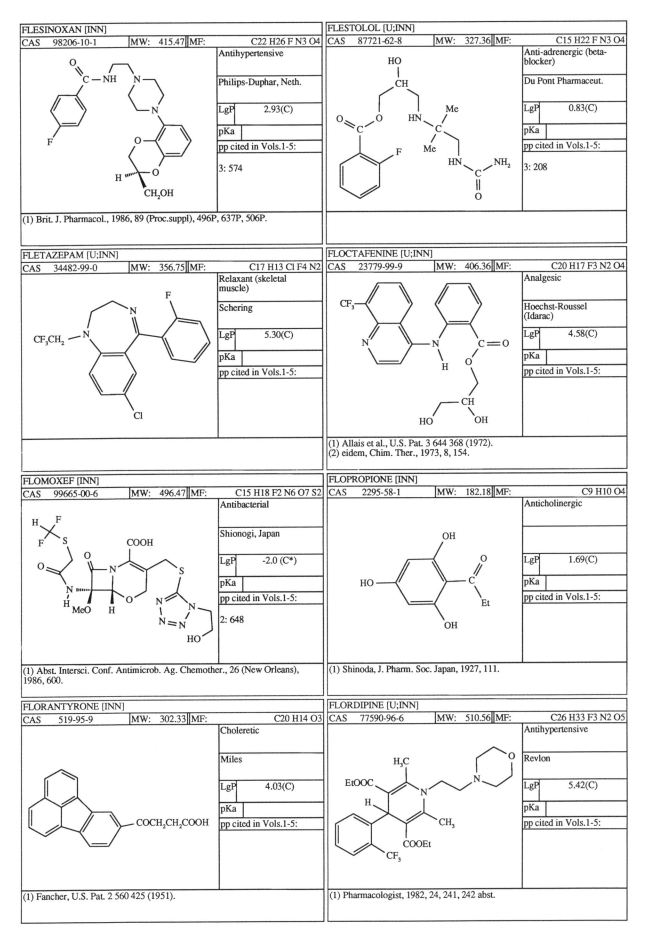

FLESINOXAN [INN]

| CAS | 98206-10-1 | MW: | 415.47 | MF: | C22 H26 F N3 O4 |

Antihypertensive

Philips-Duphar, Neth.

LgP	2.93(C)
pKa	
pp cited in Vols.1-5:	
3: 574	

(1) Brit. J. Pharmacol., 1986, 89 (Proc.suppl), 496P, 637P, 506P.

FLESTOLOL [U;INN]

| CAS | 87721-62-8 | MW: | 327.36 | MF: | C15 H22 F N3 O4 |

Anti-adrenergic (beta-blocker)

Du Pont Pharmaceut.

LgP	0.83(C)
pKa	
pp cited in Vols.1-5:	
3: 208	

FLETAZEPAM [U;INN]

| CAS | 34482-99-0 | MW: | 356.75 | MF: | C17 H13 Cl F4 N2 |

Relaxant (skeletal muscle)

Schering

LgP	5.30(C)
pKa	
pp cited in Vols.1-5:	

FLOCTAFENINE [U;INN]

| CAS | 23779-99-9 | MW: | 406.36 | MF: | C20 H17 F3 N2 O4 |

Analgesic

Hoechst-Roussel (Idarac)

LgP	4.58(C)
pKa	
pp cited in Vols.1-5:	

(1) Allais et al., U.S. Pat. 3 644 368 (1972).
(2) eidem, Chim. Ther., 1973, 8, 154.

FLOMOXEF [INN]

| CAS | 99665-00-6 | MW: | 496.47 | MF: | C15 H18 F2 N6 O7 S2 |

Antibacterial

Shionogi, Japan

LgP	-2.0 (C*)
pKa	
pp cited in Vols.1-5:	
2: 648	

(1) Abst. Intersci. Conf. Antimicrob. Ag. Chemother., 26 (New Orleans), 1986, 600.

FLOPROPIONE [INN]

| CAS | 2295-58-1 | MW: | 182.18 | MF: | C9 H10 O4 |

Anticholinergic

LgP	1.69(C)
pKa	
pp cited in Vols.1-5:	

(1) Shinoda, J. Pharm. Soc. Japan, 1927, 111.

FLORANTYRONE [INN]

| CAS | 519-95-9 | MW: | 302.33 | MF: | C20 H14 O3 |

Choleretic

Miles

LgP	4.03(C)
pKa	
pp cited in Vols.1-5:	

(1) Fancher, U.S. Pat. 2 560 425 (1951).

FLORDIPINE [U;INN]

| CAS | 77590-96-6 | MW: | 510.56 | MF: | C26 H33 F3 N2 O5 |

Antihypertensive

Revlon

LgP	5.42(C)
pKa	
pp cited in Vols.1-5:	

(1) Pharmacologist, 1982, 24, 241, 242 abst.

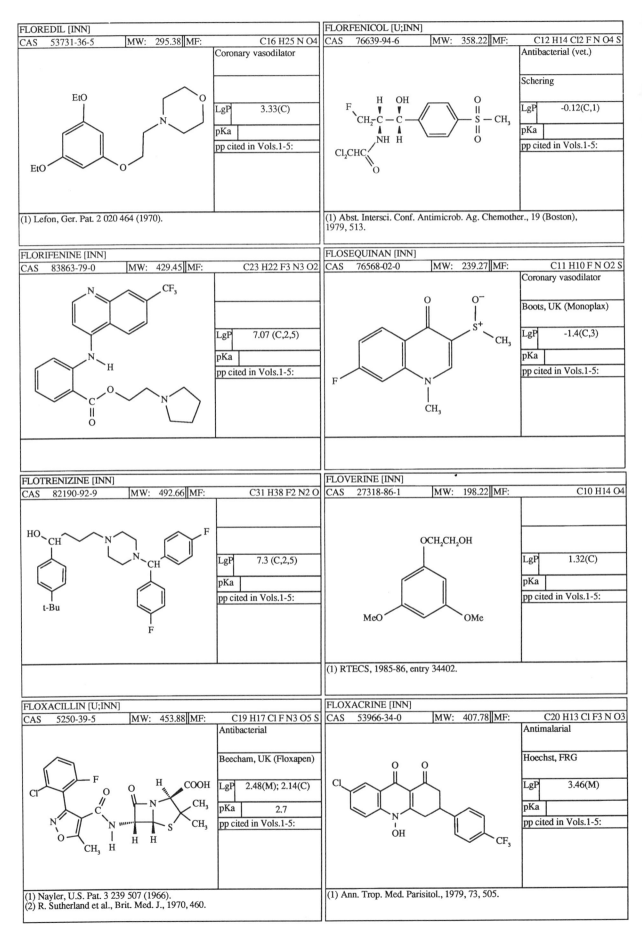

FLOREDIL [INN]

CAS 53731-36-5 | MW: 295.38 | MF: C16 H25 N O4

Coronary vasodilator

LgP 3.33(C)

pKa

pp cited in Vols.1-5:

(1) Lefon, Ger. Pat. 2 020 464 (1970).

FLORFENICOL [U;INN]

CAS 76639-94-6 | MW: 358.22 | MF: C12 H14 Cl2 F N O4 S

Antibacterial (vet.)

Schering

LgP -0.12(C,1)

pKa

pp cited in Vols.1-5:

(1) Abst. Intersci. Conf. Antimicrob. Ag. Chemother., 19 (Boston), 1979, 513.

FLORIFENINE [INN]

CAS 83863-79-0 | MW: 429.45 | MF: C23 H22 F3 N3 O2

LgP 7.07 (C,2,5)

pKa

pp cited in Vols.1-5:

FLOSEQUINAN [INN]

CAS 76568-02-0 | MW: 239.27 | MF: C11 H10 F N O2 S

Coronary vasodilator

Boots, UK (Monoplax)

LgP -1.4(C,3)

pKa

pp cited in Vols.1-5:

FLOTRENIZINE [INN]

CAS 82190-92-9 | MW: 492.66 | MF: C31 H38 F2 N2 O

LgP 7.3 (C,2,5)

pKa

pp cited in Vols.1-5:

FLOVERINE [INN]

CAS 27318-86-1 | MW: 198.22 | MF: C10 H14 O4

LgP 1.32(C)

pKa

pp cited in Vols.1-5:

(1) RTECS, 1985-86, entry 34402.

FLOXACILLIN [U;INN]

CAS 5250-39-5 | MW: 453.88 | MF: C19 H17 Cl F N3 O5 S

Antibacterial

Beecham, UK (Floxapen)

LgP 2.48(M); 2.14(C)

pKa 2.7

pp cited in Vols.1-5:

(1) Nayler, U.S. Pat. 3 239 507 (1966).
(2) R. Sutherland et al., Brit. Med. J., 1970, 460.

FLOXACRINE [INN]

CAS 53966-34-0 | MW: 407.78 | MF: C20 H13 Cl F3 N O3

Antimalarial

Hoechst, FRG

LgP 3.46(M)

pKa

pp cited in Vols.1-5:

(1) Ann. Trop. Med. Parisitol., 1979, 73, 505.

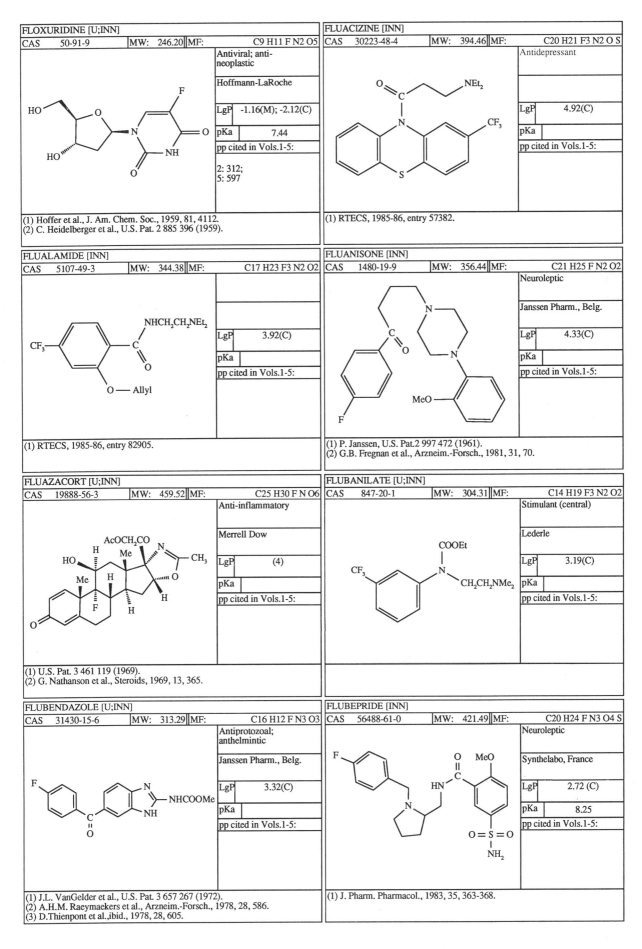

FLOXURIDINE [U;INN]

CAS	50-91-9	MW:	246.20	MF:	C9 H11 F N2 O5

Antiviral; antineoplastic

Hoffmann-LaRoche

LgP	-1.16(M); -2.12(C)
pKa	7.44

pp cited in Vols.1-5:

2: 312;
5: 597

(1) Hoffer et al., J. Am. Chem. Soc., 1959, 81, 4112.
(2) C. Heidelberger et al., U.S. Pat. 2 885 396 (1959).

FLUACIZINE [INN]

CAS	30223-48-4	MW:	394.46	MF:	C20 H21 F3 N2 O S

Antidepressant

LgP	4.92(C)
pKa	

pp cited in Vols.1-5:

(1) RTECS, 1985-86, entry 57382.

FLUALAMIDE [INN]

CAS	5107-49-3	MW:	344.38	MF:	C17 H23 F3 N2 O2

LgP	3.92(C)
pKa	

pp cited in Vols.1-5:

(1) RTECS, 1985-86, entry 82905.

FLUANISONE [INN]

CAS	1480-19-9	MW:	356.44	MF:	C21 H25 F N2 O2

Neuroleptic

Janssen Pharm., Belg.

LgP	4.33(C)
pKa	

pp cited in Vols.1-5:

(1) P. Janssen, U.S. Pat.2 997 472 (1961).
(2) G.B. Fregnan et al., Arzneim.-Forsch., 1981, 31, 70.

FLUAZACORT [U;INN]

CAS	19888-56-3	MW:	459.52	MF:	C25 H30 F N O6

Anti-inflammatory

Merrell Dow

LgP	(4)
pKa	

pp cited in Vols.1-5:

(1) U.S. Pat. 3 461 119 (1969).
(2) G. Nathanson et al., Steroids, 1969, 13, 365.

FLUBANILATE [U;INN]

CAS	847-20-1	MW:	304.31	MF:	C14 H19 F3 N2 O2

Stimulant (central)

Lederle

LgP	3.19(C)
pKa	

pp cited in Vols.1-5:

FLUBENDAZOLE [U;INN]

CAS	31430-15-6	MW:	313.29	MF:	C16 H12 F N3 O3

Antiprotozoal; anthelmintic

Janssen Pharm., Belg.

LgP	3.32(C)
pKa	

pp cited in Vols.1-5:

(1) J.L. VanGelder et al., U.S. Pat. 3 657 267 (1972).
(2) A.H.M. Raeymaekers et al., Arzneim.-Forsch., 1978, 28, 586.
(3) D.Thienpont et al.,ibid., 1978, 28, 605.

FLUBEPRIDE [INN]

CAS	56488-61-0	MW:	421.49	MF:	C20 H24 F N3 O4 S

Neuroleptic

Synthelabo, France

LgP	2.72 (C)
pKa	8.25

pp cited in Vols.1-5:

(1) J. Pharm. Pharmacol., 1983, 35, 363-368.

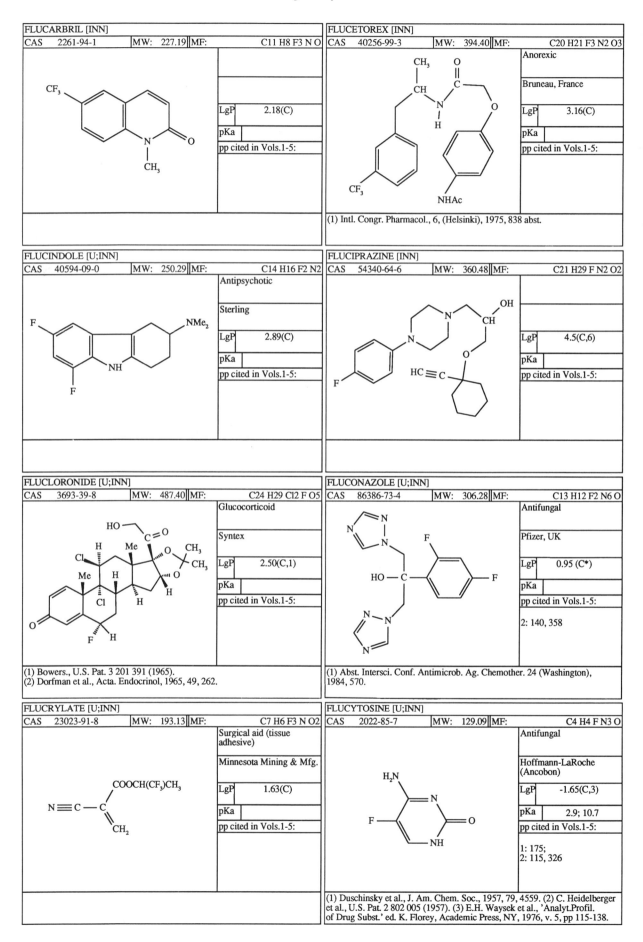

FLUCARBRIL [INN]			
CAS 2261-94-1	MW: 227.19	MF:	C11 H8 F3 N O

LgP	2.18(C)
pKa	
pp cited in Vols.1-5:	

FLUCETOREX [INN]			
CAS 40256-99-3	MW: 394.40	MF:	C20 H21 F3 N2 O3

Anorexic

Bruneau, France

LgP	3.16(C)
pKa	
pp cited in Vols.1-5:	

(1) Intl. Congr. Pharmacol., 6, (Helsinki), 1975, 838 abst.

FLUCINDOLE [U;INN]			
CAS 40594-09-0	MW: 250.29	MF:	C14 H16 F2 N2

Antipsychotic

Sterling

LgP	2.89(C)
pKa	
pp cited in Vols.1-5:	

FLUCIPRAZINE [INN]			
CAS 54340-64-6	MW: 360.48	MF:	C21 H29 F N2 O2

LgP	4.5(C,6)
pKa	
pp cited in Vols.1-5:	

FLUCLORONIDE [U;INN]			
CAS 3693-39-8	MW: 487.40	MF:	C24 H29 Cl2 F O5

Glucocorticoid

Syntex

LgP	2.50(C,1)
pKa	
pp cited in Vols.1-5:	

(1) Bowers., U.S. Pat. 3 201 391 (1965).
(2) Dorfman et al., Acta. Endocrinol, 1965, 49, 262.

FLUCONAZOLE [U;INN]			
CAS 86386-73-4	MW: 306.28	MF:	C13 H12 F2 N6 O

Antifungal

Pfizer, UK

LgP	0.95 (C*)
pKa	
pp cited in Vols.1-5:	
2: 140, 358	

(1) Abst. Intersci. Conf. Antimicrob. Ag. Chemother. 24 (Washington), 1984, 570.

FLUCRYLATE [U;INN]			
CAS 23023-91-8	MW: 193.13	MF:	C7 H6 F3 N O2

Surgical aid (tissue adhesive)

Minnesota Mining & Mfg.

LgP	1.63(C)
pKa	
pp cited in Vols.1-5:	

FLUCYTOSINE [U;INN]			
CAS 2022-85-7	MW: 129.09	MF:	C4 H4 F N3 O

Antifungal

Hoffmann-LaRoche (Ancobon)

LgP	-1.65(C,3)
pKa	2.9; 10.7
pp cited in Vols.1-5:	
1: 175; 2: 115, 326	

(1) Duschinsky et al., J. Am. Chem. Soc., 1957, 79, 4559. (2) C. Heidelberger et al., U.S. Pat. 2 802 005 (1957). (3) E.H. Waysek et al., 'Analyt.Profil. of Drug Subst.' ed. K. Florey, Academic Press, NY, 1976, v. 5, pp 115-138.

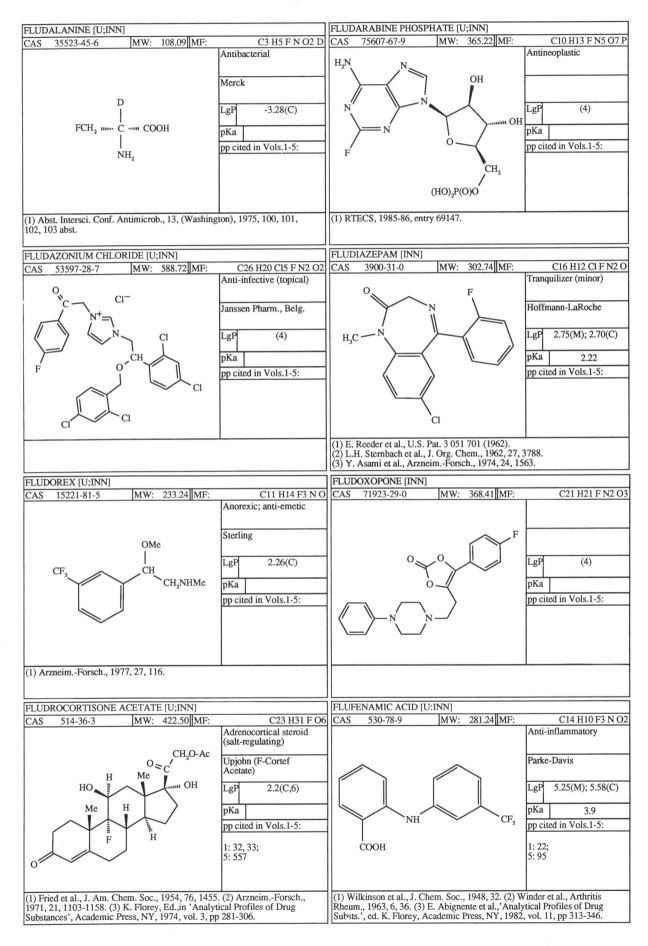

FLUDALANINE [U;INN]

CAS	35523-45-6	MW: 108.09	MF:	C3 H5 F N O2 D

Antibacterial

Merck

LgP	-3.28(C)
pKa	
pp cited in Vols.1-5:	

(1) Abst. Intersci. Conf. Antimicrob., 13, (Washington), 1975, 100, 101, 102, 103 abst.

FLUDARABINE PHOSPHATE [U;INN]

CAS	75607-67-9	MW: 365.22	MF:	C10 H13 F N5 O7 P

Antineoplastic

LgP	(4)
pKa	
pp cited in Vols.1-5:	

(1) RTECS, 1985-86, entry 69147.

FLUDAZONIUM CHLORIDE [U;INN]

CAS	53597-28-7	MW: 588.72	MF:	C26 H20 Cl5 F N2 O2

Anti-infective (topical)

Janssen Pharm., Belg.

LgP	(4)
pKa	
pp cited in Vols.1-5:	

FLUDIAZEPAM [INN]

CAS	3900-31-0	MW: 302.74	MF:	C16 H12 Cl F N2 O

Tranquilizer (minor)

Hoffmann-LaRoche

LgP	2.75(M); 2.70(C)
pKa	2.22
pp cited in Vols.1-5:	

(1) E. Reeder et al., U.S. Pat. 3 051 701 (1962).
(2) L.H. Sternbach et al., J. Org. Chem., 1962, 27, 3788.
(3) Y. Asami et al., Arzneim.-Forsch., 1974, 24, 1563.

FLUDOREX [U;INN]

CAS	15221-81-5	MW: 233.24	MF:	C11 H14 F3 N O

Anorexic; anti-emetic

Sterling

LgP	2.26(C)
pKa	
pp cited in Vols.1-5:	

(1) Arzneim.-Forsch., 1977, 27, 116.

FLUDOXOPONE [INN]

CAS	71923-29-0	MW: 368.41	MF:	C21 H21 F N2 O3

LgP	(4)
pKa	
pp cited in Vols.1-5:	

FLUDROCORTISONE ACETATE [U;INN]

CAS	514-36-3	MW: 422.50	MF:	C23 H31 F O6

Adrenocortical steroid (salt-regulating)

Upjohn (F-Cortef Acetate)

LgP	2.2(C,6)
pKa	
pp cited in Vols.1-5:	
1: 32, 33; 5: 557	

(1) Fried et al., J. Am. Chem. Soc., 1954, 76, 1455. (2) Arzneim.-Forsch., 1971, 21, 1103-1158. (3) K. Florey, Ed.,in 'Analytical Profiles of Drug Substances', Academic Press, NY, 1974, vol. 3, pp 281-306.

FLUFENAMIC ACID [U:INN]

CAS	530-78-9	MW: 281.24	MF:	C14 H10 F3 N O2

Anti-inflammatory

Parke-Davis

LgP	5.25(M); 5.58(C)
pKa	3.9
pp cited in Vols.1-5:	
1: 22; 5: 95	

(1) Wilkinson et al., J. Chem. Soc., 1948, 32. (2) Winder et al., Arthritis Rheum., 1963, 6, 36. (3) E. Abignente et al.,'Analytical Profiles of Drug Substs.', ed. K. Florey, Academic Press, NY, 1982, vol. 11, pp 313-346.

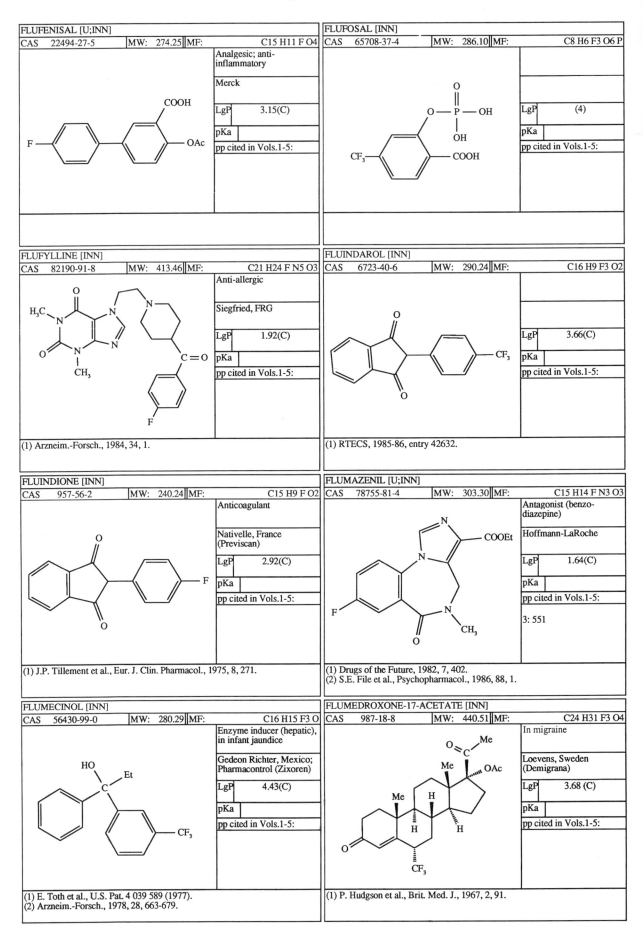

FLUFENISAL [U;INN]

CAS	22494-27-5	MW:	274.25	MF:	C15 H11 F O4

Analgesic; anti-inflammatory

Merck

LgP	3.15(C)
pKa	
pp cited in Vols.1-5:	

FLUFOSAL [INN]

CAS	65708-37-4	MW:	286.10	MF:	C8 H6 F3 O6 P

LgP	(4)
pKa	
pp cited in Vols.1-5:	

FLUFYLLINE [INN]

CAS	82190-91-8	MW:	413.46	MF:	C21 H24 F N5 O3

Anti-allergic

Siegfried, FRG

LgP	1.92(C)
pKa	
pp cited in Vols.1-5:	

(1) Arzneim.-Forsch., 1984, 34, 1.

FLUINDAROL [INN]

CAS	6723-40-6	MW:	290.24	MF:	C16 H9 F3 O2

LgP	3.66(C)
pKa	
pp cited in Vols.1-5:	

(1) RTECS, 1985-86, entry 42632.

FLUINDIONE [INN]

CAS	957-56-2	MW:	240.24	MF:	C15 H9 F O2

Anticoagulant

Nativelle, France (Previscan)

LgP	2.92(C)
pKa	
pp cited in Vols.1-5:	

(1) J.P. Tillement et al., Eur. J. Clin. Pharmacol., 1975, 8, 271.

FLUMAZENIL [U;INN]

CAS	78755-81-4	MW:	303.30	MF:	C15 H14 F N3 O3

Antagonist (benzo-diazepine)

Hoffmann-LaRoche

LgP	1.64(C)
pKa	
pp cited in Vols.1-5:	
3: 551	

(1) Drugs of the Future, 1982, 7, 402.
(2) S.E. File et al., Psychopharmacol., 1986, 88, 1.

FLUMECINOL [INN]

CAS	56430-99-0	MW:	280.29	MF:	C16 H15 F3 O

Enzyme inducer (hepatic), in infant jaundice

Gedeon Richter, Mexico; Pharmacontrol (Zixoren)

LgP	4.43(C)
pKa	
pp cited in Vols.1-5:	

(1) E. Toth et al., U.S. Pat. 4 039 589 (1977).
(2) Arzneim.-Forsch., 1978, 28, 663-679.

FLUMEDROXONE-17-ACETATE [INN]

CAS	987-18-8	MW:	440.51	MF:	C24 H31 F3 O4

In migraine

Loevens, Sweden (Demigrana)

LgP	3.68 (C)
pKa	
pp cited in Vols.1-5:	

(1) P. Hudgson et al., Brit. Med. J., 1967, 2, 91.

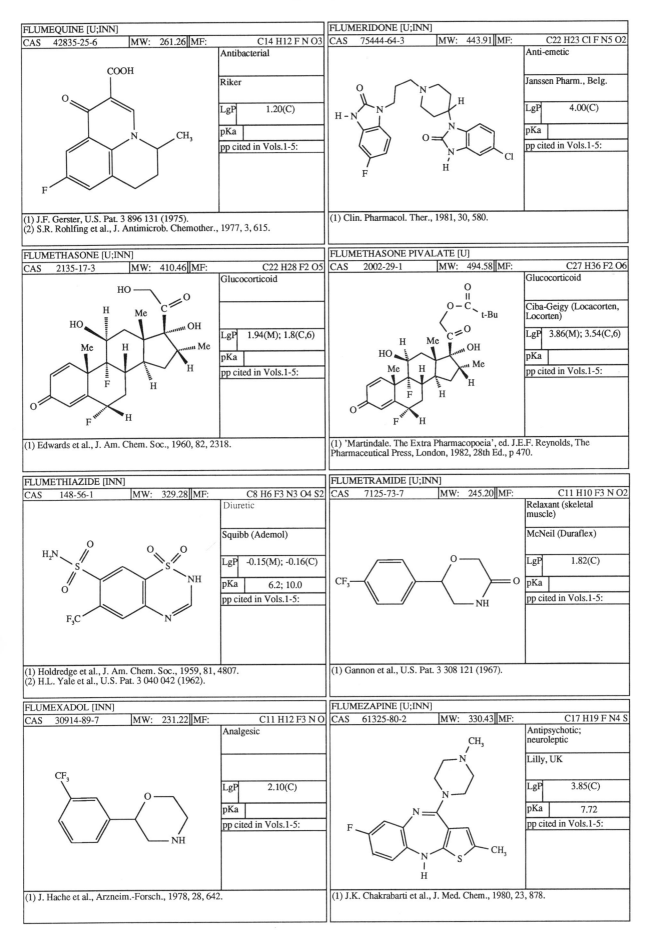

FLUMEQUINE [U;INN]

CAS	42835-25-6	MW:	261.26	MF:	C14 H12 F N O3

Antibacterial

Riker

LgP	1.20(C)

pKa	

pp cited in Vols.1-5:

(1) J.F. Gerster, U.S. Pat. 3 896 131 (1975).
(2) S.R. Rohlfing et al., J. Antimicrob. Chemother., 1977, 3, 615.

FLUMERIDONE [U;INN]

CAS	75444-64-3	MW:	443.91	MF:	C22 H23 Cl F N5 O2

Anti-emetic

Janssen Pharm., Belg.

LgP	4.00(C)

pKa	

pp cited in Vols.1-5:

(1) Clin. Pharmacol. Ther., 1981, 30, 580.

FLUMETHASONE [U;INN]

CAS	2135-17-3	MW:	410.46	MF:	C22 H28 F2 O5

Glucocorticoid

LgP	1.94(M); 1.8(C,6)

pKa	

pp cited in Vols.1-5:

(1) Edwards et al., J. Am. Chem. Soc., 1960, 82, 2318.

FLUMETHASONE PIVALATE [U]

CAS	2002-29-1	MW:	494.58	MF:	C27 H36 F2 O6

Glucocorticoid

Ciba-Geigy (Locacorten, Locorten)

LgP	3.86(M); 3.54(C,6)

pKa	

pp cited in Vols.1-5:

(1) 'Martindale. The Extra Pharmacopoeia', ed. J.E.F. Reynolds, The Pharmaceutical Press, London, 1982, 28th Ed., p 470.

FLUMETHIAZIDE [INN]

CAS	148-56-1	MW:	329.28	MF:	C8 H6 F3 N3 O4 S2

Diuretic

Squibb (Ademol)

LgP	-0.15(M); -0.16(C)

pKa	6.2; 10.0

pp cited in Vols.1-5:

(1) Holdredge et al., J. Am. Chem. Soc., 1959, 81, 4807.
(2) H.L. Yale et al., U.S. Pat. 3 040 042 (1962).

FLUMETRAMIDE [U;INN]

CAS	7125-73-7	MW:	245.20	MF:	C11 H10 F3 N O2

Relaxant (skeletal muscle)

McNeil (Duraflex)

LgP	1.82(C)

pKa	

pp cited in Vols.1-5:

(1) Gannon et al., U.S. Pat. 3 308 121 (1967).

FLUMEXADOL [INN]

CAS	30914-89-7	MW:	231.22	MF:	C11 H12 F3 N O

Analgesic

LgP	2.10(C)

pKa	

pp cited in Vols.1-5:

(1) J. Hache et al., Arzneim.-Forsch., 1978, 28, 642.

FLUMEZAPINE [U;INN]

CAS	61325-80-2	MW:	330.43	MF:	C17 H19 F N4 S

Antipsychotic; neuroleptic

Lilly, UK

LgP	3.85(C)

pKa	7.72

pp cited in Vols.1-5:

(1) J.K. Chakrabarti et al., J. Med. Chem., 1980, 23, 878.

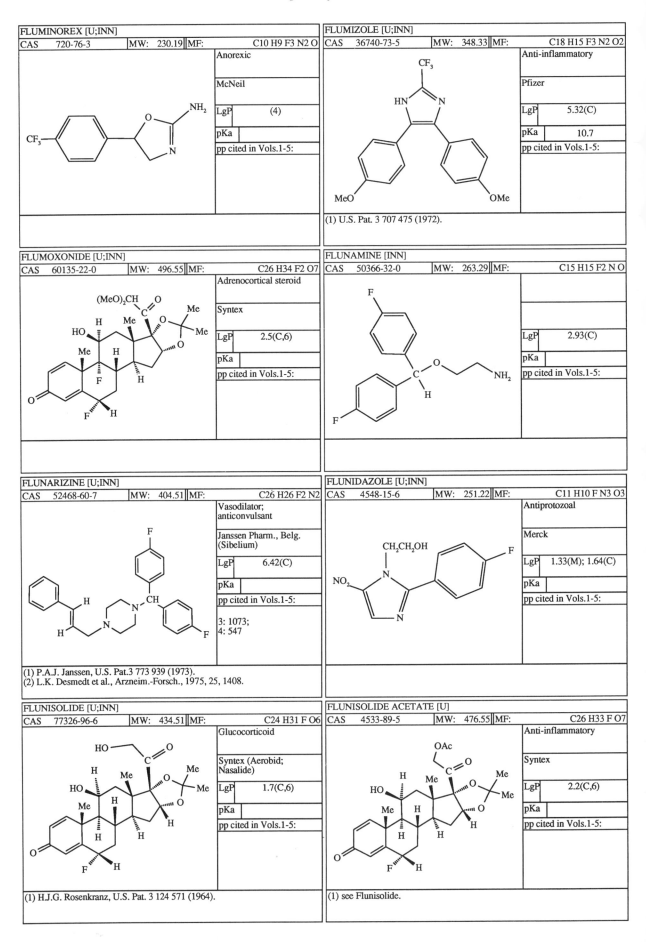

FLUMINOREX [U;INN]

CAS	720-76-3	MW:	230.19	MF:	C10 H9 F3 N2 O

Anorexic

McNeil

LgP	(4)
pKa	
pp cited in Vols.1-5:	

FLUMIZOLE [U;INN]

CAS	36740-73-5	MW:	348.33	MF:	C18 H15 F3 N2 O2

Anti-inflammatory

Pfizer

LgP	5.32(C)
pKa	10.7
pp cited in Vols.1-5:	

(1) U.S. Pat. 3 707 475 (1972).

FLUMOXONIDE [U;INN]

CAS	60135-22-0	MW:	496.55	MF:	C26 H34 F2 O7

Adrenocortical steroid

Syntex

LgP	2.5(C,6)
pKa	
pp cited in Vols.1-5:	

FLUNAMINE [INN]

CAS	50366-32-0	MW:	263.29	MF:	C15 H15 F2 N O

LgP	2.93(C)
pKa	
pp cited in Vols.1-5:	

FLUNARIZINE [U;INN]

CAS	52468-60-7	MW:	404.51	MF:	C26 H26 F2 N2

Vasodilator;
anticonvulsant

Janssen Pharm., Belg.
(Sibelium)

LgP	6.42(C)
pKa	
pp cited in Vols.1-5:	
3: 1073;	
4: 547	

(1) P.A.J. Janssen, U.S. Pat.3 773 939 (1973).
(2) L.K. Desmedt et al., Arzneim.-Forsch., 1975, 25, 1408.

FLUNIDAZOLE [U;INN]

CAS	4548-15-6	MW:	251.22	MF:	C11 H10 F N3 O3

Antiprotozoal

Merck

LgP	1.33(M); 1.64(C)
pKa	
pp cited in Vols.1-5:	

FLUNISOLIDE [U;INN]

CAS	77326-96-6	MW:	434.51	MF:	C24 H31 F O6

Glucocorticoid

Syntex (Aerobid;
Nasalide)

LgP	1.7(C,6)
pKa	
pp cited in Vols.1-5:	

(1) H.J.G. Rosenkranz, U.S. Pat. 3 124 571 (1964).

FLUNISOLIDE ACETATE [U]

CAS	4533-89-5	MW:	476.55	MF:	C26 H33 F O7

Anti-inflammatory

Syntex

LgP	2.2(C,6)
pKa	
pp cited in Vols.1-5:	

(1) see Flunisolide.

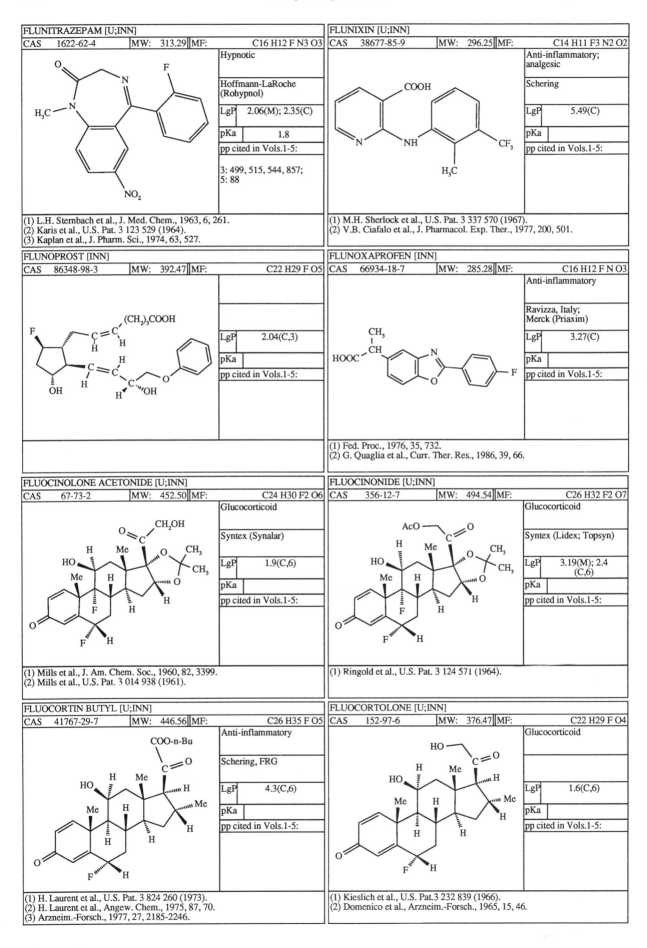

FLUNITRAZEPAM [U;INN]

CAS	1622-62-4	MW:	313.29	MF:	C16 H12 F N3 O3

Hypnotic

Hoffmann-LaRoche (Rohypnol)

LgP	2.06(M); 2.35(C)

pKa	1.8

pp cited in Vols.1-5:

3: 499, 515, 544, 857; 5: 88

(1) L.H. Sternbach et al., J. Med. Chem., 1963, 6, 261.
(2) Karis et al., U.S. Pat. 3 123 529 (1964).
(3) Kaplan et al., J. Pharm. Sci., 1974, 63, 527.

FLUNIXIN [U;INN]

CAS	38677-85-9	MW:	296.25	MF:	C14 H11 F3 N2 O2

Anti-inflammatory; analgesic

Schering

LgP	5.49(C)

pKa	

pp cited in Vols.1-5:

(1) M.H. Sherlock et al., U.S. Pat. 3 337 570 (1967).
(2) V.B. Ciafalo et al., J. Pharmacol. Exp. Ther., 1977, 200, 501.

FLUNOPROST [INN]

CAS	86348-98-3	MW:	392.47	MF:	C22 H29 F O5

LgP	2.04(C,3)

pKa	

pp cited in Vols.1-5:

FLUNOXAPROFEN [INN]

CAS	66934-18-7	MW:	285.28	MF:	C16 H12 F N O3

Anti-inflammatory

Ravizza, Italy; Merck (Priaxim)

LgP	3.27(C)

pKa	

pp cited in Vols.1-5:

(1) Fed. Proc., 1976, 35, 732.
(2) G. Quaglia et al., Curr. Ther. Res., 1986, 39, 66.

FLUOCINOLONE ACETONIDE [U;INN]

CAS	67-73-2	MW:	452.50	MF:	C24 H30 F2 O6

Glucocorticoid

Syntex (Synalar)

LgP	1.9(C,6)

pKa	

pp cited in Vols.1-5:

(1) Mills et al., J. Am. Chem. Soc., 1960, 82, 3399.
(2) Mills et al., U.S. Pat. 3 014 938 (1961).

FLUOCINONIDE [U;INN]

CAS	356-12-7	MW:	494.54	MF:	C26 H32 F2 O7

Glucocorticoid

Syntex (Lidex; Topsyn)

LgP	3.19(M); 2.4 (C,6)

pKa	

pp cited in Vols.1-5:

(1) Ringold et al., U.S. Pat. 3 124 571 (1964).

FLUOCORTIN BUTYL [U;INN]

CAS	41767-29-7	MW:	446.56	MF:	C26 H35 F O5

Anti-inflammatory

Schering, FRG

LgP	4.3(C,6)

pKa	

pp cited in Vols.1-5:

(1) H. Laurent et al., U.S. Pat. 3 824 260 (1973).
(2) H. Laurent et al., Angew. Chem., 1975, 87, 70.
(3) Arzneim.-Forsch., 1977, 27, 2185-2246.

FLUOCORTOLONE [U;INN]

CAS	152-97-6	MW:	376.47	MF:	C22 H29 F O4

Glucocorticoid

LgP	1.6(C,6)

pKa	

pp cited in Vols.1-5:

(1) Kieslich et al., U.S. Pat. 3 232 839 (1966).
(2) Domenico et al., Arzneim.-Forsch., 1965, 15, 46.

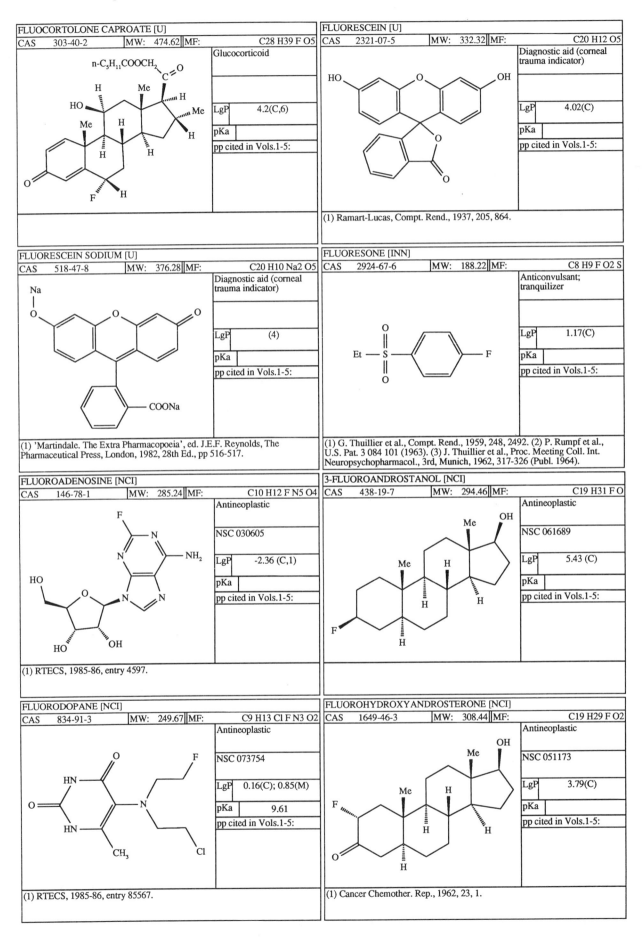

FLUOCORTOLONE CAPROATE [U]

CAS	303-40-2	MW:	474.62	MF:	C28 H39 F O5

Glucocorticoid

LgP	4.2(C,6)
pKa	
pp cited in Vols.1-5:	

n-C₅H₁₁COOCH₂

FLUORESCEIN [U]

CAS	2321-07-5	MW:	332.32	MF:	C20 H12 O5

Diagnostic aid (corneal trauma indicator)

LgP	4.02(C)
pKa	
pp cited in Vols.1-5:	

(1) Ramart-Lucas, Compt. Rend., 1937, 205, 864.

FLUORESCEIN SODIUM [U]

CAS	518-47-8	MW:	376.28	MF:	C20 H10 Na2 O5

Diagnostic aid (corneal trauma indicator)

LgP	(4)
pKa	
pp cited in Vols.1-5:	

COONa

(1) 'Martindale. The Extra Pharmacopoeia', ed. J.E.F. Reynolds, The Pharmaceutical Press, London, 1982, 28th Ed., pp 516-517.

FLUORESONE [INN]

CAS	2924-67-6	MW:	188.22	MF:	C8 H9 F O2 S

Anticonvulsant; tranquilizer

LgP	1.17(C)
pKa	
pp cited in Vols.1-5:	

(1) G. Thuillier et al., Compt. Rend., 1959, 248, 2492. (2) P. Rumpf et al., U.S. Pat. 3 084 101 (1963). (3) J. Thuillier et al., Proc. Meeting Coll. Int. Neuropsychopharmacol., 3rd, Munich, 1962, 317-326 (Publ. 1964).

FLUOROADENOSINE [NCI]

CAS	146-78-1	MW:	285.24	MF:	C10 H12 F N5 O4

Antineoplastic

NSC 030605

LgP	-2.36 (C,1)
pKa	
pp cited in Vols.1-5:	

(1) RTECS, 1985-86, entry 4597.

3-FLUOROANDROSTANOL [NCI]

CAS	438-19-7	MW:	294.46	MF:	C19 H31 F O

Antineoplastic

NSC 061689

LgP	5.43 (C)
pKa	
pp cited in Vols.1-5:	

FLUORODOPANE [NCI]

CAS	834-91-3	MW:	249.67	MF:	C9 H13 Cl F N3 O2

Antineoplastic

NSC 073754

LgP	0.16(C); 0.85(M)
pKa	9.61
pp cited in Vols.1-5:	

(1) RTECS, 1985-86, entry 85567.

FLUOROHYDROXYANDROSTERONE [NCI]

CAS	1649-46-3	MW:	308.44	MF:	C19 H29 F O2

Antineoplastic

NSC 051173

LgP	3.79(C)
pKa	
pp cited in Vols.1-5:	

(1) Cancer Chemother. Rep., 1962, 23, 1.

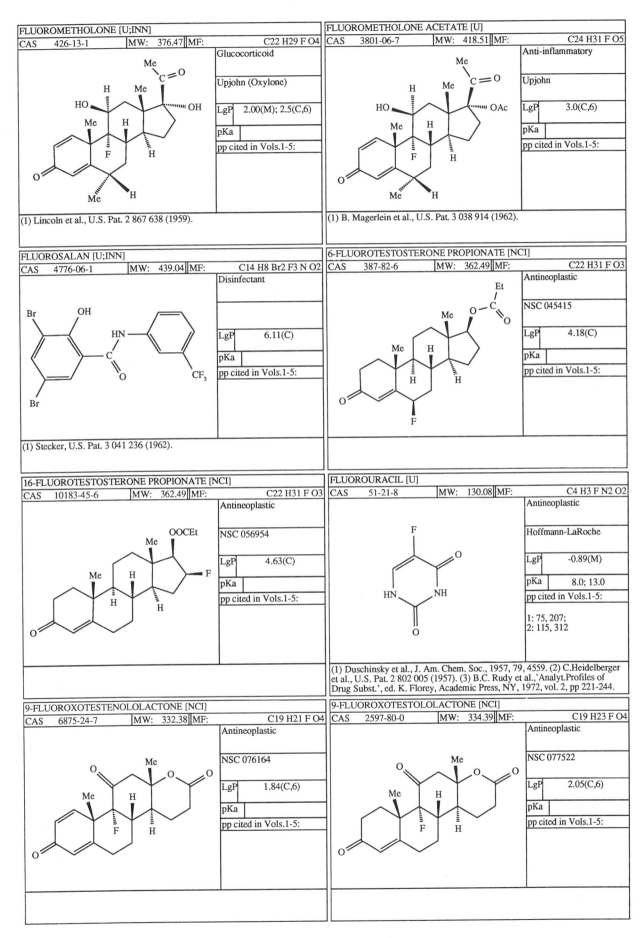

FLUOROMETHOLONE [U;INN]

CAS	426-13-1	MW:	376.47	MF:	C22 H29 F O4

Glucocorticoid

Upjohn (Oxylone)

LgP 2.00(M); 2.5(C,6)

pKa

pp cited in Vols.1-5:

(1) Lincoln et al., U.S. Pat. 2 867 638 (1959).

FLUOROMETHOLONE ACETATE [U]

CAS	3801-06-7	MW:	418.51	MF:	C24 H31 F O5

Anti-inflammatory

Upjohn

LgP 3.0(C,6)

pKa

pp cited in Vols.1-5:

(1) B. Magerlein et al., U.S. Pat. 3 038 914 (1962).

FLUOROSALAN [U;INN]

CAS	4776-06-1	MW:	439.04	MF:	C14 H8 Br2 F3 N O2

Disinfectant

LgP 6.11(C)

pKa

pp cited in Vols.1-5:

(1) Stecker, U.S. Pat. 3 041 236 (1962).

6-FLUOROTESTOSTERONE PROPIONATE [NCI]

CAS	387-82-6	MW:	362.49	MF:	C22 H31 F O3

Antineoplastic

NSC 045415

LgP 4.18(C)

pKa

pp cited in Vols.1-5:

16-FLUOROTESTOSTERONE PROPIONATE [NCI]

CAS	10183-45-6	MW:	362.49	MF:	C22 H31 F O3

Antineoplastic

NSC 056954

LgP 4.63(C)

pKa

pp cited in Vols.1-5:

FLUOROURACIL [U]

CAS	51-21-8	MW:	130.08	MF:	C4 H3 F N2 O2

Antineoplastic

Hoffmann-LaRoche

LgP -0.89(M)

pKa 8.0; 13.0

pp cited in Vols.1-5:

1: 75, 207;
2: 115, 312

(1) Duschinsky et al., J. Am. Chem. Soc., 1957, 79, 4559. (2) C.Heidelberger et al., U.S. Pat. 2 802 005 (1957). (3) B.C. Rudy et al.,'Analyt.Profiles of Drug Subst.', ed. K. Florey, Academic Press, NY, 1972, vol. 2, pp 221-244.

9-FLUOROXOTESTENOLOLACTONE [NCI]

CAS	6875-24-7	MW:	332.38	MF:	C19 H21 F O4

Antineoplastic

NSC 076164

LgP 1.84(C,6)

pKa

pp cited in Vols.1-5:

9-FLUOROXOTESTOLOLACTONE [NCI]

CAS	2597-80-0	MW:	334.39	MF:	C19 H23 F O4

Antineoplastic

NSC 077522

LgP 2.05(C,6)

pKa

pp cited in Vols.1-5:

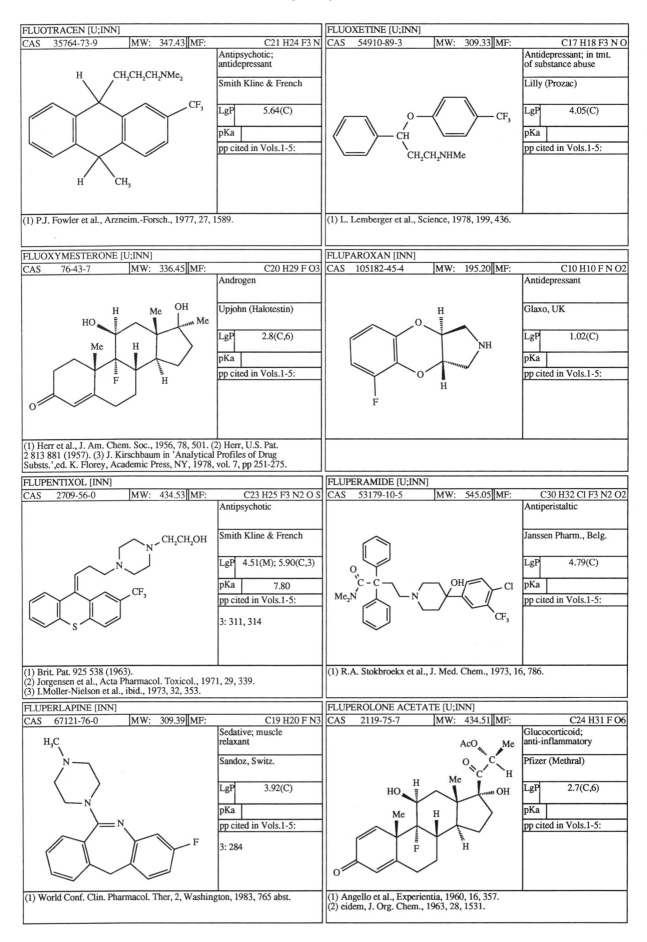

FLUOTRACEN [U;INN]

CAS	35764-73-9	MW: 347.43	MF:	C21 H24 F3 N

Antipsychotic; antidepressant

Smith Kline & French

LgP	5.64(C)

pKa	

pp cited in Vols.1-5:

(1) P.J. Fowler et al., Arzneim.-Forsch., 1977, 27, 1589.

FLUOXETINE [U;INN]

CAS	54910-89-3	MW: 309.33	MF:	C17 H18 F3 N O

Antidepressant; in tmt. of substance abuse

Lilly (Prozac)

LgP	4.05(C)

pKa	

pp cited in Vols.1-5:

(1) L. Lemberger et al., Science, 1978, 199, 436.

FLUOXYMESTERONE [U;INN]

CAS	76-43-7	MW: 336.45	MF:	C20 H29 F O3

Androgen

Upjohn (Halotestin)

LgP	2.8(C,6)

pKa	

pp cited in Vols.1-5:

(1) Herr et al., J. Am. Chem. Soc., 1956, 78, 501. (2) Herr, U.S. Pat. 2 813 881 (1957). (3) J. Kirschbaum in 'Analytical Profiles of Drug Substs.',ed. K. Florey, Academic Press, NY, 1978, vol. 7, pp 251-275.

FLUPAROXAN [INN]

CAS	105182-45-4	MW: 195.20	MF:	C10 H10 F N O2

Antidepressant

Glaxo, UK

LgP	1.02(C)

pKa	

pp cited in Vols.1-5:

FLUPENTIXOL [INN]

CAS	2709-56-0	MW: 434.53	MF:	C23 H25 F3 N2 O S

Antipsychotic

Smith Kline & French

LgP	4.51(M); 5.90(C,3)

pKa	7.80

pp cited in Vols.1-5:

3: 311, 314

(1) Brit. Pat. 925 538 (1963).
(2) Jorgensen et al., Acta Pharmacol. Toxicol., 1971, 29, 339.
(3) I.Moller-Nielson et al., ibid., 1973, 32, 353.

FLUPERAMIDE [U;INN]

CAS	53179-10-5	MW: 545.05	MF:	C30 H32 Cl F3 N2 O2

Antiperistaltic

Janssen Pharm., Belg.

LgP	4.79(C)

pKa	

pp cited in Vols.1-5:

(1) R.A. Stokbroekx et al., J. Med. Chem., 1973, 16, 786.

FLUPERLAPINE [INN]

CAS	67121-76-0	MW: 309.39	MF:	C19 H20 F N3

Sedative; muscle relaxant

Sandoz, Switz.

LgP	3.92(C)

pKa	

pp cited in Vols.1-5:

3: 284

(1) World Conf. Clin. Pharmacol. Ther, 2, Washington, 1983, 765 abst.

FLUPEROLONE ACETATE [U;INN]

CAS	2119-75-7	MW: 434.51	MF:	C24 H31 F O6

Glucocorticoid; anti-inflammatory

Pfizer (Methral)

LgP	2.7(C,6)

pKa	

pp cited in Vols.1-5:

(1) Angello et al., Experientia, 1960, 16, 357.
(2) eidem, J. Org. Chem., 1963, 28, 1531.

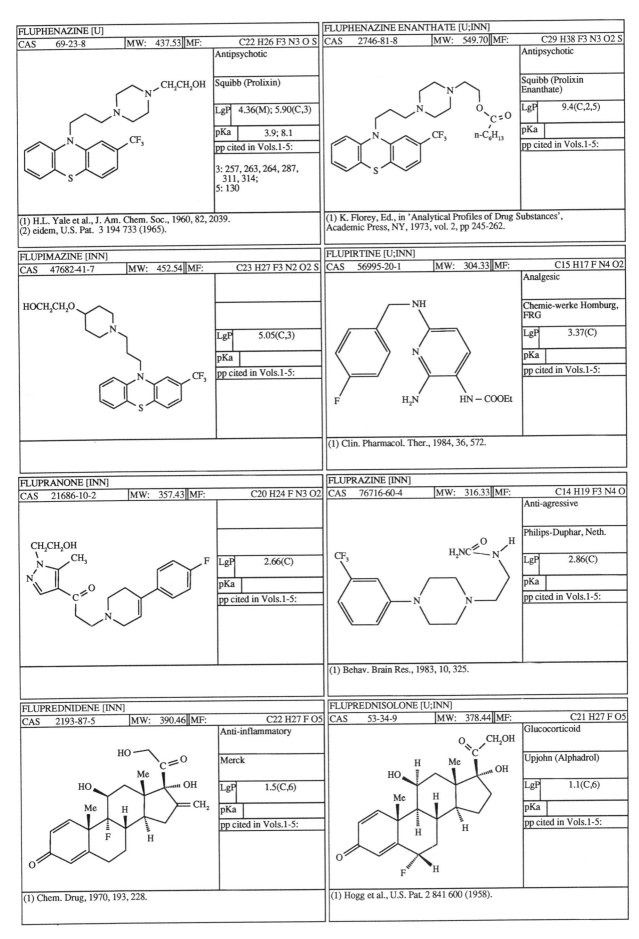

FLUPHENAZINE [U]

CAS	69-23-8	MW: 437.53	MF:	C22 H26 F3 N3 O S

Antipsychotic

Squibb (Prolixin)

LgP	4.36(M); 5.90(C,3)

pKa	3.9; 8.1

pp cited in Vols.1-5:

3: 257, 263, 264, 287, 311, 314;
5: 130

(1) H.L. Yale et al., J. Am. Chem. Soc., 1960, 82, 2039.
(2) eidem, U.S. Pat. 3 194 733 (1965).

FLUPHENAZINE ENANTHATE [U;INN]

CAS	2746-81-8	MW: 549.70	MF:	C29 H38 F3 N3 O2 S

Antipsychotic

Squibb (Prolixin Enanthate)

LgP	9.4(C,2,5)

pKa	

pp cited in Vols.1-5:

(1) K. Florey, Ed., in 'Analytical Profiles of Drug Substances',
Academic Press, NY, 1973, vol. 2, pp 245-262.

FLUPIMAZINE [INN]

CAS	47682-41-7	MW: 452.54	MF:	C23 H27 F3 N2 O2 S

LgP	5.05(C,3)

pKa	

pp cited in Vols.1-5:

FLUPIRTINE [U;INN]

CAS	56995-20-1	MW: 304.33	MF:	C15 H17 F N4 O2

Analgesic

Chemie-werke Homburg, FRG

LgP	3.37(C)

pKa	

pp cited in Vols.1-5:

(1) Clin. Pharmacol. Ther., 1984, 36, 572.

FLUPRANONE [INN]

CAS	21686-10-2	MW: 357.43	MF:	C20 H24 F N3 O2

LgP	2.66(C)

pKa	

pp cited in Vols.1-5:

FLUPRAZINE [INN]

CAS	76716-60-4	MW: 316.33	MF:	C14 H19 F3 N4 O

Anti-agressive

Philips-Duphar, Neth.

LgP	2.86(C)

pKa	

pp cited in Vols.1-5:

(1) Behav. Brain Res., 1983, 10, 325.

FLUPREDNIDENE [INN]

CAS	2193-87-5	MW: 390.46	MF:	C22 H27 F O5

Anti-inflammatory

Merck

LgP	1.5(C,6)

pKa	

pp cited in Vols.1-5:

(1) Chem. Drug, 1970, 193, 228.

FLUPREDNISOLONE [U;INN]

CAS	53-34-9	MW: 378.44	MF:	C21 H27 F O5

Glucocorticoid

Upjohn (Alphadrol)

LgP	1.1(C,6)

pKa	

pp cited in Vols.1-5:

(1) Hogg et al., U.S. Pat. 2 841 600 (1958).

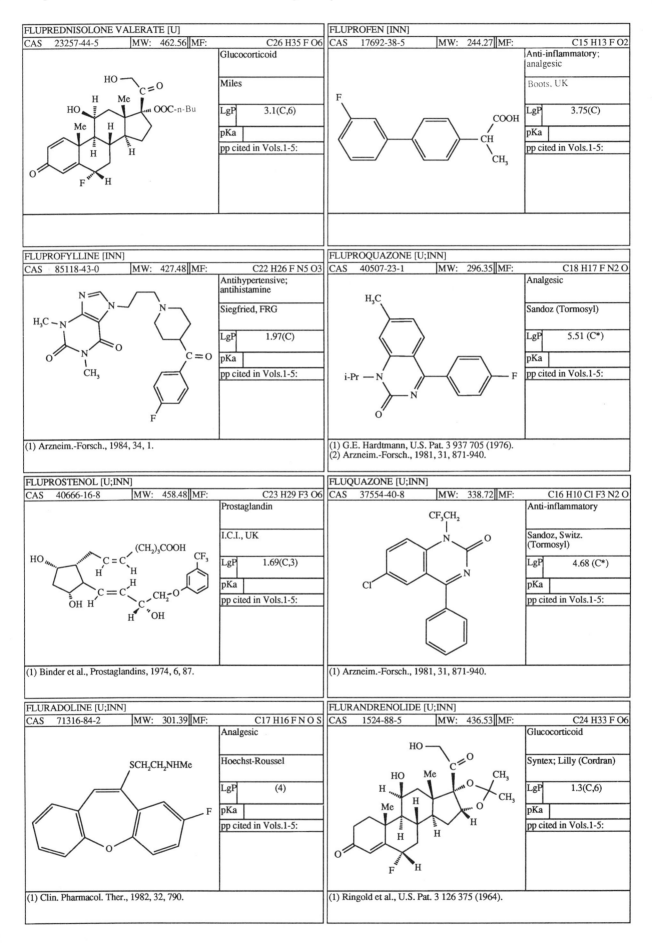

FLUPREDNISOLONE VALERATE [U]

CAS	23257-44-5	MW: 462.56	MF:	C26 H35 F O6

Glucocorticoid

Miles

LgP	3.1(C,6)
pKa	
pp cited in Vols.1-5:	

FLUPROFEN [INN]

CAS	17692-38-5	MW: 244.27	MF:	C15 H13 F O2

Anti-inflammatory; analgesic

Boots, UK

LgP	3.75(C)
pKa	
pp cited in Vols.1-5:	

FLUPROFYLLINE [INN]

CAS	85118-43-0	MW: 427.48	MF:	C22 H26 F N5 O3

Antihypertensive; antihistamine

Siegfried, FRG

LgP	1.97(C)
pKa	
pp cited in Vols.1-5:	

(1) Arzneim.-Forsch., 1984, 34, 1.

FLUPROQUAZONE [U;INN]

CAS	40507-23-1	MW: 296.35	MF:	C18 H17 F N2 O

Analgesic

Sandoz (Tormosyl)

LgP	5.51 (C*)
pKa	
pp cited in Vols.1-5:	

(1) G.E. Hardtmann, U.S. Pat. 3 937 705 (1976).
(2) Arzneim.-Forsch., 1981, 31, 871-940.

FLUPROSTENOL [U;INN]

CAS	40666-16-8	MW: 458.48	MF:	C23 H29 F3 O6

Prostaglandin

I.C.I., UK

LgP	1.69(C,3)
pKa	
pp cited in Vols.1-5:	

(1) Binder et al., Prostaglandins, 1974, 6, 87.

FLUQUAZONE [U;INN]

CAS	37554-40-8	MW: 338.72	MF:	C16 H10 Cl F3 N2 O

Anti-inflammatory

Sandoz, Switz. (Tormosyl)

LgP	4.68 (C*)
pKa	
pp cited in Vols.1-5:	

(1) Arzneim.-Forsch., 1981, 31, 871-940.

FLURADOLINE [U;INN]

CAS	71316-84-2	MW: 301.39	MF:	C17 H16 F N O S

Analgesic

Hoechst-Roussel

LgP	(4)
pKa	
pp cited in Vols.1-5:	

(1) Clin. Pharmacol. Ther., 1982, 32, 790.

FLURANDRENOLIDE [U;INN]

CAS	1524-88-5	MW: 436.53	MF:	C24 H33 F O6

Glucocorticoid

Syntex; Lilly (Cordran)

LgP	1.3(C,6)
pKa	
pp cited in Vols.1-5:	

(1) Ringold et al., U.S. Pat. 3 126 375 (1964).

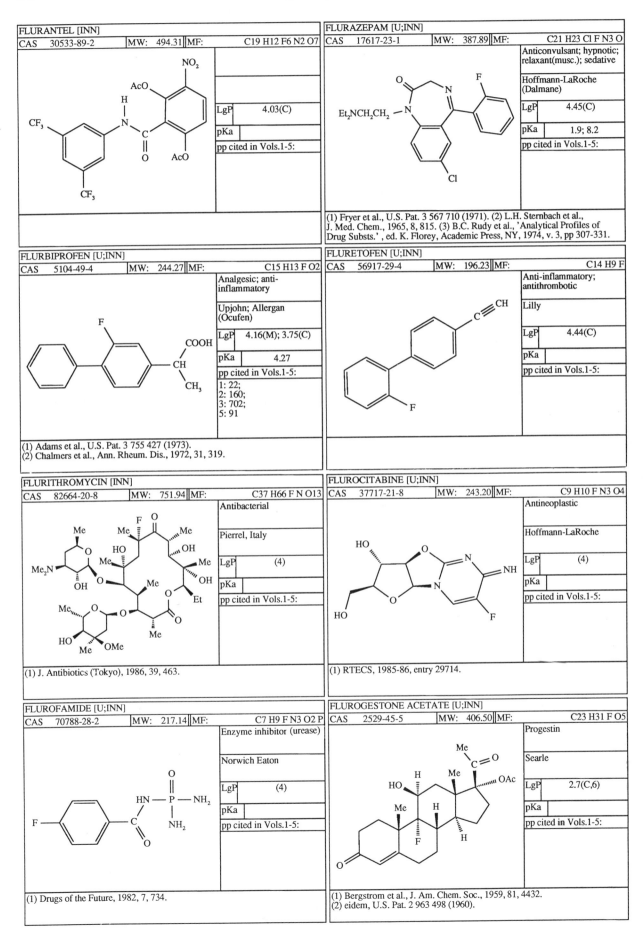

FLURANTEL [INN]

CAS	30533-89-2	MW: 494.31	MF:	C19 H12 F6 N2 O7

LgP	4.03(C)
pKa	
pp cited in Vols.1-5:	

FLURAZEPAM [U;INN]

CAS	17617-23-1	MW: 387.89	MF:	C21 H23 Cl F N3 O

Anticonvulsant; hypnotic; relaxant(musc.); sedative

Hoffmann-LaRoche (Dalmane)

LgP	4.45(C)
pKa	1.9; 8.2
pp cited in Vols.1-5:	

(1) Fryer et al., U.S. Pat. 3 567 710 (1971). (2) L.H. Sternbach et al., J. Med. Chem., 1965, 8, 815. (3) B.C. Rudy et al., 'Analytical Profiles of Drug Substs.' , ed. K. Florey, Academic Press, NY, 1974, v. 3, pp 307-331.

FLURBIPROFEN [U;INN]

CAS	5104-49-4	MW: 244.27	MF:	C15 H13 F O2

Analgesic; anti-inflammatory

Upjohn; Allergan (Ocufen)

LgP	4.16(M); 3.75(C)
pKa	4.27
pp cited in Vols.1-5:	1: 22; 2: 160; 3: 702; 5: 91

(1) Adams et al., U.S. Pat. 3 755 427 (1973).
(2) Chalmers et al., Ann. Rheum. Dis., 1972, 31, 319.

FLURETOFEN [U;INN]

CAS	56917-29-4	MW: 196.23	MF:	C14 H9 F

Anti-inflammatory; antithrombotic

Lilly

LgP	4.44(C)
pKa	
pp cited in Vols.1-5:	

FLURITHROMYCIN [INN]

CAS	82664-20-8	MW: 751.94	MF:	C37 H66 F N O13

Antibacterial

Pierrel, Italy

LgP	(4)
pKa	
pp cited in Vols.1-5:	

(1) J. Antibiotics (Tokyo), 1986, 39, 463.

FLUROCITABINE [U;INN]

CAS	37717-21-8	MW: 243.20	MF:	C9 H10 F N3 O4

Antineoplastic

Hoffmann-LaRoche

LgP	(4)
pKa	
pp cited in Vols.1-5:	

(1) RTECS, 1985-86, entry 29714.

FLUROFAMIDE [U;INN]

CAS	70788-28-2	MW: 217.14	MF:	C7 H9 F N3 O2 P

Enzyme inhibitor (urease)

Norwich Eaton

LgP	(4)
pKa	
pp cited in Vols.1-5:	

(1) Drugs of the Future, 1982, 7, 734.

FLUROGESTONE ACETATE [U;INN]

CAS	2529-45-5	MW: 406.50	MF:	C23 H31 F O5

Progestin

Searle

LgP	2.7(C,6)
pKa	
pp cited in Vols.1-5:	

(1) Bergstrom et al., J. Am. Chem. Soc., 1959, 81, 4432.
(2) eidem, U.S. Pat. 2 963 498 (1960).

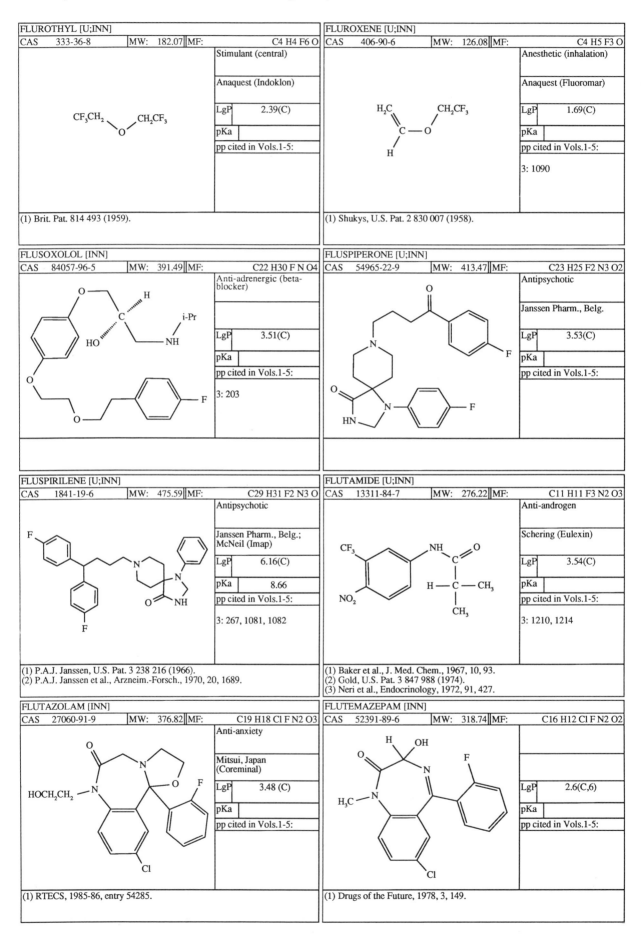

FLUROTHYL [U;INN]		
CAS 333-36-8	MW: 182.07	MF: C4 H4 F6 O

Stimulant (central)

Anaquest (Indoklon)

LgP	2.39(C)
pKa	

pp cited in Vols.1-5:

(1) Brit. Pat. 814 493 (1959).

FLUROXENE [U;INN]		
CAS 406-90-6	MW: 126.08	MF: C4 H5 F3 O

Anesthetic (inhalation)

Anaquest (Fluoromar)

LgP	1.69(C)
pKa	

pp cited in Vols.1-5:

3: 1090

(1) Shukys, U.S. Pat. 2 830 007 (1958).

FLUSOXOLOL [INN]		
CAS 84057-96-5	MW: 391.49	MF: C22 H30 F N O4

Anti-adrenergic (beta-blocker)

LgP	3.51(C)
pKa	

pp cited in Vols.1-5:

3: 203

FLUSPIPERONE [U;INN]		
CAS 54965-22-9	MW: 413.47	MF: C23 H25 F2 N3 O2

Antipsychotic

Janssen Pharm., Belg.

LgP	3.53(C)
pKa	

pp cited in Vols.1-5:

FLUSPIRILENE [U;INN]		
CAS 1841-19-6	MW: 475.59	MF: C29 H31 F2 N3 O

Antipsychotic

Janssen Pharm., Belg.; McNeil (Imap)

LgP	6.16(C)
pKa	8.66

pp cited in Vols.1-5:

3: 267, 1081, 1082

(1) P.A.J. Janssen, U.S. Pat. 3 238 216 (1966).
(2) P.A.J. Janssen et al., Arzneim.-Forsch., 1970, 20, 1689.

FLUTAMIDE [U;INN]		
CAS 13311-84-7	MW: 276.22	MF: C11 H11 F3 N2 O3

Anti-androgen

Schering (Eulexin)

LgP	3.54(C)
pKa	

pp cited in Vols.1-5:

3: 1210, 1214

(1) Baker et al., J. Med. Chem., 1967, 10, 93.
(2) Gold, U.S. Pat. 3 847 988 (1974).
(3) Neri et al., Endocrinology, 1972, 91, 427.

FLUTAZOLAM [INN]		
CAS 27060-91-9	MW: 376.82	MF: C19 H18 Cl F N2 O3

Anti-anxiety

Mitsui, Japan (Coreminal)

LgP	3.48 (C)
pKa	

pp cited in Vols.1-5:

(1) RTECS, 1985-86, entry 54285.

FLUTEMAZEPAM [INN]		
CAS 52391-89-6	MW: 318.74	MF: C16 H12 Cl F N2 O2

LgP	2.6(C,6)
pKa	

pp cited in Vols.1-5:

(1) Drugs of the Future, 1978, 3, 149.

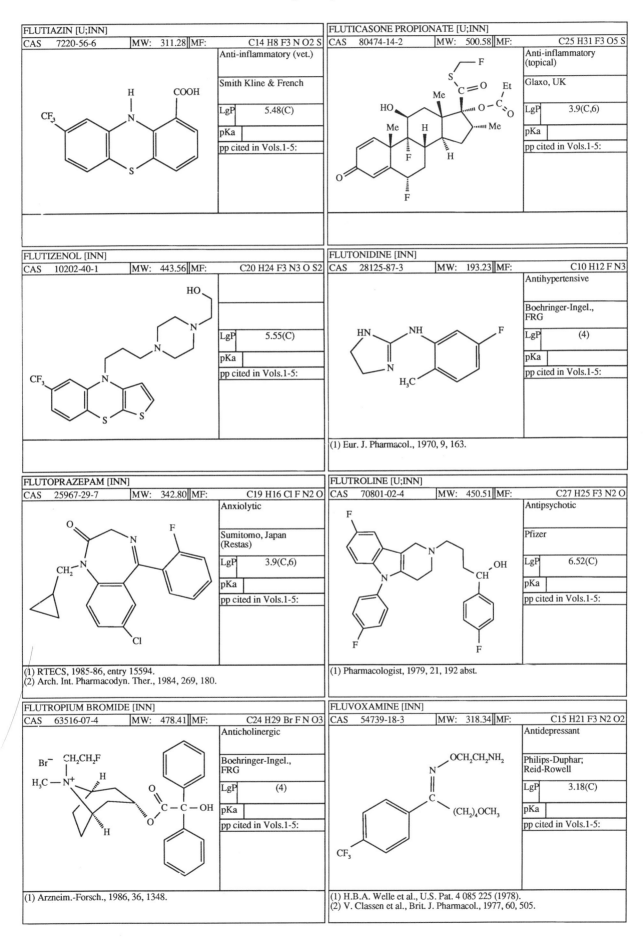

FLUTIAZIN [U;INN]

CAS	7220-56-6	MW:	311.28	MF:	C14 H8 F3 N O2 S

Anti-inflammatory (vet.)

Smith Kline & French

LgP	5.48(C)
pKa	
pp cited in Vols.1-5:	

FLUTICASONE PROPIONATE [U;INN]

CAS	80474-14-2	MW:	500.58	MF:	C25 H31 F3 O5 S

Anti-inflammatory (topical)

Glaxo, UK

LgP	3.9(C,6)
pKa	
pp cited in Vols.1-5:	

FLUTIZENOL [INN]

CAS	10202-40-1	MW:	443.56	MF:	C20 H24 F3 N3 O S2

LgP	5.55(C)
pKa	
pp cited in Vols.1-5:	

FLUTONIDINE [INN]

CAS	28125-87-3	MW:	193.23	MF:	C10 H12 F N3

Antihypertensive

Boehringer-Ingel., FRG

LgP	(4)
pKa	
pp cited in Vols.1-5:	

(1) Eur. J. Pharmacol., 1970, 9, 163.

FLUTOPRAZEPAM [INN]

CAS	25967-29-7	MW:	342.80	MF:	C19 H16 Cl F N2 O

Anxiolytic

Sumitomo, Japan (Restas)

LgP	3.9(C,6)
pKa	
pp cited in Vols.1-5:	

(1) RTECS, 1985-86, entry 15594.
(2) Arch. Int. Pharmacodyn. Ther., 1984, 269, 180.

FLUTROLINE [U;INN]

CAS	70801-02-4	MW:	450.51	MF:	C27 H25 F3 N2 O

Antipsychotic

Pfizer

LgP	6.52(C)
pKa	
pp cited in Vols.1-5:	

(1) Pharmacologist, 1979, 21, 192 abst.

FLUTROPIUM BROMIDE [INN]

CAS	63516-07-4	MW:	478.41	MF:	C24 H29 Br F N O3

Anticholinergic

Boehringer-Ingel., FRG

LgP	(4)
pKa	
pp cited in Vols.1-5:	

(1) Arzneim.-Forsch., 1986, 36, 1348.

FLUVOXAMINE [INN]

CAS	54739-18-3	MW:	318.34	MF:	C15 H21 F3 N2 O2

Antidepressant

Philips-Duphar; Reid-Rowell

LgP	3.18(C)
pKa	
pp cited in Vols.1-5:	

(1) H.B.A. Welle et al., U.S. Pat. 4 085 225 (1978).
(2) V. Classen et al., Brit. J. Pharmacol., 1977, 60, 505.

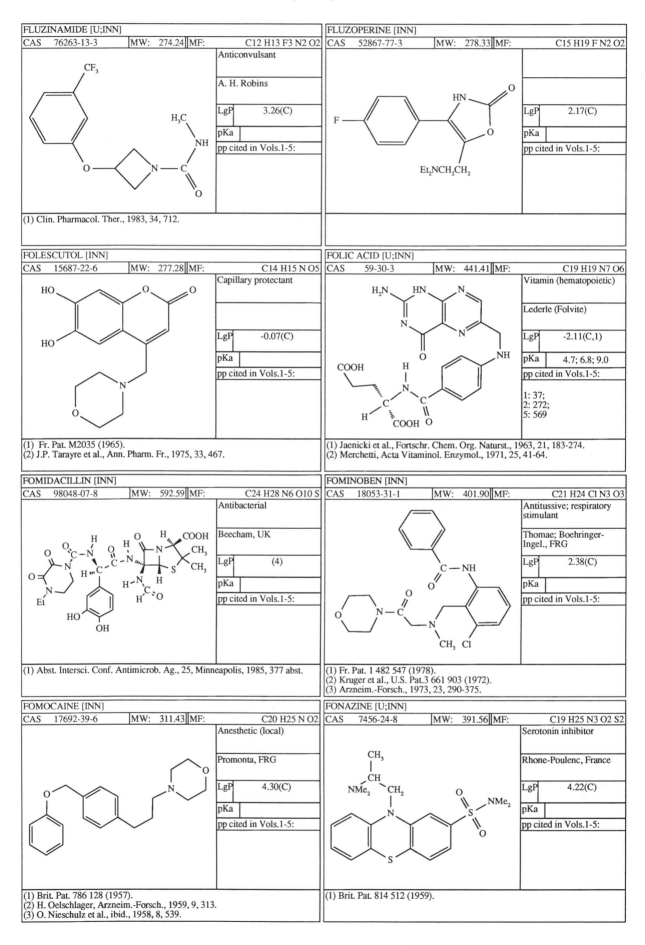

FLUZINAMIDE [U;INN]

CAS	76263-13-3	MW:	274.24	MF:	C12 H13 F3 N2 O2

Anticonvulsant

A. H. Robins

LgP	3.26(C)
pKa	
pp cited in Vols.1-5:	

(1) Clin. Pharmacol. Ther., 1983, 34, 712.

FLUZOPERINE [INN]

CAS	52867-77-3	MW:	278.33	MF:	C15 H19 F N2 O2

LgP	2.17(C)
pKa	
pp cited in Vols.1-5:	

FOLESCUTOL [INN]

CAS	15687-22-6	MW:	277.28	MF:	C14 H15 N O5

Capillary protectant

LgP	-0.07(C)
pKa	
pp cited in Vols.1-5:	

(1) Fr. Pat. M2035 (1965).
(2) J.P. Tarayre et al., Ann. Pharm. Fr., 1975, 33, 467.

FOLIC ACID [U;INN]

CAS	59-30-3	MW:	441.41	MF:	C19 H19 N7 O6

Vitamin (hematopoietic)

Lederle (Folvite)

LgP	-2.11(C,1)
pKa	4.7; 6.8; 9.0
pp cited in Vols.1-5:	

1: 37;
2: 272;
5: 569

(1) Jaenicki et al., Fortschr. Chem. Org. Naturst., 1963, 21, 183-274.
(2) Merchetti, Acta Vitaminol. Enzymol., 1971, 25, 41-64.

FOMIDACILLIN [INN]

CAS	98048-07-8	MW:	592.59	MF:	C24 H28 N6 O10 S

Antibacterial

Beecham, UK

LgP	(4)
pKa	
pp cited in Vols.1-5:	

(1) Abst. Intersci. Conf. Antimicrob. Ag., 25, Minneapolis, 1985, 377 abst.

FOMINOBEN [INN]

CAS	18053-31-1	MW:	401.90	MF:	C21 H24 Cl N3 O3

Antitussive; respiratory stimulant

Thomae; Boehringer-Ingel., FRG

LgP	2.38(C)
pKa	
pp cited in Vols.1-5:	

(1) Fr. Pat. 1 482 547 (1978).
(2) Kruger et al., U.S. Pat. 3 661 903 (1972).
(3) Arzneim.-Forsch., 1973, 23, 290-375.

FOMOCAINE [INN]

CAS	17692-39-6	MW:	311.43	MF:	C20 H25 N O2

Anesthetic (local)

Promonta, FRG

LgP	4.30(C)
pKa	
pp cited in Vols.1-5:	

(1) Brit. Pat. 786 128 (1957).
(2) H. Oelschlager, Arzneim.-Forsch., 1959, 9, 313.
(3) O. Nieschulz et al., ibid., 1958, 8, 539.

FONAZINE [U;INN]

CAS	7456-24-8	MW:	391.56	MF:	C19 H25 N3 O2 S2

Serotonin inhibitor

Rhone-Poulenc, France

LgP	4.22(C)
pKa	
pp cited in Vols.1-5:	

(1) Brit. Pat. 814 512 (1959).

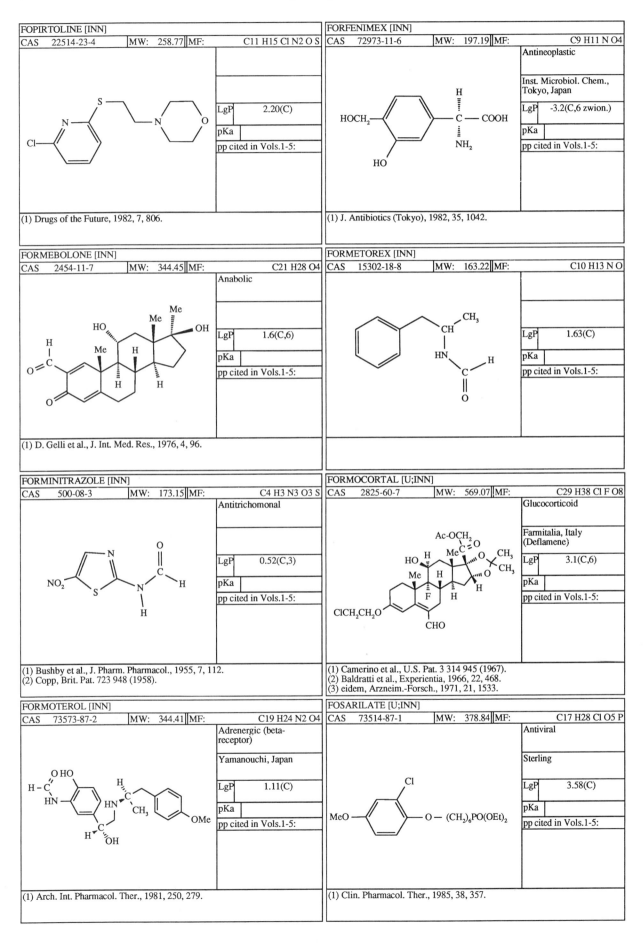

FOPIRTOLINE [INN]

CAS	22514-23-4	MW:	258.77	MF:	C11 H15 Cl N2 O S

LgP	2.20(C)
pKa	
pp cited in Vols.1-5:	

(1) Drugs of the Future, 1982, 7, 806.

FORFENIMEX [INN]

CAS	72973-11-6	MW:	197.19	MF:	C9 H11 N O4

Antineoplastic

Inst. Microbiol. Chem., Tokyo, Japan

LgP	-3.2(C,6 zwion.)
pKa	
pp cited in Vols.1-5:	

(1) J. Antibiotics (Tokyo), 1982, 35, 1042.

FORMEBOLONE [INN]

CAS	2454-11-7	MW:	344.45	MF:	C21 H28 O4

Anabolic

LgP	1.6(C,6)
pKa	
pp cited in Vols.1-5:	

(1) D. Gelli et al., J. Int. Med. Res., 1976, 4, 96.

FORMETOREX [INN]

CAS	15302-18-8	MW:	163.22	MF:	C10 H13 N O

LgP	1.63(C)
pKa	
pp cited in Vols.1-5:	

FORMINITRAZOLE [INN]

CAS	500-08-3	MW:	173.15	MF:	C4 H3 N3 O3 S

Antitrichomonal

LgP	0.52(C,3)
pKa	
pp cited in Vols.1-5:	

(1) Bushby et al., J. Pharm. Pharmacol., 1955, 7, 112.
(2) Copp, Brit. Pat. 723 948 (1958).

FORMOCORTAL [U;INN]

CAS	2825-60-7	MW:	569.07	MF:	C29 H38 Cl F O8

Glucocorticoid

Farmitalia, Italy (Deflamene)

LgP	3.1(C,6)
pKa	
pp cited in Vols.1-5:	

(1) Camerino et al., U.S. Pat. 3 314 945 (1967).
(2) Baldratti et al., Experientia, 1966, 22, 468.
(3) eidem, Arzneim.-Forsch., 1971, 21, 1533.

FORMOTEROL [INN]

CAS	73573-87-2	MW:	344.41	MF:	C19 H24 N2 O4

Adrenergic (beta-receptor)

Yamanouchi, Japan

LgP	1.11(C)
pKa	
pp cited in Vols.1-5:	

(1) Arch. Int. Pharmacol. Ther., 1981, 250, 279.

FOSARILATE [U;INN]

CAS	73514-87-1	MW:	378.84	MF:	C17 H28 Cl O5 P

Antiviral

Sterling

LgP	3.58(C)
pKa	
pp cited in Vols.1-5:	

(1) Clin. Pharmacol. Ther., 1985, 38, 357.

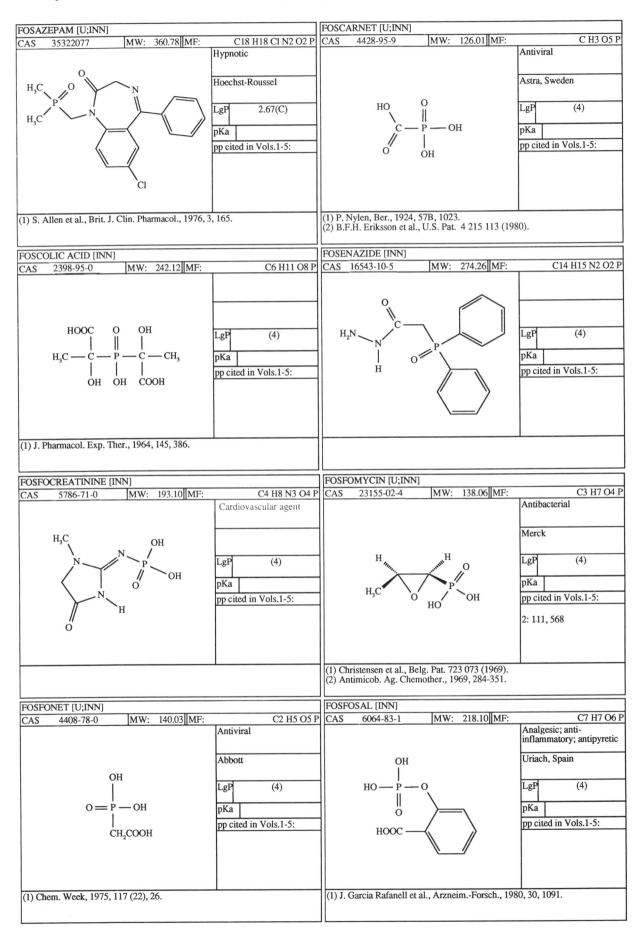

FOSAZEPAM [U;INN]

CAS	35322077	MW:	360.78	MF:	C18 H18 Cl N2 O2 P

Hypnotic

Hoechst-Roussel

LgP	2.67(C)
pKa	
pp cited in Vols.1-5:	

(1) S. Allen et al., Brit. J. Clin. Pharmacol., 1976, 3, 165.

FOSCARNET [U;INN]

CAS	4428-95-9	MW:	126.01	MF:	C H3 O5 P

Antiviral

Astra, Sweden

LgP	(4)
pKa	
pp cited in Vols.1-5:	

(1) P. Nylen, Ber., 1924, 57B, 1023.
(2) B.F.H. Eriksson et al., U.S. Pat. 4 215 113 (1980).

FOSCOLIC ACID [INN]

CAS	2398-95-0	MW:	242.12	MF:	C6 H11 O8 P

LgP	(4)
pKa	
pp cited in Vols.1-5:	

(1) J. Pharmacol. Exp. Ther., 1964, 145, 386.

FOSENAZIDE [INN]

CAS	16543-10-5	MW:	274.26	MF:	C14 H15 N2 O2 P

LgP	(4)
pKa	
pp cited in Vols.1-5:	

FOSFOCREATININE [INN]

CAS	5786-71-0	MW:	193.10	MF:	C4 H8 N3 O4 P

Cardiovascular agent

LgP	(4)
pKa	
pp cited in Vols.1-5:	

FOSFOMYCIN [U;INN]

CAS	23155-02-4	MW:	138.06	MF:	C3 H7 O4 P

Antibacterial

Merck

LgP	(4)
pKa	
pp cited in Vols.1-5:	

2: 111, 568

(1) Christensen et al., Belg. Pat. 723 073 (1969).
(2) Antimicob. Ag. Chemother., 1969, 284-351.

FOSFONET [U;INN]

CAS	4408-78-0	MW:	140.03	MF:	C2 H5 O5 P

Antiviral

Abbott

LgP	(4)
pKa	
pp cited in Vols.1-5:	

(1) Chem. Week, 1975, 117 (22), 26.

FOSFOSAL [INN]

CAS	6064-83-1	MW:	218.10	MF:	C7 H7 O6 P

Analgesic; anti-inflammatory; antipyretic

Uriach, Spain

LgP	(4)
pKa	
pp cited in Vols.1-5:	

(1) J. Garcia Rafanell et al., Arzneim.-Forsch., 1980, 30, 1091.

FOSINAPRIL SODIUM [U;INN]

CAS	88889-14-9	MW:	585.66	MF:	C30 H45 N Na O7 P

Antihypertensive (ACE inhibitor)

Squibb

LgP	(4)
pKa	
pp cited in Vols.1-5:	5: 556

(1) U.S. Pat. 4 337 201 (1982).

FOSMENIC ACID [INN]

CAS	13237-70-2	MW:	176.15	MF:	C7 H13 O3 P

Anti-arteriosclerotic

LgP	(4)
pKa	
pp cited in Vols.1-5:	

FOSMIDOMYCIN [INN]

CAS	66508-53-0	MW:	183.10	MF:	C4 H10 N O5 P

Antibacterial

Fujisawa, Japan

LgP	(4)
pKa	
pp cited in Vols.1-5:	

(1) J. Antibiotics (Tokyo), 1980, 33, 80-50.

FOSPIRATE [U;INN]

CAS	5598-52-7	MW:	306.47	MF:	C7 H7 Cl3 N O4 P

Anthelmintic (vet.)

Dow Chemical (Torelle)

LgP	2.61(C)
pKa	
pp cited in Vols.1-5:	

(1) Rigterink et al., J. Agr. Food. Chem., 1966, 14, 304.

FOSTEDIL [U;INN]

CAS	75889-62-2	MW:	361.40	MF:	C18 H20 N O3 P S

Vasodilator (calcium channel blocker)

Kanebo, Japan; Abbott

LgP	3.58(C)
pKa	
pp cited in Vols.1-5:	

(1) K. Harakawa et al., Arzneim.-Forsch., 1982, 32, 1068.

FOSTRIECIN [U;INN]

CAS	87810-56-8	MW:	430.40	MF:	C19 H27 O9 P

Antineoplastic

Parke-Davis

LgP	(4)
pKa	
pp cited in Vols.1-5:	

(1) Pharmacologist, 1986, 28, 177, 634 abst.

FOTEMUSTINE [INN]

CAS	92118-27-9	MW:	315.70	MF:	C9 H19 Cl N3 O5 P

Antineoplastic

Servier, France

LgP	0.78 (C)
pKa	
pp cited in Vols.1-5:	

(1) Proc. Ann. Mtg. Amer. Assoc. Cancer Res., 79 (New Orleans), 1988, 1318 abst.

FOTRETAMINE [INN]

CAS	37132-72-2	MW:	431.36	MF:	C14 H28 N9 O P3

Antineoplastic

LgP	(4)
pKa	
pp cited in Vols.1-5:	

(1) RTECS, 1985-86, entry 83658.

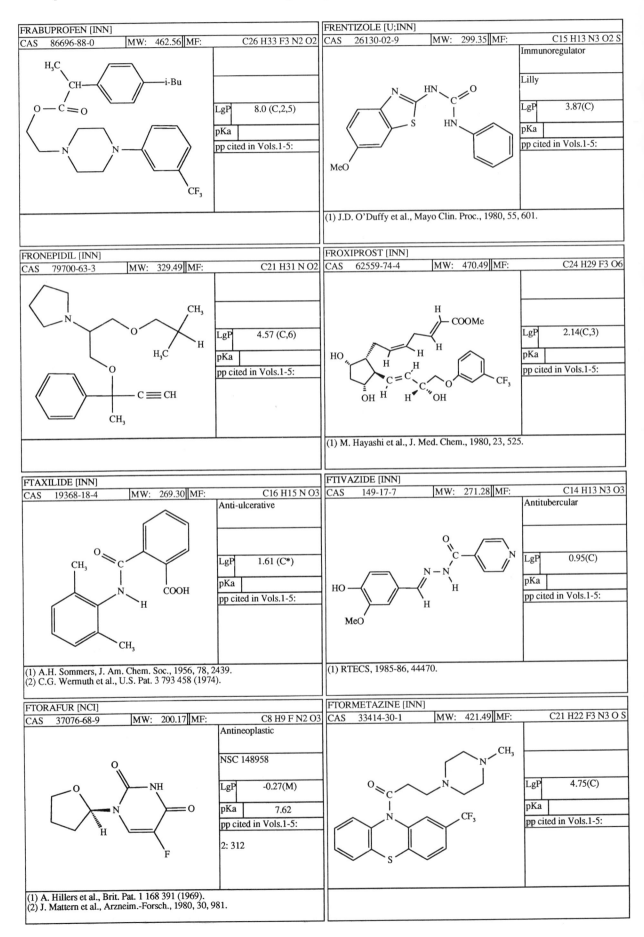

FRABUPROFEN [INN]

CAS 86696-88-0 | MW: 462.56 | MF: | C26 H33 F3 N2 O2

LgP 8.0 (C,2,5)

pKa

pp cited in Vols.1-5:

FRENTIZOLE [U;INN]

CAS 26130-02-9 | MW: 299.35 | MF: | C15 H13 N3 O2 S

Immunoregulator

Lilly

LgP 3.87(C)

pKa

pp cited in Vols.1-5:

(1) J.D. O'Duffy et al., Mayo Clin. Proc., 1980, 55, 601.

FRONEPIDIL [INN]

CAS 79700-63-3 | MW: 329.49 | MF: | C21 H31 N O2

LgP 4.57 (C,6)

pKa

pp cited in Vols.1-5:

FROXIPROST [INN]

CAS 62559-74-4 | MW: 470.49 | MF: | C24 H29 F3 O6

LgP 2.14(C,3)

pKa

pp cited in Vols.1-5:

(1) M. Hayashi et al., J. Med. Chem., 1980, 23, 525.

FTAXILIDE [INN]

CAS 19368-18-4 | MW: 269.30 | MF: | C16 H15 N O3

Anti-ulcerative

LgP 1.61 (C*)

pKa

pp cited in Vols.1-5:

(1) A.H. Sommers, J. Am. Chem. Soc., 1956, 78, 2439.
(2) C.G. Wermuth et al., U.S. Pat. 3 793 458 (1974).

FTIVAZIDE [INN]

CAS 149-17-7 | MW: 271.28 | MF: | C14 H13 N3 O3

Antitubercular

LgP 0.95(C)

pKa

pp cited in Vols.1-5:

(1) RTECS, 1985-86, 44470.

FTORAFUR [NCI]

CAS 37076-68-9 | MW: 200.17 | MF: | C8 H9 F N2 O3

Antineoplastic

NSC 148958

LgP -0.27(M)

pKa 7.62

pp cited in Vols.1-5:

2: 312

(1) A. Hillers et al., Brit. Pat. 1 168 391 (1969).
(2) J. Mattern et al., Arzneim.-Forsch., 1980, 30, 981.

FTORMETAZINE [INN]

CAS 33414-30-1 | MW: 421.49 | MF: | C21 H22 F3 N3 O S

LgP 4.75(C)

pKa

pp cited in Vols.1-5:

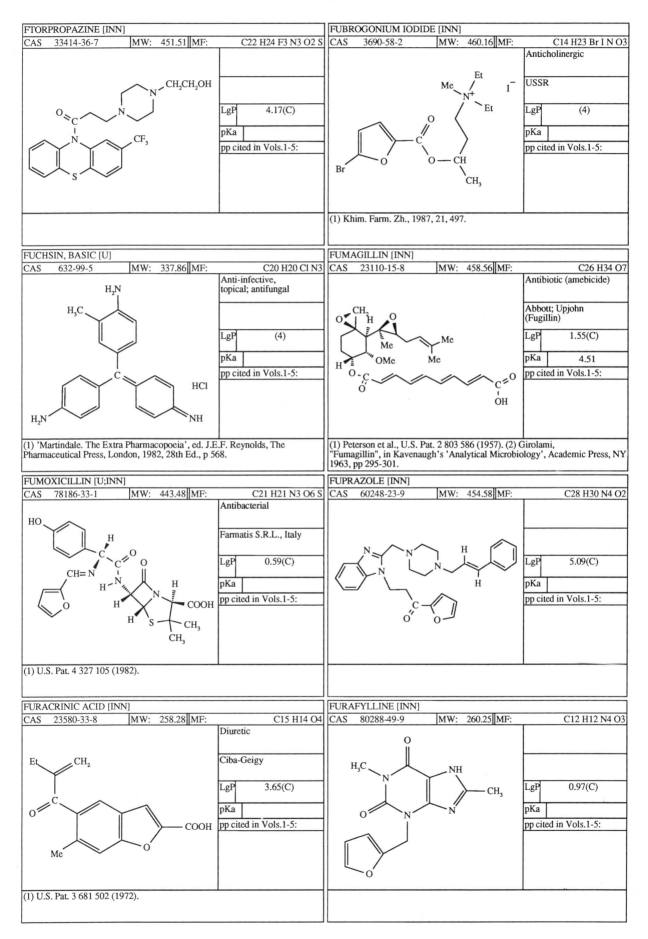

FTORPROPAZINE [INN]			
CAS 33414-36-7	MW: 451.51	MF:	C22 H24 F3 N3 O2 S

LgP   4.17(C)

pKa

pp cited in Vols.1-5:

FUBROGONIUM IODIDE [INN]			
CAS 3690-58-2	MW: 460.16	MF:	C14 H23 Br I N O3

Anticholinergic

USSR

LgP   (4)

pKa

pp cited in Vols.1-5:

(1) Khim. Farm. Zh., 1987, 21, 497.

FUCHSIN, BASIC [U]			
CAS 632-99-5	MW: 337.86	MF:	C20 H20 Cl N3

Anti-infective, topical; antifungal

LgP   (4)

pKa

pp cited in Vols.1-5:

(1) 'Martindale. The Extra Pharmacopoeia', ed. J.E.F. Reynolds, The Pharmaceutical Press, London, 1982, 28th Ed., p 568.

FUMAGILLIN [INN]			
CAS 23110-15-8	MW: 458.56	MF:	C26 H34 O7

Antibiotic (amebicide)

Abbott; Upjohn (Fugillin)

LgP   1.55(C)

pKa   4.51

pp cited in Vols.1-5:

(1) Peterson et al., U.S. Pat. 2 803 586 (1957). (2) Girolami, "Fumagillin", in Kavenaugh's 'Analytical Microbiology', Academic Press, NY, 1963, pp 295-301.

FUMOXICILLIN [U;INN]			
CAS 78186-33-1	MW: 443.48	MF:	C21 H21 N3 O6 S

Antibacterial

Farmatis S.R.L., Italy

LgP   0.59(C)

pKa

pp cited in Vols.1-5:

(1) U.S. Pat. 4 327 105 (1982).

FUPRAZOLE [INN]			
CAS 60248-23-9	MW: 454.58	MF:	C28 H30 N4 O2

LgP   5.09(C)

pKa

pp cited in Vols.1-5:

FURACRINIC ACID [INN]			
CAS 23580-33-8	MW: 258.28	MF:	C15 H14 O4

Diuretic

Ciba-Geigy

LgP   3.65(C)

pKa

pp cited in Vols.1-5:

(1) U.S. Pat. 3 681 502 (1972).

FURAFYLLINE [INN]			
CAS 80288-49-9	MW: 260.25	MF:	C12 H12 N4 O3

LgP   0.97(C)

pKa

pp cited in Vols.1-5:

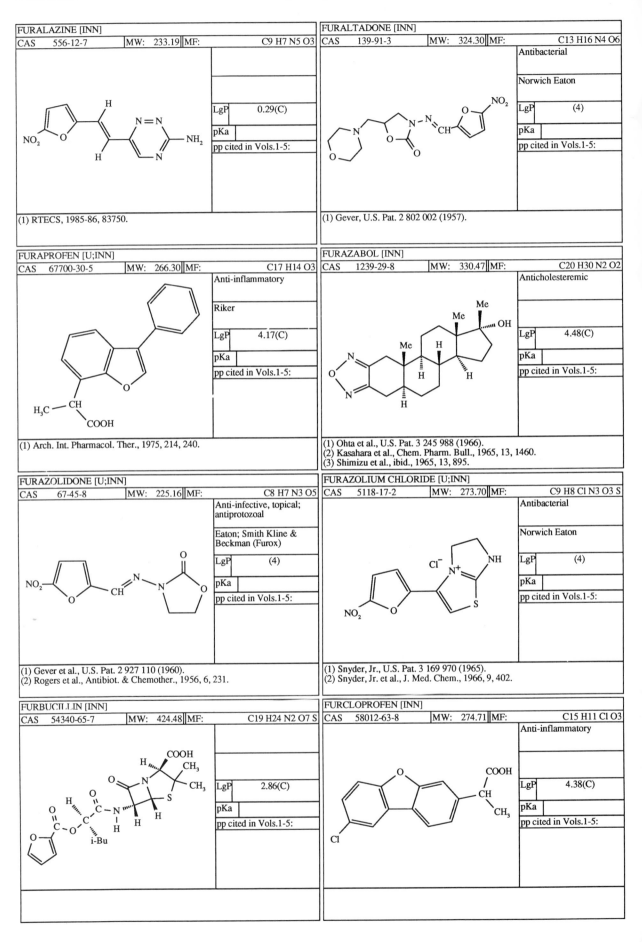

FURALAZINE [INN]

| CAS | 556-12-7 | MW: | 233.19 | MF: | C9 H7 N5 O3 |

LgP	0.29(C)
pKa	
pp cited in Vols.1-5:	

(1) RTECS, 1985-86, 83750.

FURALTADONE [INN]

| CAS | 139-91-3 | MW: | 324.30 | MF: | C13 H16 N4 O6 |

Antibacterial

Norwich Eaton

LgP	(4)
pKa	
pp cited in Vols.1-5:	

(1) Gever, U.S. Pat. 2 802 002 (1957).

FURAPROFEN [U;INN]

| CAS | 67700-30-5 | MW: | 266.30 | MF: | C17 H14 O3 |

Anti-inflammatory

Riker

LgP	4.17(C)
pKa	
pp cited in Vols.1-5:	

(1) Arch. Int. Pharmacol. Ther., 1975, 214, 240.

FURAZABOL [INN]

| CAS | 1239-29-8 | MW: | 330.47 | MF: | C20 H30 N2 O2 |

Anticholesteremic

LgP	4.48(C)
pKa	
pp cited in Vols.1-5:	

(1) Ohta et al., U.S. Pat. 3 245 988 (1966).
(2) Kasahara et al., Chem. Pharm. Bull., 1965, 13, 1460.
(3) Shimizu et al., ibid., 1965, 13, 895.

FURAZOLIDONE [U;INN]

| CAS | 67-45-8 | MW: | 225.16 | MF: | C8 H7 N3 O5 |

Anti-infective, topical; antiprotozoal

Eaton; Smith Kline & Beckman (Furox)

LgP	(4)
pKa	
pp cited in Vols.1-5:	

(1) Gever et al., U.S. Pat. 2 927 110 (1960).
(2) Rogers et al., Antibiot. & Chemother., 1956, 6, 231.

FURAZOLIUM CHLORIDE [U;INN]

| CAS | 5118-17-2 | MW: | 273.70 | MF: | C9 H8 Cl N3 O3 S |

Antibacterial

Norwich Eaton

LgP	(4)
pKa	
pp cited in Vols.1-5:	

(1) Snyder, Jr., U.S. Pat. 3 169 970 (1965).
(2) Snyder, Jr. et al., J. Med. Chem., 1966, 9, 402.

FURBUCILLIN [INN]

| CAS | 54340-65-7 | MW: | 424.48 | MF: | C19 H24 N2 O7 S |

LgP	2.86(C)
pKa	
pp cited in Vols.1-5:	

FURCLOPROFEN [INN]

| CAS | 58012-63-8 | MW: | 274.71 | MF: | C15 H11 Cl O3 |

Anti-inflammatory

LgP	4.38(C)
pKa	
pp cited in Vols.1-5:	

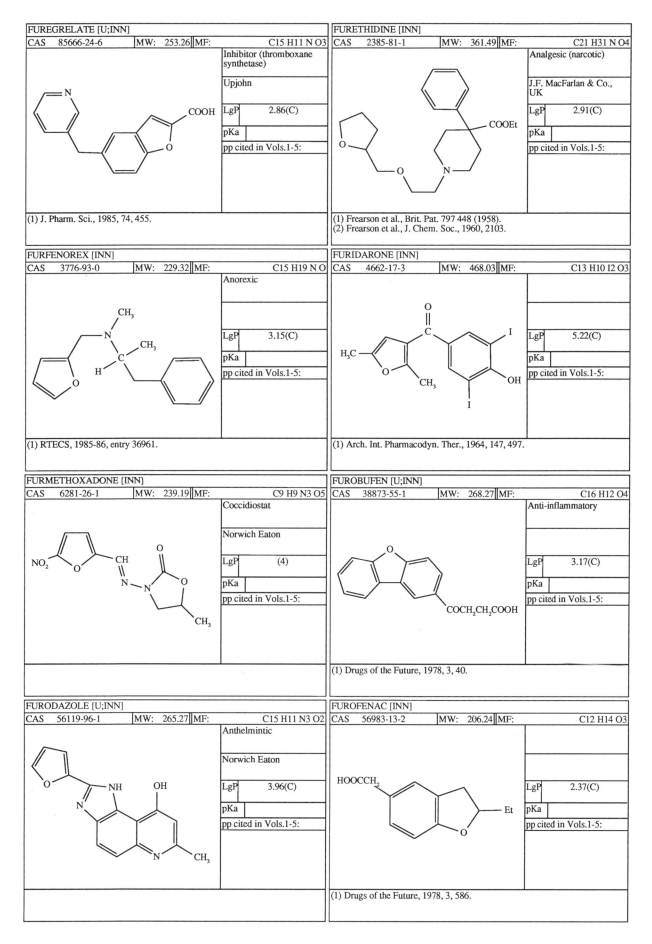

FUREGRELATE [U;INN]

CAS 85666-24-6	MW: 253.26	MF: C15 H11 N O3

Inhibitor (thromboxane synthetase)

Upjohn

LgP	2.86(C)
pKa	
pp cited in Vols.1-5:	

(1) J. Pharm. Sci., 1985, 74, 455.

FURETHIDINE [INN]

CAS 2385-81-1	MW: 361.49	MF: C21 H31 N O4

Analgesic (narcotic)

J.F. MacFarlan & Co., UK

LgP	2.91(C)
pKa	
pp cited in Vols.1-5:	

(1) Frearson et al., Brit. Pat. 797 448 (1958).
(2) Frearson et al., J. Chem. Soc., 1960, 2103.

FURFENOREX [INN]

CAS 3776-93-0	MW: 229.32	MF: C15 H19 N O

Anorexic

LgP	3.15(C)
pKa	
pp cited in Vols.1-5:	

(1) RTECS, 1985-86, entry 36961.

FURIDARONE [INN]

CAS 4662-17-3	MW: 468.03	MF: C13 H10 I2 O3

LgP	5.22(C)
pKa	
pp cited in Vols.1-5:	

(1) Arch. Int. Pharmacodyn. Ther., 1964, 147, 497.

FURMETHOXADONE [INN]

CAS 6281-26-1	MW: 239.19	MF: C9 H9 N3 O5

Coccidiostat

Norwich Eaton

LgP	(4)
pKa	
pp cited in Vols.1-5:	

FUROBUFEN [U;INN]

CAS 38873-55-1	MW: 268.27	MF: C16 H12 O4

Anti-inflammatory

LgP	3.17(C)
pKa	
pp cited in Vols.1-5:	

(1) Drugs of the Future, 1978, 3, 40.

FURODAZOLE [U;INN]

CAS 56119-96-1	MW: 265.27	MF: C15 H11 N3 O2

Anthelmintic

Norwich Eaton

LgP	3.96(C)
pKa	
pp cited in Vols.1-5:	

FUROFENAC [INN]

CAS 56983-13-2	MW: 206.24	MF: C12 H14 O3

LgP	2.37(C)
pKa	
pp cited in Vols.1-5:	

(1) Drugs of the Future, 1978, 3, 586.

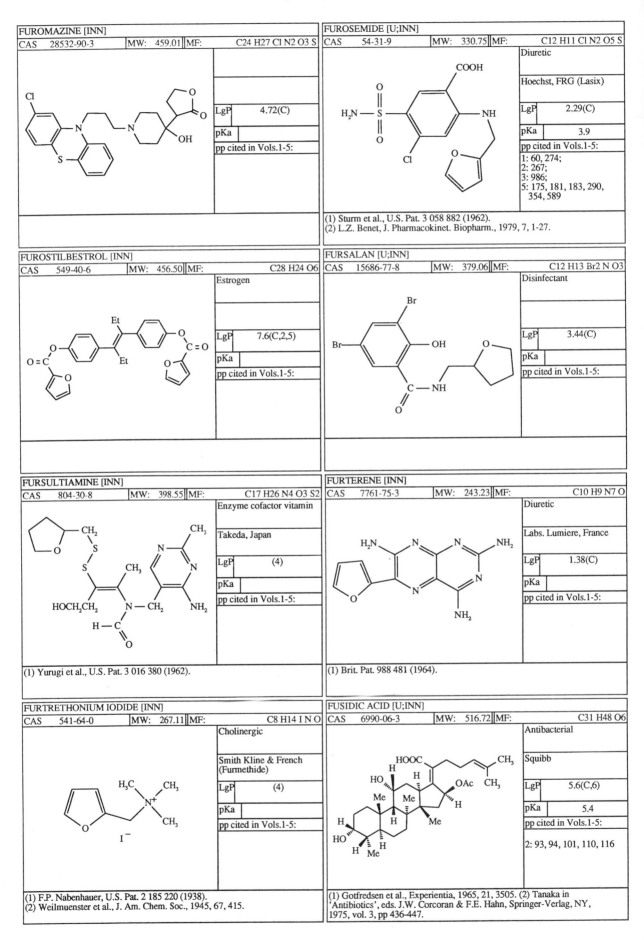

FUROMAZINE [INN]

CAS	28532-90-3	MW:	459.01	MF:	C24 H27 Cl N2 O3 S

LgP	4.72(C)
pKa	
pp cited in Vols.1-5:	

FUROSEMIDE [U;INN]

CAS	54-31-9	MW:	330.75	MF:	C12 H11 Cl N2 O5 S

Diuretic

Hoechst, FRG (Lasix)

LgP	2.29(C)
pKa	3.9
pp cited in Vols.1-5:	

1: 60, 274;
2: 267;
3: 986;
5: 175, 181, 183, 290, 354, 589

(1) Sturm et al., U.S. Pat. 3 058 882 (1962).
(2) L.Z. Benet, J. Pharmacokinet. Biopharm., 1979, 7, 1-27.

FUROSTILBESTROL [INN]

CAS	549-40-6	MW:	456.50	MF:	C28 H24 O6

Estrogen

LgP	7.6(C,2,5)
pKa	
pp cited in Vols.1-5:	

FURSALAN [U;INN]

CAS	15686-77-8	MW:	379.06	MF:	C12 H13 Br2 N O3

Disinfectant

LgP	3.44(C)
pKa	
pp cited in Vols.1-5:	

FURSULTIAMINE [INN]

CAS	804-30-8	MW:	398.55	MF:	C17 H26 N4 O3 S2

Enzyme cofactor vitamin

Takeda, Japan

LgP	(4)
pKa	
pp cited in Vols.1-5:	

(1) Yurugi et al., U.S. Pat. 3 016 380 (1962).

FURTERENE [INN]

CAS	7761-75-3	MW:	243.23	MF:	C10 H9 N7 O

Diuretic

Labs. Lumiere, France

LgP	1.38(C)
pKa	
pp cited in Vols.1-5:	

(1) Brit. Pat. 988 481 (1964).

FURTRETHONIUM IODIDE [INN]

CAS	541-64-0	MW:	267.11	MF:	C8 H14 I N O

Cholinergic

Smith Kline & French
(Furmethide)

LgP	(4)
pKa	
pp cited in Vols.1-5:	

(1) F.P. Nabenhauer, U.S. Pat. 2 185 220 (1938).
(2) Weilmuenster et al., J. Am. Chem. Soc., 1945, 67, 415.

FUSIDIC ACID [U;INN]

CAS	6990-06-3	MW:	516.72	MF:	C31 H48 O6

Antibacterial

Squibb

LgP	5.6(C,6)
pKa	5.4
pp cited in Vols.1-5:	

2: 93, 94, 101, 110, 116

(1) Gotfredsen et al., Experientia, 1965, 21, 3505. (2) Tanaka in
'Antibiotics', eds. J.W. Corcoran & F.E. Hahn, Springer-Verlag, NY,
1975, vol. 3, pp 436-447.

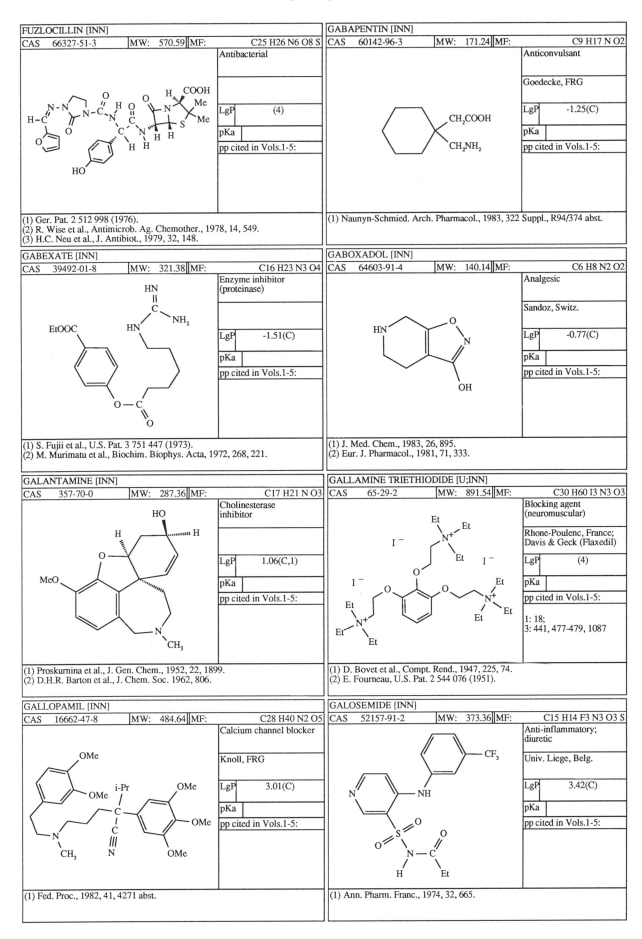

FUZLOCILLIN [INN]

CAS	66327-51-3	MW: 570.59	MF:	C25 H26 N6 O8 S

Antibacterial

LgP	(4)
pKa	
pp cited in Vols.1-5:	

(1) Ger. Pat. 2 512 998 (1976).
(2) R. Wise et al., Antimicrob. Ag. Chemother., 1978, 14, 549.
(3) H.C. Neu et al., J. Antibiot., 1979, 32, 148.

GABEXATE [INN]

CAS	39492-01-8	MW: 321.38	MF:	C16 H23 N3 O4

Enzyme inhibitor (proteinase)

LgP	-1.51(C)
pKa	
pp cited in Vols.1-5:	

(1) S. Fujii et al., U.S. Pat. 3 751 447 (1973).
(2) M. Murimatu et al., Biochim. Biophys. Acta, 1972, 268, 221.

GALANTAMINE [INN]

CAS	357-70-0	MW: 287.36	MF:	C17 H21 N O3

Cholinesterase inhibitor

LgP	1.06(C,1)
pKa	
pp cited in Vols.1-5:	

(1) Proskurnina et al., J. Gen. Chem., 1952, 22, 1899.
(2) D.H.R. Barton et al., J. Chem. Soc. 1962, 806.

GALLOPAMIL [INN]

CAS	16662-47-8	MW: 484.64	MF:	C28 H40 N2 O5

Calcium channel blocker

Knoll, FRG

LgP	3.01(C)
pKa	
pp cited in Vols.1-5:	

(1) Fed. Proc., 1982, 41, 4271 abst.

GABAPENTIN [INN]

CAS	60142-96-3	MW: 171.24	MF:	C9 H17 N O2

Anticonvulsant

Goedecke, FRG

LgP	-1.25(C)
pKa	
pp cited in Vols.1-5:	

(1) Naunyn-Schmied. Arch. Pharmacol., 1983, 322 Suppl., R94/374 abst.

GABOXADOL [INN]

CAS	64603-91-4	MW: 140.14	MF:	C6 H8 N2 O2

Analgesic

Sandoz, Switz.

LgP	-0.77(C)
pKa	
pp cited in Vols.1-5:	

(1) J. Med. Chem., 1983, 26, 895.
(2) Eur. J. Pharmacol., 1981, 71, 333.

GALLAMINE TRIETHIODIDE [U;INN]

CAS	65-29-2	MW: 891.54	MF:	C30 H60 I3 N3 O3

Blocking agent (neuromuscular)

Rhone-Poulenc, France; Davis & Geck (Flaxedil)

LgP	(4)
pKa	
pp cited in Vols.1-5:	

1: 18;
3: 441, 477-479, 1087

(1) D. Bovet et al., Compt. Rend., 1947, 225, 74.
(2) E. Fourneau, U.S. Pat. 2 544 076 (1951).

GALOSEMIDE [INN]

CAS	52157-91-2	MW: 373.36	MF:	C15 H14 F3 N3 O3 S

Anti-inflammatory; diuretic

Univ. Liege, Belg.

LgP	3.42(C)
pKa	
pp cited in Vols.1-5:	

(1) Ann. Pharm. Franc., 1974, 32, 665.

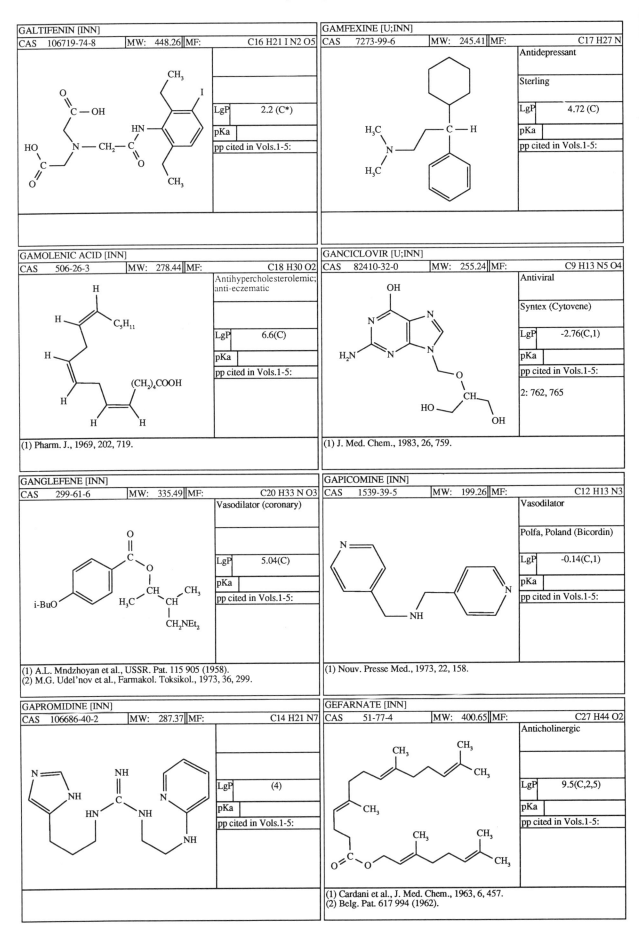

GALTIFENIN [INN]

CAS	106719-74-8	MW:	448.26	MF:	C16 H21 I N2 O5

LgP	2.2 (C*)
pKa	
pp cited in Vols.1-5:	

GAMFEXINE [U;INN]

CAS	7273-99-6	MW:	245.41	MF:	C17 H27 N

Antidepressant

Sterling

LgP	4.72 (C)
pKa	
pp cited in Vols.1-5:	

GAMOLENIC ACID [INN]

CAS	506-26-3	MW:	278.44	MF:	C18 H30 O2

Antihypercholesterolemic; anti-eczematic

LgP	6.6(C)
pKa	
pp cited in Vols.1-5:	

(1) Pharm. J., 1969, 202, 719.

GANCICLOVIR [U;INN]

CAS	82410-32-0	MW:	255.24	MF:	C9 H13 N5 O4

Antiviral

Syntex (Cytovene)

LgP	-2.76(C,1)
pKa	
pp cited in Vols.1-5:	

2: 762, 765

(1) J. Med. Chem., 1983, 26, 759.

GANGLEFENE [INN]

CAS	299-61-6	MW:	335.49	MF:	C20 H33 N O3

Vasodilator (coronary)

LgP	5.04(C)
pKa	
pp cited in Vols.1-5:	

(1) A.L. Mndzhoyan et al., USSR. Pat. 115 905 (1958).
(2) M.G. Udel'nov et al., Farmakol. Toksikol., 1973, 36, 299.

GAPICOMINE [INN]

CAS	1539-39-5	MW:	199.26	MF:	C12 H13 N3

Vasodilator

Polfa, Poland (Bicordin)

LgP	-0.14(C,1)
pKa	
pp cited in Vols.1-5:	

(1) Nouv. Presse Med., 1973, 22, 158.

GAPROMIDINE [INN]

CAS	106686-40-2	MW:	287.37	MF:	C14 H21 N7

LgP	(4)
pKa	
pp cited in Vols.1-5:	

GEFARNATE [INN]

CAS	51-77-4	MW:	400.65	MF:	C27 H44 O2

Anticholinergic

LgP	9.5(C,2,5)
pKa	
pp cited in Vols.1-5:	

(1) Cardani et al., J. Med. Chem., 1963, 6, 457.
(2) Belg. Pat. 617 994 (1962).

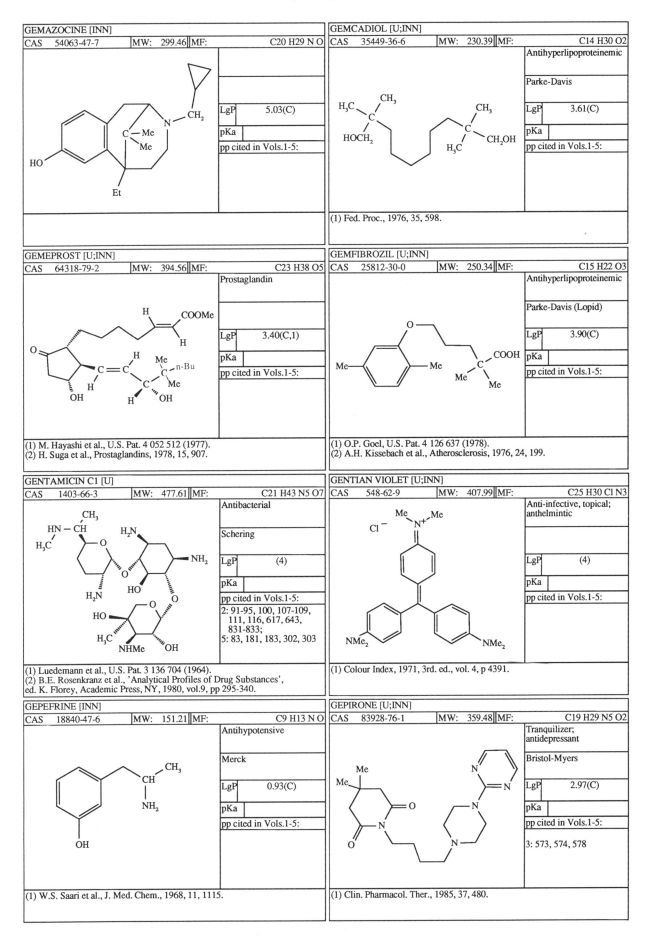

GEMAZOCINE [INN]

CAS 54063-47-7	MW: 299.46	MF:	C20 H29 N O
		LgP	5.03(C)
		pKa	
		pp cited in Vols.1-5:	

GEMCADIOL [U;INN]

CAS 35449-36-6	MW: 230.39	MF:	C14 H30 O2
			Antihyperlipoproteinemic
			Parke-Davis
		LgP	3.61(C)
		pKa	
		pp cited in Vols.1-5:	

(1) Fed. Proc., 1976, 35, 598.

GEMEPROST [U;INN]

CAS 64318-79-2	MW: 394.56	MF:	C23 H38 O5
			Prostaglandin
		LgP	3.40(C,1)
		pKa	
		pp cited in Vols.1-5:	

(1) M. Hayashi et al., U.S. Pat. 4 052 512 (1977).
(2) H. Suga et al., Prostaglandins, 1978, 15, 907.

GEMFIBROZIL [U;INN]

CAS 25812-30-0	MW: 250.34	MF:	C15 H22 O3
			Antihyperlipoproteinemic
			Parke-Davis (Lopid)
		LgP	3.90(C)
		pKa	
		pp cited in Vols.1-5:	

(1) O.P. Goel, U.S. Pat. 4 126 637 (1978).
(2) A.H. Kissebach et al., Atherosclerosis, 1976, 24, 199.

GENTAMICIN C1 [U]

CAS 1403-66-3	MW: 477.61	MF:	C21 H43 N5 O7
			Antibacterial
			Schering
		LgP	(4)
		pKa	
		pp cited in Vols.1-5:	
			2: 91-95, 100, 107-109, 111, 116, 617, 643, 831-833;
			5: 83, 181, 183, 302, 303

(1) Luedemann et al., U.S. Pat. 3 136 704 (1964).
(2) B.E. Rosenkranz et al., 'Analytical Profiles of Drug Substances',
ed. K. Florey, Academic Press, NY, 1980, vol.9, pp 295-340.

GENTIAN VIOLET [U;INN]

CAS 548-62-9	MW: 407.99	MF:	C25 H30 Cl N3
			Anti-infective, topical; anthelmintic
		LgP	(4)
		pKa	
		pp cited in Vols.1-5:	

(1) Colour Index, 1971, 3rd. ed., vol. 4, p 4391.

GEPEFRINE [INN]

CAS 18840-47-6	MW: 151.21	MF:	C9 H13 N O
			Antihypotensive
			Merck
		LgP	0.93(C)
		pKa	
		pp cited in Vols.1-5:	

(1) W.S. Saari et al., J. Med. Chem., 1968, 11, 1115.

GEPIRONE [U;INN]

CAS 83928-76-1	MW: 359.48	MF:	C19 H29 N5 O2
			Tranquilizer; antidepressant
			Bristol-Myers
		LgP	2.97(C)
		pKa	
		pp cited in Vols.1-5:	
			3: 573, 574, 578

(1) Clin. Pharmacol. Ther., 1985, 37, 480.

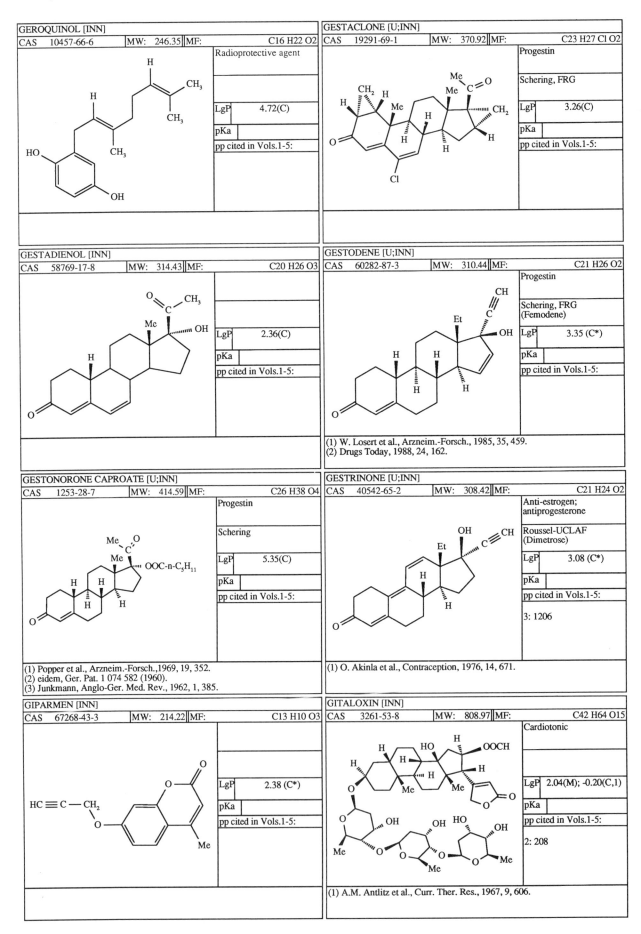

GEROQUINOL [INN]

CAS 10457-66-6 | MW: 246.35 | MF: C16 H22 O2

Radioprotective agent

LgP 4.72(C)

pKa

pp cited in Vols.1-5:

GESTACLONE [U;INN]

CAS 19291-69-1 | MW: 370.92 | MF: C23 H27 Cl O2

Progestin

Schering, FRG

LgP 3.26(C)

pKa

pp cited in Vols.1-5:

GESTADIENOL [INN]

CAS 58769-17-8 | MW: 314.43 | MF: C20 H26 O3

LgP 2.36(C)

pKa

pp cited in Vols.1-5:

GESTODENE [U;INN]

CAS 60282-87-3 | MW: 310.44 | MF: C21 H26 O2

Progestin

Schering, FRG
(Femodene)

LgP 3.35 (C*)

pKa

pp cited in Vols.1-5:

(1) W. Losert et al., Arzneim.-Forsch., 1985, 35, 459.
(2) Drugs Today, 1988, 24, 162.

GESTONORONE CAPROATE [U;INN]

CAS 1253-28-7 | MW: 414.59 | MF: C26 H38 O4

Progestin

Schering

LgP 5.35(C)

pKa

pp cited in Vols.1-5:

(1) Popper et al., Arzneim.-Forsch.,1969, 19, 352.
(2) eidem, Ger. Pat. 1 074 582 (1960).
(3) Junkmann, Anglo-Ger. Med. Rev., 1962, 1, 385.

GESTRINONE [U;INN]

CAS 40542-65-2 | MW: 308.42 | MF: C21 H24 O2

Anti-estrogen;
antiprogesterone

Roussel-UCLAF
(Dimetrose)

LgP 3.08 (C*)

pKa

pp cited in Vols.1-5:

3: 1206

(1) O. Akinla et al., Contraception, 1976, 14, 671.

GIPARMEN [INN]

CAS 67268-43-3 | MW: 214.22 | MF: C13 H10 O3

LgP 2.38 (C*)

pKa

pp cited in Vols.1-5:

GITALOXIN [INN]

CAS 3261-53-8 | MW: 808.97 | MF: C42 H64 O15

Cardiotonic

LgP 2.04(M); -0.20(C,1)

pKa

pp cited in Vols.1-5:

2: 208

(1) A.M. Antlitz et al., Curr. Ther. Res., 1967, 9, 606.

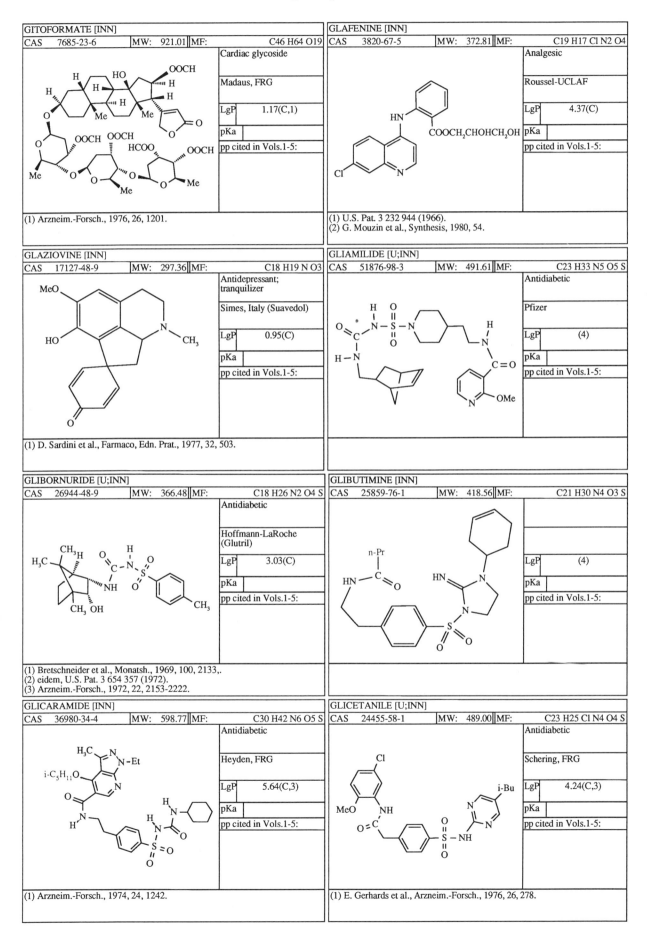

GITOFORMATE [INN]

CAS	7685-23-6	MW:	921.01	MF:	C46 H64 O19

Cardiac glycoside

Madaus, FRG

LgP   1.17(C,1)

pKa

pp cited in Vols.1-5:

(1) Arzneim.-Forsch., 1976, 26, 1201.

GLAFENINE [INN]

CAS	3820-67-5	MW:	372.81	MF:	C19 H17 Cl N2 O4

Analgesic

Roussel-UCLAF

LgP   4.37(C)

pKa

pp cited in Vols.1-5:

(1) U.S. Pat. 3 232 944 (1966).
(2) G. Mouzin et al., Synthesis, 1980, 54.

GLAZIOVINE [INN]

CAS	17127-48-9	MW:	297.36	MF:	C18 H19 N O3

Antidepressant; tranquilizer

Simes, Italy (Suavedol)

LgP   0.95(C)

pKa

pp cited in Vols.1-5:

(1) D. Sardini et al., Farmaco, Edn. Prat., 1977, 32, 503.

GLIAMILIDE [U;INN]

CAS	51876-98-3	MW:	491.61	MF:	C23 H33 N5 O5 S

Antidiabetic

Pfizer

LgP   (4)

pKa

pp cited in Vols.1-5:

GLIBORNURIDE [U;INN]

CAS	26944-48-9	MW:	366.48	MF:	C18 H26 N2 O4 S

Antidiabetic

Hoffmann-LaRoche (Glutril)

LgP   3.03(C)

pKa

pp cited in Vols.1-5:

(1) Bretschneider et al., Monatsh., 1969, 100, 2133,.
(2) eidem, U.S. Pat. 3 654 357 (1972).
(3) Arzneim.-Forsch., 1972, 22, 2153-2222.

GLIBUTIMINE [INN]

CAS	25859-76-1	MW:	418.56	MF:	C21 H30 N4 O3 S

LgP   (4)

pKa

pp cited in Vols.1-5:

GLICARAMIDE [INN]

CAS	36980-34-4	MW:	598.77	MF:	C30 H42 N6 O5 S

Antidiabetic

Heyden, FRG

LgP   5.64(C,3)

pKa

pp cited in Vols.1-5:

(1) Arzneim.-Forsch., 1974, 24, 1242.

GLICETANILE [U;INN]

CAS	24455-58-1	MW:	489.00	MF:	C23 H25 Cl N4 O4 S

Antidiabetic

Schering, FRG

LgP   4.24(C,3)

pKa

pp cited in Vols.1-5:

(1) E. Gerhards et al., Arzneim.-Forsch., 1976, 26, 278.

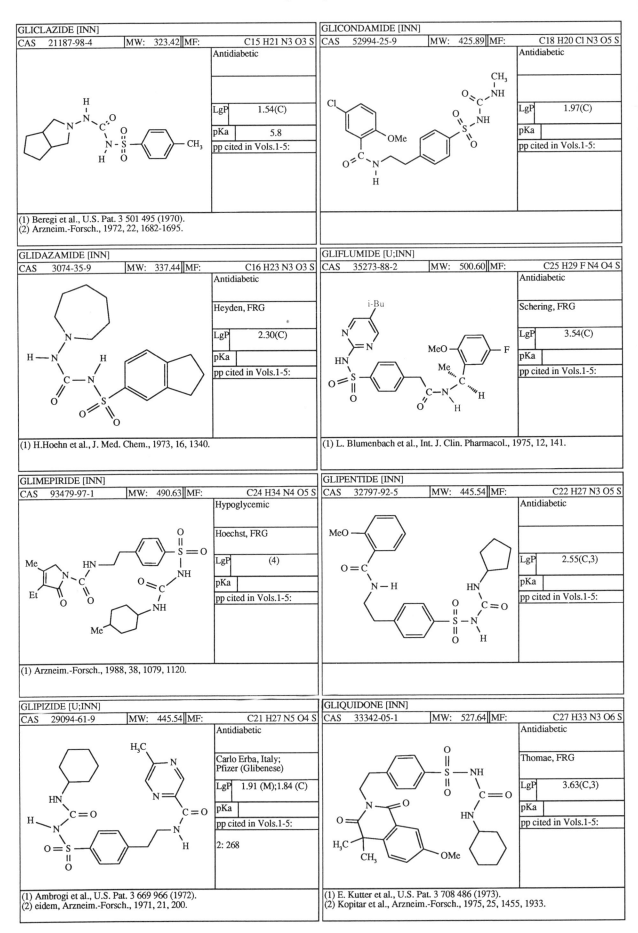

GLICLAZIDE [INN]

CAS	21187-98-4	MW: 323.42	MF:	C15 H21 N3 O3 S

Antidiabetic

LgP	1.54(C)
pKa	5.8
pp cited in Vols.1-5:	

(1) Beregi et al., U.S. Pat. 3 501 495 (1970).
(2) Arzneim.-Forsch., 1972, 22, 1682-1695.

GLICONDAMIDE [INN]

CAS	52994-25-9	MW: 425.89	MF:	C18 H20 Cl N3 O5 S

Antidiabetic

LgP	1.97(C)
pKa	
pp cited in Vols.1-5:	

GLIDAZAMIDE [INN]

CAS	3074-35-9	MW: 337.44	MF:	C16 H23 N3 O3 S

Antidiabetic

Heyden, FRG

LgP	2.30(C)
pKa	
pp cited in Vols.1-5:	

(1) H.Hoehn et al., J. Med. Chem., 1973, 16, 1340.

GLIFLUMIDE [U;INN]

CAS	35273-88-2	MW: 500.60	MF:	C25 H29 F N4 O4 S

Antidiabetic

Schering, FRG

LgP	3.54(C)
pKa	
pp cited in Vols.1-5:	

(1) L. Blumenbach et al., Int. J. Clin. Pharmacol., 1975, 12, 141.

GLIMEPIRIDE [INN]

CAS	93479-97-1	MW: 490.63	MF:	C24 H34 N4 O5 S

Hypoglycemic

Hoechst, FRG

LgP	(4)
pKa	
pp cited in Vols.1-5:	

(1) Arzneim.-Forsch., 1988, 38, 1079, 1120.

GLIPENTIDE [INN]

CAS	32797-92-5	MW: 445.54	MF:	C22 H27 N3 O5 S

Antidiabetic

LgP	2.55(C,3)
pKa	
pp cited in Vols.1-5:	

GLIPIZIDE [U;INN]

CAS	29094-61-9	MW: 445.54	MF:	C21 H27 N5 O4 S

Antidiabetic

Carlo Erba, Italy;
Pfizer (Glibenese)

LgP	1.91 (M);1.84 (C)
pKa	
pp cited in Vols.1-5:	
2: 268	

(1) Ambrogi et al., U.S. Pat. 3 669 966 (1972).
(2) eidem, Arzneim.-Forsch., 1971, 21, 200.

GLIQUIDONE [INN]

CAS	33342-05-1	MW: 527.64	MF:	C27 H33 N3 O6 S

Antidiabetic

Thomae, FRG

LgP	3.63(C,3)
pKa	
pp cited in Vols.1-5:	

(1) E. Kutter et al., U.S. Pat. 3 708 486 (1973).
(2) Kopitar et al., Arzneim.-Forsch., 1975, 25, 1455, 1933.

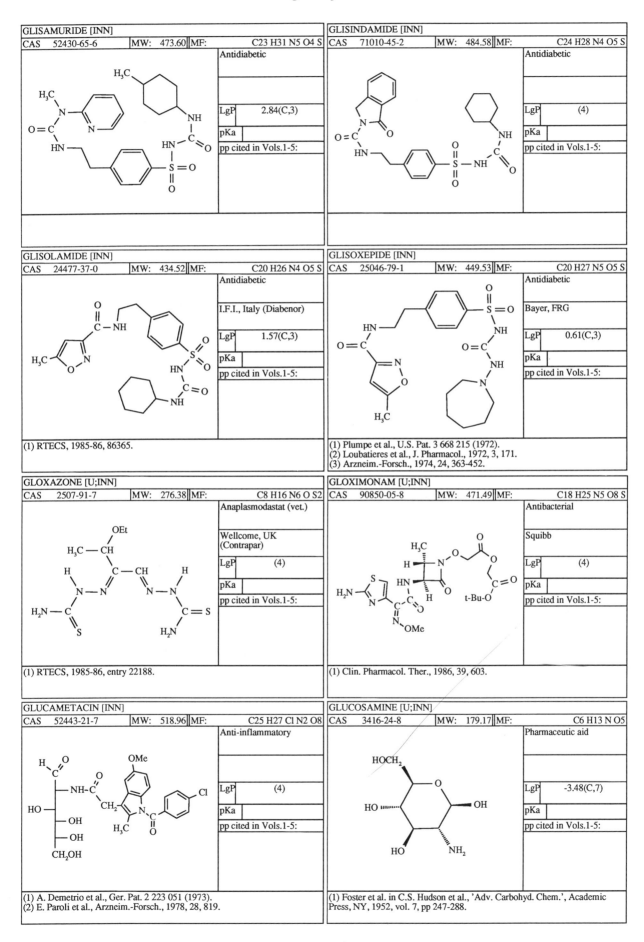

GLISAMURIDE [INN]
CAS 52430-65-6 | MW: 473.60 | MF: C23 H31 N5 O4 S
Antidiabetic
LgP 2.84(C,3)
pKa
pp cited in Vols.1-5:

GLISINDAMIDE [INN]
CAS 71010-45-2 | MW: 484.58 | MF: C24 H28 N4 O5 S
Antidiabetic
LgP (4)
pKa
pp cited in Vols.1-5:

GLISOLAMIDE [INN]
CAS 24477-37-0 | MW: 434.52 | MF: C20 H26 N4 O5 S
Antidiabetic
I.F.I., Italy (Diabenor)
LgP 1.57(C,3)
pKa
pp cited in Vols.1-5:
(1) RTECS, 1985-86, 86365.

GLISOXEPIDE [INN]
CAS 25046-79-1 | MW: 449.53 | MF: C20 H27 N5 O5 S
Antidiabetic
Bayer, FRG
LgP 0.61(C,3)
pKa
pp cited in Vols.1-5:
(1) Plumpe et al., U.S. Pat. 3 668 215 (1972).
(2) Loubatieres et al., J. Pharmacol., 1972, 3, 171.
(3) Arzneim.-Forsch., 1974, 24, 363-452.

GLOXAZONE [U;INN]
CAS 2507-91-7 | MW: 276.38 | MF: C8 H16 N6 O S2
Anaplasmodastat (vet.)
Wellcome, UK (Contrapar)
LgP (4)
pKa
pp cited in Vols.1-5:
(1) RTECS, 1985-86, entry 22188.

GLOXIMONAM [U;INN]
CAS 90850-05-8 | MW: 471.49 | MF: C18 H25 N5 O8 S
Antibacterial
Squibb
LgP (4)
pKa
pp cited in Vols.1-5:
(1) Clin. Pharmacol. Ther., 1986, 39, 603.

GLUCAMETACIN [INN]
CAS 52443-21-7 | MW: 518.96 | MF: C25 H27 Cl N2 O8
Anti-inflammatory
LgP (4)
pKa
pp cited in Vols.1-5:
(1) A. Demetrio et al., Ger. Pat. 2 223 051 (1973).
(2) E. Paroli et al., Arzneim.-Forsch., 1978, 28, 819.

GLUCOSAMINE [U;INN]
CAS 3416-24-8 | MW: 179.17 | MF: C6 H13 N O5
Pharmaceutic aid
LgP -3.48(C,7)
pKa
pp cited in Vols.1-5:
(1) Foster et al. in C.S. Hudson et al., 'Adv. Carbohyd. Chem.', Academic Press, NY, 1952, vol. 7, pp 247-288.

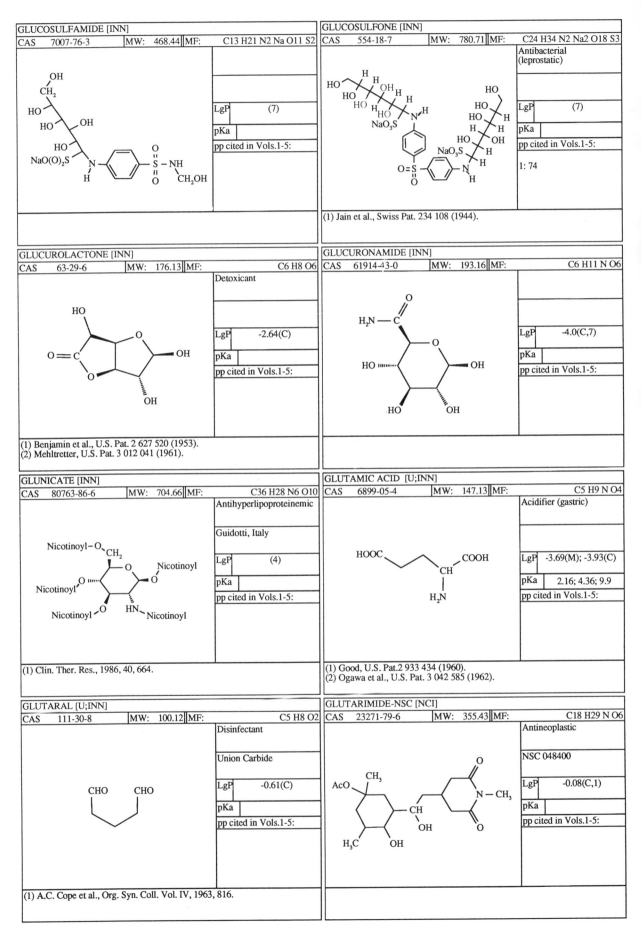

GLUCOSULFAMIDE [INN]		
CAS 7007-76-3	MW: 468.44	MF: C13 H21 N2 Na O11 S2
LgP	(7)	
pKa		
pp cited in Vols.1-5:		

GLUCOSULFONE [INN]		
CAS 554-18-7	MW: 780.71	MF: C24 H34 N2 Na2 O18 S3
Antibacterial (leprostatic)		
LgP	(7)	
pKa		
pp cited in Vols.1-5:		
1: 74		

(1) Jain et al., Swiss Pat. 234 108 (1944).

GLUCUROLACTONE [INN]		
CAS 63-29-6	MW: 176.13	MF: C6 H8 O6
Detoxicant		
LgP	-2.64(C)	
pKa		
pp cited in Vols.1-5:		

(1) Benjamin et al., U.S. Pat. 2 627 520 (1953).
(2) Mehltretter, U.S. Pat. 3 012 041 (1961).

GLUCURONAMIDE [INN]		
CAS 61914-43-0	MW: 193.16	MF: C6 H11 N O6
LgP	-4.0(C,7)	
pKa		
pp cited in Vols.1-5:		

GLUNICATE [INN]		
CAS 80763-86-6	MW: 704.66	MF: C36 H28 N6 O10
Antihyperlipoproteinemic		
Guidotti, Italy		
LgP	(4)	
pKa		
pp cited in Vols.1-5:		

(1) Clin. Ther. Res., 1986, 40, 664.

GLUTAMIC ACID [U;INN]		
CAS 6899-05-4	MW: 147.13	MF: C5 H9 N O4
Acidifier (gastric)		
LgP	-3.69(M); -3.93(C)	
pKa	2.16; 4.36; 9.9	
pp cited in Vols.1-5:		

(1) Good, U.S. Pat.2 933 434 (1960).
(2) Ogawa et al., U.S. Pat. 3 042 585 (1962).

GLUTARAL [U;INN]		
CAS 111-30-8	MW: 100.12	MF: C5 H8 O2
Disinfectant		
Union Carbide		
LgP	-0.61(C)	
pKa		
pp cited in Vols.1-5:		

(1) A.C. Cope et al., Org. Syn. Coll. Vol. IV, 1963, 816.

GLUTARIMIDE-NSC [NCI]		
CAS 23271-79-6	MW: 355.43	MF: C18 H29 N O6
Antineoplastic		
NSC 048400		
LgP	-0.08(C,1)	
pKa		
pp cited in Vols.1-5:		

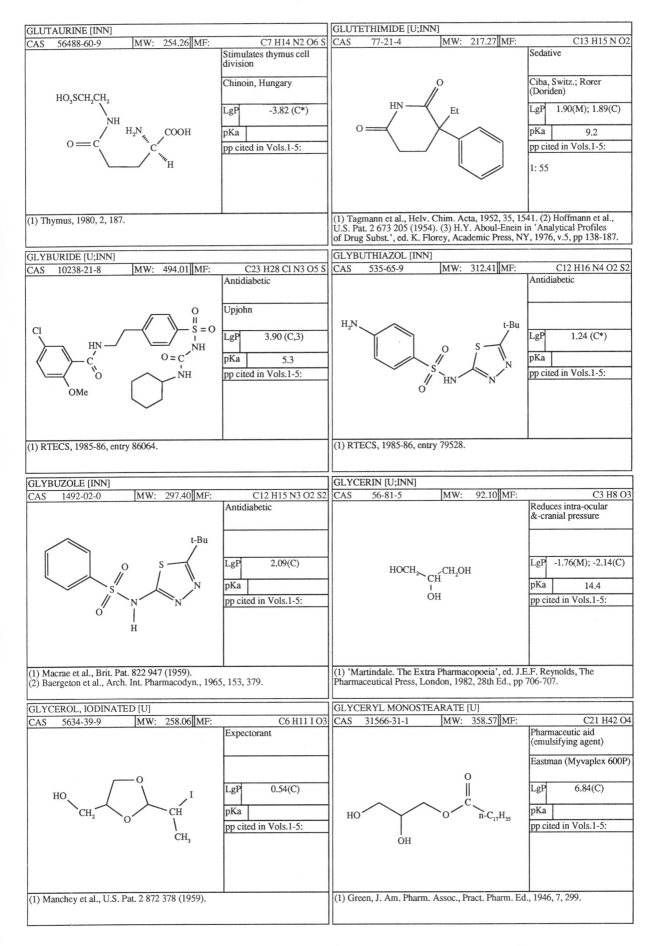

GLUTAURINE [INN]

| CAS | 56488-60-9 | MW: | 254.26 | MF: | C7 H14 N2 O6 S |

Stimulates thymus cell division

Chinoin, Hungary

LgP	-3.82 (C*)
pKa	
pp cited in Vols.1-5:	

HO₃SCH₂CH₂

(1) Thymus, 1980, 2, 187.

GLUTETHIMIDE [U;INN]

| CAS | 77-21-4 | MW: | 217.27 | MF: | C13 H15 N O2 |

Sedative

Ciba, Switz.; Rorer (Doriden)

LgP	1.90(M); 1.89(C)
pKa	9.2
pp cited in Vols.1-5:	

1: 55

(1) Tagmann et al., Helv. Chim. Acta, 1952, 35, 1541. (2) Hoffmann et al., U.S. Pat. 2 673 205 (1954). (3) H.Y. Aboul-Enein in 'Analytical Profiles of Drug Subst.', ed. K. Florey, Academic Press, NY, 1976, v.5, pp 138-187.

GLYBURIDE [U;INN]

| CAS | 10238-21-8 | MW: | 494.01 | MF: | C23 H28 Cl N3 O5 S |

Antidiabetic

Upjohn

LgP	3.90 (C,3)
pKa	5.3
pp cited in Vols.1-5:	

(1) RTECS, 1985-86, entry 86064.

GLYBUTHIAZOL [INN]

| CAS | 535-65-9 | MW: | 312.41 | MF: | C12 H16 N4 O2 S2 |

Antidiabetic

LgP	1.24 (C*)
pKa	
pp cited in Vols.1-5:	

(1) RTECS, 1985-86, entry 79528.

GLYBUZOLE [INN]

| CAS | 1492-02-0 | MW: | 297.40 | MF: | C12 H15 N3 O2 S2 |

Antidiabetic

LgP	2.09(C)
pKa	
pp cited in Vols.1-5:	

(1) Macrae et al., Brit. Pat. 822 947 (1959).
(2) Baergeton et al., Arch. Int. Pharmacodyn., 1965, 153, 379.

GLYCERIN [U;INN]

| CAS | 56-81-5 | MW: | 92.10 | MF: | C3 H8 O3 |

Reduces intra-ocular &-cranial pressure

LgP	-1.76(M); -2.14(C)
pKa	14.4
pp cited in Vols.1-5:	

HOCH₂—CH—CH₂OH
              |
             OH

(1) 'Martindale. The Extra Pharmacopoeia', ed. J.E.F. Reynolds, The Pharmaceutical Press, London, 1982, 28th Ed., pp 706-707.

GLYCEROL, IODINATED [U]

| CAS | 5634-39-9 | MW: | 258.06 | MF: | C6 H11 I O3 |

Expectorant

LgP	0.54(C)
pKa	
pp cited in Vols.1-5:	

(1) Manchey et al., U.S. Pat. 2 872 378 (1959).

GLYCERYL MONOSTEARATE [U]

| CAS | 31566-31-1 | MW: | 358.57 | MF: | C21 H42 O4 |

Pharmaceutic aid (emulsifying agent)

Eastman (Myvaplex 600P)

LgP	6.84(C)
pKa	
pp cited in Vols.1-5:	

(1) Green, J. Am. Pharm. Assoc., Pract. Pharm. Ed., 1946, 7, 299.

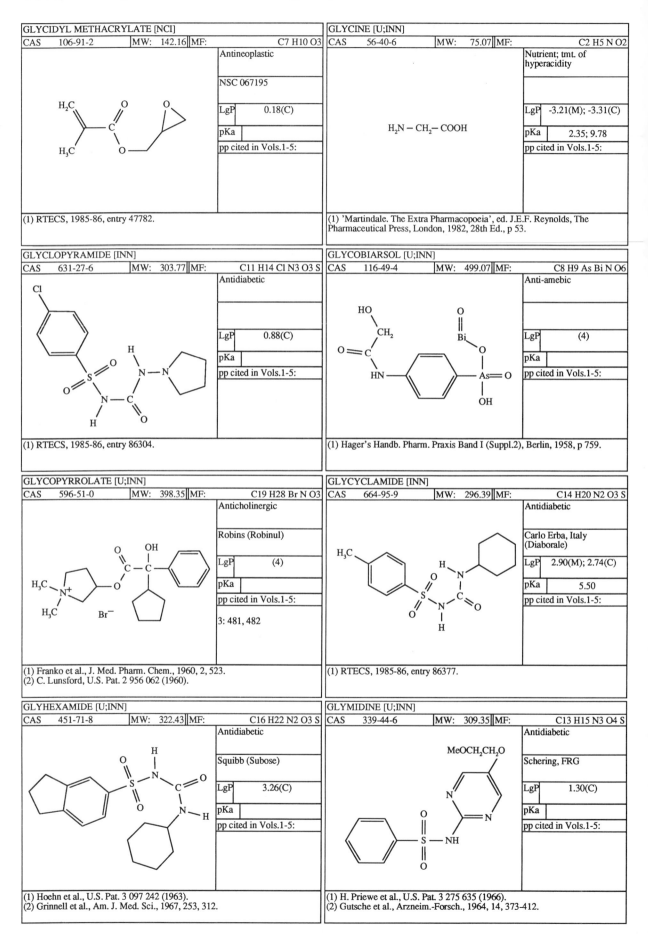

GLYCIDYL METHACRYLATE [NCI]

CAS 106-91-2	MW: 142.16	MF: C7 H10 O3

Antineoplastic

NSC 067195

LgP	0.18(C)
pKa	
pp cited in Vols.1-5:	

(1) RTECS, 1985-86, entry 47782.

GLYCINE [U;INN]

CAS 56-40-6	MW: 75.07	MF: C2 H5 N O2

Nutrient; tmt. of hyperacidity

$H_2N - CH_2 - COOH$

LgP	-3.21(M); -3.31(C)
pKa	2.35; 9.78
pp cited in Vols.1-5:	

(1) 'Martindale. The Extra Pharmacopoeia', ed. J.E.F. Reynolds, The Pharmaceutical Press, London, 1982, 28th Ed., p 53.

GLYCLOPYRAMIDE [INN]

CAS 631-27-6	MW: 303.77	MF: C11 H14 Cl N3 O3 S

Antidiabetic

LgP	0.88(C)
pKa	
pp cited in Vols.1-5:	

(1) RTECS, 1985-86, entry 86304.

GLYCOBIARSOL [U;INN]

CAS 116-49-4	MW: 499.07	MF: C8 H9 As Bi N O6

Anti-amebic

LgP	(4)
pKa	
pp cited in Vols.1-5:	

(1) Hager's Handb. Pharm. Praxis Band I (Suppl.2), Berlin, 1958, p 759.

GLYCOPYRROLATE [U;INN]

CAS 596-51-0	MW: 398.35	MF: C19 H28 Br N O3

Anticholinergic

Robins (Robinul)

LgP	(4)
pKa	
pp cited in Vols.1-5:	

3: 481, 482

(1) Franko et al., J. Med. Pharm. Chem., 1960, 2, 523.
(2) C. Lunsford, U.S. Pat. 2 956 062 (1960).

GLYCYCLAMIDE [INN]

CAS 664-95-9	MW: 296.39	MF: C14 H20 N2 O3 S

Antidiabetic

Carlo Erba, Italy (Diaborale)

LgP	2.90(M); 2.74(C)
pKa	5.50
pp cited in Vols.1-5:	

(1) RTECS, 1985-86, entry 86377.

GLYHEXAMIDE [U;INN]

CAS 451-71-8	MW: 322.43	MF: C16 H22 N2 O3 S

Antidiabetic

Squibb (Subose)

LgP	3.26(C)
pKa	
pp cited in Vols.1-5:	

(1) Hoehn et al., U.S. Pat. 3 097 242 (1963).
(2) Grinnell et al., Am. J. Med. Sci., 1967, 253, 312.

GLYMIDINE [U;INN]

CAS 339-44-6	MW: 309.35	MF: C13 H15 N3 O4 S

Antidiabetic

$MeOCH_2CH_2O$

Schering, FRG

LgP	1.30(C)
pKa	
pp cited in Vols.1-5:	

(1) H. Priewe et al., U.S. Pat. 3 275 635 (1966).
(2) Gutsche et al., Arzneim.-Forsch., 1964, 14, 373-412.

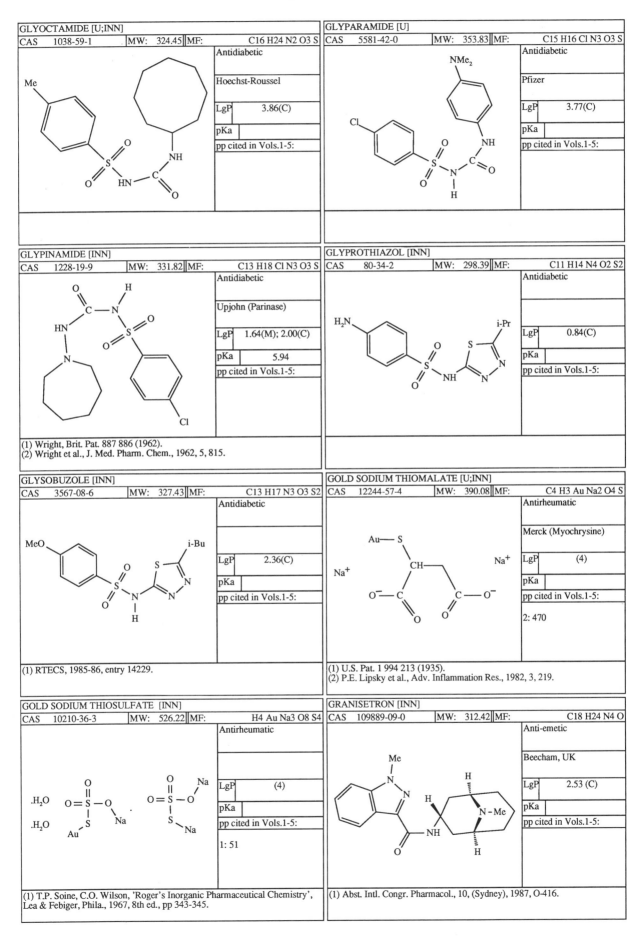

GLYOCTAMIDE [U;INN]

CAS 1038-59-1	MW: 324.45	MF:	C16 H24 N2 O3 S

Antidiabetic

Hoechst-Roussel

LgP	3.86(C)
pKa	
pp cited in Vols.1-5:	

GLYPARAMIDE [U]

CAS 5581-42-0	MW: 353.83	MF:	C15 H16 Cl N3 O3 S

Antidiabetic

Pfizer

LgP	3.77(C)
pKa	
pp cited in Vols.1-5:	

GLYPINAMIDE [INN]

CAS 1228-19-9	MW: 331.82	MF:	C13 H18 Cl N3 O3 S

Antidiabetic

Upjohn (Parinase)

LgP	1.64(M); 2.00(C)
pKa	5.94
pp cited in Vols.1-5:	

(1) Wright, Brit. Pat. 887 886 (1962).
(2) Wright et al., J. Med. Pharm. Chem., 1962, 5, 815.

GLYPROTHIAZOL [INN]

CAS 80-34-2	MW: 298.39	MF:	C11 H14 N4 O2 S2

Antidiabetic

LgP	0.84(C)
pKa	
pp cited in Vols.1-5:	

GLYSOBUZOLE [INN]

CAS 3567-08-6	MW: 327.43	MF:	C13 H17 N3 O3 S2

Antidiabetic

LgP	2.36(C)
pKa	
pp cited in Vols.1-5:	

(1) RTECS, 1985-86, entry 14229.

GOLD SODIUM THIOMALATE [U;INN]

CAS 12244-57-4	MW: 390.08	MF:	C4 H3 Au Na2 O4 S

Antirheumatic

Merck (Myochrysine)

LgP	(4)
pKa	
pp cited in Vols.1-5:	
2: 470	

(1) U.S. Pat. 1 994 213 (1935).
(2) P.E. Lipsky et al., Adv. Inflammation Res., 1982, 3, 219.

GOLD SODIUM THIOSULFATE [INN]

CAS 10210-36-3	MW: 526.22	MF:	H4 Au Na3 O8 S4

Antirheumatic

LgP	(4)
pKa	
pp cited in Vols.1-5:	
1: 51	

(1) T.P. Soine, C.O. Wilson, 'Roger's Inorganic Pharmaceutical Chemistry',
Lea & Febiger, Phila., 1967, 8th ed., pp 343-345.

GRANISETRON [INN]

CAS 109889-09-0	MW: 312.42	MF:	C18 H24 N4 O

Anti-emetic

Beecham, UK

LgP	2.53 (C)
pKa	
pp cited in Vols.1-5:	

(1) Abst. Intl. Congr. Pharmacol., 10, (Sydney), 1987, O-416.

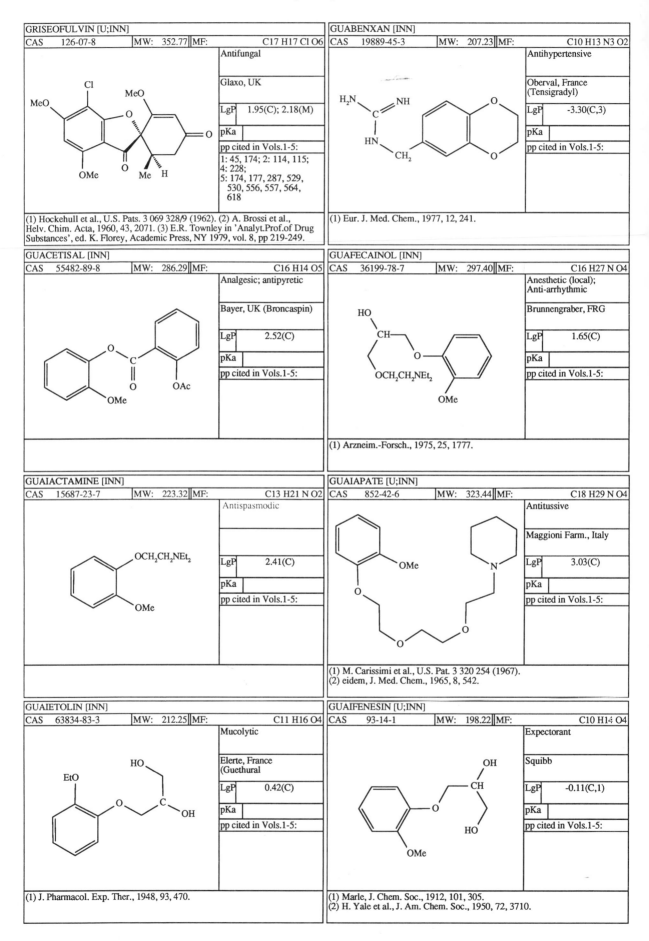

GRISEOFULVIN [U;INN]		
CAS 126-07-8	MW: 352.77	MF: C17 H17 Cl O6

Antifungal

Glaxo, UK

LgP	1.95(C); 2.18(M)
pKa	
pp cited in Vols.1-5:	

1: 45, 174; 2: 114, 115;
4: 228;
5: 174, 177, 287, 529,
  530, 556, 557, 564,
  618

(1) Hockehull et al., U.S. Pats. 3 069 328/9 (1962). (2) A. Brossi et al., Helv. Chim. Acta, 1960, 43, 2071. (3) E.R. Townley in 'Analyt.Prof.of Drug Substances', ed. K. Florey, Academic Press, NY 1979, vol. 8, pp 219-249.

GUABENXAN [INN]		
CAS 19889-45-3	MW: 207.23	MF: C10 H13 N3 O2

Antihypertensive

Oberval, France
(Tensigradyl)

LgP	-3.30(C,3)
pKa	
pp cited in Vols.1-5:	

(1) Eur. J. Med. Chem., 1977, 12, 241.

GUACETISAL [INN]		
CAS 55482-89-8	MW: 286.29	MF: C16 H14 O5

Analgesic; antipyretic

Bayer, UK (Broncaspin)

LgP	2.52(C)
pKa	
pp cited in Vols.1-5:	

GUAFECAINOL [INN]		
CAS 36199-78-7	MW: 297.40	MF: C16 H27 N O4

Anesthetic (local);
Anti-arrhythmic

Brunnengraber, FRG

LgP	1.65(C)
pKa	
pp cited in Vols.1-5:	

(1) Arzneim.-Forsch., 1975, 25, 1777.

GUAIACTAMINE [INN]		
CAS 15687-23-7	MW: 223.32	MF: C13 H21 N O2

Antispasmodic

LgP	2.41(C)
pKa	
pp cited in Vols.1-5:	

GUAIAPATE [U;INN]		
CAS 852-42-6	MW: 323.44	MF: C18 H29 N O4

Antitussive

Maggioni Farm., Italy

LgP	3.03(C)
pKa	
pp cited in Vols.1-5:	

(1) M. Carissimi et al., U.S. Pat. 3 320 254 (1967).
(2) eidem, J. Med. Chem., 1965, 8, 542.

GUAIETOLIN [INN]		
CAS 63834-83-3	MW: 212.25	MF: C11 H16 O4

Mucolytic

Elerte, France
(Guethural)

LgP	0.42(C)
pKa	
pp cited in Vols.1-5:	

(1) J. Pharmacol. Exp. Ther., 1948, 93, 470.

GUAIFENESIN [U;INN]		
CAS 93-14-1	MW: 198.22	MF: C10 H14 O4

Expectorant

Squibb

LgP	-0.11(C,1)
pKa	
pp cited in Vols.1-5:	

(1) Marle, J. Chem. Soc., 1912, 101, 305.
(2) H. Yale et al., J. Am. Chem. Soc., 1950, 72, 3710.

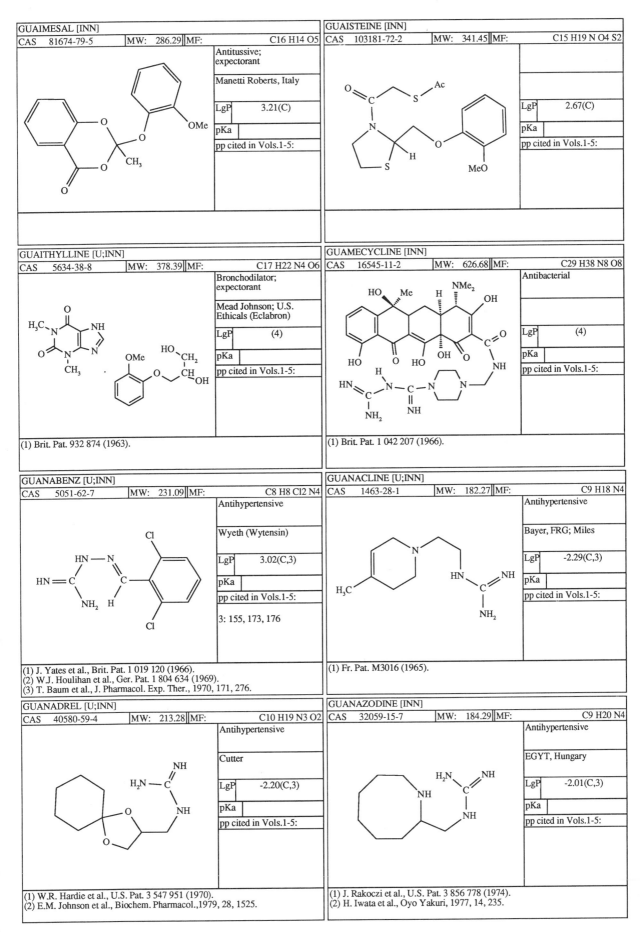

GUAIMESAL [INN]

CAS	81674-79-5	MW:	286.29	MF:	C16 H14 O5

Antitussive; expectorant

Manetti Roberts, Italy

LgP	3.21(C)
pKa	
pp cited in Vols.1-5:	

GUAISTEINE [INN]

CAS	103181-72-2	MW:	341.45	MF:	C15 H19 N O4 S2

LgP	2.67(C)
pKa	
pp cited in Vols.1-5:	

GUAITHYLLINE [U;INN]

CAS	5634-38-8	MW:	378.39	MF:	C17 H22 N4 O6

Bronchodilator; expectorant

Mead Johnson; U.S. Ethicals (Eclabron)

LgP	(4)
pKa	
pp cited in Vols.1-5:	

(1) Brit. Pat. 932 874 (1963).

GUAMECYCLINE [INN]

CAS	16545-11-2	MW:	626.68	MF:	C29 H38 N8 O8

Antibacterial

LgP	(4)
pKa	
pp cited in Vols.1-5:	

(1) Brit. Pat. 1 042 207 (1966).

GUANABENZ [U;INN]

CAS	5051-62-7	MW:	231.09	MF:	C8 H8 Cl2 N4

Antihypertensive

Wyeth (Wytensin)

LgP	3.02(C,3)
pKa	
pp cited in Vols.1-5:	

3: 155, 173, 176

(1) J. Yates et al., Brit. Pat. 1 019 120 (1966).
(2) W.J. Houlihan et al., Ger. Pat. 1 804 634 (1969).
(3) T. Baum et al., J. Pharmacol. Exp. Ther., 1970, 171, 276.

GUANACLINE [U;INN]

CAS	1463-28-1	MW:	182.27	MF:	C9 H18 N4

Antihypertensive

Bayer, FRG; Miles

LgP	-2.29(C,3)
pKa	
pp cited in Vols.1-5:	

(1) Fr. Pat. M3016 (1965).

GUANADREL [U;INN]

CAS	40580-59-4	MW:	213.28	MF:	C10 H19 N3 O2

Antihypertensive

Cutter

LgP	-2.20(C,3)
pKa	
pp cited in Vols.1-5:	

(1) W.R. Hardie et al., U.S. Pat. 3 547 951 (1970).
(2) E.M. Johnson et al., Biochem. Pharmacol.,1979, 28, 1525.

GUANAZODINE [INN]

CAS	32059-15-7	MW:	184.29	MF:	C9 H20 N4

Antihypertensive

EGYT, Hungary

LgP	-2.01(C,3)
pKa	
pp cited in Vols.1-5:	

(1) J. Rakoczi et al., U.S. Pat. 3 856 778 (1974).
(2) H. Iwata et al., Oyo Yakuri, 1977, 14, 235.

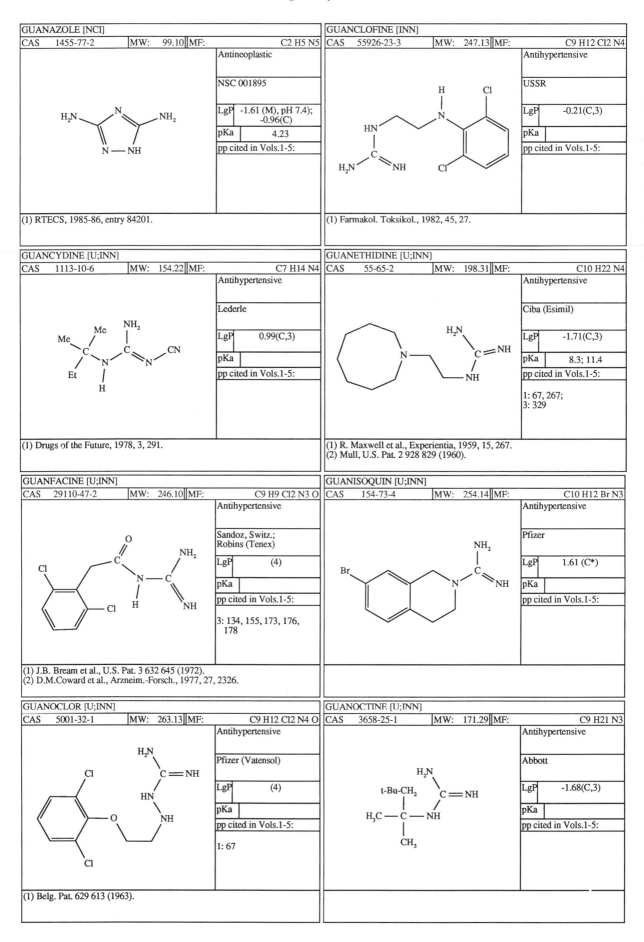

GUANAZOLE [NCI]

CAS	1455-77-2	MW:	99.10	MF:	C2 H5 N5

Antineoplastic

NSC 001895

LgP	-1.61 (M), pH 7.4); -0.96(C)
pKa	4.23
pp cited in Vols.1-5:	

(1) RTECS, 1985-86, entry 84201.

GUANCLOFINE [INN]

CAS	55926-23-3	MW:	247.13	MF:	C9 H12 Cl2 N4

Antihypertensive

USSR

LgP	-0.21(C,3)
pKa	
pp cited in Vols.1-5:	

(1) Farmakol. Toksikol., 1982, 45, 27.

GUANCYDINE [U;INN]

CAS	1113-10-6	MW:	154.22	MF:	C7 H14 N4

Antihypertensive

Lederle

LgP	0.99(C,3)
pKa	
pp cited in Vols.1-5:	

(1) Drugs of the Future, 1978, 3, 291.

GUANETHIDINE [U;INN]

CAS	55-65-2	MW:	198.31	MF:	C10 H22 N4

Antihypertensive

Ciba (Esimil)

LgP	-1.71(C,3)
pKa	8.3; 11.4
pp cited in Vols.1-5:	

1: 67, 267;
3: 329

(1) R. Maxwell et al., Experientia, 1959, 15, 267.
(2) Mull, U.S. Pat. 2 928 829 (1960).

GUANFACINE [U;INN]

CAS	29110-47-2	MW:	246.10	MF:	C9 H9 Cl2 N3 O

Antihypertensive

Sandoz, Switz.; Robins (Tenex)

LgP	(4)
pKa	
pp cited in Vols.1-5:	

3: 134, 155, 173, 176, 178

(1) J.B. Bream et al., U.S. Pat. 3 632 645 (1972).
(2) D.M.Coward et al., Arzneim.-Forsch., 1977, 27, 2326.

GUANISOQUIN [U;INN]

CAS	154-73-4	MW:	254.14	MF:	C10 H12 Br N3

Antihypertensive

Pfizer

LgP	1.61 (C*)
pKa	
pp cited in Vols.1-5:	

GUANOCLOR [U;INN]

CAS	5001-32-1	MW:	263.13	MF:	C9 H12 Cl2 N4 O

Antihypertensive

Pfizer (Vatensol)

LgP	(4)
pKa	
pp cited in Vols.1-5:	

1: 67

(1) Belg. Pat. 629 613 (1963).

GUANOCTINE [U;INN]

CAS	3658-25-1	MW:	171.29	MF:	C9 H21 N3

Antihypertensive

Abbott

LgP	-1.68(C,3)
pKa	
pp cited in Vols.1-5:	

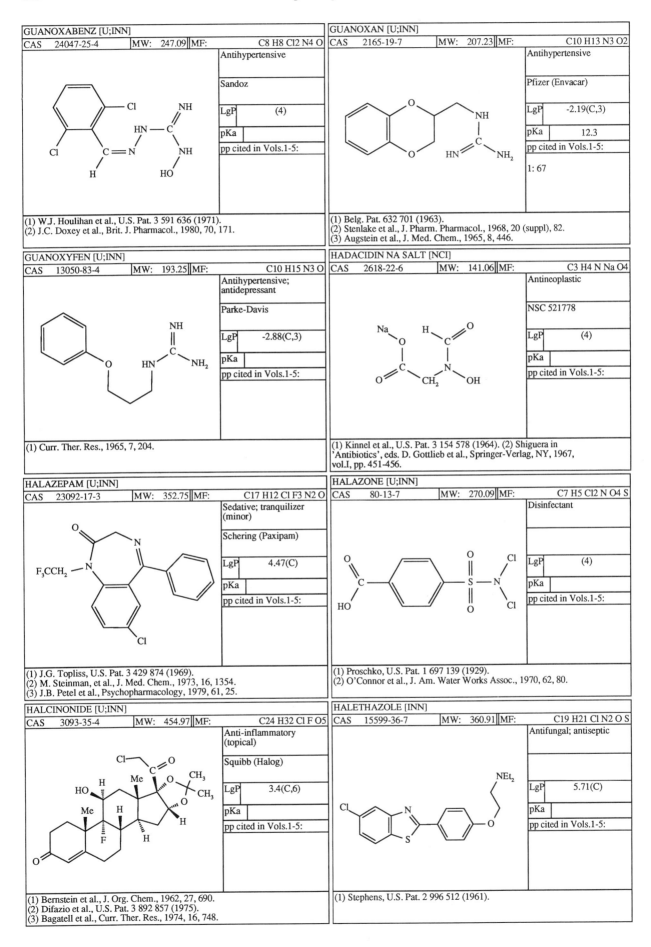

GUANOXABENZ [U;INN]

CAS	24047-25-4	MW: 247.09	MF:	C8 H8 Cl2 N4 O

Antihypertensive

Sandoz

LgP	(4)
pKa	
pp cited in Vols.1-5:	

(1) W.J. Houlihan et al., U.S. Pat. 3 591 636 (1971).
(2) J.C. Doxey et al., Brit. J. Pharmacol., 1980, 70, 171.

GUANOXAN [U;INN]

CAS	2165-19-7	MW: 207.23	MF:	C10 H13 N3 O2

Antihypertensive

Pfizer (Envacar)

LgP	-2.19(C,3)
pKa	12.3
pp cited in Vols.1-5:	

1: 67

(1) Belg. Pat. 632 701 (1963).
(2) Stenlake et al., J. Pharm. Pharmacol., 1968, 20 (suppl), 82.
(3) Augstein et al., J. Med. Chem., 1965, 8, 446.

GUANOXYFEN [U;INN]

CAS	13050-83-4	MW: 193.25	MF:	C10 H15 N3 O

Antihypertensive; antidepressant

Parke-Davis

LgP	-2.88(C,3)
pKa	
pp cited in Vols.1-5:	

(1) Curr. Ther. Res., 1965, 7, 204.

HADACIDIN NA SALT [NCI]

CAS	2618-22-6	MW: 141.06	MF:	C3 H4 N Na O4

Antineoplastic

NSC 521778

LgP	(4)
pKa	
pp cited in Vols.1-5:	

(1) Kinnel et al., U.S. Pat. 3 154 578 (1964). (2) Shiguera in
'Antibiotics', eds. D. Gottlieb et al., Springer-Verlag, NY, 1967,
vol.I, pp. 451-456.

HALAZEPAM [U;INN]

CAS	23092-17-3	MW: 352.75	MF:	C17 H12 Cl F3 N2 O

Sedative; tranquilizer (minor)

Schering (Paxipam)

LgP	4.47(C)
pKa	
pp cited in Vols.1-5:	

(1) J.G. Topliss, U.S. Pat. 3 429 874 (1969).
(2) M. Steinman, et al., J. Med. Chem., 1973, 16, 1354.
(3) J.B. Petel et al., Psychopharmacology, 1979, 61, 25.

HALAZONE [U;INN]

CAS	80-13-7	MW: 270.09	MF:	C7 H5 Cl2 N O4 S

Disinfectant

LgP	(4)
pKa	
pp cited in Vols.1-5:	

(1) Proschko, U.S. Pat. 1 697 139 (1929).
(2) O'Connor et al., J. Am. Water Works Assoc., 1970, 62, 80.

HALCINONIDE [U;INN]

CAS	3093-35-4	MW: 454.97	MF:	C24 H32 Cl F O5

Anti-inflammatory (topical)

Squibb (Halog)

LgP	3.4(C,6)
pKa	
pp cited in Vols.1-5:	

(1) Bernstein et al., J. Org. Chem., 1962, 27, 690.
(2) Difazio et al., U.S. Pat. 3 892 857 (1975).
(3) Bagatell et al., Curr. Ther. Res., 1974, 16, 748.

HALETHAZOLE [INN]

CAS	15599-36-7	MW: 360.91	MF:	C19 H21 Cl N2 O S

Antifungal; antiseptic

LgP	5.71(C)
pKa	
pp cited in Vols.1-5:	

(1) Stephens, U.S. Pat. 2 996 512 (1961).

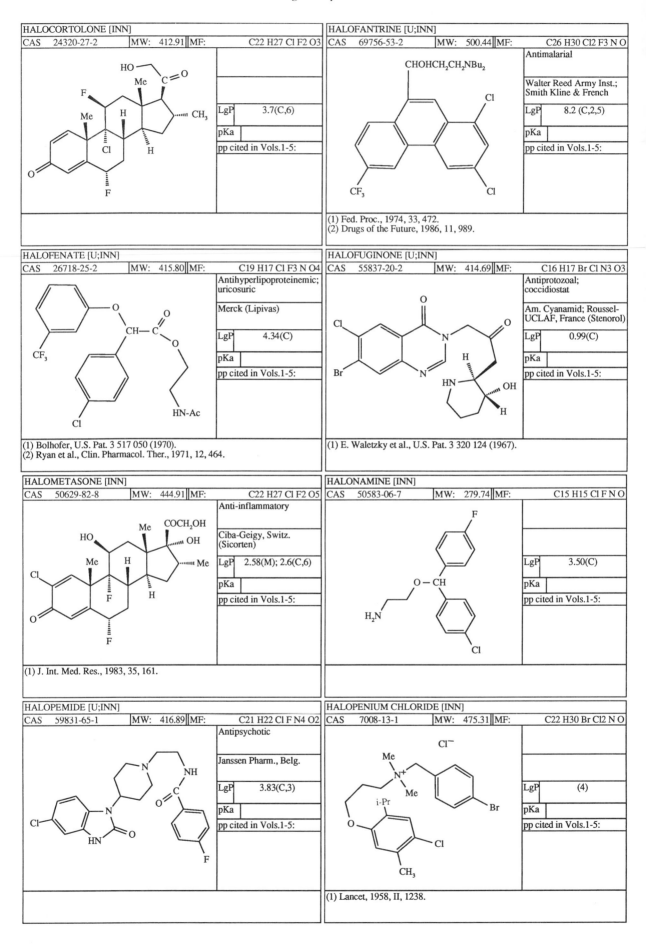

HALOCORTOLONE [INN]			
CAS 24320-27-2	MW: 412.91	MF:	C22 H27 Cl F2 O3

LgP	3.7(C,6)
pKa	
pp cited in Vols.1-5:	

HALOFANTRINE [U;INN]			
CAS 69756-53-2	MW: 500.44	MF:	C26 H30 Cl2 F3 N O

Antimalarial

Walter Reed Army Inst.; Smith Kline & French

LgP	8.2 (C,2,5)
pKa	
pp cited in Vols.1-5:	

(1) Fed. Proc., 1974, 33, 472.
(2) Drugs of the Future, 1986, 11, 989.

HALOFENATE [U;INN]			
CAS 26718-25-2	MW: 415.80	MF:	C19 H17 Cl F3 N O4

Antihyperlipoproteinemic; uricosuric

Merck (Lipivas)

LgP	4.34(C)
pKa	
pp cited in Vols.1-5:	

(1) Bolhofer, U.S. Pat. 3 517 050 (1970).
(2) Ryan et al., Clin. Pharmacol. Ther., 1971, 12, 464.

HALOFUGINONE [U;INN]			
CAS 55837-20-2	MW: 414.69	MF:	C16 H17 Br Cl N3 O3

Antiprotozoal; coccidiostat

Am. Cyanamid; Roussel-UCLAF, France (Stenorol)

LgP	0.99(C)
pKa	
pp cited in Vols.1-5:	

(1) E. Waletzky et al., U.S. Pat. 3 320 124 (1967).

HALOMETASONE [INN]			
CAS 50629-82-8	MW: 444.91	MF:	C22 H27 Cl F2 O5

Anti-inflammatory

Ciba-Geigy, Switz. (Sicorten)

LgP	2.58(M); 2.6(C,6)
pKa	
pp cited in Vols.1-5:	

(1) J. Int. Med. Res., 1983, 35, 161.

HALONAMINE [INN]			
CAS 50583-06-7	MW: 279.74	MF:	C15 H15 Cl F N O

LgP	3.50(C)
pKa	
pp cited in Vols.1-5:	

HALOPEMIDE [U;INN]			
CAS 59831-65-1	MW: 416.89	MF:	C21 H22 Cl F N4 O2

Antipsychotic

Janssen Pharm., Belg.

LgP	3.83(C,3)
pKa	
pp cited in Vols.1-5:	

HALOPENIUM CHLORIDE [INN]			
CAS 7008-13-1	MW: 475.31	MF:	C22 H30 Br Cl2 N O

LgP	(4)
pKa	
pp cited in Vols.1-5:	

(1) Lancet, 1958, II, 1238.

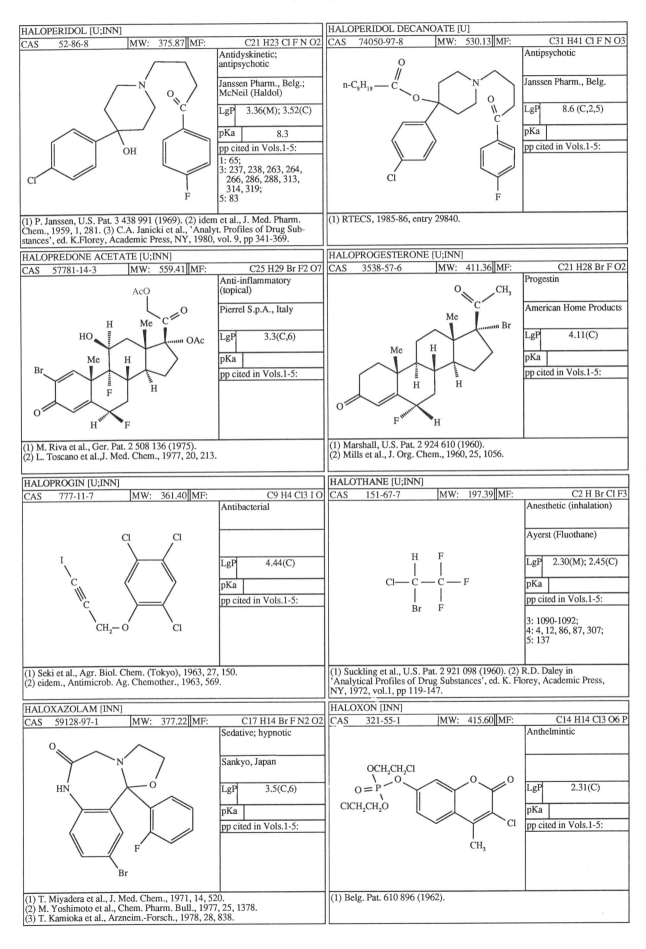

HALOPERIDOL [U;INN]

| CAS | 52-86-8 | MW: | 375.87 | MF: | C21 H23 Cl F N O2 |

Antidyskinetic; antipsychotic

Janssen Pharm., Belg.; McNeil (Haldol)

| LgP | 3.36(M); 3.52(C) |

| pKa | 8.3 |

pp cited in Vols.1-5:

1: 65;
3: 237, 238, 263, 264, 266, 286, 288, 313, 314, 319;
5: 83

(1) P. Janssen, U.S. Pat. 3 438 991 (1969). (2) idem et al., J. Med. Pharm. Chem., 1959, 1, 281. (3) C.A. Janicki et al., 'Analyt. Profiles of Drug Substances', ed. K.Florey, Academic Press, NY, 1980, vol. 9, pp 341-369.

HALOPERIDOL DECANOATE [U]

| CAS | 74050-97-8 | MW: | 530.13 | MF: | C31 H41 Cl F N O3 |

Antipsychotic

Janssen Pharm., Belg.

| LgP | 8.6 (C,2,5) |

| pKa | |

pp cited in Vols.1-5:

(1) RTECS, 1985-86, entry 29840.

HALOPREDONE ACETATE [U;INN]

| CAS | 57781-14-3 | MW: | 559.41 | MF: | C25 H29 Br F2 O7 |

Anti-inflammatory (topical)

Pierrel S.p.A., Italy

| LgP | 3.3(C,6) |

| pKa | |

pp cited in Vols.1-5:

(1) M. Riva et al., Ger. Pat. 2 508 136 (1975).
(2) L. Toscano et al.,J. Med. Chem., 1977, 20, 213.

HALOPROGESTERONE [U;INN]

| CAS | 3538-57-6 | MW: | 411.36 | MF: | C21 H28 Br F O2 |

Progestin

American Home Products

| LgP | 4.11(C) |

| pKa | |

pp cited in Vols.1-5:

(1) Marshall, U.S. Pat. 2 924 610 (1960).
(2) Mills et al., J. Org. Chem., 1960, 25, 1056.

HALOPROGIN [U;INN]

| CAS | 777-11-7 | MW: | 361.40 | MF: | C9 H4 Cl3 I O |

Antibacterial

| LgP | 4.44(C) |

| pKa | |

pp cited in Vols.1-5:

(1) Seki et al., Agr. Biol. Chem. (Tokyo), 1963, 27, 150.
(2) eidem., Antimicrob. Ag. Chemother., 1963, 569.

HALOTHANE [U;INN]

| CAS | 151-67-7 | MW: | 197.39 | MF: | C2 H Br Cl F3 |

Anesthetic (inhalation)

Ayerst (Fluothane)

| LgP | 2.30(M); 2.45(C) |

| pKa | |

pp cited in Vols.1-5:

3: 1090-1092;
4: 4, 12, 86, 87, 307;
5: 137

(1) Suckling et al., U.S. Pat. 2 921 098 (1960). (2) R.D. Daley in 'Analytical Profiles of Drug Substances', ed. K. Florey, Academic Press, NY, 1972, vol.1, pp 119-147.

HALOXAZOLAM [INN]

| CAS | 59128-97-1 | MW: | 377.22 | MF: | C17 H14 Br F N2 O2 |

Sedative; hypnotic

Sankyo, Japan

| LgP | 3.5(C,6) |

| pKa | |

pp cited in Vols.1-5:

(1) T. Miyadera et al., J. Med. Chem., 1971, 14, 520.
(2) M. Yoshimoto et al., Chem. Pharm. Bull., 1977, 25, 1378.
(3) T. Kamioka et al., Arzneim.-Forsch., 1978, 28, 838.

HALOXON [INN]

| CAS | 321-55-1 | MW: | 415.60 | MF: | C14 H14 Cl3 O6 P |

Anthelmintic

| LgP | 2.31(C) |

| pKa | |

pp cited in Vols.1-5:

(1) Belg. Pat. 610 896 (1962).

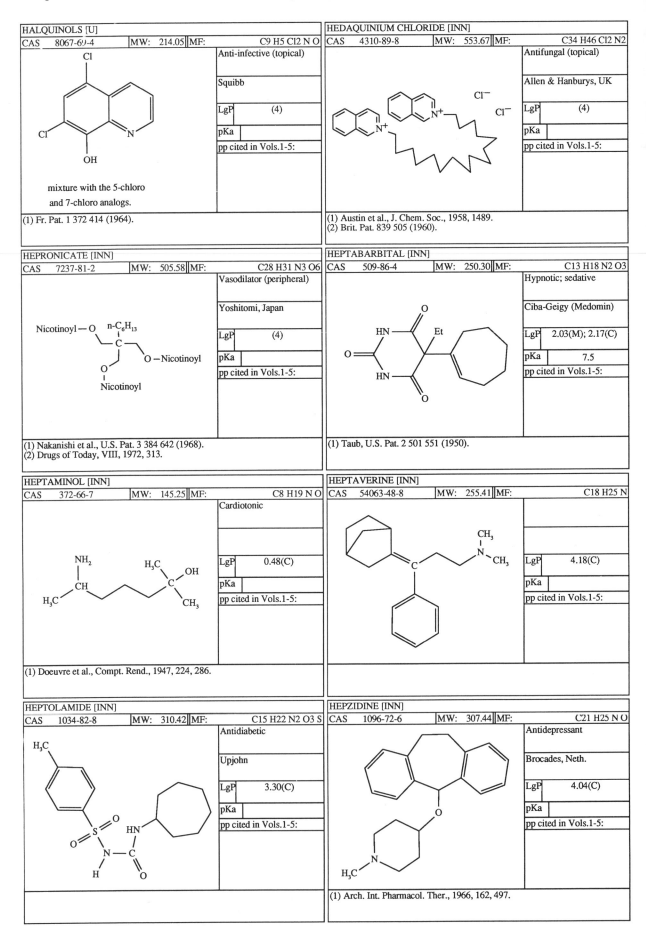

HALQUINOLS [U]

CAS	8067-69-4	MW:	214.05	MF:	C9 H5 Cl2 N O

Anti-infective (topical)

Squibb

LgP	(4)

pKa	

pp cited in Vols.1-5:

mixture with the 5-chloro
and 7-chloro analogs.

(1) Fr. Pat. 1 372 414 (1964).

HEDAQUINIUM CHLORIDE [INN]

CAS	4310-89-8	MW:	553.67	MF:	C34 H46 Cl2 N2

Antifungal (topical)

Allen & Hanburys, UK

LgP	(4)

pKa	

pp cited in Vols.1-5:

(1) Austin et al., J. Chem. Soc., 1958, 1489.
(2) Brit. Pat. 839 505 (1960).

HEPRONICATE [INN]

CAS	7237-81-2	MW:	505.58	MF:	C28 H31 N3 O6

Vasodilator (peripheral)

Yoshitomi, Japan

LgP	(4)

pKa	

pp cited in Vols.1-5:

(1) Nakanishi et al., U.S. Pat. 3 384 642 (1968).
(2) Drugs of Today, VIII, 1972, 313.

HEPTABARBITAL [INN]

CAS	509-86-4	MW:	250.30	MF:	C13 H18 N2 O3

Hypnotic; sedative

Ciba-Geigy (Medomin)

LgP	2.03(M); 2.17(C)

pKa	7.5

pp cited in Vols.1-5:

(1) Taub, U.S. Pat. 2 501 551 (1950).

HEPTAMINOL [INN]

CAS	372-66-7	MW:	145.25	MF:	C8 H19 N O

Cardiotonic

LgP	0.48(C)

pKa	

pp cited in Vols.1-5:

(1) Doeuvre et al., Compt. Rend., 1947, 224, 286.

HEPTAVERINE [INN]

CAS	54063-48-8	MW:	255.41	MF:	C18 H25 N

LgP	4.18(C)

pKa	

pp cited in Vols.1-5:

HEPTOLAMIDE [INN]

CAS	1034-82-8	MW:	310.42	MF:	C15 H22 N2 O3 S

Antidiabetic

Upjohn

LgP	3.30(C)

pKa	

pp cited in Vols.1-5:

HEPZIDINE [INN]

CAS	1096-72-6	MW:	307.44	MF:	C21 H25 N O

Antidepressant

Brocades, Neth.

LgP	4.04(C)

pKa	

pp cited in Vols.1-5:

(1) Arch. Int. Pharmacol. Ther., 1966, 162, 497.

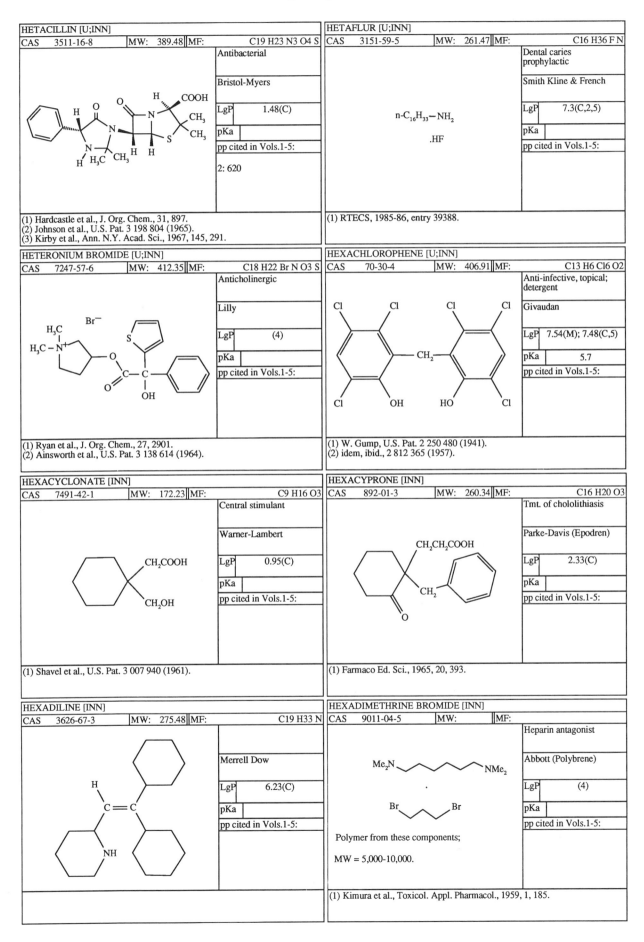

HETACILLIN [U;INN]

CAS 3511-16-8	MW: 389.48	MF:	C19 H23 N3 O4 S

Antibacterial

Bristol-Myers

LgP	1.48(C)
pKa	
pp cited in Vols.1-5:	

2: 620

(1) Hardcastle et al., J. Org. Chem., 31, 897.
(2) Johnson et al., U.S. Pat. 3 198 804 (1965).
(3) Kirby et al., Ann. N.Y. Acad. Sci., 1967, 145, 291.

HETAFLUR [U;INN]

CAS 3151-59-5	MW: 261.47	MF:	C16 H36 F N

Dental caries
prophylactic

Smith Kline & French

LgP	7.3(C,2,5)
pKa	
pp cited in Vols.1-5:	

n-C$_{16}$H$_{33}$—NH$_2$

.HF

(1) RTECS, 1985-86, entry 39388.

HETERONIUM BROMIDE [U;INN]

CAS 7247-57-6	MW: 412.35	MF:	C18 H22 Br N O3 S

Anticholinergic

Lilly

LgP	(4)
pKa	
pp cited in Vols.1-5:	

(1) Ryan et al., J. Org. Chem., 27, 2901.
(2) Ainsworth et al., U.S. Pat. 3 138 614 (1964).

HEXACHLOROPHENE [U;INN]

CAS 70-30-4	MW: 406.91	MF:	C13 H6 Cl6 O2

Anti-infective, topical;
detergent

Givaudan

LgP	7.54(M); 7.48(C,5)
pKa	5.7
pp cited in Vols.1-5:	

(1) W. Gump, U.S. Pat. 2 250 480 (1941).
(2) idem, ibid., 2 812 365 (1957).

HEXACYCLONATE [INN]

CAS 7491-42-1	MW: 172.23	MF:	C9 H16 O3

Central stimulant

Warner-Lambert

LgP	0.95(C)
pKa	
pp cited in Vols.1-5:	

(1) Shavel et al., U.S. Pat. 3 007 940 (1961).

HEXACYPRONE [INN]

CAS 892-01-3	MW: 260.34	MF:	C16 H20 O3

Tmt. of chololithiasis

Parke-Davis (Epodren)

LgP	2.33(C)
pKa	
pp cited in Vols.1-5:	

(1) Farmaco Ed. Sci., 1965, 20, 393.

HEXADILINE [INN]

CAS 3626-67-3	MW: 275.48	MF:	C19 H33 N

Merrell Dow

LgP	6.23(C)
pKa	
pp cited in Vols.1-5:	

HEXADIMETHRINE BROMIDE [INN]

CAS 9011-04-5	MW:	MF:	

Heparin antagonist

Abbott (Polybrene)

LgP	(4)
pKa	
pp cited in Vols.1-5:	

Polymer from these components;

MW = 5,000-10,000.

(1) Kimura et al., Toxicol. Appl. Pharmacol., 1959, 1, 185.

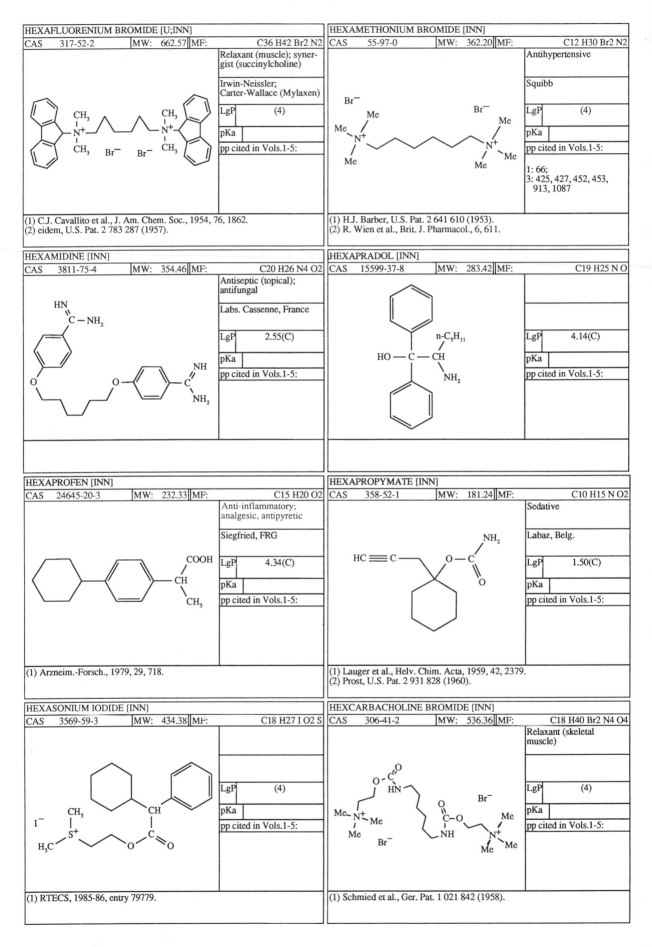

HEXAFLUORENIUM BROMIDE [U;INN]

CAS	317-52-2	MW:	662.57	MF:	C36 H42 Br2 N2

Relaxant (muscle); synergist (succinylcholine)

Irwin-Neissler; Carter-Wallace (Mylaxen)

LgP	(4)
pKa	
pp cited in Vols.1-5:	

(1) C.J. Cavallito et al., J. Am. Chem. Soc., 1954, 76, 1862.
(2) eidem, U.S. Pat. 2 783 287 (1957).

HEXAMETHONIUM BROMIDE [INN]

CAS	55-97-0	MW:	362.20	MF:	C12 H30 Br2 N2

Antihypertensive

Squibb

LgP	(4)
pKa	
pp cited in Vols.1-5:	

1: 66;
3: 425, 427, 452, 453, 913, 1087

(1) H.J. Barber, U.S. Pat. 2 641 610 (1953).
(2) R. Wien et al., Brit. J. Pharmacol., 6, 611.

HEXAMIDINE [INN]

CAS	3811-75-4	MW:	354.46	MF:	C20 H26 N4 O2

Antiseptic (topical); antifungal

Labs. Cassenne, France

LgP	2.55(C)
pKa	
pp cited in Vols.1-5:	

HEXAPRADOL [INN]

CAS	15599-37-8	MW:	283.42	MF:	C19 H25 N O

LgP	4.14(C)
pKa	
pp cited in Vols.1-5:	

HEXAPROFEN [INN]

CAS	24645-20-3	MW:	232.33	MF:	C15 H20 O2

Anti-inflammatory; analgesic, antipyretic

Siegfried, FRG

LgP	4.34(C)
pKa	
pp cited in Vols.1-5:	

(1) Arzneim.-Forsch., 1979, 29, 718.

HEXAPROPYMATE [INN]

CAS	358-52-1	MW:	181.24	MF:	C10 H15 N O2

Sedative

Labaz, Belg.

LgP	1.50(C)
pKa	
pp cited in Vols.1-5:	

(1) Lauger et al., Helv. Chim. Acta, 1959, 42, 2379.
(2) Prost, U.S. Pat. 2 931 828 (1960).

HEXASONIUM IODIDE [INN]

CAS	3569-59-3	MW:	434.38	MF:	C18 H27 I O2 S

LgP	(4)
pKa	
pp cited in Vols.1-5:	

(1) RTECS, 1985-86, entry 79779.

HEXCARBACHOLINE BROMIDE [INN]

CAS	306-41-2	MW:	536.36	MF:	C18 H40 Br2 N4 O4

Relaxant (skeletal muscle)

LgP	(4)
pKa	
pp cited in Vols.1-5:	

(1) Schmied et al., Ger. Pat. 1 021 842 (1958).

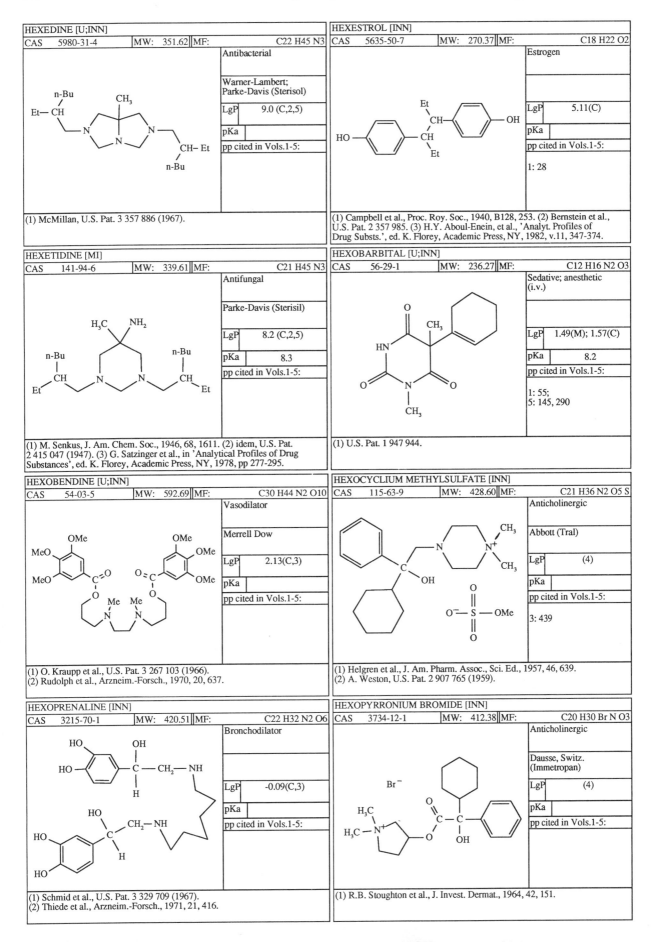

HEXEDINE [U;INN]

| CAS | 5980-31-4 | MW: | 351.62 | MF: | C22 H45 N3 |

Antibacterial

Warner-Lambert; Parke-Davis (Sterisol)

| LgP | 9.0 (C,2,5) |
| pKa | |
| pp cited in Vols.1-5: |

(1) McMillan, U.S. Pat. 3 357 886 (1967).

HEXESTROL [INN]

| CAS | 5635-50-7 | MW: | 270.37 | MF: | C18 H22 O2 |

Estrogen

| LgP | 5.11(C) |
| pKa | |
| pp cited in Vols.1-5: |
| 1: 28 |

(1) Campbell et al., Proc. Roy. Soc., 1940, B128, 253. (2) Bernstein et al., U.S. Pat. 2 357 985. (3) H.Y. Aboul-Enein, et al., 'Analyt. Profiles of Drug Substs.', ed. K. Florey, Academic Press, NY, 1982, v.11, 347-374.

HEXETIDINE [MI]

| CAS | 141-94-6 | MW: | 339.61 | MF: | C21 H45 N3 |

Antifungal

Parke-Davis (Sterisil)

| LgP | 8.2 (C,2,5) |
| pKa | 8.3 |
| pp cited in Vols.1-5: |

(1) M. Senkus, J. Am. Chem. Soc., 1946, 68, 1611. (2) idem, U.S. Pat. 2 415 047 (1947). (3) G. Satzinger et al., in 'Analytical Profiles of Drug Substances', ed. K. Florey, Academic Press, NY, 1978, pp 277-295.

HEXOBARBITAL [U;INN]

| CAS | 56-29-1 | MW: | 236.27 | MF: | C12 H16 N2 O3 |

Sedative; anesthetic (i.v.)

| LgP | 1.49(M); 1.57(C) |
| pKa | 8.2 |
| pp cited in Vols.1-5: |
| 1: 55; |
| 5: 145, 290 |

(1) U.S. Pat. 1 947 944.

HEXOBENDINE [U;INN]

| CAS | 54-03-5 | MW: | 592.69 | MF: | C30 H44 N2 O10 |

Vasodilator

Merrell Dow

| LgP | 2.13(C,3) |
| pKa | |
| pp cited in Vols.1-5: |

(1) O. Kraupp et al., U.S. Pat. 3 267 103 (1966).
(2) Rudolph et al., Arzneim.-Forsch., 1970, 20, 637.

HEXOCYCLIUM METHYLSULFATE [INN]

| CAS | 115-63-9 | MW: | 428.60 | MF: | C21 H36 N2 O5 S |

Anticholinergic

Abbott (Tral)

| LgP | (4) |
| pKa | |
| pp cited in Vols.1-5: |
| 3: 439 |

(1) Helgren et al., J. Am. Pharm. Assoc., Sci. Ed., 1957, 46, 639.
(2) A. Weston, U.S. Pat. 2 907 765 (1959).

HEXOPRENALINE [INN]

| CAS | 3215-70-1 | MW: | 420.51 | MF: | C22 H32 N2 O6 |

Bronchodilator

| LgP | -0.09(C,3) |
| pKa | |
| pp cited in Vols.1-5: |

(1) Schmid et al., U.S. Pat. 3 329 709 (1967).
(2) Thiede et al., Arzneim.-Forsch., 1971, 21, 416.

HEXOPYRRONIUM BROMIDE [INN]

| CAS | 3734-12-1 | MW: | 412.38 | MF: | C20 H30 Br N O3 |

Anticholinergic

Dausse, Switz. (Immetropan)

| LgP | (4) |
| pKa | |
| pp cited in Vols.1-5: |

(1) R.B. Stoughton et al., J. Invest. Dermat., 1964, 42, 151.

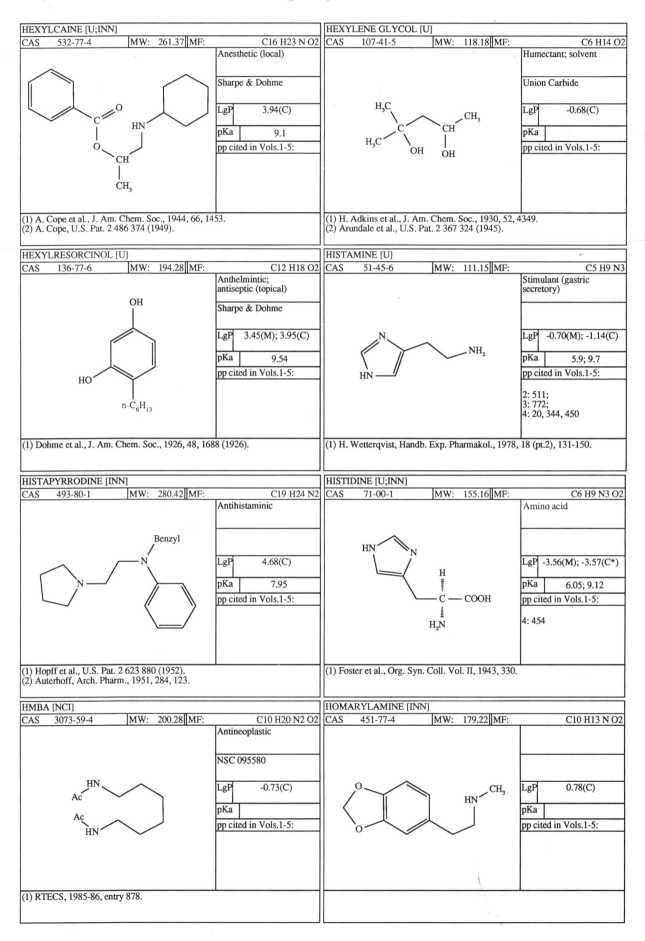

HEXYLCAINE [U;INN]

CAS	532-77-4	MW:	261.37	MF:	C16 H23 N O2

Anesthetic (local)

Sharpe & Dohme

LgP	3.94(C)
pKa	9.1
pp cited in Vols.1-5:	

(1) A. Cope et al., J. Am. Chem. Soc., 1944, 66, 1453.
(2) A. Cope, U.S. Pat. 2 486 374 (1949).

HEXYLENE GLYCOL [U]

CAS	107-41-5	MW:	118.18	MF:	C6 H14 O2

Humectant; solvent

Union Carbide

LgP	-0.68(C)
pKa	
pp cited in Vols.1-5:	

(1) H. Adkins et al., J. Am. Chem. Soc., 1930, 52, 4349.
(2) Arundale et al., U.S. Pat. 2 367 324 (1945).

HEXYLRESORCINOL [U]

CAS	136-77-6	MW:	194.28	MF:	C12 H18 O2

Anthelmintic; antiseptic (topical)

Sharpe & Dohme

LgP	3.45(M); 3.95(C)
pKa	9.54
pp cited in Vols.1-5:	

(1) Dohme et al., J. Am. Chem. Soc., 1926, 48, 1688 (1926).

HISTAMINE [U]

CAS	51-45-6	MW:	111.15	MF:	C5 H9 N3

Stimulant (gastric secretory)

LgP	-0.70(M); -1.14(C)
pKa	5.9; 9.7
pp cited in Vols.1-5:	

2: 511;
3: 772;
4: 20, 344, 450

(1) H. Wetterqvist, Handb. Exp. Pharmakol., 1978, 18 (pt.2), 131-150.

HISTAPYRRODINE [INN]

CAS	493-80-1	MW:	280.42	MF:	C19 H24 N2

Antihistaminic

LgP	4.68(C)
pKa	7.95
pp cited in Vols.1-5:	

(1) Hopff et al., U.S. Pat. 2 623 880 (1952).
(2) Auterhoff, Arch. Pharm., 1951, 284, 123.

HISTIDINE [U;INN]

CAS	71-00-1	MW:	155.16	MF:	C6 H9 N3 O2

Amino acid

LgP	-3.56(M); -3.57(C*)
pKa	6.05; 9.12
pp cited in Vols.1-5:	

4: 454

(1) Foster et al., Org. Syn. Coll. Vol. II, 1943, 330.

HMBA [NCI]

CAS	3073-59-4	MW:	200.28	MF:	C10 H20 N2 O2

Antineoplastic

NSC 095580

LgP	-0.73(C)
pKa	
pp cited in Vols.1-5:	

(1) RTECS, 1985-86, entry 878.

HOMARYLAMINE [INN]

CAS	451-77-4	MW:	179.22	MF:	C10 H13 N O2

LgP	0.78(C)
pKa	
pp cited in Vols.1-5:	

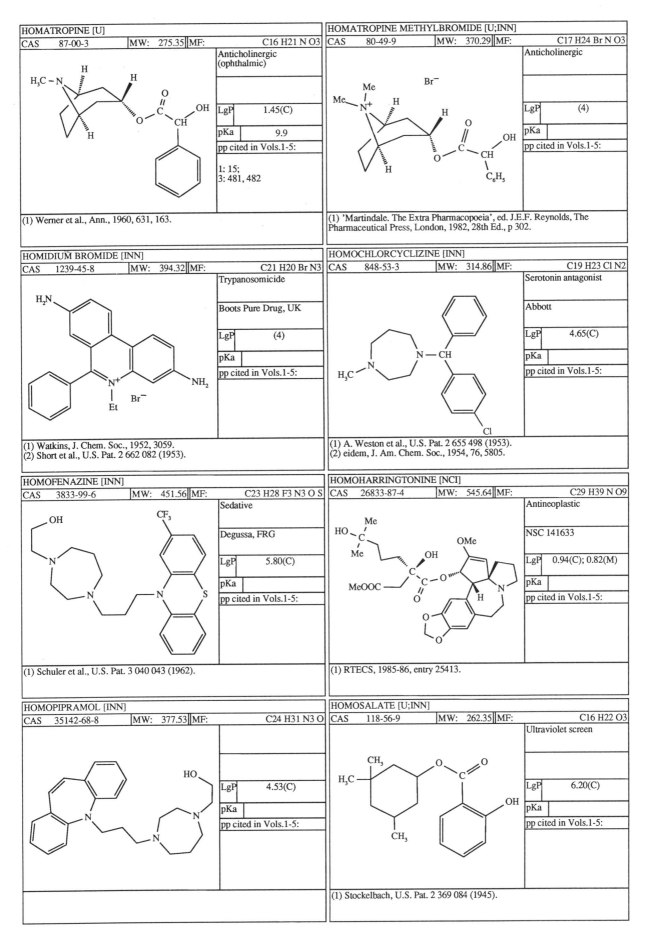

HOMATROPINE [U]

| CAS | 87-00-3 | MW: | 275.35 | MF: | C16 H21 N O3 |

Anticholinergic (ophthalmic)

| LgP | 1.45(C) |
| pKa | 9.9 |
| pp cited in Vols.1-5: |

1: 15;
3: 481, 482

(1) Werner et al., Ann., 1960, 631, 163.

HOMATROPINE METHYLBROMIDE [U;INN]

| CAS | 80-49-9 | MW: | 370.29 | MF: | C17 H24 Br N O3 |

Anticholinergic

| LgP | (4) |
| pKa | |
| pp cited in Vols.1-5: |

(1) 'Martindale. The Extra Pharmacopoeia', ed. J.E.F. Reynolds, The Pharmaceutical Press, London, 1982, 28th Ed., p 302.

HOMIDIUM BROMIDE [INN]

| CAS | 1239-45-8 | MW: | 394.32 | MF: | C21 H20 Br N3 |

Trypanosomicide

Boots Pure Drug, UK

| LgP | (4) |
| pKa | |
| pp cited in Vols.1-5: |

(1) Watkins, J. Chem. Soc., 1952, 3059.
(2) Short et al., U.S. Pat. 2 662 082 (1953).

HOMOCHLORCYCLIZINE [INN]

| CAS | 848-53-3 | MW: | 314.86 | MF: | C19 H23 Cl N2 |

Serotonin antagonist

Abbott

| LgP | 4.65(C) |
| pKa | |
| pp cited in Vols.1-5: |

(1) A. Weston et al., U.S. Pat. 2 655 498 (1953).
(2) eidem, J. Am. Chem. Soc., 1954, 76, 5805.

HOMOFENAZINE [INN]

| CAS | 3833-99-6 | MW: | 451.56 | MF: | C23 H28 F3 N3 O S |

Sedative

Degussa, FRG

| LgP | 5.80(C) |
| pKa | |
| pp cited in Vols.1-5: |

(1) Schuler et al., U.S. Pat. 3 040 043 (1962).

HOMOHARRINGTONINE [NCI]

| CAS | 26833-87-4 | MW: | 545.64 | MF: | C29 H39 N O9 |

Antineoplastic

NSC 141633

| LgP | 0.94(C); 0.82(M) |
| pKa | |
| pp cited in Vols.1-5: |

(1) RTECS, 1985-86, entry 25413.

HOMOPIPRAMOL [INN]

| CAS | 35142-68-8 | MW: | 377.53 | MF: | C24 H31 N3 O |

| LgP | 4.53(C) |
| pKa | |
| pp cited in Vols.1-5: |

HOMOSALATE [U;INN]

| CAS | 118-56-9 | MW: | 262.35 | MF: | C16 H22 O3 |

Ultraviolet screen

| LgP | 6.20(C) |
| pKa | |
| pp cited in Vols.1-5: |

(1) Stockelbach, U.S. Pat. 2 369 084 (1945).

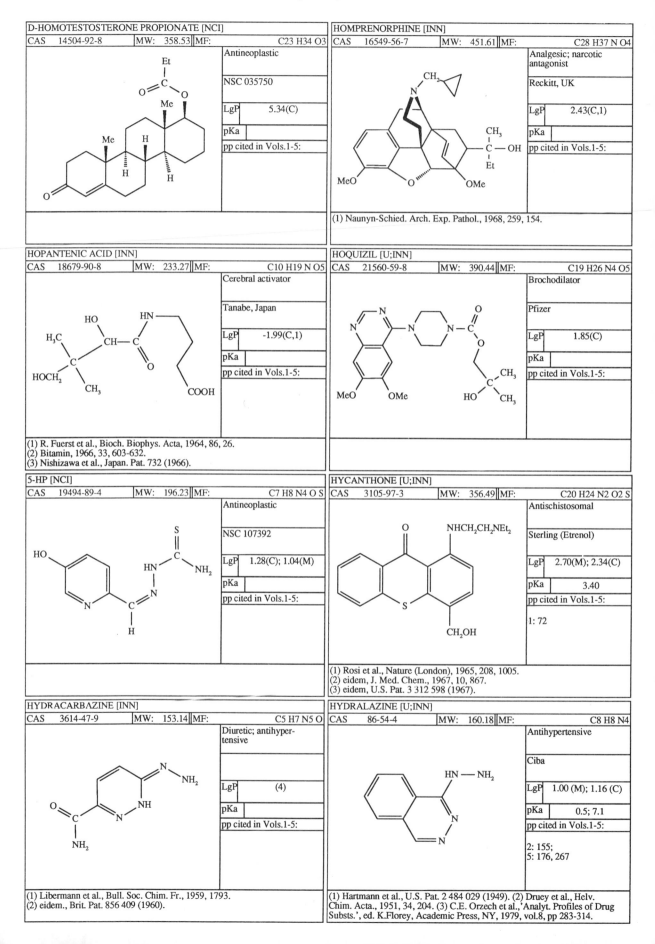

D-HOMOTESTOSTERONE PROPIONATE [NCI]

CAS	14504-92-8	MW:	358.53	MF:	C23 H34 O3

Antineoplastic

NSC 035750

LgP	5.34(C)
pKa	
pp cited in Vols.1-5:	

HOMPRENORPHINE [INN]

CAS	16549-56-7	MW:	451.61	MF:	C28 H37 N O4

Analgesic; narcotic antagonist

Reckitt, UK

LgP	2.43(C,1)
pKa	
pp cited in Vols.1-5:	

(1) Naunyn-Schied. Arch. Exp. Pathol., 1968, 259, 154.

HOPANTENIC ACID [INN]

CAS	18679-90-8	MW:	233.27	MF:	C10 H19 N O5

Cerebral activator

Tanabe, Japan

LgP	-1.99(C,1)
pKa	
pp cited in Vols.1-5:	

(1) R. Fuerst et al., Bioch. Biophys. Acta, 1964, 86, 26.
(2) Bitamin, 1966, 33, 603-632.
(3) Nishizawa et al., Japan. Pat. 732 (1966).

HOQUIZIL [U;INN]

CAS	21560-59-8	MW:	390.44	MF:	C19 H26 N4 O5

Brochodilator

Pfizer

LgP	1.85(C)
pKa	
pp cited in Vols.1-5:	

5-HP [NCI]

CAS	19494-89-4	MW:	196.23	MF:	C7 H8 N4 O S

Antineoplastic

NSC 107392

LgP	1.28(C); 1.04(M)
pKa	
pp cited in Vols.1-5:	

HYCANTHONE [U;INN]

CAS	3105-97-3	MW:	356.49	MF:	C20 H24 N2 O2 S

Antischistosomal

Sterling (Etrenol)

LgP	2.70(M); 2.34(C)
pKa	3.40
pp cited in Vols.1-5:	

1: 72

(1) Rosi et al., Nature (London), 1965, 208, 1005.
(2) eidem, J. Med. Chem., 1967, 10, 867.
(3) eidem, U.S. Pat. 3 312 598 (1967).

HYDRACARBAZINE [INN]

CAS	3614-47-9	MW:	153.14	MF:	C5 H7 N5 O

Diuretic; antihypertensive

LgP	(4)
pKa	
pp cited in Vols.1-5:	

(1) Libermann et al., Bull. Soc. Chim. Fr., 1959, 1793.
(2) eidem., Brit. Pat. 856 409 (1960).

HYDRALAZINE [U;INN]

CAS	86-54-4	MW:	160.18	MF:	C8 H8 N4

Antihypertensive

Ciba

LgP	1.00 (M); 1.16 (C)
pKa	0.5; 7.1
pp cited in Vols.1-5:	

2: 155;
5: 176, 267

(1) Hartmann et al., U.S. Pat. 2 484 029 (1949). (2) Druey et al., Helv. Chim. Acta., 1951, 34, 204. (3) C.E. Orzech et al.,'Analyt. Profiles of Drug Substs.', ed. K.Florey, Academic Press, NY, 1979, vol.8, pp 283-314.

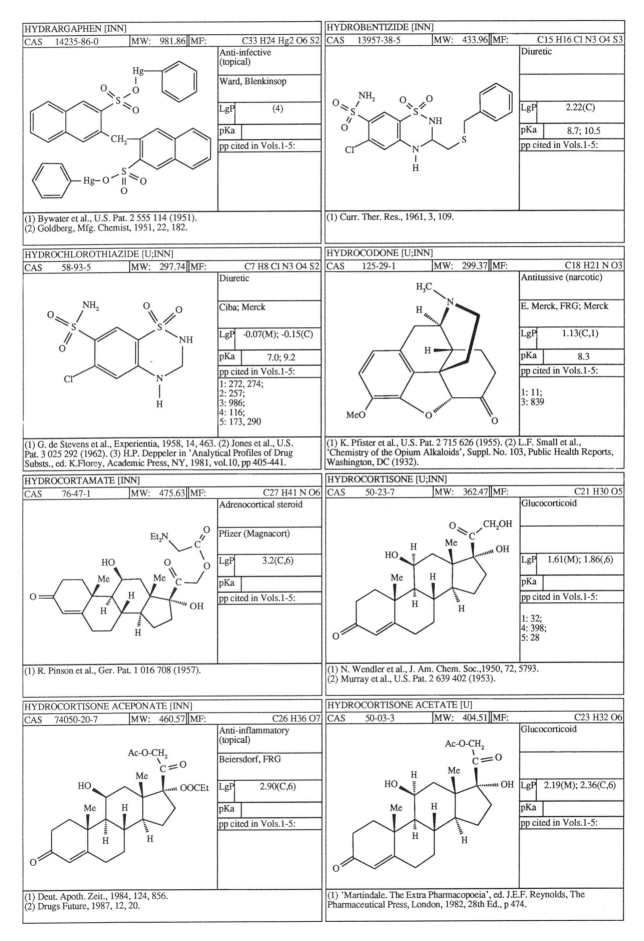

HYDRARGAPHEN [INN]		
CAS 14235-86-0	MW: 981.86	MF: C33 H24 Hg2 O6 S2

Anti-infective (topical)

Ward, Blenkinsop

LgP	(4)
pKa	
pp cited in Vols.1-5:	

(1) Bywater et al., U.S. Pat. 2 555 114 (1951).
(2) Goldberg, Mfg. Chemist, 1951, 22, 182.

HYDROBENTIZIDE [INN]		
CAS 13957-38-5	MW: 433.96	MF: C15 H16 Cl N3 O4 S3

Diuretic

LgP	2.22(C)
pKa	8.7; 10.5
pp cited in Vols.1-5:	

(1) Curr. Ther. Res., 1961, 3, 109.

HYDROCHLOROTHIAZIDE [U;INN]		
CAS 58-93-5	MW: 297.74	MF: C7 H8 Cl N3 O4 S2

Diuretic

Ciba; Merck

LgP	-0.07(M); -0.15(C)
pKa	7.0; 9.2
pp cited in Vols.1-5:	

1: 272, 274;
2: 257;
3: 986;
4: 116;
5: 173, 290

(1) G. de Stevens et al., Experientia, 1958, 14, 463. (2) Jones et al., U.S. Pat. 3 025 292 (1962). (3) H.P. Deppeler in 'Analytical Profiles of Drug Substs., ed. K.Florey, Academic Press, NY, 1981, vol.10, pp 405-441.

HYDROCODONE [U;INN]		
CAS 125-29-1	MW: 299.37	MF: C18 H21 N O3

Antitussive (narcotic)

E. Merck, FRG; Merck

LgP	1.13(C,1)
pKa	8.3
pp cited in Vols.1-5:	

1: 11;
3: 839

(1) K. Pfister et al., U.S. Pat. 2 715 626 (1955). (2) L.F. Small et al., 'Chemistry of the Opium Alkaloids', Suppl. No. 103, Public Health Reports, Washington, DC (1932).

HYDROCORTAMATE [INN]		
CAS 76-47-1	MW: 475.63	MF: C27 H41 N O6

Adrenocortical steroid

Pfizer (Magnacort)

LgP	3.2(C,6)
pKa	
pp cited in Vols.1-5:	

(1) R. Pinson et al., Ger. Pat. 1 016 708 (1957).

HYDROCORTISONE [U;INN]		
CAS 50-23-7	MW: 362.47	MF: C21 H30 O5

Glucocorticoid

LgP	1.61(M); 1.86(,6)
pKa	
pp cited in Vols.1-5:	

1: 32;
4: 398;
5: 28

(1) N. Wendler et al., J. Am. Chem. Soc.,1950, 72, 5793.
(2) Murray et al., U.S. Pat. 2 639 402 (1953).

HYDROCORTISONE ACEPONATE [INN]		
CAS 74050-20-7	MW: 460.57	MF: C26 H36 O7

Anti-inflammatory (topical)

Beiersdorf, FRG

LgP	2.90(C,6)
pKa	
pp cited in Vols.1-5:	

(1) Deut. Apoth. Zeit., 1984, 124, 856.
(2) Drugs Future, 1987, 12, 20.

HYDROCORTISONE ACETATE [U]		
CAS 50-03-3	MW: 404.51	MF: C23 H32 O6

Glucocorticoid

LgP	2.19(M); 2.36(C,6)
pKa	
pp cited in Vols.1-5:	

(1) 'Martindale. The Extra Pharmacopoeia', ed. J.E.F. Reynolds, The Pharmaceutical Press, London, 1982, 28th Ed., p 474.

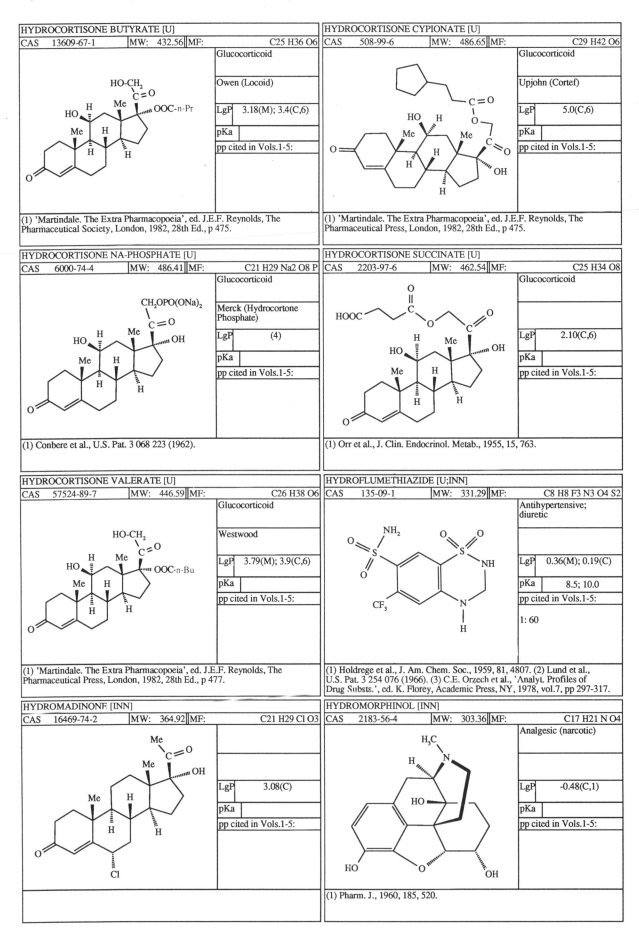

HYDROCORTISONE BUTYRATE [U]			
CAS　13609-67-1	MW:　432.56	MF:	C25 H36 O6

Glucocorticoid

Owen (Locoid)

LgP	3.18(M); 3.4(C,6)
pKa	
pp cited in Vols.1-5:	

(1) 'Martindale. The Extra Pharmacopoeia', ed. J.E.F. Reynolds, The Pharmaceutical Society, London, 1982, 28th Ed., p 475.

HYDROCORTISONE CYPIONATE [U]			
CAS　508-99-6	MW:　486.65	MF:	C29 H42 O6

Glucocorticoid

Upjohn (Cortef)

LgP	5.0(C,6)
pKa	
pp cited in Vols.1-5:	

(1) 'Martindale. The Extra Pharmacopoeia', ed. J.E.F. Reynolds, The Pharmaceutical Press, London, 1982, 28th Ed., p 475.

HYDROCORTISONE NA-PHOSPHATE [U]			
CAS　6000-74-4	MW:　486.41	MF:	C21 H29 Na2 O8 P

Glucocorticoid

Merck (Hydrocortone Phosphate)

LgP	(4)
pKa	
pp cited in Vols.1-5:	

(1) Conbere et al., U.S. Pat. 3 068 223 (1962).

HYDROCORTISONE SUCCINATE [U]			
CAS　2203-97-6	MW:　462.54	MF:	C25 H34 O8

Glucocorticoid

LgP	2.10(C,6)
pKa	
pp cited in Vols.1-5:	

(1) Orr et al., J. Clin. Endocrinol. Metab., 1955, 15, 763.

HYDROCORTISONE VALERATE [U]			
CAS　57524-89-7	MW:　446.59	MF:	C26 H38 O6

Glucocorticoid

Westwood

LgP	3.79(M); 3.9(C,6)
pKa	
pp cited in Vols.1-5:	

(1) 'Martindale. The Extra Pharmacopoeia', ed. J.E.F. Reynolds, The Pharmaceutical Press, London, 1982, 28th Ed., p 477.

HYDROFLUMETHIAZIDE [U;INN]			
CAS　135-09-1	MW:　331.29	MF:	C8 H8 F3 N3 O4 S2

Antihypertensive; diuretic

LgP	0.36(M); 0.19(C)
pKa	8.5; 10.0
pp cited in Vols.1-5:	
	1: 60

(1) Holdrege et al., J. Am. Chem. Soc., 1959, 81, 4807. (2) Lund et al., U.S. Pat. 3 254 076 (1966). (3) C.E. Orzech et al., 'Analyt. Profiles of Drug Substs.', ed. K. Florey, Academic Press, NY, 1978, vol.7, pp 297-317.

HYDROMADINONE [INN]			
CAS　16469-74-2	MW:　364.92	MF:	C21 H29 Cl O3

LgP	3.08(C)
pKa	
pp cited in Vols.1-5:	

HYDROMORPHINOL [INN]			
CAS　2183-56-4	MW:　303.36	MF:	C17 H21 N O4

Analgesic (narcotic)

LgP	-0.48(C,1)
pKa	
pp cited in Vols.1-5:	

(1) Pharm. J., 1960, 185, 520.

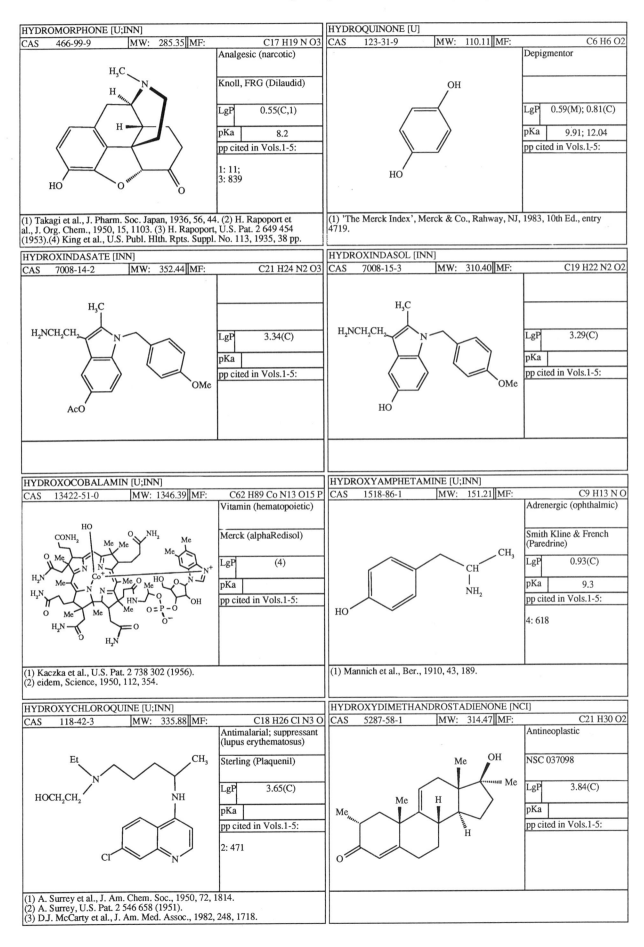

HYDROMORPHONE [U;INN]

CAS	466-99-9	MW: 285.35	MF:	C17 H19 N O3

Analgesic (narcotic)

Knoll, FRG (Dilaudid)

LgP	0.55(C,1)
pKa	8.2
pp cited in Vols.1-5:	1: 11; 3: 839

(1) Takagi et al., J. Pharm. Soc. Japan, 1936, 56, 44. (2) H. Rapoport et al., J. Org. Chem., 1950, 15, 1103. (3) H. Rapoport, U.S. Pat. 2 649 454 (1953).(4) King et al., U.S. Publ. Hlth. Rpts. Suppl. No. 113, 1935, 38 pp.

HYDROQUINONE [U]

CAS	123-31-9	MW: 110.11	MF:	C6 H6 O2

Depigmentor

LgP	0.59(M); 0.81(C)
pKa	9.91; 12.04
pp cited in Vols.1-5:	

(1) 'The Merck Index', Merck & Co., Rahway, NJ, 1983, 10th Ed., entry 4719.

HYDROXINDASATE [INN]

CAS	7008-14-2	MW: 352.44	MF:	C21 H24 N2 O3

LgP	3.34(C)
pKa	
pp cited in Vols.1-5:	

HYDROXINDASOL [INN]

CAS	7008-15-3	MW: 310.40	MF:	C19 H22 N2 O2

LgP	3.29(C)
pKa	
pp cited in Vols.1-5:	

HYDROXOCOBALAMIN [U;INN]

CAS	13422-51-0	MW: 1346.39	MF:	C62 H89 Co N13 O15 P

Vitamin (hematopoietic)

Merck (alphaRedisol)

LgP	(4)
pKa	
pp cited in Vols.1-5:	

(1) Kaczka et al., U.S. Pat. 2 738 302 (1956).
(2) eidem, Science, 1950, 112, 354.

HYDROXYAMPHETAMINE [U;INN]

CAS	1518-86-1	MW: 151.21	MF:	C9 H13 N O

Adrenergic (ophthalmic)

Smith Kline & French (Paredrine)

LgP	0.93(C)
pKa	9.3
pp cited in Vols.1-5:	4: 618

(1) Mannich et al., Ber., 1910, 43, 189.

HYDROXYCHLOROQUINE [U;INN]

CAS	118-42-3	MW: 335.88	MF:	C18 H26 Cl N3 O

Antimalarial; suppressant (lupus erythematosus)

Sterling (Plaquenil)

LgP	3.65(C)
pKa	
pp cited in Vols.1-5:	2: 471

(1) A. Surrey et al., J. Am. Chem. Soc., 1950, 72, 1814.
(2) A. Surrey, U.S. Pat. 2 546 658 (1951).
(3) D.J. McCarty et al., J. Am. Med. Assoc., 1982, 248, 1718.

HYDROXYDIMETHANDROSTADIENONE [NCI]

CAS	5287-58-1	MW: 314.47	MF:	C21 H30 O2

Antineoplastic

NSC 037098

LgP	3.84(C)
pKa	
pp cited in Vols.1-5:	

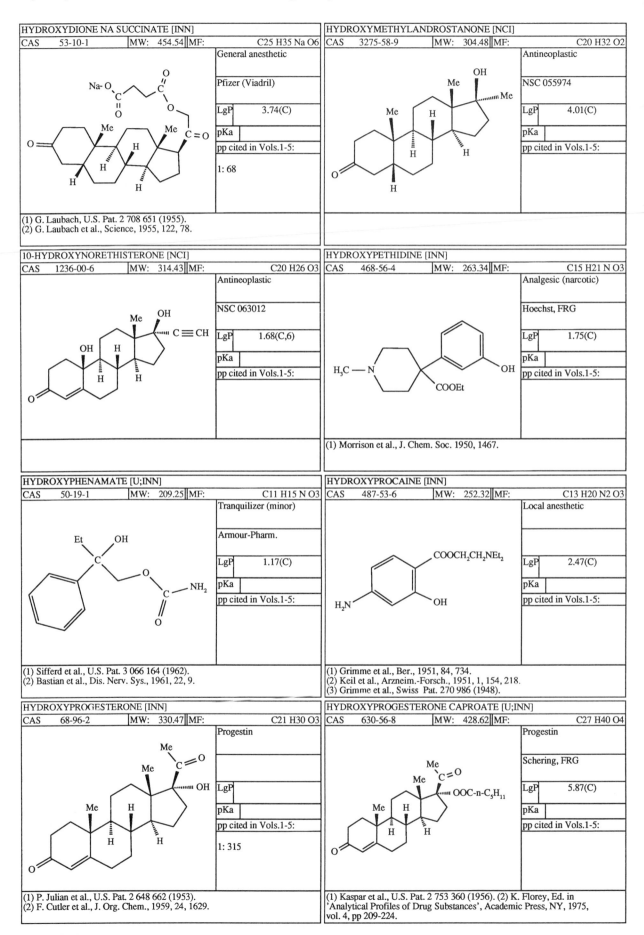

HYDROXYDIONE NA SUCCINATE [INN]

CAS	53-10-1	MW:	454.54	MF:	C25 H35 Na O6

General anesthetic

Pfizer (Viadril)

LgP	3.74(C)
pKa	

pp cited in Vols.1-5:

1: 68

(1) G. Laubach, U.S. Pat. 2 708 651 (1955).
(2) G. Laubach et al., Science, 1955, 122, 78.

HYDROXYMETHYLANDROSTANONE [NCI]

CAS	3275-58-9	MW:	304.48	MF:	C20 H32 O2

Antineoplastic

NSC 055974

LgP	4.01(C)
pKa	

pp cited in Vols.1-5:

10-HYDROXYNORETHISTERONE [NCI]

CAS	1236-00-6	MW:	314.43	MF:	C20 H26 O3

Antineoplastic

NSC 063012

LgP	1.68(C,6)
pKa	

pp cited in Vols.1-5:

HYDROXYPETHIDINE [INN]

CAS	468-56-4	MW:	263.34	MF:	C15 H21 N O3

Analgesic (narcotic)

Hoechst, FRG

LgP	1.75(C)
pKa	

pp cited in Vols.1-5:

(1) Morrison et al., J. Chem. Soc. 1950, 1467.

HYDROXYPHENAMATE [U;INN]

CAS	50-19-1	MW:	209.25	MF:	C11 H15 N O3

Tranquilizer (minor)

Armour-Pharm.

LgP	1.17(C)
pKa	

pp cited in Vols.1-5:

(1) Sifferd et al., U.S. Pat. 3 066 164 (1962).
(2) Bastian et al., Dis. Nerv. Sys., 1961, 22, 9.

HYDROXYPROCAINE [INN]

CAS	487-53-6	MW:	252.32	MF:	C13 H20 N2 O3

Local anesthetic

LgP	2.47(C)
pKa	

pp cited in Vols.1-5:

(1) Grimme et al., Ber., 1951, 84, 734.
(2) Keil et al., Arzneim.-Forsch., 1951, 1, 154, 218.
(3) Grimme et al., Swiss Pat. 270 986 (1948).

HYDROXYPROGESTERONE [INN]

CAS	68-96-2	MW:	330.47	MF:	C21 H30 O3

Progestin

LgP	
pKa	

pp cited in Vols.1-5:

1: 315

(1) P. Julian et al., U.S. Pat. 2 648 662 (1953).
(2) F. Cutler et al., J. Org. Chem., 1959, 24, 1629.

HYDROXYPROGESTERONE CAPROATE [U;INN]

CAS	630-56-8	MW:	428.62	MF:	C27 H40 O4

Progestin

Schering, FRG

LgP	5.87(C)
pKa	

pp cited in Vols.1-5:

(1) Kaspar et al., U.S. Pat. 2 753 360 (1956). (2) K. Florey, Ed. in
'Analytical Profiles of Drug Substances', Academic Press, NY, 1975,
vol. 4, pp 209-224.

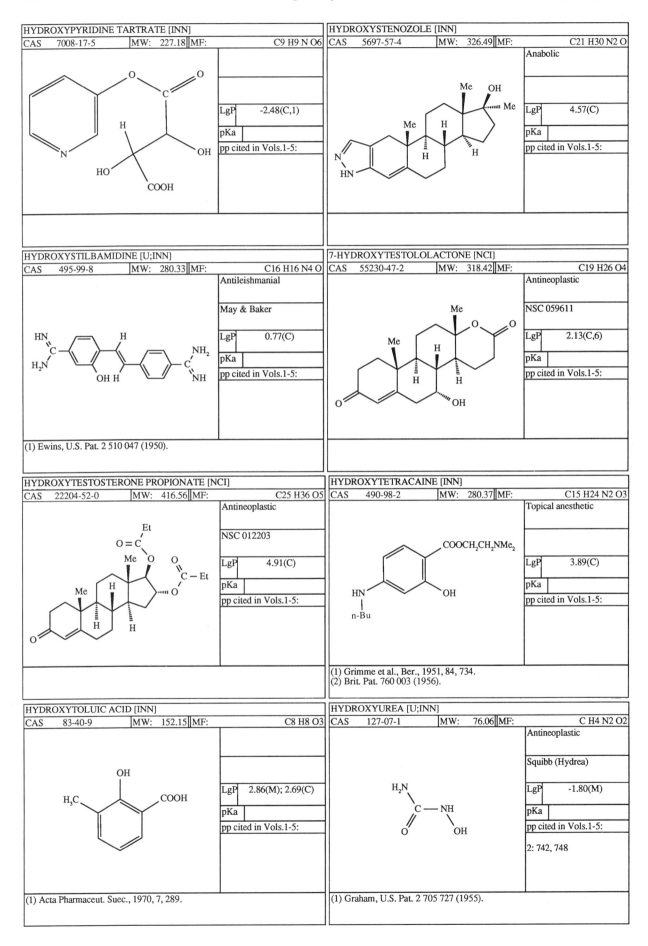

HYDROXYPYRIDINE TARTRATE [INN]

CAS	7008-17-5	MW: 227.18	MF:	C9 H9 N O6

LgP	-2.48(C,1)
pKa	
pp cited in Vols.1-5:	

HYDROXYSTENOZOLE [INN]

CAS	5697-57-4	MW: 326.49	MF:	C21 H30 N2 O

Anabolic

LgP	4.57(C)
pKa	
pp cited in Vols.1-5:	

HYDROXYSTILBAMIDINE [U;INN]

CAS	495-99-8	MW: 280.33	MF:	C16 H16 N4 O

Antileishmanial

May & Baker

LgP	0.77(C)
pKa	
pp cited in Vols.1-5:	

(1) Ewins, U.S. Pat. 2 510 047 (1950).

7-HYDROXYTESTOLOLACTONE [NCI]

CAS	55230-47-2	MW: 318.42	MF:	C19 H26 O4

Antineoplastic

NSC 059611

LgP	2.13(C,6)
pKa	
pp cited in Vols.1-5:	

HYDROXYTESTOSTERONE PROPIONATE [NCI]

CAS	22204-52-0	MW: 416.56	MF:	C25 H36 O5

Antineoplastic

NSC 012203

LgP	4.91(C)
pKa	
pp cited in Vols.1-5:	

HYDROXYTETRACAINE [INN]

CAS	490-98-2	MW: 280.37	MF:	C15 H24 N2 O3

Topical anesthetic

LgP	3.89(C)
pKa	
pp cited in Vols.1-5:	

(1) Grimme et al., Ber., 1951, 84, 734.
(2) Brit. Pat. 760 003 (1956).

HYDROXYTOLUIC ACID [INN]

CAS	83-40-9	MW: 152.15	MF:	C8 H8 O3

LgP	2.86(M); 2.69(C)
pKa	
pp cited in Vols.1-5:	

(1) Acta Pharmaceut. Suec., 1970, 7, 289.

HYDROXYUREA [U;INN]

CAS	127-07-1	MW: 76.06	MF:	C H4 N2 O2

Antineoplastic

Squibb (Hydrea)

LgP	-1.80(M)
pKa	
pp cited in Vols.1-5:	

2: 742, 748

(1) Graham, U.S. Pat. 2 705 727 (1955).

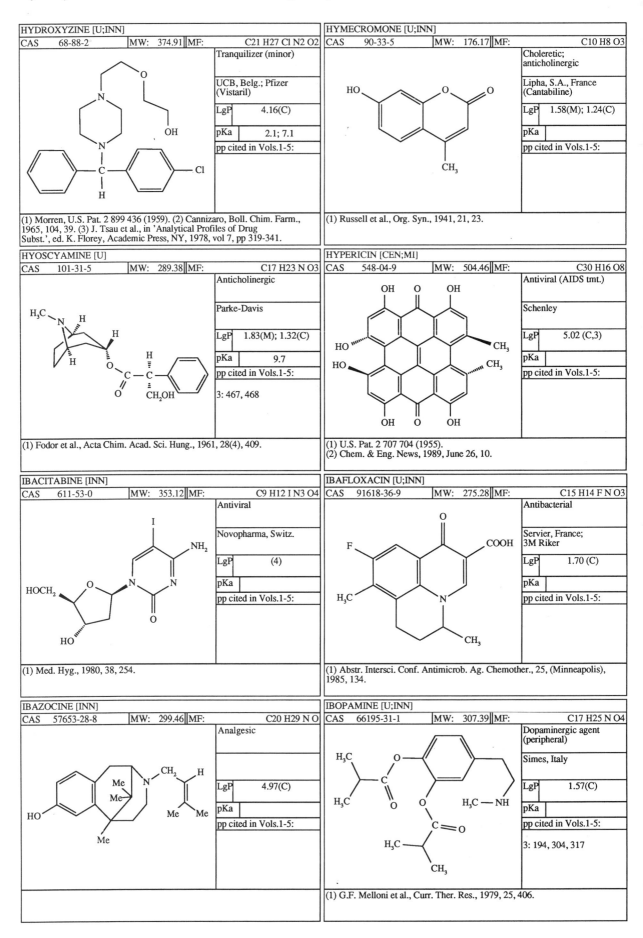

HYDROXYZINE [U;INN]		
CAS 68-88-2	MW: 374.91	MF: C21 H27 Cl N2 O2

Tranquilizer (minor)

UCB, Belg.; Pfizer (Vistaril)

LgP	4.16(C)
pKa	2.1; 7.1
pp cited in Vols.1-5:	

(1) Morren, U.S. Pat. 2 899 436 (1959). (2) Cannizaro, Boll. Chim. Farm., 1965, 104, 39. (3) J. Tsau et al., in 'Analytical Profiles of Drug Subst.', ed. K. Florey, Academic Press, NY, 1978, vol 7, pp 319-341.

HYMECROMONE [U;INN]		
CAS 90-33-5	MW: 176.17	MF: C10 H8 O3

Choleretic; anticholinergic

Lipha, S.A., France (Cantabiline)

LgP	1.58(M); 1.24(C)
pKa	
pp cited in Vols.1-5:	

(1) Russell et al., Org. Syn., 1941, 21, 23.

HYOSCYAMINE [U]		
CAS 101-31-5	MW: 289.38	MF: C17 H23 N O3

Anticholinergic

Parke-Davis

LgP	1.83(M); 1.32(C)
pKa	9.7
pp cited in Vols.1-5:	
3: 467, 468	

(1) Fodor et al., Acta Chim. Acad. Sci. Hung., 1961, 28(4), 409.

HYPERICIN [CEN;MI]		
CAS 548-04-9	MW: 504.46	MF: C30 H16 O8

Antiviral (AIDS tmt.)

Schenley

LgP	5.02 (C,3)
pKa	
pp cited in Vols.1-5:	

(1) U.S. Pat. 2 707 704 (1955).
(2) Chem. & Eng. News, 1989, June 26, 10.

IBACITABINE [INN]		
CAS 611-53-0	MW: 353.12	MF: C9 H12 I N3 O4

Antiviral

Novopharma, Switz.

LgP	(4)
pKa	
pp cited in Vols.1-5:	

(1) Med. Hyg., 1980, 38, 254.

IBAFLOXACIN [U;INN]		
CAS 91618-36-9	MW: 275.28	MF: C15 H14 F N O3

Antibacterial

Servier, France; 3M Riker

LgP	1.70 (C)
pKa	
pp cited in Vols.1-5:	

(1) Abstr. Intersci. Conf. Antimicrob. Ag. Chemother., 25, (Minneapolis), 1985, 134.

IBAZOCINE [INN]		
CAS 57653-28-8	MW: 299.46	MF: C20 H29 N O

Analgesic

LgP	4.97(C)
pKa	
pp cited in Vols.1-5:	

IBOPAMINE [U;INN]		
CAS 66195-31-1	MW: 307.39	MF: C17 H25 N O4

Dopaminergic agent (peripheral)

Simes, Italy

LgP	1.57(C)
pKa	
pp cited in Vols.1-5:	
3: 194, 304, 317	

(1) G.F. Melloni et al., Curr. Ther. Res., 1979, 25, 406.

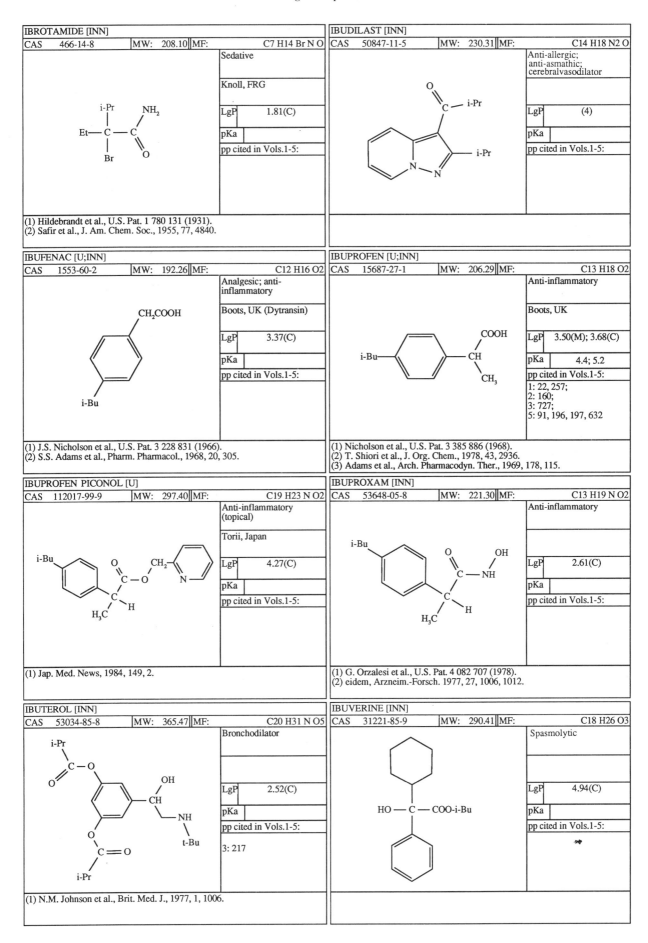

IBROTAMIDE [INN]

CAS	466-14-8	MW:	208.10	MF:	C7 H14 Br N O

Sedative

Knoll, FRG

LgP	1.81(C)
pKa	
pp cited in Vols.1-5:	

(1) Hildebrandt et al., U.S. Pat. 1 780 131 (1931).
(2) Safir et al., J. Am. Chem. Soc., 1955, 77, 4840.

IBUDILAST [INN]

CAS	50847-11-5	MW:	230.31	MF:	C14 H18 N2 O

Anti-allergic;
anti-asmathic;
cerebralvasodilator

LgP	(4)
pKa	
pp cited in Vols.1-5:	

IBUFENAC [U;INN]

CAS	1553-60-2	MW:	192.26	MF:	C12 H16 O2

Analgesic; anti-
inflammatory

Boots, UK (Dytransin)

LgP	3.37(C)
pKa	
pp cited in Vols.1-5:	

(1) J.S. Nicholson et al., U.S. Pat. 3 228 831 (1966).
(2) S.S. Adams et al., Pharm. Pharmacol., 1968, 20, 305.

IBUPROFEN [U;INN]

CAS	15687-27-1	MW:	206.29	MF:	C13 H18 O2

Anti-inflammatory

Boots, UK

LgP	3.50(M); 3.68(C)
pKa	4.4; 5.2
pp cited in Vols.1-5:	

1: 22, 257;
2: 160;
3: 727;
5: 91, 196, 197, 632

(1) Nicholson et al., U.S. Pat. 3 385 886 (1968).
(2) T. Shiori et al., J. Org. Chem., 1978, 43, 2936.
(3) Adams et al., Arch. Pharmacodyn. Ther., 1969, 178, 115.

IBUPROFEN PICONOL [U]

CAS	112017-99-9	MW:	297.40	MF:	C19 H23 N O2

Anti-inflammatory
(topical)

Torii, Japan

LgP	4.27(C)
pKa	
pp cited in Vols.1-5:	

(1) Jap. Med. News, 1984, 149, 2.

IBUPROXAM [INN]

CAS	53648-05-8	MW:	221.30	MF:	C13 H19 N O2

Anti-inflammatory

LgP	2.61(C)
pKa	
pp cited in Vols.1-5:	

(1) G. Orzalesi et al., U.S. Pat. 4 082 707 (1978).
(2) eidem, Arzneim.-Forsch. 1977, 27, 1006, 1012.

IBUTEROL [INN]

CAS	53034-85-8	MW:	365.47	MF:	C20 H31 N O5

Bronchodilator

LgP	2.52(C)
pKa	
pp cited in Vols.1-5:	

3: 217

(1) N.M. Johnson et al., Brit. Med. J., 1977, 1, 1006.

IBUVERINE [INN]

CAS	31221-85-9	MW:	290.41	MF:	C18 H26 O3

Spasmolytic

LgP	4.94(C)
pKa	
pp cited in Vols.1-5:	

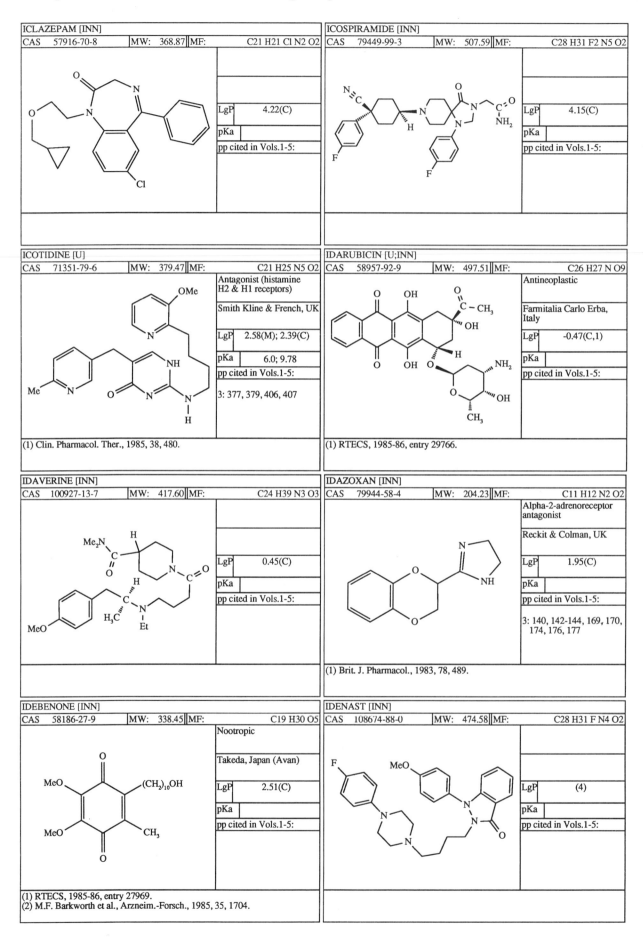

ICLAZEPAM [INN]

CAS	57916-70-8	MW:	368.87	MF:	C21 H21 Cl N2 O2

LgP	4.22(C)
pKa	
pp cited in Vols.1-5:	

ICOSPIRAMIDE [INN]

CAS	79449-99-3	MW:	507.59	MF:	C28 H31 F2 N5 O2

LgP	4.15(C)
pKa	
pp cited in Vols.1-5:	

ICOTIDINE [U]

CAS	71351-79-6	MW:	379.47	MF:	C21 H25 N5 O2

Antagonist (histamine H2 & H1 receptors)

Smith Kline & French, UK

LgP	2.58(M); 2.39(C)
pKa	6.0; 9.78
pp cited in Vols.1-5:	

3: 377, 379, 406, 407

(1) Clin. Pharmacol. Ther., 1985, 38, 480.

IDARUBICIN [U;INN]

CAS	58957-92-9	MW:	497.51	MF:	C26 H27 N O9

Antineoplastic

Farmitalia Carlo Erba, Italy

LgP	-0.47(C,1)
pKa	
pp cited in Vols.1-5:	

(1) RTECS, 1985-86, entry 29766.

IDAVERINE [INN]

CAS	100927-13-7	MW:	417.60	MF:	C24 H39 N3 O3

LgP	0.45(C)
pKa	
pp cited in Vols.1-5:	

IDAZOXAN [INN]

CAS	79944-58-4	MW:	204.23	MF:	C11 H12 N2 O2

Alpha-2-adrenoreceptor antagonist

Reckit & Colman, UK

LgP	1.95(C)
pKa	
pp cited in Vols.1-5:	

3: 140, 142-144, 169, 170, 174, 176, 177

(1) Brit. J. Pharmacol., 1983, 78, 489.

IDEBENONE [INN]

CAS	58186-27-9	MW:	338.45	MF:	C19 H30 O5

Nootropic

Takeda, Japan (Avan)

LgP	2.51(C)
pKa	
pp cited in Vols.1-5:	

(1) RTECS, 1985-86, entry 27969.
(2) M.F. Barkworth et al., Arzneim.-Forsch., 1985, 35, 1704.

IDENAST [INN]

CAS	108674-88-0	MW:	474.58	MF:	C28 H31 F N4 O2

LgP	(4)
pKa	
pp cited in Vols.1-5:	

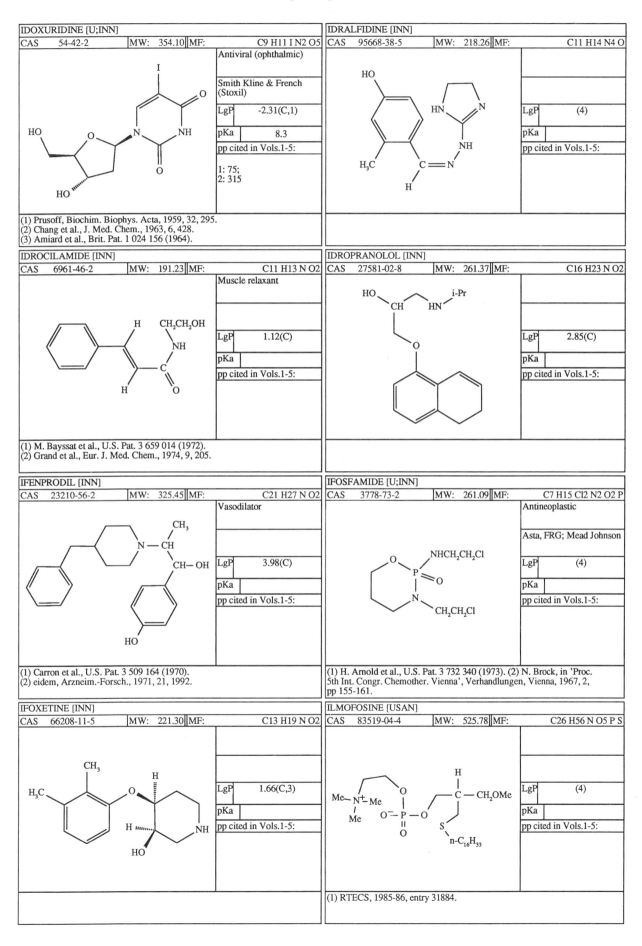

IDOXURIDINE [U;INN]

CAS	54-42-2	MW:	354.10	MF:	C9 H11 I N2 O5

Antiviral (ophthalmic)

Smith Kline & French (Stoxil)

LgP	-2.31(C,1)
pKa	8.3

pp cited in Vols.1-5:

1: 75;
2: 315

(1) Prusoff, Biochim. Biophys. Acta, 1959, 32, 295.
(2) Chang et al., J. Med. Chem., 1963, 6, 428.
(3) Amiard et al., Brit. Pat. 1 024 156 (1964).

IDRALFIDINE [INN]

CAS	95668-38-5	MW:	218.26	MF:	C11 H14 N4 O

LgP	(4)
pKa	

pp cited in Vols.1-5:

IDROCILAMIDE [INN]

CAS	6961-46-2	MW:	191.23	MF:	C11 H13 N O2

Muscle relaxant

LgP	1.12(C)
pKa	

pp cited in Vols.1-5:

(1) M. Bayssat et al., U.S. Pat. 3 659 014 (1972).
(2) Grand et al., Eur. J. Med. Chem., 1974, 9, 205.

IDROPRANOLOL [INN]

CAS	27581-02-8	MW:	261.37	MF:	C16 H23 N O2

LgP	2.85(C)
pKa	

pp cited in Vols.1-5:

IFENPRODIL [INN]

CAS	23210-56-2	MW:	325.45	MF:	C21 H27 N O2

Vasodilator

LgP	3.98(C)
pKa	

pp cited in Vols.1-5:

(1) Carron et al., U.S. Pat. 3 509 164 (1970).
(2) eidem, Arzneim.-Forsch., 1971, 21, 1992.

IFOSFAMIDE [U;INN]

CAS	3778-73-2	MW:	261.09	MF:	C7 H15 Cl2 N2 O2 P

Antineoplastic

Asta, FRG; Mead Johnson

LgP	(4)
pKa	

pp cited in Vols.1-5:

(1) H. Arnold et al., U.S. Pat. 3 732 340 (1973). (2) N. Brock, in 'Proc.
5th Int. Congr. Chemother. Vienna', Verhandlungen, Vienna, 1967, 2,
pp 155-161.

IFOXETINE [INN]

CAS	66208-11-5	MW:	221.30	MF:	C13 H19 N O2

LgP	1.66(C,3)
pKa	

pp cited in Vols.1-5:

ILMOFOSINE [USAN]

CAS	83519-04-4	MW:	525.78	MF:	C26 H56 N O5 P S

LgP	(4)
pKa	

pp cited in Vols.1-5:

(1) RTECS, 1985-86, entry 31884.

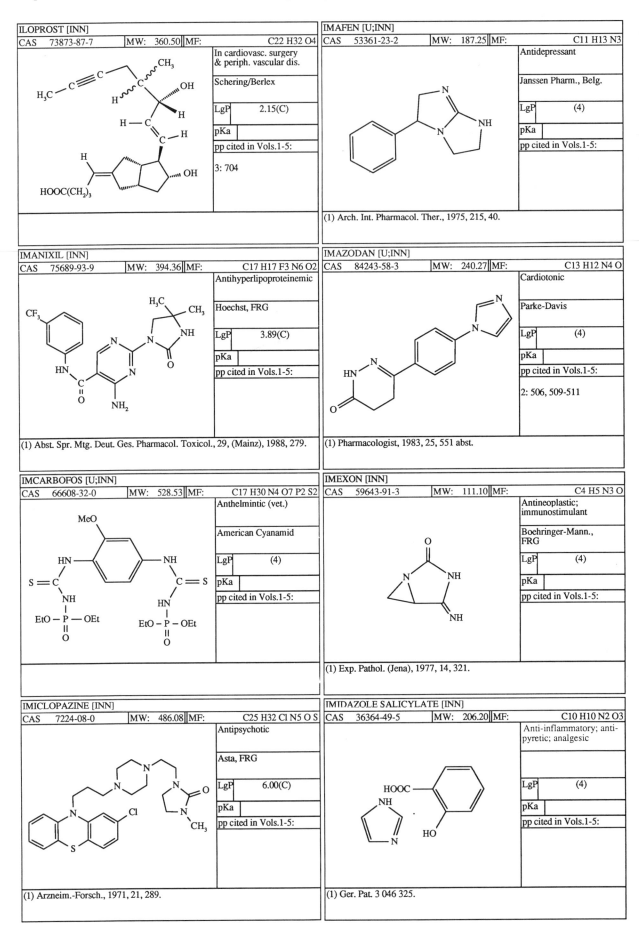

ILOPROST [INN]
CAS 73873-87-7   MW: 360.50 MF: C22 H32 O4
In cardiovasc. surgery & periph. vascular dis.
Schering/Berlex
LgP 2.15(C)
pKa
pp cited in Vols.1-5:
3: 704

IMAFEN [U;INN]
CAS 53361-23-2   MW: 187.25 MF: C11 H13 N3
Antidepressant
Janssen Pharm., Belg.
LgP (4)
pKa
pp cited in Vols.1-5:
(1) Arch. Int. Pharmacol. Ther., 1975, 215, 40.

IMANIXIL [INN]
CAS 75689-93-9   MW: 394.36 MF: C17 H17 F3 N6 O2
Antihyperlipoproteinemic
Hoechst, FRG
LgP 3.89(C)
pKa
pp cited in Vols.1-5:
(1) Abst. Spr. Mtg. Deut. Ges. Pharmacol. Toxicol., 29, (Mainz), 1988, 279.

IMAZODAN [U;INN]
CAS 84243-58-3   MW: 240.27 MF: C13 H12 N4 O
Cardiotonic
Parke-Davis
LgP (4)
pKa
pp cited in Vols.1-5:
2: 506, 509-511
(1) Pharmacologist, 1983, 25, 551 abst.

IMCARBOFOS [U;INN]
CAS 66608-32-0   MW: 528.53 MF: C17 H30 N4 O7 P2 S2
Anthelmintic (vet.)
American Cyanamid
LgP (4)
pKa
pp cited in Vols.1-5:

IMEXON [INN]
CAS 59643-91-3   MW: 111.10 MF: C4 H5 N3 O
Antineoplastic; immunostimulant
Boehringer-Mann., FRG
LgP (4)
pKa
pp cited in Vols.1-5:
(1) Exp. Pathol. (Jena), 1977, 14, 321.

IMICLOPAZINE [INN]
CAS 7224-08-0   MW: 486.08 MF: C25 H32 Cl N5 O S
Antipsychotic
Asta, FRG
LgP 6.00(C)
pKa
pp cited in Vols.1-5:
(1) Arzneim.-Forsch., 1971, 21, 289.

IMIDAZOLE SALICYLATE [INN]
CAS 36364-49-5   MW: 206.20 MF: C10 H10 N2 O3
Anti-inflammatory; anti-pyretic; analgesic
LgP (4)
pKa
pp cited in Vols.1-5:
(1) Ger. Pat. 3 046 325.

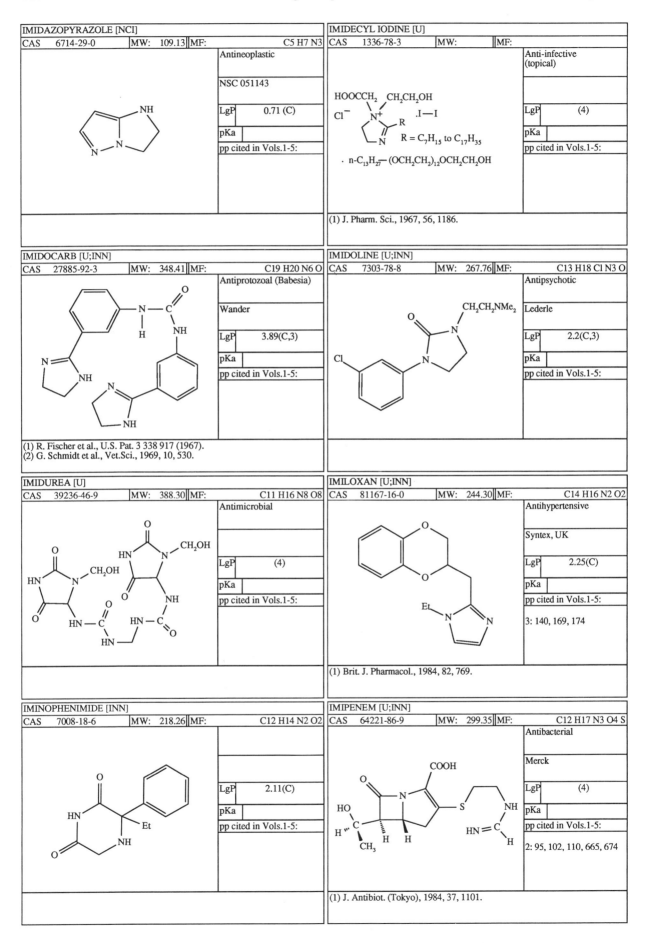

IMIDAZOPYRAZOLE [NCI]

CAS	6714-29-0	MW:	109.13	MF:	C5 H7 N3

Antineoplastic

NSC 051143

LgP	0.71 (C)
pKa	
pp cited in Vols.1-5:	

IMIDECYL IODINE [U]

CAS	1336-78-3	MW:		MF:	

Anti-infective (topical)

HOOCCH₂ CH₂CH₂OH
Cl⁻ .I—I
R = C₇H₁₅ to C₁₇H₃₅
· n-C₁₃H₂₇—(OCH₂CH₂)₁₂OCH₂CH₂OH

LgP	(4)
pKa	
pp cited in Vols.1-5:	

(1) J. Pharm. Sci., 1967, 56, 1186.

IMIDOCARB [U;INN]

CAS	27885-92-3	MW:	348.41	MF:	C19 H20 N6 O

Antiprotozoal (Babesia)

Wander

LgP	3.89(C,3)
pKa	
pp cited in Vols.1-5:	

(1) R. Fischer et al., U.S. Pat. 3 338 917 (1967).
(2) G. Schmidt et al., Vet.Sci., 1969, 10, 530.

IMIDOLINE [U;INN]

CAS	7303-78-8	MW:	267.76	MF:	C13 H18 Cl N3 O

Antipsychotic

Lederle

LgP	2.2(C,3)
pKa	
pp cited in Vols.1-5:	

IMIDUREA [U]

CAS	39236-46-9	MW:	388.30	MF:	C11 H16 N8 O8

Antimicrobial

LgP	(4)
pKa	
pp cited in Vols.1-5:	

IMILOXAN [U;INN]

CAS	81167-16-0	MW:	244.30	MF:	C14 H16 N2 O2

Antihypertensive

Syntex, UK

LgP	2.25(C)
pKa	
pp cited in Vols.1-5:	

3: 140, 169, 174

(1) Brit. J. Pharmacol., 1984, 82, 769.

IMINOPHENIMIDE [INN]

CAS	7008-18-6	MW:	218.26	MF:	C12 H14 N2 O2

LgP	2.11(C)
pKa	
pp cited in Vols.1-5:	

IMIPENEM [U;INN]

CAS	64221-86-9	MW:	299.35	MF:	C12 H17 N3 O4 S

Antibacterial

Merck

LgP	(4)
pKa	
pp cited in Vols.1-5:	

2: 95, 102, 110, 665, 674

(1) J. Antibiot. (Tokyo), 1984, 37, 1101.

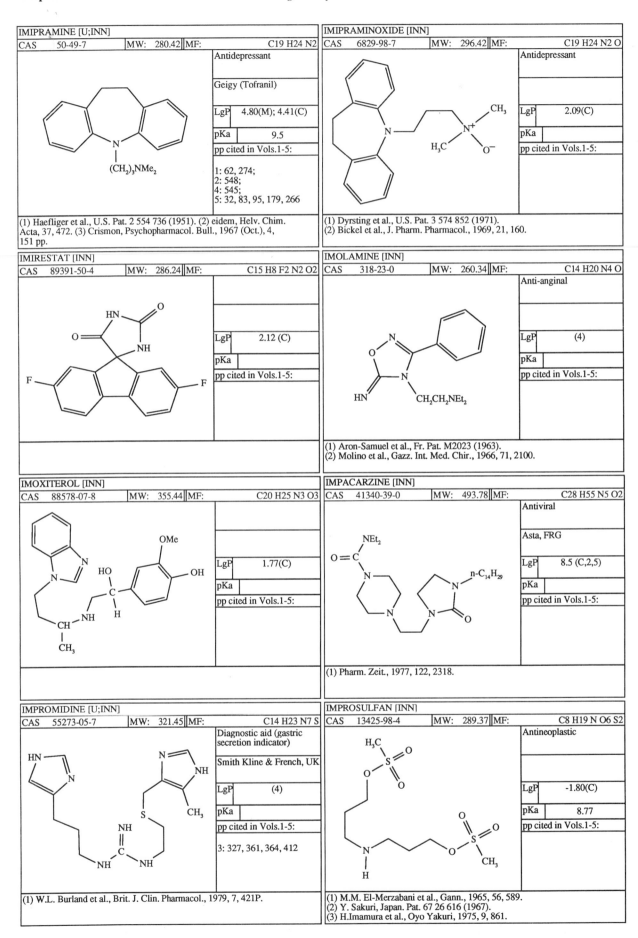

IMIPRAMINE [U;INN]

CAS 50-49-7	MW: 280.42	MF:	C19 H24 N2

Antidepressant

Geigy (Tofranil)

LgP	4.80(M); 4.41(C)
pKa	9.5

pp cited in Vols.1-5:

1: 62, 274;
2: 548;
4: 545;
5: 32, 83, 95, 179, 266

(1) Haefliger et al., U.S. Pat. 2 554 736 (1951). (2) eidem, Helv. Chim. Acta, 37, 472. (3) Crismon, Psychopharmacol. Bull., 1967 (Oct.), 4, 151 pp.

IMIPRAMINOXIDE [INN]

CAS 6829-98-7	MW: 296.42	MF:	C19 H24 N2 O

Antidepressant

LgP	2.09(C)
pKa	

pp cited in Vols.1-5:

(1) Dyrsting et al., U.S. Pat. 3 574 852 (1971).
(2) Bickel et al., J. Pharm. Pharmacol., 1969, 21, 160.

IMIRESTAT [INN]

CAS 89391-50-4	MW: 286.24	MF:	C15 H8 F2 N2 O2

LgP	2.12 (C)
pKa	

pp cited in Vols.1-5:

IMOLAMINE [INN]

CAS 318-23-0	MW: 260.34	MF:	C14 H20 N4 O

Anti-anginal

LgP	(4)
pKa	

pp cited in Vols.1-5:

(1) Aron-Samuel et al., Fr. Pat. M2023 (1963).
(2) Molino et al., Gazz. Int. Med. Chir., 1966, 71, 2100.

IMOXITEROL [INN]

CAS 88578-07-8	MW: 355.44	MF:	C20 H25 N3 O3

LgP	1.77(C)
pKa	

pp cited in Vols.1-5:

IMPACARZINE [INN]

CAS 41340-39-0	MW: 493.78	MF:	C28 H55 N5 O2

Antiviral

Asta, FRG

LgP	8.5 (C,2,5)
pKa	

pp cited in Vols.1-5:

(1) Pharm. Zeit., 1977, 122, 2318.

IMPROMIDINE [U;INN]

CAS 55273-05-7	MW: 321.45	MF:	C14 H23 N7 S

Diagnostic aid (gastric secretion indicator)

Smith Kline & French, UK

LgP	(4)
pKa	

pp cited in Vols.1-5:

3: 327, 361, 364, 412

(1) W.L. Burland et al., Brit. J. Clin. Pharmacol., 1979, 7, 421P.

IMPROSULFAN [INN]

CAS 13425-98-4	MW: 289.37	MF:	C8 H19 N O6 S2

Antineoplastic

LgP	-1.80(C)
pKa	8.77

pp cited in Vols.1-5:

(1) M.M. El-Merzabani et al., Gann., 1965, 56, 589.
(2) Y. Sakuri, Japan. Pat. 67 26 616 (1967).
(3) H.Imamura et al., Oyo Yakuri, 1975, 9, 861.

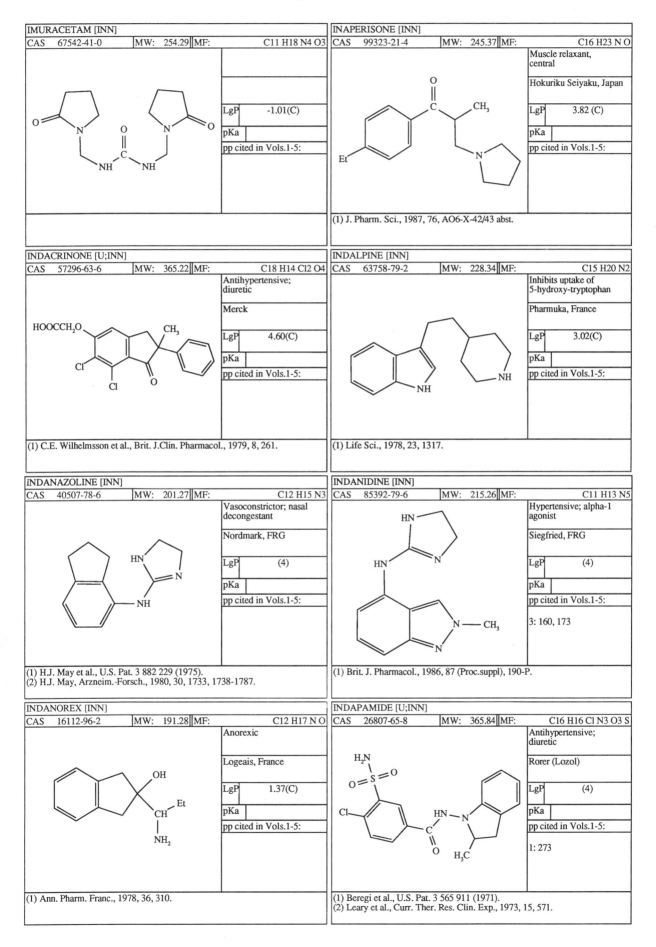

IMURACETAM [INN]

CAS	67542-41-0	MW:	254.29	MF:	C11 H18 N4 O3

LgP	-1.01(C)
pKa	
pp cited in Vols.1-5:	

INAPERISONE [INN]

CAS	99323-21-4	MW:	245.37	MF:	C16 H23 N O

Muscle relaxant, central

Hokuriku Seiyaku, Japan

LgP	3.82 (C)
pKa	
pp cited in Vols.1-5:	

(1) J. Pharm. Sci., 1987, 76, AO6-X-42/43 abst.

INDACRINONE [U;INN]

CAS	57296-63-6	MW:	365.22	MF:	C18 H14 Cl2 O4

Antihypertensive; diuretic

Merck

LgP	4.60(C)
pKa	
pp cited in Vols.1-5:	

(1) C.E. Wilhelmsson et al., Brit. J.Clin. Pharmacol., 1979, 8, 261.

INDALPINE [INN]

CAS	63758-79-2	MW:	228.34	MF:	C15 H20 N2

Inhibits uptake of 5-hydroxy-tryptophan

Pharmuka, France

LgP	3.02(C)
pKa	
pp cited in Vols.1-5:	

(1) Life Sci., 1978, 23, 1317.

INDANAZOLINE [INN]

CAS	40507-78-6	MW:	201.27	MF:	C12 H15 N3

Vasoconstrictor; nasal decongestant

Nordmark, FRG

LgP	(4)
pKa	
pp cited in Vols.1-5:	

(1) H.J. May et al., U.S. Pat. 3 882 229 (1975).
(2) H.J. May, Arzneim.-Forsch., 1980, 30, 1733, 1738-1787.

INDANIDINE [INN]

CAS	85392-79-6	MW:	215.26	MF:	C11 H13 N5

Hypertensive; alpha-1 agonist

Siegfried, FRG

LgP	(4)
pKa	
pp cited in Vols.1-5:	

3: 160, 173

(1) Brit. J. Pharmacol., 1986, 87 (Proc.suppl), 190-P.

INDANOREX [INN]

CAS	16112-96-2	MW:	191.28	MF:	C12 H17 N O

Anorexic

Logeais, France

LgP	1.37(C)
pKa	
pp cited in Vols.1-5:	

(1) Ann. Pharm. Franc., 1978, 36, 310.

INDAPAMIDE [U;INN]

CAS	26807-65-8	MW:	365.84	MF:	C16 H16 Cl N3 O3 S

Antihypertensive; diuretic

Rorer (Lozol)

LgP	(4)
pKa	
pp cited in Vols.1-5:	

1: 273

(1) Beregi et al., U.S. Pat. 3 565 911 (1971).
(2) Leary et al., Curr. Ther. Res. Clin. Exp., 1973, 15, 571.

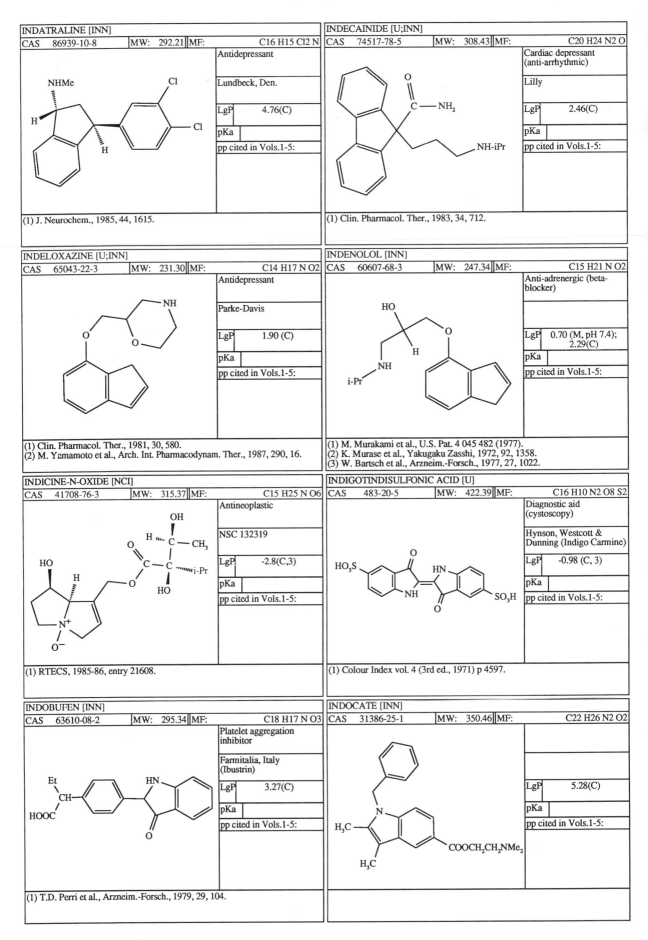

INDATRALINE [INN]

CAS	86939-10-8	MW:	292.21	MF:	C16 H15 Cl2 N

Antidepressant

Lundbeck, Den.

LgP	4.76(C)
pKa	
pp cited in Vols.1-5:	

NHMe

(1) J. Neurochem., 1985, 44, 1615.

INDECAINIDE [U;INN]

CAS	74517-78-5	MW:	308.43	MF:	C20 H24 N2 O

Cardiac depressant (anti-arrhythmic)

Lilly

LgP	2.46(C)
pKa	
pp cited in Vols.1-5:	

(1) Clin. Pharmacol. Ther., 1983, 34, 712.

INDELOXAZINE [U;INN]

CAS	65043-22-3	MW:	231.30	MF:	C14 H17 N O2

Antidepressant

Parke-Davis

LgP	1.90 (C)
pKa	
pp cited in Vols.1-5:	

(1) Clin. Pharmacol. Ther., 1981, 30, 580.
(2) M. Yamamoto et al., Arch. Int. Pharmacodynam. Ther., 1987, 290, 16.

INDENOLOL [INN]

CAS	60607-68-3	MW:	247.34	MF:	C15 H21 N O2

Anti-adrenergic (beta-blocker)

LgP	0.70 (M, pH 7.4); 2.29(C)
pKa	
pp cited in Vols.1-5:	

(1) M. Murakami et al., U.S. Pat. 4 045 482 (1977).
(2) K. Murase et al., Yakugaku Zasshi, 1972, 92, 1358.
(3) W. Bartsch et al., Arzneim.-Forsch., 1977, 27, 1022.

INDICINE-N-OXIDE [NCI]

CAS	41708-76-3	MW:	315.37	MF:	C15 H25 N O6

Antineoplastic

NSC 132319

LgP	-2.8(C,3)
pKa	
pp cited in Vols.1-5:	

(1) RTECS, 1985-86, entry 21608.

INDIGOTINDISULFONIC ACID [U]

CAS	483-20-5	MW:	422.39	MF:	C16 H10 N2 O8 S2

Diagnostic aid (cystoscopy)

Hynson, Westcott & Dunning (Indigo Carmine)

LgP	-0.98 (C, 3)
pKa	
pp cited in Vols.1-5:	

(1) Colour Index vol. 4 (3rd ed., 1971) p 4597.

INDOBUFEN [INN]

CAS	63610-08-2	MW:	295.34	MF:	C18 H17 N O3

Platelet aggregation inhibitor

Farmitalia, Italy (Ibustrin)

LgP	3.27(C)
pKa	
pp cited in Vols.1-5:	

(1) T.D. Perri et al., Arzneim.-Forsch., 1979, 29, 104.

INDOCATE [INN]

CAS	31386-25-1	MW:	350.46	MF:	C22 H26 N2 O2

LgP	5.28(C)
pKa	
pp cited in Vols.1-5:	

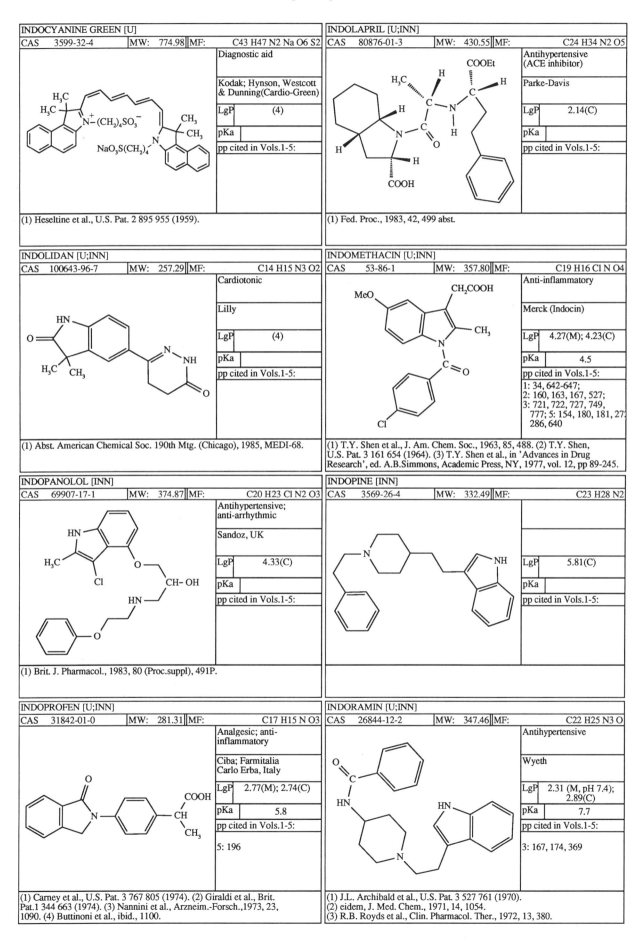

INDOCYANINE GREEN [U]

CAS	3599-32-4	MW:	774.98	MF:	C43 H47 N2 Na O6 S2

Diagnostic aid

Kodak; Hynson, Westcott & Dunning(Cardio-Green)

LgP	(4)
pKa	
pp cited in Vols.1-5:	

(1) Heseltine et al., U.S. Pat. 2 895 955 (1959).

INDOLAPRIL [U;INN]

CAS	80876-01-3	MW:	430.55	MF:	C24 H34 N2 O5

Antihypertensive (ACE inhibitor)

Parke-Davis

LgP	2.14(C)
pKa	
pp cited in Vols.1-5:	

(1) Fed. Proc., 1983, 42, 499 abst.

INDOLIDAN [U;INN]

CAS	100643-96-7	MW:	257.29	MF:	C14 H15 N3 O2

Cardiotonic

Lilly

LgP	(4)
pKa	
pp cited in Vols.1-5:	

(1) Abst. American Chemical Soc. 190th Mtg. (Chicago), 1985, MEDI-68.

INDOMETHACIN [U;INN]

CAS	53-86-1	MW:	357.80	MF:	C19 H16 Cl N O4

Anti-inflammatory

Merck (Indocin)

LgP	4.27(M); 4.23(C)
pKa	4.5
pp cited in Vols.1-5:	

1: 34, 642-647;
2: 160, 163, 167, 527;
3: 721, 722, 727, 749, 777; 5: 154, 180, 181, 27? 286, 640

(1) T.Y. Shen et al., J. Am. Chem. Soc., 1963, 85, 488. (2) T.Y. Shen, U.S. Pat. 3 161 654 (1964). (3) T.Y. Shen et al., in 'Advances in Drug Research', ed. A.B.Simmons, Academic Press, NY, 1977, vol. 12, pp 89-245.

INDOPANOLOL [INN]

CAS	69907-17-1	MW:	374.87	MF:	C20 H23 Cl N2 O3

Antihypertensive; anti-arrhythmic

Sandoz, UK

LgP	4.33(C)
pKa	
pp cited in Vols.1-5:	

(1) Brit. J. Pharmacol., 1983, 80 (Proc.suppl), 491P.

INDOPINE [INN]

CAS	3569-26-4	MW:	332.49	MF:	C23 H28 N2

LgP	5.81(C)
pKa	
pp cited in Vols.1-5:	

INDOPROFEN [U;INN]

CAS	31842-01-0	MW:	281.31	MF:	C17 H15 N O3

Analgesic; anti-inflammatory

Ciba; Farmitalia Carlo Erba, Italy

LgP	2.77(M); 2.74(C)
pKa	5.8
pp cited in Vols.1-5:	

5: 196

(1) Carney et al., U.S. Pat. 3 767 805 (1974). (2) Giraldi et al., Brit. Pat.1 344 663 (1974). (3) Nannini et al., Arzneim.-Forsch.,1973, 23, 1090. (4) Buttinoni et al., ibid., 1100.

INDORAMIN [U;INN]

CAS	26844-12-2	MW:	347.46	MF:	C22 H25 N3 O

Antihypertensive

Wyeth

LgP	2.31 (M, pH 7.4); 2.89(C)
pKa	7.7
pp cited in Vols.1-5:	

3: 167, 174, 369

(1) J.L. Archibald et al., U.S. Pat. 3 527 761 (1970).
(2) eidem, J. Med. Chem., 1971, 14, 1054.
(3) R.B. Royds et al., Clin. Pharmacol. Ther., 1972, 13, 380.

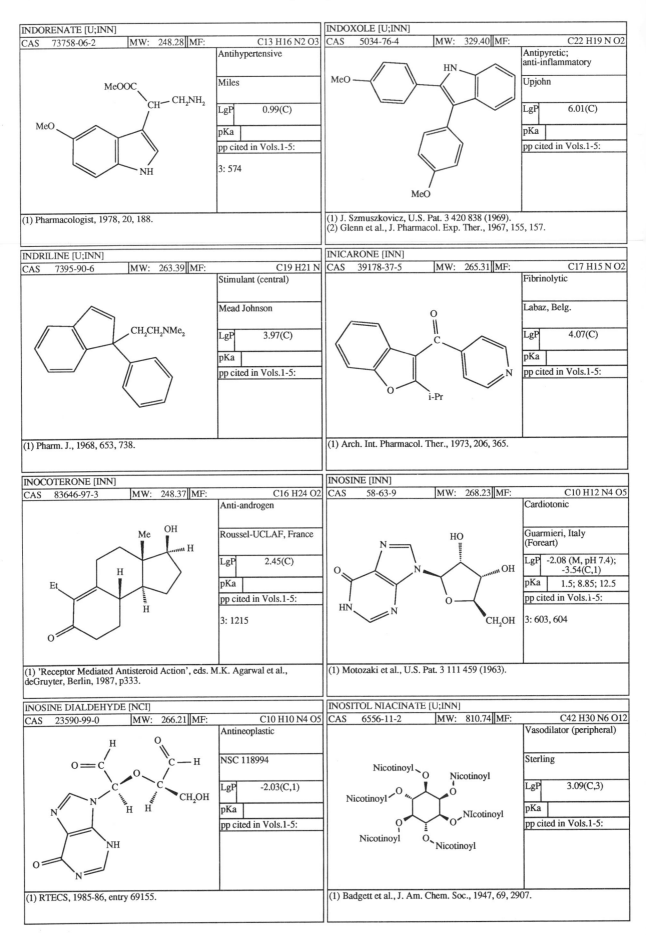

INDORENATE [U;INN]

CAS 73758-06-2　MW: 248.28　MF: C13 H16 N2 O3

Antihypertensive

Miles

LgP	0.99(C)

pKa

pp cited in Vols.1-5:

3: 574

(1) Pharmacologist, 1978, 20, 188.

INDOXOLE [U;INN]

CAS 5034-76-4　MW: 329.40　MF: C22 H19 N O2

Antipyretic;
anti-inflammatory

Upjohn

LgP	6.01(C)

pKa

pp cited in Vols.1-5:

(1) J. Szmuszkovicz, U.S. Pat. 3 420 838 (1969).
(2) Glenn et al., J. Pharmacol. Exp. Ther., 1967, 155, 157.

INDRILINE [U;INN]

CAS 7395-90-6　MW: 263.39　MF: C19 H21 N

Stimulant (central)

Mead Johnson

LgP	3.97(C)

pKa

pp cited in Vols.1-5:

(1) Pharm. J., 1968, 653, 738.

INICARONE [INN]

CAS 39178-37-5　MW: 265.31　MF: C17 H15 N O2

Fibrinolytic

Labaz, Belg.

LgP	4.07(C)

pKa

pp cited in Vols.1-5:

(1) Arch. Int. Pharmacol. Ther., 1973, 206, 365.

INOCOTERONE [INN]

CAS 83646-97-3　MW: 248.37　MF: C16 H24 O2

Anti-androgen

Roussel-UCLAF, France

LgP	2.45(C)

pKa

pp cited in Vols.1-5:

3: 1215

(1) 'Receptor Mediated Antisteroid Action', eds. M.K. Agarwal et al.,
deGruyter, Berlin, 1987, p333.

INOSINE [INN]

CAS 58-63-9　MW: 268.23　MF: C10 H12 N4 O5

Cardiotonic

Guarmieri, Italy
(Foreart)

LgP	-2.08 (M, pH 7.4); -3.54(C,1)
pKa	1.5; 8.85; 12.5

pp cited in Vols.1-5:

3: 603, 604

(1) Motozaki et al., U.S. Pat. 3 111 459 (1963).

INOSINE DIALDEHYDE [NCI]

CAS 23590-99-0　MW: 266.21　MF: C10 H10 N4 O5

Antineoplastic

NSC 118994

LgP	-2.03(C,1)

pKa

pp cited in Vols.1-5:

(1) RTECS, 1985-86, entry 69155.

INOSITOL NIACINATE [U;INN]

CAS 6556-11-2　MW: 810.74　MF: C42 H30 N6 O12

Vasodilator (peripheral)

Sterling

LgP	3.09(C,3)

pKa

pp cited in Vols.1-5:

(1) Badgett et al., J. Am. Chem. Soc., 1947, 69, 2907.

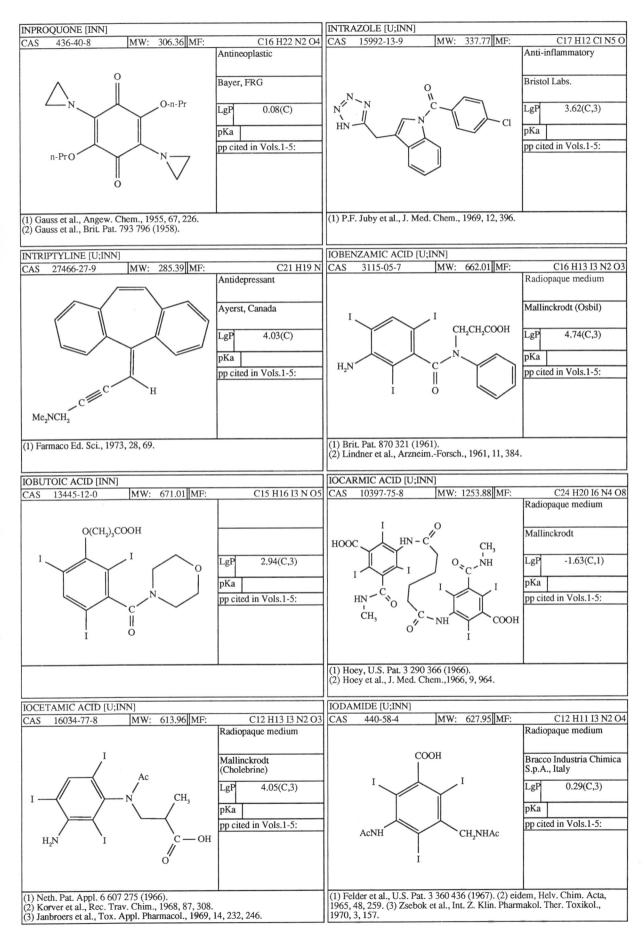

INPROQUONE [INN]

CAS 436-40-8	MW: 306.36	MF:	C16 H22 N2 O4

Antineoplastic

Bayer, FRG

LgP	0.08(C)

pKa	

pp cited in Vols.1-5:

(1) Gauss et al., Angew. Chem., 1955, 67, 226.
(2) Gauss et al., Brit. Pat. 793 796 (1958).

INTRAZOLE [U;INN]

CAS 15992-13-9	MW: 337.77	MF:	C17 H12 Cl N5 O

Anti-inflammatory

Bristol Labs.

LgP	3.62(C,3)

pKa	

pp cited in Vols.1-5:

(1) P.F. Juby et al., J. Med. Chem., 1969, 12, 396.

INTRIPTYLINE [U;INN]

CAS 27466-27-9	MW: 285.39	MF:	C21 H19 N

Antidepressant

Ayerst, Canada

LgP	4.03(C)

pKa	

pp cited in Vols.1-5:

(1) Farmaco Ed. Sci., 1973, 28, 69.

IOBENZAMIC ACID [U;INN]

CAS 3115-05-7	MW: 662.01	MF:	C16 H13 I3 N2 O3

Radiopaque medium

Mallinckrodt (Osbil)

LgP	4.74(C,3)

pKa	

pp cited in Vols.1-5:

(1) Brit. Pat. 870 321 (1961).
(2) Lindner et al., Arzneim.-Forsch., 1961, 11, 384.

IOBUTOIC ACID [INN]

CAS 13445-12-0	MW: 671.01	MF:	C15 H16 I3 N O5

LgP	2.94(C,3)

pKa	

pp cited in Vols.1-5:

IOCARMIC ACID [U;INN]

CAS 10397-75-8	MW: 1253.88	MF:	C24 H20 I6 N4 O8

Radiopaque medium

Mallinckrodt

LgP	-1.63(C,1)

pKa	

pp cited in Vols.1-5:

(1) Hoey, U.S. Pat. 3 290 366 (1966).
(2) Hoey et al., J. Med. Chem.,1966, 9, 964.

IOCETAMIC ACID [U;INN]

CAS 16034-77-8	MW: 613.96	MF:	C12 H13 I3 N2 O3

Radiopaque medium

Mallinckrodt (Cholebrine)

LgP	4.05(C,3)

pKa	

pp cited in Vols.1-5:

(1) Neth. Pat. Appl. 6 607 275 (1966).
(2) Korver et al., Rec. Trav. Chim., 1968, 87, 308.
(3) Janbroers et al., Tox. Appl. Pharmacol., 1969, 14, 232, 246.

IODAMIDE [U;INN]

CAS 440-58-4	MW: 627.95	MF:	C12 H11 I3 N2 O4

Radiopaque medium

Bracco Industria Chimica S.p.A., Italy

LgP	0.29(C,3)

pKa	

pp cited in Vols.1-5:

(1) Felder et al., U.S. Pat. 3 360 436 (1967). (2) eidem, Helv. Chim. Acta, 1965, 48, 259. (3) Zsebok et al., Int. Z. Klin. Pharmakol. Ther. Toxikol., 1970, 3, 157.

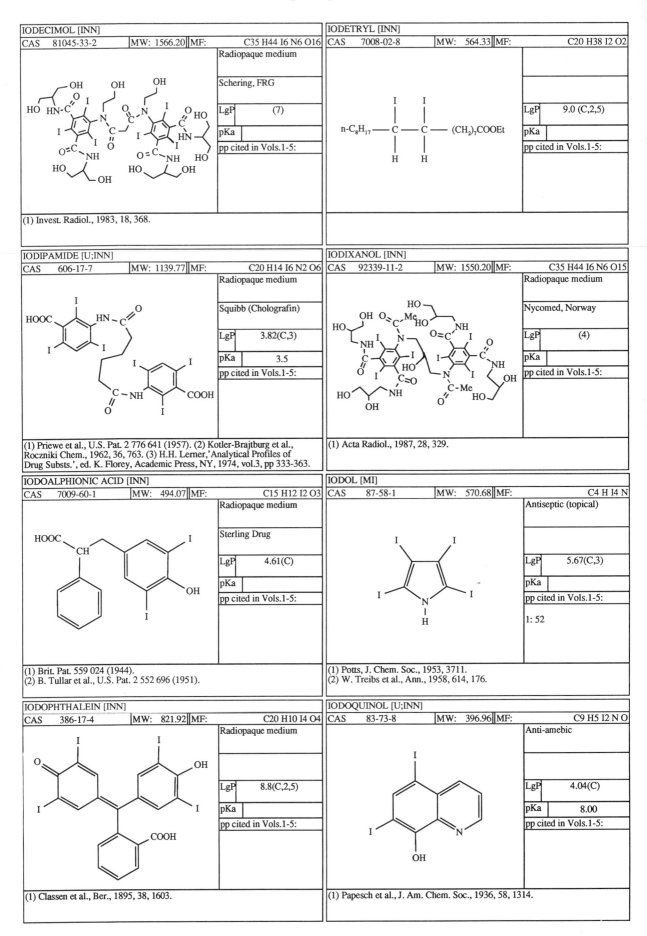

IODECIMOL [INN]		
CAS 81045-33-2	MW: 1566.20 MF:	C35 H44 I6 N6 O16

Radiopaque medium

Schering, FRG

LgP	(7)
pKa	
pp cited in Vols.1-5:	

(1) Invest. Radiol., 1983, 18, 368.

IODETRYL [INN]		
CAS 7008-02-8	MW: 564.33 MF:	C20 H38 I2 O2

n-C₈H₁₇—C—C—(CH₂)₇COOEt

LgP	9.0 (C,2,5)
pKa	
pp cited in Vols.1-5:	

IODIPAMIDE [U;INN]		
CAS 606-17-7	MW: 1139.77 MF:	C20 H14 I6 N2 O6

Radiopaque medium

Squibb (Cholografin)

LgP	3.82(C,3)
pKa	3.5
pp cited in Vols.1-5:	

(1) Priewe et al., U.S. Pat. 2 776 641 (1957). (2) Kotler-Brajtburg et al., Roczniki Chem., 1962, 36, 763. (3) H.H. Lerner,'Analytical Profiles of Drug Substs.', ed. K. Florey, Academic Press, NY, 1974, vol.3, pp 333-363.

IODIXANOL [INN]		
CAS 92339-11-2	MW: 1550.20 MF:	C35 H44 I6 N6 O15

Radiopaque medium

Nycomed, Norway

LgP	(4)
pKa	
pp cited in Vols.1-5:	

(1) Acta Radiol., 1987, 28, 329.

IODOALPHIONIC ACID [INN]		
CAS 7009-60-1	MW: 494.07 MF:	C15 H12 I2 O3

Radiopaque medium

Sterling Drug

LgP	4.61(C)
pKa	
pp cited in Vols.1-5:	

(1) Brit. Pat. 559 024 (1944).
(2) B. Tullar et al., U.S. Pat. 2 552 696 (1951).

IODOL [MI]		
CAS 87-58-1	MW: 570.68 MF:	C4 H I4 N

Antiseptic (topical)

LgP	5.67(C,3)
pKa	
pp cited in Vols.1-5:	
1: 52	

(1) Potts, J. Chem. Soc., 1953, 3711.
(2) W. Treibs et al., Ann., 1958, 614, 176.

IODOPHTHALEIN [INN]		
CAS 386-17-4	MW: 821.92 MF:	C20 H10 I4 O4

Radiopaque medium

LgP	8.8(C,2,5)
pKa	
pp cited in Vols.1-5:	

(1) Classen et al., Ber., 1895, 38, 1603.

IODOQUINOL [U;INN]		
CAS 83-73-8	MW: 396.96 MF:	C9 H5 I2 N O

Anti-amebic

LgP	4.04(C)
pKa	8.00
pp cited in Vols.1-5:	

(1) Papesch et al., J. Am. Chem. Soc., 1936, 58, 1314.

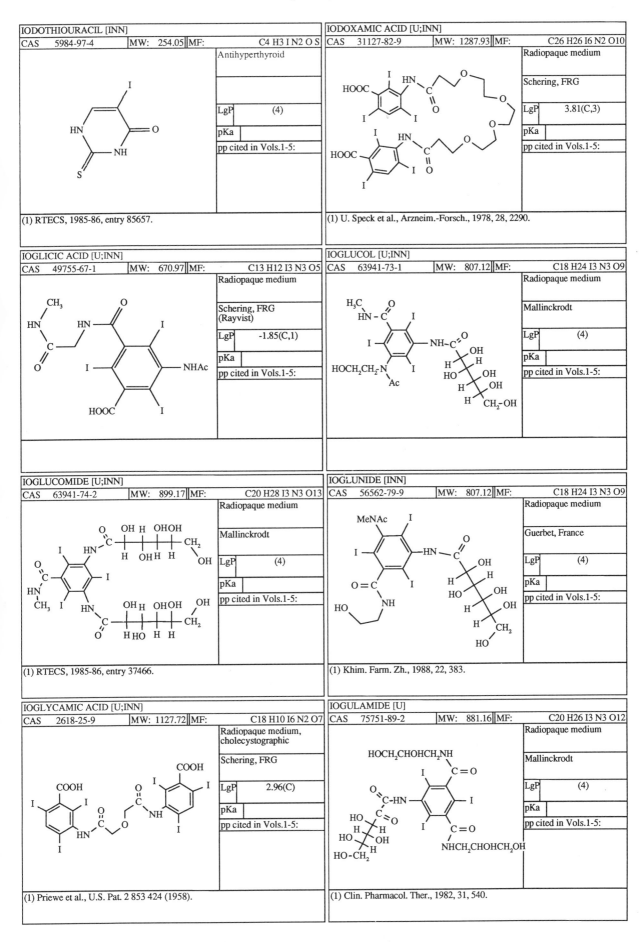

IODOTHIOURACIL [INN]

CAS	5984-97-4	MW:	254.05	MF:	C4 H3 I N2 O S

Antihyperthyroid

LgP	(4)
pKa	
pp cited in Vols.1-5:	

(1) RTECS, 1985-86, entry 85657.

IODOXAMIC ACID [U;INN]

CAS	31127-82-9	MW:	1287.93	MF:	C26 H26 I6 N2 O10

Radiopaque medium

Schering, FRG

LgP	3.81(C,3)
pKa	
pp cited in Vols.1-5:	

(1) U. Speck et al., Arzneim.-Forsch., 1978, 28, 2290.

IOGLICIC ACID [U;INN]

CAS	49755-67-1	MW:	670.97	MF:	C13 H12 I3 N3 O5

Radiopaque medium

Schering, FRG
(Rayvist)

LgP	-1.85(C,1)
pKa	
pp cited in Vols.1-5:	

IOGLUCOL [U;INN]

CAS	63941-73-1	MW:	807.12	MF:	C18 H24 I3 N3 O9

Radiopaque medium

Mallinckrodt

LgP	(4)
pKa	
pp cited in Vols.1-5:	

IOGLUCOMIDE [U;INN]

CAS	63941-74-2	MW:	899.17	MF:	C20 H28 I3 N3 O13

Radiopaque medium

Mallinckrodt

LgP	(4)
pKa	
pp cited in Vols.1-5:	

(1) RTECS, 1985-86, entry 37466.

IOGLUNIDE [INN]

CAS	56562-79-9	MW:	807.12	MF:	C18 H24 I3 N3 O9

Radiopaque medium

Guerbet, France

LgP	(4)
pKa	
pp cited in Vols.1-5:	

(1) Khim. Farm. Zh., 1988, 22, 383.

IOGLYCAMIC ACID [U;INN]

CAS	2618-25-9	MW:	1127.72	MF:	C18 H10 I6 N2 O7

Radiopaque medium,
cholecystographic

Schering, FRG

LgP	2.96(C)
pKa	
pp cited in Vols.1-5:	

(1) Priewe et al., U.S. Pat. 2 853 424 (1958).

IOGULAMIDE [U]

CAS	75751-89-2	MW:	881.16	MF:	C20 H26 I3 N3 O12

Radiopaque medium

Mallinckrodt

LgP	(4)
pKa	
pp cited in Vols.1-5:	

(1) Clin. Pharmacol. Ther., 1982, 31, 540.

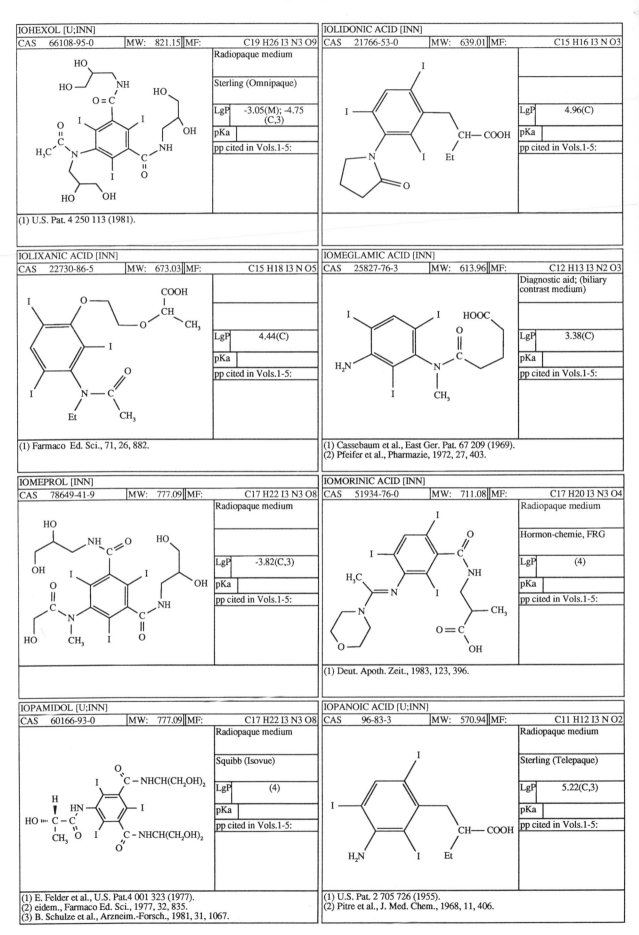

IOHEXOL [U;INN]

CAS 66108-95-0	MW: 821.15	MF:	C19 H26 I3 N3 O9

Radiopaque medium

Sterling (Omnipaque)

LgP	-3.05(M); -4.75 (C,3)
pKa	
pp cited in Vols.1-5:	

(1) U.S. Pat. 4 250 113 (1981).

IOLIDONIC ACID [INN]

CAS 21766-53-0	MW: 639.01	MF:	C15 H16 I3 N O3

LgP	4.96(C)
pKa	
pp cited in Vols.1-5:	

IOLIXANIC ACID [INN]

CAS 22730-86-5	MW: 673.03	MF:	C15 H18 I3 N O5

LgP	4.44(C)
pKa	
pp cited in Vols.1-5:	

(1) Farmaco Ed. Sci., 71, 26, 882.

IOMEGLAMIC ACID [INN]

CAS 25827-76-3	MW: 613.96	MF:	C12 H13 I3 N2 O3

Diagnostic aid; (biliary contrast medium)

LgP	3.38(C)
pKa	
pp cited in Vols.1-5:	

(1) Cassebaum et al., East Ger. Pat. 67 209 (1969).
(2) Pfeifer et al., Pharmazie, 1972, 27, 403.

IOMEPROL [INN]

CAS 78649-41-9	MW: 777.09	MF:	C17 H22 I3 N3 O8

Radiopaque medium

LgP	-3.82(C,3)
pKa	
pp cited in Vols.1-5:	

IOMORINIC ACID [INN]

CAS 51934-76-0	MW: 711.08	MF:	C17 H20 I3 N3 O4

Radiopaque medium

Hormon-chemie, FRG

LgP	(4)
pKa	
pp cited in Vols.1-5:	

(1) Deut. Apoth. Zeit., 1983, 123, 396.

IOPAMIDOL [U;INN]

CAS 60166-93-0	MW: 777.09	MF:	C17 H22 I3 N3 O8

Radiopaque medium

Squibb (Isovue)

LgP	(4)
pKa	
pp cited in Vols.1-5:	

(1) E. Felder et al., U.S. Pat.4 001 323 (1977).
(2) eidem., Farmaco Ed. Sci., 1977, 32, 835.
(3) B. Schulze et al., Arzneim.-Forsch., 1981, 31, 1067.

IOPANOIC ACID [U;INN]

CAS 96-83-3	MW: 570.94	MF:	C11 H12 I3 N O2

Radiopaque medium

Sterling (Telepaque)

LgP	5.22(C,3)
pKa	
pp cited in Vols.1-5:	

(1) U.S. Pat. 2 705 726 (1955).
(2) Pitre et al., J. Med. Chem., 1968, 11, 406.

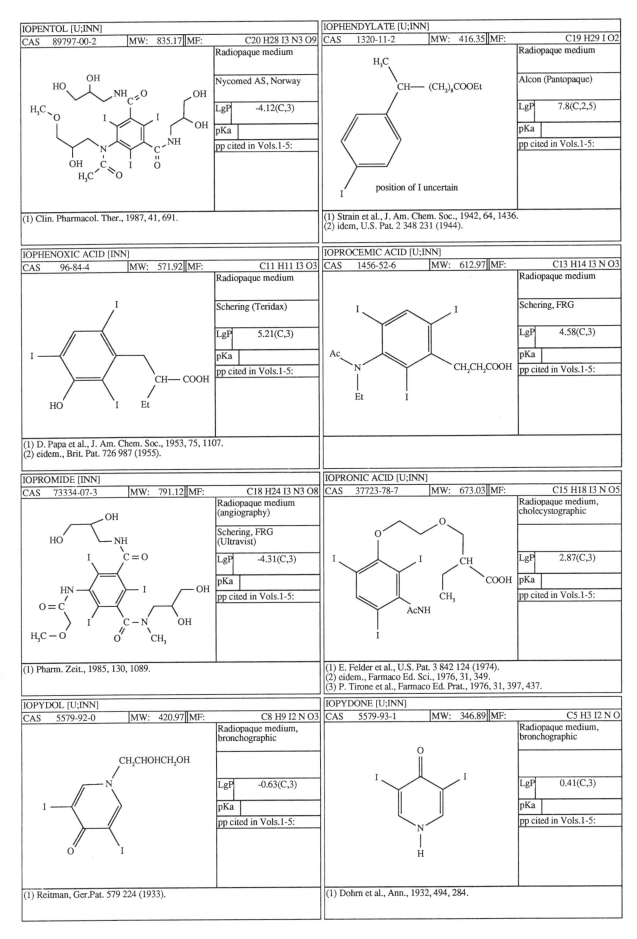

IOPENTOL [U;INN]

CAS	89797-00-2	MW:	835.17	MF:	C20 H28 I3 N3 O9

Radiopaque medium

Nycomed AS, Norway

LgP	-4.12(C,3)

pKa	

pp cited in Vols.1-5:

(1) Clin. Pharmacol. Ther., 1987, 41, 691.

IOPHENDYLATE [U;INN]

CAS	1320-11-2	MW:	416.35	MF:	C19 H29 I O2

Radiopaque medium

Alcon (Pantopaque)

LgP	7.8(C,2,5)

pKa	

pp cited in Vols.1-5:

position of I uncertain

(1) Strain et al., J. Am. Chem. Soc., 1942, 64, 1436.
(2) idem, U.S. Pat. 2 348 231 (1944).

IOPHENOXIC ACID [INN]

CAS	96-84-4	MW:	571.92	MF:	C11 H11 I3 O3

Radiopaque medium

Schering (Teridax)

LgP	5.21(C,3)

pKa	

pp cited in Vols.1-5:

(1) D. Papa et al., J. Am. Chem. Soc., 1953, 75, 1107.
(2) eidem., Brit. Pat. 726 987 (1955).

IOPROCEMIC ACID [U;INN]

CAS	1456-52-6	MW:	612.97	MF:	C13 H14 I3 N O3

Radiopaque medium

Schering, FRG

LgP	4.58(C,3)

pKa	

pp cited in Vols.1-5:

IOPROMIDE [INN]

CAS	73334-07-3	MW:	791.12	MF:	C18 H24 I3 N3 O8

Radiopaque medium (angiography)

Schering, FRG (Ultravist)

LgP	-4.31(C,3)

pKa	

pp cited in Vols.1-5:

(1) Pharm. Zeit., 1985, 130, 1089.

IOPRONIC ACID [U;INN]

CAS	37723-78-7	MW:	673.03	MF:	C15 H18 I3 N O5

Radiopaque medium, cholecystographic

LgP	2.87(C,3)

pKa	

pp cited in Vols.1-5:

(1) E. Felder et al., U.S. Pat. 3 842 124 (1974).
(2) eidem., Farmaco Ed. Sci., 1976, 31, 349.
(3) P. Tirone et al., Farmaco Ed. Prat., 1976, 31, 397, 437.

IOPYDOL [U;INN]

CAS	5579-92-0	MW:	420.97	MF:	C8 H9 I2 N O3

Radiopaque medium, bronchographic

LgP	-0.63(C,3)

pKa	

pp cited in Vols.1-5:

(1) Reitman, Ger.Pat. 579 224 (1933).

IOPYDONE [U;INN]

CAS	5579-93-1	MW:	346.89	MF:	C5 H3 I2 N O

Radiopaque medium, bronchographic

LgP	0.41(C,3)

pKa	

pp cited in Vols.1-5:

(1) Dohrn et al., Ann., 1932, 494, 284.

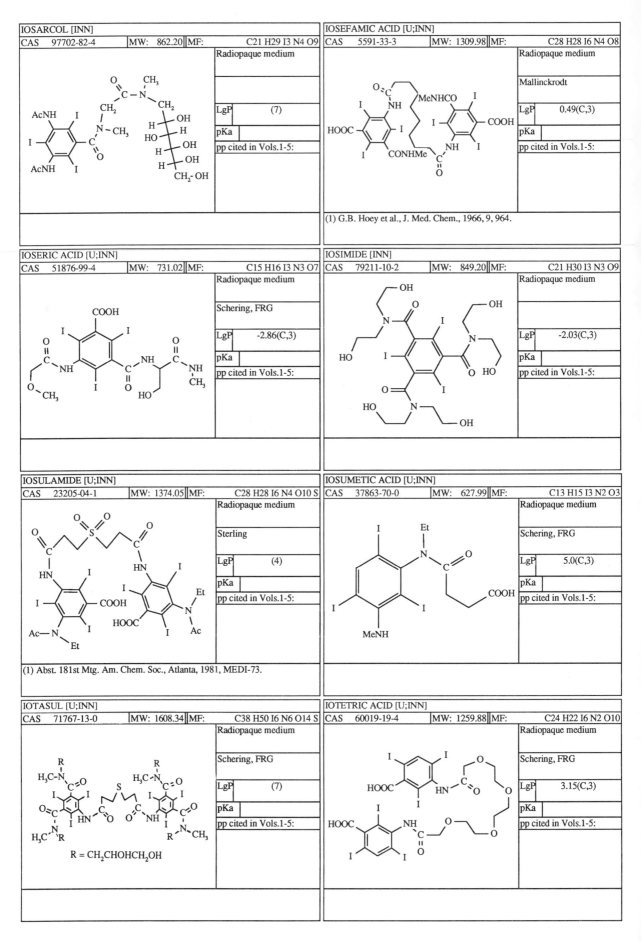

IOSARCOL [INN]

CAS	97702-82-4	MW:	862.20	MF:	C21 H29 I3 N4 O9

Radiopaque medium

LgP	(7)
pKa	
pp cited in Vols.1-5:	

IOSEFAMIC ACID [U;INN]

CAS	5591-33-3	MW:	1309.98	MF:	C28 H28 I6 N4 O8

Radiopaque medium

Mallinckrodt

LgP	0.49(C,3)
pKa	
pp cited in Vols.1-5:	

(1) G.B. Hoey et al., J. Med. Chem., 1966, 9, 964.

IOSERIC ACID [U;INN]

CAS	51876-99-4	MW:	731.02	MF:	C15 H16 I3 N3 O7

Radiopaque medium

Schering, FRG

LgP	-2.86(C,3)
pKa	
pp cited in Vols.1-5:	

IOSIMIDE [INN]

CAS	79211-10-2	MW:	849.20	MF:	C21 H30 I3 N3 O9

Radiopaque medium

LgP	-2.03(C,3)
pKa	
pp cited in Vols.1-5:	

IOSULAMIDE [U;INN]

CAS	23205-04-1	MW:	1374.05	MF:	C28 H28 I6 N4 O10 S

Radiopaque medium

Sterling

LgP	(4)
pKa	
pp cited in Vols.1-5:	

(1) Abst. 181st Mtg. Am. Chem. Soc., Atlanta, 1981, MEDI-73.

IOSUMETIC ACID [U;INN]

CAS	37863-70-0	MW:	627.99	MF:	C13 H15 I3 N2 O3

Radiopaque medium

Schering, FRG

LgP	5.0(C,3)
pKa	
pp cited in Vols.1-5:	

IOTASUL [U;INN]

CAS	71767-13-0	MW:	1608.34	MF:	C38 H50 I6 N6 O14 S

Radiopaque medium

Schering, FRG

LgP	(7)
pKa	
pp cited in Vols.1-5:	

R = CH₂CHOHCH₂OH

IOTETRIC ACID [U;INN]

CAS	60019-19-4	MW:	1259.88	MF:	C24 H22 I6 N2 O10

Radiopaque medium

Schering, FRG

LgP	3.15(C,3)
pKa	
pp cited in Vols.1-5:	

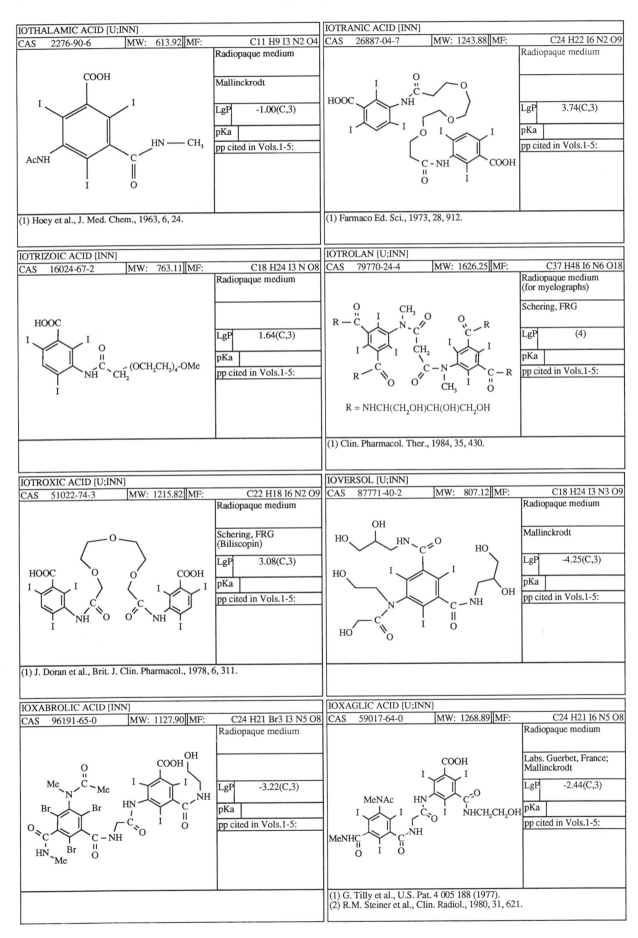

IOTHALAMIC ACID [U;INN]

CAS	2276-90-6	MW: 613.92	MF:	C11 H9 I3 N2 O4

Radiopaque medium

Mallinckrodt

LgP	-1.00(C,3)
pKa	
pp cited in Vols.1-5:	

(1) Hoey et al., J. Med. Chem., 1963, 6, 24.

IOTRANIC ACID [INN]

CAS	26887-04-7	MW: 1243.88	MF:	C24 H22 I6 N2 O9

Radiopaque medium

LgP	3.74(C,3)
pKa	
pp cited in Vols.1-5:	

(1) Farmaco Ed. Sci., 1973, 28, 912.

IOTRIZOIC ACID [INN]

CAS	16024-67-2	MW: 763.11	MF:	C18 H24 I3 N O8

Radiopaque medium

LgP	1.64(C,3)
pKa	
pp cited in Vols.1-5:	

IOTROLAN [U;INN]

CAS	79770-24-4	MW: 1626.25	MF:	C37 H48 I6 N6 O18

Radiopaque medium
(for myelographs)

Schering, FRG

LgP	(4)
pKa	
pp cited in Vols.1-5:	

R = NHCH(CH₂OH)CH(OH)CH₂OH

(1) Clin. Pharmacol. Ther., 1984, 35, 430.

IOTROXIC ACID [U;INN]

CAS	51022-74-3	MW: 1215.82	MF:	C22 H18 I6 N2 O9

Radiopaque medium

Schering, FRG
(Biliscopin)

LgP	3.08(C,3)
pKa	
pp cited in Vols.1-5:	

(1) J. Doran et al., Brit. J. Clin. Pharmacol., 1978, 6, 311.

IOVERSOL [U;INN]

CAS	87771-40-2	MW: 807.12	MF:	C18 H24 I3 N3 O9

Radiopaque medium

Mallinckrodt

LgP	-4.25(C,3)
pKa	
pp cited in Vols.1-5:	

IOXABROLIC ACID [INN]

CAS	96191-65-0	MW: 1127.90	MF:	C24 H21 Br3 I3 N5 O8

Radiopaque medium

LgP	-3.22(C,3)
pKa	
pp cited in Vols.1-5:	

IOXAGLIC ACID [U;INN]

CAS	59017-64-0	MW: 1268.89	MF:	C24 H21 I6 N5 O8

Radiopaque medium

Labs. Guerbet, France;
Mallinckrodt

LgP	-2.44(C,3)
pKa	
pp cited in Vols.1-5:	

(1) G. Tilly et al., U.S. Pat. 4 005 188 (1977).
(2) R.M. Steiner et al., Clin. Radiol., 1980, 31, 621.

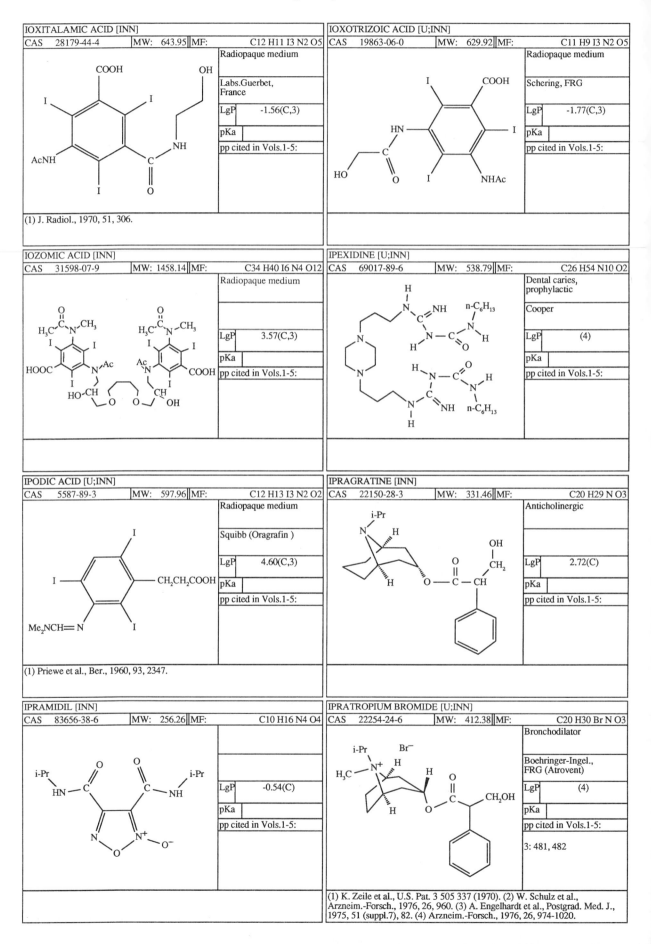

IOXITALAMIC ACID [INN]

CAS	28179-44-4	MW:	643.95	MF:	C12 H11 I3 N2 O5

Radiopaque medium

Labs.Guerbet, France

LgP	-1.56(C,3)
pKa	

pp cited in Vols.1-5:

(1) J. Radiol., 1970, 51, 306.

IOXOTRIZOIC ACID [U;INN]

CAS	19863-06-0	MW:	629.92	MF:	C11 H9 I3 N2 O5

Radiopaque medium

Schering, FRG

LgP	-1.77(C,3)
pKa	

pp cited in Vols.1-5:

IOZOMIC ACID [INN]

CAS	31598-07-9	MW:	1458.14	MF:	C34 H40 I6 N4 O12

Radiopaque medium

LgP	3.57(C,3)
pKa	

pp cited in Vols.1-5:

IPEXIDINE [U;INN]

CAS	69017-89-6	MW:	538.79	MF:	C26 H54 N10 O2

Dental caries, prophylactic

Cooper

LgP	(4)
pKa	

pp cited in Vols.1-5:

IPODIC ACID [U;INN]

CAS	5587-89-3	MW:	597.96	MF:	C12 H13 I3 N2 O2

Radiopaque medium

Squibb (Oragrafin)

LgP	4.60(C,3)
pKa	

pp cited in Vols.1-5:

(1) Priewe et al., Ber., 1960, 93, 2347.

IPRAGRATINE [INN]

CAS	22150-28-3	MW:	331.46	MF:	C20 H29 N O3

Anticholinergic

LgP	2.72(C)
pKa	

pp cited in Vols.1-5:

IPRAMIDIL [INN]

CAS	83656-38-6	MW:	256.26	MF:	C10 H16 N4 O4

LgP	-0.54(C)
pKa	

pp cited in Vols.1-5:

IPRATROPIUM BROMIDE [U;INN]

CAS	22254-24-6	MW:	412.38	MF:	C20 H30 Br N O3

Bronchodilator

Boehringer-Ingel., FRG (Atrovent)

LgP	(4)
pKa	

pp cited in Vols.1-5:

3: 481, 482

(1) K. Zeile et al., U.S. Pat. 3 505 337 (1970). (2) W. Schulz et al., Arzneim.-Forsch., 1976, 26, 960. (3) A. Engelhardt et al., Postgrad. Med. J., 1975, 51 (suppl.7), 82. (4) Arzneim.-Forsch., 1976, 26, 974-1020.

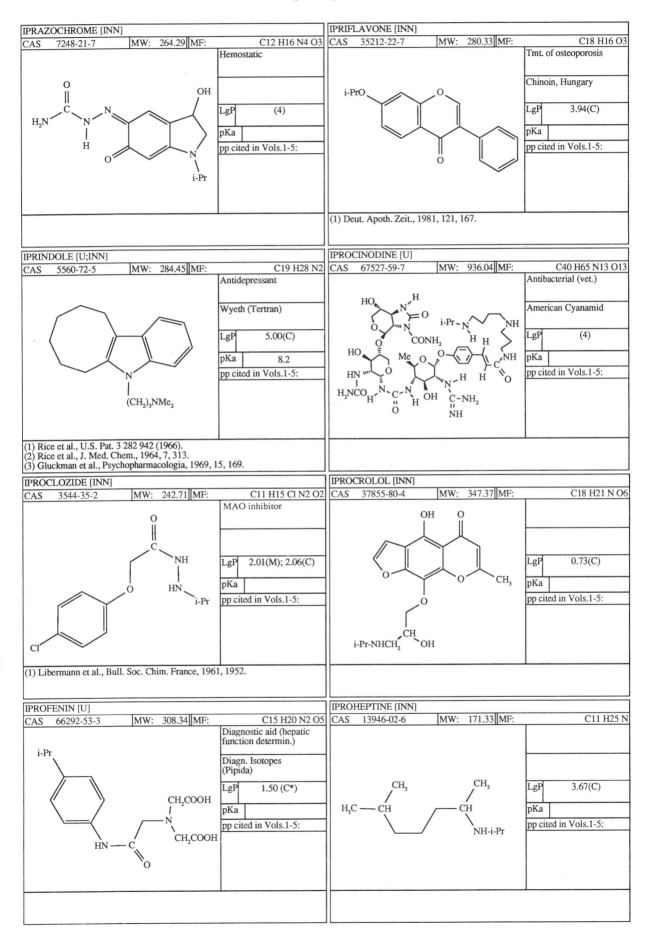

IPRAZOCHROME [INN]

CAS	7248-21-7	MW:	264.29	MF:	C12 H16 N4 O3

Hemostatic

LgP	(4)
pKa	
pp cited in Vols.1-5:	

IPRIFLAVONE [INN]

CAS	35212-22-7	MW:	280.33	MF:	C18 H16 O3

Tmt. of osteoporosis

Chinoin, Hungary

LgP	3.94(C)
pKa	
pp cited in Vols.1-5:	

(1) Deut. Apoth. Zeit., 1981, 121, 167.

IPRINDOLE [U;INN]

CAS	5560-72-5	MW:	284.45	MF:	C19 H28 N2

Antidepressant

Wyeth (Tertran)

LgP	5.00(C)
pKa	8.2
pp cited in Vols.1-5:	

(1) Rice et al., U.S. Pat. 3 282 942 (1966).
(2) Rice et al., J. Med. Chem., 1964, 7, 313.
(3) Gluckman et al., Psychopharmacologia, 1969, 15, 169.

IPROCINODINE [U]

CAS	67527-59-7	MW:	936.04	MF:	C40 H65 N13 O13

Antibacterial (vet.)

American Cyanamid

LgP	(4)
pKa	
pp cited in Vols.1-5:	

IPROCLOZIDE [INN]

CAS	3544-35-2	MW:	242.71	MF:	C11 H15 Cl N2 O2

MAO inhibitor

LgP	2.01(M); 2.06(C)
pKa	
pp cited in Vols.1-5:	

(1) Libermann et al., Bull. Soc. Chim. France, 1961, 1952.

IPROCROLOL [INN]

CAS	37855-80-4	MW:	347.37	MF:	C18 H21 N O6

LgP	0.73(C)
pKa	
pp cited in Vols.1-5:	

IPROFENIN [U]

CAS	66292-53-3	MW:	308.34	MF:	C15 H20 N2 O5

Diagnostic aid (hepatic function determin.)

Diagn. Isotopes (Pipida)

LgP	1.50 (C*)
pKa	
pp cited in Vols.1-5:	

IPROHEPTINE [INN]

CAS	13946-02-6	MW:	171.33	MF:	C11 H25 N

LgP	3.67(C)
pKa	
pp cited in Vols.1-5:	

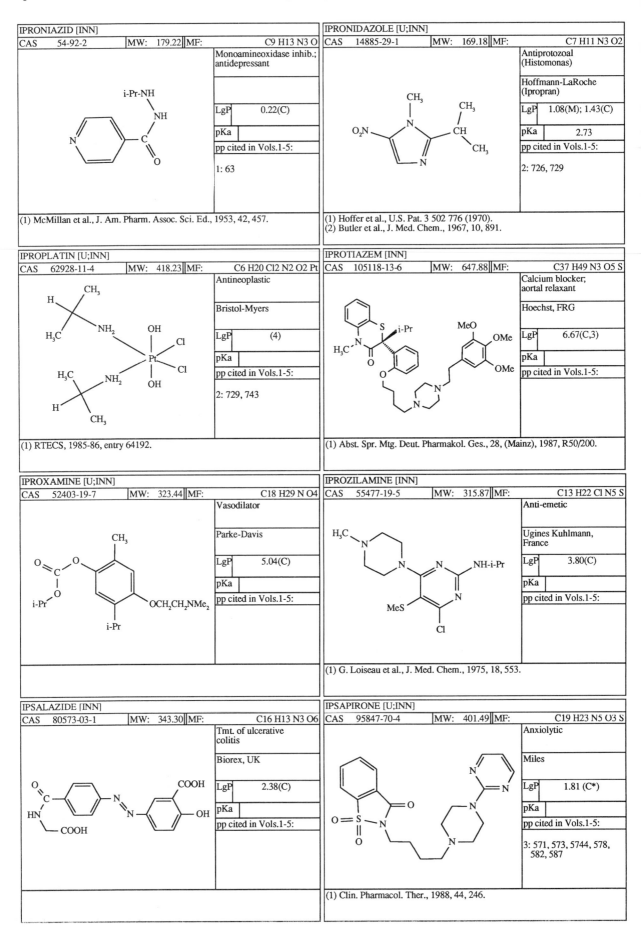

IPRONIAZID [INN]		
CAS 54-92-2	MW: 179.22 MF:	C9 H13 N3 O

Monoamineoxidase inhib.; antidepressant

LgP	0.22(C)
pKa	
pp cited in Vols.1-5:	

1: 63

(1) McMillan et al., J. Am. Pharm. Assoc. Sci. Ed., 1953, 42, 457.

IPRONIDAZOLE [U;INN]		
CAS 14885-29-1	MW: 169.18 MF:	C7 H11 N3 O2

Antiprotozoal (Histomonas)

Hoffmann-LaRoche (Ipropran)

LgP	1.08(M); 1.43(C)
pKa	2.73
pp cited in Vols.1-5:	

2: 726, 729

(1) Hoffer et al., U.S. Pat. 3 502 776 (1970).
(2) Butler et al., J. Med. Chem., 1967, 10, 891.

IPROPLATIN [U;INN]		
CAS 62928-11-4	MW: 418.23 MF:	C6 H20 Cl2 N2 O2 Pt

Antineoplastic

Bristol-Myers

LgP	(4)
pKa	
pp cited in Vols.1-5:	

2: 729, 743

(1) RTECS, 1985-86, entry 64192.

IPROTIAZEM [INN]		
CAS 105118-13-6	MW: 647.88 MF:	C37 H49 N3 O5 S

Calcium blocker; aortal relaxant

Hoechst, FRG

LgP	6.67(C,3)
pKa	
pp cited in Vols.1-5:	

(1) Abst. Spr. Mtg. Deut. Pharmakol. Ges., 28, (Mainz), 1987, R50/200.

IPROXAMINE [U;INN]		
CAS 52403-19-7	MW: 323.44 MF:	C18 H29 N O4

Vasodilator

Parke-Davis

LgP	5.04(C)
pKa	
pp cited in Vols.1-5:	

IPROZILAMINE [INN]		
CAS 55477-19-5	MW: 315.87 MF:	C13 H22 Cl N5 S

Anti-emetic

Ugines Kuhlmann, France

LgP	3.80(C)
pKa	
pp cited in Vols.1-5:	

(1) G. Loiseau et al., J. Med. Chem., 1975, 18, 553.

IPSALAZIDE [INN]		
CAS 80573-03-1	MW: 343.30 MF:	C16 H13 N3 O6

Tmt. of ulcerative colitis

Biorex, UK

LgP	2.38(C)
pKa	
pp cited in Vols.1-5:	

IPSAPIRONE [U;INN]		
CAS 95847-70-4	MW: 401.49 MF:	C19 H23 N5 O3 S

Anxiolytic

Miles

LgP	1.81 (C*)
pKa	
pp cited in Vols.1-5:	

3: 571, 573, 5744, 578, 582, 587

(1) Clin. Pharmacol. Ther., 1988, 44, 246.

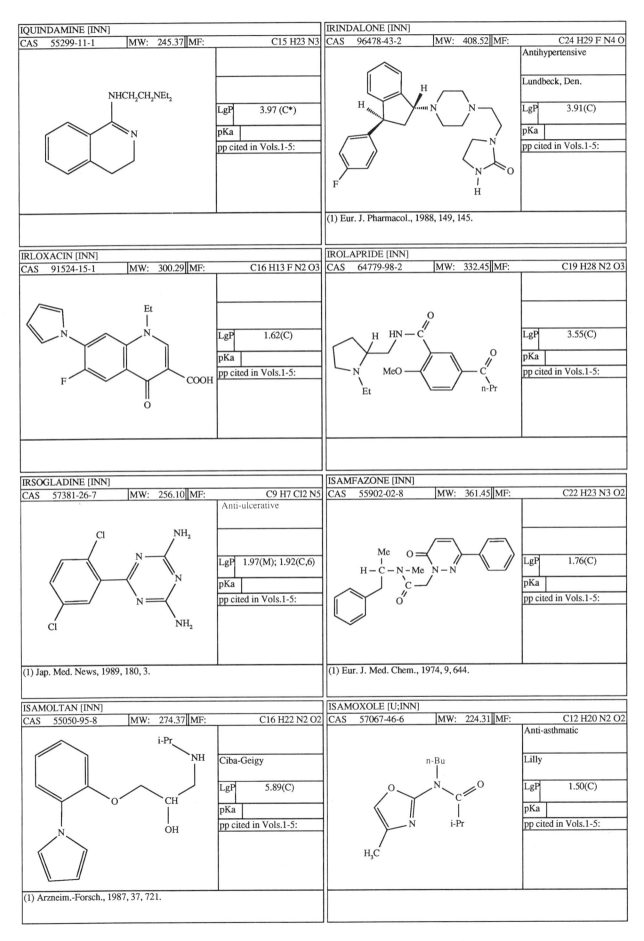

IQUINDAMINE [INN]				
CAS 55299-11-1	MW: 245.37	MF:		C15 H23 N3
			LgP	3.97 (C*)
			pKa	
			pp cited in Vols.1-5:	

IRINDALONE [INN]				
CAS 96478-43-2	MW: 408.52	MF:		C24 H29 F N4 O
		Antihypertensive		
		Lundbeck, Den.		
			LgP	3.91(C)
			pKa	
			pp cited in Vols.1-5:	
(1) Eur. J. Pharmacol., 1988, 149, 145.				

IRLOXACIN [INN]				
CAS 91524-15-1	MW: 300.29	MF:		C16 H13 F N2 O3
			LgP	1.62(C)
			pKa	
			pp cited in Vols.1-5:	

IROLAPRIDE [INN]				
CAS 64779-98-2	MW: 332.45	MF:		C19 H28 N2 O3
			LgP	3.55(C)
			pKa	
			pp cited in Vols.1-5:	

IRSOGLADINE [INN]				
CAS 57381-26-7	MW: 256.10	MF:		C9 H7 Cl2 N5
		Anti-ulcerative		
			LgP	1.97(M); 1.92(C,6)
			pKa	
			pp cited in Vols.1-5:	
(1) Jap. Med. News, 1989, 180, 3.				

ISAMFAZONE [INN]				
CAS 55902-02-8	MW: 361.45	MF:		C22 H23 N3 O2
			LgP	1.76(C)
			pKa	
			pp cited in Vols.1-5:	
(1) Eur. J. Med. Chem., 1974, 9, 644.				

ISAMOLTAN [INN]				
CAS 55050-95-8	MW: 274.37	MF:		C16 H22 N2 O2
		Ciba-Geigy		
			LgP	5.89(C)
			pKa	
			pp cited in Vols.1-5:	
(1) Arzneim.-Forsch., 1987, 37, 721.				

ISAMOXOLE [U;INN]				
CAS 57067-46-6	MW: 224.31	MF:		C12 H20 N2 O2
		Anti-asthmatic		
		Lilly		
			LgP	1.50(C)
			pKa	
			pp cited in Vols.1-5:	

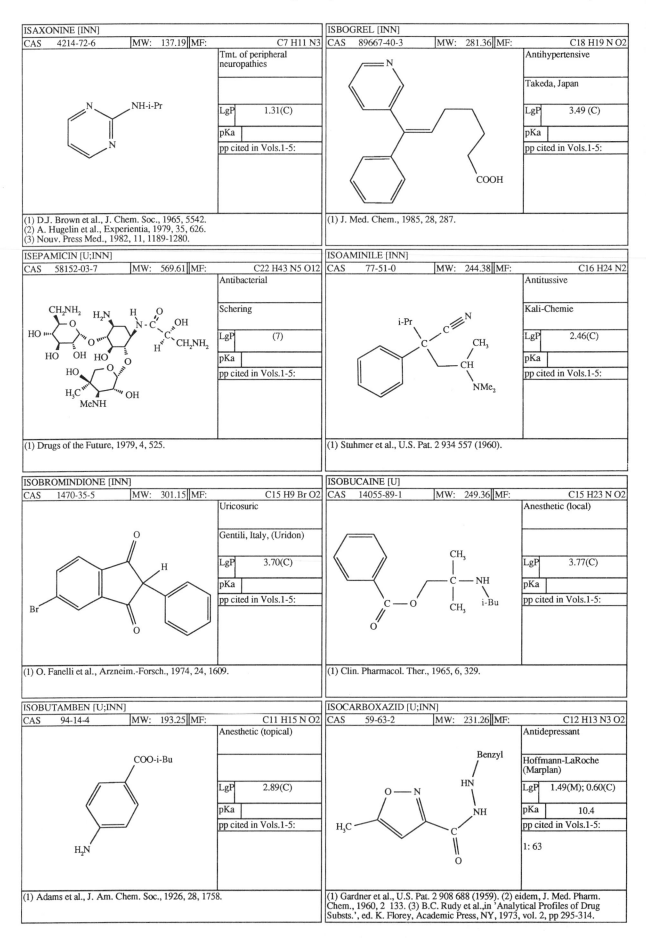

ISAXONINE [INN]

CAS	4214-72-6	MW:	137.19	MF:	C7 H11 N3

Tmt. of peripheral neuropathies

LgP	1.31(C)
pKa	
pp cited in Vols.1-5:	

(1) D.J. Brown et al., J. Chem. Soc., 1965, 5542.
(2) A. Hugelin et al., Experientia, 1979, 35, 626.
(3) Nouv. Press Med., 1982, 11, 1189-1280.

ISBOGREL [INN]

CAS	89667-40-3	MW:	281.36	MF:	C18 H19 N O2

Antihypertensive

Takeda, Japan

LgP	3.49 (C)
pKa	
pp cited in Vols.1-5:	

(1) J. Med. Chem., 1985, 28, 287.

ISEPAMICIN [U;INN]

CAS	58152-03-7	MW:	569.61	MF:	C22 H43 N5 O12

Antibacterial

Schering

LgP	(7)
pKa	
pp cited in Vols.1-5:	

(1) Drugs of the Future, 1979, 4, 525.

ISOAMINILE [INN]

CAS	77-51-0	MW:	244.38	MF:	C16 H24 N2

Antitussive

Kali-Chemie

LgP	2.46(C)
pKa	
pp cited in Vols.1-5:	

(1) Stuhmer et al., U.S. Pat. 2 934 557 (1960).

ISOBROMINDIONE [INN]

CAS	1470-35-5	MW:	301.15	MF:	C15 H9 Br O2

Uricosuric

Gentili, Italy, (Uridon)

LgP	3.70(C)
pKa	
pp cited in Vols.1-5:	

(1) O. Fanelli et al., Arzneim.-Forsch., 1974, 24, 1609.

ISOBUCAINE [U]

CAS	14055-89-1	MW:	249.36	MF:	C15 H23 N O2

Anesthetic (local)

LgP	3.77(C)
pKa	
pp cited in Vols.1-5:	

(1) Clin. Pharmacol. Ther., 1965, 6, 329.

ISOBUTAMBEN [U;INN]

CAS	94-14-4	MW:	193.25	MF:	C11 H15 N O2

Anesthetic (topical)

LgP	2.89(C)
pKa	
pp cited in Vols.1-5:	

(1) Adams et al., J. Am. Chem. Soc., 1926, 28, 1758.

ISOCARBOXAZID [U;INN]

CAS	59-63-2	MW:	231.26	MF:	C12 H13 N3 O2

Antidepressant

Hoffmann-LaRoche (Marplan)

LgP	1.49(M); 0.60(C)
pKa	10.4
pp cited in Vols.1-5:	
	1: 63

(1) Gardner et al., U.S. Pat. 2 908 688 (1959). (2) eidem, J. Med. Pharm. Chem., 1960, 2 133. (3) B.C. Rudy et al.,in 'Analytical Profiles of Drug Substs.', ed. K. Florey, Academic Press, NY, 1973, vol. 2, pp 295-314.

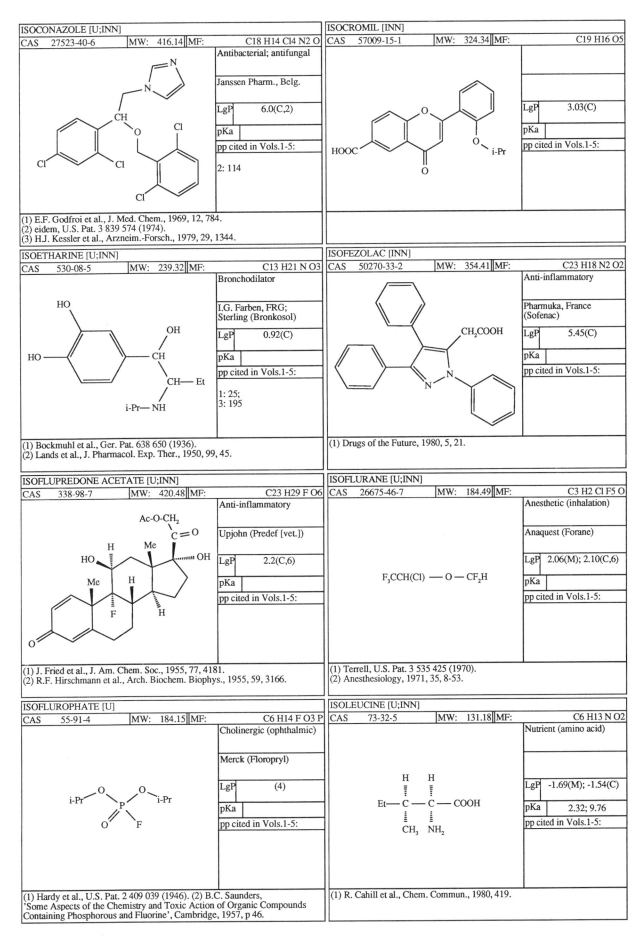

ISOCONAZOLE [U;INN]

CAS	27523-40-6	MW:	416.14	MF:	C18 H14 Cl4 N2 O

Antibacterial; antifungal

Janssen Pharm., Belg.

LgP	6.0(C,2)
pKa	

pp cited in Vols.1-5:

2: 114

(1) E.F. Godfroi et al., J. Med. Chem., 1969, 12, 784.
(2) eidem, U.S. Pat. 3 839 574 (1974).
(3) H.J. Kessler et al., Arzneim.-Forsch., 1979, 29, 1344.

ISOCROMIL [INN]

CAS	57009-15-1	MW:	324.34	MF:	C19 H16 O5

LgP	3.03(C)
pKa	

pp cited in Vols.1-5:

ISOETHARINE [U;INN]

CAS	530-08-5	MW:	239.32	MF:	C13 H21 N O3

Bronchodilator

I.G. Farben, FRG;
Sterling (Bronkosol)

LgP	0.92(C)
pKa	

pp cited in Vols.1-5:

1: 25;
3: 195

(1) Bockmuhl et al., Ger. Pat. 638 650 (1936).
(2) Lands et al., J. Pharmacol. Exp. Ther., 1950, 99, 45.

ISOFEZOLAC [INN]

CAS	50270-33-2	MW:	354.41	MF:	C23 H18 N2 O2

Anti-inflammatory

Pharmuka, France
(Sofenac)

LgP	5.45(C)
pKa	

pp cited in Vols.1-5:

(1) Drugs of the Future, 1980, 5, 21.

ISOFLUPREDONE ACETATE [U;INN]

CAS	338-98-7	MW:	420.48	MF:	C23 H29 F O6

Anti-inflammatory

Upjohn (Predef [vet.])

LgP	2.2(C,6)
pKa	

pp cited in Vols.1-5:

(1) J. Fried et al., J. Am. Chem. Soc., 1955, 77, 4181.
(2) R.F. Hirschmann et al., Arch. Biochem. Biophys., 1955, 59, 3166.

ISOFLURANE [U;INN]

CAS	26675-46-7	MW:	184.49	MF:	C3 H2 Cl F5 O

Anesthetic (inhalation)

Anaquest (Forane)

LgP	2.06(M); 2.10(C,6)
pKa	

pp cited in Vols.1-5:

$F_3CCH(Cl) — O — CF_2H$

(1) Terrell, U.S. Pat. 3 535 425 (1970).
(2) Anesthesiology, 1971, 35, 8-53.

ISOFLUROPHATE [U]

CAS	55-91-4	MW:	184.15	MF:	C6 H14 F O3 P

Cholinergic (ophthalmic)

Merck (Floropryl)

LgP	(4)
pKa	

pp cited in Vols.1-5:

(1) Hardy et al., U.S. Pat. 2 409 039 (1946). (2) B.C. Saunders,
'Some Aspects of the Chemistry and Toxic Action of Organic Compounds
Containing Phosphorous and Fluorine', Cambridge, 1957, p 46.

ISOLEUCINE [U;INN]

CAS	73-32-5	MW:	131.18	MF:	C6 H13 N O2

Nutrient (amino acid)

LgP	-1.69(M); -1.54(C)
pKa	2.32; 9.76

pp cited in Vols.1-5:

(1) R. Cahill et al., Chem. Commun., 1980, 419.

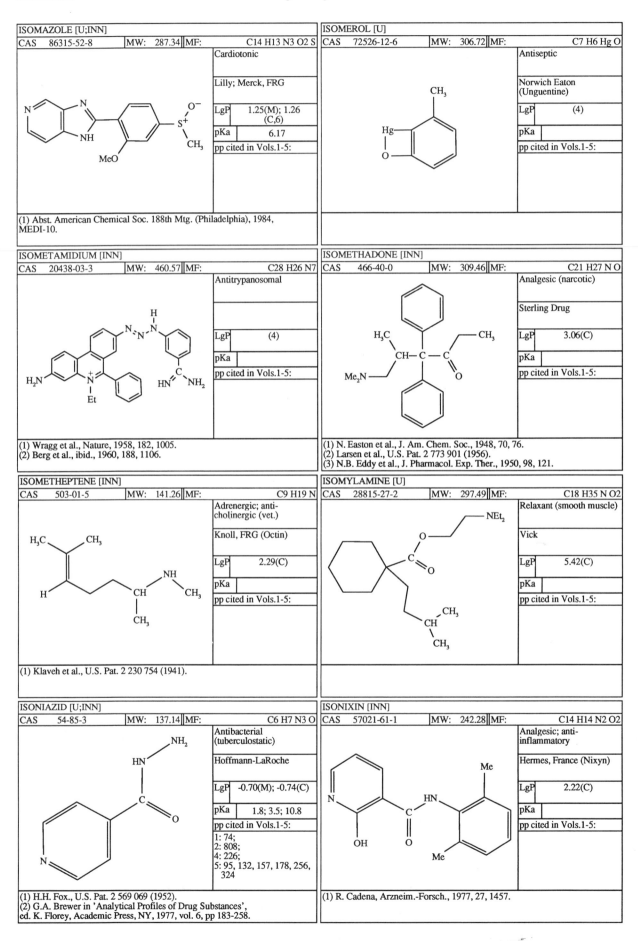

ISOMAZOLE [U;INN]			
CAS 86315-52-8	MW: 287.34	MF:	C14 H13 N3 O2 S

Cardiotonic

Lilly; Merck, FRG

LgP	1.25(M); 1.26 (C,6)
pKa	6.17
pp cited in Vols.1-5:	

(1) Abst. American Chemical Soc. 188th Mtg. (Philadelphia), 1984, MEDI-10.

ISOMEROL [U]			
CAS 72526-12-6	MW: 306.72	MF:	C7 H6 Hg O

Antiseptic

Norwich Eaton (Unguentine)

LgP	(4)
pKa	
pp cited in Vols.1-5:	

ISOMETAMIDIUM [INN]			
CAS 20438-03-3	MW: 460.57	MF:	C28 H26 N7

Antitrypanosomal

LgP	(4)
pKa	
pp cited in Vols.1-5:	

(1) Wragg et al., Nature, 1958, 182, 1005.
(2) Berg et al., ibid., 1960, 188, 1106.

ISOMETHADONE [INN]			
CAS 466-40-0	MW: 309.46	MF:	C21 H27 N O

Analgesic (narcotic)

Sterling Drug

LgP	3.06(C)
pKa	
pp cited in Vols.1-5:	

(1) N. Easton et al., J. Am. Chem. Soc., 1948, 70, 76.
(2) Larsen et al., U.S. Pat. 2 773 901 (1956).
(3) N.B. Eddy et al., J. Pharmacol. Exp. Ther., 1950, 98, 121.

ISOMETHEPTENE [INN]			
CAS 503-01-5	MW: 141.26	MF:	C9 H19 N

Adrenergic; anti-cholinergic (vet.)

Knoll, FRG (Octin)

LgP	2.29(C)
pKa	
pp cited in Vols.1-5:	

(1) Klaveh et al., U.S. Pat. 2 230 754 (1941).

ISOMYLAMINE [U]			
CAS 28815-27-2	MW: 297.49	MF:	C18 H35 N O2

Relaxant (smooth muscle)

Vick

LgP	5.42(C)
pKa	
pp cited in Vols.1-5:	

ISONIAZID [U;INN]			
CAS 54-85-3	MW: 137.14	MF:	C6 H7 N3 O

Antibacterial (tuberculostatic)

Hoffmann-LaRoche

LgP	-0.70(M); -0.74(C)
pKa	1.8; 3.5; 10.8
pp cited in Vols.1-5:	

1: 74;
2: 808;
4: 226;
5: 95, 132, 157, 178, 256, 324

(1) H.H. Fox., U.S. Pat. 2 569 069 (1952).
(2) G.A. Brewer in 'Analytical Profiles of Drug Substances', ed. K. Florey, Academic Press, NY, 1977, vol. 6, pp 183-258.

ISONIXIN [INN]			
CAS 57021-61-1	MW: 242.28	MF:	C14 H14 N2 O2

Analgesic; anti-inflammatory

Hermes, France (Nixyn)

LgP	2.22(C)
pKa	
pp cited in Vols.1-5:	

(1) R. Cadena, Arzneim.-Forsch., 1977, 27, 1457.

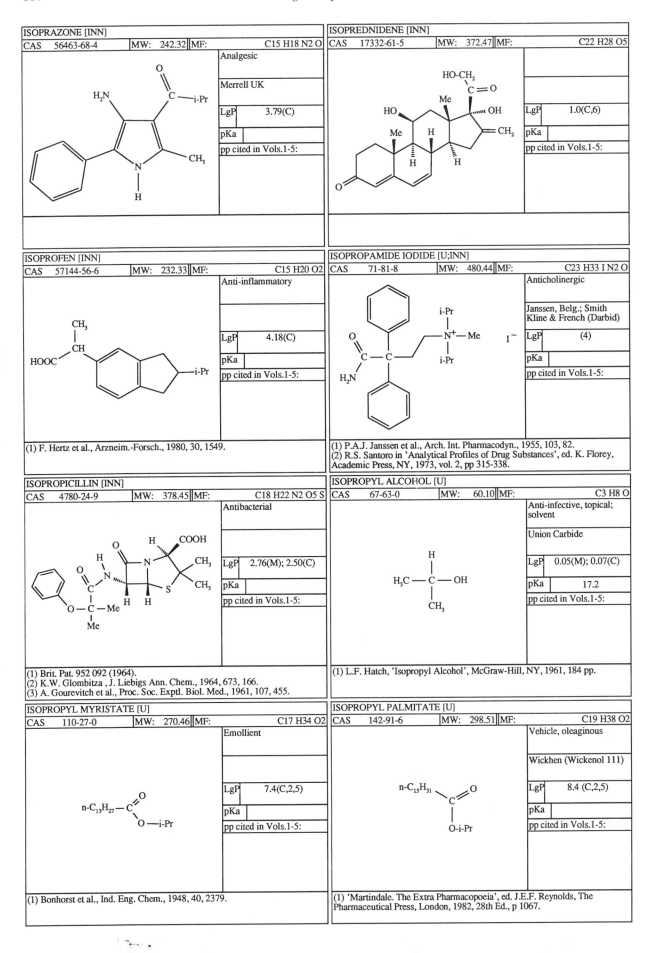

ISOPRAZONE [INN]

CAS	56463-68-4	MW:	242.32	MF:	C15 H18 N2 O

Analgesic

Merrell UK

LgP	3.79(C)
pKa	
pp cited in Vols.1-5:	

ISOPREDNIDENE [INN]

CAS	17332-61-5	MW:	372.47	MF:	C22 H28 O5

LgP	1.0(C,6)
pKa	
pp cited in Vols.1-5:	

ISOPROFEN [INN]

CAS	57144-56-6	MW:	232.33	MF:	C15 H20 O2

Anti-inflammatory

LgP	4.18(C)
pKa	
pp cited in Vols.1-5:	

(1) F. Hertz et al., Arzneim.-Forsch., 1980, 30, 1549.

ISOPROPAMIDE IODIDE [U;INN]

CAS	71-81-8	MW:	480.44	MF:	C23 H33 I N2 O

Anticholinergic

Janssen, Belg.; Smith Kline & French (Darbid)

LgP	(4)
pKa	
pp cited in Vols.1-5:	

(1) P.A.J. Janssen et al., Arch. Int. Pharmacodyn., 1955, 103, 82.
(2) R.S. Santoro in 'Analytical Profiles of Drug Substances', ed. K. Florey, Academic Press, NY, 1973, vol. 2, pp 315-338.

ISOPROPICILLIN [INN]

CAS	4780-24-9	MW:	378.45	MF:	C18 H22 N2 O5 S

Antibacterial

LgP	2.76(M); 2.50(C)
pKa	
pp cited in Vols.1-5:	

(1) Brit. Pat. 952 092 (1964).
(2) K.W. Glombitza , J. Liebigs Ann. Chem., 1964, 673, 166.
(3) A. Gourevitch et al., Proc. Soc. Exptl. Biol. Med., 1961, 107, 455.

ISOPROPYL ALCOHOL [U]

CAS	67-63-0	MW:	60.10	MF:	C3 H8 O

Anti-infective, topical; solvent

Union Carbide

LgP	0.05(M); 0.07(C)
pKa	17.2
pp cited in Vols.1-5:	

(1) L.F. Hatch, 'Isopropyl Alcohol', McGraw-Hill, NY, 1961, 184 pp.

ISOPROPYL MYRISTATE [U]

CAS	110-27-0	MW:	270.46	MF:	C17 H34 O2

Emollient

LgP	7.4(C,2,5)
pKa	
pp cited in Vols.1-5:	

(1) Bonhorst et al., Ind. Eng. Chem., 1948, 40, 2379.

ISOPROPYL PALMITATE [U]

CAS	142-91-6	MW:	298.51	MF:	C19 H38 O2

Vehicle, oleaginous

Wickhen (Wickenol 111)

LgP	8.4 (C,2,5)
pKa	
pp cited in Vols.1-5:	

(1) 'Martindale. The Extra Pharmacopoeia', ed. J.E.F. Reynolds, The Pharmaceutical Press, London, 1982, 28th Ed., p 1067.

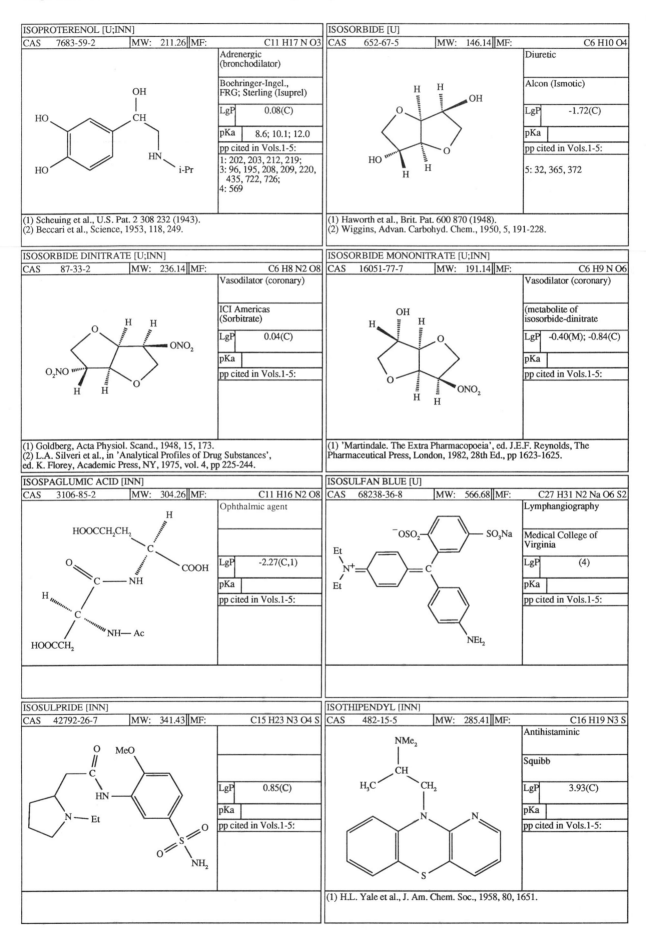

ISOPROTERENOL [U;INN]			
CAS 7683-59-2	MW: 211.26	MF:	C11 H17 N O3

Adrenergic (bronchodilator)

Boehringer-Ingel., FRG; Sterling (Isuprel)

LgP	0.08(C)

pKa	8.6; 10.1; 12.0

pp cited in Vols.1-5:
1: 202, 203, 212, 219;
3: 96, 195, 208, 209, 220, 435, 722, 726;
4: 569

(1) Scheuing et al., U.S. Pat. 2 308 232 (1943).
(2) Beccari et al., Science, 1953, 118, 249.

ISOSORBIDE [U]			
CAS 652-67-5	MW: 146.14	MF:	C6 H10 O4

Diuretic

Alcon (Ismotic)

LgP	-1.72(C)

pKa	

pp cited in Vols.1-5:

5: 32, 365, 372

(1) Haworth et al., Brit. Pat. 600 870 (1948).
(2) Wiggins, Advan. Carbohyd. Chem., 1950, 5, 191-228.

ISOSORBIDE DINITRATE [U;INN]			
CAS 87-33-2	MW: 236.14	MF:	C6 H8 N2 O8

Vasodilator (coronary)

ICI Americas (Sorbitrate)

LgP	0.04(C)

pKa	

pp cited in Vols.1-5:

(1) Goldberg, Acta Physiol. Scand., 1948, 15, 173.
(2) L.A. Silveri et al., in 'Analytical Profiles of Drug Substances', ed. K. Florey, Academic Press, NY, 1975, vol. 4, pp 225-244.

ISOSORBIDE MONONITRATE [U;INN]			
CAS 16051-77-7	MW: 191.14	MF:	C6 H9 N O6

Vasodilator (coronary)

(metabolite of isosorbide-dinitrate

LgP	-0.40(M); -0.84(C)

pKa	

pp cited in Vols.1-5:

(1) 'Martindale. The Extra Pharmacopoeia', ed. J.E.F. Reynolds, The Pharmaceutical Press, London, 1982, 28th Ed., pp 1623-1625.

ISOSPAGLUMIC ACID [INN]			
CAS 3106-85-2	MW: 304.26	MF:	C11 H16 N2 O8

Ophthalmic agent

LgP	-2.27(C,1)

pKa	

pp cited in Vols.1-5:

ISOSULFAN BLUE [U]			
CAS 68238-36-8	MW: 566.68	MF:	C27 H31 N2 Na O6 S2

Lymphangiography

Medical College of Virginia

LgP	(4)

pKa	

pp cited in Vols.1-5:

ISOSULPRIDE [INN]			
CAS 42792-26-7	MW: 341.43	MF:	C15 H23 N3 O4 S

LgP	0.85(C)

pKa	

pp cited in Vols.1-5:

ISOTHIPENDYL [INN]			
CAS 482-15-5	MW: 285.41	MF:	C16 H19 N3 S

Antihistaminic

Squibb

LgP	3.93(C)

pKa	

pp cited in Vols.1-5:

(1) H.L. Yale et al., J. Am. Chem. Soc., 1958, 80, 1651.

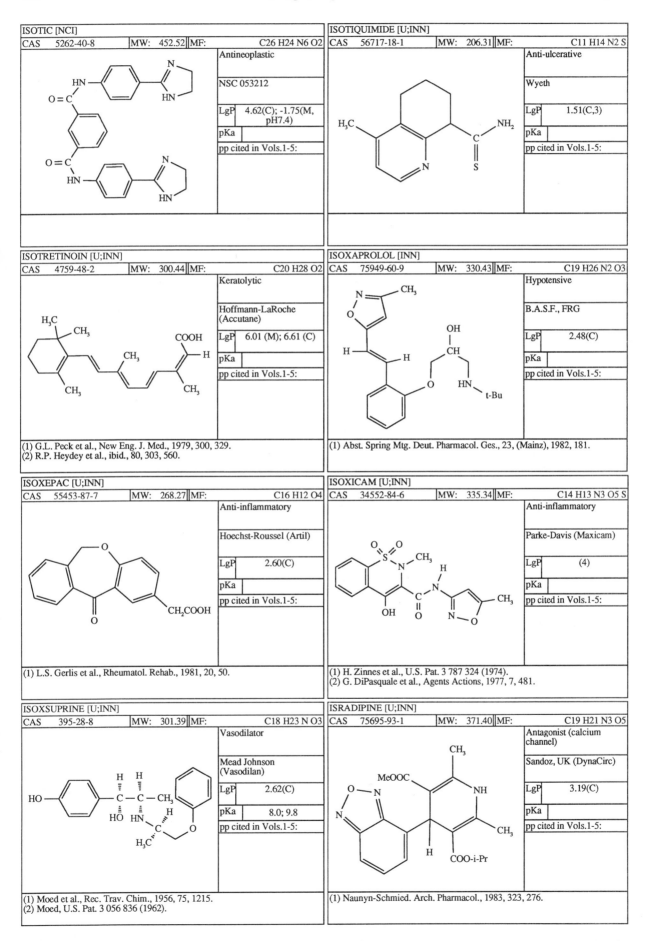

ISOTIC [NCI]				
CAS	5262-40-8	MW: 452.52	MF:	C26 H24 N6 O2

Antineoplastic

NSC 053212

LgP	4.62(C); -1.75(M, pH7.4)
pKa	
pp cited in Vols.1-5:	

ISOTIQUIMIDE [U;INN]				
CAS	56717-18-1	MW: 206.31	MF:	C11 H14 N2 S

Anti-ulcerative

Wyeth

LgP	1.51(C,3)
pKa	
pp cited in Vols.1-5:	

ISOTRETINOIN [U;INN]				
CAS	4759-48-2	MW: 300.44	MF:	C20 H28 O2

Keratolytic

Hoffmann-LaRoche (Accutane)

LgP	6.01 (M); 6.61 (C)
pKa	
pp cited in Vols.1-5:	

(1) G.L. Peck et al., New Eng. J. Med., 1979, 300, 329.
(2) R.P. Heydey et al., ibid., 80, 303, 560.

ISOXAPROLOL [INN]				
CAS	75949-60-9	MW: 330.43	MF:	C19 H26 N2 O3

Hypotensive

B.A.S.F., FRG

LgP	2.48(C)
pKa	
pp cited in Vols.1-5:	

(1) Abst. Spring Mtg. Deut. Pharmacol. Ges., 23, (Mainz), 1982, 181.

ISOXEPAC [U;INN]				
CAS	55453-87-7	MW: 268.27	MF:	C16 H12 O4

Anti-inflammatory

Hoechst-Roussel (Artil)

LgP	2.60(C)
pKa	
pp cited in Vols.1-5:	

(1) L.S. Gerlis et al., Rheumatol. Rehab., 1981, 20, 50.

ISOXICAM [U;INN]				
CAS	34552-84-6	MW: 335.34	MF:	C14 H13 N3 O5 S

Anti-inflammatory

Parke-Davis (Maxicam)

LgP	(4)
pKa	
pp cited in Vols.1-5:	

(1) H. Zinnes et al., U.S. Pat. 3 787 324 (1974).
(2) G. DiPasquale et al., Agents Actions, 1977, 7, 481.

ISOXSUPRINE [U;INN]				
CAS	395-28-8	MW: 301.39	MF:	C18 H23 N O3

Vasodilator

Mead Johnson (Vasodilan)

LgP	2.62(C)
pKa	8.0; 9.8
pp cited in Vols.1-5:	

(1) Moed et al., Rec. Trav. Chim., 1956, 75, 1215.
(2) Moed, U.S. Pat. 3 056 836 (1962).

ISRADIPINE [U;INN]				
CAS	75695-93-1	MW: 371.40	MF:	C19 H21 N3 O5

Antagonist (calcium channel)

Sandoz, UK (DynaCirc)

LgP	3.19(C)
pKa	
pp cited in Vols.1-5:	

(1) Naunyn-Schmied. Arch. Pharmacol., 1983, 323, 276.

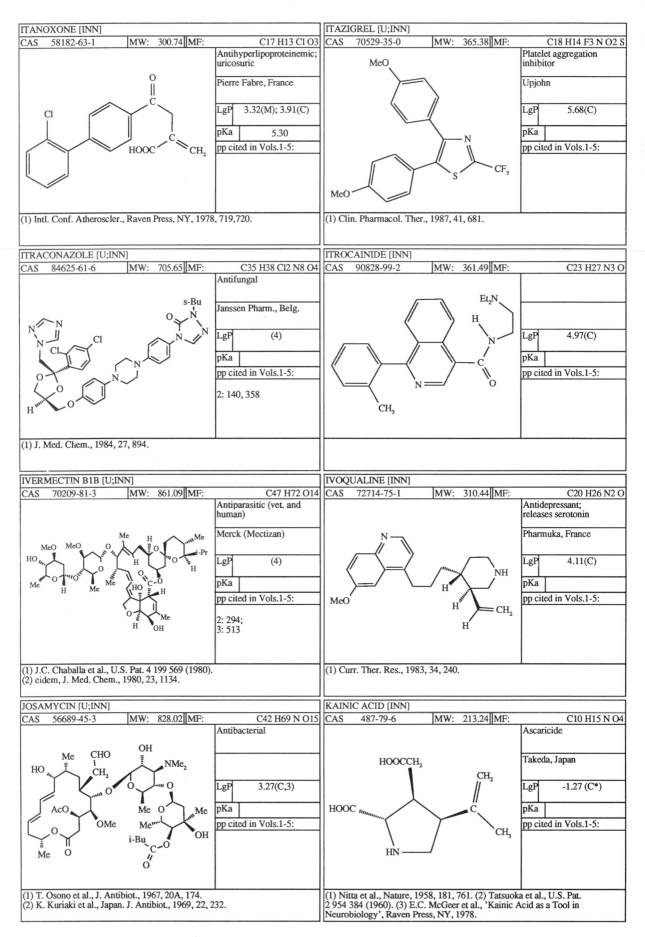

ITANOXONE [INN]

| CAS | 58182-63-1 | MW: | 300.74 | MF: | C17 H13 Cl O3 |

Antihyperlipoproteinemic; uricosuric

Pierre Fabre, France

LgP   3.32(M); 3.91(C)

pKa   5.30

pp cited in Vols.1-5:

(1) Intl. Conf. Atheroscler., Raven Press, NY, 1978, 719,720.

ITAZIGREL [U;INN]

| CAS | 70529-35-0 | MW: | 365.38 | MF: | C18 H14 F3 N O2 S |

Platelet aggregation inhibitor

Upjohn

LgP   5.68(C)

pKa

pp cited in Vols.1-5:

(1) Clin. Pharmacol. Ther., 1987, 41, 681.

ITRACONAZOLE [U;INN]

| CAS | 84625-61-6 | MW: | 705.65 | MF: | C35 H38 Cl2 N8 O4 |

Antifungal

Janssen Pharm., Belg.

LgP   (4)

pKa

pp cited in Vols.1-5:

2: 140, 358

(1) J. Med. Chem., 1984, 27, 894.

ITROCAINIDE [INN]

| CAS | 90828-99-2 | MW: | 361.49 | MF: | C23 H27 N3 O |

LgP   4.97(C)

pKa

pp cited in Vols.1-5:

IVERMECTIN B1B [U;INN]

| CAS | 70209-81-3 | MW: | 861.09 | MF: | C47 H72 O14 |

Antiparasitic (vet. and human)

Merck (Mectizan)

LgP   (4)

pKa

pp cited in Vols.1-5:

2: 294;
3: 513

(1) J.C. Chaballa et al., U.S. Pat. 4 199 569 (1980).
(2) eidem, J. Med. Chem., 1980, 23, 1134.

IVOQUALINE [INN]

| CAS | 72714-75-1 | MW: | 310.44 | MF: | C20 H26 N2 O |

Antidepressant; releases serotonin

Pharmuka, France

LgP   4.11(C)

pKa

pp cited in Vols.1-5:

(1) Curr. Ther. Res., 1983, 34, 240.

JOSAMYCIN [U;INN]

| CAS | 56689-45-3 | MW: | 828.02 | MF: | C42 H69 N O15 |

Antibacterial

LgP   3.27(C,3)

pKa

pp cited in Vols.1-5:

(1) T. Osono et al., J. Antibiot., 1967, 20A, 174.
(2) K. Kuriaki et al., Japan. J. Antibiot., 1969, 22, 232.

KAINIC ACID [INN]

| CAS | 487-79-6 | MW: | 213.24 | MF: | C10 H15 N O4 |

Ascaricide

Takeda, Japan

LgP   -1.27 (C*)

pKa

pp cited in Vols.1-5:

(1) Nitta et al., Nature, 1958, 181, 761. (2) Tatsuoka et al., U.S. Pat. 2 954 384 (1960). (3) E.C. McGeer et al., 'Kainic Acid as a Tool in Neurobiology', Raven Press, NY, 1978.

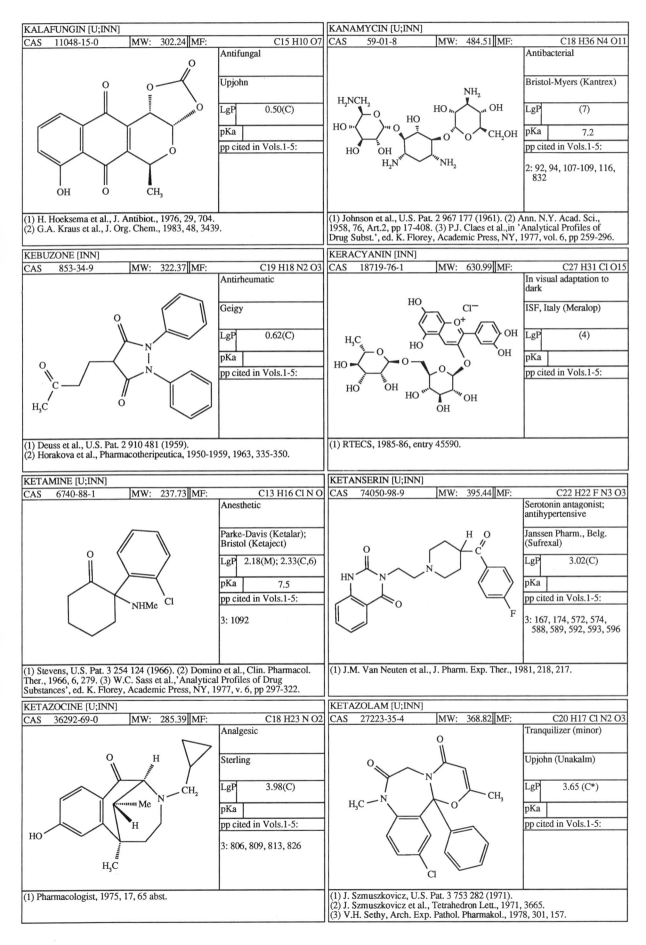

KALAFUNGIN [U;INN]

CAS	11048-15-0	MW:	302.24	MF:	C15 H10 O7

Antifungal

Upjohn

LgP	0.50(C)
pKa	
pp cited in Vols.1-5:	

(1) H. Hoeksema et al., J. Antibiot., 1976, 29, 704.
(2) G.A. Kraus et al., J. Org. Chem., 1983, 48, 3439.

KANAMYCIN [U;INN]

CAS	59-01-8	MW:	484.51	MF:	C18 H36 N4 O11

Antibacterial

Bristol-Myers (Kantrex)

LgP	(7)
pKa	7.2
pp cited in Vols.1-5:	

2: 92, 94, 107-109, 116, 832

(1) Johnson et al., U.S. Pat. 2 967 177 (1961). (2) Ann. N.Y. Acad. Sci., 1958, 76, Art.2, pp 17-408. (3) P.J. Claes et al.,in 'Analytical Profiles of Drug Subst.', ed. K. Florey, Academic Press, NY, 1977, vol. 6, pp 259-296.

KEBUZONE [INN]

CAS	853-34-9	MW:	322.37	MF:	C19 H18 N2 O3

Antirheumatic

Geigy

LgP	0.62(C)
pKa	
pp cited in Vols.1-5:	

(1) Deuss et al., U.S. Pat. 2 910 481 (1959).
(2) Horakova et al., Pharmacotheripeutica, 1950-1959, 1963, 335-350.

KERACYANIN [INN]

CAS	18719-76-1	MW:	630.99	MF:	C27 H31 Cl O15

In visual adaptation to dark

ISF, Italy (Meralop)

LgP	(4)
pKa	
pp cited in Vols.1-5:	

(1) RTECS, 1985-86, entry 45590.

KETAMINE [U;INN]

CAS	6740-88-1	MW:	237.73	MF:	C13 H16 Cl N O

Anesthetic

Parke-Davis (Ketalar); Bristol (Ketaject)

LgP	2.18(M); 2.33(C,6)
pKa	7.5
pp cited in Vols.1-5:	

3: 1092

(1) Stevens, U.S. Pat. 3 254 124 (1966). (2) Domino et al., Clin. Pharmacol. Ther., 1966, 6, 279. (3) W.C. Sass et al.,'Analytical Profiles of Drug Substances', ed. K. Florey, Academic Press, NY, 1977, v. 6, pp 297-322.

KETANSERIN [U;INN]

CAS	74050-98-9	MW:	395.44	MF:	C22 H22 F N3 O3

Serotonin antagonist; antihypertensive

Janssen Pharm., Belg. (Sufrexal)

LgP	3.02(C)
pKa	
pp cited in Vols.1-5:	

3: 167, 174, 572, 574, 588, 589, 592, 593, 596

(1) J.M. Van Neuten et al., J. Pharm. Exp. Ther., 1981, 218, 217.

KETAZOCINE [U;INN]

CAS	36292-69-0	MW:	285.39	MF:	C18 H23 N O2

Analgesic

Sterling

LgP	3.98(C)
pKa	
pp cited in Vols.1-5:	

3: 806, 809, 813, 826

(1) Pharmacologist, 1975, 17, 65 abst.

KETAZOLAM [U;INN]

CAS	27223-35-4	MW:	368.82	MF:	C20 H17 Cl N2 O3

Tranquilizer (minor)

Upjohn (Unakalm)

LgP	3.65 (C*)
pKa	
pp cited in Vols.1-5:	

(1) J. Szmuszkovicz, U.S. Pat. 3 753 282 (1971).
(2) J. Szmuszkovicz et al., Tetrahedron Lett., 1971, 3665.
(3) V.H. Sethy, Arch. Exp. Pathol. Pharmakol., 1978, 301, 157.

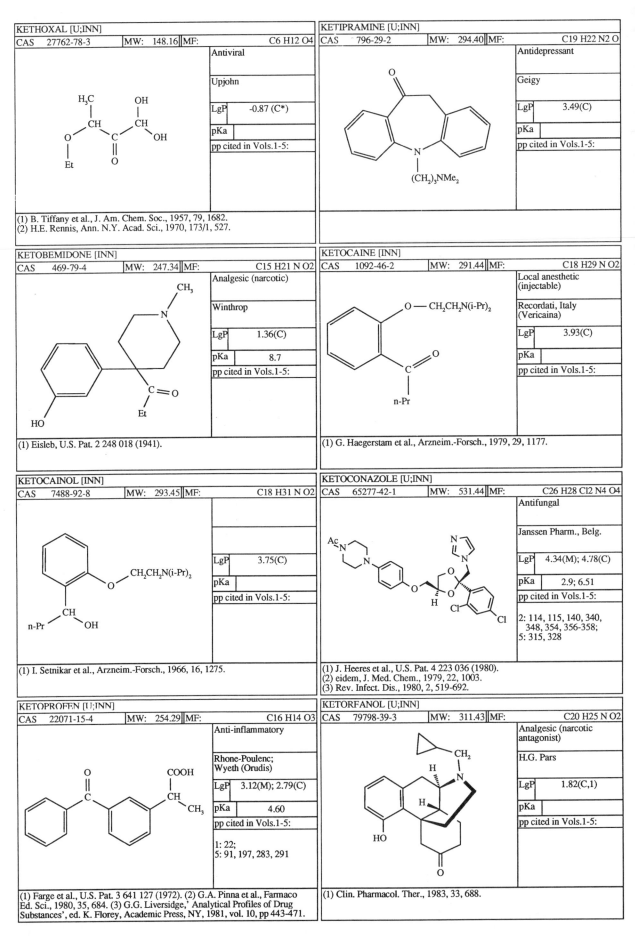

KETHOXAL [U;INN]

CAS	27762-78-3	MW: 148.16	MF:	C6 H12 O4

Antiviral

Upjohn

LgP   -0.87 (C*)

pKa

pp cited in Vols.1-5:

(1) B. Tiffany et al., J. Am. Chem. Soc., 1957, 79, 1682.
(2) H.E. Rennis, Ann. N.Y. Acad. Sci., 1970, 173/1, 527.

KETIPRAMINE [U;INN]

CAS	796-29-2	MW: 294.40	MF:	C19 H22 N2 O

Antidepressant

Geigy

LgP   3.49(C)

pKa

pp cited in Vols.1-5:

KETOBEMIDONE [INN]

CAS	469-79-4	MW: 247.34	MF:	C15 H21 N O2

Analgesic (narcotic)

Winthrop

LgP   1.36(C)

pKa   8.7

pp cited in Vols.1-5:

(1) Eisleb, U.S. Pat. 2 248 018 (1941).

KETOCAINE [INN]

CAS	1092-46-2	MW: 291.44	MF:	C18 H29 N O2

Local anesthetic (injectable)

Recordati, Italy (Vericaina)

LgP   3.93(C)

pKa

pp cited in Vols.1-5:

(1) G. Haegerstam et al., Arzneim.-Forsch., 1979, 29, 1177.

KETOCAINOL [INN]

CAS	7488-92-8	MW: 293.45	MF:	C18 H31 N O2

LgP   3.75(C)

pKa

pp cited in Vols.1-5:

(1) I. Setnikar et al., Arzneim.-Forsch., 1966, 16, 1275.

KETOCONAZOLE [U;INN]

CAS	65277-42-1	MW: 531.44	MF:	C26 H28 Cl2 N4 O4

Antifungal

Janssen Pharm., Belg.

LgP   4.34(M); 4.78(C)

pKa   2.9; 6.51

pp cited in Vols.1-5:

2: 114, 115, 140, 340, 348, 354, 356-358;
5: 315, 328

(1) J. Heeres et al., U.S. Pat. 4 223 036 (1980).
(2) eidem, J. Med. Chem., 1979, 22, 1003.
(3) Rev. Infect. Dis., 1980, 2, 519-692.

KETOPROFEN [U;INN]

CAS	22071-15-4	MW: 254.29	MF:	C16 H14 O3

Anti-inflammatory

Rhone-Poulenc; Wyeth (Orudis)

LgP   3.12(M); 2.79(C)

pKa   4.60

pp cited in Vols.1-5:

1: 22;
5: 91, 197, 283, 291

(1) Farge et al., U.S. Pat. 3 641 127 (1972). (2) G.A. Pinna et al., Farmaco Ed. Sci., 1980, 35, 684. (3) G.G. Liversidge,' Analytical Profiles of Drug Substances', ed. K. Florey, Academic Press, NY, 1981, vol. 10, pp 443-471.

KETORFANOL [U;INN]

CAS	79798-39-3	MW: 311.43	MF:	C20 H25 N O2

Analgesic (narcotic antagonist)

H.G. Pars

LgP   1.82(C,1)

pKa

pp cited in Vols.1-5:

(1) Clin. Pharmacol. Ther., 1983, 33, 688.

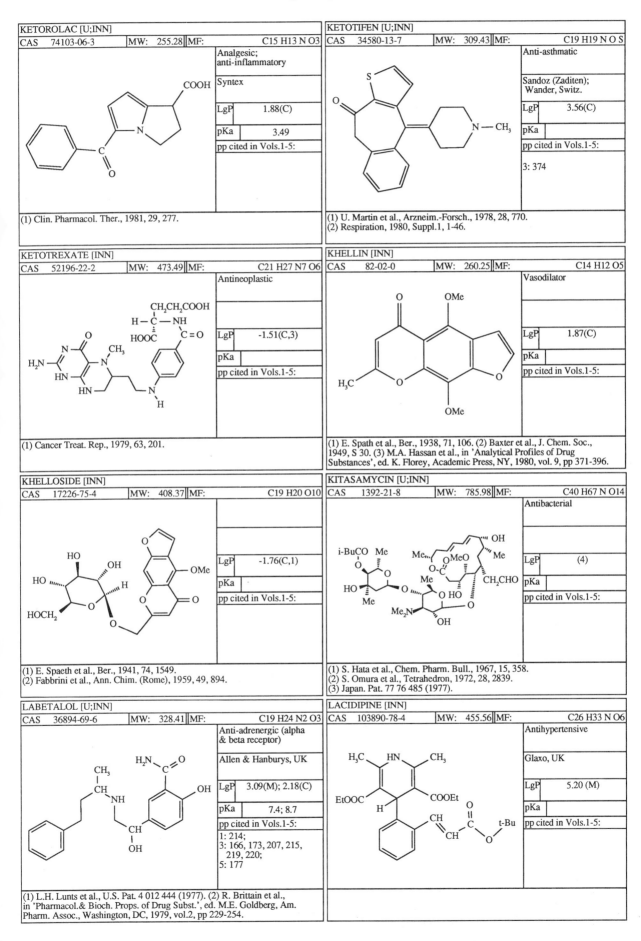

KETOROLAC [U;INN]

CAS	74103-06-3	MW:	255.28	MF:	C15 H13 N O3

Analgesic; anti-inflammatory

Syntex

LgP	1.88(C)
pKa	3.49
pp cited in Vols.1-5:	

(1) Clin. Pharmacol. Ther., 1981, 29, 277.

KETOTIFEN [U;INN]

CAS	34580-13-7	MW:	309.43	MF:	C19 H19 N O S

Anti-asthmatic

Sandoz (Zaditen); Wander, Switz.

LgP	3.56(C)
pKa	
pp cited in Vols.1-5:	

3: 374

(1) U. Martin et al., Arzneim.-Forsch., 1978, 28, 770.
(2) Respiration, 1980, Suppl.1, 1-46.

KETOTREXATE [INN]

CAS	52196-22-2	MW:	473.49	MF:	C21 H27 N7 O6

Antineoplastic

LgP	-1.51(C,3)
pKa	
pp cited in Vols.1-5:	

(1) Cancer Treat. Rep., 1979, 63, 201.

KHELLIN [INN]

CAS	82-02-0	MW:	260.25	MF:	C14 H12 O5

Vasodilator

LgP	1.87(C)
pKa	
pp cited in Vols.1-5:	

(1) E. Spath et al., Ber., 1938, 71, 106. (2) Baxter et al., J. Chem. Soc., 1949, S 30. (3) M.A. Hassan et al., in 'Analytical Profiles of Drug Substances', ed. K. Florey, Academic Press, NY, 1980, vol. 9, pp 371-396.

KHELLOSIDE [INN]

CAS	17226-75-4	MW:	408.37	MF:	C19 H20 O10

LgP	-1.76(C,1)
pKa	
pp cited in Vols.1-5:	

(1) E. Spaeth et al., Ber., 1941, 74, 1549.
(2) Fabbrini et al., Ann. Chim. (Rome), 1959, 49, 894.

KITASAMYCIN [U;INN]

CAS	1392-21-8	MW:	785.98	MF:	C40 H67 N O14

Antibacterial

LgP	(4)
pKa	
pp cited in Vols.1-5:	

(1) S. Hata et al., Chem. Pharm. Bull., 1967, 15, 358.
(2) S. Omura et al., Tetrahedron, 1972, 28, 2839.
(3) Japan. Pat. 77 76 485 (1977).

LABETALOL [U;INN]

CAS	36894-69-6	MW:	328.41	MF:	C19 H24 N2 O3

Anti-adrenergic (alpha & beta receptor)

Allen & Hanburys, UK

LgP	3.09(M); 2.18(C)
pKa	7.4; 8.7
pp cited in Vols.1-5:	

1: 214;
3: 166, 173, 207, 215, 219, 220;
5: 177

(1) L.H. Lunts et al., U.S. Pat. 4 012 444 (1977). (2) R. Brittain et al., in 'Pharmacol.& Bioch. Props. of Drug Subst.', ed. M.E. Goldberg, Am. Pharm. Assoc., Washington, DC, 1979, vol.2, pp 229-254.

LACIDIPINE [INN]

CAS	103890-78-4	MW:	455.56	MF:	C26 H33 N O6

Antihypertensive

Glaxo, UK

LgP	5.20 (M)
pKa	
pp cited in Vols.1-5:	

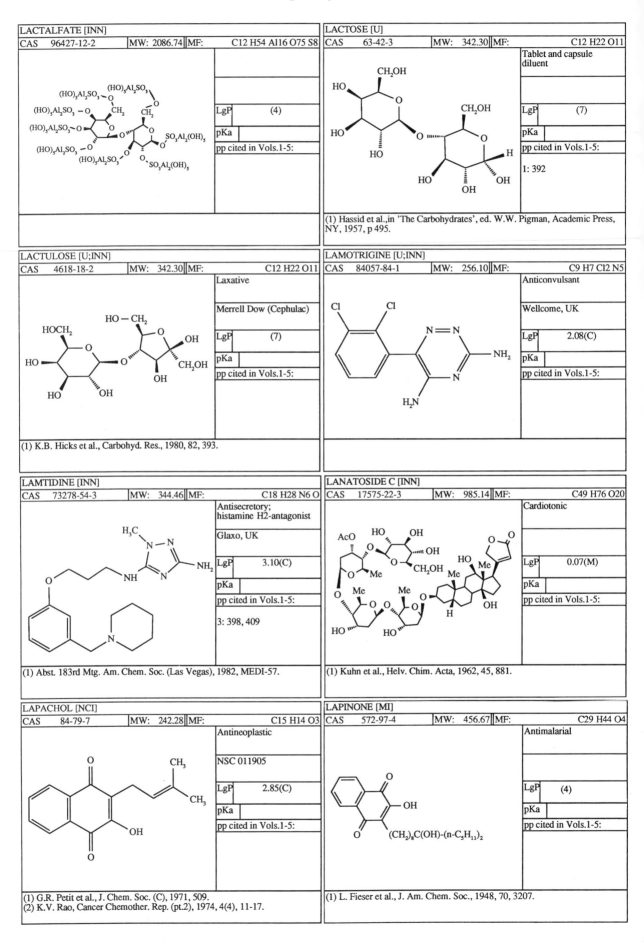

LACTALFATE [INN]

CAS	96427-12-2	MW: 2086.74	MF:	C12 H54 Al16 O75 S8

LgP	(4)
pKa	
pp cited in Vols.1-5:	

LACTOSE [U]

CAS	63-42-3	MW: 342.30	MF:	C12 H22 O11

Tablet and capsule diluent

LgP	(7)
pKa	
pp cited in Vols.1-5:	

1: 392

(1) Hassid et al.,in 'The Carbohydrates', ed. W.W. Pigman, Academic Press, NY, 1957, p 495.

LACTULOSE [U;INN]

CAS	4618-18-2	MW: 342.30	MF:	C12 H22 O11

Laxative

Merrell Dow (Cephulac)

LgP	(7)
pKa	
pp cited in Vols.1-5:	

(1) K.B. Hicks et al., Carbohyd. Res., 1980, 82, 393.

LAMOTRIGINE [U;INN]

CAS	84057-84-1	MW: 256.10	MF:	C9 H7 Cl2 N5

Anticonvulsant

Wellcome, UK

LgP	2.08(C)
pKa	
pp cited in Vols.1-5:	

LAMTIDINE [INN]

CAS	73278-54-3	MW: 344.46	MF:	C18 H28 N6 O

Antisecretory; histamine H2-antagonist

Glaxo, UK

LgP	3.10(C)
pKa	
pp cited in Vols.1-5:	

3: 398, 409

(1) Abst. 183rd Mtg. Am. Chem. Soc. (Las Vegas), 1982, MEDI-57.

LANATOSIDE C [INN]

CAS	17575-22-3	MW: 985.14	MF:	C49 H76 O20

Cardiotonic

LgP	0.07(M)
pKa	
pp cited in Vols.1-5:	

(1) Kuhn et al., Helv. Chim. Acta, 1962, 45, 881.

LAPACHOL [NCI]

CAS	84-79-7	MW: 242.28	MF:	C15 H14 O3

Antineoplastic

NSC 011905

LgP	2.85(C)
pKa	
pp cited in Vols.1-5:	

(1) G.R. Petit et al., J. Chem. Soc. (C), 1971, 509.
(2) K.V. Rao, Cancer Chemother. Rep. (pt.2), 1974, 4(4), 11-17.

LAPINONE [MI]

CAS	572-97-4	MW: 456.67	MF:	C29 H44 O4

Antimalarial

LgP	(4)
pKa	
pp cited in Vols.1-5:	

(1) L. Fieser et al., J. Am. Chem. Soc., 1948, 70, 3207.

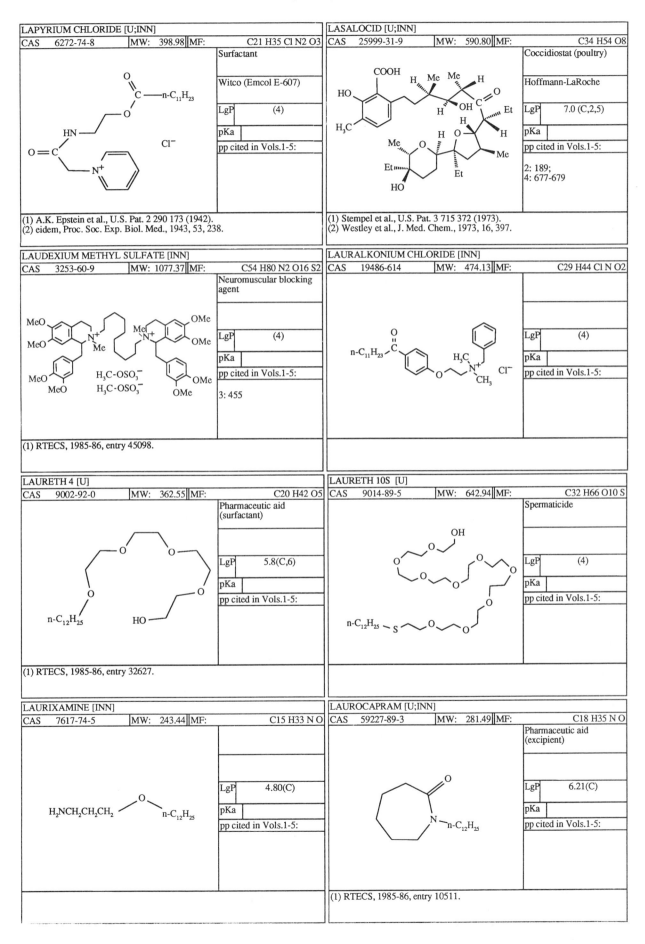

LAPYRIUM CHLORIDE [U;INN]

CAS	6272-74-8	MW:	398.98	MF:	C21 H35 Cl N2 O3

Surfactant

Witco (Emcol E-607)

LgP	(4)
pKa	
pp cited in Vols.1-5:	

(1) A.K. Epstein et al., U.S. Pat. 2 290 173 (1942).
(2) eidem, Proc. Soc. Exp. Biol. Med., 1943, 53, 238.

LASALOCID [U;INN]

CAS	25999-31-9	MW:	590.80	MF:	C34 H54 O8

Coccidiostat (poultry)

Hoffmann-LaRoche

LgP	7.0 (C,2,5)
pKa	
pp cited in Vols.1-5:	

2: 189;
4: 677-679

(1) Stempel et al., U.S. Pat. 3 715 372 (1973).
(2) Westley et al., J. Med. Chem., 1973, 16, 397.

LAUDEXIUM METHYL SULFATE [INN]

CAS	3253-60-9	MW:	1077.37	MF:	C54 H80 N2 O16 S2

Neuromuscular blocking agent

LgP	(4)
pKa	
pp cited in Vols.1-5:	

3: 455

(1) RTECS, 1985-86, entry 45098.

LAURALKONIUM CHLORIDE [INN]

CAS	19486-614	MW:	474.13	MF:	C29 H44 Cl N O2

LgP	(4)
pKa	
pp cited in Vols.1-5:	

LAURETH 4 [U]

CAS	9002-92-0	MW:	362.55	MF:	C20 H42 O5

Pharmaceutic aid (surfactant)

LgP	5.8(C,6)
pKa	
pp cited in Vols.1-5:	

(1) RTECS, 1985-86, entry 32627.

LAURETH 10S [U]

CAS	9014-89-5	MW:	642.94	MF:	C32 H66 O10 S

Spermaticide

LgP	(4)
pKa	
pp cited in Vols.1-5:	

LAURIXAMINE [INN]

CAS	7617-74-5	MW:	243.44	MF:	C15 H33 N O

LgP	4.80(C)
pKa	
pp cited in Vols.1-5:	

LAUROCAPRAM [U;INN]

CAS	59227-89-3	MW:	281.49	MF:	C18 H35 N O

Pharmaceutic aid (excipient)

LgP	6.21(C)
pKa	
pp cited in Vols.1-5:	

(1) RTECS, 1985-86, entry 10511.

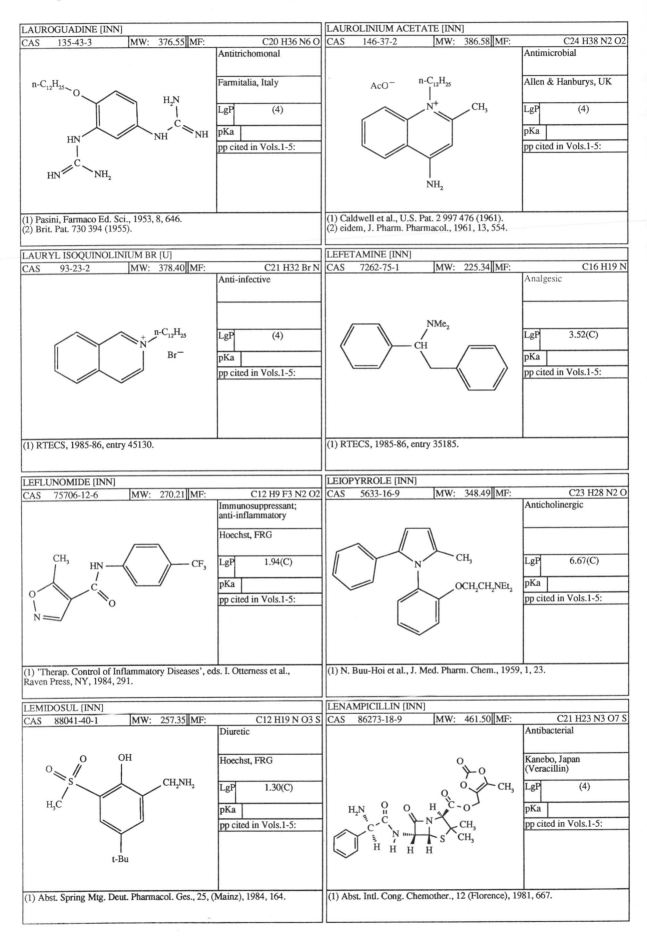

LAUROGUADINE [INN]		
CAS 135-43-3	MW: 376.55	MF: C20 H36 N6 O

Antitrichomonal

Farmitalia, Italy

LgP	(4)
pKa	
pp cited in Vols.1-5:	

(1) Pasini, Farmaco Ed. Sci., 1953, 8, 646.
(2) Brit. Pat. 730 394 (1955).

LAUROLINIUM ACETATE [INN]		
CAS 146-37-2	MW: 386.58	MF: C24 H38 N2 O2

Antimicrobial

Allen & Hanburys, UK

LgP	(4)
pKa	
pp cited in Vols.1-5:	

(1) Caldwell et al., U.S. Pat. 2 997 476 (1961).
(2) eidem, J. Pharm. Pharmacol., 1961, 13, 554.

LAURYL ISOQUINOLINIUM BR [U]		
CAS 93-23-2	MW: 378.40	MF: C21 H32 Br N

Anti-infective

LgP	(4)
pKa	
pp cited in Vols.1-5:	

(1) RTECS, 1985-86, entry 45130.

LEFETAMINE [INN]		
CAS 7262-75-1	MW: 225.34	MF: C16 H19 N

Analgesic

LgP	3.52(C)
pKa	
pp cited in Vols.1-5:	

(1) RTECS, 1985-86, entry 35185.

LEFLUNOMIDE [INN]		
CAS 75706-12-6	MW: 270.21	MF: C12 H9 F3 N2 O2

Immunosuppressant; anti-inflammatory

Hoechst, FRG

LgP	1.94(C)
pKa	
pp cited in Vols.1-5:	

(1) 'Therap. Control of Inflammatory Diseases', eds. I. Otterness et al., Raven Press, NY, 1984, 291.

LEIOPYRROLE [INN]		
CAS 5633-16-9	MW: 348.49	MF: C23 H28 N2 O

Anticholinergic

LgP	6.67(C)
pKa	
pp cited in Vols.1-5:	

(1) N. Buu-Hoi et al., J. Med. Pharm. Chem., 1959, 1, 23.

LEMIDOSUL [INN]		
CAS 88041-40-1	MW: 257.35	MF: C12 H19 N O3 S

Diuretic

Hoechst, FRG

LgP	1.30(C)
pKa	
pp cited in Vols.1-5:	

(1) Abst. Spring Mtg. Deut. Pharmacol. Ges., 25, (Mainz), 1984, 164.

LENAMPICILLIN [INN]		
CAS 86273-18-9	MW: 461.50	MF: C21 H23 N3 O7 S

Antibacterial

Kanebo, Japan
(Veracillin)

LgP	(4)
pKa	
pp cited in Vols.1-5:	

(1) Abst. Intl. Cong. Chemother., 12 (Florence), 1981, 667.

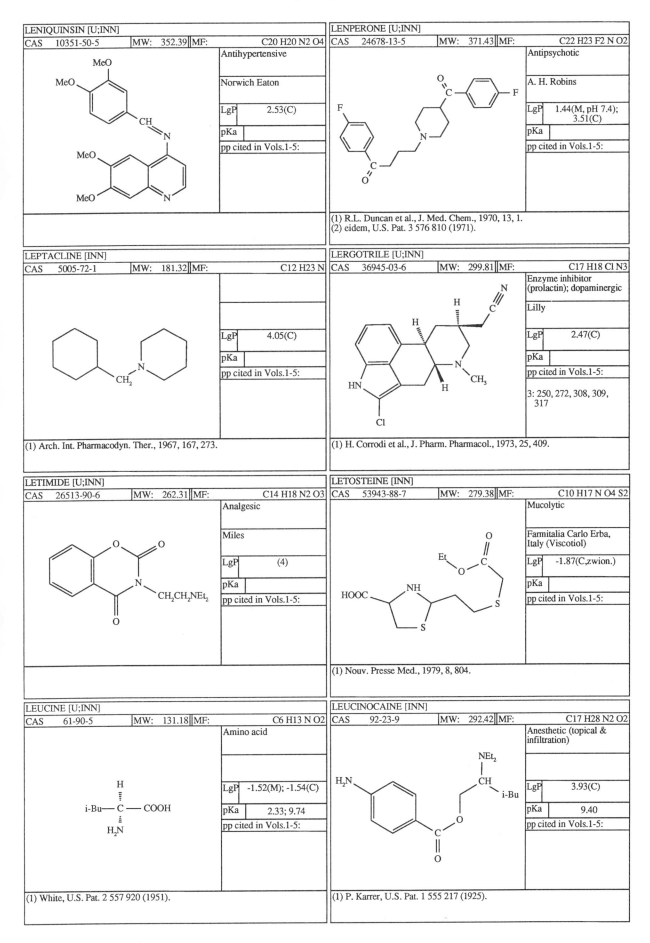

LENIQUINSIN [U;INN]

CAS	10351-50-5	MW:	352.39	MF:	C20 H20 N2 O4

Antihypertensive

Norwich Eaton

LgP	2.53(C)
pKa	
pp cited in Vols.1-5:	

LENPERONE [U;INN]

CAS	24678-13-5	MW:	371.43	MF:	C22 H23 F2 N O2

Antipsychotic

A. H. Robins

LgP	1.44(M, pH 7.4); 3.51(C)
pKa	
pp cited in Vols.1-5:	

(1) R.L. Duncan et al., J. Med. Chem., 1970, 13, 1.
(2) eidem, U.S. Pat. 3 576 810 (1971).

LEPTACLINE [INN]

CAS	5005-72-1	MW:	181.32	MF:	C12 H23 N

LgP	4.05(C)
pKa	
pp cited in Vols.1-5:	

(1) Arch. Int. Pharmacodyn. Ther., 1967, 167, 273.

LERGOTRILE [U;INN]

CAS	36945-03-6	MW:	299.81	MF:	C17 H18 Cl N3

Enzyme inhibitor (prolactin); dopaminergic

Lilly

LgP	2.47(C)
pKa	
pp cited in Vols.1-5:	

3: 250, 272, 308, 309, 317

(1) H. Corrodi et al., J. Pharm. Pharmacol., 1973, 25, 409.

LETIMIDE [U;INN]

CAS	26513-90-6	MW:	262.31	MF:	C14 H18 N2 O3

Analgesic

Miles

LgP	(4)
pKa	
pp cited in Vols.1-5:	

LETOSTEINE [INN]

CAS	53943-88-7	MW:	279.38	MF:	C10 H17 N O4 S2

Mucolytic

Farmitalia Carlo Erba, Italy (Viscotiol)

LgP	-1.87(C,zwion.)
pKa	
pp cited in Vols.1-5:	

(1) Nouv. Presse Med., 1979, 8, 804.

LEUCINE [U;INN]

CAS	61-90-5	MW:	131.18	MF:	C6 H13 N O2

Amino acid

LgP	-1.52(M); -1.54(C)
pKa	2.33; 9.74
pp cited in Vols.1-5:	

(1) White, U.S. Pat. 2 557 920 (1951).

LEUCINOCAINE [INN]

CAS	92-23-9	MW:	292.42	MF:	C17 H28 N2 O2

Anesthetic (topical & infiltration)

LgP	3.93(C)
pKa	9.40
pp cited in Vols.1-5:	

(1) P. Karrer, U.S. Pat. 1 555 217 (1925).

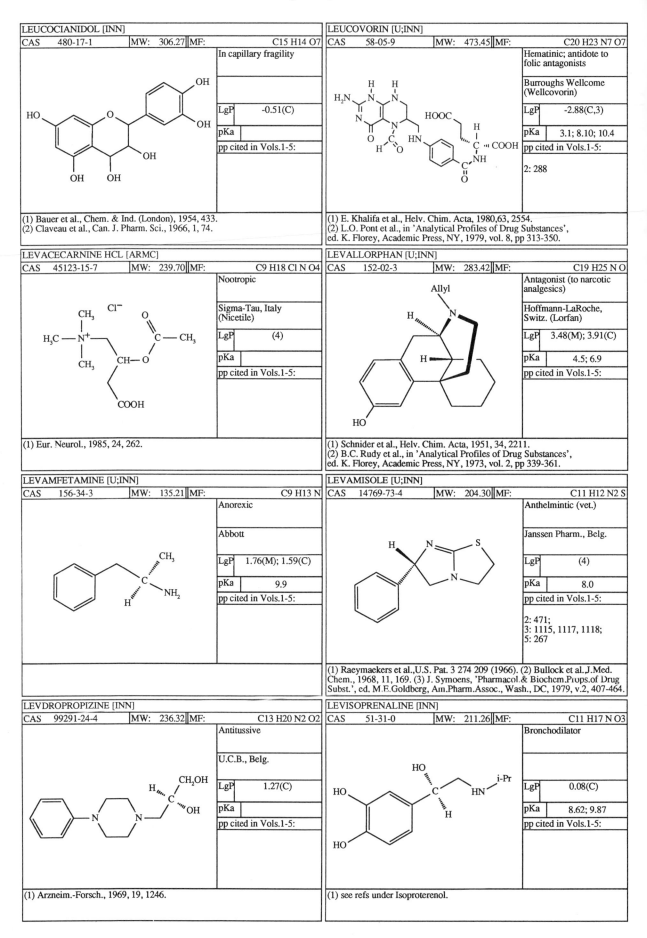

LEUCOCIANIDOL [INN]			
CAS 480-17-1	MW: 306.27	MF:	C15 H14 O7

In capillary fragility

LgP	-0.51(C)
pKa	
pp cited in Vols.1-5:	

(1) Bauer et al., Chem. & Ind. (London), 1954, 433.
(2) Claveau et al., Can. J. Pharm. Sci., 1966, 1, 74.

LEUCOVORIN [U;INN]			
CAS 58-05-9	MW: 473.45	MF:	C20 H23 N7 O7

Hematinic; antidote to folic antagonists

Burroughs Wellcome (Wellcovorin)

LgP	-2.88(C,3)
pKa	3.1; 8.10; 10.4
pp cited in Vols.1-5:	

2: 288

(1) E. Khalifa et al., Helv. Chim. Acta, 1980,63, 2554.
(2) L.O. Pont et al., in 'Analytical Profiles of Drug Substances', ed. K. Florey, Academic Press, NY, 1979, vol. 8, pp 313-350.

LEVACECARNINE HCL [ARMC]			
CAS 45123-15-7	MW: 239.70	MF:	C9 H18 Cl N O4

Nootropic

Sigma-Tau, Italy (Nicetile)

LgP	(4)
pKa	
pp cited in Vols.1-5:	

(1) Eur. Neurol., 1985, 24, 262.

LEVALLORPHAN [U;INN]			
CAS 152-02-3	MW: 283.42	MF:	C19 H25 N O

Antagonist (to narcotic analgesics)

Hoffmann-LaRoche, Switz. (Lorfan)

LgP	3.48(M); 3.91(C)
pKa	4.5; 6.9
pp cited in Vols.1-5:	

(1) Schnider et al., Helv. Chim. Acta, 1951, 34, 2211.
(2) B.C. Rudy et al., in 'Analytical Profiles of Drug Substances', ed. K. Florey, Academic Press, NY, 1973, vol. 2, pp 339-361.

LEVAMFETAMINE [U;INN]			
CAS 156-34-3	MW: 135.21	MF:	C9 H13 N

Anorexic

Abbott

LgP	1.76(M); 1.59(C)
pKa	9.9
pp cited in Vols.1-5:	

LEVAMISOLE [U;INN]			
CAS 14769-73-4	MW: 204.30	MF:	C11 H12 N2 S

Anthelmintic (vet.)

Janssen Pharm., Belg.

LgP	(4)
pKa	8.0
pp cited in Vols.1-5:	

2: 471;
3: 1115, 1117, 1118;
5: 267

(1) Raeymaekers et al.,U.S. Pat. 3 274 209 (1966). (2) Bullock et al.,J.Med. Chem., 1968, 11, 169. (3) J. Symoens, 'Pharmacol & Biochem.Props.of Drug Subst.', ed. M.E.Goldberg, Am.Pharm.Assoc., Wash., DC, 1979, v.2, 407-464.

LEVDROPROPIZINE [INN]			
CAS 99291-24-4	MW: 236.32	MF:	C13 H20 N2 O2

Antitussive

U.C.B., Belg.

LgP	1.27(C)
pKa	
pp cited in Vols.1-5:	

(1) Arzneim.-Forsch., 1969, 19, 1246.

LEVISOPRENALINE [INN]			
CAS 51-31-0	MW: 211.26	MF:	C11 H17 N O3

Bronchodilator

LgP	0.08(C)
pKa	8.62; 9.87
pp cited in Vols.1-5:	

(1) see refs under Isoproterenol.

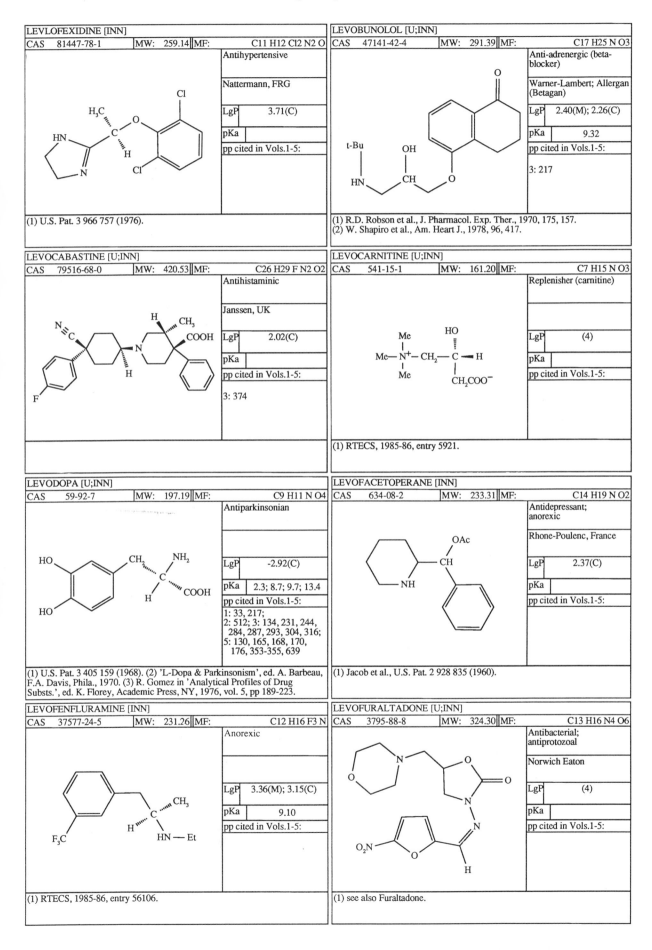

LEVLOFEXIDINE [INN]

| CAS | 81447-78-1 | MW: 259.14 | MF: | C11 H12 Cl2 N2 O |

Antihypertensive

Nattermann, FRG

LgP	3.71(C)
pKa	
pp cited in Vols.1-5:	

(1) U.S. Pat. 3 966 757 (1976).

LEVOBUNOLOL [U;INN]

| CAS | 47141-42-4 | MW: 291.39 | MF: | C17 H25 N O3 |

Anti-adrenergic (beta-blocker)

Warner-Lambert; Allergan (Betagan)

LgP	2.40(M); 2.26(C)
pKa	9.32
pp cited in Vols.1-5:	

3: 217

(1) R.D. Robson et al., J. Pharmacol. Exp. Ther., 1970, 175, 157.
(2) W. Shapiro et al., Am. Heart J., 1978, 96, 417.

LEVOCABASTINE [U;INN]

| CAS | 79516-68-0 | MW: 420.53 | MF: | C26 H29 F N2 O2 |

Antihistaminic

Janssen, UK

LgP	2.02(C)
pKa	
pp cited in Vols.1-5:	

3: 374

LEVOCARNITINE [U;INN]

| CAS | 541-15-1 | MW: 161.20 | MF: | C7 H15 N O3 |

Replenisher (carnitine)

LgP	(4)
pKa	
pp cited in Vols.1-5:	

(1) RTECS, 1985-86, entry 5921.

LEVODOPA [U;INN]

| CAS | 59-92-7 | MW: 197.19 | MF: | C9 H11 N O4 |

Antiparkinsonian

LgP	-2.92(C)
pKa	2.3; 8.7; 9.7; 13.4
pp cited in Vols.1-5:	

1: 33, 217;
2: 512; 3: 134, 231, 244,
284, 287, 293, 304, 316;
5: 130, 165, 168, 170,
176, 353-355, 639

(1) U.S. Pat. 3 405 159 (1968). (2) 'L-Dopa & Parkinsonism', ed. A. Barbeau,
F.A. Davis, Phila., 1970. (3) R. Gomez in 'Analytical Profiles of Drug
Substs.', ed. K. Florey, Academic Press, NY, 1976, vol. 5, pp 189-223.

LEVOFACETOPERANE [INN]

| CAS | 634-08-2 | MW: 233.31 | MF: | C14 H19 N O2 |

Antidepressant;
anorexic

Rhone-Poulenc, France

LgP	2.37(C)
pKa	
pp cited in Vols.1-5:	

(1) Jacob et al., U.S. Pat. 2 928 835 (1960).

LEVOFENFLURAMINE [INN]

| CAS | 37577-24-5 | MW: 231.26 | MF: | C12 H16 F3 N |

Anorexic

LgP	3.36(M); 3.15(C)
pKa	9.10
pp cited in Vols.1-5:	

(1) RTECS, 1985-86, entry 56106.

LEVOFURALTADONE [U;INN]

| CAS | 3795-88-8 | MW: 324.30 | MF: | C13 H16 N4 O6 |

Antibacterial;
antiprotozoal

Norwich Eaton

LgP	(4)
pKa	
pp cited in Vols.1-5:	

(1) see also Furaltadone.

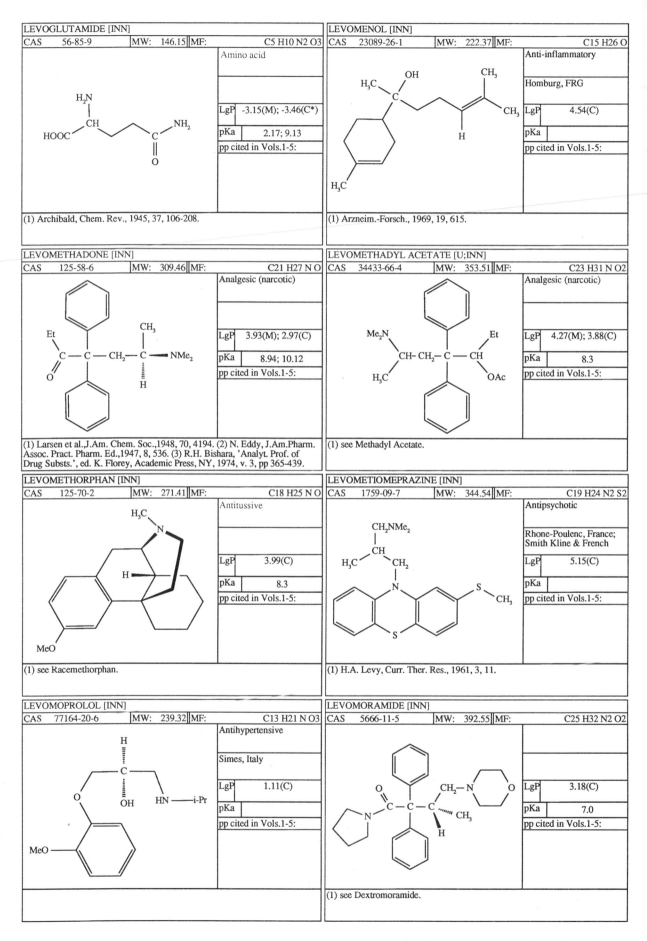

LEVOGLUTAMIDE [INN]

CAS	56-85-9	MW:	146.15	MF:	C5 H10 N2 O3

Amino acid

LgP	-3.15(M); -3.46(C*)
pKa	2.17; 9.13
pp cited in Vols.1-5:	

(1) Archibald, Chem. Rev., 1945, 37, 106-208.

LEVOMENOL [INN]

CAS	23089-26-1	MW:	222.37	MF:	C15 H26 O

Anti-inflammatory

Homburg, FRG

LgP	4.54(C)
pKa	
pp cited in Vols.1-5:	

(1) Arzneim.-Forsch., 1969, 19, 615.

LEVOMETHADONE [INN]

CAS	125-58-6	MW:	309.46	MF:	C21 H27 N O

Analgesic (narcotic)

LgP	3.93(M); 2.97(C)
pKa	8.94; 10.12
pp cited in Vols.1-5:	

(1) Larsen et al.,J.Am. Chem. Soc.,1948, 70, 4194. (2) N. Eddy, J.Am.Pharm. Assoc. Pract. Pharm. Ed.,1947, 8, 536. (3) R.H. Bishara, 'Analyt. Prof. of Drug Substs.', ed. K. Florey, Academic Press, NY, 1974, v. 3, pp 365-439.

LEVOMETHADYL ACETATE [U;INN]

CAS	34433-66-4	MW:	353.51	MF:	C23 H31 N O2

Analgesic (narcotic)

LgP	4.27(M); 3.88(C)
pKa	8.3
pp cited in Vols.1-5:	

(1) see Methadyl Acetate.

LEVOMETHORPHAN [INN]

CAS	125-70-2	MW:	271.41	MF:	C18 H25 N O

Antitussive

LgP	3.99(C)
pKa	8.3
pp cited in Vols.1-5:	

(1) see Racemethorphan.

LEVOMETIOMEPRAZINE [INN]

CAS	1759-09-7	MW:	344.54	MF:	C19 H24 N2 S2

Antipsychotic

Rhone-Poulenc, France; Smith Kline & French

LgP	5.15(C)
pKa	
pp cited in Vols.1-5:	

(1) H.A. Levy, Curr. Ther. Res., 1961, 3, 11.

LEVOMOPROLOL [INN]

CAS	77164-20-6	MW:	239.32	MF:	C13 H21 N O3

Antihypertensive

Simes, Italy

LgP	1.11(C)
pKa	
pp cited in Vols.1-5:	

LEVOMORAMIDE [INN]

CAS	5666-11-5	MW:	392.55	MF:	C25 H32 N2 O2

LgP	3.18(C)
pKa	7.0
pp cited in Vols.1-5:	

(1) see Dextromoramide.

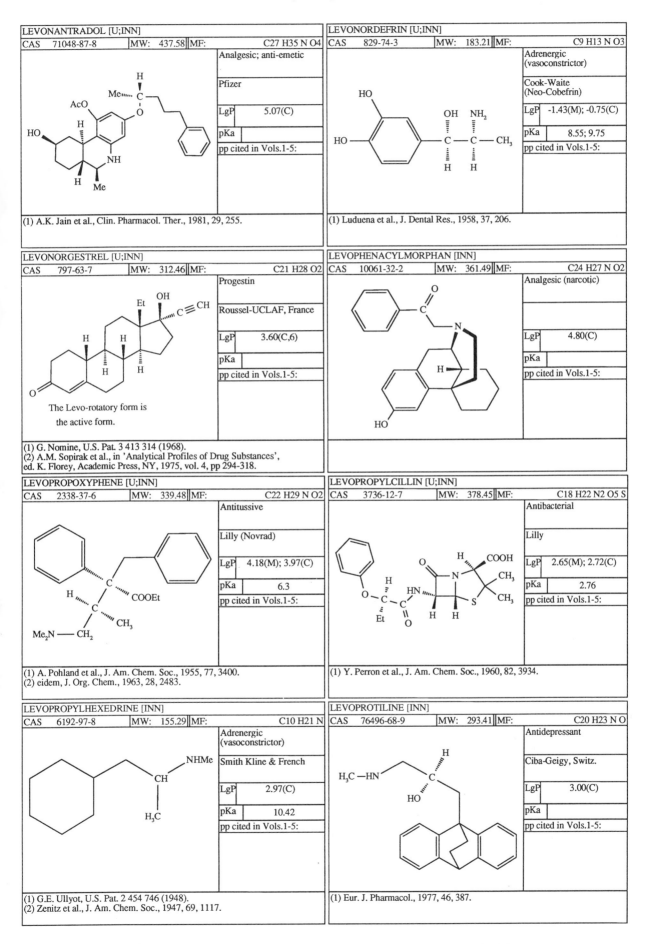

LEVONANTRADOL [U;INN]			
CAS 71048-87-8	MW: 437.58	MF:	C27 H35 N O4

Analgesic; anti-emetic

Pfizer

LgP	5.07(C)
pKa	
pp cited in Vols.1-5:	

(1) A.K. Jain et al., Clin. Pharmacol. Ther., 1981, 29, 255.

LEVONORDEFRIN [U;INN]			
CAS 829-74-3	MW: 183.21	MF:	C9 H13 N O3

Adrenergic
(vasoconstrictor)

Cook-Waite
(Neo-Cobefrin)

LgP	-1.43(M); -0.75(C)
pKa	8.55; 9.75
pp cited in Vols.1-5:	

(1) Luduena et al., J. Dental Res., 1958, 37, 206.

LEVONORGESTREL [U;INN]			
CAS 797-63-7	MW: 312.46	MF:	C21 H28 O2

Progestin

Roussel-UCLAF, France

LgP	3.60(C,6)
pKa	
pp cited in Vols.1-5:	

The Levo-rotatory form is
the active form.

(1) G. Nomine, U.S. Pat. 3 413 314 (1968).
(2) A.M. Sopirak et al., in 'Analytical Profiles of Drug Substances',
ed. K. Florey, Academic Press, NY, 1975, vol. 4, pp 294-318.

LEVOPHENACYLMORPHAN [INN]			
CAS 10061-32-2	MW: 361.49	MF:	C24 H27 N O2

Analgesic (narcotic)

LgP	4.80(C)
pKa	
pp cited in Vols.1-5:	

LEVOPROPOXYPHENE [U;INN]			
CAS 2338-37-6	MW: 339.48	MF:	C22 H29 N O2

Antitussive

Lilly (Novrad)

LgP	4.18(M); 3.97(C)
pKa	6.3
pp cited in Vols.1-5:	

(1) A. Pohland et al., J. Am. Chem. Soc., 1955, 77, 3400.
(2) eidem, J. Org. Chem., 1963, 28, 2483.

LEVOPROPYLCILLIN [U;INN]			
CAS 3736-12-7	MW: 378.45	MF:	C18 H22 N2 O5 S

Antibacterial

Lilly

LgP	2.65(M); 2.72(C)
pKa	2.76
pp cited in Vols.1-5:	

(1) Y. Perron et al., J. Am. Chem. Soc., 1960, 82, 3934.

LEVOPROPYLHEXEDRINE [INN]			
CAS 6192-97-8	MW: 155.29	MF:	C10 H21 N

Adrenergic
(vasoconstrictor)

Smith Kline & French

LgP	2.97(C)
pKa	10.42
pp cited in Vols.1-5:	

(1) G.E. Ullyot, U.S. Pat. 2 454 746 (1948).
(2) Zenitz et al., J. Am. Chem. Soc., 1947, 69, 1117.

LEVOPROTILINE [INN]			
CAS 76496-68-9	MW: 293.41	MF:	C20 H23 N O

Antidepressant

Ciba-Geigy, Switz.

LgP	3.00(C)
pKa	
pp cited in Vols.1-5:	

(1) Eur. J. Pharmacol., 1977, 46, 387.

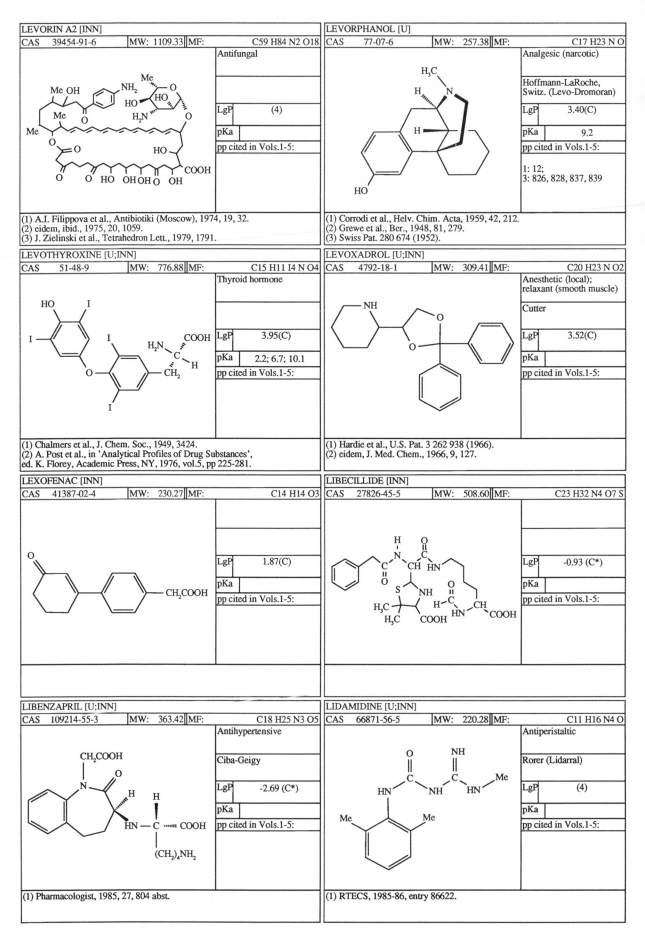

LEVORIN A2 [INN]			
CAS 39454-91-6	MW: 1109.33	MF:	C59 H84 N2 O18

Antifungal

LgP (4)

pKa

pp cited in Vols.1-5:

(1) A.I. Filippova et al., Antibiotiki (Moscow), 1974, 19, 32.
(2) eidem, ibid., 1975, 20, 1059.
(3) J. Zielinski et al., Tetrahedron Lett., 1979, 1791.

LEVORPHANOL [U]			
CAS 77-07-6	MW: 257.38	MF:	C17 H23 N O

Analgesic (narcotic)

Hoffmann-LaRoche, Switz. (Levo-Dromoran)

LgP 3.40(C)

pKa 9.2

pp cited in Vols.1-5:

1: 12;
3: 826, 828, 837, 839

(1) Corrodi et al., Helv. Chim. Acta, 1959, 42, 212.
(2) Grewe et al., Ber., 1948, 81, 279.
(3) Swiss Pat. 280 674 (1952).

LEVOTHYROXINE [U;INN]			
CAS 51-48-9	MW: 776.88	MF:	C15 H11 I4 N O4

Thyroid hormone

LgP 3.95(C)

pKa 2.2; 6.7; 10.1

pp cited in Vols.1-5:

(1) Chalmers et al., J. Chem. Soc., 1949, 3424.
(2) A. Post et al., in 'Analytical Profiles of Drug Substances',
ed. K. Florey, Academic Press, NY, 1976, vol.5, pp 225-281.

LEVOXADROL [U;INN]			
CAS 4792-18-1	MW: 309.41	MF:	C20 H23 N O2

Anesthetic (local);
relaxant (smooth muscle)

Cutter

LgP 3.52(C)

pKa

pp cited in Vols.1-5:

(1) Hardie et al., U.S. Pat. 3 262 938 (1966).
(2) eidem, J. Med. Chem., 1966, 9, 127.

LEXOFENAC [INN]			
CAS 41387-02-4	MW: 230.27	MF:	C14 H14 O3

LgP 1.87(C)

pKa

pp cited in Vols.1-5:

LIBECILLIDE [INN]			
CAS 27826-45-5	MW: 508.60	MF:	C23 H32 N4 O7 S

LgP -0.93 (C*)

pKa

pp cited in Vols.1-5:

LIBENZAPRIL [U;INN]			
CAS 109214-55-3	MW: 363.42	MF:	C18 H25 N3 O5

Antihypertensive

Ciba-Geigy

LgP -2.69 (C*)

pKa

pp cited in Vols.1-5:

(1) Pharmacologist, 1985, 27, 804 abst.

LIDAMIDINE [U;INN]			
CAS 66871-56-5	MW: 220.28	MF:	C11 H16 N4 O

Antiperistaltic

Rorer (Lidarral)

LgP (4)

pKa

pp cited in Vols.1-5:

(1) RTECS, 1985-86, entry 86622.

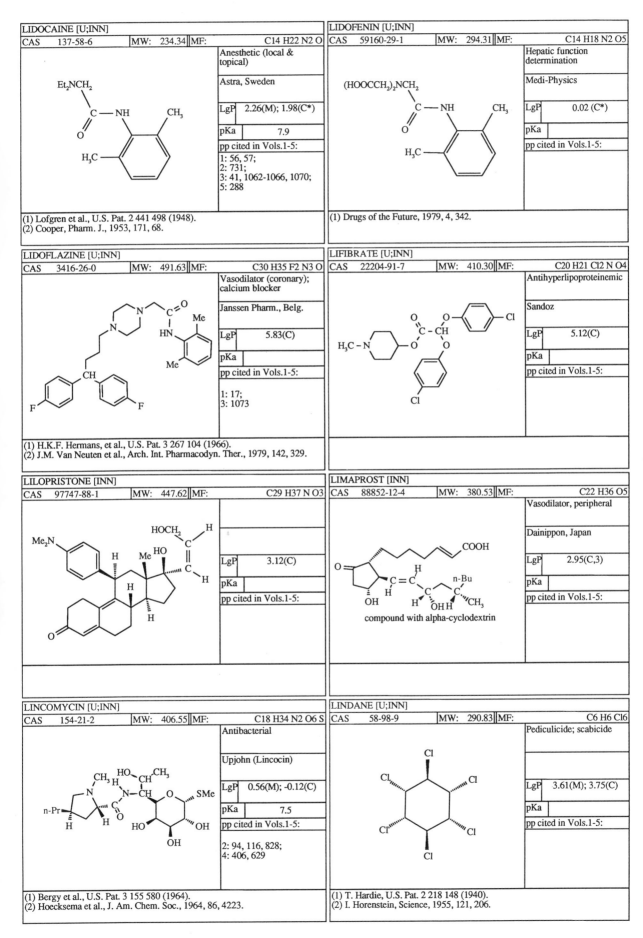

LIDOCAINE [U;INN]

CAS 137-58-6	MW: 234.34	MF: C14 H22 N2 O

Anesthetic (local & topical)

Astra, Sweden

LgP	2.26(M); 1.98(C*)
pKa	7.9

pp cited in Vols.1-5:
1: 56, 57;
2: 731;
3: 41, 1062-1066, 1070;
5: 288

(1) Lofgren et al., U.S. Pat. 2 441 498 (1948).
(2) Cooper, Pharm. J., 1953, 171, 68.

LIDOFENIN [U;INN]

CAS 59160-29-1	MW: 294.31	MF: C14 H18 N2 O5

Hepatic function determination

Medi-Physics

LgP	0.02 (C*)
pKa	

pp cited in Vols.1-5:

(1) Drugs of the Future, 1979, 4, 342.

LIDOFLAZINE [U;INN]

CAS 3416-26-0	MW: 491.63	MF: C30 H35 F2 N3 O

Vasodilator (coronary); calcium blocker

Janssen Pharm., Belg.

LgP	5.83(C)
pKa	

pp cited in Vols.1-5:

1: 17;
3: 1073

(1) H.K.F. Hermans, et al., U.S. Pat. 3 267 104 (1966).
(2) J.M. Van Neuten et al., Arch. Int. Pharmacodyn. Ther., 1979, 142, 329.

LIFIBRATE [U;INN]

CAS 22204-91-7	MW: 410.30	MF: C20 H21 Cl2 N O4

Antihyperlipoproteinemic

Sandoz

LgP	5.12(C)
pKa	

pp cited in Vols.1-5:

LILOPRISTONE [INN]

CAS 97747-88-1	MW: 447.62	MF: C29 H37 N O3

LgP	3.12(C)
pKa	

pp cited in Vols.1-5:

LIMAPROST [INN]

CAS 88852-12-4	MW: 380.53	MF: C22 H36 O5

Vasodilator, peripheral

Dainippon, Japan

LgP	2.95(C,3)
pKa	

pp cited in Vols.1-5:

compound with alpha-cyclodextrin

LINCOMYCIN [U;INN]

CAS 154-21-2	MW: 406.55	MF: C18 H34 N2 O6 S

Antibacterial

Upjohn (Lincocin)

LgP	0.56(M); -0.12(C)
pKa	7.5

pp cited in Vols.1-5:

2: 94, 116, 828;
4: 406, 629

(1) Bergy et al., U.S. Pat. 3 155 580 (1964).
(2) Hoecksema et al., J. Am. Chem. Soc., 1964, 86, 4223.

LINDANE [U;INN]

CAS 58-98-9	MW: 290.83	MF: C6 H6 Cl6

Pediculicide; scabicide

LgP	3.61(M); 3.75(C)
pKa	

pp cited in Vols.1-5:

(1) T. Hardie, U.S. Pat. 2 218 148 (1940).
(2) I. Horenstein, Science, 1955, 121, 206.

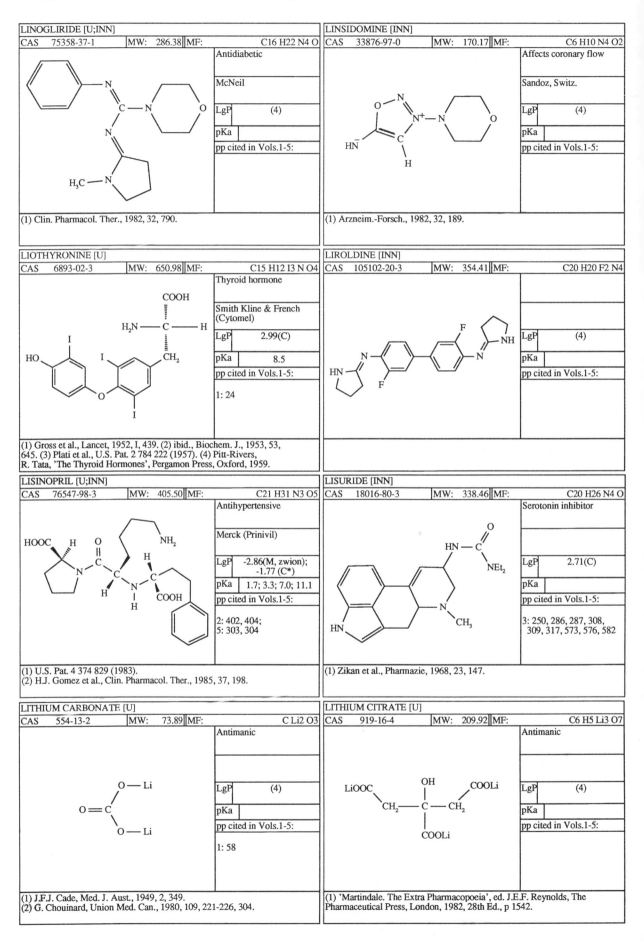

LINOGLIRIDE [U;INN]			
CAS 75358-37-1	MW: 286.38	MF:	C16 H22 N4 O
	Antidiabetic		
	McNeil		
	LgP	(4)	
	pKa		
	pp cited in Vols.1-5:		

(1) Clin. Pharmacol. Ther., 1982, 32, 790.

LINSIDOMINE [INN]			
CAS 33876-97-0	MW: 170.17	MF:	C6 H10 N4 O2
	Affects coronary flow		
	Sandoz, Switz.		
	LgP	(4)	
	pKa		
	pp cited in Vols.1-5:		

(1) Arzneim.-Forsch., 1982, 32, 189.

LIOTHYRONINE [U]			
CAS 6893-02-3	MW: 650.98	MF:	C15 H12 I3 N O4
	Thyroid hormone		
	Smith Kline & French (Cytomel)		
	LgP	2.99(C)	
	pKa	8.5	
	pp cited in Vols.1-5:		
	1: 24		

(1) Gross et al., Lancet, 1952, I, 439. (2) ibid., Biochem. J., 1953, 53, 645. (3) Plati et al., U.S. Pat. 2 784 222 (1957). (4) Pitt-Rivers, R. Tata, 'The Thyroid Hormones', Pergamon Press, Oxford, 1959.

LIROLDINE [INN]			
CAS 105102-20-3	MW: 354.41	MF:	C20 H20 F2 N4
	LgP	(4)	
	pKa		
	pp cited in Vols.1-5:		

LISINOPRIL [U;INN]			
CAS 76547-98-3	MW: 405.50	MF:	C21 H31 N3 O5
	Antihypertensive		
	Merck (Prinivil)		
	LgP	-2.86(M, zwion); -1.77 (C*)	
	pKa	1.7; 3.3; 7.0; 11.1	
	pp cited in Vols.1-5:		
	2: 402, 404; 5: 303, 304		

(1) U.S. Pat. 4 374 829 (1983).
(2) H.J. Gomez et al., Clin. Pharmacol. Ther., 1985, 37, 198.

LISURIDE [INN]			
CAS 18016-80-3	MW: 338.46	MF:	C20 H26 N4 O
	Serotonin inhibitor		
	LgP	2.71(C)	
	pKa		
	pp cited in Vols.1-5:		
	3: 250, 286, 287, 308, 309, 317, 573, 576, 582		

(1) Zikan et al., Pharmazie, 1968, 23, 147.

LITHIUM CARBONATE [U]			
CAS 554-13-2	MW: 73.89	MF:	C Li2 O3
	Antimanic		
	LgP	(4)	
	pKa		
	pp cited in Vols.1-5:		
	1: 58		

(1) J.F.J. Cade, Med. J. Aust., 1949, 2, 349.
(2) G. Chouinard, Union Med. Can., 1980, 109, 221-226, 304.

LITHIUM CITRATE [U]			
CAS 919-16-4	MW: 209.92	MF:	C6 H5 Li3 O7
	Antimanic		
	LgP	(4)	
	pKa		
	pp cited in Vols.1-5:		

(1) 'Martindale. The Extra Pharmacopoeia', ed. J.E.F. Reynolds, The Pharmaceutical Press, London, 1982, 28th Ed., p 1542.

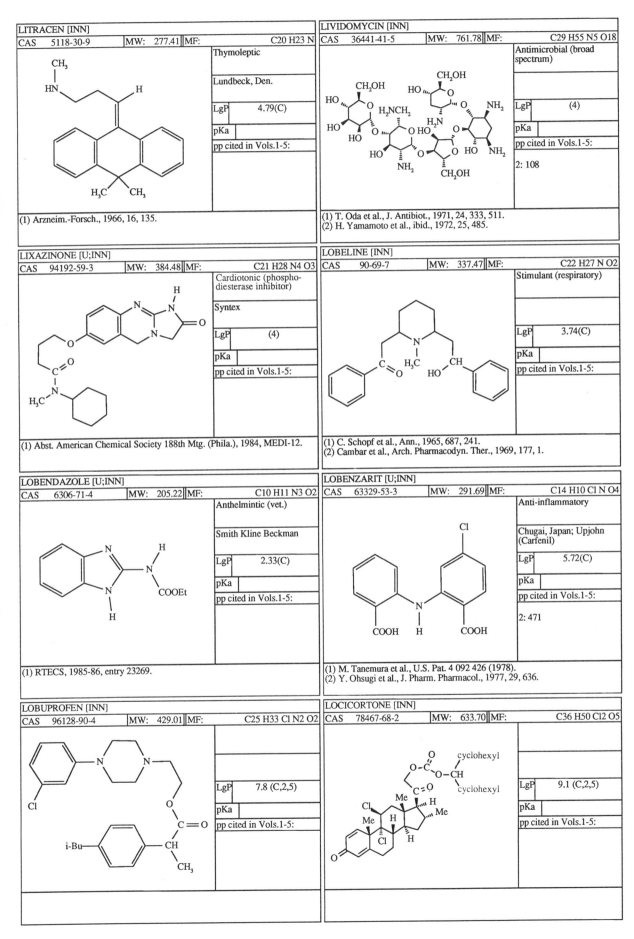

LITRACEN [INN]

CAS	5118-30-9	MW:	277.41	MF:	C20 H23 N

Thymoleptic

Lundbeck, Den.

LgP	4.79(C)

pKa

pp cited in Vols.1-5:

(1) Arzneim.-Forsch., 1966, 16, 135.

LIVIDOMYCIN [INN]

CAS	36441-41-5	MW:	761.78	MF:	C29 H55 N5 O18

Antimicrobial (broad spectrum)

LgP	(4)

pKa

pp cited in Vols.1-5:

2: 108

(1) T. Oda et al., J. Antibiot., 1971, 24, 333, 511.
(2) H. Yamamoto et al., ibid., 1972, 25, 485.

LIXAZINONE [U;INN]

CAS	94192-59-3	MW:	384.48	MF:	C21 H28 N4 O3

Cardiotonic (phospho-diesterase inhibitor)

Syntex

LgP	(4)

pKa

pp cited in Vols.1-5:

(1) Abst. American Chemical Society 188th Mtg. (Phila.), 1984, MEDI-12.

LOBELINE [INN]

CAS	90-69-7	MW:	337.47	MF:	C22 H27 N O2

Stimulant (respiratory)

LgP	3.74(C)

pKa

pp cited in Vols.1-5:

(1) C. Schopf et al., Ann., 1965, 687, 241.
(2) Cambar et al., Arch. Pharmacodyn. Ther., 1969, 177, 1.

LOBENDAZOLE [U;INN]

CAS	6306-71-4	MW:	205.22	MF:	C10 H11 N3 O2

Anthelmintic (vet.)

Smith Kline Beckman

LgP	2.33(C)

pKa

pp cited in Vols.1-5:

(1) RTECS, 1985-86, entry 23269.

LOBENZARIT [U;INN]

CAS	63329-53-3	MW:	291.69	MF:	C14 H10 Cl N O4

Anti-inflammatory

Chugai, Japan; Upjohn (Carfenil)

LgP	5.72(C)

pKa

pp cited in Vols.1-5:

2: 471

(1) M. Tanemura et al., U.S. Pat. 4 092 426 (1978).
(2) Y. Ohsugi et al., J. Pharm. Pharmacol., 1977, 29, 636.

LOBUPROFEN [INN]

CAS	96128-90-4	MW:	429.01	MF:	C25 H33 Cl N2 O2

LgP	7.8 (C,2,5)

pKa

pp cited in Vols.1-5:

LOCICORTONE [INN]

CAS	78467-68-2	MW:	633.70	MF:	C36 H50 Cl2 O5

LgP	9.1 (C,2,5)

pKa

pp cited in Vols.1-5:

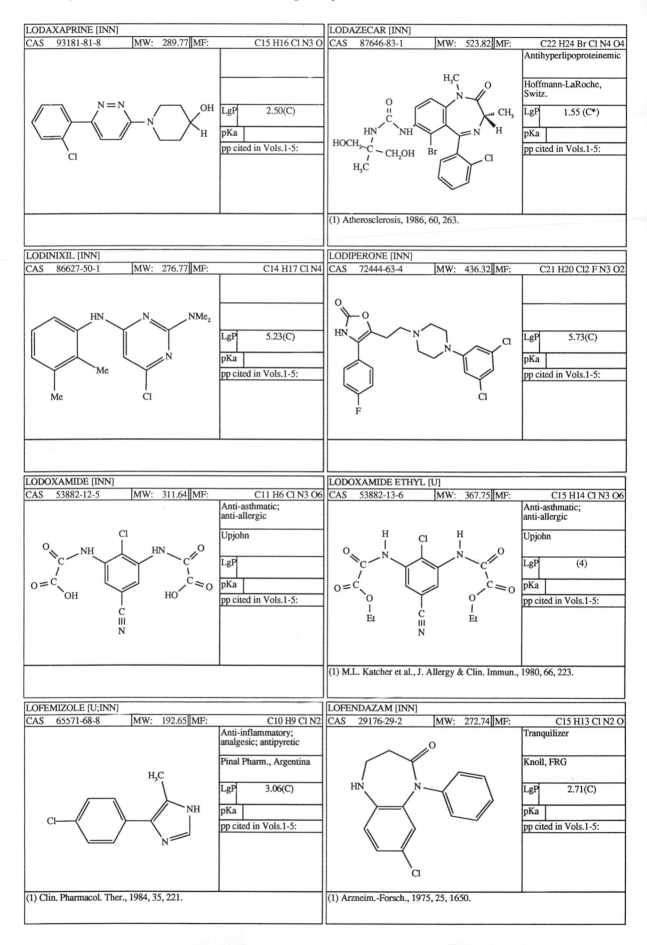

LODAXAPRINE [INN]

CAS 93181-81-8 | MW: 289.77 | MF: C15 H16 Cl N3 O

LgP 2.50(C)

pKa

pp cited in Vols.1-5:

LODAZECAR [INN]

CAS 87646-83-1 | MW: 523.82 | MF: C22 H24 Br Cl N4 O4

Antihyperlipoproteinemic

Hoffmann-LaRoche, Switz.

LgP 1.55 (C*)

pKa

pp cited in Vols.1-5:

(1) Atherosclerosis, 1986, 60, 263.

LODINIXIL [INN]

CAS 86627-50-1 | MW: 276.77 | MF: C14 H17 Cl N4

LgP 5.23(C)

pKa

pp cited in Vols.1-5:

LODIPERONE [INN]

CAS 72444-63-4 | MW: 436.32 | MF: C21 H20 Cl2 F N3 O2

LgP 5.73(C)

pKa

pp cited in Vols.1-5:

LODOXAMIDE [INN]

CAS 53882-12-5 | MW: 311.64 | MF: C11 H6 Cl N3 O6

Anti-asthmatic; anti-allergic

Upjohn

LgP

pKa

pp cited in Vols.1-5:

LODOXAMIDE ETHYL [U]

CAS 53882-13-6 | MW: 367.75 | MF: C15 H14 Cl N3 O6

Anti-asthmatic; anti-allergic

Upjohn

LgP (4)

pKa

pp cited in Vols.1-5:

(1) M.L. Katcher et al., J. Allergy & Clin. Immun., 1980, 66, 223.

LOFEMIZOLE [U;INN]

CAS 65571-68-8 | MW: 192.65 | MF: C10 H9 Cl N2

Anti-inflammatory; analgesic; antipyretic

Pinal Pharm., Argentina

LgP 3.06(C)

pKa

pp cited in Vols.1-5:

(1) Clin. Pharmacol. Ther., 1984, 35, 221.

LOFENDAZAM [INN]

CAS 29176-29-2 | MW: 272.74 | MF: C15 H13 Cl N2 O

Tranquilizer

Knoll, FRG

LgP 2.71(C)

pKa

pp cited in Vols.1-5:

(1) Arzneim.-Forsch., 1975, 25, 1650.

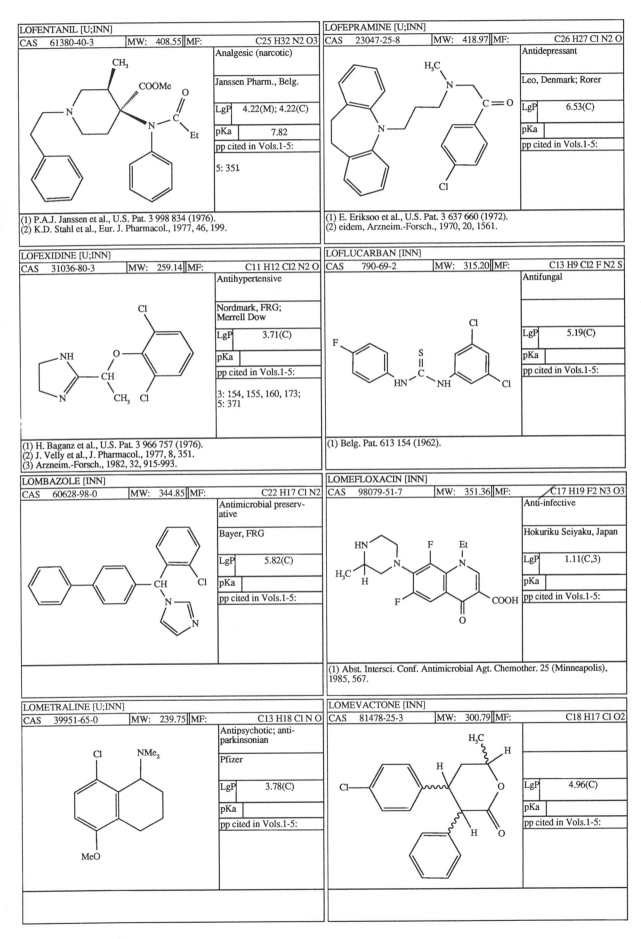

LOFENTANIL [U;INN]

CAS	61380-40-3	MW:	408.55	MF:	C25 H32 N2 O3

Analgesic (narcotic)

Janssen Pharm., Belg.

LgP	4.22(M); 4.22(C)
pKa	7.82

pp cited in Vols.1-5:

5: 351

(1) P.A.J. Janssen et al., U.S. Pat. 3 998 834 (1976).
(2) K.D. Stahl et al., Eur. J. Pharmacol., 1977, 46, 199.

LOFEPRAMINE [U;INN]

CAS	23047-25-8	MW:	418.97	MF:	C26 H27 Cl N2 O

Antidepressant

Leo, Denmark; Rorer

LgP	6.53(C)
pKa	

pp cited in Vols.1-5:

(1) E. Eriksoo et al., U.S. Pat. 3 637 660 (1972).
(2) eidem, Arzneim.-Forsch., 1970, 20, 1561.

LOFEXIDINE [U;INN]

CAS	31036-80-3	MW:	259.14	MF:	C11 H12 Cl2 N2 O

Antihypertensive

Nordmark, FRG;
Merrell Dow

LgP	3.71(C)
pKa	

pp cited in Vols.1-5:

3: 154, 155, 160, 173;
5: 371

(1) H. Baganz et al., U.S. Pat. 3 966 757 (1976).
(2) J. Velly et al., J. Pharmacol., 1977, 8, 351.
(3) Arzneim.-Forsch., 1982, 32, 915-993.

LOFLUCARBAN [INN]

CAS	790-69-2	MW:	315.20	MF:	C13 H9 Cl2 F N2 S

Antifungal

LgP	5.19(C)
pKa	

pp cited in Vols.1-5:

(1) Belg. Pat. 613 154 (1962).

LOMBAZOLE [INN]

CAS	60628-98-0	MW:	344.85	MF:	C22 H17 Cl N2

Antimicrobial preservative

Bayer, FRG

LgP	5.82(C)
pKa	

pp cited in Vols.1-5:

LOMEFLOXACIN [INN]

CAS	98079-51-7	MW:	351.36	MF:	C17 H19 F2 N3 O3

Anti-infective

Hokuriku Seiyaku, Japan

LgP	1.11(C,3)
pKa	

pp cited in Vols.1-5:

(1) Abst. Intersci. Conf. Antimicrobial Agt. Chemother. 25 (Minneapolis), 1985, 567.

LOMETRALINE [U;INN]

CAS	39951-65-0	MW:	239.75	MF:	C13 H18 Cl N O

Antipsychotic; anti-parkinsonian

Pfizer

LgP	3.78(C)
pKa	

pp cited in Vols.1-5:

LOMEVACTONE [INN]

CAS	81478-25-3	MW:	300.79	MF:	C18 H17 Cl O2

LgP	4.96(C)
pKa	

pp cited in Vols.1-5:

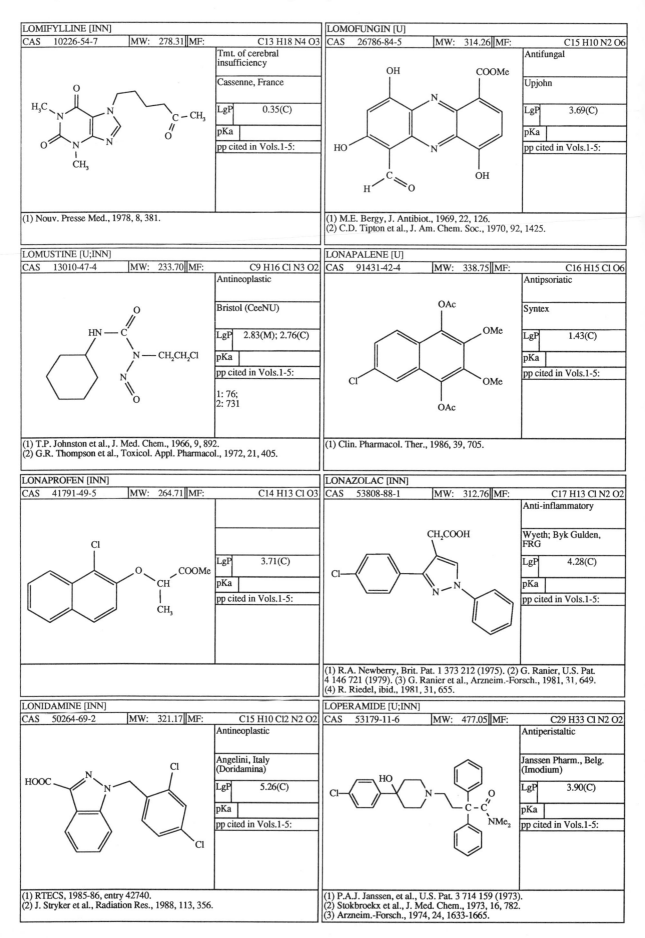

LOMIFYLLINE [INN]			
CAS 10226-54-7	MW: 278.31	MF:	C13 H18 N4 O3

Tmt. of cerebral insufficiency

Cassenne, France

LgP	0.35(C)
pKa	
pp cited in Vols.1-5:	

(1) Nouv. Presse Med., 1978, 8, 381.

LOMOFUNGIN [U]			
CAS 26786-84-5	MW: 314.26	MF:	C15 H10 N2 O6

Antifungal

Upjohn

LgP	3.69(C)
pKa	
pp cited in Vols.1-5:	

(1) M.E. Bergy, J. Antibiot., 1969, 22, 126.
(2) C.D. Tipton et al., J. Am. Chem. Soc., 1970, 92, 1425.

LOMUSTINE [U;INN]			
CAS 13010-47-4	MW: 233.70	MF:	C9 H16 Cl N3 O2

Antineoplastic

Bristol (CeeNU)

LgP	2.83(M); 2.76(C)
pKa	
pp cited in Vols.1-5:	

1: 76;
2: 731

(1) T.P. Johnston et al., J. Med. Chem., 1966, 9, 892.
(2) G.R. Thompson et al., Toxicol. Appl. Pharmacol., 1972, 21, 405.

LONAPALENE [U]			
CAS 91431-42-4	MW: 338.75	MF:	C16 H15 Cl O6

Antipsoriatic

Syntex

LgP	1.43(C)
pKa	
pp cited in Vols.1-5:	

(1) Clin. Pharmacol. Ther., 1986, 39, 705.

LONAPROFEN [INN]			
CAS 41791-49-5	MW: 264.71	MF:	C14 H13 Cl O3

LgP	3.71(C)
pKa	
pp cited in Vols.1-5:	

LONAZOLAC [INN]			
CAS 53808-88-1	MW: 312.76	MF:	C17 H13 Cl N2 O2

Anti-inflammatory

Wyeth; Byk Gulden, FRG

LgP	4.28(C)
pKa	
pp cited in Vols.1-5:	

(1) R.A. Newberry, Brit. Pat. 1 373 212 (1975). (2) G. Ranier, U.S. Pat. 4 146 721 (1979). (3) G. Ranier et al., Arzneim.-Forsch., 1981, 31, 649. (4) R. Riedel, ibid., 1981, 31, 655.

LONIDAMINE [INN]			
CAS 50264-69-2	MW: 321.17	MF:	C15 H10 Cl2 N2 O2

Antineoplastic

Angelini, Italy (Doridamina)

LgP	5.26(C)
pKa	
pp cited in Vols.1-5:	

(1) RTECS, 1985-86, entry 42740.
(2) J. Stryker et al., Radiation Res., 1988, 113, 356.

LOPERAMIDE [U;INN]			
CAS 53179-11-6	MW: 477.05	MF:	C29 H33 Cl N2 O2

Antiperistaltic

Janssen Pharm., Belg. (Imodium)

LgP	3.90(C)
pKa	
pp cited in Vols.1-5:	

(1) P.A.J. Janssen, et al., U.S. Pat. 3 714 159 (1973).
(2) Stokbroekx et al., J. Med. Chem., 1973, 16, 782.
(3) Arzneim.-Forsch., 1974, 24, 1633-1665.

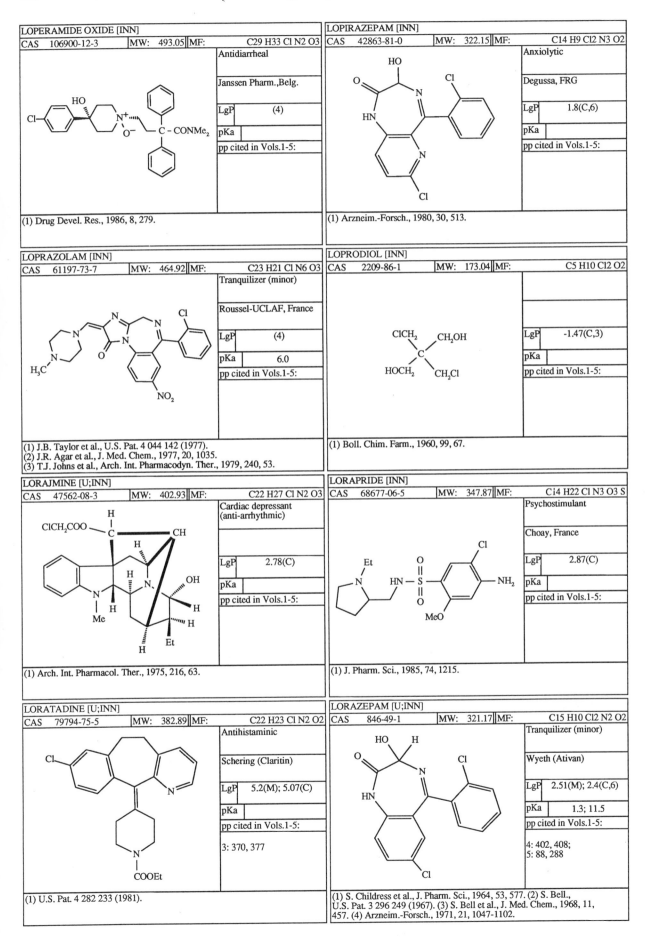

LOPERAMIDE OXIDE [INN]

CAS	106900-12-3	MW:	493.05	MF:	C29 H33 Cl N2 O3

Antidiarrheal

Janssen Pharm.,Belg.

LgP	(4)
pKa	
pp cited in Vols.1-5:	

(1) Drug Devel. Res., 1986, 8, 279.

LOPIRAZEPAM [INN]

CAS	42863-81-0	MW:	322.15	MF:	C14 H9 Cl2 N3 O2

Anxiolytic

Degussa, FRG

LgP	1.8(C,6)
pKa	
pp cited in Vols.1-5:	

(1) Arzneim.-Forsch., 1980, 30, 513.

LOPRAZOLAM [INN]

CAS	61197-73-7	MW:	464.92	MF:	C23 H21 Cl N6 O3

Tranquilizer (minor)

Roussel-UCLAF, France

LgP	(4)
pKa	6.0
pp cited in Vols.1-5:	

(1) J.B. Taylor et al., U.S. Pat. 4 044 142 (1977).
(2) J.R. Agar et al., J. Med. Chem., 1977, 20, 1035.
(3) T.J. Johns et al., Arch. Int. Pharmacodyn. Ther., 1979, 240, 53.

LOPRODIOL [INN]

CAS	2209-86-1	MW:	173.04	MF:	C5 H10 Cl2 O2

LgP	-1.47(C,3)
pKa	
pp cited in Vols.1-5:	

(1) Boll. Chim. Farm., 1960, 99, 67.

LORAJMINE [U;INN]

CAS	47562-08-3	MW:	402.93	MF:	C22 H27 Cl N2 O3

Cardiac depressant
(anti-arrhythmic)

LgP	2.78(C)
pKa	
pp cited in Vols.1-5:	

(1) Arch. Int. Pharmacol. Ther., 1975, 216, 63.

LORAPRIDE [INN]

CAS	68677-06-5	MW:	347.87	MF:	Ci4 H22 Cl N3 O3 S

Psychostimulant

Choay, France

LgP	2.87(C)
pKa	
pp cited in Vols.1-5:	

(1) J. Pharm. Sci., 1985, 74, 1215.

LORATADINE [U;INN]

CAS	79794-75-5	MW:	382.89	MF:	C22 H23 Cl N2 O2

Antihistaminic

Schering (Claritin)

LgP	5.2(M); 5.07(C)
pKa	
pp cited in Vols.1-5:	
	3: 370, 377

(1) U.S. Pat. 4 282 233 (1981).

LORAZEPAM [U;INN]

CAS	846-49-1	MW:	321.17	MF:	C15 H10 Cl2 N2 O2

Tranquilizer (minor)

Wyeth (Ativan)

LgP	2.51(M); 2.4(C,6)
pKa	1.3; 11.5
pp cited in Vols.1-5:	
	4: 402, 408; 5: 88, 288

(1) S. Childress et al., J. Pharm. Sci., 1964, 53, 577. (2) S. Bell., U.S. Pat. 3 296 249 (1967). (3) S. Bell et al., J. Med. Chem., 1968, 11, 457. (4) Arzneim.-Forsch., 1971, 21, 1047-1102.

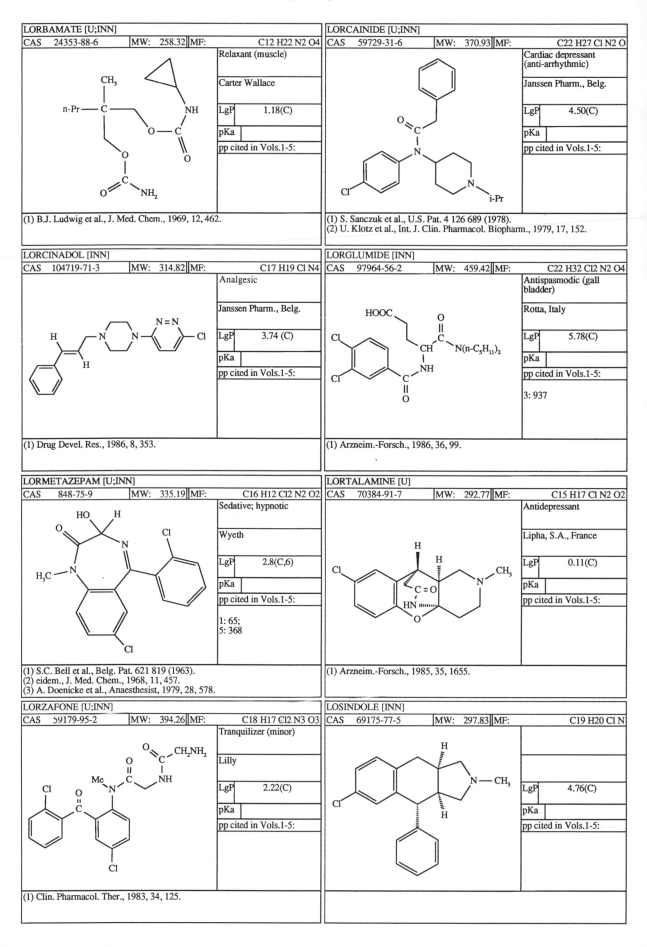

LORBAMATE [U;INN]		
CAS 24353-88-6	MW: 258.32	MF: C12 H22 N2 O4

Relaxant (muscle)

Carter Wallace

LgP 1.18(C)

pKa

pp cited in Vols.1-5:

(1) B.J. Ludwig et al., J. Med. Chem., 1969, 12, 462.

LORCAINIDE [U;INN]		
CAS 59729-31-6	MW: 370.93	MF: C22 H27 Cl N2 O

Cardiac depressant (anti-arrhythmic)

Janssen Pharm., Belg.

LgP 4.50(C)

pKa

pp cited in Vols.1-5:

(1) S. Sanczuk et al., U.S. Pat. 4 126 689 (1978).
(2) U. Klotz et al., Int. J. Clin. Pharmacol. Biopharm., 1979, 17, 152.

LORCINADOL [INN]		
CAS 104719-71-3	MW: 314.82	MF: C17 H19 Cl N4

Analgesic

Janssen Pharm., Belg.

LgP 3.74 (C)

pKa

pp cited in Vols.1-5:

(1) Drug Devel. Res., 1986, 8, 353.

LORGLUMIDE [INN]		
CAS 97964-56-2	MW: 459.42	MF: C22 H32 Cl2 N2 O4

Antispasmodic (gall bladder)

Rotta, Italy

LgP 5.78(C)

pKa

pp cited in Vols.1-5:

3: 937

(1) Arzneim.-Forsch., 1986, 36, 99.

LORMETAZEPAM [U;INN]		
CAS 848-75-9	MW: 335.19	MF: C16 H12 Cl2 N2 O2

Sedative; hypnotic

Wyeth

LgP 2.8(C,6)

pKa

pp cited in Vols.1-5:

1: 65;
5: 368

(1) S.C. Bell et al., Belg. Pat. 621 819 (1963).
(2) eidem., J. Med. Chem., 1968, 11, 457.
(3) A. Doenicke et al., Anaesthesist, 1979, 28, 578.

LORTALAMINE [U]		
CAS 70384-91-7	MW: 292.77	MF: C15 H17 Cl N2 O2

Antidepressant

Lipha, S.A., France

LgP 0.11(C)

pKa

pp cited in Vols.1-5:

(1) Arzneim.-Forsch., 1985, 35, 1655.

LORZAFONE [U;INN]		
CAS 59179-95-2	MW: 394.26	MF: C18 H17 Cl2 N3 O3

Tranquilizer (minor)

Lilly

LgP 2.22(C)

pKa

pp cited in Vols.1-5:

(1) Clin. Pharmacol. Ther., 1983, 34, 125.

LOSINDOLE [INN]		
CAS 69175-77-5	MW: 297.83	MF: C19 H20 Cl N

LgP 4.76(C)

pKa

pp cited in Vols.1-5:

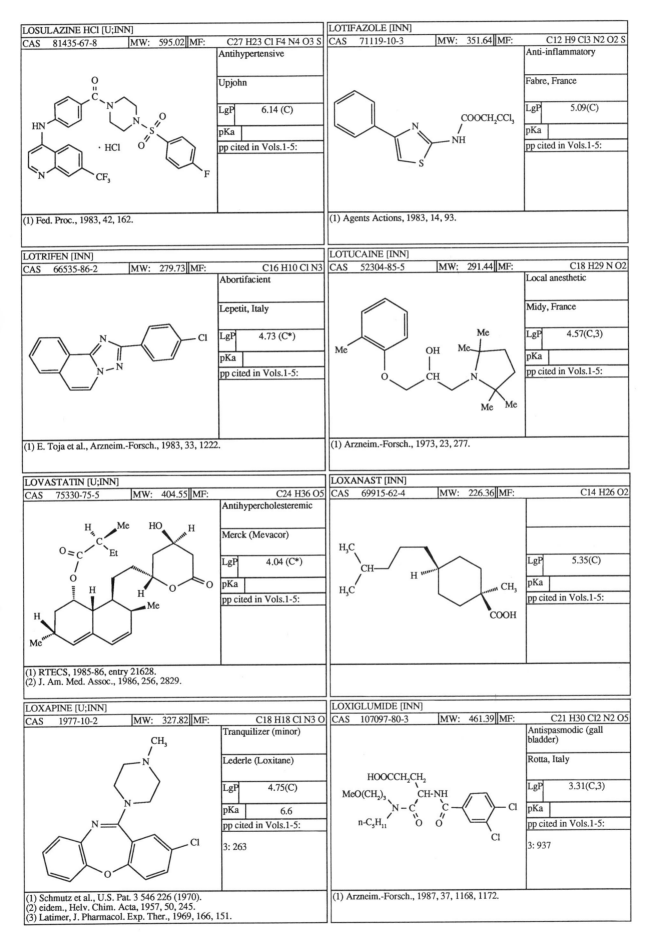

LOSULAZINE HCl [U;INN]

CAS 81435-67-8	MW: 595.02	MF:	C27 H23 Cl F4 N4 O3 S

Antihypertensive

Upjohn

LgP	6.14 (C)
pKa	
pp cited in Vols.1-5:	

(1) Fed. Proc., 1983, 42, 162.

LOTIFAZOLE [INN]

CAS 71119-10-3	MW: 351.64	MF:	C12 H9 Cl3 N2 O2 S

Anti-inflammatory

Fabre, France

LgP	5.09(C)
pKa	
pp cited in Vols.1-5:	

(1) Agents Actions, 1983, 14, 93.

LOTRIFEN [INN]

CAS 66535-86-2	MW: 279.73	MF:	C16 H10 Cl N3

Abortifacient

Lepetit, Italy

LgP	4.73 (C*)
pKa	
pp cited in Vols.1-5:	

(1) E. Toja et al., Arzneim.-Forsch., 1983, 33, 1222.

LOTUCAINE [INN]

CAS 52304-85-5	MW: 291.44	MF:	C18 H29 N O2

Local anesthetic

Midy, France

LgP	4.57(C,3)
pKa	
pp cited in Vols.1-5:	

(1) Arzneim.-Forsch., 1973, 23, 277.

LOVASTATIN [U;INN]

CAS 75330-75-5	MW: 404.55	MF:	C24 H36 O5

Antihypercholesteremic

Merck (Mevacor)

LgP	4.04 (C*)
pKa	
pp cited in Vols.1-5:	

(1) RTECS, 1985-86, entry 21628.
(2) J. Am. Med. Assoc., 1986, 256, 2829.

LOXANAST [INN]

CAS 69915-62-4	MW: 226.36	MF:	C14 H26 O2

LgP	5.35(C)
pKa	
pp cited in Vols.1-5:	

LOXAPINE [U;INN]

CAS 1977-10-2	MW: 327.82	MF:	C18 H18 Cl N3 O

Tranquilizer (minor)

Lederle (Loxitane)

LgP	4.75(C)
pKa	6.6
pp cited in Vols.1-5:	

3: 263

(1) Schmutz et al., U.S. Pat. 3 546 226 (1970).
(2) eidem., Helv. Chim. Acta, 1957, 50, 245.
(3) Latimer, J. Pharmacol. Exp. Ther., 1969, 166, 151.

LOXIGLUMIDE [INN]

CAS 107097-80-3	MW: 461.39	MF:	C21 H30 Cl2 N2 O5

Antispasmodic (gall bladder)

Rotta, Italy

LgP	3.31(C,3)
pKa	
pp cited in Vols.1-5:	

3: 937

(1) Arzneim.-Forsch., 1987, 37, 1168, 1172.

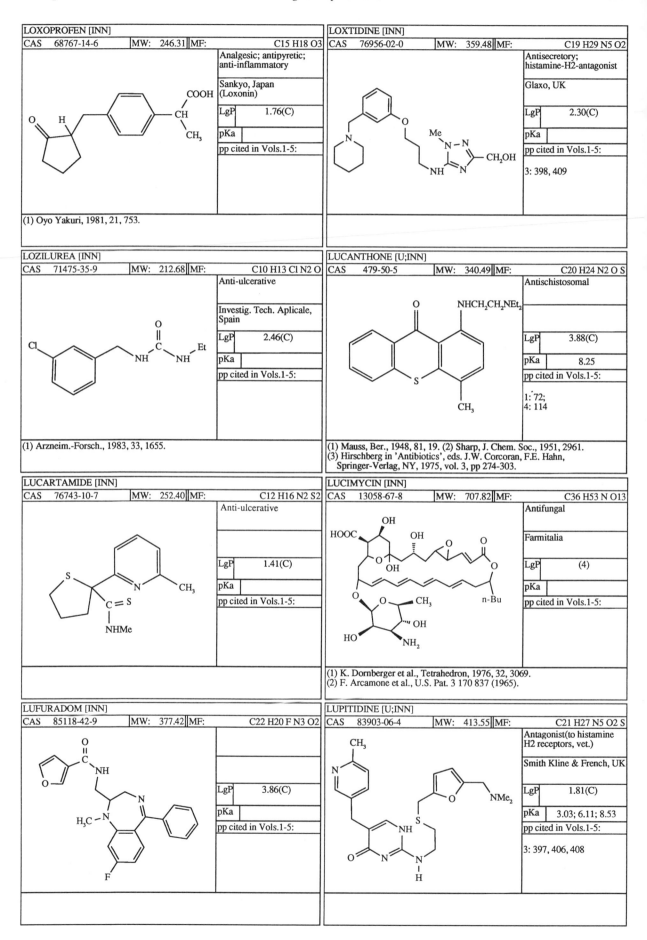

LOXOPROFEN [INN]

CAS	68767-14-6	MW: 246.31	MF:	C15 H18 O3

Analgesic; antipyretic; anti-inflammatory

Sankyo, Japan (Loxonin)

LgP	1.76(C)
pKa	
pp cited in Vols.1-5:	

(1) Oyo Yakuri, 1981, 21, 753.

LOXTIDINE [INN]

CAS	76956-02-0	MW: 359.48	MF:	C19 H29 N5 O2

Antisecretory; histamine-H2-antagonist

Glaxo, UK

LgP	2.30(C)
pKa	
pp cited in Vols.1-5:	

3: 398, 409

LOZILUREA [INN]

CAS	71475-35-9	MW: 212.68	MF:	C10 H13 Cl N2 O

Anti-ulcerative

Investig. Tech. Aplicale, Spain

LgP	2.46(C)
pKa	
pp cited in Vols.1-5:	

(1) Arzneim.-Forsch., 1983, 33, 1655.

LUCANTHONE [U;INN]

CAS	479-50-5	MW: 340.49	MF:	C20 H24 N2 O S

Antischistosomal

LgP	3.88(C)
pKa	8.25
pp cited in Vols.1-5:	

1: 72;
4: 114

(1) Mauss, Ber., 1948, 81, 19. (2) Sharp, J. Chem. Soc., 1951, 2961.
(3) Hirschberg in 'Antibiotics', eds. J.W. Corcoran, F.E. Hahn,
Springer-Verlag, NY, 1975, vol. 3, pp 274-303.

LUCARTAMIDE [INN]

CAS	76743-10-7	MW: 252.40	MF:	C12 H16 N2 S2

Anti-ulcerative

LgP	1.41(C)
pKa	
pp cited in Vols.1-5:	

LUCIMYCIN [INN]

CAS	13058-67-8	MW: 707.82	MF:	C36 H53 N O13

Antifungal

Farmitalia

LgP	(4)
pKa	
pp cited in Vols.1-5:	

(1) K. Dornberger et al., Tetrahedron, 1976, 32, 3069.
(2) F. Arcamone et al., U.S. Pat. 3 170 837 (1965).

LUFURADOM [INN]

CAS	85118-42-9	MW: 377.42	MF:	C22 H20 F N3 O2

LgP	3.86(C)
pKa	
pp cited in Vols.1-5:	

LUPITIDINE [U;INN]

CAS	83903-06-4	MW: 413.55	MF:	C21 H27 N5 O2 S

Antagonist(to histamine H2 receptors, vet.)

Smith Kline & French, UK

LgP	1.81(C)
pKa	3.03; 6.11; 8.53
pp cited in Vols.1-5:	

3: 397, 406, 408

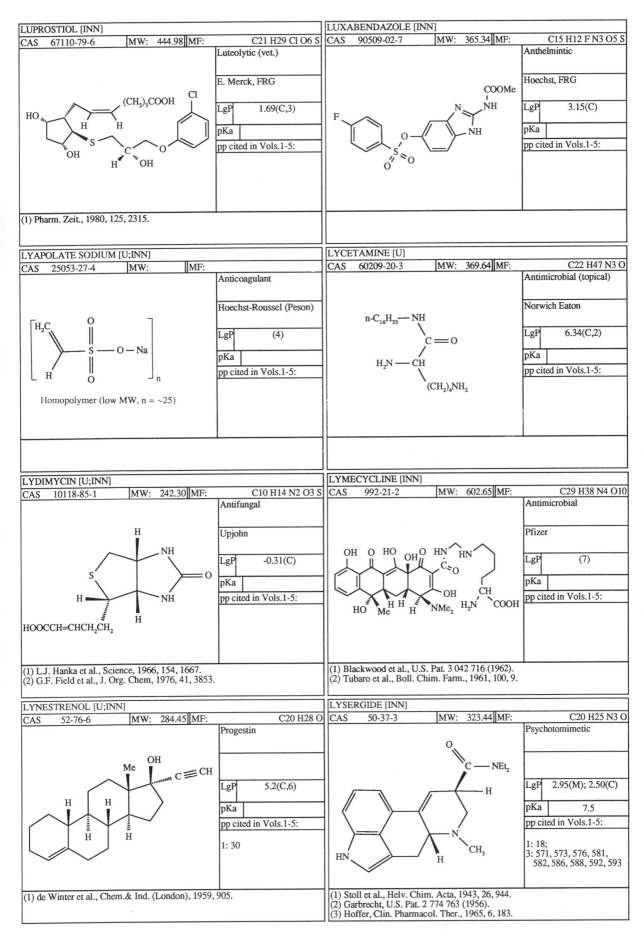

LUPROSTIOL [INN]

CAS	67110-79-6	MW:	444.98	MF:	C21 H29 Cl O6 S

Luteolytic (vet.)

E. Merck, FRG

LgP	1.69(C,3)
pKa	
pp cited in Vols.1-5:	

(1) Pharm. Zeit., 1980, 125, 2315.

LUXABENDAZOLE [INN]

CAS	90509-02-7	MW:	365.34	MF:	C15 H12 F N3 O5 S

Anthelmintic

Hoechst, FRG

LgP	3.15(C)
pKa	
pp cited in Vols.1-5:	

LYAPOLATE SODIUM [U;INN]

CAS	25053-27-4	MW:		MF:	

Anticoagulant

Hoechst-Roussel (Peson)

LgP	(4)
pKa	
pp cited in Vols.1-5:	

Homopolymer (low MW, n = ~25)

LYCETAMINE [U]

CAS	60209-20-3	MW:	369.64	MF:	C22 H47 N3 O

Antimicrobial (topical)

Norwich Eaton

LgP	6.34(C,2)
pKa	
pp cited in Vols.1-5:	

LYDIMYCIN [U;INN]

CAS	10118-85-1	MW:	242.30	MF:	C10 H14 N2 O3 S

Antifungal

Upjohn

LgP	-0.31(C)
pKa	
pp cited in Vols.1-5:	

(1) L.J. Hanka et al., Science, 1966, 154, 1667.
(2) G.F. Field et al., J. Org. Chem, 1976, 41, 3853.

LYMECYCLINE [INN]

CAS	992-21-2	MW:	602.65	MF:	C29 H38 N4 O10

Antimicrobial

Pfizer

LgP	(7)
pKa	
pp cited in Vols.1-5:	

(1) Blackwood et al., U.S. Pat. 3 042 716 (1962).
(2) Tubaro et al., Boll. Chim. Farm., 1961, 100, 9.

LYNESTRENOL [U;INN]

CAS	52-76-6	MW:	284.45	MF:	C20 H28 O

Progestin

LgP	5.2(C,6)
pKa	
pp cited in Vols.1-5:	

1: 30

(1) de Winter et al., Chem.& Ind. (London), 1959, 905.

LYSERGIDE [INN]

CAS	50-37-3	MW:	323.44	MF:	C20 H25 N3 O

Psychotomimetic

LgP	2.95(M); 2.50(C)
pKa	7.5
pp cited in Vols.1-5:	

1: 18;
3: 571, 573, 576, 581,
582, 586, 588, 592, 593

(1) Stoll et al., Helv. Chim. Acta, 1943, 26, 944.
(2) Garbrecht, U.S. Pat. 2 774 763 (1956).
(3) Hoffer, Clin. Pharmacol. Ther., 1965, 6, 183.

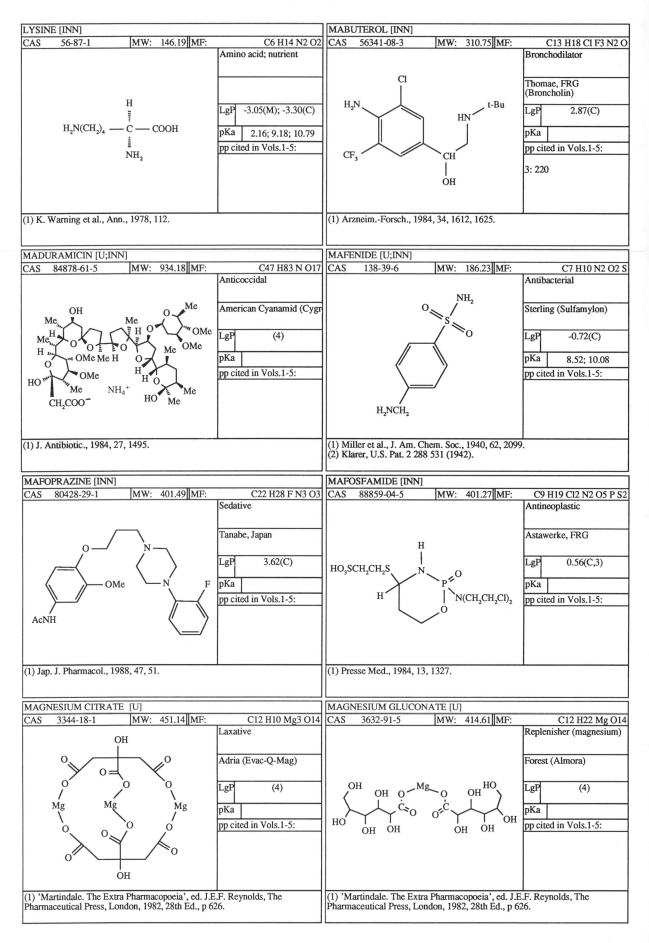

LYSINE [INN]			
CAS 56-87-1	MW: 146.19	MF:	C6 H14 N2 O2

Amino acid; nutrient

LgP	-3.05(M); -3.30(C)
pKa	2.16; 9.18; 10.79
pp cited in Vols.1-5:	

(1) K. Warning et al., Ann., 1978, 112.

MABUTEROL [INN]			
CAS 56341-08-3	MW: 310.75	MF:	C13 H18 Cl F3 N2 O

Bronchodilator

Thomae, FRG (Broncholin)

LgP	2.87(C)
pKa	
pp cited in Vols.1-5:	
3: 220	

(1) Arzneim.-Forsch., 1984, 34, 1612, 1625.

MADURAMICIN [U;INN]			
CAS 84878-61-5	MW: 934.18	MF:	C47 H83 N O17

Anticoccidal

American Cyanamid (Cygr

LgP	(4)
pKa	
pp cited in Vols.1-5:	

(1) J. Antibiotic., 1984, 27, 1495.

MAFENIDE [U;INN]			
CAS 138-39-6	MW: 186.23	MF:	C7 H10 N2 O2 S

Antibacterial

Sterling (Sulfamylon)

LgP	-0.72(C)
pKa	8.52; 10.08
pp cited in Vols.1-5:	

(1) Miller et al., J. Am. Chem. Soc., 1940, 62, 2099.
(2) Klarer, U.S. Pat. 2 288 531 (1942).

MAFOPRAZINE [INN]			
CAS 80428-29-1	MW: 401.49	MF:	C22 H28 F N3 O3

Sedative

Tanabe, Japan

LgP	3.62(C)
pKa	
pp cited in Vols.1-5:	

(1) Jap. J. Pharmacol., 1988, 47, 51.

MAFOSFAMIDE [INN]			
CAS 88859-04-5	MW: 401.27	MF:	C9 H19 Cl2 N2 O5 P S2

Antineoplastic

Astawerke, FRG

LgP	0.56(C,3)
pKa	
pp cited in Vols.1-5:	

(1) Presse Med., 1984, 13, 1327.

MAGNESIUM CITRATE [U]			
CAS 3344-18-1	MW: 451.14	MF:	C12 H10 Mg3 O14

Laxative

Adria (Evac-Q-Mag)

LgP	(4)
pKa	
pp cited in Vols.1-5:	

(1) 'Martindale. The Extra Pharmacopoeia', ed. J.E.F. Reynolds, The Pharmaceutical Press, London, 1982, 28th Ed., p 626.

MAGNESIUM GLUCONATE [U]			
CAS 3632-91-5	MW: 414.61	MF:	C12 H22 Mg O14

Replenisher (magnesium)

Forest (Almora)

LgP	(4)
pKa	
pp cited in Vols.1-5:	

(1) 'Martindale. The Extra Pharmacopoeia', ed. J.E.F. Reynolds, The Pharmaceutical Press, London, 1982, 28th Ed., p 626.

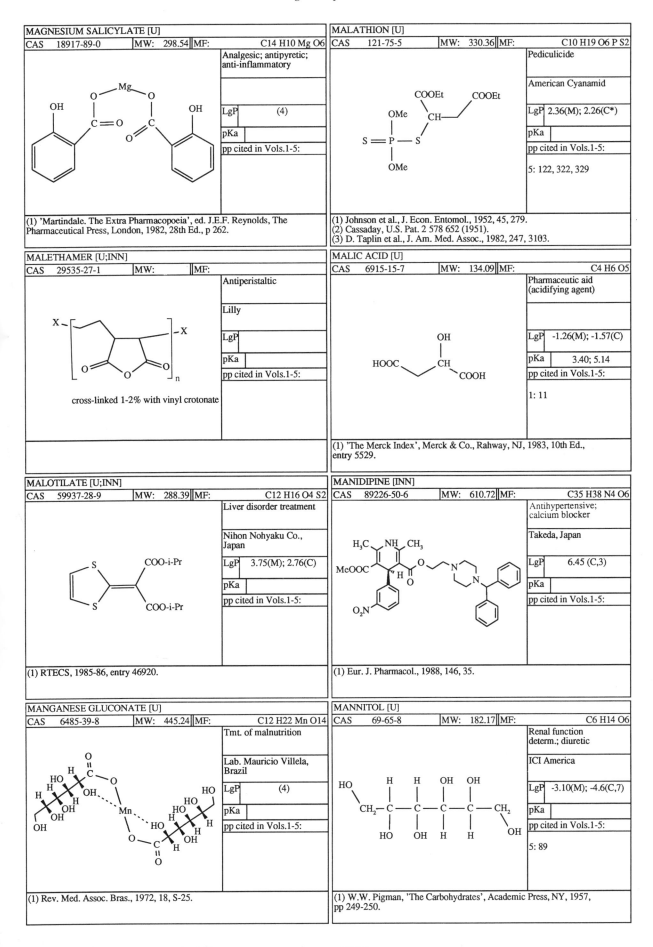

MAGNESIUM SALICYLATE [U]

CAS 18917-89-0	MW: 298.54	MF:	C14 H10 Mg O6

Analgesic; antipyretic; anti-inflammatory

LgP	(4)
pKa	
pp cited in Vols.1-5:	

(1) 'Martindale. The Extra Pharmacopoeia', ed. J.E.F. Reynolds, The Pharmaceutical Press, London, 1982, 28th Ed., p 262.

MALETHAMER [U;INN]

CAS 29535-27-1	MW:	MF:	

Antiperistaltic

Lilly

LgP	
pKa	
pp cited in Vols.1-5:	

cross-linked 1-2% with vinyl crotonate

MALOTILATE [U;INN]

CAS 59937-28-9	MW: 288.39	MF:	C12 H16 O4 S2

Liver disorder treatment

Nihon Nohyaku Co., Japan

LgP	3.75(M); 2.76(C)
pKa	
pp cited in Vols.1-5:	

(1) RTECS, 1985-86, entry 46920.

MANGANESE GLUCONATE [U]

CAS 6485-39-8	MW: 445.24	MF:	C12 H22 Mn O14

Tmt. of malnutrition

Lab. Mauricio Villela, Brazil

LgP	(4)
pKa	
pp cited in Vols.1-5:	

(1) Rev. Med. Assoc. Bras., 1972, 18, S-25.

MALATHION [U]

CAS 121-75-5	MW: 330.36	MF:	C10 H19 O6 P S2

Pediculicide

American Cyanamid

LgP	2.36(M); 2.26(C*)
pKa	
pp cited in Vols.1-5:	

5: 122, 322, 329

(1) Johnson et al., J. Econ. Entomol., 1952, 45, 279.
(2) Cassaday, U.S. Pat. 2 578 652 (1951).
(3) D. Taplin et al., J. Am. Med. Assoc., 1982, 247, 3103.

MALIC ACID [U]

CAS 6915-15-7	MW: 134.09	MF:	C4 H6 O5

Pharmaceutic aid (acidifying agent)

LgP	-1.26(M); -1.57(C)
pKa	3.40; 5.14
pp cited in Vols.1-5:	

1: 11

(1) 'The Merck Index', Merck & Co., Rahway, NJ, 1983, 10th Ed., entry 5529.

MANIDIPINE [INN]

CAS 89226-50-6	MW: 610.72	MF:	C35 H38 N4 O6

Antihypertensive; calcium blocker

Takeda, Japan

LgP	6.45 (C,3)
pKa	
pp cited in Vols.1-5:	

(1) Eur. J. Pharmacol., 1988, 146, 35.

MANNITOL [U]

CAS 69-65-8	MW: 182.17	MF:	C6 H14 O6

Renal function determ.; diuretic

ICI America

LgP	-3.10(M); -4.6(C,7)
pKa	
pp cited in Vols.1-5:	

5: 89

(1) W.W. Pigman, 'The Carbohydrates', Academic Press, NY, 1957, pp 249-250.

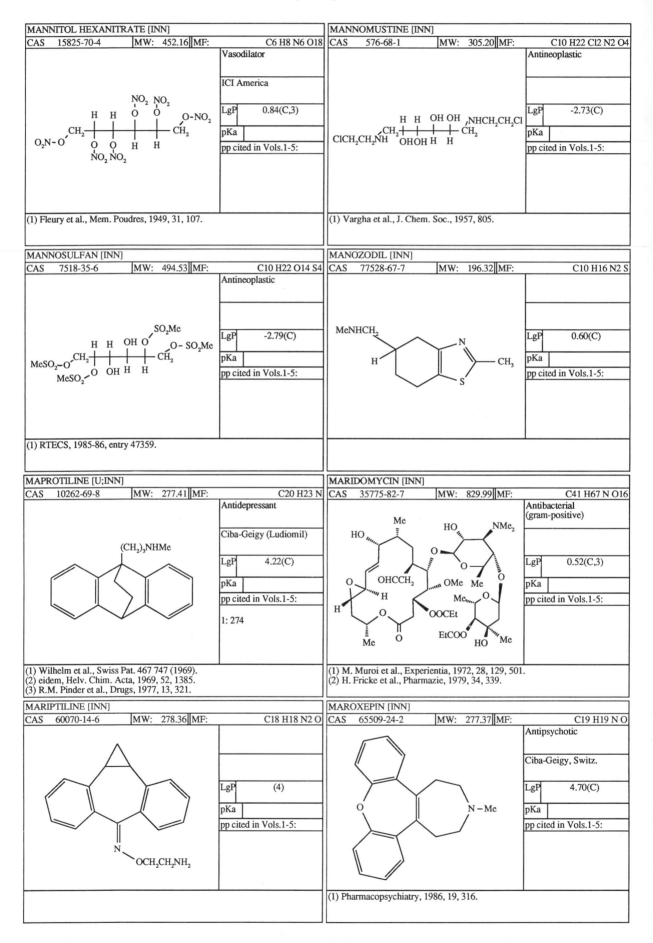

MANNITOL HEXANITRATE [INN]		
CAS 15825-70-4	MW: 452.16	MF: $C_6 H_8 N_6 O_{18}$

Vasodilator

ICI America

| LgP | 0.84(C,3) |
| pKa | |
| pp cited in Vols.1-5: |

(1) Fleury et al., Mem. Poudres, 1949, 31, 107.

MANNOMUSTINE [INN]		
CAS 576-68-1	MW: 305.20	MF: $C_{10} H_{22} Cl_2 N_2 O_4$

Antineoplastic

| LgP | -2.73(C) |
| pKa | |
| pp cited in Vols.1-5: |

(1) Vargha et al., J. Chem. Soc., 1957, 805.

MANNOSULFAN [INN]		
CAS 7518-35-6	MW: 494.53	MF: $C_{10} H_{22} O_{14} S_4$

Antineoplastic

| LgP | -2.79(C) |
| pKa | |
| pp cited in Vols.1-5: |

(1) RTECS, 1985-86, entry 47359.

MANOZODIL [INN]		
CAS 77528-67-7	MW: 196.32	MF: $C_{10} H_{16} N_2 S$

| LgP | 0.60(C) |
| pKa | |
| pp cited in Vols.1-5: |

MAPROTILINE [U;INN]		
CAS 10262-69-8	MW: 277.41	MF: $C_{20} H_{23} N$

Antidepressant

Ciba-Geigy (Ludiomil)

| LgP | 4.22(C) |
| pKa | |
| pp cited in Vols.1-5: |
| 1: 274 |

(1) Wilhelm et al., Swiss Pat. 467 747 (1969).
(2) eidem, Helv. Chim. Acta, 1969, 52, 1385.
(3) R.M. Pinder et al., Drugs, 1977, 13, 321.

MARIDOMYCIN [INN]		
CAS 35775-82-7	MW: 829.99	MF: $C_{41} H_{67} N O_{16}$

Antibacterial
(gram-positive)

| LgP | 0.52(C,3) |
| pKa | |
| pp cited in Vols.1-5: |

(1) M. Muroi et al., Experientia, 1972, 28, 129, 501.
(2) H. Fricke et al., Pharmazie, 1979, 34, 339.

MARIPTILINE [INN]		
CAS 60070-14-6	MW: 278.36	MF: $C_{18} H_{18} N_2 O$

| LgP | (4) |
| pKa | |
| pp cited in Vols.1-5: |

MAROXEPIN [INN]		
CAS 65509-24-2	MW: 277.37	MF: $C_{19} H_{19} N O$

Antipsychotic

Ciba-Geigy, Switz.

| LgP | 4.70(C) |
| pKa | |
| pp cited in Vols.1-5: |

(1) Pharmacopsychiatry, 1986, 19, 316.

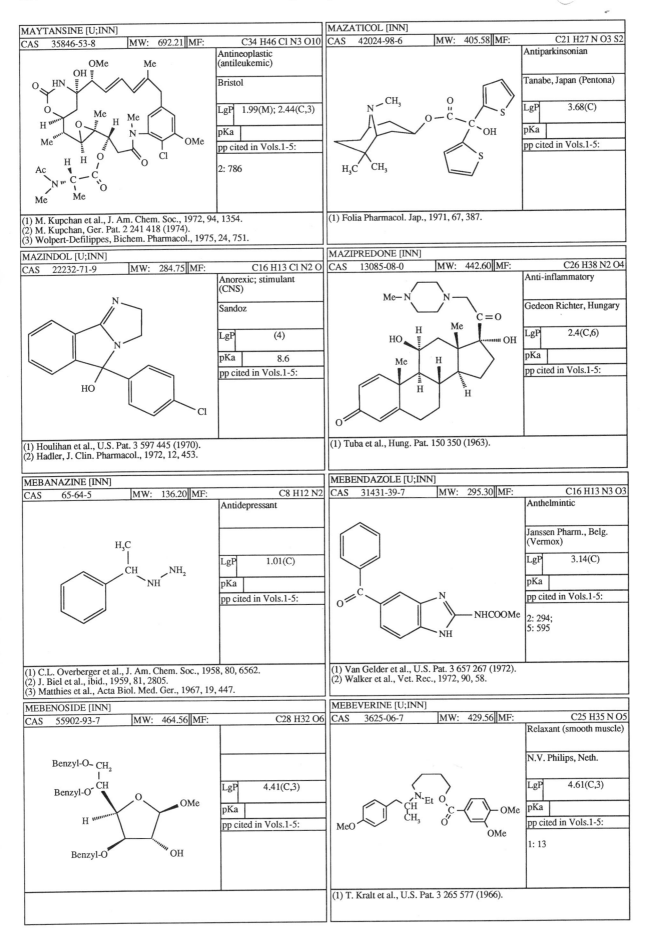

MAYTANSINE [U;INN]		
CAS 35846-53-8	MW: 692.21	MF: C34 H46 Cl N3 O10

Antineoplastic (antileukemic)

Bristol

LgP	1.99(M); 2.44(C,3)
pKa	
pp cited in Vols.1-5:	

2: 786

(1) M. Kupchan et al., J. Am. Chem. Soc., 1972, 94, 1354.
(2) M. Kupchan, Ger. Pat. 2 241 418 (1974).
(3) Wolpert-Defilippes, Bichem. Pharmacol., 1975, 24, 751.

MAZINDOL [U;INN]		
CAS 22232-71-9	MW: 284.75	MF: C16 H13 Cl N2 O

Anorexic; stimulant (CNS)

Sandoz

LgP	(4)
pKa	8.6
pp cited in Vols.1-5:	

(1) Houlihan et al., U.S. Pat. 3 597 445 (1970).
(2) Hadler, J. Clin. Pharmacol., 1972, 12, 453.

MEBANAZINE [INN]		
CAS 65-64-5	MW: 136.20	MF: C8 H12 N2

Antidepressant

LgP	1.01(C)
pKa	
pp cited in Vols.1-5:	

(1) C.L. Overberger et al., J. Am. Chem. Soc., 1958, 80, 6562.
(2) J. Biel et al., ibid., 1959, 81, 2805.
(3) Matthies et al., Acta Biol. Med. Ger., 1967, 19, 447.

MEBENOSIDE [INN]		
CAS 55902-93-7	MW: 464.56	MF: C28 H32 O6

LgP	4.41(C,3)
pKa	
pp cited in Vols.1-5:	

MAZATICOL [INN]		
CAS 42024-98-6	MW: 405.58	MF: C21 H27 N O3 S2

Antiparkinsonian

Tanabe, Japan (Pentona)

LgP	3.68(C)
pKa	
pp cited in Vols.1-5:	

(1) Folia Pharmacol. Jap., 1971, 67, 387.

MAZIPREDONE [INN]		
CAS 13085-08-0	MW: 442.60	MF: C26 H38 N2 O4

Anti-inflammatory

Gedeon Richter, Hungary

LgP	2.4(C,6)
pKa	
pp cited in Vols.1-5:	

(1) Tuba et al., Hung. Pat. 150 350 (1963).

MEBENDAZOLE [U;INN]		
CAS 31431-39-7	MW: 295.30	MF: C16 H13 N3 O3

Anthelmintic

Janssen Pharm., Belg. (Vermox)

LgP	3.14(C)
pKa	
pp cited in Vols.1-5:	

2: 294;
5: 595

(1) Van Gelder et al., U.S. Pat. 3 657 267 (1972).
(2) Walker et al., Vet. Rec., 1972, 90, 58.

MEBEVERINE [U;INN]		
CAS 3625-06-7	MW: 429.56	MF: C25 H35 N O5

Relaxant (smooth muscle)

N.V. Philips, Neth.

LgP	4.61(C,3)
pKa	
pp cited in Vols.1-5:	

1: 13

(1) T. Kralt et al., U.S. Pat. 3 265 577 (1966).

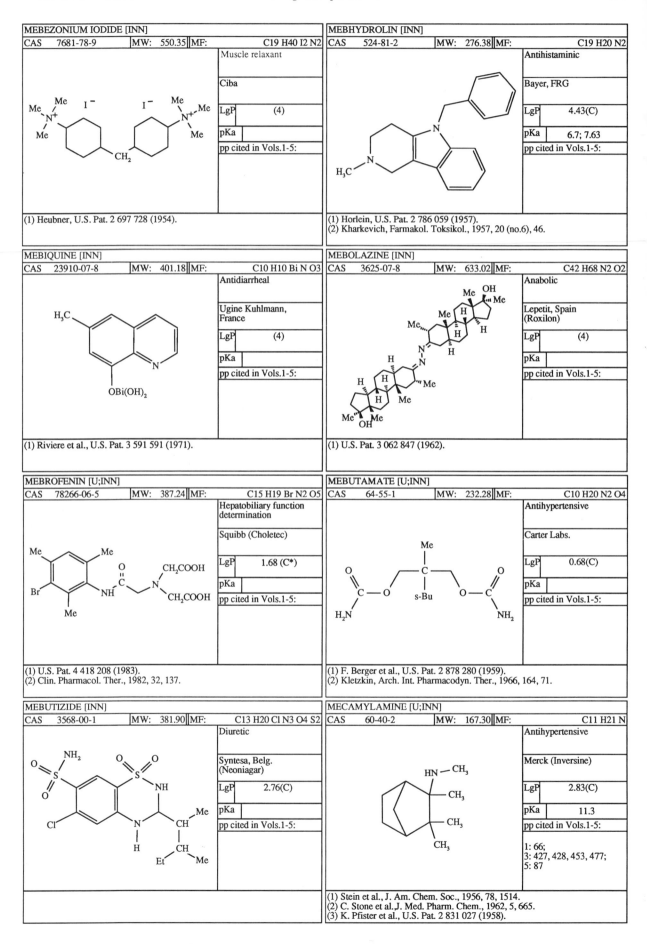

MEBEZONIUM IODIDE [INN]

CAS	7681-78-9	MW:	550.35	MF:	C19 H40 I2 N2

Muscle relaxant

Ciba

LgP	(4)
pKa	
pp cited in Vols.1-5:	

(1) Heubner, U.S. Pat. 2 697 728 (1954).

MEBHYDROLIN [INN]

CAS	524-81-2	MW:	276.38	MF:	C19 H20 N2

Antihistaminic

Bayer, FRG

LgP	4.43(C)
pKa	6.7; 7.63
pp cited in Vols.1-5:	

(1) Horlein, U.S. Pat. 2 786 059 (1957).
(2) Kharkevich, Farmakol. Toksikol., 1957, 20 (no.6), 46.

MEBIQUINE [INN]

CAS	23910-07-8	MW:	401.18	MF:	C10 H10 Bi N O3

Antidiarrheal

Ugine Kuhlmann, France

LgP	(4)
pKa	
pp cited in Vols.1-5:	

(1) Riviere et al., U.S. Pat. 3 591 591 (1971).

MEBOLAZINE [INN]

CAS	3625-07-8	MW:	633.02	MF:	C42 H68 N2 O2

Anabolic

Lepetit, Spain (Roxilon)

LgP	(4)
pKa	
pp cited in Vols.1-5:	

(1) U.S. Pat. 3 062 847 (1962).

MEBROFENIN [U;INN]

CAS	78266-06-5	MW:	387.24	MF:	C15 H19 Br N2 O5

Hepatobiliary function determination

Squibb (Choletec)

LgP	1.68 (C*)
pKa	
pp cited in Vols.1-5:	

(1) U.S. Pat. 4 418 208 (1983).
(2) Clin. Pharmacol. Ther., 1982, 32, 137.

MEBUTAMATE [U;INN]

CAS	64-55-1	MW:	232.28	MF:	C10 H20 N2 O4

Antihypertensive

Carter Labs.

LgP	0.68(C)
pKa	
pp cited in Vols.1-5:	

(1) F. Berger et al., U.S. Pat. 2 878 280 (1959).
(2) Kletzkin, Arch. Int. Pharmacodyn. Ther., 1966, 164, 71.

MEBUTIZIDE [INN]

CAS	3568-00-1	MW:	381.90	MF:	C13 H20 Cl N3 O4 S2

Diuretic

Syntesa, Belg. (Neoniagar)

LgP	2.76(C)
pKa	
pp cited in Vols.1-5:	

MECAMYLAMINE [U;INN]

CAS	60-40-2	MW:	167.30	MF:	C11 H21 N

Antihypertensive

Merck (Inversine)

LgP	2.83(C)
pKa	11.3
pp cited in Vols.1-5:	

1: 66;
3: 427, 428, 453, 477;
5: 87

(1) Stein et al., J. Am. Chem. Soc., 1956, 78, 1514.
(2) C. Stone et al. J. Med. Pharm. Chem., 1962, 5, 665.
(3) K. Pfister et al., U.S. Pat. 2 831 027 (1958).

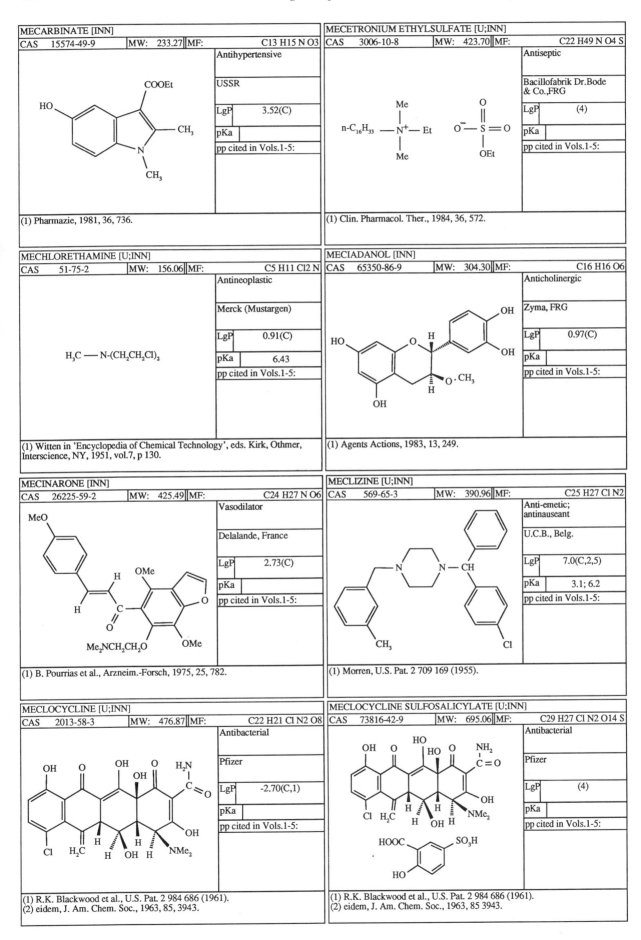

MECARBINATE [INN]

CAS 15574-49-9	MW: 233.27	MF: C13 H15 N O3

Antihypertensive

USSR

LgP	3.52(C)
pKa	
pp cited in Vols.1-5:	

(1) Pharmazie, 1981, 36, 736.

MECETRONIUM ETHYLSULFATE [U;INN]

CAS 3006-10-8	MW: 423.70	MF: C22 H49 N O4 S

Antiseptic

Bacillofabrik Dr.Bode & Co.,FRG

LgP	(4)
pKa	
pp cited in Vols.1-5:	

(1) Clin. Pharmacol. Ther., 1984, 36, 572.

MECHLORETHAMINE [U;INN]

CAS 51-75-2	MW: 156.06	MF: C5 H11 Cl2 N

Antineoplastic

Merck (Mustargen)

LgP	0.91(C)
pKa	6.43
pp cited in Vols.1-5:	

$H_3C - N-(CH_2CH_2Cl)_2$

(1) Witten in 'Encyclopedia of Chemical Technology', eds. Kirk, Othmer, Interscience, NY, 1951, vol.7, p 130.

MECIADANOL [INN]

CAS 65350-86-9	MW: 304.30	MF: C16 H16 O6

Anticholinergic

Zyma, FRG

LgP	0.97(C)
pKa	
pp cited in Vols.1-5:	

(1) Agents Actions, 1983, 13, 249.

MECINARONE [INN]

CAS 26225-59-2	MW: 425.49	MF: C24 H27 N O6

Vasodilator

Delalande, France

LgP	2.73(C)
pKa	
pp cited in Vols.1-5:	

(1) B. Pourrias et al., Arzneim.-Forsch, 1975, 25, 782.

MECLIZINE [U;INN]

CAS 569-65-3	MW: 390.96	MF: C25 H27 Cl N2

Anti-emetic; antinauseant

U.C.B., Belg.

LgP	7.0(C,2,5)
pKa	3.1; 6.2
pp cited in Vols.1-5:	

(1) Morren, U.S. Pat. 2 709 169 (1955).

MECLOCYCLINE [U;INN]

CAS 2013-58-3	MW: 476.87	MF: C22 H21 Cl N2 O8

Antibacterial

Pfizer

LgP	-2.70(C,1)
pKa	
pp cited in Vols.1-5:	

(1) R.K. Blackwood et al., U.S. Pat. 2 984 686 (1961).
(2) eidem, J. Am. Chem. Soc., 1963, 85, 3943.

MECLOCYCLINE SULFOSALICYLATE [U;INN]

CAS 73816-42-9	MW: 695.06	MF: C29 H27 Cl N2 O14 S

Antibacterial

Pfizer

LgP	(4)
pKa	
pp cited in Vols.1-5:	

(1) R.K. Blackwood et al., U.S. Pat. 2 984 686 (1961).
(2) eidem, J. Am. Chem. Soc., 1963, 85 3943.

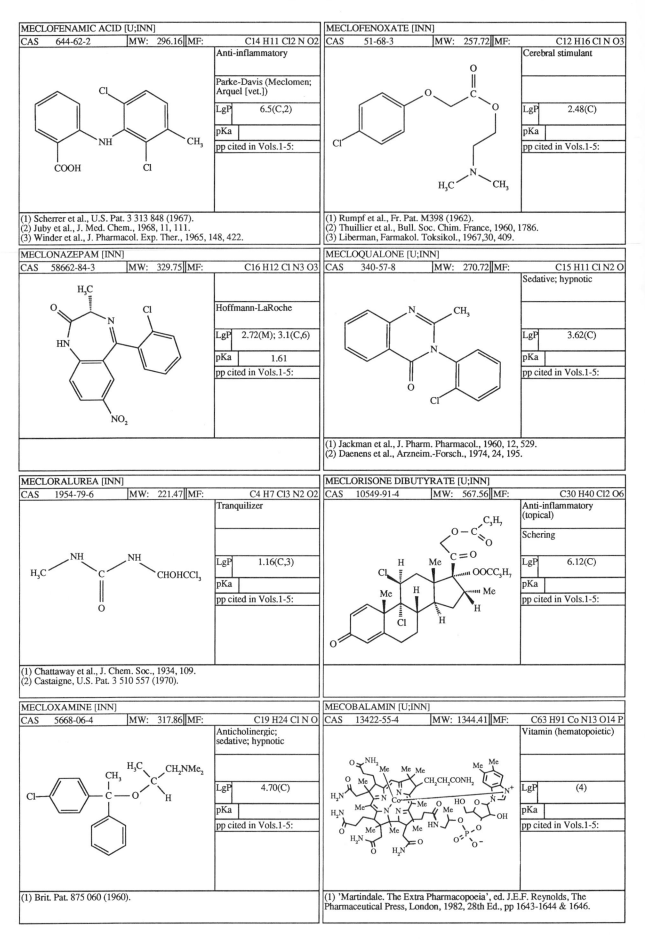

MECLOFENAMIC ACID [U;INN]			
CAS 644-62-2	MW: 296.16	MF:	C14 H11 Cl2 N O2
		Anti-inflammatory	
		Parke-Davis (Meclomen; Arquel [vet.])	
		LgP	6.5(C,2)
		pKa	
		pp cited in Vols.1-5:	

(1) Scherrer et al., U.S. Pat. 3 313 848 (1967).
(2) Juby et al., J. Med. Chem., 1968, 11, 111.
(3) Winder et al., J. Pharmacol. Exp. Ther., 1965, 148, 422.

MECLOFENOXATE [INN]			
CAS 51-68-3	MW: 257.72	MF:	C12 H16 Cl N O3
		Cerebral stimulant	
		LgP	2.48(C)
		pKa	
		pp cited in Vols.1-5:	

(1) Rumpf et al., Fr. Pat. M398 (1962).
(2) Thuillier et al., Bull. Soc. Chim. France, 1960, 1786.
(3) Liberman, Farmakol. Toksikol., 1967, 30, 409.

MECLONAZEPAM [INN]			
CAS 58662-84-3	MW: 329.75	MF:	C16 H12 Cl N3 O3
		Hoffmann-LaRoche	
		LgP	2.72(M); 3.1(C,6)
		pKa	1.61
		pp cited in Vols.1-5:	

MECLOQUALONE [U;INN]			
CAS 340-57-8	MW: 270.72	MF:	C15 H11 Cl N2 O
		Sedative; hypnotic	
		LgP	3.62(C)
		pKa	
		pp cited in Vols.1-5:	

(1) Jackman et al., J. Pharm. Pharmacol., 1960, 12, 529.
(2) Daenens et al., Arzneim.-Forsch., 1974, 24, 195.

MECLORALUREA [INN]			
CAS 1954-79-6	MW: 221.47	MF:	C4 H7 Cl3 N2 O2
		Tranquilizer	
		LgP	1.16(C,3)
		pKa	
		pp cited in Vols.1-5:	

(1) Chattaway et al., J. Chem. Soc., 1934, 109.
(2) Castaigne, U.S. Pat. 3 510 557 (1970).

MECLORISONE DIBUTYRATE [U;INN]			
CAS 10549-91-4	MW: 567.56	MF:	C30 H40 Cl2 O6
		Anti-inflammatory (topical)	
		Schering	
		LgP	6.12(C)
		pKa	
		pp cited in Vols.1-5:	

MECLOXAMINE [INN]			
CAS 5668-06-4	MW: 317.86	MF:	C19 H24 Cl N O
		Anticholinergic; sedative; hypnotic	
		LgP	4.70(C)
		pKa	
		pp cited in Vols.1-5:	

(1) Brit. Pat. 875 060 (1960).

MECOBALAMIN [U;INN]			
CAS 13422-55-4	MW: 1344.41	MF:	C63 H91 Co N13 O14 P
		Vitamin (hematopoietic)	
		LgP	(4)
		pKa	
		pp cited in Vols.1-5:	

(1) 'Martindale. The Extra Pharmacopoeia', ed. J.E.F. Reynolds, The Pharmaceutical Press, London, 1982, 28th Ed., pp 1643-1644 & 1646.

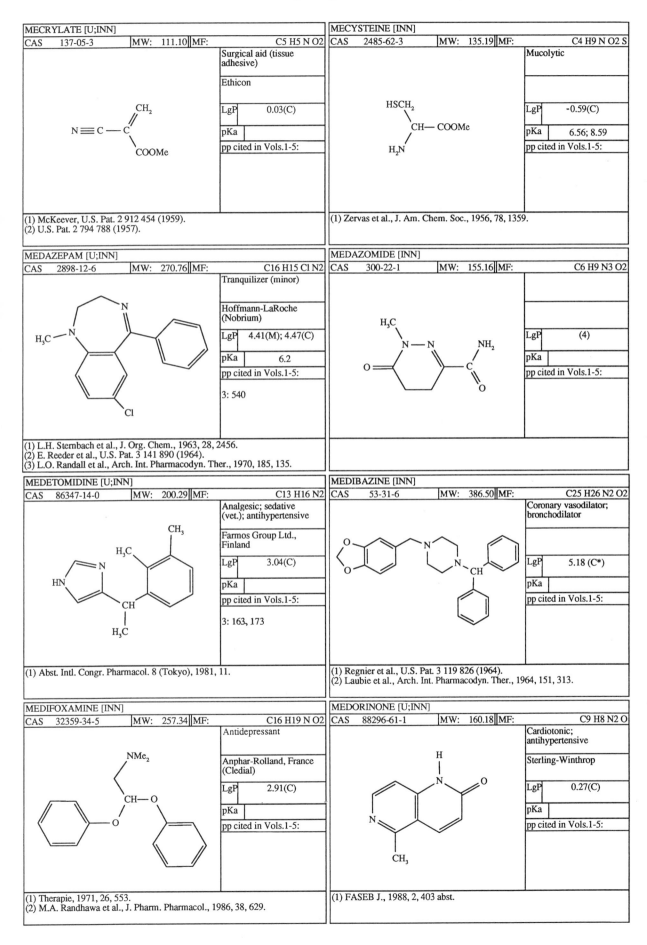

MECRYLATE [U;INN]			
CAS 137-05-3	MW: 111.10	MF:	C5 H5 N O2

Surgical aid (tissue adhesive)

Ethicon

LgP	0.03(C)
pKa	
pp cited in Vols.1-5:	

(1) McKeever, U.S. Pat. 2 912 454 (1959).
(2) U.S. Pat. 2 794 788 (1957).

MECYSTEINE [INN]			
CAS 2485-62-3	MW: 135.19	MF:	C4 H9 N O2 S

Mucolytic

LgP	-0.59(C)
pKa	6.56; 8.59
pp cited in Vols.1-5:	

(1) Zervas et al., J. Am. Chem. Soc., 1956, 78, 1359.

MEDAZEPAM [U;INN]			
CAS 2898-12-6	MW: 270.76	MF:	C16 H15 Cl N2

Tranquilizer (minor)

Hoffmann-LaRoche (Nobrium)

LgP	4.41(M); 4.47(C)
pKa	6.2
pp cited in Vols.1-5:	
3: 540	

(1) L.H. Sternbach et al., J. Org. Chem., 1963, 28, 2456.
(2) E. Reeder et al., U.S. Pat. 3 141 890 (1964).
(3) L.O. Randall et al., Arch. Int. Pharmacodyn. Ther., 1970, 185, 135.

MEDAZOMIDE [INN]			
CAS 300-22-1	MW: 155.16	MF:	C6 H9 N3 O2

LgP	(4)
pKa	
pp cited in Vols.1-5:	

MEDETOMIDINE [U;INN]			
CAS 86347-14-0	MW: 200.29	MF:	C13 H16 N2

Analgesic; sedative (vet.); antihypertensive

Farmos Group Ltd., Finland

LgP	3.04(C)
pKa	
pp cited in Vols.1-5:	
3: 163, 173	

(1) Abst. Intl. Congr. Pharmacol. 8 (Tokyo), 1981, 11.

MEDIBAZINE [INN]			
CAS 53-31-6	MW: 386.50	MF:	C25 H26 N2 O2

Coronary vasodilator; bronchodilator

LgP	5.18 (C*)
pKa	
pp cited in Vols.1-5:	

(1) Regnier et al., U.S. Pat. 3 119 826 (1964).
(2) Laubie et al., Arch. Int. Pharmacodyn. Ther., 1964, 151, 313.

MEDIFOXAMINE [INN]			
CAS 32359-34-5	MW: 257.34	MF:	C16 H19 N O2

Antidepressant

Anphar-Rolland, France (Cledial)

LgP	2.91(C)
pKa	
pp cited in Vols.1-5:	

(1) Therapie, 1971, 26, 553.
(2) M.A. Randhawa et al., J. Pharm. Pharmacol., 1986, 38, 629.

MEDORINONE [U;INN]			
CAS 88296-61-1	MW: 160.18	MF:	C9 H8 N2 O

Cardiotonic; antihypertensive

Sterling-Winthrop

LgP	0.27(C)
pKa	
pp cited in Vols.1-5:	

(1) FASEB J., 1988, 2, 403 abst.

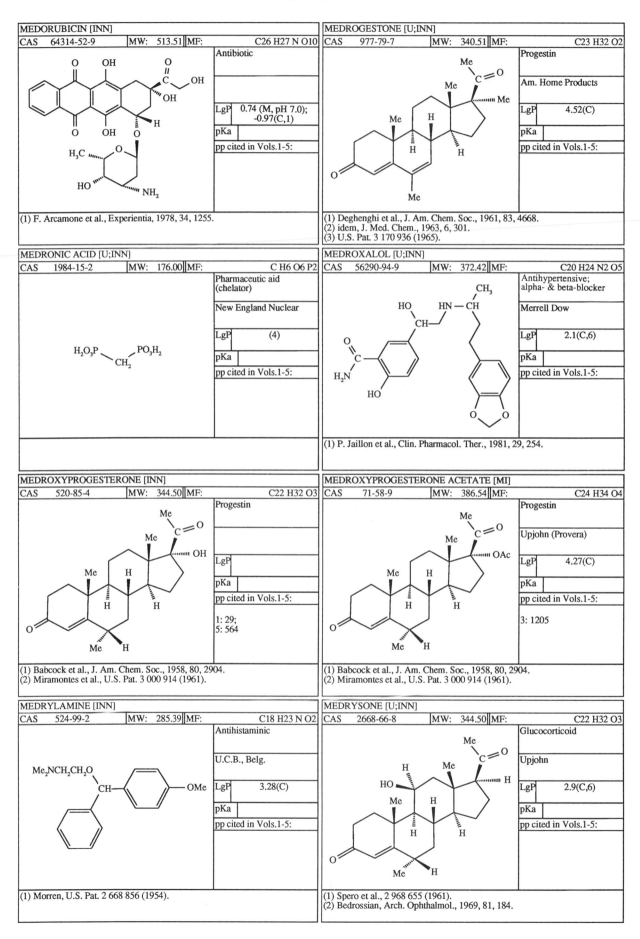

MEDORUBICIN [INN]

CAS 64314-52-9	MW: 513.51	MF: C26 H27 N O10

Antibiotic

LgP	0.74 (M, pH 7.0); -0.97(C,1)
pKa	
pp cited in Vols.1-5:	

(1) F. Arcamone et al., Experientia, 1978, 34, 1255.

MEDROGESTONE [U;INN]

CAS 977-79-7	MW: 340.51	MF: C23 H32 O2

Progestin

Am. Home Products

LgP	4.52(C)
pKa	
pp cited in Vols.1-5:	

(1) Deghenghi et al., J. Am. Chem. Soc., 1961, 83, 4668.
(2) idem, J. Med. Chem., 1963, 6, 301.
(3) U.S. Pat. 3 170 936 (1965).

MEDRONIC ACID [U;INN]

CAS 1984-15-2	MW: 176.00	MF: C H6 O6 P2

Pharmaceutic aid (chelator)

New England Nuclear

LgP	(4)
pKa	
pp cited in Vols.1-5:	

MEDROXALOL [U;INN]

CAS 56290-94-9	MW: 372.42	MF: C20 H24 N2 O5

Antihypertensive; alpha- & beta-blocker

Merrell Dow

LgP	2.1(C,6)
pKa	
pp cited in Vols.1-5:	

(1) P. Jaillon et al., Clin. Pharmacol. Ther., 1981, 29, 254.

MEDROXYPROGESTERONE [INN]

CAS 520-85-4	MW: 344.50	MF: C22 H32 O3

Progestin

LgP	
pKa	
pp cited in Vols.1-5:	
	1: 29; 5: 564

(1) Babcock et al., J. Am. Chem. Soc., 1958, 80, 2904.
(2) Miramontes et al., U.S. Pat. 3 000 914 (1961).

MEDROXYPROGESTERONE ACETATE [MI]

CAS 71-58-9	MW: 386.54	MF: C24 H34 O4

Progestin

Upjohn (Provera)

LgP	4.27(C)
pKa	
pp cited in Vols.1-5:	
	3: 1205

(1) Babcock et al., J. Am. Chem. Soc., 1958, 80, 2904.
(2) Miramontes et al., U.S. Pat. 3 000 914 (1961).

MEDRYLAMINE [INN]

CAS 524-99-2	MW: 285.39	MF: C18 H23 N O2

Antihistaminic

U.C.B., Belg.

LgP	3.28(C)
pKa	
pp cited in Vols.1-5:	

(1) Morren, U.S. Pat. 2 668 856 (1954).

MEDRYSONE [U;INN]

CAS 2668-66-8	MW: 344.50	MF: C22 H32 O3

Glucocorticoid

Upjohn

LgP	2.9(C,6)
pKa	
pp cited in Vols.1-5:	

(1) Spero et al., 2 968 655 (1961).
(2) Bedrossian, Arch. Ophthalmol., 1969, 81, 184.

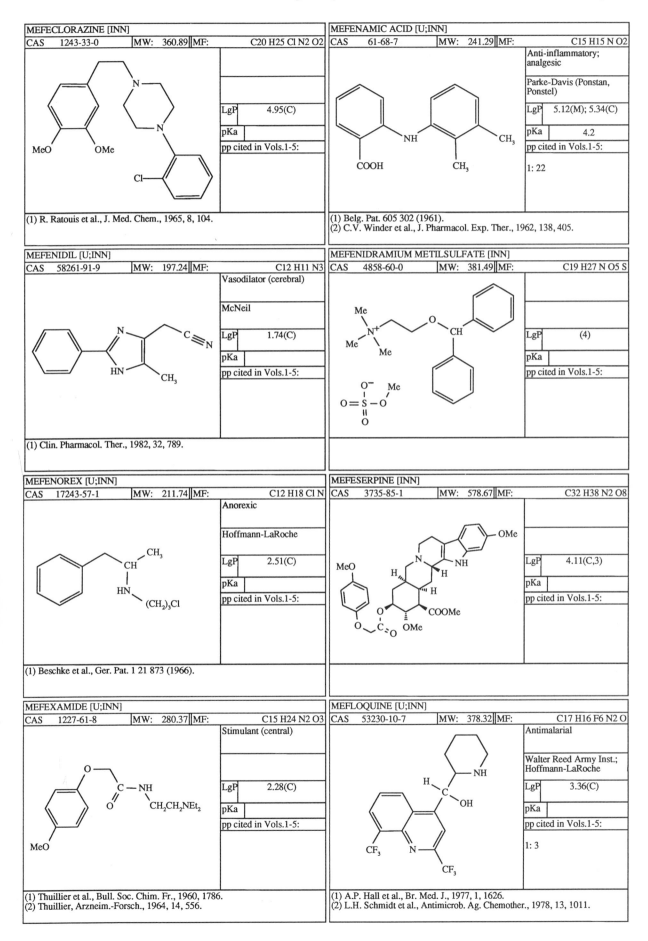

MEFECLORAZINE [INN]

CAS	1243-33-0	MW: 360.89	MF:	C20 H25 Cl N2 O2

LgP	4.95(C)
pKa	
pp cited in Vols.1-5:	

(1) R. Ratouis et al., J. Med. Chem., 1965, 8, 104.

MEFENAMIC ACID [U;INN]

CAS	61-68-7	MW: 241.29	MF:	C15 H15 N O2

Anti-inflammatory; analgesic	
Parke-Davis (Ponstan, Ponstel)	
LgP	5.12(M); 5.34(C)
pKa	4.2
pp cited in Vols.1-5:	
1: 22	

(1) Belg. Pat. 605 302 (1961).
(2) C.V. Winder et al., J. Pharmacol. Exp. Ther., 1962, 138, 405.

MEFENIDIL [U;INN]

CAS	58261-91-9	MW: 197.24	MF:	C12 H11 N3

Vasodilator (cerebral)	
McNeil	
LgP	1.74(C)
pKa	
pp cited in Vols.1-5:	

(1) Clin. Pharmacol. Ther., 1982, 32, 789.

MEFENIDRAMIUM METILSULFATE [INN]

CAS	4858-60-0	MW: 381.49	MF:	C19 H27 N O5 S

LgP	(4)
pKa	
pp cited in Vols.1-5:	

MEFENOREX [U;INN]

CAS	17243-57-1	MW: 211.74	MF:	C12 H18 Cl N

Anorexic	
Hoffmann-LaRoche	
LgP	2.51(C)
pKa	
pp cited in Vols.1-5:	

(1) Beschke et al., Ger. Pat. 1 21 873 (1966).

MEFESERPINE [INN]

CAS	3735-85-1	MW: 578.67	MF:	C32 H38 N2 O8

LgP	4.11(C,3)
pKa	
pp cited in Vols.1-5:	

MEFEXAMIDE [U;INN]

CAS	1227-61-8	MW: 280.37	MF:	C15 H24 N2 O3

Stimulant (central)	
LgP	2.28(C)
pKa	
pp cited in Vols.1-5:	

(1) Thuillier et al., Bull. Soc. Chim. Fr., 1960, 1786.
(2) Thuillier, Arzneim.-Forsch., 1964, 14, 556.

MEFLOQUINE [U;INN]

CAS	53230-10-7	MW: 378.32	MF:	C17 H16 F6 N2 O

Antimalarial	
Walter Reed Army Inst.; Hoffmann-LaRoche	
LgP	3.36(C)
pKa	
pp cited in Vols.1-5:	
1: 3	

(1) A.P. Hall et al., Br. Med. J., 1977, 1, 1626.
(2) L.H. Schmidt et al., Antimicrob. Ag. Chemother., 1978, 13, 1011.

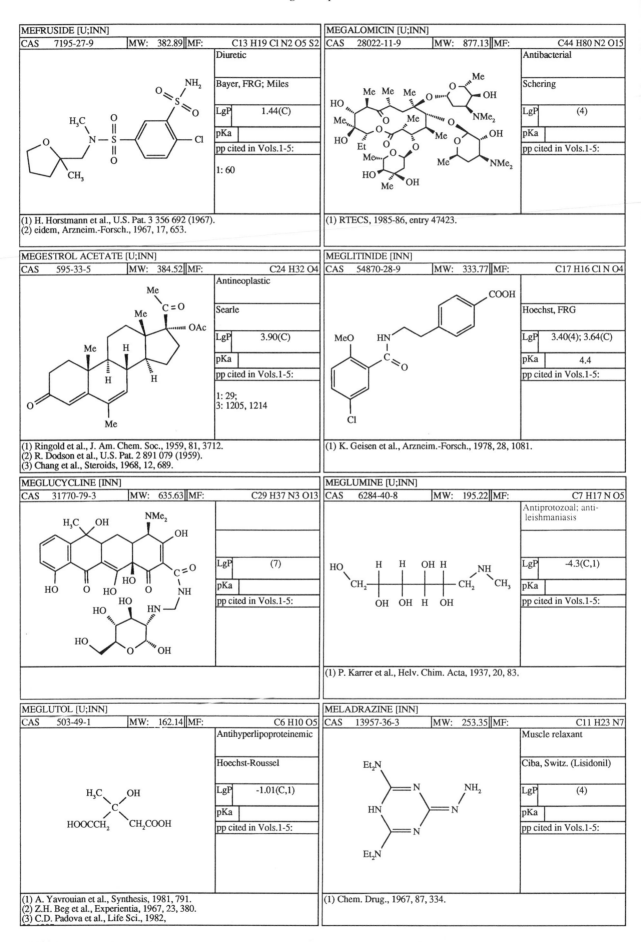

MEFRUSIDE [U;INN]

CAS	7195-27-9	MW:	382.89	MF:	C13 H19 Cl N2 O5 S2

Diuretic

Bayer, FRG; Miles

LgP	1.44(C)
pKa	
pp cited in Vols.1-5:	
1: 60	

(1) H. Horstmann et al., U.S. Pat. 3 356 692 (1967).
(2) eidem, Arzneim.-Forsch., 1967, 17, 653.

MEGALOMICIN [U;INN]

CAS	28022-11-9	MW:	877.13	MF:	C44 H80 N2 O15

Antibacterial

Schering

LgP	(4)
pKa	
pp cited in Vols.1-5:	

(1) RTECS, 1985-86, entry 47423.

MEGESTROL ACETATE [U;INN]

CAS	595-33-5	MW:	384.52	MF:	C24 H32 O4

Antineoplastic

Searle

LgP	3.90(C)
pKa	
pp cited in Vols.1-5:	
1: 29;	
3: 1205, 1214	

(1) Ringold et al., J. Am. Chem. Soc., 1959, 81, 3712.
(2) R. Dodson et al., U.S. Pat. 2 891 079 (1959).
(3) Chang et al., Steroids, 1968, 12, 689.

MEGLITINIDE [INN]

CAS	54870-28-9	MW:	333.77	MF:	C17 H16 Cl N O4

Hoechst, FRG

LgP	3.40(4); 3.64(C)
pKa	4.4
pp cited in Vols.1-5:	

(1) K. Geisen et al., Arzneim.-Forsch., 1978, 28, 1081.

MEGLUCYCLINE [INN]

CAS	31770-79-3	MW:	635.63	MF:	C29 H37 N3 O13

LgP	(7)
pKa	
pp cited in Vols.1-5:	

MEGLUMINE [U;INN]

CAS	6284-40-8	MW:	195.22	MF:	C7 H17 N O5

Antiprotozoal; anti-leishmaniasis

LgP	-4.3(C,1)
pKa	
pp cited in Vols.1-5:	

(1) P. Karrer et al., Helv. Chim. Acta, 1937, 20, 83.

MEGLUTOL [U;INN]

CAS	503-49-1	MW:	162.14	MF:	C6 H10 O5

Antihyperlipoproteinemic

Hoechst-Roussel

LgP	-1.01(C,1)
pKa	
pp cited in Vols.1-5:	

(1) A. Yavrouian et al., Synthesis, 1981, 791.
(2) Z.H. Beg et al., Experientia, 1967, 23, 380.
(3) C.D. Padova et al., Life Sci., 1982,

MELADRAZINE [INN]

CAS	13957-36-3	MW:	253.35	MF:	C11 H23 N7

Muscle relaxant

Ciba, Switz. (Lisidonil)

LgP	(4)
pKa	
pp cited in Vols.1-5:	

(1) Chem. Drug., 1967, 87, 334.

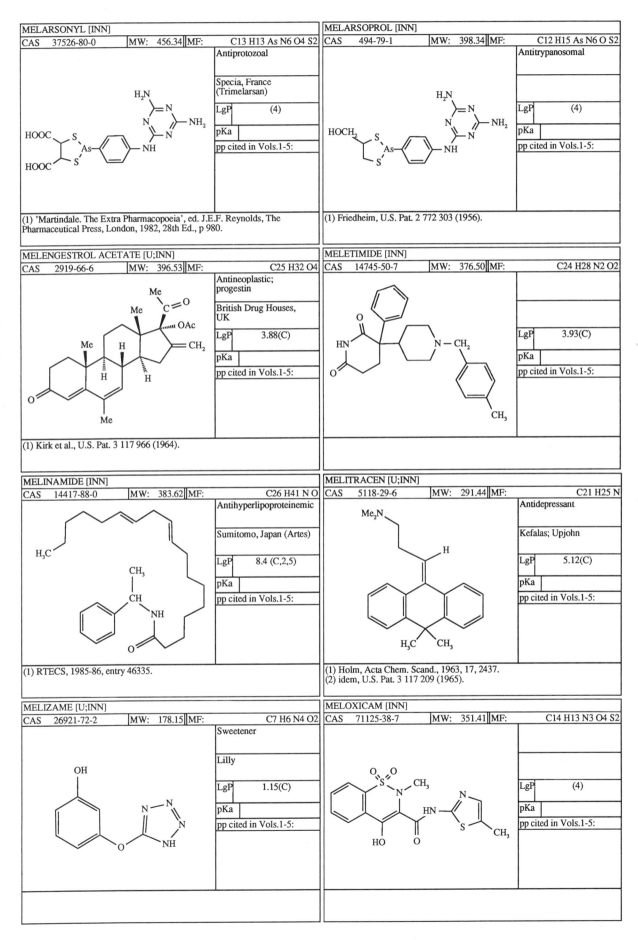

MELARSONYL [INN]

| CAS | 37526-80-0 | MW: | 456.34 | MF: | C13 H13 As N6 O4 S2 |

Antiprotozoal

Specia, France (Trimelarsan)

LgP	(4)
pKa	
pp cited in Vols.1-5:	

(1) 'Martindale. The Extra Pharmacopoeia', ed. J.E.F. Reynolds, The Pharmaceutical Press, London, 1982, 28th Ed., p 980.

MELARSOPROL [INN]

| CAS | 494-79-1 | MW: | 398.34 | MF: | C12 H15 As N6 O S2 |

Antitrypanosomal

LgP	(4)
pKa	
pp cited in Vols.1-5:	

(1) Friedheim, U.S. Pat. 2 772 303 (1956).

MELENGESTROL ACETATE [U;INN]

| CAS | 2919-66-6 | MW: | 396.53 | MF: | C25 H32 O4 |

Antineoplastic; progestin

British Drug Houses, UK

LgP	3.88(C)
pKa	
pp cited in Vols.1-5:	

(1) Kirk et al., U.S. Pat. 3 117 966 (1964).

MELETIMIDE [INN]

| CAS | 14745-50-7 | MW: | 376.50 | MF: | C24 H28 N2 O2 |

LgP	3.93(C)
pKa	
pp cited in Vols.1-5:	

MELINAMIDE [INN]

| CAS | 14417-88-0 | MW: | 383.62 | MF: | C26 H41 N O |

Antihyperlipoproteinemic

Sumitomo, Japan (Artes)

LgP	8.4 (C,2,5)
pKa	
pp cited in Vols.1-5:	

(1) RTECS, 1985-86, entry 46335.

MELITRACEN [U;INN]

| CAS | 5118-29-6 | MW: | 291.44 | MF: | C21 H25 N |

Antidepressant

Kefalas; Upjohn

LgP	5.12(C)
pKa	
pp cited in Vols.1-5:	

(1) Holm, Acta Chem. Scand., 1963, 17, 2437.
(2) idem, U.S. Pat. 3 117 209 (1965).

MELIZAME [U;INN]

| CAS | 26921-72-2 | MW: | 178.15 | MF: | C7 H6 N4 O2 |

Sweetener

Lilly

LgP	1.15(C)
pKa	
pp cited in Vols.1-5:	

MELOXICAM [INN]

| CAS | 71125-38-7 | MW: | 351.41 | MF: | C14 H13 N3 O4 S2 |

LgP	(4)
pKa	
pp cited in Vols.1-5:	

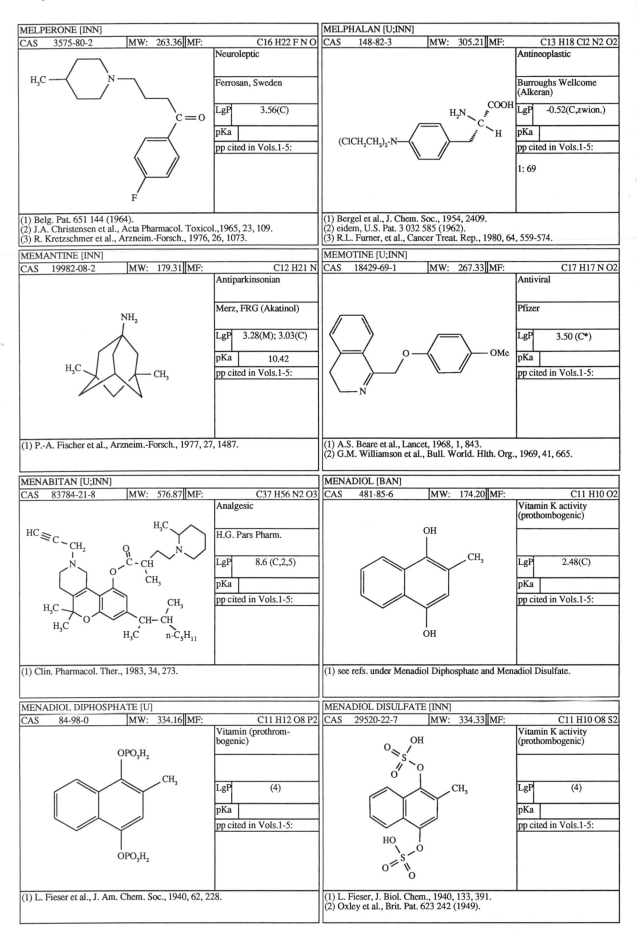

MELPERONE [INN]		
CAS 3575-80-2	MW: 263.36 MF:	C16 H22 F N O
		Neuroleptic
		Ferrosan, Sweden
		LgP 3.56(C)
		pKa
		pp cited in Vols.1-5:

(1) Belg. Pat. 651 144 (1964).
(2) J.A. Christensen et al., Acta Pharmacol. Toxicol.,1965, 23, 109.
(3) R. Kretzschmer et al., Arzneim.-Forsch., 1976, 26, 1073.

MELPHALAN [U;INN]		
CAS 148-82-3	MW: 305.21 MF:	C13 H18 Cl2 N2 O2
		Antineoplastic
		Burroughs Wellcome (Alkeran)
		LgP -0.52(C,zwion.)
		pKa
		pp cited in Vols.1-5:
		1: 69

(1) Bergel et al., J. Chem. Soc., 1954, 2409.
(2) eidem, U.S. Pat. 3 032 585 (1962).
(3) R.L. Furner, et al., Cancer Treat. Rep., 1980, 64, 559-574.

MEMANTINE [INN]		
CAS 19982-08-2	MW: 179.31 MF:	C12 H21 N
		Antiparkinsonian
		Merz, FRG (Akatinol)
		LgP 3.28(M); 3.03(C)
		pKa 10.42
		pp cited in Vols.1-5:

(1) P.-A. Fischer et al., Arzneim.-Forsch., 1977, 27, 1487.

MEMOTINE [U;INN]		
CAS 18429-69-1	MW: 267.33 MF:	C17 H17 N O2
		Antiviral
		Pfizer
		LgP 3.50 (C*)
		pKa
		pp cited in Vols.1-5:

(1) A.S. Beare et al., Lancet, 1968, 1, 843.
(2) G.M. Williamson et al., Bull. World. Hlth. Org., 1969, 41, 665.

MENABITAN [U;INN]		
CAS 83784-21-8	MW: 576.87 MF:	C37 H56 N2 O3
		Analgesic
		H.G. Pars Pharm.
		LgP 8.6 (C,2,5)
		pKa
		pp cited in Vols.1-5:

(1) Clin. Pharmacol. Ther., 1983, 34, 273.

MENADIOL [BAN]		
CAS 481-85-6	MW: 174.20 MF:	C11 H10 O2
		Vitamin K activity (prothombogenic)
		LgP 2.48(C)
		pKa
		pp cited in Vols.1-5:

(1) see refs. under Menadiol Diphosphate and Menadiol Disulfate.

MENADIOL DIPHOSPHATE [U]		
CAS 84-98-0	MW: 334.16 MF:	C11 H12 O8 P2
		Vitamin (prothrombogenic)
		LgP (4)
		pKa
		pp cited in Vols.1-5:

(1) L. Fieser et al., J. Am. Chem. Soc., 1940, 62, 228.

MENADIOL DISULFATE [INN]		
CAS 29520-22-7	MW: 334.33 MF:	C11 H10 O8 S2
		Vitamin K activity (prothombogenic)
		LgP (4)
		pKa
		pp cited in Vols.1-5:

(1) L. Fieser, J. Biol. Chem., 1940, 133, 391.
(2) Oxley et al., Brit. Pat. 623 242 (1949).

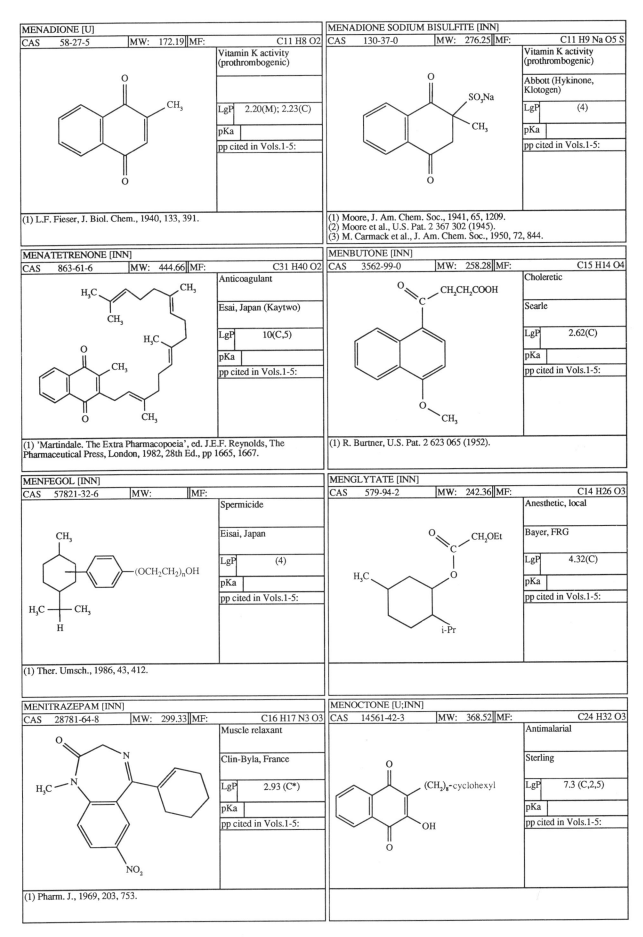

MENADIONE [U]

CAS	58-27-5	MW:	172.19	MF:	C11 H8 O2

Vitamin K activity (prothrombogenic)

LgP 2.20(M); 2.23(C)

pKa

pp cited in Vols.1-5:

(1) L.F. Fieser, J. Biol. Chem., 1940, 133, 391.

MENADIONE SODIUM BISULFITE [INN]

CAS	130-37-0	MW:	276.25	MF:	C11 H9 Na O5 S

Vitamin K activity (prothrombogenic)

Abbott (Hykinone, Klotogen)

LgP (4)

pKa

pp cited in Vols.1-5:

(1) Moore, J. Am. Chem. Soc., 1941, 65, 1209.
(2) Moore et al., U.S. Pat. 2 367 302 (1945).
(3) M. Carmack et al., J. Am. Chem. Soc., 1950, 72, 844.

MENATETRENONE [INN]

CAS	863-61-6	MW:	444.66	MF:	C31 H40 O2

Anticoagulant

Esai, Japan (Kaytwo)

LgP 10(C,5)

pKa

pp cited in Vols.1-5:

(1) 'Martindale. The Extra Pharmacopoeia', ed. J.E.F. Reynolds, The Pharmaceutical Press, London, 1982, 28th Ed., pp 1665, 1667.

MENBUTONE [INN]

CAS	3562-99-0	MW:	258.28	MF:	C15 H14 O4

Choleretic

Searle

LgP 2.62(C)

pKa

pp cited in Vols.1-5:

(1) R. Burtner, U.S. Pat. 2 623 065 (1952).

MENFEGOL [INN]

CAS	57821-32-6	MW:		MF:	

Spermicide

Eisai, Japan

LgP (4)

pKa

pp cited in Vols.1-5:

(1) Ther. Umsch., 1986, 43, 412.

MENGLYTATE [INN]

CAS	579-94-2	MW:	242.36	MF:	C14 H26 O3

Anesthetic, local

Bayer, FRG

LgP 4.32(C)

pKa

pp cited in Vols.1-5:

MENITRAZEPAM [INN]

CAS	28781-64-8	MW:	299.33	MF:	C16 H17 N3 O3

Muscle relaxant

Clin-Byla, France

LgP 2.93 (C*)

pKa

pp cited in Vols.1-5:

(1) Pharm. J., 1969, 203, 753.

MENOCTONE [U;INN]

CAS	14561-42-3	MW:	368.52	MF:	C24 H32 O3

Antimalarial

Sterling

LgP 7.3 (C,2,5)

pKa

pp cited in Vols.1-5:

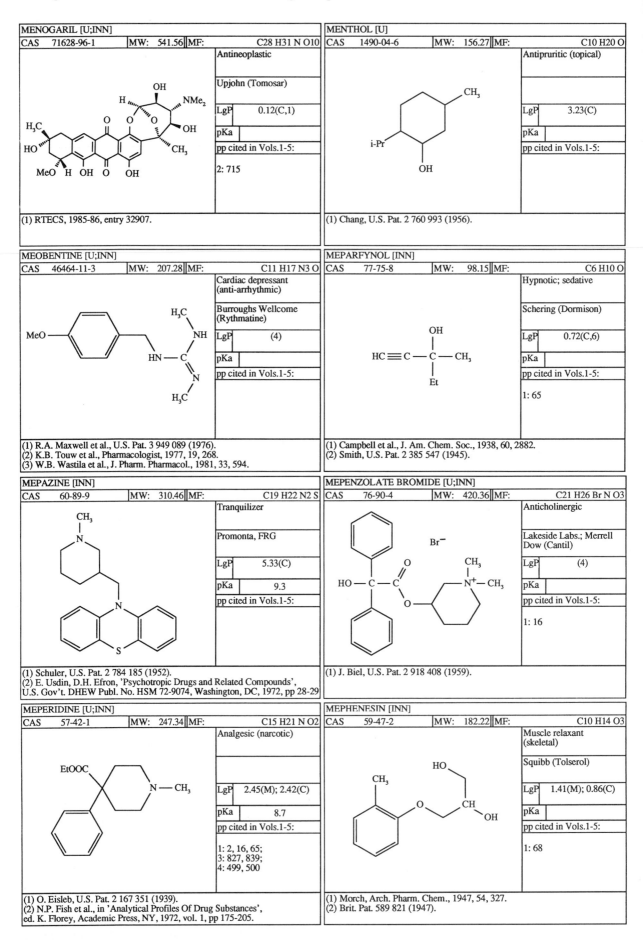

MENOGARIL [U;INN]			
CAS 71628-96-1	MW: 541.56	MF:	C28 H31 N O10

Antineoplastic

Upjohn (Tomosar)

LgP	0.12(C,1)
pKa	
pp cited in Vols.1-5:	
	2: 715

(1) RTECS, 1985-86, entry 32907.

MENTHOL [U]			
CAS 1490-04-6	MW: 156.27	MF:	C10 H20 O

Antipruritic (topical)

LgP	3.23(C)
pKa	
pp cited in Vols.1-5:	

(1) Chang, U.S. Pat. 2 760 993 (1956).

MEOBENTINE [U;INN]			
CAS 46464-11-3	MW: 207.28	MF:	C11 H17 N3 O

Cardiac depressant (anti-arrhythmic)

Burroughs Wellcome (Rythmatine)

LgP	(4)
pKa	
pp cited in Vols.1-5:	

(1) R.A. Maxwell et al., U.S. Pat. 3 949 089 (1976).
(2) K.B. Touw et al., Pharmacologist, 1977, 19, 268.
(3) W.B. Wastila et al., J. Pharm. Pharmacol., 1981, 33, 594.

MEPARFYNOL [INN]			
CAS 77-75-8	MW: 98.15	MF:	C6 H10 O

Hypnotic; sedative

Schering (Dormison)

LgP	0.72(C,6)
pKa	
pp cited in Vols.1-5:	
	1: 65

(1) Campbell et al., J. Am. Chem. Soc., 1938, 60, 2882.
(2) Smith, U.S. Pat. 2 385 547 (1945).

MEPAZINE [INN]			
CAS 60-89-9	MW: 310.46	MF:	C19 H22 N2 S

Tranquilizer

Promonta, FRG

LgP	5.33(C)
pKa	9.3
pp cited in Vols.1-5:	

(1) Schuler, U.S. Pat. 2 784 185 (1952).
(2) E. Usdin, D.H. Efron, 'Psychotropic Drugs and Related Compounds',
U.S. Gov't. DHEW Publ. No. HSM 72-9074, Washington, DC, 1972, pp 28-29

MEPENZOLATE BROMIDE [U;INN]			
CAS 76-90-4	MW: 420.36	MF:	C21 H26 Br N O3

Anticholinergic

Lakeside Labs.; Merrell Dow (Cantil)

LgP	(4)
pKa	
pp cited in Vols.1-5:	
	1: 16

(1) J. Biel, U.S. Pat. 2 918 408 (1959).

MEPERIDINE [U;INN]			
CAS 57-42-1	MW: 247.34	MF:	C15 H21 N O2

Analgesic (narcotic)

LgP	2.45(M); 2.42(C)
pKa	8.7
pp cited in Vols.1-5:	
	1: 2, 16, 65; 3: 827, 839; 4: 499, 500

(1) O. Eisleb, U.S. Pat. 2 167 351 (1939).
(2) N.P. Fish et al., in 'Analytical Profiles Of Drug Substances',
ed. K. Florey, Academic Press, NY, 1972, vol. 1, pp 175-205.

MEPHENESIN [INN]			
CAS 59-47-2	MW: 182.22	MF:	C10 H14 O3

Muscle relaxant (skeletal)

Squibb (Tolserol)

LgP	1.41(M); 0.86(C)
pKa	
pp cited in Vols.1-5:	
	1: 68

(1) Morch, Arch. Pharm. Chem., 1947, 54, 327.
(2) Brit. Pat. 589 821 (1947).

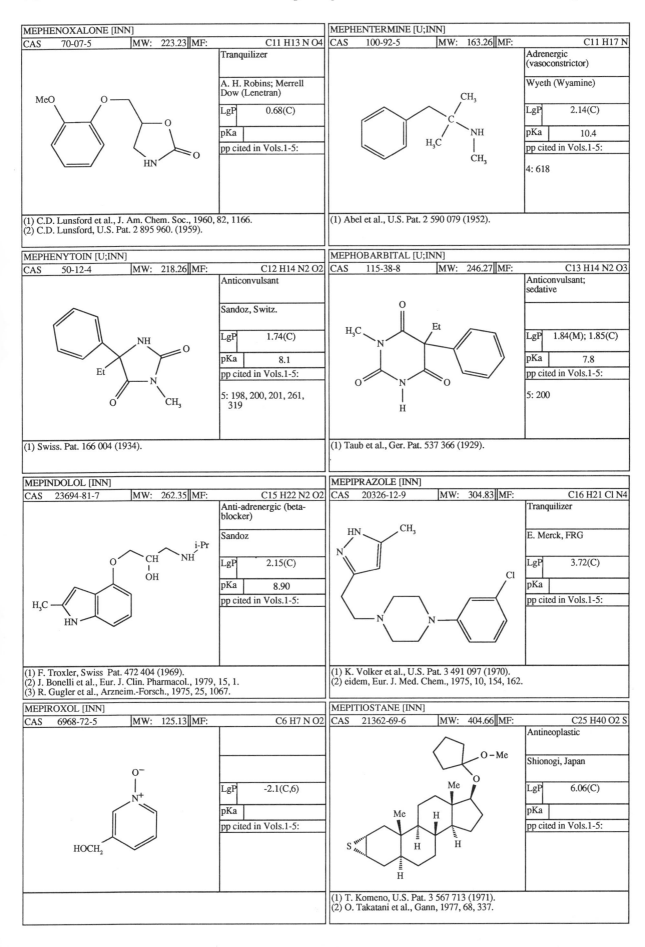

MEPHENOXALONE [INN]

| CAS | 70-07-5 | MW: | 223.23 | MF: | C11 H13 N O4 |

Tranquilizer

A. H. Robins; Merrell Dow (Lenetran)

LgP	0.68(C)
pKa	
pp cited in Vols.1-5:	

(1) C.D. Lunsford et al., J. Am. Chem. Soc., 1960, 82, 1166.
(2) C.D. Lunsford, U.S. Pat. 2 895 960. (1959).

MEPHENYTOIN [U;INN]

| CAS | 50-12-4 | MW: | 218.26 | MF: | C12 H14 N2 O2 |

Anticonvulsant

Sandoz, Switz.

LgP	1.74(C)
pKa	8.1
pp cited in Vols.1-5:	

5: 198, 200, 201, 261, 319

(1) Swiss. Pat. 166 004 (1934).

MEPINDOLOL [INN]

| CAS | 23694-81-7 | MW: | 262.35 | MF: | C15 H22 N2 O2 |

Anti-adrenergic (beta-blocker)

Sandoz

LgP	2.15(C)
pKa	8.90
pp cited in Vols.1-5:	

(1) F. Troxler, Swiss Pat. 472 404 (1969).
(2) J. Bonelli et al., Eur. J. Clin. Pharmacol., 1979, 15, 1.
(3) R. Gugler et al., Arzneim.-Forsch., 1975, 25, 1067.

MEPIROXOL [INN]

| CAS | 6968-72-5 | MW: | 125.13 | MF: | C6 H7 N O2 |

LgP	-2.1(C,6)
pKa	
pp cited in Vols.1-5:	

MEPHENTERMINE [U;INN]

| CAS | 100-92-5 | MW: | 163.26 | MF: | C11 H17 N |

Adrenergic (vasoconstrictor)

Wyeth (Wyamine)

LgP	2.14(C)
pKa	10.4
pp cited in Vols.1-5:	

4: 618

(1) Abel et al., U.S. Pat. 2 590 079 (1952).

MEPHOBARBITAL [U;INN]

| CAS | 115-38-8 | MW: | 246.27 | MF: | C13 H14 N2 O3 |

Anticonvulsant; sedative

LgP	1.84(M); 1.85(C)
pKa	7.8
pp cited in Vols.1-5:	

5: 200

(1) Taub et al., Ger. Pat. 537 366 (1929).

MEPIPRAZOLE [INN]

| CAS | 20326-12-9 | MW: | 304.83 | MF: | C16 H21 Cl N4 |

Tranquilizer

E. Merck, FRG

LgP	3.72(C)
pKa	
pp cited in Vols.1-5:	

(1) K. Volker et al., U.S. Pat. 3 491 097 (1970).
(2) eidem, Eur. J. Med. Chem., 1975, 10, 154, 162.

MEPITIOSTANE [INN]

| CAS | 21362-69-6 | MW: | 404.66 | MF: | C25 H40 O2 S |

Antineoplastic

Shionogi, Japan

LgP	6.06(C)
pKa	
pp cited in Vols.1-5:	

(1) T. Komeno, U.S. Pat. 3 567 713 (1971).
(2) O. Takatani et al., Gann, 1977, 68, 337.

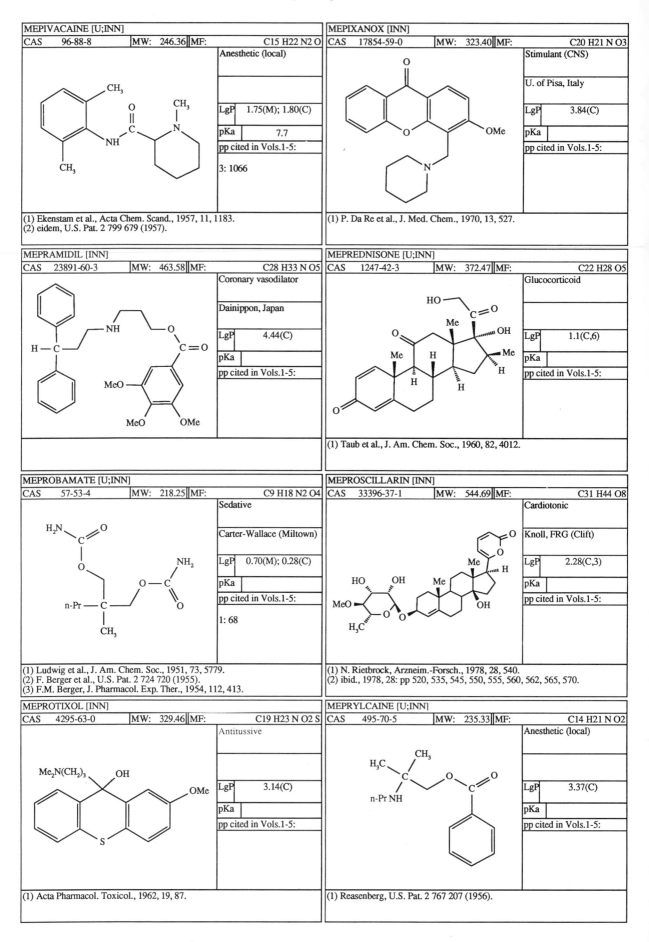

MEPIVACAINE [U;INN]

CAS 96-88-8	MW: 246.36 MF: C15 H22 N2 O

Anesthetic (local)

LgP	1.75(M); 1.80(C)
pKa	7.7
pp cited in Vols.1-5:	3: 1066

(1) Ekenstam et al., Acta Chem. Scand., 1957, 11, 1183.
(2) eidem, U.S. Pat. 2 799 679 (1957).

MEPIXANOX [INN]

CAS 17854-59-0	MW: 323.40 MF: C20 H21 N O3

Stimulant (CNS)

U. of Pisa, Italy

LgP	3.84(C)
pKa	
pp cited in Vols.1-5:	

(1) P. Da Re et al., J. Med. Chem., 1970, 13, 527.

MEPRAMIDIL [INN]

CAS 23891-60-3	MW: 463.58 MF: C28 H33 N O5

Coronary vasodilator

Dainippon, Japan

LgP	4.44(C)
pKa	
pp cited in Vols.1-5:	

MEPREDNISONE [U;INN]

CAS 1247-42-3	MW: 372.47 MF: C22 H28 O5

Glucocorticoid

LgP	1.1(C,6)
pKa	
pp cited in Vols.1-5:	

(1) Taub et al., J. Am. Chem. Soc., 1960, 82, 4012.

MEPROBAMATE [U;INN]

CAS 57-53-4	MW: 218.25 MF: C9 H18 N2 O4

Sedative

Carter-Wallace (Miltown)

LgP	0.70(M); 0.28(C)
pKa	
pp cited in Vols.1-5:	1: 68

(1) Ludwig et al., J. Am. Chem. Soc., 1951, 73, 5779.
(2) F. Berger et al., U.S. Pat. 2 724 720 (1955).
(3) F.M. Berger, J. Pharmacol. Exp. Ther., 1954, 112, 413.

MEPROSCILLARIN [INN]

CAS 33396-37-1	MW: 544.69 MF: C31 H44 O8

Cardiotonic

Knoll, FRG (Clift)

LgP	2.28(C,3)
pKa	
pp cited in Vols.1-5:	

(1) N. Rietbrock, Arzneim.-Forsch., 1978, 28, 540.
(2) ibid., 1978, 28: pp 520, 535, 545, 550, 555, 560, 562, 565, 570.

MEPROTIXOL [INN]

CAS 4295-63-0	MW: 329.46 MF: C19 H23 N O2 S

Antitussive

LgP	3.14(C)
pKa	
pp cited in Vols.1-5:	

(1) Acta Pharmacol. Toxicol., 1962, 19, 87.

MEPRYLCAINE [U;INN]

CAS 495-70-5	MW: 235.33 MF: C14 H21 N O2

Anesthetic (local)

LgP	3.37(C)
pKa	
pp cited in Vols.1-5:	

(1) Reasenberg, U.S. Pat. 2 767 207 (1956).

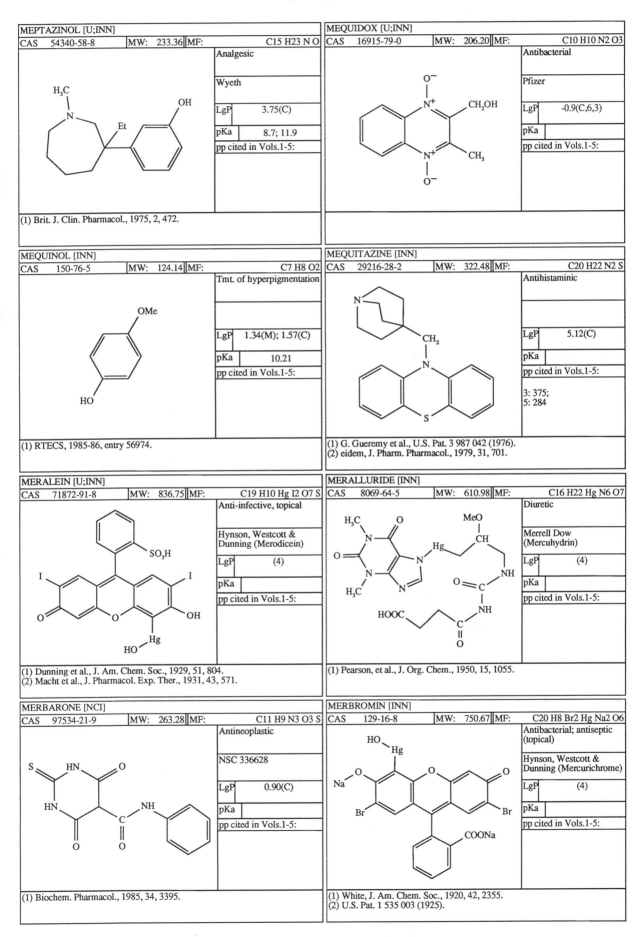

MEPTAZINOL [U;INN]

CAS	54340-58-8		MW: 233.36	MF:	C15 H23 N O

Analgesic

Wyeth

LgP	3.75(C)
pKa	8.7; 11.9
pp cited in Vols.1-5:	

(1) Brit. J. Clin. Pharmacol., 1975, 2, 472.

MEQUIDOX [U;INN]

CAS	16915-79-0		MW: 206.20	MF:	C10 H10 N2 O3

Antibacterial

Pfizer

LgP	-0.9(C,6,3)
pKa	
pp cited in Vols.1-5:	

MEQUINOL [INN]

CAS	150-76-5		MW: 124.14	MF:	C7 H8 O2

Tmt. of hyperpigmentation

LgP	1.34(M); 1.57(C)
pKa	10.21
pp cited in Vols.1-5:	

(1) RTECS, 1985-86, entry 56974.

MEQUITAZINE [INN]

CAS	29216-28-2		MW: 322.48	MF:	C20 H22 N2 S

Antihistaminic

LgP	5.12(C)
pKa	
pp cited in Vols.1-5:	

3: 375;
5: 284

(1) G. Gueremy et al., U.S. Pat. 3 987 042 (1976).
(2) eidem, J. Pharm. Pharmacol., 1979, 31, 701.

MERALEIN [U;INN]

CAS	71872-91-8		MW: 836.75	MF:	C19 H10 Hg I2 O7 S

Anti-infective, topical

Hynson, Westcott & Dunning (Merodicein)

LgP	(4)
pKa	
pp cited in Vols.1-5:	

(1) Dunning et al., J. Am. Chem. Soc., 1929, 51, 804.
(2) Macht et al., J. Pharmacol. Exp. Ther., 1931, 43, 571.

MERALLURIDE [INN]

CAS	8069-64-5		MW: 610.98	MF:	C16 H22 Hg N6 O7

Diuretic

Merrell Dow (Mercuhydrin)

LgP	(4)
pKa	
pp cited in Vols.1-5:	

(1) Pearson, et al., J. Org. Chem., 1950, 15, 1055.

MERBARONE [NCI]

CAS	97534-21-9		MW: 263.28	MF:	C11 H9 N3 O3 S

Antineoplastic

NSC 336628

LgP	0.90(C)
pKa	
pp cited in Vols.1-5:	

(1) Biochem. Pharmacol., 1985, 34, 3395.

MERBROMIN [INN]

CAS	129-16-8		MW: 750.67	MF:	C20 H8 Br2 Hg Na2 O6

Antibacterial; antiseptic (topical)

Hynson, Westcott & Dunning (Mercurichrome)

LgP	(4)
pKa	
pp cited in Vols.1-5:	

(1) White, J. Am. Chem. Soc., 1920, 42, 2355.
(2) U.S. Pat. 1 535 003 (1925).

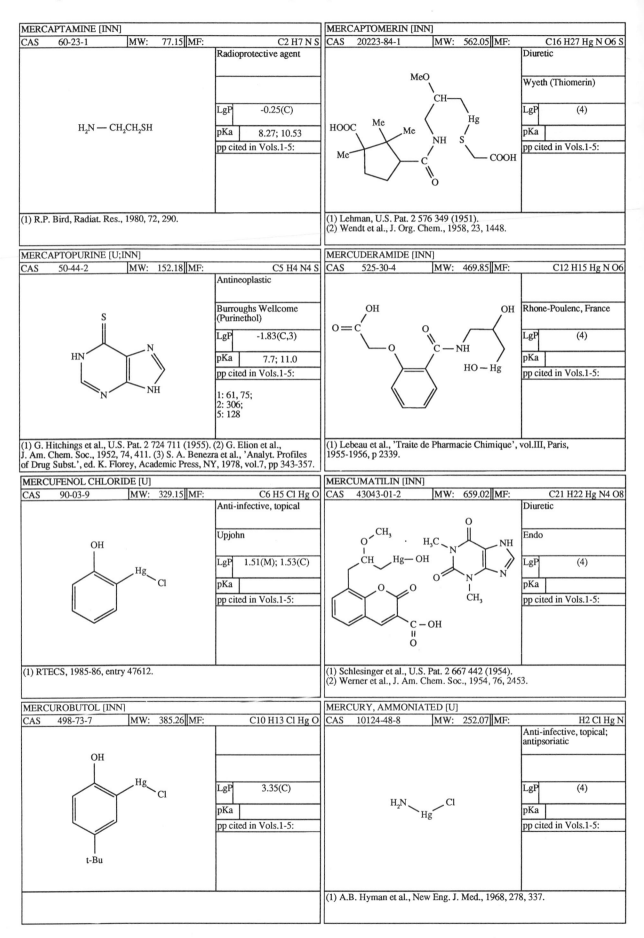

MERCAPTAMINE [INN]			
CAS 60-23-1	MW: 77.15	MF:	C2 H7 N S

Radioprotective agent

LgP	-0.25(C)
pKa	8.27; 10.53
pp cited in Vols.1-5:	

$H_2N — CH_2CH_2SH$

(1) R.P. Bird, Radiat. Res., 1980, 72, 290.

MERCAPTOMERIN [INN]			
CAS 20223-84-1	MW: 562.05	MF:	C16 H27 Hg N O6 S

Diuretic

Wyeth (Thiomerin)

LgP	(4)
pKa	
pp cited in Vols.1-5:	

(1) Lehman, U.S. Pat. 2 576 349 (1951).
(2) Wendt et al., J. Org. Chem., 1958, 23, 1448.

MERCAPTOPURINE [U;INN]			
CAS 50-44-2	MW: 152.18	MF:	C5 H4 N4 S

Antineoplastic

Burroughs Wellcome (Purinethol)

LgP	-1.83(C,3)
pKa	7.7; 11.0
pp cited in Vols.1-5:	

1: 61, 75;
2: 306;
5: 128

(1) G. Hitchings et al., U.S. Pat. 2 724 711 (1955). (2) G. Elion et al.,
J. Am. Chem. Soc., 1952, 74, 411. (3) S. A. Benezra et al., 'Analyt. Profiles
of Drug Subst.', ed. K. Florey, Academic Press, NY, 1978, vol.7, pp 343-357.

MERCUDERAMIDE [INN]			
CAS 525-30-4	MW: 469.85	MF:	C12 H15 Hg N O6

Rhone-Poulenc, France

LgP	(4)
pKa	
pp cited in Vols.1-5:	

(1) Lebeau et al., 'Traite de Pharmacie Chimique', vol.III, Paris,
1955-1956, p 2339.

MERCUFENOL CHLORIDE [U]			
CAS 90-03-9	MW: 329.15	MF:	C6 H5 Cl Hg O

Anti-infective, topical

Upjohn

LgP	1.51(M); 1.53(C)
pKa	
pp cited in Vols.1-5:	

(1) RTECS, 1985-86, entry 47612.

MERCUMATILIN [INN]			
CAS 43043-01-2	MW: 659.02	MF:	C21 H22 Hg N4 O8

Diuretic

Endo

LgP	(4)
pKa	
pp cited in Vols.1-5:	

(1) Schlesinger et al., U.S. Pat. 2 667 442 (1954).
(2) Werner et al., J. Am. Chem. Soc., 1954, 76, 2453.

MERCUROBUTOL [INN]			
CAS 498-73-7	MW: 385.26	MF:	C10 H13 Cl Hg O

LgP	3.35(C)
pKa	
pp cited in Vols.1-5:	

MERCURY, AMMONIATED [U]			
CAS 10124-48-8	MW: 252.07	MF:	H2 Cl Hg N

Anti-infective, topical;
antipsoriatic

LgP	(4)
pKa	
pp cited in Vols.1-5:	

(1) A.B. Hyman et al., New Eng. J. Med., 1968, 278, 337.

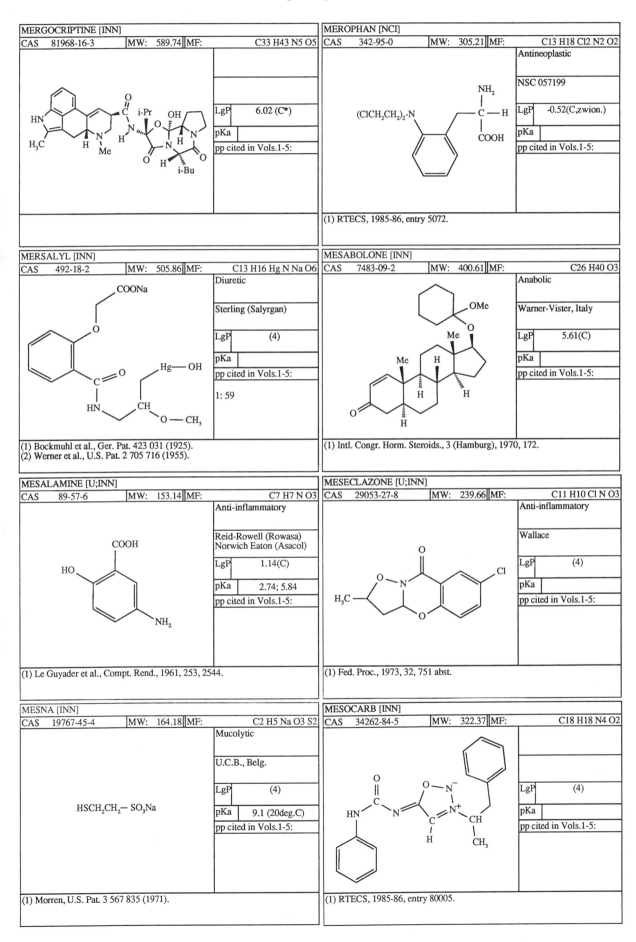

MERGOCRIPTINE [INN]

CAS	81968-16-3	MW:	589.74	MF:	C33 H43 N5 O5

LgP	6.02 (C*)
pKa	
pp cited in Vols.1-5:	

MEROPHAN [NCI]

CAS	342-95-0	MW:	305.21	MF:	C13 H18 Cl2 N2 O2

Antineoplastic

NSC 057199

LgP	-0.52(C,zwion.)
pKa	
pp cited in Vols.1-5:	

(1) RTECS, 1985-86, entry 5072.

MERSALYL [INN]

CAS	492-18-2	MW:	505.86	MF:	C13 H16 Hg N Na O6

Diuretic

Sterling (Salyrgan)

LgP	(4)
pKa	
pp cited in Vols.1-5:	

1: 59

(1) Bockmuhl et al., Ger. Pat. 423 031 (1925).
(2) Werner et al., U.S. Pat. 2 705 716 (1955).

MESABOLONE [INN]

CAS	7483-09-2	MW:	400.61	MF:	C26 H40 O3

Anabolic

Warner-Vister, Italy

LgP	5.61(C)
pKa	
pp cited in Vols.1-5:	

(1) Intl. Congr. Horm. Steroids., 3 (Hamburg), 1970, 172.

MESALAMINE [U;INN]

CAS	89-57-6	MW:	153.14	MF:	C7 H7 N O3

Anti-inflammatory

Reid-Rowell (Rowasa)
Norwich Eaton (Asacol)

LgP	1.14(C)
pKa	2.74; 5.84
pp cited in Vols.1-5:	

(1) Le Guyader et al., Compt. Rend., 1961, 253, 2544.

MESECLAZONE [U;INN]

CAS	29053-27-8	MW:	239.66	MF:	C11 H10 Cl N O3

Anti-inflammatory

Wallace

LgP	(4)
pKa	
pp cited in Vols.1-5:	

(1) Fed. Proc., 1973, 32, 751 abst.

MESNA [INN]

CAS	19767-45-4	MW:	164.18	MF:	C2 H5 Na O3 S2

Mucolytic

U.C.B., Belg.

LgP	(4)
pKa	9.1 (20deg.C)
pp cited in Vols.1-5:	

HSCH$_2$CH$_2$— SO$_3$Na

(1) Morren, U.S. Pat. 3 567 835 (1971).

MESOCARB [INN]

CAS	34262-84-5	MW:	322.37	MF:	C18 H18 N4 O2

LgP	(4)
pKa	
pp cited in Vols.1-5:	

(1) RTECS, 1985-86, entry 80005.

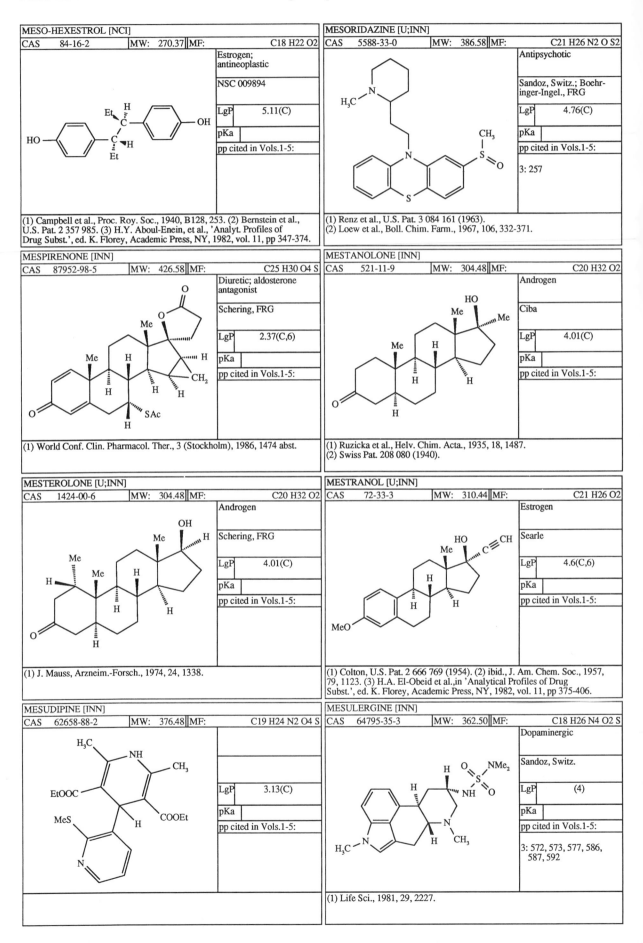

MESO-HEXESTROL [NCI]

CAS	84-16-2	MW: 270.37	MF:	C18 H22 O2

Estrogen; antineoplastic

NSC 009894

LgP	5.11(C)
pKa	
pp cited in Vols.1-5:	

(1) Campbell et al., Proc. Roy. Soc., 1940, B128, 253. (2) Bernstein et al., U.S. Pat. 2 357 985. (3) H.Y. Aboul-Enein, et al., 'Analyt. Profiles of Drug Subst.', ed. K. Florey, Academic Press, NY, 1982, vol. 11, pp 347-374.

MESORIDAZINE [U;INN]

CAS	5588-33-0	MW: 386.58	MF:	C21 H26 N2 O S2

Antipsychotic

Sandoz, Switz.; Boehringer-Ingel., FRG

LgP	4.76(C)
pKa	
pp cited in Vols.1-5:	

3: 257

(1) Renz et al., U.S. Pat. 3 084 161 (1963).
(2) Loew et al., Boll. Chim. Farm., 1967, 106, 332-371.

MESPIRENONE [INN]

CAS	87952-98-5	MW: 426.58	MF:	C25 H30 O4 S

Diuretic; aldosterone antagonist

Schering, FRG

LgP	2.37(C,6)
pKa	
pp cited in Vols.1-5:	

(1) World Conf. Clin. Pharmacol. Ther., 3 (Stockholm), 1986, 1474 abst.

MESTANOLONE [INN]

CAS	521-11-9	MW: 304.48	MF:	C20 H32 O2

Androgen

Ciba

LgP	4.01(C)
pKa	
pp cited in Vols.1-5:	

(1) Ruzicka et al., Helv. Chim. Acta., 1935, 18, 1487.
(2) Swiss Pat. 208 080 (1940).

MESTEROLONE [U;INN]

CAS	1424-00-6	MW: 304.48	MF:	C20 H32 O2

Androgen

Schering, FRG

LgP	4.01(C)
pKa	
pp cited in Vols.1-5:	

(1) J. Mauss, Arzneim.-Forsch., 1974, 24, 1338.

MESTRANOL [U;INN]

CAS	72-33-3	MW: 310.44	MF:	C21 H26 O2

Estrogen

Searle

LgP	4.6(C,6)
pKa	
pp cited in Vols.1-5:	

(1) Colton, U.S. Pat. 2 666 769 (1954). (2) ibid., J. Am. Chem. Soc., 1957, 79, 1123. (3) H.A. El-Obeid et al.,in 'Analytical Profiles of Drug Subst.', ed. K. Florey, Academic Press, NY, 1982, vol. 11, pp 375-406.

MESUDIPINE [INN]

CAS	62658-88-2	MW: 376.48	MF:	C19 H24 N2 O4 S

LgP	3.13(C)
pKa	
pp cited in Vols.1-5:	

MESULERGINE [INN]

CAS	64795-35-3	MW: 362.50	MF:	C18 H26 N4 O2 S

Dopaminergic

Sandoz, Switz.

LgP	(4)
pKa	
pp cited in Vols.1-5:	

3: 572, 573, 577, 586, 587, 592

(1) Life Sci., 1981, 29, 2227.

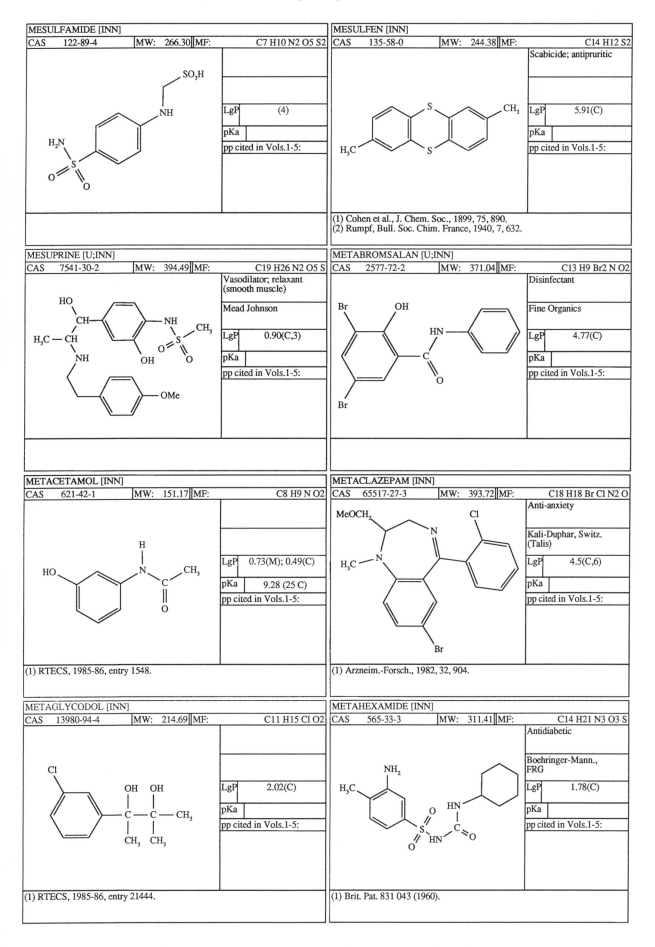

MESULFAMIDE [INN]

CAS	122-89-4	MW: 266.30	MF:	C7 H10 N2 O5 S2

LgP	(4)
pKa	
pp cited in Vols.1-5:	

MESULFEN [INN]

CAS	135-58-0	MW: 244.38	MF:	C14 H12 S2

Scabicide; antipruritic

LgP	5.91(C)
pKa	
pp cited in Vols.1-5:	

(1) Cohen et al., J. Chem. Soc., 1899, 75, 890.
(2) Rumpf, Bull. Soc. Chim. France, 1940, 7, 632.

MESUPRINE [U;INN]

CAS	7541-30-2	MW: 394.49	MF:	C19 H26 N2 O5 S

Vasodilator; relaxant (smooth muscle)

Mead Johnson

LgP	0.90(C,3)
pKa	
pp cited in Vols.1-5:	

METABROMSALAN [U;INN]

CAS	2577-72-2	MW: 371.04	MF:	C13 H9 Br2 N O2

Disinfectant

Fine Organics

LgP	4.77(C)
pKa	
pp cited in Vols.1-5:	

METACETAMOL [INN]

CAS	621-42-1	MW: 151.17	MF:	C8 H9 N O2

LgP	0.73(M); 0.49(C)
pKa	9.28 (25 C)
pp cited in Vols.1-5:	

(1) RTECS, 1985-86, entry 1548.

METACLAZEPAM [INN]

CAS	65517-27-3	MW: 393.72	MF:	C18 H18 Br Cl N2 O

Anti-anxiety

Kali-Duphar, Switz. (Talis)

LgP	4.5(C,6)
pKa	
pp cited in Vols.1-5:	

(1) Arzneim.-Forsch., 1982, 32, 904.

METAGLYCODOL [INN]

CAS	13980-94-4	MW: 214.69	MF:	C11 H15 Cl O2

LgP	2.02(C)
pKa	
pp cited in Vols.1-5:	

(1) RTECS, 1985-86, entry 21444.

METAHEXAMIDE [INN]

CAS	565-33-3	MW: 311.41	MF:	C14 H21 N3 O3 S

Antidiabetic

Boehringer-Mann., FRG

LgP	1.78(C)
pKa	
pp cited in Vols.1-5:	

(1) Brit. Pat. 831 043 (1960).

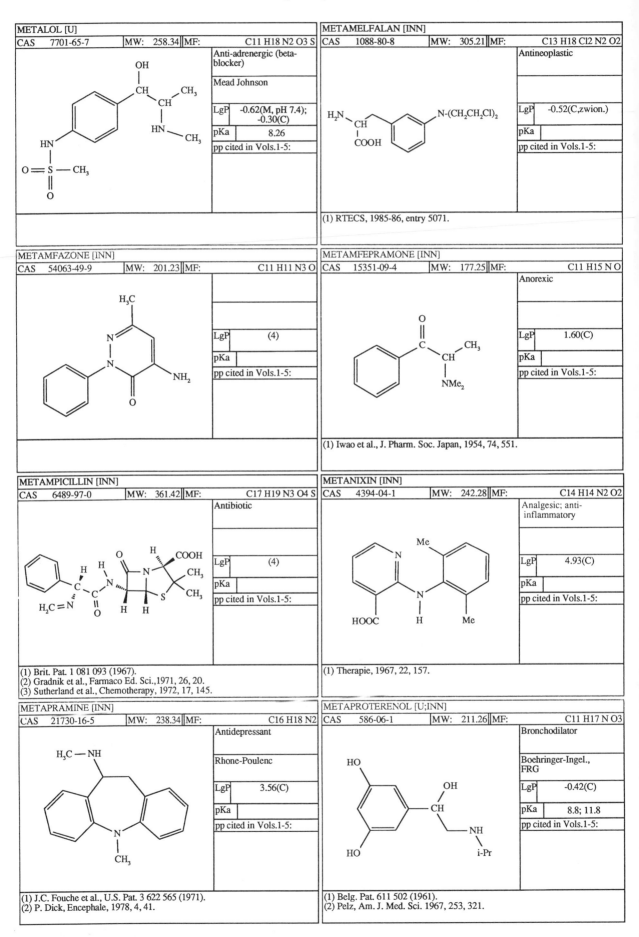

METALOL [U]			
CAS 7701-65-7	MW: 258.34	MF:	C11 H18 N2 O3 S

Anti-adrenergic (beta-blocker)

Mead Johnson

LgP	-0.62(M, pH 7.4); -0.30(C)
pKa	8.26
pp cited in Vols.1-5:	

METAMELFALAN [INN]			
CAS 1088-80-8	MW: 305.21	MF:	C13 H18 Cl2 N2 O2

Antineoplastic

LgP	-0.52(C,zwion.)
pKa	
pp cited in Vols.1-5:	

(1) RTECS, 1985-86, entry 5071.

METAMFAZONE [INN]			
CAS 54063-49-9	MW: 201.23	MF:	C11 H11 N3 O

LgP	(4)
pKa	
pp cited in Vols.1-5:	

METAMFEPRAMONE [INN]			
CAS 15351-09-4	MW: 177.25	MF:	C11 H15 N O

Anorexic

LgP	1.60(C)
pKa	
pp cited in Vols.1-5:	

(1) Iwao et al., J. Pharm. Soc. Japan, 1954, 74, 551.

METAMPICILLIN [INN]			
CAS 6489-97-0	MW: 361.42	MF:	C17 H19 N3 O4 S

Antibiotic

LgP	(4)
pKa	
pp cited in Vols.1-5:	

(1) Brit. Pat. 1 081 093 (1967).
(2) Gradnik et al., Farmaco Ed. Sci.,1971, 26, 20.
(3) Sutherland et al., Chemotherapy, 1972, 17, 145.

METANIXIN [INN]			
CAS 4394-04-1	MW: 242.28	MF:	C14 H14 N2 O2

Analgesic; anti-inflammatory

LgP	4.93(C)
pKa	
pp cited in Vols.1-5:	

(1) Therapie, 1967, 22, 157.

METAPRAMINE [INN]			
CAS 21730-16-5	MW: 238.34	MF:	C16 H18 N2

Antidepressant

Rhone-Poulenc

LgP	3.56(C)
pKa	
pp cited in Vols.1-5:	

(1) J.C. Fouche et al., U.S. Pat. 3 622 565 (1971).
(2) P. Dick, Encephale, 1978, 4, 41.

METAPROTERENOL [U;INN]			
CAS 586-06-1	MW: 211.26	MF:	C11 H17 N O3

Bronchodilator

Boehringer-Ingel., FRG

LgP	-0.42(C)
pKa	8.8; 11.8
pp cited in Vols.1-5:	

(1) Belg. Pat. 611 502 (1961).
(2) Pelz, Am. J. Med. Sci. 1967, 253, 321.

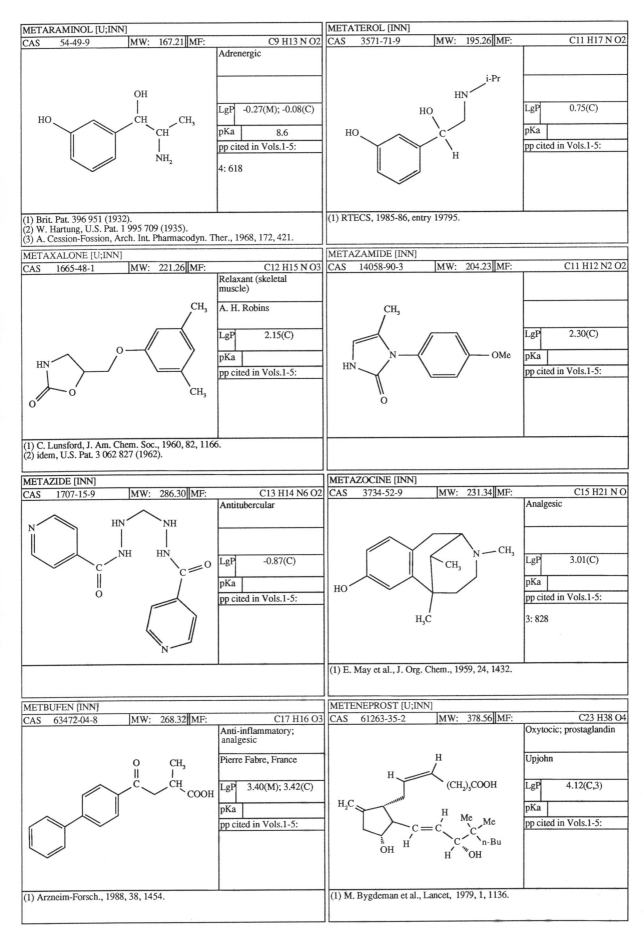

METARAMINOL [U;INN]

CAS	54-49-9	MW:	167.21	MF:	C9 H13 N O2

Adrenergic

LgP	-0.27(M); -0.08(C)
pKa	8.6
pp cited in Vols.1-5:	

4: 618

(1) Brit. Pat. 396 951 (1932).
(2) W. Hartung, U.S. Pat. 1 995 709 (1935).
(3) A. Cession-Fossion, Arch. Int. Pharmacodyn. Ther., 1968, 172, 421.

METATEROL [INN]

CAS	3571-71-9	MW:	195.26	MF:	C11 H17 N O2

LgP	0.75(C)
pKa	
pp cited in Vols.1-5:	

(1) RTECS, 1985-86, entry 19795.

METAXALONE [U;INN]

CAS	1665-48-1	MW:	221.26	MF:	C12 H15 N O3

Relaxant (skeletal muscle)

A. H. Robins

LgP	2.15(C)
pKa	
pp cited in Vols.1-5:	

(1) C. Lunsford, J. Am. Chem. Soc., 1960, 82, 1166.
(2) idem, U.S. Pat. 3 062 827 (1962).

METAZAMIDE [INN]

CAS	14058-90-3	MW:	204.23	MF:	C11 H12 N2 O2

LgP	2.30(C)
pKa	
pp cited in Vols.1-5:	

METAZIDE [INN]

CAS	1707-15-9	MW:	286.30	MF:	C13 H14 N6 O2

Antitubercular

LgP	-0.87(C)
pKa	
pp cited in Vols.1-5:	

METAZOCINE [INN]

CAS	3734-52-9	MW:	231.34	MF:	C15 H21 N O

Analgesic

LgP	3.01(C)
pKa	
pp cited in Vols.1-5:	

3: 828

(1) E. May et al., J. Org. Chem., 1959, 24, 1432.

METBUFEN [INN]

CAS	63472-04-8	MW:	268.32	MF:	C17 H16 O3

Anti-inflammatory; analgesic

Pierre Fabre, France

LgP	3.40(M); 3.42(C)
pKa	
pp cited in Vols.1-5:	

(1) Arzneim-Forsch., 1988, 38, 1454.

METENEPROST [U;INN]

CAS	61263-35-2	MW:	378.56	MF:	C23 H38 O4

Oxytocic; prostaglandin

Upjohn

LgP	4.12(C,3)
pKa	
pp cited in Vols.1-5:	

(1) M. Bygdeman et al., Lancet, 1979, 1, 1136.

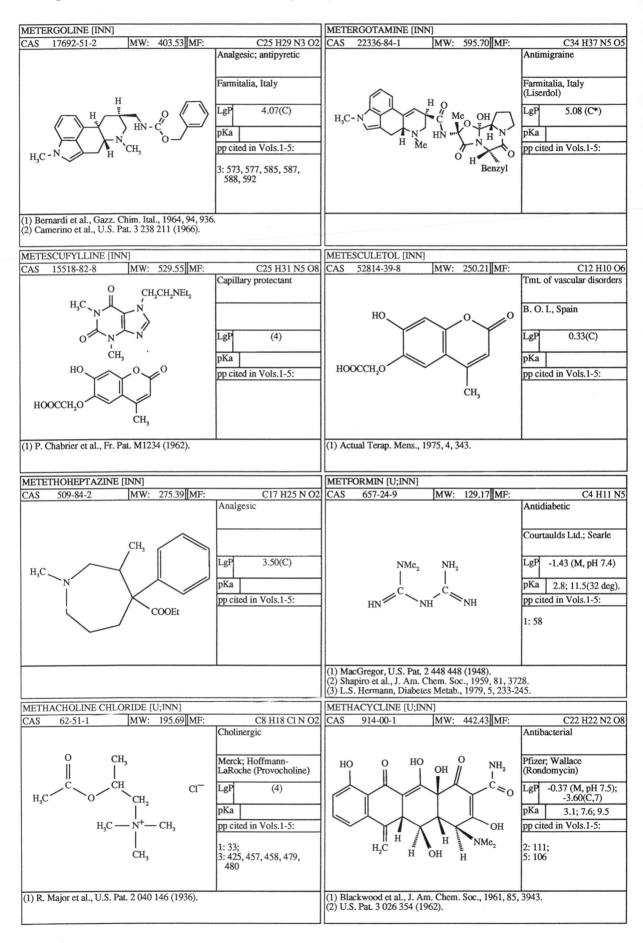

METERGOLINE [INN]

CAS	17692-51-2	MW:	403.53	MF:	C25 H29 N3 O2

Analgesic; antipyretic

Farmitalia, Italy

LgP	4.07(C)
pKa	
pp cited in Vols.1-5:	

3: 573, 577, 585, 587, 588, 592

(1) Bernardi et al., Gazz. Chim. Ital., 1964, 94, 936.
(2) Camerino et al., U.S. Pat. 3 238 211 (1966).

METERGOTAMINE [INN]

CAS	22336-84-1	MW:	595.70	MF:	C34 H37 N5 O5

Antimigraine

Farmitalia, Italy
(Liserdol)

LgP	5.08 (C*)
pKa	
pp cited in Vols.1-5:	

METESCUFYLLINE [INN]

CAS	15518-82-8	MW:	529.55	MF:	C25 H31 N5 O8

Capillary protectant

LgP	(4)
pKa	
pp cited in Vols.1-5:	

(1) P. Chabrier et al., Fr. Pat. M1234 (1962).

METESCULETOL [INN]

CAS	52814-39-8	MW:	250.21	MF:	C12 H10 O6

Tmt. of vascular disorders

B. O. I., Spain

LgP	0.33(C)
pKa	
pp cited in Vols.1-5:	

(1) Actual Terap. Mens., 1975, 4, 343.

METETHOHEPTAZINE [INN]

CAS	509-84-2	MW:	275.39	MF:	C17 H25 N O2

Analgesic

LgP	3.50(C)
pKa	
pp cited in Vols.1-5:	

METFORMIN [U;INN]

CAS	657-24-9	MW:	129.17	MF:	C4 H11 N5

Antidiabetic

Courtaulds Ltd.; Searle

LgP	-1.43 (M, pH 7.4)
pKa	2.8; 11.5(32 deg).
pp cited in Vols.1-5:	

1: 58

(1) MacGregor, U.S. Pat. 2 448 448 (1948).
(2) Shapiro et al., J. Am. Chem. Soc., 1959, 81, 3728.
(3) L.S. Hermann, Diabetes Metab., 1979, 5, 233-245.

METHACHOLINE CHLORIDE [U;INN]

CAS	62-51-1	MW:	195.69	MF:	C8 H18 Cl N O2

Cholinergic

Merck; Hoffmann-LaRoche (Provocholine)

LgP	(4)
pKa	
pp cited in Vols.1-5:	

1: 33;
3: 425, 457, 458, 479, 480

(1) R. Major et al., U.S. Pat. 2 040 146 (1936).

METHACYCLINE [U;INN]

CAS	914-00-1	MW:	442.43	MF:	C22 H22 N2 O8

Antibacterial

Pfizer; Wallace
(Rondomycin)

LgP	-0.37 (M, pH 7.5); -3.60(C,7)
pKa	3.1; 7.6; 9.5
pp cited in Vols.1-5:	

2: 111;
5: 106

(1) Blackwood et al., J. Am. Chem. Soc., 1961, 85, 3943.
(2) U.S. Pat. 3 026 354 (1962).

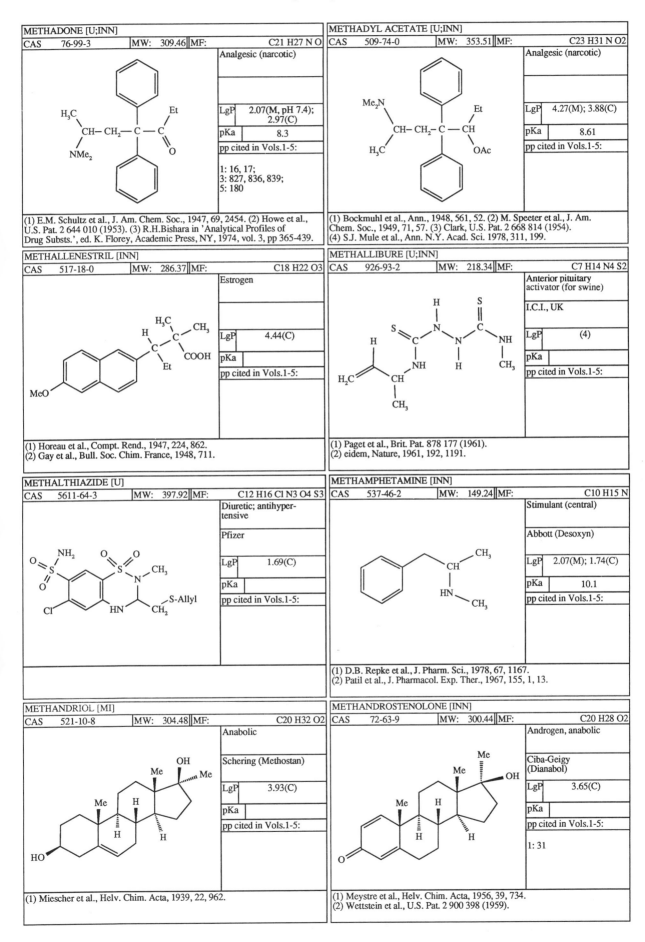

METHADONE [U;INN]

CAS	76-99-3	MW: 309.46	MF:	C21 H27 N O

Analgesic (narcotic)

LgP	2.07(M, pH 7.4); 2.97(C)
pKa	8.3
pp cited in Vols.1-5:	

1: 16, 17;
3: 827, 836, 839;
5: 180

(1) E.M. Schultz et al., J. Am. Chem. Soc., 1947, 69, 2454. (2) Howe et al., U.S. Pat. 2 644 010 (1953). (3) R.H.Bishara in 'Analytical Profiles of Drug Substs.', ed. K. Florey, Academic Press, NY, 1974, vol. 3, pp 365-439.

METHADYL ACETATE [U;INN]

CAS	509-74-0	MW: 353.51	MF:	C23 H31 N O2

Analgesic (narcotic)

LgP	4.27(M); 3.88(C)
pKa	8.61
pp cited in Vols.1-5:	

(1) Bockmuhl et al., Ann., 1948, 561, 52. (2) M. Speeter et al., J. Am. Chem. Soc., 1949, 71, 57. (3) Clark, U.S. Pat. 2 668 814 (1954). (4) S.J. Mule et al., Ann. N.Y. Acad. Sci. 1978, 311, 199.

METHALLENESTRIL [INN]

CAS	517-18-0	MW: 286.37	MF:	C18 H22 O3

Estrogen

LgP	4.44(C)
pKa	
pp cited in Vols.1-5:	

(1) Horeau et al., Compt. Rend., 1947, 224, 862.
(2) Gay et al., Bull. Soc. Chim. France, 1948, 711.

METHALLIBURE [U;INN]

CAS	926-93-2	MW: 218.34	MF:	C7 H14 N4 S2

Anterior pituitary activator (for swine)

I.C.I., UK

LgP	(4)
pKa	
pp cited in Vols.1-5:	

(1) Paget et al., Brit. Pat. 878 177 (1961).
(2) eidem, Nature, 1961, 192, 1191.

METHALTHIAZIDE [U]

CAS	5611-64-3	MW: 397.92	MF:	C12 H16 Cl N3 O4 S3

Diuretic; antihyper-tensive

Pfizer

LgP	1.69(C)
pKa	
pp cited in Vols.1-5:	

METHAMPHETAMINE [INN]

CAS	537-46-2	MW: 149.24	MF:	C10 H15 N

Stimulant (central)

Abbott (Desoxyn)

LgP	2.07(M); 1.74(C)
pKa	10.1
pp cited in Vols.1-5:	

(1) D.B. Repke et al., J. Pharm. Sci., 1978, 67, 1167.
(2) Patil et al., J. Pharmacol. Exp. Ther., 1967, 155, 1, 13.

METHANDRIOL [MI]

CAS	521-10-8	MW: 304.48	MF:	C20 H32 O2

Anabolic

Schering (Methostan)

LgP	3.93(C)
pKa	
pp cited in Vols.1-5:	

(1) Miescher et al., Helv. Chim. Acta, 1939, 22, 962.

METHANDROSTENOLONE [INN]

CAS	72-63-9	MW: 300.44	MF:	C20 H28 O2

Androgen, anabolic

Ciba-Geigy (Dianabol)

LgP	3.65(C)
pKa	
pp cited in Vols.1-5:	

1: 31

(1) Meystre et al., Helv. Chim. Acta, 1956, 39, 734.
(2) Wettstein et al., U.S. Pat. 2 900 398 (1959).

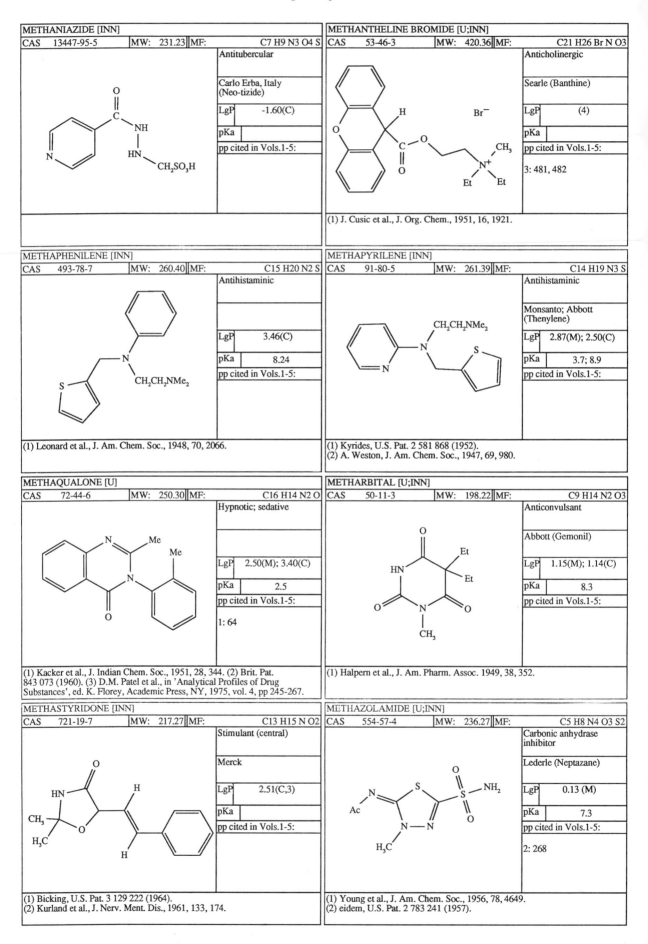

METHANIAZIDE [INN]

CAS	13447-95-5	MW:	231.23	MF:	C7 H9 N3 O4 S

Antitubercular

Carlo Erba, Italy
(Neo-tizide)

LgP	-1.60(C)
pKa	
pp cited in Vols.1-5:	

METHANTHELINE BROMIDE [U;INN]

CAS	53-46-3	MW:	420.36	MF:	C21 H26 Br N O3

Anticholinergic

Searle (Banthine)

LgP	(4)
pKa	
pp cited in Vols.1-5:	

3: 481, 482

(1) J. Cusic et al., J. Org. Chem., 1951, 16, 1921.

METHAPHENILENE [INN]

CAS	493-78-7	MW:	260.40	MF:	C15 H20 N2 S

Antihistaminic

LgP	3.46(C)
pKa	8.24
pp cited in Vols.1-5:	

(1) Leonard et al., J. Am. Chem. Soc., 1948, 70, 2066.

METHAPYRILENE [INN]

CAS	91-80-5	MW:	261.39	MF:	C14 H19 N3 S

Antihistaminic

Monsanto; Abbott
(Thenylene)

LgP	2.87(M); 2.50(C)
pKa	3.7; 8.9
pp cited in Vols.1-5:	

(1) Kyrides, U.S. Pat. 2 581 868 (1952).
(2) A. Weston, J. Am. Chem. Soc., 1947, 69, 980.

METHAQUALONE [U]

CAS	72-44-6	MW:	250.30	MF:	C16 H14 N2 O

Hypnotic; sedative

LgP	2.50(M); 3.40(C)
pKa	2.5
pp cited in Vols.1-5:	

1: 64

(1) Kacker et al., J. Indian Chem. Soc., 1951, 28, 344. (2) Brit. Pat.
843 073 (1960). (3) D.M. Patel et al., in 'Analytical Profiles of Drug
Substances', ed. K. Florey, Academic Press, NY, 1975, vol. 4, pp 245-267.

METHARBITAL [U;INN]

CAS	50-11-3	MW:	198.22	MF:	C9 H14 N2 O3

Anticonvulsant

Abbott (Gemonil)

LgP	1.15(M); 1.14(C)
pKa	8.3
pp cited in Vols.1-5:	

(1) Halpern et al., J. Am. Pharm. Assoc. 1949, 38, 352.

METHASTYRIDONE [INN]

CAS	721-19-7	MW:	217.27	MF:	C13 H15 N O2

Stimulant (central)

Merck

LgP	2.51(C,3)
pKa	
pp cited in Vols.1-5:	

(1) Bicking, U.S. Pat. 3 129 222 (1964).
(2) Kurland et al., J. Nerv. Ment. Dis., 1961, 133, 174.

METHAZOLAMIDE [U;INN]

CAS	554-57-4	MW:	236.27	MF:	C5 H8 N4 O3 S2

Carbonic anhydrase
inhibitor

Lederle (Neptazane)

LgP	0.13 (M)
pKa	7.3
pp cited in Vols.1-5:	

2: 268

(1) Young et al., J. Am. Chem. Soc., 1956, 78, 4649.
(2) eidem, U.S. Pat. 2 783 241 (1957).

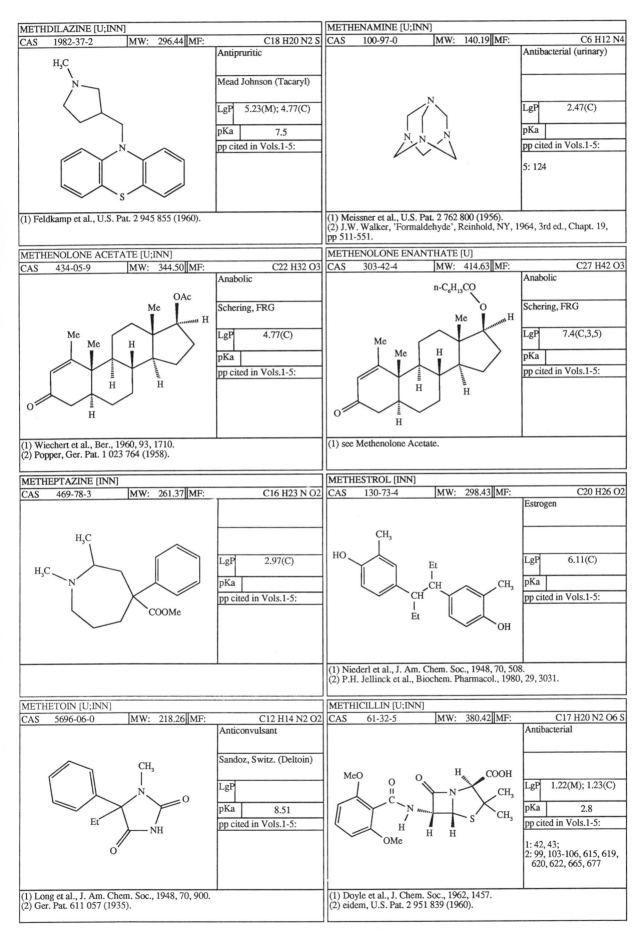

METHDILAZINE [U;INN]

CAS	1982-37-2	MW: 296.44	MF:	C18 H20 N2 S

Antipruritic

Mead Johnson (Tacaryl)

LgP	5.23(M); 4.77(C)
pKa	7.5
pp cited in Vols.1-5:	

(1) Feldkamp et al., U.S. Pat. 2 945 855 (1960).

METHENAMINE [U;INN]

CAS	100-97-0	MW: 140.19	MF:	C6 H12 N4

Antibacterial (urinary)

LgP	2.47(C)
pKa	
pp cited in Vols.1-5:	
5: 124	

(1) Meissner et al., U.S. Pat. 2 762 800 (1956).
(2) J.W. Walker, 'Formaldehyde', Reinhold, NY, 1964, 3rd ed., Chapt. 19, pp 511-551.

METHENOLONE ACETATE [U;INN]

CAS	434-05-9	MW: 344.50	MF:	C22 H32 O3

Anabolic

Schering, FRG

LgP	4.77(C)
pKa	
pp cited in Vols.1-5:	

(1) Wiechert et al., Ber., 1960, 93, 1710.
(2) Popper, Ger. Pat. 1 023 764 (1958).

METHENOLONE ENANTHATE [U]

CAS	303-42-4	MW: 414.63	MF:	C27 H42 O3

Anabolic

Schering, FRG

LgP	7.4(C,3,5)
pKa	
pp cited in Vols.1-5:	

(1) see Methenolone Acetate.

METHEPTAZINE [INN]

CAS	469-78-3	MW: 261.37	MF:	C16 H23 N O2

LgP	2.97(C)
pKa	
pp cited in Vols.1-5:	

METHESTROL [INN]

CAS	130-73-4	MW: 298.43	MF:	C20 H26 O2

Estrogen

LgP	6.11(C)
pKa	
pp cited in Vols.1-5:	

(1) Niederl et al., J. Am. Chem. Soc., 1948, 70, 508.
(2) P.H. Jellinck et al., Biochem. Pharmacol., 1980, 29, 3031.

METHETOIN [U;INN]

CAS	5696-06-0	MW: 218.26	MF:	C12 H14 N2 O2

Anticonvulsant

Sandoz, Switz. (Deltoin)

LgP	
pKa	8.51
pp cited in Vols.1-5:	

(1) Long et al., J. Am. Chem. Soc., 1948, 70, 900.
(2) Ger. Pat. 611 057 (1935).

METHICILLIN [U;INN]

CAS	61-32-5	MW: 380.42	MF:	C17 H20 N2 O6 S

Antibacterial

LgP	1.22(M); 1.23(C)
pKa	2.8
pp cited in Vols.1-5:	
1: 42, 43;	
2: 99, 103-106, 615, 619, 620, 622, 665, 677	

(1) Doyle et al., J. Chem. Soc., 1962, 1457.
(2) eidem, U.S. Pat. 2 951 839 (1960).

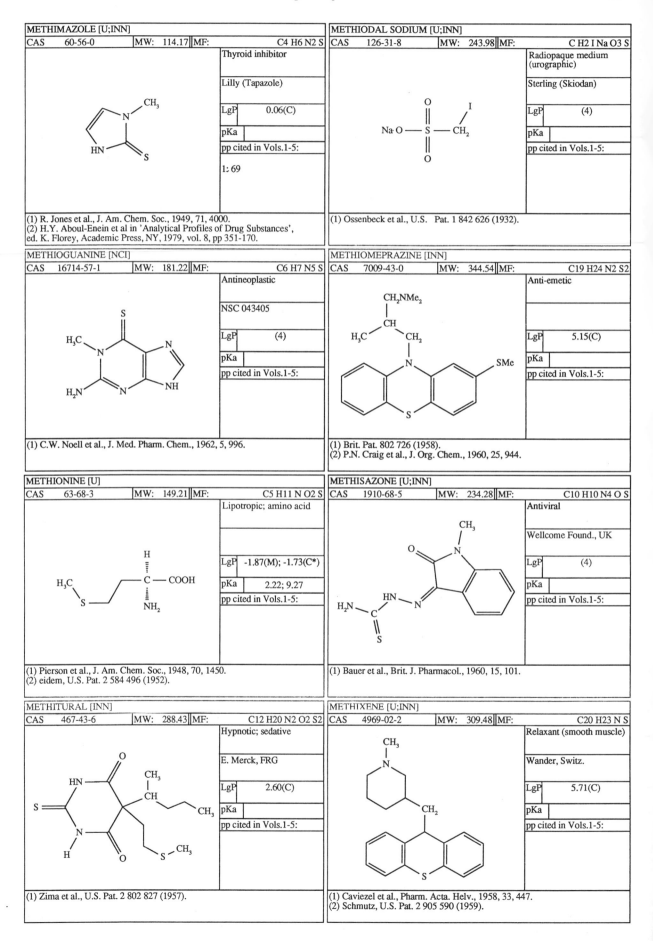

METHIMAZOLE [U;INN]

CAS	60-56-0	MW:	114.17	MF:	C4 H6 N2 S

Thyroid inhibitor

Lilly (Tapazole)

LgP	0.06(C)

pKa	

pp cited in Vols.1-5:

1: 69

(1) R. Jones et al., J. Am. Chem. Soc., 1949, 71, 4000.
(2) H.Y. Aboul-Enein et al in 'Analytical Profiles of Drug Substances',
ed. K. Florey, Academic Press, NY, 1979, vol. 8, pp 351-170.

METHIODAL SODIUM [U;INN]

CAS	126-31-8	MW:	243.98	MF:	C H2 I Na O3 S

Radiopaque medium (urographic)

Sterling (Skiodan)

LgP	(4)

pKa	

pp cited in Vols.1-5:

(1) Ossenbeck et al., U.S. Pat. 1 842 626 (1932).

METHIOGUANINE [NCI]

CAS	16714-57-1	MW:	181.22	MF:	C6 H7 N5 S

Antineoplastic

NSC 043405

LgP	(4)

pKa	

pp cited in Vols.1-5:

(1) C.W. Noell et al., J. Med. Pharm. Chem., 1962, 5, 996.

METHIOMEPRAZINE [INN]

CAS	7009-43-0	MW:	344.54	MF:	C19 H24 N2 S2

Anti-emetic

LgP	5.15(C)

pKa	

pp cited in Vols.1-5:

(1) Brit. Pat. 802 726 (1958).
(2) P.N. Craig et al., J. Org. Chem., 1960, 25, 944.

METHIONINE [U]

CAS	63-68-3	MW:	149.21	MF:	C5 H11 N O2 S

Lipotropic; amino acid

LgP	-1.87(M); -1.73(C*)

pKa	2.22; 9.27

pp cited in Vols.1-5:

(1) Pierson et al., J. Am. Chem. Soc., 1948, 70, 1450.
(2) eidem, U.S. Pat. 2 584 496 (1952).

METHISAZONE [U;INN]

CAS	1910-68-5	MW:	234.28	MF:	C10 H10 N4 O S

Antiviral

Wellcome Found., UK

LgP	(4)

pKa	

pp cited in Vols.1-5:

(1) Bauer et al., Brit. J. Pharmacol., 1960, 15, 101.

METHITURAL [INN]

CAS	467-43-6	MW:	288.43	MF:	C12 H20 N2 O2 S2

Hypnotic; sedative

E. Merck, FRG

LgP	2.60(C)

pKa	

pp cited in Vols.1-5:

(1) Zima et al., U.S. Pat. 2 802 827 (1957).

METHIXENE [U;INN]

CAS	4969-02-2	MW:	309.48	MF:	C20 H23 N S

Relaxant (smooth muscle)

Wander, Switz.

LgP	5.71(C)

pKa	

pp cited in Vols.1-5:

(1) Caviezel et al., Pharm. Acta. Helv., 1958, 33, 447.
(2) Schmutz, U.S. Pat. 2 905 590 (1959).

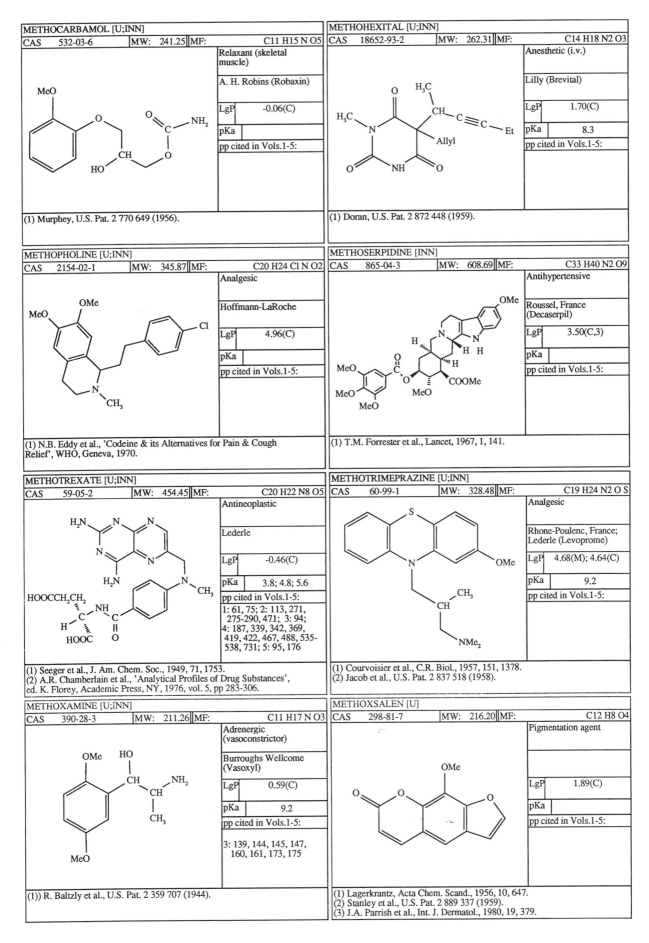

METHOCARBAMOL [U;INN]

CAS	532-03-6	MW:	241.25	MF:	C11 H15 N O5

Relaxant (skeletal muscle)

A. H. Robins (Robaxin)

LgP	-0.06(C)
pKa	
pp cited in Vols.1-5:	

(1) Murphey, U.S. Pat. 2 770 649 (1956).

METHOHEXITAL [U;INN]

CAS	18652-93-2	MW:	262.31	MF:	C14 H18 N2 O3

Anesthetic (i.v.)

Lilly (Brevital)

LgP	1.70(C)
pKa	8.3
pp cited in Vols.1-5:	

(1) Doran, U.S. Pat. 2 872 448 (1959).

METHOPHOLINE [U;INN]

CAS	2154-02-1	MW:	345.87	MF:	C20 H24 Cl N O2

Analgesic

Hoffmann-LaRoche

LgP	4.96(C)
pKa	
pp cited in Vols.1-5:	

(1) N.B. Eddy et al., 'Codeine & its Alternatives for Pain & Cough Relief', WHO, Geneva, 1970.

METHOSERPIDINE [INN]

CAS	865-04-3	MW:	608.69	MF:	C33 H40 N2 O9

Antihypertensive

Roussel, France (Decaserpil)

LgP	3.50(C,3)
pKa	
pp cited in Vols.1-5:	

(1) T.M. Forrester et al., Lancet, 1967, 1, 141.

METHOTREXATE [U;INN]

CAS	59-05-2	MW:	454.45	MF:	C20 H22 N8 O5

Antineoplastic

Lederle

LgP	-0.46(C)
pKa	3.8; 4.8; 5.6
pp cited in Vols.1-5:	

1: 61, 75; 2: 113, 271, 275-290, 471; 3: 94; 4: 187, 339, 342, 369, 419, 422, 467, 488, 535-538, 731; 5: 95, 176

(1) Seeger et al., J. Am. Chem. Soc., 1949, 71, 1753.
(2) A.R. Chamberlain et al., 'Analytical Profiles of Drug Substances', ed. K. Florey, Academic Press, NY, 1976, vol. 5, pp 283-306.

METHOTRIMEPRAZINE [U;INN]

CAS	60-99-1	MW:	328.48	MF:	C19 H24 N2 O S

Analgesic

Rhone-Poulenc, France; Lederle (Levoprome)

LgP	4.68(M); 4.64(C)
pKa	9.2
pp cited in Vols.1-5:	

(1) Courvoisier et al., C.R. Biol., 1957, 151, 1378.
(2) Jacob et al., U.S. Pat. 2 837 518 (1958).

METHOXAMINE [U;INN]

CAS	390-28-3	MW:	211.26	MF:	C11 H17 N O3

Adrenergic (vasoconstrictor)

Burroughs Wellcome (Vasoxyl)

LgP	0.59(C)
pKa	9.2
pp cited in Vols.1-5:	

3: 139, 144, 145, 147, 160, 161, 173, 175

(1)) R. Baltzly et al., U.S. Pat. 2 359 707 (1944).

METHOXSALEN [U]

CAS	298-81-7	MW:	216.20	MF:	C12 H8 O4

Pigmentation agent

LgP	1.89(C)
pKa	
pp cited in Vols.1-5:	

(1) Lagerkrantz, Acta Chem. Scand., 1956, 10, 647.
(2) Stanley et al., U.S. Pat. 2 889 337 (1959).
(3) J.A. Parrish et al., Int. J. Dermatol., 1980, 19, 379.

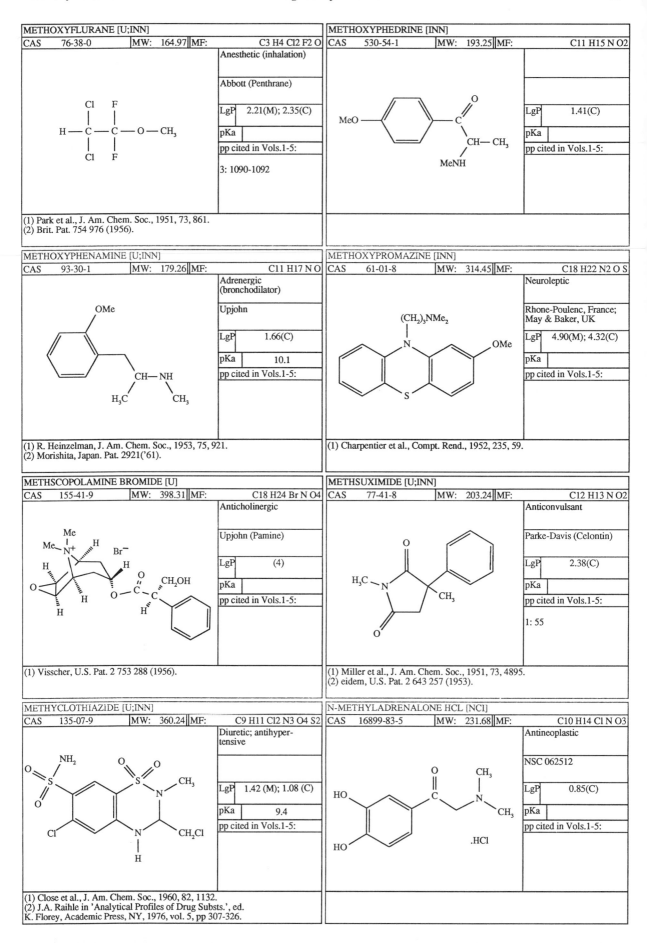

METHOXYFLURANE [U;INN]					
CAS 76-38-0	MW: 164.97	MF:			C3 H4 Cl2 F2 O

Anesthetic (inhalation)

Abbott (Penthrane)

LgP	2.21(M); 2.35(C)
pKa	
pp cited in Vols.1-5:	

3: 1090-1092

(1) Park et al., J. Am. Chem. Soc., 1951, 73, 861.
(2) Brit. Pat. 754 976 (1956).

METHOXYPHEDRINE [INN]					
CAS 530-54-1	MW: 193.25	MF:			C11 H15 N O2

LgP	1.41(C)
pKa	
pp cited in Vols.1-5:	

METHOXYPHENAMINE [U;INN]					
CAS 93-30-1	MW: 179.26	MF:			C11 H17 N O

Adrenergic
(bronchodilator)

Upjohn

LgP	1.66(C)
pKa	10.1
pp cited in Vols.1-5:	

(1) R. Heinzelman, J. Am. Chem. Soc., 1953, 75, 921.
(2) Morishita, Japan. Pat. 2921('61).

METHOXYPROMAZINE [INN]					
CAS 61-01-8	MW: 314.45	MF:			C18 H22 N2 O S

Neuroleptic

Rhone-Poulenc, France;
May & Baker, UK

LgP	4.90(M); 4.32(C)
pKa	
pp cited in Vols.1-5:	

(1) Charpentier et al., Compt. Rend., 1952, 235, 59.

METHSCOPOLAMINE BROMIDE [U]					
CAS 155-41-9	MW: 398.31	MF:			C18 H24 Br N O4

Anticholinergic

Upjohn (Pamine)

LgP	(4)
pKa	
pp cited in Vols.1-5:	

(1) Visscher, U.S. Pat. 2 753 288 (1956).

METHSUXIMIDE [U;INN]					
CAS 77-41-8	MW: 203.24	MF:			C12 H13 N O2

Anticonvulsant

Parke-Davis (Celontin)

LgP	2.38(C)
pKa	
pp cited in Vols.1-5:	

1: 55

(1) Miller et al., J. Am. Chem. Soc., 1951, 73, 4895.
(2) eidem, U.S. Pat. 2 643 257 (1953).

METHYCLOTHIAZIDE [U;INN]					
CAS 135-07-9	MW: 360.24	MF:			C9 H11 Cl2 N3 O4 S2

Diuretic; antihyper-
tensive

LgP	1.42 (M); 1.08 (C)
pKa	9.4
pp cited in Vols.1-5:	

(1) Close et al., J. Am. Chem. Soc., 1960, 82, 1132.
(2) J.A. Raihle in 'Analytical Profiles of Drug Substs.', ed.
K. Florey, Academic Press, NY, 1976, vol. 5, pp 307-326.

N-METHYLADRENALONE HCL [NCI]					
CAS 16899-83-5	MW: 231.68	MF:			C10 H14 Cl N O3

Antineoplastic

NSC 062512

LgP	0.85(C)
pKa	
pp cited in Vols.1-5:	

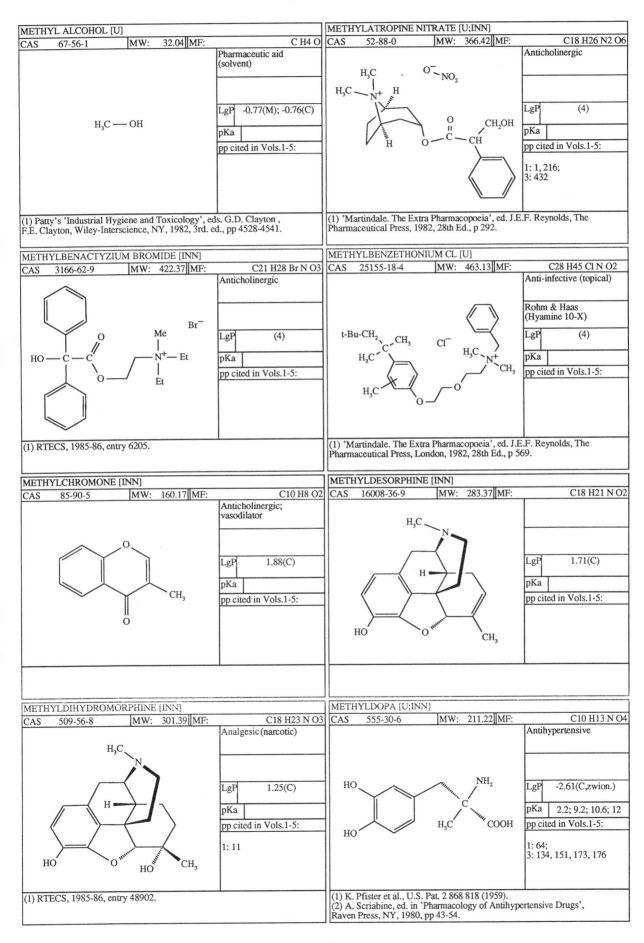

METHYL ALCOHOL [U]

CAS 67-56-1	MW: 32.04	MF: C H4 O

Pharmaceutic aid (solvent)

LgP -0.77(M); -0.76(C)

pKa

pp cited in Vols.1-5:

H₃C — OH

(1) Patty's 'Industrial Hygiene and Toxicology', eds. G.D. Clayton , F.E. Clayton, Wiley-Interscience, NY, 1982, 3rd. ed., pp 4528-4541.

METHYLATROPINE NITRATE [U;INN]

CAS 52-88-0	MW: 366.42	MF: C18 H26 N2 O6

Anticholinergic

LgP (4)

pKa

pp cited in Vols.1-5:

1: 1, 216;
3: 432

(1) 'Martindale. The Extra Pharmacopoeia', ed. J.E.F. Reynolds, The Pharmaceutical Press, 1982, 28th Ed., p 292.

METHYLBENACTYZIUM BROMIDE [INN]

CAS 3166-62-9	MW: 422.37	MF: C21 H28 Br N O3

Anticholinergic

LgP (4)

pKa

pp cited in Vols.1-5:

(1) RTECS, 1985-86, entry 6205.

METHYLBENZETHONIUM CL [U]

CAS 25155-18-4	MW: 463.13	MF: C28 H45 Cl N O2

Anti-infective (topical)

Rohm & Haas
(Hyamine 10-X)

LgP (4)

pKa

pp cited in Vols.1-5:

(1) 'Martindale. The Extra Pharmacopoeia', ed. J.E.F. Reynolds, The Pharmaceutical Press, London, 1982, 28th Ed., p 569.

METHYLCHROMONE [INN]

CAS 85-90-5	MW: 160.17	MF: C10 H8 O2

Anticholinergic; vasodilator

LgP 1.88(C)

pKa

pp cited in Vols.1-5:

METHYLDESORPHINE [INN]

CAS 16008-36-9	MW: 283.37	MF: C18 H21 N O2

LgP 1.71(C)

pKa

pp cited in Vols.1-5:

METHYLDIHYDROMORPHINE [INN]

CAS 509-56-8	MW: 301.39	MF: C18 H23 N O3

Analgesic (narcotic)

LgP 1.25(C)

pKa

pp cited in Vols.1-5:

1: 11

(1) RTECS, 1985-86, entry 48902.

METHYLDOPA [U;INN]

CAS 555-30-6	MW: 211.22	MF: C10 H13 N O4

Antihypertensive

LgP -2.61(C,zwion.)

pKa 2.2; 9.2; 10.6; 12

pp cited in Vols.1-5:

1: 64;
3: 134, 151, 173, 176

(1) K. Pfister et al., U.S. Pat. 2 868 818 (1959).
(2) A. Scriabine, ed. in 'Pharmacology of Antihypertensive Drugs', Raven Press, NY, 1980, pp 43-54.

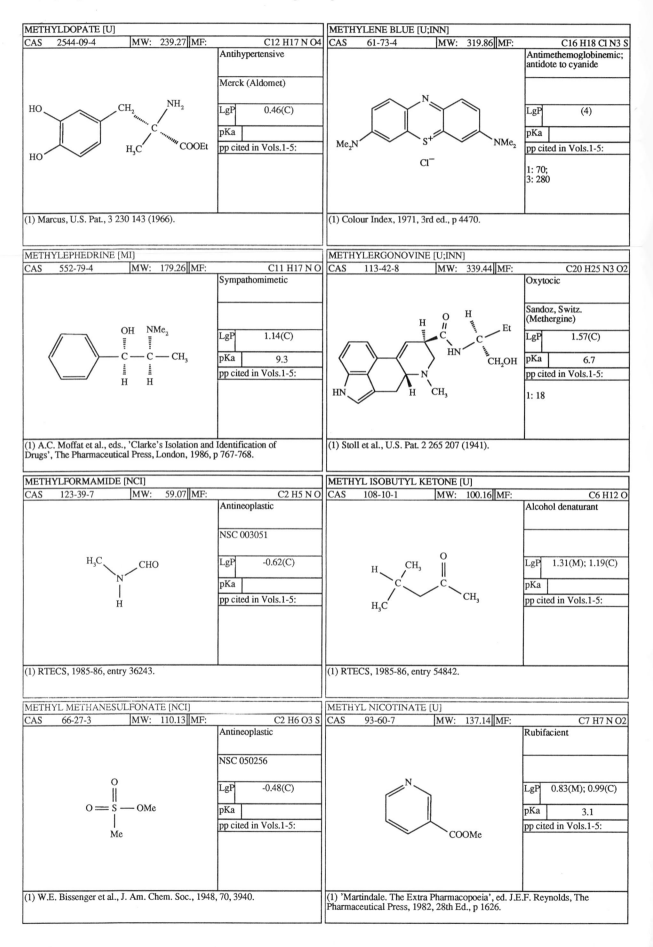

METHYLDOPATE [U]

CAS	2544-09-4	MW:	239.27	MF:	C12 H17 N O4

Antihypertensive

Merck (Aldomet)

LgP	0.46(C)
pKa	
pp cited in Vols.1-5:	

(1) Marcus, U.S. Pat., 3 230 143 (1966).

METHYLENE BLUE [U;INN]

CAS	61-73-4	MW:	319.86	MF:	C16 H18 Cl N3 S

Antimethemoglobinemic; antidote to cyanide

LgP	(4)
pKa	
pp cited in Vols.1-5:	

1: 70;
3: 280

(1) Colour Index, 1971, 3rd ed., p 4470.

METHYLEPHEDRINE [MI]

CAS	552-79-4	MW:	179.26	MF:	C11 H17 N O

Sympathomimetic

LgP	1.14(C)
pKa	9.3
pp cited in Vols.1-5:	

(1) A.C. Moffat et al., eds., 'Clarke's Isolation and Identification of Drugs', The Pharmaceutical Press, London, 1986, p 767-768.

METHYLERGONOVINE [U;INN]

CAS	113-42-8	MW:	339.44	MF:	C20 H25 N3 O2

Oxytocic

Sandoz, Switz. (Methergine)

LgP	1.57(C)
pKa	6.7
pp cited in Vols.1-5:	

1: 18

(1) Stoll et al., U.S. Pat. 2 265 207 (1941).

METHYLFORMAMIDE [NCI]

CAS	123-39-7	MW:	59.07	MF:	C2 H5 N O

Antineoplastic

NSC 003051

LgP	-0.62(C)
pKa	
pp cited in Vols.1-5:	

(1) RTECS, 1985-86, entry 36243.

METHYL ISOBUTYL KETONE [U]

CAS	108-10-1	MW:	100.16	MF:	C6 H12 O

Alcohol denaturant

LgP	1.31(M); 1.19(C)
pKa	
pp cited in Vols.1-5:	

(1) RTECS, 1985-86, entry 54842.

METHYL METHANESULFONATE [NCI]

CAS	66-27-3	MW:	110.13	MF:	C2 H6 O3 S

Antineoplastic

NSC 050256

LgP	-0.48(C)
pKa	
pp cited in Vols.1-5:	

(1) W.E. Bissenger et al., J. Am. Chem. Soc., 1948, 70, 3940.

METHYL NICOTINATE [U]

CAS	93-60-7	MW:	137.14	MF:	C7 H7 N O2

Rubifacient

LgP	0.83(M); 0.99(C)
pKa	3.1
pp cited in Vols.1-5:	

(1) 'Martindale. The Extra Pharmacopoeia', ed. J.E.F. Reynolds, The Pharmaceutical Press, 1982, 28th Ed., p 1626.

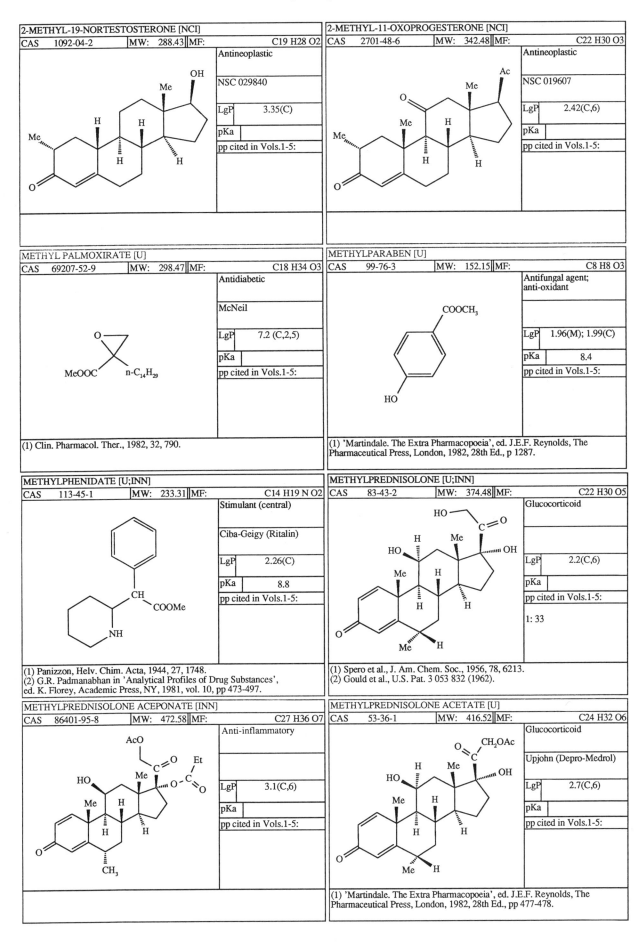

2-METHYL-19-NORTESTOSTERONE [NCI]

CAS	1092-04-2	MW: 288.43	MF:	C19 H28 O2

Antineoplastic

NSC 029840

LgP	3.35(C)
pKa	

pp cited in Vols.1-5:

2-METHYL-11-OXOPROGESTERONE [NCI]

CAS	2701-48-6	MW: 342.48	MF:	C22 H30 O3

Antineoplastic

NSC 019607

LgP	2.42(C,6)
pKa	

pp cited in Vols.1-5:

METHYL PALMOXIRATE [U]

CAS	69207-52-9	MW: 298.47	MF:	C18 H34 O3

Antidiabetic

McNeil

LgP	7.2 (C,2,5)
pKa	

pp cited in Vols.1-5:

(1) Clin. Pharmacol. Ther., 1982, 32, 790.

METHYLPARABEN [U]

CAS	99-76-3	MW: 152.15	MF:	C8 H8 O3

Antifungal agent; anti-oxidant

LgP	1.96(M); 1.99(C)
pKa	8.4

pp cited in Vols.1-5:

(1) 'Martindale. The Extra Pharmacopoeia', ed. J.E.F. Reynolds, The Pharmaceutical Press, London, 1982, 28th Ed., p 1287.

METHYLPHENIDATE [U;INN]

CAS	113-45-1	MW: 233.31	MF:	C14 H19 N O2

Stimulant (central)

Ciba-Geigy (Ritalin)

LgP	2.26(C)
pKa	8.8

pp cited in Vols.1-5:

(1) Panizzon, Helv. Chim. Acta, 1944, 27, 1748.
(2) G.R. Padmanabhan in 'Analytical Profiles of Drug Substances', ed. K. Florey, Academic Press, NY, 1981, vol. 10, pp 473-497.

METHYLPREDNISOLONE [U;INN]

CAS	83-43-2	MW: 374.48	MF:	C22 H30 O5

Glucocorticoid

LgP	2.2(C,6)
pKa	

pp cited in Vols.1-5:

1: 33

(1) Spero et al., J. Am. Chem. Soc., 1956, 78, 6213.
(2) Gould et al., U.S. Pat. 3 053 832 (1962).

METHYLPREDNISOLONE ACEPONATE [INN]

CAS	86401-95-8	MW: 472.58	MF:	C27 H36 O7

Anti-inflammatory

LgP	3.1(C,6)
pKa	

pp cited in Vols.1-5:

METHYLPREDNISOLONE ACETATE [U]

CAS	53-36-1	MW: 416.52	MF:	C24 H32 O6

Glucocorticoid

Upjohn (Depro-Medrol)

LgP	2.7(C,6)
pKa	

pp cited in Vols.1-5:

(1) 'Martindale. The Extra Pharmacopoeia', ed. J.E.F. Reynolds, The Pharmaceutical Press, London, 1982, 28th Ed., pp 477-478.

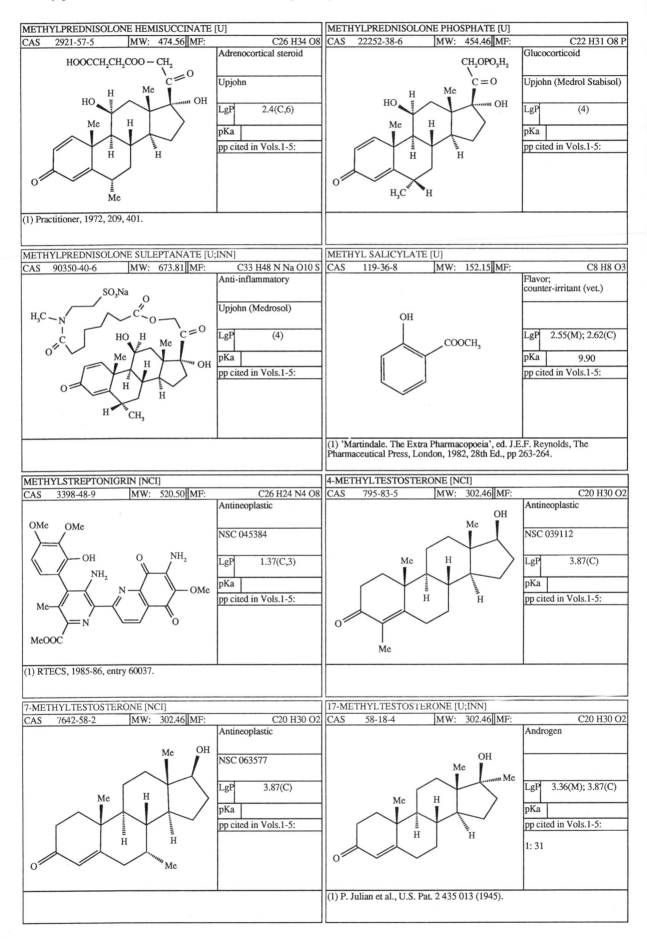

METHYLPREDNISOLONE HEMISUCCINATE [U]			
CAS 2921-57-5	MW: 474.56	MF:	C26 H34 O8

Adrenocortical steroid

Upjohn

LgP	2.4(C,6)
pKa	
pp cited in Vols.1-5:	

(1) Practitioner, 1972, 209, 401.

METHYLPREDNISOLONE PHOSPHATE [U]			
CAS 22252-38-6	MW: 454.46	MF:	C22 H31 O8 P

Glucocorticoid

Upjohn (Medrol Stabisol)

LgP	(4)
pKa	
pp cited in Vols.1-5:	

METHYLPREDNISOLONE SULEPTANATE [U;INN]			
CAS 90350-40-6	MW: 673.81	MF:	C33 H48 N Na O10 S

Anti-inflammatory

Upjohn (Medrosol)

LgP	(4)
pKa	
pp cited in Vols.1-5:	

METHYL SALICYLATE [U]			
CAS 119-36-8	MW: 152.15	MF:	C8 H8 O3

Flavor;
counter-irritant (vet.)

LgP	2.55(M); 2.62(C)
pKa	9.90
pp cited in Vols.1-5:	

(1) 'Martindale. The Extra Pharmacopoeia', ed. J.E.F. Reynolds, The Pharmaceutical Press, London, 1982, 28th Ed., pp 263-264.

METHYLSTREPTONIGRIN [NCI]			
CAS 3398-48-9	MW: 520.50	MF:	C26 H24 N4 O8

Antineoplastic

NSC 045384

LgP	1.37(C,3)
pKa	
pp cited in Vols.1-5:	

(1) RTECS, 1985-86, entry 60037.

4-METHYLTESTOSTERONE [NCI]			
CAS 795-83-5	MW: 302.46	MF:	C20 H30 O2

Antineoplastic

NSC 039112

LgP	3.87(C)
pKa	
pp cited in Vols.1-5:	

7-METHYLTESTOSTERONE [NCI]			
CAS 7642-58-2	MW: 302.46	MF:	C20 H30 O2

Antineoplastic

NSC 063577

LgP	3.87(C)
pKa	
pp cited in Vols.1-5:	

17-METHYLTESTOSTERONE [U;INN]			
CAS 58-18-4	MW: 302.46	MF:	C20 H30 O2

Androgen

LgP	3.36(M); 3.87(C)
pKa	
pp cited in Vols.1-5:	
1: 31	

(1) P. Julian et al., U.S. Pat. 2 435 013 (1945).

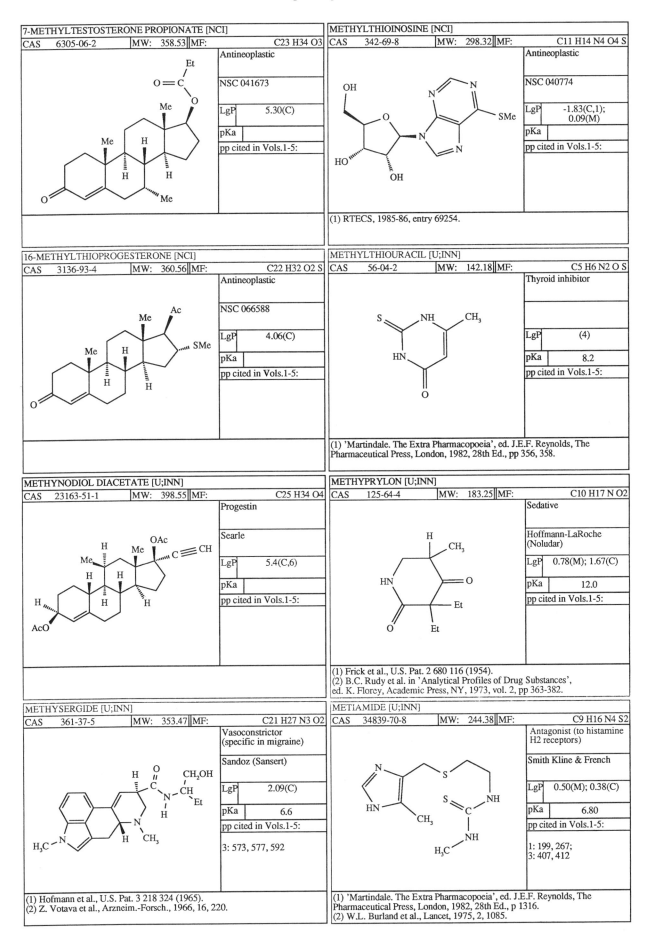

7-METHYLTESTOSTERONE PROPIONATE [NCI]

CAS	6305-06-2	MW:	358.53	MF:	C23 H34 O3

Antineoplastic
NSC 041673

LgP	5.30(C)
pKa	
pp cited in Vols.1-5:	

METHYLTHIOINOSINE [NCI]

CAS	342-69-8	MW:	298.32	MF:	C11 H14 N4 O4 S

Antineoplastic
NSC 040774

LgP	-1.83(C,1); 0.09(M)
pKa	
pp cited in Vols.1-5:	

(1) RTECS, 1985-86, entry 69254.

16-METHYLTHIOPROGESTERONE [NCI]

CAS	3136-93-4	MW:	360.56	MF:	C22 H32 O2 S

Antineoplastic
NSC 066588

LgP	4.06(C)
pKa	
pp cited in Vols.1-5:	

METHYLTHIOURACIL [U;INN]

CAS	56-04-2	MW:	142.18	MF:	C5 H6 N2 O S

Thyroid inhibitor

LgP	(4)
pKa	8.2
pp cited in Vols.1-5:	

(1) 'Martindale. The Extra Pharmacopoeia', ed. J.E.F. Reynolds, The Pharmaceutical Press, London, 1982, 28th Ed., pp 356, 358.

METHYNODIOL DIACETATE [U;INN]

CAS	23163-51-1	MW:	398.55	MF:	C25 H34 O4

Progestin
Searle

LgP	5.4(C,6)
pKa	
pp cited in Vols.1-5:	

METHYPRYLON [U;INN]

CAS	125-64-4	MW:	183.25	MF:	C10 H17 N O2

Sedative
Hoffmann-LaRoche (Noludar)

LgP	0.78(M); 1.67(C)
pKa	12.0
pp cited in Vols.1-5:	

(1) Frick et al., U.S. Pat. 2 680 116 (1954).
(2) B.C. Rudy et al. in 'Analytical Profiles of Drug Substances', ed. K. Florey, Academic Press, NY, 1973, vol. 2, pp 363-382.

METHYSERGIDE [U;INN]

CAS	361-37-5	MW:	353.47	MF:	C21 H27 N3 O2

Vasoconstrictor (specific in migraine)
Sandoz (Sansert)

LgP	2.09(C)
pKa	6.6
pp cited in Vols.1-5:	
3: 573, 577, 592	

(1) Hofmann et al., U.S. Pat. 3 218 324 (1965).
(2) Z. Votava et al., Arzneim.-Forsch., 1966, 16, 220.

METIAMIDE [U;INN]

CAS	34839-70-8	MW:	244.38	MF:	C9 H16 N4 S2

Antagonist (to histamine H2 receptors)
Smith Kline & French

LgP	0.50(M); 0.38(C)
pKa	6.80
pp cited in Vols.1-5:	
1: 199, 267; 3: 407, 412	

(1) 'Martindale. The Extra Pharmacopoeia', ed. J.E.F. Reynolds, The Pharmaceutical Press, London, 1982, 28th Ed., p 1316.
(2) W.L. Burland et al., Lancet, 1975, 2, 1085.

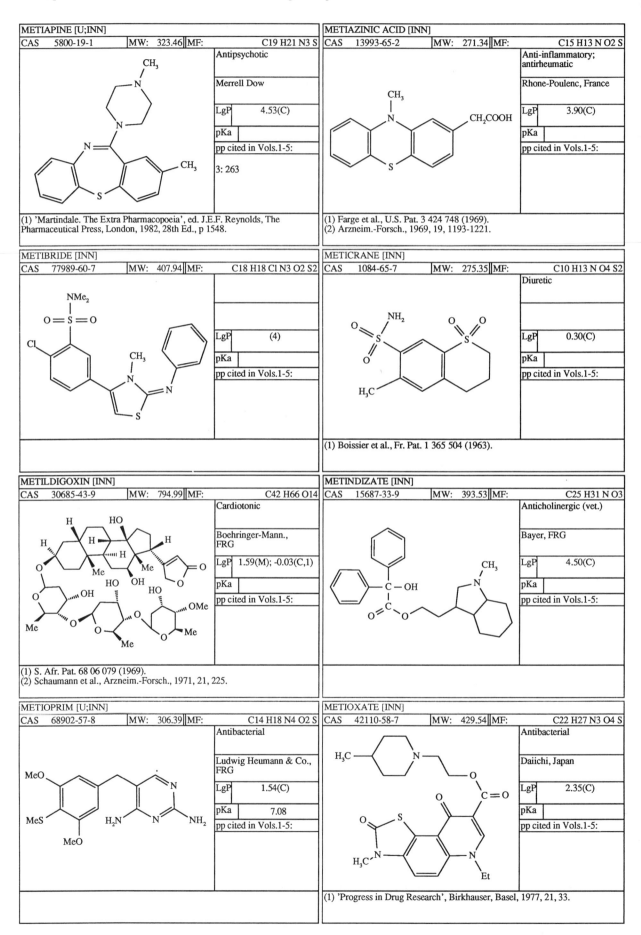

METIAPINE [U;INN]		
CAS 5800-19-1 MW: 323.46 MF: C19 H21 N3 S		
	Antipsychotic	
	Merrell Dow	
	LgP 4.53(C)	
	pKa	
	pp cited in Vols.1-5:	
	3: 263	

(1) 'Martindale. The Extra Pharmacopoeia', ed. J.E.F. Reynolds, The Pharmaceutical Press, London, 1982, 28th Ed., p 1548.

METIAZINIC ACID [INN]		
CAS 13993-65-2 MW: 271.34 MF: C15 H13 N O2 S		
	Anti-inflammatory; antirheumatic	
	Rhone-Poulenc, France	
	LgP 3.90(C)	
	pKa	
	pp cited in Vols.1-5:	

(1) Farge et al., U.S. Pat. 3 424 748 (1969).
(2) Arzneim.-Forsch., 1969, 19, 1193-1221.

METIBRIDE [INN]		
CAS 77989-60-7 MW: 407.94 MF: C18 H18 Cl N3 O2 S2		
	LgP (4)	
	pKa	
	pp cited in Vols.1-5:	

METICRANE [INN]		
CAS 1084-65-7 MW: 275.35 MF: C10 H13 N O4 S2		
	Diuretic	
	LgP 0.30(C)	
	pKa	
	pp cited in Vols.1-5:	

(1) Boissier et al., Fr. Pat. 1 365 504 (1963).

METILDIGOXIN [INN]		
CAS 30685-43-9 MW: 794.99 MF: C42 H66 O14		
	Cardiotonic	
	Boehringer-Mann., FRG	
	LgP 1.59(M); -0.03(C,1)	
	pKa	
	pp cited in Vols.1-5:	

(1) S. Afr. Pat. 68 06 079 (1969).
(2) Schaumann et al., Arzneim.-Forsch., 1971, 21, 225.

METINDIZATE [INN]		
CAS 15687-33-9 MW: 393.53 MF: C25 H31 N O3		
	Anticholinergic (vet.)	
	Bayer, FRG	
	LgP 4.50(C)	
	pKa	
	pp cited in Vols.1-5:	

METIOPRIM [U;INN]		
CAS 68902-57-8 MW: 306.39 MF: C14 H18 N4 O2 S		
	Antibacterial	
	Ludwig Heumann & Co., FRG	
	LgP 1.54(C)	
	pKa 7.08	
	pp cited in Vols.1-5:	

METIOXATE [INN]		
CAS 42110-58-7 MW: 429.54 MF: C22 H27 N3 O4 S		
	Antibacterial	
	Daiichi, Japan	
	LgP 2.35(C)	
	pKa	
	pp cited in Vols.1-5:	

(1) 'Progress in Drug Research', Birkhauser, Basel, 1977, 21, 33.

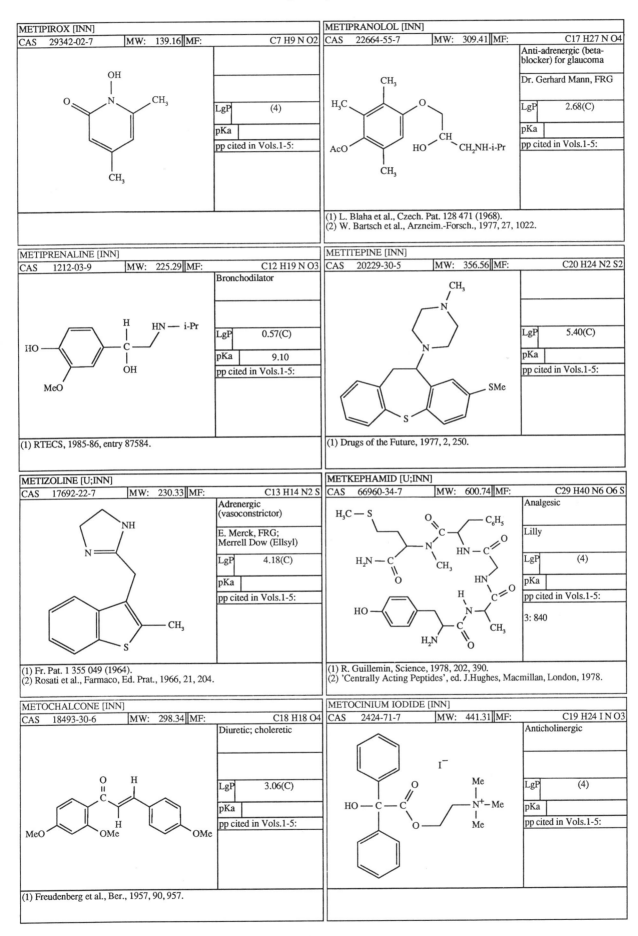

METIPIROX [INN]

CAS	29342-02-7	MW:	139.16	MF:	C7 H9 N O2

LgP	(4)
pKa	
pp cited in Vols.1-5:	

METIPRANOLOL [INN]

CAS	22664-55-7	MW:	309.41	MF:	C17 H27 N O4

Anti-adrenergic (beta-blocker) for glaucoma

Dr. Gerhard Mann, FRG

LgP	2.68(C)
pKa	
pp cited in Vols.1-5:	

(1) L. Blaha et al., Czech. Pat. 128 471 (1968).
(2) W. Bartsch et al., Arzneim.-Forsch., 1977, 27, 1022.

METIPRENALINE [INN]

CAS	1212-03-9	MW:	225.29	MF:	C12 H19 N O3

Bronchodilator

LgP	0.57(C)
pKa	9.10
pp cited in Vols.1-5:	

(1) RTECS, 1985-86, entry 87584.

METITEPINE [INN]

CAS	20229-30-5	MW:	356.56	MF:	C20 H24 N2 S2

LgP	5.40(C)
pKa	
pp cited in Vols.1-5:	

(1) Drugs of the Future, 1977, 2, 250.

METIZOLINE [U;INN]

CAS	17692-22-7	MW:	230.33	MF:	C13 H14 N2 S

Adrenergic (vasoconstrictor)

E. Merck, FRG; Merrell Dow (Ellsyl)

LgP	4.18(C)
pKa	
pp cited in Vols.1-5:	

(1) Fr. Pat. 1 355 049 (1964).
(2) Rosati et al., Farmaco, Ed. Prat., 1966, 21, 204.

METKEPHAMID [U;INN]

CAS	66960-34-7	MW:	600.74	MF:	C29 H40 N6 O6 S

Analgesic

Lilly

LgP	(4)
pKa	
pp cited in Vols.1-5:	
	3: 840

(1) R. Guillemin, Science, 1978, 202, 390.
(2) 'Centrally Acting Peptides', ed. J.Hughes, Macmillan, London, 1978.

METOCHALCONE [INN]

CAS	18493-30-6	MW:	298.34	MF:	C18 H18 O4

Diuretic; choleretic

LgP	3.06(C)
pKa	
pp cited in Vols.1-5:	

(1) Freudenberg et al., Ber., 1957, 90, 957.

METOCINIUM IODIDE [INN]

CAS	2424-71-7	MW:	441.31	MF:	C19 H24 I N O3

Anticholinergic

LgP	(4)
pKa	
pp cited in Vols.1-5:	

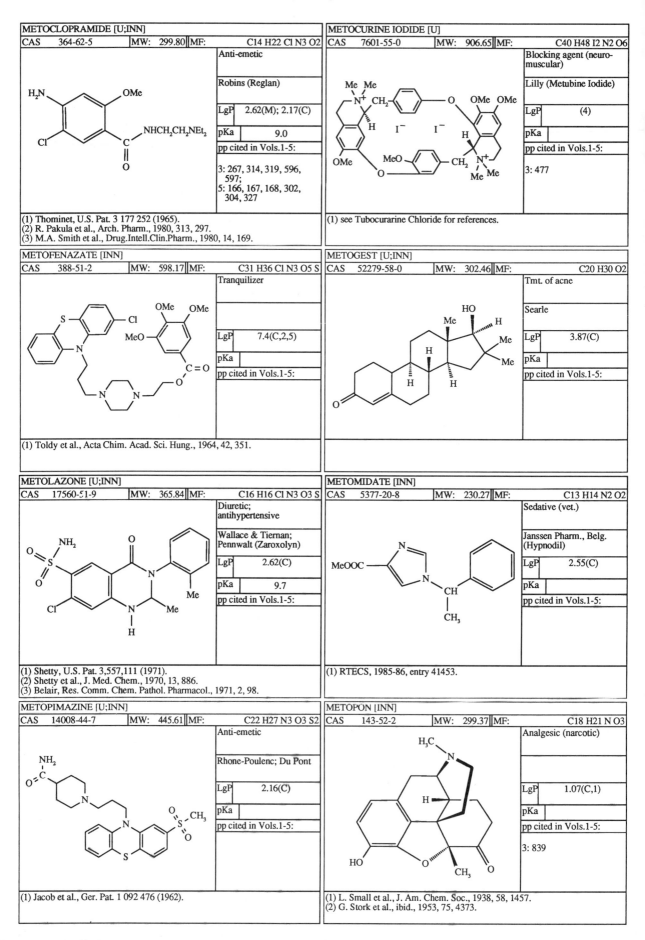

METOCLOPRAMIDE [U;INN]

CAS	364-62-5	MW:	299.80	MF:	C14 H22 Cl N3 O2

Anti-emetic

Robins (Reglan)

LgP	2.62(M); 2.17(C)
pKa	9.0
pp cited in Vols.1-5:	

3: 267, 314, 319, 596, 597;
5: 166, 167, 168, 302, 304, 327

(1) Thominet, U.S. Pat. 3 177 252 (1965).
(2) R. Pakula et al., Arch. Pharm., 1980, 313, 297.
(3) M.A. Smith et al., Drug.Intell.Clin.Pharm., 1980, 14, 169.

METOCURINE IODIDE [U]

CAS	7601-55-0	MW:	906.65	MF:	C40 H48 I2 N2 O6

Blocking agent (neuro-muscular)

Lilly (Metubine Iodide)

LgP	(4)
pKa	
pp cited in Vols.1-5:	

3: 477

(1) see Tubocurarine Chloride for references.

METOFENAZATE [INN]

CAS	388-51-2	MW:	598.17	MF:	C31 H36 Cl N3 O5 S

Tranquilizer

LgP	7.4(C,2,5)
pKa	
pp cited in Vols.1-5:	

(1) Toldy et al., Acta Chim. Acad. Sci. Hung., 1964, 42, 351.

METOGEST [U;INN]

CAS	52279-58-0	MW:	302.46	MF:	C20 H30 O2

Tmt. of acne

Searle

LgP	3.87(C)
pKa	
pp cited in Vols.1-5:	

METOLAZONE [U;INN]

CAS	17560-51-9	MW:	365.84	MF:	C16 H16 Cl N3 O3 S

Diuretic; antihypertensive

Wallace & Tiernan; Pennwalt (Zaroxolyn)

LgP	2.62(C)
pKa	9.7
pp cited in Vols.1-5:	

(1) Shetty, U.S. Pat. 3,557,111 (1971).
(2) Shetty et al., J. Med. Chem., 1970, 13, 886.
(3) Belair, Res. Comm. Chem. Pathol. Pharmacol., 1971, 2, 98.

METOMIDATE [INN]

CAS	5377-20-8	MW:	230.27	MF:	C13 H14 N2 O2

Sedative (vet.)

Janssen Pharm., Belg. (Hypnodil)

LgP	2.55(C)
pKa	
pp cited in Vols.1-5:	

(1) RTECS, 1985-86, entry 41453.

METOPIMAZINE [U;INN]

CAS	14008-44-7	MW:	445.61	MF:	C22 H27 N3 O3 S2

Anti-emetic

Rhone-Poulenc; Du Pont

LgP	2.16(C)
pKa	
pp cited in Vols.1-5:	

(1) Jacob et al., Ger. Pat. 1 092 476 (1962).

METOPON [INN]

CAS	143-52-2	MW:	299.37	MF:	C18 H21 N O3

Analgesic (narcotic)

LgP	1.07(C,1)
pKa	
pp cited in Vols.1-5:	

3: 839

(1) L. Small et al., J. Am. Chem. Soc., 1938, 58, 1457.
(2) G. Stork et al., ibid., 1953, 75, 4373.

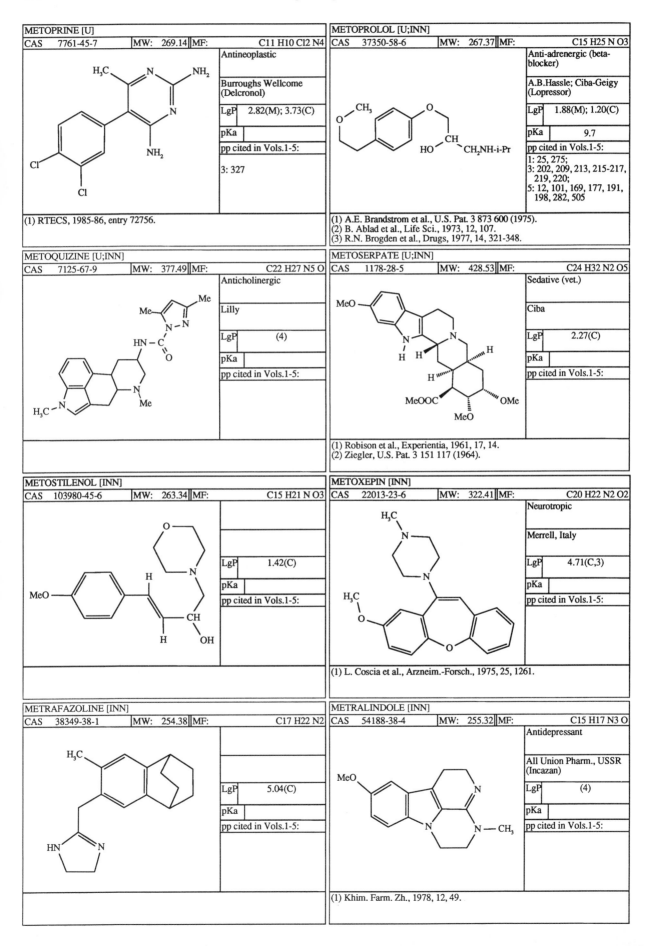

METOPRINE [U]

CAS 7761-45-7	MW: 269.14 MF: C11 H10 Cl2 N4

Antineoplastic

Burroughs Wellcome (Delcronol)

LgP 2.82(M); 3.73(C)

pKa

pp cited in Vols.1-5:

3: 327

(1) RTECS, 1985-86, entry 72756.

METOPROLOL [U;INN]

CAS 37350-58-6	MW: 267.37 MF: C15 H25 N O3

Anti-adrenergic (beta-blocker)

A.B.Hassle; Ciba-Geigy (Lopressor)

LgP 1.88(M); 1.20(C)

pKa 9.7

pp cited in Vols.1-5:

1: 25, 275;
3: 202, 209, 213, 215-217, 219, 220;
5: 12, 101, 169, 177, 191, 198, 282, 505

(1) A.E. Brandstrom et al., U.S. Pat. 3 873 600 (1975).
(2) B. Ablad et al., Life Sci., 1973, 12, 107.
(3) R.N. Brogden et al., Drugs, 1977, 14, 321-348.

METOQUIZINE [U;INN]

CAS 7125-67-9	MW: 377.49 MF: C22 H27 N5 O

Anticholinergic

Lilly

LgP (4)

pKa

pp cited in Vols.1-5:

METOSERPATE [U;INN]

CAS 1178-28-5	MW: 428.53 MF: C24 H32 N2 O5

Sedative (vet.)

Ciba

LgP 2.27(C)

pKa

pp cited in Vols.1-5:

(1) Robison et al., Experientia, 1961, 17, 14.
(2) Ziegler, U.S. Pat. 3 151 117 (1964).

METOSTILENOL [INN]

CAS 103980-45-6	MW: 263.34 MF: C15 H21 N O3

LgP 1.42(C)

pKa

pp cited in Vols.1-5:

METOXEPIN [INN]

CAS 22013-23-6	MW: 322.41 MF: C20 H22 N2 O2

Neurotropic

Merrell, Italy

LgP 4.71(C,3)

pKa

pp cited in Vols.1-5:

(1) L. Coscia et al., Arzneim.-Forsch., 1975, 25, 1261.

METRAFAZOLINE [INN]

CAS 38349-38-1	MW: 254.38 MF: C17 H22 N2

LgP 5.04(C)

pKa

pp cited in Vols.1-5:

METRALINDOLE [INN]

CAS 54188-38-4	MW: 255.32 MF: C15 H17 N3 O

Antidepressant

All Union Pharm., USSR (Incazan)

LgP (4)

pKa

pp cited in Vols.1-5:

(1) Khim. Farm. Zh., 1978, 12, 49.

METRAZIFONE [INN]

CAS 68289-14-5	MW: 349.44	MF:	C20 H23 N5 O

LgP	(4)
pKa	
pp cited in Vols.1-5:	

(1) Ann. Pharm. Franc., 1985, 43, 407.

METRENPERONE [INN]

CAS 81043-56-3	MW: 407.49	MF:	C24 H26 F N3 O2

Serotonin anatagonist

Janssen Pharm., Belg.

LgP	2.78(C)
pKa	
pp cited in Vols.1-5:	

METRIBOLONE [INN]

CAS 965-93-5	MW: 284.40	MF:	C19 H24 O2

Anabolic

Roussel-UCLAF, France

LgP	2.53(C)
pKa	
pp cited in Vols.1-5:	

(1) Velluz et al., Compt. Rend., 1963, 257, 569.
(2) Neth. Pat. 6 401 555 (1964).

METRIFONATE [INN]

CAS 52-68-6	MW: 257.44	MF:	C4 H8 Cl3 O4 P

Anthelmintic (vet.); antischistosomal (human)

Bayer, FRG (Bilarcil)

LgP	0.51(M); 0.30(C)
pKa	
pp cited in Vols.1-5:	

(1) 'Martindale. The Extra Pharmacopoeia', ed. J.E.F. Reynolds, The Pharmaceutical Press, London, 1982, 28th Ed., p 839.

METRIFUDIL [INN]

CAS 23707-33-7	MW: 371.40	MF:	C18 H21 N5 O4

Coronary vasodilator

Boehringer-Mann., FRG

LgP	0.39(C,1)
pKa	
pp cited in Vols.1-5:	

(1) Arzneim.-Forsch., 1972, 22, 783.

METRIZAMIDE [U;INN]

CAS 31112-62-6	MW: 789.10	MF:	C18 H22 I3 N3 O8

Radiopaque medium

Nyegaard, Norway; Sterling (Amipaque)

LgP	(4)
pKa	
pp cited in Vols.1-5:	

(1) H.O. Torsten et al., U.S. Pat. 3 701 771 (1972).
(2) G.F. Dibona, Proc. Soc. Exp. Biol. Med. 1978, 157, 453.

METRIZOIC ACID [U;INN]

CAS 1949-45-7	MW: 627.95	MF:	C12 H11 I3 N2 O4

Radiopaque medium

Nyegaard, Norway; Sterling

LgP	0.69(C,3)
pKa	
pp cited in Vols.1-5:	

(1) Pitre et al., Farmaco Ed. Sci., 1962, 17, 340.
(2) Brit. Pat. 973 881 (1964).

METRONIDAZOLE [U;INN]

CAS 443-48-1	MW: 171.16	MF:	C6 H9 N3 O3

Antiprotozoal (Trichomonas)

Rhone-Poulenc, France; Searle (Flagyl)

LgP	-0.02(M); -0.35(C)
pKa	2.5
pp cited in Vols.1-5:	

2: 92, 726, 728-730, 741;
5: 33, 119, 150, 195

(1) Jacob et al., U.S. Pat. 2 944 061 (1960).
(2) Cossar et al., Arzneim.-Forsch., 1966, 16, 23.
(3) R.N. Brogden et al., Drugs, 1978, 16, 387.

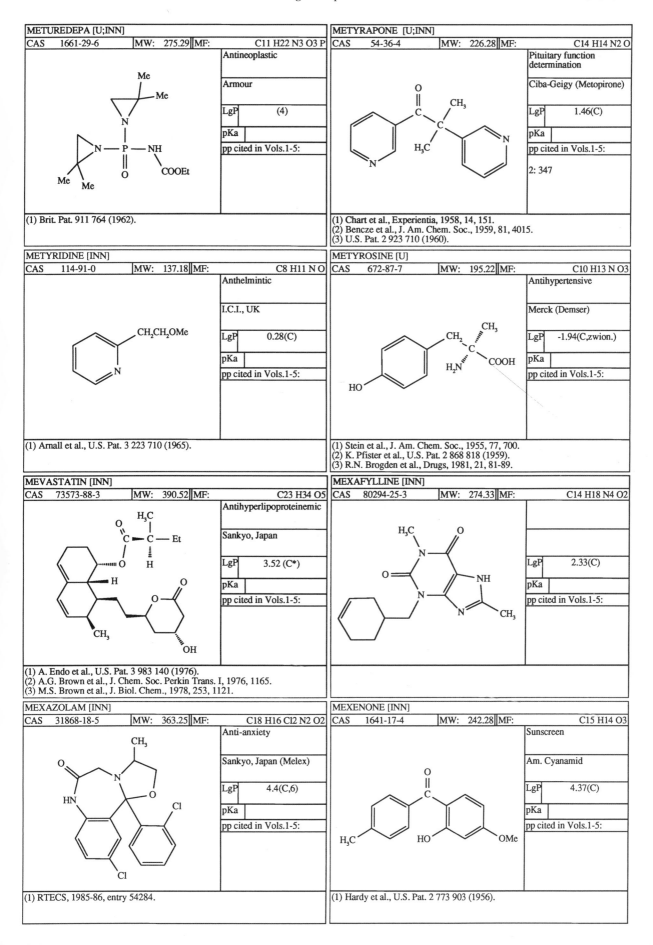

METUREDEPA [U;INN]

CAS	1661-29-6	MW:	275.29	MF:	C11 H22 N3 O3 P

Antineoplastic

Armour

LgP	(4)
pKa	
pp cited in Vols.1-5:	

(1) Brit. Pat. 911 764 (1962).

METYRAPONE [U;INN]

CAS	54-36-4	MW:	226.28	MF:	C14 H14 N2 O

Pituitary function determination

Ciba-Geigy (Metopirone)

LgP	1.46(C)
pKa	
pp cited in Vols.1-5:	

2: 347

(1) Chart et al., Experientia, 1958, 14, 151.
(2) Bencze et al., J. Am. Chem. Soc., 1959, 81, 4015.
(3) U.S. Pat. 2 923 710 (1960).

METYRIDINE [INN]

CAS	114-91-0	MW:	137.18	MF:	C8 H11 N O

Anthelmintic

I.C.I., UK

LgP	0.28(C)
pKa	
pp cited in Vols.1-5:	

(1) Arnall et al., U.S. Pat. 3 223 710 (1965).

METYROSINE [U]

CAS	672-87-7	MW:	195.22	MF:	C10 H13 N O3

Antihypertensive

Merck (Demser)

LgP	-1.94(C,zwion.)
pKa	
pp cited in Vols.1-5:	

(1) Stein et al., J. Am. Chem. Soc., 1955, 77, 700.
(2) K. Pfister et al., U.S. Pat. 2 868 818 (1959).
(3) R.N. Brogden et al., Drugs, 1981, 21, 81-89.

MEVASTATIN [INN]

CAS	73573-88-3	MW:	390.52	MF:	C23 H34 O5

Antihyperlipoproteinemic

Sankyo, Japan

LgP	3.52 (C*)
pKa	
pp cited in Vols.1-5:	

(1) A. Endo et al., U.S. Pat. 3 983 140 (1976).
(2) A.G. Brown et al., J. Chem. Soc. Perkin Trans. I, 1976, 1165.
(3) M.S. Brown et al., J. Biol. Chem., 1978, 253, 1121.

MEXAFYLLINE [INN]

CAS	80294-25-3	MW:	274.33	MF:	C14 H18 N4 O2

LgP	2.33(C)
pKa	
pp cited in Vols.1-5:	

MEXAZOLAM [INN]

CAS	31868-18-5	MW:	363.25	MF:	C18 H16 Cl2 N2 O2

Anti-anxiety

Sankyo, Japan (Melex)

LgP	4.4(C,6)
pKa	
pp cited in Vols.1-5:	

(1) RTECS, 1985-86, entry 54284.

MEXENONE [INN]

CAS	1641-17-4	MW:	242.28	MF:	C15 H14 O3

Sunscreen

Am. Cyanamid

LgP	4.37(C)
pKa	
pp cited in Vols.1-5:	

(1) Hardy et al., U.S. Pat. 2 773 903 (1956).

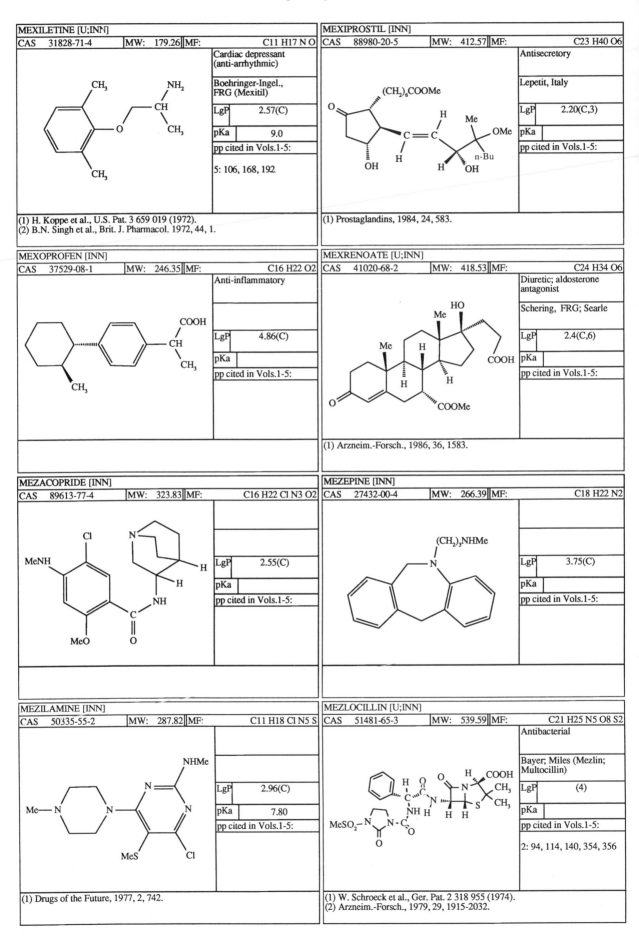

MEXILETINE [U;INN]

| CAS | 31828-71-4 | MW: | 179.26 | MF: | C11 H17 N O |

Cardiac depressant (anti-arrhythmic)

Boehringer-Ingel., FRG (Mexitil)

| LgP | 2.57(C) |
| pKa | 9.0 |
| pp cited in Vols.1-5: |

5: 106, 168, 192

(1) H. Koppe et al., U.S. Pat. 3 659 019 (1972).
(2) B.N. Singh et al., Brit. J. Pharmacol. 1972, 44, 1.

MEXIPROSTIL [INN]

| CAS | 88980-20-5 | MW: | 412.57 | MF: | C23 H40 O6 |

Antisecretory

Lepetit, Italy

| LgP | 2.20(C,3) |
| pKa | |
| pp cited in Vols.1-5: |

(1) Prostaglandins, 1984, 24, 583.

MEXOPROFEN [INN]

| CAS | 37529-08-1 | MW: | 246.35 | MF: | C16 H22 O2 |

Anti-inflammatory

| LgP | 4.86(C) |
| pKa | |
| pp cited in Vols.1-5: |

MEXRENOATE [U;INN]

| CAS | 41020-68-2 | MW: | 418.53 | MF: | C24 H34 O6 |

Diuretic; aldosterone antagonist

Schering, FRG; Searle

| LgP | 2.4(C,6) |
| pKa | |
| pp cited in Vols.1-5: |

(1) Arzneim.-Forsch., 1986, 36, 1583.

MEZACOPRIDE [INN]

| CAS | 89613-77-4 | MW: | 323.83 | MF: | C16 H22 Cl N3 O2 |

| LgP | 2.55(C) |
| pKa | |
| pp cited in Vols.1-5: |

(1) Drugs of the Future, 1977, 2, 742.

MEZEPINE [INN]

| CAS | 27432-00-4 | MW: | 266.39 | MF: | C18 H22 N2 |

| LgP | 3.75(C) |
| pKa | |
| pp cited in Vols.1-5: |

MEZILAMINE [INN]

| CAS | 50335-55-2 | MW: | 287.82 | MF: | C11 H18 Cl N5 S |

| LgP | 2.96(C) |
| pKa | 7.80 |
| pp cited in Vols.1-5: |

MEZLOCILLIN [U;INN]

| CAS | 51481-65-3 | MW: | 539.59 | MF: | C21 H25 N5 O8 S2 |

Antibacterial

Bayer; Miles (Mezlin; Multocillin)

| LgP | (4) |
| pKa | |
| pp cited in Vols.1-5: |

2: 94, 114, 140, 354, 356

(1) W. Schroeck et al., Ger. Pat. 2 318 955 (1974).
(2) Arzneim.-Forsch., 1979, 29, 1915-2032.

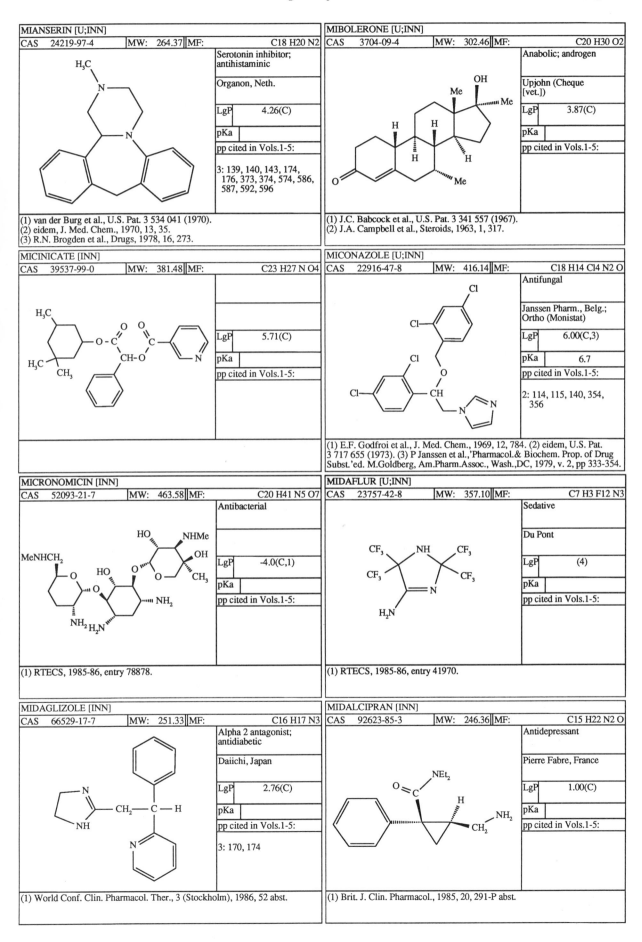

MIANSERIN [U;INN]

CAS 24219-97-4	MW: 264.37	MF:	C18 H20 N2

Serotonin inhibitor; antihistaminic

Organon, Neth.

LgP 4.26(C)

pKa

pp cited in Vols.1-5:

3: 139, 140, 143, 174, 176, 373, 374, 574, 586, 587, 592, 596

(1) van der Burg et al., U.S. Pat. 3 534 041 (1970).
(2) eidem, J. Med. Chem., 1970, 13, 35.
(3) R.N. Brogden et al., Drugs, 1978, 16, 273.

MIBOLERONE [U;INN]

CAS 3704-09-4	MW: 302.46	MF:	C20 H30 O2

Anabolic; androgen

Upjohn (Cheque [vet.])

LgP 3.87(C)

pKa

pp cited in Vols.1-5:

(1) J.C. Babcock et al., U.S. Pat. 3 341 557 (1967).
(2) J.A. Campbell et al., Steroids, 1963, 1, 317.

MICINICATE [INN]

CAS 39537-99-0	MW: 381.48	MF:	C23 H27 N O4

LgP 5.71(C)

pKa

pp cited in Vols.1-5:

MICONAZOLE [U;INN]

CAS 22916-47-8	MW: 416.14	MF:	C18 H14 Cl4 N2 O

Antifungal

Janssen Pharm., Belg.; Ortho (Monistat)

LgP 6.00(C,3)

pKa 6.7

pp cited in Vols.1-5:

2: 114, 115, 140, 354, 356

(1) E.F. Godfroi et al., J. Med. Chem., 1969, 12, 784. (2) eidem, U.S. Pat. 3 717 655 (1973). (3) P Janssen et al.,'Pharmacol.& Biochem. Prop. of Drug Subst.'ed. M.Goldberg, Am.Pharm.Assoc., Wash.,DC, 1979, v. 2, pp 333-354.

MICRONOMICIN [INN]

CAS 52093-21-7	MW: 463.58	MF:	C20 H41 N5 O7

Antibacterial

LgP -4.0(C,1)

pKa

pp cited in Vols.1-5:

(1) RTECS, 1985-86, entry 78878.

MIDAFLUR [U;INN]

CAS 23757-42-8	MW: 357.10	MF:	C7 H3 F12 N3

Sedative

Du Pont

LgP (4)

pKa

pp cited in Vols.1-5:

(1) RTECS, 1985-86, entry 41970.

MIDAGLIZOLE [INN]

CAS 66529-17-7	MW: 251.33	MF:	C16 H17 N3

Alpha 2 antagonist; antidiabetic

Daiichi, Japan

LgP 2.76(C)

pKa

pp cited in Vols.1-5:

3: 170, 174

(1) World Conf. Clin. Pharmacol. Ther., 3 (Stockholm), 1986, 52 abst.

MIDALCIPRAN [INN]

CAS 92623-85-3	MW: 246.36	MF:	C15 H22 N2 O

Antidepressant

Pierre Fabre, France

LgP 1.00(C)

pKa

pp cited in Vols.1-5:

(1) Brit. J. Clin. Pharmacol., 1985, 20, 291-P abst.

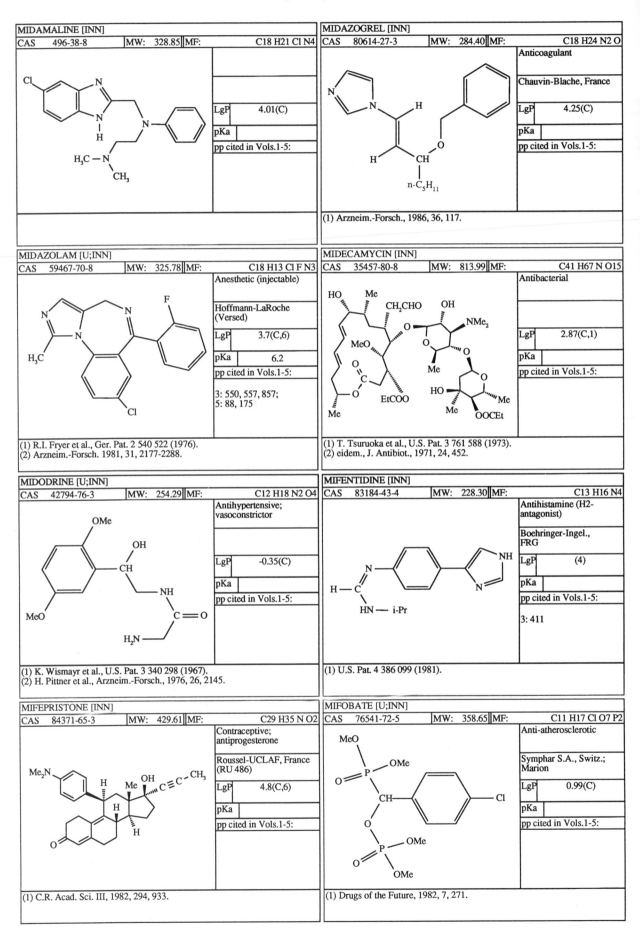

MIDAMALINE [INN]			
CAS 496-38-8	MW: 328.85	MF:	C18 H21 Cl N4

LgP	4.01(C)
pKa	
pp cited in Vols.1-5:	

MIDAZOGREL [INN]			
CAS 80614-27-3	MW: 284.40	MF:	C18 H24 N2 O

Anticoagulant

Chauvin-Blache, France

LgP	4.25(C)
pKa	
pp cited in Vols.1-5:	

(1) Arzneim.-Forsch., 1986, 36, 117.

MIDAZOLAM [U;INN]			
CAS 59467-70-8	MW: 325.78	MF:	C18 H13 Cl F N3

Anesthetic (injectable)

Hoffmann-LaRoche (Versed)

LgP	3.7(C,6)
pKa	6.2
pp cited in Vols.1-5:	

3: 550, 557, 857;
5: 88, 175

(1) R.I. Fryer et al., Ger. Pat. 2 540 522 (1976).
(2) Arzneim.-Forsch. 1981, 31, 2177-2288.

MIDECAMYCIN [INN]			
CAS 35457-80-8	MW: 813.99	MF:	C41 H67 N O15

Antibacterial

LgP	2.87(C,1)
pKa	
pp cited in Vols.1-5:	

(1) T. Tsuruoka et al., U.S. Pat. 3 761 588 (1973).
(2) eidem., J. Antibiot., 1971, 24, 452.

MIDODRINE [U;INN]			
CAS 42794-76-3	MW: 254.29	MF:	C12 H18 N2 O4

Antihypertensive; vasoconstrictor

LgP	-0.35(C)
pKa	
pp cited in Vols.1-5:	

(1) K. Wismayr et al., U.S. Pat. 3 340 298 (1967).
(2) H. Pittner et al., Arzneim.-Forsch., 1976, 26, 2145.

MIFENTIDINE [INN]			
CAS 83184-43-4	MW: 228.30	MF:	C13 H16 N4

Antihistamine (H2-antagonist)

Boehringer-Ingel., FRG

LgP	(4)
pKa	
pp cited in Vols.1-5:	

3: 411

(1) U.S. Pat. 4 386 099 (1981).

MIFEPRISTONE [INN]			
CAS 84371-65-3	MW: 429.61	MF:	C29 H35 N O2

Contraceptive; antiprogesterone

Roussel-UCLAF, France (RU 486)

LgP	4.8(C,6)
pKa	
pp cited in Vols.1-5:	

(1) C.R. Acad. Sci. III, 1982, 294, 933.

MIFOBATE [U;INN]			
CAS 76541-72-5	MW: 358.65	MF:	C11 H17 Cl O7 P2

Anti-atherosclerotic

Symphar S.A., Switz.; Marion

LgP	0.99(C)
pKa	
pp cited in Vols.1-5:	

(1) Drugs of the Future, 1982, 7, 271.

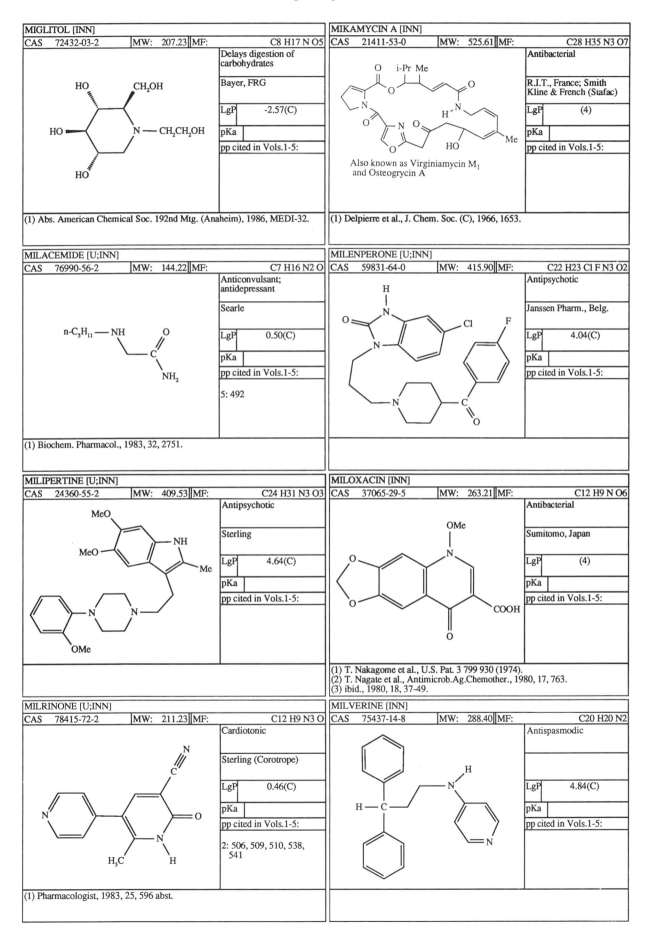

MIGLITOL [INN]
CAS 72432-03-2 | MW: 207.23 | MF: C8 H17 N O5
Delays digestion of carbohydrates
Bayer, FRG
LgP -2.57(C)
pKa
pp cited in Vols.1-5:
(1) Abs. American Chemical Soc. 192nd Mtg. (Anaheim), 1986, MEDI-32.

MIKAMYCIN A [INN]
CAS 21411-53-0 | MW: 525.61 | MF: C28 H35 N3 O7
Antibacterial
R.I.T., France; Smith Kline & French (Stafac)
LgP (4)
pKa
pp cited in Vols.1-5:
Also known as Virginiamycin M₁ and Osteogrycin A
(1) Delpierre et al., J. Chem. Soc. (C), 1966, 1653.

MILACEMIDE [U;INN]
CAS 76990-56-2 | MW: 144.22 | MF: C7 H16 N2 O
Anticonvulsant; antidepressant
Searle
LgP 0.50(C)
pKa
pp cited in Vols.1-5: 5: 492
(1) Biochem. Pharmacol., 1983, 32, 2751.

MILENPERONE [U;INN]
CAS 59831-64-0 | MW: 415.90 | MF: C22 H23 Cl F N3 O2
Antipsychotic
Janssen Pharm., Belg.
LgP 4.04(C)
pKa
pp cited in Vols.1-5:

MILIPERTINE [U;INN]
CAS 24360-55-2 | MW: 409.53 | MF: C24 H31 N3 O3
Antipsychotic
Sterling
LgP 4.64(C)
pKa
pp cited in Vols.1-5:

MILOXACIN [INN]
CAS 37065-29-5 | MW: 263.21 | MF: C12 H9 N O6
Antibacterial
Sumitomo, Japan
LgP (4)
pKa
pp cited in Vols.1-5:
(1) T. Nakagome et al., U.S. Pat. 3 799 930 (1974).
(2) T. Nagate et al., Antimicrob.Ag.Chemother., 1980, 17, 763.
(3) ibid., 1980, 18, 37-49.

MILRINONE [U;INN]
CAS 78415-72-2 | MW: 211.23 | MF: C12 H9 N3 O
Cardiotonic
Sterling (Corotrope)
LgP 0.46(C)
pKa
pp cited in Vols.1-5: 2: 506, 509, 510, 538, 541
(1) Pharmacologist, 1983, 25, 596 abst.

MILVERINE [INN]
CAS 75437-14-8 | MW: 288.40 | MF: C20 H20 N2
Antispasmodic
LgP 4.84(C)
pKa
pp cited in Vols.1-5:

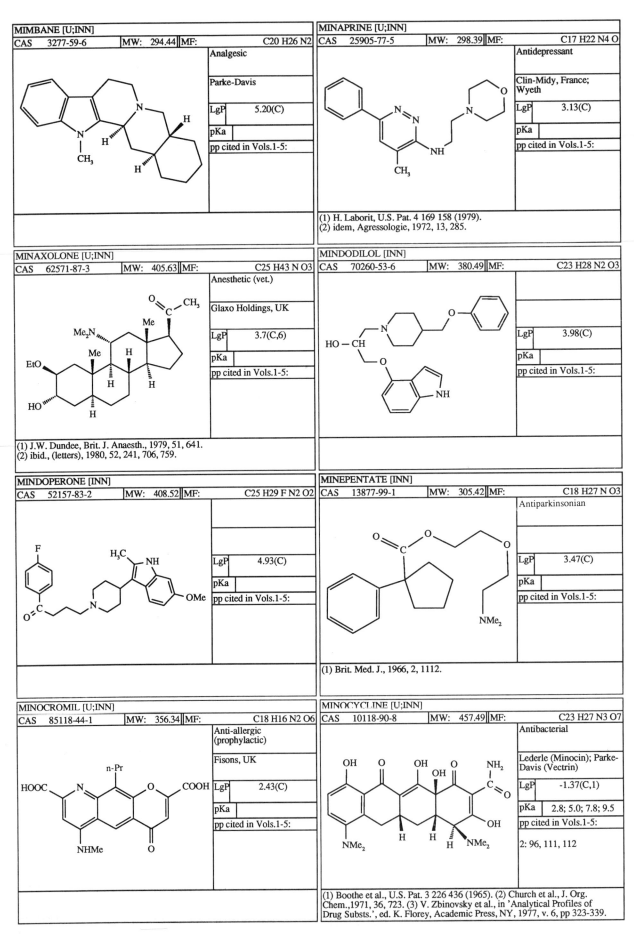

MIMBANE [U;INN]

CAS 3277-59-6	MW: 294.44	MF:	C20 H26 N2

Analgesic

Parke-Davis

LgP	5.20(C)
pKa	
pp cited in Vols.1-5:	

MINAPRINE [U;INN]

CAS 25905-77-5	MW: 298.39	MF:	C17 H22 N4 O

Antidepressant

Clin-Midy, France; Wyeth

LgP	3.13(C)
pKa	
pp cited in Vols.1-5:	

(1) H. Laborit, U.S. Pat. 4 169 158 (1979).
(2) idem, Agressologie, 1972, 13, 285.

MINAXOLONE [U;INN]

CAS 62571-87-3	MW: 405.63	MF:	C25 H43 N O3

Anesthetic (vet.)

Glaxo Holdings, UK

LgP	3.7(C,6)
pKa	
pp cited in Vols.1-5:	

(1) J.W. Dundee, Brit. J. Anaesth., 1979, 51, 641.
(2) ibid., (letters), 1980, 52, 241, 706, 759.

MINDODILOL [INN]

CAS 70260-53-6	MW: 380.49	MF:	C23 H28 N2 O3

LgP	3.98(C)
pKa	
pp cited in Vols.1-5:	

MINDOPERONE [INN]

CAS 52157-83-2	MW: 408.52	MF:	C25 H29 F N2 O2

LgP	4.93(C)
pKa	
pp cited in Vols.1-5:	

MINEPENTATE [INN]

CAS 13877-99-1	MW: 305.42	MF:	C18 H27 N O3

Antiparkinsonian

LgP	3.47(C)
pKa	
pp cited in Vols.1-5:	

(1) Brit. Med. J., 1966, 2, 1112.

MINOCROMIL [U;INN]

CAS 85118-44-1	MW: 356.34	MF:	C18 H16 N2 O6

Anti-allergic (prophylactic)

Fisons, UK

LgP	2.43(C)
pKa	
pp cited in Vols.1-5:	

MINOCYCLINE [U;INN]

CAS 10118-90-8	MW: 457.49	MF:	C23 H27 N3 O7

Antibacterial

Lederle (Minocin); Parke-Davis (Vectrin)

LgP	-1.37(C,1)
pKa	2.8; 5.0; 7.8; 9.5
pp cited in Vols.1-5:	

2: 96, 111, 112

(1) Boothe et al., U.S. Pat. 3 226 436 (1965). (2) Church et al., J. Org. Chem.,1971, 36, 723. (3) V. Zbinovsky et al., in 'Analytical Profiles of Drug Substs.', ed. K. Florey, Academic Press, NY, 1977, v. 6, pp 323-339.

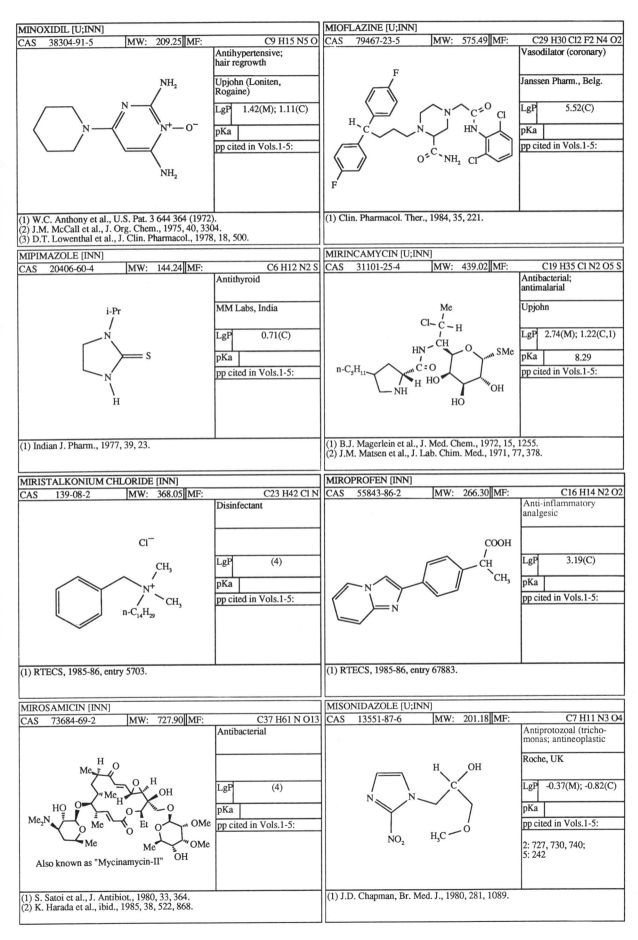

MINOXIDIL [U;INN]

CAS	38304-91-5	MW:	209.25	MF:	C9 H15 N5 O

Antihypertensive; hair regrowth

Upjohn (Loniten, Rogaine)

LgP	1.42(M); 1.11(C)
pKa	
pp cited in Vols.1-5:	

(1) W.C. Anthony et al., U.S. Pat. 3 644 364 (1972).
(2) J.M. McCall et al., J. Org. Chem., 1975, 40, 3304.
(3) D.T. Lowenthal et al., J. Clin. Pharmacol., 1978, 18, 500.

MIOFLAZINE [U;INN]

CAS	79467-23-5	MW:	575.49	MF:	C29 H30 Cl2 F2 N4 O2

Vasodilator (coronary)

Janssen Pharm., Belg.

LgP	5.52(C)
pKa	
pp cited in Vols.1-5:	

(1) Clin. Pharmacol. Ther., 1984, 35, 221.

MIPIMAZOLE [INN]

CAS	20406-60-4	MW:	144.24	MF:	C6 H12 N2 S

Antithyroid

MM Labs, India

LgP	0.71(C)
pKa	
pp cited in Vols.1-5:	

(1) Indian J. Pharm., 1977, 39, 23.

MIRINCAMYCIN [U;INN]

CAS	31101-25-4	MW:	439.02	MF:	C19 H35 Cl N2 O5 S

Antibacterial; antimalarial

Upjohn

LgP	2.74(M); 1.22(C,1)
pKa	8.29
pp cited in Vols.1-5:	

(1) B.J. Magerlein et al., J. Med. Chem., 1972, 15, 1255.
(2) J.M. Matsen et al., J. Lab. Chim. Med., 1971, 77, 378.

MIRISTALKONIUM CHLORIDE [INN]

CAS	139-08-2	MW:	368.05	MF:	C23 H42 Cl N

Disinfectant

LgP	(4)
pKa	
pp cited in Vols.1-5:	

(1) RTECS, 1985-86, entry 5703.

MIROPROFEN [INN]

CAS	55843-86-2	MW:	266.30	MF:	C16 H14 N2 O2

Anti-inflammatory analgesic

LgP	3.19(C)
pKa	
pp cited in Vols.1-5:	

(1) RTECS, 1985-86, entry 67883.

MIROSAMICIN [INN]

CAS	73684-69-2	MW:	727.90	MF:	C37 H61 N O13

Antibacterial

LgP	(4)
pKa	
pp cited in Vols.1-5:	

Also known as "Mycinamycin-II"

(1) S. Satoi et al., J. Antibiot., 1980, 33, 364.
(2) K. Harada et al., ibid., 1985, 38, 522, 868.

MISONIDAZOLE [U;INN]

CAS	13551-87-6	MW:	201.18	MF:	C7 H11 N3 O4

Antiprotozoal (trichomonas; antineoplastic

Roche, UK

LgP	-0.37(M); -0.82(C)
pKa	
pp cited in Vols.1-5:	
	2: 727, 730, 740; 5: 242

(1) J.D. Chapman, Br. Med. J., 1980, 281, 1089.

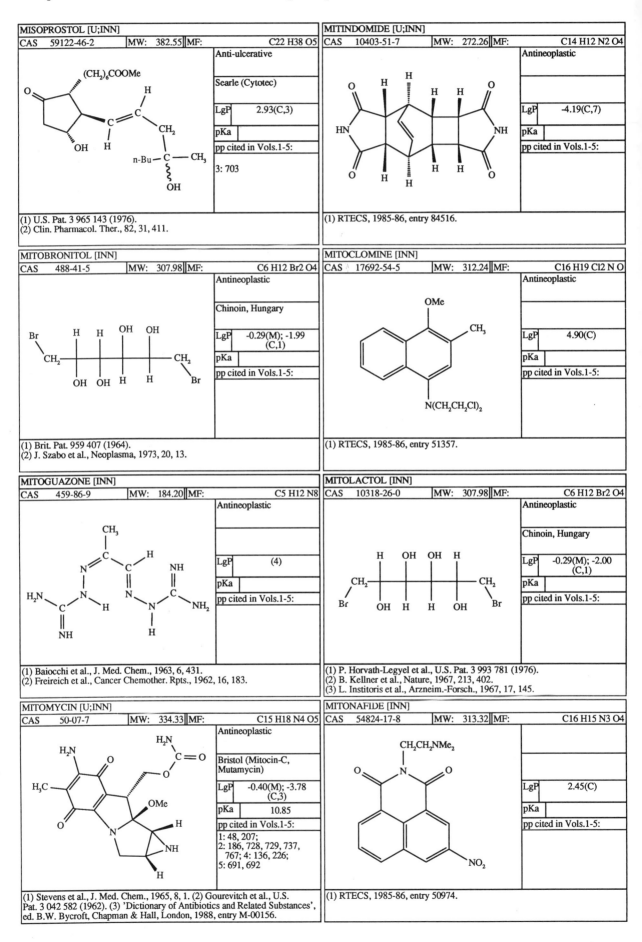

MISOPROSTOL [U;INN]

CAS 59122-46-2	MW: 382.55	MF:	C22 H38 O5

Anti-ulcerative

Searle (Cytotec)

LgP	2.93(C,3)
pKa	
pp cited in Vols.1-5:	

3: 703

(1) U.S. Pat. 3 965 143 (1976).
(2) Clin. Pharmacol. Ther., 82, 31, 411.

MITINDOMIDE [U;INN]

CAS 10403-51-7	MW: 272.26	MF:	C14 H12 N2 O4

Antineoplastic

LgP	-4.19(C,7)
pKa	
pp cited in Vols.1-5:	

(1) RTECS, 1985-86, entry 84516.

MITOBRONITOL [INN]

CAS 488-41-5	MW: 307.98	MF:	C6 H12 Br2 O4

Antineoplastic

Chinoin, Hungary

LgP	-0.29(M); -1.99 (C,1)
pKa	
pp cited in Vols.1-5:	

(1) Brit. Pat. 959 407 (1964).
(2) J. Szabo et al., Neoplasma, 1973, 20, 13.

MITOCLOMINE [INN]

CAS 17692-54-5	MW: 312.24	MF:	C16 H19 Cl2 N O

Antineoplastic

LgP	4.90(C)
pKa	
pp cited in Vols.1-5:	

(1) RTECS, 1985-86, entry 51357.

MITOGUAZONE [INN]

CAS 459-86-9	MW: 184.20	MF:	C5 H12 N8

Antineoplastic

LgP	(4)
pKa	
pp cited in Vols.1-5:	

(1) Baiocchi et al., J. Med. Chem., 1963, 6, 431.
(2) Freireich et al., Cancer Chemother. Rpts., 1962, 16, 183.

MITOLACTOL [INN]

CAS 10318-26-0	MW: 307.98	MF:	C6 H12 Br2 O4

Antineoplastic

Chinoin, Hungary

LgP	-0.29(M); -2.00 (C,1)
pKa	
pp cited in Vols.1-5:	

(1) P. Horvath-Legyel et al., U.S. Pat. 3 993 781 (1976).
(2) B. Kellner et al., Nature, 1967, 213, 402.
(3) L. Institoris et al., Arzneim.-Forsch., 1967, 17, 145.

MITOMYCIN [U;INN]

CAS 50-07-7	MW: 334.33	MF:	C15 H18 N4 O5

Antineoplastic

Bristol (Mitocin-C, Mutamycin)

LgP	-0.40(M); -3.78 (C,3)
pKa	10.85
pp cited in Vols.1-5:	

1: 48, 207;
2: 186, 728, 729, 737, 767; 4: 136, 226;
5: 691, 692

(1) Stevens et al., J. Med. Chem., 1965, 8, 1. (2) Gourevitch et al., U.S. Pat. 3 042 582 (1962). (3) 'Dictionary of Antibiotics and Related Substances', ed. B.W. Bycroft, Chapman & Hall, London, 1988, entry M-00156.

MITONAFIDE [INN]

CAS 54824-17-8	MW: 313.32	MF:	C16 H15 N3 O4

LgP	2.45(C)
pKa	
pp cited in Vols.1-5:	

(1) RTECS, 1985-86, entry 50974.

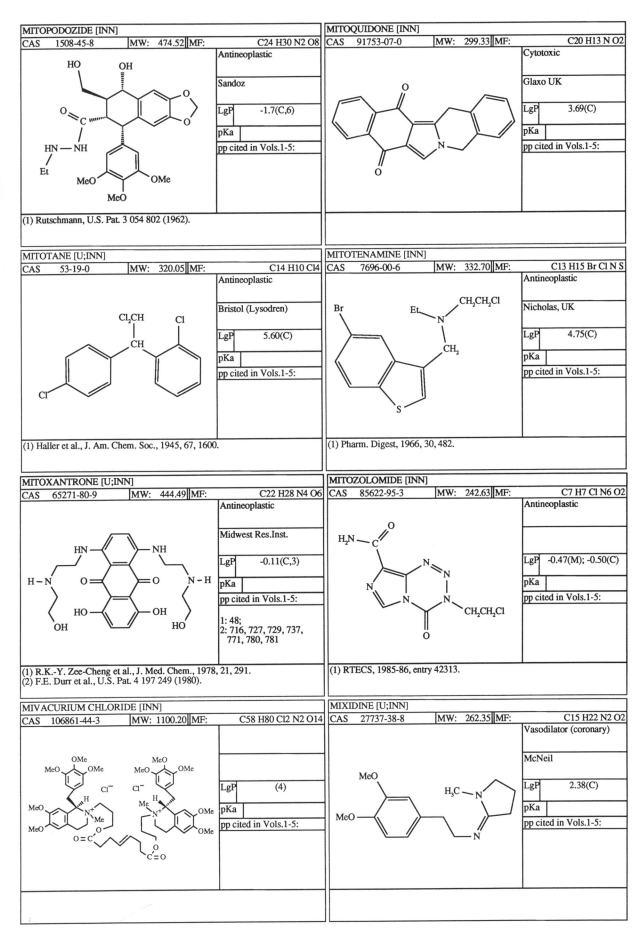

MITOPODOZIDE [INN]

CAS	1508-45-8	MW:	474.52	MF:	C24 H30 N2 O8

Antineoplastic

Sandoz

LgP	-1.7(C,6)
pKa	
pp cited in Vols.1-5:	

(1) Rutschmann, U.S. Pat. 3 054 802 (1962).

MITOQUIDONE [INN]

CAS	91753-07-0	MW:	299.33	MF:	C20 H13 N O2

Cytotoxic

Glaxo UK

LgP	3.69(C)
pKa	
pp cited in Vols.1-5:	

MITOTANE [U;INN]

CAS	53-19-0	MW:	320.05	MF:	C14 H10 Cl4

Antineoplastic

Bristol (Lysodren)

LgP	5.60(C)
pKa	
pp cited in Vols.1-5:	

(1) Haller et al., J. Am. Chem. Soc., 1945, 67, 1600.

MITOTENAMINE [INN]

CAS	7696-00-6	MW:	332.70	MF:	C13 H15 Br Cl N S

Antineoplastic

Nicholas, UK

LgP	4.75(C)
pKa	
pp cited in Vols.1-5:	

(1) Pharm. Digest, 1966, 30, 482.

MITOXANTRONE [U;INN]

CAS	65271-80-9	MW:	444.49	MF:	C22 H28 N4 O6

Antineoplastic

Midwest Res.Inst.

LgP	-0.11(C,3)
pKa	
pp cited in Vols.1-5:	

1: 48;
2: 716, 727, 729, 737, 771, 780, 781

(1) R.K.-Y. Zee-Cheng et al., J. Med. Chem., 1978, 21, 291.
(2) F.E. Durr et al., U.S. Pat. 4 197 249 (1980).

MITOZOLOMIDE [INN]

CAS	85622-95-3	MW:	242.63	MF:	C7 H7 Cl N6 O2

Antineoplastic

LgP	-0.47(M); -0.50(C)
pKa	
pp cited in Vols.1-5:	

(1) RTECS, 1985-86, entry 42313.

MIVACURIUM CHLORIDE [INN]

CAS	106861-44-3	MW:	1100.20	MF:	C58 H80 Cl2 N2 O14

LgP	(4)
pKa	
pp cited in Vols.1-5:	

MIXIDINE [U;INN]

CAS	27737-38-8	MW:	262.35	MF:	C15 H22 N2 O2

Vasodilator (coronary)

McNeil

LgP	2.38(C)
pKa	
pp cited in Vols.1-5:	

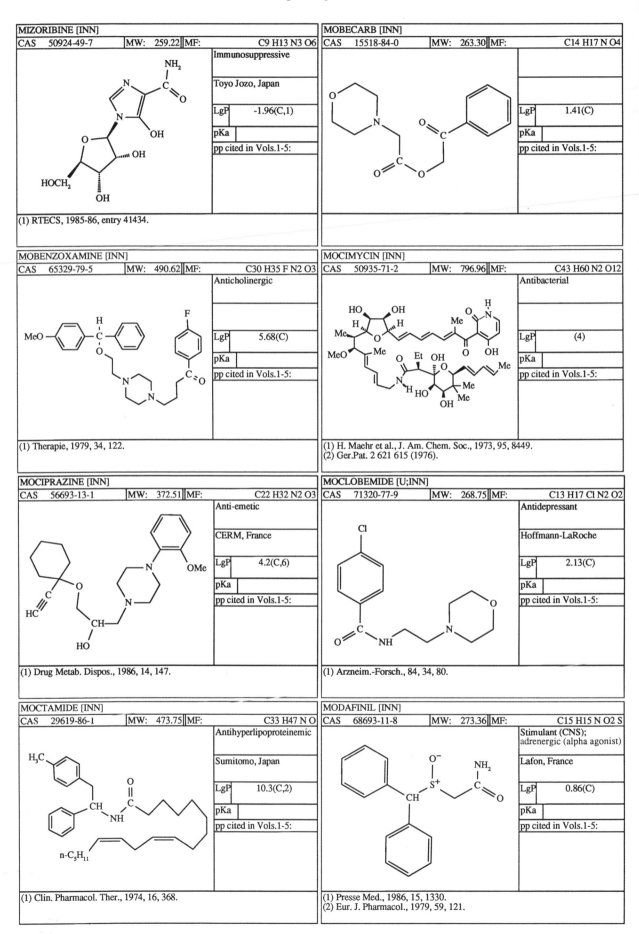

MIZORIBINE [INN]		
CAS 50924-49-7	MW: 259.22	MF: C9 H13 N3 O6

Immunosuppressive

Toyo Jozo, Japan

LgP	-1.96(C,1)
pKa	
pp cited in Vols.1-5:	

(1) RTECS, 1985-86, entry 41434.

MOBECARB [INN]		
CAS 15518-84-0	MW: 263.30	MF: C14 H17 N O4

LgP	1.41(C)
pKa	
pp cited in Vols.1-5:	

MOBENZOXAMINE [INN]		
CAS 65329-79-5	MW: 490.62	MF: C30 H35 F N2 O3

Anticholinergic

LgP	5.68(C)
pKa	
pp cited in Vols.1-5:	

(1) Therapie, 1979, 34, 122.

MOCIMYCIN [INN]		
CAS 50935-71-2	MW: 796.96	MF: C43 H60 N2 O12

Antibacterial

LgP	(4)
pKa	
pp cited in Vols.1-5:	

(1) H. Maehr et al., J. Am. Chem. Soc., 1973, 95, 8449.
(2) Ger.Pat. 2 621 615 (1976).

MOCIPRAZINE [INN]		
CAS 56693-13-1	MW: 372.51	MF: C22 H32 N2 O3

Anti-emetic

CERM, France

LgP	4.2(C,6)
pKa	
pp cited in Vols.1-5:	

(1) Drug Metab. Dispos., 1986, 14, 147.

MOCLOBEMIDE [U;INN]		
CAS 71320-77-9	MW: 268.75	MF: C13 H17 Cl N2 O2

Antidepressant

Hoffmann-LaRoche

LgP	2.13(C)
pKa	
pp cited in Vols.1-5:	

(1) Arzneim.-Forsch., 84, 34, 80.

MOCTAMIDE [INN]		
CAS 29619-86-1	MW: 473.75	MF: C33 H47 N O

Antihyperlipoproteinemic

Sumitomo, Japan

LgP	10.3(C,2)
pKa	
pp cited in Vols.1-5:	

(1) Clin. Pharmacol. Ther., 1974, 16, 368.

MODAFINIL [INN]		
CAS 68693-11-8	MW: 273.36	MF: C15 H15 N O2 S

Stimulant (CNS);
adrenergic (alpha agonist)

Lafon, France

LgP	0.86(C)
pKa	
pp cited in Vols.1-5:	

(1) Presse Med., 1986, 15, 1330.
(2) Eur. J. Pharmacol., 1979, 59, 121.

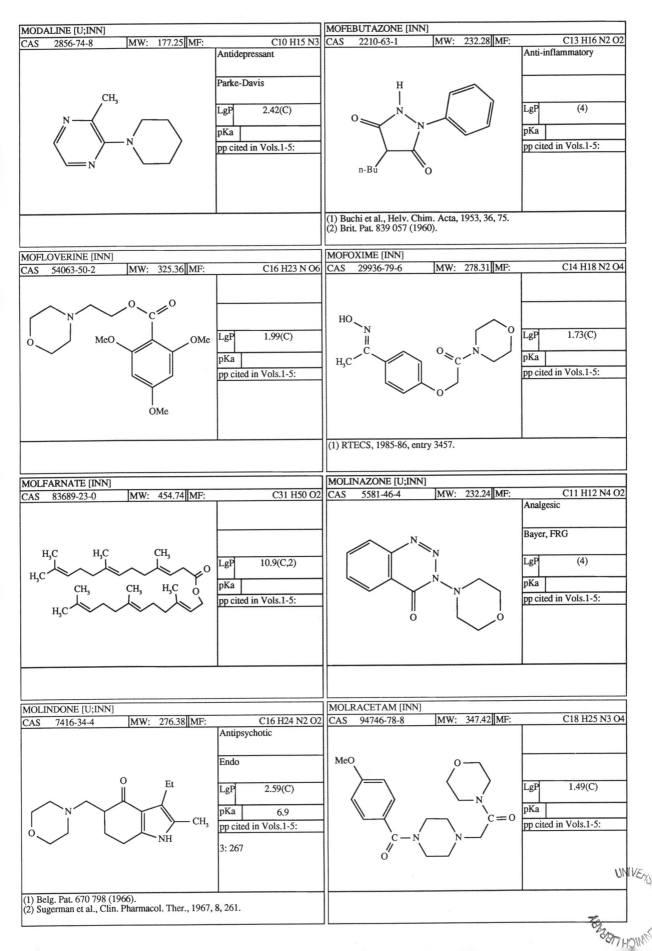

MODALINE [U;INN]

CAS	2856-74-8	MW:	177.25	MF:	C10 H15 N3

Antidepressant

Parke-Davis

LgP	2.42(C)
pKa	
pp cited in Vols.1-5:	

MOFEBUTAZONE [INN]

CAS	2210-63-1	MW:	232.28	MF:	C13 H16 N2 O2

Anti-inflammatory

LgP	(4)
pKa	
pp cited in Vols.1-5:	

(1) Buchi et al., Helv. Chim. Acta, 1953, 36, 75.
(2) Brit. Pat. 839 057 (1960).

MOFLOVERINE [INN]

CAS	54063-50-2	MW:	325.36	MF:	C16 H23 N O6

LgP	1.99(C)
pKa	
pp cited in Vols.1-5:	

MOFOXIME [INN]

CAS	29936-79-6	MW:	278.31	MF:	C14 H18 N2 O4

LgP	1.73(C)
pKa	
pp cited in Vols.1-5:	

(1) RTECS, 1985-86, entry 3457.

MOLFARNATE [INN]

CAS	83689-23-0	MW:	454.74	MF:	C31 H50 O2

LgP	10.9(C,2)
pKa	
pp cited in Vols.1-5:	

MOLINAZONE [U;INN]

CAS	5581-46-4	MW:	232.24	MF:	C11 H12 N4 O2

Analgesic

Bayer, FRG

LgP	(4)
pKa	
pp cited in Vols.1-5:	

MOLINDONE [U;INN]

CAS	7416-34-4	MW:	276.38	MF:	C16 H24 N2 O2

Antipsychotic

Endo

LgP	2.59(C)
pKa	6.9
pp cited in Vols.1-5:	
3: 267	

(1) Belg. Pat. 670 798 (1966).
(2) Sugerman et al., Clin. Pharmacol. Ther., 1967, 8, 261.

MOLRACETAM [INN]

CAS	94746-78-8	MW:	347.42	MF:	C18 H25 N3 O4

LgP	1.49(C)
pKa	
pp cited in Vols.1-5:	

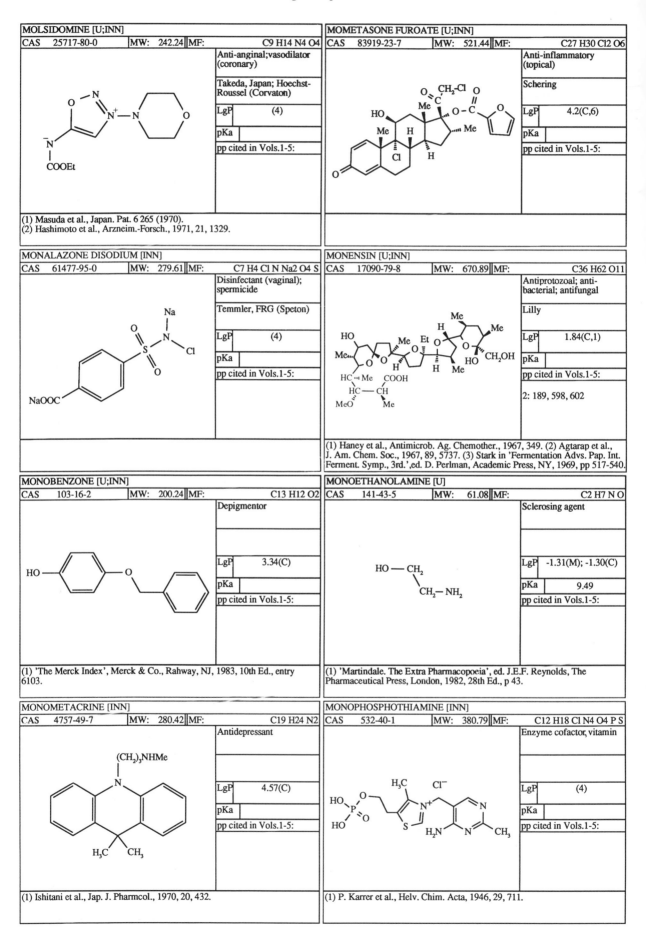

MOLSIDOMINE [U;INN]			
CAS 25717-80-0	MW: 242.24	MF:	C9 H14 N4 O4

Anti-anginal;vasodilator (coronary)

Takeda, Japan; Hoechst-Roussel (Corvaton)

LgP	(4)
pKa	
pp cited in Vols.1-5:	

(1) Masuda et al., Japan. Pat. 6 265 (1970).
(2) Hashimoto et al., Arzneim.-Forsch., 1971, 21, 1329.

MOMETASONE FUROATE [U;INN]			
CAS 83919-23-7	MW: 521.44	MF:	C27 H30 Cl2 O6

Anti-inflammatory (topical)

Schering

LgP	4.2(C,6)
pKa	
pp cited in Vols.1-5:	

MONALAZONE DISODIUM [INN]			
CAS 61477-95-0	MW: 279.61	MF:	C7 H4 Cl N Na2 O4 S

Disinfectant (vaginal); spermicide

Temmler, FRG (Speton)

LgP	(4)
pKa	
pp cited in Vols.1-5:	

MONENSIN [U;INN]			
CAS 17090-79-8	MW: 670.89	MF:	C36 H62 O11

Antiprotozoal; anti-bacterial; antifungal

Lilly

LgP	1.84(C,1)
pKa	
pp cited in Vols.1-5:	
	2: 189, 598, 602

(1) Haney et al., Antimicrob. Ag. Chemother., 1967, 349. (2) Agtarap et al., J. Am. Chem. Soc., 1967, 89, 5737. (3) Stark in 'Fermentation Advs. Pap. Int. Ferment. Symp., 3rd.',ed. D. Perlman, Academic Press, NY, 1969, pp 517-540.

MONOBENZONE [U;INN]			
CAS 103-16-2	MW: 200.24	MF:	C13 H12 O2

Depigmentor

LgP	3.34(C)
pKa	
pp cited in Vols.1-5:	

(1) 'The Merck Index', Merck & Co., Rahway, NJ, 1983, 10th Ed., entry 6103.

MONOETHANOLAMINE [U]			
CAS 141-43-5	MW: 61.08	MF:	C2 H7 N O

Sclerosing agent

LgP	-1.31(M); -1.30(C)
pKa	9.49
pp cited in Vols.1-5:	

(1) 'Martindale. The Extra Pharmacopoeia', ed. J.E.F. Reynolds, The Pharmaceutical Press, London, 1982, 28th Ed., p 43.

MONOMETACRINE [INN]			
CAS 4757-49-7	MW: 280.42	MF:	C19 H24 N2

Antidepressant

LgP	4.57(C)
pKa	
pp cited in Vols.1-5:	

(1) Ishitani et al., Jap. J. Pharmcol., 1970, 20, 432.

MONOPHOSPHOTHIAMINE [INN]			
CAS 532-40-1	MW: 380.79	MF:	C12 H18 Cl N4 O4 P S

Enzyme cofactor, vitamin

LgP	(4)
pKa	
pp cited in Vols.1-5:	

(1) P. Karrer et al., Helv. Chim. Acta, 1946, 29, 711.

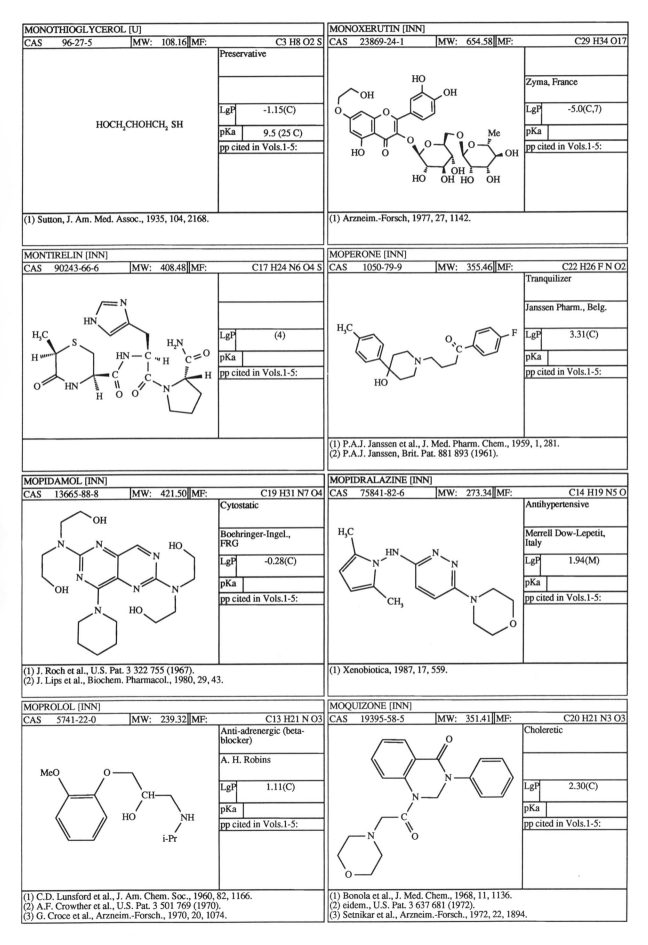

MONOTHIOGLYCEROL [U]

CAS	96-27-5	MW:	108.16	MF:	C3 H8 O2 S

Preservative

HOCH₂CHOHCH₂ SH

LgP	-1.15(C)
pKa	9.5 (25 C)
pp cited in Vols.1-5:	

(1) Sutton, J. Am. Med. Assoc., 1935, 104, 2168.

MONOXERUTIN [INN]

CAS	23869-24-1	MW:	654.58	MF:	C29 H34 O17

Zyma, France

LgP	-5.0(C,7)
pKa	
pp cited in Vols.1-5:	

(1) Arzneim.-Forsch, 1977, 27, 1142.

MONTIRELIN [INN]

CAS	90243-66-6	MW:	408.48	MF:	C17 H24 N6 O4 S

LgP	(4)
pKa	
pp cited in Vols.1-5:	

MOPERONE [INN]

CAS	1050-79-9	MW:	355.46	MF:	C22 H26 F N O2

Tranquilizer

Janssen Pharm., Belg.

LgP	3.31(C)
pKa	
pp cited in Vols.1-5:	

(1) P.A.J. Janssen et al., J. Med. Pharm. Chem., 1959, 1, 281.
(2) P.A.J. Janssen, Brit. Pat. 881 893 (1961).

MOPIDAMOL [INN]

CAS	13665-88-8	MW:	421.50	MF:	C19 H31 N7 O4

Cytostatic

Boehringer-Ingel., FRG

LgP	-0.28(C)
pKa	
pp cited in Vols.1-5:	

(1) J. Roch et al., U.S. Pat. 3 322 755 (1967).
(2) J. Lips et al., Biochem. Pharmacol., 1980, 29, 43.

MOPIDRALAZINE [INN]

CAS	75841-82-6	MW:	273.34	MF:	C14 H19 N5 O

Antihypertensive

Merrell Dow-Lepetit, Italy

LgP	1.94(M)
pKa	
pp cited in Vols.1-5:	

(1) Xenobiotica, 1987, 17, 559.

MOPROLOL [INN]

CAS	5741-22-0	MW:	239.32	MF:	C13 H21 N O3

Anti-adrenergic (beta-blocker)

A. H. Robins

LgP	1.11(C)
pKa	
pp cited in Vols.1-5:	

(1) C.D. Lunsford et al., J. Am. Chem. Soc., 1960, 82, 1166.
(2) A.F. Crowther et al., U.S. Pat. 3 501 769 (1970).
(3) G. Croce et al., Arzneim.-Forsch., 1970, 20, 1074.

MOQUIZONE [INN]

CAS	19395-58-5	MW:	351.41	MF:	C20 H21 N3 O3

Choleretic

LgP	2.30(C)
pKa	
pp cited in Vols.1-5:	

(1) Bonola et al., J. Med. Chem., 1968, 11, 1136.
(2) eidem., U.S. Pat. 3 637 681 (1972).
(3) Setnikar et al., Arzneim.-Forsch., 1972, 22, 1894.

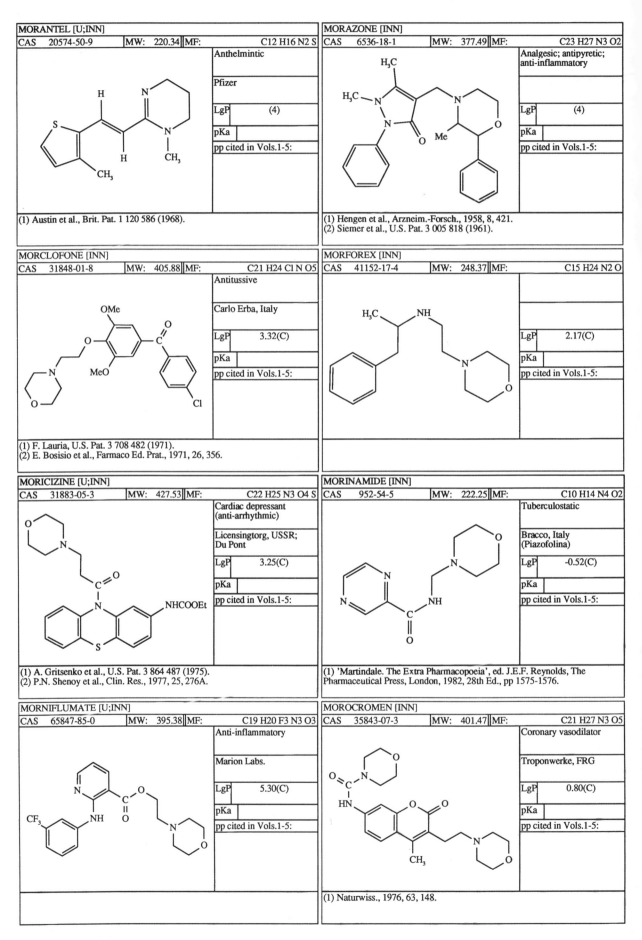

MORANTEL [U;INN]

CAS	20574-50-9	MW:	220.34	MF:	C12 H16 N2 S

Anthelmintic

Pfizer

LgP	(4)
pKa	
pp cited in Vols.1-5:	

(1) Austin et al., Brit. Pat. 1 120 586 (1968).

MORAZONE [INN]

CAS	6536-18-1	MW:	377.49	MF:	C23 H27 N3 O2

Analgesic; antipyretic; anti-inflammatory

LgP	(4)
pKa	
pp cited in Vols.1-5:	

(1) Hengen et al., Arzneim.-Forsch., 1958, 8, 421.
(2) Siemer et al., U.S. Pat. 3 005 818 (1961).

MORCLOFONE [INN]

CAS	31848-01-8	MW:	405.88	MF:	C21 H24 Cl N O5

Antitussive

Carlo Erba, Italy

LgP	3.32(C)
pKa	
pp cited in Vols.1-5:	

(1) F. Lauria, U.S. Pat. 3 708 482 (1971).
(2) E. Bosisio et al., Farmaco Ed. Prat., 1971, 26, 356.

MORFOREX [INN]

CAS	41152-17-4	MW:	248.37	MF:	C15 H24 N2 O

LgP	2.17(C)
pKa	
pp cited in Vols.1-5:	

MORICIZINE [U;INN]

CAS	31883-05-3	MW:	427.53	MF:	C22 H25 N3 O4 S

Cardiac depressant (anti-arrhythmic)

Licensingtorg, USSR; Du Pont

LgP	3.25(C)
pKa	
pp cited in Vols.1-5:	

(1) A. Gritsenko et al., U.S. Pat. 3 864 487 (1975).
(2) P.N. Shenoy et al., Clin. Res., 1977, 25, 276A.

MORINAMIDE [INN]

CAS	952-54-5	MW:	222.25	MF:	C10 H14 N4 O2

Tuberculostatic

Bracco, Italy (Piazofolina)

LgP	-0.52(C)
pKa	
pp cited in Vols.1-5:	

(1) 'Martindale. The Extra Pharmacopoeia', ed. J.E.F. Reynolds, The Pharmaceutical Press, London, 1982, 28th Ed., pp 1575-1576.

MORNIFLUMATE [U;INN]

CAS	65847-85-0	MW:	395.38	MF:	C19 H20 F3 N3 O3

Anti-inflammatory

Marion Labs.

LgP	5.30(C)
pKa	
pp cited in Vols.1-5:	

MOROCROMEN [INN]

CAS	35843-07-3	MW:	401.47	MF:	C21 H27 N3 O5

Coronary vasodilator

Troponwerke, FRG

LgP	0.80(C)
pKa	
pp cited in Vols.1-5:	

(1) Naturwiss., 1976, 63, 148.

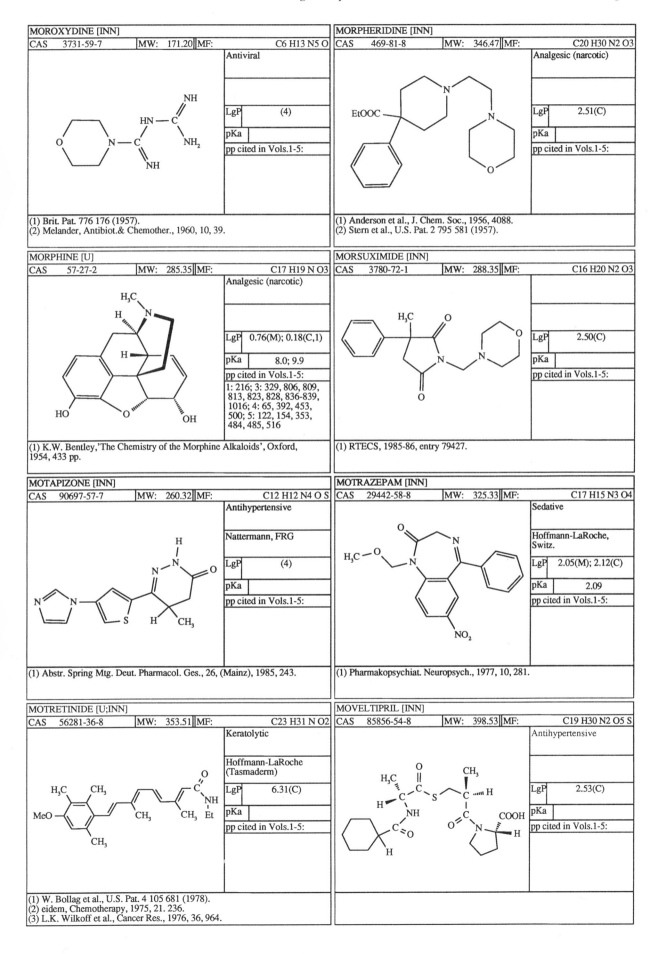

MOROXYDINE [INN]

CAS	3731-59-7	MW: 171.20	MF:	C6 H13 N5 O

Antiviral

LgP	(4)
pKa	
pp cited in Vols.1-5:	

(1) Brit. Pat. 776 176 (1957).
(2) Melander, Antibiot.& Chemother., 1960, 10, 39.

MORPHERIDINE [INN]

CAS	469-81-8	MW: 346.47	MF:	C20 H30 N2 O3

Analgesic (narcotic)

LgP	2.51(C)
pKa	
pp cited in Vols.1-5:	

(1) Anderson et al., J. Chem. Soc., 1956, 4088.
(2) Stern et al., U.S. Pat. 2 795 581 (1957).

MORPHINE [U]

CAS	57-27-2	MW: 285.35	MF:	C17 H19 N O3

Analgesic (narcotic)

LgP	0.76(M); 0.18(C,1)
pKa	8.0; 9.9
pp cited in Vols.1-5:	1: 216; 3: 329, 806, 809, 813, 823, 828, 836-839, 1016; 4: 65, 392, 453, 500; 5: 122, 154, 353, 484, 485, 516

(1) K.W. Bentley,'The Chemistry of the Morphine Alkaloids', Oxford, 1954, 433 pp.

MORSUXIMIDE [INN]

CAS	3780-72-1	MW: 288.35	MF:	C16 H20 N2 O3

LgP	2.50(C)
pKa	
pp cited in Vols.1-5:	

(1) RTECS, 1985-86, entry 79427.

MOTAPIZONE [INN]

CAS	90697-57-7	MW: 260.32	MF:	C12 H12 N4 O S

Antihypertensive

Nattermann, FRG

LgP	(4)
pKa	
pp cited in Vols.1-5:	

(1) Abstr. Spring Mtg. Deut. Pharmacol. Ges., 26, (Mainz), 1985, 243.

MOTRAZEPAM [INN]

CAS	29442-58-8	MW: 325.33	MF:	C17 H15 N3 O4

Sedative

Hoffmann-LaRoche, Switz.

LgP	2.05(M); 2.12(C)
pKa	2.09
pp cited in Vols.1-5:	

(1) Pharmakopsychiat. Neuropsych., 1977, 10, 281.

MOTRETINIDE [U;INN]

CAS	56281-36-8	MW: 353.51	MF:	C23 H31 N O2

Keratolytic

Hoffmann-LaRoche (Tasmaderm)

LgP	6.31(C)
pKa	
pp cited in Vols.1-5:	

(1) W. Bollag et al., U.S. Pat. 4 105 681 (1978).
(2) eidem, Chemotherapy, 1975, 21. 236.
(3) L.K. Wilkoff et al., Cancer Res., 1976, 36, 964.

MOVELTIPRIL [INN]

CAS	85856-54-8	MW: 398.53	MF:	C19 H30 N2 O5 S

Antihypertensive

LgP	2.53(C)
pKa	
pp cited in Vols.1-5:	

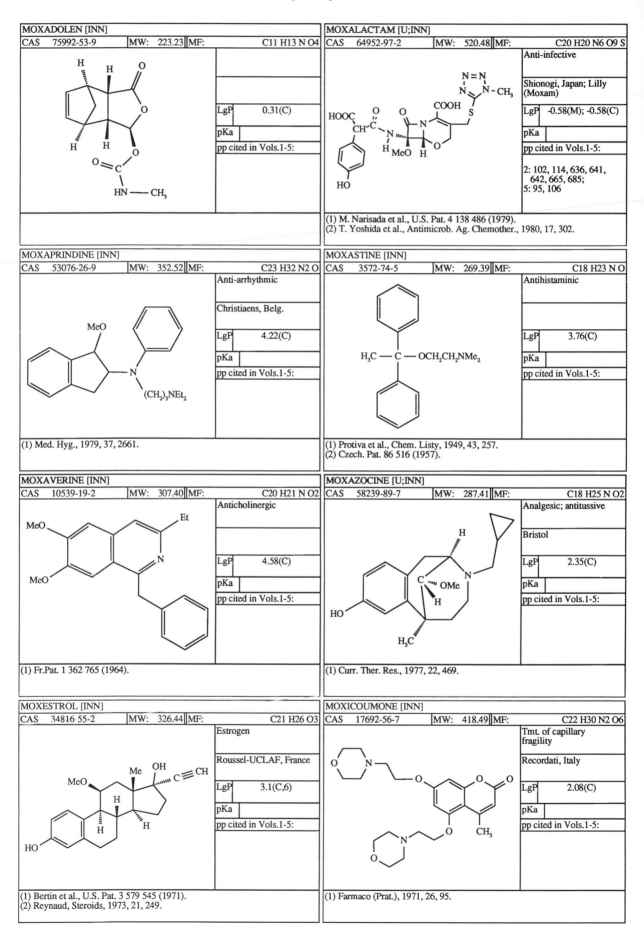

MOXADOLEN [INN]

CAS	75992-53-9	MW:	223.23	MF:	C11 H13 N O4

LgP	0.31(C)
pKa	
pp cited in Vols.1-5:	

MOXAPRINDINE [INN]

CAS	53076-26-9	MW:	352.52	MF:	C23 H32 N2 O

Anti-arrhythmic

Christiaens, Belg.

LgP	4.22(C)
pKa	
pp cited in Vols.1-5:	

(1) Med. Hyg., 1979, 37, 2661.

MOXAVERINE [INN]

CAS	10539-19-2	MW:	307.40	MF:	C20 H21 N O2

Anticholinergic

LgP	4.58(C)
pKa	
pp cited in Vols.1-5:	

(1) Fr.Pat. 1 362 765 (1964).

MOXESTROL [INN]

CAS	34816 55-2	MW:	326.44	MF:	C21 H26 O3

Estrogen

Roussel-UCLAF, France

LgP	3.1(C,6)
pKa	
pp cited in Vols.1-5:	

(1) Bertin et al., U.S. Pat. 3 579 545 (1971).
(2) Reynaud, Steroids, 1973, 21, 249.

MOXALACTAM [U;INN]

CAS	64952-97-2	MW:	520.48	MF:	C20 H20 N6 O9 S

Anti-infective

Shionogi, Japan; Lilly (Moxam)

LgP	-0.58(M); -0.58(C)
pKa	
pp cited in Vols.1-5:	

2: 102, 114, 636, 641, 642, 665, 685;
5: 95, 106

(1) M. Narisada et al., U.S. Pat. 4 138 486 (1979).
(2) T. Yoshida et al., Antimicrob. Ag. Chemother., 1980, 17, 302.

MOXASTINE [INN]

CAS	3572-74-5	MW:	269.39	MF:	C18 H23 N O

Antihistaminic

LgP	3.76(C)
pKa	
pp cited in Vols.1-5:	

(1) Protiva et al., Chem. Listy, 1949, 43, 257.
(2) Czech. Pat. 86 516 (1957).

MOXAZOCINE [U;INN]

CAS	58239-89-7	MW:	287.41	MF:	C18 H25 N O2

Analgesic; antitussive

Bristol

LgP	2.35(C)
pKa	
pp cited in Vols.1-5:	

(1) Curr. Ther. Res., 1977, 22, 469.

MOXICOUMONE [INN]

CAS	17692-56-7	MW:	418.49	MF:	C22 H30 N2 O6

Tmt. of capillary fragility

Recordati, Italy

LgP	2.08(C)
pKa	
pp cited in Vols.1-5:	

(1) Farmaco (Prat.), 1971, 26, 95.

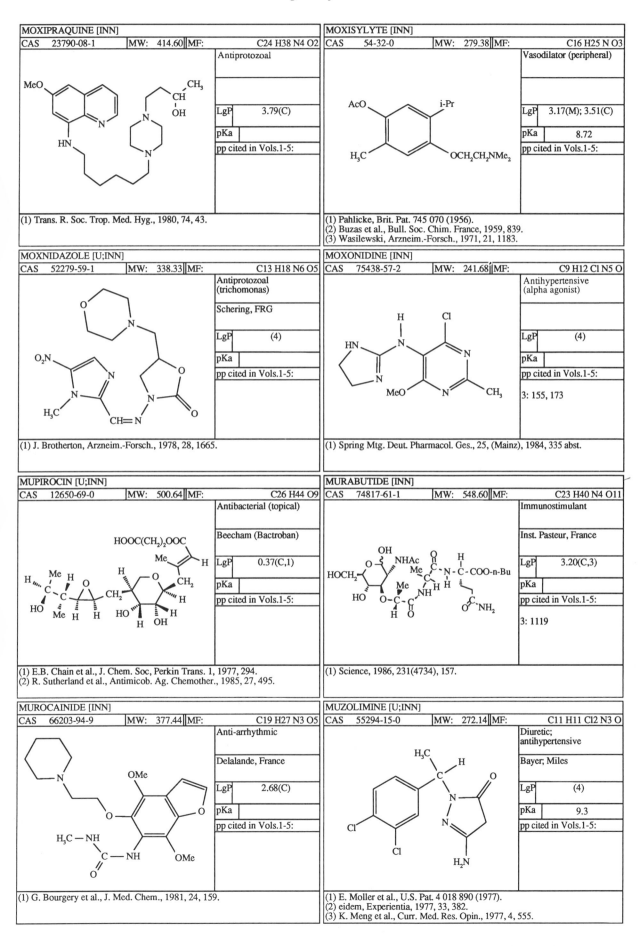

MOXIPRAQUINE [INN]

CAS	23790-08-1	MW:	414.60	MF:	C24 H38 N4 O2

Antiprotozoal

LgP	3.79(C)
pKa	
pp cited in Vols.1-5:	

(1) Trans. R. Soc. Trop. Med. Hyg., 1980, 74, 43.

MOXISYLYTE [INN]

CAS	54-32-0	MW:	279.38	MF:	C16 H25 N O3

Vasodilator (peripheral)

LgP	3.17(M); 3.51(C)
pKa	8.72
pp cited in Vols.1-5:	

(1) Pahlicke, Brit. Pat. 745 070 (1956).
(2) Buzas et al., Bull. Soc. Chim. France, 1959, 839.
(3) Wasilewski, Arzneim.-Forsch., 1971, 21, 1183.

MOXNIDAZOLE [U;INN]

CAS	52279-59-1	MW:	338.33	MF:	C13 H18 N6 O5

Antiprotozoal
(trichomonas)

Schering, FRG

LgP	(4)
pKa	
pp cited in Vols.1-5:	

(1) J. Brotherton, Arzneim.-Forsch., 1978, 28, 1665.

MOXONIDINE [INN]

CAS	75438-57-2	MW:	241.68	MF:	C9 H12 Cl N5 O

Antihypertensive
(alpha agonist)

LgP	(4)
pKa	
pp cited in Vols.1-5:	

3: 155, 173

(1) Spring Mtg. Deut. Pharmacol. Ges., 25, (Mainz), 1984, 335 abst.

MUPIROCIN [U;INN]

CAS	12650-69-0	MW:	500.64	MF:	C26 H44 O9

Antibacterial (topical)

Beecham (Bactroban)

LgP	0.37(C,1)
pKa	
pp cited in Vols.1-5:	

(1) E.B. Chain et al., J. Chem. Soc, Perkin Trans. 1, 1977, 294.
(2) R. Sutherland et al., Antimicob. Ag. Chemother., 1985, 27, 495.

MURABUTIDE [INN]

CAS	74817-61-1	MW:	548.60	MF:	C23 H40 N4 O11

Immunostimulant

Inst. Pasteur, France

LgP	3.20(C,3)
pKa	
pp cited in Vols.1-5:	

3: 1119

(1) Science, 1986, 231(4734), 157.

MUROCAINIDE [INN]

CAS	66203-94-9	MW:	377.44	MF:	C19 H27 N3 O5

Anti-arrhythmic

Delalande, France

LgP	2.68(C)
pKa	
pp cited in Vols.1-5:	

(1) G. Bourgery et al., J. Med. Chem., 1981, 24, 159.

MUZOLIMINE [U;INN]

CAS	55294-15-0	MW:	272.14	MF:	C11 H11 Cl2 N3 O

Diuretic;
antihypertensive

Bayer; Miles

LgP	(4)
pKa	9.3
pp cited in Vols.1-5:	

(1) E. Moller et al., U.S. Pat. 4 018 890 (1977).
(2) eidem, Experientia, 1977, 33, 382.
(3) K. Meng et al., Curr. Med. Res. Opin., 1977, 4, 555.

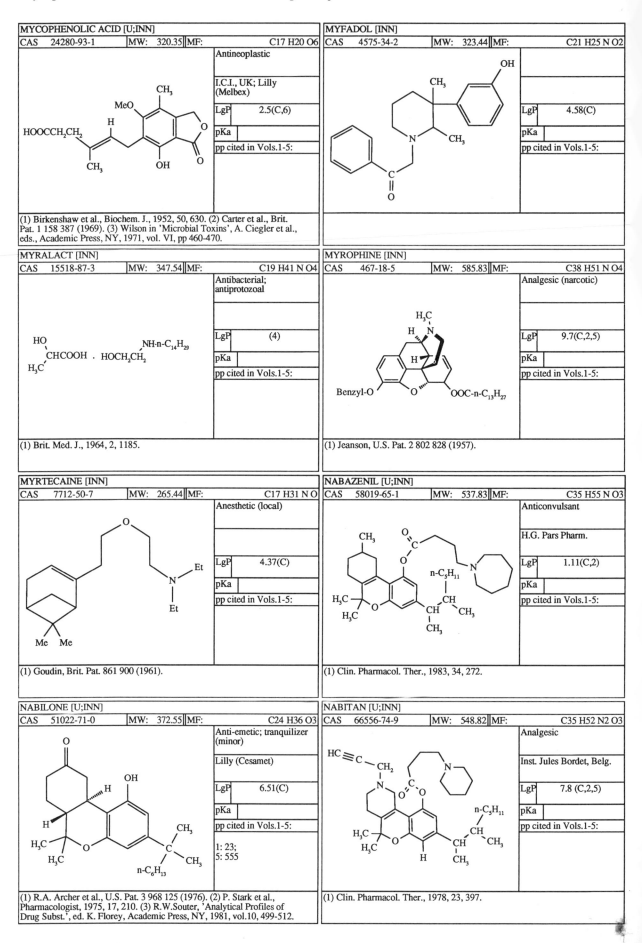

MYCOPHENOLIC ACID [U;INN]

CAS	24280-93-1	MW:	320.35	MF:	C17 H20 O6

Antineoplastic

I.C.I., UK; Lilly (Melbex)

LgP	2.5(C,6)
pKa	
pp cited in Vols.1-5:	

(1) Birkenshaw et al., Biochem. J., 1952, 50, 630. (2) Carter et al., Brit. Pat. 1 158 387 (1969). (3) Wilson in 'Microbial Toxins', A. Ciegler et al., eds., Academic Press, NY, 1971, vol. VI, pp 460-470.

MYFADOL [INN]

CAS	4575-34-2	MW:	323.44	MF:	C21 H25 N O2

LgP	4.58(C)
pKa	
pp cited in Vols.1-5:	

MYRALACT [INN]

CAS	15518-87-3	MW:	347.54	MF:	C19 H41 N O4

Antibacterial; antiprotozoal

LgP	(4)
pKa	
pp cited in Vols.1-5:	

(1) Brit. Med. J., 1964, 2, 1185.

MYROPHINE [INN]

CAS	467-18-5	MW:	585.83	MF:	C38 H51 N O4

Analgesic (narcotic)

LgP	9.7(C,2,5)
pKa	
pp cited in Vols.1-5:	

(1) Jeanson, U.S. Pat. 2 802 828 (1957).

MYRTECAINE [INN]

CAS	7712-50-7	MW:	265.44	MF:	C17 H31 N O

Anesthetic (local)

LgP	4.37(C)
pKa	
pp cited in Vols.1-5:	

(1) Goudin, Brit. Pat. 861 900 (1961).

NABAZENIL [U;INN]

CAS	58019-65-1	MW:	537.83	MF:	C35 H55 N O3

Anticonvulsant

H.G. Pars Pharm.

LgP	1.11(C,2)
pKa	
pp cited in Vols.1-5:	

(1) Clin. Pharmacol. Ther., 1983, 34, 272.

NABILONE [U;INN]

CAS	51022-71-0	MW:	372.55	MF:	C24 H36 O3

Anti-emetic; tranquilizer (minor)

Lilly (Cesamet)

LgP	6.51(C)
pKa	
pp cited in Vols.1-5:	

1: 23;
5: 555

(1) R.A. Archer et al., U.S. Pat. 3 968 125 (1976). (2) P. Stark et al., Pharmacologist, 1975, 17, 210. (3) R.W.Souter, 'Analytical Profiles of Drug Subst.', ed. K. Florey, Academic Press, NY, 1981, vol.10, 499-512.

NABITAN [U;INN]

CAS	66556-74-9	MW:	548.82	MF:	C35 H52 N2 O3

Analgesic

Inst. Jules Bordet, Belg.

LgP	7.8 (C,2,5)
pKa	
pp cited in Vols.1-5:	

(1) Clin. Pharmacol. Ther., 1978, 23, 397.

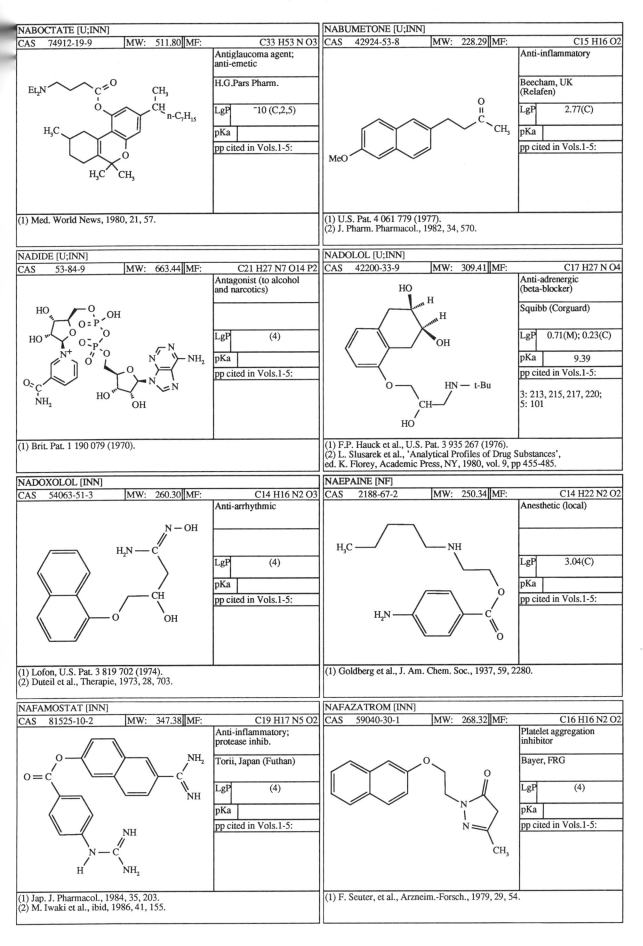

NABOCTATE [U;INN]

CAS	74912-19-9	MW:	511.80	MF:	C33 H53 N O3

Antiglaucoma agent; anti-emetic

H.G.Pars Pharm.

LgP	~10 (C,2,5)
pKa	
pp cited in Vols.1-5:	

(1) Med. World News, 1980, 21, 57.

NABUMETONE [U;INN]

CAS	42924-53-8	MW:	228.29	MF:	C15 H16 O2

Anti-inflammatory

Beecham, UK (Relafen)

LgP	2.77(C)
pKa	
pp cited in Vols.1-5:	

(1) U.S. Pat. 4 061 779 (1977).
(2) J. Pharm. Pharmacol., 1982, 34, 570.

NADIDE [U;INN]

CAS	53-84-9	MW:	663.44	MF:	C21 H27 N7 O14 P2

Antagonist (to alcohol and narcotics)

LgP	(4)
pKa	
pp cited in Vols.1-5:	

(1) Brit. Pat. 1 190 079 (1970).

NADOLOL [U;INN]

CAS	42200-33-9	MW:	309.41	MF:	C17 H27 N O4

Anti-adrenergic (beta-blocker)

Squibb (Corguard)

LgP	0.71(M); 0.23(C)
pKa	9.39
pp cited in Vols.1-5:	

3: 213, 215, 217, 220;
5: 101

(1) F.P. Hauck et al., U.S. Pat. 3 935 267 (1976).
(2) L. Slusarek et al., 'Analytical Profiles of Drug Substances', ed. K. Florey, Academic Press, NY, 1980, vol. 9, pp 455-485.

NADOXOLOL [INN]

CAS	54063-51-3	MW:	260.30	MF:	C14 H16 N2 O3

Anti-arrhythmic

LgP	(4)
pKa	
pp cited in Vols.1-5:	

(1) Lofon, U.S. Pat. 3 819 702 (1974).
(2) Duteil et al., Therapie, 1973, 28, 703.

NAEPAINE [NF]

CAS	2188-67-2	MW:	250.34	MF:	C14 H22 N2 O2

Anesthetic (local)

LgP	3.04(C)
pKa	
pp cited in Vols.1-5:	

(1) Goldberg et al., J. Am. Chem. Soc., 1937, 59, 2280.

NAFAMOSTAT [INN]

CAS	81525-10-2	MW:	347.38	MF:	C19 H17 N5 O2

Anti-inflammatory; protease inhib.

Torii, Japan (Futhan)

LgP	(4)
pKa	
pp cited in Vols.1-5:	

(1) Jap. J. Pharmacol., 1984, 35, 203.
(2) M. Iwaki et al., ibid, 1986, 41, 155.

NAFAZATROM [INN]

CAS	59040-30-1	MW:	268.32	MF:	C16 H16 N2 O2

Platelet aggregation inhibitor

Bayer, FRG

LgP	(4)
pKa	
pp cited in Vols.1-5:	

(1) F. Seuter, et al., Arzneim.-Forsch., 1979, 29, 54.

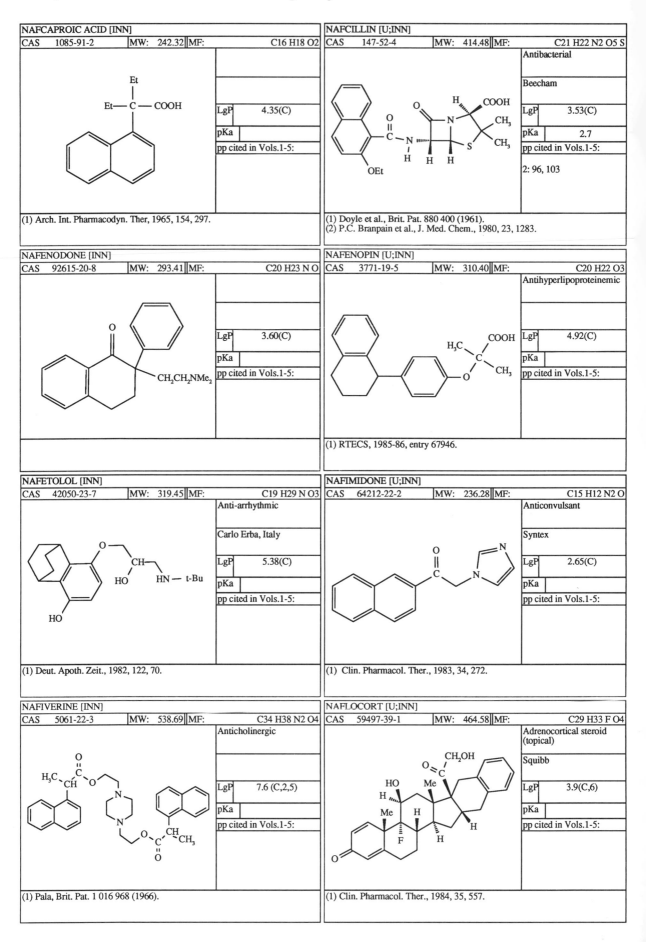

NAFCAPROIC ACID [INN]

CAS	1085-91-2	MW:	242.32	MF:	C16 H18 O2

LgP	4.35(C)
pKa	
pp cited in Vols.1-5:	

(1) Arch. Int. Pharmacodyn. Ther, 1965, 154, 297.

NAFCILLIN [U;INN]

CAS	147-52-4	MW:	414.48	MF:	C21 H22 N2 O5 S

Antibacterial

Beecham

LgP	3.53(C)
pKa	2.7
pp cited in Vols.1-5:	

2: 96, 103

(1) Doyle et al., Brit. Pat. 880 400 (1961).
(2) P.C. Branpain et al., J. Med. Chem., 1980, 23, 1283.

NAFENODONE [INN]

CAS	92615-20-8	MW:	293.41	MF:	C20 H23 N O

LgP	3.60(C)
pKa	
pp cited in Vols.1-5:	

NAFENOPIN [U;INN]

CAS	3771-19-5	MW:	310.40	MF:	C20 H22 O3

Antihyperlipoproteinemic

LgP	4.92(C)
pKa	
pp cited in Vols.1-5:	

(1) RTECS, 1985-86, entry 67946.

NAFETOLOL [INN]

CAS	42050-23-7	MW:	319.45	MF:	C19 H29 N O3

Anti-arrhythmic

Carlo Erba, Italy

LgP	5.38(C)
pKa	
pp cited in Vols.1-5:	

(1) Deut. Apoth. Zeit., 1982, 122, 70.

NAFIMIDONE [U;INN]

CAS	64212-22-2	MW:	236.28	MF:	C15 H12 N2 O

Anticonvulsant

Syntex

LgP	2.65(C)
pKa	
pp cited in Vols.1-5:	

(1) Clin. Pharmacol. Ther., 1983, 34, 272.

NAFIVERINE [INN]

CAS	5061-22-3	MW:	538.69	MF:	C34 H38 N2 O4

Anticholinergic

LgP	7.6 (C,2,5)
pKa	
pp cited in Vols.1-5:	

(1) Pala, Brit. Pat. 1 016 968 (1966).

NAFLOCORT [U;INN]

CAS	59497-39-1	MW:	464.58	MF:	C29 H33 F O4

Adrenocortical steroid (topical)

Squibb

LgP	3.9(C,6)
pKa	
pp cited in Vols.1-5:	

(1) Clin. Pharmacol. Ther., 1984, 35, 557.

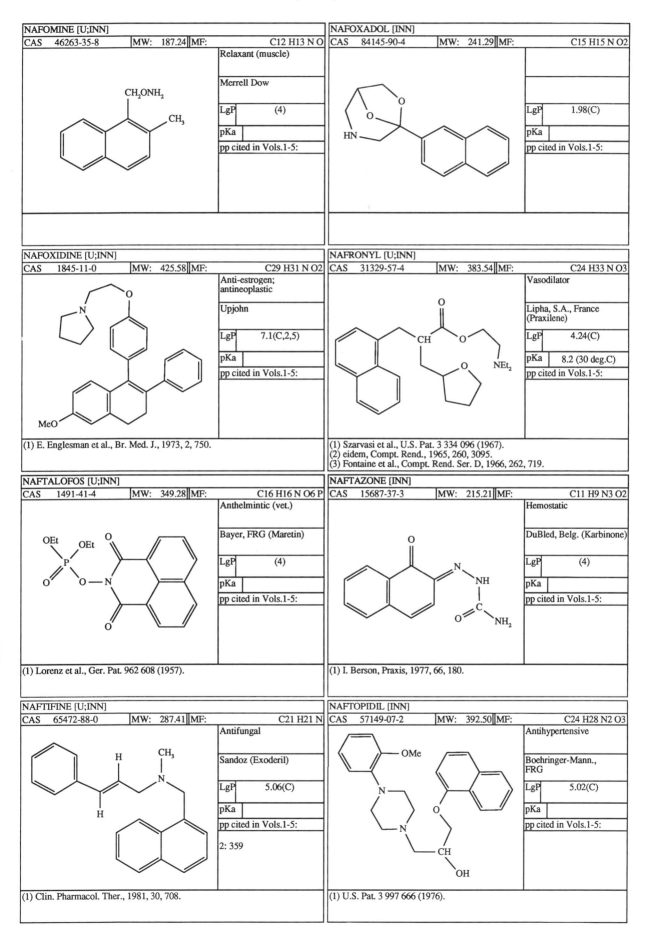

NAFOMINE [U;INN]

CAS	46263-35-8	MW:	187.24	MF:	C12 H13 N O

Relaxant (muscle)

Merrell Dow

LgP	(4)
pKa	
pp cited in Vols.1-5:	

NAFOXADOL [INN]

CAS	84145-90-4	MW:	241.29	MF:	C15 H15 N O2

LgP	1.98(C)
pKa	
pp cited in Vols.1-5:	

NAFOXIDINE [U;INN]

CAS	1845-11-0	MW:	425.58	MF:	C29 H31 N O2

Anti-estrogen; antineoplastic

Upjohn

LgP	7.1(C,2,5)
pKa	
pp cited in Vols.1-5:	

(1) E. Englesman et al., Br. Med. J., 1973, 2, 750.

NAFRONYL [U;INN]

CAS	31329-57-4	MW:	383.54	MF:	C24 H33 N O3

Vasodilator

Lipha, S.A., France (Praxilene)

LgP	4.24(C)
pKa	8.2 (30 deg.C)
pp cited in Vols.1-5:	

(1) Szarvasi et al., U.S. Pat. 3 334 096 (1967).
(2) eidem, Compt. Rend., 1965, 260, 3095.
(3) Fontaine et al., Compt. Rend. Ser. D, 1966, 262, 719.

NAFTALOFOS [U;INN]

CAS	1491-41-4	MW:	349.28	MF:	C16 H16 N O6 P

Anthelmintic (vet.)

Bayer, FRG (Maretin)

LgP	(4)
pKa	
pp cited in Vols.1-5:	

(1) Lorenz et al., Ger. Pat. 962 608 (1957).

NAFTAZONE [INN]

CAS	15687-37-3	MW:	215.21	MF:	C11 H9 N3 O2

Hemostatic

DuBled, Belg. (Karbinone)

LgP	(4)
pKa	
pp cited in Vols.1-5:	

(1) I. Berson, Praxis, 1977, 66, 180.

NAFTIFINE [U;INN]

CAS	65472-88-0	MW:	287.41	MF:	C21 H21 N

Antifungal

Sandoz (Exoderil)

LgP	5.06(C)
pKa	
pp cited in Vols.1-5:	
	2: 359

(1) Clin. Pharmacol. Ther., 1981, 30, 708.

NAFTOPIDIL [INN]

CAS	57149-07-2	MW:	392.50	MF:	C24 H28 N2 O3

Antihypertensive

Boehringer-Mann., FRG

LgP	5.02(C)
pKa	
pp cited in Vols.1-5:	

(1) U.S. Pat. 3 997 666 (1976).

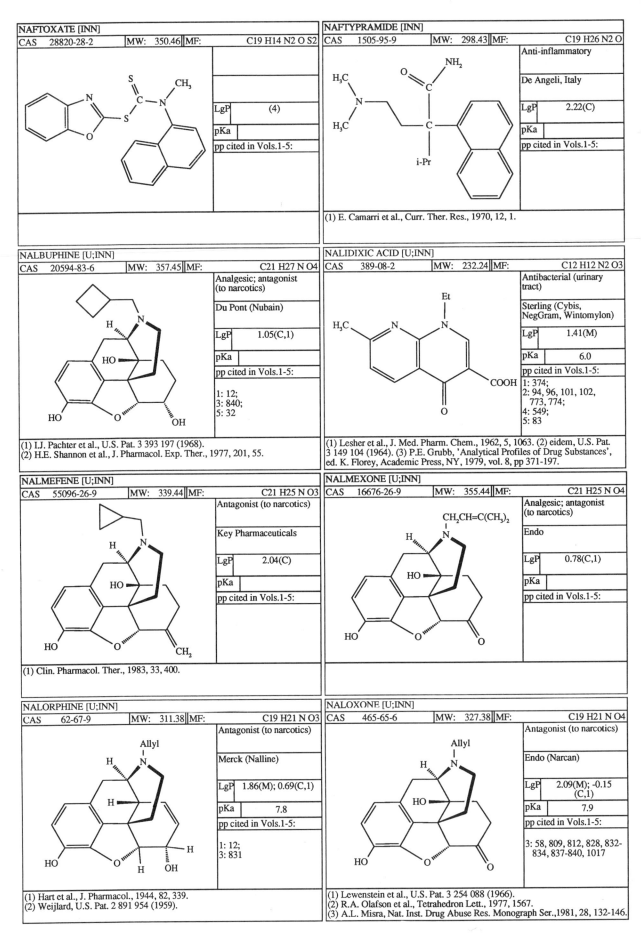

NAFTOXATE [INN]

CAS	28820-28-2	MW:	350.46	MF:	C19 H14 N2 O S2

LgP	(4)
pKa	
pp cited in Vols.1-5:	

NAFTYPRAMIDE [INN]

CAS	1505-95-9	MW:	298.43	MF:	C19 H26 N2 O

Anti-inflammatory

De Angeli, Italy

LgP	2.22(C)
pKa	
pp cited in Vols.1-5:	

(1) E. Camarri et al., Curr. Ther. Res., 1970, 12, 1.

NALBUPHINE [U;INN]

CAS	20594-83-6	MW:	357.45	MF:	C21 H27 N O4

Analgesic; antagonist (to narcotics)

Du Pont (Nubain)

LgP	1.05(C,1)
pKa	
pp cited in Vols.1-5:	

1: 12;
3: 840;
5: 32

(1) I.J. Pachter et al., U.S. Pat. 3 393 197 (1968).
(2) H.E. Shannon et al., J. Pharmacol. Exp. Ther., 1977, 201, 55.

NALIDIXIC ACID [U;INN]

CAS	389-08-2	MW:	232.24	MF:	C12 H12 N2 O3

Antibacterial (urinary tract)

Sterling (Cybis, NegGram, Wintomylon)

LgP	1.41(M)
pKa	6.0
pp cited in Vols.1-5:	

1: 374;
2: 94, 96, 101, 102, 773, 774;
4: 549;
5: 83

(1) Lesher et al., J. Med. Pharm. Chem., 1962, 5, 1063. (2) eidem, U.S. Pat. 3 149 104 (1964). (3) P.E. Grubb, 'Analytical Profiles of Drug Substances', ed. K. Florey, Academic Press, NY, 1979, vol. 8, pp 371-197.

NALMEFENE [U;INN]

CAS	55096-26-9	MW:	339.44	MF:	C21 H25 N O3

Antagonist (to narcotics)

Key Pharmaceuticals

LgP	2.04(C)
pKa	
pp cited in Vols.1-5:	

(1) Clin. Pharmacol. Ther., 1983, 33, 400.

NALMEXONE [U;INN]

CAS	16676-26-9	MW:	355.44	MF:	C21 H25 N O4

Analgesic; antagonist (to narcotics)

Endo

LgP	0.78(C,1)
pKa	
pp cited in Vols.1-5:	

NALORPHINE [U;INN]

CAS	62-67-9	MW:	311.38	MF:	C19 H21 N O3

Antagonist (to narcotics)

Merck (Nalline)

LgP	1.86(M); 0.69(C,1)
pKa	7.8
pp cited in Vols.1-5:	

1: 12;
3: 831

(1) Hart et al., J. Pharmacol., 1944, 82, 339.
(2) Weijlard, U.S. Pat. 2 891 954 (1959).

NALOXONE [U;INN]

CAS	465-65-6	MW:	327.38	MF:	C19 H21 N O4

Antagonist (to narcotics)

Endo (Narcan)

LgP	2.09(M); -0.15 (C,1)
pKa	7.9
pp cited in Vols.1-5:	

3: 58, 809, 812, 828, 832-834, 837-840, 1017

(1) Lewenstein et al., U.S. Pat. 3 254 088 (1966).
(2) R.A. Olafson et al., Tetrahedron Lett., 1977, 1567.
(3) A.L. Misra, Nat. Inst. Drug Abuse Res. Monograph Ser.,1981, 28, 132-146.

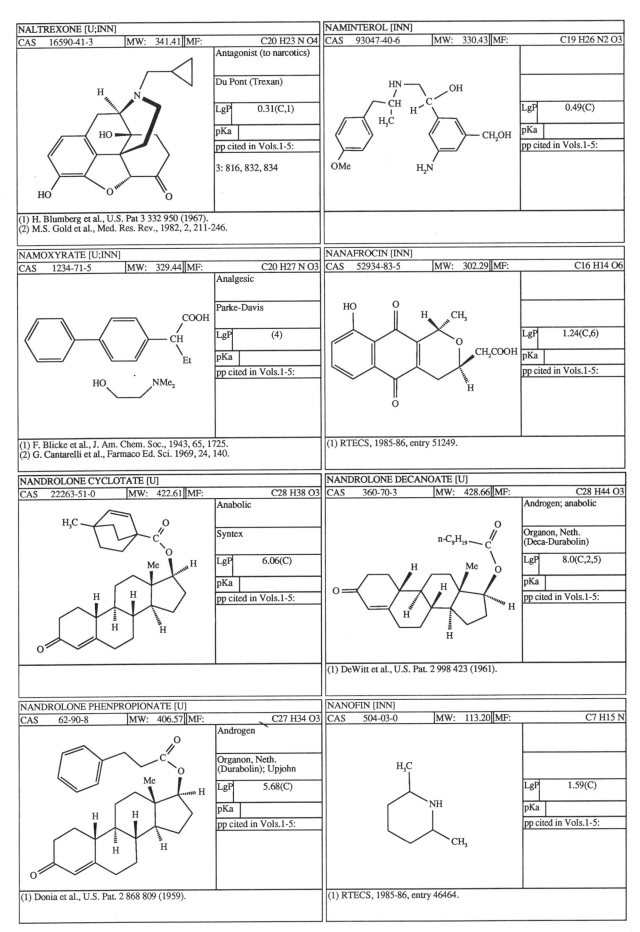

NALTREXONE [U;INN]

CAS	16590-41-3	MW:	341.41	MF:	C20 H23 N O4

Antagonist (to narcotics)

Du Pont (Trexan)

LgP	0.31(C,1)
pKa	
pp cited in Vols.1-5:	

3: 816, 832, 834

(1) H. Blumberg et al., U.S. Pat 3 332 950 (1967).
(2) M.S. Gold et al., Med. Res. Rev., 1982, 2, 211-246.

NAMINTEROL [INN]

CAS	93047-40-6	MW:	330.43	MF:	C19 H26 N2 O3

LgP	0.49(C)
pKa	
pp cited in Vols.1-5:	

NAMOXYRATE [U;INN]

CAS	1234-71-5	MW:	329.44	MF:	C20 H27 N O3

Analgesic

Parke-Davis

LgP	(4)
pKa	
pp cited in Vols.1-5:	

(1) F. Blicke et al., J. Am. Chem. Soc., 1943, 65, 1725.
(2) G. Cantarelli et al., Farmaco Ed. Sci. 1969, 24, 140.

NANAFROCIN [INN]

CAS	52934-83-5	MW:	302.29	MF:	C16 H14 O6

LgP	1.24(C,6)
pKa	
pp cited in Vols.1-5:	

(1) RTECS, 1985-86, entry 51249.

NANDROLONE CYCLOTATE [U]

CAS	22263-51-0	MW:	422.61	MF:	C28 H38 O3

Anabolic

Syntex

LgP	6.06(C)
pKa	
pp cited in Vols.1-5:	

NANDROLONE DECANOATE [U]

CAS	360-70-3	MW:	428.66	MF:	C28 H44 O3

Androgen; anabolic

Organon, Neth.
(Deca-Durabolin)

LgP	8.0(C,2,5)
pKa	
pp cited in Vols.1-5:	

(1) DeWitt et al., U.S. Pat. 2 998 423 (1961).

NANDROLONE PHENPROPIONATE [U]

CAS	62-90-8	MW:	406.57	MF:	C27 H34 O3

Androgen

Organon, Neth.
(Durabolin); Upjohn

LgP	5.68(C)
pKa	
pp cited in Vols.1-5:	

(1) Donia et al., U.S. Pat. 2 868 809 (1959).

NANOFIN [INN]

CAS	504-03-0	MW:	113.20	MF:	C7 H15 N

LgP	1.59(C)
pKa	
pp cited in Vols.1-5:	

(1) RTECS, 1985-86, entry 46464.

NANTRADOL [U;INN]		
CAS 65511-41-3 MW: 437.58 MF: C27 H35 N O4		

	Analgesic
	Pfizer
	LgP 5.07(C)
	pKa
	pp cited in Vols.1-5:

(1) Pharmacologist, 1979, 21, 326, 661 abst.

NAPACTADINE [U;INN]		
CAS 76631-45-3 MW: 212.30 MF: C14 H16 N2		

	Antidepressant
	Merrell Dow
	LgP (4)
	pKa
	pp cited in Vols.1-5:

(1) Clin. Pharmacol. Ther., 1981, 30, 580.

NAPAMEZOLE [U;INN]		
CAS 91524-14-0 MW: 212.30 MF: C14 H16 N2		

	Antidepressant; adren-ergic (alpha-2 receptor)
	Sterling
	LgP 3.71(C)
	pKa
	pp cited in Vols.1-5:

(1) Clin. Pharmacol. Ther., 1985, 38, 357.

NAPHAZOLINE [U;INN]		
CAS 835-31-4 MW: 210.28 MF: C14 H14 N2		

	Adrenergic (vasocon-strictor)
	Ciba (Privine)
	LgP 3.83(C)
	pKa 10.9
	pp cited in Vols.1-5:
	1: 26; 3: 139, 148, 151, 173

(1) U.S. Pat. 2 161 939 (1939).

NAPHTHONONE [INN]		
CAS 7114-11-6 MW: 240.30 MF: C16 H16 O2		

	Antitussive
	Clin-Byla
	LgP 3.00(C)
	pKa
	pp cited in Vols.1-5:

(1) Mousseron, U.S. Pat. 2 882 201 (1959).

NAPRODOXIME [INN]		
CAS 57925-64-1 MW: 230.27 MF: C13 H14 N2 O2		

	Anti-agressive
	Labaz, Belg.
	LgP (4)
	pKa
	pp cited in Vols.1-5:

(1) Eur. J. Med. Chem., 1975, 10, 463.

NAPROXEN [U;INN]		
CAS 22204-53-1 MW: 230.27 MF: C14 H14 O3		

	Anti-inflammatory; analgesic; antipyretic
	Syntex (Naprosyn)
	LgP 3.18(M); 2.82(C)
	pKa 4.2
	pp cited in Vols.1-5:
	1: 22, 329; 2: 468; 5: 91

(1) Fried et al., S. Afr. Pat. 67 07 597 (1968).
(2) Harrison et al., J. Med. Chem. 1970, 13, 203.
(3) Arzneim.-Forsch., 1975, 25, 278-332.

NAPROXOL [U;INN]		
CAS 26159-36-4 MW: 216.28 MF: C14 H16 O2		

	Anti-inflammatory; analgesic; antipyretic
	Syntex
	LgP 2.68(C)
	pKa
	pp cited in Vols.1-5:

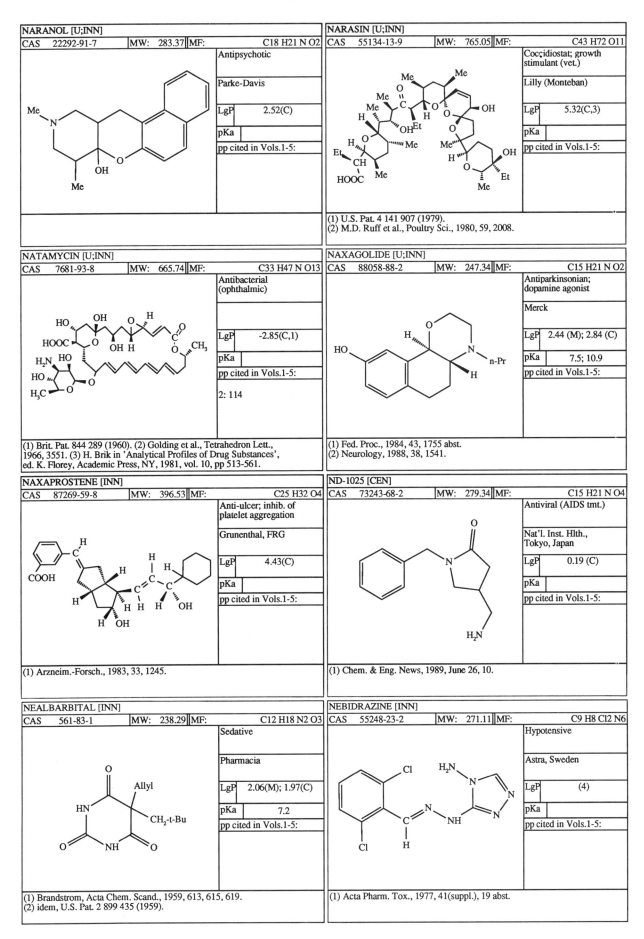

NARANOL [U;INN]

CAS	22292-91-7	MW:	283.37	MF:	C18 H21 N O2

Antipsychotic

Parke-Davis

LgP	2.52(C)

pKa	

pp cited in Vols.1-5:

NARASIN [U;INN]

CAS	55134-13-9	MW:	765.05	MF:	C43 H72 O11

Coccidiostat; growth stimulant (vet.)

Lilly (Monteban)

LgP	5.32(C,3)

pKa	

pp cited in Vols.1-5:

(1) U.S. Pat. 4 141 907 (1979).
(2) M.D. Ruff et al., Poultry Sci., 1980, 59, 2008.

NATAMYCIN [U;INN]

CAS	7681-93-8	MW:	665.74	MF:	C33 H47 N O13

Antibacterial (ophthalmic)

LgP	-2.85(C,1)

pKa	

pp cited in Vols.1-5:

2: 114

(1) Brit. Pat. 844 289 (1960). (2) Golding et al., Tetrahedron Lett., 1966, 3551. (3) H. Brik in 'Analytical Profiles of Drug Substances', ed. K. Florey, Academic Press, NY, 1981, vol. 10, pp 513-561.

NAXAGOLIDE [U;INN]

CAS	88058-88-2	MW:	247.34	MF:	C15 H21 N O2

Antiparkinsonian; dopamine agonist

Merck

LgP	2.44 (M); 2.84 (C)

pKa	7.5; 10.9

pp cited in Vols.1-5:

(1) Fed. Proc., 1984, 43, 1755 abst.
(2) Neurology, 1988, 38, 1541.

NAXAPROSTENE [INN]

CAS	87269-59-8	MW:	396.53	MF:	C25 H32 O4

Anti-ulcer; inhib. of platelet aggregation

Grunenthal, FRG

LgP	4.43(C)

pKa	

pp cited in Vols.1-5:

(1) Arzneim.-Forsch., 1983, 33, 1245.

ND-1025 [CEN]

CAS	73243-68-2	MW:	279.34	MF:	C15 H21 N O4

Antiviral (AIDS tmt.)

Nat'l. Inst. Hlth., Tokyo, Japan

LgP	0.19 (C)

pKa	

pp cited in Vols.1-5:

(1) Chem. & Eng. News, 1989, June 26, 10.

NEALBARBITAL [INN]

CAS	561-83-1	MW:	238.29	MF:	C12 H18 N2 O3

Sedative

Pharmacia

LgP	2.06(M); 1.97(C)

pKa	7.2

pp cited in Vols.1-5:

(1) Brandstrom, Acta Chem. Scand., 1959, 613, 615, 619.
(2) idem, U.S. Pat. 2 899 435 (1959).

NEBIDRAZINE [INN]

CAS	55248-23-2	MW:	271.11	MF:	C9 H8 Cl2 N6

Hypotensive

Astra, Sweden

LgP	(4)

pKa	

pp cited in Vols.1-5:

(1) Acta Pharm. Tox., 1977, 41(suppl.), 19 abst.

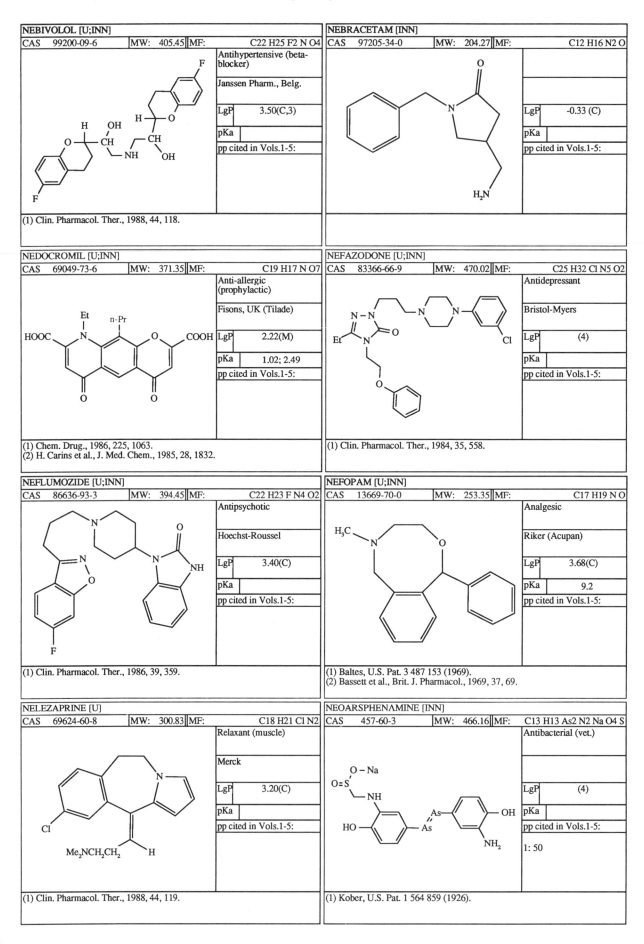

NEBIVOLOL [U;INN]

CAS	99200-09-6	MW:	405.45	MF:	C22 H25 F2 N O4

Antihypertensive (beta-blocker)

Janssen Pharm., Belg.

LgP	3.50(C,3)
pKa	
pp cited in Vols.1-5:	

(1) Clin. Pharmacol. Ther., 1988, 44, 118.

NEBRACETAM [INN]

CAS	97205-34-0	MW:	204.27	MF:	C12 H16 N2 O

LgP	-0.33 (C)
pKa	
pp cited in Vols.1-5:	

NEDOCROMIL [U;INN]

CAS	69049-73-6	MW:	371.35	MF:	C19 H17 N O7

Anti-allergic (prophylactic)

Fisons, UK (Tilade)

LgP	2.22(M)
pKa	1.02; 2.49
pp cited in Vols.1-5:	

(1) Chem. Drug., 1986, 225, 1063.
(2) H. Carins et al., J. Med. Chem., 1985, 28, 1832.

NEFAZODONE [U;INN]

CAS	83366-66-9	MW:	470.02	MF:	C25 H32 Cl N5 O2

Antidepressant

Bristol-Myers

LgP	(4)
pKa	
pp cited in Vols.1-5:	

(1) Clin. Pharmacol. Ther., 1984, 35, 558.

NEFLUMOZIDE [U;INN]

CAS	86636-93-3	MW:	394.45	MF:	C22 H23 F N4 O2

Antipsychotic

Hoechst-Roussel

LgP	3.40(C)
pKa	
pp cited in Vols.1-5:	

(1) Clin. Pharmacol. Ther., 1986, 39, 359.

NEFOPAM [U;INN]

CAS	13669-70-0	MW:	253.35	MF:	C17 H19 N O

Analgesic

Riker (Acupan)

LgP	3.68(C)
pKa	9.2
pp cited in Vols.1-5:	

(1) Baltes, U.S. Pat. 3 487 153 (1969).
(2) Bassett et al., Brit. J. Pharmacol., 1969, 37, 69.

NELEZAPRINE [U]

CAS	69624-60-8	MW:	300.83	MF:	C18 H21 Cl N2

Relaxant (muscle)

Merck

LgP	3.20(C)
pKa	
pp cited in Vols.1-5:	

(1) Clin. Pharmacol. Ther., 1988, 44, 119.

NEOARSPHENAMINE [INN]

CAS	457-60-3	MW:	466.16	MF:	C13 H13 As2 N2 Na O4 S

Antibacterial (vet.)

LgP	(4)
pKa	
pp cited in Vols.1-5:	
1: 50	

(1) Kober, U.S. Pat. 1 564 859 (1926).

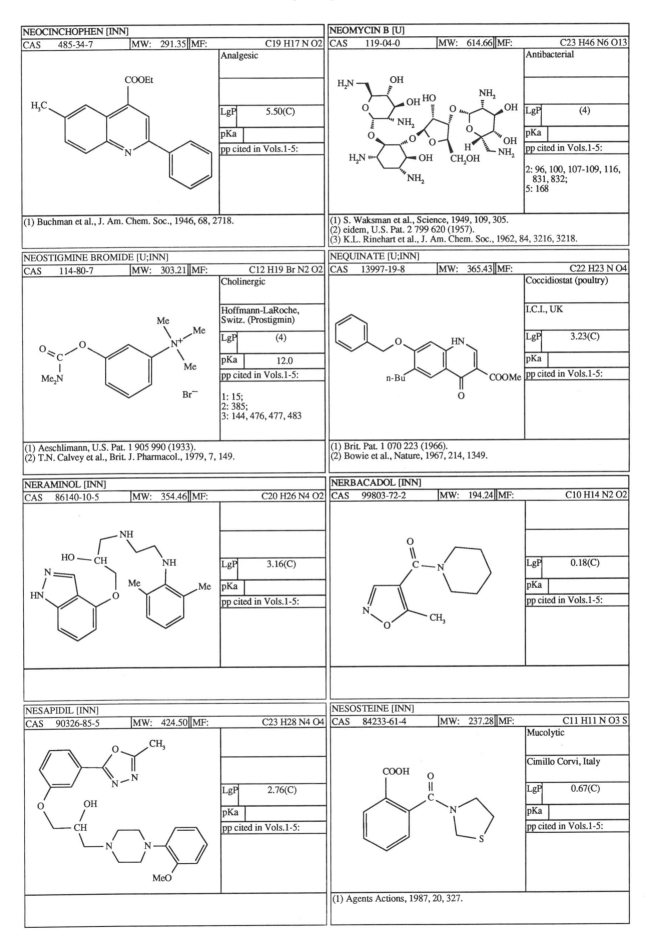

NEOCINCHOPHEN [INN]

CAS	485-34-7	MW: 291.35	MF:	C19 H17 N O2

Analgesic

LgP	5.50(C)
pKa	
pp cited in Vols.1-5:	

(1) Buchman et al., J. Am. Chem. Soc., 1946, 68, 2718.

NEOMYCIN B [U]

CAS	119-04-0	MW: 614.66	MF:	C23 H46 N6 O13

Antibacterial

LgP	(4)
pKa	
pp cited in Vols.1-5:	

2: 96, 100, 107-109, 116, 831, 832;
5: 168

(1) S. Waksman et al., Science, 1949, 109, 305.
(2) eidem, U.S. Pat. 2 799 620 (1957).
(3) K.L. Rinehart et al., J. Am. Chem. Soc., 1962, 84, 3216, 3218.

NEOSTIGMINE BROMIDE [U;INN]

CAS	114-80-7	MW: 303.21	MF:	C12 H19 Br N2 O2

Cholinergic

Hoffmann-LaRoche, Switz. (Prostigmin)

LgP	(4)
pKa	12.0
pp cited in Vols.1-5:	

1: 15;
2: 385;
3: 144, 476, 477, 483

(1) Aeschlimann, U.S. Pat. 1 905 990 (1933).
(2) T.N. Calvey et al., Brit. J. Pharmacol., 1979, 7, 149.

NEQUINATE [U;INN]

CAS	13997-19-8	MW: 365.43	MF:	C22 H23 N O4

Coccidiostat (poultry)

I.C.I., UK

LgP	3.23(C)
pKa	
pp cited in Vols.1-5:	

(1) Brit. Pat. 1 070 223 (1966).
(2) Bowie et al., Nature, 1967, 214, 1349.

NERAMINOL [INN]

CAS	86140-10-5	MW: 354.46	MF:	C20 H26 N4 O2

LgP	3.16(C)
pKa	
pp cited in Vols.1-5:	

NERBACADOL [INN]

CAS	99803-72-2	MW: 194.24	MF:	C10 H14 N2 O2

LgP	0.18(C)
pKa	
pp cited in Vols.1-5:	

NESAPIDIL [INN]

CAS	90326-85-5	MW: 424.50	MF:	C23 H28 N4 O4

LgP	2.76(C)
pKa	
pp cited in Vols.1-5:	

NESOSTEINE [INN]

CAS	84233-61-4	MW: 237.28	MF:	C11 H11 N O3 S

Mucolytic

Cimillo Corvi, Italy

LgP	0.67(C)
pKa	
pp cited in Vols.1-5:	

(1) Agents Actions, 1987, 20, 327.

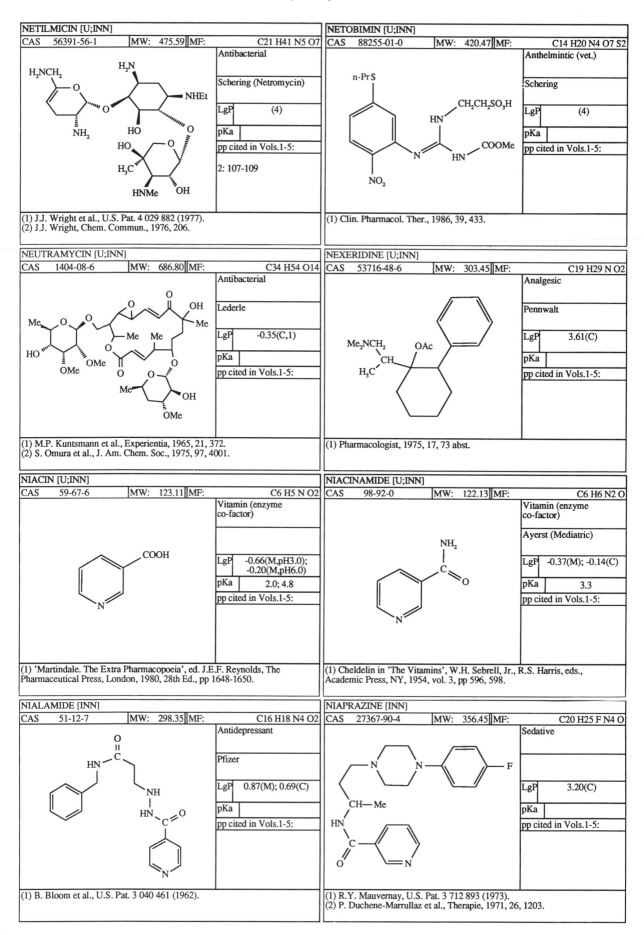

NETILMICIN [U;INN]

CAS	56391-56-1	MW:	475.59	MF:	C21 H41 N5 O7

Antibacterial

Schering (Netromycin)

LgP	(4)
pKa	
pp cited in Vols.1-5:	

2: 107-109

(1) J.J. Wright et al., U.S. Pat. 4 029 882 (1977).
(2) J.J. Wright, Chem. Commun., 1976, 206.

NETOBIMIN [U;INN]

CAS	88255-01-0	MW:	420.47	MF:	C14 H20 N4 O7 S2

Anthelmintic (vet.)

Schering

LgP	(4)
pKa	
pp cited in Vols.1-5:	

(1) Clin. Pharmacol. Ther., 1986, 39, 433.

NEUTRAMYCIN [U;INN]

CAS	1404-08-6	MW:	686.80	MF:	C34 H54 O14

Antibacterial

Lederle

LgP	-0.35(C,1)
pKa	
pp cited in Vols.1-5:	

(1) M.P. Kuntsmann et al., Experientia, 1965, 21, 372.
(2) S. Omura et al., J. Am. Chem. Soc., 1975, 97, 4001.

NEXERIDINE [U;INN]

CAS	53716-48-6	MW:	303.45	MF:	C19 H29 N O2

Analgesic

Pennwalt

LgP	3.61(C)
pKa	
pp cited in Vols.1-5:	

(1) Pharmacologist, 1975, 17, 73 abst.

NIACIN [U;INN]

CAS	59-67-6	MW:	123.11	MF:	C6 H5 N O2

Vitamin (enzyme co-factor)

LgP	-0.66(M,pH3.0); -0.20(M,pH6.0)
pKa	2.0; 4.8
pp cited in Vols.1-5:	

(1) 'Martindale. The Extra Pharmacopoeia', ed. J.E.F. Reynolds, The Pharmaceutical Press, London, 1980, 28th Ed., pp 1648-1650.

NIACINAMIDE [U;INN]

CAS	98-92-0	MW:	122.13	MF:	C6 H6 N2 O

Vitamin (enzyme co-factor)

Ayerst (Mediatric)

LgP	-0.37(M); -0.14(C)
pKa	3.3
pp cited in Vols.1-5:	

(1) Cheldelin in 'The Vitamins', W.H. Sebrell, Jr., R.S. Harris, eds., Academic Press, NY, 1954, vol. 3, pp 596, 598.

NIALAMIDE [INN]

CAS	51-12-7	MW:	298.35	MF:	C16 H18 N4 O2

Antidepressant

Pfizer

LgP	0.87(M); 0.69(C)
pKa	
pp cited in Vols.1-5:	

(1) B. Bloom et al., U.S. Pat. 3 040 461 (1962).

NIAPRAZINE [INN]

CAS	27367-90-4	MW:	356.45	MF:	C20 H25 F N4 O

Sedative

LgP	3.20(C)
pKa	
pp cited in Vols.1-5:	

(1) R.Y. Mauvernay, U.S. Pat. 3 712 893 (1973).
(2) P. Duchene-Marullaz et al., Therapie, 1971, 26, 1203.

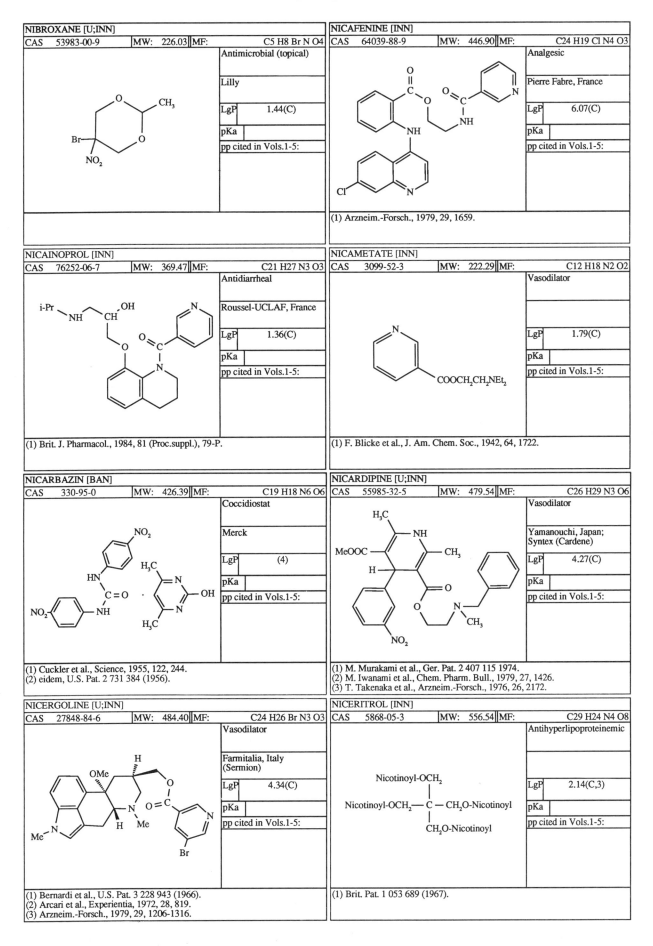

NIBROXANE [U;INN]

CAS	53983-00-9	MW: 226.03	MF:	C5 H8 Br N O4

Antimicrobial (topical)

Lilly

LgP	1.44(C)
pKa	
pp cited in Vols.1-5:	

NICAINOPROL [INN]

CAS	76252-06-7	MW: 369.47	MF:	C21 H27 N3 O3

Antidiarrheal

Roussel-UCLAF, France

LgP	1.36(C)
pKa	
pp cited in Vols.1-5:	

(1) Brit. J. Pharmacol., 1984, 81 (Proc.suppl.), 79-P.

NICARBAZIN [BAN]

CAS	330-95-0	MW: 426.39	MF:	C19 H18 N6 O6

Coccidiostat

Merck

LgP	(4)
pKa	
pp cited in Vols.1-5:	

(1) Cuckler et al., Science, 1955, 122, 244.
(2) eidem, U.S. Pat. 2 731 384 (1956).

NICERGOLINE [U;INN]

CAS	27848-84-6	MW: 484.40	MF:	C24 H26 Br N3 O3

Vasodilator

Farmitalia, Italy
(Sermion)

LgP	4.34(C)
pKa	
pp cited in Vols.1-5:	

(1) Bernardi et al., U.S. Pat. 3 228 943 (1966).
(2) Arcari et al., Experientia, 1972, 28, 819.
(3) Arzneim.-Forsch., 1979, 29, 1206-1316.

NICAFENINE [INN]

CAS	64039-88-9	MW: 446.90	MF:	C24 H19 Cl N4 O3

Analgesic

Pierre Fabre, France

LgP	6.07(C)
pKa	
pp cited in Vols.1-5:	

(1) Arzneim.-Forsch., 1979, 29, 1659.

NICAMETATE [INN]

CAS	3099-52-3	MW: 222.29	MF:	C12 H18 N2 O2

Vasodilator

LgP	1.79(C)
pKa	
pp cited in Vols.1-5:	

(1) F. Blicke et al., J. Am. Chem. Soc., 1942, 64, 1722.

NICARDIPINE [U;INN]

CAS	55985-32-5	MW: 479.54	MF:	C26 H29 N3 O6

Vasodilator

Yamanouchi, Japan;
Syntex (Cardene)

LgP	4.27(C)
pKa	
pp cited in Vols.1-5:	

(1) M. Murakami et al., Ger. Pat. 2 407 115 1974.
(2) M. Iwanami et al., Chem. Pharm. Bull., 1979, 27, 1426.
(3) T. Takenaka et al., Arzneim.-Forsch., 1976, 26, 2172.

NICERITROL [INN]

CAS	5868-05-3	MW: 556.54	MF:	C29 H24 N4 O8

Antihyperlipoproteinemic

LgP	2.14(C,3)
pKa	
pp cited in Vols.1-5:	

(1) Brit. Pat. 1 053 689 (1967).

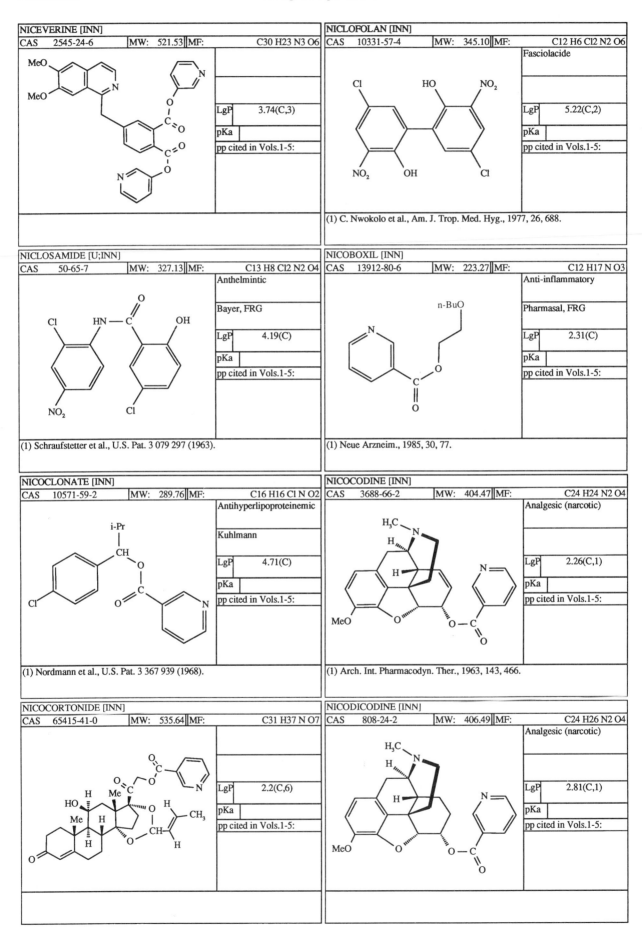

NICEVERINE [INN]

CAS	2545-24-6	MW:	521.53	MF:	C30 H23 N3 O6

LgP	3.74(C,3)
pKa	
pp cited in Vols.1-5:	

NICLOFOLAN [INN]

CAS	10331-57-4	MW:	345.10	MF:	C12 H6 Cl2 N2 O6

Fasciolacide

LgP	5.22(C,2)
pKa	
pp cited in Vols.1-5:	

(1) C. Nwokolo et al., Am. J. Trop. Med. Hyg., 1977, 26, 688.

NICLOSAMIDE [U;INN]

CAS	50-65-7	MW:	327.13	MF:	C13 H8 Cl2 N2 O4

Anthelmintic

Bayer, FRG

LgP	4.19(C)
pKa	
pp cited in Vols.1-5:	

(1) Schraufstetter et al., U.S. Pat. 3 079 297 (1963).

NICOBOXIL [INN]

CAS	13912-80-6	MW:	223.27	MF:	C12 H17 N O3

Anti-inflammatory

Pharmasal, FRG

LgP	2.31(C)
pKa	
pp cited in Vols.1-5:	

(1) Neue Arzneim., 1985, 30, 77.

NICOCLONATE [INN]

CAS	10571-59-2	MW:	289.76	MF:	C16 H16 Cl N O2

Antihyperlipoproteinemic

Kuhlmann

LgP	4.71(C)
pKa	
pp cited in Vols.1-5:	

(1) Nordmann et al., U.S. Pat. 3 367 939 (1968).

NICOCODINE [INN]

CAS	3688-66-2	MW:	404.47	MF:	C24 H24 N2 O4

Analgesic (narcotic)

LgP	2.26(C,1)
pKa	
pp cited in Vols.1-5:	

(1) Arch. Int. Pharmacodyn. Ther., 1963, 143, 466.

NICOCORTONIDE [INN]

CAS	65415-41-0	MW:	535.64	MF:	C31 H37 N O7

LgP	2.2(C,6)
pKa	
pp cited in Vols.1-5:	

NICODICODINE [INN]

CAS	808-24-2	MW:	406.49	MF:	C24 H26 N2 O4

Analgesic (narcotic)

LgP	2.81(C,1)
pKa	
pp cited in Vols.1-5:	

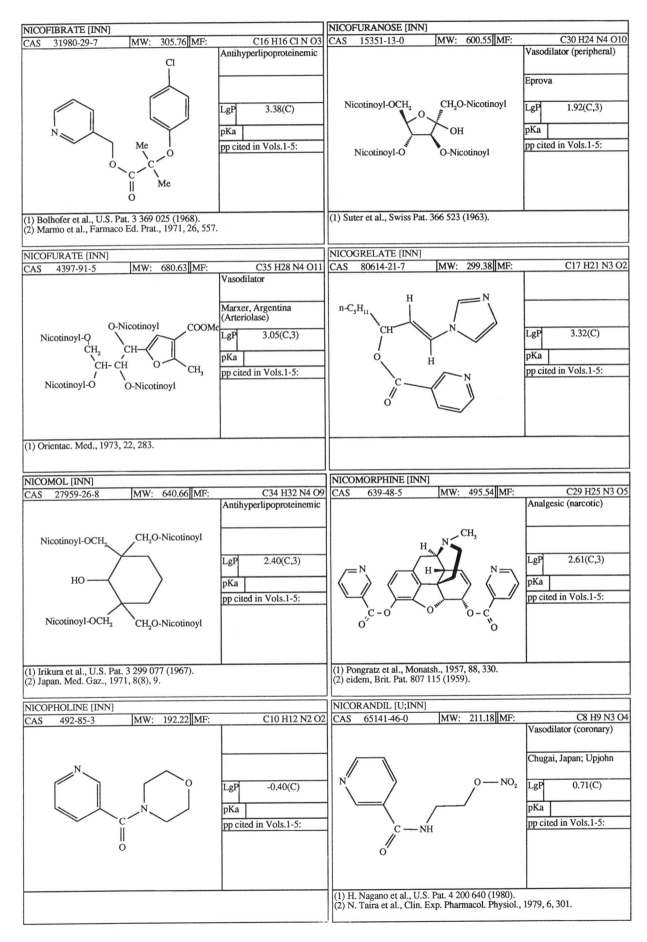

NICOFIBRATE [INN]

CAS 31980-29-7	MW: 305.76	MF:	C16 H16 Cl N O3

Antihyperlipoproteinemic

LgP	3.38(C)
pKa	
pp cited in Vols.1-5:	

(1) Bolhofer et al., U.S. Pat. 3 369 025 (1968).
(2) Marmo et al., Farmaco Ed. Prat., 1971, 26, 557.

NICOFURANOSE [INN]

CAS 15351-13-0	MW: 600.55	MF:	C30 H24 N4 O10

Vasodilator (peripheral)

Eprova

LgP	1.92(C,3)
pKa	
pp cited in Vols.1-5:	

(1) Suter et al., Swiss Pat. 366 523 (1963).

NICOFURATE [INN]

CAS 4397-91-5	MW: 680.63	MF:	C35 H28 N4 O11

Vasodilator

Marxer, Argentina (Arteriolase)

LgP	3.05(C,3)
pKa	
pp cited in Vols.1-5:	

(1) Orientac. Med., 1973, 22, 283.

NICOGRELATE [INN]

CAS 80614-21-7	MW: 299.38	MF:	C17 H21 N3 O2

LgP	3.32(C)
pKa	
pp cited in Vols.1-5:	

NICOMOL [INN]

CAS 27959-26-8	MW: 640.66	MF:	C34 H32 N4 O9

Antihyperlipoproteinemic

LgP	2.40(C,3)
pKa	
pp cited in Vols.1-5:	

(1) Irikura et al., U.S. Pat. 3 299 077 (1967).
(2) Japan. Med. Gaz., 1971, 8(8), 9.

NICOMORPHINE [INN]

CAS 639-48-5	MW: 495.54	MF:	C29 H25 N3 O5

Analgesic (narcotic)

LgP	2.61(C,3)
pKa	
pp cited in Vols.1-5:	

(1) Pongratz et al., Monatsh., 1957, 88, 330.
(2) eidem, Brit. Pat. 807 115 (1959).

NICOPHOLINE [INN]

CAS 492-85-3	MW: 192.22	MF:	C10 H12 N2 O2

LgP	-0.40(C)
pKa	
pp cited in Vols.1-5:	

NICORANDIL [U;INN]

CAS 65141-46-0	MW: 211.18	MF:	C8 H9 N3 O4

Vasodilator (coronary)

Chugai, Japan; Upjohn

LgP	0.71(C)
pKa	
pp cited in Vols.1-5:	

(1) H. Nagano et al., U.S. Pat. 4 200 640 (1980).
(2) N. Taira et al., Clin. Exp. Pharmacol. Physiol., 1979, 6, 301.

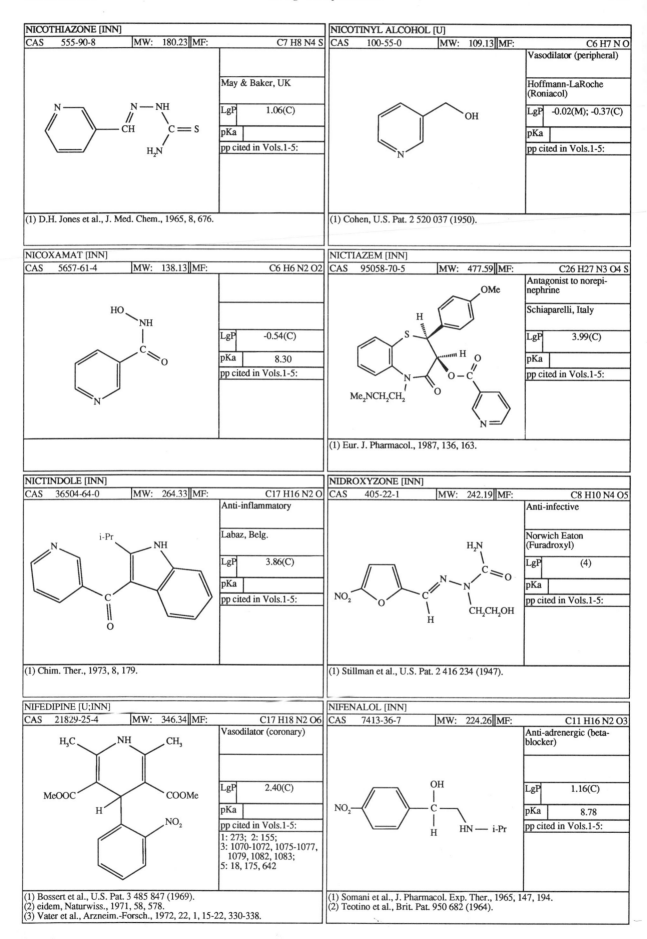

NICOTHIAZONE [INN]

CAS	555-90-8	MW:	180.23	MF:	C7 H8 N4 S

	May & Baker, UK
LgP	1.06(C)
pKa	
pp cited in Vols.1-5:	

(1) D.H. Jones et al., J. Med. Chem., 1965, 8, 676.

NICOTINYL ALCOHOL [U]

CAS	100-55-0	MW:	109.13	MF:	C6 H7 N O

Vasodilator (peripheral)
Hoffmann-LaRoche (Roniacol)
LgP	-0.02(M); -0.37(C)
pKa	
pp cited in Vols.1-5:	

(1) Cohen, U.S. Pat. 2 520 037 (1950).

NICOXAMAT [INN]

CAS	5657-61-4	MW:	138.13	MF:	C6 H6 N2 O2

LgP	-0.54(C)
pKa	8.30
pp cited in Vols.1-5:	

NICTIAZEM [INN]

CAS	95058-70-5	MW:	477.59	MF:	C26 H27 N3 O4 S

Antagonist to norepi-nephrine
Schiaparelli, Italy
LgP	3.99(C)
pKa	
pp cited in Vols.1-5:	

(1) Eur. J. Pharmacol., 1987, 136, 163.

NICTINDOLE [INN]

CAS	36504-64-0	MW:	264.33	MF:	C17 H16 N2 O

Anti-inflammatory
Labaz, Belg.
LgP	3.86(C)
pKa	
pp cited in Vols.1-5:	

(1) Chim. Ther., 1973, 8, 179.

NIDROXYZONE [INN]

CAS	405-22-1	MW:	242.19	MF:	C8 H10 N4 O5

Anti-infective
Norwich Eaton (Furadroxyl)
LgP	(4)
pKa	
pp cited in Vols.1-5:	

(1) Stillman et al., U.S. Pat. 2 416 234 (1947).

NIFEDIPINE [U;INN]

CAS	21829-25-4	MW:	346.34	MF:	C17 H18 N2 O6

Vasodilator (coronary)
LgP	2.40(C)
pKa	
pp cited in Vols.1-5:
1: 273; 2: 155; 3: 1070-1072, 1075-1077, 1079, 1082, 1083; 5: 18, 175, 642

(1) Bossert et al., U.S. Pat. 3 485 847 (1969).
(2) eidem, Naturwiss., 1971, 58, 578.
(3) Vater et al., Arzneim.-Forsch., 1972, 22, 1, 15-22, 330-338.

NIFENALOL [INN]

CAS	7413-36-7	MW:	224.26	MF:	C11 H16 N2 O3

Anti-adrenergic (beta-blocker)
LgP	1.16(C)
pKa	8.78
pp cited in Vols.1-5:	

(1) Somani et al., J. Pharmacol. Exp. Ther., 1965, 147, 194.
(2) Teotino et al., Brit. Pat. 950 682 (1964).

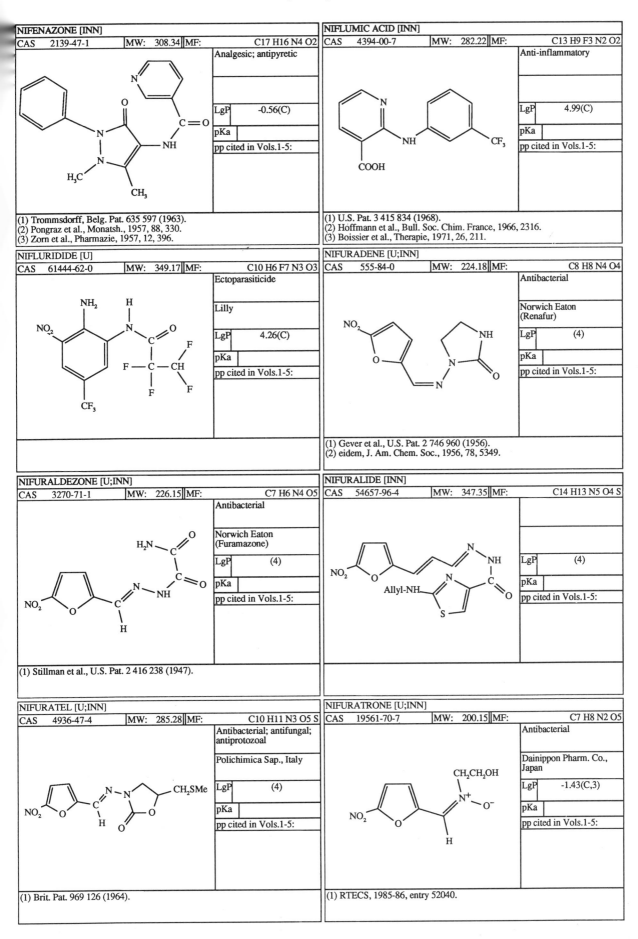

NIFENAZONE [INN]

CAS	2139-47-1	MW: 308.34	MF:	C17 H16 N4 O2

Analgesic; antipyretic

LgP	-0.56(C)
pKa	
pp cited in Vols.1-5:	

(1) Trommsdorff, Belg. Pat. 635 597 (1963).
(2) Pongraz et al., Monatsh., 1957, 88, 330.
(3) Zorn et al., Pharmazie, 1957, 12, 396.

NIFLUMIC ACID [INN]

CAS	4394-00-7	MW: 282.22	MF:	C13 H9 F3 N2 O2

Anti-inflammatory

LgP	4.99(C)
pKa	
pp cited in Vols.1-5:	

(1) U.S. Pat. 3 415 834 (1968).
(2) Hoffmann et al., Bull. Soc. Chim. France, 1966, 2316.
(3) Boissier et al., Therapie, 1971, 26, 211.

NIFLURIDIDE [U]

CAS	61444-62-0	MW: 349.17	MF:	C10 H6 F7 N3 O3

Ectoparasiticide

Lilly

LgP	4.26(C)
pKa	
pp cited in Vols.1-5:	

NIFURADENE [U;INN]

CAS	555-84-0	MW: 224.18	MF:	C8 H8 N4 O4

Antibacterial

Norwich Eaton
(Renafur)

LgP	(4)
pKa	
pp cited in Vols.1-5:	

(1) Gever et al., U.S. Pat. 2 746 960 (1956).
(2) eidem, J. Am. Chem. Soc., 1956, 78, 5349.

NIFURALDEZONE [U;INN]

CAS	3270-71-1	MW: 226.15	MF:	C7 H6 N4 O5

Antibacterial

Norwich Eaton
(Furamazone)

LgP	(4)
pKa	
pp cited in Vols.1-5:	

(1) Stillman et al., U.S. Pat. 2 416 238 (1947).

NIFURALIDE [INN]

CAS	54657-96-4	MW: 347.35	MF:	C14 H13 N5 O4 S

LgP	(4)
pKa	
pp cited in Vols.1-5:	

NIFURATEL [U;INN]

CAS	4936-47-4	MW: 285.28	MF:	C10 H11 N3 O5 S

Antibacterial; antifungal;
antiprotozoal

Polichimica Sap., Italy

LgP	(4)
pKa	
pp cited in Vols.1-5:	

(1) Brit. Pat. 969 126 (1964).

NIFURATRONE [U;INN]

CAS	19561-70-7	MW: 200.15	MF:	C7 H8 N2 O5

Antibacterial

Dainippon Pharm. Co.,
Japan

LgP	-1.43(C,3)
pKa	
pp cited in Vols.1-5:	

(1) RTECS, 1985-86, entry 52040.

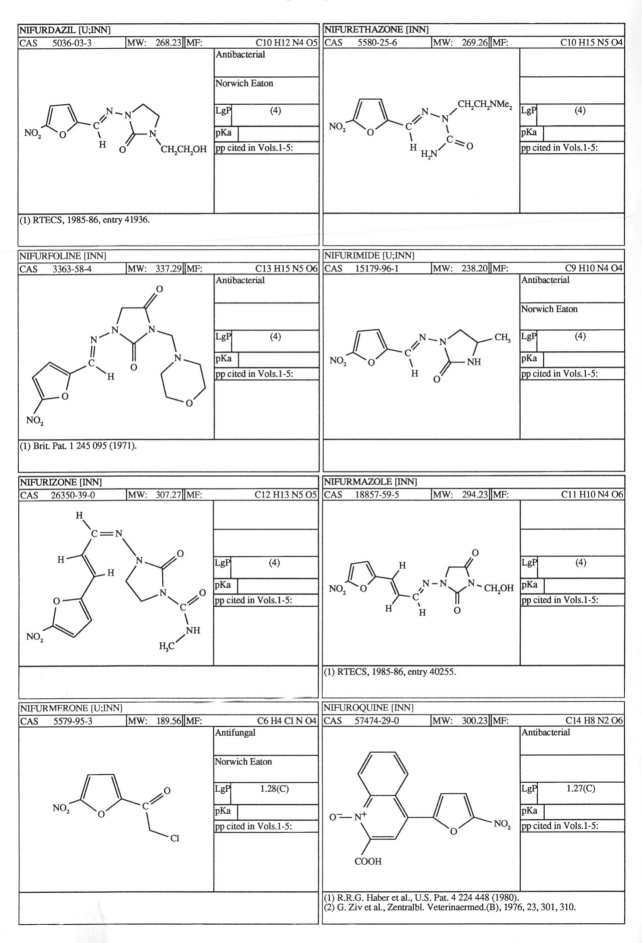

NIFURDAZIL [U;INN]

CAS	5036-03-3	MW:	268.23	MF:	C10 H12 N4 O5

Antibacterial

Norwich Eaton

LgP	(4)
pKa	
pp cited in Vols.1-5:	

(1) RTECS, 1985-86, entry 41936.

NIFURETHAZONE [INN]

CAS	5580-25-6	MW:	269.26	MF:	C10 H15 N5 O4

LgP	(4)
pKa	
pp cited in Vols.1-5:	

NIFURFOLINE [INN]

CAS	3363-58-4	MW:	337.29	MF:	C13 H15 N5 O6

Antibacterial

LgP	(4)
pKa	
pp cited in Vols.1-5:	

(1) Brit. Pat. 1 245 095 (1971).

NIFURIMIDE [U;INN]

CAS	15179-96-1	MW:	238.20	MF:	C9 H10 N4 O4

Antibacterial

Norwich Eaton

LgP	(4)
pKa	
pp cited in Vols.1-5:	

NIFURIZONE [INN]

CAS	26350-39-0	MW:	307.27	MF:	C12 H13 N5 O5

LgP	(4)
pKa	
pp cited in Vols.1-5:	

NIFURMAZOLE [INN]

CAS	18857-59-5	MW:	294.23	MF:	C11 H10 N4 O6

LgP	(4)
pKa	
pp cited in Vols.1-5:	

(1) RTECS, 1985-86, entry 40255.

NIFURMERONE [U;INN]

CAS	5579-95-3	MW:	189.56	MF:	C6 H4 Cl N O4

Antifungal

Norwich Eaton

LgP	1.28(C)
pKa	
pp cited in Vols.1-5:	

NIFUROQUINE [INN]

CAS	57474-29-0	MW:	300.23	MF:	C14 H8 N2 O6

Antibacterial

LgP	1.27(C)
pKa	
pp cited in Vols.1-5:	

(1) R.R.G. Haber et al., U.S. Pat. 4 224 448 (1980).
(2) G. Ziv et al., Zentralbl. Veterinaermed.(B), 1976, 23, 301, 310.

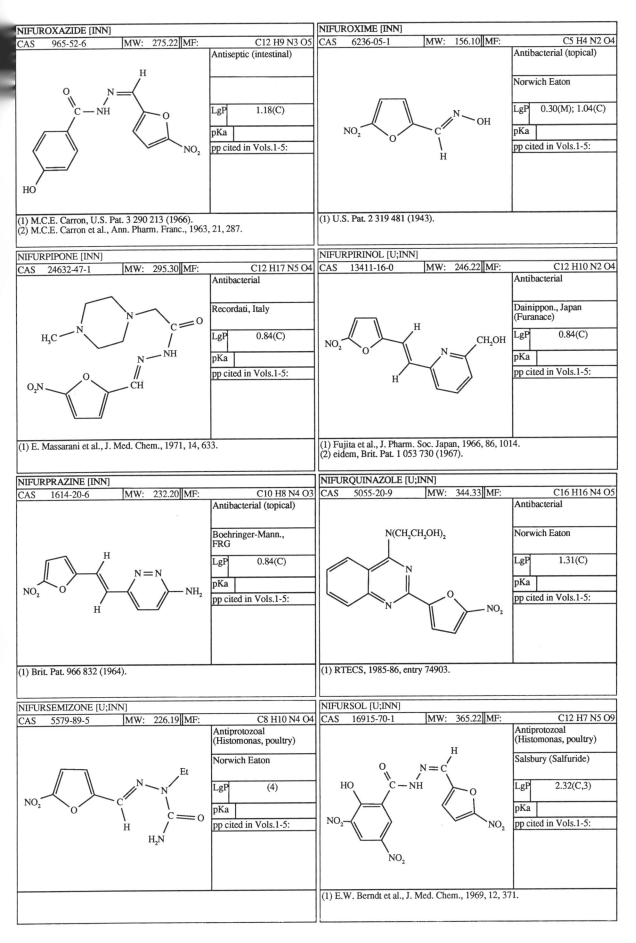

NIFUROXAZIDE [INN]

CAS	965-52-6	MW:	275.22	MF:	C12 H9 N3 O5

Antiseptic (intestinal)

LgP 1.18(C)

pKa

pp cited in Vols.1-5:

(1) M.C.E. Carron, U.S. Pat. 3 290 213 (1966).
(2) M.C.E. Carron et al., Ann. Pharm. Franc., 1963, 21, 287.

NIFUROXIME [INN]

CAS	6236-05-1	MW:	156.10	MF:	C5 H4 N2 O4

Antibacterial (topical)

Norwich Eaton

LgP 0.30(M); 1.04(C)

pKa

pp cited in Vols.1-5:

(1) U.S. Pat. 2 319 481 (1943).

NIFURPIPONE [INN]

CAS	24632-47-1	MW:	295.30	MF:	C12 H17 N5 O4

Antibacterial

Recordati, Italy

LgP 0.84(C)

pKa

pp cited in Vols.1-5:

(1) E. Massarani et al., J. Med. Chem., 1971, 14, 633.

NIFURPIRINOL [U;INN]

CAS	13411-16-0	MW:	246.22	MF:	C12 H10 N2 O4

Antibacterial

Dainippon., Japan
(Furanace)

LgP 0.84(C)

pKa

pp cited in Vols.1-5:

(1) Fujita et al., J. Pharm. Soc. Japan, 1966, 86, 1014.
(2) eidem, Brit. Pat. 1 053 730 (1967).

NIFURPRAZINE [INN]

CAS	1614-20-6	MW:	232.20	MF:	C10 H8 N4 O3

Antibacterial (topical)

Boehringer-Mann.,
FRG

LgP 0.84(C)

pKa

pp cited in Vols.1-5:

(1) Brit. Pat. 966 832 (1964).

NIFURQUINAZOLE [U;INN]

CAS	5055-20-9	MW:	344.33	MF:	C16 H16 N4 O5

Antibacterial

Norwich Eaton

LgP 1.31(C)

pKa

pp cited in Vols.1-5:

(1) RTECS, 1985-86, entry 74903.

NIFURSEMIZONE [U;INN]

CAS	5579-89-5	MW:	226.19	MF:	C8 H10 N4 O4

Antiprotozoal
(Histomonas, poultry)

Norwich Eaton

LgP (4)

pKa

pp cited in Vols.1-5:

NIFURSOL [U;INN]

CAS	16915-70-1	MW:	365.22	MF:	C12 H7 N5 O9

Antiprotozoal
(Histomonas, poultry)

Salsbury (Salfuride)

LgP 2.32(C,3)

pKa

pp cited in Vols.1-5:

(1) E.W. Berndt et al., J. Med. Chem., 1969, 12, 371.

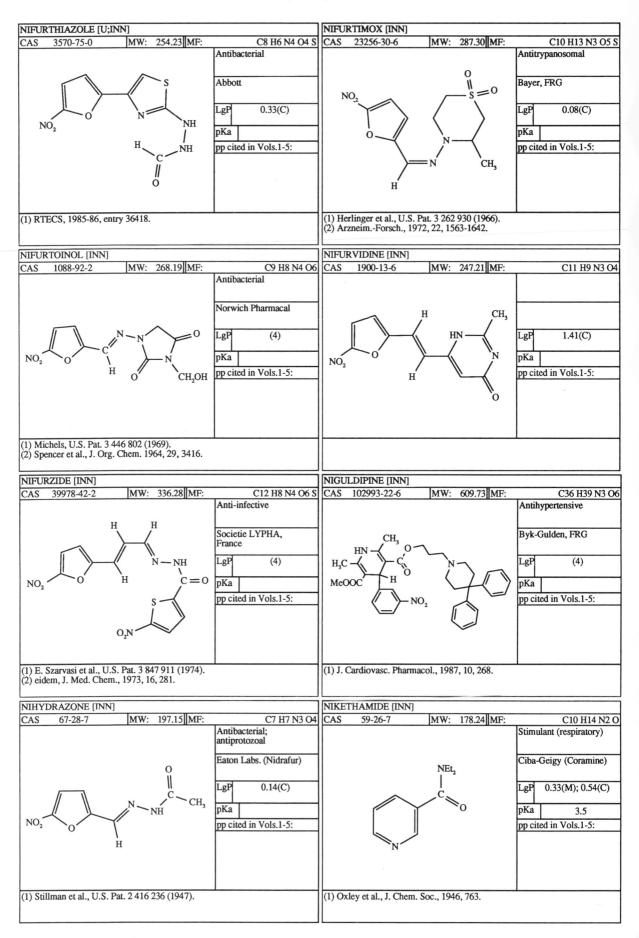

NIFURTHIAZOLE [U;INN]

CAS	3570-75-0	MW:	254.23	MF:	C8 H6 N4 O4 S

Antibacterial

Abbott

LgP	0.33(C)

pKa	

pp cited in Vols.1-5:	

(1) RTECS, 1985-86, entry 36418.

NIFURTOINOL [INN]

CAS	1088-92-2	MW:	268.19	MF:	C9 H8 N4 O6

Antibacterial

Norwich Pharmacal

LgP	(4)

pKa	

pp cited in Vols.1-5:	

(1) Michels, U.S. Pat. 3 446 802 (1969).
(2) Spencer et al., J. Org. Chem. 1964, 29, 3416.

NIFURZIDE [INN]

CAS	39978-42-2	MW:	336.28	MF:	C12 H8 N4 O6 S

Anti-infective

Societie LYPHA, France

LgP	(4)

pKa	

pp cited in Vols.1-5:	

(1) E. Szarvasi et al., U.S. Pat. 3 847 911 (1974).
(2) eidem, J. Med. Chem., 1973, 16, 281.

NIHYDRAZONE [INN]

CAS	67-28-7	MW:	197.15	MF:	C7 H7 N3 O4

Antibacterial; antiprotozoal

Eaton Labs. (Nidrafur)

LgP	0.14(C)

pKa	

pp cited in Vols.1-5:	

(1) Stillman et al., U.S. Pat. 2 416 236 (1947).

NIFURTIMOX [INN]

CAS	23256-30-6	MW:	287.30	MF:	C10 H13 N3 O5 S

Antitrypanosomal

Bayer, FRG

LgP	0.08(C)

pKa	

pp cited in Vols.1-5:	

(1) Herlinger et al., U.S. Pat. 3 262 930 (1966).
(2) Arzneim.-Forsch., 1972, 22, 1563-1642.

NIFURVIDINE [INN]

CAS	1900-13-6	MW:	247.21	MF:	C11 H9 N3 O4

Antibacterial

LgP	1.41(C)

pKa	

pp cited in Vols.1-5:	

NIGULDIPINE [INN]

CAS	102993-22-6	MW:	609.73	MF:	C36 H39 N3 O6

Antihypertensive

Byk-Gulden, FRG

LgP	(4)

pKa	

pp cited in Vols.1-5:	

(1) J. Cardiovasc. Pharmacol., 1987, 10, 268.

NIKETHAMIDE [INN]

CAS	59-26-7	MW:	178.24	MF:	C10 H14 N2 O

Stimulant (respiratory)

Ciba-Geigy (Coramine)

LgP	0.33(M); 0.54(C)

pKa	3.5

pp cited in Vols.1-5:	

(1) Oxley et al., J. Chem. Soc., 1946, 763.

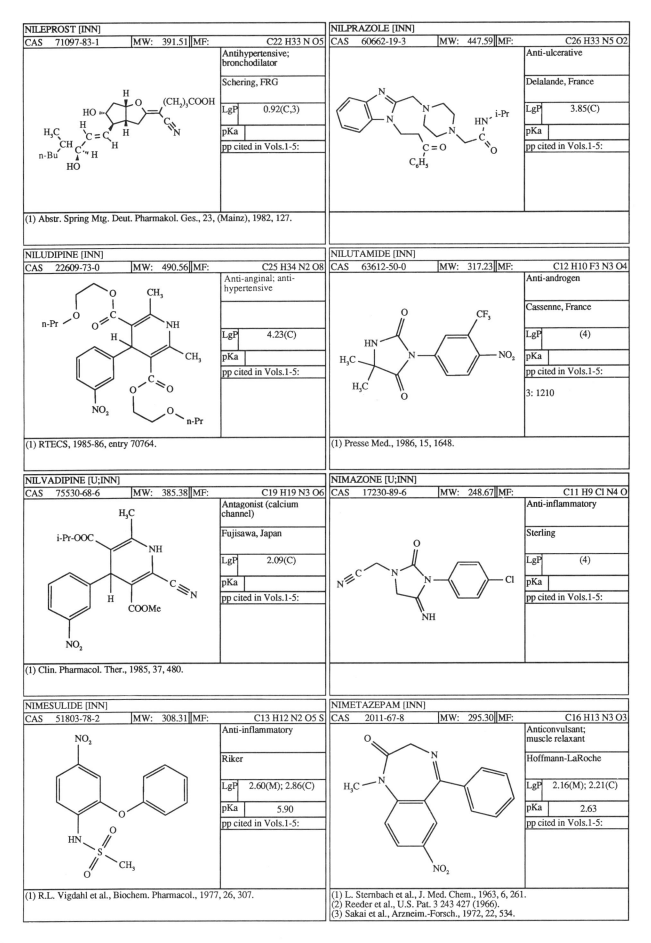

NILEPROST [INN]

CAS 71097-83-1	MW: 391.51	MF: C22 H33 N O5

Antihypertensive; bronchodilator

Schering, FRG

LgP	0.92(C,3)
pKa	
pp cited in Vols.1-5:	

(1) Abstr. Spring Mtg. Deut. Pharmakol. Ges., 23, (Mainz), 1982, 127.

NILPRAZOLE [INN]

CAS 60662-19-3	MW: 447.59	MF: C26 H33 N5 O2

Anti-ulcerative

Delalande, France

LgP	3.85(C)
pKa	
pp cited in Vols.1-5:	

NILUDIPINE [INN]

CAS 22609-73-0	MW: 490.56	MF: C25 H34 N2 O8

Anti-anginal; anti-hypertensive

LgP	4.23(C)
pKa	
pp cited in Vols.1-5:	

(1) RTECS, 1985-86, entry 70764.

NILUTAMIDE [INN]

CAS 63612-50-0	MW: 317.23	MF: C12 H10 F3 N3 O4

Anti-androgen

Cassenne, France

LgP	(4)
pKa	
pp cited in Vols.1-5:	
3: 1210	

(1) Presse Med., 1986, 15, 1648.

NILVADIPINE [U;INN]

CAS 75530-68-6	MW: 385.38	MF: C19 H19 N3 O6

Antagonist (calcium channel)

Fujisawa, Japan

LgP	2.09(C)
pKa	
pp cited in Vols.1-5:	

(1) Clin. Pharmacol. Ther., 1985, 37, 480.

NIMAZONE [U;INN]

CAS 17230-89-6	MW: 248.67	MF: C11 H9 Cl N4 O

Anti-inflammatory

Sterling

LgP	(4)
pKa	
pp cited in Vols.1-5:	

NIMESULIDE [INN]

CAS 51803-78-2	MW: 308.31	MF: C13 H12 N2 O5 S

Anti-inflammatory

Riker

LgP	2.60(M); 2.86(C)
pKa	5.90
pp cited in Vols.1-5:	

(1) R.L. Vigdahl et al., Biochem. Pharmacol., 1977, 26, 307.

NIMETAZEPAM [INN]

CAS 2011-67-8	MW: 295.30	MF: C16 H13 N3 O3

Anticonvulsant; muscle relaxant

Hoffmann-LaRoche

LgP	2.16(M); 2.21(C)
pKa	2.63
pp cited in Vols.1-5:	

(1) L. Sternbach et al., J. Med. Chem., 1963, 6, 261.
(2) Reeder et al., U.S. Pat. 3 243 427 (1966).
(3) Sakai et al., Arzneim.-Forsch., 1972, 22, 534.

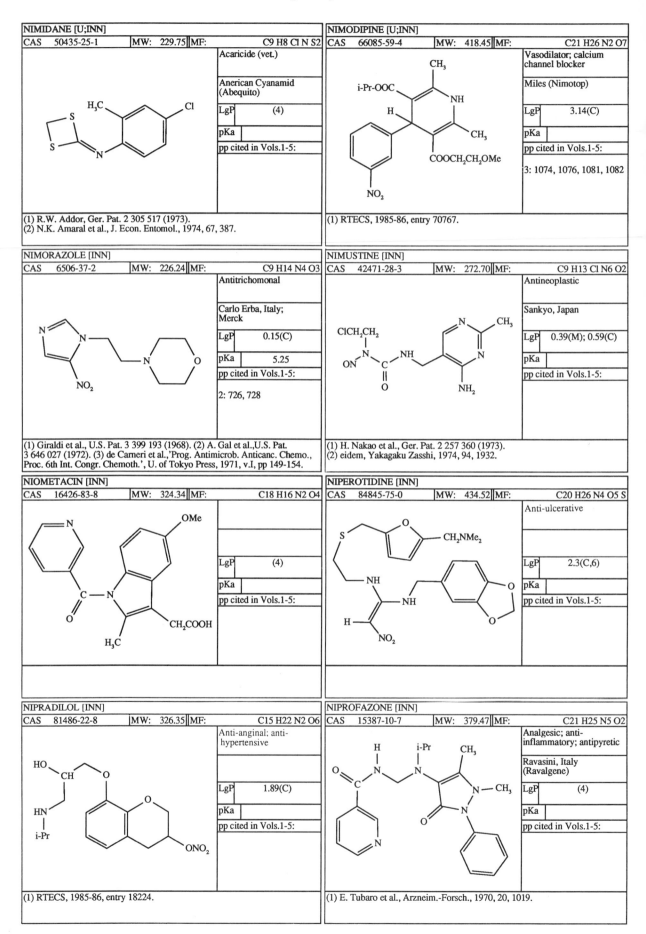

NIMIDANE [U;INN]

CAS	50435-25-1	MW:	229.75	MF:	C9 H8 Cl N S2

Acaricide (vet.)

Anerican Cyanamid (Abequito)

LgP	(4)
pKa	
pp cited in Vols.1-5:	

(1) R.W. Addor, Ger. Pat. 2 305 517 (1973).
(2) N.K. Amaral et al., J. Econ. Entomol., 1974, 67, 387.

NIMODIPINE [U;INN]

CAS	66085-59-4	MW:	418.45	MF:	C21 H26 N2 O7

Vasodilator; calcium channel blocker

Miles (Nimotop)

LgP	3.14(C)
pKa	
pp cited in Vols.1-5:	

3: 1074, 1076, 1081, 1082

(1) RTECS, 1985-86, entry 70767.

NIMORAZOLE [INN]

CAS	6506-37-2	MW:	226.24	MF:	C9 H14 N4 O3

Antitrichomonal

Carlo Erba, Italy; Merck

LgP	0.15(C)
pKa	5.25
pp cited in Vols.1-5:	

2: 726, 728

(1) Giraldi et al., U.S. Pat. 3 399 193 (1968). (2) A. Gal et al.,U.S. Pat. 3 646 027 (1972). (3) de Carneri et al.,'Prog. Antimicrob. Anticanc. Chemo., Proc. 6th Int. Congr. Chemoth.', U. of Tokyo Press, 1971, v.I, pp 149-154.

NIMUSTINE [INN]

CAS	42471-28-3	MW:	272.70	MF:	C9 H13 Cl N6 O2

Antineoplastic

Sankyo, Japan

LgP	0.39(M); 0.59(C)
pKa	
pp cited in Vols.1-5:	

(1) H. Nakao et al., Ger. Pat. 2 257 360 (1973).
(2) eidem, Yakagaku Zasshi, 1974, 94, 1932.

NIOMETACIN [INN]

CAS	16426-83-8	MW:	324.34	MF:	C18 H16 N2 O4

LgP	(4)
pKa	
pp cited in Vols.1-5:	

NIPEROTIDINE [INN]

CAS	84845-75-0	MW:	434.52	MF:	C20 H26 N4 O5 S

Anti-ulcerative

LgP	2.3(C,6)
pKa	
pp cited in Vols.1-5:	

NIPRADILOL [INN]

CAS	81486-22-8	MW:	326.35	MF:	C15 H22 N2 O6

Anti-anginal; anti-hypertensive

LgP	1.89(C)
pKa	
pp cited in Vols.1-5:	

(1) RTECS, 1985-86, entry 18224.

NIPROFAZONE [INN]

CAS	15387-10-7	MW:	379.47	MF:	C21 H25 N5 O2

Analgesic; anti-inflammatory; antipyretic

Ravasini, Italy (Ravalgene)

LgP	(4)
pKa	
pp cited in Vols.1-5:	

(1) E. Tubaro et al., Arzneim.-Forsch., 1970, 20, 1019.

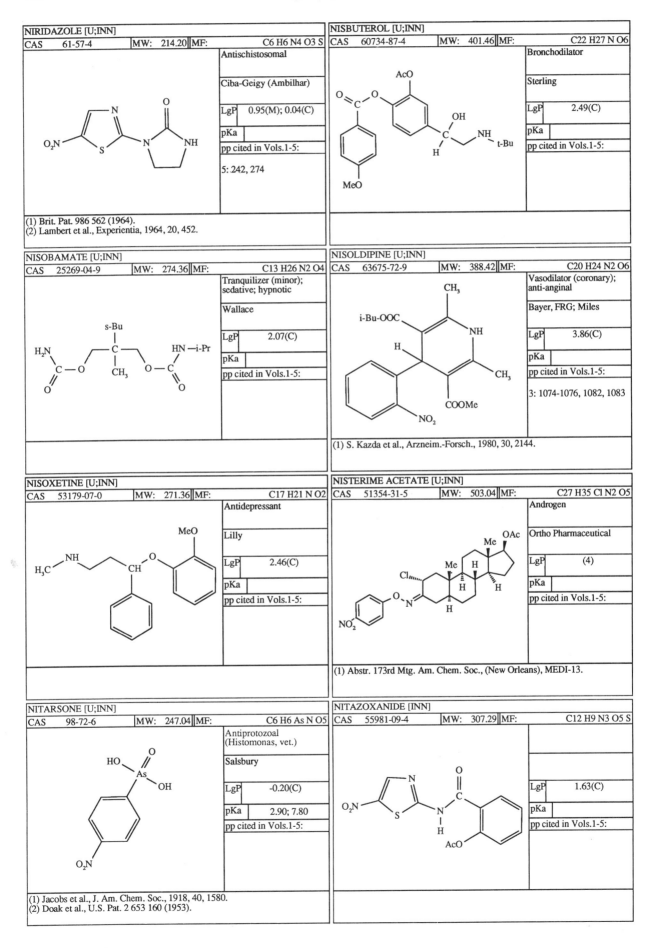

NIRIDAZOLE [U;INN]

CAS	61-57-4	MW:	214.20	MF:	C6 H6 N4 O3 S

Antischistosomal

Ciba-Geigy (Ambilhar)

LgP	0.95(M); 0.04(C)
pKa	
pp cited in Vols.1-5:	

5: 242, 274

(1) Brit. Pat. 986 562 (1964).
(2) Lambert et al., Experientia, 1964, 20, 452.

NISBUTEROL [U;INN]

CAS	60734-87-4	MW:	401.46	MF:	C22 H27 N O6

Bronchodilator

Sterling

LgP	2.49(C)
pKa	
pp cited in Vols.1-5:	

NISOBAMATE [U;INN]

CAS	25269-04-9	MW:	274.36	MF:	C13 H26 N2 O4

Tranquilizer (minor); sedative; hypnotic

Wallace

LgP	2.07(C)
pKa	
pp cited in Vols.1-5:	

NISOLDIPINE [U;INN]

CAS	63675-72-9	MW:	388.42	MF:	C20 H24 N2 O6

Vasodilator (coronary); anti-anginal

Bayer, FRG; Miles

LgP	3.86(C)
pKa	
pp cited in Vols.1-5:	

3: 1074-1076, 1082, 1083

(1) S. Kazda et al., Arzneim.-Forsch., 1980, 30, 2144.

NISOXETINE [U;INN]

CAS	53179-07-0	MW:	271.36	MF:	C17 H21 N O2

Antidepressant

Lilly

LgP	2.46(C)
pKa	
pp cited in Vols.1-5:	

NISTERIME ACETATE [U;INN]

CAS	51354-31-5	MW:	503.04	MF:	C27 H35 Cl N2 O5

Androgen

Ortho Pharmaceutical

LgP	(4)
pKa	
pp cited in Vols.1-5:	

(1) Abstr. 173rd Mtg. Am. Chem. Soc., (New Orleans), MEDI-13.

NITARSONE [U;INN]

CAS	98-72-6	MW:	247.04	MF:	C6 H6 As N O5

Antiprotozoal (Histomonas, vet.)

Salsbury

LgP	-0.20(C)
pKa	2.90; 7.80
pp cited in Vols.1-5:	

(1) Jacobs et al., J. Am. Chem. Soc., 1918, 40, 1580.
(2) Doak et al., U.S. Pat. 2 653 160 (1953).

NITAZOXANIDE [INN]

CAS	55981-09-4	MW:	307.29	MF:	C12 H9 N3 O5 S

LgP	1.63(C)
pKa	
pp cited in Vols.1-5:	

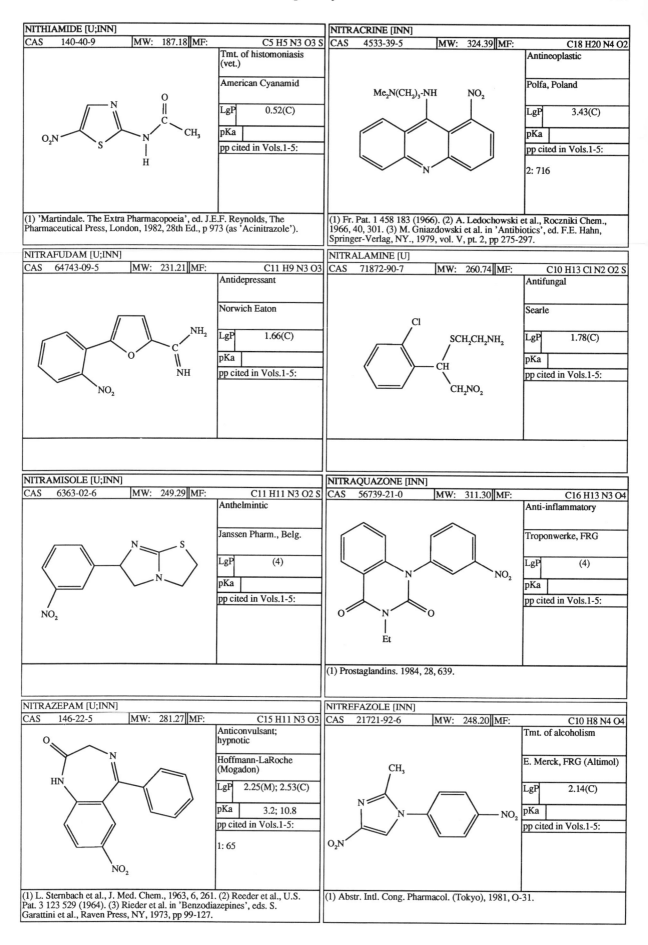

NITHIAMIDE [U;INN]

CAS	140-40-9	MW:	187.18	MF:	C5 H5 N3 O3 S

Tmt. of histomoniasis (vet.)

American Cyanamid

LgP	0.52(C)
pKa	
pp cited in Vols.1-5:	

(1) 'Martindale. The Extra Pharmacopoeia', ed. J.E.F. Reynolds, The Pharmaceutical Press, London, 1982, 28th Ed., p 973 (as 'Acinitrazole').

NITRACRINE [INN]

CAS	4533-39-5	MW:	324.39	MF:	C18 H20 N4 O2

Antineoplastic

Polfa, Poland

LgP	3.43(C)
pKa	
pp cited in Vols.1-5:	

2: 716

(1) Fr. Pat. 1 458 183 (1966). (2) A. Ledochowski et al., Roczniki Chem., 1966, 40, 301. (3) M. Gniazdowski et al. in 'Antibiotics', ed. F.E. Hahn, Springer-Verlag, NY., 1979, vol. V, pt. 2, pp 275-297.

NITRAFUDAM [U;INN]

CAS	64743-09-5	MW:	231.21	MF:	C11 H9 N3 O3

Antidepressant

Norwich Eaton

LgP	1.66(C)
pKa	
pp cited in Vols.1-5:	

NITRALAMINE [U]

CAS	71872-90-7	MW:	260.74	MF:	C10 H13 Cl N2 O2 S

Antifungal

Searle

LgP	1.78(C)
pKa	
pp cited in Vols.1-5:	

NITRAMISOLE [U;INN]

CAS	6363-02-6	MW:	249.29	MF:	C11 H11 N3 O2 S

Anthelmintic

Janssen Pharm., Belg.

LgP	(4)
pKa	
pp cited in Vols.1-5:	

NITRAQUAZONE [INN]

CAS	56739-21-0	MW:	311.30	MF:	C16 H13 N3 O4

Anti-inflammatory

Troponwerke, FRG

LgP	(4)
pKa	
pp cited in Vols.1-5:	

(1) Prostaglandins. 1984, 28, 639.

NITRAZEPAM [U;INN]

CAS	146-22-5	MW:	281.27	MF:	C15 H11 N3 O3

Anticonvulsant; hypnotic

Hoffmann-LaRoche (Mogadon)

LgP	2.25(M); 2.53(C)
pKa	3.2; 10.8
pp cited in Vols.1-5:	

1: 65

(1) L. Sternbach et al., J. Med. Chem., 1963, 6, 261. (2) Reeder et al., U.S. Pat. 3 123 529 (1964). (3) Rieder et al. in 'Benzodiazepines', eds. S. Garattini et al., Raven Press, NY, 1973, pp 99-127.

NITREFAZOLE [INN]

CAS	21721-92-6	MW:	248.20	MF:	C10 H8 N4 O4

Tmt. of alcoholism

E. Merck, FRG (Altimol)

LgP	2.14(C)
pKa	
pp cited in Vols.1-5:	

(1) Abstr. Intl. Cong. Pharmacol. (Tokyo), 1981, O-31.

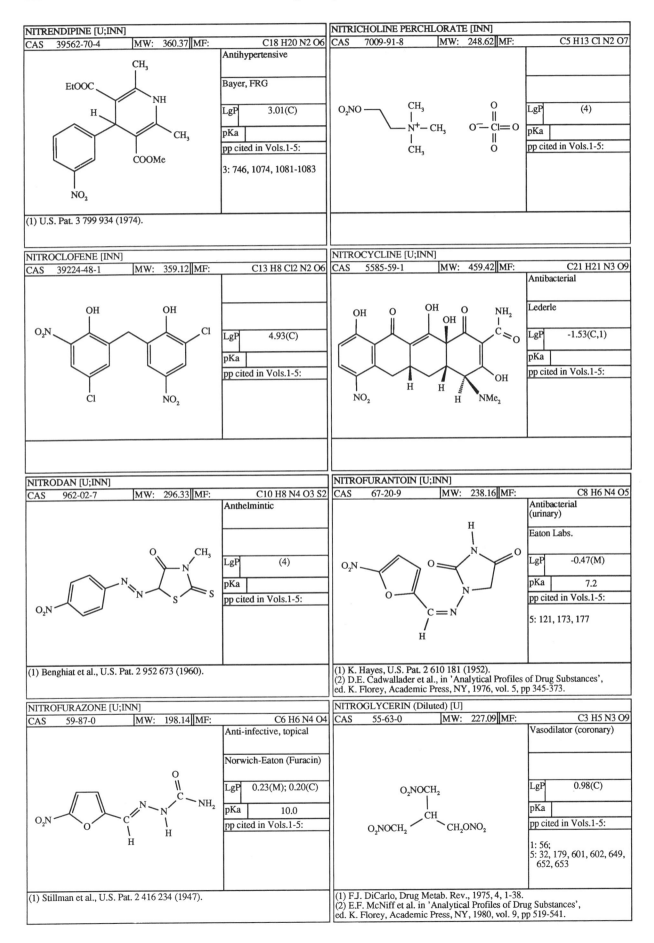

NITRENDIPINE [U;INN]

CAS	39562-70-4	MW:	360.37	MF:	C18 H20 N2 O6

Antihypertensive

Bayer, FRG

LgP	3.01(C)
pKa	
pp cited in Vols.1-5:	

3: 746, 1074, 1081-1083

(1) U.S. Pat. 3 799 934 (1974).

NITRICHOLINE PERCHLORATE [INN]

CAS	7009-91-8	MW:	248.62	MF:	C5 H13 Cl N2 O7

LgP	(4)
pKa	
pp cited in Vols.1-5:	

NITROCLOFENE [INN]

CAS	39224-48-1	MW:	359.12	MF:	C13 H8 Cl2 N2 O6

LgP	4.93(C)
pKa	
pp cited in Vols.1-5:	

NITROCYCLINE [U;INN]

CAS	5585-59-1	MW:	459.42	MF:	C21 H21 N3 O9

Antibacterial

Lederle

LgP	-1.53(C,1)
pKa	
pp cited in Vols.1-5:	

NITRODAN [U;INN]

CAS	962-02-7	MW:	296.33	MF:	C10 H8 N4 O3 S2

Anthelmintic

LgP	(4)
pKa	
pp cited in Vols.1-5:	

(1) Benghiat et al., U.S. Pat. 2 952 673 (1960).

NITROFURANTOIN [U;INN]

CAS	67-20-9	MW:	238.16	MF:	C8 H6 N4 O5

Antibacterial
(urinary)

Eaton Labs.

LgP	-0.47(M)
pKa	7.2
pp cited in Vols.1-5:	

5: 121, 173, 177

(1) K. Hayes, U.S. Pat. 2 610 181 (1952).
(2) D.E. Cadwallader et al., in 'Analytical Profiles of Drug Substances',
ed. K. Florey, Academic Press, NY, 1976, vol. 5, pp 345-373.

NITROFURAZONE [U;INN]

CAS	59-87-0	MW:	198.14	MF:	C6 H6 N4 O4

Anti-infective, topical

Norwich-Eaton (Furacin)

LgP	0.23(M); 0.20(C)
pKa	10.0
pp cited in Vols.1-5:	

(1) Stillman et al., U.S. Pat. 2 416 234 (1947).

NITROGLYCERIN (Diluted) [U]

CAS	55-63-0	MW:	227.09	MF:	C3 H5 N3 O9

Vasodilator (coronary)

LgP	0.98(C)
pKa	
pp cited in Vols.1-5:	

1: 56;
5: 32, 179, 601, 602, 649,
   652, 653

(1) F.J. DiCarlo, Drug Metab. Rev., 1975, 4, 1-38.
(2) E.F. McNiff et al. in 'Analytical Profiles of Drug Substances',
ed. K. Florey, Academic Press, NY, 1980, vol. 9, pp 519-541.

NITROMERSOL [U]

CAS 133-58-4	MW: 351.71	MF: C7 H5 Hg N O3

Anti-infective, topical

LgP (4)

pKa

pp cited in Vols.1-5:

1: 50

(1) U.S. Pat 1 544 293; reissue 17 563 (1925).

NITROMIDE [U]

CAS 121-81-3	MW: 211.14	MF: C7 H5 N3 O5

Coccidiostat (poultry); antibacterial

Salsbury

LgP 0.83(M); 0.79(C)

pKa

pp cited in Vols.1-5:

(1) Finan et al., J. Chem. Soc., 1962, 2824.

NITROMIFENE [INN]

CAS 10448-84-7	MW: 444.54	MF: C27 H28 N2 O4

Anti-estrogen

Parke-Davis

LgP 5.48(C)

pKa

pp cited in Vols.1-5:

(1) D.J. Collins et al., J. Med. Chem, 1971, 14, 952.

NITROSCANATE [U;INN]

CAS 19881-18-6	MW: 272.28	MF: C13 H8 N2 O3 S

Anthelmintic (vet.)

Agripat; Ciba-Geigy (Lopatol)

LgP 5.70(C)

pKa

pp cited in Vols.1-5:

(1) Fr. Pat. 1 491 477 (1967).
(2) Gemmell et al., Res. Vet. Sci., 1975, 19, 217.

NITROSULFATHIAZOLE [INN]

CAS 473-42-7	MW: 285.30	MF: C9 H7 N3 O4 S2

Antibacterial

Sterling (Nisulfazole)

LgP 1.59(C,3)

pKa

pp cited in Vols.1-5:

(1) Lorenz et al., U.S. Pat. 2 443 742 (1948).

NITROUS OXIDE [U]

CAS 10024-97-2	MW: 44.01	MF: N2 O

Anesthetic (inhalation)

LgP 0.43(M)

pKa

pp cited in Vols.1-5:

1: 53, 55;
3: 1090

(1) Jones in 'Comprehensive Inorganic Chemistry', eds. J.C. Bailar, Jr. et al., Pergamon Press, Oxford, 1973, vol. VIII, suppl.II, Nitrogen (pt. 2), pp 316-323.

NITROXINIL [INN]

CAS 1689-89-0	MW: 290.02	MF: C7 H3 I N2 O3

Anthelmintic (fasciolicide)

May & Baker, UK

LgP 2.58(C)

pKa

pp cited in Vols.1-5:

(1) Collins et al., U.S. Pat. 3 331 738 (1967).

NITROXOLINE [INN]

CAS 4008-48-4	MW: 190.16	MF: C9 H6 N2 O3

Antibacterial

LgP 2.02(C)

pKa

pp cited in Vols.1-5:

(1) Petrow et al., J. Chem. Soc., 1954, 570.

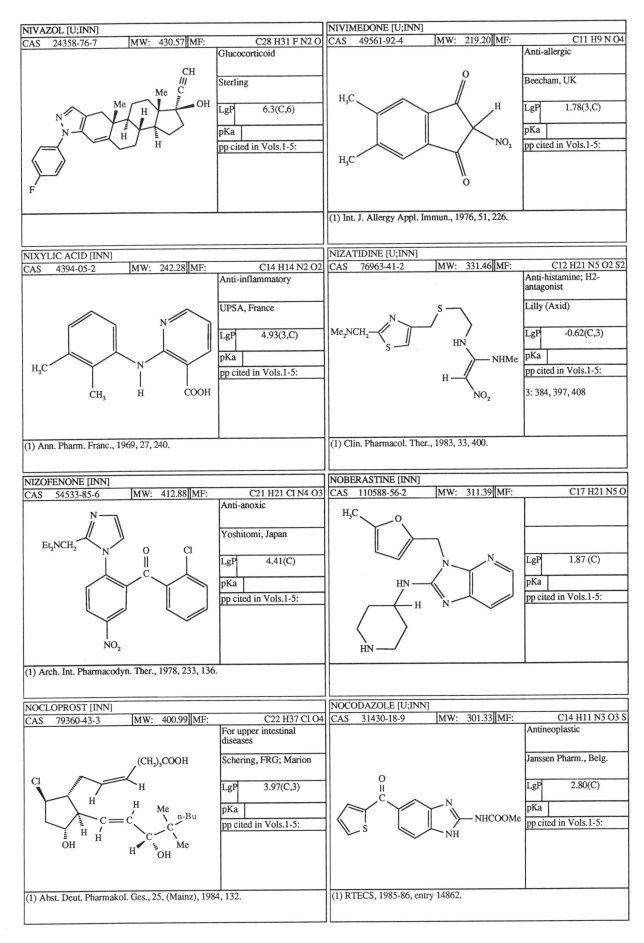

NIVAZOL [U;INN]

CAS	24358-76-7	MW:	430.57	MF:	C28 H31 F N2 O

Glucocorticoid

Sterling

LgP 6.3(C,6)

pKa

pp cited in Vols.1-5:

NIVIMEDONE [U;INN]

CAS	49561-92-4	MW:	219.20	MF:	C11 H9 N O4

Anti-allergic

Beecham, UK

LgP 1.78(3,C)

pKa

pp cited in Vols.1-5:

(1) Int. J. Allergy Appl. Immun., 1976, 51, 226.

NIXYLIC ACID [INN]

CAS	4394-05-2	MW:	242.28	MF:	C14 H14 N2 O2

Anti-inflammatory

UPSA, France

LgP 4.93(3,C)

pKa

pp cited in Vols.1-5:

(1) Ann. Pharm. Franc., 1969, 27, 240.

NIZATIDINE [U;INN]

CAS	76963-41-2	MW:	331.46	MF:	C12 H21 N5 O2 S2

Anti-histamine; H2-antagonist

Lilly (Axid)

LgP -0.62(C,3)

pKa

pp cited in Vols.1-5:

3: 384, 397, 408

(1) Clin. Pharmacol. Ther., 1983, 33, 400.

NIZOFENONE [INN]

CAS	54533-85-6	MW:	412.88	MF:	C21 H21 Cl N4 O3

Anti-anoxic

Yoshitomi, Japan

LgP 4.41(C)

pKa

pp cited in Vols.1-5:

(1) Arch. Int. Pharmacodyn. Ther., 1978, 233, 136.

NOBERASTINE [INN]

CAS	110588-56-2	MW:	311.39	MF:	C17 H21 N5 O

LgP 1.87 (C)

pKa

pp cited in Vols.1-5:

NOCLOPROST [INN]

CAS	79360-43-3	MW:	400.99	MF:	C22 H37 Cl O4

For upper intestinal diseases

Schering, FRG; Marion

LgP 3.97(C,3)

pKa

pp cited in Vols.1-5:

(1) Abst. Deut. Pharmakol. Ges., 25, (Mainz), 1984, 132.

NOCODAZOLE [U;INN]

CAS	31430-18-9	MW:	301.33	MF:	C14 H11 N3 O3 S

Antineoplastic

Janssen Pharm., Belg.

LgP 2.80(C)

pKa

pp cited in Vols.1-5:

(1) RTECS, 1985-86, entry 14862.

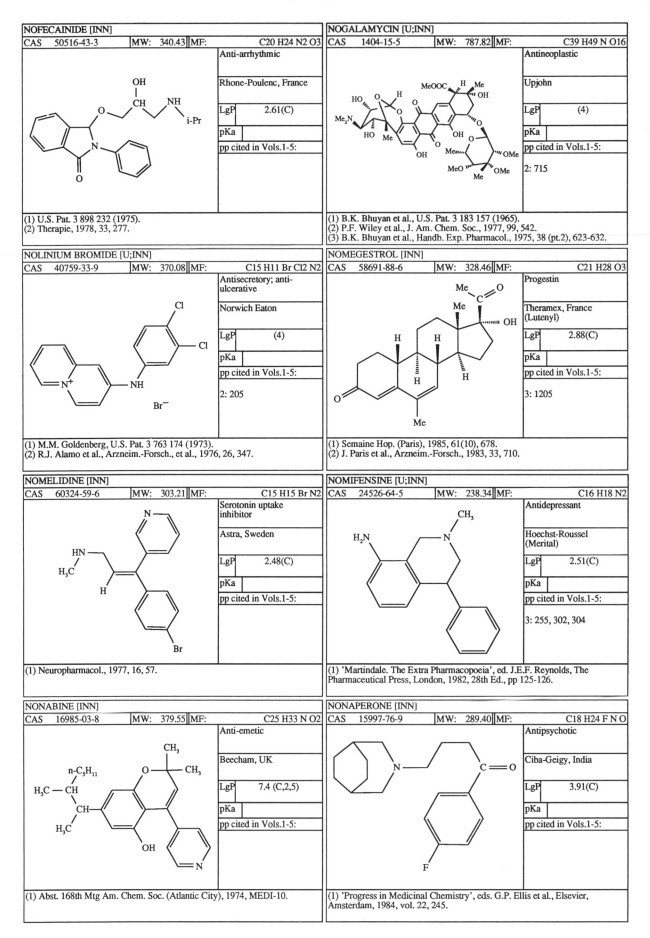

NOFECAINIDE [INN]

CAS	50516-43-3	MW:	340.43	MF:	C20 H24 N2 O3

Anti-arrhythmic

Rhone-Poulenc, France

LgP	2.61(C)
pKa	
pp cited in Vols.1-5:	

(1) U.S. Pat. 3 898 232 (1975).
(2) Therapie, 1978, 33, 277.

NOGALAMYCIN [U;INN]

CAS	1404-15-5	MW:	787.82	MF:	C39 H49 N O16

Antineoplastic

Upjohn

LgP	(4)
pKa	
pp cited in Vols.1-5:	

2: 715

(1) B.K. Bhuyan et al., U.S. Pat. 3 183 157 (1965).
(2) P.F. Wiley et al., J. Am. Chem. Soc., 1977, 99, 542.
(3) B.K. Bhuyan et al., Handb. Exp. Pharmacol., 1975, 38 (pt.2), 623-632.

NOLINIUM BROMIDE [U;INN]

CAS	40759-33-9	MW:	370.08	MF:	C15 H11 Br Cl2 N2

Antisecretory; anti-ulcerative

Norwich Eaton

LgP	(4)
pKa	
pp cited in Vols.1-5:	

2: 205

(1) M.M. Goldenberg, U.S. Pat. 3 763 174 (1973).
(2) R.J. Alamo et al., Arzneim.-Forsch., et al., 1976, 26, 347.

NOMEGESTROL [INN]

CAS	58691-88-6	MW:	328.46	MF:	C21 H28 O3

Progestin

Theramex, France (Lutenyl)

LgP	2.88(C)
pKa	
pp cited in Vols.1-5:	

3: 1205

(1) Semaine Hop. (Paris), 1985, 61(10), 678.
(2) J. Paris et al., Arzneim.-Forsch., 1983, 33, 710.

NOMELIDINE [INN]

CAS	60324-59-6	MW:	303.21	MF:	C15 H15 Br N2

Serotonin uptake inhibitor

Astra, Sweden

LgP	2.48(C)
pKa	
pp cited in Vols.1-5:	

(1) Neuropharmacol., 1977, 16, 57.

NOMIFENSINE [U;INN]

CAS	24526-64-5	MW:	238.34	MF:	C16 H18 N2

Antidepressant

Hoechst-Roussel (Merital)

LgP	2.51(C)
pKa	
pp cited in Vols.1-5:	

3: 255, 302, 304

(1) 'Martindale. The Extra Pharmacopoeia', ed. J.E.F. Reynolds, The Pharmaceutical Press, London, 1982, 28th Ed., pp 125-126.

NONABINE [INN]

CAS	16985-03-8	MW:	379.55	MF:	C25 H33 N O2

Anti-emetic

Beecham, UK

LgP	7.4 (C,2,5)
pKa	
pp cited in Vols.1-5:	

(1) Abst. 168th Mtg Am. Chem. Soc. (Atlantic City), 1974, MEDI-10.

NONAPERONE [INN]

CAS	15997-76-9	MW:	289.40	MF:	C18 H24 F N O

Antipsychotic

Ciba-Geigy, India

LgP	3.91(C)
pKa	
pp cited in Vols.1-5:	

(1) 'Progress in Medicinal Chemistry', eds. G.P. Ellis et al., Elsevier, Amsterdam, 1984, vol. 22, 245.

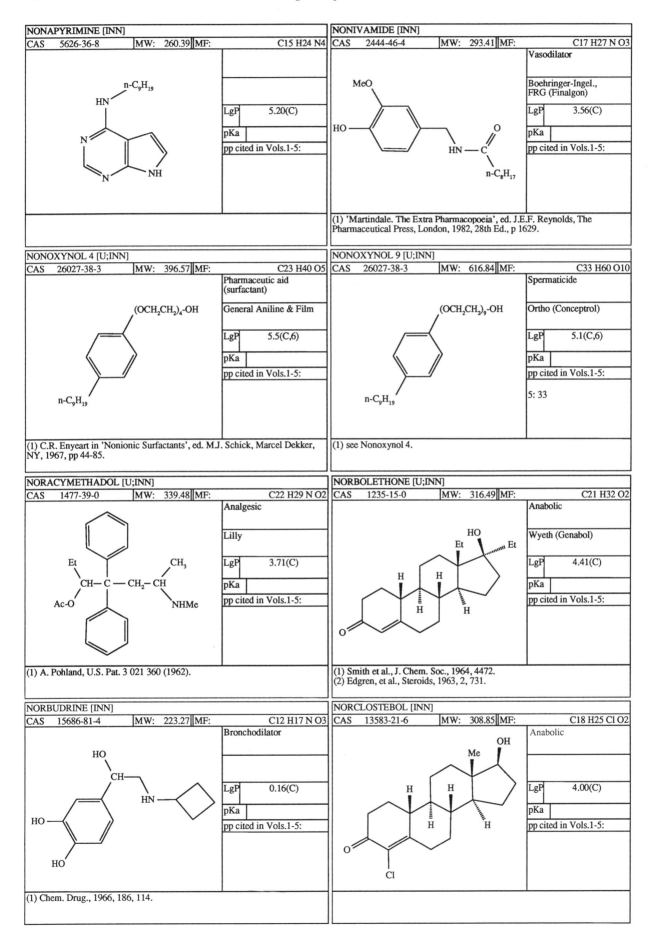

NONAPYRIMINE [INN]		
CAS 5626-36-8	MW: 260.39	MF: C15 H24 N4
	LgP	5.20(C)
	pKa	
	pp cited in Vols.1-5:	

NONIVAMIDE [INN]		
CAS 2444-46-4	MW: 293.41	MF: C17 H27 N O3
	Vasodilator	
	Boehringer-Ingel., FRG (Finalgon)	
	LgP	3.56(C)
	pKa	
	pp cited in Vols.1-5:	

(1) 'Martindale. The Extra Pharmacopoeia', ed. J.E.F. Reynolds, The Pharmaceutical Press, London, 1982, 28th Ed., p 1629.

NONOXYNOL 4 [U;INN]		
CAS 26027-38-3	MW: 396.57	MF: C23 H40 O5
	Pharmaceutic aid (surfactant)	
	General Aniline & Film	
	LgP	5.5(C,6)
	pKa	
	pp cited in Vols.1-5:	

(1) C.R. Enyeart in 'Nonionic Surfactants', ed. M.J. Schick, Marcel Dekker, NY, 1967, pp 44-85.

NONOXYNOL 9 [U;INN]		
CAS 26027-38-3	MW: 616.84	MF: C33 H60 O10
	Spermaticide	
	Ortho (Conceptrol)	
	LgP	5.1(C,6)
	pKa	
	pp cited in Vols.1-5:	
	5: 33	

(1) see Nonoxynol 4.

NORACYMETHADOL [U;INN]		
CAS 1477-39-0	MW: 339.48	MF: C22 H29 N O2
	Analgesic	
	Lilly	
	LgP	3.71(C)
	pKa	
	pp cited in Vols.1-5:	

(1) A. Pohland, U.S. Pat. 3 021 360 (1962).

NORBOLETHONE [U;INN]		
CAS 1235-15-0	MW: 316.49	MF: C21 H32 O2
	Anabolic	
	Wyeth (Genabol)	
	LgP	4.41(C)
	pKa	
	pp cited in Vols.1-5:	

(1) Smith et al., J. Chem. Soc., 1964, 4472.
(2) Edgren, et al., Steroids, 1963, 2, 731.

NORBUDRINE [INN]		
CAS 15686-81-4	MW: 223.27	MF: C12 H17 N O3
	Bronchodilator	
	LgP	0.16(C)
	pKa	
	pp cited in Vols.1-5:	

(1) Chem. Drug., 1966, 186, 114.

NORCLOSTEBOL [INN]		
CAS 13583-21-6	MW: 308.85	MF: C18 H25 Cl O2
	Anabolic	
	LgP	4.00(C)
	pKa	
	pp cited in Vols.1-5:	

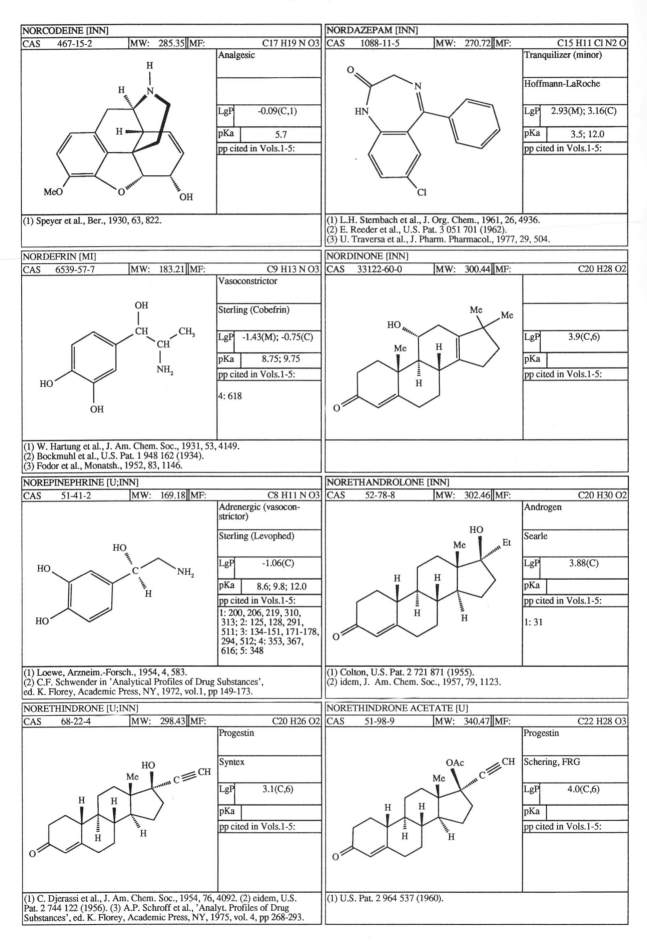

NORCODEINE [INN]

CAS	467-15-2	MW:	285.35	MF:	C17 H19 N O3

Analgesic

LgP	-0.09(C,1)
pKa	5.7
pp cited in Vols.1-5:	

(1) Speyer et al., Ber., 1930, 63, 822.

NORDAZEPAM [INN]

CAS	1088-11-5	MW:	270.72	MF:	C15 H11 Cl N2 O

Tranquilizer (minor)

Hoffmann-LaRoche

LgP	2.93(M); 3.16(C)
pKa	3.5; 12.0
pp cited in Vols.1-5:	

(1) L.H. Sternbach et al., J. Org. Chem., 1961, 26, 4936.
(2) E. Reeder et al., U.S. Pat. 3 051 701 (1962).
(3) U. Traversa et al., J. Pharm. Pharmacol., 1977, 29, 504.

NORDEFRIN [MI]

CAS	6539-57-7	MW:	183.21	MF:	C9 H13 N O3

Vasoconstrictor

Sterling (Cobefrin)

LgP	-1.43(M); -0.75(C)
pKa	8.75; 9.75
pp cited in Vols.1-5:	
4: 618	

(1) W. Hartung et al., J. Am. Chem. Soc., 1931, 53, 4149.
(2) Bockmuhl et al., U.S. Pat. 1 948 162 (1934).
(3) Fodor et al., Monatsh., 1952, 83, 1146.

NORDINONE [INN]

CAS	33122-60-0	MW:	300.44	MF:	C20 H28 O2

LgP	3.9(C,6)
pKa	
pp cited in Vols.1-5:	

NOREPINEPHRINE [U;INN]

CAS	51-41-2	MW:	169.18	MF:	C8 H11 N O3

Adrenergic (vasoconstrictor)

Sterling (Levophed)

LgP	-1.06(C)
pKa	8.6; 9.8; 12.0
pp cited in Vols.1-5:	
1: 200, 206, 219, 310, 313; 2: 125, 128, 291, 511; 3: 134-151, 171-178, 294, 512; 4: 353, 367, 616; 5: 348	

(1) Loewe, Arzneim.-Forsch., 1954, 4, 583.
(2) C.F. Schwender in 'Analytical Profiles of Drug Substances',
ed. K. Florey, Academic Press, NY, 1972, vol.1, pp 149-173.

NORETHANDROLONE [INN]

CAS	52-78-8	MW:	302.46	MF:	C20 H30 O2

Androgen

Searle

LgP	3.88(C)
pKa	
pp cited in Vols.1-5:	
1: 31	

(1) Colton, U.S. Pat. 2 721 871 (1955).
(2) idem, J. Am. Chem. Soc., 1957, 79, 1123.

NORETHINDRONE [U;INN]

CAS	68-22-4	MW:	298.43	MF:	C20 H26 O2

Progestin

Syntex

LgP	3.1(C,6)
pKa	
pp cited in Vols.1-5:	

(1) C. Djerassi et al., J. Am. Chem. Soc., 1954, 76, 4092. (2) eidem, U.S. Pat. 2 744 122 (1956). (3) A.P. Schroff et al., 'Analyt. Profiles of Drug Substances', ed. K. Florey, Academic Press, NY, 1975, vol. 4, pp 268-293.

NORETHINDRONE ACETATE [U]

CAS	51-98-9	MW:	340.47	MF:	C22 H28 O3

Progestin

Schering, FRG

LgP	4.0(C,6)
pKa	
pp cited in Vols.1-5:	

(1) U.S. Pat. 2 964 537 (1960).

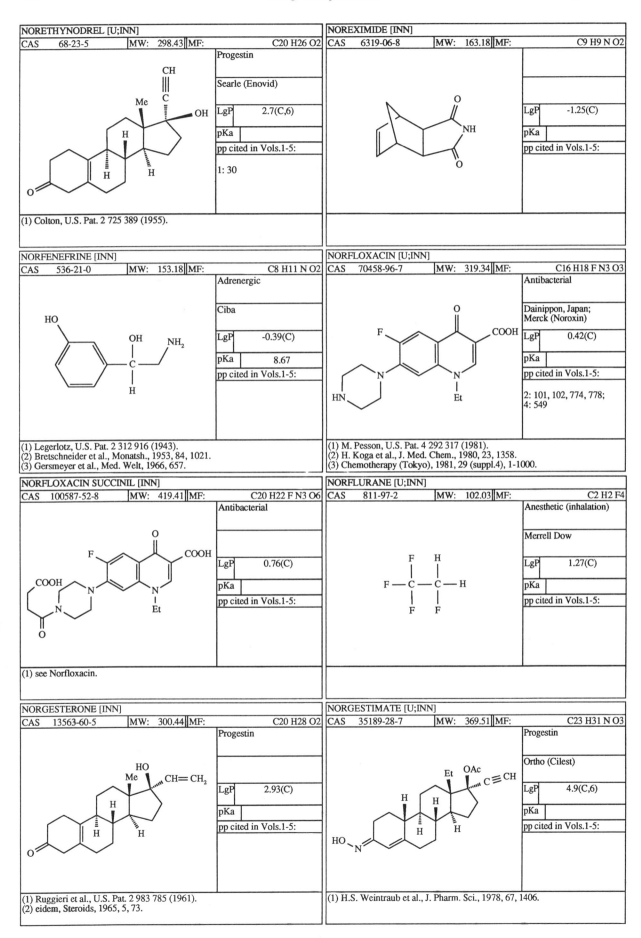

NORETHYNODREL [U;INN]

| CAS | 68-23-5 | MW: | 298.43 | MF: | C20 H26 O2 |

Progestin

Searle (Enovid)

| LgP | 2.7(C,6) |
| pKa | |
| pp cited in Vols.1-5: |
| 1: 30 |

(1) Colton, U.S. Pat. 2 725 389 (1955).

NOREXIMIDE [INN]

| CAS | 6319-06-8 | MW: | 163.18 | MF: | C9 H9 N O2 |

| LgP | -1.25(C) |
| pKa | |
| pp cited in Vols.1-5: |

NORFENEFRINE [INN]

| CAS | 536-21-0 | MW: | 153.18 | MF: | C8 H11 N O2 |

Adrenergic

Ciba

| LgP | -0.39(C) |
| pKa | 8.67 |
| pp cited in Vols.1-5: |

(1) Legerlotz, U.S. Pat. 2 312 916 (1943).
(2) Bretschneider et al., Monatsh., 1953, 84, 1021.
(3) Gersmeyer et al., Med. Welt, 1966, 657.

NORFLOXACIN [U;INN]

| CAS | 70458-96-7 | MW: | 319.34 | MF: | C16 H18 F N3 O3 |

Antibacterial

Dainippon, Japan; Merck (Noroxin)

| LgP | 0.42(C) |
| pKa | |
| pp cited in Vols.1-5: |
| 2: 101, 102, 774, 778; 4: 549 |

(1) M. Pesson, U.S. Pat. 4 292 317 (1981).
(2) H. Koga et al., J. Med. Chem., 1980, 23, 1358.
(3) Chemotherapy (Tokyo), 1981, 29 (suppl.4), 1-1000.

NORFLOXACIN SUCCINIL [INN]

| CAS | 100587-52-8 | MW: | 419.41 | MF: | C20 H22 F N3 O6 |

Antibacterial

| LgP | 0.76(C) |
| pKa | |
| pp cited in Vols.1-5: |

(1) see Norfloxacin.

NORFLURANE [U;INN]

| CAS | 811-97-2 | MW: | 102.03 | MF: | C2 H2 F4 |

Anesthetic (inhalation)

Merrell Dow

| LgP | 1.27(C) |
| pKa | |
| pp cited in Vols.1-5: |

NORGESTERONE [INN]

| CAS | 13563-60-5 | MW: | 300.44 | MF: | C20 H28 O2 |

Progestin

| LgP | 2.93(C) |
| pKa | |
| pp cited in Vols.1-5: |

(1) Ruggieri et al., U.S. Pat. 2 983 785 (1961).
(2) eidem, Steroids, 1965, 5, 73.

NORGESTIMATE [U;INN]

| CAS | 35189-28-7 | MW: | 369.51 | MF: | C23 H31 N O3 |

Progestin

Ortho (Cilest)

| LgP | 4.9(C,6) |
| pKa | |
| pp cited in Vols.1-5: |

(1) H.S. Weintraub et al., J. Pharm. Sci., 1978, 67, 1406.

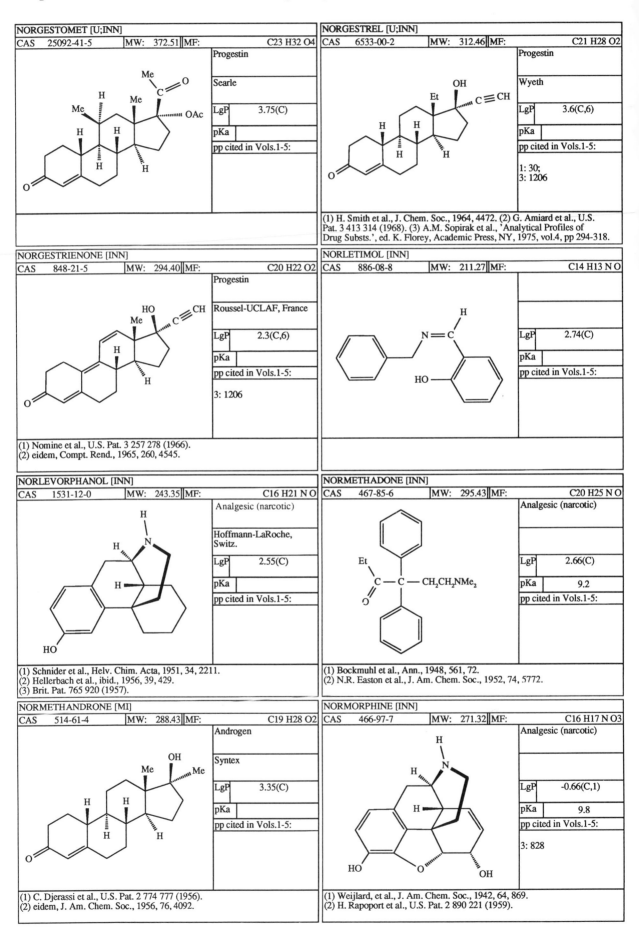

NORGESTOMET [U;INN]

CAS	25092-41-5	MW: 372.51	MF:	C23 H32 O4

Progestin

Searle

LgP	3.75(C)

pKa	

pp cited in Vols.1-5:

NORGESTREL [U;INN]

CAS	6533-00-2	MW: 312.46	MF:	C21 H28 O2

Progestin

Wyeth

LgP	3.6(C,6)

pKa	

pp cited in Vols.1-5:

1: 30;
3: 1206

(1) H. Smith et al., J. Chem. Soc., 1964, 4472. (2) G. Amiard et al., U.S. Pat. 3 413 314 (1968). (3) A.M. Sopirak et al., 'Analytical Profiles of Drug Substs.', ed. K. Florey, Academic Press, NY, 1975, vol.4, pp 294-318.

NORGESTRIENONE [INN]

CAS	848-21-5	MW: 294.40	MF:	C20 H22 O2

Progestin

Roussel-UCLAF, France

LgP	2.3(C,6)

pKa	

pp cited in Vols.1-5:

3: 1206

(1) Nomine et al., U.S. Pat. 3 257 278 (1966).
(2) eidem, Compt. Rend., 1965, 260, 4545.

NORLETIMOL [INN]

CAS	886-08-8	MW: 211.27	MF:	C14 H13 N O

LgP	2.74(C)

pKa	

pp cited in Vols.1-5:

NORLEVORPHANOL [INN]

CAS	1531-12-0	MW: 243.35	MF:	C16 H21 N O

Analgesic (narcotic)

Hoffmann-LaRoche, Switz.

LgP	2.55(C)

pKa	

pp cited in Vols.1-5:

(1) Schnider et al., Helv. Chim. Acta, 1951, 34, 2211.
(2) Hellerbach et al., ibid., 1956, 39, 429.
(3) Brit. Pat. 765 920 (1957).

NORMETHADONE [INN]

CAS	467-85-6	MW: 295.43	MF:	C20 H25 N O

Analgesic (narcotic)

LgP	2.66(C)

pKa	9.2

pp cited in Vols.1-5:

(1) Bockmuhl et al., Ann., 1948, 561, 72.
(2) N.R. Easton et al., J. Am. Chem. Soc., 1952, 74, 5772.

NORMETHANDRONE [MI]

CAS	514-61-4	MW: 288.43	MF:	C19 H28 O2

Androgen

Syntex

LgP	3.35(C)

pKa	

pp cited in Vols.1-5:

(1) C. Djerassi et al., U.S. Pat. 2 774 777 (1956).
(2) eidem, J. Am. Chem. Soc., 1956, 76, 4092.

NORMORPHINE [INN]

CAS	466-97-7	MW: 271.32	MF:	C16 H17 N O3

Analgesic (narcotic)

LgP	-0.66(C,1)

pKa	9.8

pp cited in Vols.1-5:

3: 828

(1) Weijlard, et al., J. Am. Chem. Soc., 1942, 64, 869.
(2) H. Rapoport et al., U.S. Pat. 2 890 221 (1959).

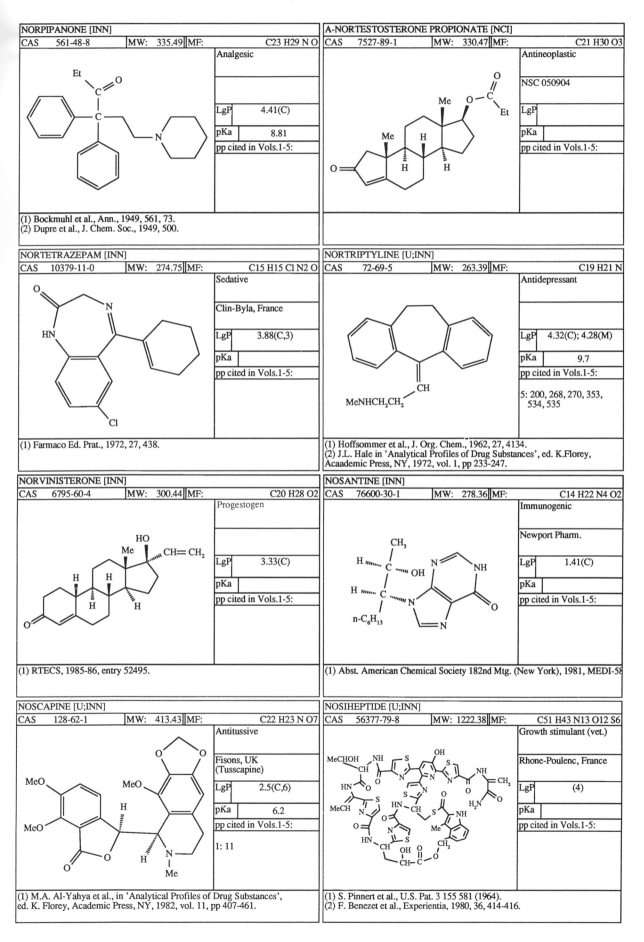

NORPIPANONE [INN]

CAS	561-48-8	MW:	335.49	MF:	C23 H29 N O

Analgesic

LgP	4.41(C)
pKa	8.81
pp cited in Vols.1-5:	

(1) Bockmuhl et al., Ann., 1949, 561, 73.
(2) Dupre et al., J. Chem. Soc., 1949, 500.

A-NORTESTOSTERONE PROPIONATE [NCI]

CAS	7527-89-1	MW:	330.47	MF:	C21 H30 O3

Antineoplastic

NSC 050904

LgP	
pKa	
pp cited in Vols.1-5:	

NORTETRAZEPAM [INN]

CAS	10379-11-0	MW:	274.75	MF:	C15 H15 Cl N2 O

Sedative

Clin-Byla, France

LgP	3.88(C,3)
pKa	
pp cited in Vols.1-5:	

(1) Farmaco Ed. Prat., 1972, 27, 438.

NORTRIPTYLINE [U;INN]

CAS	72-69-5	MW:	263.39	MF:	C19 H21 N

Antidepressant

MeNHCH₂CH₂

LgP	4.32(C); 4.28(M)
pKa	9.7
pp cited in Vols.1-5:	
	5: 200, 268, 270, 353, 534, 535

(1) Hoffsommer et al., J. Org. Chem., 1962, 27, 4134.
(2) J.L. Hale in 'Analytical Profiles of Drug Substances', ed. K.Florey, Acaademic Press, NY, 1972, vol. 1, pp 233-247.

NORVINISTERONE [INN]

CAS	6795-60-4	MW:	300.44	MF:	C20 H28 O2

Progestogen

LgP	3.33(C)
pKa	
pp cited in Vols.1-5:	

(1) RTECS, 1985-86, entry 52495.

NOSANTINE [INN]

CAS	76600-30-1	MW:	278.36	MF:	C14 H22 N4 O2

Immunogenic

Newport Pharm.

LgP	1.41(C)
pKa	
pp cited in Vols.1-5:	

(1) Abst. American Chemical Society 182nd Mtg. (New York), 1981, MEDI-58

NOSCAPINE [U;INN]

CAS	128-62-1	MW:	413.43	MF:	C22 H23 N O7

Antitussive

Fisons, UK (Tusscapine)

LgP	2.5(C,6)
pKa	6.2
pp cited in Vols.1-5:	
	1: 11

(1) M.A. Al-Yahya et al., in 'Analytical Profiles of Drug Substances', ed. K. Florey, Academic Press, NY, 1982, vol. 11, pp 407-461.

NOSIHEPTIDE [U;INN]

CAS	56377-79-8	MW:	1222.38	MF:	C51 H43 N13 O12 S6

Growth stimulant (vet.)

Rhone-Poulenc, France

LgP	(4)
pKa	
pp cited in Vols.1-5:	

(1) S. Pinnert et al., U.S. Pat. 3 155 581 (1964).
(2) F. Benezet et al., Experientia, 1980, 36, 414-416.

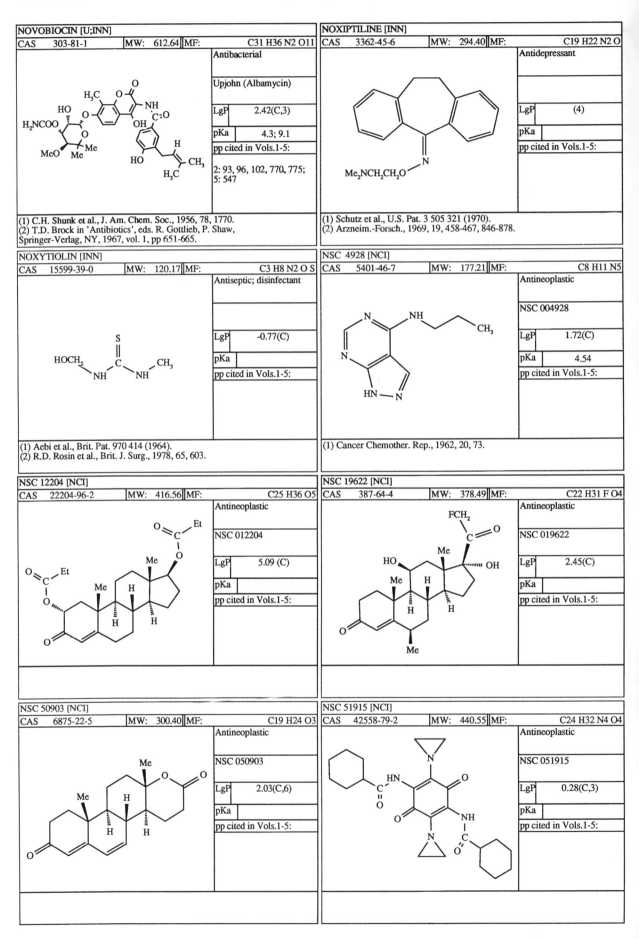

NOVOBIOCIN [U;INN]

CAS	303-81-1	MW:	612.64	MF:	C31 H36 N2 O11

Antibacterial

Upjohn (Albamycin)

LgP	2.42(C,3)
pKa	4.3; 9.1
pp cited in Vols.1-5:	

2: 93, 96, 102, 770, 775; 5: 547

(1) C.H. Shunk et al., J. Am. Chem. Soc., 1956, 78, 1770.
(2) T.D. Brock in 'Antibiotics', eds. R. Gottlieb, P. Shaw, Springer-Verlag, NY, 1967, vol. 1, pp 651-665.

NOXIPTILINE [INN]

CAS	3362-45-6	MW:	294.40	MF:	C19 H22 N2 O

Antidepressant

LgP	(4)
pKa	
pp cited in Vols.1-5:	

(1) Schutz et al., U.S. Pat. 3 505 321 (1970).
(2) Arzneim.-Forsch., 1969, 19, 458-467, 846-878.

NOXYTIOLIN [INN]

CAS	15599-39-0	MW:	120.17	MF:	C3 H8 N2 O S

Antiseptic; disinfectant

LgP	-0.77(C)
pKa	
pp cited in Vols.1-5:	

(1) Aebi et al., Brit. Pat. 970 414 (1964).
(2) R.D. Rosin et al., Brit. J. Surg., 1978, 65, 603.

NSC 4928 [NCI]

CAS	5401-46-7	MW:	177.21	MF:	C8 H11 N5

Antineoplastic

NSC 004928

LgP	1.72(C)
pKa	4.54
pp cited in Vols.1-5:	

(1) Cancer Chemother. Rep., 1962, 20, 73.

NSC 12204 [NCI]

CAS	22204-96-2	MW:	416.56	MF:	C25 H36 O5

Antineoplastic

NSC 012204

LgP	5.09 (C)
pKa	
pp cited in Vols.1-5:	

NSC 19622 [NCI]

CAS	387-64-4	MW:	378.49	MF:	C22 H31 F O4

Antineoplastic

NSC 019622

LgP	2.45(C)
pKa	
pp cited in Vols.1-5:	

NSC 50903 [NCI]

CAS	6875-22-5	MW:	300.40	MF:	C19 H24 O3

Antineoplastic

NSC 050903

LgP	2.03(C,6)
pKa	
pp cited in Vols.1-5:	

NSC 51915 [NCI]

CAS	42558-79-2	MW:	440.55	MF:	C24 H32 N4 O4

Antineoplastic

NSC 051915

LgP	0.28(C,3)
pKa	
pp cited in Vols.1-5:	

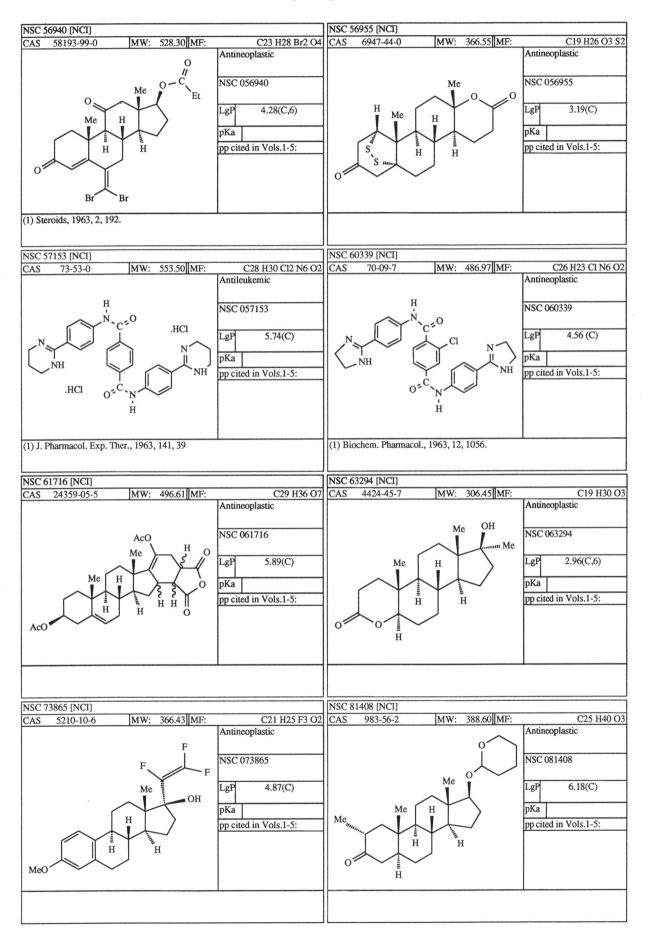

NSC 56940 [NCI]

CAS	58193-99-0	MW:	528.30	MF:	C23 H28 Br2 O4

Antineoplastic

NSC 056940

LgP	4.28(C,6)

pKa	

pp cited in Vols.1-5:

(1) Steroids, 1963, 2, 192.

NSC 56955 [NCI]

CAS	6947-44-0	MW:	366.55	MF:	C19 H26 O3 S2

Antineoplastic

NSC 056955

LgP	3.19(C)

pKa	

pp cited in Vols.1-5:

NSC 57153 [NCI]

CAS	73-53-0	MW:	553.50	MF:	C28 H30 Cl2 N6 O2

Antileukemic

NSC 057153

LgP	5.74(C)

pKa	

pp cited in Vols.1-5:

(1) J. Pharmacol. Exp. Ther., 1963, 141, 39

NSC 60339 [NCI]

CAS	70-09-7	MW:	486.97	MF:	C26 H23 Cl N6 O2

Antineoplastic

NSC 060339

LgP	4.56 (C)

pKa	

pp cited in Vols.1-5:

(1) Biochem. Pharmacol., 1963, 12, 1056.

NSC 61716 [NCI]

CAS	24359-05-5	MW:	496.61	MF:	C29 H36 O7

Antineoplastic

NSC 061716

LgP	5.89(C)

pKa	

pp cited in Vols.1-5:

NSC 63294 [NCI]

CAS	4424-45-7	MW:	306.45	MF:	C19 H30 O3

Antineoplastic

NSC 063294

LgP	2.96(C,6)

pKa	

pp cited in Vols.1-5:

NSC 73865 [NCI]

CAS	5210-10-6	MW:	366.43	MF:	C21 H25 F3 O2

Antineoplastic

NSC 073865

LgP	4.87(C)

pKa	

pp cited in Vols.1-5:

NSC 81408 [NCI]

CAS	983-56-2	MW:	388.60	MF:	C25 H40 O3

Antineoplastic

NSC 081408

LgP	6.18(C)

pKa	

pp cited in Vols.1-5:

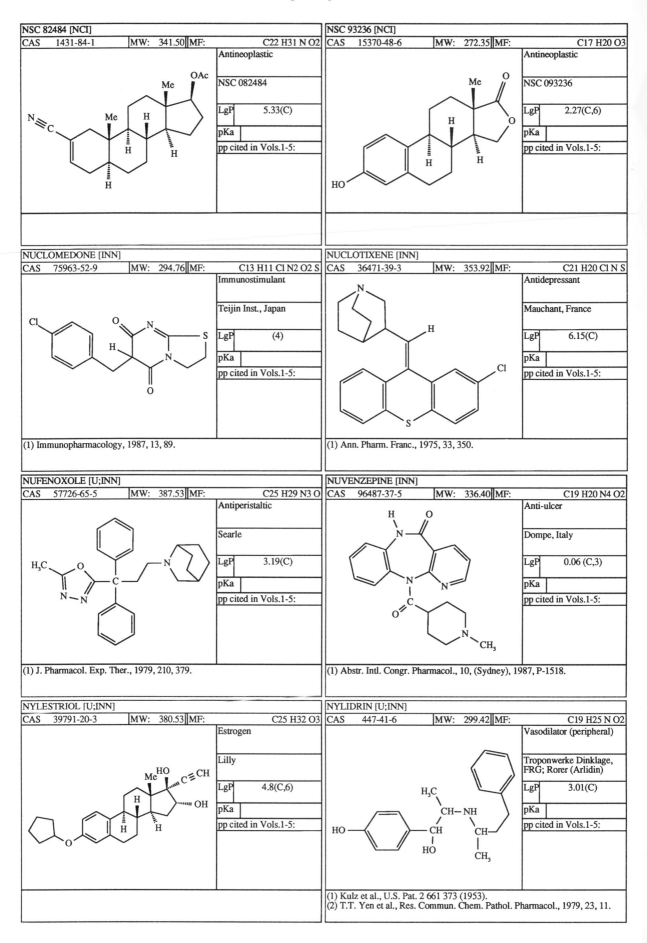

NSC 82484 [NCI]

CAS	1431-84-1	MW:	341.50	MF:	C22 H31 N O2

Antineoplastic

NSC 082484

LgP	5.33(C)
pKa	
pp cited in Vols.1-5:	

NSC 93236 [NCI]

CAS	15370-48-6	MW:	272.35	MF:	C17 H20 O3

Antineoplastic

NSC 093236

LgP	2.27(C,6)
pKa	
pp cited in Vols.1-5:	

NUCLOMEDONE [INN]

CAS	75963-52-9	MW:	294.76	MF:	C13 H11 Cl N2 O2 S

Immunostimulant

Teijin Inst., Japan

LgP	(4)
pKa	
pp cited in Vols.1-5:	

(1) Immunopharmacology, 1987, 13, 89.

NUCLOTIXENE [INN]

CAS	36471-39-3	MW:	353.92	MF:	C21 H20 Cl N S

Antidepressant

Mauchant, France

LgP	6.15(C)
pKa	
pp cited in Vols.1-5:	

(1) Ann. Pharm. Franc., 1975, 33, 350.

NUFENOXOLE [U;INN]

CAS	57726-65-5	MW:	387.53	MF:	C25 H29 N3 O

Antiperistaltic

Searle

LgP	3.19(C)
pKa	
pp cited in Vols.1-5:	

(1) J. Pharmacol. Exp. Ther., 1979, 210, 379.

NUVENZEPINE [INN]

CAS	96487-37-5	MW:	336.40	MF:	C19 H20 N4 O2

Anti-ulcer

Dompe, Italy

LgP	0.06 (C,3)
pKa	
pp cited in Vols.1-5:	

(1) Abstr. Intl. Congr. Pharmacol., 10, (Sydney), 1987, P-1518.

NYLESTRIOL [U;INN]

CAS	39791-20-3	MW:	380.53	MF:	C25 H32 O3

Estrogen

Lilly

LgP	4.8(C,6)
pKa	
pp cited in Vols.1-5:	

NYLIDRIN [U;INN]

CAS	447-41-6	MW:	299.42	MF:	C19 H25 N O2

Vasodilator (peripheral)

Troponwerke Dinklage, FRG; Rorer (Arlidin)

LgP	3.01(C)
pKa	
pp cited in Vols.1-5:	

(1) Kulz et al., U.S. Pat. 2 661 373 (1953).
(2) T.T. Yen et al., Res. Commun. Chem. Pathol. Pharmacol., 1979, 23, 11.

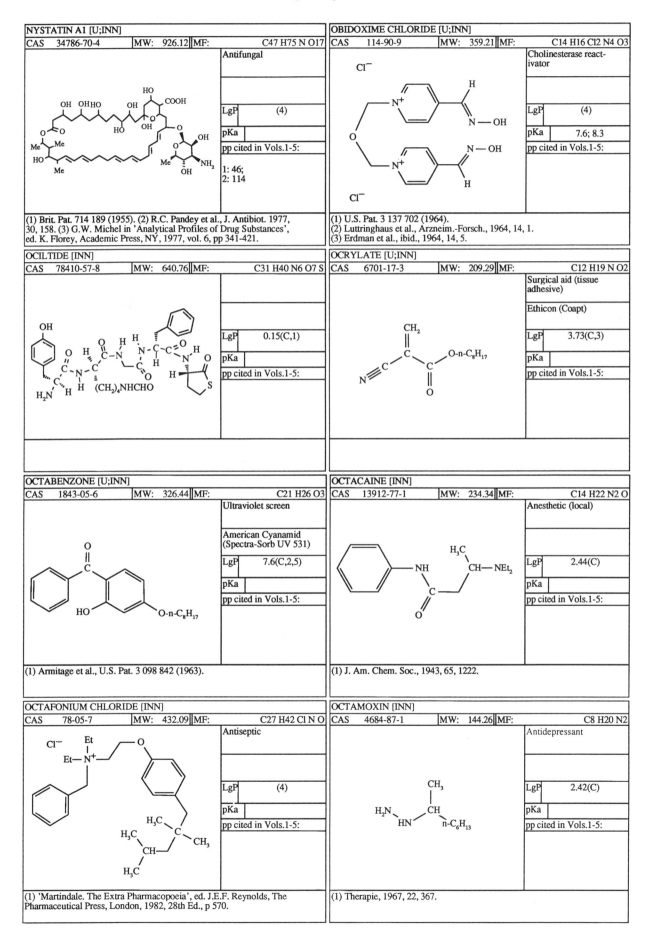

NYSTATIN A1 [U;INN]

CAS	34786-70-4	MW:	926.12	MF:	C47 H75 N O17

Antifungal

LgP   (4)

pKa

pp cited in Vols.1-5:
1: 46;
2: 114

(1) Brit. Pat. 714 189 (1955). (2) R.C. Pandey et al., J. Antibiot. 1977, 30, 158. (3) G.W. Michel in 'Analytical Profiles of Drug Substances', ed. K. Florey, Academic Press, NY, 1977, vol. 6, pp 341-421.

OBIDOXIME CHLORIDE [U;INN]

CAS	114-90-9	MW:	359.21	MF:	C14 H16 Cl2 N4 O3

Cholinesterase react-ivator

LgP   (4)

pKa   7.6; 8.3

pp cited in Vols.1-5:

(1) U.S. Pat. 3 137 702 (1964).
(2) Luttringhaus et al., Arzneim.-Forsch., 1964, 14, 1.
(3) Erdman et al., ibid., 1964, 14, 5.

OCILTIDE [INN]

CAS	78410-57-8	MW:	640.76	MF:	C31 H40 N6 O7 S

LgP   0.15(C,1)

pKa

pp cited in Vols.1-5:

OCRYLATE [U;INN]

CAS	6701-17-3	MW:	209.29	MF:	C12 H19 N O2

Surgical aid (tissue adhesive)

Ethicon (Coapt)

LgP   3.73(C,3)

pKa

pp cited in Vols.1-5:

OCTABENZONE [U;INN]

CAS	1843-05-6	MW:	326.44	MF:	C21 H26 O3

Ultraviolet screen

American Cyanamid (Spectra-Sorb UV 531)

LgP   7.6(C,2,5)

pKa

pp cited in Vols.1-5:

(1) Armitage et al., U.S. Pat. 3 098 842 (1963).

OCTACAINE [INN]

CAS	13912-77-1	MW:	234.34	MF:	C14 H22 N2 O

Anesthetic (local)

LgP   2.44(C)

pKa

pp cited in Vols.1-5:

(1) J. Am. Chem. Soc., 1943, 65, 1222.

OCTAFONIUM CHLORIDE [INN]

CAS	78-05-7	MW:	432.09	MF:	C27 H42 Cl N O

Antiseptic

LgP   (4)

pKa

pp cited in Vols.1-5:

(1) 'Martindale. The Extra Pharmacopoeia', ed. J.E.F. Reynolds, The Pharmaceutical Press, London, 1982, 28th Ed., p 570.

OCTAMOXIN [INN]

CAS	4684-87-1	MW:	144.26	MF:	C8 H20 N2

Antidepressant

LgP   2.42(C)

pKa

pp cited in Vols.1-5:

(1) Therapie, 1967, 22, 367.

OCTAMYLAMINE [INN]

CAS	502-59-0	MW:	199.38	MF:	C13 H29 N

Anticholinergic

Knoll, FRG

LgP	2.82(C)
pKa	
pp cited in Vols.1-5:	

(1) Swiss Pat. 258 452 (1942).

OCTANOIC ACID [U;INN]

CAS	124-07-2	MW:	144.22	MF:	C8 H16 O2

Antifungal

Norwich Eaton

$n-C_7H_{15}$ — COOH

LgP	3.05(M); 2.94(C)
pKa	4.88
pp cited in Vols.1-5:	

(1) K.S. Markley, 'Fatty Acids', Interscience, NY, 1960, 2nd. ed., 1960, pp 34, 38.

OCTAPINOL [INN]

CAS	71138-71-1	MW:	241.42	MF:	C15 H31 N O

LgP	2.91(C)
pKa	
pp cited in Vols.1-5:	

OCTASTINE [INN]

CAS	59767-12-3	MW:	371.95	MF:	C23 H30 Cl N O

LgP	7.0 (C,2,5)
pKa	
pp cited in Vols.1-5:	

OCTAVERINE [INN]

CAS	549-68-8	MW:	397.48	MF:	C23 H27 N O5

Anticholinergic

LgP	4.32(C)
pKa	
pp cited in Vols.1-5:	

(1) Fr. Pat. 760 825 (1934).
(2) Goldberg et al., J. Pharm. Pharmacol., 1954, 6, 171.

OCTAZAMIDE [U;INN]

CAS	56391-55-0	MW:	217.27	MF:	C13 H15 N O2

Analgesic

ICI Americas

LgP	-0.36(C)
pKa	
pp cited in Vols.1-5:	

OCTENIDINE [U;INN]

CAS	71251-02-0	MW:	550.92	MF:	C36 H62 N4

Anti-infective, topical

Sterling

LgP	(4)
pKa	
pp cited in Vols.1-5:	

OCTENIDINE SACCHARIN [U]

CAS	86767-75-1	MW:	917.30	MF:	C50 H72 N6 O6 S2

Dental plaque inhibitor

Sterling

LgP	(4)
pKa	
pp cited in Vols.1-5:	

(1) Clin. Pharmacol. Ther., 1984, 36, 281.

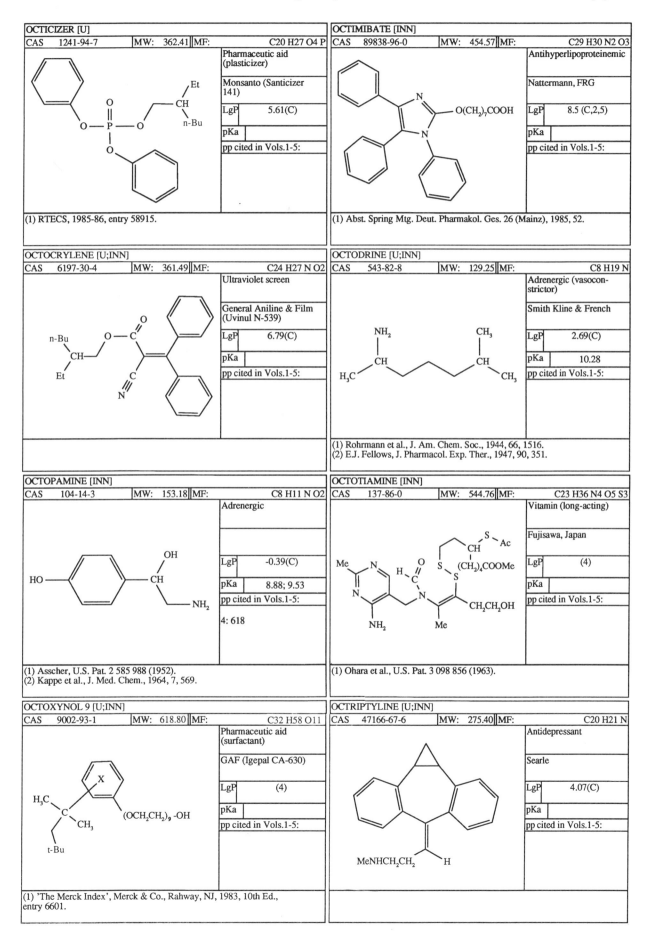

OCTICIZER [U]

CAS	1241-94-7	MW:	362.41	MF:	C20 H27 O4 P

Pharmaceutic aid (plasticizer)

Monsanto (Santicizer 141)

LgP	5.61(C)
pKa	
pp cited in Vols.1-5:	

(1) RTECS, 1985-86, entry 58915.

OCTIMIBATE [INN]

CAS	89838-96-0	MW:	454.57	MF:	C29 H30 N2 O3

Antihyperlipoproteinemic

Nattermann, FRG

LgP	8.5 (C,2,5)
pKa	
pp cited in Vols.1-5:	

$O(CH_2)_7COOH$

(1) Abst. Spring Mtg. Deut. Pharmakol. Ges. 26 (Mainz), 1985, 52.

OCTOCRYLENE [U;INN]

CAS	6197-30-4	MW:	361.49	MF:	C24 H27 N O2

Ultraviolet screen

General Aniline & Film (Uvinul N-539)

LgP	6.79(C)
pKa	
pp cited in Vols.1-5:	

OCTODRINE [U;INN]

CAS	543-82-8	MW:	129.25	MF:	C8 H19 N

Adrenergic (vasoconstrictor)

Smith Kline & French

LgP	2.69(C)
pKa	10.28
pp cited in Vols.1-5:	

(1) Rohrmann et al., J. Am. Chem. Soc., 1944, 66, 1516.
(2) E.J. Fellows, J. Pharmacol. Exp. Ther., 1947, 90, 351.

OCTOPAMINE [INN]

CAS	104-14-3	MW:	153.18	MF:	C8 H11 N O2

Adrenergic

LgP	-0.39(C)
pKa	8.88; 9.53
pp cited in Vols.1-5:	
	4: 618

(1) Asscher, U.S. Pat. 2 585 988 (1952).
(2) Kappe et al., J. Med. Chem., 1964, 7, 569.

OCTOTIAMINE [INN]

CAS	137-86-0	MW:	544.76	MF:	C23 H36 N4 O5 S3

Vitamin (long-acting)

Fujisawa, Japan

LgP	(4)
pKa	
pp cited in Vols.1-5:	

(1) Ohara et al., U.S. Pat. 3 098 856 (1963).

OCTOXYNOL 9 [U;INN]

CAS	9002-93-1	MW:	618.80	MF:	C32 H58 O11

Pharmaceutic aid (surfactant)

GAF (Igepal CA-630)

LgP	(4)
pKa	
pp cited in Vols.1-5:	

(1) 'The Merck Index', Merck & Co., Rahway, NJ, 1983, 10th Ed., entry 6601.

OCTRIPTYLINE [U;INN]

CAS	47166-67-6	MW:	275.40	MF:	C20 H21 N

Antidepressant

Searle

LgP	4.07(C)
pKa	
pp cited in Vols.1-5:	

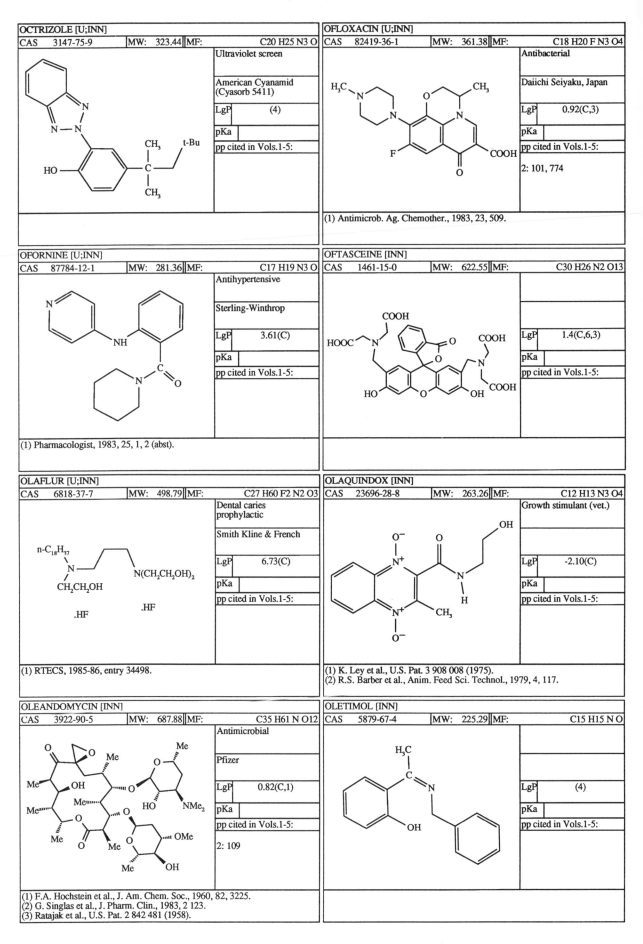

OCTRIZOLE [U;INN]		
CAS 3147-75-9	MW: 323.44	MF: C20 H25 N3 O
	Ultraviolet screen	
	American Cyanamid (Cyasorb 5411)	
	LgP (4)	
	pKa	
	pp cited in Vols.1-5:	

OFLOXACIN [U;INN]		
CAS 82419-36-1	MW: 361.38	MF: C18 H20 F N3 O4
	Antibacterial	
	Daiichi Seiyaku, Japan	
	LgP 0.92(C,3)	
	pKa	
	pp cited in Vols.1-5:	
	2: 101, 774	

(1) Antimicrob. Ag. Chemother., 1983, 23, 509.

OFORNINE [U;INN]		
CAS 87784-12-1	MW: 281.36	MF: C17 H19 N3 O
	Antihypertensive	
	Sterling-Winthrop	
	LgP 3.61(C)	
	pKa	
	pp cited in Vols.1-5:	

(1) Pharmacologist, 1983, 25, 1, 2 (abst).

OFTASCEINE [INN]		
CAS 1461-15-0	MW: 622.55	MF: C30 H26 N2 O13
	LgP 1.4(C,6,3)	
	pKa	
	pp cited in Vols.1-5:	

OLAFLUR [U;INN]		
CAS 6818-37-7	MW: 498.79	MF: C27 H60 F2 N2 O3
	Dental caries prophylactic	
	Smith Kline & French	
	LgP 6.73(C)	
	pKa	
	pp cited in Vols.1-5:	

(1) RTECS, 1985-86, entry 34498.

OLAQUINDOX [INN]		
CAS 23696-28-8	MW: 263.26	MF: C12 H13 N3 O4
	Growth stimulant (vet.)	
	LgP -2.10(C)	
	pKa	
	pp cited in Vols.1-5:	

(1) K. Ley et al., U.S. Pat. 3 908 008 (1975).
(2) R.S. Barber et al., Anim. Feed Sci. Technol., 1979, 4, 117.

OLEANDOMYCIN [INN]		
CAS 3922-90-5	MW: 687.88	MF: C35 H61 N O12
	Antimicrobial	
	Pfizer	
	LgP 0.82(C,1)	
	pKa	
	pp cited in Vols.1-5:	
	2: 109	

(1) F.A. Hochstein et al., J. Am. Chem. Soc., 1960, 82, 3225.
(2) G. Singlas et al., J. Pharm. Clin., 1983, 2 123.
(3) Ratajak et al., U.S. Pat. 2 842 481 (1958).

OLETIMOL [INN]		
CAS 5879-67-4	MW: 225.29	MF: C15 H15 N O
	LgP (4)	
	pKa	
	pp cited in Vols.1-5:	

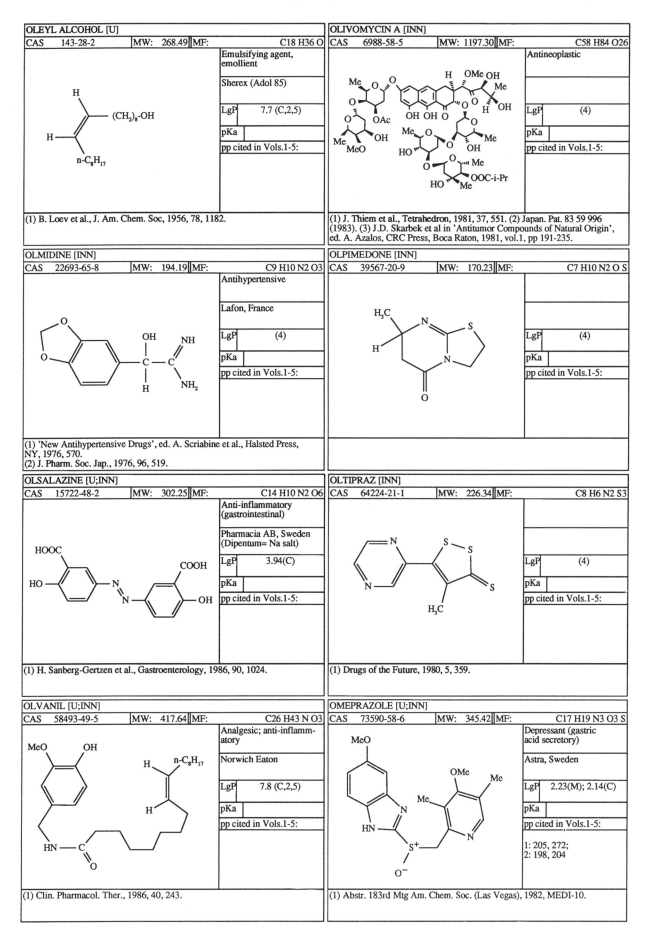

OLEYL ALCOHOL [U]

| CAS | 143-28-2 | MW: | 268.49 | MF: | C18 H36 O |

Emulsifying agent, emollient

Sherex (Adol 85)

LgP	7.7 (C,2,5)
pKa	
pp cited in Vols.1-5:	

(1) B. Loev et al., J. Am. Chem. Soc, 1956, 78, 1182.

OLIVOMYCIN A [INN]

| CAS | 6988-58-5 | MW: | 1197.30 | MF: | C58 H84 O26 |

Antineoplastic

LgP	(4)
pKa	
pp cited in Vols.1-5:	

(1) J. Thiem et al., Tetrahedron, 1981, 37, 551. (2) Japan. Pat. 83 59 996 (1983). (3) J.D. Skarbek et al in 'Antitumor Compounds of Natural Origin', ed. A. Azalos, CRC Press, Boca Raton, 1981, vol.1, pp 191-235.

OLMIDINE [INN]

| CAS | 22693-65-8 | MW: | 194.19 | MF: | C9 H10 N2 O3 |

Antihypertensive

Lafon, France

LgP	(4)
pKa	
pp cited in Vols.1-5:	

(1) 'New Antihypertensive Drugs', ed. A. Scriabine et al., Halsted Press, NY, 1976, 570.
(2) J. Pharm. Soc. Jap., 1976, 96, 519.

OLPIMEDONE [INN]

| CAS | 39567-20-9 | MW: | 170.23 | MF: | C7 H10 N2 O S |

LgP	(4)
pKa	
pp cited in Vols.1-5:	

OLSALAZINE [U;INN]

| CAS | 15722-48-2 | MW: | 302.25 | MF: | C14 H10 N2 O6 |

Anti-inflammatory (gastrointestinal)

Pharmacia AB, Sweden (Dipentum= Na salt)

LgP	3.94(C)
pKa	
pp cited in Vols.1-5:	

(1) H. Sanberg-Gertzen et al., Gastroenterology, 1986, 90, 1024.

OLTIPRAZ [INN]

| CAS | 64224-21-1 | MW: | 226.34 | MF: | C8 H6 N2 S3 |

LgP	(4)
pKa	
pp cited in Vols.1-5:	

(1) Drugs of the Future, 1980, 5, 359.

OLVANIL [U;INN]

| CAS | 58493-49-5 | MW: | 417.64 | MF: | C26 H43 N O3 |

Analgesic; anti-inflammatory

Norwich Eaton

LgP	7.8 (C,2,5)
pKa	
pp cited in Vols.1-5:	

(1) Clin. Pharmacol. Ther., 1986, 40, 243.

OMEPRAZOLE [U;INN]

| CAS | 73590-58-6 | MW: | 345.42 | MF: | C17 H19 N3 O3 S |

Depressant (gastric acid secretory)

Astra, Sweden

LgP	2.23(M); 2.14(C)
pKa	
pp cited in Vols.1-5:	

1: 205, 272;
2: 198, 204

(1) Abstr. 183rd Mtg Am. Chem. Soc. (Las Vegas), 1982, MEDI-10.

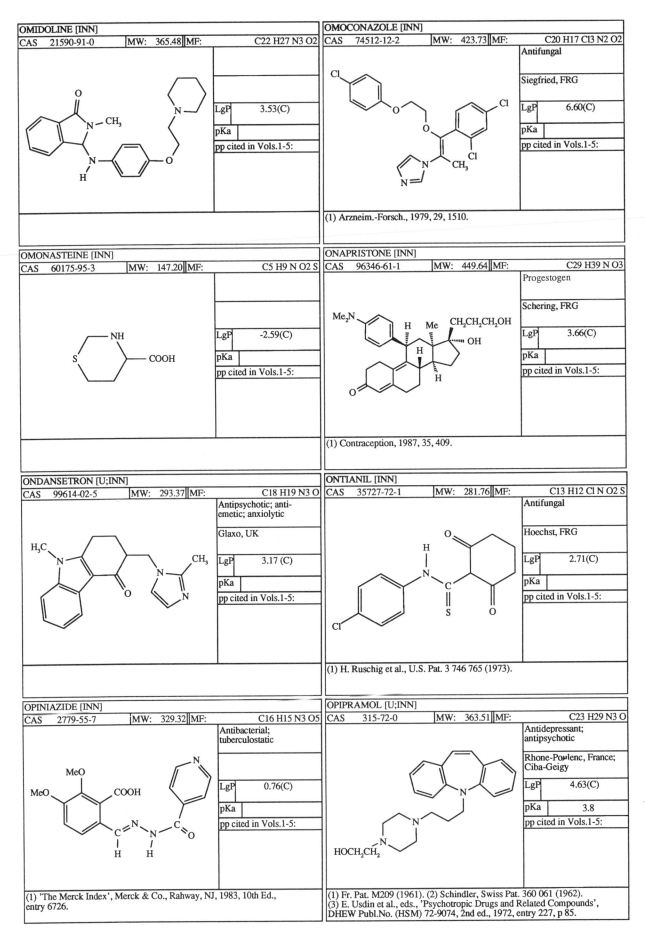

OMIDOLINE [INN]

CAS 21590-91-0	MW: 365.48	MF:	C22 H27 N3 O2

LgP	3.53(C)
pKa	
pp cited in Vols.1-5:	

OMOCONAZOLE [INN]

CAS 74512-12-2	MW: 423.73	MF:	C20 H17 Cl3 N2 O2

Antifungal

Siegfried, FRG

LgP	6.60(C)
pKa	
pp cited in Vols.1-5:	

(1) Arzneim.-Forsch., 1979, 29, 1510.

OMONASTEINE [INN]

CAS 60175-95-3	MW: 147.20	MF:	C5 H9 N O2 S

LgP	-2.59(C)
pKa	
pp cited in Vols.1-5:	

ONAPRISTONE [INN]

CAS 96346-61-1	MW: 449.64	MF:	C29 H39 N O3

Progestogen

Schering, FRG

LgP	3.66(C)
pKa	
pp cited in Vols.1-5:	

(1) Contraception, 1987, 35, 409.

ONDANSETRON [U;INN]

CAS 99614-02-5	MW: 293.37	MF:	C18 H19 N3 O

Antipsychotic; anti-emetic; anxiolytic

Glaxo, UK

LgP	3.17 (C)
pKa	
pp cited in Vols.1-5:	

ONTIANIL [INN]

CAS 35727-72-1	MW: 281.76	MF:	C13 H12 Cl N O2 S

Antifungal

Hoechst, FRG

LgP	2.71(C)
pKa	
pp cited in Vols.1-5:	

(1) H. Ruschig et al., U.S. Pat. 3 746 765 (1973).

OPINIAZIDE [INN]

CAS 2779-55-7	MW: 329.32	MF:	C16 H15 N3 O5

Antibacterial; tuberculostatic

LgP	0.76(C)
pKa	
pp cited in Vols.1-5:	

(1) 'The Merck Index', Merck & Co., Rahway, NJ, 1983, 10th Ed., entry 6726.

OPIPRAMOL [U;INN]

CAS 315-72-0	MW: 363.51	MF:	C23 H29 N3 O

Antidepressant; antipsychotic

Rhone-Poulenc, France; Ciba-Geigy

LgP	4.63(C)
pKa	3.8
pp cited in Vols.1-5:	

(1) Fr. Pat. M209 (1961). (2) Schindler, Swiss Pat. 360 061 (1962).
(3) E. Usdin et al., eds., 'Psychotropic Drugs and Related Compounds', DHEW Publ.No. (HSM) 72-9074, 2nd ed., 1972, entry 227, p 85.

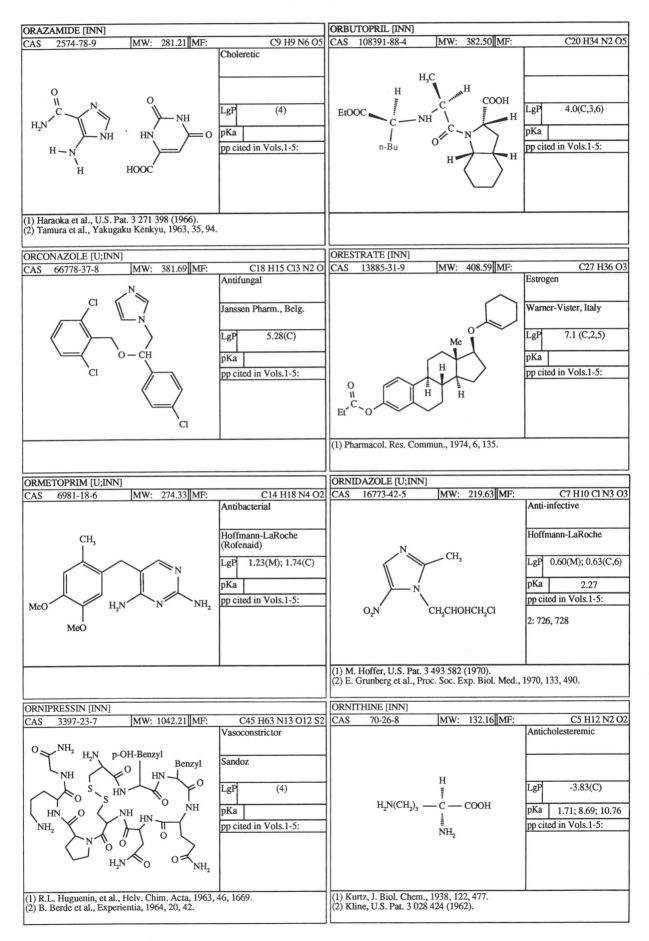

ORAZAMIDE [INN]

CAS	2574-78-9	MW: 281.21	MF:	C9 H9 N6 O5

Choleretic

LgP	(4)
pKa	
pp cited in Vols.1-5:	

(1) Haraoka et al., U.S. Pat. 3 271 398 (1966).
(2) Tamura et al., Yakugaku Kenkyu, 1963, 35, 94.

ORBUTOPRIL [INN]

CAS	108391-88-4	MW: 382.50	MF:	C20 H34 N2 O5

LgP	4.0(C,3,6)
pKa	
pp cited in Vols.1-5:	

ORCONAZOLE [U;INN]

CAS	66778-37-8	MW: 381.69	MF:	C18 H15 Cl3 N2 O

Antifungal

Janssen Pharm., Belg.

LgP	5.28(C)
pKa	
pp cited in Vols.1-5:	

ORESTRATE [INN]

CAS	13885-31-9	MW: 408.59	MF:	C27 H36 O3

Estrogen

Warner-Vister, Italy

LgP	7.1 (C,2,5)
pKa	
pp cited in Vols.1-5:	

(1) Pharmacol. Res. Commun., 1974, 6, 135.

ORMETOPRIM [U;INN]

CAS	6981-18-6	MW: 274.33	MF:	C14 H18 N4 O2

Antibacterial

Hoffmann-LaRoche (Rofenaid)

LgP	1.23(M); 1.74(C)
pKa	
pp cited in Vols.1-5:	

ORNIDAZOLE [U;INN]

CAS	16773-42-5	MW: 219.63	MF:	C7 H10 Cl N3 O3

Anti-infective

Hoffmann-LaRoche

LgP	0.60(M); 0.63(C,6)
pKa	2.27
pp cited in Vols.1-5:	

2: 726, 728

(1) M. Hoffer, U.S. Pat. 3 493 582 (1970).
(2) E. Grunberg et al., Proc. Soc. Exp. Biol. Med., 1970, 133, 490.

ORNIPRESSIN [INN]

CAS	3397-23-7	MW: 1042.21	MF:	C45 H63 N13 O12 S2

Vasoconstrictor

Sandoz

LgP	(4)
pKa	
pp cited in Vols.1-5:	

(1) R.L. Huguenin, et al., Helv. Chim. Acta, 1963, 46, 1669.
(2) B. Berde et al., Experientia, 1964, 20, 42.

ORNITHINE [INN]

CAS	70-26-8	MW: 132.16	MF:	C5 H12 N2 O2

Anticholesteremic

LgP	-3.83(C)
pKa	1.71; 8.69; 10.76
pp cited in Vols.1-5:	

(1) Kurtz, J. Biol. Chem., 1938, 122, 477.
(2) Kline, U.S. Pat. 3 028 424 (1962).

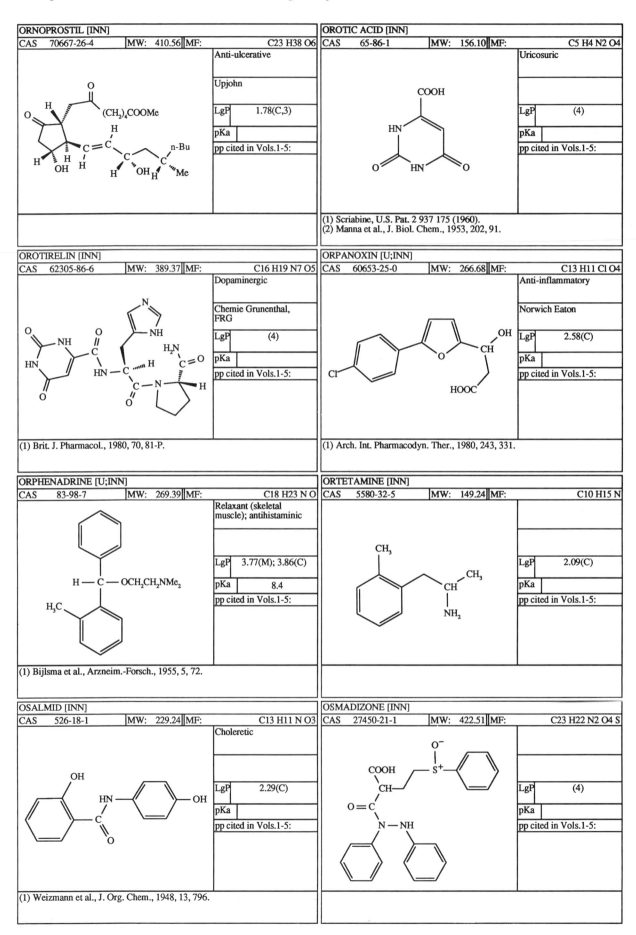

ORNOPROSTIL [INN]		
CAS 70667-26-4	MW: 410.56	MF: C23 H38 O6

Anti-ulcerative

Upjohn

LgP	1.78(C,3)
pKa	
pp cited in Vols.1-5:	

OROTIC ACID [INN]		
CAS 65-86-1	MW: 156.10	MF: C5 H4 N2 O4

Uricosuric

LgP	(4)
pKa	
pp cited in Vols.1-5:	

(1) Scriabine, U.S. Pat. 2 937 175 (1960).
(2) Manna et al., J. Biol. Chem., 1953, 202, 91.

OROTIRELIN [INN]		
CAS 62305-86-6	MW: 389.37	MF: C16 H19 N7 O5

Dopaminergic

Chemie Grunenthal, FRG

LgP	(4)
pKa	
pp cited in Vols.1-5:	

(1) Brit. J. Pharmacol., 1980, 70, 81-P.

ORPANOXIN [U;INN]		
CAS 60653-25-0	MW: 266.68	MF: C13 H11 Cl O4

Anti-inflammatory

Norwich Eaton

LgP	2.58(C)
pKa	
pp cited in Vols.1-5:	

(1) Arch. Int. Pharmacodyn. Ther., 1980, 243, 331.

ORPHENADRINE [U;INN]		
CAS 83-98-7	MW: 269.39	MF: C18 H23 N O

Relaxant (skeletal muscle); antihistaminic

LgP	3.77(M); 3.86(C)
pKa	8.4
pp cited in Vols.1-5:	

(1) Bijlsma et al., Arzneim.-Forsch., 1955, 5, 72.

ORTETAMINE [INN]		
CAS 5580-32-5	MW: 149.24	MF: C10 H15 N

LgP	2.09(C)
pKa	
pp cited in Vols.1-5:	

OSALMID [INN]		
CAS 526-18-1	MW: 229.24	MF: C13 H11 N O3

Choleretic

LgP	2.29(C)
pKa	
pp cited in Vols.1-5:	

(1) Weizmann et al., J. Org. Chem., 1948, 13, 796.

OSMADIZONE [INN]		
CAS 27450-21-1	MW: 422.51	MF: C23 H22 N2 O4 S

LgP	(4)
pKa	
pp cited in Vols.1-5:	

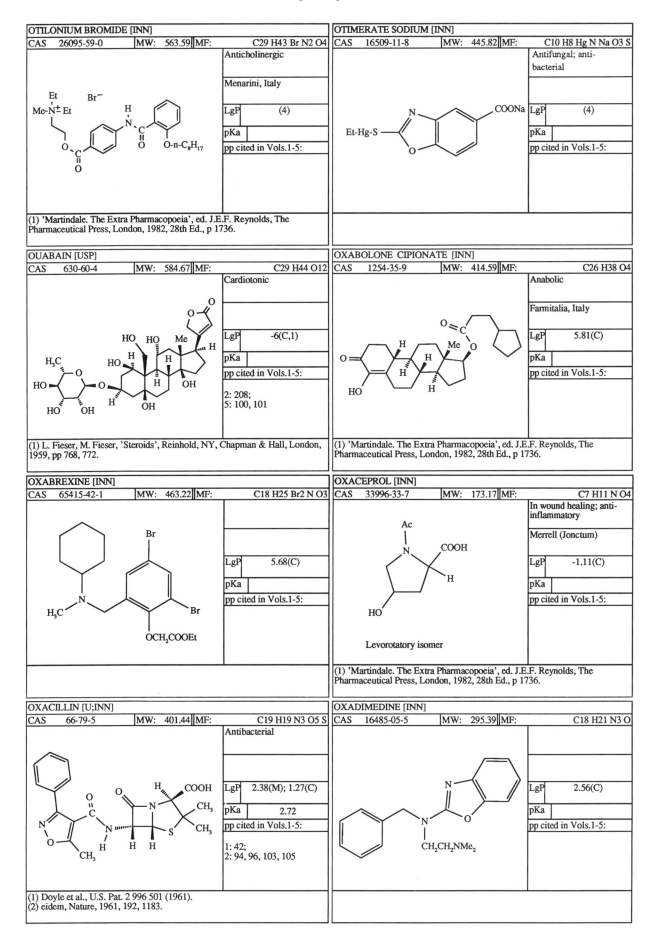

OTILONIUM BROMIDE [INN]

CAS	26095-59-0	MW:	563.59	MF:	C29 H43 Br N2 O4

Anticholinergic

Menarini, Italy

LgP	(4)
pKa	
pp cited in Vols.1-5:	

(1) 'Martindale. The Extra Pharmacopoeia', ed. J.E.F. Reynolds, The Pharmaceutical Press, London, 1982, 28th Ed., p 1736.

OTIMERATE SODIUM [INN]

CAS	16509-11-8	MW:	445.82	MF:	C10 H8 Hg N Na O3 S

Antifungal; anti-bacterial

LgP	(4)
pKa	
pp cited in Vols.1-5:	

OUABAIN [USP]

CAS	630-60-4	MW:	584.67	MF:	C29 H44 O12

Cardiotonic

LgP	-6(C,1)
pKa	
pp cited in Vols.1-5:	

2: 208;
5: 100, 101

(1) L. Fieser, M. Fieser, 'Steroids', Reinhold, NY, Chapman & Hall, London, 1959, pp 768, 772.

OXABOLONE CIPIONATE [INN]

CAS	1254-35-9	MW:	414.59	MF:	C26 H38 O4

Anabolic

Farmitalia, Italy

LgP	5.81(C)
pKa	
pp cited in Vols.1-5:	

(1) 'Martindale. The Extra Pharmacopoeia', ed. J.E.F. Reynolds, The Pharmaceutical Press, London, 1982, 28th Ed., p 1736.

OXABREXINE [INN]

CAS	65415-42-1	MW:	463.22	MF:	C18 H25 Br2 N O3

LgP	5.68(C)
pKa	
pp cited in Vols.1-5:	

OCH₂COOEt

OXACEPROL [INN]

CAS	33996-33-7	MW:	173.17	MF:	C7 H11 N O4

In wound healing; anti-inflammatory

Merrell (Jonctum)

LgP	-1.11(C)
pKa	
pp cited in Vols.1-5:	

Levorotatory isomer

(1) 'Martindale. The Extra Pharmacopoeia', ed. J.E.F. Reynolds, The Pharmaceutical Press, London, 1982, 28th Ed., p 1736.

OXACILLIN [U;INN]

CAS	66-79-5	MW:	401.44	MF:	C19 H19 N3 O5 S

Antibacterial

LgP	2.38(M); 1.27(C)
pKa	2.72
pp cited in Vols.1-5:	

1: 42;
2: 94, 96, 103, 105

(1) Doyle et al., U.S. Pat. 2 996 501 (1961).
(2) eidem, Nature, 1961, 192, 1183.

OXADIMEDINE [INN]

CAS	16485-05-5	MW:	295.39	MF:	C18 H21 N3 O

LgP	2.56(C)
pKa	
pp cited in Vols.1-5:	

CH₂CH₂NMe₂

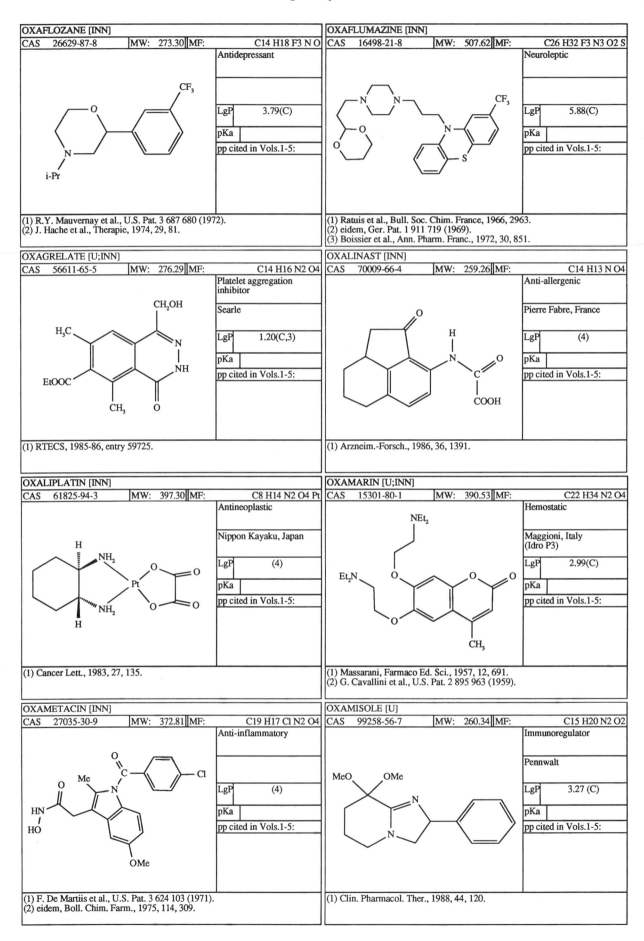

OXAFLOZANE [INN]			
CAS 26629-87-8	MW: 273.30	MF:	C14 H18 F3 N O
			Antidepressant
		LgP	3.79(C)
		pKa	
		pp cited in Vols.1-5:	

(1) R.Y. Mauvernay et al., U.S. Pat. 3 687 680 (1972).
(2) J. Hache et al., Therapie, 1974, 29, 81.

OXAFLUMAZINE [INN]			
CAS 16498-21-8	MW: 507.62	MF:	C26 H32 F3 N3 O2 S
			Neuroleptic
		LgP	5.88(C)
		pKa	
		pp cited in Vols.1-5:	

(1) Ratuis et al., Bull. Soc. Chim. France, 1966, 2963.
(2) eidem, Ger. Pat. 1 911 719 (1969).
(3) Boissier et al., Ann. Pharm. Franc., 1972, 30, 851.

OXAGRELATE [U;INN]			
CAS 56611-65-5	MW: 276.29	MF:	C14 H16 N2 O4
			Platelet aggregation inhibitor
			Searle
		LgP	1.20(C,3)
		pKa	
		pp cited in Vols.1-5:	

(1) RTECS, 1985-86, entry 59725.

OXALINAST [INN]			
CAS 70009-66-4	MW: 259.26	MF:	C14 H13 N O4
			Anti-allergenic
			Pierre Fabre, France
		LgP	(4)
		pKa	
		pp cited in Vols.1-5:	

(1) Arzneim.-Forsch., 1986, 36, 1391.

OXALIPLATIN [INN]			
CAS 61825-94-3	MW: 397.30	MF:	C8 H14 N2 O4 Pt
			Antineoplastic
			Nippon Kayaku, Japan
		LgP	(4)
		pKa	
		pp cited in Vols.1-5:	

(1) Cancer Lett., 1983, 27, 135.

OXAMARIN [U;INN]			
CAS 15301-80-1	MW: 390.53	MF:	C22 H34 N2 O4
			Hemostatic
			Maggioni, Italy (Idro P3)
		LgP	2.99(C)
		pKa	
		pp cited in Vols.1-5:	

(1) Massarani, Farmaco Ed. Sci., 1957, 12, 691.
(2) G. Cavallini et al., U.S. Pat. 2 895 963 (1959).

OXAMETACIN [INN]			
CAS 27035-30-9	MW: 372.81	MF:	C19 H17 Cl N2 O4
			Anti-inflammatory
		LgP	(4)
		pKa	
		pp cited in Vols.1-5:	

(1) F. De Martiis et al., U.S. Pat. 3 624 103 (1971).
(2) eidem, Boll. Chim. Farm., 1975, 114, 309.

OXAMISOLE [U]			
CAS 99258-56-7	MW: 260.34	MF:	C15 H20 N2 O2
			Immunoregulator
			Pennwalt
		LgP	3.27 (C)
		pKa	
		pp cited in Vols.1-5:	

(1) Clin. Pharmacol. Ther., 1988, 44, 120.

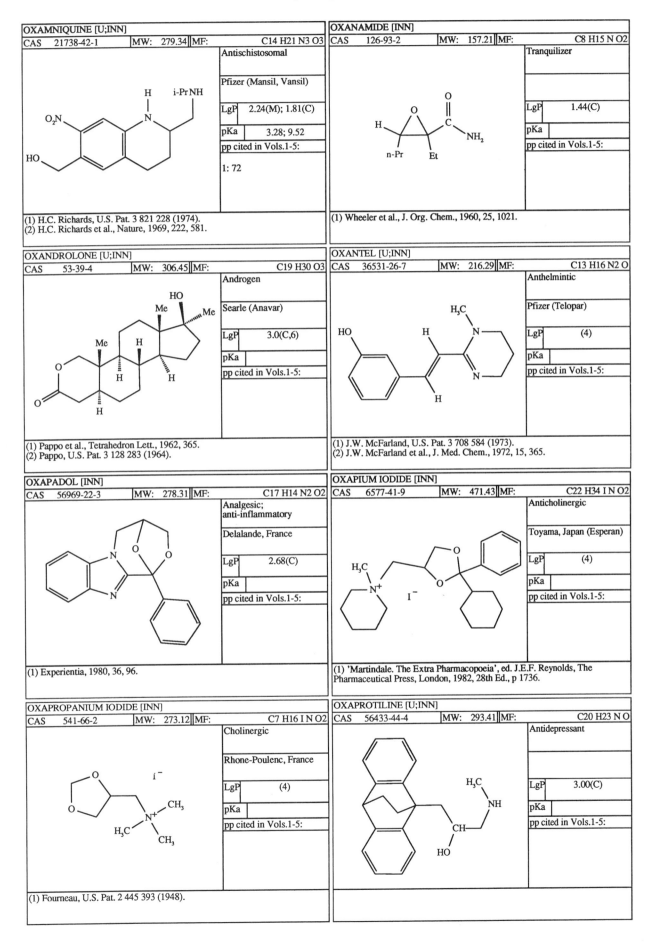

OXAMNIQUINE [U;INN]

CAS	21738-42-1	MW:	279.34	MF:	C14 H21 N3 O3

Antischistosomal

Pfizer (Mansil, Vansil)

LgP	2.24(M); 1.81(C)
pKa	3.28; 9.52
pp cited in Vols.1-5:	

1: 72

(1) H.C. Richards, U.S. Pat. 3 821 228 (1974).
(2) H.C. Richards et al., Nature, 1969, 222, 581.

OXANAMIDE [INN]

CAS	126-93-2	MW:	157.21	MF:	C8 H15 N O2

Tranquilizer

LgP	1.44(C)
pKa	
pp cited in Vols.1-5:	

(1) Wheeler et al., J. Org. Chem., 1960, 25, 1021.

OXANDROLONE [U;INN]

CAS	53-39-4	MW:	306.45	MF:	C19 H30 O3

Androgen

Searle (Anavar)

LgP	3.0(C,6)
pKa	
pp cited in Vols.1-5:	

(1) Pappo et al., Tetrahedron Lett., 1962, 365.
(2) Pappo, U.S. Pat. 3 128 283 (1964).

OXANTEL [U;INN]

CAS	36531-26-7	MW:	216.29	MF:	C13 H16 N2 O

Anthelmintic

Pfizer (Telopar)

LgP	(4)
pKa	
pp cited in Vols.1-5:	

(1) J.W. McFarland, U.S. Pat. 3 708 584 (1973).
(2) J.W. McFarland et al., J. Med. Chem., 1972, 15, 365.

OXAPADOL [INN]

CAS	56969-22-3	MW:	278.31	MF:	C17 H14 N2 O2

Analgesic;
anti-inflammatory

Delalande, France

LgP	2.68(C)
pKa	
pp cited in Vols.1-5:	

(1) Experientia, 1980, 36, 96.

OXAPIUM IODIDE [INN]

CAS	6577-41-9	MW:	471.43	MF:	C22 H34 I N O2

Anticholinergic

Toyama, Japan (Esperan)

LgP	(4)
pKa	
pp cited in Vols.1-5:	

(1) 'Martindale. The Extra Pharmacopoeia', ed. J.E.F. Reynolds, The
Pharmaceutical Press, London, 1982, 28th Ed., p 1736.

OXAPROPANIUM IODIDE [INN]

CAS	541-66-2	MW:	273.12	MF:	C7 H16 I N O2

Cholinergic

Rhone-Poulenc, France

LgP	(4)
pKa	
pp cited in Vols.1-5:	

(1) Fourneau, U.S. Pat. 2 445 393 (1948).

OXAPROTILINE [U;INN]

CAS	56433-44-4	MW:	293.41	MF:	C20 H23 N O

Antidepressant

LgP	3.00(C)
pKa	
pp cited in Vols.1-5:	

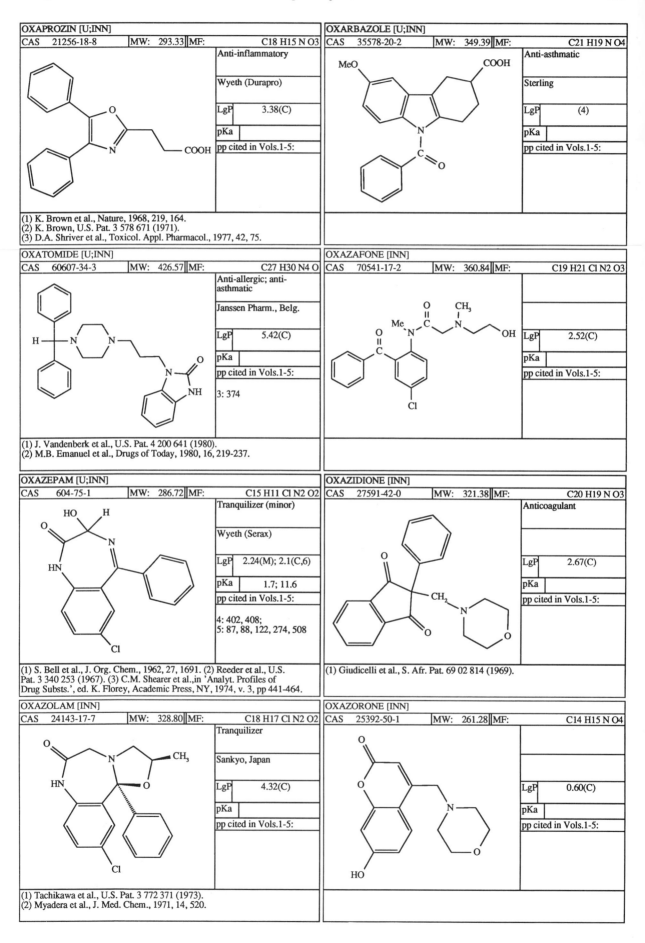

OXAPROZIN [U;INN]

CAS	21256-18-8	MW:	293.33	MF:	C18 H15 N O3

Anti-inflammatory

Wyeth (Durapro)

LgP 3.38(C)

pKa

pp cited in Vols.1-5:

(1) K. Brown et al., Nature, 1968, 219, 164.
(2) K. Brown, U.S. Pat. 3 578 671 (1971).
(3) D.A. Shriver et al., Toxicol. Appl. Pharmacol., 1977, 42, 75.

OXARBAZOLE [U;INN]

CAS	35578-20-2	MW:	349.39	MF:	C21 H19 N O4

Anti-asthmatic

Sterling

LgP (4)

pKa

pp cited in Vols.1-5:

OXATOMIDE [U;INN]

CAS	60607-34-3	MW:	426.57	MF:	C27 H30 N4 O

Anti-allergic; anti-asthmatic

Janssen Pharm., Belg.

LgP 5.42(C)

pKa

pp cited in Vols.1-5:

3: 374

(1) J. Vandenberk et al., U.S. Pat. 4 200 641 (1980).
(2) M.B. Emanuel et al., Drugs of Today, 1980, 16, 219-237.

OXAZAFONE [INN]

CAS	70541-17-2	MW:	360.84	MF:	C19 H21 Cl N2 O3

LgP 2.52(C)

pKa

pp cited in Vols.1-5:

OXAZEPAM [U;INN]

CAS	604-75-1	MW:	286.72	MF:	C15 H11 Cl N2 O2

Tranquilizer (minor)

Wyeth (Serax)

LgP 2.24(M); 2.1(C,6)

pKa 1.7; 11.6

pp cited in Vols.1-5:

4: 402, 408;
5: 87, 88, 122, 274, 508

(1) S. Bell et al., J. Org. Chem., 1962, 27, 1691. (2) Reeder et al., U.S. Pat. 3 340 253 (1967). (3) C.M. Shearer et al.,in 'Analyt. Profiles of Drug Substs.', ed. K. Florey, Academic Press, NY, 1974, v. 3, pp 441-464.

OXAZIDIONE [INN]

CAS	27591-42-0	MW:	321.38	MF:	C20 H19 N O3

Anticoagulant

LgP 2.67(C)

pKa

pp cited in Vols.1-5:

(1) Giudicelli et al., S. Afr. Pat. 69 02 814 (1969).

OXAZOLAM [INN]

CAS	24143-17-7	MW:	328.80	MF:	C18 H17 Cl N2 O2

Tranquilizer

Sankyo, Japan

LgP 4.32(C)

pKa

pp cited in Vols.1-5:

(1) Tachikawa et al., U.S. Pat. 3 772 371 (1973).
(2) Myadera et al., J. Med. Chem., 1971, 14, 520.

OXAZORONE [INN]

CAS	25392-50-1	MW:	261.28	MF:	C14 H15 N O4

LgP 0.60(C)

pKa

pp cited in Vols.1-5:

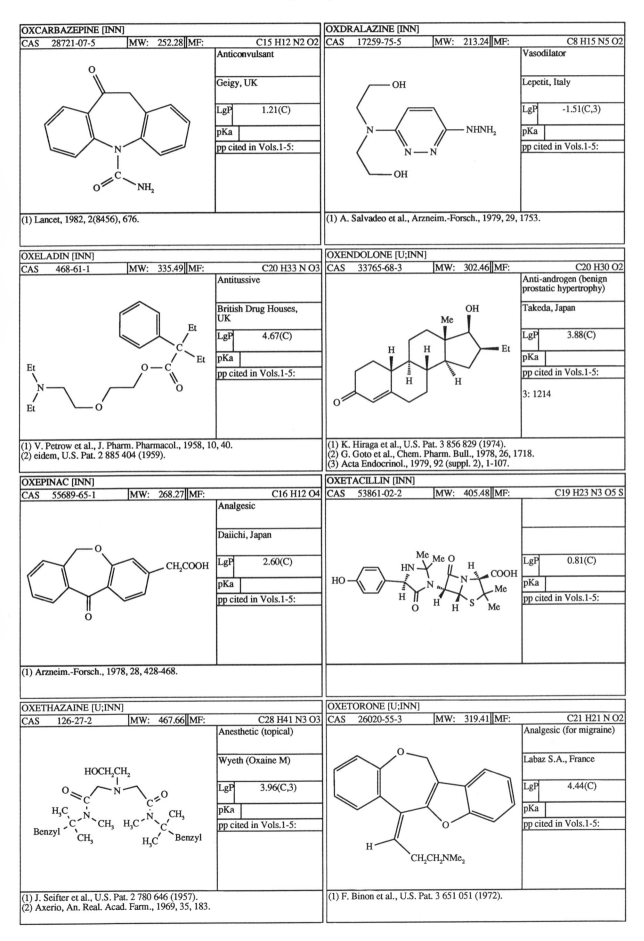

OXCARBAZEPINE [INN]

| CAS 28721-07-5 | MW: 252.28 | MF: C15 H12 N2 O2 |

Anticonvulsant

Geigy, UK

LgP 1.21(C)

pKa

pp cited in Vols.1-5:

(1) Lancet, 1982, 2(8456), 676.

OXDRALAZINE [INN]

| CAS 17259-75-5 | MW: 213.24 | MF: C8 H15 N5 O2 |

Vasodilator

Lepetit, Italy

LgP -1.51(C,3)

pKa

pp cited in Vols.1-5:

(1) A. Salvadeo et al., Arzneim.-Forsch., 1979, 29, 1753.

OXELADIN [INN]

| CAS 468-61-1 | MW: 335.49 | MF: C20 H33 N O3 |

Antitussive

British Drug Houses, UK

LgP 4.67(C)

pKa

pp cited in Vols.1-5:

(1) V. Petrow et al., J. Pharm. Pharmacol., 1958, 10, 40.
(2) eidem, U.S. Pat. 2 885 404 (1959).

OXENDOLONE [U;INN]

| CAS 33765-68-3 | MW: 302.46 | MF: C20 H30 O2 |

Anti-androgen (benign prostatic hypertrophy)

Takeda, Japan

LgP 3.88(C)

pKa

pp cited in Vols.1-5:

3: 1214

(1) K. Hiraga et al., U.S. Pat. 3 856 829 (1974).
(2) G. Goto et al., Chem. Pharm. Bull., 1978, 26, 1718.
(3) Acta Endocrinol., 1979, 92 (suppl. 2), 1-107.

OXEPINAC [INN]

| CAS 55689-65-1 | MW: 268.27 | MF: C16 H12 O4 |

Analgesic

Daiichi, Japan

LgP 2.60(C)

pKa

pp cited in Vols.1-5:

(1) Arzneim.-Forsch., 1978, 28, 428-468.

OXETACILLIN [INN]

| CAS 53861-02-2 | MW: 405.48 | MF: C19 H23 N3 O5 S |

LgP 0.81(C)

pKa

pp cited in Vols.1-5:

OXETHAZAINE [U;INN]

| CAS 126-27-2 | MW: 467.66 | MF: C28 H41 N3 O3 |

Anesthetic (topical)

Wyeth (Oxaine M)

LgP 3.96(C,3)

pKa

pp cited in Vols.1-5:

(1) J. Seifter et al., U.S. Pat. 2 780 646 (1957).
(2) Axerio, An. Real. Acad. Farm., 1969, 35, 183.

OXETORONE [U;INN]

| CAS 26020-55-3 | MW: 319.41 | MF: C21 H21 N O2 |

Analgesic (for migraine)

Labaz S.A., France

LgP 4.44(C)

pKa

pp cited in Vols.1-5:

(1) F. Binon et al., U.S. Pat. 3 651 051 (1972).

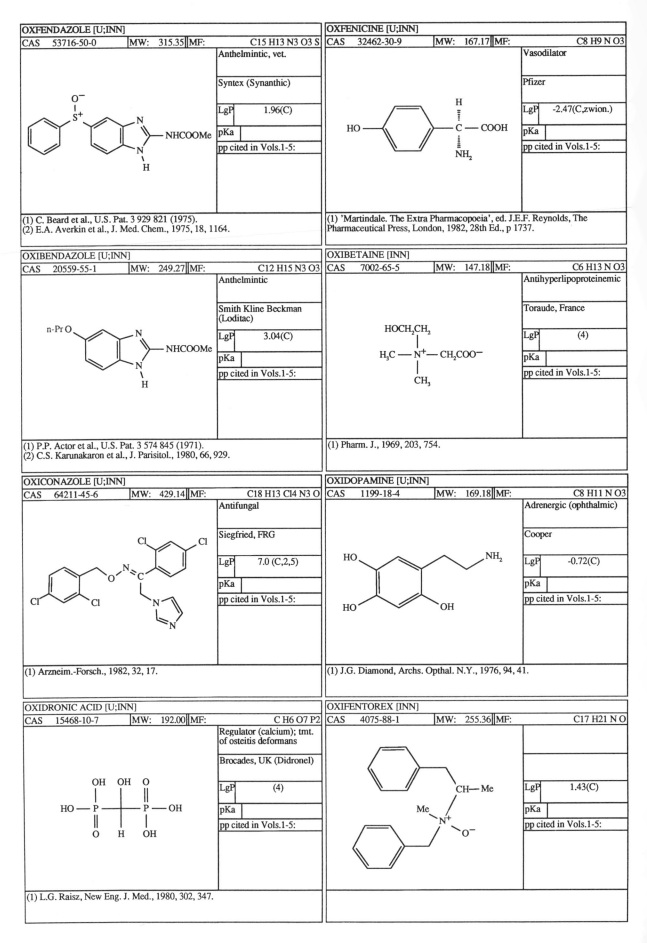

OXFENDAZOLE [U;INN]

CAS	53716-50-0	MW:	315.35	MF:	C15 H13 N3 O3 S

Anthelmintic, vet.

Syntex (Synanthic)

LgP	1.96(C)
pKa	
pp cited in Vols.1-5:	

(1) C. Beard et al., U.S. Pat. 3 929 821 (1975).
(2) E.A. Averkin et al., J. Med. Chem., 1975, 18, 1164.

OXFENICINE [U;INN]

CAS	32462-30-9	MW:	167.17	MF:	C8 H9 N O3

Vasodilator

Pfizer

LgP	-2.47(C,zwion.)
pKa	
pp cited in Vols.1-5:	

(1) 'Martindale. The Extra Pharmacopoeia', ed. J.E.F. Reynolds, The Pharmaceutical Press, London, 1982, 28th Ed., p 1737.

OXIBENDAZOLE [U;INN]

CAS	20559-55-1	MW:	249.27	MF:	C12 H15 N3 O3

Anthelmintic

Smith Kline Beckman (Loditac)

LgP	3.04(C)
pKa	
pp cited in Vols.1-5:	

(1) P.P. Actor et al., U.S. Pat. 3 574 845 (1971).
(2) C.S. Karunakaron et al., J. Parisitol., 1980, 66, 929.

OXIBETAINE [INN]

CAS	7002-65-5	MW:	147.18	MF:	C6 H13 N O3

Antihyperlipoproteinemic

Toraude, France

LgP	(4)
pKa	
pp cited in Vols.1-5:	

(1) Pharm. J., 1969, 203, 754.

OXICONAZOLE [U;INN]

CAS	64211-45-6	MW:	429.14	MF:	C18 H13 Cl4 N3 O

Antifungal

Siegfried, FRG

LgP	7.0 (C,2,5)
pKa	
pp cited in Vols.1-5:	

(1) Arzneim.-Forsch., 1982, 32, 17.

OXIDOPAMINE [U;INN]

CAS	1199-18-4	MW:	169.18	MF:	C8 H11 N O3

Adrenergic (ophthalmic)

Cooper

LgP	-0.72(C)
pKa	
pp cited in Vols.1-5:	

(1) J.G. Diamond, Archs. Opthal. N.Y., 1976, 94, 41.

OXIDRONIC ACID [U;INN]

CAS	15468-10-7	MW:	192.00	MF:	C H6 O7 P2

Regulator (calcium); tmt. of osteitis deformans

Brocades, UK (Didronel)

LgP	(4)
pKa	
pp cited in Vols.1-5:	

(1) L.G. Raisz, New Eng. J. Med., 1980, 302, 347.

OXIFENTOREX [INN]

CAS	4075-88-1	MW:	255.36	MF:	C17 H21 N O

LgP	1.43(C)
pKa	
pp cited in Vols.1-5:	

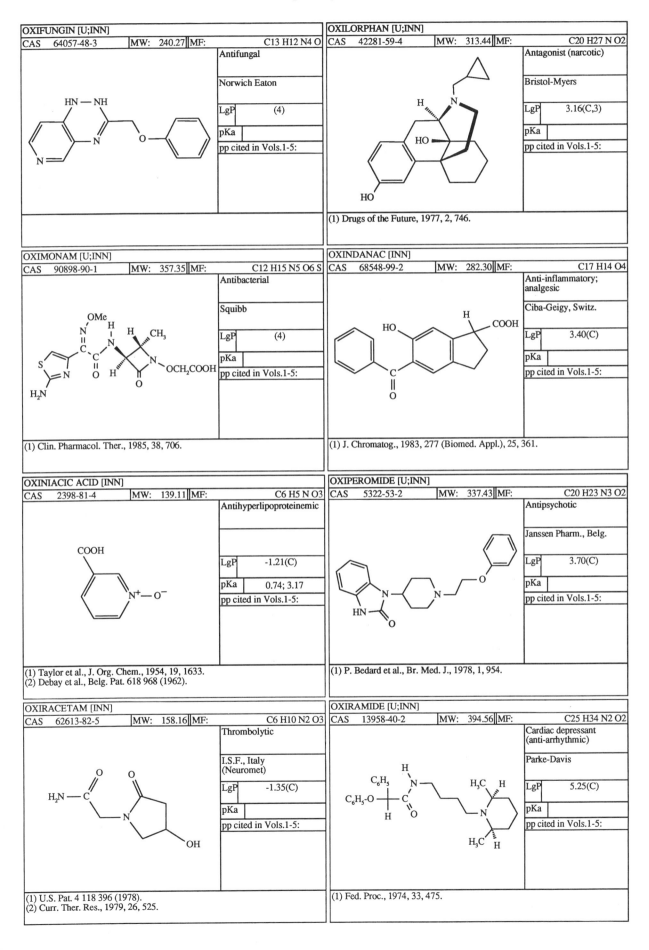

OXIFUNGIN [U;INN]

CAS	64057-48-3	MW:	240.27	MF:	C13 H12 N4 O

Antifungal

Norwich Eaton

LgP	(4)
pKa	
pp cited in Vols.1-5:	

OXILORPHAN [U;INN]

CAS	42281-59-4	MW:	313.44	MF:	C20 H27 N O2

Antagonist (narcotic)

Bristol-Myers

LgP	3.16(C,3)
pKa	
pp cited in Vols.1-5:	

(1) Drugs of the Future, 1977, 2, 746.

OXIMONAM [U;INN]

CAS	90898-90-1	MW:	357.35	MF:	C12 H15 N5 O6 S

Antibacterial

Squibb

LgP	(4)
pKa	
pp cited in Vols.1-5:	

(1) Clin. Pharmacol. Ther., 1985, 38, 706.

OXINDANAC [INN]

CAS	68548-99-2	MW:	282.30	MF:	C17 H14 O4

Anti-inflammatory; analgesic

Ciba-Geigy, Switz.

LgP	3.40(C)
pKa	
pp cited in Vols.1-5:	

(1) J. Chromatog., 1983, 277 (Biomed. Appl.), 25, 361.

OXINIACIC ACID [INN]

CAS	2398-81-4	MW:	139.11	MF:	C6 H5 N O3

Antihyperlipoproteinemic

LgP	-1.21(C)
pKa	0.74; 3.17
pp cited in Vols.1-5:	

(1) Taylor et al., J. Org. Chem., 1954, 19, 1633.
(2) Debay et al., Belg. Pat. 618 968 (1962).

OXIPEROMIDE [U;INN]

CAS	5322-53-2	MW:	337.43	MF:	C20 H23 N3 O2

Antipsychotic

Janssen Pharm., Belg.

LgP	3.70(C)
pKa	
pp cited in Vols.1-5:	

(1) P. Bedard et al., Br. Med. J., 1978, 1, 954.

OXIRACETAM [INN]

CAS	62613-82-5	MW:	158.16	MF:	C6 H10 N2 O3

Thrombolytic

I.S.F., Italy (Neuromet)

LgP	-1.35(C)
pKa	
pp cited in Vols.1-5:	

(1) U.S. Pat. 4 118 396 (1978).
(2) Curr. Ther. Res., 1979, 26, 525.

OXIRAMIDE [U;INN]

CAS	13958-40-2	MW:	394.56	MF:	C25 H34 N2 O2

Cardiac depressant (anti-arrhythmic)

Parke-Davis

LgP	5.25(C)
pKa	
pp cited in Vols.1-5:	

(1) Fed. Proc., 1974, 33, 475.

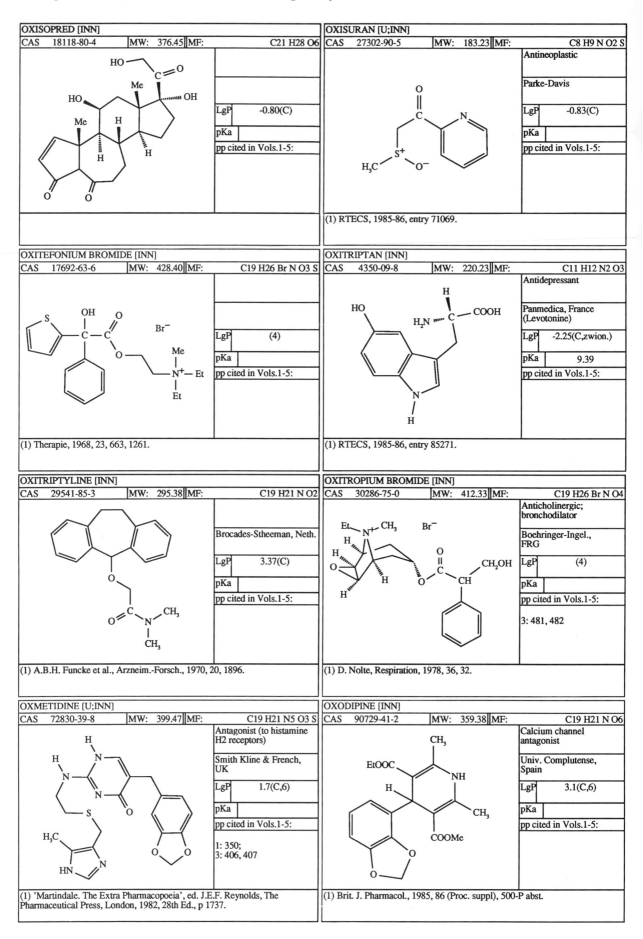

OXISOPRED [INN]

CAS	18118-80-4	MW:	376.45	MF:	C21 H28 O6

LgP	-0.80(C)
pKa	
pp cited in Vols.1-5:	

OXISURAN [U;INN]

CAS	27302-90-5	MW:	183.23	MF:	C8 H9 N O2 S

Antineoplastic

Parke-Davis

LgP	-0.83(C)
pKa	
pp cited in Vols.1-5:	

(1) RTECS, 1985-86, entry 71069.

OXITEFONIUM BROMIDE [INN]

CAS	17692-63-6	MW:	428.40	MF:	C19 H26 Br N O3 S

LgP	(4)
pKa	
pp cited in Vols.1-5:	

(1) Therapie, 1968, 23, 663, 1261.

OXITRIPTAN [INN]

CAS	4350-09-8	MW:	220.23	MF:	C11 H12 N2 O3

Antidepressant

Panmedica, France
(Levotonine)

LgP	-2.25(C,zwion.)
pKa	9.39
pp cited in Vols.1-5:	

(1) RTECS, 1985-86, entry 85271.

OXITRIPTYLINE [INN]

CAS	29541-85-3	MW:	295.38	MF:	C19 H21 N O2

Brocades-Stheeman, Neth.

LgP	3.37(C)
pKa	
pp cited in Vols.1-5:	

(1) A.B.H. Funcke et al., Arzneim.-Forsch., 1970, 20, 1896.

OXITROPIUM BROMIDE [INN]

CAS	30286-75-0	MW:	412.33	MF:	C19 H26 Br N O4

Anticholinergic;
bronchodilator

Boehringer-Ingel.,
FRG

LgP	(4)
pKa	
pp cited in Vols.1-5:	
3: 481, 482	

(1) D. Nolte, Respiration, 1978, 36, 32.

OXMETIDINE [U;INN]

CAS	72830-39-8	MW:	399.47	MF:	C19 H21 N5 O3 S

Antagonist (to histamine
H2 receptors)

Smith Kline & French,
UK

LgP	1.7(C,6)
pKa	
pp cited in Vols.1-5:	
1: 350; 3: 406, 407	

(1) 'Martindale. The Extra Pharmacopoeia', ed. J.E.F. Reynolds, The Pharmaceutical Press, London, 1982, 28th Ed., p 1737.

OXODIPINE [INN]

CAS	90729-41-2	MW:	359.38	MF:	C19 H21 N O6

Calcium channel
antagonist

Univ. Complutense,
Spain

LgP	3.1(C,6)
pKa	
pp cited in Vols.1-5:	

(1) Brit. J. Pharmacol., 1985, 86 (Proc. suppl), 500-P abst.

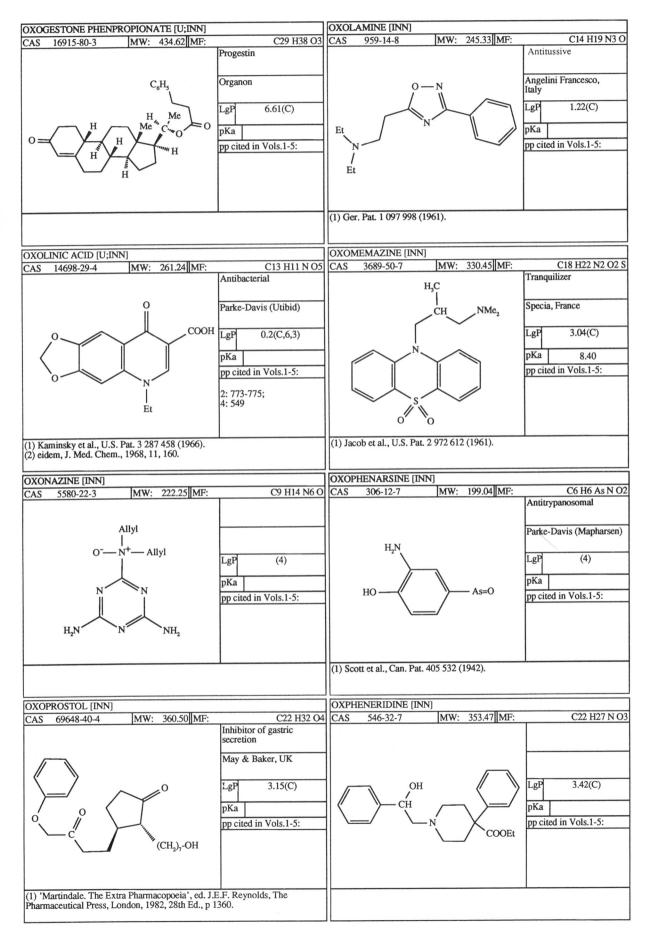

OXOGESTONE PHENPROPIONATE [U;INN]

CAS	16915-80-3	MW:	434.62	MF:	C29 H38 O3

Progestin

Organon

LgP	6.61(C)
pKa	
pp cited in Vols.1-5:	

OXOLAMINE [INN]

CAS	959-14-8	MW:	245.33	MF:	C14 H19 N3 O

Antitussive

Angelini Francesco, Italy

LgP	1.22(C)
pKa	
pp cited in Vols.1-5:	

(1) Ger. Pat. 1 097 998 (1961).

OXOLINIC ACID [U;INN]

CAS	14698-29-4	MW:	261.24	MF:	C13 H11 N O5

Antibacterial

Parke-Davis (Utibid)

LgP	0.2(C,6,3)
pKa	
pp cited in Vols.1-5:	

2: 773-775;
4: 549

(1) Kaminsky et al., U.S. Pat. 3 287 458 (1966).
(2) eidem, J. Med. Chem., 1968, 11, 160.

OXOMEMAZINE [INN]

CAS	3689-50-7	MW:	330.45	MF:	C18 H22 N2 O2 S

Tranquilizer

Specia, France

LgP	3.04(C)
pKa	8.40
pp cited in Vols.1-5:	

(1) Jacob et al., U.S. Pat. 2 972 612 (1961).

OXONAZINE [INN]

CAS	5580-22-3	MW:	222.25	MF:	C9 H14 N6 O

LgP	(4)
pKa	
pp cited in Vols.1-5:	

OXOPHENARSINE [INN]

CAS	306-12-7	MW:	199.04	MF:	C6 H6 As N O2

Antitrypanosomal

Parke-Davis (Mapharsen)

LgP	(4)
pKa	
pp cited in Vols.1-5:	

(1) Scott et al., Can. Pat. 405 532 (1942).

OXOPROSTOL [INN]

CAS	69648-40-4	MW:	360.50	MF:	C22 H32 O4

Inhibitor of gastric secretion

May & Baker, UK

LgP	3.15(C)
pKa	
pp cited in Vols.1-5:	

(1) 'Martindale. The Extra Pharmacopoeia', ed. J.E.F. Reynolds, The Pharmaceutical Press, London, 1982, 28th Ed., p 1360.

OXPHENERIDINE [INN]

CAS	546-32-7	MW:	353.47	MF:	C22 H27 N O3

LgP	3.42(C)
pKa	
pp cited in Vols.1-5:	

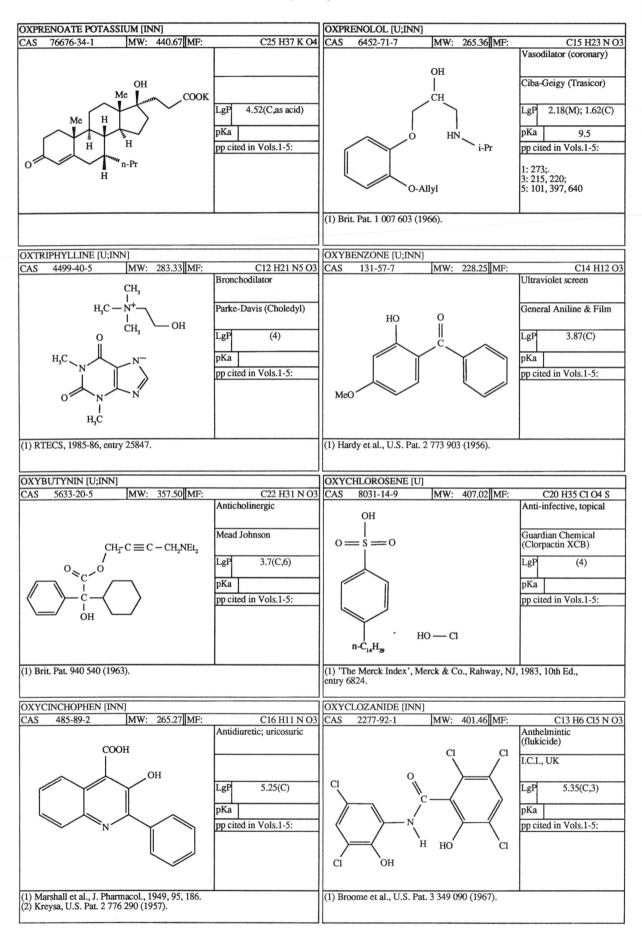

OXPRENOATE POTASSIUM [INN]			
CAS 76676-34-1	MW: 440.67	MF:	C25 H37 K O4

LgP	4.52(C,as acid)
pKa	
pp cited in Vols.1-5:	

OXPRENOLOL [U;INN]			
CAS 6452-71-7	MW: 265.36	MF:	C15 H23 N O3

Vasodilator (coronary)

Ciba-Geigy (Trasicor)

LgP	2.18(M); 1.62(C)
pKa	9.5
pp cited in Vols.1-5:	

1: 273;
3: 215, 220;
5: 101, 397, 640

(1) Brit. Pat. 1 007 603 (1966).

OXTRIPHYLLINE [U;INN]			
CAS 4499-40-5	MW: 283.33	MF:	C12 H21 N5 O3

Bronchodilator

Parke-Davis (Choledyl)

LgP	(4)
pKa	
pp cited in Vols.1-5:	

(1) RTECS, 1985-86, entry 25847.

OXYBENZONE [U;INN]			
CAS 131-57-7	MW: 228.25	MF:	C14 H12 O3

Ultraviolet screen

General Aniline & Film

LgP	3.87(C)
pKa	
pp cited in Vols.1-5:	

(1) Hardy et al., U.S. Pat. 2 773 903 (1956).

OXYBUTYNIN [U;INN]			
CAS 5633-20-5	MW: 357.50	MF:	C22 H31 N O3

Anticholinergic

Mead Johnson

LgP	3.7(C,6)
pKa	
pp cited in Vols.1-5:	

(1) Brit. Pat. 940 540 (1963).

OXYCHLOROSENE [U]			
CAS 8031-14-9	MW: 407.02	MF:	C20 H35 Cl O4 S

Anti-infective, topical

Guardian Chemical
(Clorpactin XCB)

LgP	(4)
pKa	
pp cited in Vols.1-5:	

(1) 'The Merck Index', Merck & Co., Rahway, NJ, 1983, 10th Ed., entry 6824.

OXYCINCHOPHEN [INN]			
CAS 485-89-2	MW: 265.27	MF:	C16 H11 N O3

Antidiuretic; uricosuric

LgP	5.25(C)
pKa	
pp cited in Vols.1-5:	

(1) Marshall et al., J. Pharmacol., 1949, 95, 186.
(2) Kreysa, U.S. Pat. 2 776 290 (1957).

OXYCLOZANIDE [INN]			
CAS 2277-92-1	MW: 401.46	MF:	C13 H6 Cl5 N O3

Anthelmintic
(flukicide)

I.C.I., UK

LgP	5.35(C,3)
pKa	
pp cited in Vols.1-5:	

(1) Broome et al., U.S. Pat. 3 349 090 (1967).

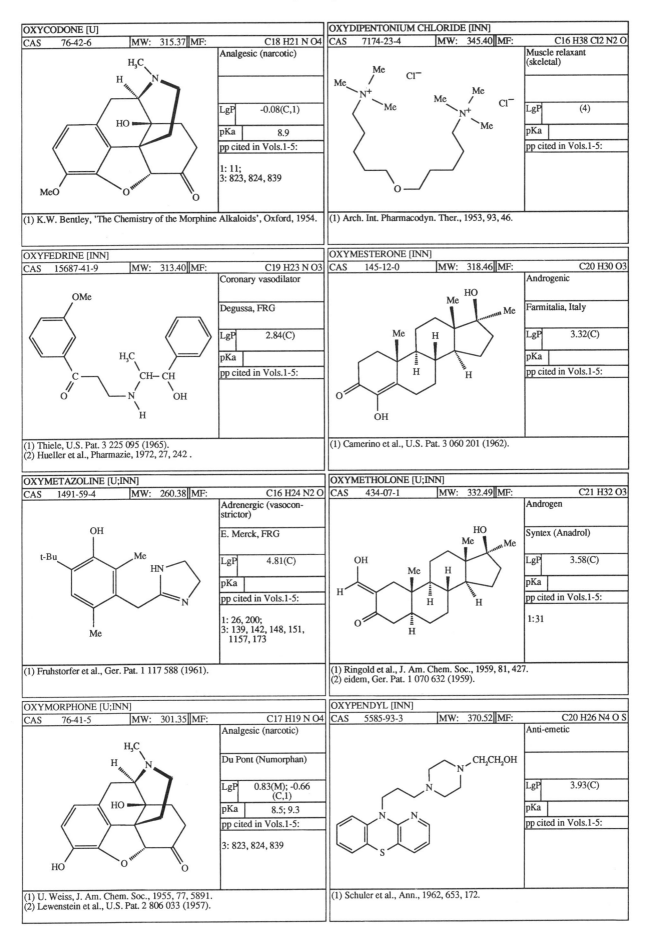

OXYCODONE [U]			
CAS 76-42-6	MW: 315.37	MF:	C18 H21 N O4

Analgesic (narcotic)

LgP	-0.08(C,1)
pKa	8.9

pp cited in Vols.1-5:

1: 11;
3: 823, 824, 839

(1) K.W. Bentley, 'The Chemistry of the Morphine Alkaloids', Oxford, 1954.

OXYDIPENTONIUM CHLORIDE [INN]			
CAS 7174-23-4	MW: 345.40	MF:	C16 H38 Cl2 N2 O

Muscle relaxant (skeletal)

LgP	(4)
pKa	

pp cited in Vols.1-5:

(1) Arch. Int. Pharmacodyn. Ther., 1953, 93, 46.

OXYFEDRINE [INN]			
CAS 15687-41-9	MW: 313.40	MF:	C19 H23 N O3

Coronary vasodilator

Degussa, FRG

LgP	2.84(C)
pKa	

pp cited in Vols.1-5:

(1) Thiele, U.S. Pat. 3 225 095 (1965).
(2) Hueller et al., Pharmazie, 1972, 27, 242 .

OXYMESTERONE [INN]			
CAS 145-12-0	MW: 318.46	MF:	C20 H30 O3

Androgenic

Farmitalia, Italy

LgP	3.32(C)
pKa	

pp cited in Vols.1-5:

(1) Camerino et al., U.S. Pat. 3 060 201 (1962).

OXYMETAZOLINE [U;INN]			
CAS 1491-59-4	MW: 260.38	MF:	C16 H24 N2 O

Adrenergic (vasocon-strictor)

E. Merck, FRG

LgP	4.81(C)
pKa	

pp cited in Vols.1-5:

1: 26, 200;
3: 139, 142, 148, 151,
   1157, 173

(1) Fruhstorfer et al., Ger. Pat. 1 117 588 (1961).

OXYMETHOLONE [U;INN]			
CAS 434-07-1	MW: 332.49	MF:	C21 H32 O3

Androgen

Syntex (Anadrol)

LgP	3.58(C)
pKa	

pp cited in Vols.1-5:

1:31

(1) Ringold et al., J. Am. Chem. Soc., 1959, 81, 427.
(2) eidem, Ger. Pat. 1 070 632 (1959).

OXYMORPHONE [U;INN]			
CAS 76-41-5	MW: 301.35	MF:	C17 H19 N O4

Analgesic (narcotic)

Du Pont (Numorphan)

LgP	0.83(M); -0.66 (C,1)
pKa	8.5; 9.3

pp cited in Vols.1-5:

3: 823, 824, 839

(1) U. Weiss, J. Am. Chem. Soc., 1955, 77, 5891.
(2) Lewenstein et al., U.S. Pat. 2 806 033 (1957).

OXYPENDYL [INN]			
CAS 5585-93-3	MW: 370.52	MF:	C20 H26 N4 O S

Anti-emetic

LgP	3.93(C)
pKa	

pp cited in Vols.1-5:

(1) Schuler et al., Ann., 1962, 653, 172.

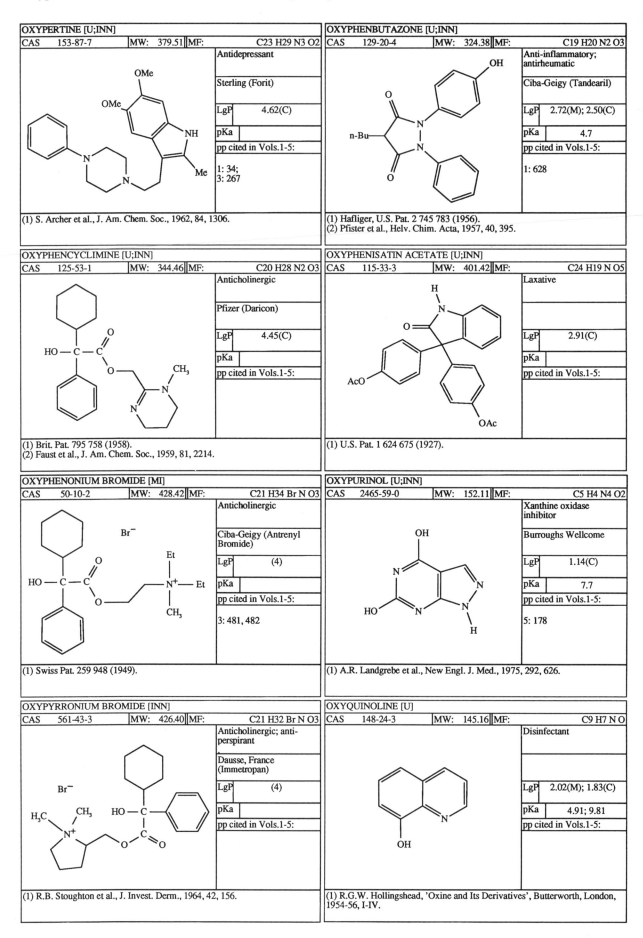

OXYPERTINE [U;INN]

CAS	153-87-7	MW:	379.51	MF:	C23 H29 N3 O2

Antidepressant

Sterling (Forit)

LgP	4.62(C)
pKa	

pp cited in Vols.1-5:

1: 34;
3: 267

(1) S. Archer et al., J. Am. Chem. Soc., 1962, 84, 1306.

OXYPHENBUTAZONE [U;INN]

CAS	129-20-4	MW:	324.38	MF:	C19 H20 N2 O3

Anti-inflammatory;
antirheumatic

Ciba-Geigy (Tandearil)

LgP	2.72(M); 2.50(C)
pKa	4.7

pp cited in Vols.1-5:

1: 628

(1) Hafliger, U.S. Pat. 2 745 783 (1956).
(2) Pfister et al., Helv. Chim. Acta, 1957, 40, 395.

OXYPHENCYCLIMINE [U;INN]

CAS	125-53-1	MW:	344.46	MF:	C20 H28 N2 O3

Anticholinergic

Pfizer (Daricon)

LgP	4.45(C)
pKa	

pp cited in Vols.1-5:

(1) Brit. Pat. 795 758 (1958).
(2) Faust et al., J. Am. Chem. Soc., 1959, 81, 2214.

OXYPHENISATIN ACETATE [U;INN]

CAS	115-33-3	MW:	401.42	MF:	C24 H19 N O5

Laxative

LgP	2.91(C)
pKa	

pp cited in Vols.1-5:

(1) U.S. Pat. 1 624 675 (1927).

OXYPHENONIUM BROMIDE [MI]

CAS	50-10-2	MW:	428.42	MF:	C21 H34 Br N O3

Anticholinergic

Ciba-Geigy (Antrenyl
Bromide)

LgP	(4)
pKa	

pp cited in Vols.1-5:

3: 481, 482

(1) Swiss Pat. 259 948 (1949).

OXYPURINOL [U;INN]

CAS	2465-59-0	MW:	152.11	MF:	C5 H4 N4 O2

Xanthine oxidase
inhibitor

Burroughs Wellcome

LgP	1.14(C)
pKa	7.7

pp cited in Vols.1-5:

5: 178

(1) A.R. Landgrebe et al., New Engl. J. Med., 1975, 292, 626.

OXYPYRRONIUM BROMIDE [INN]

CAS	561-43-3	MW:	426.40	MF:	C21 H32 Br N O3

Anticholinergic; anti-
perspirant

Dausse, France
(Immetropan)

LgP	(4)
pKa	

pp cited in Vols.1-5:

(1) R.B. Stoughton et al., J. Invest. Derm., 1964, 42, 156.

OXYQUINOLINE [U]

CAS	148-24-3	MW:	145.16	MF:	C9 H7 N O

Disinfectant

LgP	2.02(M); 1.83(C)
pKa	4.91; 9.81

pp cited in Vols.1-5:

(1) R.G.W. Hollingshead, 'Oxine and Its Derivatives', Butterworth, London,
1954-56, I-IV.

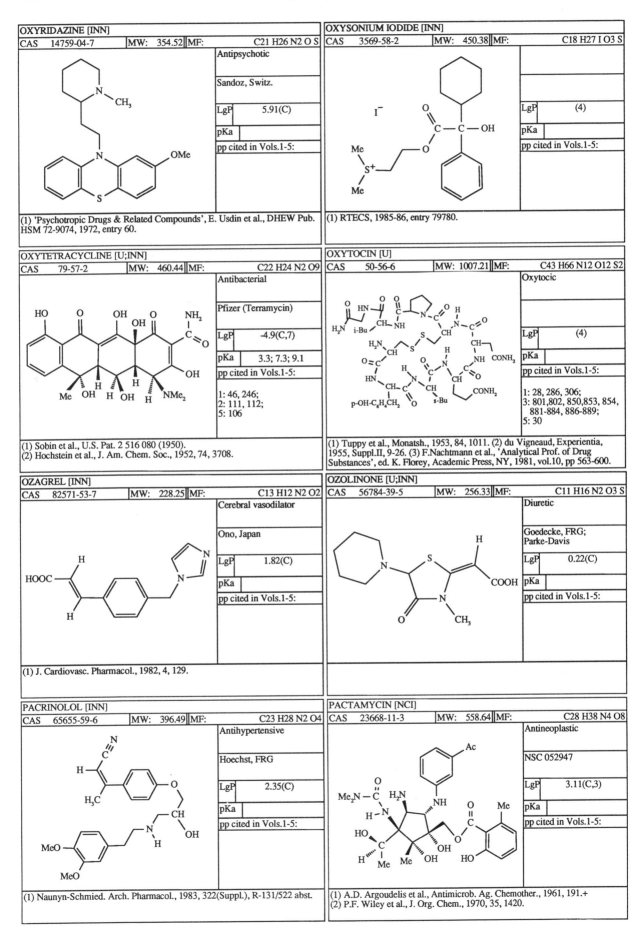

OXYRIDAZINE [INN]

CAS	14759-04-7	MW:	354.52	MF:	C21 H26 N2 O S

Antipsychotic

Sandoz, Switz.

LgP	5.91(C)

pKa	

pp cited in Vols.1-5:

(1) 'Psychotropic Drugs & Related Compounds', E. Usdin et al., DHEW Pub. HSM 72-9074, 1972, entry 60.

OXYSONIUM IODIDE [INN]

CAS	3569-58-2	MW:	450.38	MF:	C18 H27 I O3 S

LgP	(4)

pKa	

pp cited in Vols.1-5:

(1) RTECS, 1985-86, entry 79780.

OXYTETRACYCLINE [U;INN]

CAS	79-57-2	MW:	460.44	MF:	C22 H24 N2 O9

Antibacterial

Pfizer (Terramycin)

LgP	-4.9(C,7)

pKa	3.3; 7.3; 9.1

pp cited in Vols.1-5:

1: 46, 246;
2: 111, 112;
5: 106

(1) Sobin et al., U.S. Pat. 2 516 080 (1950).
(2) Hochstein et al., J. Am. Chem. Soc., 1952, 74, 3708.

OXYTOCIN [U]

CAS	50-56-6	MW:	1007.21	MF:	C43 H66 N12 O12 S2

Oxytocic

LgP	(4)

pKa	

pp cited in Vols.1-5:

1: 28, 286, 306;
3: 801,802, 850,853, 854,
   881-884, 886-889;
5: 30

(1) Tuppy et al., Monatsh., 1953, 84, 1011. (2) du Vigneaud, Experientia, 1955, Suppl.II, 9-26. (3) F.Nachtmann et al., 'Analytical Prof. of Drug Substances', ed. K. Florey, Academic Press, NY, 1981, vol.10, pp 563-600.

OZAGREL [INN]

CAS	82571-53-7	MW:	228.25	MF:	C13 H12 N2 O2

Cerebral vasodilator

Ono, Japan

LgP	1.82(C)

pKa	

pp cited in Vols.1-5:

(1) J. Cardiovasc. Pharmacol., 1982, 4, 129.

OZOLINONE [U;INN]

CAS	56784-39-5	MW:	256.33	MF:	C11 H16 N2 O3 S

Diuretic

Goedecke, FRG;
Parke-Davis

LgP	0.22(C)

pKa	

pp cited in Vols.1-5:

PACRINOLOL [INN]

CAS	65655-59-6	MW:	396.49	MF:	C23 H28 N2 O4

Antihypertensive

Hoechst, FRG

LgP	2.35(C)

pKa	

pp cited in Vols.1-5:

(1) Naunyn-Schmied. Arch. Pharmacol., 1983, 322(Suppl.), R-131/522 abst.

PACTAMYCIN [NCI]

CAS	23668-11-3	MW:	558.64	MF:	C28 H38 N4 O8

Antineoplastic

NSC 052947

LgP	3.11(C,3)

pKa	

pp cited in Vols.1-5:

(1) A.D. Argoudelis et al., Antimicrob. Ag. Chemother., 1961, 191.+
(2) P.F. Wiley et al., J. Org. Chem., 1970, 35, 1420.

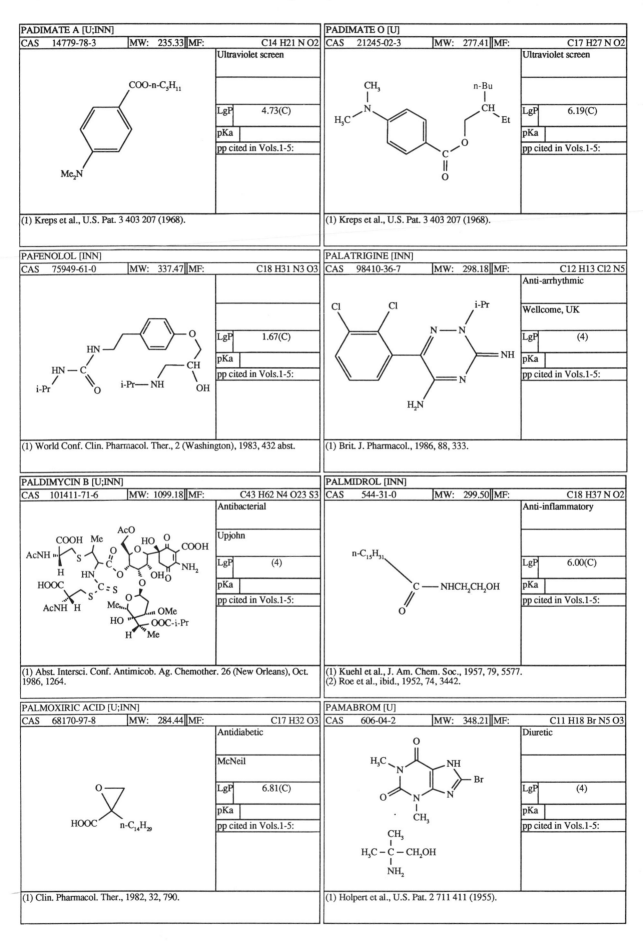

PADIMATE A [U;INN]

CAS	14779-78-3	MW:	235.33	MF:	C14 H21 N O2

Ultraviolet screen

LgP	4.73(C)
pKa	
pp cited in Vols.1-5:	

(1) Kreps et al., U.S. Pat. 3 403 207 (1968).

PADIMATE O [U]

CAS	21245-02-3	MW:	277.41	MF:	C17 H27 N O2

Ultraviolet screen

LgP	6.19(C)
pKa	
pp cited in Vols.1-5:	

(1) Kreps et al., U.S. Pat. 3 403 207 (1968).

PAFENOLOL [INN]

CAS	75949-61-0	MW:	337.47	MF:	C18 H31 N3 O3

LgP	1.67(C)
pKa	
pp cited in Vols.1-5:	

(1) World Conf. Clin. Pharmacol. Ther., 2 (Washington), 1983, 432 abst.

PALATRIGINE [INN]

CAS	98410-36-7	MW:	298.18	MF:	C12 H13 Cl2 N5

Anti-arrhythmic

Wellcome, UK

LgP	(4)
pKa	
pp cited in Vols.1-5:	

(1) Brit. J. Pharmacol., 1986, 88, 333.

PALDIMYCIN B [U;INN]

CAS	101411-71-6	MW:	1099.18	MF:	C43 H62 N4 O23 S3

Antibacterial

Upjohn

LgP	(4)
pKa	
pp cited in Vols.1-5:	

(1) Abst. Intersci. Conf. Antimicob. Ag. Chemother. 26 (New Orleans), Oct. 1986, 1264.

PALMIDROL [INN]

CAS	544-31-0	MW:	299.50	MF:	C18 H37 N O2

Anti-inflammatory

LgP	6.00(C)
pKa	
pp cited in Vols.1-5:	

(1) Kuehl et al., J. Am. Chem. Soc., 1957, 79, 5577.
(2) Roe et al., ibid., 1952, 74, 3442.

PALMOXIRIC ACID [U;INN]

CAS	68170-97-8	MW:	284.44	MF:	C17 H32 O3

Antidiabetic

McNeil

LgP	6.81(C)
pKa	
pp cited in Vols.1-5:	

(1) Clin. Pharmacol. Ther., 1982, 32, 790.

PAMABROM [U]

CAS	606-04-2	MW:	348.21	MF:	C11 H18 Br N5 O3

Diuretic

LgP	(4)
pKa	
pp cited in Vols.1-5:	

(1) Holpert et al., U.S. Pat. 2 711 411 (1955).

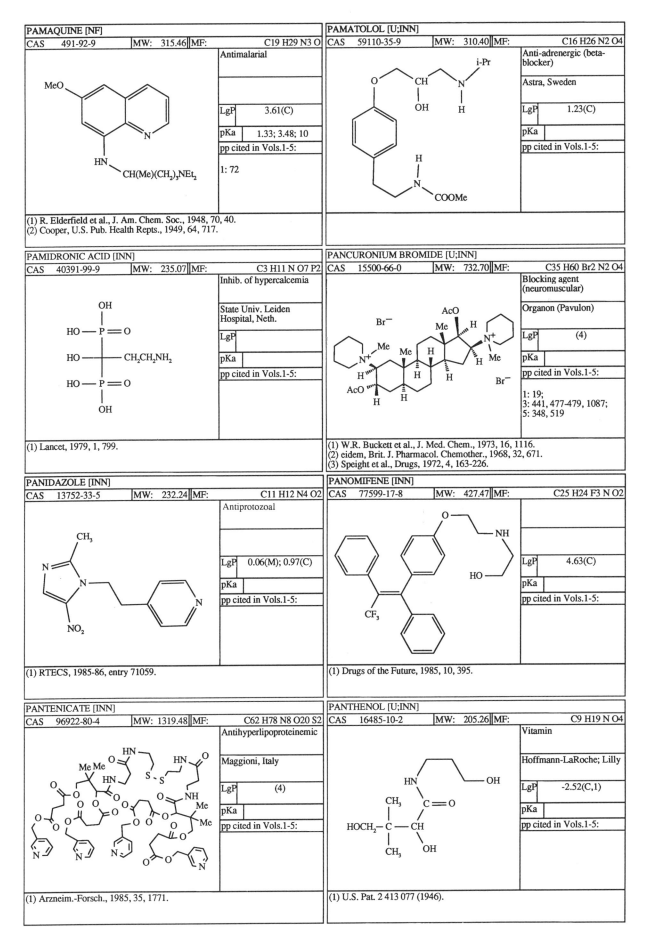

PAMAQUINE [NF]

CAS	491-92-9	MW:	315.46	MF:	C19 H29 N3 O

Antimalarial

LgP	3.61(C)
pKa	1.33; 3.48; 10
pp cited in Vols.1-5:	

1: 72

MeO

HN—CH(Me)(CH₂)₃NEt₂

(1) R. Elderfield et al., J. Am. Chem. Soc., 1948, 70, 40.
(2) Cooper, U.S. Pub. Health Repts., 1949, 64, 717.

PAMATOLOL [U;INN]

CAS	59110-35-9	MW:	310.40	MF:	C16 H26 N2 O4

Anti-adrenergic (beta-blocker)

Astra, Sweden

LgP	1.23(C)
pKa	
pp cited in Vols.1-5:	

PAMIDRONIC ACID [INN]

CAS	40391-99-9	MW:	235.07	MF:	C3 H11 N O7 P2

Inhib. of hypercalcemia

State Univ. Leiden Hospital, Neth.

LgP	
pKa	
pp cited in Vols.1-5:	

(1) Lancet, 1979, 1, 799.

PANCURONIUM BROMIDE [U;INN]

CAS	15500-66-0	MW:	732.70	MF:	C35 H60 Br2 N2 O4

Blocking agent (neuromuscular)

Organon (Pavulon)

LgP	(4)
pKa	
pp cited in Vols.1-5:	

1: 19;
3: 441, 477-479, 1087;
5: 348, 519

(1) W.R. Buckett et al., J. Med. Chem., 1973, 16, 1116.
(2) eidem, Brit. J. Pharmacol. Chemother., 1968, 32, 671.
(3) Speight et al., Drugs, 1972, 4, 163-226.

PANIDAZOLE [INN]

CAS	13752-33-5	MW:	232.24	MF:	C11 H12 N4 O2

Antiprotozoal

LgP	0.06(M); 0.97(C)
pKa	
pp cited in Vols.1-5:	

(1) RTECS, 1985-86, entry 71059.

PANOMIFENE [INN]

CAS	77599-17-8	MW:	427.47	MF:	C25 H24 F3 N O2

LgP	4.63(C)
pKa	
pp cited in Vols.1-5:	

(1) Drugs of the Future, 1985, 10, 395.

PANTENICATE [INN]

CAS	96922-80-4	MW:	1319.48	MF:	C62 H78 N8 O20 S2

Antihyperlipoproteinemic

Maggioni, Italy

LgP	(4)
pKa	
pp cited in Vols.1-5:	

(1) Arzneim.-Forsch., 1985, 35, 1771.

PANTHENOL [U;INN]

CAS	16485-10-2	MW:	205.26	MF:	C9 H19 N O4

Vitamin

Hoffmann-LaRoche; Lilly

LgP	-2.52(C,1)
pKa	
pp cited in Vols.1-5:	

(1) U.S. Pat. 2 413 077 (1946).

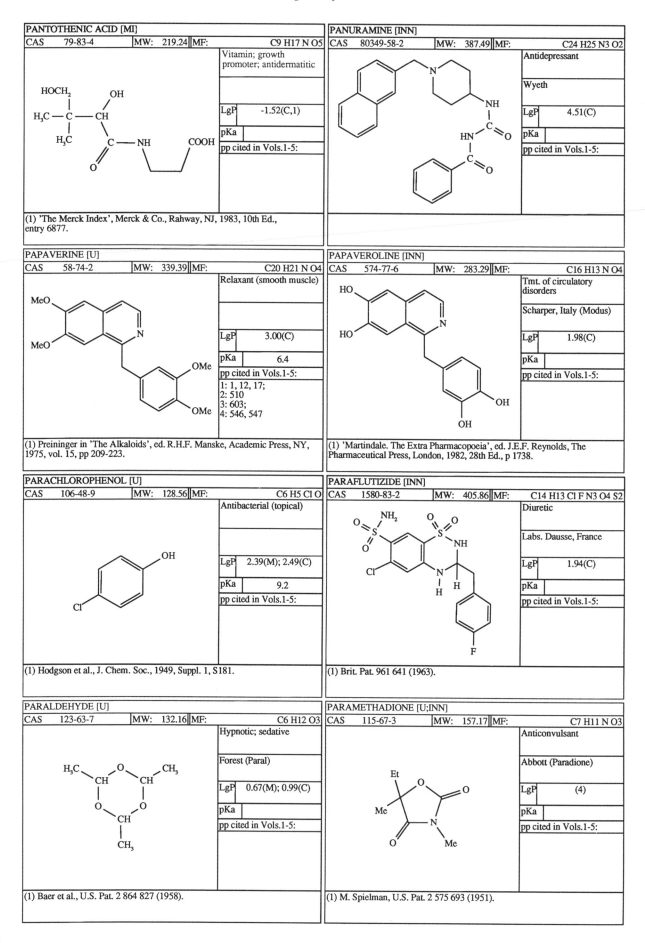

PANTOTHENIC ACID [MI]

CAS	79-83-4	MW:	219.24	MF:	C9 H17 N O5

Vitamin; growth promoter; antidermatitic

LgP	-1.52(C,1)
pKa	
pp cited in Vols.1-5:	

(1) 'The Merck Index', Merck & Co., Rahway, NJ, 1983, 10th Ed., entry 6877.

PANURAMINE [INN]

CAS	80349-58-2	MW:	387.49	MF:	C24 H25 N3 O2

Antidepressant

Wyeth

LgP	4.51(C)
pKa	
pp cited in Vols.1-5:	

PAPAVERINE [U]

CAS	58-74-2	MW:	339.39	MF:	C20 H21 N O4

Relaxant (smooth muscle)

LgP	3.00(C)
pKa	6.4
pp cited in Vols.1-5:	1: 1, 12, 17; 2: 510 3: 603; 4: 546, 547

(1) Preininger in 'The Alkaloids', ed. R.H.F. Manske, Academic Press, NY, 1975, vol. 15, pp 209-223.

PAPAVEROLINE [INN]

CAS	574-77-6	MW:	283.29	MF:	C16 H13 N O4

Tmt. of circulatory disorders

Scharper, Italy (Modus)

LgP	1.98(C)
pKa	
pp cited in Vols.1-5:	

(1) 'Martindale. The Extra Pharmacopoeia', ed. J.E.F. Reynolds, The Pharmaceutical Press, London, 1982, 28th Ed., p 1738.

PARACHLOROPHENOL [U]

CAS	106-48-9	MW:	128.56	MF:	C6 H5 Cl O

Antibacterial (topical)

LgP	2.39(M); 2.49(C)
pKa	9.2
pp cited in Vols.1-5:	

(1) Hodgson et al., J. Chem. Soc., 1949, Suppl. 1, S181.

PARAFLUTIZIDE [INN]

CAS	1580-83-2	MW:	405.86	MF:	C14 H13 Cl F N3 O4 S2

Diuretic

Labs. Dausse, France

LgP	1.94(C)
pKa	
pp cited in Vols.1-5:	

(1) Brit. Pat. 961 641 (1963).

PARALDEHYDE [U]

CAS	123-63-7	MW:	132.16	MF:	C6 H12 O3

Hypnotic; sedative

Forest (Paral)

LgP	0.67(M); 0.99(C)
pKa	
pp cited in Vols.1-5:	

(1) Baer et al., U.S. Pat. 2 864 827 (1958).

PARAMETHADIONE [U;INN]

CAS	115-67-3	MW:	157.17	MF:	C7 H11 N O3

Anticonvulsant

Abbott (Paradione)

LgP	(4)
pKa	
pp cited in Vols.1-5:	

(1) M. Spielman, U.S. Pat. 2 575 693 (1951).

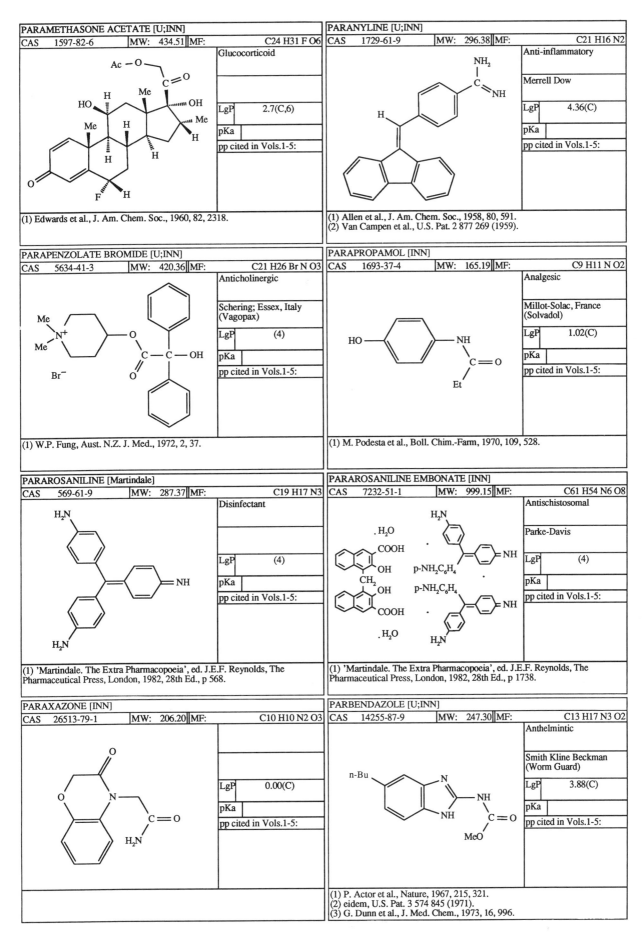

PARAMETHASONE ACETATE [U;INN]

CAS	1597-82-6	MW:	434.51	MF:	C24 H31 F O6

Glucocorticoid

LgP	2.7(C,6)
pKa	
pp cited in Vols.1-5:	

(1) Edwards et al., J. Am. Chem. Soc., 1960, 82, 2318.

PARANYLINE [U;INN]

CAS	1729-61-9	MW:	296.38	MF:	C21 H16 N2

Anti-inflammatory

Merrell Dow

LgP	4.36(C)
pKa	
pp cited in Vols.1-5:	

(1) Allen et al., J. Am. Chem. Soc., 1958, 80, 591.
(2) Van Campen et al., U.S. Pat. 2 877 269 (1959).

PARAPENZOLATE BROMIDE [U;INN]

CAS	5634-41-3	MW:	420.36	MF:	C21 H26 Br N O3

Anticholinergic

Schering; Essex, Italy (Vagopax)

LgP	(4)
pKa	
pp cited in Vols.1-5:	

(1) W.P. Fung, Aust. N.Z. J. Med., 1972, 2, 37.

PARAPROPAMOL [INN]

CAS	1693-37-4	MW:	165.19	MF:	C9 H11 N O2

Analgesic

Millot-Solac, France (Solvadol)

LgP	1.02(C)
pKa	
pp cited in Vols.1-5:	

(1) M. Podesta et al., Boll. Chim.-Farm, 1970, 109, 528.

PARAROSANILINE [Martindale]

CAS	569-61-9	MW:	287.37	MF:	C19 H17 N3

Disinfectant

LgP	(4)
pKa	
pp cited in Vols.1-5:	

(1) 'Martindale. The Extra Pharmacopoeia', ed. J.E.F. Reynolds, The Pharmaceutical Press, London, 1982, 28th Ed., p 568.

PARAROSANILINE EMBONATE [INN]

CAS	7232-51-1	MW:	999.15	MF:	C61 H54 N6 O8

Antischistosomal

Parke-Davis

LgP	(4)
pKa	
pp cited in Vols.1-5:	

(1) 'Martindale. The Extra Pharmacopoeia', ed. J.E.F. Reynolds, The Pharmaceutical Press, London, 1982, 28th Ed., p 1738.

PARAXAZONE [INN]

CAS	26513-79-1	MW:	206.20	MF:	C10 H10 N2 O3

LgP	0.00(C)
pKa	
pp cited in Vols.1-5:	

PARBENDAZOLE [U;INN]

CAS	14255-87-9	MW:	247.30	MF:	C13 H17 N3 O2

Anthelmintic

Smith Kline Beckman (Worm Guard)

LgP	3.88(C)
pKa	
pp cited in Vols.1-5:	

(1) P. Actor et al., Nature, 1967, 215, 321.
(2) eidem, U.S. Pat. 3 574 845 (1971).
(3) G. Dunn et al., J. Med. Chem., 1973, 16, 996.

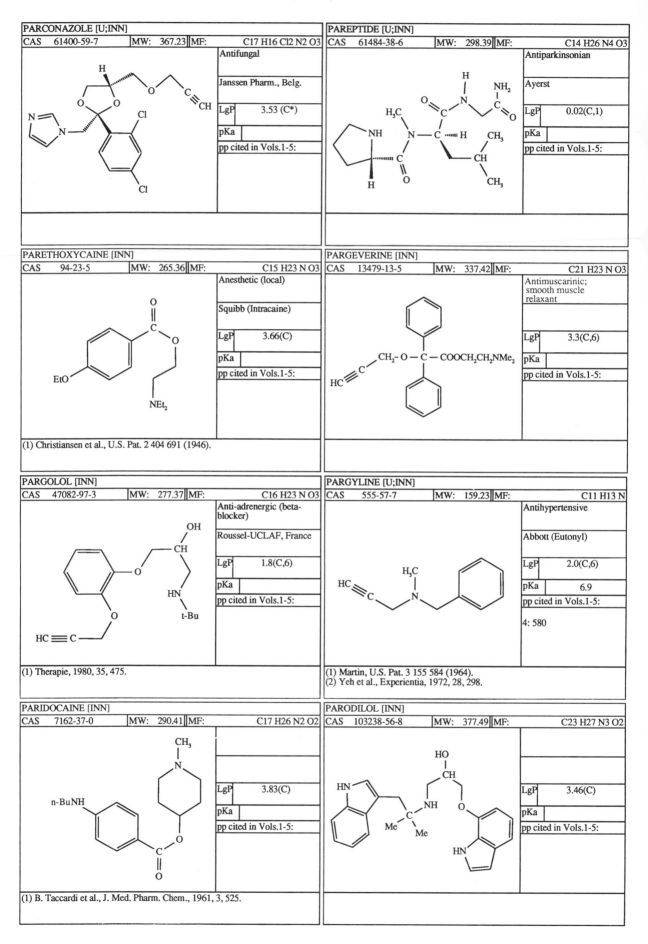

PARCONAZOLE [U;INN]		
CAS 61400-59-7	MW: 367.23	MF: C17 H16 Cl2 N2 O3

Antifungal

Janssen Pharm., Belg.

LgP 3.53 (C*)

pKa

pp cited in Vols.1-5:

PAREPTIDE [U;INN]		
CAS 61484-38-6	MW: 298.39	MF: C14 H26 N4 O3

Antiparkinsonian

Ayerst

LgP 0.02(C,1)

pKa

pp cited in Vols.1-5:

PARETHOXYCAINE [INN]		
CAS 94-23-5	MW: 265.36	MF: C15 H23 N O3

Anesthetic (local)

Squibb (Intracaine)

LgP 3.66(C)

pKa

pp cited in Vols.1-5:

(1) Christiansen et al., U.S. Pat. 2 404 691 (1946).

PARGEVERINE [INN]		
CAS 13479-13-5	MW: 337.42	MF: C21 H23 N O3

Antimuscarinic; smooth muscle relaxant

LgP 3.3(C,6)

pKa

pp cited in Vols.1-5:

PARGOLOL [INN]		
CAS 47082-97-3	MW: 277.37	MF: C16 H23 N O3

Anti-adrenergic (beta-blocker)

Roussel-UCLAF, France

LgP 1.8(C,6)

pKa

pp cited in Vols.1-5:

(1) Therapie, 1980, 35, 475.

PARGYLINE [U;INN]		
CAS 555-57-7	MW: 159.23	MF: C11 H13 N

Antihypertensive

Abbott (Eutonyl)

LgP 2.0(C,6)

pKa 6.9

pp cited in Vols.1-5:

4: 580

(1) Martin, U.S. Pat. 3 155 584 (1964).
(2) Yeh et al., Experientia, 1972, 28, 298.

PARIDOCAINE [INN]		
CAS 7162-37-0	MW: 290.41	MF: C17 H26 N2 O2

LgP 3.83(C)

pKa

pp cited in Vols.1-5:

(1) B. Taccardi et al., J. Med. Pharm. Chem., 1961, 3, 525.

PARODILOL [INN]		
CAS 103238-56-8	MW: 377.49	MF: C23 H27 N3 O2

LgP 3.46(C)

pKa

pp cited in Vols.1-5:

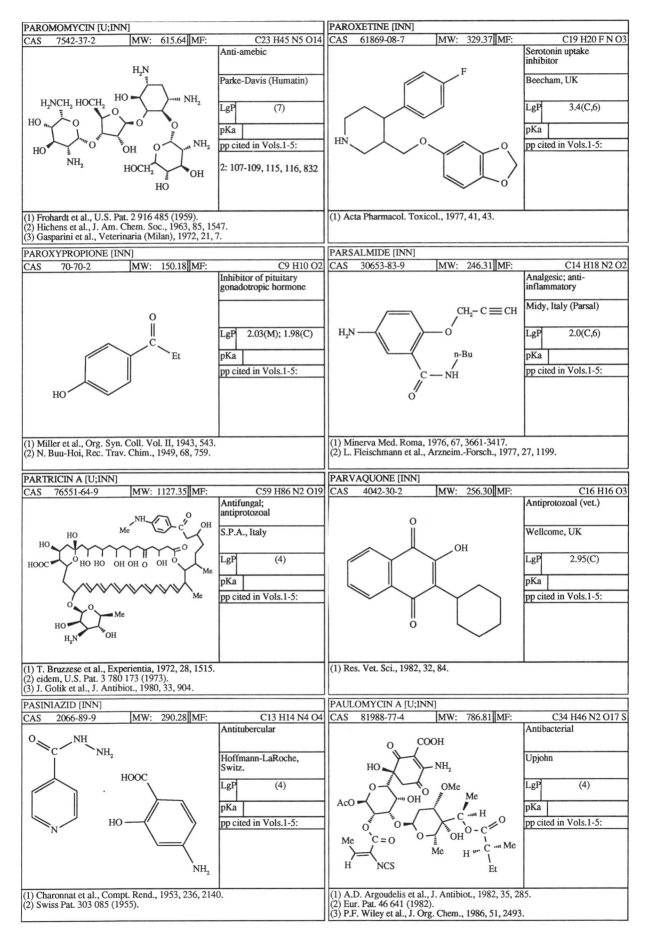

PAROMOMYCIN [U;INN]

CAS	7542-37-2	MW: 615.64	MF:	C23 H45 N5 O14

Anti-amebic

Parke-Davis (Humatin)

LgP	(7)
pKa	

pp cited in Vols.1-5:

2: 107-109, 115, 116, 832

(1) Frohardt et al., U.S. Pat. 2 916 485 (1959).
(2) Hichens et al., J. Am. Chem. Soc., 1963, 85, 1547.
(3) Gasparini et al., Veterinaria (Milan), 1972, 21, 7.

PAROXETINE [INN]

CAS	61869-08-7	MW: 329.37	MF:	C19 H20 F N O3

Serotonin uptake inhibitor

Beecham, UK

LgP	3.4(C,6)
pKa	

pp cited in Vols.1-5:

(1) Acta Pharmacol. Toxicol., 1977, 41, 43.

PAROXYPROPIONE [INN]

CAS	70-70-2	MW: 150.18	MF:	C9 H10 O2

Inhibitor of pituitary gonadotropic hormone

LgP	2.03(M); 1.98(C)
pKa	

pp cited in Vols.1-5:

(1) Miller et al., Org. Syn. Coll. Vol. II, 1943, 543.
(2) N. Buu-Hoi, Rec. Trav. Chim., 1949, 68, 759.

PARSALMIDE [INN]

CAS	30653-83-9	MW: 246.31	MF:	C14 H18 N2 O2

Analgesic; anti-inflammatory

Midy, Italy (Parsal)

LgP	2.0(C,6)
pKa	

pp cited in Vols.1-5:

(1) Minerva Med. Roma, 1976, 67, 3661-3417.
(2) L. Fleischmann et al., Arzneim.-Forsch., 1977, 27, 1199.

PARTRICIN A [U;INN]

CAS	76551-64-9	MW: 1127.35	MF:	C59 H86 N2 O19

Antifungal; antiprotozoal

S.P.A., Italy

LgP	(4)
pKa	

pp cited in Vols.1-5:

(1) T. Bruzzese et al., Experientia, 1972, 28, 1515.
(2) eidem, U.S. Pat. 3 780 173 (1973).
(3) J. Golik et al., J. Antibiot., 1980, 33, 904.

PARVAQUONE [INN]

CAS	4042-30-2	MW: 256.30	MF:	C16 H16 O3

Antiprotozoal (vet.)

Wellcome, UK

LgP	2.95(C)
pKa	

pp cited in Vols.1-5:

(1) Res. Vet. Sci., 1982, 32, 84.

PASINIAZID [INN]

CAS	2066-89-9	MW: 290.28	MF:	C13 H14 N4 O4

Antitubercular

Hoffmann-LaRoche, Switz.

LgP	(4)
pKa	

pp cited in Vols.1-5:

(1) Charonnat et al., Compt. Rend., 1953, 236, 2140.
(2) Swiss Pat. 303 085 (1955).

PAULOMYCIN A [U;INN]

CAS	81988-77-4	MW: 786.81	MF:	C34 H46 N2 O17 S

Antibacterial

Upjohn

LgP	(4)
pKa	

pp cited in Vols.1-5:

(1) A.D. Argoudelis et al., J. Antibiot., 1982, 35, 285.
(2) Eur. Pat. 46 641 (1982).
(3) P.F. Wiley et al., J. Org. Chem., 1986, 51, 2493.

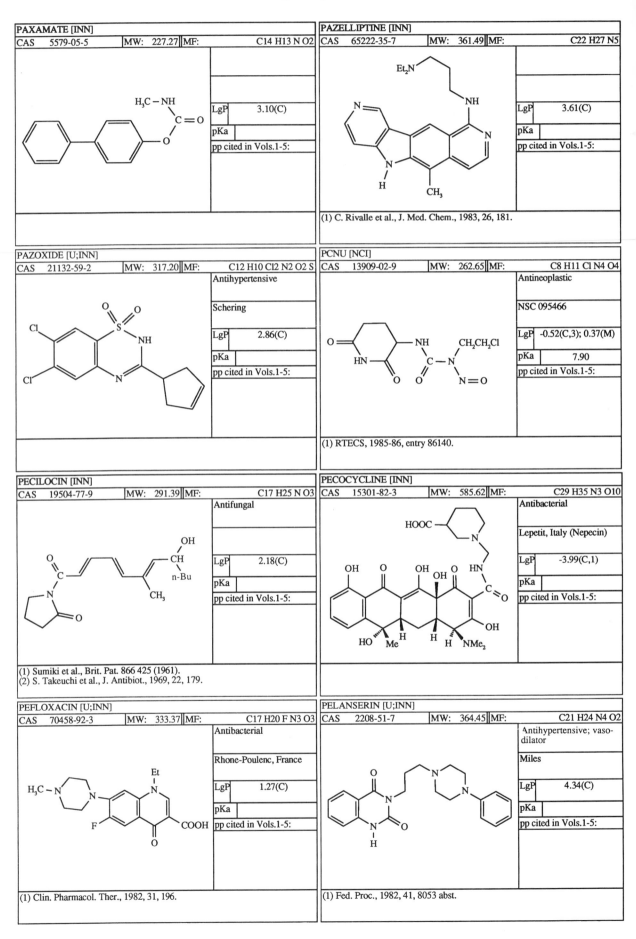

PAXAMATE [INN]

CAS	5579-05-5	MW:	227.27	MF:	C14 H13 N O2

LgP	3.10(C)
pKa	
pp cited in Vols.1-5:	

PAZELLIPTINE [INN]

CAS	65222-35-7	MW:	361.49	MF:	C22 H27 N5

LgP	3.61(C)
pKa	
pp cited in Vols.1-5:	

(1) C. Rivalle et al., J. Med. Chem., 1983, 26, 181.

PAZOXIDE [U;INN]

CAS	21132-59-2	MW:	317.20	MF:	C12 H10 Cl2 N2 O2 S

Antihypertensive	
Schering	
LgP	2.86(C)
pKa	
pp cited in Vols.1-5:	

PCNU [NCI]

CAS	13909-02-9	MW:	262.65	MF:	C8 H11 Cl N4 O4

Antineoplastic	
NSC 095466	
LgP	-0.52(C,3); 0.37(M)
pKa	7.90
pp cited in Vols.1-5:	

(1) RTECS, 1985-86, entry 86140.

PECILOCIN [INN]

CAS	19504-77-9	MW:	291.39	MF:	C17 H25 N O3

Antifungal	
LgP	2.18(C)
pKa	
pp cited in Vols.1-5:	

(1) Sumiki et al., Brit. Pat. 866 425 (1961).
(2) S. Takeuchi et al., J. Antibiot., 1969, 22, 179.

PECOCYCLINE [INN]

CAS	15301-82-3	MW:	585.62	MF:	C29 H35 N3 O10

Antibacterial	
Lepetit, Italy (Nepecin)	
LgP	-3.99(C,1)
pKa	
pp cited in Vols.1-5:	

PEFLOXACIN [U;INN]

CAS	70458-92-3	MW:	333.37	MF:	C17 H20 F N3 O3

Antibacterial	
Rhone-Poulenc, France	
LgP	1.27(C)
pKa	
pp cited in Vols.1-5:	

(1) Clin. Pharmacol. Ther., 1982, 31, 196.

PELANSERIN [U;INN]

CAS	2208-51-7	MW:	364.45	MF:	C21 H24 N4 O2

Antihypertensive; vaso-dilator	
Miles	
LgP	4.34(C)
pKa	
pp cited in Vols.1-5:	

(1) Fed. Proc., 1982, 41, 8053 abst.

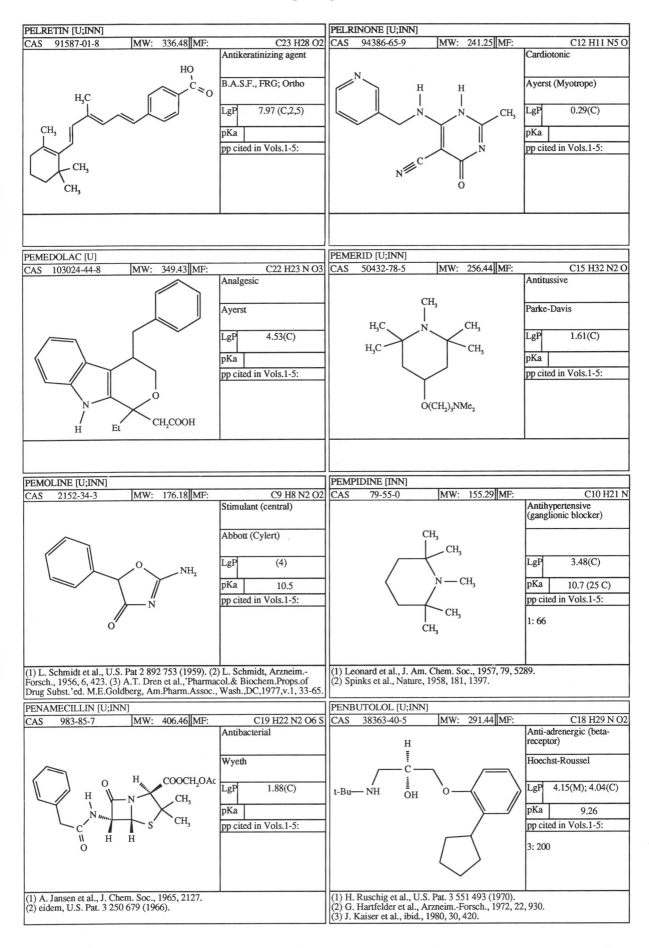

PELRETIN [U;INN]

CAS 91587-01-8	MW: 336.48	MF:	C23 H28 O2

| Antikeratinizing agent |
| B.A.S.F., FRG; Ortho |
LgP	7.97 (C,2,5)
pKa	
pp cited in Vols.1-5:	

PELRINONE [U;INN]

CAS 94386-65-9	MW: 241.25	MF:	C12 H11 N5 O

| Cardiotonic |
| Ayerst (Myotrope) |
LgP	0.29(C)
pKa	
pp cited in Vols.1-5:	

PEMEDOLAC [U]

CAS 103024-44-8	MW: 349.43	MF:	C22 H23 N O3

| Analgesic |
| Ayerst |
LgP	4.53(C)
pKa	
pp cited in Vols.1-5:	

PEMERID [U;INN]

CAS 50432-78-5	MW: 256.44	MF:	C15 H32 N2 O

| Antitussive |
| Parke-Davis |
LgP	1.61(C)
pKa	
pp cited in Vols.1-5:	

PEMOLINE [U;INN]

CAS 2152-34-3	MW: 176.18	MF:	C9 H8 N2 O2

| Stimulant (central) |
| Abbott (Cylert) |
LgP	(4)
pKa	10.5
pp cited in Vols.1-5:	

(1) L. Schmidt et al., U.S. Pat 2 892 753 (1959). (2) L. Schmidt, Arzneim.-Forsch., 1956, 6, 423. (3) A.T. Dren et al.,'Pharmacol.& Biochem.Props.of Drug Subst.'ed. M.E.Goldberg, Am.Pharm.Assoc., Wash.,DC,1977,v.1, 33-65.

PEMPIDINE [INN]

CAS 79-55-0	MW: 155.29	MF:	C10 H21 N

| Antihypertensive (ganglionic blocker) |
| |
LgP	3.48(C)
pKa	10.7 (25 C)
pp cited in Vols.1-5:	
1: 66

(1) Leonard et al., J. Am. Chem. Soc., 1957, 79, 5289.
(2) Spinks et al., Nature, 1958, 181, 1397.

PENAMECILLIN [U;INN]

CAS 983-85-7	MW: 406.46	MF:	C19 H22 N2 O6 S

| Antibacterial |
| Wyeth |
LgP	1.88(C)
pKa	
pp cited in Vols.1-5:	

(1) A. Jansen et al., J. Chem. Soc., 1965, 2127.
(2) eidem, U.S. Pat. 3 250 679 (1966).

PENBUTOLOL [U;INN]

CAS 38363-40-5	MW: 291.44	MF:	C18 H29 N O2

| Anti-adrenergic (beta-receptor) |
| Hoechst-Roussel |
LgP	4.15(M); 4.04(C)
pKa	9.26
pp cited in Vols.1-5:	
3: 200

(1) H. Ruschig et al., U.S. Pat. 3 551 493 (1970).
(2) G. Hartfelder et al., Arzneim.-Forsch., 1972, 22, 930.
(3) J. Kaiser et al., ibid., 1980, 30, 420.

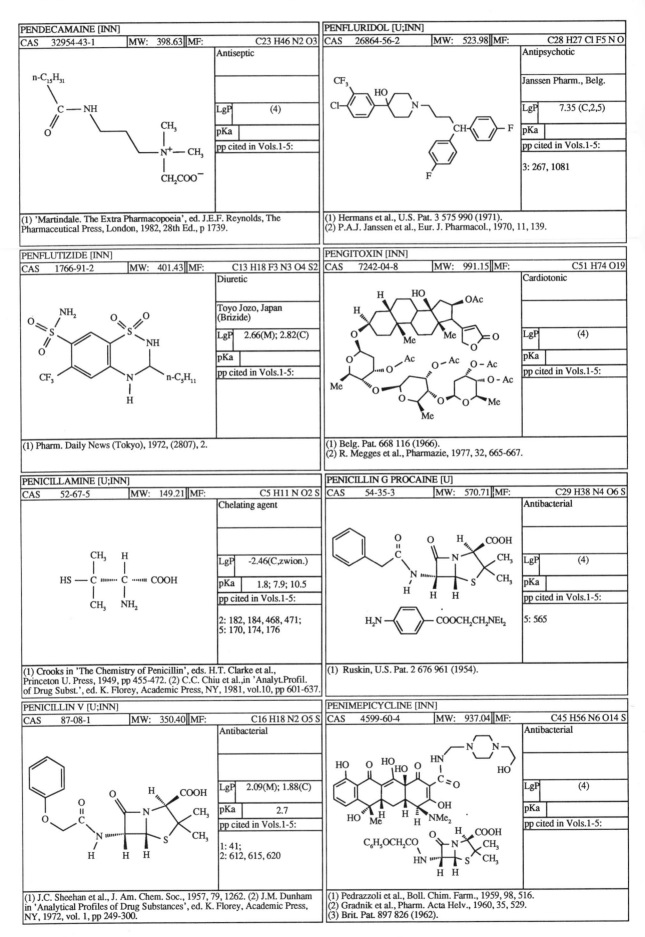

PENDECAMAINE [INN]

CAS	32954-43-1	MW:	398.63	MF:	C23 H46 N2 O3

Antiseptic

LgP	(4)
pKa	
pp cited in Vols.1-5:	

(1) 'Martindale. The Extra Pharmacopoeia', ed. J.E.F. Reynolds, The Pharmaceutical Press, London, 1982, 28th Ed., p 1739.

PENFLURIDOL [U;INN]

CAS	26864-56-2	MW:	523.98	MF:	C28 H27 Cl F5 N O

Antipsychotic

Janssen Pharm., Belg.

LgP	7.35 (C,2,5)
pKa	
pp cited in Vols.1-5:	

3: 267, 1081

(1) Hermans et al., U.S. Pat. 3 575 990 (1971).
(2) P.A.J. Janssen et al., Eur. J. Pharmacol., 1970, 11, 139.

PENFLUTIZIDE [INN]

CAS	1766-91-2	MW:	401.43	MF:	C13 H18 F3 N3 O4 S2

Diuretic

Toyo Jozo, Japan (Brizide)

LgP	2.66(M); 2.82(C)
pKa	
pp cited in Vols.1-5:	

(1) Pharm. Daily News (Tokyo), 1972, (2807), 2.

PENGITOXIN [INN]

CAS	7242-04-8	MW:	991.15	MF:	C51 H74 O19

Cardiotonic

LgP	(4)
pKa	
pp cited in Vols.1-5:	

(1) Belg. Pat. 668 116 (1966).
(2) R. Megges et al., Pharmazie, 1977, 32, 665-667.

PENICILLAMINE [U;INN]

CAS	52-67-5	MW:	149.21	MF:	C5 H11 N O2 S

Chelating agent

LgP	-2.46(C,zwion.)
pKa	1.8; 7.9; 10.5
pp cited in Vols.1-5:	

2: 182, 184, 468, 471;
5: 170, 174, 176

(1) Crooks in 'The Chemistry of Penicillin', eds. H.T. Clarke et al., Princeton U. Press, 1949, pp 455-472. (2) C.C. Chiu et al.,in 'Analyt.Profil. of Drug Subst.', ed. K. Florey, Academic Press, NY, 1981, vol.10, pp 601-637.

PENICILLIN G PROCAINE [U]

CAS	54-35-3	MW:	570.71	MF:	C29 H38 N4 O6 S

Antibacterial

LgP	(4)
pKa	
pp cited in Vols.1-5:	

5: 565

(1) Ruskin, U.S. Pat. 2 676 961 (1954).

PENICILLIN V [U;INN]

CAS	87-08-1	MW:	350.40	MF:	C16 H18 N2 O5 S

Antibacterial

LgP	2.09(M); 1.88(C)
pKa	2.7
pp cited in Vols.1-5:	

1: 41;
2: 612, 615, 620

(1) J.C. Sheehan et al., J. Am. Chem. Soc., 1957, 79, 1262. (2) J.M. Dunham in 'Analytical Profiles of Drug Substances', ed. K. Florey, Academic Press, NY, 1972, vol. 1, pp 249-300.

PENIMEPICYCLINE [INN]

CAS	4599-60-4	MW:	937.04	MF:	C45 H56 N6 O14 S

Antibacterial

LgP	(4)
pKa	
pp cited in Vols.1-5:	

(1) Pedrazzoli et al., Boll. Chim. Farm., 1959, 98, 516.
(2) Gradnik et al., Pharm. Acta Helv., 1960, 35, 529.
(3) Brit. Pat. 897 826 (1962).

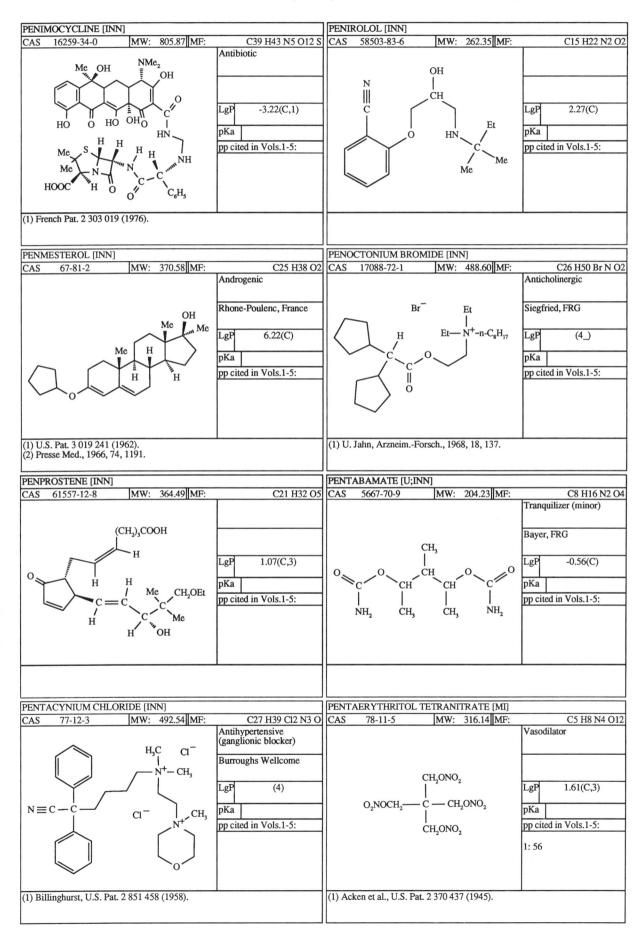

PENIMOCYCLINE [INN]

CAS	16259-34-0	MW:	805.87	MF:	C39 H43 N5 O12 S

Antibiotic

LgP	-3.22(C,1)
pKa	
pp cited in Vols.1-5:	

(1) French Pat. 2 303 019 (1976).

PENIROLOL [INN]

CAS	58503-83-6	MW:	262.35	MF:	C15 H22 N2 O2

LgP	2.27(C)
pKa	
pp cited in Vols.1-5:	

PENMESTEROL [INN]

CAS	67-81-2	MW:	370.58	MF:	C25 H38 O2

Androgenic

Rhone-Poulenc, France

LgP	6.22(C)
pKa	
pp cited in Vols.1-5:	

(1) U.S. Pat. 3 019 241 (1962).
(2) Presse Med., 1966, 74, 1191.

PENOCTONIUM BROMIDE [INN]

CAS	17088-72-1	MW:	488.60	MF:	C26 H50 Br N O2

Anticholinergic

Siegfried, FRG

LgP	(4_)
pKa	
pp cited in Vols.1-5:	

(1) U. Jahn, Arzneim.-Forsch., 1968, 18, 137.

PENPROSTENE [INN]

CAS	61557-12-8	MW:	364.49	MF:	C21 H32 O5

LgP	1.07(C,3)
pKa	
pp cited in Vols.1-5:	

PENTABAMATE [U;INN]

CAS	5667-70-9	MW:	204.23	MF:	C8 H16 N2 O4

Tranquilizer (minor)

Bayer, FRG

LgP	-0.56(C)
pKa	
pp cited in Vols.1-5:	

PENTACYNIUM CHLORIDE [INN]

CAS	77-12-3	MW:	492.54	MF:	C27 H39 Cl2 N3 O

Antihypertensive (ganglionic blocker)

Burroughs Wellcome

LgP	(4)
pKa	
pp cited in Vols.1-5:	

(1) Billinghurst, U.S. Pat. 2 851 458 (1958).

PENTAERYTHRITOL TETRANITRATE [MI]

CAS	78-11-5	MW:	316.14	MF:	C5 H8 N4 O12

Vasodilator

LgP	1.61(C,3)
pKa	
pp cited in Vols.1-5:	

1: 56

(1) Acken et al., U.S. Pat. 2 370 437 (1945).

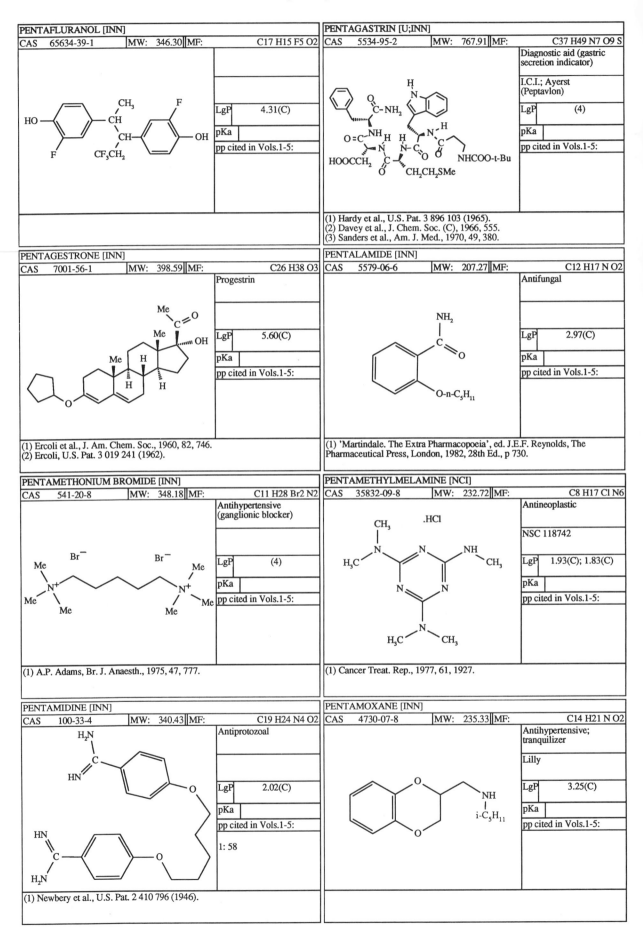

PENTAFLURANOL [INN]

CAS	65634-39-1	MW:	346.30	MF:	C17 H15 F5 O2

LgP	4.31(C)
pKa	
pp cited in Vols.1-5:	

PENTAGASTRIN [U;INN]

CAS	5534-95-2	MW:	767.91	MF:	C37 H49 N7 O9 S

Diagnostic aid (gastric secretion indicator)

I.C.I.; Ayerst (Peptavlon)

LgP	(4)
pKa	
pp cited in Vols.1-5:	

(1) Hardy et al., U.S. Pat. 3 896 103 (1965).
(2) Davey et al., J. Chem. Soc. (C), 1966, 555.
(3) Sanders et al., Am. J. Med., 1970, 49, 380.

PENTAGESTRONE [INN]

CAS	7001-56-1	MW:	398.59	MF:	C26 H38 O3

Progestrin

LgP	5.60(C)
pKa	
pp cited in Vols.1-5:	

(1) Ercoli et al., J. Am. Chem. Soc., 1960, 82, 746.
(2) Ercoli, U.S. Pat. 3 019 241 (1962).

PENTALAMIDE [INN]

CAS	5579-06-6	MW:	207.27	MF:	C12 H17 N O2

Antifungal

LgP	2.97(C)
pKa	
pp cited in Vols.1-5:	

(1) 'Martindale. The Extra Pharmacopoeia', ed. J.E.F. Reynolds, The Pharmaceutical Press, London, 1982, 28th Ed., p 730.

PENTAMETHONIUM BROMIDE [INN]

CAS	541-20-8	MW:	348.18	MF:	C11 H28 Br2 N2

Antihypertensive (ganglionic blocker)

LgP	(4)
pKa	
pp cited in Vols.1-5:	

(1) A.P. Adams, Br. J. Anaesth., 1975, 47, 777.

PENTAMETHYLMELAMINE [NCI]

CAS	35832-09-8	MW:	232.72	MF:	C8 H17 Cl N6

Antineoplastic

NSC 118742

LgP	1.93(C); 1.83(C)
pKa	
pp cited in Vols.1-5:	

(1) Cancer Treat. Rep., 1977, 61, 1927.

PENTAMIDINE [INN]

CAS	100-33-4	MW:	340.43	MF:	C19 H24 N4 O2

Antiprotozoal

LgP	2.02(C)
pKa	
pp cited in Vols.1-5:	
1: 58	

(1) Newbery et al., U.S. Pat. 2 410 796 (1946).

PENTAMOXANE [INN]

CAS	4730-07-8	MW:	235.33	MF:	C14 H21 N O2

Antihypertensive; tranquilizer

Lilly

LgP	3.25(C)
pKa	
pp cited in Vols.1-5:	

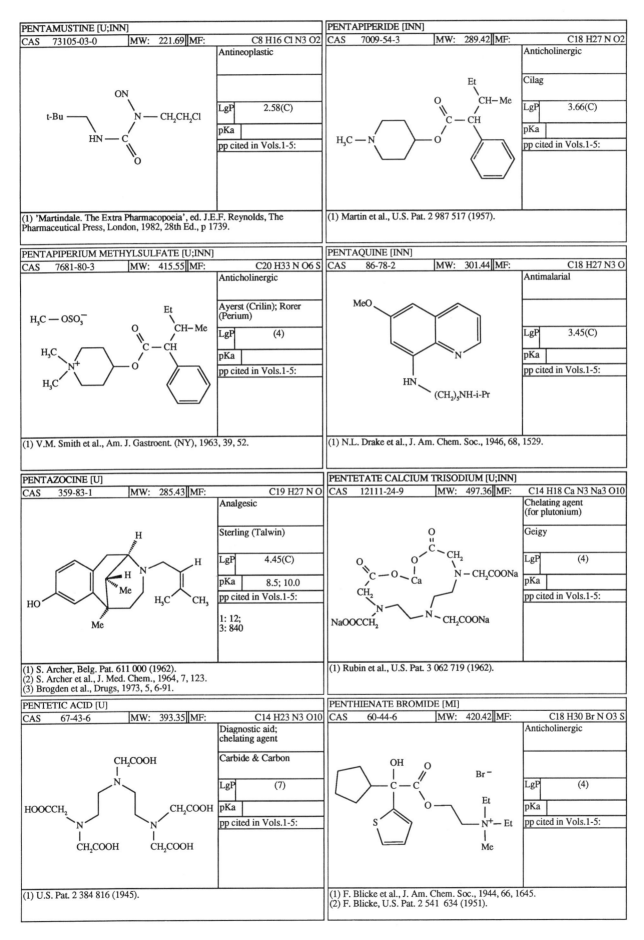

PENTAMUSTINE [U;INN]

CAS	73105-03-0	MW:	221.69	MF:	C8 H16 Cl N3 O2

Antineoplastic

LgP	2.58(C)
pKa	
pp cited in Vols.1-5:	

(1) 'Martindale. The Extra Pharmacopoeia', ed. J.E.F. Reynolds, The Pharmaceutical Press, London, 1982, 28th Ed., p 1739.

PENTAPIPERIDE [INN]

CAS	7009-54-3	MW:	289.42	MF:	C18 H27 N O2

Anticholinergic

Cilag

LgP	3.66(C)
pKa	
pp cited in Vols.1-5:	

(1) Martin et al., U.S. Pat. 2 987 517 (1957).

PENTAPIPERIUM METHYLSULFATE [U;INN]

CAS	7681-80-3	MW:	415.55	MF:	C20 H33 N O6 S

Anticholinergic

Ayerst (Crilin); Rorer (Perium)

LgP	(4)
pKa	
pp cited in Vols.1-5:	

(1) V.M. Smith et al., Am. J. Gastroent. (NY), 1963, 39, 52.

PENTAQUINE [INN]

CAS	86-78-2	MW:	301.44	MF:	C18 H27 N3 O

Antimalarial

LgP	3.45(C)
pKa	
pp cited in Vols.1-5:	

(1) N.L. Drake et al., J. Am. Chem. Soc., 1946, 68, 1529.

PENTAZOCINE [U]

CAS	359-83-1	MW:	285.43	MF:	C19 H27 N O

Analgesic

Sterling (Talwin)

LgP	4.45(C)
pKa	8.5; 10.0
pp cited in Vols.1-5:	

1: 12;
3: 840

(1) S. Archer, Belg. Pat. 611 000 (1962).
(2) S. Archer et al., J. Med. Chem., 1964, 7, 123.
(3) Brogden et al., Drugs, 1973, 5, 6-91.

PENTETATE CALCIUM TRISODIUM [U;INN]

CAS	12111-24-9	MW:	497.36	MF:	C14 H18 Ca N3 Na3 O10

Chelating agent (for plutonium)

Geigy

LgP	(4)
pKa	
pp cited in Vols.1-5:	

(1) Rubin et al., U.S. Pat. 3 062 719 (1962).

PENTETIC ACID [U]

CAS	67-43-6	MW:	393.35	MF:	C14 H23 N3 O10

Diagnostic aid; chelating agent

Carbide & Carbon

LgP	(7)
pKa	
pp cited in Vols.1-5:	

(1) U.S. Pat. 2 384 816 (1945).

PENTHIENATE BROMIDE [MI]

CAS	60-44-6	MW:	420.42	MF:	C18 H30 Br N O3 S

Anticholinergic

LgP	(4)
pKa	
pp cited in Vols.1-5:	

(1) F. Blicke et al., J. Am. Chem. Soc., 1944, 66, 1645.
(2) F. Blicke, U.S. Pat. 2 541 634 (1951).

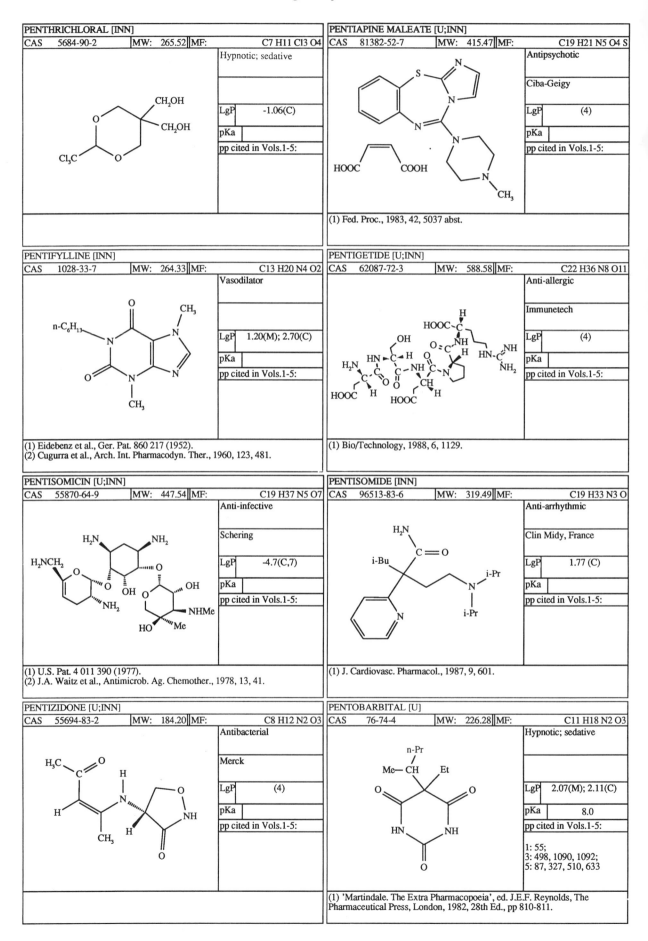

PENTHRICHLORAL [INN]

CAS	5684-90-2	MW:	265.52	MF:	C7 H11 Cl3 O4

Hypnotic; sedative

LgP	-1.06(C)
pKa	
pp cited in Vols.1-5:	

PENTIFYLLINE [INN]

CAS	1028-33-7	MW:	264.33	MF:	C13 H20 N4 O2

Vasodilator

LgP	1.20(M); 2.70(C)
pKa	
pp cited in Vols.1-5:	

(1) Eidebenz et al., Ger. Pat. 860 217 (1952).
(2) Cugurra et al., Arch. Int. Pharmacodyn. Ther., 1960, 123, 481.

PENTISOMICIN [U;INN]

CAS	55870-64-9	MW:	447.54	MF:	C19 H37 N5 O7

Anti-infective

Schering

LgP	-4.7(C,7)
pKa	
pp cited in Vols.1-5:	

(1) U.S. Pat. 4 011 390 (1977).
(2) J.A. Waitz et al., Antimicrob. Ag. Chemother., 1978, 13, 41.

PENTIZIDONE [U;INN]

CAS	55694-83-2	MW:	184.20	MF:	C8 H12 N2 O3

Antibacterial

Merck

LgP	(4)
pKa	
pp cited in Vols.1-5:	

PENTIAPINE MALEATE [U;INN]

CAS	81382-52-7	MW:	415.47	MF:	C19 H21 N5 O4 S

Antipsychotic

Ciba-Geigy

LgP	(4)
pKa	
pp cited in Vols.1-5:	

(1) Fed. Proc., 1983, 42, 5037 abst.

PENTIGETIDE [U;INN]

CAS	62087-72-3	MW:	588.58	MF:	C22 H36 N8 O11

Anti-allergic

Immunetech

LgP	(4)
pKa	
pp cited in Vols.1-5:	

(1) Bio/Technology, 1988, 6, 1129.

PENTISOMIDE [INN]

CAS	96513-83-6	MW:	319.49	MF:	C19 H33 N3 O

Anti-arrhythmic

Clin Midy, France

LgP	1.77 (C)
pKa	
pp cited in Vols.1-5:	

(1) J. Cardiovasc. Pharmacol., 1987, 9, 601.

PENTOBARBITAL [U]

CAS	76-74-4	MW:	226.28	MF:	C11 H18 N2 O3

Hypnotic; sedative

LgP	2.07(M); 2.11(C)
pKa	8.0
pp cited in Vols.1-5:	

1: 55;
3: 498, 1090, 1092;
5: 87, 327, 510, 633

(1) 'Martindale. The Extra Pharmacopoeia', ed. J.E.F. Reynolds, The
Pharmaceutical Press, London, 1982, 28th Ed., pp 810-811.

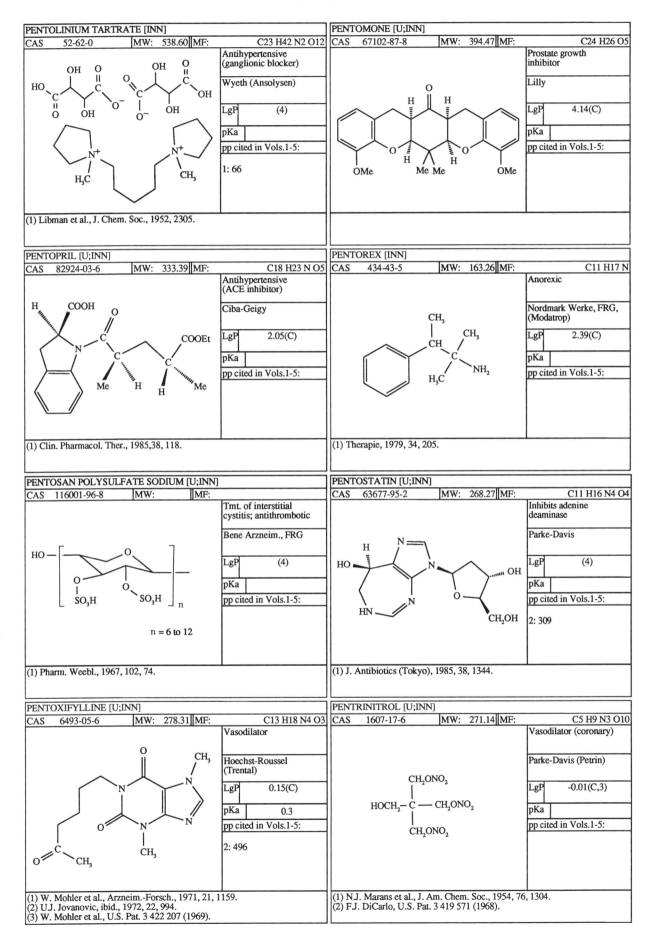

PENTOLINIUM TARTRATE [INN]

CAS 52-62-0	MW: 538.60	MF: C23 H42 N2 O12

Antihypertensive (ganglionic blocker)

Wyeth (Ansolysen)

LgP	(4)
pKa	
pp cited in Vols.1-5:	

1: 66

(1) Libman et al., J. Chem. Soc., 1952, 2305.

PENTOMONE [U;INN]

CAS 67102-87-8	MW: 394.47	MF: C24 H26 O5

Prostate growth inhibitor

Lilly

LgP	4.14(C)
pKa	
pp cited in Vols.1-5:	

PENTOPRIL [U;INN]

CAS 82924-03-6	MW: 333.39	MF: C18 H23 N O5

Antihypertensive (ACE inhibitor)

Ciba-Geigy

LgP	2.05(C)
pKa	
pp cited in Vols.1-5:	

(1) Clin. Pharmacol. Ther., 1985,38, 118.

PENTOREX [INN]

CAS 434-43-5	MW: 163.26	MF: C11 H17 N

Anorexic

Nordmark Werke, FRG, (Modatrop)

LgP	2.39(C)
pKa	
pp cited in Vols.1-5:	

(1) Therapie, 1979, 34, 205.

PENTOSAN POLYSULFATE SODIUM [U;INN]

CAS 116001-96-8	MW:	MF:

Tmt. of interstitial cystitis; antithrombotic

Bene Arzneim., FRG

LgP	(4)
pKa	
pp cited in Vols.1-5:	

n = 6 to 12

(1) Pharm. Weebl., 1967, 102, 74.

PENTOSTATIN [U;INN]

CAS 63677-95-2	MW: 268.27	MF: C11 H16 N4 O4

Inhibits adenine deaminase

Parke-Davis

LgP	(4)
pKa	
pp cited in Vols.1-5:	

2: 309

(1) J. Antibiotics (Tokyo), 1985, 38, 1344.

PENTOXIFYLLINE [U;INN]

CAS 6493-05-6	MW: 278.31	MF: C13 H18 N4 O3

Vasodilator

Hoechst-Roussel (Trental)

LgP	0.15(C)
pKa	0.3
pp cited in Vols.1-5:	

2: 496

(1) W. Mohler et al., Arzneim.-Forsch., 1971, 21, 1159.
(2) U.J. Jovanovic, ibid., 1972, 22, 994.
(3) W. Mohler et al., U.S. Pat. 3 422 207 (1969).

PENTRINITROL [U;INN]

CAS 1607-17-6	MW: 271.14	MF: C5 H9 N3 O10

Vasodilator (coronary)

Parke-Davis (Petrin)

LgP	-0.01(C,3)
pKa	
pp cited in Vols.1-5:	

(1) N.J. Marans et al., J. Am. Chem. Soc., 1954, 76, 1304.
(2) F.J. DiCarlo, U.S. Pat. 3 419 571 (1968).

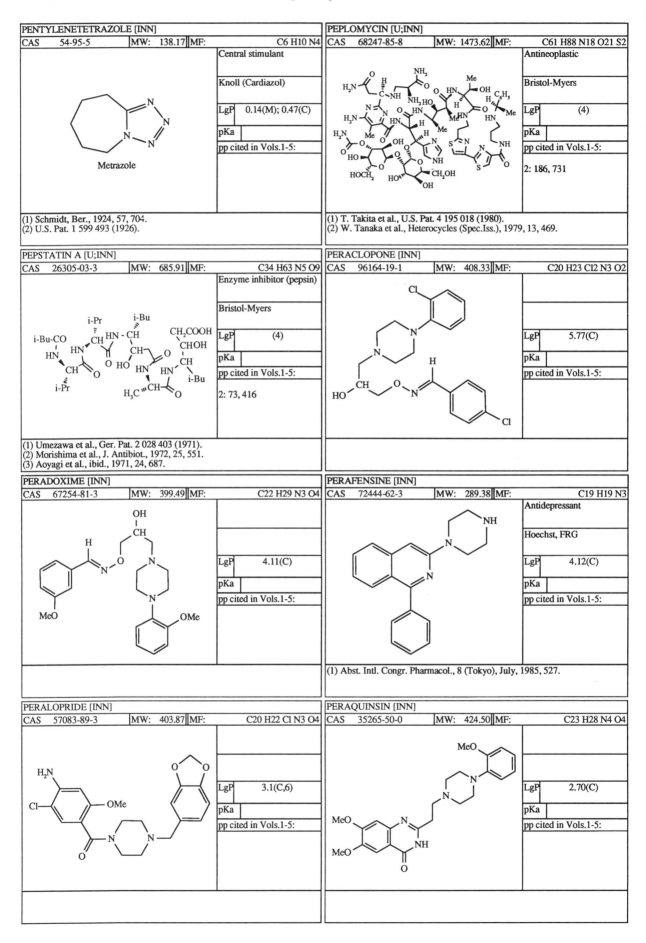

PENTYLENETETRAZOLE [INN]

CAS	54-95-5	MW:	138.17	MF:	C6 H10 N4

Central stimulant

Knoll (Cardiazol)

LgP	0.14(M); 0.47(C)
pKa	
pp cited in Vols.1-5:	

Metrazole

(1) Schmidt, Ber., 1924, 57, 704.
(2) U.S. Pat. 1 599 493 (1926).

PEPLOMYCIN [U;INN]

CAS	68247-85-8	MW:	1473.62	MF:	C61 H88 N18 O21 S2

Antineoplastic

Bristol-Myers

LgP	(4)
pKa	
pp cited in Vols.1-5:	

2: 186, 731

(1) T. Takita et al., U.S. Pat. 4 195 018 (1980).
(2) W. Tanaka et al., Heterocycles (Spec.Iss.), 1979, 13, 469.

PEPSTATIN A [U;INN]

CAS	26305-03-3	MW:	685.91	MF:	C34 H63 N5 O9

Enzyme inhibitor (pepsin)

Bristol-Myers

LgP	(4)
pKa	
pp cited in Vols.1-5:	

2: 73, 416

(1) Umezawa et al., Ger. Pat. 2 028 403 (1971).
(2) Morishima et al., J. Antibiot., 1972, 25, 551.
(3) Aoyagi et al., ibid., 1971, 24, 687.

PERACLOPONE [INN]

CAS	96164-19-1	MW:	408.33	MF:	C20 H23 Cl2 N3 O2

LgP	5.77(C)
pKa	
pp cited in Vols.1-5:	

PERADOXIME [INN]

CAS	67254-81-3	MW:	399.49	MF:	C22 H29 N3 O4

LgP	4.11(C)
pKa	
pp cited in Vols.1-5:	

PERAFENSINE [INN]

CAS	72444-62-3	MW:	289.38	MF:	C19 H19 N3

Antidepressant

Hoechst, FRG

LgP	4.12(C)
pKa	
pp cited in Vols.1-5:	

(1) Abst. Intl. Congr. Pharmacol., 8 (Tokyo), July, 1985, 527.

PERALOPRIDE [INN]

CAS	57083-89-3	MW:	403.87	MF:	C20 H22 Cl N3 O4

LgP	3.1(C,6)
pKa	
pp cited in Vols.1-5:	

PERAQUINSIN [INN]

CAS	35265-50-0	MW:	424.50	MF:	C23 H28 N4 O4

LgP	2.70(C)
pKa	
pp cited in Vols.1-5:	

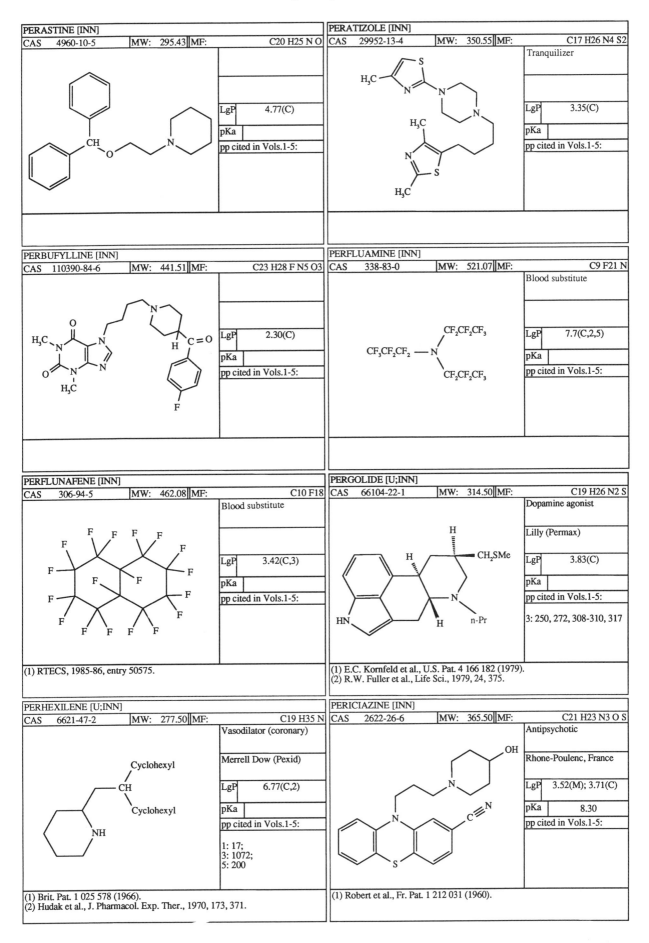

PERASTINE [INN]

CAS	4960-10-5	MW:	295.43	MF:	C20 H25 N O

LgP	4.77(C)
pKa	
pp cited in Vols.1-5:	

PERATIZOLE [INN]

CAS	29952-13-4	MW:	350.55	MF:	C17 H26 N4 S2

Tranquilizer

LgP	3.35(C)
pKa	
pp cited in Vols.1-5:	

PERBUFYLLINE [INN]

CAS	110390-84-6	MW:	441.51	MF:	C23 H28 F N5 O3

LgP	2.30(C)
pKa	
pp cited in Vols.1-5:	

PERFLUAMINE [INN]

CAS	338-83-0	MW:	521.07	MF:	C9 F21 N

Blood substitute

LgP	7.7(C,2,5)
pKa	
pp cited in Vols.1-5:	

PERFLUNAFENE [INN]

CAS	306-94-5	MW:	462.08	MF:	C10 F18

Blood substitute

LgP	3.42(C,3)
pKa	
pp cited in Vols.1-5:	

(1) RTECS, 1985-86, entry 50575.

PERGOLIDE [U;INN]

CAS	66104-22-1	MW:	314.50	MF:	C19 H26 N2 S

Dopamine agonist

Lilly (Permax)

LgP	3.83(C)
pKa	
pp cited in Vols.1-5:	
3: 250, 272, 308-310, 317	

(1) E.C. Kornfeld et al., U.S. Pat. 4 166 182 (1979).
(2) R.W. Fuller et al., Life Sci., 1979, 24, 375.

PERHEXILENE [U;INN]

CAS	6621-47-2	MW:	277.50	MF:	C19 H35 N

Vasodilator (coronary)

Merrell Dow (Pexid)

LgP	6.77(C,2)
pKa	
pp cited in Vols.1-5:	
1: 17; 3: 1072; 5: 200	

(1) Brit. Pat. 1 025 578 (1966).
(2) Hudak et al., J. Pharmacol. Exp. Ther., 1970, 173, 371.

PERICIAZINE [INN]

CAS	2622-26-6	MW:	365.50	MF:	C21 H23 N3 O S

Antipsychotic

Rhone-Poulenc, France

LgP	3.52(M); 3.71(C)
pKa	8.30
pp cited in Vols.1-5:	

(1) Robert et al., Fr. Pat. 1 212 031 (1960).

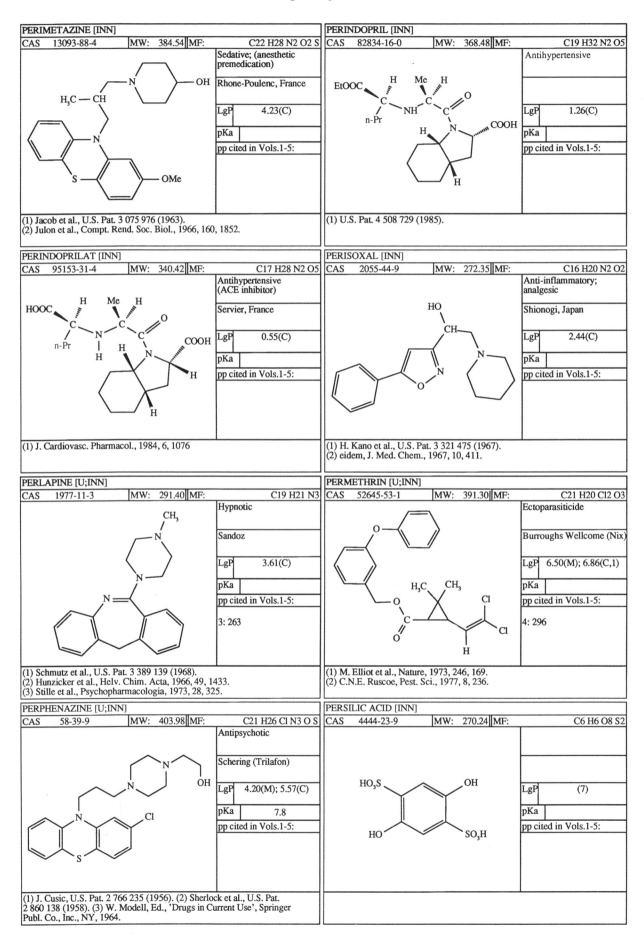

PERIMETAZINE [INN]

CAS 13093-88-4 | MW: 384.54 | MF: C22 H28 N2 O2 S

Sedative; (anesthetic premedication)

Rhone-Poulenc, France

LgP 4.23(C)

pKa

pp cited in Vols.1-5:

(1) Jacob et al., U.S. Pat. 3 075 976 (1963).
(2) Julon et al., Compt. Rend. Soc. Biol., 1966, 160, 1852.

PERINDOPRIL [INN]

CAS 82834-16-0 | MW: 368.48 | MF: C19 H32 N2 O5

Antihypertensive

LgP 1.26(C)

pKa

pp cited in Vols.1-5:

(1) U.S. Pat. 4 508 729 (1985).

PERINDOPRILAT [INN]

CAS 95153-31-4 | MW: 340.42 | MF: C17 H28 N2 O5

Antihypertensive (ACE inhibitor)

Servier, France

LgP 0.55(C)

pKa

pp cited in Vols.1-5:

(1) J. Cardiovasc. Pharmacol., 1984, 6, 1076

PERISOXAL [INN]

CAS 2055-44-9 | MW: 272.35 | MF: C16 H20 N2 O2

Anti-inflammatory; analgesic

Shionogi, Japan

LgP 2.44(C)

pKa

pp cited in Vols.1-5:

(1) H. Kano et al., U.S. Pat. 3 321 475 (1967).
(2) eidem, J. Med. Chem., 1967, 10, 411.

PERLAPINE [U;INN]

CAS 1977-11-3 | MW: 291.40 | MF: C19 H21 N3

Hypnotic

Sandoz

LgP 3.61(C)

pKa

pp cited in Vols.1-5:

3: 263

(1) Schmutz et al., U.S. Pat. 3 389 139 (1968).
(2) Hunzicker et al., Helv. Chim. Acta, 1966, 49, 1433.
(3) Stille et al., Psychopharmacologia, 1973, 28, 325.

PERMETHRIN [U;INN]

CAS 52645-53-1 | MW: 391.30 | MF: C21 H20 Cl2 O3

Ectoparasiticide

Burroughs Wellcome (Nix)

LgP 6.50(M); 6.86(C,1)

pKa

pp cited in Vols.1-5:

4: 296

(1) M. Elliot et al., Nature, 1973, 246, 169.
(2) C.N.E. Ruscoe, Pest. Sci., 1977, 8, 236.

PERPHENAZINE [U;INN]

CAS 58-39-9 | MW: 403.98 | MF: C21 H26 Cl N3 O S

Antipsychotic

Schering (Trilafon)

LgP 4.20(M); 5.57(C)

pKa 7.8

pp cited in Vols.1-5:

(1) J. Cusic, U.S. Pat. 2 766 235 (1956). (2) Sherlock et al., U.S. Pat. 2 860 138 (1958). (3) W. Modell, Ed., 'Drugs in Current Use', Springer Publ. Co., Inc., NY, 1964.

PERSILIC ACID [INN]

CAS 4444-23-9 | MW: 270.24 | MF: C6 H6 O8 S2

LgP (7)

pKa

pp cited in Vols.1-5:

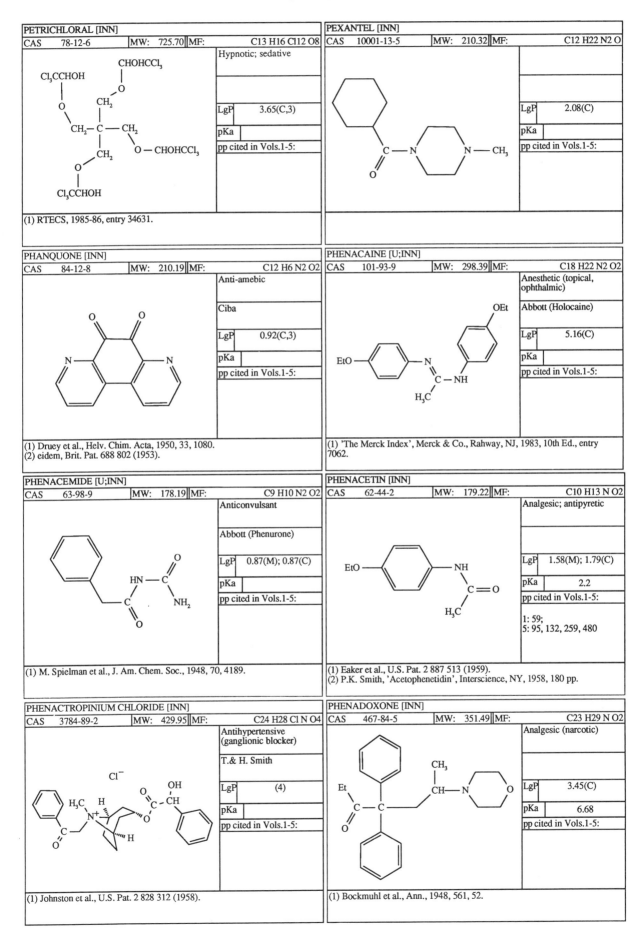

PETRICHLORAL [INN]

CAS	78-12-6	MW:	725.70	MF:	C13 H16 Cl12 O8

Hypnotic; sedative

LgP	3.65(C,3)
pKa	
pp cited in Vols.1-5:	

(1) RTECS, 1985-86, entry 34631.

PEXANTEL [INN]

CAS	10001-13-5	MW:	210.32	MF:	C12 H22 N2 O

LgP	2.08(C)
pKa	
pp cited in Vols.1-5:	

PHANQUONE [INN]

CAS	84-12-8	MW:	210.19	MF:	C12 H6 N2 O2

Anti-amebic

Ciba

LgP	0.92(C,3)
pKa	
pp cited in Vols.1-5:	

(1) Druey et al., Helv. Chim. Acta, 1950, 33, 1080.
(2) eidem, Brit. Pat. 688 802 (1953).

PHENACAINE [U;INN]

CAS	101-93-9	MW:	298.39	MF:	C18 H22 N2 O2

Anesthetic (topical, ophthalmic)

Abbott (Holocaine)

LgP	5.16(C)
pKa	
pp cited in Vols.1-5:	

(1) 'The Merck Index', Merck & Co., Rahway, NJ, 1983, 10th Ed., entry 7062.

PHENACEMIDE [U;INN]

CAS	63-98-9	MW:	178.19	MF:	C9 H10 N2 O2

Anticonvulsant

Abbott (Phenurone)

LgP	0.87(M); 0.87(C)
pKa	
pp cited in Vols.1-5:	

(1) M. Spielman et al., J. Am. Chem. Soc., 1948, 70, 4189.

PHENACETIN [INN]

CAS	62-44-2	MW:	179.22	MF:	C10 H13 N O2

Analgesic; antipyretic

LgP	1.58(M); 1.79(C)
pKa	2.2
pp cited in Vols.1-5:	

1: 59;
5: 95, 132, 259, 480

(1) Eaker et al., U.S. Pat. 2 887 513 (1959).
(2) P.K. Smith, 'Acetophenetidin', Interscience, NY, 1958, 180 pp.

PHENACTROPINIUM CHLORIDE [INN]

CAS	3784-89-2	MW:	429.95	MF:	C24 H28 Cl N O4

Antihypertensive (ganglionic blocker)

T.& H. Smith

LgP	(4)
pKa	
pp cited in Vols.1-5:	

(1) Johnston et al., U.S. Pat. 2 828 312 (1958).

PHENADOXONE [INN]

CAS	467-84-5	MW:	351.49	MF:	C23 H29 N O2

Analgesic (narcotic)

LgP	3.45(C)
pKa	6.68
pp cited in Vols.1-5:	

(1) Bockmuhl et al., Ann., 1948, 561, 52.

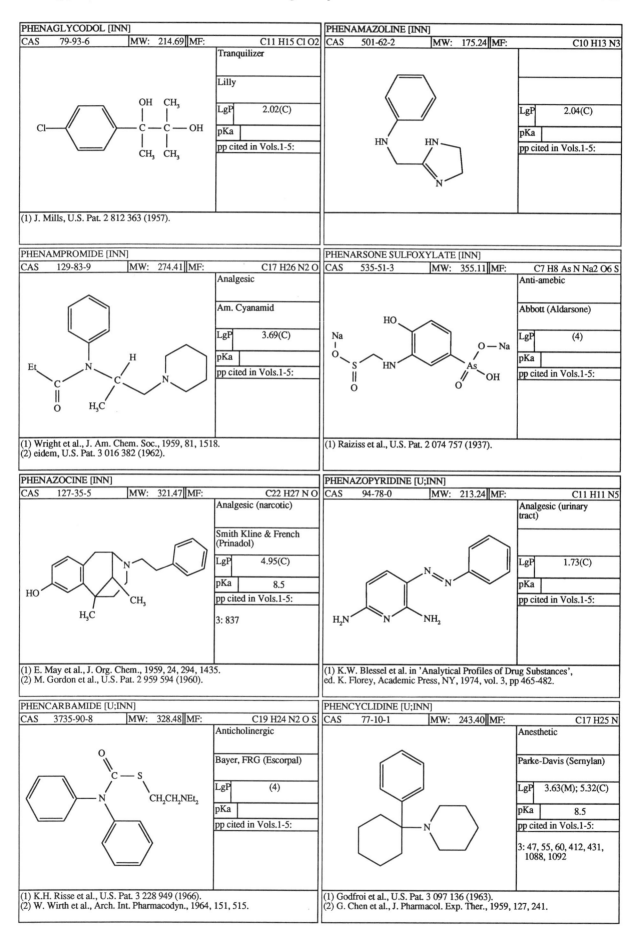

PHENAGLYCODOL [INN]

CAS	79-93-6	MW:	214.69	MF:	C11 H15 Cl O2

Tranquilizer

Lilly

LgP	2.02(C)
pKa	
pp cited in Vols.1-5:	

(1) J. Mills, U.S. Pat. 2 812 363 (1957).

PHENAMAZOLINE [INN]

CAS	501-62-2	MW:	175.24	MF:	C10 H13 N3

LgP	2.04(C)
pKa	
pp cited in Vols.1-5:	

PHENAMPROMIDE [INN]

CAS	129-83-9	MW:	274.41	MF:	C17 H26 N2 O

Analgesic

Am. Cyanamid

LgP	3.69(C)
pKa	
pp cited in Vols.1-5:	

(1) Wright et al., J. Am. Chem. Soc., 1959, 81, 1518.
(2) eidem, U.S. Pat. 3 016 382 (1962).

PHENARSONE SULFOXYLATE [INN]

CAS	535-51-3	MW:	355.11	MF:	C7 H8 As N Na2 O6 S

Anti-amebic

Abbott (Aldarsone)

LgP	(4)
pKa	
pp cited in Vols.1-5:	

(1) Raiziss et al., U.S. Pat. 2 074 757 (1937).

PHENAZOCINE [INN]

CAS	127-35-5	MW:	321.47	MF:	C22 H27 N O

Analgesic (narcotic)

Smith Kline & French (Prinadol)

LgP	4.95(C)
pKa	8.5
pp cited in Vols.1-5:	

3: 837

(1) E. May et al., J. Org. Chem., 1959, 24, 294, 1435.
(2) M. Gordon et al., U.S. Pat. 2 959 594 (1960).

PHENAZOPYRIDINE [U;INN]

CAS	94-78-0	MW:	213.24	MF:	C11 H11 N5

Analgesic (urinary tract)

LgP	1.73(C)
pKa	
pp cited in Vols.1-5:	

(1) K.W. Blessel et al. in 'Analytical Profiles of Drug Substances',
ed. K. Florey, Academic Press, NY, 1974, vol. 3, pp 465-482.

PHENCARBAMIDE [U;INN]

CAS	3735-90-8	MW:	328.48	MF:	C19 H24 N2 O S

Anticholinergic

Bayer, FRG (Escorpal)

LgP	(4)
pKa	
pp cited in Vols.1-5:	

(1) K.H. Risse et al., U.S. Pat. 3 228 949 (1966).
(2) W. Wirth et al., Arch. Int. Pharmacodyn., 1964, 151, 515.

PHENCYCLIDINE [U;INN]

CAS	77-10-1	MW:	243.40	MF:	C17 H25 N

Anesthetic

Parke-Davis (Sernylan)

LgP	3.63(M); 5.32(C)
pKa	8.5
pp cited in Vols.1-5:	

3: 47, 55, 60, 412, 431, 1088, 1092

(1) Godfroi et al., U.S. Pat. 3 097 136 (1963).
(2) G. Chen et al., J. Pharmacol. Exp. Ther., 1959, 127, 241.

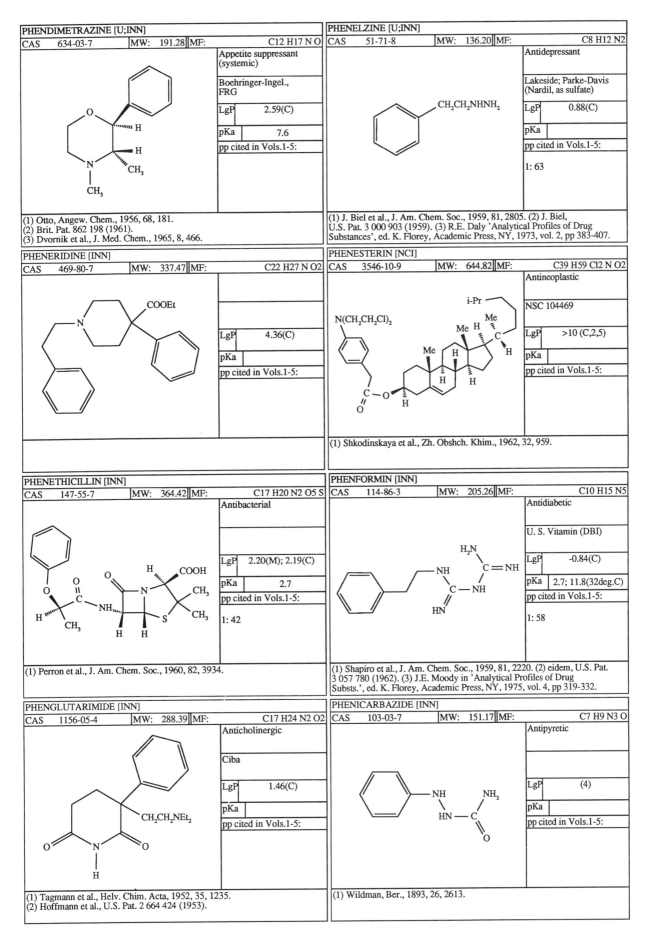

PHENDIMETRAZINE [U;INN]

CAS	634-03-7	MW:	191.28	MF:	C12 H17 N O

Appetite suppressant (systemic)

Boehringer-Ingel., FRG

LgP	2.59(C)
pKa	7.6

pp cited in Vols.1-5:

(1) Otto, Angew. Chem., 1956, 68, 181.
(2) Brit. Pat. 862 198 (1961).
(3) Dvornik et al., J. Med. Chem., 1965, 8, 466.

PHENELZINE [U;INN]

CAS	51-71-8	MW:	136.20	MF:	C8 H12 N2

Antidepressant

Lakeside; Parke-Davis (Nardil, as sulfate)

LgP	0.88(C)
pKa	

pp cited in Vols.1-5:

1: 63

(1) J. Biel et al., J. Am. Chem. Soc., 1959, 81, 2805. (2) J. Biel, U.S. Pat. 3 000 903 (1959). (3) R.E. Daly 'Analytical Profiles of Drug Substances', ed. K. Florey, Academic Press, NY, 1973, vol. 2, pp 383-407.

PHENERIDINE [INN]

CAS	469-80-7	MW:	337.47	MF:	C22 H27 N O2

LgP	4.36(C)
pKa	

pp cited in Vols.1-5:

PHENESTERIN [NCI]

CAS	3546-10-9	MW:	644.82	MF:	C39 H59 Cl2 N O2

Antineoplastic

NSC 104469

LgP	>10 (C,2,5)
pKa	

pp cited in Vols.1-5:

(1) Shkodinskaya et al., Zh. Obshch. Khim., 1962, 32, 959.

PHENETHICILLIN [INN]

CAS	147-55-7	MW:	364.42	MF:	C17 H20 N2 O5 S

Antibacterial

LgP	2.20(M); 2.19(C)
pKa	2.7

pp cited in Vols.1-5:

1: 42

(1) Perron et al., J. Am. Chem. Soc., 1960, 82, 3934.

PHENFORMIN [INN]

CAS	114-86-3	MW:	205.26	MF:	C10 H15 N5

Antidiabetic

U. S. Vitamin (DBI)

LgP	-0.84(C)
pKa	2.7; 11.8(32deg.C)

pp cited in Vols.1-5:

1: 58

(1) Shapiro et al., J. Am. Chem. Soc., 1959, 81, 2220. (2) eidem, U.S. Pat. 3 057 780 (1962). (3) J.E. Moody in 'Analytical Profiles of Drug Substs.', ed. K. Florey, Academic Press, NY, 1975, vol. 4, pp 319-332.

PHENGLUTARIMIDE [INN]

CAS	1156-05-4	MW:	288.39	MF:	C17 H24 N2 O2

Anticholinergic

Ciba

LgP	1.46(C)
pKa	

pp cited in Vols.1-5:

(1) Tagmann et al., Helv. Chim. Acta, 1952, 35, 1235.
(2) Hoffmann et al., U.S. Pat. 2 664 424 (1953).

PHENICARBAZIDE [INN]

CAS	103-03-7	MW:	151.17	MF:	C7 H9 N3 O

Antipyretic

LgP	(4)
pKa	

pp cited in Vols.1-5:

(1) Wildman, Ber., 1893, 26, 2613.

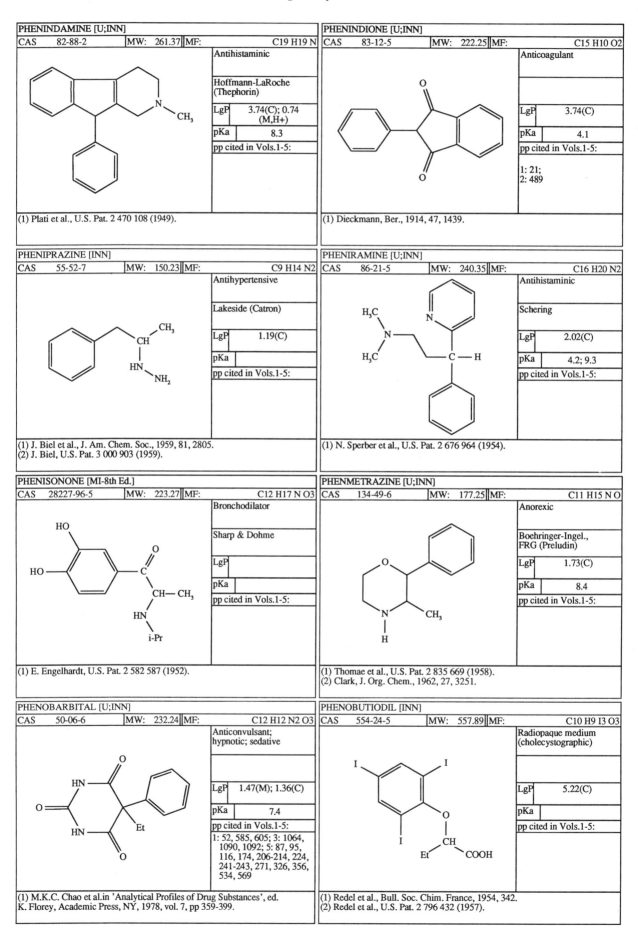

PHENINDAMINE [U;INN]

| CAS | 82-88-2 | MW: | 261.37 | MF: | C19 H19 N |

Antihistaminic

Hoffmann-LaRoche (Thephorin)

LgP	3.74(C); 0.74 (M,H+)
pKa	8.3
pp cited in Vols.1-5:	

(1) Plati et al., U.S. Pat. 2 470 108 (1949).

PHENINDIONE [U;INN]

| CAS | 83-12-5 | MW: | 222.25 | MF: | C15 H10 O2 |

Anticoagulant

LgP	3.74(C)
pKa	4.1
pp cited in Vols.1-5:	

1: 21;
2: 489

(1) Dieckmann, Ber., 1914, 47, 1439.

PHENIPRAZINE [INN]

| CAS | 55-52-7 | MW: | 150.23 | MF: | C9 H14 N2 |

Antihypertensive

Lakeside (Catron)

LgP	1.19(C)
pKa	
pp cited in Vols.1-5:	

(1) J. Biel et al., J. Am. Chem. Soc., 1959, 81, 2805.
(2) J. Biel, U.S. Pat. 3 000 903 (1959).

PHENIRAMINE [U;INN]

| CAS | 86-21-5 | MW: | 240.35 | MF: | C16 H20 N2 |

Antihistaminic

Schering

LgP	2.02(C)
pKa	4.2; 9.3
pp cited in Vols.1-5:	

(1) N. Sperber et al., U.S. Pat. 2 676 964 (1954).

PHENISONONE [MI-8th Ed.]

| CAS | 28227-96-5 | MW: | 223.27 | MF: | C12 H17 N O3 |

Bronchodilator

Sharp & Dohme

LgP	
pKa	
pp cited in Vols.1-5:	

(1) E. Engelhardt, U.S. Pat. 2 582 587 (1952).

PHENMETRAZINE [U;INN]

| CAS | 134-49-6 | MW: | 177.25 | MF: | C11 H15 N O |

Anorexic

Boehringer-Ingel., FRG (Preludin)

LgP	1.73(C)
pKa	8.4
pp cited in Vols.1-5:	

(1) Thomae et al., U.S. Pat. 2 835 669 (1958).
(2) Clark, J. Org. Chem., 1962, 27, 3251.

PHENOBARBITAL [U;INN]

| CAS | 50-06-6 | MW: | 232.24 | MF: | C12 H12 N2 O3 |

Anticonvulsant; hypnotic; sedative

LgP	1.47(M); 1.36(C)
pKa	7.4
pp cited in Vols.1-5:	

1: 52, 585, 605; 3: 1064, 1090, 1092; 5: 87, 95, 116, 174, 206-214, 224, 241-243, 271, 326, 356, 534, 569

(1) M.K.C. Chao et al.in 'Analytical Profiles of Drug Substances', ed. K. Florey, Academic Press, NY, 1978, vol. 7, pp 359-399.

PHENOBUTIODIL [INN]

| CAS | 554-24-5 | MW: | 557.89 | MF: | C10 H9 I3 O3 |

Radiopaque medium (cholecystographic)

LgP	5.22(C)
pKa	
pp cited in Vols.1-5:	

(1) Redel et al., Bull. Soc. Chim. France, 1954, 342.
(2) Redel et al., U.S. Pat. 2 796 432 (1957).

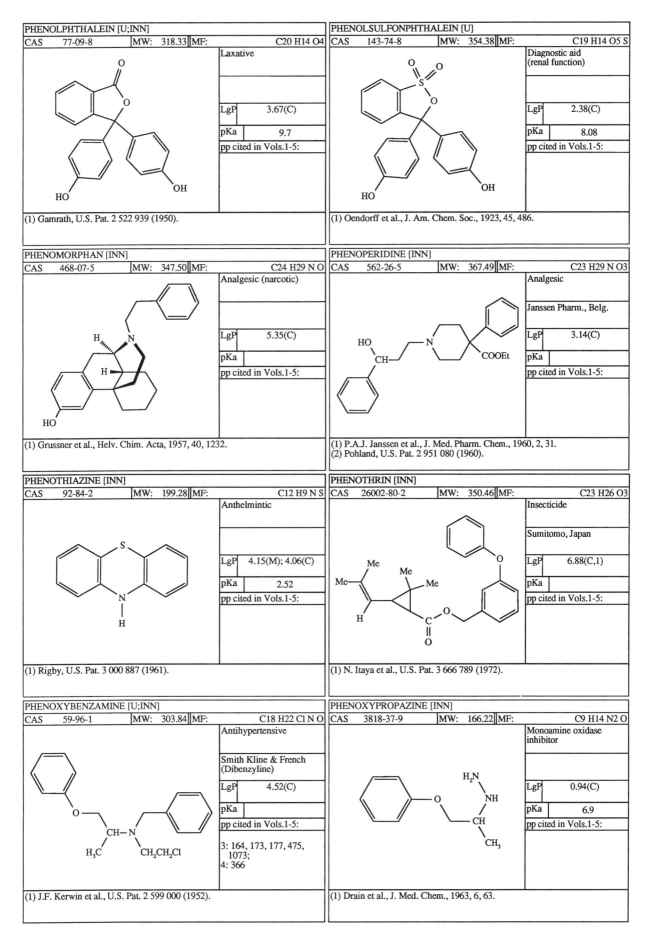

PHENOLPHTHALEIN [U;INN]

CAS	77-09-8	MW:	318.33	MF:	C20 H14 O4

Laxative

LgP	3.67(C)
pKa	9.7
pp cited in Vols.1-5:	

(1) Gamrath, U.S. Pat. 2 522 939 (1950).

PHENOLSULFONPHTHALEIN [U]

CAS	143-74-8	MW:	354.38	MF:	C19 H14 O5 S

Diagnostic aid
(renal function)

LgP	2.38(C)
pKa	8.08
pp cited in Vols.1-5:	

(1) Oendorff et al., J. Am. Chem. Soc., 1923, 45, 486.

PHENOMORPHAN [INN]

CAS	468-07-5	MW:	347.50	MF:	C24 H29 N O

Analgesic (narcotic)

LgP	5.35(C)
pKa	
pp cited in Vols.1-5:	

(1) Grussner et al., Helv. Chim. Acta, 1957, 40, 1232.

PHENOPERIDINE [INN]

CAS	562-26-5	MW:	367.49	MF:	C23 H29 N O3

Analgesic

Janssen Pharm., Belg.

LgP	3.14(C)
pKa	
pp cited in Vols.1-5:	

(1) P.A.J. Janssen et al., J. Med. Pharm. Chem., 1960, 2, 31.
(2) Pohland, U.S. Pat. 2 951 080 (1960).

PHENOTHIAZINE [INN]

CAS	92-84-2	MW:	199.28	MF:	C12 H9 N S

Anthelmintic

LgP	4.15(M); 4.06(C)
pKa	2.52
pp cited in Vols.1-5:	

(1) Rigby, U.S. Pat. 3 000 887 (1961).

PHENOTHRIN [INN]

CAS	26002-80-2	MW:	350.46	MF:	C23 H26 O3

Insecticide

Sumitomo, Japan

LgP	6.88(C,1)
pKa	
pp cited in Vols.1-5:	

(1) N. Itaya et al., U.S. Pat. 3 666 789 (1972).

PHENOXYBENZAMINE [U;INN]

CAS	59-96-1	MW:	303.84	MF:	C18 H22 Cl N O

Antihypertensive

Smith Kline & French
(Dibenzyline)

LgP	4.52(C)
pKa	
pp cited in Vols.1-5:	

3: 164, 173, 177, 475,
  1073;
4: 366

(1) J.F. Kerwin et al., U.S. Pat. 2 599 000 (1952).

PHENOXYPROPAZINE [INN]

CAS	3818-37-9	MW:	166.22	MF:	C9 H14 N2 O

Monoamine oxidase
inhibitor

LgP	0.94(C)
pKa	6.9
pp cited in Vols.1-5:	

(1) Drain et al., J. Med. Chem., 1963, 6, 63.

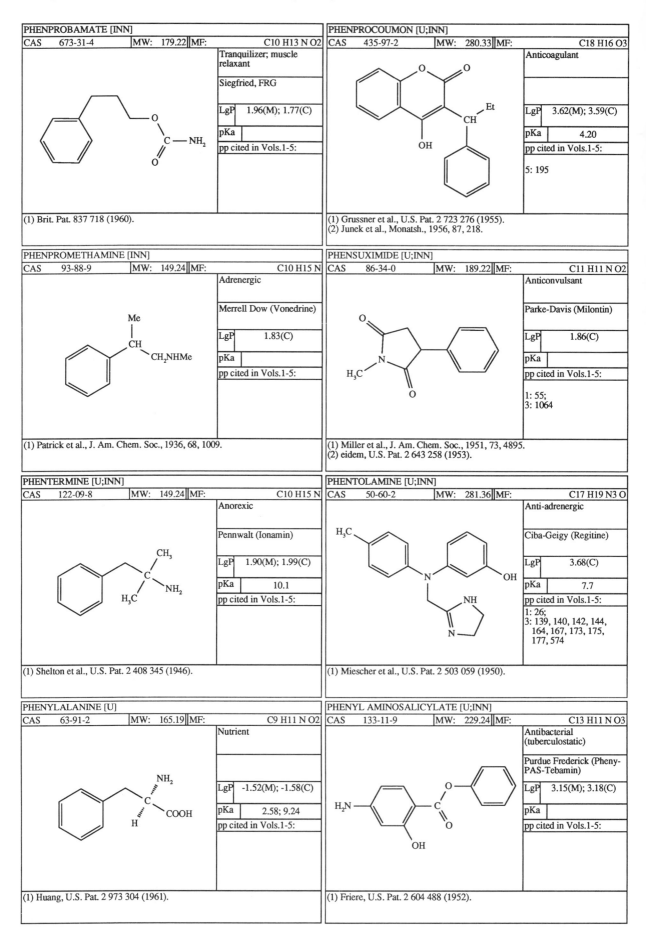

PHENPROBAMATE [INN]			
CAS 673-31-4	MW: 179.22	MF:	C10 H13 N O2

Tranquilizer; muscle relaxant

Siegfried, FRG

LgP	1.96(M); 1.77(C)
pKa	
pp cited in Vols.1-5:	

(1) Brit. Pat. 837 718 (1960).

PHENPROCOUMON [U;INN]			
CAS 435-97-2	MW: 280.33	MF:	C18 H16 O3

Anticoagulant

LgP	3.62(M); 3.59(C)
pKa	4.20
pp cited in Vols.1-5:	

5: 195

(1) Grussner et al., U.S. Pat. 2 723 276 (1955).
(2) Junek et al., Monatsh., 1956, 87, 218.

PHENPROMETHAMINE [INN]			
CAS 93-88-9	MW: 149.24	MF:	C10 H15 N

Adrenergic

Merrell Dow (Vonedrine)

LgP	1.83(C)
pKa	
pp cited in Vols.1-5:	

(1) Patrick et al., J. Am. Chem. Soc., 1936, 68, 1009.

PHENSUXIMIDE [U;INN]			
CAS 86-34-0	MW: 189.22	MF:	C11 H11 N O2

Anticonvulsant

Parke-Davis (Milontin)

LgP	1.86(C)
pKa	
pp cited in Vols.1-5:	

1: 55;
3: 1064

(1) Miller et al., J. Am. Chem. Soc., 1951, 73, 4895.
(2) eidem, U.S. Pat. 2 643 258 (1953).

PHENTERMINE [U;INN]			
CAS 122-09-8	MW: 149.24	MF:	C10 H15 N

Anorexic

Pennwalt (Ionamin)

LgP	1.90(M); 1.99(C)
pKa	10.1
pp cited in Vols.1-5:	

(1) Shelton et al., U.S. Pat. 2 408 345 (1946).

PHENTOLAMINE [U;INN]			
CAS 50-60-2	MW: 281.36	MF:	C17 H19 N3 O

Anti-adrenergic

Ciba-Geigy (Regitine)

LgP	3.68(C)
pKa	7.7
pp cited in Vols.1-5:	

1: 26;
3: 139, 140, 142, 144, 164, 167, 173, 175, 177, 574

(1) Miescher et al., U.S. Pat. 2 503 059 (1950).

PHENYLALANINE [U]			
CAS 63-91-2	MW: 165.19	MF:	C9 H11 N O2

Nutrient

LgP	-1.52(M); -1.58(C)
pKa	2.58; 9.24
pp cited in Vols.1-5:	

(1) Huang, U.S. Pat. 2 973 304 (1961).

PHENYL AMINOSALICYLATE [U;INN]			
CAS 133-11-9	MW: 229.24	MF:	C13 H11 N O3

Antibacterial (tuberculostatic)

Purdue Frederick (Pheny-PAS-Tebamin)

LgP	3.15(M); 3.18(C)
pKa	
pp cited in Vols.1-5:	

(1) Friere, U.S. Pat. 2 604 488 (1952).

PHENYLBUTAZONE [U;INN]	
CAS 50-33-9 MW: 308.38 MF: C19 H20 N2 O2	

Anti-inflammatory

Geigy, Switz.

LgP	3.16(M); 3.17(C)
pKa	4.4
pp cited in Vols.1-5:	

1: 268, 600, 628;
4: 389;
5: 83, 193, 259, 268, 270, 272, 326, 327, 328, 555

(1) Stenzl, U.S. Pat. 2 562 830 (1951).
(2) S.L. Ali in 'Analytical Profiles of Drug Substances', ed. K. Florey, Academic Press, NY, 1982, vol. 11, pp 483-521.

PHENYLEPHRINE [U;INN]	
CAS 59-42-7 MW: 167.21 MF: C9 H13 N O2	

Adrenergic (ophthalmic)

LgP	-0.31(M); -0.09(C)
pKa	8.9; 10.1
pp cited in Vols.1-5:	

3: 139, 141, 144, 147, 160, 173, 177, 178
4: 366, 618

(1) Bergmann et al., J. Org. Chem., 1951, 16, 84.
(2) C.A. Gaglia in 'Analytical Profiles of Drug Substances', ed. K.Florey, Academic Press, NY, 1974, vol. 3, pp 483-512.

PHENYLETHYL ALCOHOL [U]	
CAS 60-12-8 MW: 122.17 MF: C8 H10 O	

Antimicrobial agent, (opthalmic)

LgP	1.36(M); 1.18(C)
pKa	
pp cited in Vols.1-5:	

(1) 'Martindale. The Extra Pharmacopoeia', ed. J.E.F. Reynolds, The Pharmaceutical Press, London, 1982, 28th Ed., p 1288.

PHENYLMERCURIC ACETATE [U]	
CAS 62-38-4 MW: 336.74 MF: C8 H8 Hg O2	

Antimicrobial agent

LgP	0.71(M); 0.71(C)
pKa	
pp cited in Vols.1-5:	

(1) F.C. Whitmore, 'Organic Compounds of Mercury', Chemical Catalog Co., NY, 1921, pp 175-176.

PHENYLMERCURIC BORATE [INN]	
CAS 8017-88-7 MW: 633.23 MF: C12 H13 B Hg2 O4	

Antiseptic (topical)

Lever Bros.

LgP	(4)
pKa	
pp cited in Vols.1-5:	

(1) Christiansen, U.S. Pat. 2 196 384 (1950).

PHENYLMERCURIC CHLORIDE [MI]	
CAS 100-56-1 MW: 313.15 MF: C6 H5 Cl Hg	

Antimicrobial agent

LgP	1.78(C); 1.78(M)
pKa	
pp cited in Vols.1-5:	

see Phenylmercuric Acetate

PHENYLMERCURIC NITRATE [U]	
CAS 55-68-5 MW: 634.41 MF: C12 H11 Hg2 N O4	

Pharmaceutic aid (antimicrobial agent)

Schering-Kahlbaum

LgP	(4)
pKa	
pp cited in Vols.1-5:	

1: 50

(1) Pyman et al., Pharm. J., 1934, 133, 2269.
(2) Brit. Pat. 466 703 (1936).

PHENYLMETHYLBARBITURIC ACID [MI]	
CAS 76-94-8 MW: 218.21 MF: C11 H10 N2 O3	

Anticonvulsant; sedative

LgP	0.84(C)
pKa	7.7
pp cited in Vols.1-5:	

(1) A.C. Moffat et al., eds, 'Clarke's Isolation and Identification of Drugs', The Pharmaceutical Press, London, 1986, pp 894-895.

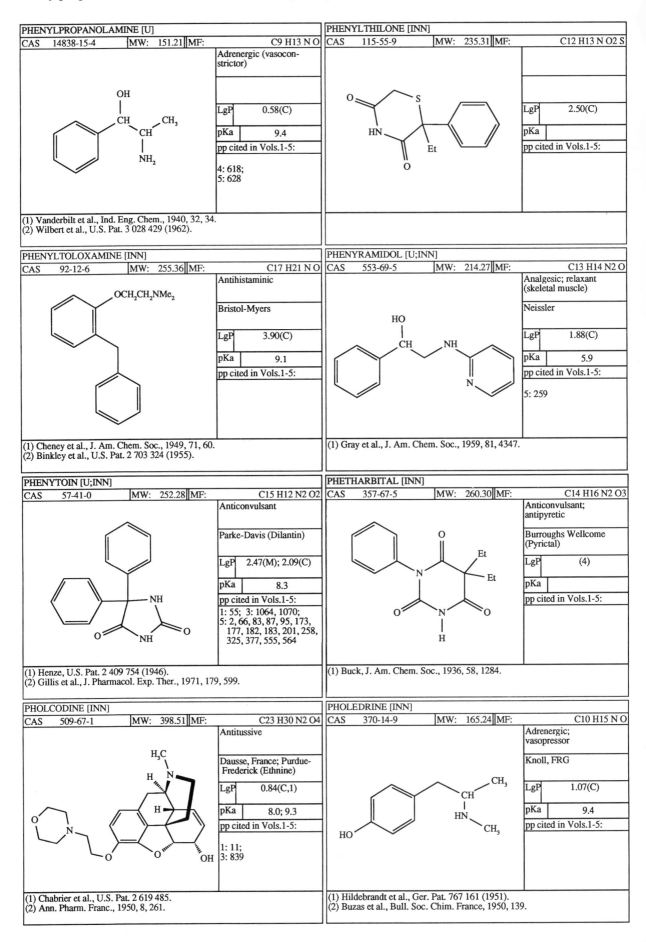

PHENYLPROPANOLAMINE [U]			
CAS 14838-15-4	MW: 151.21	MF:	C9 H13 N O

Adrenergic (vasocon-strictor)

LgP	0.58(C)
pKa	9.4

pp cited in Vols.1-5:

4: 618;
5: 628

(1) Vanderbilt et al., Ind. Eng. Chem., 1940, 32, 34.
(2) Wilbert et al., U.S. Pat. 3 028 429 (1962).

PHENYLTHILONE [INN]			
CAS 115-55-9	MW: 235.31	MF:	C12 H13 N O2 S

LgP	2.50(C)
pKa	

pp cited in Vols.1-5:

PHENYLTOLOXAMINE [INN]			
CAS 92-12-6	MW: 255.36	MF:	C17 H21 N O

Antihistaminic

Bristol-Myers

LgP	3.90(C)
pKa	9.1

pp cited in Vols.1-5:

(1) Cheney et al., J. Am. Chem. Soc., 1949, 71, 60.
(2) Binkley et al., U.S. Pat. 2 703 324 (1955).

PHENYRAMIDOL [U;INN]			
CAS 553-69-5	MW: 214.27	MF:	C13 H14 N2 O

Analgesic; relaxant (skeletal muscle)

Neissler

LgP	1.88(C)
pKa	5.9

pp cited in Vols.1-5:

5: 259

(1) Gray et al., J. Am. Chem. Soc., 1959, 81, 4347.

PHENYTOIN [U;INN]			
CAS 57-41-0	MW: 252.28	MF:	C15 H12 N2 O2

Anticonvulsant

Parke-Davis (Dilantin)

LgP	2.47(M); 2.09(C)
pKa	8.3

pp cited in Vols.1-5:

1: 55; 3: 1064, 1070;
5: 2, 66, 83, 87, 95, 173,
177, 182, 183, 201, 258,
325, 377, 555, 564

(1) Henze, U.S. Pat. 2 409 754 (1946).
(2) Gillis et al., J. Pharmacol. Exp. Ther., 1971, 179, 599.

PHETHARBITAL [INN]			
CAS 357-67-5	MW: 260.30	MF:	C14 H16 N2 O3

Anticonvulsant; antipyretic

Burroughs Wellcome (Pyrictal)

LgP	(4)
pKa	

pp cited in Vols.1-5:

(1) Buck, J. Am. Chem. Soc., 1936, 58, 1284.

PHOLCODINE [INN]			
CAS 509-67-1	MW: 398.51	MF:	C23 H30 N2 O4

Antitussive

Dausse, France; Purdue-Frederick (Ethnine)

LgP	0.84(C,1)
pKa	8.0; 9.3

pp cited in Vols.1-5:

1: 11;
3: 839

(1) Chabrier et al., U.S. Pat. 2 619 485.
(2) Ann. Pharm. Franc., 1950, 8, 261.

PHOLEDRINE [INN]			
CAS 370-14-9	MW: 165.24	MF:	C10 H15 N O

Adrenergic; vasopressor

Knoll, FRG

LgP	1.07(C)
pKa	9.4

pp cited in Vols.1-5:

(1) Hildebrandt et al., Ger. Pat. 767 161 (1951).
(2) Buzas et al., Bull. Soc. Chim. France, 1950, 139.

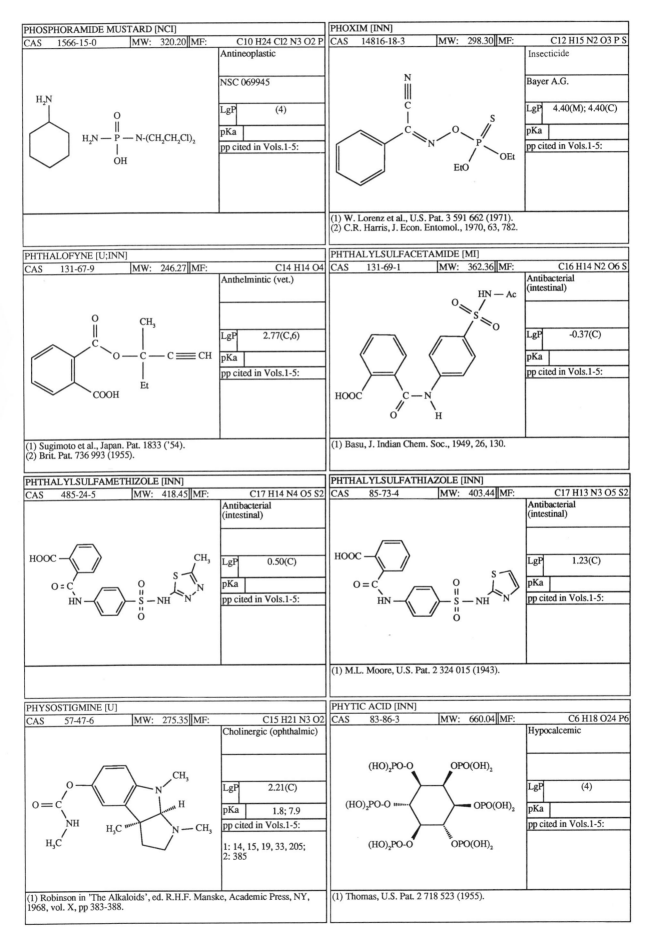

PHOSPHORAMIDE MUSTARD [NCI]

CAS	1566-15-0	MW:	320.20	MF:	C10 H24 Cl2 N3 O2 P

Antineoplastic

NSC 069945

LgP	(4)
pKa	
pp cited in Vols.1-5:	

PHOXIM [INN]

CAS	14816-18-3	MW:	298.30	MF:	C12 H15 N2 O3 P S

Insecticide

Bayer A.G.

LgP	4.40(M); 4.40(C)
pKa	
pp cited in Vols.1-5:	

(1) W. Lorenz et al., U.S. Pat. 3 591 662 (1971).
(2) C.R. Harris, J. Econ. Entomol., 1970, 63, 782.

PHTHALOFYNE [U;INN]

CAS	131-67-9	MW:	246.27	MF:	C14 H14 O4

Anthelmintic (vet.)

LgP	2.77(C,6)
pKa	
pp cited in Vols.1-5:	

(1) Sugimoto et al., Japan. Pat. 1833 ('54).
(2) Brit. Pat. 736 993 (1955).

PHTHALYLSULFACETAMIDE [MI]

CAS	131-69-1	MW:	362.36	MF:	C16 H14 N2 O6 S

Antibacterial
(intestinal)

LgP	-0.37(C)
pKa	
pp cited in Vols.1-5:	

(1) Basu, J. Indian Chem. Soc., 1949, 26, 130.

PHTHALYLSULFAMETHIZOLE [INN]

CAS	485-24-5	MW:	418.45	MF:	C17 H14 N4 O5 S2

Antibacterial
(intestinal)

LgP	0.50(C)
pKa	
pp cited in Vols.1-5:	

PHTHALYLSULFATHIAZOLE [INN]

CAS	85-73-4	MW:	403.44	MF:	C17 H13 N3 O5 S2

Antibacterial
(intestinal)

LgP	1.23(C)
pKa	
pp cited in Vols.1-5:	

(1) M.L. Moore, U.S. Pat. 2 324 015 (1943).

PHYSOSTIGMINE [U]

CAS	57-47-6	MW:	275.35	MF:	C15 H21 N3 O2

Cholinergic (ophthalmic)

LgP	2.21(C)
pKa	1.8; 7.9
pp cited in Vols.1-5:	

1: 14, 15, 19, 33, 205;
2: 385

(1) Robinson in 'The Alkaloids', ed. R.H.F. Manske, Academic Press, NY, 1968, vol. X, pp 383-388.

PHYTIC ACID [INN]

CAS	83-86-3	MW:	660.04	MF:	C6 H18 O24 P6

Hypocalcemic

LgP	(4)
pKa	
pp cited in Vols.1-5:	

(1) Thomas, U.S. Pat. 2 718 523 (1955).

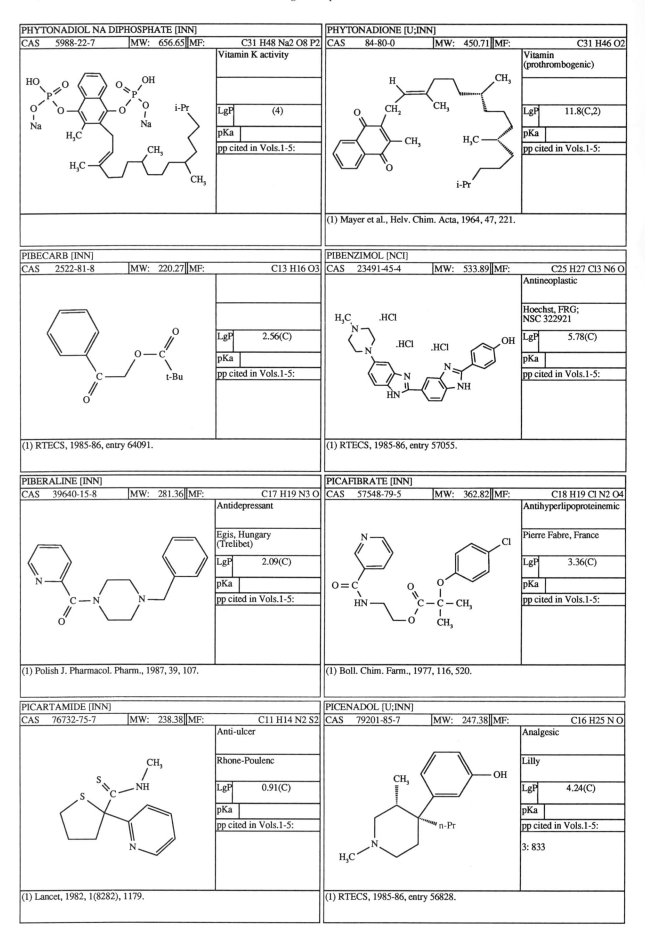

PHYTONADIOL NA DIPHOSPHATE [INN]

CAS	5988-22-7	MW:	656.65	MF:	C31 H48 Na2 O8 P2

Vitamin K activity

LgP	(4)
pKa	
pp cited in Vols.1-5:	

PHYTONADIONE [U;INN]

CAS	84-80-0	MW:	450.71	MF:	C31 H46 O2

Vitamin (prothrombogenic)

LgP	11.8(C,2)
pKa	
pp cited in Vols.1-5:	

(1) Mayer et al., Helv. Chim. Acta, 1964, 47, 221.

PIBECARB [INN]

CAS	2522-81-8	MW:	220.27	MF:	C13 H16 O3

LgP	2.56(C)
pKa	
pp cited in Vols.1-5:	

(1) RTECS, 1985-86, entry 64091.

PIBENZIMOL [NCI]

CAS	23491-45-4	MW:	533.89	MF:	C25 H27 Cl3 N6 O

Antineoplastic

Hoechst, FRG; NSC 322921

LgP	5.78(C)
pKa	
pp cited in Vols.1-5:	

(1) RTECS, 1985-86, entry 57055.

PIBERALINE [INN]

CAS	39640-15-8	MW:	281.36	MF:	C17 H19 N3 O

Antidepressant

Egis, Hungary (Trelibet)

LgP	2.09(C)
pKa	
pp cited in Vols.1-5:	

(1) Polish J. Pharmacol. Pharm., 1987, 39, 107.

PICAFIBRATE [INN]

CAS	57548-79-5	MW:	362.82	MF:	C18 H19 Cl N2 O4

Antihyperlipoproteinemic

Pierre Fabre, France

LgP	3.36(C)
pKa	
pp cited in Vols.1-5:	

(1) Boll. Chim. Farm., 1977, 116, 520.

PICARTAMIDE [INN]

CAS	76732-75-7	MW:	238.38	MF:	C11 H14 N2 S2

Anti-ulcer

Rhone-Poulenc

LgP	0.91(C)
pKa	
pp cited in Vols.1-5:	

(1) Lancet, 1982, 1(8282), 1179.

PICENADOL [U;INN]

CAS	79201-85-7	MW:	247.38	MF:	C16 H25 N O

Analgesic

Lilly

LgP	4.24(C)
pKa	
pp cited in Vols.1-5:	
3: 833	

(1) RTECS, 1985-86, entry 56828.

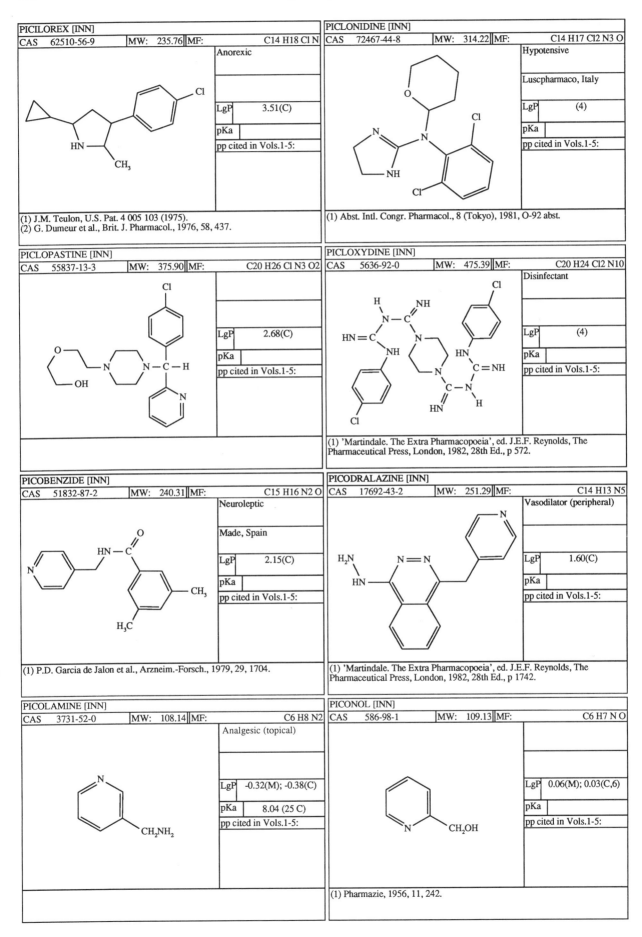

PICILOREX [INN]

CAS	62510-56-9	MW:	235.76	MF:	C14 H18 Cl N

Anorexic

LgP	3.51(C)

pKa	

pp cited in Vols.1-5:

(1) J.M. Teulon, U.S. Pat. 4 005 103 (1975).
(2) G. Dumeur et al., Brit. J. Pharmacol., 1976, 58, 437.

PICLONIDINE [INN]

CAS	72467-44-8	MW:	314.22	MF:	C14 H17 Cl2 N3 O

Hypotensive

Luscpharmaco, Italy

LgP	(4)

pKa	

pp cited in Vols.1-5:

(1) Abst. Intl. Congr. Pharmacol., 8 (Tokyo), 1981, O-92 abst.

PICLOPASTINE [INN]

CAS	55837-13-3	MW:	375.90	MF:	C20 H26 Cl N3 O2

LgP	2.68(C)

pKa	

pp cited in Vols.1-5:

PICLOXYDINE [INN]

CAS	5636-92-0	MW:	475.39	MF:	C20 H24 Cl2 N10

Disinfectant

LgP	(4)

pKa	

pp cited in Vols.1-5:

(1) 'Martindale. The Extra Pharmacopoeia', ed. J.E.F. Reynolds, The
Pharmaceutical Press, London, 1982, 28th Ed., p 572.

PICOBENZIDE [INN]

CAS	51832-87-2	MW:	240.31	MF:	C15 H16 N2 O

Neuroleptic

Made, Spain

LgP	2.15(C)

pKa	

pp cited in Vols.1-5:

(1) P.D. Garcia de Jalon et al., Arzneim.-Forsch., 1979, 29, 1704.

PICODRALAZINE [INN]

CAS	17692-43-2	MW:	251.29	MF:	C14 H13 N5

Vasodilator (peripheral)

LgP	1.60(C)

pKa	

pp cited in Vols.1-5:

(1) 'Martindale. The Extra Pharmacopoeia', ed. J.E.F. Reynolds, The
Pharmaceutical Press, London, 1982, 28th Ed., p 1742.

PICOLAMINE [INN]

CAS	3731-52-0	MW:	108.14	MF:	C6 H8 N2

Analgesic (topical)

LgP	-0.32(M); -0.38(C)

pKa	8.04 (25 C)

pp cited in Vols.1-5:

PICONOL [INN]

CAS	586-98-1	MW:	109.13	MF:	C6 H7 N O

LgP	0.06(M); 0.03(C,6)

pKa	

pp cited in Vols.1-5:

(1) Pharmazie, 1956, 11, 242.

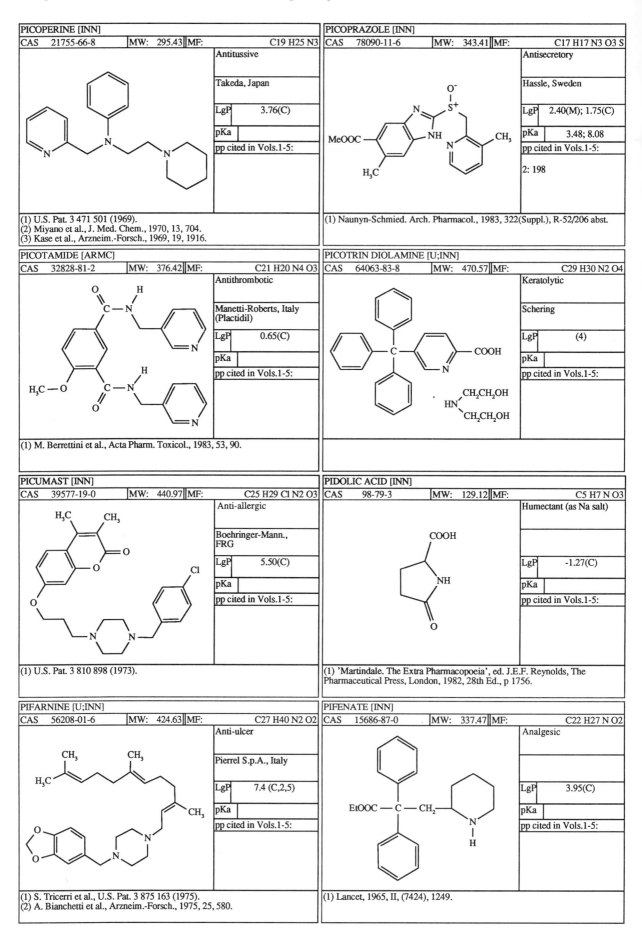

PICOPERINE [INN]		
CAS 21755-66-8	MW: 295.43	MF: C19 H25 N3

Antitussive

Takeda, Japan

LgP	3.76(C)
pKa	
pp cited in Vols.1-5:	

(1) U.S. Pat. 3 471 501 (1969).
(2) Miyano et al., J. Med. Chem., 1970, 13, 704.
(3) Kase et al., Arzneim.-Forsch., 1969, 19, 1916.

PICOPRAZOLE [INN]		
CAS 78090-11-6	MW: 343.41	MF: C17 H17 N3 O3 S

Antisecretory

Hassle, Sweden

LgP	2.40(M); 1.75(C)
pKa	3.48; 8.08
pp cited in Vols.1-5:	
	2: 198

(1) Naunyn-Schmied. Arch. Pharmacol., 1983, 322(Suppl.), R-52/206 abst.

PICOTAMIDE [ARMC]		
CAS 32828-81-2	MW: 376.42	MF: C21 H20 N4 O3

Antithrombotic

Manetti-Roberts, Italy
(Plactidil)

LgP	0.65(C)
pKa	
pp cited in Vols.1-5:	

(1) M. Berrettini et al., Acta Pharm. Toxicol., 1983, 53, 90.

PICOTRIN DIOLAMINE [U;INN]		
CAS 64063-83-8	MW: 470.57	MF: C29 H30 N2 O4

Keratolytic

Schering

LgP	(4)
pKa	
pp cited in Vols.1-5:	

PICUMAST [INN]		
CAS 39577-19-0	MW: 440.97	MF: C25 H29 Cl N2 O3

Anti-allergic

Boehringer-Mann.,
FRG

LgP	5.50(C)
pKa	
pp cited in Vols.1-5:	

(1) U.S. Pat. 3 810 898 (1973).

PIDOLIC ACID [INN]		
CAS 98-79-3	MW: 129.12	MF: C5 H7 N O3

Humectant (as Na salt)

LgP	-1.27(C)
pKa	
pp cited in Vols.1-5:	

(1) 'Martindale. The Extra Pharmacopoeia', ed. J.E.F. Reynolds, The
Pharmaceutical Press, London, 1982, 28th Ed., p 1756.

PIFARNINE [U;INN]		
CAS 56208-01-6	MW: 424.63	MF: C27 H40 N2 O2

Anti-ulcer

Pierrel S.p.A., Italy

LgP	7.4 (C,2,5)
pKa	
pp cited in Vols.1-5:	

(1) S. Tricerri et al., U.S. Pat. 3 875 163 (1975).
(2) A. Bianchetti et al., Arzneim.-Forsch., 1975, 25, 580.

PIFENATE [INN]		
CAS 15686-87-0	MW: 337.47	MF: C22 H27 N O2

Analgesic

LgP	3.95(C)
pKa	
pp cited in Vols.1-5:	

(1) Lancet, 1965, II, (7424), 1249.

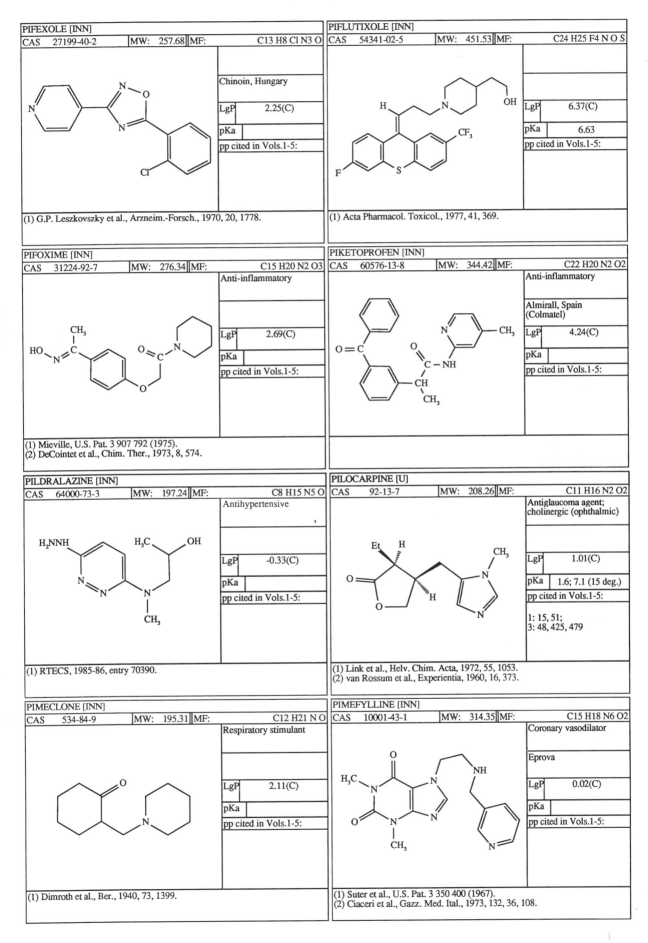

PIFEXOLE [INN]

| CAS | 27199-40-2 | MW: | 257.68 | MF: | C13 H8 Cl N3 O |

Chinoin, Hungary

LgP	2.25(C)
pKa	
pp cited in Vols.1-5:	

(1) G.P. Leszkovszky et al., Arzneim.-Forsch., 1970, 20, 1778.

PIFLUTIXOLE [INN]

| CAS | 54341-02-5 | MW: | 451.53 | MF: | C24 H25 F4 N O S |

LgP	6.37(C)
pKa	6.63
pp cited in Vols.1-5:	

(1) Acta Pharmacol. Toxicol., 1977, 41, 369.

PIFOXIME [INN]

| CAS | 31224-92-7 | MW: | 276.34 | MF: | C15 H20 N2 O3 |

Anti-inflammatory

LgP	2.69(C)
pKa	
pp cited in Vols.1-5:	

(1) Mieville, U.S. Pat. 3 907 792 (1975).
(2) DeCointet et al., Chim. Ther., 1973, 8, 574.

PIKETOPROFEN [INN]

| CAS | 60576-13-8 | MW: | 344.42 | MF: | C22 H20 N2 O2 |

Anti-inflammatory

Almirall, Spain
(Colmatel)

LgP	4.24(C)
pKa	
pp cited in Vols.1-5:	

PILDRALAZINE [INN]

| CAS | 64000-73-3 | MW: | 197.24 | MF: | C8 H15 N5 O |

Antihypertensive

LgP	-0.33(C)
pKa	
pp cited in Vols.1-5:	

(1) RTECS, 1985-86, entry 70390.

PILOCARPINE [U]

| CAS | 92-13-7 | MW: | 208.26 | MF: | C11 H16 N2 O2 |

Antiglaucoma agent;
cholinergic (ophthalmic)

LgP	1.01(C)
pKa	1.6; 7.1 (15 deg.)
pp cited in Vols.1-5:	

1: 15, 51;
3: 48, 425, 479

(1) Link et al., Helv. Chim. Acta, 1972, 55, 1053.
(2) van Rossum et al., Experientia, 1960, 16, 373.

PIMECLONE [INN]

| CAS | 534-84-9 | MW: | 195.31 | MF: | C12 H21 N O |

Respiratory stimulant

LgP	2.11(C)
pKa	
pp cited in Vols.1-5:	

(1) Dimroth et al., Ber., 1940, 73, 1399.

PIMEFYLLINE [INN]

| CAS | 10001-43-1 | MW: | 314.35 | MF: | C15 H18 N6 O2 |

Coronary vasodilator

Eprova

LgP	0.02(C)
pKa	
pp cited in Vols.1-5:	

(1) Suter et al., U.S. Pat. 3 350 400 (1967).
(2) Ciaceri et al., Gazz. Med. Ital., 1973, 132, 36, 108.

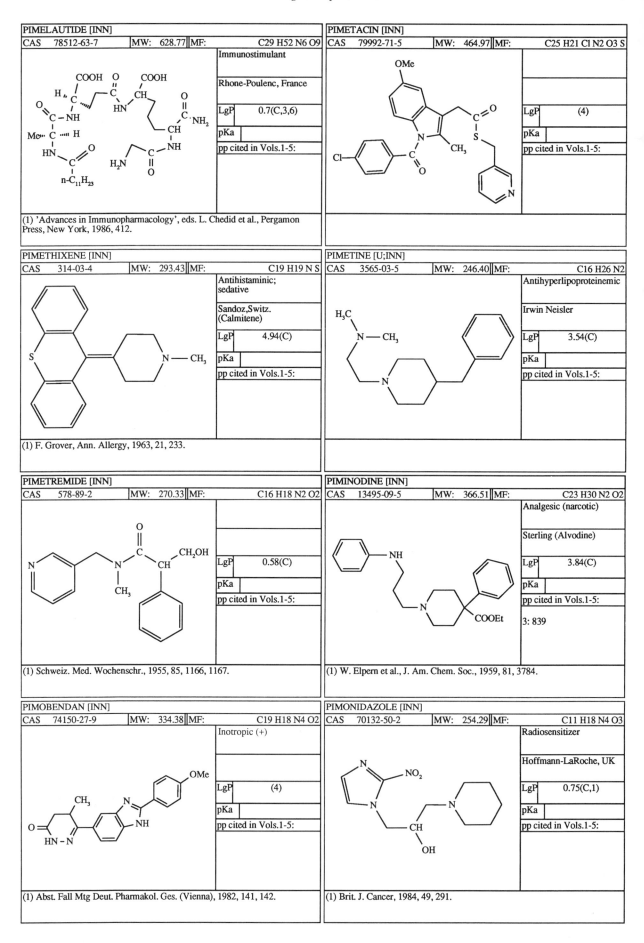

PIMELAUTIDE [INN]		
CAS 78512-63-7	MW: 628.77	MF: C29 H52 N6 O9

Immunostimulant

Rhone-Poulenc, France

LgP	0.7(C,3,6)
pKa	
pp cited in Vols.1-5:	

(1) 'Advances in Immunopharmacology', eds. L. Chedid et al., Pergamon Press, New York, 1986, 412.

PIMETACIN [INN]		
CAS 79992-71-5	MW: 464.97	MF: C25 H21 Cl N2 O3 S

LgP	(4)
pKa	
pp cited in Vols.1-5:	

PIMETHIXENE [INN]		
CAS 314-03-4	MW: 293.43	MF: C19 H19 N S

Antihistaminic; sedative

Sandoz,Switz. (Calmitene)

LgP	4.94(C)
pKa	
pp cited in Vols.1-5:	

(1) F. Grover, Ann. Allergy, 1963, 21, 233.

PIMETINE [U;INN]		
CAS 3565-03-5	MW: 246.40	MF: C16 H26 N2

Antihyperlipoproteinemic

Irwin Neisler

LgP	3.54(C)
pKa	
pp cited in Vols.1-5:	

PIMETREMIDE [INN]		
CAS 578-89-2	MW: 270.33	MF: C16 H18 N2 O2

LgP	0.58(C)
pKa	
pp cited in Vols.1-5:	

(1) Schweiz. Med. Wochenschr., 1955, 85, 1166, 1167.

PIMINODINE [INN]		
CAS 13495-09-5	MW: 366.51	MF: C23 H30 N2 O2

Analgesic (narcotic)

Sterling (Alvodine)

LgP	3.84(C)
pKa	
pp cited in Vols.1-5:	
3: 839	

(1) W. Elpern et al., J. Am. Chem. Soc., 1959, 81, 3784.

PIMOBENDAN [INN]		
CAS 74150-27-9	MW: 334.38	MF: C19 H18 N4 O2

Inotropic (+)

LgP	(4)
pKa	
pp cited in Vols.1-5:	

(1) Abst. Fall Mtg Deut. Pharmakol. Ges. (Vienna), 1982, 141, 142.

PIMONIDAZOLE [INN]		
CAS 70132-50-2	MW: 254.29	MF: C11 H18 N4 O3

Radiosensitizer

Hoffmann-LaRoche, UK

LgP	0.75(C,1)
pKa	
pp cited in Vols.1-5:	

(1) Brit. J. Cancer, 1984, 49, 291.

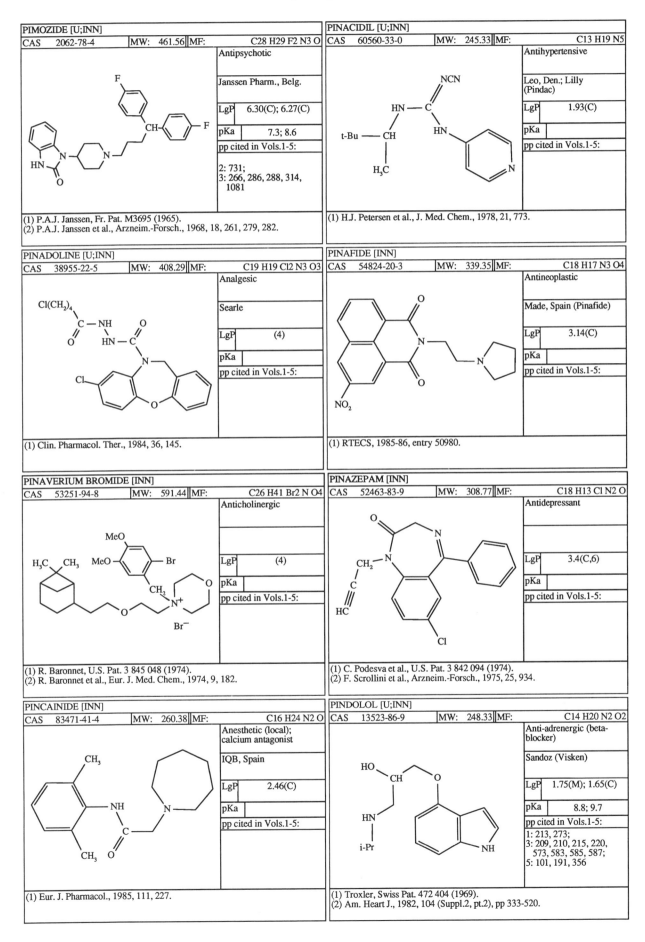

PIMOZIDE [U;INN]

CAS	2062-78-4	MW:	461.56	MF:	C28 H29 F2 N3 O

Antipsychotic

Janssen Pharm., Belg.

LgP	6.30(C); 6.27(C)
pKa	7.3; 8.6
pp cited in Vols.1-5:	

2: 731;
3: 266, 286, 288, 314, 1081

(1) P.A.J. Janssen, Fr. Pat. M3695 (1965).
(2) P.A.J. Janssen et al., Arzneim.-Forsch., 1968, 18, 261, 279, 282.

PINACIDIL [U;INN]

CAS	60560-33-0	MW:	245.33	MF:	C13 H19 N5

Antihypertensive

Leo, Den.; Lilly (Pindac)

LgP	1.93(C)
pKa	
pp cited in Vols.1-5:	

(1) H.J. Petersen et al., J. Med. Chem., 1978, 21, 773.

PINADOLINE [U;INN]

CAS	38955-22-5	MW:	408.29	MF:	C19 H19 Cl2 N3 O3

Analgesic

Searle

LgP	(4)
pKa	
pp cited in Vols.1-5:	

(1) Clin. Pharmacol. Ther., 1984, 36, 145.

PINAFIDE [INN]

CAS	54824-20-3	MW:	339.35	MF:	C18 H17 N3 O4

Antineoplastic

Made, Spain (Pinafide)

LgP	3.14(C)
pKa	
pp cited in Vols.1-5:	

(1) RTECS, 1985-86, entry 50980.

PINAVERIUM BROMIDE [INN]

CAS	53251-94-8	MW:	591.44	MF:	C26 H41 Br2 N O4

Anticholinergic

LgP	(4)
pKa	
pp cited in Vols.1-5:	

(1) R. Baronnet, U.S. Pat. 3 845 048 (1974).
(2) R. Baronnet et al., Eur. J. Med. Chem., 1974, 9, 182.

PINAZEPAM [INN]

CAS	52463-83-9	MW:	308.77	MF:	C18 H13 Cl N2 O

Antidepressant

LgP	3.4(C,6)
pKa	
pp cited in Vols.1-5:	

(1) C. Podesva et al., U.S. Pat. 3 842 094 (1974).
(2) F. Scrollini et al., Arzneim.-Forsch., 1975, 25, 934.

PINCAINIDE [INN]

CAS	83471-41-4	MW:	260.38	MF:	C16 H24 N2 O

Anesthetic (local); calcium antagonist

IQB, Spain

LgP	2.46(C)
pKa	
pp cited in Vols.1-5:	

(1) Eur. J. Pharmacol., 1985, 111, 227.

PINDOLOL [U;INN]

CAS	13523-86-9	MW:	248.33	MF:	C14 H20 N2 O2

Anti-adrenergic (beta-blocker)

Sandoz (Visken)

LgP	1.75(M); 1.65(C)
pKa	8.8; 9.7
pp cited in Vols.1-5:	

1: 213, 273;
3: 209, 210, 215, 220, 573, 583, 585, 587;
5: 101, 191, 356

(1) Troxler, Swiss Pat. 472 404 (1969).
(2) Am. Heart J., 1982, 104 (Suppl.2, pt.2), pp 333-520.

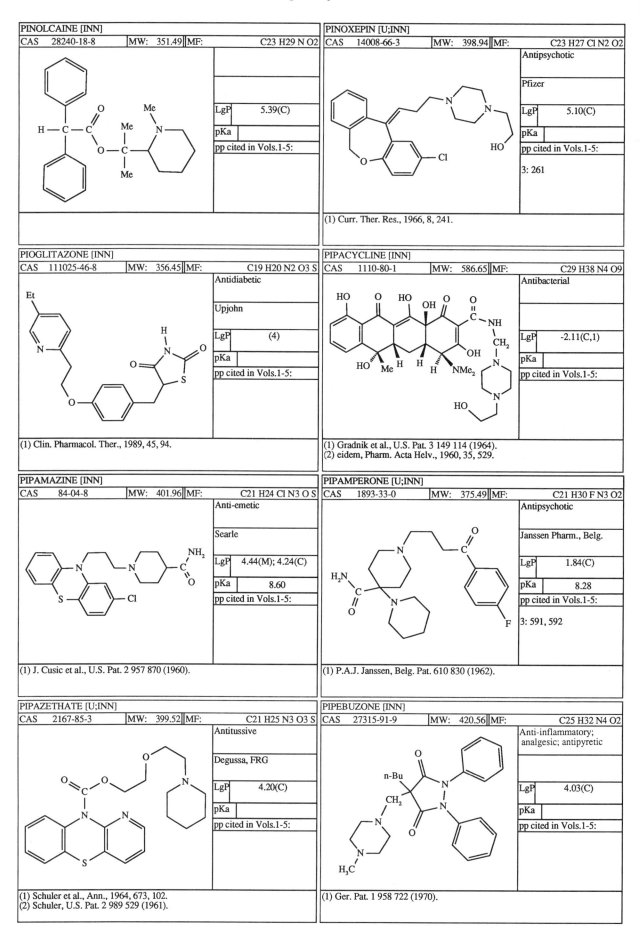

PINOLCAINE [INN]		
CAS 28240-18-8	MW: 351.49	MF: C23 H29 N O2

LgP	5.39(C)
pKa	
pp cited in Vols.1-5:	

PINOXEPIN [U;INN]		
CAS 14008-66-3	MW: 398.94	MF: C23 H27 Cl N2 O2

Antipsychotic

Pfizer

LgP	5.10(C)
pKa	
pp cited in Vols.1-5:	
3: 261	

(1) Curr. Ther. Res., 1966, 8, 241.

PIOGLITAZONE [INN]		
CAS 111025-46-8	MW: 356.45	MF: C19 H20 N2 O3 S

Antidiabetic

Upjohn

LgP	(4)
pKa	
pp cited in Vols.1-5:	

(1) Clin. Pharmacol. Ther., 1989, 45, 94.

PIPACYCLINE [INN]		
CAS 1110-80-1	MW: 586.65	MF: C29 H38 N4 O9

Antibacterial

LgP	-2.11(C,1)
pKa	
pp cited in Vols.1-5:	

(1) Gradnik et al., U.S. Pat. 3 149 114 (1964).
(2) eidem, Pharm. Acta Helv., 1960, 35, 529.

PIPAMAZINE [INN]		
CAS 84-04-8	MW: 401.96	MF: C21 H24 Cl N3 O S

Anti-emetic

Searle

LgP	4.44(M); 4.24(C)
pKa	8.60
pp cited in Vols.1-5:	

(1) J. Cusic et al., U.S. Pat. 2 957 870 (1960).

PIPAMPERONE [U;INN]		
CAS 1893-33-0	MW: 375.49	MF: C21 H30 F N3 O2

Antipsychotic

Janssen Pharm., Belg.

LgP	1.84(C)
pKa	8.28
pp cited in Vols.1-5:	
3: 591, 592	

(1) P.A.J. Janssen, Belg. Pat. 610 830 (1962).

PIPAZETHATE [U;INN]		
CAS 2167-85-3	MW: 399.52	MF: C21 H25 N3 O3 S

Antitussive

Degussa, FRG

LgP	4.20(C)
pKa	
pp cited in Vols.1-5:	

(1) Schuler et al., Ann., 1964, 673, 102.
(2) Schuler, U.S. Pat. 2 989 529 (1961).

PIPEBUZONE [INN]		
CAS 27315-91-9	MW: 420.56	MF: C25 H32 N4 O2

Anti-inflammatory; analgesic; antipyretic

LgP	4.03(C)
pKa	
pp cited in Vols.1-5:	

(1) Ger. Pat. 1 958 722 (1970).

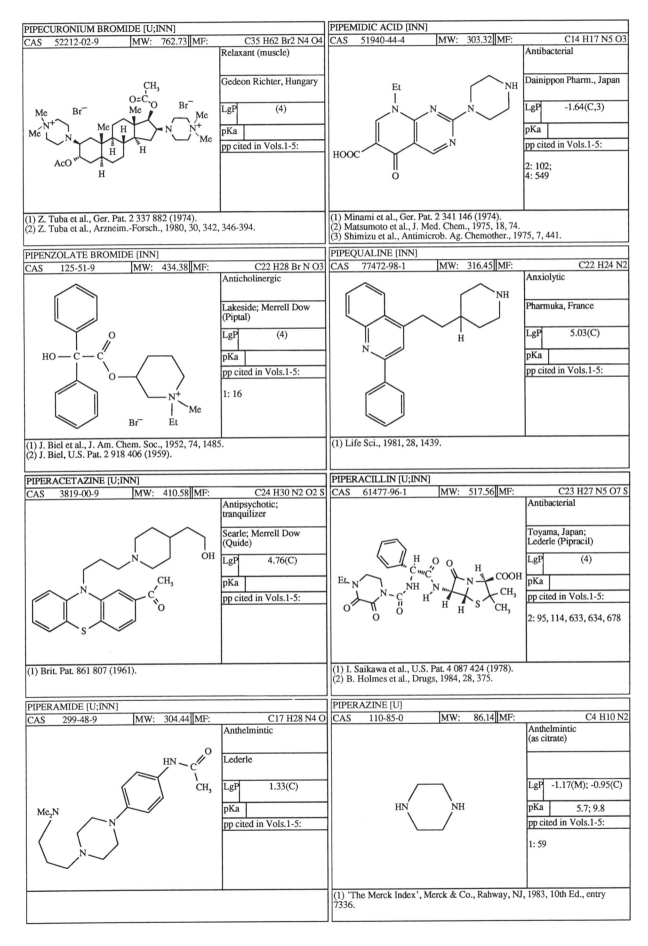

PIPECURONIUM BROMIDE [U;INN]

CAS	52212-02-9	MW:	762.73	MF:	C35 H62 Br2 N4 O4

Relaxant (muscle)

Gedeon Richter, Hungary

LgP	(4)
pKa	
pp cited in Vols.1-5:	

(1) Z. Tuba et al., Ger. Pat. 2 337 882 (1974).
(2) Z. Tuba et al., Arzneim.-Forsch., 1980, 30, 342, 346-394.

PIPEMIDIC ACID [INN]

CAS	51940-44-4	MW:	303.32	MF:	C14 H17 N5 O3

Antibacterial

Dainippon Pharm., Japan

LgP	-1.64(C,3)
pKa	
pp cited in Vols.1-5:	

2: 102;
4: 549

(1) Minami et al., Ger. Pat. 2 341 146 (1974).
(2) Matsumoto et al., J. Med. Chem., 1975, 18, 74.
(3) Shimizu et al., Antimicrob. Ag. Chemother., 1975, 7, 441.

PIPENZOLATE BROMIDE [INN]

CAS	125-51-9	MW:	434.38	MF:	C22 H28 Br N O3

Anticholinergic

Lakeside; Merrell Dow (Piptal)

LgP	(4)
pKa	
pp cited in Vols.1-5:	

1: 16

(1) J. Biel et al., J. Am. Chem. Soc., 1952, 74, 1485.
(2) J. Biel, U.S. Pat. 2 918 406 (1959).

PIPEQUALINE [INN]

CAS	77472-98-1	MW:	316.45	MF:	C22 H24 N2

Anxiolytic

Pharmuka, France

LgP	5.03(C)
pKa	
pp cited in Vols.1-5:	

(1) Life Sci., 1981, 28, 1439.

PIPERACETAZINE [U;INN]

CAS	3819-00-9	MW:	410.58	MF:	C24 H30 N2 O2 S

Antipsychotic; tranquilizer

Searle; Merrell Dow (Quide)

LgP	4.76(C)
pKa	
pp cited in Vols.1-5:	

(1) Brit. Pat. 861 807 (1961).

PIPERACILLIN [U;INN]

CAS	61477-96-1	MW:	517.56	MF:	C23 H27 N5 O7 S

Antibacterial

Toyama, Japan; Lederle (Pipracil)

LgP	(4)
pKa	
pp cited in Vols.1-5:	

2: 95, 114, 633, 634, 678

(1) I. Saikawa et al., U.S. Pat. 4 087 424 (1978).
(2) B. Holmes et al., Drugs, 1984, 28, 375.

PIPERAMIDE [U;INN]

CAS	299-48-9	MW:	304.44	MF:	C17 H28 N4 O

Anthelmintic

Lederle

LgP	1.33(C)
pKa	
pp cited in Vols.1-5:	

PIPERAZINE [U]

CAS	110-85-0	MW:	86.14	MF:	C4 H10 N2

Anthelmintic (as citrate)

LgP	-1.17(M); -0.95(C)
pKa	5.7; 9.8
pp cited in Vols.1-5:	

1: 59

(1) 'The Merck Index', Merck & Co., Rahway, NJ, 1983, 10th Ed., entry 7336.

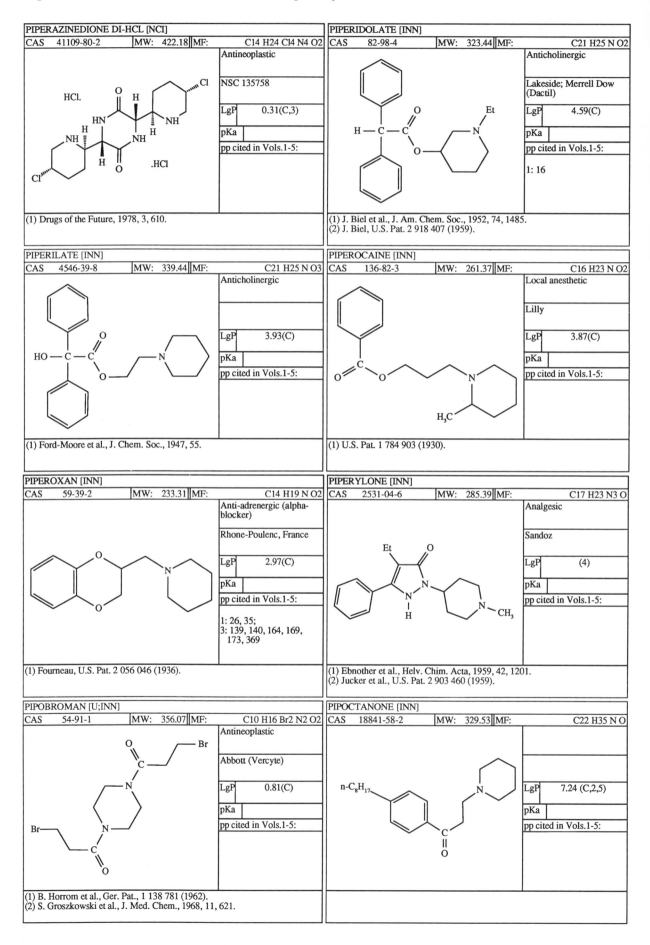

PIPERAZINEDIONE DI-HCL [NCI]

CAS	41109-80-2	MW:	422.18	MF:	C14 H24 Cl4 N4 O2

Antineoplastic

NSC 135758

LgP	0.31(C,3)
pKa	

pp cited in Vols.1-5:

(1) Drugs of the Future, 1978, 3, 610.

PIPERIDOLATE [INN]

CAS	82-98-4	MW:	323.44	MF:	C21 H25 N O2

Anticholinergic

Lakeside; Merrell Dow (Dactil)

LgP	4.59(C)
pKa	

pp cited in Vols.1-5:

1: 16

(1) J. Biel et al., J. Am. Chem. Soc., 1952, 74, 1485.
(2) J. Biel, U.S. Pat. 2 918 407 (1959).

PIPERILATE [INN]

CAS	4546-39-8	MW:	339.44	MF:	C21 H25 N O3

Anticholinergic

LgP	3.93(C)
pKa	

pp cited in Vols.1-5:

(1) Ford-Moore et al., J. Chem. Soc., 1947, 55.

PIPEROCAINE [INN]

CAS	136-82-3	MW:	261.37	MF:	C16 H23 N O2

Local anesthetic

Lilly

LgP	3.87(C)
pKa	

pp cited in Vols.1-5:

(1) U.S. Pat. 1 784 903 (1930).

PIPEROXAN [INN]

CAS	59-39-2	MW:	233.31	MF:	C14 H19 N O2

Anti-adrenergic (alpha-blocker)

Rhone-Poulenc, France

LgP	2.97(C)
pKa	

pp cited in Vols.1-5:

1: 26, 35;
3: 139, 140, 164, 169, 173, 369

(1) Fourneau, U.S. Pat. 2 056 046 (1936).

PIPERYLONE [INN]

CAS	2531-04-6	MW:	285.39	MF:	C17 H23 N3 O

Analgesic

Sandoz

LgP	(4)
pKa	

pp cited in Vols.1-5:

(1) Ebnother et al., Helv. Chim. Acta, 1959, 42, 1201.
(2) Jucker et al., U.S. Pat. 2 903 460 (1959).

PIPOBROMAN [U;INN]

CAS	54-91-1	MW:	356.07	MF:	C10 H16 Br2 N2 O2

Antineoplastic

Abbott (Vercyte)

LgP	0.81(C)
pKa	

pp cited in Vols.1-5:

(1) B. Horrom et al., Ger. Pat., 1 138 781 (1962).
(2) S. Groszkowski et al., J. Med. Chem., 1968, 11, 621.

PIPOCTANONE [INN]

CAS	18841-58-2	MW:	329.53	MF:	C22 H35 N O

LgP	7.24 (C,2,5)
pKa	

pp cited in Vols.1-5:

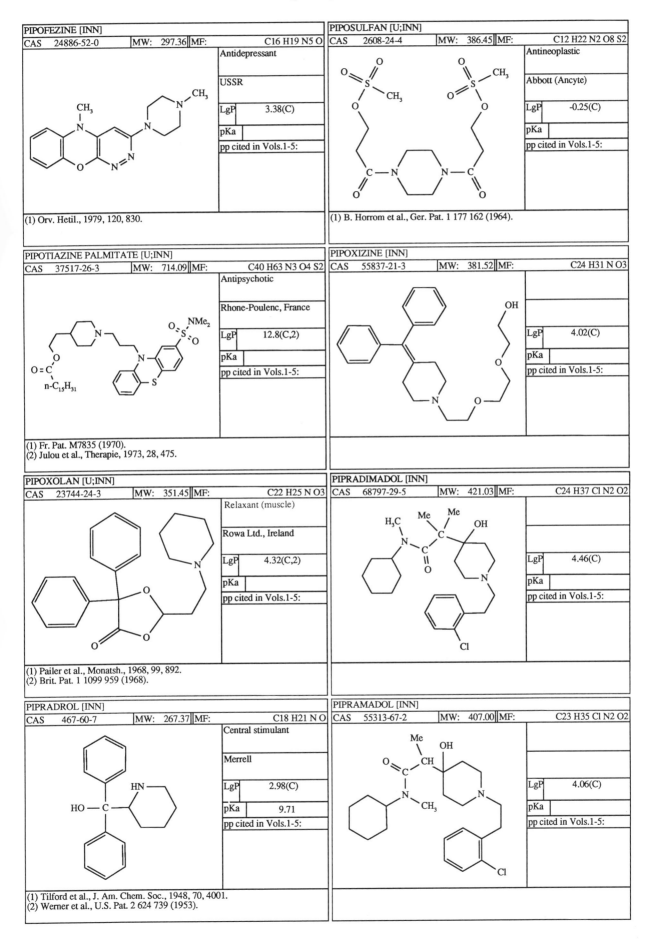

PIPOFEZINE [INN]

| CAS | 24886-52-0 | MW: | 297.36 | MF: | C16 H19 N5 O |

Antidepressant

USSR

LgP	3.38(C)
pKa	
pp cited in Vols.1-5:	

(1) Orv. Hetil., 1979, 120, 830.

PIPOSULFAN [U;INN]

| CAS | 2608-24-4 | MW: | 386.45 | MF: | C12 H22 N2 O8 S2 |

Antineoplastic

Abbott (Ancyte)

LgP	-0.25(C)
pKa	
pp cited in Vols.1-5:	

(1) B. Horrom et al., Ger. Pat. 1 177 162 (1964).

PIPOTIAZINE PALMITATE [U;INN]

| CAS | 37517-26-3 | MW: | 714.09 | MF: | C40 H63 N3 O4 S2 |

Antipsychotic

Rhone-Poulenc, France

LgP	12.8(C,2)
pKa	
pp cited in Vols.1-5:	

(1) Fr. Pat. M7835 (1970).
(2) Julou et al., Therapie, 1973, 28, 475.

PIPOXIZINE [INN]

| CAS | 55837-21-3 | MW: | 381.52 | MF: | C24 H31 N O3 |

LgP	4.02(C)
pKa	
pp cited in Vols.1-5:	

PIPOXOLAN [U;INN]

| CAS | 23744-24-3 | MW: | 351.45 | MF: | C22 H25 N O3 |

Relaxant (muscle)

Rowa Ltd., Ireland

LgP	4.32(C,2)
pKa	
pp cited in Vols.1-5:	

(1) Pailer et al., Monatsh., 1968, 99, 892.
(2) Brit. Pat. 1 1099 959 (1968).

PIPRADIMADOL [INN]

| CAS | 68797-29-5 | MW: | 421.03 | MF: | C24 H37 Cl N2 O2 |

LgP	4.46(C)
pKa	
pp cited in Vols.1-5:	

PIPRADROL [INN]

| CAS | 467-60-7 | MW: | 267.37 | MF: | C18 H21 N O |

Central stimulant

Merrell

LgP	2.98(C)
pKa	9.71
pp cited in Vols.1-5:	

(1) Tilford et al., J. Am. Chem. Soc., 1948, 70, 4001.
(2) Werner et al., U.S. Pat. 2 624 739 (1953).

PIPRAMADOL [INN]

| CAS | 55313-67-2 | MW: | 407.00 | MF: | C23 H35 Cl N2 O2 |

LgP	4.06(C)
pKa	
pp cited in Vols.1-5:	

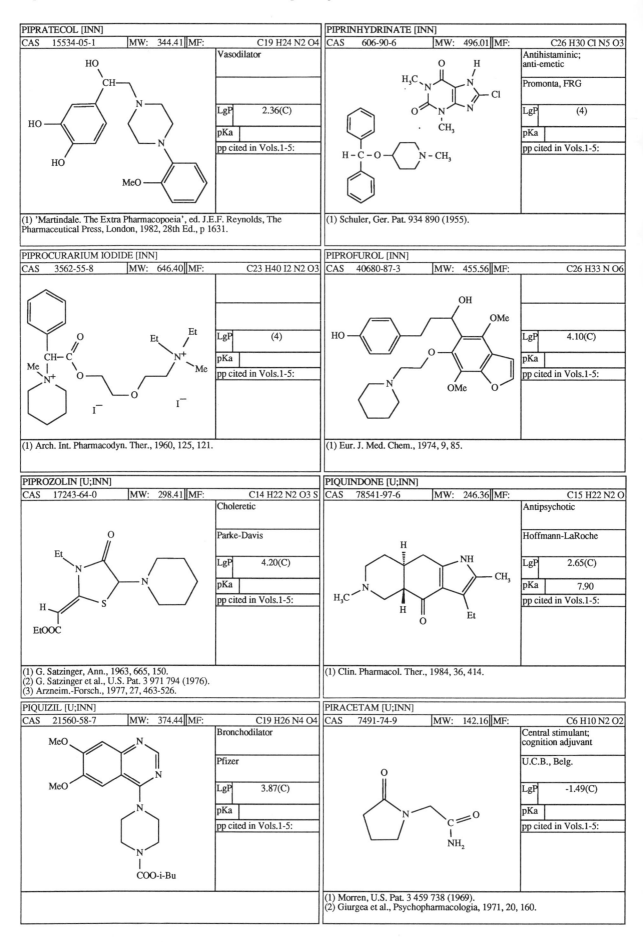

PIPRATECOL [INN]

CAS	15534-05-1	MW:	344.41	MF:	C19 H24 N2 O4

Vasodilator

LgP	2.36(C)
pKa	
pp cited in Vols.1-5:	

(1) 'Martindale. The Extra Pharmacopoeia', ed. J.E.F. Reynolds, The Pharmaceutical Press, London, 1982, 28th Ed., p 1631.

PIPRINHYDRINATE [INN]

CAS	606-90-6	MW:	496.01	MF:	C26 H30 Cl N5 O3

Antihistaminic; anti-emetic

Promonta, FRG

LgP	(4)
pKa	
pp cited in Vols.1-5:	

(1) Schuler, Ger. Pat. 934 890 (1955).

PIPROCURARIUM IODIDE [INN]

CAS	3562-55-8	MW:	646.40	MF:	C23 H40 I2 N2 O3

LgP	(4)
pKa	
pp cited in Vols.1-5:	

(1) Arch. Int. Pharmacodyn. Ther., 1960, 125, 121.

PIPROFUROL [INN]

CAS	40680-87-3	MW:	455.56	MF:	C26 H33 N O6

LgP	4.10(C)
pKa	
pp cited in Vols.1-5:	

(1) Eur. J. Med. Chem., 1974, 9, 85.

PIPROZOLIN [U;INN]

CAS	17243-64-0	MW:	298.41	MF:	C14 H22 N2 O3 S

Choleretic

Parke-Davis

LgP	4.20(C)
pKa	
pp cited in Vols.1-5:	

(1) G. Satzinger, Ann., 1963, 665, 150.
(2) G. Satzinger et al., U.S. Pat. 3 971 794 (1976).
(3) Arzneim.-Forsch., 1977, 27, 463-526.

PIQUINDONE [U;INN]

CAS	78541-97-6	MW:	246.36	MF:	C15 H22 N2 O

Antipsychotic

Hoffmann-LaRoche

LgP	2.65(C)
pKa	7.90
pp cited in Vols.1-5:	

(1) Clin. Pharmacol. Ther., 1984, 36, 414.

PIQUIZIL [U;INN]

CAS	21560-58-7	MW:	374.44	MF:	C19 H26 N4 O4

Bronchodilator

Pfizer

LgP	3.87(C)
pKa	
pp cited in Vols.1-5:	

PIRACETAM [U;INN]

CAS	7491-74-9	MW:	142.16	MF:	C6 H10 N2 O2

Central stimulant; cognition adjuvant

U.C.B., Belg.

LgP	-1.49(C)
pKa	
pp cited in Vols.1-5:	

(1) Morren, U.S. Pat. 3 459 738 (1969).
(2) Giurgea et al., Psychopharmacologia, 1971, 20, 160.

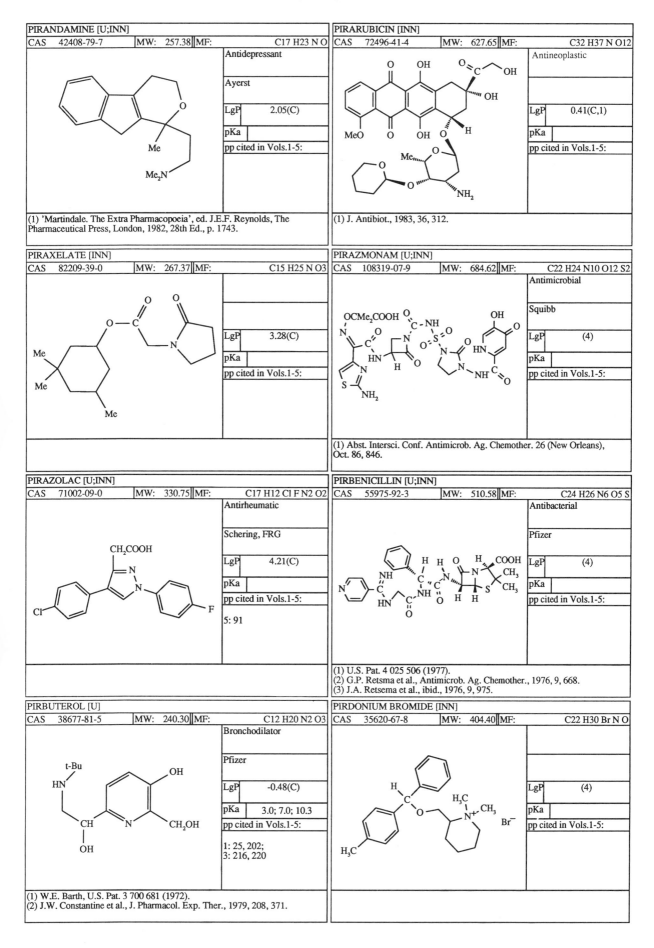

PIRANDAMINE [U;INN]

CAS	42408-79-7	MW:	257.38	MF:	C17 H23 N O

Antidepressant

Ayerst

LgP	2.05(C)
pKa	
pp cited in Vols.1-5:	

(1) 'Martindale. The Extra Pharmacopoeia', ed. J.E.F. Reynolds, The Pharmaceutical Press, London, 1982, 28th Ed., p. 1743.

PIRARUBICIN [INN]

CAS	72496-41-4	MW:	627.65	MF:	C32 H37 N O12

Antineoplastic

LgP	0.41(C,1)
pKa	
pp cited in Vols.1-5:	

(1) J. Antibiot., 1983, 36, 312.

PIRAXELATE [INN]

CAS	82209-39-0	MW:	267.37	MF:	C15 H25 N O3

LgP	3.28(C)
pKa	
pp cited in Vols.1-5:	

PIRAZMONAM [U;INN]

CAS	108319-07-9	MW:	684.62	MF:	C22 H24 N10 O12 S2

Antimicrobial

Squibb

LgP	(4)
pKa	
pp cited in Vols.1-5:	

(1) Abst. Intersci. Conf. Antimicrob. Ag. Chemother. 26 (New Orleans), Oct. 86, 846.

PIRAZOLAC [U;INN]

CAS	71002-09-0	MW:	330.75	MF:	C17 H12 Cl F N2 O2

Antirheumatic

Schering, FRG

LgP	4.21(C)
pKa	
pp cited in Vols.1-5:	
5: 91	

PIRBENICILLIN [U;INN]

CAS	55975-92-3	MW:	510.58	MF:	C24 H26 N6 O5 S

Antibacterial

Pfizer

LgP	(4)
pKa	
pp cited in Vols.1-5:	

(1) U.S. Pat. 4 025 506 (1977).
(2) G.P. Retsma et al., Antimicrob. Ag. Chemother., 1976, 9, 668.
(3) J.A. Retsema et al., ibid., 1976, 9, 975.

PIRBUTEROL [U]

CAS	38677-81-5	MW:	240.30	MF:	C12 H20 N2 O3

Bronchodilator

Pfizer

LgP	-0.48(C)
pKa	3.0; 7.0; 10.3
pp cited in Vols.1-5:	
1: 25, 202; 3: 216, 220	

(1) W.E. Barth, U.S. Pat. 3 700 681 (1972).
(2) J.W. Constantine et al., J. Pharmacol. Exp. Ther., 1979, 208, 371.

PIRDONIUM BROMIDE [INN]

CAS	35620-67-8	MW:	404.40	MF:	C22 H30 Br N O

LgP	(4)
pKa	
pp cited in Vols.1-5:	

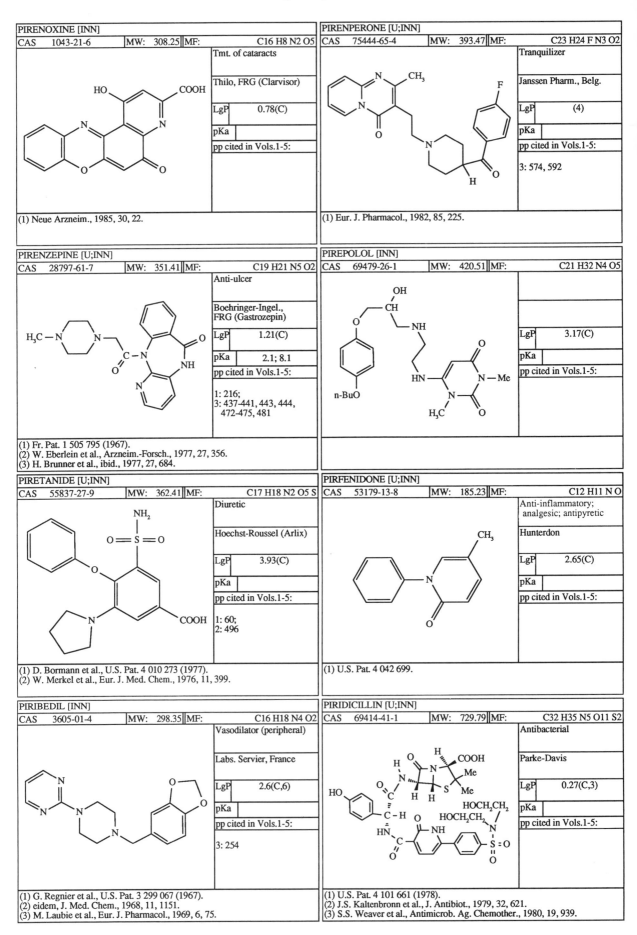

PIRENOXINE [INN]		
CAS 1043-21-6	MW: 308.25 MF:	C16 H8 N2 O5
	Tmt. of cataracts	
	Thilo, FRG (Clarvisor)	
	LgP	0.78(C)
	pKa	
	pp cited in Vols.1-5:	

(1) Neue Arzneim., 1985, 30, 22.

PIRENPERONE [U;INN]		
CAS 75444-65-4	MW: 393.47 MF:	C23 H24 F N3 O2
	Tranquilizer	
	Janssen Pharm., Belg.	
	LgP	(4)
	pKa	
	pp cited in Vols.1-5:	
	3: 574, 592	

(1) Eur. J. Pharmacol., 1982, 85, 225.

PIRENZEPINE [U;INN]		
CAS 28797-61-7	MW: 351.41 MF:	C19 H21 N5 O2
	Anti-ulcer	
	Boehringer-Ingel., FRG (Gastrozepin)	
	LgP	1.21(C)
	pKa	2.1; 8.1
	pp cited in Vols.1-5:	
	1: 216; 3: 437-441, 443, 444, 472-475, 481	

(1) Fr. Pat. 1 505 795 (1967).
(2) W. Eberlein et al., Arzneim.-Forsch., 1977, 27, 356.
(3) H. Brunner et al., ibid., 1977, 27, 684.

PIREPOLOL [INN]		
CAS 69479-26-1	MW: 420.51 MF:	C21 H32 N4 O5
	LgP	3.17(C)
	pKa	
	pp cited in Vols.1-5:	

PIRETANIDE [U;INN]		
CAS 55837-27-9	MW: 362.41 MF:	C17 H18 N2 O5 S
	Diuretic	
	Hoechst-Roussel (Arlix)	
	LgP	3.93(C)
	pKa	
	pp cited in Vols.1-5:	
	1: 60; 2: 496	

(1) D. Bormann et al., U.S. Pat. 4 010 273 (1977).
(2) W. Merkel et al., Eur. J. Med. Chem., 1976, 11, 399.

PIRFENIDONE [U;INN]		
CAS 53179-13-8	MW: 185.23 MF:	C12 H11 N O
	Anti-inflammatory; analgesic; antipyretic	
	Hunterdon	
	LgP	2.65(C)
	pKa	
	pp cited in Vols.1-5:	

(1) U.S. Pat. 4 042 699.

PIRIBEDIL [INN]		
CAS 3605-01-4	MW: 298.35 MF:	C16 H18 N4 O2
	Vasodilator (peripheral)	
	Labs. Servier, France	
	LgP	2.6(C,6)
	pKa	
	pp cited in Vols.1-5:	
	3: 254	

(1) G. Regnier et al., U.S. Pat. 3 299 067 (1967).
(2) eidem, J. Med. Chem., 1968, 11, 1151.
(3) M. Laubie et al., Eur. J. Pharmacol., 1969, 6, 75.

PIRIDICILLIN [U;INN]		
CAS 69414-41-1	MW: 729.79 MF:	C32 H35 N5 O11 S2
	Antibacterial	
	Parke-Davis	
	LgP	0.27(C,3)
	pKa	
	pp cited in Vols.1-5:	

(1) U.S. Pat. 4 101 661 (1978).
(2) J.S. Kaltenbronn et al., J. Antibiot., 1979, 32, 621.
(3) S.S. Weaver et al., Antimicrob. Ag. Chemother., 1980, 19, 939.

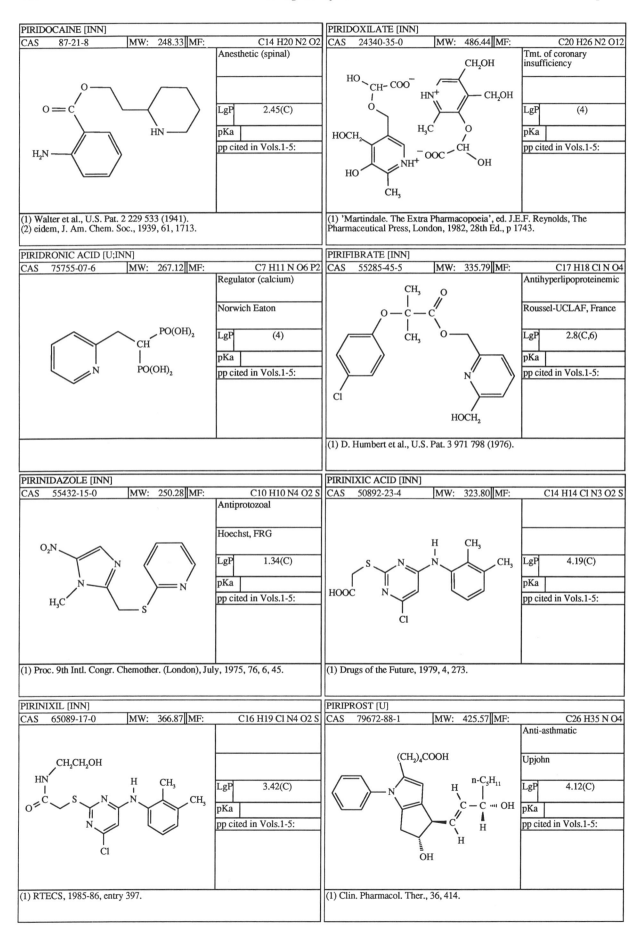

PIRIDOCAINE [INN]

CAS 87-21-8	MW: 248.33	MF:	C14 H20 N2 O2

Anesthetic (spinal)

LgP	2.45(C)
pKa	
pp cited in Vols.1-5:	

(1) Walter et al., U.S. Pat. 2 229 533 (1941).
(2) eidem, J. Am. Chem. Soc., 1939, 61, 1713.

PIRIDOXILATE [INN]

CAS 24340-35-0	MW: 486.44	MF:	C20 H26 N2 O12

Tmt. of coronary insufficiency

LgP	(4)
pKa	
pp cited in Vols.1-5:	

(1) 'Martindale. The Extra Pharmacopoeia', ed. J.E.F. Reynolds, The Pharmaceutical Press, London, 1982, 28th Ed., p 1743.

PIRIDRONIC ACID [U;INN]

CAS 75755-07-6	MW: 267.12	MF:	C7 H11 N O6 P2

Regulator (calcium)

Norwich Eaton

LgP	(4)
pKa	
pp cited in Vols.1-5:	

PIRIFIBRATE [INN]

CAS 55285-45-5	MW: 335.79	MF:	C17 H18 Cl N O4

Antihyperlipoproteinemic

Roussel-UCLAF, France

LgP	2.8(C,6)
pKa	
pp cited in Vols.1-5:	

(1) D. Humbert et al., U.S. Pat. 3 971 798 (1976).

PIRINIDAZOLE [INN]

CAS 55432-15-0	MW: 250.28	MF:	C10 H10 N4 O2 S

Antiprotozoal

Hoechst, FRG

LgP	1.34(C)
pKa	
pp cited in Vols.1-5:	

(1) Proc. 9th Intl. Congr. Chemother. (London), July, 1975, 76, 6, 45.

PIRINIXIC ACID [INN]

CAS 50892-23-4	MW: 323.80	MF:	C14 H14 Cl N3 O2 S

LgP	4.19(C)
pKa	
pp cited in Vols.1-5:	

(1) Drugs of the Future, 1979, 4, 273.

PIRINIXIL [INN]

CAS 65089-17-0	MW: 366.87	MF:	C16 H19 Cl N4 O2 S

LgP	3.42(C)
pKa	
pp cited in Vols.1-5:	

(1) RTECS, 1985-86, entry 397.

PIRIPROST [U]

CAS 79672-88-1	MW: 425.57	MF:	C26 H35 N O4

Anti-asthmatic

Upjohn

LgP	4.12(C)
pKa	
pp cited in Vols.1-5:	

(1) Clin. Pharmacol. Ther., 36, 414.

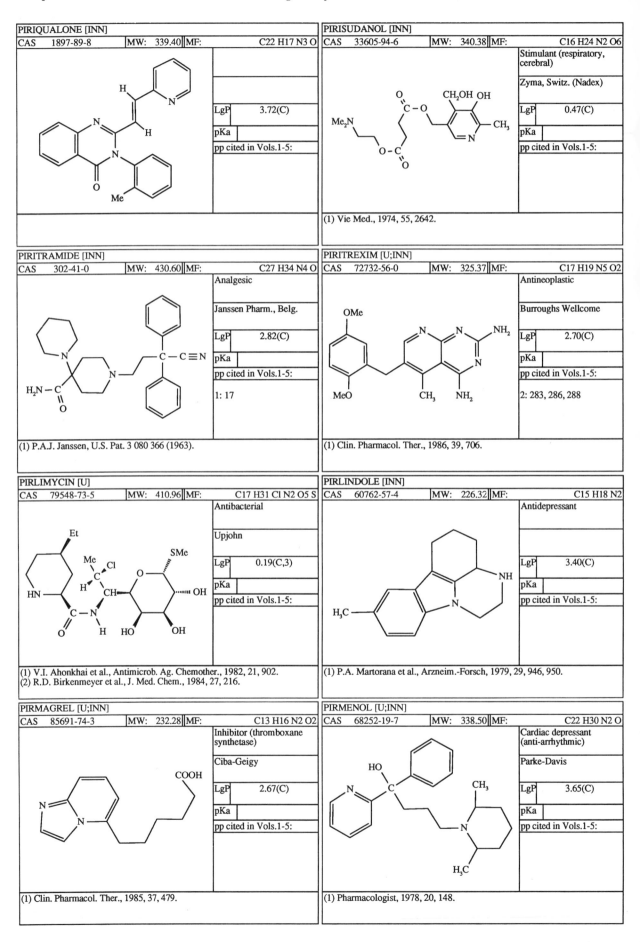

PIRIQUALONE [INN]			
CAS 1897-89-8	MW: 339.40	MF:	C22 H17 N3 O
		LgP	3.72(C)
		pKa	
		pp cited in Vols.1-5:	

PIRISUDANOL [INN]			
CAS 33605-94-6	MW: 340.38	MF:	C16 H24 N2 O6
		Stimulant (respiratory, cerebral)	
		Zyma, Switz. (Nadex)	
		LgP	0.47(C)
		pKa	
		pp cited in Vols.1-5:	

(1) Vie Med., 1974, 55, 2642.

PIRITRAMIDE [INN]			
CAS 302-41-0	MW: 430.60	MF:	C27 H34 N4 O
		Analgesic	
		Janssen Pharm., Belg.	
		LgP	2.82(C)
		pKa	
		pp cited in Vols.1-5:	
		1: 17	

(1) P.A.J. Janssen, U.S. Pat. 3 080 366 (1963).

PIRITREXIM [U;INN]			
CAS 72732-56-0	MW: 325.37	MF:	C17 H19 N5 O2
		Antineoplastic	
		Burroughs Wellcome	
		LgP	2.70(C)
		pKa	
		pp cited in Vols.1-5:	
		2: 283, 286, 288	

(1) Clin. Pharmacol. Ther., 1986, 39, 706.

PIRLIMYCIN [U]			
CAS 79548-73-5	MW: 410.96	MF:	C17 H31 Cl N2 O5 S
		Antibacterial	
		Upjohn	
		LgP	0.19(C,3)
		pKa	
		pp cited in Vols.1-5:	

(1) V.I. Ahonkhai et al., Antimicrob. Ag. Chemother., 1982, 21, 902.
(2) R.D. Birkenmeyer et al., J. Med. Chem., 1984, 27, 216.

PIRLINDOLE [INN]			
CAS 60762-57-4	MW: 226.32	MF:	C15 H18 N2
		Antidepressant	
		LgP	3.40(C)
		pKa	
		pp cited in Vols.1-5:	

(1) P.A. Martorana et al., Arzneim.-Forsch, 1979, 29, 946, 950.

PIRMAGREL [U;INN]			
CAS 85691-74-3	MW: 232.28	MF:	C13 H16 N2 O2
		Inhibitor (thromboxane synthetase)	
		Ciba-Geigy	
		LgP	2.67(C)
		pKa	
		pp cited in Vols.1-5:	

(1) Clin. Pharmacol. Ther., 1985, 37, 479.

PIRMENOL [U;INN]			
CAS 68252-19-7	MW: 338.50	MF:	C22 H30 N2 O
		Cardiac depressant (anti-arrhythmic)	
		Parke-Davis	
		LgP	3.65(C)
		pKa	
		pp cited in Vols.1-5:	

(1) Pharmacologist, 1978, 20, 148.

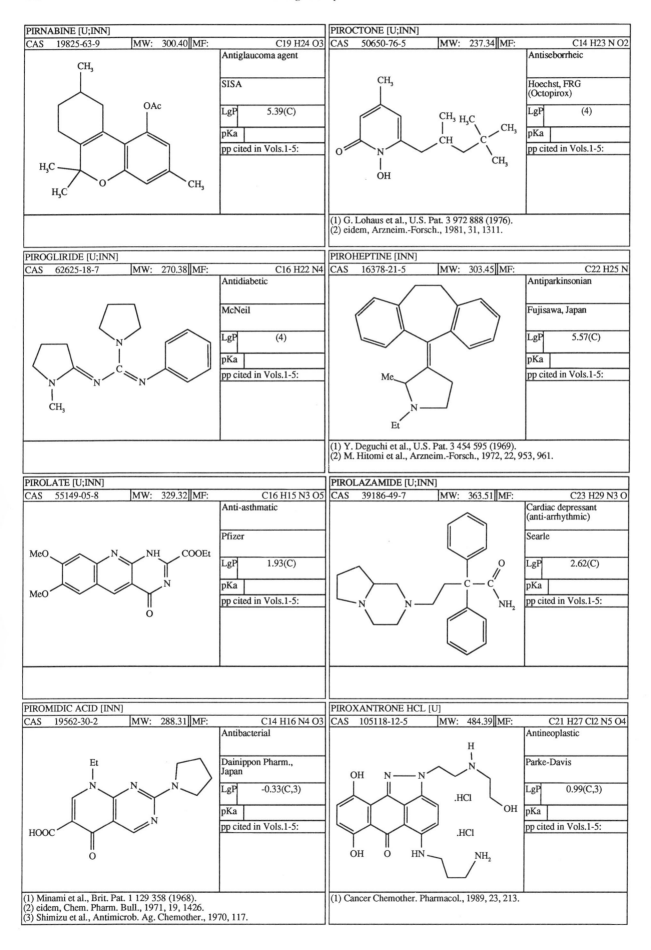

PIRNABINE [U;INN]

CAS 19825-63-9	MW: 300.40	MF:	C19 H24 O3

Antiglaucoma agent

SISA

LgP	5.39(C)
pKa	
pp cited in Vols.1-5:	

PIROCTONE [U;INN]

CAS 50650-76-5	MW: 237.34	MF:	C14 H23 N O2

Antiseborrheic

Hoechst, FRG
(Octopirox)

LgP	(4)
pKa	
pp cited in Vols.1-5:	

(1) G. Lohaus et al., U.S. Pat. 3 972 888 (1976).
(2) eidem, Arzneim.-Forsch., 1981, 31, 1311.

PIROGLIRIDE [U;INN]

CAS 62625-18-7	MW: 270.38	MF:	C16 H22 N4

Antidiabetic

McNeil

LgP	(4)
pKa	
pp cited in Vols.1-5:	

PIROHEPTINE [INN]

CAS 16378-21-5	MW: 303.45	MF:	C22 H25 N

Antiparkinsonian

Fujisawa, Japan

LgP	5.57(C)
pKa	
pp cited in Vols.1-5:	

(1) Y. Deguchi et al., U.S. Pat. 3 454 595 (1969).
(2) M. Hitomi et al., Arzneim.-Forsch., 1972, 22, 953, 961.

PIROLATE [U;INN]

CAS 55149-05-8	MW: 329.32	MF:	C16 H15 N3 O5

Anti-asthmatic

Pfizer

LgP	1.93(C)
pKa	
pp cited in Vols.1-5:	

PIROLAZAMIDE [U;INN]

CAS 39186-49-7	MW: 363.51	MF:	C23 H29 N3 O

Cardiac depressant
(anti-arrhythmic)

Searle

LgP	2.62(C)
pKa	
pp cited in Vols.1-5:	

PIROMIDIC ACID [INN]

CAS 19562-30-2	MW: 288.31	MF:	C14 H16 N4 O3

Antibacterial

Dainippon Pharm.,
Japan

LgP	-0.33(C,3)
pKa	
pp cited in Vols.1-5:	

(1) Minami et al., Brit. Pat. 1 129 358 (1968).
(2) eidem, Chem. Pharm. Bull., 1971, 19, 1426.
(3) Shimizu et al., Antimicrob. Ag. Chemother., 1970, 117.

PIROXANTRONE HCL [U]

CAS 105118-12-5	MW: 484.39	MF:	C21 H27 Cl2 N5 O4

Antineoplastic

Parke-Davis

LgP	0.99(C,3)
pKa	
pp cited in Vols.1-5:	

(1) Cancer Chemother. Pharmacol., 1989, 23, 213.

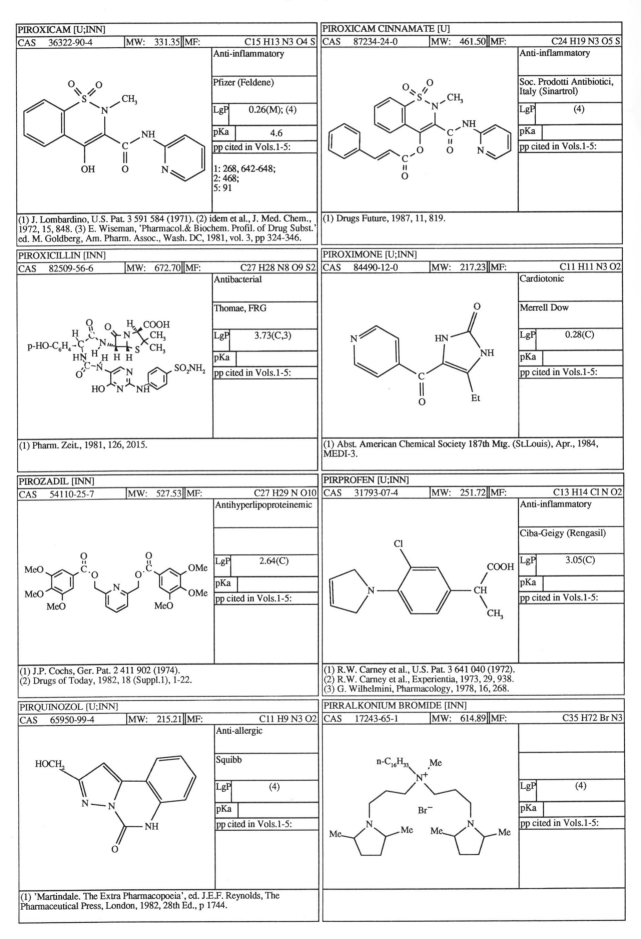

PIROXICAM [U;INN]			
CAS 36322-90-4	MW: 331.35	MF:	C15 H13 N3 O4 S

Anti-inflammatory

Pfizer (Feldene)

LgP	0.26(M); (4)
pKa	4.6
pp cited in Vols.1-5:	

1: 268, 642-648;
2: 468;
5: 91

(1) J. Lombardino, U.S. Pat. 3 591 584 (1971). (2) idem et al., J. Med. Chem., 1972, 15, 848. (3) E. Wiseman, 'Pharmacol.& Biochem. Profil. of Drug Subst.' ed. M. Goldberg, Am. Pharm. Assoc., Wash. DC, 1981, vol. 3, pp 324-346.

PIROXICAM CINNAMATE [U]			
CAS 87234-24-0	MW: 461.50	MF:	C24 H19 N3 O5 S

Anti-inflammatory

Soc. Prodotti Antibiotici, Italy (Sinartrol)

LgP	(4)
pKa	
pp cited in Vols.1-5:	

(1) Drugs Future, 1987, 11, 819.

PIROXICILLIN [INN]			
CAS 82509-56-6	MW: 672.70	MF:	C27 H28 N8 O9 S2

Antibacterial

Thomae, FRG

LgP	3.73(C,3)
pKa	
pp cited in Vols.1-5:	

(1) Pharm. Zeit., 1981, 126, 2015.

PIROXIMONE [U;INN]			
CAS 84490-12-0	MW: 217.23	MF:	C11 H11 N3 O2

Cardiotonic

Merrell Dow

LgP	0.28(C)
pKa	
pp cited in Vols.1-5:	

(1) Abst. American Chemical Society 187th Mtg. (St.Louis), Apr., 1984, MEDI-3.

PIROZADIL [INN]			
CAS 54110-25-7	MW: 527.53	MF:	C27 H29 N O10

Antihyperlipoproteinemic

LgP	2.64(C)
pKa	
pp cited in Vols.1-5:	

(1) J.P. Cochs, Ger. Pat. 2 411 902 (1974).
(2) Drugs of Today, 1982, 18 (Suppl.1), 1-22.

PIRPROFEN [U;INN]			
CAS 31793-07-4	MW: 251.72	MF:	C13 H14 Cl N O2

Anti-inflammatory

Ciba-Geigy (Rengasil)

LgP	3.05(C)
pKa	
pp cited in Vols.1-5:	

(1) R.W. Carney et al., U.S. Pat. 3 641 040 (1972).
(2) R.W. Carney et al., Experientia, 1973, 29, 938.
(3) G. Wilhelmini, Pharmacology, 1978, 16, 268.

PIRQUINOZOL [U;INN]			
CAS 65950-99-4	MW: 215.21	MF:	C11 H9 N3 O2

Anti-allergic

Squibb

LgP	(4)
pKa	
pp cited in Vols.1-5:	

(1) 'Martindale. The Extra Pharmacopoeia', ed. J.E.F. Reynolds, The Pharmaceutical Press, London, 1982, 28th Ed., p 1744.

PIRRALKONIUM BROMIDE [INN]			
CAS 17243-65-1	MW: 614.89	MF:	C35 H72 Br N3

LgP	(4)
pKa	
pp cited in Vols.1-5:	

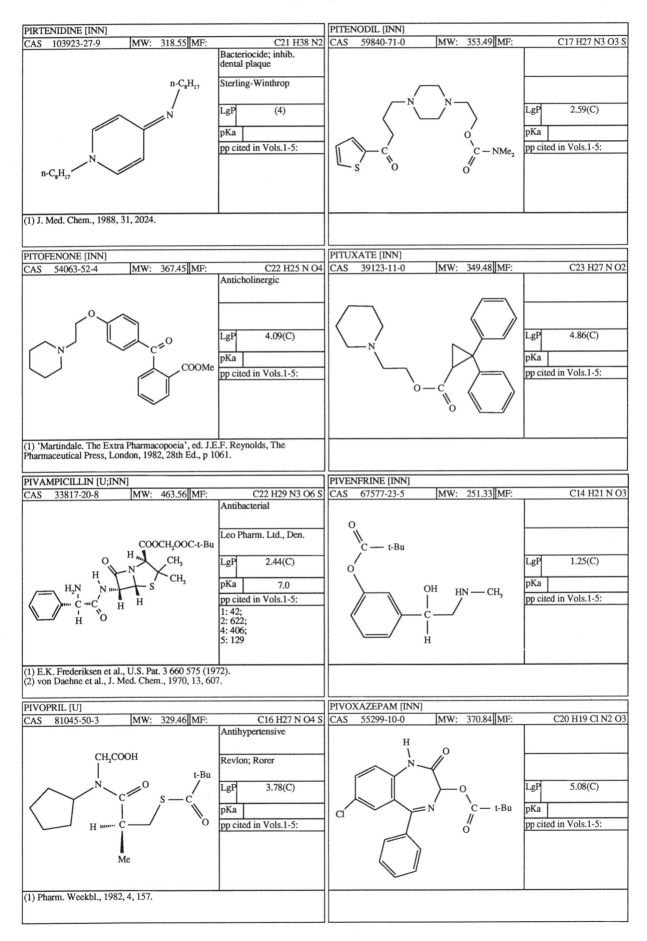

PIRTENIDINE [INN]

CAS	103923-27-9	MW:	318.55	MF:	C21 H38 N2

Bacteriocide; inhib. dental plaque

Sterling-Winthrop

LgP	(4)
pKa	
pp cited in Vols.1-5:	

(1) J. Med. Chem., 1988, 31, 2024.

PITENODIL [INN]

CAS	59840-71-0	MW:	353.49	MF:	C17 H27 N3 O3 S

LgP	2.59(C)
pKa	
pp cited in Vols.1-5:	

PITOFENONE [INN]

CAS	54063-52-4	MW:	367.45	MF:	C22 H25 N O4

Anticholinergic

LgP	4.09(C)
pKa	
pp cited in Vols.1-5:	

(1) 'Martindale. The Extra Pharmacopoeia', ed. J.E.F. Reynolds, The Pharmaceutical Press, London, 1982, 28th Ed., p 1061.

PITUXATE [INN]

CAS	39123-11-0	MW:	349.48	MF:	C23 H27 N O2

LgP	4.86(C)
pKa	
pp cited in Vols.1-5:	

PIVAMPICILLIN [U;INN]

CAS	33817-20-8	MW:	463.56	MF:	C22 H29 N3 O6 S

Antibacterial

Leo Pharm. Ltd., Den.

LgP	2.44(C)
pKa	7.0
pp cited in Vols.1-5:	
1: 42; 2: 622; 4: 406; 5: 129	

(1) E.K. Frederiksen et al., U.S. Pat. 3 660 575 (1972).
(2) von Daehne et al., J. Med. Chem., 1970, 13, 607.

PIVENFRINE [INN]

CAS	67577-23-5	MW:	251.33	MF:	C14 H21 N O3

LgP	1.25(C)
pKa	
pp cited in Vols.1-5:	

PIVOPRIL [U]

CAS	81045-50-3	MW:	329.46	MF:	C16 H27 N O4 S

Antihypertensive

Revlon; Rorer

LgP	3.78(C)
pKa	
pp cited in Vols.1-5:	

(1) Pharm. Weekbl., 1982, 4, 157.

PIVOXAZEPAM [INN]

CAS	55299-10-0	MW:	370.84	MF:	C20 H19 Cl N2 O3

LgP	5.08(C)
pKa	
pp cited in Vols.1-5:	

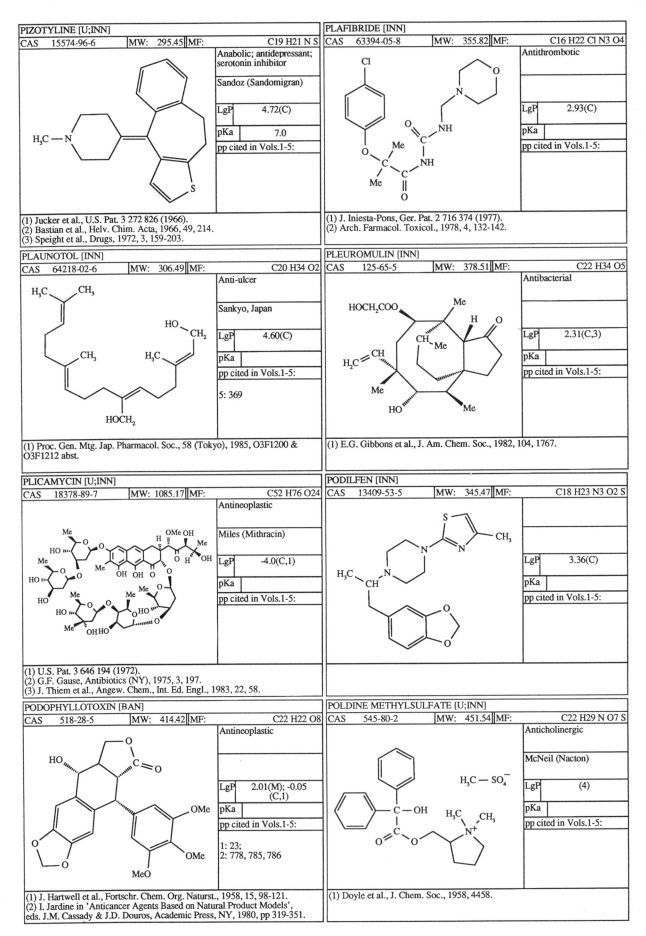

PIZOTYLINE [U;INN]			
CAS 15574-96-6	MW: 295.45	MF:	C19 H21 N S

Anabolic; antidepressant; serotonin inhibitor

Sandoz (Sandomigran)

LgP	4.72(C)
pKa	7.0
pp cited in Vols.1-5:	

(1) Jucker et al., U.S. Pat. 3 272 826 (1966).
(2) Bastian et al., Helv. Chim. Acta, 1966, 49, 214.
(3) Speight et al., Drugs, 1972, 3, 159-203.

PLAFIBRIDE [INN]			
CAS 63394-05-8	MW: 355.82	MF:	C16 H22 Cl N3 O4

Antithrombotic

LgP	2.93(C)
pKa	
pp cited in Vols.1-5:	

(1) J. Iniesta-Pons, Ger. Pat. 2 716 374 (1977).
(2) Arch. Farmacol. Toxicol., 1978, 4, 132-142.

PLAUNOTOL [INN]			
CAS 64218-02-6	MW: 306.49	MF:	C20 H34 O2

Anti-ulcer

Sankyo, Japan

LgP	4.60(C)
pKa	
pp cited in Vols.1-5:	
	5: 369

(1) Proc. Gen. Mtg. Jap. Pharmacol. Soc., 58 (Tokyo), 1985, O3F1200 & O3F1212 abst.

PLEUROMULIN [INN]			
CAS 125-65-5	MW: 378.51	MF:	C22 H34 O5

Antibacterial

LgP	2.31(C,3)
pKa	
pp cited in Vols.1-5:	

(1) E.G. Gibbons et al., J. Am. Chem. Soc., 1982, 104, 1767.

PLICAMYCIN [U;INN]			
CAS 18378-89-7	MW: 1085.17	MF:	C52 H76 O24

Antineoplastic

Miles (Mithracin)

LgP	-4.0(C,1)
pKa	
pp cited in Vols.1-5:	

(1) U.S. Pat. 3 646 194 (1972).
(2) G.F. Gause, Antibiotics (NY), 1975, 3, 197.
(3) J. Thiem et al., Angew. Chem., Int. Ed. Engl., 1983, 22, 58.

PODILFEN [INN]			
CAS 13409-53-5	MW: 345.47	MF:	C18 H23 N3 O2 S

LgP	3.36(C)
pKa	
pp cited in Vols.1-5:	

PODOPHYLLOTOXIN [BAN]			
CAS 518-28-5	MW: 414.42	MF:	C22 H22 O8

Antineoplastic

LgP	2.01(M); -0.05 (C,1)
pKa	
pp cited in Vols.1-5:	
	1: 23; 2: 778, 785, 786

(1) J. Hartwell et al., Fortschr. Chem. Org. Naturst., 1958, 15, 98-121.
(2) I. Jardine in 'Anticancer Agents Based on Natural Product Models', eds. J.M. Cassady & J.D. Douros, Academic Press, NY, 1980, pp 319-351.

POLDINE METHYLSULFATE [U;INN]			
CAS 545-80-2	MW: 451.54	MF:	C22 H29 N O7 S

Anticholinergic

McNeil (Nacton)

LgP	(4)
pKa	
pp cited in Vols.1-5:	

(1) Doyle et al., J. Chem. Soc., 1958, 4458.

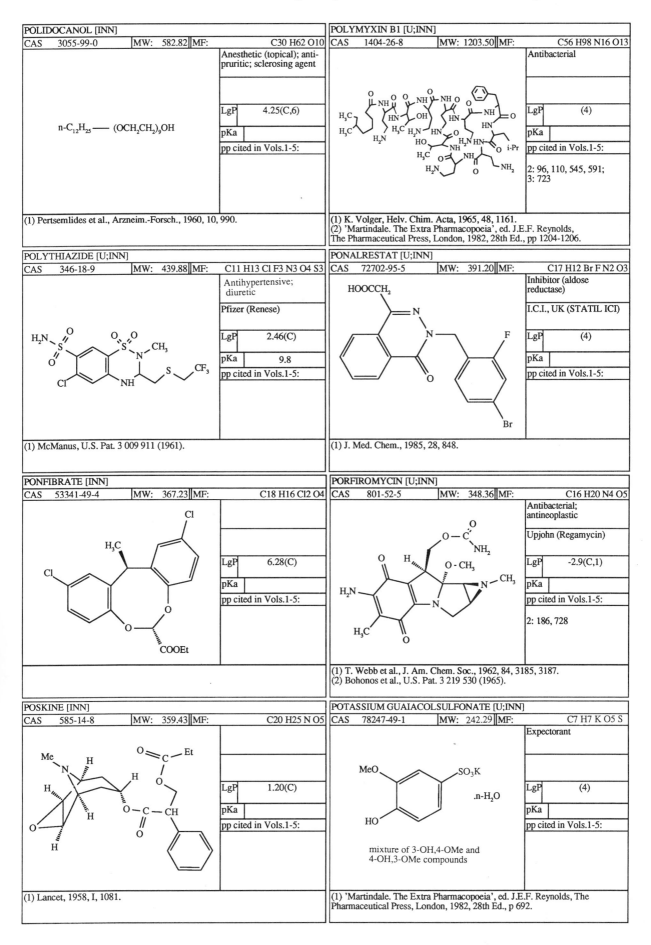

POLIDOCANOL [INN]

CAS	3055-99-0	MW:	582.82	MF:	C30 H62 O10

Anesthetic (topical); anti-pruritic; sclerosing agent

LgP	4.25(C,6)
pKa	
pp cited in Vols.1-5:	

n-C$_{12}$H$_{25}$——(OCH$_2$CH$_2$)$_9$OH

(1) Pertsemlides et al., Arzneim.-Forsch., 1960, 10, 990.

POLYMYXIN B1 [U;INN]

CAS	1404-26-8	MW:	1203.50	MF:	C56 H98 N16 O13

Antibacterial

LgP	(4)
pKa	
pp cited in Vols.1-5:	

2: 96, 110, 545, 591;
3: 723

(1) K. Volger, Helv. Chim. Acta, 1965, 48, 1161.
(2) 'Martindale. The Extra Pharmacopoeia', ed. J.E.F. Reynolds,
The Pharmaceutical Press, London, 1982, 28th Ed., pp 1204-1206.

POLYTHIAZIDE [U;INN]

CAS	346-18-9	MW:	439.88	MF:	C11 H13 Cl F3 N3 O4 S3

Antihypertensive; diuretic

Pfizer (Renese)

LgP	2.46(C)
pKa	9.8
pp cited in Vols.1-5:	

(1) McManus, U.S. Pat. 3 009 911 (1961).

PONALRESTAT [U;INN]

CAS	72702-95-5	MW:	391.20	MF:	C17 H12 Br F N2 O3

Inhibitor (aldose reductase)

I.C.I., UK (STATIL ICI)

LgP	(4)
pKa	
pp cited in Vols.1-5:	

(1) J. Med. Chem., 1985, 28, 848.

PONFIBRATE [INN]

CAS	53341-49-4	MW:	367.23	MF:	C18 H16 Cl2 O4

LgP	6.28(C)
pKa	
pp cited in Vols.1-5:	

PORFIROMYCIN [U;INN]

CAS	801-52-5	MW:	348.36	MF:	C16 H20 N4 O5

Antibacterial; antineoplastic

Upjohn (Regamycin)

LgP	-2.9(C,1)
pKa	
pp cited in Vols.1-5:	

2: 186, 728

(1) T. Webb et al., J. Am. Chem. Soc., 1962, 84, 3185, 3187.
(2) Bohonos et al., U.S. Pat. 3 219 530 (1965).

POSKINE [INN]

CAS	585-14-8	MW:	359.43	MF:	C20 H25 N O5

LgP	1.20(C)
pKa	
pp cited in Vols.1-5:	

(1) Lancet, 1958, I, 1081.

POTASSIUM GUAIACOLSULFONATE [U;INN]

CAS	78247-49-1	MW:	242.29	MF:	C7 H7 K O5 S

Expectorant

LgP	(4)
pKa	
pp cited in Vols.1-5:	

.n-H$_2$O

mixture of 3-OH,4-OMe and
4-OH,3-OMe compounds

(1) 'Martindale. The Extra Pharmacopoeia', ed. J.E.F. Reynolds, The
Pharmaceutical Press, London, 1982, 28th Ed., p 692.

POTASSIUM NITRAZEPATE [INN]		
CAS 5571-84-6	MW: 363.38 MF:	C16 H10 K N3 O5

Tranquilizer

Clin-Byla, France

LgP	(4)
pKa	
pp cited in Vols.1-5:	

(1) P. Mesnard et al., Prod. Pharm., 1967, 22, 89.

POTASSIUM SODIUM TARTRATE [U]		
CAS 304-59-6	MW: 210.16 MF:	C4 H4 K Na O6

Laxative

LgP	(4)
pKa	4.2;4.5;15.0
pp cited in Vols.1-5:	

(1) 'Martindale. The Extra Pharmacopoeia', ed. J.E.F. Reynolds, The Pharmaceutical Press, London, 1982, 28th Ed., pp 642-643.

POTASSIUM SORBATE [U]		
CAS 590-00-1	MW: 150.22 MF:	C6 H7 K O2

Antimicrobial agent

Hoechst, FRG

LgP	(4)
pKa	
pp cited in Vols.1-5:	

(1) Probst et al., U.S. Pat. 3 173 948 (1965).

POTASSIUM THIOCYANATE [MI]		
CAS 333-20-0	MW: 97.18 MF:	C K N S

Hypotensive

LgP	(4)
pKa	
pp cited in Vols.1-5:	

K — S — C ≡ N

(1) 'The Merck Index', Merck & Co., Rahway, NJ, 1983, 10th Ed., entry 7581.

PRACTOLOL [U;INN]		
CAS 6673-35-4	MW: 266.34 MF:	C14 H22 N2 O3

Anti-adrenergic (beta-blocker)

I. C. I., UK

LgP	0.79(M); 0.78(C)
pKa	9.5
pp cited in Vols.1-5:	
1: 25, 570, 571, 629, 636; 3: 201, 209, 211, 215, 216, 218, 220; 5: 101	

(1) Howe et al., U.S. Pat. 3 408 387 (1968).
(2) Danilewicz et al., J. Med. Chem., 1973, 16, 168.
(3) Dunlop et al., Brit. J. Pharmacol. Chemother., 1968, 32, 201.

PRAJMALIUM [INN]		
CAS 35080-11-6	MW: 369.53 MF:	C23 H33 N2 O2 X

Anti-arrhythmic

LgP	(4)
pKa	
pp cited in Vols.1-5:	

(1) Keck, Z. Naturforsch., 1963, 18b, 177.
(2) idem, U.S. Pat. 3 414 577 (1968).

PRALIDOXIME CHLORIDE [U]		
CAS 51-15-0	MW: 172.62 MF:	C7 H9 Cl N2 O

Cholinesterase reactivator

LgP	(4)
pKa	8.0
pp cited in Vols.1-5:	

(1) Kondritzer et al., J. Pharm. Sci., 1961, 50, 109.
(2) Ellin et al., U.S. Pat. 3 140 289 (1964).

PRAMIPEXOLE [INN]		
CAS 104632-26-0	MW: 211.33 MF:	C10 H17 N3 S

LgP	0.45(C)
pKa	
pp cited in Vols.1-5:	

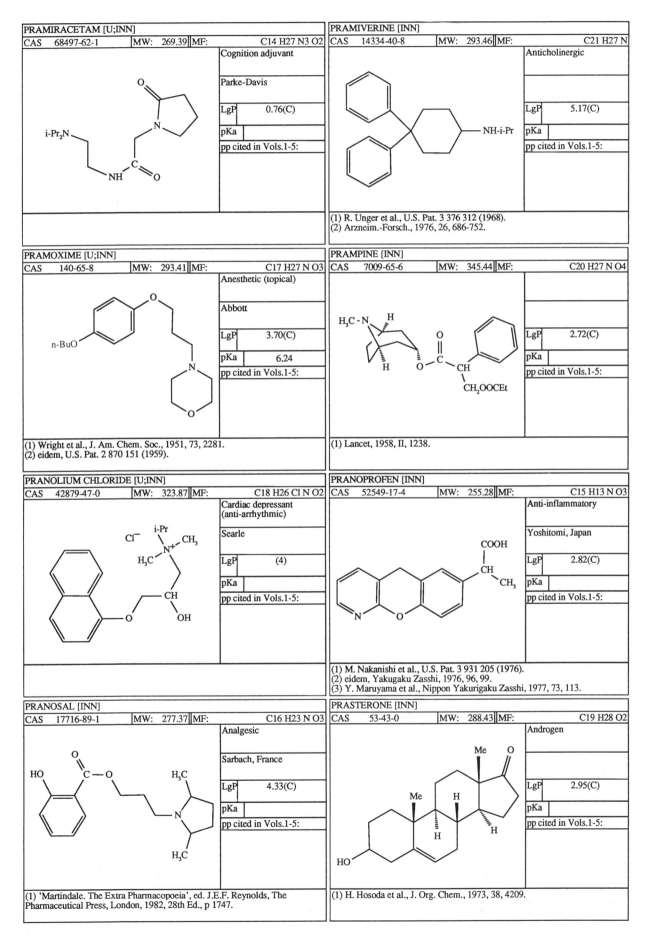

PRAMIRACETAM [U;INN]

CAS	68497-62-1	MW:	269.39	MF:	C14 H27 N3 O2

Cognition adjuvant

Parke-Davis

LgP	0.76(C)
pKa	
pp cited in Vols.1-5:	

PRAMIVERINE [INN]

CAS	14334-40-8	MW:	293.46	MF:	C21 H27 N

Anticholinergic

LgP	5.17(C)
pKa	
pp cited in Vols.1-5:	

(1) R. Unger et al., U.S. Pat. 3 376 312 (1968).
(2) Arzneim.-Forsch., 1976, 26, 686-752.

PRAMOXIME [U;INN]

CAS	140-65-8	MW:	293.41	MF:	C17 H27 N O3

Anesthetic (topical)

Abbott

LgP	3.70(C)
pKa	6.24
pp cited in Vols.1-5:	

(1) Wright et al., J. Am. Chem. Soc., 1951, 73, 2281.
(2) eidem, U.S. Pat. 2 870 151 (1959).

PRAMPINE [INN]

CAS	7009-65-6	MW:	345.44	MF:	C20 H27 N O4

LgP	2.72(C)
pKa	
pp cited in Vols.1-5:	

(1) Lancet, 1958, II, 1238.

PRANOLIUM CHLORIDE [U;INN]

CAS	42879-47-0	MW:	323.87	MF:	C18 H26 Cl N O2

Cardiac depressant (anti-arrhythmic)

Searle

LgP	(4)
pKa	
pp cited in Vols.1-5:	

PRANOPROFEN [INN]

CAS	52549-17-4	MW:	255.28	MF:	C15 H13 N O3

Anti-inflammatory

Yoshitomi, Japan

LgP	2.82(C)
pKa	
pp cited in Vols.1-5:	

(1) M. Nakanishi et al., U.S. Pat. 3 931 205 (1976).
(2) eidem, Yakugaku Zasshi, 1976, 96, 99.
(3) Y. Maruyama et al., Nippon Yakurigaku Zasshi, 1977, 73, 113.

PRANOSAL [INN]

CAS	17716-89-1	MW:	277.37	MF:	C16 H23 N O3

Analgesic

Sarbach, France

LgP	4.33(C)
pKa	
pp cited in Vols.1-5:	

(1) 'Martindale. The Extra Pharmacopoeia', ed. J.E.F. Reynolds, The Pharmaceutical Press, London, 1982, 28th Ed., p 1747.

PRASTERONE [INN]

CAS	53-43-0	MW:	288.43	MF:	C19 H28 O2

Androgen

LgP	2.95(C)
pKa	
pp cited in Vols.1-5:	

(1) H. Hosoda et al., J. Org. Chem., 1973, 38, 4209.

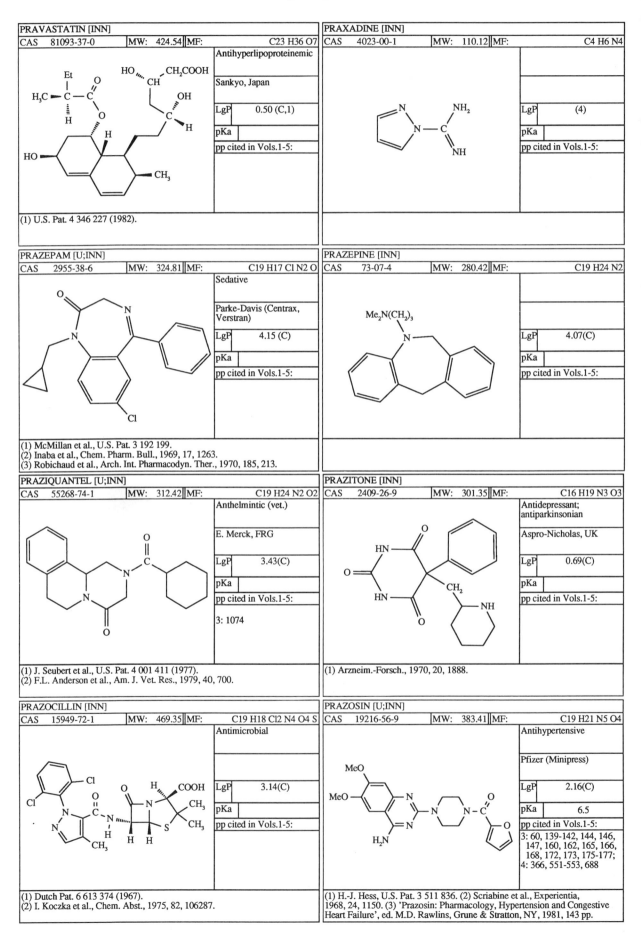

PRAVASTATIN [INN]

CAS 81093-37-0	MW: 424.54	MF:	C23 H36 O7

Antihyperlipoproteinemic

Sankyo, Japan

LgP	0.50 (C,1)
pKa	
pp cited in Vols.1-5:	

(1) U.S. Pat. 4 346 227 (1982).

PRAXADINE [INN]

CAS 4023-00-1	MW: 110.12	MF:	C4 H6 N4

LgP	(4)
pKa	
pp cited in Vols.1-5:	

PRAZEPAM [U;INN]

CAS 2955-38-6	MW: 324.81	MF:	C19 H17 Cl N2 O

Sedative

Parke-Davis (Centrax, Verstran)

LgP	4.15 (C)
pKa	
pp cited in Vols.1-5:	

(1) McMillan et al., U.S. Pat. 3 192 199.
(2) Inaba et al., Chem. Pharm. Bull., 1969, 17, 1263.
(3) Robichaud et al., Arch. Int. Pharmacodyn. Ther., 1970, 185, 213.

PRAZEPINE [INN]

CAS 73-07-4	MW: 280.42	MF:	C19 H24 N2

$Me_2N(CH_2)_3$

LgP	4.07(C)
pKa	
pp cited in Vols.1-5:	

PRAZIQUANTEL [U;INN]

CAS 55268-74-1	MW: 312.42	MF:	C19 H24 N2 O2

Anthelmintic (vet.)

E. Merck, FRG

LgP	3.43(C)
pKa	
pp cited in Vols.1-5:	
	3: 1074

(1) J. Seubert et al., U.S. Pat. 4 001 411 (1977).
(2) F.L. Anderson et al., Am. J. Vet. Res., 1979, 40, 700.

PRAZITONE [INN]

CAS 2409-26-9	MW: 301.35	MF:	C16 H19 N3 O3

Antidepressant; antiparkinsonian

Aspro-Nicholas, UK

LgP	0.69(C)
pKa	
pp cited in Vols.1-5:	

(1) Arzneim.-Forsch., 1970, 20, 1888.

PRAZOCILLIN [INN]

CAS 15949-72-1	MW: 469.35	MF:	C19 H18 Cl2 N4 O4 S

Antimicrobial

LgP	3.14(C)
pKa	
pp cited in Vols.1-5:	

(1) Dutch Pat. 6 613 374 (1967).
(2) I. Koczka et al., Chem. Abst., 1975, 82, 106287.

PRAZOSIN [U;INN]

CAS 19216-56-9	MW: 383.41	MF:	C19 H21 N5 O4

Antihypertensive

Pfizer (Minipress)

LgP	2.16(C)
pKa	6.5
pp cited in Vols.1-5:	
	3: 60, 139-142, 144, 146, 147, 160, 162, 165, 166, 168, 172, 173, 175-177; 4: 366, 551-553, 688

(1) H.-J. Hess, U.S. Pat. 3 511 836. (2) Scriabine et al., Experientia, 1968, 24, 1150. (3) 'Prazosin: Pharmacology, Hypertension and Congestive Heart Failure', ed. M.D. Rawlins, Grune & Stratton, NY, 1981, 143 pp.

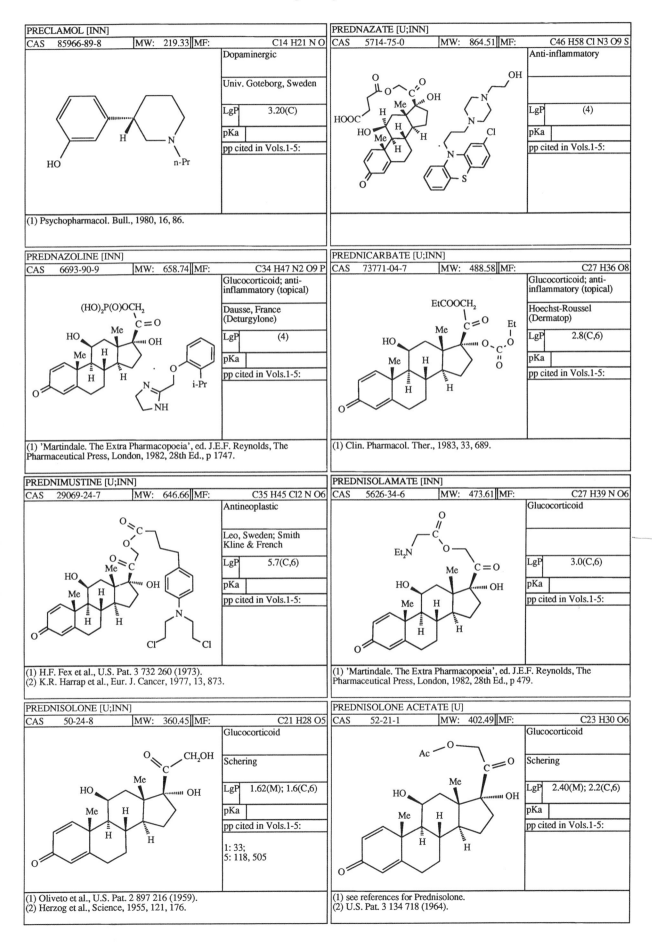

PRECLAMOL [INN]

CAS	85966-89-8	MW:	219.33	MF:	C14 H21 N O

Dopaminergic

Univ. Goteborg, Sweden

LgP	3.20(C)
pKa	
pp cited in Vols.1-5:	

(1) Psychopharmacol. Bull., 1980, 16, 86.

PREDNAZATE [U;INN]

CAS	5714-75-0	MW:	864.51	MF:	C46 H58 Cl N3 O9 S

Anti-inflammatory

LgP	(4)
pKa	
pp cited in Vols.1-5:	

PREDNAZOLINE [INN]

CAS	6693-90-9	MW:	658.74	MF:	C34 H47 N2 O9 P

Glucocorticoid; anti-inflammatory (topical)

Dausse, France (Deturgylone)

LgP	(4)
pKa	
pp cited in Vols.1-5:	

(1) 'Martindale. The Extra Pharmacopoeia', ed. J.E.F. Reynolds, The Pharmaceutical Press, London, 1982, 28th Ed., p 1747.

PREDNICARBATE [U;INN]

CAS	73771-04-7	MW:	488.58	MF:	C27 H36 O8

Glucocorticoid; anti-inflammatory (topical)

Hoechst-Roussel (Dermatop)

LgP	2.8(C,6)
pKa	
pp cited in Vols.1-5:	

(1) Clin. Pharmacol. Ther., 1983, 33, 689.

PREDNIMUSTINE [U;INN]

CAS	29069-24-7	MW:	646.66	MF:	C35 H45 Cl2 N O6

Antineoplastic

Leo, Sweden; Smith Kline & French

LgP	5.7(C,6)
pKa	
pp cited in Vols.1-5:	

(1) H.F. Fex et al., U.S. Pat. 3 732 260 (1973).
(2) K.R. Harrap et al., Eur. J. Cancer, 1977, 13, 873.

PREDNISOLAMATE [INN]

CAS	5626-34-6	MW:	473.61	MF:	C27 H39 N O6

Glucocorticoid

LgP	3.0(C,6)
pKa	
pp cited in Vols.1-5:	

(1) 'Martindale. The Extra Pharmacopoeia', ed. J.E.F. Reynolds, The Pharmaceutical Press, London, 1982, 28th Ed., p 479.

PREDNISOLONE [U;INN]

CAS	50-24-8	MW:	360.45	MF:	C21 H28 O5

Glucocorticoid

Schering

LgP	1.62(M); 1.6(C,6)
pKa	
pp cited in Vols.1-5:	

1: 33;
5: 118, 505

(1) Oliveto et al., U.S. Pat. 2 897 216 (1959).
(2) Herzog et al., Science, 1955, 121, 176.

PREDNISOLONE ACETATE [U]

CAS	52-21-1	MW:	402.49	MF:	C23 H30 O6

Glucocorticoid

Schering

LgP	2.40(M); 2.2(C,6)
pKa	
pp cited in Vols.1-5:	

(1) see references for Prednisolone.
(2) U.S. Pat. 3 134 718 (1964).

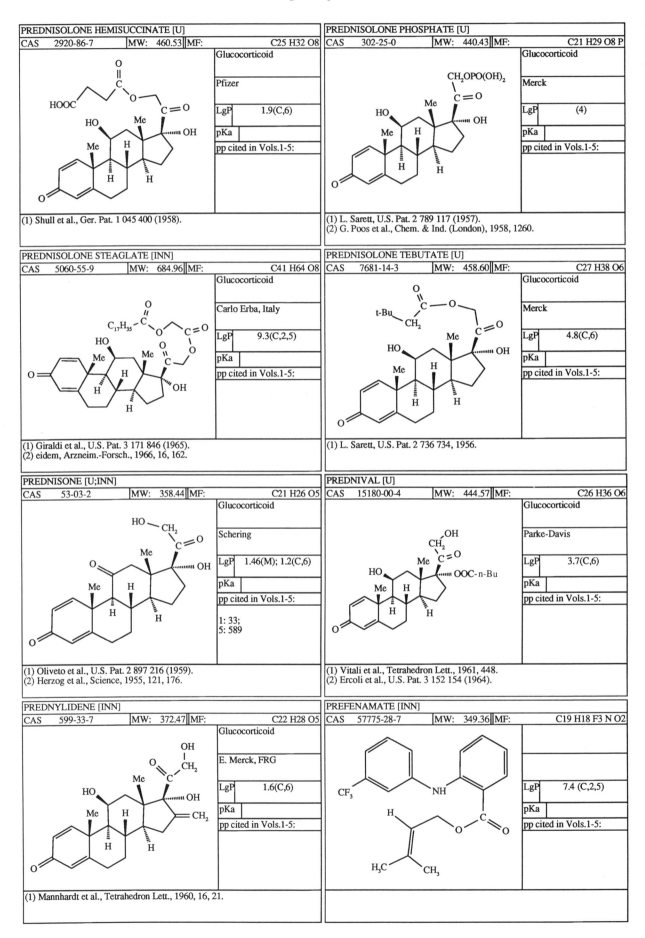

PREDNISOLONE HEMISUCCINATE [U]	
CAS 2920-86-7 MW: 460.53 MF: C25 H32 O8	Glucocorticoid
	Pfizer
	LgP 1.9(C,6)
	pKa
	pp cited in Vols.1-5:
(1) Shull et al., Ger. Pat. 1 045 400 (1958).	

PREDNISOLONE PHOSPHATE [U]	
CAS 302-25-0 MW: 440.43 MF: C21 H29 O8 P	Glucocorticoid
	Merck
	LgP (4)
	pKa
	pp cited in Vols.1-5:
(1) L. Sarett, U.S. Pat. 2 789 117 (1957). (2) G. Poos et al., Chem. & Ind. (London), 1958, 1260.	

PREDNISOLONE STEAGLATE [INN]	
CAS 5060-55-9 MW: 684.96 MF: C41 H64 O8	Glucocorticoid
	Carlo Erba, Italy
	LgP 9.3(C,2,5)
	pKa
	pp cited in Vols.1-5:
(1) Giraldi et al., U.S. Pat. 3 171 846 (1965). (2) eidem, Arzneim.-Forsch., 1966, 16, 162.	

PREDNISOLONE TEBUTATE [U]	
CAS 7681-14-3 MW: 458.60 MF: C27 H38 O6	Glucocorticoid
	Merck
	LgP 4.8(C,6)
	pKa
	pp cited in Vols.1-5:
(1) L. Sarett, U.S. Pat. 2 736 734, 1956.	

PREDNISONE [U;INN]	
CAS 53-03-2 MW: 358.44 MF: C21 H26 O5	Glucocorticoid
	Schering
	LgP 1.46(M); 1.2(C,6)
	pKa
	pp cited in Vols.1-5:
	1: 33; 5: 589
(1) Oliveto et al., U.S. Pat. 2 897 216 (1959). (2) Herzog et al., Science, 1955, 121, 176.	

PREDNIVAL [U]	
CAS 15180-00-4 MW: 444.57 MF: C26 H36 O6	Glucocorticoid
	Parke-Davis
	LgP 3.7(C,6)
	pKa
	pp cited in Vols.1-5:
(1) Vitali et al., Tetrahedron Lett., 1961, 448. (2) Ercoli et al., U.S. Pat. 3 152 154 (1964).	

PREDNYLIDENE [INN]	
CAS 599-33-7 MW: 372.47 MF: C22 H28 O5	Glucocorticoid
	E. Merck, FRG
	LgP 1.6(C,6)
	pKa
	pp cited in Vols.1-5:
(1) Mannhardt et al., Tetrahedron Lett., 1960, 16, 21.	

PREFENAMATE [INN]	
CAS 57775-28-7 MW: 349.36 MF: C19 H18 F3 N O2	
	LgP 7.4 (C,2,5)
	pKa
	pp cited in Vols.1-5:

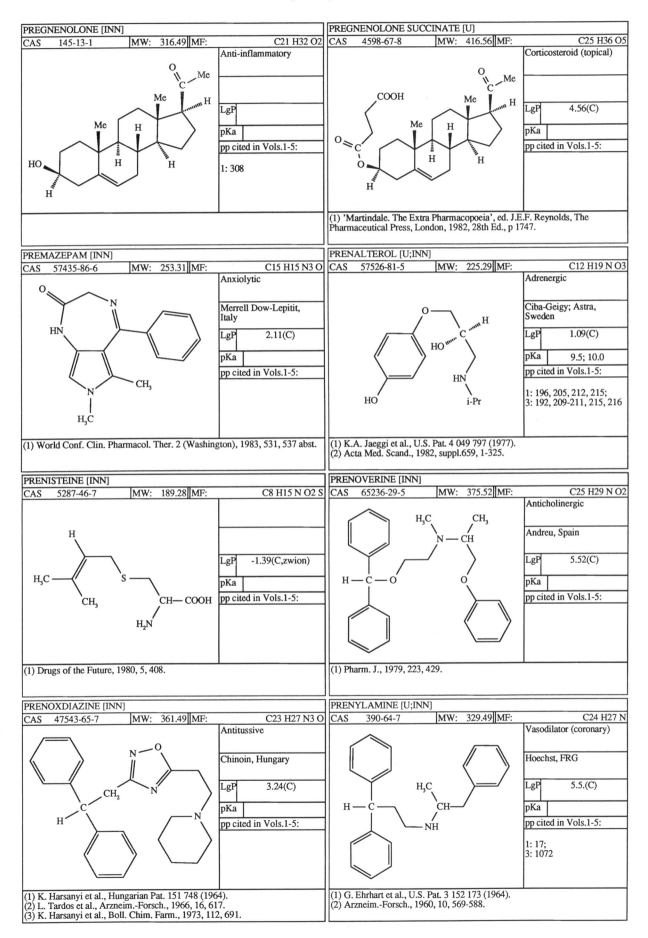

PREGNENOLONE [INN]

CAS	145-13-1	MW:	316.49	MF:	C21 H32 O2

Anti-inflammatory

LgP

pKa

pp cited in Vols.1-5:

1: 308

PREGNENOLONE SUCCINATE [U]

CAS	4598-67-8	MW:	416.56	MF:	C25 H36 O5

Corticosteroid (topical)

LgP   4.56(C)

pKa

pp cited in Vols.1-5:

(1) 'Martindale. The Extra Pharmacopoeia', ed. J.E.F. Reynolds, The Pharmaceutical Press, London, 1982, 28th Ed., p 1747.

PREMAZEPAM [INN]

CAS	57435-86-6	MW:	253.31	MF:	C15 H15 N3 O

Anxiolytic

Merrell Dow-Lepitit, Italy

LgP   2.11(C)

pKa

pp cited in Vols.1-5:

(1) World Conf. Clin. Pharmacol. Ther. 2 (Washington), 1983, 531, 537 abst.

PRENALTEROL [U;INN]

CAS	57526-81-5	MW:	225.29	MF:	C12 H19 N O3

Adrenergic

Ciba-Geigy; Astra, Sweden

LgP   1.09(C)

pKa   9.5; 10.0

pp cited in Vols.1-5:

1: 196, 205, 212, 215;
3: 192, 209-211, 215, 216

(1) K.A. Jaeggi et al., U.S. Pat. 4 049 797 (1977).
(2) Acta Med. Scand., 1982, suppl.659, 1-325.

PRENISTEINE [INN]

CAS	5287-46-7	MW:	189.28	MF:	C8 H15 N O2 S

LgP   -1.39(C,zwion)

pKa

pp cited in Vols.1-5:

(1) Drugs of the Future, 1980, 5, 408.

PRENOVERINE [INN]

CAS	65236-29-5	MW:	375.52	MF:	C25 H29 N O2

Anticholinergic

Andreu, Spain

LgP   5.52(C)

pKa

pp cited in Vols.1-5:

(1) Pharm. J., 1979, 223, 429.

PRENOXDIAZINE [INN]

CAS	47543-65-7	MW:	361.49	MF:	C23 H27 N3 O

Antitussive

Chinoin, Hungary

LgP   3.24(C)

pKa

pp cited in Vols.1-5:

(1) K. Harsanyi et al., Hungarian Pat. 151 748 (1964).
(2) L. Tardos et al., Arzneim.-Forsch., 1966, 16, 617.
(3) K. Harsanyi et al., Boll. Chim. Farm., 1973, 112, 691.

PRENYLAMINE [U;INN]

CAS	390-64-7	MW:	329.49	MF:	C24 H27 N

Vasodilator (coronary)

Hoechst, FRG

LgP   5.5.(C)

pKa

pp cited in Vols.1-5:

1: 17;
3: 1072

(1) G. Ehrhart et al., U.S. Pat. 3 152 173 (1964).
(2) Arzneim.-Forsch., 1960, 10, 569-588.

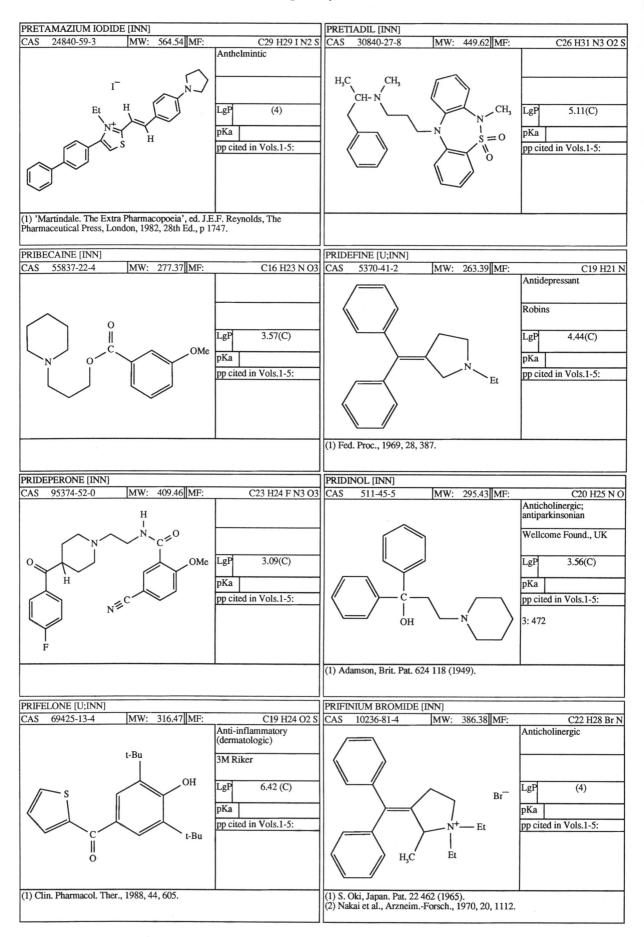

PRETAMAZIUM IODIDE [INN]

CAS	24840-59-3	MW: 564.54	MF:	C29 H29 I N2 S

Anthelmintic

LgP (4)

pKa

pp cited in Vols.1-5:

(1) 'Martindale. The Extra Pharmacopoeia', ed. J.E.F. Reynolds, The Pharmaceutical Press, London, 1982, 28th Ed., p 1747.

PRIBECAINE [INN]

CAS	55837-22-4	MW: 277.37	MF:	C16 H23 N O3

LgP 3.57(C)

pKa

pp cited in Vols.1-5:

PRIDEPERONE [INN]

CAS	95374-52-0	MW: 409.46	MF:	C23 H24 F N3 O3

LgP 3.09(C)

pKa

pp cited in Vols.1-5:

PRIFELONE [U;INN]

CAS	69425-13-4	MW: 316.47	MF:	C19 H24 O2 S

Anti-inflammatory (dermatologic)

3M Riker

LgP 6.42 (C)

pKa

pp cited in Vols.1-5:

(1) Clin. Pharmacol. Ther., 1988, 44, 605.

PRETIADIL [INN]

CAS	30840-27-8	MW: 449.62	MF:	C26 H31 N3 O2 S

LgP 5.11(C)

pKa

pp cited in Vols.1-5:

PRIDEFINE [U;INN]

CAS	5370-41-2	MW: 263.39	MF:	C19 H21 N

Antidepressant

Robins

LgP 4.44(C)

pKa

pp cited in Vols.1-5:

(1) Fed. Proc., 1969, 28, 387.

PRIDINOL [INN]

CAS	511-45-5	MW: 295.43	MF:	C20 H25 N O

Anticholinergic; antiparkinsonian

Wellcome Found., UK

LgP 3.56(C)

pKa

pp cited in Vols.1-5:

3: 472

(1) Adamson, Brit. Pat. 624 118 (1949).

PRIFINIUM BROMIDE [INN]

CAS	10236-81-4	MW: 386.38	MF:	C22 H28 Br N

Anticholinergic

LgP (4)

pKa

pp cited in Vols.1-5:

(1) S. Oki, Japan. Pat. 22 462 (1965).
(2) Nakai et al., Arzneim.-Forsch., 1970, 20, 1112.

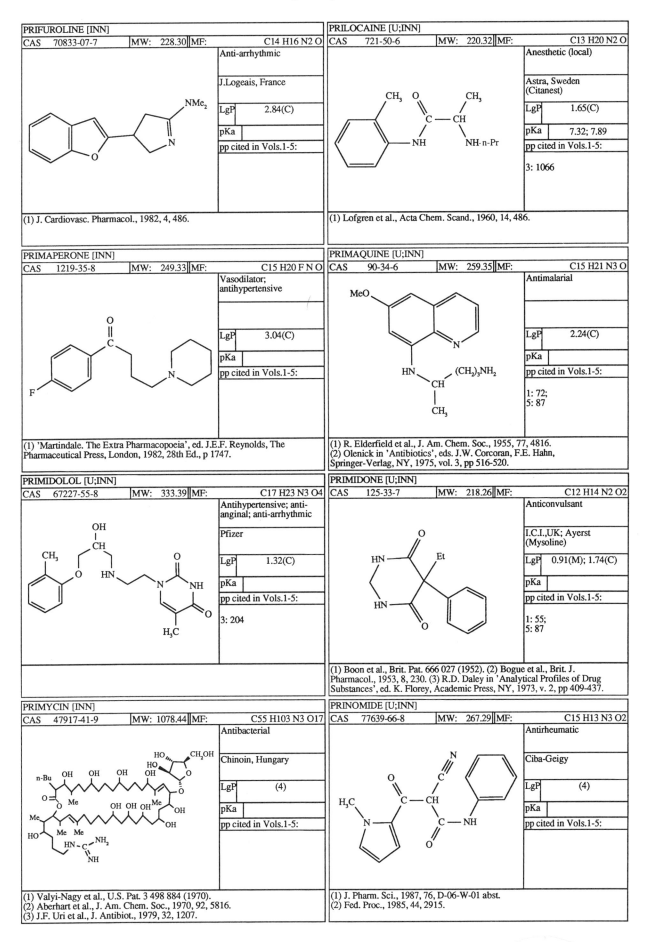

PRIFUROLINE [INN]

CAS 70833-07-7	MW: 228.30	MF:	C14 H16 N2 O

Anti-arrhythmic

J.Logeais, France

LgP	2.84(C)
pKa	
pp cited in Vols.1-5:	

(1) J. Cardiovasc. Pharmacol., 1982, 4, 486.

PRILOCAINE [U;INN]

CAS 721-50-6	MW: 220.32	MF:	C13 H20 N2 O

Anesthetic (local)

Astra, Sweden (Citanest)

LgP	1.65(C)
pKa	7.32; 7.89
pp cited in Vols.1-5:	
3: 1066	

(1) Lofgren et al., Acta Chem. Scand., 1960, 14, 486.

PRIMAPERONE [INN]

CAS 1219-35-8	MW: 249.33	MF:	C15 H20 F N O

Vasodilator; antihypertensive

LgP	3.04(C)
pKa	
pp cited in Vols.1-5:	

(1) 'Martindale. The Extra Pharmacopoeia', ed. J.E.F. Reynolds, The Pharmaceutical Press, London, 1982, 28th Ed., p 1747.

PRIMAQUINE [U;INN]

CAS 90-34-6	MW: 259.35	MF:	C15 H21 N3 O

Antimalarial

LgP	2.24(C)
pKa	
pp cited in Vols.1-5:	
1: 72; 5: 87	

(1) R. Elderfield et al., J. Am. Chem. Soc., 1955, 77, 4816.
(2) Olenick in 'Antibiotics', eds. J.W. Corcoran, F.E. Hahn, Springer-Verlag, NY, 1975, vol. 3, pp 516-520.

PRIMIDOLOL [U;INN]

CAS 67227-55-8	MW: 333.39	MF:	C17 H23 N3 O4

Antihypertensive; anti-anginal; anti-arrhythmic

Pfizer

LgP	1.32(C)
pKa	
pp cited in Vols.1-5:	
3: 204	

PRIMIDONE [U;INN]

CAS 125-33-7	MW: 218.26	MF:	C12 H14 N2 O2

Anticonvulsant

I.C.I.,UK; Ayerst (Mysoline)

LgP	0.91(M); 1.74(C)
pKa	
pp cited in Vols.1-5:	
1: 55; 5: 87	

(1) Boon et al., Brit. Pat. 666 027 (1952). (2) Bogue et al., Brit. J. Pharmacol., 1953, 8, 230. (3) R.D. Daley in 'Analytical Profiles of Drug Substances', ed. K. Florey, Academic Press, NY, 1973, v. 2, pp 409-437.

PRIMYCIN [INN]

CAS 47917-41-9	MW: 1078.44	MF:	C55 H103 N3 O17

Antibacterial

Chinoin, Hungary

LgP	(4)
pKa	
pp cited in Vols.1-5:	

(1) Valyi-Nagy et al., U.S. Pat. 3 498 884 (1970).
(2) Aberhart et al., J. Am. Chem. Soc., 1970, 92, 5816.
(3) J.F. Uri et al., J. Antibiot., 1979, 32, 1207.

PRINOMIDE [U;INN]

CAS 77639-66-8	MW: 267.29	MF:	C15 H13 N3 O2

Antirheumatic

Ciba-Geigy

LgP	(4)
pKa	
pp cited in Vols.1-5:	

(1) J. Pharm. Sci., 1987, 76, D-06-W-01 abst.
(2) Fed. Proc., 1985, 44, 2915.

PRISTINAMYCIN [INN]

CAS	11006-76-1	MW:	852.95	MF:	C44 H52 N8 O10

Antibacterial

Rhone Poulenc, France

LgP	(4)
pKa	
pp cited in Vols.1-5:	

(1) J. Preud'Homme et al., Bull. Soc. Chim. Fr., 1968, 585.

PRIZIDILOL [U;INN]

CAS	59010-44-5	MW:	331.42	MF:	C17 H25 N5 O2

Antihypertensive; vasodilator

Smith Kline & French, UK

LgP	1.41(C)
pKa	
pp cited in Vols.1-5:	

3: 206, 220

(1) A. Bell et al., Br. J. Clin. Pharmacol., 1980, 9, 299P, 300P, 301P.

PROADIFEN [U;INN]

CAS	302-33-0	MW:	353.51	MF:	C23 H31 N O2

Synergist

Smith Kline & French (SKF 525)

LgP	4.65(M); 5.64(C)
pKa	8.80
pp cited in Vols.1-5:	

n-Pr — C — COOCH₂CH₂NEt₂

(1) P.N. Craig et al., J. Am. Chem. Soc., 1951, 73, 1339.

PROBARBITAL [INN]

CAS	76-76-6	MW:	198.22	MF:	C9 H14 N2 O3

Sedative; hypnotic

LgP	0.97(M); 1.05(C)
pKa	
pp cited in Vols.1-5:	

(1) Thorp, U.S. Pat. 1 576 014 (1926).

PROBENECID [U;INN]

CAS	57-66-9	MW:	285.36	MF:	C13 H19 N O4 S

Uricosuric

Sharpe & Dohme

LgP	3.21(M); 3.37(C)
pKa	3.4
pp cited in Vols.1-5:	

2: 278, 618;
5: 180, 197

(1) C.S. Miller, U.S. Pat. 2 608 507 (1952).
(2) A.A. Al-Badr et al. in 'Analytical Profiles of Drug Substances',
ed. K. Florey, Academic Press, NY, 1981, vol. 10, pp 639-663.

PROBICROMIL [U;INN]

CAS	58805-38-2	MW:	344.28	MF:	C17 H12 O8

Anti-allergic (prophylactic)

Fisons, UK

LgP	0.42(C)
pKa	
pp cited in Vols.1-5:	

(1) 'Martindale. The Extra Pharmacopoeia', ed. J.E.F. Reynolds, The
Pharmaceutical Press, London, 1982, 28th Ed., 1748.

PROBUCOL [U;INN]

CAS	23288-49-5	MW:	516.86	MF:	C31 H48 O2 S2

Antihyperlipoproteinemic

Merrell Dow (Lorelco)

LgP	10.8(C,2)
pKa	
pp cited in Vols.1-5:	

(1) Neuworth et al., J. Med. Chem., 1970, 13, 722.
(2) Drake et al., Circulation, 1969, 40(suppl.3), 73.

PROCAINAMIDE [U;INN]

CAS	51-06-9	MW:	235.33	MF:	C13 H21 N3 O

Cardiac depressant (anti-arrhythmic)

LgP	0.88(M); 1.11(C)
pKa	9.26
pp cited in Vols.1-5:	

3: 1064, 1068, 1070;
5: 151, 180, 256

(1) M. Yamazaki et al., J. Pharm. Soc. Japan, 1953, 73, 294.
(2) R.B. Poet et al., in 'Analytical Profiles of Drug Substances',
ed. K.Florey, Academic Press, NY, 1975, vol. 4, pp 333-383.

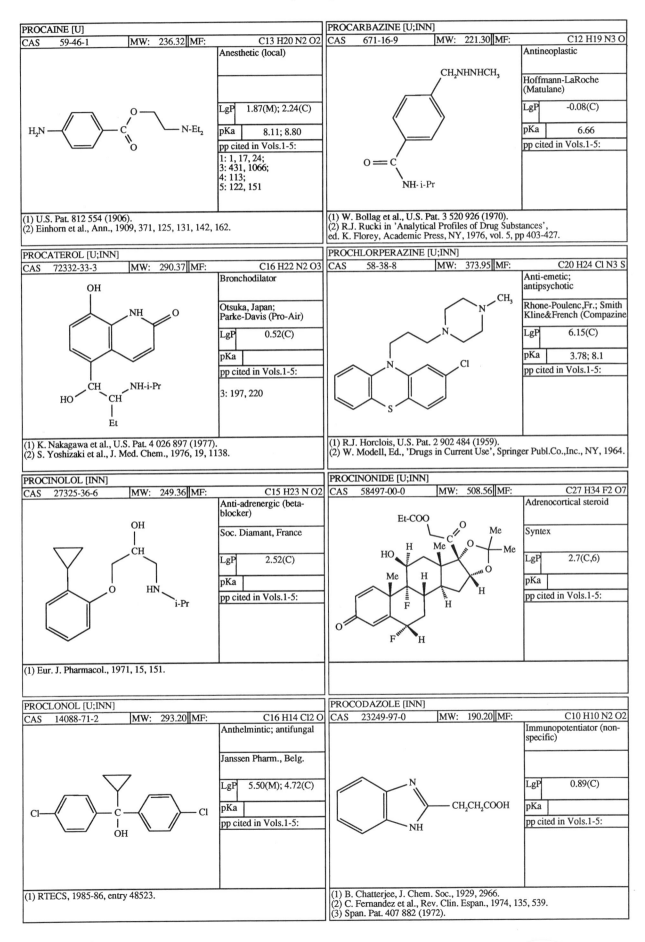

PROCAINE [U]

CAS	59-46-1	MW: 236.32	MF:	C13 H20 N2 O2

Anesthetic (local)

LgP	1.87(M); 2.24(C)
pKa	8.11; 8.80

pp cited in Vols.1-5:
1: 1, 17, 24;
3: 431, 1066;
4: 113;
5: 122, 151

(1) U.S. Pat. 812 554 (1906).
(2) Einhorn et al., Ann., 1909, 371, 125, 131, 142, 162.

PROCARBAZINE [U;INN]

CAS	671-16-9	MW: 221.30	MF:	C12 H19 N3 O

Antineoplastic

Hoffmann-LaRoche
(Matulane)

LgP	-0.08(C)
pKa	6.66

pp cited in Vols.1-5:

(1) W. Bollag et al., U.S. Pat. 3 520 926 (1970).
(2) R.J. Rucki in 'Analytical Profiles of Drug Substances',
ed. K. Florey, Academic Press, NY, 1976, vol. 5, pp 403-427.

PROCATEROL [U;INN]

CAS	72332-33-3	MW: 290.37	MF:	C16 H22 N2 O3

Bronchodilator

Otsuka, Japan;
Parke-Davis (Pro-Air)

LgP	0.52(C)
pKa	

pp cited in Vols.1-5:
3: 197, 220

(1) K. Nakagawa et al., U.S. Pat. 4 026 897 (1977).
(2) S. Yoshizaki et al., J. Med. Chem., 1976, 19, 1138.

PROCHLORPERAZINE [U;INN]

CAS	58-38-8	MW: 373.95	MF:	C20 H24 Cl N3 S

Anti-emetic;
antipsychotic

Rhone-Poulenc,Fr.; Smith
Kline&French (Compazine

LgP	6.15(C)
pKa	3.78; 8.1

pp cited in Vols.1-5:

(1) R.J. Horclois, U.S. Pat. 2 902 484 (1959).
(2) W. Modell, Ed., 'Drugs in Current Use', Springer Publ.Co.,Inc., NY, 1964.

PROCINOLOL [INN]

CAS	27325-36-6	MW: 249.36	MF:	C15 H23 N O2

Anti-adrenergic (beta-
blocker)

Soc. Diamant, France

LgP	2.52(C)
pKa	

pp cited in Vols.1-5:

(1) Eur. J. Pharmacol., 1971, 15, 151.

PROCINONIDE [U;INN]

CAS	58497-00-0	MW: 508.56	MF:	C27 H34 F2 O7

Adrenocortical steroid

Syntex

LgP	2.7(C,6)
pKa	

pp cited in Vols.1-5:

PROCLONOL [U;INN]

CAS	14088-71-2	MW: 293.20	MF:	C16 H14 Cl2 O

Anthelmintic; antifungal

Janssen Pharm., Belg.

LgP	5.50(M); 4.72(C)
pKa	

pp cited in Vols.1-5:

(1) RTECS, 1985-86, entry 48523.

PROCODAZOLE [INN]

CAS	23249-97-0	MW: 190.20	MF:	C10 H10 N2 O2

Immunopotentiator (non-
specific)

LgP	0.89(C)
pKa	

pp cited in Vols.1-5:

(1) B. Chatterjee, J. Chem. Soc., 1929, 2966.
(2) C. Fernandez et al., Rev. Clin. Espan., 1974, 135, 539.
(3) Span. Pat. 407 882 (1972).

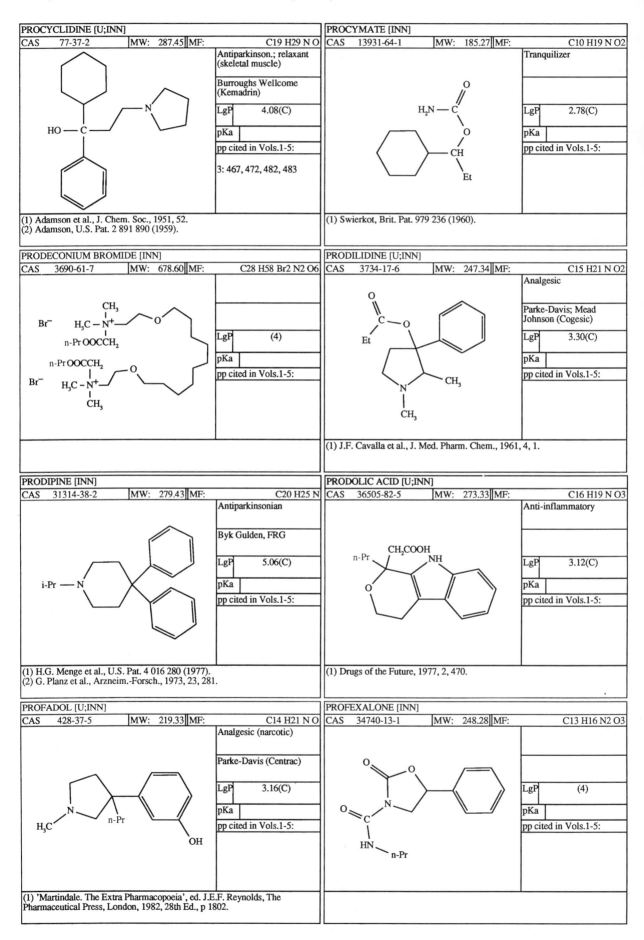

PROCYCLIDINE [U;INN]

CAS	77-37-2	MW: 287.45	MF:	C19 H29 N O

Antiparkinson.; relaxant (skeletal muscle)

Burroughs Wellcome (Kemadrin)

LgP	4.08(C)
pKa	
pp cited in Vols.1-5:	

3: 467, 472, 482, 483

(1) Adamson et al., J. Chem. Soc., 1951, 52.
(2) Adamson, U.S. Pat. 2 891 890 (1959).

PROCYMATE [INN]

CAS	13931-64-1	MW: 185.27	MF:	C10 H19 N O2

Tranquilizer

LgP	2.78(C)
pKa	
pp cited in Vols.1-5:	

(1) Swierkot, Brit. Pat. 979 236 (1960).

PRODECONIUM BROMIDE [INN]

CAS	3690-61-7	MW: 678.60	MF:	C28 H58 Br2 N2 O6

LgP	(4)
pKa	
pp cited in Vols.1-5:	

PRODILIDINE [U;INN]

CAS	3734-17-6	MW: 247.34	MF:	C15 H21 N O2

Analgesic

Parke-Davis; Mead Johnson (Cogesic)

LgP	3.30(C)
pKa	
pp cited in Vols.1-5:	

(1) J.F. Cavalla et al., J. Med. Pharm. Chem., 1961, 4, 1.

PRODIPINE [INN]

CAS	31314-38-2	MW: 279.43	MF:	C20 H25 N

Antiparkinsonian

Byk Gulden, FRG

LgP	5.06(C)
pKa	
pp cited in Vols.1-5:	

(1) H.G. Menge et al., U.S. Pat. 4 016 280 (1977).
(2) G. Planz et al., Arzneim.-Forsch., 1973, 23, 281.

PRODOLIC ACID [U;INN]

CAS	36505-82-5	MW: 273.33	MF:	C16 H19 N O3

Anti-inflammatory

LgP	3.12(C)
pKa	
pp cited in Vols.1-5:	

(1) Drugs of the Future, 1977, 2, 470.

PROFADOL [U;INN]

CAS	428-37-5	MW: 219.33	MF:	C14 H21 N O

Analgesic (narcotic)

Parke-Davis (Centrac)

LgP	3.16(C)
pKa	
pp cited in Vols.1-5:	

(1) 'Martindale. The Extra Pharmacopoeia', ed. J.E.F. Reynolds, The Pharmaceutical Press, London, 1982, 28th Ed., p 1802.

PROFEXALONE [INN]

CAS	34740-13-1	MW: 248.28	MF:	C13 H16 N2 O3

LgP	(4)
pKa	
pp cited in Vols.1-5:	

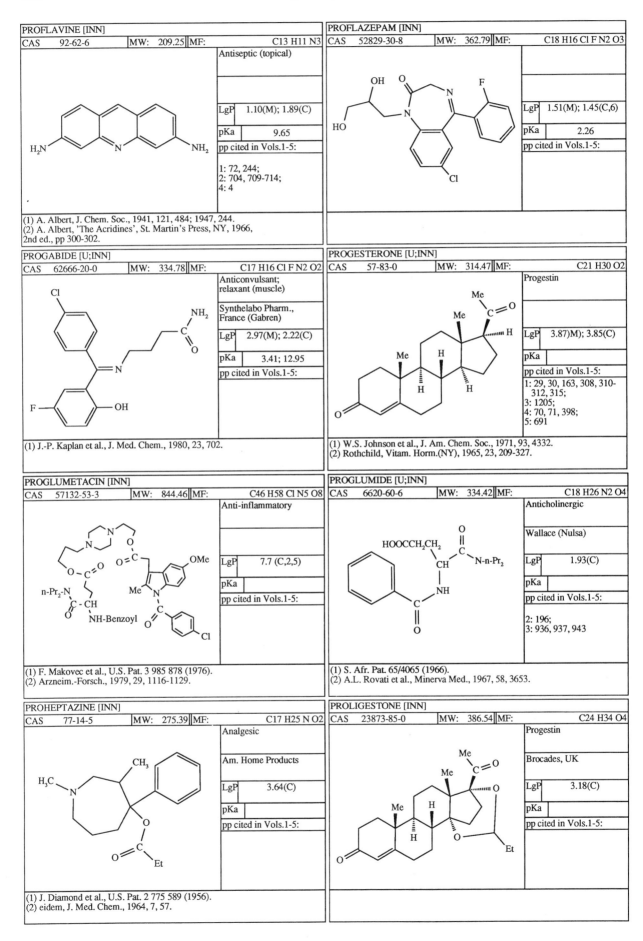

PROFLAVINE [INN]

CAS	92-62-6	MW:	209.25	MF:		C13 H11 N3

Antiseptic (topical)

LgP	1.10(M); 1.89(C)
pKa	9.65

pp cited in Vols.1-5:

1: 72, 244;
2: 704, 709-714;
4: 4

(1) A. Albert, J. Chem. Soc., 1941, 121, 484; 1947, 244.
(2) A. Albert, 'The Acridines', St. Martin's Press, NY, 1966,
2nd ed., pp 300-302.

PROFLAZEPAM [INN]

CAS	52829-30-8	MW:	362.79	MF:		C18 H16 Cl F N2 O3

LgP	1.51(M); 1.45(C,6)
pKa	2.26

pp cited in Vols.1-5:

PROGABIDE [U;INN]

CAS	62666-20-0	MW:	334.78	MF:		C17 H16 Cl F N2 O2

Anticonvulsant;
relaxant (muscle)

Synthelabo Pharm.,
France (Gabren)

LgP	2.97(M); 2.22(C)
pKa	3.41; 12.95

pp cited in Vols.1-5:

(1) J.-P. Kaplan et al., J. Med. Chem., 1980, 23, 702.

PROGESTERONE [U;INN]

CAS	57-83-0	MW:	314.47	MF:		C21 H30 O2

Progestin

LgP	3.87)M); 3.85(C)
pKa	

pp cited in Vols.1-5:

1: 29, 30, 163, 308, 310-
   312, 315;
3: 1205;
4: 70, 71, 398;
5: 691

(1) W.S. Johnson et al., J. Am. Chem. Soc., 1971, 93, 4332.
(2) Rothchild, Vitam. Horm.(NY), 1965, 23, 209-327.

PROGLUMETACIN [INN]

CAS	57132-53-3	MW:	844.46	MF:		C46 H58 Cl N5 O8

Anti-inflammatory

LgP	7.7 (C,2,5)
pKa	

pp cited in Vols.1-5:

(1) F. Makovec et al., U.S. Pat. 3 985 878 (1976).
(2) Arzneim.-Forsch., 1979, 29, 1116-1129.

PROGLUMIDE [U;INN]

CAS	6620-60-6	MW:	334.42	MF:		C18 H26 N2 O4

Anticholinergic

Wallace (Nulsa)

LgP	1.93(C)
pKa	

pp cited in Vols.1-5:

2: 196;
3: 936, 937, 943

(1) S. Afr. Pat. 65/4065 (1966).
(2) A.L. Rovati et al., Minerva Med., 1967, 58, 3653.

PROHEPTAZINE [INN]

CAS	77-14-5	MW:	275.39	MF:		C17 H25 N O2

Analgesic

Am. Home Products

LgP	3.64(C)
pKa	

pp cited in Vols.1-5:

(1) J. Diamond et al., U.S. Pat. 2 775 589 (1956).
(2) eidem, J. Med. Chem., 1964, 7, 57.

PROLIGESTONE [INN]

CAS	23873-85-0	MW:	386.54	MF:		C24 H34 O4

Progestin

Brocades, UK

LgP	3.18(C)
pKa	

pp cited in Vols.1-5:

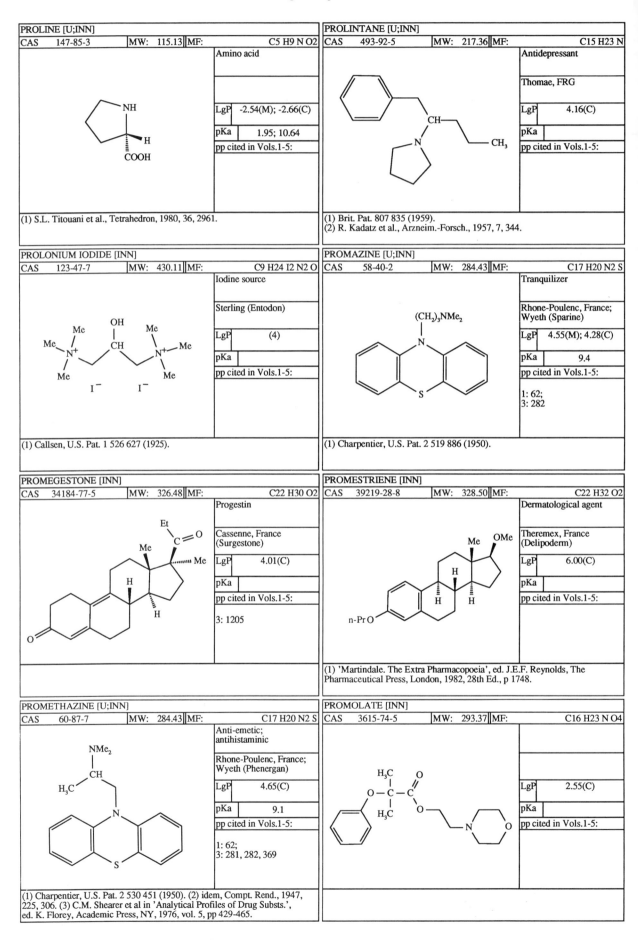

PROLINE [U;INN]

| CAS | 147-85-3 | MW: | 115.13 | MF: | C5 H9 N O2 |

Amino acid

LgP	-2.54(M); -2.66(C)
pKa	1.95; 10.64
pp cited in Vols.1-5:	

(1) S.L. Titouani et al., Tetrahedron, 1980, 36, 2961.

PROLINTANE [U;INN]

| CAS | 493-92-5 | MW: | 217.36 | MF: | C15 H23 N |

Antidepressant

Thomae, FRG

LgP	4.16(C)
pKa	
pp cited in Vols.1-5:	

(1) Brit. Pat. 807 835 (1959).
(2) R. Kadatz et al., Arzneim.-Forsch., 1957, 7, 344.

PROLONIUM IODIDE [INN]

| CAS | 123-47-7 | MW: | 430.11 | MF: | C9 H24 I2 N2 O |

Iodine source

Sterling (Entodon)

LgP	(4)
pKa	
pp cited in Vols.1-5:	

(1) Callsen, U.S. Pat. 1 526 627 (1925).

PROMAZINE [U;INN]

| CAS | 58-40-2 | MW: | 284.43 | MF: | C17 H20 N2 S |

Tranquilizer

Rhone-Poulenc, France; Wyeth (Sparine)

LgP	4.55(M); 4.28(C)
pKa	9.4
pp cited in Vols.1-5:	

1: 62;
3: 282

(1) Charpentier, U.S. Pat. 2 519 886 (1950).

PROMEGESTONE [INN]

| CAS | 34184-77-5 | MW: | 326.48 | MF: | C22 H30 O2 |

Progestin

Cassenne, France (Surgestone)

LgP	4.01(C)
pKa	
pp cited in Vols.1-5:	

3: 1205

PROMESTRIENE [INN]

| CAS | 39219-28-8 | MW: | 328.50 | MF: | C22 H32 O2 |

Dermatological agent

Theremex, France (Delipoderm)

LgP	6.00(C)
pKa	
pp cited in Vols.1-5:	

(1) 'Martindale. The Extra Pharmacopoeia', ed. J.E.F. Reynolds, The Pharmaceutical Press, London, 1982, 28th Ed., p 1748.

PROMETHAZINE [U;INN]

| CAS | 60-87-7 | MW: | 284.43 | MF: | C17 H20 N2 S |

Anti-emetic; antihistaminic

Rhone-Poulenc, France; Wyeth (Phenergan)

LgP	4.65(C)
pKa	9.1
pp cited in Vols.1-5:	

1: 62;
3: 281, 282, 369

(1) Charpentier, U.S. Pat. 2 530 451 (1950). (2) idem, Compt. Rend., 1947, 225, 306. (3) C.M. Shearer et al in 'Analytical Profiles of Drug Substs.', ed. K. Florey, Academic Press, NY, 1976, vol. 5, pp 429-465.

PROMOLATE [INN]

| CAS | 3615-74-5 | MW: | 293.37 | MF: | C16 H23 N O4 |

LgP	2.55(C)
pKa	
pp cited in Vols.1-5:	

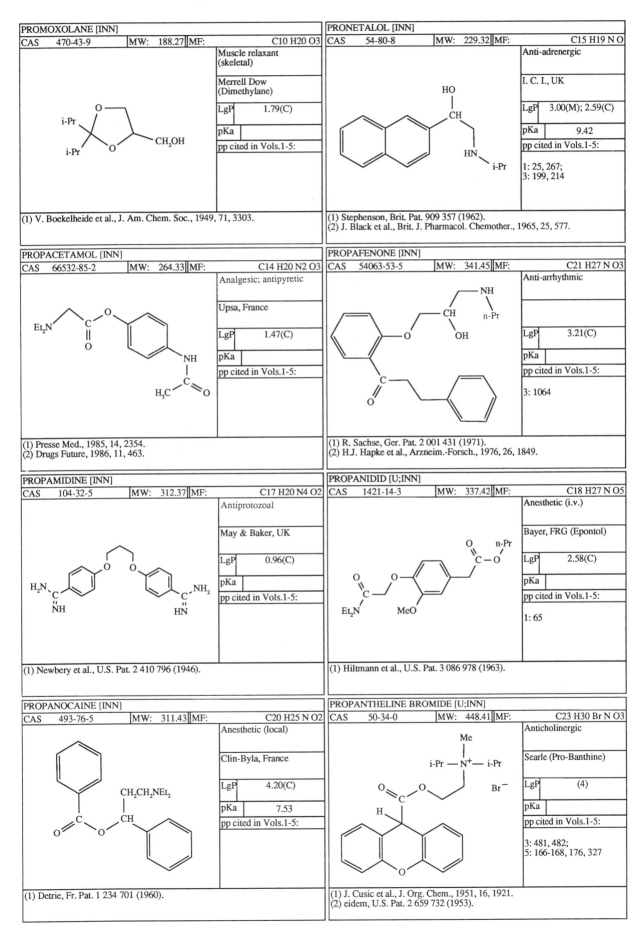

PROMOXOLANE [INN]

CAS	470-43-9	MW:	188.27	MF:	C10 H20 O3

Muscle relaxant (skeletal)

Merrell Dow (Dimethylane)

LgP	1.79(C)
pKa	
pp cited in Vols.1-5:	

(1) V. Boekelheide et al., J. Am. Chem. Soc., 1949, 71, 3303.

PRONETALOL [INN]

CAS	54-80-8	MW:	229.32	MF:	C15 H19 N O

Anti-adrenergic

I. C. I., UK

LgP	3.00(M); 2.59(C)
pKa	9.42
pp cited in Vols.1-5:	

1: 25, 267;
3: 199, 214

(1) Stephenson, Brit. Pat. 909 357 (1962).
(2) J. Black et al., Brit. J. Pharmacol. Chemother., 1965, 25, 577.

PROPACETAMOL [INN]

CAS	66532-85-2	MW:	264.33	MF:	C14 H20 N2 O3

Analgesic; antipyretic

Upsa, France

LgP	1.47(C)
pKa	
pp cited in Vols.1-5:	

(1) Presse Med., 1985, 14, 2354.
(2) Drugs Future, 1986, 11, 463.

PROPAFENONE [INN]

CAS	54063-53-5	MW:	341.45	MF:	C21 H27 N O3

Anti-arrhythmic

LgP	3.21(C)
pKa	
pp cited in Vols.1-5:	

3: 1064

(1) R. Sachse, Ger. Pat. 2 001 431 (1971).
(2) H.J. Hapke et al., Arzneim.-Forsch., 1976, 26, 1849.

PROPAMIDINE [INN]

CAS	104-32-5	MW:	312.37	MF:	C17 H20 N4 O2

Antiprotozoal

May & Baker, UK

LgP	0.96(C)
pKa	
pp cited in Vols.1-5:	

(1) Newbery et al., U.S. Pat. 2 410 796 (1946).

PROPANIDID [U;INN]

CAS	1421-14-3	MW:	337.42	MF:	C18 H27 N O5

Anesthetic (i.v.)

Bayer, FRG (Epontol)

LgP	2.58(C)
pKa	
pp cited in Vols.1-5:	

1: 65

(1) Hiltmann et al., U.S. Pat. 3 086 978 (1963).

PROPANOCAINE [INN]

CAS	493-76-5	MW:	311.43	MF:	C20 H25 N O2

Anesthetic (local)

Clin-Byla, France

LgP	4.20(C)
pKa	7.53
pp cited in Vols.1-5:	

(1) Detrie, Fr. Pat. 1 234 701 (1960).

PROPANTHELINE BROMIDE [U;INN]

CAS	50-34-0	MW:	448.41	MF:	C23 H30 Br N O3

Anticholinergic

Searle (Pro-Banthine)

LgP	(4)
pKa	
pp cited in Vols.1-5:	

3: 481, 482;
5: 166-168, 176, 327

(1) J. Cusic et al., J. Org. Chem., 1951, 16, 1921.
(2) eidem, U.S. Pat. 2 659 732 (1953).

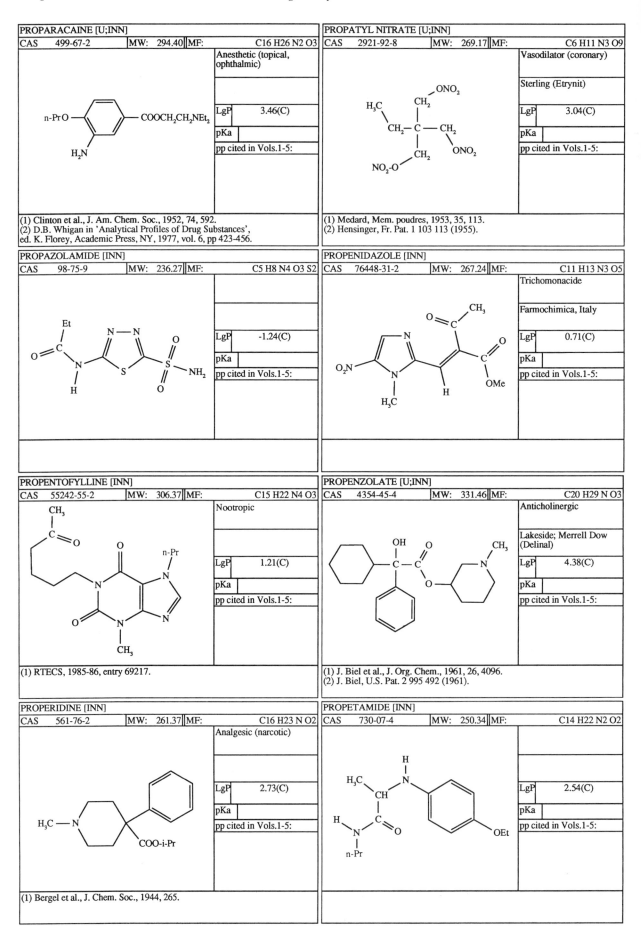

PROPARACAINE [U;INN]		
CAS 499-67-2	MW: 294.40	MF: C16 H26 N2 O3

Anesthetic (topical, ophthalmic)

LgP	3.46(C)
pKa	
pp cited in Vols.1-5:	

(1) Clinton et al., J. Am. Chem. Soc., 1952, 74, 592.
(2) D.B. Whigan in 'Analytical Profiles of Drug Substances',
ed. K. Florey, Academic Press, NY, 1977, vol. 6, pp 423-456.

PROPATYL NITRATE [U;INN]		
CAS 2921-92-8	MW: 269.17	MF: C6 H11 N3 O9

Vasodilator (coronary)

Sterling (Etrynit)

LgP	3.04(C)
pKa	
pp cited in Vols.1-5:	

(1) Medard, Mem. poudres, 1953, 35, 113.
(2) Hensinger, Fr. Pat. 1 103 113 (1955).

PROPAZOLAMIDE [INN]		
CAS 98-75-9	MW: 236.27	MF: C5 H8 N4 O3 S2

LgP	-1.24(C)
pKa	
pp cited in Vols.1-5:	

PROPENIDAZOLE [INN]		
CAS 76448-31-2	MW: 267.24	MF: C11 H13 N3 O5

Trichomonacide

Farmochimica, Italy

LgP	0.71(C)
pKa	
pp cited in Vols.1-5:	

PROPENTOFYLLINE [INN]		
CAS 55242-55-2	MW: 306.37	MF: C15 H22 N4 O3

Nootropic

LgP	1.21(C)
pKa	
pp cited in Vols.1-5:	

(1) RTECS, 1985-86, entry 69217.

PROPENZOLATE [U;INN]		
CAS 4354-45-4	MW: 331.46	MF: C20 H29 N O3

Anticholinergic

Lakeside; Merrell Dow (Delinal)

LgP	4.38(C)
pKa	
pp cited in Vols.1-5:	

(1) J. Biel et al., J. Org. Chem., 1961, 26, 4096.
(2) J. Biel, U.S. Pat. 2 995 492 (1961).

PROPERIDINE [INN]		
CAS 561-76-2	MW: 261.37	MF: C16 H23 N O2

Analgesic (narcotic)

LgP	2.73(C)
pKa	
pp cited in Vols.1-5:	

(1) Bergel et al., J. Chem. Soc., 1944, 265.

PROPETAMIDE [INN]		
CAS 730-07-4	MW: 250.34	MF: C14 H22 N2 O2

LgP	2.54(C)
pKa	
pp cited in Vols.1-5:	

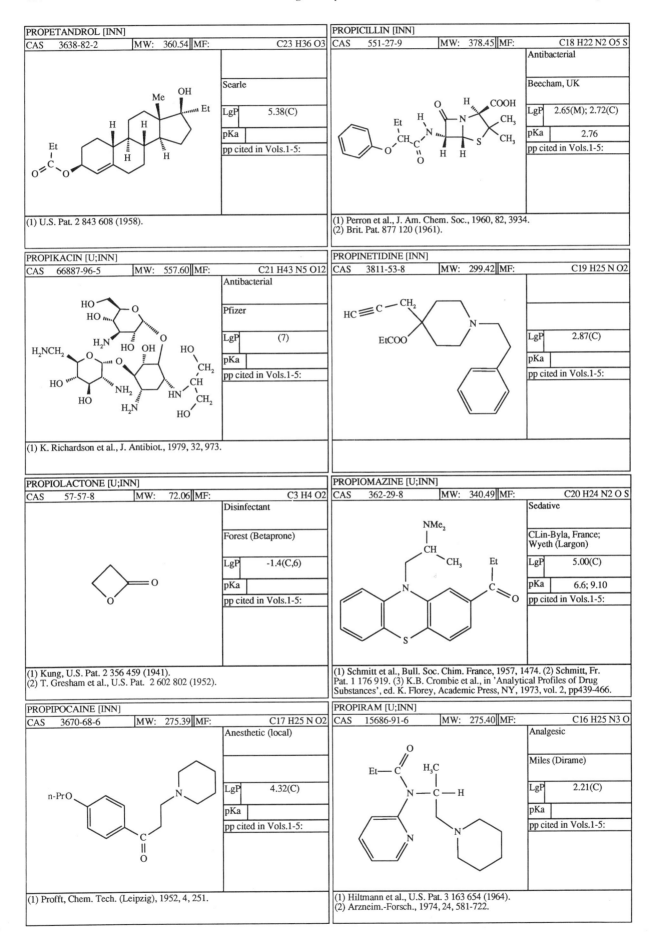

PROPETANDROL [INN]

CAS	3638-82-2	MW:	360.54	MF:	C23 H36 O3

Searle

LgP   5.38(C)

pKa

pp cited in Vols.1-5:

Me   OH   Et   H   H   H   Et   O

(1) U.S. Pat. 2 843 608 (1958).

PROPICILLIN [INN]

CAS	551-27-9	MW:	378.45	MF:	C18 H22 N2 O5 S

Antibacterial

Beecham, UK

LgP   2.65(M); 2.72(C)

pKa   2.76

pp cited in Vols.1-5:

(1) Perron et al., J. Am. Chem. Soc., 1960, 82, 3934.
(2) Brit. Pat. 877 120 (1961).

PROPIKACIN [U;INN]

CAS	66887-96-5	MW:	557.60	MF:	C21 H43 N5 O12

Antibacterial

Pfizer

LgP   (7)

pKa

pp cited in Vols.1-5:

(1) K. Richardson et al., J. Antibiot., 1979, 32, 973.

PROPINETIDINE [INN]

CAS	3811-53-8	MW:	299.42	MF:	C19 H25 N O2

LgP   2.87(C)

pKa

pp cited in Vols.1-5:

PROPIOLACTONE [U;INN]

CAS	57-57-8	MW:	72.06	MF:	C3 H4 O2

Disinfectant

Forest (Betaprone)

LgP   -1.4(C,6)

pKa

pp cited in Vols.1-5:

(1) Kung, U.S. Pat. 2 356 459 (1941).
(2) T. Gresham et al., U.S. Pat. 2 602 802 (1952).

PROPIOMAZINE [U;INN]

CAS	362-29-8	MW:	340.49	MF:	C20 H24 N2 O S

Sedative

CLin-Byla, France;
Wyeth (Largon)

LgP   5.00(C)

pKa   6.6; 9.10

pp cited in Vols.1-5:

(1) Schmitt et al., Bull. Soc. Chim. France, 1957, 1474. (2) Schmitt, Fr.
Pat. 1 176 919. (3) K.B. Crombie et al., in 'Analytical Profiles of Drug
Substances', ed. K. Florey, Academic Press, NY, 1973, vol. 2, pp439-466.

PROPIPOCAINE [INN]

CAS	3670-68-6	MW:	275.39	MF:	C17 H25 N O2

Anesthetic (local)

LgP   4.32(C)

pKa

pp cited in Vols.1-5:

(1) Profft, Chem. Tech. (Leipzig), 1952, 4, 251.

PROPIRAM [U;INN]

CAS	15686-91-6	MW:	275.40	MF:	C16 H25 N3 O

Analgesic

Miles (Dirame)

LgP   2.21(C)

pKa

pp cited in Vols.1-5:

(1) Hiltmann et al., U.S. Pat. 3 163 654 (1964).
(2) Arzneim.-Forsch., 1974, 24, 581-722.

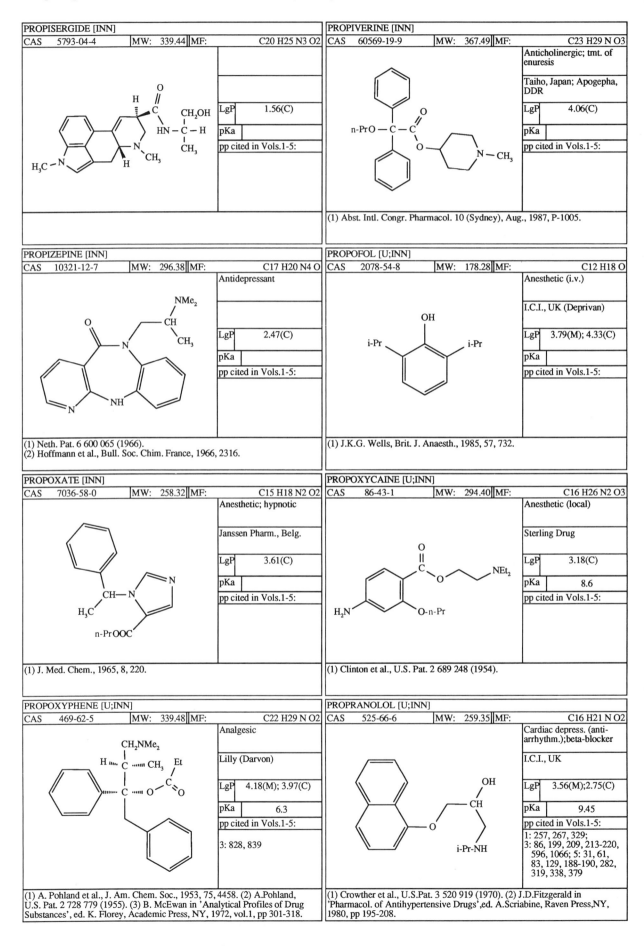

PROPISERGIDE [INN]

CAS	5793-04-4	MW: 339.44	MF:	C20 H25 N3 O2

LgP	1.56(C)
pKa	
pp cited in Vols.1-5:	

PROPIVERINE [INN]

CAS	60569-19-9	MW: 367.49	MF:	C23 H29 N O3

Anticholinergic; tmt. of enuresis

Taiho, Japan; Apogepha, DDR

LgP	4.06(C)
pKa	
pp cited in Vols.1-5:	

(1) Abst. Intl. Congr. Pharmacol. 10 (Sydney), Aug., 1987, P-1005.

PROPIZEPINE [INN]

CAS	10321-12-7	MW: 296.38	MF:	C17 H20 N4 O

Antidepressant

LgP	2.47(C)
pKa	
pp cited in Vols.1-5:	

(1) Neth. Pat. 6 600 065 (1966).
(2) Hoffmann et al., Bull. Soc. Chim. France, 1966, 2316.

PROPOFOL [U;INN]

CAS	2078-54-8	MW: 178.28	MF:	C12 H18 O

Anesthetic (i.v.)

I.C.I., UK (Deprivan)

LgP	3.79(M); 4.33(C)
pKa	
pp cited in Vols.1-5:	

(1) J.K.G. Wells, Brit. J. Anaesth., 1985, 57, 732.

PROPOXATE [INN]

CAS	7036-58-0	MW: 258.32	MF:	C15 H18 N2 O2

Anesthetic; hypnotic

Janssen Pharm., Belg.

LgP	3.61(C)
pKa	
pp cited in Vols.1-5:	

(1) J. Med. Chem., 1965, 8, 220.

PROPOXYCAINE [U;INN]

CAS	86-43-1	MW: 294.40	MF:	C16 H26 N2 O3

Anesthetic (local)

Sterling Drug

LgP	3.18(C)
pKa	8.6
pp cited in Vols.1-5:	

(1) Clinton et al., U.S. Pat. 2 689 248 (1954).

PROPOXYPHENE [U;INN]

CAS	469-62-5	MW: 339.48	MF:	C22 H29 N O2

Analgesic

Lilly (Darvon)

LgP	4.18(M); 3.97(C)
pKa	6.3
pp cited in Vols.1-5:	
3: 828, 839	

(1) A. Pohland et al., J. Am. Chem. Soc., 1953, 75, 4458. (2) A.Pohland, U.S. Pat. 2 728 779 (1955). (3) B. McEwan in 'Analytical Profiles of Drug Substances', ed. K. Florey, Academic Press, NY, 1972, vol.1, pp 301-318.

PROPRANOLOL [U;INN]

CAS	525-66-6	MW: 259.35	MF:	C16 H21 N O2

Cardiac depress. (anti-arrhythm.);beta-blocker

I.C.I., UK

LgP	3.56(M);2.75(C)
pKa	9.45
pp cited in Vols.1-5:	
1: 257, 267, 329; 3: 86, 199, 209, 213-220, 596, 1066; 5: 31, 61, 83, 129, 188-190, 282, 319, 338, 379	

(1) Crowther et al., U.S.Pat. 3 520 919 (1970). (2) J.D.Fitzgerald in 'Pharmacol. of Antihypertensive Drugs',ed. A.Scriabine, Raven Press,NY, 1980, pp 195-208.

PROPYL DOCETRIZOATE [INN]

CAS	5579-08-8	MW:	640.99	MF:	C14 H14 I3 N O4

Radiopaque medium

May & Baker, UK

LgP	5.10(C)
pKa	
pp cited in Vols.1-5:	

(1) Ashley et al., Brit. Pat., 898 780 (1962).

PROPYLENE GLYCOL [U]

CAS	57-55-6	MW:	76.10	MF:	C3 H8 O2

Humectant; solvent; suspending agent

$H_3C - CHOHCH_2OH$

LgP	-0.92(M); -1.06(C)
pKa	14.7
pp cited in Vols.1-5:	

(1) Ruddick, Toxicol. Appl. Pharmacol., 1972, 21, 102.

PROPYLENE GLYCOL MONOSTEARATE [U]

CAS	1323-39-3	MW:	342.57	MF:	C21 H42 O3

Non-ionic surfactant; food stabilizer

$n-C_{17}H_{35}$

LgP	8.3(C,2,5)
pKa	
pp cited in Vols.1-5:	

(1) 'Martindale. The Extra Pharmacopoeia', ed. J.E.F. Reynolds, The Pharmaceutical Press, London, 1982, 28th Ed., p 372

PROPYL GALLATE [U]

CAS	121-79-9	MW:	212.20	MF:	C10 H12 O5

Anti-oxidant

LgP	1.80(M); 2.11(C)
pKa	
pp cited in Vols.1-5:	

(1) M.A. Augustin et al., J. Am. Oil Chem. Soc., 1983, 60, 105.

PROPYLHEXEDRINE [U;INN]

CAS	101-40-6	MW:	155.29	MF:	C10 H21 N

Adrenergic (vasoconstrictor)

Smith Kline & French (Benzedrex)

LgP	2.97(C)
pKa	10.42
pp cited in Vols.1-5:	

(1) G.E. Ullyot, U.S. Pat. 2 454 746 (1948).
(2) Zenitz et al., J. Am. Chem. Soc., 1947, 69, 1117.

PROPYLIODONE [U;INN]

CAS	587-61-1	MW:	447.01	MF:	C10 H11 I2 N O3

Radiopaque medium

I.C.I., UK

LgP	1.90(C)
pKa	
pp cited in Vols.1-5:	

(1) Branscombe, Brit. Pat. 517 382 (1940).
(2) Tomich et al., Brit. J. Pharmacol., 1953, 8, 166.

PROPYLPARABEN [U]

CAS	94-13-3	MW:	180.21	MF:	C10 H12 O3

Antifungal; preservative

LgP	3.04(M); 3.04(C)
pKa	8.14
pp cited in Vols.1-5:	

(1) Arzneim.-Forsch., 1954, 4, 575.

PROPYLTHIOURACIL [U;INN]

CAS	51-52-5	MW:	170.23	MF:	C7 H10 N2 O S

Thyroid inhibitor

Lederle

LgP	(4)
pKa	7.8; 8.3
pp cited in Vols.1-5:	
	1: 69

(1) Anderson et al., J. Am. Chem. Soc., 1945, 67, 2197.
(2) H.Y. Aboul-Enein, 'Analytical Profiles of Drug Substances', ed. K. Florey, Academic Press, NY, 1977, vol. 6, pp 457-486.

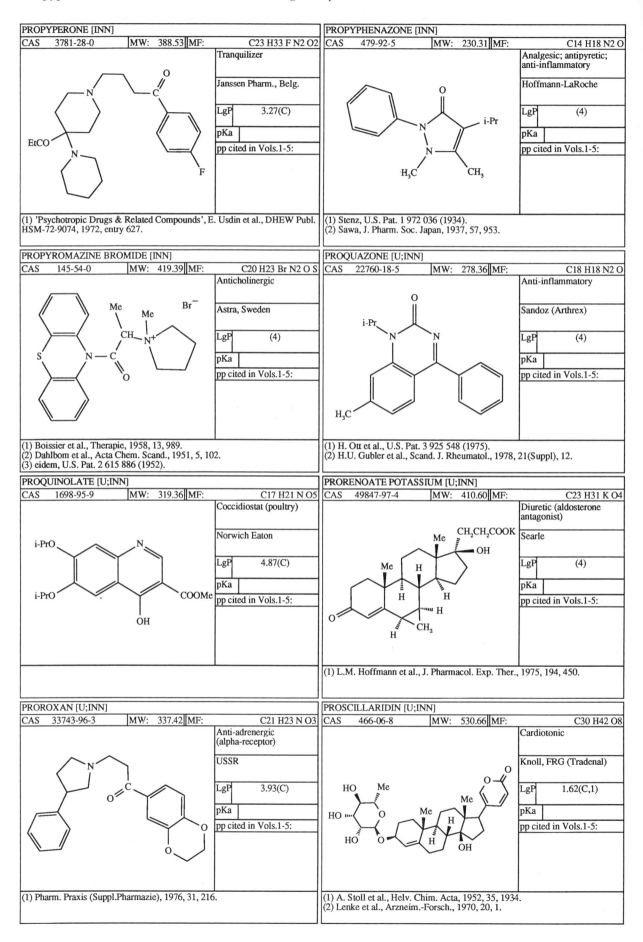

PROPYPERONE [INN]		
CAS 3781-28-0	MW: 388.53 MF:	C23 H33 F N2 O2

Tranquilizer

Janssen Pharm., Belg.

LgP	3.27(C)
pKa	
pp cited in Vols.1-5:	

(1) 'Psychotropic Drugs & Related Compounds', E. Usdin et al., DHEW Publ. HSM-72-9074, 1972, entry 627.

PROPYPHENAZONE [INN]		
CAS 479-92-5	MW: 230.31 MF:	C14 H18 N2 O

Analgesic; antipyretic; anti-inflammatory

Hoffmann-LaRoche

LgP	(4)
pKa	
pp cited in Vols.1-5:	

(1) Stenz, U.S. Pat. 1 972 036 (1934).
(2) Sawa, J. Pharm. Soc. Japan, 1937, 57, 953.

PROPYROMAZINE BROMIDE [INN]		
CAS 145-54-0	MW: 419.39 MF:	C20 H23 Br N2 O S

Anticholinergic

Astra, Sweden

LgP	(4)
pKa	
pp cited in Vols.1-5:	

(1) Boissier et al., Therapie, 1958, 13, 989.
(2) Dahlbom et al., Acta Chem. Scand., 1951, 5, 102.
(3) eidem, U.S. Pat. 2 615 886 (1952).

PROQUAZONE [U;INN]		
CAS 22760-18-5	MW: 278.36 MF:	C18 H18 N2 O

Anti-inflammatory

Sandoz (Arthrex)

LgP	(4)
pKa	
pp cited in Vols.1-5:	

(1) H. Ott et al., U.S. Pat. 3 925 548 (1975).
(2) H.U. Gubler et al., Scand. J. Rheumatol., 1978, 21(Suppl), 12.

PROQUINOLATE [U;INN]		
CAS 1698-95-9	MW: 319.36 MF:	C17 H21 N O5

Coccidiostat (poultry)

Norwich Eaton

LgP	4.87(C)
pKa	
pp cited in Vols.1-5:	

PRORENOATE POTASSIUM [U;INN]		
CAS 49847-97-4	MW: 410.60 MF:	C23 H31 K O4

Diuretic (aldosterone antagonist)

Searle

LgP	(4)
pKa	
pp cited in Vols.1-5:	

(1) L.M. Hoffmann et al., J. Pharmacol. Exp. Ther., 1975, 194, 450.

PROROXAN [U;INN]		
CAS 33743-96-3	MW: 337.42 MF:	C21 H23 N O3

Anti-adrenergic (alpha-receptor)

USSR

LgP	3.93(C)
pKa	
pp cited in Vols.1-5:	

(1) Pharm. Praxis (Suppl.Pharmazie), 1976, 31, 216.

PROSCILLARIDIN [U;INN]		
CAS 466-06-8	MW: 530.66 MF:	C30 H42 O8

Cardiotonic

Knoll, FRG (Tradenal)

LgP	1.62(C,1)
pKa	
pp cited in Vols.1-5:	

(1) A. Stoll et al., Helv. Chim. Acta, 1952, 35, 1934.
(2) Lenke et al., Arzneim.-Forsch., 1970, 20, 1.

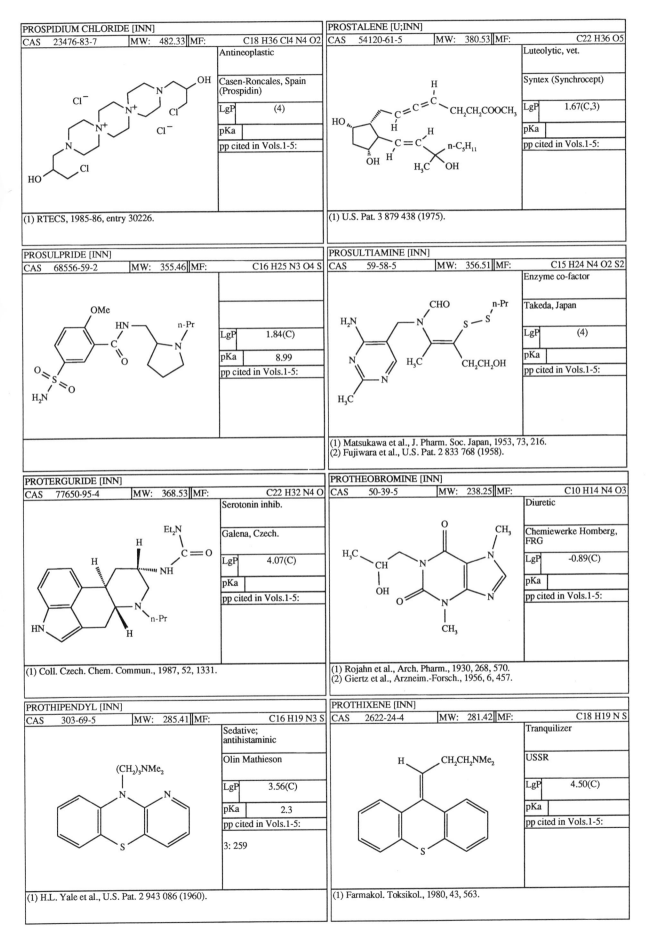

PROSPIDIUM CHLORIDE [INN]

CAS	23476-83-7	MW:	482.33	MF:	C18 H36 Cl4 N4 O2

Antineoplastic

Casen-Roncales, Spain (Prospidin)

LgP	(4)
pKa	
pp cited in Vols.1-5:	

(1) RTECS, 1985-86, entry 30226.

PROSTALENE [U;INN]

CAS	54120-61-5	MW:	380.53	MF:	C22 H36 O5

Luteolytic, vet.

Syntex (Synchrocept)

LgP	1.67(C,3)
pKa	
pp cited in Vols.1-5:	

(1) U.S. Pat. 3 879 438 (1975).

PROSULPRIDE [INN]

CAS	68556-59-2	MW:	355.46	MF:	C16 H25 N3 O4 S

LgP	1.84(C)
pKa	8.99
pp cited in Vols.1-5:	

PROSULTIAMINE [INN]

CAS	59-58-5	MW:	356.51	MF:	C15 H24 N4 O2 S2

Enzyme co-factor

Takeda, Japan

LgP	(4)
pKa	
pp cited in Vols.1-5:	

(1) Matsukawa et al., J. Pharm. Soc. Japan, 1953, 73, 216.
(2) Fujiwara et al., U.S. Pat. 2 833 768 (1958).

PROTERGURIDE [INN]

CAS	77650-95-4	MW:	368.53	MF:	C22 H32 N4 O

Serotonin inhib.

Galena, Czech.

LgP	4.07(C)
pKa	
pp cited in Vols.1-5:	

(1) Coll. Czech. Chem. Commun., 1987, 52, 1331.

PROTHEOBROMINE [INN]

CAS	50-39-5	MW:	238.25	MF:	C10 H14 N4 O3

Diuretic

Chemiewerke Homberg, FRG

LgP	-0.89(C)
pKa	
pp cited in Vols.1-5:	

(1) Rojahn et al., Arch. Pharm., 1930, 268, 570.
(2) Giertz et al., Arzneim.-Forsch., 1956, 6, 457.

PROTHIPENDYL [INN]

CAS	303-69-5	MW:	285.41	MF:	C16 H19 N3 S

Sedative; antihistaminic

Olin Mathieson

LgP	3.56(C)
pKa	2.3
pp cited in Vols.1-5:	
3: 259	

(1) H.L. Yale et al., U.S. Pat. 2 943 086 (1960).

PROTHIXENE [INN]

CAS	2622-24-4	MW:	281.42	MF:	C18 H19 N S

Tranquilizer

USSR

LgP	4.50(C)
pKa	
pp cited in Vols.1-5:	

(1) Farmakol. Toksikol., 1980, 43, 563.

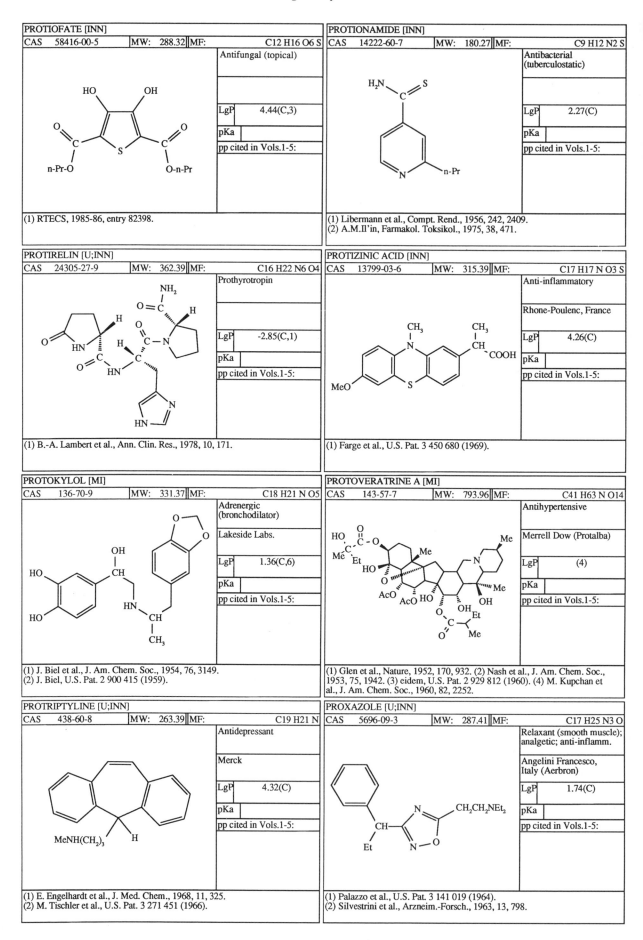

PROTIOFATE [INN]			
CAS 58416-00-5	MW: 288.32	MF:	C12 H16 O6 S
			Antifungal (topical)
		LgP	4.44(C,3)
		pKa	
		pp cited in Vols.1-5:	

(1) RTECS, 1985-86, entry 82398.

PROTIRELIN [U;INN]			
CAS 24305-27-9	MW: 362.39	MF:	C16 H22 N6 O4
			Prothyrotropin
		LgP	-2.85(C,1)
		pKa	
		pp cited in Vols.1-5:	

(1) B.-A. Lambert et al., Ann. Clin. Res., 1978, 10, 171.

PROTOKYLOL [MI]			
CAS 136-70-9	MW: 331.37	MF:	C18 H21 N O5
			Adrenergic (bronchodilator)
			Lakeside Labs.
		LgP	1.36(C,6)
		pKa	
		pp cited in Vols.1-5:	

(1) J. Biel et al., J. Am. Chem. Soc., 1954, 76, 3149.
(2) J. Biel, U.S. Pat. 2 900 415 (1959).

PROTRIPTYLINE [U;INN]			
CAS 438-60-8	MW: 263.39	MF:	C19 H21 N
			Antidepressant
			Merck
		LgP	4.32(C)
		pKa	
		pp cited in Vols.1-5:	

(1) E. Engelhardt et al., J. Med. Chem., 1968, 11, 325.
(2) M. Tischler et al., U.S. Pat. 3 271 451 (1966).

PROTIONAMIDE [INN]			
CAS 14222-60-7	MW: 180.27	MF:	C9 H12 N2 S
			Antibacterial (tuberculostatic)
		LgP	2.27(C)
		pKa	
		pp cited in Vols.1-5:	

(1) Libermann et al., Compt. Rend., 1956, 242, 2409.
(2) A.M.Il'in, Farmakol. Toksikol., 1975, 38, 471.

PROTIZINIC ACID [INN]			
CAS 13799-03-6	MW: 315.39	MF:	C17 H17 N O3 S
			Anti-inflammatory
			Rhone-Poulenc, France
		LgP	4.26(C)
		pKa	
		pp cited in Vols.1-5:	

(1) Farge et al., U.S. Pat. 3 450 680 (1969).

PROTOVERATRINE A [MI]			
CAS 143-57-7	MW: 793.96	MF:	C41 H63 N O14
			Antihypertensive
			Merrell Dow (Protalba)
		LgP	(4)
		pKa	
		pp cited in Vols.1-5:	

(1) Glen et al., Nature, 1952, 170, 932. (2) Nash et al., J. Am. Chem. Soc.,
1953, 75, 1942. (3) eidem, U.S. Pat. 2 929 812 (1960). (4) M. Kupchan et
al., J. Am. Chem. Soc., 1960, 82, 2252.

PROXAZOLE [U;INN]			
CAS 5696-09-3	MW: 287.41	MF:	C17 H25 N3 O
			Relaxant (smooth muscle); analgetic; anti-inflamm.
			Angelini Francesco, Italy (Aerbron)
		LgP	1.74(C)
		pKa	
		pp cited in Vols.1-5:	

(1) Palazzo et al., U.S. Pat. 3 141 019 (1964).
(2) Silvestrini et al., Arzneim.-Forsch., 1963, 13, 798.

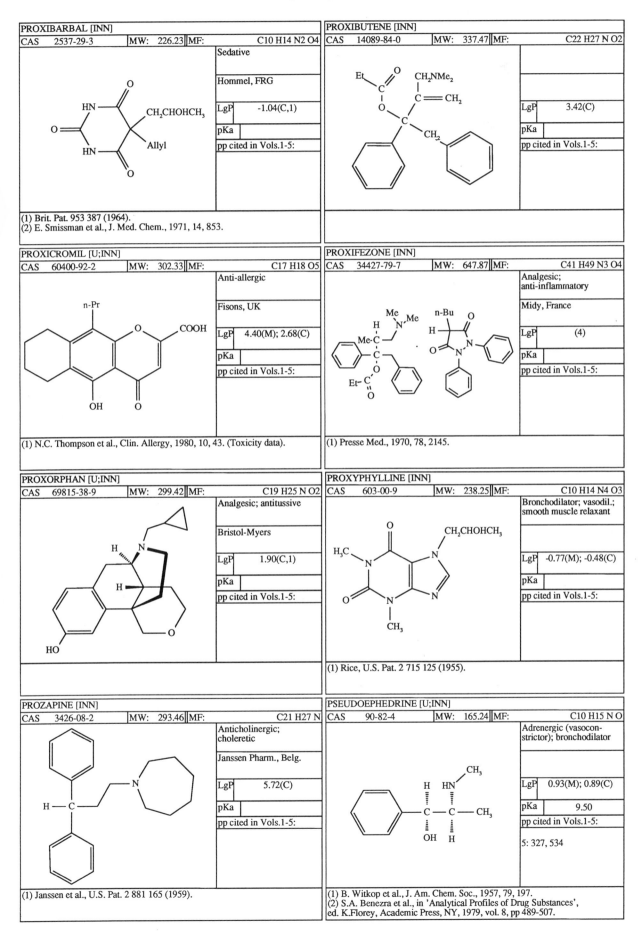

PROXIBARBAL [INN]

| CAS | 2537-29-3 | MW: | 226.23 | MF: | C10 H14 N2 O4 |

Sedative

Hommel, FRG

LgP	-1.04(C,1)
pKa	
pp cited in Vols.1-5:	

(1) Brit. Pat. 953 387 (1964).
(2) E. Smissman et al., J. Med. Chem., 1971, 14, 853.

PROXIBUTENE [INN]

| CAS | 14089-84-0 | MW: | 337.47 | MF: | C22 H27 N O2 |

LgP	3.42(C)
pKa	
pp cited in Vols.1-5:	

PROXICROMIL [U;INN]

| CAS | 60400-92-2 | MW: | 302.33 | MF: | C17 H18 O5 |

Anti-allergic

Fisons, UK

LgP	4.40(M); 2.68(C)
pKa	
pp cited in Vols.1-5:	

(1) N.C. Thompson et al., Clin. Allergy, 1980, 10, 43. (Toxicity data).

PROXIFEZONE [INN]

| CAS | 34427-79-7 | MW: | 647.87 | MF: | C41 H49 N3 O4 |

Analgesic;
anti-inflammatory

Midy, France

LgP	(4)
pKa	
pp cited in Vols.1-5:	

(1) Presse Med., 1970, 78, 2145.

PROXORPHAN [U;INN]

| CAS | 69815-38-9 | MW: | 299.42 | MF: | C19 H25 N O2 |

Analgesic; antitussive

Bristol-Myers

LgP	1.90(C,1)
pKa	
pp cited in Vols.1-5:	

PROXYPHYLLINE [INN]

| CAS | 603-00-9 | MW: | 238.25 | MF: | C10 H14 N4 O3 |

Bronchodilator; vasodil.;
smooth muscle relaxant

LgP	-0.77(M); -0.48(C)
pKa	
pp cited in Vols.1-5:	

(1) Rice, U.S. Pat. 2 715 125 (1955).

PROZAPINE [INN]

| CAS | 3426-08-2 | MW: | 293.46 | MF: | C21 H27 N |

Anticholinergic;
choleretic

Janssen Pharm., Belg.

LgP	5.72(C)
pKa	
pp cited in Vols.1-5:	

(1) Janssen et al., U.S. Pat. 2 881 165 (1959).

PSEUDOEPHEDRINE [U;INN]

| CAS | 90-82-4 | MW: | 165.24 | MF: | C10 H15 N O |

Adrenergic (vasocon-
strictor); bronchodilator

LgP	0.93(M); 0.89(C)
pKa	9.50
pp cited in Vols.1-5:	

5: 327, 534

(1) B. Witkop et al., J. Am. Chem. Soc., 1957, 79, 197.
(2) S.A. Benezra et al., in 'Analytical Profiles of Drug Substances',
ed. K.Florey, Academic Press, NY, 1979, vol. 8, pp 489-507.

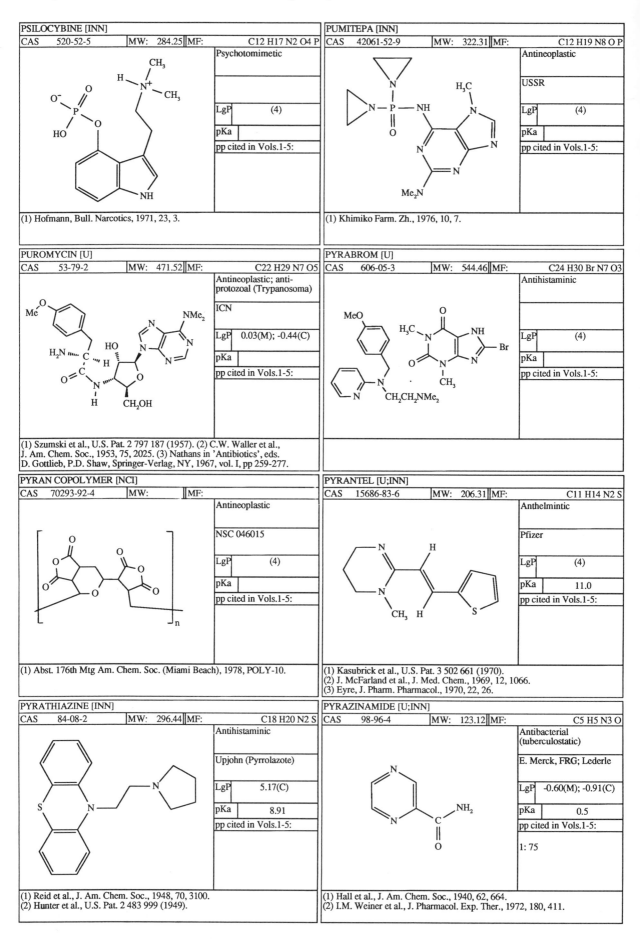

PSILOCYBINE [INN]

CAS	520-52-5	MW:	284.25	MF:	C12 H17 N2 O4 P

Psychotomimetic

LgP	(4)
pKa	
pp cited in Vols.1-5:	

(1) Hofmann, Bull. Narcotics, 1971, 23, 3.

PUMITEPA [INN]

CAS	42061-52-9	MW:	322.31	MF:	C12 H19 N8 O P

Antineoplastic

USSR

LgP	(4)
pKa	
pp cited in Vols.1-5:	

(1) Khimiko Farm. Zh., 1976, 10, 7.

PUROMYCIN [U]

CAS	53-79-2	MW:	471.52	MF:	C22 H29 N7 O5

Antineoplastic; anti-protozoal (Trypanosoma)

ICN

LgP	0.03(M); -0.44(C)
pKa	
pp cited in Vols.1-5:	

(1) Szumski et al., U.S. Pat. 2 797 187 (1957). (2) C.W. Waller et al., J. Am. Chem. Soc., 1953, 75, 2025. (3) Nathans in 'Antibiotics', eds. D. Gottlieb, P.D. Shaw, Springer-Verlag, NY, 1967, vol. I, pp 259-277.

PYRABROM [U]

CAS	606-05-3	MW:	544.46	MF:	C24 H30 Br N7 O3

Antihistaminic

LgP	(4)
pKa	
pp cited in Vols.1-5:	

PYRAN COPOLYMER [NCI]

CAS	70293-92-4	MW:		MF:	

Antineoplastic

NSC 046015

LgP	(4)
pKa	
pp cited in Vols.1-5:	

(1) Abst. 176th Mtg Am. Chem. Soc. (Miami Beach), 1978, POLY-10.

PYRANTEL [U;INN]

CAS	15686-83-6	MW:	206.31	MF:	C11 H14 N2 S

Anthelmintic

Pfizer

LgP	(4)
pKa	11.0
pp cited in Vols.1-5:	

(1) Kasubrick et al., U.S. Pat. 3 502 661 (1970).
(2) J. McFarland et al., J. Med. Chem., 1969, 12, 1066.
(3) Eyre, J. Pharm. Pharmacol., 1970, 22, 26.

PYRATHIAZINE [INN]

CAS	84-08-2	MW:	296.44	MF:	C18 H20 N2 S

Antihistaminic

Upjohn (Pyrrolazote)

LgP	5.17(C)
pKa	8.91
pp cited in Vols.1-5:	

(1) Reid et al., J. Am. Chem. Soc., 1948, 70, 3100.
(2) Hunter et al., U.S. Pat. 2 483 999 (1949).

PYRAZINAMIDE [U;INN]

CAS	98-96-4	MW:	123.12	MF:	C5 H5 N3 O

Antibacterial (tuberculostatic)

E. Merck, FRG; Lederle

LgP	-0.60(M); -0.91(C)
pKa	0.5
pp cited in Vols.1-5:	
	1: 75

(1) Hall et al., J. Am. Chem. Soc., 1940, 62, 664.
(2) I.M. Weiner et al., J. Pharmacol. Exp. Ther., 1972, 180, 411.

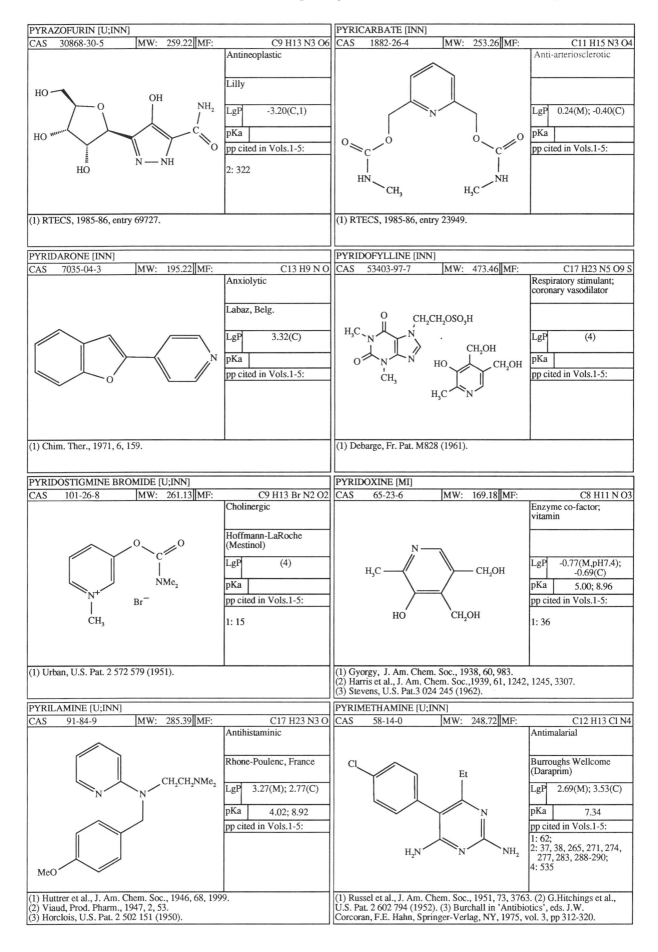

PYRAZOFURIN [U;INN]			
CAS 30868-30-5	MW: 259.22	MF:	C9 H13 N3 O6

Antineoplastic

Lilly

LgP	-3.20(C,1)
pKa	
pp cited in Vols.1-5:	

2: 322

(1) RTECS, 1985-86, entry 69727.

PYRICARBATE [INN]			
CAS 1882-26-4	MW: 253.26	MF:	C11 H15 N3 O4

Anti-arteriosclerotic

LgP	0.24(M); -0.40(C)
pKa	
pp cited in Vols.1-5:	

(1) RTECS, 1985-86, entry 23949.

PYRIDARONE [INN]			
CAS 7035-04-3	MW: 195.22	MF:	C13 H9 N O

Anxiolytic

Labaz, Belg.

LgP	3.32(C)
pKa	
pp cited in Vols.1-5:	

(1) Chim. Ther., 1971, 6, 159.

PYRIDOFYLLINE [INN]			
CAS 53403-97-7	MW: 473.46	MF:	C17 H23 N5 O9 S

Respiratory stimulant; coronary vasodilator

LgP	(4)
pKa	
pp cited in Vols.1-5:	

(1) Debarge, Fr. Pat. M828 (1961).

PYRIDOSTIGMINE BROMIDE [U;INN]			
CAS 101-26-8	MW: 261.13	MF:	C9 H13 Br N2 O2

Cholinergic

Hoffmann-LaRoche (Mestinol)

LgP	(4)
pKa	
pp cited in Vols.1-5:	

1: 15

(1) Urban, U.S. Pat. 2 572 579 (1951).

PYRIDOXINE [MI]			
CAS 65-23-6	MW: 169.18	MF:	C8 H11 N O3

Enzyme co-factor; vitamin

LgP	-0.77(M,pH7.4); -0.69(C)
pKa	5.00; 8.96
pp cited in Vols.1-5:	

1: 36

(1) Gyorgy, J. Am. Chem. Soc., 1938, 60, 983.
(2) Harris et al., J. Am. Chem. Soc.,1939, 61, 1242, 1245, 3307.
(3) Stevens, U.S. Pat.3 024 245 (1962).

PYRILAMINE [U;INN]			
CAS 91-84-9	MW: 285.39	MF:	C17 H23 N3 O

Antihistaminic

Rhone-Poulenc, France

LgP	3.27(M); 2.77(C)
pKa	4.02; 8.92
pp cited in Vols.1-5:	

(1) Huttrer et al., J. Am. Chem. Soc., 1946, 68, 1999.
(2) Viaud, Prod. Pharm., 1947, 2, 53.
(3) Horclois, U.S. Pat. 2 502 151 (1950).

PYRIMETHAMINE [U;INN]			
CAS 58-14-0	MW: 248.72	MF:	C12 H13 Cl N4

Antimalarial

Burroughs Wellcome (Daraprim)

LgP	2.69(M); 3.53(C)
pKa	7.34
pp cited in Vols.1-5:	

1: 62;
2: 37, 38, 265, 271, 274, 277, 283, 288-290;
4: 535

(1) Russel et al., J. Am. Chem. Soc., 1951, 73, 3763. (2) G.Hitchings et al., U.S. Pat. 2 602 794 (1952). (3) Burchall in 'Antibiotics', eds. J.W. Corcoran, F.E. Hahn, Springer-Verlag, NY, 1975, vol. 3, pp 312-320.

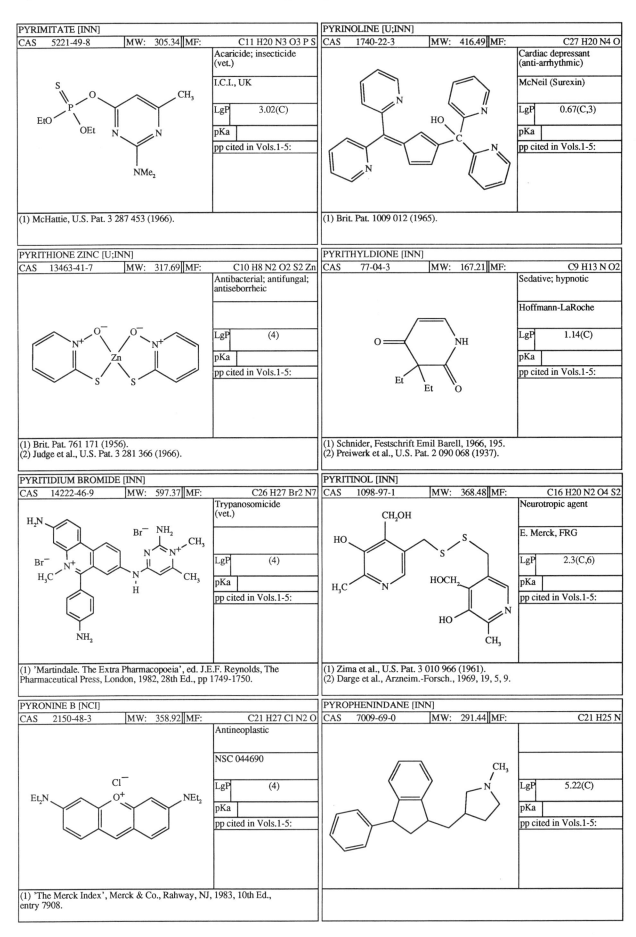

PYRIMITATE [INN]			
CAS 5221-49-8	MW: 305.34	MF:	C11 H20 N3 O3 P S

Acaricide; insecticide (vet.)

I.C.I., UK

LgP 3.02(C)

pKa

pp cited in Vols.1-5:

(1) McHattie, U.S. Pat. 3 287 453 (1966).

PYRINOLINE [U;INN]			
CAS 1740-22-3	MW: 416.49	MF:	C27 H20 N4 O

Cardiac depressant (anti-arrhythmic)

McNeil (Surexin)

LgP 0.67(C,3)

pKa

pp cited in Vols.1-5:

(1) Brit. Pat. 1009 012 (1965).

PYRITHIONE ZINC [U;INN]			
CAS 13463-41-7	MW: 317.69	MF:	C10 H8 N2 O2 S2 Zn

Antibacterial; antifungal; antiseborrheic

LgP (4)

pKa

pp cited in Vols.1-5:

(1) Brit. Pat. 761 171 (1956).
(2) Judge et al., U.S. Pat. 3 281 366 (1966).

PYRITHYLDIONE [INN]			
CAS 77-04-3	MW: 167.21	MF:	C9 H13 N O2

Sedative; hypnotic

Hoffmann-LaRoche

LgP 1.14(C)

pKa

pp cited in Vols.1-5:

(1) Schnider, Festschrift Emil Barell, 1966, 195.
(2) Preiwerk et al., U.S. Pat. 2 090 068 (1937).

PYRITIDIUM BROMIDE [INN]			
CAS 14222-46-9	MW: 597.37	MF:	C26 H27 Br2 N7

Trypanosomicide (vet.)

LgP (4)

pKa

pp cited in Vols.1-5:

(1) 'Martindale. The Extra Pharmacopoeia', ed. J.E.F. Reynolds, The Pharmaceutical Press, London, 1982, 28th Ed., pp 1749-1750.

PYRITINOL [INN]			
CAS 1098-97-1	MW: 368.48	MF:	C16 H20 N2 O4 S2

Neurotropic agent

E. Merck, FRG

LgP 2.3(C,6)

pKa

pp cited in Vols.1-5:

(1) Zima et al., U.S. Pat. 3 010 966 (1961).
(2) Darge et al., Arzneim.-Forsch., 1969, 19, 5, 9.

PYRONINE B [NCI]			
CAS 2150-48-3	MW: 358.92	MF:	C21 H27 Cl N2 O

Antineoplastic

NSC 044690

LgP (4)

pKa

pp cited in Vols.1-5:

(1) 'The Merck Index', Merck & Co., Rahway, NJ, 1983, 10th Ed., entry 7908.

PYROPHENINDANE [INN]			
CAS 7009-69-0	MW: 291.44	MF:	C21 H25 N

LgP 5.22(C)

pKa

pp cited in Vols.1-5:

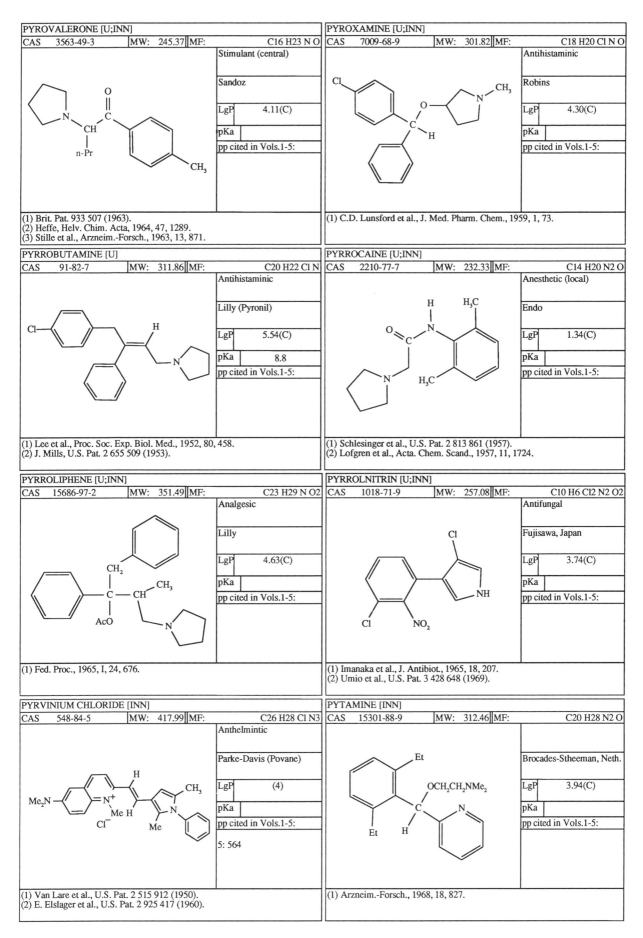

PYROVALERONE [U;INN]

CAS	3563-49-3	MW:	245.37	MF:	C16 H23 N O

Stimulant (central)

Sandoz

LgP	4.11(C)

pKa	

pp cited in Vols.1-5:

(1) Brit. Pat. 933 507 (1963).
(2) Heffe, Helv. Chim. Acta, 1964, 47, 1289.
(3) Stille et al., Arzneim.-Forsch., 1963, 13, 871.

PYROXAMINE [U;INN]

CAS	7009-68-9	MW:	301.82	MF:	C18 H20 Cl N O

Antihistaminic

Robins

LgP	4.30(C)

pKa	

pp cited in Vols.1-5:

(1) C.D. Lunsford et al., J. Med. Pharm. Chem., 1959, 1, 73.

PYRROBUTAMINE [U]

CAS	91-82-7	MW:	311.86	MF:	C20 H22 Cl N

Antihistaminic

Lilly (Pyronil)

LgP	5.54(C)

pKa	8.8

pp cited in Vols.1-5:

(1) Lee et al., Proc. Soc. Exp. Biol. Med., 1952, 80, 458.
(2) J. Mills, U.S. Pat. 2 655 509 (1953).

PYRROCAINE [U;INN]

CAS	2210-77-7	MW:	232.33	MF:	C14 H20 N2 O

Anesthetic (local)

Endo

LgP	1.34(C)

pKa	

pp cited in Vols.1-5:

(1) Schlesinger et al., U.S. Pat. 2 813 861 (1957).
(2) Lofgren et al., Acta. Chem. Scand., 1957, 11, 1724.

PYRROLIPHENE [U;INN]

CAS	15686-97-2	MW:	351.49	MF:	C23 H29 N O2

Analgesic

Lilly

LgP	4.63(C)

pKa	

pp cited in Vols.1-5:

(1) Fed. Proc., 1965, I, 24, 676.

PYRROLNITRIN [U;INN]

CAS	1018-71-9	MW:	257.08	MF:	C10 H6 Cl2 N2 O2

Antifungal

Fujisawa, Japan

LgP	3.74(C)

pKa	

pp cited in Vols.1-5:

(1) Imanaka et al., J. Antibiot., 1965, 18, 207.
(2) Umio et al., U.S. Pat. 3 428 648 (1969).

PYRVINIUM CHLORIDE [INN]

CAS	548-84-5	MW:	417.99	MF:	C26 H28 Cl N3

Anthelmintic

Parke-Davis (Povane)

LgP	(4)

pKa	

pp cited in Vols.1-5:

5: 564

(1) Van Lare et al., U.S. Pat. 2 515 912 (1950).
(2) E. Elslager et al., U.S. Pat. 2 925 417 (1960).

PYTAMINE [INN]

CAS	15301-88-9	MW:	312.46	MF:	C20 H28 N2 O

Brocades-Stheeman, Neth.

LgP	3.94(C)

pKa	

pp cited in Vols.1-5:

(1) Arzneim.-Forsch., 1968, 18, 827.

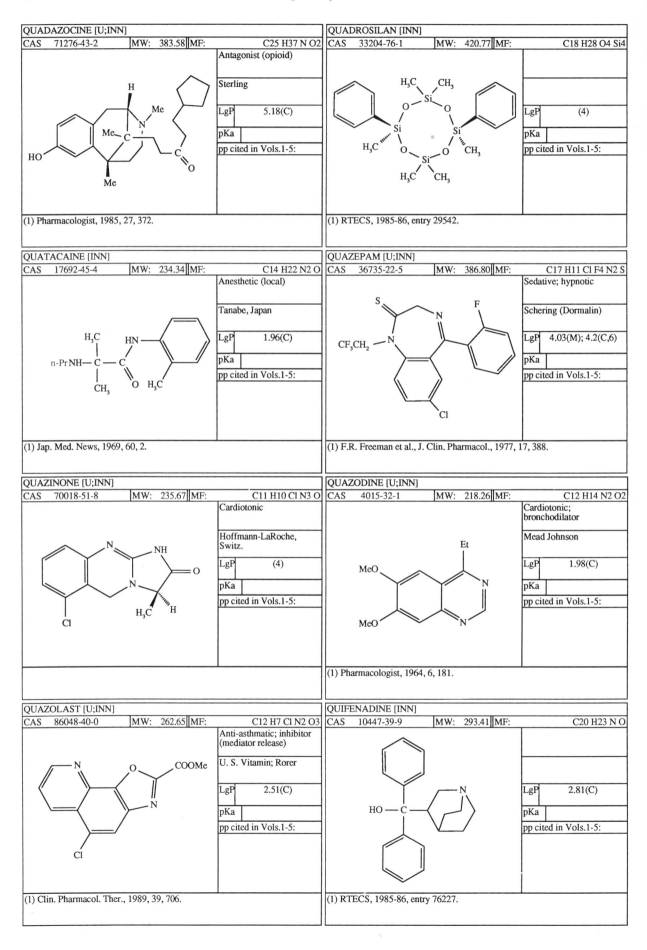

QUADAZOCINE [U;INN]		
CAS 71276-43-2	MW: 383.58	MF: C25 H37 N O2

Antagonist (opioid)

Sterling

LgP	5.18(C)
pKa	
pp cited in Vols.1-5:	

(1) Pharmacologist, 1985, 27, 372.

QUADROSILAN [INN]		
CAS 33204-76-1	MW: 420.77	MF: C18 H28 O4 Si4

LgP	(4)
pKa	
pp cited in Vols.1-5:	

(1) RTECS, 1985-86, entry 29542.

QUATACAINE [INN]		
CAS 17692-45-4	MW: 234.34	MF: C14 H22 N2 O

Anesthetic (local)

Tanabe, Japan

LgP	1.96(C)
pKa	
pp cited in Vols.1-5:	

(1) Jap. Med. News, 1969, 60, 2.

QUAZEPAM [U;INN]		
CAS 36735-22-5	MW: 386.80	MF: C17 H11 Cl F4 N2 S

Sedative; hypnotic

Schering (Dormalin)

LgP	4.03(M); 4.2(C,6)
pKa	
pp cited in Vols.1-5:	

(1) F.R. Freeman et al., J. Clin. Pharmacol., 1977, 17, 388.

QUAZINONE [U;INN]		
CAS 70018-51-8	MW: 235.67	MF: C11 H10 Cl N3 O

Cardiotonic

Hoffmann-LaRoche, Switz.

LgP	(4)
pKa	
pp cited in Vols.1-5:	

QUAZODINE [U;INN]		
CAS 4015-32-1	MW: 218.26	MF: C12 H14 N2 O2

Cardiotonic; bronchodilator

Mead Johnson

LgP	1.98(C)
pKa	
pp cited in Vols.1-5:	

(1) Pharmacologist, 1964, 6, 181.

QUAZOLAST [U;INN]		
CAS 86048-40-0	MW: 262.65	MF: C12 H7 Cl N2 O3

Anti-asthmatic; inhibitor (mediator release)

U. S. Vitamin; Rorer

LgP	2.51(C)
pKa	
pp cited in Vols.1-5:	

(1) Clin. Pharmacol. Ther., 1989, 39, 706.

QUIFENADINE [INN]		
CAS 10447-39-9	MW: 293.41	MF: C20 H23 N O

LgP	2.81(C)
pKa	
pp cited in Vols.1-5:	

(1) RTECS, 1985-86, entry 76227.

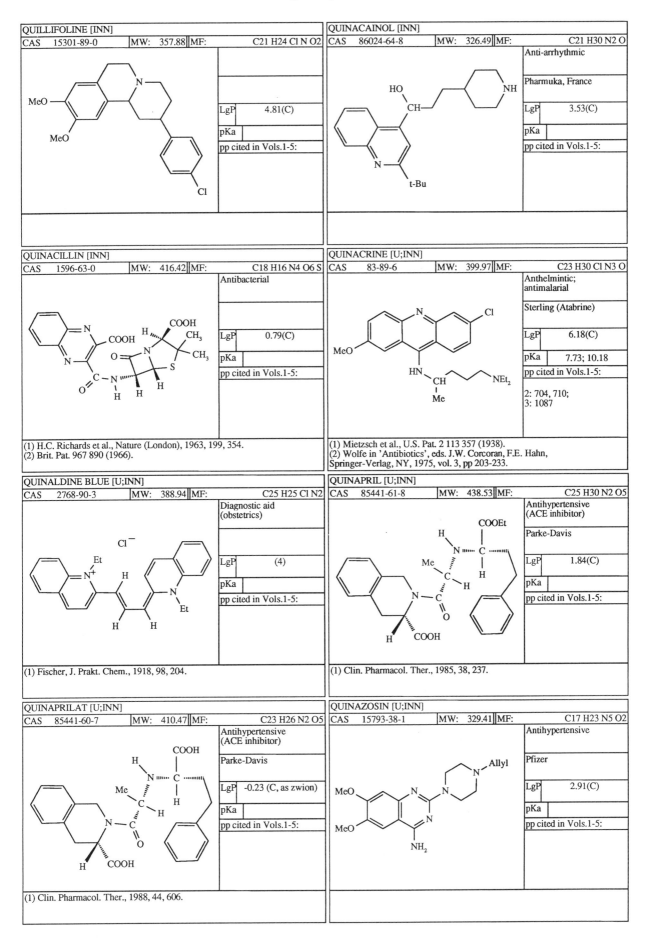

QUILLIFOLINE [INN]			
CAS 15301-89-0	MW: 357.88	MF:	C21 H24 Cl N O2

LgP 4.81(C)

pKa

pp cited in Vols.1-5:

QUINACAINOL [INN]			
CAS 86024-64-8	MW: 326.49	MF:	C21 H30 N2 O

Anti-arrhythmic

Pharmuka, France

LgP 3.53(C)

pKa

pp cited in Vols.1-5:

QUINACILLIN [INN]			
CAS 1596-63-0	MW: 416.42	MF:	C18 H16 N4 O6 S

Antibacterial

LgP 0.79(C)

pKa

pp cited in Vols.1-5:

(1) H.C. Richards et al., Nature (London), 1963, 199, 354.
(2) Brit. Pat. 967 890 (1966).

QUINACRINE [U;INN]			
CAS 83-89-6	MW: 399.97	MF:	C23 H30 Cl N3 O

Anthelmintic; antimalarial

Sterling (Atabrine)

LgP 6.18(C)

pKa 7.73; 10.18

pp cited in Vols.1-5:

2: 704, 710;
3: 1087

(1) Mietzsch et al., U.S. Pat. 2 113 357 (1938).
(2) Wolfe in 'Antibiotics', eds. J.W. Corcoran, F.E. Hahn, Springer-Verlag, NY, 1975, vol. 3, pp 203-233.

QUINALDINE BLUE [U;INN]			
CAS 2768-90-3	MW: 388.94	MF:	C25 H25 Cl N2

Diagnostic aid (obstetrics)

LgP (4)

pKa

pp cited in Vols.1-5:

(1) Fischer, J. Prakt. Chem., 1918, 98, 204.

QUINAPRIL [U;INN]			
CAS 85441-61-8	MW: 438.53	MF:	C25 H30 N2 O5

Antihypertensive (ACE inhibitor)

Parke-Davis

LgP 1.84(C)

pKa

pp cited in Vols.1-5:

(1) Clin. Pharmacol. Ther., 1985, 38, 237.

QUINAPRILAT [U;INN]			
CAS 85441-60-7	MW: 410.47	MF:	C23 H26 N2 O5

Antihypertensive (ACE inhibitor)

Parke-Davis

LgP -0.23 (C, as zwion)

pKa

pp cited in Vols.1-5:

(1) Clin. Pharmacol. Ther., 1988, 44, 606.

QUINAZOSIN [U;INN]			
CAS 15793-38-1	MW: 329.41	MF:	C17 H23 N5 O2

Antihypertensive

Pfizer

LgP 2.91(C)

pKa

pp cited in Vols.1-5:

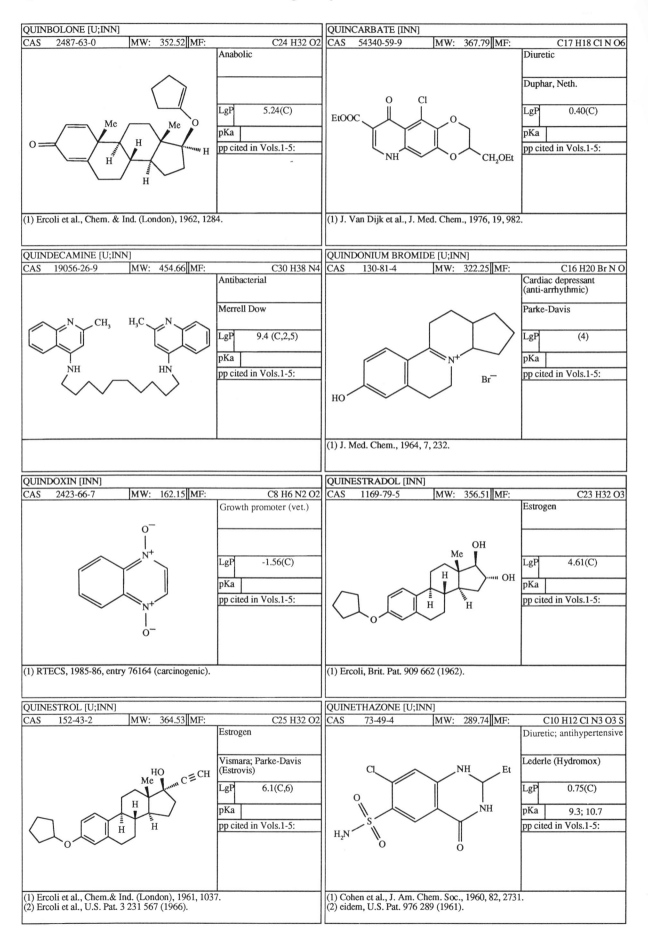

QUINBOLONE [U;INN]

CAS	2487-63-0	MW:	352.52	MF:	C24 H32 O2

Anabolic

LgP	5.24(C)
pKa	
pp cited in Vols.1-5:	

(1) Ercoli et al., Chem. & Ind. (London), 1962, 1284.

QUINCARBATE [INN]

CAS	54340-59-9	MW:	367.79	MF:	C17 H18 Cl N O6

Diuretic

Duphar, Neth.

LgP	0.40(C)
pKa	
pp cited in Vols.1-5:	

(1) J. Van Dijk et al., J. Med. Chem., 1976, 19, 982.

QUINDECAMINE [U;INN]

CAS	19056-26-9	MW:	454.66	MF:	C30 H38 N4

Antibacterial

Merrell Dow

LgP	9.4 (C,2,5)
pKa	
pp cited in Vols.1-5:	

QUINDONIUM BROMIDE [U;INN]

CAS	130-81-4	MW:	322.25	MF:	C16 H20 Br N O

Cardiac depressant (anti-arrhythmic)

Parke-Davis

LgP	(4)
pKa	
pp cited in Vols.1-5:	

(1) J. Med. Chem., 1964, 7, 232.

QUINDOXIN [INN]

CAS	2423-66-7	MW:	162.15	MF:	C8 H6 N2 O2

Growth promoter (vet.)

LgP	-1.56(C)
pKa	
pp cited in Vols.1-5:	

(1) RTECS, 1985-86, entry 76164 (carcinogenic).

QUINESTRADOL [INN]

CAS	1169-79-5	MW:	356.51	MF:	C23 H32 O3

Estrogen

LgP	4.61(C)
pKa	
pp cited in Vols.1-5:	

(1) Ercoli, Brit. Pat. 909 662 (1962).

QUINESTROL [U;INN]

CAS	152-43-2	MW:	364.53	MF:	C25 H32 O2

Estrogen

Vismara; Parke-Davis (Estrovis)

LgP	6.1(C,6)
pKa	
pp cited in Vols.1-5:	

(1) Ercoli et al., Chem.& Ind. (London), 1961, 1037.
(2) Ercoli et al., U.S. Pat. 3 231 567 (1966).

QUINETHAZONE [U;INN]

CAS	73-49-4	MW:	289.74	MF:	C10 H12 Cl N3 O3 S

Diuretic; antihypertensive

Lederle (Hydromox)

LgP	0.75(C)
pKa	9.3; 10.7
pp cited in Vols.1-5:	

(1) Cohen et al., J. Am. Chem. Soc., 1960, 82, 2731.
(2) eidem, U.S. Pat. 976 289 (1961).

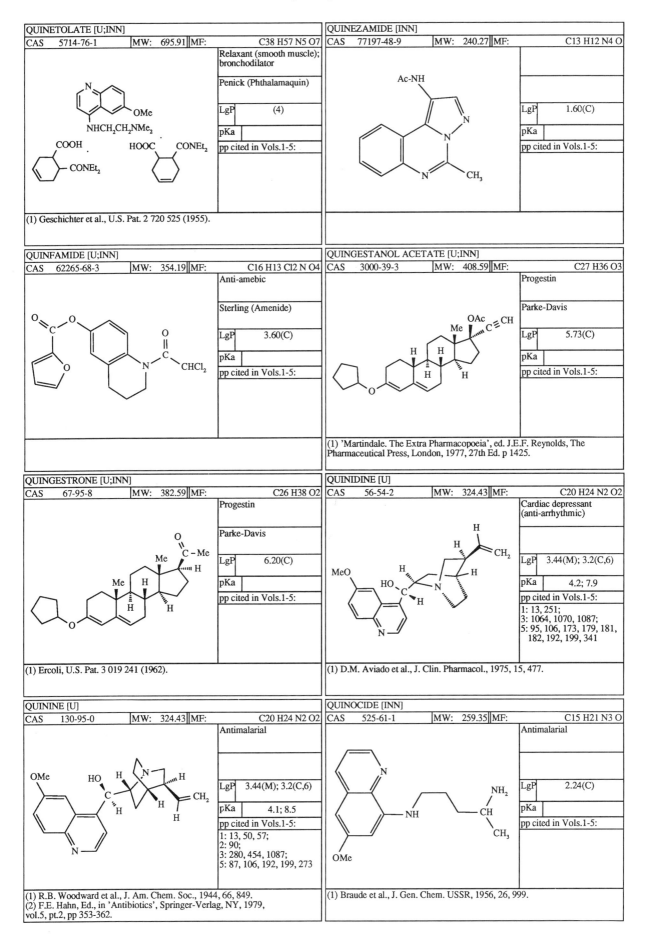

QUINETOLATE [U;INN]			
CAS 5714-76-1	MW: 695.91	MF:	C38 H57 N5 O7

Relaxant (smooth muscle); bronchodilator

Penick (Phthalamaquin)

LgP	(4)
pKa	
pp cited in Vols.1-5:	

(1) Geschichter et al., U.S. Pat. 2 720 525 (1955).

QUINEZAMIDE [INN]			
CAS 77197-48-9	MW: 240.27	MF:	C13 H12 N4 O

LgP	1.60(C)
pKa	
pp cited in Vols.1-5:	

QUINFAMIDE [U;INN]			
CAS 62265-68-3	MW: 354.19	MF:	C16 H13 Cl2 N O4

Anti-amebic

Sterling (Amenide)

LgP	3.60(C)
pKa	
pp cited in Vols.1-5:	

QUINGESTANOL ACETATE [U;INN]			
CAS 3000-39-3	MW: 408.59	MF:	C27 H36 O3

Progestin

Parke-Davis

LgP	5.73(C)
pKa	
pp cited in Vols.1-5:	

(1) 'Martindale. The Extra Pharmacopoeia', ed. J.E.F. Reynolds, The Pharmaceutical Press, London, 1977, 27th Ed. p 1425.

QUINGESTRONE [U;INN]			
CAS 67-95-8	MW: 382.59	MF:	C26 H38 O2

Progestin

Parke-Davis

LgP	6.20(C)
pKa	
pp cited in Vols.1-5:	

(1) Ercoli, U.S. Pat. 3 019 241 (1962).

QUINIDINE [U]			
CAS 56-54-2	MW: 324.43	MF:	C20 H24 N2 O2

Cardiac depressant (anti-arrhythmic)

LgP	3.44(M); 3.2(C,6)
pKa	4.2; 7.9
pp cited in Vols.1-5:	

1: 13, 251;
3: 1064, 1070, 1087;
5: 95, 106, 173, 179, 181, 182, 192, 199, 341

(1) D.M. Aviado et al., J. Clin. Pharmacol., 1975, 15, 477.

QUININE [U]			
CAS 130-95-0	MW: 324.43	MF:	C20 H24 N2 O2

Antimalarial

LgP	3.44(M); 3.2(C,6)
pKa	4.1; 8.5
pp cited in Vols.1-5:	

1: 13, 50, 57;
2: 90;
3: 280, 454, 1087;
5: 87, 106, 192, 199, 273

(1) R.B. Woodward et al., J. Am. Chem. Soc., 1944, 66, 849.
(2) F.E. Hahn, Ed., in 'Antibiotics', Springer-Verlag, NY, 1979, vol.5, pt.2, pp 353-362.

QUINOCIDE [INN]			
CAS 525-61-1	MW: 259.35	MF:	C15 H21 N3 O

Antimalarial

LgP	2.24(C)
pKa	
pp cited in Vols.1-5:	

(1) Braude et al., J. Gen. Chem. USSR, 1956, 26, 999.

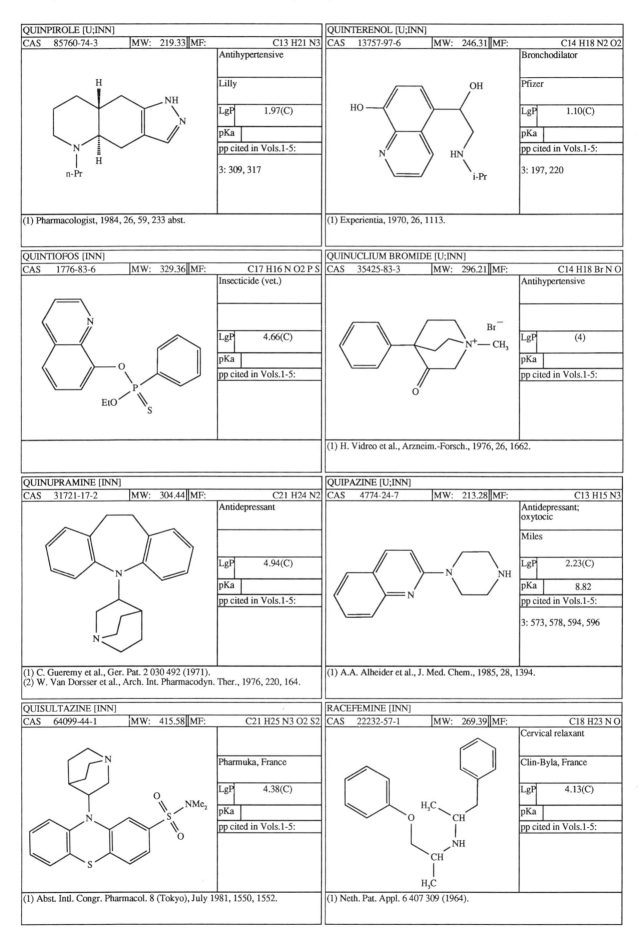

QUINPIROLE [U;INN]		
CAS 85760-74-3	MW: 219.33	MF: C13 H21 N3

Antihypertensive

Lilly

LgP	1.97(C)
pKa	
pp cited in Vols.1-5:	

3: 309, 317

n-Pr

(1) Pharmacologist, 1984, 26, 59, 233 abst.

QUINTERENOL [U;INN]		
CAS 13757-97-6	MW: 246.31	MF: C14 H18 N2 O2

Bronchodilator

Pfizer

LgP	1.10(C)
pKa	
pp cited in Vols.1-5:	

3: 197, 220

i-Pr

(1) Experientia, 1970, 26, 1113.

QUINTIOFOS [INN]		
CAS 1776-83-6	MW: 329.36	MF: C17 H16 N O2 P S

Insecticide (vet.)

LgP	4.66(C)
pKa	
pp cited in Vols.1-5:	

EtO

QUINUCLIUM BROMIDE [U;INN]		
CAS 35425-83-3	MW: 296.21	MF: C14 H18 Br N O

Antihypertensive

LgP	(4)
pKa	
pp cited in Vols.1-5:	

(1) H. Vidreo et al., Arzneim.-Forsch., 1976, 26, 1662.

QUINUPRAMINE [INN]		
CAS 31721-17-2	MW: 304.44	MF: C21 H24 N2

Antidepressant

LgP	4.94(C)
pKa	
pp cited in Vols.1-5:	

(1) C. Gueremy et al., Ger. Pat. 2 030 492 (1971).
(2) W. Van Dorsser et al., Arch. Int. Pharmacodyn. Ther., 1976, 220, 164.

QUIPAZINE [U;INN]		
CAS 4774-24-7	MW: 213.28	MF: C13 H15 N3

Antidepressant; oxytocic

Miles

LgP	2.23(C)
pKa	8.82
pp cited in Vols.1-5:	

3: 573, 578, 594, 596

(1) A.A. Alheider et al., J. Med. Chem., 1985, 28, 1394.

QUISULTAZINE [INN]		
CAS 64099-44-1	MW: 415.58	MF: C21 H25 N3 O2 S2

Pharmuka, France

LgP	4.38(C)
pKa	
pp cited in Vols.1-5:	

NMe₂

(1) Abst. Intl. Congr. Pharmacol. 8 (Tokyo), July 1981, 1550, 1552.

RACEFEMINE [INN]		
CAS 22232-57-1	MW: 269.39	MF: C18 H23 N O

Cervical relaxant

Clin-Byla, France

LgP	4.13(C)
pKa	
pp cited in Vols.1-5:	

H₃C

H₃C

(1) Neth. Pat. Appl. 6 407 309 (1964).

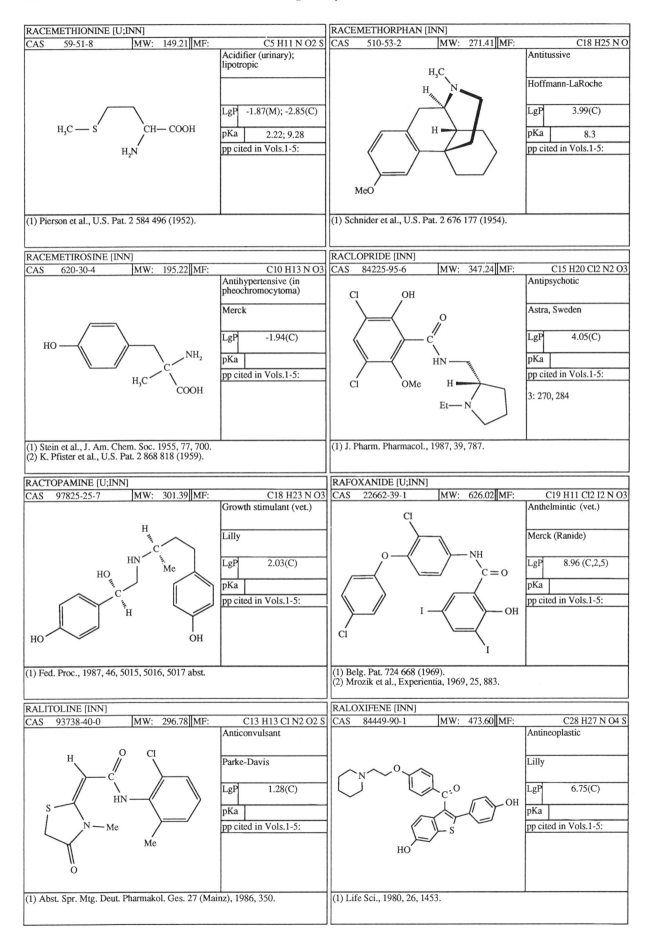

RACEMETHIONINE [U;INN]

CAS	59-51-8	MW:	149.21	MF:	C5 H11 N O2 S

Acidifier (urinary); lipotropic

LgP	-1.87(M); -2.85(C)
pKa	2.22; 9.28
pp cited in Vols.1-5:	

(1) Pierson et al., U.S. Pat. 2 584 496 (1952).

RACEMETHORPHAN [INN]

CAS	510-53-2	MW:	271.41	MF:	C18 H25 N O

Antitussive

Hoffmann-LaRoche

LgP	3.99(C)
pKa	8.3
pp cited in Vols.1-5:	

(1) Schnider et al., U.S. Pat. 2 676 177 (1954).

RACEMETIROSINE [INN]

CAS	620-30-4	MW:	195.22	MF:	C10 H13 N O3

Antihypertensive (in pheochromocytoma)

Merck

LgP	-1.94(C)
pKa	
pp cited in Vols.1-5:	

(1) Stein et al., J. Am. Chem. Soc. 1955, 77, 700.
(2) K. Pfister et al., U.S. Pat. 2 868 818 (1959).

RACLOPRIDE [INN]

CAS	84225-95-6	MW:	347.24	MF:	C15 H20 Cl2 N2 O3

Antipsychotic

Astra, Sweden

LgP	4.05(C)
pKa	
pp cited in Vols.1-5:	
3: 270, 284	

(1) J. Pharm. Pharmacol., 1987, 39, 787.

RACTOPAMINE [U;INN]

CAS	97825-25-7	MW:	301.39	MF:	C18 H23 N O3

Growth stimulant (vet.)

Lilly

LgP	2.03(C)
pKa	
pp cited in Vols.1-5:	

(1) Fed. Proc., 1987, 46, 5015, 5016, 5017 abst.

RAFOXANIDE [U;INN]

CAS	22662-39-1	MW:	626.02	MF:	C19 H11 Cl2 I2 N O3

Anthelmintic (vet.)

Merck (Ranide)

LgP	8.96 (C,2,5)
pKa	
pp cited in Vols.1-5:	

(1) Belg. Pat. 724 668 (1969).
(2) Mrozik et al., Experientia, 1969, 25, 883.

RALITOLINE [INN]

CAS	93738-40-0	MW:	296.78	MF:	C13 H13 Cl N2 O2 S

Anticonvulsant

Parke-Davis

LgP	1.28(C)
pKa	
pp cited in Vols.1-5:	

(1) Abst. Spr. Mtg. Deut. Pharmakol. Ges. 27 (Mainz), 1986, 350.

RALOXIFENE [INN]

CAS	84449-90-1	MW:	473.60	MF:	C28 H27 N O4 S

Antineoplastic

Lilly

LgP	6.75(C)
pKa	
pp cited in Vols.1-5:	

(1) Life Sci., 1980, 26, 1453.

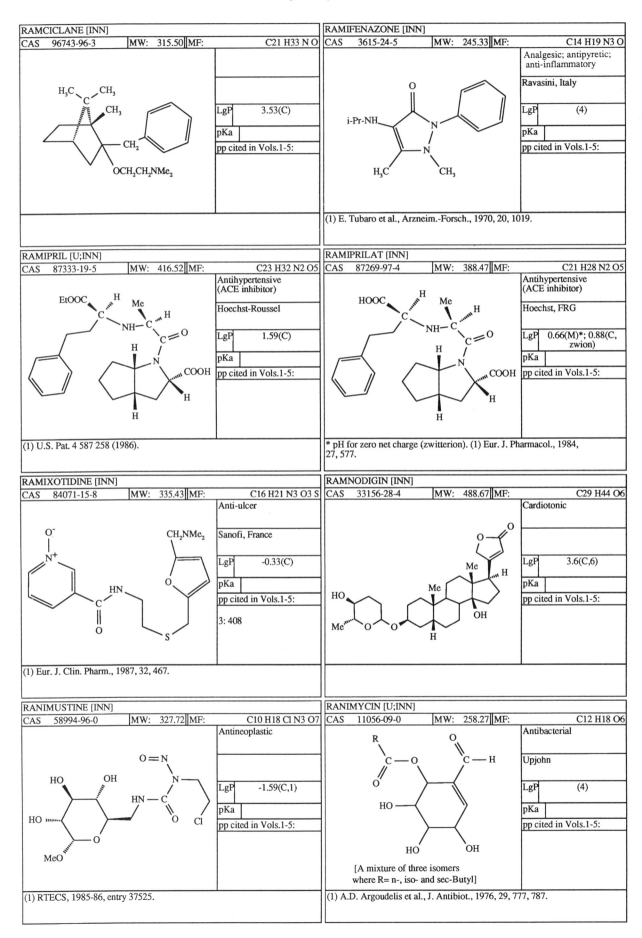

RAMCICLANE [INN]

CAS	96743-96-3	MW:	315.50	MF:	C21 H33 N O

LgP	3.53(C)
pKa	
pp cited in Vols.1-5:	

RAMIFENAZONE [INN]

CAS	3615-24-5	MW:	245.33	MF:	C14 H19 N3 O

Analgesic; antipyretic; anti-inflammatory

Ravasini, Italy

LgP	(4)
pKa	
pp cited in Vols.1-5:	

(1) E. Tubaro et al., Arzneim.-Forsch., 1970, 20, 1019.

RAMIPRIL [U;INN]

CAS	87333-19-5	MW:	416.52	MF:	C23 H32 N2 O5

Antihypertensive (ACE inhibitor)

Hoechst-Roussel

LgP	1.59(C)
pKa	
pp cited in Vols.1-5:	

(1) U.S. Pat. 4 587 258 (1986).

RAMIPRILAT [INN]

CAS	87269-97-4	MW:	388.47	MF:	C21 H28 N2 O5

Antihypertensive (ACE inhibitor)

Hoechst, FRG

LgP	0.66(M)*; 0.88(C, zwion)
pKa	
pp cited in Vols.1-5:	

* pH for zero net charge (zwitterion). (1) Eur. J. Pharmacol., 1984, 27, 577.

RAMIXOTIDINE [INN]

CAS	84071-15-8	MW:	335.43	MF:	C16 H21 N3 O3 S

Anti-ulcer

Sanofi, France

LgP	-0.33(C)
pKa	
pp cited in Vols.1-5:	
3: 408	

(1) Eur. J. Clin. Pharm., 1987, 32, 467.

RAMNODIGIN [INN]

CAS	33156-28-4	MW:	488.67	MF:	C29 H44 O6

Cardiotonic

LgP	3.6(C,6)
pKa	
pp cited in Vols.1-5:	

RANIMUSTINE [INN]

CAS	58994-96-0	MW:	327.72	MF:	C10 H18 Cl N3 O7

Antineoplastic

LgP	-1.59(C,1)
pKa	
pp cited in Vols.1-5:	

(1) RTECS, 1985-86, entry 37525.

RANIMYCIN [U;INN]

CAS	11056-09-0	MW:	258.27	MF:	C12 H18 O6

Antibacterial

Upjohn

LgP	(4)
pKa	
pp cited in Vols.1-5:	

[A mixture of three isomers where R= n-, iso- and sec-Butyl]

(1) A.D. Argoudelis et al., J. Antibiot., 1976, 29, 777, 787.

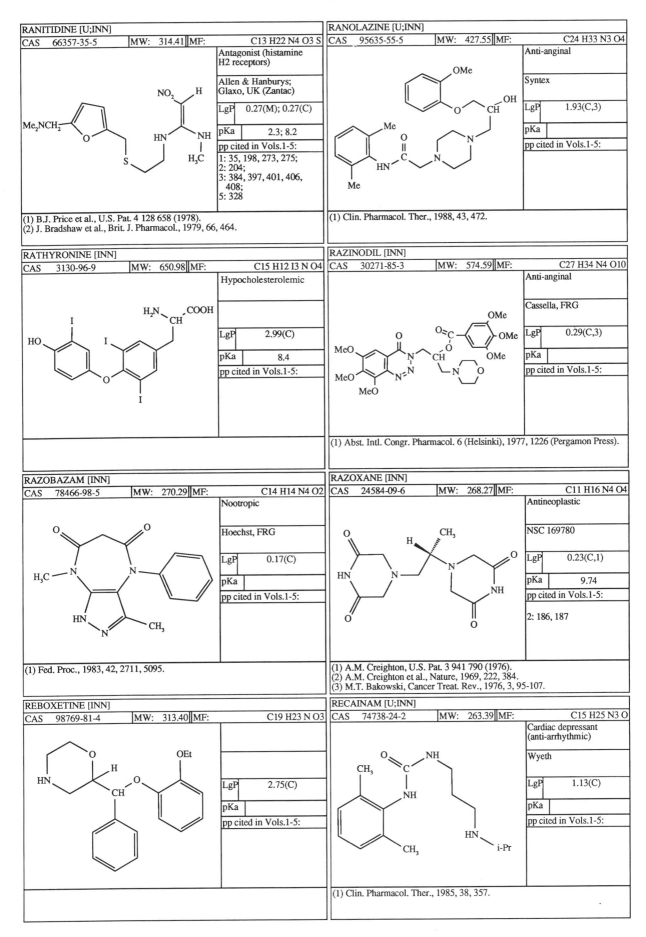

RANITIDINE [U;INN]

CAS 66357-35-5	MW: 314.41	MF:	C13 H22 N4 O3 S

Antagonist (histamine H2 receptors)

Allen & Hanburys; Glaxo, UK (Zantac)

LgP	0.27(M); 0.27(C)
pKa	2.3; 8.2

pp cited in Vols.1-5:
1: 35, 198, 273, 275;
2: 204;
3: 384, 397, 401, 406, 408;
5: 328

(1) B.J. Price et al., U.S. Pat. 4 128 658 (1978).
(2) J. Bradshaw et al., Brit. J. Pharmacol., 1979, 66, 464.

RANOLAZINE [U;INN]

CAS 95635-55-5	MW: 427.55	MF:	C24 H33 N3 O4

Anti-anginal

Syntex

LgP	1.93(C,3)
pKa	

pp cited in Vols.1-5:

(1) Clin. Pharmacol. Ther., 1988, 43, 472.

RATHYRONINE [INN]

CAS 3130-96-9	MW: 650.98	MF:	C15 H12 I3 N O4

Hypocholesterolemic

LgP	2.99(C)
pKa	8.4

pp cited in Vols.1-5:

RAZINODIL [INN]

CAS 30271-85-3	MW: 574.59	MF:	C27 H34 N4 O10

Anti-anginal

Cassella, FRG

LgP	0.29(C,3)
pKa	

pp cited in Vols.1-5:

(1) Abst. Intl. Congr. Pharmacol. 6 (Helsinki), 1977, 1226 (Pergamon Press).

RAZOBAZAM [INN]

CAS 78466-98-5	MW: 270.29	MF:	C14 H14 N4 O2

Nootropic

Hoechst, FRG

LgP	0.17(C)
pKa	

pp cited in Vols.1-5:

(1) Fed. Proc., 1983, 42, 2711, 5095.

RAZOXANE [INN]

CAS 24584-09-6	MW: 268.27	MF:	C11 H16 N4 O4

Antineoplastic

NSC 169780

LgP	0.23(C,1)
pKa	9.74

pp cited in Vols.1-5:

2: 186, 187

(1) A.M. Creighton, U.S. Pat. 3 941 790 (1976).
(2) A.M. Creighton et al., Nature, 1969, 222, 384.
(3) M.T. Bakowski, Cancer Treat. Rev., 1976, 3, 95-107.

REBOXETINE [INN]

CAS 98769-81-4	MW: 313.40	MF:	C19 H23 N O3

LgP	2.75(C)
pKa	

pp cited in Vols.1-5:

RECAINAM [U;INN]

CAS 74738-24-2	MW: 263.39	MF:	C15 H25 N3 O

Cardiac depressant (anti-arrhythmic)

Wyeth

LgP	1.13(C)
pKa	

pp cited in Vols.1-5:

(1) Clin. Pharmacol. Ther., 1985, 38, 357.

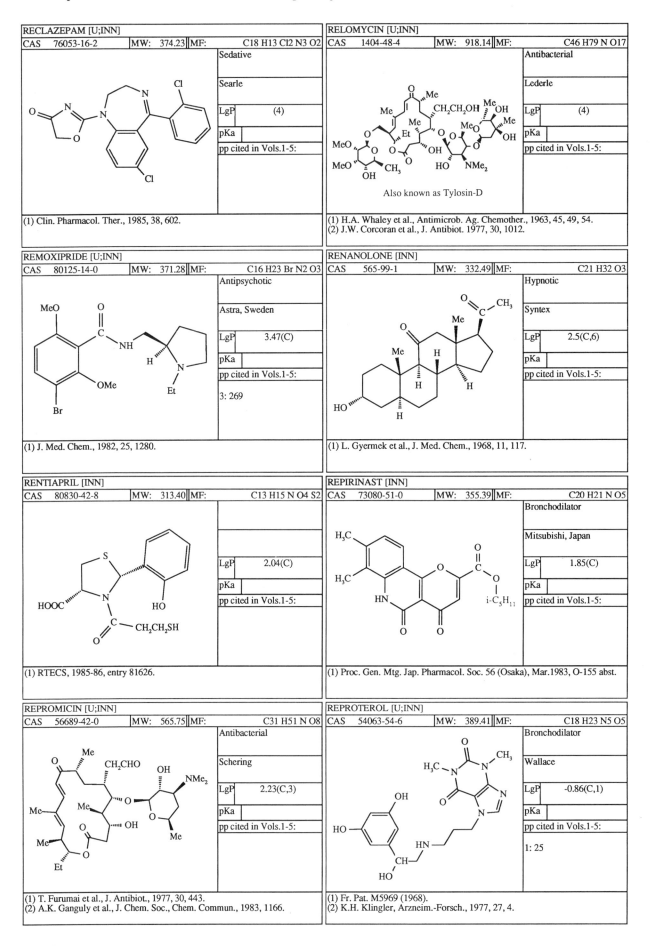

RECLAZEPAM [U;INN]

CAS	76053-16-2	MW:	374.23	MF:	C18 H13 Cl2 N3 O2

Sedative

Searle

LgP	(4)
pKa	
pp cited in Vols.1-5:	

(1) Clin. Pharmacol. Ther., 1985, 38, 602.

RELOMYCIN [U;INN]

CAS	1404-48-4	MW:	918.14	MF:	C46 H79 N O17

Antibacterial

Lederle

LgP	(4)
pKa	
pp cited in Vols.1-5:	

Also known as Tylosin-D

(1) H.A. Whaley et al., Antimicrob. Ag. Chemother., 1963, 45, 49, 54.
(2) J.W. Corcoran et al., J. Antibiot. 1977, 30, 1012.

REMOXIPRIDE [U;INN]

CAS	80125-14-0	MW:	371.28	MF:	C16 H23 Br N2 O3

Antipsychotic

Astra, Sweden

LgP	3.47(C)
pKa	
pp cited in Vols.1-5:	
3: 269	

(1) J. Med. Chem., 1982, 25, 1280.

RENANOLONE [INN]

CAS	565-99-1	MW:	332.49	MF:	C21 H32 O3

Hypnotic

Syntex

LgP	2.5(C,6)
pKa	
pp cited in Vols.1-5:	

(1) L. Gyermek et al., J. Med. Chem., 1968, 11, 117.

RENTIAPRIL [INN]

CAS	80830-42-8	MW:	313.40	MF:	C13 H15 N O4 S2

LgP	2.04(C)
pKa	
pp cited in Vols.1-5:	

(1) RTECS, 1985-86, entry 81626.

REPIRINAST [INN]

CAS	73080-51-0	MW:	355.39	MF:	C20 H21 N O5

Bronchodilator

Mitsubishi, Japan

LgP	1.85(C)
pKa	
pp cited in Vols.1-5:	

(1) Proc. Gen. Mtg. Jap. Pharmacol. Soc. 56 (Osaka), Mar.1983, O-155 abst.

REPROMICIN [U;INN]

CAS	56689-42-0	MW:	565.75	MF:	C31 H51 N O8

Antibacterial

Schering

LgP	2.23(C,3)
pKa	
pp cited in Vols.1-5:	

(1) T. Furumai et al., J. Antibiot., 1977, 30, 443.
(2) A.K. Ganguly et al., J. Chem. Soc., Chem. Commun., 1983, 1166.

REPROTEROL [U;INN]

CAS	54063-54-6	MW:	389.41	MF:	C18 H23 N5 O5

Bronchodilator

Wallace

LgP	-0.86(C,1)
pKa	
pp cited in Vols.1-5:	
1: 25	

(1) Fr. Pat. M5969 (1968).
(2) K.H. Klingler, Arzneim.-Forsch., 1977, 27, 4.

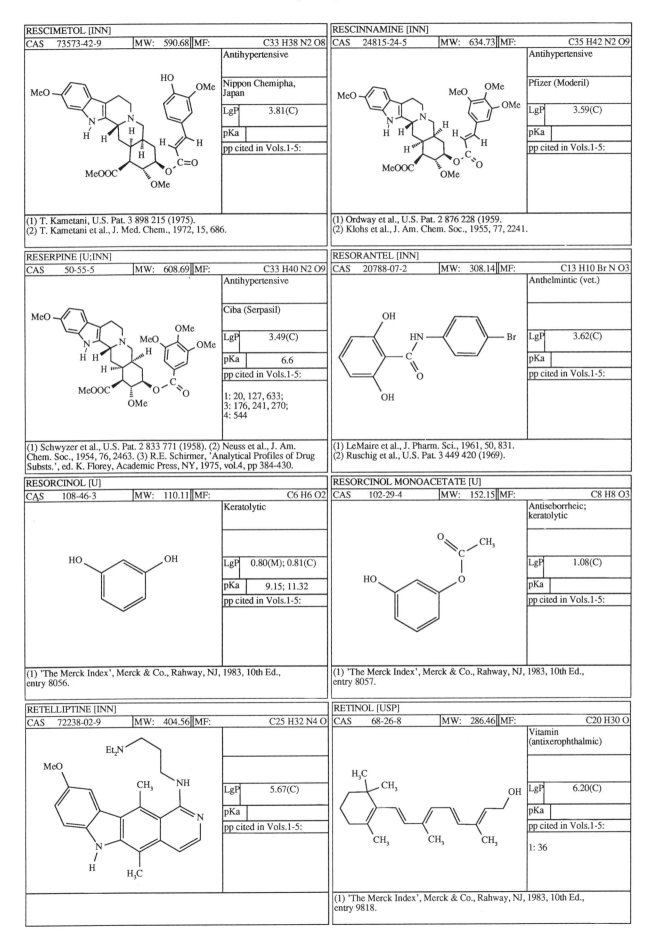

RESCIMETOL [INN]

| CAS | 73573-42-9 | MW: | 590.68 | MF: | C33 H38 N2 O8 |

Antihypertensive

Nippon Chemipha, Japan

LgP	3.81(C)
pKa	
pp cited in Vols.1-5:	

(1) T. Kametani, U.S. Pat. 3 898 215 (1975).
(2) T. Kametani et al., J. Med. Chem., 1972, 15, 686.

RESCINNAMINE [INN]

| CAS | 24815-24-5 | MW: | 634.73 | MF: | C35 H42 N2 O9 |

Antihypertensive

Pfizer (Moderil)

LgP	3.59(C)
pKa	
pp cited in Vols.1-5:	

(1) Ordway et al., U.S. Pat. 2 876 228 (1959.
(2) Klohs et al., J. Am. Chem. Soc., 1955, 77, 2241.

RESERPINE [U;INN]

| CAS | 50-55-5 | MW: | 608.69 | MF: | C33 H40 N2 O9 |

Antihypertensive

Ciba (Serpasil)

LgP	3.49(C)
pKa	6.6
pp cited in Vols.1-5:	

1: 20, 127, 633;
3: 176, 241, 270;
4: 544

(1) Schwyzer et al., U.S. Pat. 2 833 771 (1958). (2) Neuss et al., J. Am. Chem. Soc., 1954, 76, 2463. (3) R.E. Schirmer, 'Analytical Profiles of Drug Substs.', ed. K. Florey, Academic Press, NY, 1975, vol.4, pp 384-430.

RESORANTEL [INN]

| CAS | 20788-07-2 | MW: | 308.14 | MF: | C13 H10 Br N O3 |

Anthelmintic (vet.)

LgP	3.62(C)
pKa	
pp cited in Vols.1-5:	

(1) LeMaire et al., J. Pharm. Sci., 1961, 50, 831.
(2) Ruschig et al., U.S. Pat. 3 449 420 (1969).

RESORCINOL [U]

| CAS | 108-46-3 | MW: | 110.11 | MF: | C6 H6 O2 |

Keratolytic

LgP	0.80(M); 0.81(C)
pKa	9.15; 11.32
pp cited in Vols.1-5:	

(1) 'The Merck Index', Merck & Co., Rahway, NJ, 1983, 10th Ed., entry 8056.

RESORCINOL MONOACETATE [U]

| CAS | 102-29-4 | MW: | 152.15 | MF: | C8 H8 O3 |

Antiseborrheic; keratolytic

LgP	1.08(C)
pKa	
pp cited in Vols.1-5:	

(1) 'The Merck Index', Merck & Co., Rahway, NJ, 1983, 10th Ed., entry 8057.

RETELLIPTINE [INN]

| CAS | 72238-02-9 | MW: | 404.56 | MF: | C25 H32 N4 O |

LgP	5.67(C)
pKa	
pp cited in Vols.1-5:	

RETINOL [USP]

| CAS | 68-26-8 | MW: | 286.46 | MF: | C20 H30 O |

Vitamin (antixerophthalmic)

LgP	6.20(C)
pKa	
pp cited in Vols.1-5:	

1: 36

(1) 'The Merck Index', Merck & Co., Rahway, NJ, 1983, 10th Ed., entry 9818.

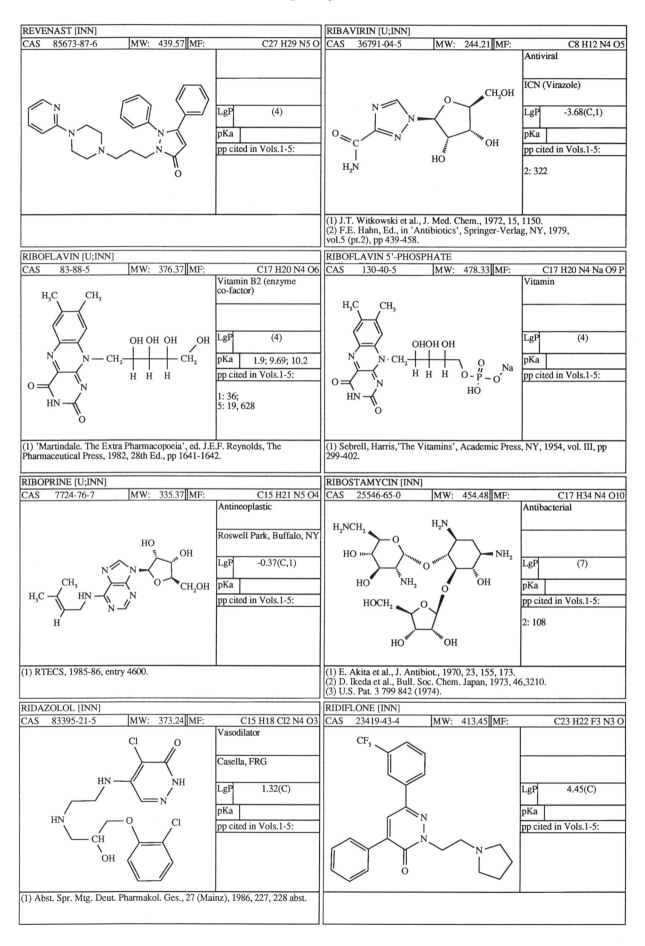

REVENAST [INN]

CAS	85673-87-6	MW:	439.57	MF:	C27 H29 N5 O

LgP	(4)
pKa	
pp cited in Vols.1-5:	

RIBAVIRIN [U;INN]

CAS	36791-04-5	MW:	244.21	MF:	C8 H12 N4 O5

Antiviral

ICN (Virazole)

LgP	-3.68(C,1)
pKa	
pp cited in Vols.1-5:	
2: 322	

(1) J.T. Witkowski et al., J. Med. Chem., 1972, 15, 1150.
(2) F.E. Hahn, Ed., in 'Antibiotics', Springer-Verlag, NY, 1979, vol.5 (pt.2), pp 439-458.

RIBOFLAVIN [U;INN]

CAS	83-88-5	MW:	376.37	MF:	C17 H20 N4 O6

Vitamin B2 (enzyme co-factor)

LgP	(4)
pKa	1.9; 9.69; 10.2
pp cited in Vols.1-5:	
1: 36;	
5: 19, 628	

(1) 'Martindale. The Extra Pharmacopoeia', ed. J.E.F. Reynolds, The Pharmaceutical Press, 1982, 28th Ed., pp 1641-1642.

RIBOFLAVIN 5'-PHOSPHATE

CAS	130-40-5	MW:	478.33	MF:	C17 H20 N4 Na O9 P

Vitamin

LgP	(4)
pKa	
pp cited in Vols.1-5:	

(1) Sebrell, Harris,'The Vitamins', Academic Press, NY, 1954, vol. III, pp 299-402.

RIBOPRINE [U;INN]

CAS	7724-76-7	MW:	335.37	MF:	C15 H21 N5 O4

Antineoplastic

Roswell Park, Buffalo, NY

LgP	-0.37(C,1)
pKa	
pp cited in Vols.1-5:	

(1) RTECS, 1985-86, entry 4600.

RIBOSTAMYCIN [INN]

CAS	25546-65-0	MW:	454.48	MF:	C17 H34 N4 O10

Antibacterial

LgP	(7)
pKa	
pp cited in Vols.1-5:	
2: 108	

(1) E. Akita et al., J. Antibiot., 1970, 23, 155, 173.
(2) D. Ikeda et al., Bull. Soc. Chem. Japan, 1973, 46,3210.
(3) U.S. Pat. 3 799 842 (1974).

RIDAZOLOL [INN]

CAS	83395-21-5	MW:	373.24	MF:	C15 H18 Cl2 N4 O3

Vasodilator

Casella, FRG

LgP	1.32(C)
pKa	
pp cited in Vols.1-5:	

(1) Abst. Spr. Mtg. Deut. Pharmakol. Ges., 27 (Mainz), 1986, 227, 228 abst.

RIDIFLONE [INN]

CAS	23419-43-4	MW:	413.45	MF:	C23 H22 F3 N3 O

LgP	4.45(C)
pKa	
pp cited in Vols.1-5:	

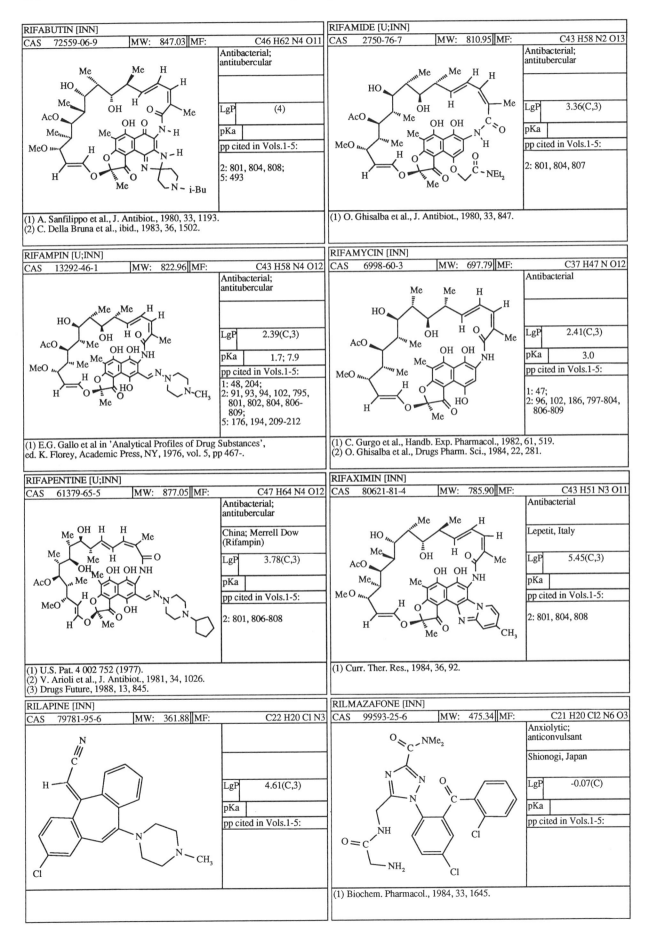

RIFABUTIN [INN]

CAS	72559-06-9	MW:	847.03	MF:	C46 H62 N4 O11

Antibacterial; antitubercular

LgP	(4)
pKa	
pp cited in Vols.1-5:	2: 801, 804, 808; 5: 493

(1) A. Sanfilippo et al., J. Antibiot., 1980, 33, 1193.
(2) C. Della Bruna et al., ibid., 1983, 36, 1502.

RIFAMIDE [U;INN]

CAS	2750-76-7	MW:	810.95	MF:	C43 H58 N2 O13

Antibacterial; antitubercular

LgP	3.36(C,3)
pKa	
pp cited in Vols.1-5:	2: 801, 804, 807

(1) O. Ghisalba et al., J. Antibiot., 1980, 33, 847.

RIFAMPIN [U;INN]

CAS	13292-46-1	MW:	822.96	MF:	C43 H58 N4 O12

Antibacterial; antitubercular

LgP	2.39(C,3)
pKa	1.7; 7.9
pp cited in Vols.1-5:	1: 48, 204; 2: 91, 93, 94, 102, 795, 801, 802, 804, 806-809; 5: 176, 194, 209-212

(1) E.G. Gallo et al in 'Analytical Profiles of Drug Substances',
ed. K. Florey, Academic Press, NY, 1976, vol. 5, pp 467-.

RIFAMYCIN [INN]

CAS	6998-60-3	MW:	697.79	MF:	C37 H47 N O12

Antibacterial

LgP	2.41(C,3)
pKa	3.0
pp cited in Vols.1-5:	1: 47; 2: 96, 102, 186, 797-804, 806-809

(1) C. Gurgo et al., Handb. Exp. Pharmacol., 1982, 61, 519.
(2) O. Ghisalba et al., Drugs Pharm. Sci., 1984, 22, 281.

RIFAPENTINE [U;INN]

CAS	61379-65-5	MW:	877.05	MF:	C47 H64 N4 O12

Antibacterial; antitubercular

China; Merrell Dow (Rifampin)	
LgP	3.78(C,3)
pKa	
pp cited in Vols.1-5:	2: 801, 806-808

(1) U.S. Pat. 4 002 752 (1977).
(2) V. Arioli et al., J. Antibiot., 1981, 34, 1026.
(3) Drugs Future, 1988, 13, 845.

RIFAXIMIN [INN]

CAS	80621-81-4	MW:	785.90	MF:	C43 H51 N3 O11

Antibacterial

Lepetit, Italy	
LgP	5.45(C,3)
pKa	
pp cited in Vols.1-5:	2: 801, 804, 808

(1) Curr. Ther. Res., 1984, 36, 92.

RILAPINE [INN]

CAS	79781-95-6	MW:	361.88	MF:	C22 H20 Cl N3

LgP	4.61(C,3)
pKa	
pp cited in Vols.1-5:	

RILMAZAFONE [INN]

CAS	99593-25-6	MW:	475.34	MF:	C21 H20 Cl2 N6 O3

Anxiolytic; anticonvulsant

Shionogi, Japan	
LgP	-0.07(C)
pKa	
pp cited in Vols.1-5:	

(1) Biochem. Pharmacol., 1984, 33, 1645.

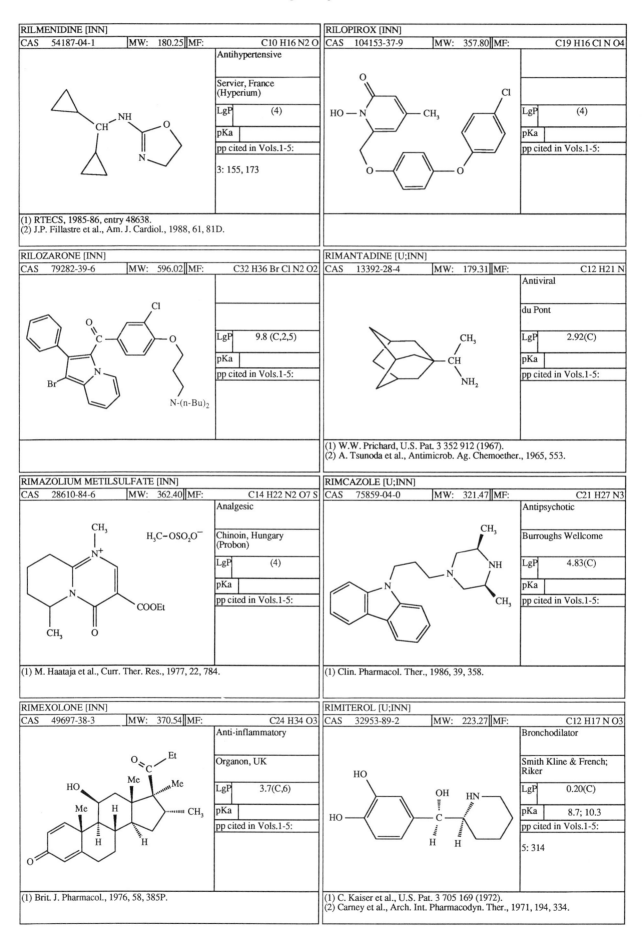

RILMENIDINE [INN]			
CAS 54187-04-1	MW: 180.25	MF:	C10 H16 N2 O
		Antihypertensive	
		Servier, France (Hyperium)	
		LgP	(4)
		pKa	
		pp cited in Vols.1-5:	
		3: 155, 173	

(1) RTECS, 1985-86, entry 48638.
(2) J.P. Fillastre et al., Am. J. Cardiol., 1988, 61, 81D.

RILOPIROX [INN]			
CAS 104153-37-9	MW: 357.80	MF:	C19 H16 Cl N O4
		LgP	(4)
		pKa	
		pp cited in Vols.1-5:	

RILOZARONE [INN]			
CAS 79282-39-6	MW: 596.02	MF:	C32 H36 Br Cl N2 O2
		LgP	9.8 (C,2,5)
		pKa	
		pp cited in Vols.1-5:	

RIMANTADINE [U;INN]			
CAS 13392-28-4	MW: 179.31	MF:	C12 H21 N
		Antiviral	
		du Pont	
		LgP	2.92(C)
		pKa	
		pp cited in Vols.1-5:	

(1) W.W. Prichard, U.S. Pat. 3 352 912 (1967).
(2) A. Tsunoda et al., Antimicrob. Ag. Chemoether., 1965, 553.

RIMAZOLIUM METILSULFATE [INN]			
CAS 28610-84-6	MW: 362.40	MF:	C14 H22 N2 O7 S
		Analgesic	
		Chinoin, Hungary (Probon)	
		LgP	(4)
		pKa	
		pp cited in Vols.1-5:	

(1) M. Haataja et al., Curr. Ther. Res., 1977, 22, 784.

RIMCAZOLE [U;INN]			
CAS 75859-04-0	MW: 321.47	MF:	C21 H27 N3
		Antipsychotic	
		Burroughs Wellcome	
		LgP	4.83(C)
		pKa	
		pp cited in Vols.1-5:	

(1) Clin. Pharmacol. Ther., 1986, 39, 358.

RIMEXOLONE [INN]			
CAS 49697-38-3	MW: 370.54	MF:	C24 H34 O3
		Anti-inflammatory	
		Organon, UK	
		LgP	3.7(C,6)
		pKa	
		pp cited in Vols.1-5:	

(1) Brit. J. Pharmacol., 1976, 58, 385P.

RIMITEROL [U;INN]			
CAS 32953-89-2	MW: 223.27	MF:	C12 H17 N O3
		Bronchodilator	
		Smith Kline & French; Riker	
		LgP	0.20(C)
		pKa	8.7; 10.3
		pp cited in Vols.1-5:	
		5: 314	

(1) C. Kaiser et al., U.S. Pat. 3 705 169 (1972).
(2) Carney et al., Arch. Int. Pharmacodyn. Ther., 1971, 194, 334.

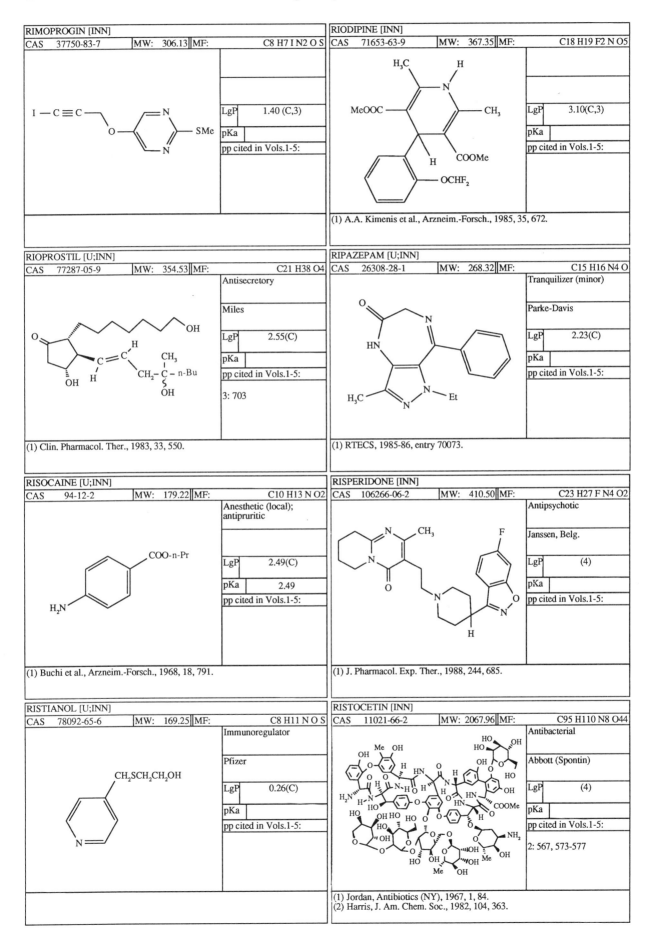

RIMOPROGIN [INN]

CAS	37750-83-7	MW:	306.13	MF:	C8 H7 I N2 O S

LgP	1.40 (C,3)
pKa	
pp cited in Vols.1-5:	

RIODIPINE [INN]

CAS	71653-63-9	MW:	367.35	MF:	C18 H19 F2 N O5

LgP	3.10(C,3)
pKa	
pp cited in Vols.1-5:	

(1) A.A. Kimenis et al., Arzneim.-Forsch., 1985, 35, 672.

RIOPROSTIL [U;INN]

CAS	77287-05-9	MW:	354.53	MF:	C21 H38 O4

Antisecretory	
Miles	
LgP	2.55(C)
pKa	
pp cited in Vols.1-5:	
3: 703	

(1) Clin. Pharmacol. Ther., 1983, 33, 550.

RIPAZEPAM [U;INN]

CAS	26308-28-1	MW:	268.32	MF:	C15 H16 N4 O

Tranquilizer (minor)	
Parke-Davis	
LgP	2.23(C)
pKa	
pp cited in Vols.1-5:	

(1) RTECS, 1985-86, entry 70073.

RISOCAINE [U;INN]

CAS	94-12-2	MW:	179.22	MF:	C10 H13 N O2

Anesthetic (local); antipruritic	
LgP	2.49(C)
pKa	2.49
pp cited in Vols.1-5:	

(1) Buchi et al., Arzneim.-Forsch., 1968, 18, 791.

RISPERIDONE [INN]

CAS	106266-06-2	MW:	410.50	MF:	C23 H27 F N4 O2

Antipsychotic	
Janssen, Belg.	
LgP	(4)
pKa	
pp cited in Vols.1-5:	

(1) J. Pharmacol. Exp. Ther., 1988, 244, 685.

RISTIANOL [U;INN]

CAS	78092-65-6	MW:	169.25	MF:	C8 H11 N O S

Immunoregulator	
Pfizer	
LgP	0.26(C)
pKa	
pp cited in Vols.1-5:	

RISTOCETIN [INN]

CAS	11021-66-2	MW:	2067.96	MF:	C95 H110 N8 O44

Antibacterial	
Abbott (Spontin)	
LgP	(4)
pKa	
pp cited in Vols.1-5:	
2: 567, 573-577	

(1) Jordan, Antibiotics (NY), 1967, 1, 84.
(2) Harris, J. Am. Chem. Soc., 1982, 104, 363.

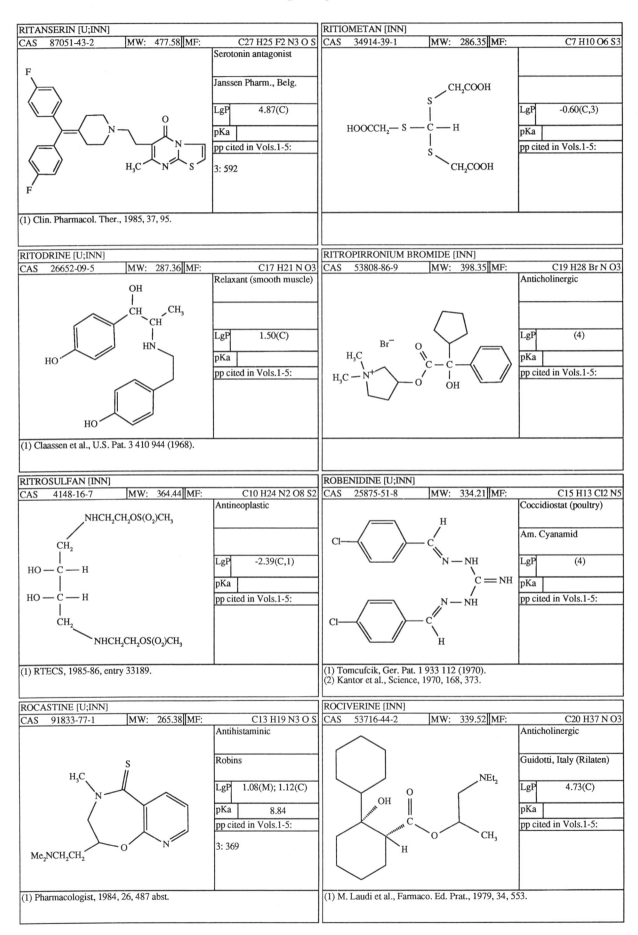

RITANSERIN [U;INN]

CAS	87051-43-2	MW: 477.58	MF:	C27 H25 F2 N3 O S

Serotonin antagonist

Janssen Pharm., Belg.

LgP	4.87(C)
pKa	
pp cited in Vols.1-5:	
3: 592	

(1) Clin. Pharmacol. Ther., 1985, 37, 95.

RITIOMETAN [INN]

CAS	34914-39-1	MW: 286.35	MF:	C7 H10 O6 S3

LgP	-0.60(C,3)
pKa	
pp cited in Vols.1-5:	

RITODRINE [U;INN]

CAS	26652-09-5	MW: 287.36	MF:	C17 H21 N O3

Relaxant (smooth muscle)

LgP	1.50(C)
pKa	
pp cited in Vols.1-5:	

(1) Claassen et al., U.S. Pat. 3 410 944 (1968).

RITROPIRRONIUM BROMIDE [INN]

CAS	53808-86-9	MW: 398.35	MF:	C19 H28 Br N O3

Anticholinergic

LgP	(4)
pKa	
pp cited in Vols.1-5:	

RITROSULFAN [INN]

CAS	4148-16-7	MW: 364.44	MF:	C10 H24 N2 O8 S2

Antineoplastic

LgP	-2.39(C,1)
pKa	
pp cited in Vols.1-5:	

(1) RTECS, 1985-86, entry 33189.

ROBENIDINE [U;INN]

CAS	25875-51-8	MW: 334.21	MF:	C15 H13 Cl2 N5

Coccidiostat (poultry)

Am. Cyanamid

LgP	(4)
pKa	
pp cited in Vols.1-5:	

(1) Tomcufcik, Ger. Pat. 1 933 112 (1970).
(2) Kantor et al., Science, 1970, 168, 373.

ROCASTINE [U;INN]

CAS	91833-77-1	MW: 265.38	MF:	C13 H19 N3 O S

Antihistaminic

Robins

LgP	1.08(M); 1.12(C)
pKa	8.84
pp cited in Vols.1-5:	
3: 369	

(1) Pharmacologist, 1984, 26, 487 abst.

ROCIVERINE [INN]

CAS	53716-44-2	MW: 339.52	MF:	C20 H37 N O3

Anticholinergic

Guidotti, Italy (Rilaten)

LgP	4.73(C)
pKa	
pp cited in Vols.1-5:	

(1) M. Laudi et al., Farmaco. Ed. Prat., 1979, 34, 553.

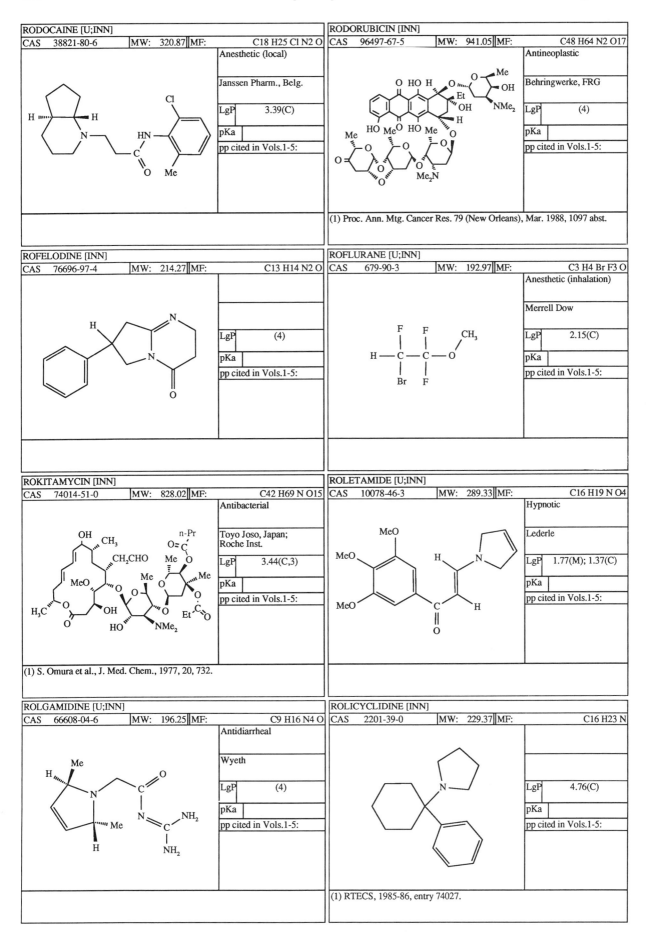

RODOCAINE [U;INN]

CAS	38821-80-6	MW:	320.87	MF:	C18 H25 Cl N2 O

Anesthetic (local)

Janssen Pharm., Belg.

LgP	3.39(C)
pKa	
pp cited in Vols.1-5:	

RODORUBICIN [INN]

CAS	96497-67-5	MW:	941.05	MF:	C48 H64 N2 O17

Antineoplastic

Behringwerke, FRG

LgP	(4)
pKa	
pp cited in Vols.1-5:	

(1) Proc. Ann. Mtg. Cancer Res. 79 (New Orleans), Mar. 1988, 1097 abst.

ROFELODINE [INN]

CAS	76696-97-4	MW:	214.27	MF:	C13 H14 N2 O

LgP	(4)
pKa	
pp cited in Vols.1-5:	

ROFLURANE [U;INN]

CAS	679-90-3	MW:	192.97	MF:	C3 H4 Br F3 O

Anesthetic (inhalation)

Merrell Dow

LgP	2.15(C)
pKa	
pp cited in Vols.1-5:	

ROKITAMYCIN [INN]

CAS	74014-51-0	MW:	828.02	MF:	C42 H69 N O15

Antibacterial

Toyo Joso, Japan; Roche Inst.

LgP	3.44(C,3)
pKa	
pp cited in Vols.1-5:	

(1) S. Omura et al., J. Med. Chem., 1977, 20, 732.

ROLETAMIDE [U;INN]

CAS	10078-46-3	MW:	289.33	MF:	C16 H19 N O4

Hypnotic

Lederle

LgP	1.77(M); 1.37(C)
pKa	
pp cited in Vols.1-5:	

ROLGAMIDINE [U;INN]

CAS	66608-04-6	MW:	196.25	MF:	C9 H16 N4 O

Antidiarrheal

Wyeth

LgP	(4)
pKa	
pp cited in Vols.1-5:	

ROLICYCLIDINE [INN]

CAS	2201-39-0	MW:	229.37	MF:	C16 H23 N

LgP	4.76(C)
pKa	
pp cited in Vols.1-5:	

(1) RTECS, 1985-86, entry 74027.

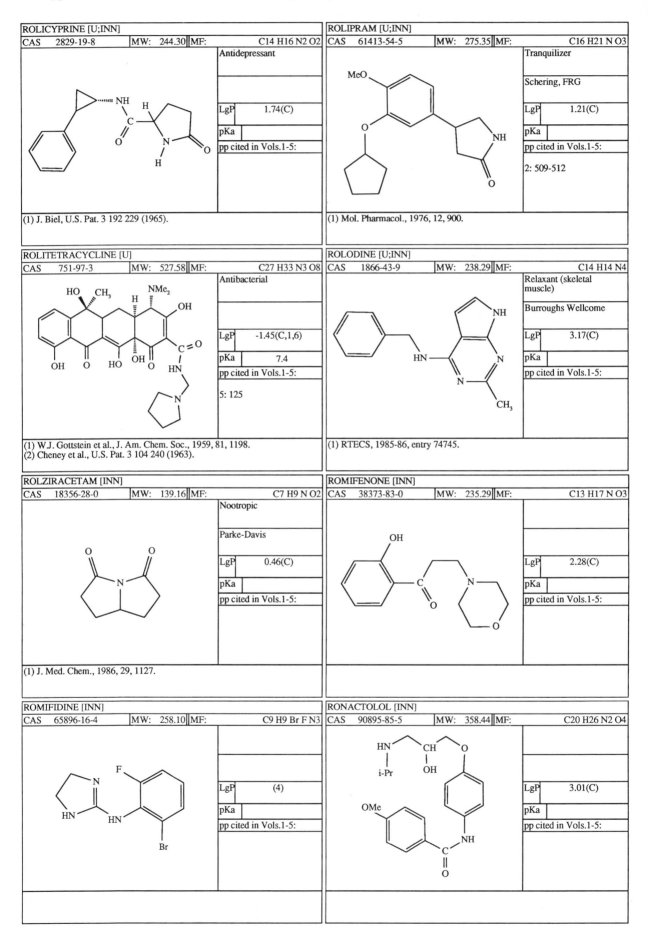

ROLICYPRINE [U;INN]

| CAS | 2829-19-8 | MW: | 244.30 | MF: | C14 H16 N2 O2 |

Antidepressant

LgP	1.74(C)
pKa	
pp cited in Vols.1-5:	

(1) J. Biel, U.S. Pat. 3 192 229 (1965).

ROLIPRAM [U;INN]

| CAS | 61413-54-5 | MW: | 275.35 | MF: | C16 H21 N O3 |

Tranquilizer

Schering, FRG

LgP	1.21(C)
pKa	
pp cited in Vols.1-5:	
2: 509-512	

(1) Mol. Pharmacol., 1976, 12, 900.

ROLITETRACYCLINE [U]

| CAS | 751-97-3 | MW: | 527.58 | MF: | C27 H33 N3 O8 |

Antibacterial

LgP	-1.45(C,1,6)
pKa	7.4
pp cited in Vols.1-5:	
5: 125	

(1) W.J. Gottstein et al., J. Am. Chem. Soc., 1959, 81, 1198.
(2) Cheney et al., U.S. Pat. 3 104 240 (1963).

ROLODINE [U;INN]

| CAS | 1866-43-9 | MW: | 238.29 | MF: | C14 H14 N4 |

Relaxant (skeletal muscle)

Burroughs Wellcome

LgP	3.17(C)
pKa	
pp cited in Vols.1-5:	

(1) RTECS, 1985-86, entry 74745.

ROLZIRACETAM [INN]

| CAS | 18356-28-0 | MW: | 139.16 | MF: | C7 H9 N O2 |

Nootropic

Parke-Davis

LgP	0.46(C)
pKa	
pp cited in Vols.1-5:	

(1) J. Med. Chem., 1986, 29, 1127.

ROMIFENONE [INN]

| CAS | 38373-83-0 | MW: | 235.29 | MF: | C13 H17 N O3 |

LgP	2.28(C)
pKa	
pp cited in Vols.1-5:	

ROMIFIDINE [INN]

| CAS | 65896-16-4 | MW: | 258.10 | MF: | C9 H9 Br F N3 |

LgP	(4)
pKa	
pp cited in Vols.1-5:	

RONACTOLOL [INN]

| CAS | 90895-85-5 | MW: | 358.44 | MF: | C20 H26 N2 O4 |

LgP	3.01(C)
pKa	
pp cited in Vols.1-5:	

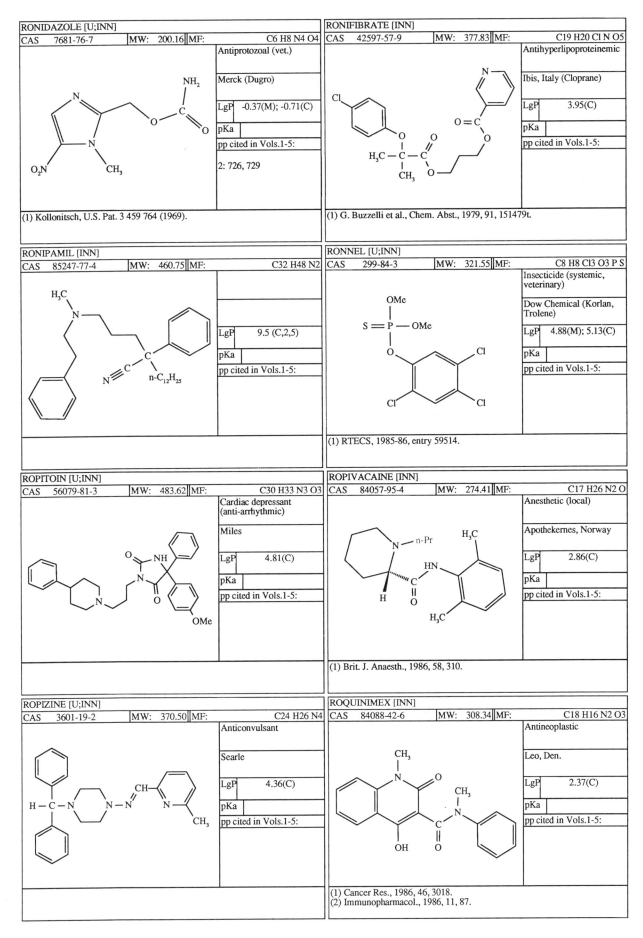

RONIDAZOLE [U;INN]

CAS	7681-76-7	MW:	200.16	MF:	C6 H8 N4 O4

Antiprotozoal (vet.)

Merck (Dugro)

LgP	-0.37(M); -0.71(C)

pKa	

pp cited in Vols.1-5:

2: 726, 729

(1) Kollonitsch, U.S. Pat. 3 459 764 (1969).

RONIFIBRATE [INN]

CAS	42597-57-9	MW:	377.83	MF:	C19 H20 Cl N O5

Antihyperlipoproteinemic

Ibis, Italy (Cloprane)

LgP	3.95(C)

pKa	

pp cited in Vols.1-5:

(1) G. Buzzelli et al., Chem. Abst., 1979, 91, 151479t.

RONIPAMIL [INN]

CAS	85247-77-4	MW:	460.75	MF:	C32 H48 N2

LgP	9.5 (C,2,5)

pKa	

pp cited in Vols.1-5:

RONNEL [U;INN]

CAS	299-84-3	MW:	321.55	MF:	C8 H8 Cl3 O3 P S

Insecticide (systemic, veterinary)

Dow Chemical (Korlan, Trolene)

LgP	4.88(M); 5.13(C)

pKa	

pp cited in Vols.1-5:

(1) RTECS, 1985-86, entry 59514.

ROPITOIN [U;INN]

CAS	56079-81-3	MW:	483.62	MF:	C30 H33 N3 O3

Cardiac depressant (anti-arrhythmic)

Miles

LgP	4.81(C)

pKa	

pp cited in Vols.1-5:

ROPIVACAINE [INN]

CAS	84057-95-4	MW:	274.41	MF:	C17 H26 N2 O

Anesthetic (local)

Apothekernes, Norway

LgP	2.86(C)

pKa	

pp cited in Vols.1-5:

(1) Brit. J. Anaesth., 1986, 58, 310.

ROPIZINE [U;INN]

CAS	3601-19-2	MW:	370.50	MF:	C24 H26 N4

Anticonvulsant

Searle

LgP	4.36(C)

pKa	

pp cited in Vols.1-5:

ROQUINIMEX [INN]

CAS	84088-42-6	MW:	308.34	MF:	C18 H16 N2 O3

Antineoplastic

Leo, Den.

LgP	2.37(C)

pKa	

pp cited in Vols.1-5:

(1) Cancer Res., 1986, 46, 3018.
(2) Immunopharmacol., 1986, 11, 87.

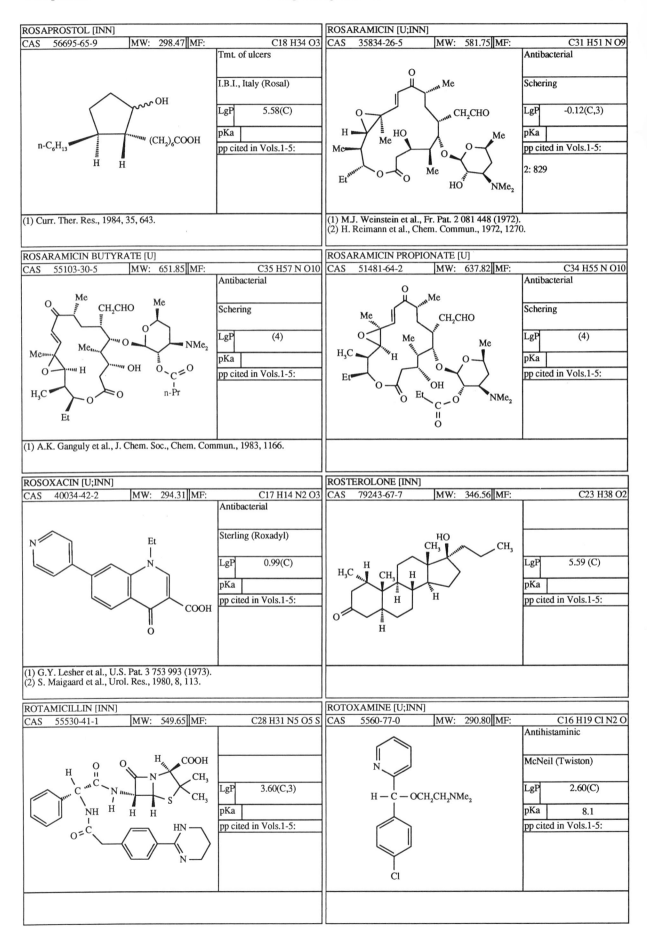

ROSAPROSTOL [INN]

CAS	56695-65-9	MW:	298.47	MF:	C18 H34 O3

Tmt. of ulcers

I.B.I., Italy (Rosal)

LgP	5.58(C)

pKa	

pp cited in Vols.1-5:

(1) Curr. Ther. Res., 1984, 35, 643.

ROSARAMICIN [U;INN]

CAS	35834-26-5	MW:	581.75	MF:	C31 H51 N O9

Antibacterial

Schering

LgP	-0.12(C,3)

pKa	

pp cited in Vols.1-5:

2: 829

(1) M.J. Weinstein et al., Fr. Pat. 2 081 448 (1972).
(2) H. Reimann et al., Chem. Commun., 1972, 1270.

ROSARAMICIN BUTYRATE [U]

CAS	55103-30-5	MW:	651.85	MF:	C35 H57 N O10

Antibacterial

Schering

LgP	(4)

pKa	

pp cited in Vols.1-5:

(1) A.K. Ganguly et al., J. Chem. Soc., Chem. Commun., 1983, 1166.

ROSARAMICIN PROPIONATE [U]

CAS	51481-64-2	MW:	637.82	MF:	C34 H55 N O10

Antibacterial

Schering

LgP	(4)

pKa	

pp cited in Vols.1-5:

ROSOXACIN [U;INN]

CAS	40034-42-2	MW:	294.31	MF:	C17 H14 N2 O3

Antibacterial

Sterling (Roxadyl)

LgP	0.99(C)

pKa	

pp cited in Vols.1-5:

(1) G.Y. Lesher et al., U.S. Pat. 3 753 993 (1973).
(2) S. Maigaard et al., Urol. Res., 1980, 8, 113.

ROSTEROLONE [INN]

CAS	79243-67-7	MW:	346.56	MF:	C23 H38 O2

LgP	5.59 (C)

pKa	

pp cited in Vols.1-5:

ROTAMICILLIN [INN]

CAS	55530-41-1	MW:	549.65	MF:	C28 H31 N5 O5 S

LgP	3.60(C,3)

pKa	

pp cited in Vols.1-5:

ROTOXAMINE [U;INN]

CAS	5560-77-0	MW:	290.80	MF:	C16 H19 Cl N2 O

Antihistaminic

McNeil (Twiston)

LgP	2.60(C)

pKa	8.1

pp cited in Vols.1-5:

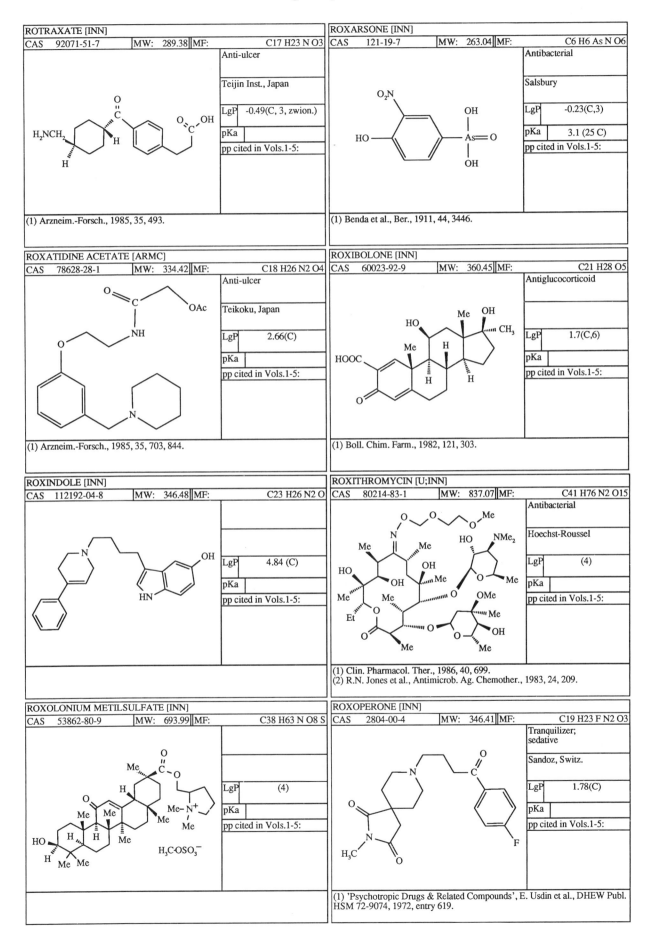

ROTRAXATE [INN]

CAS	92071-51-7	MW:	289.38	MF:	C17 H23 N O3

Anti-ulcer

Teijin Inst., Japan

LgP	-0.49(C, 3, zwion.)
pKa	
pp cited in Vols.1-5:	

(1) Arzneim.-Forsch., 1985, 35, 493.

ROXARSONE [INN]

CAS	121-19-7	MW:	263.04	MF:	C6 H6 As N O6

Antibacterial

Salsbury

LgP	-0.23(C,3)
pKa	3.1 (25 C)
pp cited in Vols.1-5:	

(1) Benda et al., Ber., 1911, 44, 3446.

ROXATIDINE ACETATE [ARMC]

CAS	78628-28-1	MW:	334.42	MF:	C18 H26 N2 O4

Anti-ulcer

Teikoku, Japan

LgP	2.66(C)
pKa	
pp cited in Vols.1-5:	

(1) Arzneim.-Forsch., 1985, 35, 703, 844.

ROXIBOLONE [INN]

CAS	60023-92-9	MW:	360.45	MF:	C21 H28 O5

Antiglucocorticoid

LgP	1.7(C,6)
pKa	
pp cited in Vols.1-5:	

(1) Boll. Chim. Farm., 1982, 121, 303.

ROXINDOLE [INN]

CAS	112192-04-8	MW:	346.48	MF:	C23 H26 N2 O

LgP	4.84 (C)
pKa	
pp cited in Vols.1-5:	

ROXITHROMYCIN [U;INN]

CAS	80214-83-1	MW:	837.07	MF:	C41 H76 N2 O15

Antibacterial

Hoechst-Roussel

LgP	(4)
pKa	
pp cited in Vols.1-5:	

(1) Clin. Pharmacol. Ther., 1986, 40, 699.
(2) R.N. Jones et al., Antimicrob. Ag. Chemother., 1983, 24, 209.

ROXOLONIUM METILSULFATE [INN]

CAS	53862-80-9	MW:	693.99	MF:	C38 H63 N O8 S

LgP	(4)
pKa	
pp cited in Vols.1-5:	

ROXOPERONE [INN]

CAS	2804-00-4	MW:	346.41	MF:	C19 H23 F N2 O3

Tranquilizer; sedative

Sandoz, Switz.

LgP	1.78(C)
pKa	
pp cited in Vols.1-5:	

(1) 'Psychotropic Drugs & Related Compounds', E. Usdin et al., DHEW Publ. HSM 72-9074, 1972, entry 619.

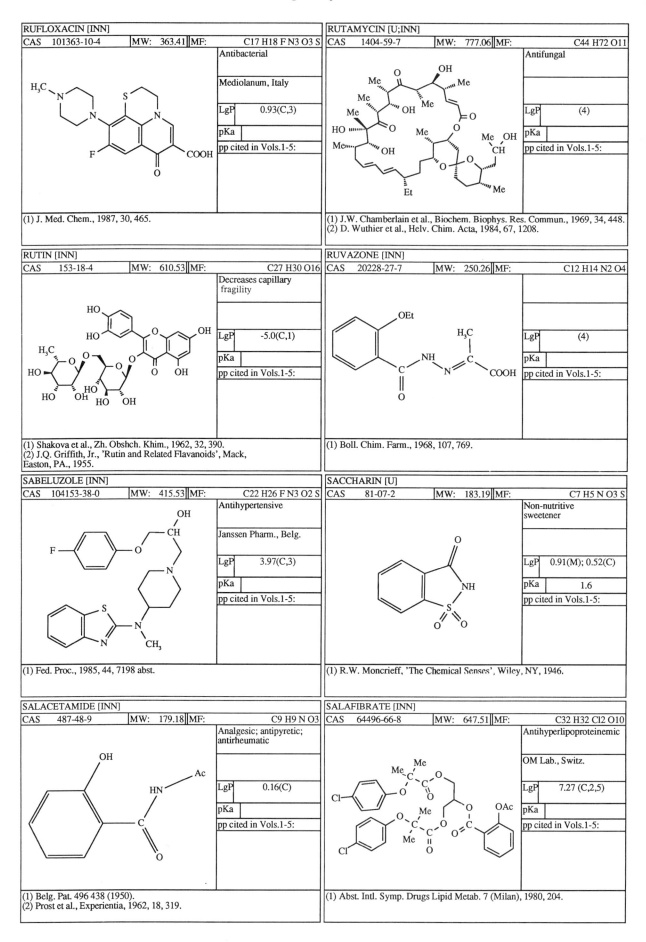

RUFLOXACIN [INN]

CAS	101363-10-4	MW:	363.41	MF:	C17 H18 F N3 O3 S

Antibacterial

Mediolanum, Italy

LgP	0.93(C,3)
pKa	
pp cited in Vols.1-5:	

(1) J. Med. Chem., 1987, 30, 465.

RUTAMYCIN [U;INN]

CAS	1404-59-7	MW:	777.06	MF:	C44 H72 O11

Antifungal

LgP	(4)
pKa	
pp cited in Vols.1-5:	

(1) J.W. Chamberlain et al., Biochem. Biophys. Res. Commun., 1969, 34, 448.
(2) D. Wuthier et al., Helv. Chim. Acta, 1984, 67, 1208.

RUTIN [INN]

CAS	153-18-4	MW:	610.53	MF:	C27 H30 O16

Decreases capillary fragility

LgP	-5.0(C,1)
pKa	
pp cited in Vols.1-5:	

(1) Shakova et al., Zh. Obshch. Khim., 1962, 32, 390.
(2) J.Q. Griffith, Jr., 'Rutin and Related Flavanoids', Mack, Easton, PA., 1955.

RUVAZONE [INN]

CAS	20228-27-7	MW:	250.26	MF:	C12 H14 N2 O4

LgP	(4)
pKa	
pp cited in Vols.1-5:	

(1) Boll. Chim. Farm., 1968, 107, 769.

SABELUZOLE [INN]

CAS	104153-38-0	MW:	415.53	MF:	C22 H26 F N3 O2 S

Antihypertensive

Janssen Pharm., Belg.

LgP	3.97(C,3)
pKa	
pp cited in Vols.1-5:	

(1) Fed. Proc., 1985, 44, 7198 abst.

SACCHARIN [U]

CAS	81-07-2	MW:	183.19	MF:	C7 H5 N O3 S

Non-nutritive sweetener

LgP	0.91(M); 0.52(C)
pKa	1.6
pp cited in Vols.1-5:	

(1) R.W. Moncrieff, 'The Chemical Senses', Wiley, NY, 1946.

SALACETAMIDE [INN]

CAS	487-48-9	MW:	179.18	MF:	C9 H9 N O3

Analgesic; antipyretic; antirheumatic

LgP	0.16(C)
pKa	
pp cited in Vols.1-5:	

(1) Belg. Pat. 496 438 (1950).
(2) Prost et al., Experientia, 1962, 18, 319.

SALAFIBRATE [INN]

CAS	64496-66-8	MW:	647.51	MF:	C32 H32 Cl2 O10

Antihyperlipoproteinemic

OM Lab., Switz.

LgP	7.27 (C,2,5)
pKa	
pp cited in Vols.1-5:	

(1) Abst. Intl. Symp. Drugs Lipid Metab. 7 (Milan), 1980, 204.

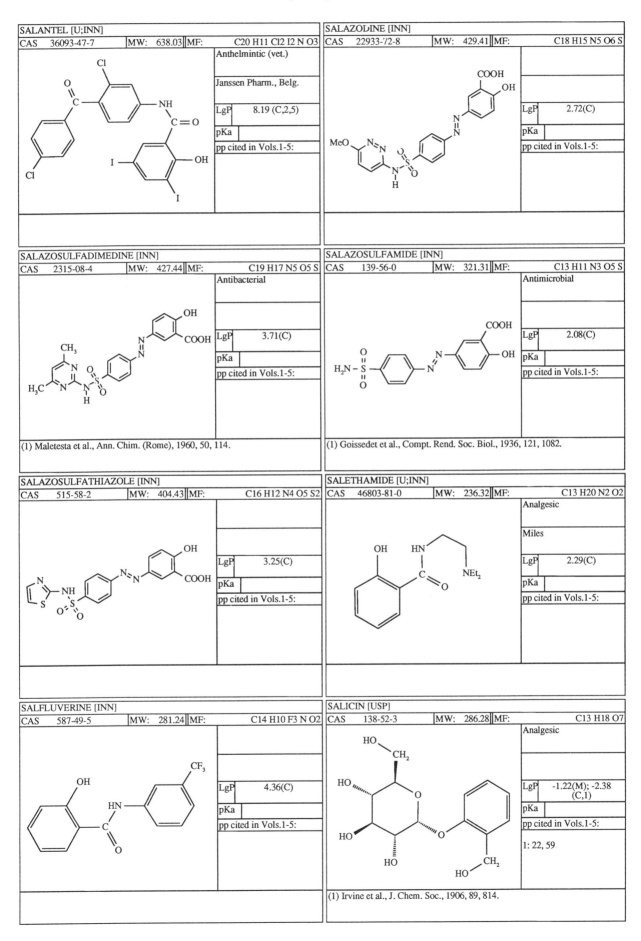

SALANTEL [U;INN]

| CAS | 36093-47-7 | MW: | 638.03 | MF: | C20 H11 Cl2 I2 N O3 |

Anthelmintic (vet.)

Janssen Pharm., Belg.

LgP	8.19 (C,2,5)
pKa	
pp cited in Vols.1-5:	

SALAZODINE [INN]

| CAS | 22933-72-8 | MW: | 429.41 | MF: | C18 H15 N5 O6 S |

LgP	2.72(C)
pKa	
pp cited in Vols.1-5:	

SALAZOSULFADIMEDINE [INN]

| CAS | 2315-08-4 | MW: | 427.44 | MF: | C19 H17 N5 O5 S |

Antibacterial

LgP	3.71(C)
pKa	
pp cited in Vols.1-5:	

(1) Maletesta et al., Ann. Chim. (Rome), 1960, 50, 114.

SALAZOSULFAMIDE [INN]

| CAS | 139-56-0 | MW: | 321.31 | MF: | C13 H11 N3 O5 S |

Antimicrobial

LgP	2.08(C)
pKa	
pp cited in Vols.1-5:	

(1) Goissedet et al., Compt. Rend. Soc. Biol., 1936, 121, 1082.

SALAZOSULFATHIAZOLE [INN]

| CAS | 515-58-2 | MW: | 404.43 | MF: | C16 H12 N4 O5 S2 |

LgP	3.25(C)
pKa	
pp cited in Vols.1-5:	

SALETHAMIDE [U;INN]

| CAS | 46803-81-0 | MW: | 236.32 | MF: | C13 H20 N2 O2 |

Analgesic

Miles

LgP	2.29(C)
pKa	
pp cited in Vols.1-5:	

SALFLUVERINE [INN]

| CAS | 587-49-5 | MW: | 281.24 | MF: | C14 H10 F3 N O2 |

LgP	4.36(C)
pKa	
pp cited in Vols.1-5:	

SALICIN [USP]

| CAS | 138-52-3 | MW: | 286.28 | MF: | C13 H18 O7 |

Analgesic

LgP	-1.22(M); -2.38 (C,1)
pKa	
pp cited in Vols.1-5:	

1: 22, 59

(1) Irvine et al., J. Chem. Soc., 1906, 89, 814.

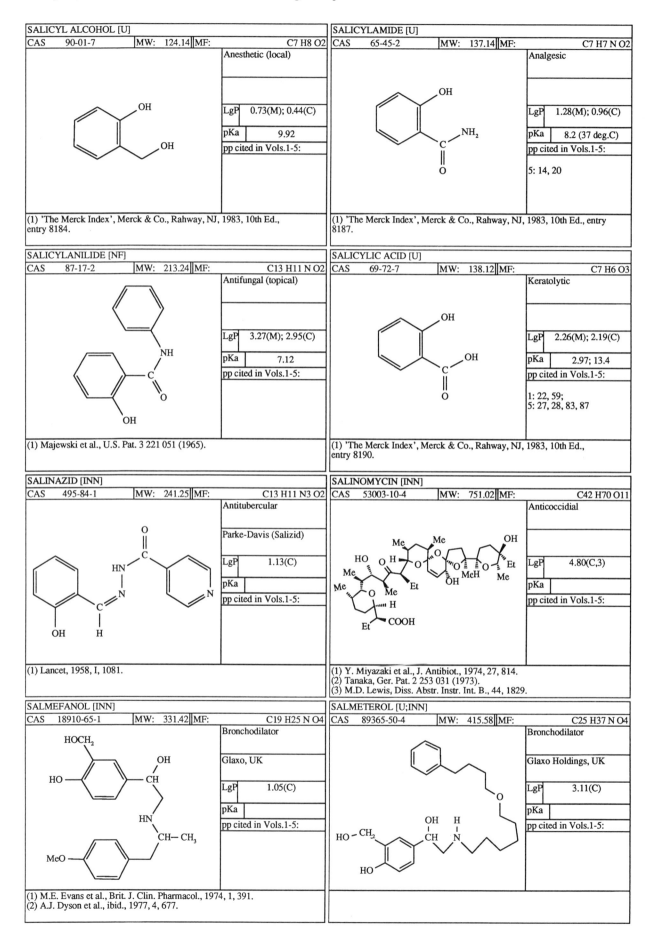

SALICYL ALCOHOL [U]

CAS	90-01-7	MW:	124.14	MF:	C7 H8 O2

Anesthetic (local)

LgP	0.73(M); 0.44(C)
pKa	9.92
pp cited in Vols.1-5:	

(1) 'The Merck Index', Merck & Co., Rahway, NJ, 1983, 10th Ed., entry 8184.

SALICYLAMIDE [U]

CAS	65-45-2	MW:	137.14	MF:	C7 H7 N O2

Analgesic

LgP	1.28(M); 0.96(C)
pKa	8.2 (37 deg.C)
pp cited in Vols.1-5:	

5: 14, 20

(1) 'The Merck Index', Merck & Co., Rahway, NJ, 1983, 10th Ed., entry 8187.

SALICYLANILIDE [NF]

CAS	87-17-2	MW:	213.24	MF:	C13 H11 N O2

Antifungal (topical)

LgP	3.27(M); 2.95(C)
pKa	7.12
pp cited in Vols.1-5:	

(1) Majewski et al., U.S. Pat. 3 221 051 (1965).

SALICYLIC ACID [U]

CAS	69-72-7	MW:	138.12	MF:	C7 H6 O3

Keratolytic

LgP	2.26(M); 2.19(C)
pKa	2.97; 13.4
pp cited in Vols.1-5:	

1: 22, 59;
5: 27, 28, 83, 87

(1) 'The Merck Index', Merck & Co., Rahway, NJ, 1983, 10th Ed., entry 8190.

SALINAZID [INN]

CAS	495-84-1	MW:	241.25	MF:	C13 H11 N3 O2

Antitubercular

Parke-Davis (Salizid)

LgP	1.13(C)
pKa	
pp cited in Vols.1-5:	

(1) Lancet, 1958, I, 1081.

SALINOMYCIN [INN]

CAS	53003-10-4	MW:	751.02	MF:	C42 H70 O11

Anticoccidial

LgP	4.80(C,3)
pKa	
pp cited in Vols.1-5:	

(1) Y. Miyazaki et al., J. Antibiot., 1974, 27, 814.
(2) Tanaka, Ger. Pat. 2 253 031 (1973).
(3) M.D. Lewis, Diss. Abstr. Instr. Int. B., 44, 1829.

SALMEFANOL [INN]

CAS	18910-65-1	MW:	331.42	MF:	C19 H25 N O4

Bronchodilator

Glaxo, UK

LgP	1.05(C)
pKa	
pp cited in Vols.1-5:	

(1) M.E. Evans et al., Brit. J. Clin. Pharmacol., 1974, 1, 391.
(2) A.J. Dyson et al., ibid., 1977, 4, 677.

SALMETEROL [U;INN]

CAS	89365-50-4	MW:	415.58	MF:	C25 H37 N O4

Bronchodilator

Glaxo Holdings, UK

LgP	3.11(C)
pKa	
pp cited in Vols.1-5:	

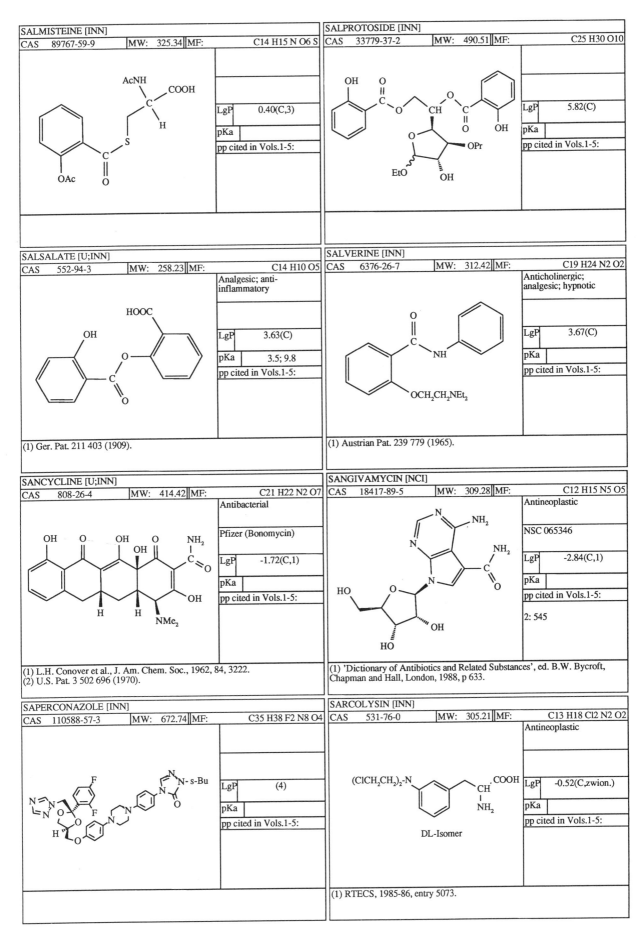

SALMISTEINE [INN]

CAS	89767-59-9	MW:	325.34	MF:	C14 H15 N O6 S

LgP	0.40(C,3)
pKa	
pp cited in Vols.1-5:	

SALPROTOSIDE [INN]

CAS	33779-37-2	MW:	490.51	MF:	C25 H30 O10

LgP	5.82(C)
pKa	
pp cited in Vols.1-5:	

SALSALATE [U;INN]

CAS	552-94-3	MW:	258.23	MF:	C14 H10 O5

Analgesic; anti-inflammatory

LgP	3.63(C)
pKa	3.5; 9.8
pp cited in Vols.1-5:	

(1) Ger. Pat. 211 403 (1909).

SALVERINE [INN]

CAS	6376-26-7	MW:	312.42	MF:	C19 H24 N2 O2

Anticholinergic; analgesic; hypnotic

LgP	3.67(C)
pKa	
pp cited in Vols.1-5:	

(1) Austrian Pat. 239 779 (1965).

SANCYCLINE [U;INN]

CAS	808-26-4	MW:	414.42	MF:	C21 H22 N2 O7

Antibacterial

Pfizer (Bonomycin)

LgP	-1.72(C,1)
pKa	
pp cited in Vols.1-5:	

(1) L.H. Conover et al., J. Am. Chem. Soc., 1962, 84, 3222.
(2) U.S. Pat. 3 502 696 (1970).

SANGIVAMYCIN [NCI]

CAS	18417-89-5	MW:	309.28	MF:	C12 H15 N5 O5

Antineoplastic

NSC 065346

LgP	-2.84(C,1)
pKa	
pp cited in Vols.1-5:	
2: 545	

(1) 'Dictionary of Antibiotics and Related Substances', ed. B.W. Bycroft, Chapman and Hall, London, 1988, p 633.

SAPERCONAZOLE [INN]

CAS	110588-57-3	MW:	672.74	MF:	C35 H38 F2 N8 O4

LgP	(4)
pKa	
pp cited in Vols.1-5:	

SARCOLYSIN [INN]

CAS	531-76-0	MW:	305.21	MF:	C13 H18 Cl2 N2 O2

Antineoplastic

LgP	-0.52(C,zwion.)
pKa	
pp cited in Vols.1-5:	

DL-Isomer

(1) RTECS, 1985-86, entry 5073.

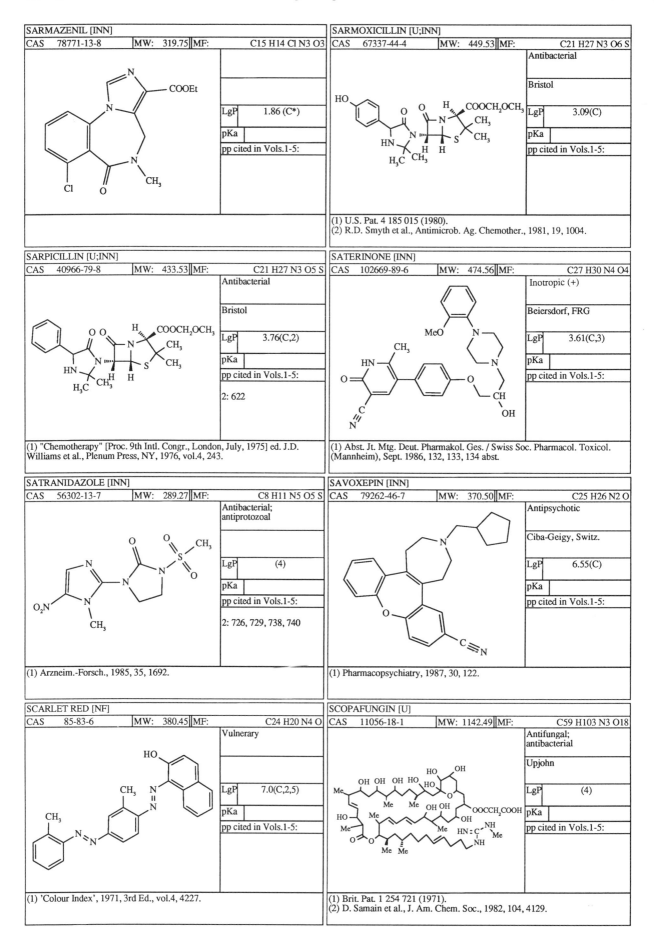

Sarmazenil   *Drug Compendium*   844

SARMAZENIL [INN]

CAS 78771-13-8	MW: 319.75	MF:	C15 H14 Cl N3 O3

LgP 1.86 (C*)

pKa

pp cited in Vols.1-5:

SARMOXICILLIN [U;INN]

CAS 67337-44-4	MW: 449.53	MF:	C21 H27 N3 O6 S

Antibacterial

Bristol

LgP 3.09(C)

pKa

pp cited in Vols.1-5:

(1) U.S. Pat. 4 185 015 (1980).
(2) R.D. Smyth et al., Antimicrob. Ag. Chemother., 1981, 19, 1004.

SARPICILLIN [U;INN]

CAS 40966-79-8	MW: 433.53	MF:	C21 H27 N3 O5 S

Antibacterial

Bristol

LgP 3.76(C,2)

pKa

pp cited in Vols.1-5:

2: 622

(1) "Chemotherapy" [Proc. 9th Intl. Congr., London, July, 1975] ed. J.D. Williams et al., Plenum Press, NY, 1976, vol.4, 243.

SATERINONE [INN]

CAS 102669-89-6	MW: 474.56	MF:	C27 H30 N4 O4

Inotropic (+)

Beiersdorf, FRG

LgP 3.61(C,3)

pKa

pp cited in Vols.1-5:

(1) Abst. Jt. Mtg. Deut. Pharmakol. Ges. / Swiss Soc. Pharmacol. Toxicol. (Mannheim), Sept. 1986, 132, 133, 134 abst.

SATRANIDAZOLE [INN]

CAS 56302-13-7	MW: 289.27	MF:	C8 H11 N5 O5 S

Antibacterial; antiprotozoal

LgP (4)

pKa

pp cited in Vols.1-5:

2: 726, 729, 738, 740

(1) Arzneim.-Forsch., 1985, 35, 1692.

SAVOXEPIN [INN]

CAS 79262-46-7	MW: 370.50	MF:	C25 H26 N2 O

Antipsychotic

Ciba-Geigy, Switz.

LgP 6.55(C)

pKa

pp cited in Vols.1-5:

(1) Pharmacopsychiatry, 1987, 30, 122.

SCARLET RED [NF]

CAS 85-83-6	MW: 380.45	MF:	C24 H20 N4 O

Vulnerary

LgP 7.0(C,2,5)

pKa

pp cited in Vols.1-5:

(1) 'Colour Index', 1971, 3rd Ed., vol.4, 4227.

SCOPAFUNGIN [U]

CAS 11056-18-1	MW: 1142.49	MF:	C59 H103 N3 O18

Antifungal; antibacterial

Upjohn

LgP (4)

pKa

pp cited in Vols.1-5:

(1) Brit. Pat. 1 254 721 (1971).
(2) D. Samain et al., J. Am. Chem. Soc., 1982, 104, 4129.

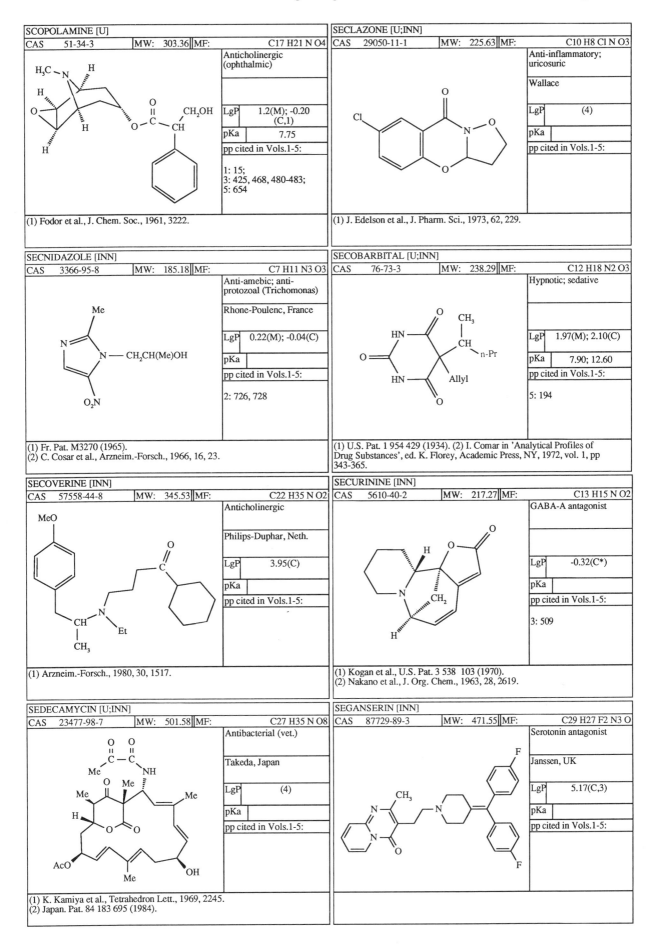

SCOPOLAMINE [U]

CAS	51-34-3	MW:	303.36	MF:	C17 H21 N O4

Anticholinergic (ophthalmic)

LgP	1.2(M); -0.20 (C,1)
pKa	7.75
pp cited in Vols.1-5:	

1: 15;
3: 425, 468, 480-483;
5: 654

(1) Fodor et al., J. Chem. Soc., 1961, 3222.

SECLAZONE [U;INN]

CAS	29050-11-1	MW:	225.63	MF:	C10 H8 Cl N O3

Anti-inflammatory; uricosuric

Wallace

LgP	(4)
pKa	
pp cited in Vols.1-5:	

(1) J. Edelson et al., J. Pharm. Sci., 1973, 62, 229.

SECNIDAZOLE [INN]

CAS	3366-95-8	MW:	185.18	MF:	C7 H11 N3 O3

Anti-amebic; anti-protozoal (Trichomonas)

Rhone-Poulenc, France

LgP	0.22(M); -0.04(C)
pKa	
pp cited in Vols.1-5:	

2: 726, 728

(1) Fr. Pat. M3270 (1965).
(2) C. Cosar et al., Arzneim.-Forsch., 1966, 16, 23.

SECOBARBITAL [U;INN]

CAS	76-73-3	MW:	238.29	MF:	C12 H18 N2 O3

Hypnotic; sedative

LgP	1.97(M); 2.10(C)
pKa	7.90; 12.60
pp cited in Vols.1-5:	

5: 194

(1) U.S. Pat. 1 954 429 (1934). (2) I. Comar in 'Analytical Profiles of Drug Substances', ed. K. Florey, Academic Press, NY, 1972, vol. 1, pp 343-365.

SECOVERINE [INN]

CAS	57558-44-8	MW:	345.53	MF:	C22 H35 N O2

Anticholinergic

Philips-Duphar, Neth.

LgP	3.95(C)
pKa	
pp cited in Vols.1-5:	

(1) Arzneim.-Forsch., 1980, 30, 1517.

SECURININE [INN]

CAS	5610-40-2	MW:	217.27	MF:	C13 H15 N O2

GABA-A antagonist

LgP	-0.32(C*)
pKa	
pp cited in Vols.1-5:	

3: 509

(1) Kogan et al., U.S. Pat. 3 538 103 (1970).
(2) Nakano et al., J. Org. Chem., 1963, 28, 2619.

SEDECAMYCIN [U;INN]

CAS	23477-98-7	MW:	501.58	MF:	C27 H35 N O8

Antibacterial (vet.)

Takeda, Japan

LgP	(4)
pKa	
pp cited in Vols.1-5:	

(1) K. Kamiya et al., Tetrahedron Lett., 1969, 2245.
(2) Japan. Pat. 84 183 695 (1984).

SEGANSERIN [INN]

CAS	87729-89-3	MW:	471.55	MF:	C29 H27 F2 N3 O

Serotonin antagonist

Janssen, UK

LgP	5.17(C,3)
pKa	
pp cited in Vols.1-5:	

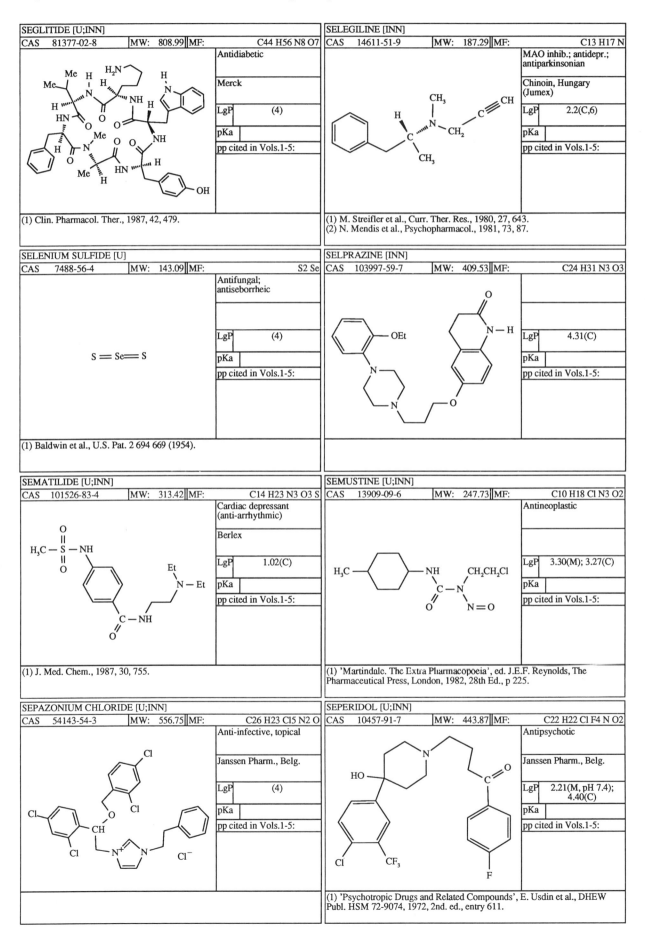

SEGLITIDE [U;INN]			
CAS 81377-02-8	MW: 808.99	MF:	C44 H56 N8 O7

Antidiabetic

Merck

LgP	(4)
pKa	
pp cited in Vols.1-5:	

(1) Clin. Pharmacol. Ther., 1987, 42, 479.

SELEGILINE [INN]			
CAS 14611-51-9	MW: 187.29	MF:	C13 H17 N

MAO inhib.; antidepr.; antiparkinsonian

Chinoin, Hungary (Jumex)

LgP	2.2(C,6)
pKa	
pp cited in Vols.1-5:	

(1) M. Streifler et al., Curr. Ther. Res., 1980, 27, 643.
(2) N. Mendis et al., Psychopharmacol., 1981, 73, 87.

SELENIUM SULFIDE [U]			
CAS 7488-56-4	MW: 143.09	MF:	S2 Se

Antifungal; antiseborrheic

LgP	(4)
pKa	
pp cited in Vols.1-5:	

S = Se = S

(1) Baldwin et al., U.S. Pat. 2 694 669 (1954).

SELPRAZINE [INN]			
CAS 103997-59-7	MW: 409.53	MF:	C24 H31 N3 O3

LgP	4.31(C)
pKa	
pp cited in Vols.1-5:	

SEMATILIDE [U;INN]			
CAS 101526-83-4	MW: 313.42	MF:	C14 H23 N3 O3 S

Cardiac depressant (anti-arrhythmic)

Berlex

LgP	1.02(C)
pKa	
pp cited in Vols.1-5:	

(1) J. Med. Chem., 1987, 30, 755.

SEMUSTINE [U;INN]			
CAS 13909-09-6	MW: 247.73	MF:	C10 H18 Cl N3 O2

Antineoplastic

LgP	3.30(M); 3.27(C)
pKa	
pp cited in Vols.1-5:	

(1) 'Martindale. The Extra Pharmacopoeia', ed. J.E.F. Reynolds, The Pharmaceutical Press, London, 1982, 28th Ed., p 225.

SEPAZONIUM CHLORIDE [U;INN]			
CAS 54143-54-3	MW: 556.75	MF:	C26 H23 Cl5 N2 O

Anti-infective, topical

Janssen Pharm., Belg.

LgP	(4)
pKa	
pp cited in Vols.1-5:	

SEPERIDOL [U;INN]			
CAS 10457-91-7	MW: 443.87	MF:	C22 H22 Cl F4 N O2

Antipsychotic

Janssen Pharm., Belg.

LgP	2.21(M, pH 7.4); 4.40(C)
pKa	
pp cited in Vols.1-5:	

(1) 'Psychotropic Drugs and Related Compounds', E. Usdin et al., DHEW Publ. HSM 72-9074, 1972, 2nd. ed., entry 611.

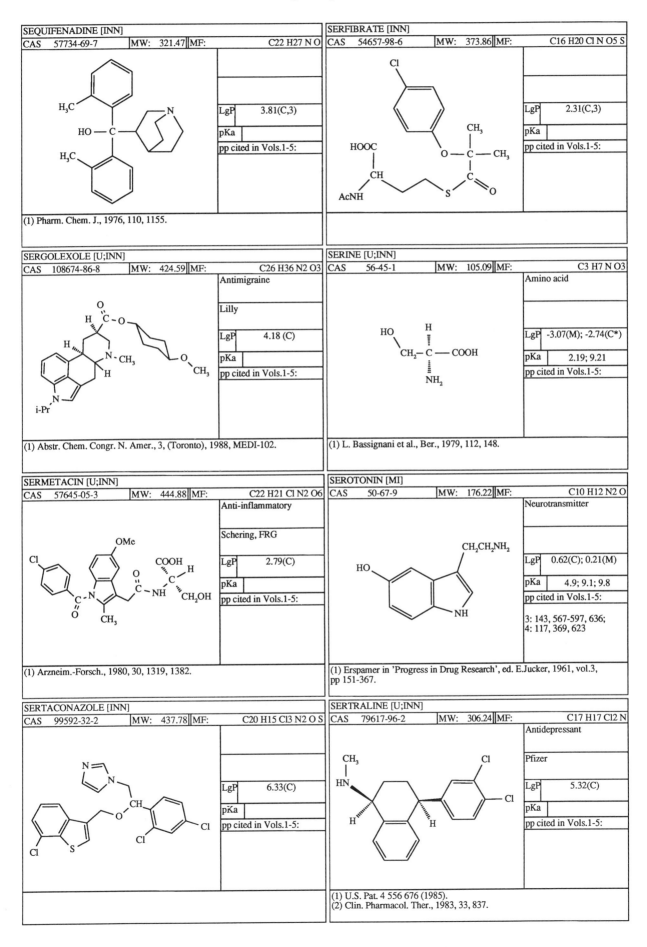

SEQUIFENADINE [INN]

CAS	57734-69-7	MW:	321.47	MF:	C22 H27 N O

LgP	3.81(C,3)
pKa	
pp cited in Vols.1-5:	

(1) Pharm. Chem. J., 1976, 110, 1155.

SERGOLEXOLE [U;INN]

CAS	108674-86-8	MW:	424.59	MF:	C26 H36 N2 O3

Antimigraine

Lilly

LgP	4.18 (C)
pKa	
pp cited in Vols.1-5:	

(1) Abstr. Chem. Congr. N. Amer., 3, (Toronto), 1988, MEDI-102.

SERMETACIN [U;INN]

CAS	57645-05-3	MW:	444.88	MF:	C22 H21 Cl N2 O6

Anti-inflammatory

Schering, FRG

LgP	2.79(C)
pKa	
pp cited in Vols.1-5:	

(1) Arzneim.-Forsch., 1980, 30, 1319, 1382.

SERTACONAZOLE [INN]

CAS	99592-32-2	MW:	437.78	MF:	C20 H15 Cl3 N2 O S

LgP	6.33(C)
pKa	
pp cited in Vols.1-5:	

SERFIBRATE [INN]

CAS	54657-98-6	MW:	373.86	MF:	C16 H20 Cl N O5 S

LgP	2.31(C,3)
pKa	
pp cited in Vols.1-5:	

SERINE [U;INN]

CAS	56-45-1	MW:	105.09	MF:	C3 H7 N O3

Amino acid

LgP	-3.07(M); -2.74(C*)
pKa	2.19; 9.21
pp cited in Vols.1-5:	

(1) L. Bassignani et al., Ber., 1979, 112, 148.

SEROTONIN [MI]

CAS	50-67-9	MW:	176.22	MF:	C10 H12 N2 O

Neurotransmitter

LgP	0.62(C); 0.21(M)
pKa	4.9; 9.1; 9.8
pp cited in Vols.1-5:	

3: 143, 567-597, 636;
4: 117, 369, 623

(1) Erspamer in 'Progress in Drug Research', ed. E.Jucker, 1961, vol.3, pp 151-367.

SERTRALINE [U;INN]

CAS	79617-96-2	MW:	306.24	MF:	C17 H17 Cl2 N

Antidepressant

Pfizer

LgP	5.32(C)
pKa	
pp cited in Vols.1-5:	

(1) U.S. Pat. 4 556 676 (1985).
(2) Clin. Pharmacol. Ther., 1983, 33, 837.

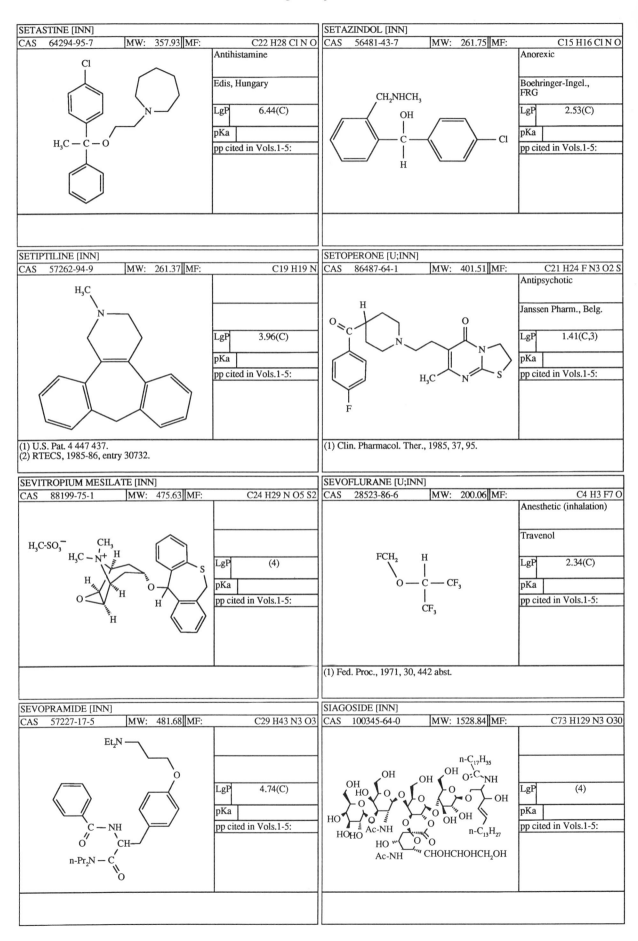

SETASTINE [INN]			
CAS 64294-95-7	MW: 357.93	MF:	C22 H28 Cl N O

Antihistamine

Edis, Hungary

LgP	6.44(C)
pKa	
pp cited in Vols.1-5:	

SETAZINDOL [INN]			
CAS 56481-43-7	MW: 261.75	MF:	C15 H16 Cl N O

Anorexic

Boehringer-Ingel., FRG

LgP	2.53(C)
pKa	
pp cited in Vols.1-5:	

SETIPTILINE [INN]			
CAS 57262-94-9	MW: 261.37	MF:	C19 H19 N

LgP	3.96(C)
pKa	
pp cited in Vols.1-5:	

(1) U.S. Pat. 4 447 437.
(2) RTECS, 1985-86, entry 30732.

SETOPERONE [U;INN]			
CAS 86487-64-1	MW: 401.51	MF:	C21 H24 F N3 O2 S

Antipsychotic

Janssen Pharm., Belg.

LgP	1.41(C,3)
pKa	
pp cited in Vols.1-5:	

(1) Clin. Pharmacol. Ther., 1985, 37, 95.

SEVITROPIUM MESILATE [INN]			
CAS 88199-75-1	MW: 475.63	MF:	C24 H29 N O5 S2

LgP	(4)
pKa	
pp cited in Vols.1-5:	

SEVOFLURANE [U;INN]			
CAS 28523-86-6	MW: 200.06	MF:	C4 H3 F7 O

Anesthetic (inhalation)

Travenol

LgP	2.34(C)
pKa	
pp cited in Vols.1-5:	

(1) Fed. Proc., 1971, 30, 442 abst.

SEVOPRAMIDE [INN]			
CAS 57227-17-5	MW: 481.68	MF:	C29 H43 N3 O3

LgP	4.74(C)
pKa	
pp cited in Vols.1-5:	

SIAGOSIDE [INN]			
CAS 100345-64-0	MW: 1528.84	MF:	C73 H129 N3 O30

LgP	(4)
pKa	
pp cited in Vols.1-5:	

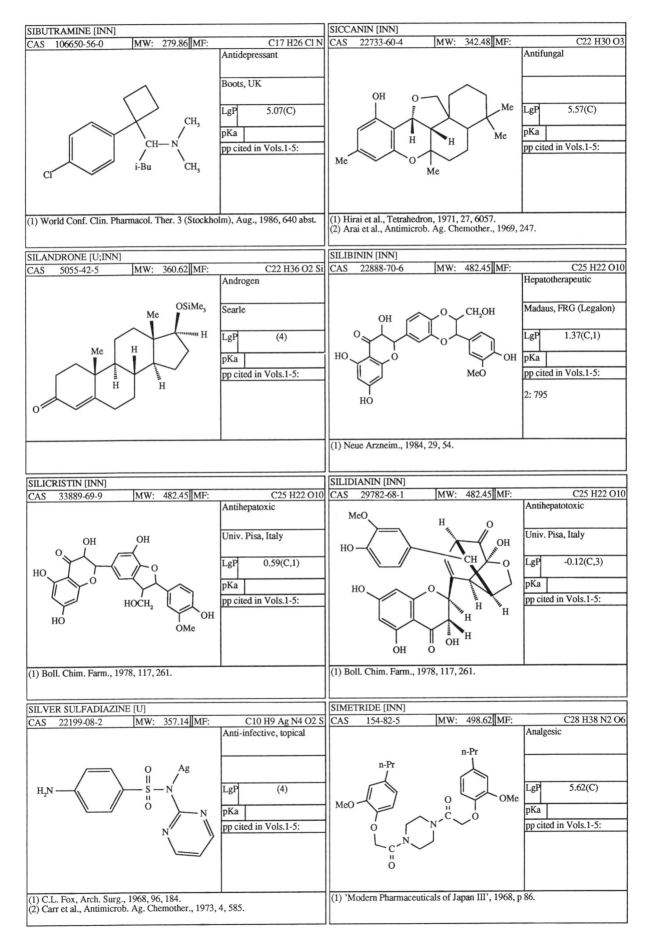

SIBUTRAMINE [INN]

CAS 106650-56-0	MW: 279.86	MF:	C17 H26 Cl N

Antidepressant

Boots, UK

LgP	5.07(C)
pKa	
pp cited in Vols.1-5:	

(1) World Conf. Clin. Pharmacol. Ther. 3 (Stockholm), Aug., 1986, 640 abst.

SICCANIN [INN]

CAS 22733-60-4	MW: 342.48	MF:	C22 H30 O3

Antifungal

LgP	5.57(C)
pKa	
pp cited in Vols.1-5:	

(1) Hirai et al., Tetrahedron, 1971, 27, 6057.
(2) Arai et al., Antimicrob. Ag. Chemother., 1969, 247.

SILANDRONE [U;INN]

CAS 5055-42-5	MW: 360.62	MF:	C22 H36 O2 Si

Androgen

Searle

LgP	(4)
pKa	
pp cited in Vols.1-5:	

SILIBININ [INN]

CAS 22888-70-6	MW: 482.45	MF:	C25 H22 O10

Hepatotherapeutic

Madaus, FRG (Legalon)

LgP	1.37(C,1)
pKa	
pp cited in Vols.1-5:	
	2: 795

(1) Neue Arzneim., 1984, 29, 54.

SILICRISTIN [INN]

CAS 33889-69-9	MW: 482.45	MF:	C25 H22 O10

Antihepatoxic

Univ. Pisa, Italy

LgP	0.59(C,1)
pKa	
pp cited in Vols.1-5:	

(1) Boll. Chim. Farm., 1978, 117, 261.

SILIDIANIN [INN]

CAS 29782-68-1	MW: 482.45	MF:	C25 H22 O10

Antihepatotoxic

Univ. Pisa, Italy

LgP	-0.12(C,3)
pKa	
pp cited in Vols.1-5:	

(1) Boll. Chim. Farm., 1978, 117, 261.

SILVER SULFADIAZINE [U]

CAS 22199-08-2	MW: 357.14	MF:	C10 H9 Ag N4 O2 S

Anti-infective, topical

LgP	(4)
pKa	
pp cited in Vols.1-5:	

(1) C.L. Fox, Arch. Surg., 1968, 96, 184.
(2) Carr et al., Antimicrob. Ag. Chemother., 1973, 4, 585.

SIMETRIDE [INN]

CAS 154-82-5	MW: 498.62	MF:	C28 H38 N2 O6

Analgesic

LgP	5.62(C)
pKa	
pp cited in Vols.1-5:	

(1) 'Modern Pharmaceuticals of Japan III', 1968, p 86.

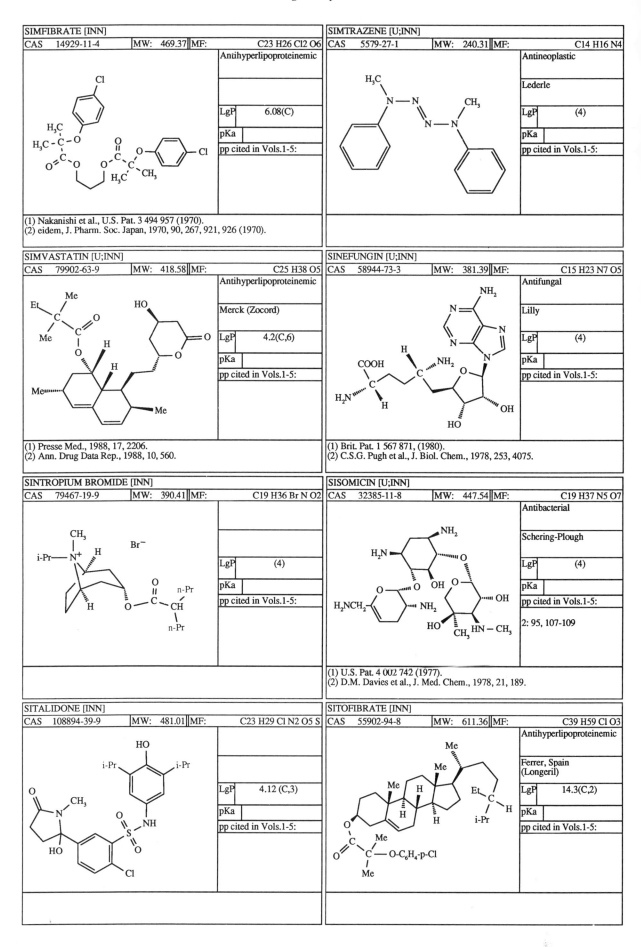

SIMFIBRATE [INN]		
CAS 14929-11-4	MW: 469.37	MF: C23 H26 Cl2 O6

Antihyperlipoproteinemic

LgP 6.08(C)

pKa

pp cited in Vols.1-5:

(1) Nakanishi et al., U.S. Pat. 3 494 957 (1970).
(2) eidem, J. Pharm. Soc. Japan, 1970, 90, 267, 921, 926 (1970).

SIMTRAZENE [U;INN]		
CAS 5579-27-1	MW: 240.31	MF: C14 H16 N4

Antineoplastic

Lederle

LgP (4)

pKa

pp cited in Vols.1-5:

SIMVASTATIN [U;INN]		
CAS 79902-63-9	MW: 418.58	MF: C25 H38 O5

Antihyperlipoproteinemic

Merck (Zocord)

LgP 4.2(C,6)

pKa

pp cited in Vols.1-5:

(1) Presse Med., 1988, 17, 2206.
(2) Ann. Drug Data Rep., 1988, 10, 560.

SINEFUNGIN [U;INN]		
CAS 58944-73-3	MW: 381.39	MF: C15 H23 N7 O5

Antifungal

Lilly

LgP (4)

pKa

pp cited in Vols.1-5:

(1) Brit. Pat. 1 567 871, (1980).
(2) C.S.G. Pugh et al., J. Biol. Chem., 1978, 253, 4075.

SINTROPIUM BROMIDE [INN]		
CAS 79467-19-9	MW: 390.41	MF: C19 H36 Br N O2

LgP (4)

pKa

pp cited in Vols.1-5:

SISOMICIN [U;INN]		
CAS 32385-11-8	MW: 447.54	MF: C19 H37 N5 O7

Antibacterial

Schering-Plough

LgP (4)

pKa

pp cited in Vols.1-5:
2: 95, 107-109

(1) U.S. Pat. 4 002 742 (1977).
(2) D.M. Davies et al., J. Med. Chem., 1978, 21, 189.

SITALIDONE [INN]		
CAS 108894-39-9	MW: 481.01	MF: C23 H29 Cl N2 O5 S

LgP 4.12 (C,3)

pKa

pp cited in Vols.1-5:

SITOFIBRATE [INN]		
CAS 55902-94-8	MW: 611.36	MF: C39 H59 Cl O3

Antihyperlipoproteinemic

Ferrer, Spain
(Longeril)

LgP 14.3(C,2)

pKa

pp cited in Vols.1-5:

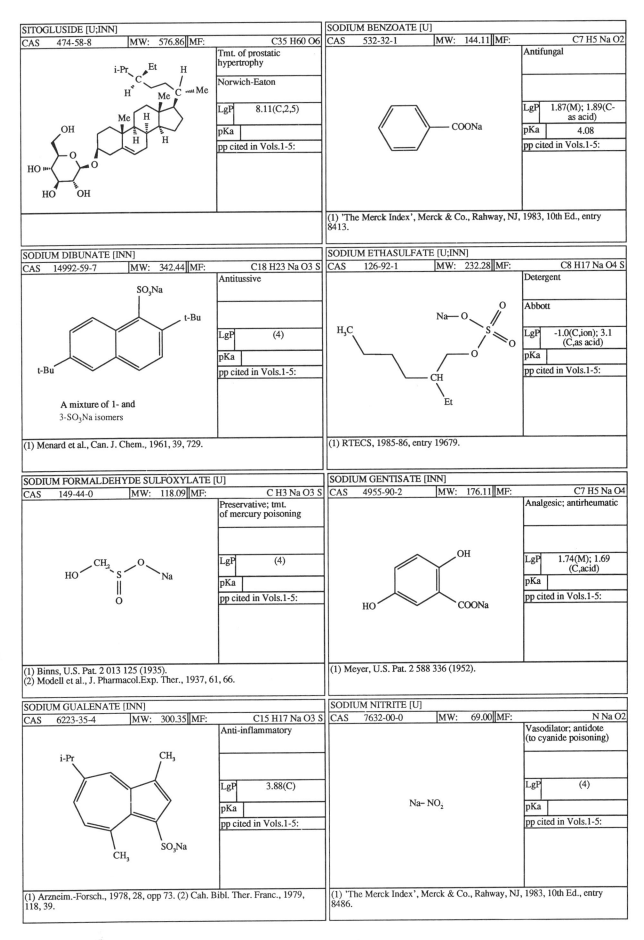

SITOGLUSIDE [U;INN]

CAS	474-58-8	MW:	576.86	MF:	C35 H60 O6

Tmt. of prostatic hypertrophy

Norwich-Eaton

LgP	8.11(C,2,5)
pKa	
pp cited in Vols.1-5:	

SODIUM BENZOATE [U]

CAS	532-32-1	MW:	144.11	MF:	C7 H5 Na O2

Antifungal

LgP	1.87(M); 1.89(C-as acid)
pKa	4.08
pp cited in Vols.1-5:	

(1) 'The Merck Index', Merck & Co., Rahway, NJ, 1983, 10th Ed., entry 8413.

SODIUM DIBUNATE [INN]

CAS	14992-59-7	MW:	342.44	MF:	C18 H23 Na O3 S

Antitussive

LgP	(4)
pKa	
pp cited in Vols.1-5:	

A mixture of 1- and 3-SO₃Na isomers

(1) Menard et al., Can. J. Chem., 1961, 39, 729.

SODIUM ETHASULFATE [U;INN]

CAS	126-92-1	MW:	232.28	MF:	C8 H17 Na O4 S

Detergent

Abbott

LgP	-1.0(C,ion); 3.1 (C,as acid)
pKa	
pp cited in Vols.1-5:	

(1) RTECS, 1985-86, entry 19679.

SODIUM FORMALDEHYDE SULFOXYLATE [U]

CAS	149-44-0	MW:	118.09	MF:	C H3 Na O3 S

Preservative; tmt. of mercury poisoning

LgP	(4)
pKa	
pp cited in Vols.1-5:	

(1) Binns, U.S. Pat. 2 013 125 (1935).
(2) Modell et al., J. Pharmacol.Exp. Ther., 1937, 61, 66.

SODIUM GENTISATE [INN]

CAS	4955-90-2	MW:	176.11	MF:	C7 H5 Na O4

Analgesic; antirheumatic

LgP	1.74(M); 1.69 (C,acid)
pKa	
pp cited in Vols.1-5:	

(1) Meyer, U.S. Pat. 2 588 336 (1952).

SODIUM GUALENATE [INN]

CAS	6223-35-4	MW:	300.35	MF:	C15 H17 Na O3 S

Anti-inflammatory

LgP	3.88(C)
pKa	
pp cited in Vols.1-5:	

(1) Arzneim.-Forsch., 1978, 28, opp 73. (2) Cah. Bibl. Ther. Franc., 1979, 118, 39.

SODIUM NITRITE [U]

CAS	7632-00-0	MW:	69.00	MF:	N Na O2

Vasodilator; antidote (to cyanide poisoning)

Na– NO₂

LgP	(4)
pKa	
pp cited in Vols.1-5:	

(1) 'The Merck Index', Merck & Co., Rahway, NJ, 1983, 10th Ed., entry 8486.

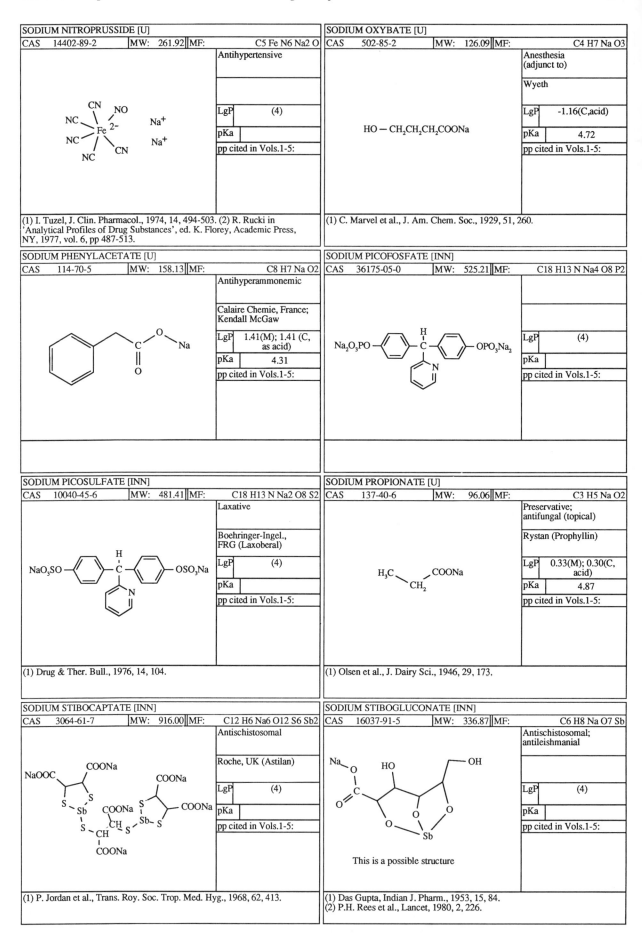

SODIUM NITROPRUSSIDE [U]	
CAS 14402-89-2	MW: 261.92 MF: C5 Fe N6 Na2 O

Antihypertensive

LgP (4)
pKa
pp cited in Vols.1-5:

(1) I. Tuzel, J. Clin. Pharmacol., 1974, 14, 494-503. (2) R. Rucki in 'Analytical Profiles of Drug Substances', ed. K. Florey, Academic Press, NY, 1977, vol. 6, pp 487-513.

SODIUM OXYBATE [U]	
CAS 502-85-2	MW: 126.09 MF: C4 H7 Na O3

Anesthesia (adjunct to)
Wyeth

HO — CH2CH2CH2COONa

LgP -1.16(C,acid)
pKa 4.72
pp cited in Vols.1-5:

(1) C. Marvel et al., J. Am. Chem. Soc., 1929, 51, 260.

SODIUM PHENYLACETATE [U]	
CAS 114-70-5	MW: 158.13 MF: C8 H7 Na O2

Antihyperammonemic
Calaire Chemie, France; Kendall McGaw

LgP 1.41(M); 1.41 (C, as acid)
pKa 4.31
pp cited in Vols.1-5:

SODIUM PICOFOSFATE [INN]	
CAS 36175-05-0	MW: 525.21 MF: C18 H13 N Na4 O8 P2

LgP (4)
pKa
pp cited in Vols.1-5:

SODIUM PICOSULFATE [INN]	
CAS 10040-45-6	MW: 481.41 MF: C18 H13 N Na2 O8 S2

Laxative
Boehringer-Ingel., FRG (Laxoberal)

LgP (4)
pKa
pp cited in Vols.1-5:

(1) Drug & Ther. Bull., 1976, 14, 104.

SODIUM PROPIONATE [U]	
CAS 137-40-6	MW: 96.06 MF: C3 H5 Na O2

Preservative; antifungal (topical)
Rystan (Prophyllin)

LgP 0.33(M); 0.30(C, acid)
pKa 4.87
pp cited in Vols.1-5:

(1) Olsen et al., J. Dairy Sci., 1946, 29, 173.

SODIUM STIBOCAPTATE [INN]	
CAS 3064-61-7	MW: 916.00 MF: C12 H6 Na6 O12 S6 Sb2

Antischistosomal
Roche, UK (Astilan)

LgP (4)
pKa
pp cited in Vols.1-5:

(1) P. Jordan et al., Trans. Roy. Soc. Trop. Med. Hyg., 1968, 62, 413.

SODIUM STIBOGLUCONATE [INN]	
CAS 16037-91-5	MW: 336.87 MF: C6 H8 Na O7 Sb

Antischistosomal; antileishmanial

LgP (4)
pKa
pp cited in Vols.1-5:

This is a possible structure

(1) Das Gupta, Indian J. Pharm., 1953, 15, 84.
(2) P.H. Rees et al., Lancet, 1980, 2, 226.

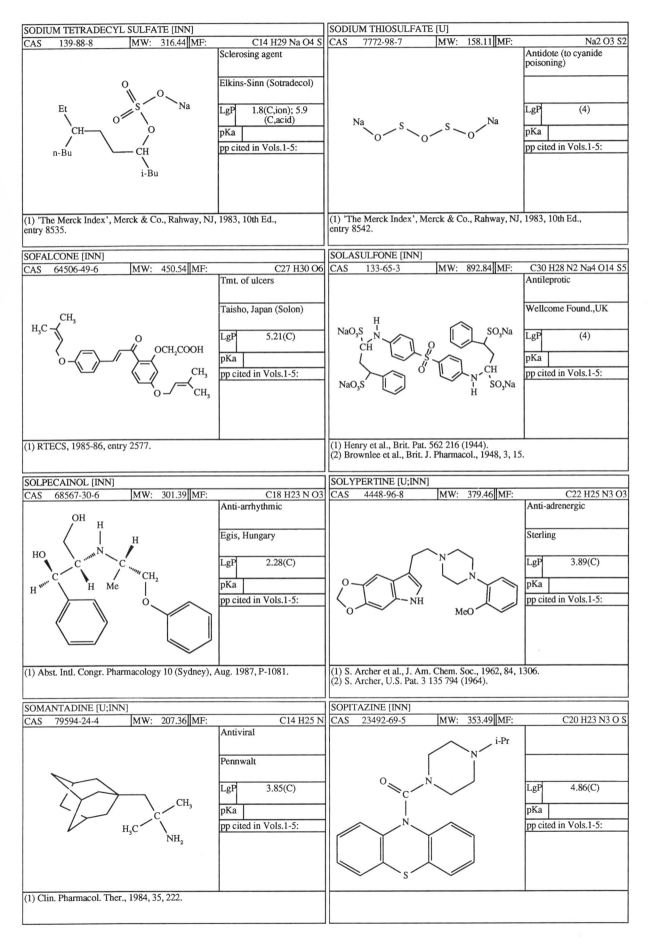

SODIUM TETRADECYL SULFATE [INN]		
CAS 139-88-8	MW: 316.44	MF: C14 H29 Na O4 S

Sclerosing agent	
Elkins-Sinn (Sotradecol)	
LgP	1.8(C,ion); 5.9 (C,acid)
pKa	
pp cited in Vols.1-5:	

(1) 'The Merck Index', Merck & Co., Rahway, NJ, 1983, 10th Ed., entry 8535.

SODIUM THIOSULFATE [U]		
CAS 7772-98-7	MW: 158.11	MF: Na2 O3 S2

Antidote (to cyanide poisoning)	
LgP	(4)
pKa	
pp cited in Vols.1-5:	

(1) 'The Merck Index', Merck & Co., Rahway, NJ, 1983, 10th Ed., entry 8542.

SOFALCONE [INN]		
CAS 64506-49-6	MW: 450.54	MF: C27 H30 O6

Tmt. of ulcers	
Taisho, Japan (Solon)	
LgP	5.21(C)
pKa	
pp cited in Vols.1-5:	

(1) RTECS, 1985-86, entry 2577.

SOLASULFONE [INN]		
CAS 133-65-3	MW: 892.84	MF: C30 H28 N2 Na4 O14 S5

Antileprotic	
Wellcome Found.,UK	
LgP	(4)
pKa	
pp cited in Vols.1-5:	

(1) Henry et al., Brit. Pat. 562 216 (1944).
(2) Brownlee et al., Brit. J. Pharmacol., 1948, 3, 15.

SOLPECAINOL [INN]		
CAS 68567-30-6	MW: 301.39	MF: C18 H23 N O3

Anti-arrhythmic	
Egis, Hungary	
LgP	2.28(C)
pKa	
pp cited in Vols.1-5:	

(1) Abst. Intl. Congr. Pharmacology 10 (Sydney), Aug. 1987, P-1081.

SOLYPERTINE [U;INN]		
CAS 4448-96-8	MW: 379.46	MF: C22 H25 N3 O3

Anti-adrenergic	
Sterling	
LgP	3.89(C)
pKa	
pp cited in Vols.1-5:	

(1) S. Archer et al., J. Am. Chem. Soc., 1962, 84, 1306.
(2) S. Archer, U.S. Pat. 3 135 794 (1964).

SOMANTADINE [U;INN]		
CAS 79594-24-4	MW: 207.36	MF: C14 H25 N

Antiviral	
Pennwalt	
LgP	3.85(C)
pKa	
pp cited in Vols.1-5:	

(1) Clin. Pharmacol. Ther., 1984, 35, 222.

SOPITAZINE [INN]		
CAS 23492-69-5	MW: 353.49	MF: C20 H23 N3 O S

LgP	4.86(C)
pKa	
pp cited in Vols.1-5:	

Sopromidine   *Drug Compendium*   854

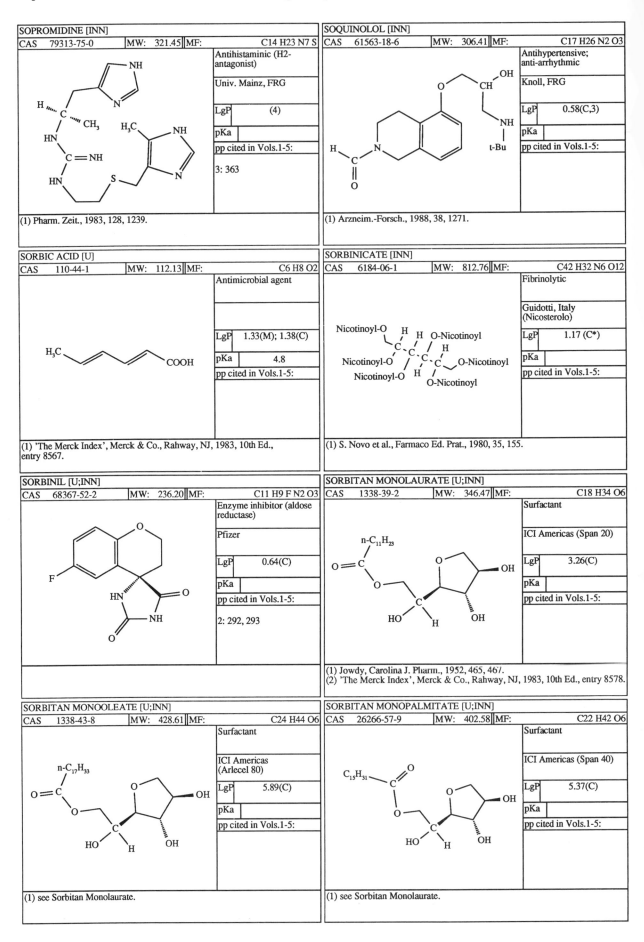

SOPROMIDINE [INN]	
CAS 79313-75-0	MW: 321.45	MF: C14 H23 N7 S

Antihistaminic (H2-antagonist)

Univ. Mainz, FRG

LgP (4)

pKa

pp cited in Vols.1-5:

3: 363

(1) Pharm. Zeit., 1983, 128, 1239.

SOQUINOLOL [INN]	
CAS 61563-18-6	MW: 306.41	MF: C17 H26 N2 O3

Antihypertensive; anti-arrhythmic

Knoll, FRG

LgP 0.58(C,3)

pKa

pp cited in Vols.1-5:

(1) Arzneim.-Forsch., 1988, 38, 1271.

SORBIC ACID [U]	
CAS 110-44-1	MW: 112.13	MF: C6 H8 O2

Antimicrobial agent

LgP 1.33(M); 1.38(C)

pKa 4.8

pp cited in Vols.1-5:

(1) 'The Merck Index', Merck & Co., Rahway, NJ, 1983, 10th Ed., entry 8567.

SORBINICATE [INN]	
CAS 6184-06-1	MW: 812.76	MF: C42 H32 N6 O12

Fibrinolytic

Guidotti, Italy (Nicosterolo)

LgP 1.17 (C*)

pKa

pp cited in Vols.1-5:

(1) S. Novo et al., Farmaco Ed. Prat., 1980, 35, 155.

SORBINIL [U;INN]	
CAS 68367-52-2	MW: 236.20	MF: C11 H9 F N2 O3

Enzyme inhibitor (aldose reductase)

Pfizer

LgP 0.64(C)

pKa

pp cited in Vols.1-5:

2: 292, 293

SORBITAN MONOLAURATE [U;INN]	
CAS 1338-39-2	MW: 346.47	MF: C18 H34 O6

Surfactant

ICI Americas (Span 20)

LgP 3.26(C)

pKa

pp cited in Vols.1-5:

(1) Jowdy, Carolina J. Pharm., 1952, 465, 467.
(2) 'The Merck Index', Merck & Co., Rahway, NJ, 1983, 10th Ed., entry 8578.

SORBITAN MONOOLEATE [U;INN]	
CAS 1338-43-8	MW: 428.61	MF: C24 H44 O6

Surfactant

ICI Americas (Arlecel 80)

LgP 5.89(C)

pKa

pp cited in Vols.1-5:

(1) see Sorbitan Monolaurate.

SORBITAN MONOPALMITATE [U;INN]	
CAS 26266-57-9	MW: 402.58	MF: C22 H42 O6

Surfactant

ICI Americas (Span 40)

LgP 5.37(C)

pKa

pp cited in Vols.1-5:

(1) see Sorbitan Monolaurate.

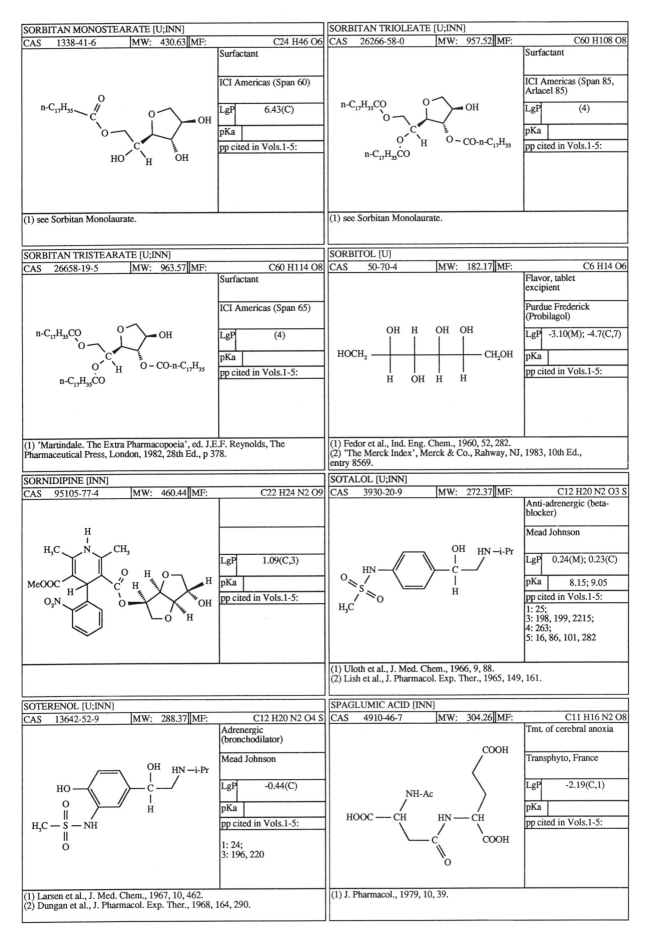

SORBITAN MONOSTEARATE [U;INN]

CAS	1338-41-6	MW:	430.63	MF:	C24 H46 O6

Surfactant

ICI Americas (Span 60)

LgP	6.43(C)
pKa	
pp cited in Vols.1-5:	

(1) see Sorbitan Monolaurate.

SORBITAN TRIOLEATE [U;INN]

CAS	26266-58-0	MW:	957.52	MF:	C60 H108 O8

Surfactant

ICI Americas (Span 85, Arlacel 85)

LgP	(4)
pKa	
pp cited in Vols.1-5:	

(1) see Sorbitan Monolaurate.

SORBITAN TRISTEARATE [U;INN]

CAS	26658-19-5	MW:	963.57	MF:	C60 H114 O8

Surfactant

ICI Americas (Span 65)

LgP	(4)
pKa	
pp cited in Vols.1-5:	

(1) 'Martindale. The Extra Pharmacopoeia', ed. J.E.F. Reynolds, The Pharmaceutical Press, London, 1982, 28th Ed., p 378.

SORBITOL [U]

CAS	50-70-4	MW:	182.17	MF:	C6 H14 O6

Flavor, tablet excipient

Purdue Frederick (Probilagol)

LgP	-3.10(M); -4.7(C,7)
pKa	
pp cited in Vols.1-5:	

(1) Fedor et al., Ind. Eng. Chem., 1960, 52, 282.
(2) 'The Merck Index', Merck & Co., Rahway, NJ, 1983, 10th Ed., entry 8569.

SORNIDIPINE [INN]

CAS	95105-77-4	MW:	460.44	MF:	C22 H24 N2 O9

LgP	1.09(C,3)
pKa	
pp cited in Vols.1-5:	

SOTALOL [U;INN]

CAS	3930-20-9	MW:	272.37	MF:	C12 H20 N2 O3 S

Anti-adrenergic (beta-blocker)

Mead Johnson

LgP	0.24(M); 0.23(C)
pKa	8.15; 9.05
pp cited in Vols.1-5:	

1: 25;
3: 198, 199, 2215;
4: 263;
5: 16, 86, 101, 282

(1) Uloth et al., J. Med. Chem., 1966, 9, 88.
(2) Lish et al., J. Pharmacol. Exp. Ther., 1965, 149, 161.

SOTERENOL [U;INN]

CAS	13642-52-9	MW:	288.37	MF:	C12 H20 N2 O4 S

Adrenergic (bronchodilator)

Mead Johnson

LgP	-0.44(C)
pKa	
pp cited in Vols.1-5:	

1: 24;
3: 196, 220

(1) Larsen et al., J. Med. Chem., 1967, 10, 462.
(2) Dungan et al., J. Pharmacol. Exp. Ther., 1968, 164, 290.

SPAGLUMIC ACID [INN]

CAS	4910-46-7	MW:	304.26	MF:	C11 H16 N2 O8

Tmt. of cerebral anoxia

Transphyto, France

LgP	-2.19(C,1)
pKa	
pp cited in Vols.1-5:	

(1) J. Pharmacol., 1979, 10, 39.

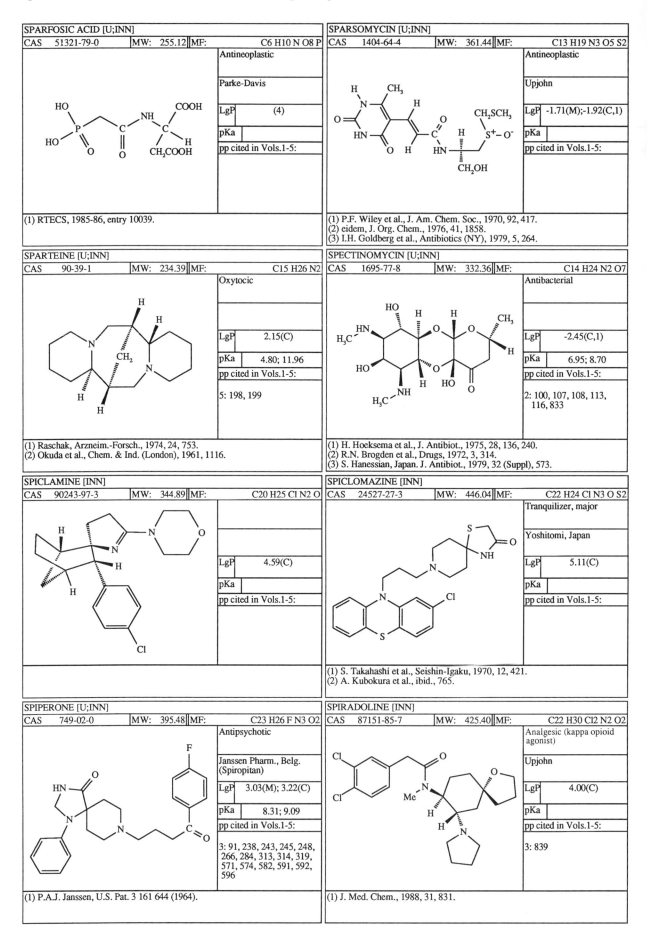

SPARFOSIC ACID [U;INN]

CAS	51321-79-0	MW:	255.12	MF:	C6 H10 N O8 P

Antineoplastic

Parke-Davis

LgP	(4)
pKa	
pp cited in Vols.1-5:	

(1) RTECS, 1985-86, entry 10039.

SPARTEINE [U;INN]

CAS	90-39-1	MW:	234.39	MF:	C15 H26 N2

Oxytocic

LgP	2.15(C)
pKa	4.80; 11.96
pp cited in Vols.1-5:	

5: 198, 199

(1) Raschak, Arzneim.-Forsch., 1974, 24, 753.
(2) Okuda et al., Chem. & Ind. (London), 1961, 1116.

SPICLAMINE [INN]

CAS	90243-97-3	MW:	344.89	MF:	C20 H25 Cl N2 O

LgP	4.59(C)
pKa	
pp cited in Vols.1-5:	

SPIPERONE [U;INN]

CAS	749-02-0	MW:	395.48	MF:	C23 H26 F N3 O2

Antipsychotic

Janssen Pharm., Belg.
(Spiropitan)

LgP	3.03(M); 3.22(C)
pKa	8.31; 9.09
pp cited in Vols.1-5:	

3: 91, 238, 243, 245, 248, 266, 284, 313, 314, 319, 571, 574, 582, 591, 592, 596

(1) P.A.J. Janssen, U.S. Pat. 3 161 644 (1964).

SPARSOMYCIN [U;INN]

CAS	1404-64-4	MW:	361.44	MF:	C13 H19 N3 O5 S2

Antineoplastic

Upjohn

LgP	-1.71(M);-1.92(C,1)
pKa	
pp cited in Vols.1-5:	

(1) P.F. Wiley et al., J. Am. Chem. Soc., 1970, 92, 417.
(2) eidem, J. Org. Chem., 1976, 41, 1858.
(3) I.H. Goldberg et al., Antibiotics (NY), 1979, 5, 264.

SPECTINOMYCIN [U;INN]

CAS	1695-77-8	MW:	332.36	MF:	C14 H24 N2 O7

Antibacterial

LgP	-2.45(C,1)
pKa	6.95; 8.70
pp cited in Vols.1-5:	

2: 100, 107, 108, 113, 116, 833

(1) H. Hoeksema et al., J. Antibiot., 1975, 28, 136, 240.
(2) R.N. Brogden et al., Drugs, 1972, 3, 314.
(3) S. Hanessian, Japan. J. Antibiot., 1979, 32 (Suppl), 573.

SPICLOMAZINE [INN]

CAS	24527-27-3	MW:	446.04	MF:	C22 H24 Cl N3 O S2

Tranquilizer, major

Yoshitomi, Japan

LgP	5.11(C)
pKa	
pp cited in Vols.1-5:	

(1) S. Takahashi et al., Seishin-Igaku, 1970, 12, 421.
(2) A. Kubokura et al., ibid., 765.

SPIRADOLINE [INN]

CAS	87151-85-7	MW:	425.40	MF:	C22 H30 Cl2 N2 O2

Analgesic (kappa opioid agonist)

Upjohn

LgP	4.00(C)
pKa	
pp cited in Vols.1-5:	

3: 839

(1) J. Med. Chem., 1988, 31, 831.

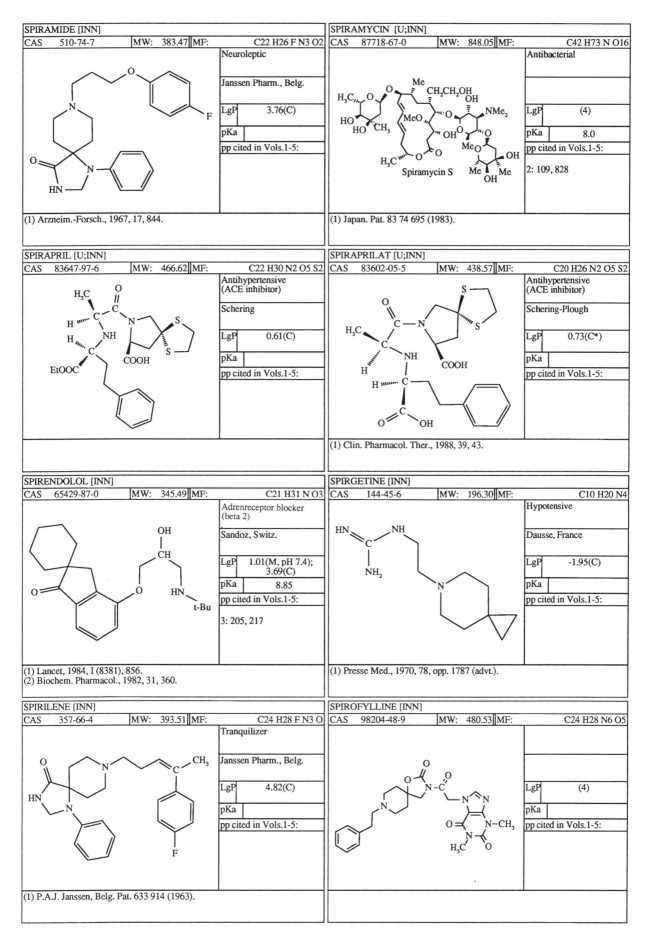

SPIRAMIDE [INN]

| CAS | 510-74-7 | MW: | 383.47 | MF: | C22 H26 F N3 O2 |

Neuroleptic

Janssen Pharm., Belg.

LgP	3.76(C)
pKa	
pp cited in Vols.1-5:	

(1) Arzneim.-Forsch., 1967, 17, 844.

SPIRAMYCIN [U;INN]

| CAS | 87718-67-0 | MW: | 848.05 | MF: | C42 H73 N O16 |

Antibacterial

Spiramycin S

LgP	(4)
pKa	8.0
pp cited in Vols.1-5:	
2: 109, 828	

(1) Japan. Pat. 83 74 695 (1983).

SPIRAPRIL [U;INN]

| CAS | 83647-97-6 | MW: | 466.62 | MF: | C22 H30 N2 O5 S2 |

Antihypertensive
(ACE inhibitor)

Schering

LgP	0.61(C)
pKa	
pp cited in Vols.1-5:	

SPIRAPRILAT [U;INN]

| CAS | 83602-05-5 | MW: | 438.57 | MF: | C20 H26 N2 O5 S2 |

Antihypertensive
(ACE inhibitor)

Schering-Plough

LgP	0.73(C*)
pKa	
pp cited in Vols.1-5:	

(1) Clin. Pharmacol. Ther., 1988, 39, 43.

SPIRENDOLOL [INN]

| CAS | 65429-87-0 | MW: | 345.49 | MF: | C21 H31 N O3 |

Adrenreceptor blocker
(beta 2)

Sandoz, Switz.

LgP	1.01(M, pH 7.4); 3.69(C)
pKa	8.85
pp cited in Vols.1-5:	
3: 205, 217	

(1) Lancet, 1984, I (8381), 856.
(2) Biochem. Pharmacol., 1982, 31, 360.

SPIRGETINE [INN]

| CAS | 144-45-6 | MW: | 196.30 | MF: | C10 H20 N4 |

Hypotensive

Dausse, France

LgP	-1.95(C)
pKa	
pp cited in Vols.1-5:	

(1) Presse Med., 1970, 78, opp. 1787 (advt.).

SPIRILENE [INN]

| CAS | 357-66-4 | MW: | 393.51 | MF: | C24 H28 F N3 O |

Tranquilizer

Janssen Pharm., Belg.

LgP	4.82(C)
pKa	
pp cited in Vols.1-5:	

(1) P.A.J. Janssen, Belg. Pat. 633 914 (1963).

SPIROFYLLINE [INN]

| CAS | 98204-48-9 | MW: | 480.53 | MF: | C24 H28 N6 O5 |

LgP	(4)
pKa	
pp cited in Vols.1-5:	

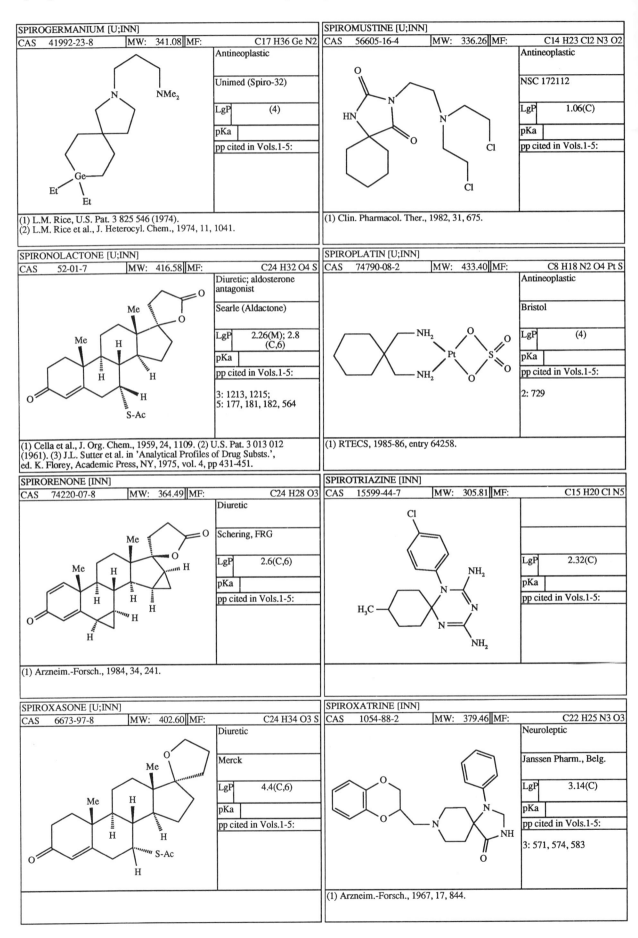

SPIROGERMANIUM [U;INN]

CAS	41992-23-8	MW:	341.08	MF:	C17 H36 Ge N2

Antineoplastic

Unimed (Spiro-32)

LgP   (4)

pKa

pp cited in Vols.1-5:

(1) L.M. Rice, U.S. Pat. 3 825 546 (1974).
(2) L.M. Rice et al., J. Heterocyl. Chem., 1974, 11, 1041.

SPIROMUSTINE [U;INN]

CAS	56605-16-4	MW:	336.26	MF:	C14 H23 Cl2 N3 O2

Antineoplastic

NSC 172112

LgP   1.06(C)

pKa

pp cited in Vols.1-5:

(1) Clin. Pharmacol. Ther., 1982, 31, 675.

SPIRONOLACTONE [U;INN]

CAS	52-01-7	MW:	416.58	MF:	C24 H32 O4 S

Diuretic; aldosterone antagonist

Searle (Aldactone)

LgP   2.26(M); 2.8 (C,6)

pKa

pp cited in Vols.1-5:

3: 1213, 1215;
5: 177, 181, 182, 564

(1) Cella et al., J. Org. Chem., 1959, 24, 1109. (2) U.S. Pat. 3 013 012
(1961). (3) J.L. Sutter et al. in 'Analytical Profiles of Drug Substs.',
ed. K. Florey, Academic Press, NY, 1975, vol. 4, pp 431-451.

SPIROPLATIN [U;INN]

CAS	74790-08-2	MW:	433.40	MF:	C8 H18 N2 O4 Pt S

Antineoplastic

Bristol

LgP   (4)

pKa

pp cited in Vols.1-5:

2: 729

(1) RTECS, 1985-86, entry 64258.

SPIRORENONE [INN]

CAS	74220-07-8	MW:	364.49	MF:	C24 H28 O3

Diuretic

Schering, FRG

LgP   2.6(C,6)

pKa

pp cited in Vols.1-5:

(1) Arzneim.-Forsch., 1984, 34, 241.

SPIROTRIAZINE [INN]

CAS	15599-44-7	MW:	305.81	MF:	C15 H20 Cl N5

LgP   2.32(C)

pKa

pp cited in Vols.1-5:

SPIROXASONE [U;INN]

CAS	6673-97-8	MW:	402.60	MF:	C24 H34 O3 S

Diuretic

Merck

LgP   4.4(C,6)

pKa

pp cited in Vols.1-5:

SPIROXATRINE [INN]

CAS	1054-88-2	MW:	379.46	MF:	C22 H25 N3 O3

Neuroleptic

Janssen Pharm., Belg.

LgP   3.14(C)

pKa

pp cited in Vols.1-5:

3: 571, 574, 583

(1) Arzneim.-Forsch., 1967, 17, 844.

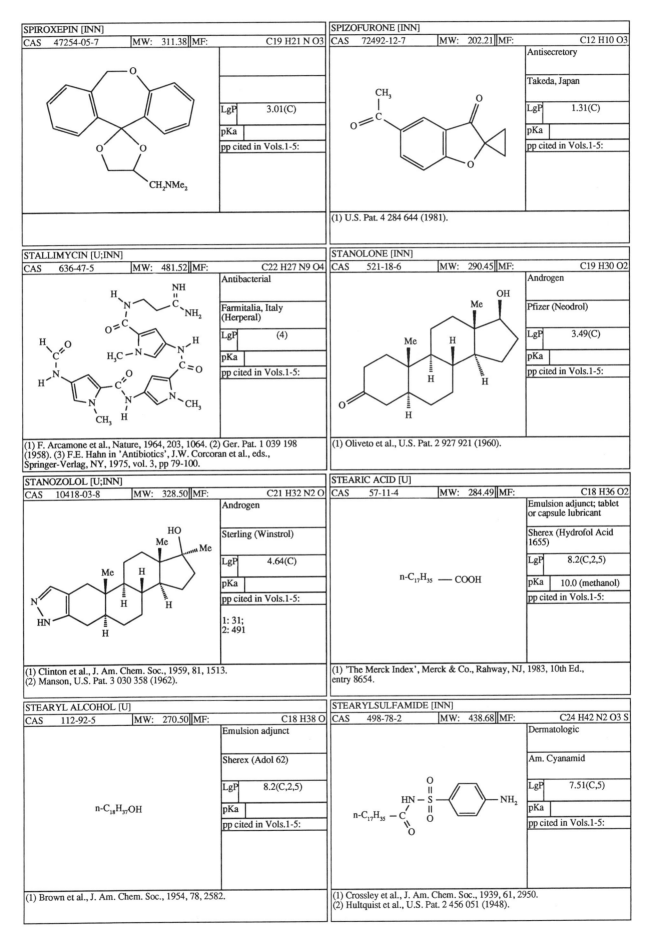

SPIROXEPIN [INN]

CAS	47254-05-7	MW:	311.38	MF:	C19 H21 N O3

LgP	3.01(C)
pKa	
pp cited in Vols.1-5:	

SPIZOFURONE [INN]

CAS	72492-12-7	MW:	202.21	MF:	C12 H10 O3

Antisecretory

Takeda, Japan

LgP	1.31(C)
pKa	
pp cited in Vols.1-5:	

(1) U.S. Pat. 4 284 644 (1981).

STALLIMYCIN [U;INN]

CAS	636-47-5	MW:	481.52	MF:	C22 H27 N9 O4

Antibacterial

Farmitalia, Italy (Herperal)

LgP	(4)
pKa	
pp cited in Vols.1-5:	

(1) F. Arcamone et al., Nature, 1964, 203, 1064. (2) Ger. Pat. 1 039 198 (1958). (3) F.E. Hahn in 'Antibiotics', J.W. Corcoran et al., eds., Springer-Verlag, NY, 1975, vol. 3, pp 79-100.

STANOLONE [INN]

CAS	521-18-6	MW:	290.45	MF:	C19 H30 O2

Androgen

Pfizer (Neodrol)

LgP	3.49(C)
pKa	
pp cited in Vols.1-5:	

(1) Oliveto et al., U.S. Pat. 2 927 921 (1960).

STANOZOLOL [U;INN]

CAS	10418-03-8	MW:	328.50	MF:	C21 H32 N2 O

Androgen

Sterling (Winstrol)

LgP	4.64(C)
pKa	
pp cited in Vols.1-5:	
	1: 31; 2: 491

(1) Clinton et al., J. Am. Chem. Soc., 1959, 81, 1513.
(2) Manson, U.S. Pat. 3 030 358 (1962).

STEARIC ACID [U]

CAS	57-11-4	MW:	284.49	MF:	C18 H36 O2

Emulsion adjunct; tablet or capsule lubricant

Sherex (Hydrofol Acid 1655)

LgP	8.2(C,2,5)
pKa	10.0 (methanol)
pp cited in Vols.1-5:	

$n\text{-}C_{17}H_{35}$ — COOH

(1) 'The Merck Index', Merck & Co., Rahway, NJ, 1983, 10th Ed., entry 8654.

STEARYL ALCOHOL [U]

CAS	112-92-5	MW:	270.50	MF:	C18 H38 O

Emulsion adjunct

Sherex (Adol 62)

LgP	8.2(C,2,5)
pKa	
pp cited in Vols.1-5:	

$n\text{-}C_{18}H_{37}OH$

(1) Brown et al., J. Am. Chem. Soc., 1954, 78, 2582.

STEARYLSULFAMIDE [INN]

CAS	498-78-2	MW:	438.68	MF:	C24 H42 N2 O3 S

Dermatologic

Am. Cyanamid

LgP	7.51(C,5)
pKa	
pp cited in Vols.1-5:	

$n\text{-}C_{17}H_{35}$ — C

(1) Crossley et al., J. Am. Chem. Soc., 1939, 61, 2950.
(2) Hultquist et al., U.S. Pat. 2 456 051 (1948).

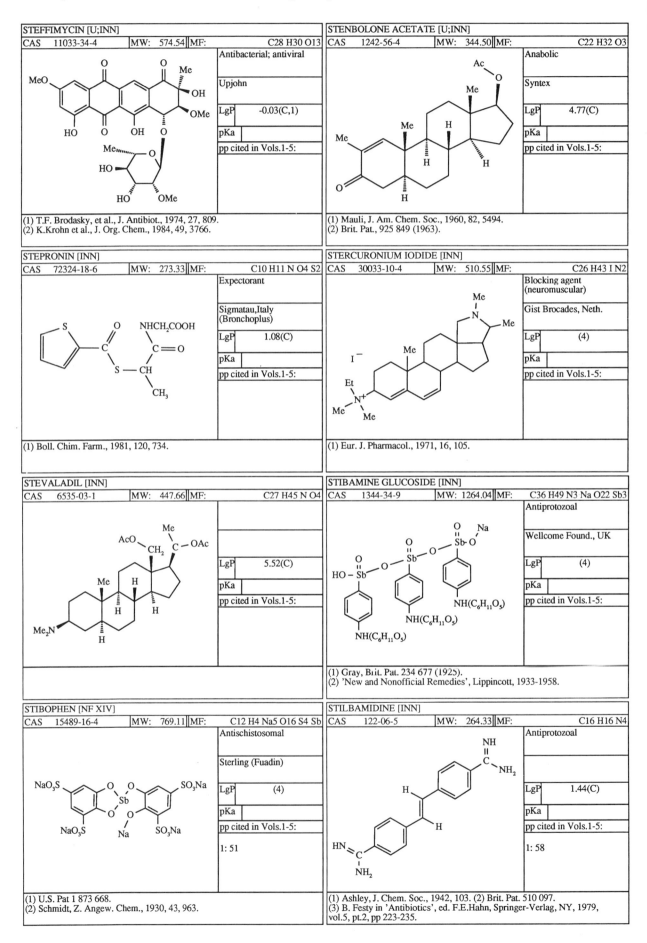

STEFFIMYCIN [U;INN]

CAS	11033-34-4	MW:	574.54	MF:	C28 H30 O13

Antibacterial; antiviral

Upjohn

LgP	-0.03(C,1)
pKa	
pp cited in Vols.1-5:	

(1) T.F. Brodasky, et al., J. Antibiot., 1974, 27, 809.
(2) K.Krohn et al., J. Org. Chem., 1984, 49, 3766.

STENBOLONE ACETATE [U;INN]

CAS	1242-56-4	MW:	344.50	MF:	C22 H32 O3

Anabolic

Syntex

LgP	4.77(C)
pKa	
pp cited in Vols.1-5:	

(1) Mauli, J. Am. Chem. Soc., 1960, 82, 5494.
(2) Brit. Pat., 925 849 (1963).

STEPRONIN [INN]

CAS	72324-18-6	MW:	273.33	MF:	C10 H11 N O4 S2

Expectorant

Sigmatau,Italy
(Bronchoplus)

LgP	1.08(C)
pKa	
pp cited in Vols.1-5:	

(1) Boll. Chim. Farm., 1981, 120, 734.

STERCURONIUM IODIDE [INN]

CAS	30033-10-4	MW:	510.55	MF:	C26 H43 I N2

Blocking agent
(neuromuscular)

Gist Brocades, Neth.

LgP	(4)
pKa	
pp cited in Vols.1-5:	

(1) Eur. J. Pharmacol., 1971, 16, 105.

STEVALADIL [INN]

CAS	6535-03-1	MW:	447.66	MF:	C27 H45 N O4

LgP	5.52(C)
pKa	
pp cited in Vols.1-5:	

STIBAMINE GLUCOSIDE [INN]

CAS	1344-34-9	MW:	1264.04	MF:	C36 H49 N3 Na O22 Sb3

Antiprotozoal

Wellcome Found., UK

LgP	(4)
pKa	
pp cited in Vols.1-5:	

(1) Gray, Brit. Pat. 234 677 (1925).
(2) 'New and Nonofficial Remedies', Lippincott, 1933-1958.

STIBOPHEN [NF XIV]

CAS	15489-16-4	MW:	769.11	MF:	C12 H4 Na5 O16 S4 Sb

Antischistosomal

Sterling (Fuadin)

LgP	(4)
pKa	
pp cited in Vols.1-5:	
1: 51	

(1) U.S. Pat 1 873 668.
(2) Schmidt, Z. Angew. Chem., 1930, 43, 963.

STILBAMIDINE [INN]

CAS	122-06-5	MW:	264.33	MF:	C16 H16 N4

Antiprotozoal

LgP	1.44(C)
pKa	
pp cited in Vols.1-5:	
1: 58	

(1) Ashley, J. Chem. Soc., 1942, 103. (2) Brit. Pat. 510 097.
(3) B. Festy in 'Antibiotics', ed. F.E.Hahn, Springer-Verlag, NY, 1979,
vol.5, pt.2, pp 223-235.

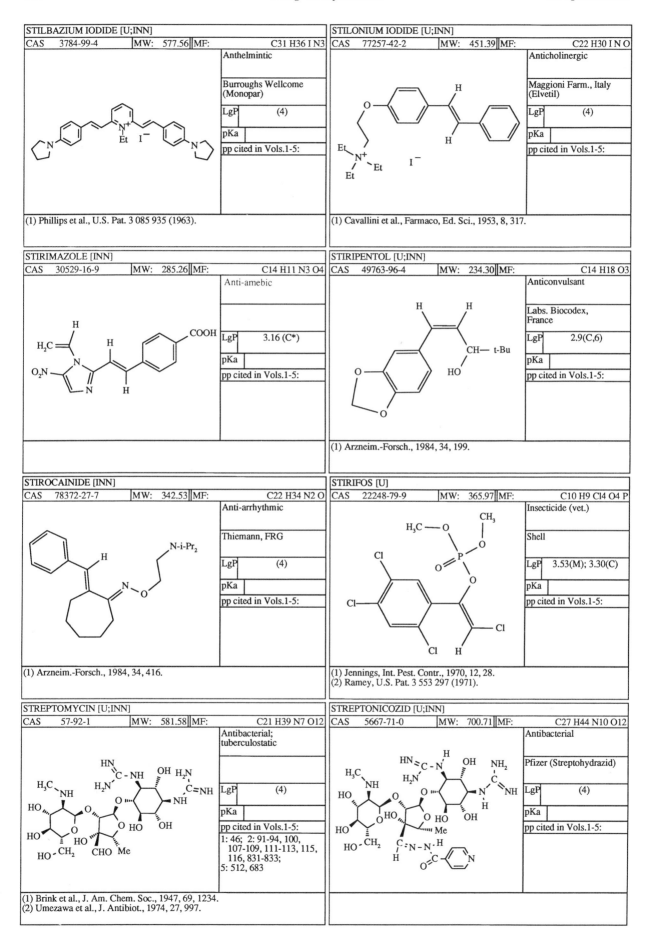

STILBAZIUM IODIDE [U;INN]

CAS	3784-99-4	MW: 577.56	MF:	C31 H36 I N3

Anthelmintic

Burroughs Wellcome (Monopar)

LgP	(4)
pKa	
pp cited in Vols.1-5:	

(1) Phillips et al., U.S. Pat. 3 085 935 (1963).

STILONIUM IODIDE [U;INN]

CAS	77257-42-2	MW: 451.39	MF:	C22 H30 I N O

Anticholinergic

Maggioni Farm., Italy (Elvetil)

LgP	(4)
pKa	
pp cited in Vols.1-5:	

(1) Cavallini et al., Farmaco, Ed. Sci., 1953, 8, 317.

STIRIMAZOLE [INN]

CAS	30529-16-9	MW: 285.26	MF:	C14 H11 N3 O4

Anti-amebic

LgP	3.16 (C*)
pKa	
pp cited in Vols.1-5:	

STIRIPENTOL [U;INN]

CAS	49763-96-4	MW: 234.30	MF:	C14 H18 O3

Anticonvulsant

Labs. Biocodex, France

LgP	2.9(C,6)
pKa	
pp cited in Vols.1-5:	

(1) Arzneim.-Forsch., 1984, 34, 199.

STIROCAINIDE [INN]

CAS	78372-27-7	MW: 342.53	MF:	C22 H34 N2 O

Anti-arrhythmic

Thiemann, FRG

LgP	(4)
pKa	
pp cited in Vols.1-5:	

(1) Arzneim.-Forsch., 1984, 34, 416.

STIRIFOS [U]

CAS	22248-79-9	MW: 365.97	MF:	C10 H9 Cl4 O4 P

Insecticide (vet.)

Shell

LgP	3.53(M); 3.30(C)
pKa	
pp cited in Vols.1-5:	

(1) Jennings, Int. Pest. Contr., 1970, 12, 28.
(2) Ramey, U.S. Pat. 3 553 297 (1971).

STREPTOMYCIN [U;INN]

CAS	57-92-1	MW: 581.58	MF:	C21 H39 N7 O12

Antibacterial; tuberculostatic

LgP	(4)
pKa	
pp cited in Vols.1-5:	1: 46; 2: 91-94, 100, 107-109, 111-113, 115, 116, 831-833; 5: 512, 683

(1) Brink et al., J. Am. Chem. Soc., 1947, 69, 1234.
(2) Umezawa et al., J. Antibiot., 1974, 27, 997.

STREPTONICOZID [U;INN]

CAS	5667-71-0	MW: 700.71	MF:	C27 H44 N10 O12

Antibacterial

Pfizer (Streptohydrazid)

LgP	(4)
pKa	
pp cited in Vols.1-5:	

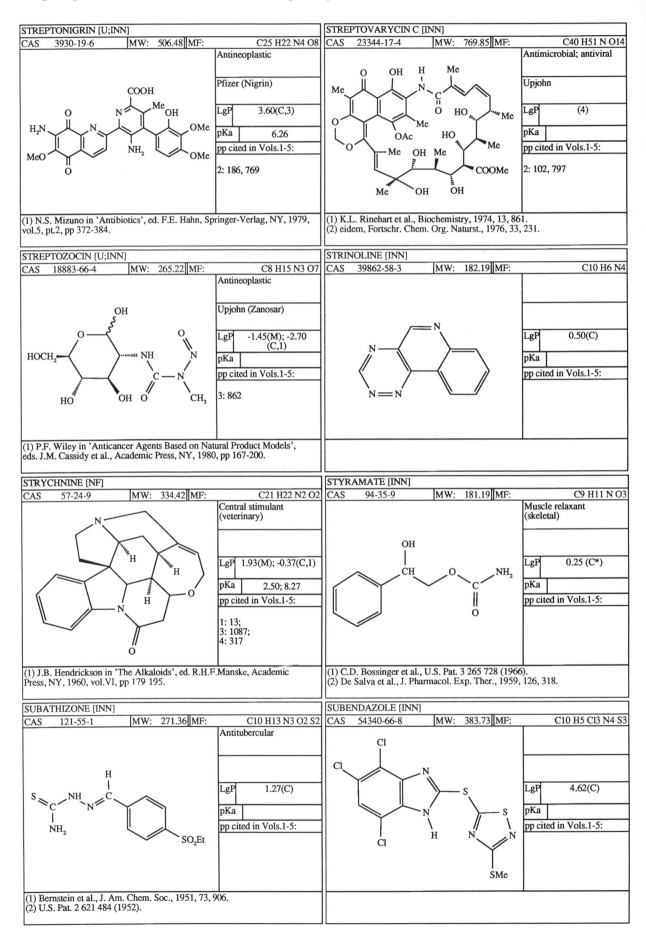

STREPTONIGRIN [U;INN]

CAS	3930-19-6	MW:	506.48	MF:	C25 H22 N4 O8

Antineoplastic

Pfizer (Nigrin)

LgP	3.60(C,3)
pKa	6.26
pp cited in Vols.1-5:	
2: 186, 769	

(1) N.S. Mizuno in 'Antibiotics', ed. F.E. Hahn, Springer-Verlag, NY, 1979, vol.5, pt.2, pp 372-384.

STREPTOVARYCIN C [INN]

CAS	23344-17-4	MW:	769.85	MF:	C40 H51 N O14

Antimicrobial; antiviral

Upjohn

LgP	(4)
pKa	
pp cited in Vols.1-5:	
2: 102, 797	

(1) K.L. Rinehart et al., Biochemistry, 1974, 13, 861.
(2) eidem, Fortschr. Chem. Org. Naturst., 1976, 33, 231.

STREPTOZOCIN [U;INN]

CAS	18883-66-4	MW:	265.22	MF:	C8 H15 N3 O7

Antineoplastic

Upjohn (Zanosar)

LgP	-1.45(M); -2.70 (C,1)
pKa	
pp cited in Vols.1-5:	
3: 862	

(1) P.F. Wiley in 'Anticancer Agents Based on Natural Product Models', eds. J.M. Cassidy et al., Academic Press, NY, 1980, pp 167-200.

STRINOLINE [INN]

CAS	39862-58-3	MW:	182.19	MF:	C10 H6 N4

LgP	0.50(C)
pKa	
pp cited in Vols.1-5:	

STRYCHNINE [NF]

CAS	57-24-9	MW:	334.42	MF:	C21 H22 N2 O2

Central stimulant (veterinary)

LgP	1.93(M); -0.37(C,1)
pKa	2.50; 8.27
pp cited in Vols.1-5:	
1: 13; 3: 1087; 4: 317	

(1) J.B. Hendrickson in 'The Alkaloids', ed. R.H.F.Manske, Academic Press, NY, 1960, vol.VI, pp 179 195.

STYRAMATE [INN]

CAS	94-35-9	MW:	181.19	MF:	C9 H11 N O3

Muscle relaxant (skeletal)

LgP	0.25 (C*)
pKa	
pp cited in Vols.1-5:	

(1) C.D. Bossinger et al., U.S. Pat. 3 265 728 (1966).
(2) De Salva et al., J. Pharmacol. Exp. Ther., 1959, 126, 318.

SUBATHIZONE [INN]

CAS	121-55-1	MW:	271.36	MF:	C10 H13 N3 O2 S2

Antitubercular

LgP	1.27(C)
pKa	
pp cited in Vols.1-5:	

(1) Bernstein et al., J. Am. Chem. Soc., 1951, 73, 906.
(2) U.S. Pat. 2 621 484 (1952).

SUBENDAZOLE [INN]

CAS	54340-66-8	MW:	383.73	MF:	C10 H5 Cl3 N4 S3

LgP	4.62(C)
pKa	
pp cited in Vols.1-5:	

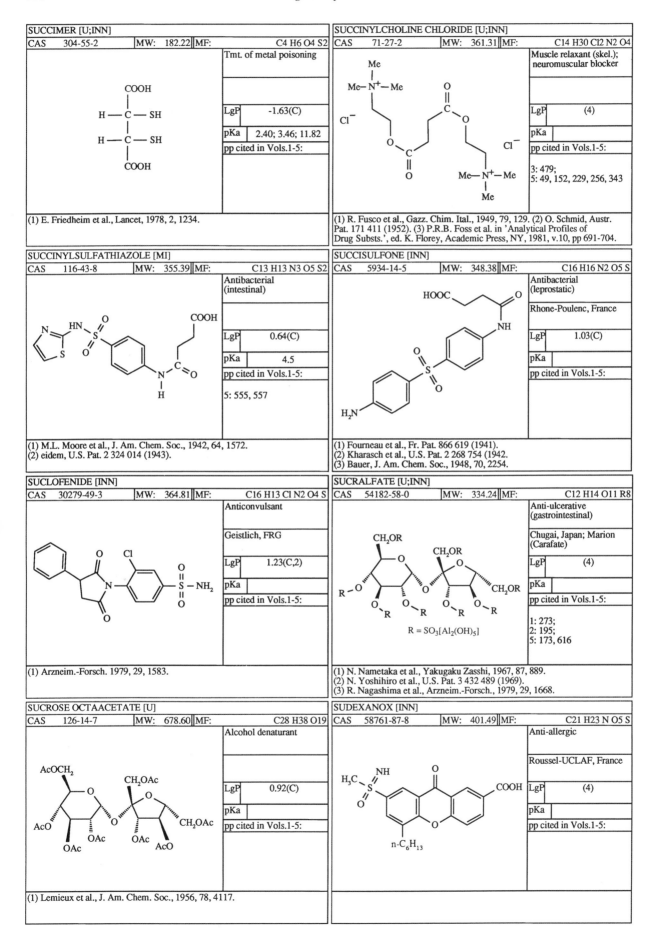

SUCCIMER [U;INN]

CAS	304-55-2	MW:	182.22	MF:	C4 H6 O4 S2

Tmt. of metal poisoning

LgP	-1.63(C)
pKa	2.40; 3.46; 11.82
pp cited in Vols.1-5:	

(1) E. Friedheim et al., Lancet, 1978, 2, 1234.

SUCCINYLCHOLINE CHLORIDE [U;INN]

CAS	71-27-2	MW:	361.31	MF:	C14 H30 Cl2 N2 O4

Muscle relaxant (skel.); neuromuscular blocker

LgP	(4)
pKa	
pp cited in Vols.1-5:	

3: 479;
5: 49, 152, 229, 256, 343

(1) R. Fusco et al., Gazz. Chim. Ital., 1949, 79, 129. (2) O. Schmid, Austr. Pat. 171 411 (1952). (3) P.R.B. Foss et al. in 'Analytical Profiles of Drug Substs.', ed. K. Florey, Academic Press, NY, 1981, v.10, pp 691-704.

SUCCINYLSULFATHIAZOLE [MI]

CAS	116-43-8	MW:	355.39	MF:	C13 H13 N3 O5 S2

Antibacterial (intestinal)

LgP	0.64(C)
pKa	4.5
pp cited in Vols.1-5:	

5: 555, 557

(1) M.L. Moore et al., J. Am. Chem. Soc., 1942, 64, 1572.
(2) eidem, U.S. Pat. 2 324 014 (1943).

SUCCISULFONE [INN]

CAS	5934-14-5	MW:	348.38	MF:	C16 H16 N2 O5 S

Antibacterial (leprostatic)

Rhone-Poulenc, France

LgP	1.03(C)
pKa	
pp cited in Vols.1-5:	

(1) Fourneau et al., Fr. Pat. 866 619 (1941).
(2) Kharasch et al., U.S. Pat. 2 268 754 (1942.
(3) Bauer, J. Am. Chem. Soc., 1948, 70, 2254.

SUCLOFENIDE [INN]

CAS	30279-49-3	MW:	364.81	MF:	C16 H13 Cl N2 O4 S

Anticonvulsant

Geistlich, FRG

LgP	1.23(C,2)
pKa	
pp cited in Vols.1-5:	

(1) Arzneim.-Forsch. 1979, 29, 1583.

SUCRALFATE [U;INN]

CAS	54182-58-0	MW:	334.24	MF:	C12 H14 O11 R8

Anti-ulcerative (gastrointestinal)

Chugai, Japan; Marion (Carafate)

LgP	(4)
pKa	
pp cited in Vols.1-5:	

$R = SO_3[Al_2(OH)_5]$

1: 273;
2: 195;
5: 173, 616

(1) N. Nametaka et al., Yakugaku Zasshi, 1967, 87, 889.
(2) N. Yoshihiro et al., U.S. Pat. 3 432 489 (1969).
(3) R. Nagashima et al., Arzneim.-Forsch., 1979, 29, 1668.

SUCROSE OCTAACETATE [U]

CAS	126-14-7	MW:	678.60	MF:	C28 H38 O19

Alcohol denaturant

LgP	0.92(C)
pKa	
pp cited in Vols.1-5:	

(1) Lemieux et al., J. Am. Chem. Soc., 1956, 78, 4117.

SUDEXANOX [INN]

CAS	58761-87-8	MW:	401.49	MF:	C21 H23 N O5 S

Anti-allergic

Roussel-UCLAF, France

LgP	(4)
pKa	
pp cited in Vols.1-5:	

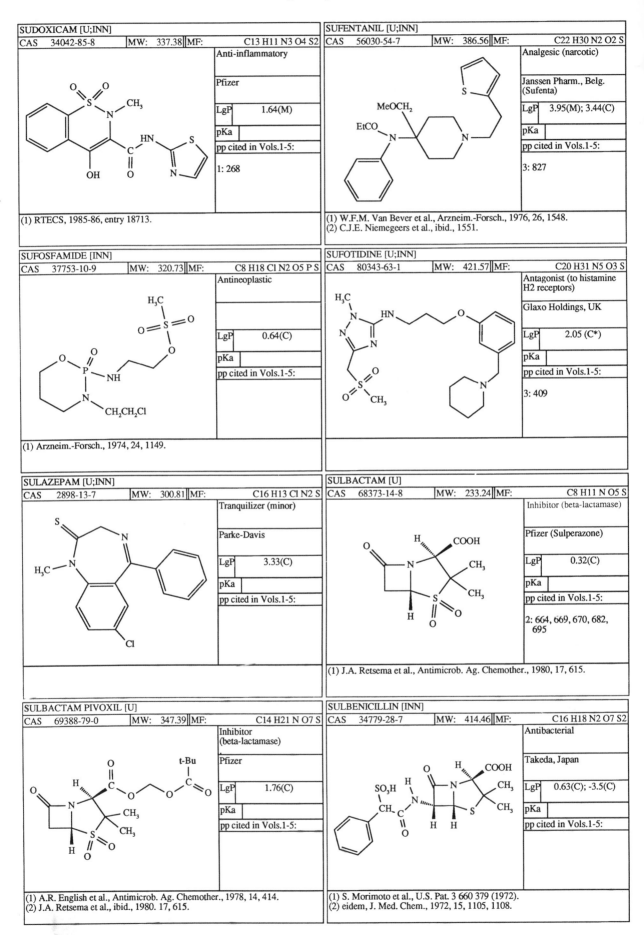

SUDOXICAM [U;INN]			
CAS 34042-85-8	MW: 337.38	MF:	C13 H11 N3 O4 S2
		Anti-inflammatory	
		Pfizer	
		LgP 1.64(M)	
		pKa	
		pp cited in Vols.1-5:	
		1: 268	

(1) RTECS, 1985-86, entry 18713.

SUFENTANIL [U;INN]			
CAS 56030-54-7	MW: 386.56	MF:	C22 H30 N2 O2 S
		Analgesic (narcotic)	
		Janssen Pharm., Belg. (Sufenta)	
		LgP 3.95(M); 3.44(C)	
		pKa	
		pp cited in Vols.1-5:	
		3: 827	

(1) W.F.M. Van Bever et al., Arzneim.-Forsch., 1976, 26, 1548.
(2) C.J.E. Niemegeers et al., ibid., 1551.

SUFOSFAMIDE [INN]			
CAS 37753-10-9	MW: 320.73	MF:	C8 H18 Cl N2 O5 P S
		Antineoplastic	
		LgP 0.64(C)	
		pKa	
		pp cited in Vols.1-5:	

(1) Arzneim.-Forsch., 1974, 24, 1149.

SUFOTIDINE [U;INN]			
CAS 80343-63-1	MW: 421.57	MF:	C20 H31 N5 O3 S
		Antagonist (to histamine H2 receptors)	
		Glaxo Holdings, UK	
		LgP 2.05 (C*)	
		pKa	
		pp cited in Vols.1-5:	
		3: 409	

SULAZEPAM [U;INN]			
CAS 2898-13-7	MW: 300.81	MF:	C16 H13 Cl N2 S
		Tranquilizer (minor)	
		Parke-Davis	
		LgP 3.33(C)	
		pKa	
		pp cited in Vols.1-5:	

SULBACTAM [U]			
CAS 68373-14-8	MW: 233.24	MF:	C8 H11 N O5 S
		Inhibitor (beta-lactamase)	
		Pfizer (Sulperazone)	
		LgP 0.32(C)	
		pKa	
		pp cited in Vols.1-5:	
		2: 664, 669, 670, 682, 695	

(1) J.A. Retsema et al., Antimicrob. Ag. Chemother., 1980, 17, 615.

SULBACTAM PIVOXIL [U]			
CAS 69388-79-0	MW: 347.39	MF:	C14 H21 N O7 S
		Inhibitor (beta-lactamase)	
		Pfizer	
		LgP 1.76(C)	
		pKa	
		pp cited in Vols.1-5:	

(1) A.R. English et al., Antimicrob. Ag. Chemother., 1978, 14, 414.
(2) J.A. Retsema et al., ibid., 1980. 17, 615.

SULBENICILLIN [INN]			
CAS 34779-28-7	MW: 414.46	MF:	C16 H18 N2 O7 S2
		Antibacterial	
		Takeda, Japan	
		LgP 0.63(C); -3.5(C)	
		pKa	
		pp cited in Vols.1-5:	

(1) S. Morimoto et al., U.S. Pat. 3 660 379 (1972).
(2) eidem, J. Med. Chem., 1972, 15, 1105, 1108.

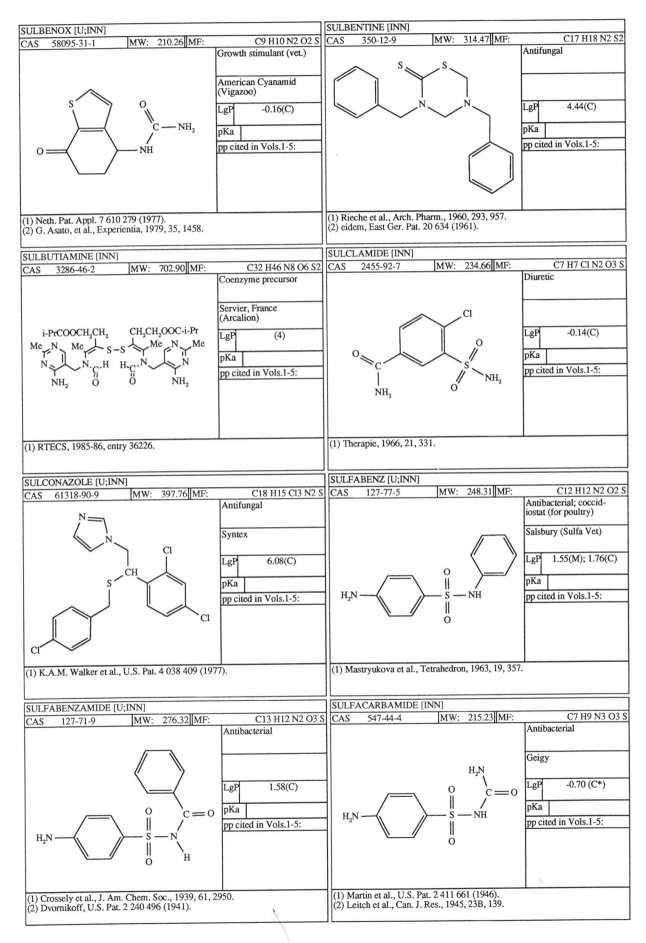

SULBENOX [U;INN]

CAS 58095-31-1	MW: 210.26	MF:	C9 H10 N2 O2 S

Growth stimulant (vet.)

American Cyanamid (Vigazoo)

LgP	-0.16(C)
pKa	
pp cited in Vols.1-5:	

(1) Neth. Pat. Appl. 7 610 279 (1977).
(2) G. Asato, et al., Experientia, 1979, 35, 1458.

SULBENTINE [INN]

CAS 350-12-9	MW: 314.47	MF:	C17 H18 N2 S2

Antifungal

LgP	4.44(C)
pKa	
pp cited in Vols.1-5:	

(1) Rieche et al., Arch. Pharm., 1960, 293, 957.
(2) eidem, East Ger. Pat. 20 634 (1961).

SULBUTIAMINE [INN]

CAS 3286-46-2	MW: 702.90	MF:	C32 H46 N8 O6 S2

Coenzyme precursor

Servier, France (Arcalion)

LgP	(4)
pKa	
pp cited in Vols.1-5:	

(1) RTECS, 1985-86, entry 36226.

SULCLAMIDE [INN]

CAS 2455-92-7	MW: 234.66	MF:	C7 H7 Cl N2 O3 S

Diuretic

LgP	-0.14(C)
pKa	
pp cited in Vols.1-5:	

(1) Therapie, 1966, 21, 331.

SULCONAZOLE [U;INN]

CAS 61318-90-9	MW: 397.76	MF:	C18 H15 Cl3 N2 S

Antifungal

Syntex

LgP	6.08(C)
pKa	
pp cited in Vols.1-5:	

(1) K.A.M. Walker et al., U.S. Pat. 4 038 409 (1977).

SULFABENZ [U;INN]

CAS 127-77-5	MW: 248.31	MF:	C12 H12 N2 O2 S

Antibacterial; coccidiostat (for poultry)

Salsbury (Sulfa Vet)

LgP	1.55(M); 1.76(C)
pKa	
pp cited in Vols.1-5:	

(1) Mastryukova et al., Tetrahedron, 1963, 19, 357.

SULFABENZAMIDE [U;INN]

CAS 127-71-9	MW: 276.32	MF:	C13 H12 N2 O3 S

Antibacterial

LgP	1.58(C)
pKa	
pp cited in Vols.1-5:	

(1) Crossely et al., J. Am. Chem. Soc., 1939, 61, 2950.
(2) Dvornikoff, U.S. Pat. 2 240 496 (1941).

SULFACARBAMIDE [INN]

CAS 547-44-4	MW: 215.23	MF:	C7 H9 N3 O3 S

Antibacterial

Geigy

LgP	-0.70 (C*)
pKa	
pp cited in Vols.1-5:	

(1) Martin et al., U.S. Pat. 2 411 661 (1946).
(2) Leitch et al., Can. J. Res., 1945, 23B, 139.

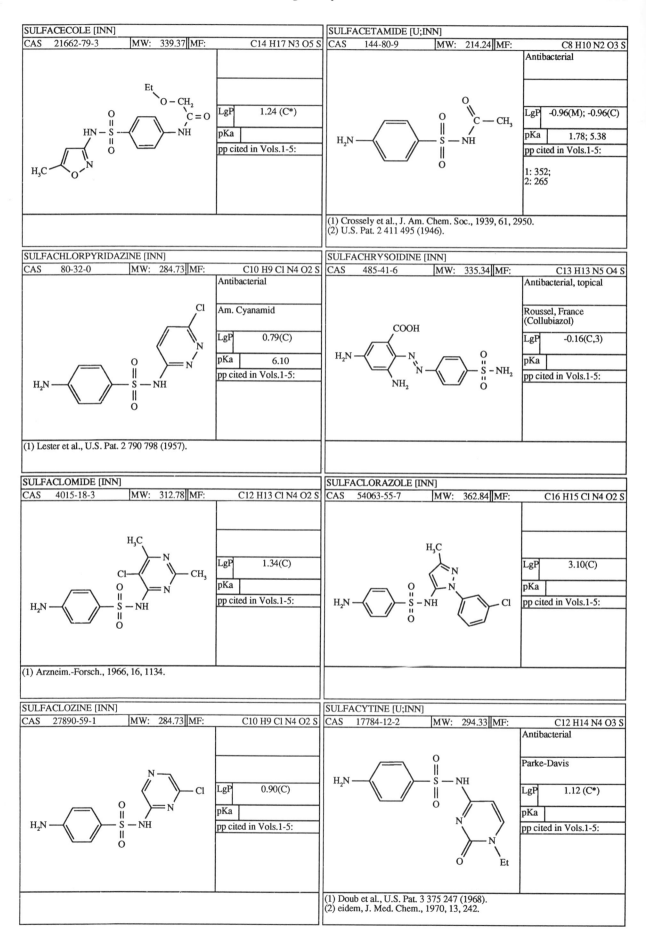

SULFACECOLE [INN]		
CAS 21662-79-3	MW: 339.37	MF: C14 H17 N3 O5 S
	LgP	1.24 (C*)
	pKa	
	pp cited in Vols.1-5:	

SULFACETAMIDE [U;INN]		
CAS 144-80-9	MW: 214.24	MF: C8 H10 N2 O3 S
Antibacterial		
	LgP	-0.96(M); -0.96(C)
	pKa	1.78; 5.38
	pp cited in Vols.1-5: 1: 352; 2: 265	

(1) Crossely et al., J. Am. Chem. Soc., 1939, 61, 2950.
(2) U.S. Pat. 2 411 495 (1946).

SULFACHLORPYRIDAZINE [INN]		
CAS 80-32-0	MW: 284.73	MF: C10 H9 Cl N4 O2 S
Antibacterial Am. Cyanamid		
	LgP	0.79(C)
	pKa	6.10
	pp cited in Vols.1-5:	

(1) Lester et al., U.S. Pat. 2 790 798 (1957).

SULFACHRYSOIDINE [INN]		
CAS 485-41-6	MW: 335.34	MF: C13 H13 N5 O4 S
Antibacterial, topical Roussel, France (Collubiazol)		
	LgP	-0.16(C,3)
	pKa	

SULFACLOMIDE [INN]		
CAS 4015-18-3	MW: 312.78	MF: C12 H13 Cl N4 O2 S
	LgP	1.34(C)
	pKa	

(1) Arzneim.-Forsch., 1966, 16, 1134.

SULFACLORAZOLE [INN]		
CAS 54063-55-7	MW: 362.84	MF: C16 H15 Cl N4 O2 S
	LgP	3.10(C)
	pKa	

SULFACLOZINE [INN]		
CAS 27890-59-1	MW: 284.73	MF: C10 H9 Cl N4 O2 S
	LgP	0.90(C)
	pKa	

SULFACYTINE [U;INN]		
CAS 17784-12-2	MW: 294.33	MF: C12 H14 N4 O3 S
Antibacterial Parke-Davis		
	LgP	1.12 (C*)
	pKa	

(1) Doub et al., U.S. Pat. 3 375 247 (1968).
(2) eidem, J. Med. Chem., 1970, 13, 242.

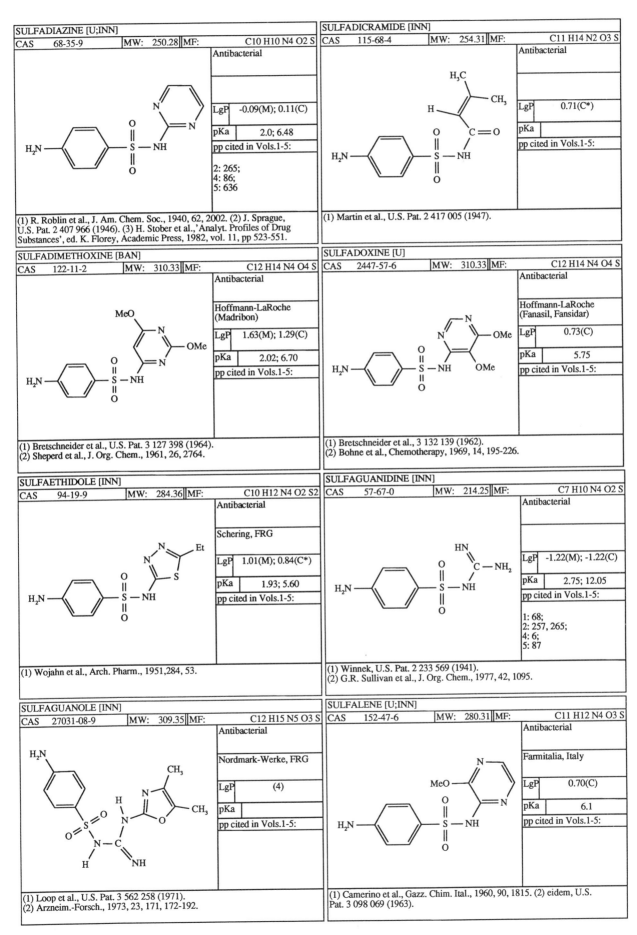

SULFADIAZINE [U;INN]

CAS	68-35-9	MW:	250.28	MF:	C10 H10 N4 O2 S

Antibacterial

LgP	-0.09(M); 0.11(C)
pKa	2.0; 6.48

pp cited in Vols.1-5:

2: 265;
4: 86;
5: 636

(1) R. Roblin et al., J. Am. Chem. Soc., 1940, 62, 2002. (2) J. Sprague, U.S. Pat. 2 407 966 (1946). (3) H. Stober et al.,'Analyt. Profiles of Drug Substances', ed. K. Florey, Academic Press, 1982, vol. 11, pp 523-551.

SULFADICRAMIDE [INN]

CAS	115-68-4	MW:	254.31	MF:	C11 H14 N2 O3 S

Antibacterial

LgP	0.71(C*)
pKa	

pp cited in Vols.1-5:

(1) Martin et al., U.S. Pat. 2 417 005 (1947).

SULFADIMETHOXINE [BAN]

CAS	122-11-2	MW:	310.33	MF:	C12 H14 N4 O4 S

Antibacterial

Hoffmann-LaRoche (Madribon)

LgP	1.63(M); 1.29(C)
pKa	2.02; 6.70

pp cited in Vols.1-5:

(1) Bretschneider et al., U.S. Pat. 3 127 398 (1964).
(2) Sheperd et al., J. Org. Chem., 1961, 26, 2764.

SULFADOXINE [U]

CAS	2447-57-6	MW:	310.33	MF:	C12 H14 N4 O4 S

Antibacterial

Hoffmann-LaRoche (Fanasil, Fansidar)

LgP	0.73(C)
pKa	5.75

pp cited in Vols.1-5:

(1) Bretschneider et al., 3 132 139 (1962).
(2) Bohne et al., Chemotherapy, 1969, 14, 195-226.

SULFAETHIDOLE [INN]

CAS	94-19-9	MW:	284.36	MF:	C10 H12 N4 O2 S2

Antibacterial

Schering, FRG

LgP	1.01(M); 0.84(C*)
pKa	1.93; 5.60

pp cited in Vols.1-5:

(1) Wojahn et al., Arch. Pharm., 1951,284, 53.

SULFAGUANIDINE [INN]

CAS	57-67-0	MW:	214.25	MF:	C7 H10 N4 O2 S

Antibacterial

LgP	-1.22(M); -1.22(C)
pKa	2.75; 12.05

pp cited in Vols.1-5:

1: 68;
2: 257, 265;
4: 6;
5: 87

(1) Winnek, U.S. Pat. 2 233 569 (1941).
(2) G.R. Sullivan et al., J. Org. Chem., 1977, 42, 1095.

SULFAGUANOLE [INN]

CAS	27031-08-9	MW:	309.35	MF:	C12 H15 N5 O3 S

Antibacterial

Nordmark-Werke, FRG

LgP	(4)
pKa	

pp cited in Vols.1-5:

(1) Loop et al., U.S. Pat. 3 562 258 (1971).
(2) Arzneim.-Forsch., 1973, 23, 171, 172-192.

SULFALENE [U;INN]

CAS	152-47-6	MW:	280.31	MF:	C11 H12 N4 O3 S

Antibacterial

Farmitalia, Italy

LgP	0.70(C)
pKa	6.1

pp cited in Vols.1-5:

(1) Camerino et al., Gazz. Chim. Ital., 1960, 90, 1815. (2) eidem, U.S. Pat. 3 098 069 (1963).

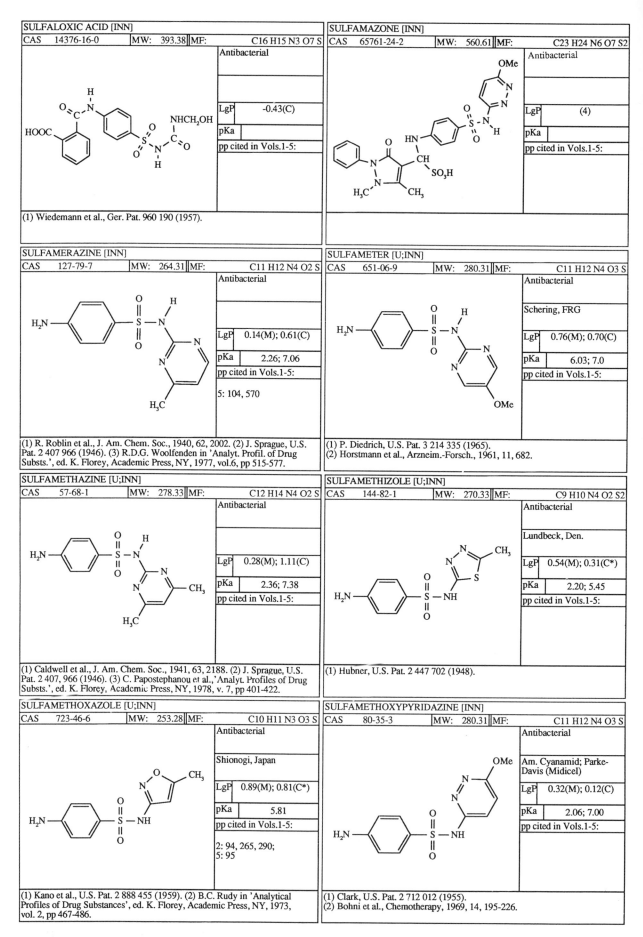

SULFALOXIC ACID [INN]

CAS	14376-16-0	MW:	393.38	MF:	C16 H15 N3 O7 S

Antibacterial

LgP	-0.43(C)
pKa	
pp cited in Vols.1-5:	

(1) Wiedemann et al., Ger. Pat. 960 190 (1957).

SULFAMAZONE [INN]

CAS	65761-24-2	MW:	560.61	MF:	C23 H24 N6 O7 S2

Antibacterial

LgP	(4)
pKa	
pp cited in Vols.1-5:	

SULFAMERAZINE [INN]

CAS	127-79-7	MW:	264.31	MF:	C11 H12 N4 O2 S

Antibacterial

LgP	0.14(M); 0.61(C)
pKa	2.26; 7.06
pp cited in Vols.1-5:	5: 104, 570

(1) R. Roblin et al., J. Am. Chem. Soc., 1940, 62, 2002. (2) J. Sprague, U.S. Pat. 2 407 966 (1946). (3) R.D.G. Woolfenden in 'Analyt. Profil. of Drug Substs.', ed. K. Florey, Academic Press, NY, 1977, vol.6, pp 515-577.

SULFAMETER [U;INN]

CAS	651-06-9	MW:	280.31	MF:	C11 H12 N4 O3 S

Antibacterial

Schering, FRG

LgP	0.76(M); 0.70(C)
pKa	6.03; 7.0
pp cited in Vols.1-5:	

(1) P. Diedrich, U.S. Pat. 3 214 335 (1965).
(2) Horstmann et al., Arzneim.-Forsch., 1961, 11, 682.

SULFAMETHAZINE [U;INN]

CAS	57-68-1	MW:	278.33	MF:	C12 H14 N4 O2 S

Antibacterial

LgP	0.28(M); 1.11(C)
pKa	2.36; 7.38
pp cited in Vols.1-5:	

(1) Caldwell et al., J. Am. Chem. Soc., 1941, 63, 2188. (2) J. Sprague, U.S. Pat. 2 407, 966 (1946). (3) C. Papostephanou et al.,'Analyt. Profiles of Drug Substs.', ed. K. Florey, Academic Press, NY, 1978, v. 7, pp 401-422.

SULFAMETHIZOLE [U;INN]

CAS	144-82-1	MW:	270.33	MF:	C9 H10 N4 O2 S2

Antibacterial

Lundbeck, Den.

LgP	0.54(M); 0.31(C*)
pKa	2.20; 5.45
pp cited in Vols.1-5:	

(1) Hubner, U.S. Pat. 2 447 702 (1948).

SULFAMETHOXAZOLE [U;INN]

CAS	723-46-6	MW:	253.28	MF:	C10 H11 N3 O3 S

Antibacterial

Shionogi, Japan

LgP	0.89(M); 0.81(C*)
pKa	5.81
pp cited in Vols.1-5:	2: 94, 265, 290; 5: 95

(1) Kano et al., U.S. Pat. 2 888 455 (1959). (2) B.C. Rudy in 'Analytical Profiles of Drug Substances', ed. K. Florey, Academic Press, NY, 1973, vol. 2, pp 467-486.

SULFAMETHOXYPYRIDAZINE [INN]

CAS	80-35-3	MW:	280.31	MF:	C11 H12 N4 O3 S

Antibacterial

Am. Cyanamid; Parke-Davis (Midicel)

LgP	0.32(M); 0.12(C)
pKa	2.06; 7.00
pp cited in Vols.1-5:	

(1) Clark, U.S. Pat. 2 712 012 (1955).
(2) Bohni et al., Chemotherapy, 1969, 14, 195-226.

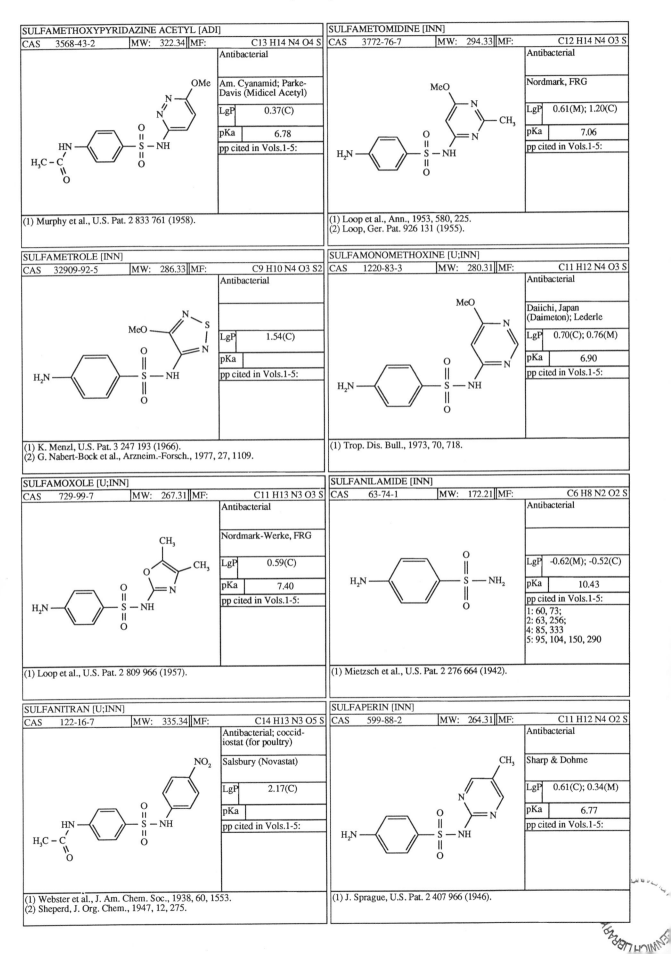

SULFAMETHOXYPYRIDAZINE ACETYL [ADI]

CAS 3568-43-2	MW: 322.34	MF:	C13 H14 N4 O4 S

Antibacterial

Am. Cyanamid; Parke-Davis (Midicel Acetyl)

LgP	0.37(C)
pKa	6.78

pp cited in Vols.1-5:

(1) Murphy et al., U.S. Pat. 2 833 761 (1958).

SULFAMETOMIDINE [INN]

CAS 3772-76-7	MW: 294.33	MF:	C12 H14 N4 O3 S

Antibacterial

Nordmark, FRG

LgP	0.61(M); 1.20(C)
pKa	7.06

pp cited in Vols.1-5:

(1) Loop et al., Ann., 1953, 580, 225.
(2) Loop, Ger. Pat. 926 131 (1955).

SULFAMETROLE [INN]

CAS 32909-92-5	MW: 286.33	MF:	C9 H10 N4 O3 S2

Antibacterial

LgP	1.54(C)
pKa	

pp cited in Vols.1-5:

(1) K. Menzl, U.S. Pat. 3 247 193 (1966).
(2) G. Nabert-Bock et al., Arzneim.-Forsch., 1977, 27, 1109.

SULFAMONOMETHOXINE [U;INN]

CAS 1220-83-3	MW: 280.31	MF:	C11 H12 N4 O3 S

Antibacterial

Daiichi, Japan (Daimeton); Lederle

LgP	0.70(C); 0.76(M)
pKa	6.90

pp cited in Vols.1-5:

(1) Trop. Dis. Bull., 1973, 70, 718.

SULFAMOXOLE [U;INN]

CAS 729-99-7	MW: 267.31	MF:	C11 H13 N3 O3 S

Antibacterial

Nordmark-Werke, FRG

LgP	0.59(C)
pKa	7.40

pp cited in Vols.1-5:

(1) Loop et al., U.S. Pat. 2 809 966 (1957).

SULFANILAMIDE [INN]

CAS 63-74-1	MW: 172.21	MF:	C6 H8 N2 O2 S

Antibacterial

LgP	-0.62(M); -0.52(C)
pKa	10.43

pp cited in Vols.1-5:
1: 60, 73;
2: 63, 256;
4: 85, 333
5: 95, 104, 150, 290

(1) Mietzsch et al., U.S. Pat. 2 276 664 (1942).

SULFANITRAN [U;INN]

CAS 122-16-7	MW: 335.34	MF:	C14 H13 N3 O5 S

Antibacterial; coccidiostat (for poultry)

Salsbury (Novastat)

LgP	2.17(C)
pKa	

pp cited in Vols.1-5:

(1) Webster et al., J. Am. Chem. Soc., 1938, 60, 1553.
(2) Sheperd, J. Org. Chem., 1947, 12, 275.

SULFAPERIN [INN]

CAS 599-88-2	MW: 264.31	MF:	C11 H12 N4 O2 S

Antibacterial

Sharp & Dohme

LgP	0.61(C); 0.34(M)
pKa	6.77

pp cited in Vols.1-5:

(1) J. Sprague, U.S. Pat. 2 407 966 (1946).

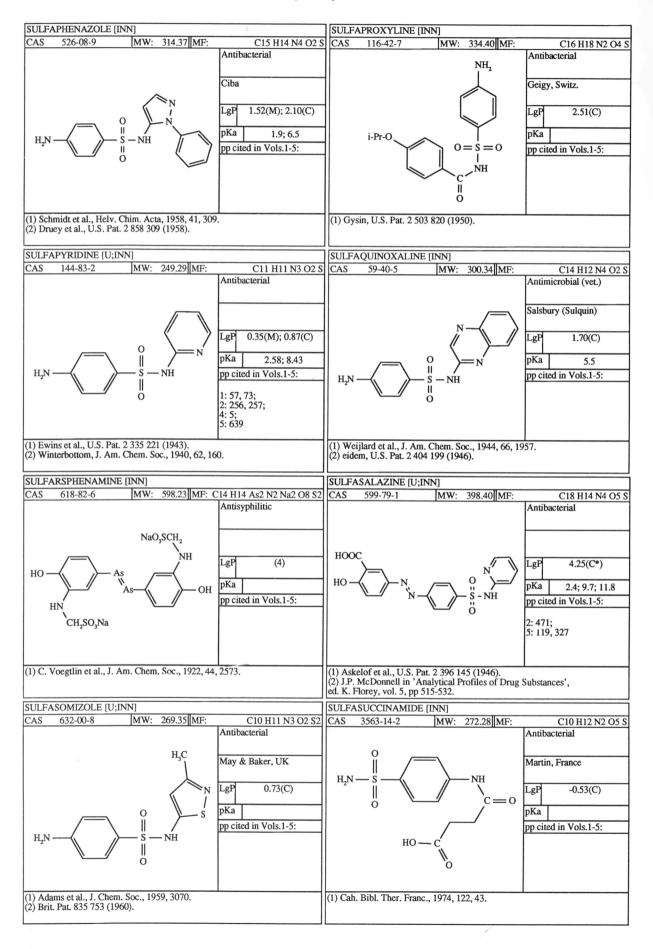

SULFAPHENAZOLE [INN]

| CAS | 526-08-9 | MW: | 314.37 | MF: | C15 H14 N4 O2 S |

Antibacterial

Ciba

LgP	1.52(M); 2.10(C)
pKa	1.9; 6.5
pp cited in Vols.1-5:	

(1) Schmidt et al., Helv. Chim. Acta, 1958, 41, 309.
(2) Druey et al., U.S. Pat. 2 858 309 (1958).

SULFAPROXYLINE [INN]

| CAS | 116-42-7 | MW: | 334.40 | MF: | C16 H18 N2 O4 S |

Antibacterial

Geigy, Switz.

LgP	2.51(C)
pKa	
pp cited in Vols.1-5:	

(1) Gysin, U.S. Pat. 2 503 820 (1950).

SULFAPYRIDINE [U;INN]

| CAS | 144-83-2 | MW: | 249.29 | MF: | C11 H11 N3 O2 S |

Antibacterial

LgP	0.35(M); 0.87(C)
pKa	2.58; 8.43
pp cited in Vols.1-5:	

1: 57, 73;
2: 256, 257;
4: 5;
5: 639

(1) Ewins et al., U.S. Pat. 2 335 221 (1943).
(2) Winterbottom, J. Am. Chem. Soc., 1940, 62, 160.

SULFAQUINOXALINE [INN]

| CAS | 59-40-5 | MW: | 300.34 | MF: | C14 H12 N4 O2 S |

Antimicrobial (vet.)

Salsbury (Sulquin)

LgP	1.70(C)
pKa	5.5
pp cited in Vols.1-5:	

(1) Weijlard et al., J. Am. Chem. Soc., 1944, 66, 1957.
(2) eidem, U.S. Pat. 2 404 199 (1946).

SULFARSPHENAMINE [INN]

| CAS | 618-82-6 | MW: | 598.23 | MF: | C14 H14 As2 N2 Na2 O8 S2 |

Antisyphilitic

LgP	(4)
pKa	
pp cited in Vols.1-5:	

(1) C. Voegtlin et al., J. Am. Chem. Soc., 1922, 44, 2573.

SULFASALAZINE [U;INN]

| CAS | 599-79-1 | MW: | 398.40 | MF: | C18 H14 N4 O5 S |

Antibacterial

LgP	4.25(C*)
pKa	2.4; 9.7; 11.8
pp cited in Vols.1-5:	

2: 471;
5: 119, 327

(1) Askelof et al., U.S. Pat. 2 396 145 (1946).
(2) J.P. McDonnell in 'Analytical Profiles of Drug Substances',
ed. K. Florey, vol. 5, pp 515-532.

SULFASOMIZOLE [U;INN]

| CAS | 632-00-8 | MW: | 269.35 | MF: | C10 H11 N3 O2 S2 |

Antibacterial

May & Baker, UK

LgP	0.73(C)
pKa	
pp cited in Vols.1-5:	

(1) Adams et al., J. Chem. Soc., 1959, 3070.
(2) Brit. Pat. 835 753 (1960).

SULFASUCCINAMIDE [INN]

| CAS | 3563-14-2 | MW: | 272.28 | MF: | C10 H12 N2 O5 S |

Antibacterial

Martin, France

LgP	-0.53(C)
pKa	
pp cited in Vols.1-5:	

(1) Cah. Bibl. Ther. Franc., 1974, 122, 43.

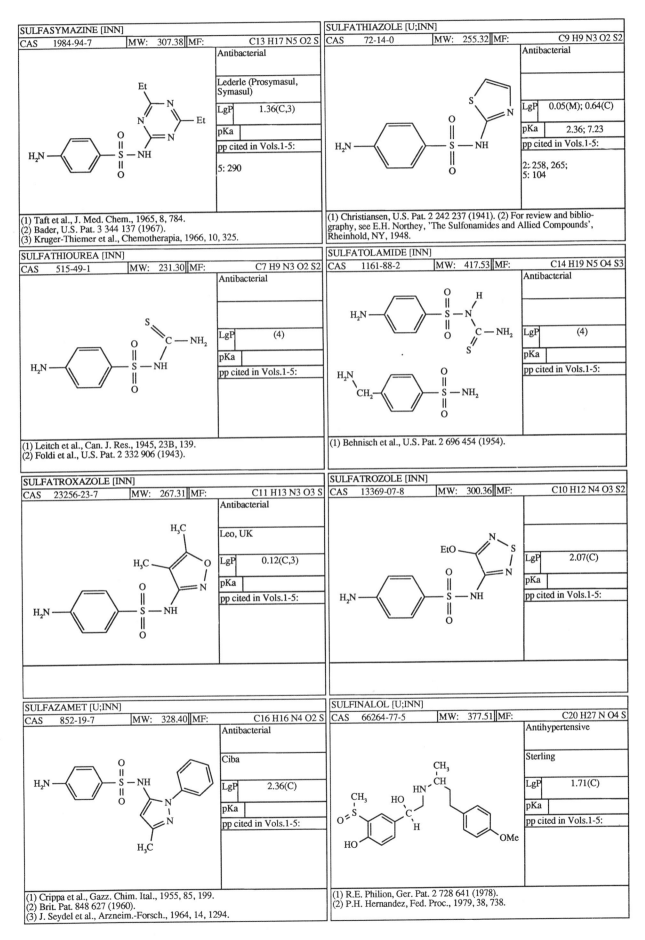

SULFASYMAZINE [INN]

CAS	1984-94-7	MW:	307.38	MF:	C13 H17 N5 O2 S

Antibacterial

Lederle (Prosymasul, Symasul)

LgP	1.36(C,3)
pKa	
pp cited in Vols.1-5:	

5: 290

(1) Taft et al., J. Med. Chem., 1965, 8, 784.
(2) Bader, U.S. Pat. 3 344 137 (1967).
(3) Kruger-Thiemer et al., Chemotherapia, 1966, 10, 325.

SULFATHIAZOLE [U;INN]

CAS	72-14-0	MW:	255.32	MF:	C9 H9 N3 O2 S2

Antibacterial

LgP	0.05(M); 0.64(C)
pKa	2.36; 7.23
pp cited in Vols.1-5:	

2: 258, 265;
5: 104

(1) Christiansen, U.S. Pat. 2 242 237 (1941). (2) For review and bibliography, see E.H. Northey, 'The Sulfonamides and Allied Compounds', Rheinhold, NY, 1948.

SULFATHIOUREA [INN]

CAS	515-49-1	MW:	231.30	MF:	C7 H9 N3 O2 S2

Antibacterial

LgP	(4)
pKa	
pp cited in Vols.1-5:	

(1) Leitch et al., Can. J. Res., 1945, 23B, 139.
(2) Foldi et al., U.S. Pat. 2 332 906 (1943).

SULFATOLAMIDE [INN]

CAS	1161-88-2	MW:	417.53	MF:	C14 H19 N5 O4 S3

Antibacterial

LgP	(4)
pKa	
pp cited in Vols.1-5:	

(1) Behnisch et al., U.S. Pat. 2 696 454 (1954).

SULFATROXAZOLE [INN]

CAS	23256-23-7	MW:	267.31	MF:	C11 H13 N3 O3 S

Antibacterial

Leo, UK

LgP	0.12(C,3)
pKa	
pp cited in Vols.1-5:	

SULFATROZOLE [INN]

CAS	13369-07-8	MW:	300.36	MF:	C10 H12 N4 O3 S2

LgP	2.07(C)
pKa	
pp cited in Vols.1-5:	

SULFAZAMET [U;INN]

CAS	852-19-7	MW:	328.40	MF:	C16 H16 N4 O2 S

Antibacterial

Ciba

LgP	2.36(C)
pKa	
pp cited in Vols.1-5:	

(1) Crippa et al., Gazz. Chim. Ital., 1955, 85, 199.
(2) Brit. Pat. 848 627 (1960).
(3) J. Seydel et al., Arzneim.-Forsch., 1964, 14, 1294.

SULFINALOL [U;INN]

CAS	66264-77-5	MW:	377.51	MF:	C20 H27 N O4 S

Antihypertensive

Sterling

LgP	1.71(C)
pKa	
pp cited in Vols.1-5:	

(1) R.E. Philion, Ger. Pat. 2 728 641 (1978).
(2) P.H. Hernandez, Fed. Proc., 1979, 38, 738.

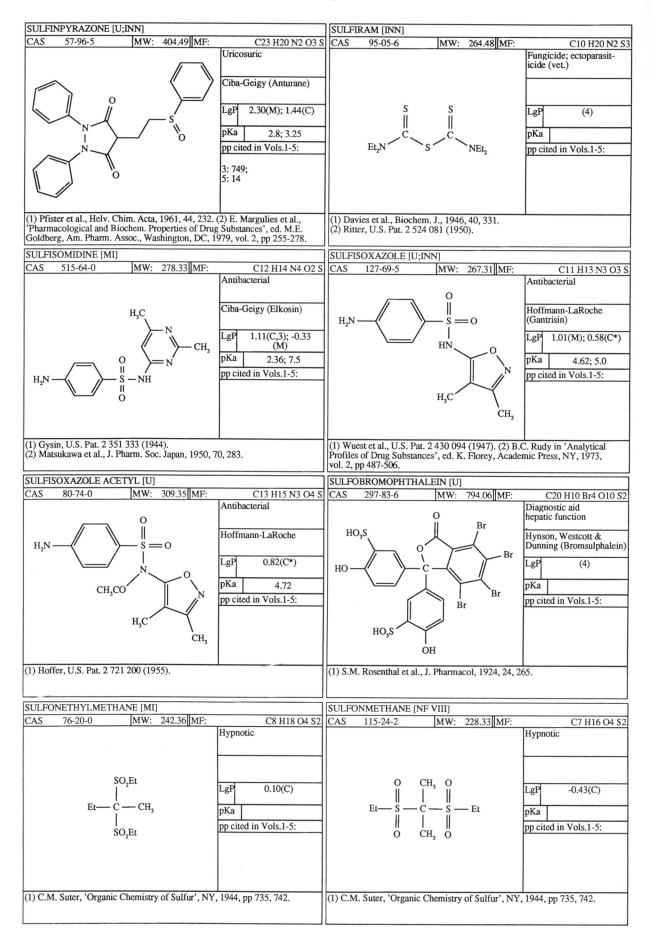

SULFINPYRAZONE [U;INN]		
CAS 57-96-5	MW: 404.49	MF: C23 H20 N2 O3 S
	Uricosuric	
	Ciba-Geigy (Anturane)	
	LgP	2.30(M); 1.44(C)
	pKa	2.8; 3.25
	pp cited in Vols.1-5:	
	3: 749; 5: 14	

(1) Pfister et al., Helv. Chim. Acta, 1961, 44, 232. (2) E. Margulies et al., 'Pharmacological and Biochem. Properties of Drug Substances', ed. M.E. Goldberg, Am. Pharm. Assoc., Washington, DC, 1979, vol. 2, pp 255-278.

SULFIRAM [INN]		
CAS 95-05-6	MW: 264.48	MF: C10 H20 N2 S3
	Fungicide; ectoparasiticide (vet.)	
	LgP	(4)
	pKa	
	pp cited in Vols.1-5:	

(1) Davies et al., Biochem. J., 1946, 40, 331.
(2) Ritter, U.S. Pat. 2 524 081 (1950).

SULFISOMIDINE [MI]		
CAS 515-64-0	MW: 278.33	MF: C12 H14 N4 O2 S
	Antibacterial	
	Ciba-Geigy (Elkosin)	
	LgP	1.11(C,3); -0.33 (M)
	pKa	2.36; 7.5
	pp cited in Vols.1-5:	

(1) Gysin, U.S. Pat. 2 351 333 (1944).
(2) Matsukawa et al., J. Pharm. Soc. Japan, 1950, 70, 283.

SULFISOXAZOLE [U;INN]		
CAS 127-69-5	MW: 267.31	MF: C11 H13 N3 O3 S
	Antibacterial	
	Hoffmann-LaRoche (Gantrisin)	
	LgP	1.01(M); 0.58(C*)
	pKa	4.62; 5.0
	pp cited in Vols.1-5:	

(1) Wuest et al., U.S. Pat. 2 430 094 (1947). (2) B.C. Rudy in 'Analytical Profiles of Drug Substances', ed. K. Florey, Academic Press, NY, 1973, vol. 2, pp 487-506.

SULFISOXAZOLE ACETYL [U]		
CAS 80-74-0	MW: 309.35	MF: C13 H15 N3 O4 S
	Antibacterial	
	Hoffmann-LaRoche	
	LgP	0.82(C*)
	pKa	4.72
	pp cited in Vols.1-5:	

(1) Hoffer, U.S. Pat. 2 721 200 (1955).

SULFOBROMOPHTHALEIN [U]		
CAS 297-83-6	MW: 794.06	MF: C20 H10 Br4 O10 S2
	Diagnostic aid hepatic function	
	Hynson, Westcott & Dunning (Bromsulphalein)	
	LgP	(4)
	pKa	
	pp cited in Vols.1-5:	

(1) S.M. Rosenthal et al., J. Pharmacol, 1924, 24, 265.

SULFONETHYLMETHANE [MI]		
CAS 76-20-0	MW: 242.36	MF: C8 H18 O4 S2
	Hypnotic	
	LgP	0.10(C)
	pKa	
	pp cited in Vols.1-5:	

(1) C.M. Suter, 'Organic Chemistry of Sulfur', NY, 1944, pp 735, 742.

SULFONMETHANE [NF VIII]		
CAS 115-24-2	MW: 228.33	MF: C7 H16 O4 S2
	Hypnotic	
	LgP	-0.43(C)
	pKa	
	pp cited in Vols.1-5:	

(1) C.M. Suter, 'Organic Chemistry of Sulfur', NY, 1944, pp 735, 742.

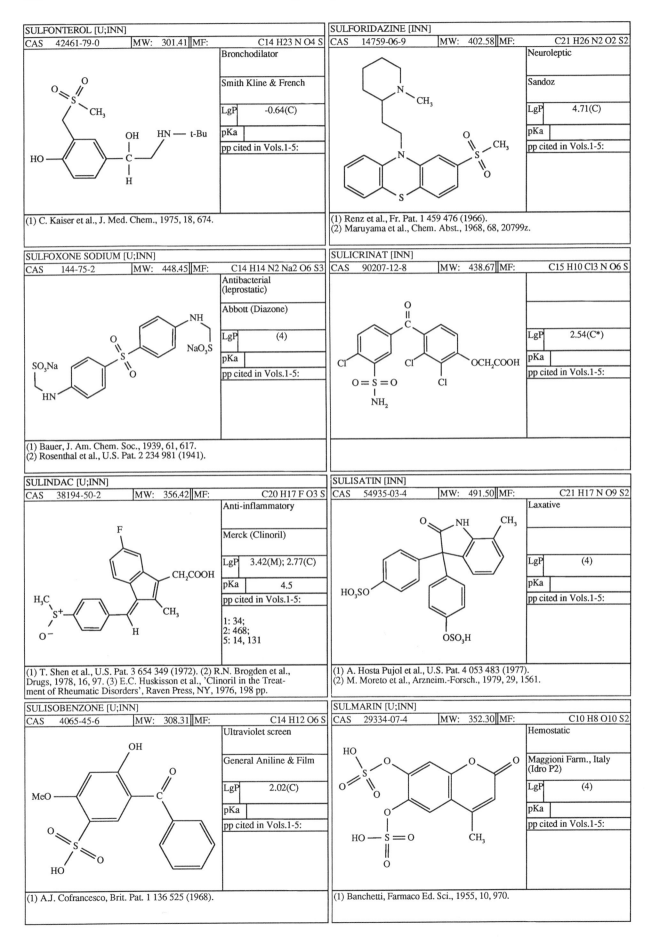

SULFONTEROL [U;INN]

CAS	42461-79-0	MW:	301.41	MF:	C14 H23 N O4 S

Bronchodilator

Smith Kline & French

LgP	-0.64(C)
pKa	
pp cited in Vols.1-5:	

(1) C. Kaiser et al., J. Med. Chem., 1975, 18, 674.

SULFORIDAZINE [INN]

CAS	14759-06-9	MW:	402.58	MF:	C21 H26 N2 O2 S2

Neuroleptic

Sandoz

LgP	4.71(C)
pKa	
pp cited in Vols.1-5:	

(1) Renz et al., Fr. Pat. 1 459 476 (1966).
(2) Maruyama et al., Chem. Abst., 1968, 68, 20799z.

SULFOXONE SODIUM [U;INN]

CAS	144-75-2	MW:	448.45	MF:	C14 H14 N2 Na2 O6 S3

Antibacterial
(leprostatic)

Abbott (Diazone)

LgP	(4)
pKa	
pp cited in Vols.1-5:	

(1) Bauer, J. Am. Chem. Soc., 1939, 61, 617.
(2) Rosenthal et al., U.S. Pat. 2 234 981 (1941).

SULICRINAT [INN]

CAS	90207-12-8	MW:	438.67	MF:	C15 H10 Cl3 N O6 S

LgP	2.54(C*)
pKa	
pp cited in Vols.1-5:	

SULINDAC [U;INN]

CAS	38194-50-2	MW:	356.42	MF:	C20 H17 F O3 S

Anti-inflammatory

Merck (Clinoril)

LgP	3.42(M); 2.77(C)
pKa	4.5
pp cited in Vols.1-5:	

1: 34;
2: 468;
5: 14, 131

(1) T. Shen et al., U.S. Pat. 3 654 349 (1972). (2) R.N. Brogden et al., Drugs, 1978, 16, 97. (3) E.C. Huskisson et al., 'Clinoril in the Treatment of Rheumatic Disorders', Raven Press, NY, 1976, 198 pp.

SULISATIN [INN]

CAS	54935-03-4	MW:	491.50	MF:	C21 H17 N O9 S2

Laxative

LgP	(4)
pKa	
pp cited in Vols.1-5:	

(1) A. Hosta Pujol et al., U.S. Pat. 4 053 483 (1977).
(2) M. Moreto et al., Arzneim.-Forsch., 1979, 29, 1561.

SULISOBENZONE [U;INN]

CAS	4065-45-6	MW:	308.31	MF:	C14 H12 O6 S

Ultraviolet screen

General Aniline & Film

LgP	2.02(C)
pKa	
pp cited in Vols.1-5:	

(1) A.J. Cofrancesco, Brit. Pat. 1 136 525 (1968).

SULMARIN [U;INN]

CAS	29334-07-4	MW:	352.30	MF:	C10 H8 O10 S2

Hemostatic

Maggioni Farm., Italy
(Idro P2)

LgP	(4)
pKa	
pp cited in Vols.1-5:	

(1) Banchetti, Farmaco Ed. Sci., 1955, 10, 970.

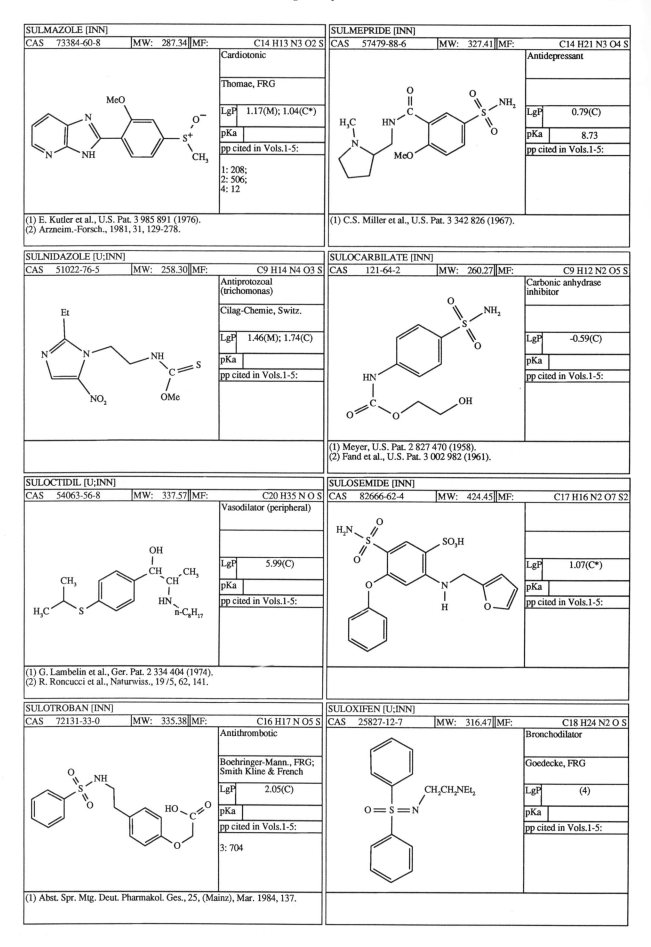

SULMAZOLE [INN]		
CAS 73384-60-8	MW: 287.34	MF: C14 H13 N3 O2 S

Cardiotonic

Thomae, FRG

LgP	1.17(M); 1.04(C*)
pKa	

pp cited in Vols.1-5:

1: 208;
2: 506;
4: 12

(1) E. Kutler et al., U.S. Pat. 3 985 891 (1976).
(2) Arzneim.-Forsch., 1981, 31, 129-278.

SULMEPRIDE [INN]		
CAS 57479-88-6	MW: 327.41	MF: C14 H21 N3 O4 S

Antidepressant

LgP	0.79(C)
pKa	8.73

pp cited in Vols.1-5:

(1) C.S. Miller et al., U.S. Pat. 3 342 826 (1967).

SULNIDAZOLE [U;INN]		
CAS 51022-76-5	MW: 258.30	MF: C9 H14 N4 O3 S

Antiprotozoal
(trichomonas)

Cilag-Chemie, Switz.

LgP	1.46(M); 1.74(C)
pKa	

pp cited in Vols.1-5:

SULOCARBILATE [INN]		
CAS 121-64-2	MW: 260.27	MF: C9 H12 N2 O5 S

Carbonic anhydrase
inhibitor

LgP	-0.59(C)
pKa	

pp cited in Vols.1-5:

(1) Meyer, U.S. Pat. 2 827 470 (1958).
(2) Fand et al., U.S. Pat. 3 002 982 (1961).

SULOCTIDIL [U;INN]		
CAS 54063-56-8	MW: 337.57	MF: C20 H35 N O S

Vasodilator (peripheral)

LgP	5.99(C)
pKa	

pp cited in Vols.1-5:

(1) G. Lambelin et al., Ger. Pat. 2 334 404 (1974).
(2) R. Roncucci et al., Naturwiss., 19 /5, 62, 141.

SULOSEMIDE [INN]		
CAS 82666-62-4	MW: 424.45	MF: C17 H16 N2 O7 S2

LgP	1.07(C*)
pKa	

pp cited in Vols.1-5:

SULOTROBAN [INN]		
CAS 72131-33-0	MW: 335.38	MF: C16 H17 N O5 S

Antithrombotic

Boehringer-Mann., FRG;
Smith Kline & French

LgP	2.05(C)
pKa	

pp cited in Vols.1-5:

3: 704

(1) Abst. Spr. Mtg. Deut. Pharmakol. Ges., 25, (Mainz), Mar. 1984, 137.

SULOXIFEN [U;INN]		
CAS 25827-12-7	MW: 316.47	MF: C18 H24 N2 O S

Bronchodilator

Goedecke, FRG

LgP	(4)
pKa	

pp cited in Vols.1-5:

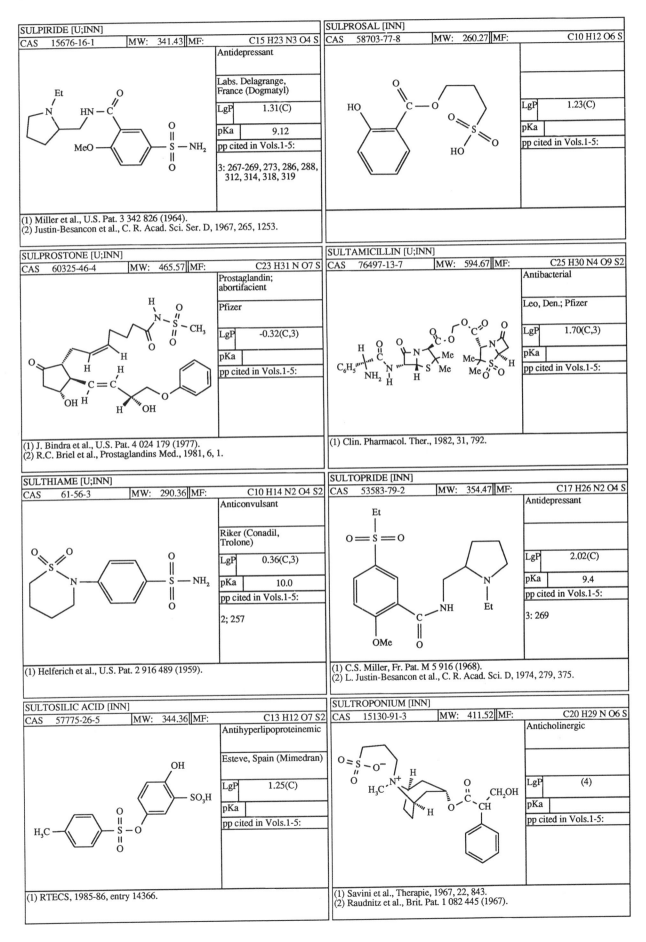

SULPIRIDE [U;INN]

CAS	15676-16-1	MW: 341.43	MF: C15 H23 N3 O4 S

Antidepressant

Labs. Delagrange, France (Dogmatyl)

LgP	1.31(C)
pKa	9.12

pp cited in Vols.1-5:

3: 267-269, 273, 286, 288, 312, 314, 318, 319

(1) Miller et al., U.S. Pat. 3 342 826 (1964).
(2) Justin-Besancon et al., C. R. Acad. Sci. Ser. D, 1967, 265, 1253.

SULPROSAL [INN]

CAS	58703-77-8	MW: 260.27	MF: C10 H12 O6 S

LgP	1.23(C)
pKa	

pp cited in Vols.1-5:

SULPROSTONE [U;INN]

CAS	60325-46-4	MW: 465.57	MF: C23 H31 N O7 S

Prostaglandin; abortifacient

Pfizer

LgP	-0.32(C,3)
pKa	

pp cited in Vols.1-5:

(1) J. Bindra et al., U.S. Pat. 4 024 179 (1977).
(2) R.C. Briel et al., Prostaglandins Med., 1981, 6, 1.

SULTAMICILLIN [U;INN]

CAS	76497-13-7	MW: 594.67	MF: C25 H30 N4 O9 S2

Antibacterial

Leo, Den.; Pfizer

LgP	1.70(C,3)
pKa	

pp cited in Vols.1-5:

(1) Clin. Pharmacol. Ther., 1982, 31, 792.

SULTHIAME [U;INN]

CAS	61-56-3	MW: 290.36	MF: C10 H14 N2 O4 S2

Anticonvulsant

Riker (Conadil, Trolone)

LgP	0.36(C,3)
pKa	10.0

pp cited in Vols.1-5:

2; 257

(1) Helferich et al., U.S. Pat. 2 916 489 (1959).

SULTOPRIDE [INN]

CAS	53583-79-2	MW: 354.47	MF: C17 H26 N2 O4 S

Antidepressant

LgP	2.02(C)
pKa	9.4

pp cited in Vols.1-5:

3: 269

(1) C.S. Miller, Fr. Pat. M 5 916 (1968).
(2) L. Justin-Besancon et al., C. R. Acad. Sci. D, 1974, 279, 375.

SULTOSILIC ACID [INN]

CAS	57775-26-5	MW: 344.36	MF: C13 H12 O7 S2

Antihyperlipoproteinemic

Esteve, Spain (Mimedran)

LgP	1.25(C)
pKa	

pp cited in Vols.1-5:

(1) RTECS, 1985-86, entry 14366.

SULTROPONIUM [INN]

CAS	15130-91-3	MW: 411.52	MF: C20 H29 N O6 S

Anticholinergic

LgP	(4)
pKa	

pp cited in Vols.1-5:

(1) Savini et al., Therapie, 1967, 22, 843.
(2) Raudnitz et al., Brit. Pat. 1 082 445 (1967).

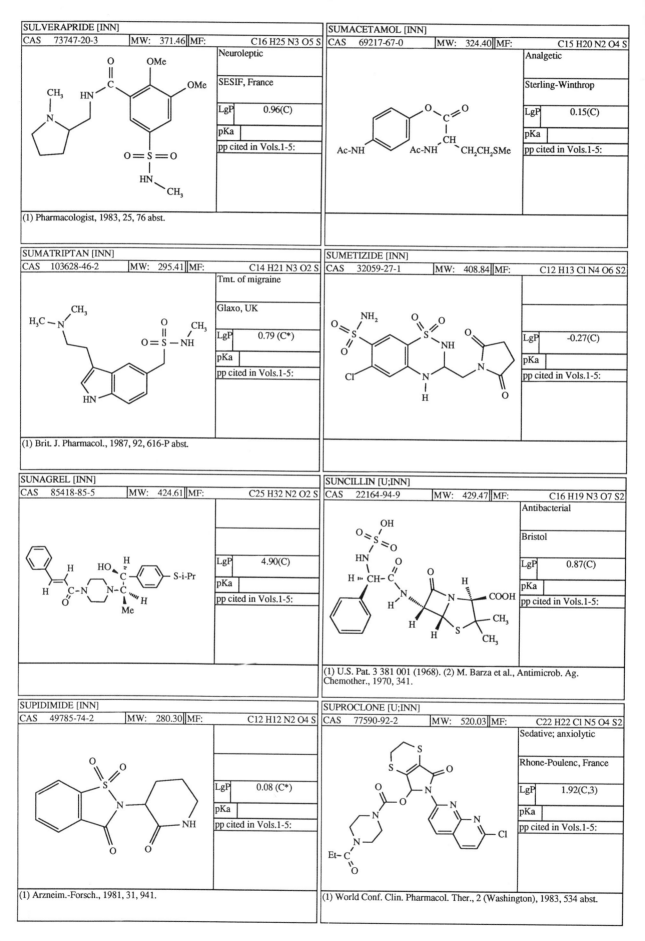

SULVERAPRIDE [INN]

CAS 73747-20-3　　MW: 371.46　MF: C16 H25 N3 O5 S

Neuroleptic

SESIF, France

LgP 0.96(C)

pKa

pp cited in Vols.1-5:

(1) Pharmacologist, 1983, 25, 76 abst.

SUMACETAMOL [INN]

CAS 69217-67-0　　MW: 324.40　MF: C15 H20 N2 O4 S

Analgetic

Sterling-Winthrop

LgP 0.15(C)

pKa

pp cited in Vols.1-5:

SUMATRIPTAN [INN]

CAS 103628-46-2　　MW: 295.41　MF: C14 H21 N3 O2 S

Tmt. of migraine

Glaxo, UK

LgP 0.79 (C*)

pKa

pp cited in Vols.1-5:

(1) Brit. J. Pharmacol., 1987, 92, 616-P abst.

SUMETIZIDE [INN]

CAS 32059-27-1　　MW: 408.84　MF: C12 H13 Cl N4 O6 S2

LgP -0.27(C)

pKa

pp cited in Vols.1-5:

SUNAGREL [INN]

CAS 85418-85-5　　MW: 424.61　MF: C25 H32 N2 O2 S

LgP 4.90(C)

pKa

pp cited in Vols.1-5:

SUNCILLIN [U;INN]

CAS 22164-94-9　　MW: 429.47　MF: C16 H19 N3 O7 S2

Antibacterial

Bristol

LgP 0.87(C)

pKa

pp cited in Vols.1-5:

(1) U.S. Pat. 3 381 001 (1968). (2) M. Barza et al., Antimicrob. Ag. Chemother., 1970, 341.

SUPIDIMIDE [INN]

CAS 49785-74-2　　MW: 280.30　MF: C12 H12 N2 O4 S

LgP 0.08 (C*)

pKa

pp cited in Vols.1-5:

(1) Arzneim.-Forsch., 1981, 31, 941.

SUPROCLONE [U;INN]

CAS 77590-92-2　　MW: 520.03　MF: C22 H22 Cl N5 O4 S2

Sedative; anxiolytic

Rhone-Poulenc, France

LgP 1.92(C,3)

pKa

pp cited in Vols.1-5:

(1) World Conf. Clin. Pharmacol. Ther., 2 (Washington), 1983, 534 abst.

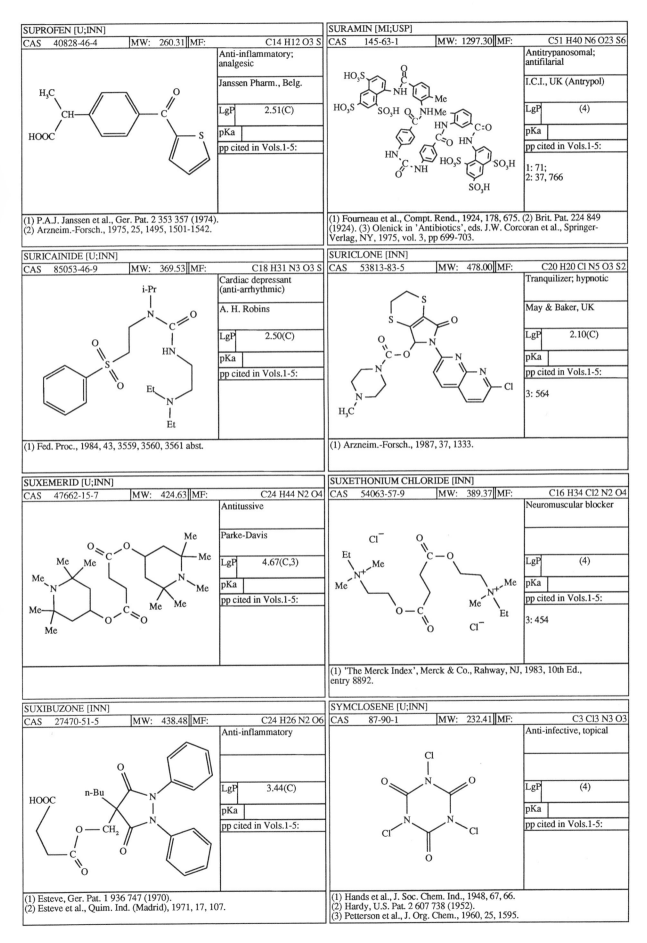

SUPROFEN [U;INN]

CAS 40828-46-4	MW: 260.31	MF:	C14 H12 O3 S

Anti-inflammatory; analgesic

Janssen Pharm., Belg.

LgP	2.51(C)

pKa

pp cited in Vols.1-5:

(1) P.A.J. Janssen et al., Ger. Pat. 2 353 357 (1974).
(2) Arzneim.-Forsch., 1975, 25, 1495, 1501-1542.

SURAMIN [MI;USP]

CAS 145-63-1	MW: 1297.30	MF:	C51 H40 N6 O23 S6

Antitrypanosomal; antifilarial

I.C.I., UK (Antrypol)

LgP	(4)

pKa

pp cited in Vols.1-5:

1: 71;
2: 37, 766

(1) Fourneau et al., Compt. Rend., 1924, 178, 675. (2) Brit. Pat. 224 849
(1924). (3) Olenick in 'Antibiotics', eds. J.W. Corcoran et al., Springer-
Verlag, NY, 1975, vol. 3, pp 699-703.

SURICAINIDE [U;INN]

CAS 85053-46-9	MW: 369.53	MF:	C18 H31 N3 O3 S

Cardiac depressant (anti-arrhythmic)

A. H. Robins

LgP	2.50(C)

pKa

pp cited in Vols.1-5:

(1) Fed. Proc., 1984, 43, 3559, 3560, 3561 abst.

SURICLONE [INN]

CAS 53813-83-5	MW: 478.00	MF:	C20 H20 Cl N5 O3 S2

Tranquilizer; hypnotic

May & Baker, UK

LgP	2.10(C)

pKa

pp cited in Vols.1-5:

3: 564

(1) Arzneim.-Forsch., 1987, 37, 1333.

SUXEMERID [U;INN]

CAS 47662-15-7	MW: 424.63	MF:	C24 H44 N2 O4

Antitussive

Parke-Davis

LgP	4.67(C,3)

pKa

pp cited in Vols.1-5:

SUXETHONIUM CHLORIDE [INN]

CAS 54063-57-9	MW: 389.37	MF:	C16 H34 Cl2 N2 O4

Neuromuscular blocker

LgP	(4)

pKa

pp cited in Vols.1-5:

3: 454

(1) 'The Merck Index', Merck & Co., Rahway, NJ, 1983, 10th Ed.,
entry 8892.

SUXIBUZONE [INN]

CAS 27470-51-5	MW: 438.48	MF:	C24 H26 N2 O6

Anti-inflammatory

LgP	3.44(C)

pKa

pp cited in Vols.1-5:

(1) Esteve, Ger. Pat. 1 936 747 (1970).
(2) Esteve et al., Quim. Ind. (Madrid), 1971, 17, 107.

SYMCLOSENE [U;INN]

CAS 87-90-1	MW: 232.41	MF:	C3 Cl3 N3 O3

Anti-infective, topical

LgP	(4)

pKa

pp cited in Vols.1-5:

(1) Hands et al., J. Soc. Chem. Ind., 1948, 67, 66.
(2) Hardy, U.S. Pat. 2 607 738 (1952).
(3) Petterson et al., J. Org. Chem., 1960, 25, 1595.

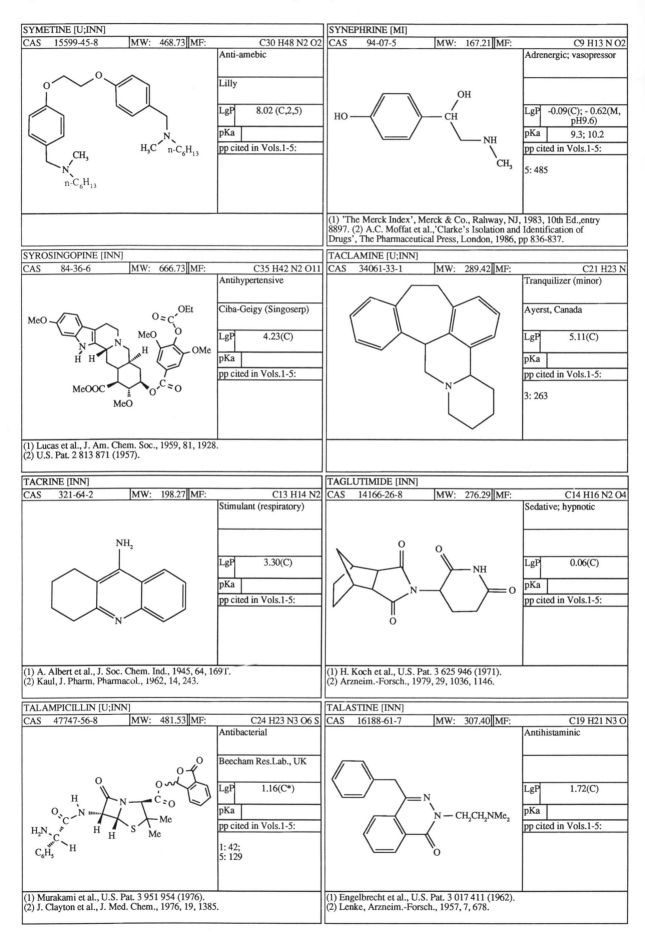

SYMETINE [U;INN]

CAS	15599-45-8	MW:	468.73	MF:	C30 H48 N2 O2

Anti-amebic

Lilly

LgP	8.02 (C,2,5)
pKa	
pp cited in Vols.1-5:	

SYNEPHRINE [MI]

CAS	94-07-5	MW:	167.21	MF:	C9 H13 N O2

Adrenergic; vasopressor

LgP	-0.09(C); - 0.62(M, pH9.6)
pKa	9.3; 10.2
pp cited in Vols.1-5:	

5: 485

(1) 'The Merck Index', Merck & Co., Rahway, NJ, 1983, 10th Ed.,entry 8897. (2) A.C. Moffat et al.,'Clarke's Isolation and Identification of Drugs', The Pharmaceutical Press, London, 1986, pp 836-837.

SYROSINGOPINE [INN]

CAS	84-36-6	MW:	666.73	MF:	C35 H42 N2 O11

Antihypertensive

Ciba-Geigy (Singoserp)

LgP	4.23(C)
pKa	
pp cited in Vols.1-5:	

(1) Lucas et al., J. Am. Chem. Soc., 1959, 81, 1928.
(2) U.S. Pat. 2 813 871 (1957).

TACLAMINE [U;INN]

CAS	34061-33-1	MW:	289.42	MF:	C21 H23 N

Tranquilizer (minor)

Ayerst, Canada

LgP	5.11(C)
pKa	
pp cited in Vols.1-5:	

3: 263

TACRINE [INN]

CAS	321-64-2	MW:	198.27	MF:	C13 H14 N2

Stimulant (respiratory)

LgP	3.30(C)
pKa	
pp cited in Vols.1-5:	

(1) A. Albert et al., J. Soc. Chem. Ind., 1945, 64, 169T.
(2) Kaul, J. Pharm. Pharmacol., 1962, 14, 243.

TAGLUTIMIDE [INN]

CAS	14166-26-8	MW:	276.29	MF:	C14 H16 N2 O4

Sedative; hypnotic

LgP	0.06(C)
pKa	
pp cited in Vols.1-5:	

(1) H. Koch et al., U.S. Pat. 3 625 946 (1971).
(2) Arzneim.-Forsch., 1979, 29, 1036, 1146.

TALAMPICILLIN [U;INN]

CAS	47747-56-8	MW:	481.53	MF:	C24 H23 N3 O6 S

Antibacterial

Beecham Res.Lab., UK

LgP	1.16(C*)
pKa	
pp cited in Vols.1-5:	

1: 42;
5: 129

(1) Murakami et al., U.S. Pat. 3 951 954 (1976).
(2) J. Clayton et al., J. Med. Chem., 1976, 19, 1385.

TALASTINE [INN]

CAS	16188-61-7	MW:	307.40	MF:	C19 H21 N3 O

Antihistaminic

LgP	1.72(C)
pKa	
pp cited in Vols.1-5:	

(1) Engelbrecht et al., U.S. Pat. 3 017 411 (1962).
(2) Lenke, Arzneim.-Forsch., 1957, 7, 678.

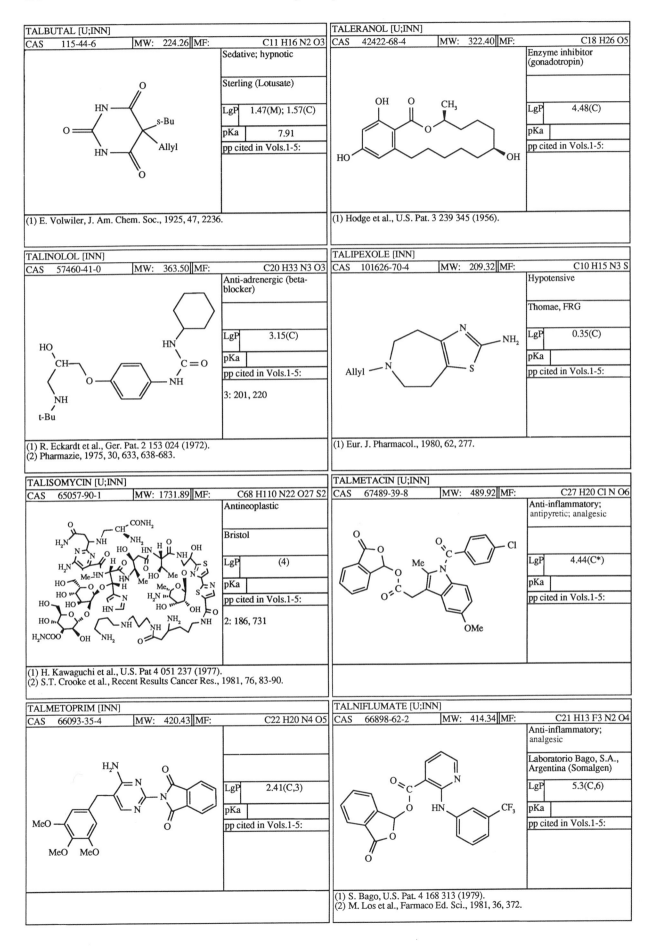

TALBUTAL [U;INN]

CAS	115-44-6	MW:	224.26	MF:	C11 H16 N2 O3

Sedative; hypnotic

Sterling (Lotusate)

LgP	1.47(M); 1.57(C)
pKa	7.91
pp cited in Vols.1-5:	

(1) E. Volwiler, J. Am. Chem. Soc., 1925, 47, 2236.

TALERANOL [U;INN]

CAS	42422-68-4	MW:	322.40	MF:	C18 H26 O5

Enzyme inhibitor (gonadotropin)

LgP	4.48(C)
pKa	
pp cited in Vols.1-5:	

(1) Hodge et al., U.S. Pat. 3 239 345 (1956).

TALINOLOL [INN]

CAS	57460-41-0	MW:	363.50	MF:	C20 H33 N3 O3

Anti-adrenergic (beta-blocker)

LgP	3.15(C)
pKa	
pp cited in Vols.1-5:	
3: 201, 220	

(1) R. Eckardt et al., Ger. Pat. 2 153 024 (1972).
(2) Pharmazie, 1975, 30, 633, 638-683.

TALIPEXOLE [INN]

CAS	101626-70-4	MW:	209.32	MF:	C10 H15 N3 S

Hypotensive

Thomae, FRG

LgP	0.35(C)
pKa	
pp cited in Vols.1-5:	

(1) Eur. J. Pharmacol., 1980, 62, 277.

TALISOMYCIN [U;INN]

CAS	65057-90-1	MW:	1731.89	MF:	C68 H110 N22 O27 S2

Antineoplastic

Bristol

LgP	(4)
pKa	
pp cited in Vols.1-5:	
2: 186, 731	

(1) H. Kawaguchi et al., U.S. Pat 4 051 237 (1977).
(2) S.T. Crooke et al., Recent Results Cancer Res., 1981, 76, 83-90.

TALMETACIN [U;INN]

CAS	67489-39-8	MW:	489.92	MF:	C27 H20 Cl N O6

Anti-inflammatory; antipyretic; analgesic

LgP	4.44(C*)
pKa	
pp cited in Vols.1-5:	

TALMETOPRIM [INN]

CAS	66093-35-4	MW:	420.43	MF:	C22 H20 N4 O5

LgP	2.41(C,3)
pKa	
pp cited in Vols.1-5:	

TALNIFLUMATE [U;INN]

CAS	66898-62-2	MW:	414.34	MF:	C21 H13 F3 N2 O4

Anti-inflammatory; analgesic

Laboratorio Bago, S.A., Argentina (Somalgen)

LgP	5.3(C,6)
pKa	
pp cited in Vols.1-5:	

(1) S. Bago, U.S. Pat. 4 168 313 (1979).
(2) M. Los et al., Farmaco Ed. Sci., 1981, 36, 372.

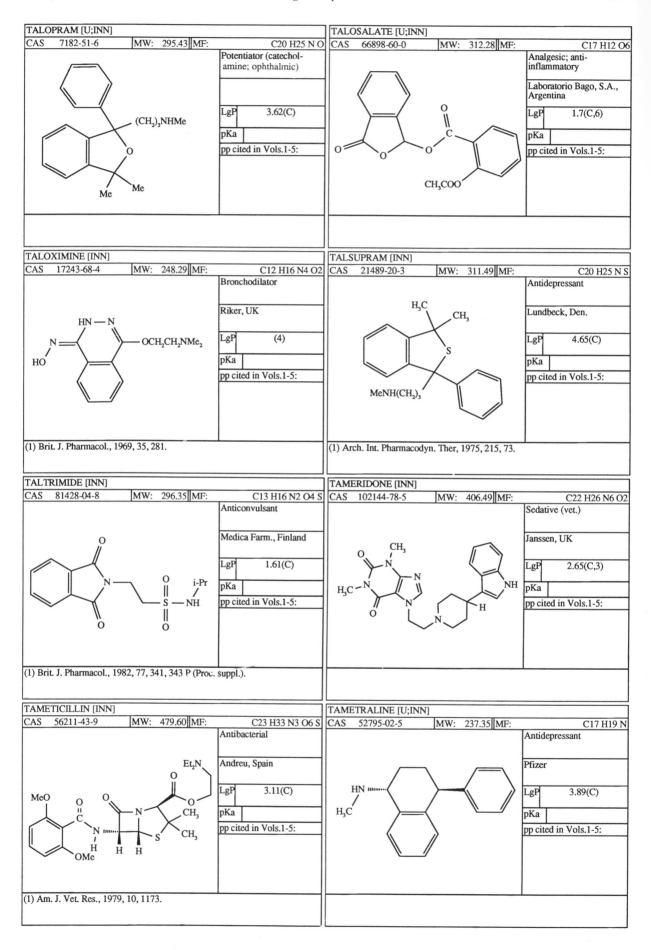

TALOPRAM [U;INN]

CAS	7182-51-6	MW:	295.43	MF:	C20 H25 N O

Potentiator (catechol-amine; ophthalmic)

LgP	3.62(C)
pKa	
pp cited in Vols.1-5:	

TALOSALATE [U;INN]

CAS	66898-60-0	MW:	312.28	MF:	C17 H12 O6

Analgesic; anti-inflammatory

Laboratorio Bago, S.A., Argentina

LgP	1.7(C,6)
pKa	
pp cited in Vols.1-5:	

TALOXIMINE [INN]

CAS	17243-68-4	MW:	248.29	MF:	C12 H16 N4 O2

Bronchodilator

Riker, UK

LgP	(4)
pKa	
pp cited in Vols.1-5:	

(1) Brit. J. Pharmacol., 1969, 35, 281.

TALSUPRAM [INN]

CAS	21489-20-3	MW:	311.49	MF:	C20 H25 N S

Antidepressant

Lundbeck, Den.

LgP	4.65(C)
pKa	
pp cited in Vols.1-5:	

(1) Arch. Int. Pharmacodyn. Ther, 1975, 215, 73.

TALTRIMIDE [INN]

CAS	81428-04-8	MW:	296.35	MF:	C13 H16 N2 O4 S

Anticonvulsant

Medica Farm., Finland

LgP	1.61(C)
pKa	
pp cited in Vols.1-5:	

(1) Brit. J. Pharmacol., 1982, 77, 341, 343 P (Proc. suppl.).

TAMERIDONE [INN]

CAS	102144-78-5	MW:	406.49	MF:	C22 H26 N6 O2

Sedative (vet.)

Janssen, UK

LgP	2.65(C,3)
pKa	
pp cited in Vols.1-5:	

TAMETICILLIN [INN]

CAS	56211-43-9	MW:	479.60	MF:	C23 H33 N3 O6 S

Antibacterial

Andreu, Spain

LgP	3.11(C)
pKa	
pp cited in Vols.1-5:	

(1) Am. J. Vet. Res., 1979, 10, 1173.

TAMETRALINE [U;INN]

CAS	52795-02-5	MW:	237.35	MF:	C17 H19 N

Antidepressant

Pfizer

LgP	3.89(C)
pKa	
pp cited in Vols.1-5:	

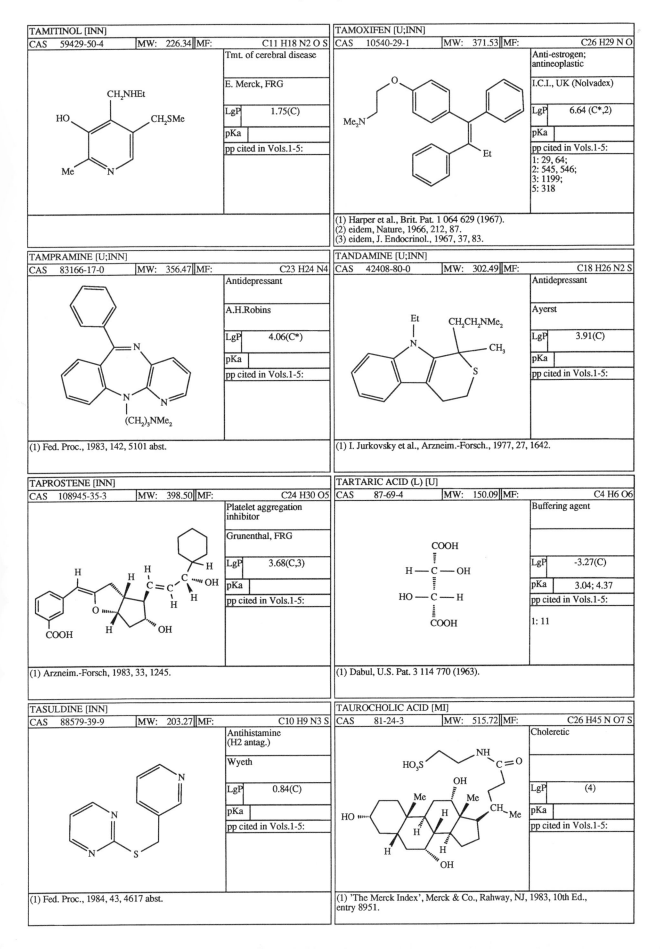

TAMITINOL [INN]

| CAS | 59429-50-4 | MW: | 226.34 | MF: | C11 H18 N2 O S |

Tmt. of cerebral disease

E. Merck, FRG

LgP	1.75(C)
pKa	
pp cited in Vols.1-5:	

TAMOXIFEN [U;INN]

| CAS | 10540-29-1 | MW: | 371.53 | MF: | C26 H29 N O |

Anti-estrogen; antineoplastic

I.C.I., UK (Nolvadex)

LgP	6.64 (C*,2)
pKa	
pp cited in Vols.1-5:	

1: 29, 64;
2: 545, 546;
3: 1199;
5: 318

(1) Harper et al., Brit. Pat. 1 064 629 (1967).
(2) eidem, Nature, 1966, 212, 87.
(3) eidem, J. Endocrinol., 1967, 37, 83.

TAMPRAMINE [U;INN]

| CAS | 83166-17-0 | MW: | 356.47 | MF: | C23 H24 N4 |

Antidepressant

A.H.Robins

LgP	4.06(C*)
pKa	
pp cited in Vols.1-5:	

(1) Fed. Proc., 1983, 142, 5101 abst.

TANDAMINE [U;INN]

| CAS | 42408-80-0 | MW: | 302.49 | MF: | C18 H26 N2 S |

Antidepressant

Ayerst

LgP	3.91(C)
pKa	
pp cited in Vols.1-5:	

(1) I. Jurkovsky et al., Arzneim.-Forsch., 1977, 27, 1642.

TAPROSTENE [INN]

| CAS | 108945-35-3 | MW: | 398.50 | MF: | C24 H30 O5 |

Platelet aggregation inhibitor

Grunenthal, FRG

LgP	3.68(C,3)
pKa	
pp cited in Vols.1-5:	

(1) Arzneim.-Forsch, 1983, 33, 1245.

TARTARIC ACID (L) [U]

| CAS | 87-69-4 | MW: | 150.09 | MF: | C4 H6 O6 |

Buffering agent

LgP	-3.27(C)
pKa	3.04; 4.37
pp cited in Vols.1-5:	

1: 11

(1) Dabul, U.S. Pat. 3 114 770 (1963).

TASULDINE [INN]

| CAS | 88579-39-9 | MW: | 203.27 | MF: | C10 H9 N3 S |

Antihistamine (H2 antag.)

Wyeth

LgP	0.84(C)
pKa	
pp cited in Vols.1-5:	

(1) Fed. Proc., 1984, 43, 4617 abst.

TAUROCHOLIC ACID [MI]

| CAS | 81-24-3 | MW: | 515.72 | MF: | C26 H45 N O7 S |

Choleretic

LgP	(4)
pKa	
pp cited in Vols.1-5:	

(1) 'The Merck Index', Merck & Co., Rahway, NJ, 1983, 10th Ed., entry 8951.

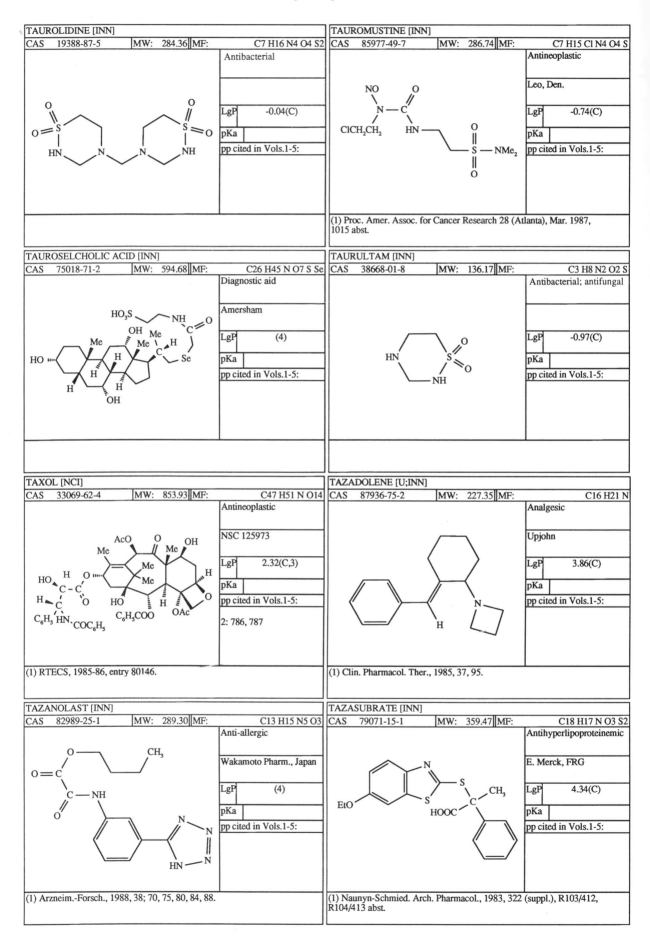

TAUROLIDINE [INN]			
CAS 19388-87-5	MW: 284.36	MF:	C7 H16 N4 O4 S2

Antibacterial

LgP	-0.04(C)
pKa	
pp cited in Vols.1-5:	

TAUROMUSTINE [INN]			
CAS 85977-49-7	MW: 286.74	MF:	C7 H15 Cl N4 O4 S

Antineoplastic

Leo, Den.

LgP	-0.74(C)
pKa	
pp cited in Vols.1-5:	

(1) Proc. Amer. Assoc. for Cancer Research 28 (Atlanta), Mar. 1987, 1015 abst.

TAUROSELCHOLIC ACID [INN]			
CAS 75018-71-2	MW: 594.68	MF:	C26 H45 N O7 S Se

Diagnostic aid

Amersham

LgP	(4)
pKa	
pp cited in Vols.1-5:	

TAURULTAM [INN]			
CAS 38668-01-8	MW: 136.17	MF:	C3 H8 N2 O2 S

Antibacterial; antifungal

LgP	-0.97(C)
pKa	
pp cited in Vols.1-5:	

TAXOL [NCI]			
CAS 33069-62-4	MW: 853.93	MF:	C47 H51 N O14

Antineoplastic

NSC 125973

LgP	2.32(C,3)
pKa	
pp cited in Vols.1-5:	
2: 786, 787	

(1) RTECS, 1985-86, entry 80146.

TAZADOLENE [U;INN]			
CAS 87936-75-2	MW: 227.35	MF:	C16 H21 N

Analgesic

Upjohn

LgP	3.86(C)
pKa	
pp cited in Vols.1-5:	

(1) Clin. Pharmacol. Ther., 1985, 37, 95.

TAZANOLAST [INN]			
CAS 82989-25-1	MW: 289.30	MF:	C13 H15 N5 O3

Anti-allergic

Wakamoto Pharm., Japan

LgP	(4)
pKa	
pp cited in Vols.1-5:	

(1) Arzneim.-Forsch., 1988, 38; 70, 75, 80, 84, 88.

TAZASUBRATE [INN]			
CAS 79071-15-1	MW: 359.47	MF:	C18 H17 N O3 S2

Antihyperlipoproteinemic

E. Merck, FRG

LgP	4.34(C)
pKa	
pp cited in Vols.1-5:	

(1) Naunyn-Schmied. Arch. Pharmacol., 1983, 322 (suppl.), R103/412, R104/413 abst.

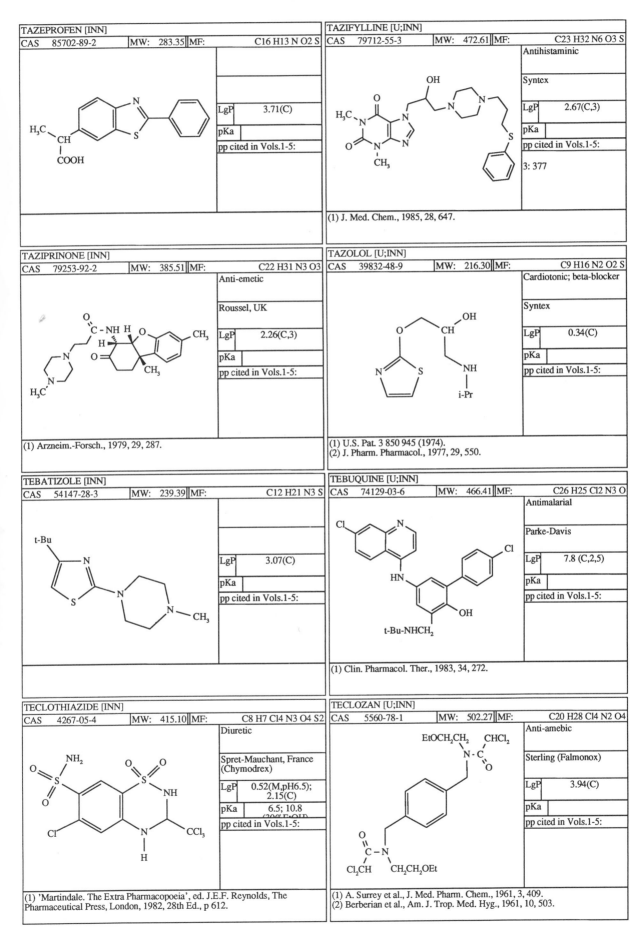

TAZEPROFEN [INN]

CAS	85702-89-2	MW:	283.35	MF:	C16 H13 N O2 S

LgP	3.71(C)
pKa	
pp cited in Vols.1-5:	

TAZIFYLLINE [U;INN]

CAS	79712-55-3	MW:	472.61	MF:	C23 H32 N6 O3 S

Antihistaminic

Syntex

LgP	2.67(C,3)
pKa	
pp cited in Vols.1-5:	
3: 377	

(1) J. Med. Chem., 1985, 28, 647.

TAZIPRINONE [INN]

CAS	79253-92-2	MW:	385.51	MF:	C22 H31 N3 O3

Anti-emetic

Roussel, UK

LgP	2.26(C,3)
pKa	
pp cited in Vols.1-5:	

(1) Arzneim.-Forsch., 1979, 29, 287.

TAZOLOL [U;INN]

CAS	39832-48-9	MW:	216.30	MF:	C9 H16 N2 O2 S

Cardiotonic; beta-blocker

Syntex

LgP	0.34(C)
pKa	
pp cited in Vols.1-5:	

(1) U.S. Pat. 3 850 945 (1974).
(2) J. Pharm. Pharmacol., 1977, 29, 550.

TEBATIZOLE [INN]

CAS	54147-28-3	MW:	239.39	MF:	C12 H21 N3 S

LgP	3.07(C)
pKa	
pp cited in Vols.1-5:	

TEBUQUINE [U;INN]

CAS	74129-03-6	MW:	466.41	MF:	C26 H25 Cl2 N3 O

Antimalarial

Parke-Davis

LgP	7.8 (C,2,5)
pKa	
pp cited in Vols.1-5:	

(1) Clin. Pharmacol. Ther., 1983, 34, 272.

TECLOTHIAZIDE [INN]

CAS	4267-05-4	MW:	415.10	MF:	C8 H7 Cl4 N3 O4 S2

Diuretic

Spret-Mauchant, France (Chymodrex)

LgP	0.52(M,pH6.5); 2.15(C)
pKa	6.5; 10.8 (20%EtOH)
pp cited in Vols.1-5:	

(1) 'Martindale. The Extra Pharmacopoeia', ed. J.E.F. Reynolds, The Pharmaceutical Press, London, 1982, 28th Ed., p 612.

TECLOZAN [U;INN]

CAS	5560-78-1	MW:	502.27	MF:	C20 H28 Cl4 N2 O4

Anti-amebic

Sterling (Falmonox)

LgP	3.94(C)
pKa	
pp cited in Vols.1-5:	

(1) A. Surrey et al., J. Med. Pharm. Chem., 1961, 3, 409.
(2) Berberian et al., Am. J. Trop. Med. Hyg., 1961, 10, 503.

TEDISAMIL [INN]

CAS 90961-53-8	MW: 288.48	MF:	C19 H32 N2

Anti-ischemic; bradycardic

Carlo Erba, Italy

LgP 3.70 (C)

pKa

pp cited in Vols.1-5:

(1) Abst. Spr. Mtg. Deut. Ges. Pharmakol.-Toxicol., 29, (Mainz), 1988, 272.

TEFAZOLINE [INN]

CAS 1082-56-0	MW: 214.31	MF:	C14 H18 N2

Vasoconstrictor

U.P.B., Belg. (Tenaphto)

LgP 4.22(C)

pKa

pp cited in Vols.1-5:

TEFENPERATE [INN]

CAS 77342-26-8	MW: 534.53	MF:	C29 H37 Cl2 N O4

LgP 8.5 (C,2,5)

pKa

pp cited in Vols.1-5:

TEFLUDAZINE [INN]

CAS 80680-06-4	MW: 408.44	MF:	C22 H24 F4 N2 O

Neuroleptic

Lundbeck, Den.

LgP 4.39(M); 4.80(C)

pKa 7.39

pp cited in Vols.1-5:

(1) J. Med. Chem., 1983, 26, 936.

TEFLURANE [U;INN]

CAS 124-72-1	MW: 180.93	MF:	C2 H Br F4

Anesthetic (inhalation)

LgP 2.01(C)

pKa

pp cited in Vols.1-5:

(1) Larsen, U.S. Pat. 2 971 990 (1961).
(2) Black et al., Brit. J. Anaesth., 1969, 41, 288.

TEFLUTIXOL [INN]

CAS 55837-23-5	MW: 454.53	MF:	C23 H26 F4 N2 O S

Neuroleptic

Lundbeck, Den.

LgP 6.24(C)

pKa 6.94

pp cited in Vols.1-5:

(1) J. Pharmacol., 1975, 6, 277.

TEGAFUR [U;INN]

CAS 17902-23-7	MW: 200.17	MF:	C8 H9 F N2 O3

Antineoplastic

Taiho, Japan

LgP -0.27(M); 0.08(C)

pKa 7.62

pp cited in Vols.1-5:

2: 312

(1) S.A. Hillers et al., Dokl. Akad. Nauk. SSSR, 1967, 176, 332.
(2) eidem, Brit. Pat. 1 168 391 (1969).
(3) M. Yasumoto et al., J. Med. Chem., 1978, 21, 738.

TELENZEPINE [INN]

CAS 80880-90-6	MW: 370.48	MF:	C19 H22 N4 O2 S

Antisecretory

Byk Gulden, FRG

LgP 1.97(C)

pKa

pp cited in Vols.1-5:

3: 438, 474, 475

(1) Eur. J. Pharmacol., 1985, 112, 211.

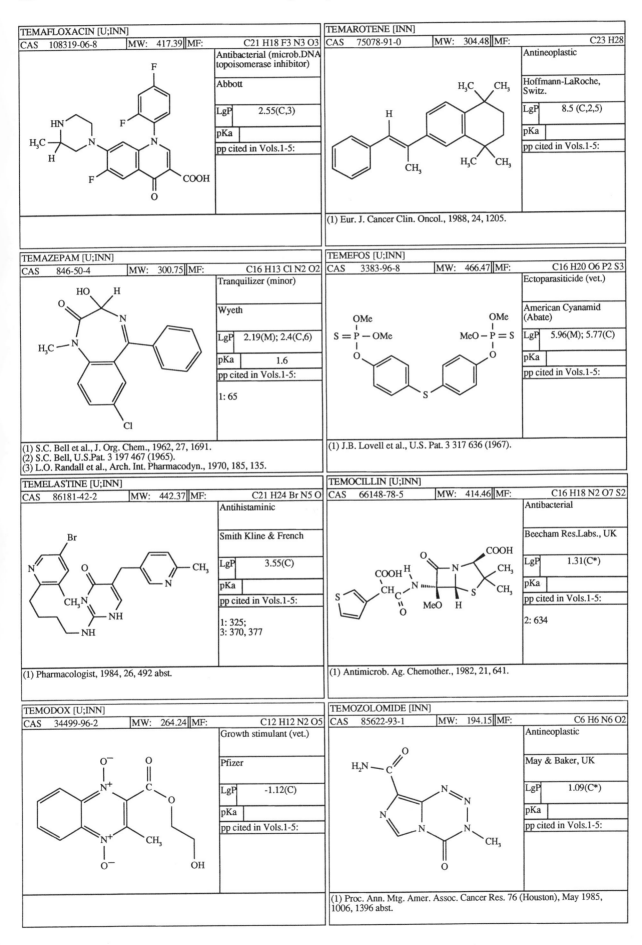

TEMAFLOXACIN [U;INN]

CAS	108319-06-8	MW:	417.39	MF:	C21 H18 F3 N3 O3

Antibacterial (microb.DNA topoisomerase inhibitor)

Abbott

LgP	2.55(C,3)
pKa	
pp cited in Vols.1-5:	

TEMAROTENE [INN]

CAS	75078-91-0	MW:	304.48	MF:	C23 H28

Antineoplastic

Hoffmann-LaRoche, Switz.

LgP	8.5 (C,2,5)
pKa	
pp cited in Vols.1-5:	

(1) Eur. J. Cancer Clin. Oncol., 1988, 24, 1205.

TEMAZEPAM [U;INN]

CAS	846-50-4	MW:	300.75	MF:	C16 H13 Cl N2 O2

Tranquilizer (minor)

Wyeth

LgP	2.19(M); 2.4(C,6)
pKa	1.6
pp cited in Vols.1-5:	

1: 65

(1) S.C. Bell et al., J. Org. Chem., 1962, 27, 1691.
(2) S.C. Bell, U.S.Pat. 3 197 467 (1965).
(3) L.O. Randall et al., Arch. Int. Pharmacodyn., 1970, 185, 135.

TEMEFOS [U;INN]

CAS	3383-96-8	MW:	466.47	MF:	C16 H20 O6 P2 S3

Ectoparasiticide (vet.)

American Cyanamid (Abate)

LgP	5.96(M); 5.77(C)
pKa	
pp cited in Vols.1-5:	

(1) J.B. Lovell et al., U.S. Pat. 3 317 636 (1967).

TEMELASTINE [U;INN]

CAS	86181-42-2	MW:	442.37	MF:	C21 H24 Br N5 O

Antihistaminic

Smith Kline & French

LgP	3.55(C)
pKa	
pp cited in Vols.1-5:	

1: 325;
3: 370, 377

(1) Pharmacologist, 1984, 26, 492 abst.

TEMOCILLIN [U;INN]

CAS	66148-78-5	MW:	414.46	MF:	C16 H18 N2 O7 S2

Antibacterial

Beecham Res.Labs., UK

LgP	1.31(C*)
pKa	
pp cited in Vols.1-5:	

2: 634

(1) Antimicrob. Ag. Chemother., 1982, 21, 641.

TEMODOX [U;INN]

CAS	34499-96-2	MW:	264.24	MF:	C12 H12 N2 O5

Growth stimulant (vet.)

Pfizer

LgP	-1.12(C)
pKa	
pp cited in Vols.1-5:	

TEMOZOLOMIDE [INN]

CAS	85622-93-1	MW:	194.15	MF:	C6 H6 N6 O2

Antineoplastic

May & Baker, UK

LgP	1.09(C*)
pKa	
pp cited in Vols.1-5:	

(1) Proc. Ann. Mtg. Amer. Assoc. Cancer Res. 76 (Houston), May 1985, 1006, 1396 abst.

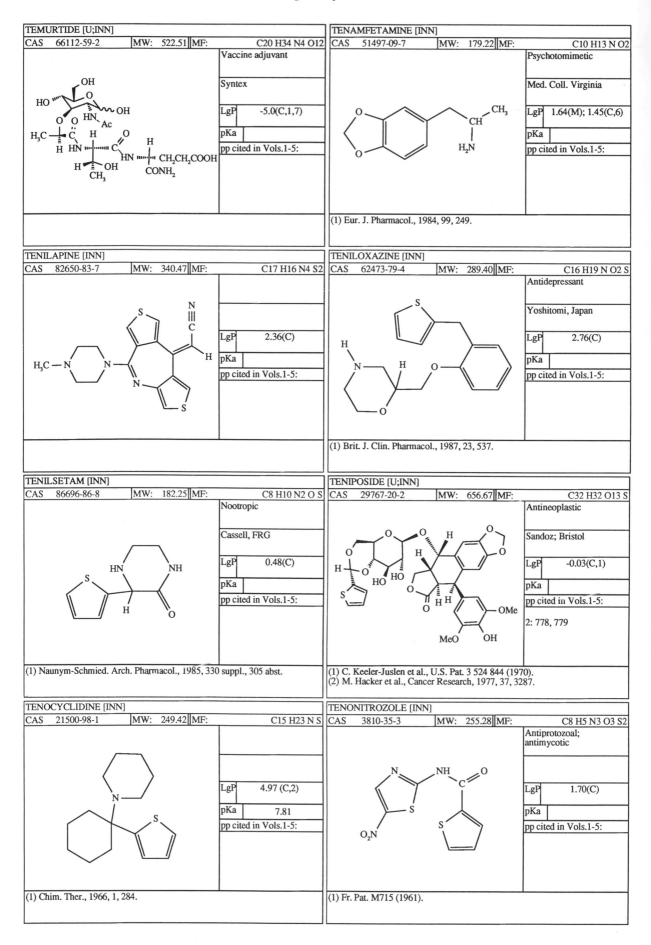

TEMURTIDE [U;INN]

CAS 66112-59-2	MW: 522.51	MF: C20 H34 N4 O12

Vaccine adjuvant

Syntex

LgP   -5.0(C,1,7)

pKa

pp cited in Vols.1-5:

TENAMFETAMINE [INN]

CAS 51497-09-7	MW: 179.22	MF: C10 H13 N O2

Psychotomimetic

Med. Coll. Virginia

LgP   1.64(M); 1.45(C,6)

pKa

pp cited in Vols.1-5:

(1) Eur. J. Pharmacol., 1984, 99, 249.

TENILAPINE [INN]

CAS 82650-83-7	MW: 340.47	MF: C17 H16 N4 S2

LgP   2.36(C)

pKa

pp cited in Vols.1-5:

TENILOXAZINE [INN]

CAS 62473-79-4	MW: 289.40	MF: C16 H19 N O2 S

Antidepressant

Yoshitomi, Japan

LgP   2.76(C)

pKa

pp cited in Vols.1-5:

(1) Brit. J. Clin. Pharmacol., 1987, 23, 537.

TENILSETAM [INN]

CAS 86696-86-8	MW: 182.25	MF: C8 H10 N2 O S

Nootropic

Cassell, FRG

LgP   0.48(C)

pKa

pp cited in Vols.1-5:

(1) Naunym-Schmied. Arch. Pharmacol., 1985, 330 suppl., 305 abst.

TENIPOSIDE [U;INN]

CAS 29767-20-2	MW: 656.67	MF: C32 H32 O13 S

Antineoplastic

Sandoz; Bristol

LgP   -0.03(C,1)

pKa

pp cited in Vols.1-5:

2: 778, 779

(1) C. Keeler-Juslen et al., U.S. Pat. 3 524 844 (1970).
(2) M. Hacker et al., Cancer Research, 1977, 37, 3287.

TENOCYCLIDINE [INN]

CAS 21500-98-1	MW: 249.42	MF: C15 H23 N S

LgP   4.97 (C,2)

pKa   7.81

pp cited in Vols.1-5:

(1) Chim. Ther., 1966, 1, 284.

TENONITROZOLE [INN]

CAS 3810-35-3	MW: 255.28	MF: C8 H5 N3 O3 S2

Antiprotozoal;
antimycotic

LgP   1.70(C)

pKa

pp cited in Vols.1-5:

(1) Fr. Pat. M715 (1961).

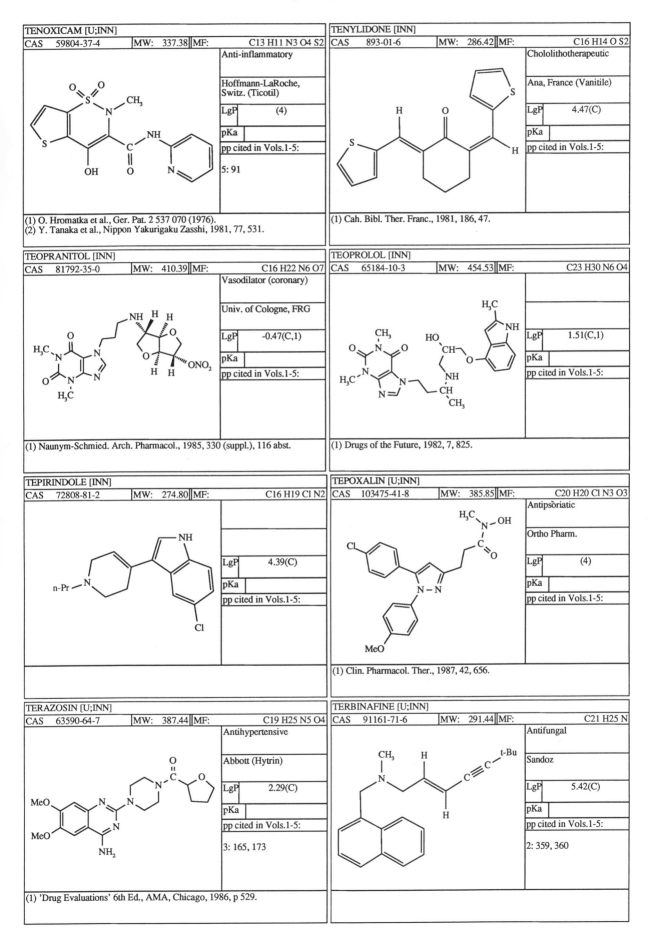

TENOXICAM [U;INN]

| CAS | 59804-37-4 | MW: | 337.38 | MF: | C13 H11 N3 O4 S2 |

Anti-inflammatory

Hoffmann-LaRoche, Switz. (Ticotil)

LgP (4)

pKa

pp cited in Vols.1-5:

5: 91

(1) O. Hromatka et al., Ger. Pat. 2 537 070 (1976).
(2) Y. Tanaka et al., Nippon Yakurigaku Zasshi, 1981, 77, 531.

TENYLIDONE [INN]

| CAS | 893-01-6 | MW: | 286.42 | MF: | C16 H14 O S2 |

Chololithotherapeutic

Ana, France (Vanitile)

LgP 4.47(C)

pKa

pp cited in Vols.1-5:

(1) Cah. Bibl. Ther. Franc., 1981, 186, 47.

TEOPRANITOL [INN]

| CAS | 81792-35-0 | MW: | 410.39 | MF: | C16 H22 N6 O7 |

Vasodilator (coronary)

Univ. of Cologne, FRG

LgP -0.47(C,1)

pKa

pp cited in Vols.1-5:

(1) Naunym-Schmied. Arch. Pharmacol., 1985, 330 (suppl.), 116 abst.

TEOPROLOL [INN]

| CAS | 65184-10-3 | MW: | 454.53 | MF: | C23 H30 N6 O4 |

LgP 1.51(C,1)

pKa

pp cited in Vols.1-5:

(1) Drugs of the Future, 1982, 7, 825.

TEPIRINDOLE [INN]

| CAS | 72808-81-2 | MW: | 274.80 | MF: | C16 H19 Cl N2 |

LgP 4.39(C)

pKa

pp cited in Vols.1-5:

TEPOXALIN [U;INN]

| CAS | 103475-41-8 | MW: | 385.85 | MF: | C20 H20 Cl N3 O3 |

Antipsoriatic

Ortho Pharm.

LgP (4)

pKa

pp cited in Vols.1-5:

(1) Clin. Pharmacol. Ther., 1987, 42, 656.

TERAZOSIN [U;INN]

| CAS | 63590-64-7 | MW: | 387.44 | MF: | C19 H25 N5 O4 |

Antihypertensive

Abbott (Hytrin)

LgP 2.29(C)

pKa

pp cited in Vols.1-5:

3: 165, 173

(1) 'Drug Evaluations' 6th Ed., AMA, Chicago, 1986, p 529.

TERBINAFINE [U;INN]

| CAS | 91161-71-6 | MW: | 291.44 | MF: | C21 H25 N |

Antifungal

Sandoz

LgP 5.42(C)

pKa

pp cited in Vols.1-5:

2: 359, 360

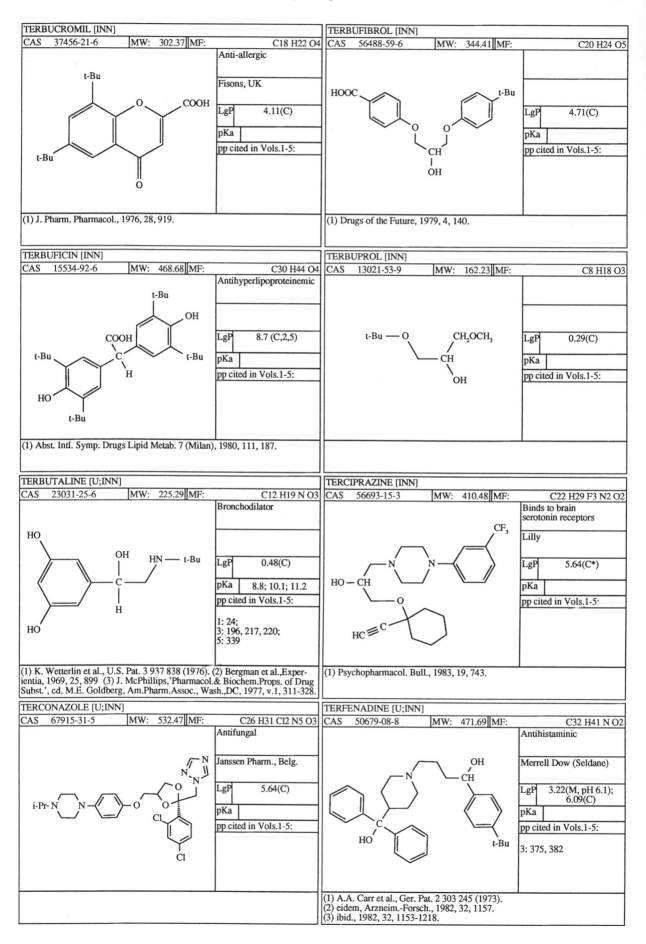

TERBUCROMIL [INN]

CAS	37456-21-6	MW:	302.37	MF:	C18 H22 O4

Anti-allergic

Fisons, UK

LgP	4.11(C)
pKa	
pp cited in Vols.1-5:	

t-Bu

COOH

t-Bu

(1) J. Pharm. Pharmacol., 1976, 28, 919.

TERBUFIBROL [INN]

CAS	56488-59-6	MW:	344.41	MF:	C20 H24 O5

HOOC   t-Bu

LgP	4.71(C)
pKa	
pp cited in Vols.1-5:	

(1) Drugs of the Future, 1979, 4, 140.

TERBUFICIN [INN]

CAS	15534-92-6	MW:	468.68	MF:	C30 H44 O4

Antihyperlipoproteinemic

t-Bu

OH

COOH

t-Bu   t-Bu

H

HO

t-Bu

LgP	8.7 (C,2,5)
pKa	
pp cited in Vols.1-5:	

(1) Abst. Intl. Symp. Drugs Lipid Metab. 7 (Milan), 1980, 111, 187.

TERBUPROL [INN]

CAS	13021-53-9	MW:	162.23	MF:	C8 H18 O3

t-Bu — O CH_2OCH_3

CH

OH

LgP	0.29(C)
pKa	
pp cited in Vols.1-5:	

TERBUTALINE [U;INN]

CAS	23031-25-6	MW:	225.29	MF:	C12 H19 N O3

Bronchodilator

HO

OH   HN — t-Bu

C

H

HO

LgP	0.48(C)
pKa	8.8; 10.1; 11.2
pp cited in Vols.1-5:	

1: 24;
3: 196, 217, 220;
5: 339

(1) K. Wetterlin et al., U.S. Pat. 3 937 838 (1976). (2) Bergman et al.,Experientia, 1969, 25, 899. (3) J. McPhillips,'Pharmacol.& Biochem.Props. of Drug Subst.', cd. M.E. Goldberg, Am.Pharm.Assoc., Wash.,DC, 1977, v.1, 311-328.

TERCIPRAZINE [INN]

CAS	56693-15-3	MW:	410.48	MF:	C22 H29 F3 N2 O2

Binds to brain serotonin receptors

Lilly

CF_3

HO — CH

N   N

O

HC≡C

LgP	5.64(C*)
pKa	
pp cited in Vols.1-5:	

(1) Psychopharmacol. Bull., 1983, 19, 743.

TERCONAZOLE [U;INN]

CAS	67915-31-5	MW:	532.47	MF:	C26 H31 Cl2 N5 O3

Antifungal

Janssen Pharm., Belg.

i-Pr-N   N   O   N

N

N

O

O

Cl

Cl

LgP	5.64(C)
pKa	
pp cited in Vols.1-5:	

TERFENADINE [U;INN]

CAS	50679-08-8	MW:	471.69	MF:	C32 H41 N O2

Antihistaminic

Merrell Dow (Seldane)

OH

N   CH

C

HO

t-Bu

LgP	3.22(M, pH 6.1); 6.09(C)
pKa	
pp cited in Vols.1-5:	

3: 375, 382

(1) A.A. Carr et al., Ger. Pat. 2 303 245 (1973).
(2) eidem, Arzneim.-Forsch., 1982, 32, 1157.
(3) ibid., 1982, 32, 1153-1218.

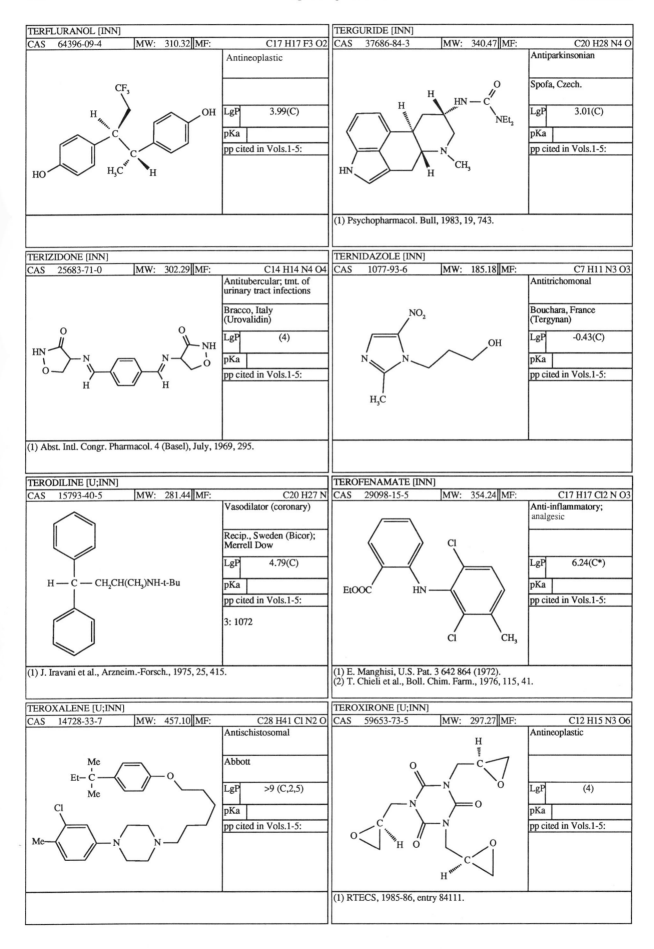

TERFLURANOL [INN]

CAS	64396-09-4	MW:	310.32	MF:	C17 H17 F3 O2

Antineoplastic

LgP	3.99(C)
pKa	
pp cited in Vols.1-5:	

TERGURIDE [INN]

CAS	37686-84-3	MW:	340.47	MF:	C20 H28 N4 O

Antiparkinsonian

Spofa, Czech.

LgP	3.01(C)
pKa	
pp cited in Vols.1-5:	

(1) Psychopharmacol. Bull, 1983, 19, 743.

TERIZIDONE [INN]

CAS	25683-71-0	MW:	302.29	MF:	C14 H14 N4 O4

Antitubercular; tmt. of urinary tract infections

Bracco, Italy (Urovalidin)

LgP	(4)
pKa	
pp cited in Vols.1-5:	

(1) Abst. Intl. Congr. Pharmacol. 4 (Basel), July, 1969, 295.

TERNIDAZOLE [INN]

CAS	1077-93-6	MW:	185.18	MF:	C7 H11 N3 O3

Antitrichomonal

Bouchara, France (Tergynan)

LgP	-0.43(C)
pKa	
pp cited in Vols.1-5:	

TERODILINE [U;INN]

CAS	15793-40-5	MW:	281.44	MF:	C20 H27 N

Vasodilator (coronary)

Recip., Sweden (Bicor); Merrell Dow

LgP	4.79(C)
pKa	
pp cited in Vols.1-5:	
3: 1072	

(1) J. Iravani et al., Arzneim.-Forsch., 1975, 25, 415.

TEROFENAMATE [INN]

CAS	29098-15-5	MW:	354.24	MF:	C17 H17 Cl2 N O3

Anti-inflammatory; analgesic

LgP	6.24(C*)
pKa	
pp cited in Vols.1-5:	

(1) E. Manghisi, U.S. Pat. 3 642 864 (1972).
(2) T. Chieli et al., Boll. Chim. Farm., 1976, 115, 41.

TEROXALENE [U;INN]

CAS	14728-33-7	MW:	457.10	MF:	C28 H41 Cl N2 O

Antischistosomal

Abbott

LgP	>9 (C,2,5)
pKa	
pp cited in Vols.1-5:	

TEROXIRONE [U;INN]

CAS	59653-73-5	MW:	297.27	MF:	C12 H15 N3 O6

Antineoplastic

LgP	(4)
pKa	
pp cited in Vols.1-5:	

(1) RTECS, 1985-86, entry 84111.

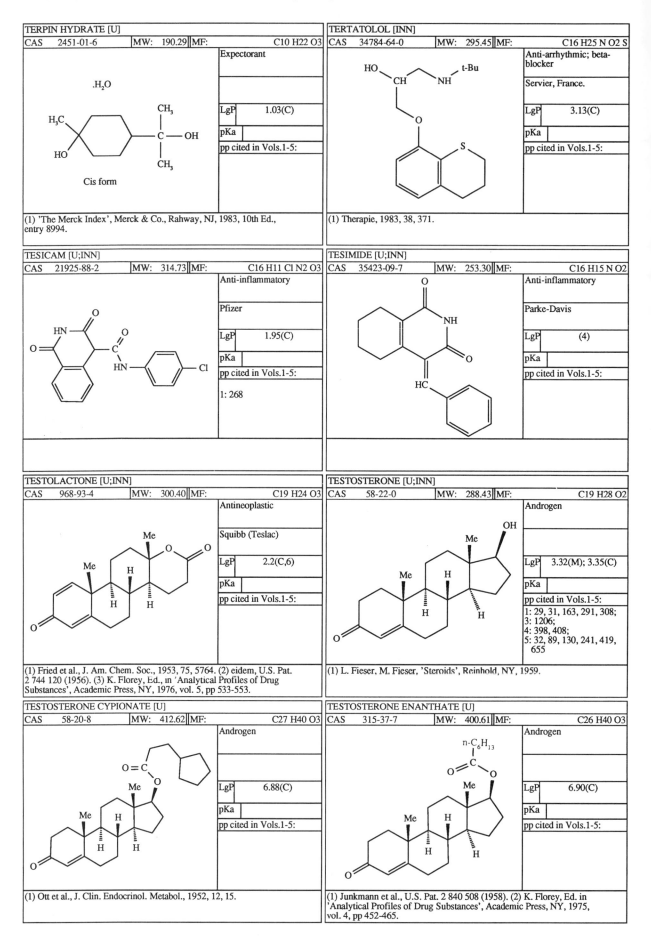

TERPIN HYDRATE [U]			
CAS 2451-01-6	MW: 190.29	MF:	C10 H22 O3
		Expectorant	
		LgP	1.03(C)
		pKa	
		pp cited in Vols.1-5:	

.H₂O

Cis form

(1) 'The Merck Index', Merck & Co., Rahway, NJ, 1983, 10th Ed., entry 8994.

TERTATOLOL [INN]			
CAS 34784-64-0	MW: 295.45	MF:	C16 H25 N O2 S
		Anti-arrhythmic; beta-blocker	
		Servier, France.	
		LgP	3.13(C)
		pKa	
		pp cited in Vols.1-5:	

(1) Therapie, 1983, 38, 371.

TESICAM [U;INN]			
CAS 21925-88-2	MW: 314.73	MF:	C16 H11 Cl N2 O3
		Anti-inflammatory	
		Pfizer	
		LgP	1.95(C)
		pKa	
		pp cited in Vols.1-5:	
		1: 268	

TESIMIDE [U;INN]			
CAS 35423-09-7	MW: 253.30	MF:	C16 H15 N O2
		Anti-inflammatory	
		Parke-Davis	
		LgP	(4)
		pKa	
		pp cited in Vols.1-5:	

TESTOLACTONE [U;INN]			
CAS 968-93-4	MW: 300.40	MF:	C19 H24 O3
		Antineoplastic	
		Squibb (Teslac)	
		LgP	2.2(C,6)
		pKa	
		pp cited in Vols.1-5:	

(1) Fried et al., J. Am. Chem. Soc., 1953, 75, 5764. (2) eidem, U.S. Pat. 2 744 120 (1956). (3) K. Florey, Ed., in 'Analytical Profiles of Drug Substances', Academic Press, NY, 1976, vol. 5, pp 533-553.

TESTOSTERONE [U;INN]			
CAS 58-22-0	MW: 288.43	MF:	C19 H28 O2
		Androgen	
		LgP	3.32(M); 3.35(C)
		pKa	
		pp cited in Vols.1-5:	
		1: 29, 31, 163, 291, 308; 3: 1206; 4: 398, 408; 5: 32, 89, 130, 241, 419, 655	

(1) L. Fieser, M. Fieser, 'Steroids', Reinhold, NY, 1959.

TESTOSTERONE CYPIONATE [U]			
CAS 58-20-8	MW: 412.62	MF:	C27 H40 O3
		Androgen	
		LgP	6.88(C)
		pKa	
		pp cited in Vols.1-5:	

(1) Ott et al., J. Clin. Endocrinol. Metabol., 1952, 12, 15.

TESTOSTERONE ENANTHATE [U]			
CAS 315-37-7	MW: 400.61	MF:	C26 H40 O3
		Androgen	
		LgP	6.90(C)
		pKa	
		pp cited in Vols.1-5:	

(1) Junkmann et al., U.S. Pat. 2 840 508 (1958). (2) K. Florey, Ed. in 'Analytical Profiles of Drug Substances', Academic Press, NY, 1975, vol. 4, pp 452-465.

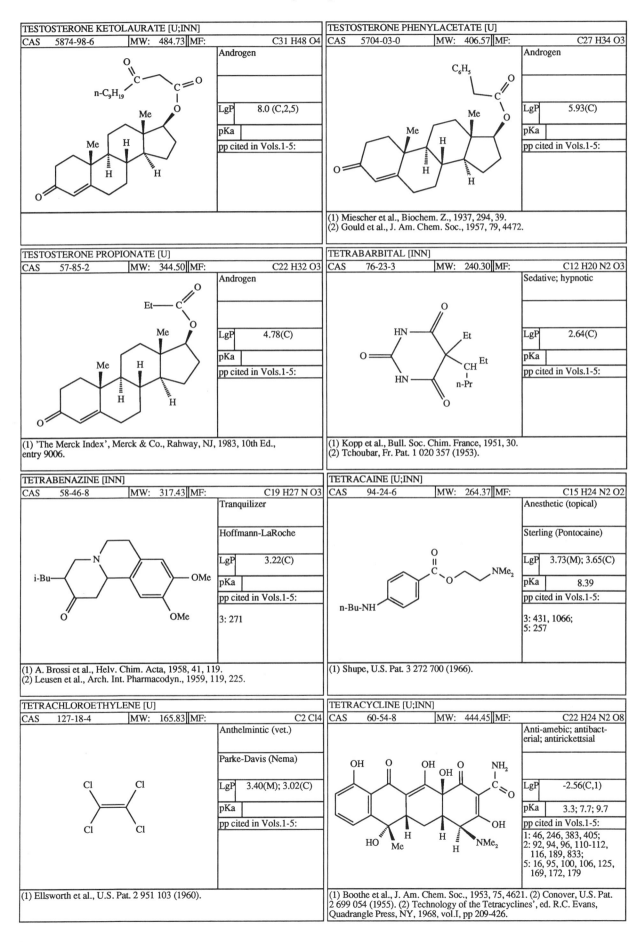

TESTOSTERONE KETOLAURATE [U;INN]			
CAS 5874-98-6	MW: 484.73	MF:	C31 H48 O4

Androgen

LgP	8.0 (C,2,5)
pKa	
pp cited in Vols.1-5:	

TESTOSTERONE PHENYLACETATE [U]			
CAS 5704-03-0	MW: 406.57	MF:	C27 H34 O3

Androgen

LgP	5.93(C)
pKa	
pp cited in Vols.1-5:	

(1) Miescher et al., Biochem. Z., 1937, 294, 39.
(2) Gould et al., J. Am. Chem. Soc., 1957, 79, 4472.

TESTOSTERONE PROPIONATE [U]			
CAS 57-85-2	MW: 344.50	MF:	C22 H32 O3

Androgen

LgP	4.78(C)
pKa	
pp cited in Vols.1-5:	

(1) 'The Merck Index', Merck & Co., Rahway, NJ, 1983, 10th Ed., entry 9006.

TETRABARBITAL [INN]			
CAS 76-23-3	MW: 240.30	MF:	C12 H20 N2 O3

Sedative; hypnotic

LgP	2.64(C)
pKa	
pp cited in Vols.1-5:	

(1) Kopp et al., Bull. Soc. Chim. France, 1951, 30.
(2) Tchoubar, Fr. Pat. 1 020 357 (1953).

TETRABENAZINE [INN]			
CAS 58-46-8	MW: 317.43	MF:	C19 H27 N O3

Tranquilizer

Hoffmann-LaRoche

LgP	3.22(C)
pKa	
pp cited in Vols.1-5:	
3: 271	

(1) A. Brossi et al., Helv. Chim. Acta, 1958, 41, 119.
(2) Leusen et al., Arch. Int. Pharmacodyn., 1959, 119, 225.

TETRACAINE [U;INN]			
CAS 94-24-6	MW: 264.37	MF:	C15 H24 N2 O2

Anesthetic (topical)

Sterling (Pontocaine)

LgP	3.73(M); 3.65(C)
pKa	8.39
pp cited in Vols.1-5:	
3: 431, 1066; 5: 257	

(1) Shupe, U.S. Pat. 3 272 700 (1966).

TETRACHLOROETHYLENE [U]			
CAS 127-18-4	MW: 165.83	MF:	C2 Cl4

Anthelmintic (vet.)

Parke-Davis (Nema)

LgP	3.40(M); 3.02(C)
pKa	
pp cited in Vols.1-5:	

(1) Ellsworth et al., U.S. Pat. 2 951 103 (1960).

TETRACYCLINE [U;INN]			
CAS 60-54-8	MW: 444.45	MF:	C22 H24 N2 O8

Anti-amebic; antibacterial; antirickettsial

LgP	-2.56(C,1)
pKa	3.3; 7.7; 9.7
pp cited in Vols.1-5:	
1: 46, 246, 383, 405; 2: 92, 94, 96, 110-112, 116, 189, 833; 5: 16, 95, 100, 106, 125, 169, 172, 179	

(1) Boothe et al., J. Am. Chem. Soc., 1953, 75, 4621. (2) Conover, U.S. Pat. 2 699 054 (1955). (2) Technology of the Tetracyclines', ed. R.C. Evans, Quadrangle Press, NY, 1968, vol.I, pp 209-426.

TETRADONIUM BROMIDE [INN]			
CAS 1119-97-7	MW: 336.41	MF:	C17 H38 Br N
		Disinfectant	
		LgP	(4)
		pKa	
		pp cited in Vols.1-5:	

n-C14H29 — N+ — CH3 (with Me, Me substituents) Br−

(1) 'Martindale. The Extra Pharmacopoeia', ed. J.E.F. Reynolds, The Pharmaceutical Press, London, 1982, 28th Ed., pp 551-553.

TETRAETHYLAMMONIUM CHLORIDE [MI]			
CAS 56-34-8	MW: 165.71	MF:	C8 H20 Cl N
		Ganglionic blocker	
		LgP	(4)
		pKa	
		pp cited in Vols.1-5:	
		3: 1087	

(1) Moe et al., Pharmacol. Rev., 1950, 2, 61-95.

TETRAHYDROZOLINE [U;INN]			
CAS 84-22-0	MW: 200.29	MF:	C13 H16 N2
		Adrenergic (vasoconstrictor)	
		Sahyun Labs.	
		LgP	3.54(C)
		pKa	10.51
		pp cited in Vols.1-5:	
		3: 151, 173	

(1) M. Synerholm et al., U.S. Pat. 2 731 471 (1956).

TETRAMETHRIN [INN]			
CAS 7696-12-0	MW: 331.42	MF:	C19 H25 N O4
		Insecticide (vet.)	
		Sumitomo, Japan	
		LgP	4.73(M); 4.20(C)
		pKa	
		pp cited in Vols.1-5:	
		3: 1062	

(1) T. Kato et al., U.S. Pat. 3 268 398 (1966).

TETRAMISOLE [U;INN]			
CAS 5036-02-2	MW: 204.30	MF:	C11 H12 N2 S
		Anthelmintic	
		Janssen Pharm., Belg.	
		LgP	(4)
		pKa	
		pp cited in Vols.1-5:	

(1) Raeymaekers et al., U.S. Pat. 3 274 209 (1966). (2) eidem., J.Med.Chem. 1966, 9, 545. (3) J.Symoens et al.,'Pharmacol.& Biochem.Props. of Drug Subst.',ed.M.E.Goldberg, Am.Pharm.Assoc.,Wash.,DC, 1979, v.2, 407-464.

TETRANDRINE [NCI]			
CAS 518-34-3	MW: 622.77	MF:	C38 H42 N2 O6
		Analgesic; antipyretic; antineoplastic	
		NSC 077037	
		LgP	8.4(C,2,5)
		pKa	
		pp cited in Vols.1-5:	

(1) Fujita et al., J. Pharm. Soc. Japan, 1951, 71, 1039.
(2) Inubushi et al., Tetrahedron Lett., 1968, 3399.

TETRANTOIN [MI]			
CAS 52094-70-9	MW: 216.24	MF:	C12 H12 N2 O2
		Anticonvulsant	
		Cutter Labs.	
		LgP	0.66(C)
		pKa	
		pp cited in Vols.1-5:	

(1) Faust et al., J. Am. Pharm. Assoc., 1957, 46, 118.
(2) Jules et al., U.S. Pat. 2 716 648 (1955).

TETRAZEPAM [INN]			
CAS 10379-14-3	MW: 288.78	MF:	C16 H17 Cl N2 O
		Tranquilizer; muscle relaxant	
		Clin-Byla, France	
		LgP	3.20(M); 3.90(C*)
		pKa	
		pp cited in Vols.1-5:	

(1) Schmitt et al., Chim. Ther., 1967, 2, 254.

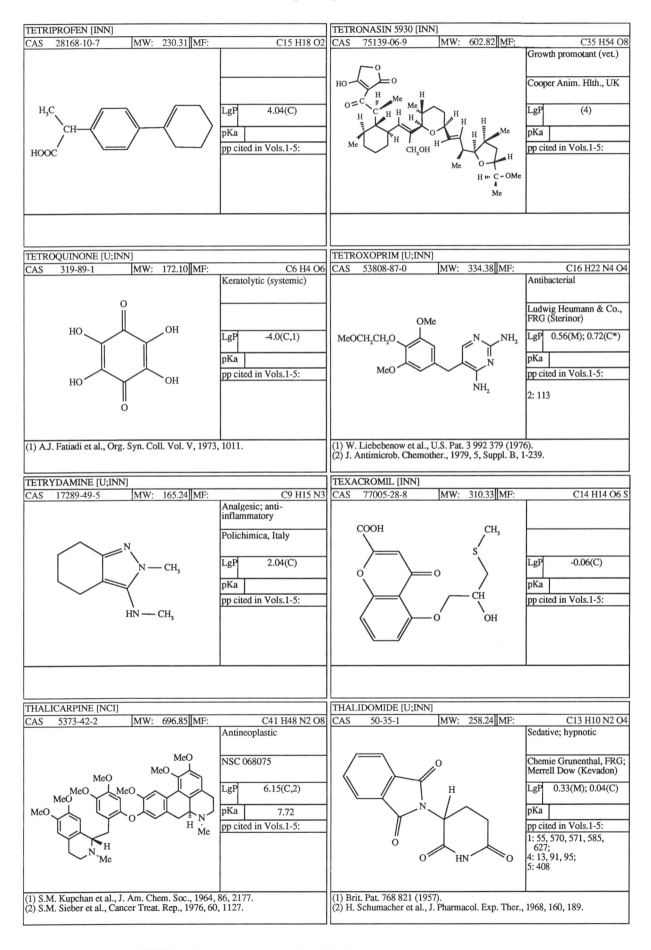

TETRIPROFEN [INN]

CAS	28168-10-7	MW:	230.31	MF:	C15 H18 O2

LgP	4.04(C)
pKa	
pp cited in Vols.1-5:	

TETRONASIN 5930 [INN]

CAS	75139-06-9	MW:	602.82	MF:	C35 H54 O8

Growth promotant (vet.)

Cooper Anim. Hlth., UK

LgP	(4)
pKa	
pp cited in Vols.1-5:	

TETROQUINONE [U;INN]

CAS	319-89-1	MW:	172.10	MF:	C6 H4 O6

Keratolytic (systemic)

LgP	-4.0(C,1)
pKa	
pp cited in Vols.1-5:	

(1) A.J. Fatiadi et al., Org. Syn. Coll. Vol. V, 1973, 1011.

TETROXOPRIM [U;INN]

CAS	53808-87-0	MW:	334.38	MF:	C16 H22 N4 O4

Antibacterial

Ludwig Heumann & Co., FRG (Sterinor)

LgP	0.56(M); 0.72(C*)
pKa	
pp cited in Vols.1-5:	
2: 113	

(1) W. Liebebenow et al., U.S. Pat. 3 992 379 (1976).
(2) J. Antimicrob. Chemother., 1979, 5, Suppl. B, 1-239.

TETRYDAMINE [U;INN]

CAS	17289-49-5	MW:	165.24	MF:	C9 H15 N3

Analgesic; anti-inflammatory

Polichimica, Italy

LgP	2.04(C)
pKa	
pp cited in Vols.1-5:	

TEXACROMIL [INN]

CAS	77005-28-8	MW:	310.33	MF:	C14 H14 O6 S

LgP	-0.06(C)
pKa	
pp cited in Vols.1-5:	

THALICARPINE [NCI]

CAS	5373-42-2	MW:	696.85	MF:	C41 H48 N2 O8

Antineoplastic

NSC 068075

LgP	6.15(C,2)
pKa	7.72
pp cited in Vols.1-5:	

(1) S.M. Kupchan et al., J. Am. Chem. Soc., 1964, 86, 2177.
(2) S.M. Sieber et al., Cancer Treat. Rep., 1976, 60, 1127.

THALIDOMIDE [U;INN]

CAS	50-35-1	MW:	258.24	MF:	C13 H10 N2 O4

Sedative; hypnotic

Chemie Grunenthal, FRG; Merrell Dow (Kevadon)

LgP	0.33(M); 0.04(C)
pKa	
pp cited in Vols.1-5:	
1: 55, 570, 571, 585, 627;	
4: 13, 91, 95;	
5: 408	

(1) Brit. Pat. 768 821 (1957).
(2) H. Schumacher et al., J. Pharmacol. Exp. Ther., 1968, 160, 189.

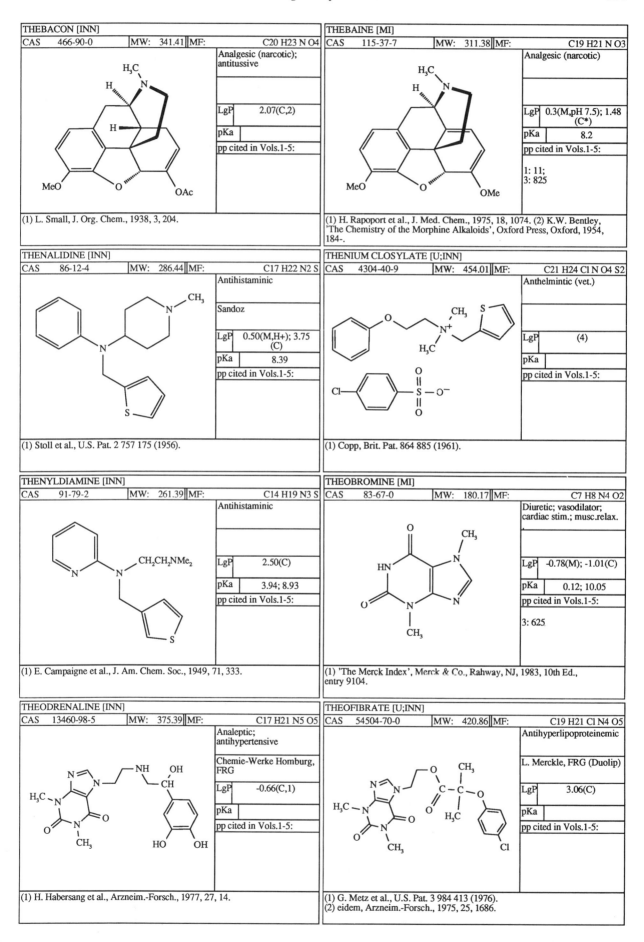

THEBACON [INN]

CAS	466-90-0	MW:	341.41	MF:	C20 H23 N O4

Analgesic (narcotic); antitussive

LgP	2.07(C,2)
pKa	
pp cited in Vols.1-5:	

(1) L. Small, J. Org. Chem., 1938, 3, 204.

THEBAINE [MI]

CAS	115-37-7	MW:	311.38	MF:	C19 H21 N O3

Analgesic (narcotic)

LgP	0.3(M,pH 7.5); 1.48 (C*)
pKa	8.2
pp cited in Vols.1-5:	

1: 11;
3: 825

(1) H. Rapoport et al., J. Med. Chem., 1975, 18, 1074. (2) K.W. Bentley, 'The Chemistry of the Morphine Alkaloids', Oxford Press, Oxford, 1954, 184-.

THENALIDINE [INN]

CAS	86-12-4	MW:	286.44	MF:	C17 H22 N2 S

Antihistaminic

Sandoz

LgP	0.50(M,H+); 3.75 (C)
pKa	8.39
pp cited in Vols.1-5:	

(1) Stoll et al., U.S. Pat. 2 757 175 (1956).

THENIUM CLOSYLATE [U;INN]

CAS	4304-40-9	MW:	454.01	MF:	C21 H24 Cl N O4 S2

Anthelmintic (vet.)

LgP	(4)
pKa	
pp cited in Vols.1-5:	

(1) Copp, Brit. Pat. 864 885 (1961).

THENYLDIAMINE [INN]

CAS	91-79-2	MW:	261.39	MF:	C14 H19 N3 S

Antihistaminic

LgP	2.50(C)
pKa	3.94; 8.93
pp cited in Vols.1-5:	

(1) E. Campaigne et al., J. Am. Chem. Soc., 1949, 71, 333.

THEOBROMINE [MI]

CAS	83-67-0	MW:	180.17	MF:	C7 H8 N4 O2

Diuretic; vasodilator; cardiac stim.; musc.relax.

LgP	-0.78(M); -1.01(C)
pKa	0.12; 10.05
pp cited in Vols.1-5:	

3: 625

(1) 'The Merck Index', Merck & Co., Rahway, NJ, 1983, 10th Ed., entry 9104.

THEODRENALINE [INN]

CAS	13460-98-5	MW:	375.39	MF:	C17 H21 N5 O5

Analeptic; antihypertensive

Chemie-Werke Homburg, FRG

LgP	-0.66(C,1)
pKa	
pp cited in Vols.1-5:	

(1) H. Habersang et al., Arzneim.-Forsch., 1977, 27, 14.

THEOFIBRATE [U;INN]

CAS	54504-70-0	MW:	420.86	MF:	C19 H21 Cl N4 O5

Antihyperlipoproteinemic

L. Merckle, FRG (Duolip)

LgP	3.06(C)
pKa	
pp cited in Vols.1-5:	

(1) G. Metz et al., U.S. Pat. 3 984 413 (1976).
(2) eidem, Arzneim.-Forsch., 1975, 25, 1686.

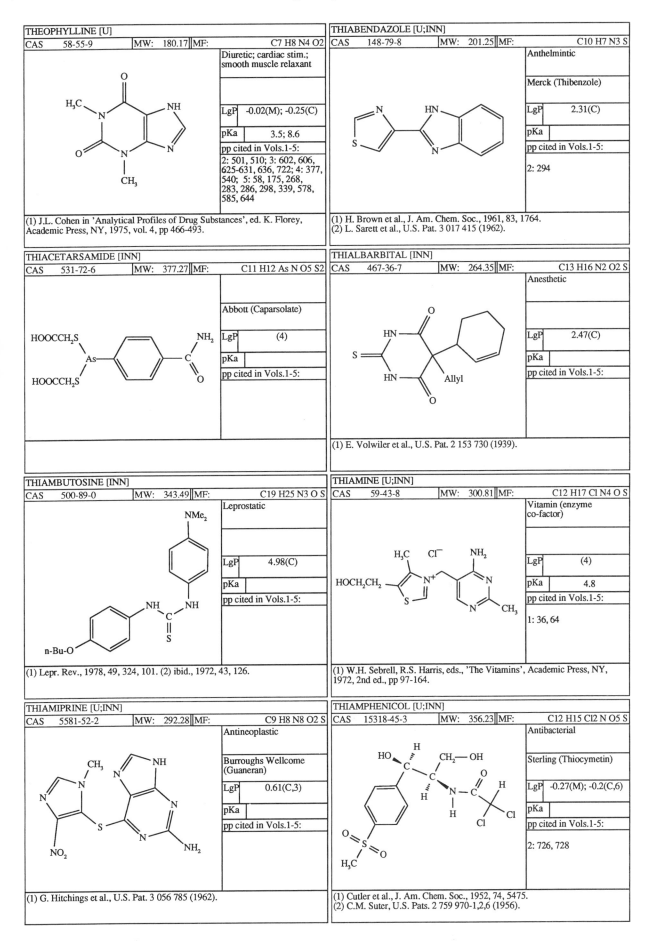

THEOPHYLLINE [U]

CAS	58-55-9	MW:	180.17	MF:	C7 H8 N4 O2

Diuretic; cardiac stim.; smooth muscle relaxant

LgP	-0.02(M); -0.25(C)
pKa	3.5; 8.6
pp cited in Vols.1-5:	

2: 501, 510; 3: 602, 606, 625-631, 636, 722; 4: 377, 540; 5: 58, 175, 268, 283, 286, 298, 339, 578, 585, 644

(1) J.L. Cohen in 'Analytical Profiles of Drug Substances', ed. K. Florey, Academic Press, NY, 1975, vol. 4, pp 466-493.

THIABENDAZOLE [U;INN]

CAS	148-79-8	MW:	201.25	MF:	C10 H7 N3 S

Anthelmintic

Merck (Thibenzole)

LgP	2.31(C)
pKa	
pp cited in Vols.1-5:	

2: 294

(1) H. Brown et al., J. Am. Chem. Soc., 1961, 83, 1764.
(2) L. Sarett et al., U.S. Pat. 3 017 415 (1962).

THIACETARSAMIDE [INN]

CAS	531-72-6	MW:	377.27	MF:	C11 H12 As N O5 S2

Abbott (Caparsolate)

LgP	(4)
pKa	
pp cited in Vols.1-5:	

THIALBARBITAL [INN]

CAS	467-36-7	MW:	264.35	MF:	C13 H16 N2 O2 S

Anesthetic

LgP	2.47(C)
pKa	
pp cited in Vols.1-5:	

(1) E. Volwiler et al., U.S. Pat. 2 153 730 (1939).

THIAMBUTOSINE [INN]

CAS	500-89-0	MW:	343.49	MF:	C19 H25 N3 O S

Leprostatic

LgP	4.98(C)
pKa	
pp cited in Vols.1-5:	

(1) Lepr. Rev., 1978, 49, 324, 101. (2) ibid., 1972, 43, 126.

THIAMINE [U;INN]

CAS	59-43-8	MW:	300.81	MF:	C12 H17 Cl N4 O S

Vitamin (enzyme co-factor)

LgP	(4)
pKa	4.8
pp cited in Vols.1-5:	

1: 36, 64

(1) W.H. Sebrell, R.S. Harris, eds., 'The Vitamins', Academic Press, NY, 1972, 2nd ed., pp 97-164.

THIAMIPRINE [U;INN]

CAS	5581-52-2	MW:	292.28	MF:	C9 H8 N8 O2 S

Antineoplastic

Burroughs Wellcome (Guaneran)

LgP	0.61(C,3)
pKa	
pp cited in Vols.1-5:	

(1) G. Hitchings et al., U.S. Pat. 3 056 785 (1962).

THIAMPHENICOL [U;INN]

CAS	15318-45-3	MW:	356.23	MF:	C12 H15 Cl2 N O5 S

Antibacterial

Sterling (Thiocymetin)

LgP	-0.27(M); -0.2(C,6)
pKa	
pp cited in Vols.1-5:	

2: 726, 728

(1) Cutler et al., J. Am. Chem. Soc., 1952, 74, 5475.
(2) C.M. Suter, U.S. Pats. 2 759 970-1,2,6 (1956).

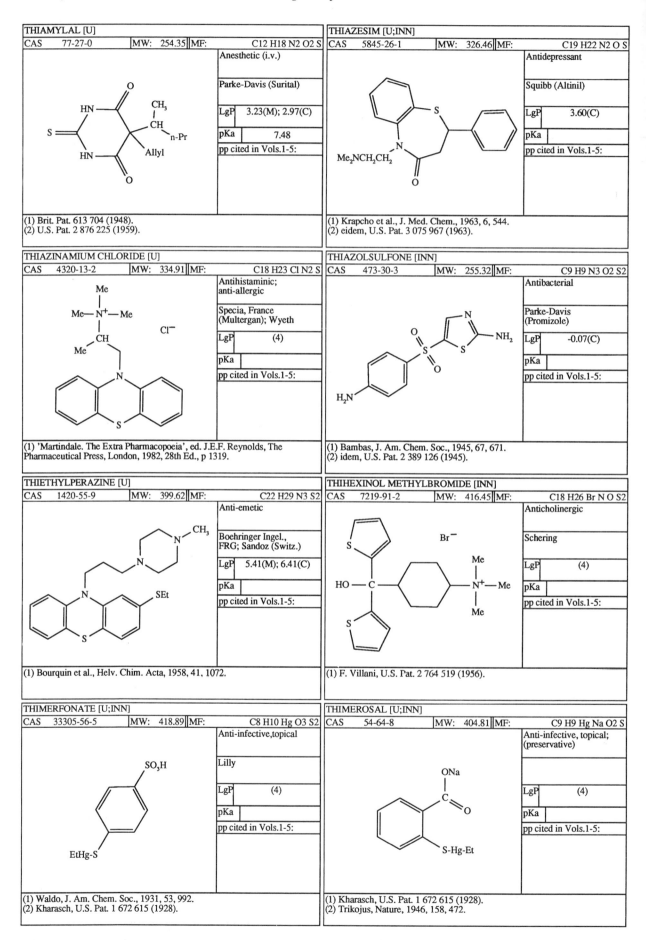

THIAMYLAL [U]

| CAS | 77-27-0 | MW: | 254.35 | MF: | C12 H18 N2 O2 S |

Anesthetic (i.v.)

Parke-Davis (Surital)

LgP	3.23(M); 2.97(C)
pKa	7.48
pp cited in Vols.1-5:	

(1) Brit. Pat. 613 704 (1948).
(2) U.S. Pat. 2 876 225 (1959).

THIAZESIM [U;INN]

| CAS | 5845-26-1 | MW: | 326.46 | MF: | C19 H22 N2 O S |

Antidepressant

Squibb (Altinil)

LgP	3.60(C)
pKa	
pp cited in Vols.1-5:	

(1) Krapcho et al., J. Med. Chem., 1963, 6, 544.
(2) eidem, U.S. Pat. 3 075 967 (1963).

THIAZINAMIUM CHLORIDE [U]

| CAS | 4320-13-2 | MW: | 334.91 | MF: | C18 H23 Cl N2 S |

Antihistaminic; anti-allergic

Specia, France (Multergan); Wyeth

LgP	(4)
pKa	
pp cited in Vols.1-5:	

(1) 'Martindale. The Extra Pharmacopoeia', ed. J.E.F. Reynolds, The Pharmaceutical Press, London, 1982, 28th Ed., p 1319.

THIAZOLSULFONE [INN]

| CAS | 473-30-3 | MW: | 255.32 | MF: | C9 H9 N3 O2 S2 |

Antibacterial

Parke-Davis (Promizole)

LgP	-0.07(C)
pKa	
pp cited in Vols.1-5:	

(1) Bambas, J. Am. Chem. Soc., 1945, 67, 671.
(2) idem, U.S. Pat. 2 389 126 (1945).

THIETHYLPERAZINE [U]

| CAS | 1420-55-9 | MW: | 399.62 | MF: | C22 H29 N3 S2 |

Anti-emetic

Boehringer Ingel., FRG; Sandoz (Switz.)

LgP	5.41(M); 6.41(C)
pKa	
pp cited in Vols.1-5:	

(1) Bourquin et al., Helv. Chim. Acta, 1958, 41, 1072.

THIHEXINOL METHYLBROMIDE [INN]

| CAS | 7219-91-2 | MW: | 416.45 | MF: | C18 H26 Br N O S2 |

Anticholinergic

Schering

LgP	(4)
pKa	
pp cited in Vols.1-5:	

(1) F. Villani, U.S. Pat. 2 764 519 (1956).

THIMERFONATE [U;INN]

| CAS | 33305-56-5 | MW: | 418.89 | MF: | C8 H10 Hg O3 S2 |

Anti-infective, topical

Lilly

LgP	(4)
pKa	
pp cited in Vols.1-5:	

(1) Waldo, J. Am. Chem. Soc., 1931, 53, 992.
(2) Kharasch, U.S. Pat. 1 672 615 (1928).

THIMEROSAL [U;INN]

| CAS | 54-64-8 | MW: | 404.81 | MF: | C9 H9 Hg Na O2 S |

Anti-infective, topical; (preservative)

LgP	(4)
pKa	
pp cited in Vols.1-5:	

(1) Kharasch, U.S. Pat. 1 672 615 (1928).
(2) Trikojus, Nature, 1946, 158, 472.

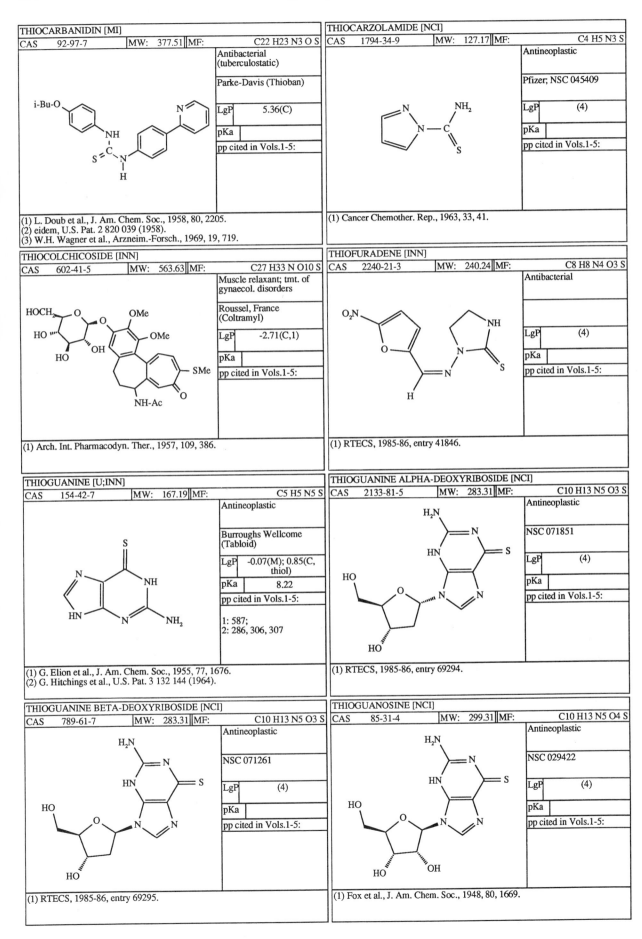

THIOCARBANIDIN [MI]

CAS	92-97-7	MW:	377.51	MF:	C22 H23 N3 O S

Antibacterial (tuberculostatic)

Parke-Davis (Thioban)

LgP	5.36(C)
pKa	
pp cited in Vols.1-5:	

i-Bu-O

(1) L. Doub et al., J. Am. Chem. Soc., 1958, 80, 2205.
(2) eidem, U.S. Pat. 2 820 039 (1958).
(3) W.H. Wagner et al., Arzneim.-Forsch., 1969, 19, 719.

THIOCARZOLAMIDE [NCI]

CAS	1794-34-9	MW:	127.17	MF:	C4 H5 N3 S

Antineoplastic

Pfizer; NSC 045409

LgP	(4)
pKa	
pp cited in Vols.1-5:	

(1) Cancer Chemother. Rep., 1963, 33, 41.

THIOCOLCHICOSIDE [INN]

CAS	602-41-5	MW:	563.63	MF:	C27 H33 N O10 S

Muscle relaxant; tmt. of gynaecol. disorders

Roussel, France (Coltramyl)

LgP	-2.71(C,1)
pKa	
pp cited in Vols.1-5:	

(1) Arch. Int. Pharmacodyn. Ther., 1957, 109, 386.

THIOFURADENE [INN]

CAS	2240-21-3	MW:	240.24	MF:	C8 H8 N4 O3 S

Antibacterial

LgP	(4)
pKa	
pp cited in Vols.1-5:	

(1) RTECS, 1985-86, entry 41846.

THIOGUANINE [U;INN]

CAS	154-42-7	MW:	167.19	MF:	C5 H5 N5 S

Antineoplastic

Burroughs Wellcome (Tabloid)

LgP	-0.07(M); 0.85(C, thiol)
pKa	8.22
pp cited in Vols.1-5:	

1: 587;
2: 286, 306, 307

(1) G. Elion et al., J. Am. Chem. Soc., 1955, 77, 1676.
(2) G. Hitchings et al., U.S. Pat. 3 132 144 (1964).

THIOGUANINE ALPHA-DEOXYRIBOSIDE [NCI]

CAS	2133-81-5	MW:	283.31	MF:	C10 H13 N5 O3 S

Antineoplastic

NSC 071851

LgP	(4)
pKa	
pp cited in Vols.1-5:	

(1) RTECS, 1985-86, entry 69294.

THIOGUANINE BETA-DEOXYRIBOSIDE [NCI]

CAS	789-61-7	MW:	283.31	MF:	C10 H13 N5 O3 S

Antineoplastic

NSC 071261

LgP	(4)
pKa	
pp cited in Vols.1-5:	

(1) RTECS, 1985-86, entry 69295.

THIOGUANOSINE [NCI]

CAS	85-31-4	MW:	299.31	MF:	C10 H13 N5 O4 S

Antineoplastic

NSC 029422

LgP	(4)
pKa	
pp cited in Vols.1-5:	

(1) Fox et al., J. Am. Chem. Soc., 1948, 80, 1669.

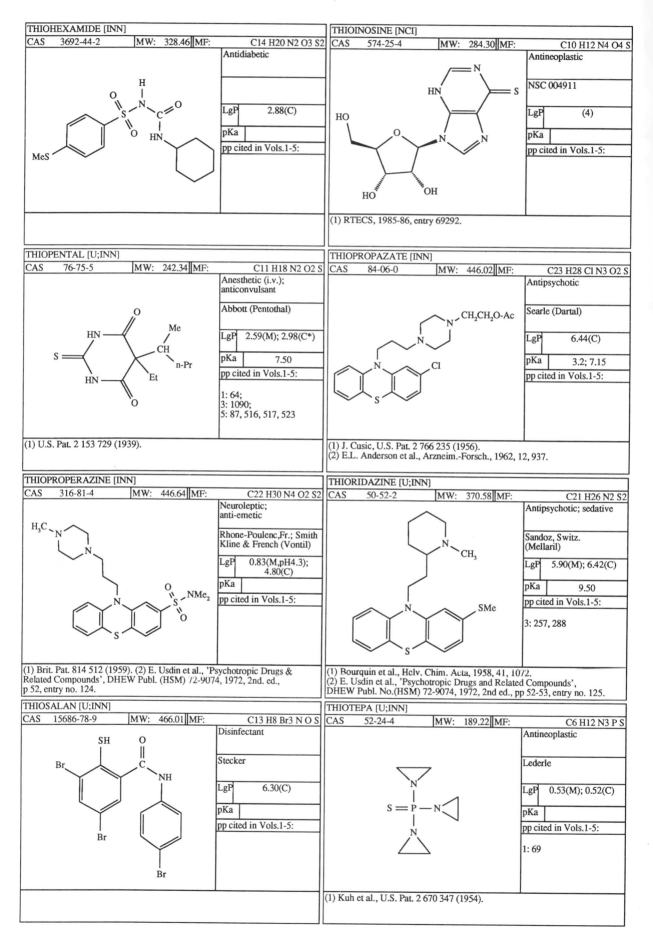

THIOHEXAMIDE [INN]

CAS	3692-44-2	MW:	328.46	MF:	C14 H20 N2 O3 S2

Antidiabetic

LgP	2.88(C)
pKa	
pp cited in Vols.1-5:	

THIOINOSINE [NCI]

CAS	574-25-4	MW:	284.30	MF:	C10 H12 N4 O4 S

Antineoplastic

NSC 004911

LgP	(4)
pKa	
pp cited in Vols.1-5:	

(1) RTECS, 1985-86, entry 69292.

THIOPENTAL [U;INN]

CAS	76-75-5	MW:	242.34	MF:	C11 H18 N2 O2 S

Anesthetic (i.v.); anticonvulsant

Abbott (Pentothal)

LgP	2.59(M); 2.98(C*)
pKa	7.50
pp cited in Vols.1-5:	

1: 64;
3: 1090;
5: 87, 516, 517, 523

(1) U.S. Pat. 2 153 729 (1939).

THIOPROPAZATE [INN]

CAS	84-06-0	MW:	446.02	MF:	C23 H28 Cl N3 O2 S

Antipsychotic

Searle (Dartal)

LgP	6.44(C)
pKa	3.2; 7.15
pp cited in Vols.1-5:	

(1) J. Cusic, U.S. Pat. 2 766 235 (1956).
(2) E.L. Anderson et al., Arzneim.-Forsch., 1962, 12, 937.

THIOPROPERAZINE [INN]

CAS	316-81-4	MW:	446.64	MF:	C22 H30 N4 O2 S2

Neuroleptic; anti-emetic

Rhone-Poulenc,Fr.; Smith Kline & French (Vontil)

LgP	0.83(M,pH4.3); 4.80(C)
pKa	
pp cited in Vols.1-5:	

(1) Brit. Pat. 814 512 (1959). (2) E. Usdin et al., 'Psychotropic Drugs & Related Compounds', DHEW Publ. (HSM) 72-9074, 1972, 2nd. ed., p 52, entry no. 124.

THIORIDAZINE [U;INN]

CAS	50-52-2	MW:	370.58	MF:	C21 H26 N2 S2

Antipsychotic; sedative

Sandoz, Switz. (Mellaril)

LgP	5.90(M); 6.42(C)
pKa	9.50
pp cited in Vols.1-5:	

3: 257, 288

(1) Bourquin et al., Helv. Chim. Acta, 1958, 41, 1072.
(2) E. Usdin et al., 'Psychotropic Drugs and Related Compounds', DHEW Publ. No.(HSM) 72-9074, 1972, 2nd ed., pp 52-53, entry no. 125.

THIOSALAN [U;INN]

CAS	15686-78-9	MW:	466.01	MF:	C13 H8 Br3 N O S

Disinfectant

Stecker

LgP	6.30(C)
pKa	
pp cited in Vols.1-5:	

THIOTEPA [U;INN]

CAS	52-24-4	MW:	189.22	MF:	C6 H12 N3 P S

Antineoplastic

Lederle

LgP	0.53(M); 0.52(C)
pKa	
pp cited in Vols.1-5:	

1: 69

(1) Kuh et al., U.S. Pat. 2 670 347 (1954).

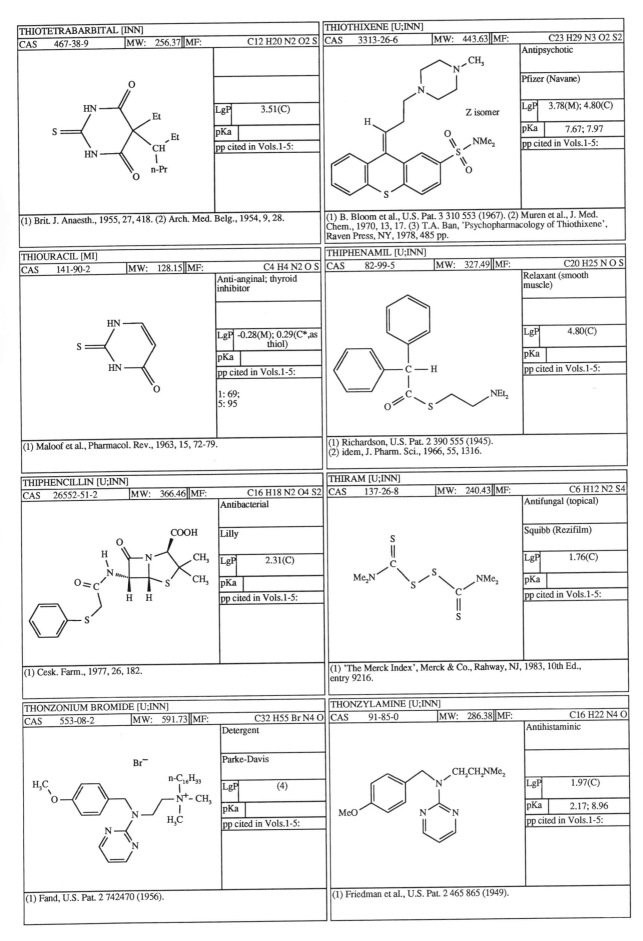

THIOTETRABARBITAL [INN]

CAS	467-38-9	MW:	256.37	MF:	C12 H20 N2 O2 S

LgP	3.51(C)
pKa	
pp cited in Vols.1-5:	

(1) Brit. J. Anaesth., 1955, 27, 418. (2) Arch. Med. Belg., 1954, 9, 28.

THIOTHIXENE [U;INN]

CAS	3313-26-6	MW:	443.63	MF:	C23 H29 N3 O2 S2

Antipsychotic

Pfizer (Navane)

Z isomer

LgP	3.78(M); 4.80(C)
pKa	7.67; 7.97
pp cited in Vols.1-5:	

(1) B. Bloom et al., U.S. Pat. 3 310 553 (1967). (2) Muren et al., J. Med. Chem., 1970, 13, 17. (3) T.A. Ban, 'Psychopharmacology of Thiothixene', Raven Press, NY, 1978, 485 pp.

THIOURACIL [MI]

CAS	141-90-2	MW:	128.15	MF:	C4 H4 N2 O S

Anti-anginal; thyroid inhibitor

LgP	-0.28(M); 0.29(C*,as thiol)
pKa	
pp cited in Vols.1-5:	

1: 69;
5: 95

(1) Maloof et al., Pharmacol. Rev., 1963, 15, 72-79.

THIPHENAMIL [U;INN]

CAS	82-99-5	MW:	327.49	MF:	C20 H25 N O S

Relaxant (smooth muscle)

LgP	4.80(C)
pKa	
pp cited in Vols.1-5:	

(1) Richardson, U.S. Pat. 2 390 555 (1945).
(2) idem, J. Pharm. Sci., 1966, 55, 1316.

THIPHENCILLIN [U;INN]

CAS	26552-51-2	MW:	366.46	MF:	C16 H18 N2 O4 S2

Antibacterial

Lilly

LgP	2.31(C)
pKa	
pp cited in Vols.1-5:	

(1) Cesk. Farm., 1977, 26, 182.

THIRAM [U;INN]

CAS	137-26-8	MW:	240.43	MF:	C6 H12 N2 S4

Antifungal (topical)

Squibb (Rezifilm)

LgP	1.76(C)
pKa	
pp cited in Vols.1-5:	

(1) 'The Merck Index', Merck & Co., Rahway, NJ, 1983, 10th Ed., entry 9216.

THONZONIUM BROMIDE [U;INN]

CAS	553-08-2	MW:	591.73	MF:	C32 H55 Br N4 O

Detergent

Parke-Davis

LgP	(4)
pKa	
pp cited in Vols.1-5:	

(1) Fand, U.S. Pat. 2 742470 (1956).

THONZYLAMINE [U;INN]

CAS	91-85-0	MW:	286.38	MF:	C16 H22 N4 O

Antihistaminic

LgP	1.97(C)
pKa	2.17; 8.96
pp cited in Vols.1-5:	

(1) Friedman et al., U.S. Pat. 2 465 865 (1949).

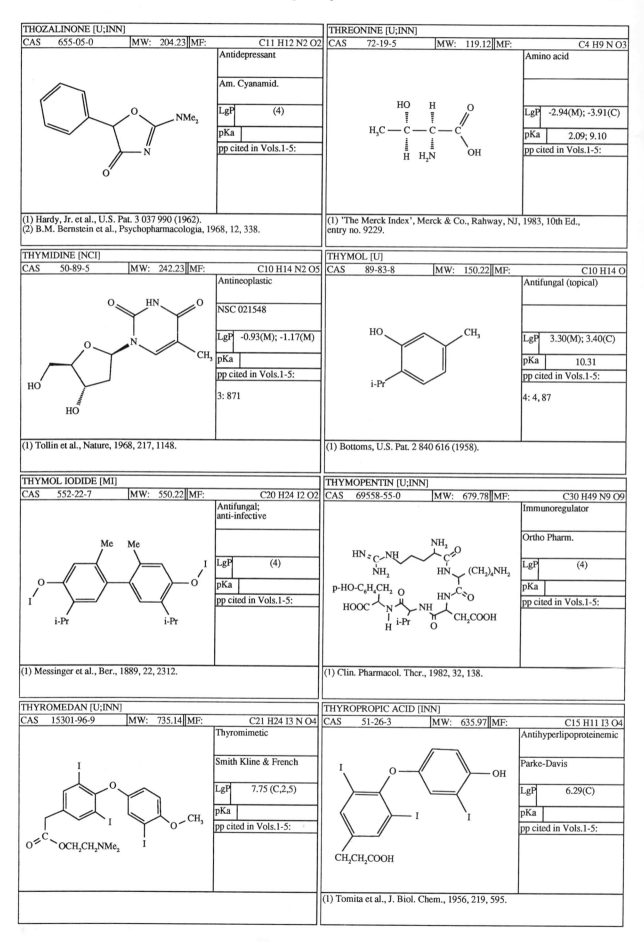

THOZALINONE [U;INN]

CAS	655-05-0	MW:	204.23	MF:	C11 H12 N2 O2

Antidepressant

Am. Cyanamid.

LgP	(4)
pKa	
pp cited in Vols.1-5:	

(1) Hardy, Jr. et al., U.S. Pat. 3 037 990 (1962).
(2) B.M. Bernstein et al., Psychopharmacologia, 1968, 12, 338.

THREONINE [U;INN]

CAS	72-19-5	MW:	119.12	MF:	C4 H9 N O3

Amino acid

LgP	-2.94(M); -3.91(C)
pKa	2.09; 9.10
pp cited in Vols.1-5:	

(1) 'The Merck Index', Merck & Co., Rahway, NJ, 1983, 10th Ed., entry no. 9229.

THYMIDINE [NCI]

CAS	50-89-5	MW:	242.23	MF:	C10 H14 N2 O5

Antineoplastic

NSC 021548

LgP	-0.93(M); -1.17(M)
pKa	
pp cited in Vols.1-5:	
3: 871	

(1) Tollin et al., Nature, 1968, 217, 1148.

THYMOL [U]

CAS	89-83-8	MW:	150.22	MF:	C10 H14 O

Antifungal (topical)

LgP	3.30(M); 3.40(C)
pKa	10.31
pp cited in Vols.1-5:	
4: 4, 87	

(1) Bottoms, U.S. Pat. 2 840 616 (1958).

THYMOL IODIDE [MI]

CAS	552-22-7	MW:	550.22	MF:	C20 H24 I2 O2

Antifungal; anti-infective

LgP	(4)
pKa	
pp cited in Vols.1-5:	

(1) Messinger et al., Ber., 1889, 22, 2312.

THYMOPENTIN [U;INN]

CAS	69558-55-0	MW:	679.78	MF:	C30 H49 N9 O9

Immunoregulator

Ortho Pharm.

LgP	(4)
pKa	
pp cited in Vols.1-5:	

(1) Clin. Pharmacol. Ther., 1982, 32, 138.

THYROMEDAN [U;INN]

CAS	15301-96-9	MW:	735.14	MF:	C21 H24 I3 N O4

Thyromimetic

Smith Kline & French

LgP	7.75 (C,2,5)
pKa	
pp cited in Vols.1-5:	

THYROPROPIC ACID [INN]

CAS	51-26-3	MW:	635.97	MF:	C15 H11 I3 O4

Antihyperlipoproteinemic

Parke-Davis

LgP	6.29(C)
pKa	
pp cited in Vols.1-5:	

(1) Tomita et al., J. Biol. Chem., 1956, 219, 595.

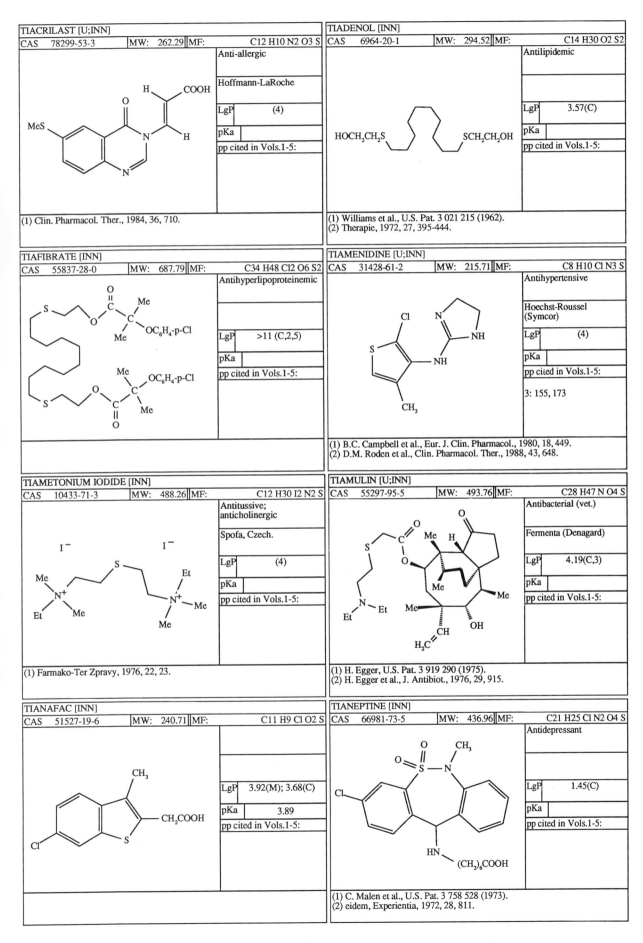

TIACRILAST [U;INN]

CAS	78299-53-3	MW:	262.29	MF:	C12 H10 N2 O3 S

Anti-allergic

Hoffmann-LaRoche

LgP	(4)
pKa	
pp cited in Vols.1-5:	

(1) Clin. Pharmacol. Ther., 1984, 36, 710.

TIADENOL [INN]

CAS	6964-20-1	MW:	294.52	MF:	C14 H30 O2 S2

Antilipidemic

LgP	3.57(C)
pKa	
pp cited in Vols.1-5:	

HOCH₂CH₂S   SCH₂CH₂OH

(1) Williams et al., U.S. Pat. 3 021 215 (1962).
(2) Therapie, 1972, 27, 395-444.

TIAFIBRATE [INN]

CAS	55837-28-0	MW:	687.79	MF:	C34 H48 Cl2 O6 S2

Antihyperlipoproteinemic

LgP	>11 (C,2,5)
pKa	
pp cited in Vols.1-5:	

TIAMENIDINE [U;INN]

CAS	31428-61-2	MW:	215.71	MF:	C8 H10 Cl N3 S

Antihypertensive

Hoechst-Roussel (Symcor)

LgP	(4)
pKa	
pp cited in Vols.1-5:	

3: 155, 173

(1) B.C. Campbell et al., Eur. J. Clin. Pharmacol., 1980, 18, 449.
(2) D.M. Roden et al., Clin. Pharmacol. Ther., 1988, 43, 648.

TIAMETONIUM IODIDE [INN]

CAS	10433-71-3	MW:	488.26	MF:	C12 H30 I2 N2 S

Antitussive; anticholinergic

Spofa, Czech.

LgP	(4)
pKa	
pp cited in Vols.1-5:	

(1) Farmako-Ter Zpravy, 1976, 22, 23.

TIAMULIN [U;INN]

CAS	55297-95-5	MW:	493.76	MF:	C28 H47 N O4 S

Antibacterial (vet.)

Fermenta (Denagard)

LgP	4.19(C,3)
pKa	
pp cited in Vols.1-5:	

(1) H. Egger, U.S. Pat. 3 919 290 (1975).
(2) H. Egger et al., J. Antibiot., 1976, 29, 915.

TIANAFAC [INN]

CAS	51527-19-6	MW:	240.71	MF:	C11 H9 Cl O2 S

LgP	3.92(M); 3.68(C)
pKa	3.89
pp cited in Vols.1-5:	

TIANEPTINE [INN]

CAS	66981-73-5	MW:	436.96	MF:	C21 H25 Cl N2 O4 S

Antidepressant

LgP	1.45(C)
pKa	
pp cited in Vols.1-5:	

(1) C. Malen et al., U.S. Pat. 3 758 528 (1973).
(2) eidem, Experientia, 1972, 28, 811.

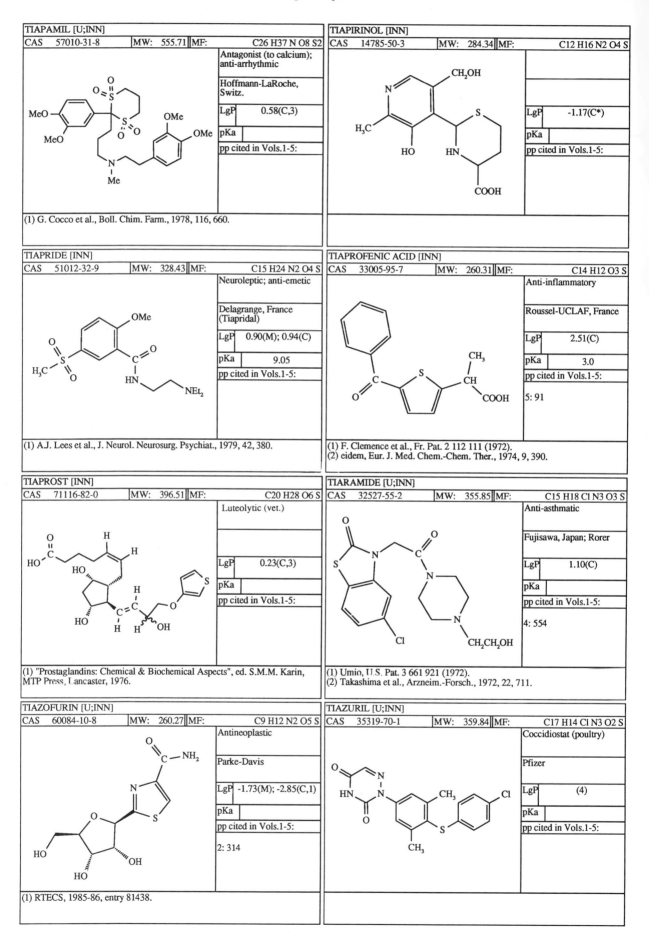

TIAPAMIL [U;INN]

CAS	57010-31-8	MW:	555.71	MF:	C26 H37 N O8 S2

Antagonist (to calcium); anti-arrhythmic

Hoffmann-LaRoche, Switz.

LgP	0.58(C,3)
pKa	
pp cited in Vols.1-5:	

MeO / MeO — OMe / OMe — N — Me

(1) G. Cocco et al., Boll. Chim. Farm., 1978, 116, 660.

TIAPIRINOL [INN]

CAS	14785-50-3	MW:	284.34	MF:	C12 H16 N2 O4 S

CH$_2$OH / H$_3$C / HO / HN / COOH

LgP	-1.17(C*)
pKa	
pp cited in Vols.1-5:	

TIAPRIDE [INN]

CAS	51012-32-9	MW:	328.43	MF:	C15 H24 N2 O4 S

Neuroleptic; anti-emetic

Delagrange, France (Tiapridal)

LgP	0.90(M); 0.94(C)
pKa	9.05
pp cited in Vols.1-5:	

H$_3$C / OMe / HN / NEt$_2$

(1) A.J. Lees et al., J. Neurol. Neurosurg. Psychiat., 1979, 42, 380.

TIAPROFENIC ACID [INN]

CAS	33005-95-7	MW:	260.31	MF:	C14 H12 O3 S

Anti-inflammatory

Roussel-UCLAF, France

LgP	2.51(C)
pKa	3.0
pp cited in Vols.1-5:	
5: 91	

CH$_3$ / CH / COOH

(1) F. Clemence et al., Fr. Pat. 2 112 111 (1972).
(2) eidem, Eur. J. Med. Chem.-Chem. Ther., 1974, 9, 390.

TIAPROST [INN]

CAS	71116-82-0	MW:	396.51	MF:	C20 H28 O6 S

Luteolytic (vet.)

LgP	0.23(C,3)
pKa	
pp cited in Vols.1-5:	

HO / HO / HO / OH

(1) "Prostaglandins: Chemical & Biochemical Aspects", ed. S.M.M. Karin, MTP Press, Lancaster, 1976.

TIARAMIDE [U;INN]

CAS	32527-55-2	MW:	355.85	MF:	C15 H18 Cl N3 O3 S

Anti-asthmatic

Fujisawa, Japan; Rorer

LgP	1.10(C)
pKa	
pp cited in Vols.1-5:	
4: 554	

Cl / CH$_2$CH$_2$OH

(1) Umio, U.S. Pat. 3 661 921 (1972).
(2) Takashima et al., Arzneim.-Forsch., 1972, 22, 711.

TIAZOFURIN [U;INN]

CAS	60084-10-8	MW:	260.27	MF:	C9 H12 N2 O5 S

Antineoplastic

Parke-Davis

LgP	-1.73(M); -2.85(C,1)
pKa	
pp cited in Vols.1-5:	
2: 314	

C—NH$_2$ / HO / OH / HO

(1) RTECS, 1985-86, entry 81438.

TIAZURIL [U;INN]

CAS	35319-70-1	MW:	359.84	MF:	C17 H14 Cl N3 O2 S

Coccidiostat (poultry)

Pfizer

LgP	(4)
pKa	
pp cited in Vols.1-5:	

HN / CH$_3$ / Cl / S / CH$_3$

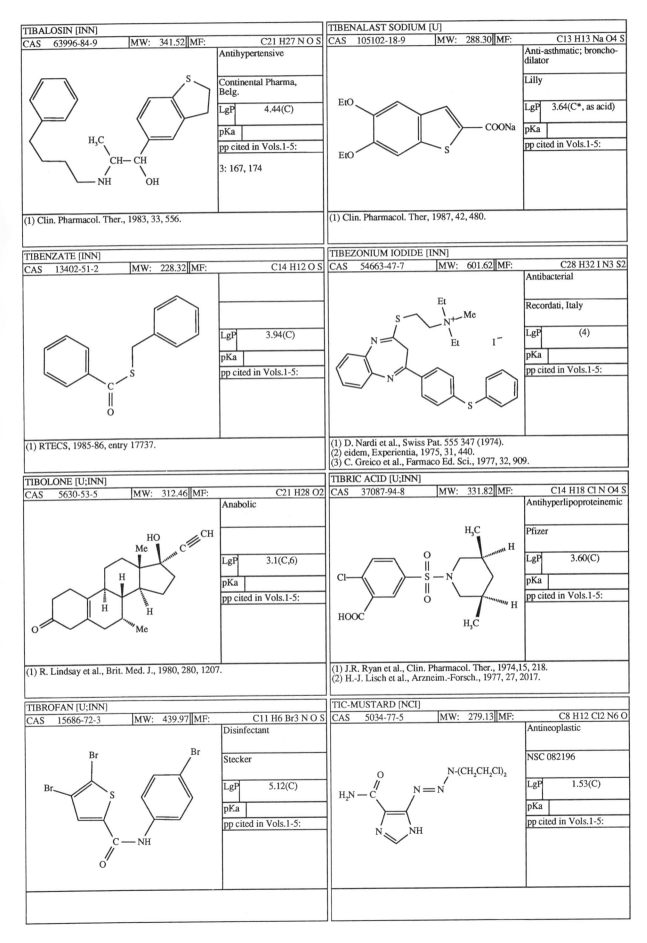

TIBALOSIN [INN]

CAS	63996-84-9	MW: 341.52	MF:	C21 H27 N O S

Antihypertensive

Continental Pharma, Belg.

LgP	4.44(C)
pKa	

pp cited in Vols.1-5:

3: 167, 174

(1) Clin. Pharmacol. Ther., 1983, 33, 556.

TIBENZATE [INN]

CAS	13402-51-2	MW: 228.32	MF:	C14 H12 O S

LgP	3.94(C)
pKa	

pp cited in Vols.1-5:

(1) RTECS, 1985-86, entry 17737.

TIBOLONE [U;INN]

CAS	5630-53-5	MW: 312.46	MF:	C21 H28 O2

Anabolic

LgP	3.1(C,6)
pKa	

pp cited in Vols.1-5:

(1) R. Lindsay et al., Brit. Med. J., 1980, 280, 1207.

TIBROFAN [U;INN]

CAS	15686-72-3	MW: 439.97	MF:	C11 H6 Br3 N O S

Disinfectant

Stecker

LgP	5.12(C)
pKa	

pp cited in Vols.1-5:

TIBENALAST SODIUM [U]

CAS	105102-18-9	MW: 288.30	MF:	C13 H13 Na O4 S

Anti-asthmatic; broncho-dilator

Lilly

LgP	3.64(C*, as acid)
pKa	

pp cited in Vols.1-5:

(1) Clin. Pharmacol. Ther, 1987, 42, 480.

TIBEZONIUM IODIDE [INN]

CAS	54663-47-7	MW: 601.62	MF:	C28 H32 I N3 S2

Antibacterial

Recordati, Italy

LgP	(4)
pKa	

pp cited in Vols.1-5:

(1) D. Nardi et al., Swiss Pat. 555 347 (1974).
(2) eidem, Experientia, 1975, 31, 440.
(3) C. Greico et al., Farmaco Ed. Sci., 1977, 32, 909.

TIBRIC ACID [U;INN]

CAS	37087-94-8	MW: 331.82	MF:	C14 H18 Cl N O4 S

Antihyperlipoproteinemic

Pfizer

LgP	3.60(C)
pKa	

pp cited in Vols.1-5:

(1) J.R. Ryan et al., Clin. Pharmacol. Ther., 1974, 15, 218.
(2) H.-J. Lisch et al., Arzneim.-Forsch., 1977, 27, 2017.

TIC-MUSTARD [NCI]

CAS	5034-77-5	MW: 279.13	MF:	C8 H12 Cl2 N6 O

Antineoplastic

NSC 082196

LgP	1.53(C)
pKa	

pp cited in Vols.1-5:

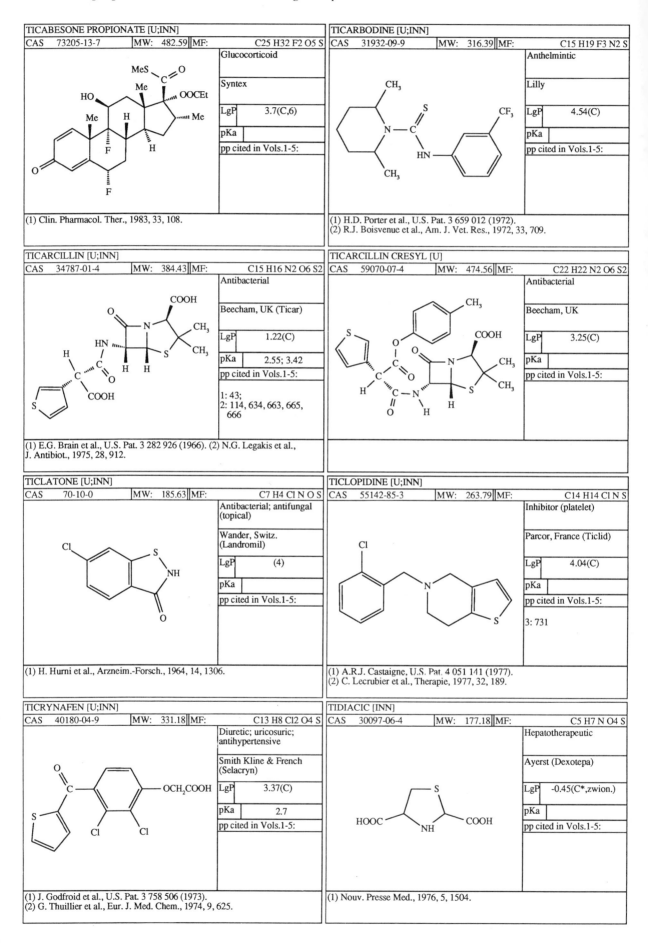

TICABESONE PROPIONATE [U;INN]

CAS	73205-13-7	MW:	482.59	MF:	C25 H32 F2 O5 S

Glucocorticoid

Syntex

LgP	3.7(C,6)
pKa	
pp cited in Vols.1-5:	

(1) Clin. Pharmacol. Ther., 1983, 33, 108.

TICARBODINE [U;INN]

CAS	31932-09-9	MW:	316.39	MF:	C15 H19 F3 N2 S

Anthelmintic

Lilly

LgP	4.54(C)
pKa	
pp cited in Vols.1-5:	

(1) H.D. Porter et al., U.S. Pat. 3 659 012 (1972).
(2) R.J. Boisvenue et al., Am. J. Vet. Res., 1972, 33, 709.

TICARCILLIN [U;INN]

CAS	34787-01-4	MW:	384.43	MF:	C15 H16 N2 O6 S2

Antibacterial

Beecham, UK (Ticar)

LgP	1.22(C)
pKa	2.55; 3.42
pp cited in Vols.1-5:	
	1: 43;
2: 114, 634, 663, 665, 666

(1) E.G. Brain et al., U.S. Pat. 3 282 926 (1966). (2) N.G. Legakis et al.,
J. Antibiot., 1975, 28, 912.

TICARCILLIN CRESYL [U]

CAS	59070-07-4	MW:	474.56	MF:	C22 H22 N2 O6 S2

Antibacterial

Beecham, UK

LgP	3.25(C)
pKa	
pp cited in Vols.1-5:	

TICLATONE [U;INN]

CAS	70-10-0	MW:	185.63	MF:	C7 H4 Cl N O S

Antibacterial; antifungal
(topical)

Wander, Switz.
(Landromil)

LgP	(4)
pKa	
pp cited in Vols.1-5:	

(1) H. Hurni et al., Arzneim.-Forsch., 1964, 14, 1306.

TICLOPIDINE [U;INN]

CAS	55142-85-3	MW:	263.79	MF:	C14 H14 Cl N S

Inhibitor (platelet)

Parcor, France (Ticlid)

LgP	4.04(C)
pKa	
pp cited in Vols.1-5:	
	3: 731

(1) A.R.J. Castaigne, U.S. Pat. 4 051 141 (1977).
(2) C. Lecrubier et al., Therapie, 1977, 32, 189.

TICRYNAFEN [U;INN]

CAS	40180-04-9	MW:	331.18	MF:	C13 H8 Cl2 O4 S

Diuretic; uricosuric;
antihypertensive

Smith Kline & French
(Selacryn)

LgP	3.37(C)
pKa	2.7
pp cited in Vols.1-5:	

(1) J. Godfroid et al., U.S. Pat. 3 758 506 (1973).
(2) G. Thuillier et al., Eur. J. Med. Chem., 1974, 9, 625.

TIDIACIC [INN]

CAS	30097-06-4	MW:	177.18	MF:	C5 H7 N O4 S

Hepatotherapeutic

Ayerst (Dexotepa)

LgP	-0.45(C*,zwion.)
pKa	
pp cited in Vols.1-5:	

(1) Nouv. Presse Med., 1976, 5, 1504.

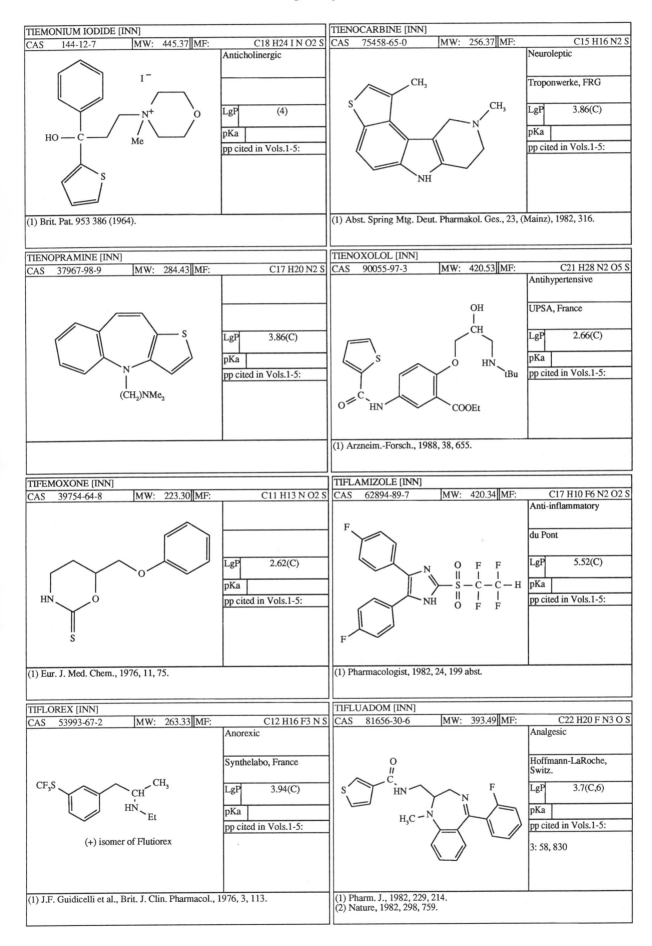

TIEMONIUM IODIDE [INN]

CAS	144-12-7	MW:	445.37	MF:	C18 H24 I N O2 S

Anticholinergic

LgP	(4)
pKa	
pp cited in Vols.1-5:	

(1) Brit. Pat. 953 386 (1964).

TIENOCARBINE [INN]

CAS	75458-65-0	MW:	256.37	MF:	C15 H16 N2 S

Neuroleptic

Troponwerke, FRG

LgP	3.86(C)
pKa	
pp cited in Vols.1-5:	

(1) Abst. Spring Mtg. Deut. Pharmakol. Ges., 23, (Mainz), 1982, 316.

TIENOPRAMINE [INN]

CAS	37967-98-9	MW:	284.43	MF:	C17 H20 N2 S

LgP	3.86(C)
pKa	
pp cited in Vols.1-5:	

TIENOXOLOL [INN]

CAS	90055-97-3	MW:	420.53	MF:	C21 H28 N2 O5 S

Antihypertensive

UPSA, France

LgP	2.66(C)
pKa	
pp cited in Vols.1-5:	

(1) Arzneim.-Forsch., 1988, 38, 655.

TIFEMOXONE [INN]

CAS	39754-64-8	MW:	223.30	MF:	C11 H13 N O2 S

LgP	2.62(C)
pKa	
pp cited in Vols.1-5:	

(1) Eur. J. Med. Chem., 1976, 11, 75.

TIFLAMIZOLE [INN]

CAS	62894-89-7	MW:	420.34	MF:	C17 H10 F6 N2 O2 S

Anti-inflammatory

du Pont

LgP	5.52(C)
pKa	
pp cited in Vols.1-5:	

(1) Pharmacologist, 1982, 24, 199 abst.

TIFLOREX [INN]

CAS	53993-67-2	MW:	263.33	MF:	C12 H16 F3 N S

Anorexic

Synthelabo, France

LgP	3.94(C)
pKa	
pp cited in Vols.1-5:	

(+) isomer of Flutiorex

(1) J.F. Guidicelli et al., Brit. J. Clin. Pharmacol., 1976, 3, 113.

TIFLUADOM [INN]

CAS	81656-30-6	MW:	393.49	MF:	C22 H20 F N3 O S

Analgesic

Hoffmann-LaRoche, Switz.

LgP	3.7(C,6)
pKa	
pp cited in Vols.1-5:	
3: 58, 830	

(1) Pharm. J., 1982, 229, 214.
(2) Nature, 1982, 298, 759.

TIFLUCARBINE [INN]

CAS 89875-86-5	MW: 288.39	MF:	C16 H17 F N2 S

Antidepressant

Troponwerke, FRG

LgP	4.66(C)
pKa	
pp cited in Vols.1-5:	

(1) Abst. Spr. Mtg. Deut. Pharmakol. Ges., 23, (Mainz), 1982, 316.

TIFORMIN [INN]

CAS 4210-97-3	MW: 144.18	MF:	C5 H12 N4 O

Tmt. of uremic diabetes

LgP	(7)
pKa	
pp cited in Vols.1-5:	

(1) W.J.H. Butterfield et al., Lancet, 1969, 2, 381.

TIFURAC [U;INN]

CAS 97483-17-5	MW: 326.37	MF:	C18 H14 O4 S

Analgesic

Syntex

LgP	3.75(C)
pKa	
pp cited in Vols.1-5:	

(1) Clin. Pharmacol. Ther., 1987, 41, 694.

TIGEMONAM [U;INN]

CAS 102507-71-1	MW: 437.41	MF:	C12 H15 N5 O9 S2

Antimicrobial

Squibb (Tigemen)

LgP	(4)
pKa	
pp cited in Vols.1-5:	

(1) Intersci. Conf. Antimicrob. Ag. Chemother. 27 (New York City), Oct., 1987, 1224 abst.

TIGESTOL [U;INN]

CAS 896-71-9	MW: 284.45	MF:	C20 H28 O

Progestin

Organon, Neth.

LgP	5.52(C*)
pKa	
pp cited in Vols.1-5:	

TIGLOIDINE [INN]

CAS 495-83-0	MW: 223.32	MF:	C13 H21 N O2

CNS depressant; antiparkinsonian

LgP	1.98(C)
pKa	
pp cited in Vols.1-5:	

(1) Evans et al., J. Chem. Soc., 1937, 1820.
(2) Sanghvi et al., Eur. J. Pharmacol., 1968, 4, 246.

TILBROQUINOL [INN]

CAS 7175-09-9	MW: 238.09	MF:	C10 H8 Br N O

Anti-amebic; anti-fungal; antibacterial

LgP	3.08(C)
pKa	
pp cited in Vols.1-5:	

TILETAMINE [U;INN]

CAS 14176-49-9	MW: 223.34	MF:	C12 H17 N O S

Anesthetic; anticonvulsant

Parke-Davis (Telazol)

LgP	2.39(C)
pKa	
pp cited in Vols.1-5:	

(1) Pharm. J., 1969, 203, 457.

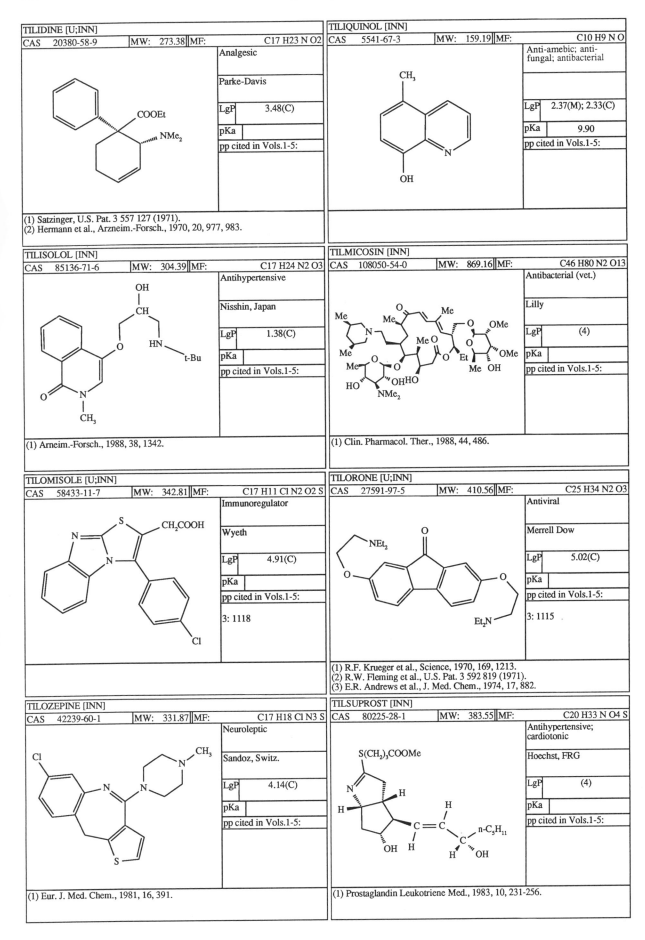

TILIDINE [U;INN]

CAS 20380-58-9	MW: 273.38	MF:	C17 H23 N O2

Analgesic

Parke-Davis

LgP	3.48(C)
pKa	
pp cited in Vols.1-5:	

(1) Satzinger, U.S. Pat. 3 557 127 (1971).
(2) Hermann et al., Arzneim.-Forsch., 1970, 20, 977, 983.

TILIQUINOL [INN]

CAS 5541-67-3	MW: 159.19	MF:	C10 H9 N O

Anti-amebic; anti-fungal; antibacterial

LgP	2.37(M); 2.33(C)
pKa	9.90
pp cited in Vols.1-5:	

TILISOLOL [INN]

CAS 85136-71-6	MW: 304.39	MF:	C17 H24 N2 O3

Antihypertensive

Nisshin, Japan

LgP	1.38(C)
pKa	
pp cited in Vols.1-5:	

(1) Arneim.-Forsch., 1988, 38, 1342.

TILMICOSIN [INN]

CAS 108050-54-0	MW: 869.16	MF:	C46 H80 N2 O13

Antibacterial (vet.)

Lilly

LgP	(4)
pKa	
pp cited in Vols.1-5:	

(1) Clin. Pharmacol. Ther., 1988, 44, 486.

TILOMISOLE [U;INN]

CAS 58433-11-7	MW: 342.81	MF:	C17 H11 Cl N2 O2 S

Immunoregulator

Wyeth

LgP	4.91(C)
pKa	
pp cited in Vols.1-5:	
3: 1118	

TILORONE [U;INN]

CAS 27591-97-5	MW: 410.56	MF:	C25 H34 N2 O3

Antiviral

Merrell Dow

LgP	5.02(C)
pKa	
pp cited in Vols.1-5:	
3: 1115	

(1) R.F. Krueger et al., Science, 1970, 169, 1213.
(2) R.W. Fleming et al., U.S. Pat. 3 592 819 (1971).
(3) E.R. Andrews et al., J. Med. Chem., 1974, 17, 882.

TILOZEPINE [INN]

CAS 42239-60-1	MW: 331.87	MF:	C17 H18 Cl N3 S

Neuroleptic

Sandoz, Switz.

LgP	4.14(C)
pKa	
pp cited in Vols.1-5:	

(1) Eur. J. Med. Chem., 1981, 16, 391.

TILSUPROST [INN]

CAS 80225-28-1	MW: 383.55	MF:	C20 H33 N O4 S

Antihypertensive; cardiotonic

Hoechst, FRG

LgP	(4)
pKa	
pp cited in Vols.1-5:	

(1) Prostaglandin Leukotriene Med., 1983, 10, 231-256.

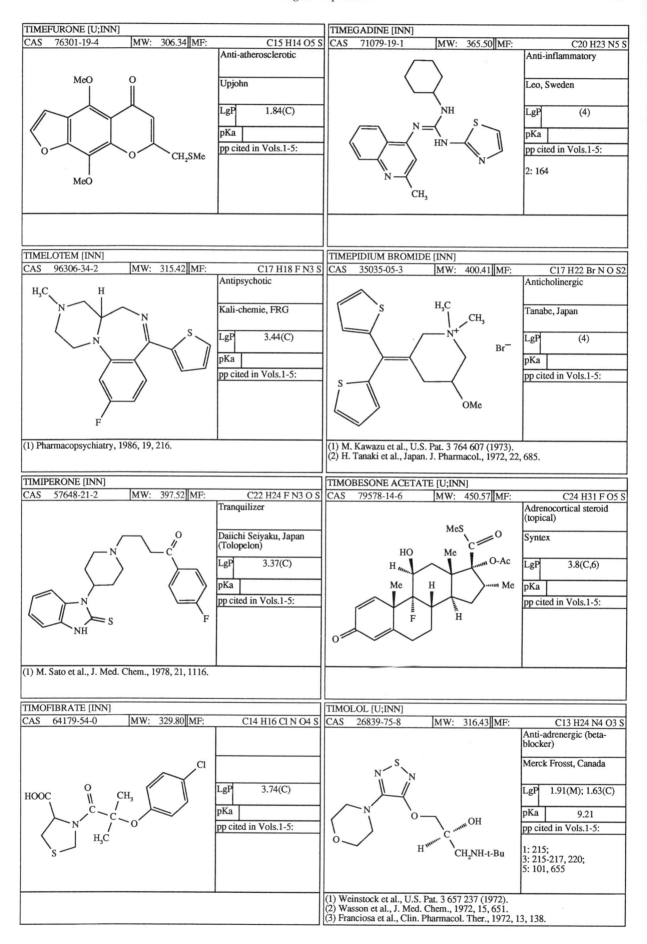

TIMEFURONE [U;INN]

CAS	76301-19-4	MW:	306.34	MF:	C15 H14 O5 S

Anti-atherosclerotic

Upjohn

LgP	1.84(C)
pKa	

pp cited in Vols.1-5:

TIMEGADINE [INN]

CAS	71079-19-1	MW:	365.50	MF:	C20 H23 N5 S

Anti-inflammatory

Leo, Sweden

LgP	(4)
pKa	

pp cited in Vols.1-5:

2: 164

TIMELOTEM [INN]

CAS	96306-34-2	MW:	315.42	MF:	C17 H18 F N3 S

Antipsychotic

Kali-chemie, FRG

LgP	3.44(C)
pKa	

pp cited in Vols.1-5:

(1) Pharmacopsychiatry, 1986, 19, 216.

TIMEPIDIUM BROMIDE [INN]

CAS	35035-05-3	MW:	400.41	MF:	C17 H22 Br N O S2

Anticholinergic

Tanabe, Japan

LgP	(4)
pKa	

pp cited in Vols.1-5:

(1) M. Kawazu et al., U.S. Pat. 3 764 607 (1973).
(2) H. Tanaki et al., Japan. J. Pharmacol., 1972, 22, 685.

TIMIPERONE [INN]

CAS	57648-21-2	MW:	397.52	MF:	C22 H24 F N3 O S

Tranquilizer

Daiichi Seiyaku, Japan
(Tolopelon)

LgP	3.37(C)
pKa	

pp cited in Vols.1-5:

(1) M. Sato et al., J. Med. Chem., 1978, 21, 1116.

TIMOBESONE ACETATE [U;INN]

CAS	79578-14-6	MW:	450.57	MF:	C24 H31 F O5 S

Adrenocortical steroid
(topical)

Syntex

LgP	3.8(C,6)
pKa	

pp cited in Vols.1-5:

TIMOFIBRATE [INN]

CAS	64179-54-0	MW:	329.80	MF:	C14 H16 Cl N O4 S

LgP	3.74(C)
pKa	

pp cited in Vols.1-5:

TIMOLOL [U;INN]

CAS	26839-75-8	MW:	316.43	MF:	C13 H24 N4 O3 S

Anti-adrenergic (beta-blocker)

Merck Frosst, Canada

LgP	1.91(M); 1.63(C)
pKa	9.21

pp cited in Vols.1-5:

1: 215;
3: 215-217, 220;
5: 101, 655

(1) Weinstock et al., U.S. Pat. 3 657 237 (1972).
(2) Wasson et al., J. Med. Chem., 1972, 15, 651.
(3) Franciosa et al., Clin. Pharmacol. Ther., 1972, 13, 138.

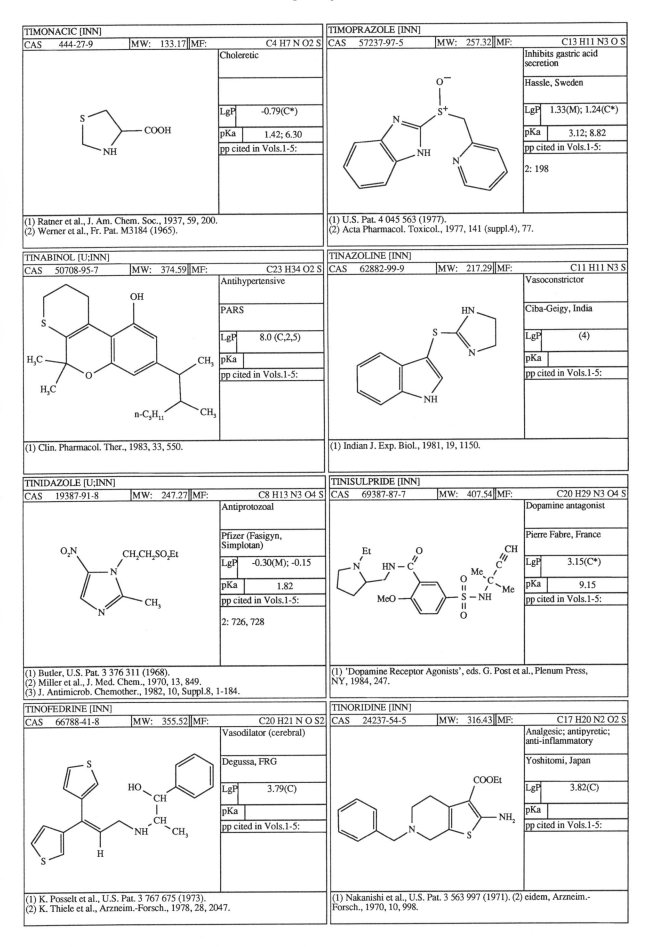

TIMONACIC [INN]

CAS	444-27-9	MW:	133.17	MF:	C4 H7 N O2 S

Choleretic

LgP	-0.79(C*)
pKa	1.42; 6.30
pp cited in Vols.1-5:	

(1) Ratner et al., J. Am. Chem. Soc., 1937, 59, 200.
(2) Werner et al., Fr. Pat. M3184 (1965).

TIMOPRAZOLE [INN]

CAS	57237-97-5	MW:	257.32	MF:	C13 H11 N3 O S

Inhibits gastric acid secretion

Hassle, Sweden

LgP	1.33(M); 1.24(C*)
pKa	3.12; 8.82
pp cited in Vols.1-5:	

2: 198

(1) U.S. Pat. 4 045 563 (1977).
(2) Acta Pharmacol. Toxicol., 1977, 141 (suppl.4), 77.

TINABINOL [U;INN]

CAS	50708-95-7	MW:	374.59	MF:	C23 H34 O2 S

Antihypertensive

PARS

LgP	8.0 (C,2,5)
pKa	
pp cited in Vols.1-5:	

(1) Clin. Pharmacol. Ther., 1983, 33, 550.

TINAZOLINE [INN]

CAS	62882-99-9	MW:	217.29	MF:	C11 H11 N3 S

Vasoconstrictor

Ciba-Geigy, India

LgP	(4)
pKa	
pp cited in Vols.1-5:	

(1) Indian J. Exp. Biol., 1981, 19, 1150.

TINIDAZOLE [U;INN]

CAS	19387-91-8	MW:	247.27	MF:	C8 H13 N3 O4 S

Antiprotozoal

Pfizer (Fasigyn, Simplotan)

LgP	-0.30(M); -0.15
pKa	1.82
pp cited in Vols.1-5:	

2: 726, 728

(1) Butler, U.S. Pat. 3 376 311 (1968).
(2) Miller et al., J. Med. Chem., 1970, 13, 849.
(3) J. Antimicrob. Chemother., 1982, 10, Suppl.8, 1-184.

TINISULPRIDE [INN]

CAS	69387-87-7	MW:	407.54	MF:	C20 H29 N3 O4 S

Dopamine antagonist

Pierre Fabre, France

LgP	3.15(C*)
pKa	9.15
pp cited in Vols.1-5:	

(1) 'Dopamine Receptor Agonists', eds. G. Post et al., Plenum Press, NY, 1984, 247.

TINOFEDRINE [INN]

CAS	66788-41-8	MW:	355.52	MF:	C20 H21 N O S2

Vasodilator (cerebral)

Degussa, FRG

LgP	3.79(C)
pKa	
pp cited in Vols.1-5:	

(1) K. Posselt et al., U.S. Pat. 3 767 675 (1973).
(2) K. Thiele et al., Arzneim.-Forsch., 1978, 28, 2047.

TINORIDINE [INN]

CAS	24237-54-5	MW:	316.43	MF:	C17 H20 N2 O2 S

Analgesic; antipyretic; anti-inflammatory

Yoshitomi, Japan

LgP	3.82(C)
pKa	
pp cited in Vols.1-5:	

(1) Nakanishi et al., U.S. Pat. 3 563 997 (1971). (2) eidem, Arzneim.-Forsch., 1970, 10, 998.

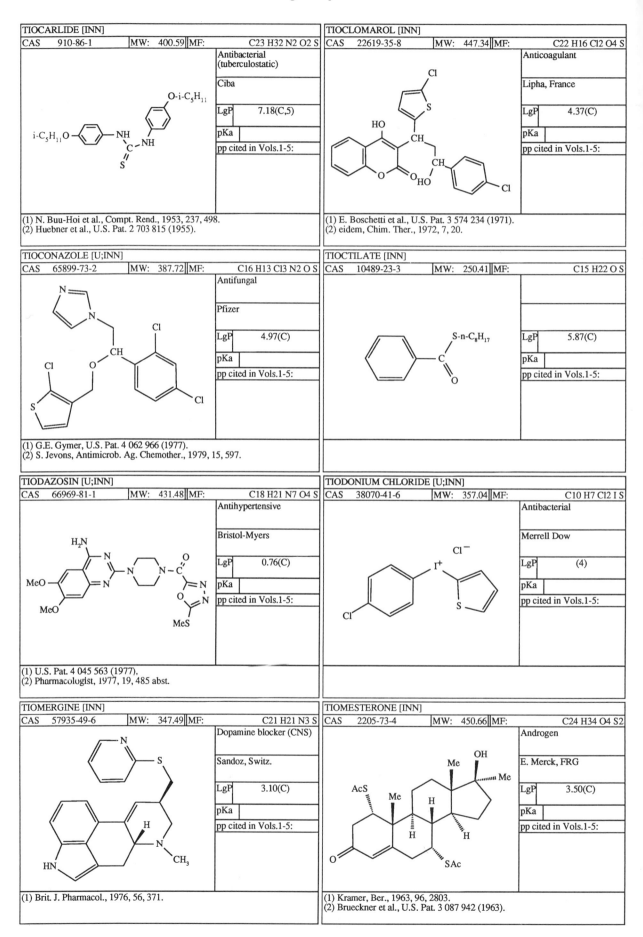

TIOCARLIDE [INN]			
CAS 910-86-1	MW: 400.59	MF:	C23 H32 N2 O2 S

Antibacterial (tuberculostatic)

Ciba

LgP 7.18(C,5)

pKa

pp cited in Vols.1-5:

(1) N. Buu-Hoi et al., Compt. Rend., 1953, 237, 498.
(2) Huebner et al., U.S. Pat. 2 703 815 (1955).

TIOCLOMAROL [INN]			
CAS 22619-35-8	MW: 447.34	MF:	C22 H16 Cl2 O4 S

Anticoagulant

Lipha, France

LgP 4.37(C)

pKa

pp cited in Vols.1-5:

(1) E. Boschetti et al., U.S. Pat. 3 574 234 (1971).
(2) eidem, Chim. Ther., 1972, 7, 20.

TIOCONAZOLE [U;INN]			
CAS 65899-73-2	MW: 387.72	MF:	C16 H13 Cl3 N2 O S

Antifungal

Pfizer

LgP 4.97(C)

pKa

pp cited in Vols.1-5:

(1) G.E. Gymer, U.S. Pat. 4 062 966 (1977).
(2) S. Jevons, Antimicrob. Ag. Chemother., 1979, 15, 597.

TIOCTILATE [INN]			
CAS 10489-23-3	MW: 250.41	MF:	C15 H22 O S

LgP 5.87(C)

pKa

pp cited in Vols.1-5:

TIODAZOSIN [U;INN]			
CAS 66969-81-1	MW: 431.48	MF:	C18 H21 N7 O4 S

Antihypertensive

Bristol-Myers

LgP 0.76(C)

pKa

pp cited in Vols.1-5:

(1) U.S. Pat. 4 045 563 (1977),
(2) Pharmacologist, 1977, 19, 485 abst.

TIODONIUM CHLORIDE [U;INN]			
CAS 38070-41-6	MW: 357.04	MF:	C10 H7 Cl2 I S

Antibacterial

Merrell Dow

LgP (4)

pKa

pp cited in Vols.1-5:

TIOMERGINE [INN]			
CAS 57935-49-6	MW: 347.49	MF:	C21 H21 N3 S

Dopamine blocker (CNS)

Sandoz, Switz.

LgP 3.10(C)

pKa

pp cited in Vols.1-5:

(1) Brit. J. Pharmacol., 1976, 56, 371.

TIOMESTERONE [INN]			
CAS 2205-73-4	MW: 450.66	MF:	C24 H34 O4 S2

Androgen

E. Merck, FRG

LgP 3.50(C)

pKa

pp cited in Vols.1-5:

(1) Kramer, Ber., 1963, 96, 2803.
(2) Brueckner et al., U.S. Pat. 3 087 942 (1963).

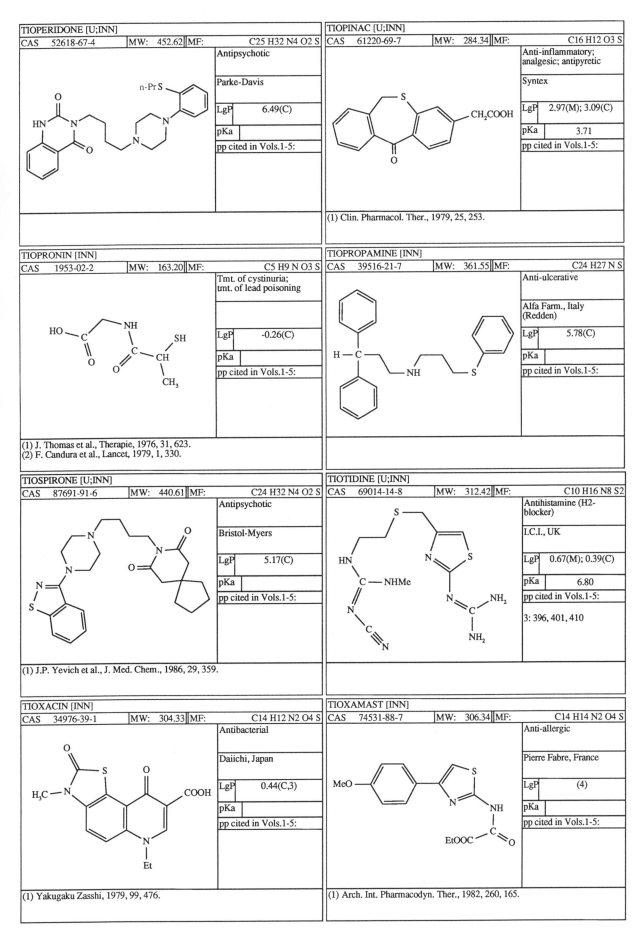

TIOPERIDONE [U;INN]

CAS	52618-67-4	MW:	452.62	MF:	C25 H32 N4 O2 S

Antipsychotic

Parke-Davis

LgP	6.49(C)
pKa	
pp cited in Vols.1-5:	

TIOPINAC [U;INN]

CAS	61220-69-7	MW:	284.34	MF:	C16 H12 O3 S

Anti-inflammatory; analgesic; antipyretic

Syntex

LgP	2.97(M); 3.09(C)
pKa	3.71
pp cited in Vols.1-5:	

(1) Clin. Pharmacol. Ther., 1979, 25, 253.

TIOPRONIN [INN]

CAS	1953-02-2	MW:	163.20	MF:	C5 H9 N O3 S

Tmt. of cystinuria; tmt. of lead poisoning

LgP	-0.26(C)
pKa	
pp cited in Vols.1-5:	

(1) J. Thomas et al., Therapie, 1976, 31, 623.
(2) F. Candura et al., Lancet, 1979, 1, 330.

TIOPROPAMINE [INN]

CAS	39516-21-7	MW:	361.55	MF:	C24 H27 N S

Anti-ulcerative

Alfa Farm., Italy (Redden)

LgP	5.78(C)
pKa	
pp cited in Vols.1-5:	

TIOSPIRONE [U;INN]

CAS	87691-91-6	MW:	440.61	MF:	C24 H32 N4 O2 S

Antipsychotic

Bristol-Myers

LgP	5.17(C)
pKa	
pp cited in Vols.1-5:	

(1) J.P. Yevich et al., J. Med. Chem., 1986, 29, 359.

TIOTIDINE [U;INN]

CAS	69014-14-8	MW:	312.42	MF:	C10 H16 N8 S2

Antihistamine (H2-blocker)

I.C.I., UK

LgP	0.67(M); 0.39(C)
pKa	6.80
pp cited in Vols.1-5:	
	3: 396, 401, 410

TIOXACIN [INN]

CAS	34976-39-1	MW:	304.33	MF:	C14 H12 N2 O4 S

Antibacterial

Daiichi, Japan

LgP	0.44(C,3)
pKa	
pp cited in Vols.1-5:	

(1) Yakugaku Zasshi, 1979, 99, 476.

TIOXAMAST [INN]

CAS	74531-88-7	MW:	306.34	MF:	C14 H14 N2 O4 S

Anti-allergic

Pierre Fabre, France

LgP	(4)
pKa	
pp cited in Vols.1-5:	

(1) Arch. Int. Pharmacodyn. Ther., 1982, 260, 165.

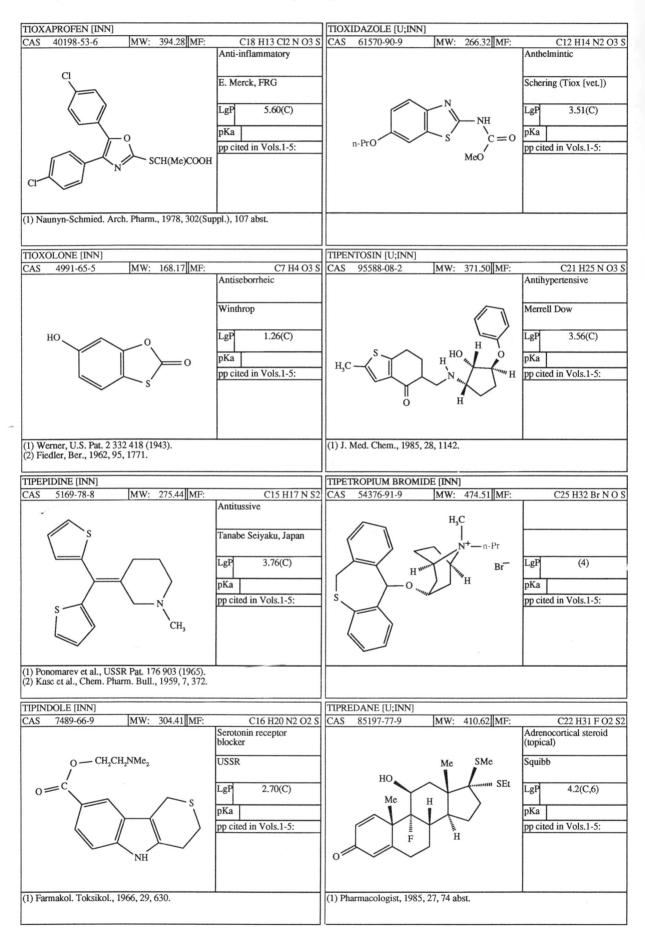

TIOXAPROFEN [INN]

CAS	40198-53-6	MW:	394.28	MF:	C18 H13 Cl2 N O3 S

Anti-inflammatory

E. Merck, FRG

LgP	5.60(C)
pKa	
pp cited in Vols.1-5:	

(1) Naunyn-Schmied. Arch. Pharm., 1978, 302(Suppl.), 107 abst.

TIOXIDAZOLE [U;INN]

CAS	61570-90-9	MW:	266.32	MF:	C12 H14 N2 O3 S

Anthelmintic

Schering (Tiox [vet.])

LgP	3.51(C)
pKa	
pp cited in Vols.1-5:	

TIOXOLONE [INN]

CAS	4991-65-5	MW:	168.17	MF:	C7 H4 O3 S

Antiseborrheic

Winthrop

LgP	1.26(C)
pKa	
pp cited in Vols.1-5:	

(1) Werner, U.S. Pat. 2 332 418 (1943).
(2) Fiedler, Ber., 1962, 95, 1771.

TIPENTOSIN [U;INN]

CAS	95588-08-2	MW:	371.50	MF:	C21 H25 N O3 S

Antihypertensive

Merrell Dow

LgP	3.56(C)
pKa	
pp cited in Vols.1-5:	

(1) J. Med. Chem., 1985, 28, 1142.

TIPEPIDINE [INN]

CAS	5169-78-8	MW:	275.44	MF:	C15 H17 N S2

Antitussive

Tanabe Seiyaku, Japan

LgP	3.76(C)
pKa	
pp cited in Vols.1-5:	

(1) Ponomarev et al., USSR Pat. 176 903 (1965).
(2) Kase et al., Chem. Pharm. Bull., 1959, 7, 372.

TIPETROPIUM BROMIDE [INN]

CAS	54376-91-9	MW:	474.51	MF:	C25 H32 Br N O S

LgP	(4)
pKa	
pp cited in Vols.1-5:	

TIPINDOLE [INN]

CAS	7489-66-9	MW:	304.41	MF:	C16 H20 N2 O2 S

Serotonin receptor blocker

USSR

LgP	2.70(C)
pKa	
pp cited in Vols.1-5:	

(1) Farmakol. Toksikol., 1966, 29, 630.

TIPREDANE [U;INN]

CAS	85197-77-9	MW:	410.62	MF:	C22 H31 F O2 S2

Adrenocortical steroid (topical)

Squibb

LgP	4.2(C,6)
pKa	
pp cited in Vols.1-5:	

(1) Pharmacologist, 1985, 27, 74 abst.

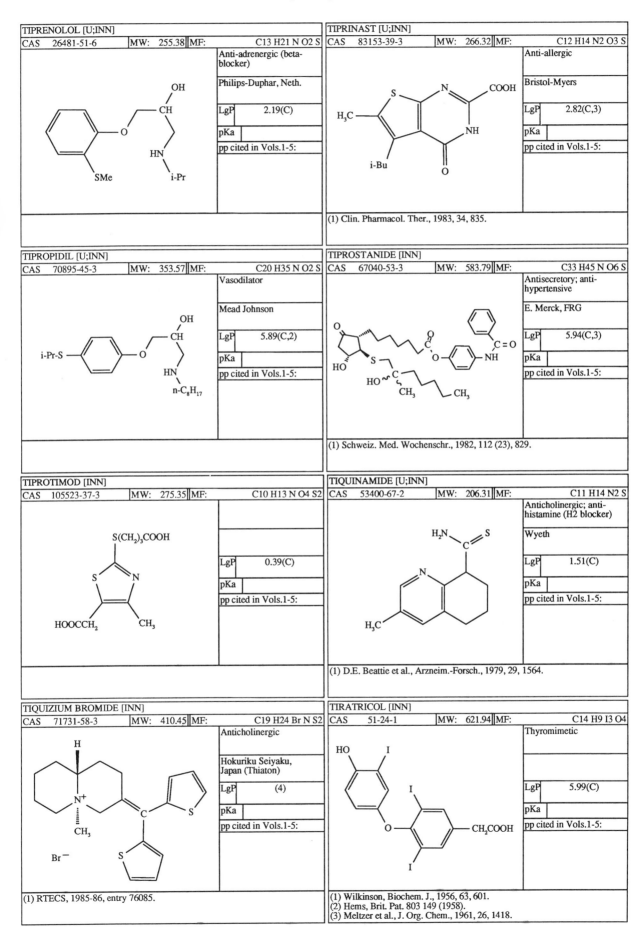

TIPRENOLOL [U;INN]

CAS	26481-51-6	MW:	255.38	MF:	C13 H21 N O2 S

Anti-adrenergic (beta-blocker)

Philips-Duphar, Neth.

LgP	2.19(C)
pKa	
pp cited in Vols.1-5:	

TIPRINAST [U;INN]

CAS	83153-39-3	MW:	266.32	MF:	C12 H14 N2 O3 S

Anti-allergic

Bristol-Myers

LgP	2.82(C,3)
pKa	
pp cited in Vols.1-5:	

(1) Clin. Pharmacol. Ther., 1983, 34, 835.

TIPROPIDIL [U;INN]

CAS	70895-45-3	MW:	353.57	MF:	C20 H35 N O2 S

Vasodilator

Mead Johnson

LgP	5.89(C,2)
pKa	
pp cited in Vols.1-5:	

TIPROSTANIDE [INN]

CAS	67040-53-3	MW:	583.79	MF:	C33 H45 N O6 S

Antisecretory; anti-hypertensive

E. Merck, FRG

LgP	5.94(C,3)
pKa	
pp cited in Vols.1-5:	

(1) Schweiz. Med. Wochenschr., 1982, 112 (23), 829.

TIPROTIMOD [INN]

CAS	105523-37-3	MW:	275.35	MF:	C10 H13 N O4 S2

LgP	0.39(C)
pKa	
pp cited in Vols.1-5:	

TIQUINAMIDE [U;INN]

CAS	53400-67-2	MW:	206.31	MF:	C11 H14 N2 S

Anticholinergic; anti-histamine (H2 blocker)

Wyeth

LgP	1.51(C)
pKa	
pp cited in Vols.1-5:	

(1) D.E. Beattie et al., Arzneim.-Forsch., 1979, 29, 1564.

TIQUIZIUM BROMIDE [INN]

CAS	71731-58-3	MW:	410.45	MF:	C19 H24 Br N S2

Anticholinergic

Hokuriku Seiyaku, Japan (Thiaton)

LgP	(4)
pKa	
pp cited in Vols.1-5:	

(1) RTECS, 1985-86, entry 76085.

TIRATRICOL [INN]

CAS	51-24-1	MW:	621.94	MF:	C14 H9 I3 O4

Thyromimetic

LgP	5.99(C)
pKa	
pp cited in Vols.1-5:	

(1) Wilkinson, Biochem. J., 1956, 63, 601.
(2) Hems, Brit. Pat. 803 149 (1958).
(3) Meltzer et al., J. Org. Chem., 1961, 26, 1418.

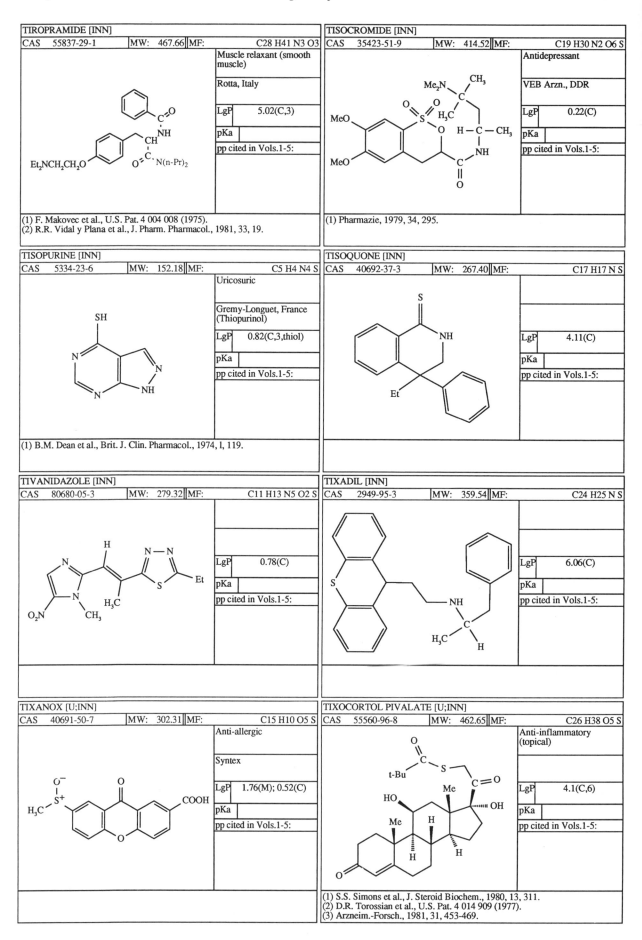

TIROPRAMIDE [INN]

CAS 55837-29-1	MW: 467.66	MF: C28 H41 N3 O3

Muscle relaxant (smooth muscle)

Rotta, Italy

LgP	5.02(C,3)
pKa	
pp cited in Vols.1-5:	

(1) F. Makovec et al., U.S. Pat. 4 004 008 (1975).
(2) R.R. Vidal y Plana et al., J. Pharm. Pharmacol., 1981, 33, 19.

TISOCROMIDE [INN]

CAS 35423-51-9	MW: 414.52	MF: C19 H30 N2 O6 S

Antidepressant

VEB Arzn., DDR

LgP	0.22(C)
pKa	
pp cited in Vols.1-5:	

(1) Pharmazie, 1979, 34, 295.

TISOPURINE [INN]

CAS 5334-23-6	MW: 152.18	MF: C5 H4 N4 S

Uricosuric

Gremy-Longuet, France (Thiopurinol)

LgP	0.82(C,3,thiol)
pKa	
pp cited in Vols.1-5:	

(1) B.M. Dean et al., Brit. J. Clin. Pharmacol., 1974, 1, 119.

TISOQUONE [INN]

CAS 40692-37-3	MW: 267.40	MF: C17 H17 N S

LgP	4.11(C)
pKa	
pp cited in Vols.1-5:	

TIVANIDAZOLE [INN]

CAS 80680-05-3	MW: 279.32	MF: C11 H13 N5 O2 S

LgP	0.78(C)
pKa	
pp cited in Vols.1-5:	

TIXADIL [INN]

CAS 2949-95-3	MW: 359.54	MF: C24 H25 N S

LgP	6.06(C)
pKa	
pp cited in Vols.1-5:	

TIXANOX [U;INN]

CAS 40691-50-7	MW: 302.31	MF: C15 H10 O5 S

Anti-allergic

Syntex

LgP	1.76(M); 0.52(C)
pKa	
pp cited in Vols.1-5:	

TIXOCORTOL PIVALATE [U;INN]

CAS 55560-96-8	MW: 462.65	MF: C26 H38 O5 S

Anti-inflammatory (topical)

LgP	4.1(C,6)
pKa	
pp cited in Vols.1-5:	

(1) S.S. Simons et al., J. Steroid Biochem., 1980, 13, 311.
(2) D.R. Torossian et al., U.S. Pat. 4 014 909 (1977).
(3) Arzneim.-Forsch., 1981, 31, 453-469.

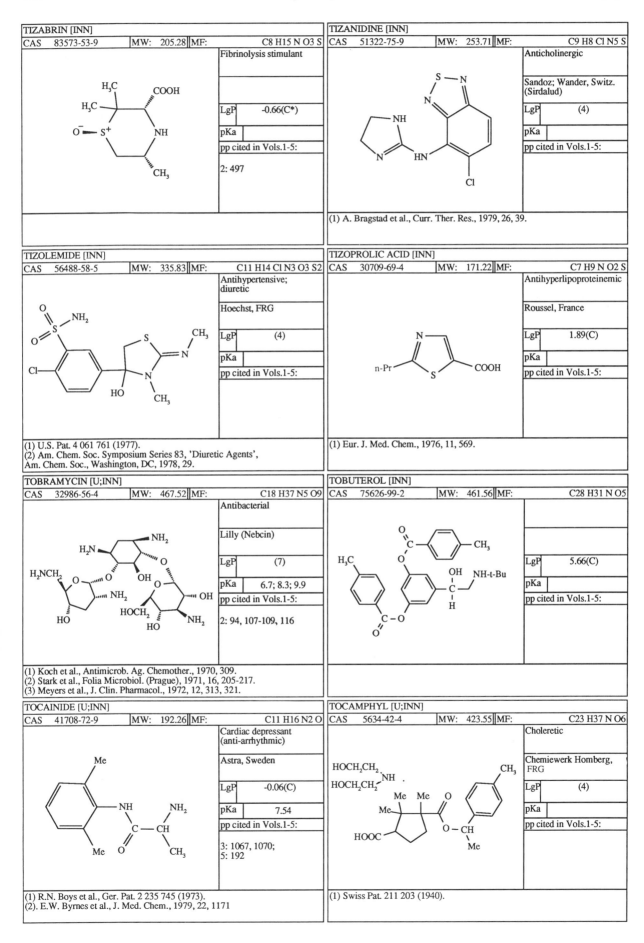

TIZABRIN [INN]

CAS	83573-53-9	MW:	205.28	MF:	C8 H15 N O3 S

Fibrinolysis stimulant

LgP	-0.66(C*)
pKa	
pp cited in Vols.1-5:	

2: 497

TIZANIDINE [INN]

CAS	51322-75-9	MW:	253.71	MF:	C9 H8 Cl N5 S

Anticholinergic

Sandoz; Wander, Switz. (Sirdalud)

LgP	(4)
pKa	
pp cited in Vols.1-5:	

(1) A. Bragstad et al., Curr. Ther. Res., 1979, 26, 39.

TIZOLEMIDE [INN]

CAS	56488-58-5	MW:	335.83	MF:	C11 H14 Cl N3 O3 S2

Antihypertensive; diuretic

Hoechst, FRG

LgP	(4)
pKa	
pp cited in Vols.1-5:	

(1) U.S. Pat. 4 061 761 (1977).
(2) Am. Chem. Soc. Symposium Series 83, 'Diuretic Agents', Am. Chem. Soc., Washington, DC, 1978, 29.

TIZOPROLIC ACID [INN]

CAS	30709-69-4	MW:	171.22	MF:	C7 H9 N O2 S

Antihyperlipoproteinemic

Roussel, France

LgP	1.89(C)
pKa	
pp cited in Vols.1-5:	

(1) Eur. J. Med. Chem., 1976, 11, 569.

TOBRAMYCIN [U;INN]

CAS	32986-56-4	MW:	467.52	MF:	C18 H37 N5 O9

Antibacterial

Lilly (Nebcin)

LgP	(7)
pKa	6.7; 8.3; 9.9
pp cited in Vols.1-5:	

2: 94, 107-109, 116

(1) Koch et al., Antimicrob. Ag. Chemother., 1970, 309.
(2) Stark et al., Folia Microbiol. (Prague), 1971, 16, 205-217.
(3) Meyers et al., J. Clin. Pharmacol., 1972, 12, 313, 321.

TOBUTEROL [INN]

CAS	75626-99-2	MW:	461.56	MF:	C28 H31 N O5

LgP	5.66(C)
pKa	
pp cited in Vols.1-5:	

TOCAINIDE [U;INN]

CAS	41708-72-9	MW:	192.26	MF:	C11 H16 N2 O

Cardiac depressant (anti-arrhythmic)

Astra, Sweden

LgP	-0.06(C)
pKa	7.54
pp cited in Vols.1-5:	

3: 1067, 1070;
5: 192

(1) R.N. Boys et al., Ger. Pat. 2 235 745 (1973).
(2). E.W. Byrnes et al., J. Med. Chem., 1979, 22, 1171

TOCAMPHYL [U;INN]

CAS	5634-42-4	MW:	423.55	MF:	C23 H37 N O6

Choleretic

Chemiewerk Homberg, FRG

LgP	(4)
pKa	
pp cited in Vols.1-5:	

(1) Swiss Pat. 211 203 (1940).

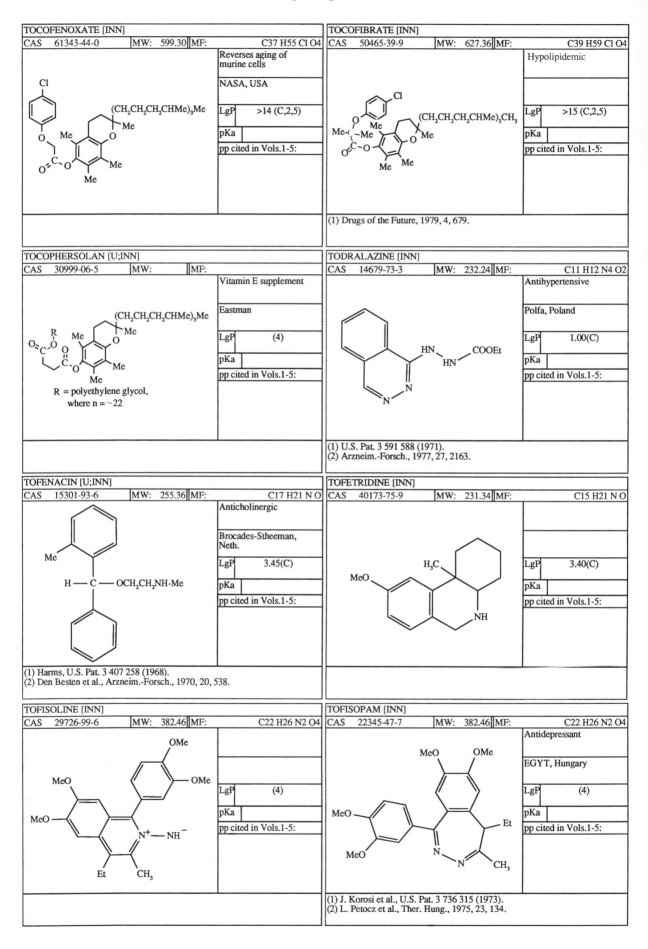

TOCOFENOXATE [INN]

CAS 61343-44-0 | MW: 599.30 | MF: C37 H55 Cl O4

Reverses aging of murine cells

NASA, USA

LgP	>14 (C,2,5)
pKa	
pp cited in Vols.1-5:	

TOCOFIBRATE [INN]

CAS 50465-39-9 | MW: 627.36 | MF: C39 H59 Cl O4

Hypolipidemic

LgP	>15 (C,2,5)
pKa	
pp cited in Vols.1-5:	

(1) Drugs of the Future, 1979, 4, 679.

TOCOPHERSOLAN [U;INN]

CAS 30999-06-5 | MW: | MF:

Vitamin E supplement

Eastman

LgP	(4)
pKa	
pp cited in Vols.1-5:	

R = polyethylene glycol, where n = ~22

TODRALAZINE [INN]

CAS 14679-73-3 | MW: 232.24 | MF: C11 H12 N4 O2

Antihypertensive

Polfa, Poland

LgP	1.00(C)
pKa	
pp cited in Vols.1-5:	

(1) U.S. Pat. 3 591 588 (1971).
(2) Arzneim.-Forsch., 1977, 27, 2163.

TOFENACIN [U;INN]

CAS 15301-93-6 | MW: 255.36 | MF: C17 H21 N O

Anticholinergic

Brocades-Stheeman, Neth.

LgP	3.45(C)
pKa	
pp cited in Vols.1-5:	

(1) Harms, U.S. Pat. 3 407 258 (1968).
(2) Den Besten et al., Arzneim.-Forsch., 1970, 20, 538.

TOFETRIDINE [INN]

CAS 40173-75-9 | MW: 231.34 | MF: C15 H21 N O

LgP	3.40(C)
pKa	
pp cited in Vols.1-5:	

TOFISOLINE [INN]

CAS 29726-99-6 | MW: 382.46 | MF: C22 H26 N2 O4

LgP	(4)
pKa	
pp cited in Vols.1-5:	

TOFISOPAM [INN]

CAS 22345-47-7 | MW: 382.46 | MF: C22 H26 N2 O4

Antidepressant

EGYT, Hungary

LgP	(4)
pKa	
pp cited in Vols.1-5:	

(1) J. Korosi et al., U.S. Pat. 3 736 315 (1973).
(2) L. Petocz et al., Ther. Hung., 1975, 23, 134.

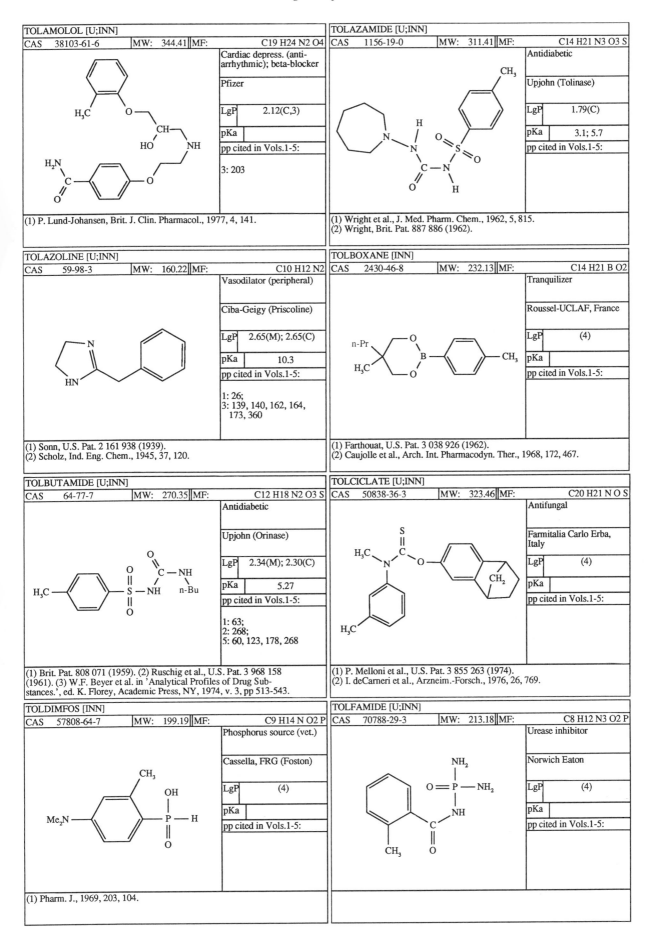

TOLAMOLOL [U;INN]

| CAS | 38103-61-6 | MW: | 344.41 | MF: | C19 H24 N2 O4 |

Cardiac depress. (anti-arrhythmic); beta-blocker

Pfizer

| LgP | 2.12(C,3) |
| pKa | |
| pp cited in Vols.1-5: |
| 3: 203 |

(1) P. Lund-Johansen, Brit. J. Clin. Pharmacol., 1977, 4, 141.

TOLAZAMIDE [U;INN]

| CAS | 1156-19-0 | MW: | 311.41 | MF: | C14 H21 N3 O3 S |

Antidiabetic

Upjohn (Tolinase)

| LgP | 1.79(C) |
| pKa | 3.1; 5.7 |
| pp cited in Vols.1-5: |

(1) Wright et al., J. Med. Pharm. Chem., 1962, 5, 815.
(2) Wright, Brit. Pat. 887 886 (1962).

TOLAZOLINE [U;INN]

| CAS | 59-98-3 | MW: | 160.22 | MF: | C10 H12 N2 |

Vasodilator (peripheral)

Ciba-Geigy (Priscoline)

| LgP | 2.65(M); 2.65(C) |
| pKa | 10.3 |
| pp cited in Vols.1-5: |
| 1: 26; |
| 3: 139, 140, 162, 164, |
| 173, 360 |

(1) Sonn, U.S. Pat. 2 161 938 (1939).
(2) Scholz, Ind. Eng. Chem., 1945, 37, 120.

TOLBOXANE [INN]

| CAS | 2430-46-8 | MW: | 232.13 | MF: | C14 H21 B O2 |

Tranquilizer

Roussel-UCLAF, France

| LgP | (4) |
| pKa | |
| pp cited in Vols.1-5: |

(1) Farthouat, U.S. Pat. 3 038 926 (1962).
(2) Caujolle et al., Arch. Int. Pharmacodyn. Ther., 1968, 172, 467.

TOLBUTAMIDE [U;INN]

| CAS | 64-77-7 | MW: | 270.35 | MF: | C12 H18 N2 O3 S |

Antidiabetic

Upjohn (Orinase)

| LgP | 2.34(M); 2.30(C) |
| pKa | 5.27 |
| pp cited in Vols.1-5: |
| 1: 63; |
| 2: 268; |
| 5: 60, 123, 178, 268 |

(1) Brit. Pat. 808 071 (1959). (2) Ruschig et al., U.S. Pat. 3 968 158 (1961). (3) W.F. Beyer et al. in 'Analytical Profiles of Drug Substances.', ed. K. Florey, Academic Press, NY, 1974, v. 3, pp 513-543.

TOLCICLATE [U;INN]

| CAS | 50838-36-3 | MW: | 323.46 | MF: | C20 H21 N O S |

Antifungal

Farmitalia Carlo Erba, Italy

| LgP | (4) |
| pKa | |
| pp cited in Vols.1-5: |

(1) P. Melloni et al., U.S. Pat. 3 855 263 (1974).
(2) I. deCarneri et al., Arzneim.-Forsch., 1976, 26, 769.

TOLDIMFOS [INN]

| CAS | 57808-64-7 | MW: | 199.19 | MF: | C9 H14 N O2 P |

Phosphorus source (vet.)

Cassella, FRG (Foston)

| LgP | (4) |
| pKa | |
| pp cited in Vols.1-5: |

(1) Pharm. J., 1969, 203, 104.

TOLFAMIDE [U;INN]

| CAS | 70788-29-3 | MW: | 213.18 | MF: | C8 H12 N3 O2 P |

Urease inhibitor

Norwich Eaton

| LgP | (4) |
| pKa | |
| pp cited in Vols.1-5: |

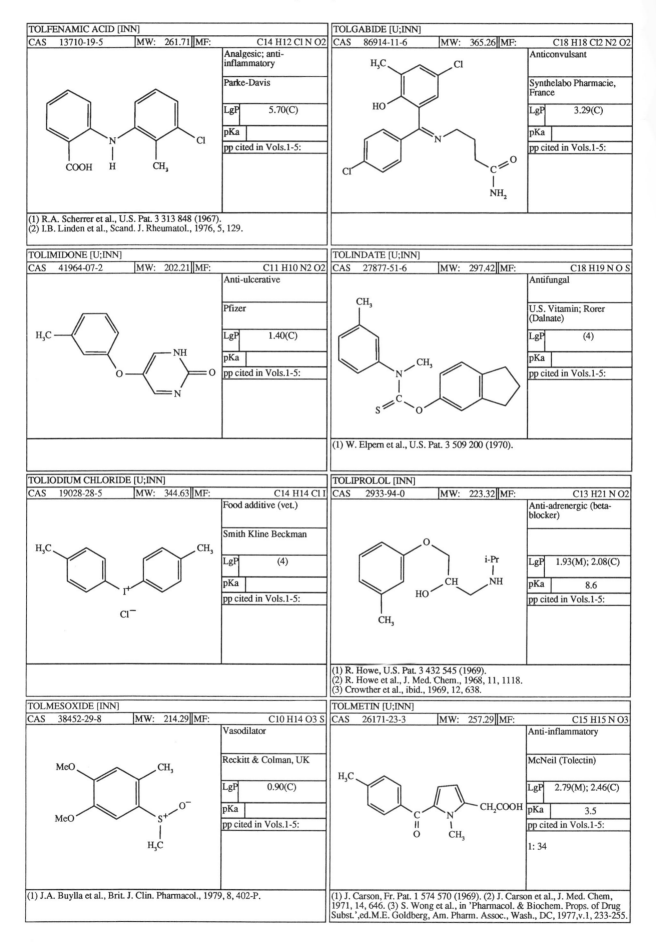

TOLFENAMIC ACID [INN]			
CAS 13710-19-5	MW: 261.71	MF:	C14 H12 Cl N O2

Analgesic; anti-inflammatory

Parke-Davis

LgP	5.70(C)
pKa	
pp cited in Vols.1-5:	

(1) R.A. Scherrer et al., U.S. Pat. 3 313 848 (1967).
(2) I.B. Linden et al., Scand. J. Rheumatol., 1976, 5, 129.

TOLGABIDE [U;INN]			
CAS 86914-11-6	MW: 365.26	MF:	C18 H18 Cl2 N2 O2

Anticonvulsant

Synthelabo Pharmacie, France

LgP	3.29(C)
pKa	
pp cited in Vols.1-5:	

TOLIMIDONE [U;INN]			
CAS 41964-07-2	MW: 202.21	MF:	C11 H10 N2 O2

Anti-ulcerative

Pfizer

LgP	1.40(C)
pKa	
pp cited in Vols.1-5:	

TOLINDATE [U;INN]			
CAS 27877-51-6	MW: 297.42	MF:	C18 H19 N O S

Antifungal

U.S. Vitamin; Rorer (Dalnate)

LgP	(4)
pKa	
pp cited in Vols.1-5:	

(1) W. Elpern et al., U.S. Pat. 3 509 200 (1970).

TOLIODIUM CHLORIDE [U;INN]			
CAS 19028-28-5	MW: 344.63	MF:	C14 H14 Cl I

Food additive (vet.)

Smith Kline Beckman

LgP	(4)
pKa	
pp cited in Vols.1-5:	

TOLIPROLOL [INN]			
CAS 2933-94-0	MW: 223.32	MF:	C13 H21 N O2

Anti-adrenergic (beta-blocker)

LgP	1.93(M); 2.08(C)
pKa	8.6
pp cited in Vols.1-5:	

(1) R. Howe, U.S. Pat. 3 432 545 (1969).
(2) R. Howe et al., J. Med. Chem., 1968, 11, 1118.
(3) Crowther et al., ibid., 1969, 12, 638.

TOLMESOXIDE [INN]			
CAS 38452-29-8	MW: 214.29	MF:	C10 H14 O3 S

Vasodilator

Reckitt & Colman, UK

LgP	0.90(C)
pKa	
pp cited in Vols.1-5:	

(1) J.A. Buylla et al., Brit. J. Clin. Pharmacol., 1979, 8, 402-P.

TOLMETIN [U;INN]			
CAS 26171-23-3	MW: 257.29	MF:	C15 H15 N O3

Anti-inflammatory

McNeil (Tolectin)

LgP	2.79(M); 2.46(C)
pKa	3.5
pp cited in Vols.1-5:	

1: 34

(1) J. Carson, Fr. Pat. 1 574 570 (1969). (2) J. Carson et al., J. Med. Chem, 1971, 14, 646. (3) S. Wong et al., in 'Pharmacol. & Biochem. Props. of Drug Subst.',ed.M.E. Goldberg, Am. Pharm. Assoc., Wash., DC, 1977,v.1, 233-255.

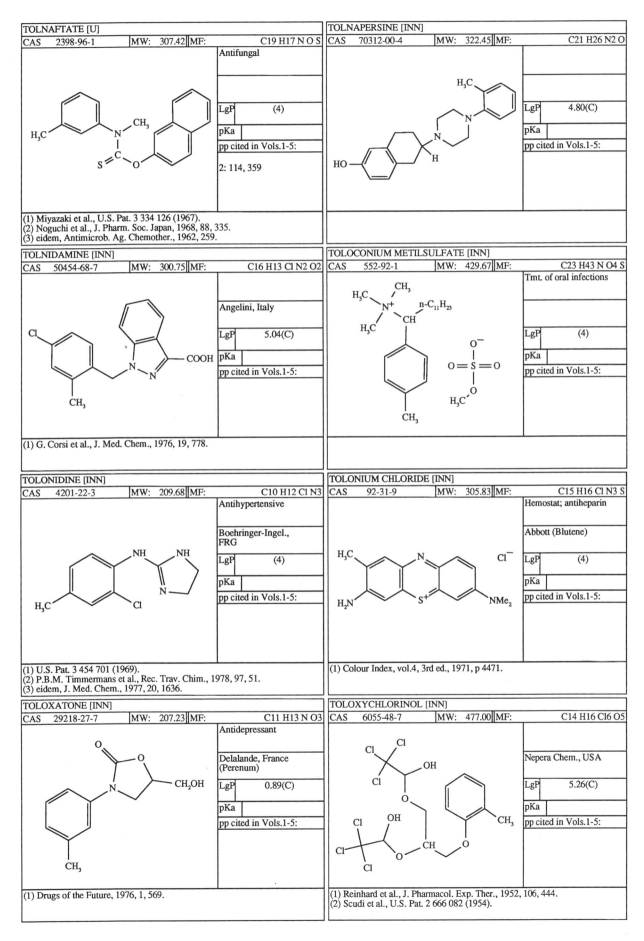

TOLNAFTATE [U]

CAS	2398-96-1	MW: 307.42	MF:	C19 H17 N O S

Antifungal

LgP	(4)
pKa	
pp cited in Vols.1-5:	

2: 114, 359

(1) Miyazaki et al., U.S. Pat. 3 334 126 (1967).
(2) Noguchi et al., J. Pharm. Soc. Japan, 1968, 88, 335.
(3) eidem, Antimicrob. Ag. Chemother., 1962, 259.

TOLNAPERSINE [INN]

CAS	70312-00-4	MW: 322.45	MF:	C21 H26 N2 O

LgP	4.80(C)
pKa	
pp cited in Vols.1-5:	

TOLNIDAMINE [INN]

CAS	50454-68-7	MW: 300.75	MF:	C16 H13 Cl N2 O2

Angelini, Italy

LgP	5.04(C)
pKa	
pp cited in Vols.1-5:	

(1) G. Corsi et al., J. Med. Chem., 1976, 19, 778.

TOLOCONIUM METILSULFATE [INN]

CAS	552-92-1	MW: 429.67	MF:	C23 H43 N O4 S

Tmt. of oral infections

LgP	(4)
pKa	
pp cited in Vols.1-5:	

TOLONIDINE [INN]

CAS	4201-22-3	MW: 209.68	MF:	C10 H12 Cl N3

Antihypertensive

Boehringer-Ingel., FRG

LgP	(4)
pKa	
pp cited in Vols.1-5:	

(1) U.S. Pat. 3 454 701 (1969).
(2) P.B.M. Timmermans et al., Rec. Trav. Chim., 1978, 97, 51.
(3) eidem, J. Med. Chem., 1977, 20, 1636.

TOLONIUM CHLORIDE [INN]

CAS	92-31-9	MW: 305.83	MF:	C15 H16 Cl N3 S

Hemostat; antiheparin

Abbott (Blutene)

LgP	(4)
pKa	
pp cited in Vols.1-5:	

(1) Colour Index, vol.4, 3rd ed., 1971, p 4471.

TOLOXATONE [INN]

CAS	29218-27-7	MW: 207.23	MF:	C11 H13 N O3

Antidepressant

Delalande, France (Perenum)

LgP	0.89(C)
pKa	
pp cited in Vols.1-5:	

(1) Drugs of the Future, 1976, 1, 569.

TOLOXYCHLORINOL [INN]

CAS	6055-48-7	MW: 477.00	MF:	C14 H16 Cl6 O5

Nepera Chem., USA

LgP	5.26(C)
pKa	
pp cited in Vols.1-5:	

(1) Reinhard et al., J. Pharmacol. Exp. Ther., 1952, 106, 444.
(2) Scudi et al., U.S. Pat. 2 666 082 (1954).

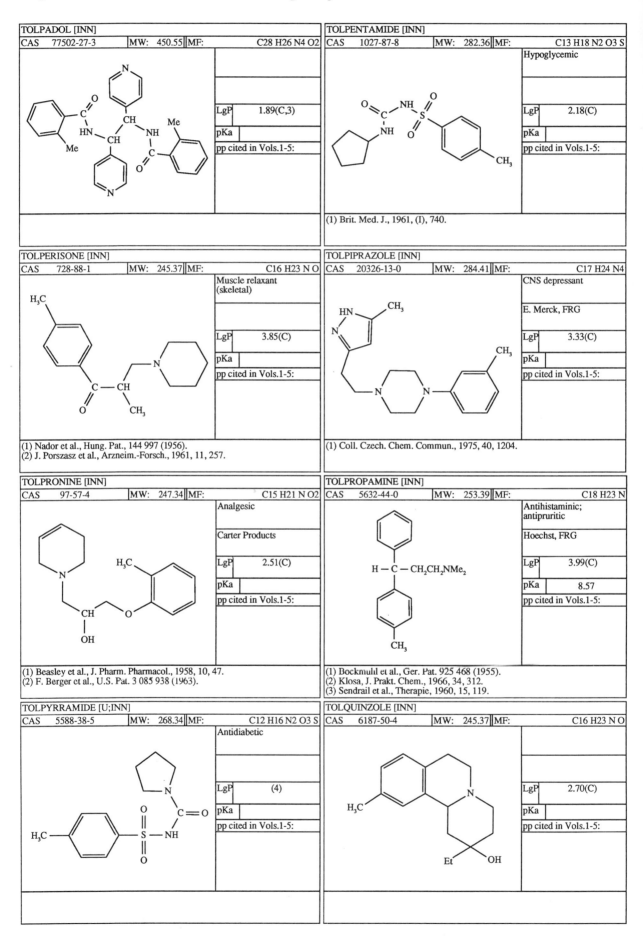

TOLPADOL [INN]		
CAS 77502-27-3	MW: 450.55 MF:	C28 H26 N4 O2
LgP	1.89(C,3)	
pKa		
pp cited in Vols.1-5:		

TOLPENTAMIDE [INN]		
CAS 1027-87-8	MW: 282.36 MF:	C13 H18 N2 O3 S
Hypoglycemic		
LgP	2.18(C)	
pKa		
pp cited in Vols.1-5:		

(1) Brit. Med. J., 1961, (I), 740.

TOLPERISONE [INN]		
CAS 728-88-1	MW: 245.37 MF:	C16 H23 N O
Muscle relaxant (skeletal)		
LgP	3.85(C)	
pKa		
pp cited in Vols.1-5:		

(1) Nador et al., Hung. Pat., 144 997 (1956).
(2) J. Porszasz et al., Arzneim.-Forsch., 1961, 11, 257.

TOLPIPRAZOLE [INN]		
CAS 20326-13-0	MW: 284.41 MF:	C17 H24 N4
CNS depressant		
E. Merck, FRG		
LgP	3.33(C)	
pKa		
pp cited in Vols.1-5:		

(1) Coll. Czech. Chem. Commun., 1975, 40, 1204.

TOLPRONINE [INN]		
CAS 97-57-4	MW: 247.34 MF:	C15 H21 N O2
Analgesic		
Carter Products		
LgP	2.51(C)	
pKa		
pp cited in Vols.1-5:		

(1) Beasley et al., J. Pharm. Pharmacol., 1958, 10, 47.
(2) F. Berger et al., U.S. Pat. 3 085 938 (1963).

TOLPROPAMINE [INN]		
CAS 5632-44-0	MW: 253.39 MF:	C18 H23 N
Antihistaminic; antipruritic		
Hoechst, FRG		
LgP	3.99(C)	
pKa	8.57	
pp cited in Vols.1-5:		

(1) Bockmuhl et al., Ger. Pat. 925 468 (1955).
(2) Klosa, J. Prakt. Chem., 1966, 34, 312.
(3) Sendrail et al., Therapie, 1960, 15, 119.

TOLPYRRAMIDE [U;INN]		
CAS 5588-38-5	MW: 268.34 MF:	C12 H16 N2 O3 S
Antidiabetic		
LgP	(4)	
pKa		
pp cited in Vols.1-5:		

TOLQUINZOLE [INN]		
CAS 6187-50-4	MW: 245.37 MF:	C16 H23 N O
LgP	2.70(C)	
pKa		
pp cited in Vols.1-5:		

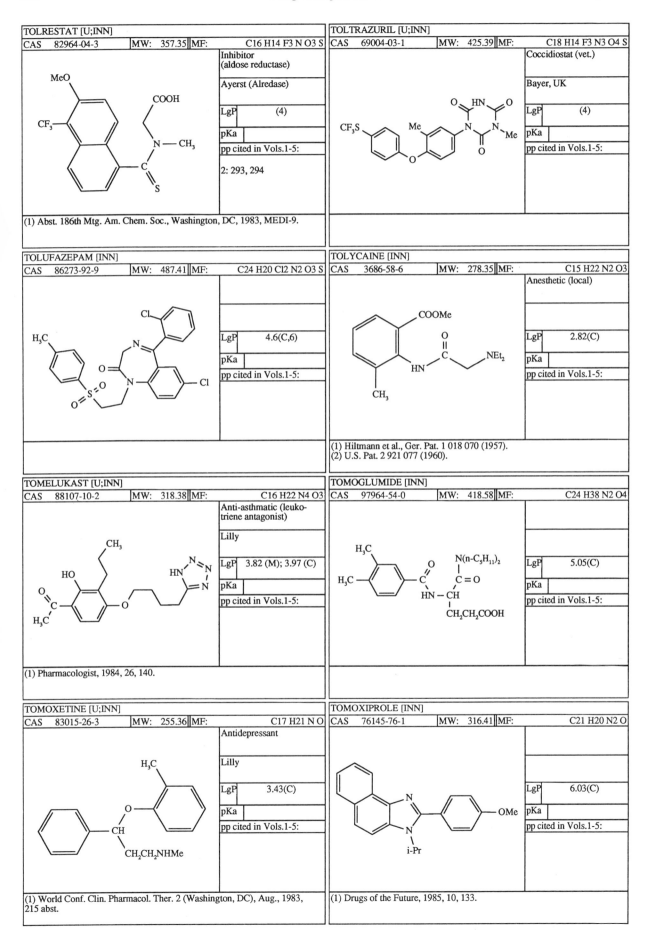

TOLRESTAT [U;INN]

CAS	82964-04-3	MW:	357.35	MF:	C16 H14 F3 N O3 S

Inhibitor (aldose reductase)

Ayerst (Alredase)

LgP	(4)
pKa	
pp cited in Vols.1-5:	

2: 293, 294

(1) Abst. 186th Mtg. Am. Chem. Soc., Washington, DC, 1983, MEDI-9.

TOLTRAZURIL [U;INN]

CAS	69004-03-1	MW:	425.39	MF:	C18 H14 F3 N3 O4 S

Coccidiostat (vet.)

Bayer, UK

LgP	(4)
pKa	
pp cited in Vols.1-5:	

TOLUFAZEPAM [INN]

CAS	86273-92-9	MW:	487.41	MF:	C24 H20 Cl2 N2 O3 S

LgP	4.6(C,6)
pKa	
pp cited in Vols.1-5:	

TOLYCAINE [INN]

CAS	3686-58-6	MW:	278.35	MF:	C15 H22 N2 O3

Anesthetic (local)

LgP	2.82(C)
pKa	
pp cited in Vols.1-5:	

(1) Hiltmann et al., Ger. Pat. 1 018 070 (1957).
(2) U.S. Pat. 2 921 077 (1960).

TOMELUKAST [U;INN]

CAS	88107-10-2	MW:	318.38	MF:	C16 H22 N4 O3

Anti-asthmatic (leuko-triene antagonist)

Lilly

LgP	3.82 (M); 3.97 (C)
pKa	
pp cited in Vols.1-5:	

(1) Pharmacologist, 1984, 26, 140.

TOMOGLUMIDE [INN]

CAS	97964-54-0	MW:	418.58	MF:	C24 H38 N2 O4

LgP	5.05(C)
pKa	
pp cited in Vols.1-5:	

TOMOXETINE [U;INN]

CAS	83015-26-3	MW:	255.36	MF:	C17 H21 N O

Antidepressant

Lilly

LgP	3.43(C)
pKa	
pp cited in Vols.1-5:	

(1) World Conf. Clin. Pharmacol. Ther. 2 (Washington, DC), Aug., 1983, 215 abst.

TOMOXIPROLE [INN]

CAS	76145-76-1	MW:	316.41	MF:	C21 H20 N2 O

LgP	6.03(C)
pKa	
pp cited in Vols.1-5:	

(1) Drugs of the Future, 1985, 10, 133.

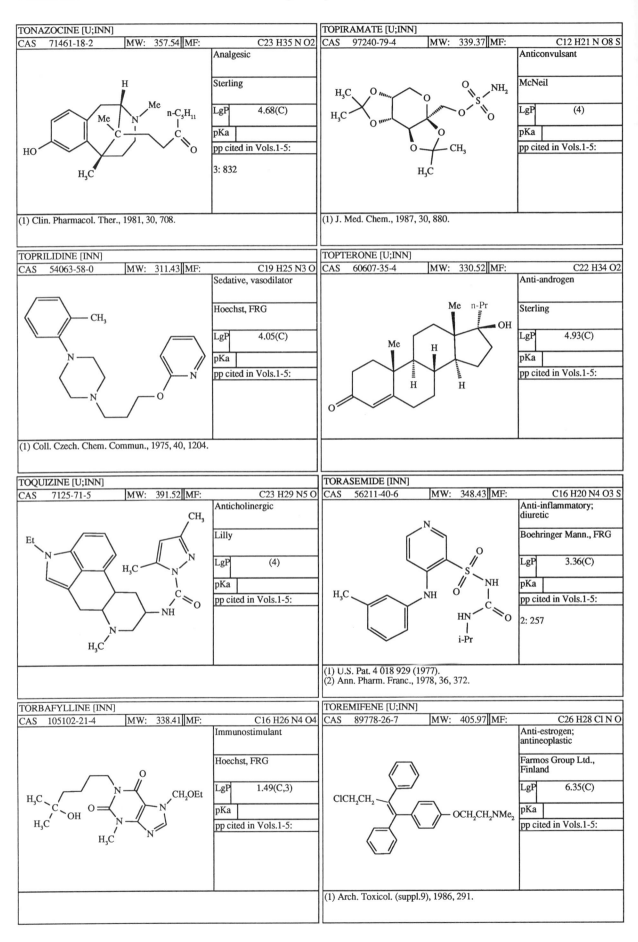

TONAZOCINE [U;INN]

| CAS | 71461-18-2 | MW: | 357.54 | MF: | C23 H35 N O2 |

Analgesic

Sterling

LgP 4.68(C)

pKa

pp cited in Vols.1-5:

3: 832

(1) Clin. Pharmacol. Ther., 1981, 30, 708.

TOPIRAMATE [U;INN]

| CAS | 97240-79-4 | MW: | 339.37 | MF: | C12 H21 N O8 S |

Anticonvulsant

McNeil

LgP (4)

pKa

pp cited in Vols.1-5:

(1) J. Med. Chem., 1987, 30, 880.

TOPRILIDINE [INN]

| CAS | 54063-58-0 | MW: | 311.43 | MF: | C19 H25 N3 O |

Sedative, vasodilator

Hoechst, FRG

LgP 4.05(C)

pKa

pp cited in Vols.1-5:

(1) Coll. Czech. Chem. Commun., 1975, 40, 1204.

TOPTERONE [U;INN]

| CAS | 60607-35-4 | MW: | 330.52 | MF: | C22 H34 O2 |

Anti-androgen

Sterling

LgP 4.93(C)

pKa

pp cited in Vols.1-5:

TOQUIZINE [U;INN]

| CAS | 7125-71-5 | MW: | 391.52 | MF: | C23 H29 N5 O |

Anticholinergic

Lilly

LgP (4)

pKa

pp cited in Vols.1-5:

TORASEMIDE [INN]

| CAS | 56211-40-6 | MW: | 348.43 | MF: | C16 H20 N4 O3 S |

Anti-inflammatory; diuretic

Boehringer Mann., FRG

LgP 3.36(C)

pKa

pp cited in Vols.1-5:

2: 257

(1) U.S. Pat. 4 018 929 (1977).
(2) Ann. Pharm. Franc., 1978, 36, 372.

TORBAFYLLINE [INN]

| CAS | 105102-21-4 | MW: | 338.41 | MF: | C16 H26 N4 O4 |

Immunostimulant

Hoechst, FRG

LgP 1.49(C,3)

pKa

pp cited in Vols.1-5:

TOREMIFENE [U;INN]

| CAS | 89778-26-7 | MW: | 405.97 | MF: | C26 H28 Cl N O |

Anti-estrogen; antineoplastic

Farmos Group Ltd., Finland

LgP 6.35(C)

pKa

pp cited in Vols.1-5:

(1) Arch. Toxicol. (suppl.9), 1986, 291.

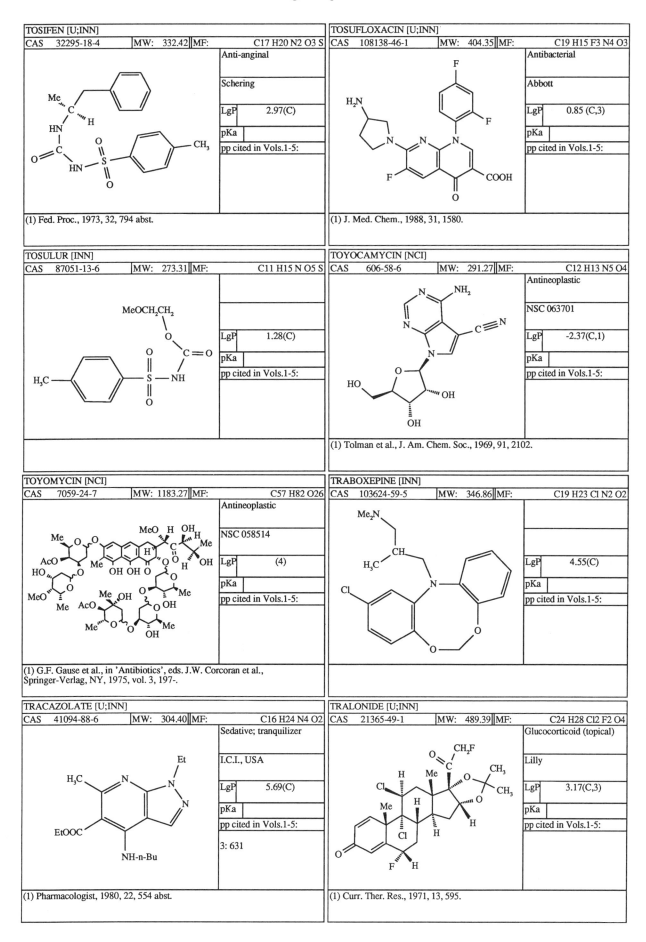

TOSIFEN [U;INN]

CAS	32295-18-4	MW: 332.42	MF:	C17 H20 N2 O3 S

Anti-anginal

Schering

LgP	2.97(C)
pKa	
pp cited in Vols.1-5:	

(1) Fed. Proc., 1973, 32, 794 abst.

TOSUFLOXACIN [U;INN]

CAS	108138-46-1	MW: 404.35	MF:	C19 H15 F3 N4 O3

Antibacterial

Abbott

LgP	0.85 (C,3)
pKa	
pp cited in Vols.1-5:	

(1) J. Med. Chem., 1988, 31, 1580.

TOSULUR [INN]

CAS	87051-13-6	MW: 273.31	MF:	C11 H15 N O5 S

LgP	1.28(C)
pKa	
pp cited in Vols.1-5:	

TOYOCAMYCIN [NCI]

CAS	606-58-6	MW: 291.27	MF:	C12 H13 N5 O4

Antineoplastic

NSC 063701

LgP	-2.37(C,1)
pKa	
pp cited in Vols.1-5:	

(1) Tolman et al., J. Am. Chem. Soc., 1969, 91, 2102.

TOYOMYCIN [NCI]

CAS	7059-24-7	MW: 1183.27	MF:	C57 H82 O26

Antineoplastic

NSC 058514

LgP	(4)
pKa	
pp cited in Vols.1-5:	

(1) G.F. Gause et al., in 'Antibiotics', eds. J.W. Corcoran et al., Springer-Verlag, NY, 1975, vol. 3, 197-.

TRABOXEPINE [INN]

CAS	103624-59-5	MW: 346.86	MF:	C19 H23 Cl N2 O2

LgP	4.55(C)
pKa	
pp cited in Vols.1-5:	

TRACAZOLATE [U;INN]

CAS	41094-88-6	MW: 304.40	MF:	C16 H24 N4 O2

Sedative; tranquilizer

I.C.I., USA

LgP	5.69(C)
pKa	
pp cited in Vols.1-5:	
3: 631	

(1) Pharmacologist, 1980, 22, 554 abst.

TRALONIDE [U;INN]

CAS	21365-49-1	MW: 489.39	MF:	C24 H28 Cl2 F2 O4

Glucocorticoid (topical)

Lilly

LgP	3.17(C,3)
pKa	
pp cited in Vols.1-5:	

(1) Curr. Ther. Res., 1971, 13, 595.

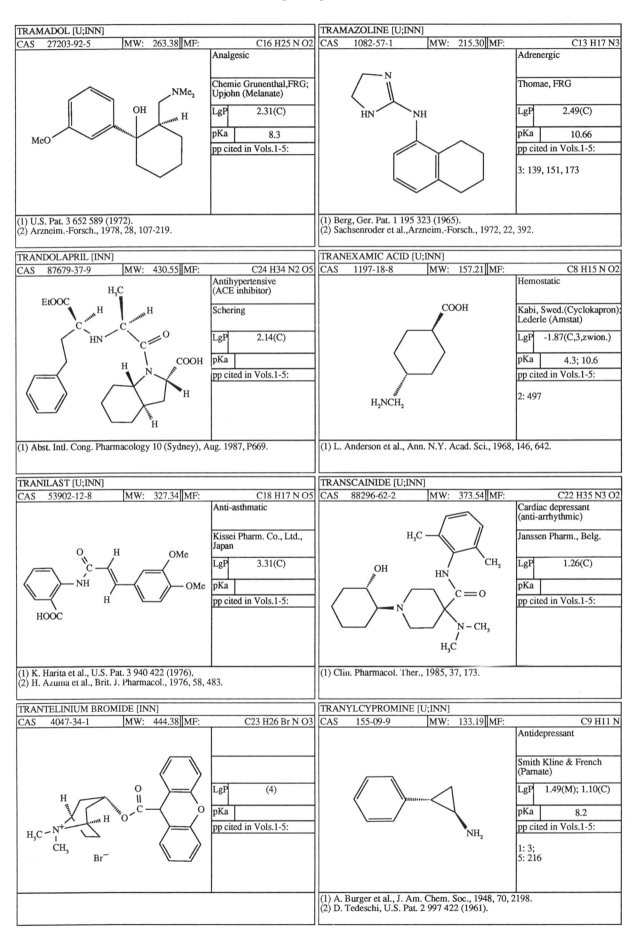

TRAMADOL [U;INN]

CAS	27203-92-5	MW:	263.38	MF:	C16 H25 N O2

Analgesic

Chemie Grunenthal,FRG; Upjohn (Melanate)

LgP	2.31(C)
pKa	8.3
pp cited in Vols.1-5:	

(1) U.S. Pat. 3 652 589 (1972).
(2) Arzneim.-Forsch., 1978, 28, 107-219.

TRAMAZOLINE [U;INN]

CAS	1082-57-1	MW:	215.30	MF:	C13 H17 N3

Adrenergic

Thomae, FRG

LgP	2.49(C)
pKa	10.66
pp cited in Vols.1-5:	

3: 139, 151, 173

(1) Berg, Ger. Pat. 1 195 323 (1965).
(2) Sachsenroder et al.,Arzneim.-Forsch., 1972, 22, 392.

TRANDOLAPRIL [INN]

CAS	87679-37-9	MW:	430.55	MF:	C24 H34 N2 O5

Antihypertensive (ACE inhibitor)

Schering

LgP	2.14(C)
pKa	
pp cited in Vols.1-5:	

(1) Abst. Intl. Cong. Pharmacology 10 (Sydney), Aug. 1987, P669.

TRANEXAMIC ACID [U;INN]

CAS	1197-18-8	MW:	157.21	MF:	C8 H15 N O2

Hemostatic

Kabi, Swed.(Cyclokapron); Lederle (Amstat)

LgP	-1.87(C,3,zwion.)
pKa	4.3; 10.6
pp cited in Vols.1-5:	

2: 497

(1) L. Anderson et al., Ann. N.Y. Acad. Sci., 1968, 146, 642.

TRANILAST [U;INN]

CAS	53902-12-8	MW:	327.34	MF:	C18 H17 N O5

Anti-asthmatic

Kissei Pharm. Co., Ltd., Japan

LgP	3.31(C)
pKa	
pp cited in Vols.1-5:	

(1) K. Harita et al., U.S. Pat. 3 940 422 (1976).
(2) H. Azuma et al., Brit. J. Pharmacol., 1976, 58, 483.

TRANSCAINIDE [U;INN]

CAS	88296-62-2	MW:	373.54	MF:	C22 H35 N3 O2

Cardiac depressant (anti-arrhythmic)

Janssen Pharm., Belg.

LgP	1.26(C)
pKa	
pp cited in Vols.1-5:	

(1) Clin. Pharmacol. Ther., 1985, 37, 173.

TRANTELINIUM BROMIDE [INN]

CAS	4047-34-1	MW:	444.38	MF:	C23 H26 Br N O3

LgP	(4)
pKa	
pp cited in Vols.1-5:	

TRANYLCYPROMINE [U;INN]

CAS	155-09-9	MW:	133.19	MF:	C9 H11 N

Antidepressant

Smith Kline & French (Parnate)

LgP	1.49(M); 1.10(C)
pKa	8.2
pp cited in Vols.1-5:	

1: 3;
5: 216

(1) A. Burger et al., J. Am. Chem. Soc., 1948, 70, 2198.
(2) D. Tedeschi, U.S. Pat. 2 997 422 (1961).

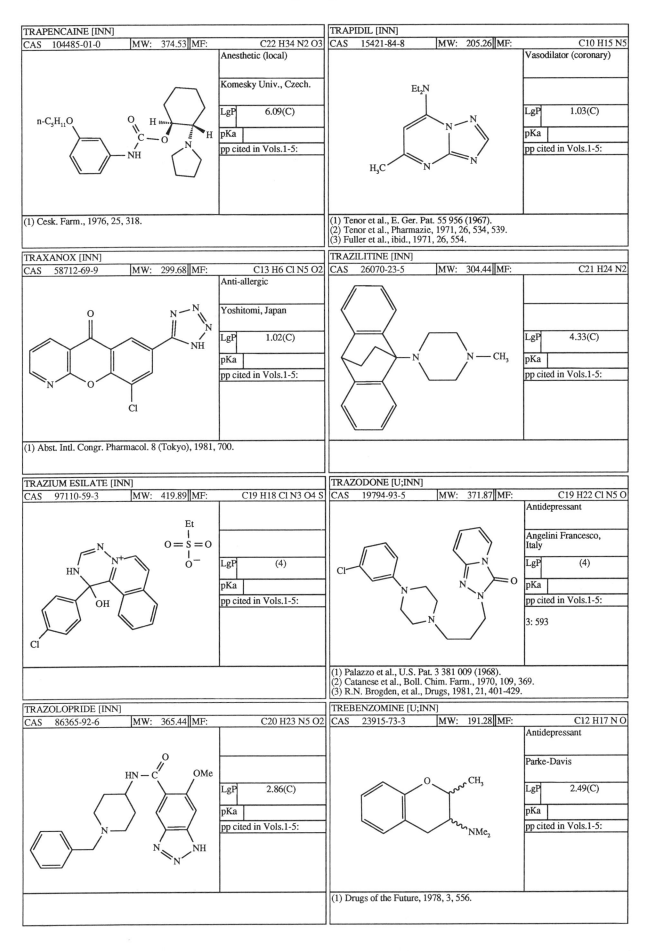

TRAPENCAINE [INN]

CAS	104485-01-0	MW:	374.53	MF:	C22 H34 N2 O3

Anesthetic (local)

Komesky Univ., Czech.

LgP	6.09(C)
pKa	
pp cited in Vols.1-5:	

(1) Cesk. Farm., 1976, 25, 318.

TRAPIDIL [INN]

CAS	15421-84-8	MW:	205.26	MF:	C10 H15 N5

Vasodilator (coronary)

LgP	1.03(C)
pKa	
pp cited in Vols.1-5:	

(1) Tenor et al., E. Ger. Pat. 55 956 (1967).
(2) Tenor et al., Pharmazie, 1971, 26, 534, 539.
(3) Fuller et al., ibid., 1971, 26, 554.

TRAXANOX [INN]

CAS	58712-69-9	MW:	299.68	MF:	C13 H6 Cl N5 O2

Anti-allergic

Yoshitomi, Japan

LgP	1.02(C)
pKa	
pp cited in Vols.1-5:	

(1) Abst. Intl. Congr. Pharmacol. 8 (Tokyo), 1981, 700.

TRAZILITINE [INN]

CAS	26070-23-5	MW:	304.44	MF:	C21 H24 N2

LgP	4.33(C)
pKa	
pp cited in Vols.1-5:	

TRAZIUM ESILATE [INN]

CAS	97110-59-3	MW:	419.89	MF:	C19 H18 Cl N3 O4 S

LgP	(4)
pKa	
pp cited in Vols.1-5:	

TRAZODONE [U;INN]

CAS	19794-93-5	MW:	371.87	MF:	C19 H22 Cl N5 O

Antidepressant

Angelini Francesco, Italy

LgP	(4)
pKa	
pp cited in Vols.1-5:	
3: 593	

(1) Palazzo et al., U.S. Pat. 3 381 009 (1968).
(2) Catanese et al., Boll. Chim. Farm., 1970, 109, 369.
(3) R.N. Brogden, et al., Drugs, 1981, 21, 401-429.

TRAZOLOPRIDE [INN]

CAS	86365-92-6	MW:	365.44	MF:	C20 H23 N5 O2

LgP	2.86(C)
pKa	
pp cited in Vols.1-5:	

TREBENZOMINE [U;INN]

CAS	23915-73-3	MW:	191.28	MF:	C12 H17 N O

Antidepressant

Parke-Davis

LgP	2.49(C)
pKa	
pp cited in Vols.1-5:	

(1) Drugs of the Future, 1978, 3, 556.

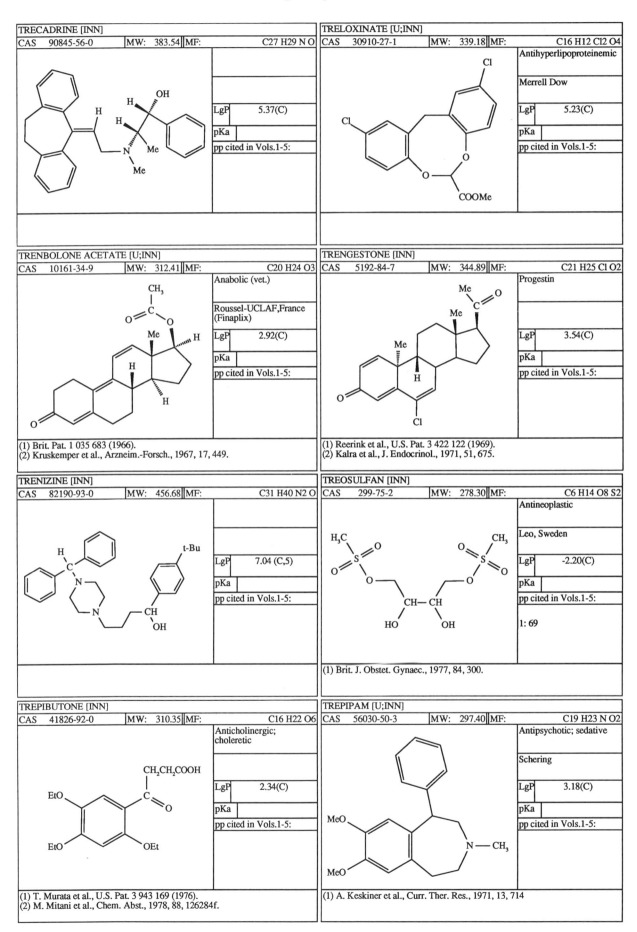

TRECADRINE [INN]			
CAS 90845-56-0	MW: 383.54	MF:	C27 H29 N O
		LgP	5.37(C)
		pKa	
		pp cited in Vols.1-5:	

TRELOXINATE [U;INN]			
CAS 30910-27-1	MW: 339.18	MF:	C16 H12 Cl2 O4
Antihyperlipoproteinemic			
Merrell Dow			
LgP	5.23(C)		
pKa			
pp cited in Vols.1-5:			

TRENBOLONE ACETATE [U;INN]			
CAS 10161-34-9	MW: 312.41	MF:	C20 H24 O3
Anabolic (vet.)			
Roussel-UCLAF,France (Finaplix)			
LgP	2.92(C)		
pKa			
pp cited in Vols.1-5:			

(1) Brit. Pat. 1 035 683 (1966).
(2) Kruskemper et al., Arzneim.-Forsch., 1967, 17, 449.

TRENGESTONE [INN]			
CAS 5192-84-7	MW: 344.89	MF:	C21 H25 Cl O2
Progestin			
LgP	3.54(C)		
pKa			
pp cited in Vols.1-5:			

(1) Reerink et al., U.S. Pat. 3 422 122 (1969).
(2) Kalra et al., J. Endocrinol., 1971, 51, 675.

TRENIZINE [INN]			
CAS 82190-93-0	MW: 456.68	MF:	C31 H40 N2 O
		LgP	7.04 (C,5)
		pKa	
		pp cited in Vols.1-5:	

TREOSULFAN [INN]			
CAS 299-75-2	MW: 278.30	MF:	C6 H14 O8 S2
Antineoplastic			
Leo, Sweden			
LgP	-2.20(C)		
pKa			
pp cited in Vols.1-5:			
1: 69			

(1) Brit. J. Obstet. Gynaec., 1977, 84, 300.

TREPIBUTONE [INN]			
CAS 41826-92-0	MW: 310.35	MF:	C16 H22 O6
Anticholinergic; choleretic			
LgP	2.34(C)		
pKa			
pp cited in Vols.1-5:			

(1) T. Murata et al., U.S. Pat. 3 943 169 (1976).
(2) M. Mitani et al., Chem. Abst., 1978, 88, 126284f.

TREPIPAM [U;INN]			
CAS 56030-50-3	MW: 297.40	MF:	C19 H23 N O2
Antipsychotic; sedative			
Schering			
LgP	3.18(C)		
pKa			
pp cited in Vols.1-5:			

(1) A. Keskiner et al., Curr. Ther. Res., 1971, 13, 714

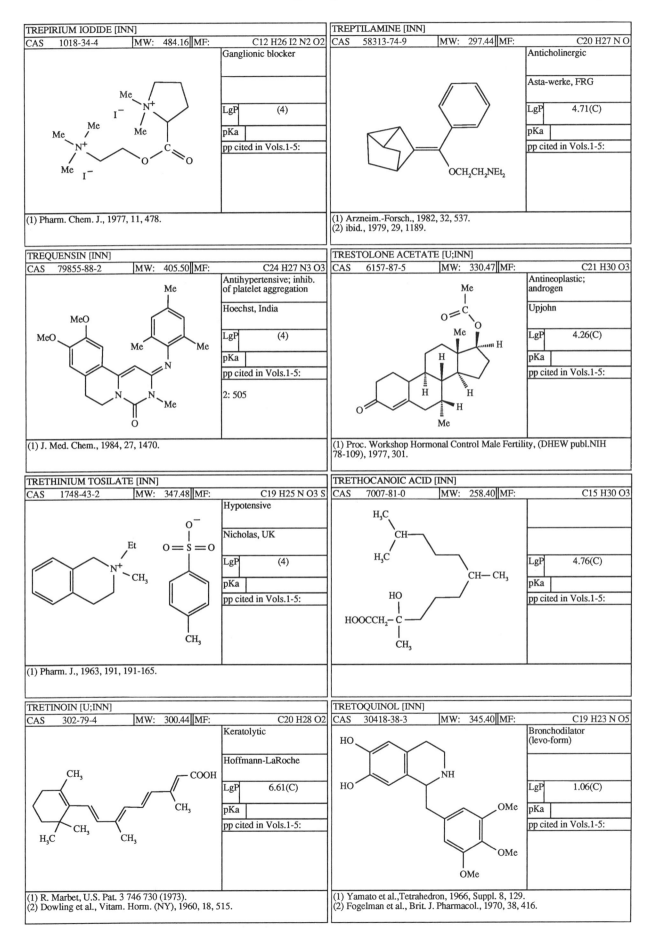

TREPIRIUM IODIDE [INN]

CAS	1018-34-4	MW:	484.16	MF:	C12 H26 I2 N2 O2

Ganglionic blocker

LgP	(4)
pKa	
pp cited in Vols.1-5:	

(1) Pharm. Chem. J., 1977, 11, 478.

TREPTILAMINE [INN]

CAS	58313-74-9	MW:	297.44	MF:	C20 H27 N O

Anticholinergic

Asta-werke, FRG

LgP	4.71(C)
pKa	
pp cited in Vols.1-5:	

(1) Arzneim.-Forsch., 1982, 32, 537.
(2) ibid., 1979, 29, 1189.

TREQUENSIN [INN]

CAS	79855-88-2	MW:	405.50	MF:	C24 H27 N3 O3

Antihypertensive; inhib. of platelet aggregation

Hoechst, India

LgP	(4)
pKa	
pp cited in Vols.1-5:	2: 505

(1) J. Med. Chem., 1984, 27, 1470.

TRESTOLONE ACETATE [U;INN]

CAS	6157-87-5	MW:	330.47	MF:	C21 H30 O3

Antineoplastic; androgen

Upjohn

LgP	4.26(C)
pKa	
pp cited in Vols.1-5:	

(1) Proc. Workshop Hormonal Control Male Fertility, (DHEW publ.NIH 78-109), 1977, 301.

TRETHINIUM TOSILATE [INN]

CAS	1748-43-2	MW:	347.48	MF:	C19 H25 N O3 S

Hypotensive

Nicholas, UK

LgP	(4)
pKa	
pp cited in Vols.1-5:	

(1) Pharm. J., 1963, 191, 191-165.

TRETHOCANOIC ACID [INN]

CAS	7007-81-0	MW:	258.40	MF:	C15 H30 O3

LgP	4.76(C)
pKa	
pp cited in Vols.1-5:	

TRETINOIN [U;INN]

CAS	302-79-4	MW:	300.44	MF:	C20 H28 O2

Keratolytic

Hoffmann-LaRoche

LgP	6.61(C)
pKa	
pp cited in Vols.1-5:	

(1) R. Marbet, U.S. Pat. 3 746 730 (1973).
(2) Dowling et al., Vitam. Horm. (NY), 1960, 18, 515.

TRETOQUINOL [INN]

CAS	30418-38-3	MW:	345.40	MF:	C19 H23 N O5

Bronchodilator (levo-form)

LgP	1.06(C)
pKa	
pp cited in Vols.1-5:	

(1) Yamato et al.,Tetrahedron, 1966, Suppl. 8, 129.
(2) Fogelman et al., Brit. J. Pharmacol., 1970, 38, 416.

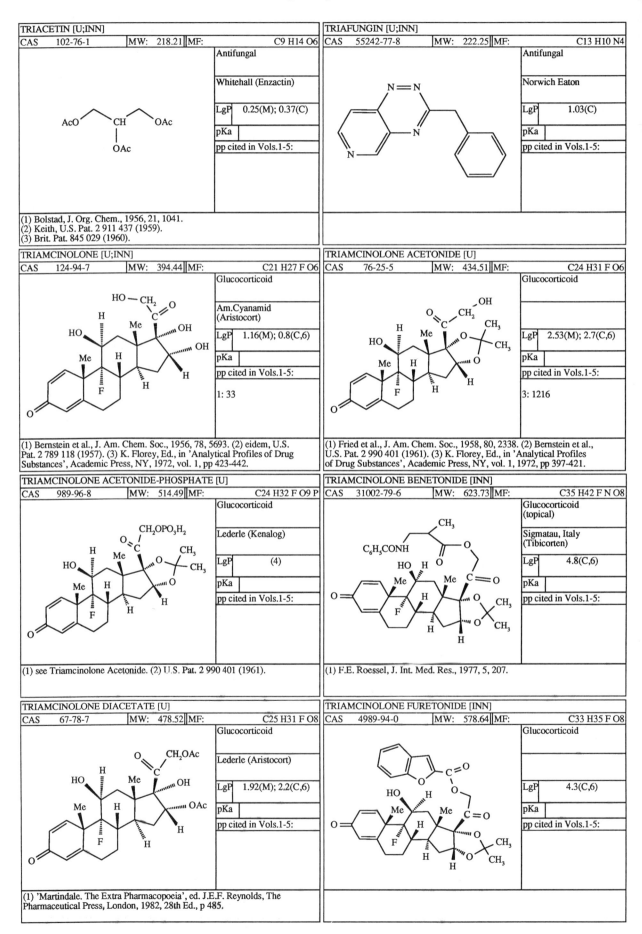

TRIACETIN [U;INN]

CAS	102-76-1	MW:	218.21	MF:	C9 H14 O6

Antifungal

Whitehall (Enzactin)

LgP	0.25(M); 0.37(C)
pKa	
pp cited in Vols.1-5:	

(1) Bolstad, J. Org. Chem., 1956, 21, 1041.
(2) Keith, U.S. Pat. 2 911 437 (1959).
(3) Brit. Pat. 845 029 (1960).

TRIAFUNGIN [U;INN]

CAS	55242-77-8	MW:	222.25	MF:	C13 H10 N4

Antifungal

Norwich Eaton

LgP	1.03(C)
pKa	
pp cited in Vols.1-5:	

TRIAMCINOLONE [U;INN]

CAS	124-94-7	MW:	394.44	MF:	C21 H27 F O6

Glucocorticoid

Am.Cyanamid (Aristocort)

LgP	1.16(M); 0.8(C,6)
pKa	
pp cited in Vols.1-5:	
1: 33	

(1) Bernstein et al., J. Am. Chem. Soc., 1956, 78, 5693. (2) eidem, U.S. Pat. 2 789 118 (1957). (3) K. Florey, Ed., in 'Analytical Profiles of Drug Substances', Academic Press, NY, 1972, vol. 1, pp 423-442.

TRIAMCINOLONE ACETONIDE [U]

CAS	76-25-5	MW:	434.51	MF:	C24 H31 F O6

Glucocorticoid

LgP	2.53(M); 2.7(C,6)
pKa	
pp cited in Vols.1-5:	
3: 1216	

(1) Fried et al., J. Am. Chem. Soc., 1958, 80, 2338. (2) Bernstein et al., U.S. Pat. 2 990 401 (1961). (3) K. Florey, Ed., in 'Analytical Profiles of Drug Substances', Academic Press, NY, vol. 1, 1972, pp 397-421.

TRIAMCINOLONE ACETONIDE-PHOSPHATE [U]

CAS	989-96-8	MW:	514.49	MF:	C24 H32 F O9 P

Glucocorticoid

Lederle (Kenalog)

LgP	(4)
pKa	
pp cited in Vols.1-5:	

(1) see Triamcinolone Acetonide. (2) U.S. Pat. 2 990 401 (1961).

TRIAMCINOLONE BENETONIDE [INN]

CAS	31002-79-6	MW:	623.73	MF:	C35 H42 F N O8

Glucocorticoid (topical)

Sigmatau, Italy (Tibicorten)

LgP	4.8(C,6)
pKa	
pp cited in Vols.1-5:	

(1) F.E. Roessel, J. Int. Med. Res., 1977, 5, 207.

TRIAMCINOLONE DIACETATE [U]

CAS	67-78-7	MW:	478.52	MF:	C25 H31 F O8

Glucocorticoid

Lederle (Aristocort)

LgP	1.92(M); 2.2(C,6)
pKa	
pp cited in Vols.1-5:	

(1) 'Martindale. The Extra Pharmacopoeia', ed. J.E.F. Reynolds, The Pharmaceutical Press, London, 1982, 28th Ed., p 485.

TRIAMCINOLONE FURETONIDE [INN]

CAS	4989-94-0	MW:	578.64	MF:	C33 H35 F O8

Glucocorticoid

LgP	4.3(C,6)
pKa	
pp cited in Vols.1-5:	

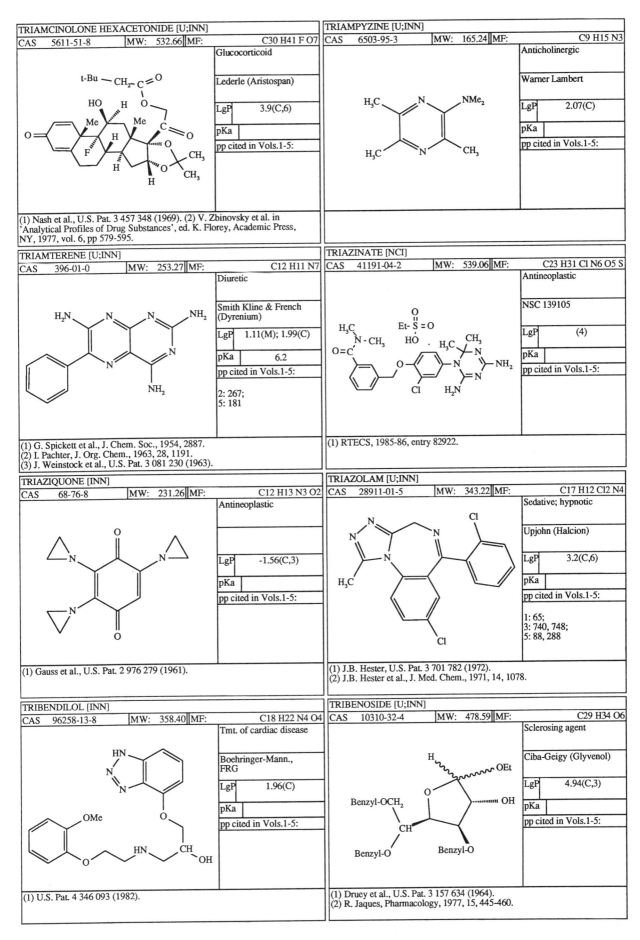

TRIAMCINOLONE HEXACETONIDE [U;INN]

CAS	5611-51-8	MW:	532.66	MF:	C30 H41 F O7

Glucocorticoid

Lederle (Aristospan)

LgP	3.9(C,6)
pKa	
pp cited in Vols.1-5:	

(1) Nash et al., U.S. Pat. 3 457 348 (1969). (2) V. Zbinovsky et al. in 'Analytical Profiles of Drug Substances', ed. K. Florey, Academic Press, NY, 1977, vol. 6, pp 579-595.

TRIAMTERENE [U;INN]

CAS	396-01-0	MW:	253.27	MF:	C12 H11 N7

Diuretic

Smith Kline & French (Dyrenium)

LgP	1.11(M); 1.99(C)
pKa	6.2
pp cited in Vols.1-5:	

2: 267;
5: 181

(1) G. Spickett et al., J. Chem. Soc., 1954, 2887.
(2) I. Pachter, J. Org. Chem., 1963, 28, 1191.
(3) J. Weinstock et al., U.S. Pat. 3 081 230 (1963).

TRIAZIQUONE [INN]

CAS	68-76-8	MW:	231.26	MF:	C12 H13 N3 O2

Antineoplastic

LgP	-1.56(C,3)
pKa	
pp cited in Vols.1-5:	

(1) Gauss et al., U.S. Pat. 2 976 279 (1961).

TRIBENDILOL [INN]

CAS	96258-13-8	MW:	358.40	MF:	C18 H22 N4 O4

Tmt. of cardiac disease

Boehringer-Mann., FRG

LgP	1.96(C)
pKa	
pp cited in Vols.1-5:	

(1) U.S. Pat. 4 346 093 (1982).

TRIAMPYZINE [U;INN]

CAS	6503-95-3	MW:	165.24	MF:	C9 H15 N3

Anticholinergic

Warner Lambert

LgP	2.07(C)
pKa	
pp cited in Vols.1-5:	

TRIAZINATE [NCI]

CAS	41191-04-2	MW:	539.06	MF:	C23 H31 Cl N6 O5 S

Antineoplastic

NSC 139105

LgP	(4)
pKa	
pp cited in Vols.1-5:	

(1) RTECS, 1985-86, entry 82922.

TRIAZOLAM [U;INN]

CAS	28911-01-5	MW:	343.22	MF:	C17 H12 Cl2 N4

Sedative; hypnotic

Upjohn (Halcion)

LgP	3.2(C,6)
pKa	
pp cited in Vols.1-5:	

1: 65;
3: 740, 748;
5: 88, 288

(1) J.B. Hester, U.S. Pat. 3 701 782 (1972).
(2) J.B. Hester et al., J. Med. Chem., 1971, 14, 1078.

TRIBENOSIDE [U;INN]

CAS	10310-32-4	MW:	478.59	MF:	C29 H34 O6

Sclerosing agent

Ciba-Geigy (Glyvenol)

LgP	4.94(C,3)
pKa	
pp cited in Vols.1-5:	

(1) Druey et al., U.S. Pat. 3 157 634 (1964).
(2) R. Jaques, Pharmacology, 1977, 15, 445-460.

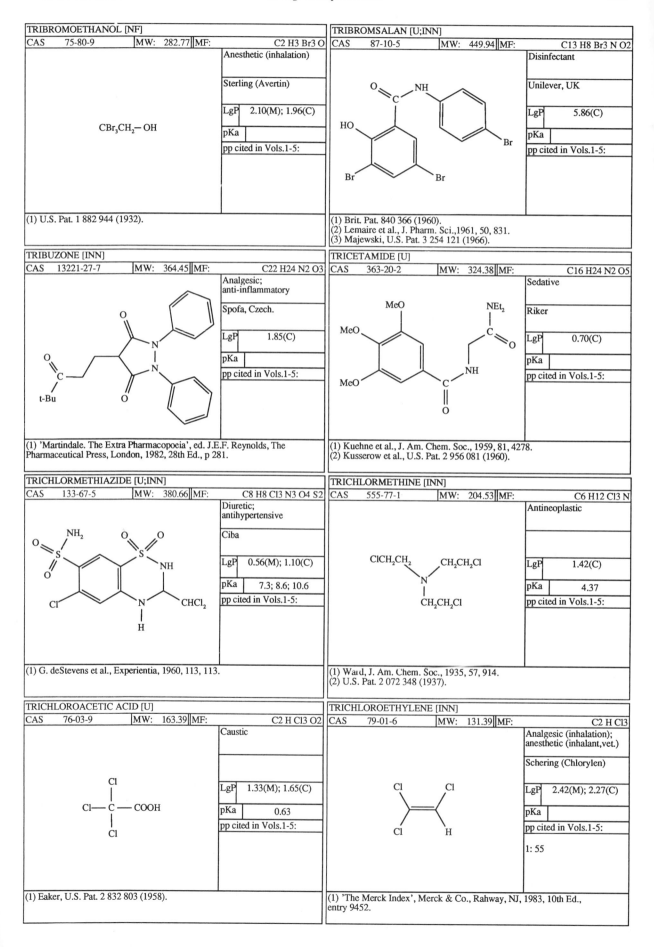

TRIBROMOETHANOL [NF]			
CAS 75-80-9	MW: 282.77	MF:	C2 H3 Br3 O
		Anesthetic (inhalation)	
		Sterling (Avertin)	
CBr₃CH₂—OH		LgP 2.10(M); 1.96(C)	
		pKa	
		pp cited in Vols.1-5:	
(1) U.S. Pat. 1 882 944 (1932).			

CBr₃CH₂—OH rendered as CBr_3CH_2-OH

TRIBROMSALAN [U;INN]			
CAS 87-10-5	MW: 449.94	MF:	C13 H8 Br3 N O2
		Disinfectant	
		Unilever, UK	
		LgP 5.86(C)	
		pKa	
		pp cited in Vols.1-5:	
(1) Brit. Pat. 840 366 (1960). (2) Lemaire et al., J. Pharm. Sci.,1961, 50, 831. (3) Majewski, U.S. Pat. 3 254 121 (1966).			

TRIBUZONE [INN]			
CAS 13221-27-7	MW: 364.45	MF:	C22 H24 N2 O3
		Analgesic; anti-inflammatory	
		Spofa, Czech.	
		LgP 1.85(C)	
		pKa	
		pp cited in Vols.1-5:	
(1) 'Martindale. The Extra Pharmacopoeia', ed. J.E.F. Reynolds, The Pharmaceutical Press, London, 1982, 28th Ed., p 281.			

TRICETAMIDE [U]			
CAS 363-20-2	MW: 324.38	MF:	C16 H24 N2 O5
		Sedative	
		Riker	
		LgP 0.70(C)	
		pKa	
		pp cited in Vols.1-5:	
(1) Kuehne et al., J. Am. Chem. Soc., 1959, 81, 4278. (2) Kusserow et al., U.S. Pat. 2 956 081 (1960).			

TRICHLORMETHIAZIDE [U;INN]			
CAS 133-67-5	MW: 380.66	MF:	C8 H8 Cl3 N3 O4 S2
		Diuretic; antihypertensive	
		Ciba	
		LgP 0.56(M); 1.10(C)	
		pKa 7.3; 8.6; 10.6	
		pp cited in Vols.1-5:	
(1) G. deStevens et al., Experientia, 1960, 113, 113.			

TRICHLORMETHINE [INN]			
CAS 555-77-1	MW: 204.53	MF:	C6 H12 Cl3 N
		Antineoplastic	
		LgP 1.42(C)	
		pKa 4.37	
		pp cited in Vols.1-5:	
(1) Ward, J. Am. Chem. Soc., 1935, 57, 914. (2) U.S. Pat. 2 072 348 (1937).			

TRICHLOROACETIC ACID [U]			
CAS 76-03-9	MW: 163.39	MF:	C2 H Cl3 O2
		Caustic	
		LgP 1.33(M); 1.65(C)	
		pKa 0.63	
		pp cited in Vols.1-5:	
(1) Eaker, U.S. Pat. 2 832 803 (1958).			

TRICHLOROETHYLENE [INN]			
CAS 79-01-6	MW: 131.39	MF:	C2 H Cl3
		Analgesic (inhalation); anesthetic (inhalant,vet.)	
		Schering (Chlorylen)	
		LgP 2.42(M); 2.27(C)	
		pKa	
		pp cited in Vols.1-5:	
		1: 55	
(1) 'The Merck Index', Merck & Co., Rahway, NJ, 1983, 10th Ed., entry 9452.			

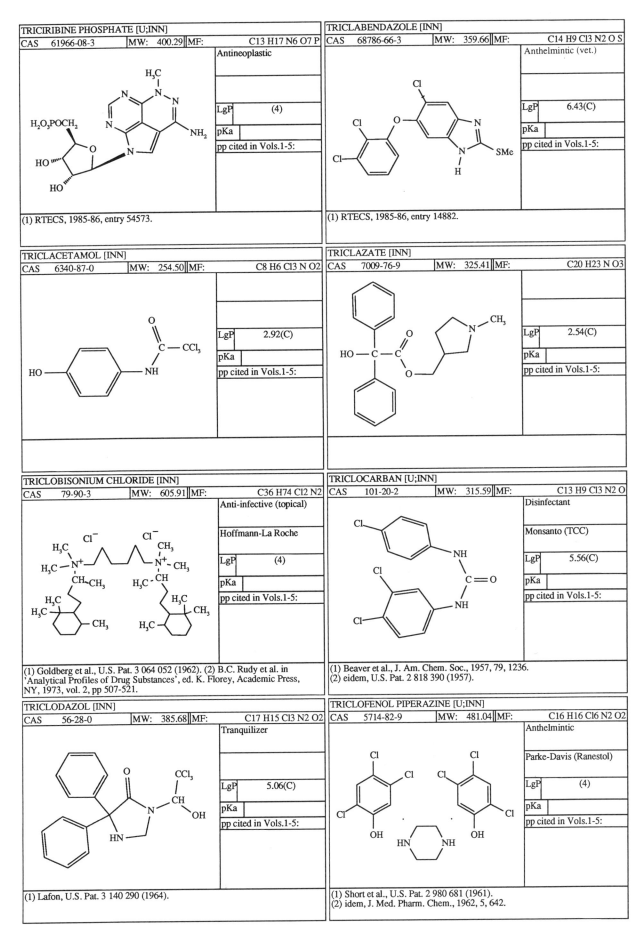

TRICIRIBINE PHOSPHATE [U;INN]

CAS	61966-08-3	MW:	400.29	MF:	C13 H17 N6 O7 P

Antineoplastic

LgP (4)

pKa

pp cited in Vols.1-5:

(1) RTECS, 1985-86, entry 54573.

TRICLABENDAZOLE [INN]

CAS	68786-66-3	MW:	359.66	MF:	C14 H9 Cl3 N2 O S

Anthelmintic (vet.)

LgP 6.43(C)

pKa

pp cited in Vols.1-5:

(1) RTECS, 1985-86, entry 14882.

TRICLACETAMOL [INN]

CAS	6340-87-0	MW:	254.50	MF:	C8 H6 Cl3 N O2

LgP 2.92(C)

pKa

pp cited in Vols.1-5:

TRICLAZATE [INN]

CAS	7009-76-9	MW:	325.41	MF:	C20 H23 N O3

LgP 2.54(C)

pKa

pp cited in Vols.1-5:

TRICLOBISONIUM CHLORIDE [INN]

CAS	79-90-3	MW:	605.91	MF:	C36 H74 Cl2 N2

Anti-infective (topical)

Hoffmann-La Roche

LgP (4)

pKa

pp cited in Vols.1-5:

(1) Goldberg et al., U.S. Pat. 3 064 052 (1962). (2) B.C. Rudy et al. in 'Analytical Profiles of Drug Substances', ed. K. Florey, Academic Press, NY, 1973, vol. 2, pp 507-521.

TRICLOCARBAN [U;INN]

CAS	101-20-2	MW:	315.59	MF:	C13 H9 Cl3 N2 O

Disinfectant

Monsanto (TCC)

LgP 5.56(C)

pKa

pp cited in Vols.1-5:

(1) Beaver et al., J. Am. Chem. Soc., 1957, 79, 1236.
(2) eidem, U.S. Pat. 2 818 390 (1957).

TRICLODAZOL [INN]

CAS	56-28-0	MW:	385.68	MF:	C17 H15 Cl3 N2 O2

Tranquilizer

LgP 5.06(C)

pKa

pp cited in Vols.1-5:

(1) Lafon, U.S. Pat. 3 140 290 (1964).

TRICLOFENOL PIPERAZINE [U;INN]

CAS	5714-82-9	MW:	481.04	MF:	C16 H16 Cl6 N2 O2

Anthelmintic

Parke-Davis (Ranestol)

LgP (4)

pKa

pp cited in Vols.1-5:

(1) Short et al., U.S. Pat. 2 980 681 (1961).
(2) idem, J. Med. Pharm. Chem., 1962, 5, 642.

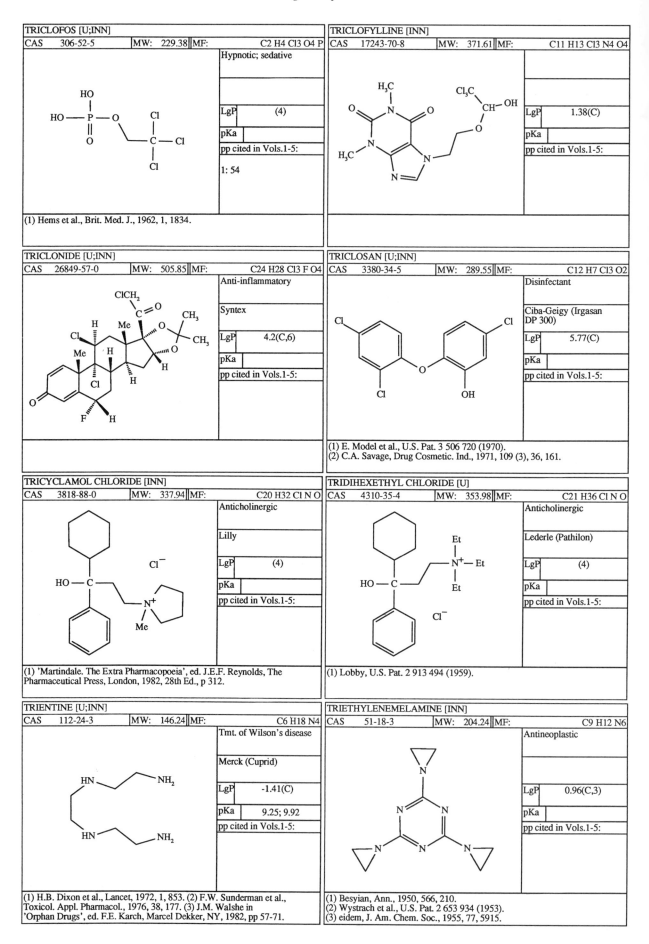

TRICLOFOS [U;INN]

CAS	306-52-5	MW: 229.38	MF:	C2 H4 Cl3 O4 P

Hypnotic; sedative

LgP (4)

pKa

pp cited in Vols.1-5:

1: 54

(1) Hems et al., Brit. Med. J., 1962, 1, 1834.

TRICLOFYLLINE [INN]

CAS	17243-70-8	MW: 371.61	MF:	C11 H13 Cl3 N4 O4

LgP 1.38(C)

pKa

pp cited in Vols.1-5:

TRICLONIDE [U;INN]

CAS	26849-57-0	MW: 505.85	MF:	C24 H28 Cl3 F O4

Anti-inflammatory

Syntex

LgP 4.2(C,6)

pKa

pp cited in Vols.1-5:

TRICLOSAN [U;INN]

CAS	3380-34-5	MW: 289.55	MF:	C12 H7 Cl3 O2

Disinfectant

Ciba-Geigy (Irgasan DP 300)

LgP 5.77(C)

pKa

pp cited in Vols.1-5:

(1) E. Model et al., U.S. Pat. 3 506 720 (1970).
(2) C.A. Savage, Drug Cosmetic. Ind., 1971, 109 (3), 36, 161.

TRICYCLAMOL CHLORIDE [INN]

CAS	3818-88-0	MW: 337.94	MF:	C20 H32 Cl N O

Anticholinergic

Lilly

LgP (4)

pKa

pp cited in Vols.1-5:

(1) 'Martindale. The Extra Pharmacopoeia', ed. J.E.F. Reynolds, The Pharmaceutical Press, London, 1982, 28th Ed., p 312.

TRIDIHEXETHYL CHLORIDE [U]

CAS	4310-35-4	MW: 353.98	MF:	C21 H36 Cl N O

Anticholinergic

Lederle (Pathilon)

LgP (4)

pKa

pp cited in Vols.1-5:

(1) Lobby, U.S. Pat. 2 913 494 (1959).

TRIENTINE [U;INN]

CAS	112-24-3	MW: 146.24	MF:	C6 H18 N4

Tmt. of Wilson's disease

Merck (Cuprid)

LgP -1.41(C)

pKa 9.25; 9.92

pp cited in Vols.1-5:

(1) H.B. Dixon et al., Lancet, 1972, 1, 853. (2) F.W. Sunderman et al., Toxicol. Appl. Pharmacol., 1976, 38, 177. (3) J.M. Walshe in 'Orphan Drugs', ed. F.E. Karch, Marcel Dekker, NY, 1982, pp 57-71.

TRIETHYLENEMELAMINE [INN]

CAS	51-18-3	MW: 204.24	MF:	C9 H12 N6

Antineoplastic

LgP 0.96(C,3)

pKa

pp cited in Vols.1-5:

(1) Besyian, Ann., 1950, 566, 210.
(2) Wystrach et al., U.S. Pat. 2 653 934 (1953).
(3) eidem, J. Am. Chem. Soc., 1955, 77, 5915.

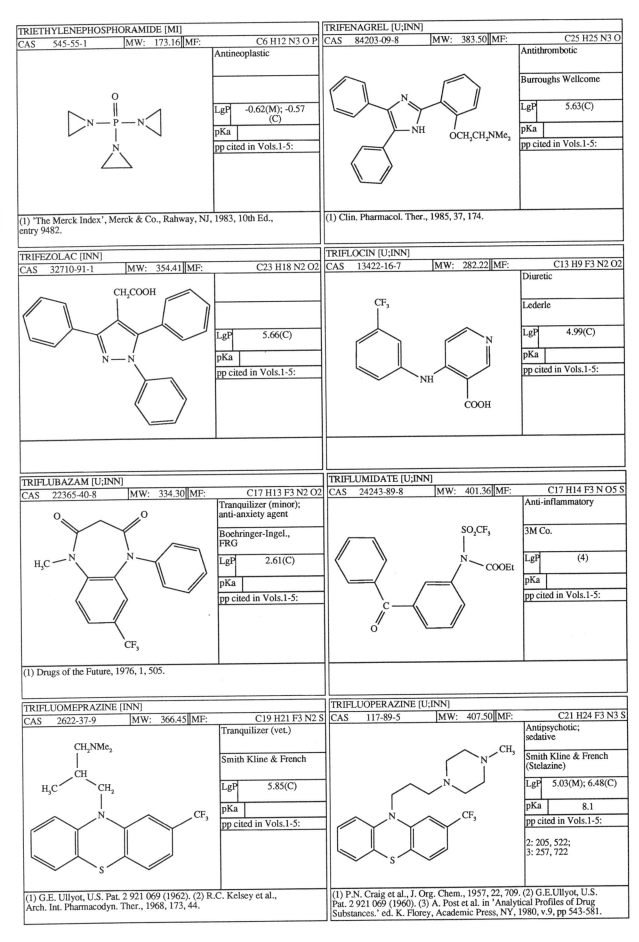

TRIETHYLENEPHOSPHORAMIDE [MI]

CAS	545-55-1	MW: 173.16	MF:	C6 H12 N3 O P

Antineoplastic

LgP	-0.62(M); -0.57 (C)
pKa	
pp cited in Vols.1-5:	

(1) 'The Merck Index', Merck & Co., Rahway, NJ, 1983, 10th Ed., entry 9482.

TRIFENAGREL [U;INN]

CAS	84203-09-8	MW: 383.50	MF:	C25 H25 N3 O

Antithrombotic

Burroughs Wellcome

LgP	5.63(C)
pKa	
pp cited in Vols.1-5:	

OCH₂CH₂NMe₂

(1) Clin. Pharmacol. Ther., 1985, 37, 174.

TRIFEZOLAC [INN]

CAS	32710-91-1	MW: 354.41	MF:	C23 H18 N2 O2

CH₂COOH

LgP	5.66(C)
pKa	
pp cited in Vols.1-5:	

TRIFLOCIN [U;INN]

CAS	13422-16-7	MW: 282.22	MF:	C13 H9 F3 N2 O2

Diuretic

Lederle

LgP	4.99(C)
pKa	
pp cited in Vols.1-5:	

TRIFLUBAZAM [U;INN]

CAS	22365-40-8	MW: 334.30	MF:	C17 H13 F3 N2 O2

Tranquilizer (minor); anti-anxiety agent

Boehringer-Ingel., FRG

LgP	2.61(C)
pKa	
pp cited in Vols.1-5:	

(1) Drugs of the Future, 1976, 1, 505.

TRIFLUMIDATE [U;INN]

CAS	24243-89-8	MW: 401.36	MF:	C17 H14 F3 N O5 S

Anti-inflammatory

3M Co.

LgP	(4)
pKa	
pp cited in Vols.1-5:	

SO₂CF₃   N—COOEt

TRIFLUOMEPRAZINE [INN]

CAS	2622-37-9	MW: 366.45	MF:	C19 H21 F3 N2 S

Tranquilizer (vet.)

Smith Kline & French

LgP	5.85(C)
pKa	
pp cited in Vols.1-5:	

(1) G.E. Ullyot, U.S. Pat. 2 921 069 (1962). (2) R.C. Kelsey et al., Arch. Int. Pharmacodyn. Ther., 1968, 173, 44.

TRIFLUOPERAZINE [U;INN]

CAS	117-89-5	MW: 407.50	MF:	C21 H24 F3 N3 S

Antipsychotic; sedative

Smith Kline & French (Stelazine)

LgP	5.03(M); 6.48(C)
pKa	8.1
pp cited in Vols.1-5:	

2: 205, 522;
3: 257, 722

(1) P.N. Craig et al., J. Org. Chem., 1957, 22, 709. (2) G.E.Ullyot, U.S. Pat. 2 921 069 (1960). (3) A. Post et al. in 'Analytical Profiles of Drug Substances.' ed. K. Florey, Academic Press, NY, 1980, v.9, pp 543-581.

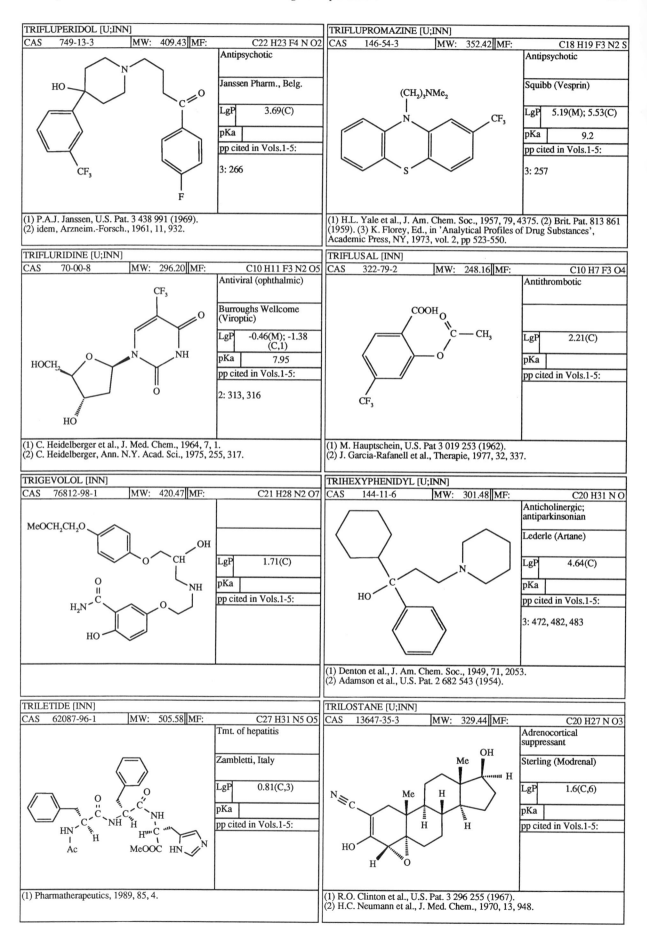

TRIFLUPERIDOL [U;INN]

CAS 749-13-3	MW: 409.43	MF: C22 H23 F4 N O2

Antipsychotic

Janssen Pharm., Belg.

LgP	3.69(C)
pKa	
pp cited in Vols.1-5:	
3: 266	

(1) P.A.J. Janssen, U.S. Pat. 3 438 991 (1969).
(2) idem, Arzneim.-Forsch., 1961, 11, 932.

TRIFLUPROMAZINE [U;INN]

CAS 146-54-3	MW: 352.42	MF: C18 H19 F3 N2 S

Antipsychotic

Squibb (Vesprin)

LgP	5.19(M); 5.53(C)
pKa	9.2
pp cited in Vols.1-5:	
3: 257	

(1) H.L. Yale et al., J. Am. Chem. Soc., 1957, 79, 4375. (2) Brit. Pat. 813 861 (1959). (3) K. Florey, Ed., in 'Analytical Profiles of Drug Substances', Academic Press, NY, 1973, vol. 2, pp 523-550.

TRIFLURIDINE [U;INN]

CAS 70-00-8	MW: 296.20	MF: C10 H11 F3 N2 O5

Antiviral (ophthalmic)

Burroughs Wellcome (Viroptic)

LgP	-0.46(M); -1.38 (C,1)
pKa	7.95
pp cited in Vols.1-5:	
2: 313, 316	

(1) C. Heidelberger et al., J. Med. Chem., 1964, 7, 1.
(2) C. Heidelberger, Ann. N.Y. Acad. Sci., 1975, 255, 317.

TRIFLUSAL [INN]

CAS 322-79-2	MW: 248.16	MF: C10 H7 F3 O4

Antithrombotic

LgP	2.21(C)
pKa	
pp cited in Vols.1-5:	

(1) M. Hauptschein, U.S. Pat 3 019 253 (1962).
(2) J. Garcia-Rafanell et al., Therapie, 1977, 32, 337.

TRIGEVOLOL [INN]

CAS 76812-98-1	MW: 420.47	MF: C21 H28 N2 O7

LgP	1.71(C)
pKa	
pp cited in Vols.1-5:	

TRIHEXYPHENIDYL [U;INN]

CAS 144-11-6	MW: 301.48	MF: C20 H31 N O

Anticholinergic; antiparkinsonian

Lederle (Artane)

LgP	4.64(C)
pKa	
pp cited in Vols.1-5:	
3: 472, 482, 483	

(1) Denton et al., J. Am. Chem. Soc., 1949, 71, 2053.
(2) Adamson et al., U.S. Pat. 2 682 543 (1954).

TRILETIDE [INN]

CAS 62087-96-1	MW: 505.58	MF: C27 H31 N5 O5

Tmt. of hepatitis

Zambletti, Italy

LgP	0.81(C,3)
pKa	
pp cited in Vols.1-5:	

(1) Pharmatherapeutics, 1989, 85, 4.

TRILOSTANE [U;INN]

CAS 13647-35-3	MW: 329.44	MF: C20 H27 N O3

Adrenocortical suppressant

Sterling (Modrenal)

LgP	1.6(C,6)
pKa	
pp cited in Vols.1-5:	

(1) R.O. Clinton et al., U.S. Pat. 3 296 255 (1967).
(2) H.C. Neumann et al., J. Med. Chem., 1970, 13, 948.

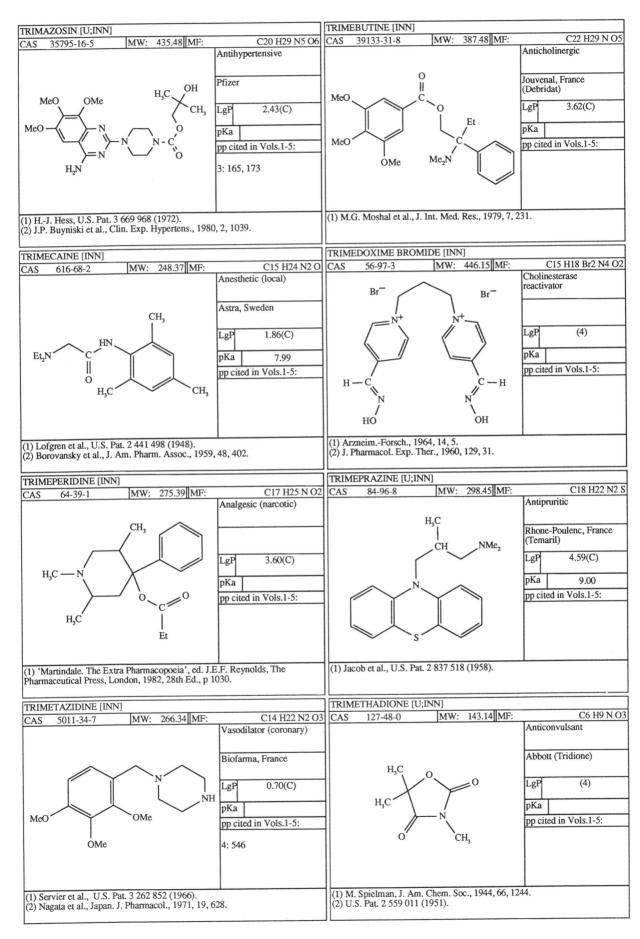

TRIMAZOSIN [U;INN]

CAS 35795-16-5	MW: 435.48	MF: C20 H29 N5 O6

Antihypertensive

Pfizer

LgP	2.43(C)

pKa	

pp cited in Vols.1-5:

3: 165, 173

(1) H.-J. Hess, U.S. Pat. 3 669 968 (1972).
(2) J.P. Buyniski et al., Clin. Exp. Hypertens., 1980, 2, 1039.

TRIMEBUTINE [INN]

CAS 39133-31-8	MW: 387.48	MF: C22 H29 N O5

Anticholinergic

Jouvenal, France
(Debridat)

LgP	3.62(C)

pKa	

pp cited in Vols.1-5:

(1) M.G. Moshal et al., J. Int. Med. Res., 1979, 7, 231.

TRIMECAINE [INN]

CAS 616-68-2	MW: 248.37	MF: C15 H24 N2 O

Anesthetic (local)

Astra, Sweden

LgP	1.86(C)

pKa	7.99

pp cited in Vols.1-5:

(1) Lofgren et al., U.S. Pat. 2 441 498 (1948).
(2) Borovansky et al., J. Am. Pharm. Assoc., 1959, 48, 402.

TRIMEDOXIME BROMIDE [INN]

CAS 56-97-3	MW: 446.15	MF: C15 H18 Br2 N4 O2

Cholinesterase
reactivator

LgP	(4)

pKa	

pp cited in Vols.1-5:

(1) Arzneim.-Forsch., 1964, 14, 5.
(2) J. Pharmacol. Exp. Ther., 1960, 129, 31.

TRIMEPERIDINE [INN]

CAS 64-39-1	MW: 275.39	MF: C17 H25 N O2

Analgesic (narcotic)

LgP	3.60(C)

pKa	

pp cited in Vols.1-5:

(1) 'Martindale. The Extra Pharmacopoeia', ed. J.E.F. Reynolds, The
Pharmaceutical Press, London, 1982, 28th Ed., p 1030.

TRIMEPRAZINE [U;INN]

CAS 84-96-8	MW: 298.45	MF: C18 H22 N2 S

Antipruritic

Rhone-Poulenc, France
(Temaril)

LgP	4.59(C)

pKa	9.00

pp cited in Vols.1-5:

(1) Jacob et al., U.S. Pat. 2 837 518 (1958).

TRIMETAZIDINE [INN]

CAS 5011-34-7	MW: 266.34	MF: C14 H22 N2 O3

Vasodilator (coronary)

Biofarma, France

LgP	0.70(C)

pKa	

pp cited in Vols.1-5:

4: 546

(1) Servier et al., U.S. Pat. 3 262 852 (1966).
(2) Nagata et al., Japan. J. Pharmacol., 1971, 19, 628.

TRIMETHADIONE [U;INN]

CAS 127-48-0	MW: 143.14	MF: C6 H9 N O3

Anticonvulsant

Abbott (Tridione)

LgP	(4)

pKa	

pp cited in Vols.1-5:

(1) M. Spielman, J. Am. Chem. Soc., 1944, 66, 1244.
(2) U.S. Pat. 2 559 011 (1951).

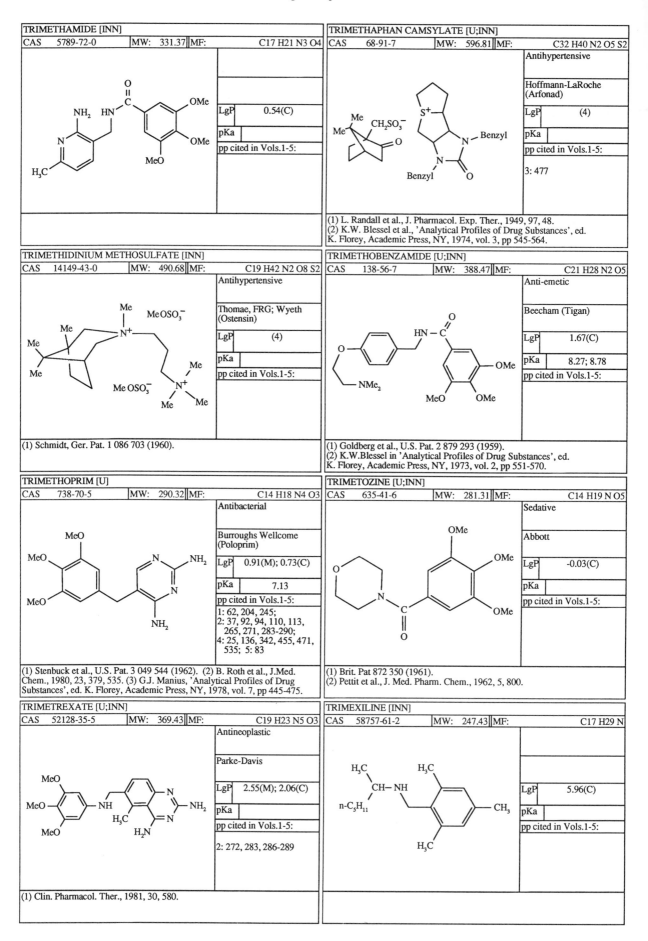

TRIMETHAMIDE [INN]

CAS	5789-72-0	MW:	331.37	MF:	C17 H21 N3 O4

LgP	0.54(C)
pKa	
pp cited in Vols.1-5:	

TRIMETHAPHAN CAMSYLATE [U;INN]

CAS	68-91-7	MW:	596.81	MF:	C32 H40 N2 O5 S2

Antihypertensive

Hoffmann-LaRoche (Arfonad)

LgP	(4)
pKa	
pp cited in Vols.1-5:	

3: 477

(1) L. Randall et al., J. Pharmacol. Exp. Ther., 1949, 97, 48.
(2) K.W. Blessel et al., 'Analytical Profiles of Drug Substances', ed.
K. Florey, Academic Press, NY, 1974, vol. 3, pp 545-564.

TRIMETHIDINIUM METHOSULFATE [INN]

CAS	14149-43-0	MW:	490.68	MF:	C19 H42 N2 O8 S2

Antihypertensive

Thomae, FRG; Wyeth (Ostensin)

LgP	(4)
pKa	
pp cited in Vols.1-5:	

(1) Schmidt, Ger. Pat. 1 086 703 (1960).

TRIMETHOBENZAMIDE [U;INN]

CAS	138-56-7	MW:	388.47	MF:	C21 H28 N2 O5

Anti-emetic

Beecham (Tigan)

LgP	1.67(C)
pKa	8.27; 8.78
pp cited in Vols.1-5:	

(1) Goldberg et al., U.S. Pat. 2 879 293 (1959).
(2) K.W.Blessel in 'Analytical Profiles of Drug Substances', ed.
K. Florey, Academic Press, NY, 1973, vol. 2, pp 551-570.

TRIMETHOPRIM [U]

CAS	738-70-5	MW:	290.32	MF:	C14 H18 N4 O3

Antibacterial

Burroughs Wellcome (Poloprim)

LgP	0.91(M); 0.73(C)
pKa	7.13
pp cited in Vols.1-5:	

1: 62, 204, 245;
2: 37, 92, 94, 110, 113, 265, 271, 283-290;
4: 25, 136, 342, 455, 471, 535; 5: 83

(1) Stenbuck et al., U.S. Pat. 3 049 544 (1962). (2) B. Roth et al., J.Med. Chem., 1980, 23, 379, 535. (3) G.J. Manius, 'Analytical Profiles of Drug Substances', ed. K. Florey, Academic Press, NY, 1978, vol. 7, pp 445-475.

TRIMETOZINE [U;INN]

CAS	635-41-6	MW:	281.31	MF:	C14 H19 N O5

Sedative

Abbott

LgP	-0.03(C)
pKa	
pp cited in Vols.1-5:	

(1) Brit. Pat 872 350 (1961).
(2) Pettit et al., J. Med. Pharm. Chem., 1962, 5, 800.

TRIMETREXATE [U;INN]

CAS	52128-35-5	MW:	369.43	MF:	C19 H23 N5 O3

Antineoplastic

Parke-Davis

LgP	2.55(M); 2.06(C)
pKa	
pp cited in Vols.1-5:	

2: 272, 283, 286-289

(1) Clin. Pharmacol. Ther., 1981, 30, 580.

TRIMEXILINE [INN]

CAS	58757-61-2	MW:	247.43	MF:	C17 H29 N

LgP	5.96(C)
pKa	
pp cited in Vols.1-5:	

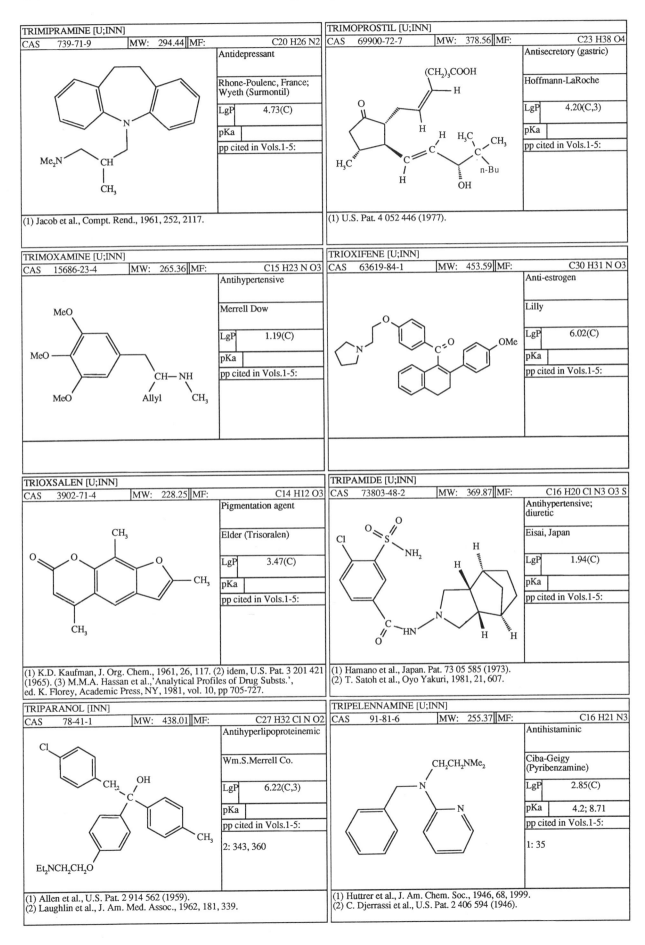

TRIMIPRAMINE [U;INN]

CAS	739-71-9	MW:	294.44	MF:	C20 H26 N2

Antidepressant

Rhone-Poulenc, France; Wyeth (Surmontil)

LgP	4.73(C)
pKa	
pp cited in Vols.1-5:	

(1) Jacob et al., Compt. Rend., 1961, 252, 2117.

TRIMOPROSTIL [U;INN]

CAS	69900-72-7	MW:	378.56	MF:	C23 H38 O4

Antisecretory (gastric)

Hoffmann-LaRoche

LgP	4.20(C,3)
pKa	
pp cited in Vols.1-5:	

(1) U.S. Pat. 4 052 446 (1977).

TRIMOXAMINE [U;INN]

CAS	15686-23-4	MW:	265.36	MF:	C15 H23 N O3

Antihypertensive

Merrell Dow

LgP	1.19(C)
pKa	
pp cited in Vols.1-5:	

TRIOXIFENE [U;INN]

CAS	63619-84-1	MW:	453.59	MF:	C30 H31 N O3

Anti-estrogen

Lilly

LgP	6.02(C)
pKa	
pp cited in Vols.1-5:	

TRIOXSALEN [U;INN]

CAS	3902-71-4	MW:	228.25	MF:	C14 H12 O3

Pigmentation agent

Elder (Trisoralen)

LgP	3.47(C)
pKa	
pp cited in Vols.1-5:	

(1) K.D. Kaufman, J. Org. Chem., 1961, 26, 117. (2) idem, U.S. Pat. 3 201 421 (1965). (3) M.M.A. Hassan et al.,'Analytical Profiles of Drug Substs.', ed. K. Florey, Academic Press, NY, 1981, vol. 10, pp 705-727.

TRIPAMIDE [U;INN]

CAS	73803-48-2	MW:	369.87	MF:	C16 H20 Cl N3 O3 S

Antihypertensive; diuretic

Eisai, Japan

LgP	1.94(C)
pKa	
pp cited in Vols.1-5:	

(1) Hamano et al., Japan. Pat. 73 05 585 (1973). (2) T. Satoh et al., Oyo Yakuri, 1981, 21, 607.

TRIPARANOL [INN]

CAS	78-41-1	MW:	438.01	MF:	C27 H32 Cl N O2

Antihyperlipoproteinemic

Wm.S.Merrell Co.

LgP	6.22(C,3)
pKa	
pp cited in Vols.1-5:	
	2: 343, 360

(1) Allen et al., U.S. Pat. 2 914 562 (1959). (2) Laughlin et al., J. Am. Med. Assoc., 1962, 181, 339.

TRIPELENNAMINE [U;INN]

CAS	91-81-6	MW:	255.37	MF:	C16 H21 N3

Antihistaminic

Ciba-Geigy (Pyribenzamine)

LgP	2.85(C)
pKa	4.2; 8.71
pp cited in Vols.1-5:	
	1: 35

(1) Huttrer et al., J. Am. Chem. Soc., 1946, 68, 1999. (2) C. Djerrassi et al., U.S. Pat. 2 406 594 (1946).

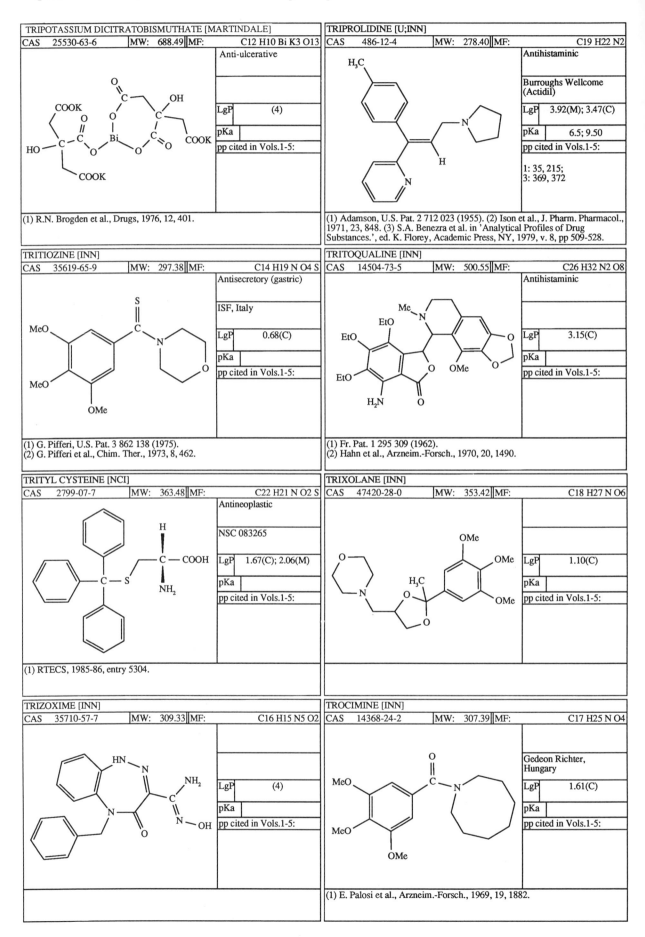

TRIPOTASSIUM DICITRATOBISMUTHATE [MARTINDALE]

CAS	25530-63-6	MW:	688.49	MF:	C12 H10 Bi K3 O13

Anti-ulcerative

LgP	(4)
pKa	
pp cited in Vols.1-5:	

(1) R.N. Brogden et al., Drugs, 1976, 12, 401.

TRIPROLIDINE [U;INN]

CAS	486-12-4	MW:	278.40	MF:	C19 H22 N2

Antihistaminic

Burroughs Wellcome (Actidil)

LgP	3.92(M); 3.47(C)
pKa	6.5; 9.50
pp cited in Vols.1-5:	

1: 35, 215;
3: 369, 372

(1) Adamson, U.S. Pat. 2 712 023 (1955). (2) Ison et al., J. Pharm. Pharmacol., 1971, 23, 848. (3) S.A. Benezra et al. in 'Analytical Profiles of Drug Substances.', ed. K. Florey, Academic Press, NY, 1979, v. 8, pp 509-528.

TRITIOZINE [INN]

CAS	35619-65-9	MW:	297.38	MF:	C14 H19 N O4 S

Antisecretory (gastric)

ISF, Italy

LgP	0.68(C)
pKa	
pp cited in Vols.1-5:	

(1) G. Pifferi, U.S. Pat. 3 862 138 (1975).
(2) G. Pifferi et al., Chim. Ther., 1973, 8, 462.

TRITOQUALINE [INN]

CAS	14504-73-5	MW:	500.55	MF:	C26 H32 N2 O8

Antihistaminic

LgP	3.15(C)
pKa	
pp cited in Vols.1-5:	

(1) Fr. Pat. 1 295 309 (1962).
(2) Hahn et al., Arzneim.-Forsch., 1970, 20, 1490.

TRITYL CYSTEINE [NCI]

CAS	2799-07-7	MW:	363.48	MF:	C22 H21 N O2 S

Antineoplastic

NSC 083265

LgP	1.67(C); 2.06(M)
pKa	
pp cited in Vols.1-5:	

(1) RTECS, 1985-86, entry 5304.

TRIXOLANE [INN]

CAS	47420-28-0	MW:	353.42	MF:	C18 H27 N O6

LgP	1.10(C)
pKa	
pp cited in Vols.1-5:	

TRIZOXIME [INN]

CAS	35710-57-7	MW:	309.33	MF:	C16 H15 N5 O2

LgP	(4)
pKa	
pp cited in Vols.1-5:	

TROCIMINE [INN]

CAS	14368-24-2	MW:	307.39	MF:	C17 H25 N O4

Gedeon Richter, Hungary

LgP	1.61(C)
pKa	
pp cited in Vols.1-5:	

(1) E. Palosi et al., Arzneim.-Forsch., 1969, 19, 1882.

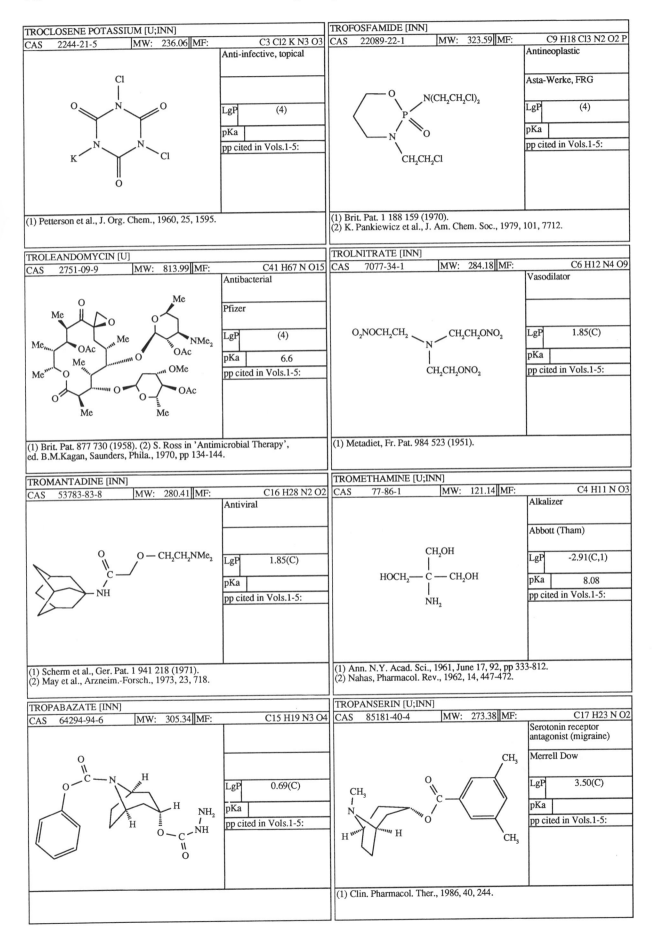

TROCLOSENE POTASSIUM [U;INN]

CAS	2244-21-5	MW:	236.06	MF:	C3 Cl2 K N3 O3

Anti-infective, topical

LgP	(4)
pKa	
pp cited in Vols.1-5:	

(1) Petterson et al., J. Org. Chem., 1960, 25, 1595.

TROFOSFAMIDE [INN]

CAS	22089-22-1	MW:	323.59	MF:	C9 H18 Cl3 N2 O2 P

Antineoplastic

Asta-Werke, FRG

LgP	(4)
pKa	
pp cited in Vols.1-5:	

(1) Brit. Pat. 1 188 159 (1970).
(2) K. Pankiewicz et al., J. Am. Chem. Soc., 1979, 101, 7712.

TROLEANDOMYCIN [U]

CAS	2751-09-9	MW:	813.99	MF:	C41 H67 N O15

Antibacterial

Pfizer

LgP	(4)
pKa	6.6
pp cited in Vols.1-5:	

(1) Brit. Pat. 877 730 (1958). (2) S. Ross in 'Antimicrobial Therapy',
ed. B.M.Kagan, Saunders, Phila., 1970, pp 134-144.

TROLNITRATE [INN]

CAS	7077-34-1	MW:	284.18	MF:	C6 H12 N4 O9

Vasodilator

LgP	1.85(C)
pKa	
pp cited in Vols.1-5:	

(1) Metadiet, Fr. Pat. 984 523 (1951).

TROMANTADINE [INN]

CAS	53783-83-8	MW:	280.41	MF:	C16 H28 N2 O2

Antiviral

LgP	1.85(C)
pKa	
pp cited in Vols.1-5:	

(1) Scherm et al., Ger. Pat. 1 941 218 (1971).
(2) May et al., Arzneim.-Forsch., 1973, 23, 718.

TROMETHAMINE [U;INN]

CAS	77-86-1	MW:	121.14	MF:	C4 H11 N O3

Alkalizer

Abbott (Tham)

LgP	-2.91(C,1)
pKa	8.08
pp cited in Vols.1-5:	

(1) Ann. N.Y. Acad. Sci., 1961, June 17, 92, pp 333-812.
(2) Nahas, Pharmacol. Rev., 1962, 14, 447-472.

TROPABAZATE [INN]

CAS	64294-94-6	MW:	305.34	MF:	C15 H19 N3 O4

LgP	0.69(C)
pKa	
pp cited in Vols.1-5:	

TROPANSERIN [U;INN]

CAS	85181-40-4	MW:	273.38	MF:	C17 H23 N O2

Serotonin receptor
antagonist (migraine)

Merrell Dow

LgP	3.50(C)
pKa	
pp cited in Vols.1-5:	

(1) Clin. Pharmacol. Ther., 1986, 40, 244.

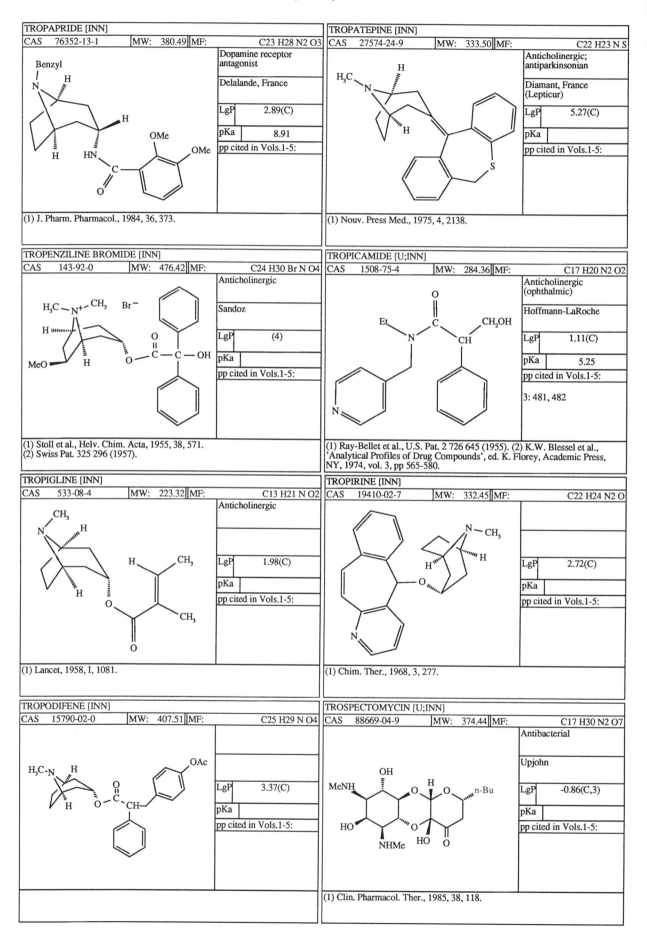

TROPAPRIDE [INN]			
CAS 76352-13-1	MW: 380.49	MF:	C23 H28 N2 O3

Dopamine receptor antagonist

Delalande, France

LgP	2.89(C)
pKa	8.91
pp cited in Vols.1-5:	

Benzyl · OMe · OMe · HN · C · O

(1) J. Pharm. Pharmacol., 1984, 36, 373.

TROPATEPINE [INN]			
CAS 27574-24-9	MW: 333.50	MF:	C22 H23 N S

Anticholinergic; antiparkinsonian

Diamant, France (Lepticur)

LgP	5.27(C)
pKa	
pp cited in Vols.1-5:	

H₃C · N · H · S

(1) Nouv. Press Med., 1975, 4, 2138.

TROPENZILINE BROMIDE [INN]			
CAS 143-92-0	MW: 476.42	MF:	C24 H30 Br N O4

Anticholinergic

Sandoz

LgP	(4)
pKa	
pp cited in Vols.1-5:	

H₃C—N⁺—CH₃ Br⁻ · MeO · H · O · C · C · OH

(1) Stoll et al., Helv. Chim. Acta, 1955, 38, 571.
(2) Swiss Pat. 325 296 (1957).

TROPICAMIDE [U;INN]			
CAS 1508-75-4	MW: 284.36	MF:	C17 H20 N2 O2

Anticholinergic (ophthalmic)

Hoffmann-LaRoche

LgP	1.11(C)
pKa	5.25
pp cited in Vols.1-5:	
3: 481, 482	

O · Et · N · C · CH · CH₂OH · N

(1) Ray-Bellet et al., U.S. Pat. 2 726 645 (1955). (2) K.W. Blessel et al., 'Analytical Profiles of Drug Compounds', ed. K. Florey, Academic Press, NY, 1974, vol. 3, pp 565-580.

TROPIGLINE [INN]			
CAS 533-08-4	MW: 223.32	MF:	C13 H21 N O2

Anticholinergic

LgP	1.98(C)
pKa	
pp cited in Vols.1-5:	

CH₃ · N · H · H · CH₃ · O · CH₃ · O

(1) Lancet, 1958, I, 1081.

TROPIRINE [INN]			
CAS 19410-02-7	MW: 332.45	MF:	C22 H24 N2 O

LgP	2.72(C)
pKa	
pp cited in Vols.1-5:	

N—CH₃ · H · O · N

(1) Chim. Ther., 1968, 3, 277.

TROPODIFENE [INN]			
CAS 15790-02-0	MW: 407.51	MF:	C25 H29 N O4

LgP	3.37(C)
pKa	
pp cited in Vols.1-5:	

H₃C-N · H · OAc · O · C · CH · O

TROSPECTOMYCIN [U;INN]			
CAS 88669-04-9	MW: 374.44	MF:	C17 H30 N2 O7

Antibacterial

Upjohn

LgP	-0.86(C,3)
pKa	
pp cited in Vols.1-5:	

OH · MeNH · H · n-Bu · HO · O · NHMe · HO · O

(1) Clin. Pharmacol. Ther., 1985, 38, 118.

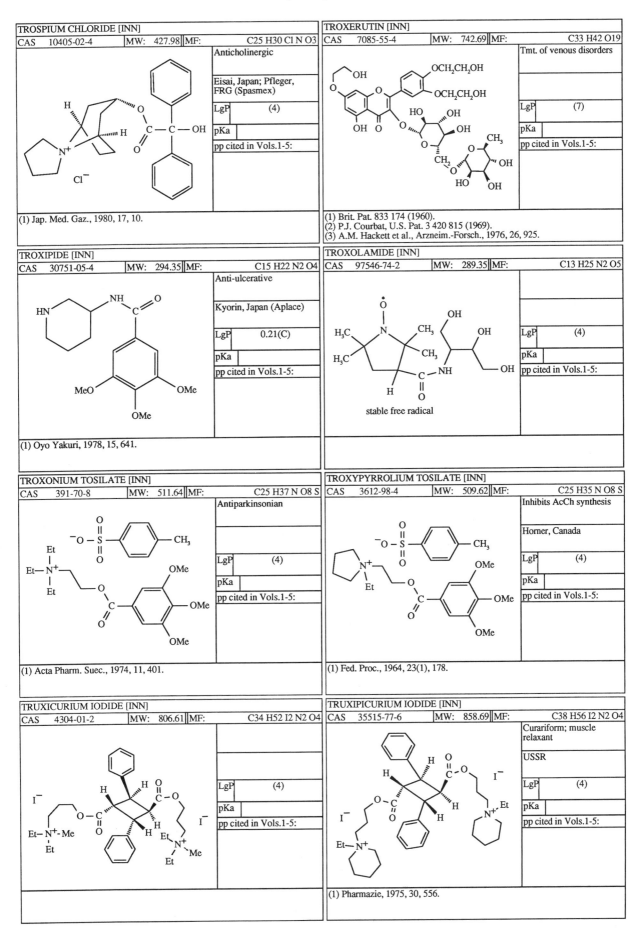

TROSPIUM CHLORIDE [INN]

CAS	10405-02-4	MW:	427.98	MF:	C25 H30 Cl N O3

Anticholinergic

Eisai, Japan; Pfleger, FRG (Spasmex)

LgP	(4)

pKa	

pp cited in Vols.1-5:

(1) Jap. Med. Gaz., 1980, 17, 10.

TROXERUTIN [INN]

CAS	7085-55-4	MW:	742.69	MF:	C33 H42 O19

Tmt. of venous disorders

LgP	(7)

pKa	

pp cited in Vols.1-5:

(1) Brit. Pat. 833 174 (1960).
(2) P.J. Courbat, U.S. Pat. 3 420 815 (1969).
(3) A.M. Hackett et al., Arzneim.-Forsch., 1976, 26, 925.

TROXIPIDE [INN]

CAS	30751-05-4	MW:	294.35	MF:	C15 H22 N2 O4

Anti-ulcerative

Kyorin, Japan (Aplace)

LgP	0.21(C)

pKa	

pp cited in Vols.1-5:

(1) Oyo Yakuri, 1978, 15, 641.

TROXOLAMIDE [INN]

CAS	97546-74-2	MW:	289.35	MF:	C13 H25 N2 O5

stable free radical

LgP	(4)

pKa	

pp cited in Vols.1-5:

TROXONIUM TOSILATE [INN]

CAS	391-70-8	MW:	511.64	MF:	C25 H37 N O8 S

Antiparkinsonian

LgP	(4)

pKa	

pp cited in Vols.1-5:

(1) Acta Pharm. Suec., 1974, 11, 401.

TROXYPYRROLIUM TOSILATE [INN]

CAS	3612-98-4	MW:	509.62	MF:	C25 H35 N O8 S

Inhibits AcCh synthesis

Horner, Canada

LgP	(4)

pKa	

pp cited in Vols.1-5:

(1) Fed. Proc., 1964, 23(1), 178.

TRUXICURIUM IODIDE [INN]

CAS	4304-01-2	MW:	806.61	MF:	C34 H52 I2 N2 O4

LgP	(4)

pKa	

pp cited in Vols.1-5:

TRUXIPICURIUM IODIDE [INN]

CAS	35515-77-6	MW:	858.69	MF:	C38 H56 I2 N2 O4

Curariform; muscle relaxant

USSR

LgP	(4)

pKa	

pp cited in Vols.1-5:

(1) Pharmazie, 1975, 30, 556.

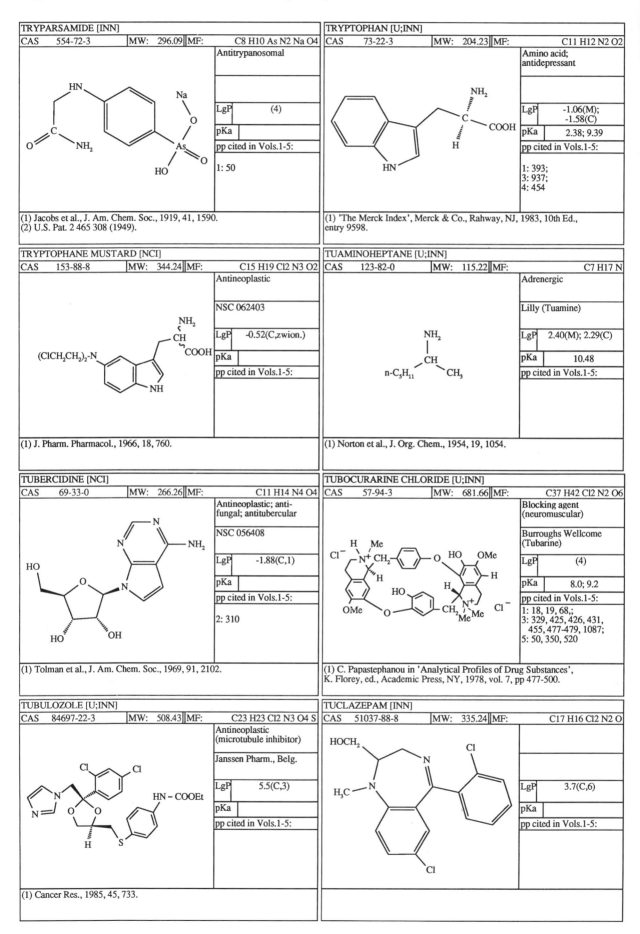

TRYPARSAMIDE [INN]

CAS	554-72-3	MW:	296.09	MF:	C8 H10 As N2 Na O4

Antitrypanosomal

LgP	(4)
pKa	
pp cited in Vols.1-5:	
1: 50	

(1) Jacobs et al., J. Am. Chem. Soc., 1919, 41, 1590.
(2) U.S. Pat. 2 465 308 (1949).

TRYPTOPHAN [U;INN]

CAS	73-22-3	MW:	204.23	MF:	C11 H12 N2 O2

Amino acid; antidepressant

LgP	-1.06(M); -1.58(C)
pKa	2.38; 9.39
pp cited in Vols.1-5:	
1: 393; 3: 937; 4: 454	

(1) 'The Merck Index', Merck & Co., Rahway, NJ, 1983, 10th Ed., entry 9598.

TRYPTOPHANE MUSTARD [NCI]

CAS	153-88-8	MW:	344.24	MF:	C15 H19 Cl2 N3 O2

Antineoplastic

NSC 062403

LgP	-0.52(C,zwion.)
pKa	
pp cited in Vols.1-5:	

(1) J. Pharm. Pharmacol., 1966, 18, 760.

TUAMINOHEPTANE [U;INN]

CAS	123-82-0	MW:	115.22	MF:	C7 H17 N

Adrenergic

Lilly (Tuamine)

LgP	2.40(M); 2.29(C)
pKa	10.48
pp cited in Vols.1-5:	

(1) Norton et al., J. Org. Chem., 1954, 19, 1054.

TUBERCIDINE [NCI]

CAS	69-33-0	MW:	266.26	MF:	C11 H14 N4 O4

Antineoplastic; antifungal; antitubercular

NSC 056408

LgP	-1.88(C,1)
pKa	
pp cited in Vols.1-5:	
2: 310	

(1) Tolman et al., J. Am. Chem. Soc., 1969, 91, 2102.

TUBOCURARINE CHLORIDE [U;INN]

CAS	57-94-3	MW:	681.66	MF:	C37 H42 Cl2 N2 O6

Blocking agent (neuromuscular)

Burroughs Wellcome (Tubarine)

LgP	(4)
pKa	8.0; 9.2
pp cited in Vols.1-5:	
1: 18, 19, 68,; 3: 329, 425, 426, 431, 455, 477-479, 1087; 5: 50, 350, 520	

(1) C. Papastephanou in 'Analytical Profiles of Drug Substances', K. Florey, ed., Academic Press, NY, 1978, vol. 7, pp 477-500.

TUBULOZOLE [U;INN]

CAS	84697-22-3	MW:	508.43	MF:	C23 H23 Cl2 N3 O4 S

Antineoplastic (microtubule inhibitor)

Janssen Pharm., Belg.

LgP	5.5(C,3)
pKa	
pp cited in Vols.1-5:	

(1) Cancer Res., 1985, 45, 733.

TUCLAZEPAM [INN]

CAS	51037-88-8	MW:	335.24	MF:	C17 H16 Cl2 N2 O

LgP	3.7(C,6)
pKa	
pp cited in Vols.1-5:	

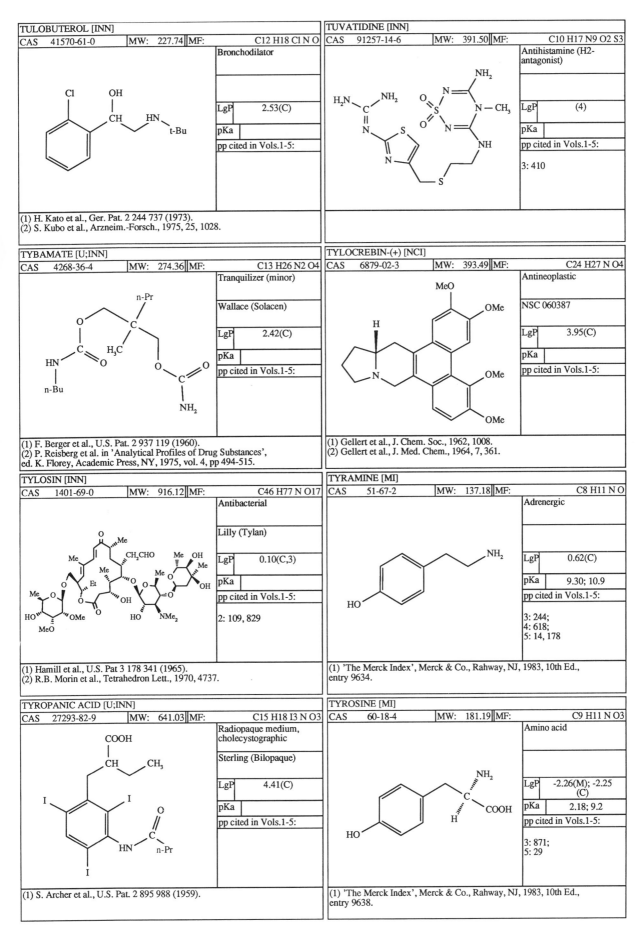

TULOBUTEROL [INN]

CAS	41570-61-0	MW:	227.74	MF:	C12 H18 Cl N O

Bronchodilator

LgP	2.53(C)

pKa	

pp cited in Vols.1-5:

(1) H. Kato et al., Ger. Pat. 2 244 737 (1973).
(2) S. Kubo et al., Arzneim.-Forsch., 1975, 25, 1028.

TUVATIDINE [INN]

CAS	91257-14-6	MW:	391.50	MF:	C10 H17 N9 O2 S3

Antihistamine (H2-antagonist)

LgP	(4)

pKa	

pp cited in Vols.1-5:

3: 410

TYBAMATE [U;INN]

CAS	4268-36-4	MW:	274.36	MF:	C13 H26 N2 O4

Tranquilizer (minor)

Wallace (Solacen)

LgP	2.42(C)

pKa	

pp cited in Vols.1-5:

(1) F. Berger et al., U.S. Pat. 2 937 119 (1960).
(2) P. Reisberg et al. in 'Analytical Profiles of Drug Substances',
ed. K. Florey, Academic Press, NY, 1975, vol. 4, pp 494-515.

TYLOCREBIN-(+) [NCI]

CAS	6879-02-3	MW:	393.49	MF:	C24 H27 N O4

Antineoplastic

NSC 060387

LgP	3.95(C)

pKa	

pp cited in Vols.1-5:

(1) Gellert et al., J. Chem. Soc., 1962, 1008.
(2) Gellert et al., J. Med. Chem., 1964, 7, 361.

TYLOSIN [INN]

CAS	1401-69-0	MW:	916.12	MF:	C46 H77 N O17

Antibacterial

Lilly (Tylan)

LgP	0.10(C,3)

pKa	

pp cited in Vols.1-5:

2: 109, 829

(1) Hamill et al., U.S. Pat 3 178 341 (1965).
(2) R.B. Morin et al., Tetrahedron Lett., 1970, 4737.

TYRAMINE [MI]

CAS	51-67-2	MW:	137.18	MF:	C8 H11 N O

Adrenergic

LgP	0.62(C)

pKa	9.30; 10.9

pp cited in Vols.1-5:

3: 244;
4: 618;
5: 14, 178

(1) 'The Merck Index', Merck & Co., Rahway, NJ, 1983, 10th Ed.,
entry 9634.

TYROPANIC ACID [U;INN]

CAS	27293-82-9	MW:	641.03	MF:	C15 H18 I3 N O3

Radiopaque medium, cholecystographic

Sterling (Bilopaque)

LgP	4.41(C)

pKa	

pp cited in Vols.1-5:

(1) S. Archer et al., U.S. Pat. 2 895 988 (1959).

TYROSINE [MI]

CAS	60-18-4	MW:	181.19	MF:	C9 H11 N O3

Amino acid

LgP	-2.26(M); -2.25 (C)

pKa	2.18; 9.2

pp cited in Vols.1-5:

3: 871;
5: 29

(1) 'The Merck Index', Merck & Co., Rahway, NJ, 1983, 10th Ed.,
entry 9638.

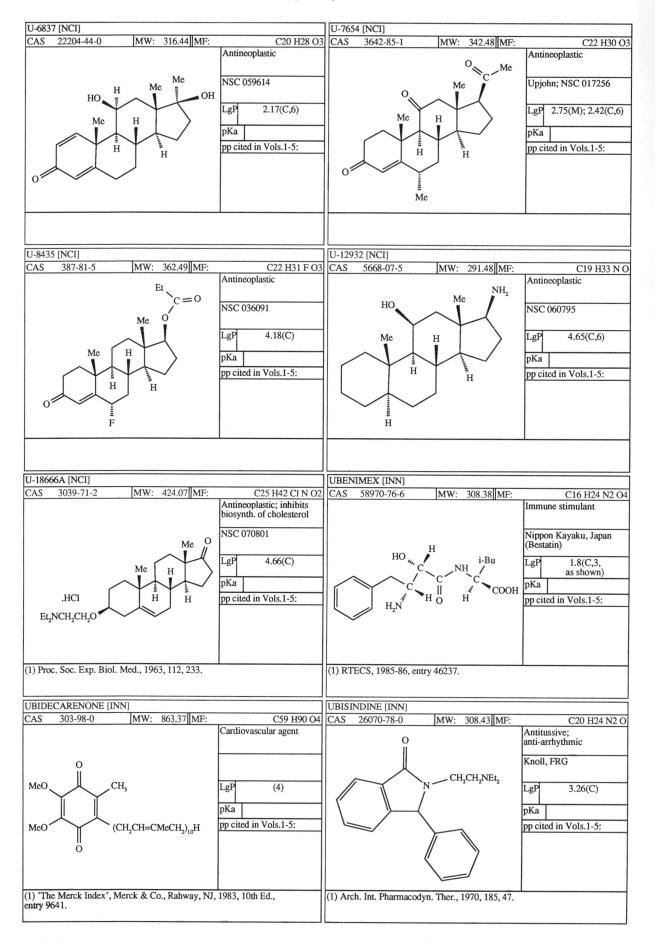

U-6837 [NCI]				
CAS	22204-44-0	MW: 316.44	MF:	C20 H28 O3

Antineoplastic

NSC 059614

LgP	2.17(C,6)
pKa	
pp cited in Vols.1-5:	

U-7654 [NCI]				
CAS	3642-85-1	MW: 342.48	MF:	C22 H30 O3

Antineoplastic

Upjohn; NSC 017256

LgP	2.75(M); 2.42(C,6)
pKa	
pp cited in Vols.1-5:	

U-8435 [NCI]				
CAS	387-81-5	MW: 362.49	MF:	C22 H31 F O3

Antineoplastic

NSC 036091

LgP	4.18(C)
pKa	
pp cited in Vols.1-5:	

U-12932 [NCI]				
CAS	5668-07-5	MW: 291.48	MF:	C19 H33 N O

Antineoplastic

NSC 060795

LgP	4.65(C,6)
pKa	
pp cited in Vols.1-5:	

U-18666A [NCI]				
CAS	3039-71-2	MW: 424.07	MF:	C25 H42 Cl N O2

Antineoplastic; inhibits biosynth. of cholesterol

NSC 070801

LgP	4.66(C)
pKa	
pp cited in Vols.1-5:	

.HCl

Et2NCH2CH2O

(1) Proc. Soc. Exp. Biol. Med., 1963, 112, 233.

UBENIMEX [INN]				
CAS	58970-76-6	MW: 308.38	MF:	C16 H24 N2 O4

Immune stimulant

Nippon Kayaku, Japan (Bestatin)

LgP	1.8(C,3, as shown)
pKa	
pp cited in Vols.1-5:	

(1) RTECS, 1985-86, entry 46237.

UBIDECARENONE [INN]				
CAS	303-98-0	MW: 863.37	MF:	C59 H90 O4

Cardiovascular agent

LgP	(4)
pKa	
pp cited in Vols.1-5:	

(1) 'The Merck Index', Merck & Co., Rahway, NJ, 1983, 10th Ed., entry 9641.

UBISINDINE [INN]				
CAS	26070-78-0	MW: 308.43	MF:	C20 H24 N2 O

Antitussive; anti-arrhythmic

Knoll, FRG

LgP	3.26(C)
pKa	
pp cited in Vols.1-5:	

(1) Arch. Int. Pharmacodyn. Ther., 1970, 185, 47.

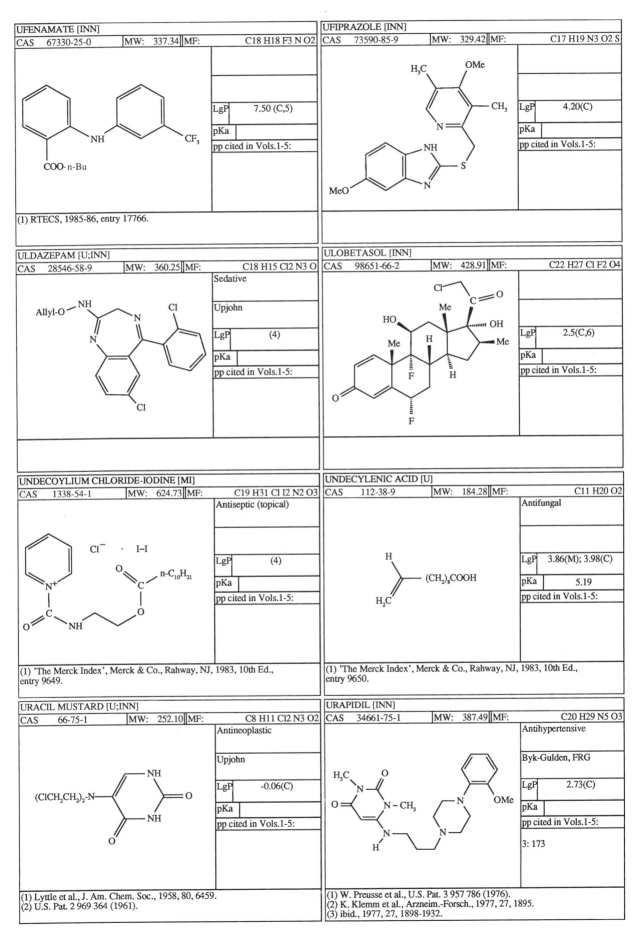

UFENAMATE [INN]

CAS	67330-25-0	MW:	337.34	MF:	C18 H18 F3 N O2

LgP	7.50 (C,5)
pKa	
pp cited in Vols.1-5:	

(1) RTECS, 1985-86, entry 17766.

UFIPRAZOLE [INN]

CAS	73590-85-9	MW:	329.42	MF:	C17 H19 N3 O2 S

LgP	4.20(C)
pKa	
pp cited in Vols.1-5:	

ULDAZEPAM [U;INN]

CAS	28546-58-9	MW:	360.25	MF:	C18 H15 Cl2 N3 O

Sedative

Upjohn

LgP	(4)
pKa	
pp cited in Vols.1-5:	

ULOBETASOL [INN]

CAS	98651-66-2	MW:	428.91	MF:	C22 H27 Cl F2 O4

LgP	2.5(C,6)
pKa	
pp cited in Vols.1-5:	

UNDECOYLIUM CHLORIDE-IODINE [MI]

CAS	1338-54-1	MW:	624.73	MF:	C19 H31 Cl I2 N2 O3

Antiseptic (topical)

LgP	(4)
pKa	
pp cited in Vols.1-5:	

(1) 'The Merck Index', Merck & Co., Rahway, NJ, 1983, 10th Ed., entry 9649.

UNDECYLENIC ACID [U]

CAS	112-38-9	MW:	184.28	MF:	C11 H20 O2

Antifungal

LgP	3.86(M); 3.98(C)
pKa	5.19
pp cited in Vols.1-5:	

(1) 'The Merck Index', Merck & Co., Rahway, NJ, 1983, 10th Ed., entry 9650.

URACIL MUSTARD [U;INN]

CAS	66-75-1	MW:	252.10	MF:	C8 H11 Cl2 N3 O2

Antineoplastic

Upjohn

LgP	-0.06(C)
pKa	
pp cited in Vols.1-5:	

(1) Lyttle et al., J. Am. Chem. Soc., 1958, 80, 6459.
(2) U.S. Pat. 2 969 364 (1961).

URAPIDIL [INN]

CAS	34661-75-1	MW:	387.49	MF:	C20 H29 N5 O3

Antihypertensive

Byk-Gulden, FRG

LgP	2.73(C)
pKa	
pp cited in Vols.1-5:	
3: 173	

(1) W. Preusse et al., U.S. Pat. 3 957 786 (1976).
(2) K. Klemm et al., Arzneim.-Forsch., 1977, 27, 1895.
(3) ibid., 1977, 27, 1898-1932.

UREA [U]				UREDEPA [U;INN]			
CAS 57-13-6	MW: 60.06	MF:	C H4 N2 O	CAS 302-49-8	MW: 219.18	MF:	C7 H14 N3 O3 P

	Diuretic			Antineoplastic
$H_2N-CO-NH_2$			(EtO-CO-HN-P(O)(N-aziridine)$_2$)	
	LgP	-2.11(M); -2.11 (C)		LgP	(4)
	pKa	0.2		pKa	
	pp cited in Vols.1-5:			pp cited in Vols.1-5:	
	1: 58;\n3: 603;\n5: 89				

(1) 'The Merck Index', Merck & Co., Rahway, NJ, 1983, 10th Ed., entry 9671.

(1) Bardos et al., Nature, 1959, 183, 399.
(2) Bardos et al., U.S. Pat. 3 201 313 (1965).

UREDOFOS [U;INN]				UREFIBRATE [INN]			
CAS 52406-01-6	MW: 500.54	MF:	C19 H25 N4 O6 P S2	CAS 38647-79-9	MW: 355.18	MF:	C15 H12 Cl2 N2 O4

	Anthelmintic (vet.)			Antihyperlipoproteinemic
	Rohm & Haas			EGYT, Hungary
	LgP	(4)		LgP	4.13(C)
	pKa			pKa	
	pp cited in Vols.1-5:			pp cited in Vols.1-5:	

(1) Am. J. Vet. Res., 1976, 37, 1483.

URETHANE [INN]				URIDINE [NCI]			
CAS 51-79-6	MW: 89.09	MF:	C3 H7 N O2	CAS 58-96-8	MW: 244.21	MF:	C9 H12 N2 O6

	Antineoplastic			Antineoplastic
$H_2N-COOEt$				NSC 020256
	LgP	-0.15(M); -0.18 (C)		LgP	-2.71(C,1); -1.98 (M)
	pKa			pKa	
	pp cited in Vols.1-5:			pp cited in Vols.1-5:	
	3: 1092				

(1) 'The Merck Index', Merck & Co., Rahway, NJ, 1983, 10th Ed., entry 9681.

(1) 'The Merck Index', Merck & Co., Rahway, NJ, 1983, 10th Ed., entry 9685.

URSODEOXYCHOLIC ACID [INN]				URSULCHOLIC ACID [INN]			
CAS 128-13-2	MW: 392.58	MF:	C24 H40 O4	CAS 88426-32-8	MW: 552.71	MF:	C24 H40 O10 S2

	Cholelitholytic			
	Erregierre			
	LgP	4.47(C)		LgP	5.05(C)
	pKa			pKa	
	pp cited in Vols.1-5:			pp cited in Vols.1-5:	

(1) Iwasaki, Z. Physiol. Chem., 1929, 185, 151.
(2) U.S. Pat. 4 404 199 (1983).

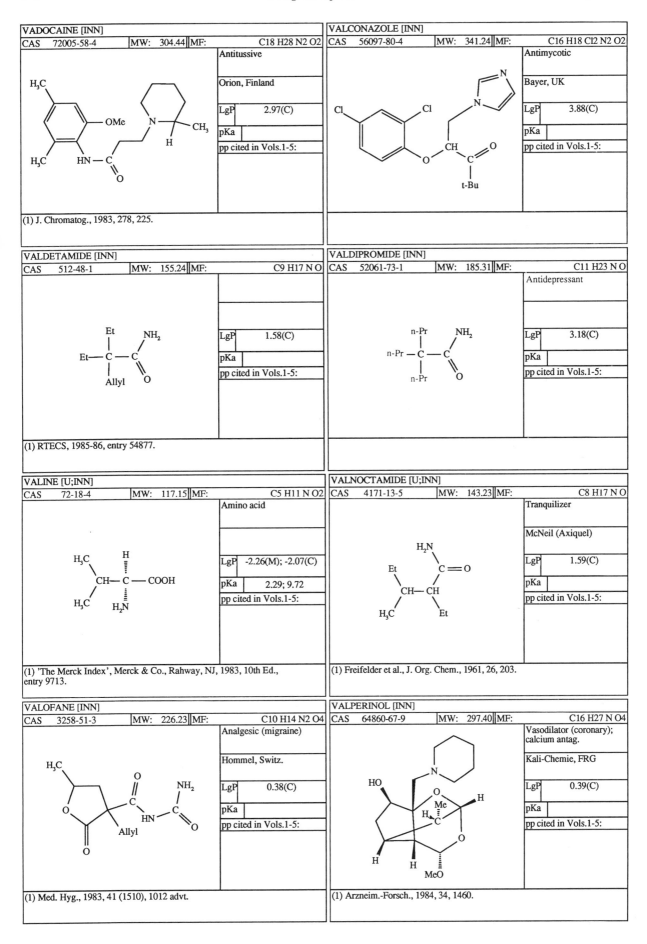

VADOCAINE [INN]

CAS	72005-58-4	MW:	304.44	MF:	C18 H28 N2 O2

Antitussive

Orion, Finland

LgP	2.97(C)
pKa	
pp cited in Vols.1-5:	

(1) J. Chromatog., 1983, 278, 225.

VALCONAZOLE [INN]

CAS	56097-80-4	MW:	341.24	MF:	C16 H18 Cl2 N2 O2

Antimycotic

Bayer, UK

LgP	3.88(C)
pKa	
pp cited in Vols.1-5:	

VALDETAMIDE [INN]

CAS	512-48-1	MW:	155.24	MF:	C9 H17 N O

LgP	1.58(C)
pKa	
pp cited in Vols.1-5:	

(1) RTECS, 1985-86, entry 54877.

VALDIPROMIDE [INN]

CAS	52061-73-1	MW:	185.31	MF:	C11 H23 N O

Antidepressant

LgP	3.18(C)
pKa	
pp cited in Vols.1-5:	

VALINE [U;INN]

CAS	72-18-4	MW:	117.15	MF:	C5 H11 N O2

Amino acid

LgP	-2.26(M); -2.07(C)
pKa	2.29; 9.72
pp cited in Vols.1-5:	

(1) 'The Merck Index', Merck & Co., Rahway, NJ, 1983, 10th Ed., entry 9713.

VALNOCTAMIDE [U;INN]

CAS	4171-13-5	MW:	143.23	MF:	C8 H17 N O

Tranquilizer

McNeil (Axiquel)

LgP	1.59(C)
pKa	
pp cited in Vols.1-5:	

(1) Freifelder et al., J. Org. Chem., 1961, 26, 203.

VALOFANE [INN]

CAS	3258-51-3	MW:	226.23	MF:	C10 H14 N2 O4

Analgesic (migraine)

Hommel, Switz.

LgP	0.38(C)
pKa	
pp cited in Vols.1-5:	

(1) Med. Hyg., 1983, 41 (1510), 1012 advt.

VALPERINOL [INN]

CAS	64860-67-9	MW:	297.40	MF:	C16 H27 N O4

Vasodilator (coronary); calcium antag.

Kali-Chemie, FRG

LgP	0.39(C)
pKa	
pp cited in Vols.1-5:	

(1) Arzneim.-Forsch., 1984, 34, 1460.

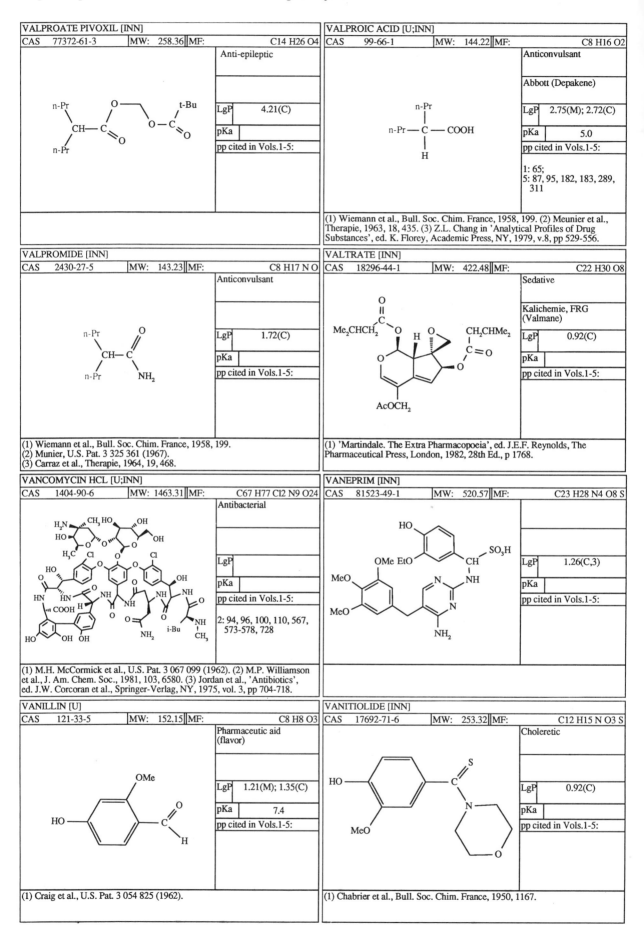

VALPROATE PIVOXIL [INN]				VALPROIC ACID [U;INN]		
CAS 77372-61-3	MW: 258.36	MF:	C14 H26 O4	CAS 99-66-1	MW: 144.22	MF:	C8 H16 O2

VALPROATE PIVOXIL [INN]
- CAS 77372-61-3 | MW: 258.36 | MF: C14 H26 O4
- Anti-epileptic
- LgP 4.21(C)
- pKa
- pp cited in Vols.1-5:

VALPROIC ACID [U;INN]
- CAS 99-66-1 | MW: 144.22 | MF: C8 H16 O2
- Anticonvulsant
- Abbott (Depakene)
- LgP 2.75(M); 2.72(C)
- pKa 5.0
- pp cited in Vols.1-5:
- 1: 65;
- 5: 87, 95, 182, 183, 289, 311

(1) Wiemann et al., Bull. Soc. Chim. France, 1958, 199. (2) Meunier et al., Therapie, 1963, 18, 435. (3) Z.L. Chang in 'Analytical Profiles of Drug Substances', ed. K. Florey, Academic Press, NY, 1979, v.8, pp 529-556.

VALPROMIDE [INN]
- CAS 2430-27-5 | MW: 143.23 | MF: C8 H17 N O
- Anticonvulsant
- LgP 1.72(C)
- pKa
- pp cited in Vols.1-5:

(1) Wiemann et al., Bull. Soc. Chim. France, 1958, 199.
(2) Munier, U.S. Pat. 3 325 361 (1967).
(3) Carraz et al., Therapie, 1964, 19, 468.

VALTRATE [INN]
- CAS 18296-44-1 | MW: 422.48 | MF: C22 H30 O8
- Sedative
- Kalichemie, FRG (Valmane)
- LgP 0.92(C)
- pKa
- pp cited in Vols.1-5:

(1) 'Martindale. The Extra Pharmacopoeia', ed. J.E.F. Reynolds, The Pharmaceutical Press, London, 1982, 28th Ed., p 1768.

VANCOMYCIN HCL [U;INN]
- CAS 1404-90-6 | MW: 1463.31 | MF: C67 H77 Cl2 N9 O24
- Antibacterial
- LgP
- pKa
- pp cited in Vols.1-5:
- 2: 94, 96, 100, 110, 567, 573-578, 728

(1) M.H. McCormick et al., U.S. Pat. 3 067 099 (1962). (2) M.P. Williamson et al., J. Am. Chem. Soc., 1981, 103, 6580. (3) Jordan et al., 'Antibiotics', ed. J.W. Corcoran et al., Springer-Verlag, NY, 1975, vol. 3, pp 704-718.

VANEPRIM [INN]
- CAS 81523-49-1 | MW: 520.57 | MF: C23 H28 N4 O8 S
- LgP 1.26(C,3)
- pKa
- pp cited in Vols.1-5:

VANILLIN [U]
- CAS 121-33-5 | MW: 152.15 | MF: C8 H8 O3
- Pharmaceutic aid (flavor)
- LgP 1.21(M); 1.35(C)
- pKa 7.4
- pp cited in Vols.1-5:

(1) Craig et al., U.S. Pat. 3 054 825 (1962).

VANITIOLIDE [INN]
- CAS 17692-71-6 | MW: 253.32 | MF: C12 H15 N O3 S
- Choleretic
- LgP 0.92(C)
- pKa
- pp cited in Vols.1-5:

(1) Chabrier et al., Bull. Soc. Chim. France, 1950, 1167.

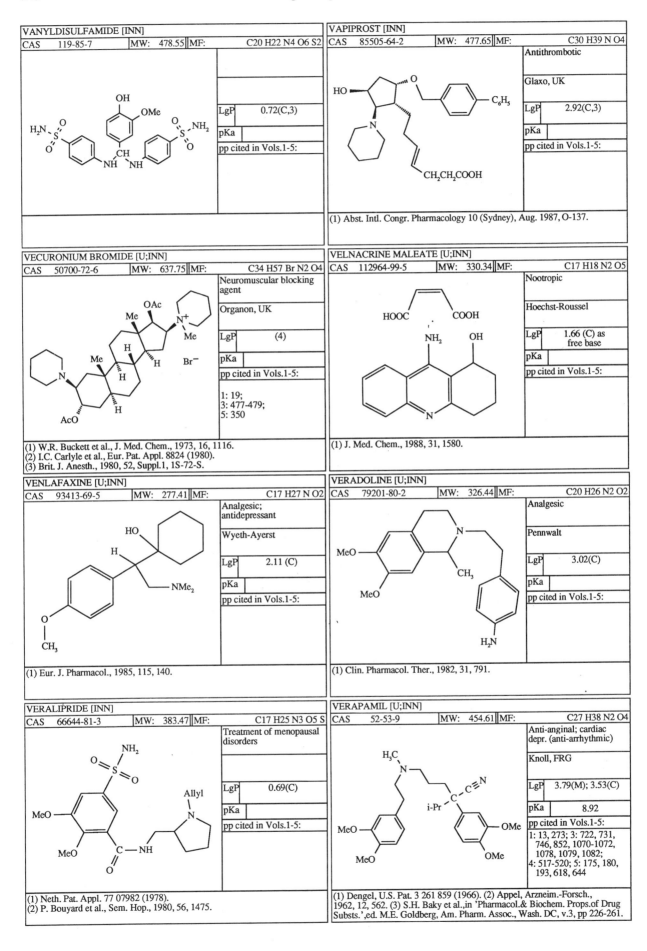

VANYLDISULFAMIDE [INN]

CAS	119-85-7	MW:	478.55	MF:	C20 H22 N4 O6 S2

LgP	0.72(C,3)
pKa	
pp cited in Vols.1-5:	

VAPIPROST [INN]

CAS	85505-64-2	MW:	477.65	MF:	C30 H39 N O4

Antithrombotic

Glaxo, UK

LgP	2.92(C,3)
pKa	
pp cited in Vols.1-5:	

(1) Abst. Intl. Congr. Pharmacology 10 (Sydney), Aug. 1987, O-137.

VECURONIUM BROMIDE [U;INN]

CAS	50700-72-6	MW:	637.75	MF:	C34 H57 Br N2 O4

Neuromuscular blocking agent

Organon, UK

LgP	(4)
pKa	
pp cited in Vols.1-5:	

1: 19;
3: 477-479;
5: 350

(1) W.R. Buckett et al., J. Med. Chem., 1973, 16, 1116.
(2) I.C. Carlyle et al., Eur. Pat. Appl. 8824 (1980).
(3) Brit. J. Anesth., 1980, 52, Suppl.1, 1S-72-S.

VELNACRINE MALEATE [U;INN]

CAS	112964-99-5	MW:	330.34	MF:	C17 H18 N2 O5

Nootropic

Hoechst-Roussel

LgP	1.66 (C) as free base
pKa	
pp cited in Vols.1-5:	

(1) J. Med. Chem., 1988, 31, 1580.

VENLAFAXINE [U;INN]

CAS	93413-69-5	MW:	277.41	MF:	C17 H27 N O2

Analgesic; antidepressant

Wyeth-Ayerst

LgP	2.11 (C)
pKa	
pp cited in Vols.1-5:	

(1) Eur. J. Pharmacol., 1985, 115, 140.

VERADOLINE [U;INN]

CAS	79201-80-2	MW:	326.44	MF:	C20 H26 N2 O2

Analgesic

Pennwalt

LgP	3.02(C)
pKa	
pp cited in Vols.1-5:	

(1) Clin. Pharmacol. Ther., 1982, 31, 791.

VERALIPRIDE [INN]

CAS	66644-81-3	MW:	383.47	MF:	C17 H25 N3 O5 S

Treatment of menopausal disorders

LgP	0.69(C)
pKa	
pp cited in Vols.1-5:	

(1) Neth. Pat. Appl. 77 07982 (1978).
(2) P. Bouyard et al., Sem. Hop., 1980, 56, 1475.

VERAPAMIL [U;INN]

CAS	52-53-9	MW:	454.61	MF:	C27 H38 N2 O4

Anti-anginal; cardiac depr. (anti-arrhythmic)

Knoll, FRG

LgP	3.79(M); 3.53(C)
pKa	8.92
pp cited in Vols.1-5:	

1: 13, 273; 3: 722, 731, 746, 852, 1070-1072, 1078, 1079, 1082;
4: 517-520; 5: 175, 180, 193, 618, 644

(1) Dengel, U.S. Pat. 3 261 859 (1966). (2) Appel, Arzneim.-Forsch., 1962, 12, 562. (3) S.H. Baky et al.,in 'Pharmacol.& Biochem. Props.of Drug Substs.',ed. M.E. Goldberg, Am. Pharm. Assoc., Wash. DC, v.3, pp 226-261.

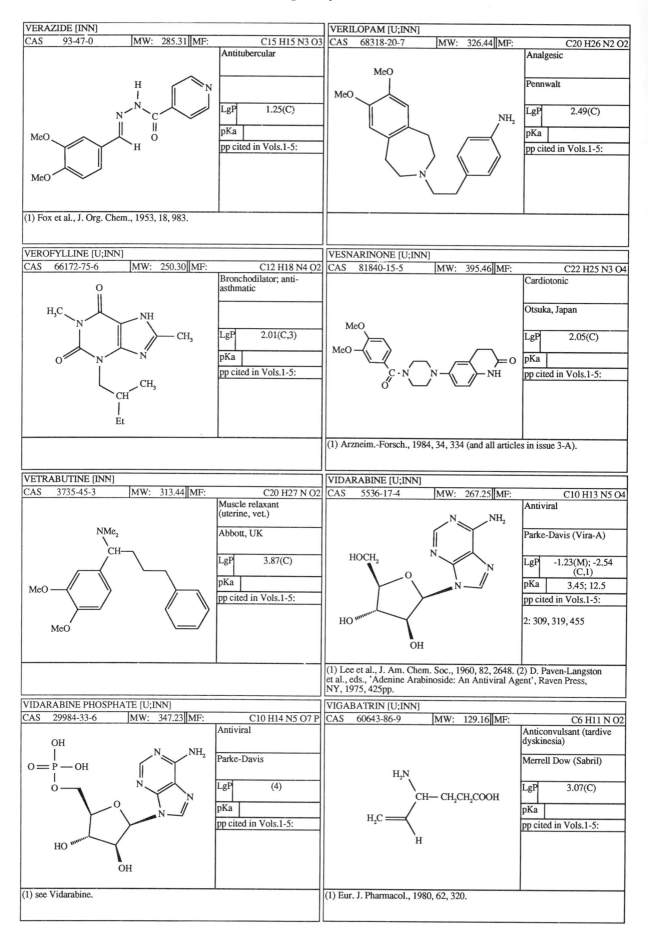

VERAZIDE [INN]

CAS	93-47-0	MW:	285.31	MF:	C15 H15 N3 O3

Antitubercular

LgP	1.25(C)
pKa	
pp cited in Vols.1-5:	

(1) Fox et al., J. Org. Chem., 1953, 18, 983.

VERILOPAM [U;INN]

CAS	68318-20-7	MW:	326.44	MF:	C20 H26 N2 O2

Analgesic

Pennwalt

LgP	2.49(C)
pKa	
pp cited in Vols.1-5:	

VEROFYLLINE [U;INN]

CAS	66172-75-6	MW:	250.30	MF:	C12 H18 N4 O2

Bronchodilator; anti-asthmatic

LgP	2.01(C,3)
pKa	
pp cited in Vols.1-5:	

VESNARINONE [U;INN]

CAS	81840-15-5	MW:	395.46	MF:	C22 H25 N3 O4

Cardiotonic

Otsuka, Japan

LgP	2.05(C)
pKa	
pp cited in Vols.1-5:	

(1) Arzneim.-Forsch., 1984, 34, 334 (and all articles in issue 3-A).

VETRABUTINE [INN]

CAS	3735-45-3	MW:	313.44	MF:	C20 H27 N O2

Muscle relaxant (uterine, vet.)

Abbott, UK

LgP	3.87(C)
pKa	
pp cited in Vols.1-5:	

VIDARABINE [U;INN]

CAS	5536-17-4	MW:	267.25	MF:	C10 H13 N5 O4

Antiviral

Parke-Davis (Vira-A)

LgP	-1.23(M); -2.54 (C,1)
pKa	3.45; 12.5
pp cited in Vols.1-5:	
2: 309, 319, 455	

(1) Lee et al., J. Am. Chem. Soc., 1960, 82, 2648. (2) D. Paven-Langston et al., eds., 'Adenine Arabinoside: An Antiviral Agent', Raven Press, NY, 1975, 425pp.

VIDARABINE PHOSPHATE [U;INN]

CAS	29984-33-6	MW:	347.23	MF:	C10 H14 N5 O7 P

Antiviral

Parke-Davis

LgP	(4)
pKa	
pp cited in Vols.1-5:	

(1) see Vidarabine.

VIGABATRIN [U;INN]

CAS	60643-86-9	MW:	129.16	MF:	C6 H11 N O2

Anticonvulsant (tardive dyskinesia)

Merrell Dow (Sabril)

LgP	3.07(C)
pKa	
pp cited in Vols.1-5:	

(1) Eur. J. Pharmacol., 1980, 62, 320.

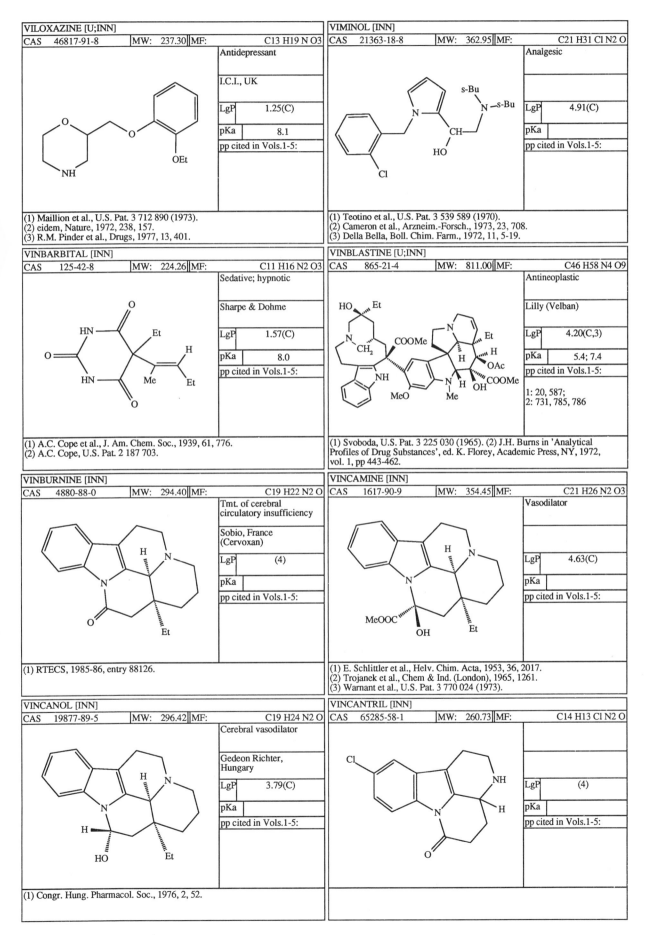

VILOXAZINE [U;INN]

CAS	46817-91-8	MW:	237.30	MF:	C13 H19 N O3

Antidepressant

I.C.I., UK

LgP	1.25(C)
pKa	8.1
pp cited in Vols.1-5:	

(1) Maillion et al., U.S. Pat. 3 712 890 (1973).
(2) eidem, Nature, 1972, 238, 157.
(3) R.M. Pinder et al., Drugs, 1977, 13, 401.

VIMINOL [INN]

CAS	21363-18-8	MW:	362.95	MF:	C21 H31 Cl N2 O

Analgesic

LgP	4.91(C)
pKa	
pp cited in Vols.1-5:	

(1) Teotino et al., U.S. Pat. 3 539 589 (1970).
(2) Cameron et al., Arzneim.-Forsch., 1973, 23, 708.
(3) Della Bella, Boll. Chim. Farm., 1972, 11, 5-19.

VINBARBITAL [INN]

CAS	125-42-8	MW:	224.26	MF:	C11 H16 N2 O3

Sedative; hypnotic

Sharpe & Dohme

LgP	1.57(C)
pKa	8.0
pp cited in Vols.1-5:	

(1) A.C. Cope et al., J. Am. Chem. Soc., 1939, 61, 776.
(2) A.C. Cope, U.S. Pat. 2 187 703.

VINBLASTINE [U;INN]

CAS	865-21-4	MW:	811.00	MF:	C46 H58 N4 O9

Antineoplastic

Lilly (Velban)

LgP	4.20(C,3)
pKa	5.4; 7.4
pp cited in Vols.1-5:	

1: 20, 587;
2: 731, 785, 786

(1) Svoboda, U.S. Pat. 3 225 030 (1965). (2) J.H. Burns in 'Analytical Profiles of Drug Substances', ed. K. Florey, Academic Press, NY, 1972, vol. 1, pp 443-462.

VINBURNINE [INN]

CAS	4880-88-0	MW:	294.40	MF:	C19 H22 N2 O

Tmt. of cerebral circulatory insufficiency

Sobio, France (Cervoxan)

LgP	(4)
pKa	
pp cited in Vols.1-5:	

(1) RTECS, 1985-86, entry 88126.

VINCAMINE [INN]

CAS	1617-90-9	MW:	354.45	MF:	C21 H26 N2 O3

Vasodilator

LgP	4.63(C)
pKa	
pp cited in Vols.1-5:	

(1) E. Schlittler et al., Helv. Chim. Acta, 1953, 36, 2017.
(2) Trojanek et al., Chem & Ind. (London), 1965, 1261.
(3) Warnant et al., U.S. Pat. 3 770 024 (1973).

VINCANOL [INN]

CAS	19877-89-5	MW:	296.42	MF:	C19 H24 N2 O

Cerebral vasodilator

Gedeon Richter, Hungary

LgP	3.79(C)
pKa	
pp cited in Vols.1-5:	

(1) Congr. Hung. Pharmacol. Soc., 1976, 2, 52.

VINCANTRIL [INN]

CAS	65285-58-1	MW:	260.73	MF:	C14 H13 Cl N2 O

LgP	(4)
pKa	
pp cited in Vols.1-5:	

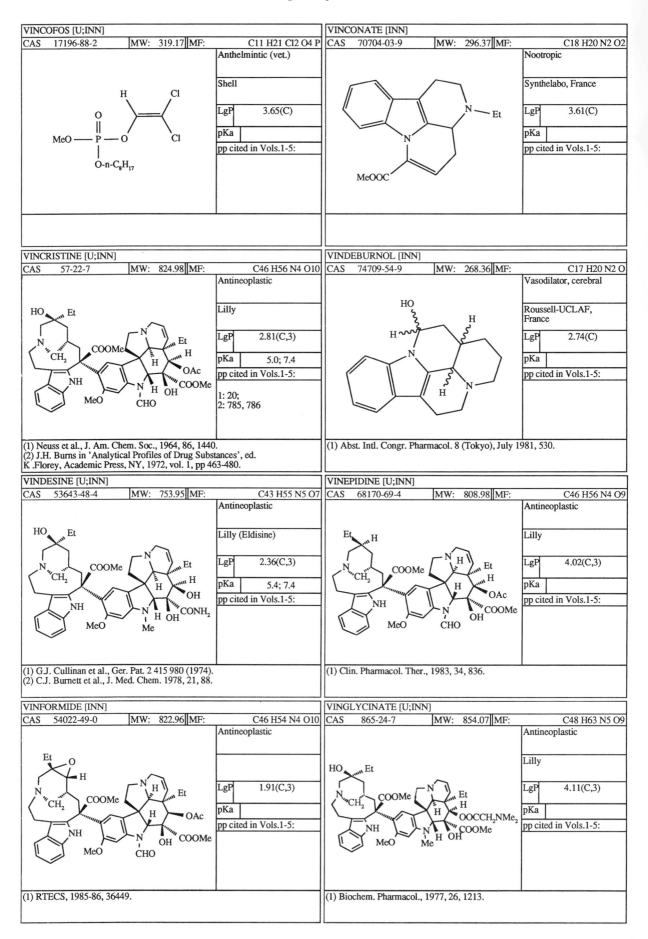

VINCOFOS [U;INN]

CAS	17196-88-2	MW:	319.17	MF:	C11 H21 Cl2 O4 P

Anthelmintic (vet.)

Shell

LgP	3.65(C)
pKa	

pp cited in Vols.1-5:

VINCONATE [INN]

CAS	70704-03-9	MW:	296.37	MF:	C18 H20 N2 O2

Nootropic

Synthelabo, France

LgP	3.61(C)
pKa	

pp cited in Vols.1-5:

VINCRISTINE [U;INN]

CAS	57-22-7	MW:	824.98	MF:	C46 H56 N4 O10

Antineoplastic

Lilly

LgP	2.81(C,3)
pKa	5.0; 7.4

pp cited in Vols.1-5:

1: 20;
2: 785, 786

(1) Neuss et al., J. Am. Chem. Soc., 1964, 86, 1440.
(2) J.H. Burns in 'Analytical Profiles of Drug Substances', ed.
K .Florey, Academic Press, NY, 1972, vol. 1, pp 463-480.

VINDEBURNOL [INN]

CAS	74709-54-9	MW:	268.36	MF:	C17 H20 N2 O

Vasodilator, cerebral

Roussell-UCLAF,
France

LgP	2.74(C)
pKa	

pp cited in Vols.1-5:

(1) Abst. Intl. Congr. Pharmacol. 8 (Tokyo), July 1981, 530.

VINDESINE [U;INN]

CAS	53643-48-4	MW:	753.95	MF:	C43 H55 N5 O7

Antineoplastic

Lilly (Eldisine)

LgP	2.36(C,3)
pKa	5.4; 7.4

pp cited in Vols.1-5:

(1) G.J. Cullinan et al., Ger. Pat. 2 415 980 (1974).
(2) C.J. Burnett et al., J. Med. Chem. 1978, 21, 88.

VINEPIDINE [U;INN]

CAS	68170-69-4	MW:	808.98	MF:	C46 H56 N4 O9

Antineoplastic

Lilly

LgP	4.02(C,3)
pKa	

pp cited in Vols.1-5:

(1) Clin. Pharmacol. Ther., 1983, 34, 836.

VINFORMIDE [INN]

CAS	54022-49-0	MW:	822.96	MF:	C46 H54 N4 O10

Antineoplastic

LgP	1.91(C,3)
pKa	

pp cited in Vols.1-5:

(1) RTECS, 1985-86, 36449.

VINGLYCINATE [U;INN]

CAS	865-24-7	MW:	854.07	MF:	C48 H63 N5 O9

Antineoplastic

Lilly

LgP	4.11(C,3)
pKa	

pp cited in Vols.1-5:

(1) Biochem. Pharmacol., 1977, 26, 1213.

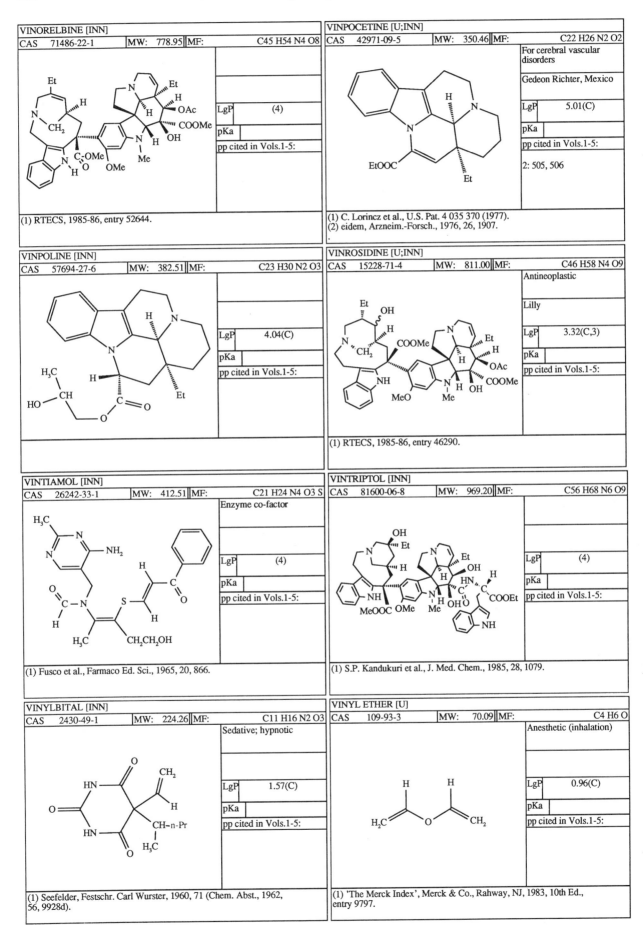

VINORELBINE [INN]

CAS	71486-22-1	MW:	778.95	MF:	C45 H54 N4 O8

LgP	(4)
pKa	
pp cited in Vols.1-5:	

(1) RTECS, 1985-86, entry 52644.

VINPOCETINE [U;INN]

CAS	42971-09-5	MW:	350.46	MF:	C22 H26 N2 O2

For cerebral vascular disorders

Gedeon Richter, Mexico

LgP	5.01(C)
pKa	
pp cited in Vols.1-5:	

2: 505, 506

(1) C. Lorincz et al., U.S. Pat. 4 035 370 (1977).
(2) eidem, Arzneim.-Forsch., 1976, 26, 1907.

VINPOLINE [INN]

CAS	57694-27-6	MW:	382.51	MF:	C23 H30 N2 O3

LgP	4.04(C)
pKa	
pp cited in Vols.1-5:	

VINROSIDINE [U;INN]

CAS	15228-71-4	MW:	811.00	MF:	C46 H58 N4 O9

Antineoplastic

Lilly

LgP	3.32(C,3)
pKa	
pp cited in Vols.1-5:	

(1) RTECS, 1985-86, entry 46290.

VINTIAMOL [INN]

CAS	26242-33-1	MW:	412.51	MF:	C21 H24 N4 O3 S

Enzyme co-factor

LgP	(4)
pKa	
pp cited in Vols.1-5:	

(1) Fusco et al., Farmaco Ed. Sci., 1965, 20, 866.

VINTRIPTOL [INN]

CAS	81600-06-8	MW:	969.20	MF:	C56 H68 N6 O9

LgP	(4)
pKa	
pp cited in Vols.1-5:	

(1) S.P. Kandukuri et al., J. Med. Chem., 1985, 28, 1079.

VINYLBITAL [INN]

CAS	2430-49-1	MW:	224.26	MF:	C11 H16 N2 O3

Sedative; hypnotic

LgP	1.57(C)
pKa	
pp cited in Vols.1-5:	

(1) Seefelder, Festschr. Carl Wurster, 1960, 71 (Chem. Abst., 1962, 56, 9928d).

VINYL ETHER [U]

CAS	109-93-3	MW:	70.09	MF:	C4 H6 O

Anesthetic (inhalation)

LgP	0.96(C)
pKa	
pp cited in Vols.1-5:	

(1) 'The Merck Index', Merck & Co., Rahway, NJ, 1983, 10th Ed., entry 9797.

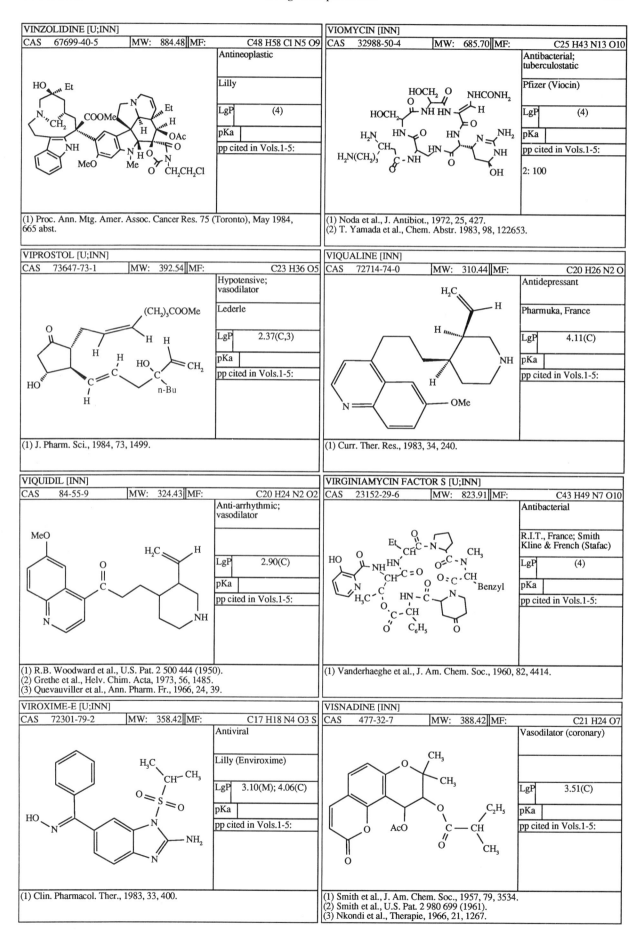

VINZOLIDINE [U;INN]			
CAS 67699-40-5	MW: 884.48	MF:	C48 H58 Cl N5 O9

Antineoplastic

Lilly

LgP	(4)
pKa	
pp cited in Vols.1-5:	

(1) Proc. Ann. Mtg. Amer. Assoc. Cancer Res. 75 (Toronto), May 1984, 665 abst.

VIOMYCIN [INN]			
CAS 32988-50-4	MW: 685.70	MF:	C25 H43 N13 O10

Antibacterial; tuberculostatic

Pfizer (Viocin)

LgP	(4)
pKa	
pp cited in Vols.1-5:	
2: 100	

(1) Noda et al., J. Antibiot., 1972, 25, 427.
(2) T. Yamada et al., Chem. Abstr. 1983, 98, 122653.

VIPROSTOL [U;INN]			
CAS 73647-73-1	MW: 392.54	MF:	C23 H36 O5

Hypotensive; vasodilator

Lederle

LgP	2.37(C,3)
pKa	
pp cited in Vols.1-5:	

(1) J. Pharm. Sci., 1984, 73, 1499.

VIQUALINE [INN]			
CAS 72714-74-0	MW: 310.44	MF:	C20 H26 N2 O

Antidepressant

Pharmuka, France

LgP	4.11(C)
pKa	
pp cited in Vols.1-5:	

(1) Curr. Ther. Res., 1983, 34, 240.

VIQUIDIL [INN]			
CAS 84-55-9	MW: 324.43	MF:	C20 H24 N2 O2

Anti-arrhythmic; vasodilator

LgP	2.90(C)
pKa	
pp cited in Vols.1-5:	

(1) R.B. Woodward et al., U.S. Pat. 2 500 444 (1950).
(2) Grethe et al., Helv. Chim. Acta, 1973, 56, 1485.
(3) Quevauviller et al., Ann. Pharm. Fr., 1966, 24, 39.

VIRGINIAMYCIN FACTOR S [U;INN]			
CAS 23152-29-6	MW: 823.91	MF:	C43 H49 N7 O10

Antibacterial

R.I.T., France; Smith Kline & French (Stafac)

LgP	(4)
pKa	
pp cited in Vols.1-5:	

(1) Vanderhaeghe et al., J. Am. Chem. Soc., 1960, 82, 4414.

VIROXIME-E [U;INN]			
CAS 72301-79-2	MW: 358.42	MF:	C17 H18 N4 O3 S

Antiviral

Lilly (Enviroxime)

LgP	3.10(M); 4.06(C)
pKa	
pp cited in Vols.1-5:	

(1) Clin. Pharmacol. Ther., 1983, 33, 400.

VISNADINE [INN]			
CAS 477-32-7	MW: 388.42	MF:	C21 H24 O7

Vasodilator (coronary)

LgP	3.51(C)
pKa	
pp cited in Vols.1-5:	

(1) Smith et al., J. Am. Chem. Soc., 1957, 79, 3534.
(2) Smith et al., U.S. Pat. 2 980 699 (1961).
(3) Nkondi et al., Therapie, 1966, 21, 1267.

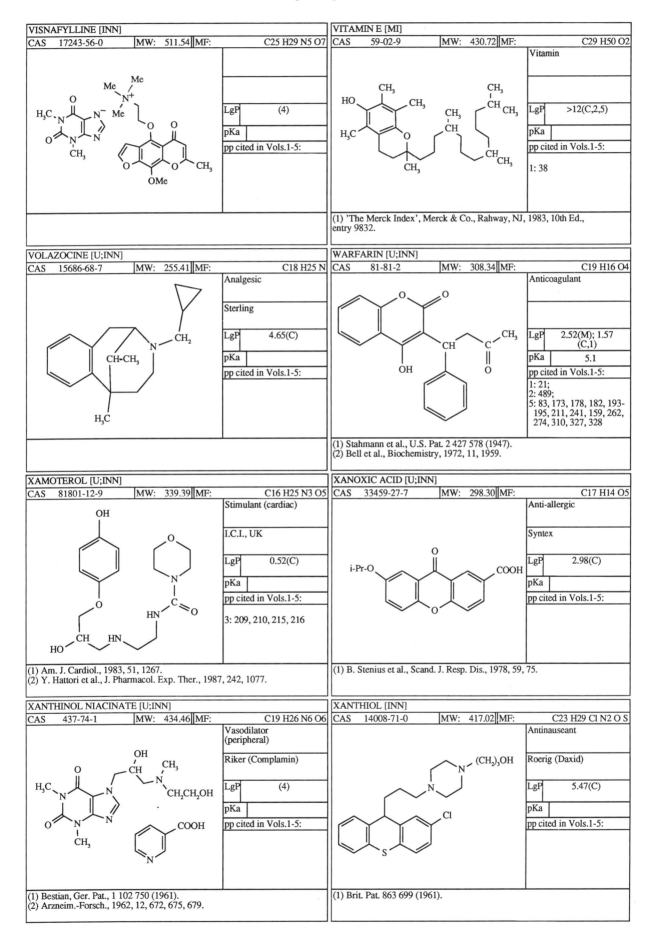

VISNAFYLLINE [INN]

CAS	17243-56-0	MW:	511.54	MF:	C25 H29 N5 O7

LgP	(4)
pKa	
pp cited in Vols.1-5:	

VITAMIN E [MI]

CAS	59-02-9	MW:	430.72	MF:	C29 H50 O2

Vitamin

LgP	>12(C,2,5)
pKa	
pp cited in Vols.1-5:	

1: 38

(1) 'The Merck Index', Merck & Co., Rahway, NJ, 1983, 10th Ed., entry 9832.

VOLAZOCINE [U;INN]

CAS	15686-68-7	MW:	255.41	MF:	C18 H25 N

Analgesic

Sterling

LgP	4.65(C)
pKa	
pp cited in Vols.1-5:	

WARFARIN [U;INN]

CAS	81-81-2	MW:	308.34	MF:	C19 H16 O4

Anticoagulant

LgP	2.52(M); 1.57 (C,1)
pKa	5.1
pp cited in Vols.1-5:	

1: 21;
2: 489;
5: 83, 173, 178, 182, 193-195, 211, 241, 159, 262, 274, 310, 327, 328

(1) Stahmann et al., U.S. Pat. 2 427 578 (1947).
(2) Bell et al., Biochemistry, 1972, 11, 1959.

XAMOTEROL [U;INN]

CAS	81801-12-9	MW:	339.39	MF:	C16 H25 N3 O5

Stimulant (cardiac)

I.C.I., UK

LgP	0.52(C)
pKa	
pp cited in Vols.1-5:	

3: 209, 210, 215, 216

(1) Am. J. Cardiol., 1983, 51, 1267.
(2) Y. Hattori et al., J. Pharmacol. Exp. Ther., 1987, 242, 1077.

XANOXIC ACID [U;INN]

CAS	33459-27-7	MW:	298.30	MF:	C17 H14 O5

Anti-allergic

Syntex

LgP	2.98(C)
pKa	
pp cited in Vols.1-5:	

(1) B. Stenius et al., Scand. J. Resp. Dis., 1978, 59, 75.

XANTHINOL NIACINATE [U;INN]

CAS	437-74-1	MW:	434.46	MF:	C19 H26 N6 O6

Vasodilator (peripheral)

Riker (Complamin)

LgP	(4)
pKa	
pp cited in Vols.1-5:	

(1) Bestian, Ger. Pat., 1 102 750 (1961).
(2) Arzneim.-Forsch., 1962, 12, 672, 675, 679.

XANTHIOL [INN]

CAS	14008-71-0	MW:	417.02	MF:	C23 H29 Cl N2 O S

Antinauseant

Roerig (Daxid)

LgP	5.47(C)
pKa	
pp cited in Vols.1-5:	

(1) Brit. Pat. 863 699 (1961).

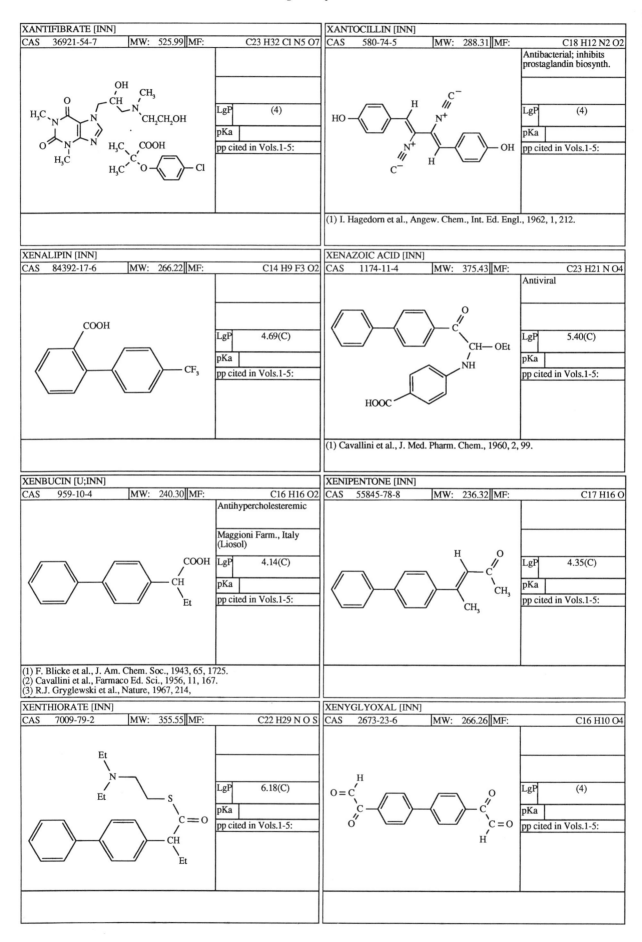

XANTIFIBRATE [INN]

CAS	36921-54-7	MW:	525.99	MF:	C23 H32 Cl N5 O7

LgP	(4)
pKa	
pp cited in Vols.1-5:	

XANTOCILLIN [INN]

CAS	580-74-5	MW:	288.31	MF:	C18 H12 N2 O2

Antibacterial; inhibits prostaglandin biosynth.

LgP	(4)
pKa	
pp cited in Vols.1-5:	

(1) I. Hagedorn et al., Angew. Chem., Int. Ed. Engl., 1962, 1, 212.

XENALIPIN [INN]

CAS	84392-17-6	MW:	266.22	MF:	C14 H9 F3 O2

LgP	4.69(C)
pKa	
pp cited in Vols.1-5:	

XENAZOIC ACID [INN]

CAS	1174-11-4	MW:	375.43	MF:	C23 H21 N O4

Antiviral

LgP	5.40(C)
pKa	
pp cited in Vols.1-5:	

(1) Cavallini et al., J. Med. Pharm. Chem., 1960, 2, 99.

XENBUCIN [U;INN]

CAS	959-10-4	MW:	240.30	MF:	C16 H16 O2

Antihypercholesteremic

Maggioni Farm., Italy (Liosol)

LgP	4.14(C)
pKa	
pp cited in Vols.1-5:	

(1) F. Blicke et al., J. Am. Chem. Soc., 1943, 65, 1725.
(2) Cavallini et al., Farmaco Ed. Sci., 1956, 11, 167.
(3) R.J. Gryglewski et al., Nature, 1967, 214,

XENIPENTONE [INN]

CAS	55845-78-8	MW:	236.32	MF:	C17 H16 O

LgP	4.35(C)
pKa	
pp cited in Vols.1-5:	

XENTHIORATE [INN]

CAS	7009-79-2	MW:	355.55	MF:	C22 H29 N O S

LgP	6.18(C)
pKa	
pp cited in Vols.1-5:	

XENYGLYOXAL [INN]

CAS	2673-23-6	MW:	266.26	MF:	C16 H10 O4

LgP	(4)
pKa	
pp cited in Vols.1-5:	

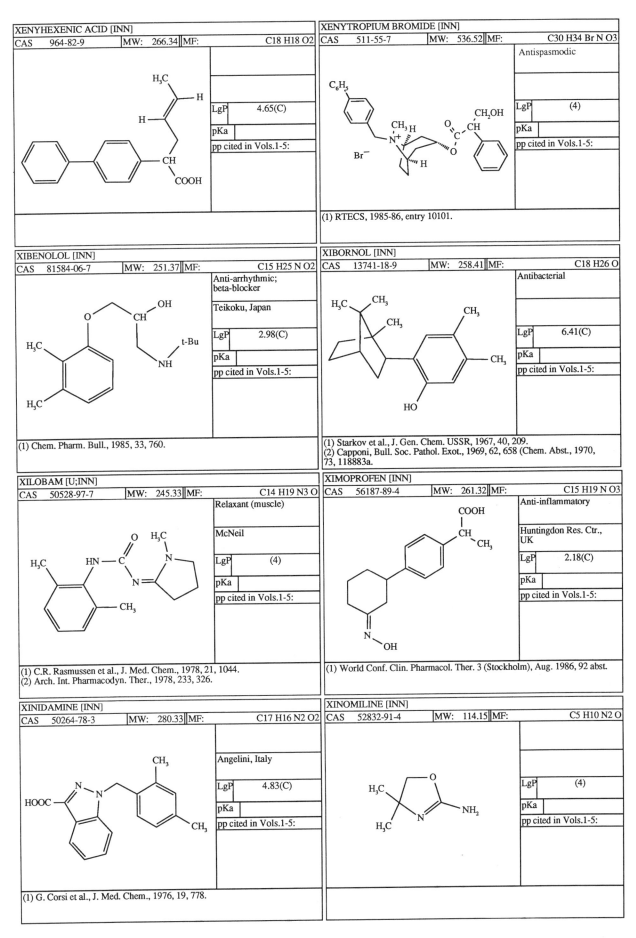

XENYHEXENIC ACID [INN]

CAS	964-82-9	MW:	266.34	MF:	C18 H18 O2

LgP	4.65(C)
pKa	
pp cited in Vols.1-5:	

XENYTROPIUM BROMIDE [INN]

CAS	511-55-7	MW:	536.52	MF:	C30 H34 Br N O3

Antispasmodic

LgP	(4)
pKa	
pp cited in Vols.1-5:	

(1) RTECS, 1985-86, entry 10101.

XIBENOLOL [INN]

CAS	81584-06-7	MW:	251.37	MF:	C15 H25 N O2

Anti-arrhythmic; beta-blocker

Teikoku, Japan

LgP	2.98(C)
pKa	
pp cited in Vols.1-5:	

(1) Chem. Pharm. Bull., 1985, 33, 760.

XIBORNOL [INN]

CAS	13741-18-9	MW:	258.41	MF:	C18 H26 O

Antibacterial

LgP	6.41(C)
pKa	
pp cited in Vols.1-5:	

(1) Starkov et al., J. Gen. Chem. USSR, 1967, 40, 209.
(2) Capponi, Bull. Soc. Pathol. Exot., 1969, 62, 658 (Chem. Abst., 1970, 73, 118883a.

XILOBAM [U;INN]

CAS	50528-97-7	MW:	245.33	MF:	C14 H19 N3 O

Relaxant (muscle)

McNeil

LgP	(4)
pKa	
pp cited in Vols.1-5:	

(1) C.R. Rasmussen et al., J. Med. Chem., 1978, 21, 1044.
(2) Arch. Int. Pharmacodyn. Ther., 1978, 233, 326.

XIMOPROFEN [INN]

CAS	56187-89-4	MW:	261.32	MF:	C15 H19 N O3

Anti-inflammatory

Huntingdon Res. Ctr., UK

LgP	2.18(C)
pKa	
pp cited in Vols.1-5:	

(1) World Conf. Clin. Pharmacol. Ther. 3 (Stockholm), Aug. 1986, 92 abst.

XINIDAMINE [INN]

CAS	50264-78-3	MW:	280.33	MF:	C17 H16 N2 O2

Angelini, Italy

LgP	4.83(C)
pKa	
pp cited in Vols.1-5:	

(1) G. Corsi et al., J. Med. Chem., 1976, 19, 778.

XINOMILINE [INN]

CAS	52832-91-4	MW:	114.15	MF:	C5 H10 N2 O

LgP	(4)
pKa	
pp cited in Vols.1-5:	

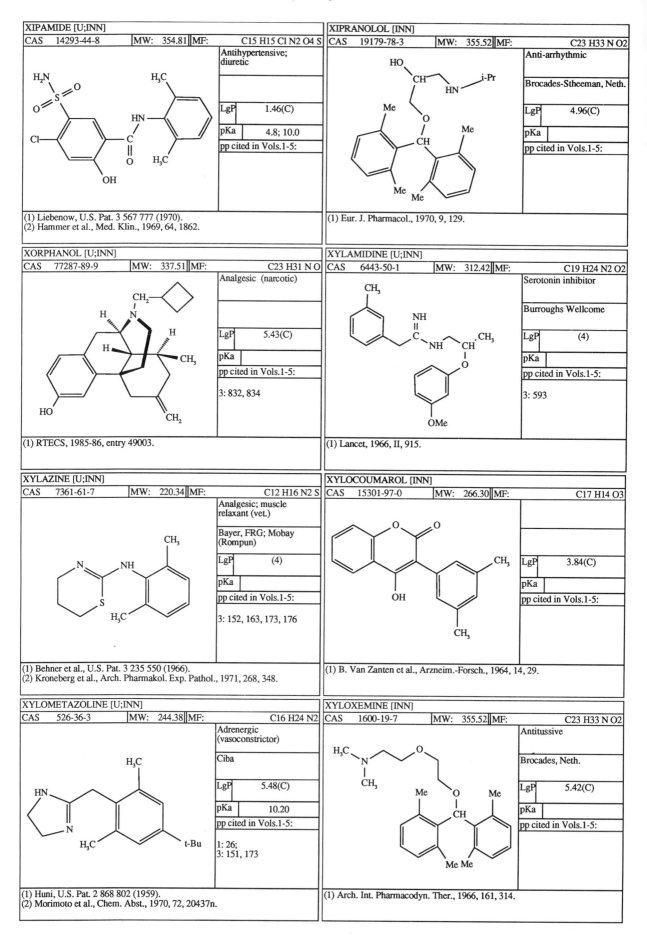

XIPAMIDE [U;INN]

CAS	14293-44-8	MW:	354.81	MF:	C15 H15 Cl N2 O4 S

Antihypertensive; diuretic

LgP	1.46(C)
pKa	4.8; 10.0

pp cited in Vols.1-5:

(1) Liebenow, U.S. Pat. 3 567 777 (1970).
(2) Hammer et al., Med. Klin., 1969, 64, 1862.

XIPRANOLOL [INN]

CAS	19179-78-3	MW:	355.52	MF:	C23 H33 N O2

Anti-arrhythmic

Brocades-Stheeman, Neth.

LgP	4.96(C)
pKa	

pp cited in Vols.1-5:

(1) Eur. J. Pharmacol., 1970, 9, 129.

XORPHANOL [U;INN]

CAS	77287-89-9	MW:	337.51	MF:	C23 H31 N O

Analgesic (narcotic)

LgP	5.43(C)
pKa	

pp cited in Vols.1-5:

3: 832, 834

(1) RTECS, 1985-86, entry 49003.

XYLAMIDINE [U;INN]

CAS	6443-50-1	MW:	312.42	MF:	C19 H24 N2 O2

Serotonin inhibitor

Burroughs Wellcome

LgP	(4)
pKa	

pp cited in Vols.1-5:

3: 593

(1) Lancet, 1966, II, 915.

XYLAZINE [U;INN]

CAS	7361-61-7	MW:	220.34	MF:	C12 H16 N2 S

Analgesic; muscle relaxant (vet.)

Bayer, FRG; Mobay (Rompun)

LgP	(4)
pKa	

pp cited in Vols.1-5:

3: 152, 163, 173, 176

(1) Behner et al., U.S. Pat. 3 235 550 (1966).
(2) Kroneberg et al., Arch. Pharmakol. Exp. Pathol., 1971, 268, 348.

XYLOCOUMAROL [INN]

CAS	15301-97-0	MW:	266.30	MF:	C17 H14 O3

LgP	3.84(C)
pKa	

pp cited in Vols.1-5:

(1) B. Van Zanten et al., Arzneim.-Forsch., 1964, 14, 29.

XYLOMETAZOLINE [U;INN]

CAS	526-36-3	MW:	244.38	MF:	C16 H24 N2

Adrenergic (vasoconstrictor)

Ciba

LgP	5.48(C)
pKa	10.20

pp cited in Vols.1-5:

1: 26;
3: 151, 173

(1) Huni, U.S. Pat. 2 868 802 (1959).
(2) Morimoto et al., Chem. Abst., 1970, 72, 20437n.

XYLOXEMINE [INN]

CAS	1600-19-7	MW:	355.52	MF:	C23 H33 N O2

Antitussive

Brocades, Neth.

LgP	5.42(C)
pKa	

pp cited in Vols.1-5:

(1) Arch. Int. Pharmacodyn. Ther., 1966, 161, 314.

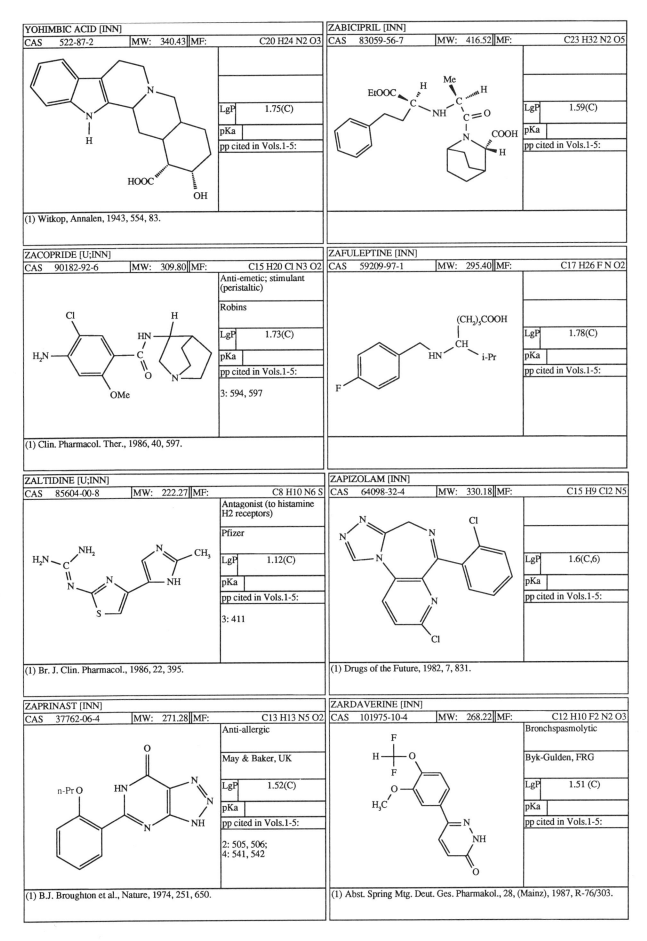

YOHIMBIC ACID [INN]			
CAS 522-87-2	MW: 340.43	MF:	C20 H24 N2 O3

LgP 1.75(C)
pKa
pp cited in Vols.1-5:

(1) Witkop, Annalen, 1943, 554, 83.

ZABICIPRIL [INN]			
CAS 83059-56-7	MW: 416.52	MF:	C23 H32 N2 O5

LgP 1.59(C)
pKa
pp cited in Vols.1-5:

ZACOPRIDE [U;INN]			
CAS 90182-92-6	MW: 309.80	MF:	C15 H20 Cl N3 O2

Anti-emetic; stimulant (peristaltic)

Robins

LgP 1.73(C)
pKa
pp cited in Vols.1-5:

3: 594, 597

(1) Clin. Pharmacol. Ther., 1986, 40, 597.

ZAFULEPTINE [INN]			
CAS 59209-97-1	MW: 295.40	MF:	C17 H26 F N O2

LgP 1.78(C)
pKa
pp cited in Vols.1-5:

ZALTIDINE [U;INN]			
CAS 85604-00-8	MW: 222.27	MF:	C8 H10 N6 S

Antagonist (to histamine H2 receptors)

Pfizer

LgP 1.12(C)
pKa
pp cited in Vols.1-5:

3: 411

(1) Br. J. Clin. Pharmacol., 1986, 22, 395.

ZAPIZOLAM [INN]			
CAS 64098-32-4	MW: 330.18	MF:	C15 H9 Cl2 N5

LgP 1.6(C,6)
pKa
pp cited in Vols.1-5:

(1) Drugs of the Future, 1982, 7, 831.

ZAPRINAST [INN]			
CAS 37762-06-4	MW: 271.28	MF:	C13 H13 N5 O2

Anti-allergic

May & Baker, UK

LgP 1.52(C)
pKa
pp cited in Vols.1-5:

2: 505, 506;
4: 541, 542

(1) B.J. Broughton et al., Nature, 1974, 251, 650.

ZARDAVERINE [INN]			
CAS 101975-10-4	MW: 268.22	MF:	C12 H10 F2 N2 O3

Bronchspasmolytic

Byk-Gulden, FRG

LgP 1.51 (C)
pKa
pp cited in Vols.1-5:

(1) Abst. Spring Mtg. Deut. Ges. Pharmakol., 28, (Mainz), 1987, R-76/303.

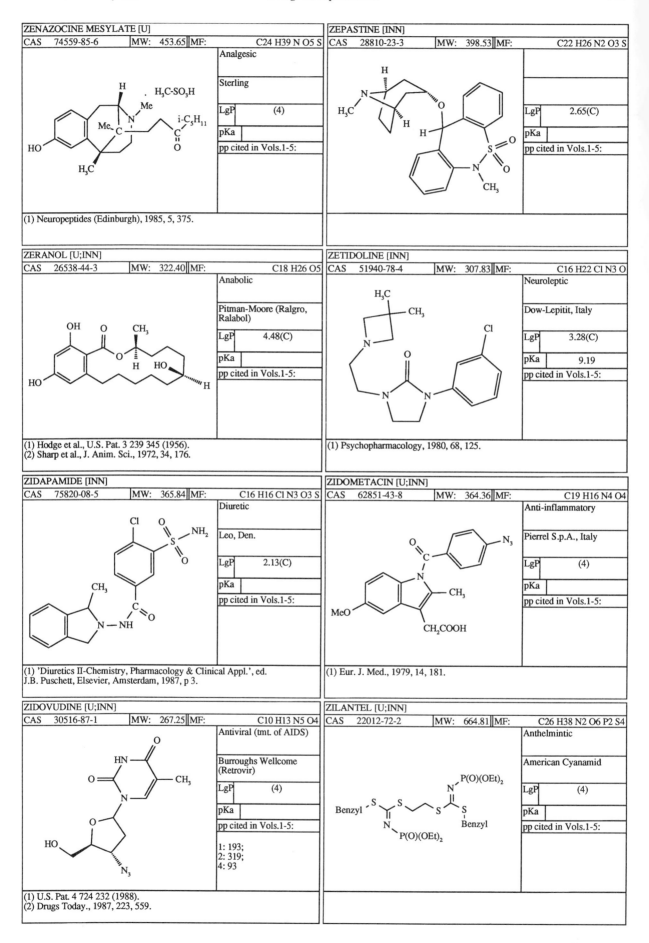

ZENAZOCINE MESYLATE [U]

CAS	74559-85-6	MW:	453.65	MF:	C24 H39 N O5 S

Analgesic

Sterling

LgP	(4)
pKa	
pp cited in Vols.1-5:	

(1) Neuropeptides (Edinburgh), 1985, 5, 375.

ZEPASTINE [INN]

CAS	28810-23-3	MW:	398.53	MF:	C22 H26 N2 O3 S

LgP	2.65(C)
pKa	
pp cited in Vols.1-5:	

ZERANOL [U;INN]

CAS	26538-44-3	MW:	322.40	MF:	C18 H26 O5

Anabolic

Pitman-Moore (Ralgro, Ralabol)

LgP	4.48(C)
pKa	
pp cited in Vols.1-5:	

(1) Hodge et al., U.S. Pat. 3 239 345 (1956).
(2) Sharp et al., J. Anim. Sci., 1972, 34, 176.

ZETIDOLINE [INN]

CAS	51940-78-4	MW:	307.83	MF:	C16 H22 Cl N3 O

Neuroleptic

Dow-Lepitit, Italy

LgP	3.28(C)
pKa	9.19
pp cited in Vols.1-5:	

(1) Psychopharmacology, 1980, 68, 125.

ZIDAPAMIDE [INN]

CAS	75820-08-5	MW:	365.84	MF:	C16 H16 Cl N3 O3 S

Diuretic

Leo, Den.

LgP	2.13(C)
pKa	
pp cited in Vols.1-5:	

(1) 'Diuretics II-Chemistry, Pharmacology & Clinical Appl.', ed. J.B. Puschett, Elsevier, Amsterdam, 1987, p 3.

ZIDOMETACIN [U;INN]

CAS	62851-43-8	MW:	364.36	MF:	C19 H16 N4 O4

Anti-inflammatory

Pierrel S.p.A., Italy

LgP	(4)
pKa	
pp cited in Vols.1-5:	

(1) Eur. J. Med., 1979, 14, 181.

ZIDOVUDINE [U;INN]

CAS	30516-87-1	MW:	267.25	MF:	C10 H13 N5 O4

Antiviral (tmt. of AIDS)

Burroughs Wellcome (Retrovir)

LgP	(4)
pKa	
pp cited in Vols.1-5:	

1: 193;
2: 319;
4: 93

(1) U.S. Pat. 4 724 232 (1988).
(2) Drugs Today., 1987, 223, 559.

ZILANTEL [U;INN]

CAS	22012-72-2	MW:	664.81	MF:	C26 H38 N2 O6 P2 S4

Anthelmintic

American Cyanamid

LgP	(4)
pKa	
pp cited in Vols.1-5:	

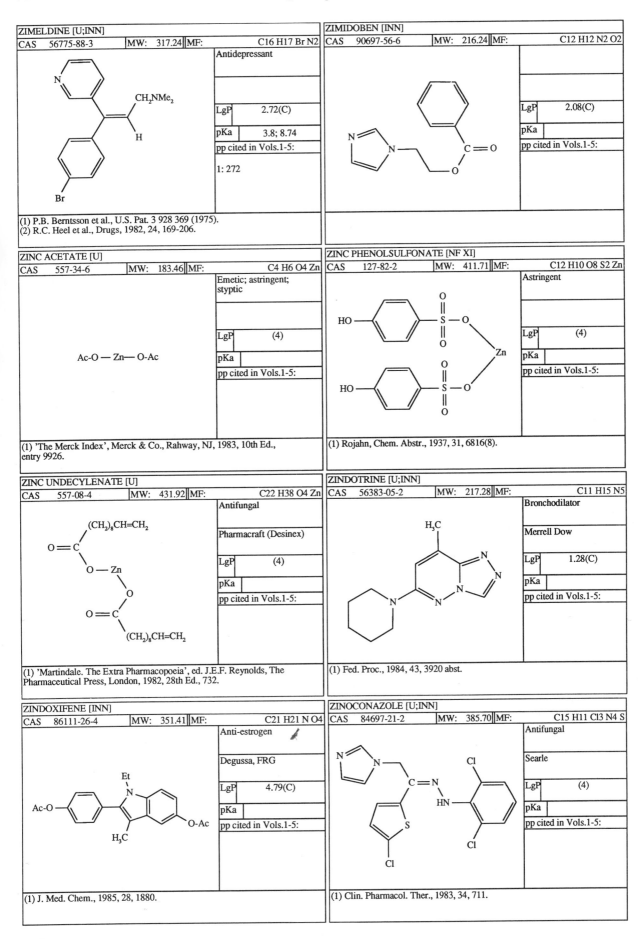

ZIMELDINE [U;INN]

CAS	56775-88-3	MW:	317.24	MF:	C16 H17 Br N2

Antidepressant

LgP	2.72(C)
pKa	3.8; 8.74

pp cited in Vols.1-5:

1: 272

(1) P.B. Berntsson et al., U.S. Pat. 3 928 369 (1975).
(2) R.C. Heel et al., Drugs, 1982, 24, 169-206.

ZIMIDOBEN [INN]

CAS	90697-56-6	MW:	216.24	MF:	C12 H12 N2 O2

LgP	2.08(C)
pKa	

pp cited in Vols.1-5:

ZINC ACETATE [U]

CAS	557-34-6	MW:	183.46	MF:	C4 H6 O4 Zn

Emetic; astringent; styptic

LgP	(4)
pKa	

pp cited in Vols.1-5:

Ac-O — Zn— O-Ac

(1) 'The Merck Index', Merck & Co., Rahway, NJ, 1983, 10th Ed., entry 9926.

ZINC PHENOLSULFONATE [NF XI]

CAS	127-82-2	MW:	411.71	MF:	C12 H10 O8 S2 Zn

Astringent

LgP	(4)
pKa	

pp cited in Vols.1-5:

(1) Rojahn, Chem. Abstr., 1937, 31, 6816(8).

ZINC UNDECYLENATE [U]

CAS	557-08-4	MW:	431.92	MF:	C22 H38 O4 Zn

Antifungal

Pharmacraft (Desinex)

LgP	(4)
pKa	

pp cited in Vols.1-5:

(1) 'Martindale. The Extra Pharmacopoeia', ed. J.E.F. Reynolds, The Pharmaceutical Press, London, 1982, 28th Ed., 732.

ZINDOTRINE [U;INN]

CAS	56383-05-2	MW:	217.28	MF:	C11 H15 N5

Bronchodilator

Merrell Dow

LgP	1.28(C)
pKa	

pp cited in Vols.1-5:

(1) Fed. Proc., 1984, 43, 3920 abst.

ZINDOXIFENE [INN]

CAS	86111-26-4	MW:	351.41	MF:	C21 H21 N O4

Anti-estrogen

Degussa, FRG

LgP	4.79(C)
pKa	

pp cited in Vols.1-5:

(1) J. Med. Chem., 1985, 28, 1880.

ZINOCONAZOLE [U;INN]

CAS	84697-21-2	MW:	385.70	MF:	C15 H11 Cl3 N4 S

Antifungal

Searle

LgP	(4)
pKa	

pp cited in Vols.1-5:

(1) Clin. Pharmacol. Ther., 1983, 34, 711.

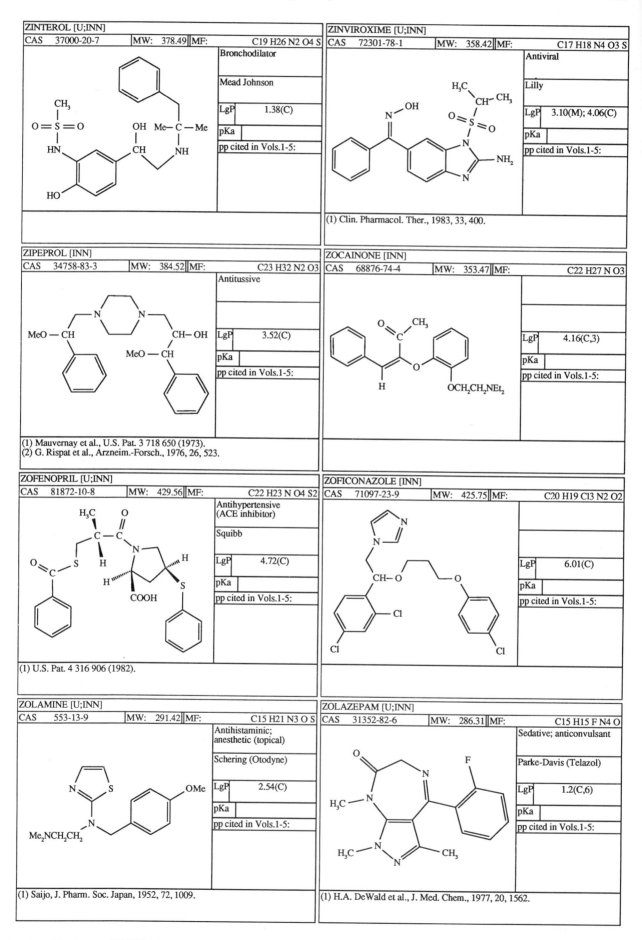

ZINTEROL [U;INN]

CAS	37000-20-7	MW: 378.49	MF:	C19 H26 N2 O4 S

Bronchodilator

Mead Johnson

LgP	1.38(C)
pKa	
pp cited in Vols.1-5:	

ZINVIROXIME [U;INN]

CAS	72301-78-1	MW: 358.42	MF:	C17 H18 N4 O3 S

Antiviral

Lilly

LgP	3.10(M); 4.06(C)
pKa	
pp cited in Vols.1-5:	

(1) Clin. Pharmacol. Ther., 1983, 33, 400.

ZIPEPROL [INN]

CAS	34758-83-3	MW: 384.52	MF:	C23 H32 N2 O3

Antitussive

LgP	3.52(C)
pKa	
pp cited in Vols.1-5:	

(1) Mauvernay et al., U.S. Pat. 3 718 650 (1973).
(2) G. Rispat et al., Arzneim.-Forsch., 1976, 26, 523.

ZOCAINONE [INN]

CAS	68876-74-4	MW: 353.47	MF:	C22 H27 N O3

LgP	4.16(C,3)
pKa	
pp cited in Vols.1-5:	

ZOFENOPRIL [U;INN]

CAS	81872-10-8	MW: 429.56	MF:	C22 H23 N O4 S2

Antihypertensive
(ACE inhibitor)

Squibb

LgP	4.72(C)
pKa	
pp cited in Vols.1-5:	

(1) U.S. Pat. 4 316 906 (1982).

ZOFICONAZOLE [INN]

CAS	71097-23-9	MW: 425.75	MF:	C20 H19 Cl3 N2 O2

LgP	6.01(C)
pKa	
pp cited in Vols.1-5:	

ZOLAMINE [U;INN]

CAS	553-13-9	MW: 291.42	MF:	C15 H21 N3 O S

Antihistaminic;
anesthetic (topical)

Schering (Otodyne)

LgP	2.54(C)
pKa	
pp cited in Vols.1-5:	

(1) Saijo, J. Pharm. Soc. Japan, 1952, 72, 1009.

ZOLAZEPAM [U;INN]

CAS	31352-82-6	MW: 286.31	MF:	C15 H15 F N4 O

Sedative; anticonvulsant

Parke-Davis (Telazol)

LgP	1.2(C,6)
pKa	
pp cited in Vols.1-5:	

(1) H.A. DeWald et al., J. Med. Chem., 1977, 20, 1562.

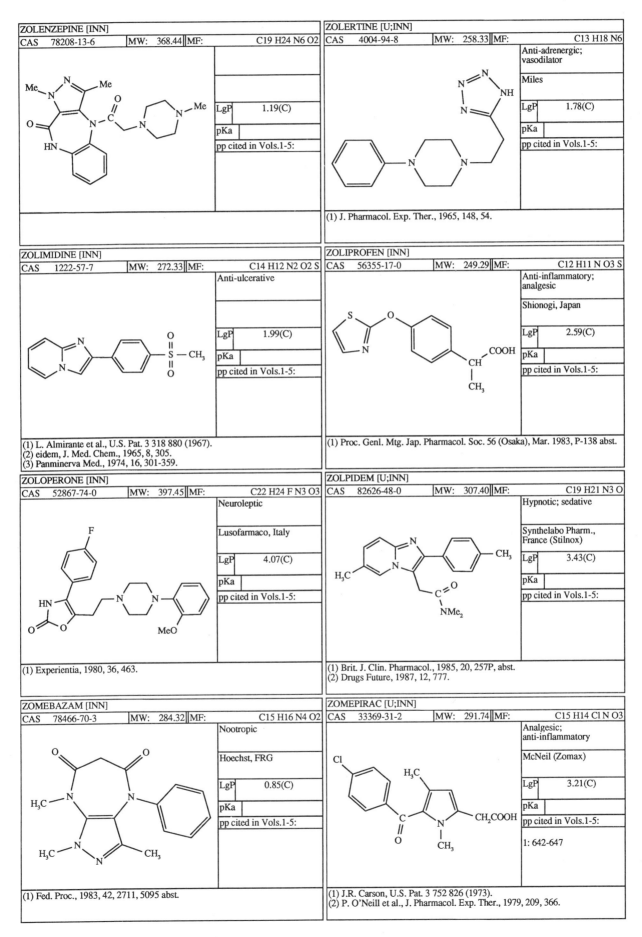

ZOLENZEPINE [INN]

CAS	78208-13-6	MW: 368.44	MF:	C19 H24 N6 O2

LgP	1.19(C)
pKa	
pp cited in Vols.1-5:	

ZOLERTINE [U;INN]

CAS	4004-94-8	MW: 258.33	MF:	C13 H18 N6

Anti-adrenergic; vasodilator

Miles

LgP	1.78(C)
pKa	
pp cited in Vols.1-5:	

(1) J. Pharmacol. Exp. Ther., 1965, 148, 54.

ZOLIMIDINE [INN]

CAS	1222-57-7	MW: 272.33	MF:	C14 H12 N2 O2 S

Anti-ulcerative

LgP	1.99(C)
pKa	
pp cited in Vols.1-5:	

(1) L. Almirante et al., U.S. Pat. 3 318 880 (1967).
(2) eidem, J. Med. Chem., 1965, 8, 305.
(3) Panminerva Med., 1974, 16, 301-359.

ZOLIPROFEN [INN]

CAS	56355-17-0	MW: 249.29	MF:	C12 H11 N O3 S

Anti-inflammatory; analgesic

Shionogi, Japan

LgP	2.59(C)
pKa	
pp cited in Vols.1-5:	

(1) Proc. Genl. Mtg. Jap. Pharmacol. Soc. 56 (Osaka), Mar. 1983, P-138 abst.

ZOLOPERONE [INN]

CAS	52867-74-0	MW: 397.45	MF:	C22 H24 F N3 O3

Neuroleptic

Lusofarmaco, Italy

LgP	4.07(C)
pKa	
pp cited in Vols.1-5:	

(1) Experientia, 1980, 36, 463.

ZOLPIDEM [U;INN]

CAS	82626-48-0	MW: 307.40	MF:	C19 H21 N3 O

Hypnotic; sedative

Synthelabo Pharm., France (Stilnox)

LgP	3.43(C)
pKa	
pp cited in Vols.1-5:	

(1) Brit. J. Clin. Pharmacol., 1985, 20, 257P, abst.
(2) Drugs Future, 1987, 12, 777.

ZOMEBAZAM [INN]

CAS	78466-70-3	MW: 284.32	MF:	C15 H16 N4 O2

Nootropic

Hoechst, FRG

LgP	0.85(C)
pKa	
pp cited in Vols.1-5:	

(1) Fed. Proc., 1983, 42, 2711, 5095 abst.

ZOMEPIRAC [U;INN]

CAS	33369-31-2	MW: 291.74	MF:	C15 H14 Cl N O3

Analgesic; anti-inflammatory

McNeil (Zomax)

LgP	3.21(C)
pKa	
pp cited in Vols.1-5:	

1: 642-647

(1) J.R. Carson, U.S. Pat. 3 752 826 (1973).
(2) P. O'Neill et al., J. Pharmacol. Exp. Ther., 1979, 209, 366.

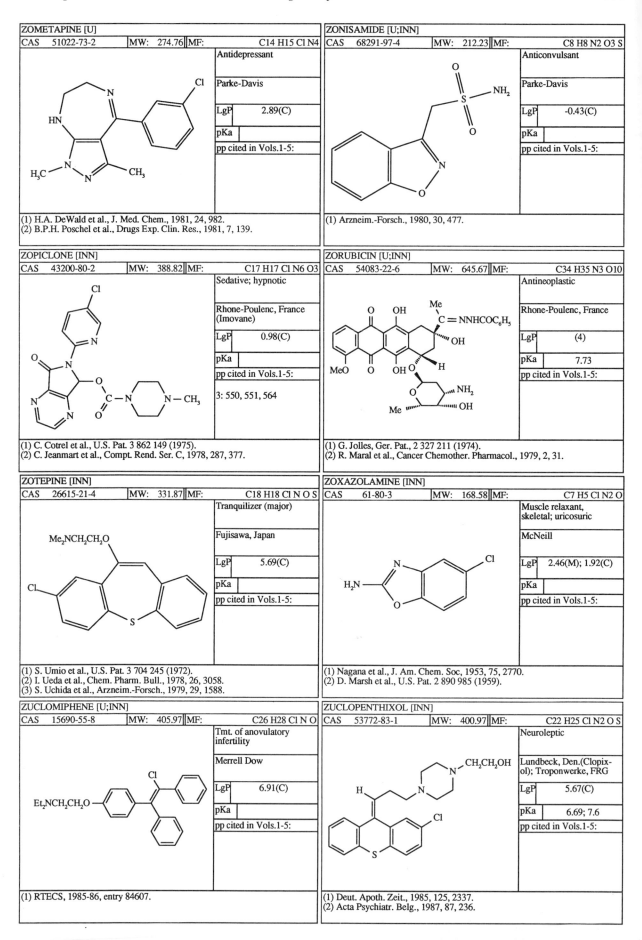

ZOMETAPINE [U]			
CAS 51022-73-2	MW: 274.76	MF:	C14 H15 Cl N4

Antidepressant

Parke-Davis

LgP	2.89(C)
pKa	

pp cited in Vols.1-5:

(1) H.A. DeWald et al., J. Med. Chem., 1981, 24, 982.
(2) B.P.H. Poschel et al., Drugs Exp. Clin. Res., 1981, 7, 139.

ZONISAMIDE [U;INN]			
CAS 68291-97-4	MW: 212.23	MF:	C8 H8 N2 O3 S

Anticonvulsant

Parke-Davis

LgP	-0.43(C)
pKa	

pp cited in Vols.1-5:

(1) Arzneim.-Forsch., 1980, 30, 477.

ZOPICLONE [INN]			
CAS 43200-80-2	MW: 388.82	MF:	C17 H17 Cl N6 O3

Sedative; hypnotic

Rhone-Poulenc, France (Imovane)

LgP	0.98(C)
pKa	

pp cited in Vols.1-5:

3: 550, 551, 564

(1) C. Cotrel et al., U.S. Pat. 3 862 149 (1975).
(2) C. Jeanmart et al., Compt. Rend. Ser. C, 1978, 287, 377.

ZORUBICIN [U;INN]			
CAS 54083-22-6	MW: 645.67	MF:	C34 H35 N3 O10

Antineoplastic

Rhone-Poulenc, France

LgP	(4)
pKa	7.73

pp cited in Vols.1-5:

(1) G. Jolles, Ger. Pat., 2 327 211 (1974).
(2) R. Maral et al., Cancer Chemother. Pharmacol., 1979, 2, 31.

ZOTEPINE [INN]			
CAS 26615-21-4	MW: 331.87	MF:	C18 H18 Cl N O S

Tranquilizer (major)

Fujisawa, Japan

LgP	5.69(C)
pKa	

pp cited in Vols.1-5:

(1) S. Umio et al., U.S. Pat. 3 704 245 (1972).
(2) I. Ueda et al., Chem. Pharm. Bull., 1978, 26, 3058.
(3) S. Uchida et al., Arzneim.-Forsch., 1979, 29, 1588.

ZOXAZOLAMINE [INN]			
CAS 61-80-3	MW: 168.58	MF:	C7 H5 Cl N2 O

Muscle relaxant, skeletal; uricosuric

McNeill

LgP	2.46(M); 1.92(C)
pKa	

pp cited in Vols.1-5:

(1) Nagana et al., J. Am. Chem. Soc, 1953, 75, 2770.
(2) D. Marsh et al., U.S. Pat. 2 890 985 (1959).

ZUCLOMIPHENE [U;INN]			
CAS 15690-55-8	MW: 405.97	MF:	C26 H28 Cl N O

Tmt. of anovulatory infertility

Merrell Dow

LgP	6.91(C)
pKa	

pp cited in Vols.1-5:

(1) RTECS, 1985-86, entry 84607.

ZUCLOPENTHIXOL [INN]			
CAS 53772-83-1	MW: 400.97	MF:	C22 H25 Cl N2 O S

Neuroleptic

Lundbeck, Den.(Clopixol); Troponwerke, FRG

LgP	5.67(C)
pKa	6.69; 7.6

pp cited in Vols.1-5:

(1) Deut. Apoth. Zeit., 1985, 125, 2337.
(2) Acta Psychiatr. Belg., 1987, 87, 236.

ZYLOFURAMINE [INN]				
CAS 3563-92-6	MW: 219.33	MF:		C14 H21 N O

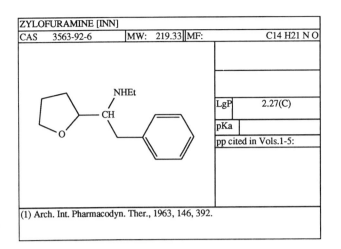

LgP	2.27(C)
pKa	
pp cited in Vols.1-5:	

(1) Arch. Int. Pharmacodyn. Ther., 1963, 146, 392.

CLASSIFICATION INDEX

Acaricide

Dioxation
Pyrimitate
Nimidane

Adrenergic

Aidimolol
Adrenalone
Aganodine
Amidephrine
Apraclonidine
Cimaterol
Cirazoline
Clorprenaline
Coumazoline
Cyclopentamine
Deterenol
Dexpropanolol
Dilevalol
Dioxethedrine
Dipivefrin
Dopamine
Draquinolol
Ephedrine
Epinephrine
Epinephryl borate
Etafedrine
Fenspiride
Flerobuterol
Formoterol
Hydroxyamphetamine
Isometheptene
Isoproterenol
Levonordefrin
Levopropylhexedrine
Mephentermine
Metaraminol
Methoxamine
Methoxyphenamine
Metizoline
Modafinil
Naphazoline
Norepinephrine
Norfenefrine
Octodrine
Octopamine
Oxidopamine
Oxymetazoline
Phenpromethamine
Phenylephrine
Phneylpropanolamine
Pholedrine
Prenalterol
Propylhexedrine
Protokylol
Pseudoephedrine
Soterenol
Synephrine
Tetrahydrozoline
Tramazoline
Tuaminoheptane
Tyramine
Xylometazoline

Aldosterone Antagonist

Canrenoic acid

Canrenone
Dicirenone
Mespirenone
Mexrenoate
Prorenoate potassium
Spironolactone

Anabolic

Bolandiol dipropionate
Bolasterone
Bolazine
Boldenone undecylenate
Bolenol
Bolmantalate
Clostebol
Ethylestrenol
Formebolone
Hydroxystenozole
Mebolazine
Mesabolone
Methandriol
Methandrostenolone
Methenolone acetate
Methenolone enanthate
Metribolone
Mibolerone
Nandrolone cyclotate
Nandrolone decanoate
Norbolethone
Norchlostebol acetate
Oxabolone cypionate
Pizotyline
Quinbolone
Stenbolone acetate
Tibolone
Trenbolone acetate
Zeranol

Analgesic

Aceclofenac
Acetaminophen
Acetaminosalol
Acetanilide
Acetorphine
Acetylsalicylic acid
Alclofenac
Aletamine
Alfaxalone
Alfentanil
Allylprodine
Alminoprofen
Alphacetylmethadol
Alphameprodine
Alphamethadol
Alphaprodine
Aminopyrine
Anidoxime
Anileridine
Anilopam
Anirolac
Antipyrine
Antrafenine
Apazone
Aprofene
Benorilate
Benoxaprofen

Benzydamine
Bermoprofen
Betacetylmethadol
Betameprodine
Betamethadol
Betaprodine
Bezitramide
Bicifadine
Bromadoline
Bromfenac
Bucetin
Bufexamac
Bumadizone
Buprenorphine
Butacetin
Butixirate
Butorphanol
Carbamazepine
Carbaspirin calcium
Carbiphene
Carfentanil
Carsalam
Chlorthenoxazine
Cinchophen
Ciprefadol
Ciramadol
Clofexamide
Clometacin
Clonitazene
Clonixeril
Clonixin
Cloponone
Codeine
Cogazocine
Conorphone
Cropropamide
Crotetamide
Cyclazocine
Desomorphine
Detomidine
Dexoxadrol
Dextrorphan
Dezocine
Diacetamate
Diacetylmorphine
Diampromide
Dibupyrone
Dibusadol
Diethylthiambutene
Difenamizole
Diflunisal
Diftalone
Dimefadane
Dimenoxadol
Dimepheptanol
Dimethylthiambutene
Dioxaphetyl butyrate
Dipipanone
Diproqualone
Dipyrone
Dizatrifone
Doxpicomine
Drindene
Drocode
Duometacin
Embutramide
Emorfazone
Epirizole
Eptazocine
Ergotamine

Analgesic (*contd*)

Eflurbiprofen
Etersalate
Ethenzamide
Ethoheptazine
Ethoxazene
Ethylmethylthiambutene
Ethylmorphine
Etofenamate
Etonitazene
Etorphine
Etosalamide
Etoxeridine
Eugenol
Famprofazone
Fenoprofen
Fentanyl
Fepradinol
Floctafenine
Flufenisal
Flumexadol
Flunixin
Flupirtine
Fluproquazone
Fluradoline
Flurbiprofen
Fosfosal
Furethidine
Gaboxadol
Glafenine
Guacetisal
Hexaprofen
Homprenorphine
Hydromorphinol
Hydromorphone
Hydroxypethidine
Ibazocine
Ibufenac
Imidazole salicylate
Indoprofen
Isomethadone
Isonixin
Isoprazone
Ketazocine
Ketobemidone
Ketorfanol
Ketorolac
Lefetamine
Letimide
Levallorphan
Levomethadone
Levomethadyl acetate
Levonantradol
Levophenacylmorphan
Levorphanol
Lofemizole
Lofentanil
Lorcinadol
Loxoprofen
Magnesium salicylate
Medetomidine
Mefenamic acid
Menabitan
Meperidine
Meptazinol
Metanixin
Metazocine
Metbufen
Metergoline
Metethoheptazine
Methadone
Methadyl acetate
Methopholine

Methotrimeprazine
Methyldihydromorphine
Metkephamid
Metopon
Mimbane
Miroprofen
Molinazone
Morazone
Morpheridine
Morphine
Moxazocine
Myrophine
Nabitan
Nalbuphine
Nalmexone
Namoxyrate
Nantradol
Naproxen
Naproxol
Nefopam
Neocinchophen
Nexeridine
Nicafenine
Nicocodine
Nicodicodine
Nicomorphine
Nifenazone
Niprofazone
Noracymethadol
Norcodine
Norlevorphanol
Normethadone
Normorphine
Norpipanone
Octazamide
Olvanil
Oxapadol
Oxepinac
Oxetorone
Oxindanac
Oxycodone
Oxymorphone
Parapropamol
Parsalmide
Pemedolac
Pentazocine
Perisoxal
Phenacetin
Phenadoxone
Phenampromide
Phenazocine
Phenazopyridine
Phenomorphan
Phenoperidine
Phenylramidol
Picenadol
Picolamine
Pifenate
Pimonodine
Pinadoline
Pipebuzone
Piperylone
Pirfenidone
Piritramide
Pranosal
Prodilidine
Profadol
Proheptazine
Propacetamol
Properidine
Propiram
Propoxyphene
Propyphenazone
Proxazole

Proxifezone
Proxorphan
Pyrroliphene
Ramifenazone
Rimazolium metilsulfate
Salacetamide
Salethamide
Salicin
Salicylamide
Salsalate
Salverine
Simetride
Sodium gentisate
Sufentanil
Sumacetamol
Suprofen
Talmetacin
Talniflumate
Talosalate
Tazadolene
Terofenamate
Tetrandrine
Tetrydamine
Thebacon
Thebaine
Tifluadom
Tifurac
Tilidine
Tinoridine
Tiopinac
Tolfenamic acid
Tolpronine
Tonazocine
Tramadol
Tribuzone
Trichloroethylene
Trimeperidine
Valofane
Venlafaxine
Veradoline
Verilopam
Viminol
Volazocine
Xorphanol
Xylazine
Zenazocine mesylate
Zoliprofen
Zomepirac

Androgen

Fluoxymesterone
Mestanolone
Mesterolone
Methandrostenolone
17-Methyltestosterone
Mibolerone
Nandrolone decanoate
Nandrolone phenpropionate
Nisterime acetate
Norethandrolone
Normethandrone
Oxandrolone
Oxymesterone
Oxymetholone
Penmesterol
Prasterone
Silandrone
Stanolone
Stanozolol
Testosterone
Testosterone cypionate
Testosterone enanthate

Androgen (*contd*)

Testosterone ketolaurate
Testosterone phenylacetate
Testosterone propionate
Tiomesterone
Trestolone acetate

Anesthetic (inhalation)

Aliflurane
Cyclopropane
Enflurane
Ether
Fluroxene
Halothane
Isoflurane
Methoxyflurane
Nitrous oxide
Norflurane
Roflurane
Sevoflurane
Teflurane
Tribromoethanol
Trichloroethylene
Vinyl ether

Anesthetic (local)

Ambucaine
Amoxecaine
Aptocaine
Betoxycaine
Brufacain
Bupivacaine
Butacaine
Butanilicaine
Butethamine
Chloroprocaine
Clodacaine
Diamocaine
Dibucaine
Etidocaine
Fenalcomine
Fomocaine
Hexylcaine
Hydroxyprocaine
Isobucaine
Ketocaine
Levoxadrol
Lidocaine
Lotucaine
Menglytate
Mepivacaine
Meprylcaine
Myrtecaine
Naepaine
Octacaine
Parethoxycaine
Pincainide
Piperocaine
Prilocaine
Procaine
Propanocaine
Propipocaine
Propoxycaine
Pyrrocaine
Quatacaine
Risocaine
Rodocaine
Ropivacaine
Salicyl alcohol

Tolycaine
Trapencaine
Trimecaine

Anesthetic (topical)

Benoxinate
Benzocaine
Biphenamine
Butamben
Cocaine
Cyclomethycaine
Dimethisoquin
Diperodon
Dycylonine
Ethyl chloride
Euprocin
Hydroxytetracaine
Isobutamben
Leucinocaine
Lidocaine
Oxethazaine
Phenacaine
Polidocanol
Pramoxine
Proparacaine
Tetracaine
Zolamine

Anesthetic (other)

Alfadolone
Amolanone
Articaine
Chloroform
Dexivacaine
Etoxadrol
Hexobarbital
Hydroxydione sodium succinate
Ketamine
Methohexital
Midazolam
Minaxolone
Perimetazine
Phencyclidine
Piridocaine
Propanidid
Propofol
Propoxate
Sodium oxybate
Thialbarbital
Thiamylal
Thiopental
Tiletamine

Anorexic

Aminorex
Amphecloral
Benzphetamine
Cathine
Cathinone
Chlorophentermine
Ciclazindol
Clobenzorex
Cloforex
Clominorex
Clortermine
Dexfenfluramine
Diethylpropion
Difemetorex

Etilamfetamine
Fenbutrazate
Fenfluramine
Fenisorex
Fenproporex
Flucetorex
Fludorex
Fluminorex
Furfenorex
Indanorex
Levamfetamine
Levofacetoperane
Mazindol
Mefenorex
Metamfepramone
Pentorex
Phenmetrazine
Phentermine
Picilorex
Setazindol
Tiflorex

Antacid

Almadrate sulfate
Almasilate
Bismuth subgallate
Carbaldrate
Dihydroxyaluminum aminoacetate
Dihydroxyaluminum sodium carbonate

Anthelmintic

Abamectin B1b
Albendazole
Albendazole oxide
Amoscanate
Antafenite
Anthelmycin
Anthiolimine
Antienite
Bephenium hydroxynaphthoate
Bisbendazole
Bitoscanate
Bromoxanide
Brotianide
Bunamidine
Butamisole
Butonate
Calcium carbimide
Cambendazole
Carbantel lauryl sulfate
Carbon tetrachloride
Clioxanide
Closantel
Coumaphos
Crufomate
Cyacetacide
Cyclobendazole
Desaspidin
Dexamisole
Diamfenetide
Dichlorophen
Dichlorvos
Diethylcarbamazine
Diphenan
Dithiazanine iodide
Dribendazole
Dymanthine
Epsiprantel
Etibendazole
Febantel

Anthelmintic (*contd*)

Fenbendazole
Flubendazole
Fospirate
Furodazole
Gentian violet
Haloxon
Hexylresorcinol
Imcarbofos
Levamisole
Lobendazole
Luxabendazole
Mebendazole
Metrifonate
Metyridine
Morantel
Naftalofos
Netobimin
Niclosamide
Nitramisole
Nitrodan
Nitroscanate
Nitroxinil
Oxantel
Oxfendazole
Oxibendazole
Oxyclozanide
Parbendazole
Phenothiazine
Phthalofyne
Piperamide
Piperazine
Praziquantel
Pretamzium iodide
Proclonol
Pyrantel
Pyrvinium chloride
Quinacrine
Rafoxanide
Resorantel
Salantel
Stilbazium iodide
Tetrachloroethylene
Tetramisole
Thenium closylate
Thiabendazole
Thiacetarsamide
Ticarbodine
Tioxidazole
Triclabendazole
Triclofenol piperazine
Uredofos
Vincofos
Zilantel

Antiadrenergic (α receptor)

Fenspiride
Labetalol
Piperoxan
Proroxan

Antiadrenergic (β receptor)

Acebutolol
Afurolol
Alprenolol
Atenolol
Befunolol
Bisoprolol

Bometolol
Bopindolol
Bornaprolol
Bucindolol
Bucumolol
Bufuralol
Bunolol
Bupranolol
Butocrolol
Butofilolol
Carazolol
Carteolol
Celiprolol
Cetamolol
Cicloprolol
Cloranolol
Dexpropanolol
Diacetolol
Dilevalol
Epanolol
Esmolol
Exaprolol
Flestolol
Indenolol
Labetalol
Levobunolol
Medroxalol
Mepindolol
Metalol
Metipranolol
Metoprolol
Moprolol
Nadolol
Nifenalol
Pamatolol
Pargolol
Penbutolol
Pindolol
Practolol
Procinolol
Propranolol
Sotalol
Talinolol
Tazolol
Tartatolol
Timolol
Tiprenolol
Tolamolol
Toliprolol
Xibenolol

Antiadrenergic (nonspecific)

Bretylium tosylate
Butidrine
Dihydroergotamine
Phentolamine
Pronetalol
Solypertine
Zolertine

Antiallergic

Astemizole
Azelastine
Bufrolin
Cetirizine
Eclazolast
Enoxamast
Fenpiprane
Fenprinast

Flufylline
Ibudilast
Lodoxamide
Lodoxamide ethyl
Minocromil
Nedocromil
Nivimedone
Oxalinast
Oxatomide
Pentigetide
Picumast
Pirquinozol
Probicromil
Proxicromil
Sudexanox
Tazanolast
Terbucromil
Thiazinamium chloride
Tiacrilast
Tioxamast
Tiprinast
Tixanox
Traxanox
Xanoxic acid
Zaprinast

Antiamebic

Berythromycin
Bialamicol
Carbarsone
Chiniofon
Chlorbetamide
Chloroquine
Clamoxyquin
Clefamide
Clioquinol
Cloquinate
Conessine
Dehydroemetine
Difetarsone
Diloxanide
Emetine
Etofamide
Fumagillin
Glycobiarsol
Iodoquinol
Paromomycin
Phanquone
Phenarsone sulfoxylate
Quinfamide
Secnidazole
Stirimazole
Symetine
Teclozan
Tetracycline
Tilbroquinol
Tiliquinol

Antiandrogen

Benorterone
Bifluranol
Cyproterone acetate
Delmadinone acetate
Flutamide
Inocoterone
Nilutamide
Oxendolone
Topterone

Antianginal

Amlodipine
Bepridil
Betaxolol
Bevantolol
Bunitrolol
Butoprozine
Carvedilol
Cinepazet
Cinepazic acid
Diproteverine
Epanolol
Imolamine
Molsidomine
Niludipine
Nipradilol
Nisoldipine
Primidolol
Ranolazine
Razinodil
Thiouracil
Tosifen
Verapamil

Antiarrhythmic

Acecainide
Alinidine
Amafolone
Amoproxan
Aprindine
Asocainol
Barucainide
Bevantolol
Bretylium tosylate
Bucainide
Bucromarone
Bufetolol
Bunaftine
Butobendine
Butoprozine
Capobenic acid
Carcainium chloride
Cifenline
Clofilium phosphate
Dazolicine
Dexpropranolol
Diprafenone
Disobutamide
Disopyramide
Drobuline
Droxicainide
Edifolone
Emilium Tosylate
Encainide
Eproxindine
Flecainide
Guafecainol
Indecainide
Indopanolol
Lorajmine
Lorcainide
Meobentine
Mexiletine
Moricizine
Moxaprindine
Murocainide
Nadoxolol
Nafetolol
Nofecainide
Oxiramide
Palatrigine

Pentisomide
Pirmenol
Pirolazamide
Prajmalium
Pranolium chloride
Prifuroline
Primidolol
Procainamide
Propafenone
Propranolol
Pyrinoline
Quinacainol
Quindonium bromide
Quidine
Recainam
Ropitoin
Sematilide
Solpecainol
Soquinolol
Stirocainide
Suricainide
Tertatolol
Tiapamil
Tocainide
Tolamolol
Transcainide
Ubisindine
Verapamil
Viquidil
Xibenolol
Xipranolol

Antiasthmatic

Amlexanox
Azelastine
Bunaprolast
Clenbuterol
Cromitrile sodium
Cromolyn
Ibudilast
Isamoxole
Ketotifen
Lodoxamide
Lodoxamide ethyl
Oxarbazole
Oxatomide
Piriprost
Pirolate
Quazolast
Tiaramide
Tibenalast sodium
Tomelukast
Tranilast
Verofylline

Antibacterial

Acetosulfone
Aconiazide
Adicillin
Aditoprim
Alafosfalin
Alexidine
Almecillin
Amdinocillin
Amdinocillin pivoxil
Amicycline
Amifloxacin
Amikacin
Aminosalicylic acid
Amoxicillin

Ampicillin
Apalcillin
Apicycline
Apramycin
Arbekacin
Asperlin
Aspoxicillin
Avilamycin-A
Azamulin
Azidocillin
Azithromycin
Azlocillin
Bacampicillin
Bacmecillinam
Baquiloprim
Bekanamycin
Benethamine penicillin
Benzoylpas
Benzyl alcohol
Benzylpenicillin
Benzylsulfamide
Berberine sulfate
Berythromycin
Betamicin
Biclotymol
Binfloxacin
Biphenamine
Bispyrithione magsulfex
Bluensomycin
Boric acid
Brodimorprim
Broxaldine
Butikacin
Butirosin
Capreomycin 1B
Carbadox
Carbenicillin
Carbenicillin indanyl
Carbenicillin phenyl
Carbomycin
Carumonam
Cefaclor
Cefadroxil
Cefalonium
Cefaloram
Cefamandole
Cefamandole nafate
Cefaparole
Cefatrizine
Cefazaflur
Cefazedone
Cefazolin
Cefbuperazone
Cefcanel
Cefcanel daloxate
Cefedrolor
Cefempidone
Cefepime
Cefetamet
Cefetrizole
Cefivitril
Cefixime
Cefmenoxime
Cefmepidium chloride
Cefmetazole
Cefminox
Cefodizime
Cefonicid
Cefoperazone
Ceforanide
Cefotaxime
Cefotetan
Cefotiam
Cefoxazole

Antibacterial (*contd*)

Cefoxitin
Cefpimizole
Cefpiramide
Cefpirome
Cefpodoxime
Cefpodoxime proxetil
Cefrotil
Cefroxadine
Cefsulodin
Cefsumide
Ceftazidime
Cefteram
Ceftezole
Ceftiofur
Ceftiolene
Ceftioxide
Ceftizoxime
Ceftriaxone
Cefuracetime
Cefuroxime
Cefuroxime axetil
Cefuzonam
Cephacetrile
Cephalexin
Cephaloglycin
Cephaloridine
Cephalothin
Cephapirin
Cephradine
Cetocycline
Cetophenicol
Chaulmosulfone
Chloramine-T
Chloramphenicol
Chloramphenicol palmitate
Chloramphenicol sodium succinate
Chlorhexidine phosphanilate
Chloroazodin
Chloroxylenol
Chlorquinaldol
Chlortetracycline
Cinodine
Cinoxacin
Ciprofloxacin
Clarithromycin
Clindamycin
Clindamycin palmitate
Clindamycin phosphate
Clofazimine
Clofoctol
Clometocillin
Clomocycline
Cloxacillin
Cloxacillin benzathine
Cloxyquin
Coumermycin A1
Cuprimyxin
Cyclacillin
Cycloserine
Dapsone
Daptomycin
Demeclocycline
Demecycline
Diathymosulfone
Diaveridine
Dibekacin
Dicloxacillin
Dihydrostreptomycin
Diproleandomycin
Dipyrithione
Dirithromycin
Doxycycline

Droxacin
Enoxacin
Enrofloxacin
Enviomycin
Epicillin
Erythromycin
Erythromycin acistrate
Erythromycin ethylsuccinate
Erythromycin propionate
Ethambutol
Ethionamide
Etisomicin
Fenbenicillin
Fibracillin
Fleroxacin
Flomoxef
Florfenicol
Floxacillin
Fludalanine
Flumequine
Flurithromycin
Fomidacillin
Fosfomycin
Fosmidomycin
Fumoxicillin
Furaltadone
Furazolium chloride
Fusidic acid
Fuzlocillin
Gentamicin chloride
Gloximonam
Glucosulfone
Guamecycline
Haloprogin
Hetacillin
Hexedine
Ibafloxacin
Imipenem
Iprocinodine
Isepamicin
Isoconazole
Isoniazid
Isopropicillin
Josamycin
Kanamycin
Kitasamycin
Lenampicillin
Levofuraltadone
Levopropylcillin
Lincomycin
Mafenide
Maridomycin
Meclocycline
Meclocycline sulfosalicylate
Megalomicin
Mequidox
Merbromin
Methacycline
Methenamine
Methicilin
Metioprim
Metioxate
Mezlocillin
Micronomicin
Midecamycin
Mikamycin A
Miloxacin
Minocycline
Mirincamycin
Mirosamicin
Mocimycin
Monensin
Mupirocin
Myralact

Nafcillin
Nalidixic acid
Natamycin
Neoarsphenamine
Neomycin B
Netilmicin
Neutramycin
Nifuradene
Nifuraldezone
Nifuratel
Nifuratrone
Nifurdazil
Nifurfoline
Nifurimide
Nifuroquine
Nifuroxime
Nifurpipone
Nifurpirinol
Nifurprazine
Nifurquinazole
Nifurthiazole
Nifurtoinol
Nihydrazone
Nitrocycline
Nitrofurantoin
Nitromide
Nitrosulfathiazole
Nitroxoline
Norfloxacin
Norfloxacin succinil
Novobiocin
Ofloxacin
Opiniazide
Ormetoprim
Oxacillin
Oximonam
Oxolinic acid
Oxytetracyline
Paldimycin B
Parachlorophenol
Paulomycin A
Pecocycline
Pefloxacin
Penamecillin
Penicillin G procaine
Penicillin V
Penimepicycline
Pentizidone
Phenethicillin
Phenyl aminosalicylate
Phthalylsulfacetamide
Phthalylsulfamethizole
Phthalylsulfathiazole
Pipacycline
Pipemidic Acid
Piperacillin
Pirbenicillin
Piridicillin
Pirlimycin
Piromidic acid
Piroxicillin
Pivampicillin
Pleuromulin
Polymyxin B1
Porfiromycin
Primycin
Pristinamycin
Propicillin
Propikacin
Protionamide
Pyrazinamide
Pyrithione zinc
Quinacillin
Quindecamine

Antibacterial (*contd*)

Ranimycin
Relomycin
Repromicin
Ribostamycin
Rifabutin
Rifamide
Rifampin
Rifamycin
Rifapentine
Rifaximin
Ristocetin
Rokitamycin
Rolitetracycline
Rosaramicin
Rosaramicin butyrate
Rosaramicin propionate
Rosoxacin
Roxarsone
Roxithromycin
Rufloxacin
Salazosulfadimedine
Sancycline
Sarmoxicillin
Sarpicillin
Satranidazole
Scopafungin
Sedecamycin
Sisomicin
Spectinomycin
Spiramycin
Stallimycin
Steffimycin
Streptomycin
Streptonicozid
Succinylsulfathiazole
Succisulfone
Sulbenicillin
Sulfabenz
Sulfabenzamide
Sulfacarbamide
Sulfacetamide
Sulfachlorpyridazine
Sulfachrysoidine
Sulfacytine
Sulfadiazine
Sulfadicramide
Sulfadimethoxine
Sulfadoxine
Sulfaethidole
Sulfaguanidine
Sulfaguanole
Sulfalene
Sulfaloxic acid
Sulfamazone
Sulfamerazine
Sulfameter
Sulfamethazine
Sulfamethizole
Sulfamethoxazole
Sulfamethoxypyridazine
Sulfamethoxypyridazine acetyl
Sulfametomidine
Sulfametrole
Sulfamonomethoxine
Sulfamoxole
Sulfanilamide
Sulfanitran
Sulfaperin
Sulfaphenazole
Sulfaproxyline
Sulfapyridine
Sulfasalazine

Sulfasomizole
Sulfasuccinamide
Sulfasymazine
Sulfathiazole
Sulfathiourea
Sulfatolamide
Sulfatroxazole
Sulazamet
Sulfisomidine
Sulfisoxazole
Sulfisoxazole acetyl
Sulfoxone soduim
Sultamicillin
Suncillin
Tamampicillin
Tameticillin
Taurolidine
Taurultam
Temafloxacin
Temocillin
Tetroxoprim
Thiamphenicol
Thiazolsulfone
Thiocarbanidin
Thiofuradene
Thiphencillin
Tiamulin
Tibezonium iodide
Ticarcillin
Ticarcillin cresyl
Ticlatone
Tilbroquinol
Tiliquinol
Tilmicosin
Tiocarlide
Tiodonium chloride
Tioxacin
Tobramycin
Tosufloxacin
Trimethoprim
Troleandomycin
Trospectomycin
Tylosin
Vancomycin hydrochloride
Viomycin
Virginiamycin factor S
Xantocillin
Xibornol

Anticholinergic (Antispasmodic)

Alibenidol
Alverine
Ambucetamide
Ambutonium bromide
Amikhelline
Aminopentamide
Aminopromazine
Amixetrine
Anisotropine methyl bromide
Atromepine
Atropine
Atropine oxide
Benactyzine
Benapryzine
Bencyclane
Benzetimide
Benzilonium bromide
Bevonium metilsulfate
Bietamiverine
Biperiden
Butaverine
Butetamate

Butinoline
Butropium bromide
Buzepide metiodide
Camylofin
Caroverine
Chlorbenzoxamine
Chlorphenoxamine
Ciclotropium bromide
Cimetropium bromide
Clebopride
Clidinium bromide
Clofeverine
Cyclandelate
Cyclarbamate
Cyclopentolate
Cyclopyrronium bromide
Cycrimine
Demelverine
Denaverine
Deptropine
Dexetimide
Dexsecoverine
Dextromoramide
Dicyclomine
Diethazine
Difemerine
Dihexyverine
Diisopromine
Dipenine bromide
Diphemanil methylsulfate
Dipiproverine
Diprofene
Domazoline
Doxofylline
Drofenine
Drotaverine
Elantrine
Elucaine
Emepronium bromide
Ethaverine
Ethopropazine
Ethybenztropine
Eucatropine
Fenclexonium metilsulfate
Fenocinol
Fenoverine
Fenipiprane
Fenpipramide
Fenpiverinium bromide
Fentonium bromide
Flopropione
Flutropium bromide
Fubrogonium iodide
Gefarnate
Glycopyrrolate
Guaiactamine
Heteronium bromide
Hexocyclium methylsulfate
Hexopyrronium bromide
Homatropine
Homatropine methylbromide
Hymecromone
Hyoscyamine
Ibuverine
Ipragratine
Isometheptene
Isopropamide iodide
Leiopyrrole
Lorglumide
Loxiglumide
Meciadanol
Mecloxamine
Mepenzolate bromide
Methantheline bromide

Anticholinergic (*contd*)

Methscopolamine bromide
Methylatropine nitrate
Methylbenactyzium bromide
Methylchromone
Metindizate
Metocinium iodide
Metoquizine
Milverine
Mobenzoxamine
Moxaverine
Nafiverine
Octamylamine
Octaverine
Otilonium bromide
Oxapium iodide
Oxitropium bromide
Oxybutynin
Oxyphencyclimine
Oxyphenonium bromide
Oxypyrronium bromide
Parapenzolate bromide
Penoctonium bromide
Pentapiperide
Pentapiperium methylsulfate
Penthienate bromide
Phencarbamide
Phenglutarimide
Pinaverium bromide
Pipenzolate bromide
Piperidolate
Piperilate
Pitofenone
Poldine methylsulfate
Pramiverine
Prenoverine
Pridinol
Prifinium bromide
Proglumide
Propantheline bromide
Propenzolate
Propiverine
Propyromazine bromide
Prozapine
Ritropirronium bromide
Rociverine
Salverine
Scopolamine
Secoverine
Stilonium iodide
Sultroponium
Thihexinol methylbromide
Tiametonium iodide
Tiemonium iodide
Timepidium bromide
Tiquinamide
Tiquizium bromide
Tizanidine
Tofenacin
Toquizine
Trepibutone
Treptilamine
Triampyzine
Tricyclamol chloride
Tridihexethyl chloride
Trihexyphenidyl
Trimebutine
Tropatepine
Tropenziline bromide
Tropicamide
Tropigline
Trospium chloride
Xenytropium bromide

Anticoagulant

Acenocoumarol
Acitemate
Anisindione
Bromindione
Clocoumarol
Clorindione
Coumetarol
Daltroban
Dicumarol
Diphenadione
Ethyl biscoumacetate
Fluindione
Lyapolate soduim
Menatetrenone
Midazogrel
Oxazidione
Phenindione
Phenprocoumon
Tioclomarol
Warfarin

Anticonvulsant

Albutoin
Allomethadione
Arfendazam
Atolide
Barbexaclone
Beclamide
Benzobarbital
Buramate
Carbamazepine
Cinromide
Citenamide
Clomethiazole
Clonazepam
Cyheptamide
Decimemide
Denzimol
Dimethadione
Dioxamate
Diphoxazide
Dizocilpine
Doxenitoin
Eterobarb
Ethosuximide
Ethotoin
Ethylphenacemide
Felbamate
Flunarizine
Fluoresone
Flurazepam
Fluzinamide
Gabapentin
Lamotrigine
Mephenytoin
Mephobarbital
Metharbital
Methetoin
Methsuximide
Milacemide
Nabazenil
Nafimidone
Nimetazepam
Nitrazepam
Oxcarbazepine
Paramethadione
Phenacemide
Phenobarbital
Phensuximide
Phenylmethylbarbituric acid

Phenytoin
Phetharbital
Primidone
Progabide
Ralitoline
Rilmazofone
Ropizine
Stiripentol
Suclofenide
Sulthiame
Taltrimide
Tetrantion
Thiopental
Tiletamine
Tolgabide
Topiramate
Trimethadione
Valproate pivoxil
Valproic acid
Valpromide
Vigabatrin
Zolazepam
Zonisamide

Antidepressant

Acecainide
Adinazolam
Adrafinil
Alaproclate
Aletamine
Almoxatone
Amedalin
Amiodarone
Amitriptyline
Amitriptylinoxide
Amixetrine
Amoxapine
Aprindine
Aptazapine
Azaloxan
Azepindole
Azipramine
Befuraline
Benmoxin
Benzaprinoxide
Binedaline
Bipenamol
Bretylium tosylate
Bucainide
Bucromarone
Bupropion
Butacetin
Butoprozine
Butriptyline
Capobenic acid
Caproxamine
Caroxazone
Cartazolate
Ciclazindol
Cidoxepin
Cifenline
Cilobamine
Citalopram
Clemeprol
Clodazon
Clofilium phosphate
Clomipramine
Clorgiline
Clovoxamine
Cotinine
Cyclindole
Cypenamine

Antidepressant (*contd*)

Cyprolidol
Cyproximide
Daledalin
Dazadrol
Dazepinil
Deanol aceglumate
Deanol acetaminobenzoate
Demexiptiline
Depramine
Desipramine
Deximafen
Dibenzepin
Diclofensine
Dimetacrine
Dimethazan
Dioxadrol
Disobutamide
Disopyramide
Dothiepin
Doxepin
Drobuline
Eclanamine
Edifolone
Emilium tosylate
Encainide
Encyprate
Etoperidone
Fantridone
Femoxetine
Fengabine
Fenmetozole
Fenmetramide
Fezolamine
Fipexide
Flecainide
Fluacizine
Fluotracen
Fluoxetine
Fluparoxan
Fluvoxamine
Gamfexine
Genpirone
Glaziovine
Guanoxyfen
Hepzidine
Imafen
Imipramine
Imipraminoxide
Indatraline
Indecainide
Indeloxazine
Intriptyline
Iprindole
Iproniazid
Isocarboxazid
Ivoqualine
Ketipramine
Levofacetoperane
Levoprotiline
Lofepramine
Lorajmine
Lorcainide
Lortalamine
Maprotiline
Mebanazine
Medifoxamine
Melitracen
Meobentine
Metapramine
Metralindole
Mexiletine

Midalcipran
Milacemide
Minaprine
Moclobemide
Modaline
Monometacrine
Moricizine
Napactadine
Napamezole
Nefazodone
Nialamide
Nisoxetine
Nitrafudam
Nomifensine
Nortriptyline
Noxiptiline
Nuclotixene
Octamoxin
Octriptyline
Opipramol
Oxaflozane
Oxaprotiline
Oxiramide
Oxitriptan
Oxypertine
Panuramine
Perafensine
Phenelzine
Piberaline
Pinazepam
Pipofezine
Pirandamine
Pirlindole
Pirmenol
Pirolazamide
Pizotyline
Pranolium chloride
Prazitone
Pridefine
Procainamide
Prolintane
Propizepine
Propranolol
Protriptyline
Pyrinoline
Quindonium bromide
Quinidine
Quinupramine
Quipazine
Recainam
Rolicyprine
Ropitoin
Sematilide
Sertraline
Sibutramine
Sulmepride
Sulpiride
Sultopride
Suricainide
Talsupram
Tametraline
Tampramine
Tandamine
Teniloxazine
Thiazesim
Thozalinone
Tianeptine
Tiflucarbine
Tigloidine
Tisocromide
Tocainide
Tofisopam
Tolamolol

Toloxatone
Tomoxetine
Transcainide
Tranylcypromine
Trazodone
Trebenzomine
Trimipramine
Tryptophan
Valdipromide
Venlafaxine
Viloxazine
Viqualine
Zimeldine
Zometapine

Antidiabetic (Hypoglycemic)

Acetohexamide
Buformin
Butoxamine
Calcium dobesilate
Carbutamide
Chlorpropamide
Ciglitazone
Epalrestat
Etoformin
Gilamilide
Glibornuride
Glicaramide
Glicetanile
Gliclazide
Glicondamide
Glidazamide
Gliflumide
Glimepiride
Glipentide
Glipizide
Gliquidone
Glisamuride
Glisindamide
Glisolamide
Glisoxepide
Glyburide
Glybuthiazol
Glybuzole
Glyclopyramide
Glycyclamide
Glyhexamide
Glymidine
Glyoctamide
Glyparamide
Glypinamide
Glyprothiazol
Glysobuzole
Heptolamide
Linogliride
Metahexamide
Metformin
Methyl palmoxirate
Midaglizole
Palmoxiric acid
Phenformin
Pioglitazone
Pirogliride
Seglitide
Thiohexamide
Tiformin
Tolazamide
Tolbutamide
Tolpentamide
Tolpyrramide

Antiemetic

Alizapride
Benzquinamide
Bromopride
Chlorpromazine
Cipropride
Clebopride
Cyclizine
Dimenhydrinate
Diphenidol
Domperidone
Dronabinol
Fludorex
Flumeridone
Granisetron
Iprozilamine
Levonantradol
Meclizine
Methiomeprazine
Metoclopramide
Metopimazine
Mociprazine
Nabilone
Naboctate
Nonabine
Ondansetron
Oxypendyl
Pipamazine
Piprinhydrinate
Prochlorperazine
Promethazine
Taziprinone
Thiethylperazine
Thioproperazine
Tiapride
Trimethobenzamide
Zacopride

Antiestrogen

Acefluranol
Clometherone
Delmadinone acetate
Dimepregnen
Droloxifene
Gestrinone
Nafoxidine
Nitromifene
Tamoxifen
Toremifene
Trioxifene
Zindoxifene

Antifungal

Acrisorcin
Aliconazole
Ambruticin
Amorolfine
Amphotericin B
Anthralin
Azaconazole
Azaserine
Bensuldazic acid
Benzoic acid
Bifonazole
Biphenamine
Bispyrithione magsulfex
Buclosamide
Butoconazole
Butylparaben

Carbol-fuschin
Cethexonium chloride
Chlordantoin
Chlormidazole
Chlorphenesin
Ciclopirox
Cilofungin
Cisconazole
Climbazole
Clotioxone
Clotrimazole
Croconazole
Cuprimyxin
Diamthazole
Dipyrithione
Doconazole
Econazole
Enilconazole
Ethonam
Ethylparaben
Etisazole
Exalamide
Fenticonazole
Fezatione
Filipin
Fluconazole
Flucytosine
Fuchsin, basic
Griseofulvin
Halethazole
Hedaquinium chloride
Hezamidine
Hexetidine
Isoconazole
Itraconazole
Kalafungin
Ketoconazole
Levorin A2
Loflucarban
Lomofungin
Lucimycin
Lydimycin
Methylparaben
Miconazole
Monensin
Naftifine
Nifuratel
Nifurmerone
Nitralamine
Nystatin A1
Octanoic acid
Octimerate sodium
Omoconazole
Ontianil
Orconazole
Oxiconazole
Oxifungin
Parconazole
Partricin A
Pecilocin
Pentalamide
Proclonol
Propylparaben
Protiofate
Pyrithione zinc
Pyrrolnitrin
Rutamycin
Salicylanilide
Scopafungin
Selenium sulfide
Siccanin
Sinefungin
Sodium benzoate
Sodium propionate

Sulbentine
Sulconazole
Sulfiram
Taurultam
Terbinafine
Terconazole
Thiram
Thymol iodide
Thymol
Ticlatone
Tilbroquinol
Tiliquinol
Tioconazole
Tolciclate
Tolindate
Tolnaftate
Triacetin
Triafungin
Tubercidine
Undecylenic acid
Zinc undecylenate
Zinoconazole

Antihistaminic (*see also* Histamine H$_2$ Antagonist)

Acrivastine
Antazoline
Astemizole
Azatadine
Bamipine
Bromodiphenhydramine
Brompheniramine
Cabastine
Carbinoxamine
Carebastine
Cetirizine
Cetoxime
Chlorcyclizine
Chloropyramine
Chlorothen
Chlorpheniramine
Cinnarizine
Cinnarizine clofibrate
Clemastine
Clemizole
Clobenzepam
Clobenztropine
Clocinizine
Closiramine
Cycliramine
Cyclizine
Cyproheptadine
Dametralast
Danitracin
Deptropine
Dexbrompheniramine
Dexchlorpheniramine
Dimelazine
Dimethindene
Diphenhydramine
Diphenylpyraline
Dorastine
Doxylamine
Ebastine
Embramine
Epinastine
Esaprazole
Etoloxamine
Ethymemazine
Fenethazine
Fluprofylline
Histapyrrodine

Antihistaminic (*contd*)

Isothipendyl
Lamtidine
Levocabastine
Loratadine
Loxtidine
Mebhydrolin
Medrylamine
Mequitazine
Methaphenilene
Methapyrilene
Mianserin
Mifentidine
Moxastine
Nizatidine
Orphenadrine
Phenindamine
Pheniramine
Phenyltoloxamine
Pimethixene
Piprinhydrinate
Promethazine
Prothipendyl
Pyrabrom
Pyrathiazine
Pyrilamine
Pyroxamine
Pyrrobutamine
Rocastine
Rotoxamine
Setastine
Sopromidine
Talastine
Tasuldine
Tazifylline
Temelastine
Terfenadine
Thenalidine
Thenyldiamine
Thiazinamium chloride
Thonzylamine
Tiotidine
Tiquinamide
Tolpropamine
Tripelennamine
Triprolidine
Tritoqualine
Tuvatidine
Zolamine

Antihyperlipoproteinemic

Acetiromate
Acifran
Acipimox
Aluminum clofibrate
Beclobrate
Beloxamide
Bezafibrate
Binifibrate
Boxidine
Butoxamine
Carnitine
Cetaben
Ciprofibrate
Clinofibrate
Clinolamide
Clofibrate
Clofibric acid
Cloridarol
Colestolone
Detrothyronine

Dextrothyroxine
Eniclobrate
Etiroxate
Etofibrate
Fenirofibrate
Fenofibrate
Fosmenic acid
Furazabol
Gamolenic acid
Gemcadiol
Gemfibrozil
Glunicate
Halofenate
Imanixil
Itanoxone
Lifibrate
Lodazecar
Lovastatin
Meglutol
Melinamide
Mevastatin
Moctamide
Nafenopin
Niceritrol
Nicoclonate
Nicofibrate
Nicomol
Octimibate
Ornithine
Oxibetaine
Oxiniacic acid
Pantenicate
Picafibrate
Pimetine
Pirfibrate
Priozadil
Pravastatin
Probucol
Pyricarbate
Rathyronine
Ronifibrate
Salafibrate
Simfibrate
Simvastatin
Sitofibrate
Sultosilic acid
Tazasubrate
Teruficin
Theofibrate
Thyropropic acid
Tiafibrate
Tibric acid
Tizoprolic acid
Tocofibrate
Treloxinate
Triparanol
U-18666A
Urefibrate
Xenbucin

Antihypertensive

Acetryptine
Ajmaline
Alacepril
Alfuzosin
Althiazide
Amiquinsin
Amosulalol
Arotinolol
Atirprosin
Azamethonium bromide
Azepexole

Bemitradine
Benidipine
Bethanidine
Biclodil
Bietaserpine
Bisoprolol
Brefonalol
Bucindolol
Budralazine
Bufeniode
Bumepidil
Bunazosin
Bupicomide
Butynamine
Cadralazine
Cafedrine
Captopril
Carvedilol
Chlorisondamine chloride
Cianergoline
Cicaprost
Cicletanine
Ciclosidomine
Cilazapril
Cilazaprilat
Clonidine
Clopamide
Cromakalim
Cyclopenthiazide
Darodipine
Debrisoquin
Delapril
Deserpidine
Dexlofexidine
Dexmedetomidine
Diapamide
Diazoxide
Dihydralazine
Dilevalol
Doxazosin
Enalapril
Enalaprilat
Endralazine
Epithiazide
Epoprostenol
Fenoldopam
Flavodil
Flesinoxan
Flordipine
Fluprofylline
Flutonidine
Fosinapril sodium
Guabenxan
Guanabenz
Guanacline
Guanadrel
Guanazodine
Guanclofine
Guancydine
Guanethidine
Guanfacine
Guanisoquin
Guanoclor
Guanoctine
Guanoxabenz
Guanoxan
Guanoxyfen
Hexamethonium bromide
Hydralazine
Hydroflumethiazide
Imiloxan
Indacrinone
Indanidine
Indapamide

Antihypertensive (*contd*)

Indolapril
Indopanolol
Indoramin
Indorenate
Irindalone
Isbogrel
Ketanserin
Lacidipine
Leniquinsin
Levlofexidine
Levomoprolol
Libenzapril
Lisinopril
Lofexidine
Losulazine hydrochloride
Manidipine
Mebutamate
Mecamylamine
Mecarbinate
Medetomidine
Medorinone
Medroxalol
Methoserpidine
Methyldopa
Methyldopate
Metolazone
Metyrosine
Midodrine
Minoxidil
Mopidralazine
Motapizone
Moveltipril
Moxonidine
Muzolimine
Naftopidil
Nebivolol
Niguldipine
Nileprost
Niludipine
Nipradilol
Nitrendipine
Ofornine
Olmidine
Pacrinolol
Pargyline
Pazocide
Pelanserin
Pempidine
Pentacynium chloride
Pentamethonium bromide
Pentamoxane
Pentolinium tartrate
Pentopril
Perindopril
Perindoprilat
Phenactropinium chloride
Pheniprazine
Phenoxybenzamine
Pildralazine
Pinacidil
Pivopril
Polythiazide .
Prazosin
Primaperone
Primidolol
Prizidilol
Protoveratrine
Quinapril
Quinaprilat
Quinazosin
Quinpirole
Quinuclium bromide

Racemetirosine
Ramipril
Ramiprilat
Rescimetol
Rescinnamine
Reserpine
Rilmenidine
Sabeluzole
Sodium nitroprusside
Soquinolol
Spirapril
Spiraprilat
Sulfinalol
Syrosingopine
Terazosin
Theodrenaline
Tiamenidine
Tibalosin
Ticrynafen
Tienoxolol
Tilisolol
Tilsuprost
Tinabinol
Tiodazosin
Tipentosin
Tiprostanide
Tizolemide
Todralazine
Tolonidine
Trandolapril
Trequensin
Trichlormethiazide
Trimazosin
Trimethaphan camsylate
Trimethidinium methosulfate
Trimoxamine
Tripamide
Urapidil
Xipamide
Zofenopril

Anti-infective (Disinfectant)

Acediasulfone
Alcohol
Aminacrine
Aminoquinuride
Bensalan
Benzethonium chloride
Benzododecinium chloride
Benzoxiquine
Benzoxonium chloride
Bithionol
Bithionoloxide
Bromchlorenone
Broxyquinoline
Camphor
Carbamide peroxide
Cetalkonium chloride
Cetylpyridinium Chloride
Chlorhexidine
Chlorocresol
Clioquinol
Cloflucarban
Clorophene
Cresol
Dequalinium chloride
Dibromsalan
Difloxacin
Dofamium chloride
Domiphen bromide
Fenticlor
Fludazonium chloride

Fluorosalan
Fuchsin, basic
Furazolidone
Fursalan
Gentian violet
Glutaral
Halazone
Halquinols
Hexachlorophene
Hydrargaphen
Imidecyl iodine
Isopropyl alcohol
Lauryl Isoquinolinium bromide
Lomefloxacin
Meralein
Mercufenol chloride
Mercury, ammoniated
Metabromsalan
Methylbenzethonium chloride
Miristalkonium chloride
Monalazone disodium
Moxalactam
Nidroxyzone
Nifurzide
Nitrofurazone
Nitromersol
Noxytiolin
Octenidine
Ornidazole
Oxychlorosene
Oxyquinoline
Pararosaniline
Pentisomicin
Picloxydine
Propiolactone
Sepazonium chloride
Silver sulfadiazine
Symclosene
Terizidone
Tetradonium bromide
Thimerfonate
Thimerosal
Thiosalan
Thymol iodide
Tibrofan
Toloconium metilsulfate
Tribromsalan
Triclobisonium chloride
Triclocarban
Triclosan
Troclosene potassium

Anti-inflammatory

Aceclofenac
Acemetacin
Acetaminosalol
Acexamic acid
Aclantate
Ademetionine
AF-2259
Alclofenac
Alclometasone dipropionate
Algestone acetonide
Alminoprofen
Amcinafal
Amcinafide
Amfenac
Amixetrine
Amprioxicam
Anirolac
Anitrazafen
Apazone

Anti-inflammatory (*contd*)

Beclomethasone dipropionate
Bendazac
Benorilate
Benoxaprofen
Benzydamine
Bermorprofen
Brefezil
Broperamole
Bucloxic Acid
Bucolome
Budesonide
Bufexamac
Butibufen
Carprofen
Cicloprofen
Cinfenoac
Cinmetacin
Cintazone
Clamidoxic acid
Clantifen
Clidanac
Cliprofen
Clobetasol propionate
Clobetasone Butyrate
Clofexamide
Clofezone
Clonixeril
Clonixin
Clopirac
Cloticasone propionate
Cloximate
Colfenamate
Cormethasone acetate
Cortodoxone
Deflazacort
Desonide
Desoximetasone
Dexamethasone dipropionate
Dexindoprofen
Diacerein
Diclofenac
Diclonixin
Difenamizole
Diflorasone diacetate
Diflumidone
Diflunisal
Difluprednate
Diftalone
Dimesone
Dimethyl sulfoxide
Ditazole
Drocinonide
Droxicam
Ebselen
Emorfazone
Endrysone
Enfenamic acid
Enolicam
Enocolone
Epirizole
Esflurbiprofen
Etersalate
Etodolac
Etofenamate
Feclobuzone
Felbinac
Fenamole
Fenbufen
Fenclofenac
Fenclorac
Fenclozic acid
Fendosal

Fenflumizole
Fenoprofen
Fenpipalone
Fentiazac
Fepradinol
Feprazone
Flazalone
Fluazacort
Flufenamic acid
Flufenisal
Flumizole
Flunisolide acetate
Flunixin
Flunoxaprofen
Fluocortin butyl
Fluorometholone acetate
Fluperolone acetate
Fluprednidene
Fluquazone
Flurbiprofen
Fluretofen
Flutiazin
Fluticasone propionate
Fosfosal
Furaprofen
Furcloprofen
Furobufen
Galosemide
Glucametacin
Halcinonide
Halometasone
Halopredone acetate
Hexaprofen
Hydrocortisone aceponate
Ibufenac
Ibuprofen
Ibuprofen piconol
Ibuproxam
Imidazole salicylate
Indomethacin
Indoprofen
Indoxole
Intrazole
Isofezolac
Isoflupredone acetate
Isonixin
Isoprofen
Isoxepac
Isoxicam
Ketoprofen
Ketorolac
Leflunomide
Levomenol
Lobenzarit
Lofemizole
Lonazolac
Lotifazole
Loxoprofen
Magnesium salicylate
Mazipredone
Meclofenamic acid
Meclorisone dibutyrate
Mefenamic acid
Mesalamine
Meseclazone
Metanixin
Metbufen
Methylprednisolone aceponate
Methylprednisolone suleptanate
Metiazinic acid
Metoprofen
Miroprofen
Mofebutazone
Mometasone furoate

Morazone
Morniflumate
Nabumetone
Nafamostat
Naftypramide
Naproxen
Naproxol
Nicoboxil
Nictindole
Niflumic acid
Nimazone
Nimesulide
Niprofazone
Nitraquazone
Nixylic Acid
Olsalazine
Olvanil
Orpanoxin
Oxaceprol
Oxametacin
Oxapadol
Oxaprozin
Oxindanac
Oxyphenbutazone
Palmidrol
Paranyline
Parsalmide
Perisoxal
Phenyylbutazone
Pifoxime
Piketoprofen
Pipebuzone
Pirfenidone
Piroxicam
Piroxicam cinnamate
Pirprofen
Pranoprofen
Prednazate
Prednazoline
Prednicarbate
Pregnenolone
Prifelone
Prodolic acid
Proglumetacin
Propyphenazone
Proquazone
Protizinic acid
Proxazole
Proxifezone
Ramifenazone
Rimexolone
Salsalate
Seclazone
Sermetacin
Sodium gualenate
Sudoxicam
Sulindac
Suprofen
Suxibuzone
Talmetacin
Talniflumate
Talosalate
Tenoxicam
Terofenamate
Tesicam
Tesimide
Tetrydamine
Tiaprofenic acid
Tiflamizole
Timegadine
Tinoridine
Tiopinac
Tioxaprofen
Tixocortol pivalate

Anti-inflammatory (*contd*)

Tolfenamic acid
Tolmetin
Torasemide
Tribuzone
Triclonide
Triflumidate
Ximoprofen
Zidometacin
Zoliprofen
Zomepirac

Antileishmanial

Aminoquinol
Hydroxystilbamidine
Meglumine
Sodium stibogluconate

Antimalarial

Acedapsone
Amopyroquin
Amquinate
Artemisinin
Berberine sulfate
Chloroguanide
Chloroquine
Chlorproguanil
Cinchonine
Cycloguanil
Enpiroline
Ethylhydrocupreine
Floxacrine
Halofantrine
Hydroxychloroquine
Lapinone
Mefloquine
Menoctone
Mirincamcycin
Pamaquine
Pentaquine
Primaquine
Pyrimethamine
Quinacrine
Quinine
Quinocide
Tebuquine

Antimigraine

Alpiropride
Dihydroergotamine
Ergotamine
Flumedroxone-17-acetate
Metergotamine
Methysergide
Oxetorone
Sergolexole
Sumatriptan
Tropanserin
Valofane

Antineoplastic

Aceglatone
Acetylcolchinol
Acivicin
Aclarubicin

Acodazole
Acronine
Alanine mustard
Alanosine
Alpha-vinylaziridinoethyl acetate
Altretamine
Ametantrone
Amidino TIC
Aminoglutethimide
Aminothiadiazole
Amonafide
Amsacrine
Amygdalin (D)
Anaxirone
Ancitabine
Androstanediol
Androstanol propionate
Androstenetrione
Androstenonol propionate
Anguidine
Aniline mustard
Anisoylbromacrylic acid, sodium salt
Anthramycin
Arabinosylmercaptopurine
Asaley
Asperlin
ATPU
Azacitidine
Azaguanidine
Azapicyl
Azastreptonigrin
Azauridine
Azetepa
Azimexon
Azotomycin
Benaxibine
Bendamustine
Benzodepa
Bisantrene
Bis(aziridinyl)butanediol
Bleomycin A2
Bofumustine
Brequinar
Bromebric acid
Bropirimine
Broxuridine
Bruceantin
Budotitane
Busulfan
Butanediol cyclic sulfite
Calusterone
Camptothecin
Caracemide
Carbestrol
Carboplatin
Carboquone
Carmofur
Carmustine
Carubicin
Carzolamide
Chlorambucil
Chlornaphazine
Chlorodihydroxyandrostenone
Chloroethyl mesylate
Chlorozotocin
Cisplatin
Corticosterone
Crisnatol
Cycloleucine
Cyclophosphamide
Cytarabine
Dacarbazine
Dactinomycin
Daunorubicin

Deazauridine
Defosfamide
Dehydro-7-methyltestosterone
Demecolcine
Deoxyspergualin
Desacetylcolchicine tartrate
Desmethylcolchicine
Desmethylmisonidazole
Desoxypryridoxine
Detorubicin
Dezaguanine
Diaminomethylphenazinium chloride
Dianhydrogalactitol
Diaziquone
Diazoacetylglycine hydrazide
Diazouracil
Dichlorallyl lawsone
Dichloromethotrexate
Didemnin B
Dideoxycytidine
Dihydroazacytidine
Dihydrolenperone
Dihydroxyfluoroprogesterone
Dimethaminostyrylquinoline
Dimethylhydroxytestosterone
Dimethylnorandrostadienone
Dimethylnortestosterone
Dimethylstilbestrol
Ditercalinium chloride
Ditiomustine
DON
Doxifluridine
Doxorubicin
Droloxifene
Dromostanolone
Dromostanolone propionate
Echinomycin
Edelfosine
Eflornithine
Elderfield's pyrimidine mustard
Elliptinium acetate
Elmustine
Enocitabine
Enpromate
Epipropidine
Epirubicin
Epitiostanol
Esorubicin
Estramustine
Estramustine phosphate sodium
Etanidazole
Ethidium chloride
9-Ethyl-6-mercaptopurine
Etiocholanolone
Etoglucid
Etoposide
Etoprine
Fazaribine
Fenretinide
Flavoneacetic acid
Floxuridine
Fludarabine phosphate
Fluoroadenosine
3-Fluoroandrostanol
Fluorodopane
Fluorohydroxyandrosterone
6-Fluorotestosterone propionate
Fluorouracil
9-Fluoroxotestenololactone
9-Fluoroxotestololactone
Flurocitabine
Forfenimex
Fostriecin
Fotemustine

Antineoplastic (*contd*)

Fotretamine
Ftorafur
Glutarimide-NSC
Glycidyl methacrylate
Guanazole
Hadacidin sodium salt
HMBA
Homoharringtonine
D-Homostesterone propionate
5-HP
Hydroxydimethandrostadienone
Hydroxymethylandrostanone
10-Hydroxynorethisterone
7-Hydroxytestololactone
Hydroxytestosterone propionate
Hydroxyurea
Idarubicin
Ifosfamide
Imexon
Imidazopyrazole
Improsulfan
Indicine-*N*-oxide
Inosine dialdehyde
Inproquone
Iproplatin
Isotic
Ketotrexate
Lapachol
Lomustine
Lonidamine
Mafosfamide
Mannomustine
Mannosulfan
Maytansine
Mechlorethamine
Megestrol acetate
Melengestrol acetate
Melphalan
Menogaril
Mepitiostane
Merbarone
Mercaptopurine
Merophan
Meso-hexestrol
Metamelfalan
Methioguanine
Methotrexate
N-Methyladrenalone hydrochloride
Methylformamide
Methyl methanesulfonate
2-Methyl-19-nortestosterone
2-Methyl-11-oxoprogesterone
Methylstreptonigrin
4-Methyltestosterone
7-Methyltestosterone
7-Methyltestosterone propionate
Methylthioinosine
16-Methylthioprogesterone
Metoprine
Meturedepa
Misonidazole
Mitindomide
Mitobronitol
Mitoclomine
Mitoguazone
Mitolactol
Mitomycin
Mitopodozide
Mitotane
Mitotenamine
Mitoxantrone
Mitozolomide

Mycophenolic acid
Nafoxidine
Nimustine
Nitracrine
Nocodazole
Nogalamycin
A-Norprogesterone
A-Nortestosterone propionate
NSC 4928
NSC 12204
NSC 19622
NSC 50903
NSC 51915
NSC 56940
NSC 56955
NSC 60339
NSC 61716
NSC 63294
NSC 73865
NSC 81408
NSC 82484
NSC 92326
Olivomycin A
Oxaliplatin
Oxisuran
Pactamycin
PCNU
Pentamethylmelamine
Pentamustine
Peplomycin
Phenesterin
Phosphoramide mustard
Pibenzimol
Pinafide
Piperazinedione dihydrochloride
Pipobraman
Piposulfan
Pirarubicin
Piritrexim
Piroxantrone hydrochloride
Plicamycin
Podophyllotoxin
Porfiromycin
Prednimustine
Procarbazine
Prospidium chloride
Pumitepa
Puromycin
Pyran copolymer
Pyrazofurin
Pyronine B
Raloxifene
Ranimustine
Razoxane
Riboprine
Ritrosulfan
Rodorubicin
Roquinimex
Sangivamycin
Sarcolysin
Semustine
Simtrazene
Sparfosic acid
Sparsomycin
Spirogermanium
Spiromustine
Spiroplatin
Streptonigrin
Streptozocin
Sufosfamide
Talisomycin
Tamoxifen
Tauromustine
Taxol

Tegafur
Temarotene
Temozolomide
Teniposide
Terfluranol
Teroxirone
Testolactone
Tetrandrine
Thalicarpine
Thiamiprine
Thiocarzolamide
Thioguanine
Thioguanine alpha-deoxyriboside
Thioguanine beta-deoxyriboside
Thioguanosine
Thioinosine
Thiotepa
Thymidine
Tiazofurin
TIC-mustard
Toremifene
Toyocamycin
Toyomycin
Treosulfan
Trestolone acetate
Trianzinate
Triaziquone
Trichlormethine
Triciribine phosphate
Triethylenemelamine
Triethylenephosphoramide
Trimetrexate
Trityl cysteine
Trofosfamide
Trypotophane mustard
Tubercidine
Tubulozole
Tylocrebin - (+)
U-6837
U-7654
U-8435
U-12932
U-18666A
Uracil mustard
Uredepa
Urethane
Uridine
Vinblastine
Vincristine
Vindesine
Vinepidine
Vinformide
Vinglycinate
Vinrosidine
Vinzolidine
Zorubicin

Antiparkinsonian

Benztropine
Biperiden
Bornaprine
Botiacrine
Budipine
Carmantadine
Chlorphenoxamine
Ciladopa
Diclofensine
Diethazine
Dopamantine
Droxidopa
Epicriptine
Ethopropazine

Antiparkinsonian (*contd*)

Ethybenztropine
Levodopa
Lometraline
Mazaticol
Memantine
Minepentate
Naxagolide
Pareptide
Piroheptine
Prazitone
Pridinol
Procyclidine
Prodipine
Selegiline
Terguride
Tigloidine
Trihexyphenidyl
Tropatepine
Troxonium tosilate ,

Antiperistaltic

Difenoximide
Difenoxin
Diphenoxylate
Fluperamide
Lidamidine
Loperamide
Malethamer
Nufenoxole

Antiprotozoal

Acetarsone
Aminoquinol
Amodiaquine
Arsthinol
Azanidazole
Bamnidazole
Benznidazole
Carnidazole
Chlortetracycline
Dimetridazole
Diminazene
Ethylstibamine
Fexinidazole
Flubendazole
Flunidazole
Furazolidone
Halofuginone
Imidocarb
Ipronidazole
Levofuraltadone
Meglumine
Melarsonyl
Metronidazole
Misonidazole
Monensin
Moxipraquine
Moxnidazole
Myralact
Nifuratel
Nifursemizone
Nifursol
Nihydrazone
Nitarsone
Panidazole
Partricin A
Parvaquone
Pentamidine

Pirinidazole
Propamidine
Puromycin
Ronidazole
Satranidazole
Secnidazole
Stibamine glucoside
Stilbamidine
Sulnidazole
Tenonitrozole
Tinidazole

Antipruritic

Camphor
Cyproheptadine
Dichlorisone acetate
Menthol
Mesulfen
Methdilazine
Polidocanol
Risocaine
Tolpropamine
Trimeprazine

Antipsoriatic

Acitretin
Anthralin
Azaribine
Butantrone
Cycloheximide
Etretinate
Lonapalene
Mercury, ammoniated
Tepoxalin

Antipsychotic

Acetophenazine
Alpertine
Amisulpride
Azaperone
Benperidol
Benzindopyrine
Brofoxine
Bromperidol
Bromperidol decanoate
Buramate
Butaclamol
Butaperazine
Carperone
Carphenazine
Chlorpromazine
Chlorprothixene
Cinperene
Cintriamide
Clomacran
Clopenthixol
Clopimozide
Clopipazan
Cloroperone
Clothiapine
Clothixamide
Clozapine
Cyclophenazine
Cyproximide
Dicarbine
Droperidol
Etazolate
Fenimide

Flucindole
Flumezapine
Fluotracen
Flupentixol
Fluphenazine
Fluphenazine enanthate
Fluspiperone
Fluspirilene
Flutroline
Halopemide
Haloperidol
Haloperidol decanoate
Imiclopazine
Imidoline
Lenperone
Levometiomeprazine
Lometraline
Maroxepin
Mesoridazine
Metiapine
Milenperone
Milipertine
Molindone
Naranol
Neflumozide
Nonaperone
Ondansetron
Opirpamol
Oxiperomide
Oxyridazine
Penfluridol
Pentiapine maleate
Periciazine
Perphenazine
Pimozide
Pinoxepin
Pipamperone
Piperacetazine
Pipotiazine palmitate
Piquindone
Prochlorperazine
Raclopride
Remoxipride
Rimcazole
Risperidone
Savoxepin
Seperidol
Setoperone
Spiperone
Thiopropazate
Thioridazine
Thiothixene
Timelotem
Tioperidone
Tiospirone
Trepipam
Trifluoperazine
Trifluperidol
Triflupromazine

Antipyretic

Acetaminophen
Acetaminosalol
Acetanilide
Acetylsalicylic acid
Aconitine
Alclofenac
Aminopyrine
Benorilate
Benzydamine
Berberine sulfate
Bermoprofen

Antipyretic (contd)

Bufexamac
Bumadizone
Chlorthenoxazine
Ciproquazone
Clidanac
Dipyrone
Etersalate
Etosalamide
Famprofazone
Fenacetinol
Fosfosal
Guacetisal
Hexaprofen
Imidazole salicylate
Indoxole
Lofemizole
Loxoprofen
Magnesium salicylate
Metergoline
Morazone
Naproxen
Naproxol
Nifenazone
Niprofazone
Phenacetin
Phenicarbazide
Phetharbital
Pipebuzone
Pirfenidone
Propacetamol
Propyphenazone
Ramifenazone
Salacetamide
Talmetacin
Tetrandrine
Tinoridine
Tiopinac

Antirheumatic

Acetylsalicylic acid
Auranofin
Aurothioglucose
Bumadizone
Butixirate
Clobuzarit
Enolicam
Gold sodium thiomalate
Gold sodium thiosulfate
Kebuzone
Metiazinic acid
Oxyphenbutazone
Pirazolac
Prinomide
Salacetamide
Sodium gentisate

Antirickettsial

Aminobenzoic Acid
Chloramphenicol
Chloramphenicol palmitate
Chloramphenicol sodium succinate
Tetracycline

Antischistosomal

Amphotalide
Antimony potassium tartrate
Antimony thioglycollate
Becanthone
Hycanthone
Lucanthone
Metrifonate
Niridazole
Oxamniquine
Pararosaniline embonate
Sodium stibocaptate
Sodium stibogluconate
Stibophen
Teroxalene

Antiseborrheic

Chloroxine
Cystine
Piroctone
Pyrithione zinc
Resorcinol monoacetate
Selenium sulfide
Tioxolone

Antisecretory

Arbaprostil
Cinprazole
Deprostil
Enprostil
Fenoctimine
Lamtidine
Loxtidine
Mexiprostil
Nolinium bromide
Omeprazole
Oxoprostol
Picoprazole
Rioprostil
Spizofurone
Telenzepine
Timoprazole
Tiprostanide
Trimoprostil
Tritiozine

Antithrombotic

Anagrelide
Beraprost
Cilostazol
Dazoxiben
Fenflumizole
Fluretofen
Pentosan polysulfate sodium
Picotamide
Plafibride
Sulotroban
Trifenagrel
Triflusal
Vapiprost

Antitubercular (Antimycotic)

Aconiazide
Aminosalicylic acid
Benzoylpas
Capreomycin 1B
Clofazimine
Cycloserine
Enviomycin
Ethambutol
Ethionamide
Isoniazid
Opiniazide
Phenyl aminosalicylate
Protionamide
Pyrazinamide
Rifabutin
Rifamide
Rifampin
Rifapentine
Streptomycin
Thiocarbanidin
Tiocarlide
Viomycin

Antitussive

Alloclamide
Amicibone
Benproperine
Benzonatate
Bibenzonium bromide
Bromoform
Butamirate
Butopiprine
Butorphanol
Caramiphen
Carbetapentane
Chlophedianol
Clobutinol
Cloperastine
Codeine
Codoxime
Cyclexanone
Dextromethorphan
Dimemorfan
Dimethoxanate
Dioxethedrin
Dropropizine
Drotebanol
Eprazinone
Ethyl dibunate
Ethylmorphine
Fedrilate
Fominoben
Guaiapate
Guaimesal
Hydrocodone
Isoaminile
Levdropropizine
Levomethorphan
Levopropoxyphene
Meprotixol
Morclofone
Moxazocine
Naphthonone
Noscapine
Oxeladin
Oxolamine
Pemerid
Pholcodine
Picoperine
Pipazethate
Prenoxdiazine
Proxorphan
Racemethorphan
Sodium dibunate
Suxemerid
Thebacon
Tiametonium iodide
Tipepidine
Ubisindine

Antitussive (*contd*)

Vadocaine
Xyloxemine
Zipeprol

Antiulcerative

Aceglutamide aluminum
Benexate
Beperidium iodide
Cetraxate
Clocanfamide
Dimoxaprost
Enisoprost
Enprostil
Ftaxilide
Irsogladine
Isotiquimide
Lozilurea
Lucartamide
Misoprostol
Naxaprostene
Nilprazole
Niperotidine
Nolinium bromide
Nuvenzepine
Ornoprostil
Picartamide
Pifarnine
Pirenzepine
Plaunotol
Ramixotidine
Rotraxate
Roxatidine acetate
Sucralfate
Tiopropamine
Tolimidone
Tripotassium dicitrato bismuthate
Troxipide
Zolimidine

Antiviral

Acyclovir
Amantadine
Amidapsone
Amiprilose
Aranotin
Arildone
Avridine
BCH-189
Brivudine
Bropirimine
Buciclovir
5-Chloro-3'-fluoro-2',3'-
       dideoxyuridine
Cicloxolone
Citenazone
Cyclobut-G
Cytarabine
Desciclovir
Dimepranol
Dimyristoylphosphatidyl-AZT
Disoxaril
Edoxudine
Enviradene
Famotine
Floxuridine
Fosarilate
Foscarnet
Fosfonet

Ganciclovir
Hypericin
Ibacitabine
Idoxuridine
Impacarzine
Kethoxal
Memotine
Methisazone
Moroxydine
ND-1025
Ribavirin
Rimantadine
Somantadine
Steffimycin
Streptovarycin C
Tilorone
Trifluridine
Tromantadine
Vidarabine
Vidarabine phosphate
Viroxime-E
Xenazoic acid
Zidovudine
Zinviroxime

Anxiolytic

Alozafone
Alpidem
Cartazolate
Enciprazine
Ethomoxane
Flutazolam
Flutoprazepam
Ipsapirone
Lopirazepam
Metaclazepam
Mexazolam
Ondansetron
Pipequaline
Premazepam
Pyridarone
Rilmazafone
Suproclone
Triflubazam

Bronchodilator

Acefylline
Albuterol
Azanator
Bambuterol
Bamifylline
Bevonium metilsulfate
Bitolterol
Butaprost
Carbuterol
Clorprenaline
Colterol
Dametralast
Darodipine
Dioxifedrine
Doxaprost
Dyphylline
Enprofylline
Epinephrine
Eprozinol
Ethylnorepinephrine
Fenoterol
Fenprinast
Fenspiride
Guaithylline

Hexoprenaline
Ibuterol
Ipratropium bromide
Isoetharine
Isoproterenol
Levisoprenaline
Mabuterol
Medibazine
Metaproterenol
Methoxyphenamine
Metiprenaline
Nileprost
Nisbuterol
Norbudrine
Oxitropium bromide
Oxtriphylline
Phenisonone
Piquizil
Pirbuterol
Procaterol
Protokylol
Proxyphylline
Pseudoephedrine
Quazodine
Quinetolate
Quinerenol
Repirinast
Reproterol
Rimiterol
Salmefanol
Salmeterol
Soterenol
Sulfonterol
Suloxifen
Taloximine
Terbutaline
Tretoquinol
Tulobuterol
Verofylline
Zindotrine
Zinterol

Calcium Channel Blocker

Anipamil
Benidipine
Gallopamil
Isradipine
Nilvadipine
Oxodipine

Calcium Regulator

APD
Calcifediol
Calcitriol
Clodronic acid
Dihydrotachysterol
Etidronic acid
Oxidronic acid
Piridronic acid

Calcium Replenisher

Calcium glubionate
Calcium gluceptate
Calcium gluconate
Calcium lactate
Calcium levulinate
Calcium phosphate, dibasic
Calcium phosphate, tribasic

Carbonic Anhydrase Inhibitor

Acetazolamide
Carzenide
Dichlorphenamide
Ethoxzolamide
Methazolamide
Sulocarbilate

Cardiotonic

Acefylline piperazine
Acetyldigitoxin
Acrihellin
Actodigin
Adibendan
Amrinone
Bemarinone
Benfurodil hemisuccinate
Bucladesine
Butopamine
Carbazeran
Denopamine
Deslanoside
Digitoxin
Digoxin
Dobutamine
Enoximone
Fosfocreatinine
Gitaloxin
Heptaminol
Imazodan
Indolidan
Inosine
Isomazole
Lanatoside
Lixazinone
Medorinone
Meproscillarin
Metildigoxin
Milrinone
Ouabain
Pelrinone
Pengitoxin
Pimobendan
Piroximone
Proscillaridin
Quazinone
Quazodine
Ramnodigin
Sulmazole
Tazolol
Tilsuprost
Vesnarinone

Chelating Agent

Deferoxamine
Ditocarb sodium
Edetic acid
Egtazic acid
Medronic acid
Penicillamine
Pentetate calcium trisodium
Pentetic acid

Choleretic

Alibendol
Azintamide
Cholic acid

Cicloxilic acid
Cicrotic acid
Clanobutin
Cyclobutyrol
Cyclovalone
Cynarine
Dehydrocholic acid
Dimecrotic acid
Etamiphyllin
Exiproben
Februpol
Fencibutirol
Fenipentol
Florantyrone
Hymecromone
Menbutone
Metochalcone
Moquizone
Orazamide
Osalmid
Piprozolin
Prozapine
Taurocholic acid
Timonacic
Tocamphyl
Trepibutone
Vanitiolide

Cholinergic

Aceclidine
Acetylcholine chloride
Aclatonium napadisilate
Ambenonium chloride
Benzopyrronium bromide
Benzpyrinium bromide
Bethanechol chloride
Carbachol
Carpronium chloride
Demecarium bromide
Dexpanthenol
Echothiophate iodide
Eseridine
Furtrethonium iodide
Isoflurophate
Methacholine chloride
Neostigmine bromide
Oxapropanium iodide
Physostigmine
Pilocarpine
Pyridostigmine bromide

Cholinesterase Reactivator

Obidoxime chloride
Pralidoxime chloride
Trimedoxime bromide

Coccidiostat

Aklomide
Amprolium
Arprinocid
Beclotiamine
Buquinolate
Clazuril
Clopidol
Cyproquinate
Decoquinate
Diclazuril
Dinitolmide

Dinsed
Furmethoxadone
Halofuginone
Lasalocid
Narasin
Nequinate
Nicarbazin
Nitromide
Proquinolate
Robenidine
Salinomycin
Tiazuril
Toltrazuril

Dental Caries Prophylactic

Dectaflur
Hetaflur
Ipexidine
Olaflur

Depigmentor

Captamine
Hydroquinone
Monobenzone

Disinfectant

Amantanium bromide
Bensalan
Benzododecinium chloride
Benzoxiquine
Benzoxonium chloride
Broxyquinoline
Chlorocresol
Cloflucarban
Clorophene
Cresol
Dequalinium chloride
Dibromsalan
Dofamium chloride
Drazidox
Fluorosalan
Fursalan
Glutaral
Halazone
Metabromsalan
Miristalkonium chloride
Monalazone disodium
Noxytiolin
Oxyquinoline
Pararosaniline
Picloxydine
Propiolactone
Tetradonium bromide
Thiosalan
Tibrofan
Tribromsalan
Triclocarban
Triclosan

Diuretic

Acefylline
Acefylline piperazine
Alipamide
Amanoxine
Ambuphylline
Ambuside

Diuretic (*contd*)

Amiloride
Aminometradine
Amisometradine
Ampyrimine
Azolimine
Azosemide
Bemetizide
Bemitradine
Bendroflumethiazide
Benzthiazide
Besunide
Brocrinat
Bumetanide
Buthiazide
Caffeine
Canrenoic acid
Chlorazanal
Chlormerodrin
Chlorothiazide
Chlorthalidone
Clazolimine
Clofenamide
Clopamide
Clorexolone
Cyclothiazide
Diapamide
Disulfamide
Epithiazide
Ethacrynic acid
Ethiazide
Ethoxzolamide
Ethyl nitrite
Etofylline
Etozolin
Fenquizone
Flumethiazide
Furacrinic acid
Furosemide
Furterene
Galosemide
Hydracarbazine
Hydrabentizide
Hydrochlorothiazide
Hydroflumethiazide
Indacrinone
Indapamide
Isosorbide
Lemidosul
Mannitol
Mebutizide
Mefruside
Meralluride
Mercaptomerin
Mercumatilin
Mersalyl
Mespirenone
Methalthiazide
Methyclothiazide
Meticrane
Metochalcone
Metolazone
Mexrenoate
Muzolimine
Ozolinone
Pamabrom
Paraflutizide
Penflutizide
Piretanide
Polythiazide
Prorenoate potassium
Protheobromine
Quincarbate

Quinethazone
Spironolactone
Spirorenone
Spiroxasone
Sulclamide
Teclothiazide
Theobromine
Theophylline
Ticrynafen
Tizolemide
Torasemide
Triamterene
Trichlormethiazide
Triflocin
Tripamide
Urea
Xipamide
Zidapamide

Dopamine Agonist or Dopaminergic

Ciladopa
Delergotrile
Etisulergine
Etrabamine
Fenoldopam
Ibopamine
Lergotrile
Mesulergine
Naxagolide
Orotirelin
Pergolide
Preclamol

Dopamine Antagonist

Alepride
Broclepride
Cinitapride
Eticlopride
Tinisulpride
Tropapride

Estrogen

Benzestrol
Broparestrol
Chlorotrianisene
Dienestrol
Diethylstilbestrol
Diethylstilbestrol diphosphate
Diethylstilbestrol dipropionate
Equilin
Estradiol
Estradiol benzoate
Estradiol cypionate
Estradiol dipropionate
Estradiol enanthate
Estradiol undecylate
Estradiol valerate
Estrapronicate
Estrazinol
Estiol
Estrofurate
Estrone
Estrone hydrogen sulfate
Estropipate
Ethinyl estradiol
Fenestrel
Furostilbestrol
Hexestrol

Meso-hexestrol
Mestranol
Methallenestril
Methestrol
Moxestrol
Nylestriol
Orestrate
Quinestradol
Quinestrol

Expectorant

Adamexine
Ambroxol
Bromhexine
Cistinexine
Glycerol, iodinated
Guaifenesin
Guaimesal
Guaithylline
Potassium guaiacolsulfonate
Stepronin
Terpin hydrate

Fibrinolytic

Bisobrin
Inicarone
Sorbinicate

Glucocorticoid

Amcinonide
Aminoglutethimide
Beclomethasone dipropionate
Betamethasone
Betamethasone acetate
Betamethasone acibutate
Betamethasone benzoate
Betamethasone dipropionate
Betamethasone sodium phosphate
Betamethasone valerate
Carbenoxolone
Chloroprednisone acetate
Ciprocinonide
Clocortolone acetate
Clocortolone pivalate
Cloprednol
Corticosterone
Cortisone acetate
Cortivazol
Deprodone
Descinolone acetonide
Desoxycorticosterone acetate
Desoxycorticosterone pivalate
Dexamethasone
Dexamethasone acetate
Dexamethasone phosphate
Diflucortolone
Diflucortolone pivalate
Doxibetasol
Flucloronide
Fludrocortisone acetate
Flumethasone
Flumethasone pivalate
Flumoxonide
Flunisolide
Fluocinolone acetonide
Fluocinonide
Fluocortolone
Fluocortolone caproate

Glucocorticoid (*contd*)

Fluorometholone
Fluperolone acetate
Fluprednisolone
Fluprednisolone valerate
Flurandrenolide
Formocortal
Hydrocortamate
Hydrocortisone
Hydrocortisone acetate
Hydrocortisone buyrate
Hydrocortisone cypionate
Hydrocortisone hemisuccinate
Hydrocortisone sodium phosphate
Hydrocortisone succinate
Hydrocortisone valerate
Medrysone
Meprednisone
Methylprednisolone
Methylprednisolone acetate
Methylprednisolone hemisuccinate
Methylprednisolone phosphate
Naflocort
Nivazol
Paramethasone acetate
Prednazoline
Prednicarbate
Prednisolamate
Prenisolone
Prednisone
Prednisolone acetate
Prednisolone hemisuccinate
Prednisolone phosphate
Prednisolone steaglate
Prednisolone tebutate
Prednival
Prednylidene
Procinonide
Roxibolone
Ticabesone propionate
Timobesone acetate
Tipredane
Tralonide
Trimcinolone
Triamcinolone acetonide
Trimcinolone acetonide sodium phosphate
Trimacinolone benetonide
Triamcinolone diacetate
Trimcinolone furetonide
Trimcinolone hexacetonide
Trilostane

Gout Suppressant

Amflutizole
Colchicine

Hematinic

Ferriclate calcium sodium
Ferrotrenine
Ferrous fumarate
Ferrous gluconate
Leucovorin

Hemostatic

Aminocaproic acid
Carbazochrome

Carbazochrome salicylate
Carbazachrome sodium sulfonate
Cotarnine chloride
Dicresulene
Disogluside
Ellagic acid
Ethamsylate
Iprazochrome
Naftazone
Oxamarin
Sulmarin
Tranexamic acid

Hypnotic

Acebrochol
Allobarbital
Alonimid
Amphenidone
Aprobarbital
Barbital
Brallobarbital
Bromisovalum
Brotizolam
Butabarbital
Butoctamide
Capuride
Carbocloral
Carbromal
Carbubarb
Carfimate
Cetohexazine
Chloral hydrate
Chlorhexadol
Clorethate
Cyclobarbital
Dipipanone
Estazolam
Etaqualone
Etizolam
Etodroxizine
Etomidate
Fenadiazole
Flunitrazepam
Flurazepam
Fosazepam
Haloxazolam
Heptabarbital
Lormetazepam
Mecloqualone
Mecloxamine
Meparfynol
Methaqualone
Methitural
Nisobamate
Nitrazepam
Paraldehyde
Penthrichloral
Pentobarbital
Perlapine
Petrichloral
Phenobarbital
Probarbital
Propoxate
Pyrithyldione
Quazepam
Renanolone
Roletamide
Salverine
Secobarbital
Sulfonethylmethane
Sulfonmethane
Suriclone

Taglutimide
Talbutal
Tetrabarbital
Thalidomide
Triazolam
Triclofos
Vinbarbital
Vinylbital
Zolpidem
Zopiclone

Histamine H₂ Antagonist

Cimetidine
Donetidine
Etindine
Famotidine
Icotidine
Lamtidine
Loxtidine
Lupitidine
Metiamide
Nizatidine
Oxmetidine
Ranitidine
Sopromidine
Sufotidine
Zaltidine

Hypotensive (*see also* Antihypertensive)

Apovincamine
Dicirenone
Isoxaprolol
Nebidrazine
Piclonidine
Potassium thiocyanate
Spirgetine
Talipexole
Trethinium tosilate
Viprostol

Immunoregulator or Immuno-suppressive

Azarole
Fanetizole
Frentizole
Oxamisole
Ristianol
Thymopentin
Tilomisole

Immunostimulant

Acedoben
Ciamexon
Imexon
Murabutide
Nosantine
Nuclomedone
Pimelautide
Torbafylline

Insecticide

Benoxafos
Bioresmethrin
Bromociclen

Insecticide (*contd*)

Bromofos
Carbofenotion
Chlorophenothane
Clenpirin
Clofenvinfos
Cypothrin
Dieldrin
Dimpylate
Dioxation
Phenothrin
Pyrimitate
Quintiofos
Ronnel
Stirifos
Tetramethrin

Keratolytic

Alcloxa
Aldioxa
Benzoyl peroxide
Dibenzothiophene
Epiestriol
Isotretinoin
Motretinide
Picotrin diolamine
Resorcinol
Resorcinol monoacetate
Salicylic acid
Tetroquinone
Tretinoin

Laxative

Bisacodyl
Bisoxatin acetate
Danthron
Lactulose
Magnesium citrate
Oxyphenisatin acetate
Phenolphthalein
Potassium sodium tartrate
Sodium picosulfate
Sulisatin

Luteolytic

Etiproston
Luprostiol
Prostalene
Tiaprost

MAO Inhibitors (*see also* Antidepressants)

Almoxatone
Amiflamine
Cimoxatone
Clorgiline
Selegiline

Mental Performance Enhancer

Aniracetam
Dimoxamine

Mucolytic

Acetylcysteine
Carbocysteine
Mecysteine
Bromehexine
Mesna
Letosteine
Adamexine
Domiodol
Guaietolin
Nesosteine
Erdosteine

Muscle Relaxant (general)

Afloqualone
Ambenoxan
Baclofen
Benzoctamine
Chlorproethazine
Cinflumide
Climazolam
Cyclobenzaprine
Dacuronium bromide
Dextrofemine
Eperisone
Etomidoline
Fluperlapine
Hexafluorenium bromide
Idrocilamide
Inaperisone
Lorbamate
Mebezonium iodide
Meladrazine
Menitrazepam
Nafomine
Nelezaprine
Nimetazepam
Phenprobamate
Pipecuronium bromide
Pipoxolan
Progabide
Tetrazepam
Thiocolchicoside
Truxipecurium iodide
Vetrabutine
Xilobam
Xylazine

Muscle Relaxant (skeletal)

Alcuronium chloride
Atracurium besilate
Azumolene
Benzoquinonium chloride
Carisoprodol
Chlorphenesin carbamate
Chlorzoxazone
Clodanolene
Dantrolene
Decamethonium bromide
Dimethyltubocurarinium chloride
Doxacurium chloride
Fazadinium bromide
Fenyripol
Fletazepam
Flumetramide
Hexcarbacholine bromide
Mephenesin
Metaxalone
Methocarbamol

Orphenadrine
Oxydipentonium chloride
Phenyramidol
Procyclidine
Promoxolane
Rolodine
Styramate
Succinylcholine chloride
Tolperisone
Zoxazolamine

Muscle Relaxant (smooth muscle)

Acefylline piperazine
Adiphenine
Ambuphylline
Aminophylline
Cinnamedrine
Etamiphyllin
Fenalamide
Fetoxylate
Flavoxate
Isomylamine
Levoxadrol
Mebeverine
Mesuprine
Methixene
Papaverine
Pargeverine
Proxazole
Proxyphylline
Quinetolate
Ritodrine
Theophylline
Thiphenamil
Tiropramide

Narcotic Antagonists

Alletorphine
Amiphenazole
Conorphone
Cyprenorphine
Dezocine
Diprenorphine
Fenmetozole
Homprenorphine
Ketorfanol
Levallorphan
Nadide
Nalbuphine
Nalmefene
Nalmexone
Nalorphine
Naloxone
Naltrexone
Oxilorphan

Neuroleptic

Aceprometazine
Alizapride
Cinuperone
Clocapramine
Duoperone
Erizepine
Fluanisone
Flubepride
Flumezapine
Melperone
Methoxypromazine

Neuroleptic (*contd*)

Oxaflumazine
Picobenzide
Spiramide
Spiroxatrine
Sulforidazine
Sulverapride
Tefludazine
Teflutixol
Thioproperazine
Tiapride
Tienocarbine
Tilozepine
Zetidoline
Zoloperone
Zuclopenthixol

Neuromuscular Blocker

Gallamine triethiodide
Laudexium methyl sulfate
Metocurine iodide
Pancuronium bromide
Stercuronium iodide
Succinylcholine chloride
Suxethonium chloride
Tubocurarine chloride
Vecuronium bromide

Nootropic

Bifemelane
Idebenone
Levacecarnine hydrochloride
Propentofylline
Razobazam
Rolziracetam
Tenilsetam
Velnacrine maleate
Vinconate
Zomebazam

Oxytocic

Carboprost
Carboprost methyl
Cloquinozine
Dinoprost
Dinoprostone
Ergonovine
Meteneprost
Methylergonovine
Oxytocin
Quipazine
Sparteine

Pediculicide

Benzyl benzoate
Chlorophenothane
Lindane
Malathion

Preservative

Benzalkonium chloride
Benzethonium chloride
Cetylpyridinium chloride
Dehydroacetic acid

Monothioglycerol
Propylparaben
Sodium formaldehyde sulfoxylate
Sodium propionate
Thimerosal

Progestin

Algestone acetophenide
Allylestrenol
Altrenogest
Amadinone acetate
Anagestone acetate
Chlormadinone acetate
Cingestol
Clogestone acetate
Clomegestone acetate
Delmadinone acetate
Demegestone
Desogestrel
Dimethisterone
Domoprednate
Dydrogesterone
Edogestrone
Ethisterone
Ethynerone
Ethynodiol diacetate
Flurogestone acetate
Gestaclone
Gestodene
Gestonorone caproate
Haloprogesterone
Hydroxyprogesterone
Hydroxyprogesterone caproate
Levonorgestrel
Lynestrenol
Medrogestone
Medroxyprogesterone
Medroxyprogesterone acetate
Melengestrol acetate
Methynodiol diacetate
Nomegestrol
Norethindrone
Norethindrone acetate
Norethynodrel
Norgesterone
Norgestimate
Norgestomet
Norgestrel
Norgestrienone
Norvinisterone
Onapristone
Oxogestone phenpropionate
Progesterone
Proligestone
Promegestone
Quingestanol acetate
Quingestrone
Tigestol
Trengestone

Prostaglandins

Alfaprostol
Cloprostenol
Dinoprost
Dinoprostone
Fluprostenol
Gemeprost
Meteneprost
Prostalene
Sulprostone

Prostate Growth Inhibitor

Pentomone

Radiopaque

Acetrizoate
Bunamiodyl
Diatrizoic acid
Dimethiodal sodium
Diprotrizoate
Ethyl cartrizoate
Iobenzamic acid
Iocarmic acid
Iocetamic acid
Iodamide
Iodecimol
Iodipamide
Iodixanol
Iodoalphionic acid
Iodophthalein
Iodoxamic acid
Ioglicic acid
Ioglucol
Ioglucomide
Ioglunide
Ioglycamic acid
Iogulamide
Iohexol
Iomeprol
Iomorinic acid
Iopamidol
Iopanoic acid
Iopentol
Iophendylate
Iophenoxic acid
Ioprocemic acid
Iopromide
Iopronic acid
Iopydol
Iopydone
Iosarcol
Iosefamic acid
Ioseric acid
Iosimide
Iosulamide
Iosumetic acid
Iotasul
Iotetric acid
Iothalamic acid
Iotranic acid
Iotrizoic acid
Iotrolan
Iotroxic acid
Ioversol
Ioxabrolic acid
Ioxaglic acid
Ioxitalamic acid
Ioxotrizoic acid
Iozomic acid
Ipodic acid
Methiodal sodium
Metrizamide
Metrizoic acid
Phenobutiodil
Propyl docetrizoate
Propyliodone
Tyropanic acid

Radioprotective

Ethiofos

Radioprotective (*contd*)

Geroquinol
Mercaptamine

Repellent (Arthropod) or Arthropod Repellent

Diethyltoluamide

Scabicide

Amitraz
Benzyl benzoate
Crotamiton
Lindane
Mesulfen

Sedative

Acebrochol
Acecarbromal
Acepromazine
Acevaltrate
Adinazolam
Alfaxalone
Alonimid
Alprazolam
Amobarbital
Amphenidone
Aprobarbital
Barbital
Bentazepam
Benzoclidine
Benzoctamine
Bromisovalum
Bromoform
Butabarbital
Butalbital
Butoctamide
Cannabinol
Captodiame
Carbromal
Carbubarb
Carburazepam
Carfimate
Chloral betaine
Chloral hydrate
Chloralose
Clomethiazole
Cloperidone
Clorethate
Clozapine
Cyclobarbital
Cyprazepam
Declenperone
Delorazepam
Detomidine
Dexclamol
Diazepam
Didrovaltrate
Dimelazine
Dipipanone
Doxefazepam
Ectylurea
Estazolam
Etaqualone
Ethchlorvynol
Ethinamate
Fenobam
Fluperlapine

Flurazepam
Glutethimide
Halazepam
Haloxazolam
Heptabarbital
Hexapropymate
Hexobarbital
Homofenazine
Ibrotamide
Lormetazepam
Mafoprazine
Mecloqualone
Mecloxamine
Medetomidine
Meparfynol
Mephobarbital
Meprobamate
Methaqualone
Methitural
Methyprylon
Metomidate
Metoserpate
Midaflur
Motrazepam
Nealbarbital
Niaprazine
Nisobamate
Nortetrazepam
Paraldehyde
Penthrichloral
Pentobarbital
Perimetazine
Petrichloral
Phenobarbital
Phenylmethylbarbituric acid
Pimethixene
Prazepam
Probarbital
Propiomazine
Prothipendyl
Proxibarbal
Pyrithyldione
Quazepam
Reclazepam
Roxoperone
Secobarbital
Suproclone
Taglutimide
Talbutal
Tameridone
Tetrabarbital
Thalidomide
Thioridazine
Toprilidine
Tracazolate
Trepipam
Triazolam
Tricetamide
Triclofos
Trifluoperazine
Trimetozine
Uldazepam
Valtrate
Vinbarbital
Vinylbital
Zolazepam
Zolpidem
Zopiclone

Serotonin Antagonist or Inhibitor

Altanserin
Benanserin

Cianopramine
Cinanserin
Danitracin
Fenclonine
Fonazine
Homochlorcyclizine
Ivoqualine
Ketanserin
Lisuride
Metrenperone
Mianserin
Nomelidine
Paroxetine
Pizotyline
Proterguride
Ritanserin
Seganserin
Terciprazine
Tipindole
Tropanserin
Xylamidine

Spermaticide

Chlorindanol
Laureth 10S
Nonoxynol 9

Stimulant (Cardiac)

Etofylline
Fenalcomine
Xamoterol

Stimulant (Central)

Amfonelic acid
Amineptine
Amphetamine
Ampyzine
Azabon
Bemegride
Brucine
Caffeine
Cyprodenate
Dexoxadrol
Dextroamphetamine
Difluanine
Ethamivan
Etifelmine
Etryptamine
Fencamfamine
Fenethylline
Fenozolone
Flubanilate
Flurothyl
Hexacyclonate
Indriline
Mazindol
Mefexamide
Mepixanox
Methamphetamine
Methastyridone
Methylphenidate
Modafinil
Pemoline
Pentylenetetrazole
Pipradrol
Piracetam
Pyrovalerone
Strychnine

Stimulant (Cerebral)

Citicoline
Meclofenoxate
Pirisudanol

Stimulant (Respiratory)

Almitrine
Dimefline
Doxapram
Ethamivan
Fominoben
Lobeline
Nikethamide
Pimeclone
Pirisudanol
Pyridofylline
Tacrine

Suppressant (*Lupus erythematosus*)

Bismuth sodium triglycollamate
Bismuth subsalicylate
Chloroquine
Hydroxychloroquine

Sweetener

Acesulfame
Aspartame
Cyclamic acid
Melizame
Saccharin

Thyroid Hormone or Thyromimetic

Levothyroxine
Liothyronine
Protirelin
Thyromedan
Tiratricol

Thyroid Inhibitor

Aminothiazole
Carbimazole
Methimazole
Methylthiouracil
Mipimazole
Propylthiouracil
Thiouracil

Tranquilizer (*see also* Antipsychotic)

Ajmaline
Arfendazam
Azacyclonol
Benactyzine
Bentazepam
Camazepam
Carburazepam
Cetohexazine
Chlormezanone
Chlorproethazine
Ciclotizolam
Clofenetamine
Clomethiazole

Clotiazepam
Cyamemazine
Cyclarbamate
Difencloxazine
Emylcamate
Etifoxine
Etymemazine
Febarbamate
Fenpentadiol
Fluoresone
Gepirone
Glaziovine
Lofendazam
Mecloralurea
Mepazine
Mephenoxalone
Mepiprazole
Metofenazate
Moperone
Oxanamide
Oxazolam
Oxomemazine
Pentamoxane
Peratizole
Phenaglycodol
Phenprobamate
Piperacetazine
Pirenperone
Potassium nitrazepate
Procymate
Promazine
Propyperone
Prothixene
Rolipram
Roxoperone
Spiclomazine
Spirilene
Suriclone
Tetrabenazine
Tetrazepam
Timiperone
Tolboxane
Tracazolate
Triclodazol
Trifluomeprazine
Valnoctamide
Zotepine

Tranquilizer (Minor)

Alprazolam
Bromazepam
Buspirone
Chlordiazepoxide
Clazolam
Clobazam
Cloxazolam
Demoxepam
Ethyl loflazepate
Fludiazepam
Halazepam
Hydroxyphenamate
Hydroxyzine
Ketazolam
Loprazolam
Lorazepam
Lorzafone
Loxapine
Medazepam
Nabilone
Nisobamate
Nordazepam
Oxazepam

Pentabamate
Ripazepam
Sulazepam
Taclamine
Temazepam
Triflubazam
Tybamate

Ultraviolet Screen

Actinoquinol
Aminobenzoic acid
Beta carotene
Bornelone
Bumetrizole
Cinoxate
Dioxybenzone
Drometrizole
Etocrylene
Homosalate
Octabenzone
Octocrylene
Octrizole
Oxybenzone
Padimate A
Padimate O
Sulisobenzone

Uricosuric

Benzbromarone
Benzmalecene
Etebenecid
Halofenate
Isobromindione
Itanoxone
Orotic acid
Oxycinchophen
Probenecid
Seclazone
Sulfinpyrazone
Ticrynafen
Tisopurine
Zoxazolamine

Vasoconstrictor

Coumazoline
Cyclopentamine
Indanazoline
Levonordefrin
Levopropylhexedrine
Mephentermine
Methoxamine
Methysergide
Metizoline
Midodrine
Nordefrin
Ornipressin
Propylhexedrine
Tefazoline
Tetrahydrozoline
Tinazoline
Xylometazoline

Vasodilator

Acetylcholine chloride
Alprostadil
Aminoethyl nitrate

Vasodilator (*contd*)

Amyl nitrite
Bamethan
Bemarinone
Bencianol
Bencyclane
Bendazol
Benfurodil hemisuccinate
Bepridil
Betahistine
Biclodil
Bufeniode
Cicletanine
Ciclonicate
Cinepazic acid
Cinnarizine clofibrate
Creatinolfosfate
Daltroban
Dimetofrine
Dimoxyline
Elziverine
Etofylline
Felodipine
Fenoxedil
Flunarizine
Fostedil
Gapicomine
Hexobendine
Ifenprodil
Iproxamine
Isoxsuprine
Khellin
Mannitol hexanitrate
Mecinarone
Mesuprine
Methylchromone
Nafronyl
Nicametate
Nicardipine
Nicergoline
Nicofurate
Nimodipine
Nonivamide
Oxdralazine
Oxfenicine
Pentaerythritol tetranitrate
Pentifylline
Pentoxifylline
Pipratecol
Primaperone
Prizidilol
Ridazolol
Sodium nitrite
Theobromine
Tipropidil
Tolmesoxide
Toprilidine
Trolnitrate
Vincamine
Viprostol
Viquidil
Zolertine

Vasodilator (Cerebral)

Brovincamine
Ibudilast

Mefenidil
Ozagrel
Tinofedrine
Vincanol
Vindeburnol

Vasodilator (Coronary)

Amotriphene
Azaclorzine
Benziodarone
Bumepidil
Chloracyzine
Chromonar
Cinpropazide
Clonitrate
Cloricromen
Devapamil
Dilazep
Diltiazem
Diniprofylline
Dipyridamole
Droprenilamine
Efloxate
Erythrityl tetranitrate
Etafenone
Fendiline
Floredil
Flosequinan
Ganglefene
Isosorbide dinitrate
Isosorbide mononitrate
Lidoflazine
Medibazine
Mepramidil
Metrifudil
Mioflazine
Mixidine
Molsidomine
Morocromen
Nicorandil
Nifedipine
Nisoldipine
Nitroglycerin
Oxprenolol
Oxyfedrine
Pentrinitrol
Perhexilene
Pimefylline
Prenylamine
Propatyl nitrate
Pyridofylline
Teopranitol
Terodiline
Trapidil
Trimetazidine
Valperinol
Visnadine

Vasodilator (Peripheral)

Buflomedil
Butalamine
Buterizine
Cetiedil
Cinepazide
Hepronicate

Inositol niacinate
Limaprost
Moxisylyte
Nicofuranose
Nicotinyl alcohol
Nylidrin
Picodralazine
Piribedil
Suloctidil
Tolazoline
Xanthinol niacinate

Vitamin (Provitamin or Cofactor)

Alfacalcidol
Ascorbic acid
Benfotiamine
Biotin
Bisbentiamine
Calcium pantothenate
Cetotiamine
Cholecalciferol
Cobamide
Cocarboxylase
Cyanocobalamine
Cycotiamine
Ergocalciferol
Ergosterol
Folic acid
Fursultiamine
Hydroxocobalamin
Mecobalamin
Menadiol
Menadiol diphosphate
Menadiol disulfate
Menadione
Menadione sodium bisulfite
Monophosphothiamine
Niacin
Niacinamide
Octotiamine
Panthenol
Pantothenic acid
Phytonadiol sodium diphosphate
Phytonadione
Pyridoxine
Retinol
Riboflavin
Riboflavin 5′-phosphate
Thiamine
Tocophersolan
Vitamin E

Vulnerary

Allantoin
Allopurinol
Allylthiourea
Oxypurinol
Scarlet red

Xanthine Oxidase Inhibitor

Allopurinol
Oxypurinol